AP* Edition

BIOLOGY

Sylvia S. Mader

Michael Windelspecht
Appalachian State University

With contributions by

April Cognato
Michigan State University

David Cox
Lincoln Land Community College

Jeffrey Isaacson
Nebraska Wesleyan University

Ian Quitadamo
Central Washington University

Reinforced Binding

What does it mean?

This textbook is widely adopted by colleges and universities yet it is frequently used in high school for teaching Advanced Placement*, honors and electives courses. Since high schools frequently adopt for several years, it is important that a textbook can withstand the wear and tear of usage by multiple students. To ensure durability, McGraw-Hill has elected to manufacture this textbook with a reinforced binding.

* Advanced Placement program and AP are registered trademarks of the College Board, which was not involved in the production of, and does not endorse, this product.

Mc Graw Hill

Connect Learn Succeed™

BIOLOGY, AP EDITION

Published by McGraw-Hill, a business unit of The McGraw-Hill Companies, Inc., 1221 Avenue of the Americas, New York, NY 10020. Copyright © 2013 by The McGraw-Hill Companies, Inc. All rights reserved. Printed in the United States of America. Previous editions © 2010, 2007, and 2004. No part of this publication may be reproduced or distributed in any form or by any means, or stored in a database or retrieval system, without the prior written consent of The McGraw-Hill Companies, Inc., including, but not limited to, in any network or other electronic storage or transmission, or broadcast for distance learning.

Some ancillaries, including electronic and print components, may not be available to customers outside the United States.

This book is printed on acid-free paper.

4 5 6 7 8 9 0 QVR/LEH 16 15 14 13 12

ISBN 978-0-07-662004-3
MHID 0-07-662004-2

Vice President, Editor-in-Chief: *Marty Lange*
Vice President, EDP: *Kimberly Meriwether David*
Senior Director of Development: *Kristine Tibbetts*
Publisher: *Michael S. Hackett*
Sponsoring Editor: *Eric Weber*
Senior Developmental Editor: *Rose M. Koos*
Director of Digital Content Development: *Tod Duncan, Ph.D.*
Senior Marketing Manager: *Tamara Maury*
Senior Project Manager: *Jayne L. Klein*
Senior Buyer: *Sandy Ludovissy*
Senior Media Project Manager: *Jodi K. Banowetz*
Senior Designer: *Laurie B. Janssen*
Cover Designer: *Elise Lansdon*
Cover Image: *Corbis RF/Alamy*
Senior Photo Research Coordinator: *Lori Hancock*
Photo Research: *Evelyn Jo Johnson*
Compositor: *Lachina Publishing Services*
Typeface: *10/12 Slimbach*
Printer: *Quad/Graphics*

All credits appearing on page or at the end of the book are considered to be an extension of the copyright page.

Library of Congress Cataloging-in-Publication Data

Biology / Sylvia S. Mader, Michael Windelspecht with contributions by April Cognato ... [et al.]. — 11th ed.
   p. cm.
     Rev. ed. of: Biology / Sylvia S. Mader. c2010.
     Includes index.
     ISBN 978-0-07-662004-3 — ISBN 0-07-662004-2 (hard copy : alk. paper) 1. Biology.
I. Mader, Sylvia S. II. Windelspecht, Michael.
   QH308.2.M23 2013
   570—dc23

                 2011030846

www.mheonline.com

About the Authors

Sylvia S. Mader

Sylvia S. Mader has authored several nationally recognized biology texts published by McGraw-Hill. Educated at Bryn Mawr College, Harvard University, Tufts University, and Nova Southeastern University, she holds degrees in both Biology and Education. Over the years she has taught at University of Massachusetts, Lowell; Massachusetts Bay Community College; Suffolk University; and Nathan Mayhew Seminars. Her ability to reach out to science-shy students led to the writing of her first text, *Inquiry into Life*, that is now in its thirteenth edition. Highly acclaimed for her crisp and entertaining writing style, her books have become models for others who write in the field of biology.

Although her writing schedule is always quite demanding, Dr. Mader enjoys taking time to visit and explore the various ecosystems of the biosphere. Her several trips to the Florida Everglades and Caribbean coral reefs resulted in talks she has given to various groups around the country. She has visited the tundra in Alaska, the taiga in the Canadian Rockies, the Sonoran Desert in Arizona, and tropical rain forests in South America and Australia. A photo safari to the Serengeti in Kenya resulted in a number of photographs for her texts. She was thrilled to think of walking in Darwin's steps when she journeyed to the Galápagos Islands with a group of biology educators. Dr. Mader was also a member of a group of biology educators who traveled to China to meet with their Chinese counterparts and exchange ideas about the teaching of modern-day biology.

Michael Windelspecht

As an educator, Dr. Windelspecht has taught introductory biology, genetics, and human genetics in the online, traditional, and hybrid environments at community colleges, comprehensive universities, and military institutions. For over a decade he served as the Introductory Biology Coordinator at Appalachian State University, where he directed a program that enrolled over 4,500 students annually. He was educated at Michigan State University and the University of South Florida. Dr. Windelspecht is also active in promoting the scientific literacy of secondary school educators. He has led multiple workshops on integrating water quality research into the science curriculum, and has spent several summers teaching Pakistani middle school teachers.

As an author, Dr. Windelspecht has published five reference textbooks and multiple print and online lab manuals. He served as the series editor for a ten-volume work on the human body. For years Dr. Windelspecht has been active in the development of multimedia resources for online and hybrid science classrooms. Along with his wife, Sandra, he owns a multimedia production company that actively develops and assesses the use of new technologies in the classroom.

Contributors

April Cognato serves as an Assistant Professor in the Department of Zoology at Michigan State University. She was educated at University of California–Davis, and at Texas A&M University where she earned a master's degree and Ph.D. in evolutionary biology. Dr. Cognato is an accomplished research biologist and educator. As an educator, Dr. Cognato designs and teaches introductory non-majors biology, majors biology, and genetics. In addition to her teaching assignments, Dr. Cognato has expertise in the integration of digital resources into education, and she authors fully online courses in genetics and evolution.

Dave Cox serves as Associate Professor of Biology at Lincoln Land Community College, in Springfield, Illinois. He was educated at Illinois College and Western Illinois University. As an educator, Professor Cox teaches introductory biology for non-majors in the traditional classroom format as well as in a hybrid format. He also teaches biology for majors, and marine biology and biological field studies as study-abroad courses in Belize. He serves as the Educational Director for the Sibun Education and Adventure Lodge, located in Belmopan, Belize.

Jeffrey Isaacson is an Associate Professor of Biology at Nebraska Wesleyan University, where he teaches courses in microbiology, immunology, and cell biology. He currently serves as Chair of the Undergraduate Curriculum Committee and Co-Chair of the General Education Revision Team at NWU. Dr. Isaacson was educated at Nebraska Wesleyan, Kansas State College of Veterinary Medicine, and Iowa State University. Previously, he worked as a small-animal veterinarian, and completed a postdoctoral fellowship in the Department of Immunology at the Mayo Clinic. As an author, Dr. Isaacson has served as a significant contributor and coauthor for the twelfth and thirteenth editions of Mader's *Inquiry Into Life*.

Ian Quitadamo has a dual appointment in Biological Sciences and Science Education at Central Washington University where he teaches courses in cell biology, genetics, biotechnology, and non-majors biology. In addition, he facilitates courses in teaching methodology for future science teachers and interdisciplinary content courses focused on alternative energy and sustainability. Dr. Quitadamo was educated at Washington State University where he earned an interdisciplinary Ph.D. in science, education, and technology. Dr. Quitadamo is a recipient of the Crystal Apple award for teaching excellence. He is a fourth-degree black belt in Kyokushin Karate and owns an academic consulting company.

Guided Chapter Tour

Pedagogy

Chapter Opener Each chapter opener now includes two to three questions designed to integrate the subject matter into the chapter and stimulate class discussions.

Chapter Outline The major sections that will be discussed in the chapter are listed for easy reference.

Before You Begin The content of each chapter is linked with material from earlier in the text. The questions designate important topics that students should understand before proceeding into the chapter.

Following the Big Ideas This section highlights the Big Ideas from the AP Curriculum Framework covered in the chapter.

Connecting the Concepts with the Big Ideas Located at the end of each chapter these boxes include the most important items from the AP Curriculum Framework found within that chapter's pages. They also include citations for the illustrative examples.

Media Study Tools This feature provides a link to the Mader Online Learning Center, which contains animations, videos, and multimedia assets that can help the student succeed in their study of biology. In many chapters, icons direct teachers and students to the presence of 3D animations that may be used to further student comprehension of difficult topics.

The McGraw-Hill ConnectPlus® platform provides a media-rich eBook, interactive learning tools, and access to the LearnSmart system. ConnectPlus® also contains AP chapter banks and an AP Practice Test to assess student comprehension of the course content.

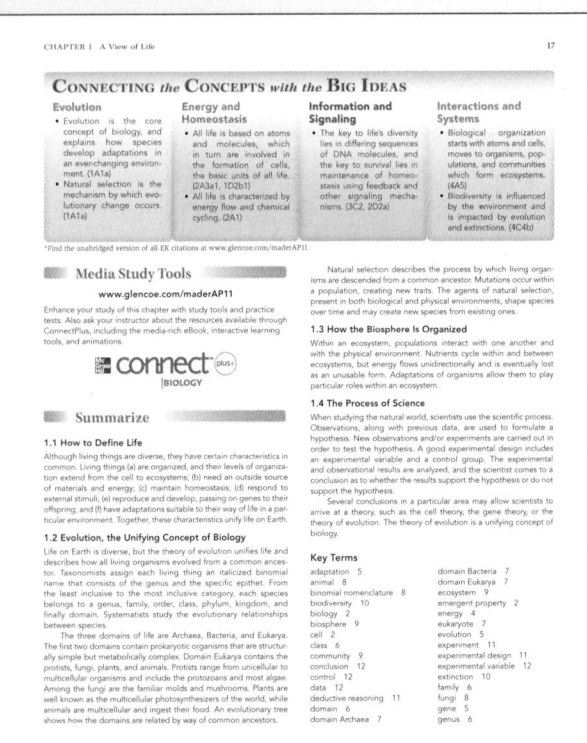

McGraw-Hill Education and Blackboard Have Teamed Up

The Best of Both Worlds

Now you and your students can access McGraw-Hill's Connect™ and Create™ right from with your Blackboard® course—all with one single sign-on. This allows for deep integration of McGraw-Hill content and content engines right into Blackboard. When a student completes an integrated Connect assignment, the grade is automatically fed into your Blackboard grade center. Contact your McGraw-Hill representative for details.

McGraw-Hill LearnSmart™

McGraw-Hill LearnSmart™ is available as an integrated feature of Connect® Biology. Using artificial intelligence, LearnSmart™ intelligently assesses a student's knowledge of course content through a series of adaptive questions. It pinpoints concepts the student does not understand and maps out a personalized study plan for success. This innovative study tool allows teachers to see exactly what students have accomplished.

McGraw-Hill Connect Biology provides online presentation, assignment, and assessment solutions. It connects your students with the tools and resources they'll need to achieve success.

With Connect Biology, you can deliver assignments, quizzes, AP chapter tests, and an AP Practice Test online. As a teacher, you can edit existing questions, author new questions, and track individual student performance—by question, assignment, or in relation to the class overall—with detailed grade reports.

ConnectPlus Biology provides students with all the advantages of Connect Biology, plus 24/7 online access to an eBook. This media-rich version of the book allows seamless integration of text, media, and assessments.

To learn more, visit
www.mcgrawhillconnect.com/advplac

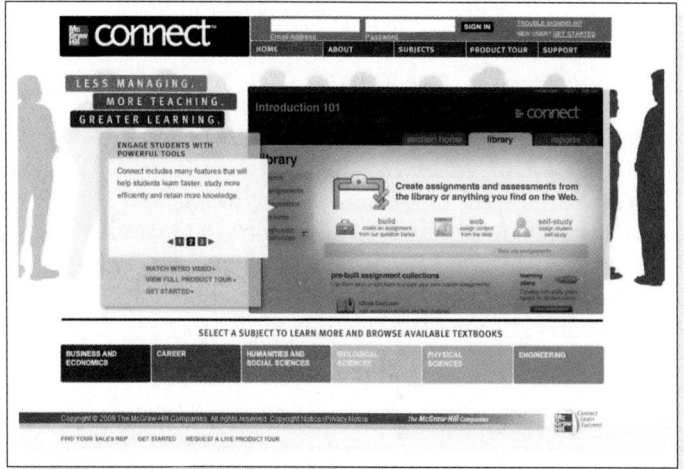

McGraw-Hill Create™

With **McGraw-Hill Create™**, www.mcgrawhillcreate.com, you can easily rearrange chapters, combine material from other content sources, and quickly upload content you have written like your course syllabus or teaching notes. Find the content you need in Create by searching through thousands of leading McGraw-Hill textbooks. Arrange your book to fit your teaching style. Create even allows you to personalize your book's appearance by selecting the cover and adding your name, school, and course information. Contact your McGraw-Hill sales representative to receive a complimentary copy.

Laboratory Manual

The *Biology Laboratory Manual* is written by Dr. Syvia Mader. Every laboratory has been written to help students learn the fundamental concepts of biology and the specific content of the chapter, as well as a better understanding of the scientific method.

Companion Website

www.mheonline.com/maderAP11
or
www.glencoe.com/maderAP11

The Mader *Biology* companion website allows students to access a variety of free digital learning tools that include

- Chapter-level quizzing
- Animations and videos
- Presentation Tools
- Vocabulary flashcards
- Virtual labs
- Presentation Center

Presentation Tools

Everything you need for outstanding presentations in one place.

FlexArt Image Powerpoints—including every piece of art that has been sized and cropped specifically for superior presentations, as well as labels that can be edited and flexible art that can be picked up and moved on key figures. Also included are tables, photographs, and unlabeled art pieces.

Lecture PowerPoints with Animations—animations illustrating important processes are embedded in the lecture material.

Animation PowerPoints—animations only are provided in PowerPoint.

Labeled JPEG Images—full-color digital files of all illustrations that can be readily incorporated into presentations, exams, or custom-made classroom materials.

Base Art Image Files—unlabeled digital files of all illustrations.

Acknowledgments

Dr. Sylvia Mader represents one of the icons of science education. Her dedication to her students, coupled to her clear, concise writing style, has benefited the education of thousands of students over the past three decades. It is an honor to continue her legacy, and to bring her message to the next generation of students. I have been privileged to work with the ultimate team of science educators—April Cognato, Dave Cox, Jeff Isaacson, and Ian Quitadamo. These dedicated and talented individuals have made a significant contribution to this text, and my abilities as a science educator have been enhanced by our teamwork. Together, we have striven to ensure that the material was written and illustrated in the familiar Mader style.

Many dedicated and talented individuals assisted in the development of *Biology*. I am very grateful for the help of so many professionals at McGraw-Hill who were involved in bringing this book to fruition. In particular, let me thank Rose Koos, the developmental editor who lent her exemplary talents, project management skills, advice, and most of all, patience, to all those who worked on this text. The biology publisher is Michael Hackett, who steadfastly encouraged and supported all aspects of this project. The desire of my editor, Eric Weber, to impact the lives of our students, is evident throughout this text. The project manager, Jayne Klein, faithfully and carefully steered the book through the publication process. Tamara Maury, the marketing manager, tirelessly promoted the text and educated the sales reps on its message.

The design of the book is the result of the creative talents of Laurie Janssen and many others who assisted in deciding the appearance of each element in the text. I was very lucky to have Jody Larson as my copyeditor, and Lauren Timmer and Dawnelle Krouse as proofreaders. Lachina Publishing Services produced this textbook, in the process emphasizing pedagogy and beauty to arrive at the best presentation on the page. Lori Hancock and Evelyn Jo Johnson did a superb job of finding just the right photographs and micrographs.

Who I am, as an educator and an author, is a direct reflection of what I have learned from my students. Education is a two-way street, and it is my honest opinion that both my professional and personal life have been enriched by my interactions with my students. They have encouraged me to learn more, teach better, and never stop questioning the world around me. I would also like to acknowledge my wife, Sandra. She has never wavered in her patience and support of my endeavors.

Michael Windelspecht, Ph.D.
Blowing Rock, NC

The excellent quality of the eleventh edition of *Biology* would not have been possible without all of the contributors involved in its development, including several highly-respected AP teachers, table leaders, and table readers. Below is a list of the many contributors and reviewers involved in the development of this edition.

AP Contributors

AP-specific Student Edition Content: Carolyn Schofield Bronston; **AP Teacher's Manual:** Lynn Meldru; **AP Practice Test Booklet:** G. Roger Pratt; **AP ExamView:** Claire Vollmar; **AP Connect Chapter Banks:** Jennifer Pfannerstill

Ancillary Authors

Connect Question Bank: Krissy Johnson, Alex James, and Betsy Harris, *Appalachian State University;* Sandy Windelspecht, *Ricochet Creative Productions;* **Test Bank:** Dave Cox, *Lincoln Land Community College;* **Lecture Outlines:** Felicia Scott, *Macomb Community College;* **Instructor's Manual:** Andrea Thomason, *Roxbury Community College;* **Practice Tests:** Deborah Dardis, *Southeastern Louisiana University;* **eBook Quizzes:** Eric Rabitoy, *Citrus College;* **LearnSmart Authors:** Patrick Galliart, *North Iowa Area Community College;* Sylvester Allred, *Northern Arizona University;* Tammy Atchison, *Pitt Community College;* Dena Berg, *Tarrant County College;* Joy Brookshire, *Kennesaw State University;* Jeffrey Isaacson, *Nebraska Wesleyan University;* **LearnSmart Reviewers:** Jill Nugent, *University of North Texas;* Murad Odeh, *South Texas College*

AP Reviewers

Pat Mote, Lynn Meldru, Cheryl Hollinger

Eleventh Edition Reviewers

Nina Abubakari, *Wayne County Community College District*
LaQuetta B. Anderson, *Grambling State University*
Rachele Arrigoni-Restrepo, *New York City College of Technology*
Dennis Bakewicz, *New York City College of Technology*
Sarah Bales, *Moraine Valley Community College*
Marilyn Banta, *Texas State University–San Marcos*
Isaac Barjis, *New York City College of Technology*
Morgan Benowitz-Fredericks, *Bucknell University*
Gretchen Bernard, *Moraine Valley Community College*
Karen Bledsoe, *Western Oregon University*
Lois Borek, *Georgia State University*

Anthony Botyrius, *York College of Pennsylvania*

Denise Chung, *Long Island University–Brooklyn*

Pamela Anderson Cole, *Shelton State Community College*

David Cox, *Lincoln Land Community College*

Deborah Dardis, *Southeastern Louisiana University*

Diane Day, *Georgia State University*

Lewis Deaton, *University of Louisiana at Lafayette*

Helen Donis-Keller, *Franklin W. Olin College of Engineering*

C. Craig Farquhar, *University of Texas–Austin*

Jennifer Foulk, *Montgomery County Community College*

Cynthia M. Galloway, *Texas A&M University–Kingsville*

Raul Galvan, *South Texas College*

Kristine Garner, *University of Arkansas–Fort Smith*

Michele B. Garrett, *Guilford Technical Community College–Jamestown*

Sandra Gibbons, *Moraine Valley Community College*

Melanie Glasscock, *Wallace State Community College*

Andrew Goliszek, *North Carolina A&T State University*

Lula (Gwen) Gordon, *Wayne County Community College–Downtown*

Susan Michele Green, *Texas State University–San Marcos*

Robert Greene, *Cranbrook Kingswood School*

Bradley L. Griggs, *Piedmont Technical College*

Tray Hamil, *University of South Alabama*

Jerrie Hanible, *Southeastern Louisiana University*

Chris Haynes, *Shelton State Community College*

Jennifer (Wearly) Hooks, *Winthrop University*

Brenda Hunzinger, *Lake Land College*

Felix Ifeanyi, *Grambling State University*

David Jarrell, *Armstrong Atlantic State University*

Ragupathy Kannan, *University of Arkansas–Fort Smith*

Carolyn Lebsack, *Linn-Benton Community College*

Stephen G. Lebsack, *Linn-Benton Community College*

Julian Lee, *University of Miami*

Tammy J. Liles, *Bluegrass Community and Technical College*

Lynne Lohmeier, *Mississippi Gulf Coast Community College*

Chintamani S. Manish, *Midland Lutheran College*

Jessica Mayfield, *Southeastern Louisiana University*

Mark Meade, *Jacksonville State University*

Sandra L. Millward, *The Christ College of Nursing and Health Sciences*

Scott Murdoch, *Moraine Valley Community College*

Joseph Murray, *Blue Ridge Community College*

Necia M. Nicholas, *Calhoun Community College*

Therese Poole, *Georgia State University*

Michelle Priest, *Los Angeles Valley College*

Kirstin Purcell, *Saint Paul College*

Jason Raymond, *University of California–Merced*

Pamela Riddell, *Macomb Community College–Center Campus*

Abraham Saraya, *New York City College of Technology*

Dale Smoak, *Piedmont Technical College*

Phillip Snider, *Gadsden State Community College*

Lisa Strain, *Northeast Lakeview College*

Diane Teter, *South Texas College*

Mark X. VanCura, *Cape Fear Community College*

Van Wheat, *South Texas College*

Leslie Whiteman-Richardson, *Virginia State University*

Ann R. Witham, *University of Cincinnati, Raymond Walters College*

Frank Wray, *University of Cincinnati, Raymond Walters College*

AP Course Information

To the AP Biology Teacher

As someone who was lucky enough to teach AP Biology for 29 years, I know it to be the best thing that has ever happened to advanced high school science curricula. Starting in the 1950s with the idea of allowing bright and motivated students to test for college credit, it has evolved dramatically over the years. Originally a simple 3-page course outline, twelve laboratory experiences were eventually added, and later themes and concepts were overlaid on the course outline to emphasize the unity and connections among all aspects of the study of life.

Now, in a huge evolutionary leap, AP is about to embark in a totally new direction, one that attempts to pare down the voluminous amount of material normally covered in a beginning college course to the truly basic bones, focusing on the major concepts and essential skills which will prepare and equip students for continuing biological studies at the university level. The Redesign Committee, composed of dynamic classroom teachers and college professors, came up with four Big Ideas that every student of life sciences should deeply and completely understand. To flesh out each Big Idea, Enduring Understanding (EU) and Essential Knowledge (EK) pieces detail the information students need to internalize while Learning Objectives (LOs) specify skills and thinking patterns essential in "doing" science. An important note: all the questions on the AP Biology Exam will be based on these LOs. And to keep study confined and focused, explicit exclusion statements have been added to steer teachers away from excessive detail or from material that does not further conceptual understanding. With a very specific framework and detailed exclusions, the Committee hopes to be very transparent about and take the guess work out of what is going to be fair game for the exam.

This textbook will be an invaluable asset as you begin to navigate the new curriculum. At the beginning of each chapter, there are color-coded *Following the Big Ideas* boxes that highlight which concepts will appear within the chapter.

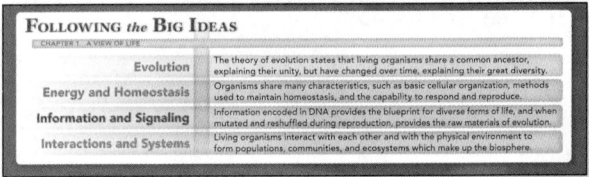

Unit Openers cite the major themes that flow through the material in each unit.

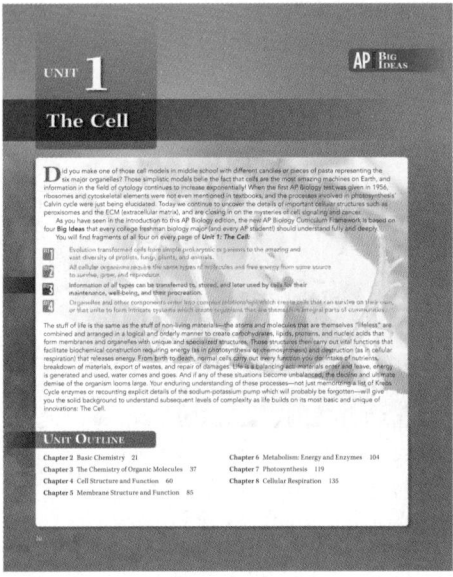

At the end of each chapter you will find *Connecting the Concepts with the Big Ideas* boxes. These boxes include the most important items from the framework found within each chapter's pages including citations for the illustrative examples that you might choose to use.

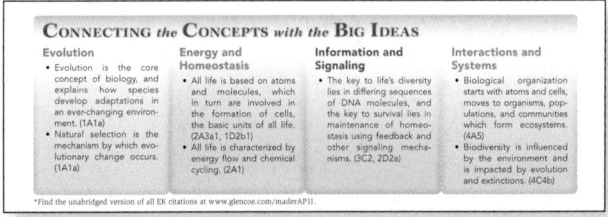

Look at the example below. This will help you decipher the correlations within these boxes.

Big   Enduring   Essential   Illustrative
Idea   Understanding   Knowledge   Example

Size and number of tropic levels are impacted by availability of free energy. (2A1f*IE*)

A complete listing of these Essential Knowledge correlations can be found on the Mader *Biology* Online Learning Center at www.glencoe.com/maderAP11.

The Mader AP Edition has many wonderful resources which will help you address the new curriculum framework.

- **Mader AP Teacher's Manual**—provides teaching strategies, activities, and teaching tips to help you teach the new AP Biology course.
- **Mader AP Practice Test Booklet**—includes one complete practice test. This booklet will help students prepare for the May exam.
- **Mader AP ExamView**—includes hundreds of questions correlated to the new curriculum framework.
- **Biology ConnectPlus**—provides students with online access to the student edition, as well as chapter banks correlated to the new curriculum framework. There is also a complete practice test available online.

To the AP Biology Student

Welcome to the world of Advanced Placement! You are about to be dazzled, enthralled, and wowed by college biology taught in the smaller and more personal environment of your high school classroom. What you get out of this course depends upon you! Doing college work at the high school level is not easy, but is deeply rewarding: completion of AP classes has come to be an extremely reliable predictor of college success.

The College Board has long required that AP Biology students do laboratory work which will enlarge their skill set and scientific enterprise. Rather than being "canned" experiments with known outcomes, these labs are open-ended, inquiry-based, and often student-designed so that curiosity, creativity, and real discovery are back in the picture. At least 25% of your AP class time will likely be devoted to laboratory investigations.

The most important tips to doing well academically in AP Biology are simple and should serve you well in most college classes:

- Keep the class calendar/syllabus front and center. Mark all important test and due dates and work ahead of deadlines— DO NOT PROCRASTINATE!
- Show up for class prepared everyday. Read your assignment, write down questions you have about the current lesson or lab, and try to clarify them at the beginning of class the next day.

- Take notes in whatever style suits you best and keep all your notes and papers organized by date! Bad note taker? Get better! Borrow someone's notes and see what they record and how they do it.
- Find a study partner or group. When you talk over biology with others, questions will come up that can be clarified quickly (or recorded for checking with the teacher); you get reinforcement for what you think you know and fill in what you don't.
- Try different study aids—note cards, concept maps, outlines—and find out what works for you.
- Excellent essay writing is an important skill for any student of science and one at which you can become very proficient. Brainstorm and sketch out a "key words" outline to organize your thoughts. Always read the question twice; do not restate the question, but answer it directly in complete sentences with details and at least one example where possible. Answer all parts of the question, labeling them and keeping them in order. The verbs will tell you what is required of you: analyze, compare, contrast, calculate, describe, discuss, evaluate, explain, interpret, etc. Do what they say! Be concise and precise; length is not scored, content is. Grammar and spelling do not detract from your essay score, but bad penmanship can be a real negative. A grader must be able to decipher your valuable thoughts.
- Here's a tip most students don't often utilize. Learn from your mistakes. Go over the first exam when you get it back. Why did you miss what you did? Was the information not in your notes? Did you not read the question carefully? Did you not spend enough time studying so you were confident rather than tentative on your multiple choice? Did you not directly answer the essay question, being precise and concise in your answer and using good examples to prove your point? Make an appointment with your teacher, tell him/her that you would like an opinion on where you fell down and suggestions on what to do differently.

Wishing you a class that will give you big ideas, hone your skills, pique your curiosity, fill you with knowledge both essential and enduring, remind you that no one yet has all the answers, and be one of the most memorable, challenging, and enjoyable experiences of your life.

Carolyn Schofield Bronston

Advanced Placement Correlation

Big Idea 1: The process of evolution drives the diversity and unity of life.

Essential Knowledge	Pages	Illustrative Examples Covered
1.A.1 Natural selection is a major mechanism of evolution	6–8, 272–286, 290–303	• Graphical analysis of allele frequencies in a population (293) • Application of Hardy-Weinberg Equation (291)
1.A.2 Natural selection acts on phenotypic variations in populations	174–175, 278–280, 290–292, 296–303, 376–377, 679, 908	• Peppered moth (279–280, 290–292) • Sickle cell anemia (302–303) • Artificial selection (278–279) • Overuse of antibiotics (279–280, 376–377, 679) • Loss of genetic diversity within a crop species (908)
1.A.3 Evolutionary change is also driven by random processes	290–295	
1.A.4 Biological evolution is supported by scientific evidence from many disciplines, including mathematics.	275–286, 290–295, 333–337, 354–359	• Graphical analyses of allele frequencies in a population (293) • Analysis of sequence data sets (285) • Analysis of phylogenetic trees (355) • Construction of phylogenetic trees based on sequence data (358)
1.B.1 Organisms share many conserved core processes and features that evolved and are widely distributed among organisms today.	67–68, 72–74, 76–81, 216–217, 224, 328–337	• Cytoskeleton (is a network of structural proteins that facilitate cell movement, morphological integrity and organelle transport) (68, 78–81) • Membrane-bound organelles (mitochondria and/or chloroplasts) (67, 76–77) • Endomembrane systems, including the nuclear envelope (67, 72–74)
1.B.2 Phylogenetic trees and cladograms are graphical representations (models) of evolutionary history that can be tested.	348–359, 420–422, 550–551, 571–575	• mammalian evolution including opposable thumb (355) (plus other examples) • evolutionary plant history (420–422) • chordate development (356, 550–551) • primate evolution (358–359, 571–575)
1.C.1 Speciation and extinction have occurred throughout the Earth's history.	307–318, 342–344, 913–917	• Five major extinctions (343–344) • Human impact on ecosystems and species extinction rates (913–914)
1.C.2 Speciation may occur when two populations become reproductively isolated from each other.	313–318	
1.C.3 Populations of organisms continue to evolve.	278–280, 307–323, 369–371, 376–377, 608–609, 694–696	• Chemical resistance (mutations for resistance to antibiotics, pesticides, herbicides or chemotherapy drugs occur in the absence of the chemical) (376–377) • Emergent diseases (369–371) • Observed directional phenotypic change in a population (Grants' observations of Darwin's finches in the Galapagos) (278–280) • A eukaryotic example that describes evolution of a structure of process such as heart chambers, limbs, brain, and immune system (608–609, 694–696)
1.D.1 There are several hypotheses about the natural origin of life on Earth, each with supporting scientific evidence.	328–332	
1.D.2 Scientific evidence from many different disciplines supports models of the origin of life.	333–341	

Big Idea 2: Biological systems utilize free energy and molecular building blocks to grow, to reproduce and to maintain dynamic homeostasis.

Essential Knowledge	Pages	Illustrative Examples Covered
2.A.1 All living systems require constant input of free energy.	105–115, 122–131, 136–148, 553, 556, 558, 563, 600–601, 870–879	• Krebs Cycle (142) • Glycolysis (138–139) • Calvin Cycle (128–129) • Fermentation (140–141) • Endothermy is the use of thermal energy generated by metabolism to maintain homeostatic body temperatures. (563) • Ectothermy is the use of external thermal energy to help regulate and maintain body temperature. (553, 556, 558) • Change in the producer level can affect the number and size of other trophic levels. (870, 873) • Change in energy resources levels such as sunlight can affect the number and size of the trophic levels. (873)
2.A.2 Organisms capture and store free energy for use in biological processes.	113–115, 120–131, 136–148, 335–337, 375	• $NADP^+$ in photosynthesis (125–126) • Oxygen in cellular respiration (144)
2.A.3 Organisms must exchange matter with the environment to grow, reproduce, and maintain organization.	28–31, 61–64, 467, 469–471, 652–653, 668, 870–879	• Cohesion (30–31) • Adhesion (30–31) • High specific heat capacity (29) • Universal solvent supports reactions (29–30) • Heat of vaporization (29) • Root hairs (467, 469–471) • Cells of the alveoli (668) • Cells of the villi (652–653) • Microvilli (652–653)
2.B.1 Cell membranes are selectively permeable due to their structure.	86–91, 98–100	
2.B.2 Growth and dynamic homeostasis are maintained by the constant movement of molecules across membranes.	91–100	• Glucose transport (94) • Na^+/K^+ transport (95–96)
2.B.3 Eukaryotic cells maintain internal membranes that partition the cell into specialized regions.	65–81	• Endoplasmic Reticulum (72) • Mitochondria (77) • Chloroplasts (76–77) • Golgi (72–73) • Nuclear Envelope (70–71)
2.C.1 Organisms use negative feedback mechanisms to maintain their internal environments and respond to external environmental changes.	238–240, 473–477, 489, 599–601, 620–621, 688, 757, 761, 765–767, 782–783	• Operons in gene regulation (238–240) • Temperature regulation in animals (600–601) • Plant responses to water limitations (473–477) • Lactation in mammals (757, 782–783) • Onset of labor in childbirth (601) • Ripening of fruit (489) • Diabetes mellitus in response to decreased insulin (765–767) • Dehydration in response to decreased ADH (688, 757) • Graves' disease (761) (hyperthyroidism) • Blood clotting (620–621)

Essential Knowledge	Pages	Illustrative Examples Covered
2.C.2 Organisms respond to changes in their external environments.	405–408, 490–498, 600–601, 768, 820–834	• Photoperiodism and phototropism in plants (494–496) • Hibernation and migration in animals (825) • Chemotaxis in bacteria, sexual reproduction in fungi (405–408) • Nocturnal and diurnal activity: circadian rhythms (494–495, 768) • Shivering and sweating in humans (600–601)
2.D.1 All biological systems from cells and organisms to populations, communities, and ecosystems are affected by complex biotic and abiotic interactions involving exchange of matter and free energy	163, 377, 392, 839–850, 858–879, 884–904, 910–913	• Cell density (163) • Temperature (840) • Water availability (875) • Sunlight (840) • Symbiosis (mutualism, commensalism, parasitism) (864–866) • Predator-prey relationships (860–864) • Water and nutrient availability, temperature, salinity, pH (870) • Food chains and food webs (870–874) • Species diversity (910–913) • Population density (840, 847) • Algal blooms (377, 392)
2.D.2 Homeostatic mechanisms reflect both common ancestry and divergence due to adaptation in different environments.	372, 395, 465–479, 599–601, 609–615, 647–656, 664–671, 681–690	• Gas exchange in aquatic and terrestrial plants (476–478) • Digestive mechanisms in animals such as food vacuoles, gastrovascular cavities, one-way digestive systems (647–649) • Respiratory systems of aquatic and terrestrial animals (664–668) • Nitrogenous waste production and elimination in aquatic and terrestrial animals (681–690) • Excretory systems in flatworms, earthworms, and vertebrates (681–683) • Osmoregulation in bacteria, fish, and protists (372, 395, 683–684) • Osmoregulation in aquatic and terrestrial plants (475–478) • Circulatory systems in fish, amphibians, and mammals (609)
2.D.3 Biological systems are affected by disruptions to their dynamic homeostasis.	631–642, 683–690, 757, 847, 908–918	• Physiological responses to toxic substances (632) • Dehydration (683–684, 688, 757) • Immunological responses to pathogens, toxins and allergens (631–639, 641–642) • Invasive and/or eruptive species (913–915) • Human Impact (913–918) • Hurricanes, floods, earthquakes, volcanoes, fires (847) • Water limitation (912)
2.D.4 Plants and animals have a variety of chemical defenses against infections that affect dynamic homeostasis.	448, 496–498, 627–640	• Invertebrate immune systems have nonspecific response mechanisms, but they lack pathogen-specific defense responses. (627–628) • Plant defenses against pathogens include molecular recognition systems with systemic responses; infection triggers chemical responses that destroy infected and adjacent cells by apoptosis, thus localizing the effects. (448) • Vertebrate immune systems have non-specific and non-heritable defense mechanisms against pathogens. (630–633)
2.E.1 Timing and coordination of specific events are necessary for the normal development of an organism, and these events are regulated by a variety of mechanisms.	154–156, 241–247, 484, 502–505, 633–639, 796–810	• Morphogenesis of fingers and toes (155) • Immune function (633–639) • *C. elegans* development (802) • Flower development (502–505)

Essential Knowledge	Pages	Illustrative Examples Covered
2.E.2 Timing and coordination of physiological events are regulated by multiple mechanisms.	407–409, 484–498, 753–769, 825–834	• Circadian rhythms, or the physiological cycle of about 24 hours that is present in all eukaryotes and persists even in the absence of external cues (768) • Diurnal/nocturnal and sleep/awake cycles (491, 768) • Seasonal responses, such as hibernation, estivation, and migration (825) • Release and reaction to pheromones (826) • Visual displays in reproductive cycle (828–829) • Fruiting body formation in fungi, slime molds, and certain types of bacteria (407–409)
2.E.3 Timing and coordination of behavior are regulated by various mechanisms and are important in natural selection.	375, 377, 407–409, 412–415, 472, 490–498, 506–507, 820–826, 831–833, 858–867	• Migration (825) • Courtship (831–833) • Availability of resources leading to fruiting body formation in fungi and certain types of bacteria (407–409, 412–413) • Niche and resource partitioning (860) • Mutualistic relationships (lichens; bacteria in digestive tracts of animals; mycorrhizae) (375, 377, 414–415, 472, 498, 866) • Biology of pollination (506–507)

Big Idea 3: Living systems store, retrieve, transmit and respond to information essential to life processes.

Essential Knowledge	Pages	Illustrative Examples Covered
3.A.1 DNA, and in some cases RNA, is the primary source of heritable information.	49–53, 88–89, 109–113, 215–234, 255–266, 368, 656–659	• Addition of a poly-A tail (226) • Addition of a GTP cap (226) • Excision of introns (226–227) • Enzymatic reactions (109–113) • Transport by proteins (88–89) • Synthesis (49–53) • Degradation (656–659) • Electrophoresis (257) • Plasmid-based transformation (255–256) • Restriction enzyme analysis of DNA (256–257) • Polymerase Chain Reaction (PCR) (256) • Genetically-modified foods (258) • Transgenic animals (259) • Cloned animals (259) • Pharmaceuticals, such as human insulin or factor X (259)
3.A.2 In eukaryotes, heritable information is passed to the next generation via processes that include the cell cycle and mitosis, or meiosis plus fertilization.	154–167, 172–182	• Cancer results from disruptions in cell cycle control (163–165)
3.A.3 The chromosomal basis of inheritance provides an understanding of the pattern of passage (transmission) of genes from parent to offspring	73, 184–185, 193–210	• Sickle cell anemia (206–207) • Cystic fibrosis (202–203) • Tay-Sachs (73) • Huntington's Disease (204) • X-linked Color Blindness (209) • Trisomy 21/Down Syndrome (184) • Klinefelter Syndrome (185) • Reproduction issues (201–204, 209–210) • Civic issues such as ownership of genetic information, privacy, historical contexts, etc. (198–201)

Essential Knowledge	Pages	Illustrative Examples Covered
3.A.4 The inheritance pattern of many traits cannot be explained by simple Mendelian genetics.	180–181, 205–210, 337, 768	• Sex-linked genes reside on sex chromosomes (X in humans). (208–209) • In mammals and flies, the Y chromosome is very small and carries very few genes. (209) • In mammals and flies, females are XX and males are XY; as such, X-linked recessive traits are always expressed in males. (209) • Some traits are sex limited, and expression depends on the sex of the individual, such as milk production in female mammals and pattern baldness in males. (768)
3.B.1 Gene regulation results in differential gene expression, leading to cell specialization.	225–227, 238–250	• Promoters (225–226, 238–244) • Terminators (226) • Enhancers (244)
3.B.2 A variety of intercellular and intracellular signal transmissions mediate gene expression.	90, 155, 164, 184, 208, 238–240, 320–322, 484–489, 524, 634, 637, 800–804	• Cytokines regulate gene expression to allow for cell replication and division. (90, 155, 484–485, 634, 637) • Levels of cAMP regulate metabolic gene expression in bacteria. (240) • Expression of the SRY gene triggers the male sexual development pathway in animals. (184, 208) • Ethylene levels cause changes in the production of different enzymes, allowing fruit ripening. (488–489) • Gibberellin promotes seed germination in plants. (486–488) • Morphogens stimulate cell differentiation and development. (800–804) • Changes in p53 activity can result in cancer. (164) • HOX genes play a role in development. (320–322, 524)
3.C.1 Changes in genotype can result in changes in phenotype.	183–188, 222, 246–250, 280, 302–303	• Antibiotic resistance mutations (280) • Pesticide resistance mutations (280) • Sickle cell disorder and heterozygote advantage (302–303)
3.C.2 Biological systems have multiple processes that increase genetic variation.	172–175, 373	
3.C.3 Viral replication results in genetic variation, and viral infection can introduce genetic variation into the hosts.	262, 364–373	• Transduction in bacteria (373) • Transposons present in incoming DNA (262)
3.D.1 Cell communication processes share common features that reflect a shared evolutionary history.	90, 248, 378, 484, 713, 755–756, 762–763, 826–828	• Use of chemical messengers by microbes to communicate with nearby cells and to regulate specific pathways in response to population density (Quorum sensing) (378) • Use of pheromones to trigger reproduction and developmental pathways (826–828) • Epinephrine stimulation of glycogen breakdown in mammals (713, 755–756, 762–763) • DNA repair mechanisms (248)
3.D.2 Cells communicate with each other through direct contact with other cells or from a distance via chemical signaling.	86–90, 99–100, 378, 496–498, 630–640, 697–701, 757–761, 765–768, 802	• Immune cells interact by cell-cell contact, antigen-presenting-cells (APCs), helper T-cells, killer T-cells. (633–636) • Plasmodesmata between plant cells that allow material to be transported from cell to cell (100) • Neurotransmitters (700–701) • Plant immune response (498) • Quorum sensing in bacteria (378) • Morphogens in embryonic development (802) • Insulin (765–767) • Human Growth Hormone (757–759) • Thyroid hormones (760–761) • Testosterone (767) • Estrogen (767–768)

Essential Knowledge	Pages	Illustrative Examples Covered
3.D.3 Signal transduction pathways link signal reception with cellular response.	90, 164–165, 484–497, 698–699, 753–756	• G-protein linked receptors (756) • Ligand-gated ion channels (698–699) • Receptor tyrosine kinases (164–165) • Second messengers such as: cyclic GMP, cyclic AMP calcium ions (Ca^{2+}), and inositol triphosphate (IP_3) (756)
3.D.4 Changes in signal transduction pathways can alter cellular response.	163, 374, 376, 612–613, 633–640, 642–643, 706–707, 766–767, 783–784	• Diabetes, heart disease, neurological disease, autoimmune disease, cancer, cholera (163, 374, 612–613, 642–643, 706–707, 766–767) • Effects of neurotoxins, poisons, pesticides (376) • Drugs (Antihistamines, and Birth Control) (641–642, 783–784)
3.E.1 Individuals can act on information and communicate it to others.	506–507, 713, 824–834, 863–864	• Fight or flight response (713) • Predator warnings (828) • Protection of young (830) • Avoidance responses (824) • Territorial marking in mammals (830) • Coloration in flowers (506–507) • Birds songs (828) • Pack behavior in animals (833) • Herd, flock, and schooling behavior in animals (834) • Colony and swarming behavior in insects (833) • Coloration (863–864) • Parent and offspring interactions (834) • Migration patterns (825) • Courtship and mating behaviors (831–833) • Foraging in bees and other animals (830–831)
3.E.2 Animals have nervous systems that detect external and internal signals, transmit and integrate information, and produce responses.	694–713, 717–731, 744–749, 757	• Acetylcholine (700, 713) • Epinephrine (713) • Norepinephrine (700, 713) • Dopamine (700) • Serotonin (700) • Vision (720–725) • Hearing (726–730) • Muscle movement (744–749) • Abstract thought and emotions (703–705) • Neuro-hormone production (757) • Forebrain (cerebrum), midbrain (brainstem), and hindbrain (cerebellum) (703–705) • Right and left cerebral hemispheres in humans (703–704)

Big Idea 4: Biological systems interact, and these systems and their interactions possess complex properties.

Essential Knowledge	Pages	Illustrative Examples Covered
4.A.1 The subcomponents of biological molecules and their sequence determine the properties of that molecule.	26–27, 38–56, 215–222	
4.A.2 The structure and function of subcellular components, and their interactions, provide essential cellular processes.	61–81	

Essential Knowledge	Pages	Illustrative Examples Covered
4.A.3 Interactions between external stimuli and regulated gene expression result in specialization of cells, tissues and organs.	800–804	
4.A.4 Organisms exhibit complex properties due to interactions between their constituent parts.	444–459, 608–615, 618–623, 650–656, 664–672, 684–690, 707–713, 744–749	• Stomach and small intestines (652–654) • Kidney and bladder (684) • Root, stem, and leaf (444–459) • Respiratory and circulatory (608–615, 621, 671–672) • Nervous and muscular (707–708, 744–749) • Plant vascular and leaf (449–459)
4.A.5 Communities are composed of populations of organisms that interact in complex ways.	840–853, 858–879, 913–918	• Predator/prey relationships spreadsheet model (863) • Symbiotic relationship (864–866) • Graphical representation of field data (860) • Introduction of species (913–915) • Global climate change models (879, 915–916)
4.A.6 Interactions among living systems and with their environment result in the movement of matter and energy.	858–866, 870–879, 884–904, 913–918	
4.B.1 Interactions between molecules affect their structure and function.	109–113	
4.B.2 Cooperative interactions within organisms promote efficiency in the use of energy and matter.	67–81, 375, 379, 449–450, 459, 469–479, 606–623, 650–657, 664–672, 681–690, 694–713, 736–749, 753–769, 775–783	• Exchange of gases (477–478, 664–672) • Circulation of fluids (449–450, 474–479, 606–623) • Digestion of food (650–657) • Excretion of wastes (681–690) • Bacterial community in the rumen of animals (375) • Bacterial community in and around deep sea vents (379)
4.B.3 Interactions between and within populations influence patterns of species distribution and abundance.	410, 843–848, 858–869, 913–918	• Loss of keystone species (918) • Kudzu (914–915) • Dutch Elm Disease (410)
4.B.4 Distribution of local and global ecosystems changes over time.	342–344, 391, 410, 559, 868, 904, 913–920	• Logging, slash and burn agriculture, urbanization, mono-cropping, infrastructure development (dams, transmission lines, roads), and global climate change threaten ecosystems and life on Earth. (913–918) • An introduced species can exploit a new niche free of predators or competitors, thus exploiting new resources. (913–915) • Dutch Elm disease (410) • Potato blight (391) • El Niño (904) • Continental Drift (342–343, 868) • Meteor Impact on Dinosaurs (343–344, 559)

Essential Knowledge	Pages	Illustrative Examples Covered
4.C.1 Variation in molecular units provides cells with a wider range of functions.	88–89, 124, 203, 300–303, 634–636, 643	• Different types of hemoglobin (203) • MHC Proteins (635–636, 643) • Chlorophylls (124) • Molecular diversity of antibodies in response to an antigen (634–635)
4.C.2 Environmental factors influence the expression of the genotype in an organism.	207, 239–240, 494–496, 598	• Effect of adding lactose to a Lac + bacterial culture (239–240) • Effect of increased UV on melanin production in animals (598) • Alterations in timing of flowering due to climate changes (494–496)
4.C.3 The level of variation in a population affects population dynamics.	391, 413, 630–632, 908–913	• California condor (908) • Black-footed ferrets (909) • Potato blight causing the potato famine (391) • Corn rust affects on agricultural crops (413) • Not all individuals in a population in a disease outbreak are equally affected; some may not show symptoms, have mild symptoms or may be naturally immune and resistant to the disease. (630–632)
4.C.4 The diversity of species within an ecosystem may influence the stability of the ecosystem.	908–918	

*The chapter correlation is available at **www.mheonline.com/maderAP11** or **www.glencoe.com/maderAP11**

Contents

Contents

The tuatara (*Sphenodon punctatus*).

A View of Life

At first glance, the tuatara of New Zealand appears to be just another lizard. However, recently, scientists have realized that this reptile, which may live as long as 80 years, is something very special. The tuatara represents a living fossil, a remnant of the type of reptiles that lived before the age of the dinosaurs hundreds of millions of years ago. Yet, despite being very old, the tuatara continues to mystify scientists. Some studies suggest that some traits in tuataras are evolving at rates faster than any other vertebrate animal, allowing scientists to use this organism as a test case for how species evolve over long periods of time.

The Earth hosts a wide variety of ecosystems, from which spring a mind-boggling diversity of life, including the tuataras. Even so, all Earth's organisms, regardless of form, are united by a number of common characteristics, such as the need to acquire nutrients, the ability to respond to a changing environment, and to reproduce their own kind. Incredibly, even organisms as diverse as the tuatara and a human being share similar characteristics, including a common chemistry and genetic code. As you read this chapter, reflect on the staggering diversity of life on Earth and on the many ties that bind even the most diverse organisms, from bacteria to the titan arum to humans. It is through these ties that our fates are linked together in the web of life.

As you read through this chapter, think about the following questions

1. What are the general characteristics shared by all living organisms?

2. What is the relationship between evolutionary change and the study of biology?

3. How do scientists use the scientific method to study living organisms?

FOLLOWING *the* BIG IDEAS

CHAPTER 1 A VIEW OF LIFE	
Evolution	The theory of evolution states that living organisms share a common ancestor, explaining their unity, but have changed over time, explaining their great diversity.
Energy and Homeostasis	Organisms share many characteristics, such as basic cellular organization, methods used to maintain homeostasis, and the capability to respond and reproduce.
Information and Signaling	Information encoded in DNA provides the blueprint for diverse forms of life, and when mutated and reshuffled during reproduction, provides the raw materials of evolution.
Interactions and Systems	Living organisms interact with each other and with the physical environment to form populations, communities, and ecosystems which make up the biosphere.

1.1 How to Define Life

> **Learning Outcomes**
>
> Upon completion of this section, you should be able to
> 1. Distinguish between the levels of biological organization.
> 2. Identify the basic characteristics of life.

Biology is the scientific study of life. Life on Earth takes on a staggering variety of forms, often functioning and behaving in ways strange to humans. For example, gastric-brooding frogs swallow their embryos and give birth to them later by throwing them up! Some species of puffballs, a type of fungus, are capable of producing trillions of spores when they reproduce. Fetal sand sharks kill and eat their siblings while still inside their mother. Some *Ophrys* orchids look so much like female bees that male bees try to mate with them. Octopuses and squid have remarkable problem-solving abilities despite a small brain. Some bacteria live their entire life in 15 minutes, while bristlecone pine trees outlive ten generations of humans. Simply put, from the deepest oceanic trenches to the upper reaches of the atmosphere, life is plentiful and diverse.

Figure 1.1 illustrates the major groups of living organisms. From left to right, bacteria are widely distributed, tiny, microscopic organisms with a very simple structure. A *Paramecium* is an example of a microscopic protist. Protists are larger in size and more complex than bacteria. The other organisms in Figure 1.1 are easily seen with the naked eye. They can be distinguished by how they get their food. A morel is a fungus that digests its food externally. A sunflower is a photosynthetic plant that makes its own food, and an octopus is an aquatic animal that ingests its food.

Although life is tremendously diverse, it may be defined by several basic characteristics that are shared by all organisms. Like nonliving things, organisms are composed of chemical elements. Also, organisms obey the same laws of chemistry and physics that govern everything within the universe. The characteristics of life, however, will provide great insight into the unique nature of organisms and will help us distinguish living things from nonliving things.

MP3 Life Characteristics

Living Things Are Organized

The levels of organization depicted in Figure 1.2 begin with atoms, which are the basic units of matter. Atoms combine with other atoms of the same or different elements to form molecules. The **cell,** which is composed of a variety of molecules working together, is the basic unit of structure and function of all living things. Some cells, such as **unicellular** paramecia, live independently. Other cells, for example, the colonial alga *Volvox*, cluster together in microscopic colonies.

Many living things are **multicellular,** meaning they contain more than one cell. In multicellular organisms, similar cells combine to form a tissue; nerve tissue is a common tissue in animals. Tissues make up organs, as when various tissues combine to form the brain. Organs work together in systems; for example, the brain works with the spinal cord and a network of nerves to form the nervous system. Organ systems are joined together to form a complete living thing, or organism, such as an elephant.

The levels of biological organization extend beyond the individual organism. All the members of one species in a particular area belong to a population. A nearby forest may have a population of gray squirrels and a population of white oaks, for example. The populations of various animals and plants in the forest make up a community. The community of populations interacts with the physical environment (water, land, climate) to form an ecosystem. Collectively, all the Earth's ecosystems make up the biosphere.

Emergent Properties

Each level of biological organization builds upon the previous level and is more complex. Moving up the hierarchy, each level acquires new **emergent properties** that are determined by the interactions between the individual parts. When cells are broken down into bits of membrane and liquids, these parts themselves cannot carry out the business of living. For example, you can take apart a lump of coal, rearrange the pieces in any order, and still have a lump of coal with the same function as the original one. But, if you slice apart a living plant and rearrange the pieces, the plant is no longer functional as a complete plant, because it depends on the exact order of those pieces.

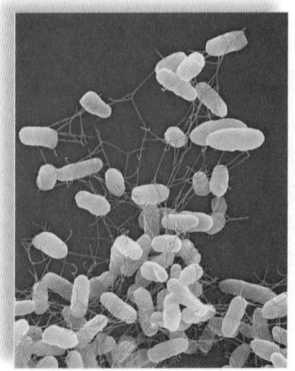   

Bacteria   *Paramecium*   Morel   Sunflower   Octopus

Figure 1.1 Diversity of life.   Biology is the scientific study of life. Many diverse forms of life are found on planet Earth.

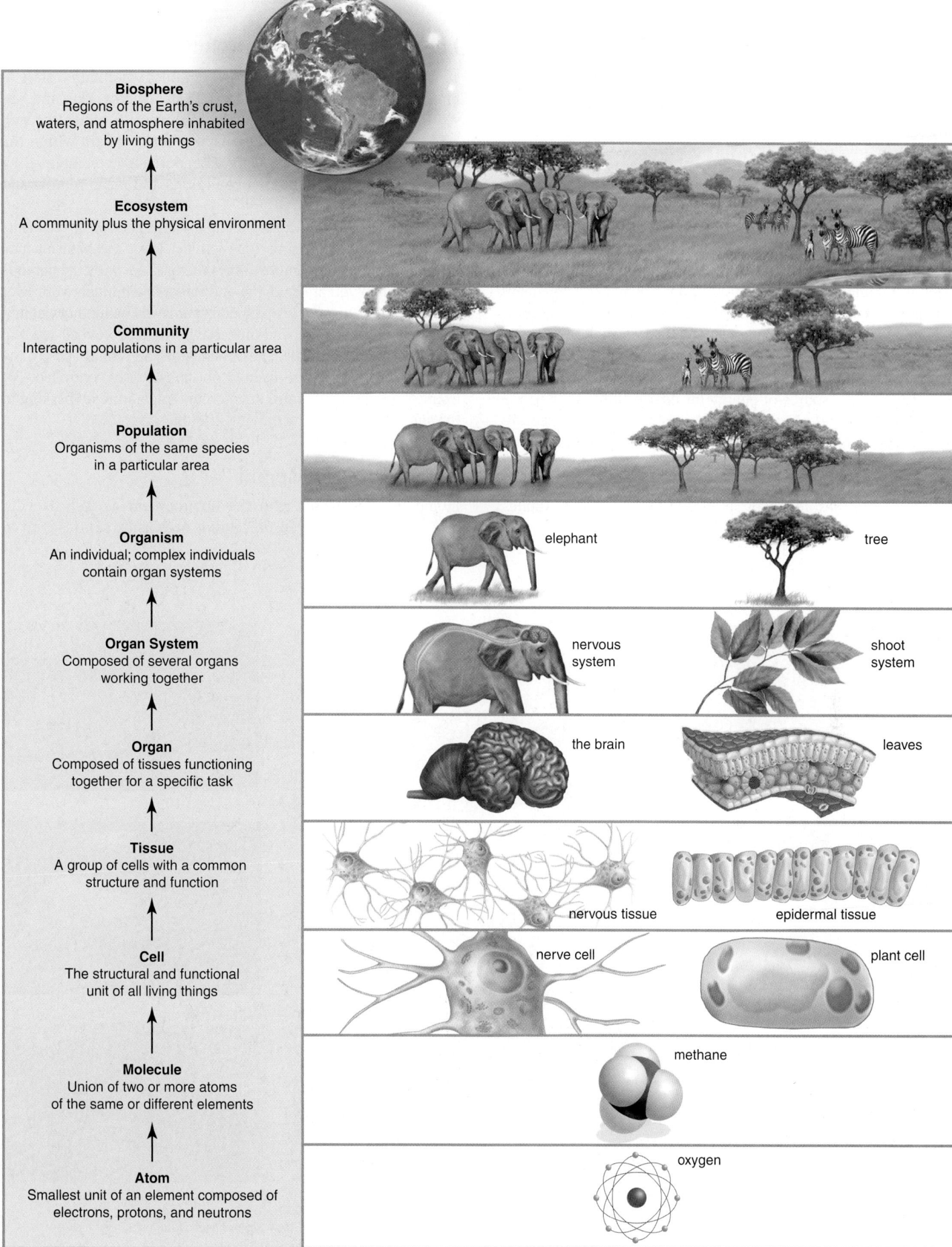

Biosphere
Regions of the Earth's crust, waters, and atmosphere inhabited by living things

Ecosystem
A community plus the physical environment

Community
Interacting populations in a particular area

Population
Organisms of the same species in a particular area

Organism
An individual; complex individuals contain organ systems

elephant

tree

Organ System
Composed of several organs working together

nervous system

shoot system

Organ
Composed of tissues functioning together for a specific task

the brain

leaves

Tissue
A group of cells with a common structure and function

nervous tissue

epidermal tissue

Cell
The structural and functional unit of all living things

nerve cell

plant cell

Molecule
Union of two or more atoms of the same or different elements

methane

Atom
Smallest unit of an element composed of electrons, protons, and neutrons

oxygen

Figure 1.2 Levels of biological organization.

In the living world, the whole is indeed more than the sum of its parts. The emergent properties created by the interactions between levels of biological organization are new, unique characteristics. These properties are governed by the laws of chemistry and physics.

Living Things Acquire Materials and Energy

Living things cannot maintain their organization or carry on life's activities without an outside source of nutrients and energy (Fig. 1.3). Food provides nutrients, which are used as building blocks or for energy. **Energy** is the capacity to do work, and it takes work to maintain the organization of the cell and the organism. When cells use nutrient molecules to make their parts and products, they carry out a sequence of chemical reactions. The term **metabolism** [Gk. *meta*, change] encompasses all the chemical reactions that occur in a cell.

The ultimate source of energy for nearly all life on Earth is the Sun. Plants and certain other organisms are able to capture solar energy and carry on **photosynthesis,** a process that transforms solar energy into the chemical energy of organic nutrient molecules. All life on Earth acquires energy by metabolizing nutrient molecules made by photosynthesizers. This applies even to plants themselves.

Living Things Maintain Homeostasis

To survive, it is imperative that an organism maintain a state of biological balance, or **homeostasis** [Gk. *homoios*, like, and *stasis*, the same]. For life to continue, temperature, moisture level, acidity, and other physiological factors must remain within the tolerance range of the organism. Homeostasis is maintained by systems that monitor internal conditions and make routine and necessary adjustments.

Organisms have intricate feedback and control mechanisms that do not require any conscious activity. These mechanisms may be controlled by one or more tissues themselves, or by the nervous system. When you are studying and forget to eat lunch, your liver releases stored sugar to keep blood sugar levels within normal limits. Many organisms depend on behavior to regulate their internal environment. In animals, these behaviors are controlled by the nervous system, and are usually not consciously controlled. For example, a lizard may raise its internal temperature by basking in the sun, or cool down by moving into the shade.

Living Things Respond

Living things interact with the environment as well as with other living things. Even unicellular organisms can respond to

Figure 1.3 Acquiring nutrients and energy. **a.** An eagle ingesting fish. **b.** A human eating an apple. **c.** A cypress tree capturing sunlight. **d.** An amoeba engulfing food. **e.** A fungus feeding on a tree. **f.** A bison eating grass.

their environment. In some, the beating of microscopic hairs or, in others, the snapping of whiplike tails moves them toward or away from light or chemicals. Multicellular organisms can manage more complex responses. A vulture can detect a carcass a kilometer away and soar toward dinner. A monarch butterfly can sense the approach of Fall and begin its flight south where resources are still abundant.

The ability to respond often results in movement: the leaves of a land plant turn toward the Sun, and animals dart toward safety. Appropriate responses help ensure survival of the organism and allow it to carry on its daily activities. All together, these activities are termed the behavior of the organism. Organisms display a variety of behaviors as they maintain homeostasis and search and compete for energy, nutrients, shelter, and mates. Many organisms display complex communication, hunting, and defense behaviors.

Living Things Reproduce and Develop

Life comes only from life. Every type of living thing can **reproduce,** or make another organism like itself. Bacteria, protists, and other unicellular organisms simply split in two. In most multicellular organisms, the reproductive process begins with the pairing of a sperm from one partner and an egg from the other partner. The union of sperm and egg, followed by many cell divisions, results in an immature stage, which grows and develops through various stages to become the adult.

An embryo develops into a humpback whale or a purple iris because of a blueprint inherited from its parents. The instructions, or blueprint, for an organism's metabolism and organization are encoded in genes. The **genes,** which contain specific information for how the organism is to be ordered, are made of long molecules of DNA (deoxyribonucleic acid). DNA has a shape resembling a spiral staircase with millions of steps. Housed within this spiral staircase is the genetic code that is shared by all living things.

When living things reproduce, their genes are passed on to the next generation. Random combinations of sperm and egg, each of which contains a unique collection of genes, ensure that the new individual has new and different characteristics. The DNA of organisms, over time, also undergoes mutations (changes) that may be passed on to the next generation. These events help to create a staggering diversity of life, even within a group of otherwise identical organisms. Sometimes, organisms inherit characteristics that allow them to be more suited to their way of life.

Living Things Have Adaptations

Adaptations [L. *ad*, toward, and *aptus*, suitable] are modifications that make organisms better able to function in a particular environment. For example, penguins are adapted to an aquatic existence in the Antarctic. An extra layer of downy feathers is covered by short, thick feathers that form a waterproof coat. Layers of blubber also keep the birds warm in cold water. Most birds have forelimbs proportioned for

Figure 1.4 Living Things Have Adaptations.   Penguins have evolved complex behaviors, such as sliding across ice to conserve energy, to adapt to their environment.

flying, but penguins have stubby, flattened wings suitable for swimming. Their feet and tails serve as rudders in the water, but the flat feet also allow them to walk on land. Penguins also have many behavioral adaptations to living in the Antarctic. Penguins often slide on their bellies across the snow in order to conserve energy when moving quickly (Fig. 1.4). Their eggs—one or at most two—are carried on the feet, where they are protected by a pouch of skin. This also allows the birds to huddle together for warmth while standing erect and incubating eggs.

From penguins to fire ants, life on Earth is very diverse because over long periods of time, organisms respond to ever-changing environments by developing new adaptations. These adaptations are unintentional, but they provide the framework for evolutionary change. **Evolution** [L. *evolutio*, an unrolling] includes the way in which populations of organisms change over the course of many generations to become more suited to their environments. All living things have the capacity to evolve, and the process of evolution constantly reshapes every species on the planet, potentially providing a way for organisms to persist, despite a changing environment.

Check Your Progress   1.1

1. Distinguish between an ecosystem and a population in the levels of biological organization.
2. List the common characteristics of all living organisms.
3. Explain how adaptations relate to evolutionary change.

1.2 Evolution, the Unifying Concept of Biology

Learning Outcomes

Upon completion of this section you should be able to

1. Distinguish between the three domains of life.
2. Explain the relationship between the process of natural selection and evolutionary change.

Despite diversity in form, function, and lifestyle, organisms share the same basic characteristics. As mentioned, they are all composed of cells organized in a similar manner. Their genes are composed of DNA, and they carry out the same metabolic reactions to acquire energy and maintain their organization. The unity of living things suggests that they are descended from a common ancestor—the first cell or cells.

An evolutionary tree is like a family tree (Fig. 1.5). Just as a family tree shows how a group of people have descended from one couple, an evolutionary tree traces the ancestry of life on Earth to a common ancestor. One couple can have diverse children, and likewise a population can be a common ancestor to several other groups, each adapted to a particular set of environmental conditions. In this way, over time, diverse life-forms have arisen. Evolution may be considered the unifying concept of biology because it explains so many aspects of biology, including how living organisms arose from a single ancestor.

Organizing Diversity

Because life is so diverse, it is helpful to group organisms into categories. **Taxonomy** [Gk. *tasso*, arrange, and *nomos*, usage] is the discipline of identifying and grouping organisms according to certain rules. Taxonomy makes sense out of the bewildering variety of life on Earth and is meant to provide valuable insight into evolution. **Systematics** is the study of the evolutionary relationships between organisms. As systematists learn more about living things, the taxonomy often changes. DNA technology is now widely used

Table 1.1 Levels of Classification

Category	Human	Corn
Domain	Eukarya	Eukarya
Kingdom	Animalia	Plantae
Phylum	Chordata	Anthophyta
Class	Mammalia	Monocotyledones
Order	Primates	Commelinales
Family	Hominidae	Poaceae
Genus	*Homo*	*Zea*
Species*	*H. sapiens*	*Z. mays*

*To specify an organism, you must use the full binomial name, such as Homo sapiens.

by systematists to revise current information and to discover previously unknown relationships between organisms.

Several of the basic classification categories, or *taxa*, going from least inclusive to most inclusive, are **species, genus, family, order, class, phylum, kingdom,** and **domain** (Table 1.1). The least inclusive category, species [L. *species*, model, kind], is defined as a group of interbreeding individuals. Each successive classification category above species contains more types of organisms than the preceding one. Species placed within one genus share many specific characteristics and are the most closely related, while species placed in the same kingdom share only general characteristics with one another. For example, all species in the genus *Pisum* look

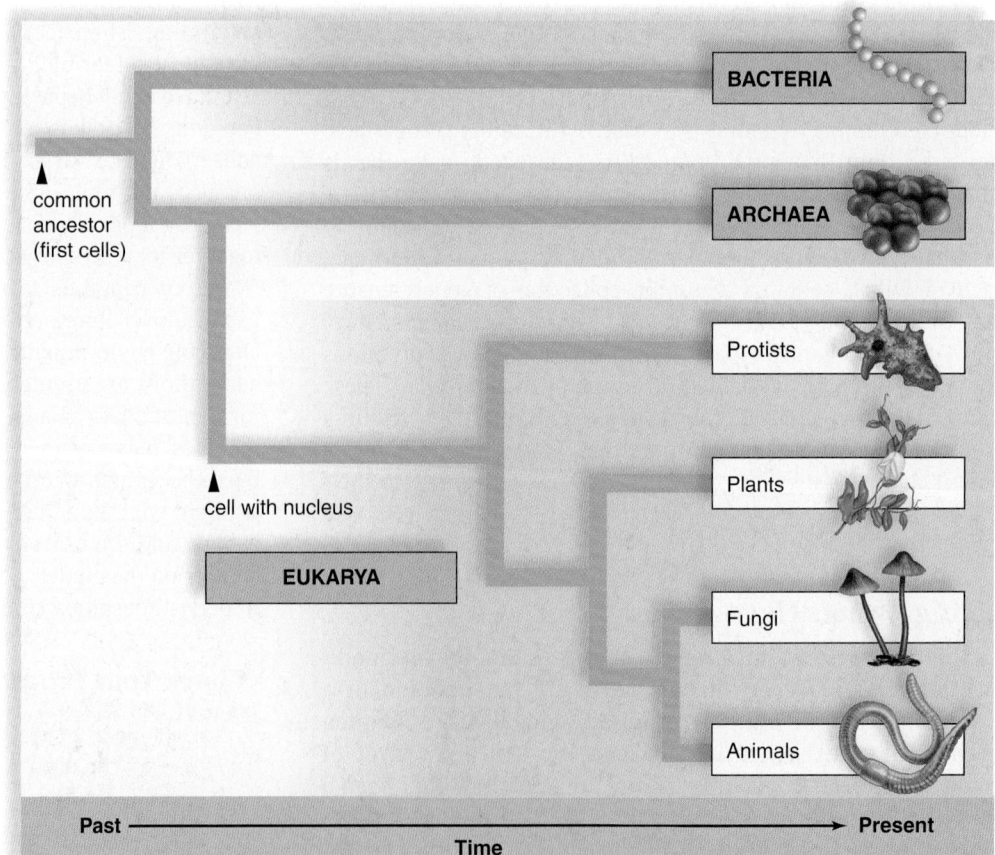

Figure 1.5 Evolutionary tree of life. As existing organisms change over time, they give rise to new species. Evolutionary studies show that all living organisms arose from a common ancestor about 4 billion years ago. Domain Archaea and domain Bacteria include the prokaryotes. Domain Eukarya includes both unicellular and multicellular organisms that possess a membrane-bounded nucleus.

pretty much the same—that is, like pea plants—but species in the plant kingdom can be quite varied, as is evident when we compare grasses to trees. Species placed in different domains are the most distantly related.

Domains

Biochemical evidence suggests that there are only three domains: **domain Bacteria, domain Archaea,** and **domain Eukarya.** Figure 1.5 shows how the domains are believed to be related. Both domain Bacteria and domain Archaea may have evolved from the first common ancestor soon after life began. These two domains contain the **prokaryotes,** which lack the membrane-bounded nucleus found in the **eukaryotes** of domain Eukarya. However, archaea organize their DNA differently than bacteria, and their cell walls and membranes are chemically more similar to eukaryotes than to bacteria. So, the conclusion is that eukarya split off from the archaeal line of descent.

Animation
Three Domains

Prokaryotes are structurally simple but metabolically complex. Archaea (Fig. 1.6) can live in aquatic environments that lack oxygen or are too salty, too hot, or too acidic for most other organisms. Perhaps these environments are similar to those of the primitive Earth, and archaea (Gk. *archae*, ancient) are the least evolved forms of life, as their name implies. Bacteria (Fig. 1.7) are variously adapted to living almost anywhere—in the water, soil, and atmosphere, as well as on our skin and in our mouths and large intestines.

Taxonomists are in the process of deciding how to categorize archaea and bacteria into kingdoms. Domain Eukarya, on the other hand, contains four major groups of organisms (Fig. 1.8). **Protists,** which now comprise a number of kingdoms, range from unicellular forms to a few multicellular ones. Some are photosynthesizers, and some must acquire their food. Common protists include algae, the protozoans, and the water molds. Figure 1.5 shows that plants, fungi, and animals most likely evolved from

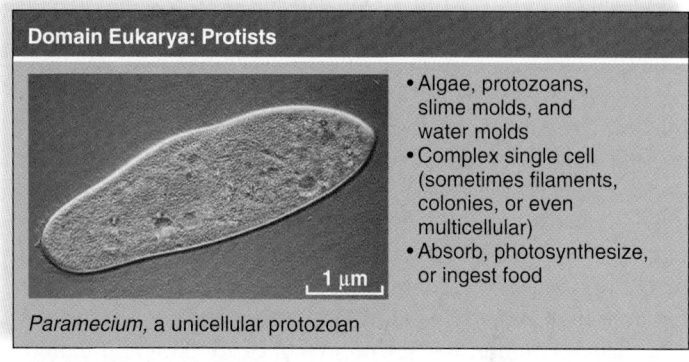

Domain Eukarya: Protists

- Algae, protozoans, slime molds, and water molds
- Complex single cell (sometimes filaments, colonies, or even multicellular)
- Absorb, photosynthesize, or ingest food

1 μm

Paramecium, a unicellular protozoan

Domain Eukarya: Kingdom Fungi

- Molds, mushrooms, yeasts, and ringworms
- Mostly multicellular filaments with specialized, complex cells
- Absorb food

Amanita, a mushroom

Domain Eukarya: Kingdom Plantae

- Certain algae, mosses, ferns, conifers, and flowering plants
- Multicellular, usually with specialized tissues, containing complex cells
- Photosynthesize food

Passiflora, passion flower, a flowering plant

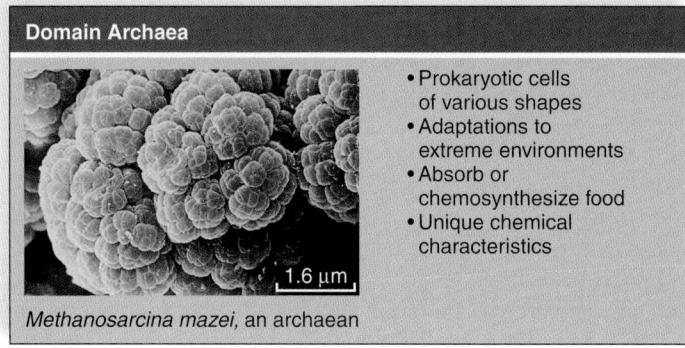

Domain Archaea

- Prokaryotic cells of various shapes
- Adaptations to extreme environments
- Absorb or chemosynthesize food
- Unique chemical characteristics

1.6 μm

Methanosarcina mazei, an archaean

Figure 1.6 Domain Archaea.

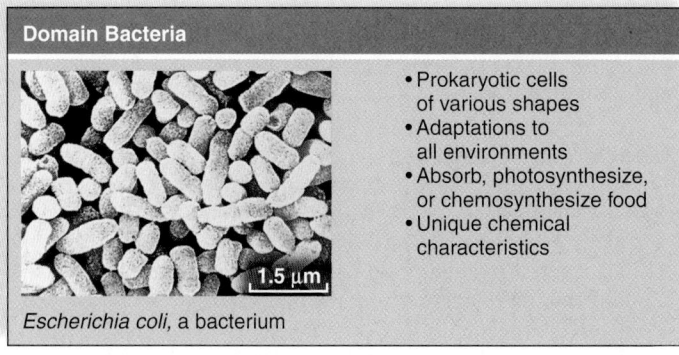

Domain Bacteria

- Prokaryotic cells of various shapes
- Adaptations to all environments
- Absorb, photosynthesize, or chemosynthesize food
- Unique chemical characteristics

1.5 μm

Escherichia coli, a bacterium

Figure 1.7 Domain Bacteria.

Domain Eukarya: Kingdom Animalia

- Sponges, worms, insects, fishes, frogs, turtles, birds, and mammals
- Multicellular with specialized tissues containing complex cells
- Ingest food

Vulpes, a red fox

Figure 1.8 Domain Eukarya.

protists. **Plants** (kingdom Plantae) are multicellular photosynthetic organisms. Example plants include azaleas, zinnias, and pines. Among the **fungi** (kingdom Fungi) are the familiar molds and mushrooms that, along with bacteria, help decompose dead organisms. **Animals** (kingdom Animalia) are multicellular organisms that must ingest and process their food. Aardvarks, jellyfish, and zebras are representative animals.

Scientific Name

Biologists use **binomial nomenclature** [L. *bi*, two, and *nomen*, name] to assign each living thing a two-part name called a scientific name. For example, the scientific name for mistletoe is *Phoradendron tomentosum*. The first word is the genus, and the second word is the species designation (or *specific epithet*) of each species within a genus. The genus may be abbreviated (e.g., *P. tomentosum*) and if the species has not been determined, it may simply be indicated with a generic abbreviation (e.g., *Phoradendron* sp.). Scientific names are universally used by biologists to avoid confusion. Common names tend to overlap and often differ depending on locality and the language of a particular country. But scientific names are based on Latin, a universally used language that not too long ago was well known by most scholars.

Evolution Is Common Descent with Modification

The phrase "common descent with modification" sums up the process of evolution because it means that as descent occurs from common ancestors, so do modifications that cause organisms to be adapted to their environment. Through many observations and experiments, Charles Darwin came to the conclusion that **natural selection** was the process that made modification—that is, adaptation—possible.

Natural Selection

During the process of natural selection, some aspect of the environment selects which traits are more apt to be passed on to the next generation. The selective agent can be an abiotic agent (part of the physical environment, such as altitude) or it can be a biotic agent (part of the living environment, such as a deer). Figure 1.9 shows how the dietary habits of deer might eventually affect the characteristics of the leaves of a particular land plant.

Mutations fuel natural selection because mutation introduces variations among the members of a population. In Figure 1.9, a plant species generally produces smooth leaves, but a mutation occurs that causes one plant to have leaves that are covered with small extensions or "hairs." The plant with hairy leaves has an advantage because the deer (the selective agent) prefer to eat smooth leaves and not hairy leaves. Therefore, the plant with hairy leaves survives best and produces more seeds than most of its neighbors. As a result, generations later most plants of this species produce hairy leaves.

As with this example, Darwin realized that although all individuals within a population have the potential to reproduce, not all do so with the same success. Prevention of reproduction can run the gamut from an inability to capture resources, as when

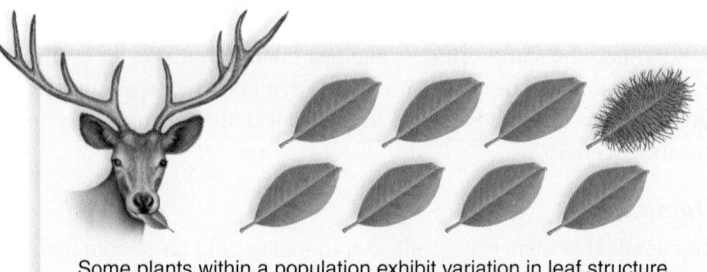

Some plants within a population exhibit variation in leaf structure.

Deer prefer a diet of smooth leaves over hairy leaves. Plants with hairy leaves reproduce more than other plants in the population.

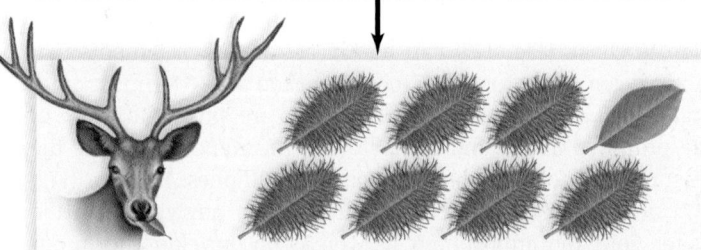

Generations later, most plants within the population have hairy leaves, as smooth leaves are selected against.

Figure 1.9 Natural selection. Natural selection selects for or against new traits introduced into a population by mutations. Over many generations, selective forces such as competition, predation, and the physical environment alter the makeup of a population, favoring those more suited to the environment and lifestyle.

long-necked, but not short-necked, giraffes can reach their food source, to an inability to escape being eaten because long legs, but not short legs, can carry an animal to safety.

Whatever the example, it can be seen that living things with advantageous traits can produce more offspring than those that lack them. In this way, living things change over time, and these changes are passed on from one generation to the next. Over long periods of time, the introduction of newer, more advantageous traits into a population may drastically reshape a species. Natural selection tends to sculpt a species to fit its environment and lifestyle and can create new species from existing ones. The end result is the diversity of life classified into the three domains of life (see Fig. 1.5).

Video
Finches—Natural Selection

Check Your Progress   1.2

1. List the levels of taxonomic classification from most inclusive to least inclusive.
2. Describe the differences that might be used to distinguish the various kingdoms of domain Eukarya.
3. Explain how natural selection results in new adaptations within a species.

1.3 How the Biosphere Is Organized

Learning Outcomes

Upon completion of this section you should be able to

1. Distinguish among populations, communities, ecosystems, and the biosphere.
2. Recognize the importance of maintaining biodiversity.

The organization of life extends beyond the individual organism to the **biosphere,** the zone of air, land, and water at the surface of the Earth where organisms exist (see Fig. 1.2). Individual organisms belong to a **population,** which is all the members of a species within a particular area. The populations of a **community** interact among themselves and with the physical environment (e.g., soil, atmosphere, and chemicals), thereby forming an **ecosystem.**

Figure 1.10 depicts a grassland inhabited by populations of rabbits, mice, snakes, hawks, and various types of land plants. These populations exchange gases with and give off heat to the atmosphere. They also take in water from and give off water to the physical environment. In addition, populations interact by forming food chains in which one population feeds on another. Mice feed on plants and seeds, snakes feed on mice, and hawks feed on rabbits and snakes, for example. Interactions between the various food chains make up a food web.

Ecosystems are characterized by chemical cycling and energy flow, both of which begin when photosynthetic plants, aquatic algae, and some bacteria take in solar energy and inorganic nutrients to produce food in the form of organic nutrients. The gray arrows in Figure 1.10 represent chemical cycling—chemicals move from one population to another in a food chain, until with death and decomposition, inorganic nutrients are returned to living plants once again. The yellow to red arrows represent energy flow. Energy flows from the Sun through plants and other members of the food chain as one population feeds on another. With each transfer some energy is lost as heat. Eventually, all the energy taken in by photosynthesizers has dissipated into the atmosphere. Because energy flows and does not cycle, ecosystems could not stay in existence without a constant input of solar energy and the ability of photosynthesizers to absorb it.

Video Tallgrass Prairie Ecology

The Human Population

Humans possess the unique ability to modify existing ecosystems, which can greatly upset their natural nutrient cycles. When an ecosystem's natural energy flow has been disrupted by eliminating food sources for other animal populations even the human population can eventually suffer harm. Humans clear forests or grasslands to grow crops; later, they build houses on what was once farmland; and finally, they convert small towns into cities. Coastal ecosystems are most vulnerable. As

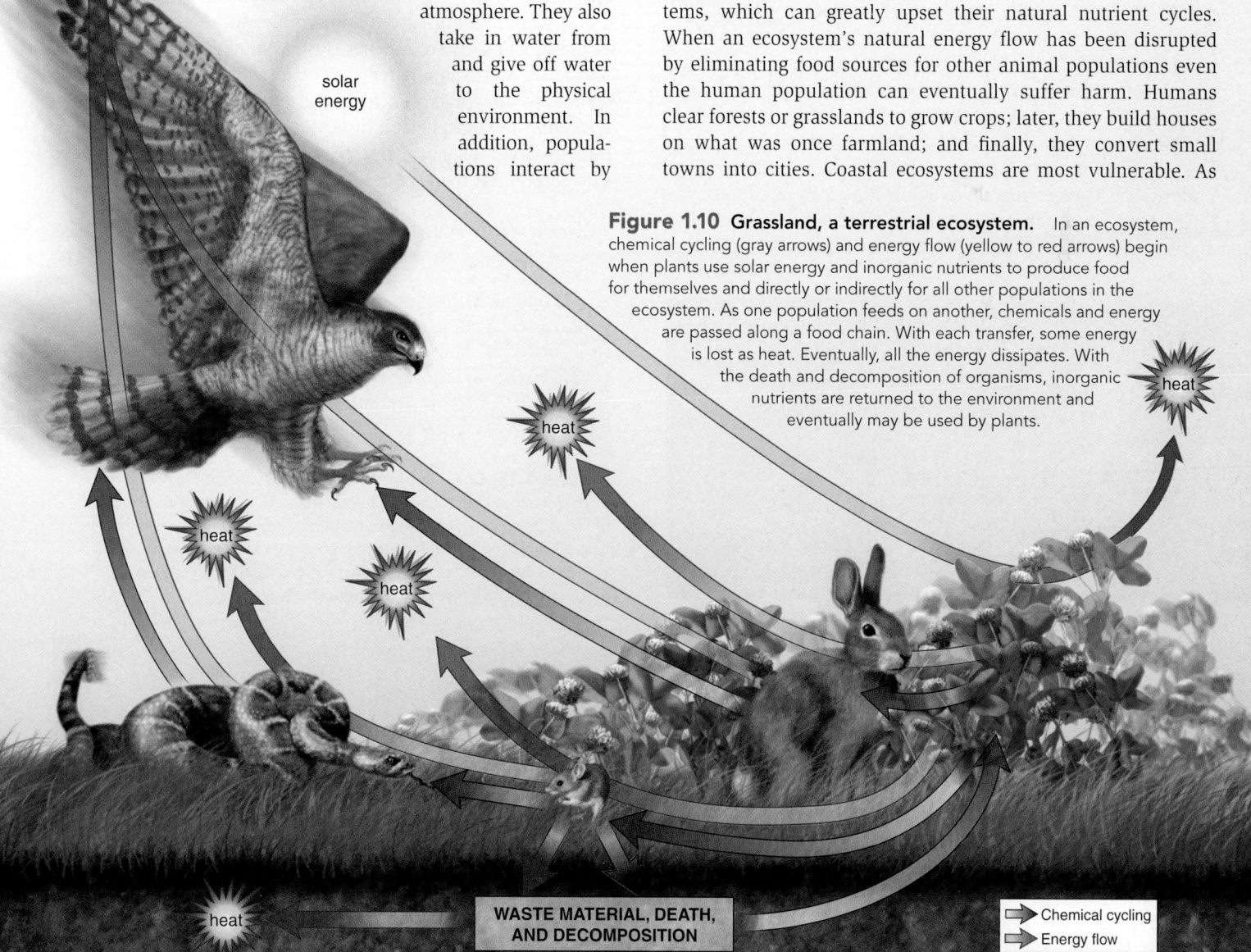

Figure 1.10 Grassland, a terrestrial ecosystem. In an ecosystem, chemical cycling (gray arrows) and energy flow (yellow to red arrows) begin when plants use solar energy and inorganic nutrients to produce food for themselves and directly or indirectly for all other populations in the ecosystem. As one population feeds on another, chemicals and energy are passed along a food chain. With each transfer, some energy is lost as heat. Eventually, all the energy dissipates. With the death and decomposition of organisms, inorganic nutrients are returned to the environment and eventually may be used by plants.

heat

solar energy

heat

heat

heat

heat

heat

heat

WASTE MATERIAL, DEATH, AND DECOMPOSITION

⇨ Chemical cycling
⇨ Energy flow

they are developed, humans send sediments, sewage, and other pollutants into the sea. Human activities destroy valuable coastal wetlands, which serve as protection against storms and as nurseries for a myriad of invertebrates and vertebrates.

Biodiversity

The two most biologically diverse ecosystems—tropical rain forests and coral reefs—are home to many organisms. These ecosystems are also threatened by human activities. The canopy of the tropical rain forest alone supports a variety of organisms including orchids, insects, and monkeys. Coral reefs, which are found just offshore of the continents and islands of the Southern Hemisphere, are built up from calcium carbonate skeletons of sea animals called corals. Reefs provide a habitat for many animals, including jellyfish, sponges, snails, crabs, lobsters, sea turtles, moray eels, and some of the world's most colorful fishes (Fig. 1.11a). Like tropical rain forests, coral reefs are severely threatened as the human population increases in size. Some reefs are 50 million years old, and yet in just a few decades, human activities have destroyed an estimated 25% of all coral reefs and seriously degraded another 30% (Fig. 1.11b). At this rate, nearly three-quarters could be destroyed within 40 years. Similar statistics are available for tropical rain forests.

Video Coral Reef Ecosystems

Destruction of healthy ecosystems has many unintended effects. For example, we depend on them for food, medicines, and various raw materials. Draining the natural wetlands of the Mississippi and Ohio rivers and the construction of levees has worsened flooding problems, making once fertile farmland undesirable. The destruction of South American rain forests has killed many species that may have yielded the next miracle drug and has also decreased the availability of many types of lumber.

We are only now beginning to realize that we depend on ecosystems even more for the services they provide. Just as chemical cycling occurs within a single ecosystem, so all ecosystems keep chemicals cycling throughout the entire biosphere. The workings of ecosystems ensure that the environmental conditions of the biosphere are suitable for the continued existence of humans. And several studies show that ecosystems cannot function properly unless they remain biologically diverse.

Biodiversity is the total number and relative abundance of species, the variability of their genes, and the different ecosystems in which they live. The present biodiversity of our planet has been estimated to be as high as 15 million species, and so far, less than 2 million have been identified and named. **Extinction** is the death of a species or larger classification category. It is estimated that presently we are losing as many as 400 species per day due to human activities and that as much as 38% of all species, including most primates, birds, and amphibians, may be in danger of extinction before the end of the century. Many biologists are alarmed about the present rate of extinction and hypothesize it may eventually rival the rates of the five mass extinctions that have occurred during our planet's history. The last mass extinction, about 65 million years ago, caused many plant and animal species, including the dinosaurs, to become extinct.

It would seem that the primary bioethical issue of our time is preservation of ecosystems. Just as a native fisherman who assists in overfishing a reef is doing away with his own food source, so are we as a society contributing to the destruction of our home, the biosphere. If instead we adopt a conservation ethic that preserves the biosphere, we would help ensure the continued existence of our own species.

Check Your Progress   1.3

1. Explain the relationship between a population, a community, and an ecosystem.
2. Describe some unintentional ways in which human activities affect ecosystems.
3. Discuss why ecosystems with high biodiversity might be more vulnerable to destruction by human activities.

a. Healthy coral reef

Figure 1.11 Coral reef, a marine ecosystem.   a. Coral reefs, a type of ecosystem found in tropical seas, contain many diverse forms of life, a few of which are shown here. **b.** Various human activities have caused catastrophic damage to this coral reef off the coast of Florida, as shown over the course of 19 years. Preserving biodiversity is a modern-day challenge of great proportions.

1975 Minimal coral death | 1985 Some coral death with no fish present | 1995 Coral bleaching with limited chance of recovery | 2004 Coral is black from sedimentation; bleaching still evident

b.

1.4 The Process of Science

Learning Outcomes

Upon completion of this section you should be able to
1. Identify the components of the scientific method.
2. Distinguish between a theory and a hypothesis.
3. Analyze a scientific experiment and identify the hypothesis, experiment, control groups, and conclusions.

The process of science pertains to the study of biology. Biology consists of many disciplines and areas of specialty because life has numerous aspects. Some biological disciplines are cytology, the study of cells; anatomy, the study of structure; physiology, the study of function; botany, the study of plants; zoology, the study of animals; genetics, the study of heredity; and ecology, the study of the interrelationships between organisms and their environment.

Religion, aesthetics, ethics, and science are all ways in which human beings seek order in the natural world. Science differs from these other ways of knowing and learning because the scientific process uses the **scientific method,** a standard series of steps used in gaining new knowledge that is widely accepted among scientists. The steps of the scientific method are often applicable to other situations, and begin with observation (Fig. 1.12).

Observation

Scientists believe that nature is orderly and measurable—that natural laws, such as the law of gravity, do not change with time, and that a natural event, or **phenomenon,** can be understood more fully through **observation**—a formal way of "seeing what happens."

Scientists use all of their senses in making observations. The behavior of chimpanzees can be observed through visual means, the disposition of a skunk can be observed through olfactory means, and the warning rattles of a rattlesnake provide auditory information of imminent danger. Scientists also extend the ability of their senses by using instruments; for example, the microscope enables us to see objects that could never be seen by the naked eye. Finally, scientists may expand their understanding even further by taking advantage of the knowledge and experiences of other scientists. For instance, they may look up past studies at the library or on the Internet, or they may write or speak to others who are researching similar topics.

Hypothesis

After making observations and gathering knowledge about a phenomenon, a scientist uses inductive reasoning to formulate a possible explanation. **Inductive reasoning** occurs whenever a person uses creative thinking to combine isolated facts into a cohesive whole. In some cases, chance alone may help a scientist arrive at an idea.

One famous case pertains to the antibiotic penicillin, which was discovered in 1928. While examining a petri dish of bacteria that had become contaminated with the mold *Penicillium,*

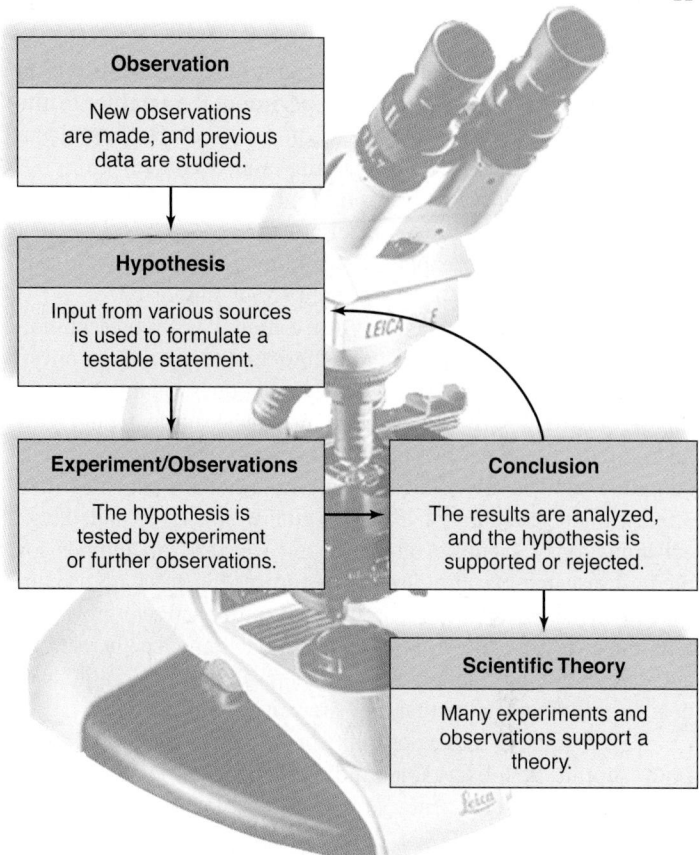

Figure 1.12 Flow diagram for the scientific method. On the basis of new and/or previous observations, a scientist formulates a hypothesis, which is then tested by further observations and/or experiments. Experiments provide data that either support or do not support the hypothesis. The return arrow indicates that a scientist often chooses to retest the same hypothesis or to test a related hypothesis. Conclusions from many different but related experiments may lead to the development of a scientific theory.

Alexander Flemming observed an area that was free of bacteria. Flemming, an early expert on antibacterial substances, reasoned that the mold might be producing an antibacterial compound.

We call such a possible explanation for a natural event a **hypothesis.** A hypothesis is not merely a guess; rather, it is an informed statement that can be tested in a manner suited to the processes of science.

All of a scientist's past experiences, no matter what they might be, have the potential to influence the formation of a hypothesis. But a scientist considers only hypotheses that can be tested. Moral and religious beliefs, while very important in the lives of many people, differ between cultures and through time, and these may not be scientifically testable.

Experiments, Observations, and Data

Scientists often perform an **experiment,** which is a series of procedures, to test a hypothesis. To determine how to test a hypothesis, a scientist uses deductive reasoning. **Deductive reasoning** involves "if, then" logic. In designing the experiment, the scientist may make a **prediction,** or an expected outcome, based on knowledge of the factors in the experiment.

The manner in which a scientist intends to conduct an experiment is called the **experimental design.** A good experimental design ensures that scientists are examining the contribution of a specific variable, called the **experimental variable**, to the observation. To ensure that the results will be meaningful, an experiment contains both test groups and a **control** group. A test group is exposed to the environmental variable, but the control group is not. If the control groups and test groups show the same results, the experimenter knows that the hypothesis predicting a difference between them is not supported.

In some cases, scientists may use a **model** as a representation of the actual object because altering the actual object may be physically impossible, very expensive, or morally questionable. Later in this section, a scientist uses bluebird models because it would have been impossible to get live birds to cooperate. As other examples, computer models are used to decide how human activities may affect climate, because of expense, ethical concerns, and physical limitations. Scientists often use mice instead of humans for medical research because of ethical concerns. Bacteria are used in much genetic research because they are inexpensive to grow and reproduce very quickly. Although these models are usually relevant and give useful information, they are themselves still hypotheses in need of testing to ensure that they are valid representations.

The results of an experiment are referred to as the **data.** Data should be observable and objective, rather than subjective. Mathematical data are often displayed in the form of a graph or table. Often, these results include a **standard deviation**, a statistical analysis that is a measure of how much the data in the experiment varies. Many studies, such as the one discussed in the Nature of Science feature on page 13, rely on statistical data.

An Example

As a hypothetical example, let's say an investigator wants to know if eating onions can prevent women from getting osteoporosis (weak bones). The scientist conducts a survey asking women about their onion-eating habits and then correlates this data with the condition of their bones. Other scientists critiquing this study would want to know: How many women were surveyed? How old were the women? What were their exercise habits? What proportion of the diet consisted of onions? And what criteria were used to determine the condition of their bones?

If the investigators conclude that eating onions does protect a woman from osteoporosis, then other scientists might want to know the statistical probability of error. The probability of error is a mathematical calculation based on the conditions and methods of the experiment. If the results are significant at a 0.30 level, then the probability that the correlation is incorrect is 30% or less. (This would be considered a high probability of error.) The greater the variance in the data, the greater the probability of error.

Even if this study had a low probability of error, it would be considered hypothetical until scientists identified one or more active components in onions that have a direct biochemical or physiological effect on bones. Therefore, scientists must be skeptics who always pressure one another to continue investigating a particular topic.

Conclusion

Scientists must analyze the data in order to reach a **conclusion** as to whether the hypothesis is supported or not (see Fig. 1.12). Because science progresses, the conclusion of one experiment can lead to the hypothesis for another experiment, as represented by the return arrow in Figure 1.12. Results that do not support one hypothesis can often help a scientist formulate another hypothesis to be tested.

Scientists report their findings in scientific journals so that their methodology and data are available to other scientists for critique. Experiments and observations must be repeatable—that is, the reporting scientist and any scientist who repeats the experiment must get the same results, or else the data are suspect.

Scientific Theory

The ultimate goal of science is to understand the natural world in terms of **scientific theories,** which are concepts that join together well-supported and related hypotheses. In ordinary speech, the word *theory* refers to a speculative idea. In contrast, a scientific theory is supported by a broad range of observations, experiments, and data often from a variety of disciplines. Some of the basic theories of biology are:

Theory	*Concept*
Cell	All organisms are composed of cells, and new cells come only from preexisting cells.
Homeostasis	The internal environment of an organism stays relatively constant—within a range that is protective of life.
Gene	Organisms contain coded information that dictates their form, function, and behavior.
Ecosystem	Organisms are members of populations, which interact with each other and the physical environment within a particular locale.
Evolution	All living things have a common ancestor, but each is adapted to a particular way of life.

As stated earlier, the theory of evolution is the unifying concept of biology because it pertains to many different aspects of living things. For example, the theory of evolution enables scientists to understand the history of living things, and the anatomy, physiology, and embryological development of organisms. Even behavior can be described through evolution, as we shall see in a study discussed later in this chapter.

The theory of evolution has been a fruitful scientific theory, meaning that it has helped scientists generate new hypotheses. Because this theory has been supported by so many observations and experiments for over 100 years, some biologists refer to the **principle** of evolution, a term sometimes used for theories that are generally accepted by an overwhelming number of scientists. The term **law** instead of principle is preferred by some. For instance, in a subsequent chapter concerning energy relationships, we will examine the laws of thermodynamics.

Nature of Science

The Benefits and Limitations of Statistical Studies

Many of the studies published in scientific journals and reported in the news are statistical studies, so it behooves us to be aware of their benefits and limitations. At the start, you should know that a statistical study will gather numerical information from various sources and then try to make sense out of it, for the purpose of coming to a conclusion.

Example of a Statistical Study

Let's take a look at a study that allows us to conclude that babies conceived 18 months to five years after a previous birth are healthier than those conceived at shorter or longer intervals. In other words, spacing children about two to five years apart is a good idea (Fig. 1A). Here is how the authors collected their data and the results they published in the *Journal of the American Medical Association*.[1]

Objective. To determine whether there is an association between birth spacing and a healthy baby when data are corrected for maternal characteristics or socioeconomic status.

Data. The authors collected data from studies performed around the world in 1966 through January 2006. The studies were published in various journals, reported on at professional meetings, or were known to the authors by personal contact. The authors gathered a very large pool of data that included over 11 million pregnancies from 67 individual studies. Twenty of the studies were from the United States, with the remaining 47 coming from 61 different countries. The authors attempted to adjust the data (by elimination of certain data) for factors such as mother's age, wealth, access to prenatal care, and breast-feeding. These adjustments allow the findings to be applied to both developed and developing countries.

Conclusion.

1. A pregnancy that begins less than six months after a previous birth has a 77% higher chance of being preterm and a 39% higher chance of lower birth weight.
2. For up to 18 months between pregnancies, the chance of a preterm birth decreases by 2% per month, and the chance of a low-weight birth decreases

Figure 1A Does spacing pregnancies lead to healthier children? A recent statistical study suggests that it does. If so, which mother, left or right, may have a healthier younger child?

by 3% per month as the 18-month time period is approached.
3. Babies conceived after 59 months have the same risk as those conceived in the less-than-six-months group.
4. The optimum spacing between pregnancies appears to be 18 months to five years after a previous birth.

The study leader, Agustin Conde-Agudelo, said, "Health officials should counsel women who have just given birth to delay their next conception by 18 to 59 months."

Limitations of Experimental Studies

The expression "statistical study" is a bit of a misnomer because most scientists collect quantitative data and use them to come to a conclusion. However, if we compare this study to experimental studies, we can see that the experimental studies include both a control group and test groups. The groups are treated the same except for the experimental variable. Obviously, you wouldn't be able to divide women of the same childbearing age into various groups and tell each group when they will conceive their children for the purpose of deciding the best interval between pregnancies for the health of the newborn. So, what is the next best thing? Do a statistical study utilizing data already available about women who became pregnant at different intervals.

A statistical study is really a correlation study. In our example, the authors studied the correlation between birth spacing and the health of a newborn. The more data collected from more varied sources make a correlation study more reliable. The study by Conde-Agudelo has a very large sample size, which goes a long way to validating the results. Even so, a correlation does not necessarily translate to causation. So, it is not surprising that Dr. Mark A. Klebanoff, director of the National Institute of Child Health and Human Development, commented that many factors will affect birth spacing and that the study is not detailed enough to take all factors into consideration. Is any statistical study detailed enough? Most likely not.

Benefits of Statistical Studies

Before we give up on statistical studies, let's consider that they do provide us with information not attainable otherwise. Regardless of whether we understand the intricacies of statistical analysis, statistical studies do allow scientists to gain information and insights into many problems. True, further study is needed to find out if the observed correlation does mean causation, but science is always a work in progress, with additional findings being published every day.

Questions to Consider

1. Why is a large sample size needed in a statistical study? Why might this be a challenge for scientists?
2. How might both statistical and experimental studies be combined to enhance the validity of a scientific study?

1. Conde-Agudelo, A., Rosas-Bermudez, A., and Kafury-Goeta, A. C. 2006. Birth spacing and risk of adverse perinatal outcomes. Abstract. *JAMA* 295: 1809–23.

Using the Scientific Method

This section demonstrates the use of the scientific method in a controlled study to ensure that the outcome is due to the experimental variable, or independent variable—the component or factor being tested. The result is termed the **responding variable,** or dependent variable, because it is due to the experimental variable:

Experimental Variable (Independent Variable)	**Responding Variable** (Dependent Variable)
Factor of the experiment being tested	Result or change that occurs due to the experimental variable

Observation

Researchers doing this study knew that in the short run, nitrogen fertilizer enhances yield and increases food supplies. However, excessive nitrogen fertilizer application can cause pollution by adding toxic levels of nitrates to water supplies. Also, applying nitrogen fertilizer year after year may alter soil properties to the point that crop yields may decrease instead of increase. At that point, the only solution is to let the land remain unplanted for several years until the soil recovers naturally.

An alternative to the use of nitrogen fertilizers is the use of legumes, plants such as peas and beans, that increase soil nitrogen. Legumes provide a home for bacteria that convert atmospheric nitrogen to a form usable by the plant. The bacteria live in nodules on the roots (Fig. 1.13). The bacteria supply the plant with nitrogen compounds, and in turn, the plant passes the product of photosyntheis to the nodules.

Numerous legume crops can be rotated (planted every other season) with any number of cereal crops. The nitrogen added to the soil by the legume crop is a natural fertilizer that increases the yield of cereal crops. The particular rotation used by farmers tends to depend on the location, climate, and market demand. In this study, researchers performed an experiment in which method of fertilization is the experimental variable and enhanced yield is the responding variable.

Figure 1.13 Root nodules.   Bacteria that live in nodules on the roots of legumes, such as pea plants, convert nitrogen in the air to a form that land plants can use to make proteins and other nitrogen-containing molecules.

Hypothesis

Researchers doing this study knew that the pigeon pea plant is a legume with a high rate of atmospheric nitrogen conversion. This plant is widely grown as a food crop in India, Kenya, Uganda, Pakistan, and other subtropical countries. Researchers formulated the hypothesis that a pigeon pea/winter wheat rotation would be a reasonable alternative to the use of nitrogen fertilizer to increase the yield of winter wheat.

HYPOTHESIS: A pigeon pea/winter wheat rotation will cause winter wheat production to increase as well as or better than the use of nitrogen fertilizer.

PREDICTION: Wheat biomass following the growth of pigeon peas will surpass wheat biomass following nitrogen fertilizer treatment.

Experiment and Data

In this study, the investigators decided on the following experimental design (Fig. 1.14a):

CONTROL POTS
- Winter wheat was planted in pots of soil that received no fertilization treatment—that is, no nitrogen fertilizer and no preplanting of pigeon peas.

TEST POTS
- Winter wheat was grown in clay pots in soil treated with nitrogen fertilizer equivalent to 45 kilograms (kg)/hectare (ha).
- Winter wheat was grown in clay pots in soil treated with nitrogen fertilizer equivalent to 90 kg/ha.
- Pigeon pea plants were grown in clay pots in the summer. The pigeon pea plants were then tilled into the soil and winter wheat was planted in the same pots.

To ensure a controlled experiment, the conditions for the control pots and the test pots were identical; the plants were exposed to the same environmental conditions and watered equally. During the following spring, the wheat plants were dried and weighed to determine wheat biomass production in each of the pots.

After the first year, wheat biomass was higher in certain test pots than in the control pots (Fig. 1.14b). Specifically, test pots with 45 kg/ha of nitrogen fertilizer (orange) had only slightly more wheat biomass production than the control pots, but test pots that received 90 kg/ha treatment (green) demonstrated nearly twice the biomass production of the control pots. To the surprise of investigators, wheat production following summer planting of pigeon peas (brown) did not demonstrate as high a biomass production as the control pots.

Conclusion and Further Investigation

Wheat biomass following the growth of pigeon peas is not as great as that obtained with nitrogen fertilizer treatments, meaning that the data from the experiment did not support the investigators' hypothesis. This is not an uncommon event in scientific investigations. However, the investigators decided to continue the experiment using the same design and the same pots as before, to see if the buildup of residual soil nitrogen from pigeon peas would eventually increase wheat biomass. So they proposed a new hypothesis.

Figure 1.14 Pigeon pea/winter wheat rotation study.
a. Experiment involves control pots and test pots of three types: test pots that received 45 kg/ha of nitrogen; test pots that received 90 kg/ha of nitrogen; and test pots in which pigeon peas rotated with winter wheat. **b.** The graph compares wheat biomass for each of three years. Wheat biomass in test pots that received the most nitrogen fertilizer (green) declined while wheat biomass in test pots with pigeon pea/winter wheat rotation (brown) increased dramatically.

b. Results

HYPOTHESIS: A sustained pigeon pea/winter wheat rotation will eventually cause an increase in winter wheat production.

PREDICTION: Wheat biomass following two years of pigeon pea/winter wheat rotation will surpass wheat biomass following nitrogen fertilizer treatment.

After two years, the yield following 90 kg/ha nitrogen treatment (green) was not as much as it was the first year (Fig. 1.14b). Indeed, wheat biomass following summer planting of pigeon peas (brown) was the highest of all treatments, suggesting that buildup of residual nitrogen from pigeon peas had the potential to provide fertilization for winter wheat growth.

CONCLUSION: The hypothesis is supported. At the end of two years, the yield of winter wheat following a pigeon pea/winter wheat rotation was better than for the other type pots.

The researchers continued their experiment for still another year. After three years, winter wheat biomass production had decreased in the control pots and in the pots treated with nitrogen fertilizer. Pots treated with nitrogen fertilizer still had increased wheat biomass production compared with the control

pots, but not nearly as much as pots following summer planting of pigeon peas. Compared to the first year, wheat biomass increased almost fourfold in pots having a pigeon pea/winter wheat rotation (brown, Fig. 1.14b). The researchers suggested that the soil was improved by the organic matter as well as the addition of nitrogen from the pigeon peas. The researchers published their results in a scientific journal.[1]

A Field Study

Researcher David Barash observed the mating behavior of mountain bluebirds (Fig. 1.15a, b) and noted that males perform aggressive behavior during the mating season. He was interested in whether this behavior changed during the events of mating and reproduction.

1 Bidlack, J. E., Rao, S. C., and Demezas, D. H. 2001. Nodulation, nitrogenase activity, and dry weight of chickpea and pigeon pea cultivars using different *Bradyrhizobium* strains. *Journal of Plant Nutrition* 24: 549–60.

Hypothesis

Barash formulated the hypothesis that aggression of the male varies during the reproductive cycle. To test this hypothesis, he reasoned that he should evaluate the intensity of male aggression at three stages: after the nest is built, after the first egg is laid, and after the eggs hatch.

> HYPOTHESIS: Male bluebird aggression varies during the reproductive cycle.
>
> PREDICTION: Aggression intensity will change after the nest is built, after the first egg is laid, and after hatching.

Experiment and Data

For his experiment, Barash decided to measure aggression intensity by recording the number of approaches per minute a male made toward a rival male and his own female mate. To provide a rival, Barash posted a male bluebird model near the nests while resident males were out foraging. The aggressive behavior (approaches) of the resident male was noted and counted during the first 10 minutes of the male's return (Fig. 1.15c).

To give his results validity, Barash included a control group. For his control, Barash posted a male robin model instead of a male bluebird near certain nests.

Resident males of the control group did not exhibit any aggressive behavior, but resident males of the experimental groups did exhibit aggressive behavior. Barash graphed his mathematical data on number of approaches (Fig. 1.15d). By examining the graph, you can see that the resident male was more aggressive toward the rival male model than toward his female mate, and that he was most aggressive while the nest was under construction, less aggressive after the first egg was laid, and least aggressive after the eggs hatched.

Conclusion

The results allowed Barash to conclude that aggression in male bluebirds is related to their reproductive cycle. Therefore, his hypothesis was supported. If male bluebirds were always aggressive, even toward male robin models, his hypothesis would not have been supported.

> CONCLUSION: The hypothesis is supported. Male bluebird aggression does vary during the reproductive cycle.

Barash reported his experiment in *The American Naturalist*.[2] In this article, Barash gave an evolutionary interpretation to his results. It is adaptive, he said, for male bluebirds to be less aggressive after the first egg is laid because by then the male bird is "sure the offspring is his own." It is maladaptive for the male bird to waste energy being aggressive after hatching, because his offspring are already present.

Check Your Progress   1.4

1. Identify the role of the experimental variable in an experiment.
2. Distinguish between the roles of the test group and the control group in an experiment.
3. Describe the process by which a scientist may test a hypothesis about an observation.

---

2 Barash, D. P. 1976. Male response to apparent female adultery in the mountain bluebird (*Sialia currucoides*): an evolutionary interpretation. *The American Naturalist* 110: 1097–1101.

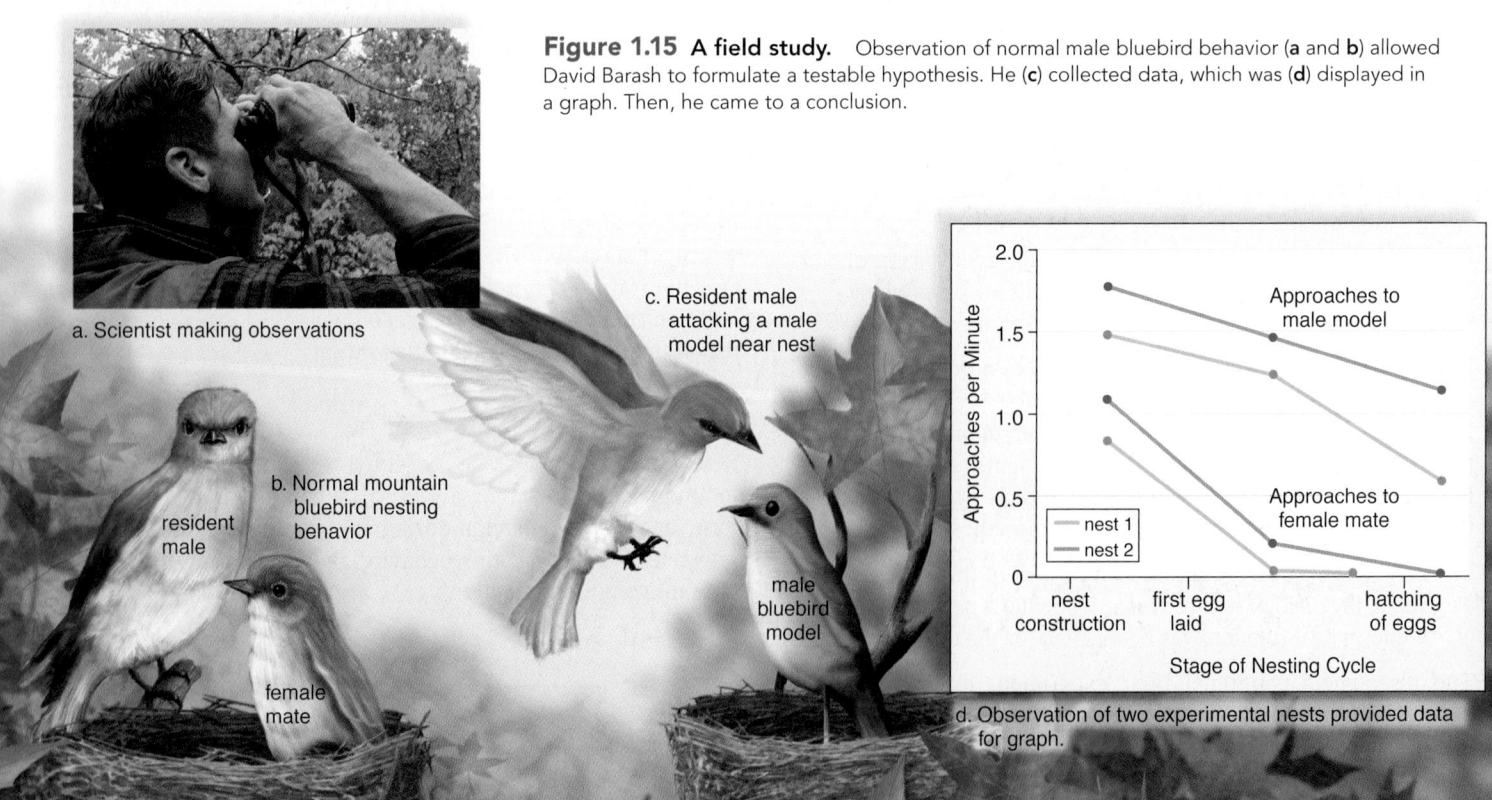

Figure 1.15 A field study. Observation of normal male bluebird behavior (**a** and **b**) allowed David Barash to formulate a testable hypothesis. He (**c**) collected data, which was (**d**) displayed in a graph. Then, he came to a conclusion.

a. Scientist making observations

b. Normal mountain bluebird nesting behavior

resident male

female mate

c. Resident male attacking a male model near nest

male bluebird model

d. Observation of two experimental nests provided data for graph.

Approaches per Minute

Approaches to male model

Approaches to female mate

nest 1
nest 2

nest construction   first egg laid   hatching of eggs

Stage of Nesting Cycle

CONNECTING *the* CONCEPTS *with the* BIG IDEAS

Evolution

- Evolution is the core concept of biology, and explains how species develop adaptations in an ever-changing environment. (1A1a)
- Natural selection is the mechanism by which evolutionary change occurs. (1A1a)

Energy and Homeostasis

- All life is based on atoms and molecules, which in turn are involved in the formation of cells, the basic units of all life. (2A3a1, 1D2b1)
- All life is characterized by energy flow and chemical cycling. (2A1)

Information and Signaling

- The key to life's diversity lies in differing sequences of DNA molecules, and the key to survival lies in maintenance of homeostasis using feedback and other signaling mechanisms. (3C2, 2D2a)

Interactions and Systems

- Biological organization starts with atoms and cells, moves to organisms, populations, and communities which form ecosystems. (4A5)
- Biodiversity is influenced by the environment and is impacted by evolution and extinctions. (4C4b)

*Find the unabridged version of all EK citations at www.glencoe.com/maderAP11.

Media Study Tools

www.glencoe.com/maderAP11

Enhance your study of this chapter with study tools and practice tests. Also ask your instructor about the resources available through ConnectPlus, including the media-rich eBook, interactive learning tools, and animations.

Summarize

1.1 How to Define Life

Although living things are diverse, they have certain characteristics in common. Living things (a) are organized, and their levels of organization extend from the cell to ecosystems; (b) need an outside source of materials and energy; (c) maintain homeostasis; (d) respond to external stimuli; (e) reproduce and develop, passing on genes to their offspring; and (f) have adaptations suitable to their way of life in a particular environment. Together, these characteristics unify life on Earth.

1.2 Evolution, the Unifying Concept of Biology

Life on Earth is diverse, but the theory of evolution unifies life and describes how all living organisms evolved from a common ancestor. Taxonomists assign each living thing an italicized binomial name that consists of the genus and the specific epithet. From the least inclusive to the most inclusive category, each species belongs to a genus, family, order, class, phylum, kingdom, and finally domain. Systematists study the evolutionary relationships between species.

The three domains of life are Archaea, Bacteria, and Eukarya. The first two domains contain prokaryotic organisms that are structurally simple but metabolically complex. Domain Eukarya contains the protists, fungi, plants, and animals. Protists range from unicellular to multicellular organisms and include the protozoans and most algae. Among the fungi are the familiar molds and mushrooms. Plants are well known as the multicellular photosynthesizers of the world, while animals are multicellular and ingest their food. An evolutionary tree shows how the domains are related by way of common ancestors.

Natural selection describes the process by which living organisms are descended from a common ancestor. Mutations occur within a population, creating new traits. The agents of natural selection, present in both biological and physical environments, shape species over time and may create new species from existing ones.

1.3 How the Biosphere Is Organized

Within an ecosystem, populations interact with one another and with the physical environment. Nutrients cycle within and between ecosystems, but energy flows unidirectionally and is eventually lost as an unusable form. Adaptations of organisms allow them to play particular roles within an ecosystem.

1.4 The Process of Science

When studying the natural world, scientists use the scientific process. Observations, along with previous data, are used to formulate a hypothesis. New observations and/or experiments are carried out in order to test the hypothesis. A good experimental design includes an experimental variable and a control group. The experimental and observational results are analyzed, and the scientist comes to a conclusion as to whether the results support the hypothesis or do not support the hypothesis.

Several conclusions in a particular area may allow scientists to arrive at a theory, such as the cell theory, the gene theory, or the theory of evolution. The theory of evolution is a unifying concept of biology.

Key Terms

adaptation 5	domain Bacteria 7
animal 8	domain Eukarya 7
binomial nomenclature 8	ecosystem 9
biodiversity 10	emergent property 2
biology 2	energy 4
biosphere 9	eukaryote 7
cell 2	evolution 5
class 6	experiment 11
community 9	experimental design 11
conclusion 12	experimental variable 12
control 12	extinction 10
data 12	family 6
deductive reasoning 11	fungi 8
domain 6	gene 5
domain Archaea 7	genus 6

homeostasis   4
hypothesis   11
inductive reasoning   11
kingdom   6
law   12
metabolism   4
model   12
multicellular   2
natural selection   8
observation   11
order   6
phenomenon   11
photosynthesis   4
phylum   6
plant   8

population   9
prediction   11
principle   12
prokaryote   7
protist   7
reproduce   5
responding variable   14
scientific method   11
scientific theory   12
species   6
standard deviation 12
systematics 6
taxonomy   6
unicellular   2

Assess

Reviewing This Chapter

1. What are the common characteristics of life?   2–5
2. Describe the levels of biological organization.   2
3. Why do living things require an outside source of nutrients and energy? Describe these sources.   4
4. What is passed from generation to generation when organisms reproduce? What has to happen to the hereditary material DNA for evolution to occur?   5
5. How does evolution explain both the unity and the diversity of life?   5–6
6. What are the categories of classification? How does the domain Eukarya differ from domain Bacteria and domain Archaea?   6
7. Explain the scientific name of an organism.   6
8. How does natural selection result in adaptation to the environment?   8
9. What is an ecosystem, and why should human beings preserve ecosystems?   9–10
10. Describe the series of steps involved in the scientific method.   11–12
11. Give an example of a controlled study. Name the experimental variable and the responding variable.   14–15

Testing Yourself

Choose the best answer for each question.

1. Which of these is not a property of all living organisms?
   a. organization
   b. acquisition of materials and energy
   c. care for their offspring
   d. reproduction
   e. responding to the environment

2. Describe an emergent property that might arise when moving from a single neuron (nerve cell) to nervous tissue.

3. The level of organization that includes cells of similar structure and function would be
   a. an organ.
   b. a tissue.
   c. an organ system.
   d. an organism.

4. Which of the following is an example of adaptation?
   a. In a very wet year, some plants grow unusually tall stalks and large leaves.
   b. Over millions of years, the eyes of cave salamanders lose their function.
   c. An escaped dog joins a pack of wild dogs and begins interbreeding with them.
   d. A harsh winter kills many birds within a population, especially the smallest ones.

5. Energy is brought into ecosystems by which of the following?
   a. fungi and other decomposers
   b. cows and other organisms that graze on grass
   c. meat-eating animals
   d. organisms that photosynthesize, such as plants
   e. All of these are correct.

6. Which of the following statements is a hypothesis?
   a. Will increasing my cat's food increase her weight?
   b. Increasing my cat's food consumption will result in a 25% increase in her weight.
   c. I will feed my cat more food.
   d. My cat has gained weight; therefore, she is eating more food.

7. After formulating a hypothesis, a scientist
   a. proves the hypothesis true or false.
   b. tests the hypothesis.
   c. decides how to best avoid having a control.
   d. makes sure environmental conditions are just right.
   e. formulates a scientific theory.

8. The experimental variable in the bluebird experiment (pages 15-16) was the
   a. use of a model male bluebird.
   b. observations of the experimenter.
   c. various behavior of the males.
   d. identification of what bluebirds to study.
   e. All of these are correct.

9. The control group in the pigeon pea/winter wheat experiment (pages 14-15) was the pots that were
   a. planted with pigeon peas.
   b. treated with nitrogen fertilizer.
   c. not treated.
   d. not watered.
   e. Both c and d are correct.

10. Which of the following are agents of natural selection?
    a. changes in the environment
    b. competition among individuals for food and water
    c. predation by another species
    d. competition among members of a population for prime nesting sites
    e. All of these are correct.

11. Which of the following is an example of natural selection?
    a. In a very wet year, some plants grow unusually tall stalks and large leaves.
    b. After several unusually cold winters, squirrels with an extra layer of fat have more offspring.
    c. Squirrels may have long or short tails.
    d. Dogs with longer legs are able to run faster than dogs with shorter legs.

12. Which of the following statements regarding evolution is false?
    a. Adaptations may be physical or behavioral.
    b. Natural selection always results in organisms becoming more adapted to the environment.
    c. A trait selected for, may suddenly become selected against when the environment changes.
    d. Some traits are neither selected for nor against.

For questions 13–15, write a brief answer.

13. Why is it said that all energy used by living organisms originates from the Sun?

14. Carbon dioxide emissions have been blamed for climate change by many scientists. How might excessive amounts of carbon dioxide affect nutrient cycling?

15. Would the accidental introduction of a new species to an ecosystem necessarily have a negative effect on biodiversity? Why or why not?

Engage

Virtual Lab
Dependent and Independent Variables

The virtual lab "Dependent and Independent Variables" provides an interactive exploration of how scientists construct scientific experiments.

Thinking Scientifically

1. An investigator spills dye on a culture plate and notices that the bacteria live despite exposure to sunlight. He decides to test if the dye is protective against ultraviolet (UV) light. He exposes one group of culture plates containing bacteria and dye and another group containing only bacteria to UV light. The bacteria on all plates die. Complete the following diagram.

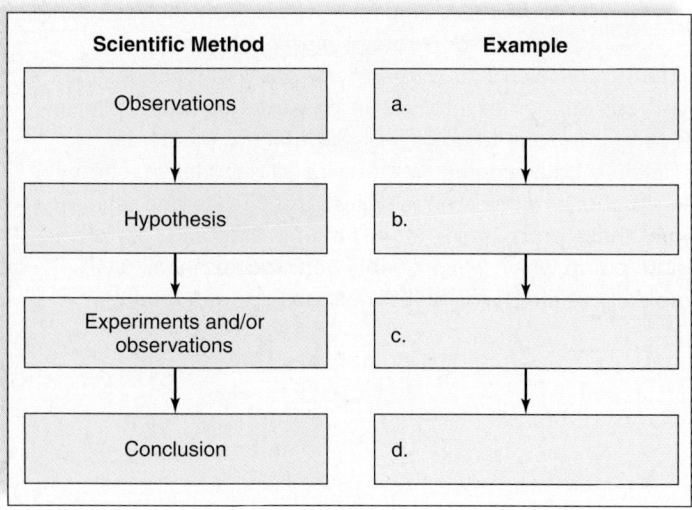

2. You want to grow large tomatoes and notice that a name-brand fertilizer claims to produce larger produce than a generic brand. How would you test this claim?

3. A scientist wishes to test her hypothesis that a commonly used drug causes heart attacks in some individuals. What kind of study should she initiate? What would you expect her experimental and responding variables to be?

Bioethical Issue

Oil Drilling in the Arctic

Established by an act of Congress in 1980, the Arctic National Wildlife Refuge (ANWR) covers a total of 19 million acres of northernmost Alaska far above the Arctic Circle. ANWR is home to a variety of wildlife, such as caribou, migratory birds, grizzly and polar bears, wolves, and musk oxen. But it is also home to substantial oil reserves, which has led to an ongoing contentious debate over its future: Should Congress allow development of ANWR for oil exploration and drilling?

Those who favor oil drilling in ANWR insist that first and foremost, the impact on the land would be minimal. The affected area would be roughly the size of an airport in a total area roughly the size of the state of South Carolina. They contend that the effect would mainly be underground because new techniques allow us to go lower and spread out beneath the surface to get the oil. Waste treatment and disposal methods have also improved. Acquiring the oil, advocates say, would also protect jobs and national security in the United States by lessening dependence on often hostile foreign countries for oil, and would have the added benefit of insulating the U.S. economy from oil price spikes and supply shocks.

Those who do not favor oil drilling in ANWR are eager to point out that at current levels of consumption, the oil coming from ANWR would hardly have a noticeable impact on prices and supply levels. Furthermore, they believe that the best solution to the current energy crunch would be for U.S. citizens to adopt simple energy conservation measures and invest in research on alternative fuels. They suggest that this would save many times the oil that could come from drilling in the Arctic refuge and that, by using a renewable energy resource, the environment in the lower 48 states would be protected, in addition to protecting the wildlife in the Arctic National Wildlife Refuge. Using renewable energy would lessen the need for foreign oil, and would also protect our national security.

Should Congress approve oil drilling in ANWR? Or should Congress invest in alternative and renewable energy forms, and insist that citizens adopt energy conservation measures? Should public tax monies be made available to Congress for oil exploration or for investment in alternative energy sources?

UNIT 1

The Cell

AP | BIG IDEAS

Did you make one of those cell models in middle school with different candies or pieces of pasta representing the six major organelles? Those simplistic models belie the fact that cells are the most amazing machines on Earth, and information in the field of cytology continues to increase exponentially! When the first AP Biology test was given in 1956, ribosomes and cytoskeletal elements were not even mentioned in textbooks, and the processes involved in photosynthesis' Calvin cycle were just being elucidated. Today we continue to uncover the details of important cellular structures such as peroxisomes and the ECM (extracellular matrix), and are closing in on the mysteries of cell signaling and cancer.

As you have seen in the introduction to this AP Biology edition, the new AP Biology Curriculum Framework is based on four **Big Ideas** that every college freshman biology major (and every AP student!) should understand fully and deeply.

You will find fragments of all four on every page of *Unit 1: The Cell:*

BI 1 Evolution transformed cells from simple prokaryotic organisms to the amazing and vast diversity of protists, fungi, plants, and animals.

BI 2 All cellular organisms require the same types of molecules and free energy from some source to survive, grow, and reproduce.

BI 3 Information of all types can be transferred to, stored, and later used by cells for their maintenance, well-being, and their procreation.

BI 4 Organelles and other components enter into complex relationships which create cells that can survive on their own, or that unite to form intricate systems which create organisms that are themselves integral parts of communities.

The stuff of life is the same as the stuff of non-living materials—the atoms and molecules that are themselves "lifeless" are combined and arranged in a logical and orderly manner to create carbohydrates, lipids, proteins, and nucleic acids that form membranes and organelles with unique and specialized structures. Those structures then carry out vital functions that facilitate biochemical construction requiring energy (as in photosynthesis or chemosynthesis) and destruction (as in cellular respiration) that releases energy. From birth to death, normal cells carry out every function you do: intake of nutrients, breakdown of materials, export of wastes, and repair of damages. Life is a balancing act: materials enter and leave, energy is generated and used, water comes and goes. And if any of these situations become unbalanced, the decline and ultimate demise of the organism looms large. Your enduring understanding of these processes—not just memorizing a list of Krebs Cycle enzymes or recounting explicit details of the sodium-potassium pump which will probably be forgotten—will give you the solid background to understand subsequent levels of complexity as life builds on its most basic and unique of innovations: The Cell.

UNIT OUTLINE

Bottle-nosed dolphin (*Tursiops truncatus*).

2

Basic Chemistry

A bottle-nosed dolphin, an amazing and intelligent creature, can tolerate a certain salinity, can stay underwater for only so long, and must eat a particular diet to keep its complex organ systems functioning. Chemistry affects every aspect of the dolphin's life, whether it is playing in the Gulf of Mexico or performing at Sea World. Without molecular chemistry, a dolphin wouldn't be able to live at all.

At one time, people believed that organisms contained a "vital force" that allowed them to live. Over time, science has shown us that living and nonliving things are all composed of the same elements. It is true, though, that living and nonliving things differ as to which elements are most common, as we shall see. This chapter reviews inorganic chemistry, which largely pertains to nonliving things, and also explores the composition and chemistry of water, an inorganic substance that is intimately connected to the life of organisms on planet Earth. Our search for water on other planets further emphasizes the essential role of water in all life.

As you read through the chapter, think about the following questions:

1. What unique properties do chemical elements have?
2. How do elements interact with one another?
3. How is it possible for inanimate chemical elements to produce a living organism?

BEFORE YOU BEGIN

Before beginning this chapter, take a few moments to review the following discussions.

Figure 1.2 Why is it important to understand scale when studying biology?

Figures 1.5-1.8 Why do we organize biological concepts into systems?

Section 1.4 How does scientific study help us to understand the natural world?

FOLLOWING *the* BIG IDEAS

CHAPTER 2   BASIC CHEMISTRY

Energy and Homeostasis	Knowledge of the properties of atoms, the bonds that they make, and the unique properties they possess help explain the structures and functions of life systems.
Interactions and Systems	The chemistry of life, its atomic and molecular structure, will explain life workings at all levels of organization.

2.1 Chemical Elements

Throw a ball, pat your dog, rake leaves, turn a page; everything we touch—from the water we drink to the air we breathe—is composed of matter. **Matter** refers to anything that takes up space and has mass. Although matter has many diverse forms—anything from molten lava to kidney stones—it only exists in three distinct states: solid, liquid, or gas.

Elements

All matter, both nonliving and living, is composed of certain basic substances called **elements.** An element is a substance that cannot be broken down to simpler substances with different properties (a property is a physical or chemical characteristic, such as density, solubility, melting point, and reactivity) by ordinary chemical means. It is quite remarkable that, in the known universe, there are only 92 naturally occurring elements that serve as the building blocks of matter. Other elements have been artificially constructed by physicists and are not biologically important.

Both the Earth's crust and all organisms are composed of elements, but they differ as to which ones are common. Only six elements—carbon, hydrogen, nitrogen, oxygen, phosphorus, and sulfur—are basic to life and make up about 95% of the body weight of organisms. The acronym CHNOPS helps us remember these six elements. The properties of these elements are essential to the uniqueness of cells and organisms, such as the macaws in Figure 2.1. The macaws have gathered on a salt lick in South America. Salt contains the elements sodium and chlorine and is commonly sought after by many forms of life. Potassium, calcium, iron, and magnesium are still other elements found in living things.

Atoms

In the early 1800s, the English scientist John Dalton (1776-1844) developed the atomic theory, which says that elements consist of tiny particles called **atoms** [Gk. *atomos*, uncut, indivisible]. An atom is the smallest part of an element that displays the properties of the element. An element and its atoms share the same name. One or two letters create the **atomic symbol,** which stands for this name. For example, the symbol H means a hydrogen atom, the symbol Rn stands for radon, and the symbol Na (for *natrium* in Latin) is used for a sodium atom.

Physicists have identified a number of subatomic particles that make up atoms. The three best known subatomic particles include positively charged **protons,** uncharged **neutrons,** and negatively charged **electrons** [Gk. *elektron*, electricity]. Protons and neutrons are located within the nucleus of an atom, and electrons move about the nucleus. Figure 2.2 shows the arrangement of the subatomic particles in a helium atom, which has only two electrons. In Figure 2.2*a*, the stippling shows the

Figure 2.1 Elements that make up the Earth's crust and its organisms. Scarlet macaws gather on a salt lick in South America. The graph inset shows the Earth's crust primarily contains the elements silicon (Si), aluminum (Al) and oxygen (O). Living organisms primarily contain the elements oxygen (O), nitrogen (N), carbon (C), and hydrogen (H). Biological molecules also contain the elements sulfur (S) and phosphorus (P).

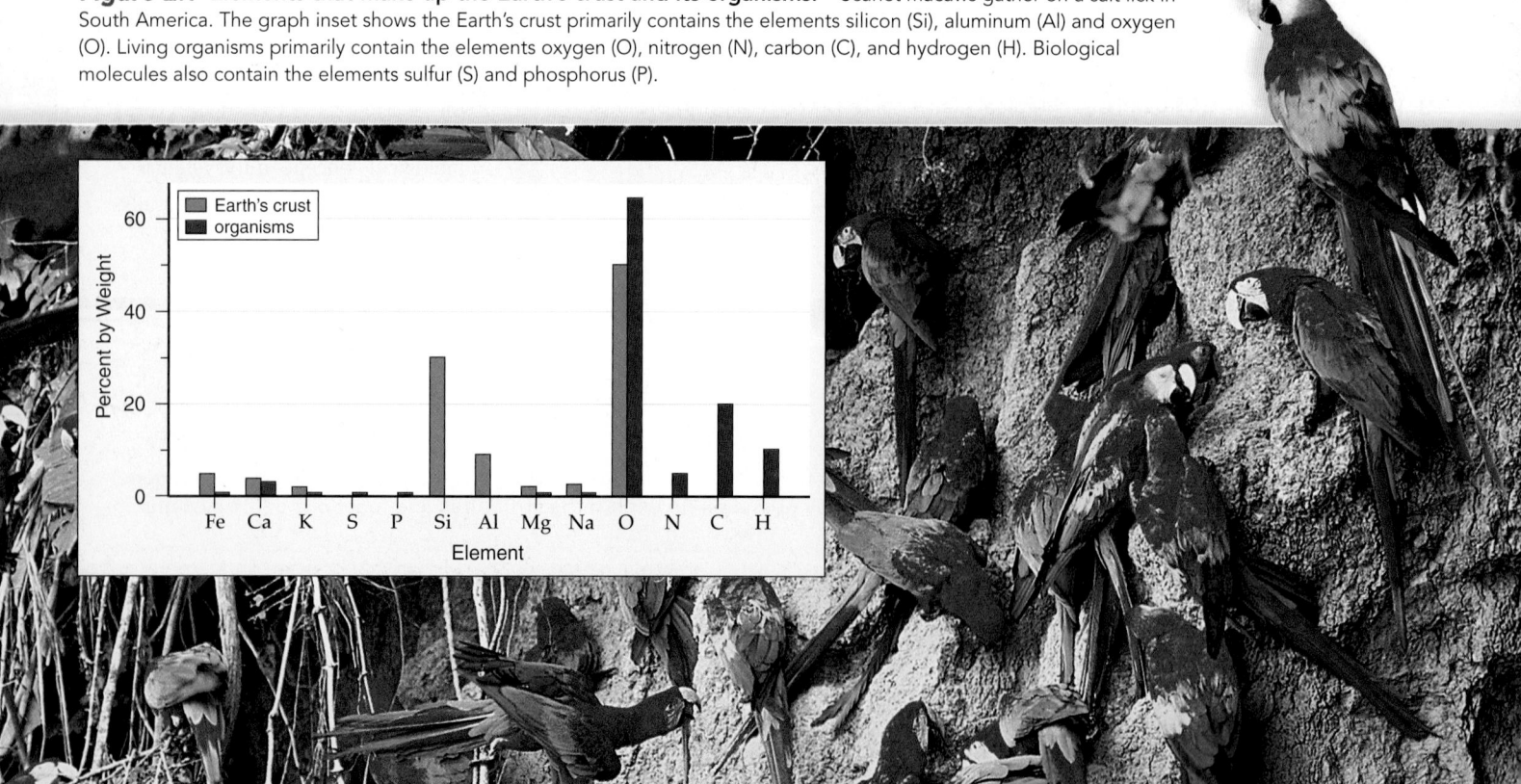

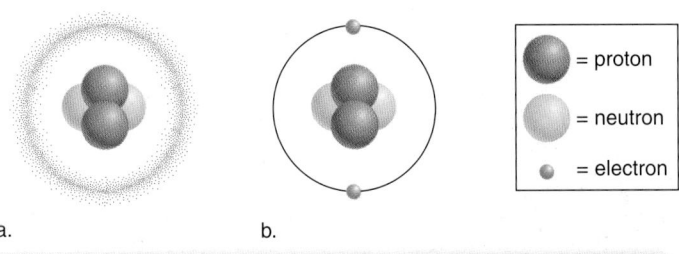

a.   b.

	= proton
	= neutron
	= electron

Subatomic Particles

Particle	Electric Charge	Atomic Mass Unit (AMU)	Location
Proton	+1	1	Nucleus
Neutron	0	1	Nucleus
Electron	−1	0	Electron shell

c.

Figure 2.2 Model of helium (He). Atoms contain subatomic particles, which are located as shown. Protons and neutrons are found within the nucleus, and electrons are outside the nucleus. **a.** The stippling shows the probable location of the electrons in the helium atom. **b.** The average location of an electron is sometimes represented by a circle termed an electron shell. **c.** The electric charge and the atomic mass units (AMU) of the subatomic particles vary as shown.

Figure 2.3 A portion of the periodic table. In the periodic table, elements are listed in the order of their atomic numbers, but arranged so each element is placed in a group (vertical column) and period (horizontal row). All the atoms in a particular group have the same number of valence electrons and therefore share common chemical characteristics. Each period shows the number of electron shells for an element. This abbreviated periodic table contains elements most important in biology; the complete periodic table is in Appendix D.

probable location of electrons, and in Figure 2.2*b*, the circle represents an **electron shell,** the average location of electrons.

The concept of an atom has changed greatly since Dalton's day. If an atom could be drawn the size of a football field, the nucleus would be like a gumball in the center of the field, and the electrons would be tiny specks whirling about in the upper stands. Most of an atom is empty space. We should also realize that we can only indicate where the electrons are expected to be most of the time. In our analogy, the electrons might very well stray outside the stadium at times.

Atomic Number and Mass Number

Atoms not only have an atomic symbol, they also have an atomic number and mass number. All atoms of an element have the same number of protons housed in the nucleus. This is called the **atomic number,** which accounts for the unique properties of this type of atom.

Each atom also has its own mass number dependent on the number of subatomic particles in that atom. Protons and neutrons are assigned one atomic mass unit (AMU) each. Electrons are so small that their AMU is considered zero in most calculations (Fig. 2.2*c*). Therefore, the **mass number** of an atom is the sum of protons and neutrons in the nucleus.

The term mass is used, and not *weight,* because mass is constant, while weight changes according to the gravitational force of a body. The gravitational force of the Earth is greater than that of the moon; therefore, substances weigh less on the moon, even though their mass has not changed.

By convention, when an atom stands alone (and not in the periodic table, discussed next), the atomic number is written as a subscript to the lower left of the atomic symbol. The mass number is written as a superscript to the upper left of the atomic

symbol. Regardless of position, the smaller number is always the atomic number, as shown here for carbon.

$$^{12}_{6}\text{C}$$

mass number ——— atomic symbol
atomic number ———

The Periodic Table

Once chemists discovered a number of the elements, they began to realize that even though each element consists of a different atom, certain chemical and physical characteristics recur. The periodic table, developed by the Russian chemist Dmitri Mendeleev (1834–1907), was constructed as a way to group the elements, and therefore atoms, according to these characteristics.

Figure 2.3 is a portion of the periodic table, which is shown in total in Appendix D. The atoms shown in the periodic table are assumed to be electrically neutral. Therefore, the atomic number not only tells you the number of protons, it also tells you the number of electrons. The **atomic mass** is the average of the AMU for all the isotopes (discussed next) of that atom. To determine the number of neutrons, subtract the number of protons from the atomic mass, and take the closest whole number.

In the periodic table, every atom is in a particular period (the horizontal rows) and in a particular group (the vertical columns). The atomic number of every atom in a period increases by one if you read from left to right. All the atoms in a group share the same binding characteristics. For example, all the atoms in group VII react with one atom at a time, for reasons we will soon explore. The atoms in group VIII are called the noble gases because they are inert and rarely react with another atom. Notice that helium, neon, argon, and krypton are noble gases.

Isotopes

Isotopes [Gk. *isos,* equal, and *topos,* place] are atoms of the same element that differ in the number of neutrons. Isotopes have the same number of protons, but they have different atomic masses. For example, the element carbon has three common isotopes:

$$^{12}_{6}C \qquad\qquad ^{13}_{6}C \qquad\qquad ^{14}_{6}C*$$

*radioactive

Carbon 12 has six neutrons, carbon 13 has seven neutrons, and carbon 14 has eight neutrons. Unlike the other two isotopes of carbon, carbon 14 is unstable; it changes over time into nitrogen 14, which is a stable isotope of the element nitrogen. As carbon 14 decays, it releases various types of energy in the form of rays and subatomic particles, and therefore it is termed a radioactive isotope. The radiation given off by radioactive isotopes can be detected in various ways. The Geiger counter is an instrument that is commonly used to detect radiation. In 1860, the French physicist Antoine-Henri Becquerel (1852-1908) discovered that a sample of uranium would produce a bright image on a photographic plate even in the dark, and a similar method of detecting radiation is still in use today. Marie Curie (1867-1934), who worked with Becquerel, coined the term "radioactivity" and contributed much to its study. Today, radiation is used by biologists to date objects from our distant past, create images, and trace the movement of substances in the body.

🎞 **Animation** Half-Life

Low Levels of Radiation

The chemical behavior of a radioactive isotope is essentially the same as that of the stable isotopes of an element. This means that you can put a small amount of radioactive isotope in a sample and it becomes a **tracer** by which to detect molecular changes. Melvin Calvin and his co-workers used carbon 14 to detect all the various reactions that occur during the process of photosynthesis (see Chapter 7).

The importance of chemistry to medicine is nowhere more evident than in the many medical uses of radioactive isotopes. Specific tracers are used in imaging the body's organs and tissues. For example, after a patient drinks a solution containing a minute amount of ^{131}I, it becomes concentrated in the thyroid—the only organ to take it up. A subsequent image of the thyroid indicates whether it is healthy in structure and function (Fig. 2.4*a*).

Positron-emission tomography (PET) is a way to determine the comparative activity of tissues. Radioactively labeled glucose, which emits a subatomic particle known as a positron, is injected into the body. The radiation given off is detected by sensors and analyzed by a computer. The result is a color image that shows which tissues took up glucose and are therefore metabolically active. The red areas surrounded by green in Figure 2.4*b* indicate which areas of the brain are most active. PET scans of the brain are used to evaluate patients who have memory disorders of an undetermined cause or suspected brain tumors or seizure disorders that could possibly benefit from surgery. PET scans, utilizing radioactive thallium, can detect signs of coronary artery disease and low blood flow to the heart.

📹 **Video** Nuclear Medicine

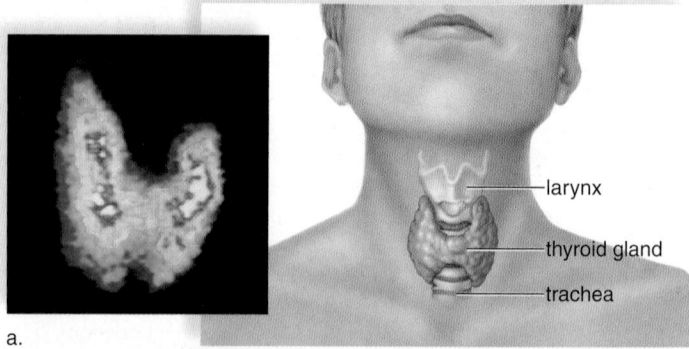
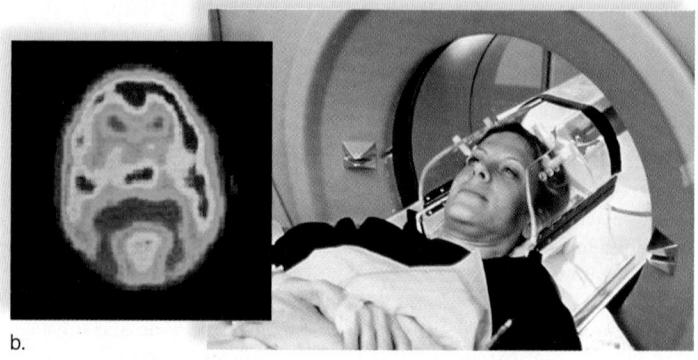

Figure 2.4 Low levels of radiation.   a. Medical scan of the thyroid gland (colored image) indicates the presence of a tumor that does not take up the radioactive iodine. **b.** A PET (positron-emission tomography) scan reveals which portions of the brain are most active (green and red colors).

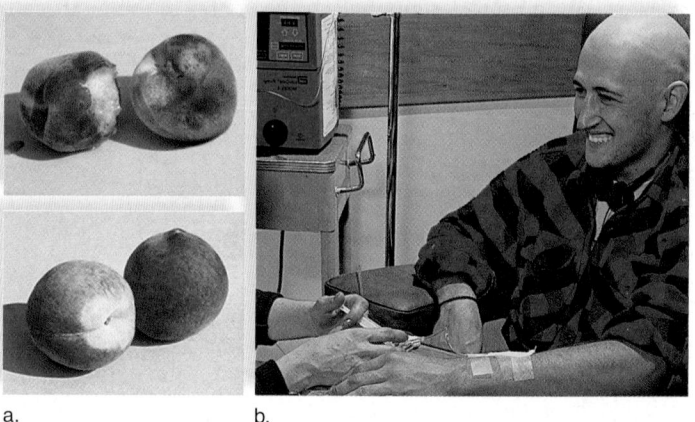

Figure 2.5 High levels of radiation.   a. Radiation used to kill bacteria and fungi on peaches helps them spoil less quickly and keep for a longer period of time. **b.** Physicians use targeted radiation therapy to kill cancer cells.

High Levels of Radiation

Radioactive substances in the environment can harm cells, damage DNA, and cause cancer. When Marie Curie was studying radiation, its harmful effects were not known, and she and many of her co-workers developed cancer. The release of radioactive particles following a nuclear power plant accident can have far-reaching and long-lasting effects on human health. The harmful effects of radiation can be put to good use, however (Fig. 2.5). Radiation from radioactive isotopes has been used for many years to sterilize medical and dental products. Now it can be used to sterilize the U.S. mail and other packages to free them of possible pathogens, such as anthrax spores. High radiation is often used to kill cancer cells. Targeted radioisotopes can be introduced into the body so that the subatomic particles emitted destroy only cancer cells, with little risk to the rest of the body.

Electrons and Energy

In an electrically neutral atom, the positive charges of the protons in the nucleus are balanced by the negative charges of electrons moving about the nucleus. Various models in years past have attempted to illustrate the precise location of electrons. Figure 2.6 uses the Bohr model, which is named after the physicist Niels Bohr (1885-1962). The Bohr model is useful as a way to visualize electron location, but we need to realize that today's physicists tell us it is not possible to determine the precise location of any individual electron at any given moment.

In the Bohr model, the electron shells about the nucleus also represent energy levels. Because negatively charged electrons are attracted to the positively charged nucleus, it takes energy to push them away and keep them in their own shell. The more distant the shell, the more energy it takes. Therefore, it is more accurate to speak of electrons as being at particular energy levels in relation to the nucleus. When you study photosynthesis, you will learn that when atoms absorb the energy of the Sun, electrons are boosted to a higher energy level. Later, as the electrons return to their original energy level, energy is released and transformed into chemical energy. This chemical energy supports all life on Earth and therefore our very existence is dependent on the energy of electrons.

You will want to learn to draw a Bohr model for each of the elements that occurs in the periodic table shown in Figure 2.3. Let's begin by examining the models depicted in Figure 2.6. Notice that the first shell (closest to the nucleus) can contain two electrons; thereafter, each additional shell can contain eight electrons. Also, each lower level is filled with electrons before the next higher level contains any electrons.

Animation
Atomic Structure

The sulfur atom, with an atomic number of 16, has two electrons in the first shell, eight electrons in the second shell, and six electrons in the third, or outer, shell. Revisit the periodic table (see Fig. 2.3), and note that sulfur is in the third period. In other words, the period tells you how many shells an atom has. Also note that sulfur is in group VI. The group tells you how many electrons an atom has in its outer shell.

Regardless of how many shells an atom has, the outermost shell is called the valence shell. The valence shell is important because it determines many of an atom's chemical properties. If an atom has only one shell, the outermost valence shell is complete when it has two electrons. But if an atom has more than one shell, the **octet rule** holds. This rule states that the outermost shell is most stable when it has eight electrons. As mentioned previously, atoms in group VIII of the periodic table are called the noble gases because they do not ordinarily react. Stability exists because the valence shell is filled with eight electrons and therefore has less energy. In general, lower energy states represent stability, as we will have an opportunity to point out again in Chapter 6.

Just as you sometimes communicate with and react to other people by using your hands, so atoms use the electrons in their valence shells to undergo reactions. Atoms with fewer than eight electrons in the outer shell react with other atoms in such a way that after the reaction, each has a stable outer shell. As we shall see, the number of electrons in an atom's **valence shell** determines whether the atom gives up, accepts, or shares electrons to acquire eight electrons in the outer shell. Each atom in a group within the periodic table has the same number of electrons in its valence shell.

Check Your Progress   2.1

1. Contrast atomic number and mass number.
2. Examine the periods and groups from the periodic table to determine the electron configuration of chlorine.
3. Explain what would happen if you used low radiation to medically treat cancer.

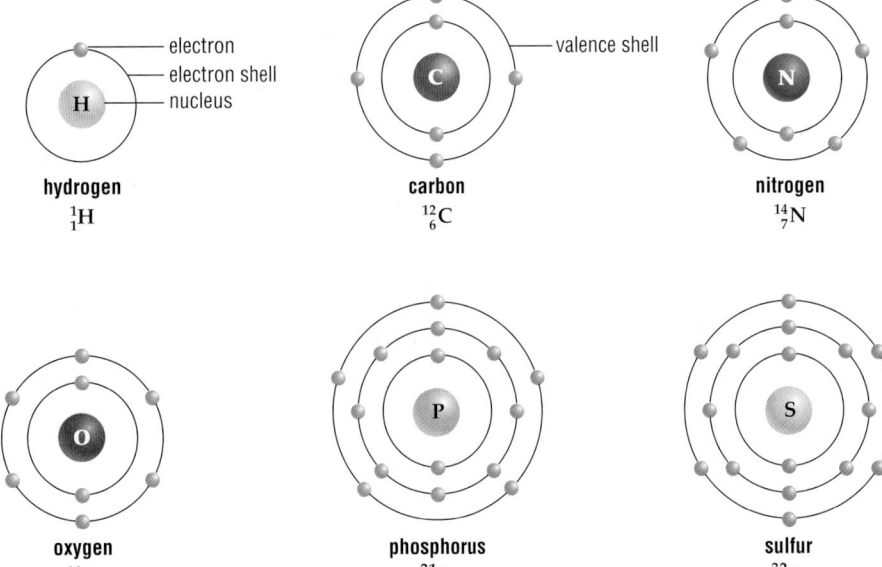

Figure 2.6 Bohr models of atoms.
Electrons orbit the nucleus at particular energy levels (electron shells). The first shell contains up to two electrons, and each shell thereafter can contain up to eight electrons as long as we consider only atoms with an atomic number of 20 or below. Each shell is filled before electrons are placed in the next shell. The outermost, or valence, shell helps determine the atom's chemical properties and how many other elements it can interact with.

2.2 Molecules and Compounds

Learning Outcomes

Upon completion of this section, you should be able to

1. Describe how elements are combined into molecules and compounds.
2. List different types of bonds that occur between elements.
3. Compare the relative strengths of ionic, covalent, and hydrogen bonds.

A **molecule** [L. *moles,* mass] exists when two or more elements bond together, and is the smallest part of a compound that retains its chemical properties. A **compound** is a molecule containing at least two different elements. In practice, these two terms are used interchangeably, but in biology, we usually speak of molecules. Water (H_2O) is a molecule that contains atoms of hydrogen and oxygen. A **formula** tells you the number of each kind of atom in a molecule. For example, in glucose:

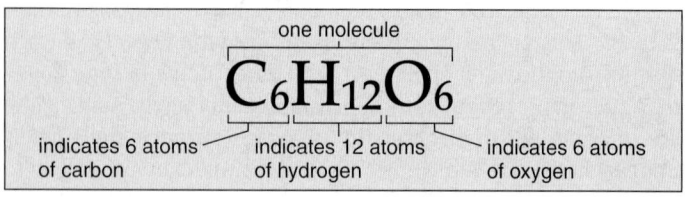

one molecule

$$C_6H_{12}O_6$$

indicates 6 atoms of carbon — indicates 12 atoms of hydrogen — indicates 6 atoms of oxygen

Electrons possess energy, and the bonds that exist between atoms also contain energy. Organisms are directly dependent on chemical-bond energy to maintain their organization. As you may know, organisms routinely break down glucose, the sugar shown to the left, to obtain energy. When a chemical reaction occurs, as when glucose is broken down, electrons shift in their relationship to one another, and energy is released. Spontaneous reactions, which are ones that occur freely, always release energy.

MP3
Chemical Bonding

Ionic Bonding

Sodium (Na), with only one electron in its third shell, tends to be an electron donor (Fig. 2.7a). Once it gives up this electron, the second shell, with eight electrons, becomes its outer shell. Chlorine (Cl), on the other hand, tends to be an electron acceptor. Its outer shell has seven electrons, so if it acquires only one more electron it has a completed outer shell. When a sodium atom and a chlorine atom come together, an electron is transferred from the sodium atom to the chlorine atom. Now both atoms have eight electrons in their outer shells.

This electron transfer, however, causes a charge imbalance in each atom. After giving up an electron, the sodium atom has one more proton than it has electrons; therefore, it has a net charge of +1 (symbolized by Na^+). After accepting an electron, the chlorine atom has one more electron than it has protons; therefore, it has a net charge of −1 (symbolized by Cl^-). Such charged particles are called **ions.** Sodium (Na^+) and chloride (Cl^-) are not the only biologically important ions. Some, such as potassium (K^+), are formed by the transfer of a single electron to another atom; others, such as calcium (Ca^{2+}) and magnesium (Mg^{2+}), are formed by the transfer of two electrons.

Ionic compounds are held together by a strong attraction between negatively and positively charged ions called an **ionic bond.** When sodium reacts with chlorine, an ionic compound called sodium chloride (NaCl) results. Sodium chloride is

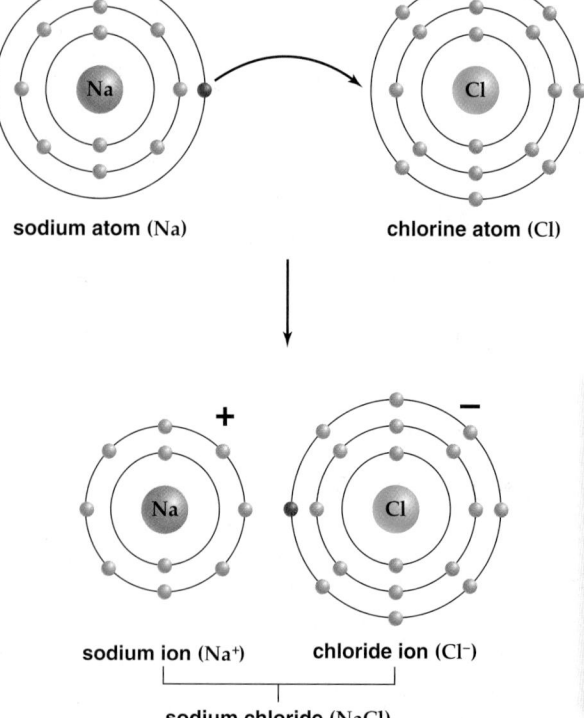

sodium atom (Na)   chlorine atom (Cl)

+   −

sodium ion (Na^+)   chloride ion (Cl^-)

sodium chloride (NaCl)

a.

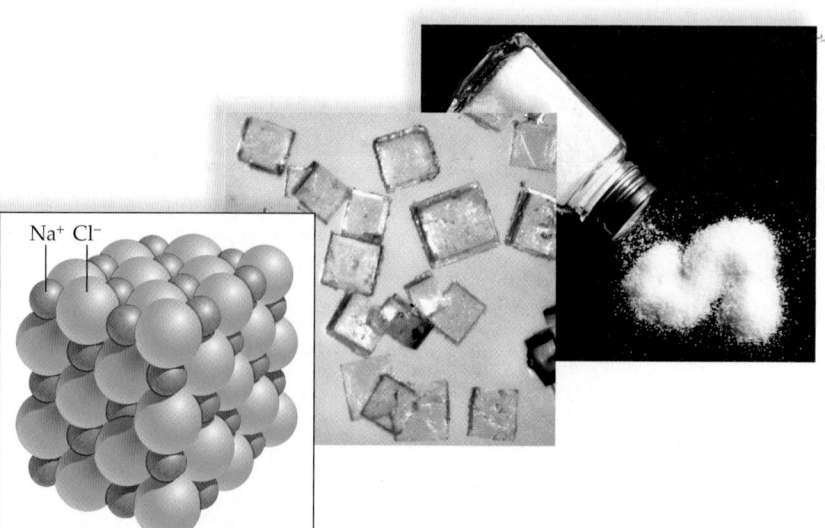

Na^+ Cl^-

b.

Figure 2.7 Formation of sodium chloride (table salt). a. During the formation of sodium chloride, an electron is transferred from the sodium atom to the chlorine atom. At the completion of the reaction, each atom has eight electrons in the outer shell, but each also carries a charge as shown. **b.** In a sodium chloride crystal, ionic bonding between Na^+ and Cl^- causes the atoms to assume a three-dimensional lattice in which each sodium ion is surrounded by six chloride ions, and each chloride ion is surrounded by six sodium ions. The result is crystals of salt as in table salt.

a salt, commonly known as table salt, because it is used to season our food (Fig. 2.7b). **Salts** are solid substances that usually separate and exist as individual ions in water, as discussed on page 30.

Animation
Ionic Bonds

Covalent Bonding

A **covalent bond** [L. *co*, together, with, and *valens,* strength] results when two atoms share electrons in such a way that each atom has an octet of electrons in the outer shell (or two electrons, in the case of hydrogen). In a hydrogen atom, the outer shell is complete when it contains two electrons. If hydrogen is in the presence of a strong electron acceptor, it gives up its electron to become a hydrogen ion (H^+). But if this is not possible, hydrogen can share with another atom and thereby have a completed outer shell. For example, one hydrogen atom will share with another hydrogen atom. Their two electron shells overlap and the electrons are shared between them (Fig. 2.8a). Because they share the electron pair, each atom has a completed outer shell.

A more common way to symbolize that atoms are sharing electrons is to draw a line between the two atoms, as in the structural formula H—H. Just as a handshake requires two hands, one from each person, a covalent bond between two atoms requires two electrons, one from each atom. In a molecular formula, the line is omitted and the molecule is simply written as H_2.

Sometimes, atoms share more than one pair of electrons to complete their octets. A double covalent bond occurs when two atoms share two pairs of electrons (Fig. 2.8b). To show that oxygen gas (O_2) contains a double bond, the molecule can be written as O = O. It is also possible for atoms to form triple covalent bonds, as in nitrogen gas (N_2), which can be written as N≡N. Single covalent bonds between atoms are quite strong, but double and triple bonds are even stronger.

Nonpolar and Polar Covalent Bonds

When the sharing of electrons between two atoms is equal, the covalent bond is said to be a **nonpolar covalent bond.** If one atom is able to attract electrons to a greater degree than the other atom, it is the more electronegative atom. **Electronegativity** is dependent on the number of protons—the greater the number of protons, the greater the electronegativity. When electrons are not shared equally, the covalent bond is a **polar covalent bond.**

Animation
Electronegativity

You can readily see that the bonds in methane (Fig. 2.8c) must be polar because carbon has more protons than a hydrogen atom. However, methane is a symmetrical molecule and the polarities cancel each other out—methane is a nonpolar molecule. Not so in water, which has this shape:

Oxygen is partially negative (δ^-)

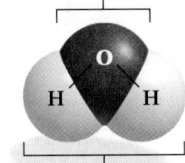

Hydrogens are partially positive (δ^+)

In water, the oxygen atom is more electronegative than the hydrogen atoms and the bonds are polar. Moreover, because of its

Electron Model	Structural Formula	Molecular Formula
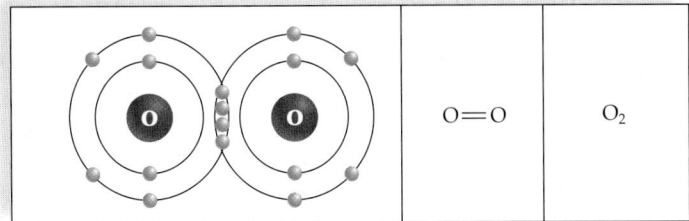	H—H	H_2

a. Hydrogen gas

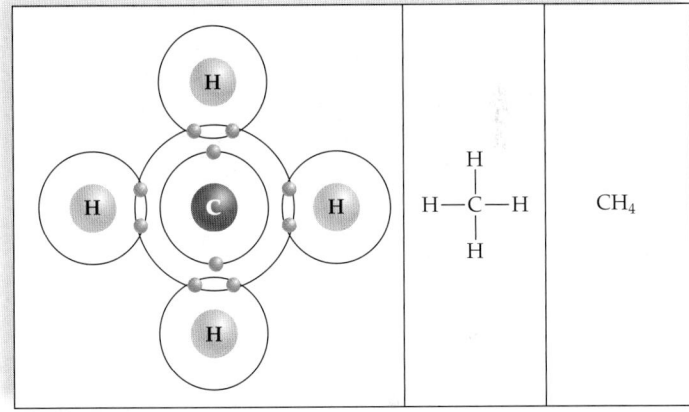	O=O	O_2

b. Oxygen gas

	$H-\overset{\displaystyle H}{\underset{\displaystyle H}{C}}-H$	CH_4

c. Methane

Figure 2.8 Covalently bonded molecules.   In a covalent bond, atoms share electrons, allowing each atom to have a completed outer shell. **a.** A molecule of hydrogen (H_2) contains two hydrogen atoms sharing a pair of electrons. This single covalent bond can be represented in any of the three ways shown. **b.** A molecule of oxygen (O_2) contains two oxygen atoms sharing two pairs of electrons. This results in a double covalent bond. **c.** A molecule of methane (CH_4) contains one carbon atom bonded to four hydrogen atoms.

nonsymmetrical shape, the polar bonds cannot cancel each other, and water is a polar molecule. The more electronegative end of the molecule is designated slightly negative (δ^-), and the hydrogens are designated slightly positive (δ^+).

Water is not the only polar molecule in living things. For example, the amine group (—NH_2) is polar, and this causes amino acids and nucleic acids to exhibit polarity, as we shall see in the next chapter. The polarity of molecules affects how they interact with other molecules.

Animation
Ionic vs. Covalent Bonding

Check Your Progress   2.2

1. Compare and contrast an ionic bond with a covalent bond.
2. Explain why calcium ion carries two positive charges.
3. Describe how the atoms in methane (CH_4) produce a complete outer shell.

2.3 Chemistry of Water

Learning Outcomes

Upon completion of this section, you should be able to

1. Describe how water associates with other molecules in solution.
2. Evaluate which property of water is important for biological life.
3. Analyze how water's solid, liquid, and vapor states allow life to exist on Earth.

Figure 2.9a recaps what we know about the water molecule. The structural formula on the far left shows that when water forms, an oxygen atom is sharing electrons with two hydrogen atoms. The ball-and-stick model in the center shows that the covalent bonds between oxygen and each of the hydrogens are at an angle of 104.5°. Finally, the space-filling molecule gives us the three-dimensional shape of the molecule and indicates its polarity.

The shape of water and of all organic molecules is necessary to the structural and functional roles they play in living things. For example, hormones have specific shapes that allow them to be recognized by the cells in the body. We can stay well only when antibodies recognize the shape of disease-causing agents, like a key fits a lock, and are able to remove them. Similarly, homeostasis is only maintained when enzymes have the proper shape to carry out their particular reactions in cells.

The shape of a water molecule and its polarity make hydrogen bonding possible. A **hydrogen bond** is the attraction of a slightly positive hydrogen to a slightly negative atom in the vicinity. In carbon dioxide, $O = C = O$, a slight difference in polarity between carbon and the oxygens is present, but because carbon dioxide is symmetrical, the opposing charges cancel one another and hydrogen bonding does not occur.

Hydrogen Bonding

The dotted lines in Figure 2.9b indicate that the hydrogen atoms in one water molecule are attracted to the oxygen atoms in other water molecules. Each of these hydrogen bonds is weaker than an ionic or covalent bond. The dotted lines indicate that hydrogen bonds are more easily broken than the other bonds.

Hydrogen bonding is not unique to water. Other biological molecules, such as DNA, have polar covalent bonds involving an electropositive hydrogen and usually an electronegative oxygen or nitrogen. In these instances, a hydrogen bond can occur within the same molecule or between nearby molecules.

Although a single hydrogen bond is more easily broken than a single covalent bond, multiple hydrogen bonds are collectively quite strong. Hydrogen bonds between cellular molecules help maintain their proper structure and function. For example, hydrogen bonds hold the two strands of DNA together.

Figure 2.9 Water molecule.   **a.** Three models for the structure of water. The electron model does not indicate the shape of the molecule. The ball-and-stick model shows that the two bonds in a water molecule are angled at 104.5°. The space-filling model also shows the V shape of a water molecule. **b.** Hydrogen bonding between water molecules. Each water molecule can hydrogen bond with up to four other molecules, in three dimensions. When in a liquid state, water is constantly forming and breaking hydrogen bonds.

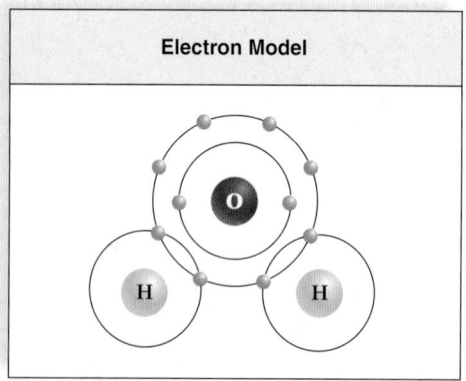

Electron Model

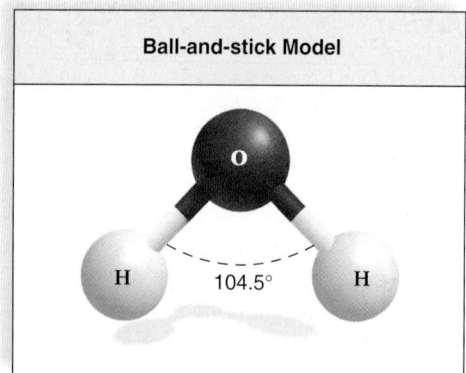

Ball-and-stick Model

104.5°

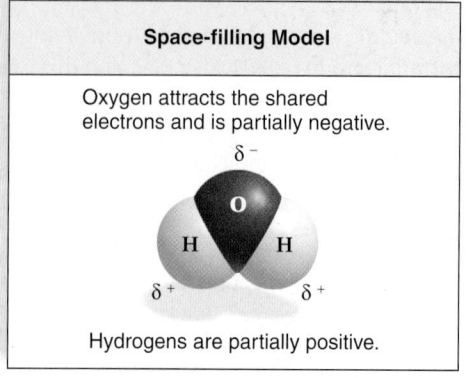

Space-filling Model

Oxygen attracts the shared electrons and is partially negative.

δ^-

O

H   H

δ^+ δ^+

Hydrogens are partially positive.

a. Water (H_2O)

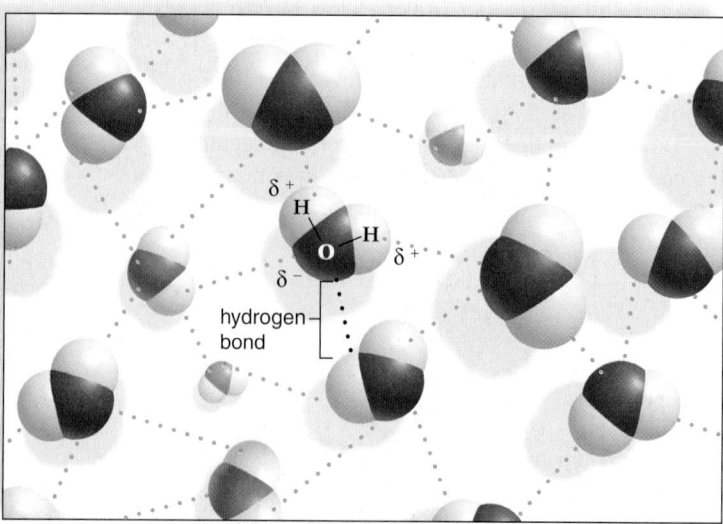

δ^+
H
O — H δ^+
δ^-
hydrogen bond

b. Hydrogen bonding between water molecules

When DNA makes a copy of itself, hydrogen bonds easily break, allowing DNA to unzip. But normally, the hydrogen bonds add stability to the DNA molecule. Similarly, the shape of protein molecules is often maintained by hydrogen bonding between different parts of the same molecule. As we shall see, many of the important properties of water are the result of hydrogen bonding.

Properties of Water

The first cell(s) evolved in water, and all living things are 70–90% water. Because of hydrogen bonding, water molecules cling together, and this association gives water its unique chemical properties. Without hydrogen bonding between molecules, water would freeze at –100°C and boil at –91°C, making most of the water on Earth steam, and life unlikely. Hydrogen bonding is responsible for water being a liquid at temperatures typically found on the Earth's surface. It freezes at 0°C and boils at 100°C. These and other unique properties of water make it essential to the existence of life as we know it. When scientists examine other planets with the hope of finding life, they first look for signs of water.

Animation
Properties
of Water

Water Has a High Heat Capacity.   A **calorie** is the amount of heat energy needed to raise the temperature of 1 g of water 1°C. In comparison, other covalently bonded liquids require input of only about half this amount of energy to rise in temperature 1°C. The many hydrogen bonds that link water molecules together help water absorb heat without a great change in temperature. Converting 1 g of the coldest liquid water to ice requires the loss of 80 calories of heat energy (Fig. 2.10a). Water holds onto its heat, and its temperature falls more slowly than that of other liquids. This property of water is important not only for aquatic organisms but also for all living things.

Because the temperature of water rises and falls slowly, organisms are better able to maintain their normal internal temperatures and are protected from rapid temperature changes.

Water Has a High Heat of Evaporation.   When water boils, it evaporates—that is, vaporizes into the environment. Converting 1 g of the hottest water to a gas requires an input of 540 calories of energy. Water has a high heat of evaporation because hydrogen bonds must be broken before water boils.

Water's high heat of vaporization gives animals in a hot environment an efficient way to release excess body heat. When an animal sweats, or gets splashed, body heat is used to vaporize water, thus cooling the animal (Fig. 2.10b). Because of water's high heat of vaporization and ability to hold onto its heat, temperatures along the coasts are moderate. During the summer, the ocean absorbs and stores solar heat, and during the winter, the ocean releases it slowly. In contrast, the interior regions of continents experience abrupt changes in temperatures.

Water Is a Solvent.   Due to its polarity, water facilitates chemical reactions, both outside and within living systems. As

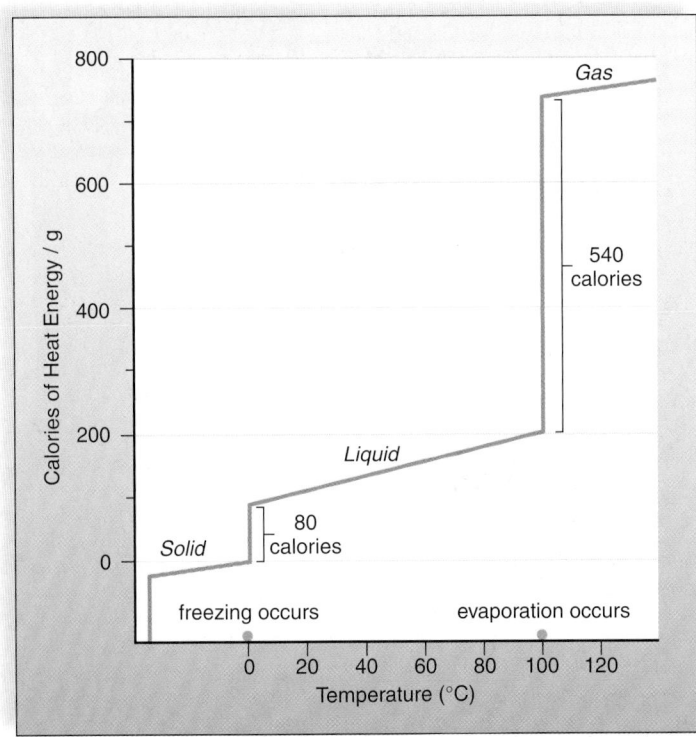

a. Calories lost when 1 g of liquid water freezes and calories required when 1 g of liquid water evaporates.

b. Bodies of organisms cool when their heat is used to evaporate water.

Figure 2.10 Temperature and water.   **a.** Water can be a solid, a liquid, or a gas at naturally occurring environmental temperatures. At room temperature and pressure, water is a liquid. When water freezes and becomes a solid (ice), it gives off heat, and this heat can help keep the environmental temperature higher than expected. On the other hand, when water evaporates, it takes up a large amount of heat as it changes from a liquid to a gas.
b. This means that splashing water on the body will help keep body temperature within a normal range. Can you also see why water's properties help keep the coasts moderate in both winter and summer?

a solvent, it dissolves a great number of substances, especially those that are also polar. A **solution** contains dissolved substances, which are then called **solutes**. When ionic salts—for example, sodium chloride (NaCl)—are put into water, the negative ends of the water molecules are attracted to the sodium ions, and the positive ends of the water molecules are attracted to the chloride ions. This attraction causes the sodium ions and the chloride ions to separate, or dissociate, in water.

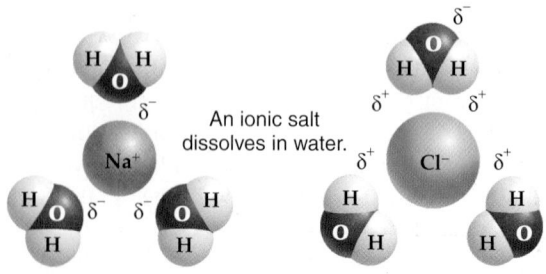

An ionic salt dissolves in water.

Water is also a solvent for larger polar molecules, such as ammonia (NH_3).

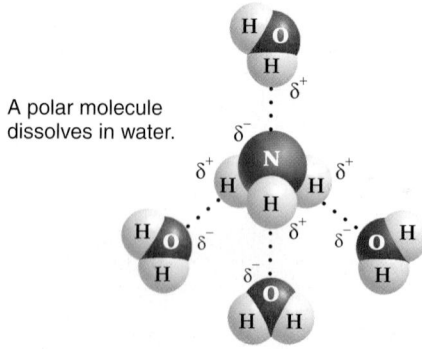

A polar molecule dissolves in water.

Molecules that can attract water are said to be **hydrophilic** [Gk. *hydrias,* of water, and *phileo,* love]. When ions and molecules disperse in water, they move about and collide, allowing reactions to occur. Nonionized and nonpolar molecules that cannot attract water are said to be **hydrophobic** [Gk. *hydrias,* of water, and *phobos,* fear]. Hydrophilic molecules tend to attract other polar molecules; similarly, hydrophobic substances usually associate with other nonpolar molecules. Gasoline contains nonpolar molecules, and therefore does not mix with water and is hydrophobic.

Water Molecules Are Cohesive and Adhesive. Cohesion refers to the ability of water molecules to cling to each other due to hydrogen bonding. At any moment in time, a water molecule can form hydrogen bonds with at most four other water molecules. Because of cohesion, water exists as a liquid under the conditions of temperature and pressure present at the Earth's surface. The strong cohesion of water molecules is apparent because water flows freely, yet water molecules do not separate from each other.

Adhesion refers to the ability of water molecules to cling to other polar surfaces. This is a result of water's polarity. Multicellular animals often contain internal vessels in which water assists the transport of nutrients and wastes because the cohesion and adhesion of water allows blood to fill the tubular

vessels of the cardiovascular system. For example, the liquid portion of our blood, which transports dissolved and suspended substances about the body, is 90% water.

Cohesion and adhesion also contribute to the transport of water in plants. Plants have their roots anchored in the soil, where they absorb water, but the leaves are uplifted and exposed to solar energy. Water evaporating from the leaves is immediately replaced with water molecules from transport vessels that extend from the roots to the leaves (Fig. 2.11). Because water molecules are cohesive, a tension is created that pulls the water column up from the roots. Adhesion of water to the walls of the transport vessels also helps prevent the water column

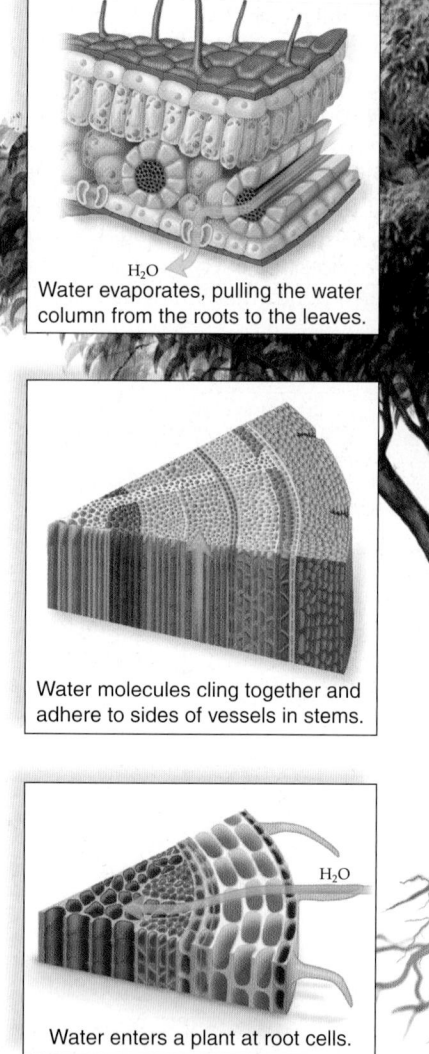

H_2O
Water evaporates, pulling the water column from the roots to the leaves.

Water molecules cling together and adhere to sides of vessels in stems.

H_2O
Water enters a plant at root cells.

Figure 2.11 Water as a transport medium.
How does water rise to the top of tall trees when they have no heart to pump it? Plants have water-filled pipelines that run from the roots to the leaves. When water evaporates from the leaves, the water column is pulled upward due to the cohesion of water molecules to one another and the adhesion of water molecules to the sides of the vessels. This capillary action is critical for plants to function.

from breaking apart. This capillary action is essential to plant life, as will be discussed in chapter 25.

Because water molecules are attracted to each other, they cling together where the liquid surface is exposed to air. The stronger the force between molecules in a liquid, the greater the **surface tension.** Water's high surface tension makes it possible for humans to skip rocks on water. Water striders, a common insect, can even walk on the surface of a pond without breaking the surface.

Frozen Water (Ice) Is Less Dense than Liquid Water.   As liquid water cools, the molecules come closer together. Water is most dense at 4°C, but the water molecules are still moving about (Fig. 2.12). At temperatures below 4°C, only vibrational movement occurs, and hydrogen bonding becomes more rigid but also more open. This means that water expands as it reaches 0°C and freezes, which is why cans of soda burst when placed in a freezer, or why frost heaves make northern roads bumpy in the winter. It also means that ice is less dense than liquid water, and therefore ice floats on liquid water.

If ice did not float on water, it would sink to the bottom, and ponds, lakes, and perhaps even the ocean would freeze solid, making life impossible in the water and also on land. Instead, bodies of water always freeze from the top down. When a body of water freezes on the surface, the ice acts as an insulator to prevent the water below it from freezing. This allows aquatic organisms to survive the winter. As ice melts in the spring, it draws heat from the environment, helping to prevent a sudden change in temperature that might be harmful to life.

Check Your Progress   2.3

1. Explain how water's high heat of vaporization allows coastal cities to have a consistent temperature throughout the year.
2. Analyze which property of water helps children cool off during summertime by playing in a sprinkler.
3. Evaluate what would happen to life if ice sank instead of floating in the winter.

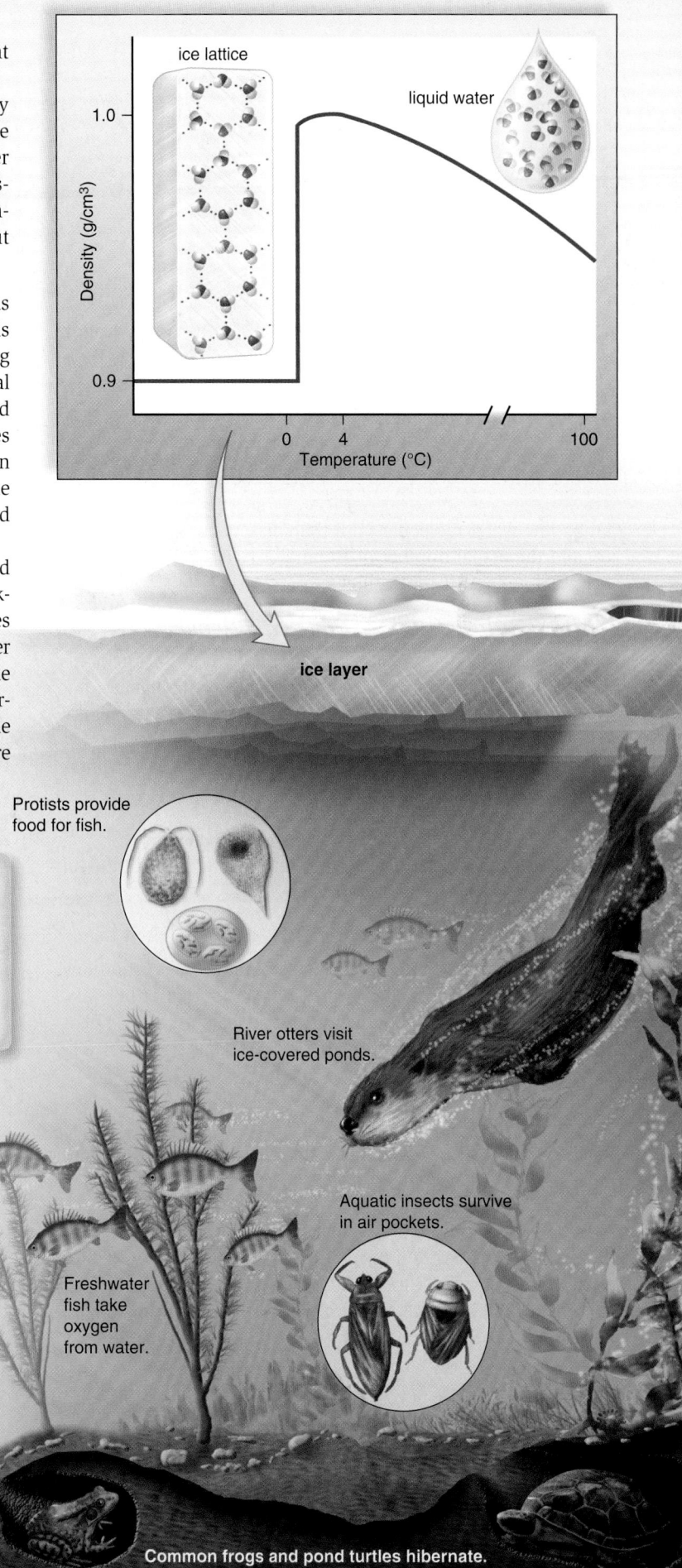

Protists provide food for fish.

River otters visit ice-covered ponds.

Aquatic insects survive in air pockets.

Freshwater fish take oxygen from water.

Common frogs and pond turtles hibernate.

Figure 2.12 A pond in winter.   *Above:* Remarkably, water is more dense at 4°C than at 0°C. Most substances contract when they solidify, but water expands when it freezes because in ice, water molecules form a lattice in which the hydrogen bonds are farther apart than in liquid water. *Below:* The layer of ice that forms at the top of a pond shields the water and protects the protists, plants, and animals so that they can survive the winter. These animals, except for the otter, are ectothermic, which means that they take on the temperature of the outside environment. This might seem disadvantageous until you realize that water remains relatively warm because of its high heat capacity. During the winter, frogs and turtles hibernate and in this way, lower their oxygen needs. Insects survive in air pockets. Fish, as you will learn later in this text, have an efficient means of extracting oxygen from the water, and they need less oxygen than the endothermic otter, which depends on muscle activity to warm its body.

2.4 Acids and Bases

Learning Outcomes

Upon completion of this section, you should be able to

1. Identify common acidic and basic substances.
2. Determine pH from a known H^+ or OH^- concentration.
3. Analyze how buffers prevent large pH changes in solutions.

When water ionizes, it releases an equal number of **hydrogen ions (H^+)** (also called protons[1]) and **hydroxide ions (OH^-)**:

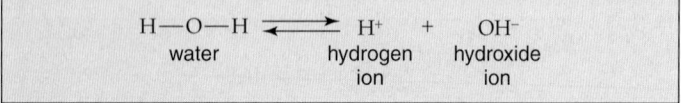

$$H\text{—}O\text{—}H \rightleftharpoons H^+ + OH^-$$

water   hydrogen   hydroxide
                   ion   ion

Only a few water molecules at a time dissociate, and the actual number of H^+ and OH^- is very small (1×10^{-7} moles/liter).[2]

MP3
Water and pH

Acidic Solutions (High H⁺ Concentrations)

Lemon juice, vinegar, tomatoes, and coffee are all acidic solutions. What do they have in common? **Acids** are substances that dissociate in water, releasing hydrogen ions (H^+). The acidity of a substance depends on how fully it dissociates in water. For example, hydrochloric acid (HCl) is a strong acid that dissociates almost completely in this manner:

$$HCl \longrightarrow H^+ + Cl^-$$

If hydrochloric acid is added to a beaker of water, the number of hydrogen ions (H^+) increases greatly.

Basic Solutions (Low H⁺ Concentration)

Milk of magnesia and ammonia are common basic solutions familiar to most people. **Bases** are substances that either take up hydrogen ions (H^+) or release hydroxide ions (OH^-). For example, sodium hydroxide (NaOH) is a strong base that dissociates almost completely in this manner:

$$NaOH \longrightarrow Na^+ + OH^-$$

If sodium hydroxide is added to a beaker of water, the number of hydroxide ions increases.

pH Scale

The **pH scale** is used to indicate the acidity or basicity (alkalinity) of a solution.[3] The pH scale (Fig. 2.13) ranges from 0 to 14. A pH of 7 represents a neutral state in which the hydrogen ion and hydroxide ion concentrations are equal. A pH below 7

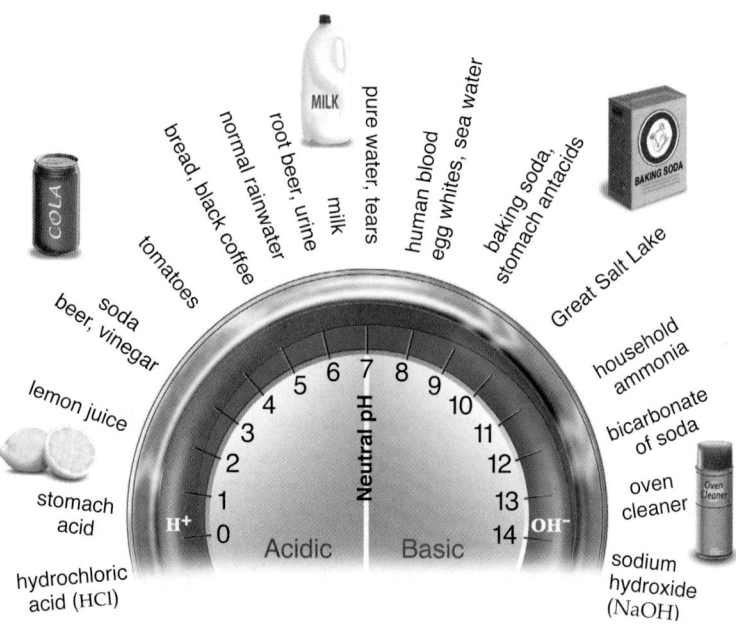

Figure 2.13 The pH scale. The pH scale ranges from 0 to 14 with 0 being the most acidic and 14 being the most basic. pH 7 (neutral pH) has equal amounts of hydrogen ions (H^+) and hydroxide ions (OH^-). An acidic pH has more H^+ than OH^- and a basic pH has more OH^- than H^+.

is an acidic solution because the hydrogen ion concentration is greater than the hydroxide concentration. A pH above 7 is basic because the $[OH^-]$ is greater than the $[H^+]$. Further, as we move down the pH scale from pH 14 to pH 0, each unit is 10 times more acidic than the previous unit. As we move up the scale from 0 to 14, each unit is 10 times more basic than the previous unit. Therefore pH 5 is 100 times more acidic than pH 7 and a 100 times more basic than pH 3.

The pH scale was devised to eliminate the use of cumbersome numbers. For example, the possible hydrogen ion concentrations of a solution are on the left of this listing and the pH is on the right:

	[H⁺] (moles per liter)	pH
0.000001	$= 1 \times 10^{-6}$	6
0.0000001	$= 1 \times 10^{-7}$	7
0.00000001	$= 1 \times 10^{-8}$	8

To further illustrate the relationship between hydrogen ion concentration and pH, consider the following question. Which of the pH values listed indicates a higher hydrogen ion concentration $[H^+]$ than pH 7, and therefore would be an acidic solution? A number with a smaller negative exponent indicates a greater quantity of hydrogen ions than one with a larger negative exponent. Therefore, pH 6 is an acidic solution.

The Biological Systems feature on page 33 describes detrimental environmental consequences to nonliving and living things as rain and snow have become more acidic. In humans, pH needs to be maintained within a narrow range or there are health consequences. The pH of blood is around 7.4, and blood is buffered in the manner described next to keep the pH within a normal range.

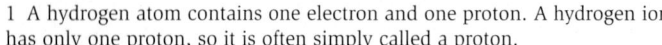

1 A hydrogen atom contains one electron and one proton. A hydrogen ion has only one proton, so it is often simply called a proton.

2 In chemistry, a mole is defined as 6.02×10^{23} of any atom, molecule, or ion. For example, 6.02×10^{23} atoms of 12C would have a mass of exactly 12 g. The same number of glucose molecules (1 mole) would have a mass of 180 g.

3 pH is defined as the negative log of the hydrogen ion concentration $[H^+]$. A log is the power to which 10 must be raised to produce a given number.

Biological Systems

The Interconnectedness of Water, Plants, and People

Acid Deposition

Normally, rainwater has a pH of about 5.6 because the carbon dioxide in the air combines with water to give a weak solution of carbonic acid. Acid deposition includes rain or snow that has a pH of less than 5, as well as dry acidic particles that fall to Earth from the atmosphere.

When fossil fuels such as coal, oil, and gasoline are burned, sulfur dioxide and nitrogen oxides combine with water to produce sulfuric and nitric acids. These pollutants are generally found eastward of where they originated because of wind patterns. The use of very tall smokestacks causes them to be carried even hundreds of miles away. For example, acid rain in southeastern Canada results from the burning of fossil fuels in factories and power plants in the midwestern United States.

Impact on Lakes

Acid rain adversely affects many aspects of biological systems. Aluminum may leach from the soil of lakes, particularly in areas where the soil is thin and lacks limestone (calcium carbonate, or $CaCO_3$) as a buffer. Acid rain may convert mercury in lake bottom sediments to toxic methyl mercury. Methyl mercury accumulates in fish, which wildlife and people eat. Over time, methyl mercury can accumulate in body tissues and cause serious sensory and muscular health problems. Acid rain in Canada and New England has caused hundreds of lakes to be devoid of fish, and in some cases, any life at all.

Impact on Forests

The leaves of plants damaged by acid rain can no longer carry on photosynthesis as before. When plants are under stress, they become susceptible to diseases and pests of all types. Forests on mountaintops re-

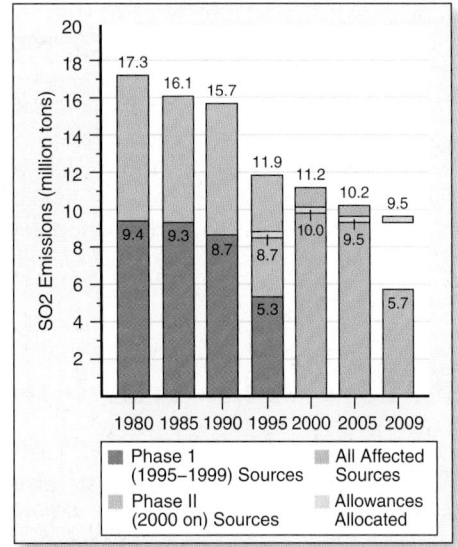
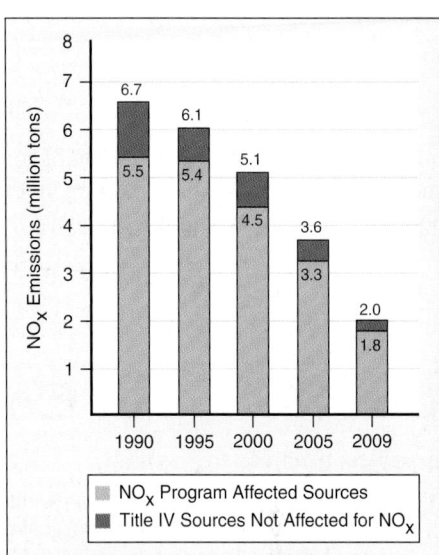

Figure 2A   Trends in U.S. acid rain emissions.   The burning of fossil fuels in factories, automobiles, and other industrial processes produces chemicals like SO_2 and NO_x that lead to acid deposition and destruction of the environment, Clean air legislation and stricter emission standards over the past 20 years have resulted in steady decreases in SO_2 and NO_x, chemicals that lead to acid rain.

ceive more rain than those at lower levels; therefore, they are more affected by acid rain. Forests are also damaged when toxic chemicals such as aluminum are leached from the soil. These kill soil fungi that assist roots in acquiring the nutrients trees need. In New England, 1.3 million acres of high-elevation forests have been devastated. Sulfur dioxide and nitrogen oxides, the main precursors of acid rain, have been steadily decreasing in the United States due to clean air legislation and strict emission limits (see Figure 2A).

Impact on Humans and Structures

Humans may be affected by acid rain. Inhaling dry sulfate and nitrate particles appears to increase the occurrence of respiratory illnesses, such as asthma. Buildings and monu-

ments made of limestone and marble break down when exposed to acid rain. The paint on homes and automobiles is likewise degraded. However, damage to natural systems and human structures due to acid rain is likely to decrease if we continue efforts to reduce chemicals that contribute to acid rain.

Questions to Consider

1. What acid rain trends are evident from the EPA data?
2. Considering that manufacturing is essential to our national interests, how might we modify industrial processing to reduce sulfur dioxide and nitrogen oxide contamination?
3. How might we prevent methyl mercury from entering biological systems and reduce the amount already present?

Buffers and pH

A **buffer** is a chemical or a combination of chemicals that keeps pH within normal limits. Many commercial products such as shampoos or deodorants are buffered as an added incentive for us to buy them.

In living things, the pH of body fluids is maintained within a narrow range, or else molecules don't function correctly and our

health suffers. The pH of our blood when we are healthy is always about 7.4—that is, just slightly basic (alkaline). If the blood pH drops to about 7, acidosis results. If the blood pH rises to about 7.8, alkalosis results. Both conditions can be life threatening, so the blood pH must be kept around 7.4. Normally, pH stability is possible because the body has built-in mechanisms to prevent pH changes. Buffers are one of these important mechanisms.

Buffers help keep the pH within normal limits because they are chemicals or combinations of chemicals that take up excess hydrogen ions (H⁺) or hydroxide ions (OH⁻). For example, carbonic acid (H_2CO_3) is a weak acid that minimally dissociates and then re-forms in the following manner:

$$H_2CO_3 \underset{\text{re-forms}}{\overset{\text{dissociates}}{\rightleftharpoons}} H^+ + HCO_3^-$$

carbonic acid   bicarbonate ion

Blood always contains a combination of some carbonic acid and some bicarbonate ions. When hydrogen ions (H⁺) are added to blood, the following reaction reduces acidity:

$$H^+ + HCO_3^- \longrightarrow H_2CO_3$$

When hydroxide ions (OH⁻) are added to blood, this reaction reduces basicity:

$$OH^- + H_2CO_3 \longrightarrow HCO_3^- + H_2O$$

These reactions prevent any significant change in blood pH.

Check Your Progress 2.4

1. Explain the difference in H⁺ concentration between an acid and a base.
2. Predict what would happen to plant and animal life where you live if the pH of rain dropped to 6.
3. Interpret what would happen to your blood pH if the amount of buffer decreased by a factor of 10.

CONNECTING the CONCEPTS with the BIG IDEAS

Energy and Homeostasis

- As in all systems, chemistry involves inputs and outputs, with resulting changes in energy; biological systems must increase inputs of energy to maintain order. (2A1a3)
- The movement of electrons between molecules allows for molecular bonding and is a primary driver of energy transfer in living organisms. (2A2d3, 2A2g2)
- Water is a polar molecule whose strong electronegativity forms hydrogen bonds, the basis of many of water's unique properties including performance as a universal solvent, high specific heat capacity, cohesion, adhesions, as well as life-supporting thermal conductivity and heats of vaporization and fusion. (2A3a3*IE*)

Interactions and Systems

- Life consists of interconnected systems, beginning with the atomic and molecular interactions which will eventually build to show emergent properties at organismal, population, community, and ecosystem level. (4A commentary)

*Find the unabridged version of all EK citations at www.glencoe.com/maderAP11.

Media Study Tools

www.glencoe.com/maderAP11

Enhance your study of this chapter with study tools and practice tests. Also ask your instructor about the resources available through ConnectPlus, including the media-rich eBook, interactive learning tools, and animations.

Summarize

2.1 Chemical Elements

Both living and nonliving things are composed of matter consisting of elements. The most significant elements (atoms) are found in living things: carbon, hydrogen, nitrogen, oxygen, phosphorus, and sulfur (CHNOPS).

Elements contain atoms, and atoms contain subatomic particles. Protons have positive charges, neutrons are uncharged, and electrons have negative charges. Protons and neutrons in the nucleus determine the mass number of an atom. The atomic number indicates the number of protons and the number of electrons in electrically neutral atoms.

Isotopes are atoms of a single element that differ in their numbers of neutrons. Radioactive isotopes have many uses, including serving as tracers in biological experiments and medical procedures.

Electrons occupy energy levels (electron shells) at discrete distances from the nucleus. The number of electrons in the outer shell determines the reactivity of an atom. The first shell is complete when it is occupied by two electrons. In atoms up through calcium, number 20, every shell beyond the first shell is complete when eight electrons are present. The octet rule states that atoms react with one another in order to have a completed outer shell. Most atoms, including those common to living things, do not have filled outer shells and this causes them to react with one another to form compounds and/or molecules. Following the reaction, the atoms have completed outer shells.

2.2 Molecules and Compounds

Compounds and molecules are formed when elements associate with each other. Ions, which are created when atoms lose or gain one or more electrons to achieve a completed outer shell, form ionic bonds between oppositely charged ions.

Covalent compounds form when atoms share one or more pairs of electrons. There are single, double, and triple covalent bonds.

In polar covalent bonds, the sharing of electrons is not equal. If the molecule is polar, the more electronegative atom carries a slightly negative charge and the other atom carries a slightly positive charge.

2.3 Chemistry of Water

Water is an essential molecule for life. The polarity of water molecules allows hydrogen bonding to occur between water molecules. A hydrogen bond is a weak attraction between a slightly positive hydrogen atom and a slightly negative oxygen or nitrogen atom within the same or a different molecule. Hydrogen bonds help maintain the structure and function of cellular molecules.

Water's polarity and hydrogen bonding account for its unique properties. These features allow living things to exist and carry on cellular activities.

2.4 Acids and Bases

pH is a measure of how acidic or basic a substance is. A small fraction of water molecules dissociate to produce an equal number of hydrogen ions and hydroxide ions. Solutions with equal numbers of H^+ and OH^- are termed neutral.

Acidic solutions contain more hydrogen ions than hydroxide ions; these solutions have a pH less than 7. Basic solutions have more hydroxide ions than hydrogen ions; these solutions have a pH greater than 7. Cells are sensitive to pH changes. Biological systems often contain buffers that help keep the pH within a normal range.

Key Terms

acid 32	electron shell 23
atom 22	element 22
atomic mass 23	formula 26
atomic number 23	hydrogen bond 28
atomic symbol 22	hydrogen ion (H^+) 32
base 32	hydrophilic 30
buffer 33	hydrophobic 30
calorie 29	hydroxide ion (OH^-) 32
compound 26	ion 26
covalent bond 27	ionic bond 26
electron 22	isotope 24
electronegativity 27	mass number 23

matter 22	proton 22
molecule 26	salt 27
neutron 22	solute 30
nonpolar covalent bond 27	solution 30
octet rule 25	surface tension 31
pH scale 32	valence shell 25
polar covalent bond 27	

Assess

Reviewing This Chapter

1. Name the kinds of subatomic particles studied. What is their atomic mass unit, charge, and location in an atom?   21–23
2. What is an isotope? A radioactive isotope? Radioactivity is always considered dangerous. Why?   24–26
3. Using the Bohr model, draw an atomic structure for a carbon that has six protons and six neutrons.   26
4. Draw an atomic representation for $MgCl_2$. Using the octet rule, explain the structure of the compound.   27
5. Explain whether CO_2 ($O = C = O$) is an ionic or a covalent compound. Why does this arrangement satisfy all atoms involved?   27
6. Of what significance is the shape of molecules in organisms?   28
7. Explain why water is a polar molecule. What does the polarity and shape of water have to do with its ability to form hydrogen bonds?   28
8. Name five properties of water, and relate them to the structure of water, including its polarity and hydrogen bonding between molecules.   28–31
9. On the pH scale, which numbers indicate a solution is acidic? Basic? Neutral?   32
10. What are buffers, and why are they important to life?   32–33

Testing Yourself

Choose the best answer for each question.

1. Which of the subatomic particles contributes almost no weight to an atom?
   a. protons in the electron shells
   b. electrons in the nucleus
   c. neutrons in the nucleus
   d. electrons at various energy levels

2. The atomic number tells you the
   a. number of neutrons in the nucleus.
   b. number of protons in the atom.
   c. atomic mass of the atom.
   d. number of its electrons if the atom is neutral.
   e. Both b and d are correct.

3. An atom that has two electrons in the outer shell, such as magnesium, would most likely
   a. share to acquire a completed outer shell.
   b. lose these two electrons and become a negatively charged ion.
   c. lose these two electrons and become a positively charged ion.
   d. bind with carbon by way of hydrogen bonds.
   e. bind with another calcium atom to satisfy its energy needs.

4. Isotopes differ in their
   a. number of protons.   c. number of neutrons.
   b. atomic number.   d. number of electrons.

5. When an atom gains electrons, it
   a. forms a negatively charged ion.
   b. forms a positively charged ion.
   c. forms covalent bonds.
   d. forms ionic bonds.
   e. gains atomic mass.

6. A covalent bond is indicated by
   a. plus and minus charges attached to atoms.
   b. dotted lines between hydrogen atoms.
   c. concentric circles about a nucleus.
   d. overlapping electron shells or a straight line between atomic symbols.
   e. the touching of atomic nuclei.

7. The shape of a molecule
   a. is dependent in part on the angle of bonds between its atoms.
   b. influences its biological function.
   c. is dependent on its electronegativity.
   d. is dependent on its place in the periodic table.
   e. Both a and b are correct.

8. In the molecule

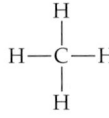

   a. all atoms have eight electrons in the outer shell.
   b. all atoms are sharing electrons.
   c. carbon could accept more hydrogen atoms.
   d. the bonds point to the corners of a square.
   e. All of these are correct.

9. Which of these properties of water cannot be attributed to hydrogen bonding between water molecules?
   a. Water stabilizes temperature inside and outside the cell.
   b. Water molecules are cohesive.
   c. Water is a solvent for many molecules.
   d. Ice floats on liquid water.
   e. Both b and c are correct.

10. Complete this diagram by placing an O for oxygen or an H for hydrogen on the appropriate atoms. Place partial charges where they belong.

11. H_2CO_3/$NaHCO_3$ is a buffer system in the body. What effect will the addition of an acid have on the pH of a solution that is buffered?
    a. The pH will rise.   c. The pH will not change.
    b. The pH will lower.   d. All of these are correct.

12. Acids
    a. release hydrogen ions in solution.
    b. cause the pH of a solution to rise above 7.
    c. take up hydroxide ions and become neutral.
    d. increase the number of water molecules.
    e. Both a and b are correct.

13. If a chemical accepted H^+ from the surrounding solution, the chemical could be
    a. a base.   d. None of the above are correct.
    b. an acid.   e. Both a and c are correct.
    c. a buffer.

14. The periodic table tells us
    a. the atomic number, symbol, and mass.
    b. how many shells an atom has.
    c. how many electrons are in the outer shell.
    d. whether the atom will react or not.
    e. All of these are correct.

15. Which of these best describes the changes that occur when a solution goes from pH 5 to pH 7?
    a. The solution is now 100 times more acidic.
    b. The solution is now 100 times more basic.
    c. The hydrogen ion concentration decreases by only a factor of 20, as the solution goes from basic to acidic.
    d. The hydrogen ion concentration changes by only a factor of 20, as the solution goes from acidic to basic.

For questions 16–19, match the statements with a property of water in the key.

KEY:
   a. Water flows because it is cohesive.
   b. Water holds its heat.
   c. Water has neutral pH.
   d. Water has a high heat of vaporization.

16. Sweating helps cool us off.

17. Our blood is composed mostly of water and cells.

18. Our blood is just about pH 7.

19. We usually maintain a normal body temperature.

Engage

Thinking Scientifically

1. Natural phenomena often require an explanation. Based on how sodium chloride dissociates in water (see pages 29–30) and Figure 2.12, explain why the oceans don't freeze.

2. Melvin Calvin used radioactive carbon (as a tracer) to discover a series of molecules that form during photosynthesis. Explain why carbon behaves chemically the same, even when radioactive.

Bioethical Issue
The Right to Refuse an IV

When a person gets sick or endures physical stress—as, for example, during childbirth—pH levels may dip or rise too far, endangering that person's life. In most U.S. hospitals, doctors routinely administer IVs, or intravenous infusions, of certain fluids to maintain a patient's pH level. Some people who oppose IVs for philosophical reasons may refuse an IV. That's relatively safe, as long as the person is healthy.

Problems arise when hospital policy dictates an IV, even though a patient does not want one. Should a patient be allowed to refuse an IV? Or does a hospital have the right to insist, for health reasons, that patients accept IV fluids? And what role should doctors play—patient advocates or hospital representatives?

Granddaughter Sylvia with azalea plant.

3

The Chemistry of Organic Molecules

All life is interconnected because it uses a common set of chemicals to build larger, more complex molecules that are used as the building blocks of every living thing. In fact, the same organic molecules are used to build a bacterium, a flower, or a little girl. These macromolecules—namely, carbohydrates, lipids, proteins, and nucleic acids—are combined to produce different structures which lead them to have different functions. The similarities in molecules makes it possible for vegetarians to eat only plants and be healthy, as long as they include enough different plants in their diet. When we consume other things, we break down their macromolecules to smaller molecules, and then we use these smaller molecules to build our own types of carbohydrates, lipids, proteins, and nucleic acids. Both the breakdown and building processes are made possible using special catalysts called enzymes.

A bacterium, plant, and human are different because of their genes. All genes are made of DNA, and the way genes function in cells is essentially the same in all organisms. In this chapter, we extend our knowledge of chemistry and consider macromolecules, how their structure and function are related, and how these support life on Earth.

As you read through the chapter, think about the following questions:

1. How can a relatively small number of chemicals be used to build many large, complex, and diverse molecules?
2. Why can one organism use molecules from another organism to build their own?
3. How are organic molecules combined to create the structures and functions necessary to support life?

CHAPTER OUTLINE

3.1 Organic Molecules 38
3.2 Carbohydrates 41
3.3 Lipids 45
3.4 Proteins 49
3.5 Nucleic Acids 54

BEFORE YOU BEGIN

Before beginning this chapter, take a few moments to review the following discussions.

Figures 1.5–1.8 What are the the major kingdoms of life?

Figure 2.3 What chemical elements are most important for producing organic molecules?

Sections 2.2–2.3 What types of bonds can form between elements in an organic molecule?

FOLLOWING *the* BIG IDEAS

CHAPTER 3   THE CHEMISTRY OF ORGANIC MOLECULES

Evolution	The diversity of biological life is the result of changes in DNA sequences and the biomolecules for which they code.
Energy and Homeostasis	Carbon's stable and versatile covalent bonding leads to tremendous variety in the carbohydrates, lipids, proteins, and nucleic acids observed in living organisms.
Information and Signaling	Nucleic acids are unique among the biomolecules in their ability to store and transmit genetic information.
Interactions and Systems	Macromolecules form the functional basis for all cellular systems.

3.1 Organic Molecules

Learning Outcomes

Upon completion of this section, you should be able to

1. Explain how the properties of carbon enable it to produce diverse organic molecules.
2. Describe how functional groups affect a carbon molecule's chemical reactivity.
3. Compare what is added and what is produced during biomolecule synthesis and degradation reactions.

Table 3.1 Inorganic Versus Organic Molecules

Inorganic Molecules	Organic Molecules
Usually contain positive and negative ions	Always contain carbon and hydrogen
Usually ionic bonding	Always covalent bonding
Always contain a small number of atoms	Often quite large, with many atoms
Often associated with nonliving matter	Usually associated with living organisms

Chemists of the nineteenth century thought that the molecules of cells must contain a vital force, so they divided chemistry into **organic chemistry,** the chemistry of living organisms, and **inorganic chemistry,** the chemistry of nonliving matter. We still use this terminology, even though many types of organic molecules can now be synthesized in the laboratory. Today, we simply define **organic molecules** as molecules that contain both carbon and hydrogen atoms (Table 3.1).

There are only four classes of organic compounds in any living thing: carbohydrates, lipids, proteins, and nucleic acids. Collectively, these are called the **biomolecules** and despite the limited number of types, their functions in a cell are quite diverse. A bacterial cell contains some 5,000 different organic molecules, and a plant or animal cell has twice that number. The diversity of life is possible because of this diversity of organic molecules (Fig. 3.1). Ultimately, the variety of organic molecules is based on the unique chemical properties of the carbon atom.

The Carbon Atom

What is there about carbon that makes organic molecules the same and also different? Carbon is quite small, with only a total of six electrons: two electrons in the first shell and four electrons in the outer shell. To acquire four electrons to complete its outer shell, a carbon atom almost always shares electrons with—you guessed it—CHNOPS, the elements that make up most of the weight of living things (see Fig. 2.1).

Because carbon needs four electrons to complete its outer shell, it can share with as many as four other elements. This flexibility makes carbon an ideal building block for biomolecules, and makes the diversity of molecules we see in nature possible. But even more significant to the shape, and therefore the function, of biomolecules, is the ability of carbon to share electrons with other carbon atoms. The C−C

Figure 3.1 Carbon and life. Carbon (**a**) is the basis for many compounds in living things, including (**b**) an apple and other foods, (**c**) plants, and (**d**) animals, as well as nonliving things like (**e**) tires, (**f**) diamonds, and (**g**) airplanes.

bond that is formed is quite stable, and allows formation of long carbon chains. The molecules termed hydrocarbons are chains of carbon atoms that have additional bonds exclusively with hydrogen atoms.

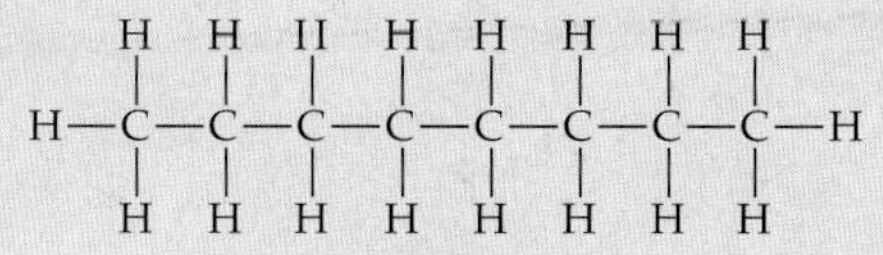

octane

Branching of the chain at any carbon atom is possible, and a hydrocarbon can also turn back on itself to form a ring compound when placed in water:

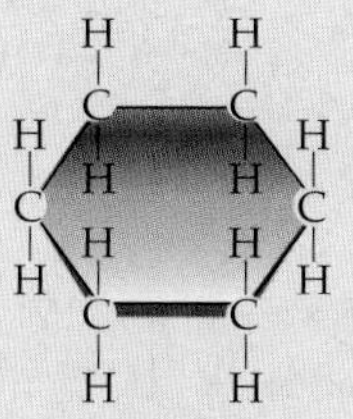

cyclohexane

In addition to forming single bonds, carbon can form double bonds with itself and other atoms. Double bonds aren't as flexible as single bonds, and they restrict the movement of bonded atoms. Double bonds affect a molecule's shape and therefore influence its function. The presence of double bonds is one way to distinguish between saturated and unsaturated fats, which are important to heart health. Carbon is also capable of forming triple bonds with itself, as in acetylene, $H-C\equiv C-H$, a gas used in industrial welding.

The diversity of organic molecules is further enhanced by the presence of particular functional groups, chemical groups attached to the carbon skeleton or backbone, as discussed next. Contrast the structure of cyclohexane, above, with the structure of glucose in Figure 3.6. Notice how the difference in structure can be attributed to the functional groups added to the same number of carbons.

The Carbon Skeleton and Functional Groups

The carbon chain of an organic molecule is called its skeleton or backbone. Just as a skeleton accounts for your body's shape, so does the carbon skeleton of an organic molecule account for its shape. The diversity of vertebrates, species with a backbone, results from the overall shapes of the organisms and the types of appendages (fins, wings, limbs) they have developed. Likewise, the diversity of organic molecules comes from the attachment of different functional groups to the carbon skeleton.

Table 3.2 Functional groups.

Functional Groups			
Group	**Structure**	**Compound**	**Significance**
Hydroxyl	$R-OH$	Alcohol as in ethanol	Polar, forms hydrogen bond Present in sugars, some amino acids
Carbonyl	$R-C\overset{O}{\underset{H}{\diagdown}}$	Aldehyde as in formaldehyde	Polar Present in sugars
	$R-\overset{O}{\overset{\|}{C}}-R$	Ketone as in acetone	Polar Present in sugars
Carboxyl (acidic)	$R-C\overset{O}{\underset{OH}{\diagdown}}$	Carboxylic acid as in acetic acid	Polar, acidic Present in fatty acids, amino acids
Amino	$R-N\overset{H}{\underset{H}{\diagdown}}$	Amine as in tryptophan	Polar, basic, forms hydrogen bonds Present in amino acids
Sulfhydryl	$R-SH$	Thiol as in ethanethiol	Forms disulfide bonds Present in some amino acids
Phosphate	$R-O-\overset{O}{\underset{OH}{\overset{\|}{P}}}-OH$	Organic phosphate as in phosphorylated molecules	Polar, acidic Present in nucleotides, phospholipids

R = remainder of molecule

A **functional group** is a specific combination of bonded atoms that always reacts in the same way, regardless of the particular carbon skeleton. Much of a biomolecule's chemistry can be attributed to its functional groups, rather than to the carbon skeleton to which they are attached. As in Table 3.2, it is even acceptable to use an R to stand for the "remainder" of the molecule, which is the carbon skeleton, because only the functional group is involved in a chemical reaction.

When a particular functional group is added to a carbon skeleton, the molecule becomes a certain type of compound. For example, the addition of an $-OH$ (hydroxyl group) to a carbon skeleton turns that molecule into an alcohol. When an −OH replaces one of the hydrogens in ethane, a 2-carbon hydrocarbon, it becomes ethanol, a type of alcohol that is familiar because humans can consume it. Whereas ethane, like other hydrocarbons, is **hydrophobic** (not soluble in water), ethanol is **hydrophilic** (soluble in water) because the −OH functional group makes the otherwise nonpolar carbon skeleton polar. Because cells are 70–90% water, the ability to interact with and be soluble in water profoundly affects the function of organic molecules in cells.

Organic molecules containing carboxyl (acidic) groups (−COOH) are highly polar. They tend to ionize and release hydrogen ions in solution:

$$-COOH \longrightarrow -COO^- + H^+$$

The attached functional groups determine not only the polarity of an organic molecule, but also the types of reactions it will undergo. You will see that alcohols react with carboxyl groups when a fat forms, and that carboxyl groups react with amino groups during protein formation.

Isomers

Isomers [Gk. *isos,* equal, and *meros,* part, portion] are organic molecules that have identical molecular formulas but a different arrangement of atoms. In essence, isomers are variations in the molecular architecture of a molecule. Isomers are another example of how the chemistry of carbon leads to variations in organic molecules.

The two molecules in Figure 3.2 are isomers of one another; they have the same molecular formula but different functional groups. Therefore, we would expect them to have different properties and react differently in chemical reactions.

The Biomolecules of Cells

You are familiar with the names of biomolecules—carbohydrates, lipids, proteins, and nucleic acids—because certain foods are known to be rich in them, as shown in Figure 3.3. For example, bread is rich in carbohydrates, and meat is rich in proteins. When you digest food, it gets broken down into smaller molecules that are subunits for biomolecules that your body uses. Digestion of bread releases glucose molecules, and digestion of

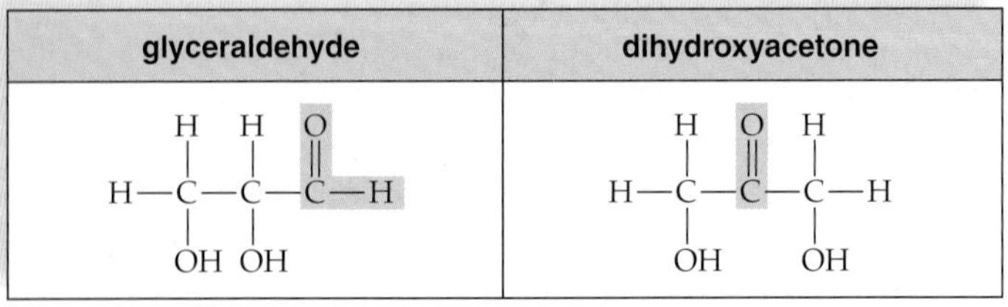

glyceraldehyde	dihydroxyacetone
H H O | | || H—C—C—C—H | | OH OH	H O H | || | H—C—C—C—H | | OH OH

Figure 3.2 Isomers. Isomers have the same molecular formula but different atomic configurations. Both of these compounds have the formula $C_3H_6O_3$. In glyceraldehyde, a colorless crystalline solid, oxygen is double-bonded to an end carbon. In dihydroxyacetone, a white crystalline solid, oxygen is double-bonded to the middle carbon.

Biomolecules		
Category	**Subunit(s)**	**Polymer**
Carbohydrates*	Monosaccharide ⟶	Polysaccharide
Lipids	Glycerol and fatty acids ⟶	Fat
Proteins*	Amino acids ⟶	Polypeptide
Nucleic acids*	Nucleotide ⟶	DNA, RNA

*Polymers

Figure 3.3 Common foods. Carbohydrates in bread and pasta are digested to sugars; lipids such as oils are digested to glycerol and fatty acids; and proteins in meat are digested to amino acids. Cells use these subunit molecules as a source of energy and to build their own biomolecules.

meat releases amino acids. Your body then takes these subunits and builds from them the particular carbohydrates and proteins that make up your cells (Fig. 3.3).

Polymers

The largest of the biomolecules are called **polymers,** and like all biomolecules, polymers are constructed by linking together a large number of the same type of subunit. However, in the case of polymers, the subunits are called **monomers.** A polysaccharide, a protein, and a nucleic acid are all polymers that contain many monomers. Just as a train increases in length when boxcars are hitched together one by one, so a polymer gets longer as monomers bond to one another.

Synthesis and Degradation

A cell uses a condensation reaction to synthesize (build up) any type of biomolecule. It's called a **dehydration reaction** because the equivalent of a water molecule—that is, an −OH (hydroxyl group) and an −H (hydrogen atom), is removed as subunits are

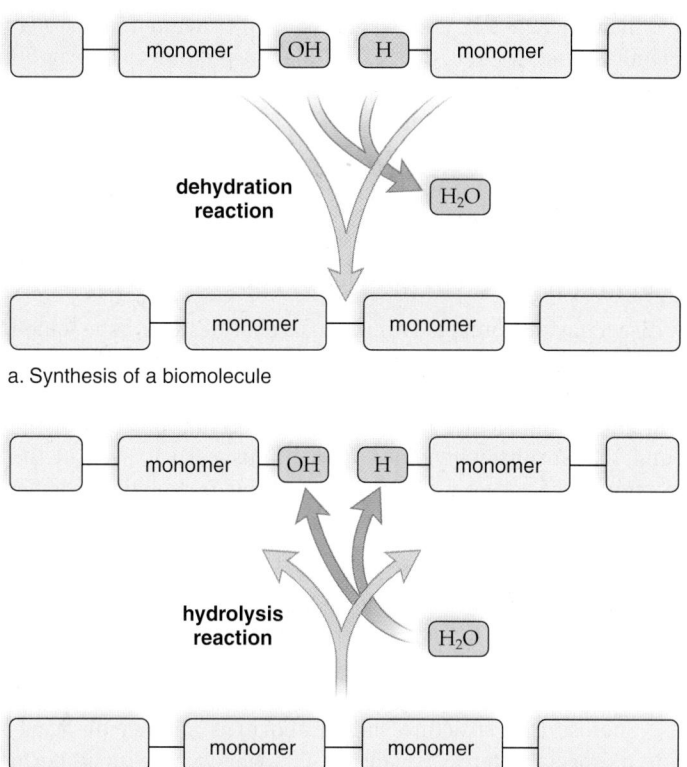

Figure 3.4 Synthesis and degradation of biomolecules.
a. In cells, synthesis often occurs when subunits bond during a dehydration reaction (removal of H_2O). **b.** Degradation occurs when the subunits separate during a hydrolysis reaction (the addition of H_2O).

a. Synthesis of a biomolecule

b. Degradation of a biomolecule

joined. Therefore, water molecules are formed as biomolecules are synthesized (Fig. 3.4*a*).

To break down biomolecules, a cell uses an opposite type of reaction. During a **hydrolysis** [Gk. *hydro,* water, and *lyse,* break] **reaction,** an −OH group from water attaches to one subunit, and an −H from water attaches to the other subunit. In other words, biomolecules are broken down by adding water to them (Fig. 3.4*b*).

Enzymes are required for cells to carry out dehydration and hydrolysis reactions. An **enzyme** is a molecule that speeds a reaction by bringing reactants together and helping them to form new molecules. The enzyme partici-pates in the reaction but is unchanged by it.

MP3
Organic Molecules

Check Your Progress   3.1

1. Describe the properties of a carbon atom that make it ideally suited to produce varied carbon skeletons.
2. Compare solubility in water of a 2-carbon alcohol and a 2-carbon carboxylic acid biomolecule.
3. Discuss what would happen if no water was present during degradation of a biomolecule.

3.2 Carbohydrates

Learning Outcomes

Upon completion of this section, you should be able to

1. List several examples of important monosaccharides and polysaccharides.
2. Compare the energy and structural uses of starch, glycogen, and cellulose.

Carbohydrates are almost universally used as an immediate energy source in living things, but they also play structural roles in a variety of organisms (Fig. 3.5). The majority of carbohydrates have a carbon to hydrogen to oxygen ratio of 1:2:1. The term *carbohydrate* (literally, carbon-water) includes single sugar molecules and chains of sugars. Chain length varies from a few sugars to hundreds of sugars. The monomer subunits, called monosaccharides, are assembled into long polymer chains called polysaccharides.

b. Shell contains chitin.

a. Cell walls contain cellulose.   c. Cell walls contain peptidoglycan.

Figure 3.5 Carbohydrates as structural materials. **a.** Plants, such as the cactus shown here, have the carbohydrate cellulose in their cell walls. **b.** The shell of a crab contains chitin, a different carbohydrate. **c.** The cell walls of bacteria contain another type of carbohydrate known as peptidoglycan.

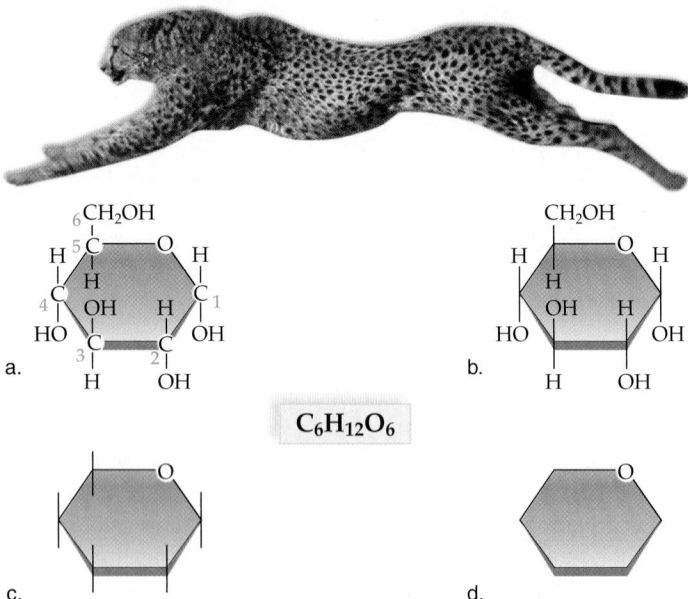

Figure 3.6 Glucose. Glucose provides energy for organisms, such as this cheetah. Each of these structural formulas is glucose. **a.** The carbon skeleton and all attached groups are shown. **b.** The carbon skeleton is omitted. **c.** The carbon skeleton and attached groups are omitted. **d.** Only the ring shape, which includes one oxygen atom, remains.

Monosaccharides: Ready Energy

Monosaccharides [Gk. *monos*, single, and *sacchar*, sugar] consist of only a single sugar molecule and are called simple sugars. A simple sugar can have a carbon backbone of three to seven carbons. The molecular formula for a simple sugar is some multiple of CH_2O, suggesting that every carbon atom is bonded to an $-H$ and an $-OH$. This is not strictly correct, as you can see by examining the structural formula for glucose (Fig. 3.6). Still, sugars do have many hydroxyl groups, and this polar functional group makes them soluble in water.

Glucose, with six carbon atoms, is a **hexose** [Gk. *hex*, six] sugar and has a molecular formula of $C_6H_{12}O_6$. Despite the fact that glucose has several isomers, such as fructose and galactose, we usually think of $C_6H_{12}O_6$ as glucose. Glucose is critical to biological function, and is the major source of cellular fuel for all living things. Glucose is transported in the blood of animals, and

it is the molecule that is broken down and converted into stored chemical energy (ATP) during cellular respiration in nearly all types of organisms.

Ribose and **deoxyribose,** with five carbon atoms, are **pentose** [Gk. *pent*, five] sugars that are significant because they are found respectively in the nucleic acids RNA and DNA. RNA and DNA are discussed later in the chapter.

Disaccharides: Varied Uses

A **disaccharide** contains two monosaccharides that have joined during a dehydration reaction. Figure 3.7 shows how the disaccharide maltose (an ingredient used in brewing) arises when two glucose molecules bond together. Note the position of the bond that results when the $-OH$ groups participating in the reaction project below the ring. When our hydrolytic digestive juices break this bond, the result is two glucose molecules.

Sucrose (the structure shown above) is another disaccharide of special interest because it is sugar we use at home to sweeten our food. Sucrose is also the form in which sugar is transported in plants. We acquire sucrose from plants such as sugarcane and sugar beets. You may also have heard of lactose, a disaccharide found in milk. Lactose is glucose combined with galactose. Individuals that are lactose intolerant cannot break this disaccharide down and have subsequent medical problems. To prevent problems they can buy foods in which lactose has been broken down into its subunit monosaccharides.

Polysaccharides: Energy Storage Molecules

Polysaccharides are polymers of monosaccharides. Some types of polysaccharides function as short-term energy storage molecules. When an organism requires energy, the polysaccharide is broken down to release sugar molecules. The helical shape of the polysaccharides in Figure 3.8 exposes the sugar linkages to the hydrolytic enzymes that can break them down.

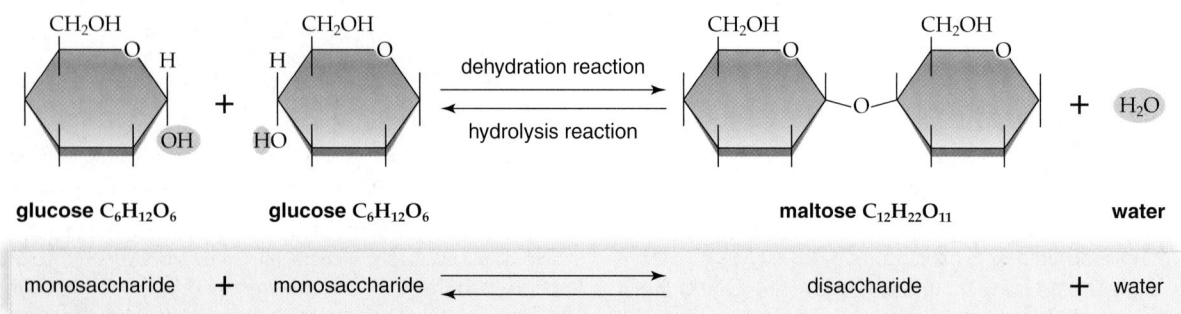

Figure 3.7 Synthesis and degradation of maltose, a disaccharide. Synthesis of maltose occurs following a dehydration reaction when a bond forms between two glucose molecules, and water is removed. Degradation of maltose occurs following a hydrolysis reaction when this bond is broken by the addition of water.

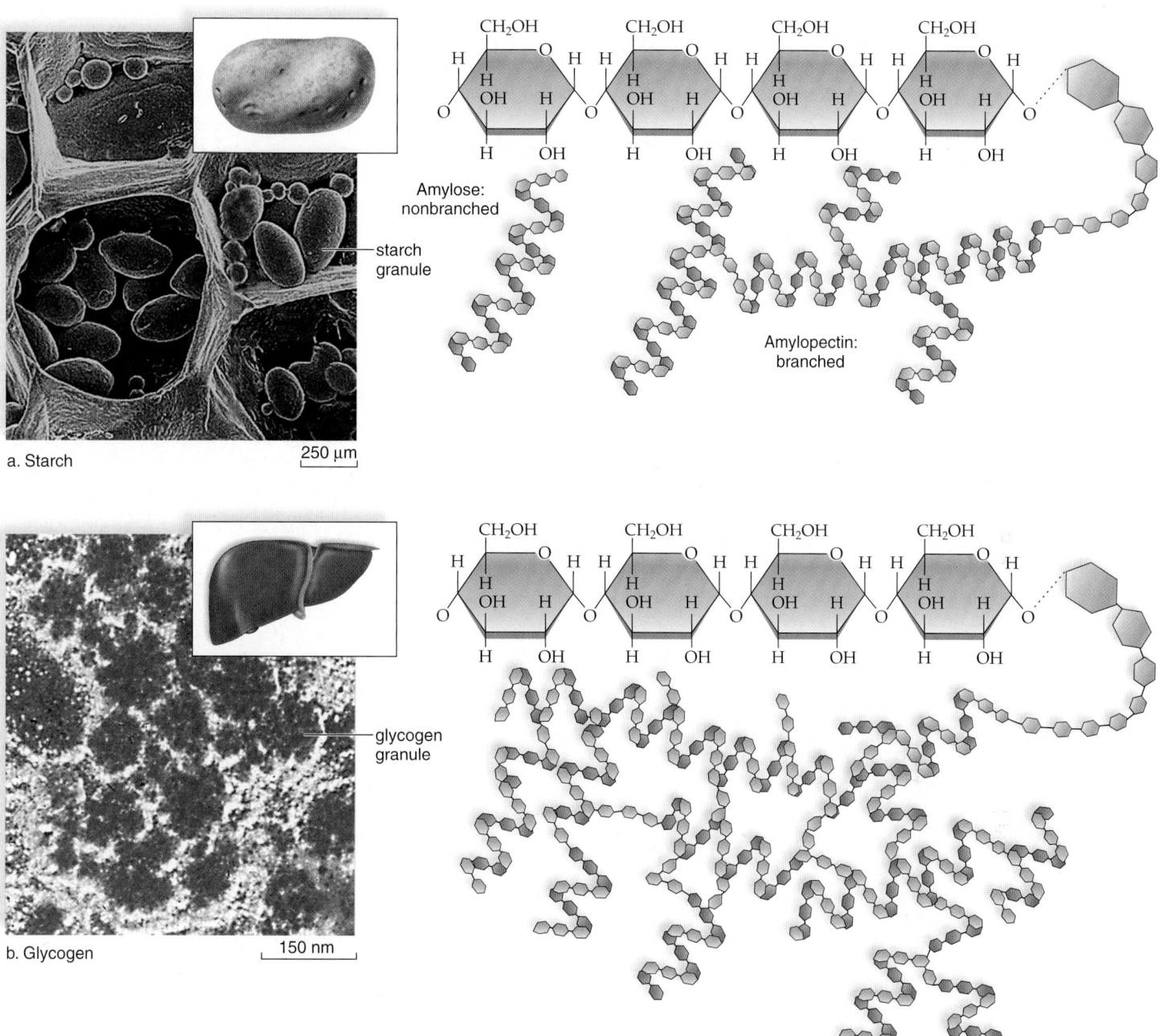

a. Starch   250 µm

Amylose: nonbranched

Amylopectin: branched

starch granule

b. Glycogen   150 nm

glycogen granule

Figure 3.8 Starch and glycogen structure and function. **a.** The electron micrograph shows the location of starch in plant cells. Starch is a chain of glucose molecules that can be branched or unbranched. **b.** The electron micrograph shows glycogen deposits in a portion of a liver cell. Glycogen is a highly branched polymer of glucose molecules.

Plants store glucose as **starch.** The cells of a potato contain granules where starch resides during winter until energy is needed for growth in the spring. Notice in Figure 3.8a that starch exists in two forms: One form (amylose) is nonbranched and the other (amylopectin) is branched. When a polysaccharide is branched, there is no main carbon chain because new chains occur at regular intervals, always at the sixth carbon of the monomer.

Animals store glucose as **glycogen.** In our bodies and those of other vertebrates, liver cells contain granules where glycogen

is stored until needed. The storage and release of glucose from liver cells is controlled by hormones. After we eat, the release of the hormone insulin from the pancreas promotes the storage of glucose as glycogen. Notice in Figure 3.8b that glycogen is even more branched than starch.

Polysaccharides serve as storage molecules because they are not as soluble in water, and are much larger than a simple sugar. Therefore, polysaccharides cannot easily pass through the plasma membrane, a sheetlike structure that encloses cells.

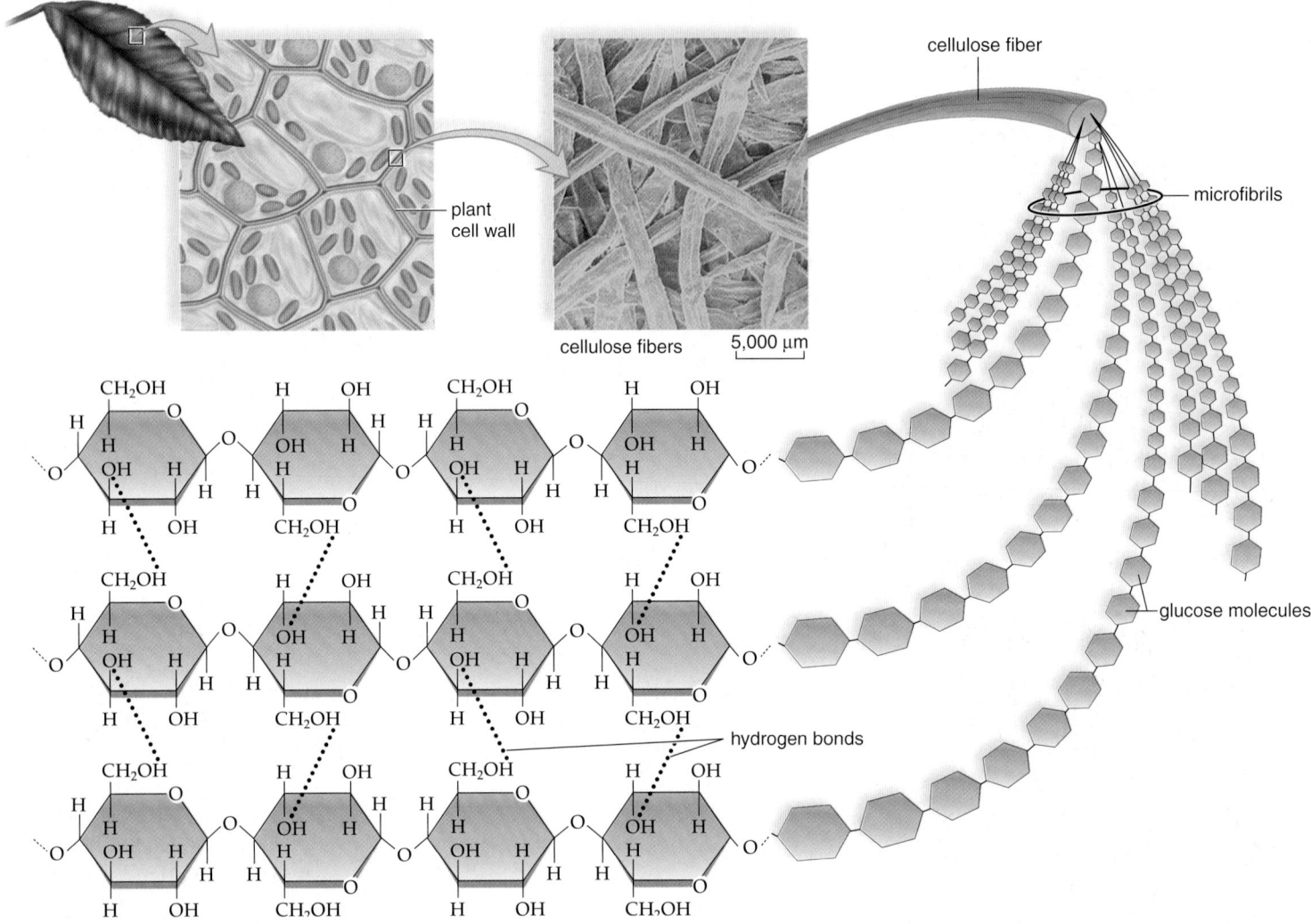

Figure 3.9 Cellulose fibrils. Cellulose fibers criss-cross in plant cell walls for added strength. A cellulose fiber contains several microfibrils, each a polymer of glucose molecules—notice that the linkage bonds differ from those of starch. Every other glucose is flipped, permitting hydrogen bonding and greater strength between the microfibrils.

Polysaccharides: Structural Molecules

Structural polysaccharides include **cellulose** in plants, **chitin** in animals and fungi, and **peptidoglycan** in bacteria (see Fig. 3.5). In all three, monomers are joined by the type of bond shown for cellulose in Figure 3.9. The cellulose monomer is simply glucose, but in chitin, the monomer has an attached amino group. The structure of peptidoglycan is even more complex because each monomer also has an amino acid chain. In both cases, the addition of a functional group to the glucose monomer changes its chemical properties.

Cellulose is the most abundant carbohydrate, and in fact, the most abundant organic molecule on Earth—over 100 billion tons of cellulose is produced by plants each year. Wood, a cellulose plant product, is used for construction, and cotton is used for cloth. Microorganisms, but not animals, are able to digest the bond between glucose monomers in cellulose. The protozoans in the gut of termites enable these insects to digest wood. In cows and other ruminants, microorganisms break down cellulose in a special digestive-tract pouch before the "cud" is returned to the mouth for more chewing and reswallowing. In rabbits, microorganisms digest cellulose in a pouch where it is packaged into pellets. In order to make use of these nutrient pellets, rabbits have to reswallow them as soon as they pass out at the anus. For other animals, including humans, that have no means of digesting cellulose, cellulose serves as dietary fiber, which maintains regularity of fecal elimination.

Chitin [Gk. *chiton,* tunic] is found in fungal cell walls and in the exoskeletons of crabs and related animals, such as lobsters, scorpions, and insects. Chitin, like cellulose, cannot be digested by animals; however, humans have found many other good uses for chitin. Seeds are coated with chitin, and this protects them from attack by soil fungi. Because chitin also has antibacterial and antiviral properties, it is processed and used in medicine as a wound dressing and suture material. Chitin is even useful during the production of cosmetics and various foods.

MP3
Carbohydrates

Check Your Progress   3.2

1. Explain why humans cannot utilize the glucose in cellulose as a nutrient source.
2. Compare and contrast the structure and function of cellulose with chitin.

3.3 Lipids

Learning Outcomes

Upon completion of this section, you should be able to

1. Describe why lipids are essential to living organisms.
2. Explain where fats and oils are produced.
3. Contrast the structures of fats, phospholipids, and steroids.
4. Compare the functions of phospholipids and steroids in cells.

Table 3.3 Lipids

Type	Functions	Human Uses
Fats	Long-term energy storage and insulation in animals	Butter, lard
Oils	Long-term energy storage in plants and their seeds	Cooking oils
Phospholipids	Component of plasma membrane	—
Steroids	Component of plasma membrane (cholesterol), sex hormones	Medicines
Waxes	Protection, prevent water loss (cuticle of plant surfaces), beeswax, earwax	Candles, polishes

A variety of organic compounds are classified as **lipids** [Gk. *lipos*, fat] (Table 3.3). These compounds are insoluble in water due to their hydrocarbon chains. Hydrogens bonded only to carbon are nonpolar and have no tendency to form hydrogen bonds with water molecules. Fat, a well-known lipid, is used by animals for both insulation and long-term energy storage. Fat below the skin of marine mammals is called blubber; in humans, it is given slang expressions such as "spare tire" and "love handles." Plants use oil instead of fat for long-term energy storage. We are familiar with fats and oils because we use them as foods and for cooking.

Phospholipids and steroids are also important lipids found in living things. They serve as major components of the plasma membrane in cells. Steroids are also involved in cell communication. Waxes, which are sticky, not greasy like fats and oils, tend to have a protective function in living things.

Nature of Science

Everyone Needs a Little Fat, Right?

Fats are an essential part of the diet. They provide lots of energy, having more than twice the caloric density of carbohydrates. Fats are needed to build and maintain cell membranes, which are critical to biological life, as well as hormones like testosterone and estrogen, and they provide essential padding for our internal organs. Fats are also a major reason why some foods taste good.

A short time ago, scientists and nutritionists thought that too much fat was bad for you. Now we know that saturated fats, which come from animals and are solid at room temperature, have effects in the body that are different from those of unsaturated fats, which come from plants and are liquid at room temperature. Saturated fats are flat molecules that easily stick together in the blood, and too much saturated fat has been shown by scientists to negatively affect heart health, contributing to clogging of arteries and coronary heart disease (CHD). By comparison, unsaturated fats seem to help prevent CHD because they don't stick together in the blood, and therefore don't clog arteries.

A Food Revolution

Unsaturated fats might be healthier for you, but plant oils can easily go rancid and aren't solid at room temperature, which makes them more difficult to cook with and to use in solid food products. To get around this problem, food manufacturers hydrogenated unsaturated fatty acids by heating the oil and exposing it to hydrogen gas. This treatment made the otherwise liquid plant oils semi-solid at room temperature, and gave foods containing partially hydrogenated oils better shelf life.

An unintended consequence of hydrogenation, however, was the formation of **trans-fats**. Many commercially packaged foods contain trans-fats, which recently have been shown to increase LDL or "bad" cholesterol and lower HDL or "good" cholesterol in the blood. Trans-fat consumption also appears to increase risk of CHD and heart attack.

Scientific Evidence Changes Our Perceptions

Initially, investigators thought that the total amount of lipid in the diet caused coronary and other heart-related diseases. As scientific evidence accumulated showing a distinction between the effects of saturated and unsaturated fats, public perception changed. Until recently, trans-fats were of little concern to the general public, and people readily consumed them without much thought. As science has brought the negative effects of trans-fats to light, perception has changed once again. Public outcry has prompted changes in the food services industry, with clear labeling of trans-fats on all food products and with more restaurants using trans-fat-free oils during cooking.

This example shows how our perceptions change over time based on scientific evidence, and it illustrates the essential role that science plays in the common good. Science constantly refines what we know as new evidence provides greater insights into how we function and live.

Questions to Consider

1. How much trans-fat do you consume daily?
2. What is the chemical structure of a trans-fat compared to a non-trans-fat?
3. How might your new scientific knowledge of fats affect your nutritional decision-making?

Triglycerides: Long-Term Energy Storage

Fats and **oils** contain two types of subunit molecules: fatty acids and glycerol. Each **fatty acid** consists of a long hydrocarbon chain with an even number of carbons and a −COOH (carboxyl) group at one end. Most of the fatty acids in cells contain 16 or 18 carbon atoms per molecule, although smaller ones are also found. Fatty acids are either saturated or unsaturated. **Saturated fatty acids** have no double bonds between the carbon atoms and contain as many hydrogens as they can potentially hold. **Unsaturated fatty acids** have double bonds in the carbon chain, which reduces the number of bonded hydrogen atoms. In addition, double bonds in unsaturated fatty acids may have chemical groups arranged on the same side (termed *cis* configuration) or on opposite sides (termed *trans* configuration) of the double bond. The *cis* or *trans* configuration of an unsaturated fatty acid affects its biological activity.

Glycerol is a 3-carbon compound with three −OH groups. The −OH groups are polar, making glycerol soluble in water. When a fat or oil forms, the −COOH functional groups of three fatty acids react with the −OH groups of glycerol during a dehydration reaction (Fig. 3.10a), resulting in a fat molecule and three molecules of water. Fats and oils are degraded during a hydrolysis reaction. Because three fatty acids are attached to each glycerol molecule, fats and oils are sometimes called **triglycerides.** Notice that triglycerides have many nonpolar C−H bonds; therefore, they do not mix with water. Even though cooking oils and water are both liquid, they do not mix, even after shaking, because the nonpolar oil and polar water are chemically incompatible.

Triglycerides containing fatty acids with unsaturated bonds melt at a lower temperature than those containing only saturated fatty acids. The reason is that a double bond creates a kink in the fatty acid chain that prevents close packing between the hydrocarbon chains (Fig. 3.10a). We can infer that butter, a fat that is solid at room temperature, must contain primarily saturated fatty acids, while corn oil, which is a liquid even when placed in the refrigerator, must contain primarily unsaturated fatty acids (Fig. 3.10b). This difference is useful to living things. For example, the feet of reindeer and penguins contain unsaturated triglycerides, and this helps protect those exposed parts from freezing.

In general, fats, which most often come from animals, are solid at room temperature, whereas oils, which come from plants, are liquid at room temperature. Diets high in animal fat have been associated with circulatory disorders because saturated fats and other molecules can accumulate inside the lining of blood vessels and block blood flow. Health organizations have recommended replacing fat with oils such as olive oil and canola oil in our diet whenever possible.

Nearly all animals use fat rather than glycogen for long-term energy storage. Gram for gram, fat stores more energy than glycogen. The C−H bonds of fatty acids make them a richer source of chemical energy than glycogen, because more bonds with stored energy are present in fatty acids; in contrast, glycogen has many C−OH bonds, which are less energetic bonds. Also, fat droplets do not contain water because they are nonpolar. Small birds, like the broad-tailed hummingbird, store a great deal of fat before they start their long spring and fall migratory flights. About 0.15 g of fat per gram of body weight is accumulated each day. If the same amount of energy were stored as glycogen, a bird would be so heavy it would not be able to fly.

Phospholipids: Membrane Components

Phospholipids [Gk. *phos,* light, and *lipos,* fat], as implied by their name, contain a phosphate group. Essentially, a phospholipid is constructed like a fat, except that in place of the third fatty acid attached to glycerol, there is a polar phosphate group. The phosphate group is usually bonded to another organic group, indicated by *R* in Figure 3.11a. This portion of the molecule becomes the polar head, while the hydrocarbon chains of the fatty acids become the nonpolar tails. Notice that a double bond causes a tail to kink.

Phospholipids have hydrophilic heads and hydrophobic tails. When exposed to water, phospholipids tend to arrange themselves so that the polar heads are oriented toward water, and the nonpolar fatty acid tails are oriented away from water. In living things, which are made mostly of water, phospholipids tend to become a bilayer (double layer) because the polar heads prefer to interact with other polar molecules like water. Conversely, the nonpolar tails associate together and stay away from polar water molecules. So, phospholipids arrange themselves like a "sandwich," with the polar heads facing to the the outside (the bread slices) and the fatty acid tails on the inside (the filling). This phospholipid bilayer is a key component used to keep cells and the biological compartments within cells separate.

The plasma membrane that surrounds cells consists primarily of a phospholipid bilayer (Fig. 3.11b). The presence of kinks in the tails causes the plasma membrane to be fluid across a range of temperatures found in nature. A plasma membrane is absolutely essential to the structure and function of a cell, and this signifies the importance of phospholipids to living things.

Steroids: Four Fused Rings

Steroids are lipids that have entirely different structures from those of fats. Steroid molecules have skeletons of four fused carbon rings (Fig. 3.12a). Each type of steroid differs primarily by the types of functional groups attached to the carbon skeleton.

Cholesterol is an essential component of an animal cell's plasma membrane, where it provides physical stability. Cholesterol is the precursor of several other steroids, such as the sex hormones testosterone and estrogen (Fig. 3.12b, c). The male sex hormone, testosterone, is formed primarily in the testes, and the female sex hormone, estrogen, is formed primarily in

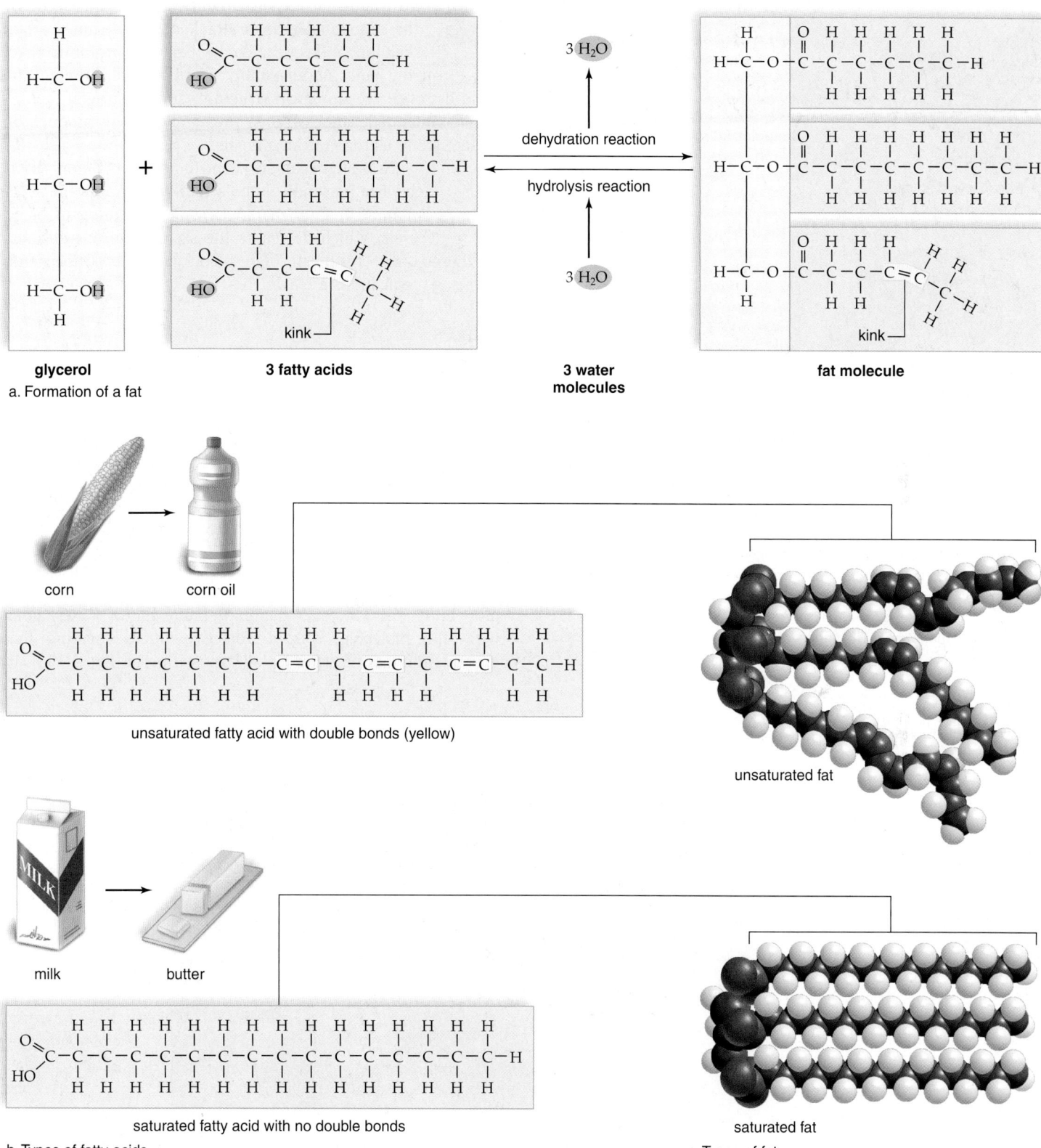

glycerol
a. Formation of a fat

3 fatty acids

3 water molecules

fat molecule

corn

corn oil

unsaturated fatty acid with double bonds (yellow)

unsaturated fat

milk

butter

saturated fatty acid with no double bonds

saturated fat

b. Types of fatty acids

c. Types of fats

Figure 3.10 Fat and fatty acids. **a.** Following a dehydration reaction, glycerol is bonded to three fatty acid molecules as fat forms and water is given off. Following a hydrolysis reaction, the bonds are broken due to the addition of water. **b.** A fatty acid has a carboxyl group attached to a long hydrocarbon chain. If there are double bonds between some of the carbons in the chain, the fatty acid is unsaturated and a kink occurs in the chain. If there are no double bonds, the fatty acid is saturated. **c.** Space-filling models of an unsaturated fat and a saturated fat.

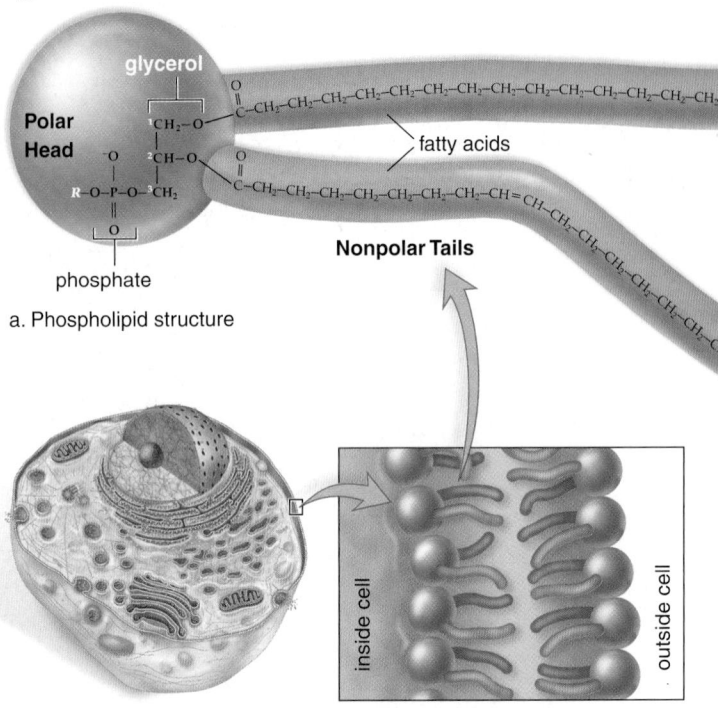

a. Phospholipid structure

b. Plasma membrane of a cell

Figure 3.11 Phospholipids form membranes.   a. Phospholipids are constructed like fats, except that in place of the third fatty acid, they have a polar phosphate group. The hydrophilic (polar) head is soluble in water, whereas the two hydrophobic (nonpolar) tails are not. A tail has a kink wherever there is an unsaturated bond. **b.** Because of their structure, phospholipids form a bilayer that serves as the major component of a cell's plasma membrane. The fluidity of the plasma membrane is affected by kinks in the phospholipids' tails.

the ovaries. Testosterone and estrogen differ only by the functional groups attached to the same carbon skeleton, and yet each have their own profound effect on the body and sexuality of an animal. Human and plant estrogen are similar in structure, and if estrogen therapy is recommended, some women prefer taking soy products in preference to estrogen from animals.

Cholesterol can also contribute to circulatory disorders. The presence of cholesterol encourages the accumulation of fatty material inside the lining of blood vessels, which decreases the size of the opening and thereby can result in high blood pressure. Cholesterol-lowering medications are available.

Waxes

In **waxes,** long-chain fatty acids bond with long-chain alcohols:

long-chain fatty acid

$O = C - CH_2 - CH_2 - CH_2 - CH_2 - CH_2 - CH_2 - CH_2 - CH_2 - CH_2 - CH_3$

$\underset{H}{\overset{O}{|}}$

$\underset{H}{\overset{|}{C}} - CH_2 - CH_2 - CH_2 - CH_2 - CH_2 - CH_2 - CH_2 - CH_2 - CH_3$

long-chain alcohol

Waxes are solid at normal temperatures because they have a high melting point. Being hydrophobic, they are also waterproof and resistant to degradation. In many plants, waxes, along with other molecules, form a protective cuticle (covering) that prevents the loss of water for all exposed parts (Fig. 3.13a). In

b. Testosterone

a. Cholesterol

c. Estrogen

Figure 3.12 Steroid diversity.   a. Built like cholesterol, (**b**) testosterone and (**c**) estrogen have different effects on the body due to different functional groups attached to the same carbon skeleton. Testosterone is the male sex hormone active in peacocks (*left*), and estrogen is the female sex hormone active in peahens (*right*). These hormones are present in many living creatures.

a.

b.

Figure 3.13 Waxes. Waxes are a type of lipid. **a.** Fruits are protected by a waxy coating that is visible on these plums. **b.** Bees secrete the wax that allows them to build a comb where they store honey. This bee has collected pollen (yellow) to feed growing larvae.

many animals, waxes are involved in skin and fur maintenance. In humans, wax is produced by glands in the outer ear canal. Earwax contains cerumin, an organic compound that at the very least repels insects, and in some cases even kills them. It also traps dust and dirt, preventing these contaminants from reaching the eardrum.

A honeybee produces beeswax in glands on the underside of its abdomen. Beeswax is used to make the six-sided cells of the comb where honey is stored (Fig. 3.13b). Honey contains the sugars fructose and glucose, breakdown products of the sugar sucrose.

Humans have found a myriad of uses for waxes, from making candles to polishing cars, furniture, and floors.

MP3 Lipids

Check Your Progress   3.3

1. Evaluate why lipids and water do not mix.
2. Contrast a saturated fatty acid with an unsaturated fatty acid. Which of these is preferred in the diet, and why?
3. Explain why phospholipids form a bilayer in water.

3.4 Proteins

Learning Outcomes

Upon completion of this section, you should be able to

1. Describe functions of proteins in cells.
2. Explain how a polypeptide is constructed from amino acids.
3. Compare the four levels of protein structure.
4. Analyze the factors that affect protein structure and function.

Proteins [Gk. *proteios,* first place], as their Greek derivation implies, are of primary importance to the structure and function of cells. As much as 50% of the dry weight of most cells consists of proteins. Presently, several hundred thousand proteins have been identified. Here are some of their many functions in animals:

Metabolism Enzyme proteins bring reactants together and thereby speed chemical reactions in cells. They are specific for one particular type of reaction and function best at specific body temperatures and pH.

Support Some proteins have a structural function. For example, keratin makes up hair and nails, while collagen gives strength to ligaments, tendons, and skin.

Transport Channel and carrier proteins in the plasma membrane regulate what substances enter and exit cells. Other proteins transport molecules in the blood of animals; **hemoglobin** is a complex protein that transports oxygen to tissues and cells.

Defense Antibodies are proteins of our immune system that combine with foreign substances, called antigens. Antibodies bind and prevent antigens from destroying cells and upsetting homeostasis.

Regulation Some hormones are proteins that regulate how cells behave. They serve as intercellular messengers that influence cell metabolism. The hormone insulin regulates how much glucose is in the blood and in cells; the presence of growth hormone during childhood and adolescence determines the height of an individual.

Motion The contractile proteins actin and myosin allow parts of cells to move and cause muscles to contract. Muscle contraction allows animals to travel from place to place. All cells contain proteins that move cell components to different internal locations. Without such proteins, cells would not be able to function.

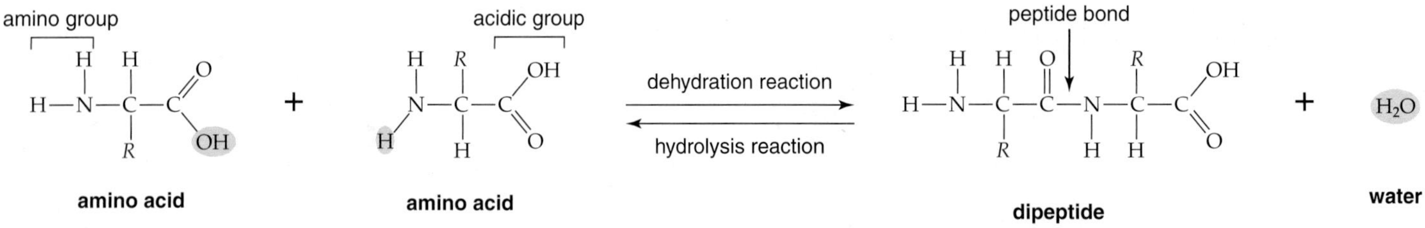

Figure 3.14 Synthesis and degradation of a peptide. Following a dehydration reaction, a peptide bond joins two amino acids and a water molecule is released. Following a hydrolysis reaction, the bond is broken due to the addition of water.

Proteins are such a major part of living organisms that tissues and cells of the body can sometimes be characterized by the proteins they contain or produce. For example, muscle cells contain large amounts of actin and myosin for contraction; red blood cells are filled with hemoglobin for oxygen transport; and support tissues, such as ligaments and tendons, contain the protein collagen, which is composed of tough fibers.

Peptides

Proteins are polymers constructed from amino acid monomers. Figure 3.14 shows how a dehydration reaction joins the carboxyl group of one amino acid to the amino group of another amino acid. The resulting covalent bond between two amino acids is called a **peptide bond.** The atoms associated with the peptide bond share the electrons unevenly because oxygen is more electronegative than nitrogen. Therefore, the hydrogen attached to the nitrogen has a slightly positive charge, while the oxygen has a slightly negative charge:

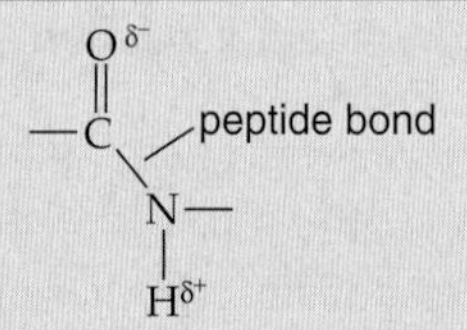

The polarity of the peptide bond means that hydrogen bonding is possible between the $-CO$ of one amino acid and the $-NH$ of another amino acid in a polypeptide. This hydrogen bonding influences the structure, or shape, of a protein.

A **peptide** is two or more amino acids bonded together, and a **polypeptide** is a chain of many amino acids joined by peptide bonds. A protein is a polypeptide that has been folded into a particular shape and has function. Some proteins may consist of more than one polypeptide chain, making it possible for some proteins to have a very large number of amino acids.

In 1953, Frederick Sanger developed a method to determine the sequence of amino acids in a polypeptide. Now that we know the sequences of many thousands of polypeptides found in nature, it is clear that the amino acid sequence greatly influences the final three-dimensional shape and function of a protein. Proteins that have an abnormal sequence often have a three-dimensional shape that causes them to function improperly. From an evolutionary perspective, we also know that, for a particular protein, the sequences of amino acids are highly similar within a species and are different across species.

Amino Acids: Protein Monomers

The name **amino acid** is appropriate because one of these groups is an $-NH_2$ (amino group) and another is a $-COOH$ (an acid group). The third group is called an R group for an amino acid:

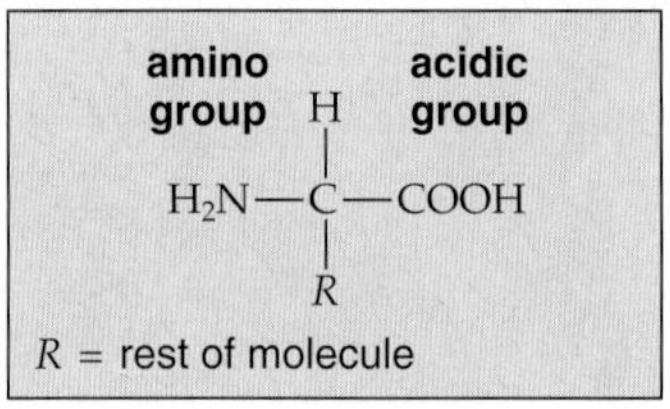

Note that the central carbon atom in an amino acid bonds to a hydrogen atom and also to three other groups of atoms, one of which is the R group (Fig. 3.14). Amino acids differ according to their particular R group (Fig. 3.15). The R groups range in complexity from a single hydrogen atom to complicated ring compounds. Some R groups are polar and associate with water, whereas others are nonpolar and do not. Also, the amino acid cysteine has an R group that ends with an $-SH$ (sulfide) group, which often serves to covalently connect one chain of amino acids to another by a disulfide bond, $-S-S-$. Several other amino acids commonly found in cells are shown in Figure 3.15.

Each protein has a sequence of amino acids that is defined by information contained within a gene. This amino acid sequence forms the basis for all levels of protein structure, which directly affect protein function.

Shape of Proteins

A protein can have up to four levels of structure, termed primary, secondary, tertiary, and quaternary; however, not all proteins have all four levels.

Primary Structure

The primary structure of a protein is the sequence of amino acids defined by a gene. Just as millions of different words can be constructed from just 26 letters in the English alphabet, so too can hundreds of thousands of different polypeptides be built from just 20 amino acids. To make a new word in English, all

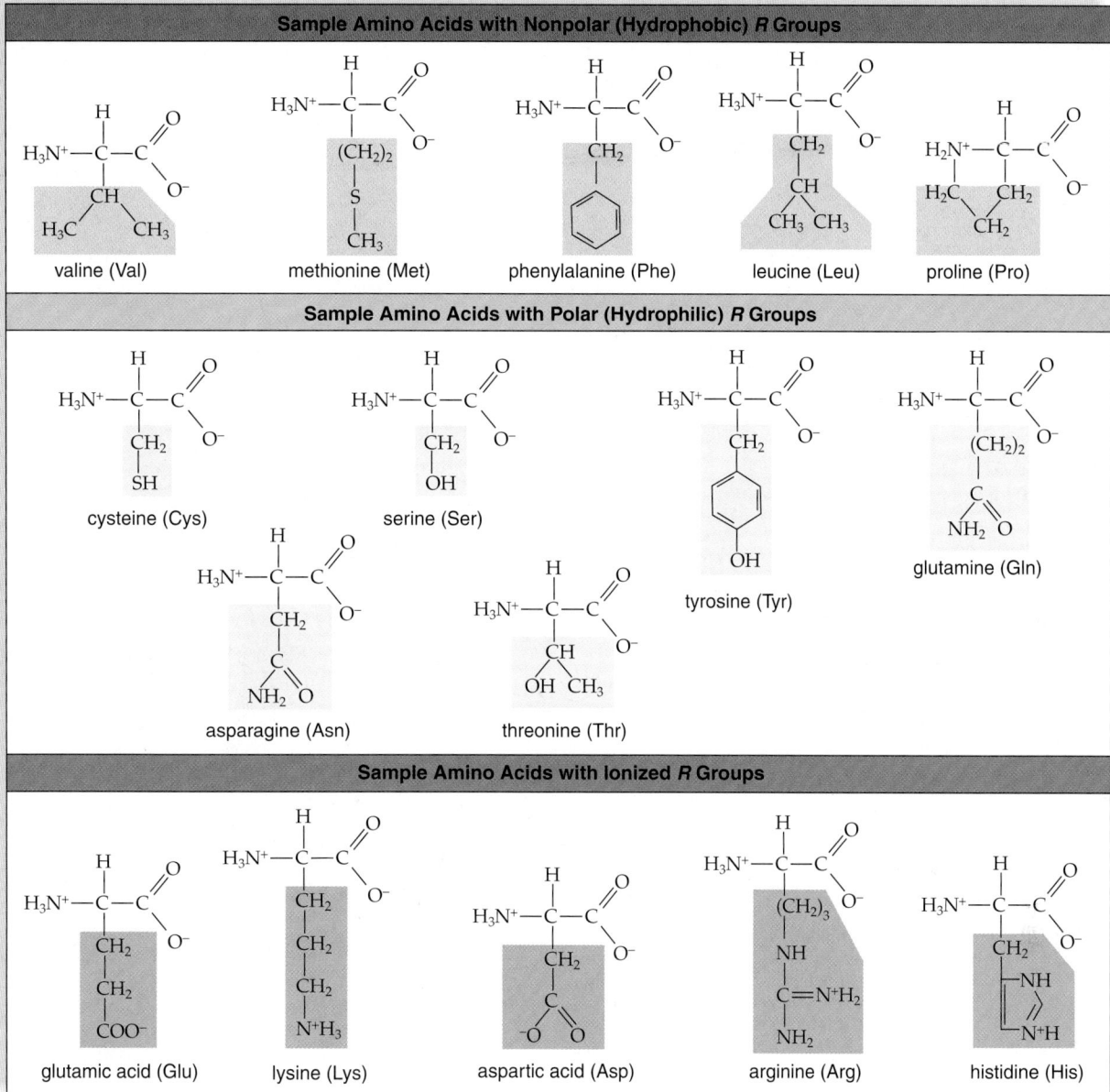

Figure 3.15 Amino acids. Polypeptides contain 20 different kinds of amino acids, some of which are shown here. Amino acids differ by the particular *R* group (blue) attached to the central carbon. Some *R* groups are nonpolar and hydrophobic (red), some are polar and hydrophilic (light blue), and some are ionized and hydrophilic (dark blue). The amino acids are shown in ionized form.

that is required is to vary the number and sequence of a few letters. Likewise, changing the sequence of 20 amino acids in a polypeptide can produce a huge array of different proteins.

Secondary Structure

The secondary structure of a protein occurs when the polypeptide coils or folds in a particular way (Fig. 3.16).

Linus Pauling and Robert Corey, who began studying the structure of amino acids in the late 1930s, concluded that a coiling they called an α (alpha) helix and a pleated sheet they called the β (beta) sheet were two basic patterns of structure amino acids assumed within a polypeptide. The names came from the fact that the α helix was the first, and the β sheet the second, pattern they discovered. Each polypeptide can have multiple α helices and β pleated sheets.

The spiral shape of α helices is formed by hydrogen bonding between every fourth amino acid within the polypeptide chain, whereas β sheets are formed when the polypeptide turns back upon itself, allowing hydrogen bonding to occur between extended lengths of the polypeptide. **Fibrous proteins,** which are structural proteins, exist only as helices or pleated sheets that hydrogen-bond to each other. Examples are keratin, a protein in hair, and silk, a protein that forms spider webs. Both of these proteins have only a secondary structure (Fig. 3.17).

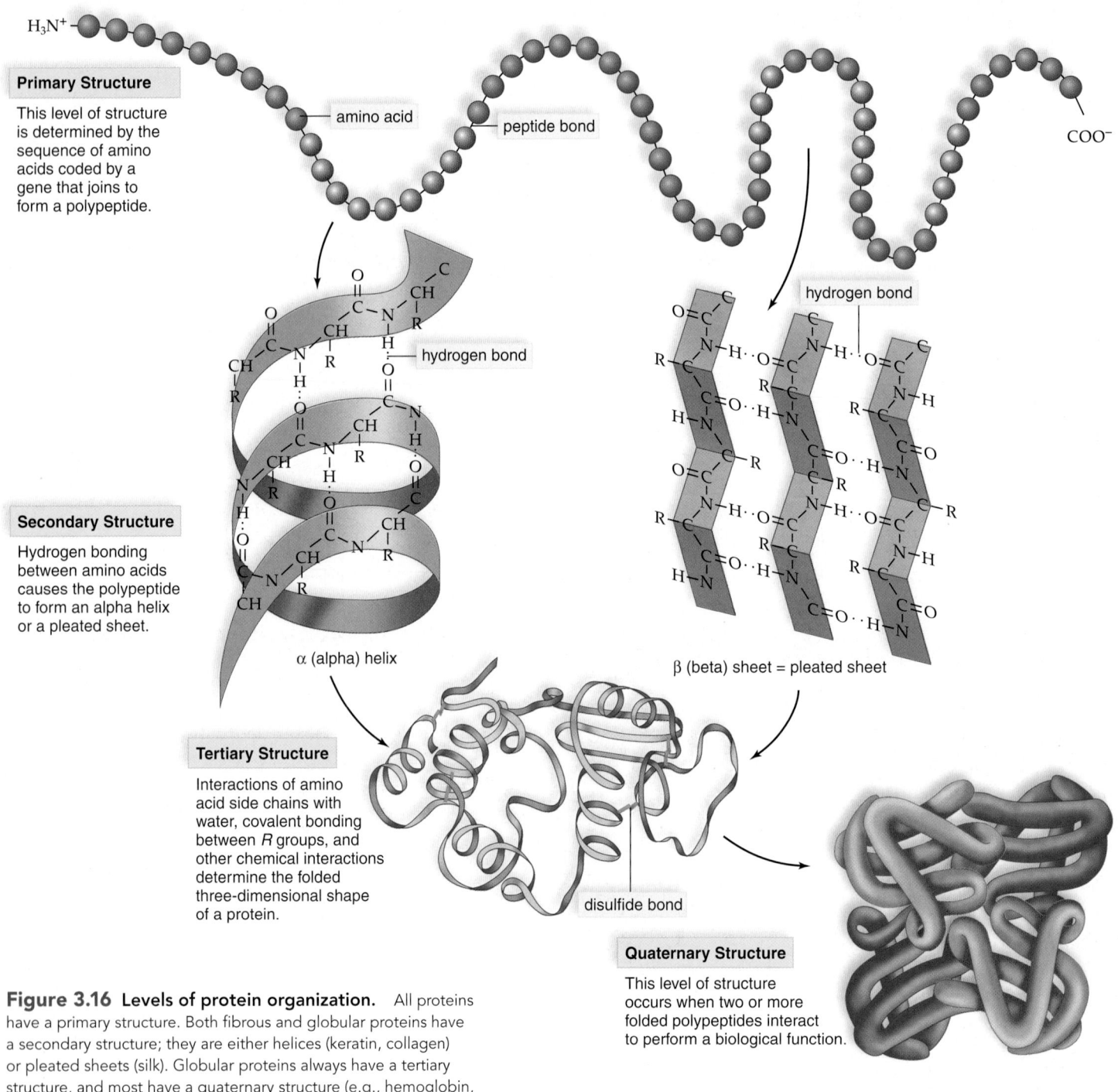

Primary Structure

This level of structure is determined by the sequence of amino acids coded by a gene that joins to form a polypeptide.

amino acid

peptide bond

COO⁻

Secondary Structure

Hydrogen bonding between amino acids causes the polypeptide to form an alpha helix or a pleated sheet.

hydrogen bond

α (alpha) helix

hydrogen bond

β (beta) sheet = pleated sheet

Tertiary Structure

Interactions of amino acid side chains with water, covalent bonding between *R* groups, and other chemical interactions determine the folded three-dimensional shape of a protein.

disulfide bond

Quaternary Structure

This level of structure occurs when two or more folded polypeptides interact to perform a biological function.

Figure 3.16 Levels of protein organization.   All proteins have a primary structure. Both fibrous and globular proteins have a secondary structure; they are either helices (keratin, collagen) or pleated sheets (silk). Globular proteins always have a tertiary structure, and most have a quaternary structure (e.g., hemoglobin, and enzymes such as RNAase).

Tertiary Structure

A tertiary structure is the folding that results in the final three-dimensional shape of a polypeptide. **Globular proteins,** which tend to ball up into rounded shapes, have tertiary structure.

The interaction of hydrophobic amino acids in the polypeptide chain with the surrounding water is a major factor in how proteins fold into, and maintain, their final shape. These nonpolar amino acids tend to group together in the interior of a protein, to be as far away from water as possible. In contrast, the polar hydrophilic and ionic amino acids interact well with water, and tend to orient themselves on the protein's surface. These chemical interactions, along with hydrogen

Figure 3.17 Fibrous proteins. Fibrous proteins are structural proteins. **a.** Keratin—found, for example, in hair, horns, and hoofs—exemplifies fibrous proteins that are helical for most of their length. Keratin is a hydrogen-bonded triple helix. In this photo, Drew Barrymore has straight hair. **b.** In order to give her curly hair, water was used to disrupt the hydrogen bonds, and when the hair dried, new hydrogen bonding allowed it to take on the shape of a curler. A permanent-wave lotion induces new covalent bonds within the helix. **c.** Silk made by spiders and silkworms exemplifies fibrous proteins that are pleated sheets for most of their length. Hydrogen bonding between parts of the molecule occurs as the pleated sheet doubles back on itself.

bonds, ionic bonds, and covalent bonds between R groups, all contribute to the tertiary structure of a protein. Strong disulfide linkages ($-S-S-$) in particular help maintain the tertiary shape.

Enzymes are globular proteins. Enzymes work best at body temperature, and each one also has an optimal pH at which the rate of the reaction is highest. At this temperature and pH, the enzyme can maintain its normal shape. A high temperature and change in pH can disrupt the interactions that maintain the shape of the enzyme. When a protein loses its natural shape, it is said to be **denatured.** An organism can die if too many proteins become denatured because it can no longer maintain the metabolic processes necessary for life.

Animation
Protein Denaturation

Quaternary Structure

Some proteins have a quaternary structure because they consist of more than one polypeptide. Hemoglobin is a much-studied globular protein that consists of four polypeptides, and therefore it has a quaternary structure. Each polypeptide in hemoglobin has a primary, secondary, and tertiary structure. However, a protein can have only two polypeptides and still have quaternary structure.

MP3
Proteins

Protein-Folding Diseases

Proteins cannot function properly unless they fold into their correct shape. In recent years investigators have found that the cell contains **chaperone proteins,** which help new proteins fold into their normal shape. At first it seemed as if chaperone proteins ensured that proteins folded properly, but now it appears that they might correct any misfolding of a new protein. These new findings serve as an example of how scientific research helps correct previous misunderstandings about how life works.

In any case, without fully functioning chaperone proteins, a cell's proteins may not be functional because they have misfolded. Several diseases in humans, such as cystic fibrosis and Alzheimer disease, are associated with misshapen proteins. The possibility exists that the diseases are due to missing or malfunctioning chaperone proteins.

Other diseases in humans are due to misfolded proteins, but the cause may be different. For years, investigators have been studying fatal brain diseases, known as TSEs,[1] that have no cure because no infective agent can be found. Mad cow disease is a well-known example of a TSE disease. Now it appears that TSE diseases could be due to misfolded proteins, called **prions,** that cause other proteins of the same type to fold the wrong way too. A possible relationship between prions and the functioning of chaperone proteins is now under investigation.

Animation
How Prions Arise

Check Your Progress 3.4

1. Explain where the information that specifies amino acid sequence in a polypeptide comes from.
2. Examine which types of amino acids are most likely to be found in the interior of a protein and why.
3. Evaluate which factors are most important to protein folding.

---

1 TSEs: transmissible spongiform encephalopathies.

3.5 Nucleic Acids

Learning Outcomes

Upon completion of this section, you should be able to

1. Distinguish between a nucleotide and nucleic acid.
2. Compare the structure and function of DNA and RNA nucleic acids.
3. Examine why purines and pyrimidines pair together.
4. Explain how ATP is able to store energy.

Each cell has a storehouse of information that specifies how a cell should behave, respond to the environment, and divide to make new cells. **Nucleic acids,** which are polymers of nucleotides, store information, include instructions for life, and conduct chemical reactions. **DNA (deoxyribonucleic acid)** is one type of nucleic acid that not only stores information about how to copy, or replicate, itself, but also specifies the order in which amino acids are to be joined to make a protein.

RNA (ribonucleic acid) is another diverse type of nucleic acid that has multiple uses. Messenger RNA (mRNA) is a temporary copy of a gene in the DNA that specifies what the amino acid sequence will be during the process of protein synthesis. Transfer RNA (tRNA) is also necessary in synthesizing proteins, and helps to translate the sequence of nucleic acids in a gene into the correct sequence of amino acid during protein synthesis. Ribosomal RNA (rRNA) works as an enzyme to form the peptide bonds between amino acids in a polypeptide. A wide range of other RNA molecules also perform important functions within the cell.

Not all nucleotides are made into DNA or RNA polymers. Some nucleotides are directly involved in metabolic functions in cells. For example, some are components of **coenzymes,** nonprotein organic molecules that help regulate enzymatic reactions. **ATP (adenosine triphosphate)** is a special nucleotide that stores large amounts of energy needed for synthetic reactions and for various other energy-requiring processes in cells.

Structure of DNA and RNA

Every **nucleotide** is comprised of three types of molecules: a pentose sugar, a phosphate (phosphoric acid), and a nitrogen-containing base (Fig. 3.18a). In DNA, the pentose sugar is deoxyribose, and in RNA the pentose sugar is ribose. A difference in the structure of these 5-carbon sugars accounts for their respective names because, as you might guess, deoxyribose lacks an oxygen atom found in ribose (Fig. 3.18b).

Both DNA and RNA contain combinations of four nucleotides, but these differ somewhat between the two nucleic acids (Fig. 3.18c). Nucleotides that have a base with a single ring are called pyrimidines, and nucleotides with a double ring are called purines. In DNA, the pyrimidine bases are cytosine and thymine; in RNA, the pyrimidine bases are cytosine and uracil. Both DNA and RNA contain the purine bases adenine and guanine. These molecules are called bases because their presence raises the pH of a solution.

Nucleotides are joined into a DNA or RNA polymer by a series of dehydration reactions. The resulting polymer is a linear molecule called a strand, in which the backbone is made up of an alternating series of sugar-phosphate-sugar-phosphate molecules. The bases project to one side of the backbone. Nucleotides are joined in an order specified by the strand they are copied from. DNA is double-stranded, and RNA is single-stranded (Fig. 3.19).

The two strands in double-stranded DNA usually twist around each other to form a double helix (Fig. 3.20a, b). The

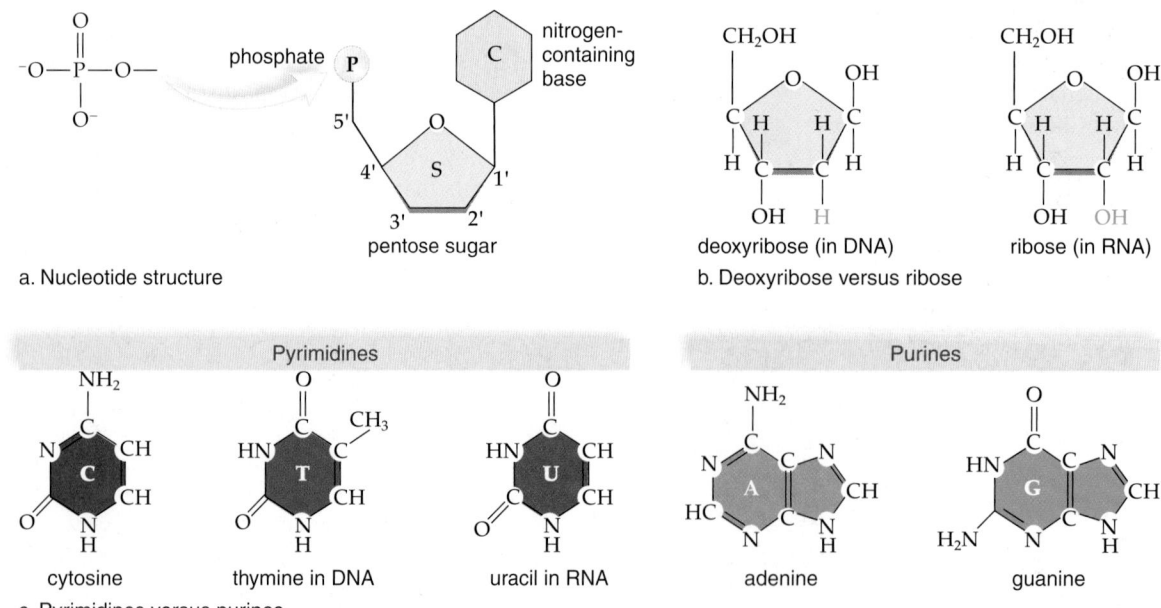

a. Nucleotide structure

b. Deoxyribose versus ribose

c. Pyrimidines versus purines

Figure 3.18 Nucleotides. **a.** A nucleotide consists of a pentose sugar, a phosphate molecule, and a nitrogen-containing base. **b.** DNA contains the sugar deoxyribose, and RNA contains the sugar ribose. **c.** DNA contains the pyrimidines C and T and the purines A and G. RNA contains the pyrimidines C and U and the purines A and G.

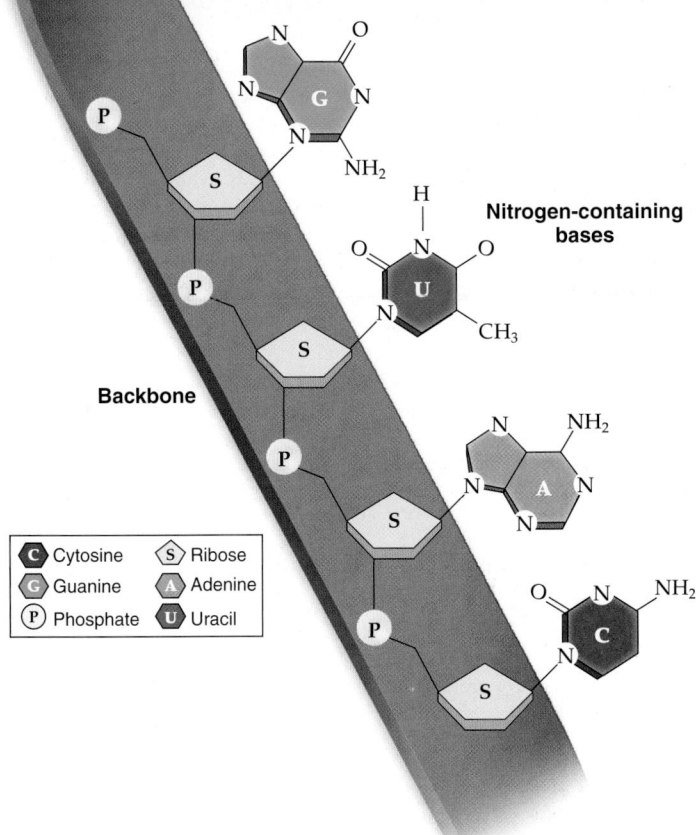

Figure 3.19 RNA structure. RNA is a single-stranded polymer of nucleotides. When the nucleotides join, the phosphate group of one is bonded to the sugar of the next. The bases project out to the side of the resulting sugar-phosphate backbone.

two strands are held together by hydrogen bonds between pyrimidine and purine base pairs. The bases can be in any order within a strand, but between strands, thymine (T) is always paired with adenine (A), and guanine (G) is always paired with cytosine (C). This is called **complementary base pairing.** Therefore, regardless of the order or the quantity of any particular base pair, the number of purine bases (A + G) always equals the number of pyrimidine bases (T + C) (Fig. 3.20c).

Animation
DNA Structure

Table 3.4 summarizes the differences between DNA and RNA.

ATP (Adenosine Triphosphate)

ATP is a special nucleotide comprised of adenine and ribose (adenosine) and three phosphates (triphosphate). The three phosphate groups are attached together and to ribose, the pentose sugar (Fig. 3.21).

Table 3.4 DNA Structure Compared to RNA Structure

	DNA	RNA
Sugar	Deoxyribose	Ribose
Bases	Adenine, guanine, thymine, cytosine	Adenine, guanine, uracil, cytosine
Strands	Double stranded with base pairing	Single stranded
Helix	Yes	No

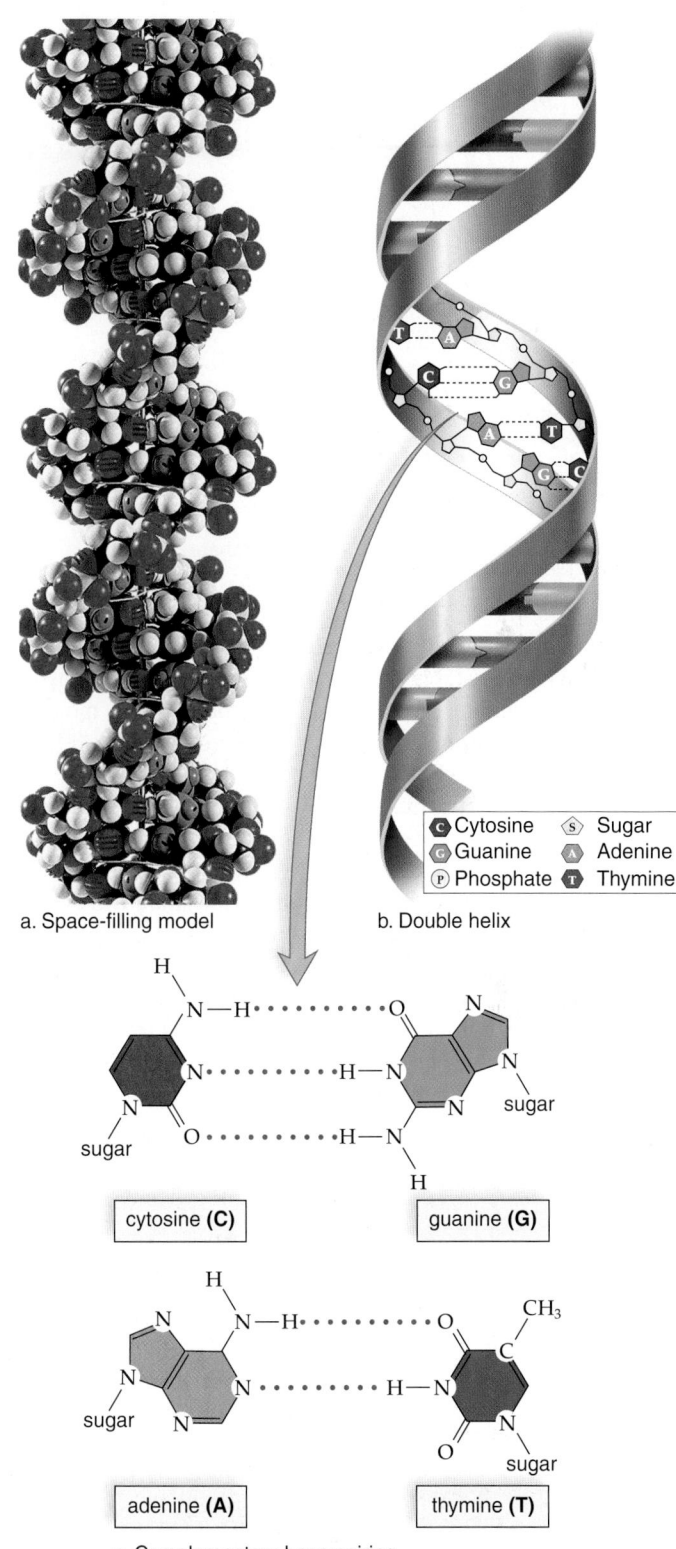

a. Space-filling model   b. Double helix

c. Complementary base pairing

Figure 3.20 DNA structure. **a.** Space-filling model of DNA. **b.** DNA is a double helix in which the two polynucleotide strands twist about each other. **c.** Hydrogen bonds (dotted lines) occur between the complementarily paired bases: C is always paired with G, and A is always paired with T.

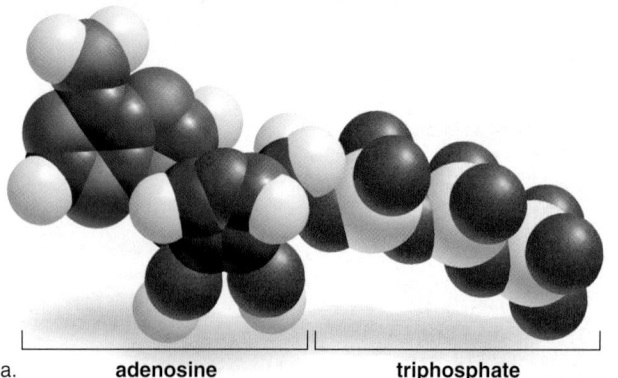

a. adenosine   triphosphate

c.

Figure 3.21 ATP. ATP, the universal energy currency of cells, is composed of adenosine and three phosphate groups. **a.** Space-filling model of ATP. **b.** When cells require energy, ATP becomes ADP + ⓟ, and energy is released. **c.** The breakdown of ATP provides the energy that an animal, such as a chipmunk, uses to acquire food and make more ATP.

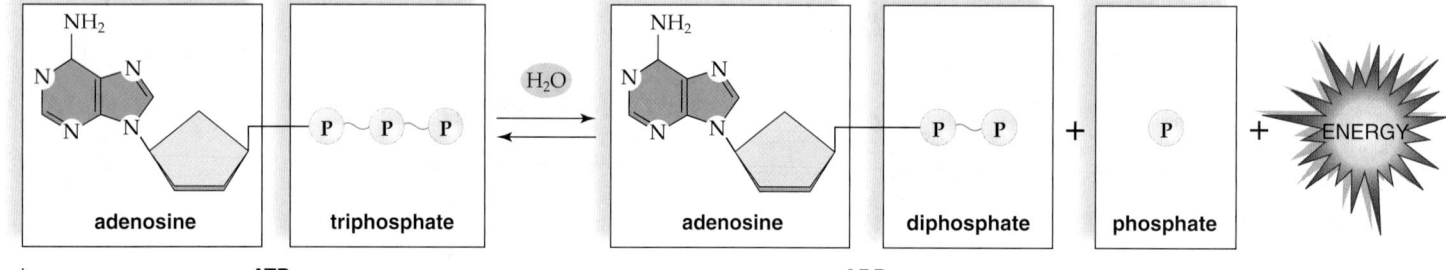

b. **ATP**   **ADP**

ATP is a high-energy molecule because the last two phosphate bonds are unstable and are easily broken. In cells, hydrolysis of the terminal phosphate bond produces the molecule **ADP (adenosine diphosphate)**, a phosphate molecule ⓟ, and lots of energy to do cellular work.

The energy that is released by ATP hydrolysis is used to power many cellular processes, including enzyme reactions, cell communication, and cell division. ATP hydrolysis is chemically favored because ADP and ⓟ are more stable than the original ATP molecule. Even though the third phosphate bond is broken, it is the whole molecule that releases energy.

In many cases, the hydrolysis of the ATP nucleotide is coupled to chemically unfavorable reactions in cells to allow these reactions to proceed. For example, key steps in the synthesis of macromolecules, such as carbohydrates and proteins, are able

to proceed because the energy from ATP breakdown is used to pay the energy costs of the chemical reaction. ATP also supplies the energy for muscle contraction and nerve impulse conduction. Just as you spend money when you pay for a product or a service, cells "spend" ATP when they need something. That's why ATP is called the energy currency of cells.

MP3
Nucleic Acids

Check Your Progress   3.5

1. Examine how a nucleic acid stores information.
2. Describe the three components of a nucleotide.
3. Evaluate the properties of ATP that make it an ideal carrier of energy.

CONNECTING *the* CONCEPTS *with the* BIG IDEAS

Evolution

- Molecules undergo evolution. A change in nucleic acid sequence often affects amino acid sequence change which may alter the structure and function of the protein produced, leading to a change within a species. (3C1a, 3C1d)
- Specific molecular building blocks seem to be universally used in all life forms. (1D2b1)

Energy and Homeostasis

- Carbon from the environment is the basis of the carbohydrates, lipids, proteins and nucleic acids that make up living things. (2A3a1)

Information and Signaling

- Genetic information is stored and transmitted in the form of specific sequences of DNA and RNA nucleotides. (3A1a1)

Interactions and Systems

- Nucleotide sequences encode biological information; while amino acid sequences dictate the shape and function of proteins. (4A1a1-2)
- Lipids and carbohydrates serve in structural and energy storage roles. (4A1a3-4)
- The orientation of biomolecules influences their structure, replication, and bond formation. (4A1b1-3)

Media Study Tools

www.glencoe.com/maderAP11

Enhance your study of this chapter with study tools and practice tests. Also ask your instructor about the resources available through ConnectPlus, including the media-rich eBook, interactive learning tools, and animations.

Summarize

3.1 Organic Molecules

The chemistry of carbon accounts for the diversity of organic molecules found in living things. Carbon can bond with as many as four other atoms. It can also bond with itself to form both chains and rings. Differences in the carbon skeleton and attached functional groups cause organic molecules to have different chemical properties. The chemical properties of a molecule determine how it interacts with other molecules and its role in the cell. Some functional groups are hydrophobic and others are hydrophilic.

There are four classes of biomolecules in cells: carbohydrates, lipids, proteins, and nucleic acids. Polysaccharides, the largest of the carbohydrates, are polymers of simple sugars called monosaccharides. Lipids are diverse hydrophobic molecules made of long chains or ring carbon structures. The polypeptides of proteins are polymers of amino acids, and nucleic acids are polymers of nucleotides. Polymers are formed by the joining together of monomers. For each bond formed during a dehydration reaction, a molecule of water is removed, and for each bond broken during a hydrolysis reaction, a molecule of water is added.

3.2 Carbohydrates

Monosaccharides, disaccharides, and polysaccharides are all carbohydrates. Therefore, the term *carbohydrate* includes both the monomers (e.g., glucose) and the polymers (e.g., starch, glycogen, and cellulose). Glucose is the immediate energy source of cells. Polysaccharides such as starch, glycogen, and cellulose are all polymers of glucose that have different types of chemical bonds. Starch in plants and glycogen in animals are used to store energy, whereas cellulose in plants and chitin in crabs and related animals, as well as fungi, provides structure.

3.3 Lipids

Lipids include a wide variety of compounds that are insoluble in water. Fats and oils, which allow long-term energy storage, contain one glycerol and three fatty acids. Both glycerol and fatty acids have polar groups, but fats and oils are nonpolar, and this accounts for their insolubility in water. Fats tend to contain saturated fatty acids, and oils tend to contain unsaturated fatty acids. Saturated fatty acids do not have carbon–carbon double bonds, but unsaturated fatty acids do have double bonds in their hydrocarbon chain. The double bond causes a kink in the molecule that accounts for the liquid nature of oils at room temperature.

A phospholipid replaces one of the fatty acids with a phosphate group. In water, phospholipids form a bilayer because the head of each molecule is hydrophilic and the tails are hydrophobic. Steroids have the same four-ring structure as cholesterol, but each differs by the attached functional groups. Waxes are composed of a fatty acid with a long hydrocarbon chain bonded to an alcohol.

3.4 Proteins

Proteins carry out many diverse functions in cells and organisms, including support, metabolism, transport, defense, regulation, and motion. Proteins are polymers of amino acids.

A polypeptide is a long chain of amino acids joined by peptide bonds. There are 20 different amino acids in cells, and they differ only by their *R* groups. Whether or not the *R* groups are hydrophilic or hydrophobic helps determine the structure, and therefore the function, of the protein. A polypeptide has up to four levels of structure: The primary level is the linear sequence of the amino acids, which is determined by the DNA; the secondary level contains α helices and β (pleated) sheets held in place by hydrogen bonding between amino acids along the polypeptide chain; and the tertiary level is the final folded polypeptide, which is held in place by internal bonding and hydrophobic interactions between *R* groups. Proteins that contain more than one polypeptide have a quaternary level of structure as well.

Some proteins serve as enzymes, which regulate and carry out body functions. As with other proteins, the shape of an enzyme is important to its function. Both high temperatures and drastic pH change can cause proteins to denature, lose their shape, and decrease their function.

3.5 Nucleic Acids

The nucleic acids DNA and RNA are linear polymers of nucleotides. Nucleotides can be sequenced in any order. Changes in nucleotide sequence produce the diversity of life seen on Earth. Each nucleotide has three components: a 5-carbon sugar, a phosphate (phosphoric acid), and a nitrogen-containing base.

DNA, which contains the sugar deoxyribose, phosphate, and nitrogen-containing bases, is the genetic material that stores information for its own replication and specifies the order in which amino acids are sequenced in proteins. DNA uses mRNA to direct protein synthesis. DNA is a double-stranded helix in which A pairs with T and C pairs with G through hydrogen bonding. RNA is single stranded, contains the sugar ribose and phosphate, and has the same bases as DNA except for uracil. There are many different types of RNA.

ATP is a special nucleotide that, with its unstable phosphate bonds, stores energy to do cellular work. Hydrolysis of ATP to ADP + Ⓟ releases energy needed by the cell to make a product or conduct metabolism.

Key Terms

ADP (adenosine diphosphate) 56
amino acid 50
ATP (adenosine triphosphate) 54
biomolecule 38
carbohydrate 41
cellulose 44
chaperone protein 53
chitin 44
coenzyme 54
complementary base pairing 55
dehydration reaction 40
denatured 53
deoxyribose 42
disaccharide 42

DNA (deoxyribonucleic acid) 54
enzyme 41
fat 46
fatty acid 46
fibrous protein 51
functional group 39
globular protein 52
glucose 42
glycerol 46
glycogen 43
hemoglobin 49
hexose 42
hydrolysis reaction 41
hydrophilic 40
hydrophobic 40
inorganic chemistry 38

Assess

Reviewing This Chapter

1. How do the chemical characteristics of carbon affect the structure of organic molecules?   38–39
2. Give examples of functional groups, and discuss the importance of these groups being hydrophobic or hydrophilic.   39–40
3. What biomolecules are monomers of the polymers studied in this chapter? How do monomers join to produce polymers, and how are polymers broken down to monomers?   40
4. Name several monosaccharides, disaccharides, and polysaccharides, and give a function of each. How are these molecules structurally distinguishable?   42–44
5. What is the difference between a saturated and an unsaturated fatty acid? Explain the structure of a fat molecule by stating its components and how they join together.   46
6. How does the structure of a phospholipid differ from that of a fat? How do phospholipids form a bilayer in the presence of water?   46
7. Describe the structure of a generalized steroid. How does one steroid differ from another?   46–48
8. Draw the structure of an amino acid and a peptide, pointing out the peptide bond.   50
9. Discuss the four possible levels of protein structure, and relate each level to particular bonding patterns.   50–53
10. How do nucleotides bond to form nucleic acids? State and explain several differences between the structure of DNA and that of RNA.   54–55
11. Discuss the structure and function of ATP.   55–56

Testing Yourself

Choose the best answer for each question.

1. Which of these is not a characteristic of carbon?
   a. forms four covalent bonds
   b. bonds with other carbon atoms
   c. is sometimes ionic
   d. can form long chains
   e. sometimes shares two pairs of electrons with another atom

2. A hydrophilic group is
   a. attracted to water.
   b. a polar and/or ionized group.
   c. found at the end of fatty acids.
   d. the opposite of a hydrophobic group.
   e. All of these are correct.

3. Which of these is an example of a hydrolysis reaction?
   a. amino acid + amino acid $\longrightarrow$ dipeptide + H_2O
   b. dipeptide + $H_2O \longrightarrow$ amino acid + amino acid
   c. denaturation of a polypeptide
   d. Both a and b are correct.
   e. Both b and c are correct.

4. Which of these makes cellulose nondigestible in humans?
   a. a polymer of glucose subunits
   b. a fibrous protein
   c. the linkage between the glucose molecules
   d. the peptide linkage between the amino acid molecules
   e. the carboxyl groups ionize

5. A fatty acid is unsaturated if it
   a. contains hydrogen.
   b. contains carbon–carbon double bonds.
   c. contains a carboxyl (acidic) group.
   d. bonds to glycogen.
   e. bonds to a nucleotide.

6. The difference between one amino acid and another is found in the
   a. amino group.
   b. carboxyl group.
   c. R group.
   d. peptide bond.
   e. carbon atoms.

7. The shape of a polypeptide is
   a. maintained by bonding between parts of the polypeptide.
   b. ultimately dependent on the primary structure.
   c. necessary to its function.
   d. All of these are correct.

8. Nucleotides
   a. contain a sugar, a nitrogen-containing base, and a phosphate group.
   b. are the monomers of fats and polysaccharides.
   c. join together by covalent bonding between the bases.
   d. are present in both DNA and RNA.
   e. Both a and d are correct.

9. ATP
   a. is an amino acid.
   b. has a helical structure.
   c. is a high-energy molecule that can break down to ADP and phosphate.
   d. provides enzymes for metabolism.
   e. is most energetic when in the ADP state.

10. Label the following diagram using the terms H_2O, monomer, hydrolysis reaction, dehydration reaction, and polymer. Terms can be used more than once and a term need not be used.

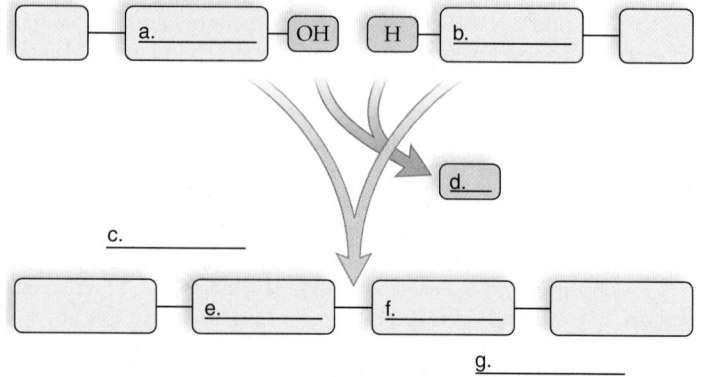

11. The monomer of a carbohydrate is
    a. an amino acid.
    b. a nucleic acid.
    c. a monosaccharide.
    d. a fatty acid.

12. The joining of two adjacent amino acids is called
    a. a peptide bond.
    b. a dehydration reaction.
    c. a covalent bond.
    d. All of these are correct.

13. The shape of a polypeptide
    a. is maintained by bonding between parts of the polypeptide.
    b. is ultimately dependent on the primary structure.
    c. involves hydrogen bonding.
    d. All of these are correct.

14. Which of the following pertains to an RNA nucleotide and not to a DNA nucleotide?
    a. contains the sugar ribose
    b. contains a nitrogen-containing base
    c. contains a phosphate molecule
    d. becomes bonded to other nucleotides following a dehydration reaction

For questions 15–18, match the items to those in the key. Some answers are used more than once.

KEY:
    a. carbohydrate
    b. fats and oils
    c. protein
    d. nucleic acid

15. the 6-carbon sugar, glucose

16. polymer of amino acids

17. glycerol and fatty acids

18. genes

19. Which is a correct statement about carbohydrates?
    a. All polysaccharides serve as energy storage molecules.
    b. Glucose is broken down for immediate energy.
    c. Glucose is not a carbohydrate, only polysaccharides are.
    d. Starch, glycogen, and cellulose have different monomers.
    e. Both a and c are correct.

20. In phospholipids,
    a. heads are polar.
    b. tails are nonpolar.
    c. heads contain phosphate.
    d. All of these are correct.

Engage

Thinking Scientifically

1. The seeds of temperate plants tend to contain unsaturated fatty acids, while the seeds of tropical plants tend to have saturated fatty acids. **a.** How would you test your hypothesis. **b.** Assuming your hypothesis is supported, give an explanation.

2. Chemical analysis reveals that an abnormal form of an enzyme contains a polar amino acid at the location where the normal form has a nonpolar amino acid. Formulate a testable hypothesis concerning the abnormal enzyme.

Bioethical Issue
Organic Pollutants

Organic compounds include the carbohydrates, proteins, lipids, and nucleic acids that make up our bodies. Modern industry also uses all sorts of organic compounds that are synthetically produced. Indeed, our modern way of life wouldn't be possible without synthetic organic compounds.

Pesticides, herbicides, disinfectants, plastics, and textiles contain organic substances that are termed pollutants when they enter the natural environment and cause harm to living things. Global use of pesticides has increased dramatically since the 1950s, and modern pesticides are ten times more toxic than those of the 1950s. The Centers for Disease Control and Prevention in Atlanta reports that 40% of children working in agricultural fields now show signs of pesticide poisoning. The U.S. Geological Survey estimates that 32 million people in urban areas and 10 million people in rural areas are using groundwater that contains organic pollutants. J. Charles Fox, an official of the Environmental Protection Agency, says that "over the life of a person, ingestion of these chemicals has been shown to have adverse health effects such as cancer, reproductive problems, and developmental effects."

At one time, people failed to realize that everything in the environment is connected to everything else. In other words, they didn't know that an organic chemical can wander far from the site of its entry into the environment and that eventually these chemicals can enter our own bodies and cause harm. Now that we are aware of this outcome, we have to decide as a society how to proceed. We might decide to do nothing if the percentage of people dying from exposure to organic pollutants is small. Or we might decide to regulate the use of industrial compounds more strictly than has been done in the past. We could also decide that we need better ways of purifying public and private water supplies so that they do not contain organic pollutants.

4

Cell Structure and Function

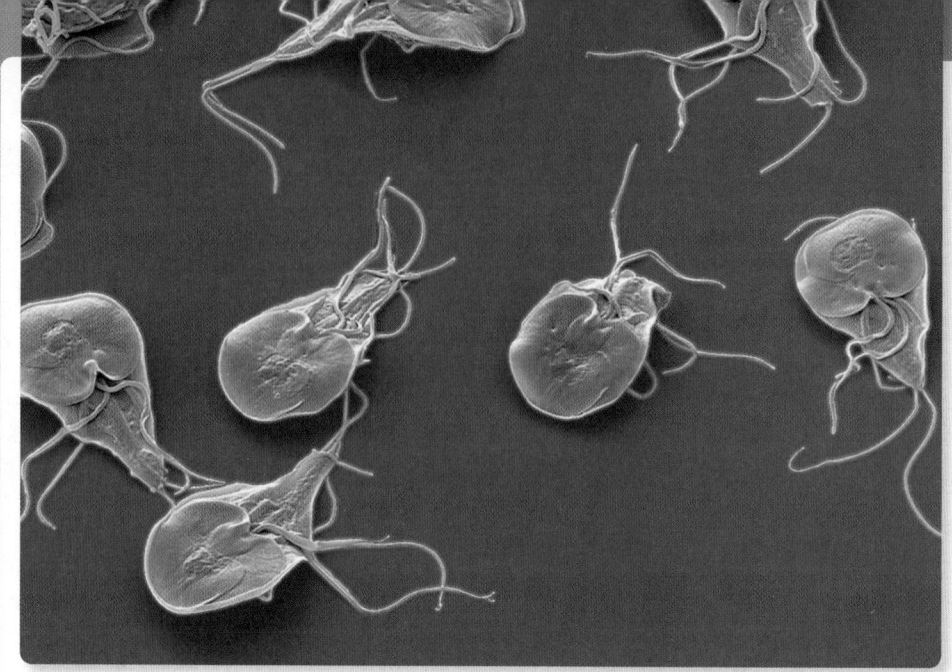

Electron micrograph of *Giardia lamblia*, a cause of diarrhea.

BEFORE YOU BEGIN

Before beginning this chapter, take a few moments to review the following discussions.

Figures 3.1 and 3.2 What role do carbon and functional groups serve in biological molecules?

Sections 3.2 to 3.5 What macromolecules are needed to construct a cell?

Figures 3.8, 3.12, 3.17, and 3.21 How does the structure of a macromolecule affect its function?

The Dutch shopkeeper Antonie van Leeuwenhoek (1632–1723) may have been the first person to see living cells. Using a microscope he built himself, he looked at everything possible, including his own stool. He wrote, "I have usually of a morning a well-formed stool. But, hitherto, I have had sometimes a looseness of bowels, so I went to stool some twice, thrice, or four times a day. My excrement being so thin, I was at diverse times constrained to examine it. Wherein I have sometimes seen animalcules a moving prettily. Their bodies were somewhat longer than broad, and the belly, which was flat-lie, furnished with sundry little paws. . . ." [November 9, 1681]

In this way, Antonie van Leeuwenhoek reported seeing the parasite *Giardia lamblia* in his feces. *Giardia* are unicellular organisms, while humans are multicellular organisms. In this chapter, you will see that cells are the fundamental building blocks of organisms, organized to carry out basic metabolic functions and adapt to changing environmental conditions. The presentation concentrates on the generalized bacterial, animal, or plant cell; however, all cells are specialized in particular ways.

As you read through the chapter, think about the following questions:

1. How is a cell more than simply the sum of its macromolecular parts?
2. What characteristics enable cells to be alive and allow them to self-replicate?
3. How are cells able to metabolize and respond to environmental changes so quickly?

FOLLOWING *the* BIG IDEAS

CHAPTER 4 CELL STRUCTURE AND FUNCTION

Evolution	All cells are produced from existing cells, creating an unbroken lineage back to the first cells almost 4 billion years ago.
Energy and Homeostasis	Eukaryotic cells contain multiple cooperating and specialized organelles which produce the structure and accomplish the functions necessary for life.
Information and Signaling	Every cell contains DNA (with or without a nuclear cover) which encrypts the information for all of its structures and functional molecules.
Interactions and Systems	Cells' systems metabolize and adapt to changing environmental conditions.

4.1 Cellular Level of Organization

Learning Outcomes

Upon completion of this section, you should be able to

1. Explain why cells are the basic unit of life.
2. List the tenets of cell theory.
3. Compare surface-area-to-volume ratios for large and small cells.

Cells are the basic units of life. All of the chemistry and bio-molecules we have discussed to this point are necessary but insufficient on their own to support life. It is only when these components are brought together and organized into a cell that life is possible.

All organisms are made up of cells. When we observe plants, animals, and other organisms, it is important to realize that what we are seeing is a collection of cells that work together in a highly organized, regulated manner, and thus conduct the business of life. Figure 4.1 shows the connection between whole organisms and their component cells. Although the cellular basis of life is clear to us now, scientists were unaware of this fact as recently as two hundred years ago. The link between cells and life became clear to microscopists during the 1830s.

The **cell** is the smallest unit of living matter. The collective work of the 19th century scientists Robert Brown (1773-1858), Matthais Schleiden (1804-1881), and Theodor Schwann (1810-1882) helped determine that plants and animals are composed of cells. Further work by the German physician Rudolph Virchow (1821-1902) showed that cells self-reproduce and "every cell comes from a preexisting cell." Today, we know that various illnesses of the body, such as diabetes and prostate cancer, are due to cellular malfunction. Countless scientific investigations since that time verify these initial findings. From these results, we can infer that all life on Earth today came from cells in ancient times, and that all cells are related in some way. In reality, a continuity of cells has been present from generation to generation, even back to the very first cell (or cells) in the history of life.

Today, some life-forms exist as single cells, whereas others are complex, interconnected systems of cells. When unicellular organisms reproduce, a single cell divides and becomes two new organisms. When multicellular organisms grow, many cells divide. The presence of many cells allows some to specialize to do particular jobs within the multicellular organism, including the cells that create genetic variation through sexual reproduction.

The work of Schleiden, Schwann, and Virchow helped created the **cell theory.** It states that:

1. all organisms are composed of cells,
2. cells are the basic units of structure and function in organisms, and
3. cells come only from preexisting cells because cells are self-reproducing.

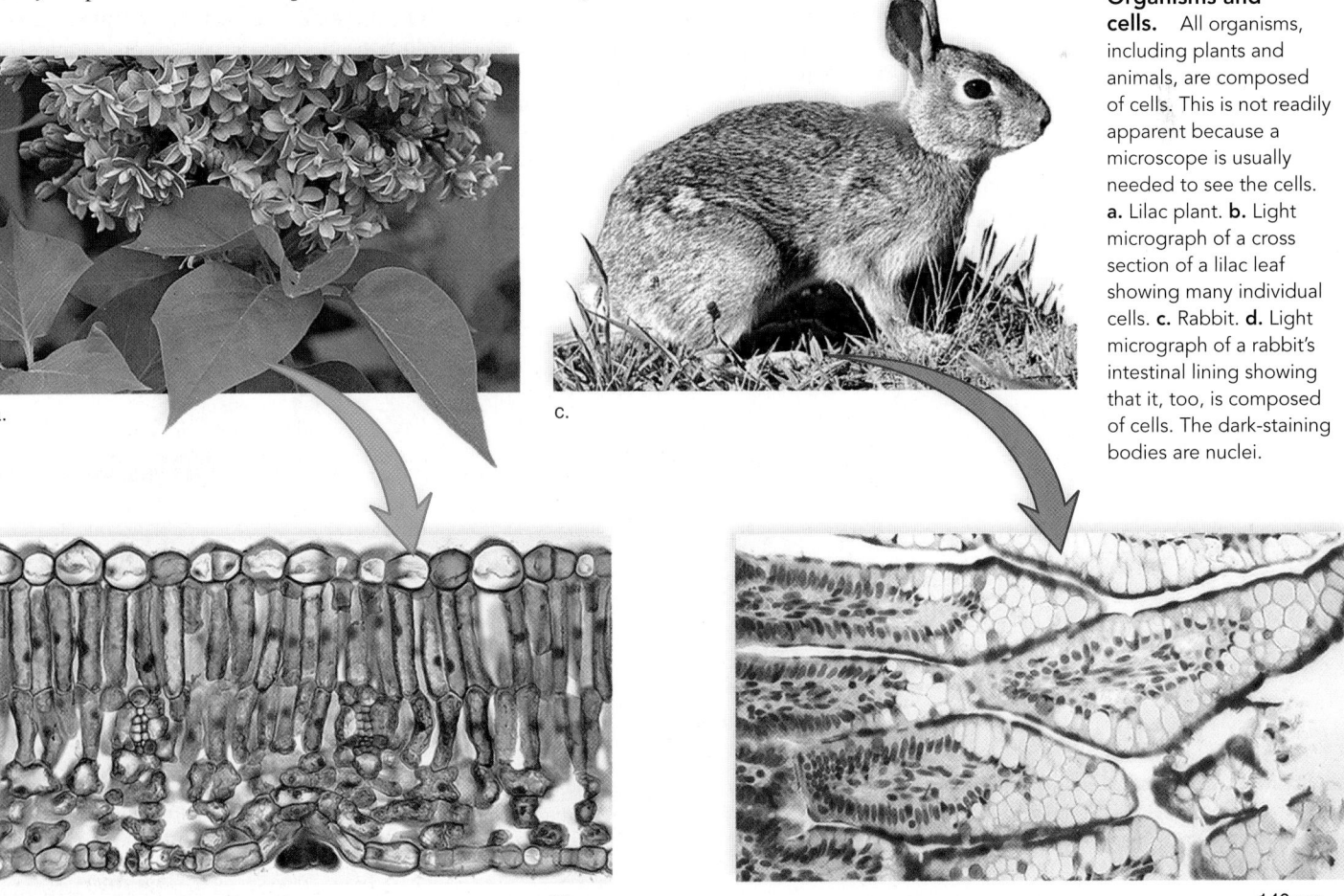

Figure 4.1 Organisms and cells. All organisms, including plants and animals, are composed of cells. This is not readily apparent because a microscope is usually needed to see the cells. **a.** Lilac plant. **b.** Light micrograph of a cross section of a lilac leaf showing many individual cells. **c.** Rabbit. **d.** Light micrograph of a rabbit's intestinal lining showing that it, too, is composed of cells. The dark-staining bodies are nuclei.

a.

b.   50 µm

c.

d.   140 µm

Cell Size

Although they range in size, cells are generally quite small. A frog's egg, at about 1 millimeter (mm) in diameter, is large enough to be seen by the human eye. But most cells are far smaller than 1 mm; some are even as small as 1 micrometer (μm)—one thousandth of a millimeter. Cell inclusions and macromolecules are smaller than a micrometer and are measured in terms of nanometers (nm).

Because of their size, very small biological structures can only be viewed with microscopes, which magnify a visual image. Figure 4.2 shows the visual range of the eye, light microscope, and electron microscope; the discussion of microscopy in the Nature of Science feature on pages 63–64 explains why the electron microscope allows us to see so much more detail than the light microscope does.

Why are cells so small? To answer this question, consider that a cell is a system by itself, and as such needs a surface area large enough to allow adequate nutrients to enter and for wastes to be eliminated. Small cells, not large cells, are likely to have an adequate surface area for exchanging wastes for nutrients. As cells increase in size, the surface area becomes inadequate to exchange the materials that the volume of the cell requires.

Figure 4.3 visually demonstrates this relationship; dividing a large cube into smaller cubes provides a lot more surface area per volume. Calculations show that a 1-cm cube has a **surface-area-to-volume ratio** of 6:1, whereas a 4-cm cube has a surface-area-to-volume ratio of 1.5:1. The former ratio allows sufficient transport in and out of the cell to support life; the latter ratio does not.

A mental image might help you visualize the importance of surface-area-to-volume-ratios and why this relationship favors smaller cell size. Imagine a small room and a large room filled with people. The small room, which holds 20 people, has only two doors, and the large room, which holds 80 people, has four doors. If a fire occurred in both rooms, it would be faster to

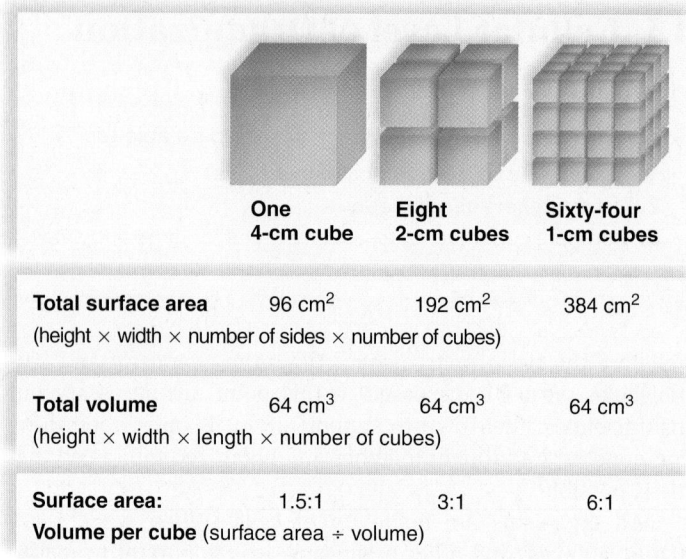

	One 4-cm cube	Eight 2-cm cubes	Sixty-four 1-cm cubes
Total surface area (height × width × number of sides × number of cubes)	96 cm^2	192 cm^2	384 cm^2
Total volume (height × width × length × number of cubes)	64 cm^3	64 cm^3	64 cm^3
Surface area: **Volume per cube** (surface area ÷ volume)	1.5:1	3:1	6:1

Figure 4.3 Surface-area-to-volume relationships.   As cell size decreases from 4 cm^3 to 1 cm^3, the surface-area-to-volume ratio increases.

get the people out of the smaller room because it has the more favorable ratio of doors to people. Similarly, a small cell size is more advantageous for exchanging molecules because of its greater surface-area-to-volume ratio.

Check Your Progress   4.1

1. Describe the major components of cell theory.
2. Evaluate why a cell and a whole organism are both examples of biological systems.
3. Explain why a large surface-area-to-volume ratio is needed for the proper functioning of cells.

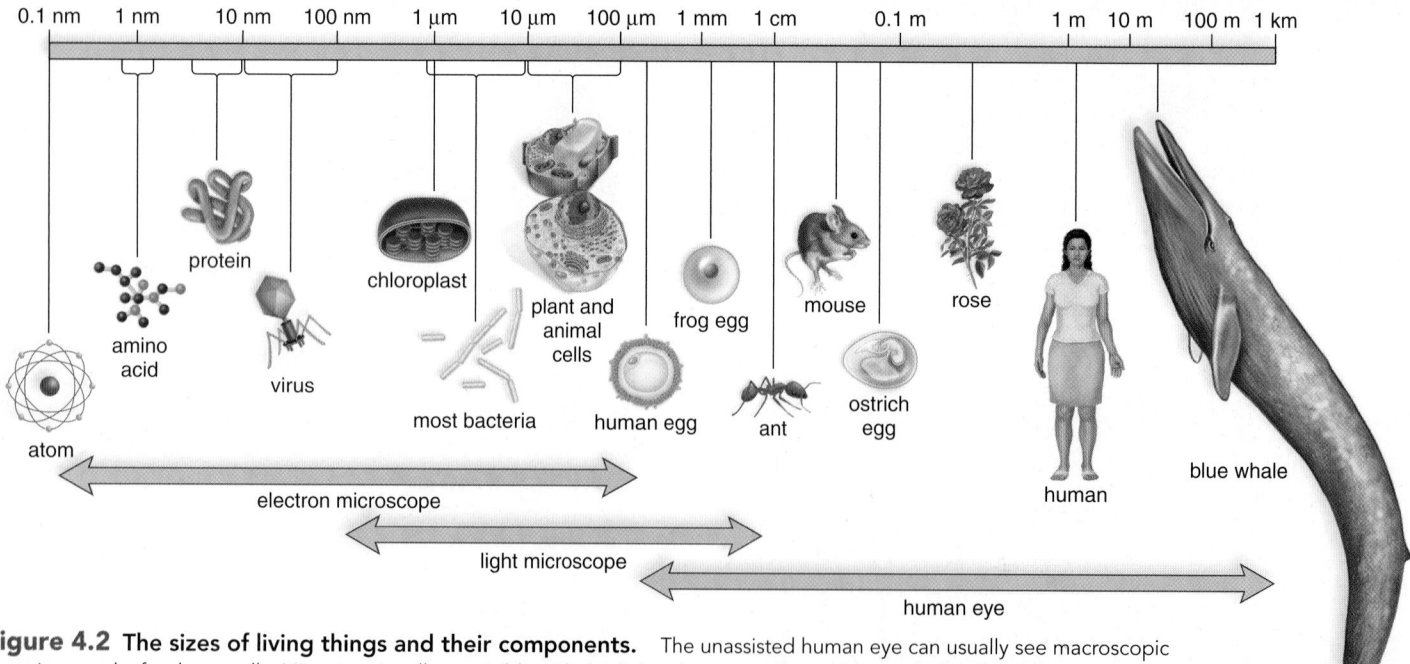

Figure 4.2 The sizes of living things and their components.   The unassisted human eye can usually see macroscopic organisms and a few large cells. Microscopic cells are visible with the light microscope, but not in much detail. An electron microscope is necessary to see organelles in detail and to observe viruses and molecules. In the metric system (see back endsheet), each higher unit is ten times greater than the preceding unit. (1 meter = 10^2 cm = 10^3 mm = 10^6 μm = 10^9 nm)

Nature of Science

Microscopy Today

Since cells are the basic unit of life, the more we learn about cells, the more we understand life. Cells were not discovered until the seventeenth century, when the microscope was invented. Since that time, various types of microscopes have been developed for studying cells and their components.

Many times when scientists don't have suitable tools to investigate natural phenomena, they invent them. Microscopes have given scientists a deeper look into how life works than is possible with the naked eye. Today, there are many types of microscopes. A *compound light microscope* uses a set of glass lenses to focus light rays passing through a specimen to produce an image that is viewed by the human eye. A *transmission electron microscope (TEM)* uses a set of electromagnetic lenses to focus electrons passing through a specimen to produce an image that is projected onto a fluorescent screen or photographic film. A *scanning electron microscope (SEM)* uses a narrow beam of electrons to scan over the surface of a specimen that is coated with a thin metal layer. Secondary electrons given off by the metal are detected and used to produce a three-dimensional image on a television screen. Figure 4A shows these three types of microscopic images.

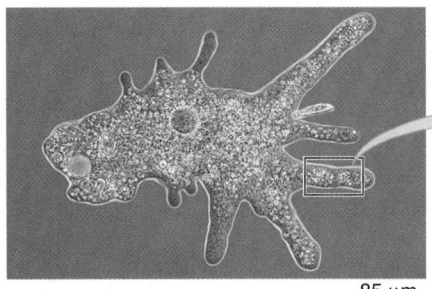

amoeba, light micrograph

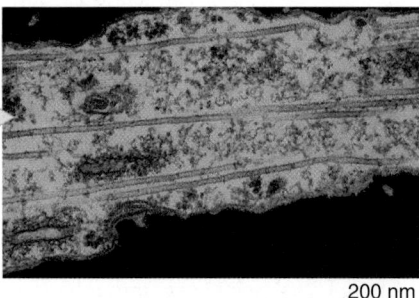

pseudopod segment, transmission electron micrograph

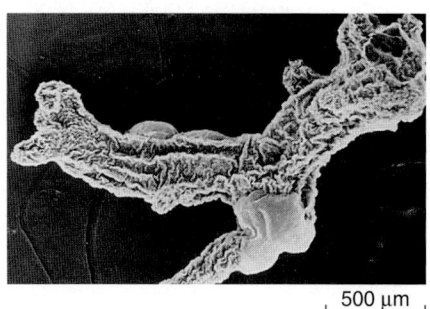

amoeba, scanning electron micrograph

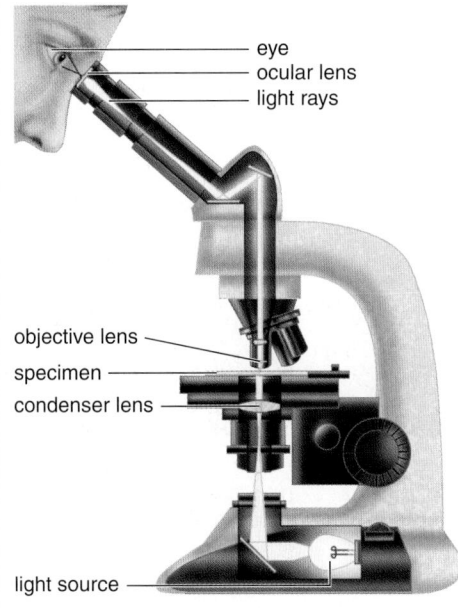

a. Compound light microscope

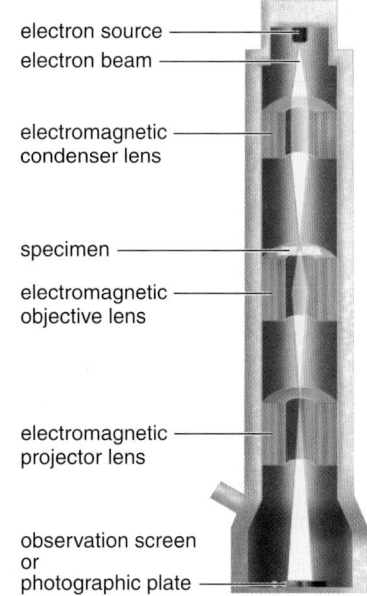

b. Transmission electron microscope

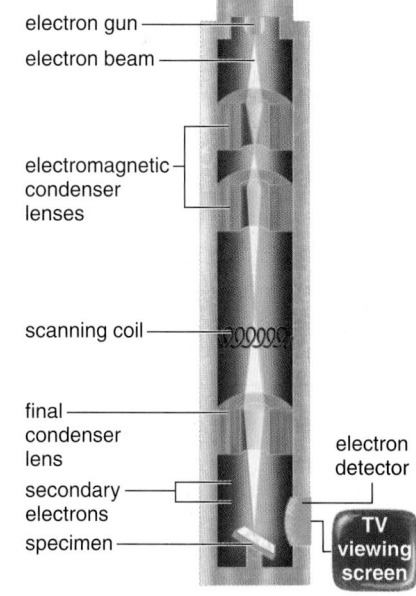

c. Scanning electron microscope

Figure 4A Diagram of microscopes with accompanying micrographs of *Amoeba proteus*. a. The compound light microscope and **(b)** the transmission electron microscope provide an internal view of an organism. **c.** The scanning electron microscope provides an external view of an organism.

Magnification, Resolution, and Contrast

Magnification is the ratio between the size of an image and its actual size. Electron microscopes magnify to a greater extent than do compound light microscopes. A light microscope can magnify objects about a thousand times, but an electron microscope can magnify them hundreds of thousands of times. The difference lies in the means of illumination. The path of light rays and electrons moving through space is wavelike, but the wavelength of electrons is much shorter than the wavelength of light. This difference in wavelength accounts for the electron microscope's greater magnifying capability and its greater ability to distinguish between two points (resolving power).

Resolution is the minimum distance between two objects that allows them to be seen as two separate objects. A microscope with poor resolution might enable a student to see only one cellular granule, while the microscope with the better resolution would show two granules next to each other. The greater the resolving power, the greater the detail seen.

If oil is placed between the sample and the objective lens of the compound light microscope, the resolving power is increased, and if ultraviolet light is used instead of visible light, it is also increased. But typically, a light microscope can resolve down to 0.2 μm, while the transmission electron microscope can resolve down to 0.0002 μm. If the resolving power of the average human eye is set at 1.0, then the typical compound light microscope is about 500, and the transmission electron microscope is 100,000 (Fig. 4A*b*).

The ability to make out, or resolve, a particular object can depend on *contrast*, a difference in the shading of an object compared to its background. Higher contrast is often achieved by staining cells with colored dyes (light microscopy) or with electron-dense metals (electron microscopy), which make them easier to see. Optical methods such as phase contrast and differential interference contrast (Fig. 4B) can also be used to improve contrast. Using fluorescently tagged antibodies can also help us visualize subcellular components like specific proteins (see Fig. 4.18).

Illumination, Viewing, and Recording

Light rays can be bent (refracted) and brought to focus as they pass through glass lenses, but electrons do not pass through glass. Electrons have a charge that allows them to be brought into focus by electromagnetic lenses. The human eye uses light to see an object but cannot use electrons for the same purpose. Therefore, electrons leaving the specimen in the electron microscope are directed toward a screen or a photographic plate that is sensitive to their presence. Humans can view the image on the screen or photograph.

A major advancement in illumination has been the introduction of *confocal microscopy*, which uses a laser beam scanned across the specimen to focus on a single shallow plane within the cell. The microscopist can "optically section" the specimen by focusing up and down, and a series of optical sections can be combined in a computer to create a three-dimensional image, which can be displayed and rotated on the computer screen.

An image from a microscope may be recorded by placing a television camera where the eye would view the image. The television camera converts the light image into an electronic image, which can be entered into a computer. In *video-enhanced contrast microscopy*, the computer makes the darkest areas of the original image much darker and the lightest areas of the original much lighter. The result is a high-contrast image with deep blacks and bright whites. Even more contrast can be introduced by the computer if shades of gray are replaced by colors.

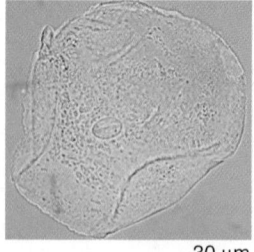 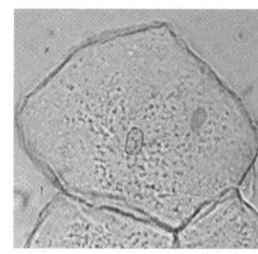 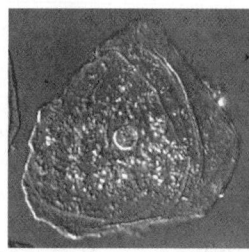 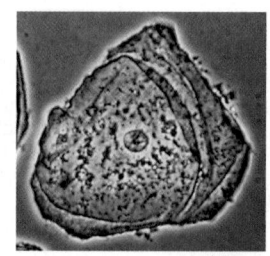 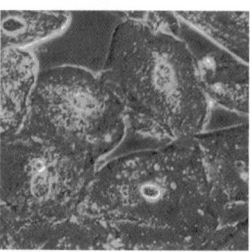

30 μm

Bright-field. Light passing through the specimen is brought directly into focus. Usually, the low level of contrast within the specimen interferes with viewing all but its largest components.

30 μm

Bright-field (stained). Dyes are used to stain the specimen. Certain components take up the dye more than other components, and therefore contrast is enhanced.

25 μm

Differential interference contrast. Optical methods are used to enhance density differences within the specimen so that certain regions appear brighter than others. This technique is used to view living cells, chromosomes, and organelle masses.

25 μm

Phase contrast. Density differences in the specimen cause light rays to come out of "phase." The microscope enhances these phase differences so that some regions of the specimen appear brighter or darker than others. The technique is widely used to observe living cells and organelles.

25 μm

Dark-field. Light is passed through the specimen at an oblique angle so that the objective lens receives only light diffracted and scattered by the object. This technique is used to view organelles, which appear quite bright against a dark field.

Figure 4B Photomicrographs of cheek cells.   Bright-field microscopy is the most common form used with a compound light microscope. Other types of microscopy include differential interference contrast, phase contrast, and dark-field.

4.2 Prokaryotic Cells

Learning Outcomes

Upon completion of this section, you should be able to

1. Examine the evolutionary relatedness of prokaryotes, eukaryotes, and archaeans.
2. Describe the fundamental components of a bacterial cell.

Fundamentally, three different types of cells exist in nature. **Prokaryotic cells** [Gk. *pro,* before, and *karyon,* kernel, nucleus] are so named because they lack a membrane-enclosed nucleus. Another type of cell, called a **eukaryotic cell** [Gk. *eu,* true, and *karyon,* kernel, nucleus], has a nucleus (see Figs. 4.6 and 4.7). A third group of cells, called **archaeans** [Gk. *arkhaios,* ancient], possess qualities of both prokaryotes and eukaryotes. Prokaryotic cells and archaeans were once thought to be closely related because of their similar size and shape. Comparisons of DNA and RNA sequences now show archaeans to be biochemically distinct from prokaryotes and eukaryotes. Although we have learned more about archaeans than was previously known, we have studied prokaryotes and eukaryotes in much greater detail.

Prokaryotes as a group are one of the most abundant and diverse life-forms on Earth, and they are present in great numbers in the air, water, and soil, as well as living in and on other organisms. Although they are structurally less complicated than eukaryotes, their metabolic capabilities as a group far exceed those of eukaryotes. Prokaryotes are an extremely successful group of organisms whose evolutionary history dates back to the first cells on Earth.

Bacteria are well known because they cause some serious diseases, such as tuberculosis, anthrax, tetanus, throat infections, and gonorrhea. But many species of bacteria are important to the environment because they decompose the remains of dead organisms and contribute to ecological cycles. Bacteria also assist humans in still another way—we use them to manufacture all sorts of products, from industrial chemicals to foodstuffs and drugs. For example, today we know how to place human genes in certain cultured bacteria so that they can produce human insulin, a necessary hormone for the treatment of diabetes.

Video
E. coli **Wars**

The Structure of Prokaryotes

Prokaryotes are quite small; an average size is 1.1–1.5 μm wide and 2.0–6.0 μm long. While other prokaryote shapes have been identified, three basic shapes are most common:

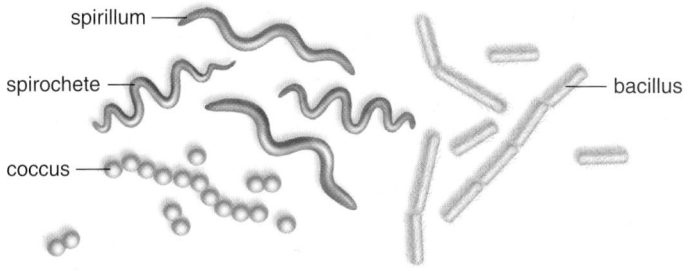

A rod-shaped bacterium is called a **bacillus,** while a spherical-shaped bacterium is a **coccus.** Both of these can occur

as pairs or chains, and in addition, cocci can occur as clusters. Some long rods are twisted into spirals, in which case they are **spirilla** if they are rigid or **spirochetes** if they are flexible.

Animation
Prokaryotic Cell Shapes

Figure 4.4 shows the generalized structure of a bacterium. "Generalized" means that not all bacteria have all the structures depicted, and some have more than one of each. Also, for the sake of discussion, we divide the organization of bacteria into the cell envelope, the cytoplasm, and the appendages.

Cell Envelope

In bacteria, the **cell envelope** includes the plasma membrane, the cell wall, and the glycocalyx. The **plasma membrane** is a phospholipid bilayer with embedded proteins:

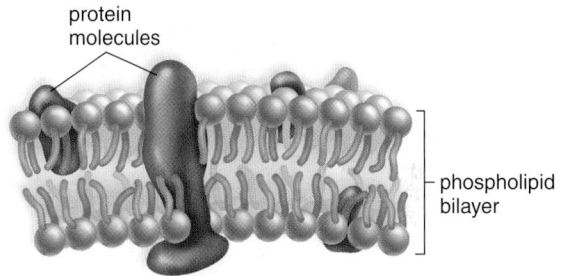

The plasma membrane has the important function of regulating the entrance and exit of substances into and out of the cytoplasm. Regulating the flow of materials in and out of the cytoplasm is necessary in order to maintain its normal composition.

In prokaryotes, the plasma membrane can form internal pouches called mesosomes. **Mesosomes** most likely increase the internal surface area for the attachment of enzymes that are carrying on metabolic activities.

The **cell wall,** when present, maintains the shape of the cell, even if the cytoplasm should happen to take up an abundance of water. You may recall that the cell wall of a plant cell is strengthened by cellulose; in contrast, the cell wall of a bacterium contains peptidoglycan, a complex molecule containing a unique amino disaccharide and peptide fragments.

Animation
Cell Wall Antibiotics

The **glycocalyx** is a layer of polysaccharides that lies outside the cell wall in some bacteria. When the layer is well organized and not easily washed off, it is called a **capsule.** A slime layer, on the other hand, is not well organized and is easily removed. The glycocalyx aids against drying out and helps bacteria resist a host's immune system. It also helps bacteria attach to almost any surface.

Cytoplasm

The **cytoplasm** is a semifluid solution composed of water and inorganic and organic molecules encased by a plasma membrane. Among the organic molecules are a variety of enzymes, which speed the many types of chemical reactions involved in metabolism.

The DNA of a prokaryote is found in a single, coiled chromosome that is located in a region called the **nucleoid.** Many bacteria also have extrachromosomal pieces of circular DNA called **plasmids.** Plasmids are routinely used in biotechnology

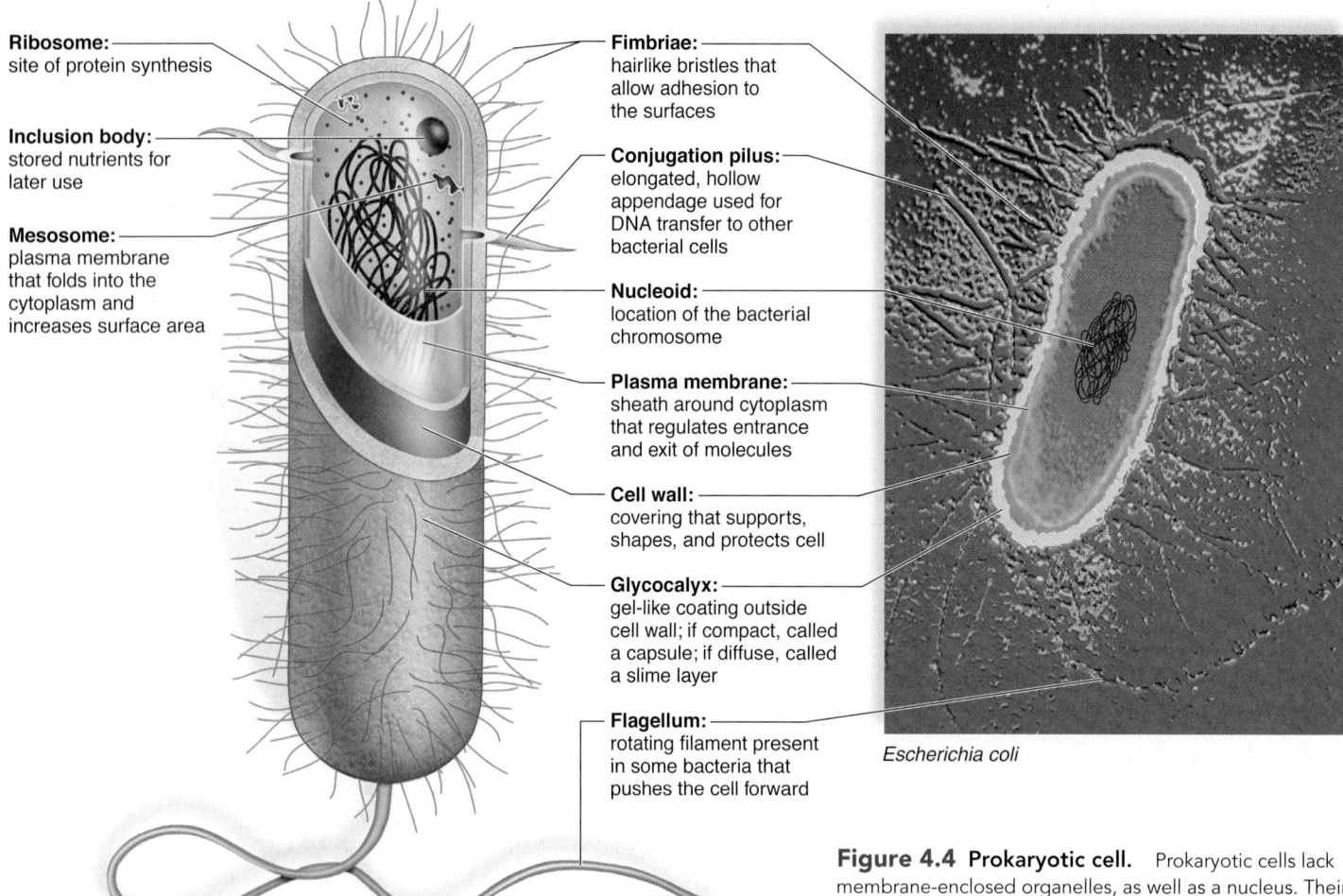

Ribosome: site of protein synthesis

Inclusion body: stored nutrients for later use

Mesosome: plasma membrane that folds into the cytoplasm and increases surface area

Fimbriae: hairlike bristles that allow adhesion to the surfaces

Conjugation pilus: elongated, hollow appendage used for DNA transfer to other bacterial cells

Nucleoid: location of the bacterial chromosome

Plasma membrane: sheath around cytoplasm that regulates entrance and exit of molecules

Cell wall: covering that supports, shapes, and protects cell

Glycocalyx: gel-like coating outside cell wall; if compact, called a capsule; if diffuse, called a slime layer

Flagellum: rotating filament present in some bacteria that pushes the cell forward

Escherichia coli

Figure 4.4 Prokaryotic cell. Prokaryotic cells lack membrane-enclosed organelles, as well as a nucleus. Their DNA is located in a region called a nucleoid.

laboratories as a molecular vehicle, also called a **vector**, to transport DNA from different organisms, including humans, into a bacterium. This technique is possible because all life on Earth is constructed from the same four DNA nucleotides: A, G, C, and T. The ability to combine genes from different life-forms is important in the production of new medicines and many commerical products we use every day.

The many proteins encoded by the bacterial DNA are synthesized on tiny structures called **ribosomes.** A bacterial cell contains thousands of ribosomes that are similar in shape and function, but are smaller than eukaryotic ribosomes. Like their eukaryotic counterparts, bacterial ribosomes still contain RNA and protein in two subunits.

Most bacteria carry out metabolism in the same manner as animals, but the **cyanobacteria** are bacteria that are capable of photosynthesis in the same manner as plants. These organisms live in water, in ditches, on buildings, and on the bark of trees. Their cytoplasm contains extensive internal membranes called **thylakoids** [Gk. *thylakon,* small sac], where chlorophyll and other pigments absorb solar energy for the production of carbohydrates. Cyanobacteria are called the blue-green bacteria because some have a pigment that adds a shade of blue to the cell, in addition to the green color of chlorophyll. The cyanobacteria release oxygen as a by-product of photosynthesis, which is one reason why scientists think that ancestral cyanobacteria were the earliest photosynthesizers on Earth. Many sources of evidence show that the composition of the early Earth's atmosphere was changed by addition of oxygen.

Appendages

The appendages of a bacterium, namely the flagella, fimbriae, and conjugation pili, are made of protein. Motile bacteria can propel themselves in water by the means of appendages called **flagella** (usually 20 nm in diameter and 1–70 nm long). The bacterial flagellum is one of the great wonders of nature, and it consists of a filament, a hook, and a basal body. The basal body is a series of rings anchored in the cell wall and membrane. The hook rotates 360° within the basal body, and this rotation propels bacteria—the bacterial flagellum does not move back and forth like a whip. Sometimes flagella occur only at the two ends of a cell, and sometimes they are dispersed randomly over the surface. The number and location of flagella can be used to help distinguish different types of bacteria.

Animation Bacterial Locomotion

Fimbriae are small, bristlelike fibers that sprout from the cell surface. They are not involved in locomotion; instead, fimbriae attach bacteria to a surface. **Conjugation pili** are rigid tubular structures used by bacteria to pass DNA from cell to cell. Bacteria reproduce asexually by binary fission, but they can exchange DNA by way of the conjugation pili. They can also take up DNA from the external medium or by way of viruses.

Check Your Progress 4.2

1. Explain the major differences between a prokaryotic and eukaryotic cell.
2. Describe the functions of the bacterial cell envelope, cytoplasm, and appendages.

4.3 Introducing Eukaryotic Cells

Learning Outcomes

Upon completion of this section, you should be able to

1. Explain how membranes compartmentalize a cell.
2. Examine how organelles divide cellular work.
3. Apply the endosymbiosis theory to eukaryotic cell structure.

Eukaryotic cells, like prokaryotic cells and archaeans, have a plasma membrane that separates the contents of the cell from the environment and regulates the passage of molecules into and out of the cytoplasm. The plasma membrane is a phospholipid bilayer with embedded proteins. Some scientists have suggested that the nucleus evolved as the result of the invagination, or pocketing, of the plasma membrane (Fig. 4.5). The resulting compartment may have allowed specific functions to occur in the nucleus, freeing up cellular resources in the cytoplasm for other work. An increased ability to metabolize would have given the new cell a comparative advantage over other cells.

Origin of the Eukaryotic Cell

Much scientific evidence also shows that mitochondria and chloroplasts, the two energy-related organelles, arose when a large eukaryotic cell engulfed independent prokaryotes. This relationship is referred to as the **endosymbiotic theory**. This theory explains why mitochondria and chloroplasts are bounded by a double membrane and contain their own genetic material separate from that of the nucleus.

Figure 4.5 shows conceptually how endosymbiosis is thought to have occurred. Figures 4.6 and 4.7 show general features of fully evolved, present day animal and plant cells. Specialized cells, as opposed to generalized cells, do not necessarily contain all the structures depicted and may have more or fewer copies of any particular organelle, depending on their particular function. These generalized depictions of plant and animal cells are useful for study purposes. A baseline understanding of cell structure and function will be helpful when you study the function of specialized cells later in this text.

Structure of a Eukaryotic Cell

As shown in Figures 4.6 and 4.7, eukaryotic cells are compartmentalized. Membranes create internal spaces that divide the labor necessary to conduct life functions. The compartments of a eukaryotic cell, typically called **organelles,** carry out specialized functions that together allow the cell to be more efficient and successful. Nearly all organelles are surrounded by a membrane with embedded proteins, many of which are enzymes. These enzymes make products specific to that organelle, but their action benefits the whole cell system. The cell can be seen as a system of interconnected organelles that work together to metabolize, regulate, and conduct life processes. For example, the nucleus is a compartment that houses the genetic material within eukaryotic chromosomes and contains hereditary information. The nucleus communicates with ribosomes in the cytoplasm, and the organelles of the endomembrane

system—notably the endoplasmic reticulum and the Golgi apparatus—communicate with one another.

Production of specific molecules takes place inside or on the surface of organelles. As mentioned, enzymes embedded in the organelles' membranes make these molecules. These products are then transported around the cell by transport **vesicles,** membranous sacs that enclose the molecules and keep them separate from the cytoplasm. For example, the endoplasmic reticulum communicates with the Golgi apparatus by means of transport vesicles. Communication with the energy-related organelles—mitochondria and chloroplasts—is less obvious, but it does occur because they import particular molecules from the cytoplasm.

MP3 Cellular Organelles

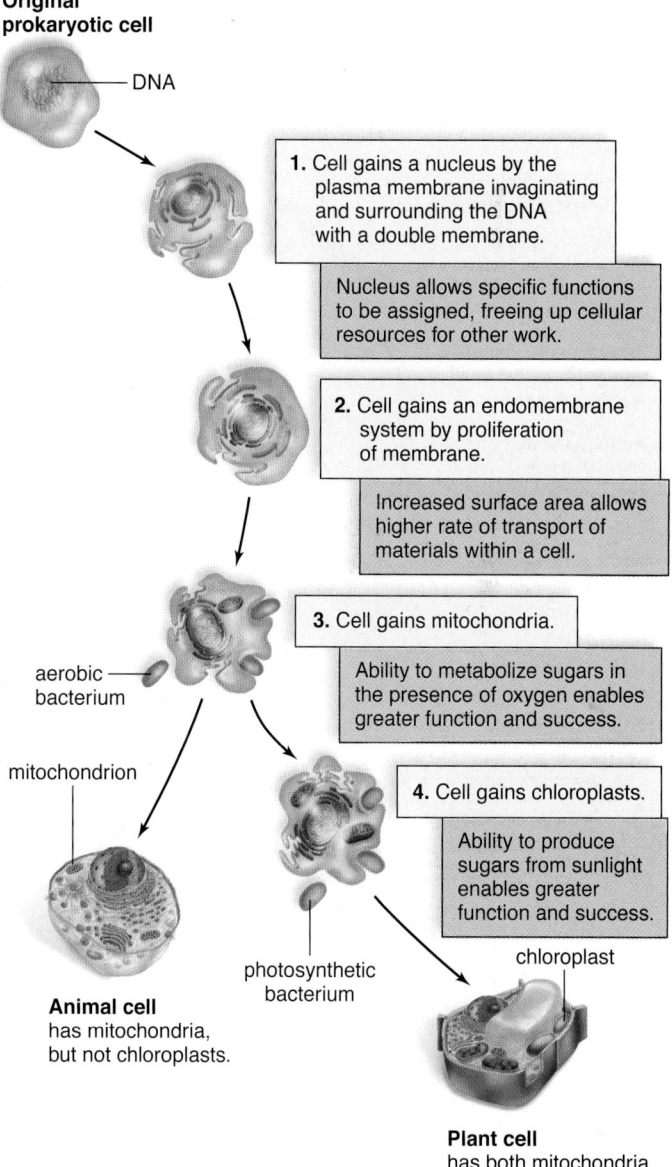

Original prokaryotic cell

—— DNA

1. Cell gains a nucleus by the plasma membrane invaginating and surrounding the DNA with a double membrane.

Nucleus allows specific functions to be assigned, freeing up cellular resources for other work.

2. Cell gains an endomembrane system by proliferation of membrane.

Increased surface area allows higher rate of transport of materials within a cell.

3. Cell gains mitochondria.

Ability to metabolize sugars in the presence of oxygen enables greater function and success.

aerobic bacterium

mitochondrion

4. Cell gains chloroplasts.

Ability to produce sugars from sunlight enables greater function and success.

photosynthetic bacterium

chloroplast

Animal cell has mitochondria, but not chloroplasts.

Plant cell has both mitochondria and chloroplasts.

Figure 4.5 Origin of organelles.   Invagination of the plasma membrane could have created the nuclear envelope and an endomembrane system that involves several organelles. The endosymbiotic theory states that mitochondria and chloroplasts were independent prokaryotes that took up residence in a eukaryotic cell. Endosymbiosis was a first step toward the origin of the eukaryotic cell during the evolutionary history of life.

Figure 4.6 Animal cell anatomy. Micrograph of an insect cell (*right*) and drawing of a generalized animal cell (*below*).

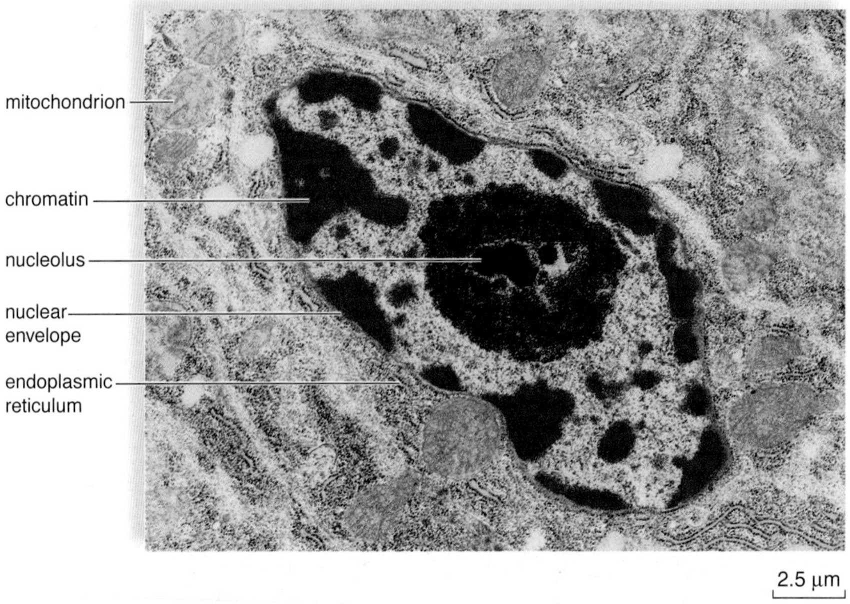

mitochondrion

chromatin

nucleolus

nuclear envelope

endoplasmic reticulum

2.5 μm

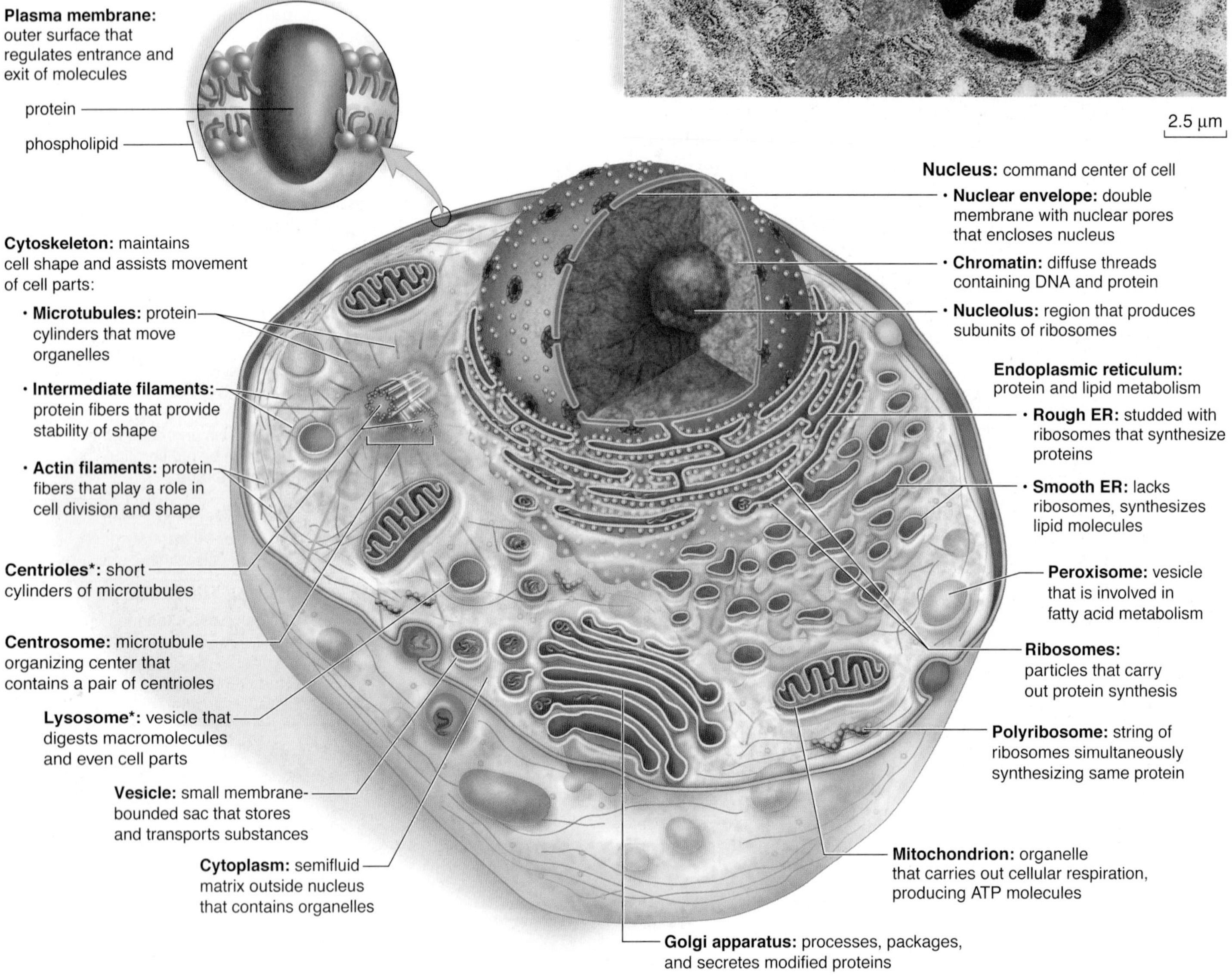

Plasma membrane: outer surface that regulates entrance and exit of molecules

protein

phospholipid

Cytoskeleton: maintains cell shape and assists movement of cell parts:

• **Microtubules:** protein cylinders that move organelles

• **Intermediate filaments:** protein fibers that provide stability of shape

• **Actin filaments:** protein fibers that play a role in cell division and shape

Centrioles*: short cylinders of microtubules

Centrosome: microtubule organizing center that contains a pair of centrioles

Lysosome*: vesicle that digests macromolecules and even cell parts

Vesicle: small membrane-bounded sac that stores and transports substances

Cytoplasm: semifluid matrix outside nucleus that contains organelles

Nucleus: command center of cell

• **Nuclear envelope:** double membrane with nuclear pores that encloses nucleus

• **Chromatin:** diffuse threads containing DNA and protein

• **Nucleolus:** region that produces subunits of ribosomes

Endoplasmic reticulum: protein and lipid metabolism

• **Rough ER:** studded with ribosomes that synthesize proteins

• **Smooth ER:** lacks ribosomes, synthesizes lipid molecules

Peroxisome: vesicle that is involved in fatty acid metabolism

Ribosomes: particles that carry out protein synthesis

Polyribosome: string of ribosomes simultaneously synthesizing same protein

Mitochondrion: organelle that carries out cellular respiration, producing ATP molecules

Golgi apparatus: processes, packages, and secretes modified proteins

*not in plant cells

Vesicles move around by means of an extensive network or lattice of protein fibers called the **cytoskeleton,** which also maintains cell shape and assists with cell movement. The protein fibers serve as tracks for the transport vesicles that are taking molecules from one organelle to another. Organelles are also moved from place to place using this transport system. Think of the cytoskeleton as a three-dimensional road system inside cells used to transport important cargo from place to place. The cytoskeleton is discussed in detail later in this chapter.

In addition to the plasma membrane, some eukaryotic cells, notably plant cells and those of fungi and many protists, have a cell wall. A plant cell wall contains cellulose fibrils and, therefore, has a different composition from the bacterial cell wall.

Cells can vary the proportion of organelles they have, depending on the specialized function of the cell. For example, a liver cell whose function is partly to detoxify drugs and other ingested compounds contains a greater proportion of smooth

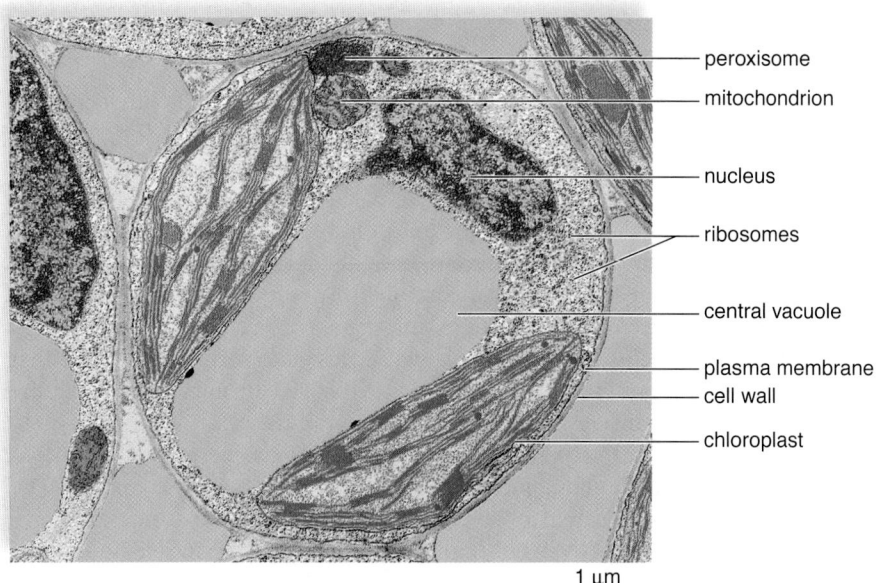

peroxisome
mitochondrion
nucleus
ribosomes
central vacuole
plasma membrane
cell wall
chloroplast

1 μm

Figure 4.7 Plant cell anatomy. False-colored micrograph of a young plant cell (*above*) and drawing of a generalized plant cell (*below*).

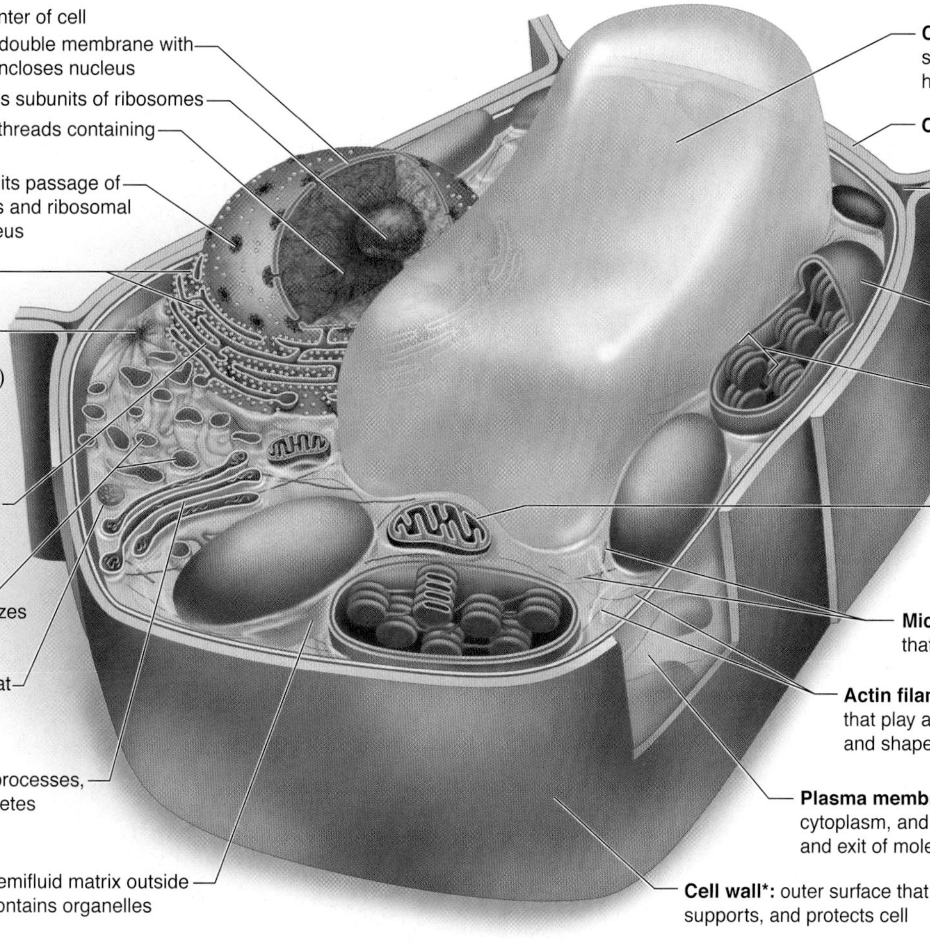

Nucleus: command center of cell
• **Nuclear envelope:** double membrane with nuclear pores that encloses nucleus
• **Nucleolus:** produces subunits of ribosomes
• **Chromatin:** diffuse threads containing DNA and protein
• **Nuclear pore:** permits passage of proteins into nucleus and ribosomal subunits out of nucleus

Ribosomes: carry out protein synthesis

Centrosome: microtubule organizing center (lacks centrioles)

Endoplasmic reticulum: protein and lipid metabolism
• **Rough ER:** studded with ribosomes that synthesize proteins
• **Smooth ER:** lacks ribosomes, synthesizes lipid molecules

Peroxisome: vesicle that is involved in fatty acid metabolism

Golgi apparatus: processes, packages, and secretes modified proteins

Cytoplasm: semifluid matrix outside nucleus that contains organelles

Central vacuole*: large, fluid-filled sac that stores metabolites and helps maintain turgor pressure

Cell wall of adjacent cell

Middle lamella: cements together the primary cell walls of adjacent plant cells

Chloroplast*: carries out photosynthesis, producing sugars

Granum*: a stack of chlorophyll-containing thylakoids in a chloroplast

Mitochondrion: organelle that carries out cellular respiration, producing ATP molecules

Microtubules: protein cylinders that aid movement of organelles

Actin filaments: protein fibers that play a role in cell division and shape

Plasma membrane: surrounds cytoplasm, and regulates entrance and exit of molecules

Cell wall*: outer surface that shapes, supports, and protects cell

*not in animal cells

endoplasmic reticulum, the organelle that accomplishes that task. A nerve cell, whose job is to conduct electrical signals across long distances, contains more plasma membrane relative to other cells. Other cells may specialize so extensively that they completely lose an organelle, like a red blood cell that ejects its nucleus to increase the surface area needed to carry oxygen in the blood. The Nature of Science feature describes the process by which investigators were able to discover the structure and function of various organelles.

Check Your Progress 4.3

1. Describe three benefits of compartmentalization found in cells.
2. Examine why organelles increase cell efficiency and function.
3. Infer how the proportion of organelles might differ between a muscle cell and a nerve cell.

4.4 The Nucleus and Ribosomes

Learning Outcomes

Upon completion of this section, you should be able to

1. Describe the structure and function of the nucleus.

2. Distinguish the flow of information from DNA to a protein.

3. Explain the role of ribosomes in protein synthesis.

The nucleus is essential to the life of a cell. It contains the genetic information that is passed on from cell to cell and from generation to generation. It specifies the information that ribosomes use to carry out protein synthesis. It also contains instructions for copying itself.

The Nucleus

The nucleus, which has a diameter of about 5 μm, is a prominent structure in the eukaryotic cell (Fig. 4.8). It generally appears as an oval structure located near the center of most cells. Some cells, such as skeletal muscle cells, can have more than one nucleus. The nucleus contains **chromatin** [Gk. *chroma*, color, and *teino*, stretch] in a semifluid matrix called the **nucleoplasm.** Chromatin looks grainy, but actually it is a network of strands that condenses and undergoes coiling into rodlike structures called **chromosomes** [Gk. *chroma*, color, and *soma*, body], just before the cell divides. All the cells of an individual contain the same number of chromosomes, and the mechanics of nuclear division ensure that each daughter cell receives the normal number of chromosomes, except for the egg and sperm, which usually have half this number. This alone suggested to early investigators that the chromosomes are the carriers of genetic information and that the nucleus is the command center of the cell.

Chromatin, and therefore chromosomes, contains DNA, protein, and some RNA (ribonucleic acid). **Genes,** composed of DNA, are units of heredity located on the chromosomes.

Three types of RNA are produced in the nucleus: *ribosomal RNA (rRNA), messenger RNA (mRNA),* and *transfer RNA (tRNA).* Ribosomal RNA is produced in the **nucleolus,** a dark region of chromatin where rRNA joins with proteins to form the subunits of ribosomes. Ribosomes are small bodies in the cytoplasm that facilitate protein synthesis. Messenger RNA, a mobile molecule, acts as an intermediary for DNA, a sedentary molecule, which specifies the sequence of amino acids in a protein. Transfer RNA participates in the assembly of amino acids into a polypeptide by recognizing both mRNA and amino acids during protein synthesis.

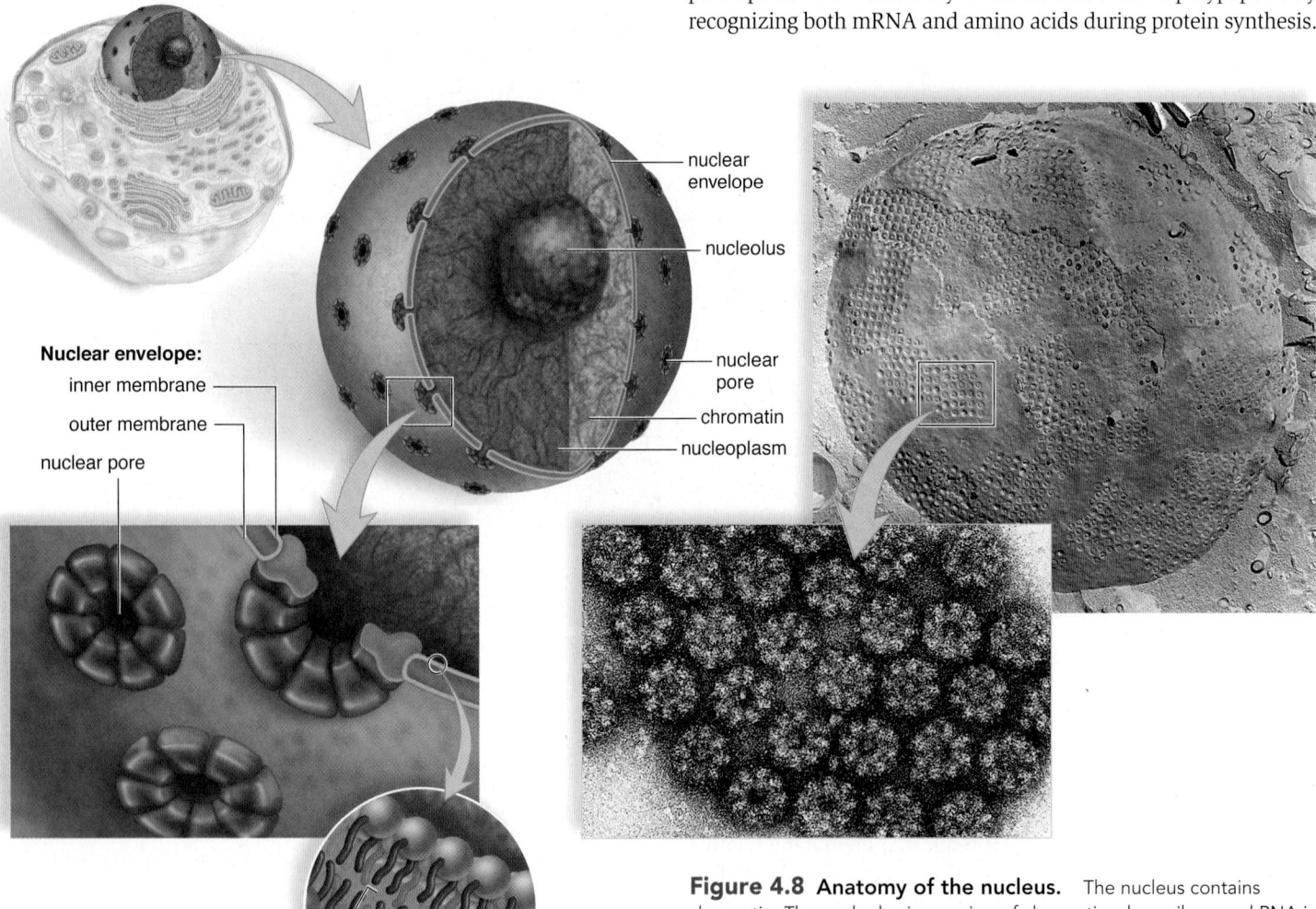

Nuclear envelope:
inner membrane
outer membrane
nuclear pore

phospholipid

nuclear envelope
nucleolus
nuclear pore
chromatin
nucleoplasm

Figure 4.8 Anatomy of the nucleus. The nucleus contains chromatin. The nucleolus is a region of chromatin where ribosomal RNA is produced, and ribosomal subunits are assembled. The nuclear envelope contains pores, as shown in the larger micrograph of a freeze-fractured nuclear envelope. Each pore is lined by a complex of eight proteins, as shown in the smaller micrograph and drawing. Nuclear pore complexes serve as passageways for substances to pass into and out of the nucleus.

The nucleus is important to cell structure and function because it specifies the code to make proteins. Although the nucleus is physically separated from the cytoplasm by a double membrane known as the **nuclear envelope,** it is still able to communicate with the cytoplasm through nuclear pores. **Nuclear pores** are of sufficient size (100 nm) to permit the passage of ribosomal subunits and mRNA out of the nucleus into the cytoplasm, and the passage of proteins from the cytoplasm into the nucleus. High-resolution electron micrographs show that nonmembrane components associated with the pores form a nuclear pore complex. Nuclear pore complexes act like gatekeepers to regulate what comes in and goes out of a nucleus.

Ribosomes

Ribosomes are particles where protein synthesis occurs. A large and small ribosomal subunit, each comprised of a mix of proteins and rRNA, are necessary components of a functional ribosome. In eukaryotes, ribosomes are 20 nm by 30 nm, and in prokaryotes they are slightly smaller. The number of ribosomes in a cell varies depending on its functions; for example, pancreatic cells and those of other glands have many ribosomes because they produce secretions that contain proteins.

In eukaryotic cells, some ribosomes occur freely within the cytoplasm, either singly or in groups called **polyribosomes,** whereas others are attached to the endoplasmic reticulum (ER), a membranous system of flattened saccules (small sacs) and tubules,

which is discussed more fully in the next section. In the nucleus, a gene is copied into mRNA, which is exported through a nuclear pore complex into the cytoplasm. Ribosomes receive the mRNA, which carries a coded message from DNA indicating the correct sequence of amino acids in a particular protein. Proteins synthesized by cytoplasmic ribosomes are used in the cytoplasm, and those synthesized by attached ribosomes end up in the ER.

What causes a ribosome to bind to the endoplasmic reticulum? Binding occurs only if the protein being synthesized by a ribosome begins with a sequence of amino acids called a **signal peptide.** The signal peptide binds a particle (signal recognition particle, SRP), which then binds to a receptor on the ER. Once the protein enters the ER, an enzyme cleaves off the signal peptide, and the protein ends up within the lumen (interior) of the ER, where it folds into its final shape (Fig. 4.9).

The sequence of DNA being transcribed into mRNA, and this in turn being translated into a protein, occurs in all living cells, at least during some point in their lifespan. Because of its universality, the DNA—mRNA—protein sequence of events is termed the *central dogma of molecular biology.*

Check Your Progress   4.4

1. Explain the importance of nuclear pore complexes.
2. Describe the sequence of events from a gene to a functional protein.

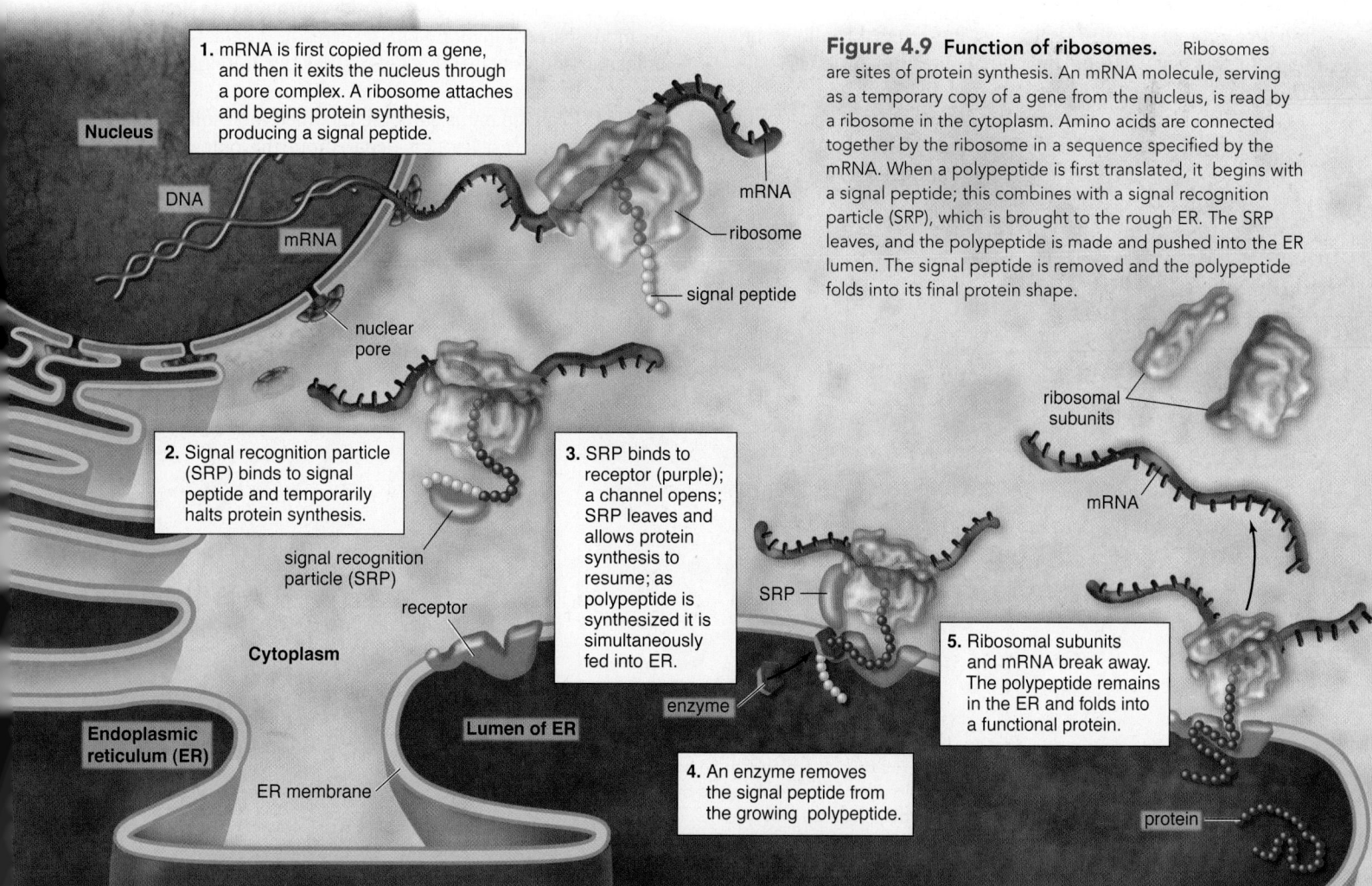

1. mRNA is first copied from a gene, and then it exits the nucleus through a pore complex. A ribosome attaches and begins protein synthesis, producing a signal peptide.

Nucleus

DNA

mRNA

nuclear pore

2. Signal recognition particle (SRP) binds to signal peptide and temporarily halts protein synthesis.

signal recognition particle (SRP)

receptor

Cytoplasm

Endoplasmic reticulum (ER)

ER membrane

3. SRP binds to receptor (purple); a channel opens; SRP leaves and allows protein synthesis to resume; as polypeptide is synthesized it is simultaneously fed into ER.

enzyme

Lumen of ER

4. An enzyme removes the signal peptide from the growing polypeptide.

mRNA

ribosome

signal peptide

ribosomal subunits

mRNA

SRP

5. Ribosomal subunits and mRNA break away. The polypeptide remains in the ER and folds into a functional protein.

protein

Figure 4.9 Function of ribosomes.   Ribosomes are sites of protein synthesis. An mRNA molecule, serving as a temporary copy of a gene from the nucleus, is read by a ribosome in the cytoplasm. Amino acids are connected together by the ribosome in a sequence specified by the mRNA. When a polypeptide is first translated, it begins with a signal peptide; this combines with a signal recognition particle (SRP), which is brought to the rough ER. The SRP leaves, and the polypeptide is made and pushed into the ER lumen. The signal peptide is removed and the polypeptide folds into its final protein shape.

4.5 The Endomembrane System

Learning Outcomes

Upon completion of this section, you should be able to

1. Explain the importance of the endomembrane system in cellular function.
2. Examine how ER, Golgi, and lysosome membranes differ from one another.
3. Describe how endomembrane vesicles are able to fuse with organelles.

The **endomembrane system** consists of the nuclear envelope, the membranes of the endoplasmic reticulum, the Golgi apparatus, and several types of vesicles. This system compartmentalizes the cell so that particular enzymatic reactions are restricted to specific regions and overall cell efficiency is increased. The vesicles transport molecules from one part of the system to another.

Endoplasmic Reticulum

The **endoplasmic reticulum (ER)** [Gk. *endon*, within; *plasma*, something molded; L. *reticulum*, net], consisting of a complicated system of membranous channels and saccules (flattened vesicles), is physically continuous with the nuclear envelope (Fig. 4.10). The ER consists of rough ER and smooth ER, which have different structures and functions.

Rough ER is studded with ribosomes on the side of the membrane that faces the cytoplasm, giving it the capacity to produce proteins. Inside its lumen, the rough ER allows proteins to fold and take on their final three-dimensional shape. The rough ER also contains enzymes that can add carbohydrate (sugar) chains to proteins, forming glycoproteins, that are important in many cell functions.

Smooth ER, which is continuous with the nuclear envelope and the rough ER, does not have attached ribosomes. Certain organs contain cells with an abundance of smooth ER, depending on the organ's function. In some organs, increased smooth ER helps produce more lipids. For example, in the testes, smooth ER produces testosterone, a steroid hormone. In the liver, smooth ER helps detoxify drugs. The smooth ER of the liver increases in quantity when a person consumes alcohol or takes barbiturates on a regular basis. Regardless of functional differences, both rough and smooth ER form vesicles that transport molecules to other parts of the cell, notably the Golgi apparatus.

The Golgi Apparatus

The **Golgi apparatus** is named for Camillo Golgi (1843-1926), who discovered its presence in cells in 1898. The Golgi apparatus typically consists of a stack of three to twenty slightly curved, flattened saccules whose appearance can be compared to a stack of pancakes (Fig. 4.11). In animal cells, one side of the stack (the *cis* or inner face) is directed toward the ER, and the other side of the stack (the *trans* or outer face) is directed toward the plasma membrane. Vesicles can frequently be seen at the edges of the saccules.

Protein-filled vesicles that bud from the rough ER and lipid-filled vesicles that bud from the smooth ER are received by the Golgi apparatus at its inner face. These substances are altered as they move through the saccules. For example, the Golgi apparatus contains enzymes that modify the carbohydrate chains first attached to proteins in the rough ER. It can change or modify one sugar into another sugar on glycoproteins. In some cases, the modified carbohydrate chain serves as a signal molecule or molecular address label that determines the protein's final destination in the cell.

The Golgi apparatus sorts the modified molecules and packages them into vesicles that depart from the outer face. These vesicles may be transported to various locations within the cell,

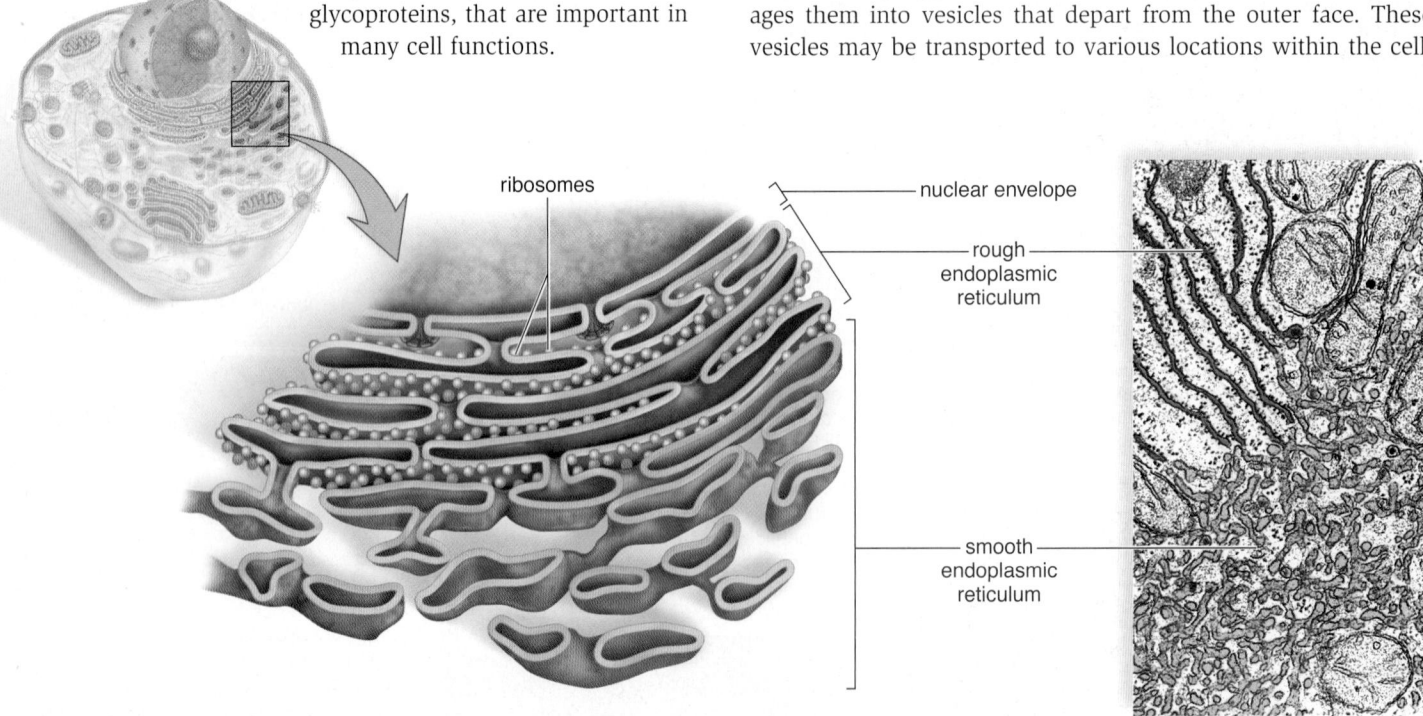

ribosomes

nuclear envelope

rough endoplasmic reticulum

smooth endoplasmic reticulum

0.08 μm

Figure 4.10 Endoplasmic reticulum (ER).   Ribosomes are present on rough ER, which consists of flattened saccules, but not on smooth ER, which is more tubular. Proteins are synthesized by rough ER. Smooth ER is involved in lipid synthesis, detoxification reactions, and several other possible functions.

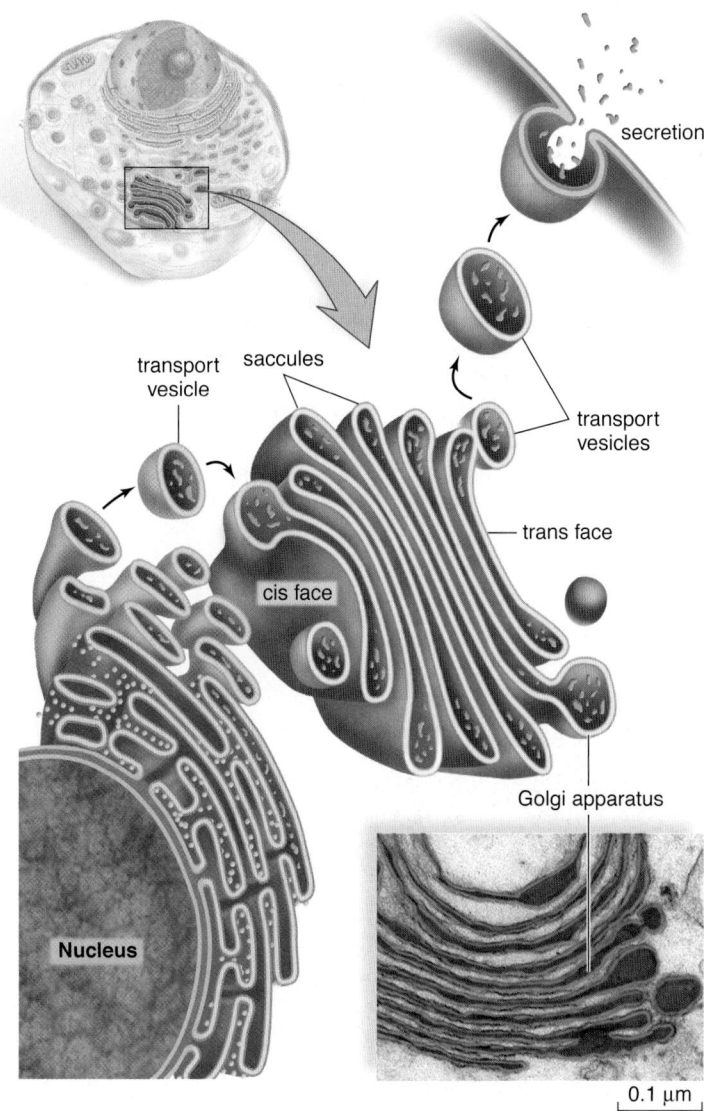

secretion

transport vesicle

saccules

transport vesicles

trans face

cis face

Golgi apparatus

Nucleus

0.1 µm

Figure 4.11 Golgi apparatus. The Golgi apparatus is a stack of flattened, curved saccules. It processes proteins and lipids and packages them in transport vesicles that distribute these molecules to various locations within the cell, or secrete them externally.

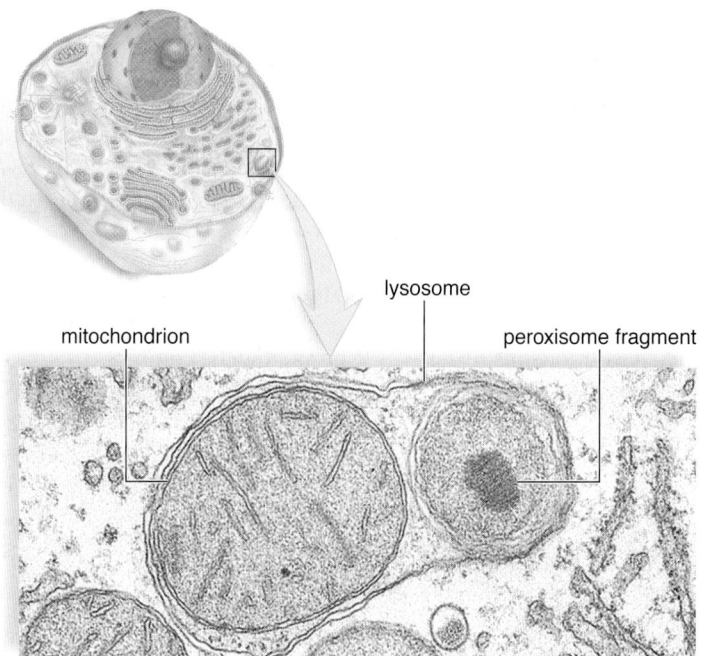

lysosome

mitochondrion

peroxisome fragment

a. Mitochondrion and a peroxisome in a lysosome

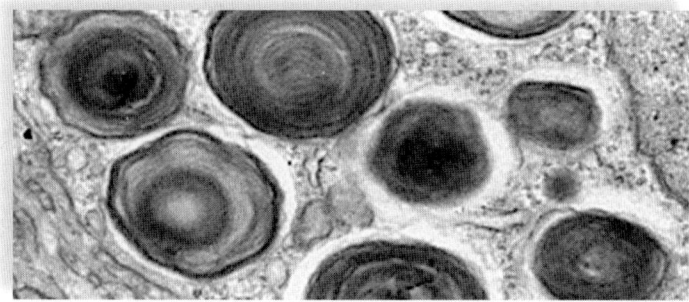

b. Storage bodies in a cell with defective lysosomes

Figure 4.12 Lysosomes. **a.** Lysosomes, which bud off the Golgi apparatus in cells, are filled with hydrolytic enzymes that digest molecules and parts of the cell. Here a lysosome digests a worn mitochondrion and a peroxisome. **b.** The nerve cells of a person with Tay-Sachs disease are filled with membranous cytoplasmic bodies storing a fat that lysosomes are unable to digest.

depending on their molecular address label. In animal cells, some of these vesicles are lysosomes, which are discussed next. Other vesicles may return to the ER or proceed to the plasma membrane, where they merge and discharge their contents to the outside of the cell during **secretion.** Secretion is termed exocytosis because the substance exits the cytoplasm.

Lysosomes

Lysosomes [Gk. *lyo,* loose, and *soma,* body] are membrane-bounded vesicles produced by the Golgi apparatus. They have a very low pH and store powerful hydrolytic-digestive enzymes in an inactive state. Lysosomes act much like your stomach in that they assist in digesting material taken into the cell. They also destroy nonfunctional organelles and portions of cytoplasm (Fig. 4.12).

Materials can be brought into a cell by vesicle or vacuole formation at the plasma membrane. When a lysosome fuses with either, the lysosomal enzymes are activated and digest the material into simpler subunits that are exported into the cytoplasm

and recycled by other cell processes. White blood cells, specialized to protect the body from foreign entities, are well known for engulfing pathogens (e.g., disease-causing viruses and bacteria) that are then broken down in lysosomes. White blood cells have a greater proportion of lysosomes than other cells because their specialized function is digestion of foreign bodies.

A number of human lysosomal storage diseases are due to a missing lysosomal enzyme. In Tay-Sachs disease, the missing enzyme digests a fatty substance that helps insulate nerve cells and increases their efficiency. The fatty substance accumulates in so many storage bodies that nerve cells die off. Affected individuals appear normal at birth but begin to develop neurological problems at four to six months of age. Eventually, the child suffers cerebral degeneration, slow paralysis, blindness, and loss of motor function. Children with Tay-Sachs disease live only about three to four years. In the future, it may be possible to provide the missing enzyme and, in that way, prevent lysosomal storage diseases.

Animation
Lysosomes

Endomembrane System Summary

You have seen that the endomembrane system is a series of membranous organelles that work together and communicate by means of transport vesicles. The endoplasmic reticulum (ER) and the Golgi apparatus are essentially flattened saccules, and lysosomes are specialized vesicles.

Organelles within the endomembrane system can interact because their membranes readily fuse together, and because membrane-associated proteins enable communication and specialized functions. Figure 4.13 shows how the components of the endomembrane system work together. Products of both rough ER and smooth ER are carried in transport vesicles to the Golgi apparatus, where they are further modified. Using signaling sequences and molecular address labels, the Golgi apparatus sorts these products and packages them into vesicles that transport them to various cellular destinations. Secretory vesicles take the proteins to the plasma membrane, where they exit the cell by exocytosis. For example, secretion into ducts occurs when the mammary glands produce milk or the pancreas produces digestive enzymes.

In animal cells, the Golgi apparatus also produces lysosomes that contain stored hydrolytic enzymes. Lysosomes fuse with incoming vesicles from the plasma membrane and digest macromolecules brought into a cell.

Check Your Progress   4.5

1. Contrast the structure and functions of rough and smooth endoplasmic reticulum.
2. Describe the relationship between the components of the endomembrane system.
3. Examine how cellular function would be affected if the Golgi apparatus ceased to function.

Figure 4.13
Endomembrane system.
The organelles in the endomembrane system work together to carry out the functions noted. Plant cells do not have lysosomes, nor do they have incoming and outgoing (secretory) vesicles.

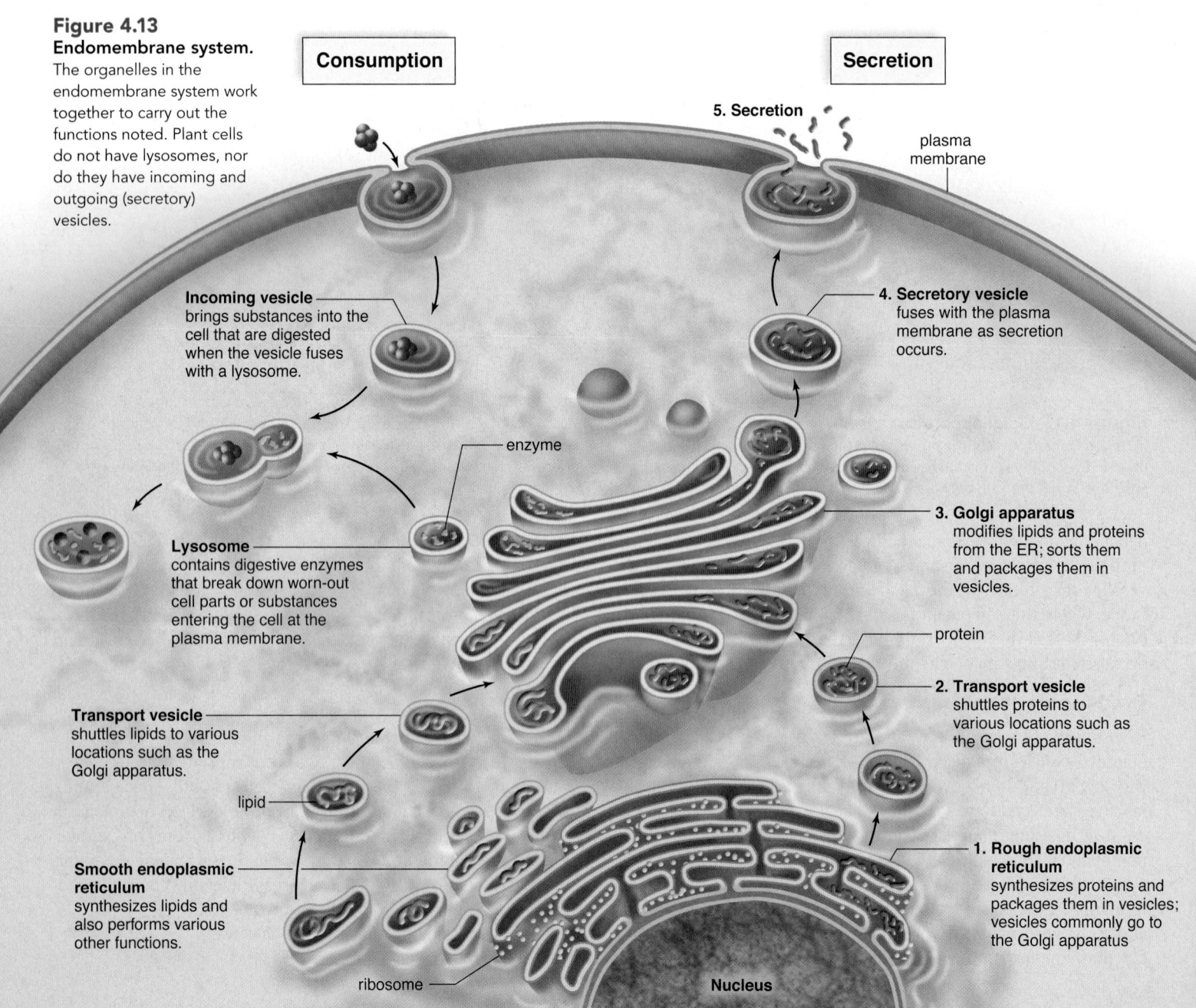

Consumption

Secretion

5. Secretion

plasma membrane

Incoming vesicle brings substances into the cell that are digested when the vesicle fuses with a lysosome.

4. Secretory vesicle fuses with the plasma membrane as secretion occurs.

enzyme

Lysosome contains digestive enzymes that break down worn-out cell parts or substances entering the cell at the plasma membrane.

3. Golgi apparatus modifies lipids and proteins from the ER; sorts them and packages them in vesicles.

protein

Transport vesicle shuttles lipids to various locations such as the Golgi apparatus.

2. Transport vesicle shuttles proteins to various locations such as the Golgi apparatus.

lipid

Smooth endoplasmic reticulum synthesizes lipids and also performs various other functions.

1. Rough endoplasmic reticulum synthesizes proteins and packages them in vesicles; vesicles commonly go to the Golgi apparatus

ribosome

Nucleus

4.6 Other Vesicles and Vacuoles

Learning Outcomes

Upon completion of this section, you should be able to

1. Describe the role of peroxisomes and vacuoles in cell function.
2. Contrast peroxisomes and vacuoles with endomembrane organelles.

Peroxisomes and the vacuoles of cells do not communicate with the organelles of the endomembrane system, and therefore are not part of it.

Peroxisomes

Peroxisomes, similar to lysosomes, are membrane-bounded vesicles that enclose enzymes. However, the enzymes in peroxisomes are synthesized by free ribosomes and transported into a peroxisome from the cytoplasm. All peroxisomes contain enzymes whose actions result in hydrogen peroxide (H_2O_2):

$$RH_2 + O_2 \longrightarrow R + H_2O_2$$

Hydrogen peroxide, a toxic molecule, is immediately broken down to water and oxygen by another peroxisomal enzyme called catalase. When hydrogen peroxide is applied to a wound, bubbling occurs as catalase breaks it down.

Peroxisomes are metabolic assistants to the other organelles. They have varied functions but are especially prevalent in cells that synthesize and break down lipids. In the liver, some peroxisomes produce bile salts from cholesterol, and others break down fats. The 1992 film *Lorenzo's Oil* is based on a medical case of a boy with defective peroxisomes. Lorenzo Odone's peroxisomes lacked a membrane protein

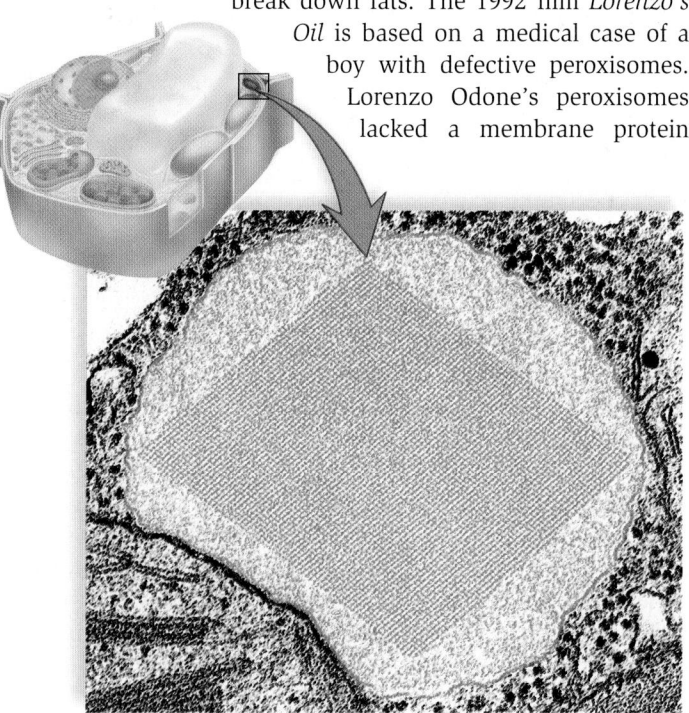

Figure 4.14 Peroxisomes.   Peroxisomes contain one or more enzymes that can oxidize various organic substances.

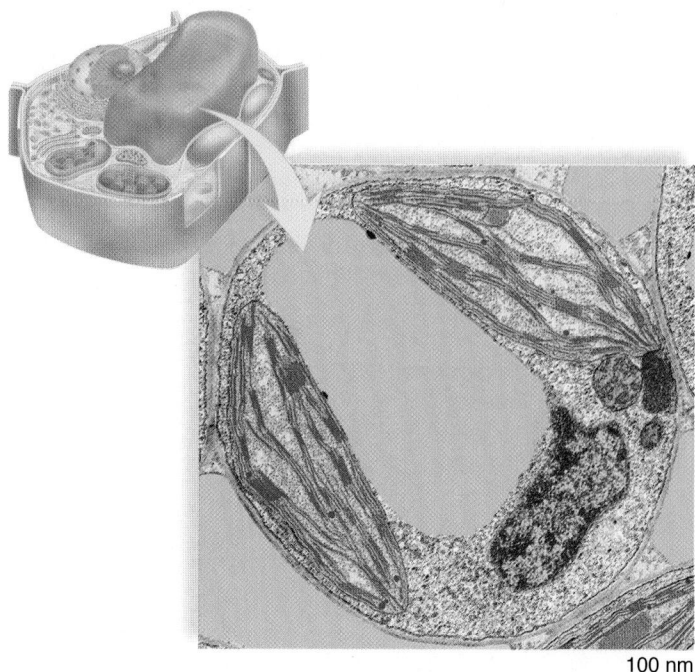

100 nm

Figure 4.15 Plant cell central vacuole.   The large central vacuole of plant cells has numerous functions, from storing molecules to helping the cell increase in size.

needed to import a specific enzyme and/or long chain fatty acids from the cytoplasm. As a result, long-chain fatty acids accumulated in his brain, and he suffered neurological damage. This disorder is known as adrenoleukodystrophy (ALD).

Plant cells also have peroxisomes (Fig. 4.14). In germinating seeds, they oxidize fatty acids into molecules that can be converted to sugars needed by the growing plant. In leaves, peroxisomes can carry out a reaction that is opposite to photosynthesis—the reaction uses up oxygen and releases carbon dioxide.

Vacuoles

Like vesicles, **vacuoles** are membranous sacs, but vacuoles are larger than vesicles. The vacuoles of some protists are quite specialized, including contractile vacuoles for ridding the cell of excess water and digestive vacuoles for breaking down nutrients. Vacuoles usually store substances. In general, few animal cells contain vacuoles; however, fat cells contain a very large lipid-engorged vacuole that takes up nearly two-thirds of the volume of the cell!

Vacuoles are essential to plant function. Plant vacuoles contain not only water, sugars, and salts but also water-soluble pigments and toxic molecules. The pigments are responsible for many of the red, blue, or purple colors of flowers and some leaves. The toxic substances help protect a land plant from herbivorous animals.

Plant Cell Central Vacuole

Typically, plant cells have a large **central vacuole** that may take up to 90% of the volume of the cell. The vacuole is filled with a watery fluid called cell sap that gives added support to the cell (Fig. 4.15). The central vacuole maintains hydrostatic pressure or turgor pressure in plant cells, which provides structural support. A plant cell can rapidly increase in size by enlarging its vacuole. Eventually, a plant cell also produces more cytoplasm.

The central vacuole functions in storage of both nutrients and waste products. Metabolic waste products are pumped across the vacuole membrane and stored permanently in the central vacuole. As organelles age and become nonfunctional, they fuse with the vacuole, where digestive enzymes break them down. This is an analogous function to that carried out by lysosomes in animal cells.

Check Your Progress 4.6

1. Compare the structure and functions of a peroxisome with a lysosome.
2. Distinguish between where peroxisome and lysosome proteins are produced.

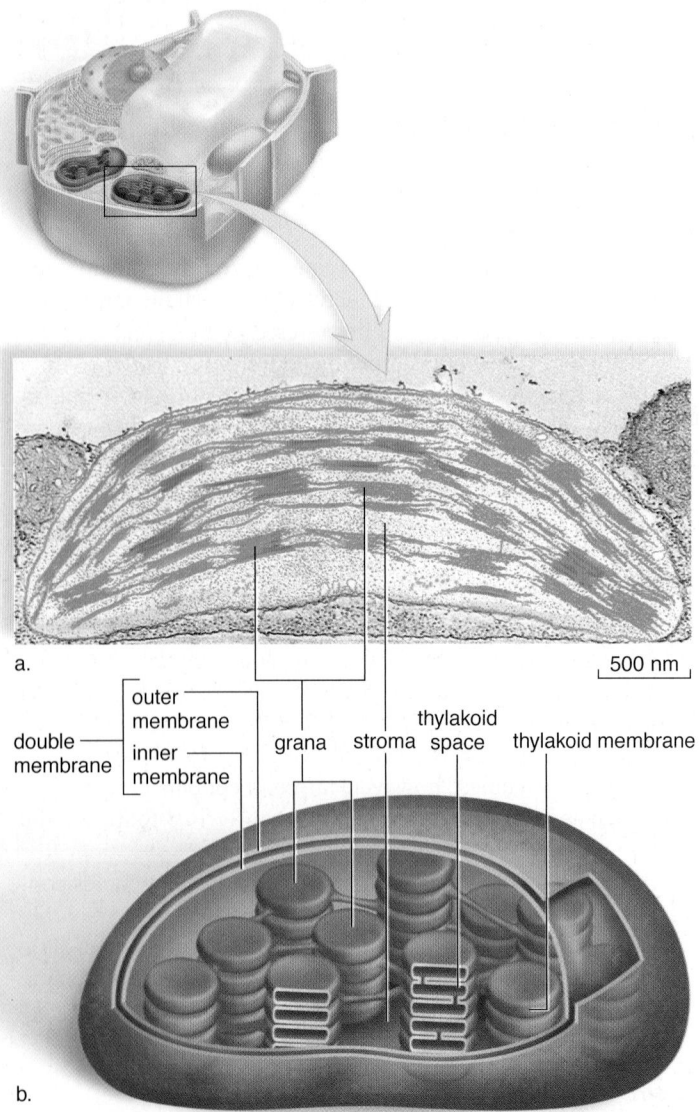

a.

500 nm

double membrane — outer membrane / inner membrane — grana — stroma — thylakoid space — thylakoid membrane

b.

Figure 4.16 Chloroplast structure. Chloroplasts carry out photosynthesis. **a.** Electron micrograph of a longitudinal section of a chloroplast. **b.** Generalized drawing of a chloroplast in which the outer and inner membranes have been cut away to reveal the grana, each of which is a stack of membranous sacs called thylakoids. In some grana, but not all, thylakoid spaces are interconnected.

4.7 The Energy-Related Organelles

Learning Outcomes

Upon completion of this section, you should be able to

1. Compare the energy management of cellular respiration versus photosynthesis.
2. Describe the evidence that suggests chloroplasts and mitochondria are derived from ancient bacteria and endosymbiosis.
3. Explain why increased membrane surface area is necessary for chloroplast and mitochondrial function.

Life is possible only because a constant input of energy maintains the structure of cells. Chloroplasts and mitochondria are the two eukaryotic membranous organelles that specialize in converting energy to a form that can be used by the cell. Although animal cells contain only mitochodria, plant cells contain both mitochondria and chloroplasts.

During photosynthesis, **chloroplasts** [Gk. *chloros,* green, and *plastos,* formed, molded] use solar energy to synthesize carbohydrates, which serve as organic nutrient molecules for plants and all living things on Earth. *Photosynthesis* can be represented by this equation:

solar energy + carbon dioxide + water → carbohydrate + oxygen

Plants, algae, and cyanobacteria are capable of conducting photosynthesis in this manner, but only plants and algae have chloroplasts because they are eukaryotes.

In cellular respiration, **mitochondria** (sing., mitochondrion) break down carbohydrate-derived products to produce ATP (adenosine triphosphate). *Cellular respiration* can be represented by this equation:

carbohydrate + oxygen → carbon dioxide + water + energy

Here the word *energy* stands for ATP molecules. When a cell needs energy, ATP supplies it. The energy of ATP is used to drive synthetic reactions, active transport, and all energy-requiring processes in cells.

Chloroplasts

Some algal cells have only one chloroplast, while some plant cells have as many as a hundred. Chloroplasts can be quite large, being twice as wide and as much as five times the length of a mitochondrion.

Chloroplasts have a three-membrane system (Fig. 4.16). They are bounded by a double membrane, which includes an outer membrane and an inner membrane. The double membrane encloses the semifluid **stroma,** which contains enzymes and **thylakoids,** disk-like sacs formed from a third chloroplast membrane. A stack of thylakoids is a **granum.** The lumens of the thylakoids are believed to form a large internal compartment called the thylakoid space. Chlorophyll and the other pigments that capture solar energy are located in the thylakoid membrane, and the enzymes that synthesize carbohydrates are located outside the thylakoid in the fluid of the stroma.

The endosymbiotic theory holds that chloroplasts are derived from a photosynthetic bacterium that was engulfed by a eukaryotic cell. This certainly explains why a chloroplast is bounded by a double membrane—one membrane is derived from the vesicle that brought the prokaryote into the cell, while the inner membrane is derived from the prokaryote. The endosymbiotic theory is also supported by the finding that chloroplasts have their own prokaryotic-type chromosome and ribosomes, and they produce some of their own enzymes even today.

Other Types of Plastids

A chloroplast is a type of plastid. **Plastids** are plant organelles that are surrounded by a double membrane and have varied functions. **Chromoplasts** contain pigments that result in a yellow, orange, or red color. Chromoplasts are responsible for the color of autumn leaves, fruits, carrots, and some flowers. **Leucoplasts** are generally colorless plastids that synthesize and store starches and oils. A microscopic examination of potato tissue reveals a number of leucoplasts.

Mitochondria

Nearly all eukaryotic cells, and certainly all plant and algal cells in addition to animal cells, contain mitochondria. Even though mitochondria are smaller than chloroplasts, they can usually be seen using a light microscope. The number of mitochondria can vary depending on the metabolic activities and energy needed within a cell. Some cells, such as liver cells, may have as many as 1,000 mitochondria.

We think of mitochondria as having a shape like that shown in Figure 4.17, but actually they often change shape to be longer and thinner or shorter and broader. Mitochondria can form long, moving chains, or they can remain fixed in one location—typically where energy is most needed. For example, they are packed between the contractile elements of cardiac cells and wrapped around the interior of a sperm's flagellum. In contrast, fat cells contain few mitochondria—they function in fat storage, which does not require energy.

Mitochondria have two membranes, the outer membrane and the inner membrane. The inner membrane is highly convoluted into folds called **cristae** that project into the matrix. These cristae increase the surface area of the inner membrane so much that in a liver cell they account for about one-third the total membrane in the cell. The inner membrane encloses a semifluid **matrix,** which contains mitochondrial DNA and ribosomes. Again, the presence of a double membrane and mitochondrial genes is consistent with the endosymbiotic theory regarding the origin of mitochondria, which was illustrated in Figure 4.5.

Mitochondria are often called the powerhouses of the cell because they produce most of the ATP utilized by the cell. The procedure described in the Nature of Science feature on page 63 allowed investigators to separate the inner membrane, the outer membrane, and the matrix from each other. Then they discovered that the matrix is a highly concentrated mixture of enzymes that break down carbohydrates and other nutrient molecules. These reactions supply the chemical energy needed for a chain of proteins on the inner membrane to create the conditions that allow ATP synthesis to take place. The entire process, which also involves the cytoplasm, is called cellular respiration because oxygen is used and carbon dioxide is given off, as shown at the beginning of this section.

▶ Animation Endosymbiosis

Figure 4.17
Mitochondrion structure. Mitochondria are involved in cellular respiration. **a.** Electron micrograph of a longitudinal section of a mitochondrion. **b.** Generalized drawing in which the outer membrane and portions of the inner membrane have been cut away to reveal the cristae.

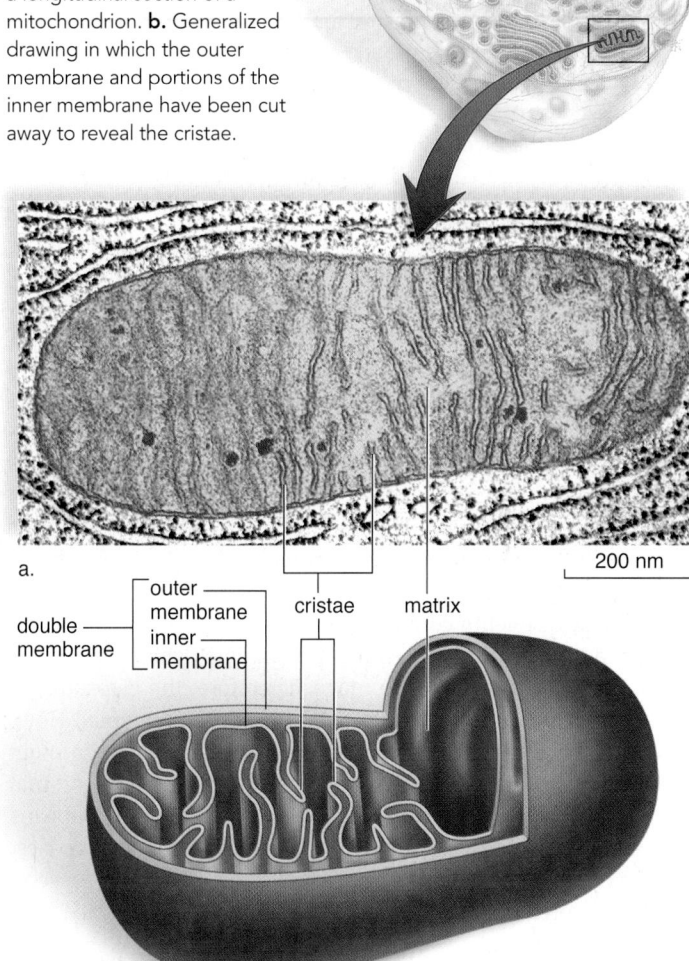

a.   200 nm

double membrane { outer membrane / inner membrane }   cristae   matrix

b.

Mitochondrial Diseases

So far, more than 40 different mitochondrial diseases that affect the brain, muscles, kidneys, heart, liver, eyes, ears, or pancreas have been identified. The common factor among these genetic diseases is that the patient's mitochondria are unable to completely metabolize organic molecules to produce ATP. As a result, toxins accumulate inside the mitochondria and the body. The toxins can be free radicals (substances that readily form harmful compounds when they react with other molecules), and these compounds damage mitochondria over time. In the United States, between 1,000 and 4,000 children per year are born with a mitochondrial disease. In addition, it is possible that many diseases of aging are due to malfunctioning mitochondria.

▶ Video Aging Secret

Check Your Progress   4.7

1. Discuss the evidence that chloroplasts and mitochondria are derived from ancient bacteria.
2. Explain the role of ATP in photosynthesis and cellular respiration.

4.8 The Cytoskeleton

Learning Outcomes

Upon completion of this section, you should be able to

1. Compare the structure and function of actin filaments, intermediate filaments, and microtubules.
2. Describe how motor molecules interact with cytoskeletal elements to produce movement.
3. Explain the diverse roles of microtubules within the cell.

Cells are exposed to many physical forces. Cell shape, movement, and internal transport all require structural support provided by the cytoskeleton. The protein components of the cytoskeleton [Gk. *kytos,* cell, and *skeleton,* dried body] interconnect and extend from the nucleus to the plasma membrane in eukaryotic cells. Prior to the 1970s, it was believed that the cytoplasm was an unorganized mixture of organic molecules. Then, high-voltage electron microscopes, which can penetrate thicker specimens, showed instead that the cytoplasm was highly organized. The technique of immunofluorescence microscopy identified the makeup of the protein components within the cytoskeletal network (Fig. 4.18).

The cytoskeleton contains actin filaments, intermediate filaments, and microtubules, which maintain cell shape and allow the cell and its organelles to move. Therefore, the cytoskeleton is often compared to the bones and muscles of an animal. However, the cytoskeleton is dynamic, and can rearrange its protein components as necessary in response to changes in internal and external environments. A number of different mechanisms appear to regulate this process, including protein phosphatases, which remove phosphates from proteins and bring about assembly, and protein kinases, which phosphorylate proteins and lead to disassembly.

Actin Filaments

Actin filaments (formerly called microfilaments) are long, extremely thin, flexible fibers (about 7 nm in diameter) that occur in bundles or meshlike networks. Each actin filament contains two chains of globular actin monomers twisted about one another in a helical manner.

Actin filaments provide structural support as a dense, complex web just under the plasma membrane, to which they are anchored by special proteins. Sometimes, actin filaments can dynamically rearrange themselves and facilitate cellular movement, such as when an amoeba moves over a surface with pseudopods [L. *pseudo,* false, and *pod,* feet], or when intestinal cell microvilli lengthen and shorten into the gut lumen (the space where ingested food is processed). In plant cells, actin filaments form the tracks along which chloroplasts circulate in a particular direction in a process called cytoplasmic streaming.

Actin filaments move the cell and its organelles by interacting with **motor molecules,** which are proteins that can attach, detach, and reattach farther along an actin filament. The motor molecule myosin uses ATP to pull actin filaments along in this way. Myosin has both a head and a tail. In muscle cells, the tails of several myosin molecules are joined to form a thick filament. In non-muscle cells, cytoplasmic myosin tails are bound to membranes, but the heads still interact with actin:

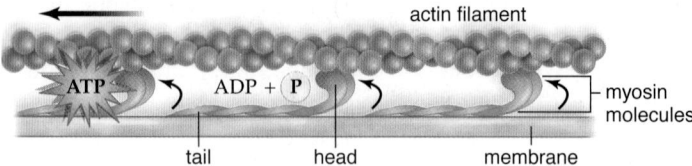

During animal cell division, the two new cells form when actin, in conjunction with myosin, pinches off the cells from one another.

Intermediate Filaments

Intermediate filaments (8–11 nm in diameter) are so named because they are intermediate in size between actin filaments and microtubules. They form a ropelike assembly of fibrous polypeptides, but the specific filament type varies according to the tissue. Some intermediate filaments support the nuclear envelope, whereas others support the plasma membrane and take part in the formation of cell-to-cell junctions. In the skin, intermediate filaments made of the protein keratin give great mechanical strength to skin cells. Like other cytoskeletal components, intermediate filaments are highly dynamic and disassemble when phosphate is added to them by a kinase.

Microtubules

Microtubules [Gk. *mikros,* small, little; L. *tubus,* tube] are small, hollow cylinders about 25 nm in diameter and from 0.2 to 25 μm in length.

Microtubules are made of a globular protein called tubulin, which is of two types called α and β. Alpha tubulin has a slightly different amino acid sequence than β tubulin. When assembly occurs, α and β tubulin molecules come together as dimers, and the dimers arrange themselves in rows. Microtubules have 13 rows of tubulin dimers, surrounding what appears in electron micrographs to be an empty central core.

Microtubule assembly is under the regulatory control of a microtubule organizing center (MTOC). In most eukaryotic cells, the main MTOC is in the **centrosome** [Gk. *centrum,* center, and *soma,* body], which lies near the nucleus. Microtubules radiate from the centrosome, helping to maintain the shape of the cell and acting as tracks along which organelles can be moved. Whereas the motor molecule myosin is associated with actin filaments, the motor molecules kinesin and dynein are associated with microtubules:

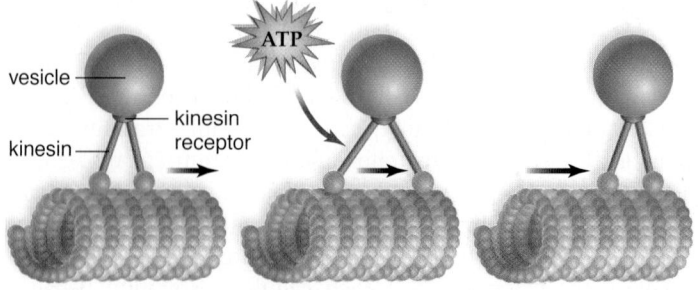

vesicle moves, not microtubule

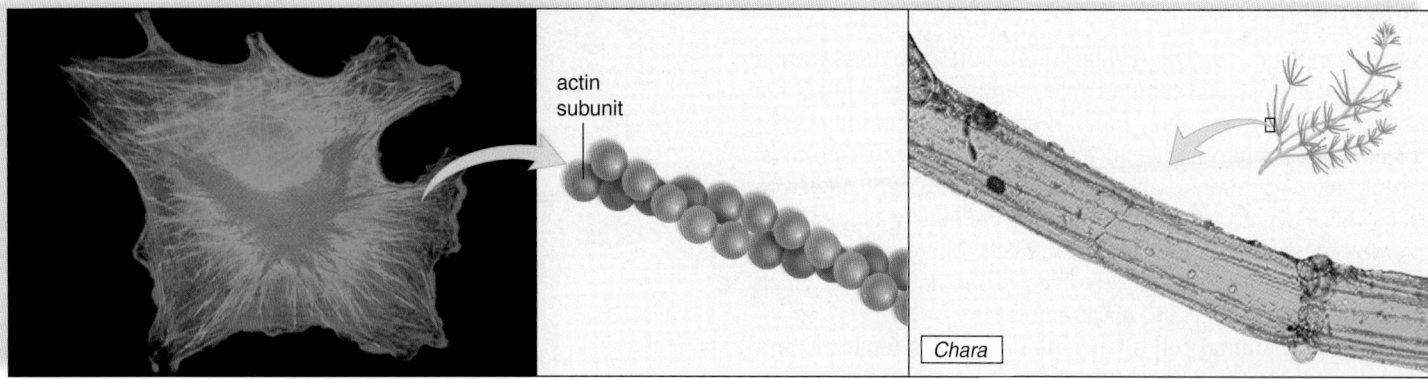

a. Actin filaments

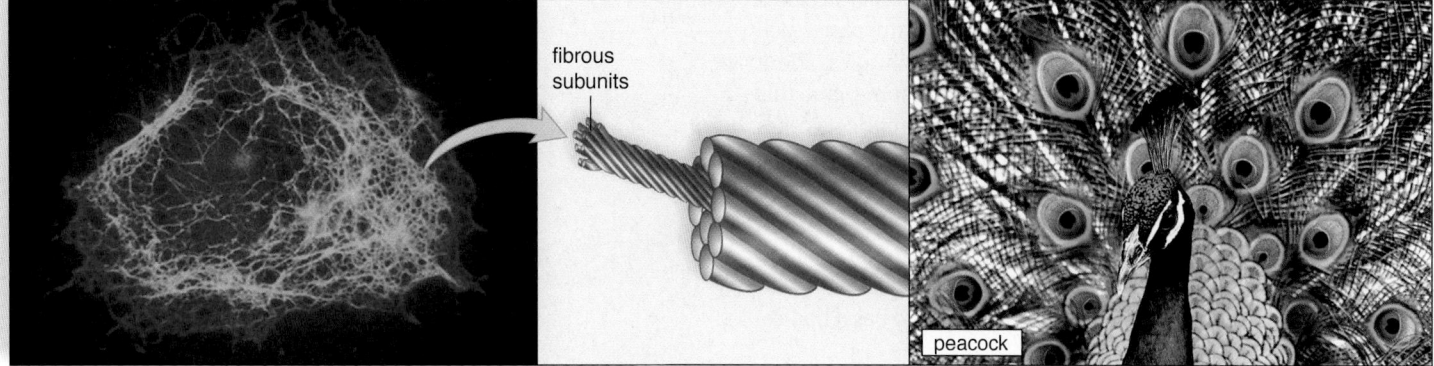

b. Intermediate filaments

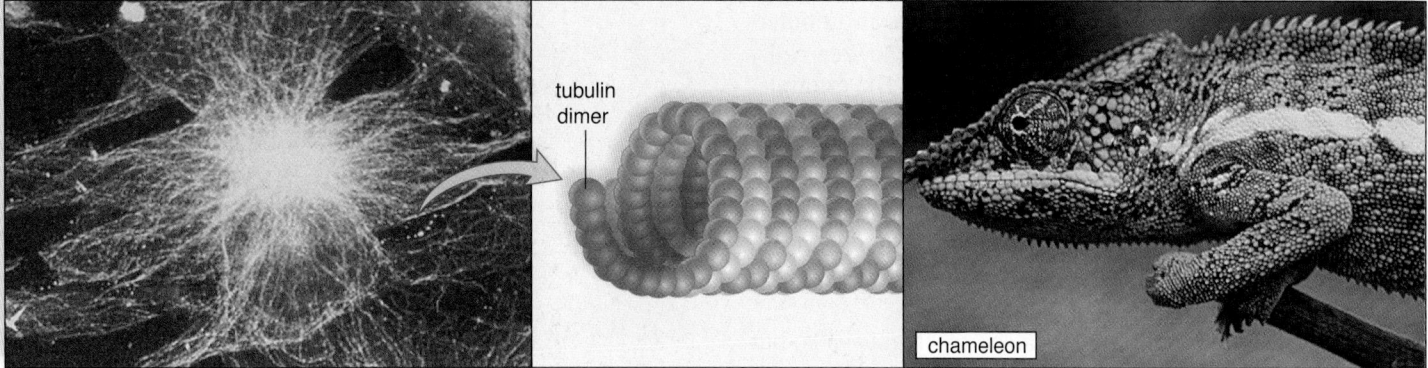

c. Microtubules

Figure 4.18 The cytoskeleton. The cytoskeleton maintains a cell's shape and allows its parts to move. Three types of protein components make up the cytoskeleton. They can be detected in cells by using labeling and fluorescence microscopy. **a.** *Left to right:* Animal cells showing a twisted double chain of actin filaments (green fibers). The giant cells of the green alga *Chara* use actin filaments to move organelles within the cell. **b.** *Left to right:* Animal cells showing fibrous, ropelike intermediate filaments (blue fibers). A peacock's colorful feathers are strengthened by intermediate filaments. **c.** *Left to right:* Animal cells showing hollow microtubules made of tubulin dimers (orange fibers). A chameleon's skin cells use microtubules to move pigment granules around so that they take on the color of their environment.

There are different types of kinesin proteins, each specialized to move one kind of vesicle or cellular organelle. Kinesin moves vesicles or organelles in an opposite direction from dynein. Cytoplasmic dynein is closely related to the molecule dynein found in flagella.

Before a cell divides, microtubules disassemble and then reassemble into a structure called a spindle that distributes chromosomes in an orderly manner. At the end of cell division, the spindle disassembles, and microtubules reassemble once again into their former array. In the "arms race" between plants and herbivores, plants have evolved various types of poisons that prevent them from being eaten. Colchicine is a plant poison that binds tubulin and blocks the assembly of microtubules.

Centrioles

Centrioles [Gk. *centrum,* center] are short cylinders with a 9 + 0 pattern of microtubule triplets—nine sets of triplets are arranged in an outer ring, but the center of a centriole does not contain a microtubule. In animal cells and most protists, a centrosome contains two centrioles lying at right angles to each other. A centrosome, as mentioned previously, is the major microtubule-organizing center for the cell. Therefore, it is possible that centrioles are also involved in the process by which microtubules assemble and disassemble.

Before an animal cell divides, the centrioles replicate, and the members of each pair are at right angles to one another (Fig. 4.19). Then each pair becomes part of a separate centrosome. During cell division, the centrosomes move apart and most likely function to organize the mitotic spindle. In any case, each new cell has its own centrosome and pair of centrioles. Plant and fungal cells have the equivalent of a centrosome, but this structure does not contain centrioles, suggesting that centrioles are not necessary to the assembly of cytoplasmic microtubules.

A **basal body** is an organelle that lies at the base of cilia and flagella and may direct the organization of microtubules within these structures. In other words, a basal body may do for a cilium or flagellum what the centrosome does for the cell. In cells with cilia and flagella, centrioles are believed to give rise to basal bodies.

Cilia and Flagella

Cilia [L. *cilium,* eyelash, hair] and **flagella** [L. *flagello,* whip] are hairlike projections that can move either in an undulating fashion, like a whip, or stiffly, like an oar. In free cells, cilia (or flagella) move the cell through liquid. For example, unicellular paramecia are organisms that move by means of cilia, whereas sperm cells move by means of flagella. If the cell is attached to other cells, cilia (or flagella) are capable of moving liquid over the cell. The cells that line our upper respiratory tract have cilia that sweep debris trapped within mucus back up into the throat, where it can be swallowed or expelled. This action helps keep the lungs clean.

In eukaryotic cells, cilia are much shorter than flagella, but they have a similar construction. Both are membrane-bounded cylinders enclosing a matrix area. In the matrix are nine microtubule doublets arranged in a circle around two central microtubules; this is called the 9 + 2 pattern of microtubules (Fig. 4.20). Cilia and flagella move when the microtubule doublets slide past one another using motor molecules.

As mentioned, each cilium and flagellum has a basal body lying in the cytoplasm at its base. Basal bodies have the same circular arrangement of microtubule triplets as centrioles and are believed to be derived from them. It is possible that basal bodies organize the microtubules within cilia and flagella, but this idea is not supported by the observation that cilia and flagella grow by the addition of tubulin dimers to their tips.

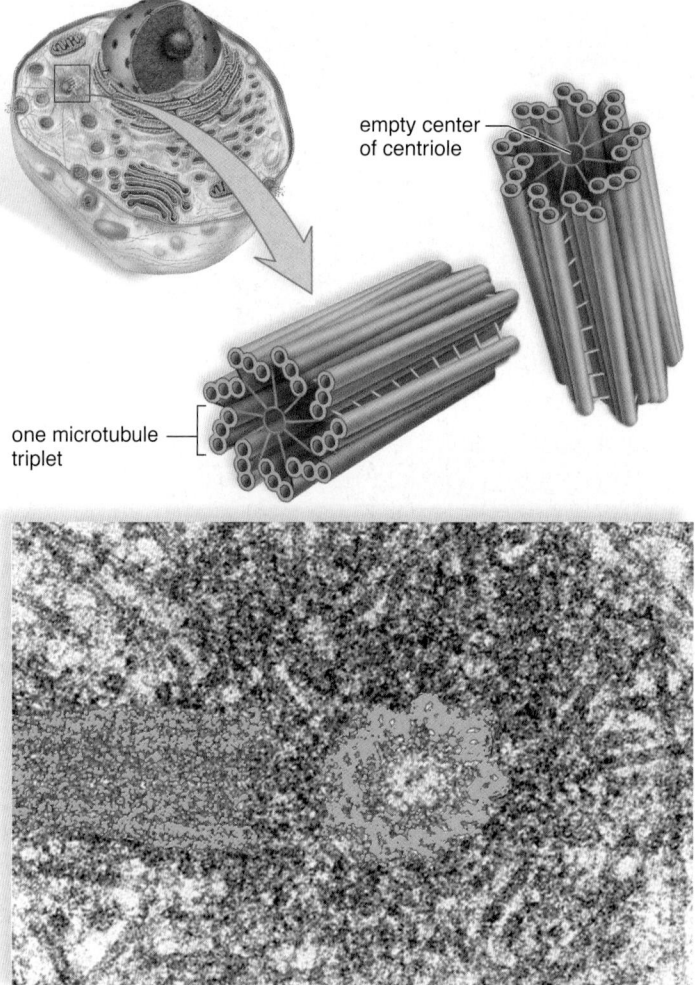

empty center of centriole

one microtubule triplet

one centrosome: one pair of centrioles

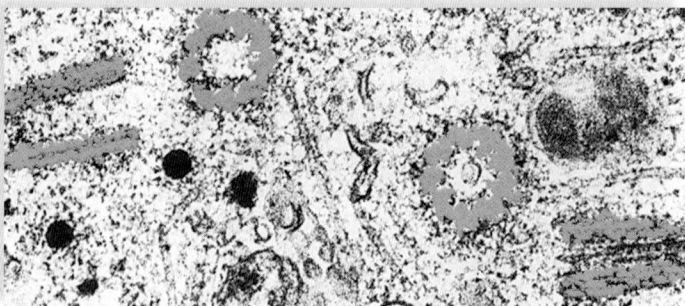

two centrosomes: two pairs of centrioles   200 nm

Figure 4.19 Centrioles.   In a nondividing animal cell, a single pair of centrioles is present in the centrosome located just outside the nucleus. Just before a cell divides, the centrioles replicate, producing two centrosomes. During cell division, the centrosomes separate so that each new cell has one centrosome containing one pair of centrioles.

Check Your Progress   4.8

1. Differentiate between the components of the cytoskeleton and how they provide support to the cell.
2. Explain how ATP is used to produce movement in a cell.
3. Describe the role of motor molecules in cilia and flagella.

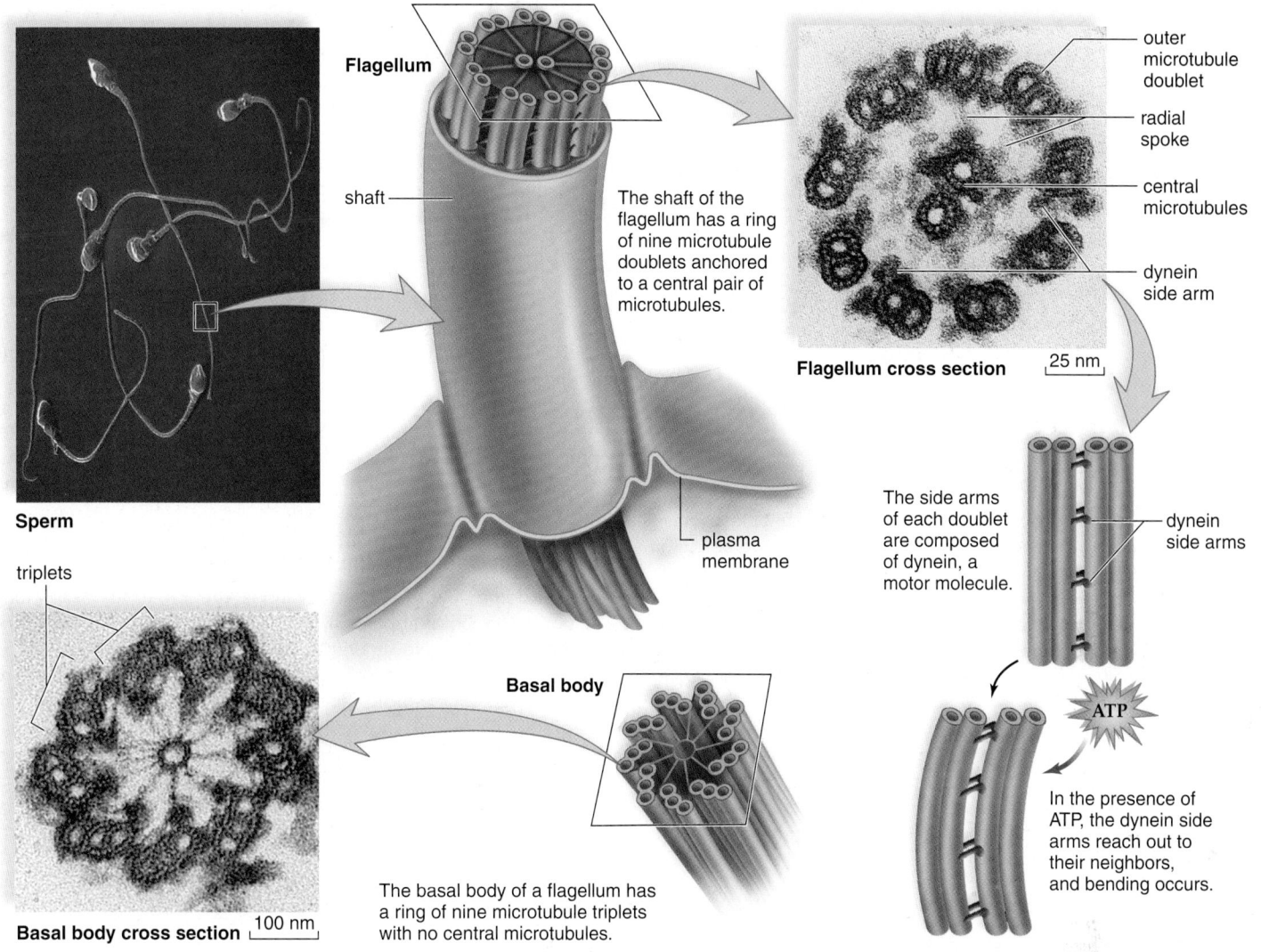

Sperm

Flagellum

shaft

The shaft of the flagellum has a ring of nine microtubule doublets anchored to a central pair of microtubules.

outer microtubule doublet

radial spoke

central microtubules

dynein side arm

Flagellum cross section 25 nm

The side arms of each doublet are composed of dynein, a motor molecule.

dynein side arms

ATP

In the presence of ATP, the dynein side arms reach out to their neighbors, and bending occurs.

plasma membrane

triplets

Basal body

Basal body cross section 100 nm

The basal body of a flagellum has a ring of nine microtubule triplets with no central microtubules.

Figure 4.20 Structure of a flagellum. *Below, left:* The basal body of a flagellum has a 9 + 0 pattern of microtubule triplets. Notice the ring of nine triplets, with no central microtubules. *Above, left:* In sperm, the shaft of the flagellum has a 9 + 2 pattern (a ring of nine microtubule doublets surrounds a central pair of microtubules). *Middle, right:* In place of the triplets seen in a basal body, a flagellum's outer doublets have side arms of dynein, a motor molecule. *Below, right:* In the presence of ATP, the dynein side arms reach out and attempt to move along their neighboring doublet. Because of the radial spokes connecting the doublets to the central microtubules and motor molecules, bending occurs.

CONNECTING *the* CONCEPTS *with the* BIG IDEAS

Evolution

- Fossil records show the first cells on Earth were primitive prokaryotes. (1D2a1)

Energy and Homeostasis

- All cells have membranes, cytoplasm, ribosomes, and DNA. (2B3c, 4B2a1)
- Cells specialize by modifying surfaces, architecture, and organelle assortment, compartmentalizing chemical reactions and storage. (4A2g, 4B2a1)
- Large surface area to volume ratios promote cells' favorable material exchange. (2A3b1, 2A3b2)

Information and Signaling

- The genetic material of the cell is stored in chromosomes composed of DNA. (3A1a2)

Interactions and Systems

- Specialized organelles allow eukaryotic cells to accomplish vital functions, often by compartmentalizing enzymes and metabolic pathways. (4A2a-g, 2B3b*IE*)
- The endomembrane system connects membrane-bounded organelles for more efficient delivery and processing of materials. (4A commentary, 4A2b, c, e, f)

Media Study Tools

Summarize

4.1 Cellular Level of Organization

All organisms are composed of cells, the smallest units of living matter. Cells self-reproduce, and existing cells come only from preexisting cells. Cells are very small (measured in micrometers) and must remain small in order to have an adequate surface area to volume ratio. The plasma membrane regulates exchange of materials between the cell interior and the external environment.

4.2 Prokaryotic Cells

Prokaryotic cells lack the nucleus of eukaryotic cells. The cell envelope of bacteria includes a plasma membrane, a cell wall, and an outer glycocalyx. The cytoplasm contains ribosomes, inclusion bodies, and a nucleoid without a nuclear envelope. The cytoplasm of cyanobacteria also includes thylakoids. The appendages of a bacterium are flagella, fimbriae, and conjugation pili.

4.3 Introducing Eukaryotic Cells

Eukaryotic cells are much larger than prokaryotic cells, and contain compartmentalized structures called organelles that each have a specific structure and function (Table 4.1) that increases cell efficiency. Endosymbiont theory helps explain the evolutionary origins of many membrane-enclosed organelles. Most membranous organelles are in constant communication.

4.4 The Nucleus and Ribosomes

The nucleus of eukaryotic cells is bounded by a nuclear envelope containing pores that regulate transport between the cytoplasm and the nucleoplasm. The nucleus contains DNA that is organized into chromosomes.

Ribosomes are organelles that function in protein synthesis. In order to make a protein, mRNA is copied exactly from the DNA, processed, and exits the nucleus through a nuclear pore. After a ribosome attaches to an mRNA, most of the time this assembly goes to the rough ER to make a protein.

4.5 The Endomembrane System

The endomembrane system includes the ER (both rough and smooth), the Golgi apparatus, the lysosomes (in animal cells), and transport vesicles. Newly produced proteins made in the rough ER are modified before they are packaged in transport vesicles, many of which go to the Golgi apparatus. The smooth ER has various metabolic functions, depending on the cell type, but it generally makes lipids that are carried by vesicles to different locations, particularly the Golgi apparatus. The Golgi apparatus modifies, sorts, and repackages proteins and also processes lipids. Some proteins are tagged for transport to different cellular destinations; others are secreted from the cell.

Table 4.1 Comparison of Prokaryotic Cells and Eukaryotic Cells

	Prokaryotic Cells (1–20 μm in diameter)	Eukaryotic Cells (10–100 μm in diameter)	
		Animal	Plant
Cell wall	Usually (peptidoglycan)	No	Yes (cellulose)
Plasma membrane	Yes	Yes	Yes
Nucleus	No	Yes	Yes
Nucleolus	No	Yes	Yes
Ribosomes	Yes (smaller)	Yes	Yes
Endoplasmic reticulum	No	Yes	Yes
Golgi apparatus	No	Yes	Yes
Lysosomes	No	Yes	No
Mitochondria	No	Yes	Yes
Chloroplasts	No	No	Yes
Peroxisomes	No	Usually	Usually
Cytoskeleton	No	Yes	Yes
Centrioles	No	Yes	No
9 + 2 cilia or flagella	No	Often	No (in flowering plants) Yes (sperm of bryophytes, ferns, and cycads)

4.6 Other Vesicles and Vacuoles

Cells contain numerous vesicles and vacuoles, some of which, such as lysosomes, have already been discussed. Peroxisomes are vesicles that are involved in the metabolism of long chain fatty acids. The large central vacuole in plant cells functions in storage and also in the breakdown of molecules and cell parts.

4.7 The Energy-Related Organelles

Cells require a constant input of energy to maintain their structure. Chloroplasts capture the energy of the sun and conduct photosynthesis, which produces carbohydrates. Carbohydrate-derived products are broken down in mitochondria in the presence of oxygen via cellular respiration, and ATP is produced as a result.

4.8 The Cytoskeleton

The cytoskeleton contains actin filaments, intermediate filaments, and microtubules. These maintain cell shape and help transport orgenelles from place to place within the cell. Actin, the thinnest filaments, interact with motor proteins to allow a range of functions from muscular contraction to cellular division. Intermediate filaments support the nuclear and plasma membranes and participate in the cell-to-cell junctions that produce tissues. Microtubules radiate out from the centrosome and are present in centrioles, cilia, and flagella. They serve as an internal transport system along which vesicles and other organelles move.

Key Terms

actin filament 78	leucoplast 77
archaean 65	lysosome 73
bacillus 65	matrix 77
basal body 80	mesosome 65
capsule 65	microtubule 78
cell 61	mitochondrion 76
cell envelope 65	motor molecule 78
cell theory 61	nuclear envelope 71
cell wall 65	nuclear pore 71
central vacuole 75	nucleoid 65
centriole 79	nucleolus 70
centrosome 78	nucleoplasm 70
chloroplast 76	organelle 67
chromatin 70	peroxisome 75
chromoplast 77	plasma membrane 65
chromosome 70	plasmid 65
cilium 80	plastid 77
coccus 65	polyribosome 71
conjugation pili 66	prokaryotic cell 65
cristae 77	ribosome 66, 71
cyanobacteria 66	rough ER 72
cytoplasm 65	secretion 73
cytoskeleton 68	signal peptide 71
endomembrane system 72	smooth ER 72
endoplasmic reticulum (ER) 72	spirillum 65
endosymbiotic theory 67	spirochete 65
eukaryotic cell 65	stroma 76
fimbriae 66	surface-area-to-volume ratio 62
flagellum (pl., flagella) 66, 80	thylakoid 66, 76
gene 70	vector 66
glycocalyx 65	vacuole 75
Golgi apparatus 72	vesicle 67
granum 76	
intermediate filament 78	

Assess

Reviewing This Chapter

1. What are the three basic principles of the cell theory?   60
2. Why is it advantageous for cells to be small?   61
3. Roughly sketch a bacterial (prokaryotic) cell, label its parts, and state a function for each of these.   65
4. How do eukaryotic and prokaryotic cells differ?   66
5. Describe how the nucleus, the chloroplast, and the mitochondrion may have become a part of the eukaryotic cell.   66
6. What does it mean to say that the eukaryotic cell is compartmentalized?   66–67
7. Describe the structure and the function of the nuclear envelope and the nuclear pores.   70–71
8. Distinguish between the nucleolus, rRNA, and ribosomes.   70–71
9. Name organelles that are a part of the endomembrane system and explain the term.   72
10. Trace the path of a protein from rough ER to the plasma membrane.   74
11. Give the overall equations for photosynthesis and cellular respiration, contrast the two, and tell how they are related.   76
12. Describe the structure and function of chloroplasts and mitochondria. How are these two organelles related to one another?   76–77
13. What are the three components of the cytoskeleton? What are their structures and functions?   78–79
14. Relate the structure of flagella (and cilia) to centrioles, and discuss the function of both.   80

Testing Yourself

Choose the best answer for each question.

1. The small size of cells best correlates with
   a. the fact that they are self-reproducing.
   b. their prokaryotic versus eukaryotic nature.
   c. an adequate surface area for exchange of materials.
   d. the fact that they come in multiple sizes.
   e. All of these are correct.

2. Which of these is not a true comparison of the compound light microscope and the transmission electron microscope?

	LIGHT	ELECTRON
a.	Uses light to "view" object	Uses electrons to "view" object
b.	Uses glass lenses for focusing	Uses magnetic lenses for focusing
c.	Specimen must be killed and stained	Specimen may be alive and nonstained
d.	Magnification is not as great	Magnification is greater
e.	Resolution is not as great	Resolution is greater

3. Which of these best distinguishes a prokaryotic cell from a eukaryotic cell?
   a. Prokaryotic cells have a cell wall, but eukaryotic cells never do.
   b. Prokaryotic cells are much larger than eukaryotic cells.
   c. Prokaryotic cells have flagella, but eukaryotic cells do not.
   d. Prokaryotic cells do not have a membrane-bounded nucleus, but eukaryotic cells do have such a nucleus.
   e. Prokaryotic cells have ribosomes, but eukaryotic cells do not have ribosomes.

4. Which of these is not found in the nucleus?
   a. functioning ribosomes
   b. chromatin that condenses to chromosomes
   c. nucleolus that produces rRNA
   d. nucleoplasm instead of cytoplasm
   e. all forms of RNA

5. Vesicles from the ER most likely are on their way to
   a. the rough ER.
   b. the lysosomes.
   c. the Golgi apparatus.
   d. the plant cell vacuole only.
   e. the location suitable to their size.

6. Lysosomes function in
   a. protein synthesis.
   b. processing and packaging.
   c. intracellular digestion.
   d. lipid synthesis.
   e. production of hydrogen peroxide.

7. Mitochondria
   a. are involved in cellular respiration.
   b. break down ATP to release energy for cells.
   c. contain grana and cristae.
   d. are present in animal cells but not plant cells.
   e. All of these are correct.

8. Which organelle releases oxygen?
   a. ribosome
   b. Golgi apparatus
   c. chloroplast
   d. smooth ER

9. Which of these is not true?
   a. Actin filaments are found in muscle cells.
   b. Microtubules radiate out from the ER.
   c. Intermediate filaments sometimes contain keratin.
   d. Motor molecules use microtubules as tracks.

10. Cilia and flagella
    a. have a 9 + 0 pattern of microtubules, same as basal bodies.
    b. contain myosin that pulls on actin filaments.
    c. are organized by basal bodies derived from centrioles.
    d. are constructed similarly in prokaryotes and eukaryotes.
    e. Both a and c are correct.

11. Which of the following organelles contains its (their) own DNA, suggesting they were once independent prokaryotes?
    a. Golgi apparatus
    b. mitochondria
    c. chloroplasts
    d. ribosomes
    e. Both b and c are correct.

12. Which organelle most likely originated by invagination of the plasma membrane?
    a. mitochondria
    b. flagella
    c. nucleus
    d. chloroplasts
    e. All of these are correct.

13. Which structures are found in a prokaryotic cell?
    a. cell wall, ribosomes, thylakoids, chromosome
    b. cell wall, plasma membrane, nucleus, flagellum
    c. nucleoid, ribosomes, chloroplasts, capsule
    d. plasmid, ribosomes, enzymes, DNA, mitochondria
    e. chlorophyll, enzymes, Golgi apparatus, plasmids

14. Study the example given in (a) below. Then for each other organelle listed, state another that is structurally and functionally related. Tell why you paired these two organelles.
    a. The nucleus can be paired with nucleoli because nucleoli are found in the nucleus. Nucleoli occur where chromatin is producing rRNA.
    b. mitochondria
    c. centrioles
    d. ER

Engage

Thinking Scientifically

1. The protists that cause malaria contribute to infections associated with AIDS. Scientists have discovered that an antibiotic that inhibits prokaryotic enzymes will kill the parasite because it is effective against the plastids in the protist. What can you conclude about the origin of the plastids?

2. For your cytology study, you have decided to label and, thereby, detect the presence of the base uracil in an animal cell. In what parts of the cell do you expect to find your radioactive tracer?

Bioethical Issue

Stem Cells

A stem cell is an immature cell capable of producing many different types of differentiated mature cells. Stem cells exist in the various organs of the human body; however, they are difficult to obtain, except for those that reside in red bone marrow and produce all types of blood cells. A rich source of stem cells is an umbilical cord; another, less controversial source is adult stem cells from, say, skin. Since the nucleus contains the genetic information, it is possible to take a 2n adult nucleus, manipulate it genetically, and put it in an enucleated egg cell. Under the right conditions, the new cells can be grown and used to make neurological tissues that could possibly cure Alzheimer or Parkinson disease or any other type of neurological disorder. However, if development were to continue, a clone of the human that donated the 2n nucleus could possibly result.

   Is it bioethical to continue investigating such research? Especially when you consider that the "embryo" that provided the stem cells was not produced by the normal method of having a sperm fertilize an egg? Or, is it wrong to produce an embryo only to serve as a source of stem cells?

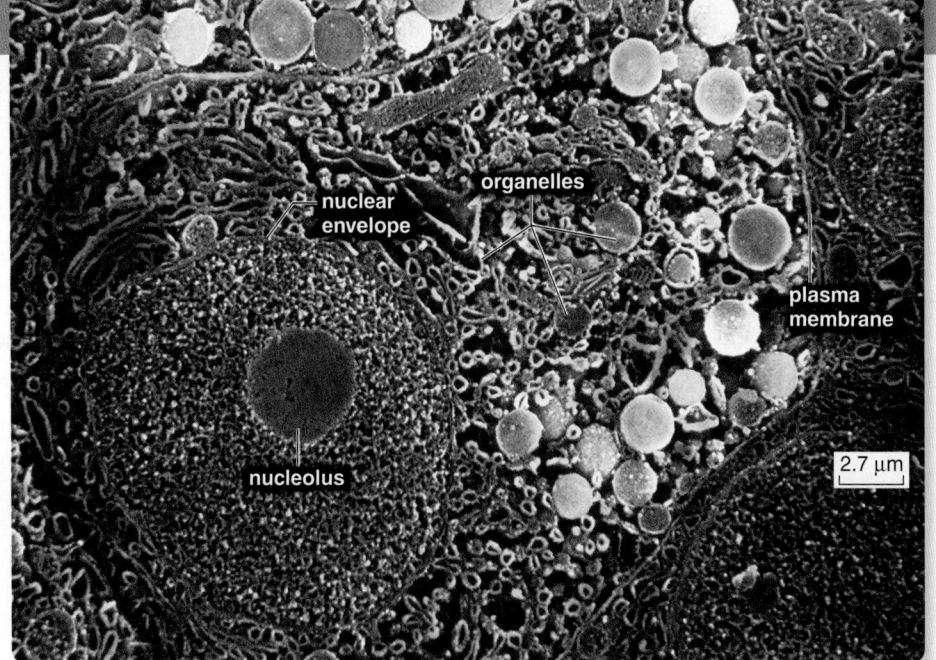

nuclear envelope

organelles

plasma membrane

nucleolus

2.7 μm

A eukaryotic cell is surrounded by a plasma membrane, and membrane also compartmentalizes the cell into various organelles with diverse functions.

Membrane Structure and Function

An overweight diabetic, an African pygmy, and a young child with cystic fibrosis all suffer from a defect in their cells' plasma membrane. The diabetic's plasma membrane does not respond properly to insulin, growth hormone does not bind to the pygmy's membrane, and the child's membrane does not transport chloride from the cells.

Every living cell is enclosed by a plasma membrane, which creates a dynamic barrier that allows life functions to occur separate from the external environment. The plasma membrane regulates what goes into and out of the cell and enables cells to communicate with one another. Internal cellular membranes create compartments such as organelles, which together with membrane-associated enzymes, allow many, sometimes incompatible chemical processes to occur simultaneously within the cell. This "division of labor" is an essential cell feature that has undergone many evolutionary adaptations, including giving rise to the distinctions between plant and animal cells. This chapter describes the plasma membrane and its core functions. It also discusses various ways cells communicate so that the activities of tissues and organs are coordinated.

As you read through the chapter, think about the following questions:

1. Why is creating compartments with membranes necessary for cellular life?
2. How does compartmentalization increase cell efficiency and use of energy?
3. In what ways have membranes enabled cells to specialize within an organism?

BEFORE YOU BEGIN

Before beginning this chapter, take a few moments to review the following discussions.

Figures 4.6 and 4.7 What are the key features of animal and plant cells?

Figure 4.13 How do membranes work together in cellular systems?

Figure 4.18 How is the cytoskeleton related to cellular membranes?

FOLLOWING *the* BIG IDEAS

CHAPTER 5   MEMBRANE STRUCTURE AND FUNCTION

Energy and Homeostasis	The plasma membrane is appropriately called the gatekeeper of the cell because it maintains the identity and the integrity of the cell as it "stands guard" over what enters and leaves.
Information and Signaling	Membrane receptor proteins act as intercellular signal receivers.
Interactions and Systems	Membranes are an integral part of an interconnected cellular system of communication and response to environment.

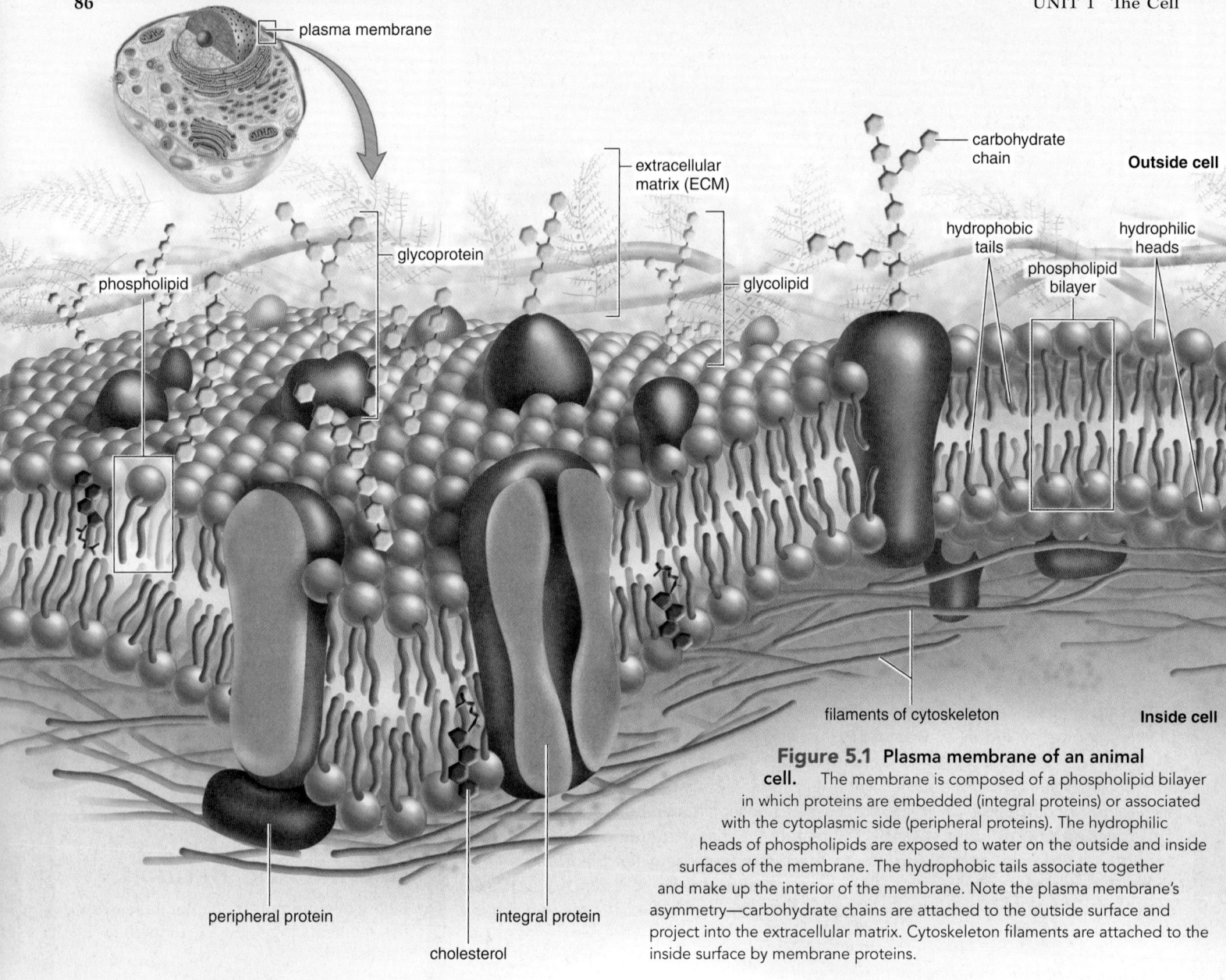

phospholipid

plasma membrane

glycoprotein

extracellular matrix (ECM)

glycolipid

carbohydrate chain

Outside cell

hydrophobic tails

hydrophilic heads

phospholipid bilayer

peripheral protein

cholesterol

integral protein

filaments of cytoskeleton

Inside cell

Figure 5.1 Plasma membrane of an animal cell.   The membrane is composed of a phospholipid bilayer in which proteins are embedded (integral proteins) or associated with the cytoplasmic side (peripheral proteins). The hydrophilic heads of phospholipids are exposed to water on the outside and inside surfaces of the membrane. The hydrophobic tails associate together and make up the interior of the membrane. Note the plasma membrane's asymmetry—carbohydrate chains are attached to the outside surface and project into the extracellular matrix. Cytoskeleton filaments are attached to the inside surface by membrane proteins.

5.1 Plasma Membrane Structure and Function

Learning Outcomes

Upon completion of this section, you should be able to

1. Distinguish between the different structural components of membranes.
2. Explain the functional benefits of using membrane-bound cellular compartments.
3. Describe the diverse role of proteins in membranes.
4. Compare membrane permeability for polar and nonpolar molecules.

The ability to create compartments is a key feature of cells. Membranes, made of a phospholipid bilayer, create separation between the cell and the external environment as well as compartments within the cell itself. Having separate spaces allows multiple, sometimes incompatible, chemical processes to occur simultaneously. This "division of labor" allows cells to operate more efficiently and respond to changing environmental conditions.

Components of the Plasma Membrane

The structure of an animal cell's plasma membrane is depicted in Figure 5.1. In addition to the phospholipid bilayer, membrane components include protein molecules that are either partially or wholly embedded in the bilayer. **Cholesterol** is another lipid found in the animal plasma membrane; related steroids are found in the plasma membrane of plants. As discussed later in this chapter, cholesterol helps modify the fluidity of the membrane over a range of temperatures.

Recall that a phospholipid is an *amphipathic molecule*, meaning that it has both a hydrophilic (water-loving) region and a hydrophobic (water-fearing) region. The amphipathic nature of phospholipids largely explains why they form a bilayer in water.

Because similar substances associate with one another, the hydrophilic polar heads of the phospholipid molecules naturally associate with the polar water molecules found on the outside and inside of the cell. Likewise, the hydrophobic nonpolar tails associate with each other because they want to "get away" from the polar water.

Cell membranes are highly similar in the types of molecules they contain, which makes them interchageable and allows them to fuse together fairly easily. What makes one membrane different from another are the types of proteins integrated into the membrane. As shown in Figure 5.1, proteins are scattered throughout the membrane in an irregular pattern, and this pattern can vary from membrane to membrane.

Electron micrographs verify the embedded nature of many membrane proteins. A research method called freeze-fracture first freezes, and then splits, the membrane so that the upper and lower layers separate. The proteins remain intact and go with one layer or the other. The embedded proteins are termed integral proteins, whereas the proteins that occur only on the cytoplasmic side of the membrane are termed peripheral proteins.

Some integral proteins protrude from only one surface of the bilayer, but most span the membrane, with a hydrophobic core region that associates with the nonpolar core of the membrane. Hydrophilic ends of integral proteins protrude from both surfaces of the bilayer, interacting with polar water molecules. Integral proteins can be held in place by attachments to protein fibers of the cytoskeleton (inside) and fibers of the extracellular matrix (outside). Only animal cells have an **extracellular matrix (ECM),** which contains various protein fibers and very large, complex carbohydrate molecules. The ECM, which is discussed in greater detail at the end of the chapter, has a number of functions, from lending external support to the plasma membrane to assisting in communication between cells.

bilayer is fluid. The fluidity of the membrane also prevents it from solidifying as external temperatures drop.

The lipid content of the membrane is responsible for its fluidity. At body temperature, the phospholipid bilayer of the plasma membrane has the consistency of olive oil. The greater the concentration of unsaturated fatty acid residues, the more fluid the bilayer. In each monolayer, the fatty acid tails jostle around, and an entire phospholipid molecule can move sideways at a rate averaging about 2 μm—the length of a prokaryotic cell—per second. Although it is possible for phospholipid molecules to flip-flop from one monolayer to the other, they rarely do so because this would require the hydrophilic head to move through the hydrophobic center of the membrane. However, at times special proteins help the phospholipids flip.

The presence of cholesterol molecules prevents the plasma membrane from becoming too fluid at higher temperatures and too solid at lower temperatures. At higher temperatures, cholesterol stiffens the membrane and makes it less fluid than it would otherwise be. At lower temperatures, cholesterol helps prevent the membrane from freezing by not allowing contact between certain phospholipid tails.

A plasma membrane is considered a mosaic because of the presence of many proteins. The number and kinds of proteins can vary in the plasma membrane and in the membranes of the various organelles. The position of these proteins can shift over time, unless they are anchored to another structure, such as the cytoskeleton. Figure 5.2 describes an experiment in which the proteins were tagged prior to allowing mouse and human cells to fuse. An hour after fusion, the proteins were completely

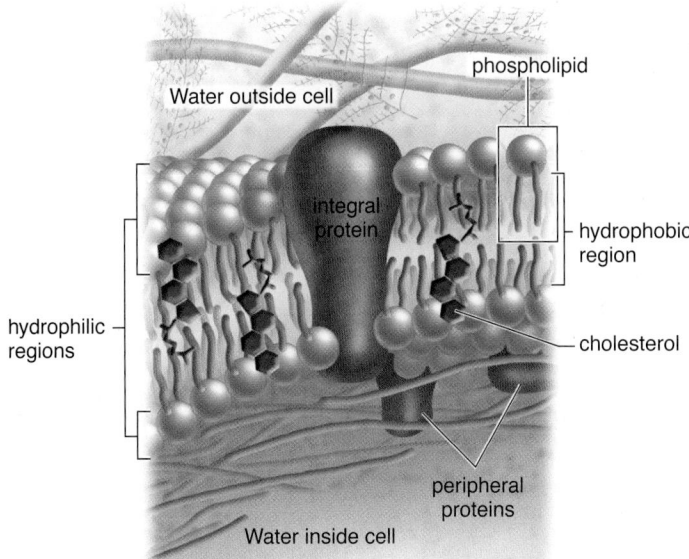

Fluid-Mosaic Model

Membranes are not rigid, but rather are flexible structures. One model used to describe the plasma membrane is called the **fluid-mosaic model.** Cells are pliable because the phospholipid

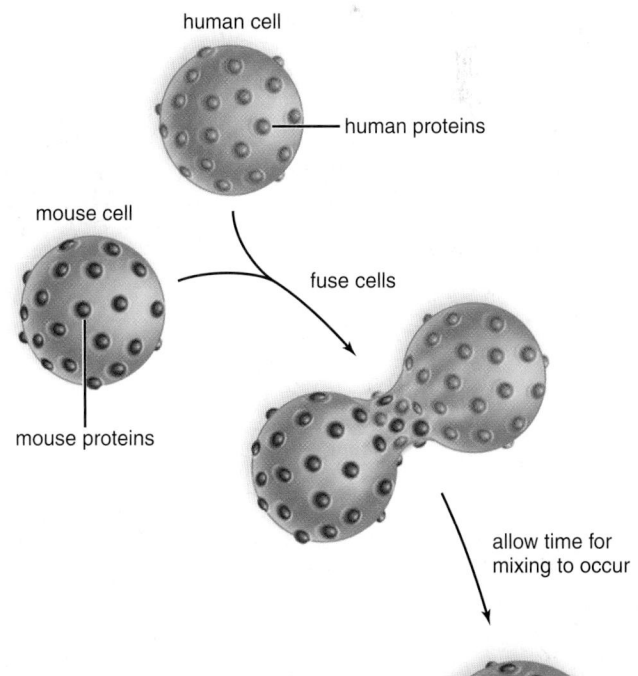

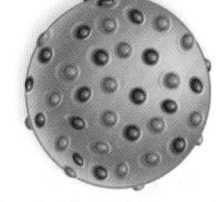

Figure 5.2 Drifting of plasma membrane proteins. After human and mouse cells fuse, the plasma membrane proteins of the mouse (purple circles) and human cell (orange circles) mix within a short time.

mixed, suggesting that at least some proteins are able to move sideways in the membrane.

Scientists once thought that all membrane proteins could freely move sideways within the fluid bilayer. Today, however, we know that membrane proteins are often associated with the ECM, the cytoskeleton, or both. These connections hold a protein in place and serve to partially anchor the otherwise fluid phospholipid bilayer.

3D Animation Lipid Bilayer   **MP3** Membrane Structure

Carbohydrate Chains

Phospholipids and proteins that have attached carbohydrate (sugar) chains are called **glycolipids** and **glycoproteins,** respectively. Because the carbohydrate chains occur only on the outside surface, and peripheral proteins occur on one surface or the other, the two sides of the membrane are not identical, and the membrane is said to be asymmetrical.

In animal cells, the carbohydrate chains attached to proteins give the cell a "sugar coat," more properly called the glycocalyx. The glycocalyx protects the cell and has various other functions, including cell-to-cell adhesion, reception of signaling molecules, and cell-to-cell recognition.

The carbohydrate (sugar) chains on a cell's exterior can be highly diverse. The chains can vary in the number (15 is usual, but there can be several hundred) and sequence of sugars, and in whether the chain is branched. Each cell within an individual has its own particular "fingerprint" because of these chains.

As you probably know, transplanted tissues are often rejected by the recipient. Rejection occurs because the immune system is able to detect that the foreign tissue's cells do not have the appropriate carbohydrate chains to be recognized as self. In humans, carbohydrate chains are also the basis for the A, B, and O blood groups.

The Functions of the Proteins

Although the protein components of cell membranes differ depending on the type of cell and the processes it is undergoing, several types of proteins are likely to be routinely present:

Channel proteins   Channel proteins are involved in passing molecules through the membrane. They form a channel that allows a substance to simply move from one side to the other (Fig. 5.3a). For example, a channel protein allows hydrogen ions to flow across the inner mitochondrial membrane. Without this movement of hydrogen ions, ATP would never be produced.

Animation Receptors Linked to a Channel Protein

Carrier proteins   Carrier proteins are also involved in passing molecules through the membrane. They receive a substance and change their shape, and this change serves to move the substance across the membrane (Fig. 5.3b). A carrier protein transports sodium and potassium ions across the plasma membrane of a nerve cell. Without this carrier protein, nerve impulse conduction would be impossible.

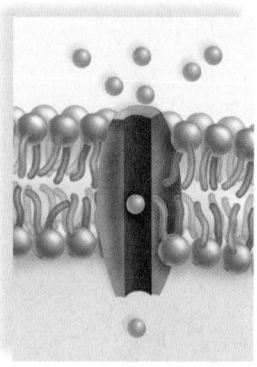

Channel Protein: Allows a particular molecule or ion to cross the plasma membrane freely. Cystic fibrosis, an inherited disorder, is caused by a faulty chloride (Cl^-) channel; a thick mucus collects in airways and in pancreatic and liver ducts.

a.

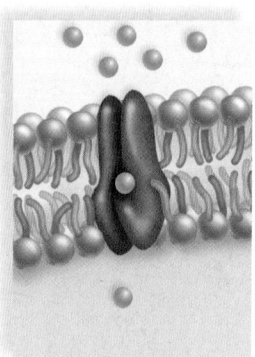

Carrier Protein: Selectively interacts with a specific molecule or ion so that it can cross the plasma membrane. The inability of some persons to use energy for sodium-potassium (Na^+–K^+) transport has been suggested as the cause of their obesity.

b.

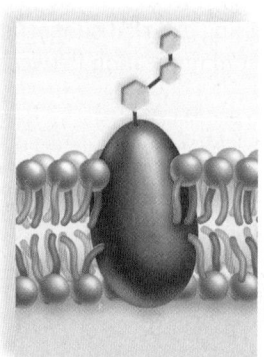

Cell Recognition Protein: The MHC (major histocompatibility complex) glycoproteins are different for each person, so organ transplants are difficult to achieve. Cells with foreign MHC glycoproteins are attacked by white blood cells responsible for immunity.

c.

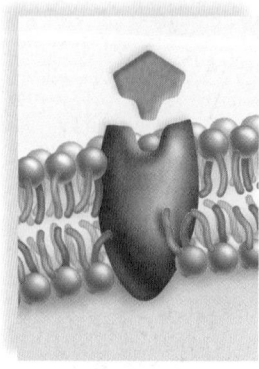

Receptor Protein: Is shaped in such a way that a specific molecule can bind to it. Pygmies are short, not because they do not produce enough growth hormone, but because their plasma membrane growth hormone receptors are faulty and cannot interact with growth hormone.

d.

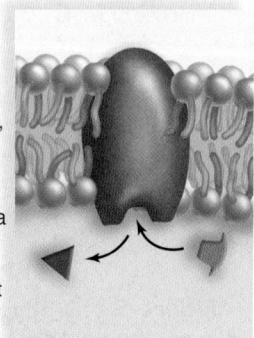

Enzymatic Protein: Catalyzes a specific reaction. The membrane protein, adenylate cyclase, is involved in ATP metabolism. Cholera bacteria release a toxin that interferes with the proper functioning of adenylate cyclase; sodium (Na^+) and water leave intestinal cells, and the individual may die from severe diarrhea.

e.

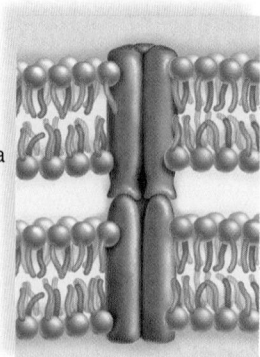

Junction Proteins: Tight junctions join cells so that a tissue can fulfill a function, as when a tissue pinches off the neural tube during development. Without this cooperation between cells, an animal embryo would have no nervous system.

f.

Figure 5.3 Membrane protein diversity.   These are some of the functions performed by proteins found in the plasma membrane.

Table 5.1 Passage of Molecules into and out of the Cell

Name	Direction	Requirement	Examples
Diffusion	Toward lower concentration	Concentration gradient	Lipid-soluble molecules, and gases
Facilitated transport	Toward lower concentration	Channels or carrier and concentration gradient	Some sugars, and amino acids
Active transport	Toward higher concentration	Carrier plus energy	Sugars, amino acids, and ions
Bulk transport	Toward outside or inside	Vesicle utilization	Macromolecules

Cell recognition proteins Cell recognition proteins are glycoproteins (Fig. 5.3c). Among other functions, these proteins help the body recognize when it is being invaded by pathogens so that an immune response can occur. Without this recognition, pathogens would be able to freely invade the body and hinder its function.

Receptor proteins Receptor proteins have a shape that allows only a specific molecule to bind to it (Fig. 5.3d). The binding of this molecule causes the protein to change its shape and thereby bring about a cellular response. The coordination of the body's organs is totally dependent on such signaling molecules. For example, the liver stores glucose after it is signaled to do so by insulin.

Enzymatic proteins Some plasma membrane proteins are enzymes that carry out metabolic reactions directly (Fig. 5.3e). Without these enzymes, some of which are attached to the various membranes of the cell, a cell would never be able to perform the chemical reactions needed to maintain its metabolism.

Junction proteins As discussed on page 88, proteins are involved in forming various types of junctions between animal cells (Fig. 5.3f). Signaling molecules that pass through gap junctions allow the cilia of cells that line your respiratory tract to beat in unison.

Permeability of the Plasma Membrane

The plasma membrane regulates the passage of molecules into and out of the cell. This function is critical because the cell must maintain its normal composition under changing environmental conditions. The plasma membrane is essential because it is **selectively permeable,** allowing only certain substances into the cell while keeping others out.

Molecules that can freely cross a membrane generally require no energy to do so. Substances that are hydrophobic and therefore similar to the phospholipid center of the membrane are able to diffuse across membranes at no energy cost. Polar molecules, however, are chemically incompatible with the center of the membrane, and so require an expenditure of energy to drive their transport.

Table 5.1 lists, and Figure 5.4 illustrates, which types of molecules can passively cross a membrane (no energy required), and which may require transport by a carrier protein and/or an expenditure of energy. In general, small, noncharged molecules, such as carbon dioxide, oxygen, glycerol, and alcohol, can freely cross the membrane. They are able to slip between the hydrophilic heads of the phospholipids and pass through the hydrophobic tails of the membrane because they are similarly nonpolar.

These molecules follow their **concentration gradient** as they move from an area where their concentration is high, to an area where their concentration is low. Consider that a cell is always using oxygen when it carries on cellular respiration. The internal consumption of oxygen results in a low cellular concentration. Because oxygen concentration is higher outside than inside the cell, oxygen tends to move across the membrane into the cell. The concentration of carbon dioxide, on the other hand, is highest inside the cell because it is produced during cellular respiration. Therefore, carbon dioxide tends to move with its concentration gradient from inside to outside the cell.

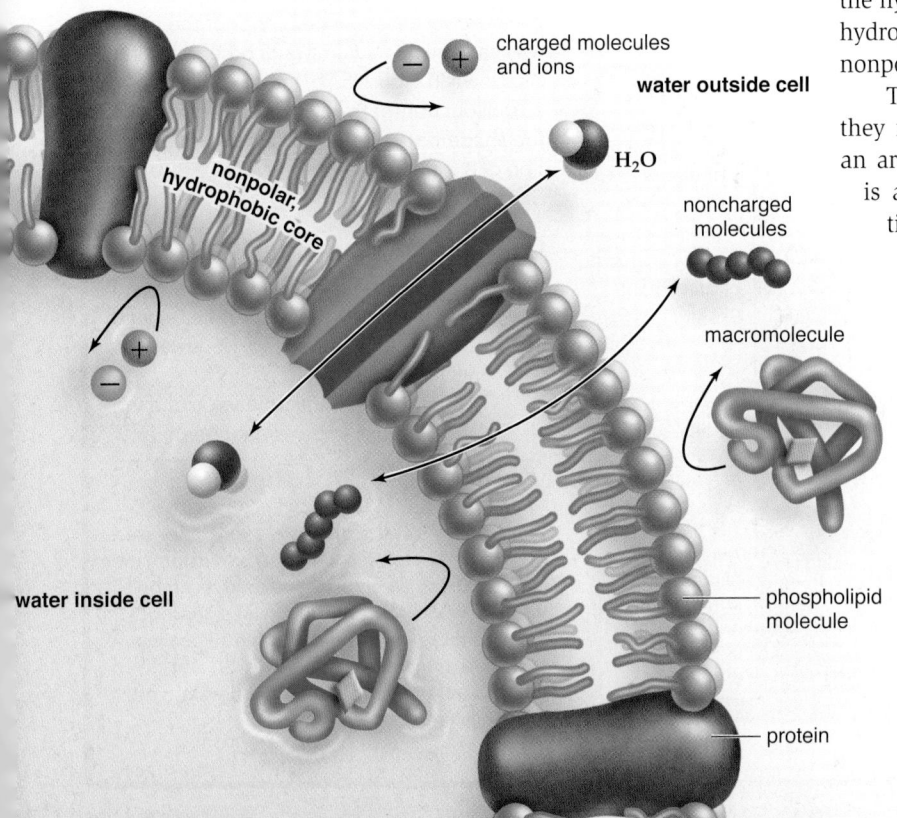

charged molecules and ions

water outside cell

H_2O

nonpolar, hydrophobic core

noncharged molecules

macromolecule

water inside cell

phospholipid molecule

protein

Figure 5.4 How molecules cross the plasma membrane. The curved arrows indicate that these substances cannot passively cross the plasma membrane, and the long back-and-forth arrows indicate that these substances can diffuse across the plasma membrane.

Biological Systems

How Cells Talk to One Another

All organisms are comprised of cells that are able to sense and respond to specific signals in their environment. A bacterium that lives in your body responds to signaling molecules when it finds food and escapes immune cells in order to stay alive. Signaling helps the bread mold that grows on stale bread detect an opposite mating strain to begin its sexual life cycle. Similarly, the cells of a developing embryo respond to signaling molecules as they move to specific locations and become specific tissues (Fig. 5Aa).

In newborn animals, internal signals like hormones are essential to ensure specific tissues develop when and how they should. In plants, external signals, such as a change in the amount of light, tells them when it is time to resume growth or flower. Internal signaling molecules enable animals and plants to coordinate their cellular activities, to metabolize, and better respond in a changing environment. The ability of cells to communicate with one another is an essential part of all biological systems.

Cell Signaling

The cells of a multicellular organism "talk" to one another by using signaling molecules, sometimes called chemical messengers. Some messengers are produced in one location and, in animals, are carried by the circulatory system to various target sites around the body. For example, the pancreas releases a hormone called insulin, which is transported in blood vessels to the liver, and this signal causes the liver to store glucose as glycogen. Failure of the liver to respond appropriately results in a medical condition called diabetes.

In Chapter 9, we are particularly interested in growth factors, which act locally as signaling molecules and cause cells to divide. Overproduction of growth factors can disrupt the balance in cellular systems. If left uncorrected, uncontrolled cell growth and formation of a tumor can result. The importance of cell signaling in regulating cell systems is the focus of much research in cell biology.

Cells respond to only certain signaling molecules. Why? Because they must bind to a receptor protein, and only cells that possess matching receptors can respond to certain signaling molecules. Each cell has a mix of receptors, which gives them the ability to respond differently to a variety of external and internal stimuli. Each cell is also able to balance the relative strength of incoming signals in order to change cellular structure or function. If a minimum level of signaling is not met, the cell dies.

Signaling molecules interacting with their receptor is only the beginning of a complex process of communication that tells the cell how to respond. Once a signaling molecule and receptor interact, a cascade of events occurs that increase, decrease, or otherwise change the signal to elicit a cellular response. This process is called a signal transduction pathway. This pathway is analogous to television transmission: a TV camera (the receptor) views a scene, converts the picture into electrical signals (transduction pathway) that are understood by the TV receiver in your house, which converts these signals to a picture on your screen (the response). The process in cells is more complicated because each member of the pathway can activate a number of other proteins. As shown in Figure 5Ab, the cell response to a transduction pathway can be a change in the shape or movement of a cell, the activation of a particular enzyme, or the activation of a specific gene.

Questions to Consider

1. If your cells needed to rapidly respond to a changing environment, would you want their effect to be short- or long-lived?
2. Given the essential role of signaling in cellular and organismal health, how might diseases arise from signaling errors?

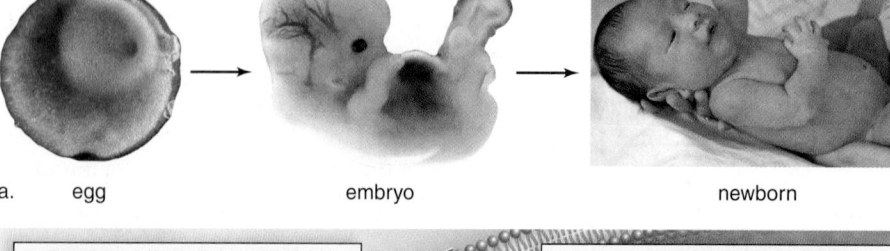

a.   egg   embryo   newborn

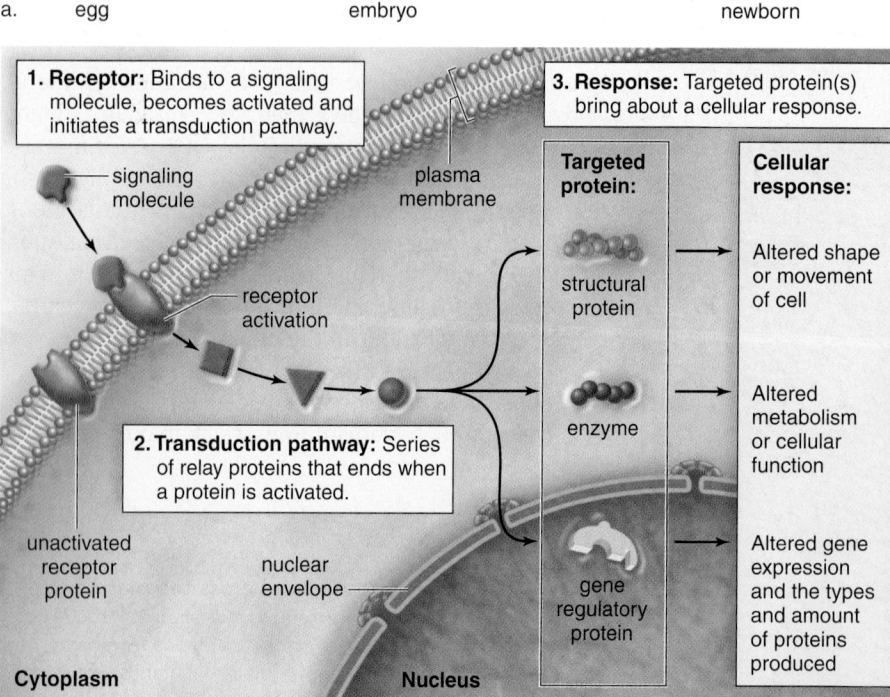

1. Receptor: Binds to a signaling molecule, becomes activated and initiates a transduction pathway.

signaling molecule

plasma membrane

receptor activation

2. Transduction pathway: Series of relay proteins that ends when a protein is activated.

unactivated receptor protein

nuclear envelope

Cytoplasm

3. Response: Targeted protein(s) bring about a cellular response.

Targeted protein: **Cellular response:**

structural protein → Altered shape or movement of cell

enzyme → Altered metabolism or cellular function

gene regulatory protein → Altered gene expression and the types and amount of proteins produced

Nucleus

b.

Figure 5A   Cell signaling.   a. The process of signaling helps account for the transformation of an egg into an embryo and then an embryo into a newborn. **b.** The process of signaling involves three steps: binding of the signaling molecule, transduction of the signal, and response of the cell depending on what type protein is targeted.

Water, a polar molecule, would not be expected to readily cross the primarily nonpolar membrane. However, scientists have discovered that some, perhaps all, cells have channel proteins called **aquaporins** that allow water to cross a membrane more quickly than expected. Aquaporins also allow cells to equalize water pressure differences between their interior and exterior environments so their membranes don't burst from environmental pressure changes.

Ions and polar molecules, such as glucose and amino acids, can slowly cross a membrane. To move as quickly as is necessary, they are often assisted across the plasma membrane by carrier proteins. Each carrier protein recognizes particular shapes of molecules, and must combine with an ion, such as sodium (Na^+), or a molecule, such as glucose, before changing its shape and transporting the molecule across the membrane. Therefore, carrier proteins are specific for the substances they transport across the plasma membrane.

Bulk transport is a way that large particles can exit or enter a cell. During exocytosis, fusion of a vesicle with the plasma membrane moves a particle to outside the membrane. During endocytosis, vesicle formation moves a particle to inside the plasma membrane. Vesicle formation is reserved for movement of macromolecules or even for something larger, such as a virus. As with many other processes, a cell is selective about what enters by endocytosis.

Check Your Progress   5.1

1. Examine the effect of reducing cholesterol in cellular membranes.
2. Explain the role of proteins in the fluid mosaic model.
3. Compare how cells transport polar and nonpolar molecules across a membrane.

5.2 Passive Transport Across a Membrane

Learning Outcomes

Upon completion of this section, you should be able to

1. Describe how molecules move from high to low concentration.
2. Compare diffusion and osmosis across a membrane.
3. Differentiate between hypotonic, isotonic, and hypertonic solutions for animal and plant cells.

Diffusion is the movement of molecules from a higher to a lower concentration—that is, down their concentration gradient—until equilibrium is achieved and the molecules are distributed equally. Diffusion is a physical process that results from random molecular motion that can be observed with any type of molecule. For example, when a crystal of dye is placed in water (Fig. 5.5), the dye and water molecules move in various directions, but their net movement, which is the sum of their motion, is toward the region of lower concentration. Eventually, the dye is dissolved evenly in the water, resulting in equilibrium and a uniform colored solution.

A **solution** contains both a solute, usually a solid, and a solvent, usually a liquid. In this case, the **solute** is the dye and the **solvent** is the water molecules. Once the solute and solvent are evenly distributed, they continue to move about, but there is no net movement of either one in any direction.

The chemical and physical properties of the plasma membrane allow only a few types of molecules to enter and exit a cell simply by diffusion. Gases can freely diffuse through the lipid bilayer because they are small and nonpolar; this is the

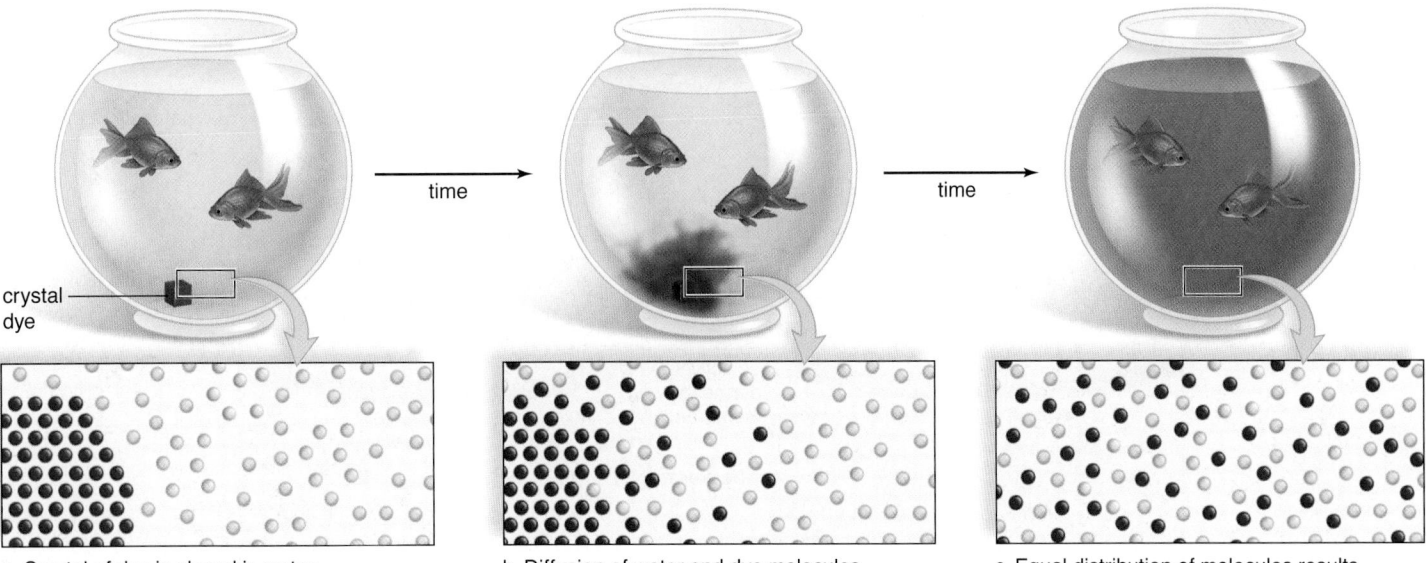

crystal dye

a. Crystal of dye is placed in water   b. Diffusion of water and dye molecules   c. Equal distribution of molecules results

Figure 5.5 Process of diffusion. Diffusion is spontaneous, and no chemical energy is required to bring it about. **a.** When a dye crystal is placed in water, it is concentrated in one area. **b.** The dye dissolves in the water, and over time a net movement of dye molecules from a higher to a lower concentration occurs. There is also a net movement of water molecules from a higher to a lower concentration. **c.** Eventually, the water and the dye molecules are equally distributed throughout the container.

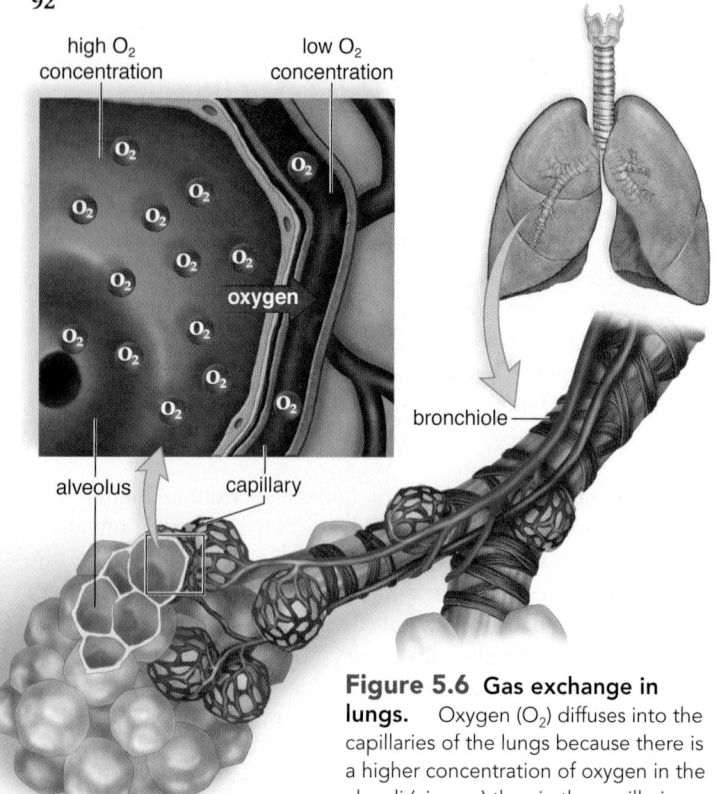

high O₂ concentration

low O₂ concentration

oxygen

alveolus   capillary

bronchiole

Figure 5.6 Gas exchange in lungs. Oxygen (O₂) diffuses into the capillaries of the lungs because there is a higher concentration of oxygen in the alveoli (air sacs) than in the capillaries.

mechanism by which oxygen enters cells and carbon dioxide exits cells. This is also how oxygen diffuses from the alveoli (air sacs) of the lungs into the blood in the lung capillaries (Fig. 5.6). After inhalation (breathing in), the concentration of oxygen in the alveoli is higher than that in the blood; therefore, oxygen diffuses into the blood along its concentration gradient.

Several factors influence the rate of diffusion, including temperature, pressure, electrical currents, and molecular size. For example, as temperature increases, the rate of diffusion increases. The movement of fishes in the tank would also speed the rate of diffusion (Fig. 5.5).

3D Animation Diffusion   **MP3** Diffusion   **Animation** How Diffusion Works

Osmosis

The diffusion of water across a selectively permeable membrane from high to low concentration is called **osmosis**. To illustrate osmosis, a thistle tube containing a 10% solute solution[1] is covered at one end by a selectively permeable membrane and then placed in a beaker containing a 5% solute solution (Fig. 5.7a). The beaker has a higher concentration of water molecules (lower percentage of solute), and the thistle tube has a lower concentration of water molecules (higher percentage of solute). Diffusion

1 Percent solutions are grams of solute per 100 mL of solvent. Therefore, a 10% solution is 10 g of sugar with water added to make 100 mL of solution.

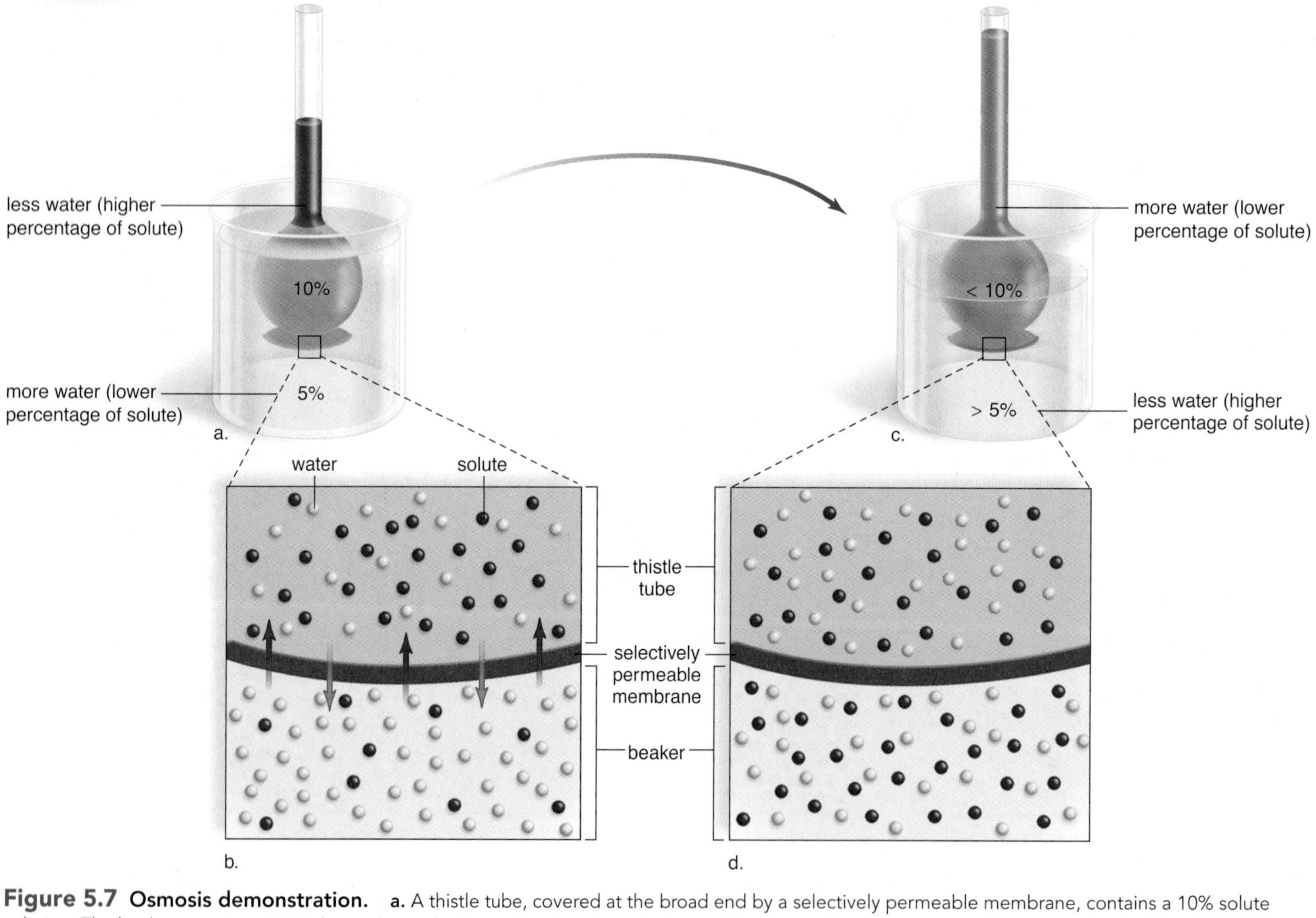

less water (higher percentage of solute)

more water (lower percentage of solute)

10%

5%

a.

more water (lower percentage of solute)

< 10%

> 5%

c.

less water (higher percentage of solute)

water   solute

thistle tube

selectively permeable membrane

beaker

b.

d.

Figure 5.7 Osmosis demonstration. **a.** A thistle tube, covered at the broad end by a selectively permeable membrane, contains a 10% solute solution. The beaker contains a 5% solute solution. **b.** The solute (purple circles) is unable to pass through the membrane, but the water (blue circles) passes through in both directions. There is a net movement of water toward the inside of the thistle tube, where there is a lower percentage of water molecules. **c.** Due to the incoming water molecules, the level of the solution rises in the thistle tube. **d.** Eventually, the concentration of water across the membrane equalizes.

always occurs from higher to lower concentration. Therefore, a net movement of water takes place, moving across the membrane from the beaker to the inside of the thistle tube (Fig. 5.7*b*).

MP3 Osmosis

The solute does not diffuse out of the thistle tube. Why not? Because the membrane is not permeable to the solute. As water enters and the solute does not exit, the level of the solution within the thistle tube rises (Fig. 5.7*c*). In the end, the concentration of solute in the thistle tube is less than 10%. Why? Because there is now less solute per unit volume. And the concentration of solute in the beaker is greater than 5%. Why? Because there is now more solute per unit volume.

Water enters the thistle tube due to the osmotic pressure of the solution within the thistle tube until it reaches equilibrium (Fig 5.7*d*). **Osmotic pressure** is the pressure that develops in a system due to osmosis.[2] In other words, the greater the possible osmotic pressure, the more likely it is that water will diffuse in that direction. Due to osmotic pressure, water is absorbed by the kidneys and taken up by capillaries in the tissues. Osmosis also occurs across the plasma membrane, as we'll see next (Fig. 5.8).

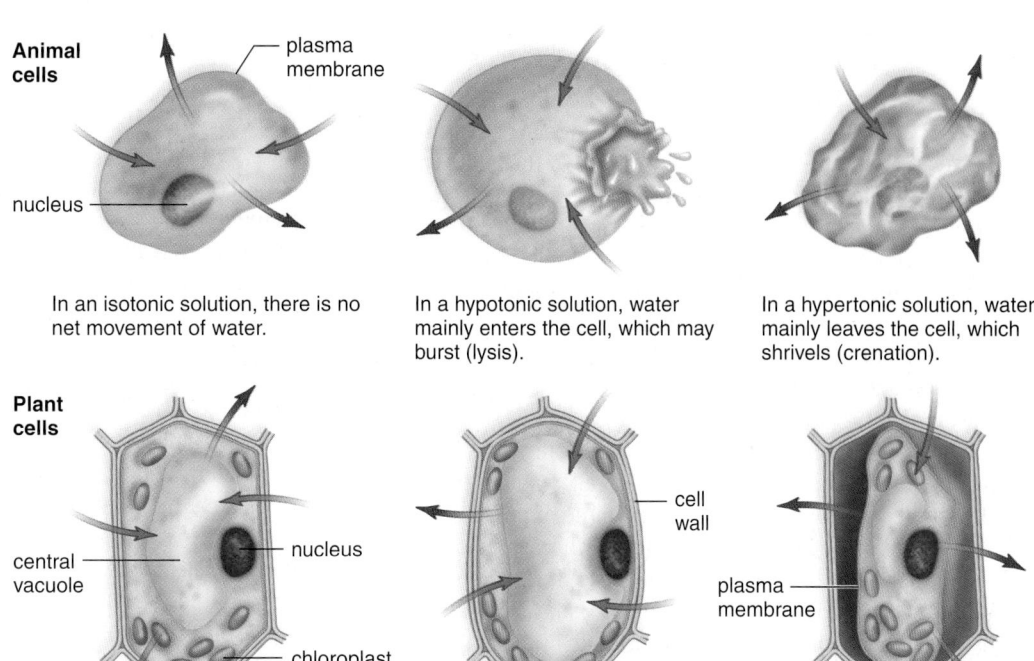

In an isotonic solution, there is no net movement of water.

In a hypotonic solution, water mainly enters the cell, which may burst (lysis).

In a hypertonic solution, water mainly leaves the cell, which shrivels (crenation).

In an isotonic solution, there is no net movement of water.

In a hypotonic solution, vacuoles fill with water, turgor pressure develops, and chloroplasts are seen next to the cell wall.

In a hypertonic solution, vacuoles lose water, the cytoplasm shrinks (plasmolysis), and chloroplasts are seen in the center of the cell.

Figure 5.8 Osmosis in animal and plant cells. The arrows indicate the net movement of water molecules. To determine the net movement of water, compare the number of blue arrows that are taking water molecules into the cell versus the number of red arrows that are taking water out of the cell. In an isotonic solution, a cell neither gains nor loses water; in a hypotonic solution, a cell gains water; and in a hypertonic solution, a cell loses water.

Isotonic Solution

In the laboratory, cells are normally placed in **isotonic solutions**—that is, the solute concentration and the water concentration both inside and outside the cell are equal, and therefore there is no net gain or loss of water. The prefix *iso* means "the same as," and the term **tonicity** refers to the strength of the solution. A 0.9% solution of the salt sodium chloride (NaCl) is known to be isotonic to red blood cells. Therefore, intravenous solutions medically administered usually have this tonicity. Terrestrial animals can usually take in either water or salt as needed to maintain the tonicity of their internal environment. Many animals living in an estuary, such as oysters, blue crabs, and some fishes, are able to cope with changes in the salinity (salt concentrations) of their environment using specialized kidneys, gills, and other structures.

Hypotonic Solution

Solutions that cause cells to swell, or even to burst, due to an intake of water are said to be **hypotonic solutions.** The prefix *hypo* means "less than" and refers to a solution with a lower concentration of solute (higher concentration of water) than inside the cell. If a cell is placed in a hypotonic solution, water enters the cell because the lower cellular concentration of water prompts a net movement of water from the outside to the inside of the cell.

Any concentration of a salt solution lower than 0.9% is hypotonic to red blood cells. Animal cells placed in such a solution expand and sometimes burst because of the buildup of pressure. The term *cytolysis* is used to refer to disrupted cells; hemolysis, then, is disrupted red blood cells.

The swelling of a plant cell in a hypotonic solution creates **turgor pressure.** When a plant cell is placed in a hypotonic solution, the cytoplasm expands because the large central vacuole gains water and the plasma membrane pushes against the rigid cell wall. Unlike animal cells that have no cell wall, the plant cell does not burst because the cell wall does not give way. Turgor pressure in plant cells is extremely important to the maintenance of the plant's erect position. If you forget to water your plants, they wilt due to decreased turgor pressure.

Organisms that live in fresh water have to avoid taking in too much water. Many protozoans, such as paramecia, have contractile vacuoles that rid the body of excess water. Freshwater fishes have well-developed kidneys that excrete a large volume of dilute urine. These fish still have to take in salts through their gills. Even though freshwater fishes are good osmoregulators, they would not be able to survive in either distilled water or a salty marine environment.

Video Contractile Vacuoles

---

2 Osmotic pressure is measured by placing a solution in an osmometer and then immersing the osmometer in pure water. The pressure that develops is the osmotic pressure of a solution.

Hypertonic Solution

Solutions that cause cells to shrink or shrivel due to loss of water are said to be **hypertonic solutions.** The prefix *hyper* means "more than" and refers to a solution with a higher percentage of solute (lower concentration of water) outside of the cell. If a cell is placed in a hypertonic solution, water leaves the cell; the net movement of water is from the inside to the outside of the cell.

Any concentration of a salt solution higher than 0.9% is hypertonic to red blood cells. If animal cells are placed in this solution, they shrink. The term **crenation** refers to red blood cells in this condition. Meats are sometimes preserved by salting them. The bacteria are not killed by the salt but by the lack of water in the meat.

When a plant cell is placed in a hypertonic solution, the plasma membrane pulls away from the cell wall as the large central vacuole loses water. This is an example of **plasmolysis,** a shrinking of the cytoplasm due to osmosis. The dead plants you may see along a salted roadside died because they were exposed to a hypertonic solution during the winter. Also, when salt water invades coastal marshes due to storms and human activities, coastal plants die. Without roots to hold the soil, it washes into the sea, doing away with many acres of valuable wetlands.

Marine animals cope with their hypertonic environment in various ways that prevent them from losing excess water to the environment. Sharks increase or decrease urea in their blood until their blood is isotonic with the environment, and in this way do not lose too much water. Marine fishes and other types of animals drink no water but excrete salts across their gills. Have you ever seen a marine turtle cry? It is ridding its body of salt by means of glands near the eye.

 **Animation** Plasmolysis

 **3D Animation** Osmosis

 **Animation** Hemolysis and Crenation

Facilitated Transport

The plasma membrane impedes the passage of all but a few substances. Yet, biologically useful molecules are able to rapidly enter and exit the cell either by way of a channel protein or because of carrier proteins in the membrane. These transport proteins are specific; each can transport only a certain type of molecule or ion across the membrane. How carrier proteins function is not completely understood, but after a carrier combines with a molecule, the carrier is believed to undergo a conformational change in shape that moves the molecule across the membrane. Carrier proteins are utilized for both facilitated transport (movement with concentration gradient; requires no energy) and active transport (movement against concentration gradient; requires energy)(see Table 5.1).

Facilitated transport explains how molecules such as glucose and amino acids are rapidly transported across the plasma membrane. Whereas water moves through a channel protein, the passage of glucose and amino acids is facilitated by their reversible combination with carrier proteins, which transport them through the plasma membrane. These carrier proteins are specific. For example, various sugar molecules of identical size might be present inside or outside the cell, but glucose can cross the membrane hundreds of times faster than the other sugars. As stated earlier, this is the reason the membrane can be called selectively permeable.

A model for facilitated transport (Fig. 5.9) shows that after a carrier has assisted the movement of a molecule to the other side of the membrane, it is free to assist the passage of other solute molecules. Neither diffusion nor facilitated transport requires an expenditure of energy because the molecules are moving down their concentration gradient.

Animation How Facilitated Diffusion Works

Check Your Progress   5.2

1. Explain how polar water can rapidly move across the nonpolar plasma membrane.
2. Contrast diffusion with facilitated transport.

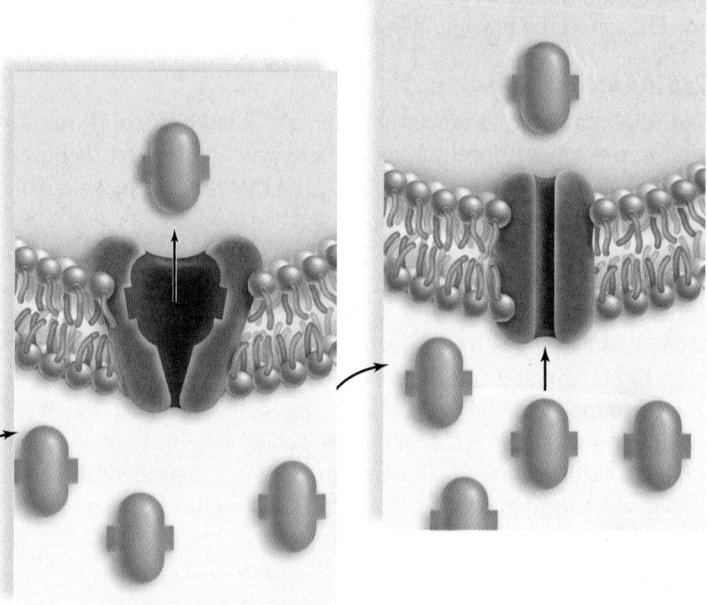

Figure 5.9 Facilitated transport. Facilitated transport by a carrier protein. A carrier protein can speed the rate at which a solute crosses the plasma membrane toward a lower concentration. Note that the carrier protein undergoes a change in shape as it moves a solute across the membrane.

Inside

plasma membrane

carrier protein

solute

Outside

5.3 Active Transport Across a Membrane

Learning Outcomes

Upon completion of this section, you should be able to

1. Explain how active transport moves substances across a membrane.
2. Compare the energy requirements of passive and active transport.
3. Contrast the active transport of large and small substances into a cell.

At times, a cell may need to further increase a concentration gradient across a membrane in order to do more work. As you might imagine, transporting a molecule against its concentration gradient, also called **active transport**, requires energy. During active transport, molecules or ions move through the plasma membrane, accumulating either inside or outside the cell. For example, iodine collects in the cells of the thyroid gland; glucose is completely absorbed from the gut by the cells lining the digestive tract; and sodium can be almost completely withdrawn from urine by cells lining the kidney tubules. In each of these instances, molecules have moved from a lower to a higher concentration, exactly opposite to the process of diffusion.

Carrier proteins and an expenditure of energy are both needed to transport molecules against their concentration gradient. In this case, chemical energy (ATP molecules usually) is required for the carrier to combine with the substance to be transported. Therefore, it is not surprising that cells involved primarily in active transport, such as kidney cells, have a large number of mitochondria near membranes where active transport is occurring.

Proteins involved in active transport often are called pumps because, just as a water pump uses energy to move water against the force of gravity, proteins use energy to move a substance against its concentration gradient. One type of pump that is active in all animal cells, but is especially associated with nerve and muscle cells, moves sodium ions (Na^+) to the outside of the cell and potassium ions (K^+) to the inside of the cell. The transport of sodium and potassium are linked together through the same carrier protein, called a **sodium-potassium pump.**

The sodium-potassium carrier protein has an initial shape that allows it to bind three sodium ions. Phosphate from an ATP molecule is added to the carrier protein, and it changes shape; this shape change moves sodium across the membrane. The new shape is no longer compatible with binding to the sodium, which falls away.

The new shape, however, is compatible with picking up two potassium ions, which bind to their sites. As the phosphate that was added from ATP in an earlier step leaves, the carrier protein assumes its original shape, and the two potassium ions are released inside the cell (Fig. 5.10). This bidirectional transport of three sodium and two potassium creates a solute gradient and electrical gradient across the plasma membrane.

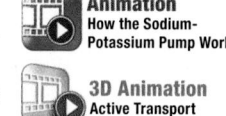

Animation
How the Sodium-Potassium Pump Works

3D Animation
Active Transport

The passage of salt (NaCl) across a plasma membrane is of primary importance to most cells. The chloride ion (Cl^-) usually crosses the plasma membrane because it is attracted by positively charged sodium ions (Na^+). First sodium ions are pumped across a membrane, and then chloride ions simply diffuse through channels that allow their passage.

Animation
Cotransport

As noted in Figure 5.3a, the genetic disorder cystic fibrosis results from a faulty chloride channel protein. When chloride is unable to exit a cell, water stays behind. The lack of water outside the cells causes abnormally thick mucus in the bronchial tubes and pancreatic ducts, thus interfering with the function of the lungs and pancreas.

Video
Good Poison

Bulk Transport

How do large molecules such as proteins, polysaccharides, or nucleic acids enter and exit a cell? These molecules are too large to be transported by carrier proteins, so they are instead transported into and out of the cell by vesicles. Membrane vesicles formed around macromolecules require an expenditure of cellular energy, but the cost is worth it because each vesicle keeps its cargo from mixing with molecules within the cytoplasm that could alter the cell's function. Generally, substances can exit a cell through exocytosis, and enter a cell through endocytosis.

Exocytosis

During **exocytosis**, an intracellular vesicle fuses with the plasma membrane as secretion occurs (Fig. 5.11). Hormones, neurotransmitters, and digestive enzymes are secreted from cells in this manner. The Golgi body often produces the vesicles that carry these cell products to the membrane. During exocytosis, the membrane of the vesicle becomes a part of the plasma membrane, because both are nonpolar. Adding additional vesicle membrane to the plasma membrane can enlarge the cell, and is a part of growth in some cells. The proteins released from the vesicle may adhere to the cell surface or become incorporated into an extracellular matrix.

Cells of particular organs are specialized to produce and export molecules. For example, pancreatic cells produce digestive enzymes or insulin, and anterior pituitary cells produce growth hormone, among other hormones. In these cells, secretory vesicles accumulate near the plasma membrane, and the vesicles release their contents only when the cell is stimulated by a signal received at the plasma membrane. A rise in blood sugar, for example, signals pancreatic cells to release the hormone insulin. This is called regulated secretion, because vesicles fuse with the plasma membrane only when the needs of the body trigger it to do so.

Figure 5.10 The sodium-potassium pump. The same carrier protein transports sodium ions (Na^+) to the outside of the cell and potassium ions (K^+) to the inside of the cell because it undergoes an ATP-dependent change in shape. Three sodium ions are carried outward for every two potassium ions carried inward; therefore, the inside of the cell is less positively charged compared to the outside.

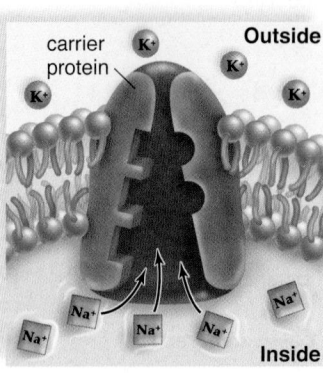

1. Carrier has a shape that allows it to take up 3 Na^+.

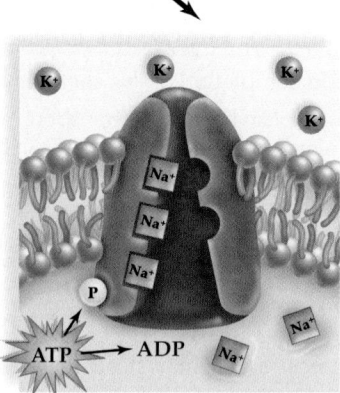

2. ATP is split, and phosphate group attaches to carrier.

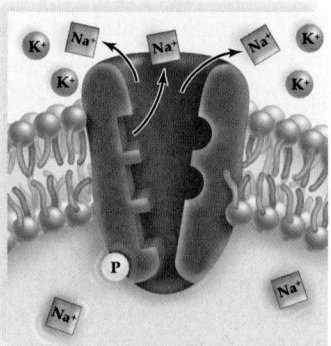

3. Change in shape results and causes carrier to release 3 Na^+ outside the cell.

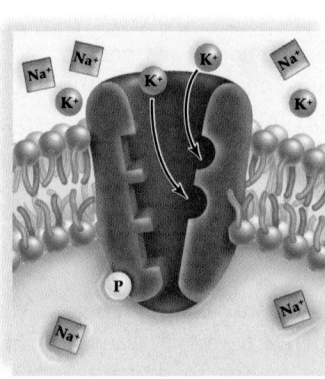

4. Carrier has a shape that allows it to take up 2 K^+.

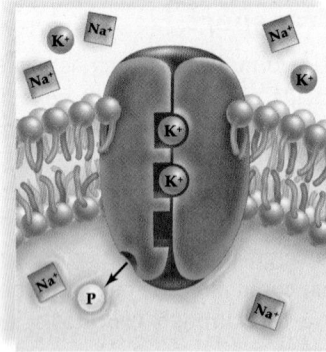

5. Phosphate group is released from carrier.

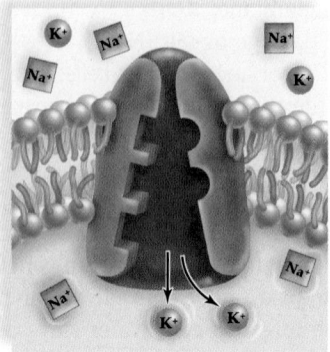

6. Change in shape results and causes carrier to release 2 K^+ inside the cell.

Figure 5.11 Exocytosis. Exocytosis secretes or deposits substances on the outside of the cell.

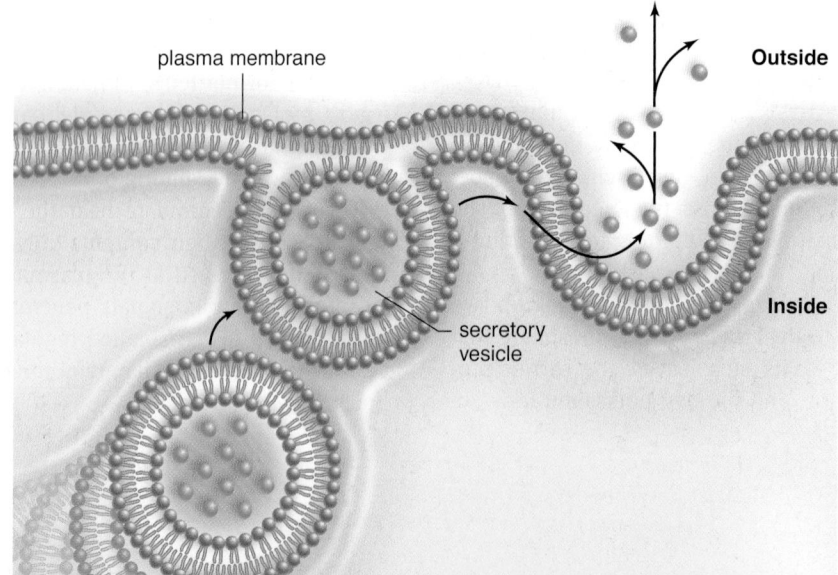

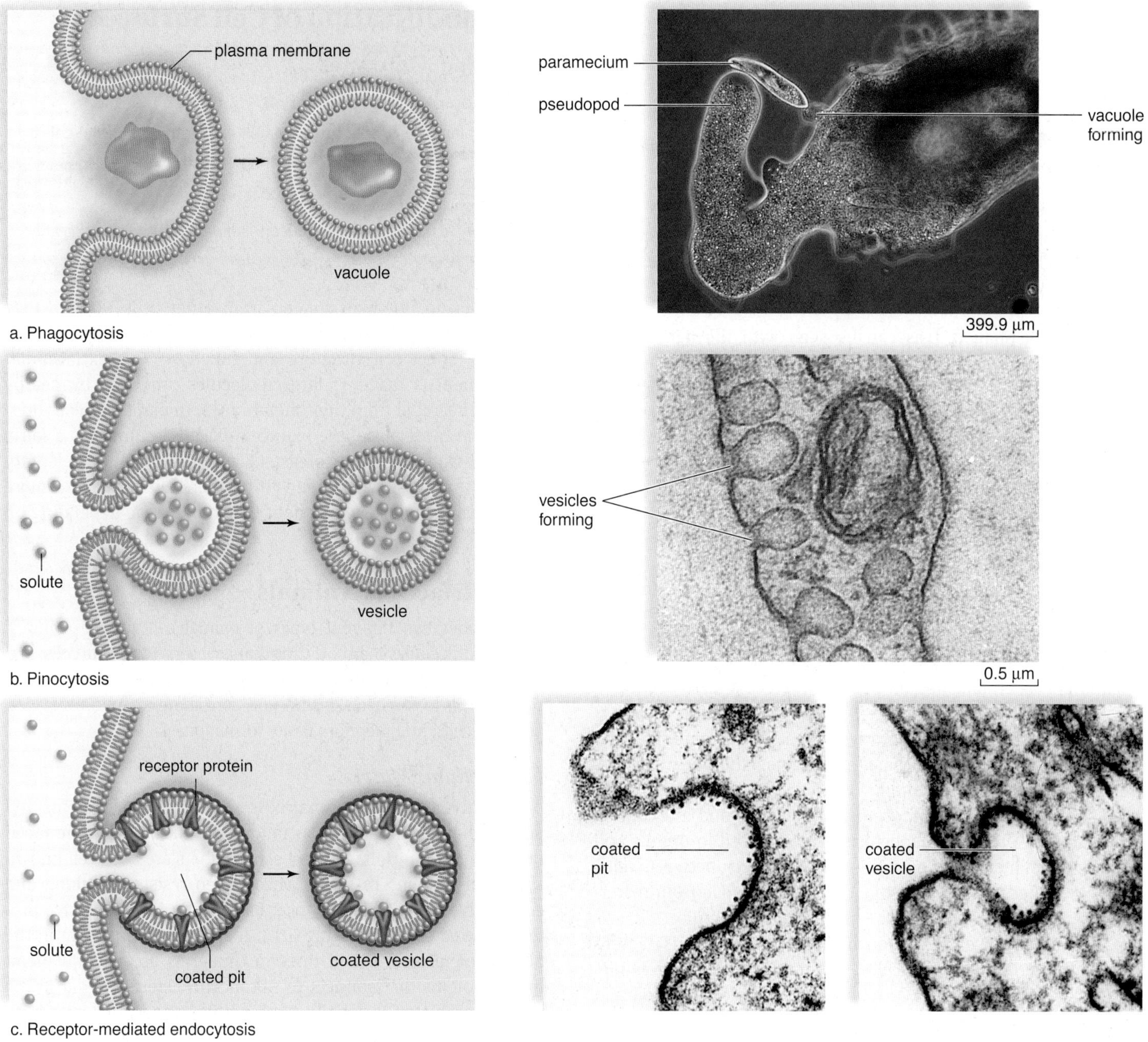

a. Phagocytosis

b. Pinocytosis

c. Receptor-mediated endocytosis

Figure 5.12 Three methods of endocytosis. a. Phagocytosis occurs when the substance to be transported into the cell is large; amoebas ingest by phagocytosis. Digestion occurs when the resulting vacuole fuses with a lysosome. **b.** Pinocytosis occurs when a macromolecule such as a polypeptide is transported into the cell. The result is a vesicle (small vacuole). **c.** Receptor-mediated endocytosis is a form of pinocytosis. Molecules first bind to specific receptor proteins, which migrate to or are already in a coated pit. The coated vesicle that forms contains the molecules and their receptors.

Endocytosis

During **endocytosis,** cells take in substances by forming vesicles around the material. A portion of the plasma membrane invaginates to envelop the substance, and then the membrane pinches off to form an intracellular vesicle. Endocytosis occurs in one of three ways, as illustrated in Figure 5.12. Phagocytosis transports large substances, such as a virus, and pinocytosis transports small substances, such as a macromolecule, into a cell. Receptor-mediated endocytosis is a special form of pinocytosis.

 **Animation**
Endocytosis and
Exocytosis

Phagocytosis. When the material taken in by endocytosis is large, such as a food particle or another cell, the process is called **phagocytosis** [Gk. *phagein,* to eat]. Phagocytosis is common in unicellular organisms such as amoebas (Fig. 5.12*a*). It also occurs in humans. Certain types of human white blood cells are amoeboid—that is, they are mobile like an amoeba, and they can engulf debris such as worn-out red blood cells or viruses. When an endocytic vesicle fuses with a lysosome, digestion occurs. Later in this text you will see that this process is a necessary and preliminary step toward the development of our immunity to bacterial diseases.

Pinocytosis.   Pinocytosis [Gk. *pinein*, to drink] occurs when vesicles form around a liquid or around very small particles (Fig. 5.12*b*). Blood cells, cells that line the kidney tubules or the intestinal wall, and plant root cells all use pinocytosis to ingest substances.

Whereas phagocytosis can be seen with the light microscope, the electron microscope must be used to observe pinocytic vesicles, which are no larger than 0.1–0.2 μm. Still, pinocytosis involves a significant amount of the plasma membrane because it occurs continuously. Cells do not shrink in size because the loss of plasma membrane due to pinocytosis is balanced by the occurrence of exocytosis.

Receptor-Mediated Endocytosis.   Receptor-mediated endocytosis is a form of pinocytosis that is quite specific because it uses a receptor protein to recognize compatible molecules and bring them into the cell. Molecules such as vitamins, peptide hormones, or lipoproteins can bind to specific receptors, found in special locations in the plasma membrane (Fig. 5.12*c*). This location is called a coated pit because there is a layer of protein on the cytoplasmic side of the pit. Once formed, the vesicle is uncoated and may fuse with a lysosome. When empty, a used vesicle fuses with the plasma membrane, and the receptors return to their former location.

Receptor-mediated endocytosis is selective and much more efficient than ordinary pinocytosis. It is involved in uptake and also in the transfer and exchange of substances between cells. Such exchanges take place when substances move from maternal blood into fetal blood at the placenta, for example.

The importance of receptor-mediated endocytosis is demonstrated by a genetic disorder called familial hypercholesterolemia. Cholesterol is transported in blood by a complex of lipids and proteins called low-density lipoprotein (LDL). Ordinarily, body cells take up LDL when LDL receptors gather in a coated pit. But in some individuals, the LDL receptor is unable to properly bind to the coated pit, and the cells are unable to take up cholesterol. Instead, cholesterol accumulates in the walls of arterial blood vessels, leading to high blood pressure, occluded (blocked) arteries, and heart attacks.

Check Your Progress   5.3

1. Compare facilitated transport with active transport.
2. Examine how exocytosis and endocytosis change membrane surface area.

5.4 Modification of Cell Surfaces

Learning Outcomes

Upon completion of this section, you should be able to

1. Explain the role of the extracellular matrix in animal cell behavior.
2. Compare the structure and function of adhesion, tight, and gap junctions in animals.
3. Contrast cell-to-cell junctions between animals and plants.

Most cells do not live isolated from other cells. Rather, they live and interact within an external environment that can dramatically affect cell structure and function. This extracellular environment is made of large molecules produced by nearby cells and secreted from their membranes. In plants, prokaryotes, fungi, and most algae, the extracellular environment is a fairly rigid cell wall, which is consistent with a somewhat sedentary lifestyle. Animals, which tend to be more active, have a more varied extracellular environment that can change depending on the tissue type.

Cell Surfaces in Animals

We consider two different types of animal cell surface features: (1) the extracellular matrix (ECM) that is observed outside cells, and (2) junctions that occur between some types of cells. Both of these can connect to the cytoskeleton and contribute to communication between cells, and therefore tissue formation.

Extracellular Matrix

A protective extracellular matrix is a meshwork of proteins and polysaccharides in close association with the cell that produced them (Fig. 5.13). Collagen and elastin fibers are two well-known structural proteins in the ECM; collagen resists stretching and elastin gives the ECM resilience. Fibronectin is an adhesive protein, colored green in Figure 5.13, that binds to a protein in the plasma membrane called integrin. Integrins are integral membrane proteins that connect to fibronectin externally and to the actin cytoskeleton internally. Through its connections with both the ECM and the cytoskeleton, integrin plays a role in cell signaling, permitting the ECM to influence the activities of the cytoskeleton and, therefore, the shape and activities of the cell.

Amino sugars in the ECM form multiple polysaccharides that attach to a protein and are, therefore, called proteoglycans. Proteoglycans, in turn, attach to a very long, centrally placed polysaccharide. The entire structure, which looks like an enormous bottle brush, resists compression of the extracellular matrix. Proteoglycans assist cell signaling when they regulate the passage of molecules through the ECM to the plasma membrane, where receptors are located. During development, they help bring about differentiation by guiding cell migration along collagen fibers to specific locations. Thus, the ECM has a dynamic role in all aspects of a cell's behavior.

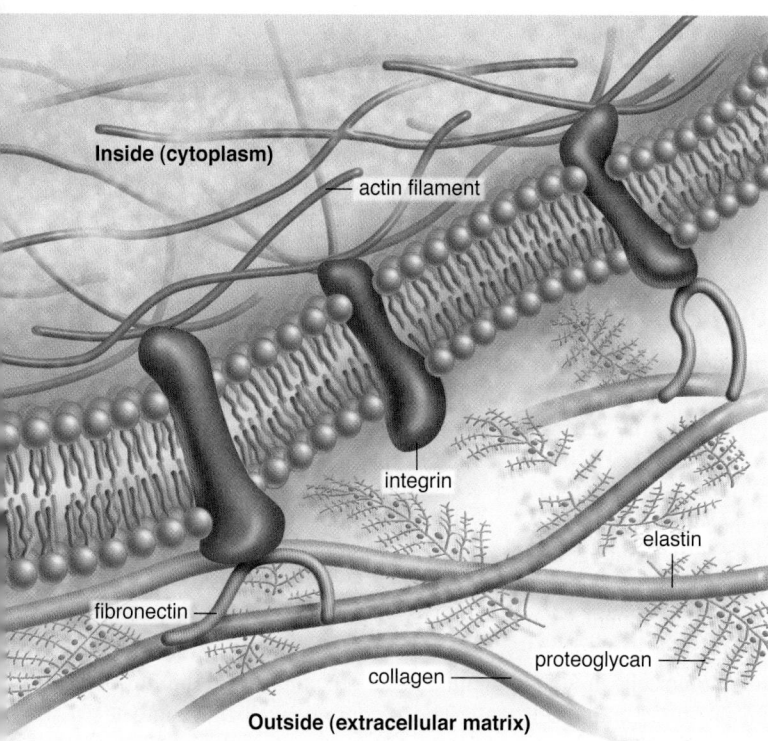

Figure 5.13 Animal cell extracellular matrix. In the extracellular matrix, collagen and elastin have a support function, while fibronectins bind to integrin, and in this way, assist communication between ECM and the cytoskeleton.

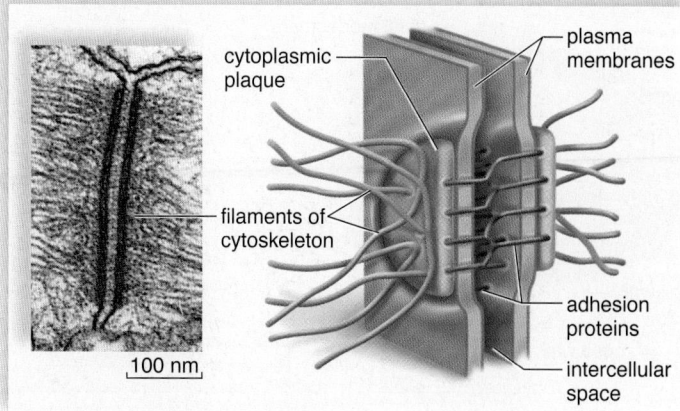

a. Adhesion junction

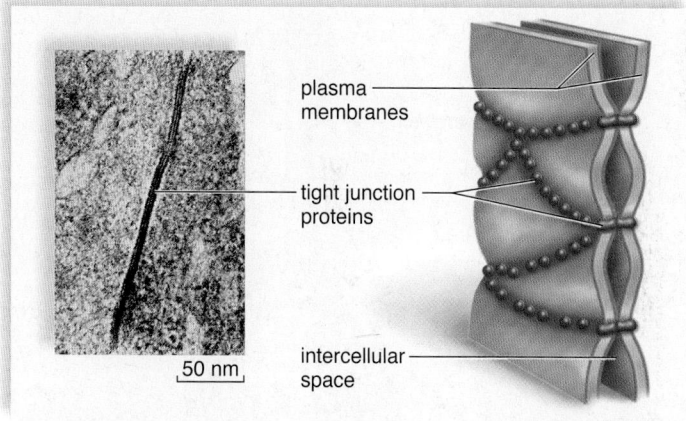

b. Tight junction

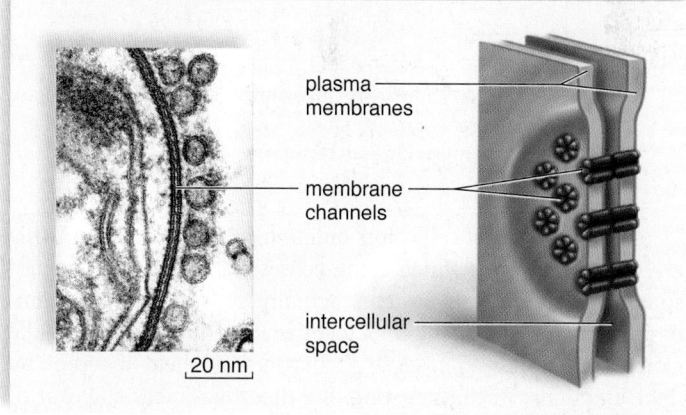

c. Gap junction

Figure 5.14 Junctions between cells of the intestinal wall.
a. In adhesion junctions such as a desmosome, adhesive proteins connect two cells. **b.** Tight junctions between cells form an impermeable barrier because their adjacent plasma membranes are joined and don't allow molecules to pass. **c.** Gap junctions allow communication between two cells because adjacent plasma membrane channels are joined.

Later on, in the discussion of tissues, you'll see that the extracellular matrix varies in quantity and in consistency from being quite flexible, as in loose connective tissue; semiflexible, as in cartilage; and rock solid, as in bone. The extracellular matrix of bone is hard because, in addition to the components mentioned, mineral salts, notably calcium salts, are deposited outside the cell.

The proportion of cells to ECM also varies. In the small intestine, for example, epithelial cells comprise the majority of the tissue, and the ECM is a thin sheet beneath the cells. In bone, the ECM makes up most of the tissue, with comparatively fewer cells.

Junctions Between Cells

Certain tissues of vertebrate animals are known to have junctions between their cells that allow them to behave in a coordinated manner. Three types of junctions are shown in Figure 5.14.

Adhesion junctions serve to mechanically attach adjacent cells. Two types of adhesion junctions are described here. In **desmosomes,** internal cytoplasmic plaques, firmly attached to the intermediate filament cytoskeleton within each cell, are joined by integral membrane proteins called cadherins between cells. The result is a sturdy but flexible sheet of cells. In some organs—such as the heart, stomach, and bladder, where tissues get stretched—desmosomes hold the cells together. At a *hemidesmosome,* the intermediate filaments of the cytoskeleton

are attached to the ECM through integrin proteins. Adhesion junctions are the most common type of intercellular junction between skin cells.

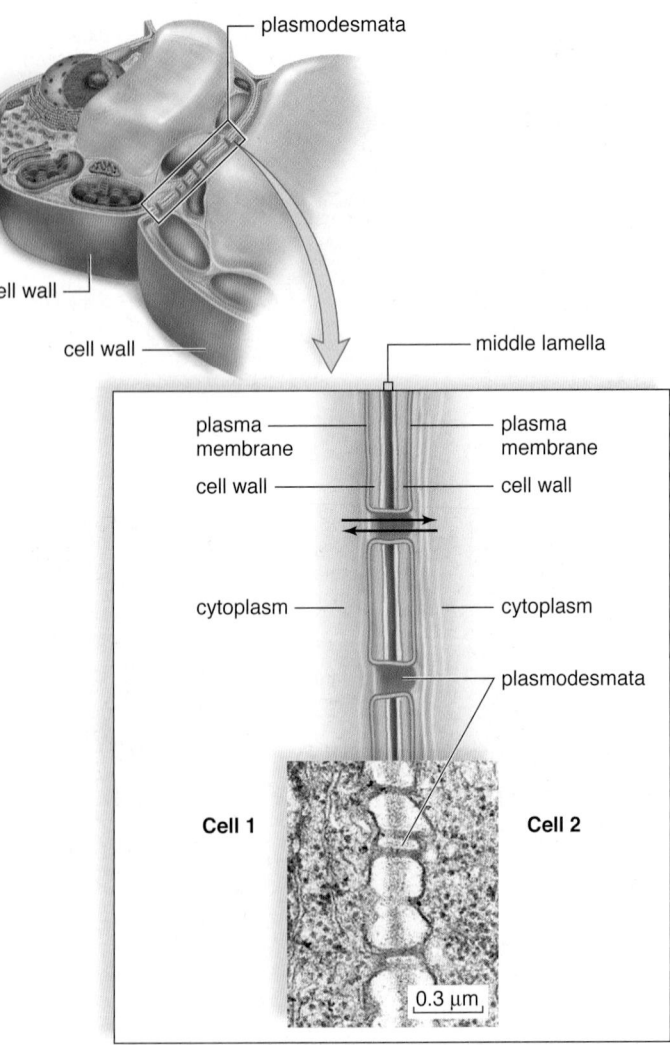

Figure 5.15 Plasmodesmata.   Plant cells are joined by membrane-lined channels that contain cytoplasm. Water and small molecules can pass from cell to cell.

Another type of adhesion junction between adjacent cells are **tight junctions,** which bring cells even closer than desmosomes. Tight junction proteins actually connect plasma membranes between adjacent cells together, producing a zipperlike fastening. Tissues that serve as barriers are held together by tight junctions; in the intestine, the digestive juices stay out of the rest of the body, and in the kidneys the urine stays within kidney tubules, because the cells are joined by tight junctions.

A **gap junction** allows cells to communicate. A gap junction is formed when two identical plasma membrane channels join. The channel of each cell is lined by six plasma membrane proteins. A gap junction lends strength to the cells, but it also allows small molecules and ions to pass between them. Gap junctions are important in heart muscle and smooth muscle because they permit a flow of ions that is required for the cells to contract as a unit.

Plant Cell Walls

In addition to a plasma membrane, plant cells are surrounded by a porous **cell wall** that varies in thickness, depending on the function of the cell.

All plant cells have a primary cell wall. The primary cell wall contains cellulose fibrils in which microfibrils are held together by noncellulose substances. Pectins allow the wall to stretch when the cell is growing, and noncellulose polysaccharides harden the wall when the cell is mature. Pectins are especially abundant in the middle lamella, which is a layer of adhesive substances that holds the cells together.

Some cells in woody plants have a secondary wall that forms inside the primary cell wall. The secondary wall has a greater quantity of cellulose fibrils than the primary wall, and layers of cellulose fibrils are laid down at right angles to one another. Lignin, a substance that adds strength, is a common ingredient of secondary cell walls in woody plants.

In a plant, the cytoplasm of living cells is connected by **plasmodesmata** (sing., plasmodesma), numerous narrow, membrane-lined channels that pass through the cell wall (Fig. 5.15). Cytoplasmic strands within these channels allow direct exchange of some materials between adjacent plant cells and eventually connect all the cells within a plant. The plasmodesmata allow only water and small solutes to pass freely from cell to cell. This limitation means that plant cells can maintain their own concentrations of larger substances and differentiate into particular cell types.

Check Your Progress   5.4

1. Describe the molecular composition of the extracellular matrix of an animal cell.
2. Contrast a plant's primary cell wall with its secondary cell wall.
3. Compare the size of molecules allowed passage in a gap junction vs. plasmodesmata.

CONNECTING *the* CONCEPTS *with the* BIG IDEAS

Energy and Homeostasis

- The fluid mosaic model combines phospholipids and proteins to form a flexible, asymmetric, amphipathic, responsive, and semi-permeable membrane which often groups and expands surface area. (2B1b1, 2B3a-b)
- Molecules cross cell membranes based on their size, charge, and polarity; the diffusion of water is a major concern for living cells. (2B1b4, 2B2a3, 2D3a)
- Diffusion, facilitated diffusion, active transport, and endocytosis/exocytosis differ in their "direction" of molecular movement, assistance by specific membrane proteins, and need for free energy input. (2B2a-c)

Information and Signaling

- Embedded membrane proteins provide structural, enzymatic, passage, recognition, and reception functions for the cell. (3B2b, 3D3a2)

Interactions and Systems

- Membranes that surround various organelles in the cell often interconnect, allowing sequential transport and storage, bringing greater efficiency to the entire cell system. (4A2a-g, 4B2a1)

*Find the unabridged version of all EK citations at www.glencoe.com/maderAP11.

Media Study Tools

www.glencoe.com/maderAP11

Enhance your study of this chapter with study tools and practice tests. Also ask your instructor about the resources available through ConnectPlus, including the media-rich eBook, interactive learning tools, and animations.

3D Animation
Membrane
Transport

For a detailed examination of the processes involved in the movement of molecules across the plasma membrane, watch McGraw-Hill's new 3D animation "Membrane Transport."

Summarize

5.1 Plasma Membrane Structure and Function

Two components of the plasma membrane are lipids and proteins. In the lipid bilayer, phospholipids are arranged with their hydrophilic (polar) heads adjacent to water and their hydrophobic (nonpolar) tails buried in the interior. The lipid bilayer has the consistency of oil but acts as a barrier to the entrance and exit of most biological molecules. Membrane glycolipids and glycoproteins are involved in marking the cell as belonging to a particular individual and tissue.

The hydrophobic portion of an integral protein lies in the lipid bilayer of the plasma membrane, and the hydrophilic portion lies at the surfaces. Proteins act as receptors, carry on enzymatic reactions, join cells together, form channels, or act as carriers to move substances across the membrane. Some of these proteins make contact with the extracellular matrix (ECM) outside and with the cytoskeleton inside. Thus, the ECM can influence the happenings inside the cell.

5.2 Passive Transport Across a Membrane

The plasma membrane is selectively permeable. Some molecules (lipid-soluble compounds, water, and gases) simply diffuse across the membrane from the area of higher concentration to the area of lower concentration. No metabolic energy is required for diffusion to occur.

The diffusion of water across a selectively permeable membrane is called osmosis. Water moves across the membrane into the area of higher solute (less water) content per volume. When cells are in an isotonic solution, they neither gain nor lose water. When cells are in a hypotonic solution, they gain water, and when they are in a hypertonic solution, they lose water (Table 5.2).

Other molecules are transported across the membrane either by a channel protein or by carrier proteins that span the membrane. During facilitated transport, a substance moves down its concentration gradient. No energy is required.

Table 5.2 Effect of Osmosis on a Cell

	Concentrations			
Tonicity of Solution	Solute	Water	Net Movement of Water	Effect on Cell
Isotonic	Same as cell	Same as cell	None	None
Hypotonic	Less than cell	More than cell	Cell gains water	Swells, turgor pressure
Hypertonic	More than cell	Less than cell	Cell loses water	Shrinks, plasmolysis

5.3 Active Transport Across a Membrane

During active transport, a carrier protein acts as a pump that causes a substance to move against its concentration gradient. One example is the sodium-potassium pump that carries Na^+ to the outside of the cell and K^+ to the inside of the cell. Energy in the form of ATP molecules is required for active transport to occur.

Larger substances can enter and exit a membrane by exocytosis and endocytosis. Exocytosis involves secretion. Endocytosis includes phagocytosis, pinocytosis, and receptor-mediated endocytosis. Receptor-mediated endocytosis makes use of receptor proteins in the plasma membrane. Once a specific solute binds to receptors, a coated pit becomes a coated vesicle. After losing the coat, the vesicle can join with the lysosome, or after discharging the substance, the receptor-containing vesicle can fuse with the plasma membrane.

5.4 Modification of Cell Surfaces

Animal cells have an extracellular matrix (ECM) that influences their shape and behavior. The amount and character of the ECM varies by tissue type. Some animal cells have junction proteins that join them to other cells of the same tissue. Adhesion junctions and tight junctions help hold cells together; gap junctions allow passage of small molecules between cells.

Plant cells have a freely permeable cell wall, with cellulose as its main component. Also, plant cells are joined by narrow, membrane-lined channels called plasmodesmata that span the cell wall and contain strands of cytoplasm that allow materials to pass from one cell to another.

Key Terms

active transport 95	hypertonic solution 94
adhesion junction 99	hypotonic solution 93
aquaporin 91	isotonic solution 93
bulk transport 91	junction protein 89
carrier protein 88	osmosis 92
cell recognition protein 89	osmotic pressure 93
cell wall 100	phagocytosis 97
channel protein 88	pinocytosis 98
cholesterol 86	plasmodesmata 100
concentration gradient 89	plasmolysis 94
crenation 94	receptor-mediated
desmosome 99	endocytosis 98
diffusion 91	receptor protein 89
endocytosis 97	selectively permeable 89
enzymatic protein 89	sodium-potassium pump 95
exocytosis 95	solute 91
extracellular matrix (ECM) 87	solution 91
facilitated transport 94	solvent 91
fluid-mosaic model 87	tight junction 100
gap junction 100	tonicity 93
glycolipid 88	turgor pressure 93
glycoprotein 88	

▌ Assess

Reviewing This Chapter

1. Describe the fluid-mosaic model of membrane structure. 87–88
2. Tell how the phospholipids are arranged in the plasma membrane. What other lipid is present in the membrane, and what functions does it serve? 86–87
3. Describe the possible functions of proteins in the plasma membrane. 86–88

4. What is cell signaling and how does it occur? 90
5. Define diffusion. What factors can influence the rate of diffusion? What substances can diffuse through a differentially permeable membrane? 91–92
6. Define osmosis. Describe verbally and with drawings what happens to an animal cell and a plant cell when placed in isotonic, hypotonic, and hypertonic solutions. 92–94
7. Why do most substances have to be assisted through the plasma membrane? Contrast movement by facilitated transport with movement by active transport. 94–95
8. Draw and explain a diagram that shows how the sodium-potassium pump works. 95–96
9. Describe and contrast three methods of endocytosis. 97
10. Describe the structure and function of animal and plant cell modifications. 98–100

Testing Yourself

Choose the best answer for each question.

1. Write hypotonic solution or hypertonic solution beneath each cell. Justify your conclusions.

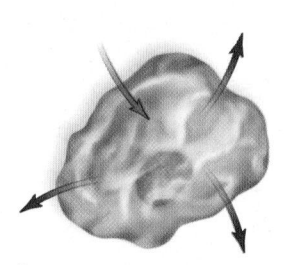 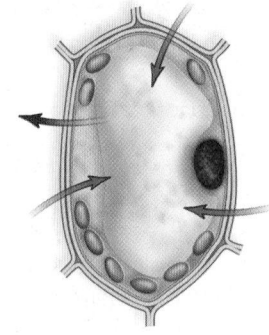

a. _____   b. _____

2. Electron micrographs following freeze-fracture of the plasma membrane indicate that
   a. the membrane is a phospholipid bilayer.
   b. some proteins span the membrane.
   c. protein is found only on the surfaces of the membrane.
   d. glycolipids and glycoproteins are antigenic.
   e. there are receptors in the membrane.

3. A phospholipid molecule has a head and two tails. The tails are found
   a. at the surfaces of the membrane.
   b. in the interior of the membrane.
   c. spanning the membrane.
   d. where the environment is hydrophilic.
   e. Both a and b are correct.

4. During diffusion,
   a. solvents move from the area of higher to lower concentration, but solutes do not.
   b. there is a net movement of molecules from the area of higher to lower concentration.
   c. a cell must be present for any movement of molecules to occur.
   d. molecules move against their concentration gradient if they are small and charged.
   e. All of these are correct.

5. When a cell is placed in a hypotonic solution,
   a. solute exits the cell to equalize the concentration on both sides of the membrane.

b. water exits the cell toward the area of lower solute concentration.

c. water enters the cell toward the area of higher solute concentration.

d. solute exits and water enters the cell.

e. Both c and d are correct.

6. When a cell is placed in a hypertonic solution,
   a. solute exits the cell to equalize the concentration on both sides of the membrane.
   b. water exits the cell toward the area of lower solute concentration.
   c. water exits the cell toward the area of higher solute concentration.
   d. solute exits and water enters the cell.
   e. Both a and c are correct.

7. Active transport
   a. requires a carrier protein.
   b. moves a molecule against its concentration gradient.
   c. requires a supply of chemical energy.
   d. does not occur during facilitated transport.
   e. All of these are correct.

8. The sodium-potassium pump
   a. helps establish an electrochemical gradient across the membrane.
   b. concentrates sodium on the outside of the membrane.
   c. uses a carrier protein and chemical energy.
   d. is present in the plasma membrane.
   e. All of these are correct.

9. Receptor-mediated endocytosis
   a. is no different from phagocytosis.
   b. brings specific solutes into the cell.
   c. helps concentrate proteins in vesicles.
   d. results in high osmotic pressure.
   e. All of these are correct.

10. Plant cells
    a. always have a secondary cell wall, even though the primary one may disappear.
    b. have channels between cells that allow strands of cytoplasm to pass from cell to cell.
    c. develop turgor pressure when water enters the nucleus.
    d. do not have cell-to-cell junctions like animal cells.
    e. All of these are correct.

11. Label this diagram of the plasma membrane.

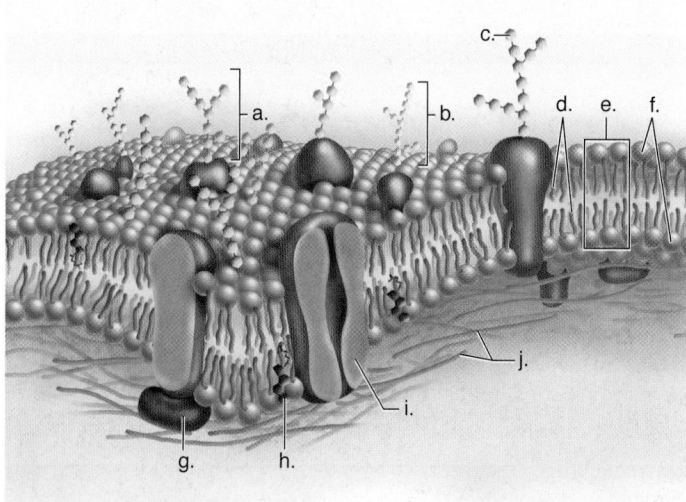

12. The fluid-mosaic model of membrane structure refers to
    a. the fluidity of proteins and the pattern of phospholipids in the membrane.
    b. the fluidity of phospholipids and the pattern of proteins in the membrane.
    c. the fluidity of cholesterol and the pattern of carbohydrate chains outside the membrane.
    d. the lack of fluidity of internal membranes compared to the plasma membrane, and the ability of the proteins to move laterally in the membrane.
    e. the fluidity of hydrophobic regions, proteins, and the mosaic pattern of hydrophilic regions.

13. Which of the following is not a function of proteins present in the plasma membrane? Proteins
    a. assist the passage of materials into the cell.
    b. interact and recognize other cells.
    c. bind with specific hormones.
    d. carry out specific metabolic reactions.
    e. produce lipid molecules.

14. The carbohydrate chains projecting from the plasma membrane are involved in
    a. adhesion between cells.
    b. reception of molecules.
    c. cell-to-cell recognition.
    d. All of these are correct.

15. Plants wilt on a hot summer day because of a decrease in
    a. turgor pressure.
    b. evaporation.
    c. condensation.
    d. diffusion.

16. The extracellular matrix
    a. assists in the movement of substances across the plasma membrane.
    b. prevents the loss of water when cells are placed in a hypertonic solution.
    c. has numerous functions that affect the shape and activities of the cell that produced it.
    d. contains the junctions that sometimes occur between cells.
    e. All of these are correct.

Engage

Thinking Scientifically

1. The mucus in bronchial tubes must be thin enough for cilia to move bacteria and viruses up into the throat away from the lungs. Which way would Cl^- normally cross the plasma membrane of bronchial tube cells in order for mucus to be thin (see Fig. 5.3a)? Use the concept of osmosis to explain your answer.

2. Winter wheat is planted in the early fall, grows over the winter when the weather is colder, and is harvested in the spring. As the temperature drops, the makeup of the plasma membrane of winter wheat changes. Unsaturated fatty acids replace saturated fatty acids in the phospholipids of the membrane. Why is this a suitable adaptation?

6

Metabolism: Energy and Enzymes

The cheetah, and more directly the impala, depend on solar energy captured by photosynthesizers.

CHAPTER OUTLINE

BEFORE YOU BEGIN

Before beginning this chapter, take a few moments to review the following discussions.

Figures 5.1 and 5.3 How are proteins embedded in biological membranes important to cellular function?

Figure 5.4 What membrane characteristics allow some biological molecules to freely pass and not others?

Section 5.3 What is necessary to move molecules against their concentration gradient?

Photosynthesizing grasses on an African plain provide impalas with organic building blocks and the energy they need to evade being caught by a cheetah. Eating impalas provides cheetahs with food and the energy they need to be quick enough to catch impalas.

All life on Earth depends on the flow of energy coming from the Sun. You, like the cheetah, consume plants and animals that get their energy either directly or indirectly from the Sun. Solar energy is concentrated enough to allow plants to photosynthesize and make biological molecules that, in turn, provide a continual supply of food for you and other creatures within the biosphere.

As you digest food and break the bonds in vegetables and meat, some of the energy is used for work, and some energy escapes into the environment as heat. Energy is critical to metabolism and enzymatic reactions, so it is the first topic we consider in this chapter. Without enzymes, you and the cheetah would not be able to use energy to maintain your bodies, nor to carry on any type of activity.

As you read through the chapter, think about the following questions:

1. What forms of energy are used by a cell?
2. How is energy used to drive biological processes within cells?
3. What might happen if insufficient energy is available for cells to function?

FOLLOWING *the* BIG IDEAS

CHAPTER 6 METABOLISM: ENERGY AND ENZYMES

Energy and Homeostasis	The production of cellular energy requires specific sequences of enzymes and membranes embedded with specific sequences of carrier molecules to power a proton pump.
Interactions and Systems	Energy flows through all biological systems, creating the ability to do work, although eventually all energy is lost as heat.

6.1 Cells and the Flow of Energy

Learning Outcomes

Upon completion of this section, you should be able to

1. Compare potential and kinetic energy.
2. Describe the first and second laws of thermodynamics.
3. Examine how organization and structure of living things is related to heat and entropy.

To maintain their structural organization and carry out metabolic activities, cells—and organisms comprised of cells—need a constant supply of energy. **Energy**, defined as the ability to do work or bring about a change, allows living things to carry on the processes of life, including growth, development, metabolism, and reproduction.

Organic nutrients, made by photosynthesizing producers (algae, plants, and some bacteria), directly provide organisms with energy by capturing energy from sunlight. Considering that producers use light energy to produce organic nutrients, the majority of life on Earth is ultimately dependent on solar energy.

Forms of Energy

Energy occurs in two forms: kinetic and potential energy. **Kinetic energy** is the energy of motion, as when water flows over a waterfall, a ball rolls down a hill, or a moose walks through grass. **Potential energy** is stored energy whose capacity to accomplish work is not being used at the moment. The food we eat has potential energy because the energy stored in chemical bonds can be converted into various types of kinetic energy. Food is specifically called **chemical energy** because it is composed of organic molecules such as carbohydrates, proteins, and fat. When a moose walks, it converts chemical energy into a type of kinetic energy called **mechanical energy** (Fig. 6.1).

Animation
Energy
Conversion

Two Laws of Thermodynamics

In nature, energy flows in biological systems. Figure 6.1 illustrates the flow of energy in a terrestrial ecosystem. Plants capture only a small portion of solar energy, and much of it dissipates as **heat.** When plants photosynthesize and then make use of the food they produce, more heat results. Even with this considerable heat loss, there is enough remaining to sustain a moose and the other organisms in an ecosystem. As they metabolize nutrient molecules, all the captured solar energy eventually dissipates as heat. Therefore, energy flows and does not cycle.

Two **laws of thermodynamics,** formulated by early energy researchers, explain why energy flows through ecosystems and through cells:

> The first law of thermodynamics—the law of conservation of energy—states energy cannot be created or destroyed, but it can be changed from one form to another.

When leaf cells photosynthesize, they use solar energy to form carbohydrate molecules from carbon dioxide gas and water. (Carbohydrates are energy-rich molecules because they have many bonds that store energy; carbon dioxide and water are energy-poor molecules because of the relative lack of bonds.) Not all of the captured solar energy becomes carbohydrates; some becomes heat:

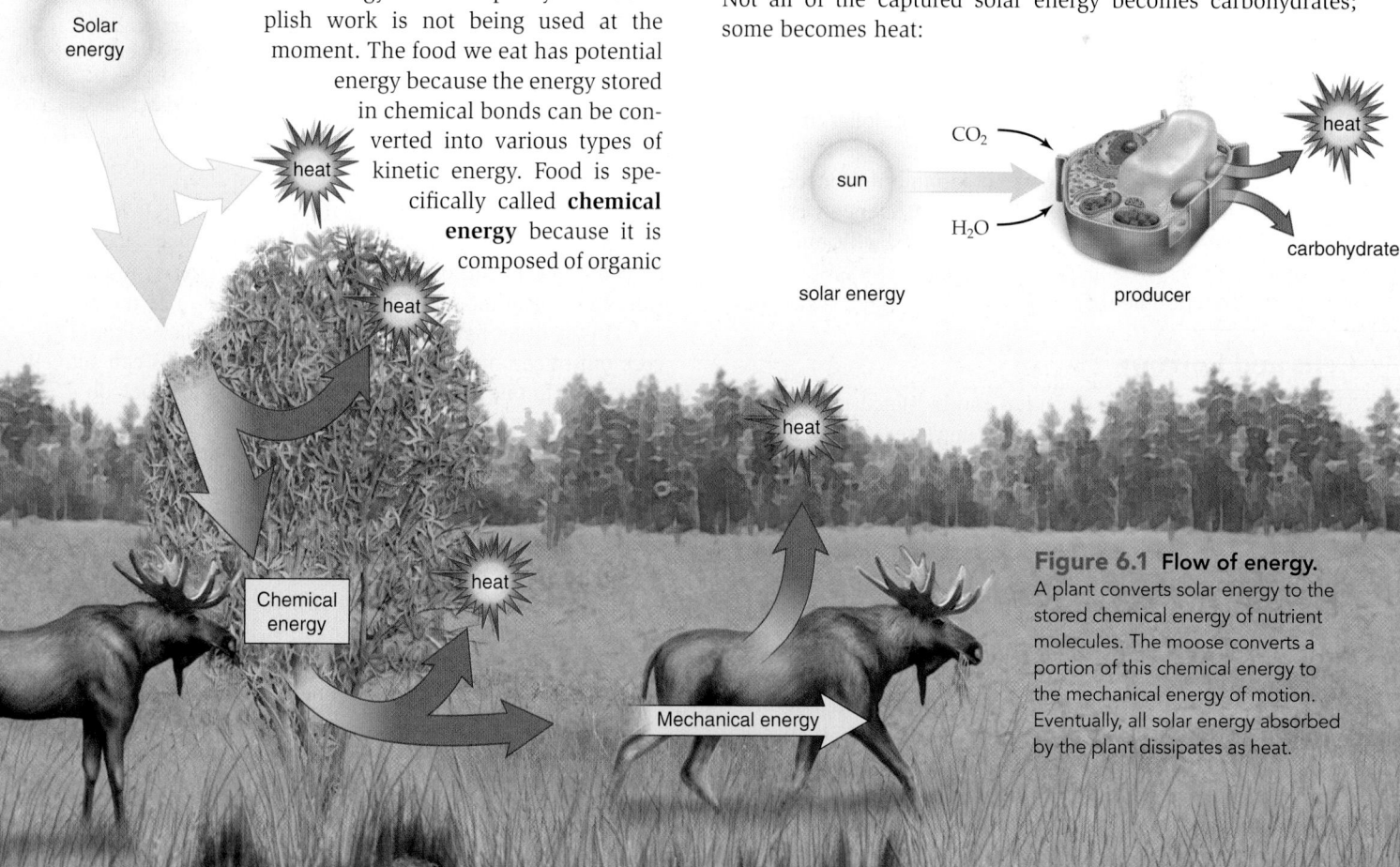

Figure 6.1 **Flow of energy.**
A plant converts solar energy to the stored chemical energy of nutrient molecules. The moose converts a portion of this chemical energy to the mechanical energy of motion. Eventually, all solar energy absorbed by the plant dissipates as heat.

Obviously, plant cells do not create the energy they use to produce carbohydrate molecules; that energy comes from the Sun. Is any energy destroyed? No, because the heat they give off is also a form of energy. Similarly, as a moose walks, it uses the potential energy stored in carbohydrates to kinetically power its muscles. As its cells use this energy, none is destroyed, but each energy exchange produces some heat, which dissipates into the environment:

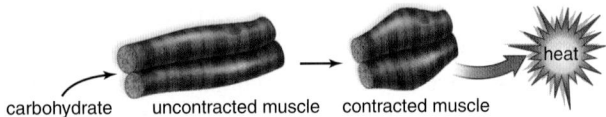

The second law of thermodynamics therefore applies to living systems:

> The second law of thermodynamics states energy cannot be changed from one form to another without a loss of usable energy.

In our example, this law is upheld because some of the solar energy taken in by the plant and some of the chemical energy within the nutrient molecules taken in by the moose become heat. When heat dissipates into the environment, it is no longer usable—that is, it is not available to do work. Each energy transformation moves us closer to a condition where all usable forms of energy become heat that is lost to the environment. Heat that dissipates into the environment cannot be captured and converted to one of the other forms of energy.

As a result of the second law of thermodynamics, no process requiring a conversion of energy is ever 100% efficient. Much of the energy is lost in the form of heat. In automobiles, the internal combustion engine is between 20% and 30% efficient in converting chemical energy stored in gasoline into mechanical energy used to drive the wheels. The majority of energy is lost as dissipated heat. Cells are capable of about 40% efficiency, with the remaining energy being given off to the surrounding environment as heat.

MP3
Laws of
Thermodynamics

Cells and Entropy

The second law of thermodynamics can be stated another way: Every energy transformation makes the universe less organized, or structured, and more disordered, or chaotic. The term **entropy** [Gk. *entrope*, a turning inward] is used to indicate the relative amount of disorganization. Because the processes that occur in cells are energy transformations, the second law means that every process that occurs in cells always does so in a way that increases the total entropy of the universe. The second law means that each cellular process makes less energy available to do useful work in the future.

Figure 6.2 shows two processes that occur in cells. The second law of thermodynamics tells us that glucose tends to break apart into carbon dioxide and water over time. Why? Because glucose is more organized and structured, and therefore less stable, than its breakdown products. Also, hydrogen ions on one side of a membrane tend to move to the other side unless they are prevented from doing so. Why? Because when they are distributed randomly, entropy has increased. As an analogy, you know from

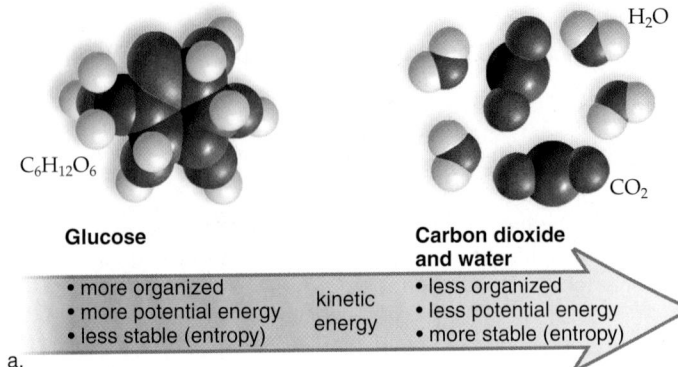

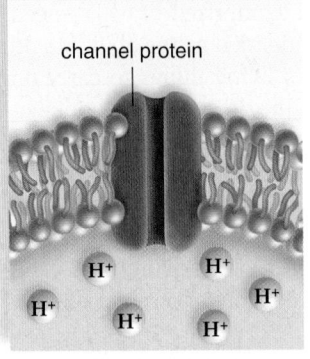

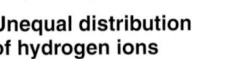

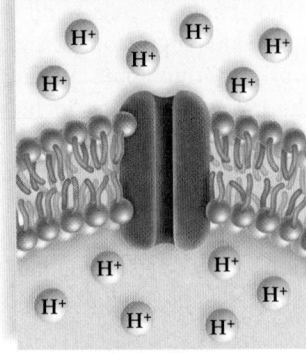

Figure 6.2 Cells and entropy.   The second law of thermodynamics tells us that **(a)** glucose, which is more organized, tends to break down to carbon dioxide and water, which are less organized. **b.** Similarly, hydrogen ions (H⁺) on one side of a membrane tend to move to the other side so that the ions are randomly distributed. Both processes result in an increase in entropy.

experience that a neat room is more organized but less stable than a messy room, which is disorganized but more stable. How do you know a neat room is less stable than a messy room? Consider that a neat room always tends to become more messy.

On the other hand, you know that some cells can make glucose out of carbon dioxide and water, and all cells can actively move ions to one side of the membrane. How do they do it? By the input of energy from an outside source. Photosynthesizing producers use energy from sunlight to create organized structure in biological molecules. Organisms that consume producers are then able to use this potential energy to kinetically drive their own metabolic processes. Thus, living things depend on a constant supply of energy ultimately provided by the Sun. The ultimate fate of all solar energy in the biosphere is to become randomized in the universe as heat. A living cell can function because it serves as a temporary repository of order, purchased at the cost of a constant flow of energy.

Check Your Progress   6.1

> 1. Discuss where glucose stores its potential energy.
> 2. Appraise how the second law of thermodynamics and entropy may be related to room cleanliness.

6.2 Metabolic Reactions and Energy Transformations

Learning Outcomes

Upon completion of this section, you should be able to

1. Compare the energy associated with endergonic and exergonic reactions.
2. Describe how energy is stored in a molecule of ATP.
3. Examine how cells use ATP to drive energetically unfavorable reactions.

All living things maintain their structure and function through chemical reactions. **Metabolism** is the sum of all the chemical reactions that occur in a cell. **Reactants** are substances that participate in a reaction, while **products** are substances that form as a result of a reaction. In the reaction A + B ⟶ C + D, A and B are the reactants while C and D are the products. Whether a reaction occurs spontaneously—that is, without an input of energy—depends on how much energy is left over after the reaction. Using the concept of entropy or disorder, a reaction occurs spontaneously if it increases the entropy of the universe.

In cell biology, which occurs on a small scale, we are less concerned about the entire universe, which is vast. In such specific instances, cell biologists use the concept of free energy instead of entropy. **Free energy** is the amount of energy left to do work after a chemical reaction has occurred. The change in free energy (also called "delta G," or ΔG) after a reaction occurs is determined by subtracting the free energy content of the reactants from that of the products. A negative result (−ΔG) means that the products have less free energy than the reactants, and the reaction will occur spontaneously. In our reaction, if C and D have less free energy than A and B, then the reaction occurs without additional input of energy.

Metabolism includes both spontaneous reactions and energy-requiring reactions. **Exergonic reactions** are spontaneous and release energy, while **endergonic reactions** require an input of energy to occur. In the body, many reactions, such as protein synthesis, nerve conduction, or muscle contraction, are endergonic. For these nonspontaneous reactions to occur during metabolism, they must be coupled with exergonic reactions, such that a net spontaneous reaction results. Many biological processes use ATP as an energy carrier between exergonic and endergonic reactions.

ATP: Energy for Cells

ATP (adenosine triphosphate) is the common energy currency of cells; when cells require energy, they use ATP. A sedentary oak tree, a flying bat, and a human being require vast amounts of ATP. The more active the organism, the greater the demand for ATP. However, the amount on hand at any one moment is minimal because ATP is constantly being generated from **ADP (adenosine diphosphate)** and a molecule of inorganic phosphate, Ⓟ (Fig. 6.3). A cell is constantly creating a supply of ATP from breakdown of glucose and other biomolecules during cellular respiration. Of all the free energy stored in the bonds of glucose, only 39% of it is transformed to ATP; the rest is lost as heat.

There are many biological advantages to the use of ATP as an energy carrier in living systems. ATP provides a common and universal energy currency because it can be used in many different types of reactions. Also, when ATP is converted to energy, ADP, and Ⓟ, the amount of energy released is sufficient to drive particular biological functions with little waste of energy. In addition, ATP breakdown can be coupled to endergonic reactions in such a way that it minimizes energy loss.

Structure of ATP

ATP is a nucleotide composed of the nitrogen-containing base adenine and the 5-carbon sugar ribose (together called adenosine) and three phosphate groups. The three phosphates of ATP repel each other, creating instability and potential energy (Fig. 6.3). ATP is called a "high-energy" molecule because a phosphate group can be easily removed. Under cellular conditions, the amount of energy released when ATP is hydrolyzed to ADP + Ⓟ is about 7.3 kcal per **mole** (a unit of measurement in chemistry—equal to the molecular weight of a molecule expressed in grams).

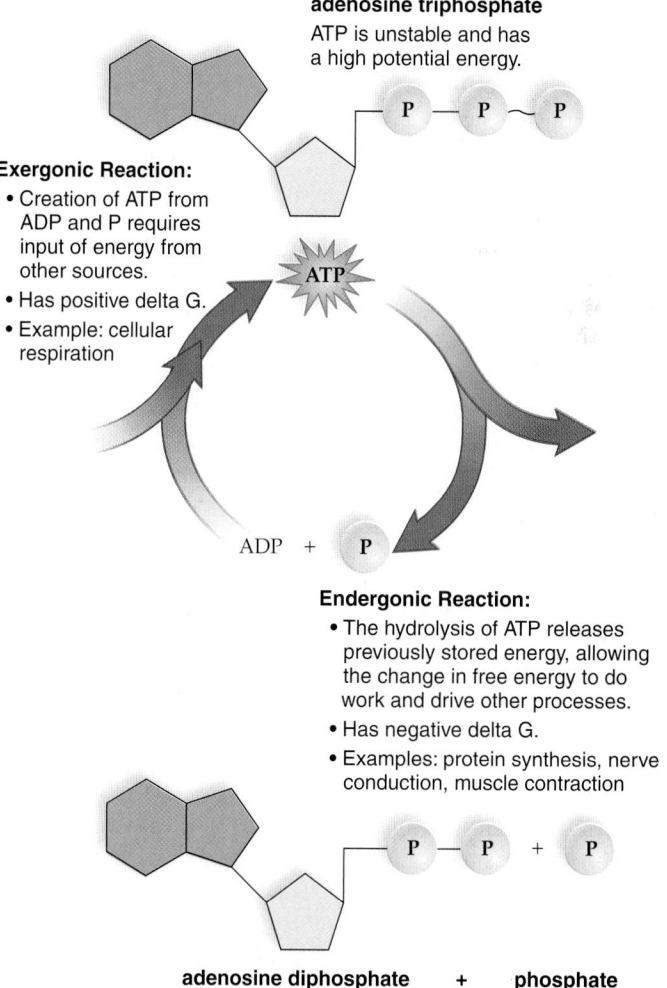

adenosine triphosphate
ATP is unstable and has a high potential energy.

Exergonic Reaction:
- Creation of ATP from ADP and P requires input of energy from other sources.
- Has positive delta G.
- Example: cellular respiration

ATP

ADP + P

Endergonic Reaction:
- The hydrolysis of ATP releases previously stored energy, allowing the change in free energy to do work and drive other processes.
- Has negative delta G.
- Examples: protein synthesis, nerve conduction, muscle contraction

adenosine diphosphate   +   phosphate
ADP is more stable and has lower potential energy than ATP.

Figure 6.3 The ATP cycle. In cells, ATP carries energy between exergonic reactions and endergonic reactions. When a phosphate group is removed by hydrolysis, ATP releases the appropriate amount of energy for most metabolic reactions.

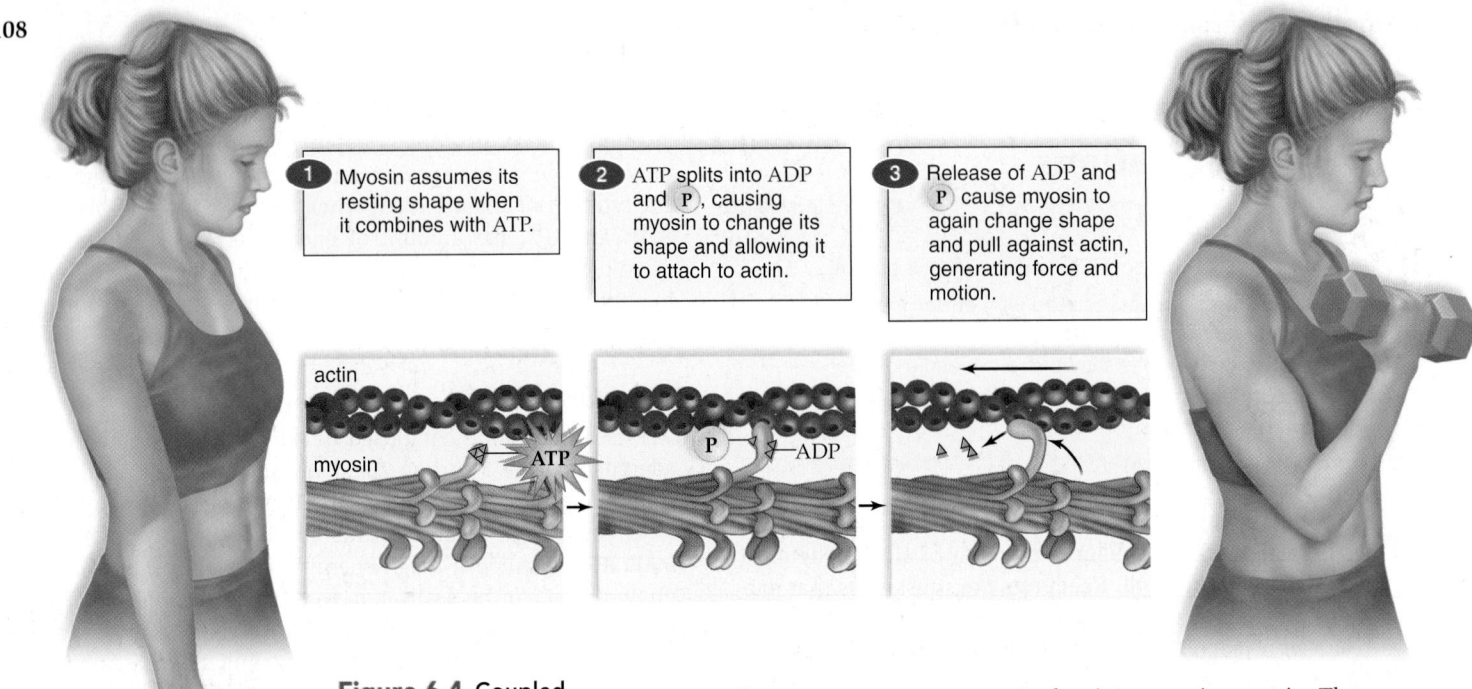

1. Myosin assumes its resting shape when it combines with ATP.

2. ATP splits into ADP and ⓟ, causing myosin to change its shape and allowing it to attach to actin.

3. Release of ADP and ⓟ cause myosin to again change shape and pull against actin, generating force and motion.

actin

myosin

Figure 6.4 Coupled reactions. Muscle contraction occurs only when it is coupled to ATP breakdown.

Coupled Reactions

How can the energy released by ATP hydrolysis be transferred to an endergonic reaction that requires energy and therefore would not ordinarily occur? In other words, how does ATP act as a carrier of chemical energy? How can that energy be transferred efficiently to an energetically unfavorable reaction?

The answer is that ATP breakdown is *coupled* to the energy-requiring reaction, such that both the energetically favorable and unfavorable reactions occur in the same place, at the same time. Usually the energy-releasing reaction is the hydrolysis of ATP. Because the cleavage of ATP's phosphate groups releases more energy than the amount consumed by the energy-requiring reaction, the net reaction is exergonic, entropy increases, and both reactions proceed. The simplest way to represent a coupled reaction is like this:

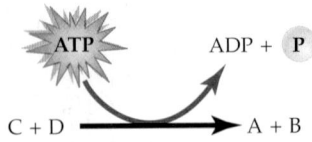

This reaction tells you that coupling occurs, but it does not show how coupling is achieved. A cell has two main ways to couple ATP hydrolysis to an energy-requiring reaction: ATP is used to energize a reactant, or ATP is used to change the shape of a reactant. Both can be achieved by transferring a phosphate group to the reactant so that the product is *phosphorylated.*

For example, when a polar ion moves across the nonpolar plasma membrane of a cell, it requires a carrier protein. In order to make the carrier protein assume a shape conducive to the ion, ATP is hydrolyzed, and then, instead of the last phosphate group floating away, an enzyme attaches it to a carrier protein. The negatively charged phosphate causes the protein to undergo a change in shape that allows it to interact with and move the ion across the membrane. Another example of a coupled reaction is when a polypeptide is synthesized at a ribosome. There, an enzyme transfers a phosphate group from ATP to each amino acid in turn, and this transfer supplies the energy needed to overcome the energy cost associated with bonding one amino acid to another.

Figure 6.4 shows how ATP hydrolysis provides the necessary energy for muscle contraction. During muscle contraction, myosin filaments pull actin filaments to the center of the cell, and the muscle shortens. ❶ A myosin head combines with ATP (three connected green triangles) and takes on its resting shape. ❷ ATP breaks down to ADP (two connected green triangles) plus ⓟ (one green triangle). A resulting change in shape allows myosin to attach to actin. ❸ The release of ADP and ⓟ from the myosin head causes it to change its shape again and pull on the actin filament. The cycle begins again at ❶, when the myosin head combines with ATP and takes on its resting shape. During this cycle, chemical energy has been transformed to mechanical energy, and entropy has increased.

Through coupled reactions, ATP drives forward energetically unfavorable processes that must occur to create the high degree of order and structure essential for life. Macromolecules must be made and organized to form cells and tissues; the internal composition of the cell and the organism must be maintained; and movement of cellular organelles and the organism must occur if life is to continue.

Animation
Breakdown of ATP and Cross-Bridge Movement During Muscle Contraction

Check Your Progress 6.2

1. Explain why ATP is an effective short-term energy storage molecule.
2. Examine how transferring a phosphate from ATP changes a molecule's structure and function.

6.3 Metabolic Pathways and Enzymes

Learning Outcomes

Upon completion of this section, you should be able to

1. Explain the purpose of a metabolic pathway and how enzymes help to regulate it.
2. Examine how enzymes lower activation energy and increase reaction rate.
3. Distinguish between conditions and factors that affect an enzyme's rate of reaction.

The chemical reactions that comprise metabolism would not easily occur without the use of organic catalysts called enzymes. An **enzyme** is a protein molecule that functions to speed a chemical reaction without itself being affected by the reaction. Enzymes allow reactions to occur under mild conditions, and they regulate metabolism, partly by eliminating nonspecific side reactions.

Not all enzymes are proteins. **Ribozymes,** which are made of RNA instead of proteins, can also serve as biological catalysts. Ribozymes are involved in the synthesis of RNA and the synthesis of proteins at ribosomes.

Chemical reactions do not occur haphazardly in healthy cells; they are usually part of a **metabolic pathway,** a series of linked reactions. Metabolic pathways begin with a particular reactant and end with a final product. Many specific steps can be involved in a metabolic pathway, and each step is a chemical reaction catalyzed by an enzyme. The reactants for the first reaction are converted into products, and those products then serve as the reactants for the next enzyme-catalyzed reaction. One reaction leads to the next reaction in an organized, highly regulated manner.

This arrangement makes it possible for one pathway to interact with several others, because different pathways may have several molecules in common. Also, metabolic pathways are useful for releasing and capturing small increments of molecular energy rather than releasing it all at once. Ultimately, enzymes in metabolic pathways enable cells to regulate and respond to changing environmental conditions.

 **Animation** Biochemical Pathways

A metabolic pathway can be represented by the following diagram:

$$\begin{array}{ccccccccccccc} & E_1 & & E_2 & & E_3 & & E_4 & & E_5 & & E_6 & \\ A & \rightarrow & B & \rightarrow & C & \rightarrow & D & \rightarrow & E & \rightarrow & F & \rightarrow & G \end{array}$$

In this diagram, reactant A is converted by enzyme E_1 to product B. Product B then becomes the reactant for enzyme E_2 and is converted to product C. Any one of the molecules (A–G) in this metabolic pathway could also be a reactant in another pathway. A diagram showing all the possibilities would be highly branched.

Each step in the metabolic chain can be regulated because each step requires an enzyme. This gives our cells fine control over how they respond in a changing environment, and helps maximize cell efficiency.

Energy of Activation

Molecules frequently do not react with one another unless they are activated in some way. In the lab, for example, in the absence of an enzyme, molecules may be heated in order to increase the number of effective collisions. The energy that must be added to cause molecules to react with one another is called the **energy of activation (E_a).** Activation energy is essential to keep molecules from spontaneously degrading within the cell. Figure 6.5 shows that an enzyme effectively lowers E_a, thus reducing the energy needed for a chemical reaction to begin. The enzyme has no effect on the energy content of the product; rather, only the reactants are affected.

Animation Energy of Activation

Enzymes allow reactions to occur under mild conditions by bringing reactants into contact with one another in such a way that the chemistry occurs more readily than it otherwise would have.

Enzyme-Substrate Complex

The reactants in an enzymatic reaction are called the **substrates** for that enzyme. Considering the metabolic pathway shown previously, A is the substrate for E_1, and B is the product. Now B becomes the substrate for E_2, and C is the product. This process continues until the final product G forms.

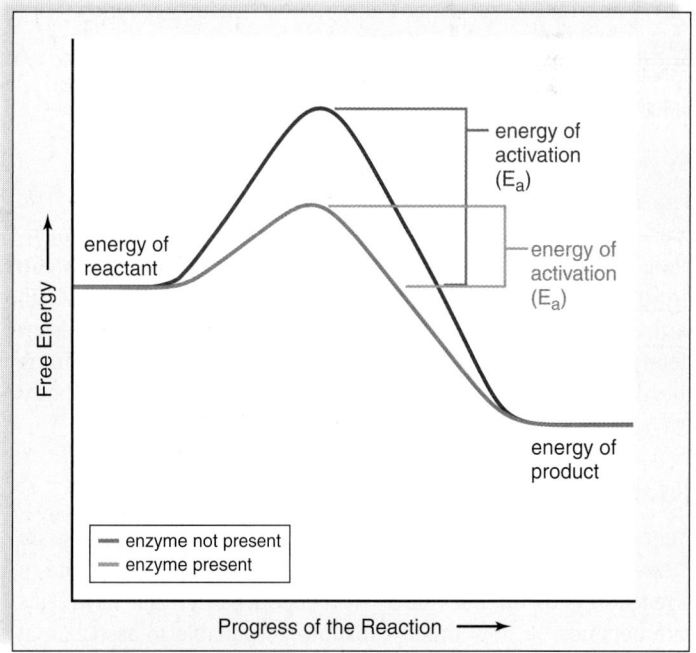

Figure 6.5 Energy of activation (E_a). Enzymes speed the rate of reactions because they lower the amount of energy required for the reactants to activate. Even spontaneous reactions like this one, in which the energy of the product is less than the energy of the reactant, speed up when an enzyme is present.

The following equation, which is illustrated in Figure 6.6, is often used to indicate that an enzyme forms a complex with its substrate:

$$\text{E} + \text{S} \rightarrow \text{ES} \rightarrow \text{E} + \text{P}$$

| enzyme | substrate | enzyme-substrate complex | enzyme | product |

In most instances, only one small part of the enzyme, called the **active site,** associates directly with the substrate(s). In the active site, the enzyme and substrate are positioned in such a way that they more easily fit together, seemingly like a key fits a lock. However, an active site differs from a lock and key because it undergoes a slight change in shape to accommodate the substrate(s). This is called the **induced fit model** because the enzyme is induced to undergo a slight alteration to achieve optimum fit for the substrates.

 **Animation** How Enzymes Work

The change in shape of the active site facilitates the reaction that now occurs. After the reaction has been completed, the product(s) is released, and the active site returns to its original state, ready to bind to another substrate molecule. Only a small amount of enzyme is actually needed in a cell because enzymes are not used up by the reaction; they merely enable it to happen more quickly.

Some enzymes do more than simply form a complex with their substrate(s); they participate in the reaction. Trypsin digests protein by breaking peptide bonds. The active site of trypsin contains three amino acids with *R* groups that actually interact with members of the peptide bond—first to break the bond and then to introduce the components of water. This illustrates that the formation of the enzyme-substrate complex is very important in speeding the reaction. Because enzymes bind only with their substrates, they are sometimes named for their substrates, and usually end in *-ase.* For example, lipase is involved in hydrolyzing lipids.

MP3 Enzymes

Regulation of Metabolism

The specificity of enzymes allows regulation of metabolism. The presence of particular enzymes helps determine which metabolic pathways are operative. In addition, some reactants can produce more than one type of product, depending upon which pathway is open to them. Therefore, which enzyme is present determines which product is produced as well as determining the direction of metabolism, without several alternative pathways being activated.

Factors Affecting Enzymatic Speed

Generally, enzymes work quickly, and in some instances they can increase the reaction rate more than 10 million times. The rate of a reaction is the amount of product produced per unit time. This rate depends on how much substrate is available to associate at the active sites of enzymes. Therefore, increasing the amount of substrate and also the amount of enzyme can increase the rate of the reaction. Any factor that alters the shape of the active site, such as pH or temperature or an inhibitor, can decrease the rate of a reaction. Thus, enzymes require specific conditions to be met in order to be fully operational. In fact, some enzymes require

additional molecules called cofactors that help speed the rate of the reaction because they help bind the substrate to the active site, or they participate in the reaction at the active site.

Substrate Concentration

Molecules must collide to react. Generally, enzyme activity increases as substrate concentration increases because there are more collisions between substrate molecules and the enzyme. As more substrate molecules fill active sites, more product results per unit time. But when the active sites are filled almost continuously with substrate, the rate of the reaction can no longer increase. Maximum rate has been reached.

Just as the amount of substrate can increase or limit the rate of an enzymatic reaction, so the amount of active enzyme can also increase or limit the rate of an enzymatic reaction. Therefore, sufficient concentrations of substrate and enzymes are necessary to achieve maximum reaction rate.

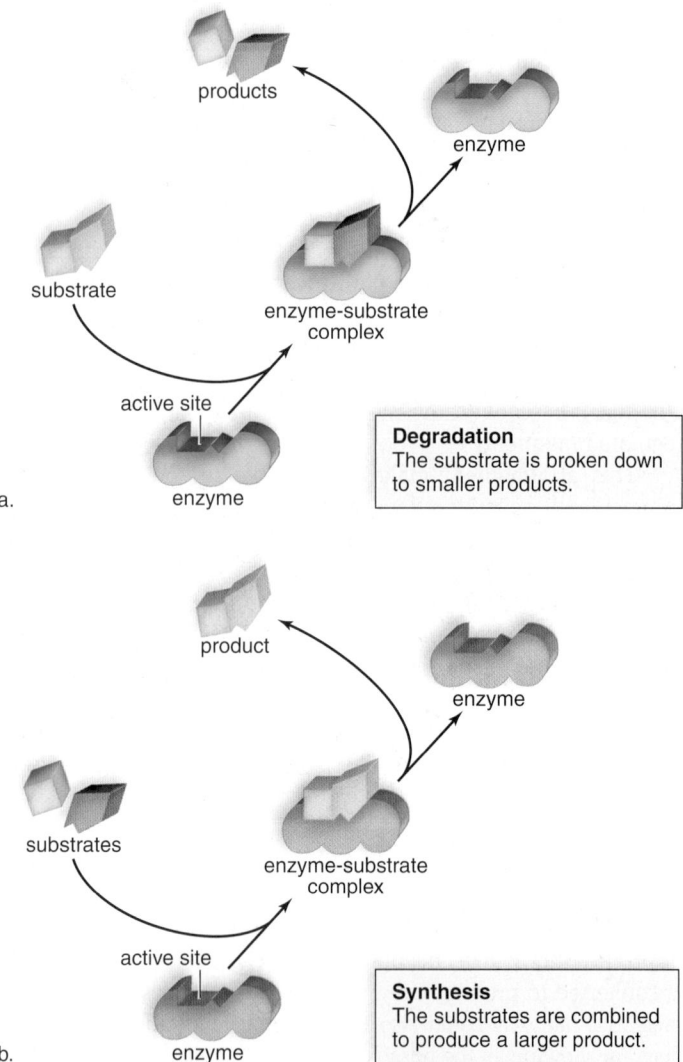

Degradation The substrate is broken down to smaller products.

Synthesis The substrates are combined to produce a larger product.

Figure 6.6 Enzymatic actions. Enzymes have an active site where the substrate(s) specifically fit together so the reaction will occur. Following the reaction, the product(s) is released, and the enzyme is free to act again. Certain enzymes carry out **(a)** degradation and others carry out **(b)** synthesis.

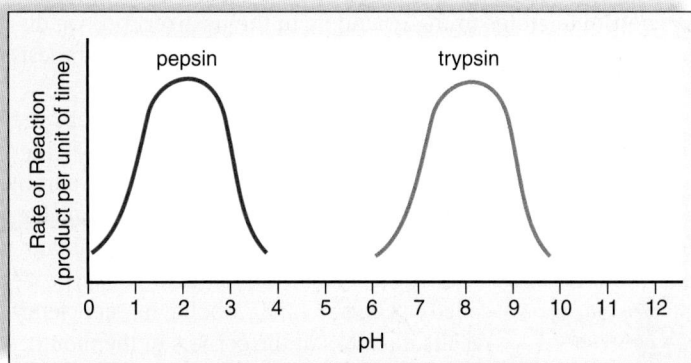

Figure 6.7 The effect of pH on rate of reaction. The optimal pH for pepsin, an enzyme that acts in the stomach, is about 2, while the optimal pH for trypsin, an enzyme that acts in the small intestine, is about 8. Enzyme shape is best maintained at the optimal pH, which allows it to best function and bind with its substrates.

Optimal pH

Each enzyme also has an optimal pH at which the reaction rate is highest. Figure 6.7 shows the optimal pH for the enzymes pepsin and trypsin. At their respective pH values, each enzyme can maintain its normal structural configuration, which enables optimum function. The globular structure of an enzyme is dependent on interactions, such as hydrogen bonding, between R groups. A change in pH can alter the ionization of these side chains and disrupt normal interactions, and under extreme conditions of pH, the enzyme loses its structure and becomes inactive. Inactivity occurs because the enzyme has an altered shape and is then unable to combine efficiently with its substrate.

Temperature

Typically, as temperature rises, enzyme activity increases (Fig. 6.8a). This occurs because warmer temperatures cause more effective collisions between enzyme and substrate. The body temperature of an animal seems to affect whether it is normally active or inactive (Fig. 6.8b, c). It has been suggested that mammals are more prevalent today than reptiles because they maintain a warm internal temperature that allows their enzymes to work at a rapid rate.

In the laboratory and in your body, if the temperature rises beyond a certain point, enzyme activity eventually levels out and then declines rapidly because the enzyme is **denatured.** An enzyme's shape changes during denaturation, and then it can no longer bind its substrate(s) efficiently.

Exceptions to this generalization do occur. For example, some prokaryotes can live in hot springs because their enzymes do not denature. These organisms are responsible for the brilliant colors of the hot springs. Another exception involves the coat color of animals. Siamese cats have inherited a mutation that causes an enzyme to be active only at cooler body temperatures. The enzyme's activity causes the cooler regions of the body—the face, ears, legs, and tail—to be dark in color (Fig. 6.9). The coat color pattern in several other animals can be explained similarly.

Figure 6.9 The effect of temperature on enzymes. Siamese cats have inherited a mutation that causes an enzyme to be active only at cooler body temperatures. Therefore, only certain regions of the body are dark in color.

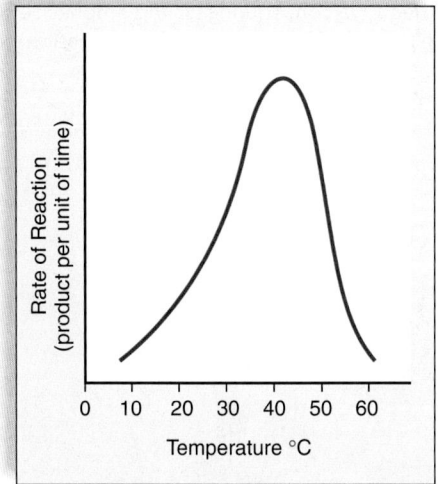

a. Rate of reaction as a function of temperature

b. Body temperature of ectothermic animals often limits rates of reactions.

c. Body temperature of endothermic animals promotes rates of reactions.

Figure 6.8 The effect of temperature on rate of reaction. **a.** Usually, the rate of an enzymatic reaction doubles with every 10°C rise in temperature. This enzymatic reaction is maximum at about 40°C; then it decreases until the reaction stops altogether, because the enzyme has become denatured. **b.** The body temperature of ectothermic animals, such as an iguana, which take on the temperature of their environment, often limits rates of reactions. **c.** The body temperature of endothermic animals, such as a polar bear, promotes rates of reaction.

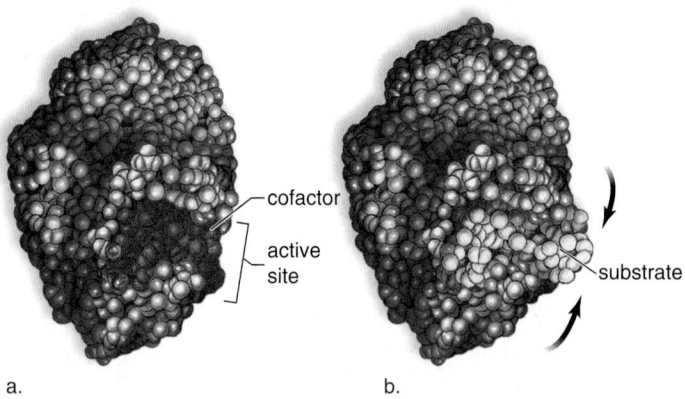

a.   b.

Figure 6.10 Cofactors at active site.   a. Cofactors, including inorganic ions and organic coenzymes, may participate in the reaction at the active site (**b**).

Enzyme Cofactors

Many enzymes require the presence of an inorganic ion or nonprotein organic molecule at the active site in order to work properly; these necessary ions or molecules are called **cofactors** (Fig. 6.10). The inorganic ions include metals such as copper, zinc, or iron. The nonprotein organic molecules are called **coenzymes.** These cofactors participate in the reaction and may even accept or contribute atoms to the reactions. In the next section, we discuss two coenzymes that play significant roles in photosynthesis and cellular respiration, respectively.

Vitamins are relatively small organic molecules that are required in trace amounts in our diet and in the diets of other animals for synthesis of coenzymes. The vitamin becomes part of a coenzyme's molecular structure. If a vitamin is not available, enzymatic activity will decrease, and the result will be a vitamin-deficiency disorder: Niacin deficiency results in a skin disease called pellagra, and riboflavin deficiency results in cracks at the corners of the mouth.

**Animation
B Vitamins**

Enzyme Inhibition

Sometimes it is necessary to limit enzyme activity. **Enzyme inhibition** occurs when a molecule (the inhibitor) binds to an enzyme and decreases its activity. ❶ In Figure 6.11, F is the end product of a metabolic pathway that can act as an inhibitor. This type of inhibition is beneficial because once sufficient end product is present, inhibiting further production can conserve raw materials and energy.

❷ Figure 6.11 also illustrates **noncompetitive inhibition** because the inhibitor (F, the end product) binds to the enzyme E_1 at a location other than the active site. The site is called an **allosteric site.** When an inhibitor is at the allosteric site, the active site of the enzyme changes shape, which in turn changes its function.

Nature of Science

Enzyme Inhibitors Can Spell Death

Cyanide gas was formerly used to execute people. How did it work? Cyanide can be fatal because it binds to a mitochondrial enzyme necessary for the production of ATP. MPTP (1-methyl-4-phenyl-1,2,3.6-tetrahydropyridine) is another enzyme inhibitor that stops mitochondria from producing ATP. The toxic nature of MPTP was discovered in the early 1980s, when a group of intravenous drug users in California suddenly developed symptoms of Parkinson disease, including uncontrollable tremors and rigidity. All of the drug users had injected a synthetic form of heroin that was contaminated with MPTP. Parkinson disease is characterized by the death of brain cells, the very ones that are also destroyed by MPTP.

Sarin is a chemical that inhibits an enzyme at neuromuscular junctions, where nerves stimulate muscles. When the enzyme is inhibited, the signal for muscle contraction cannot be turned off, so the muscles are unable to relax and become paralyzed. Sarin can be fatal if the muscles needed for breathing become paralyzed. In 1995, terrorists released sarin gas on a subway in Japan (Fig. 6A). Although many people developed symptoms, only 17 died.

A fungus that contaminates and causes spoilage of sweet clover produces a chemical called warfarin. Cattle that eat the spoiled feed die from internal bleeding because warfarin inhibits a crucial enzyme for blood clotting. Today, warfarin is widely used as a rat poison. Unfortunately, it is not uncommon for warfarin to be mistakenly eaten by pets and even very small children, with tragic results.

Many people are prescribed a medicine called Coumadin to prevent inappropriate blood clotting. For example, those who have received an artificial heart valve need such a medication. Coumadin contains a nonlethal dose of warfarin.

These examples all show how our understanding of science can have positive or negative consequences, and emphasize the role of ethics in scientific investigation.

Questions to Consider

1. Do you feel it is ethical to use dangerous chemicals to treat diseases?
2. Under what conditions do you think they should be used?

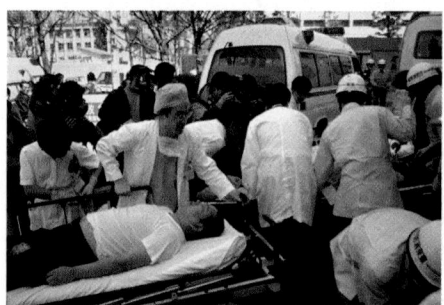

Figure 6A Sarin gas.   The aftermath when sarin, a nerve gas that results in the inability to breathe, was released by terrorists in a Japanese subway in 1995.

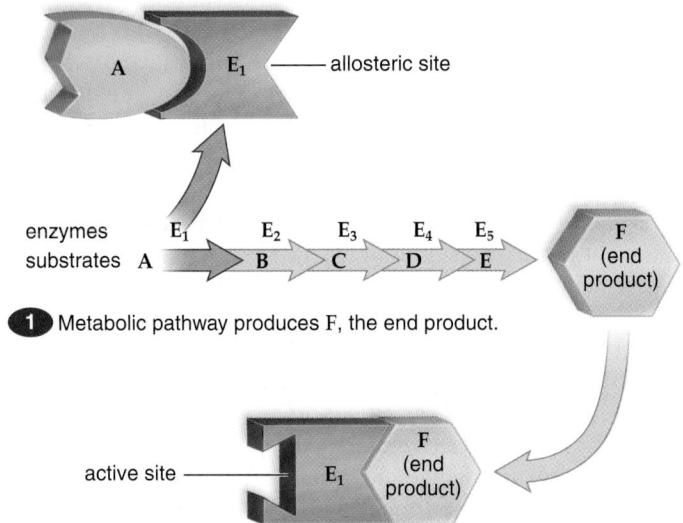

① Metabolic pathway produces F, the end product.

② F binds to allosteric site and the active site of E_1 changes shape.

③ A cannot bind to E_1; the enzyme has been inhibited by F.

Figure 6.11 Noncompetitive inhibition of an enzyme.
In the pathway, A–E are substrates, E_1–E_5 are enzymes, and F is the end product of the pathway that inhibits enzyme E_1. This negative feedback is useful because it prevents wasteful production of product F when it is not needed.

③ The enzyme E_1 is inhibited because it is unable to bind to A, its substrate. The inhibition of E_1 means that the metabolic pathway is inhibited and no more end product is produced, until conditions change and more end product is needed.

In contrast to noncompetitive inhibition, **competitive inhibition** occurs when an inhibitor and the substrate compete for the active site of an enzyme. Product forms only when the substrate, not the inhibitor, is at the active site. In this way, the amount of product is regulated.

Normally, enzyme inhibition is reversible, and the enzyme is not damaged by being inhibited. When enzyme inhibition is irreversible, the inhibitor permanently inactivates or destroys an enzyme.

Animation
Feedback Inhibition of Biochemical Pathways

Check Your Progress   6.3

1. Explain how enzymes maintain specificity in metabolic pathways.
2. Evaluate the usefulness of cofactors in enzyme regulation.

6.4 Organelles and the Flow of Energy

Learning Outcomes

Upon completion of this section, you should be able to
1. Compare the role of carbon dioxide, water, and oxygen in photosynthesis and cellular respiration.
2. Describe how high energy electrons are captured by NAD^+ and used to do work.
3. Explain how electrochemical gradients are used to produce ATP.

Two organelles are particularly involved in the flow of energy from the Sun through all living things. Photosynthesis, a process that captures solar energy to produce carbohydrates, takes place in chloroplasts. Cellular respiration, which breaks down carbohydrates and produces chemical energy, takes place in mitochondria.

Photosynthesis

The overall reaction for photosynthesis can be written like this:

$$6\ CO_2\ +\ 6\ H_2O\ +\ energy \longrightarrow C_6H_{12}O_6\ +\ 6\ O_2$$
carbon dioxide   water   glucose   oxygen

This equation shows that hydrogen atoms ($H^+ + e^-$) are transferred from water to carbon dioxide, when glucose is formed. This transfer of electrons from one molecule to another is called an **oxidation-reduction reaction. Oxidation** is defined as the loss of electrons and **reduction** is the gain of electrons. Therefore, water has been oxidized and carbon dioxide has been reduced. This reaction is also referred to as a **redox reaction.**

Animation
Redox

The creation of a glucose molecule requires input of a lot of energy. Chloroplasts are able to capture solar energy and convert it by way of an electron transport chain (discussed on next page) to the chemical energy of ATP molecules. ATP is then used along with hydrogen atoms to reduce carbon dioxide to glucose.

Glucose production also requires a high-energy electron-carrier molecule, or coenzyme, of oxidation-reduction called **NADP$^+$ (nicotinamide adenine dinucleotide phosphate)**. This molecule carries a positive charge, and therefore is written as $NADP^+$. During photosynthesis, $NADP^+$ accepts electrons plus a hydrogen ion derived from water, and it later passes them along by way of a metabolic pathway to reduce carbon dioxide and form glucose. The reaction that reduces $NADP^+$ is:

$$NADP^+ + 2\ e^- + H^+ \longrightarrow NADPH$$

Cellular Respiration

The overall equation for cellular respiration is opposite to that for photosynthesis:

$$C_6H_{12}O_6 \quad + \quad 6\,O_2 \quad \longrightarrow \quad 6\,CO_2 \quad + \quad 6\,H_2O \quad + \quad \text{energy}$$

$$\text{glucose} \qquad \text{oxygen} \qquad \text{carbon} \qquad \text{water}$$
$$\text{dioxide}$$

In this reaction, glucose has lost hydrogen atoms (been oxidized), and oxygen has gained hydrogen atoms (been reduced). The hydrogen atoms that were formerly bonded to carbon are now bonded to oxygen. Glucose is a high-energy molecule, whereas its breakdown products, carbon dioxide and water, are low-energy molecules; therefore, energy is released. Mitochondria use the energy released from glucose breakdown to build ATP molecules by way of an electron transport chain, as depicted in Figure 6.12.

In metabolic pathways, most oxidations such as those that occur during cellular respiration involve a coenzyme called **NAD$^+$ (nicotinamide adenine dinucleotide).** This molecule carries a positive charge, and therefore is represented as NAD$^+$. During oxidation reactions, NAD$^+$ accepts two electrons but only one hydrogen ion. The reaction that reduces NAD$^+$ is:

Animation
How NAD$^+$ Works

$$NAD^+ + 2\,e^- + H^+ \longrightarrow NADH$$

Electron Transport Chain

As previously mentioned, chloroplasts use solar energy to generate ATP, and mitochondria use energy stored in glucose and other biomolecules to generate ATP by way of an electron

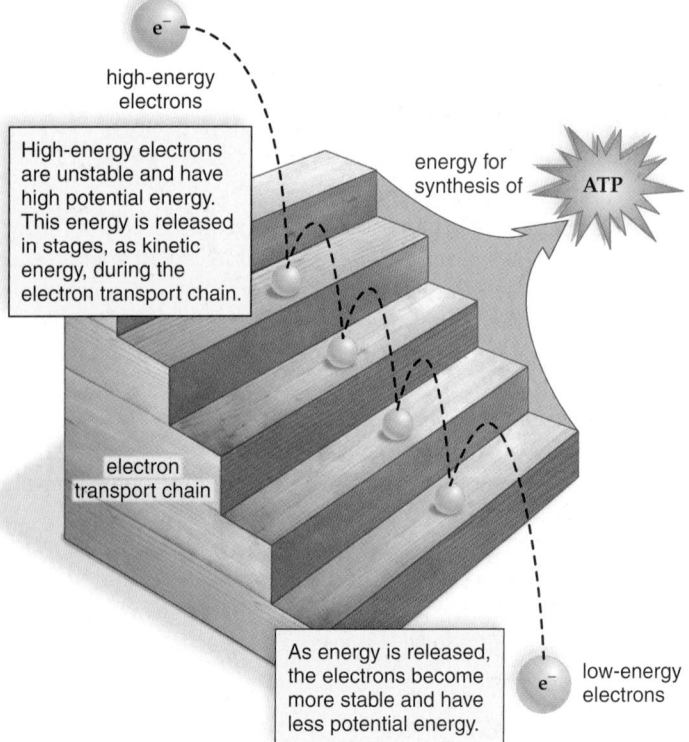

high-energy
electrons

High-energy electrons
are unstable and have
high potential energy.
This energy is released
in stages, as kinetic
energy, during the
electron transport chain.

energy for
synthesis of   **ATP**

electron
transport chain

As energy is released,
the electrons become
more stable and have
less potential energy.

e$^-$ low-energy
electrons

Figure 6.12 Electron transport chain. With each redox step, high energy electrons release energy that is used for ATP production.

transport chain. An **electron transport chain (ETC)** is a series of membrane-bound carriers that pass electrons from one carrier to another via redox reactions. High-energy electrons are delivered to the chain, and low-energy electrons leave it. As an analogy, if a hot potato is passed from one person to another, it loses heat with each transfer. In the same manner, each time electrons are transferred to a new carrier, energy is released. However, unlike in the hot potato transfer example, the cell is able to capture the released energy and use it to produce ATP molecules (Fig. 6.12).

In certain redox reactions, the result is release of energy, and in others, energy is required. In an ETC, each carrier is reduced and then oxidized in turn. The overall effect of oxidation-reduction as electrons are passed from carrier to carrier of the electron transport chain is the release of energy that is ultimately used for ATP production.

Animation
Electron Transport
Chain and ATP
Synthesis

ATP Production

For many years, scientists knew that ATP synthesis was somehow coupled to the ETC, but the exact mechanism could not be determined. Peter Mitchell, a British biochemist, received a Nobel Prize in 1978 for his theory of ATP production in both mitochondria and chloroplasts.

In chloroplasts and mitochondria, the proteins and other molecules of the ETC are located within a membrane: thylakoid membranes in chloroplasts, and cristae in mitochondria. Hydrogen ions (H$^+$), which are often referred to as protons in this context, are in low concentration on one side of the membrane and are pumped to the other side of the membrane. This process requires energy because certain carriers of the electron transport chain move these protons against their diffusion gradient. This establishes a larger electrochemical gradient across the membrane, and the potential energy of this gradient can then be used to do work, becoming kinetic energy for ATP production. Protons can move back across the membrane by interacting with other enzyme complexes in the membrane called **ATP synthases.** Each ATP synthase contains a channel that allows hydrogen ions to flow down their electrochemical gradient. The flow of hydrogen ions through the channel provides the kinetic energy for the ATP synthase enzyme to produce ATP from ADP + Ⓟ (Fig. 6.13). The production of ATP due to a hydrogen ion gradient across a membrane is called **chemiosmosis** [Gk. *osmos*, push].

Consider this analogy to understand chemiosmosis: The Sun's rays evaporate water from the seas and help create the winds that blow clouds to the mountains, where water falls in the form of rain and snow. The water in a mountain reservoir has a higher potential energy than water in the ocean. The potential energy is converted to electrical energy when water is released and used to turn turbines in an electrochemical dam before it makes its way to the ocean. The continual release of water results in a continual production of electricity.

Likewise, during photosynthesis, the solar energy collected by chloroplasts leads to continuous ATP production. Energized electrons lead to the pumping of hydrogen ions across a thylakoid membrane, which acts like a dam to retain them. The

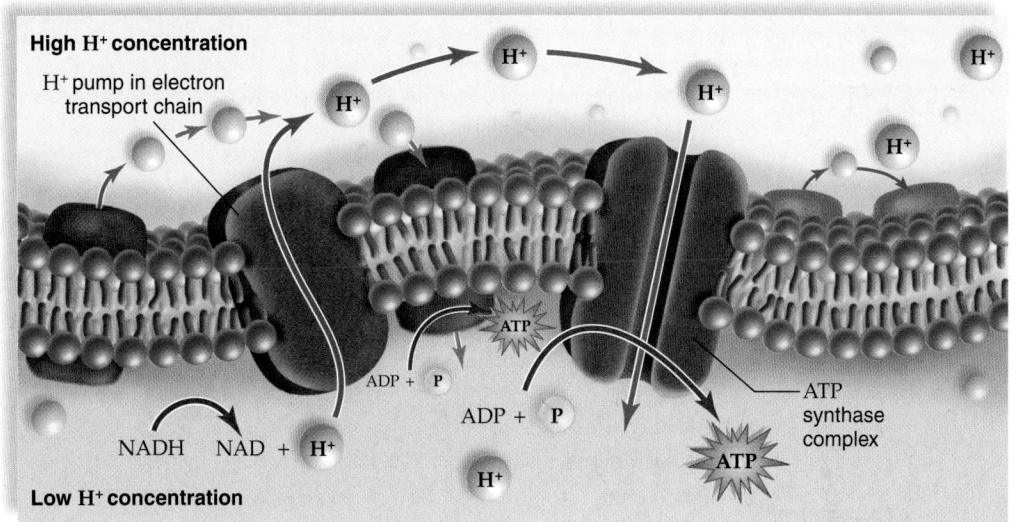

Figure 6.13 Chemiosmosis.
Carriers in the electron transport chain use energy from redox reactions to pump hydrogen ions (H⁺) across a membrane, creating greater potential energy. When the hydrogen ions flow back across the membrane through an ATP synthase complex, ATP is synthesized. This gradient-based chemiosmosis occurs in chloroplasts and mitochondria.

hydrogen ions flow through the channel of an ATP synthase complex. This complex couples the flow of hydrogen ions to the formation of ATP, just as the turbines in a hydroelectric dam system couple the flow of water to the production of electricity.

Similarly during cellular respiration, glucose breakdown provides the energy to establish a hydrogen ion gradient across the cristae of mitochondria. And again, hydrogen ions flow through the channel within an ATP synthase complex that couples the flow of hydrogen ions to the production of ATP.

Check Your Progress 6.4

1. Compare the role of carbon dioxide in photosynthesis and cellular respiration.
2. Distinguish how energy from electrons is used to establish an electrochemical gradient in chloroplasts and mitochondria.

CONNECTING *the* CONCEPTS *with the* BIG IDEAS

Energy and Homeostasis

- The workings of the laws of thermodynamics are observed throughout the biosphere as solar energy is transformed to chemical energy which later provides free energy to run all aspects of cellular workings. (2A commentary, 2A2)
- The move toward entropy (disorder) is constant, so living organisms must try to maintain order and homeostasis by free energy input. Excess energy can be used for growth or storage, while deficits of free energy result in deterioration and eventually death. (2A1a-b, 2A1b2, 2A1d4-5)
- Free energy for food production can be captured from sunlight by photosynthesizers while chemosynthetizers use energy from inorganic molecules. Heterotrophs must metabolize food produced by these autotrophs in cellular respiration to obtain their free energy. (2A2a1-2, 2A2b)
- Biological pathways dealing with energy have specific sequences with multiple-entry points, as seen in glycolysis, fermentation, the Krebs and Calvin cycles. (2A1c*IE*)

Interactions and Systems

- An enzyme's shape dictates the substrate it will "induced fit" with, lowering the activation energy for the particular reaction; cofactors and coenzymes may enhance enzyme function by favorably adjusting its shape. Enhancing or inhibiting molecules can also attach reversibly or irreversibly to enzymes, affecting their activity. (4B1b1-2, 4B1c)
- The work of enzymes can be tracked by analysis of amount of substrate or product present vs. time. (4B1d)

*Find the unabridged version of all EK citations at www.glencoe.com/maderAP11.

www.glencoe.com/maderAP11

Enhance your study of this chapter with study tools and practice tests. Also ask your instructor about the resources available through ConnectPlus, including the media-rich eBook, interactive learning tools, and animations.

Summarize

6.1 Cells and the Flow of Energy

Two energy laws are basic to understanding patterns of energy throughout biology. The first law of thermodynamics states that energy cannot be created or destroyed, but can only be transferred or transformed. The second law of thermodynamics states that one usable form of energy cannot be completely converted into another usable form. As a result of these laws, we know that the entropy of the universe is increasing and that only a flow of energy from the Sun maintains the organization of living things.

6.2 Metabolic Reactions and Energy Transformations

The term *metabolism* encompasses all the chemical reactions occurring in a cell. Considering individual reactions, only those that result in a negative free-energy difference—that is, the products have less usable energy than the reactants—occur spontaneously. Such reactions, called exergonic reactions, release energy.

An endergonic reaction, which requires an input of energy, occurs only when coupled with an exergonic process. The energy released from the exergonic reaction must exceed the net energy cost of the endergonic reaction in order for the overall reaction to proceed. Hydrolysis of ATP, an exergonic reaction, is commonly used to drive energetically unfavorable metabolic reactions.

6.3 Metabolic Pathways and Enzymes

A metabolic pathway is a series of reactions that proceed in an orderly, step-by-step manner. Enzymes speed reactions by lowering the energy of activation when they form a complex with their substrates. Enzymes regulate metabolism because, in general, no reaction occurs unless its enzyme is present. Which enzymes are present determine which metabolic pathways will be utilized.

Generally, enzyme activity increases as substrate concentration increases; once all active sites are filled, maximum reaction rate has been achieved. Environmental factors such as temperature or pH can affect the shape of an enzyme, and therefore its function. Many enzymes need cofactors or coenzymes to carry out their reactions. The activity of most metabolic pathways is regulated by feedback inhibition.

6.4 Organelles and the Flow of Energy

A flow of energy occurs through organisms because (1) photosynthesis in chloroplasts captures solar energy and produces carbohydrates, and (2) cellular respiration in mitochondria breaks down these carbohydrates to produce ATP molecules, which (3) provide energy

for metabolic reactions. The overall equation for photosynthesis is opposite that for cellular respiration.

Both processes use an electron transport chain in which electrons are transferred from one carrier to the next. Each transfer releases energy that is ultimately used to produce ATP molecules. Chemiosmosis explains how the electron transport chain produces ATP. The carriers of this system actively pump hydrogen ions (H^+) from one side of a membrane to the other, thereby creating a larger gradient. This larger electrochemical gradient pushes hydrogen ions through an ATP synthase complex, which makes ATP from ADP and $\textcircled{P}$.

Key Terms

Assess

Reviewing This Chapter

1. State the first law of thermodynamics, and give an example.   105
2. State the second law of thermodynamics, and give an example.   106
3. Explain why the entropy of the universe is always increasing and why an organized system such as an organism requires a constant input of useful energy.   106
4. What is the difference between exergonic reactions and endergonic reactions? Why can exergonic but not endergonic reactions occur spontaneously?   107
5. Why is ATP called the energy currency of cells? What is the ATP cycle?   107
6. Define coupling, and write an equation that shows an endergonic reaction being coupled to ATP breakdown.   108
7. Diagram a metabolic pathway. Label the reactants, products, and enzymes. Explain how enzymes regulate metabolism.   109–10

8. Why is less energy needed for a reaction to occur when an enzyme is present?   109
9. Why are enzymes specific, and why can't each one speed many different reactions?   110
10. Name and explain the manner in which at least three environmental factors can influence the speed of an enzymatic reaction. How do cells regulate the activity of enzymes?   111–12
11. What are cofactors and coenzymes?   112
12. Compare and contrast competitive and noncompetitive inhibition.   112–113
13. How do chloroplasts and mitochondria permit a flow of energy through all organisms. What role is played by oxidation and reduction?   113–14
14. Describe an electron transport chain.   114
15. Tell how cells form ATP during chemiosmosis.   114–15

Testing Yourself

Choose the best answer for each question.

1. A form of potential energy is
   a. a boulder at the top of a hill.
   b. the bonds of a glucose molecule.
   c. a starch molecule.
   d. stored fat tissue.
   e. All of these are correct.

2. Consider this reaction: A + B ⟶ C + D + energy.
   a. This reaction is exergonic.
   b. An enzyme could still speed the reaction.
   c. ATP is not needed to make the reaction go.
   d. A and B are reactants; C and D are products.
   e. All of these are correct.

3. The active site of an enzyme
   a. is similar to that of any other enzyme.
   b. is the part of the enzyme where its substrate can fit.
   c. can be used over and over again.
   d. is not affected by environmental factors, such as pH and temperature.
   e. Both b and c are correct.

4. If you want to increase the amount of product per unit time of an enzymatic reaction, do not increase the
   a. amount of substrate.
   b. amount of enzyme.
   c. temperature somewhat.
   d. pH.
   e. All of these are correct.

5. An allosteric site on an enzyme is
   a. the same as the active site.
   b. nonprotein in nature.
   c. where ATP attaches and gives up its energy.
   d. often involved in feedback inhibition.
   e. All of these are correct.

6. During photosynthesis, carbon dioxide
   a. is oxidized to oxygen.
   b. is reduced to glucose.
   c. gives up water to the environment.
   d. is a coenzyme of oxidation-reduction.
   e. All of these are correct.

7. Use these terms to label the following diagram: substrates, enzyme (used twice), active site, product, and enzyme-substrate complex. Explain the importance of an enzyme's shape to its activity.

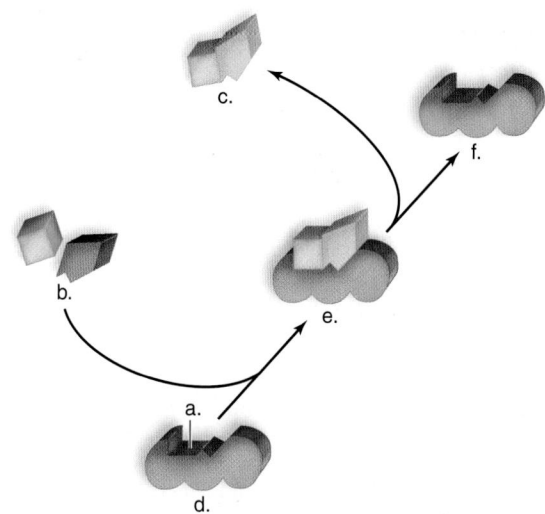

8. Coenzymes
   a. have specific functions in reactions.
   b. have an active site just as enzymes do.
   c. can be carriers for proteins.
   d. always have a phosphate group.
   e. are used in photosynthesis, but not in cellular respiration.

For questions 9–14 match each description to a process in the key.

KEY:

   a. photosynthesis
   b. cellular respiration
   c. Both
   d. Neither

9. captures solar energy
10. requires enzymes and coenzymes
11. releases CO_2 and H_2O
12. utilizes an electron transport chain
13. transforms one form of energy into another form with the release of heat
14. creates energy for the living world

15. Oxidation
   a. is the opposite of reduction.
   b. sometimes uses NAD^+.
   c. is involved in cellular respiration.
   d. occurs when ATP goes to ADP + ⓅP.
   e. All of these but d are correct.

16. NAD^+ is the _____ form, and when it later becomes NADH, it is said to be _____.
   a. reduced, oxidized
   b. neutral, a coenzyme
   c. oxidized, reduced
   d. active, denatured

17. Electron transport chains
    a. are found in both mitochondria and chloroplasts.
    b. release energy as electrons are transferred.
    c. are involved in the production of ATP.
    d. are located in a membrane.
    e. All of these are correct.

18. Chemiosmosis is dependent on
    a. the diffusion of water across a differentially permeable membrane.
    b. an outside supply of phosphate and other chemicals.
    c. the establishment of an electrochemical hydrogen ion (H^+) gradient.
    d. the ability of ADP to join with $\textcircled{P}$ even in the absence of a supply of energy.
    e. All of these are correct.

19. The difference between NAD^+ and $NADP^+$ is that
    a. only NAD^+ production requires niacin in the diet.
    b. one is an organic molecule, and the other is inorganic because it contains phosphate.
    c. one carries electrons to the electron transport chain, and the other carries them to synthetic reactions.
    d. one is involved in cellular respiration, and the other is involved in photosynthesis.
    e. Both c and d are correct.

20. Label this diagram describing chemiosmosis.

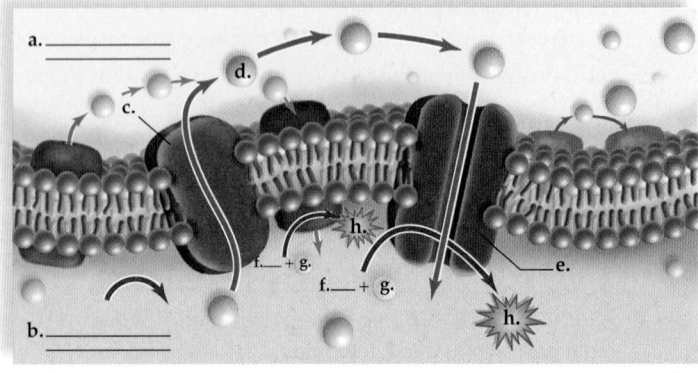

Engage

Virtual Lab
Enzyme Controlled Reactions

The virtual lab "Enzyme Controlled Reactions" provides an interactive examination of many of the factors that regulate enzymatic activity.

Thinking Scientifically

1. A flower generates heat in order to attract pollinating insects. Why might the flower break down a sugar and not ATP to produce heat?

2. You decide to calculate how much energy is released when sucrose is broken down by a flower and run into complications because you have to first heat the sucrose before it breaks down. Explain why this complication is not a problem for the flower.

Plants appear green because chlorophyll reflects green light (*left*). Otherwise, plants would be black (*right*).

Photosynthesis

Sunlight, the kind that shines down on us every day, contains many different colors, from violet to green, yellow, orange, and red. Plants use all the colors of light, except green, when they photosynthesize—that's why most plants are green! They are reflecting the green light back to our eyes. Does this mean that if plants used green light, in addition to all the other colors, they would appear black to us? In fact, yes. As a result natural areas like the one pictured above would be black, as shown on the right.

How did it happen that plants do not use green light for photosynthesis? When the ancestors of plants evolved in the ocean, green light was already being absorbed by other photosynthesizers, so natural selection favored the evolution of a pigment such as chlorophyll, which does not absorb green light. When plants moved onto land, chlorophyll remained as the primary pigment, but because there is plentiful sunlight on land, the selective pressure relaxed, allowing for the evolution of additional pigments.

Photosynthesis, as discussed in this chapter, consists of two interconnected pathways that allow chloroplasts to produce carbohydrate while releasing oxygen. This remarkable process is the basis for much life on Earth.

As you read through the chapter, think about the following questions:

1. Why are plants so important to most life on Earth?

2. What relationships do heterotrophs have with the photosynthesizing autotrophs?

3. How might the increased destruction of plants affect our environment, climate, and lives in the future?

BEFORE YOU BEGIN

Before beginning this chapter, take a few moments to review the following discussions.

Figure 6.1 How does energy flow in biological systems?

Section 6.3 What role do enzymes play in regulating metabolic processes?

Section 6.4 How are redox reactions and membranes used to conduct cellular work?

FOLLOWING *the* BIG IDEAS

CHAPTER 7 PHOTOSYNTHESIS

Evolution	Prokaryotes were the first to develop photosynthesis, adding oxygen to the ancient atmosphere. Some of these autotrophic creatures may survive as the chloroplasts of modern eukaryotes.
Energy and Homeostasis	Energy does not cycle, and therefore all life is dependent on the ability of autotrophs to capture solar energy and fix CO_2 into carbohydrates.
Interactions and Systems	Most ecosystems on Earth depend on photosynthesizing autotrophs as the basis of food webs.

7.1 Photosynthetic Organisms

Learning Outcomes

Upon completion of this section, you should be able to

1. Explain how autotrophs are able to produce their own food.
2. Describe the components of a chloroplast.
3. Compare the role of carbon dioxide in autotrophs and heterotrophs.

Photosynthesis converts solar energy into the chemical energy of a carbohydrate. Photosynthetic organisms, including land plants, algae, and cyanobacteria, are called **autotrophs** because they produce their own food (Fig. 7.1). Globally, photosynthesis produces an enormous amount of carbohydrate—so much that, if it were instantly converted to coal and the coal were loaded into standard railroad cars (each car holding about 50 tons), the photosynthesizers of the biosphere would fill more than 100 cars per second with coal.

No wonder photosynthetic organisms are able to sustain themselves and all other living things on Earth. With few exceptions, it is possible to trace any food chain back to plants

and algae. In other words, producers, which have the ability to synthesize carbohydrates, feed not only themselves but also consumers, which must take in preformed organic molecules. Collectively, consumers are called **heterotrophs.** Both autotrophs and heterotrophs use organic molecules produced by photosynthesis as a source of building blocks for growth and repair and as a source of chemical energy for cellular work.

Photosynthesizers also produce copious amounts of oxygen gas (O_2) as a by-product. Oxygen, which is required by organisms when they carry on cellular respiration, rises high into the atmosphere, where it forms an ozone shield that filters out ultraviolet radiation and makes terrestrial life possible.

Our analogy about photosynthetic products becoming coal is apt because the bodies of many ancient plants did become the coal we burn today, usually to produce electricity. Coal formation happened several hundred million years ago, and that is why coal is called a fossil fuel. Today's trees are also commonly used as fuel. Fermentation of plant materials produces ethanol, which can be used to fuel automobiles directly or as a gasoline additive.

The products of photosynthesis are critical to humankind in a number of other ways. They serve as a source of building materials, fabrics, paper, and pharmaceuticals. Of course, we also appreciate green plants for the simple beauty of a magnolia blossom, the scent of a rose, or the majesty of the Earth's forests.

Figure 7.1 Photosynthetic organisms. Photosynthetic organisms include plants, such as trees, garden plants, and mosses, which typically live on land; photosynthetic protists, such as *Euglena*, diatoms, and kelp, which typically live in water; and cyanobacteria, a type of bacterium that lives in water, damp soil, and rocks.

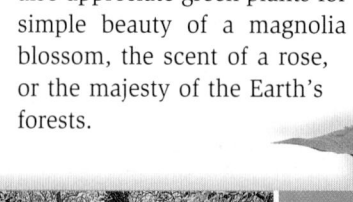

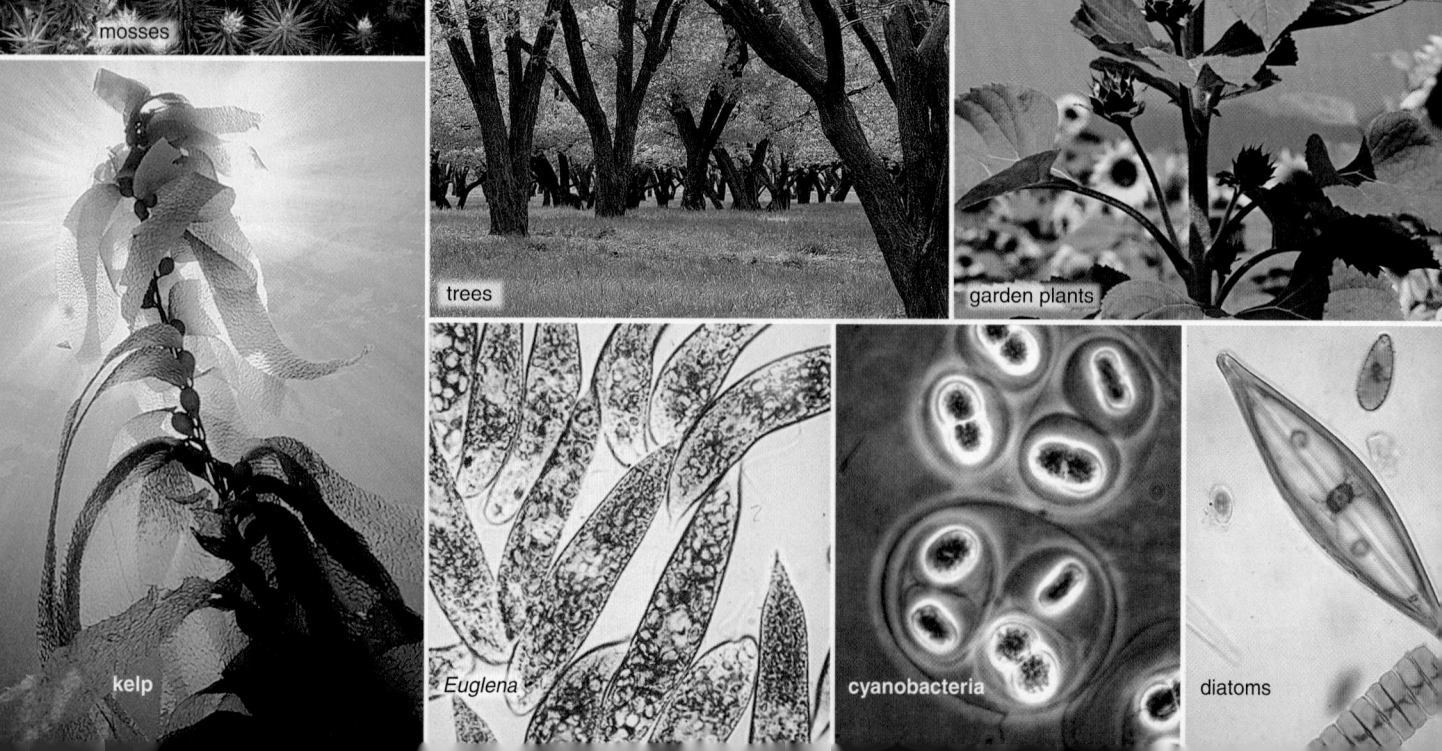

mosses

trees

garden plants

kelp

Euglena

cyanobacteria

diatoms

Flowering Plants as Photosynthesizers

Photosynthesis takes place in the green portions of plants. The leaves of a flowering plant contain mesophyll tissue in which cells are specialized for photosynthesis (Fig. 7.2). The raw materials for photosynthesis are water and carbon dioxide. The roots of a plant absorb water, which then moves in vascular tissue up the stem to a leaf by way of the leaf veins. Carbon dioxide in the air enters a leaf through small openings called **stomata** (sing., stoma). After entering a leaf, carbon dioxide and water diffuse into **chloroplasts** [Gk. *chloros,* green, and *plastos,* formed, molded], the organelles that carry on photosynthesis.

A double membrane surrounds a chloroplast, and its semi-fluid interior called the **stroma** [Gk. *stroma,* bed, mattress]. A different membrane system within the stroma forms flattened sacs called **thylakoids** [Gk. *thylakos,* sack, and *eides,* like, resembling], which in some places are stacked to form **grana** (sing., granum), so called because they looked like piles of seeds to early microscopists. The space of each thylakoid is thought to be connected to the space of every other thylakoid within a chloroplast, thereby forming an inner compartment within chloroplasts called the thylakoid space. Overall, chloroplast membranes provide tremendous surface area for photosynthesis to occur.

The thylakoid membrane contains **chlorophyll** and other pigments that are capable of absorbing solar energy, the type of energy that drives photosynthesis. The stroma contains an enzyme-rich solution where carbon dioxide is first attached to an organic compound and is then reduced to a carbohydrate.

Therefore, it is proper to associate the absorption of solar energy with the thylakoid membranes making up the grana and to associate the reduction of carbon dioxide to a carbohydrate with the stroma of a chloroplast.

Human beings, and indeed nearly all organisms, release carbon dioxide into the air when they respire. Some of these same carbon dioxide molecules enter a leaf through the stoma and are converted to carbohydrate. Carbohydrate, in the form of glucose, is the chief source of chemical energy for most organisms. Thus, an interdependent relationship exists between organisms that make their own food (autotrophs) and those that consume their food (heterotrophs).

Check Your Progress   7.1

1. Describe three major groups of photosynthetic organisms.
2. Distinguish the part of a chloroplast that absorbs solar energy from the part that forms a carbohydrate.

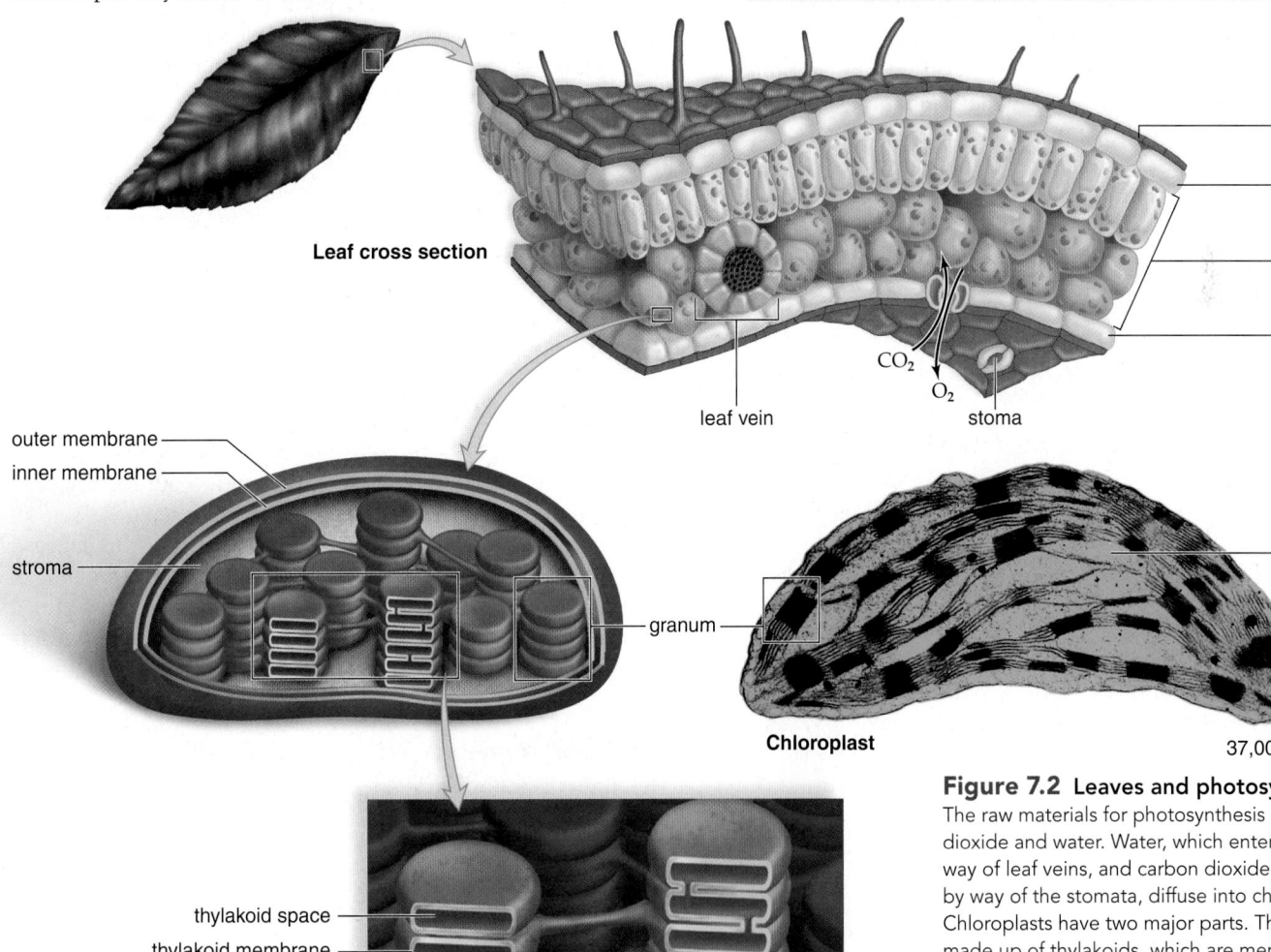

Leaf cross section

cuticle

upper epidermis

mesophyll

lower epidermis

CO_2

O_2

stoma

leaf vein

outer membrane
inner membrane

stroma

granum

stroma

Chloroplast   37,000×

thylakoid space
thylakoid membrane

independent thylakoid in a granum

overlapping thylakoid in a granum

Grana

Figure 7.2 Leaves and photosynthesis.
The raw materials for photosynthesis are carbon dioxide and water. Water, which enters a leaf by way of leaf veins, and carbon dioxide, which enters by way of the stomata, diffuse into chloroplasts. Chloroplasts have two major parts. The grana are made up of thylakoids, which are membranous disks. Their membrane contains photosynthetic pigments such as chlorophylls *a* and *b*. These pigments absorb solar energy. The stroma is a semifluid interior where carbon dioxide is enzymatically reduced to a carbohydrate.

7.2 The Process of Photosynthesis

Learning Outcomes

Upon completion of this section, you should be able to:

1. Describe the overall process of photosynthesis.
2. Compare energy input and output of the light reaction.
3. Compare carbon input and output of the Calvin cycle reaction.

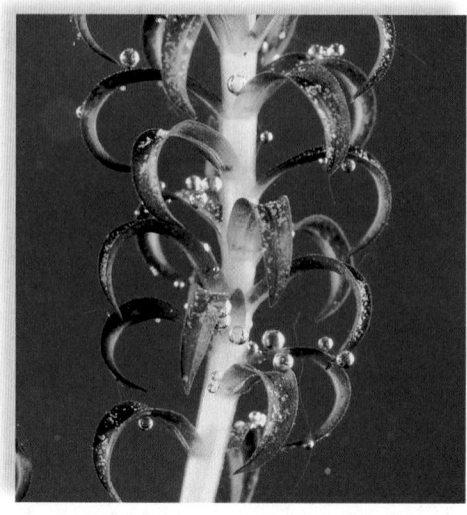

Figure 7.3 Photosynthesis releases oxygen. Bubbling indicates that the aquatic plant *Elodea* releases O_2 gas when it photosynthesizes.

The overall process of photosynthesis can be represented by an equation:

$$6CO_2 + 12H_2O \xrightarrow{\text{solar energy}} 6(CH_2O) + 6H_2O + 6O_2$$

In this equation, (CH$_2$O) represents carbohydrate. If the equation were multiplied by six, the carbohydrate would be $C_6H_{12}O_6$, or glucose.

The overall equation implies that photosynthesis involves oxidation-reduction (redox) and the movement of electrons from one molecule to another. Recall that oxidation is the loss of electrons, and reduction is the gain of electrons. In living things, as discussed in Chapter 6, the electrons are very often accompanied by hydrogen ions so that oxidation is the loss of hydrogen atoms ($H^+ + e^-$), and reduction is the gain of hydrogen atoms. This simplified rewrite of the above equation makes it clear that carbon dioxide has been reduced, and water has been oxidized:

$$CO_2 + H_2O \xrightarrow{\text{solar energy}} (CH_2O) + O_2$$

with Reduction and Oxidation labeled.

It takes hydrogen atoms and a lot of energy to reduce carbon dioxide. From your study of energy and enzymes in Chapter 6, you probably expect that solar energy is not used directly during photosynthesis; rather, it is converted to ATP molecules. ATP is the energy currency of cells and, when cells need something, they spend ATP. In this case, solar energy is used to generate the ATP needed to reduce carbon dioxide to a carbohydrate. Of course, we always want to keep in mind that this carbohydrate represents the food produced by land plants, algae, and cyanobacteria that feeds the biosphere.

The Role of NADP⁺/NADPH

A review of Section 6.4 will also lead you to suspect that the electrons needed to reduce carbon dioxide are carried by a coenzyme. NADP$^+$ is the coenzyme of oxidation-reduction (redox coenzyme) active during photosynthesis. When NADP$^+$ is reduced, it has accepted two electrons and one hydrogen atom, and when NADPH is oxidized, it gives up its electrons:

$$NADP^+ + 2\ e^- + H^+ \longrightarrow NADPH$$

What molecule supplies the electrons that reduce NADP$^+$ during photosynthesis? Put a sprig of *Elodea* in a beaker, and supply it with light, and you will observe a bubbling (Fig. 7.3). The bubbling occurs because the plant is releasing oxygen as it photosynthesizes.

A very famous experiment performed by C. B. van Niel of Stanford University found that the oxygen given off by photosynthesizers comes from water. Van Niel performed two separate experiments. When an isotope of oxygen, ^{18}O, was a part of water, the O_2 given off by the plant contained ^{18}O. When ^{18}O was a part of carbon dioxide supplied to a plant, the O_2 given off by a plant did not contain the ^{18}O. Why not? Because the oxygen in carbon dioxide doesn't come from water, it comes from the air. This was the first step toward discovering that water splits during photosynthesis. When water splits, oxygen is released, and the hydrogen atoms ($H^+ + e^-$) are taken up by NADP$^+$. Later, NADPH reduces carbon dioxide to a carbohydrate.

Two Sets of Reactions

Many investigators have contributed to our understanding of the overall equation of photosynthesis and to our current realization that photosynthesis consists of two separate sets of reactions. F. F. Blackman (1866-1947) was the first to suggest, in 1905, that enzymes must be involved in the reduction of carbon dioxide to a carbohydrate and that the process must consist of two separate sets of reactions. We call the two sets of reactions the light reactions and the Calvin cycle reactions.

Light Reactions

The **light reactions** are so named because they only occur when solar energy is available (during daylight hours). The overall equation for photosynthesis gives no hint that the green pigment chlorophyll, present in thylakoid membranes, is largely responsible for absorbing the solar energy that drives photosynthesis.

During the light reactions, solar energy energizes electrons that move down an electron transport chain (see Fig. 6.12). As the electrons move down the chain, energy is released and captured to produce ATP molecules. Energized electrons are also taken up by $NADP^+$, which is reduced and becomes NADPH. This equation can be used to summarize the light reactions because, during the light reactions, solar energy is converted to chemical energy:

$$\text{solar energy} \longrightarrow \text{chemical energy}$$
$$\text{(ATP, NADPH)}$$

Calvin Cycle Reactions

The **Calvin cycle reactions** are named for Melvin Calvin, who received a Nobel Prize for discovering the enzymatic reactions that reduce carbon dioxide to a carbohydrate in the stroma of chloroplasts (Fig. 7.4). The enzymes that speed the reduction of carbon dioxide during both day and night are located in the semifluid substance of the chloroplast stroma.

During the Calvin cycle reactions, CO_2 is taken up and then reduced to a carbohydrate that can later be converted to glucose. This equation can be used to summarize the Calvin cycle reactions because during these reactions, the ATP and NADPH formed during the light reactions are used to reduce carbon dioxide:

$$\text{chemical energy} \longrightarrow \text{chemical energy}$$
$$\text{(ATP, NADPH)} \qquad \text{(carbohydrate)}$$

Summary

Figure 7.5 summarizes our discussion so far and shows that during the light reactions, (1) solar energy is absorbed, (2) water is split so that oxygen is released, and (3) ATP and NADPH are produced.

During the Calvin cycle reactions, (1) CO_2 is absorbed and (2) reduced to a carbohydrate (CH_2O) by utilizing ATP and

Figure 7.4
Melvin Calvin in the laboratory.
Melvin Calvin used tracers to discover the cycle of reactions that reduce CO_2 to a carbohydrate.

NADPH from the light reactions (see bottom set of red arrows). The top set of red arrows takes ADP + $\textcircled{P}$ and $NADP^+$ back to light reactions, where they become ATP and NADPH once more so that carbohydrate production can continue.

Check Your Progress   7.2

1. Explain how redox reactions are used in photosynthesis.
2. Describe the role of enzymes during photosynthesis.

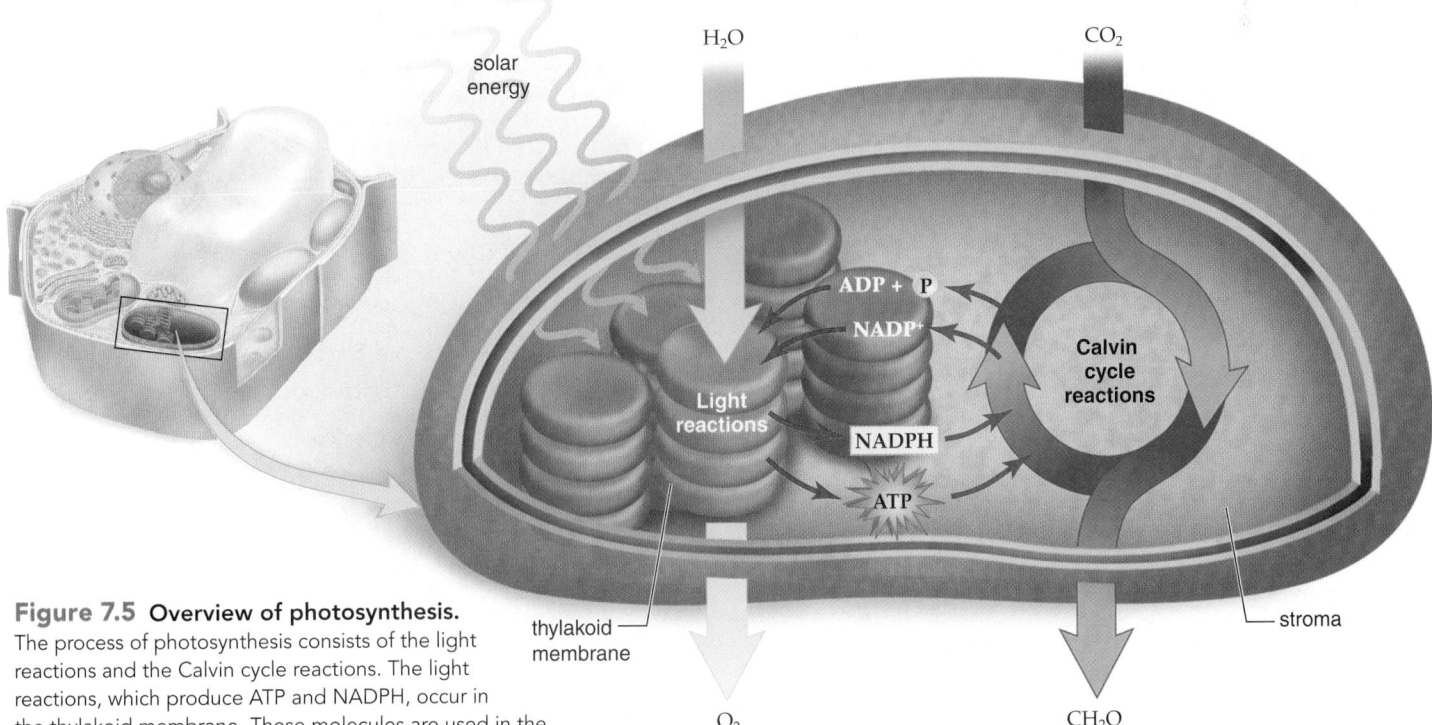

Figure 7.5 Overview of photosynthesis.
The process of photosynthesis consists of the light reactions and the Calvin cycle reactions. The light reactions, which produce ATP and NADPH, occur in the thylakoid membrane. These molecules are used in the Calvin cycle reactions which take place in the stroma. The Calvin cycle reactions reduce carbon dioxide to a carbohydrate.

7.3 Plants as Solar Energy Converters

Learning Outcomes

Upon completion of this section, you should be able to

1. Describe the relationship between wavelength and energy in the electromagnetic spectrum.
2. Explain the role of photosynthetic pigments in harnessing solar energy.
3. Examine how ATP and NADPH are produced from redox reactions and membrane gradients.

Solar energy can be described in terms of its wavelength and its energy content. Figure 7.6a lists the different types of radiant energy from the shortest wavelength, gamma rays, to the longest, radio waves. Most of the radiation reaching the Earth is within the visible-light range. Higher-energy wavelengths are screened out by the ozone layer in the atmosphere, and lower-energy wavelengths are screened out by water vapor and carbon dioxide, before they reach the Earth's surface. Because visible light is the most prevalent in the environment, we can conclude that organic molecules and processes within organisms, such as vision and photosynthesis, are chemically adapted to the radiation associated with visible light (Fig. 7.6a).

 **3D Animation** Properties of Light

Pigments and Photosystems

Pigment molecules absorb wavelengths of light. Most pigments absorb only some wavelengths; they reflect or transmit the other wavelengths. The pigments found in chloroplasts are capable of absorbing various portions of visible light. This is called their **absorption spectrum.**

Photosynthetic organisms differ in the type of chlorophyll they contain. In plants, chlorophyll *a* and chlorophyll *b* play prominent roles in photosynthesis. **Carotenoids** play an accessory role. Both chlorophylls *a* and *b* absorb violet, blue, and red light better than the light of other colors. Because green light is transmitted and reflected by chlorophyll, plant leaves appear green to us. The carotenoids, which are shades of yellow and orange, are able to absorb light in the violet-blue-green range. These pigments become noticeable in the fall when chlorophyll breaks down.

Animation Absorption of Light

How do you determine the absorption spectrum of pigments? To identify the absorption spectrum of a particular pigment, a purified sample is exposed to different wavelengths of light inside an instrument called a spectrophotometer. A spectrophotometer measures the amount of light that passes through the sample, and from this it is possible to calculate how much was absorbed. The amount of light absorbed at each wavelength is plotted on a graph, and the result is a record of the pigment's absorption spectrum (Fig. 7.6b).

A **photosystem** consists of a pigment complex (molecules of chlorophyll *a*, chlorophyll *b*, and the carotenoids) and electron acceptor molecules within the thylakoid membrane. The pigment complex serves as an "antenna" for gathering solar energy.

The Electron Flow in the Light Reactions

The light reactions utilize two photosystems, called photosystem I (PS I) and photosystem II (PS II). The photosystems are named for the order in which they were discovered, not for the order in which they occur in the thylakoid membrane or participate in the photosynthetic process.

During the light reactions, electrons usually, but not always, follow a **noncyclic pathway** that begins with photosystem II (Fig. 7.7). The pigment complex absorbs solar energy, which is then passed from one pigment to the other until it is concentrated in a particular pair of chlorophyll *a* molecules, called the *reaction center*. Electrons (e^-) in the reaction center become so energized that they escape from the reaction center and move to nearby electron acceptor molecules.

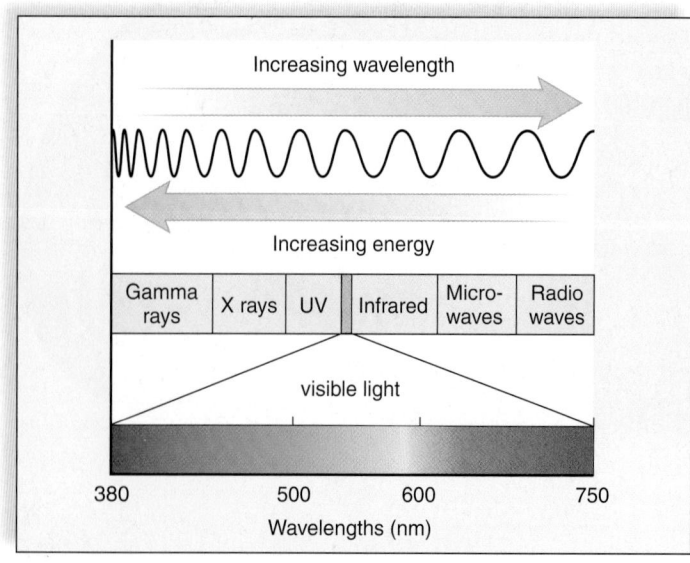

a. The electromagnetic spectrum includes visible light.

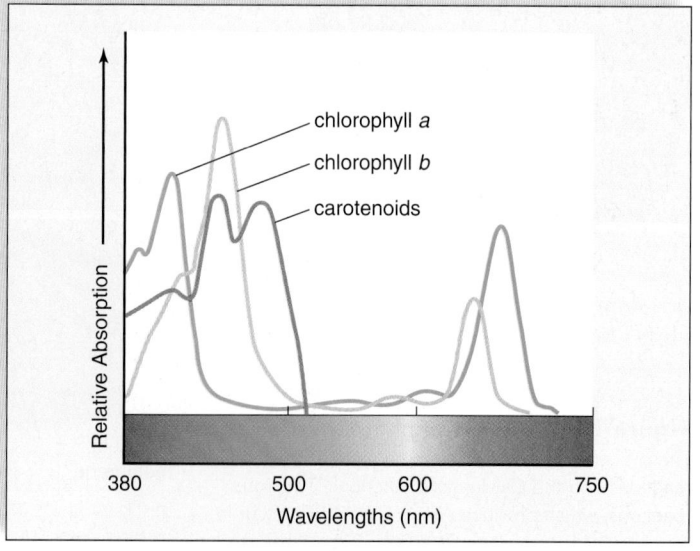

b. Absorption spectrum of photosynthetic pigments.

Figure 7.6 Photosynthetic pigments and photosynthesis.   **a.** The wavelengths in visible light differ according to energy content and color. **b.** The photosynthetic pigments in chlorophylls *a* and *b* and the carotenoids absorb certain wavelengths within visible light. This is their absorption spectrum.

PS II would disintegrate without replacement electrons, and these are removed from water, which splits, releasing oxygen to the atmosphere. Notice that with the loss of electrons, water has been oxidized, and that indeed, the oxygen released during photosynthesis does come from water. Many organisms, including plants themselves and human beings, use this oxygen within their mitochondria to make ATP. The hydrogen ions (H^+) stay in the thylakoid space and contribute to the formation of a hydrogen ion gradient.

An electron acceptor sends energized electrons, received from the reaction center, down an **electron transport chain (ETC)**, a series of carriers that pass electrons from one to the other (see Fig. 6.13). As the electrons pass from one carrier to the next, energy is captured and stored in the form of a hydrogen ion (H^+) gradient. When these hydrogen ions flow down their electrochemical gradient through ATP synthase complexes, ATP production occurs (see Fig. 7.8). Notice that this ATP is then used by the Calvin cycle reactions in the stroma to reduce carbon dioxide to a carbohydrate.

When the PS I pigment complex absorbs solar energy, energized electrons leave its reaction center and are captured by electron acceptors. (Low-energy electrons from the *electron transport chain* adjacent to PS II replace those lost by PS I.) The electron acceptors in PS I pass their electrons to $NADP^+$ molecules. Each $NADP^+$ accepts two electrons and an H^+ to become reduced and forms NADPH. This NADPH is then used by the Calvin cycle reactions in the stroma along with ATP in the reduction of carbon dioxide to a carbohydrate.

3D Animation
Light-Dependent Reactions

ATP and NADPH are not made in equal amounts during the light reactions, and more ATP than NADPH is required during the Calvin cycle. Where does this extra ATP come from? Every so often, an electron moving down the noncyclic pathway is rerouted back to an earlier point in the electron transport chain. This cyclic pathway enables electrons to participate in additional redox reactions, moving more H^+ across the thylakoid membrane and through ATP synthase, and ultimately producing more ATP.

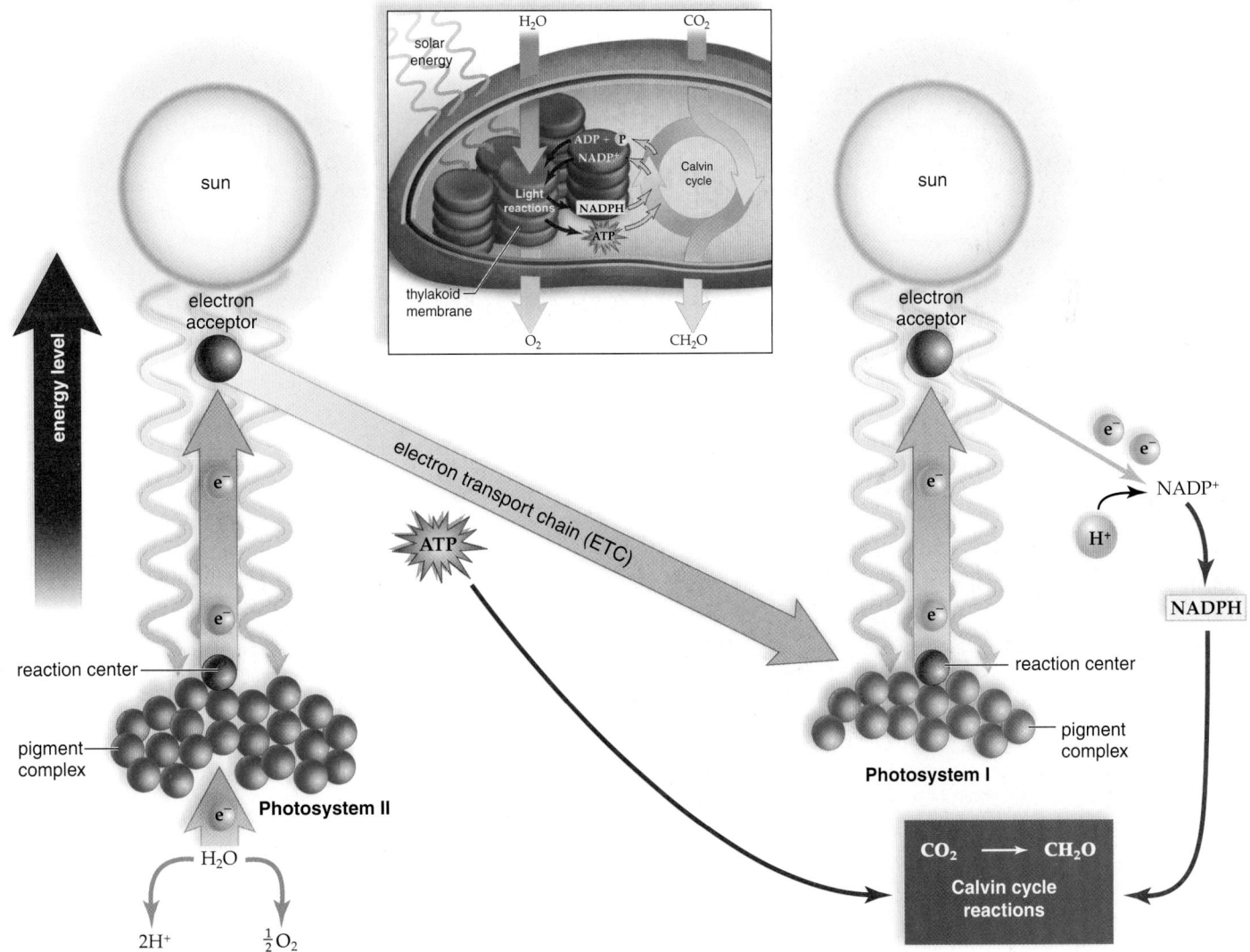

Figure 7.7 Noncyclic pathway: Electrons move from water to $NADP^+$.   Energized electrons (replaced from water, which splits, releasing oxygen) leave photosystem II and pass down an electron transport chain, leading to the formation of ATP. Energized electrons (replaced by photosystem II by way of the ETC) leave photosystem I and pass to $NADP^+$, which then combines with H^+, becoming NADPH.

The Organization of the Thylakoid Membrane

As we have discussed, the following molecular complexes are present in the thylakoid membrane (Fig. 7.8):

PS II, which consists of a pigment complex and electron-acceptor molecules, receives electrons from water as water splits, releasing oxygen.

The electron transport chain (ETC), consisting of Pq (plastoquinone) and cytochrome complexes, carries electrons from PS II to PS I via redox reactions. Pq also pumps H^+ from the stroma into the thylakoid space.

PS I, which also consists of a pigment complex and electron-acceptor molecules, is adjacent to NADP reductase, which reduces $NADP^+$ to NADPH.

The **ATP synthase** complex, which has a channel and a protruding ATP synthase, is an enzyme that joins ADP + ⓟ.

Animation
Photosynthetic Electron and ATP Synthesis

ATP Production

The thylakoid space acts as a reservoir for many hydrogen ions (H^+). First, each time water is oxidized, two H^+ remain in the thylakoid space. Second, as the electrons move from carrier to carrier via redox reactions along the electron transport chain, the electrons give up energy, which is used to pump H^+ from the stroma into the thylakoid space. Therefore, there are more H^+ in the thylakoid space than in the stroma. This difference and the resulting flow of H^+ (often referred to as protons in this context) from high to low concentration provides kinetic energy that allows an ATP synthase complex enzyme to enzymatically produce ATP from ADP + ⓟ. This method of producing ATP is called **chemiosmosis** because ATP production is tied to the establishment of an H^+ gradient (see Fig. 6.13).

Animation
Proton Pump

Video
Spinach Battery

Check Your Progress 7.3

1. Distinguish visible light from the electromagnetic spectrum.
2. Evaluate the energy level of molecules that go in and come out of the light reaction.

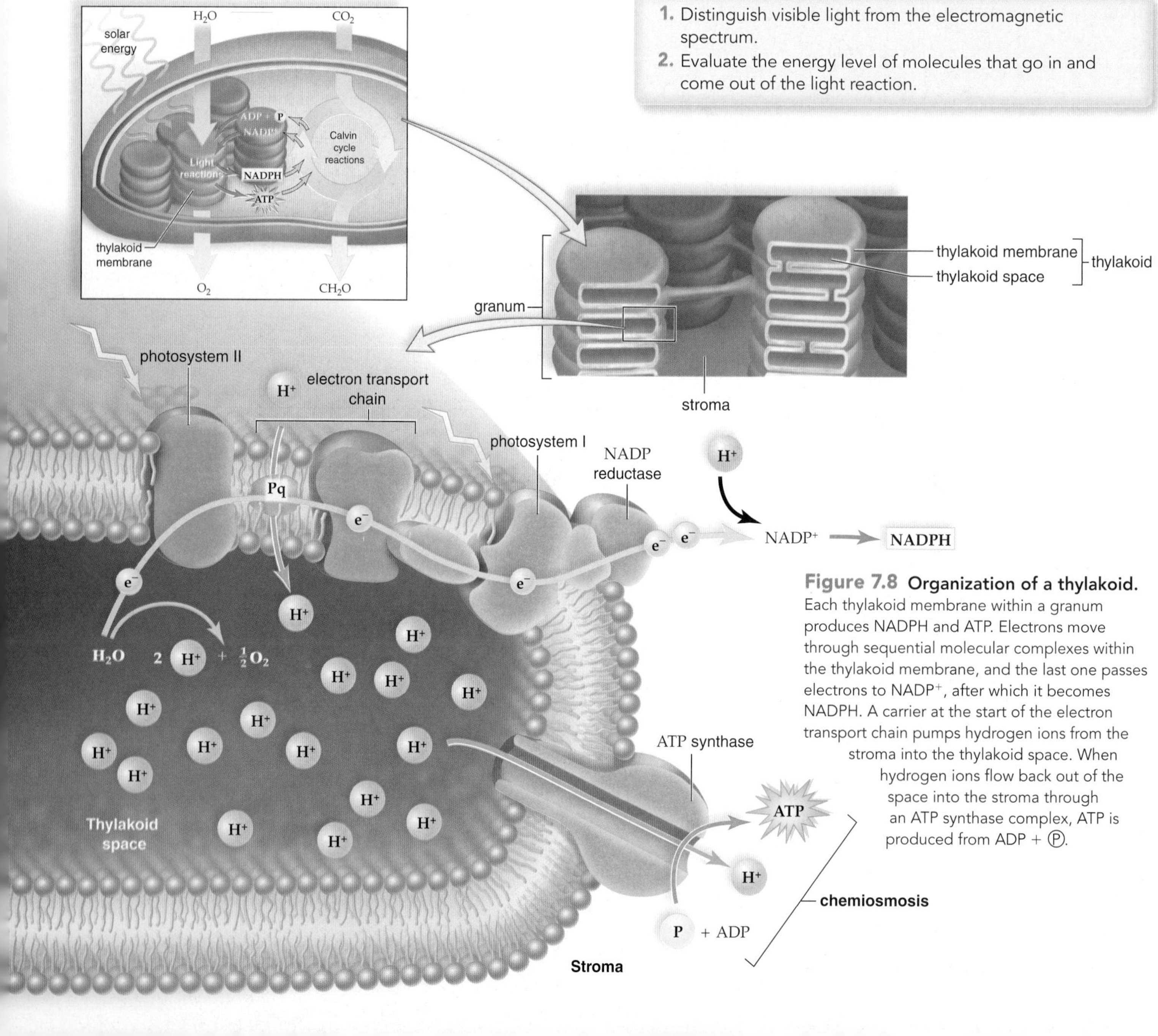

Figure 7.8 Organization of a thylakoid. Each thylakoid membrane within a granum produces NADPH and ATP. Electrons move through sequential molecular complexes within the thylakoid membrane, and the last one passes electrons to $NADP^+$, after which it becomes NADPH. A carrier at the start of the electron transport chain pumps hydrogen ions from the stroma into the thylakoid space. When hydrogen ions flow back out of the space into the stroma through an ATP synthase complex, ATP is produced from ADP + ⓟ.

Biological Systems

Tropical Rain Forest Destruction and Climate Change

Al Gore, former presidential candidate, won the 2007 Nobel Peace Prize for raising public awareness concerning climate change. The Nobel Committee said that "global warming could induce large-scale migrations and lead to greater competition for the Earth's resources. As such, it may increase the danger of violent conflicts and wars, within and between countries."

Climate change refers to an expected rise in the average global temperature during the twenty-first century due to the introduction of certain gases into the atmosphere. For at least a thousand years prior to 1850, atmospheric carbon dioxide (CO_2) levels remained fairly constant at 0.028%. Since the 1850s, when industrialization began, the amount of CO_2 in the atmosphere has increased to 0.038% (Fig. 7A).

Role of Carbon Dioxide

In much the same way as the panes of a greenhouse, CO_2 and other gases in our atmosphere trap radiant heat from the Sun. Therefore, these gases are called greenhouse gases. Without any greenhouse gases, the Earth's temperature would be about 33°C cooler than it is now. Likewise, increasing the concentration of greenhouse gases is predicted to affect climate change.

Certainly, the burning of fossil fuels adds CO_2 to the atmosphere. But another factor that contributes to an increase in atmospheric CO_2 is tropical rain forest destruction.

Role of Tropical Rain Forests

Tropical rain forests are considered by many scientists to be the "lungs" of the Earth. Between 10 and 30 million hectares of rain forests are lost every year to ranching, logging, mining, and otherwise developing areas of the forest for human needs.

Each year, deforestation in tropical rain forests accounts for 20–30% of all CO_2 in the atmosphere. With your body, if you lose lung capacity, you lose body function. Similarly, the consequence of losing forests is greater trouble for climate change because burning a forest adds CO_2 to the atmosphere and removes trees that would ordinarily absorb CO_2.

The Earth Is a System

Carbon dioxide is removed from the air via photosynthesis, which takes place in forests, oceans, and other terrestrial and marine ecosystems. In fact, photosynthesis produces organic matter that is 300 to 600 times the mass of people currently living on Earth this year. Thus, these environments act as a sink for CO_2, preventing too much from accumulating in the atmosphere where CO_2 can affect global temperatures and bring about climate change.

Despite their reduction in size from an original 14% to 6% of land surface today, tropical rain forests make a substantial contribution to global CO_2 removal. They are a critical element of the Earth's systems, and, like any biological system, are essential for normal, healthy function. Tropical rain forests contribute greatly to the uptake of CO_2 and the productivity of photosynthesis because they are the most efficient of all terrestrial ecosystems.

Tropical rain forests occur near the equator. They can exist wherever temperatures are above 26°C and rainfall is heavy (from 100–200 cm) and regular. Huge trees with buttressed trunks and broad, undivided, dark-green leaves predominate. Nearly all land plants in a tropical rain forest are woody, and woody vines are also abundant.

It might be hypothesized that an increased amount of CO_2 in the atmosphere would cause photosynthesis to increase in the remaining portion of the forest. To study this possibility, investigators measured atmospheric CO_2 levels, daily temperature levels, and tree girth in La Selva, Costa Rica, for 16 years. The results demonstrated relatively lower forest productivity at higher temperatures. These findings suggest that, as temperatures rise, tropical rain forests may add to ongoing atmospheric CO_2 accumulation and accelerated climate change rather than the reverse.

These and other studies show that, as with any biological system, equilibrium and balance are necessary for healthy function. As a biological system, the Earth is sensitive to environmental change. Learning how to properly balance human activity with the needs of the biosphere requires that people become educated about how the Earth functions.

Questions to Consider

1. Are you concerned about climate change?
2. What actions can a person take to learn more about the facts of climate change?

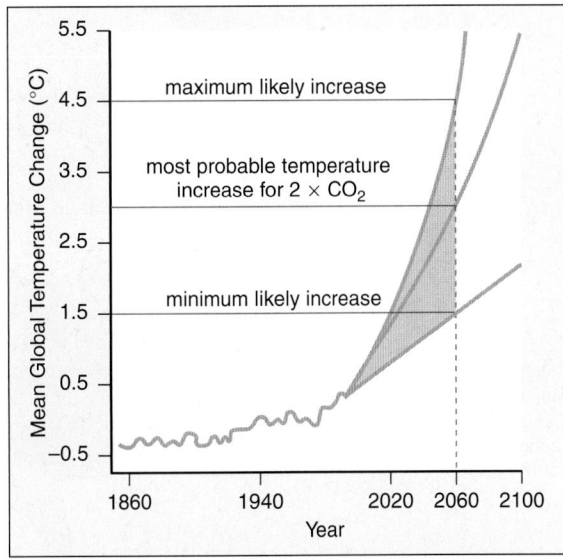

Figure 7A Climate change. Mean global temperature change is expected to rise due to the introduction of greenhouse gases into the atmosphere.

7.4 Plants as Carbon Dioxide Fixers

The Calvin cycle is a series of reactions that occur after the light reactions. The Calvin cycle produces carbohydrate before returning to its starting point, ready to accept more carbon dioxide (CO_2) (Fig. 7.9). The cycle is named for Melvin Calvin, who, with colleagues, used the radioactive isotope ^{14}C as a tracer to discover the reactions making up the cycle.

This series of reactions uses carbon dioxide from the atmosphere to produce carbohydrate. How does carbon dioxide get into the atmosphere? We and most other organisms take in oxygen from the atmosphere and release carbon dioxide to the atmosphere. The Calvin cycle includes (1) carbon dioxide fixation, (2) carbon dioxide reduction, and (3) regeneration of RuBP (pronounced "ruby-P"; described shortly).

3D Animation
Calvin Cycle

Fixation of Carbon Dioxide

Carbon dioxide fixation is the first step of the Calvin cycle. During this reaction, a molecule of carbon dioxide from the atmosphere is attached to RuBP (ribulose-1,5-bisphosphate), a 5-carbon molecule. The result is one 6-carbon molecule, which splits into two 3-carbon molecules.

The enzyme that speeds this reaction, called **RuBP carboxylase,** is a protein that makes up about 20–50% of the protein content in chloroplasts. The reason for its abundance may be

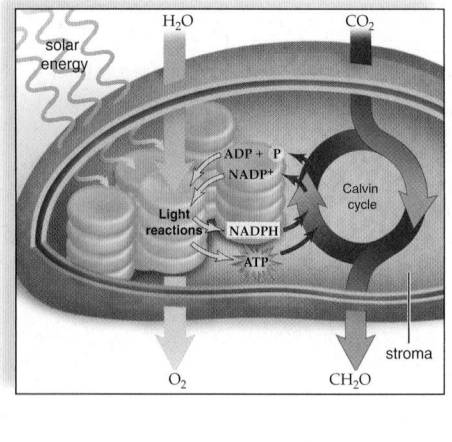

Figure 7.9 The Calvin cycle reactions.   The Calvin cycle is divided into three portions: CO_2 fixation, CO_2 reduction, and regeneration of RuBP. Because five G3P are needed to re-form three RuBP, it takes three turns of the cycle to have a net gain of one G3P. Two G3P molecules are needed to form glucose.

Metabolites of the Calvin Cycle	
RuBP	ribulose-1,5-bisphosphate
3PG	3-phosphoglycerate
BPG	1,3-bisphosphoglycerate
G3P	glyceraldehyde-3-phosphate

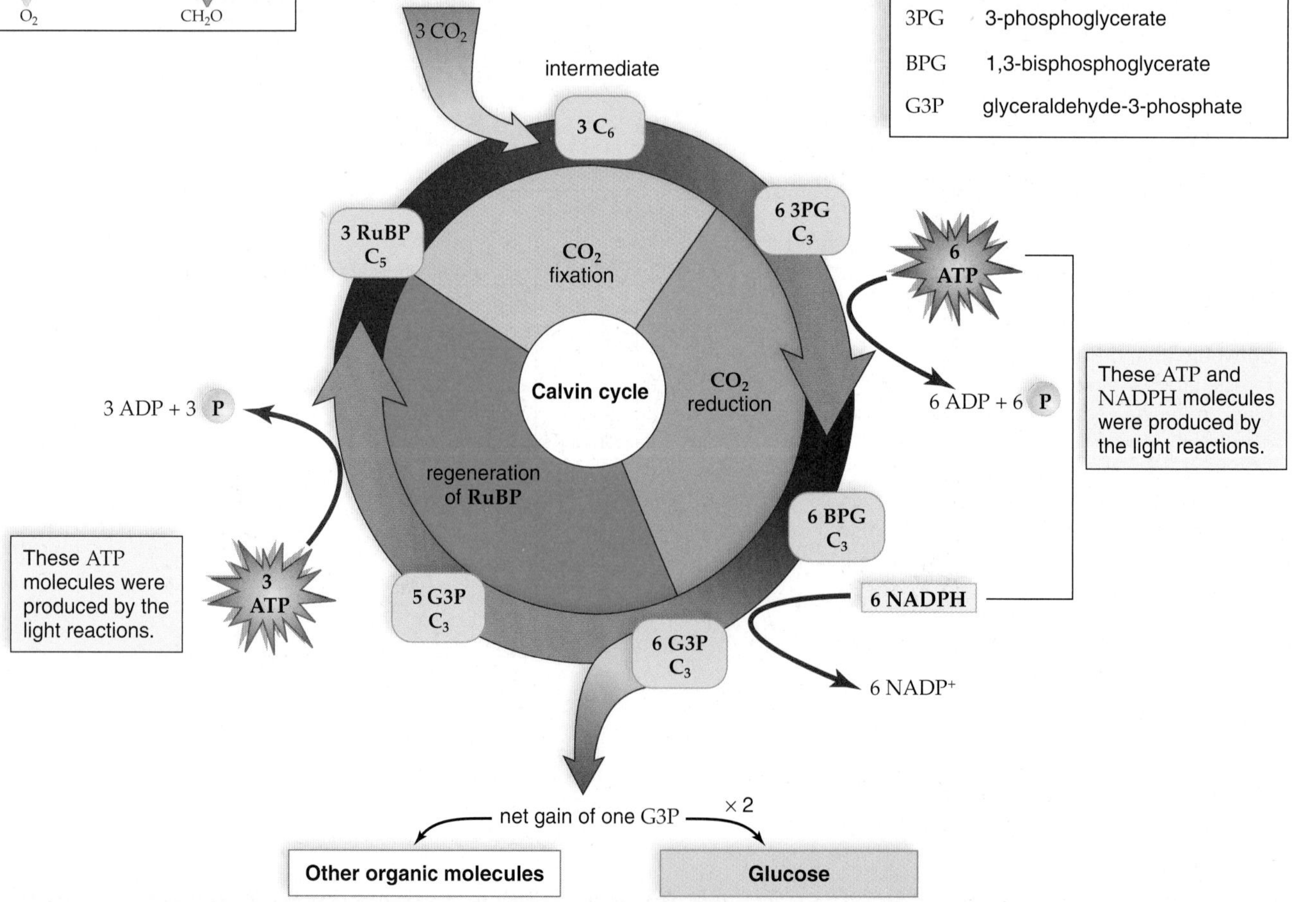

3 CO_2

intermediate

3 C_6

3 RuBP
C_5

6 3PG
C_3

CO_2 fixation

Calvin cycle

6 ATP

These ATP and NADPH molecules were produced by the light reactions.

3 ADP + 3 P

CO_2 reduction

6 ADP + 6 P

regeneration of RuBP

6 BPG
C_3

These ATP molecules were produced by the light reactions.

3 ATP

5 G3P
C_3

6 NADPH

6 G3P
C_3

6 NADP$^+$

net gain of one G3P ——— ×2

Other organic molecules   **Glucose**

that it is unusually slow—it processes only a few molecules of substrate per second compared to thousands per second for a typical enzyme—and so there has to be a lot of it to keep the Calvin cycle going.

Reduction of Carbon Dioxide

The first 3-carbon molecule in the Calvin cycle is called 3PG (3-phosphoglycerate). Each of two 3PG molecules undergoes reduction to G3P (glyceraldehyde-3-phosphate) in two steps:

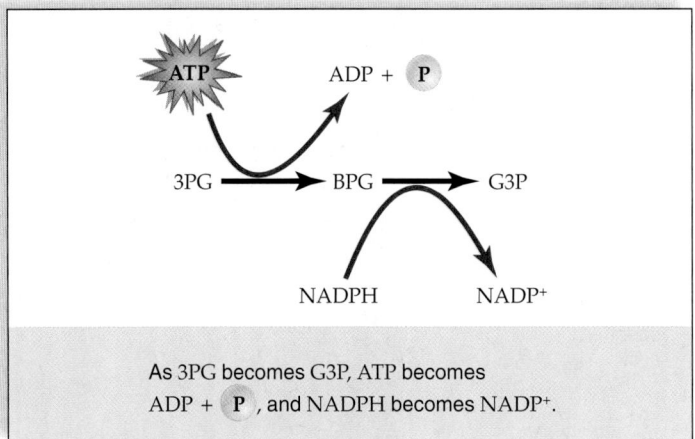

As 3PG becomes G3P, ATP becomes ADP + P, and NADPH becomes NADP+.

This is the sequence of reactions that uses some ATP and NADPH from the light reactions. This sequence signifies the reduction of carbon dioxide to a carbohydrate because $R—CO_2$ has become $R—CH_2O$. Energy and electrons are needed for this reduction reaction, and these are supplied by ATP and NADPH.

Regeneration of RuBP

Notice that the Calvin cycle reactions in Figure 7.9 are multiplied by three because it takes three turns of the Calvin cycle to allow one G3P to exit. Why? Because, for every three turns of the Calvin cycle, five molecules of G3P are used to re-form three molecules of RuBP, and the cycle continues. Notice that 5×3 (carbons in G3P) = 3×5 (carbons in RuBP):

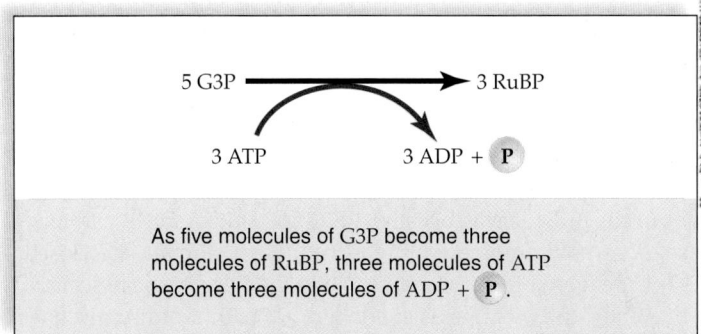

As five molecules of G3P become three molecules of RuBP, three molecules of ATP become three molecules of ADP + P.

This reaction also uses some of the ATP produced by the light reactions.

 **Animation** How the Calvin Cycle Works

The Importance of the Calvin Cycle

G3P is the product of the Calvin cycle that can be converted to other molecules a plant needs. Notice that glucose phosphate is

among the organic molecules that result from G3P metabolism (Fig. 7.10). This is of interest to us because glucose is the molecule that plants and animals most often metabolize to produce the ATP molecules they require for their energy needs.

Glucose phosphate can be combined with fructose (and the phosphate removed) to form sucrose, the molecule that plants use to transport carbohydrates from one part of the plant to the other.

Glucose phosphate is also the starting point for the synthesis of starch and cellulose. Starch is the storage form of glucose. Some starch is stored in chloroplasts, but most starch is stored in amyloplasts in roots. Cellulose is a structural component of plant cell walls and becomes fiber in our diet because we are unable to digest it.

A plant can use the hydrocarbon skeleton of G3P to form fatty acids and glycerol, which are combined in plant oils. We are all familiar with corn oil, sunflower oil, or olive oil used in cooking. Also, when nitrogen is added to the hydrocarbon skeleton derived from G3P, amino acids are formed, allowing the plant to produce protein.

Check Your Progress 7.4

1. Describe the three major steps of the Calvin cycle.
2. Illustrate why it takes 3 turns of the Calvin cycle to produce 1 G3P.

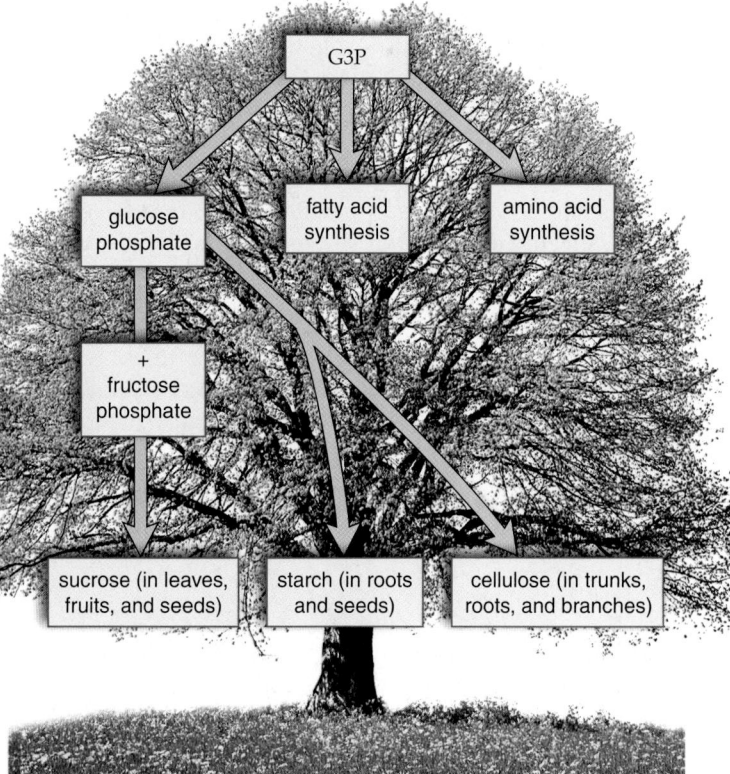

Figure 7.10 Fate of G3P. G3P is the first reactant in a number of plant cell metabolic pathways. Two G3Ps are needed to form glucose phosphate; glucose is often considered the end product of photosynthesis. Sucrose is the transport sugar in plants; starch is the storage form of glucose; and cellulose is a major constituent of plant cell walls.

7.5 Other Types of Photosynthesis

Learning Outcomes

Upon completion of this section, you should be able to

1. Compare the internal location of photosynthesis in C_3 and C_4 plants.
2. Contrast C_3/C_4 modes of photosynthesis with CAM photosynthesis.
3. Explain how different photosynthesis modes allow plants to adapt to particular environments.

The majority of land plants, such as azaleas, maples, and tulips, carry on photosynthesis as described and are called **C_3 plants** (Fig. 7.11a). C_3 plants use the enzyme RuBP carboxylase to fix CO_2 to RuBP in mesophyll cells. The first detected molecule following fixation is the 3-carbon molecule 3PG:

$$\text{RuBP} + CO_2 \xrightarrow{\text{RuBP carboxylase}} 2 \ 3PG$$

As shown in Figure 7.2, leaves have small openings called stomata through which water can leave and carbon dioxide (CO_2) can enter. If the weather is hot and dry, the stomata close, conserving water. (Water loss might cause the plant to wilt and die.) Now the concentration of CO_2 decreases in leaves, while O_2, a by-product of photosynthesis, increases. When O_2 rises in C_3 plants, RuBP carboxylase combines it with RuBP instead of CO_2. The result is one molecule of 3PG and the eventual release of CO_2. This is called **photorespiration** because in the presence of light (*photo*), oxygen is taken up and CO_2 is released (*respiration*).

An adaptation called C_4 photosynthesis enables some plants to avoid photorespiration.

C_4 Photosynthesis

In a C_3 plant, the mesophyll cells contain well-formed chloroplasts and are arranged in parallel layers. In a C_4 leaf, the bundle sheath cells, as well as the mesophyll cells, contain chloroplasts. Further, the mesophyll cells are arranged concentrically around the bundle sheath cells:

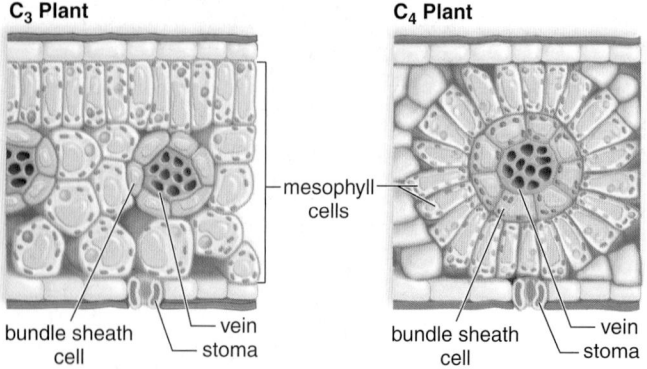

C_4 plants fix CO_2 to PEP (phosphoenolpyruvate, a C_3 molecule) using the enzyme PEP carboxylase (PEPCase). The result is oxaloacetate, a C_4 molecule:

$$\text{PEP} + CO_2 \xrightarrow{\text{PEPCase}} \text{oxaloacetate}$$

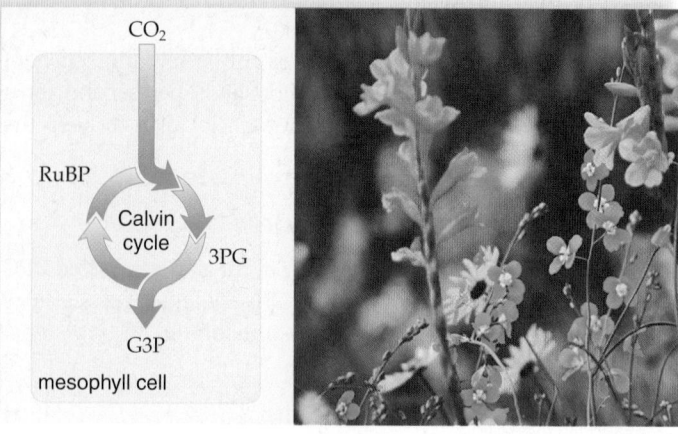

a. CO_2 fixation in a C_3 plant, wildflowers

b. CO_2 fixation in a C_4 plant, corn, *Zea mays*

Figure 7.11 Carbon dioxide fixation in C_3 and C_4 plants.
a. In C_3 plants, CO_2 is taken up by the Calvin cycle directly in mesophyll cells. **b.** C_4 plants form a C_4 molecule in mesophyll cells prior to releasing CO_2 to the Calvin cycle in bundle sheath cells.

In a C_4 plant, CO_2 is taken up in mesophyll cells, and then malate, a reduced form of oxaloacetate, is pumped into the bundle sheath cells (Fig. 7.11b). Here, and only here, does CO_2 enter the Calvin cycle.

Because it takes energy to pump molecules, you would think that the C_4 pathway would be disadvantageous. Yet in hot, dry climates, the net photosynthetic rate of C_4 plants, such as sugarcane, corn, and Bermuda grass, is about two to three times that of C_3 plants (e.g., wheat, rice, and oats). Why do C_4 plants enjoy such an advantage? The answer is that they can avoid photorespiration, discussed previously. Photorespiration is wasteful because it is not part of the Calvin cycle. Photorespiration does not occur in C_4 leaves because PEPCase, unlike RuBP carboxylase, does not combine with O_2. Even when stomata are closed, CO_2 is delivered to the Calvin cycle in the bundle sheath cells.

When the weather is moderate, C_3 plants ordinarily have the advantage, but when the weather becomes hot and dry, C_4 plants have the advantage, and we can expect them to predominate. In the early summer, C_3 plants such as Kentucky bluegrass and creeping bent grass predominate in lawns in the cooler parts of the United States, but by midsummer, crabgrass, a C_4 plant, begins to take over.

CO$_2$ fixation in a CAM plant, pineapple, *Ananas comosus*

Figure 7.12 Carbon dioxide fixation in a CAM plant. CAM plants, such as pineapple, fix CO$_2$ at night, forming a C$_4$ molecule that is released to the Calvin cycle during the day.

CAM Photosynthesis

CAM stands for crassulacean-acid metabolism; the Crassulaceae is a family of flowering succulent (water-containing) plants that live in warm, dry regions of the world. CAM was first discovered in these plants, but now it is known to be prevalent among other groups of plants.

Whereas a C$_4$ plant represents partitioning in space—carbon dioxide fixation occurs in mesophyll cells while the Calvin cycle occurs in bundle sheath cells—CAM is partitioning by the use of time. During the night, CAM plants use PEPCase to fix some CO$_2$, forming C$_4$ molecules, which are stored in large vacuoles in mesophyll cells. During the day, C$_4$ molecules (malate) release CO$_2$ to the Calvin cycle when NADPH and ATP are available from the light reactions (Fig. 7.12). The primary advantage for this partitioning again has to do with the conservation of water. CAM plants open their stomata only at night, and therefore only at that time does atmospheric CO$_2$ enter the plant. During the day, the stomata close; this conserves water, but CO$_2$ cannot enter the plant.

Photosynthesis in a CAM plant is minimal because a limited amount of CO$_2$ is fixed at night, but it does allow CAM plants to live under stressful conditions.

Photosynthesis and Adaptation to the Environment

The different types of photosynthesis give us an opportunity to consider that organisms are metabolically adapted to their environment. Each method of photosynthesis has its advantages and disadvantages, depending on the climate.

C$_4$ plants most likely evolved in, and are adapted to, areas of high light intensities, high temperatures, and limited rainfall. C$_4$ plants, however, are more sensitive to cold, and C$_3$ plants do better than C$_4$ plants below 25°C. CAM plants, on the other hand, compete well with either type of plant when the environment is extremely arid. Surprisingly, CAM is quite widespread and has evolved in 23 families of flowering plants, including some lilies and orchids! And it is found among nonflowering plants, including some ferns and cone-bearing trees.

Check Your Progress 7.5

1. Describe some plants that use a method of photosynthesis other than C$_3$ photosynthesis.
2. Explain why C$_4$ photosynthesis is advantageous in hot, dry conditions.

CONNECTING *the* CONCEPTS *with the* BIG IDEAS

Evolution

- The success of prokaryotic photosynthesis leads the way for its transfer to eukaryotic cells. (2A2e)

Energy and Homeostasis

- Light energy boosts chlorophyll's electrons energetically; passing through the ETC, they power proton pumps that produce ATP. (2A2d1-4, 4A2g2)
- ATP and NADPH are used in the Calvin cycle to produce carbohydrates from carbon dioxide. (2A2D5)
- Chloroplast's structure enhances grouping of Calvin cycle enzymes and the ETC chain. (2B3b, 4A2g1, 4A2g3)
- Photosystems I and II trap solar energy which power ATP production; water is also split, providing H$^+$ ions and releasing O$_2$. (2A2d1-2)

Interactions and Systems

- Only autotrophs are able to create food from inorganic molecules; photosynthesizing organisms form the basis of most food chains on Earth. (4A6d)
- In photosynthesis, carbon dioxide is reduced and carbohydrates are produced while water is oxidized; the "waste products" are the raw materials of cellular respiration. Energy flows but molecules recycle. (4A6a)
- Chlorophyll is the most common light-trapping molecule; it functions in many ways, increasing photosynthesis efficiency. (4A2g2, 4C1a*IE*)

*Find the unabridged version of all EK citations at www.glencoe.com/maderAP11.

▇ Media Study Tools ▇

www.glencoe.com/maderAP11

Enhance your study of this chapter with study tools and practice tests. Also ask your instructor about the resources available through ConnectPlus, including the media-rich eBook, interactive learning tools, and animations.

 **3D Animation** For a detailed examination of the process
Photosynthesis of photosynthesis, including a description
of the properties of light, watch McGraw-Hill's new 3D animation
"Photosynthesis."

Summarize

7.1 Photosynthetic Organisms

Photosynthesis produces carbohydrates and releases oxygen, both
of which are used by the majority of living things. Cyanobacteria,
algae, and land plants carry on photosynthesis. In plants, photosynthesis takes place in chloroplasts. A chloroplast is bounded by a
double membrane and contains two main components: the semifluid
stroma and the membranous grana made up of thylakoids.

7.2 The Process of Photosynthesis

The overall equation for photosynthesis shows that it is a redox reaction. Carbon dioxide is reduced, and water is oxidized. During photosynthesis, the light reactions take place in the thylakoid membranes,
and the Calvin cycle reactions take place in the stroma.

7.3 Plants as Solar Energy Converters

Photosynthesis uses solar energy in the visible-light range. Specifically, chlorophylls a and b absorb violet, blue, and red wavelengths
best and reflect green light, whereas the carotenoids absorb violet-
blue-green light and reflect yellow-to-orange light.

In the light reactions, noncyclic electron flow begins when solar
energy enters PS II and energizes chlorophyll a electrons. The oxidation
(splitting) of water replaces these electrons in the reaction-center chlorophyll a molecules. Oxygen is released to the atmosphere, and hydrogen ions (H^+) remain in the thylakoid space. Electrons are ultimately
passed to PS I via an electron transport chain, which pumps hydrogen
ions across the thylakoid membrane and contributes to the chemiosmotic gradient used to make ATP via ATP synthetase. Light-energized
electrons from PS I are captured by $NADP^+$, which combines with H^+
from the stroma to become NADPH. Cyclic electron flow in the light
reactions pumps additional hydrogen ions and also contributes to ATP
production.

7.4 Plants as Carbon Dioxide Fixers

The energy yield of the light reactions is stored in ATP and NADPH.
These molecules are used by the Calvin cycle reactions to reduce
CO_2 to carbohydrate, namely G3P, which is then converted to all the
organic molecules a plant needs.

During the first stage of the Calvin cycle, the enzyme RuBP carboxylase fixes CO_2 to RuBP, producing a 6-carbon molecule that
immediately breaks down to two C_3 molecules. During the second
stage, CO_2 (incorporated into an organic molecule) is reduced to carbohydrate (CH_2O). This step requires the NADPH and some of the ATP
from the light reactions. For every three turns of the Calvin cycle, the
net gain is one G3P molecule; the other five G3P molecules are used to
re-form three molecules of RuBP, which also requires ATP. It takes two
G3P molecules to make one glucose molecule.

7.5 Other Types of Photosynthesis

Plants have adapted ways other than the C3 process just described to
photosynthesize in various environments. In C_4 plants, carbon dioxide
is first fixed in mesophyll cells via PEPCase, is transported to a different location in bundle sheath cells, and then released to the Calvin
cycle. PEPCase has an advantage over RuBP carboxylase because it
doesn't participate in photorespiration. C_4 plants avoid the photores-

piration complication by dividing where carbon fixation occurs from
where the Calvin cycle occurs.

CAM plants, which live in hot, dry environments, cannot leave
their stomata open during the day, or they will die from loss of water.
CAM plants fix carbon only at night, conserving water. Stores of CO_2
are released to the Calvin cycle during the day, when photosynthesis
is possible. CAM plants avoid drying out by dividing when they bring
in carbon dioxide from when they release it to the Calvin cycle.

Key Terms

absorption spectrum 124	chloroplast 121
ATP synthase 126	electron transport chain 125
autotroph 120	grana (sing., granum) 121
C_3 plant 130	heterotroph 120
C_4 plant 130	light reactions 122
Calvin cycle reactions 123	noncyclic pathway 124
CAM 131	photorespiration 130
carbon dioxide (CO_2)	photosynthesis 120
fixation 128	photosystem 124
carotenoid 124	RuBP carboxylase 128
chemiosmosis 126	stomata 121
climate change 127	stroma 121
chlorophyll 121	thylakoid 121

 Assess

Reviewing This Chapter

1. Why is it proper to say that almost all living things are
   dependent on solar energy?   120
2. Name the two major components of chloroplasts, and
   associate each with one of two sets of reactions that occur
   during photosynthesis. How are the two sets of reactions
   related?   121–23
3. Write the overall equation of photosynthesis and associate
   each participant with either the light reactions or the Calvin
   cycle reactions.   122–23
4. Discuss the electromagnetic spectrum and the combined
   absorption spectrum of chlorophylls a and b and the
   carotenoids. Why is chlorophyll a green pigment, and the
   carotenoids a yellow-orange pigment?   124
5. Trace the noncyclic electron pathway, naming and explaining
   all the events that occur as the electrons move from water to
   $NADP^+$.   124–25
6. How is the thylakoid membrane organized? Name the main
   complexes in the membrane. Give a function for each.   126
7. Explain what is meant by chemiosmosis, and relate this
   process to the electron transport chain present in the thylakoid
   membrane.   126
8. Describe the three stages of the Calvin cycle. Which stage uses
   the ATP and NADPH from the light reactions?   128–29
9. Compare C_3 and C_4 photosynthesis, contrasting the actions of
   RuBP carboxylase and PEPCase.   130
10. Explain CAM photosynthesis, contrasting it to C_4
    photosynthesis in terms of partitioning a pathway.   131

accepts two electrons and two hydrogen ions (H^+) to become $FADH_2$.

 **Animation** How the NAD⁺Works

Phases of Cellular Respiration

Cellular respiration involves four phases: glycolysis, the preparatory reaction, the citric acid cycle, and the electron transport chain (Fig. 8.2). Glycolysis takes place outside the mitochondria and does not require the presence of oxygen. Therefore, glycolysis is **anaerobic.** The other phases of cellular respiration take place inside the mitochondria, where oxygen is the final acceptor of electrons. Because they require oxygen, these phases are called **aerobic.**

During these phases, notice where CO_2 and H_2O, the end products of cellular respiration, and ATP, the main outcome of respiration, are produced.

- **Glycolysis** [Gk. *glycos*, sugar, and *lysis*, splitting] is the breakdown of glucose (a 6-carbon molecule) to two molecules of pyruvate (two 3-carbon molecules). Oxidation results in NADH and provides enough energy for the net gain of two ATP molecules.
- The **preparatory (prep) reaction** takes place in the matrix of the mitochondrion. Pyruvate is broken down from a 3-carbon (C_3) to a 2-carbon (C_2) acetyl group, and a 1-carbon CO_2 molecule is released. Since glycolysis ends with two molecules of pyruvate, the prep reaction occurs twice per glucose molecule.

- The **citric acid cycle** also takes place in the matrix of the mitochondrion. Each 2-carbon acetyl group matches up with a 4-carbon molecule, forming two 6-carbon citrate molecules. As citrate bonds are broken and oxidation occurs, NADH and $FADH_2$ are formed, and two CO_2 per citrate are released. The citric acid cycle is able to produce one ATP per turn. Because two acetyl groups enter the cycle per glucose molecule, the cycle turns twice.
- The **electron transport chain (ETC)** is a series of carriers on the cristae of the mitochondria. NADH and $FADH_2$ give up their high-energy electrons to the chain. Energy is released and captured as the electrons move from a higher-energy to a lower-energy state during each redox reaction. Later, this energy is used for the production of ATP by chemiosmosis. After oxygen receives electrons at the end of the chain, it combines with hydrogen ions (H^+) and becomes water (H_2O).

Pyruvate, the end product of glycolysis, is a pivotal metabolite; its further treatment depends on whether oxygen is available. If oxygen is available, pyruvate enters a mitochondrion and is broken down completely to CO_2 and H_2O as shown in the cellular respiration equation (p. 136). If oxygen is not available, pyruvate is further metabolized in the cytoplasm by an anaerobic process called **fermentation.** Fermentation results in a net gain of only two ATP per glucose molecule.

MP3 Cellular Respiration

Check Your Progress   8.1

1. Explain the benefit of slow glucose breakdown rather than rapid breakdown during cellular respiration.
2. Describe the four phases of complete glucose breakdown, including which release CO_2 and which produce H_2O.

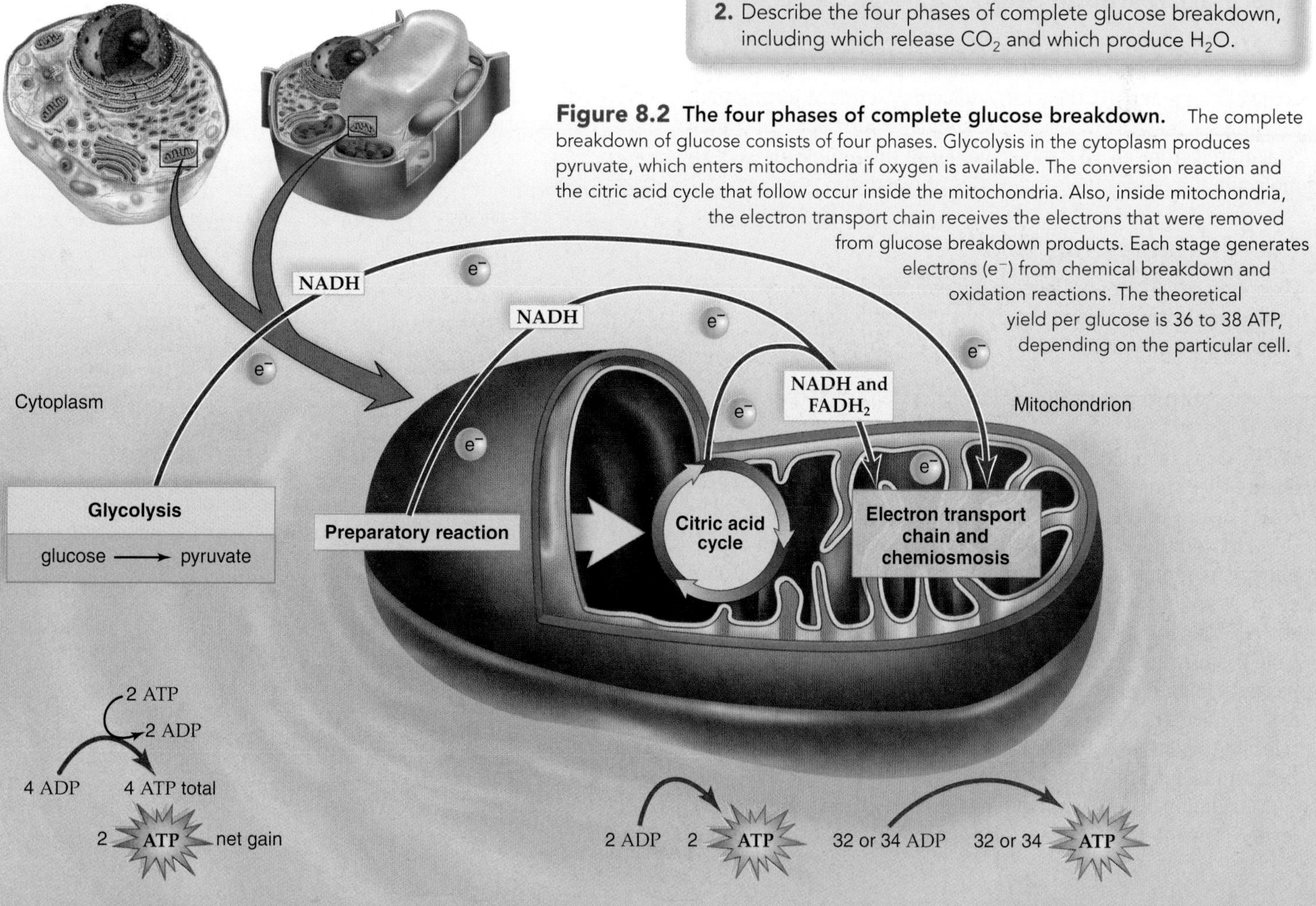

Figure 8.2 The four phases of complete glucose breakdown. The complete breakdown of glucose consists of four phases. Glycolysis in the cytoplasm produces pyruvate, which enters mitochondria if oxygen is available. The conversion reaction and the citric acid cycle that follow occur inside the mitochondria. Also, inside mitochondria, the electron transport chain receives the electrons that were removed from glucose breakdown products. Each stage generates electrons (e^-) from chemical breakdown and oxidation reactions. The theoretical yield per glucose is 36 to 38 ATP, depending on the particular cell.

NADH

NADH

e^-

e^-

e^-

NADH and $FADH_2$

e^-

e^-

e^-

e^-

Cytoplasm

Mitochondrion

Glycolysis

glucose ⟶ pyruvate

Preparatory reaction

Citric acid cycle

Electron transport chain and chemiosmosis

2 ATP

2 ADP

4 ADP   4 ATP total

2 ATP net gain

2 ADP   2 ATP   32 or 34 ADP   32 or 34 ATP

8.2 Outside the Mitochondria: Glycolysis

Glycolysis, which takes place within the cytoplasm outside the mitochondria, is the breakdown of C_6 (6-carbon) glucose to two C_3 (3-carbon) pyruvate molecules. Since glycolysis occurs universally in organisms, it most likely evolved before the citric acid cycle and the electron transport chain. This may be why glycolysis occurs in the cytoplasm and does not require oxygen. There was no free oxygen in the early atmosphere of the Earth.

Glycolysis is series of ten reactions, and just as you would expect for a metabolic pathway, each step has its own enzyme. The pathway can be conveniently divided into the energy-investment step and the energy-harvesting steps. During the energy-investment step, ATP is used to "jump-start" glycolysis. During the energy-harvesting steps, four total ATP are made, producing 2 net ATP overall.

Energy-Investment Step

As glycolysis begins, two ATP are used to activate glucose by adding phosphate. Glucose eventually splits into two C_3 molecules known as G3P, the same molecule produced during photosynthesis. Each G3P has a phosphate group, each of which is acquired from an ATP molecule. From this point on, each C_3 molecule undergoes the same series of reactions.

Energy-Harvesting Step

Oxidation of G3P now occurs by the removal of electrons accompanied by hydrogen ions. In duplicate reactions, electrons are picked up by coenzyme NAD^+, which becomes

$$\text{NADH: } 2\,NAD^+ + 4\,e^- + 2\,H^+ \longrightarrow 2\,NADH$$

When O_2 is available, each NADH molecule carries two high-energy electrons to the electron transport chain and becomes NAD^+ again. In this way, NAD^+ is recycled and used again.

The addition of inorganic phosphate results in a high-energy phosphate group on each C_3 molecule. These phosphate groups are used to directly synthesize two ATP in the later steps of glycolysis. This is called **substrate-level ATP synthesis** (sometimes called substrate-level phosphorylation) because an enzyme passes a high-energy phosphate to ADP, and ATP results (Fig. 8.3). Notice that this is an example of coupling: An energy-releasing reaction is driving forward an energy-requiring reaction on the surface of the enzyme.

Oxidation occurs again, but by the removal of H_2O. Substrate-level ATP synthesis occurs again per each C_3, and two molecules

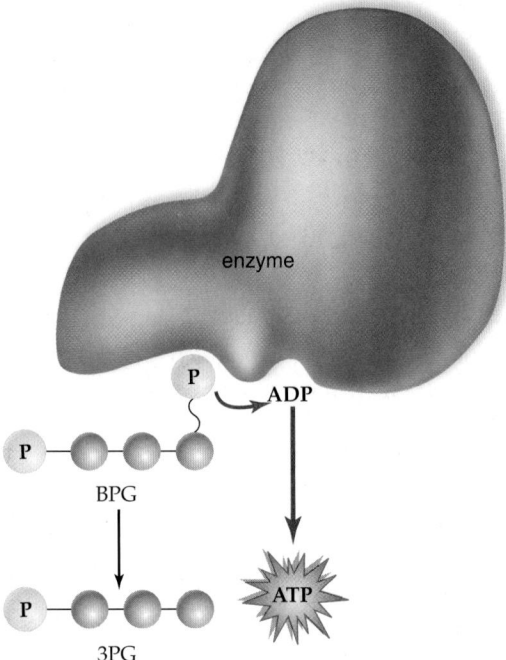

Figure 8.3 Substrate-level ATP synthesis. Substrates participating in the reaction are oriented on the enzyme. A phosphate group is transferred to ADP, producing one ATP molecule. During glycolysis (see Fig. 8.4), BPG is a C_3 substrate (each gray ball is a carbon atom) that gives up a phosphate group to ADP. This reaction occurs twice per glucose molecule.

of pyruvate result. Subtracting the two ATP that were used to get started, and the four ATP produced overall, there is a net gain of two ATP from glycolysis (Fig. 8.4).

Animation How Glycolysis Works

Inputs and Outputs of Glycolysis

All together, the inputs and outputs of glycolysis are as follows:

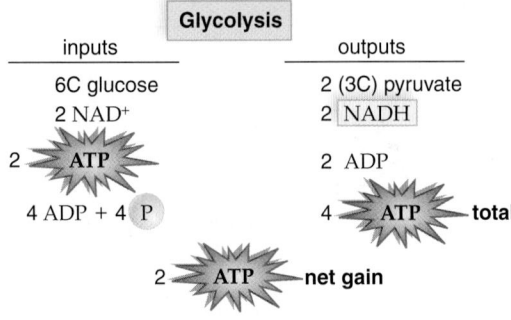

Notice that, so far, we have accounted for only two of the 36 to 38 ATP molecules that are theoretically possible when glucose is completely broken down to CO_2 and H_2O. When O_2 is available, the end product of glycolysis, pyruvate, enters the mitochondria, where it is metabolized. If O_2 is not available, fermentation, which is discussed next, occurs.

3D Animation Glycolysis

Check Your Progress   8.2

1. Examine where ATP is used and is produced in glycolysis.
2. Explain how ATP is produced from ADP and phosphate during glycolysis.

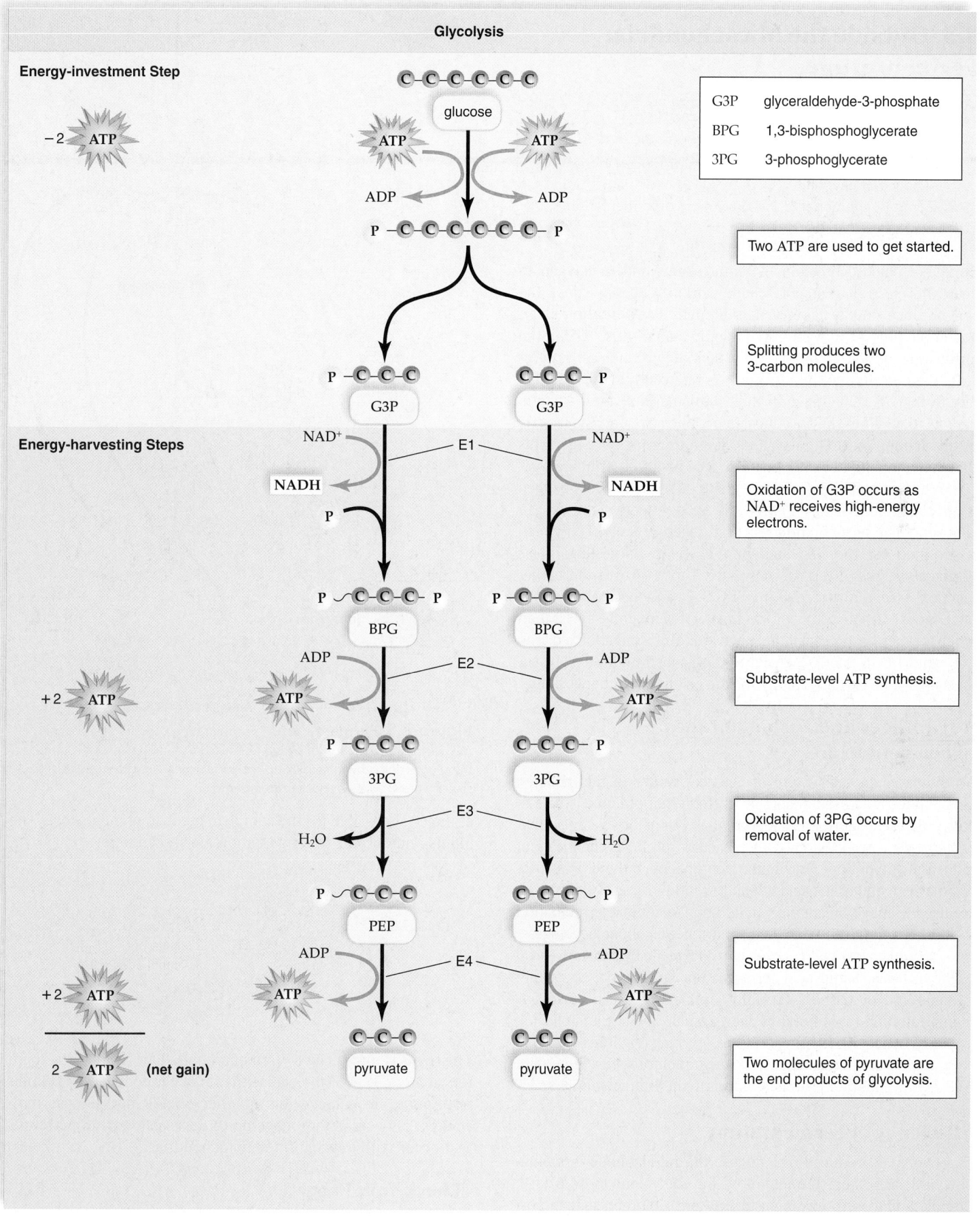

Figure 8.4 Glycolysis. This metabolic pathway begins with C$_6$ glucose (each gray ball is a carbon atom) and ends with two C$_3$ pyruvate molecules. Net gain of two ATP molecules can be calculated by subtracting those expended during the energy-investment step from those produced during the energy-harvesting steps. Each step is catalyzed by a specialized enzyme (E).

8.3 Outside the Mitochondria: Fermentation

Complete glucose breakdown requires an input of oxygen to keep the electron transport chain working. So how does the cell produce energy if oxygen is limited? **Fermentation** is an anaerobic process that produces a limited amount of ATP in the absence of oxygen. In animal cells, including human cells, pyruvate, the end product of glycolysis, is reduced by NADH to lactate (Fig. 8.5). Depending on their particular enzymes, bacteria vary as to whether they produce an organic acid, such as lactate, or an alcohol and CO_2. Yeasts are good examples of organisms that generate ethyl alcohol and CO_2 as a result of fermentation.

Why is it beneficial for pyruvate to be reduced when oxygen is not available? Because the cell still needs energy when oxygen is absent. The fermentation reaction regenerates NAD^+, which is required for the first step in the energy-harvesting phase of glycolyis. This NAD^+ is now "free" to return to the earlier reaction (see return arrow in Fig. 8.5) and become reduced once more. Although this process generates much less ATP than when oxygen is present and glucose is fully metabolized into CO_2 and H_2O in the ETC, glycolysis and substrate-level ATP synthesis produce enough energy for the cell to continue working.

Advantages and Disadvantages of Fermentation

As discussed in the Nature of Science feature in this chapter, people have long used anaerobic bacteria that produce lactate to create cheese, yogurt, and sauerkraut—even before we knew that bacteria were responsible! Other bacteria produce chemicals of industrial importance, including isopropanol, butyric acid, propionic acid, and acetic acid when they ferment. Yeasts, of course, are used to make breads rise. In addition, alcoholic fermentation is utilized to produce wine, beer, and other alcoholic beverages.

Despite its low yield of only two ATP made by substrate-level ATP synthesis, lactic acid fermentation is essential to certain animals and/or tissues. Typically, animals use lactic acid fermentation for a rapid burst of energy, such as a cheetah chasing a gazelle. Also, when muscles are working vigorously over a short period of time, lactic acid fermentation provides them with ATP, even though oxygen is temporarily in limited supply.

Efficiency of Fermentation

The two ATP produced per glucose during alcoholic fermentation and lactic acid fermentation are equivalent to 14.6 kcal. Complete glucose breakdown to CO_2 and H_2O represents a possible energy yield of 686 kcal per molecule. Therefore, the efficiency of fermentation is only 14.6 kcal/686 kcal × 100, or 2.1%

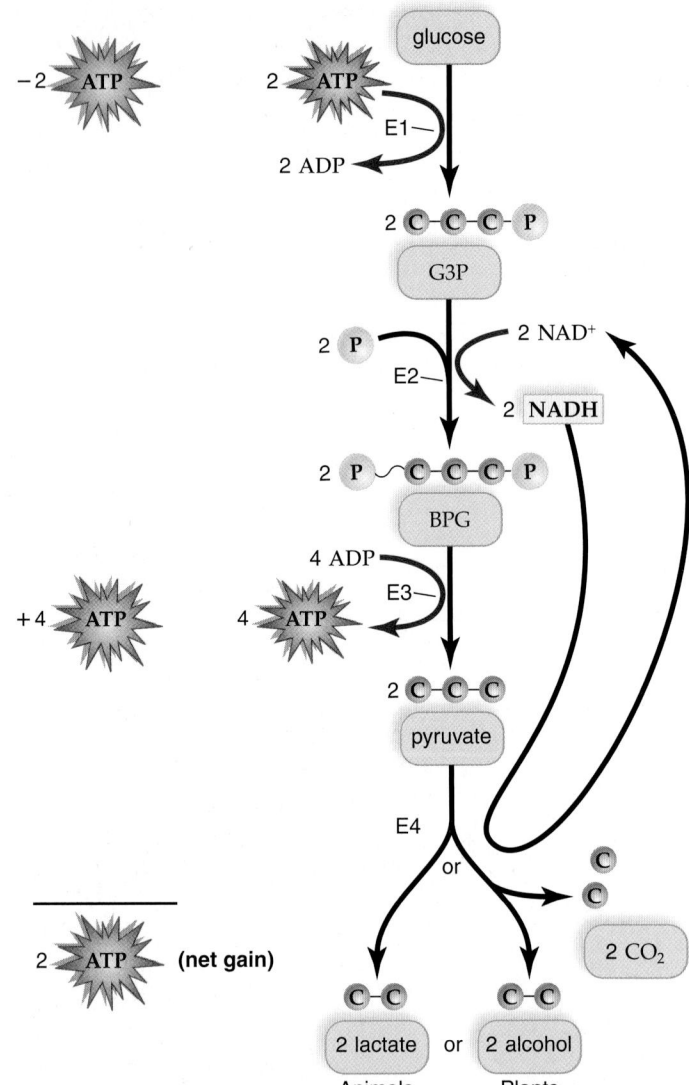

Figure 8.5 Fermentation. Fermentation consists of glycolysis followed by a reduction of pyruvate. This "frees" NAD^+ and it returns to the glycolytic pathway to pick up more electrons. As with glycolysis, each step is catalyzed by a specialized enzyme (E).

of the total possible for the complete breakdown of glucose. The inputs and outputs of fermentation are shown here:

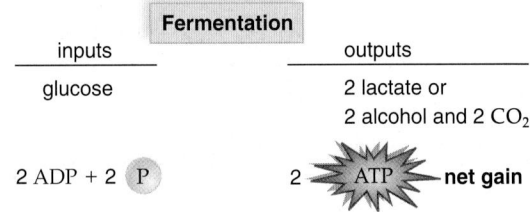

The two ATP produced by fermentation fall far short of the theoretical 36 or 38 ATP molecules that may be produced by cellular respiration. To achieve this number of ATP per glucose molecule, it is necessary to move on to the reactions and pathways that occur with oxygen in the mitochondria.

Check Your Progress 8.3

1. Describe the role of NADH in fermentation.

Nature of Science

Fermentation Helps Produce Numerous Food Products

At the grocery store, you will find such items as bread, yogurt, soy sauce, pickles, and maybe even beer or wine (**Fig. 8A**). These are just a few of the many foods that are produced when microorganisms ferment (break down sugar in the absence of oxygen). Foods produced by fermentation last longer because the fermenting organisms have removed many of the nutrients that would attract other organisms. The products of fermentation can even be dangerous to the very organisms that produced them, as when yeasts are killed by the alcohol they produce.

Yeast Fermentation

Baker's yeast, *Saccharomyces cerevisiae,* is added to bread for the purpose of leavening—the dough rises when the yeasts give off CO_2. The ethyl alcohol produced by the fermenting yeast evaporates during baking. The many different varieties of sourdough breads obtain their leavening from a starter

composed of fermenting yeasts along with bacteria from the environment. Depending on the community of microorganisms in the starter, the flavor of the bread may range from sour and tangy, as in San Francisco–style sourdough, to a milder taste, such as that produced by most Amish friendship bread recipes.

Ethyl alcohol in beer and wine is produced when yeasts ferment carbohydrates. When yeasts ferment fruit carbohydrates, the end result is wine. If they ferment grain, beer results. A few specialized varieties of beer, such as traditional wheat beers, have a distinctive sour taste because they are produced with the assistance of lactic acid–producing bacteria, such as those of the genus *Lactobacillus.* Stronger alcoholic drinks (e.g., whiskey and vodka) require distillation to concentrate the alcohol content.

Bacteria that produce acetic acid, including *Acetobacter aceti,* spoil wine. These bacteria convert the alcohol in wine or cider to acetic acid (vinegar). Until the renowned nineteenth-century scientist Louis Pasteur invented the process of pasteurization, acetic acid bacteria commonly caused wine to spoil. Although today we generally associate the process of pasteurization with making milk safe to drink, it was originally developed to reduce bacterial contamination in wine so that limited acetic acid would be produced. The discovery of pasteurization is another example of how the pursuit of scientific knowledge can positively affect our lives.

Bacterial Fermentation

Yogurt, sour cream, and cheese are produced through the action of various lactic acid bacteria that cause milk to sour. Milk contains lactose, which these bacteria use as a carbohydrate source for fermentation. Yogurt, for example, is made by adding lactic acid bacteria, such as *Streptococcus thermophilus* and *Lactobacillus bulgaricus,* to milk and then incubating it to encourage the bacteria to convert the lactose. During the production of cheese, an enzyme called rennin must also be added to the milk to cause it to coagulate and become solid.

Old-fashioned brine cucumber pickles, sauerkraut, and kimchi are pickled vegetables produced by the action of acid-producing, fermenting bacteria that can survive in high-salt environments. Salt is used to draw liquid out of

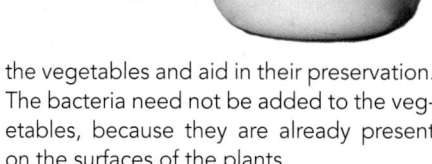

the vegetables and aid in their preservation. The bacteria need not be added to the vegetables, because they are already present on the surfaces of the plants.

Soy Sauce Production

Soy sauce is traditionally made by adding a mold, *Aspergillus,* and a combination of yeasts and fermenting bacteria to soybeans and wheat. The mold breaks down starch, supplying the fermenting microorganisms with sugar they can use to produce alcohol and organic acids.

As you can see from each of these examples, fermentation is a biologically and economically important process that scientists use for the betterment of our lives.

Questions to Consider:

1. How many products of fermentation do you consume daily?
2. What might everyday life be like without fermentation?

Figure 8A Products from fermentation. *Fermentation of different carbohydrates by microorganisms like bacteria and yeast helps produce the products shown.*

8.4 Inside the Mitochondria

The preparatory (prep) reaction, the citric acid cycle, and the electron transport chain, which are needed for the complete breakdown of glucose, take place within the mitochondria. A **mitochondrion** has a double membrane with an intermembrane space (between the outer and inner membrane). Cristae are folds of inner membrane that jut out into the matrix, the innermost compartment, which is filled with a gel-like fluid (Fig. 8.6). Like a chloroplast, a mitochondrion is highly structured, and as such we would expect reactions to be located in particular parts of this organelle.

The enzymes that speed the prep reaction and the citric acid cycle are arranged in the matrix, and the electron transport chain is located in the cristae in a very organized manner.

Most of the ATP from cellular respiration is produced in mitochondria; therefore, mitochondria are often called the powerhouses of the cell.

The Preparatory Reaction

The **preparatory (prep) reaction** is so called because it converts products from glycolysis into products that enter the citric acid cycle. In this reaction, the C_3 pyruvate is converted to a C_2 acetyl group and CO_2 is given off. This is an oxidation-reaction in which electrons are removed from pyruvate by NAD^+, and NADH is formed. One prep reaction occurs per pyruvate, so altogether, the prep reaction occurs twice per glucose molecule:

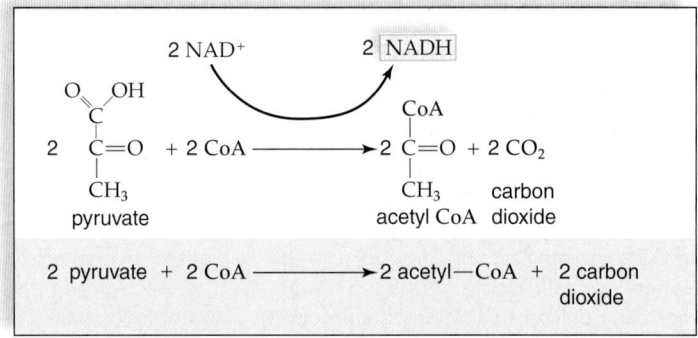

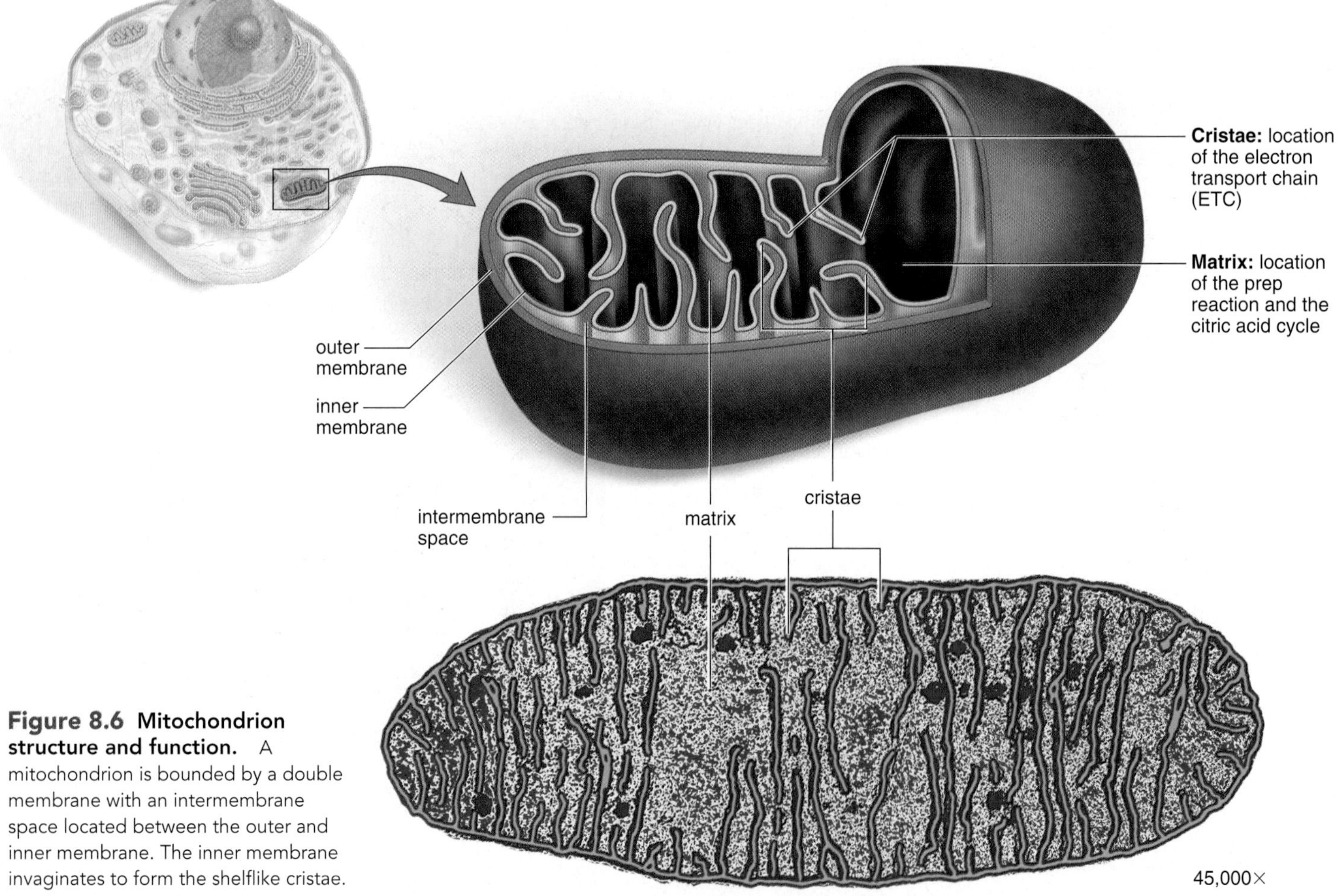

Figure 8.6 Mitochondrion structure and function. A mitochondrion is bounded by a double membrane with an intermembrane space located between the outer and inner membrane. The inner membrane invaginates to form the shelflike cristae.

Cristae: location of the electron transport chain (ETC)

Matrix: location of the prep reaction and the citric acid cycle

outer membrane

inner membrane

intermembrane space

matrix

cristae

45,000×

The C_2 acetyl group is combined with a molecule known as CoA. CoA will carry the acetyl group to the citric acid cycle in the mitochondrial matrix. The two NADH carry electrons to the electron transport chain. What about the CO_2? In vertebrates, such as ourselves, CO_2 freely diffuses out of cells into the blood, which transports it to the lungs where it is exhaled.

Citric Acid Cycle

The **citric acid cycle** is a cyclical metabolic pathway located in the matrix of mitochondria (Fig. 8.7). The citric acid cycle is also known as the Krebs cycle, after Hans Krebs, the chemist who worked out the fundamentals of the process in the 1930s.

Animation
How the Krebs
Cycle Works

At the start of the citric acid cycle, the (C_2) acetyl group carried by CoA joins with a C_4 molecule, and a C_6 citrate molecule results. During the cycle, oxidation occurs when electrons are accepted by NAD^+ in three instances and by FAD in one instance. Therefore, three NADH and one $FADH_2$ are formed as a result of one turn of the citric acid cycle. Also, the acetyl group received from the prep reaction is oxidized to two CO_2 molecules. Substrate-level ATP synthesis is also an important event of the citric acid cycle. In substrate-level ATP synthesis, you will recall, an enzyme passes a high-energy phosphate to ADP, and ATP results.

Because the citric acid cycle turns twice for each original glucose molecule, the inputs and outputs of the citric acid cycle per glucose molecule are as follows:

Citric acid cycle	
inputs	outputs
2 (2C) acetyl groups	4 CO_2
6 NAD^+	6 NADH
2 FAD	2 $FADH_2$
2 ADP + 2 P	2 ATP

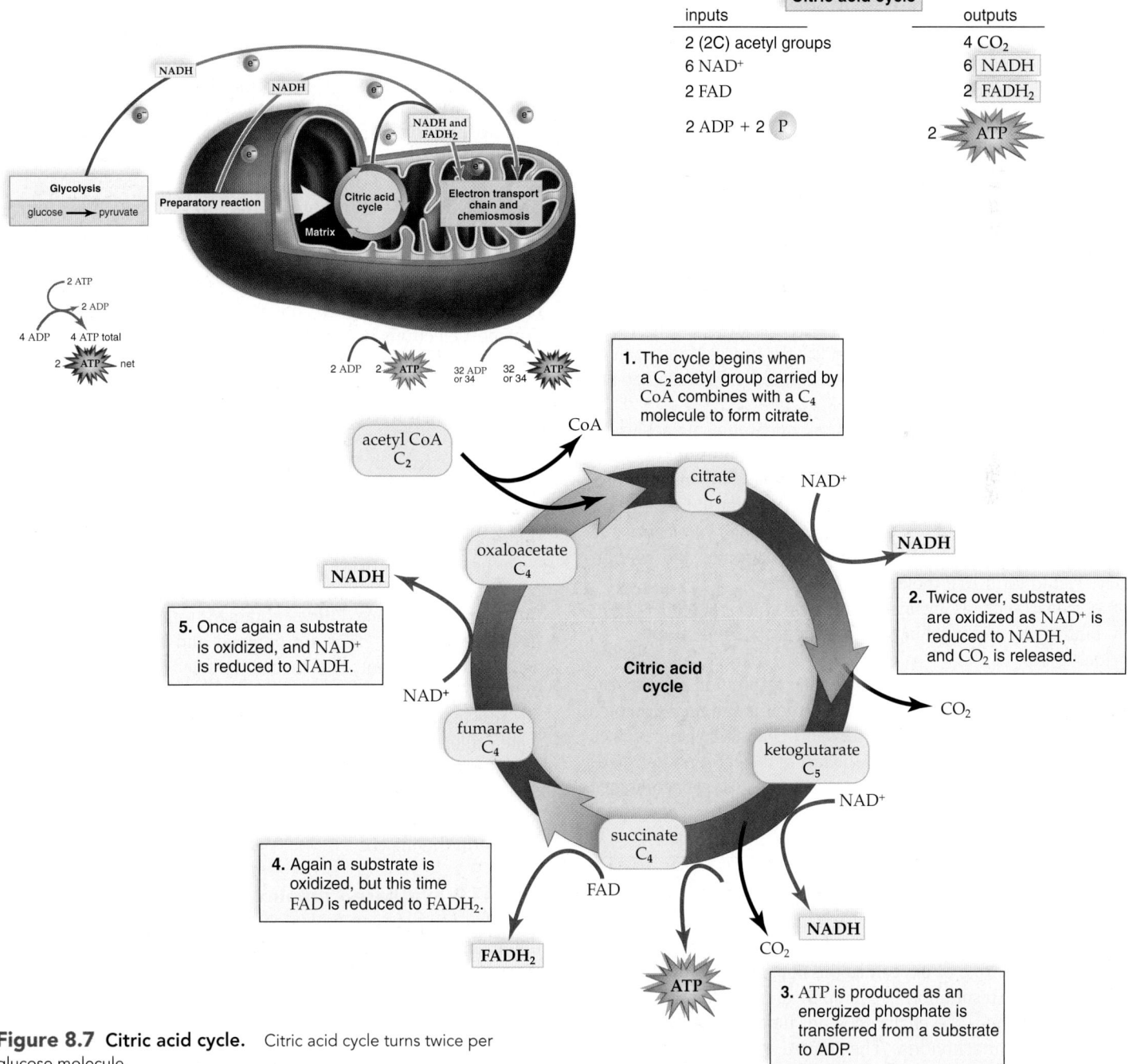

Figure 8.7 Citric acid cycle. Citric acid cycle turns twice per glucose molecule.

Production of CO₂

The six carbon atoms originally located in a glucose molecule have now become CO_2. The prep reaction produces the first two CO_2, and the citric acid cycle produces the last four CO_2 per glucose molecule. We have already mentioned that this is the CO_2 humans and other animals breathe out.

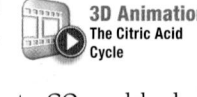
3D Animation
The Citric Acid Cycle

Thus far, we have broken down glucose to CO_2 and hydrogen atoms. Recall that, as bonds are broken and glucose gets converted to CO_2, energy in the form of high energy electrons is released. NADH and $FADH_2$ capture those high-energy electrons and carry them to the electron transport chain, as discussed next.

Electron Transport Chain

The **electron transport chain (ETC),** located in the cristae of the mitochondria and the plasma membrane of aerobic prokaryotes, is a series of carriers that pass electrons from one to the other. The high-energy electrons that enter the electron transport chain are carried by NADH and $FADH_2$. Figure 8.8 is arranged to show that high-energy electrons enter the chain, and low-energy electrons leave the chain.

Members of the Chain

When NADH gives up its electrons, it becomes oxidized to NAD^+, and when $FADH_2$ gives up its electrons, it becomes oxidized to FAD. The next carrier gains the electrons and is reduced. This oxidation-reduction reaction starts the process, and each of the carriers, in turn, becomes reduced and then oxidized as the electrons move down the chain.

Many of the redox carriers are cytochrome molecules. A **cytochrome** is a protein that has a tightly bound heme group with a central atom of iron, the same as hemoglobin does. When the iron accepts electrons, it becomes reduced, and when iron gives them up, it becomes oxidized. As the pair of electrons is passed from carrier to carrier, energy is captured and eventually used to form ATP molecules. A number of poisons, such as cyanide, cause death by binding to and blocking the function of cytochromes.

Animation
Electron Transport
System and ATP
Synthesis

What is the role of oxygen in cellular respiration and the reason we breathe to take in oxygen? Oxygen is the final acceptor of electrons from the electron transport chain. Oxygen receives the energy-spent electrons from the last of the carriers (i.e., cytochrome oxidase). After receiving electrons, oxygen combines with hydrogen ions, and water forms:

$$\tfrac{1}{2}O_2 + 2\,e^- + 2\,H^+ \longrightarrow H_2O$$

The critical role of oxygen as the final acceptor of electrons during cellular respiration is exemplified by noting that if oxygen is not present, the chain does not function, and no ATP is produced by mitochondria. The limited capacity of the body

to form ATP in a way that does not involve the electron transport chain means that death eventually results if oxygen is not available.

Cycling of Carriers

When NADH delivers high energy electrons to the first carrier of the electron transport chain, enough energy is captured by the time the electrons are received by O_2 to permit the production of three ATP molecules. When $FADH_2$ delivers high-energy electrons to the electron transport chain, two ATP are produced.

Once NADH has delivered electrons to the electron transport chain and becomes NAD^+, it is able to return and pick up more hydrogen atoms. The reuse of coenzymes increases cellular efficiency because the cell does not have to constantly make new NAD^+; it simply recycles what is already there.

The Cristae of a Mitochondrion and Chemiosmosis

The carriers of the electron transport chain and the proteins involved with ATP synthesis are spatially arranged on the cristae of mitochondria. Their sequential arrangement on the cristae allows the production of ATP to occur.

The ETC Pumps Hydrogen Ions.   Essentially, the electron transport chain consists of three protein complexes and two carriers. The three protein complexes include NADH-Q reductase complex, the cytochrome reductase complex, and cytochrome oxidase complex. The two other carriers that transport electrons between the complexes are coenzyme Q and cytochrome c (Fig. 8.8).

Animation
Proton Pump

We have already seen that the members of the electron transport chain accept electrons, which they pass from one to the other via redox reactions. So what happens to the hydrogen ions (H^+) carried by NADH and $FADH_2$? The complexes of the electron transport chain use the energy released during redox reactions to pump these hydrogen ions from the matrix into the intermembrane space of a mitochondrion.

The vertical arrows in Figure 8.8 show that the protein complexes of the electron transport chain all pump H^+ into the intermembrane space. Energy obtained from electron passage is needed because H^+ ions are pumped and actively transported against their gradient. This means the few H^+ ions in the matrix will be moved to the intermembrane space, where there are already many H^+ ions. Just as the walls of a dam hold back water, allowing it to collect, so do cristae hold back hydrogen ions. Eventually, a strong electrochemical gradient develops; about ten times as many H^+ are found in the intermembrane space as are present in the matrix.

Animation
Electron Transport System
and ATP Synthesis

The ATP Synthase Complex Produces ATP.   The ATP synthase complex can be likened to the gates of a dam. When the gates of a hydroelectric dam are opened, water rushes through, and electricity (energy) is produced. Similarly, when H^+ flows down a gradient from the intermembrane space into the matrix, the enzyme ATP synthase synthesizes ATP from ADP + Ⓟ. This process

3D Animation
Electron Transport
Chain

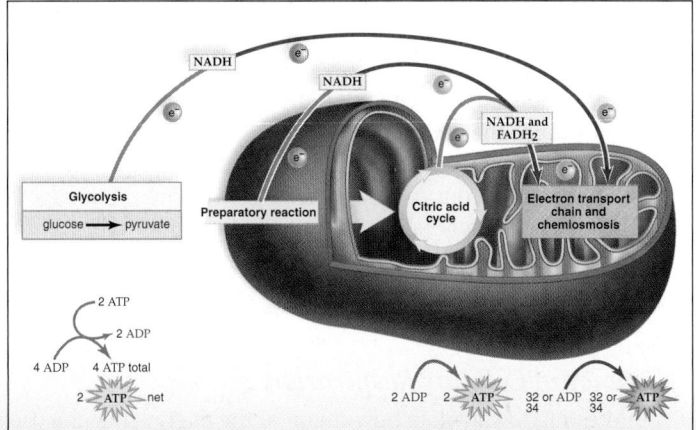

Figure 8.8 Organization and function of the electron transport chain. The electron transport chain is located in the mitochondrial cristae. NADH and FADH$_2$ bring electrons to the electron transport chain. As electrons move from one protein complex to the other via redox reactions, energy is used to pump hydrogen ions (H$^+$) from the matrix into the intermembrane space. As hydrogen ions flow down a concentration gradient from the intermembrane space into the mitochondrial matrix, ATP is synthesized by the enzyme ATP synthase. For every pair of electrons that enters by way of NADH, three ATP result. For every pair of electrons that enters by way of FADH$_2$, two ATP result. Oxygen, the final acceptor of the electrons, becomes a part of water. ATP leaves the matrix by way of a channel protein.

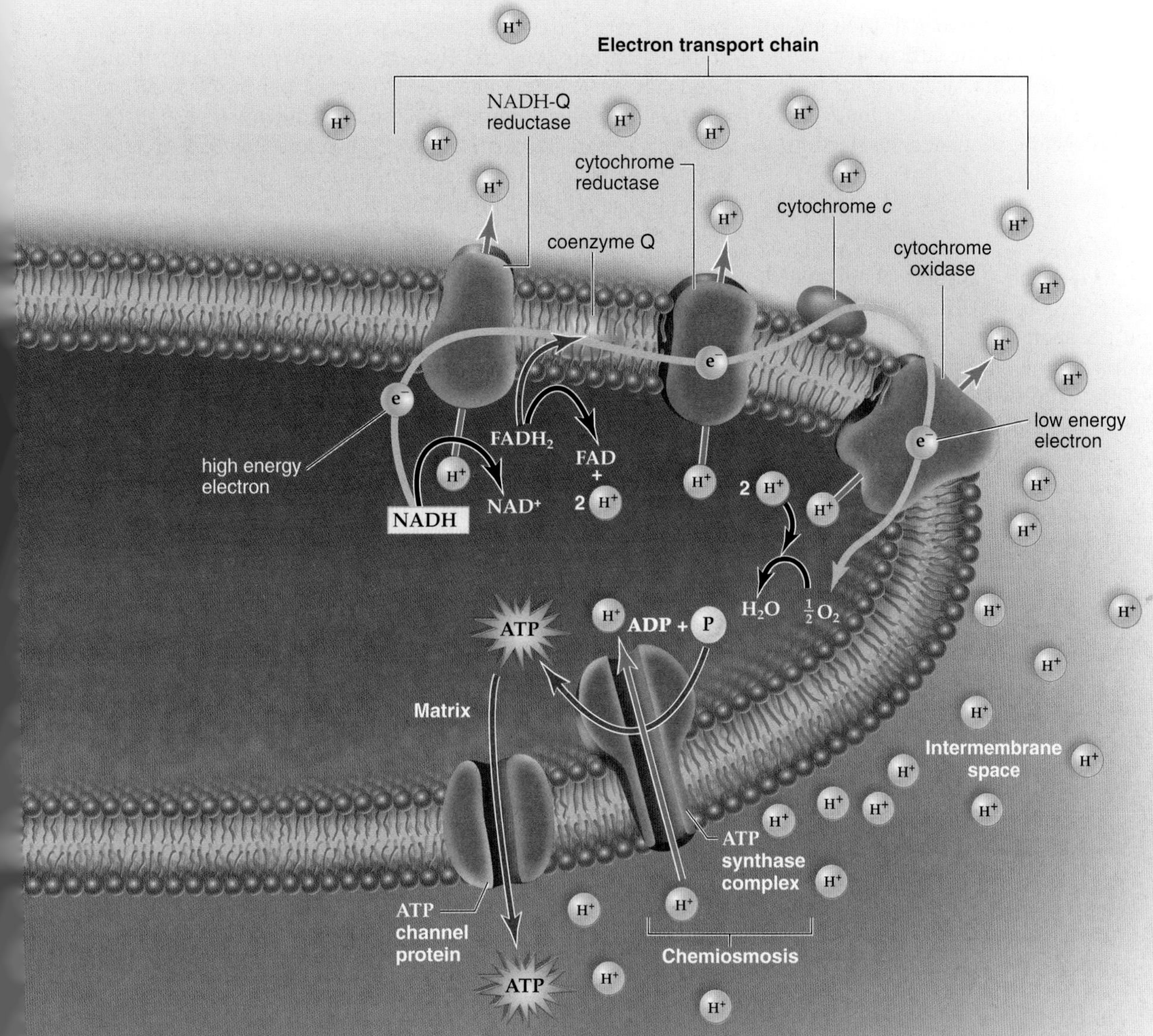

is called **chemiosmosis** because ATP production is tied to the establishment of an H$^+$ gradient.

Once formed, ATP moves out of mitochondria and is used to perform cellular work, during which it breaks down to ADP and Ⓟ. These molecules are then returned to mitochondria for recycling. At any given time, the amount of ATP in a human would sustain life for only about a minute; therefore, ATP synthase must constantly produce ATP. It is estimated that mitochondria produce our body weight in ATP every day.

Active Tissues Contain More Mitochondria. Active tissues, such as muscles, require greater amounts of ATP and have more mitochondria than less active cells. When a burst of energy is required, however, muscles still utilize fermentation.

As an example of the relative amounts of ATP, consider that the dark meat of chickens, namely the thigh meat, contains more mitochondria than the white meat of the breast. This suggests that chickens mainly walk or run, rather than fly, about the barnyard.

Energy Yield from Glucose Metabolism

Figure 8.9 calculates the theoretical ATP yield for the complete breakdown of glucose to CO_2 and H_2O during cellular respiration. Notice that the diagram includes the number of ATP produced directly by glycolysis and the citric acid cycle (to the left), as well as the number produced as a result of electrons passing down the electron transport chain (to the right). A maximum of between 32 to 34 ATP molecules may be produced by the electron transport chain.

Substrate-Level ATP Synthesis

Per glucose molecule, there is a net gain of two ATP from glycolysis, which takes place in the cytoplasm. The citric acid cycle, which occurs in the matrix of mitochondria, accounts for two ATP per glucose molecule. This means that a total of four ATP are formed by substrate-level ATP synthesis outside the electron transport chain.

ETC and Chemiosmosis

Most ATP is produced by the electron transport chain and chemiosmosis. Per glucose molecule, ten NADH and two $FADH_2$ take electrons to the electron transport chain. For each NADH formed *inside* the mitochondria by the citric acid cycle, three ATP result, but for each $FADH_2$, only two ATP are produced. Figure 8.8 explains the reason for this difference: $FADH_2$ delivers its electrons to the transport chain after NADH, and therefore these electrons do not participate in as many redox reactions and don't pump as many H^+ as NADH. Therefore, $FADH_2$ cannot account for as much ATP production.

What about the ATP yield per NADH generated *outside* the mitochondria by the glycolytic pathway? In some cells, NADH cannot cross mitochondrial membranes, but a "shuttle" mechanism allows its electrons to be delivered to the electron transport chain inside the mitochondria. The cost to the cell is one ATP for each NADH that is shuttled to the ETC. This reduces the overall count of ATP produced as a result of glycolysis, in some cells, to four instead of six ATP.

Efficiency of Cellular Respiration

It is interesting to calculate how much of the energy in a glucose molecule eventually becomes available to the cell. The difference in energy content between the reactants (glucose and O_2) and the products (CO_2 and H_2O) is 686 kcal. An ATP phosphate bond has an energy content of 7.3 kcal, and 36 of these are potentially produced during glucose breakdown; 36 phosphates are equivalent to a total of 263 kcal. Therefore, 263/686, or 39%, of the available energy is usually transferred from glucose to ATP. The rest of the energy is lost in the form of heat.

3D Animation Summary of Cellular Respiration

In the next section, we consider how cellular respiration fits into metabolism as a whole.

Check Your Progress 8.4

1. Explain when carbon is converted from glucose into carbon dioxide during cellular respiration.
2. Examine which processes during glucose breakdown produce the most ATP.
3. Compare the function of the mitochondrial inner membrane to a hydroelectric dam.

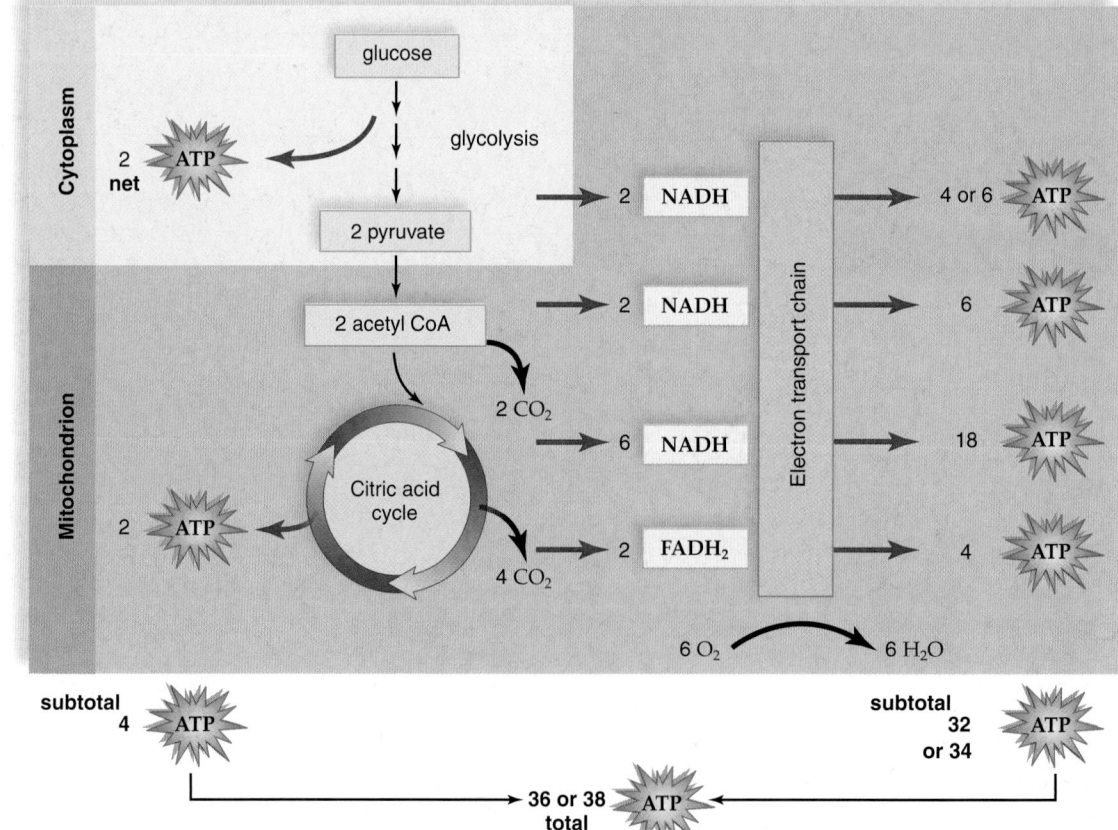

Figure 8.9 Accounting of energy yield per glucose molecule breakdown. Substrate-level ATP synthesis during glycolysis and the citric acid cycle accounts for 4 ATP. The electron transport chain accounts for 32 or 34 ATP, making the theoretical grand total of ATP between 36 and 38 ATP. Other factors may reduce the efficiency of cellular respiration. For example, cells differ as to the delivery of the electrons from NADH generated outside the mitochondria. If they are delivered by a shuttle mechanism to the start of the electron transport chain, 6 ATP result; otherwise, 4 ATP result.

8.5 Metabolic Pool

Key metabolic pathways routinely draw from pools of particular substrates needed to synthesize or degrade larger molecules. Substrates like the end product of glycolysis, pyruvate, exist as a pool that is continuously affected by changes in cellular and environmental conditions (Fig. 8.10). Degradative reactions, termed **catabolism,** that break down molecules must be dynamically balanced with constructive reactions, or **anabolism.** For example, catabolic breakdown of fats will occur when insufficient carbodydrate is present; this breakdown adds to the **metabolic pool** of pyruvate. When energy needs to be stored as fat, pyruvate is taken from the pool. This dynamic balance of catabolism and anabolism is essential to optimal cellular function.

Catabolism

We already know that glucose is broken down during cellular respiration. However, other molecules like fats and proteins can also be broken down as necessary. When a fat is used as an energy source, it breaks down to glycerol and three fatty acids. As Figure 8.10 indicates, glycerol can be converted to pyruvate and enter glycolysis. The fatty acids are converted to 2-carbon acetyl CoA that enters the citric acid cycle. An 18-carbon fatty acid results in nine acetyl CoA molecules. Calculation shows that respiration of these can produce a total of 108 ATP molecules. This is why fats are an efficient form of stored energy—the three long fatty acid chains per fat molecule can produce considerable ATP when needed.

Proteins are less frequently used as an energy source, but are available as necessary. The carbon skeleton of amino acids can enter glycolysis, be converted to acetyl groups, or enter the citric acid cycle at some other juncture. The carbon skeleton is produced in the liver when an amino acid undergoes **deamination,** or the removal of the amino group. The amino group becomes ammonia (NH_3), which enters the urea cycle and becomes part of urea, the primary excretory product of humans. Just where the carbon skeleton begins degradation depends on the length of the R group, since this determines the number of carbons left after deamination.

Anabolism

We have already mentioned that the building of new molecules requires ATP produced during breakdown of molecules. These catabolic reactions also provide the basic components used to build new molecules. For example, excessive carbohydrate intake can result in the formation of fat. Extra G3P from glycolysis can be converted to glycerol, and acetyl groups from

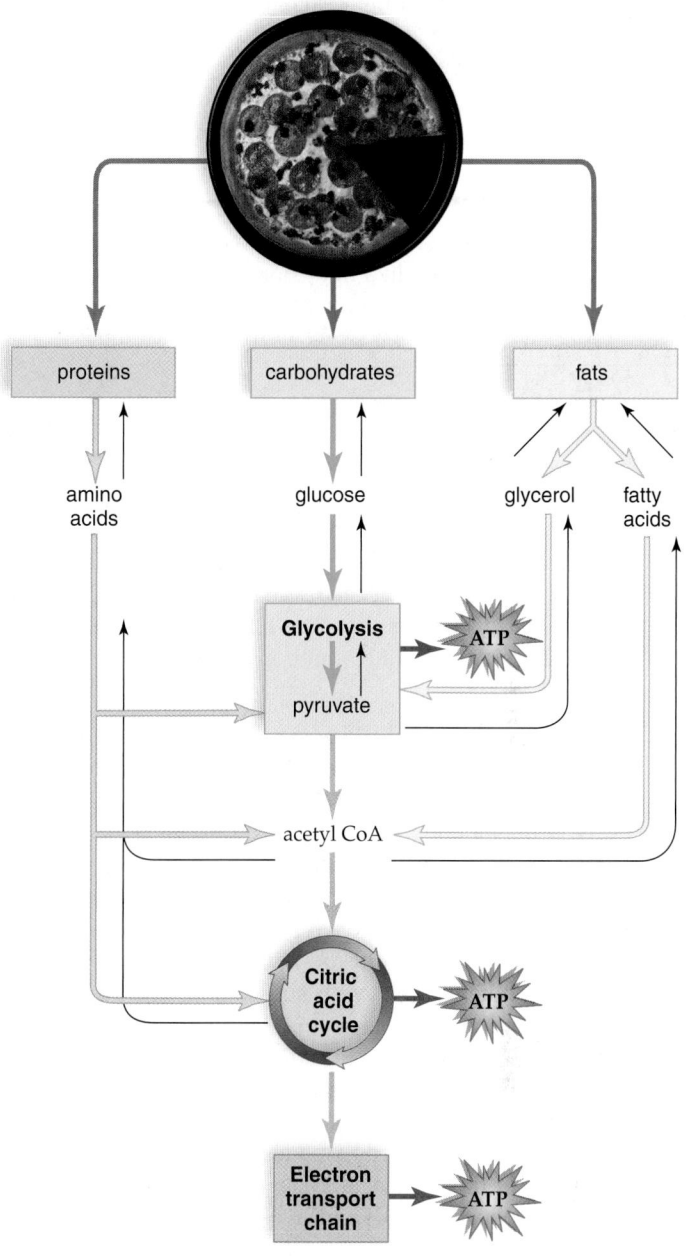

Figure 8.10 The metabolic pool concept. Carbohydrates, fats, and proteins can be used as energy sources, and their monomers (carbohydrates and proteins) or subunits (fats) enter degradative pathways at specific points. Catabolism produces molecules that can also be used for anabolism of other compounds.

glycolysis can be joined to form fatty acids, which in turn are used to synthesize fat. This explains why you gain weight from eating too much candy, ice cream, or cake.

Some substrates of the citric acid cycle can be converted to amino acids through transamination—the transfer of an amino group to an organic acid, forming a different amino acid. Plants are able to synthesize all of the amino acids they need. Animals, however, lack some of the enzymes necessary for synthesis of all amino acids. Adult humans, for example, can synthesize 11 of the common amino acids, but they cannot synthesize the other 9. The amino acids that cannot be synthesized must be supplied by the diet; they are called the essential amino acids. (The amino acids that can be synthesized are called nonessential.) It is quite possible for animals to suffer from protein deficiency if their diets do not contain adequate quantities of all the essential amino acids.

The Energy Organelles Revisited

The equation for photosynthesis in a chloroplast is opposite to that of cellular respiration in a mitochondrion (Fig. 8.11):

$$\text{energy} + 6\,CO_2 + 6\,H_2O \underset{\text{cellular respiration}}{\overset{\text{photosynthesis}}{\rightleftharpoons}} C_6H_{12}O_6 + 6\,O_2$$

While you were studying photosynthesis and cellular respiration, you may have noticed a remarkable similarity in the structural organization of chloroplasts and mitochondria. Through evolution, all organisms are related, and the similar organization of these organelles suggests that they may be related also. The two organelles carry out related but opposite processes:

1. *Use of membrane.* In a chloroplast, an inner membrane forms the thylakoids of the grana. In a mitochondrion, an inner membrane forms the convoluted cristae.
2. *Electron transport chain (ETC).* An ETC is located on the thylakoid membrane of chloroplasts and the cristae of mitochondria. In chloroplasts, the electrons passed down the ETC have been energized by the Sun; in mitochondria, energized electrons have been removed from glucose and glucose products. In both, the ETC establishes an electrochemical gradient of H^+ with subsequent ATP production by chemiosmosis.
3. *Enzymes.* In a chloroplast, the stroma contains the enzymes of the Calvin cycle and in mitochondria, the matrix contains the enzymes of the citric acid cycle. In the Calvin cycle, NADPH and ATP are used to reduce carbon dioxide to a carbohydrate. In the citric acid cycle, the oxidation of glucose products produces NADH and ATP.

Flow of Energy

The ultimate source of energy for producing a carbohydrate in chloroplasts is the Sun; the ultimate goal of cellular respiration in a mitochondrion is the conversion of carbohydrate energy into that of ATP molecules. Therefore, energy flows from the Sun, through chloroplasts to carbohydrates, and then through mitochondria to ATP molecules.

This flow of energy maintains biological organization at all levels from molecules, organisms, and ultimately the biosphere. In keeping with the energy laws, some energy is lost with each chemical transformation, and eventually, the solar energy captured by plants is lost in the form of heat. Therefore, living things depend on a continual input of solar energy.

Although energy flows through organisms, chemicals cycle within natural systems. Aerobic organisms utilize the carbohydrate and oxygen produced by chloroplasts to generate energy

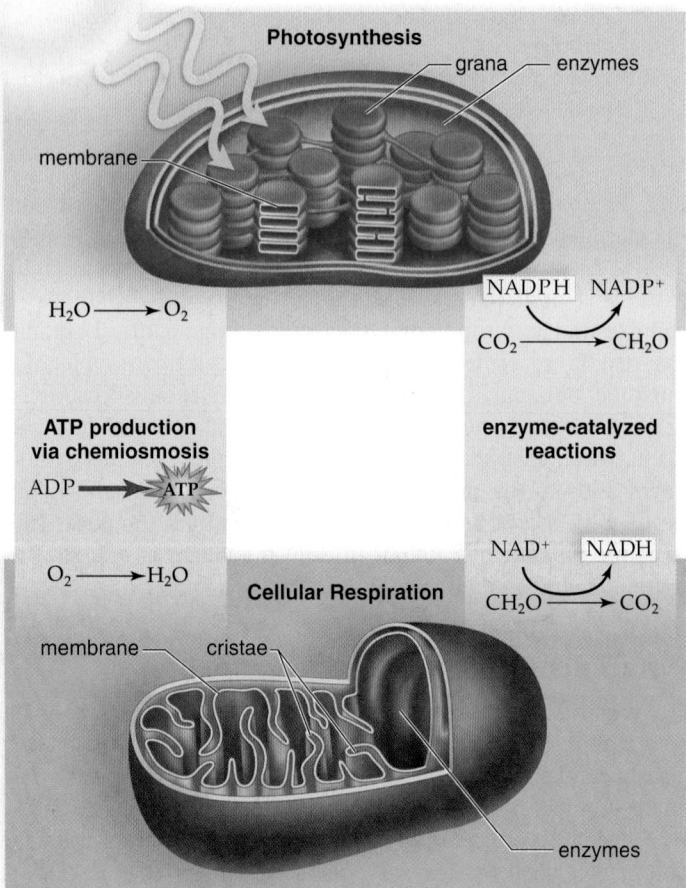

Figure 8.11 Photosynthesis versus cellular respiration.
In photosynthesis (top), water is oxidized and oxygen is released; in cellular respiration (bottom), oxygen is reduced to water. Both processes have an electron transport chain located within membranes (the grana of chloroplasts and the cristae of mitochondria), where ATP is produced by chemiosmosis. Both have enzyme-catalyzed reactions within the semifluid interior. In photosynthesis, CO_2 is reduced to a carbohydrate; in cellular respiration, a carbohydrate is oxidized to CO_2.

within the mitochondria to sustain life. Likewise, the carbon dioxide produced by mitochondria returns to chloroplasts to be used in the manufacture of carbohydrates, producing oxygen as a byproduct. Therefore, chloroplasts and mitochondria are instrumental in not only allowing a flow of energy through living things, but also permitting a cycling of chemicals.

Check Your Progress   8.5

1. Evaluate how catabolism and anabolism are balanced within a cell.
2. Compare the structure and function of chloroplasts and mitochondria.

CONNECTING *the* CONCEPTS *with the* BIG IDEAS

Energy and Homeostasis

- While the first steps of cellular respiration require no oxygen, aerobic conditions are essential for the function of the electron transport chain. In the absence of O_2, fermentations occur to eke out small amounts of ATP but create toxic byproducts like alcohol or lactic acid. (2A2b2)
- The slow, step-by-step enzymatic processes of glycolysis and Krebs cycle metabolize carbohydrates to water and CO_2, allowing slow release of free energy and partial capture in the creation of ATP by the electron transport chain; the remaining energy is lost as heat. (2A2f1-5, 2A2g1-4)
- NAD and FAD constantly ferry electrons and H^+ ions to the electron transport chain where they power ATP production by chemiosmosis. (2A2f4, 2A2g2-4)
- The final electron acceptor in respiration is oxygen. (2A2c*IE*)

Interactions and Systems

- The double-membraned structure of the mitochondrion provides increased surface area and enables compartmentalization of Krebs cycle enzyme and ETC functions. (4A2d1-3)
- In cellular respiration, carbohydrates or other classes of organic molecules are oxidized while oxygen is reduced, producing water; the "waste products" of respiration are the raw materials of photosynthesis. Energy flows but molecules recycle. (4A6a)

*Find the unabridged version of all EK citations at www.glencoe.com/maderAP11.

Media Study Tools

www.glencoe.com/maderAP11

Enhance your study of this chapter with study tools and practice tests. Also ask your instructor about the resources available through ConnectPlus, including the media-rich eBook, interactive learning tools, and animations.

3D Animation
Cellular
Respiration

For an interactive exploration of the processes of cellular respiration, take a moment to watch McGraw-Hill's new 3D animation on cellular respiration.

Summarize

8.1 Cellular Respiration

Cellular respiration, during which glucose is completely broken down to CO_2 and H_2O, consists of four phases: glycolysis, the prep reaction, the citric acid cycle, and the passage of electrons along the electron transport chain. Oxidation of substrates involves the removal of hydrogen atoms ($H^+ + e^-$), usually by redox coenzymes. NAD^+ becomes NADH, and FAD becomes $FADH_2$.

8.2 Outside the Mitochondria: Glycolysis

Glycolysis, the breakdown of glucose to two molecules of pyruvate, is a series of enzymatic reactions that occurs in the cytoplasm and is anaerobic. Breakdown releases enough energy to immediately give a net gain of two ATP by substrate-level ATP synthesis and the production of 2 NADH.

8.3 Outside the Mitochondria: Fermentation

Fermentation involves glycolysis followed by the reduction of pyruvate by NADH either to lactate (animals) or to alcohol (yeast) and carbon dioxide (CO_2). The reduction process "frees" NAD^+ so that it can accept more hydrogen atoms from glycolysis.

Although fermentation results in only two ATP molecules, it still serves a purpose. Many of the products of fermentation are used in the baking and brewing industries. In vertebrates, it provides a quick burst of ATP energy for short-term, strenuous muscular activity. The accumulation of lactate puts the individual in oxygen debt because oxygen is needed when lactate is completely metabolized to CO_2 and H_2O.

8.4 Inside the Mitochondria

When oxygen is available, pyruvate from glycolysis enters the mitochondrion, where the prep reaction takes place. During this reaction, oxidation occurs as CO_2 is removed from pyruvate. NAD^+ is reduced, and CoA receives the C_2 acetyl group that remains. Because the reaction must take place twice per glucose molecule, two NADH result.

The acetyl group enters the citric acid cycle, a cyclical series of reactions located in the mitochondrial matrix. Complete oxidation follows, as two CO_2 molecules, three NADH molecules, and one $FADH_2$ molecule are formed. The cycle also produces one ATP molecule. The entire cycle must turn twice per glucose molecule.

The final stage of glucose breakdown involves the electron transport chain located in the cristae of the mitochondria. The electrons received from NADH and $FADH_2$ are passed down a chain of carriers until they are finally received by oxygen, which combines with H^+ to produce water. As the electrons pass down the chain, energy is captured and stored for ATP production.

The cristae of mitochondria contain complexes of the electron transport chain that not only pass electrons from one to the other but also pump H^+ into the intermembrane space, setting up an electrochemical gradient. When H^+ flows down this gradient through an ATP synthase complex, energy is captured and used to form ATP molecules from ADP and Ⓟ. This is ATP synthesis by chemiosmosis.

Of the 36 or 38 ATP formed by complete glucose breakdown, four are the result of substrate-level ATP synthesis and the rest are produced as a result of the electron transport chain. For most NADH molecules that donate electrons to the electron transport chain, three ATP molecules are produced. However, in some cells, each NADH formed in the cytoplasm results in only two ATP molecules because a shuttle, rather than NADH, takes electrons through the mitochondrial membrane. $FADH_2$ results in the formation of only two ATP because its electrons enter the electron transport chain at a lower energy level.

8.5 Metabolic Pool

Carbohydrate, protein, and fat can be metabolized by entering the degradative pathways at different locations. These pathways also provide metabolites needed for the anabolism of various important substances. Therefore, catabolism and anabolism both use the same pools of metabolites.

Similar to the metabolic pool concept, photosynthesis and cellular respiration can be compared. For example, both utilize an ETC and chemiosmosis. As a result of the ETC in chloroplasts, water is split, while in mitochondria, water is formed. The enzymatic reactions in chloroplasts reduce CO_2 to a carbohydrate, while the enzymatic reactions in mitochondria oxidize carbohydrate with the release of CO_2.

Key Terms

aerobic 137	FAD 136
anabolism 147	fermentation 137, 140
anaerobic 137	glycolysis 137, 138
catabolism 147	metabolic pool 147
cellular respiration 136	mitochondrion 142
chemiosmosis 145	NAD^+ 136
citric acid cycle 137, 143	preparatory (prep)
cytochrome 144	reaction 137, 142
deamination 147	substrate-level
electron transport chain	ATP synthesis 138
(ETC) 137, 144	

 **Assess**

Reviewing This Chapter

1. What is the overall chemical equation for the complete breakdown of glucose to CO_2 and H_2O? Explain how this is an oxidation-reduction reaction.   136
2. What are NAD^+ and FAD? What are their functions?   136–37
3. Briefly describe the four phases of cellular respiration.   137
4. What are the main events of glycolysis? How is ATP formed?   138–39
5. What is fermentation, and how does it differ from glycolysis? Mention the benefit of pyruvate reduction during fermentation. What types of organisms carry out lactic acid fermentation, and what types carry out alcoholic fermentation?   140–41

6. Give the substrates and products of the prep reaction. Where does it take place?   142–43
7. What are the main events of the citric acid cycle?   143
8. What is the electron transport chain, and what are its functions?   144–46
9. Describe the organization of protein complexes within the cristae. Explain how the complexes are involved in ATP production.   145
10. Calculate the theoretical energy yield of glycolysis and complete glucose breakdown. Compare the yields from substrate-level ATP synthesis and from the electron transport chain.   146
11. Give examples to support the concept of the metabolic pool.   147
12. Compare the structure and function of chloroplasts and mitochondria. Explain the flow of energy concept.   148

Testing Yourself

Choose the best answer for each question.
For questions 1–8, identify the pathway involved by matching each description to the terms in the key.

KEY:
   a. glycolysis
   b. citric acid cycle
   c. electron transport chain

1. carbon dioxide (CO_2) given off
2. water (H_2O) formed
3. G3P
4. NADH becomes NAD^+
5. pump H^+
6. cytochrome carriers
7. pyruvate
8. FAD becomes $FADH_2$
9. The prep reaction
   a. connects glycolysis to the citric acid cycle.
   b. gives off CO_2.
   c. uses NAD^+.
   d. results in an acetyl group.
   e. All of these are correct.
10. The greatest contributor of electrons to the electron transport chain is
   a. oxygen.
   b. glycolysis.
   c. the citric acid cycle.
   d. the prep reaction.
   e. fermentation.
11. Substrate-level ATP synthesis takes place in
   a. glycolysis and the citric acid cycle.
   b. the electron transport chain and the prep reaction.
   c. glycolysis and the electron transport chain.
   d. the citric acid cycle and the prep reaction.
   e. Both b and d are correct.
12. Which of these is not true of fermentation?
   a. net gain of only two ATP
   b. occurs in cytoplasm
   c. NADH donates electrons to electron transport chain
   d. begins with glucose
   e. carried on by yeast

13. Fatty acids are broken down to
    a. pyruvate molecules, which take electrons to the electron transport chain.
    b. acetyl groups, which enter the citric acid cycle.
    c. amino acids, which excrete ammonia.
    d. glycerol, which is found in fats.
    e. All of these are correct.

14. How many ATP molecules are usually produced per NADH?
    a. 1
    b. 3
    c. 36
    d. 10

15. How many NADH molecules are produced during the complete breakdown of one molecule of glucose?
    a. 5
    b. 30
    c. 10
    d. 6

16. What is the name of the process that adds the third phosphate to an ADP molecule using the flow of hydrogen ions?
    a. substrate-level ATP synthesis
    b. fermentation
    c. reduction
    d. chemiosmosis

17. The metabolic process that produces the most ATP molecules is
    a. glycolysis.
    b. citric acid cycle.
    c. electron transport chain.
    d. fermentation.

18. Which of these is not true of the citric acid cycle? The citric acid cycle
    a. includes the prep reaction.
    b. produces ATP by substrate-level ATP synthesis.
    c. occurs in the mitochondria.
    d. is a metabolic pathway, as is glycolysis.

19. Which of these is not true of the electron transport chain? The electron transport chain
    a. is located on the cristae.
    b. produces more NADH than any metabolic pathway.
    c. contains cytochrome molecules.
    d. ends when oxygen accepts electrons.

20. The oxygen required by cellular respiration is reduced and becomes part of which molecule?
    a. ATP
    b. H_2O
    c. pyruvate
    d. CO_2

Engage

Thinking Scientifically

1. You are able to extract mitochondria from the cell and remove the outer membrane. You want to show that the mitochondria can still produce ATP if placed in the right solution. The solution should be isotonic, but at what pH? Why?

2. You are working with acetyl CoA molecules that contain only radioactive carbon. They are incubated with all the components of the citric acid cycle long enough for one turn of the cycle. Examine Figure 8.7 and explain why the carbon dioxide given off is radioactive.

Bioethical Issue

Alternative Medicine

Feeling tired and run-down? Want to jump-start your mitochondria? If you seem to have no specific ailment, you might be tempted to turn to what is now called alternative medicine. Alternative medicine includes such nonconventional therapies as herbal supplements, acupuncture, chiropractic therapy, homeopathy, osteopathy, and therapeutic touch (e.g., laying on of hands).

Advocates of alternative medicine have made some headway in having alternative medicine practices accepted by almost anyone. For example, Congress has established the National Center for Complementary and Alternative Medicine. It has also passed the Dietary Supplement Health and Education Act, which allows vitamins, minerals, and herbs to be marketed without first being approved by the Food and Drug Administration (FDA).

But is this a mistake? Many physicians believe controlled studies are needed to test the efficacy of alternative medications and practices. Do you agree? Should every food supplement or approach to health be subject to scientific testing, or are there other ways to evaluate successful treatment? Explain your reasoning.

UNIT 2

Genetic Basis of Life

Of all the topics in the world of science, the fields of genetics and molecular biology are changing the most rapidly and dramatically. When the AP Biology course first started, the momentous Watson-Crick paper titled "A Structure for Deoxyribose Nucleic Acid" was just 3 years old. **BI3** Today, the DNA molecule and all its ancillaries and offshoots, such as genomics and biotechnology, are the prime focus of the third of the Big Ideas in the AP Biology Curriculum Framework: "Living systems store, retrieve, transmit and respond to information essential to life processes." You will also find pieces of the other Big Ideas throughout the chapters:

 Evolutionarily, meiosis and the genotypes it produces provide the raw material for natural selection.

 Populations able to respond to change usually exhibit great genetic diversity.

Well before the importance of the double helix was realized, Gregor Mendel pieced together the basics of the mechanism by which many traits are passed to offspring. Once a genotype is set by meiosis and fertilization, the process of mitosis can take over, creating exact nuclear copies so that each cell produced has a complete set of plans. Ironically, identical divisions still allows for differentiation and specialization because of complex regulatory mechanisms that turn some genes on and others off in a variety of ways. RNA molecules that were always considered the "supporting cast" for the DNA stars have proved to be essential in editing and in sometimes silencing messages on their way to translation. Then there is the stuff of science fiction—the ever-enlarging field of biotechnology. From the movies *GATTACA* to *Jurassic Park*, things that were once considered impossible don't seem so far-fetched anymore. The news is filled with reports about the ability to clone yet another species, crops modified to double the protein content, gene therapy to cure dreadful diseases, development of a new CSI biotech tool...Once we understood the coding system for life, all things became possible. Perhaps the largest question for society now is "Just because we can do something, should we?"

Take time to see the "big picture" of genetics. Once you do, the stages of mitosis and meiosis that produce very specific DNA outcomes suddenly make sense: the elegant process of DNA replication, coupled with the opportunity to modify and regulate the messages it transcribes, becomes so logical; the new abilities to decipher and tinker with the alphabet of life become so compelling. The greatest secrets of life are inscribed in a 4-letter code: understand it and the future is yours.

UNIT OUTLINE

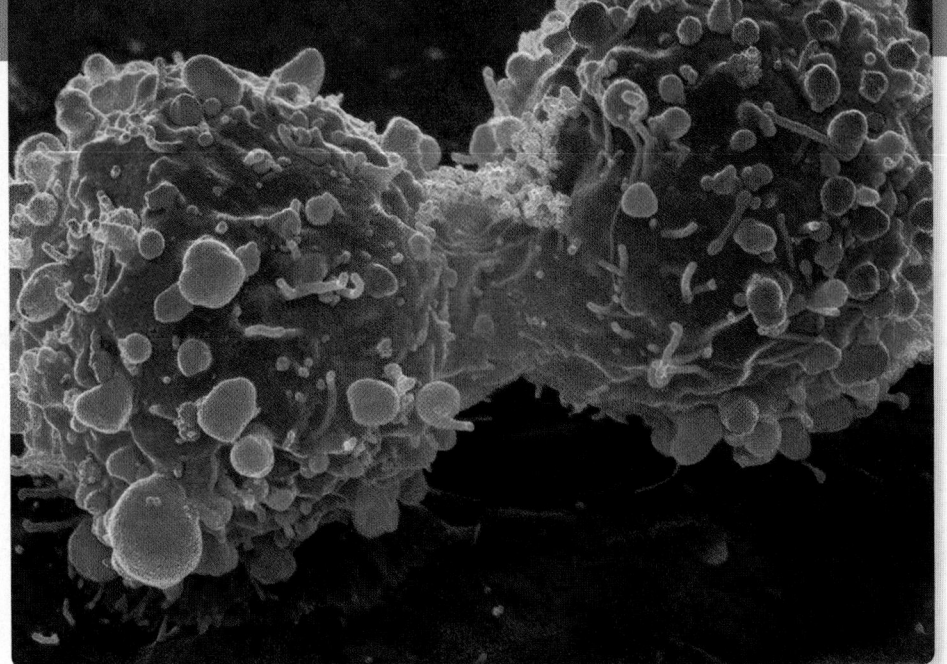

A cell may become cancerous when regulation of cell division fails.

9

The Cell Cycle and Cellular Reproduction

The process of cell division is highly regulated. In humans, life begins as a single cell, yet in a very short period of time the process of cell division produces trillions of cells, each specialized for a special function. Over 200 different types of cells are found in the human body, and although each is specialized, they all work together in harmony.

But what happens when the regulation of cell division fails? In the United States this year, over 68,000 individuals will be diagnosed with melanoma, a form of skin cancer, and around 8,000 people will die from this disease. In many instances of melanoma, exposure to ultraviolet radiation (UV) from the Sun has caused a mutation in the regulatory mechanisms of the cell cycle. Without proper regulation, cell division occurs continuously, a characteristic of cancer. For melanoma, this loss of cell cycle control results from a mutation in a gene known as *CDKN2A*. This gene is an example of a tumor suppressor gene, one of the key regulatory mechanisms of the cell cycle. In this chapter we describe the process of cell division, how it is regulated, and how cancer may develop when regulatory mechanisms malfunction.

As you read through the chapter, think about the following questions:

1. What is the normal sequence of events in the process of cellular reproduction?

2. What are the roles of the checkpoints in a cell cycle?

3. How do tumor suppressor genes regulate the cell cycle?

BEFORE YOU BEGIN

Before beginning this chapter, take a few moments to review the following discussions.

Section 3.5 What is the role of the DNA in a cell?

Sections 4.2 and 4.3 What are the major differences between prokaryotic and eukaryotic cells?

Section 4.8 What is the role of the cytoskeleton in a eukaryotic cell?

FOLLOWING *the* BIG IDEAS

CHAPTER 9   THE CELL CYCLE AND CELLULAR REPRODUCTION

Information and Signaling

For unicellular organisms, cell division results in the formation of two new organisms, while in multicellular organisms it is the basis of growth and repair.

9.1 The Cell Cycle

The **cell cycle** is an orderly set of stages that take place between the time a eukaryotic cell divides and the time the resulting daughter cells also divide. When a cell is going to divide, it grows larger, the number of organelles doubles, and the amount of DNA doubles as DNA replication occurs. The two portions of the cell cycle are interphase, which includes a number of stages, and the mitotic stage when mitosis and cytokinesis occur.

Animation Overview of Cell Division

Interphase

As Figure 9.1 shows, most of the cell cycle is spent in **interphase.** This is the time when a cell performs its usual functions, depending on its location in the body. The amount of time the cell takes for interphase varies widely. Embryonic cells complete the entire cell cycle in just a few hours. For adult mammalian cells, interphase lasts for about 20 hours, which is 90% of the cell cycle. In the past, interphase was known as the resting stage. However, today it is known that interphase is very busy, and that preparations are being made for mitosis. Interphase consists of three stages, referred to as G_1, S, and G_2.

3D Animation Interphase

G_1 Stage

Cell biologists named the stage before DNA replication G_1, and they named the stage after DNA replication G_2. G stood for "gap," but now that we know how metabolically active the cell is, it is better to think of G as standing for "growth." During G_1, the cell recovers from the previous division. The cell grows in size, increases the number of organelles (such as mitochondria and ribosomes), and accumulates materials that will be used for DNA synthesis. Otherwise, cells are constantly performing their normal daily functions during G_1, including communicating with other cells, secreting substances, and carrying out cellular respiration.

Some cells, such as nerve and muscle cells, typically do not complete the cell cycle and are permanently arrested. These cells exit interphase and enter a stage called G_0. While in the G_0 stage, the cells continue to perform normal everyday processes, but no preparations are being made for cell division. Cells may not leave the G_0 stage without proper signals from other cells and other parts of the body. Thus, completion of the cell cycle is very tightly controlled.

S Stage

Following G_1, the cell enters the S stage, when DNA synthesis or replication occurs. At the beginning of the S stage, each chromosome is composed of one DNA double helix. Following DNA replication, each chromosome is composed of two identical DNA double helix molecules. Each double helix is called a **chromatid,** and the two identical chromatids are referred to as **sister chromatids.** The sister chromatids remain attached until they are separated during mitosis.

G_2 Stage

Following the S stage, G_2 is the stage from the completion of DNA replication to the onset of mitosis. During this stage, the cell synthesizes proteins that will assist cell division. For example, it makes the proteins that form microtubules. Microtubules are used during the mitotic stage to form the mitotic spindle that is critical during M stage.

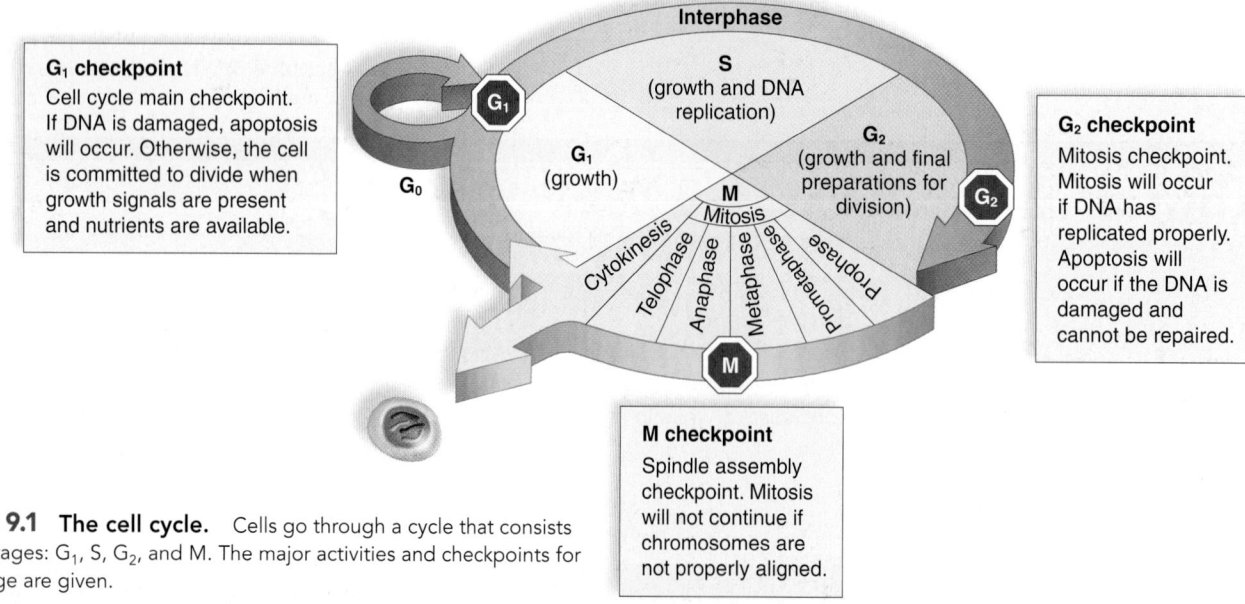

G_1 checkpoint
Cell cycle main checkpoint. If DNA is damaged, apoptosis will occur. Otherwise, the cell is committed to divide when growth signals are present and nutrients are available.

G_2 checkpoint
Mitosis checkpoint. Mitosis will occur if DNA has replicated properly. Apoptosis will occur if the DNA is damaged and cannot be repaired.

M checkpoint
Spindle assembly checkpoint. Mitosis will not continue if chromosomes are not properly aligned.

Interphase

S (growth and DNA replication)

G_1 (growth)

G_2 (growth and final preparations for division)

M

Mitosis

Cytokinesis Telophase Anaphase Metaphase Prometaphase Prophase

G_0

Figure 9.1 The cell cycle. Cells go through a cycle that consists of four stages: G_1, S, G_2, and M. The major activities and checkpoints for each stage are given.

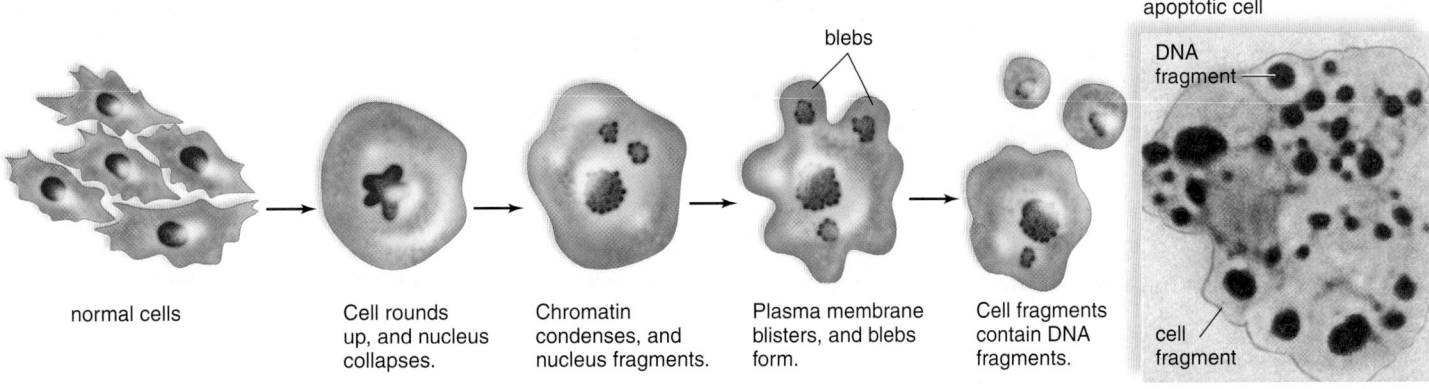

Figure 9.2 Apoptosis. Apoptosis is a sequence of events that results in a fragmented cell. The fragments are phagocytized (engulfed) by white blood cells and neighboring tissue cells.

M (Mitotic) Stage

Following interphase, the cell enters the M (for mitotic) stage. This cell division stage includes **mitosis** (nuclear division) and **cytokinesis** (division of the cytoplasm). During mitosis, daughter chromosomes are distributed by the **mitotic spindle** to two daughter nuclei. When division of the cytoplasm is complete, two daughter cells are present.

Control of the Cell Cycle

A **signal** is an agent that influences the activities of a cell. **Growth factors** are signaling proteins received at the plasma membrane. Even cells arrested in G_0 will finish the cell cycle if stimulated to do so by growth factors. In general, signals ensure that the cell cycle stages follow one another in the normal sequence.

> **Animation**
> Cell Proliferation
> Signaling Pathway

Cell Cycle Checkpoints

The red barriers in Figure 9.1 represent three checkpoints at which the cell cycle either stops or continues on, depending on the internal signals received. Researchers have identified a family of internal signaling proteins called **cyclins** that increase and decrease as the cell cycle continues. Specific cyclins must be present for the cell to proceed from the G_1 stage to the S stage and from the G_2 stage to the M stage.

> **3D Animation**
> Checkpoints

As discussed in the Nature of Science feature "The G_1 Checkpoint," the primary checkpoint of the cell cycle is the G_1 checkpoint. In mammalian cells, the signaling protein **p53** stops the cycle at the G_1 checkpoint when DNA is damaged. (In the name p53, p stands for *protein* and 53 represents its molecular weight in kilodaltons.) First, p53 attempts to initiate DNA repair, but rising levels of p53 can bring about **apoptosis**, which is programmed cell death (Fig. 9.2). Another protein, called **RB**, is responsible for interpreting growth signals and also nutrient availability signals. RB stands for *retinoblastoma*, a cancer of the retina that occurs when the *RB* gene undergoes a mutation.

The cell cycle may also stop at the G_2 checkpoint if DNA has not finished replicating. This checkpoint prevents the initiation of the M stage before completion of the S stage. If DNA is physically damaged, such as from exposure to solar radiation or X-rays, the G_2 checkpoint also offers the opportunity for DNA to be repaired.

Another cell cycle checkpoint occurs during the mitotic stage. The cycle stops if the chromosomes are not properly attached to the mitotic spindle. Normally, the mitotic spindle ensures that the chromosomes are distributed accurately to the daughter cells.

> **Animation**
> Control of the Cell
> Cycle

Apoptosis

Apoptosis is often defined as programmed cell death because the cell progresses through a typical series of events that bring about its destruction (Fig. 9.2). The cell rounds up, causing it to lose contact with its neighbors. The nucleus fragments, and the plasma membrane develops blisters. Finally, the cell fragments are engulfed by white blood cells and/or neighboring cells.

A remarkable finding of the past few years is that the enzymes that bring about apoptosis, called *caspases,* are always present in the cell. The enzymes are ordinarily held in check by inhibitors, but they can be unleashed by either internal or external signals.

Apoptosis and Cell Division. In living systems, opposing events keep the body in balance and maintain homeostasis. cell division and apoptosis are two opposing processes that keep the number of cells in the body at an appropriate level. Cell division increases and apoptosis decreases the number of **somatic** (body) **cells.** Both are normal parts of growth and development. An organism begins as a single cell that repeatedly divides to produce many cells, but eventually some cells must die for the organism to take shape. For example, when a tadpole becomes a frog, the tail disappears as apoptosis occurs. In a human embryo, the fingers and toes are at first webbed, but then they are usually freed from one another as a result of apoptosis.

Cell division occurs during your entire life. Even now, your body is producing thousands of new red blood cells, skin cells, and cells that line your respiratory and digestive tracts. Also, if you suffer a cut, cell division repairs the injury. Apoptosis occurs all the time too, particularly if an abnormal cell that could become cancerous appears, or a cell becomes infected with a virus. Death through apoptosis prevents a tumor from developing and helps to limit the spread of viruses.

Check Your Progress 9.1

1. List, in order, the four stages of the cell cycle and briefly summarize what is happening at each stage.
2. Explain what conditions might cause a cell to halt the cell cycle.
3. Discuss how apoptosis represents a regulatory event of the cell cycle.

Nature of Science

The G₁ Checkpoint

Cell division is very tightly regulated so that only certain cells in an adult body are actively dividing. After cell division occurs, cells enter the G_1 stage. Upon completing G_1, they will divide again, but before this happens they have to pass through the G_1 checkpoint.

The G_1 checkpoint ensures that conditions are right for making the commitment to divide by evaluating the meaning of growth signals, determining the availability of nutrients, and assessing the integrity of DNA. Failure to meet any one of these criteria results in a cell's halting the cell cycle and entering G_0 stage, or undergoing apoptosis if the problems are severe.

Evaluating Growth Signals

Multicellular organisms tightly control cell division so that it occurs only when needed. Signaling molecules, such as hormones, may be sent from nearby cells or distant tissues to encourage or discourage cells from entering the cell cycle. Such signals may cause a cell to enter a G_0 stage, or complete G_1 and enter the S stage. Growth signals that promote cell division cause a cyclin-dependent-kinase (CDK) to add a phosphate group to the RB protein, a major regulator of the G_1 checkpoint.

Ordinarily, a protein called E2F is bound to RB, but when RB is phosphorylated, its shape changes and it releases E2F. Now, E2F binds to DNA, activating certain genes whose products are needed to complete the cell cycle (Fig. 9Aa). Likewise, growth signals prompt cells that are in G_0 stage to reenter the G_1 stage, complete it, and enter the S stage. If growth signals are sufficient, a cell passes through the G_1 checkpoint and cell division occurs.

Determining Nutrient Availability

Just as experienced hikers ensure that they have sufficient food for their journey, a cell ensures that nutrient levels are adequate before committing to cell division. For example, scientists know that starving cells in culture enter G_0. At that time, phosphate groups are removed from RB (see reverse arrows in Figure 9Aa); RB does not release E2F; and the proteins needed to complete the cell cycle are not produced. When nutrients become available, CDKs bring about the phosphorylation of RB, which then releases E2F (see forward arrows in Figure 9Aa). After E2F binds to DNA, proteins needed to complete the cell cycle are produced. Therefore, you can see that cells do not commit to divide until conditions are conducive for them to do so.

Assessing DNA Integrity

For cell division to occur, DNA must be free of errors and damage. The p53 protein is involved in this quality control function. Ordinarily, p53 is broken down because it has no job to do. In response to DNA damage, CDK phosphorylates p53 (Fig. 9Ab). Now, the molecule is not broken down as usual, and instead its level in the nucleus begins to rise. Phosphorylated p53 binds to DNA; certain genes are activated; and DNA repair proteins are produced. If the DNA damage cannot be repaired, p53 levels continue to rise, and apoptosis is triggered. If the damage is successfully repaired, p53 levels fall, and the cell is allowed to complete G_1 stage—as long as growth signals and nutrients are present, for example.

Actually, many criteria must be met for a cell to commit to cell division, and the failure to meet any one of them may cause the cell cycle to be halted and/or apoptosis to be initiated. The G_1 checkpoint is currently an area of intense research because understanding it holds the key to possibly curing cancer, and for unleashing the power of normal, healthy cells to regenerate tissues, which could be used to cure many other human conditions.

Questions to Consider

1. What potentially could be the effect of an abnormally high level of a growth hormone on the regulation of the cell cycle?
2. Why might some cancers be associated with a mutation in the gene encoding the p53 protein?

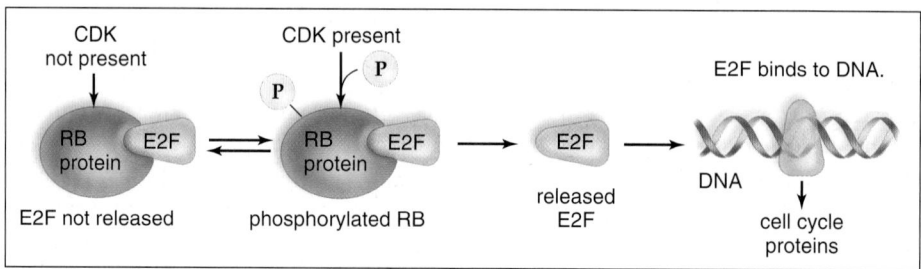

a.

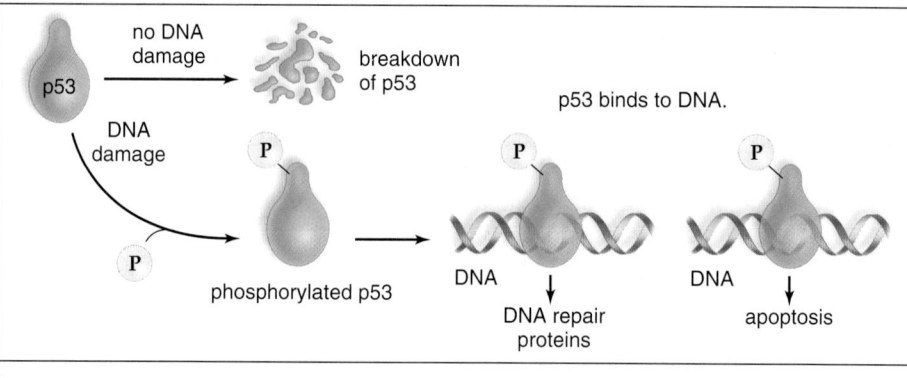

b.

Figure 9A Regulation of the G1 checkpoint.   a. When CDK (cyclin-dependent-kinase) is not present, RB retains E2F. When CDK is present, a phosphorylated RB releases E2F, and after it binds to DNA, proteins necessary to completing cell division are produced. **b.** If DNA is damaged, p53 is not broken down, and instead is involved in the production of DNA repair enzymes and in triggering apoptosis when repair is impossible.

9.2 Mitosis and Cytokinesis

Learning Outcomes

Upon completion of this section you should be able to

1. Explain how the cell prepares the chromosomes and centrosomes prior to nuclear division.
2. Summarize the major events that occur during mitosis and cytokinesis.
3. Discuss why human stem cells continuously conduct mitosis.

As mentioned, cell division in eukaryotes involves mitosis, which is nuclear division, and cytokinesis, which is division of the cytoplasm. During mitosis, the sister chromatids are separated and distributed to two daughter cells.

Eukaryotic Chromosomes

The DNA in the chromosomes of eukaryotes is associated with various proteins, including **histones** that are especially involved in organizing chromosomes. When a eukaryotic cell is not undergoing division, the DNA (and associated proteins) are located within **chromatin,** which has the appearance of a tangled mass of thin threads. Before mitosis begins, chromatin becomes highly coiled and condensed, and it is easy to see the individual chromosomes.

When the chromosomes are visible, it is possible to photograph and count them. Each species has a characteristic chromosome number (Table 9.1). This is the full or **diploid (2n)** number [Gk. *diplos,* twofold, and *-eides,* like] of chromosomes that is found in all cells of the individual. The diploid number includes two chromosomes of each kind. Half the diploid number, called the **haploid (n)** number [Gk. *haplos,* single, and *-eides,* like] of chromosomes, contains only one chromosome of each kind. Typically, only sperm and eggs have the haploid number of chromosomes in the life cycle of animals.

Preparations for Mitosis

During interphase, a cell must make preparations for cell division. These arrangements include replicating the chromosomes and duplicating most cellular organelles, including the centrosome, which will organize the spindle apparatus necessary for movement of chromosomes.

Chromosome Duplication

During mitosis, a 2n nucleus divides to produce daughter nuclei that are also 2n. The dividing cell is called the *parent cell,* and the resulting cells are called the *daughter cells.* Before nuclear division takes place, DNA replicates, duplicating the chromosomes in the parent cell. This occurs during the S stage of interphase. Now each chromosome has two identical double helical molecules. Each double helix is a *chromatid,* and the two identical chromatids are called *sister chromatids* (Fig. 9.3). Sister chromatids are constricted and attached to each other at a region called the **centromere.** Protein complexes called **kinetochores** develop on either side of the centromere during cell division.

Table 9.1 Diploid Chromosome Numbers of Some Eukaryotes

Type of Organism	Name of Chromosome	Chromosome Number
Fungi	*Saccharomyces cerevisiae* (yeast)	32
Plants	*Pisum sativum* (garden pea)	14
	Solanum tuberosum (potato)	48
	Ophioglossum vulgatum (Southern adder's tongue fern)	1,320
Animals	*Drosophila melanogaster* (fruit fly)	8
	Homo sapiens (human)	46
	Carassius auratus (goldfish)	94

During nuclear division, the two sister chromatids separate at the centromere, and in this way each duplicated chromosome gives rise to two daughter chromosomes. Each daughter chromosome has only one double helix molecule. The daughter chromosomes are distributed equally to the daughter cells. In this way, each daughter nucleus gets a copy of each chromosome that was in the parent cell.

Division of the Centrosome

The **centrosome** [Gk. *centrum,* center, and *soma,* body], the main microtubule-organizing center of the cell, also divides before mitosis begins. Each centrosome in an animal cell contains a pair

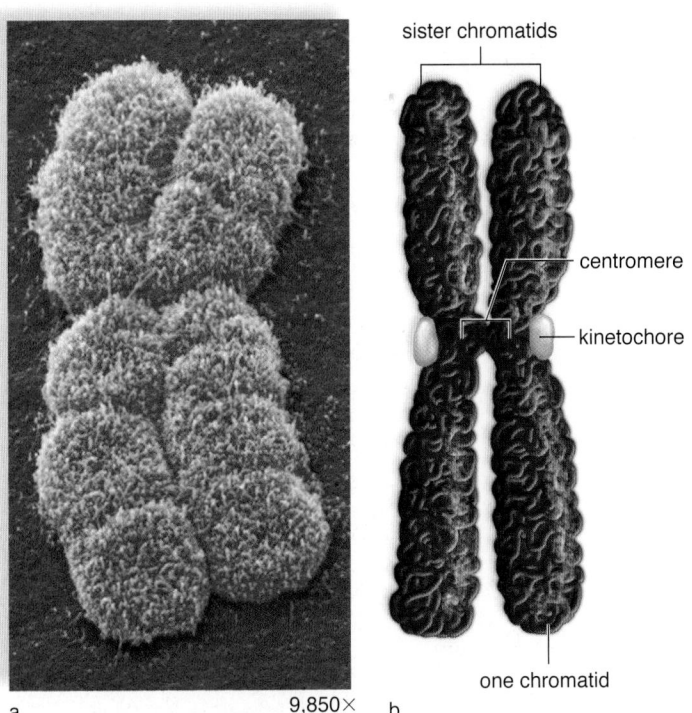

Figure 9.3 Duplicated chromosomes. A duplicated chromosome contains two sister chromatids, each with a copy of the same genes. **a.** Electron micrograph of a highly coiled and condensed chromosome, typical of a nucleus about to divide. **b.** Diagrammatic drawing of a condensed chromosome. The chromatids are held together at a region called the centromere.

of barrel-shaped organelles called **centrioles.** Centrioles are not found in plant cells.

The centrosomes organize the mitotic spindle, which contains many fibers, each of which is composed of a bundle of microtubules. Microtubules are hollow cylinders made up of the protein tubulin. They assemble when tubulin subunits join, and when they disassemble, tubulin subunits become free once more. The microtubules of the cytoskeleton disassemble when spindle fibers begin forming. Most likely, this provides tubulin for the formation of the spindle fibers, or it may allow the cell to change shape as needed for cell division.

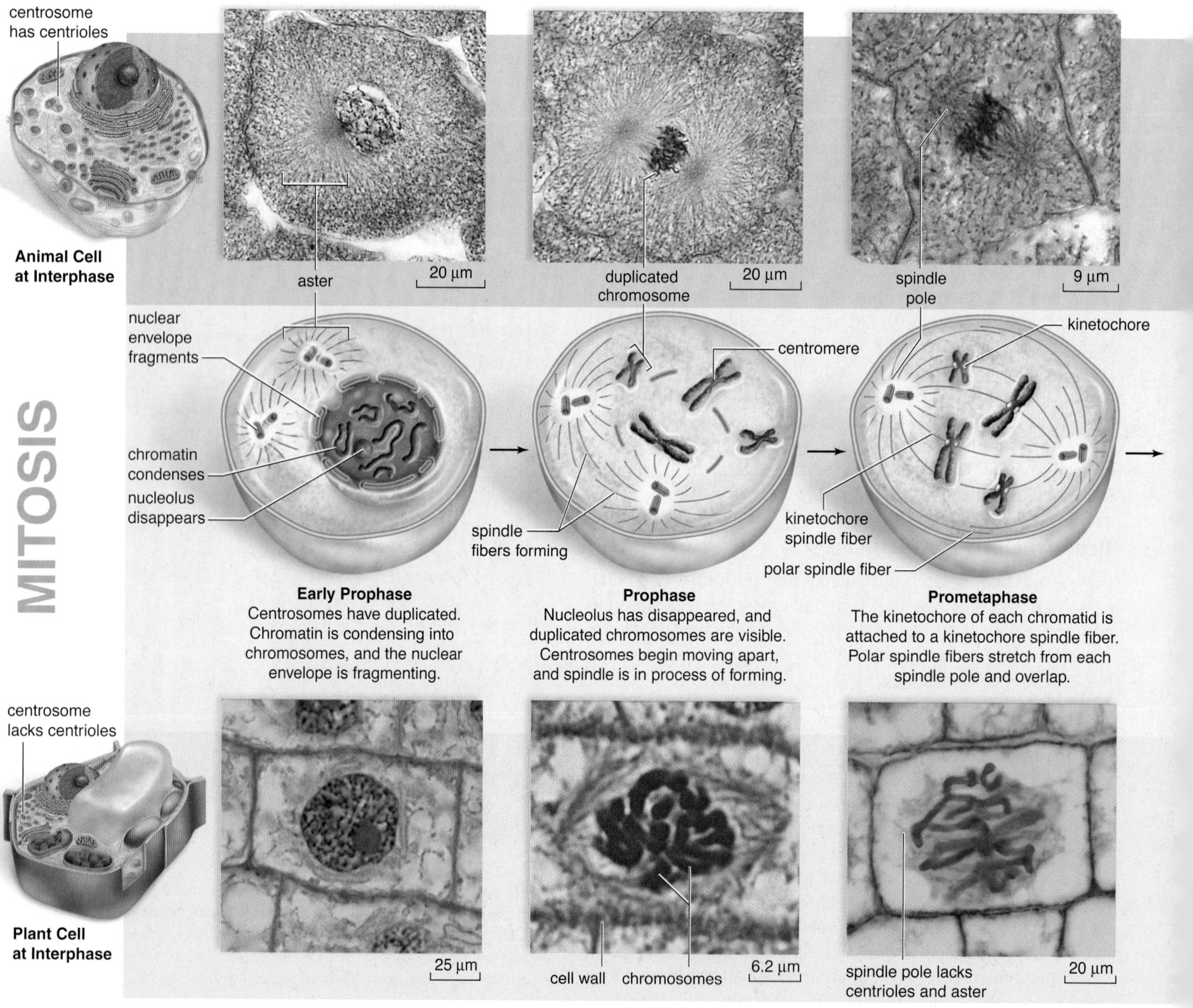

Figure 9.4 Phases of mitosis in animal and plant cells. The blue chromosomes were inherited from one parent and the red from the other parent.

Phases of Mitosis

Mitosis is a continuous process that is arbitrarily divided into five phases for convenience of description: prophase, prometaphase, metaphase, anaphase, and telophase (Fig. 9.4).

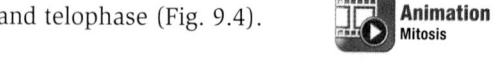

Prophase

It is apparent during **prophase** that nuclear division is about to occur because chromatin has condensed and the chromosomes are visible. Recall that DNA replication occurred during interphase, and therefore the *parental chromosomes are already duplicated and composed of two sister chromatids held together at a centromere.* Counting the number of centromeres in diagrammatic drawings gives the number of chromosomes for the cell depicted.

MITOSIS

Animal Cell at Interphase
centrosome has centrioles

aster   20 μm

duplicated chromosome   20 μm

spindle pole   9 μm

nuclear envelope fragments
chromatin condenses
nucleolus disappears
spindle fibers forming

centromere
kinetochore
kinetochore spindle fiber
polar spindle fiber

Early Prophase
Centrosomes have duplicated. Chromatin is condensing into chromosomes, and the nuclear envelope is fragmenting.

Prophase
Nucleolus has disappeared, and duplicated chromosomes are visible. Centrosomes begin moving apart, and spindle is in process of forming.

Prometaphase
The kinetochore of each chromatid is attached to a kinetochore spindle fiber. Polar spindle fibers stretch from each spindle pole and overlap.

Plant Cell at Interphase
centrosome lacks centrioles

25 μm

cell wall   chromosomes   6.2 μm

spindle pole lacks centrioles and aster   20 μm

During prophase, the nucleolus disappears and the nuclear envelope fragments. The spindle begins to assemble as the two centrosomes migrate away from one another. In animal cells, an array of microtubules radiates toward the plasma membrane from the centrosomes. These structures are called **asters.** It is thought that asters serve to brace the centrioles during later stages of cell division. Notice that the chromosomes have no particular orientation because the spindle has not yet formed.

Prometaphase (Late Prophase)

During **prometaphase,** preparations for sister chromatid separation are evident. Kinetochores appear on each side of the centromere, and these attach sister chromatids to the *kinetochore spindle fibers.* These fibers extend from the poles to the chromosomes, which will soon be located at the center of the spindle.

The kinetochore fibers attach the sister chromatids to opposite poles of the spindle, and the chromosomes are pulled first toward one pole and then toward the other before the chromosomes come into alignment. Notice that even though the chromosomes are attached to the spindle fibers in prometaphase, they are still not in alignment.

Metaphase

During **metaphase,** the centromeres of chromosomes are now in alignment on a single plane at the center of the cell. The chromosomes usually appear as a straight line across the middle of the cell when viewed under a light microscope. An imaginary plane that is perpendicular and passes through this circle is called the **metaphase plate.** It indicates the future axis of cell division.

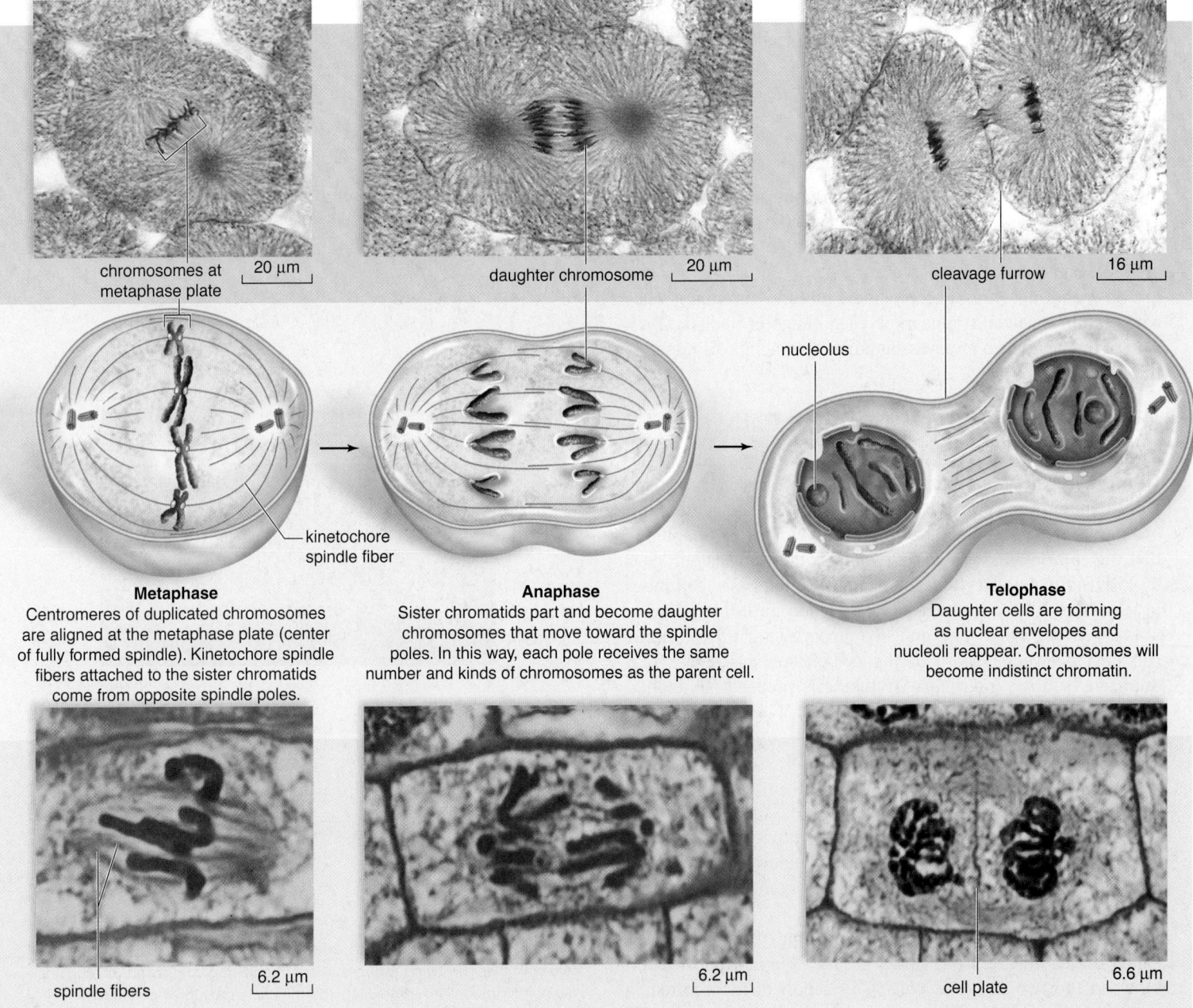

chromosomes at metaphase plate   20 μm

daughter chromosome   20 μm

cleavage furrow   16 μm

nucleolus

kinetochore spindle fiber

Metaphase
Centromeres of duplicated chromosomes are aligned at the metaphase plate (center of fully formed spindle). Kinetochore spindle fibers attached to the sister chromatids come from opposite spindle poles.

Anaphase
Sister chromatids part and become daughter chromosomes that move toward the spindle poles. In this way, each pole receives the same number and kinds of chromosomes as the parent cell.

Telophase
Daughter cells are forming as nuclear envelopes and nucleoli reappear. Chromosomes will become indistinct chromatin.

spindle fibers   6.2 μm

6.2 μm

cell plate   6.6 μm

Several nonattached spindle fibers called *polar spindle fibers* reach beyond the metaphase plate and overlap. A cell cycle checkpoint, the M checkpoint, delays the start of anaphase until the kinetochores of each chromosome are attached properly to spindle fibers and the chromosomes are properly aligned along the metaphase plate.

Anaphase

At the start of **anaphase,** the two sister chromatids of each duplicated chromosome separate at the centromere, giving rise to two daughter chromosomes. Daughter chromosomes, each with a centromere and single chromatid composed of a single double helix, appear to move toward opposite poles. Actually, the daughter chromosomes are being pulled to the opposite poles as the kinetochore spindle fibers disassemble at the region of the kinetochores.

Even as the daughter chromosomes move toward the spindle poles, the poles themselves are moving farther apart because the polar spindle fibers are sliding past one another. Microtubule-associated proteins such as the motor molecules kinesin and dynein are involved in the sliding process. Anaphase is the shortest phase of mitosis.

Telophase

During **telophase,** the spindle disappears as new nuclear envelopes form around the daughter chromosomes. Each daughter nucleus contains the same number and kinds of chromosomes as the original parent cell. Remnants of the polar spindle fibers are still visible between the two nuclei.

The chromosomes become more diffuse chromatin once again, and a nucleolus appears in each daughter nucleus. Division of the cytoplasm requires cytokinesis, which is discussed in the next section.

Cytokinesis in Animal and Plant Cells

As mentioned previously, cytokinesis is division of the cytoplasm. Cytokinesis accompanies mitosis in most cells but not all. When mitosis occurs but cytokinesis doesn't occur, the result is a multinucleated cell. For example, you will see in Chapter 27 that the embryo sac in flowering plants is multinucleated.

Division of the cytoplasm begins in anaphase, continues in telophase, but does not reach completion until the following interphase begins. By the end of mitosis each newly forming cell has received a share of the cytoplasmic organelles that duplicated during interphase. Cytokinesis proceeds differently in plant and animal cells because of differences in cell structure.

Animation
Cytokinesis

Cytokinesis in Animal Cells

In animal cells a **cleavage furrow,** which is an indentation of the membrane between the two daughter nuclei, forms just as anaphase draws to a close. By that time, the newly forming cells have received a share of the cytoplasmic organelles that duplicated during the previous interphase.

The cleavage furrow deepens when a band of actin filaments, called the contractile ring, slowly forms a circular constriction between the two daughter cells. The action of the contractile ring can be likened to pulling a drawstring ever tighter about the middle of a balloon. As the drawstring is pulled tight, the balloon constricts in the middle as the material on either side of the constriction gathers in folds. These folds are represented by the longitudinal lines in Figure 9.5.

A narrow bridge between the two cells can be seen during telophase, and then the contractile ring continues to separate the cytoplasm until there are two independent daughter cells (Fig. 9.5).

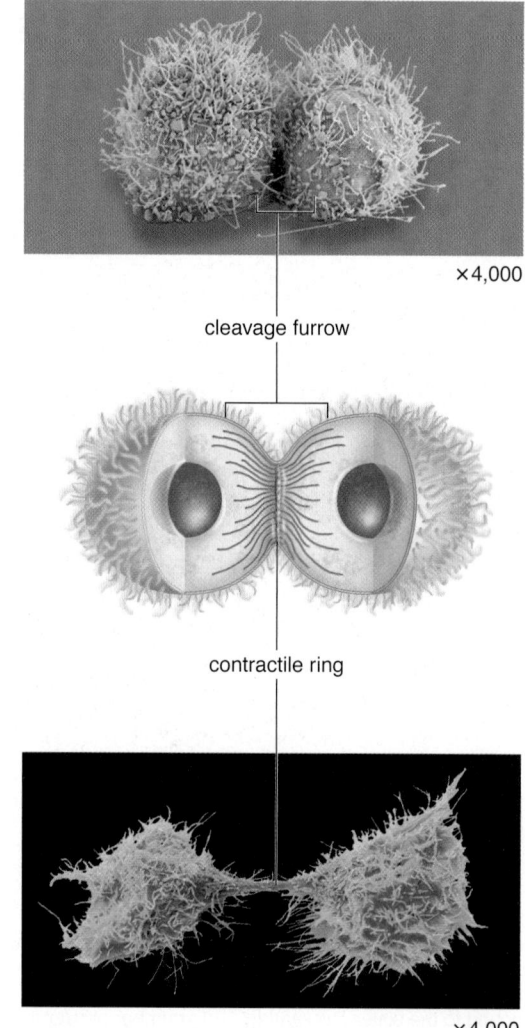

×4,000

cleavage furrow

contractile ring

×4,000

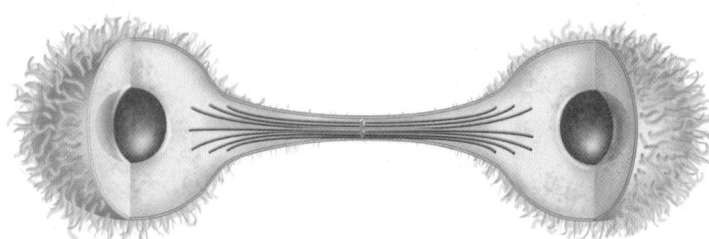

Figure 9.5 Cytokinesis in animal cells. A single cell becomes two cells by a furrowing process. A contractile ring composed of actin filaments gradually gets smaller, and the cleavage furrow pinches the cell into two cells.

Cytokinesis in Plant Cells

Cytokinesis in plant cells occurs by a process different from that seen in animal cells (Fig. 9.6). The rigid cell wall that surrounds plant cells does not permit cytokinesis by furrowing. Instead, cytokinesis in plant cells involves the building of new cell walls between the daughter cells.

Cytokinesis is apparent when a small, flattened disk appears between the two daughter plant cells near the site where the metaphase plate once was. In electron micrographs, it is possible to see that the disk is at right angles to a set of microtubules that radiate outward from the forming nuclei. The Golgi apparatus produces vesicles, which move along the microtubules to the region of the disk. As more vesicles arrive and fuse, a cell plate can be seen. The **cell plate** is simply newly formed plasma membrane that expands outward until it reaches the old plasma membrane and fuses with this membrane.

The new membrane releases molecules that form the new plant cell walls. These cell walls, known as primary cell walls, are later strengthened by the addition of cellulose fibrils. The space between the daughter cells becomes filled with middle lamella, which cements the primary cell walls together.

3D Animation Cytokinesis

The Functions of Mitosis

Mitosis permits growth and repair. In both plants and animals, mitosis is required during development as a single cell develops into an individual. In plants, the individual could be a fern or daisy, while in animals, the individual could be a grasshopper or a human being.

In flowering plants, meristematic tissue retains the ability to divide throughout the life of a plant. Meristematic tissue at the shoot tip accounts for an increase in the height of a plant for as long as it lives. Then, too, lateral meristem accounts for the ability of trees to increase their girth each growing season.

In human beings and other mammals, mitosis is necessary as a fertilized egg becomes an embryo and as the embryo becomes a fetus. Mitosis also occurs after birth as a child becomes an adult. Throughout life, mitosis allows a cut to heal or a broken bone to mend.

Video Mitosis

Stem Cells

Earlier, you learned that the cell cycle is tightly controlled, and that most cells of the body at adulthood are permanently arrested in the G_0 stage. However, mitosis is needed to repair injuries, such as a cut or a broken bone. Many mammalian organs contain stem cells (often called adult stem cells) that retain the ability to divide. As one example, red bone marrow stem cells repeatedly divide to produce millions of cells that go on to become various types of blood cells.

Researchers are learning to manipulate the production of various types of tissues from adult stem cells in the laboratory. If successful, these tissues could be used to cure illnesses. As discussed in the Nature of Science feature "Reproductive and Therapeutic Cloning," **therapeutic cloning,** which is used to produce human tissues, can begin with either adult stem cells or embryonic stem cells. Embryonic stem cells can also be used for **reproductive cloning,** the production of a new individual.

Video Heart Stem Cells

Video Stem Cells

Check Your Progress   9.2

1. Describe the major events that occur during each phase of mitosis.
2. Summarize the differences between cytokinesis in animal and plant cells and explain why the differences are necessary.
3. Discuss the importance of stem cells in the human body.

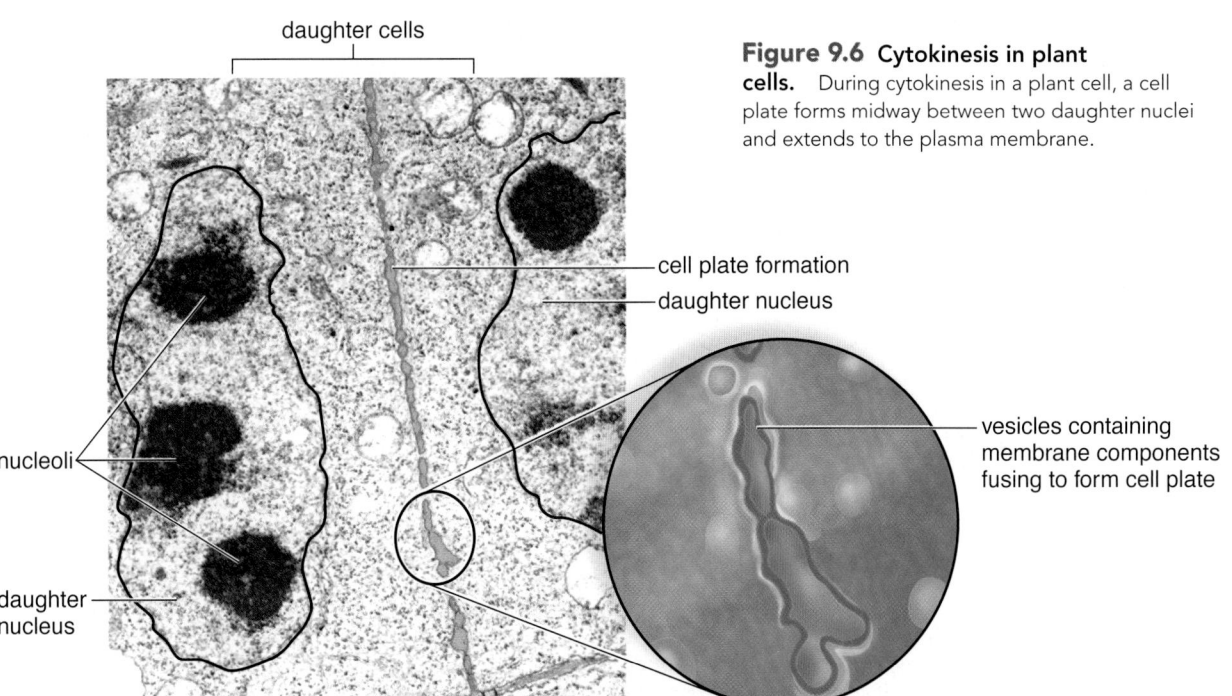

daughter cells

cell plate formation

daughter nucleus

nucleoli

daughter nucleus

vesicles containing membrane components fusing to form cell plate

Figure 9.6 Cytokinesis in plant cells. During cytokinesis in a plant cell, a cell plate forms midway between two daughter nuclei and extends to the plasma membrane.

Nature of Science

Reproductive and Therapeutic Cloning

Our knowledge of how the cell cycle is controlled has yielded major technological breakthroughs, including reproductive cloning—the ability to clone an adult animal from a normal body cell, and therapeutic cloning, which allows the rapid production of mature cells of a specific type. Both types of cloning are a direct result of recent discoveries about how the cell cycle is controlled.

Reproductive cloning, or the cloning of adult animals, was once thought to be impossible because investigators found it difficult to have the nucleus of an adult cell "start over" with the cell cycle, even when it was placed in an egg cell that had its own nucleus removed.

In 1997, Dolly the sheep demonstrated that reproductive cloning is indeed possible. The donor cells were starved before the cell's nucleus was placed in an enucleated egg. This caused them to stop dividing and go into a G_0 (resting) stage, and this made the nuclei amenable to cytoplasmic signals for initiation of development (Fig. 9Ba). This advance has made it possible to clone all sorts of farm animals that have desirable traits and even to clone rare animals that might otherwise become extinct. Despite the encouraging results, however, there are still obstacles to be overcome, and a ban on the use of federal funds in experiments to clone human beings remains firmly in place.

In therapeutic cloning, however, the objective is to produce mature cells of various cell types rather than an individual organism. The purpose of therapeutic cloning is (1) to learn more about how specialization of cells occurs and (2) to provide cells and tissues that could be used to treat human illnesses, such as diabetes, or major injuries like strokes or spinal cord injuries.

There are two possible ways to carry out therapeutic cloning. The first way is to use the exact same procedure as reproductive cloning, except that *embryonic stem cells (ESCs)* are separated and each is subjected to a treatment that causes it to develop into a particular type of cell, such as red blood cells, muscle cells, or nerve cells (Fig. 9Bb). Some have ethical concerns about this type of therapeutic cloning, which is still experimental, because if the embryo were allowed to continue development, it would become an individual.

The second way to carry out therapeutic cloning is to use *adult stem cells*. Stem cells are found in many organs of the adult's body; for example, the bone marrow has stem cells that produce new blood cells. However, adult stem cells are limited in the possible number of cell types that they may become. Nevertheless, scientists are beginning to overcome this obstacle. In 2006, by adding just four genes to adult skin stem cells, Japanese scientists were able to coax the cells, called fibroblasts, into becoming induced pluripotent stem cells (iPS), a type of stem cell that is similar to an ESC. The researchers were then able to create heart and brain cells from the adult stem cells. Other researchers have used this technique to reverse Parkinson-like symptoms in rats.

Although questions exist on the benefits of iPS cells, these advances demonstrate that scientists are actively investigating methods of overcoming the current limitations and ethical concerns of using embryonic stem cells.

Questions to Consider

1. How might the study of therapeutic cloning benefit scientific studies of reproductive cloning?
2. What types of diseases might not be treatable using therapeutic cloning?

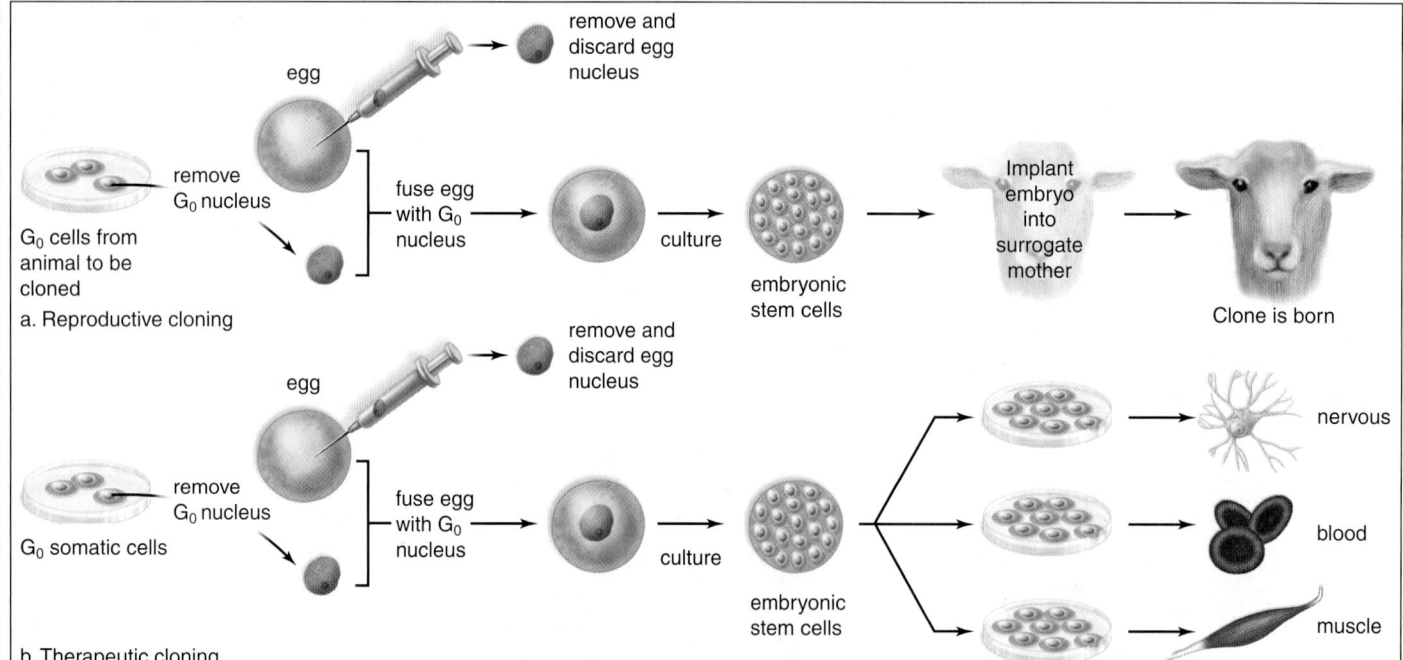

Figure 9B **Two types of cloning.** **a.** The purpose of somatic cell cloning is to produce an individual that is genetically identical to the one that donated a nucleus. The nucleus is placed in an enucleated egg, and, after several mitotic divisions, the embryo is implanted into a surrogate mother for further development. **b.** The purpose of therapeutic cloning is to produce specialized tissue cells. A nucleus is placed in an enucleated egg, and, after several mitotic divisions, the embryonic cells (called embryonic stem cells) are separated and treated to become specialized cells.

9.3 The Cell Cycle and Cancer

Learning Outcomes

Upon completion of this section, you should be able to

1. Describe the basic characteristics of cancer cells.
2. Distinguish between the roles of the tumor-suppressor genes and proto-oncogenes in the regulation of the cell cycle.

Cancer is a cellular growth disorder that occurs when cells divide uncontrollably. Although causes widely differ, most cancers are the result of accumulating mutations that ultimately cause a loss of control of the cell cycle.

Although cancers vary greatly, they usually follow a common multi-step progression (Fig. 9.7). Most cancers begin as an abnormal cell growth that is **benign,** or not cancerous, and usually does not grow larger. However, additional mutations may occur, causing the abnormal cells to fail to respond to inhibiting signals that control the cell cycle. When this occurs, the growth becomes **malignant,** meaning that it is cancerous and possesses the ability to spread.

Characteristics of Cancer Cells

The development of cancer is gradual. A mutation in a cell may cause it to become precancerous, but many other regulatory processes within the body prevent it from becoming cancerous. In fact, it may be decades before a cell possesses most or all of the characteristics of a cancer cell (Table 9.2 and Fig. 9.7). Although cancers vary greatly, cells that possess the following characteristics are generally recognized as cancerous:

Cancer cells lack differentiation. Cancer cells are not specialized and do not contribute to the functioning of a tissue. Although cancer cells may still possess many of the characteristics of surrounding normal cells, they usually look distinctly abnormal. Normal cells can enter the cell cycle about 50 times before they are incapable of dividing again. Cancer cells can enter the cell cycle an indefinite number of times, and in this way seem immortal.

Cancer cells have abnormal nuclei. The nuclei of cancer cells are enlarged and may contain an abnormal number of chromosomes. Often, extra copies of one or more chromosomes may be present. Often, there are also duplicated portions of some chromosomes present, which causes gene amplification,

New mutations arise, and one cell (brown) has the ability to start a tumor.

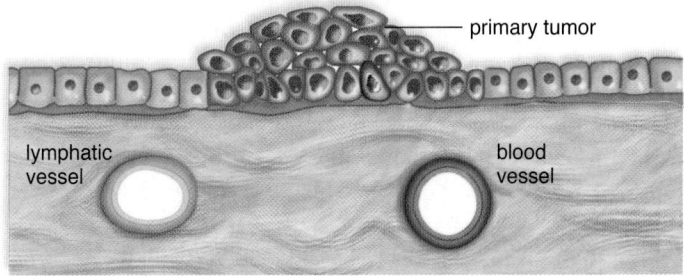

Cancer in situ. The tumor is at its place of origin. One cell (purple) mutates further.

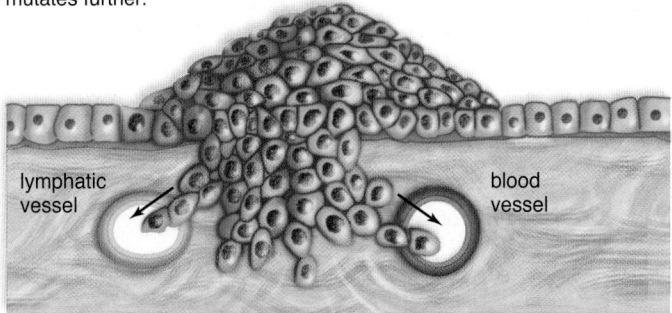

Cancer cells now have the ability to invade lymphatic and blood vessels and travel throughout the body.

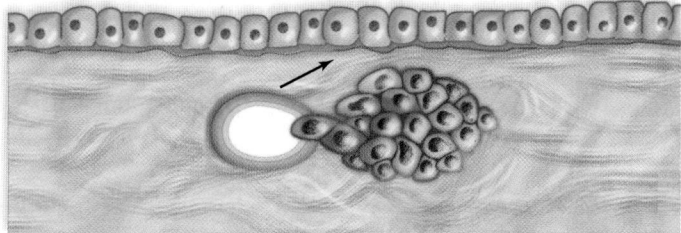

New metastatic tumors are found some distance from the primary tumor.

Figure 9.7 Progression of cancer. The development of cancer requires a series of mutations leading first to a localized tumor and then to metastatic tumors. With each successive step toward cancer, the most genetically altered and aggressive cell becomes the dominant type of tumor. The cells take on characteristics of embryonic cells; they are not differentiated, they can divide uncontrollably; and they are able to metastasize and spread to other tissues.

or extra copies of specific genes. Some chromosomes may also possess deleted portions.

Cancer cells do not undergo apoptosis. Ordinarily, cells with damaged DNA undergo apoptosis, or programmed cell death. The immune system can also recognize abnormal cells and trigger apoptosis, which normally prevents tumors from developing. Cancer cells fail to undergo apoptosis even though they are abnormal cells.

Cancer cells form tumors. Normal cells anchor themselves to a substratum and/or adhere to their neighbors. They exhibit contact inhibition—in other words, when they come in contact with a neighbor, they stop dividing. Cancer cells have lost all restraint and do not exhibit contact inhibition. The abnormal cancer cells pile on top of one another and grow in multiple

Table 9.2 Cancer Cells Versus Normal Cells

Cancer Cells	Normal Cells
Nondifferentiated cells	Differentiated cells
Abnormal nuclei	Normal nuclei
Do not undergo apoptosis	Undergo apoptosis
No contact inhibition	Contact inhibition
Disorganized, multilayered	One organized layer
Undergo metastasis	

layers, forming a **tumor.** During carcinogenesis, the most aggressive cell becomes the dominant cell of the tumor.

Cancer cells undergo metastasis and angiogenesis. Additional mutations may cause a benign tumor, which is usually contained within a capsule and cannot invade adjacent tissue, to become malignant, and spread throughout the body, forming new tumors distant from the primary tumor. These cells now produce enzymes that they normally do not express, allowing tumor cells to invade underlying tissues. Then, they travel through the blood and lymph, to start tumors elsewhere in the body. This process is known as **metastasis.**

Tumors that are actively growing soon encounter another obstacle—the blood vessels supplying nutrients to the tumor cells become insufficient to support the rapid growth of the tumor. In order to grow further, the cells of the tumor must receive additional nutrition. Thus, the formation of new blood vessels is required to bring nutrients and oxygen to support further growth. Additional mutations occurring in tumor cells allow them to direct the growth of new blood vessels into the tumor in a process called **angiogenesis.** Some modes of cancer treatment are aimed at preventing angiogenesis from occurring.

Origin of Cancer

Normal growth and maintenance of body tissues depend on a balance between signals that promote and inhibit cell division. When this balance is upset, conditions such as cancer may occur. Thus, cancer is usually caused by mutations affecting genes that directly or indirectly affect this balance, such as those shown in Figure 9.8. These two types of genes are usually affected:

1. **Proto-oncogenes** code for proteins that promote the cell cycle and prevent apoptosis. They are often likened to the gas pedal of a car because they cause the cell cycle to speed up.
2. **Tumor suppressor genes** code for proteins that inhibit the cell cycle and promote apoptosis. They are often likened to the brakes of a car because they cause the cell cycle to go more slowly or even stop.

Proto-oncogenes Become Oncogenes

Proto-oncogenes are normal genes that promote progression through the cell cycle. They are often at the end of a *stimulatory pathway* extending from the plasma membrane to the nucleus. A stimulus, such as an injury, results in the release of a growth factor that binds to a receptor protein in the plasma membrane. This sets in motion a whole series of enzymatic reactions leading to the activation of genes that promote the cell cycle, both directly and indirectly. Proto-oncogenes include the receptors and signal molecules that make up these pathways.

When mutations occur in proto-oncogenes, they become **oncogenes,** or cancer-causing genes. Oncogenes are under constant stimulation and keep on promoting the cell cycle regardless of circumstances. For example, an oncogene may code for a faulty receptor in the stimulatory pathway such that the cell cycle is stimulated, even when no growth factor is present! Or, an oncogene may specify either an abnormal protein product or

produce abnormally high levels of a normal product that stimulate the cell cycle to begin or to go to completion. As a result, uncontrolled cell division may occur.

Researchers have identified perhaps 100 oncogenes that can cause increased growth and lead to tumors. The oncogenes most frequently involved in human cancers belong to the *ras* gene family. Mutant forms of the *BRCA1* oncogene (breast cancer predisposition gene 1) are associated with certain hereditary forms of breast and ovarian cancer.

Tumor Suppressor Genes Become Inactive

Tumor suppressor genes, on the other hand, directly or indirectly inhibit the cell cycle and prevent cells from dividing uncontrollably. Some tumor suppressor genes prevent progression of the cell cycle when DNA is damaged. Other tumor suppressor genes may promote apoptosis as a last resort.

A mutation in a tumor suppressor gene is much like brake failure in a car; when the mechanism that slows down and stops cell division does not function, the cell cycle accelerates and does not halt. Researchers have identified about a half-dozen tumor suppressor genes. Among these are the *RB* and *p53* genes that code for the RB and p53 proteins. The Nature of Science feature "The G_1 Checkpoint" discusses the function of these proteins in controlling the cell cycle. The *RB* tumor suppressor gene was discovered when the inherited condition retinoblastoma was being studied, but malfunctions of this gene have now been identified in many other cancers as well, including breast, prostate, and bladder cancers. The *p53* gene turns on the expression of other genes that inhibit the cell cycle. The p53 protein can also stimulate apoptosis. It is estimated that over half of human cancers involve an abnormal or deleted *p53* gene.

Animation
How Tumor Suppressor Genes Block Cell Division

Other Causes of Cancer

As mentioned previously, cancer develops when the delicate balance between promotion and inhibition of cell division is tilted toward uncontrolled cell division. Other mutations may occur within a cell that affect this balance. For example, while a mutation affecting the cell's DNA repair system will not immediately cause cancer, it leads to a much greater chance of a mutation occurring within a proto-oncogene or tumor suppressor gene. And in some cancer cells, mutation of the telomerase enzyme that regulates the length of **telomeres,** or the ends of chromosomes, causes the telomeres to remain at a constant length. Because cells with shortened telomeres normally stop dividing, keeping the telomeres at a constant length allows the cancer cells to continue dividing over and over again.

Animation
Telomerase Function

Check Your Progress   9.3

1. List the major characteristics of cancer cells that distinguish them from normal cells.
2. Distinguish between a malignant and benign tumor.
3. Compare and contrast the effect on the cell cycle of **(a)** a mutation in a proto-oncogene; **(b)** a mutation in a tumor suppressor gene.

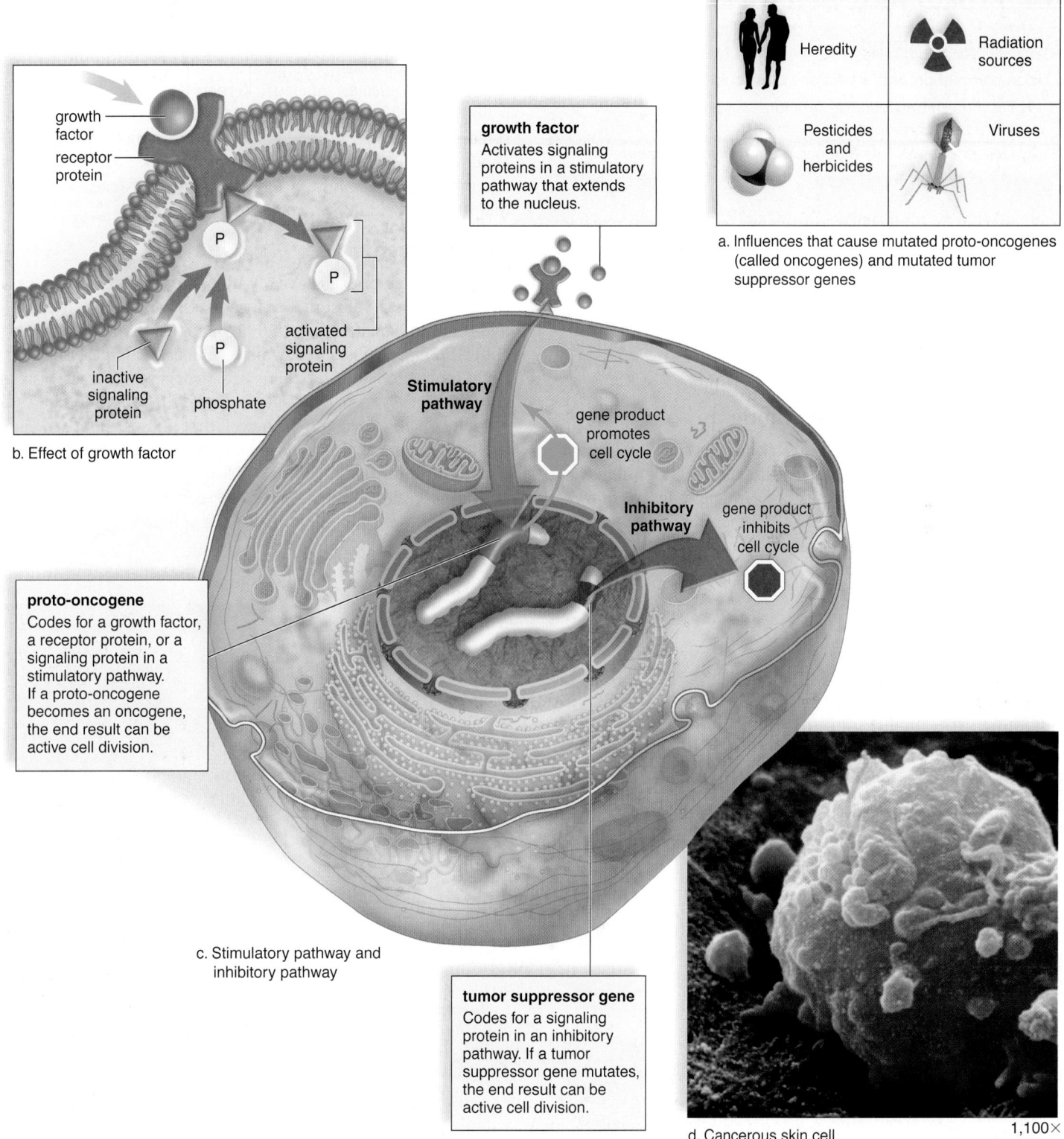

growth factor

receptor protein

P

P

inactive signaling protein

phosphate

activated signaling protein

b. Effect of growth factor

growth factor
Activates signaling proteins in a stimulatory pathway that extends to the nucleus.

Heredity

Radiation sources

Pesticides and herbicides

Viruses

a. Influences that cause mutated proto-oncogenes (called oncogenes) and mutated tumor suppressor genes

Stimulatory pathway

gene product promotes cell cycle

Inhibitory pathway

gene product inhibits cell cycle

proto-oncogene
Codes for a growth factor, a receptor protein, or a signaling protein in a stimulatory pathway. If a proto-oncogene becomes an oncogene, the end result can be active cell division.

c. Stimulatory pathway and inhibitory pathway

tumor suppressor gene
Codes for a signaling protein in an inhibitory pathway. If a tumor suppressor gene mutates, the end result can be active cell division.

d. Cancerous skin cell

1,100×

Figure 9.8 Causes of cancer.   a. Mutated genes that cause cancer can be due to the influences noted. **b.** A growth factor that binds to a receptor protein initiates a reaction that triggers a stimulatory pathway. **c.** A stimulatory pathway that begins at the plasma membrane turns on proto-oncogenes. The products of these genes promote the cell cycle and double back to become part of the stimulatory pathway. When proto-oncogenes become oncogenes, they are turned on all the time. An inhibitory pathway begins with tumor suppressor genes whose products inhibit the cell cycle. When tumor suppressor genes mutate, the cell cycle is no longer inhibited. **d.** Cancerous skin cell.

9.4 Prokaryotic Cell Division

Learning Outcomes

Upon completion of this section you should be able to

1. Distinguish between the structure of a prokaryotic and eukaryotic chromosome.
2. Describe the events that occur during binary fission.

Cell division in unicellular organisms, such as prokaryotes, produces two new individuals. This is **asexual reproduction** in which the offspring are genetically identical to the parent. In prokaryotes, reproduction consists of duplicating the single chromosome and distributing a copy to each of the daughter cells. Unless a mutation has occurred, the daughter cells are genetically identical to the parent cell.

The Prokaryotic Chromosome

Prokaryotes (bacteria and archaea) lack a nucleus and other membranous organelles found in eukaryotic cells. Still, they do have a chromosome, which is composed of DNA and a limited number of associated proteins. The single chromosome of prokaryotes contains just a few proteins and is organized differently from eukaryotic chromosomes. A eukaryotic chromosome has many more associated proteins than does a prokaryotic chromosome.

In electron micrographs, the bacterial chromosome appears as an electron-dense, irregularly shaped region called the **nucleoid** [L. *nucleus*, nucleus, kernel; Gk. *-eides*, like], which is not enclosed by membrane. When stretched out, the chromosome is seen to be a circular loop with a length that is up to about a thousand times the length of the cell. Special enzymes and proteins help coil the chromosome so that it will fit within the prokaryotic cell.

Animation Bacterial Chromosome Compaction

Binary Fission

Prokaryotes reproduce asexually by binary fission. The process is termed **binary fission** because division (fission) produces two (binary) daughter cells that are identical to the original parent cell. Before division takes place, the cell enlarges, and after DNA replication occurs, there are two chromosomes. These chromosomes attach to a special plasma membrane site and separate by an elongation of the cell that pulls them apart. During this period, new plasma membrane and cell wall develop and grow inward to divide the cell. When the cell is approximately twice its original length, the new cell wall and plasma membrane for each cell are complete (Fig. 9.9).

Animation Binary Fission

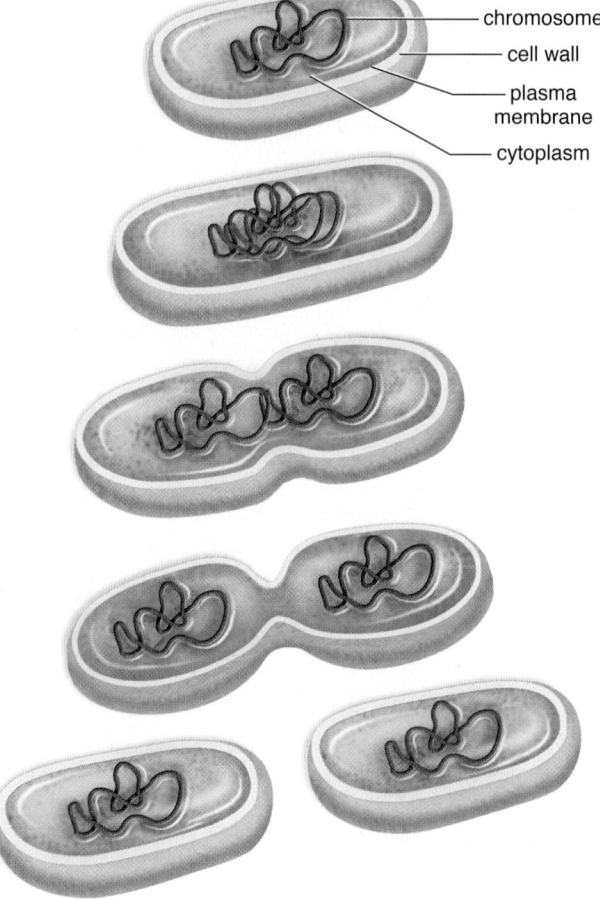

1. Attachment of chromosome to a special plasma membrane site indicates that this bacterium is about to divide.

2. The cell is preparing for binary fission by enlarging its cell wall, plasma membrane, and overall volume.

3. DNA replication has produced two identical chromosomes. Cell wall and plasma membrane begin to grow inward.

4. As the cell elongates, the chromosomes are pulled apart. Cytoplasm is being distributed evenly.

5. New cell wall and plasma membrane has divided the daughter cells.

chromosome
cell wall
plasma membrane
cytoplasm

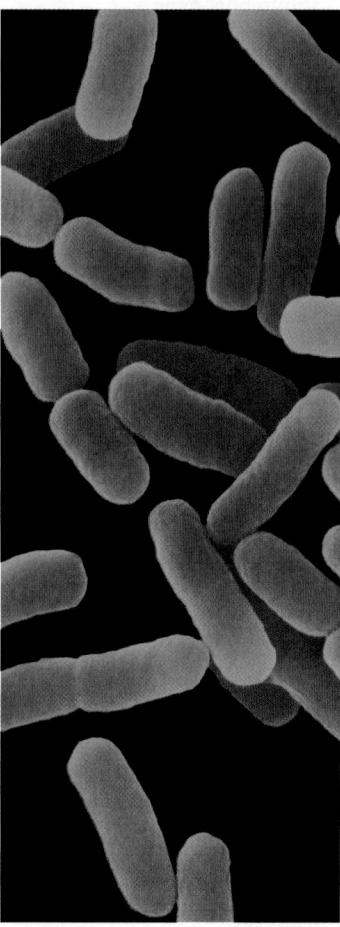

SEM 2,345×

Figure 9.9 Binary fission. First, DNA replicates, and as the cell lengthens, the two chromosomes separate, and the cells become divided. The two resulting bacteria are identical.

Escherichia coli, which lives in our intestines, has a generation time (the time it takes the cell to divide) of about 20 minutes under favorable conditions. In about seven hours, a single cell can increase to over 1 million cells! The division rate of other bacteria varies depending on the species and conditions.

Comparing Prokaryotes and Eukaryotes

Both binary fission and mitosis ensure that each daughter cell is genetically identical to the parent cell. The genes are portions of DNA found in the chromosomes. Prokaryotes (bacteria and archaea), protists (many algae and protozoans), and some fungi (yeasts) are unicellular. Cell division in unicellular organisms produces two new individuals:

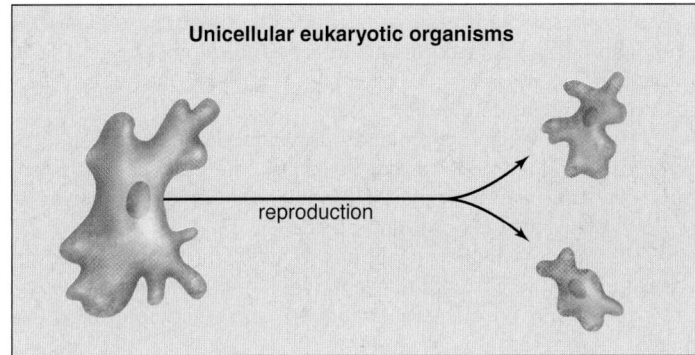

Unicellular eukaryotic organisms

reproduction

This is a form of asexual reproduction because one parent has produced identical offspring (Table 9.3).

Table 9.3 Functions of Cell Division

Type of Organism	Cell Division	Function
Prokaryotes		
Bacteria and archaea	Binary fission	Asexual reproduction
Eukaryotes		
Protists, and some fungi (yeast)	Mitosis and cytokinesis	Asexual reproduction
Other fungi, plants, and animals	Mitosis and cytokinesis	Development, growth, and repair

In multicellular fungi (molds and mushrooms), plants, and animals, cell division is part of the growth process. It produces the multicellular form we recognize as the mature organism. Cell division is also important in multicellular forms for renewal and repair:

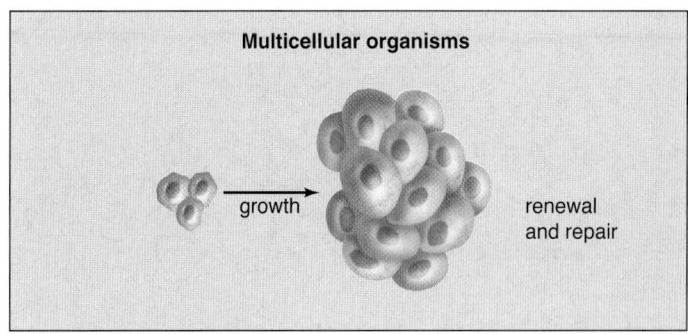

Multicellular organisms

growth

renewal and repair

The chromosomes of eukaryotic cells are composed of DNA and many associated proteins. The histone proteins organize a chromosome, allowing it to extend as chromatin during interphase and to coil and condense just prior to mitosis. Each species of multicellular eukaryotes has a characteristic number of chromosomes in the nuclei. As a result of mitosis, each daughter cell receives the same number and kinds of chromosomes as the parent cell. The spindle, which appears during mitosis, is involved in distributing the daughter chromosomes to the daughter nuclei. Cytokinesis, either by the formation of a cell plate (plant cells) or by furrowing (animal cells), is division of the cytoplasm.

In prokaryotes, the single chromosome consists largely of DNA with a few associated proteins. During binary fission, this chromosome duplicates, and each daughter cell receives one copy as the parent cell elongates, and a new cell wall and plasma membrane form between the daughter cells. No spindle is involved in binary fission.

Check Your Progress   9.4

1. Explain how binary fission in prokaryotes differs from mitosis and cytokinesis in eukaryotes.
2. Distinguish between the structure of a prokaryotic and eukaryotic chromosome.

CONNECTING *the* CONCEPTS *with the* BIG IDEAS

Information and Signaling

- The omnipresence of mitosis suggests a common cellular evolutionary lineage. (3A2)
- Prokaryotes duplicate a single circular chromosome before binary fission and may transmit extra-chromosomal plasmids. (3A1a2-3)
- The cell cycle includes growth and cell division relying on checkpoints and chemical signals such as MPF, PDGF, and cyclins to regulate its rate or stopping point. (3A2a2*IE*, 3A2a5)

- Mitosis is preceded by interphase during which cells grow, replicate DNA, and prepare for division. (3A2a1, 4)
- Apoptosis is programmed cell death. (2 narrative)
- Chromatid connection and alignment after DNA replication assures equal partitioning of genetic material into new nuclei during mitotic separation. (3A2b4)
- Mitosis and cytokinesis produce genetically identical cells for repair, growth, and asexual reproduction. (3A2b2-3)
- Abnormal cell cycle regulatory mechanisms may explain cancer. (3A2a2*IE*)

Media Study Tools

www.glencoe.com/maderAP11

Enhance your study of this chapter with study tools and practice tests. Also ask your instructor about the resources available through ConnectPlus, including the media-rich eBook, interactive learning tools, and animations.

3D Animation
Cell Cycle and
Mitosis

For an interactive exploration of the cell cycle and mitosis, take a moment to watch McGraw-Hill's new 3D animation.

Summarize

9.1 The Cell Cycle

The cell cycle of a eukaryotic cell includes (1) interphase and (2) a mitotic stage that consists of mitosis and cytokinesis. Interphase, in turn, is composed of three stages: G_1 (growth as certain organelles double), S (the synthesis stage, where the chromosomes are duplicated), and G_2 (growth as the cell prepares to divide). Cells of the body that are no longer dividing are said to be arrested in a G_0 state. During the mitotic stage (M), the chromosomes are sorted into two daughter cells so that each receives a full complement of chromosomes.

The cell cycle is regulated by three well-known checkpoints—the G_1 checkpoint, the G_2 checkpoint prior to the M stage, and the M stage checkpoint, or spindle assembly checkpoint, immediately before anaphase. The G_1 checkpoint ensures that conditions are favorable and that the proper signals are present, and also checks the DNA for damage. If the DNA is damaged beyond repair, apoptosis may occur. Cell division and apoptosis are two opposing processes that keep the number of healthy cells in balance.

9.2 Mitosis and Cytokinesis

Interphase represents the portion of the cell cycle between nuclear divisions, and during this time, preparations are made for cell division. These preparations include duplication of most cellular contents, including the centrosome, which organizes the mitotic spindle. The DNA is duplicated during S stage, at which time the chromosomes, which consisted of a single chromatid each, are duplicated. The G_2 checkpoint ensures that DNA has replicated properly. This results in a nucleus containing the same number of chromosomes, with each now consisting of two chromatids attached at the centromere. During interphase, the chromosomes are not distinct and are collectively called chromatin. Each eukaryotic species has a characteristic number of chromosomes. The total number is called the diploid number, and half this number is the haploid number.

Among eukaryotes, cell division involves both mitosis (nuclear division) and division of the cytoplasm (cytokinesis). As a result of mitosis, the chromosome number stays constant because each chromosome is duplicated and gives rise to two daughter chromosomes that consist of a single chromatid each.

Mitosis consists of five phases:

Prophase—The nucleolus disappears, the nuclear envelope fragments, and the spindle forms between centrosomes. The chromosomes condense and become visible under a light microscope. In animal cells, asters radiate from the centrioles within the centrosomes. Plant cells lack centrioles and, therefore, asters. Even so, the mitotic spindle forms.

Prometaphase (late prophase)—The kinetochores of sister chromatids attach to kinetochore spindle fibers extending from opposite poles. The chromosomes move back and forth until they are aligned at the metaphase plate.

Metaphase—The spindle is fully formed, and the duplicated chromosomes are aligned at the metaphase plate. The spindle consists of polar spindle fibers that overlap at the metaphase plate and kinetochore spindle fibers that are attached to chromosomes. The M stage checkpoint, or spindle assembly checkpoint, must be satisfied before progressing to the next phase.

Anaphase—Sister chromatids separate, becoming daughter chromosomes that move toward the poles. The polar spindle fibers slide past one another, and the kinetochore spindle fibers disassemble. Cytokinesis by furrowing begins.

Telophase—Nuclear envelopes re-form, chromosomes begin changing back to chromatin, the nucleoli reappear, and the spindle disappears. Cytokinesis continues, and is complete by the end of telophase.

Cytokinesis in animal cells is a furrowing process that divides the cytoplasm. Cytokinesis in plant cells involves the formation of a cell plate from which the plasma membrane and cell wall are completed.

9.3 The Cell Cycle and Cancer

The development of cancer is primarily due to the mutation of genes involved in control of the cell cycle. Cancer cells lack differentiation, have abnormal nuclei, do not undergo apoptosis, form tumors, and undergo metastasis and angiogenesis. Cancer often follows a progression in which mutations accumulate, gradually causing uncontrolled growth and the development of a tumor.

Proto-oncogenes stimulate the cell cycle after they are turned on by environmental signals such as growth factors. Oncogenes are mutated proto-oncogenes that stimulate the cell cycle without need of environmental signals. Tumor suppressor genes inhibit the cell cycle. Mutated tumor suppressor genes no longer inhibit the cell cycle, allowing unchecked cell division.

9.4 Prokaryotic Cell Division

Binary fission (in prokaryotes) and mitosis (in unicellular eukaryotic protists and fungi) allow organisms to reproduce asexually. Mitosis in multicellular eukaryotes is primarily for the purpose of development, growth, and repair of tissues.

The prokaryotic chromosome has a few proteins and a single, long loop of DNA. When binary fission occurs, the chromosome attaches to the inside of the plasma membrane and replicates. As the cell elongates, the chromosomes are pulled apart. Inward growth of the plasma membrane and formation of new cell wall material divide the cell in two.

Key Terms

anaphase   160
angiogenesis   164
apoptosis   155
asexual reproduction   166
aster   159
benign   163
binary fission   166
cancer   163
cell cycle   154
cell plate   161
centriole   158
centromere   157
centrosome   157
chromatid   154
chromatin   157
cleavage furrow   160
cyclin   155
cytokinesis   155
diploid (2n)   157
growth factor   155
haploid (n)   157
histone   157
interphase   154

kinetochore   157
malignant   163
metaphase   159
metaphase plate   159
metastasis   164
mitosis   155
mitotic spindle   155
nucleoid   166
oncogene   164
p53   155
prometaphase   159
prophase   158
proto-oncogene   164
RB   155
reproductive cloning   161
signal   155
sister chromatids   154
somatic cell   155
telomere   164
telophase   160
therapeutic cloning   161
tumor   164
tumor suppressor gene   164

Assess

Reviewing This Chapter

1. Describe the cell cycle, including its different stages.   154–55
2. Describe three checkpoints of the cell cycle.   155
3. What is apoptosis, and what are its functions?   155
4. Distinguish between chromosome, chromatin, chromatid, centromere, and kinetochore.   154–57
5. Describe the events that occur during the phases of mitosis.   158–60
6. Contrast cytokinesis in animal cells and plant cells.   160–61
7. List and discuss characteristics of cancer cells that distinguish them from normal cells.   163–64
8. Compare and contrast the functions of proto-oncogenes and tumor suppressor genes in controlling the cell cycle.   164–65
9. Describe the prokaryotic chromosome and the process of binary fission.   166
10. Contrast the function of cell division in prokaryotic and eukaryotic cells.   167

Testing Yourself

Choose the best answer for each question.

1. In contrast to a eukaryotic chromosome, a prokaryotic chromosome
   a. is shorter and fatter.
   b. has a single loop of DNA.
   c. never replicates.
   d. contains many histones.
   e. All of these are correct.
2. The diploid number of chromosomes
   a. is the 2n number.
   b. is in a parent cell and therefore in the two daughter cells following mitosis.
   c. varies according to the particular organism.
   d. is in every somatic cell.
   e. All of these are correct.

For questions 3–5, match the descriptions that follow to the terms in the key.

KEY:
   a. centrosome
   b. chromosome
   c. centromere
   d. cyclin

3. Point of attachment for sister chromatids
4. Found at a spindle pole in the center of an aster
5. Coiled and condensed chromatin
6. If a parent cell has 14 chromosomes prior to mitosis, how many chromosomes will each daughter cell have?
   a. 28 because each chromatid is a chromosome
   b. 14 because the chromatids separate
   c. only 7 after mitosis is finished
   d. any number between 7 and 28
   e. 7 in the nucleus and 7 in the cytoplasm, for a total of 14
7. In which phase of mitosis are the kinetochores of the chromosomes being attached to spindle fibers?
   a. prophase
   b prometaphase
   c. metaphase
   d. anaphase
   e. telophase
8. Interphase
   a. is the same as prophase, metaphase, anaphase, and telophase.
   b. is composed of G_1, S, and G_2 stages.
   c. requires the use of polar spindle fibers and kinetochore spindle fibers.
   d. is the majority of the cell cycle.
   e. Both b and d are correct.
9. At the metaphase plate during metaphase of mitosis, there are
   a. single chromosomes.
   b. duplicated chromosomes.
   c. G_1 stage chromosomes.
   d. always 23 chromosomes.
10. During which mitotic phases are duplicated chromosomes present?
   a. all but telophase
   b. prophase and anaphase
   c. all but anaphase and telophase
   d. only during metaphase at the metaphase plate
   e. Both a and b are correct.
11. Which of these is paired incorrectly?
   a. prometaphase—the kinetochores become attached to spindle fibers
   b. anaphase—daughter chromosomes are located at the spindle poles
   c. prophase—the nucleolus disappears and the nuclear envelope disintegrates
   d. metaphase—the chromosomes are aligned in the metaphase plate
   e. telophase—a resting phase between cell division cycles
12. When cancer occurs,
   a. cells cannot pass the G_1 checkpoint.
   b. control of the cell cycle is impaired.
   c. apoptosis has occurred.
   d. the cells can no longer enter the cell cycle.
   e. All of these are correct.

13. Which of the following is not characteristic of cancer cells?
    a. Cancer cells often undergo angiogenesis.
    b. Cancer cells tend to be nonspecialized.
    c. Cancer cells undergo apoptosis.
    d. Cancer cells often have abnormal nuclei.
    e. Cancer cells can metastasize.

14. Which of the following statements is true?
    a. Proto-oncogenes cause a loss of control of the cell cycle.
    b. The products of oncogenes may inhibit the cell cycle.
    c. Tumor-suppressor-gene products inhibit the cell cycle.
    d. A mutation in a tumor suppressor gene may inhibit the cell cycle.
    e. A mutation in a proto-oncogene may convert it into a tumor suppressor gene.

For questions 15–18, match the descriptions to a stage in the key.

KEY:
    a. G_1 stage
    b. S stage
    c. G_2 stage
    d. M (mitotic) stage

15. At the end of this stage, each chromosome consists of two attached chromatids.

16. During this stage, daughter chromosomes are distributed to two daughter nuclei.

17. The cell doubles its organelles and accumulates the materials needed for DNA synthesis.

18. The cell synthesizes the proteins needed for cell division.

19. Which is not true of the cell cycle?
    a. The cell cycle is controlled by internal/external signals.
    b. Cyclin is a signaling molcule that increases and decreases as the cycle continues.
    c. DNA damage can stop the cell cycle at the G_1 checkpoint.
    d. Apoptosis occurs frequently during the cell cycle.

20. Label this diagram. What phase of mitosis does it represent?

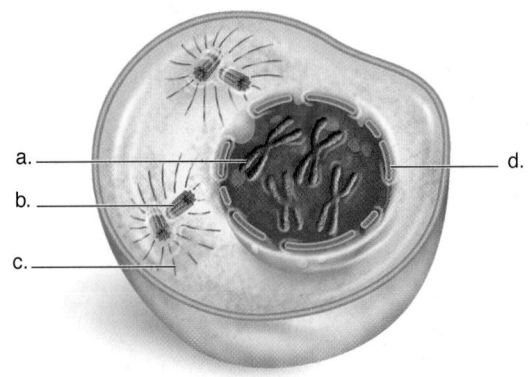

a. _____   d. _____

b. _____

c. _____

Engage

Virtual Lab
Cell Reproduction

The virtual lab "Cell Reproduction" provides an interactive examination of the cell cycle and provides for a comparison of normal and cancer cells.

Thinking Scientifically

1. After DNA is duplicated in eukaryotes, it must be bound to histones. This requires the synthesis of hundreds of millions of new protein molecules. With reference to Figure 9.1, when in the cell cycle would histones be made?

2. The survivors of the atomic bombs that were dropped on Hiroshima and Nagasaki have been the subjects of long-term studies of the effects of ionizing radiation on cancer incidence. The frequencies of different types of cancer in these individuals varied across the decades. In the 1950s, high levels of leukemia and cancers of the lung and thyroid gland were observed. The 1960s and 1970s brought high levels of breast and salivary gland cancers. In the 1980s, rates of colon cancer were especially high. Why do you suppose the rates of different types of cancer varied across time?

Bioethical Issue

Paying for Cancer Treatment

The risk factors for developing cancer are generally well known. Many lifestyle factors, such as smoking, poor dietary habits, obesity, physical inactivity, risky sexual behavior, and alcohol abuse, among others, have all been linked to higher risks of developing cancer. The greatly increasing rates of cancer over the past few decades have been decried as a public health epidemic. But aside from the cost in human life, the rising tide of cancer is causing a major crisis in today's society—how to pay for it all.

Despite increasing cure rates, effective new drugs, and novel treatments for various types of cancer, the costs of treatment continue to skyrocket. Nowhere is this more apparent than in the pharmaceutical industry. For example, new cancer drugs, while effective, are extremely expensive. Drug companies claim that it costs them between $500 million and $1 billion to bring a single new medicine to market. This cost may seem overblown, especially when you consider that the National Cancer Institute funds basic research into cancer biology and that drug companies often benefit indirectly from the findings. But the drug companies tell us that they need one successful drug to pay for the many drugs they try to develop that do not pay off.

Still, it does seem as if successful drug companies try to keep lower-cost competitors out of the market. The question of how much drug companies can charge for drugs and who should pay for them is a thorny one. If drug companies don't show a profit, they may go out of business and there will be no new drugs. The same is true for insurance companies if they can't raise the cost of insurance to pay for expensive drugs. If the government buys drugs for Medicare patients, taxes may go up dramatically.

But how should the cost of treatment be met? Cancer is an illness that can be the direct result of poor lifestyle choices, but it can also occur in otherwise healthy individuals who make proper choices. And with increasing life spans, the incidence of cancer can only be expected to increase in future years. Should people who develop cancer due to poor lifestyle choices be held fully or partly responsible for paying for treatment? And if so, how? And how should the cost of developing new drugs and treatments be borne? There are no easy answers for any of these questions, but as cancer continues to extract a high toll in both human life and financial resources, future generations may face some difficult choices.

Nanu Ram Jogi, one of the world's oldest fathers.

10

Meiosis and Sexual Reproduction

BEFORE YOU BEGIN

Before beginning this chapter, take a few moments to review the following discussions.

Figures 3.19 and 3.20 How is genetic information stored in nucleic acids?

Section 4.8 What are microtubules and how do they interact with chromosomes?

Section 9.2 How are eukaryotic chromosomes organized and replicated prior to cell division?

anu Ram Jogi, at 90 years old, recently became a new father. As he hoisted his newborn daughter into the air amid a throng of cameras, microphones, and reporters, he boasted that he plans to continue fathering children with his wife, Saburi, now 50, until he is 100. He cannot even recall how many children he has fathered over the many years of his life, but it is estimated that he has at least twelve sons, nine daughters, and twenty grandchildren. Ramjit Raghavanu, at 94 years old, is now thought to be the world's oldest new father. Nadya Suleman, the so-called Octo-mom, gave birth to 8 babies in 2009. Extreme cases such as these remind us of the huge reproductive potential of most species.

This chapter discusses meiosis, the process that occurs during sexual reproduction and ensures that offspring will have a different combination of genes than their parents. This genetic diversity is essential for survival of a species, but occasionally, offspring inherit a detrimental combination of genes and chromosomes. Such events do not detract from the success of a species overall because they allow a species to evolve and become adapted to an ever-changing environment.

As you read through the chapter, think about the following questions:

1. What is the role of meiosis in introducing new variation?

2. Why is genetic variation necessary for species survival?

3. What effects are seen when different combinations of chromosomes are produced?

FOLLOWING *the* BIG IDEAS

CHAPTER 10 MEIOSIS AND SEXUAL REPRODUCTION

Evolution	Variation introduced during meiosis produces the new genetic combinations natural selection can act upon, leading to evolution. (1A2b)
Information and Signaling	Some cells undergo meiosis to produce spores or gametes, resulting in mixed versions of genes and increase genetic diversity.

10.1 Halving the Chromosome Number

Learning Outcomes

Upon completion of this section, you should be able to

1. Contrast haploid and diploid chromosome numbers.
2. Explain what is meant by homologous chromosomes.
3. Describe the central concept of meiosis.

In sexually reproducing organisms, **meiosis** [Gk. *mio*, less, and *-sis*, act or process of] is the type of nuclear division that reduces the chromosome number from the diploid (2n) number [Gk. *diplos*, twofold, and *-eides*, like] to the haploid (n) number [Gk. *haplos*, single, and *-eides*, like]. The **diploid (2n) number** refers to the total number of chromosomes, which exists in two sets. The **haploid (n) number** of chromosomes is half the diploid number, or a single set of chromosomes. In humans, the diploid number of 46 is reduced to the haploid number of 23.

Gametes, or reproductive cells, (often the sperm and egg) usually have the haploid number of chromosomes. In **sexual reproduction**, haploid gametes are produced during meiosis that subsequently merge into a diploid cell called a **zygote.** In plants and animals, the zygote undergoes development to become an adult organism.

Meiosis is necessary in sexually reproducing organisms because the diploid number of chromosomes has to be reduced by half in each of the parents in order to produce diploid offspring. Otherwise, the number of chromosomes would double with each new generation. Within a few generations, the cells of an animal would be nothing but chromosomes! For example, in humans with a diploid number of 46 chromosomes, in five generations the chromosome number would increase to 1,472 chromosomes (46×2^5). In 10 generations this number would increase to a staggering 47,104 chromosomes (46×2^{10}). The early cytologists (biologists who study cells) realized this, and Pierre-Joseph van Beneden (1809-1894), a Belgian, was gratified to find in 1883 that the sperm and the egg of the roundworm *Ascaris* each contain only two chromosomes, while the zygote and subsequent embryonic cells always have four chromosomes.

Homologous Pairs of Chromosomes

In diploid body cells, the chromosomes occur in pairs. Figure 10.1*a*, a pictorial display of human chromosomes called a karyotype, shows the chromosomes arranged according to pairs. The members of each pair are called homologous chromosomes. **Homologous chromosomes** or **homologues** [Gk. *homologos*, agreeing, corresponding] look alike; they have the same length and centromere position. When stained, homologues have a similar banding pattern because they contain genes for the same traits in the same order in the same locations on both chromosomes in the homologous pair. But while homologous chromosomes have genes for the same traits, such as finger length, the DNA (deoxyribonucleic acid) sequence for the gene on one homologue may code for short fingers and the gene at the same

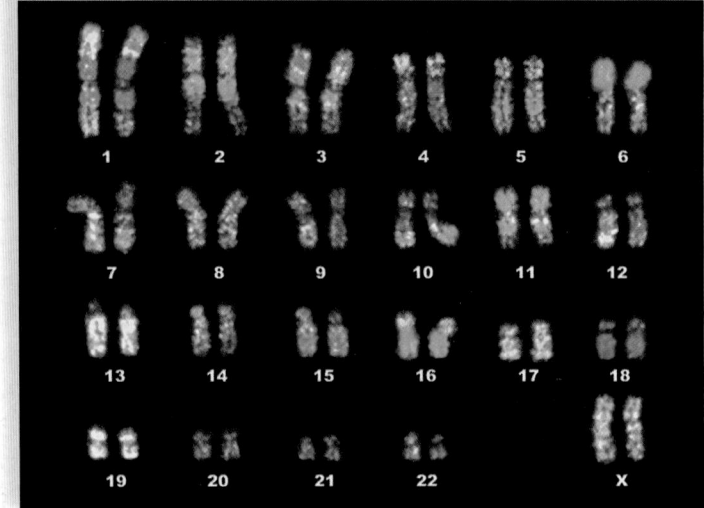

a.

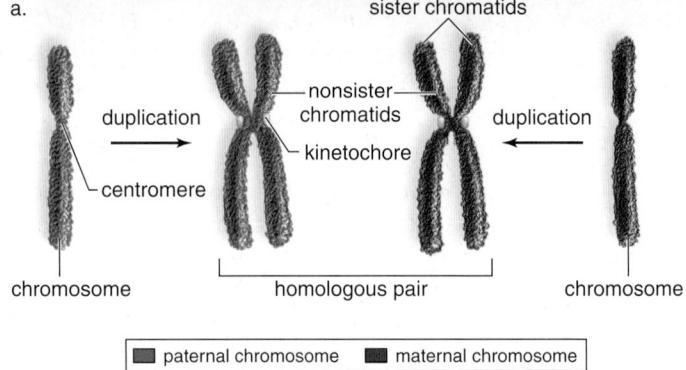

b.

Figure 10.1 Homologous chromosomes. In diploid body cells, the chromosomes occur in pairs called homologous chromosomes. **a.** In this micrograph of stained chromosomes from a human cell, the pairs have been numbered 1–23. Note that chromosome pairs 1–22 are autosomes, coding for non-sex traits, whereas pair 23 include the sex chromosomes, and help determine human gender. **b.** These chromosomes are duplicated, and each chromosome in the homologous pair is composed of two chromatids. The sister chromatids contain the exact same genes; the nonsister chromatids contain genes for the same traits (e.g., type of hair, color of eyes), but they may differ in that one could have DNA that codes for trait variations, such as dark hair versus light hair.

location on the other homologue may code for long fingers. Alternate forms of a gene (as for long fingers and short fingers) are called **alleles.** The DNA sequences of alleles are highly similar, but different enough to produce alternative physical traits like long or short fingers.

To properly produce a haploid number of chromsomes in gametes, you first have to double the amount of DNA. The chromosomes in Figure 10.1*a* are duplicated as they would be just before nuclear division. Recall that during the S stage of the cell cycle, DNA replicates and the chromosomes become duplicated. The results of the duplication process are depicted in Figure 10.1*b*. When duplicated, a chromosome is composed of two identical parts called sister chromatids, each containing one DNA double helix molecule. The sister chromatids are held together at a common region called the centromere.

Why does the zygote have paired chromosomes? One member of a homologous pair was inherited from the male parent, and the other was inherited from the female parent when the haploid sperm and egg fused together. In Figure 10.1*b* and throughout this chapter, the paternal chromosome is colored blue, and the maternal chromosome is colored red. *However, this is simply a conventional distinction. Geneticists generally use chromosome length and centromere location, not color, to recognize homologues.* You will see shortly how meiosis reduces the chromosome number. Whereas the zygote and body cells have homologous pairs of chromosomes, the gametes have only one chromosome of each kind—derived from either the paternal or maternal homologue.

Overview of Meiosis

The central purpose of meiosis is to reduce the chromosome number from 2n to n. Meiosis requires two nuclear divisions and produces four haploid daughter cells, each having one of each kind of chromosome. The process begins by replicating the chrosomosomes, then splitting the matched homologous pairs to go from 2n to n chromosomes during the first division. The second division reduces the amount of DNA in n chromosomes to an amount appropriate for each gamete. Once the DNA has been replicated and chromosomes become a pair, they may exchange genes, creating a genetic mixture different from the parent. The first nuclear division separates each homologous pair, reducing the chromosome number from 2n to n. Even though each daughter cell now has n chromosomes, each chromosome still has a sister chromatid, making a second nuclear division necessary. The end result of meiosis is four gametes with n chromosomes.

Figure 10.2 presents an overview of meiosis, indicating the two nuclear divisions, meiosis I and meiosis II. Prior to meiosis I, DNA replication has occurred; therefore, each chromosome has two sister chromatids. During meiosis I, something new happens that does not occur in mitosis. The homologous chromosomes come together and line up side by side, forming a **synaptonemal complex**. This process is called **synapsis** [Gk. *synaptos*, united, joined together] and results in a **bivalent** [L. *bis*, two, and *valens*, strength]—that is, two homologous chromosomes that stay in close association during the first two phases of meiosis I. Sometimes the term tetrad [Gk. *tetra*, four] is used instead of bivalent because, as you can see, a bivalent contains four chromatids. Chromosomes may recombine or exchange genetic information during this association (see Section 10.2).

Following synapsis, homologous pairs align at the metaphase plate, and then the members of each pair separate. This separation means that only one duplicated chromosome from each homologous pair reaches a daughter nucleus, reducing the chromosome number from 2n to n. It is important for each daughter nucleus to have a member from each pair of homologous chromosomes because only in that way can there be a copy of each *kind* of chromosome in the daughter nuclei. Notice in Figure 10.2 that two possible combinations of chromosomes in the daughter cells are shown: short red with long blue and short blue with long red. Knowing that all daughter cells have to have one short chromosome and one long chromosome, what are the other two possible combinations of chromosomes for these particular cells?

Notice that DNA replication occurs only once during meiosis; no replication is needed between meiosis I and meiosis II because the chromosomes are already duplicated; they already

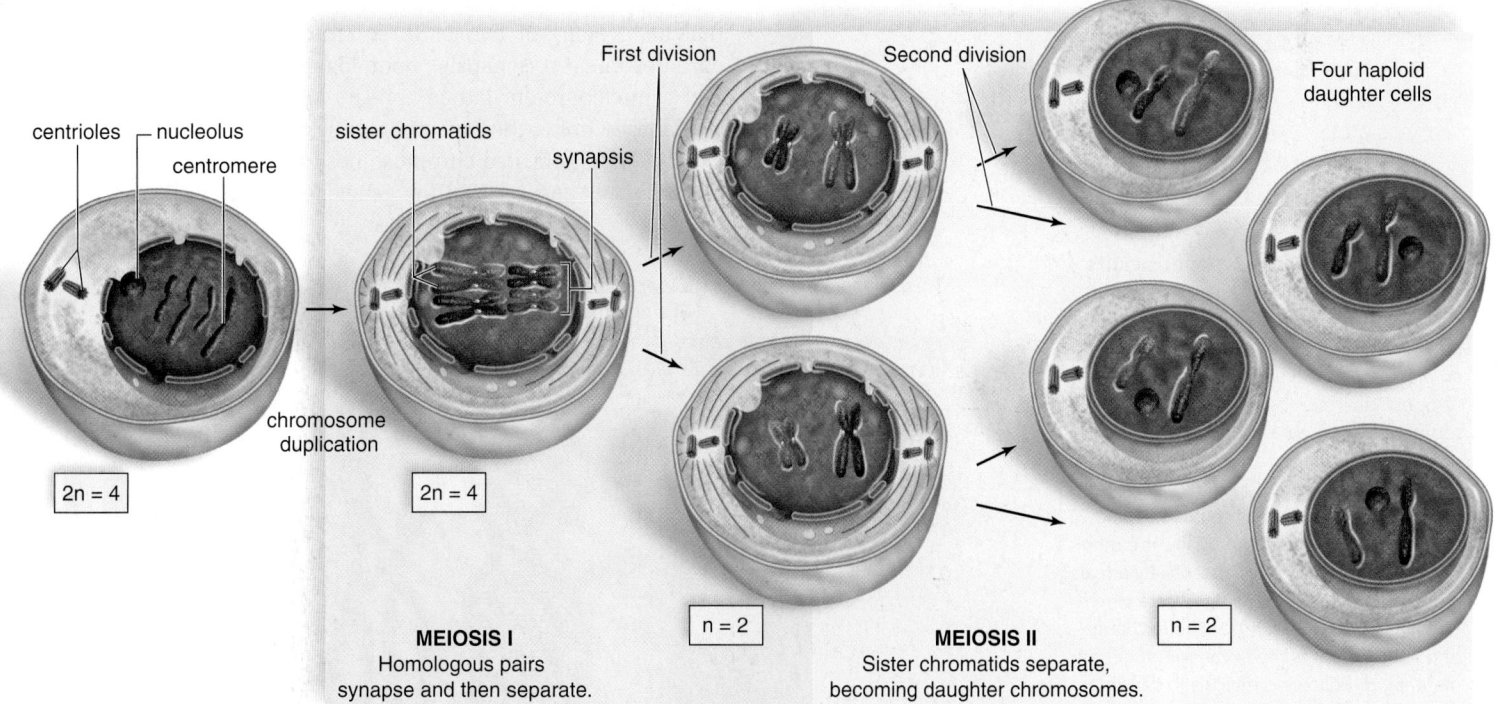

MEIOSIS I
Homologous pairs synapse and then separate.

MEIOSIS II
Sister chromatids separate, becoming daughter chromosomes.

Figure 10.2 Overview of meiosis. Following DNA replication, each chromosome is duplicated and consists of two chromatids. During meiosis I, homologous chromosomes pair and separate. During meiosis II, the sister chromatids of each duplicated chromosome separate. At the completion of meiosis, there are four haploid daughter cells. Each daughter cell has one of each kind of chromosome.

have two sister chromatids. During meiosis II, the sister chromatids separate, becoming daughter chromosomes that move to opposite poles. The chromosomes in each of the four daughter cells now contain only one DNA double helix molecule in the form of a haploid chromosome.

The number of centromeres can be counted to verify that the parent cell has the diploid number of chromosomes. At the end of meiosis I, the chromosome number has been reduced because there are half as many centromeres present, even though each chromosome still consists of two chromatids each. Each daughter cell that forms has the haploid number of chromosomes. At the end of meiosis II, sister chromatids separate, and each daughter cell that forms still contains the haploid number of chromosomes, each consisting of a single chromatid.

 **Animation**
How Meiosis Works

 **MP3**
Meiosis

Fate of Daughter Cells

In the plant life cycle, the daughter cells become haploid spores that germinate to become a haploid generation. This generation produces the gametes by mitosis. The plant life cycle is studied in Chapter 27. In the animal life cycle, the daughter cells become the gametes, either sperm or eggs. The body cells of an animal normally contain the diploid number of chromosomes due to the fusion of sperm and egg during fertilization. If meiotic events go wrong, the gametes can contain the wrong number of chromosomes or altered chromosomes. This possibility and its consequences are discussed in Section 10.6.

Check Your Progress   10.1

1. Describe what is meant by a homologous pair of chromosomes.
2. Examine how chromosome number changes during meiosis I and meiosis II.
3. Explain the purpose of a bivalent in chromosome pairing.

10.2 Genetic Variation

Learning Outcomes

Upon completion of this section, you should be able to

1. Contrast the effects of asexual and sexual reproduction on genetic variation.
2. Explain how crossing-over contributes to genetic variation.
3. Examine how independent assortment contributes to genetic variation in the offspring.

We have seen that meiosis provides a way to keep the chromosome number constant generation after generation. Without meiosis, the chromosome number of the next generation would continually increase. The events of meiosis also help ensure that genetic variation occurs with each generation.

Genetic variation is essential for a species to be able to evolve and adapt in a changing environment. Asexually reproducing organisms, such as the prokaryotes, depend primarily on mutations to generate variation among offspring. This is sufficient for their survival because they produce great numbers of offspring very quickly. Although mutations also occur among sexually reproducing organisms, the reshuffling of genetic material during sexual reproduction ensures that offspring will have a different combination of genes from their parents. Meiosis brings about genetic variation in two key ways: crossing-over and independent assortment of homologous chromosomes.

Genetic Recombination

Crossing-over is an exchange of genetic material between nonsister chromatids of a bivalent during meiosis I. It is estimated that an average of two or three crossovers occur per human chromosome. At synapsis, homologues line up side by side, and a nucleoprotein lattice appears between them (Fig. 10.3). This lattice holds the bivalent together in such a way that the DNA of the duplicated chromosomes of each homologue pair is aligned. This ensures that the genes contained on the nonsister

Figure 10.3 Crossing-over occurs during meiosis I. **a.** The homologous chromosomes pair up, and a nucleoprotein lattice develops between them. This is an electron micrograph of the lattice. It "zippers" the members of the bivalent together so that corresponding genes on paired chromosomes are in alignment. **b.** This visual representation shows only two places where nonsister chromatids 1 and 3 have come into contact. Actually, the other two nonsister chromatids most likely are also crossing-over. **c.** Chiasmata indicate where crossing-over has occurred. The exchange of color represents the exchange of genetic material. **d.** Following meiosis II, daughter chromosomes have a new combination of genetic material due to crossing-over, which occurred between nonsister chromatids during meiosis I.

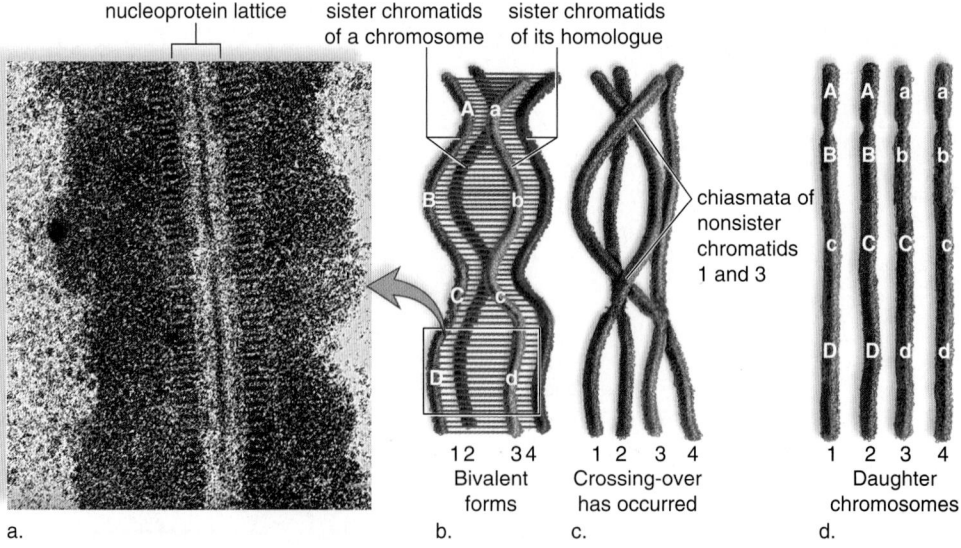

nucleoprotein lattice

sister chromatids of a chromosome

sister chromatids of its homologue

chiasmata of nonsister chromatids 1 and 3

1 2 3 4
Bivalent forms

1 2   3 4
Crossing-over has occurred

1 2 3 4
Daughter chromosomes

a.   b.   c.   d.

chromatids are directly aligned. Now crossing-over may occur. As the lattice breaks down, homologues are temporarily held together by *chiasmata* (sing., chiasma), regions where the non-sister chromatids are attached due to DNA strand exchange and crossing-over. After exchange of genetic information between the mother's and father's chromosomes, the homologues separate and are distributed to different daughter cells.

Crossing-over has been shown to be essential for the normal segregation of chromosomes during meiosis. For example, reduced levels of crossing-over have been linked to Down syndrome, which is caused by an extra copy of chromosome 21.

To appreciate the significance of crossing-over, keep in mind that the members of a homologous pair can carry slightly different instructions for the same genetic traits. In the end, due to a swapping of genetic material during crossing-over, the chromatids held together by a centromere are no longer identical. Therefore, when the chromatids separate during meiosis II, some of the daughter cells receive daughter chromosomes with recombined alleles. Due to **genetic recombination,** the offspring have a different set of alleles, and therefore genes, than their parents. This increases the genetic variation of the offspring.

Animation
Meiosis with Crossing-Over

Independent Assortment of Homologous Chromosomes

During **independent assortment,** the homologous chromosome pairs separate independently, or in a random manner. When homologues align at the metaphase plate, the maternal or paternal homologue may be oriented toward either pole. Figure 10.4 shows the possible chromosome orientations for a cell that contains only three pairs of homologous chromosomes. Once all possible alignments of independent assortment are considered for these three pairs, the result will be 2^3, or eight, combinations of maternal and paternal chromosomes in the resulting gametes from this cell, simply due to independent assortment of homologues.

Animation
Random Orientation of Chromosomes During Meiosis

Significance of Genetic Variation

In humans, who have 23 pairs of chromosomes, the possible chromosomal combinations in the gametes is a staggering 2^{23}, or 8,388,608. The variation that results from meiosis is enhanced by **fertilization,** the union of the male and female gametes. The chromosomes donated by the parents are combined, and in humans, this means that there are $(2^{23})^2$, or 70,368,744,000,000, chromosomally different zygotes possible, even assuming no crossing-over. If crossing-over occurs once, then $(4^{23})^2$, or 4,951,760,200,000,000,000,000,000, genetically different zygotes are possible for every couple. Keep in mind that crossing-over can occur several times in each chromosome!

Animation
Genetic Diversity

The staggering amount of genetic variation achieved through meiosis is particularly important to the long-term survival of a species because it increases genetic variation within a population. (Asexual reproduction passes on exactly the same combination of chromosomes and genes.) The process of sexual reproduction brings about genetic recombinations among members of a population.

If a parent is already successful in a particular environment, is asexual reproduction advantageous? It would seem so as long as the environment remains unchanged. However, if the environment changes, genetic variability among offspring introduced by sexual reproduction may be advantageous. Under the new conditions, some offspring may have a better chance of survival and reproductive success than others in a population. For example, suppose the ambient temperature were to rise due to climate change. Perhaps a dog with genes for the least amount of fur may have an advantage over other dogs of its generation.

In a changing environment, sexual reproduction, with its reshuffling of genes due to meiosis and fertilization, might give a few offspring a better chance to survive and reproduce, thereby increasing the possibility of passing on their genes to the next generation.

Check Your Progress   10.2

1. Describe the two main ways in which meiosis contributes to genetic variation.
2. Examine how many combinations of chromosomes are possible in the gametes in a cell with four pairs of homologous chromosomes.
3. Evaluate why meiosis and sexual reproduction are important in responding to the changing environment.

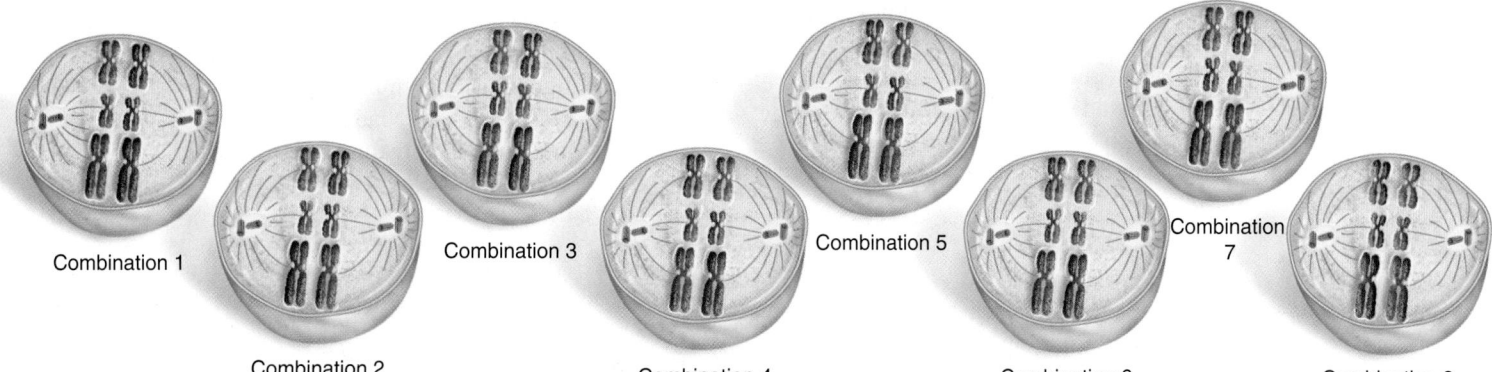

Combination 1
Combination 2
Combination 3
Combination 4
Combination 5
Combination 6
Combination 7
Combination 8

Figure 10.4 Independent assortment. When a parent cell has three pairs of homologous chromosomes, there are 2^3, or 8, possible chromosome alignments at the metaphase plate due to independent assortment. Among the 16 daughter nuclei resulting from these alignments, there are 8 different combinations of chromosomes. Each possible combination is shown, one in each cell.

10.3 The Phases of Meiosis

Learning Outcomes

Upon completion of this section, you should be able to

1. Describe the phases of meiosis and the major events that occur during each phase.
2. Identify the phase of meiosis where chromosome number is reduced from diploid to haploid.

Meiosis consists of two unique, consecutive cell divisions, meiosis I and meiosis II. DNA is replicated in S phase of the cell cycle prior to meiosis I but not meiosis II. Both meiosis I and meiosis II each contain a prophase, metaphase, anaphase, and telophase.

Animation Stages of Meiosis

Prophase I

It is apparent during prophase I that nuclear division is about to occur because a spindle forms as the centrosomes migrate away from one another. The nuclear envelope fragments, and the nucleolus disappears.

The homologous chromosomes, each having replicated during S phase of the cell cycle, consist of two sister chromatids. The homologous chromsomes undergo synapsis to form bivalents. At this time crossing-over may occur between the non-sister chromatids (Fig. 10.3). As described earlier, crossing-over increases the genetic diversity of the daughter cells, because after crossing-over, the sister chromatids are no longer identical.

Throughout prophase I, the homologous chromosomes have been condensing so that by now they have the appearance of compacted metaphase chromosomes.

Metaphase I

During metaphase I, the bivalents held together by chiasmata (see Fig. 10.3) have moved toward the metaphase plate (equator of the spindle). Metaphase I is characterized by a fully formed spindle and alignment of the bivalents at the metaphase plate. As in mitosis, kinetochores are seen, but the two kinetochores of a duplicated chromosome are attached to the same kinetochore spindle fiber.

Bivalents independently align themselves at the metaphase plate of the spindle. Either the maternal or paternal homologue of each bivalent may be oriented toward either pole of the cell. The orientation of one bivalent is not dependent on the orientation of the other bivalents. This independent assortment of chromosomes contributes to the genetic variability of the daughter cells because all possible combinations of chromosomes can occur in the daughter cells.

Anaphase I

During anaphase I, the homologues of each bivalent separate and move to opposite poles, but sister chromatids do not separate. This splitting of the homologous pair reduces the chromosome number from 2n to n. However, each chromosome still has two chromatids (see Fig. 10.5).

Telophase I

Completion of telophase I is not necessary during meiosis. That is, the spindle disappears, but new nuclear envelopes need not form before the daughter cells proceed to meiosis II. Also, this phase may or may not be accompanied by cytokinesis, which is separation of the cytoplasm. Figure 10.5 shows only two of the four possible combinations of haploid chromosomes when the parent cell has two homologous pairs of chromosomes. Can you determine what the other two possible combinations of chromosomes are?

Animation Meiosis I

Interkinesis

Following telophase, the cells enter interkinesis, a short rest period prior to beginning the second nuclear division, meiosis II. The process of **interkinesis** is similar to interphase between mitotic divisions except that DNA replication does not occur because the chromosomes are already duplicated.

Meiosis II and Gamete Formation

At the beginning of meiosis II, the two daughter cells contain the haploid number of chromosomes, or one chromosome from each homologous pair. Note that these chromosomes still consist of duplicated sister chromatids at this point. During metaphase II, the chromosomes align at the metaphase plate, but do not align in homologous pairs as in meiosis I because only one chromosome of each homologous pair is present (see Fig. 10.5). Thus, the alignment of the chromosomes at the metaphase plate is similar to what is observed during mitosis.

During anaphase II, the sister chromatids separate, becoming daughter chromosomes that are not duplicated. These daughter chromosomes move toward the poles. At the end of telophase II and cytokinesis, there are four haploid cells. Because of crossing-over of chromatids during meiosis I, each gamete most likely contains chromosomes with a mixture of maternal and paternal genes.

Animation Meiosis II

As mentioned, following meiosis II, the haploid cells become gametes in animals (see Section 10.5). In plants, they become **spores,** reproductive cells that develop into new multicellular structures without the need to fuse with another reproductive cell. The multicellular structure is the haploid generation, which produces gametes. The resulting zygote develops into a diploid generation. Therefore, plants have both haploid and diploid phases in their life cycle, and plants are said to exhibit an **alternation of generations.** In most fungi and algae, the zygote undergoes meiosis, and the daughter cells develop into new individuals. Therefore, the organism is always haploid.

Check Your Progress   10.3

1. Explain what would cause daughter cells following meiosis II to contain identical chromosomes or nonidentical chromosomes.
2. Examine what might happen if homologous chromosomes lined up top to bottom instead of side by side during meiosis I.

10.4 Meiosis Compared to Mitosis

Learning Outcomes

Upon completion of this section, you should be able to

1. Contrast changes in chromosome number, genetic variability, and number of daughter cells between meiosis and mitosis.
2. Distinguish the events that occur during prophase I of meiosis that do not occur during prophase of mitosis.
3. Compare chromosome alignment during meiosis I to mitosis.

Figure 10.6 graphically compares meiosis and mitosis. Several of the fundamental differences between the two processes include:

- Meiosis requires two nuclear divisions, but mitosis requires only one nuclear division.
- Meiosis produces four daughter nuclei. Following cytokinesis there are four daughter cells. Mitosis followed by cytokinesis results in two daughter cells.
- Following meiosis, the four daughter cells are haploid and have half the chromosome number as the diploid parent cell. Following mitosis, the daughter cells have the same chromosome number as the parent cell.
- Following meiosis, the daughter cells are neither genetically identical to each other nor to the parent cell. Following mitosis, the daughter cells are genetically identical to each other and to the parent cell.

In addition to the fundamental differences between meiosis and mitosis, two specific differences between the two types of nuclear divisions can be categorized. These differences involve occurrence and process.

Occurrence

Meiosis occurs only at certain times in the life cycle of sexually reproducing organisms. In humans, meiosis occurs only in the reproductive organs and produces the gametes. Mitosis is more common because it occurs in all tissues during growth and repair.

Animation
Comparison of Meiosis and Mitosis

Process

We now compare the processes of both meiosis I and meiosis II to mitosis.

Meiosis I Compared to Mitosis

Notice that these events distinguish meiosis I from mitosis:

- During prophase I, bivalents form and crossing-over occurs. These events do not occur during mitosis.
- During metaphase I of meiosis, bivalents independently align at the metaphase plate. The paired chromosomes have a total of four chromatids each. During metaphase in mitosis, individual chromosomes align at the metaphase plate. They each have two chromatids.

Table 10.1 Meiosis I Compared to Mitosis

Meiosis I	Mitosis
Prophase I	**Prophase**
Pairing of homologous chromosomes	No pairing of chromosomes
Metaphase I	**Metaphase**
Bivalents at metaphase plate	Duplicated chromosomes at metaphase plate
Anaphase I	**Anaphase**
Homologues of each bivalent separate and duplicated chromosomes move to poles	Sister chromatids separate, becoming daughter chromosomes that move to the poles
Telophase I	**Telophase**
Two haploid daughter cells, not identical to the parent cell	Two diploid daughter cells, identical to the parent cell

Table 10.2 Meiosis II Compared to Mitosis

Meiosis II	Mitosis
Prophase II	**Prophase**
No pairing of chromosomes	No pairing of chromosomes
Metaphase II	**Metaphase**
Haploid number of duplicated chromosomes at metaphase plate	Diploid number of duplicated chromosomes at metaphase plate
Anaphase II	**Anaphase**
Sister chromatids separate, becoming daughter chromosomes that move to the poles	Sister chromatids separate, becoming daughter chromosomes that move to the poles
Telophase II	**Telophase**
Four haploid daughter cells, not genetically identical	Two diploid daughter cells, identical to the parent cell

- During anaphase I of meiosis, homologues of each bivalent separate and duplicated chromosomes (with centromeres intact) move to opposite poles. During anaphase of mitosis, sister chromatids separate, becoming daughter chromosomes that move to opposite poles.

Meiosis II Compared to Mitosis

The events of meiosis II are similar to those of mitosis except that in meiosis II, the nuclei contain the haploid number of chromosomes. In mitosis, the original number of chromosomes is maintained. Meiosis II produces two daughter cells from each parent cell that completes meiosis I, for a total of four daughter cells. These daughter cells contain the same number of chromosomes as they did at the end of meiosis I. Tables 10.1 and 10.2 compare meiosis I and II to mitosis.

Check Your Progress   10.4

1. Compare chromosome alignment between metaphase I of meiosis and metaphase of mitosis.
2. Explain how meiosis II is more similar to mitosis than to meiosis I.

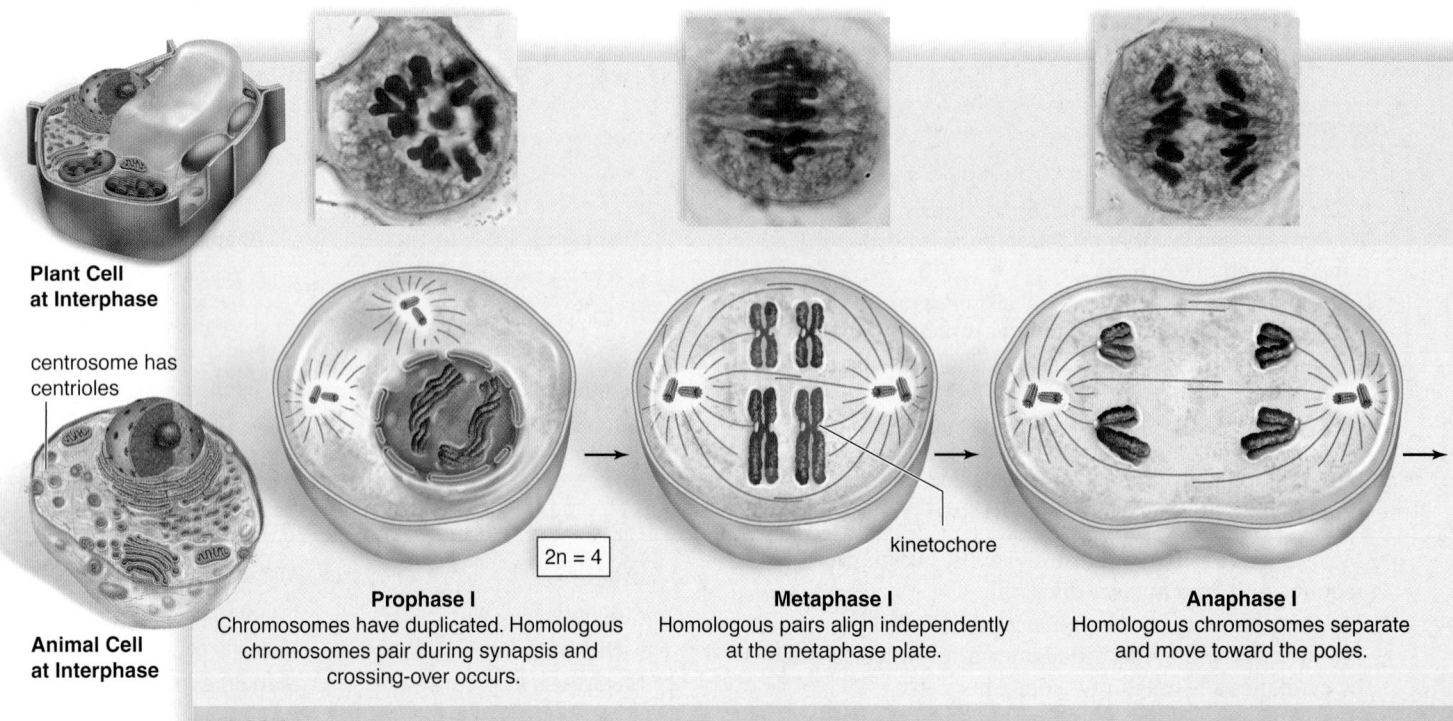

Plant Cell at Interphase

centrosome has centrioles

Animal Cell at Interphase

2n = 4

kinetochore

Prophase I
Chromosomes have duplicated. Homologous chromosomes pair during synapsis and crossing-over occurs.

Metaphase I
Homologous pairs align independently at the metaphase plate.

Anaphase I
Homologous chromosomes separate and move toward the poles.

MEIOSIS I

Figure 10.5 Meiosis I and II in plant cell micrographs and animal cell drawings. When diploid homologous chromosomes pair during meiosis I, crossing-over occurs as represented by the exchange of color. Pairs of homologous chromosomes separate during meiosis I, and chromatids separate, becoming haploid daughter chromosomes with two copies of each during meiosis II. Following meiosis II and the separation of sister chromatids, four haploid daughter cells are produced.

n = 2

n = 2

Prophase II
Cells have one chromosome from each homologous pair.

Metaphase II
Chromosomes align at the metaphase plate.

Anaphase II
Sister chromatids separate and become daughter chromosomes.

MEIOSIS II

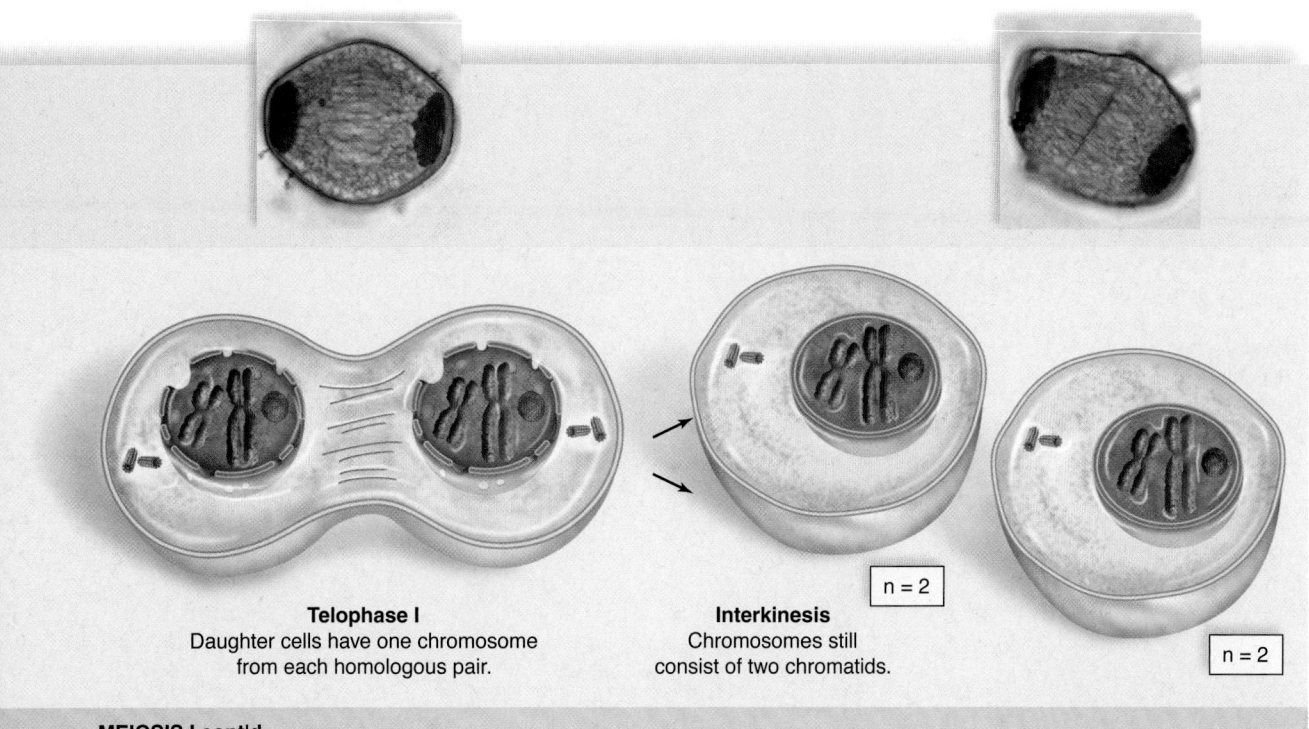

Telophase I
Daughter cells have one chromosome
from each homologous pair.

n = 2

Interkinesis
Chromosomes still
consist of two chromatids.

n = 2

MEIOSIS I cont'd

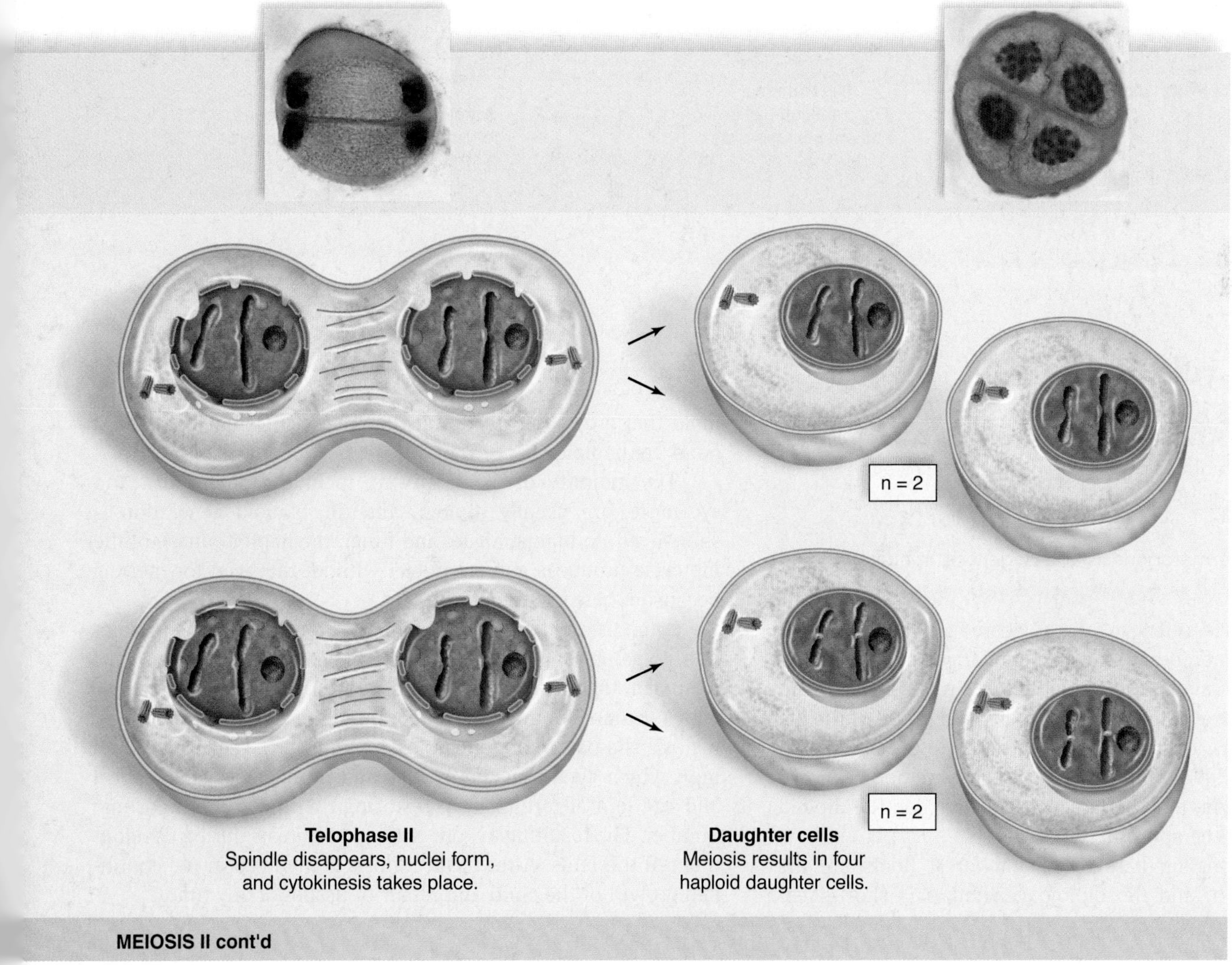

Telophase II
Spindle disappears, nuclei form,
and cytokinesis takes place.

n = 2

Daughter cells
Meiosis results in four
haploid daughter cells.

n = 2

MEIOSIS II cont'd

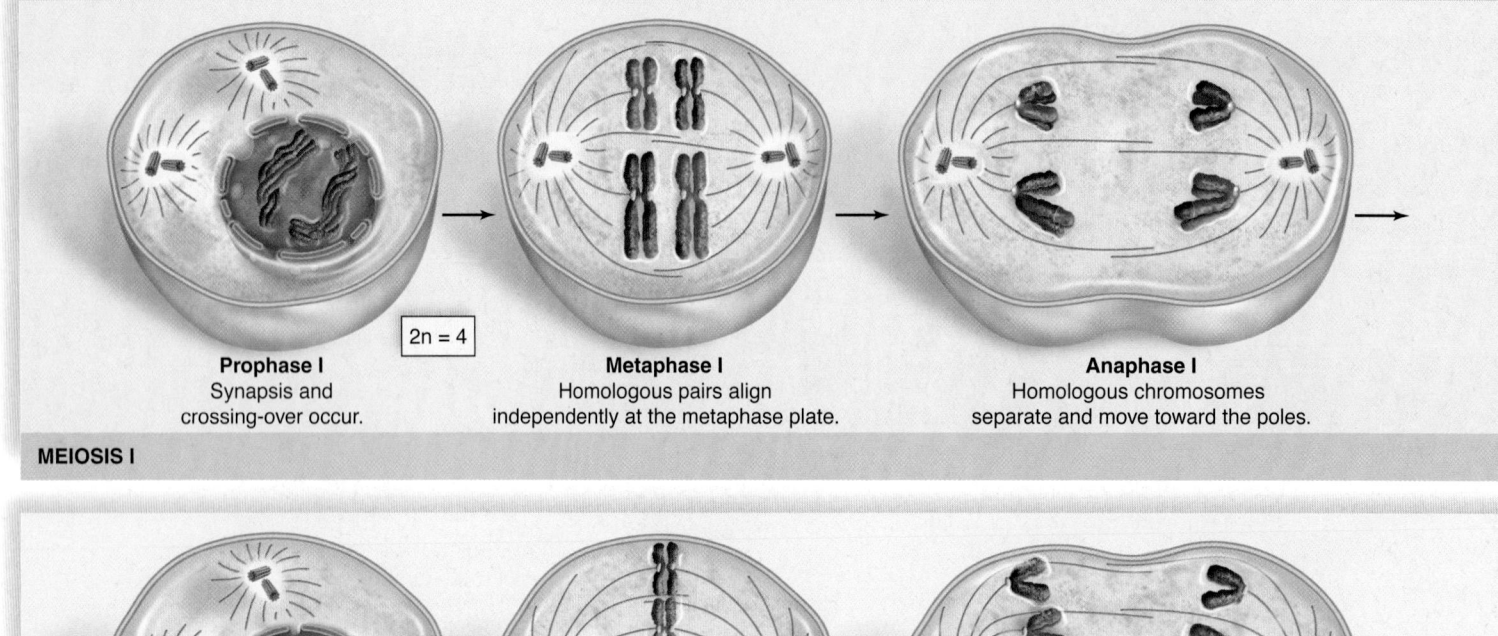

Prophase I
Synapsis and
crossing-over occur.

2n = 4

Metaphase I
Homologous pairs align
independently at the metaphase plate.

Anaphase I
Homologous chromosomes
separate and move toward the poles.

MEIOSIS I

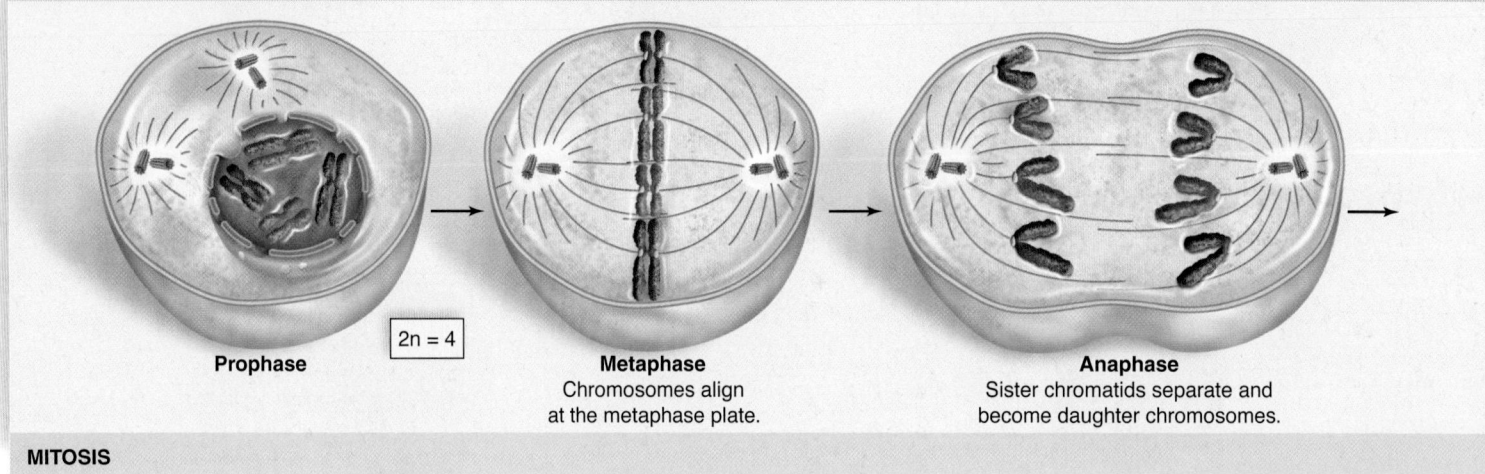

Prophase

2n = 4

Metaphase
Chromosomes align
at the metaphase plate.

Anaphase
Sister chromatids separate and
become daughter chromosomes.

MITOSIS

10.5 The Cycle of Life

Learning Outcomes

Upon completion of this section, you should be able to
1. Contrast the life cycle of plants with the life cycle of animals.
2. Describe spermatogenesis and oogenesis in humans.

The term **life cycle** refers to all the reproductive events that occur from one generation to the next similar generation. In animals, including humans, the individual is always diploid, and meiosis produces the gametes, the only haploid phase of the life cycle (Fig. 10.7). In contrast, plants have a haploid phase that alternates with a diploid phase. The haploid generation, known as the **gametophyte,** may be larger or smaller than the diploid generation, called the **sporophyte.**

Mosses growing on bare rocks and forest floors are the haploid generation, and the diploid generation is short-lived.

In most fungi and algae, the zygote is the only diploid portion of the life cycle, and it undergoes meiosis. Therefore, the black mold that grows on bread and the green scum that floats on a pond are haploid.

The majority of plant species, including pine, corn, and sycamore, are usually diploid, and the haploid generation is short-lived. In plants, algae, and fungi, the haploid phase of the life cycle produces gamete nuclei without the need for meiosis because it has occurred earlier.

Animals are diploid, and meiosis occurs during the production of gametes **(gametogenesis).** In males, meiosis is a part of **spermatogenesis** [Gk. *sperma,* seed; L. *genitus,* producing], which occurs in the testes and produces sperm. In females, meiosis is a part of **oogenesis** [Gk. *oon,* egg; L. *genitus,* producing], which occurs in the ovaries and produces eggs. A sperm and egg join at fertilization, restoring the diploid chromosome number. The resulting zygote undergoes mitosis during development of the fetus. After birth, mitosis is involved in the continued growth of the child and repair of tissues at any time.

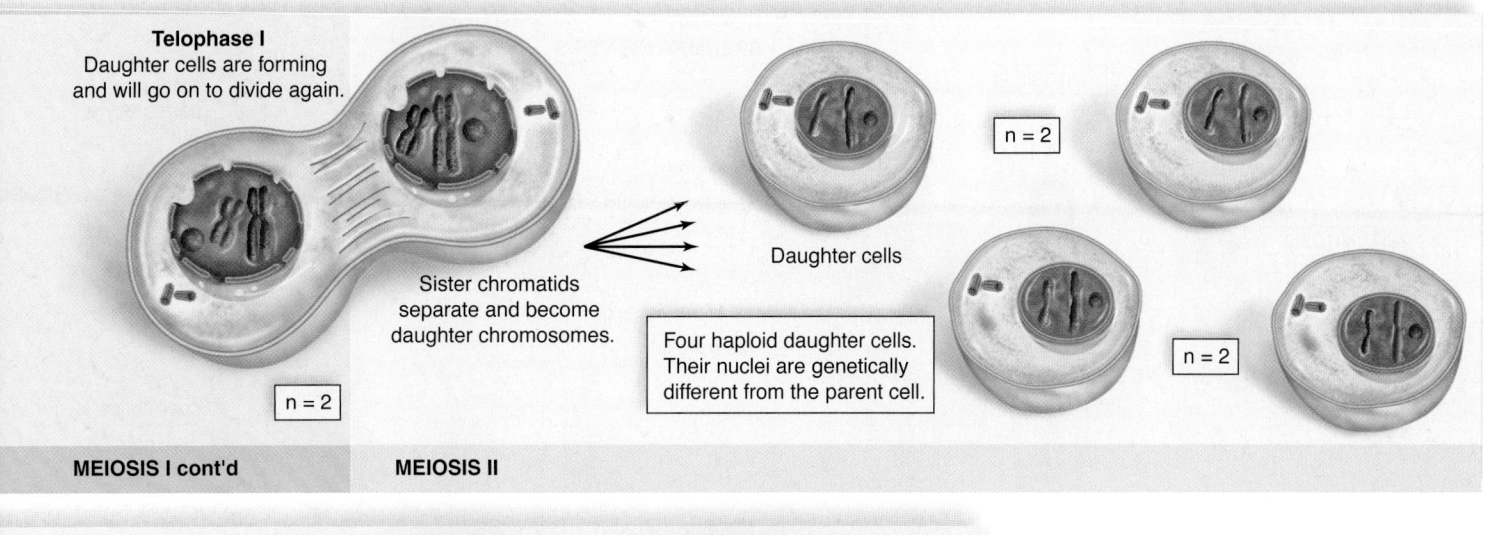

Telophase I
Daughter cells are forming and will go on to divide again.

Sister chromatids separate and become daughter chromosomes.

n = 2

n = 2

n = 2

Daughter cells

Four haploid daughter cells. Their nuclei are genetically different from the parent cell.

MEIOSIS I cont'd   MEIOSIS II

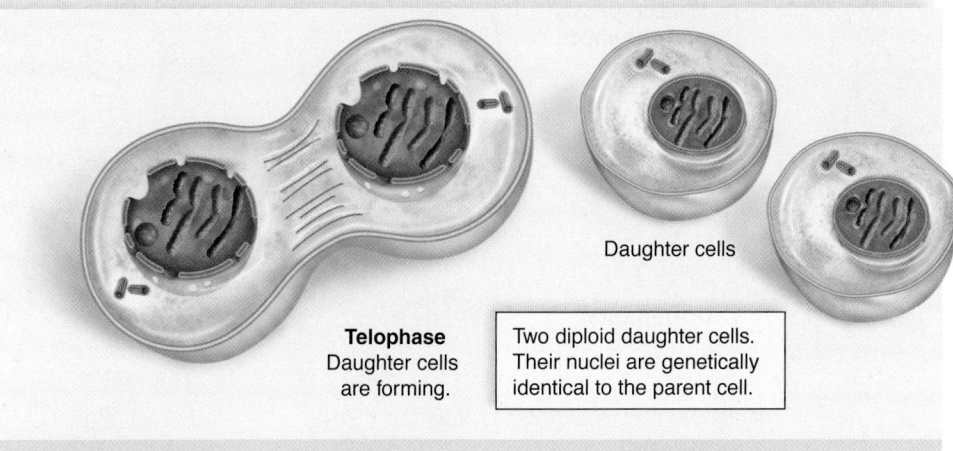

Daughter cells

Telophase
Daughter cells are forming.

Two diploid daughter cells. Their nuclei are genetically identical to the parent cell.

MITOSIS cont'd

Figure 10.6 Meiosis compared to mitosis.   Why does meiosis produce daughter cells with half the number of chromosomes, while mitosis produces daughter cells with the same number of chromosomes as the parent cell? Compare metaphase I of meiosis to metaphase of mitosis. Only in metaphase I of meiosis are the homologous chromosomes paired at the metaphase plate. Members of homologous chromosome pairs separate during anaphase I, and therefore the daughter cells are haploid. The exchange of color between nonsister chromatids represents the crossing-over that occurred during meiosis I. The blue chromosomes were inherited from the paternal parent, and the red chromosomes were inherited from the maternal parent.

Spermatogenesis and Oogenesis in Humans

In human males, spermatogenesis occurs within the testes, and in females, oogenesis occurs within the ovaries.

Spermatogenesis

The testes contain stem cells called spermatogonia, and these cells keep the testes supplied with primary spermatocytes that undergo spermatogenesis as described in Figure 10.8, *top*. Primary spermatocytes with 46 chromosomes undergo meiosis I to form two secondary spermatocytes, each with 23 duplicated chromosomes. Secondary spermatocytes undergo meiosis II to produce four spermatids with 23 daughter chromosomes. Spermatids then differentiate into viable sperm (spermatozoa). Upon sexual arousal, the sperm enter ducts and exit the penis upon ejaculation.

Animation Spermatogenesis

Oogenesis

The ovaries contain stem cells called oogonia that produce many primary oocytes with 46 chromosomes during fetal development.

They even begin oogenesis, but only a few continue when a female has become sexually mature. The result of meiosis I is two haploid cells with 23 chromosomes each (Fig. 10.8, *bottom*). One of these cells, termed the **secondary oocyte** [Gk, *oon,* egg, and *kytos,* cell], receives almost all the cytoplasm. The other is a **polar body** that may either disintegrate or divide again.

The secondary oocyte begins meiosis II but stops at metaphase II. Then the secondary oocyte leaves the ovary and enters an oviduct, where sperm may be present. If no sperm are in the oviduct, or if a sperm does not enter the secondary oocyte, it eventually disintegrates without completing meiosis. If a sperm does enter the oocyte, some of its contents trigger the completion of meiosis II in the secondary oocyte, and another polar body forms.

At the completion of oogenesis, following entrance of a sperm, there is one egg and two to three polar bodies. The polar bodies are a way to "dispose" of chromosomes while retaining much of the cytoplasm in the egg. Cytoplasmic molecules and organelles are needed by a developing embryo following

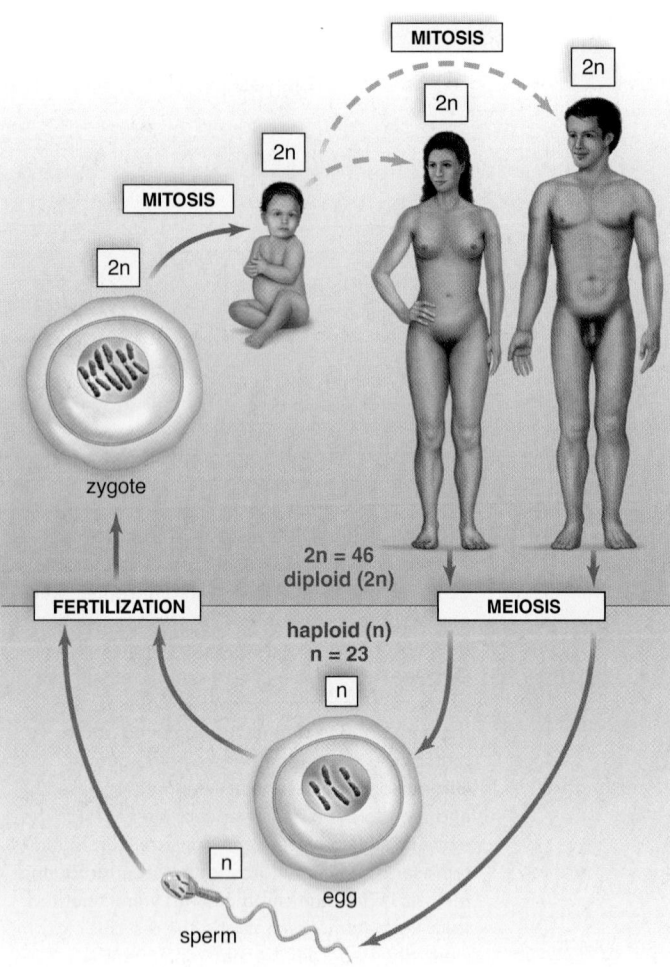

Figure 10.7 Life cycle of humans. Meiosis in males is a part of sperm production, and meiosis in females is a part of egg production. When a haploid sperm fertilizes a haploid egg, the zygote is diploid. The zygote undergoes mitosis as it develops into a newborn child. Mitosis continues throughout life during growth and repair.

fertilization. Some zygote components, such as the centrosome, are contributed by the sperm.

The mature egg has 23 chromosomes, but the zygote formed when the sperm and egg nuclei fuse has 46 chromosomes. Therefore, fertilization restores the diploid number of chromosomes. The production of haploid gametes and subsequent fusion of those gametes into a diploid zygote completes a human life cycle.

Check Your Progress   10.5

1. Describe where cells that undergo meiosis are located in humans.
2. Compare the number of gametes produced during oogenesis and spermatogenesis in humans.

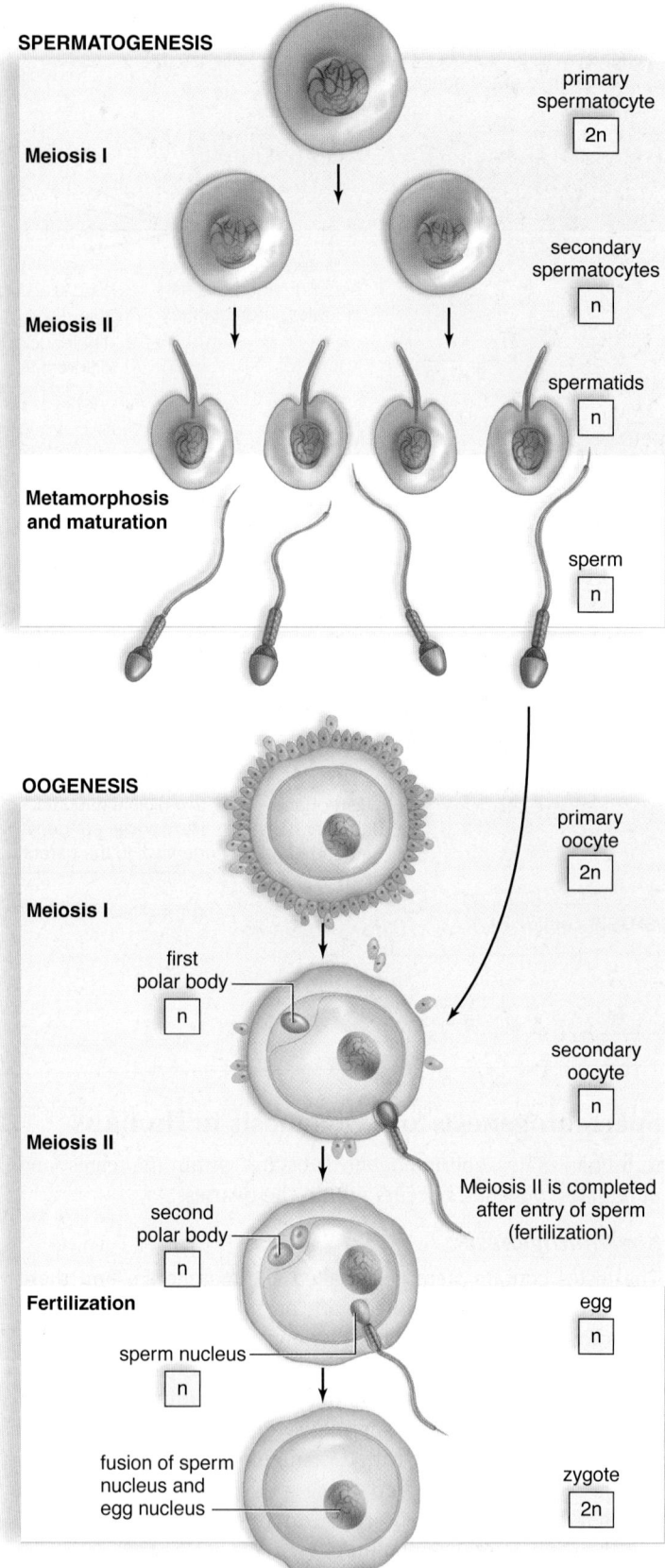

Figure 10.8 Spermatogenesis and oogenesis in mammals. Spermatogenesis produces four viable sperm, whereas oogenesis produces one egg and at least two polar bodies. In humans, both sperm and egg have 23 chromosomes each; therefore, following fertilization, the zygote has 46 chromosomes.

10.6 Changes in Chromosome Number and Structure

Learning Outcomes

Upon completion of this section, you should be able to

1. Distinguish between euploidy and aneuploidy.
2. Explain how nondisjunction can cause monosomy and trisomy aneuploidy.
3. Describe human diseases caused by changes in the number of sex chromosomes.
4. Examine how changes in chromosome structure can lead to human diseases.

We have seen that crossing-over creates variation within a population and is essential for the normal separation of chromosomes during meiosis. Furthermore, the proper separation of homologous chromosomes during meiosis I and the separation of sister chromatids during meiosis II are essential for the maintenance of normal chromosome numbers in living organisms. Although meiosis almost always proceeds normally, failure of chromosomes to separate, or **nondisjunction,** may occur, resulting in gain or loss of chromosomes. Errors in crossing-over may result in extra or missing parts of chromosomes.

Aneuploidy

The correct number of chromosomes in a species is known as **euploidy.** A change in the chromosome number resulting from nondisjunction during meiosis is called **aneuploidy.** Aneuploidy is seen in both plants and animals. Monosomy and trisomy are two aneuploid states.

Monosomy (2n − 1) occurs when an individual has only one of a particular type of chromosome when they should have two, and **trisomy** (2n + 1) occurs when an individual has three of a particular type of chromosome when they should have two. Both monosomy and trisomy are the result of nondisjunction during mitosis or meiosis. *Primary nondisjunction* occurs during meiosis I when both members of a homologous pair go into the same daughter cell (Fig. 10.9a). *Secondary nondisjunction* occurs during meiosis II when the sister chromatids fail to separate and both daughter chromosomes go into the same gamete (Fig. 10.9b).

Notice that when secondary nondisjunction occurs, there are two normal gametes and two aneuploid gametes. In contrast, when primary nondisjunction occurs, no normal gametes are produced. Therefore, primary nondisjunction tends to have more deleterious effects than secondary nondisjunction.

In animals, monosomies and trisomies of non-sex, or autosomal, chromosomes are generally lethal, but a trisomic individual is more likely to survive than a monosomic one. In humans,

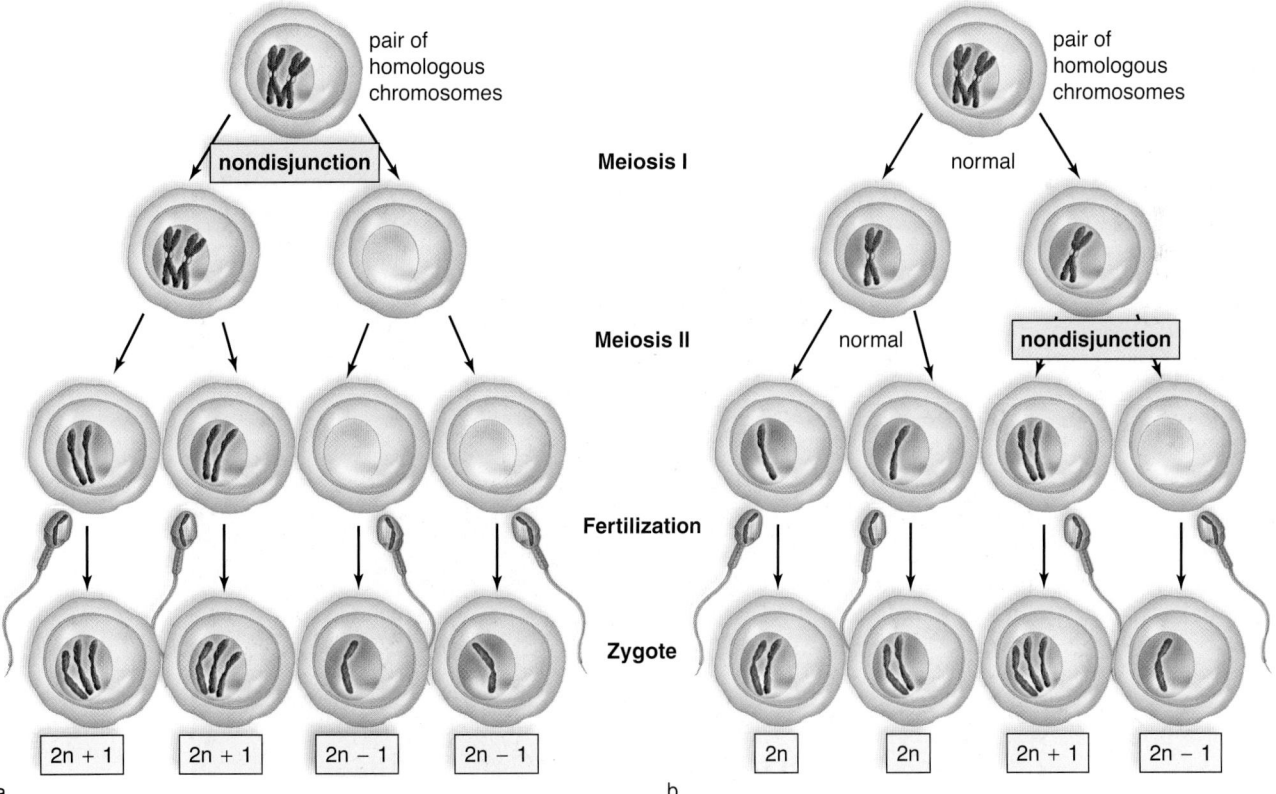

Figure 10.9 Nondisjunction of chromosomes during oogenesis, followed by fertilization with normal sperm. **a.** Nondisjunction can occur during meiosis I (primary nondisjunction) and results in abnormal eggs that also have one more or one less than the normal number of chromosomes. Fertilization of these abnormal eggs with normal sperm results in a zygote with abnormal chromosome numbers. 2n = diploid number of chromosomes. **b.** Nondisjunction can also occur during meiosis II (secondary nondisjunction) if the sister chromatids separate but the resulting daughter chromosomes go into the same daughter cell. Then the egg will have one more or one less than the usual number of chromosomes. Fertilization of these abnormal eggs with normal sperm produces a zygote with abnormal chromosome numbers.

only three autosomal trisomic conditions are known to be viable beyond birth: trisomy 13, 18, and 21. Only trisomy 21 is viable beyond early childhood, and is characterized by a distinctive set of physical and mental abnormalities. In comparison, sex chromosome aneuploids are better tolerated in animals and have a better chance of producing survivors.

Trisomy 21

The most common autosomal trisomy seen among humans is trisomy 21, also called Down syndrome. This syndrome is easily recognized by these characteristics: short stature; an eyelid fold; a flat face; stubby fingers; a wide gap between the first and second toes; a large, fissured tongue; a round head; a distinctive palm crease; heart problems; and some degree of mental retardation, which can sometimes be severe. Individuals with Down syndrome also have a greatly increased risk of developing leukemia and tend to age rapidly, resulting in a shortened life expectancy. In addition, these individuals have an increased chance of developing Alzheimer disease later in life.

Many scientists agree that the symptoms of Down syndrome are caused by gene dosage effects resulting from the presence of the extra chromosome. Recent studies indicate that not all of the genes on the chromosome are expressed at a level of 150%, challenging this theory; however, scientists have identified several genes that have been linked to increased risk of leukemia, cataracts, aging, and mental retardation.

The chances of a woman having a child with Down syndrome increase rapidly with age. In women ages 20 to 30, the incidence of trisomy 21 is 1 in 1,400 births, and in women 30 to 35, the incidence is about 1 in 750 births. It is thought that the longer the oocytes are stored in the female, the greater the chances of nondisjunction occurring. However, even though an older woman is more likely to have a Down syndrome child, most babies with Down syndrome are born to women younger than age 40 because this is the age group having the most babies. Furthermore, some recent research also indicate that in 23% of the cases studied, the sperm contributed the extra chromosome. A **karyotype,** a visual display of the chromosomes arranged by size, shape, and banding pattern, may be performed to identify babies with Down syndrome and other aneuploid conditions (Fig. 10.10).

Changes in Sex Chromosome Number

An abnormal sex chromosome number is the result of inheriting too many or too few X or Y chromosomes. Nondisjunction during oogenesis or spermatogenesis can result in gametes with an abnormal number of sex chromosomes. However, extra copies of the sex chromosomes are much more easily tolerated in humans than are extra copies of autosomes.

A person with Turner syndrome (XO) is a female, and a person with Klinefelter syndrome (XXY) is a male. However, deletion of the *SRY* gene on the short arm of the Y chromosome results in Swyer syndrome, or an "XY female." Individuals with Swyer syndrome lack a hormone called testis-determining factor, which plays a critical role in the development of male genitals. Furthermore, movement of this same gene onto the X chromosome may result in de la Chapelle syndrome, or an "XX male." Men with de la Chapelle syndrome exhibit undersized testes, sterility, and rudimentary breast development. Together, these observations suggest that in humans, the presence of the *SRY* gene, not the number of X chromosomes, determines maleness. In its absence, a person develops as a female.

a.

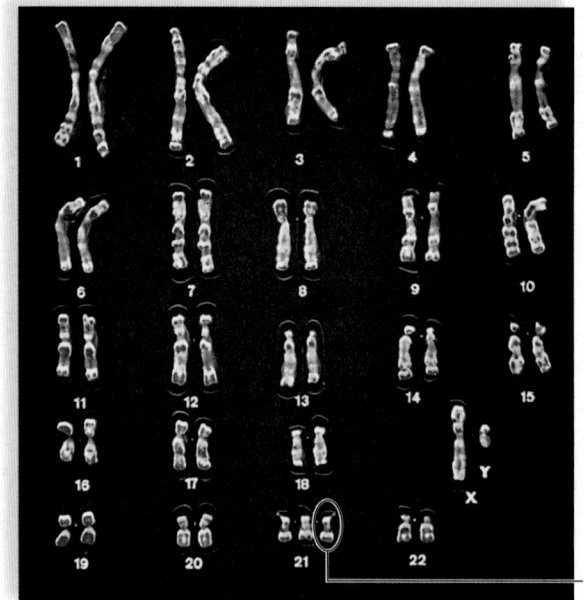

b.

extra chromosome 21

Figure 10.10 Trisomy 21.   Persons with Down syndrome, or trisomy 21, have an extra chromosome 21. **a.** Common characteristics of the syndrome include a wide, rounded face and a fold on the upper eyelids. Mental disabilities, along with an enlarged tongue, may make it difficult for a person with Down syndrome to speak distinctly. **b.** The karyotype of an individual with Down syndrome shows three copies of chromosome 21. Therefore, the individual has three copies instead of two copies of each gene on chromosome 21. Researchers are using new techniques to discover which genes on chromosome 21 are causing the syndrome's disabilities.

Why are newborns with an abnormal sex chromosome number more likely to survive than those with an abnormal autosome number? Because females have two X chromosomes and males have only one, we might expect females to produce twice the amount of each gene from this chromosome, but both males and females produce roughly the same amount. In reality, both males and females only have one functioning X chromosome. In females, and in males with extra X chromosomes, any additional X chromosomes become an inactive mass called a **Barr body,** named after Murray Barr, the person who discovered it. This inactivation provides a natural method for gene dosage compensation of the sex chromosomes and explains why extra sex chromosomes are more easily tolerated than extra autosomes.

Turner Syndrome.   From birth, an XO individual with Turner syndrome has only one sex chromosome, an X; the O signifies the absence of a second sex chromosome (Fig. 10.11a). Therefore, the nucleus does not contain a Barr body. The approximate incidence is 1 in 10,000 females.

Turner females are short, with a broad chest and widely spaced nipples. These individuals also have a low posterior hairline and neck webbing. The ovaries, oviducts, and uterus are very small and underdeveloped. Turner females do not undergo puberty or menstruate, and their breasts do not develop. However, some have given birth following in vitro fertilization using donor eggs. They usually are of normal intelligence and can lead fairly normal lives if they receive hormone supplements.

Klinefelter Syndrome.   A male with Klinefelter syndrome has two or more X chromosomes in addition to a Y chromosome (Fig. 10.11b). The extra X chromosomes become Barr bodies. The approximate incidence for Klinefelter syndrome is 1 in 500 to 1,000 males.

In Klinefelter males, the testes and prostate gland are underdeveloped and facial hair is lacking. They may exhibit some breast development. Affected individuals have large hands and feet and very long arms and legs. They are usually slow to learn but not mentally retarded unless they inherit more than two X chromosomes. No matter how many X chromosomes there are, an individual with a Y chromosome is a male.

While males with Klinefelter syndrome exhibit no other major health abnormalities, they have an increased risk of some disorders, including breast cancer, osteoporosis, and lupus, which disproportionately affect females. Although men with

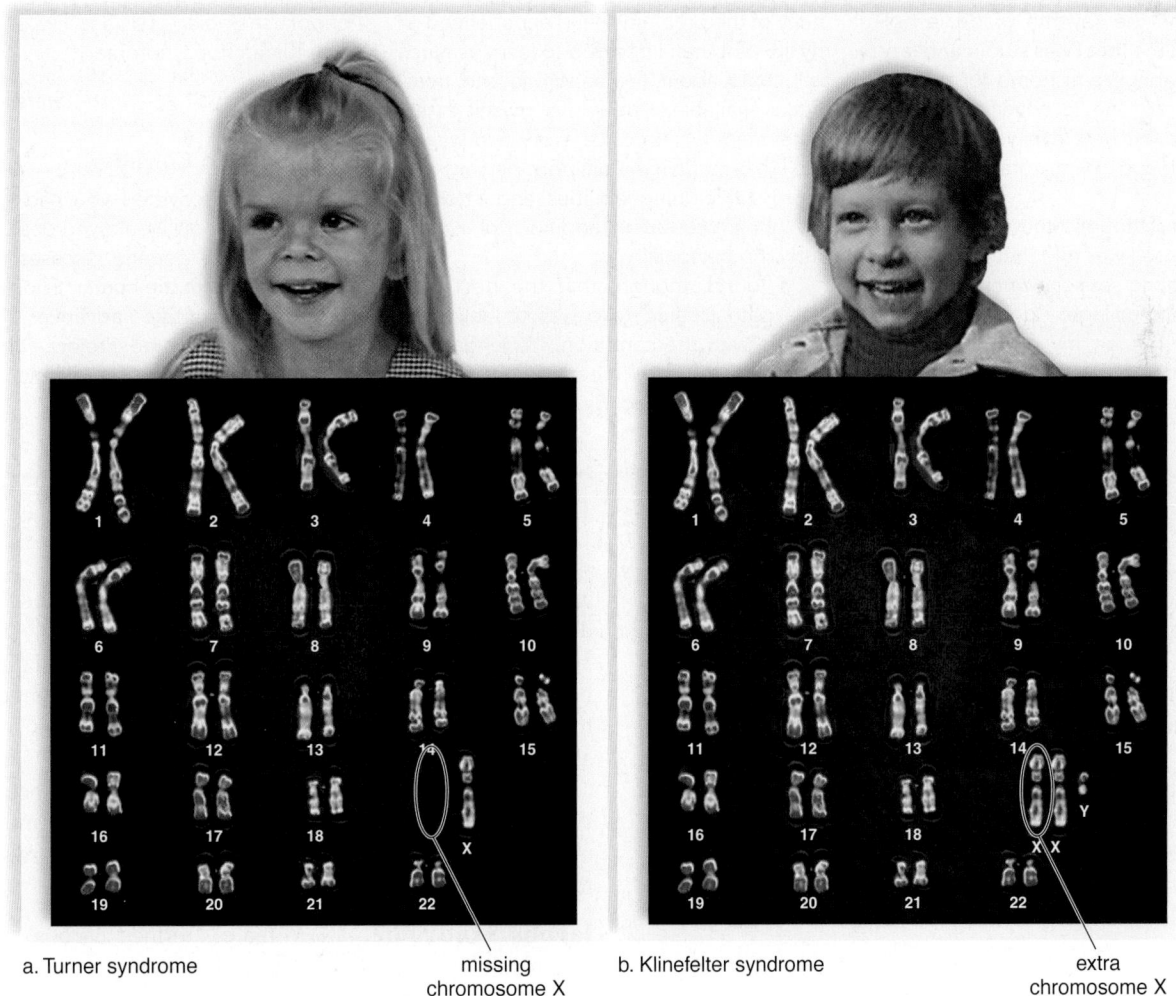

a. Turner syndrome   missing
                                                          chromosome X

b. Klinefelter syndrome   extra
                                                          chromosome X

Figure 10.11   Abnormal sex chromosome number.   Nondisjunction of sex chromosomes is tolerated better than with autosomes. People with (a) Turner syndrome, who have only one X chromosome, as shown, and (b) Klinefelter syndrome, who have more than one X chromosome plus a Y chromosome, as shown, can look relatively normal (especially as children) and can lead relatively normal lives.

Evolution

Living with Klinefelter Syndrome

In 1996, at the age of 25, I was diagnosed with Klinefelter syndrome (KS). Being diagnosed has changed my life for the better.

I was a happy baby, but when I was still very young, my parents began to believe that there was something wrong with me. I knew something was different about me, too, as early on as five years old. I was very shy and had trouble making friends. One minute I'd be well behaved, and the next I'd be picking fights and flying into a rage. Many psychologists, therapists, and doctors tested me because of school and social problems and severe mood changes. Their only diagnosis was "learning disabilities" in such areas as reading comprehension, abstract thinking, word retrieval, and auditory processing.

No one could figure out what the real problem was, and I hated the tutoring sessions I had. In the seventh grade, a psychologist told me that I was stupid and lazy, I would probably live at home for the rest of my life, and I would never amount to anything. For the next five years, he was basically right, and I barely graduated from high school.

I believe, though, that I have succeeded because I was told that I would fail. I quit the tutoring sessions when I enrolled at a community college; I decided I could figure things out on my own. I received an associate degree there, then transferred to a small liberal arts college. I never told anyone about my learning disabilities and never sought special help. However, I never had a semester below a 3.0, and I graduated with two B.S. degrees. I was accepted into a graduate program but decided instead to accept a job as a software engineer even though I did not have an educational background in this field. As I later learned, many KS'ers excel in computer skills. I had been using a computer for many years and had learned everything I needed to know on my own, through trial and error.

Around the time I started the computer job, I went to my physician for a physical. He sent me for blood tests because he noticed that my testes were smaller than usual. The results were conclusive: Klinefelter syndrome with sex chromosomes XXY. I initially felt denial, depression, and anger, even though I now had an explanation for many of the problems I had experienced all my life. But then I decided to learn as much as I could about the condition and treatments available. I now give myself a testosterone injection once every two weeks, and it has made me a different person, with improved learning abilities and stronger thought processes in addition to a more outgoing personality.

I found, though, that the best possible path I could take was to help others live with the condition. I attended my first support group meeting four months after I was diagnosed. By spring 1997, I had developed an interest in KS that was more than just a part-time hobby. I wanted to be able to work with this condition and help people forever. I have been very involved in KS conferences and have helped to start support groups in the United States, Spain, and Australia.

Since my diagnosis, it has been my dream to have a son with KS, although when I was diagnosed, I found out it was unlikely that I could have biological children. Through my work with KS, I had the opportunity to meet my wife, Chris. She has two wonderful children: a daughter, and a son who has the same condition that I do. There are a lot of similarities between my stepson and me, and I am happy I will be able to help him get the head start in coping with KS that I never had. I also look forward to many more years of helping other people seek diagnosis and live a good life with Klinefelter syndrome.

—Stefan Schwarz

Questions to Consider

1. If you discovered you had a genetic disease, how would you deal with it?
2. Why have genetic diseases persisted over time in the human population?
3. With scientific advances like the Human Genome Project, what treatments might be available to treat genetic diseases?

Klinefelter syndrome typically do not need medical treatment, some have found that testosterone therapy may help increase muscle strength, sex drive, and concentration ability. Testosterone treatment, however, does not reverse the sterility associated with Klinefelter syndrome due to the incomplete testicle development.

The Evolution feature in this chapter describes the personal experiences of a person with Klinefelter syndrome. The essay suggests that it is best for parents to know right away that they have a child with this abnormality because much can be done to help the child lead a normal life.

Poly-X Females.   A poly-X female, sometimes called a superfemale, has more than two X chromosomes and, therefore, extra Barr bodies in the nucleus. Females with three X chromosomes have no distinctive phenotype aside from a tendency to be tall and thin. Although some have delayed motor and language development, as well as learning problems, most poly-X females are not mentally retarded. Some may have menstrual difficulties, but many menstruate regularly and are fertile. Children usually have a normal karyotype. The incidence for poly-X females is about 1 in 1,500 females.

Females with more than three X chromosomes occur rarely. Unlike XXX females, XXXX females are usually tall and severely mentally retarded. Various physical abnormalities are seen, but they may menstruate normally.

Jacobs Syndrome.   XYY males, termed Jacobs syndrome, can result only from nondisjunction during spermatogenesis. These individuals are sometimes called supermales. Among all live male births, the frequency of the XYY karyotype is about 1 in 1,000. Affected males are usually taller than average, suffer from persistent acne, and tend to have speech and reading problems, but are fertile and may have children. Based upon the number of XYY individuals in prisons and mental facilities, it was suggested at one time that these men were likely to be criminally aggressive, but it has since been shown that the incidence of such behavior among them may be no greater than among XY males.

Changes in Chromosome Structure

Changes in chromosome structure are another type of chromosomal mutation. Some, but not all, changes in chromosome structure can be detected microscopically. Various agents in the environment, such as radiation, certain organic chemicals, or even viruses, can cause chromosomes to break. Ordinarily, when breaks occur in chromosomes, the two broken ends reunite to give the same sequence of genes. Sometimes, however, the broken ends of one or more chromosomes do not rejoin in the same pattern as before, and the result is various types of chromosomal mutations.

Changes in chromosome structure include deletions, duplications, translocations, and inversions of chromosome segments. A **deletion** occurs when an end of a chromosome breaks off or when two simultaneous breaks lead to the loss of an internal segment (Fig. 10.12*a*). Even when only one member of a pair of chromosomes is affected, a deletion often causes abnormalities.

Animation
Changes in Chromosome Structure

A **duplication** is the presence of a chromosomal segment more than once in the same chromosome (Fig. 10.12*b*). Duplications may or may not cause visible abnormalities, depending on the size of the duplicated region. An **inversion** has occurred when a segment of a chromosome is turned around 180° (Fig. 10.12*c*). Most individuals with inversions exhibit no abnormalities, but this reversed sequence of genes can result in duplications or deletions being passed on to their children, as described in Figure 10.13.

A **translocation** is the movement of a chromosome segment from one chromosome to another, nonhomologous chromosome. The translocation shown in Figure 10.12*d* is *balanced*, meaning that there is a reciprocal swap of one piece of the chromosome

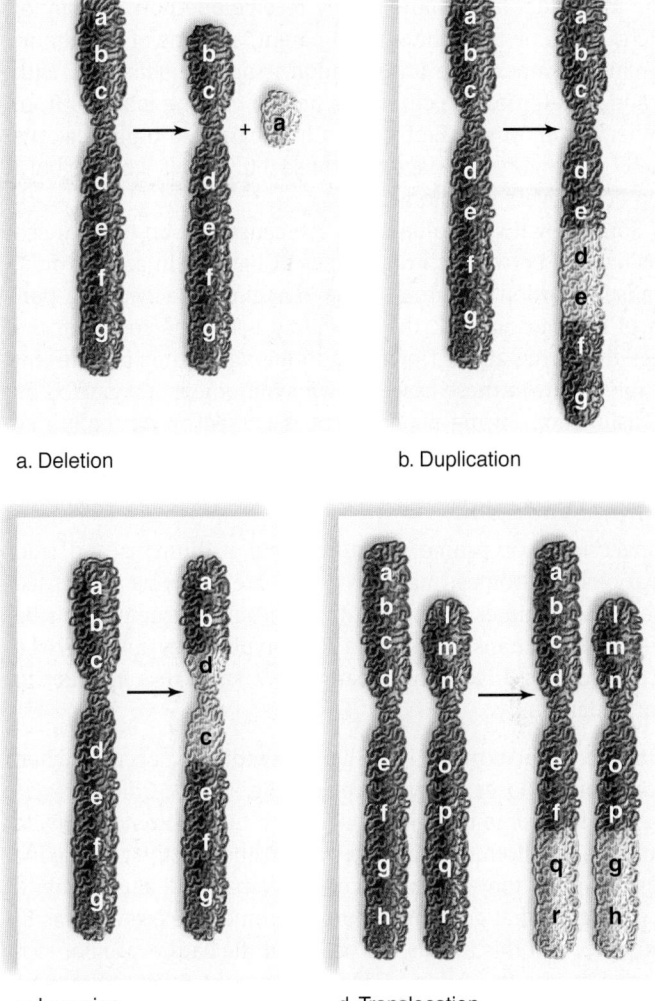

a. Deletion

b. Duplication

c. Inversion

d. Translocation

Figure 10.12 Types of chromosomal mutations. **a.** Deletion is the loss of a chromosome piece. **b.** Duplication occurs when the same piece is repeated within the chromosome. **c.** Inversion occurs when a piece of chromosome breaks loose and then rejoins in the reversed direction. **d.** Translocation is the exchange of chromosome pieces between nonhomologous pairs.

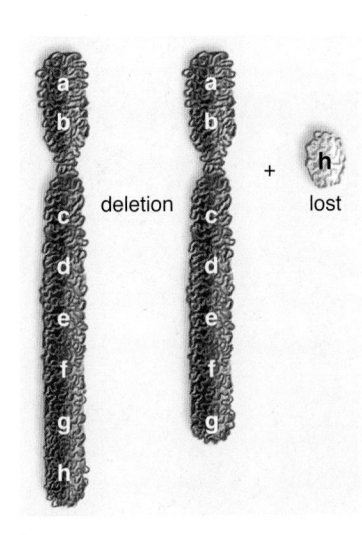

deletion

lost

Figure 10.13 Deletion.
a. When chromosome 7 loses an end piece, the result is Williams syndrome. **b.** These children, although unrelated, have the same appearance, health, and behavioral problems.

a.

b.

for the other. Often, there are no visible effects of the swap, but if the individual has children, they receive one normal copy of the chromosome from the normal parent, and one of the abnormal chromosomes. The translocation is now *unbalanced*, with extra material from one chromosome and missing material from another chromosome. Embryos with unbalanced translocations usually result in miscarriage, but those individuals who are born often have severe symptoms.

Some Down syndrome cases are caused by an unbalanced translocation between chromosomes 21 and 14. In other words, because a portion of chromosome 21 is now attached to a portion of chromosome 14, the individual has three copies of the genes that bring about Down syndrome when they are present in triplet copy. In these cases, Down syndrome is not caused by nondisjunction during meiosis, but is passed on normally like any other genetic trait as described in Chapter 11.

Human Syndromes

Changes in chromosome structure occur in humans and lead to various syndromes, many of which are just now being discovered. Sometimes changes in chromosome structure can be detected in humans by doing a karyotype. They may also be discovered by studying the inheritance pattern of a disorder in a particular family.

Deletion Syndromes.   Williams syndrome occurs when chromosome 7 loses a tiny end piece (Fig. 10.13). Children who have this syndrome look like pixies, with turned-up noses, wide mouths, a small chin, and large ears. Although their academic skills are poor, they exhibit excellent verbal and musical abilities. The gene that governs the production of the protein elastin is missing, and this affects the health of the cardiovascular system and causes their skin to age prematurely. Such individuals are very friendly but need an ordered life, perhaps because of the loss of a gene for a protein that is normally active in the brain.

Cri du chat (cat's cry) syndrome is seen when chromosome 5 is missing an end piece. The affected individual has a small head, is mentally retarded, and has facial abnormalities. Abnormal development of the glottis and larynx results in the most characteristic symptom—the infant's cry resembles that of a cat.

Translocation Syndromes.   A person who has both of the chromosomes involved in a translocation has the normal amount of genetic material and is healthy, unless the chromosome exchange breaks an allele into two pieces. The person who inherits only one of the translocated chromosomes no doubt has only one copy of certain alleles and three copies of certain other alleles. A genetic counselor begins to suspect a translocation has occurred when spontaneous abortions are commonplace and family members suffer from various syndromes. A special microscopic technique allows a technician to determine that a translocation has occurred.

Figure 10.14 shows a person who has a translocation between chromosomes 2 and 20. Although they have the normal amount of genetic material, they have the distinctive face, abnormalities of the eyes and internal organs, and severe itching characteristic of Alagille syndrome. People with this syndrome ordinarily have a deletion on chromosome 20 (Fig. 10.14a), which can lead to a congenital heart condition called Tetralogy of Fallot that produces digital clubbing of the fingers (Fig. 10.14b). The symptoms of Alagille syndrome range from mild to severe, so some people may not be aware they have the syndrome until after they've had children

Translocations can also be responsible for a variety of other disorders including certain types of cancer. In the 1970s, new staining techniques identified that a translocation from a portion of chromosome 22 to chromosome 9 was responsible for many cases of chronic myelogenous leukemia. This translocated chromosome was called Philadelphia chromosome. In Burkitt lymphoma, a cancer common in children in equatorial Africa, a large tumor develops from lymph glands in the region of the jaw. This disorder involves a translocation from a portion of chromosome 8 to chromosome 14.

Check Your Progress   10.6

1. Explain the kinds of changes in chromosome number that can be caused by nondisjunction in meiosis.
2. Examine why sex chromosome aneuploidy is more common than autosome aneuploidy.
3. Compare structural changes between an inversion and a translocation.

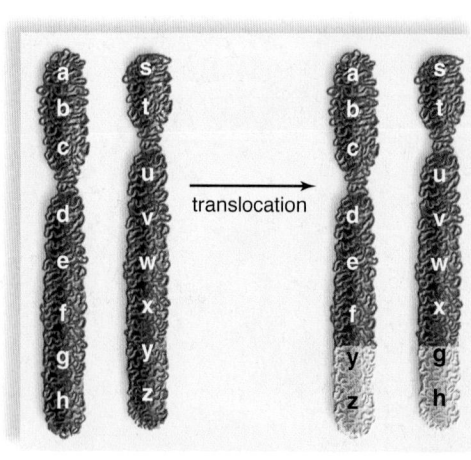

Figure 10.14
Translocation.

a. When chromosomes 2 and 20 exchange segments, (**b**) Alagille syndrome, with distinctive body features, sometimes results because of organ malfunction caused by the chromosome 20 translocation.

translocation

a.

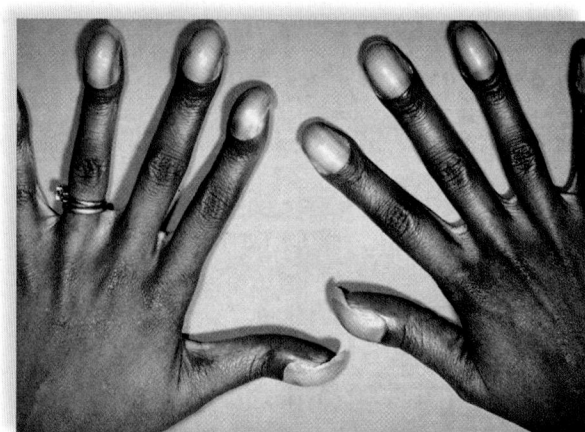

b.

CONNECTING *the* CONCEPTS *with the* BIG IDEAS

Evolution

- Meiosis provides the variation in zygotes upon which natural selection can act. (1A2b)

Information and Signaling

- Meiotic regulatory mechanisms ensure homologous pairing so maternal and paternal copies separate into four haploid cells. Also, crossing over and independent assortment exponentially increase gamete variation. (3A2c1-4)
- Fertilization restores diploidy, creating unique new combinations of genetic material in the zygote. (3A2c5, 3C2c)
- Single gene traits include many genetic diseases including Huntington's, Tay-Sachs, etc. Nondisjunction creates cells with extra or missing chromosomes, causing syndromes like Down and Kleinfelter's. (3A3c*IE*)
- Meiotic chromosomal behavior with segregation of homologs allows prediction of patterns of inheritance. (3A2c2-3)

*Find the unabridged version of all EK citations at www.glencoe.com/maderAP11.

Media Study Tools

www.glencoe.com/maderAP11

Enhance your study of this chapter with study tools and practice tests. Also ask your instructor about the resources available through ConnectPlus, including the media-rich eBook, interactive learning tools, and animations.

Summarize

10.1 Halving the Chromosome Number

Meiosis ensures that the chromosome number in offspring stays constant generation after generation. The nucleus contains pairs of chromosomes, called homologous chromosomes (homologues).

Meiosis requires two cell divisions and results in four daughter cells. Replication of DNA takes place before meiosis begins. During meiosis I, the homologues undergo synapsis (resulting in a bivalent) and align independently at the metaphase plate. The daughter cells receive one member of each pair of homologous chromosomes. There is no replication of DNA during interkinesis, the pause between meiosis I and II. During meiosis II, the sister chromatids separate, becoming daughter chromosomes that move to opposite poles as they do in mitosis. The four daughter cells contain the haploid number of chromosomes and only one of each kind.

10.2 Genetic Variation

Sexual reproduction ensures that the offspring have a different genetic makeup than the parents, and increases the ability of a species to survive. Meiosis contributes to genetic variability in two ways: crossing-over and independent assortment of the homologous chromosomes. When homologous chromosomes lie side by side during synapsis, nonsister chromatids may exchange genetic material. Due to crossing-over, the chromatids that separate during meiosis II have a different combination of genes.

When the homologous chromosomes align at the metaphase plate during metaphase I, either the maternal or the paternal chromosome can be facing either pole. Therefore, there will be all possible combinations of chromosomes in the gametes.

10.3 The Phases of Meiosis

Meiosis I, which splits pairs of homologous chromosomes and reduces the chromosome number from 2n to n, is divided into four phases:

Prophase I—Bivalents form, and crossing-over occurs as chromosomes condense; the nuclear envelope fragments.

Metaphase I—Bivalents independently align at the metaphase plate.

Anaphase I—Homologous chromosomes separate, and duplicated chromosomes move to poles.

Telophase I—Nuclei become haploid, having received one duplicated chromosome from each homologous pair.

Meiosis II, which reduces the amount of DNA in half from previously replicated 1n chromosomes, is divided into four phases:

Prophase II—Chromosomes condense, and the nuclear envelope fragments.

Metaphase II—The haploid number of still duplicated chromosomes align at the metaphase plate.

Anaphase II—Sister chromatids separate, becoming daughter chromosomes that move to the poles.

Telophase II—Four haploid daughter cells are genetically different from the parent cell.

10.4 Meiosis Compared to Mitosis

Mitosis and meiosis can be compared in this manner:

Meiosis I	Mitosis
Prophase	
Pairing of homologous chromosomes	No pairing of chromosomes
Metaphase	
Bivalents at metaphase plate	Duplicated chromosomes at metaphase plate
Anaphase	
Homologous chromosomes separate and move to poles	Sister chromatids separate, becoming daughter chromosomes that move to the poles
Telophase	
Daughter nuclei have the haploid number of chromosomes	Daughter nuclei have the parent cell chromosome number

Meiosis II is like mitosis except the nuclei are haploid.

10.5 The Cycle of Life

Meiosis occurs in any life cycle that involves sexual reproduction. In the animal life cycle, only the gametes are haploid; in plants, meiosis produces spores that develop into a multicellular haploid adult that produces the gametes. In unicellular protists and fungi, the zygote undergoes meiosis, and spores become a haploid adult that gives rise to gametes.

During the life cycle of humans and other animals, meiosis is involved in spermatogenesis and oogenesis. Whereas spermatogenesis produces four sperm per meiosis, oogenesis produces one egg and two to three nonfunctional polar bodies. Spermatogenesis occurs in males, and oogenesis occurs in females. When a sperm fertilizes an egg, the zygote has the diploid number of chromosomes. Mitosis, which is involved in growth and repair, also occurs during the life cycle of all animals.

10.6 Changes in Chromosome Number and Structure

Nondisjunction during meiosis I or meiosis II may result in aneuploidy (extra or missing copies of chromosomes). Monosomy occurs when an individual has only one of a particular type of chromosome (2n − 1) and is usually lethal; trisomy occurs when an individual has three of a particular type of chromosome (2n + 1). Down syndrome is a well-known trisomy in human beings resulting from an extra copy of chromosome 21.

Aneuploidy of the sex chromosomes is tolerated more easily than aneuploidy of the autosomes. Turner syndrome, Klinefelter syndrome, poly-X females, and Jacobs syndrome are examples of sex chromosome aneuploidy.

Abnormalities in crossing-over may result in deletions, duplications, inversions, and translocations within chromosomes. Many human syndromes, including Williams syndrome, cri du chat syndrome, and Alagille syndrome, result from changes in chromosome structure.

Key Terms

allele 172	interkinesis 176
alternation of generations 176	inversion 187
aneuploidy 183	karyotype 184
Barr body 185	life cycle 180
bivalent 173	meiosis 172
crossing-over 174	monosomy 183
deletion 187	nondisjunction 183
diploid (2n) number 172	oogenesis 180
duplication 187	polar body 181
euploidy 183	secondary oocyte 181
fertilization 175	sexual reproduction 172
gamete 172	spermatogenesis 180
gametogenesis 180	spore 176
gametophyte 180	sporophyte 180
genetic recombination 175	synapsis 173
haploid (n) number 172	synaptonemal complex 173
homologous	translocation 187
chromosome 172	trisomy 183
homologue 172	zygote 172
independent assortment 175	

Assess

Reviewing This Chapter

1. Why did early investigators predict that there must be a reduction division in the sexual reproduction process?   172
2. What are homologous chromosomes? Contrast the genetic makeup of sister chromatids with that of nonsister chromatids.   172–74
3. Draw and explain a diagram that illustrates crossing-over and another that shows all possible results from independent assortment of homologous pairs. How do these events ensure genetic variation among the gametes?   174–75
4. Draw and explain a series of diagrams that illustrate the stages of meiosis I and meiosis II.   177–81
5. What accounts for **(a)** the genetic similarity between daughter cells and the parent cell following mitosis, and **(b)** the genetic dissimilarity between daughter cells and the parent cell following meiosis?   180–81
6. Explain the human (animal) life cycle and the roles of meiosis and mitosis.   180–81
7. Compare spermatogenesis in males to oogenesis in females.   181
8. How does aneuploidy occur? Why is sex chromosome aneuploidy more common than autosomal aneuploidy? What are some human syndromes associated with aneuploidy?   183–87
9. Name and explain four types of changes in chromosome structure.   187–88
10. Name some syndromes that occur in humans due to changes in chromosome structure.   188

Testing Yourself

Choose the best answer for each question.

1. A bivalent is
   a. a homologous chromosome.
   b. the paired homologous chromosomes.
   c. a duplicated chromosome composed of sister chromatids.
   d. the two daughter cells after meiosis I.
   e. the two centrioles in a centrosome.

2. If a parent cell has 16 chromosomes, then each of the daughter cells following meiosis will have
   a. 48 chromosomes.   c. 16 chromosomes.
   b. 32 chromosomes.   d. 8 chromosomes.

3. At the metaphase plate during metaphase I of meiosis, there are
   a. chromosomes consisting of one chromatid.
   b. unpaired duplicated chromosomes.
   c. bivalents.
   d. homologous pairs of chromosomes.
   e. Both c and d are correct.

4. At the metaphase plate during metaphase II of meiosis, there are
   a. chromosomes consisting of one chromatid.
   b. unpaired duplicated chromosomes.
   c. bivalents.
   d. homologous pairs of chromosomes.
   e. Both c and d are correct.

5. Crossing-over occurs between
   a. sister chromatids of the same chromosome.
   b. two different kinds of bivalents.
   c. two different kinds of chromosomes.
   d. nonsister chromatids of a bivalent.
   e. two daughter nuclei.

6. During which phase of meiosis do homologous chromosomes separate?
   a. prophase I   c. anaphase I
   b. telophase I   d. anaphase II

7. Nondisjunction during meiosis I of oogenesis will result in eggs that have
   a. the normal number of chromosomes.
   b. one too many chromosomes.
   c. one less than the normal number of chromosomes.
   d. Both b and c are correct.

8. Which two of these chromosomal mutations are most likely to occur when an inverted chromosome is undergoing synapsis?
   a. deletion and translocation
   b. deletion and duplication
   c. duplication and translocation
   d. inversion and duplication

9. A male with underdeveloped testes and some breast development most likely has
   a. Down syndrome.   c. Turner syndrome.
   b. Jacobs syndrome.   d. Klinefelter syndrome.

For questions 10–13, fill in the blanks.

10. If the parent cell has 24 chromosomes, the daughter cells following mitosis will have _____ chromosomes and following meiosis will have _____ chromosomes.

11. Meiosis in males is a part of _____, and meiosis in females is a part of _____.

12. Oogenesis will not go to completion unless _____ occurs.

13. During oogenesis, the primary oocyte has the _____ and the secondary oocyte has the _____ number of chromosomes.

For questions 14–19, match the statements that follow to the items in the key. Answers may be used more than once, and more than one answer may be used.

KEY:
   a. mitosis
   b. meiosis I
   c. meiosis II
   d. Both meiosis I and meiosis II are correct.
   e. All of these are correct.

14. A parent cell with ten duplicated chromosomes will produce daughter cells with five duplicated chromosomes each.

15. Involves pairing of duplicated homologous chromosomes.

16. A parent cell with five duplicated chromosomes will produce daughter cells with five chromosomes consisting of one chromatid each.

17. Nondisjunction may occur, causing abnormal gametes to form.

18. A parent cell with ten duplicated chromosomes will produce daughter cells with ten chromosomes consisting of one chromatid each.

19. Involved in growth and repair of tissues.

Engage

Thinking Scientifically

1. Why is the first meiotic division considered to be the reduction division for chromosome number?

2. Recall that during interphase, the G_2 checkpoint ensures that the DNA has been faithfully replicated before the cell is allowed to divide by mitosis. Would you expect this checkpoint to be active during interkinesis? How might you set up an experiment to test your hypothesis?

3. A man has a balanced translocation between chromosome 2 and 6. If he reproduces with a normal woman could the child have the same translocation? Why or why not?

Bioethical Issue

The Risks of Advanced Maternal Age

In today's society, it is commonplace for women to embark on careers and pursue higher education, delaying marriage and childbirth until later years. Between 1991 and 2001, the birthrate among women aged 35 to 39 increased over 30%, while the birthrate among women aged 40 to 44 leaped by almost 70%. The U.S. Census Bureau indicates the average age at which a woman first gives birth is now 25.1 years, as compared to 21.1 years in 1970. These increases have occurred as society has changed, spurred by the elimination of the social stigmas, better prenatal care, and new medical technologies that can overcome the decline in fertility associated with age and treat at-risk children.

The decision to delay childbirth does carry risks. Although the reasons are not well understood, the risk of many disorders associated with meiotic nondisjunction, such as Down syndrome, increase greatly with age, rising from nearly 1 in 900 at age 30 to 1 in 109 by age 40. The risk of complications to the mother, such as gestational diabetes, are also much higher in women over 30. Thus, the medical community has embarked on a campaign to ensure that women who are pregnant and over age 35 are offered more intensive prenatal care. Many people are concerned about the ultimate cost to society, through increased insurance premiums and increased costs to governments to pay for it.

Although definite risks are associated with advanced maternal age, some people contend that having children later in life provides many advantages. Women over age 35 are usually at a later stage in their careers and have higher salaries, lessening the need for many social welfare programs. Furthermore, women over 35 often have a more stable living situation, are less likely to experience unplanned pregnancy, and are often able to devote more time to the child than are younger women. Therefore, while older mothers may require more medical attention, the overall costs to society are lower.

Considering both the benefits and the disadvantages, are we as a society obligated to fund intense screening and prenatal care for women of advanced maternal age, and to pay for treating the maladies associated with it? As birthrates among women over age 30 continue to soar, the debate over advanced maternal age is not likely to abate any time soon.

11

Mendelian Patterns of Inheritance

Trimethylaminuria is a genetic disorder that produces a fishy body odor.

CHAPTER OUTLINE

BEFORE YOU BEGIN

Before beginning this chapter, take a few moments to review the following discussions.

Figure 10.1 How is DNA-based genotype related to protein-based phenotype?

Section 10.3 How are chromosomes segregated during meiosis I and II?

Figure 10.9 What genetic changes are possible in gametes when chromosomes fail to segregate properly?

Camille was painfully aware of her foul body odor because children teased her relentlessly, calling her "Miss Fishy" and other nasty names. She had no idea that she suffers from trimethylaminuria, or "fish odor syndrome," an extremely rare genetic disorder she shares with only 1 in 10,000 people. People with this syndrome have a defective gene that makes a nonfunctional protein unable to break down the smelly chemical trimethylamine, which accumulates in the body and ends up in their urine, sweat, and sometimes even in their breath.

Like the rest of us, you are the product of your family tree. The DNA you inherit from your parents directly affects the proteins that enable your body to function properly. Rare genetic disorders like Camille's pique our curiosity about how traits are inherited from one generation to the next. In this chapter, you will learn that the process of meiosis can be used to predict the inheritance of a trait, and that the genetic diversity produced through meiosis can sometimes create cases like Camille's. Through patterns of inheritance discovered by Mendel, you will also learn that certain traits, such as trimethylaminuria, are recessive and it takes two nonfunctional copies of that gene before you are affected. This chapter will introduce you to observable patterns of inheritance, including human genetic disorders that are linked to specific genes on the chromosomes.

As you read through the chapter, think about the following questions:

1. How does the collection of chromosomes we inherit from our parents affect our body's appearance and function?

2. What patterns of inheritance can be observed across generations?

3. How does meiosis help predict the probability of producing gametes and inheriting a trait?

FOLLOWING *the* BIG IDEAS

CHAPTER 11   MENDELIAN PATTERNS OF INHERITANCE

Evolution	Inheritance of randomly shuffled genes within a population is the cornerstone of a species' ability to change over time.
Information and Signaling	Mendel's simple but eloquent explanations concerning inheritance of characteristics in diploid organisms have been substantiated by details of the meiotic process.
Interactions and Systems	The activity of genes may vary depending upon the environment in which they find themselves.

11.1 Gregor Mendel

The science of genetics explains the stability of inheritance (why you are human, as are your parents) and also variations between offspring from one generation to the next (why you have a different combination of traits than your parents). Virtually every culture in history has attempted to explain observed inheritance patterns. An understanding of these patterns has always been important to agriculture, animal husbandry (the science of breeding animals), and medicine.

The Blending Concept of Inheritance

Until the late nineteenth century, most plant and animal breeders believed that traits were inherited by the blending concept of inheritance, which stated that an offspring's genetic makeup was intermediate to that of its parents. While they acknowledged that both sexes contribute equally to a new individual, they believed that parents of contrasting appearance always produce offspring of intermediate appearance.

If the blending concept were true, a cross between plants with red flowers and plants with white flowers would yield only plants with pink flowers. However, scientists who supported the blending theory could not explain, for example, why pink flowers were able to produce offspring with red and white flowers in later generations. The breeders mistakenly attributed this to instability of the genetic material.

The blending concept of inheritance offered little help to Charles Darwin, the father of evolution, whose treatise on natural selection lacked a strong genetic basis. If populations contained only intermediate individuals and normally lacked variations, how could diverse forms evolve?

Mendel's Particulate Theory of Inheritance

Gregor Mendel was an Austrian monk who developed a particulate theory of inheritance after performing a series of ingenious experiments in the 1860s (Fig. 11.1). Mendel studied science and mathematics at the University of Vienna, and at the time of his research in genetics, he was a substitute natural science teacher at a local high school.

Mendel was a successful scientist for several reasons. First, he was one of the first scientists to apply mathematics to biology. Most likely his background in mathematics prompted him to apply statistical methods and the laws of probability to his breeding experiments. He was also a careful, deliberate scientist who followed the scientific method very closely and kept

Figure 11.1 Gregor Mendel, 1822–84. Mendel grew and tended the pea plants he used for his experiments. For each experiment, he observed as many offspring as possible. For a cross that required him to calculate the ratio of round seeds to wrinkled seeds, he observed and counted a total of 7,324 peas!

very detailed, accurate records. He prepared for his experiments carefully and conducted many preliminary studies with various animals and plants.

Mendel's theory of inheritance is called a particulate theory because it is based on the existence of minute particles or hereditary units we now call genes. Inheritance involves the reshuffling of the same genes from generation to generation. The two laws he proposed, the law of segregation and the law of independent assortment, which we will discuss shortly, describe the behavior of these particulate units of heredity as they are passed from one generation to the next. Much of modern genetics is based upon Mendel's theories, which have withstood the test of time and have been supported by innumerable experiments.

Mendel Worked with the Garden Pea

Mendel's preliminary experiments prompted him to choose the garden pea, *Pisum sativum* (Fig. 11.2*a*), as his experimental organism. The garden pea was a good choice for many reasons. The plants were easy to cultivate and had a short generation time. Although peas normally self-pollinate (pollen only goes to the same flower), they could be cross-pollinated by hand by

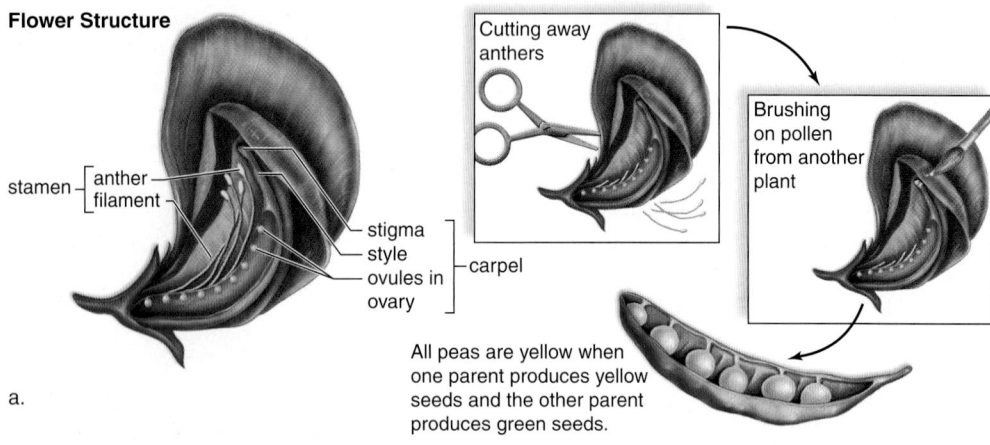

Flower Structure

a.

b.

Figure 11.2 Garden pea anatomy and a few traits. **a.** In the garden pea, *Pisum sativum,* pollen grains produced in the anther contain sperm, and ovules in the ovary contain eggs. When Mendel performed crosses, he brushed pollen from one plant onto the stigma of another plant. This cross-pollination allowed sperm to fertilize eggs, and ovules to develop into seeds (peas). The open pod shows the seed color trait that resulted from a cross between plants with yellow seeds and plants with green seeds. **b.** Mendel selected traits to study that were one form or another, and not a blend of two traits. In his research, Mendel observed a dominant, or more prevalent form, as well as a recessive or less prevalent form, in the pea's physical appearance. The text and Figure 11.3 explain the approach Mendel used and how he interpreted his results to discover basic patterns of inheritance.

| Trait | Characteristics | |
	Dominant	Recessive
Stem length	Tall	Short
Pod shape	Inflated	Constricted
Seed shape	Round	Wrinkled
Seed color	Yellow	Green
Flower position	Axial	Terminal
Flower color	Purple	White
Pod color	Green	Yellow

transferring pollen from the anther (male part of a flower) to the stigma (female part of a flower).

Many varieties of peas were available, and Mendel chose 22 for his experiments. When these varieties self-pollinated, over generations they became *true-breeding*—meaning that all the offspring were the same and exactly like the parent plants. Unlike his predecessors, Mendel studied the inheritance of relatively simple and discrete traits that were not subjective and were easy to observe, such as seed shape, seed color, and flower

color. In his crosses, Mendel observed either dominant or recessive characteristics but no intermediate ones (Fig. 11.2*b*).

Check Your Progress   11.1

1. Describe Gregor Mendel's scientific approach and how it helped make his experiments successful.
2. Explain why the garden pea was a good choice for Mendel's experiments.

11.2 Mendel's Laws

After ensuring that his pea plants were true-breeding—for example, that his tall plants always had tall offspring and his short plants always had short offspring—Mendel was ready to perform cross-pollination experiments (see Fig. 11.2a). These crosses allowed Mendel to formulate his law of segregation.

Law of Segregation

For these initial experiments, Mendel chose varieties that differed in only one trait (e.g. plant height). If the blending theory of inheritance were correct, the cross should yield plants with an intermediate appearance of medium height compared to the parents that were all tall or all short.

Mendel's Experimental Design and Results

Mendel called the original, true-breeding all tall or all short parents the *P generation*. The first generation of offspring were called the F_1, or *filial* [L. *filius,* sons and daughters] *generation* (Fig. 11.3). He performed *reciprocal crosses:* First he dusted the pollen of tall plants onto the stigmas of short plants, and then he dusted the pollen of short plants onto the stigmas of tall plants. In both cases, all F_1 offspring resembled the tall parent.

Certainly, these results were contrary to those predicted by the blending theory of inheritance. Rather than being intermediate, the F_1 plants were all tall and resembled only one parent. Did these results mean that the other characteristic (i.e., shortness) had disappeared permanently? Apparently not, because when Mendel allowed the F_1 plants to self-pollinate, $^3/_4$ of the next generation of offspring, or F_2 *generation*, were tall and $^1/_4$ were short, a 3:1 ratio (Fig. 11.3).

Mendel inferred that the F_1 plants were able to pass on a factor for shortness—it didn't disappear, it just skipped a generation. Because the F_1 plants were tall but clearly still contained the shortness characteristic, Mendel deduced that tallness was dominant to shortness (Fig 11.2b and 11.3).

Mendel counted many plants in his plant height and other experiments. When he allowed the F_1 pea plants (which were all tall but carried the characteristic for shortness) to self-fertilize and produce offspring, he counted a total of 1,064 plants, of which 787 were tall and 277 were short. This type of experiment is called a **monohybrid cross** [L. *mono,* single, and *hybrida,* mixture] because it is a cross of a single trait (i.e., plant height)

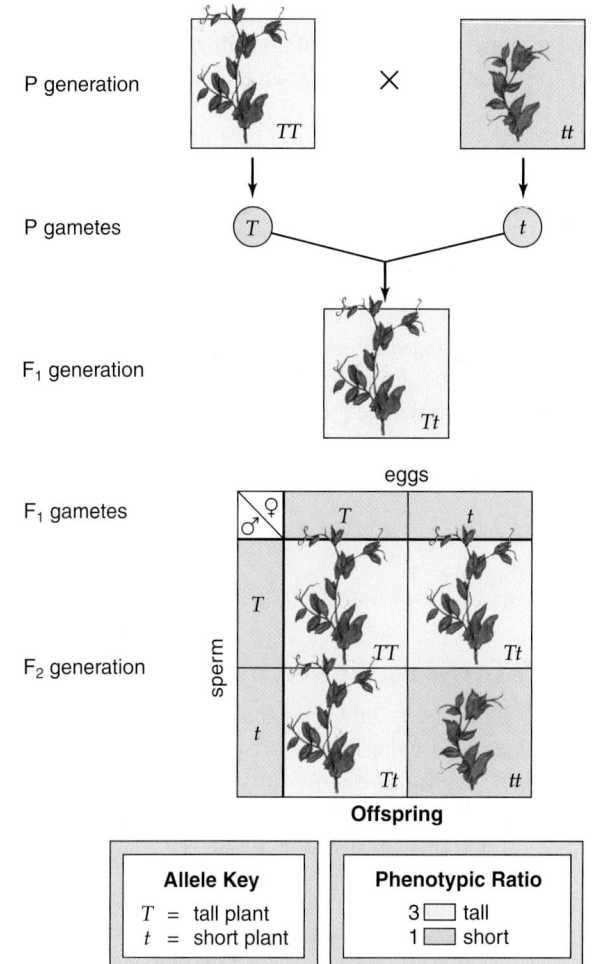

Figure 11.3 Monohybrid cross done by Mendel. The P generation pea plants differ in only one trait—length of the stem. The F_1 generation plants are all tall, but the factor for short has not disappeared because $^1/_4$ of the F_2 generation plants are short. The 3:1 ratio allowed Mendel to deduce that individuals have two discrete and separate genetic factors for each trait.

with organisms that are a hybrid (i.e., tall and short characteristics). In fact, in all monohybrid crosses that he performed for the traits shown in Figure 11.2b, he found a 3:1 ratio in the F_2 generation. The characteristic for shortness that had disappeared in the F_1 generation reappeared in $^1/_4$ of the F_2 offspring. *Today, we know that the expected phenotypic results of a monohybrid cross are always 3:1.*

Mendel's Conclusion

Mendel's mathematical approach led him to interpret his results differently from previous breeders. He knew that the same ratio was obtained among the F_2 generation time and time again when he did a monohybrid cross involving one of the seven traits he was studying. Eventually Mendel arrived at this explanation: A 3:1 ratio among the F_2 offspring was possible if (1) the

F$_1$ parents contained two separate copies of each hereditary factor, one of these being dominant and the other recessive; (2) the factors separated when the gametes were formed, and each gamete carried only one copy of each factor; and (3) random fusion of all possible gametes occurred upon fertilization. Only in this way could shortness reoccur in the F$_2$ generation. Thinking this, Mendel arrived at the first of his laws of inheritance—the law of segregation. The law of segregation is a cornerstone of his particulate theory of inheritance.

> The **law of segregation** states the following:
> • Each individual has two factors for each trait.
> • The factors segregate (separate) during the formation of the gametes.
> • Each gamete contains only one factor from each pair of factors.
> • Fertilization gives each new individual two factors for each trait.

Mendel's Cross as Viewed by Classical Genetics

Figure 11.3 also shows how classical scientists interpreted the results of Mendel's experiments on inheritance of stem length in peas. Stem length in peas is controlled by a single gene. This gene occurs on a homologous pair of chromosomes at a particular location that is called the gene **locus** (Fig. 11.4). Alternative versions of a gene are called **alleles** [Gk. *allelon*, reciprocal, parallel]. The **dominant allele** is so named because the DNA sequence that comprises that gene makes a fully functional protein. The **recessive allele** contains a slightly different DNA sequence and produces a protein with little or no function.

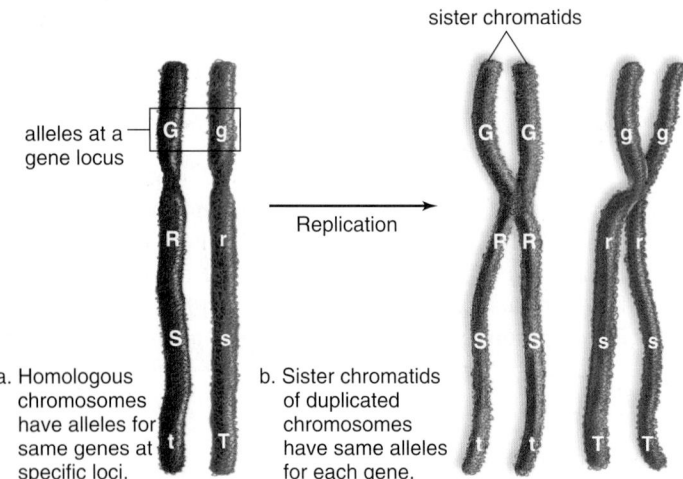

Figure 11.4 **Classical view of homologous chromosomes.**
a. The letters represent alleles; that is, alternative forms of a gene. Each allelic pair, such as *Gg* or *Tt*, is located on homologous chromosomes at a particular physical location called a gene locus. **b.** Sister chromatids carry the same alleles in the same order. Proteins made from each allele determine the observable traits.

When dominant and recessive alleles for a gene are present in a cell, the resulting mixture of functional and nonfunctional protein allows the dominant physical trait to be shown and the recessive trait to be masked. The dominant allele is identified by a capital letter, and the recessive allele by the same letter but lowercase. Usually, the first letter designating a trait is chosen to identify the allele. Using the plant height example, there is an allele for tallness (*T*) and an allele for shortness (*t*).

As described in the preceding chapter, meiosis is the type of cell division that reduces the chromosome number from diploid (2n) to haploid (n). During meiosis I, the members of bivalents (homologous chromosomes each having sister chromatids) separate. This means that the two alleles for each gene separate from each other during meiosis (see Fig. 11.7). Therefore, the process of meiosis gives an explanation for Mendel's law of segregation, and why only one allele for each trait is in a gamete.

In Mendel's cross, the original parents (P generation) were true-breeding; therefore, the tall plants had two alleles for tallness (*TT*), and the short plants had two alleles for shortness (*tt*). When an organism has two identical alleles, as these had, we say it is **homozygous** [Gk. *homo*, same, and *zygos*, balance, yoke]. Because the parents were homozygous, all gametes produced by the tall plant contained the allele for tallness (*T*), and all gametes produced by the short plant contained an allele for shortness (*t*).

After cross-pollination between different pea plants, all the individuals of the resulting F$_1$ generation had one allele for tallness and one for shortness (*Tt*). When an organism has two different alleles at a gene locus, we say that it is **heterozygous** [Gk. *hetero*, different, and *zygos*, balance, yoke]. Although the plants of the F$_1$ generation had one of each type of allele, they were all tall. The allele that is expressed in a heterozygous individual is the dominant allele. The allele that is not expressed in a heterozygote is the recessive allele. This explains why shortness, the recessive trait, skipped a generation in Mendel's experiment.

Continuing with the discussion of Mendel's cross (see Fig. 11.3), the F$_1$ plants produce gametes in which 50% have the dominant allele *T* and 50% have the recessive allele *t*. During the process of fertilization, we assume that all types of sperm (i.e., *T* or *t*) have an equal chance to fertilize all types of eggs (i.e., *T* or *t*). When this occurs, such a monohybrid cross always produces a 3:1 (dominant to recessive) ratio among the offspring. Figure 11.5 gives Mendel's results for several monohybrid crosses, and you can see that the results were always close to 3:1.

Genotype Versus Phenotype

It is obvious from our discussion that two organisms with different allelic combinations for a trait can have the same outward appearance. (That is, *TT* and *Tt* pea plants are both tall.) For this reason, it is necessary to distinguish between the alleles present in an organism and the appearance of that organism.

The word **genotype** [Gk. *genos*, birth, origin, race, and *typos*, image, shape] refers to the alleles an individual receives

Trait	Characteristics			F₂ Results		
	Dominant		Recessive	Dominant	Recessive	Ratio
Stem length	Tall 		Short 	787	277	2.84:1
Pod shape	Inflated 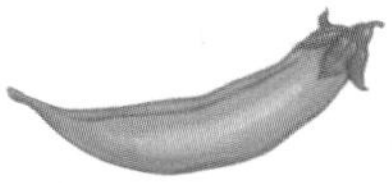		Constricted 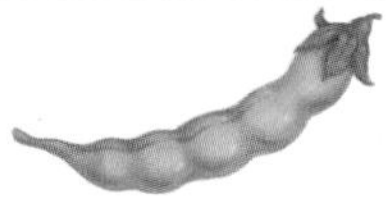	882	299	2.95:1
Seed shape	Round 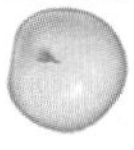		Wrinkled 	5,474	1,850	2.96:1
Seed color	Yellow 		Green 	6,022	2,001	3.01:1
Flower position	Axial 		Terminal 	651	207	3.14:1
Flower color	Purple 		White 	705	224	3.15:1
Pod color	Green 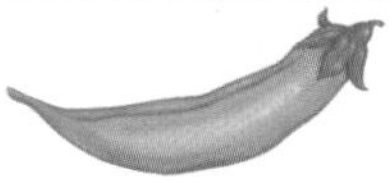		Yellow 	428	152	2.82:1
			Totals:	14,949	5,010	2.98:1

Figure 11.5 Relationship between observed phenotype and F₂ offspring.   Mendel was fortunate in choosing the pea plant because the traits he observed were quite distinct and easily classified. After crossing F₁ hybrids and counting hundreds of F₂ pea plants for each trait, Mendel discovered that each showed a 3:1 ratio.

at fertilization. Genotype may be indicated by letters or by short, descriptive phrases, and represents the DNA sequence for a particular gene. Genotype TT is called homozygous dominant, and genotype tt is called homozygous recessive. Genotype Tt is called heterozygous. These refer to the different ways that alleles can be combined in a cell.

The word **phenotype** [Gk. *phaino,* appear, and *typos,* image, shape] refers to the physical appearance of the individual, which is made from the proteins produced by the corresponding alleles. The homozygous dominant (TT) individual and the heterozygous (Tt) individual both show the dominant phenotype and are tall because they make fully functional proteins that build the tall trait, while the homozygous recessive individual that shows the recessive phenotype and makes less or nonfunctional protein for that trait is short. Thus, the DNA

that makes up the genotype produces the proteins that make up the phenotype.

Check Your Progress   11.2A

1. State all possible gametes, noting the proportion of each for the individual, for these genotypes: **a.** *WW* **b.** *Ww* **c.** *Tt* **d.** *TT*

2. Interpret which of these genotypes (*Bb, BB, bb*) a white rabbit would have if *B* = black and *b* = white.

3. What would be the expected phenotypic ratio of the offspring if two heterozygous rabbits mate? If they produced 20 offspring, how many would you expect to be white?

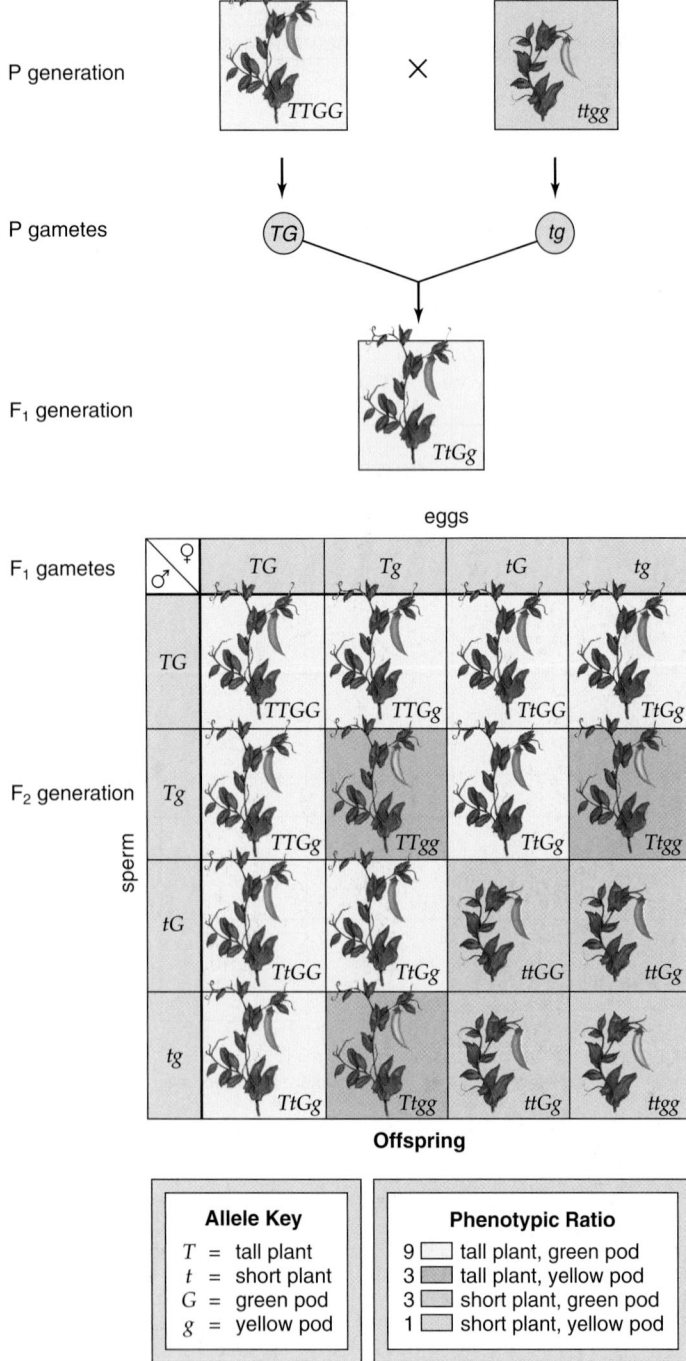

Figure 11.6 Dihybrid cross done by Mendel.   P generation plants differ in two traits: —length of the stem, and color of the pod. The F₁ generation shows only the dominant traits, but all possible phenotypes appear among the F₂ generation because the F₁ parents are hybrids. The 9:3:3:1 ratio allowed Mendel to deduce that factors segregate into gametes independently of other factors.

Mendel's Law of Independent Assortment

Mendel performed a second series of crosses in which true-breeding pea plants differed in two traits. For example, he crossed tall plants having green pods with short plants having yellow pods (Fig. 11.6). The F₁ plants showed both dominant characteristics. As before, Mendel then allowed the F₁ plants to self-pollinate. This F₁ cross is known as a **dihybrid cross** [L. *di*, two, and *hybrida*, mixture] because the plants are hybrid in two ways. Two possible results could occur in the F₂ generation:

1. If the dominant factors (*TG*) always segregate into the F₁ gametes together, and the recessive factors (*tg*) always stay together, then there would be two phenotypes among the F₂ plants—tall plants with green pods and short plants with yellow pods.
2. If the four factors segregate into the F₁ gametes independently, then there would be four phenotypes among the F₂ plants—tall plants with green pods, tall plants with yellow pods, short plants with green pods, and short plants with yellow pods.

Figure 11.6 shows that Mendel observed four phenotypes among the F₂ plants, supporting the second hypothesis. This is how Mendel formulated his second law of heredity—the law of independent assortment.

> The **law of independent assortment** states the following:
> - Each pair of factors segregates (assorts) independently of the other pairs.
> - All possible combinations of factors can occur in the gametes.

The law of independent assortment applies only to alleles on different chromosomes. Each chromosome carries a large number of alleles.

We know that the process of meiosis explains why the F₁ plants produced every possible type of gamete and, therefore, four phenotypes appear among the F₂ generation of plants. Figure 11.7 shows a parent cell with two homologous pairs of chromosomes, with alleles *A, a* on one pair and *B, b* on the other pair. Following duplication of the chromosomes during interphase, the parent cell undergoes meiosis I. At metaphase I, the homologous pairs line up independently of one another, such that the chromosomes with *A* alleles have an equal chance of lining up with the *B* alleles or the *b* alleles. The subsequent segregation of the homologous pairs during anaphase I reduces the chromosome number from 2n to n. Because *A* alleles can be sorted with *B* or *b*, and so can the *a* allele, it is possible to create gametes with *AB*, *Ab*, *aB*, and *ab* allele combinations with equal probability.

Animation
Random Orientation of Chromosomes During Meiosis

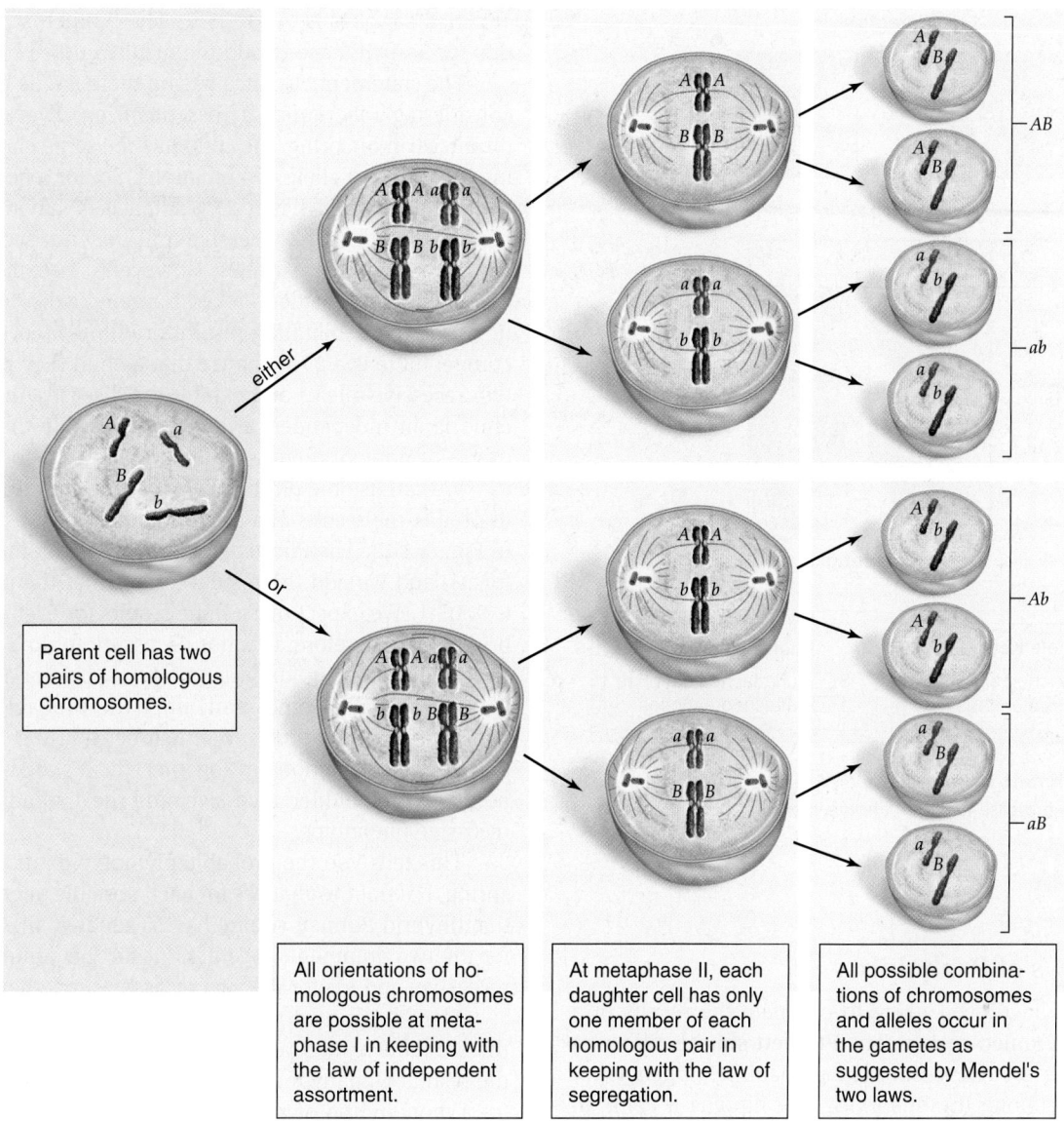

All orientations of homologous chromosomes are possible at metaphase I in keeping with the law of independent assortment.

At metaphase II, each daughter cell has only one member of each homologous pair in keeping with the law of segregation.

All possible combinations of chromosomes and alleles occur in the gametes as suggested by Mendel's two laws.

Figure 11.7 Independent assortment and segregation during meiosis. Mendel's laws hold because of the events of meiosis. The homologous pairs of chromosomes line up randomly at the metaphase plate during meiosis I. It doesn't matter which member of a homologous pair faces which spindle pole. In this example, A alleles can segregate with B or with b alleles. Likewise, a alleles can segregate with B or with b alleles. Therefore, the homologous chromosomes, and alleles they carry, segregate independently during gamete formation. All possible combinations of chromosomes and alleles, that is, AB, Ab, aB, and ab, occur in the gametes.

The same rule of independent assortment applies for the pea plant example in Figure 11.6. In that case, the possible gametes are the two dominants (such as *TG*), the two recessives (such as *tg*), and the ones that have a dominant and a recessive (such as *Tg* and *tG*). Regardless of whether we are using the A and B chromosome or the T and G chromosome examples, when all possible sperm have an opportunity to fertilize all possible eggs, *the expected phenotypic ratio of a dihybrid cross is always 9:3:3:1.*

Check Your Progress   **11.2B**

1. Identify all possible gametes for a heterozygote in fruit flies where *L* = long wings and *l* = short wings; *G* = gray body and *g* = black body.
2. Describe the expected phenotypic ratio when two heterozygous dihybrids reproduce.

Parents

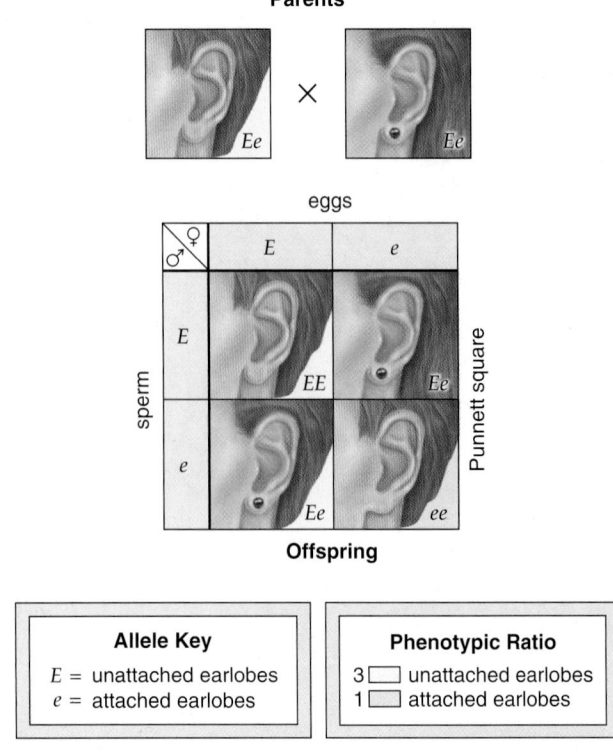

Figure 11.8 Punnett square. Use of Punnett square to calculate probable results; in this case, a 3:1 phenotypic ratio.

Mendel's Laws of Probability

The diagram we have been using to calculate the results of a cross is called a Punnett square. The **Punnett square** allows us to easily calculate the chances, or the probability, of genotypes and phenotypes among the offspring. Like flipping a coin, an offspring of the cross illustrated in the Punnett square in Figure 11.8 has a 50% (or ½) chance of receiving an E for unattached earlobe or an e for attached earlobe from each parent:

The chance of E = ½
The chance of e = ½

How likely is it that an offspring will inherit a specific set of two alleles, one from each parent? The *product rule* of probability tells us that we have to multiply the chances of independent events to get the answer:

1. The chance of EE = ½ × ½ = ¼
2. The chance of Ee = ½ × ½ = ¼
3. The chance of eE = ½ × ½ = ¼
4. The chance of ee = ½ × ½ = ¼

The Punnett square does this for us because we can easily see that each of these is ¼ of the total number of squares.

How do we get the phenotypic results? The *sum rule* of probability tells us that when the same event can occur in more than one way, we can add the results. Because 1, 2, and 3 all result in unattached earlobes, we add them up to know that the chance of unattached earlobes is ¾, or 75%. The chance of

attached earlobes is ¼, or 25%. The Punnett square doesn't do this for us—we have to add the results ourselves.

The statement "Chance has no memory" is important when considering inheritance across offspring. Every time a couple produces an offspring, the child has the same chances of inheriting the different allele combinations. So, for a heterozygous (Ee) couple, each child has a 25% chance of having attached (ee) earlobes. Inheriting a recessive trait may not seem significant if we are considering earlobes. However, it becomes quite significant when we consider a recessive genetic disorder such as cystic fibrosis, a debilitating respiratory illness. For a heterozygous couple, there is a 25% chance that a child they have will inherit two recessive alleles and exhibit the disease. And because each child is an independent event, it is possible that all their children—or none of them—could exhibit cystic fibrosis.

We can use the product rule and the sum rule of probability to predict the results of a dihybrid cross, such as the one shown in Figure 11.6. The Punnett square carries out the multiplication for us, and we add the results to find that the phenotypic ratio is 9:3:3:1. We expect these same results for each and every dihybrid cross. Therefore, it is not necessary to do a Punnett square over and over again for either a monohybrid or a dihybrid cross. Instead, we can simply remember the probable results of 3:1 and 9:3:3:1. But we have to remember that the 9 represents the two dominant phenotypes together, the 3's are a dominant phenotype with a hidden recessive, and the 1 stands for the double recessive phenotype.

This tells you the probable phenotypic ratio among the offspring, but not the chances for each possible phenotype. Because the dihybrid Punnett square has 16 squares, the chances are $9/16$ for the two dominants together, $3/16$ for the dominants with each recessive, and $1/16$ for the two recessives together.

Mendel counted the results of many similar crosses to get the probable results, and in the laboratory, we too have to count the results of many individual crosses to get the probable results for a monohybrid or a dihybrid cross. Why? Consider that each time you toss a coin, you have a 50% chance of getting heads or tails. If you tossed the coin only a couple of times, you might very well have heads or tails both times. However, if you toss the coin many times, your results are more likely to approach 50% heads and 50% tails.

Testcrosses

To confirm that the F_1 plants of his one-trait crosses were in fact heterozygous, Mendel crossed his F_1 generation tall pea plants with true-breeding short (homozygous recessive) plants; such a mating is termed a **testcross.** These crosses provided Mendel with further support for his law of segregation.

For the cross in Figure 11.9, Mendel reasoned that half the offspring should be tall and half should be short, producing a 1:1 phenotypic ratio. His results supported the hypothesis that alleles segregate when gametes are formed. In Figure 11.9*a*, the homozygous recessive parent can produce only one type of gamete—*t*—and so the Punnett square has only one column. The use of one column signifies that all the gametes carry a *t*. *The expected phenotypic ratio for this type of one-trait cross (heterozygous × recessive) is always 1:1.*

One-Trait Testcross

Today, a one-trait testcross is used to determine if an individual with the dominant phenotype is homozygous dominant (e.g., *TT*) or heterozygous (e.g., *Tt*). Because both of these genotypes produce the dominant phenotype, it is not possible to determine the genotype by observation. Figure 11.9*b* shows that if the individual is homozygous dominant, all the offspring will be tall. Each parent has only one type of gamete and, therefore, a Punnett square is not required to determine the results.

Two-Trait Testcross

When doing a two-trait testcross, an individual with the dominant phenotype is crossed with one having the recessive phenotype. Suppose you are working with fruit flies in which:

L = long wings   G = gray bodies
l = vestigial (short) wings   g = black bodies

You wouldn't know by examination whether the fly on the left was homozygous or heterozygous for wing and body color. To find out the genotype of the test fly, you cross it with the one on the right. You know by examination that this vestigial-winged and black-bodied fly is homozygous recessive for both traits.

If the test fly is homozygous dominant for both traits with the genotype *LLGG*, it will form only one gamete: *LG*. Therefore, all the offspring from the proposed cross would have long wings and a gray body.

However, if the test fly is heterozygous for both traits with the genotype *LlGg*, it will form four different types of gametes:

Gametes: *LG Lg lG lg*

and could have four different offspring:

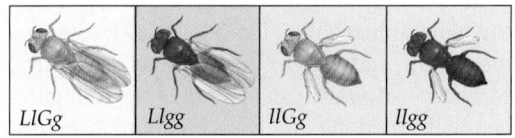

| LlGg | Llgg | llGg | llgg |

The presence of the offspring with vestigial wings and a black body shows that the test fly is heterozygous for both traits and has the genotype *LlGg*. Otherwise, it could not produce this offspring. In general, you want to remember that *the expected phenotypic ratio for this type of two-trait cross (heterozygous for two traits × recessive for both traits) is always 1:1:1:1.*

Check Your Progress   11.2C

1. State the percentage of pea plants that would have
   a. yellow seeds, and **b.** green seeds in a cross of two heterozygous plants where yellow seed color is dominant over green seed color.
2. State the chances of producing offspring with long wings and a black body from a testcross of a heterozygous (*LlGg*) fruit fly and a homozygous recessive (*llgg*) fruit fly.
3. State the genotype of all the horses if trotter (*T*) is dominant over pacer (*t*), and a pacer is produced when a trotter is mated to a pacer.

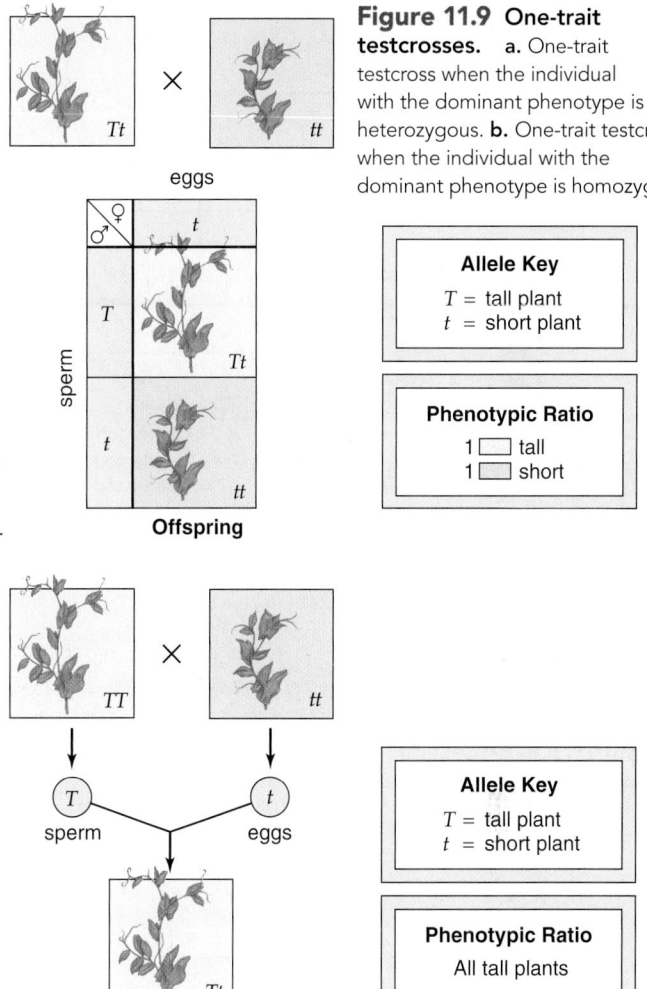

Figure 11.9 One-trait testcrosses. a. One-trait testcross when the individual with the dominant phenotype is heterozygous. **b.** One-trait testcross when the individual with the dominant phenotype is homozygous.

Allele Key
T = tall plant
t = short plant

Phenotypic Ratio
1 ▢ tall
1 ▨ short

a. **Offspring**

Allele Key
T = tall plant
t = short plant

Phenotypic Ratio
All tall plants

b. **Offspring**

Mendel's Laws and Human Genetic Disorders

Many traits and disorders in humans, and other organisms also, are genetic in origin and follow Mendel's laws. These traits are controlled by a single pair of alleles on the autosomal chromosomes. An **autosome** is any chromosome other than a sex (X or Y) chromosome.

Autosomal Patterns of Inheritance

When a genetic disorder is autosomal dominant, the normal allele (*a*) is recessive, and an individual with the alleles *AA* or *Aa* has the disorder. When a genetic disorder is autosomal recessive, the normal allele (*A*) is dominant, and only individuals with the alleles *aa* have the disorder. A pedigree shows the pattern of inheritance for a particular condition and can be used by genetic counselors to determine whether a condition is dominant or recessive. Consider these two possible patterns of inheritance:

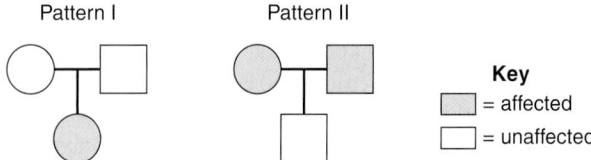

Pattern I   Pattern II

Key
▨ = affected
▢ = unaffected

In a pedigree, males are designated by squares and females by circles. Shaded circles and squares are the affected individuals. The shaded boxes do not indicate whether the condition is

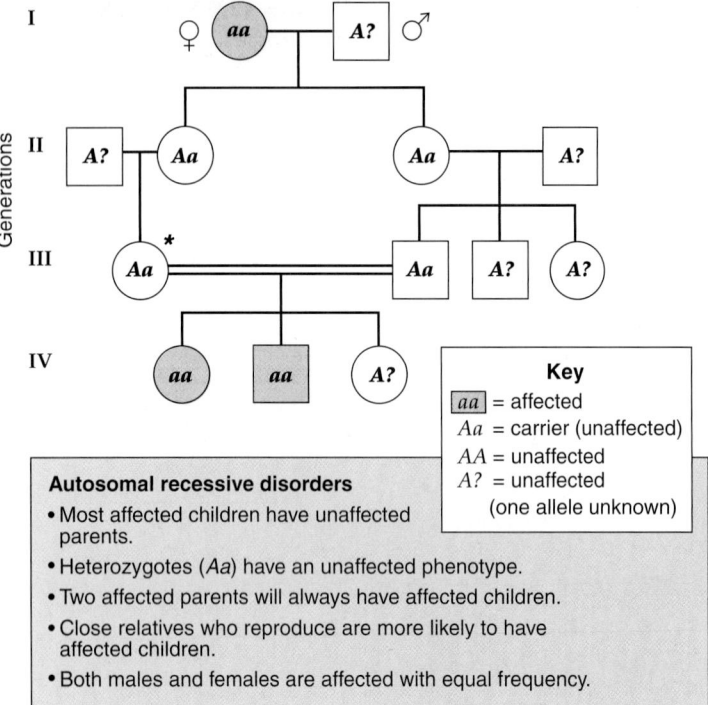

Autosomal recessive disorders

- Most affected children have unaffected parents.
- Heterozygotes (*Aa*) have an unaffected phenotype.
- Two affected parents will always have affected children.
- Close relatives who reproduce are more likely to have affected children.
- Both males and females are affected with equal frequency.

Key

- aa = affected
- *Aa* = carrier (unaffected)
- *AA* = unaffected
- *A?* = unaffected (one allele unknown)

Figure 11.10 Autosomal recessive pedigree. The list gives ways to recognize an autosomal recessive disorder. How would you know the individual at the asterisk is heterozygous? (See Appendix A for answer.)

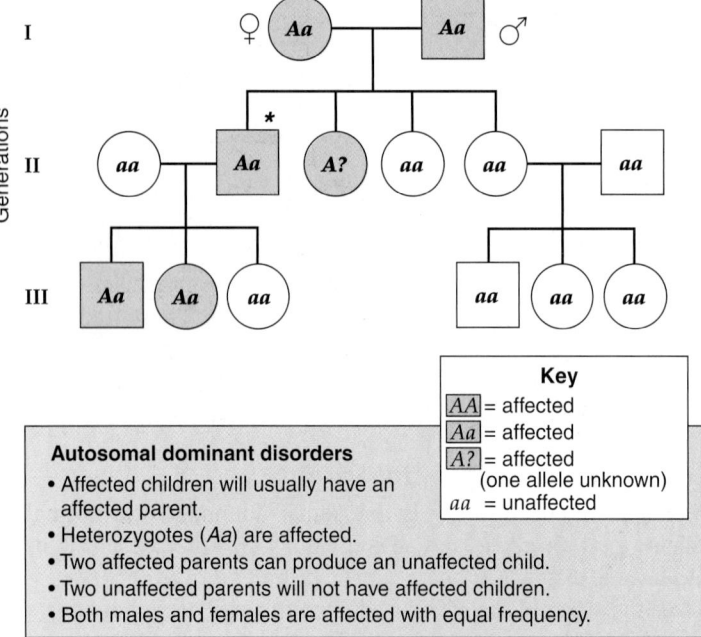

Autosomal dominant disorders

- Affected children will usually have an affected parent.
- Heterozygotes (*Aa*) are affected.
- Two affected parents can produce an unaffected child.
- Two unaffected parents will not have affected children.
- Both males and females are affected with equal frequency.

Key

- AA = affected
- Aa = affected
- A? = affected (one allele unknown)
- aa = unaffected

Figure 11.11 Autosomal dominant pedigree. The list gives ways to recognize an autosomal dominant disorder. How would you know the individual at the asterisk is heterozygous? (See Appendix A for answer.)

dominant or recessive, only that the individual exhibits the trait. A line between a square and a circle represents a union. In the patterns on page 201, a vertical line going downward leads to a single child. (If there are more children, they are lined up horizontally.) From page 201, which pattern of inheritance (I or II) do you suppose represents an autosomal dominant characteristic, and which represents an autosomal recessive characteristic?

In pattern I, the child is affected, but neither parent is; this can happen if the condition is recessive and both parents are *Aa*.

Notice that the parents are **carriers** because they appear normal (do not express the trait) but are capable of having a child with the genetic disorder. In pattern II, the child is unaffected, but the parents are affected. This can happen if the condition is dominant and the parents are *Aa*.

Figure 11.10 shows other ways to recognize an autosomal recessive pattern of inheritance, and Figure 11.11 shows other ways to recognize an autosomal dominant pattern of inheritance. In these pedigrees, generations are indicated by Roman numerals placed on the left side. Notice in the third generation of Figure 11.10 that two closely related individuals have produced three children, two of which have the affected phenotype. In this case, a double line denotes consanguineous reproduction, or inbreeding, which is reproduction between two closely related individuals. This illustrates that inbreeding significantly increases the chances of children inheriting two copies of a potentially harmful recessive allele.

Autosomal Recessive Disorders

In humans, a number of autosomal recessive disorders have been identified. Here, we discuss methemoglobinemia and cystic fibrosis.

Methemoglobinemia

Methemoglobinemia is a relatively harmless disorder that results from an accumulation of methemoglobin in the blood. This disorder has been documented for centuries, but the exact cause and genetic link had remained mysterious. Although rarely mentioned, hemoglobin, the main oxygen-carrying protein in the blood, is usually converted at a slow rate to an alternate form called methemoglobin. Unlike hemoglobin, which is bright red when carrying oxygen, methemoglobin has a bluish color, similar to that of oxygen-poor blood. Although this process is harmless, individuals with methemoglobinemia are unable to clear the abnormal blue protein from their blood, causing their skin to appear bluish-purple in color (Fig. 11.12).

A persistent and determined physician finally solved the age-old mystery of what causes methemoglobinemia by doing blood tests and pedigree analysis involving a family known as the "blue Fugates" of Troublesome Creek, Kentucky. Enzyme tests indicated that the blue Fugates lacked the enzyme diaphorase, coded for by a gene on chromosome 22. The enzyme normally converts methemoglobin back to hemoglobin.

The physician treated the disorder in a simple, but rather unconventional manner. He injected the Fugates with a dye called methylene blue! This unusual dye can donate electrons to other compounds, successfully converting the excess methemoglobin back into normal hemoglobin. The results were striking but immediate—the patient's skin quickly turned pink after treatment.

A pedigree analysis of the Fugate family indicated that the trait is common in the family because so many members carried the recessive allele.

Cystic Fibrosis

Cystic fibrosis (CF) is the most common lethal genetic disease among Caucasians in the United States (Fig. 11.13). About 1 in 20 Caucasians is a carrier, and about 1 in 2,000 newborns has the disorder. CF patients exhibit a number of characteristic symptoms, the most obvious being extremely salty sweat. In

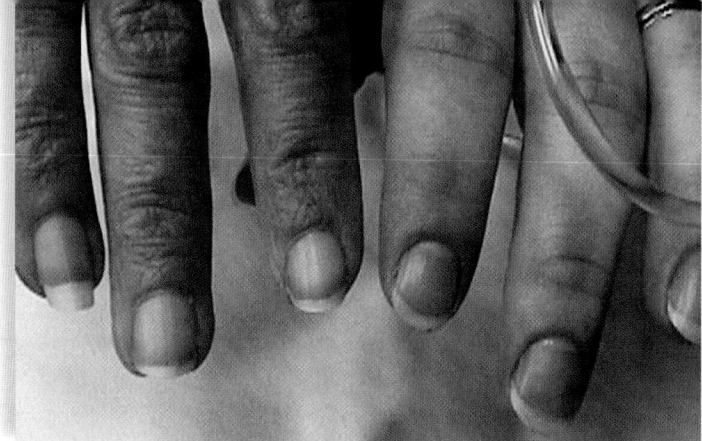

Figure 11.12 Methemoglobinemia. The hands of the woman on the right appear blue due to chemically induced methemoglobinemia.

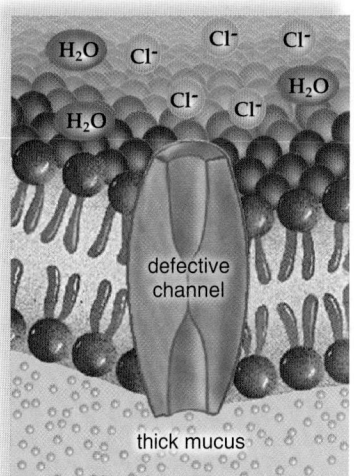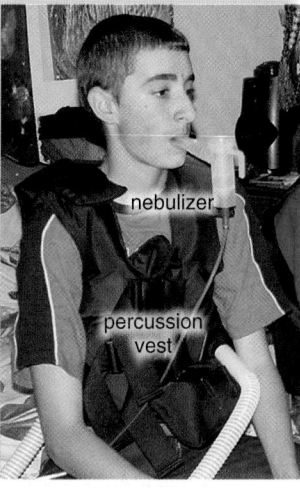

Figure 11.13 Cystic fibrosis. Cystic fibrosis is due to a faulty protein that is supposed to regulate the flow of chloride ions into and out of cells through a channel protein.

children with CF, the mucus in the bronchial tubes and pancreatic ducts is particularly thick and viscous, interfering with the function of the lungs and pancreas. To ease breathing, the thick mucus in the lungs has to be loosened periodically, but still the lungs frequently become infected. The clogged pancreatic ducts prevent digestive enzymes from reaching the small intestine, and to improve digestion, patients take digestive enzymes mixed with applesauce before every meal.

Cystic fibrosis is caused by a defective chloride ion channel that is encoded by the *CFTR* allele on chromosome 7. Research has demonstrated that chloride ions (Cl^-) fail to pass through the defective version of the CFTR chloride ion channel, which is located on the plasma membrane. Ordinarily, after chloride ions have passed through the channel to the other side of the membrane, sodium ions (Na^+) and water follow. It is believed that lack of water is the cause of the abnormally thick mucus in the bronchial tubes and pancreatic ducts.

In the past few years, new treatments have raised the average life expectancy for CF patients to as much as 35 years of age. It is hoped that other novel treatments, such as gene therapy, may be able to correct the defect by replacing a faulty copy of the gene with a normal one. Some scientists have suggested that the mutated *CFTR* allele has persisted in the human population as a means of surviving potentially fatal diseases, such as cholera.

Autosomal Dominant Disorders

A number of autosomal dominant disorders have been identified in humans. Two relatively well-known autosomal dominant disorders include osteogenesis imperfecta and hereditary spherocytosis.

Osteogenesis Imperfecta

Osteogenesis [L. *os*, bone, and *genesis*, origin] imperfecta is an autosomal dominant genetic disorder that results in weakened, brittle bones. Although at least nine types of the disorder are known, most are linked to mutations in two genes necessary to the synthesis of a type I collagen—one of the most abundant proteins in the human body. Collagen has many roles, including providing strength and rigidity to bone and forming the framework for most of the body's tissues. Osteogenesis imperfecta leads to a defective collagen I that causes the bones to be brittle and weak. Because the mutant collagen can cause structural defects even when combined with normal collagen I, osteogenesis imperfecta is generally considered to be dominant.

Osteogenesis imperfecta, which has an incidence of approximately 1 in 5,000 live births, affects all racial groups similarly

and has been documented as long as 300 years ago. Some historians think that the Viking chieftain Ivar Ragnarsson, who was known as Ivar the Boneless and was often carried into battle on a shield, had this condition. In most cases, the diagnosis is made in young children who visit the emergency room frequently due to broken bones. Some children with the disorder have an unusual blue tint in the sclera, the white portion of the eye; reduced skin elasticity; weakened teeth; and occasionally heart valve abnormalities. Currently, the disorder is treatable with a number of drugs that help to increase bone mass, but these drugs must be taken long-term.

Hereditary Spherocytosis

Hereditary spherocytosis is an autosomal dominant genetic blood disorder that results from a defective copy of the ankyrin-1 gene found on chromosome 8. The protein encoded by this gene serves as a structural component of red blood cells, and is responsible for maintaining their disklike shape. The abnormal spherocytosis protein is unable to perform its usual function, causing the affected person's red blood cells to adopt a spherical rather than disklike shape. As a result, the abnormal cells are fragile and burst easily, especially under osmotic stress. Enlargement of the spleen is also commonly seen in people with the disorder.

With an incidence of approximately 1 in 5,000, hereditary spherocytosis is one of the most common hereditary blood disorders. Roughly one-fourth of these cases result from new mutations and are not inherited from either parent. Hereditary spherocytosis exhibits incomplete penetrance, so not all individuals who inherit the mutant allele will actually show the trait. The cause of incomplete penetrance in these cases and others remains poorly understood.

Check Your Progress   11.2D

1. State the genotype of the child in Figure 11.13 and the genotypes of his parents if neither parent has cystic fibrosis. (Use this key: C = normal; c = cystic fibrosis)
2. Identify the chance that the parents in the above problem will have a child with cystic fibrosis.
3. Construct a pedigree of Ivar Ragnarsson's family tree assuming that his mother, and both her parents, were normal and that Ivar's father's father had osteogenesis imperfecta (mother was normal).

Nature of Science

Testing for Genetic Disorders

Many human genetic disorders such as Huntington disease and cystic fibrosis are the result of inheriting faulty genes. Huntington disease is a devastating neurological disease caused by the inheritance of a single dominant allele; in contrast, cystic fibrosis, being a recessive disorder, requires the inheritance of two recessive alleles. In each case, mutated sequences in these genes that have been inherited lead to defective proteins, which disrupt normal biological function.

Genetic tests have been developed that can detect a particular sequence of bases for all your genes, and these sequences tell whether you have a particular genetic disorder. When researchers set out to develop a test for Huntington disease, they first obtained multiple **family pedigrees,** such as the one shown in Figure 11A. This pedigree meets the requirements for a dominant allele and autosomal inheritance: Every individual who is affected (shaded box or circle) has a parent who is also affected, heterozygotes are affected, and both males and females are affected in equal numbers. Each offspring of an affected individual has a 50% chance of having received the faulty gene and developing Huntington disease, which doesn't appear until later in life. Thus, a person could already have produced children before they know about the disease.

The letters under the square or circle mean the individual has undergone a blood test that resulted in an analysis of their DNA. A computer analysis of these individuals' DNA found that a large number of them had a sequence designated as J, K, or L. Only the sequence of bases designated as L appears in all the individuals with Huntington disease. A closer look at the pedigree indicates that sequence L is not in the gene for Huntington because at least one individual has the sequence but does not have Huntington disease.

When genes occur on the same chromosome in close proximity, their alleles are said to be linked. The closer linked alleles are on a chromosome, the greater the chance that they will be inherited together. This is the reason that alleles must be on separate chromosomes for the law of independent assortment to hold. Still, even genes that are closely linked can undergo crossing-over and become unlinked on occasion. Testable sequences that are closely linked to that of the faulty gene are called genetic markers, and these may be used as indicators of genetic disorders, such as Huntington disease. Association studies are another way for researchers to find possible sequences that indicate someone has a genetic disorder. During an association study, the DNA of a diverse sample of the general population is tested to find similar DNA sequences. The use of genetic markers and association studies has made it possible to successfully identify many genes involved in human disease.

The mapping of disorders to genes within the human genome has yielded much valuable information to the scientific and medical communities. The information has a variety of uses including prenatal genetic testing, diagnosis of disorders in individuals before symptoms occur, carrier testing of recessive disorders, and to further understand the origin, progression, and pathology of a disorder. This may lead to the development of novel treatment methods. New techniques and technologies have greatly accelerated this process, but the tried and true methods of family pedigrees and association studies are still the primary techniques used by geneticists in pursuing the cure for many human genetic ailments.

Questions to Consider

1. Considering how quickly cell and molecular technology is emerging, what implications does this have for curing genetic diseases?
2. If you or someone you knew suffered from a genetic disease, would you choose to use emerging technologies to treat the disease?

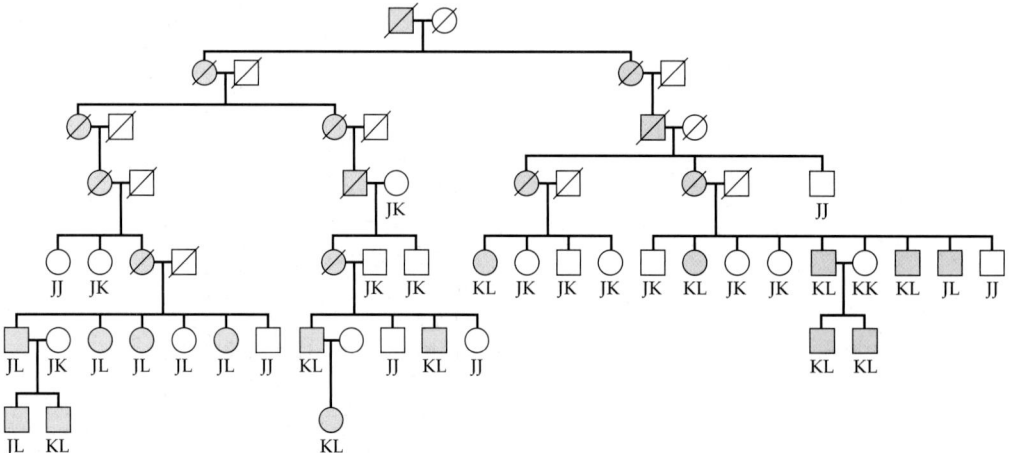

Figure 11A Blood sample testing and pedigree analysis.   In individuals with Huntingon disease, a particular sequence of DNA bases (L) is always present. The pedigree chart shows that the L sequence is not present unless an ancestor exhibited Huntington disease. A slash indicates that the ancestor is deceased.

11.3 Extending the Range of Mendelian Genetics

Learning Outcomes

Upon completion of this section you should be able to

1. Explain the inheritance pattern of traits where more than two alleles for the trait exist.
2. Contrast incomplete dominance and incomplete penetrance.
3. Describe the effects of pleiotropy on phenotypic traits.
4. Distinguish the inheritance of polygenic traits.
5. Examine X-linked inheritance and its gender-based effects.

Mendelian genetics can also be applied to complex patterns of inheritance, such as multiple alleles, incomplete dominance, pleiotropy, and polygenic inheritance.

Multiple Allelic Traits

When a trait is controlled by **multiple alleles,** the gene exists in several allelic forms within a population. For example, although a person's ABO blood type is controlled by a single gene pair, there are three possible alleles within the human population that determine the blood type. Each person receives two of these alleles (one from each parent) to determine the presence or absence of antigens on their red blood cells.

I^A = A antigen on red blood cells
I^B = B antigen on red blood cells
i = Neither A nor B antigen on red blood cells

The possible phenotypes and genotypes for blood type are as follows:

Phenotype	Genotype
A	I^AI^A, I^Ai
B	I^BI^B, I^Bi
AB	I^AI^B
O	ii

The inheritance of the ABO blood group in humans is also an example of **codominance** because both I^A and I^B are fully expressed in the presence of the other. A person who inherits chromosomes with I^A and I^B alleles will make fully functional A and B protein, and because these alleles are codominant, the resulting mixture of AB protein will give the red blood cell an AB phenotype. On the other hand, both I^A and I^B are dominant over i. Therefore, two genotypes are possible for type A blood, and two genotypes are possible for type B blood.

Use a Punnett square to confirm that reproduction between a heterozygote with type A blood and a heterozygote with type B blood can result in any one of the four blood types. Such a cross makes it clear that an offspring can have a different blood type from either parent, and for this reason, DNA fingerprinting is now used to identify the parents of an individual, instead of blood type.

Incomplete Dominance and Incomplete Penetrance

Incomplete dominance is exhibited when the heterozygote has an intermediate phenotype between that of either homozygote. In a cross between a true-breeding, red-flowered four-o'clock plant strain and a true-breeding, white-flowered strain, the offspring have pink flowers. Although this outcome might appear to be an example of the blending theory of inheritance, it is not. Here's why. When the pink plants self-pollinate, the offspring plants have a phenotypic ratio of 1 red-flowered : 2 pink-flowered : 1 white-flowered. The reappearance of the three phenotypes in this generation makes it clear that we are still dealing with a single pair of alleles (Fig. 11.14).

Incomplete dominance in four-o'clocks actually has more to do with the amount of pigment protein produced in the plant cells: A double dose of pigment results in red flowers; a single dose of pigment results in pink flowers; and a lack of any pigment produces white flowers.

Human Examples of Incomplete Dominance

In humans, familial hypercholesterolemia (FH) is an example of incomplete dominance. An individual with two alleles for this disorder develops fatty deposits in the skin and tendons and may have a heart attack as a child. An individual with one normal allele and one *FH* allele may suffer a heart attack as a young adult, and an individual with two normal alleles does not have the disorder.

Perhaps the inheritance pattern of other human disorders should be considered one of incomplete dominance. To detect the carriers of cystic fibrosis, for example, it is customary to determine the amount of cellular activity of the gene. When the activity is one-half that of the dominant homozygote, the individual is a carrier, even though the individual does not exhibit the genetic disease. In other words, at the level of gene

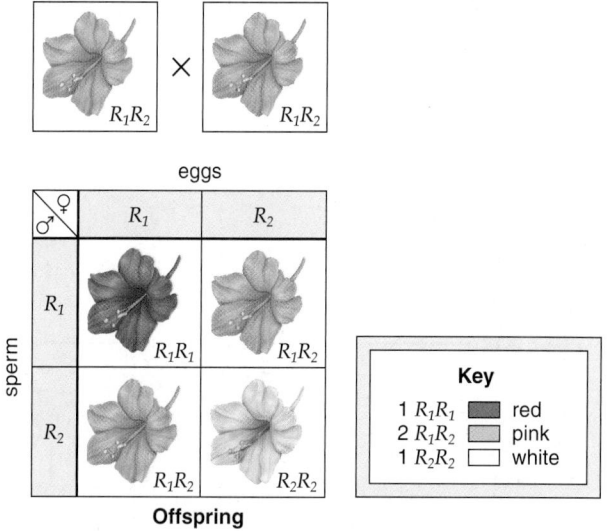

Figure 11.14 Incomplete dominance. When pink four-o'clocks self-pollinate, the results show three phenotypes. This is only possible if the pink parents had an allele for red pigment (R_1) and an allele for no pigment (R_2). Note that alleles involved in incomplete dominance are both given a capital letter.

expression, the homozygotes and heterozygotes do differ in the same manner as four-o'clock plants.

In some cases, a dominant allele may not always lead to the dominant phenotype in a heterozygote, even when the alleles show a true dominant/recessive relationship. The dominant allele in this case does not always determine the phenotype of the individual, so we describe these traits as showing **incomplete penetrance.** In other words, just because a person inherits a dominant allele doesn't mean they will fully express the gene or show the dominant phenotype. Many dominant alleles exhibit varying degrees of penetrance.

The best-known example of incomplete penetrance is polydactyly, the presence of one or more extra digits on hands, feet, or both. Polydactyly is inherited in an autosomal dominant manner; however, not all individuals who inherit the dominant allele exhibit the trait. The reasons for this are not clear, but expression of polydactyly may require additional environmental factors or be influenced by other genes, as discussed later.

Pleiotropic Effects

Pleiotropy occurs when a single mutant gene affects two or more distinct and seemingly unrelated traits. For example, persons with Marfan syndrome have disproportionately long arms, legs, hands, and feet; a weakened aorta; poor eyesight; and other characteristics (Fig. 11.15). All of these characteristics are due to the production of abnormal connective tissue.

Marfan syndrome has been linked to a mutated gene (*FBN1*) on chromosome 15 that ordinarily specifies a functional protein called fibrillin. Fibrillin is essential for the formation of elastic fibers in connective tissue. Without the structural support of normal connective tissue, the aorta can burst, particularly if the person is engaged in a strenuous sport, such as volleyball or basketball. Flo Hyman may have been the best American woman volleyball player ever, but she fell to the floor and died at the age of only 31 because her aorta gave way during a game. Now that coaches are aware of Marfan syndrome, they are on the lookout for it among very tall basketball players. Chris Weisheit, whose career was cut short after he was diagnosed with Marfan syndrome, said, "I don't want to die playing basketball."

Many other disorders, including porphyria and sickle-cell disease, are examples of pleiotropic traits. Porphyria is caused by a chemical insufficiency in the production of hemoglobin, the pigment that makes red blood cells red. The symptoms of porphyria are photosensitivity, strong abdominal pain, port-wine-colored urine, and paralysis in the arms and legs. Many members of the British royal family in the late 1700s and early 1800s suffered from this disorder, which can lead to epileptic convulsions, bizarre behavior, and coma.

In a person suffering from sickle-cell disease (Hb^SHb^S), the cells are sickle-shaped. The underlying mutation is in a gene that codes for a type of polypeptide chain in hemoglobin. Of 146 amino acids, the gene mutation changes only one amino acid, but the result is a less soluble polypeptide chain that stacks up and causes

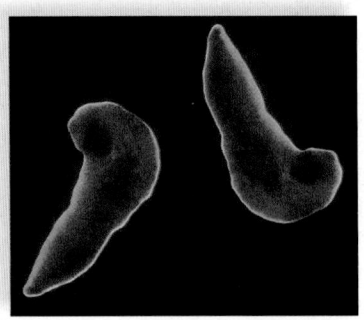

1,600×, colorized SEM

Sickled red blood cell

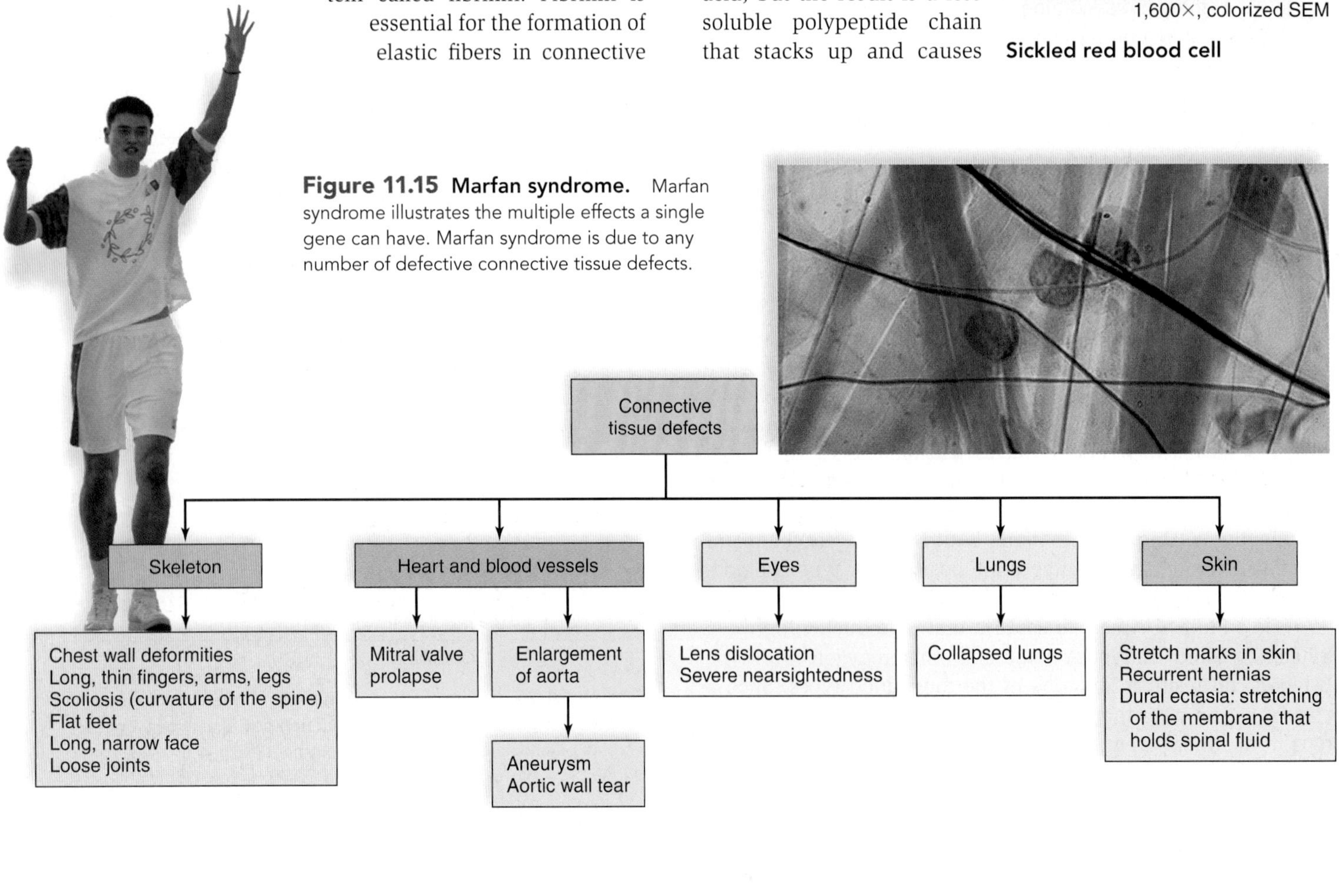

Figure 11.15 Marfan syndrome. Marfan syndrome illustrates the multiple effects a single gene can have. Marfan syndrome is due to any number of defective connective tissue defects.

Connective tissue defects

Skeleton | Heart and blood vessels | Eyes | Lungs | Skin

Chest wall deformities
Long, thin fingers, arms, legs
Scoliosis (curvature of the spine)
Flat feet
Long, narrow face
Loose joints

Mitral valve prolapse

Enlargement of aorta

Lens dislocation
Severe nearsightedness

Collapsed lungs

Stretch marks in skin
Recurrent hernias
Dural ectasia: stretching
of the membrane that
holds spinal fluid

Aneurysm
Aortic wall tear

red blood cells to be sickle-shaped. The abnormally shaped sickle cells slow down blood flow and clog small blood vessels. In addition, sickled red blood cells have a shorter life span than normal red blood cells. Affected individuals may exhibit a number of symptoms, including severe anemia, physical weakness, poor circulation, impaired mental function, pain and high fever, rheumatism, paralysis, spleen damage, low resistance to disease, and kidney and heart failure. All of these effects are due to the tendency of sickled red blood cells to break down and to the resulting decreased oxygen-carrying capacity of the blood and the damage the body suffers as a result of the condition.

Although sickle-cell disease is a devastating disorder, from an evolutionary perspective it provides heterozygous individuals with a survival advantage. People who have sickle-cell trait are resistant to the protozoan parasite that causes malaria. The parasite spends part of its life cycle in red blood cells feeding on hemoglobin, but it cannot complete its life cycle when sickle-shaped cells form and break down earlier than usual. Because of this survival benefit, the sickle-cell allele has been maintained in the human population over evolutionary time.

Polygenic Inheritance

Polygenic inheritance [Gk. *poly*, many; L. *genitus,* producing] occurs when a trait is governed by two or more sets of alleles. Examples include human height and prevalence of diabetes. The individual has a copy of all allelic pairs, possibly located on many different pairs of chromosomes. Each dominant allele has a quantitative effect on the phenotype, and these effects are additive. Therefore, a population is expected to exhibit continuous phenotypic variations, such as a wide variation in human height and weight. In Figure 11.16, a cross between genotypes *AABBCC* and *aabbcc* yields F₁ hybrids with the genotype *AaBbCc*. A range of genotypes and phenotypes results in the F₂ generation that can be depicted as a bell-shaped curve (Fig. 11.16).

Polygenic traits are controlled by many genes and may be influenced by environmental factors. We observed previously (see Fig. 6.9) that the coat color of a Siamese cat is darker in color at the ears, nose, paws, and tail because an enzyme involved in the production of melanin is active only at a low temperature. Similarly, polygenic traits like prevalence of diabetes are influenced by environmental factors like nutrition.

Human Examples of Multifactorial Inheritance

Human skin color and height are examples of polygenic traits affected by the environment. For example, exposure to the sun can affect skin color, and nutrition can affect human height. Just how many pairs of alleles control skin color is not known, but a range in colors can be explained on the basis of just two pairs when *each capital letter contributes equally to the pigment in the skin.*

Genotypes	Phenotypes
AABB	Very dark
AABb or *AaBB*	Dark
AaBb or *AAbb* or *aaBB*	Medium brown
Aabb or *aaBb*	Light
aabb	Very light

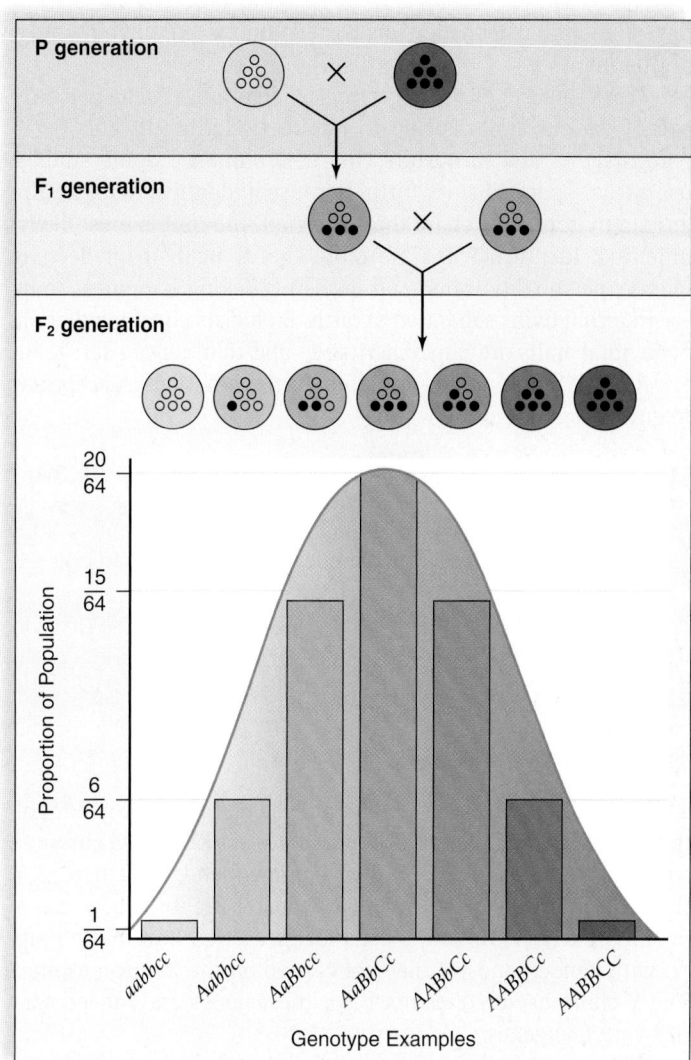

Figure 11.16 Polygenic inheritance. In polygenic inheritance, a number of pairs of genes control the trait. *Above:* Black dots and intensity of blue shading stand for the number of dominant alleles. *Below:* Orange shading shows the degree of environmental influences.

Eye color is also a polygenic trait. The amount of melanin deposited in the iris increases the darker color of the eye. Different eye colors from the brightest of blue to nearly black eyes are thought to be the result of two genes with alleles each interacting in an additive manner.

Many human disorders, such as cleft lip and/or palate, clubfoot, congenital dislocations of the hip, hypertension, diabetes, schizophrenia, and even allergies and cancers, are most likely due to the combined action of many genes plus environmental influences. In recent years, reports have surfaced that all sorts of behavioral traits, such as alcoholism, phobias, and even suicide, can be associated with particular genes.

The relative importance of genetic and environmental influences on the phenotype can vary, but in some instances the role of the environment is clear. For example, cardiovascular disease is more prevalent among those whose biological or adoptive parents have cardiovascular disease. Can you suggest environmental reasons for this correlation, based on your study of Chapter 3?

Many investigators are trying to determine what percentage of various traits is due to nature (inheritance) and what percentage is due to nurture (the environment). Some studies use twins separated since birth, because if identical twins in different environments share the same trait, the trait is most likely inherited. Identical twins are more similar in their intellectual talents, personality traits, and levels of lifelong happiness than are fraternal twins separated at birth. Biologists conclude that all behavioral traits are partly heritable, and that genes exert their effects by acting together in complex combinations susceptible to environmental influences.

Check Your Progress   11.3A

1. Interpret the genotype of the heterozygote if the inheritance pattern for a genetic disorder is shown to be incompletely dominant.
2. Examine the genotype of the child, the mother, and the possible genotypes of the father for a child with type O blood who is born to a mother with type A blood.

X-Linked Inheritance

The X and Y chromosomes in mammals determine the gender of the individual. Females are XX and males are XY. These chromosomes carry genes that control development and, in particular, if the Y chromosome contains an *SRY* gene, the embryo becomes a male. The term **X-linked** is used for genes that have nothing to do with gender, and yet they are carried on the X chromosome. The Y chromosome does not carry these genes and indeed carries very few genes.

This type of inheritance was discovered in the early 1900s by a group at Columbia University headed by Thomas Hunt Morgan. Morgan performed experiments with fruit flies, whose scientific name is *Drosophila melanogaster*. Fruit flies are even better subjects for genetic studies than garden peas. They can be easily and inexpensively raised in simple laboratory glassware; after mating, females lay hundreds of eggs during their lifetimes; and, the generation time is short, taking only about ten days from egg to adult. Fruit flies have a sex chromosome pattern similar to that of humans, and therefore Morgan's experiments with X-linked genes apply directly to humans.

Video
Why a Guy Is a Guy

Morgan's Experiment

Morgan took a newly discovered mutant male with white eyes and crossed it with a red-eyed female:

	♀		♂
P	red-eyed	×	white-eyed
F₁	red-eyed		red-eyed

From these results, he knew that red eyes are the dominant characteristic and white eyes are the recessive characteristic. He then crossed the F₁ flies. In the F₂ generation, there was the expected 3 red-eyed : 1 white-eyed ratio, but it struck him as odd that all of the white-eyed flies were males:

	♀		♂
F₁ × F₁	red-eyed	×	red-eyed
F₂	red-eyed		1 red-eyed : 1 white-eyed

Obviously, a major difference between the male flies and the female flies was their sex chromosomes. Could it be possible that an allele for eye color was on the Y chromosome but not on the X? This idea could be quickly discarded because usually females have red eyes, and they have no Y chromosome. Perhaps an allele for eye color was on the X, but not on the Y, chromosome. Figure 11.17 indicates that this explanation would match the results obtained in the experiment. These results support the chromosome theory of inheritance by showing that the

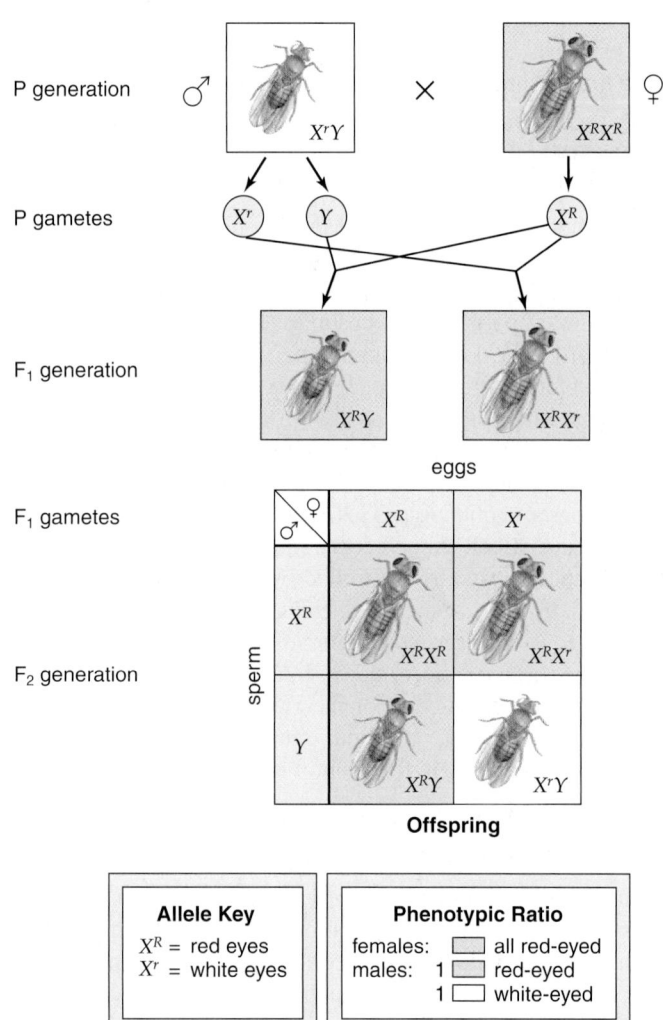

Figure 11.17 X-linked inheritance. Once researchers deduced that the alleles for red/white eye color are on the X chromosome in *Drosophila*, they were able to explain their experimental results. Males with white eyes in the F₂ generation inherit the recessive allele only from the female parent; they receive a Y chromosome lacking the allele for eye color from the male parent.

behavior of a specific allele corresponds exactly with that of a specific chromosome—the X chromosome in *Drosophila*.

Notice that X-linked alleles have a different pattern of inheritance than alleles that are on the autosomes because the Y chromosome is lacking for these alleles, and the inheritance of a Y chromosome cannot offset the inheritance of an X-linked recessive allele. For the same reason, males always receive an X-linked recessive mutant allele from the female parent—they receive only the Y chromosome from the male parent, and therefore sex-linked recessive traits appear much more frequently in males than in females.

Solving X-Linked Genetics Problems

Recall that when solving autosomal genetics problems, the allele key and genotypes can be represented as follows:

Allele key	Genotypes
L = long wings	LL, Ll
l = short wings	ll

When predicting inheritance of sex-linked traits, however, it is necessary to indicate the sex chromosomes of each individual. As noted in Figure 11.17, however, the allele key for an X-linked gene shows an allele attached to the X:

Allele key

X^R = red eyes
X^r = white eyes

The possible genotypes and phenotypes in both males and females are as follows:

Genotype	Phenotype
$X^R X^R$	red-eyed female
$X^R X^r$	red-eyed female
$X^r X^r$	white-eyed female
$X^R Y$	red-eyed male
$X^r Y$	white-eyed male

Notice that there are three possible genotypes for females but only two for males. Females can be heterozygous $X^R X^r$, in which case they are carriers. Carriers usually do not show a recessive abnormality, but they are capable of passing on a recessive allele for an abnormality. But unlike autosomal traits, males cannot be carriers for X-linked traits; if the dominant allele is on the single X chromosome, they show the dominant phenotype, and if the recessive allele is on the single X chromosome, they show the recessive phenotype. For this reason, males are considered **hemizygous** for X-linked traits, because a male only possesses one allele for the trait and, therefore, expresses whatever allele is present on the X chromosome.

We know that male fruit flies have white eyes when they receive the mutant recessive allele from the female parent. What is the inheritance pattern when females have white eyes? Females can have white eyes only when they receive a recessive allele from both parents.

Human X-Linked Disorders

Several X-linked recessive disorders occur in humans, including color blindness, Menkes syndrome, muscular dystrophy, adrenoleukodystrophy, and hemophilia.

Color Blindness. In humans, the receptors for color vision in the retina of the eyes are three different classes of cone cells. Only one type of pigment protein is present in each class of cone cell; there are blue-sensitive, red-sensitive, and green-sensitive cone cells. The allele for the blue-sensitive protein is autosomal, but the alleles for the red- and green-sensitive pigments are on the X chromosome. About 8% of Caucasian men have red-green color blindness. Most of these see brighter greens as tans, olive greens as browns, and reds as reddish browns. A few cannot tell reds from greens at all. They see only yellows, blues, blacks, whites, and grays.

Pedigrees can also reveal the unusual inheritance pattern seen in sex-linked traits. For example, the pedigree in Figure 11.18 shows the usual pattern of inheritance for color blindness. More males than females have the trait because recessive alleles on the X chromosome are expressed in males. The disorder often passes from grandfather to grandson through a carrier daughter.

Menkes Syndrome. Menkes syndrome, or kinky hair syndrome, is caused by a defective allele on the X chromosome. Normally, the gene product controls the movement of the metal

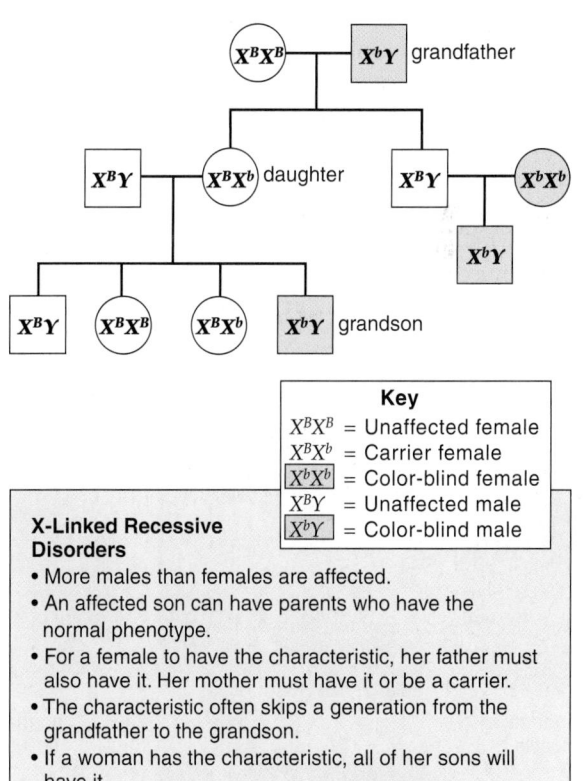

X-Linked Recessive Disorders
- More males than females are affected.
- An affected son can have parents who have the normal phenotype.
- For a female to have the characteristic, her father must also have it. Her mother must have it or be a carrier.
- The characteristic often skips a generation from the grandfather to the grandson.
- If a woman has the characteristic, all of her sons will have it.

Key	
$X^B X^B$	= Unaffected female
$X^B X^b$	= Carrier female
$X^b X^b$	= Color-blind female
$X^B Y$	= Unaffected male
$X^b Y$	= Color-blind male

Figure 11.18 X-linked recessive pedigree. This pedigree for color blindness exemplifies the inheritance pattern of an X-linked recessive disorder. The list gives various ways of recognizing the X-linked recessive pattern of inheritance.

copper in and out of cells. The symptoms of Menkes syndrome are due to accumulation of copper in some parts of the body, and the lack of the metal in other parts.

Symptoms of Menkes syndrome include poor muscle tone, seizures, abnormally low body temperature, skeletal anomalies, and the characteristic brittle, steely hair associated with the disorder. Although the condition is relatively rare, affecting approximately 1 in 100,000, mostly males, the prognosis for people with Menkes syndrome is poor, and most individuals die within the first few years of life. In recent years, some people with Menkes syndrome have been treated with injections of copper directly underneath the skin, but with mixed results, and treatment must begin very early in life to be effective.

Muscular Dystrophy.   Muscular dystrophy, as the name implies, is characterized by a wasting away of the muscles. The most common form, Duchenne muscular dystrophy, is X-linked and occurs in about 1 out of every 3,600 male births (Fig. 11.19). Symptoms, such as waddling gait, toe walking, frequent falls, and difficulty in rising, may appear as soon as the child starts to walk. Muscle weakness intensifies until the individual is confined to a wheelchair. Death usually occurs by age 20; therefore, affected males are rarely fathers. The recessive allele remains in the population through passage from carrier mother to carrier daughter.

The allele for Duchenne muscular dystrophy has been isolated, and it has been discovered that the absence of a protein called dystrophin causes the disorder. Much investigative work has determined that dystrophin is involved in the release of calcium from the sarcoplasmic reticulum in muscle fibers. The lack of dystrophin causes calcium to leak into the cell, which promotes the action of an enzyme that dissolves muscle fibers. When the body attempts to repair the tissue, fibrous tissue forms, and this cuts off the blood supply so that more and more cells die.

A test is now available to detect carriers of Duchenne muscular dystrophy. Also, various treatments have been tried. Immature muscle cells can be injected into muscles, and for every 100,000 cells injected, dystrophin production occurs in 30–40% of muscle fibers. The allele for dystrophin has been inserted into thigh muscle cells, and about 1% of these cells then produced dystrophin.

Adrenoleukodystrophy.   Adrenoleukodystrophy, or ALD, is an X-linked recessive disorder due to the failure of a carrier protein to move either an enzyme or very long chain fatty acid (24–30 carbon atoms) into peroxisomes. As a result, these fatty acids are not broken down, and they accumulate inside the cell; the result is severe nervous system damage.

Children with ALD fail to develop properly after age 5, lose adrenal gland function, exhibit very poor coordination, and show a progressive loss of hearing, speech, and vision. The condition is usually fatal, with no known cure, but the onset and severity of symptoms in patients not yet showing symptoms may be mitigated by treatment with a mixture of lipids derived from olive oil. The disease was made well-known by the 1992 movie *Lorenzo's Oil*, detailing a mother's and father's determination to devise a treatment for their son who was suffering from ALD.

Hemophilia.   About 1 in 10,000 males is a hemophiliac. There are two common types of hemophilia: Hemophilia A is due to the absence or minimal presence of a clotting factor known as factor VIII, and hemophilia B is due to the absence of clotting factor IX. Hemophilia is called the bleeder's disease because the affected person's blood either does not clot or clots very slowly. Although hemophiliacs bleed externally after an injury, they also bleed internally, particularly around joints. Hemorrhages can be stopped with transfusions of fresh blood (or plasma) or concentrates of the clotting protein. Also, clotting factors are now available as biotechnology products.

At the turn of the century, hemophilia was prevalent among the royal families of Europe, and all of the affected males could trace their ancestry to Queen Victoria of England. Of Queen Victoria's 26 grandchildren, four grandsons had hemophilia and four granddaughters were carriers. Because none of Queen Victoria's relatives were affected, it seems that the faulty allele she carried arose by mutation either in Victoria or in one of her parents. Her carrier daughters Alice and Beatrice introduced the allele into the ruling houses of Russia and Spain, respectively. Alexis, the last heir to the Russian throne before the Russian Revolution, was a hemophiliac. There are no hemophiliacs in the present British royal family because Victoria's eldest son, King Edward VII, did not receive the allele.

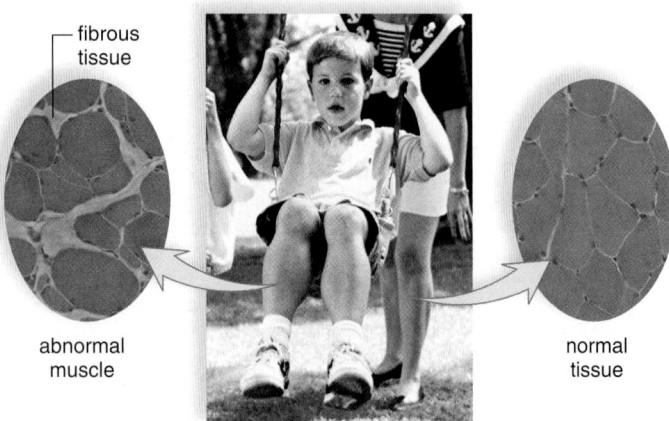

Figure 11.19 Muscular dystrophy.   In muscular dystrophy, an X-linked recessive disorder, calves enlarge because fibrous tissue develops as muscles waste away, due to lack of the protein dystrophin.

fibrous tissue

abnormal muscle

normal tissue

Check Your Progress   11.3B

1. Describe the possible genotypes if a homozygous red-eyed *Drosophila* female is crossed with a red-eyed male.
2. Explain the chances that a color-blind woman who is married to a man with normal vision will produce **a.** color-blind sons **b.** color-blind daughters **c.** carrier daughters.
3. Identify the genotype of all individuals involved in a cross between a brown-haired woman and a black-haired man that produce all brown-haired male offspring and all black-haired female offspring, assuming X-linkage.

CONNECTING *the* CONCEPTS *with the* BIG IDEAS

Evolution

- Variations in genotypes which specify phenotypes provide the material upon which natural selection acts. (1A2b-c)

Information and Signaling

- Patterns of inheritance were deciphered by Gregor Mendel using garden peas which allowed prediction about traits of offspring from particular parents based on rules of probability. (3A3a, 3A3b3)
- Mendel's laws of segregation and independent assortment are based on meiosis I events. (3A3b1)
- Sex-linked, sex-limited, polygenic, and non-nuclear inheritance are some of many situations which do not obey Mendel's laws. (3A4a-c)
- Discovery of the genetic causes of many human disorders has lead to controversial societal issues concerning privacy, genetic ownerships, prenatal testing, gene therapy, etc. (3A3d*IE*)

Interactions and Systems

- Environmental factors may affect the phenotype produced by certain genotypes, as seen in reptile sex determination, seasonal changes in body temperature and variation in fur color, effect of light on melanin production, soil influence on flower color, increase in plant hairs when subjected to herbivores, etc. (4C2a-b)

*Find the unabridged version of all EK citations at www.glencoe.com/maderAP11.

Media Study Tools

www.glencoe.com/maderAP11

Enhance your study of this chapter with study tools and practice tests. Also ask your instructor about the resources available through ConnectPlus, including the media-rich eBook, interactive learning tools, and animations.

Summarize

11.1 Gregor Mendel

Gregor Mendel used the garden pea as the subject in his genetic studies. In contrast to preceding plant breeders, his study involved nonblending traits of the garden pea. Mendel applied mathematics, followed the scientific method very closely, and kept careful records. His results supported a particulate theory of inheritance, effectively disproving the blending theory of inheritance.

11.2 Mendel's Laws

When Mendel crossed heterozygous plants with other heterozygous plants, he found that the recessive phenotype reappeared in about $1/4$ of the F_2 plants; there was a 3:1 phenotypic ratio. This allowed Mendel to propose his law of segregation, which states that the individual has two factors for each trait, and the factors segregate with equal probability into the gametes.

Mendel conducted two-trait crosses, in which the F_1 individuals showed both dominant characteristics, but there were four phenotypes among the F_2 offspring. (The actual phenotypic ratio was 9:3:3:1.) This allowed Mendel to deduce the law of independent assortment, which states that the members of one pair of factors separate independently of those of another pair. Therefore, all possible combinations of parental factors can occur in the gametes.

The laws of probability can be used to calculate the expected phenotypic ratio of a cross. A large number of offspring must be counted in order to observe the expected results, and to ensure that all possible types of sperm have fertilized all possible types of eggs, as is done in a Punnett square. The Punnett square uses the product law of probability to arrive at possible genotypes among the offspring, and then the sum law can be used to arrive at the phenotypic ratio.

Mendel also crossed the F_1 plants having the dominant phenotype with homozygous recessive plants. The 1:1 results indicated that the recessive factor was present in these F_1 plants (i.e., that they were heterozygous). Today, we call this a testcross, because it is used to test whether an individual showing the dominant characteristic is homozygous dominant or heterozygous. The two-trait testcross allows an investigator to test whether an individual showing two dominant characteristics is homozygous dominant for both traits or for one trait only, or is heterozygous for both traits.

Studies have shown that many human traits and genetic disorders can be explained on the basis of simple Mendelian inheritance. When studying human genetic disorders, biologists often construct pedigrees to show the pattern of inheritance of a characteristic within a family. The particular pattern indicates the manner in which a characteristic is inherited. Sample pedigrees for autosomal recessive and autosomal dominant patterns appear in Figures 11.10 and 11.11.

11.3 Extending the Range of Mendelian Genetics

Other patterns of inheritance have been discovered since Mendel's original contribution. For example, some genes have multiple alleles, although each individual organism has only two alleles, as in the inheritance of blood type in human beings. Inheritance of blood type also illustrates codominance. With incomplete dominance, the phenotype of F_1 individuals are intermediate between the parent phenotypes; this does not support the blending theory because the parent phenotypes reappear in F_2. With incomplete penetrance, some traits that are dominant may not be expressed due to unknown reasons.

In pleiotropy, one gene has multiple effects as with Marfan syndrome and sickle-cell disease. Polygenic traits are controlled by several genes that have an additive effect on the phenotype, resulting in quantitative variations within a population. A bell-shaped curve is seen because environmental influences bring about many intervening phenotypes, as in the inheritance of height in human beings. Skin color and eye color are also examples of polygenic inheritance (multiple genes plus the environment).

In *Drosophila*, as in humans, the sex chromosomes determine the sex of the individual, with XX being female and XY being male. Experimental support for the chromosome theory of inheritance came when Morgan and his group were able to determine that the gene for a trait unrelated to sex determination, the white-eyed allele in *Drosophila*, is on the X chromosome.

Alleles on the X chromosome are called X-linked alleles. Therefore, when doing X-linked genetics problems, it is the custom to indicate the sexes by using sex chromosomes and to indicate the alleles by superscripts attached to the X. The Y is blank because it does not carry these genes. Color blindness, Menkes syndrome, adrenoleukodystrophy, and hemophilia are X-linked recessive disorders in humans.

Key Terms

allele 196	law of segregation 196
autosome 201	locus 196
carrier 202	monohybrid cross 195
codominance 205	multiple alleles 205
dihybrid cross 198	phenotype 197
dominant allele 196	pleiotropy 206
family pedigree 204	polygenic inheritance 207
genotype 196	polygenic trait 207
hemizygous 209	Punnett square 200
heterozygous 196	recessive allele 196
homozygous 196	testcross 200
incomplete dominance 205	X-linked 208
incomplete penetrance 206	
law of independent	
assortment 198	

▓▓ Assess

Reviewing This Chapter

1. How did Mendel's procedure differ from that of his predecessors? What is his theory of inheritance called?   193
2. How does the F_2 of Mendel's one-trait cross refute the blending concept of inheritance? Using Mendel's one-trait cross as an example, trace his reasoning to arrive at the law of segregation.   195–96
3. Using Mendel's two-trait cross as an example, trace his reasoning to arrive at the law of independent assortment.   198

4. What are the two laws of probability, and how do they apply to a Punnett square?   200
5. What is a testcross, and when is it used?   200–01
6. How might you distinguish an autosomal dominant trait from an autosomal recessive trait when viewing a pedigree?   201–02
7. For autosomal recessive disorders, what are the chances of two carriers having an affected child?   202
8. For most autosomal dominant disorders, what are the chances of a heterozygote and a normal individual having an affected child?   202
9. Explain inheritance by multiple alleles. List the human blood types, and give the possible genotypes for each.   205
10. Explain the inheritance of incompletely dominant alleles and why this is not an example of blending inheritance.   205–06
11. Explain why traits controlled by polygenes show continuous variation and produce a distribution in the F_2 generation that follows a bell-shaped curve.   207
12. How do you recognize a pedigree for an X-linked recessive allele in human beings?   208–09

Testing Yourself

Choose the best answer for each question. For questions 1–4, match each item to those in the key.

KEY:
   a. 3:1
   b. 9:3:3:1
   c. 1:1
   d. 1:1:1:1
   e. 3:1:3:1

1. *TtYy* × *TtYy*
2. *Tt* × *Tt*
3. *Tt* × *tt*
4. *TtYy* × *ttyy*
5. Which of these could be a normal gamete?
   a. *GgRr*
   b. *GRr*
   c. *Gr*
   d. *GgR*
   e. None of these are correct.
6. Which of these properly describes a cross between an individual who is homozygous dominant for hairline but heterozygous for finger length, and an individual who is recessive for both characteristics? (*W* = widow's peak, *w* = straight hairline, *S* = short fingers, *s* = long fingers)
   a. *WwSs* × *WwSs*
   b. *WWSs* × *wwSs*
   c. *Ws* × *ws*
   d. *WWSs* × *wwss*
7. In peas, yellow seed (*Y*) is dominant over green seed (*y*). In the F_2 generation of a monohybrid cross that begins when a dominant homozygote is crossed with a recessive homozygote, you would expect
   a. three plants with yellow seeds to every plant with green seeds.
   b. plants with one yellow seed for every green seed.
   c. only plants with the genotype *Yy*.
   d. only plants that produce yellow seeds.
   e. Both c and d are correct.

8. In humans, pointed eyebrows (*B*) are dominant over smooth eyebrows (*b*). Mary's father has pointed eyebrows, but she and her mother have smooth. What is the genotype of the father?
   a. *BB*
   b. *Bb*
   c. *bb*
   d. *BbBb*
   e. Any one of these is correct.

9. In guinea pigs, smooth coat (*S*) is dominant over rough coat (*s*), and black coat (*B*) is dominant over white coat (*b*). In the cross *SsBb* × *SsBb*, how many of the offspring will have a smooth black coat on average?
   a. 9 only
   b. about ⁹⁄₁₆
   c. ¹⁄₁₆
   d. ⁶⁄₁₆
   e. ²⁄₆

10. In horses, *B* = black coat, *b* = brown coat, *T* = trotter, and *t* = pacer. A black trotter that has a brown pacer offspring would have which of the following genotypes?
    a. *BT*
    b. *BbTt*
    c. *bbtt*
    d. *BBtt*
    e. *BBTT*

11. In tomatoes, red fruit (*R*) is dominant over yellow fruit (*r*), and tallness (*T*) is dominant over shortness (*t*). A plant that is *RrTT* is crossed with a plant that is *rrTt*. What are the chances of an offspring possessing both recessive traits?
    a. none
    b. ½
    c. ¼
    d. ¾

12. In the cross *RrTt* × *rrtt*,
    a. all the offspring will be tall with red fruit.
    b. 75% (¾) will be tall with red fruit.
    c. 50% (½) will be tall with red fruit.
    d. 25% (¼) will be tall with red fruit.

13. A boy is color-blind (X-linked recessive) and has a straight hairline (autosomal recessive). Which could be the genotype of his mother?
    a. *bbww*   d. *XBXbWw*
    b. *XbYWw*   e. *XwXwBb*
    c. *bbXwXw*

14. Which of the following would you *not* find in a pedigree when a male has an X-linked recessive disorder?
    a. Neither parent has the disorder.
    b. Only males in the pedigree have the disorder.
    c. Only females in the pedigree have the disorder.
    d. The sons of a female with the disorder all have the disorder.
    e. Both a and c would not be seen.

For questions 15–17, match the statements to the items in the key.

KEY:
   a. multiple alleles
   b. polygenes
   c. pleiotropic gene

15. People with sickle-cell disease have many cardiovascular complications.

16. Although most people have an IQ of about 100, IQ generally ranges from about 50 to 150.

17. In humans, there are three possible alleles at the chromosomal locus that determine blood type.

18. Alice and Henry are at the opposite extremes for a polygenic trait. Their children will
    a. be bell-shaped.
    b. be a phenotype typical of a 9:3:3:1 ratio.
    c. have the middle phenotype between their two parents.
    d. look like one parent or the other.

19. Determine whether the characteristic possessed by the shaded squares (males) and circles (females) is an autosomal dominant, autosomal recessive, or X-linked recessive.

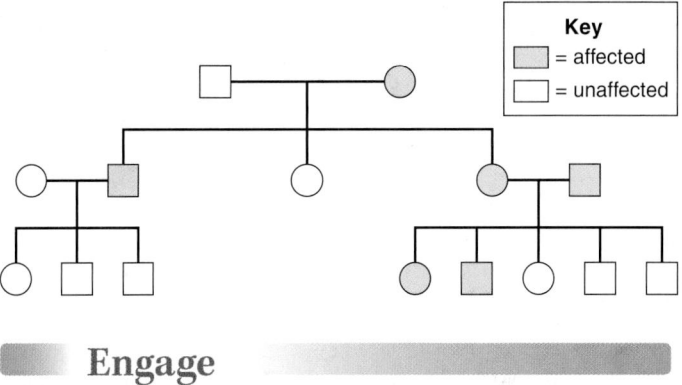

Key
■ = affected
□ = unaffected

Engage

Virtual Labs
Punnett Squares
Sex-Linked Traits

The virtual labs "Punnett Squares" and "Sex-linked Traits" both provide you with the ability to test your knowledge of Mendelian and sex-linked patterns of inheritance.

Thinking Scientifically

1. You want to determine whether a newly found *Drosophila* characteristic is dominant or recessive. Would you wait to cross this male fly with another of its own kind or cross it now with a fly that lacks the characteristic?

2. You want to test whether the leaf pattern of a plant is influenced by the amount of fertilizer in the environment. What would you do?

12

Molecular Biology of the Gene

The diversity of life is dependent on gene activity.

BEFORE YOU BEGIN

Before beginning this chapter, take a few moments to review the following discussions.

Figures 3.18 and 3.20 What are the components of a nucleotide and the structure of the DNA molecule?

Figure 11.4 How are genes organized on chromosomes, and how are alleles related to genes?

Section 11.2 What is the relationship between genotype and phenotype?

All life on Earth has the four bases of DNA—A, G, C, and T—in common. Considering up to 1.8 million different species have been discovered and named, with only a fraction of the total discovered, what makes one species different from another? It is largely due to differences in the DNA sequences that make up genes within each species.

How do DNA sequence differences determine the uniqueness of a species—for example, whether an individual is a snow leopard, a crab, or a flower? Or the variation within a species, such as whether a human has blue, brown, or hazel eyes? The diversity of life is based on the flow of genetic information from genes to proteins to observable traits, that is, from genotype to phenotype. This expression of genes is responsible for producing proteins, the molecules that carry out life functions every day.

As DNA sequences change via mutation and other mechanisms over evolutionary time, so do the proteins that are made from genes. Likewise, as proteins change, so do the life forms and diversity we see on Earth. Given that different mixtures of genes and alleles can be inherited over generations, and that life has been in existence for millions of generations, it is not surprising to see how diverse life has become.

As you read through the chapter, think about the following questions:

1. How does the flow of genetic information from DNA to protein to trait work?

2. What mechanisms are in place to ensure that genetic information is accurately expressed?

3. How might the expression of a gene change in response to environmental conditions?

FOLLOWING *the* BIG IDEAS

CHAPTER 12   MOLECULAR BIOLOGY OF THE GENE

Evolution	DNA is the transmitter of genetic information found in all life forms.
Information and Signaling	Genetic information in the form of a coded nucleotide sequence dictates the sequence of amino acids which will make a protein.

12.1 The Genetic Material

Learning Outcomes

Upon completion of this section, you should be able to

1. Describe the properties a substance must possess in order to serve as the genetic material.
2. Examine how historical researchers demonstrated that DNA was the genetic material.
3. Explain the chemical structure of DNA as defined by the Watson and Crick model.

The middle of the twentieth century was an exciting period of scientific discovery. On one hand, geneticists were busy determining that *DNA (deoxyribonucleic acid)* is the genetic material of living things. On the other hand, biochemists were in a frantic race to describe the structure of DNA. The classic experiments performed during this era set the stage for an explosion in our knowledge of modern molecular biology.

When researchers began their work, they knew that the genetic material must be

1. able to *store information* that pertains to the development, structure, and metabolic activities of the cell or organism;
2. stable so that it *can be replicated* with high accuracy during cell division and be transmitted from generation to generation;
3. able to *undergo rare changes* called mutations [L. *muta*, change] that provide the genetic variability required for evolution to occur.

This chapter will show, as the researchers of the twentieth century did, that DNA can fulfill these functions.

Transformation of Bacteria

During the late 1920s, the bacteriologist Frederick Griffith (1879-1941) was attempting to develop a vaccine against *Streptococcus pneumoniae* (pneumococcus), which causes pneumonia in mammals. In 1931, he performed a classic experiment with the bacterium. He noticed that when these bacteria are grown on culture plates, some, called S strain bacteria, produce shiny, smooth colonies, and others, called R strain bacteria, produce colonies that have a rough appearance. Under the microscope, S strain bacteria have a capsule (mucous coat) that makes them smooth but R strain bacteria do not.

When Griffith injected mice with the S strain of bacteria, the mice died, and when he injected mice with the R strain, the mice did not die (Fig. 12.1). In an effort to determine whether the capsule alone was responsible for the virulence (ability to kill) of the S strain bacteria, he injected mice with heat-killed S strain bacteria. The mice did not die.

Finally, Griffith injected the mice with a mixture of heat-killed S strain and live R strain bacteria. Most unexpectedly, the mice died—and living S strain bacteria were recovered from the bodies! Griffith concluded that some substance necessary for the bacteria to produce a capsule and be virulent must have passed from the dead S strain bacteria to the living R strain bacteria so that the R strain bacteria were *transformed* (Fig. 12.1*d*). This change in the phenotype of the R strain bacteria must be due to a change in their genotype. Indeed, couldn't the transforming substance that passed from S strain to R strain be genetic material? Reasoning such as this prompted investigators at the time to begin looking for the transforming substance to determine the chemical nature of the genetic material.

DNA: The Transforming Substance

By the time the next group of investigators, led by Oswald Avery (1877-1955) in the 1940s, began their work, it was known that the genes are on the chromosomes and that the chromosomes contain both proteins and nucleic acids. Investigators were having a much heated debate about whether protein or DNA was the genetic material. Many thought that the protein component of chromosomes must be the genetic material because proteins contain up to 20 different amino acids that can be sequenced in

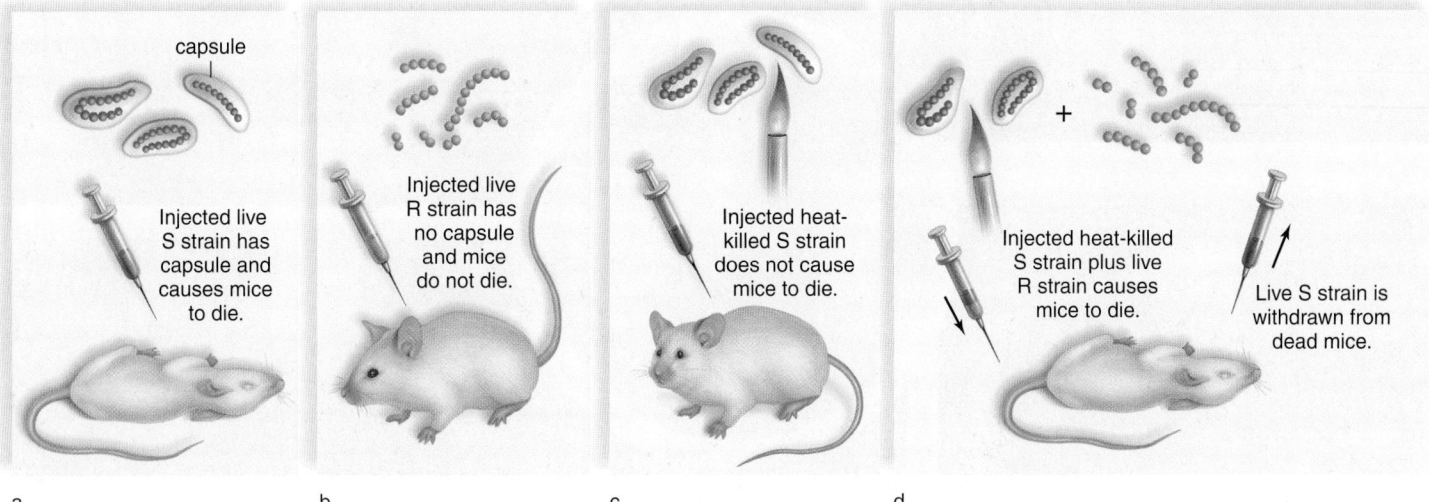

a.   b.   c.   d.

Figure 12.1 Griffith's transformation experiment. **a.** Encapsulated S strain is virulent and kills mice. **b.** Nonencapsulated R strain is not virulent and does not kill mice. **c.** Heat-killed S strain bacteria do not kill mice. **d.** If heat-killed S strain and R strain are both injected into mice, they die because the R strain bacteria have been transformed into the virulent S strain.

any particular way. On the other hand, nucleic acids—DNA and RNA—contain only four types of nucleotides as basic building blocks. Some argued that DNA did not have enough variability to be able to store information and be the genetic material.

In 1944, after 16 years of research, Oswald Avery and his coinvestigators, Colin MacLeod and Maclyn McCarty, published a paper demonstrating that the transforming substance that allows *Streptococcus* to produce a capsule and be virulent is DNA. This meant that DNA is the genetic material. Here is what they found out:

1. DNA from S strain bacteria causes R strain bacteria to be transformed so that they can produce a capsule and be virulent.
2. The addition of DNase, an enzyme that digests DNA, prevents transformation from occurring. This supports the hypothesis that DNA is the genetic material.
3. The molecular weight of the transforming substance is large. This suggests the possibility of genetic variability.
4. The addition of enzymes that degrade proteins has no effect on the transforming substance nor does RNase, an enzyme that digests RNA. This shows that neither protein nor RNA is the genetic material.

These experiments showed that DNA is the transforming substance and, therefore, the genetic material. Although some scientists remained skeptical, many felt that the evidence for DNA being the genetic material was overwhelming.

An experiment by Alfred Hershey and Martha Chase in the early 1950s helped to firmly establish DNA as the genetic material. Hershey and Chase used a virus called a T phage, composed of radioactively labeled DNA and capsid coat proteins, to infect *E. coli* bacteria. They discovered that the radioactive tracers for DNA, but not protein, ended up inside the bacterial cells, causing them to become transformed. Since only the genetic material could have caused this transformation, Hershey and Chase determined that DNA must be the genetic material.

Animation
Hershey and Chase Experiment

Transformation of Organisms Today

Because the code for living things is based on the same four nucleic acid bases of A, G, C, and T, and genes are made from this code, it is conceivable to take genes from one organism and put them into another. Transformation of organisms, resulting in *genetically modified organisms* (GMOs), is an invaluable tool in modern biotechnology today. As discussed further in the next chapter, transformation of bacteria and other organisms has resulted in commercial products that are used every day.

Early biotechnologists seeking a dramatic way to show the possibility of gene transfer between different organisms took a jellyfish gene that codes for a green fluorescent protein (GFP) and started transforming different organisms with it. When this gene is transferred to another organism, the organism glows in the dark! (Fig. 12.2.) The basic technique is relatively simple. First, isolate the jellyfish gene and then transfer it to a bacterium, or the embryo of a plant, pig, or mouse. The result is a bioluminescent organism.

Figure 12.2 Transformation of organisms.   When bacteria, plants, pigs, and mice are genetically transformed with a gene from the jellyfish *Aequorea victoria* for green fluorescent protein (GFP), these organisms glow in the dark.

Because living organisms are coded by the same four bases, genes should theoretically have no difficulty crossing the species barrier. Mammalian genes can potentially be transferred to bacteria, and an invertebrate gene, such as the GFP gene, can be transferred to a bacterium, plant, or animal. Although it is possible and relatively easy to perform, cross-species gene transfer does not always result in the host organism producing the new protein because of the innate complexity of each species.

Animation
Bacterial
Transformation

The Structure of DNA

By the early 1950s, DNA was widely accepted as the genetic material of living things. However, the structure of DNA was not known. How can a molecule with only four different nucleotides produce the great diversity of life on Earth?

To understand the structure of DNA, we need to understand how the bases in DNA are composed. Investigators knew that DNA contains four different types of nucleotides: two with *purine* bases, **adenine (A)** and **guanine (G),** which have a double ring; and two with *pyrimidine* bases, **thymine (T)** and **cytosine (C),** which have a single ring (Fig. 12.3a, b). Erwin Chargaff used new chemical techniques developed in the 1940s to analyze in detail the base content of DNA.

A sample of Chargaff's data is seen in Figure 12.3c. You can see that while some species—*E. coli* and *Zea mays* (corn), for example—do have approximately 25% of each type of nucleotide, most do not. Further, the percentage of each type of nucleotide differs from species to species. Therefore, the nucleotide content of DNA is not fixed across species, and DNA does have the *variability* between species required for it to be the genetic material.

Within each species, however, DNA was found to have the *constancy* required of the genetic material—that is, all members of a species have the same base composition. Also, the percentage of A always equals the percentage of T, and the percentage of G equals the percentage of C. It follows that if the percentage of A + G equals 40%, then the percentage of T + C would equal 60%. (Do you see why?) These relationships are called Chargaff's rules.

Chargaff's rules:

1. The amount of A, T, G, and C in DNA varies from species to species.
2. In each species, the amount of A = T and the amount of G = C.

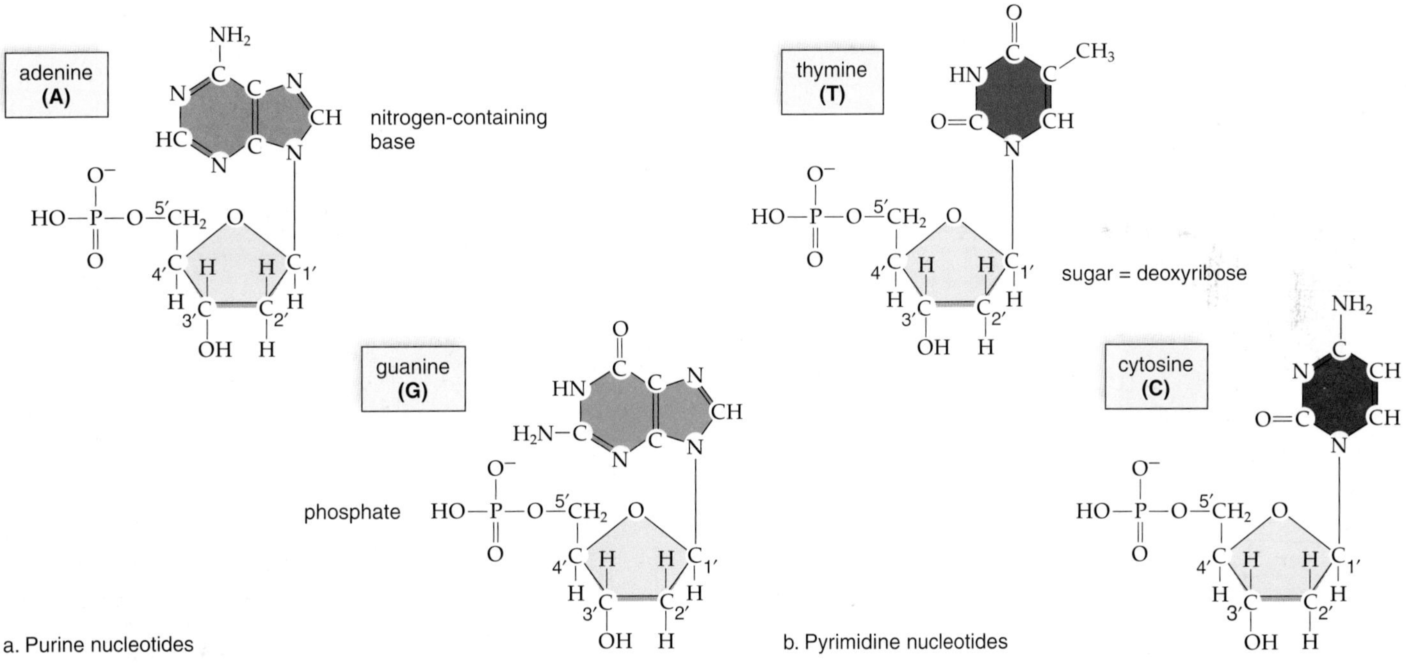

a. Purine nucleotides

b. Pyrimidine nucleotides

DNA Composition in Various Species (%)				
Species	**A**	**T**	**G**	**C**
Homo sapiens (human)	31.0	31.5	19.1	18.4
Drosophila melanogaster (fruit fly)	27.3	27.6	22.5	22.5
Zea mays (corn)	25.6	25.3	24.5	24.6
Neurospora crassa (fungus)	23.0	23.3	27.1	26.6
Escherichia coli (bacterium)	24.6	24.3	25.5	25.6
Bacillus subtilis (bacterium)	28.4	29.0	21.0	21.6

c. Chargaff's data

Figure 12.3 Nucleotide composition of DNA. All nucleotides contain phosphate, a 5-carbon sugar, and a nitrogen-containing base. In DNA, the sugar is called deoxyribose because it lacks an oxygen atom in the 2′ position, compared to ribose. The nitrogen-containing bases are **(a)** the purines adenine and guanine, which have a double ring, and **(b)** the pyrimidines thymine and cytosine, which have a single ring. **c.** Chargaff's data show that the DNA of various species differs. For example, in humans the A and T percentages are about 31%, but in fruit flies these percentages are about 27%. Note that by convention, the carbon atoms in the sugar rings are labeled with a number and a prime symbol to distinguish them from the carbon atoms in the base, which are labeled with a number only (numbers in bases not shown).

Although only one of four bases is possible at each nucleotide position in DNA, the sheer number of bases and the length of most DNA molecules is more than sufficient to provide for variability. For example, it has been calculated that each human chromosome typically contains about 140 million base pairs. This provides for a staggering number of possible sequences of nucleotides. Because any of the four possible nucleotides can be present at each nucleotide position, the total number of possible nucleotide sequences is $4^{(140 \times 10^6)}$ or $4^{140,000,000}$. No wonder each species has its own unique base percentages!

X-Ray Diffraction of DNA

Rosalind Franklin (Fig. 12.4a), a researcher at King's College in London, studied the structure of DNA using X-rays. She found that if a concentrated, viscous solution of DNA is made, it can be separated into fibers. Under the right conditions, the fibers are enough like a crystal (a solid substance whose atoms are arranged in a definite manner) that when X-rayed, an X-ray diffraction pattern results (Fig. 12.4b).

The X-ray diffraction pattern of DNA shows that DNA is a double helix. The helical shape is indicated by the crossed (X) pattern in the center of the photograph in Figure 12.4c. The dark portions at the top and bottom of the photograph indicate that some portion of the helix is repeated. Maurice H. F. Wilkins, a colleague of Franklin's, showed one of her crystallographic patterns to James Watson, who immediately grasped its significance.

Video
DNA Dark Lady

The Watson and Crick Model

James Watson, an American, was on a postdoctoral fellowship at Cavendish Laboratories in Cambridge, England, and while there he began to work with the biophysicist Francis H. C. Crick. Using the data provided from X-ray diffraction and other sources, they constructed a model of DNA for which they received a Nobel Prize in 1962.

Based on previous work of other scientists, Watson and Crick knew that DNA is a polymer of nucleotides, but they did not know how the nucleotides were arranged within the molecule. However, they deduced that DNA is a **double helix** with sugar-phosphate backbones on the outside and paired bases on the inside. This arrangement fits the mathematical measurements provided by Franklin's X-ray diffraction data for the spacing between the base pairs (0.34 nm) and for a complete turn of the double helix (3.4 nm).

According to Watson and Crick's model, the two DNA strands of the double helix are *antiparallel*, meaning the sugar-phosphate groups that are chained together to make each strand are oriented in opposite directions. As explained on page 217, each nucleotide possesses a phosphate group located at the 5′ position of the sugar. Nucleotides are joined together by linking the 5′ phosphate of one nucleotide to a free hydroxyl (–OH) located at the 3′ position on the sugar of the preceding nucleotide, giving the molecule directionality. Antiparallel simply means that while one DNA strand runs 5′ to 3′, the other strand runs in a parallel but opposite direction.

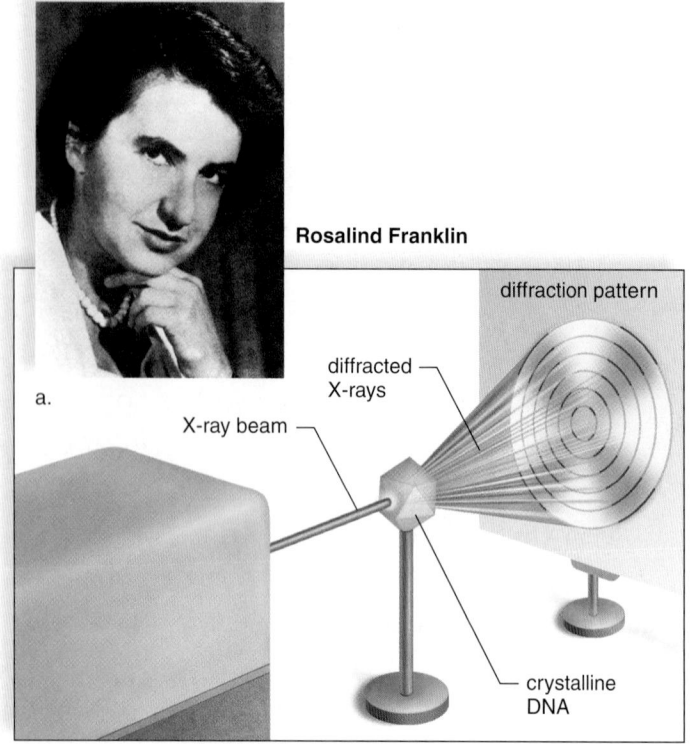

Rosalind Franklin

a.

diffraction pattern

diffracted X-rays

X-ray beam

crystalline DNA

b.

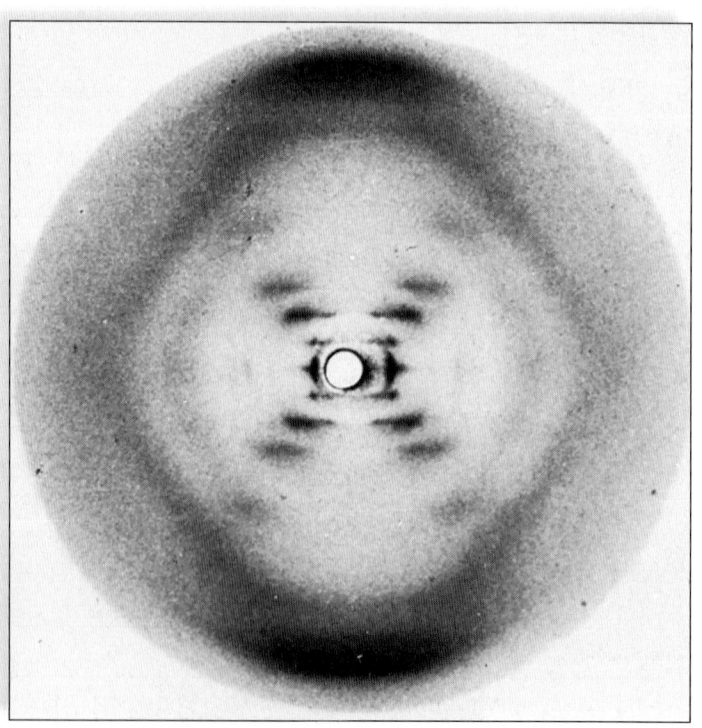

c.

Figure 12.4 X-ray diffraction of DNA.   **a.** Rosalind Franklin, 1920–1958. **b.** When a crystal is X-rayed, the way in which the beam is diffracted reflects the pattern of the molecules in the crystal. The closer together two repeating structures are in the crystal, the farther from the center the beam is diffracted. **c.** The diffraction pattern of DNA produced by Rosalind Franklin. The crossed (X) pattern in the center told investigators that DNA is a helix, and the dark portions at the top and the bottom told them that some feature is repeated over and over. Watson and Crick determined that this feature was the hydrogen-bonded bases.

This model also agreed with Chargaff's rules, which states that A = T and G = C. Figure 12.5 shows that A is hydrogen-bonded to T, and G is hydrogen-bonded to C. This so-called **complementary base pairing** means that a purine (large, two-ring base) is always bonded to a pyrimidine (smaller, one-ring base). This antiparallel pairing arrangement of the two strands ensures that the bases are oriented properly so that they can interact. The consistent spacing between the two strands of the DNA was detected by Franklin's X-ray diffraction pattern, because two pyrimidines together are too narrow, and two purines together are too wide (Fig. 12.5).

The information stored within DNA must always be read in the 5′ to 3′ direction. Thus, a DNA strand is usually replicated in a 5′ to 3′ direction.

Animation
DNA Structure

3D Animation
DNA Structure

Check Your Progress   12.1

1. Discuss how it is possible to take a gene from one organism and express it in another organism.
2. Explain the major features of DNA structure.

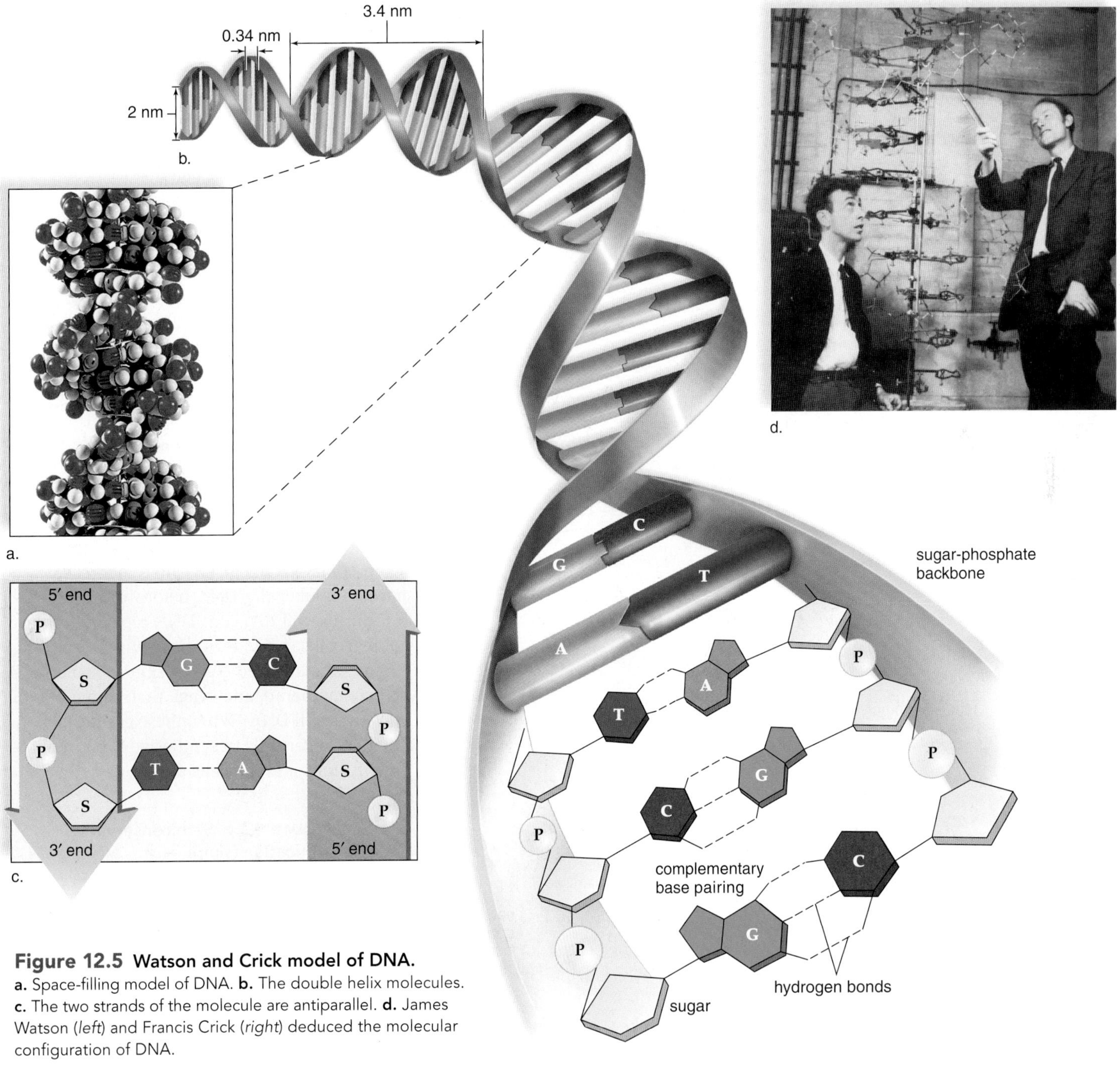

Figure 12.5 Watson and Crick model of DNA.
a. Space-filling model of DNA. **b.** The double helix molecules. **c.** The two strands of the molecule are antiparallel. **d.** James Watson (*left*) and Francis Crick (*right*) deduced the molecular configuration of DNA.

12.2 Replication of DNA

The term **DNA replication** refers to the process of copying a DNA molecule. Following replication, there is usually an exact copy of the parental DNA double helix. As soon as Watson and Crick developed their double-helix model, they commented, "It has not escaped our notice that the specific pairing we have postulated immediately suggests a possible copying mechanism for the genetic material."

A **template** is most often a mold used to produce a shape complementary to itself. During DNA replication, each DNA strand of the parental double helix serves as a template for a new strand in a daughter molecule (Fig. 12.6). DNA replication is termed **semiconservative replication** because each daughter DNA double helix contains an old strand from the parental DNA double helix and a new strand.

Animation
Meselson and
Stahl Experiment

Replication requires the following steps:

1. *Unwinding.* The old strands that make up the parental DNA molecule are unwound and "unzipped" (i.e., the weak hydrogen bonds between the paired bases are broken). A special enzyme called helicase unwinds the molecule.
2. *Complementary base pairing.* New free nucleotides, always present in the nucleus, are paired with nucleotides on the parental strands, A with T, and G with C.
3. *Joining.* The complementary nucleotides paired with the parental strands are connected to each other to form a connected chain. Each daughter DNA molecule now contains an old strand and a newly synthesized strand.

Steps 2 and 3 are carried out by an enzyme complex called **DNA polymerase.** DNA polymerase works in the test tube as well as in cells.

In Figure 12.6, the backbones of the parental DNA molecule are blue, and each base is given a particular color. Following replication, the daughter molecules each have a green backbone (new strand) and a blue backbone (old strand). Because A pairs with T, and G pairs with C, A daughter DNA double helix has the same sequence of bases as the parental DNA double helix had originally. Although we have described DNA replication simply here, it is actually a complicated process. Some of the more precise molecular events are discussed in the Biological Systems feature on page 221.

Animation
DNA Replication

You may recall from Chapters 10 and 11 that DNA must be copied before mitosis or meiosis can begin. Because the goal of these processes is to create either an exact cell copy (mitosis) or to make a gamete for reproduction (meiosis), in either case you have to double the DNA before you can separate it during cell division. DNA replication must occur before a cell can divide. Cancer, which is characterized by rapid, uncontrolled cell division, is sometimes treated with chemotherapeutic drugs that mimic one of the four nucleotides in DNA. When these are mistakenly used by the cancer cells to synthesize DNA, replication stops and the cells die off.

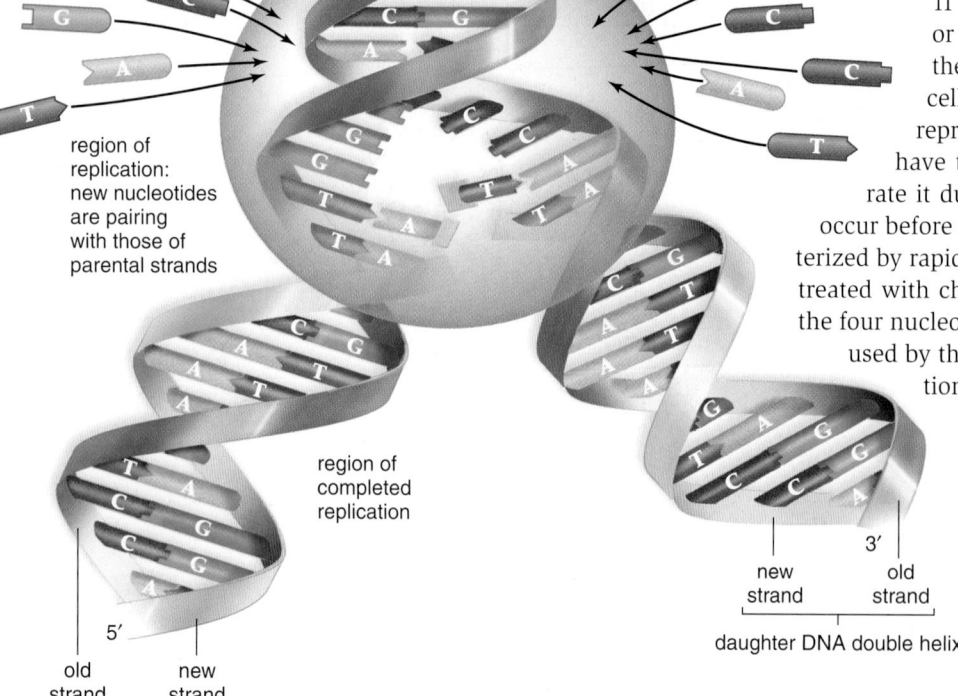

region of parental
DNA double helix

DNA
polymerase
enzyme

region of
replication:
new nucleotides
are pairing
with those of
parental strands

region of
completed
replication

old   new
strand   strand

daughter DNA double helix

new   old
strand   strand

daughter DNA double helix

Figure 12.6 Semiconservative replication (simplified). After the DNA double helix unwinds, each parental strand serves as a template for the formation of the new daughter strands. Complementary free nucleotides hydrogen bond to a matching base (e.g., A with T; G with C) in each parental strand, and are joined to form a complete daughter strand. Two helices, each with a daughter and parental strand, are produced following replication.

Biological Systems

Aspects of DNA Replication

DNA replication is an example of a complex, highly regulated biological system that requires many parts in order to function properly.

During replication, DNA polymerase needs a place to start joining new nucleotides together. In this case, it recognizes the –OH chemical group at the 3′ end of an existing nucleic acid chain, which can be DNA or RNA, and it begins synthesizing from there. During DNA replication, an RNA-producing enzyme makes a short primer that has the necessary 3′ –OH group on the end. DNA polymerase recognizes that target and begins DNA synthesis, allowing new nucleotides to form complementary base pairs with the old strand and connecting the new nucleotides together in a chain.

3 As a helicase enzyme unwinds DNA, it creates two replication forks that move away from each other. Each of the parental strands in a fork is accessible for complementary base pairing with new nucleotides and therefore synthesis of a new strand. (Binding proteins coat the newly formed, single-stranded regions and prevent them from reattaching to each other.)

3D Animation DNA Replication Fork

The parental strands are antiparallel to each other, and each of the new daughter strands must also be antiparallel to their matching parental strand—which creates a problem. **4** The new strand that gets made in the same direction as the fork is moving is called the leading strand. The other new strand in the fork must be synthesized in a direction opposite fork movement, which requires DNA polymerase to periodically start and stop. This strand is called the lagging strand. **5** Replication of this lagging strand is therefore made in segments called **6** Okazaki fragments, after the Japanese scientist Reiji Okazaki, who discovered them.

Replication is complete only when the RNA primers are removed. This works out well for the lagging strand. While checking to make sure bases are properly matched up (proofreading), DNA polymerase removes the RNA primers and replaces them with the proper DNA nucleotides. **7** Another enzyme, called DNA ligase, joins the fragments, creating a seamless DNA molecule.

3D Animation Discontinuous Synthesis

However, in eukaryotic organisms, which have linear chromosomes, there is no way for DNA polymerase to replicate all the way to the 5′ ends of both new strands after RNA primers are removed. This means that the DNA in each chromosome can get shorter for each cycle of replication. The ends of the DNA in eukaryotic chromosomes have a special nucleotide sequence called a telomere that is repeated a number of times. **Telomeres** do not code for proteins and use a repeat sequence such as TTAGGG.

Mammalian cells grown in a culture have a built-in lifespan; they can divide about 50 times before they stop. The loss of telomeres apparently signals the cell to stop dividing. Ordinarily, telomeres are maintained at their proper length by an enzyme called telomerase. Misfunctional telomerase can result in chromosome shortening, leading to premature aging and loss of cellular function. Thus, controlling telomerase activity is important to normal cell function. As you might expect, telomerase is often mistakenly turned on in cancer cells, which in effect enables them to divide indefinitely.

Questions to Consider

1. What are some potential mechanisms of regulating DNA replication?
2. How might we repair DNA when errors in replication are made?

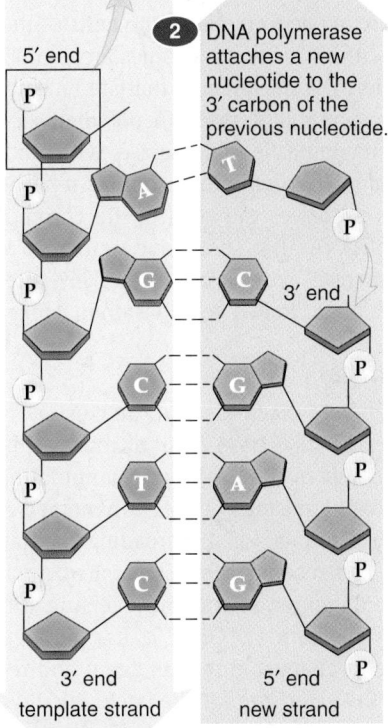

OH ← P is attached here
base is attached here

Deoxyribose molecule

2 DNA polymerase attaches a new nucleotide to the 3′ carbon of the previous nucleotide.

5′ end

3′ end

3′ end
template strand

5′ end
new strand

Direction of replication

3D Animation DNA Synthesis

5′
3′

template strand

5 lagging strand
5′

template strand
6 Okazaki fragment

3′

7 DNA ligase

Replication fork introduces complications

4 leading new strand
3′

RNA primer

DNA polymerase

3 helicase at replication fork

3′
5′

parental DNA helix

DNA polymerase

Figure 12A More detailed model of DNA replication.

Prokaryotic Versus Eukaryotic Replication

The process of DNA replication is distinctly different in prokaryotic and eukaryotic cells, although many of these organisms' basic functions are similar (Fig. 12.7).

Prokaryotic DNA Replication

Bacteria have a single circular loop chromosome whose DNA must be replicated before the cell divides. In some circular DNA molecules, replication moves around the DNA molecule in one direction only. In others, as shown in Figure 12.7a, replication occurs in two directions. The process always occurs in the 5′ to 3′ direction.

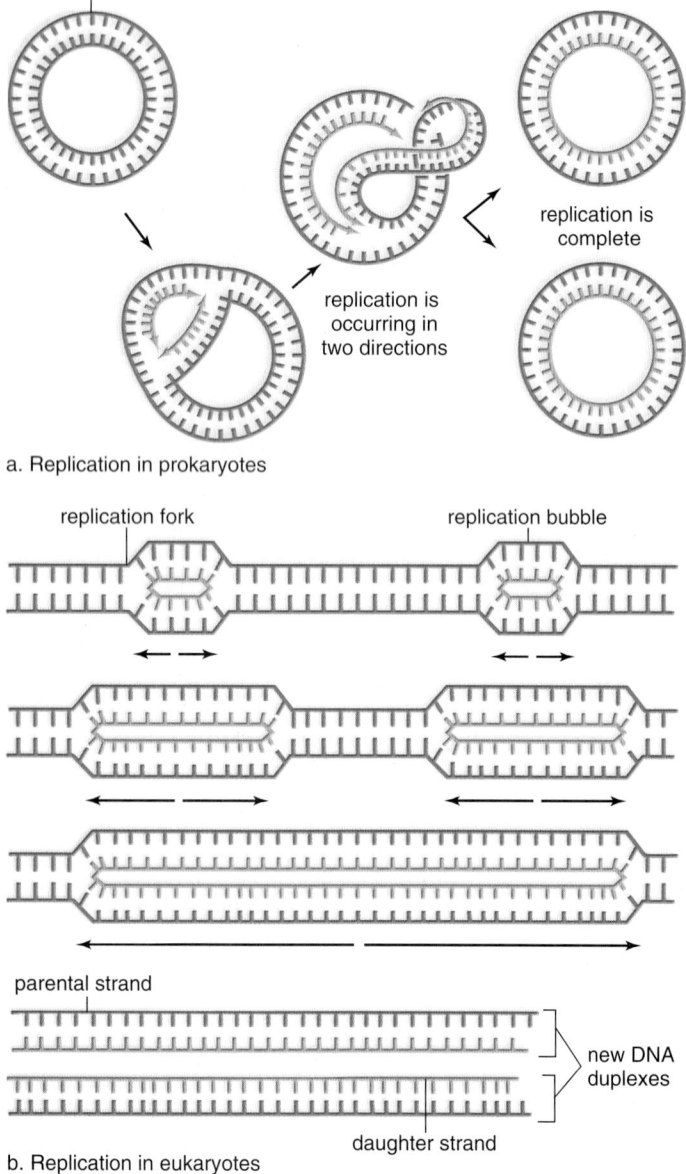

a. Replication in prokaryotes

replication fork   replication bubble

parental strand

new DNA duplexes

b. Replication in eukaryotes

Figure 12.7 Prokaryotic versus eukaryotic replication. **a.** In prokaryotes, replication can occur in two directions at once because the DNA molecule is circular. **b.** In eukaryotes, replication occurs at numerous replication bubbles, each with two forks. The forks move away from each other until they meet again and the two new daughter helices have been completed.

The process begins at the *origin of replication,* a specific site on the bacterial chromosome. The strands are separated and unwound, and a DNA polymerase enzyme binds to each side of the opening and begins the copying process. When the two DNA polymerases meet at a termination region, replication is halted, and the two copies of the chromosome are separated.

Bacterial cells require about 40 minutes to replicate the complete chromosome. Because bacterial cells are able to divide as often as once every 20 minutes, it is possible for a new round of DNA replication to begin even before the previous round is completed!

Eukaryotic DNA Replication

In eukaryotes, DNA replication begins at numerous origins of replication along the length of the linear chromosome, and the so-called replication bubbles spread bidirectionally until they meet. Notice in Figure 12.7b that there is a V shape wherever DNA is being replicated. This is called a **replication fork.**

The chromosomes of eukaryotes are long, making replication a more time-consuming process. Eukaryotes replicate their DNA at a slower rate—500 to 5,000 base pairs per minute—but there are many individual origins of replication to accelerate the process. Therefore, eukaryotic cells complete the replication of the diploid amount of DNA (in humans, over 6 billion base pairs) in a matter of hours!

The linear chromosomes of eukaryotes also pose another problem: DNA polymerase is unable to replicate the ends of the chromosomes. The ends of eukaryotic chromosomes are composed of telomeres, which are short DNA sequences that are repeated over and over. Telomeres are not copied by DNA polymerase; rather, they are added by an enzyme called telomerase, which adds the correct number of repeats after the chromosome is replicated. In stem cells, this process preserves the ends of the chromosomes and prevents the loss of DNA after successive rounds of replication. Unregulated telomerase activity can negatively affect cell function, as is seen with uncontrolled cell division in cancer cells.

Accuracy of Replication

A DNA polymerase is very accurate and makes a mistake approximately once per 100,000 base pairs at most. This error rate, however, would result in many errors accumulating over the course of several cell divisions. DNA polymerase is also capable of checking for accuracy, or proofreading the daughter strand it is making. It can recognize a mismatched nucleotide and remove it from a daughter strand by reversing direction and removing several nucleotides. Once it has removed the mismatched nucleotide, it changes direction again and resumes making DNA. Overall, the error rate for the bacterial DNA polymerase is only one in 100 million base pairs!

Check Your Progress   12.2

1. Explain the three major steps in DNA replication.
2. Compare DNA replication in prokaryotes and eukaryotes.
3. Examine how eukaryotic cells use telomerase to fully copy linear chromosomes.

12.3 The Genetic Code of Life

Learning Outcomes

Upon completion of this section, you should be able to

1. Explain the central dogma of molecular biology.
2. Determine the amino acid sequence specified by an mRNA sequence.

Table 12.1 RNA Structure Compared to DNA Structure

	RNA	DNA
Sugar	Ribose	Deoxyribose
Bases	Adenine, guanine, uracil, cytosine	Adenine, guanine, thymine, cytosine
Strands	Single stranded	Double stranded with base pairing
Helix	No	Yes

Evidence began to mount in the 1900s that metabolic disorders can be inherited. An English physician, Sir Archibald Garrod, called them "inborn errors of metabolism." Investigators George Beadle and Edward Tatum, working with red bread mold, proposed what they called the "one gene, one enzyme hypothesis," based on the observation that a defective gene caused a defective enzyme.

This and many other examples illustrate the flow of genetic information from DNA to RNA to protein to an observed trait. We now turn our attention to the transfer of information from DNA to RNA, the next component in the system.

RNA Carries the Information

Like DNA, *RNA (ribonucleic acid)* is a polymer composed of nucleotides. The nucleotides in RNA, however, contain the sugar ribose and the bases adenine (A), cytosine (C), guanine (G), and **uracil (U).** In RNA, the base uracil replaces the thymine found in DNA. Finally, RNA is single stranded and does not form a double helix in the same manner as DNA (Table 12.1 and Fig. 12.8).

There are three major classes of RNA. Each class of RNA has its own unique size, shape, and function in protein synthesis.

Messenger RNA (mRNA) takes a message from DNA in the nucleus to the ribosomes in the cytoplasm.
Transfer RNA (tRNA) transfers amino acids to the ribosomes.
Ribosomal RNA (rRNA), along with ribosomal proteins, makes up the ribosomes, where polypeptides are synthesized.

The Genetic Code

In the genetic flow of information, two major steps are needed to convert the information stored in DNA into a protein that supports body function (Fig. 12.9). First, the DNA undergoes **transcription** [L. *trans*, across, and *scriptio*, a writing], a process by which an RNA molecule is produced based on a DNA template. DNA is transcribed, or copied base by base, into mRNA, tRNA, and rRNA.

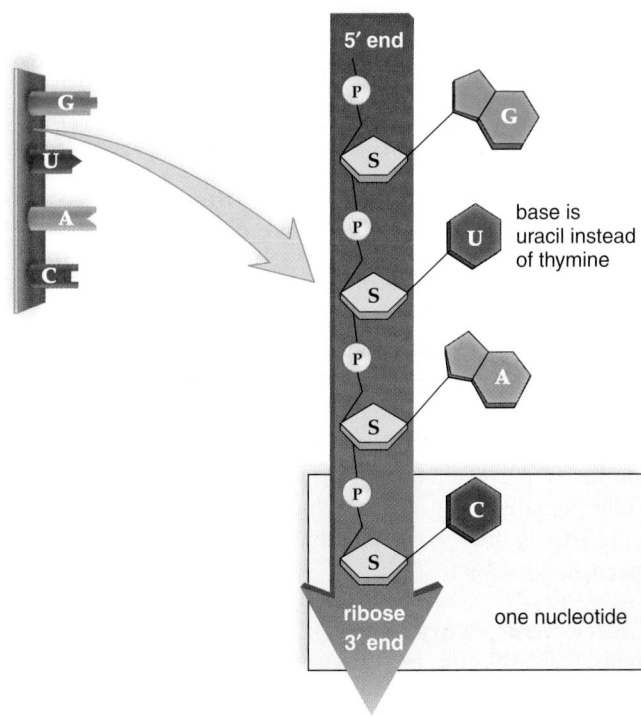

Figure 12.8 Structure of RNA. Like DNA, RNA is a polymer of nucleotides. RNA, however, is single stranded, the pentose sugar (S) is ribose, and uracil replaces thymine as one of the bases.

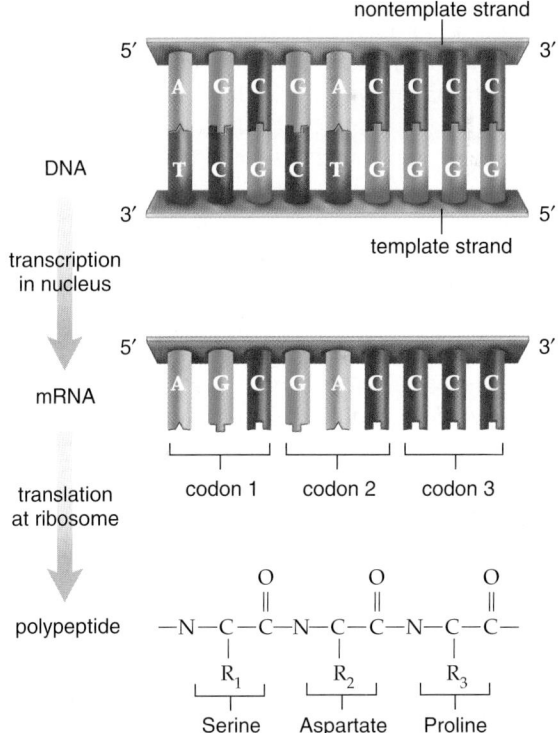

Figure 12.9 The central dogma of molecular biology. One strand of DNA acts as a template for mRNA synthesis, and the sequence of bases in mRNA determines the sequence of amino acids in a polypeptide.

Second, during **translation** [L. *trans*, across, and *latus*, carry or bear], the mRNA transcript is read by a ribosome and converted into the sequence of amino acids in a polypeptide. Like a translator who understands two languages, the cell changes a nucleotide sequence into an amino acid sequence. Together, the flow of information from DNA to RNA to protein to trait is known as the **central dogma** of molecular biology.

Now that we know that the DNA sequence within a gene is transcribed into an RNA molecule, and, for genes that code for proteins, the mRNA sequence determines the sequence of amino acids in a protein, it becomes necessary to identify the specific **genetic code** for each of the 20 amino acids found in proteins. Although scientists knew that DNA somehow directed protein production, they did not initially know specifically how the code was translated. This discovery was made in the 1960s.

Finding the Genetic Code

Logically, the genetic code would have to be at least a **triplet code;** that is, each coding unit, or **codon,** would need to be made up of three nucleotides. The reason is that fewer nucleotides would not provide sufficient variety to encode 20 different amino acids.

In 1961, Marshall Nirenberg and J. Heinrich Matthei performed an experiment that laid the groundwork for cracking the genetic code. First, they found that a cellular enzyme could be used to construct a synthetic RNA (one that does not occur in cells), and then they found that the synthetic RNA polymer could be translated in a test tube that contains the cytoplasmic contents of a cell. Their first synthetic RNA was composed only of uracil, and the protein that resulted was composed only of the amino acid phenylalanine. Therefore, the mRNA codon for phenylalanine was known to be UUU. Later, they were able to translate just three nucleotides at a time; in that way, it was possible to assign an amino acid to each of the mRNA codons (Fig. 12.10).

Like the periodic table and other major works, the genetic code seen in Figure 12.10 is a masterpiece of scientific discovery because it is a key that unlocks the very basis of biological life. Here are some of its features:

1. The genetic code is *degenerate*. This term means that most amino acids have more than one codon; leucine, serine, and arginine have six different codons, for example. The degeneracy of the code helps protect against potentially harmful mutations.
2. The genetic code is *unambiguous*. Each triplet codon has only one meaning.
3. The code has *start and stop signals*. There is only one start signal, but there are three stop signals.

The Code Is Universal

With a few exceptions, the genetic code (Fig. 12.10) is universal to all living things. In 1979, however, researchers discovered that the genetic code used by mammalian mitochondria and chloroplasts differs slightly from the more familiar genetic code.

First Base	Second Base				Third Base
	U	**C**	**A**	**G**	
U	UUU phenylalanine	UCU serine	UAU tyrosine	UGU cysteine	U
	UUC phenylalanine	UCC serine	UAC tyrosine	UGC cysteine	C
	UUA leucine	UCA serine	UAA *stop*	UGA *stop*	A
	UUG leucine	UCG serine	UAG *stop*	UGG tryptophan	G
C	CUU leucine	CCU proline	CAU histidine	CGU arginine	U
	CUC leucine	CCC proline	CAC histidine	CGC arginine	C
	CUA leucine	CCA proline	CAA glutamine	CGA arginine	A
	CUG leucine	CCG proline	CAG glutamine	CGG arginine	G
A	AUU isoleucine	ACU threonine	AAU asparagine	AGU serine	U
	AUC isoleucine	ACC threonine	AAC asparagine	AGC serine	C
	AUA isoleucine	ACA threonine	AAA lysine	AGA arginine	A
	AUG *(start)* methionine	ACG threonine	AAG lysine	AGG arginine	G
G	GUU valine	GCU alanine	GAU aspartate	GGU glycine	U
	GUC valine	GCC alanine	GAC aspartate	GGC glycine	C
	GUA valine	GCA alanine	GAA glutamate	GGA glycine	A
	GUG valine	GCG alanine	GAG glutamate	GGG glycine	G

Figure 12.10 Messenger RNA codons. Notice that in this chart, each of the codons (in boxes) is composed of three letters representing the first base, second base, and third base. For example, find the box where C for the first base and A for the second base intersect. You will see that U, C, A, or G can be the third base. The bases CAU and CAC are codons for histidine; the bases CAA and CAG are codons for glutamine.

The universal nature of the genetic code provides strong evidence that all living things share a common evolutionary heritage. Because the same genetic code is used by all living things, it is possible to transfer genes from one organism to another. Many commercial and medicinal products, such as human insulin, can be produced in this manner. Earlier, we showed that the gene for GFP could be transferred from jellyfish to a number of other organisms to cause a fluorescent green color (see Fig. 12.2). This is only made possible because the genetic code is universal.

Check Your Progress   12.3

1. Examine the flow of genetic information in a cell.
2. Describe the three major classes of RNA; what is the function of each class?
3. Explain why the genetic code is said to be degenerate.

12.4 First Step: Transcription

Learning Outcomes

Upon completion of this section, you should be able to

1. Distinguish the events of transcription that occur during formation of an mRNA molecule.
2. Describe how eukaryotic mRNA molecules are processed and exported to the cytoplasm.

During *transcription*, a segment of the DNA serves as a template for the production of an RNA molecule. Although mRNA, tRNA, and rRNA are all produced by transcription, we focus here on transcription to make mRNA, the type of RNA that eventually leads to building a protein.

MP3
Protein Synthesis

Messenger RNA Is Produced

The sequences of bases in a gene are transcribed into an mRNA molecule based on complementary base pairing: the T base in the DNA pairs with A in the mRNA, G with C, and A with U (note that uracil replaces T in the newly formed mRNA) (Fig. 12.11). When a gene is transcribed, a segment of the DNA helix unwinds and unzips, and complementary RNA nucleotides pair with DNA nucleotides of the strand opposite the gene. This strand is known as the *template strand*; the other strand is the gene strand. An **RNA polymerase** joins the nucleotides together in the 5′ ⟶ 3′ direction. Like DNA polymerase, an RNA polymerase adds a nucleotide only to the 3′ end of the polymer under construction.

Transcription begins when RNA polymerase attaches to a region of DNA called a promoter (see Fig. 12.11). A **promoter** defines the start of transcription, the direction of transcription,

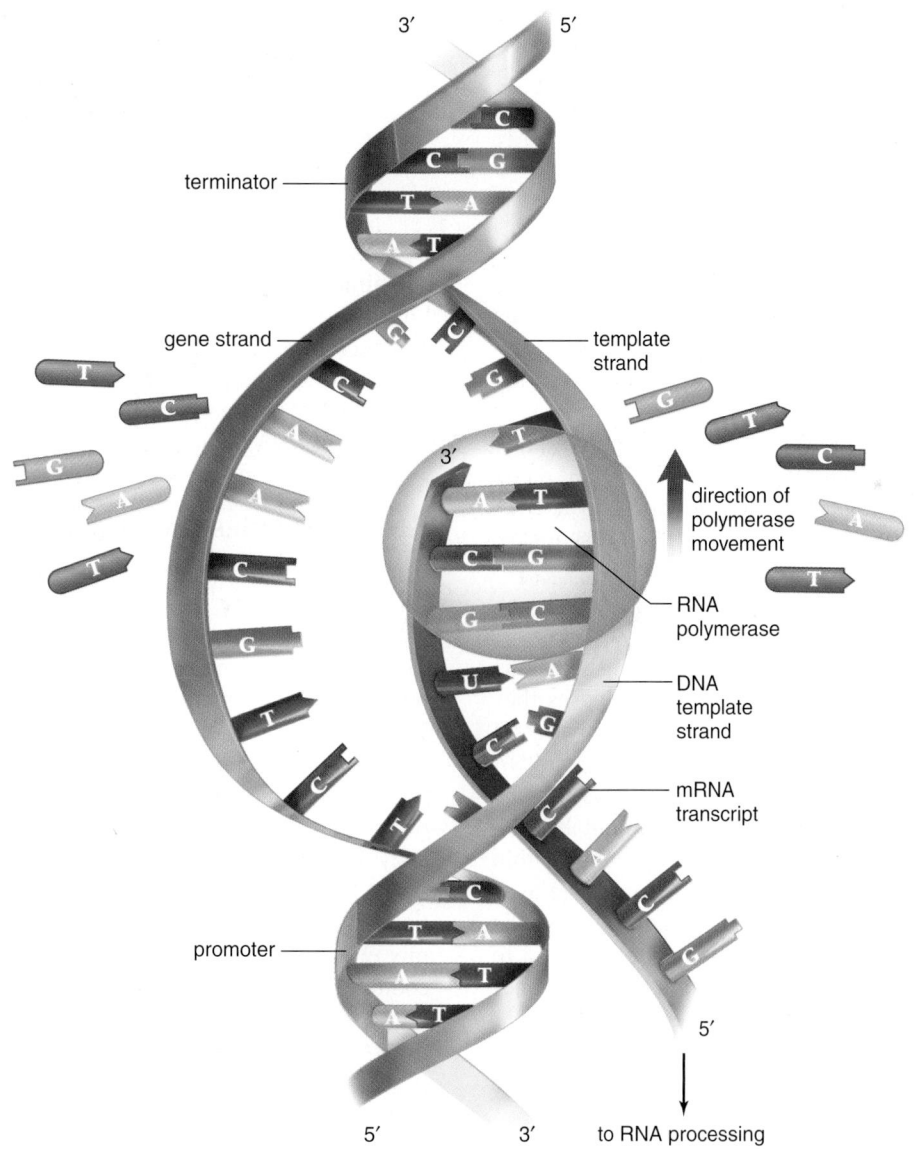

Figure 12.11 Transcription.   During transcription, complementary RNA is made from a DNA template. At the point of attachment of RNA polymerase, the DNA helix unwinds and unzips, and complementary RNA nucleotides are joined together. After RNA polymerase has passed by, the DNA strands rejoin and the mRNA transcript dangles to the side.

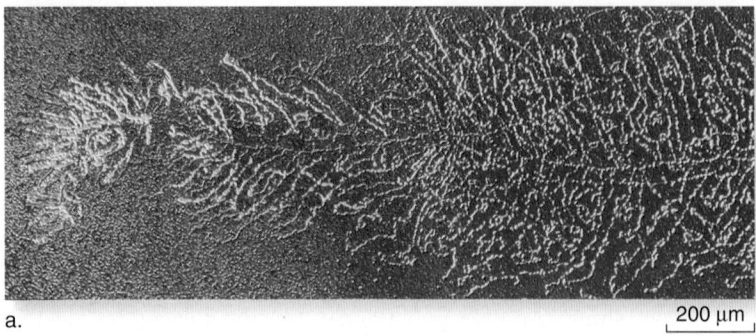

a.

200 μm

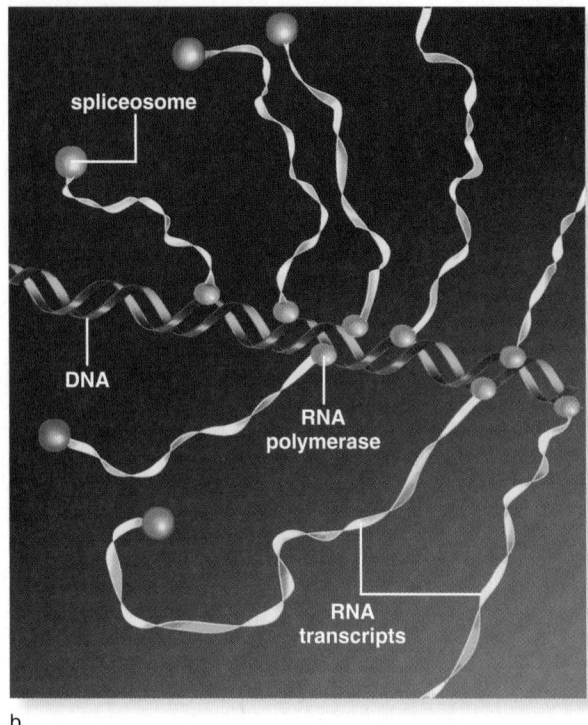

b.

Figure 12.12 RNA polymerase. **a.** Numerous RNA transcripts extend from a horizontally oriented gene in an amphibian egg cell. **b.** The strands get progressively longer because transcription begins to the left. The dots along the DNA are RNA polymerase molecules. The dots at the end of the strands are spliceosomes involved in RNA processing (see Fig. 12.13).

and the strand to be transcribed. The binding of RNA polymerase to the promoter is the *initiation* of transcription. The RNA-DNA association is not as stable as the two strands in the DNA helix. Therefore, only the newest portion of an RNA molecule that is associated with RNA polymerase is bound to the DNA, and the rest dangles off to the side.

Elongation of the mRNA molecule occurs as the RNA polymerase reads down the DNA template strand in a 5′ to 3′ direction, and continues until RNA polymerase comes to a DNA stop sequence, where *termination* occurs. The stop sequence causes RNA polymerase to stop transcribing the DNA and to release the mRNA molecule, now called an **mRNA transcript.**

Animation Stages of Transcription

It is not necessary for RNA polymerase to finish making one mRNA transcript before it starts another. As long as they have access to the gene's promoter, many RNA polymerase molecules can be working one after the other to produce mRNA transcripts at the same time (Fig. 12.12). This allows the cell to produce many thousands of copies of the same mRNA molecule, and eventually many copies of the same protein, within a shorter period of time than if a single mRNA copy were used to direct protein synthesis. This ability to rapidly express the gene enables the cell (and the organism) to better respond to changing environmental conditions and have a greater chance at survival.

Note that, for a given gene, either strand of the DNA can be a template strand. The example above uses one strand as the

template, but for another gene, the opposite strand may be the template. Assuming both genes are on the same chromosome, and therefore the same piece of DNA, can you think what the orientation of template strand in the second gene might be? (Hint: Consider the directionality of RNA polymerase.)

3D Animation Transcription

MP3 Transcription

RNA Molecules Undergo Processing

A newly formed RNA transcript, called a pre-mRNA, is modified before leaving the eukaryotic nucleus. For example, the molecule receives a cap at the 5′ end and a tail at the 3′ end (Fig. 12.13). The *cap* is a modified guanine (G) nucleotide that helps tell a ribosome where to attach when translation begins. The tail consists of a chain of 150–200 adenine (A) nucleotides. This *poly-A tail* facilitates the transport of mRNA out of the nucleus, helps initiate loading of ribosomes and the start of translation, and also delays degradation of mRNA by hydrolytic enzymes.

When the mRNA is first made by RNA polymerase from the gene, it is in a rough form. Called pre-mRNA, it contains a mix of **exons** (protein-coding regions) and **introns** (non-protein coding regions), particularly in multicellular eukaryotes. Because only the exons of the pre-mRNA will be contained in the mature mRNA, the introns, which occur in between the exons, must be spliced out.

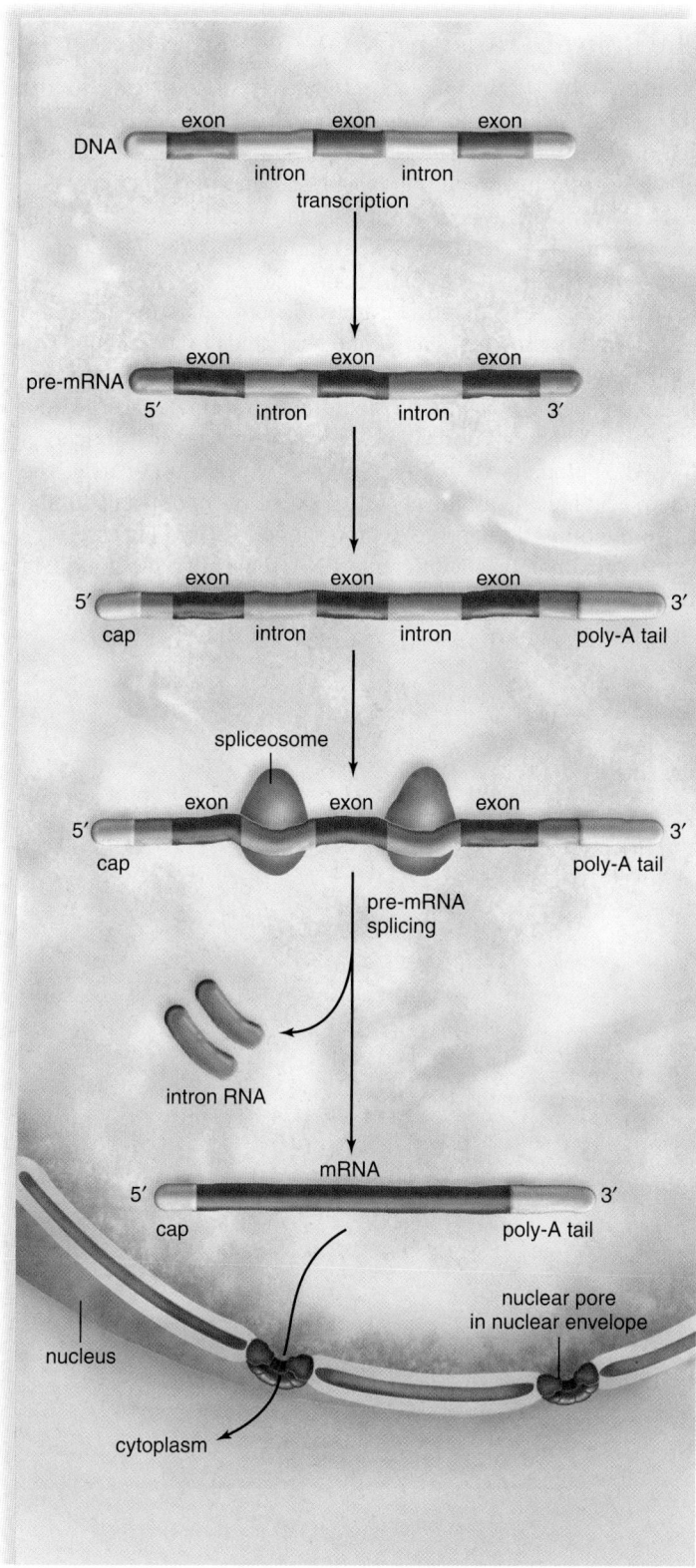

Figure 12.13 Messenger RNA (mRNA) processing in eukaryotes. DNA contains both exons (protein-coding sequences) and introns (non-protein-coding sequences). Both of these are transcribed and are present in pre-mRNA. During processing, a cap and a poly-A tail (a series of adenine nucleotides) are added to the molecule. Also, introns get cut out and the exons get spliced together by complexes called spliceosomes. Once processing is complete the mRNA molecule is ready to leave the nucleus.

In lower eukaryotes, introns are removed by "self-splicing"—that is, the intron itself has the capability of enzymatically splicing itself out of a pre-mRNA. In higher eukaryotes, the RNA splicing is done by spliceosomes, which contain *small nuclear RNAs (snRNAs)*. By means of complementary base pairing, snRNAs are capable of identifying the introns to be removed. A spliceosome utilizes a **ribozyme** (enzyme made of RNA rather than just protein) to cut and remove the introns. Following splicing of the exons together and the addition of the 5′ cap and 3′ poly-A tail, an mRNA is ready to leave the nucleus and be translated into a protein.  **Animation** How Spliceosomes Process RNA

Function of Introns

For many years, scientists thought that introns were simply wasted space within genes. Now, we realize they serve several key functions in the cell. The presence of introns allows a cell to pick and choose which exons will go into a particular mRNA (see Chapter 13). Just because an mRNA has all the exons in its pre-mRNA doesn't mean they will all make it to the final product. For example, if a gene has 3 exons, then depending on cell need and environmental conditions, it may produce an mRNA with exons 1 and 2 only, or 1 and 3 only, or 1, 2, and 3. This ability is called *alternative mRNA splicing*, and it increases the flexibility and efficiency of the cell. The snRNAs of the spliceosomes that excise the introns play an important role in alternative splicing in eukaryotes.

Some introns give rise to *microRNAs (miRNAs)*, which are small molecules involved in regulating the translation of mRNAs. These molecules bind with the mRNA through complementary base pairing and, in that way, prevent translation from occurring.  **Video** Drug Discovery / **Video** New Cancer Clue

It is also possible that the presence of introns encourages crossing-over during meiosis, and this permits a phenomenon termed *exon shuffling*, which can play a role in the evolution of new genes.  **Animation** Exon Shuffling / **3D Animation** mRNA Modifications

Check Your Progress   12.4

1. Explain the orientation of the DNA template strand relative to both the DNA gene strand and the new mRNA molecule.
2. Describe the three major modifications that occur during the processing of an mRNA.
3. Distinguish between the introns and exons of a gene.
4. Predict the sequence of nucleotides given a strand of DNA.
5. Explain the potential evolutionary benefits of alternative mRNA splicing.

12.5 Second Step: Translation

Learning Outcomes

Upon completion of this section, you should be able to

1. Describe the roles of mRNA, tRNA, and rRNA in translating the genetic code.
2. Examine the stages of translation and the events that occur during each stage.

Translation, which takes place in the cytoplasm of eukaryotic cells, is the second step needed to express a gene into a protein. During translation, the sequence of codons (nucleotide triplets) in the mRNA is read by a ribosome, which connects the sequence of amino acids dictated by the mRNA into a polypeptide. The process is called translation because it requires the conversion of information from a nucleic acid language (DNA and RNA) into an amino acid language (protein).

The Role of Transfer RNA

Transfer RNA (tRNA) molecules transfer amino acids to the ribosomes. A tRNA molecule is a single-stranded nucleic acid that doubles back on itself to create regions where complementary bases are hydrogen-bonded to one another. The structure of a tRNA molecule is generally drawn as a flat cloverleaf (Fig 12.14a), but a space-filling model shows the molecule's actual three-dimensional shape (Fig. 12.14b).

There is at least one tRNA molecule for each of the 20 amino acids found in proteins. The amino acid binds to the 3′ end. The opposite end of the molecule contains an **anticodon,** a group of three bases that is complementary and antiparallel to a specific mRNA codon. For example, a tRNA that has the anticodon 5′ AAG 3′ binds to the mRNA codon 5′ CUU 3′ and carries the amino acid leucine. In the genetic code, 61 codons specify amino acids; the other three serve as stop sequences (see Fig. 12.10).

Approximately 40 different tRNA molecules are found in most cells. There are fewer tRNAs than codons because some

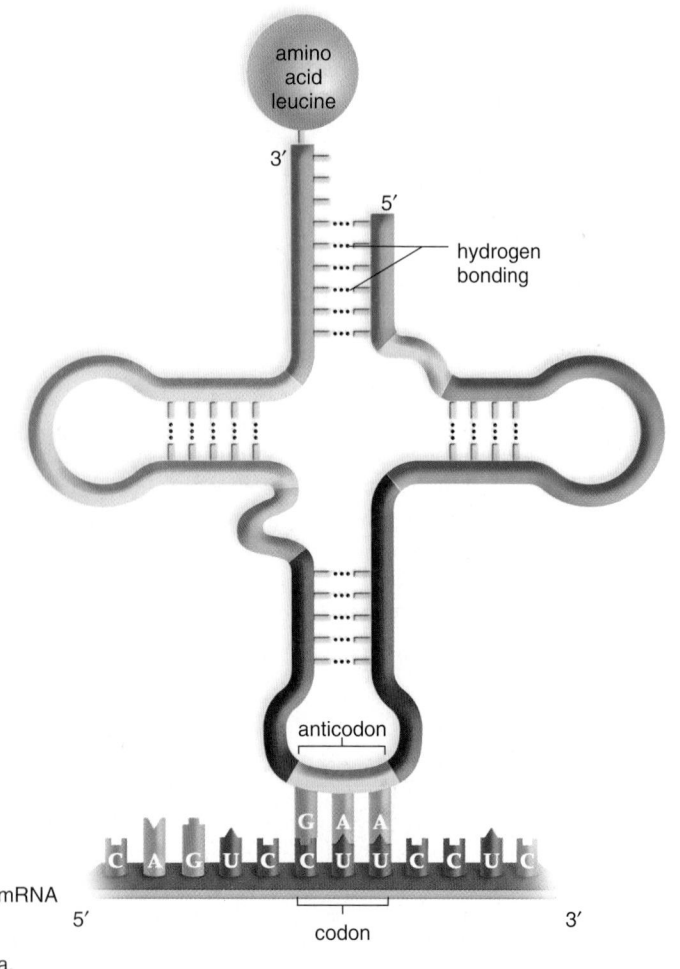

a.

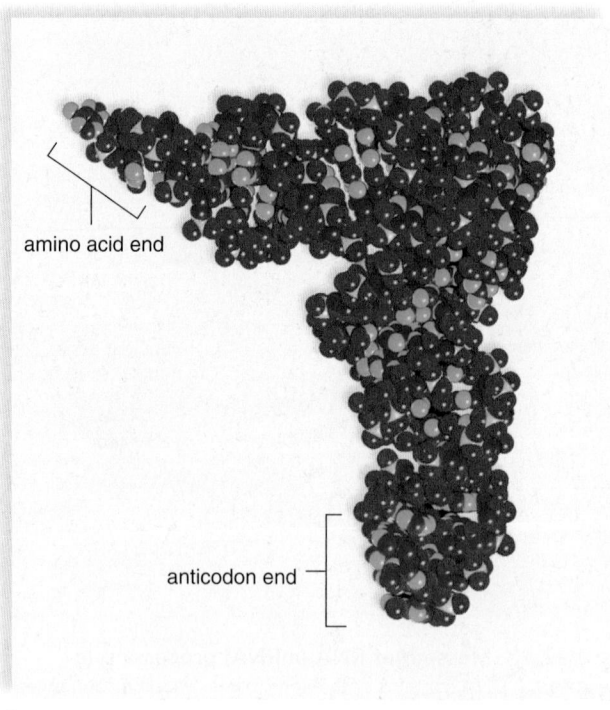

b.

Figure 12.14 Structure of a transfer RNA (tRNA) molecule.   a. Complementary base pairing indicated by hydrogen bonding occurs between nucleotides within the molecule, and this causes it to form its characteristic loops. The anticodon that base-pairs with a particular messenger RNA (mRNA) codon occurs at one end of the folded molecule; the other two loops help hold the molecule at the ribosome. An appropriate amino acid is attached at the 3′ end of the molecule in the cytoplasm by a tRNA charging enzyme. For this mRNA codon and tRNA anticodon, the specific amino acid is leucine. **b.** Space-filling model of tRNA molecule.

tRNAs can pair with more than one codon. In 1966, Francis Crick observed this phenomenon and called it the **wobble hypothesis.** He stated that the first two positions in a tRNA anticodon pair obey the A–U/G–C configuration rule. However, the third position can be variable. Some tRNA molecules can recognize as many as four separate codons differing only in the third nucleotide. The wobble effect helps ensure that despite changes in DNA base sequences, the resulting sequence of amino acids will produce a correct protein. This is one of the reasons the genetic code is said to be degenerate.

How does the correct amino acid become attached to the correct tRNA molecule? This task is carried out by amino acid–charging enzymes, generically called aminoacyl-tRNA synthetases. Just as a key fits a lock, each enzyme has a recognition site for a particular amino acid to be joined to a specific tRNA. For example, leucine-tRNA synthetase attaches the leucine amino acid to a tRNA with the correct anticodon. This is an energy-requiring process that uses ATP. A tRNA with its amino acid attached is termed a *charged tRNA*. Once the amino acid–tRNA

complex is formed, it is added to the large pool of charged tRNAs that exist in the cytoplasm, where it can now be accessed by a ribosome during protein synthesis.

The Role of Ribosomal RNA

As with so many cellular structures, the structure of a ribosome is essential to its function.

Structure of a Ribosome

In eukaryotes, ribosomal RNA (rRNA) is produced from a DNA template in the nucleolus of a nucleus. The rRNA is packaged with a variety of proteins into two ribosomal subunits, one of which is larger than the other. The subunits then move separately through nuclear envelope pores into the cytoplasm, where they join together at the start of translation (Fig. 12.15a). Once translation begins, ribosomes can remain in the cytoplasm, or they can become attached to endoplasmic reticulum.

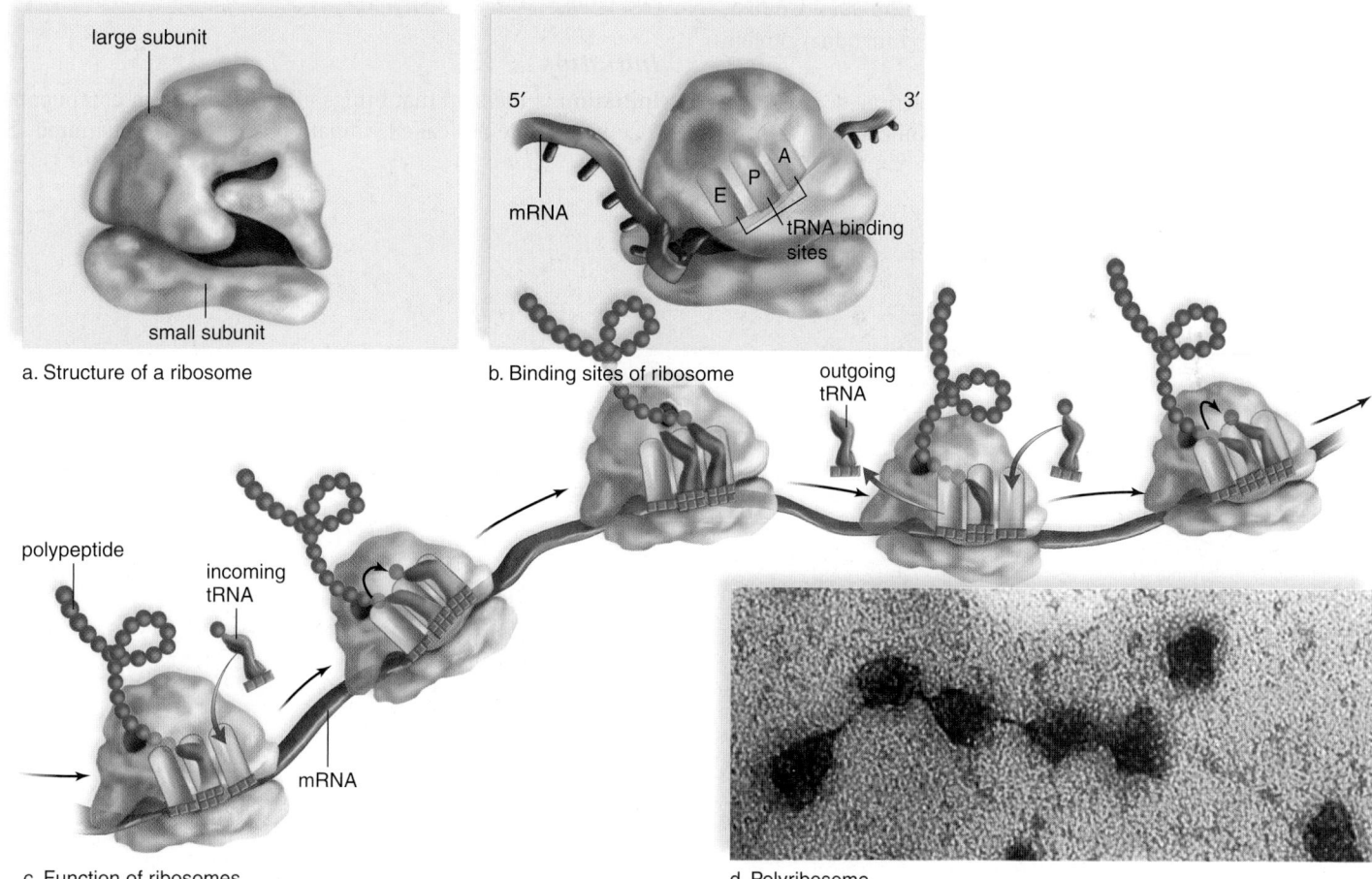

a. Structure of a ribosome

b. Binding sites of ribosome

c. Function of ribosomes

d. Polyribosome

Figure 12.15 Ribosome structure and function.   **a.** Side view of a ribosome shows a small subunit and a large subunit. **b.** Frontal view of a ribosome shows its binding sites. mRNA is bound to the small subunit, and the large subunit has three binding sites for tRNAs. **c.** Overview of protein synthesis. The tRNA bearing the growing polypeptide passes the entire chain to the new amino acid carried by the tRNA occupying the A site. The ribosome shifts, and freed of its burden, the "empty" tRNA exits. The new peptide-bearing tRNA moves over one binding site, making the A site accessible once again to a new tRNA. This cycle is repeated until the ribosome reaches the termination codon. **d.** Electron micrograph of a polyribosome, a number of ribosomes all translating the same mRNA molecule.

Function of a Ribosome

Both prokaryotic and eukaryotic cells contain thousands of ribosomes per cell because they play such a significant role in protein synthesis. Ribosomes have a binding site for mRNA and three binding sites for transfer RNA (tRNA) molecules (Fig. 12.15b). The tRNA binding sites facilitate complementary base pairing between tRNA anticodons and mRNA codons. The large ribosomal subunit has enzyme activity from rRNA (i.e., a ribozyme) that creates the peptide bond between adjacent amino acids. This peptide bond is created many times to produce a polypeptide, which in turn folds into its three-dimensional shape and becomes a protein.

When a ribosome moves down an mRNA molecule, the polypeptide increases by one amino acid at a time (Fig. 12.15c). Translation terminates at a stop codon. Once translation is complete, the polypeptide dissociates from the translation complex and folds into its normal shape. Recall from Chapter 3 that a polypeptide twists and bends into a definite shape based on the makeup of its amino acids. This folding process begins as soon as the polypeptide emerges from a ribosome. Chaperone molecules that are often present in the cytoplasm and the ER ensure protein folding proceeds as it should. For proteins that contain more than one polypeptide, each subunit is folded first, and then subunits join together into a final, functional protein complex.

Like RNA polymerase during transcription, multiple ribosomes often attach and translate the same mRNA at one time. As soon as the initial portion of mRNA has been translated by one ribosome, and the ribosome has begun to move down the mRNA, another ribosome can attach to the mRNA. The entire complex of mRNA and multiple ribosomes is called a **polyribosome** (Fig. 12.15d) and it greatly increases the efficiency of translation.

Translation Requires Three Steps

During translation, the codons of an mRNA base-pair with the anticodons of tRNA molecules carrying specific amino acids. The order of the codons determines the order of the tRNA molecules at a ribosome and the corresponding sequence of amino acids in a polypeptide. The process of translation must be extremely orderly so that the amino acids of a polypeptide are sequenced correctly. Even a single amino acid change has the potential to dramatically affect a protein's function, as is the case with individuals who carry the alleles for sickle cell anemia.

Animation
How Translation Works

Protein synthesis involves three steps: initiation, elongation, and termination. Enzymes are required for each of the three steps to function properly. The first two steps, initiation and elongation, require energy.

Initiation

Initiation is the step that brings all the translation components together. Proteins called initiation factors are required to

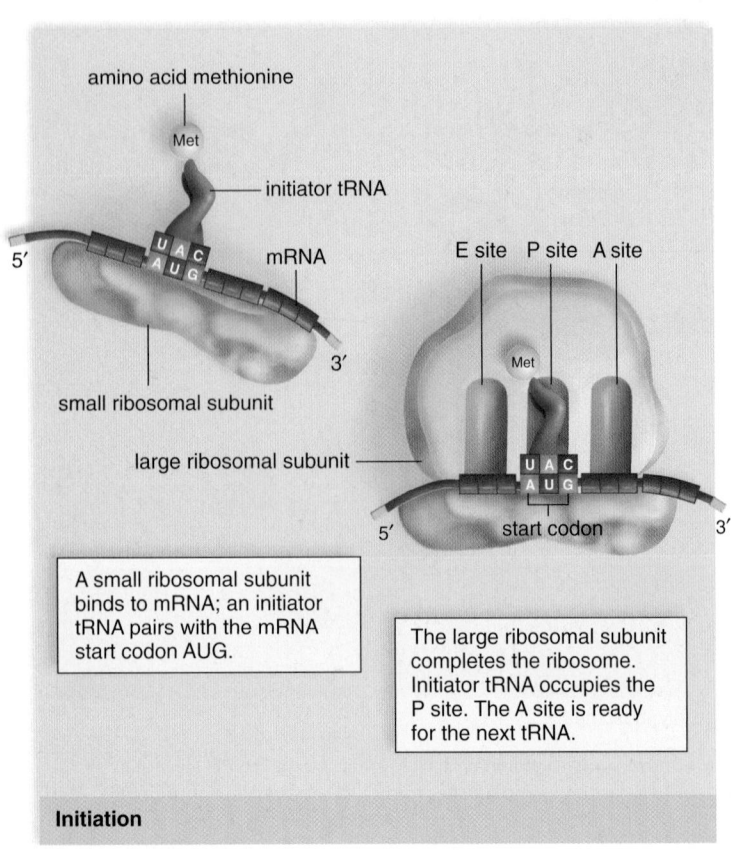

Initiation

Figure 12.16 Initiation.   In prokaryotes, participants in the translation process assemble as shown. The first amino acid is typically a special form of methionine.

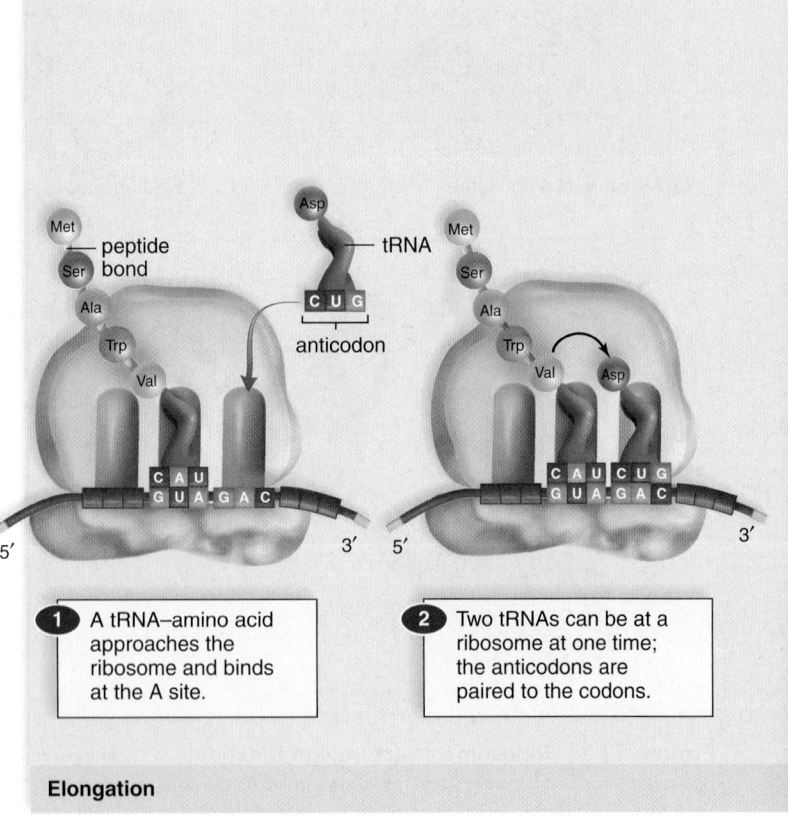

Elongation

Figure 12.17 Elongation.   Note that a polypeptide is already at the P site. During elongation, polypeptide synthesis occurs as amino acids are added one at a time to the growing chain.

assemble the small ribosomal subunit, mRNA, initiator tRNA, and the large ribosomal subunit for the start of protein synthesis.

Initiation is shown in Figure 12.16. In prokaryotes, a small ribosomal subunit attaches to the mRNA in the vicinity of the *start codon* (AUG). The first or initiator tRNA pairs with this codon. Then, a large ribosomal subunit joins to the small subunit (Fig. 12.16). Although similar in many ways, initiation in eukaryotes is much more complex.

As already discussed, a ribosome has three binding sites for tRNAs. One of these is called the E (for exit) site, second is the P (for peptide) site, and the third is the A (for amino acid) site. The initiator tRNA binds to the P site, even though it carries only the amino acid methionine (see Fig. 12.10). The A site is where tRNA carrying the next amino acid enter the ribosome, and the E site is for any tRNAs that are leaving a ribosome. Following initiation, translation continues with elongation and then termination.

Elongation

Elongation is the stage during protein synthesis when a polypeptide increases in length one amino acid at a time. In addition to the necessary tRNAs, elongation requires elongation factors, which facilitate the binding of tRNA anticodons to mRNA codons within a ribosome.

❶ Elongation is shown in Figure 12.17, where a tRNA with an attached peptide is already at the P site, and a tRNA carrying its appropriate amino acid is just arriving at the A site. **❷** Once a ribosome has verified that the incoming tRNA matches the codon and is firmly in place at the A site, the entire growing peptide will be transferred to the amino acid on the tRNA in the A site. A ribozyme, an rRNA-based enzyme which is a part of the large ribosomal subunit, uses energy to transfer the growing peptide and create a new peptide bond. **❸** Following peptide bond formation, the peptide is one amino acid longer than it was before. **❹** Next, **translocation** occurs: The ribosome moves forward, and the peptide-bearing tRNA is now in the P site of the ribosome. The spent tRNA, now at the E site, exits the ribosome. A new codon is now exposed at the A site and is ready to receive another tRNA.

The complete cycle—complementary base pairing of new tRNA, transfer of peptide chain, and translocation—is repeated at a rapid rate (about 15 times each second in the bacterium *Escherichia coli*).

Eventually, the ribosome reaches a stop codon, and termination occurs, during which the polypeptide is released.

Termination

Termination is the final step in protein synthesis. During termination, as shown in Figure 12.18, the polypeptide and the assembled components that carried out protein synthesis are separated from one another.

Termination of polypeptide synthesis occurs at a *stop codon*—that is, a codon that does not code for an amino acid.

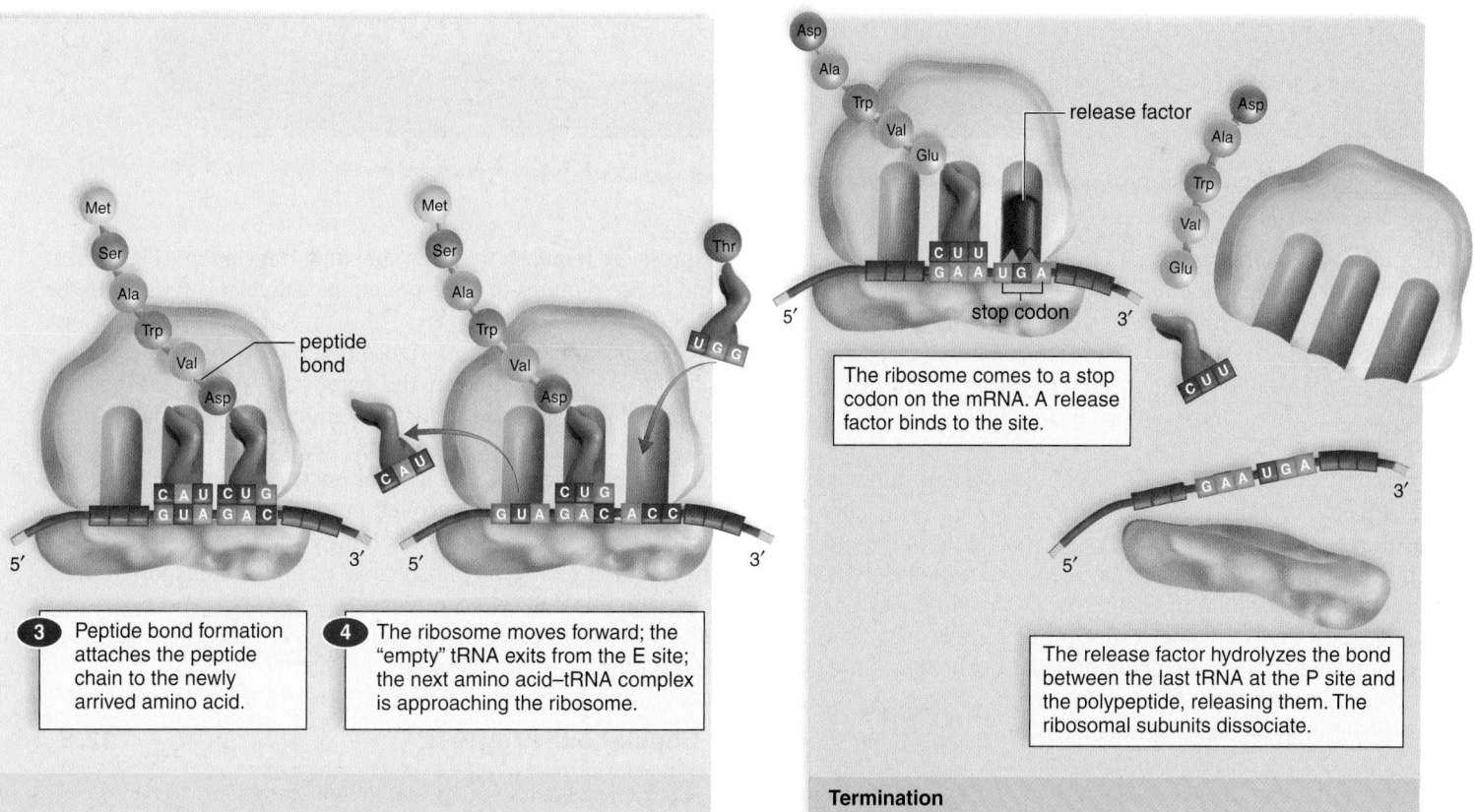

❸ Peptide bond formation attaches the peptide chain to the newly arrived amino acid.

❹ The ribosome moves forward; the "empty" tRNA exits from the E site; the next amino acid–tRNA complex is approaching the ribosome.

release factor

stop codon

The ribosome comes to a stop codon on the mRNA. A release factor binds to the site.

The release factor hydrolyzes the bond between the last tRNA at the P site and the polypeptide, releasing them. The ribosomal subunits dissociate.

Termination

Figure 12.18 Termination. During termination, the finished polypeptide is released, as is the mRNA and the last tRNA.

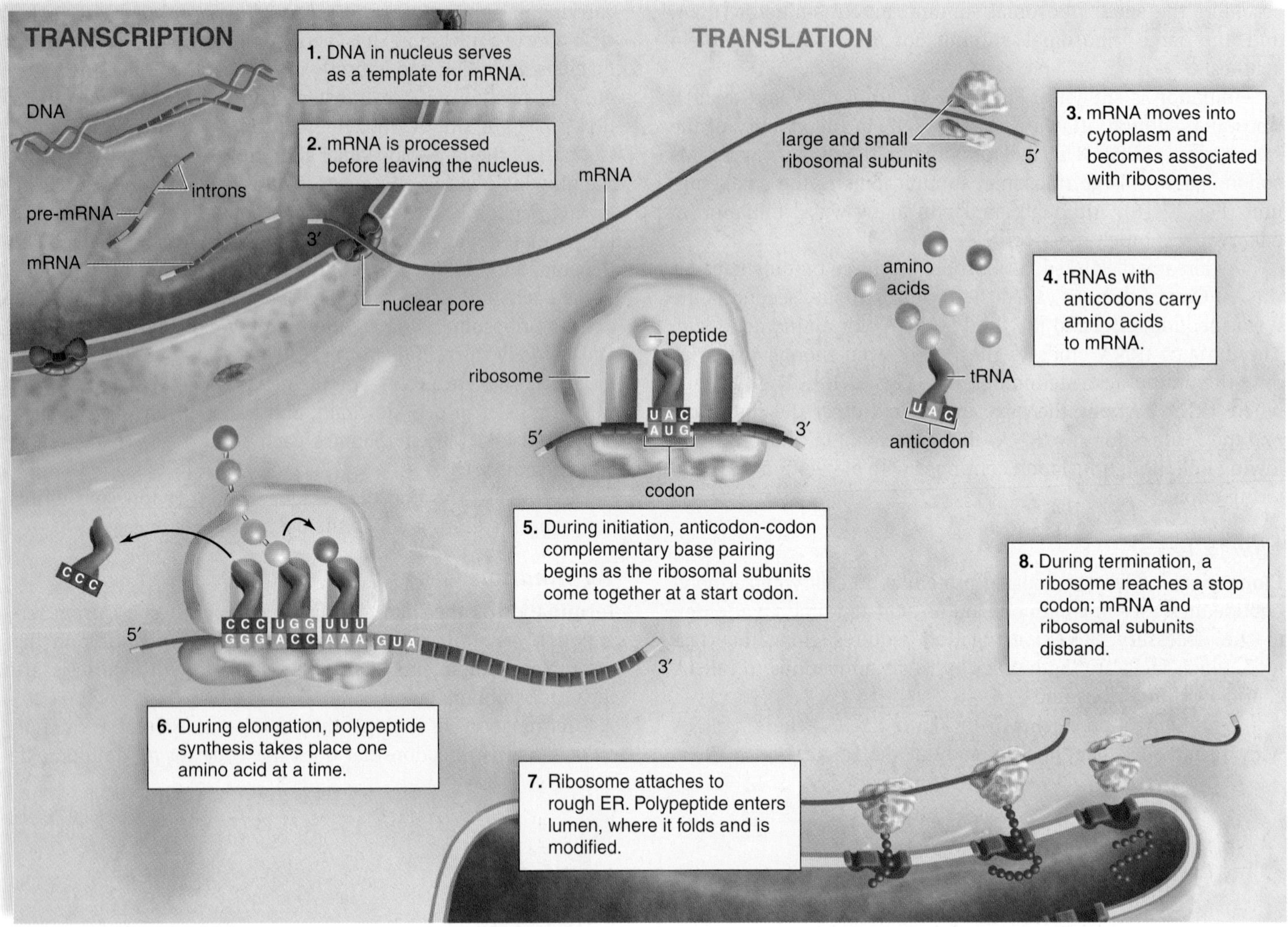

TRANSCRIPTION

TRANSLATION

1. DNA in nucleus serves as a template for mRNA.

DNA

2. mRNA is processed before leaving the nucleus.

pre-mRNA

introns

mRNA

3'

nuclear pore

mRNA

large and small ribosomal subunits

5'

3. mRNA moves into cytoplasm and becomes associated with ribosomes.

amino acids

4. tRNAs with anticodons carry amino acids to mRNA.

ribosome

peptide

U A C
A U G

5'   3'

codon

tRNA

U A C

anticodon

5. During initiation, anticodon-codon complementary base pairing begins as the ribosomal subunits come together at a start codon.

C C C

8. During termination, a ribosome reaches a stop codon; mRNA and ribosomal subunits disband.

5'

C C C U G G U U U
G G G A C C A A A G U A

3'

6. During elongation, polypeptide synthesis takes place one amino acid at a time.

7. Ribosome attaches to rough ER. Polypeptide enters lumen, where it folds and is modified.

Figure 12.19 Summary of protein synthesis in eukaryotes.

Termination requires a protein called a release factor, which can bind to a stop codon and also cleave the polypeptide from the last tRNA. After this occurs, the polypeptide is set free and begins to fold and take on its three-dimensional shape. The ribosome dissociates into its two subunits, which are returned to the cytoplasmic pool of large and small subunits, to be used again as necessary.

The next section reviews the entire process of protein synthesis (recall that a protein contains one or more polypeptides) and the role of the rough endoplasmic reticulum in the production of a polypeptide. Proteins do the work of the cell, whether they reside in a cellular membrane or are free in the cytoplasm. A whole new field of biology called **proteomics** is now dedicated to understanding the structure of proteins and how they function in metabolic pathways. One of the important goals of proteomics is to understand how proteins are modified in the endoplasmic reticulum and the Golgi apparatus.

MP3
Translation

3D Animation
Translation

Gene Expression

A gene has been expressed once its product, a protein (or an RNA), is made and is operating in the cell. For a protein, gene

expression requires transcription and translation (Fig. 12.19) and it also requires that the protein be active as discussed in the next chapter.

Translation occurs at ribosomes. Some ribosomes (polyribosomes) remain free in the cytoplasm, and some become attached to rough ER. The first few amino acids of a polypeptide act as a signal peptide that indicates where the polypeptide belongs in the cell or if it is to be secreted from the cell. Polypeptides that are to be secreted enter the lumen of the ER by way of a channel, and are then folded and further processed by the addition of sugars, phosphates, or lipids. Transport vesicles carry the proteins between organelles and to the plasma membrane as appropriate for that protein.

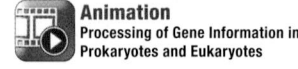
Animation
Processing of Gene Information in Prokaryotes and Eukaryotes

Check Your Progress   **12.5**

1. Explain the role of transfer RNA in translation.
2. Describe how the structure of a ribosome contributes to polypeptide synthesis.
3. Examine the events that occur during the three major steps of translation.

12.6 Structure of the Eukaryotic Chromosome

Only in recent years have investigators been able to produce models suggesting how chromosomes are organized. A eukaryotic chromosome contains a single double helix DNA molecule, but it is composed of more than 50% protein. Some of these proteins are concerned with DNA and RNA synthesis, but a large majority, termed **histones,** play primarily a structural role.

The five primary types of histone molecules are designated H1, H2A, H2B, H3, and H4 (see Fig. 13.5*b*). Remarkably, the amino acid sequences of H3 and H4 vary little between organisms. For example, the H4 of peas is only two amino acids different from the H4 of cattle. This similarity suggests that few mutations in the histone proteins have occurred during the course of evolution and that the histones, therefore, have essential functions for survival.

A human cell contains at least 2 m of DNA. Yet, all of this DNA is packed into a nucleus that is about 5 μm in diameter. The histones are responsible for packaging the DNA so that it can fit into such a small space. First, the DNA double helix is wound at intervals around a core of eight histone molecules (two copies each of H2A, H2B, H3, and H4), giving the appearance of a string of beads (Fig. 12.20*a*). Each bead is called a **nucleosome,** and the nucleosomes are said to be joined by "linker" DNA.

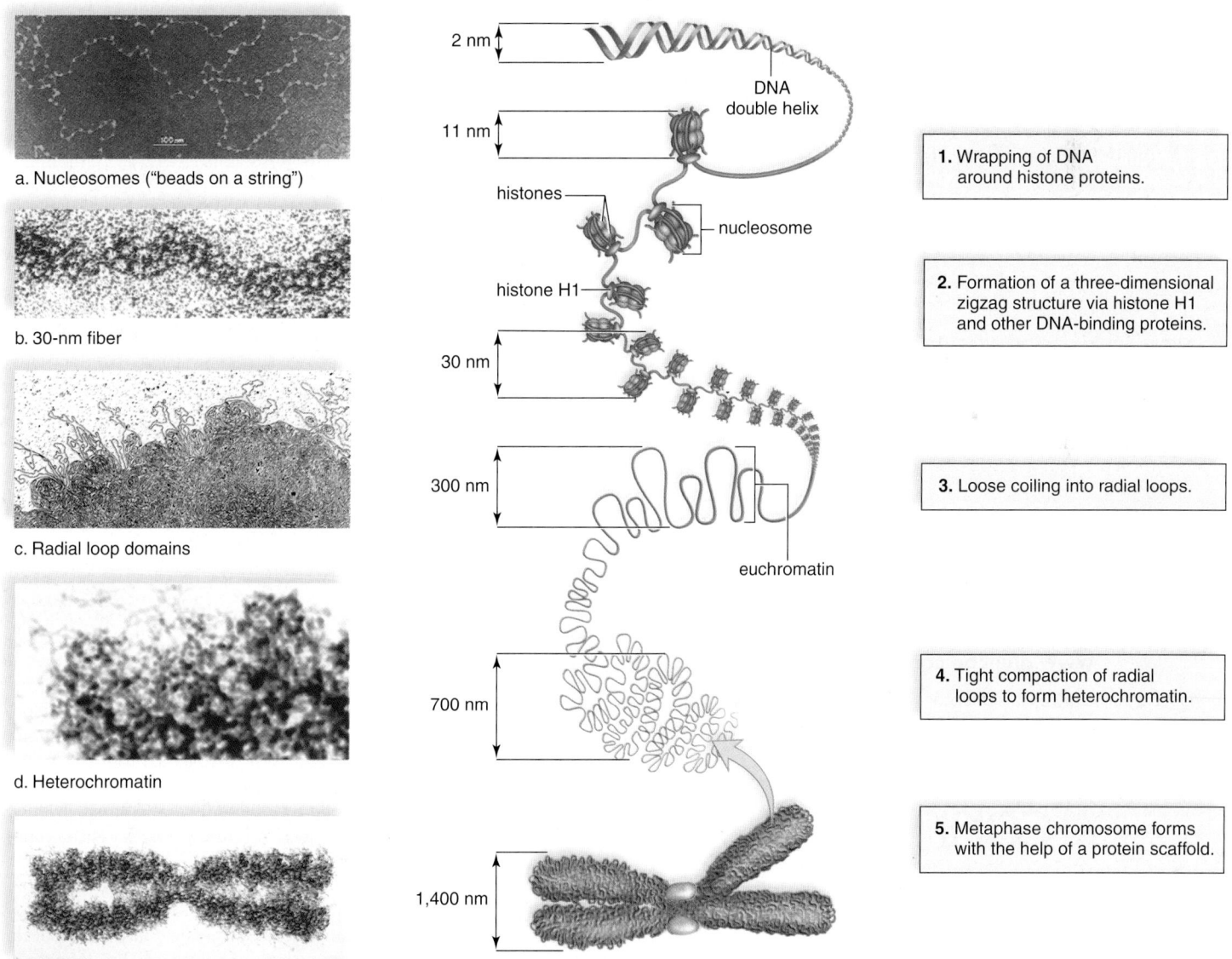

a. Nucleosomes ("beads on a string")

b. 30-nm fiber

c. Radial loop domains

d. Heterochromatin

e. Metaphase chromosome

2 nm

DNA double helix

11 nm

histones

nucleosome

histone H1

30 nm

300 nm

euchromatin

700 nm

1,400 nm

1. Wrapping of DNA around histone proteins.

2. Formation of a three-dimensional zigzag structure via histone H1 and other DNA-binding proteins.

3. Loose coiling into radial loops.

4. Tight compaction of radial loops to form heterochromatin.

5. Metaphase chromosome forms with the help of a protein scaffold.

Figure 12.20   Structure of eukaryotic chromosomes.   The DNA molecule of a chromosome is compacted at several levels. Structural changes in the DNA influence its ability to express genes. **a.** The DNA strand is wound around histones to form nucleosomes. **b.** The strand is further shortened by folding it into a zigzag structure. **c.** The fiber loops back and forth into radial loops. **d.** In heterochromatin, additional proteins further compact the radial loops. **e.** A metaphase chromosome forms.

This string is compacted by folding into a zigzag structure, further shortening the DNA strand (Fig. 12.20*b*). Histone H1 appears to mediate this coiling process. The fiber then loops back and forth into radial loops (Fig. 12.20*c*). This loosely coiled **euchromatin** represents the active chromatin containing genes that are being transcribed. The DNA of euchromatin may be accessed by RNA polymerase and other factors that are needed to promote transcription. In fact, recent research seems to indicate that regulating the level of compaction of the DNA is an important method of controlling gene expression in the cell.

Under a microscope, one often observes dark-stained fibers within the nucleus of the cell. These areas within the nucleus represent a more highly compacted form of the chromosome called **heterochromatin** (Fig. 12.20*d*). Most chromosomes

exhibit both levels of compaction in a living cell, depending on which portions of the chromosome are being used more frequently. Heterochromatin is considered inactive chromatin because the genes contained on it are infrequently transcribed, if at all.

Prior to cell division, a protein scaffold helps to further condense the chromosome into a form that is characteristic of metaphase chromosomes (Fig. 12.20*e*). No doubt, compact chromosomes are easier to move about than extended chromatin.

Check Your Progress 12.6

1. Examine how regulation of gene expression might be accomplished by changing the compaction of chromatin.

CONNECTING *the* CONCEPTS *with the* BIG IDEAS

Evolution

- The use of DNA and RNA as sources of genetic information and machinery molecules to effect protein synthesis is seen universally in all domains; a common genetic code points to a common ancestor for all life forms. (1B1a1, 1D2b2)

Information and Signaling

- Scientists worked to prove DNA was the genetic material of life and to discover its double helix structure. (3A1a4i-ii)
- DNA and RNA have differing functions, but complimentary base pairing and enzyme facilitation are essential in replication, transcription, and translation. (3A1, 3A1b1-3)
- DNA replicates semi-conservatively and bidirectionally. (3A1a5i-ii)
- RNA molecules differ in structure and sequence, facilitating their roles in the creation of proteins. (3A1b4i-iv)
- Transcribed messenger RNA molecules are often modified before leaving the nucleus. (3A1c1-2)
- mRNA carries a triplet code for an amino acid sequence to a ribosome. (3A1c3i-vii)
- Translation of mRNA information occurs when codons are matched with tRNA molecules carrying specific amino acids, creating a protein chain. (3A1c4)
- Activities of translated proteins create the physical phenotypes of organisms. (3A1d*IE*)

*Find the unabridged version of all EK citations at www.glencoe.com/maderAP11.

Media Study Tools

www.glencoe.com/maderAP11

Enhance your study of this chapter with study tools and practice tests. Also ask your instructor about the resources available through ConnectPlus, including the media-rich eBook, interactive learning tools, and animations.

3D Animations
DNA Replication
Molecular Biology of the Gene

For an interactive examination of the processes of DNA replication and gene expression, review the 3D animations "DNA Replication" and "Molecular Biology of the Gene."

Summarize

12.1 The Genetic Material

Early work illustrated that DNA was the hereditary material. Griffith injected strains of pneumococcus into mice and observed that when heat-killed S strain bacteria were injected along with live R strain bacteria, virulent S strain bacteria were recovered from the dead mice. Griffith said that the R strain had been transformed by some substance passing from the dead S strain to the live R strain. Twenty years later, Avery and his colleagues reported that the transforming substance is DNA.

To study the structure of DNA, Chargaff performed a chemical analysis of DNA and found that A = T and G = C, and that the amount of purine equals the amount of pyrimidine. Franklin prepared an X-ray photograph of DNA that showed it is helical, has repeating structural features, and has certain dimensions. Watson and Crick built a model of DNA in which the sugar-phosphate molecules made up the sides of a twisted ladder, and the complementary-paired bases were the rungs of the ladder.

12.2 Replication of DNA

The Watson and Crick model immediately suggested a method by which DNA could be replicated. Basically, the two strands unwind and unzip, and each parental strand acts as a template for a new (daughter) strand. In the end, each new helix is like the other and like the parental helix.

The enzyme DNA polymerase joins the nucleotides together and proofreads them to make sure the bases have been paired correctly. Incorrect base pairs that survive the process are a mutation. Replication in prokaryotes typically proceeds in both directions from one point of origin to a termination region until there are two copies of the circular chromosome. Replication in eukaryotes has many points of origin and many bubbles (places where the DNA strands are separating and replication is occurring). Replication occurs at the ends of the bubbles—at replication forks. Since eukaryotes have linear chromosomes, they cannot replicate the very ends of them. Therefore, the ends (telomeres) get shorter with each replication. Telomerase enzyme helps to ensure that chromosomes maintain their proper length.

12.3 The Genetic Code of Life

The central dogma of molecular biology says that the flow of genetic information is from DNA to RNA to protein to traits. More specifically, (1) DNA is a template for its own replication and also for RNA formation during transcription, and (2) the sequence of nucleotides in mRNA directs the correct sequence of amino acids of a polypeptide during translation.

The genetic code is a triplet code, and each codon (code word) consists of three bases. The code is degenerate—that is, more than one codon exists for most amino acids. There are also one start and three stop codons. The genetic code is considered universal, but there are a few exceptions.

12.4 First Step: Transcription

Transcription to produce messenger RNA (mRNA) begins when RNA polymerase attaches to the promoter of a gene. Elongation occurs until RNA polymerase reaches a stop sequence. The mRNA is processed following transcription. A cap is put onto the 5′ end, a poly-A tail is put onto the 3′ end, and introns are removed in eukaryotes by spliceosomes.

Small nuclear RNAs (snRNAs) present in spliceosomes help identify the introns to be removed. These snRNAs play a role in alternative mRNA splicing, which allows a single eukaryotic gene to code for different proteins, depending on which segments of the gene serve as introns and which serve as exons.

Some introns serve as microRNAs (miRNAs), which help regulate the translation of mRNAs. Research is now directed to discovering the many ways small RNAs influence the production of proteins in a cell.

12.5 Second Step: Translation

Translation requires mRNA, transfer RNA (tRNA), and ribosomal RNA (rRNA). Each tRNA has an anticodon at one end and an amino acid at the other; amino acid–charging enzymes ensure that the correct amino acid is attached to the correct tRNA. When tRNAs bind with their codon at a ribosome, the amino acids are correctly sequenced in a polypeptide according to the order predetermined by DNA.

In the cytoplasm, many ribosomes move along the same mRNA at a time. Collectively, these are called a polyribosome.

Translation requires these steps: During initiation, mRNA, the first (initiator) tRNA, and the two subunits of a ribosome all come together in the proper orientation at a start codon. During elongation, as the tRNA anticodons bind to their codons, the growing pep-

tide chain is transferred by peptide bonding to the next amino acid in a polypeptide. During termination at a stop codon, the polypeptide is cleaved from the last tRNA. The ribosome now dissociates.

12.6 Structure of the Eukaryotic Chromosome

Eukaryotic cells contain nearly 2 m of DNA, yet must pack it all into a nucleus no more than 20 μm in diameter. Thus, the DNA is compacted by winding it around DNA-binding proteins called histones to make nucleosomes. The nucleosomes are further compacted into a zigzag structure, which is then folded upon itself many times to form radial loops, which is the usual compaction state of euchromatin. Heterochromatin is further compacted by scaffold proteins, and further compaction can be achieved prior to mitosis and meiosis.

Key Terms

adenine (A) 217	nucleosome 233
anticodon 228	polyribosome 230
central dogma 224	promoter 225
codon 224	proteomics 232
complementary base pairing 219	replication fork 222
	ribosomal RNA (rRNA) 223
cytosine (C) 217	ribozyme 227
DNA polymerase 220	RNA polymerase 225
DNA replication 220	semiconservative replication 220
double helix 218	
elongation 231	telomere 221
euchromatin 234	template 220
exon 226	termination 231
genetic code 224	thymine (T) 217
guanine (G) 217	transcription 223
heterochromatin 234	transfer RNA (tRNA) 223
histone 233	translation 224
initiation 230	translocation 231
intron 226	triplet code 224
messenger RNA (mRNA) 223	uracil (U) 223
mRNA transcript 226	wobble hypothesis 229

Assess

Reviewing This Chapter

1. List and discuss the requirements for genetic material. 215
2. How did Avery and his colleagues demonstrate that the transforming substance is DNA? 215–16
3. Describe the Watson and Crick model of DNA structure. How did it fit the data provided by Chargaff and the X-ray diffraction patterns of Franklin? 218–19
4. Explain how DNA replicates semiconservatively. What role does DNA polymerase play? What role does helicase play? 220–21
5. List and discuss differences between prokaryotic and eukaryotic replication of DNA. 222
6. How did investigators reason that the code must be a triplet code, and in what manner was the code cracked? Why is it said that the code is degenerate, unambiguous, and almost universal? 224
7. What two steps are required for the expression of a gene? 225, 228
8. What specific steps occur during transcription of RNA off a DNA template? 225–26
9. How is messenger RNA (mRNA) processed before leaving the eukaryotic nucleus? 226–27

10. What is the role of snRNAs in the nucleus?   227
11. Compare the functions of mRNA, transfer RNA (tRNA), and ribosomal RNA (rRNA) during protein synthesis. What are the specific events of translation?   228–32
12. What are the various levels of chromosome structure?   233–34

Testing Yourself

Choose the best answer for each question.

1. If 30% of an organism's DNA is thymine, then
   a. 70% is purine.   d. 70% is pyrimidine.
   b. 20% is guanine.   e. Both c and d are correct.
   c. 30% is adenine.
2. The double-helix model of DNA resembles a twisted ladder in which the rungs of the ladder are
   a. a purine paired with a pyrimidine.
   b. A paired with G and C paired with T.
   c. sugar-phosphate paired with sugar-phosphate.
   d. a 5' end paired with a 3' end.
   e. Both a and b are correct.
3. DNA replication is said to be semiconservative because
   a. one of the new molecules conserves both of the original DNA strands.
   b. the new DNA molecule contains two new DNA strands.
   c. both of the new molecules contain one new strand and one old strand.
   d. DNA polymerase conserves both of the old strands.
4. If the sequence of bases in one strand of DNA is 5' TAGCCT 3', then the sequence of bases in the other strand will be
   a. 3' TCCGAT 5'.   c. 3' TAGCCT 5'.
   b. 3' ATCGGA 5'.   d. 3' AACGGUA 5'.
5. Transformation occurs when
   a. DNA is transformed into RNA.
   b. DNA is transformed into protein.
   c. bacteria cannot grow on penicillin.
   d. organisms receive foreign DNA and thereby acquire a new characteristic.
6. Pyrimidines
   a. are always paired with a purine.
   b. are thymine and cytosine.
   c. keep DNA from replicating too often.
   d. are adenine and guanine.
   e. Both a and b are correct.
7. A nucleotide
   a. is smaller than a base.
   b. is a subunit of nucleic acids.
   c. has a lot of variable parts.
   d. has at least four phosphates.
   e. always joins with other nucleotides.
8. This is a segment of a DNA molecule. What are (a) the RNA codons, (b) the matching tRNA anticodons, and (c) the sequence of amino acids in the eventual protein?

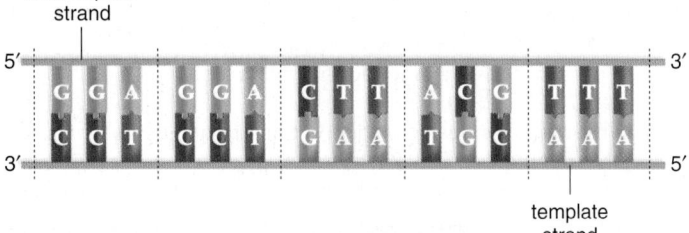

9. In prokaryotes,
   a. replication can occur in two directions at once because their DNA molecule is circular.
   b. bubbles thereby created spread out until they meet.
   c. replication occurs at numerous replication forks.
   d. a new round of DNA replication cannot begin before the previous round is complete.
   e. Both a and b are correct.
10. The central dogma of molecular biology
    a. states that DNA is a template for all RNA production.
    b. states that DNA is a template only for DNA replication.
    c. states that translation precedes transcription.
    d. states that RNA is a template for DNA replication.
    e. All of these are correct.
11. Because there are more codons than amino acids,
    a. some amino acids are specified by more than one codon.
    b. some codons specify more than one amino acid.
    c. some codons do not specify any amino acid.
    d. some amino acids do not have codons.
12. If the sequence of bases in the coding strand of a DNA is TAGC, then the sequence of bases in the mRNA will be
    a. AUCG.   c. UAGC.
    b. TAGC.   d. CGAU.
13. During protein synthesis, an anticodon on transfer RNA (tRNA) pairs with
    a. DNA nucleotide bases.
    b. ribosomal RNA (rRNA) nucleotide bases.
    c. messenger RNA (mRNA) nucleotide bases.
    d. other tRNA nucleotide bases.
    e. Any one of these can occur.
14. If the sequence of DNA on the template strand of a gene is AAA, the mRNA codon produced by transcription will be _____ and will specify the amino acid

    _____ .
    a. AAA, lysine   d. UUU, phenylalanine
    b. AAA, phenylalanine   e. TTT, lysine
    c. TTT, arginine
15. Euchromatin
    a. is organized into radial loops.
    b. is less condensed than heterochromatin.
    c. contains nucleosomes.
    d. All of these are correct.

Engage

Thinking Scientifically

1. How would you test a hypothesis that a genetic condition, such as cancer, is due to mistakes in transcription and translation?
2. Knowing that a plant will grow from a single cell in tissue culture, how could you transform a plant so that it glows in the dark?

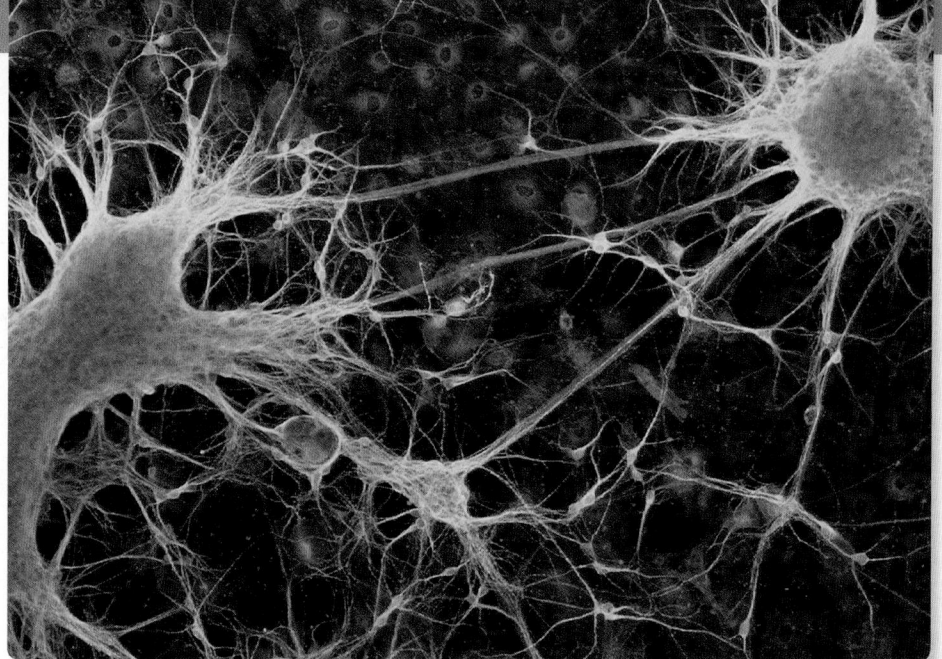

Individual neurons, shown in pink in this electron micrograph, may use unique forms of the DSCAM proteins in their plasma membranes to identify themselves.

13

Regulation of Gene Expression

It was once estimated that the human genome may contain as many as a million genes. At the start of the Human Genome Project, that number was considered to be closer to 100,000. We now know that the human genome probably contains less than 23,000 genes. So how do a mere 23,000 genes account for the variety of proteins found in human cells? The answer, surprisingly, may lie in the regulation of pre-mRNA splicing to allow the production of a myriad of proteins from a single gene. *DSCAM* is a gene associated with Down Syndrome that is present in the brain of many animals. In fruit flies *DSCAM* has four regions of alternative exons. The result is over 38,000 different possible combinations of functional mRNAs, and therefore, the *DSCAM* gene is able to specify 38,000 different proteins. Such a huge number of proteins is sufficient to provide each nerve cell with a unique identity as it communicates with others within a brain.

Complex alternative splicing, and other regulatory mechanisms, could very well account for how humans generate so many different proteins from so few genes. This chapter introduces you to regulatory mechanisms in both prokaryotes and eukaryotes, allowing you to see how these mechanisms influence the processes of transcription and translation that you learned about in the previous chapter.

As you read through this chapter, think about the following questions:

1. How does gene regulation differ between prokaryotes and eukaryotes?
2. Where in the process of gene expression does regulation occur in a eukaryotic organism?
3. How might mutations influence the ability of a cell to regulate gene expression?

BEFORE YOU BEGIN

Before starting this chapter, take a few moments to review these earlier concepts

Figure 12.9 What is the central dogma of biology?

Section 12.4 What is the purpose of transcription and where does it occur in both prokaryotic and eukaryotic cells?

Section 12.5 What is the role of translation in gene expression?

FOLLOWING *the* BIG IDEAS

CHAPTER 13   REGULATION OF GENE EXPRESSION

Evolution	Random changes in DNA can lead to phenotypic variation which can be acted upon by natural selection.
Information and Signaling	Regulation of prokaryotic genes is often at the operon level while eukaryotes fine tune gene expression and modify gene products.
Interactions and Systems	Regulation of gene activity involves intricate interactions with internal and external factors.

13.1 Prokaryotic Regulation

Learning Outcomes

Upon completion of this section you should be able to

1. Describe the structure of an operon and state the role of each component of the operon.
2. Explain how the *trp* and *lac* operons of prokaryotes are regulated.
3. Distinguish between positive and negative regulation of gene expression in prokaryotes.

Because their environment is ever changing, bacteria do not always need to express their entire complement of enzymes and proteins. In 1961, French microbiologists François Jacob and Jacques Monod showed that *Escherichia coli* is capable of regulating the expression of its genes. They observed that the genes for a metabolic pathway, called **structural genes,** are grouped on a chromosome and subsequently are transcribed at the same time. Jacob and Monod, therefore, proposed the **operon**

[L. *opera,* works] model to explain gene regulation in prokaryotes. They later received a Nobel Prize for their investigations.

An operon (Fig. 13.1) typically includes the following elements:

A **regulator gene**—Normally located outside the operon, this codes for a DNA-binding protein that acts as a **repressor.** The repressor controls whether the operon is active or not.

Promoter—A short sequence of DNA where RNA polymerase first attaches to begin transcription of the grouped genes.

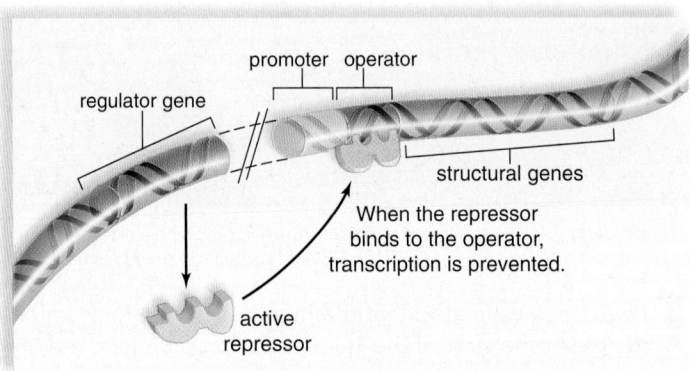

a. **Tryptophan absent.** Enzymes needed to synthesize tryptophan are produced.

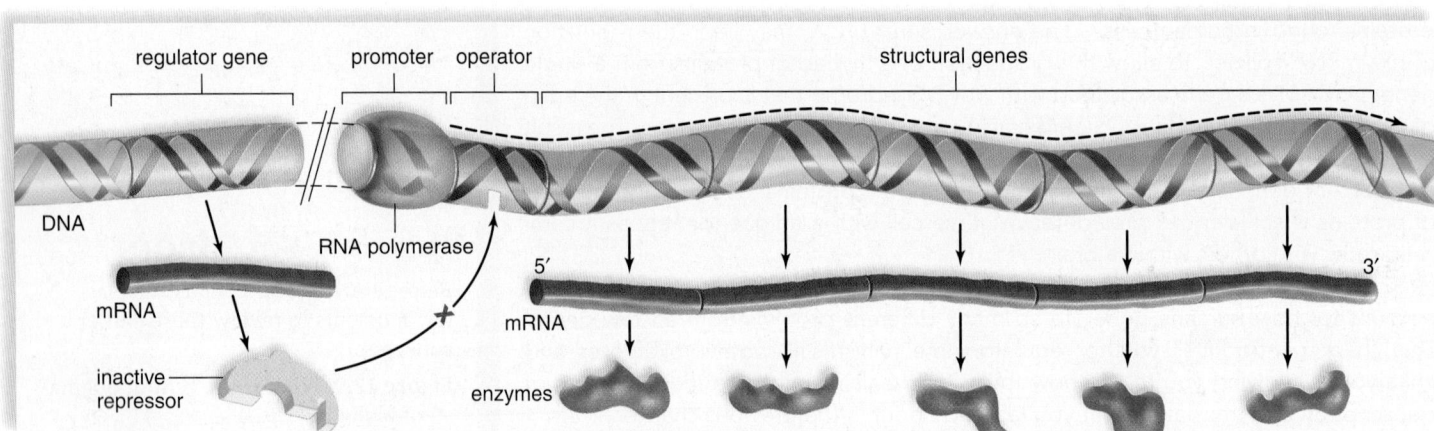

b. **Tryptophan present.** Presence of tryptophan prevents production of enzymes used to synthesize tryptophan.

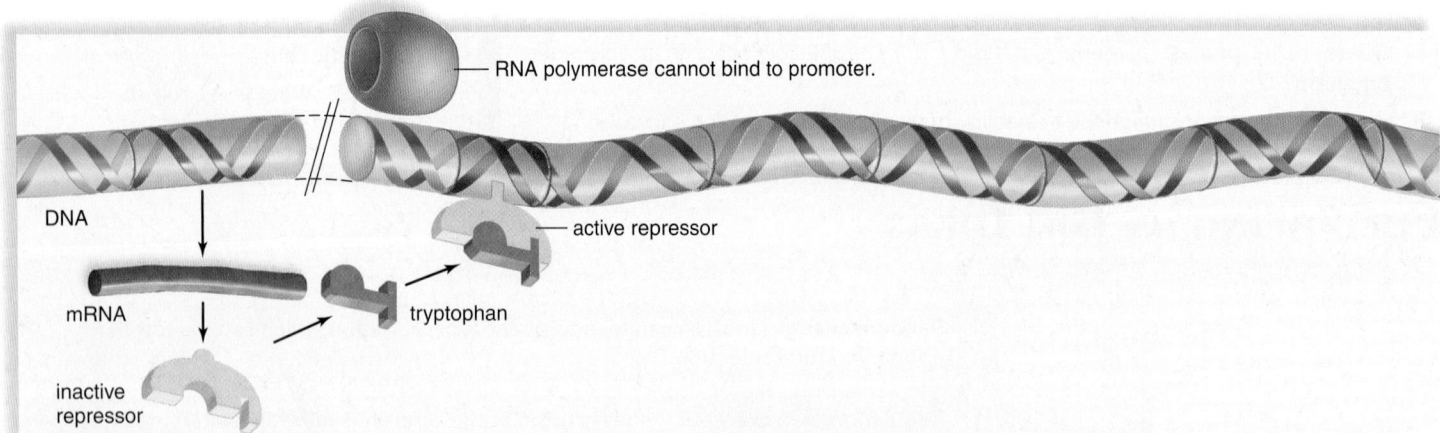

Figure 13.1 The *trp* operon.   a. The regulator gene codes for a repressor protein that is normally inactive. RNA polymerase attaches to the promoter, and the structural genes are expressed. **b.** When the nutrient tryptophan is present, it binds to the repressor, changing its shape. Now the repressor is active and can bind to the operator. RNA polymerase cannot attach to the promoter, and the structural genes are not expressed.

Basically, a promoter signals the start of the operon and the location where transcription begins.

Operator—A short portion of DNA where an active repressor binds. When an active repressor binds to the operator, RNA polymerase cannot attach to the promoter, and transcription cannot occur. In this way, the operator controls transcription of structural genes.

Structural genes—These genes code for the enzymes and proteins that are involved in the metabolic pathway of the operon. The structural genes are transcribed as a unit.

Next, we will briefly review the findings of Jacob and Monod in their studies of two *E. coli* operons: the *trp* operon and the *lac* operon.

The *trp* Operon

Many investigators, including Jacob and Monod, found that some operons in *E. coli* usually exist in the "on" rather than "off" condition. For example, in the *trp* operon, the regulator codes for a repressor that ordinarily is unable to attach to the operator. Therefore, RNA polymerase can bind to the promoter, and the structural genes of the operon are ordinarily expressed (Fig. 13.1). Their products, five different enzymes, are part of an anabolic pathway for the synthesis of the amino acid tryptophan.

If tryptophan happens to be already present in the medium, these enzymes are not needed by the cell, and the operon is turned off by the following method. Tryptophan binds to the repressor. A change in shape now allows the repressor to bind to the operator and prevent RNA polymerase from binding to the promoter, and the structural genes are not expressed. The enzymes are said to be repressible, and the entire unit is called a *repressible operon*. Tryptophan is called the **corepressor.** Repressible operons are usually involved in anabolic pathways that synthesize a substance needed by the cell.

🎬 **Animation**
The Tryptophan Repressor

The *lac* Operon

Bacteria metabolism is remarkably efficient; when there is no need for certain proteins or enzymes, the genes that are used to make them are usually inactive. For example, if the milk sugar lactose is not present, there is no need to express genes for enzymes involved in lactose catabolism. But when *E. coli* is denied glucose and is instead given lactose, the cell immediately begins to make the three enzymes needed for lactose metabolism.

The enzymes that break down lactose are encoded by three genes (Fig. 13.2): One gene is for an enzyme called β-galactosidase, which breaks down the disaccharide lactose to glucose and galactose; a second gene codes for a permease that facilitates the entry of lactose into the cell; and a third gene codes for an enzyme called transacetylase, which has an accessory function in lactose metabolism.

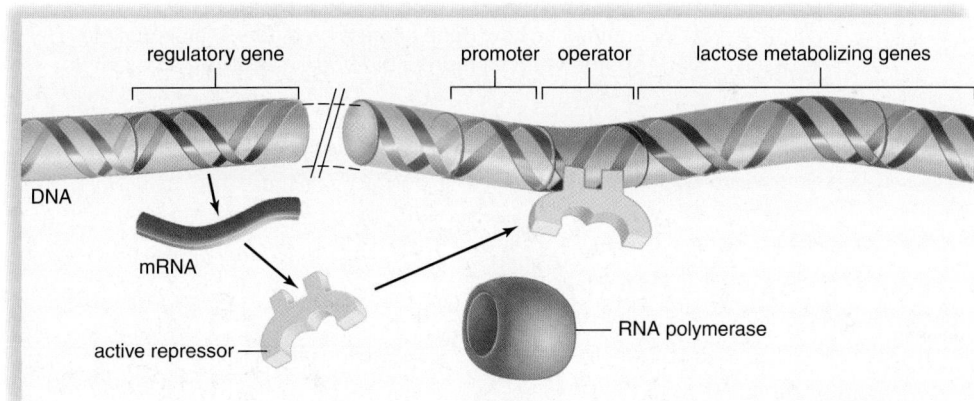

a.

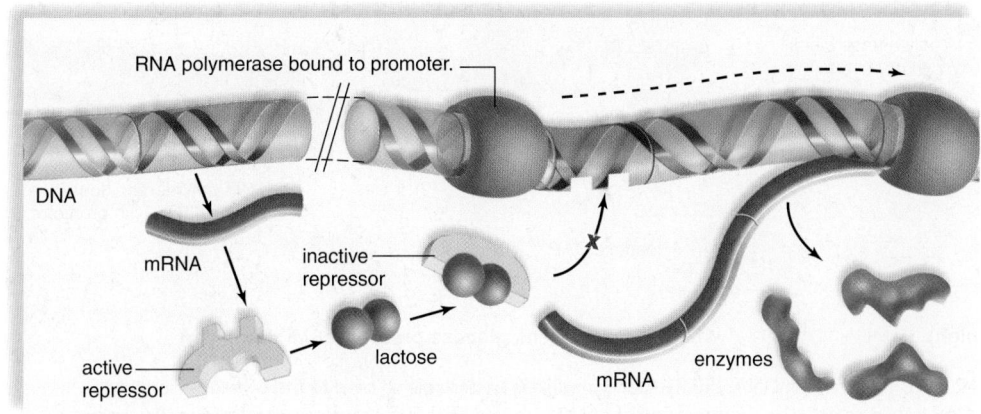

b.

Figure 13.2 The *lac* operon.
a. The regulator gene codes for a repressor that is normally active. When it binds to the operator, RNA polymerase cannot attach to the promoter, and structural genes are not expressed. **b.** When lactose is present, it binds to the repressor, changing its shape so that it is inactive and cannot bind to the operator. Now, RNA polymerase binds to the promoter, and the structural genes are expressed.

The three structural genes are adjacent to one another on the chromosome and are under the control of a single promoter and a single operator. The regulator gene codes for a *lac* operon repressor that ordinarily binds to the operator and prevents transcription of the three genes. But when glucose is absent and lactose (or more correctly, allolactose, an isomer formed from lactose) is present, lactose binds to the repressor, and the repressor undergoes a change in shape that prevents it from binding to the operator. Because the repressor is unable to bind to the operator, RNA polymerase is better able to bind to the promoter. After RNA polymerase carries out transcription, the three enzymes of lactose metabolism are synthesized.

Because the presence of lactose brings about expression of genes, it is called an **inducer** of the *lac* operon: The enzymes are said to be inducible enzymes, and the entire unit is called an *inducible operon*. Inducible operons are usually found in catabolic pathways that break down a nutrient. Why is that beneficial? Because these enzymes need to be active only when the nutrient is present.

Further Control of the lac Operon

E. coli preferentially breaks down glucose, and the bacterium has a way to ensure that the lactose operon is maximally turned on only when glucose is absent. A molecule called *cyclic AMP (cAMP)* accumulates when glucose is absent. Cyclic AMP, which is derived from ATP, has only one phosphate group, which is attached to ribose at two locations:

cyclic AMP
(cAMP)

Cyclic AMP binds to a molecule called a *catabolite activator protein (CAP)*, and the complex attaches to a CAP binding site next to the *lac* promoter. When CAP binds to DNA, DNA bends, exposing the promoter to RNA polymerase. RNA polymerase is now better able to bind to the promoter so that the *lac* operon structural genes are transcribed, leading to their expression (Fig. 13.3).

When glucose is present, there is little cAMP in the cell; CAP is inactive, and the lactose operon does not function maximally. CAP affects other operons as well and takes its name for activating the catabolism of various other metabolites when glucose is absent. A cell's ability to encourage the metabolism of lactose and other metabolites when glucose is absent provides a backup system for survival when the preferred energy source glucose is absent.

Animation
Combination of Switches:
The *lac* Operon

The CAP protein's regulation of the *lac* operon is an example of positive control. Why? Because when this molecule is active, it promotes the activity of an operon. The use of repressors, on the other hand, is an example of negative control because when active they shut down an operon. A positive control mechanism allows the cell to fine-tune its response. In the case of the *lac* operon, the operon is only maximally active when glucose is absent and lactose is present. If both glucose and lactose are present, the cell preferentially metabolizes glucose.

Check Your Progress   13.1

1. Explain the difference between the roles of the promoter and operator of an operon.
2. Summarize how gene expression differs in an inducible operon versus a repressible operon and give an example of each.
3. Describe the difference between positive control and negative control of gene expression.

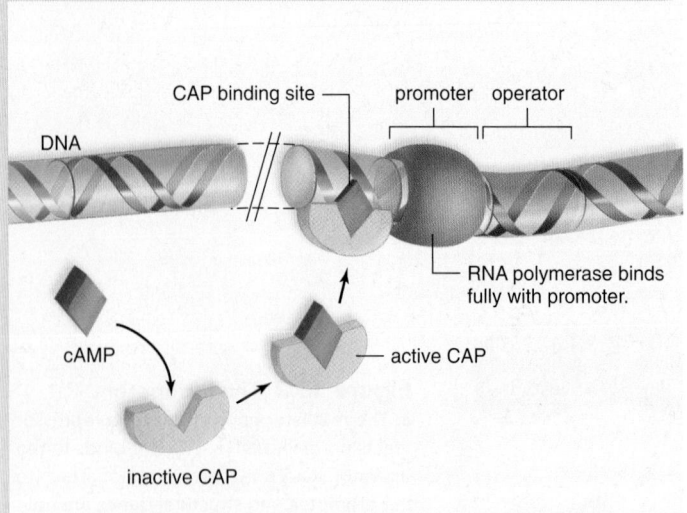

a. **Lactose present, glucose absent (cAMP level high)**

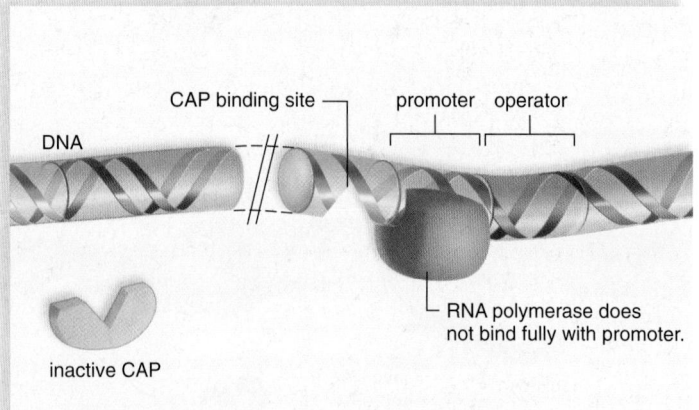

b. **Lactose present, glucose present (cAMP level low)**

Figure 13.3 Action of CAP.   When active CAP binds to its site on DNA, the RNA polymerase is better able to bind to the promoter so that the structural genes of the *lac* operon are expressed. **a.** CAP becomes active in the presence of cAMP, a molecule that is prevalent when glucose is absent. Therefore, transcription of lactose enzymes increases, and lactose is metabolized. **b.** If glucose is present, CAP is inactive, and RNA polymerase does not completely bind to the promoter. Therefore, transcription of lactose enzymes decreases, and less metabolism of lactose occurs.

13.2 Eukaryotic Regulation

Learning Outcomes

Upon completion of this section, you should be able to

1. List the levels of control of gene expression in eukaryotes.

2. Summarize how chromatin structure may be involved in regulation of gene expression in eukaryotes.

3. Identify the mechanisms of transcriptional, posttranscriptional, and translational control of gene expression.

With a few minor exceptions, each cell of a multicellular eukaryote has a complete complement of genes; the differences in cell types are determined by the different genes that are actively expressed in each cell. For example, in muscle cells a different set of genes is turned on in the nucleus and a different set of proteins is active in the cytoplasm compared to nerve cells.

Like prokaryotic cells, a variety of mechanisms regulate gene expression in eukaryotic cells. These mechanisms can be grouped under five primary levels of control; three of them pertain to the nucleus, and two pertain to the cytoplasm (Fig. 13.4). In other words, control of gene activity in eukaryotes extends from transcription to protein activity. These are the types of control in eukaryotic cells that can modify the amount of the gene product:

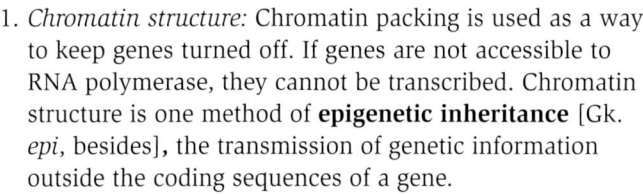 **Animation**
Control of Gene Expression in Eukaryotes

1. *Chromatin structure:* Chromatin packing is used as a way to keep genes turned off. If genes are not accessible to RNA polymerase, they cannot be transcribed. Chromatin structure is one method of **epigenetic inheritance** [Gk. *epi*, besides], the transmission of genetic information outside the coding sequences of a gene.

2. *Transcriptional control:* The degree to which a gene is transcribed into mRNA determines the amount of gene product. In the nucleus, transcription factors may promote or repress transcription, the first step in gene expression.

3. *Posttranscriptional control:* Posttranscriptional control involves mRNA processing and how fast mRNA leaves the nucleus.

4. *Translational control:* Translational control occurs in the cytoplasm and affects when translation begins and how long it continues. Any condition that can cause the persistence of the 5′ cap and 3′ poly-A tail can affect the length of translation. Excised introns may also have effects on the life span of mRNA.

5. *Posttranslational control:* Posttranslational control, which also takes place in the cytoplasm, occurs after protein synthesis. Only a functional protein is an active gene product.

We now explore each of these types of control in greater depth.

Chromatin Structure

The DNA in eukaryotes is always associated with a variety of proteins, and together they make up a stringy material called **chromatin.** Chromatin is most evident in the nucleus during interphase of the cell cycle.

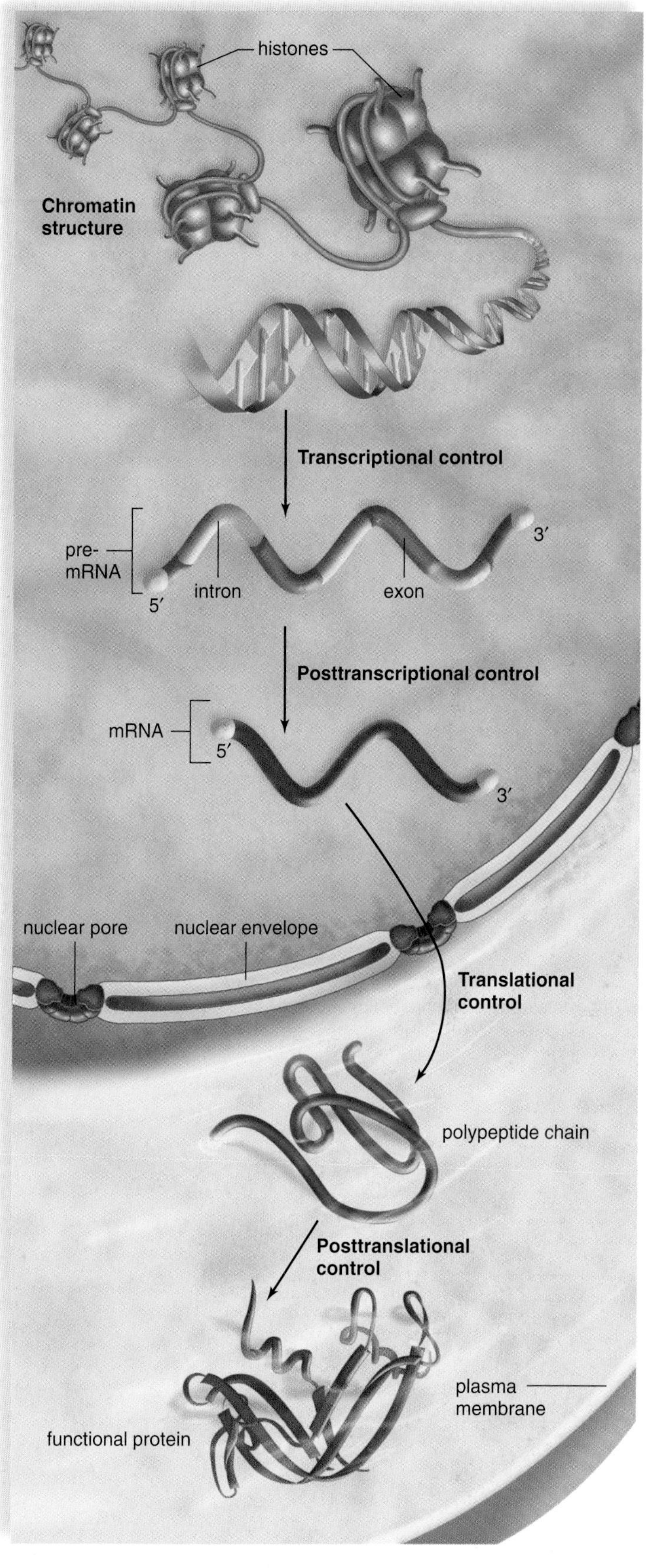

Figure 13.4 Levels at which control of gene expression occurs in eukaryotic cells. The five levels of control are (1) chromatin structure, (2) transcriptional control, and (3) posttranscriptional control, which occur in the nucleus; and (4) translational and (5) posttranslational control, which occur in the cytoplasm.

In Chapter 12 you learned that one class of these DNA-associated proteins are the histones. Histones play an important role in the compaction of the DNA (see Figure 12.20) as well as in eukaryotic gene regulation. Without histones, the DNA would not fit inside the nucleus. Each human cell contains around 2 meters of DNA, yet the nucleus is only 5 to 8 micrometers (μm) in diameter.

The degree to which chromatin is compacted greatly affects the accessibility of the chromatin to the transcriptional machinery of the cell, and thus the expression levels of the genes. Active genes in eukaryotic cells are associated with more loosely packed chromatin called *euchromatin,* while the more tightly packed DNA, called *heterochromatin,* contains mostly inactive genes. Under a microscope, the more densely compacted heterochromatin stains darker than euchromatin (Fig. 13.5*a*).

What regulates whether chromatin exists as heterochromatin or euchromatin? In Chapter 12 you learned that a *nucleosome* consists of a portion of DNA wrapped around a group of histone molecules. Histone molecules have *tails,* strings of amino acids that extend beyond the main portion of a nucleosome (Fig. 13.5*b*). In heterochromatin, the histone tails tend to bear methyl groups ($—CH_3$); in euchromatin, the histone tails tend to be acetylated and have attached acetyl groups ($—COCH_3$).

Histones regulate accessibility to DNA; euchromatin becomes genetically active when histones no longer bar access to DNA. When DNA in euchromatin is transcribed, a so-called *chromatin remodeling complex* pushes aside the histone portion of a nucleosome so that access to DNA is not barred and transcription can begin (Fig. 13.5*c*). After *unpacking* occurs,

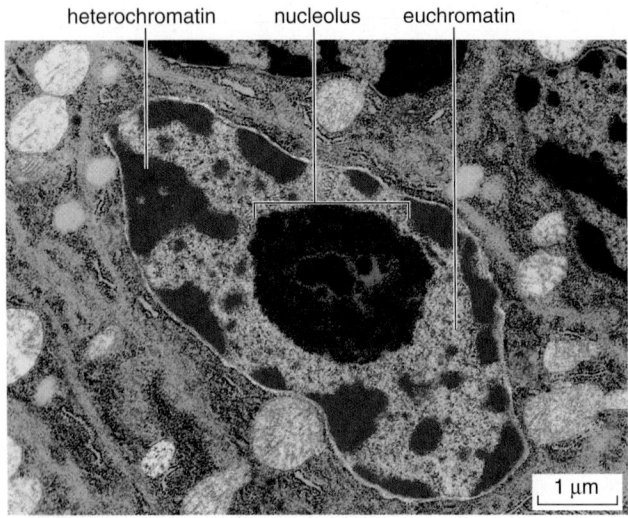

a. Darkly stained heterochromatin and lightly stained euchromatin

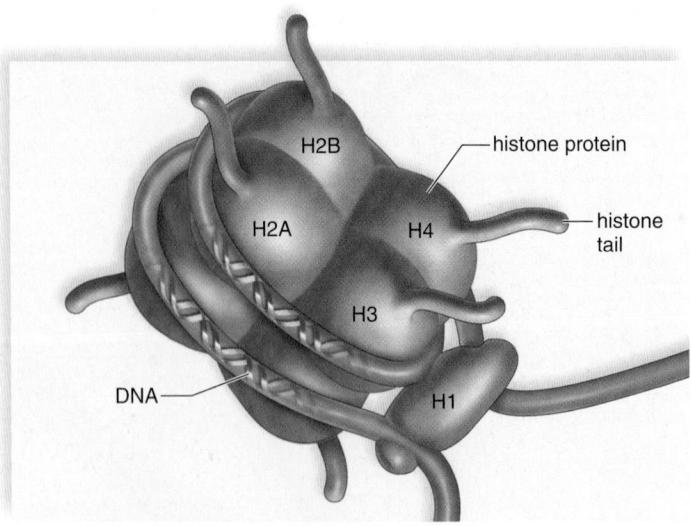

b. A nucleosome

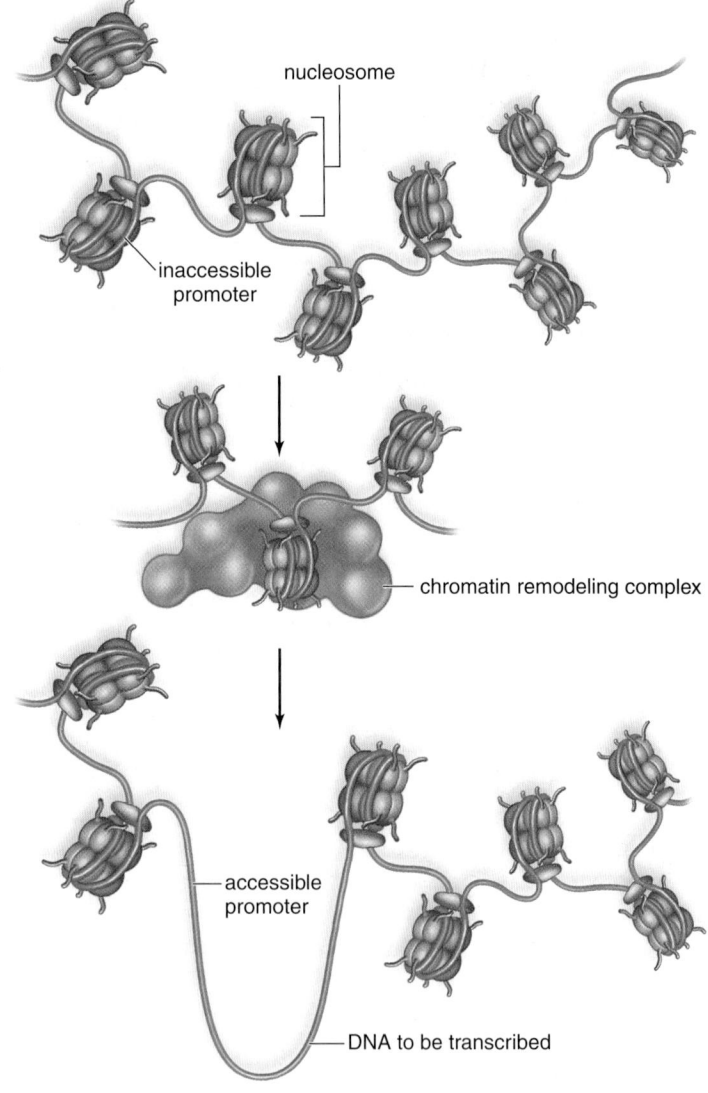

c. DNA unpacking

Figure 13.5 Chromatin structure regulates gene expression.   **a.** A eukaryotic nucleus contains highly condensed heterochromatin (darkly stained) and euchromatin (lightly stained), which is not as condensed. **b.** Nucleosomes ordinarily prevent access to DNA so that transcription cannot take place. If histone tails are acetylated, access can be achieved; if the tails are methylated, access is more difficult. **c.** A chromatin remodeling complex works on euchromatin to make a promoter accessible for transcription.

many decondensed loops radiate from the central axis of the chromosome. These chromosomes have been named lampbrush chromosomes because their feathery appearance resembles the brushes that were once used to clean kerosene lamps.

In addition to physically moving nucleosomes aside to expose promoters, chromatin remodeling complexes may also affect gene expression by adding acetyl or methyl groups to histone tails.

Heterochromatin Is Not Transcribed

In general, highly condensed heterochromatin is inaccessible to RNA polymerase, and the genes contained within are seldom or never transcribed. A dramatic example of heterochromatin is the **Barr body** in mammalian females, first mentioned in Chapter 10. This small, darkly staining mass of condensed chromatin adhering to the inner edge of the nuclear membrane is an inactive X chromosome. To compensate for the fact that female mammals have two X chromosomes (XX), while males only have one (XY), one of the X chromosomes in the cells of female embryos undergoes inactivation. The inactive X chromosome does not produce gene products, allowing both males and females to produce the same amount of gene product from a single X chromosome.

Animation
X-Inactivation

How do we know that Barr bodies are inactive X chromosomes that are not producing gene products? In a heterozygous female, 50% of the cells have one X chromosome active and 50% have the other X chromosome active. The body of a heterozygous female would therefore be a mosaic, with "patches" of genetically different cells. Investigators have discovered that human females who are heterozygous for an X-linked recessive

form of ocular albinism have patches of pigmented and nonpigmented cells at the back of the eye.

As other examples, women who are heterozygous for X-linked hereditary absence of sweat glands have patches of skin lacking sweat glands. And, the female tortoiseshell cat exhibits a difference in X-inactivation in its cells. In these cats, an allele for black coat color is on one X chromosome, and a corresponding allele for orange coat color is on the other. The patches of black and orange in the coat can be related to which X chromosome is in the Barr bodies of the cells found in the patches (Fig. 13.6).

Epigenetic Inheritance

Histone modification is sometimes linked to a phenomenon termed *epigenetic inheritance,* in which variations in the the pattern of inheritance is not due to changes in the sequence of the DNA nucleotides. For example, when histones are methylated, sometimes the DNA itself becomes methylated as well.

During *genomic imprinting,* either the mother's or the father's gene (but not both) is methylated during gamete formation. If an inherited allele is highly methylated, the gene is not expressed, even if it is a normal gene in every other respect. For traits that exhibit genomic imprinting, the expression of the gene depends on whether the unmethylated allele was inherited from the mother or the father.

The term *epigenetic inheritance* is now used broadly for other inheritance patterns that do not depend on the genes themselves. Epigenetic inheritance explains unusual inheritance patterns and also may play an important role in growth, aging, and cancer. Researchers are even hopeful that it will be easier to develop drugs to modify this level of inheritance rather than trying to change the DNA itself.

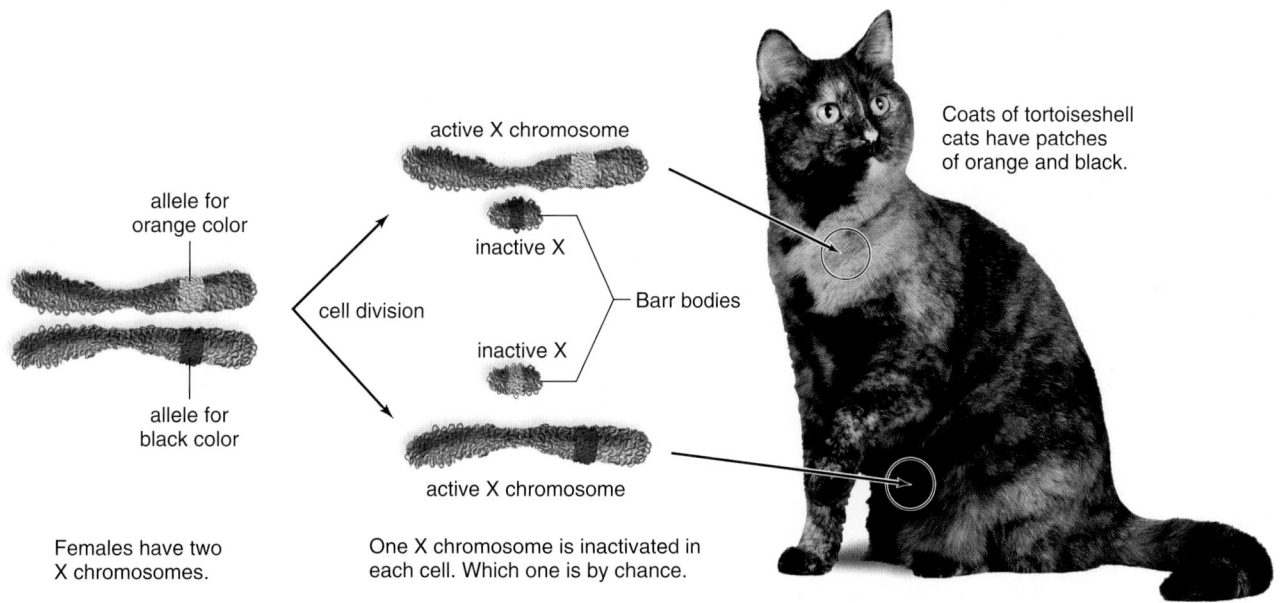

allele for orange color

active X chromosome

inactive X

cell division

Barr bodies

inactive X

allele for black color

active X chromosome

Females have two X chromosomes.

One X chromosome is inactivated in each cell. Which one is by chance.

Coats of tortoiseshell cats have patches of orange and black.

Figure 13.6 X-inactivation in mammalian females.   In cats, the alleles for black or orange coat color are carried on the X chromosomes. Random X-inactivation occurs in females. Therefore, in heterozygous females, 50% of the cells have an allele for black coat color and 50% of cells have an allele for orange coat color. The result is tortoiseshell cats that have coats with patches of both black and orange.

Transcriptional Control

Although eukaryotes have various levels of genetic control (see Fig. 13.4), **transcriptional control** remains the most critical of these levels. The first step toward transcription is availability of DNA, which involves chromatin structure. Transcriptional control also involves the participation of transcription factors, activators, and repressors.

Transcription Factors, Activators, and Repressors

Although some operons like those of prokaryotic cells have been found in eukaryotic cells, transcription in eukaryotes is still controlled by DNA-binding proteins. Every cell contains many different types of **transcription factors,** proteins that help regulate transcription by assisting the binding of the RNA polymerase to the promotor. A cell has many different types of transcription factors, and often a variety of transcription factors may be active at a single promoter. Thus, the absence of one can prevent transcription from occurring.

Animation Transcription Factors

Even if all of the transcription factors are present, transcription may not begin without the assistance of a DNA-binding protein called a **transcription activator.** These bind to regions of DNA called **enhancers,** which may be located some distance from the promoter. A hairpin loop in the DNA brings the transcription activators attached to the enhancer into contact with the transcription factor complex (Fig. 13.7). Likewise, the binding of repressors to silencers within the promoter may prohibit the transcription of certain genes. Most genes are subject to regulation by both activators and repressors also.

Animation Transcription Complex and Enhancers

The promoter structure of eukaryotic genes is often very complex, and a large variety of regulatory proteins may interact with each other and with transcription factors to affect a gene's transcription level. Mediator proteins act as a bridge between transcription factors and transcription activators at the promoter. Now RNA polymerase can begin the transcription process (Fig. 13.7). Such protein-to-protein interactions are a hallmark of eukaryotic gene regulation. Together, these mechanisms can fine-tune a gene's transcription level in response to a large variety of conditions.

Transcription factors, activators, and repressors are always present in the nucleus of a cell, but they most likely have to be activated in some way before they will bind to DNA. Activation often occurs when they are phosphorylated by a kinase. *Kinases,* which add a phosphate group to molecules, and *phosphatases,* which remove a phosphate group, are known to be signaling proteins involved in a growth regulatory network that reaches from receptors in the plasma membrane to the genes in the nucleus.

Posttranscriptional Control

Posttranscriptional control of gene expression occurs in the nucleus and includes alternative mRNA splicing and controlling the speed with which mRNA leaves the nucleus.

During pre-mRNA splicing, introns (noncoding regions) are excised, and exons (expressed regions) are joined together to

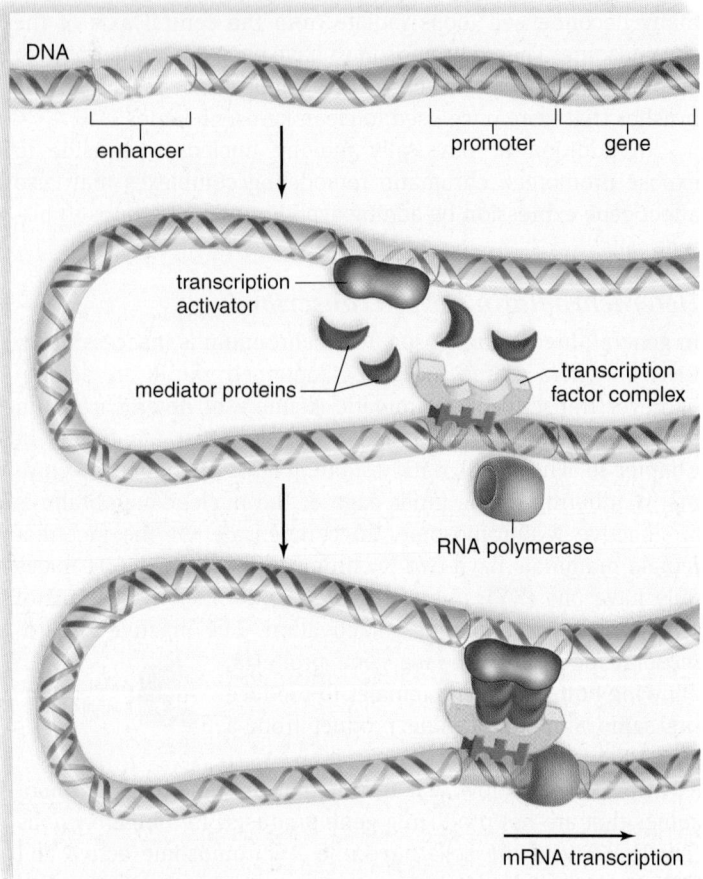

Figure 13.7 Eukaryotic transcription factors. Transcription in eukaryotic cells requires that transcription factors bind to the promoter and transcription activators bind to an enhancer. The enhancer may be far from the promoter, but the DNA loops and mediator proteins act as a bridge joining activators to factors. Only then does transcription begin.

form an mRNA (see Fig. 12.13). When introns are removed from pre-mRNA, differential splicing of exons can occur, and this affects gene expression. For example, an exon that is normally included in an mRNA transcript may be skipped, and it is excised along with the flanking introns (Fig. 13.8). The resulting mature mRNA has an altered sequence, and the protein it encodes is altered. Sometimes introns remain in an mRNA transcript; when this occurs, the protein-coding sequence is also changed.

Animation Exon-Shuffling

Examples of alternative pre-mRNA splicing abound. Both the hypothalamus and the thyroid gland produce a protein hormone called calcitonin, but the mRNA that leaves the nucleus is not the same in both types of cells. This results in the thyroid's releasing a slightly different version of calcitonin than does the hypothalamus. Evidence of alternative mRNA splicing is found in other cells, such as those that produce neurotransmitters, muscle regulatory proteins, and antibodies.

Alternative pre-mRNA splicing allows humans and other complex organisms to recombine their genes in novel ways to create the great variety of proteins found in these organisms. Researchers are busy determining how small nuclear RNAs (snRNAs) affect the splicing of pre-mRNA. They also know that, sometimes, alternative mRNA splicing can result in the inclusion

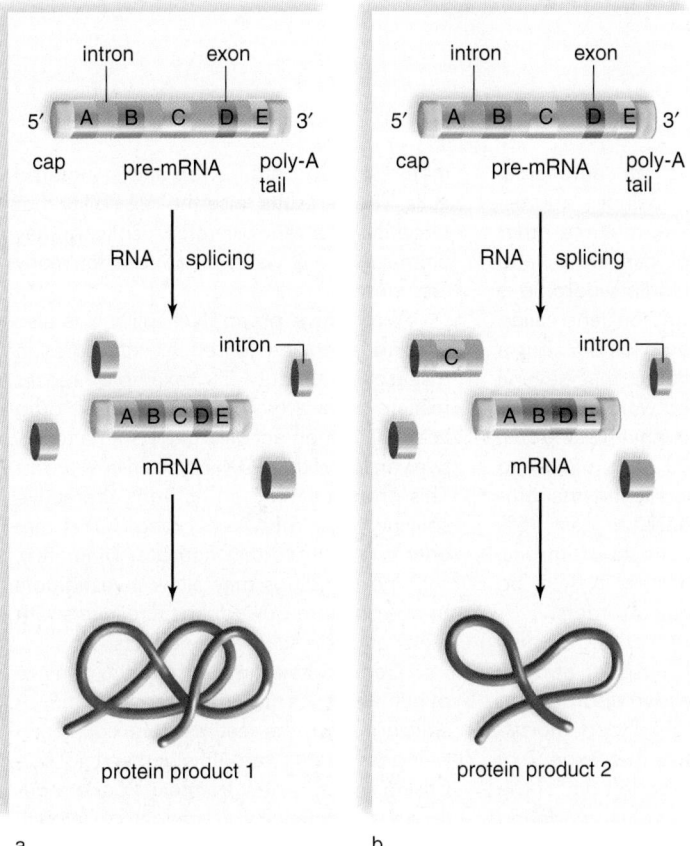

Figure 13.8 Alternative processing of pre-mRNA. Because the pre-mRNAs are processed differently in these two cells (**a** and **b**), distinct proteins result. This is a form of posttranscriptional control of gene expression.

of an intron that brings about destruction of the mRNA before it leaves the nucleus.

Further posttranscriptional control of gene expression is achieved by modifying the speed of transport of mRNA from the nucleus into the cytoplasm. Evidence indicates there is a difference in the length of time it takes various mRNA molecules to pass through a nuclear pore, affecting the amount of gene product realized per unit time following transcription.

Small RNA (sRNA) Molecules Regulate Gene Expression

For a long time, scientists were faced with a mystery: a cell appeared to contain vastly more DNA than was needed to account for the number of expressed proteins. The DNA that was not transcribed into proteins was initially termed "junk" DNA, but recently scientists have begun to understand the role of this DNA in the cell. Although only about 1.5% of the transcribed DNA codes for protein, the remainder is used to form small RNA (sRNA) molecules. We now know that these sRNA molecules represent an important form of gene regulation that functions at multiple levels of gene expression.

How do these RNA molecules regulate gene expression? Notice in Figure 13.9 that transcribed RNA can form loops as hydrogen bonding occurs between its bases. The double-stranded RNA (dsRNA) is diced up by enzymes in the cell to form sRNA molecules. Some of these sRNA molecules regulate transcription, while others are involved in the regulation of translation. Three ways have been found by which sRNA may regulate gene expression:

1. sRNA molecules have been known to alter the compaction of DNA so that some genes are inaccessible to the transcription machinery of the cell.

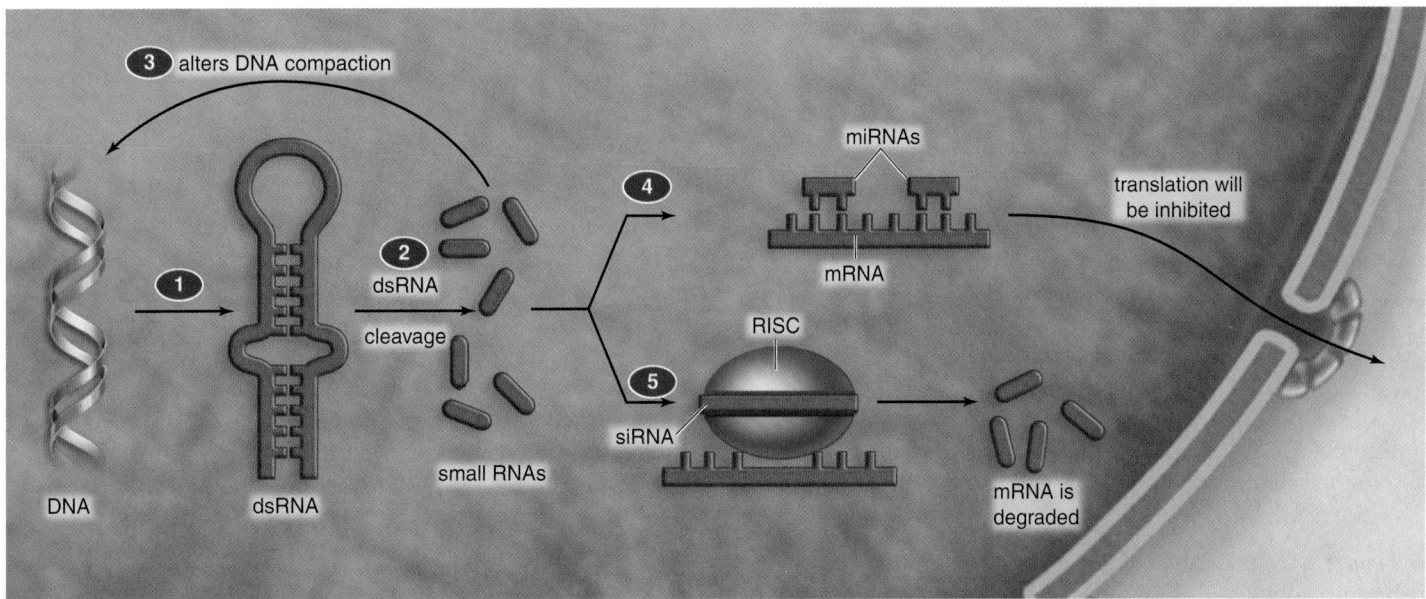

Figure 13.9 Function of small RNA molecules. Transcription of the DNA ❶ may lead to looped and double-stranded RNA (dsRNA). ❷ The cleavage of the dsRNA produces many small RNA (sRNA) molecules. ❸ An sRNA can double-back to increase DNA compaction, or may become an miRNA or siRNA. ❹ miRNA reduces translation by binding to complementary mRNA molecules. ❺ siRNA forms a complex with RISC, which then degrades any mRNA with a sequence of bases that are complementary to the siRNA.

Nature of Science

Alternative mRNA Splicing in Disease

The ability to combine the exons and introns of genes into new and novel combinations through alternative mRNA splicing is one of the mechanisms that allow humans to achieve a higher degree of complexity than simpler organisms without a huge increase in the number of genes. In more advanced organisms, the number of alternatively spliced mRNAs increases greatly. Recently, medical science has discovered that when this process goes awry, disease may result.

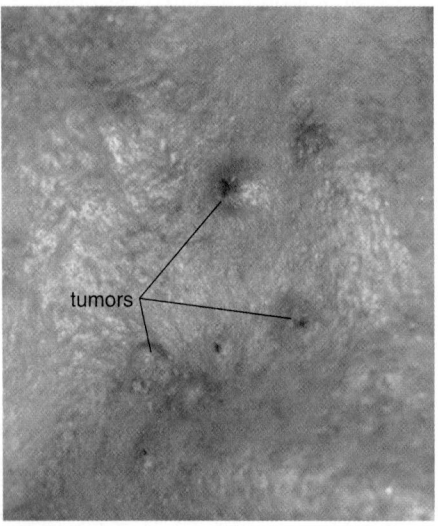

Figure 13A Skin tumors in patient with Gorlin syndrome.

Gorlin syndrome is an autosomal dominant syndrome that includes aggressive skin cancer (Fig. 13A), multiple other tumors, either benign or cancerous, and cysts in various organs. Gorlin syndrome is linked to the tumor suppressor gene called *patched* located on chromosome 9. But a recent finding demonstrated that several new mutations in *patched* were not within the gene's exons, but within its introns. These mutations caused the mRNA to be spliced incorrectly, rendering the protein nonfunctional. Because at least 95% of most human genes consists of introns, many more such mutations are likely to be discovered in other genetic disorders.

Defective pre-mRNA splicing is also a major cause of spinal muscular atrophy (SMA), an autosomal recessive disorder that is a common cause of childhood mortality. Recent research shows that exon 7 of the *SMN2* gene tends to be left out of the mature mRNA in SMA patients, rendering the protein nonfunctional. The end result is a progressive loss of spinal cord motor neurons, and eventually paralysis and skeletal muscle atrophy. Scientists at Cold Spring Harbor Laboratories turned to antisense oligonucleotide technology, a relatively new technique, in an attempt to reverse the defect. The results were stunning—several of the oligonucleotides tested were able to promote the inclusion of exon 7 in the mature mRNA both in vitro and in cultured cells. These results raise the possibility that targeting aberrant pre-mRNA splicing may ultimately be a viable treatment for many disorders.

Alternative pre-mRNA splicing is also causing scientists to rethink strategies in disease treatment. For example, recent research indicates that the common drug acetaminophen actually targets an alternative version of the COX-1 protein in neurons. This protein variant arises from alternative splicing of the mRNA encoding COX-1 that only occurs in certain neurons. Ultimately, such new findings may allow investigators to design more powerful pain relievers with fewer and less severe side effects.

Geneticists estimate that 80% or more of human genes undergo alternative mRNA splicing, and the estimate is constantly being revised upward. It is perhaps not surprising that this new frontier in gene regulation is redefining the standard approach to identifying the causes of illness and presenting new targets for the development of therapeutics.

Questions to Consider

1. What is the normal role of a tumor-suppressor gene in a cell?
2. How might the use of small RNA molecules be used to treat Gorlin syndrome?

2. Small RNAs are the source of **microRNAs (miRNAs),** small snippets of RNA that can bind to and dampen the translation of mRNA in the cytoplasm.

   Video Drug Discovery

3. Small RNAs are also the source of **small-interfering RNAs (siRNAs)** that join with an enzyme (an RNA-induced silencing complex, or RISC) to form an active silencing complex. This activated complex targets specific mRNAs in the cell for breakdown, preventing them from being expressed.

By using a combination of miRNA and siRNA molecules, a cell can fine-tune the amount of gene product being expressed, much like the way in which a dimmer switch on a light regulates the brightness of the room. Because both miRNA and siRNA molecules interfere with the normal gene expression pathways, the process is often referred to as **RNA interference.**

The first scientists to artifically construct miRNA and siRNA molecules to supress the expression of a specific gene were Andrew Fire and Craig Mello. Following this discovery, medical scientists recognized that it may be possible to use sRNA molecules as therapeutic agents to supress the expression of disease-causing genes. For their discovery, Fire and Mello received the 2006 Nobel Prize in Physiology and Medicine.

Animation RNA Interference

Video Halting Hepatitis

Video Tiny Genes Big Role

Translational Control

Translational control begins when the processed mRNA molecule reaches the cytoplasm and before there is a protein product. Translational control involves the activity of mRNA for translation at the ribosome.

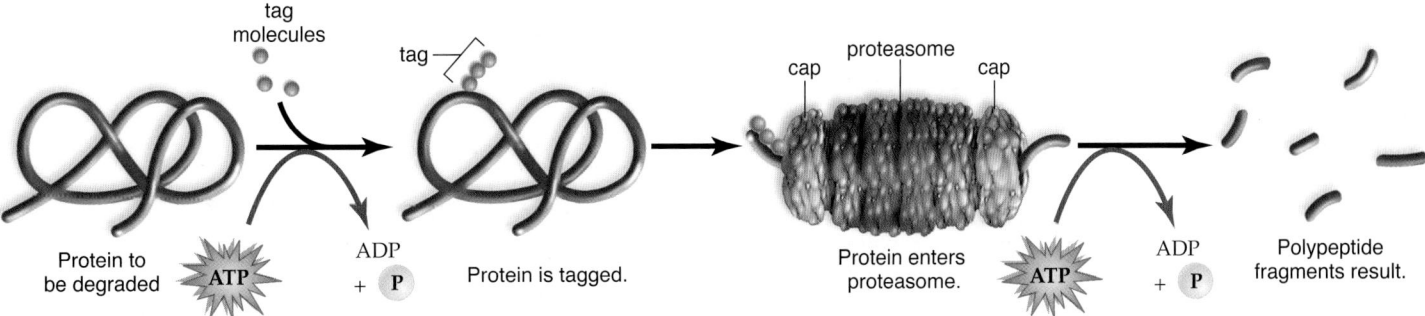

Figure 13.10 Proteasomes degrade proteins in a cell.   Proteins to be degraded are first tagged with a signaling molecule. They then enter the proteasome where they are broken down to polypeptide fragments.

Presence or absence of the 5′ cap and the length of the poly-A (adenine nucleotide) tail at the 3′ end of a mature mRNA transcript can determine whether translation takes place and how long the mRNA is active. The long life of mRNAs that code for hemoglobin in mammalian red blood cells is attributed to the persistence of their 5′ end caps and their long 3′ poly-A tails. Therefore any condition that affects the length of the poly-A tail or leads to removal of the cap may trigger the destruction of an mRNA.

Posttranslational Control

Posttranslational control begins once a protein has been synthesized and has become active. Posttranslational control represents the last chance a cell has for influencing gene expression.

Some proteins are not immediately active after synthesis. For example, at first bovine proinsulin is a single, long polypeptide that folds into a three-dimensional structure. Cleavage results in two smaller chains that are bonded together by disulfide (S—S) bonds. Only then is active insulin present. This ensures that some proteins become active only when it is appropriate for them to do so.

Just how long a protein remains active in a cell is usually regulated by the use of **proteases**, enzymes that breakdown proteins. To protect the cell, proteases are typically confined to the lysosomes or special structures called **proteasomes**. For a protein to enter a proteasome, it has to be tagged with a signaling protein that is recognized by the proteasome cap (Fig. 13.10). When the cap recognizes the tag, it opens and allows the protein to enter the core of the structure, where it is digested to peptide fragments. Notice that proteasomes help regulate gene expression because they help control the amount of protein product in the cytoplasm.

Check Your Progress   13.2

1. Describe the five levels of genetic control in eukaryotes.
2. Explain how chromatin structure influences gene expression.
3. Discuss how small RNA molecules and proteasomes regulate gene expression.

13.3 Gene Mutations

Learning Outcomes

Upon completion of this section you should be able to
1. Distinguish between spontaneous and induced mutations.
2. Identify how mutations influence protein structure.
3. Summarize how mutations may cause cancer.

A **gene mutation** is a permanent change in the sequence of bases in DNA. The effect of a DNA base sequence change on protein activity can range from no effect to complete inactivity. Germ-line mutations are those that occur in sex cells and can be passed to subsequent generations. Somatic mutations occur in body cells and, therefore, they may affect only a small number of cells in a tissue. Somatic mutations are not passed on to future generations, but they can lead to the development of cancer.

Causes of Mutations

Some mutations are spontaneous—they happen for no apparent reason—while others are induced by environmental influences. In most cases, **spontaneous mutations** arise as a result of abnormalities in normal biological processes. **Induced mutations** may result from exposure to toxic chemicals or radiation, which induce (cause) changes in the base sequence of DNA.

Spontaneous Mutations

Spontaneous mutations can be associated with any number of normal processes. For example, a movable piece of DNA, termed a *transposon*, may jump from one location to another, disrupting one or more genes and leading to an abnormal product (see Chapter 14). On rare occasions, a base in DNA can undergo a chemical change that leads to a mispairing during replication. A subsequent base pair change may be carried forth in future generations. Spontaneous mutations due to DNA replication errors, however, are rare. DNA polymerase, the enzyme that carries out replication, proofreads the new strand against the old strand and detects any mismatched nucleotides, and each is usually replaced with a correct nucleotide. In the end, only about one mistake occurs for every 1 billion nucleotide pairs replicated.

Induced Mutations

Induced mutations are caused by **mutagens,** environmental factors that can alter the base composition of DNA. Among the best-known mutagens are radiation and organic chemicals. Many mutagens are also **carcinogens** (cancer-causing).

Chemical mutagens are present in many sources, including some of the food we eat and many industrial chemicals. The mutagenic potential of AF-2, a food additive once widely used in Japan, and of safrole, a flavoring agent once used to flavor root beer, caused them to be banned. Surprisingly, many naturally occurring substances like aflatoxin, produced in moldy grain and peanuts (and present in peanut butter at an average level of 2 parts per billion), and acrylamide, a natural product found in French fries, are also suspected mutagens.

Tobacco smoke contains a number of organic chemicals that are known carcinogens, and it is estimated that one-third of all cancer deaths can be attributed to smoking. Lung cancer is the most frequent lethal cancer in the United States, and smoking is also implicated in the development of cancers of the mouth, larynx, bladder, kidney, and pancreas. The greater the number of cigarettes smoked per day, the earlier the habit starts, and the higher the tar content, the greater is the possibility of these cancers. When smoking is combined with drinking alcohol, the risk of these cancers increases even more.

Scientists use the Ames test for mutagenicity to hypothesize that a chemical can be carcinogenic (Fig. 13.11). In the Ames test, a histidine-requiring strain of bacteria is exposed to a chemical. If the chemical is mutagenic, the bacteria can grow without histidine. A large number of chemicals used in agriculture and industry give a positive Ames test. Examples are ethylene dibromide (EDB), which is added to leaded gasoline (to vaporize lead deposits in the engine and send them out the exhaust), and ziram, which is used to prevent fungus disease on crops. Some drugs, such as isoniazed (used to prevent tuberculosis), are mutagenic according to the Ames test.

Aside from chemicals, certain forms of radiation, such as X-rays and gamma rays, are called ionizing radiation because they create free radicals, ionized atoms with unpaired electrons. Free radicals react with and alter the structure of other molecules, including DNA. Ultraviolet (UV) radiation is easily absorbed by the pyrimidines in DNA. Wherever there are two thymine molecules next to one another, ultraviolet radiation may cause them to bond together, forming *thymine dimers.* A kink results in the DNA. Usually, these dimers are removed by **DNA repair enzymes,** which constantly monitor DNA and fix any irregularities. One enzyme excises a portion of DNA that contains the dimer; another makes a new section by using the other strand as a template; and still another seals the new section in place.

The importance of these repair enzymes is exemplified by individuals with the condition known as xeroderma pigmentosum. They lack some of the repair enzymes, and as a consequence, these individuals have a high incidence of skin cancer because of the large number of mutations that accumulate over time. Also, repair enzymes can fail, as when skin cancer develops because of excessive sunbathing or prolonged exposure to X-rays.

Animation
Thymine Dimers:
Formation and Repair

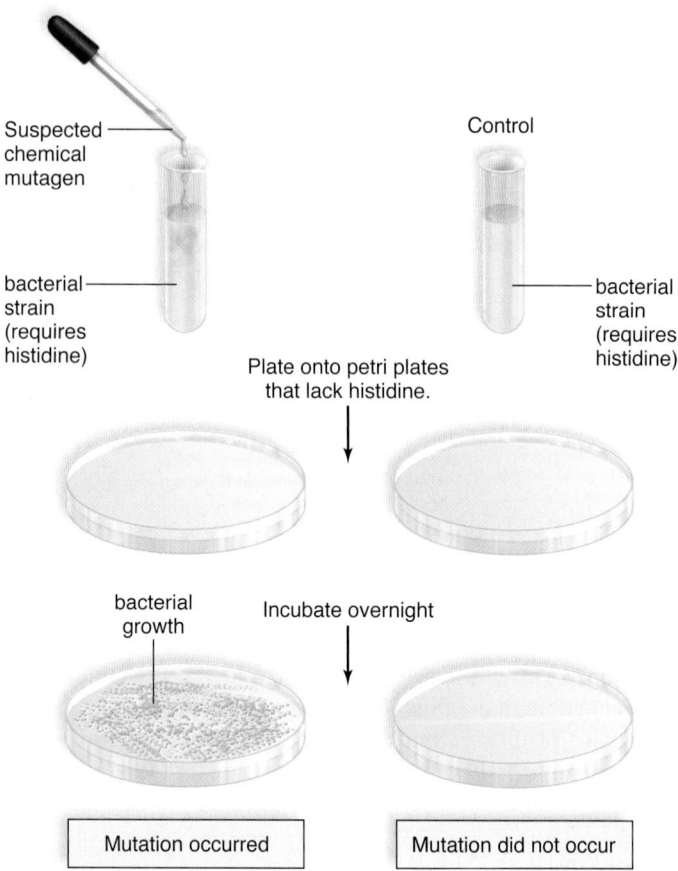

Figure 13.11 The Ames test for mutagenicity.   A bacterial strain that requires histidine as a nutrient is exposed to a suspected chemical mutagen, but a control is not exposed. The bacteria are plated on a medium that lacks histidine and only the bacteria exposed to the chemical show growth. A mutation allowed the bacteria to grow; therefore, the chemical can be carcinogenic.

Effect of Mutations on Protein Activity

Point mutations involve a change in a single DNA nucleotide and, therefore, a possible change in a specific amino acid. The base change in the second row of Figure 13.12a has no effect on the resulting amino acid in hemoglobin; the change in the third row, however, codes for the amino acid valine instead of glutamic acid. This base change accounts for the genetic disorder sickle-cell disease because the incorporation of valine instead of glutamic acid causes hemoglobin molecules to form semirigid rods, and the red blood cells to become sickle shaped. (Compare Figure 13.12b to Figure 13.12c.) Sickle-shaped cells clog blood vessels and die off more quickly than normal-shaped cells. The base change in the fourth row of Figure 13.12a may also have drastic results because the DNA now codes for a stop codon.

Animation
Mutation by Base
Substitution

Frameshift mutations occur most often when one or more nucleotides are either inserted or deleted from DNA. The result of a frameshift mutation can be a completely new sequence of codons and nonfunctional protein. Here is how this occurs: The sequence of codons is read from a specific starting point, as in this sentence, THE CAT ATE THE RAT. If the letter C is deleted from this sentence, shifting the reading frame, we read THE ATA TET HER AT—something that doesn't make sense.

Animation
Addition and
Deletion Mutations

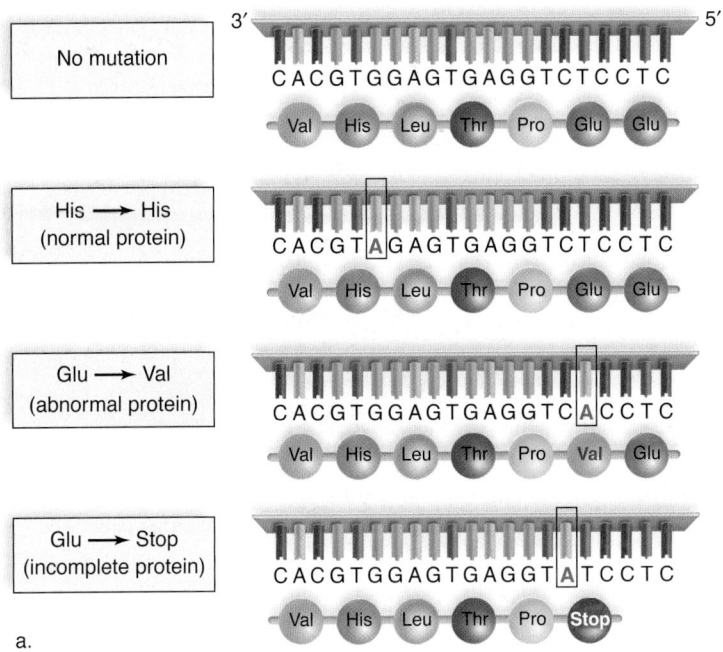

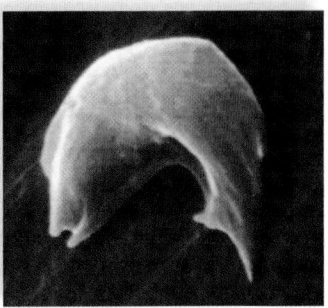

b. Normal red blood cell   c. Sickled red blood cell

Figure 13.12 Point mutations in hemoglobin.   The effect of a point mutation can vary. **a.** Starting at the *top:* Normal sequence of bases in hemoglobin; next, the base change has no effect; next, due to base change, DNA now codes for valine instead of glutamic acid, and the result is that normal red blood cells (**b**) become sickle shaped (**c**); next, a different base change causes DNA to code for termination, and the protein will be incomplete.

Nonfunctional Proteins

A single nonfunctioning protein can have a dramatic effect on the phenotype, because enzymes are often a part of metabolic pathways. One particular metabolic pathway in cells is as follows:

$$A \xrightarrow{E_A} B \xrightarrow{E_B} C$$
$$\text{(phenylalanine)} \quad \text{(tyrosine)} \quad \text{(melanin)}$$

If a faulty code for enzyme E_A is inherited, a person is unable to convert the molecule A to B. Phenylalanine builds up in the system, and the excess causes mental retardation and the other symptoms of the genetic disorder phenylketonuria (PKU). In the same pathway, if a person inherits a faulty code for enzyme E_B, then B cannot be converted to C, and the individual is an albino.

A rare condition called androgen insensitivity is due to a faulty receptor for androgens, which are male sex hormones such as testosterone. In a male with this condition, plenty of testosterone is present in the blood, but the cells are unable to respond to it. Female instead of male external genitals form, and female instead of male secondary sex characteristics occur at puberty. The individual, who appears to be a normal female, may be prompted to seek medical advice when menstruation never occurs. The karyotype is that of a male rather than a female, and the individual does not have the internal sexual organs of a female.

Mutations Can Cause Cancer

It is estimated that one in three people will develop cancer at some time in their lives. Of these affected individuals, one-third of the females and one-fourth of the males will die due to cancer. In the United States, the three deadliest forms of cancer are lung cancer, colon and rectal cancer, and breast cancer.

The development of cancer involves a series of accumulating mutations that can be different for each type of cancer. As discussed in Chapter 9, tumor suppressor genes ordinarily act as brakes on cell division, especially when it begins to occur abnormally. Proto-oncogenes stimulate cell division but are usually turned off in fully differentiated, nondividing cells. When proto-oncogenes mutate, they become oncogenes that are active all the time. Carcinogenesis begins with the loss of tumor suppressor gene activity and/or the gain of oncogene activity. When tumor suppressor genes are inactive and oncogenes are active, cell division occurs uncontrollably because a cell signaling pathway that reaches from the plasma membrane to the nucleus no longer functions as it should (Fig. 13.13 and Fig. 13.14).

It often happens that tumor suppressor genes and proto-oncogenes code for transcription factors or proteins that control transcription factors. As we have seen, transcription factors are a part of the rich and diverse types of mechanisms that control gene expression in cells. They are of fundamental importance to DNA replication and repair, cell growth and division, control of apoptosis, and cellular differentiation. Therefore, it is not surprising that inherited or acquired defects in transcription factor structure and function contribute to the development of cancer.

As examples, the major tumor suppressor gene called *p53* is more frequently mutated in human cancers than any other known gene. It has been found that the p53 protein acts as a transcription factor, and as such is involved in turning on the expression of genes whose products are cell cycle inhibitors (see Chapter 9). *p53* also promotes apoptosis (programmed cell death) when it is needed. The retinoblastoma protein (RB) controls the activity of a transcription factor for cyclin D and other genes whose products promote entry into the S stage of the cell cycle. When the tumor suppressor gene *p16* mutates, the RB

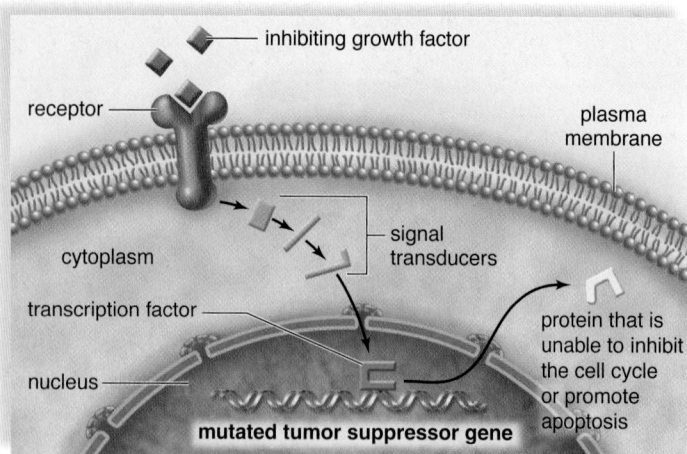

Figure 13.13 Cell signaling pathway that stimulates a mutated tumor suppressor gene. A mutated tumor suppressor gene codes for a product that directly or indirectly stimulates the cell cycle.

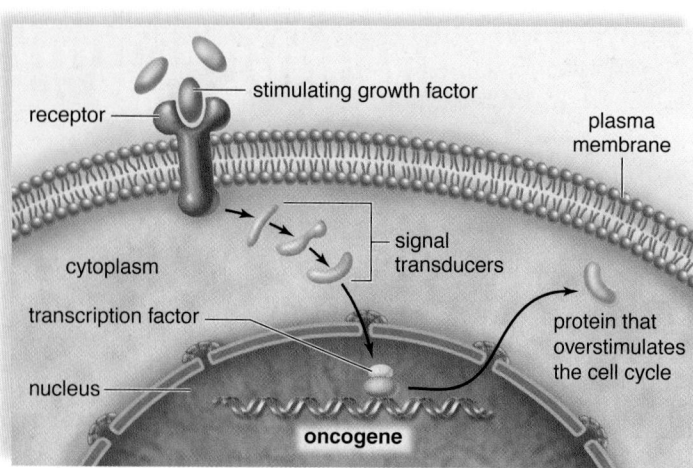

Figure 13.14 Cell signaling pathway that stimulates an oncogene. An oncogene codes for a product that either directly or indirectly overstimulates the cell cycle.

protein is always available, and the result is too much active cyclin D in the cell.

Mutations in many other genes also contribute to the development of cancer. Several proto-oncogenes code for ras proteins, which are needed for cells to grow, to make new DNA, and to not grow out of control. A point mutation is sufficient to turn a normally functioning *ras* proto-oncogene into an oncogene. Abnormal growth results.

Check Your Progress   13.3

1. List some common causes of spontaneous and induced mutations.
2. Explain how a frameshift mutation may disrupt a gene's function.
3. Discuss how a mutation in a tumor suppressor gene and in proto-oncogenes disrupts the cell cycle.

CONNECTING *the* CONCEPTS *with the* BIG IDEAS

Evolution

- All living things are subject to genetic mutations or changes in the base sequence of DNA; this is a source of genetic variability that is acted on by natural selection, allowing for evolutionary change over time. (3C1d)
- It now appears that RNA may play a prominent role in the regulation of the genome; some believe this is evidence that RNA may have preceded DNA in the evolutionary history of cells. (1D1a5)

Information and Signaling

- Prokaryotes use direct mechanisms to control their operons and gene expression; eukaryotes employ many levels of regulation. (3B1a-c)
- Regulatory DNA sequences, molecules, and transcription factors function to control gene expression. (3B1a1*IE*, 3B1c1-3)
- Specialized RNA regulates by alternative splicing and message silencing. Identical genes may produce different phenotypes due to differing gene regulation. (3B1a2, 3B1d)
- DNA repair enzymes correct irregularities and environmental damage; correction system errors lead to mutations with many effects. (3C1a-b)

Interactions and Systems

- Environmental and internal cues may affect gene regulation which ultimately determines cell differentiation and function. (4A3a-c)

*Find the unabridged version of all EK citations at www.glencoe.com/maderAP11.

Media Study Tools

www.glencoe.com/maderAP11

Enhance your study of this chapter with study tools and practice tests. Also ask your instructor about the resources available through ConnectPlus, including the media-rich eBook, interactive learning tools, and animations.

Summarize

13.1 Prokaryotic Regulation

Prokaryotes often organize genes that are involved in a common process or pathway into operons in which the genes are coordinately regulated. Gene expression in prokaryotes is usually regulated at the level of transcription. The operon model states that a regulator gene codes for a repressor. When the repressor binds to the operator RNA polymerase is unable to bind to the promoter, and transcription of the structural genes of the operon cannot take place. Operons may also be regulated by both activators and repressors.

The *trp* operon is an example of a repressible operon because when tryptophan, the corepressor, is present, it binds to the repressor. The repressor is then able to bind to the operator, and transcription of structural genes does not take place.

The *lac* operon is an example of an inducible operon because when lactose, the inducer, is present, it binds to the repressor. The repressor is unable to bind to the operator, and transcription of structural genes takes place if glucose is absent.

Both the *lac* and *trp* operons exhibit negative control, because a repressor is involved; however, some pathways also provide examples of positive control. The structural genes in the *lac* operon are not maximally expressed unless glucose is absent and lactose is present. At that time, cAMP attaches to a molecule called CAP, and then CAP binds to a site next to the promoter. Now RNA polymerase is better able to bind to the promoter, and transcription occurs.

13.2 Eukaryotic Regulation

The following levels of control of gene expression are possible in eukaryotes: chromatin structure, transcriptional control, posttranscriptional control, translational control, and posttranslational control.

Chromatin structure helps regulate transcription. Highly condensed heterochromatin is genetically inactive, as exemplified by Barr bodies. Less-condensed euchromatin is genetically active, as exemplified by lampbrush chromosomes in vertebrates.

Regulatory proteins called transcription factors, as well as DNA sequences called enhancers and silencers, play a role in controlling transcription in eukaryotes. Transcription factors bind to the promoter, and transcription activators bind to an enhancer. Small RNA molecules, such as microRNAs and small-interferring RNAs, are involved in RNA interference and play a role in gene expression.

Posttranscriptional control is achieved by creating variations in messenger RNA (mRNA) splicing, which may yield multiple mRNA messages from the same gene, and by altering the speed with which a particular mRNA molecule leaves the nucleus.

Translational control affects mRNA translation and the length of time it is translated, primarily by altering the stability of an mRNA.

Posttranslational control affects whether or not an enzyme is active and how long it is active. Proteasomes are specialized structures containing proteases that break down protein molecules and participate in the regulation of gene expression.

13.3 Gene Mutations

In molecular terms, a gene is a sequence of DNA nucleotide bases, and a genetic mutation is a change in this sequence. Mutations can be spontaneous or due to environmental mutagens such as radiation and organic chemicals. Carcinogens are mutagens that cause cancer.

Point mutations can have a range of effects, depending on the particular codon change. Sickle-cell disease is an example of a point mutation that greatly changes the activity of the affected gene. Frameshift mutations result when one or more bases are added or deleted, and the result is usually a nonfunctional protein. Most cases of cystic fibrosis are due to a frameshift mutation. Nonfunctional proteins can affect the phenotype drastically, as in a form of albinism that is due to a single faulty enzyme, and androgen insensitivity, which is due to a faulty receptor for testosterone.

Cancer is often due to an accumulation of genetic mutations among genes that code for regulatory proteins. The cell cycle occurs inappropriately when proto-oncogenes become oncogenes and tumor suppressor genes are no longer effective. Mutations that affect transcription factors and other regulators of gene expression are frequent causes of cancer.

Key Terms

Barr body 243	posttranscriptional control 244
carcinogen 248	posttranslational control 247
chromatin 241	promoter 238
corepressor 239	protease 247
DNA repair enzyme 248	proteasome 247
enhancers 244	regulator gene 238
epigenetic inheritance 241	repressor 238
frameshift mutation 248	RNA interference 246
gene mutation 247	small-interfering RNA
induced mutation 247	(siRNA) 246
inducer 240	spontaneous mutation 247
microRNA (miRNA) 246	structural gene 238
mutagen 248	transcription activator 244
operator 239	transcriptional control 244
operon 238	transcription factor 244
point mutation 248	translational control 246

Assess

Reviewing This Chapter

1. Name and state the function of the three components of operons.   238–39
2. Explain the operation of the *trp* operon, and note why it is considered a repressible operon.   239
3. Explain the operation of the *lac* operon, and note why it is considered an inducible operon.   239–40
4. What are the five levels of genetic regulatory control in eukaryotes?   241

5. Relate heterochromatin and euchromatin to levels of chromatin organization.   242–43
6. With regard to transcriptional control in eukaryotes, explain how Barr bodies show that heterochromatin is genetically inactive.   243
7. What do transcription factors do in eukaryotic cells? What are enhancers?   244
8. Explain how alternative mRNA processing may create multiple mRNAs from a single gene.   244–45
9. Describe the role of small RNA molecules in gene regulation.   245–46
10. Give examples of translational and posttranslational control in eukaryotes.   246–47
11. Name some causes of mutations.   247–48
12. What are two major types of mutations, and what effect can they have on protein activity?   247–48
13. Mutations in what types of genes, in particular, can cause cancer?   249

Testing Yourself

Choose the best answer for each question.

1. Which of the following illustrates negative control?
   a. A repressor that becomes active when bound to a corepressor and inhibits transcription.
   b. A gene that binds a repressor and becomes active.
   c. An activator that becomes active when bound to a coactivator and activates transcription.
   d. A repressor that binds a gene and becomes inactive.

2. In regulation of the *lac* operon, when lactose is present and glucose is absent,
   a. the repressor is able to bind to the operator.
   b. the repressor is unable to bind to the operator.
   c. transcription of structural genes occurs.
   d. transcription of lactose occurs.
   e. Both b and c are correct.

3. In regulation of the *trp* operon, when tryptophan is present,
   a. the repressor is able to bind to the operator.
   b. the repressor is unable to bind to the operator.
   c. transcription of the repressor in inhibited.
   d. transcription of the structural genes, operator, and promoter occurs.

4. In operon models, the function of the promoter is to
   a. code for the repressor protein.
   b. bind with RNA polymerase.
   c. bind to the repressor.
   d. code for the regulator gene.

5. Which of the following statements is/are true regarding operons?
   a. The regulator gene is transcribed with the structural genes.
   b. The structural genes are always transcribed.
   c. All genes are always transcribed.
   d. The regulator gene has its own promoter.

6. Which of the following regulate gene expression in the eukaryotic nucleus?
   a. posttranslational control
   b. transcriptional control
   c. translational control
   d. posttranscriptional control
   e. Both b and d are correct.

7. Which of the following mechanisms may create multiple mRNAs from the same gene?
   a. posttranslational control
   b. alternative mRNA splicing
   c. binding of a transcription factor
   d. chromatin remodeling
   e. miRNAs

8. Translational control of gene expression occurs within the
   a. nucleus.
   b. cytoplasm.
   c. nucleolus.
   d. mitochondria.

9. Alternative mRNA splicing is an example of which type of regulation of gene expression?
   a. transcriptional
   b. posttranscriptional
   c. translational
   d. posttranslational

10. A scientist adds radioactive uridine (label for RNA) to a culture of cells and examines an autoradiograph. Which type of chromatin is apt to show the label?
    a. heterochromatin
    b. euchromatin
    c. the histones, not the DNA
    d. the DNA, not the histones
    e. Both a and d are correct.

11. Barr bodies are
    a. genetically active X chromosomes in males.
    b. genetically inactive X chromosomes in females.
    c. genetically active Y chromosomes in males.
    d. genetically inactive Y chromosomes in females.

12. Which of these might cause a proto-oncogene to become an oncogene?
    a. exposure of the cell to radiation
    b. exposure of the cell to certain chemicals
    c. viral infection of the cell
    d. exposure of the cell to pollutants
    e. All of these are correct.

13. A cell is cancerous. You might find an abnormality in
    a. a proto-oncogene.
    b. a tumor suppressor gene.
    c. regulation of the cell cycle.
    d. tumor cells.
    e. All of these are correct.

14. A tumor suppressor gene
    a. inhibits cell division.
    b. opposes oncogenes.
    c. prevents cancer.
    d. is subject to mutations.
    e. All of these are correct.

15. Label this diagram of an operon.

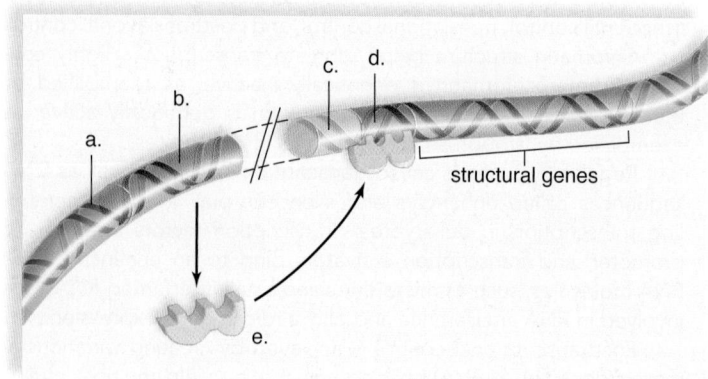

structural genes

16. If the DNA codons are CAT CAT CAT, and a guanine base is added at the beginning, which would result?
    a. CAT CAT CAT G   c. GCA TCA TCA T
    b. G CAT CAT CAT   d. GC ATC ATC AT

17. A mutation in a DNA molecule involving the replacement of one nucleotide base pair with another is called a(n)
    a. frameshift mutation.
    b. transposon.
    c. deletion mutation.
    d. point mutation.
    e. insertion mutation.

18. Which of these is characteristic of cancer?
    a. It may involve a lack of mutations over a length of time.
    b. It cannot be tied to particular environmental factors.
    c. Apoptosis is one of the first developmental effects.
    d. Mutations in certain types of genes.
    e. It typically develops within a short period of time.

19. Which is not evidence that eukaryotes control transcription?
    a. euchromatin/heterochromatin
    b. existence of transcription factors
    c. lampbrush chromosomes
    d. occurrence of mutations
    e. All of these are correct.

Engage

Virtual Lab
DNA and Genes

The virtual lab "DNA and Genes" provides an interactive tutorial for understanding the influence of mutations on protein structure.

Thinking Scientifically

1. In patients with chronic myelogenous leukemia, an odd chromosome is seen in all the cancerous cells. A small piece of chromosome 9 is connected to chromosome 22. This 9:22 translocation has been termed the Philadelphia chromosome. How could a translocation cause genetic changes that result in cancer?

2. New findings indicate that mutations outside of genes may cause disease, such as in some cases of Hirschsprung disease and multiple endocrine neoplasia. Explain how such a mutation might alter the expression of a gene.

Bioethical Issue

Environmental Estrogens and Mutation

You have learned from this chapter that many types of carcinogens, such as those found in cigarette smoke, may alter the base sequence of DNA. However, environmental estrogens are a recently identified type of carcinogen that is generating much attention and concern in recent years. Environmental estrogens are estrogen-like compounds that can disrupt normal endocrine system function in animals by competing with normal sex hormones for receptors, inadvertently activating and inactivating transcription factors and greatly affecting gene expression. They have been linked to increased mutation rates, to deformed genitals in alligators and fish, to promotion of cell division in cultured breast cancer cells, and to inhibition of sperm development in humans.

Environmental estrogens are sometimes found naturally at low concentrations in foods such as soybeans and flax seeds. However, many of these compounds are artificial, originating from chemicals such as polychlorinated biphenyls (PCBs), phthalates (found extensively in many plastics), and atrazine, a compound found in many commercial weed killers. Many people, including scientists at the EPA, contend that these artificial compounds, even at very low doses, are a major threat to the environment, to many animal species, and to human health.

However, some critics contend that the concentrations of these compounds in the soil, air, and water are far below concentrations necessary to cause problems in most animal species, including humans. They also tout studies showing high concentrations of environmental estrogens in many grains, fruits, and vegetables, and that many of these compounds are rendered harmless by the body before they have a chance to cause mutations.

Should known environmental estrogens, such as those found in plastics, herbicides, and insecticides, be closely monitored by the government, and maximal permissible levels set for their emission into the environment? And where should money to fund these regulations be derived? Or, as some critics insist, are we worried about a problem that simply does not exist?

14

Biotechnology and Genomics

A biotechnology product derived from corn may someday replace plastics produced from oil.

BEFORE YOU BEGIN

Before starting this chapter, take a few moments to review these earlier concepts.

Section 12.1 What is the basic structure of a DNA molecule?

Section 12.4 What is the difference between an intron and an exon in a gene?

Section 13.2 What is the role of microRNA molecules in a cell?

Biotechnology is the study and application of living organisms and processes to manufacture products, or improves the characteristics of bacteria, plants or animals. As a few examples, biotechnologists are actively investigating ways of making our world less dependent on fossil fuels, such as oil. Some bacteria produce biodegradable plastic granules inside of their cells. By introducing these genes into plants, such as the corn plants above, it may be possible someday to produce large quantities of biodegradable plastic. In addition to plastic-producing plants, biotechnology has made it possible for bioengineers and medical scientists to alter the genotype, and subsequently the phenotype, of other organisms—bacteria, animals, and humans. Genetically modified crops are resistant to disease and able to grow under stressful conditions. Farm animals can be made to grow larger than usual, and humans may be supplied with normal genes to make up for ones that do not function as they should.

But many people worry that genetically modified bacteria and plants might harm the environment and that the products produced by altered organisms might not be healthy for humans. Other ethical concerns abound. Is it ethical to give a cat a gene that makes it glow in the dark? To what extent would it be proper to improve the human genome? Everyone should be knowledgeable about modern genetics and biotechnology so they can participate in deciding these issues.

As you read through this chapter, think about the following questions:

1. What procedures are used to introduce a bacterial gene into a plant?
2. What is the difference between a genetically-modified and transgenic organism?
3. How are animals being genetically modified?

FOLLOWING *the* BIG IDEAS

CHAPTER 14   BIOTECHNOLOGY AND GENOMICS

Evolution	The field of comparative genomics is yielding valuable new insights into the relationships between species, impacting taxonomy and evolutionary biology.
Information and Signaling	Genetic engineering allows beneficial changes to be made in DNA and RNA sequences, improving the products or the actions of these molecules.

14.1 DNA Cloning

In biology, **cloning** is the production of genetically identical copies of DNA, cells, or organisms through some asexual means. When an underground stem or root sends up new shoots, the resulting plants are clones of one another. The members of a bacterial colony on a petri dish are clones because they all came from the division of a single original cell. Human identical twins are also considered clones. Early in embryonic development the cells separate, and each becomes a complete individual.

DNA cloning can be done to produce many identical copies of the same gene; that is, for the purpose of **gene cloning.** Scientists clone genes for a number of reasons. They might want to determine the difference in base sequence between a normal gene and a mutated gene. Or they might use the genes to genetically modify organisms in a beneficial way. When cloned genes are used to modify a human, the process is called **gene therapy.** Otherwise, the organisms are called **transgenic organisms** [L. *trans,* across, through; Gk. *genic,* producing]. Transgenic organisms are frequently used today to produce a product desired by humans.

Recombinant DNA (rDNA) technology and the polymerase chain reaction (PCR) are two procedures that scientists can use to clone DNA.

Recombinant DNA Technology

Recombinant DNA (rDNA) contains DNA from two or more different sources, such as a human cell and a bacterial cell, as shown in Figure 14.1. To make rDNA, a technician needs a **vector** [L. *vehere,* to carry] by which rDNA will be introduced into a host cell. One common vector is a plasmid. **Plasmids** are small accessory rings of DNA found in bacteria that were first discovered in the bacterium *Escherichia coli* (*E. coli*). The ring is not part of the main bacterial chromosome, and it replicates on its own.

Animation Construction of a Plasmid Vector

Two enzymes are needed to introduce foreign DNA into vector DNA: (1) a **restriction enzyme,** which cleaves DNA, and (2) an enzyme called **DNA ligase** [L. *ligo,* bind], which seals DNA into an opening created by the restriction enzyme. Hundreds of restriction enzymes occur naturally in bacteria, where they cut up any viral DNA that enters the cell. They are called restriction enzymes because they *restrict* the growth of viruses, but they also act as molecular scissors to cleave any piece of DNA at a specific site.

Notice that the restriction enzyme creates a gap in the DNA (Fig. 14.2), into which a piece of foreign DNA can be placed if it ends in bases complementary to those exposed by the restriction enzyme. The single-stranded, but complementary, ends of the two DNA molecules are called "sticky ends" because they can bind a piece of foreign DNA by complementary base-pairing. Sticky ends

Animation Restriction Endonucleases

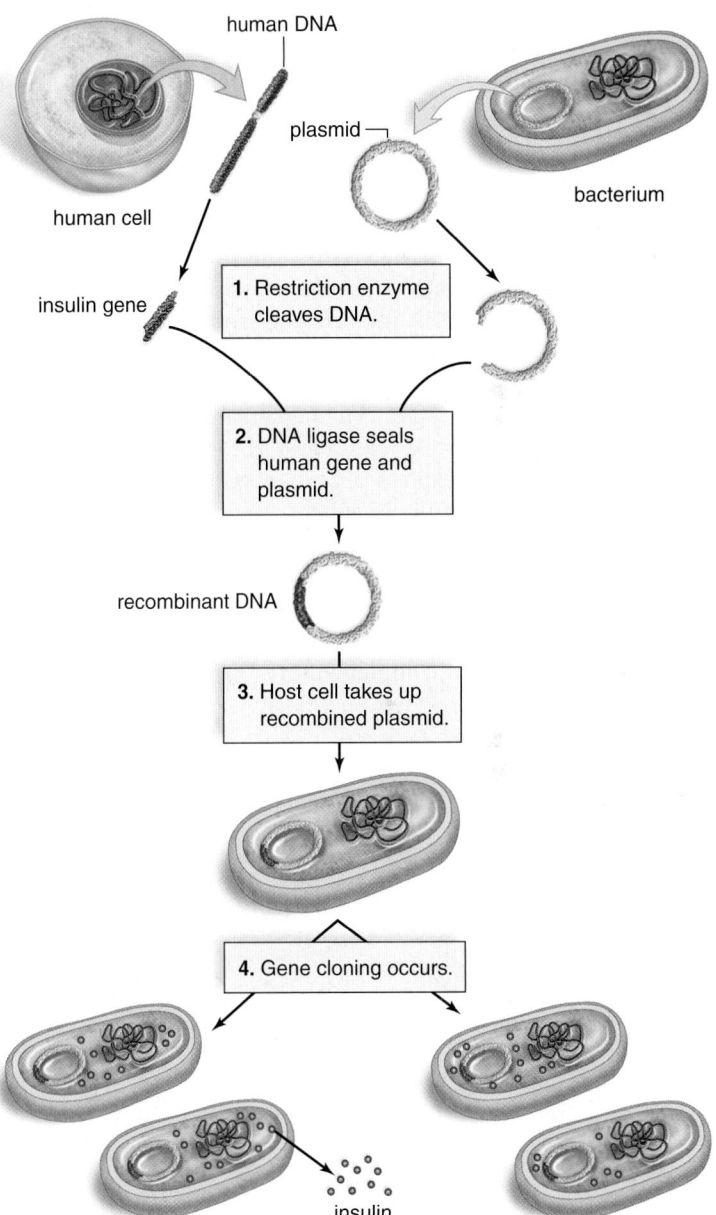

Figure 14.1 **Cloning a human gene.** This figure shows the basic steps in the cloning of a human gene. Human DNA and plasmid DNA are cleaved by a specific type of restriction enzyme. Then the human DNA, perhaps containing the insulin gene, is spliced into a plasmid by the enzyme DNA ligase. Gene cloning is achieved after a bacterium takes up the plasmid. If the gene functions normally as expected, the product (e.g., insulin) may also be retrieved.

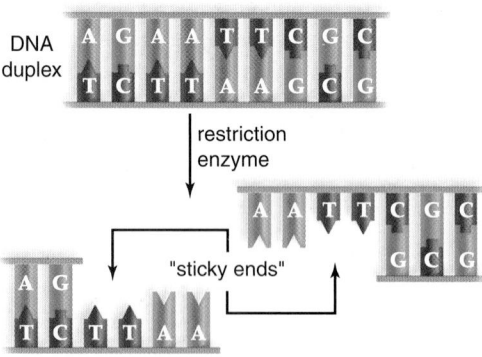

Figure 14.2 Restriction enzymes cut DNA at specific locations. Each restriction enzyme recognizes a specific sequence of nucleotides. After the enzyme cuts the DNA, "sticky ends" may be formed that are useful in the cloning of DNA sequences.

facilitate the insertion of foreign DNA into vector DNA as long as both are cleaved by the same restriction enzyme.

Next, genetic engineers use the enzyme DNA ligase to seal the foreign piece of DNA into the vector. DNA splicing is now complete; an rDNA molecule has been prepared (see Fig. 14.1). Bacterial cells take up recombinant plasmids, especially if they are treated to make their cell membranes more permeable. Thereafter, as the plasmid replicates, DNA is cloned.

For bacteria to express a human gene, the cloned gene has to be accompanied by regulatory regions unique to bacteria. Also, the gene should not contain introns because bacteria don't have introns. However, it is possible to make a human gene that lacks introns. The enzyme called reverse transcriptase can be used to make a DNA copy of human mRNA. The DNA molecule, called **complementary DNA (cDNA)**, does not

contain introns. Bacteria may then transcribe and translate the cloned cDNA to produce a human protein because the genetic code is the same in humans and bacteria.

The Polymerase Chain Reaction

The **polymerase chain reaction (PCR),** developed by Kary Mullis in 1985, is widely used in research laboratories to create copies of a segment of DNA quickly in a test tube. The process mimics DNA replication in the cell (see Section 12.2), except that PCR is very specific—it amplifies (makes copies of) only a targeted DNA sequence. The targeted sequence can be less than one part in a million of the total DNA sample!

PCR requires the use of DNA polymerase, the enzyme that carries out DNA replication, and a supply of nucleotides for the new DNA strands. The DNA polymerase used in the reaction is a heat-stable (thermostable) polymerase that has been extracted from the bacterium *Thermus aquaticus,* which lives in hot springs. The enzyme can withstand the high temperature used to separate double-stranded DNA; therefore, replication does not have to be interrupted by the need to add more enzyme. PCR is a chain reaction because the targeted DNA is repeatedly replicated as long as the process continues. The colors in Figure 14.3 distinguish the old strand from the new DNA strand. Notice that the amount of DNA doubles with each replication cycle.

Analyzing DNA

DNA amplified by PCR can be analyzed for various purposes. For example, mitochondrial DNA taken from modern living populations was used to decipher the evolutionary history of

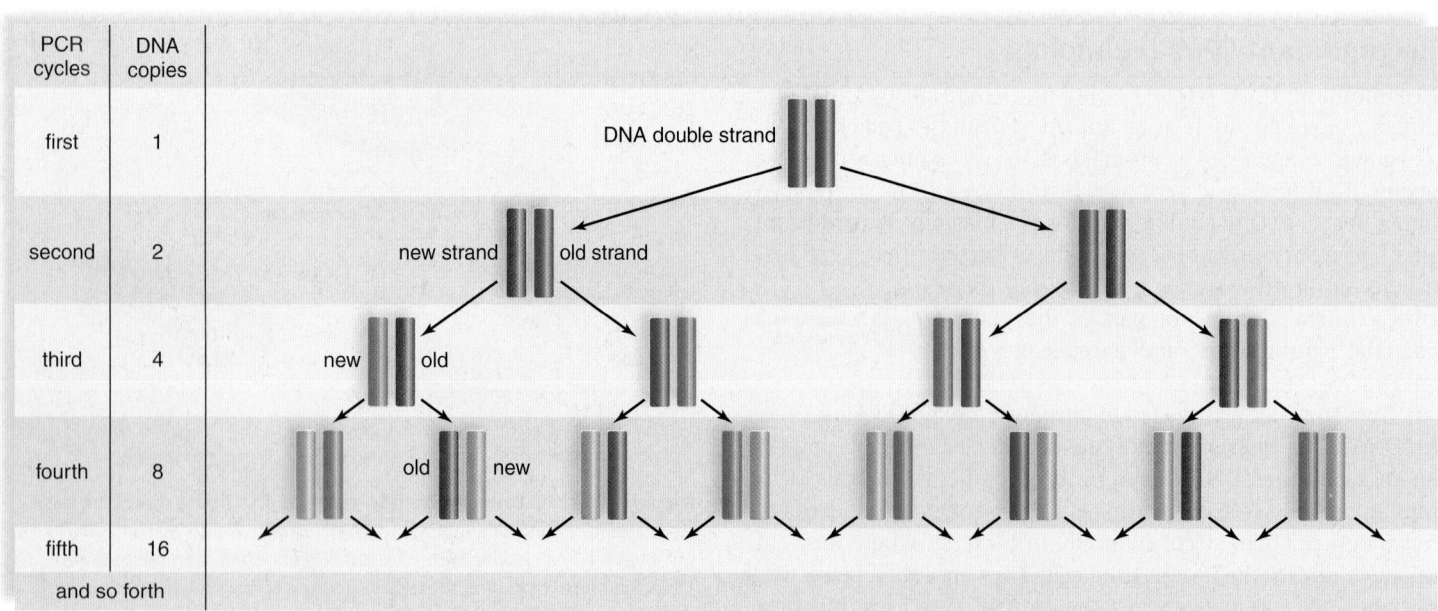

Figure 14.3 Polymerase chain reaction (PCR).   PCR allows the production of many identical copies of DNA in a laboratory setting.

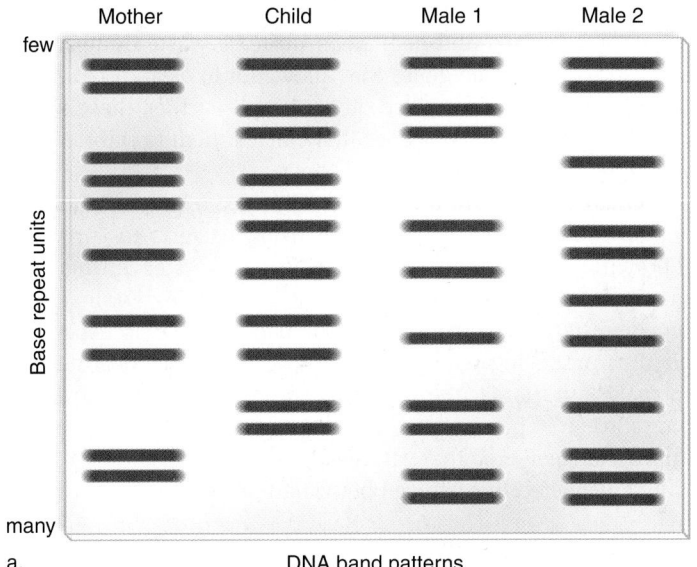

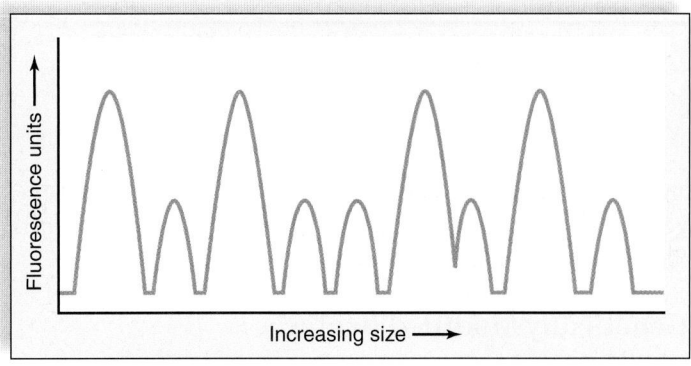

Figure 14.4 The use of STR profiling to establish paternity. **a.** In this method, DNA fragments containing STRs are separated by gel electrophoresis. Male 1 is the father. **b.** Each person's fingerprint pattern (only one is shown) can also be printed out by a machine that detects fluorescence.

human populations. For identification purposes, DNA taken from a corpse burned beyond recognition can be matched to that on the bristles of the person's toothbrush!

Analysis of DNA following PCR has undergone improvements over the years. At first, the entire genome was treated with restriction enzymes, resulting in a unique collection of different-sized fragments because each person has their own restriction enzyme sites. A process called **gel electrophoresis,** which separates DNA fragments according to their size, was then employed; the result of fragment sorting was a pattern of distinctive bands that identified the person.

Now, **short tandem repeat (STR) profiling** is the method of choice. STRs are short sequences of DNA bases that recur several times, as in TCGTCGTCG. STR profiling is advantageous because it doesn't require the use of restriction enzymes. The chromosomal locations for STRs are known, and therefore it is possible to subject only these locations to PCR and use gel electrophoresis to arrive at a band pattern that is different for each person. The band patterns are different because each person has their own number of repeats at the different locations. The greater the number of STRs at a location, the longer the DNA fragment amplified by PCR. The more STR locations employed, the more confident scientists can be of distinctive results for each person (Fig. 14.4a).

The newest method of producing DNA profiles, or "fingerprints," does away with the need to use gel electrophoresis: the DNA fragments are fluorescently labeled. A laser then excites the fluorescent STRs, and a detector records the amount of emission for each DNA fragment in terms of peaks and valleys. Therefore, the greater the fluorescence, the greater the number of repeats at a location. The printout, such as the one shown in

Figure 14.4b, is the DNA fingerprint, and each person has their own unique printout.

Other Applications

Applications of PCR are limited only by the imagination.

- A viral infection, a genetic disorder, or cancer can be confirmed when the DNA tested matches that of a known virus or mutated gene.
- DNA fingerprinted from blood or tissues at a crime scene has been successfully used in screening suspects, convicting criminals, and in exonerating those wrongly convicted.
- DNA fingerprinting through STR profiling has been extensively used to identify the victims of the natural disasters, such as the tsunamis in Indonesia and Japan.

  Video World Trade Center DNA

- Relatives can be found, paternity suits can be settled (Fig. 14.4a), genetic disorders can be detected, and illegally poached ivory and fish can be recognized using this technology (see Chapter 19).
- PCR has also shed new light on evolutionary studies by comparing extracted DNA from ancient specimens with that of living organisms.

Check Your Progress   14.1

1. Describe the process for creating a rDNA molecule.
2. List the scientific benefits of using the polymerase chain reaction.
3. Explain how DNA profiling can be used to analyze DNA molecules.

14.2 Biotechnology Products

Learning Outcomes

Upon completion of this section, you should be able to

1. Identify the benefits of genetically modified bacteria, plants, and animals to human society.
2. Describe the steps involved in the production of a transgenic animal.
3. Explain what is meant by the term xenotransplantation.

Today, transgenic bacteria, plants, and animals are often called **genetically modified organisms (GMOs),** and the products they produce are called **biotechnology products.**

Genetically Modified Bacteria

Many uses have been found for genetically modified bacteria, besides the production of proteins. Biotechnology products from bacteria include insulin, clotting factor VIII, human growth hormone, t-PA (tissue plasminogen activator), and hepatitis B vaccine.

Transgenic bacteria have many other uses as well. Some have been produced to promote the health of plants. For example, bacteria that normally live on plants and encourage the formation of ice crystals have been changed from frost-plus to frost-minus bacteria. As a result, new crops such as frost-resistant strawberries are being developed. Also, a bacterium that normally colonizes the roots of corn plants has now been endowed with genes (from another bacterium) that code for an insect toxin. The toxin protects the roots from insects.

 **Animation**
Early Genetic
Engineering Experiment

Bacteria can be selected for their ability to degrade a particular substance, and this ability can then be enhanced by bioengineering. For instance, naturally occurring bacteria that eat oil have been genetically engineered to clean up beaches (Fig. 14.5) after oil spills, such as the 2010 Deep Water Horizon spill

Figure 14.5 Genetically modified bacteria. Bacteria capable of decomposing oil have been engineered and patented. Recently, scientists used these bacteria extensively in treating the environmental impacts of the Deep Water Horizon oil spill in the Gulf of Mexico.

in the Gulf of Mexico. Bacteria can also remove sulfur from coal before it is burned and help clean up toxic waste dumps. One such strain was given genes that allowed it to clean up levels of toxins that would have killed other strains. Further, these bacteria were given "suicide" genes that caused them to self-destruct when the job had been accomplished.

Organic chemicals are often synthesized by having catalysts act on precursor molecules or by using bacteria to carry out the synthesis. Today, it is possible to go one step further and manipulate the genes that code for these enzymes. For instance, biochemists discovered a strain of bacteria that is especially good at producing phenylalanine, an organic chemical needed to make aspartame, the dipeptide sweetener better known as NutraSweet®. They isolated, altered, and formed a vector for the appropriate genes so that various other bacteria could be genetically engineered to produce phenylalanine.

Genetically Modified Plants

Techniques have been developed to introduce foreign genes into immature plant embryos or into plant cells called *protoplasts* that have had the cell wall removed. The protoplasts are treated with an electric current while they are suspended in a liquid containing foreign DNA. The current creates tiny, self-sealing holes in the plasma membrane through which the DNA can enter. These treated protoplasts go on to develop into mature plants.

Foreign genes transferred to strains of cotton, corn, potato, and even bananas, have made these plants resistant to pests such as fungi and insects, because their cells now produce a chemical that is toxic to the pest species. Similarly, soybeans have been made resistant to a common herbicide. Some corn and cotton plants are both pest- and herbicide-resistant. These and other genetically modified crops that are expected to have increased yields are now sold commercially.

Like bacteria, plants are also being engineered to produce human proteins, such as hormones, clotting factors, and antibodies, in their seeds. One type of antibody made by corn can deliver radioisotopes to tumor cells, and another made by soybeans can be used to treat genital herpes.

 **Video**
Good Tobacco

Video
Potato Vaccine

Genetically Modified Animals

Techniques have been developed to insert genes into the eggs of animals. It is possible to microinject foreign genes into eggs by hand, but another method uses vortex mixing. The eggs are placed in an agitator with DNA and silicon-carbide needles, and the needles make tiny holes through which the DNA can enter. When these eggs are fertilized, the resulting offspring are transgenic animals. Using this technique, many types of animal eggs have taken up the gene for bovine growth hormone (bGH). The procedure has been used to produce larger fishes, cows, pigs, rabbits, and sheep.

 **Video**
Cloned Milk

 Video
Firefly Rx

Gene pharming, the use of transgenic farm animals to produce pharmaceuticals, is being pursued by a number of firms. Genes that code for therapeutic and diagnostic proteins are

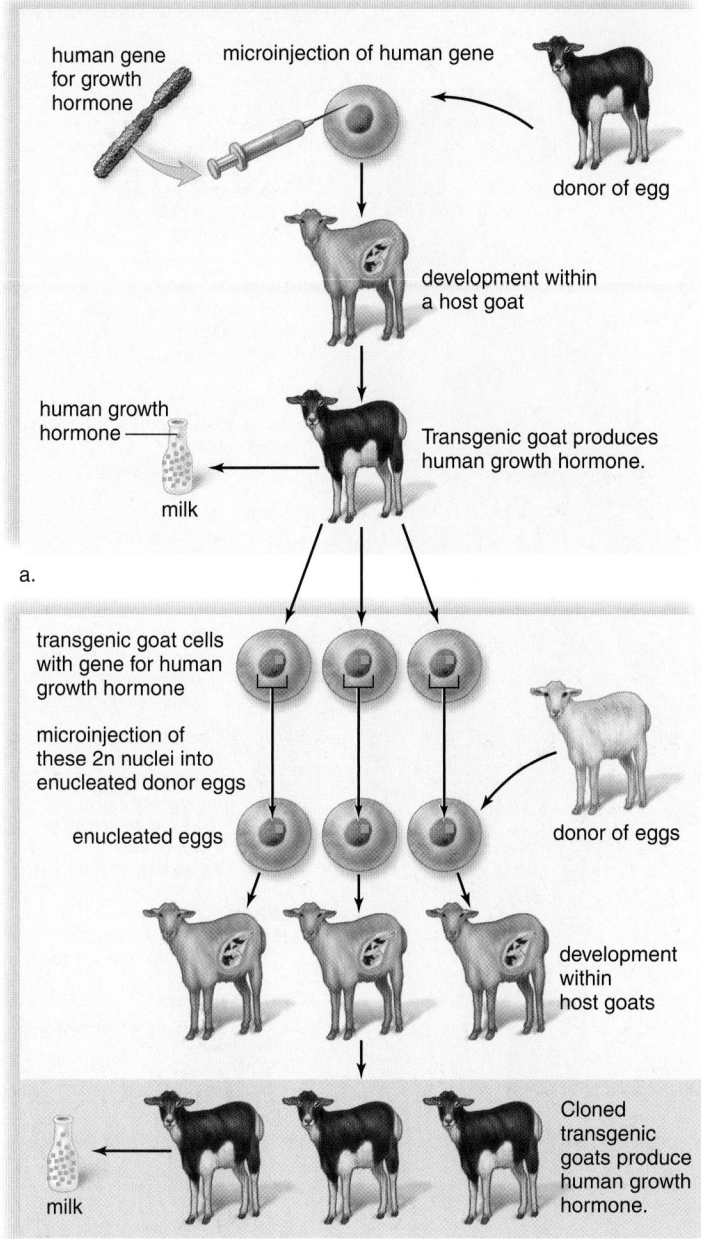

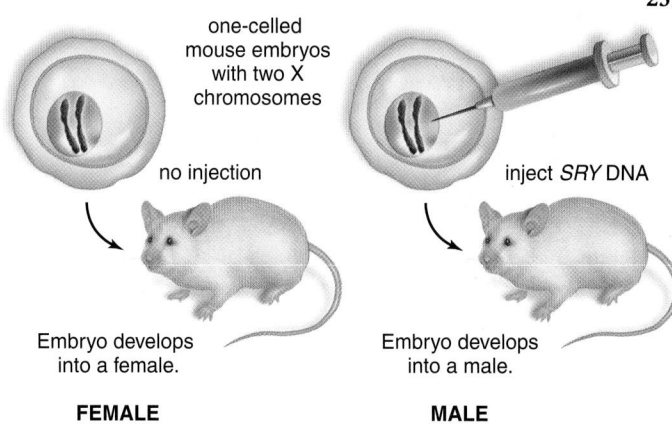

Figure 14.7 Experimental use of mice. Bioengineered mice showed that maleness is due to *SRY* DNA.

a.

Figure 14.6 Transgenic mammals produce a product. This figure illustrates the basic procedure for generating a transgenic animal. **a.** A bioengineered egg develops in a host to create a transgenic goat that produces a biotechnology product in its milk. **b.** Nuclei from the transgenic goat are transferred into donor eggs, which develop into cloned transgenic goats.

incorporated into an animal's DNA, and the proteins appear in the animal's milk. Trials are under way for drugs that treat cystic fibrosis, cancer, blood diseases, and other disorders. Figure 14.6a outlines the procedure for producing transgenic mammals: DNA containing the gene of interest is injected into donor eggs. Following in vitro fertilization, the zygotes are placed in host females, where they develop. After female offspring mature, the product is secreted in their milk.

Cloning Transgenic Animals

For many years, researchers believed that adult vertebrate animals could not be cloned because cloning requires that all the

genes of an adult cell be turned on if development is to proceed normally. This had long been thought impossible.

In 1997, however, Scottish scientists announced that they had produced a cloned sheep, which they called Dolly. Since then, calves, goats, pigs, rabbits, and even cats have also been cloned. The techniques can be applied to produce populations of transgenic animals.

As shown in Figure 14.6b, after enucleated eggs from a donor are microinjected with 2n nuclei from a single transgenic animal, they are coaxed to begin development in vitro. Development continues in host females until the clones are born. The female clones have the same product in their milk as does the donor of the eggs. Now that scientists have a way to clone animals, this procedure will undoubtedly be used routinely to procure biotechnology products. However, animal cloning is a difficult process with a low success rate (usually 1–2 viable embryos per 100 attempts). The vast majority of cloning attempts are unsuccessful, resulting in the early death of the clone.

Applications of Transgenic Animals

Researchers are using transgenic mice for many different research projects. Figure 14.7 shows how this technology has demonstrated that a section of DNA called *SRY* (sex determining region of the Y chromosome) produces a male animal. The *SRY* gene was cloned, and then one copy was injected into single-celled mouse embryos. Injected embryos developed into males, but any that were not injected developed into females.

Eliminating a gene is another way to study a gene's function. A *knockout mouse* has had both alleles of a gene removed or made nonfunctional. For example, scientists have constructed a knockout mouse lacking the *CFTR* gene, the same gene mutated in cystic fibrosis patients. The mutant mouse has a phenotype similar to a human with cystic fibrosis and can be used to test new drugs for the treatment of the disease.

Check Your Progress 14.2

1. List some of the beneficial applications of transgenic bacteria, animals, and plants.
2. Distinguish between a transgenic animal and a cloned animal.

14.3 Gene Therapy

Learning Outcomes

Upon completion of this section, you should be able to

1. Distinguish between in vivo and ex vivo gene therapy in humans.

2. List examples of how in vivo and ex vivo gene therapy has been used to treat human disease.

The manipulation of an organism's genes can be extended to humans in a process called **gene therapy.** Gene therapy is an accepted therapy for the treatment of a disorder and has been used to cure inborn errors of metabolism, as well as to treat more generalized disorders such as cardiovascular disease and cancer (Fig. 14.8).

Viruses genetically modified to be safe can be used to ferry a normal gene into the body, and so can liposomes, which are microscopic globules of lipids specially prepared to enclose the normal gene. Sometimes the gene may be injected directly into a particular region of the body. Below we give examples of *ex vivo gene therapy*, in which the gene is inserted into cells that have been removed and then returned to the body, and *in vivo gene therapy*, in which the gene is delivered directly into the body.

Ex Vivo Gene Therapy

Children who have SCID (severe combined immunodeficiency) lack the enzyme ADA (adenosine deaminase), which is involved in the maturation of immune cells. Therefore, these children are prone to constant infections and may die unless they receive treatment. To carry out gene therapy, bone marrow stem cells are removed from the bone marrow of the patient and are infected with a virus that carries a normal gene for the enzyme into their DNA. Then the cells are returned to the patient, where it is hoped they will divide to produce more blood cells with the same genes.

One of the earliest uses of ex vivo gene therapy was for familial hypercholesterolemia, a condition that develops when liver cells lack a receptor protein for removing cholesterol from the blood. The high levels of blood cholesterol make the patient subject to fatal heart attacks at a young age. In this procedure, a small portion of the liver was surgically excised and then infected with a virus containing a normal gene for the receptor before being returned to the patient. Patients experienced lowered serum cholesterol levels following this procedure. Scientists are investigating using ex vivo gene therapy to treat other human diseases, including some forms of hemophilia.

In Vivo Gene Therapy

Cystic fibrosis patients lack a gene that codes for a transmembrane carrier of the chloride ion. They often suffer from numerous and potentially deadly infections of the respiratory tract. In gene therapy trials, the gene needed to cure cystic fibrosis is sprayed into the nose or delivered to the lower respiratory tract

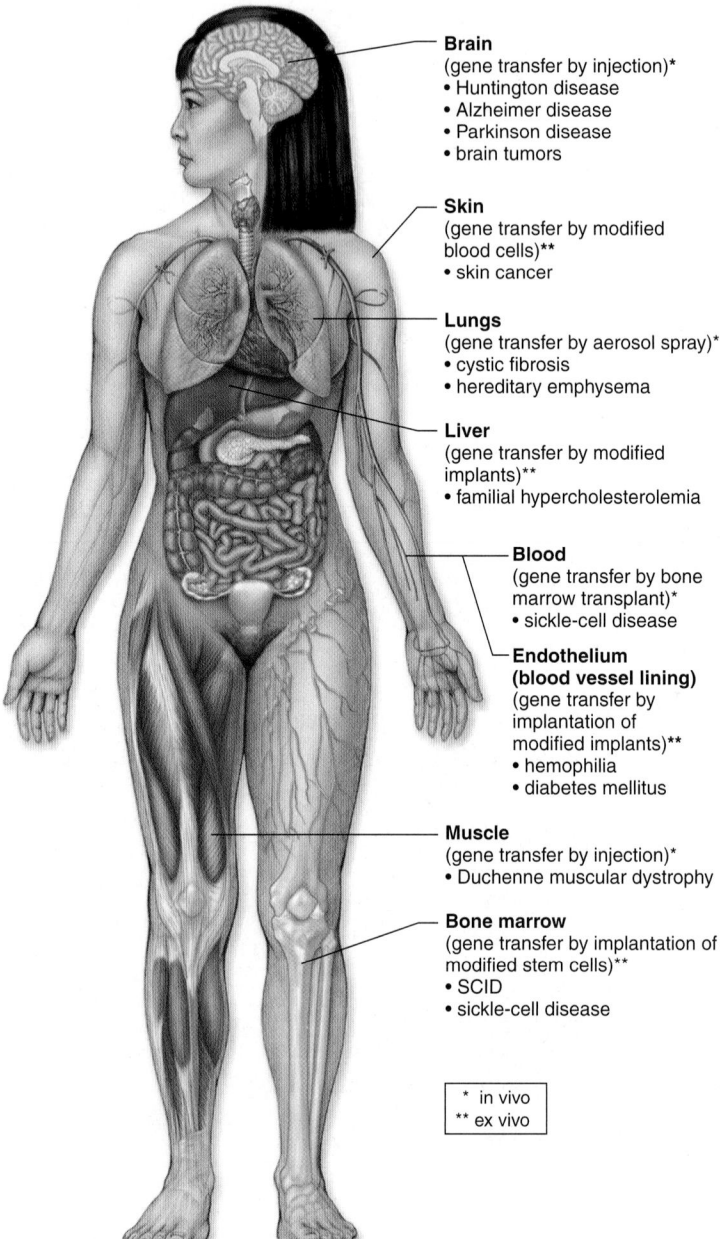

Brain
(gene transfer by injection)*
• Huntington disease
• Alzheimer disease
• Parkinson disease
• brain tumors

Skin
(gene transfer by modified blood cells)**
• skin cancer

Lungs
(gene transfer by aerosol spray)*
• cystic fibrosis
• hereditary emphysema

Liver
(gene transfer by modified implants)**
• familial hypercholesterolemia

Blood
(gene transfer by bone marrow transplant)*
• sickle-cell disease

Endothelium (blood vessel lining)
(gene transfer by implantation of modified implants)**
• hemophilia
• diabetes mellitus

Muscle
(gene transfer by injection)*
• Duchenne muscular dystrophy

Bone marrow
(gene transfer by implantation of modified stem cells)**
• SCID
• sickle-cell disease

| * in vivo |
| ** ex vivo |

Figure 14.8 Gene therapy.   Sites of ex vivo and in vivo gene therapy to cure the conditions noted.

by adenoviruses or by the use of a liposome. So far, this treatment has resulted in limited success.

In cancer patients, genes are being used to make healthy cells more tolerant of chemotherapy and to make tumors more vulnerable to chemotherapy. The gene *p53* brings about apoptosis, and there is much interest in introducing it into cancer cells that no longer have the gene and in that way killing them off.

Check Your Progress   14.3

1. Describe the methods that are being used to introduce genes into human beings for gene therapy.

2. Discuss an example of ex vivo and of in vivo gene therapy.

14.4 Genomics

Learning Outcomes

Upon completion of this section, you should be able to

1. Distinguish between the sciences of genomics, proteomics, and bioinformatics.
2. Identify the function of repetitive elements, transposons, and unique noncoding RNA sequences in the human genome.
3. Explain how DNA microarrays are used in the study of genomics.

In the preceding century, researchers discovered the structure of DNA, how DNA replicates, and how DNA and RNA are involved in the process of protein synthesis. Genetics in the twenty-first century concerns **genomics,** the study of genomes—our complete genetic makeup and also that of other organisms. Knowing the sequence of bases in genomes is the first step, and thereafter we want to understand the function of our genes and their introns, as well as the intergenic sequences. The enormity of the task can be appreciated by knowing that there are approximately 6 billion base pairs in the 2n human genome. Many other organisms have a larger number of protein-coding genes but fewer noncoding regions compared to the human genome.

Sequencing the Genome

We now know the order of the base pairs in the human genome. This feat, which has been likened to arriving at the periodic table of the elements in chemistry, was accomplished by the **Human Genome Project (HGP),** a 13-year effort that involved both university and private laboratories around the world.

In the beginning, investigators developed a laboratory procedure that would allow them to decipher a short sequence of base pairs, and then instruments became available that could carry out sequencing automatically. Over the 13-year span, DNA sequencers were constantly improved, and now modern instruments can automatically analyze up to 2 million base pairs of DNA in a 24-hour period.

Sperm DNA was the material of choice for analysis because it has a much higher ratio of DNA to protein than other types of cells. (Recall that sperm do provide both X and Y chromosomes.) However, white cells from the blood of female donors were also used in order to include female-originated samples. The male and female donors were of European, African, American (both North and South), and Asian ancestry.

Many small regions of DNA that vary among individuals, termed polymorphisms, were identified during the HGP. Most of these are *single nucleotide polymorphisms (SNPs);* these vary by only one nucleotide. Many SNPs have no effect; others may contribute to enzymatic differences affecting the phenotype. It's possible that certain SNP patterns change an individual's susceptibility to disease and alter their response to medical treatments (see Chapter 16).

Determining the number of genes in the human genome required a number of techniques, many of which relied on identifying RNAs in cells and then working backward to find the DNA that can pair with that RNA. **Structural genomics**—knowing the sequence of the bases and how many genes we have—is now being followed by functional genomics.

Estimates place the number of human genes between 23,000 and 25,000. The majority of these genes are expected to code for proteins. However, much of the human genome was formerly described as "junk" because it does not specify the order of amino acids in a polypeptide. However, recall from Chapter 12 that it is possible for RNA molecules to have a regulatory effect in cells. We examine this in more detail in the next section.

Structure of the Eukaryotic Genome

Historically, genes were defined as discrete units of heredity that corresponded to a locus on a chromosome (see Fig. 11.4). Prokaryotes typically possess a single circular chromosome with genes that are packed together very closely; eukaryotic chromosomes, in contrast, are much more complex: The genes are seemingly randomly distributed along the length of a chromosome and are fragmented into exons, with intervening sequences called introns scattered throughout the length of the gene (Fig. 14.9).

Figure 14.9 Chromosomal DNA.   A genome contains protein-coding DNA (exons) and noncoding DNA, including introns (light blue), and other intergenic sequences (red). Only the exons are present in mRNA and specify protein synthesis.

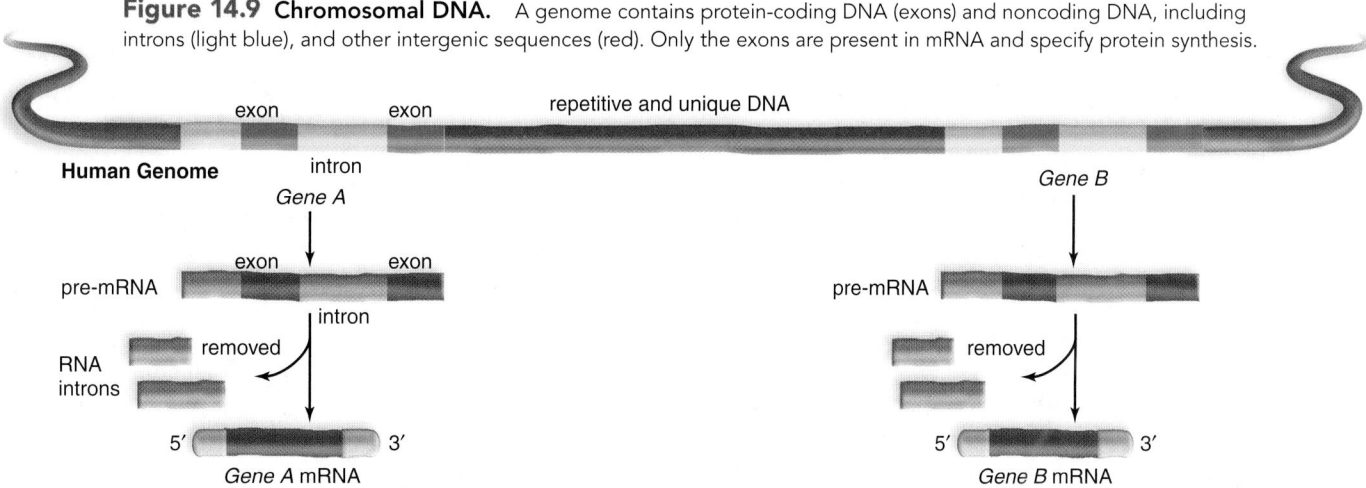

In general, more complex organisms have more complex genes with more and larger introns. In humans, 95% or more of the average protein-coding gene is introns. Once a gene is transcribed, the introns must be removed and the exons joined together to form a functional mRNA transcript (see Fig. 12.13).

Once regarded as merely intervening sequences, introns are now attracting attention as regulators of gene expression. The presence of introns allows exons to be put together in various sequences so that different mRNAs and proteins can result from a single gene. Introns might also function to regulate gene expression and help determine which genes are to be expressed and how they are to be spliced. In fact, entire genes have been found embedded within the introns of other genes.

Intergenic Sequences

DNA sequences occur between genes and are referred to as **intergenic sequences.** In general, as the complexity of an organism increases, so does the proportion of its noncoding DNA sequences. Intergenic sequences are now known to comprise the vast majority of human chromosomes, and protein-coding genes represent only about 1.5 to 2% of our total DNA. The remainder of this DNA, once dismissed as "junk DNA," is now thought to serve many important functions. Several basic types of intergenic sequences are found in the human genome, including (1) repetitive elements, (2) transposons, and (3) unique noncoding DNA. The majority of intergenic sequences belong to this last class.

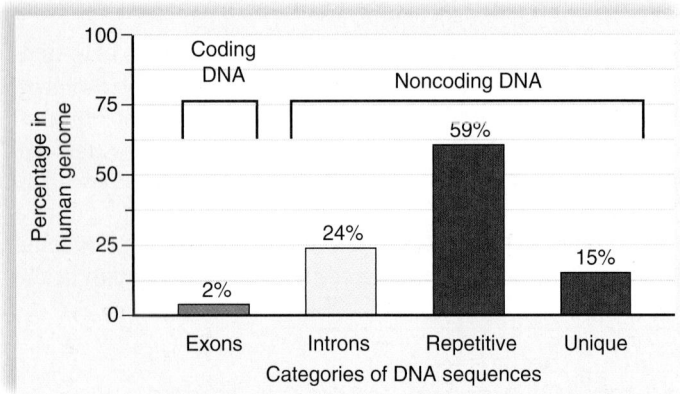

Repetitive DNA Elements

Repetitive DNA elements occur when a sequence of two or more nucleotides is repeated many times along the length of one or more chromosomes. Repetitive elements are very common—comprising nearly half of the human genome—and therefore, many scientists believe that their true significance has yet to be discovered. Although many scientists still dismiss them as having no function, others point out that the centromeres and telomeres of chromosomes are composed of repetitive elements, suggesting that repetitive DNA elements may not be as useless as once thought. For one thing, those of the centromere could possibly help with segregating the chromosomes during cell division.

Repetitive DNA elements include tandem repeats and interspersed repeats. **Tandem repeat** means that the repeated sequences are next to each other on the chromosome. Tandem repeats are often referred to as satellite DNA, because they have a different density than the rest of the DNA within the chromosome. The number and types of tandem repeats may vary significantly from one individual to another, making them invaluable as indicators of heritage. One type of tandem repeat sequence, referred to as *short tandem repeats,* or STRs, has become a standard method in forensic science for distinguishing one individual from another and for determining familial relationships (see page 257).

The second type of repetitive DNA element is called an **interspersed repeat,** meaning that the repetitions may be placed intermittently along a single chromosome, or across multiple chromosomes. For example, a repetitive DNA element, known as the *Alu* sequence, is interspersed every 5,000 base pairs in human DNA and comprises nearly 5–6% of total human DNA. Because of their common occurrence, interspersed repeats are thought to play a role in the evolution of new genes.

Transposons

Transposons are specific DNA sequences that have the remarkable ability to move within and between chromosomes. Their movement to a new location sometimes alters neighboring genes, particularly decreasing their expression. In other words, a transposon sometimes acts like a regulator gene. The movement of transposons throughout the genome is thought to be a driving force in the evolution of living things. The *Alu* repetitive element is an example of a transposon. In fact, many scientists now think that many repetitive DNA elements were originally derived from transposons.

Although Barbara McClintock first described these "movable elements" in corn over 60 years ago, it took time for the scientific community to fully appreciate this revolutionary idea. In fact, their significance was only realized within the past few decades. Transposons, sometimes termed "jumping genes," have now been discovered in bacteria, fruit flies, humans, and many other organisms. McClintock received a Nobel Prize in 1983 for her discovery of transposons and for her pioneering work in genetics.

Animation
Transposons

Unique Noncoding DNA

Genes comprise an estimated 1.5% of the human genome, and repetitive DNA elements make up about 44%; the function of the remaining half, or *unique noncoding DNA*, remains a mystery. Even though this DNA does not appear to contain any protein-coding genes, it has been highly conserved through evolution. In the many millions of years that separate humans from mice, large tracts of this mysterious DNA have remained almost unchanged. But if this DNA has no relevant function, then why has it been so meticulously maintained?

Recently, scientists observed that between 74% and 93% of the genome is transcribed into RNA, including many of these unknown sequences. Thus, what was once thought to be a vast "junk DNA" wasteland may be much more important than once thought and may play active roles in the cell. Small-sized

RNAs may be able to carry out regulatory functions more easily than proteins at times. Therefore, a previously overlooked RNA signaling network may be what allows humans, for example, to achieve structural complexity far beyond anything seen in the unicellular world. Together, these findings have revealed a much more complex, dynamic genome than was envisioned merely a few decades ago.

Revisiting the Definition of a Gene

Perhaps the modern definition of a gene should take the emphasis away from the chromosome and place it on the results of transcription. Previously, molecular genetics considered a gene to be a nucleic acid sequence that codes for the sequence of amino acids in a protein. In contrast to this definition, geneticists have known for some time that all three types of RNA are transcribed from DNA, and that these RNAs are useful products. We also know that protein-coding regions can be interrupted by regions that do not code for a protein but do produce RNAs with various functions. Taking this into consideration, an alternative definition was suggested by Mark Gerstein and associates in 2007: "A gene is a genomic sequence (either DNA or RNA) directly encoding functional products, either RNA or protein."[1]

This definition merely expands on the central dogma of genetics and recognizes that a gene product need not be a protein, and a gene need not be a particular locus on a chromosome. The DNA sequence that results in a gene product can be split and be present on one or several chromosomes. Also, any DNA sequence can result in one or more products. Furthermore, this definition recognizes that some prokaryotes have RNA genes. In other words, the genetic material need not be DNA. Again, we can view this as a simple expansion of the central dogma of genetics.

However, the definition does not spell out what is meant by functional product. It would seem, then, that sequences of DNA resulting in regulatory RNAs or proteins could be considered

1 Gerstein, M. B., Bruce, C., Rozowsky, J. S., et al. 2007. What is a gene, post-ENCODE? History and updated definition. *Genomic Research* 17: 669–81.

genes. According to this definition, "coding" no longer necessarily indicates a DNA sequence that codes for a sequence of amino acids. Instead, it simply means a sequence of DNA bases that are transcribed. The gene product can be RNA molecules, or it can be a protein.

Functional and Comparative Genomics

Since we now know the structure of our genome, the emphasis today is on functional genomics and also on comparative genomics. The aim of **functional genomics** is to understand the exact role of the genome in cells or organisms.

The Nature of Science feature on page 264 discusses the importance of a new technology that can be used to monitor the expression of thousands of genes simultaneously. **DNA microarrays,** also known as DNA chips or genome chips, contain microscopic amounts of known DNA sequences fixed onto a small glass slide or silicon chip in known locations (see Fig. 14A). The use of a microarray can tell you what genes are turned on in a specific cell or organism at a particular time and under what particular environmental circumstances. When mRNA molecules of a cell or organism bind through complementary base pairing with the various DNA sequences on an array, then that gene is active in the cell.

DNA microarrays are increasingly available that rapidly identify all the mutations in the genome of an individual. This information is called the person's **genetic profile.** The genetic profile can indicate if any genetic illnesses are likely and what type of drug therapy for an illness might be most appropriate for that individual.

Animation
Microarray

Aside from the protein-coding regions, researchers also want to know how SNPs and non-protein-coding regions, including repeats, affect which proteins are active in cells. As already discussed at length in Chapter 12, much research is now devoted to knowing the function of DNA regions that do not code for proteins.

The aim of **comparative genomics** is to compare the human genome to the genome of other organisms, such as the model organisms listed in Table 14.1. Model organisms are used in

Table 14.1 Comparison of Sequenced Genomes

Organism	Homo sapiens (human)	Mus musculus (mouse)	Drosophila melanogaster (fruit fly)	Arabidopsis thaliana (flowering plant)	Caenorhabditis elegans (roundworm)	Saccharomyces cerevisiae (yeast)
Estimated Size	3,200 million bases	2,500 million bases	180 million bases	125 million bases	97 million bases	12 million bases
Estimated Number of Genes	~25,000	~25,000	13,600	25,500	19,100	6,300
Chromosome Number	46	40	8	10	12	32

Nature of Science

DNA Microarray Technology

With advances in robotic technology, it is now possible to place the entire human genome onto a single microarray (Fig. 14A). The mRNA from the organism or the cell to be tested is labeled with a fluorescent dye and added to the chip. When the mRNAs bind to the microarray, a fluorescent pattern results that is recorded by a computer. Now the investigator knows what DNA is active in that cell or organism. A researcher can use this method to determine the difference in gene expression between two different cell types, such as between liver cells and muscle cells.  **Animation** DNA Microarray

Genetic Profiles

A mutation microarray, the most common type, can be used to generate a person's genetic profile. The microarray contains hundreds to thousands of known disease-associated mutant gene alleles. Genomic DNA from the individual to be tested is labeled with a fluorescent dye, and then added to the microarray. The spots on the microarray fluoresce if the individual's DNA binds to the mutant genes on the chip, indicating that the individual may have a particular disorder or is at risk for developing it later in life. This technique can generate a genetic profile much more quickly and inexpensively than older methods involving DNA sequencing.

Diseased Tissues

DNA microarrays also promise to hasten the identification of genes associated with diseased tissues. In the first instance, mRNA derived from diseased tissue and normal tissue is labeled with different fluorescent dyes. The normal tissue serves as a control.

The investigator applies the mRNA from both normal and abnormal tissue to the microarray. The relative intensities of fluorescence from a spot on the microarray indicate the amount of mRNA originating from that gene in the diseased tissue relative to the normal tissue. If a gene is activated in the disease, more copies of mRNA will bind to the microarray than from the control tissue, and the spot will appear more red than green.

Genomic microarrays are also used to identify links between disease and chromosomal variations. In this instance, the chip contains genomic DNA that is cut into fragments. Each spot on the microarray corresponds to a known chromosomal location. Labeled genomic DNA from diseased and control tissues bind to the DNA on the chip,

and the relative fluorescence from both dyes is determined. If the number of copies of any particular target DNA has increased, more sample DNA will bind to that spot on the microarray relative to the control DNA, and a difference in fluorescence of the two dyes will be detected. Researchers are currently using this technique to identify disease-associated copy number variations, such as those discussed in the Evolution feature on page 266.

Questions to Consider

1. Why might a researcher want to know what genes are being expressed in different cell types?
2. How might the information from a DNA microarray be used to develop new drugs to treat disease?

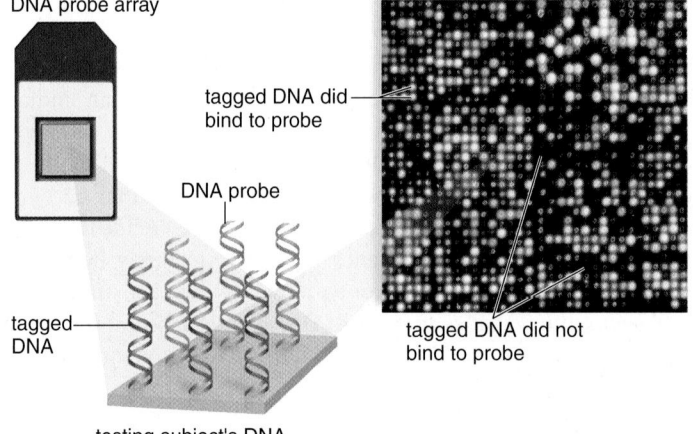

DNA probe array

tagged DNA did bind to probe

DNA probe

tagged DNA

tagged DNA did not bind to probe

testing subject's DNA

Figure 14A DNA microarray technology.
A DNA microarray contains many microscopic samples of DNA bound to known locations on a silicon chip. A fluorescently labeled mRNA from a tissue or organism binds to the DNA on the chip by complementary base pairing. The fluorescent spots indicate that binding has occurred and that the gene functions in that cell.

genetic analysis because they have many genetic mechanisms and cellular pathways in common with each other and with humans. Functional genomics has also been advanced through the study of these genomes.

Much has been learned by genetically modifying mice; however, other model organisms can also sometimes be used. Scientists inserted a human gene associated with early-onset Parkinson disease into *Drosophila melanogaster,* and the flies

showed symptoms similar to those seen in humans with the disorder. This outcome suggested that we might be able to use these organisms instead of mice to test therapies for Parkinson disease.

Comparative genomics also offers a way to study changes in a genome over time because the model organisms have a shorter generation time than humans. Comparing genomes can also help us understand the evolutionary relationships between organisms. One surprising discovery is that the genomes of all vertebrates are highly similar. Researchers were not surprised to find that the genes of chimpanzees and humans are 98% alike, but they did not expect to find that our sequence is also 85% like that of a mouse. Genomic comparisons will likely reveal evolutionary relationships between organisms never previously considered.

Proteomics

The entire collection of a species' proteins is the **proteome.** At first, it may be surprising to learn that the proteome is larger than the genome until we consider all the many regulatory mechanisms, such as alternative pre-mRNA splicing, that increase the number of possible proteins in an organism.

Proteomics is the study of the structure, function, and interaction of cellular proteins. Specific regulatory mechanisms differ between cells, and these differences account for the specialization of cells. One goal of proteomics is to identify and determine the function of the proteins within a particular cell type. Each cell produces thousands of different proteins that can vary not only between cells but also within the same cell, depending on circumstances. Therefore, the goal of proteomics is an overwhelming endeavor. Microarray technology can assist with this project and so can today's supercomputers.

Computer modeling of the three-dimensional shape of cellular proteins is also an important part of proteomics. If the primary structure of a protein is known, it should be possible to predict its final three-dimensional shape, and even the effects of DNA mutations on the protein's shape and function.

The study of protein function is viewed as essential to the discovery of new and better drugs. Also, it may be possible in the future to correlate drug treatment to the particular proteome of the individual to increase efficiency and decrease side effects. Proteomics will be a critical field of endeavor for many years to come.

Bioinformatics

Bioinformatics is the application of computer technologies, specially developed software, and statistical techniques to the study of biological information, particularly databases that contain much genomic and proteomic information (Fig. 14.10). The new, raw data produced by structural genomics and proteomics is stored in databases that are readily available to research scientists. It is called raw data because, as yet, it has little meaning. Functional genomics and proteomics are dependent on computer analysis to find significant patterns in the raw data. For example, BLAST, which stands for *basic local alignment search tool*, is a computer program that can identify homologous genes among the genomic sequences of model organisms. **Homologous genes** are genes that code for the same proteins, although the base sequences may be slightly different. Finding these differences can help trace the history of evolution among a group of organisms.

Figure 14.10 Bioinformatics. New computer programs are being used to make sense out of the raw data generated by genomics and proteomics. Bioinformatics allows researchers to study both functional and comparative genomics in a meaningful way.

Bioinformatics also has various applications in human genetics. For example, researchers found the function of the protein that causes cystic fibrosis by using the computer to search for genes in model organisms that have the same sequence. Because they knew the function of this gene in model organisms, they could deduce the function in humans. This was a necessary step toward possibly developing specific treatments for cystic fibrosis.

The human genome has 3 billion known base pairs, and without the computer it would be almost impossible to make sense of these data. For example, it is now known that an individual's genome often contains multiple copies of a gene. But individuals may differ as to the number of copies—called *copy number variations,* as discussed in the Evolution feature on page 266. Now it seems that the number of copies in a genome can be associated with specific diseases. The computer can help make correlations between genomic differences among large numbers of people and disease.

It is safe to say that without bioinformatics, our progress in determining the function of DNA sequences, comparing our genome to model organisms, knowing how genes and proteins interact in cells, and so forth, would be extremely slow. Instead, with the help of bioinformatics, progress should proceed rapidly in these and other areas.

Check Your Progress   14.4

1. Distinguish between the genome and the proteome of a cell.
2. Summarize the difference between a short tandem repeat and a transposon.
3. Explain how the use of microarrays and bioinformatics aids in the study of genomics and proteomics.

Evolution

Copy Number Variations

Geneticists have long been aware of large chromosomal duplications, deletions, and rearrangements detectable microscopically (see Fig. 10.13a, b). However, over the past decade scientists have become aware of duplications and deletions in the DNA called *copy number variations* (CNVs). Unlike shorter duplications, CNVs are sequences of DNA, usually greater than one kilobase (1,000 nucleotides) in size, that are repeated in varying numbers compared to a reference genome.

This repetition may arise from errors by the DNA polymerase during replication. For example, DNA damage or some other difficulty may cause the replication fork to stall. To continue, the replication machinery can switch to nearby chromosomal material of the same sequence. The replication fork is soon transferred back to the normal template, but the end result is extra or missing copies of small DNA segments. The fact that repetitive elements facilitate template switching suggests a new function for such sequences in our genome.

Some CNVs have known links to disease. Research shows that individuals with fewer copies of the *CCL3L1* gene are more susceptible to HIV infection than those with more copies. The disease Lupus is much more common among people with fewer copies of the complement component *C4* gene. But more surprising was a study that suggested at least some cases of autism can be linked to CNVs. The scientists who published the study examined the total chromosomal content of 1,441 autistic children and compared their DNA to more than 2,800 normal individuals. They found that in some of the autistic children, a 25-gene region of chromosome 16 was missing. Furthermore, analysis of other DNA databases revealed the same result: Approximately 1% of autism cases could be directly linked to the same deletion.[1]

CNVs are also emerging as a possible driving force in evolution. One study utilized DNA microarrays to examine the chromosomal structure of 47 individuals from many ethnic backgrounds, and found 119 regions where copy number variations existed. More surprising, none of the CNVs were found exclusively in one ethnic group, suggesting that these variants existed well before the human population spread across the Earth. Perhaps they contributed to the phenotypic variations that developed thereafter. Furthermore, many scientists are suggesting that it may be advantageous for a species to have multiple copies of genes; if one or both normal copies of an allele fail to function properly, having a third allele available might be advantageous because it could restore normal function.

Conversely, an organism's two normal alleles would free an extra gene copy from having to maintain normal function. This situation would allow the gene to accumulate mutations without major consequence, which could ultimately lead to the formation of a new, unique gene. Copy number variations may contribute to evolution because they are yet another mechanism for organisms to achieve genetic innovation.

Questions to Consider

1. Why might a CNV in the coding region of a gene change the function of the protein?
2. From an evolutionary perspective, how might variations in CNV allow an organism to adapt to a new environment?

1 Weiss, L. A., et al. 2008. Association between microdeletion and microduplication at 16p11.2 and Autism. *New England Journal of Medicine*, 358: 667–75.

CONNECTING *the* CONCEPTS *with the* BIG IDEAS

Evolution

- Genomes of organisms from every domain are being compared, revealing interesting and unexpected evolutionary relationships. (1A4b3)

Information and Signaling

- Genes from virtually any organism can be cloned using in vivo plasmid-based transformation technology; PCR allows multiple copies of a DNA sequence to be produced in vitro. (3A1e*IE*)
- Electrophoresis and restriction enzymes are widely used for DNA analysis. (3A1e*IE*)
- Genetically modified organisms (GMOs) including transgenic and cloned animals and plants have been engineered to add beneficial characteristics or make needed products such as pharmaceuticals. (3A1f*IE*)

*Find the unabridged version of all EK citations at www.glencoe.com/maderAP11.

Media Study Tools

www.glencoe.com/maderAP11

Enhance your study of this chapter with study tools and practice tests. Also ask your instructor about the resources available through ConnectPlus, including the media-rich eBook, interactive learning tools, and animations.

Summarize

14.1 DNA Cloning

DNA cloning can isolate a gene and produce many copies of it. The gene can be studied in the laboratory or inserted into a bacterium, plant, or animal. Then, this gene may be transcribed and translated to produce a protein, which can become a commercial product or used as a medicine.

Two methods are currently available for making copies of DNA: recombinant DNA technology and the polymerase chain reaction (PCR). Recombinant DNA contains DNA from two different sources. A restriction enzyme is used to cleave plasmid DNA and to cleave foreign DNA. The resulting "sticky ends" facilitate the insertion of foreign DNA into vector DNA. The foreign gene is sealed into the vector DNA by DNA ligase. Both bacterial plasmids and viruses can be used as vectors to carry foreign genes into bacterial host cells.

PCR uses the enzyme DNA polymerase to quickly make multiple copies of a specific piece (target) of DNA. PCR is a chain reaction because the targeted DNA is replicated over and over again. Analysis of DNA segments following PCR has all sorts of uses from assisting genomic research to doing DNA fingerprinting for the purpose of identifying individuals and confirming paternity.

14.2 Biotechnology Products

Transgenic organisms have had a foreign gene inserted into them. Genetically modified bacteria, agricultural plants, and farm animals now produce commercial products of interest to humans, such as hormones and vaccines. Bacteria usually secrete the product. The seeds of plants and the milk of animals contain the product.

Transgenic bacteria have also been engineered to promote the health of plants, perform bioremediation, extract minerals, and produce chemicals. Transgenic crops, engineered to resist herbicides and pests, are commercially available. Transgenic animals have been given various genes, in particular the one for bovine growth hormone (bGH). Cloning of animals is now possible.

14.3 Gene Therapy

Gene therapy, by either ex vivo or in vivo methods, is used to correct the genotype of humans and to cure various human ills. Ex vivo gene therapy has been used to treat diseases such as SCID and cystic fibrosis. A number of in vivo therapies are being employed in the fight against cancer and other human illnesses, such as cardiovascular disease.

14.4 Genomics

Researchers now know the sequence of all the base pairs along the length of the human chromosomes. So far, researchers have identi-fied around 25,000 human genes that code for proteins; the rest of our DNA consists of regions that do not code for a protein. Currently, researchers are placing an emphasis on functional and comparative genomics.

Genes comprise only 1.5% of the human genome. The rest of this DNA is surprisingly more active than once thought. About half of this DNA consists of repetitive DNA elements, which may be in tandem or interspersed throughout several chromosomes. Some of this DNA is made up of mobile DNA sequences called transposons, which are a driving evolutionary force within the genome. The role of the remaining portion of the genome is actively being investigated, but it is believed that these DNA sequences may play an important role in regulation of gene expression, thus challenging the classical definition of the gene. Functional genomics aims to understand the function of protein-coding regions and noncoding regions of our genome. To that end, researchers are utilizing tools such as DNA microarrays. Microarrays can also be used to create an individual's genetic profile, which is becoming helpful in predicting illnesses and how a person will react to particular medications.

Comparative genomics has revealed that little difference exists between the DNA sequence of our bases and those of many other organisms. Genome comparisons have revolutionized our understanding of evolutionary relationships by revealing previously unknown similarities between organisms.

Proteomics is the study of which genes are active in producing proteins in which cells under which circumstances. Bioinformatics is the use of computer technology to assist proteomics and functional and comparative genomics.

Key Terms

bioinformatics 265	intergenic sequence 262
biotechnology products 258	interspersed repeat 262
cloning 255	plasmid 255
comparative genomics 263	polymerase chain reaction
complementary DNA	(PCR) 256
(cDNA) 256	proteome 265
DNA ligase 255	proteomics 265
DNA microarray 263	recombinant DNA (rDNA) 255
functional genomics 263	repetitive DNA element 262
gel electrophoresis 257	restriction enzyme 255
gene cloning 255	short tandem repeat (STR)
gene pharming 258	profiling 257
gene therapy 255, 260	structural genomics 261
genetic profile 263	tandem repeat 262
genetically modified organism	transgenic organism 255
(GMO) 258	transposon 262
genomics 261	vector 255
homologous gene 265	
Human Genome Project	
(HGP) 261	

Assess

Reviewing This Chapter

1. What is the methodology for producing recombinant DNA? 255
2. What is the polymerase chain reaction (PCR), and how is it carried out to produce multiple copies of a DNA segment? 256
3. How does STR profiling produce a DNA fingerprint? 257

4. What are some practical applications of DNA segment analysis following PCR?   257
5. For what purposes have bacteria, plants, and animals been genetically altered?   258–59
6. Explain and give examples of ex vivo and in vivo gene therapies in humans.   260
7. What was the purpose of the Human Genome Project? What is the goal of functional genomics?   261–63
8. Describe the various types of intergenic DNA sequences found within the genome.   262–63
9. What insights into evolutionary relationships between organisms are arising from comparative genomics?   263–65
10. What are the goals of proteomics and bioinformatics?   265

Testing Yourself

Choose the best answer for each question.

1. Using this key, put the phrases in the correct order to form a plasmid-carrying recombinant DNA.

KEY:
   (1) use restriction enzymes
   (2) use DNA ligase
   (3) remove plasmid from parent bacterium
   (4) introduce plasmid into new host bacterium
   a. 1, 2, 3, 4   c. 3, 1, 2, 4
   b. 4, 3, 2, 1   d. 2, 3, 1, 4

2. Restriction enzymes found in bacterial cells are ordinarily used
   a. during DNA replication.
   b. to degrade the bacterial cell's DNA.
   c. to degrade viral DNA that enters the cell.
   d. to attach pieces of DNA together.

3. A genetic profile can
   a. assist in maintaining good health.
   b. be accomplished utilizing bioinformatics.
   c. show how many genes are normal.
   d. be accomplished utilizing a microarray.
   e. Both a and d are correct.

4. Bacteria are able to successfully transcribe and translate human genes because
   a. both bacteria and humans contain plasmid vectors.
   b. bacteria can replicate their DNA, but humans cannot.
   c. human and bacterial ribosomes are vastly different.
   d. the genetic code is nearly universal.

5. Bioinformatics can
   a. assist genomics and proteomics.
   b. compare our genome to that of a monkey.
   c. depend on computer technology.
   d. match up genes with proteins.
   e. All of these are correct.

6. The polymerase chain reaction
   a. uses RNA polymerase.
   b. takes place in huge bioreactors.
   c. uses a temperature-insensitive enzyme.
   d. makes lots of nonidentical copies of DNA.
   e. All of these are correct.

7. Which is a true statement?
   a. Genomics would be slow going without bioinformatics.
   b. Genomics is related to the field of proteomics.
   c. Genomics has now moved on to functional and comparative genomics.
   d. Genomics shows that we are related to all other organisms tested so far.
   e. All of these are correct.

8. DNA amplified by PCR and then used for fingerprinting could come from
   a. any diploid or haploid cell.
   b. only white blood cells that have been karyotyped.
   c. only skin cells after they are dead.
   d. only purified animal cells.
   e. Both b and d are correct.

9. Which was used to find the function of the cystic fibrosis gene?
   a. microarray
   b. proteomics
   c. comparative genomics and bioinformatics
   d. sequencing the gene

10. Which of these pairs is incorrectly matched?
   a. DNA ligase—mapping human chromosomes
   b. protoplast—plant cell engineering
   c. DNA fragments—DNA fingerprinting
   d. DNA polymerase—PCR

11. Which matches best to proteomics?
   a. Start with known gene sequences and build proteins.
   b. Use a microarray to discover what proteins are active in particular cells.
   c. Use bioinformatics to discover the proteins in the cells of other organisms.
   d. Match up known proteins with known genes.

12. Which is not a correct association with regard to bioengineering?
   a. plasmid as a vector—bacteria
   b. protoplast as a vector—plants
   c. RNA virus as a vector—human stem cells
   d. All of these are correct.

13. Proteomics is used to discover
   a. what genes are active in what cells.
   b. what proteins are active in what cells.
   c. the structure and function of proteins.
   d. how proteins interact.
   e. All but a are correct.

14. Which of these is an incorrect statement?
   a. Bacteria usually secrete the biotechnology product into the medium.
   b. Plants are being engineered to have human proteins in their seeds.
   c. Animals are engineered to have a human protein in their milk.
   d. Animals can be cloned, but plants and bacteria cannot.

15. Repetitive DNA elements
   a. may be tandem or spread across several chromosomes.
   b. are found in centromeres and telomeres.
   c. make up nearly half of human chromosomes.
   d. may be present just a few to many thousands of copies.
   e. All of these are correct.

16. Because of the Human Genome Project, we now know
    a. the sequence of the base pairs of our DNA.
    b. the sequence of all genes along the human chromosomes.
    c. all the mutations that lead to genetic disorders.
    d. All of these are correct.
    e. Only a and c are correct.
17. Which of the following delivery methods is not used in gene therapy?
    a. virus   c. liposomes
    b. nasal sprays   d. electric currents

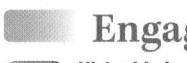
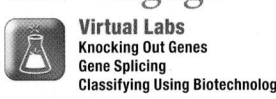

Engage

Virtual Labs
Knocking Out Genes
Gene Splicing
Classifying Using Biotechnology

The virtual labs "Knocking Out Genes," "Gene Splicing," and "Classifying Using Biotechnology" all provide an interactive examination of the material found in this chapter.

Thinking Scientifically

1. Transposons are considered by many researchers to have played a major role in the evolution of life on Earth. Explain how the movement of genetic material within the genome may produce some organisms with a selective advantage.

2. The Evolution feature on page 266 describes copy number variations within the genome. Copy number variations do not always contain genes. How might having extra or missing copies of intergenic DNA sequences be beneficial? How might it be harmful?

Bioethical Issue

Transgenic Crops

Transgenic plants can possibly allow crop yields to keep up with the ever-increasing worldwide demand for food. And some of these plants have the added benefit of requiring less fertilizer and/or pesticides, which are harmful to human health and the environment.

Some scientists believe transgenic crops pose their own threat to the environment, however, and many activists believe transgenic plants are themselves dangerous to our health. Studies have shown that wind-carried pollen can cause transgenic crops to hybridize with nearby weedy relatives. Although it has not happened yet, some fear that characteristics acquired in this way might cause weeds to become uncontrollable pests. Or perhaps a toxin produced by transgenic crops could possibly hurt other organisms in the habitat. Many researchers are conducting tests to see if this might occur. And, although transgenic crops have not caused any illnesses in humans so far, some scientists concede the possibility that people could be allergic to a transgene's protein product.

Already, transgenic plants must be approved by the Food and Drug Administration before they are considered safe for human consumption, and they must meet certain Environmental Protection Administration standards. Some people believe safety standards for transgenic crops should be further strengthened, while others fear stricter standards would result in less food produced. Another possibility is to retain the current standards but require all biotech foods to be clearly labeled so the buyer can choose whether or not to eat them. Which approach do you prefer?

UNIT 3

Evolution

When Charles Darwin embarked upon his trip around the world on the HMS *Beagle*, no one could imagine the impact that journey would have on science and the way we think about the world. Evolution, the idea that life changes with time, turned conventional explanations of the diversity of life upside-down. The evidence is overwhelming that organisms alive today are different from their ancestors, and that all organisms ultimately arose from one original form of life. The elegance and simplicity of the idea is grand: the environment in which living things find themselves determines who will perservere, and those that have the adaptations which allow survival pass those genetically-determined traits on to their offspring. Worldwide, environments are so varied and constantly changing that natural selection continues to hone the fittest to a particular situation. When conditions again alter, that honing will take life forms in a slightly different direction, creating even more biodiversity. **BI 1** Therefore, it is no coincidence that in the College Board's AP Biology Curriculum Framework, the first and most overarching principle is Big Idea 1: "The process of evolution drives the diversity and unity of life."

All around you, things evolve. Products from cars to cell phones are constantly changing to better accommodate the market; cars move toward increased fuel efficiency while cell phones expand their internet capabilities. Evolution is not a dirty word, but an expression of moving forward, adapting with time. Sometimes things cannot adapt fast enough; the record album and the cassette tape are all but gone, left in the dust of improved digital technology, yet they provided the "bones" of the new medium. So Darwin saw, and most of today's scientists see, the parade of life: new forms arising and becoming more successful due to slight modifications to the old ancestral plan. Reptiles with wings were more successful with feathers replacing heavier, less specialized scales, and eventually evolved into birds. Those creatures with modifications that do not lead to reproductive success, such as the extravagant and heavy rack of the Irish elk, eventually become extinct. Today, humans often exert inadvertent selection as they attempt to eradicate pests or accidentally import exotic species. And with the explosion of biotechnological capabilities, it is even possible to direct evolution at the genetic level, hopefully with a moral and sober hand.

Theodosius Dobzhansky, one of the first scientists to provide laboratory evidence for natural selection, said, "Nothing in biology makes sense except in the light of evolution." Understanding the theory of evolution by natural selection will not only help you make sense of the great diversity of life on Earth, but will also help you envision the "change with time" going on all around you, where mutation, adaptation, and reproductive success continue to inexorably change the face of life.

UNIT OUTLINE

Darwin and Evolution

Tiktaalik had a wrist and other features that tell us it was an early intermediate between fish and four-limbed animals.

I n 2004, a fossil "fishapod" was discovered in the Canadian Arctic. *Tiktaalik roseae* is a 375-million-year-old fossil that looks like a cross between an ancient fish and the first four-legged animals, called *tetrapods*. This unique fossil exhibits a number of transitional features that are both fishlike and tetrapod-like. It had fins, scales, and gills like a fish; but it also had a flexible neck, a flat head, and a forelimb with a wrist much like a modern tetrapod. *Tiktaalik* lived in wet, swampy areas, and a wrist may have been advantageous for moving along the bottom of shallow pools and rivers. Transitional fossils such as *Tiktaalik* support Darwin's theory that all animals descended from a common ancestor.

The evidence for evolution is not limited to fossils such as *Tiktaalik*—Darwin's theory of evolution is supported by over 150 years of biogeographical, biochemical, developmental, and genetic evidence. Evolution is not evident only in the study of fossils over long periods of time, but can be witnessed, in action, over very short periods of time such as days, weeks, and years. In this chapter we first take a look at the history of evolutionary thought, beginning with the history of ideas that influenced Darwin as he made his observations. Then, we trace Darwin's trip around the world, and present the evidence that allowed him to develop his theory of evolution by natural selection. Finally, we explore modern evidence that supports Darwin's theory.

As you read through this chapter, think about the following questions:

1. How do the features of fossil organisms tell us something about an organism's behavior and the environment in which it lived?
2. How does Darwin's theory of natural selection explain the intermediate features of modern and fossil organisms such as *Tiktaalik*?

CHAPTER OUTLINE

BEFORE YOU BEGIN

Before beginning this chapter, take a few moments to review the following discussions.

Section 1.2 What does Darwin mean by evolution as "descent with modification?"

Section 1.4 What is a scientific theory?

Section 12.3 What is the genetic basis of inheritance?

FOLLOWING *the* BIG IDEAS

CHAPTER 15 DARWIN AND EVOLUTION

Evolution	Darwin's theory of natural selection states organisms survive because they have superior adaptations, and those heritable traits are passed on to their offspring.
Interactions and Systems	Evolution by natural selection comes about from interaction between the organism and its environment.

15.1 History of Evolutionary Thought

In December 1831, a new chapter in the history of biology had its humble origins. A 22-year-old naturalist, Charles Darwin (1809–82), set sail on a journey of a lifetime aboard the British naval vessel HMS *Beagle* (Fig. 15.1). Darwin's primary mission on his journey around the world was to serve as the ship's naturalist—to collect and record the geological and biological diversity he saw during the voyage.

As Darwin set sail on the HMS *Beagle*, he was a supporter of the long-held idea that species had remained unchanged since the time of creation. Prior to Darwin, this view of the fixity of species was forged from deep-seated religious beliefs, and not by experimentation and observation of the natural world. During the five-year voyage of the *Beagle*, Darwin's observations challenged his belief that species do not change over time—in fact, his observations of geological formations and species variation led him to propose a new process by which species arise and change. This process—**evolution** [L. *evolutio*, an unrolling]—proposed that species arise, change, and become extinct due to natural, not supernatural, forces.

This new view was not readily accepted by Darwin's peers, but it gained gradual credibility as a result of a scientific and intellectual revolution that began in Europe in the late 1800s. Today, 150 years since Darwin first published his idea of natural selection, the principle he proposed has been subjected to rigorous scientific tests—so much so that it is now considered one of the unifying theories of biology. Darwin's theory of evolution by natural selection explains both the unity and diversity of life on Earth, how living things all share a common ancestor, and how species adapt to various habitats and ways of life.

Although many have believed that Darwin forged this change in worldview by himself, several biologists during the preceding century and some of Darwin's contemporaries had a large influence on Darwin as he developed his theory. The European scientists of the 18th and 19th centuries were keenly interested in understanding the nature of biological diversity. This was a time of exploration and discovery as the natural history of new lands was mapped and documented. Shipments of new plants and animals from newly explored regions were arriving in England to be identified and described by biologists—it was a time of rapid expansion of our understanding of the Earth's biological diversity. In this atmosphere of discovery, Darwin's theory first took root and grew.

Mid-Eighteenth-Century Influences

Taxonomy, the science of classifying organisms, was an important endeavor during the mid-eighteenth century. Biologists of this time used comparative anatomy, the evaluation of similar structures across a variety of species, to classify organisms into groups. By the late eighteenth century, scientists had discovered fossils and knew that they were the remains of plants and animals from the past. Explorers traveled the world and brought back newly discovered **extant** (still in existence) and fossil organisms to be compared to known living species. At first, scientists believed that each type of fossil had a living descendant, but eventually some fossils did not seem to match well with known species. Baron Georges Cuvier (1769–1832) was the first to suggest that some species known only from the fossil record had become extinct.

Chief among taxonomists was Carolus Linnaeus (1707–78), who developed the binomial system of nomenclature (a two-part name for species, such as *Homo sapiens*) and a system of classification for living things. Linnaeus, like other taxonomists of his time, believed in the fixity of species, that is, each species had an "ideal" form. He also believed in the *scala naturae*, a sequential ladder of life where the simplest beings occupy the lowest rungs, and the most complex and spiritual beings—the angels, humans, and then God—occupy the two highest rungs.

These ideas, although consistent with Judeo-Christian teachings about special creation, can be traced to the works of the Greek philosophers Plato (427–347 BC) and Aristotle (384–322 BC). Plato said that every species on Earth has a perfect, or "essential," form, and species variation is imperfection of the ideal type. Aristotle saw that organisms vary in complexity. He proposed that all organisms could be arranged in order of increasing complexity on a *scala naturae*.

Georges-Louis Leclerc (1707–88), better known as Count Buffon, was a naturalist who worked most of his life writing a 44-volume natural history series that described all known plants and animals. He provided evidence of evolution, and proposed various causes, such as environmental influence and the struggle for existence. Buffon's support of evolution seemed to waver, and often he professed to believe in special creation and the fixity of species.

Erasmus Darwin (1731–1802), Charles Darwin's grandfather, was a physician and a naturalist. His writings on both botany and zoology contained comments and footnotes that suggested the possibility of evolution. He based his conclusions on changes in animals during development, animal breeding by humans, and the presence of **vestigial structures** [L. *vestigium*, trace, footprint]—anatomical structures that apparently functioned in an ancestor but have since lost most or all of their function in a descendant. Like Buffon, Erasmus Darwin thought that species might evolve, but offered no mechanism by which this change might occur.

Late Eighteenth/Early Nineteenth–Century Influences

Baron Georges Cuvier, a distinguished zoologist, used comparative anatomy to develop a system of classifying animals. He also founded the science of **paleontology** [Gk. *palaios*, old; *ontos*, having existed; *-logy*, study of], the study of fossils, and was quite skilled at using fossil bones to deduce the structure of an animal.

Cuvier was a staunch advocate of the fixity of species and special creation, but his studies revealed that the assembly of

Figure 15.1 Voyage of the HMS *Beagle*. **a.** A young Charles Darwin in 1831 at 22 years old. He did not publish his authoritative book, *On the Origin of Species*, until 1859. **b.** A map of Darwin's journey aboard the HMS *Beagle*. **c.** Along the east coast of South America, he noted a bird called the rhea, which looks like an African ostrich. **d.** In the Patagonian Desert, he observed a rodent that resembles the European rabbit. **e.** In the Andes Mountains he observed fossils in rock layers. **f.** In the tropical rain forest he found an abundant diversity of life. **g.** On the Galápagos Islands, he saw marine iguanas with blunt snouts suited for eating algae growing on rocks.

fossil varieties changed suddenly between different layers of sediment, or **strata,** within a geographic region. He reconciled his beliefs with his observations by proposing that sudden changes in fossil variation could be explained by a series of local catastrophes, or mass extinctions, followed by repopulation by species from surrounding areas. The result of these catastrophes was a turnover in the assembly of life-forms that occupied a particular region over time. Some of Cuvier's followers suggested that there had been worldwide catastrophes, and God created new sets of species to repopulate the world. This explanation of the history of life came to be known as **catastrophism.**

Jean-Baptiste de Lamarck (1744–1829) was the first biologist to offer a testable hypothesis that explained how evolution occurs via adaptation to the environment. Lamarck's ideas about descent were entirely different from those of Cuvier. After studying the succession of fossilized life-forms in the Earth's strata, Lamarck proposed that more complex organisms are descended from less complex organisms. He mistakenly concluded, however, that increasing complexity is the result of a natural motivating force—a striving for perfection—that is inherent in all living things.

To explain the process of adaptation to the environment, Lamarck proposed the idea of **inheritance of acquired characteristics**—that the environment can produce physical changes in an organism during its lifetime that are inheritable. One example that he gave—and for which he is most famous—is that the long neck of a giraffe developed over time because their necks grew longer as they stretched to reach food in tall trees and this longer neck was then passed onto their offspring (Fig. 15.2). His hypothesis of inheritance of acquired characteristics has never been supported by experimentation. The molecular mechanism of inheritance explains why—phenotypic changes acquired during an organism's lifetime do not result in genetic changes that can be passed to subsequent generations.

In the eighteenth century, geologist James Hutton (1726–1797) proposed a theory of slow, uniform geological change. Charles Lyell (1797–1875), the foremost geologist of Darwin's time, made Hutton's ideas popular in his book, *Principles of Geology,* published in 1830. Hutton explained that the Earth was subject to slow but continuous cycles of rock formation and erosion, and not shaped by sudden catastrophes. He proposed that erosion produces dirt and rock debris that is washed into the rivers, transported to the oceans, and deposited in thick layers that are converted over time into sedimentary rock. These layers of sedimentary rocks, which often contain fossils, are then uplifted from below sea level to form land during geological upheavals.

Hutton concluded that extreme geological changes can be explained by slow, natural processes, given enough time. Lyell went on to propose the theory of **uniformitarianism,** which stated that the natural processes witnessed today are the same processes that occurred in the past. Hutton's general ideas about slow and continual geological change are still accepted today, although modern geologists realize that rates of change have not always been uniform through history. Darwin was not taken by the idea of uniform change, but he was convinced, as was Lyell, that the Earth's massive geological changes are the result of extremely slow processes, and that the Earth, therefore, must be very old.

Thomas Malthus (1766–1834) was an economist who studied the factors that influence the growth and decline of human populations. In 1798 Malthus published *An Essay on the Principle of Population,* in which he proposed that the size of human populations is limited only by the quantity of resources, such as food, water, and shelter, available to support it. He related famine, war, and epidemics to the problem of populations overstretching their limited resources. Darwin, after reading Malthus's essay in 1838, applied similar principles to animal populations—that is, animals tend to produce more offspring than can survive, and competition for limited resources in the environment is the element that determines survival. Darwin thus used Malthus's principle to formulate his idea of natural selection.

Check Your Progress 15.1

1. Define catastrophism and identify who proposed this idea.
2. Evaluate Lamarck's idea of "inheritance of acquired characteristics" as an explanation of biological diversity.
3. Construct a timeline of the history of evolutionary thought. Include major contributors and a brief description of each contribution along the timeline.

Early giraffes probably had short necks that they stretched to reach food.

Their offspring had longer necks that they stretched to reach food.

Eventually, the continued stretching of the neck resulted in today's giraffe.

Figure 15.2 Lamarck's inheritance of acquired characteristics. Lamarck proposed that the neck of a giraffe would grow longer as it strived to reach tall leaves, and this longer neck could then be passed on to its offspring. We now know that only traits encoded in genes are heritable.

15.2 Darwin's Theory of Evolution

Learning Outcomes

Upon completion of this section, you should be able to

1. Summarize the stages of evolution by natural selection.
2. List examples of the evidence Darwin gathered from fossils and biogeography that supported his growing idea of shared ancestry.
3. Give examples of how the mechanisms of evolutionary change can be identified and studied.

When Darwin signed on as the naturalist aboard the HMS *Beagle,* he possessed a suitable background for the position. Since childhood, he had been a devoted student of nature and a collector of insects. At age 16, Darwin was sent to medical school to follow in the footsteps of his grandfather and father. However, he did not take to the study of medicine, so his father encouraged him to enroll in the School of Divinity at Christ's College at Cambridge, with the intent of his becoming a clergyman.

While at Christ's College, Darwin attended many lectures on biology and geology to satisfy his interest in natural science. During this time, he became the protégé and friend of the botanist John Henslow (1796–1861) from whom he gained skills in the identification and collection of plants. Darwin gained valuable experience in geology in the summer of 1831 by conducting field-work with Adam Sedgewick (1785–1873), one of the founders of modern geology. Shortly after Darwin was awarded his BA, Henslow recommended him to serve, without pay, as the ship's naturalist aboard the HMS *Beagle* which was to explore the Southern Hemisphere.

The voyage was to take two years—but ended up taking five years—and the ship was to traverse the Southern Hemisphere (see Fig. 15.1). Along the way, Darwin encountered forms of life very different from those of his native England. As part of his duties as the ship's naturalist, Darwin began to gather evidence during the voyage that organisms are related through descent with modification from a common ancestor, and that adaptation to various environments results in diversity. Darwin also began contemplating the "mystery of mysteries," the origin of new species.

Observations of Change Over Time

On his trip, Darwin observed massive geological changes first-hand. When he explored what is now Argentina, he saw raised beaches for great distances along the coast. Many of the raised beaches had exposed layers of sediment that contained a variety of fossilized shells and bones of extinct mammals. Darwin collected fossil remains of an armadillo-like animal (*Glyptodon*), the size of a small modern-day car, and a giant ground sloth, *Mylodon darwinii,* the largest of which stood nearly 3 m tall (Fig. 15.3). Darwin also observed marine shells high in the cliffs of the impressive Andes Mountains, which suggested to him that the Earth is very old. Once Darwin accepted the possibility that the Earth must be very old, he began to think that there

a. *Glyptodon*

b. *Mylodon*

Figure 15.3 Darwin discovered fossils of extinct mammals during his exploration of South America. **a.** A giant armadillo-like glyptodont, *Glyptodon*, is known only by the study of its fossil remains. Darwin found such fossils and came to the conclusion that this extinct animal must be related to living armadillos. The glyptodont weighed 2,000 kg. **b.** Darwin also observed the fossil remains of an extinct giant ground sloth, *Mylodon*.

would have been enough time for descent with modification to occur. Therefore, living forms could be descended from extinct forms known only from the fossil record. It would seem that species were not fixed; instead, they changed over time.

Biogeographical Observations

Biogeography [Gk. *bios*, life, *geo*, earth, and *grapho*, writing] is the study of the range and geographic distribution of life-forms in different places throughout the world, as well as how and when they came to be distributed as they are today. The distribution of species and the makeup of species groups in different regions provide hints about past geological events, such as the movement of continents and the formation of volcanic islands, or about ecological change, such as glaciation and river formation.

As Darwin explored the Southern Hemisphere, he compared the animals of South America to those with which he was familiar. He noticed that although the animals in South America were different from those in Europe, similar environments on each continent had similar-looking animals. For example, instead of rabbits, he found the Patagonian cavy in the grasslands of South

America. The Patagonian cavy has long legs and ears but the face of a guinea pig, a rodent also native to South America (Fig. 15.4). Did the Patagonian cavy resemble a rabbit because the two types of animals were adapted to the same type of environment? Both animals ate grass, hid in bushes, and moved rapidly using long hind legs. Did the Patagonian cavy have the face of a guinea pig because of having an ancestor in common with guinea pigs?

As he sailed southward along the eastern coast of South America, Darwin saw how similar species replaced one another. For example, the greater rhea (an ostrichlike bird) found in the north was replaced by the lesser rhea in the south. Therefore, Darwin reasoned that related species could be modified according to environmental differences (i.e., northern vs. southern latitudes). When he explored the Galápagos Islands, he found further evidence of this phenomenon.

The Galápagos Islands are a small group of volcanic islands formed 965 km off the western coast of South America. These islands are too far from the mainland for most terrestrial animals and plants to colonize, yet life is present there. The types of

Lepus europaeus

Dolichotis patagonum

Figure 15.4 **The European hare (head only), and the Patagonian cavy.**

plants and animals Darwin found there were slightly different from species he had observed on the mainland, and even more important, they also varied from island to island. Where did animals and plants inhabiting these islands come from? Why were these species different from those on the mainland, and why were different species found on each island?

For example, each of the Galápagos Islands seemed to have its own type of tortoise, and Darwin began to wonder whether this difference was correlated with variation in vegetation among the islands (Fig. 15.5). Long-necked tortoises seemed to inhabit only dry areas where low-growing vegetation was scarce, but tall cacti were abundant. In moist regions with relatively abundant ground foliage, short-necked tortoises were found. Had an ancestral tortoise from the mainland of South America given rise to these different types, each adapted to take advantage of food sources in different environments?

One of Darwin's most famous observations from the Galápagos Islands was his study of the finches. Darwin almost overlooked the finches because of their nondescript nature compared with many of the other animals in the Galápagos. At the time, Darwin did not recognize that these birds were all finches because they were very different from the familiar finches from England. However, these birds would eventually play a major role in the formation of his thoughts about geographic barriers and their contribution to the origin of new species. Upon returning to England the birds were identified as "a series of ground finches . . . an entirely new group," and they exhibited significant variation in beak size and shape (see Fig. 15.9).

Today, many more Galápagos finches have been identified—there are ground-dwelling finches with beaks adapted to eating seeds, tree-dwelling finches with beaks sized according to their insect prey, and a cactus-eating finch with a more pointed beak used to punch holes in cactus fruit to extract pulp (see Fig. 15.9). The most unusual of the finches is a woodpecker-type finch. This bird has a sharp beak to chisel through tree bark but lacks the long tongue characteristic of a true woodpecker, which probes for insects. To compensate for this, the bird carries a

a.

b.

Figure 15.5 **Galápagos tortoises.** Darwin wondered whether the Galápagos tortoises were descended from a common ancestor. **a.** The tortoises with dome shells and short necks feed at ground level on islands with enough rainfall to support grasses. **b.** Those with shells that flare up in the front have long necks and occur on arid islands where they feed on tall, treelike cacti.

twig or cactus spine in its beak and uses it to poke into crevices. Once an insect emerges, the finch drops this tool and seizes the insect with its beak (for more information on the shape of finch beaks, see the Chapter 17 Nature of Science feature, "Genetic Basis of Beak Size in Darwin's Finches").

> **Video**
> **Galápagos Finches**

Later, Darwin speculated whether these different species of finches could have descended from a mainland finch species. In other words, he wondered if a finch from South America was the common ancestor to all the types on the Galápagos Islands. Perhaps new species had arisen because the geographic distance between the islands isolated populations of birds long enough for them to evolve independently. And perhaps the present-day species had resulted from accumulated changes occurring within each of these isolated populations.

Natural Selection and Adaptation

Upon returning to England, Darwin began to reflect on the voyage of the HMS *Beagle* and to collect additional evidence in support of his ideas about how organisms adapt to the environment. Darwin concluded early on that species change over time, and are not fixed entities crafted by a creator. However, he did not yet have a mechanism to explain how change could happen in existing species, and how new species could arise.

By 1842, Darwin had fully developed his idea of natural selection as a mechanism for evolutionary change. In 1858, Alfred Russel Wallace (1823–1913) sent an essay to Darwin in which he proposed a similar concept based on observations from the other side of the globe. The theory of natural selection was first presented to the Linnean Society of London in 1858 as a pair of essays by Darwin and Wallace.

Natural selection is a process based on the following observations:

- Organisms exhibit variation that can be passed from one generation to the next—that is, they have heritable variation.
- Organisms compete for available resources.
- Individuals within a population differ in terms of their reproductive success.
- Organisms become adapted to conditions as the environment changes.

We consider each of these characteristics in detail in the sections that follow.

Organisms Have Heritable Variation

Darwin emphasized that the members of a population vary in their functional, physical, and behavioral characteristics (Fig. 15.6). Before Darwin, variations were viewed as imperfections that should be ignored because they were not important to the description of "fixed" species (see Section 15.1). In contrast, Darwin emphasized that variation is required for the process of natural selection to operate. He suspected that a mechanism of inheritance existed, but he did not have the evidence we have today.

Figure 15.6 Variation in a population. Variation in populations, such as that seen in human populations, is required for natural selection to result in adaptation to the environment.

Now we know that genes are the unit of heredity and along with the environment, determine the phenotype of an organism, and that random mutations are a source of new genetic variation in a population. Genetic variation can be harmful, helpful, or neutral (have no effect at all) to survival and reproduction. Genetic variation arises by chance and for no particular purpose, and new variation is as likely to be harmful as helpful or neutral to the organism. However, harmful variation is removed from the population by natural selection because individuals with these mutations often do not live to reproduce. Beneficial or neutral variation can be maintained in a population. Natural selection ignores neutral variation. But a beneficial mutation increases the probability that an individual with this mutation will survive and produce more offspring. Biologists have abundant evidence that natural selection operates on heritable variation already present in a population's gene pool, and that this selection process is random, that is, it has no goal of "improvement" in anticipation of future environmental changes.

Organisms Compete for Resources

Darwin applied Malthus's treatise on human population growth to animal populations. He realized that if all offspring born to a population were to survive, insufficient resources would be available to support the growing population. He calculated the reproductive potential of elephants, assuming an average life span of 100 years and a breeding span of 30–90 years. Given these assumptions, a single female probably bears no fewer than six young, and if all these young survive and continue to reproduce at the same rate, then after only 750 years, the descendants of a single pair of elephants would number about 19 million! Obviously, no environment has the resources to support an elephant population of this magnitude, and no such elephant population has ever existed. This overproduction potential of a species is often referred to as the geometric ratio of increase.

Organisms Differ in Reproductive Success

Some individuals have favorable traits that enable them to better compete for limited resources. The individuals with favorable traits acquire more resources than the individuals with less favorable traits and can devote more energy to reproduction. Darwin called this ability to have more offspring *differential reproductive success.*

Fitness is the reproductive success of an individual relative to other members of a population. The most fit individuals are the ones that capture a larger amount of resources and convert these resources into a larger number of viable offspring. Because organisms vary, as do the challenges of local environments, fitness is influenced by different factors for different populations. For example, among western diamondback rattlesnakes (*Crotalus atrox*) living on dark-colored lava flows, the most fit are those that are black in color. But among those living on desert soil, the most fit are those with the typical light coloring with brown blotching. Background matching helps an animal both capture prey and avoid being captured; therefore, it is expected to lead to survival and increased fitness.

Natural selection occurs because certain members of a population happen to have a variation that allows them to survive and reproduce to a greater extent than do other members. For example, a variation in a desert plant that reduces water loss is beneficial; and a mutation in a wild dog that increases its sense of smell helps it find prey.

Organisms Become Adapted

An **adaptation** is any evolved trait that helps an organism be more suited to its environment. Adaptations are especially recognizable when unrelated organisms living in a particular environment display similar characteristics. For example, manatees, penguins, and sea turtles all have flippers, which help them move through the water. In Chapter 1, we described other ways in which penguins are adapted to their environment. Similarly, a Venus flytrap, a plant that lives in the nitrogen-poor soil of a bog, is able to obtain nitrogen-containing nutrients because it has specialized leaves adapted to catching and digesting flies.

Such adaptations to specific environments result from natural selection. Differential reproduction generation after generation can cause adaptive traits to be increasingly represented in each succeeding generation. Evolution includes other processes in addition to natural selection (see Chapter 16), but natural selection is the only process that results in adaptation to the environment.

We Can Observe Selection at Work

Darwin noted that humans could artificially modify desired traits in plants and animals by selecting individuals with those preferred traits for breeding. For example, the diversity of domestic dogs has resulted from prehistoric humans selectively breeding wolves with particular traits, such as hair length, height, and guarding behavior. This type of human-controlled breeding to increase the frequency of desired traits is called **artificial selection.** Artificial selection, like natural selection, is possible only because the original population exhibits a variety of characteristics, allowing humans to select traits they prefer. For dogs, the result of artificial selection is the existence of many breeds of dogs all descended from the wolf (Fig. 15.7).

Figure 15.7 Artificial selection. All dogs, *Canis lupus familiaris,* are descended from the gray wolf, *Canis lupus,* which began to be domesticated about 14,000 years ago. The process of selective breeding by humans has led to extreme phenotypic differences among breeds.

Figure 15.8 Artificial selection of plants. The vegetables Chinese cabbage, brussels sprouts, and kohlrabi are derived from wild mustard, *Brassica oleracea.* Darwin described artificial selection as a model by which to understand natural selection. With natural selection, however, the environment, and not human selection, provides the selective force.

As another example, several varieties of vegetables can be traced to a single ancestor. Chinese cabbage, brussels sprouts, and kohlrabi are all derived from a single species, *Brassica oleracea* (Fig. 15.8). Modern corn, or maize, has a wild ancestor called teosinte. Teosinte looks very different from the corn we grow for food—teosinte has 5–12 kernels in a single row, and a

a. Large, ground-dwelling finch

b. Warbler-finch

c. Cactus-finch

Figure 15.9 Galápagos finches.   Each of the 13 species of finches has a beak adapted to a particular way of life. For example, **(a)** the heavy beak of the large ground-dwelling finch (*Geospiza magnirostris*) is suited to a diet of large seeds; **(b)** the beak of the warbler-finch (*Certhidea olivacea*) is suited to feeding on insects found among ground vegetation or caught in the air; and **(c)** the longer beak, somewhat decurved, and the split tongue of the cactus-finch (*Cactornis scandens*) are suited to extracting the flesh of cactus fruit.

hard, thick outer shell encases each kernel, making it difficult to use as a food source (see Chapter 24). Strong evidence from archaeology and genetics supports the hypothesis that prehistoric humans, selecting for softer shells, more kernels, and other desirable traits, produced modern corn. Darwin surmised that if humans could create such a wide variety of organisms by artificial selection, then natural selection could also produce diversity, but with the environment, not humans, as the force selecting for particular traits.

The Galápagos finches have beaks adapted to the food they eat, with different species of finches on each island (Fig. 15.9) Today, many investigators, including Peter and Rosemary Grant of Princeton University, are documenting natural selection as it occurs on the Galápagos Islands. In 1973, the Grants began a study of the various finches on Daphne Major, an island near the center of the Galápagos Islands. The weather on this island swings widely back and forth from wet years to dry years, and

the Grants found that the beak size of the medium ground finch, *Geospiza fortis*, adapted to each weather swing, generation after generation (Fig. 15.10). These finches like to eat small, tender seeds that require a smaller beak, but when the weather turns dry, they must eat larger, drier seeds, which are harder to crush. The birds that have a larger beak depth have an advantage during the dry periods, and have more offspring. Therefore, among the next generation of *G. fortis* birds, the mean, or average, beak size has more depth than the previous generation (for more information on the shape of finch beaks, see the Chapter 17 Nature of Science feature, "Genetic Basis of Beak Size in Darwin's Finches"). The Grants' research demonstrates that evolutionary change can sometimes be observed within the timeframe of a human lifespan, rather than over thousands of years.

Recent advances in biotechnology have produced a set of new and revolutionary tools to document phenotype evolution at the level of the gene. As one example, Sean Carroll of the University of Wisconsin, Madison, studies the genes that determine variation in the color patterns on the wings of fruit flies. In one species of fly, *Drosophila biarmipes*, a close relative of the common fruit fly, *D. melanogaster*, males have a black spot on the top, forward, edge of the wing (Fig. 15.11). This spot is part of the male fly's courtship dance. Carrol's research shows that the spot on the wing of the male fly has evolved from a few simple mutations that have changed how a wing gene is switched on and off during development. A few simple mutations in the DNA code were enough to produce a change in the wing color pattern of *D. biarmipes*. This study and others like it demonstrate how new traits can evolve as a result of only a few changes in the DNA code that regulate a gene. In the case of *D. biarmipes*, natural selection, in the form of female mate choice, favors the evolution of males with spotted wings.

Industrial melanism is a common example of how natural selection can shape a trait in a population. Prior to the Industrial Revolution in Great Britain, light-colored peppered moths, *Biston betularia*, were more common than dark-colored peppered moths. It was estimated that only 10% of the moth population was dark at this time. With the advent of industry

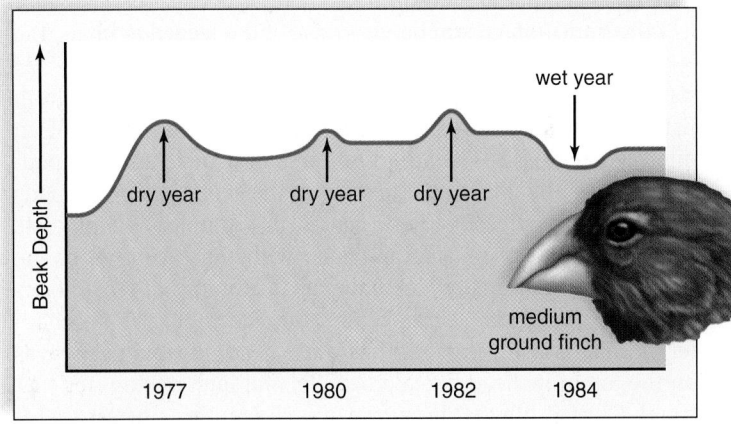

Figure 15.10 Evolution in action.   The average beak depth of medium ground finches varies from generation to generation, according to the weather. The weather affects the hardness and size of seeds on the islands, and different beak depths were better suited to eating different types of seeds. Average beak features were observed to change many times over a period of a decade. This is one way in which evolution by natural selection has been observable over a short period of time.

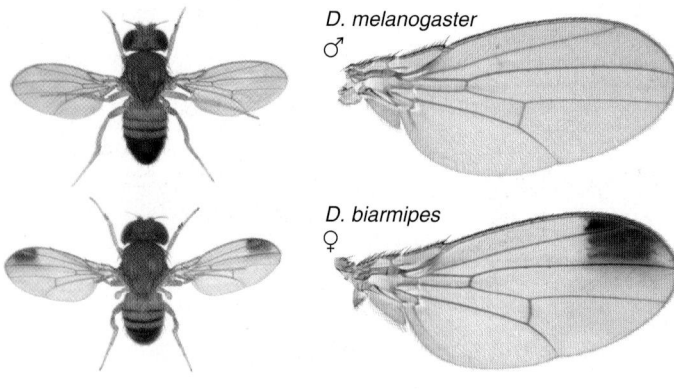

a.

b.

Figure 15.11 Wing spot in *Drosophila biarmipes*. a. Male *D. biarmipes* have a spot on the leading edge of their wings that is not present in its close relative, *D. melanogaster*. **b.** *D. biarmipes* males flash the spot on their wings to attract the attention of females during a courtship dance.

and an increase in pollution, the number of dark-colored moths exceeded 80% of the moth population. After legislation to reduce pollution, a dramatic reversal in the ratio of light-colored moths to dark-colored moths occurred. In 1994, one collecting site recorded a drop in the frequency of dark-colored moths to 19%, from a high of 94% in 1960. (We revisit this example in Chapter 16, where we discuss evolution of populations.)

The rise in bacterial resistance to antibiotics has occurred within the past 30 years or so. Resistance is an expected way of life now, not only in medicine, but also in agriculture. New chemotherapeutic and HIV drugs are required because of the resistance of cancer cells and HIV, respectively. Also, pesticides and herbicides have created resistant insects and weeds.

Check Your Progress   15.2

1. List the three categories of observations of evolution by natural selection that Darwin gathered while traveling on the HMS *Beagle*.
2. Summarize the components of Darwin's theory of evolution by natural selection.
3. Criticize this statement: "Evolution only occurs over millions of years." Provide evidence to support your position.

15.3 Evidence for Evolution

Learning Outcomes

Upon completion of this section, you should be able to

1. Interpret one example from each area of study—fossil, anatomical, biogeographical, and biochemical—as evidence supporting the descent of all life from a common ancestor.
2. Interpret misconceptions of evolution proposed by Darwin's critics.

Many different lines of evidence support the concept that organisms are related through descent from a common ancestor. This is significant because the more varied and abundant the evidence supporting a hypothesis, the more certain it becomes.

Fossil Evidence

Fossils are the remains and traces of past life or any other direct evidence of past life. Traces include trails, footprints, burrows, worm casts, or even preserved droppings. Usually when an organism dies, the soft parts are either consumed by scavengers or decomposed by bacteria. Occasionally, the organism is buried quickly and in such a way that decomposition is never completed or is completed so slowly that the soft parts leave an imprint of their structure; for example, animals or plants trapped in a landslide or mud flow. Most fossils, however, consist only of hard parts, such as shells, bones, or teeth, because these are usually not consumed or destroyed.

Transitional fossils represent the intermediate evolutionary forms of life in transition from one type to another, or a common ancestor of these types. Transitional fossils allow us to retrace the evolution of organisms over relatively long periods of time.

In 2004, a team of paleontologists discovered fossilized remains of *Tiktaalik roseae*, nicknamed the "fishapod" because it is the transitional form between fish and 4-legged animals, the tetrapods (Fig. 15.12). *Tiktaalik* fossils are estimated to be 375 million years old, and are from a time when the transition from fish to tetrapods is likely to have occurred. As expected of an intermediate fossil, *Tiktaalik* has a mix of fishlike and tetrapod-like features that illustrate the steps in the evolution of tetrapods from a fishlike ancestor (Fig. 15.12). For example, *Tiktaalik* has a very fishlike set of gills and fins, with the exception of the pectoral, or front fins, which have the beginnings of wrist bones similar to a tetrapod (see the figure at the beginning of this chapter). Unlike a fish, *Tiktaalik* has a flat head, flexible neck, eyes on the top of its head like a crocodile, and interlocking ribs that suggest it had lungs. These transitional features suggest that it had the ability to push itself along the bottom of shallow rivers and see above the surface of the water—features that would come in handy in the river habitat where it lived.

Even in Darwin's day, scientists knew of *Archaeopteryx*, which is an intermediate between dinosaurs and birds. Progressively younger fossils than *Archaeopteryx* have been found: The skeletal remains of *Sinornis* suggest it had wings that could fold against its body like those of modern birds, and its grasping

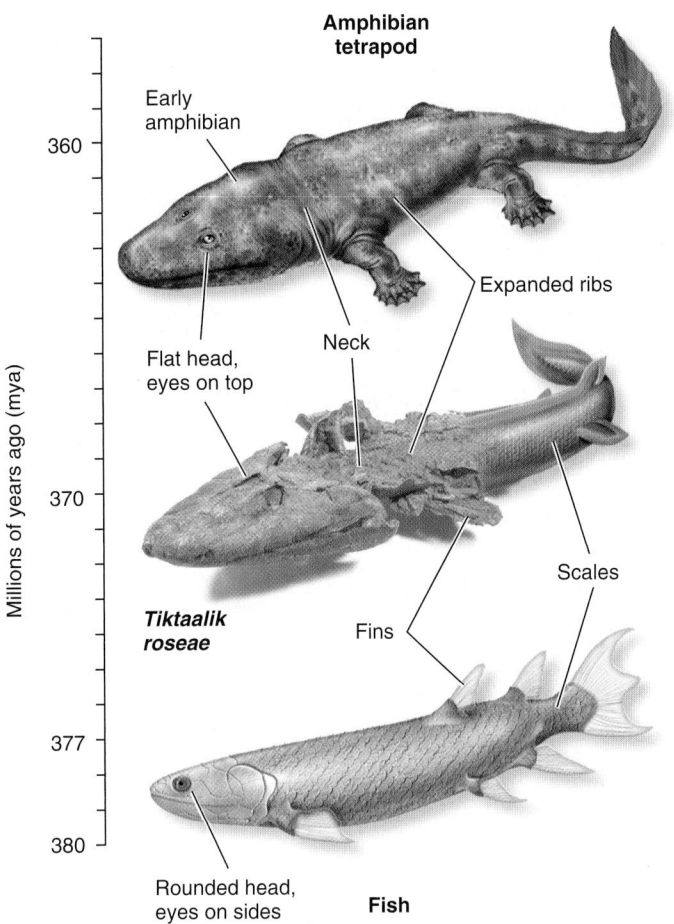

Figure 15.12 Transitional fossils. *Tiktaalik roseae* has a mix of fishlike and tetrapod-like features. Fossils such as *Tiktaalik* provide evidence that the evolution of new groups involves the modification of preexisting features in older groups. The evolutionary transition from one form to another, such as from a fish to a tetrapod, can be gradual, with intermediate forms having a suite of adapted, fully functional features.

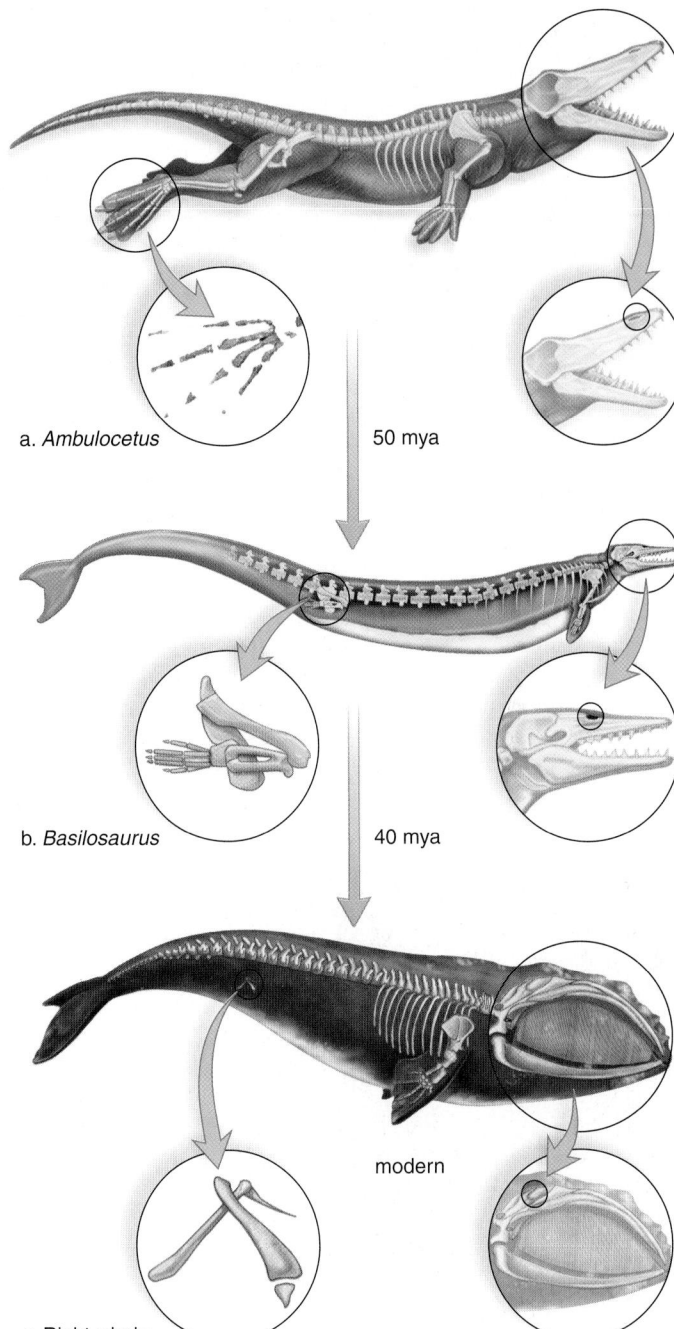

Figure 15.13 Anatomical transitions during the evolution of whales. Transitional fossils such as *Ambulocetus* and *Basilosaurus* support the hypothesis that modern whales evolved from terrestrial ancestors that walked on four limbs. These fossils show a gradual reduction in the hindlimb and a movement of the nasal opening from the tip of the nose to the top of the head—both adaptations to living in water.

feet had an opposable toe—but it still had a tail. Another fossil, *Confuciusornis*, had the first toothless beak. A third fossil, called *Iberomesornis*, had a breastbone to which powerful flight muscles could attach. Such fossils show how the species of today evolved.

Fossils have been discovered that support the hypothesis that whales had terrestrial ancestors. *Ambulocetus natans* (meaning the walking whale that swims) was the size of a large sea lion, with broad, webbed feet on its forelimbs and hindlimbs that enabled it to both walk and swim. It also had tiny hoofs on its toes and the primitive skull and teeth of early whales (Fig. 15.13). Modern whales still have a vestigial hindlimb consisting of only a few bones that are very reduced in size. As the ancestors of whales adopted an increasingly aquatic lifestyle, the location of the nasal opening underwent a transition, from the tip of the snout as in *Ambulocetus*, to midway between the tip of the snout and the skull in *Basilosaurus*, to the very top of the head in modern whales (Fig. 15.13). An older fossil, *Pakicetus*, was primarily terrestrial, and yet had the dentition of an early whale. A younger fossil, *Rodhocetus*, had reduced hindlimbs that would have been no help for either walking or swimming, but may have been used for stabilization during mating.

The origin of mammals is also well documented. The synapsids, an early amniote group, gave rise to the premammals. Slowly, mammal-like fossils acquired features that enabled them to breathe and eat at the same time, a muscular diaphragm and rib cage that helped them breathe efficiently, and so forth. The earliest true mammals were shrew-sized creatures that have been unearthed in fossil beds about 200 million years old. (We return to the topic of mammalian evolution in Chapter 29).

Nature of Science

The Tree of Life: 150 Years of Support for the Theory of Evolution by Natural Selection

Darwin spent his adult life striving to answer the question "Where do species come from?" Prior to Darwin's publication of *On the Origin of Species* in 1859, the prevailing answer to this question was that all species are "fixed" in their current state, as God created them.

Darwin challenged the idea of the fixed nature of species by proposing that species change, or evolve, in response to forces in nature. His hypothesis of evolution by natural selection explained how nature shapes variation in populations. However, Darwin admitted that he could not provide a mechanism to explain how diversity arises in the first place. It was not until the rediscovery of Gregor Mendel's work in 1900 that the concept of the genetic basis of trait inheritance became widely accepted, providing the missing mechanism to explain how new variation in populations could arise, and then be susceptible to the forces of natural selection.

Today, we know a lot more about the links between genes, inheritance, and traits. Over the last 150 years since Darwin published his book, scientists have amassed huge amounts of support for his ideas—so much evidence, in fact, that we now refer to his hypothesis as the theory of evolution by natural selection. A lot of the evidence in support of Darwin's theory has come from biomolecules—such as DNA, chromosomes, and proteins—that are compared among different species to look for a signature of evolution.

This molecular evidence has provided strong support for Darwin's proposal that all life on Earth can be traced to a single ancestor. Early in the development of his theory, Darwin kept notebooks of his thoughts. One notebook (Notebook B) contains the first known representation of life on Earth as a tree (Figure 15A). This was a revolutionary concept at the time, but today evolutionary biologists have constructed thousands

and thousands of trees similar to Darwin's from evidence provided by biomolecules and fossils.

Recently, a group of scientists has initiated a project to construct the largest evolutionary tree of all—the Tree of Life (Figure 15B). The Tree of Life project is a collaborative project to determine how all life on Earth is related, as Darwin proposed, and descended from a common ancestor. To date, the Tree of Life contains hundreds of species from all domains of life, and is growing as more species are added. What greater support of Darwin's theory than a Tree of Life that demonstrates the relatedness of all life on Earth?

Questions to Consider

1. Why was the idea of life as a tree so controversial during Darwin's time?
2. How does the tree of life support Darwin's theory that all life on Earth is descended from a common ancestor?

Figure 15A Life as a tree. *Left:* A photograph of Darwin in his later years. *Below:* A page from Darwin's notebook with his sketch of a branching "tree" connecting living organisms.

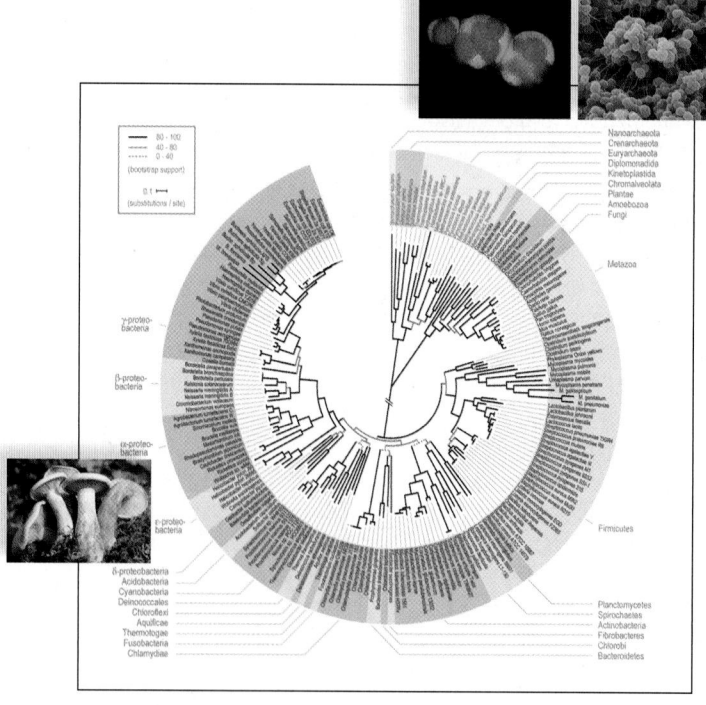

Figure 15B The Tree of Life project. This level of resolution shows the division of life into the three domains: Archaea (in green), Bacteria (in red), and Eukarya (in purple). Notice how all life can trace its descent to a single common ancestor.

Biogeographical Evidence

We described in Section 15.2 the biogeographical observations that Darwin made during his voyage on HMS *Beagle*. We noted that in cases where geography separates continents, islands, and seas, we might expect a different mix of plants and animals. For example, during his travels, Darwin observed that South America lacked rabbits, even though the environment was quite suitable to them. He concluded there were no rabbits in South America because rabbits evolved somewhere else and had no means of reaching South America. Instead, a different animal, the Patagonian cavy, occupied the environmental niche that rabbits held elsewhere.

In addition, Darwin noted that the many different species of finches on the Galápagos Islands were not found on mainland South America. One reasonable explanation is that immigrant finches of a single species from the mainland reached the Galápagos islands and over time evolved into different species on each isolated island.

In the history of the Earth, South America, Antarctica, and Australia were originally connected (see Fig. 18.16). Marsupials, mammals in which females have an external body pouch where their young complete development, had evolved from egg-laying mammalian ancestors, and today they are endemic to South America and Australia. What is now Australia separated and drifted away from the other landmasses, and the marsupials diversified into many different forms suited to various environments (Fig. 15.14).

Marsupials were free to diversify because few, if any, placental mammals were present in Australia. In placental mammals, young complete their development inside the mother's uterus, nourished by the placenta (see Chapter 29). Where placental mammals are abundant, marsupials are not as diverse due to competition. After the formation of the Isthmus of Panama, placental mammals were able to migrate into South America. As a result, marsupial mammals were outcompeted by placental mammals, and the diversity of marsupials in South America declined greatly.

Biogeographical differences, therefore, provided evidence that variability in a single, ancestral population can lead to adaptation to different environments through the forces of natural selection. Competition for resources appears to provide the pressure that leads to diversification.

Anatomical Evidence

Darwin was able to show how descent from a common ancestor can explain anatomical similarities among organisms. Vertebrate forelimbs are used for flight (birds and bats), orientation during swimming (whales and seals), running (horses), climbing (arboreal lizards), or swinging from tree branches (monkeys). Yet all vertebrate forelimbs contain the same sets of bones organized in similar ways, despite their dissimilar functions (Fig. 15.15). The most plausible explanation for this unity is that this basic forelimb plan was present in a common vertebrate ancestor, and this plan was modified independently in all descendants as each continued along its own evolutionary pathway.

Structures that are anatomically similar because they are inherited from a common ancestor are called **homologous.** In

Sugar glider, *Petaurus breviceps*, is a tree-dweller and resembles the placental flying squirrel.

The Australian wombat, *Vombatus*, is nocturnal and lives in burrows. It resembles the placental woodchuck.

Kangaroo, *Macropus*, is an herbivore that inhabits plains and forests. It resembles the placental Patagonian cavy of South America.

Figure 15.14 Biogeography. Each type of marsupial in Australia is adapted to a different way of life. All of the marsupials in Australia presumably evolved from a common ancestor that entered Australia some 60 million years ago.

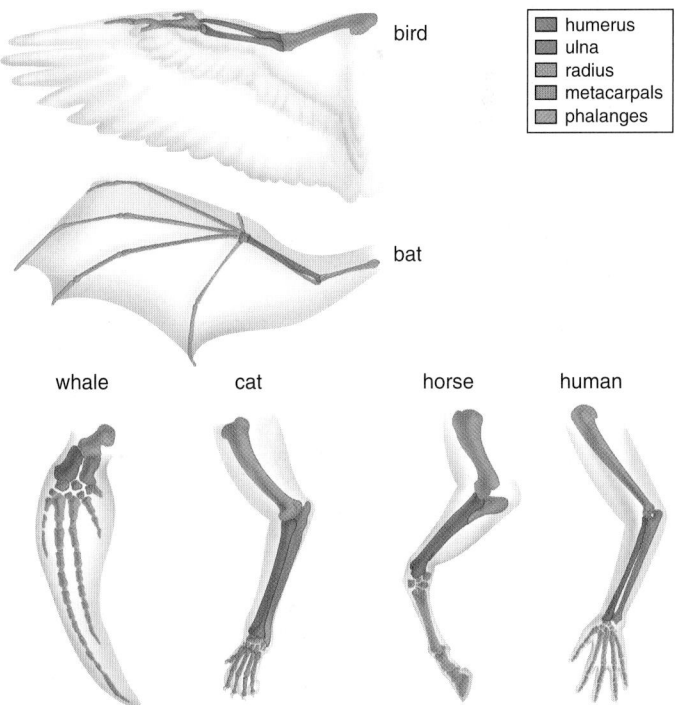

Figure 15.15 Significance of homologous structures. Although the specific design details of vertebrate forelimbs are different, the same bones are present (they are color-coded). Homologous structures provide evidence of a common ancestor.

contrast, **analogous** structures serve the same function, but originated independently in different groups of organisms that do not share a common ancestor. The wings of birds and insects are analogous structures. Thus homologous, not analogous, structures are evidence for a shared common ancestry of particular groups of organisms.

As mentioned earlier, *vestigial structures* are anatomical features that are fully developed in one group of organisms but are reduced and may have no function in related groups. Most birds, for example, have well-developed wings used for flight, while some species have greatly reduced wings and do not fly. Similarly, snakes and whales have no use for hindlimbs, and yet some species have remnants of a pelvic girdle and hindlimbs. Humans have a tailbone but no tail. Vestigial structures occur because organisms inherit their anatomy from their ancestors, and therefore their anatomy carries traces of their evolutionary history.

The homology shared by vertebrates is observable during their embryological development (Fig. 15.16). At some time during development, all vertebrates have a postanal tail and exhibit paired pharyngeal (throat) pouches supported by cartilaginous arches. In fishes and amphibian larvae, these pouches develop

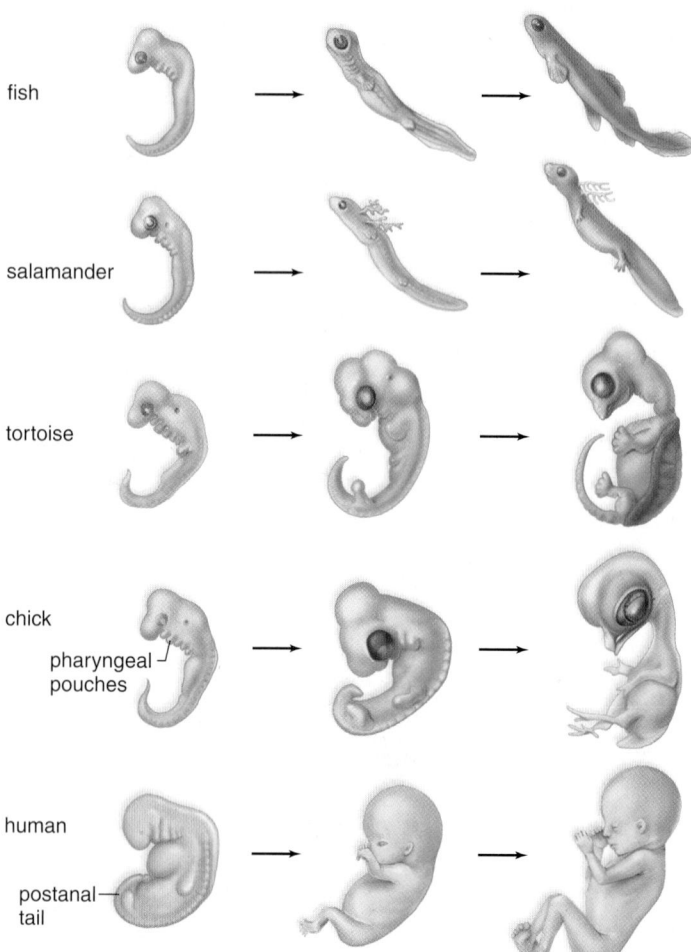

fish

salamander

tortoise

chick

pharyngeal
pouches

human

postanal
tail

Figure 15.16 Significance of developmental similarities.
At these comparable developmental stages, vertebrate embryos have many features in common, which suggests they evolved from a common ancestor. (These embryos are not drawn to scale.)

into functioning gills. In humans, the first pair of pouches and arches becomes the jawbones, the cavity of the middle ear, and the auditory tube. The second pair of pouches becomes the tonsils and facial muscle and nerves, while the third and fourth pairs become the thymus and parathyroid glands.

Why do structures like pharyngeal pouches develop in all vertebrate embryos, but then become very different structures with vastly different functions in adults of different groups? New structures or novel functions can originate only through modification of the preexisting structures in one's ancestors. All vertebrates inherited the same developmental pattern from their common ancestor, but each vertebrate group now has a specific set of modifications to this original ancestral pattern.

Biochemical Evidence

All living organisms use the same basic biochemical molecules, including DNA (deoxyribonucleic acid), RNA (ribonucleic acid), and ATP (adenosine triphosphate). We can deduce from this that these molecules were present in the first living cell or cells from which life as we know it today has arisen.

Organisms have the same triplet nucleic-acid code in their DNA that encodes the same 20 amino acids that form their proteins. Because the sequences of DNA bases in the genomes of many organisms are now known, clear evidence is available that humans have some genes in common with much simpler organisms, such as prokaryotes. Because the genetic code is universal in living things, it is possible to insert a human gene into the genome of a bacterium, and the bacterium will produce the human protein that the gene encodes.

Also, the sequence of amino acids of some proteins is similar across the tree of life. The sequence of amino acids in the human version of cytochrome *c*, a protein essential to cellular respiration, is remarkably similar to that of yeast (Fig. 15.17). The number of differences between the cytochrome *c* amino acid sequence in humans and other organisms increases with the distance in time to their common ancestor—monkey cytochrome *c* differs from that of humans by only one amino acid, from that of a duck by 11 amino acids, and from that of yeast by 51 amino acids.

Video
Molecular Clock

Evidence from Developmental Biology

The study of the evolution of development has discovered that many developmental genes are shared in common among all animals ranging from worms to humans. It appears that life's vast diversity has come about by the same set of regulatory genes that control the activity of other genes involved in development.

For example, *Hox*, or **homeobox**, genes orchestrate the development of the body plan in all animals, from invertebrates such as sea anemones and fruit flies to humans (see the Evolution feature, "Evolution of the Animal Body Plan," in Chapter 28). All animals share a *Hox* gene common ancestor, but the number and type of *Hox* genes varies among animal groups. This variation in *Hox* genes is responsible, at least in part, for the wide range of body plans seen in animals. For example, a change in the timing and duration of the expression of *Hox* genes that control the number and type of vertebrae can produce the spinal column of a chicken or the longer spinal column of a snake. Thus simple

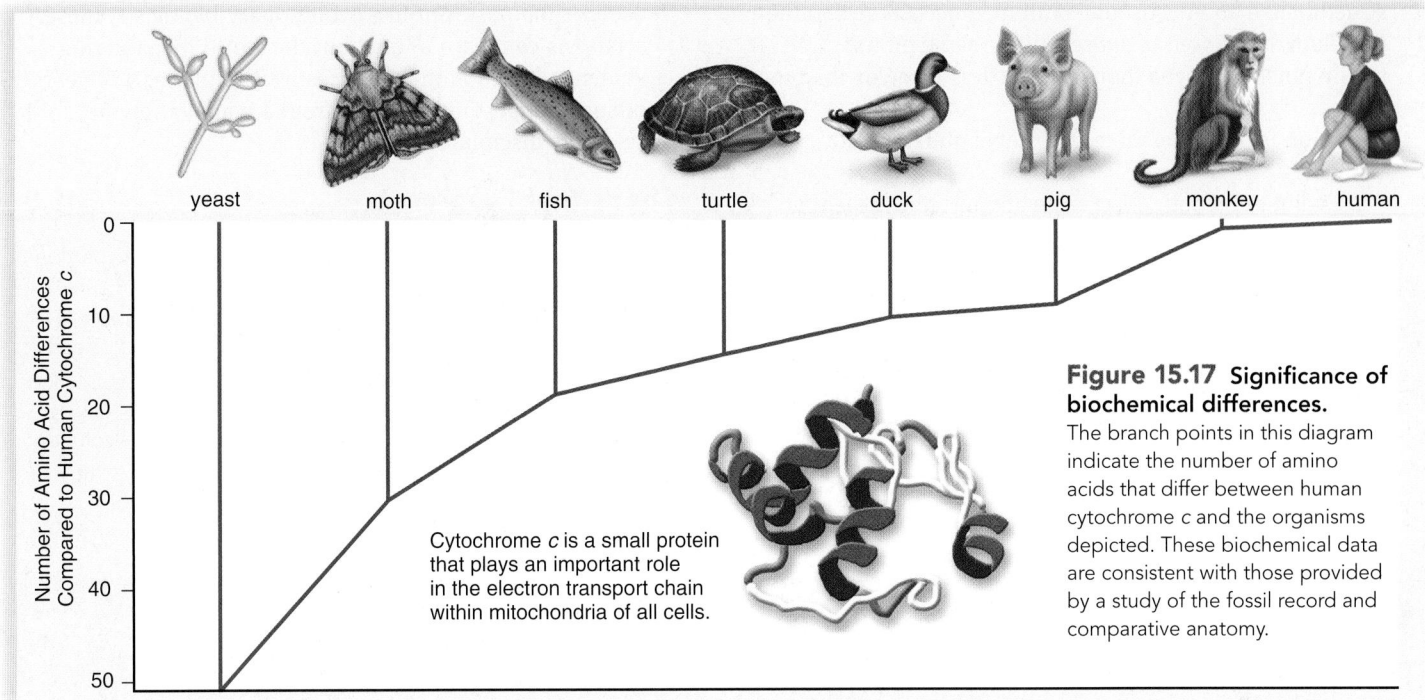

yeast moth fish turtle duck pig monkey human

Number of Amino Acid Differences Compared to Human Cytochrome *c*

0

10

20

30

40

50

Cytochrome *c* is a small protein that plays an important role in the electron transport chain within mitochondria of all cells.

Figure 15.17 Significance of biochemical differences.
The branch points in this diagram indicate the number of amino acids that differ between human cytochrome *c* and the organisms depicted. These biochemical data are consistent with those provided by a study of the fossil record and comparative anatomy.

changes in how genes are controlled can have profound effects on the phenotype of organisms.

Animation
Evolution of Homologous Genes

Criticisms of Evolution

Evolution is no longer considered a hypothesis. It is one of the great unifying theories of biology. Evolution is not "just a theory"—in science, the word *theory* is reserved for those concepts that are supported by a large number of observations (see Section 1.4). The theory of evolution has the same status in biology that the theory of heredity has in genetics. However, some people propose mechanisms other than evolution to explain the origin of new species. These alternatives, founded in religious philosophy, are untestable, and thus they are unscientific.

Many misconceptions about evolution, and the scientific process in general, are commonly used to challenge the legitimacy of the theory of evolution. Below are a few examples of misconceptions about evolution, followed by a brief scientific explanation.

1. Evolution is a theory about how life originated.
    Evolutionary biologists are concerned with how the diversity of life emerged *following* the origin of life. Certainly the study of the origin of life is interesting to evolutionary biologists, but this is not the focus of their research.
2. There are no transitional fossils.
    Biologists do not expect that all transitional forms have been preserved in the fossil record. In fact, scientists *predict* that not all transitional forms will be discovered. The reason is that a series of events must occur before a fossil can be found. First, the organism must have perished in an area that favors the preservation of its skeletal

remains. Soft tissue remains are rarely fossilized, thus, many species will not leave fossil remains at all. Second, scientists have to locate and uncover those remains. This is like finding a needle in a haystack!

Despite this, scientists have unearthed an array of transitional fossils. For example, the fossil record clearly demonstrates transitional forms in the evolution of the whale from its terrestrial ancestor (Fig. 15.13). Also, a series of intermediate fossils illustrates the transition from fish to tetrapods, such as *Tiktaalik* (Fig. 15.12).

3. Evolution proposes life changed as a result of random events; clearly traits are too complex to have originated 'by chance.'
    "Chance" does play a role in evolution, but this is only part of the story. Mutation, the process that *creates* new variation in populations, occurs randomly. However, natural selection, the process that *shapes* variation, is not random. Natural selection can act only on the variation that is present in a population, and it is thus constrained by changes that have occurred in the past. Complex structures, such as the vertebrate eye, did not evolve as a single, functioning unit with all parts intact. Complexity is the result of millions of years of modifications to preexisting traits, each of which provided a useful function at the time. For example, the bacterial flagellum—the long tail-like propeller of some bacteria—contains a complex, microscopic, rotary motor made from an assembly of many proteins. In a different group of bacteria, a simpler version of this protein assembly does not function as a rotary motor, but as a "syringe," called an injectisome, that bacteria use to "inject" eukaryotic cells with toxins. Scientists hypothesize that the flagellum evolved over time via the addition of proteins to a preexisting, simpler

structure like an injectisome. Both the injectisome and the flagellum, composed of subsets of the same proteins, are totally functional even though one is less complex than the other.

4. Evolution is not observable or testable, thus it is not 'science.'

Evolution is both observable and testable. Recently scientists discovered that there are genes that encode more than one type of trait. New variation can arise from small changes to single genes. Several studies show how traits in populations change in response to environmental changes (see Section 15.2). Other branches of science figure out how things work by accumulating a lot of evidence from the real world. Particle physicists cannot see the electrons in an atom. Geologists cannot directly observe the past. But like evolutionary biologists, these scientists can learn a lot about the world by gathering evidence from multiple sources. Evolution has 150 years of such supporting evidence from a wide variety of scientific disciplines.

Check Your Progress   15.3

1. Explain how biomolecules support the theory of evolution by natural selection.
2. Summarize the differences between homologous and analogous vestigial structures, and what each tells us about common ancestry.
3. Define transitional fossils and provide one example.
4. Evaluate one misconception about the theory of evolution.

CONNECTING *the* CONCEPTS *with the* BIG IDEAS

Evolution

- Darwin's theory of natural selection was based on observations that organisms competed for resources and those best adapted survived and reproduced. The more successful the reproduction, the higher the fitness in evolutionary terms. (1A1a, 1A1b)
- Natural selection works upon phenotype differences in organisms caused by random mutations and genetic variation. (1A1c, 1A2b)
- Artificial selection allows humans to choose the traits they deem favorable, impacting variation in many species. (1A2d)
- Darwin's scientific theory of evolution is supported by abundant evidence from the fields of paleontology, biogeography, anatomy, embryology, and biochemistry. (1A4a, b1-3)
- Transitional fossils support the theory of evolution. (1a4b1)
- All life on Earth shares the same genetic code and major metabolic pathways. (1B1a2-3, 1D2b1-2)
- Evolution can be witnessed over a short period of time, as shown by the Grant's 30-year documentation of changes in the beak depth of Galápagos finches. (1C3b*IE*)

Interactions and Systems

- Changes in the environment often drive natural selection; populations with greater genetic diversity have the best chances of adapting to those changes. (4C3a)

*Find the unabridged version of all EK citations at www.glencoe.com/maderAP11.

Media Study Tools

www.glencoe.com/maderAP11

Enhance your study of this chapter with study tools and practice tests. Also ask your instructor about the resources available through ConnectPlus, including the media-rich eBook, interactive learning tools, and animations.

Summarize

15.1 History of Evolutionary Thought

In general, the pre-Darwinian worldview was different from the post-Darwinian worldview.

A century before Darwin's trip, biologists believed in the fixity, or unchanging, nature of species. Linnaeus, the originator of taxonomy, thought that each species had a place in the *scala naturae* and that classification should describe the fixed features of species. Some naturalists, such as Count Buffon and Erasmus Darwin, put forth tentative suggestions that species change over time.

Georges Cuvier and Jean-Baptiste de Lamarck, contemporaries of Darwin in the late eighteenth century, differed sharply on evolution. To explain the fossil record of a region, Cuvier proposed that changes in the makeup of fossils in the strata of the Earth could be explained by regional catastrophes, or extinctions, followed by repopulation from other regions. Lamarck was in support of the idea that species can change as they become adapted to their environments. However, he suggested the inheritance of acquired characteristics as a mechanism for evolutionary change. We now know that only traits encoded in our genes are heritable.

Charles Lyell, the foremost geologist of Darwin's time, made popular James Hutton's theory of slow, uniform, geological change. Darwin's observations of geology led him to support Lyell's proposal that the Earth's massive geological changes are the result of extremely slow processes, and that the Earth, therefore, must be very old.

Thomas Malthus, in his influential work *Essay on the Principle of Population*, proposed that population growth is limited by the availability of resources. Darwin applied this idea to his theory of natural selection.

15.2 Darwin's Theory of Evolution

Charles Darwin formulated hypotheses concerning evolution after taking a five-year voyage as a naturalist aboard the ship HMS *Beagle*. His hypotheses were that descent with modification from a common ancestor does occur, and that natural selection results in adaptation to the environment.

Darwin's study of biogeography, including the animals of the Galápagos Islands, led him to conclude that biological diversity arises from adaptation to the environment, which can eventually lead to the formation of new species.

Natural selection is the mechanism Darwin proposed for how adaptation comes about. Members of a population exhibit random, but inherited, variations. Darwin stressed that there was a struggle for existence. The most-fit organisms are those possessing characteristics that allow them to acquire more resources and to survive and reproduce more than those less fit. In this way, natural selection can result in adaptation to an environment.

15.3 Evidence for Evolution

The theory that all organisms share a common ancestor is supported by many lines of evidence, including fossils, anatomy, biochemistry, and development. The fossil record gives us a snapshot of the history of life that allows us to trace the descent of a particular group.

Biogeography is the study of the range and distribution of plants and animals in different places throughout the world and how, and when, they got to be distributed as they are today. Therefore, a different mix of plants and animals might be expected in cases where geography separates continents, islands, and seas.

A comparison of the anatomy and the development of organisms suggests that all life on Earth is closely related. All organisms have certain biochemical molecules and a body plan encoded in genes shared in common, suggesting relatedness.

The theory of evolution has the same status in biology that the theory of heredity has in genetics. However, alternatives to evolution have been proposed to explain the origin of new species. These alternatives are untestable, and thus are unscientific, explanations founded in religious or spiritual philosophy.

Today, the theory of evolution is one of the great unifying theories of biology because it has been supported by 150 years of scientific evidence.

Key Terms

adaptation 278
analogous 284
artificial selection 278
biogeography 275
catastrophism 273
evolution 272
extant 272
fitness 278
fossil 280
homeobox 284

homologous 283
inheritance of acquired
   characteristics 273
natural selection 277
paleontology 272
strata (stratum) 273
transitional fossil 280
uniformitarianism 274
vestigial structures 272

Assess

Reviewing This Chapter

1. In general, contrast the pre-Darwinian worldview with the post-Darwinian worldview. 272–74
2. Cite naturalists who made contributions to biology in the mid-eighteenth century, and state their understandings about evolutionary descent. 272
3. How did Cuvier explain the succession of life-forms in the Earth's strata? 272–73
4. What is meant by the inheritance of acquired characteristics, a hypothesis that Lamarck used to explain adaptation to the environment? 273–74
5. What were Darwin's views on geology, and what observations did he make regarding geology? 275
6. What observations did Darwin make regarding biogeography? How did these influence his conclusions about the origin of new species? 275–77
7. What are the steps of the natural selection process as proposed by Darwin? 277–78
8. Distinguish between the concepts of fitness and adaptation to the environment. 278
9. How do transitional fossils support descent with modifications? Give an example. 280–81
10. How does biogeography support the concept of descent from a common ancestor? Explain why a diverse assemblage of marsupials evolved in Australia. 283
11. How does anatomical evidence support the concept of descent from a common ancestor? Explain why vertebrate forelimbs are similar despite different functions. 283–84
12. How does biochemical evidence support the concept of descent from a common ancestor? Explain why the sequence of amino acids in cytochrome *c* differs between two organisms. 284

Testing Yourself

Choose the best answer for each question.

1. Which of these pairs is mismatched?
   a. Charles Darwin—natural selection
   b. Linnaeus—classified organisms according to the *scala naturae*
   c. Cuvier—series of catastrophes explains the fossil record
   d. Lamarck—uniformitarianism
   e. All of these are correct.

2. According to the theory of inheritance of acquired characteristics,
   a. if a man loses his hand, then his children will also be missing a hand.
   b. changes in phenotype are passed on by way of the genotype to the next generation.
   c. organisms are able to bring about a change in their phenotype.
   d. evolution is striving toward improving particular traits.
   e. All of these are correct.

3. Why was it helpful to Darwin to learn that Lyell thought the Earth was very old?
   a. An old Earth has more fossils than a new Earth.
   b. It meant there was enough time for evolution to have occurred slowly.
   c. There was enough time for the same species to spread out into all continents.
   d. Darwin said that artificial selection occurs slowly.
   e. All of these are correct.

4. Organisms
   a. compete with other members of their species.
   b. differ in fitness.
   c. are adapted to their environment.
   d. are related by descent from common ancestors.
   e. All of these are correct.

5. DNA nucleotide similarities between organisms
   a. indicate the degree of relatedness among organisms.
   b. may reflect phenotypic (morphological) similarities.
   c. explain why there are phenotypic similarities.
   d. are to be expected if the organisms are related due to common ancestry.
   e. All of these are correct.

6. If evolution occurs, we would expect different biogeographical regions with similar environments to
   a. all contain the same mix of plants and animals.
   b. each have its own specific mixes of plants and animals.
   c. have plants and animals with similar adaptations.
   d. have plants and animals with different adaptations.
   e. Both b and c are correct.

For questions 7–14, match the evolutionary evidence in the key to the description. Choose more than one answer if correct.

KEY:
   a. biogeographical evidence
   b. fossil evidence
   c. biochemical evidence
   d. anatomical evidence
   e. developmental evidence

7. It's possible to trace the evolutionary ancestry of a species.

8. A group of related species have homologous structures.

9. The same types of molecules are found in all living things.

10. All vertebrate embryos have pharyngeal pouches.

11. Transitional fossils have been found between some major groups of organisms.

For questions 12–15, offer an explanation for each of these observations based on information in the section indicated. Write out your answer.

12. Transitional fossils serve as links between groups of organisms. See Fossil Evidence (page 280).

13. Rabbits and Patagonian cavies exist on different continents, but both have long hindlimbs and feed on grasses. See Biogeographical Evidence (page 283).

14. Amphibians, reptiles, birds, and mammals all have pharyngeal pouches at some time during development. See Anatomical Evidence (page 283–84).

15. The base sequence of DNA differs among species but yet maintains some degree of similarity. See Biochemical Evidence (page 284).

Engage

Thinking Scientifically

1. Mutations occur at random and increase the variation within a population for no particular purpose. Our immune system is capable of detecting and killing certain viruses. Would a virus that has a frequent rate of mutation be less or more able to avoid the immune system? Explain.

2. A cotton farmer applies a new insecticide against the boll weevil to his crop for several years. At first, the treatment was successful, but then the insecticide became ineffective and the boll weevil rebounded. Did evolution occur? Explain.

Bioethical Issue

Theory of Evolution

People are often confused by the terminology "theory of evolution." They believe that the word "theory" is being used in an everyday sense, such as "I have a theory about the win-loss record of the Boston Red Sox." But after studying this text, you realize that the word "theory" in science refers to a major scientific concept that has been so supported by observation and experiments that it is widely accepted by scientists as explaining many phenomena in the natural world. The theory of evolution is often referred to as the unifying theory of biology because it explains so many different aspects of living things.

Nonscientists, who may misunderstand the term "theory," have been known to suggest that other theories, aside from the theory of evolution, should be taught in school to explain the diversity of life. Do you think that the curriculum of a science course should be restricted to content that is traditionally considered scientific, or do you believe that other "theories" or ideas that may not have been tested by experimentation should also be presented in a science course? If so, how and by whom? If not, why not?

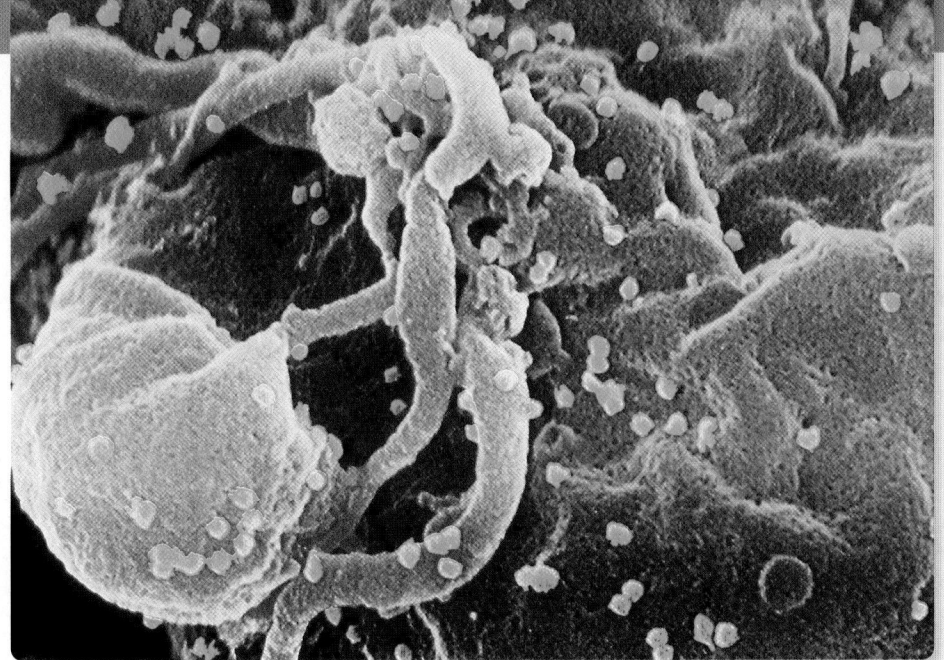

Scanning electron micrograph of HIV virions (green spheres) on the surface of a human immune cell. An average of 10.3 billion new virions (viral particles) are produced in an infected individual each day.

16

How Populations Evolve

I n 1983, two research groups independently discovered that a new virus, the human immunodeficiency virus (HIV), was responsible for the suppressed immune system in patients contracting rare forms of infections. HIV leads to the eventual development of Acquired Immune Deficiency Syndrome, or AIDS. There is no cure for HIV/AIDS, and if left untreated, the mortality rate is near 100%.

New types of HIV emerge from mutations in the virus's genes. An infected individual can have dozens of different types of HIV in their bodies. The large amount of variation in populations of HIV makes it difficult for the human immune system to adapt quickly enough to recognize, and fight against, HIV. Recent advances in drug therapies for HIV/AIDS have come about from an understanding of the way in which HIV evolves.

This chapter is about microevolution, or how a population's gene pool can change over time. An understanding of how populations evolve has helped humans to combat pesticide resistance of agricultural pests, to conserve the Earth's biodiversity, and to treat diseases such as HIV/AIDS.

As you read through the chapter, think about the following questions:

1. What is the link between genes, populations, and evolution?
2. How does a population geneticist determine whether a population is evolving?
3. How is a population defined? Can a human contain his/her own population of microorganisms?
4. With regard to populations, how does natural selection work to shape genetic diversity?

BEFORE YOU BEGIN

Before you begin reading the chapter, take a moment to review the following discussions.

Section 1.3 What is a population?

Section 11.2 How is a Punnett square used to estimate genotype frequencies?

Section 14.4 What is an allele?

FOLLOWING *the* BIG IDEAS

CHAPTER 16 HOW POPULATIONS EVOLVE	
Evolution	Microevolution, or evolution within populations, is measured as a change in allele frequencies over generations.

16.1 Genes, Populations, and Evolution

Learning Outcomes

Upon completion of this section, you should be able to

1. Explain how evolution in populations is related to a change in allele frequencies.
2. List the five conditions necessary to maintain Hardy-Weinberg equilibrium.
3. Apply the Hardy-Weinberg principle to estimate equilibrium genotype frequencies.
4. Describe the agents of evolutionary change.

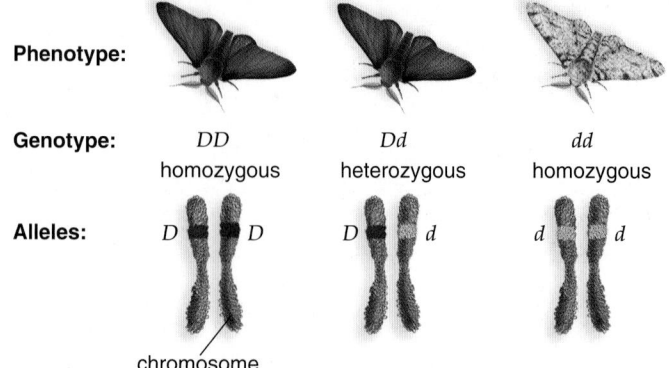

Figure 16.1 The genetic basis of body color in the peppered moth. Light or dark body color in the peppered moth is determined by a gene with two alleles, *D* and *d*. Genotypes *DD* and *Dd* produce dark body color, and *dd* produces light body color. The *D* and *d* alleles are variants of a gene at a particular locus on a chromosome.

A hiker who spends months in the Himalayan mountains gradually gets used to being at high altitude. Part of the reason is that the number of oxygen-carrying red blood cells has increased in response to the oxygen-poor environment. Many traits can change temporarily in response to a varying environment. The color change in the fur of an Arctic fox from brown to white in winter, the increased thickness of your dog's fur in cold weather, or the bronzing of your skin when exposed to the Sun last only for a season.

These are not evolutionary changes. Changes to traits over an individual's lifetime are not evidence that an individual has evolved, because these traits are not heritable. In order for traits to evolve, they must have the ability to be passed on to subsequent generations. Evolution is about change in a trait within a population, not within individuals, over many generations. A **population** is defined as a group of organisms of a single species living together in the same geographic area.

Darwin observed that populations, not individuals, evolve, but he could not explain *how* traits change over time. Now we know that genes interact with the environment to determine traits—the diversity of a population is linked to the genetic diversity of individuals within that population. Because genes and traits are linked, evolution is really about genetic change—or more specifically, *evolution is the change in allele frequencies in a population over time*.

Several mechanisms can cause a population to evolve, or to change allele frequencies, over generations. **Microevolution** pertains to evolutionary change within populations. In this chapter, we use the peppered moth example from Chapter 15 to examine how populations evolve over time.

Microevolution in the Peppered Moth

Population genetics, as its name implies, is the field of biology that studies the diversity of populations at the level of the gene. Population geneticists are interested in how genetic diversity in populations changes over generations, and in the forces that cause populations to evolve. Population geneticists study microevolution by measuring the diversity of a population in terms of allele and genotype frequencies over generations.

You may recall from Chapter 10 that diploid organisms, such as moths, carry two copies of each chromosome, with one copy of each gene on each chromosome. A single gene can come in many forms, or alleles, that encode variations of a single trait. In the peppered moth, a single gene for body color has two alleles, *D* (dark color) and *d* (light color), with *D* dominant to *d* (Fig. 16.1).

We know that with only two alleles, there are three possible genotypes (the combinations of alleles in an individual) for the color gene in the peppered moth: *DD* (homozygous dominant), *Dd* (heterozygous), or *dd* (homozygous recessive). *DD* or *Dd* genotypes produce dark moths, and the *dd* genotype produces light moths (Fig. 16.1).

Allele Frequencies

Suppose that a population geneticist collected a population of 25 moths, some dark and some light, from a forest outside London (Fig. 16.2). In this population you would expect to find a mixture of *D* and *d* alleles in the **gene pool**—the alleles of all genes in all individuals in a population (Fig. 16.2). The population geneticist ran tests to determine the alleles present in each moth. Of the 50 alleles in the peppered moth gene pool (2 alleles × 25 moths), 10 were *D* and 40 were *d*. Thus the frequency of the *D* and the *d* alleles would be 10/50, or 0.20, and 40/50, or 0.80 (Fig. 16.2). The **allele frequency,** as illustrated in this example, is the proportion of each allele in a population's gene pool.

Notice that the frequencies of *D* and *d* add up to 1. This relationship is true of the sum of allele frequencies in a population for any gene of any diploid organism. This relationship is described by the expression $p + q = 1$, where p is the frequency of one allele, in this case *D*, and q is the frequency of the other allele, *d* (Fig. 16.2).

For the next three seasons, samples of 25 moths were collected from the same forest, and the allele frequencies were always the same: 0.20 *D*, and 0.80 *d*. Because allele frequencies in this population did not change over generations, we could conclude, with regards to color, that this population has not evolved.

Hardy-Weinberg Equilibrium

A population in which allele frequencies do not change over time, such as in the moth population just described, is said to be in *genetic equilibrium* or **Hardy-Weinberg equilibrium**—a stable, non-evolving state. Hardy-Weinberg equilibrium is derived from the work of British mathematician Godfrey H. Hardy and German physician Wilhelm Weinberg, who in 1908 developed

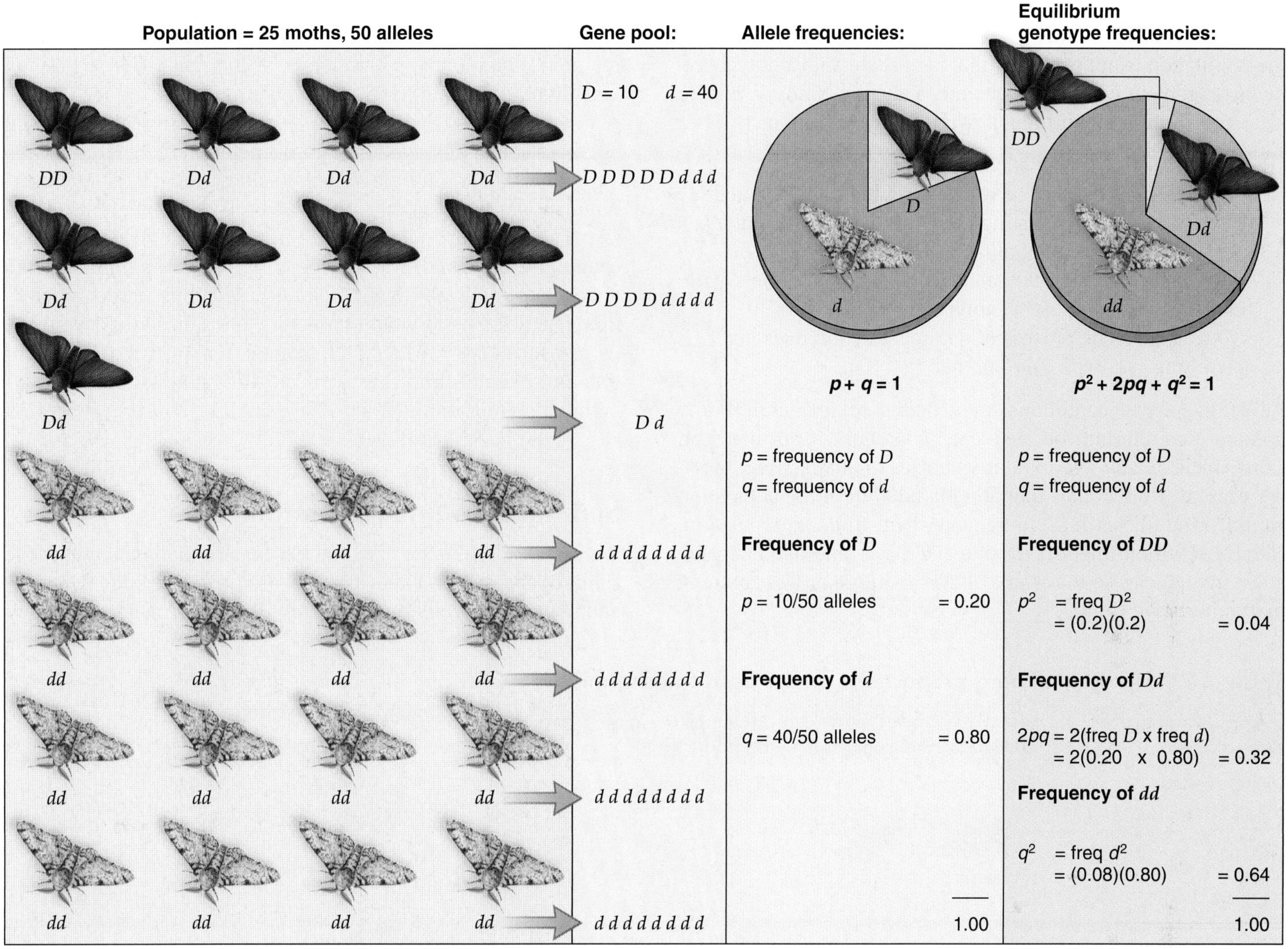

Population = 25 moths, 50 alleles | **Gene pool:** | **Allele frequencies:** | Equilibrium **genotype frequencies:**

$D = 10 \quad d = 40$

DD → D D D D D d d d

Dd Dd Dd

Dd Dd Dd Dd → D D D D d d d d

Dd → D d

dd dd dd dd → d d d d d d d d

dd dd dd dd

dd dd dd dd → d d d d d d d d

dd dd dd dd

dd dd dd dd → d d d d d d d d

Allele frequencies:

$$p + q = 1$$

p = frequency of D
q = frequency of d

Frequency of D

p = 10/50 alleles = 0.20

Frequency of d

q = 40/50 alleles = 0.80

1.00

Equilibrium genotype frequencies:

$$p^2 + 2pq + q^2 = 1$$

p = frequency of D
q = frequency of d

Frequency of DD

p^2 = freq D^2
 = (0.2)(0.2) = 0.04

Frequency of Dd

$2pq$ = 2(freq D x freq d)
 = 2(0.20 x 0.80) = 0.32

Frequency of dd

q^2 = freq d^2
 = (0.08)(0.80) = 0.64

1.00

Figure 16.2 How Hardy-Weinberg equilibrium is estimated. A population of 25 moths contains a gene pool of D and d alleles. The frequencies of the D and d alleles can be estimated from the gene pool. Under Hardy-Weinberg equilibrium, the frequencies of D and d alleles should produce in the next generation a predictable frequency of genotypes that can be calculated with the Hardy-Weinberg principle.

a mathematical model to estimate genotype frequencies of a population that is in genetic equilibrium. Their mathematical model, called the **Hardy-Weinberg principle,** proposes that the genotype frequencies of a non-evolving population can be described by the expression $p^2 + 2pq + q^2$, again with p and q representing the frequency of alleles D and d. Recall that in our moth population, $D = p$ and $d = q$, so that D^2 is the frequency of the DD genotype, $2Dd$ is the frequency of the Dd genotype, and d^2 is the frequency of the dd genotype (Fig. 16.2).

A simple Punnett square, first described in Chapter 11, is another way to illustrate how the Hardy-Weinberg principle explains the genotype frequencies of a population. The frequency of the D and d alleles in the gametes (sperm and egg) in this population would be the same as the allele frequencies, so that 20% of alleles in eggs and sperm will be D, and 80% of alleles in eggs and sperm will be d (Fig. 16.3). The genotype frequencies from the Punnett square match those predicted by the Hardy-Weinberg principle, that is, the frequency of DD, Dd, and dd is explained by the expression $D^2 + 2Dd + d^2$ (Figs. 16.2, 16.3).

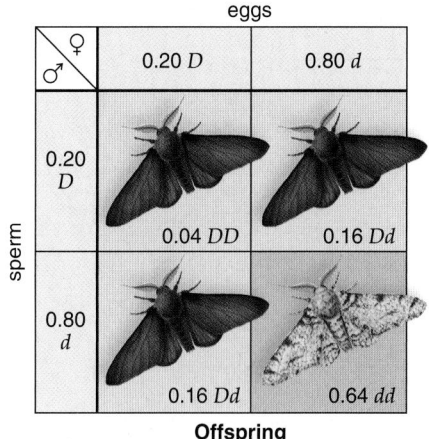

Figure 16.3 Calculation of Hardy-Weinberg equilibrium from gamete frequencies. Notice that the results of the Punnett square match those shown in Figure 16.2.

The allele and genotype frequencies of the moth population we are considering follow the Hardy-Weinberg principle only if the population is not evolving, that is, if allele and genotype frequencies remain constant over time. Thus the Hardy-Weinberg principle applies only if the following conditions are met:

1. *No mutation:* No new alleles arise by mutation.
2. *No migration:* No new members (and their alleles) join the population, and no existing members leave the population.
3. *Large gene pool:* The population is very large.
4. *Random mating:* Individuals select mates at random; mate choice is not biased by genotypes or phenotypes.
5. *No selection:* The process of natural selection does not favor one genotype over another.

All of the above conditions are required to maintain Hardy-Weinberg equilibrium because each condition, if not met, can cause allele and/or genotype frequencies to change (Fig. 16.4). For example, individuals migrating in and out of the population would bring alleles into, or remove them from, a population. Natural selection might favor one allele over another, changing allele frequencies so that some alleles are more or less common than others (Fig. 16.4).

Although possible in theory, Hardy-Weinberg equilibrium is never achieved in wild populations because all of the five required conditions are never met in the real world. The peppered moth population we are using as an example obeys all five of the conditions for genetic equilibrium, but in reality, populations are constantly evolving from generation to generation.

The Hardy-Weinberg principle does not describe natural populations, but it is an important tool for population geneticists because the violation of one or more of the five conditions causes the allele and/or genotype frequencies of a population to change in predictable ways. These predictions permit population geneticists to identify the factors that cause microevolution by measuring how allele and genotype frequencies of a population are different from those expected of a population in Hardy-Weinberg equilibrium (Table 16.1).

Microevolution and Hardy-Weinberg Equilibrium

We can modify the peppered moth scenario to illustrate how the Hardy-Weinberg principle can be used to determine whether, and how, microevolution has occurred. Population genetic theory

Table 16.1 Hardy-Weinberg proportions can be used to determine if evolution has occurred.

Hardy-Weinberg Condition	Deviation from condition	Effect of deviation on population	Expected deviation from HWE	Evolution occurred?
Random mating	Nonrandom Mating DD/Dd X DD/Dd	Allelles do not assort randomly	Change in genotype frequencies	NO
No selection	Selection	Certain alleles are seleted for or against	Change in allele frequencies	YES
No mutation	Mutation	Addition of new alleles	Change in allele frequencies	YES
No migration	Immigration or emigration	Individuals carry alleles into, or out of, the population	Change in allele frequencies	YES
Large population (no genetic drift)	Small population (genetic drift) 1. bottleneck effect 2. founder effect	Loss of allele diversity; some alleles may disappear	Change in allele frequencies	YES

F₁ generation

Allele frequencies: **Genotype frequencies:**

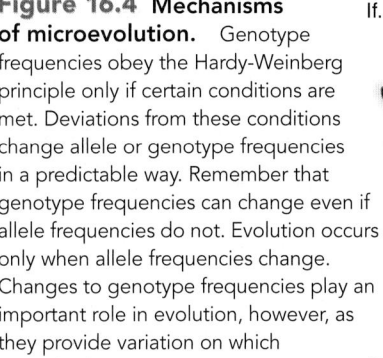

0.20 D
0.80 d

$p^2 + 2pq + q^2 = 1$

DD = 0.04
Dd = 0.32
dd = 0.64

If...
Random mating
No selection
No migration
No mutation
...then we *expect*

F₂ generation

Allele frequencies: **Genotype frequencies:** **Conclusion:**

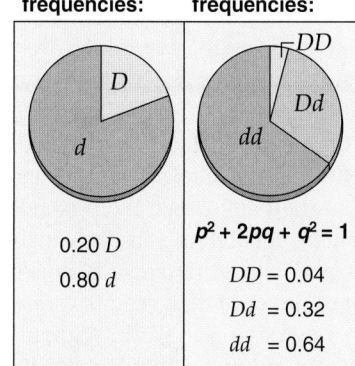

0.20 D
0.80 d

$p^2 + 2pq + q^2 = 1$

DD = 0.04
Dd = 0.32
dd = 0.64

No change in allele frequencies
No change in genotype frequencies
Evolution has not occurred

Figure 16.4 Mechanisms of microevolution. Genotype frequencies obey the Hardy-Weinberg principle only if certain conditions are met. Deviations from these conditions change allele or genotype frequencies in a predictable way. Remember that genotype frequencies can change even if allele frequencies do not. Evolution occurs only when allele frequencies change. Changes to genotype frequencies play an important role in evolution, however, as they provide variation on which natural selection can act.

If...
Nonrandom mating
...then we *observe*

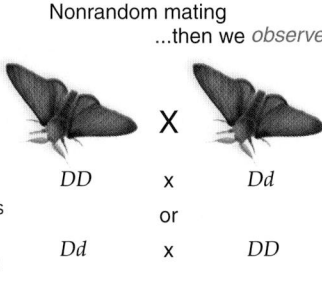

DD   x   Dd
or
Dd   x   DD

0.20 D
0.80 d

$p^2 + 2pq + q^2 = 1$

DD = 0.10
Dd = 0.20
dd = 0.70

No change in allele frequencies
Genotype frequencies change
Evolution has not occurred

If...
Selection
...then we *observe*

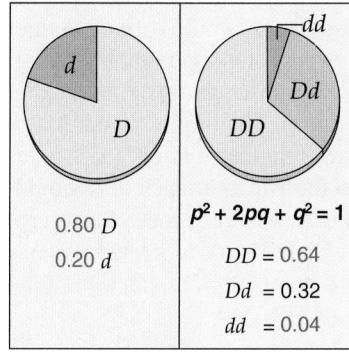

0.80 D
0.20 d

$p^2 + 2pq + q^2 = 1$

DD = 0.64
Dd = 0.32
dd = 0.04

Allele frequencies change
Genotype frequencies change
Evolution has occurred

predicts that Hardy-Weinberg equilibrium can be interrupted by deviation from any of its five conditions. Therefore, we can predict that evolutionary change can be caused by mutation, migration, small population size, nonrandom mating, or natural selection.

It is important to remember throughout these examples that we are attempting to measure evolution as a change in allele frequencies from one generation to another. Thus, we start with allele and genotype frequencies of an F₁ generation, and then remeasure the same frequencies in the F₂ generation.

Mutation

A *mutation* is a change to the DNA sequence, and it serves as a source of new genetic variation. Most mutations occur because of errors made to the DNA sequence during DNA replication. The rate of mutations is generally very low, on the order of one mutation per 100,000 cell divisions. Mutations may also occur because of exposure to *mutagens,* chemical or physical agents that cause DNA mutations, as described in Chapter 13.

Not all mutations affect the genetic equilibrium of a population. Most mutations that occur during DNA replication and from mutagens are repaired by cellular repair mechanisms. Those mutations that are not repaired can affect the gene pool of a population only if they *can be*, and *are*, transmitted from the F₁ to subsequent generations. In other words, the mutations must be carried by the gametes of individuals that successfully reproduce. In addition, mutation is a random process, and the mutation must occur in a gene such that the result is a change in the frequency of the gene's alleles. Thus, it is safe to say that *for any particular gene*, mutation, although possible, is not a major force for evolutionary change because it generally results in only small changes in the allele frequency of a single allele.

In the peppered moth, a hypothetical example of evolution by mutation could be a mutation that changes a single *D* allele to a *d* allele in gametes of a member of the F₁ generation. This change would alter the frequency of *D* and *d* alleles in the F₂ generation gene pool (Fig. 16.4).

Although inherited mutations are rare, the sum of the effect of mutations is essential to evolution. The reason is that a single organism has many thousands of genes, and each gene can have many different alleles, so that even if the frequency of heritable mutation in a particular allele of a single gene is very low, *over thousands of genes and alleles,* mutation can have a significant impact on the evolution of a population.

For example, suppose that the peppered moth has 30,000 genes in its genome; if each gene has two alleles—and genes often have more than two alleles—then any single moth could carry 60,000 different alleles. In a large population, it is likely that a moth would carry at least one new allele that arose because of mutation. These new alleles are sources of new genetic variation that is essential for evolutionary processes to work. Without new mutations, evolution could not occur, because natural selection must have variation on which to act.

Animation
Mutation by Base Substitution

Migration

Gene flow is the movement of alleles between populations. Gene flow occurs when plants or animals migrate, or more specifically, their gametes move, between populations. When gene flow brings a new or rare allele into a population, the allele frequency in the next generation changes. Gene flow in plants may result when the pollen from one plant fertilizes a plant in another population (Figure 16.5)

In the peppered moth example, the movement of a dark- or light-colored individual into a population would introduce one of the three genotypes: *DD, Dd,* or *dd.* Thus the numbers of the *D* and *d* alleles would change. In the sample group of moths, the addition of a *Dd* moth would change the number of moths to 26, the total number of alleles to 52, the number of *D* alleles to 11, and the number of *d* alleles to 41. The frequency of *D* would now be 11/52, or 0.21, instead of 0.20; the frequency of *d* would change accordingly, to 0.79. A similar change in allele frequencies would occur if a moth left the population, taking with it its *D* or *d* gametes (see Fig. 16.4).

The amount of gene flow between populations depends on several factors, including the distance between populations, the ability of individuals or their gametes to move between populations, and behavior that determines whether an individual will migrate and mate. When gene flow continuously occurs between populations, the gene pools of each population become more and more similar over time until they appear as if they are a single population. In contrast, if migration between populations does not occur, the gene pools of the populations become more and more different over time. The differences in the genetic makeup of these populations can eventually become so large that they become **reproductively isolated**—or incapable of interbreeding. Reproductive isolation is the first stage in the formation of new species, which is covered in Chapter 17.

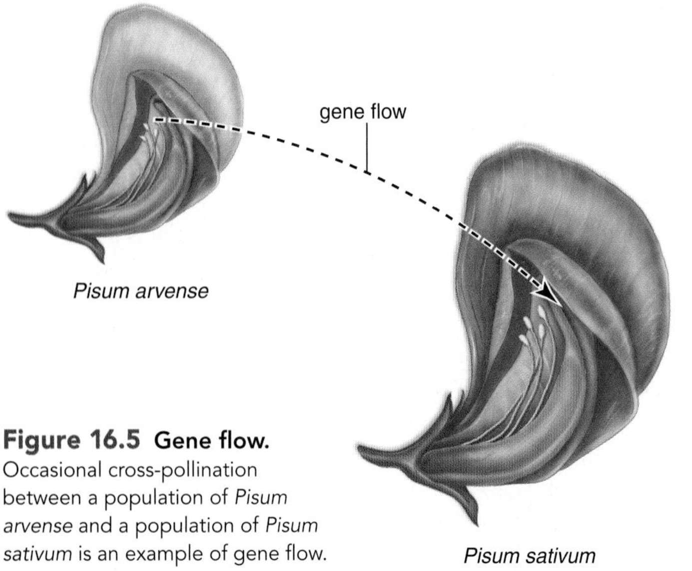

Small Population Size

Genetic drift refers to changes in the allele frequencies of a gene pool due to chance events. Such events remove individuals, and their genes, from a population at random—without regard for genotype or phenotype. Suppose the allele *B* (for brown) occurs in 10% of the members in a population of frogs (Fig. 16.6). If a storm kills half of the frogs, there is a good possibility that some of the *B* alleles would be removed from the population by chance. However, it is also possible that only green frogs would be killed, and then the allele frequency of *B* would be increased after the storm.

In the equilibrium peppered moth population of 25 individuals, 20% of the alleles are *D,* and 80% are *d.* Suppose that a heavy storm kills five of the 25 moths, or 20% of the alleles in the gene pool. To simulate the random loss of individuals from the storm, imagine that all of the 25 moths are in a canvas bag. To simulate the storm, five moths are removed at random from the bag. The remaining 20 moths in the bag represent the population after the storm. The phenotype, genotype, and allele frequencies are then determined for the remaining 20 moths, the new population after the storm. Overall, suppose we find that the loss of the five moths has increased the frequency of the *D* allele by 2.5%, and decreased the frequency of the *d* allele by 2.5%. The *D* and *d* alleles have changed frequency, and we would conclude that the population has evolved due to genetic drift.

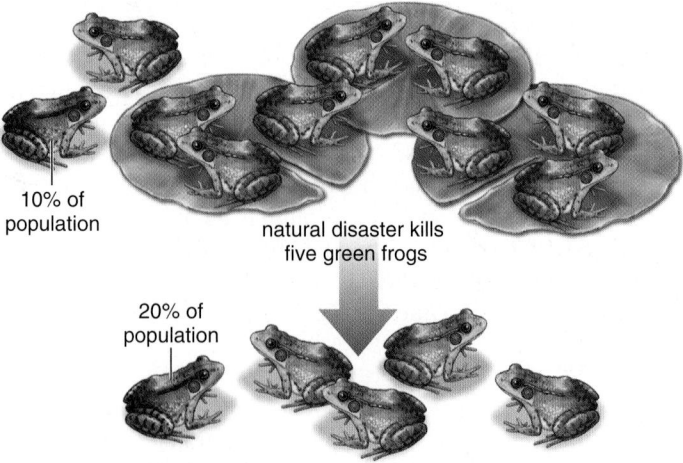

gene flow

Pisum arvense

Pisum sativum

Figure 16.5 Gene flow.
Occasional cross-pollination between a population of *Pisum arvense* and a population of *Pisum sativum* is an example of gene flow.

10% of population

natural disaster kills five green frogs

20% of population

Figure 16.6 Genetic drift. Genetic drift occurs when, by chance, only certain members of a population (in this case, green frogs) reproduce and pass on their alleles to the next generation. A natural disaster can cause the allele frequencies of the next generation's gene pool to be different from those of the previous generation. Genetic drift can be a powerful force for evolutionary change, especially in small populations.

Although genetic drift occurs in populations of all sizes, the gene pool of a smaller population is likely to be more affected by genetic drift. For example, in the population of 25 moths, removing five moths reduced the size of the gene pool by 20% (10 alleles from a gene pool of 50). But, if the storm had removed five moths from a population of 500, or ten alleles out of a pool of 1,000, only 1% of the gene pool would have been eliminated—with less effect on allele frequency. Thus, the smaller the popula-tion, the more genetic drift impacts allele frequencies.

Animation
Simulation of
Genetic Drift

In some cases, a large population can suddenly become very small, such as a *bottleneck,* when natural disasters strike and significantly reduce a population. When this occurs, the effects of genetic drift can be large. A **bottleneck effect** is a special type of genetic drift where the loss of genetic diversity is from natural disasters (e.g., hurricane, earthquake, or fire) or because of disease, overhunting, overharvesting, or habitat loss. A **founder effect**, another type of genetic drift, is similar to a bottleneck effect except that genetic variation is lost when a few individuals break away from a large population to found a new population.

The outcome of a founder and bottleneck effect is a small population with a gene pool made up of a random assortment of alleles from the original, large population. In some cases, alleles can disappear altogether. This newly assembled gene pool in the small population can be, and often is, much different from the gene pool of the population before the bottleneck or founder event occurred (Fig. 16.7). The greater the reduction in popu-lation size, the greater the effects on allele frequencies. Thus genetic drift is one of the more powerful forces for evolution of small populations.

Another byproduct of a very small population is a higher than normal occurrence of **inbreeding**, or mating between rela-tives, because after a few generations only a few, if any, unre-lated mates are available. Unlike genetic drift, *inbreeding alone does not affect the frequency of alleles* and thus does not cause a population to evolve. Nevertheless, inbreeding can have a significant impact on the genotypes, and thus the phenotypes, of individuals, sometimes with unfortunate consequences. Of particular concern are rare recessive disorders, which emerge more frequently in inbred populations (see the Nature of Science feature on page 300).

Random Mating

Hardy-Weinberg equilibrium requires individuals in a population to mate randomly, but nonrandom mating alone does not cause allele frequencies to change. Nonrandom mating, however, does affect *how the alleles in the gene pool assort into genotypes,* and thus affect the phenotypes in a population. Inbreeding, such as that in the Pingelapese described in the Nature of Science feature, is an example of nonrandom mating.

In a randomly mating popuation, the alleles in the gene pool, carried by gametes, assort at random, because individu-als mate at random. When mating is nonrandom, gametes, and thus alleles, assort according to mating behavior. For example, another type of **nonrandom mating,** called **assortative mating,** occurs when individuals choose a mate with a preferred trait,

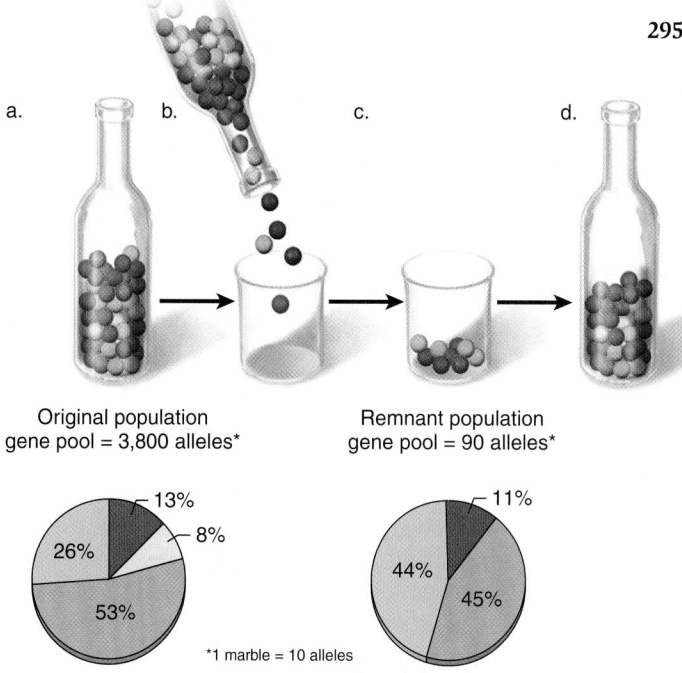

Original population
gene pool = 3,800 alleles*

Remnant population
gene pool = 90 alleles*

*1 marble = 10 alleles

Figure 16.7 Bottleneck and founder effects change allele frequencies.   a. The gene pool of a large population contains four different alleles, represented by colored marbles in a bottle, each with a different frequency. **b.** A population bottleneck occurs. The marbles, or alleles, that exit the bottle must pass through the narrow neck into the cup. The new gene pool will have a fraction of the alleles from the original population. **c.** The gene pool of the new population has changed from the original. Some alleles are in high frequency, while some are not present. **d.** A founder event is the same as a bottleneck except that in the founder event, the original population still exists.

such as a particular coat color, feather length, or body size. Assortative mating brings together alleles for these traits more often than would happen by chance. The result is certain geno-types are more frequent than others.

Although nonrandom mating does not, in itself, cause a pop-ulation to evolve, it can play an important role in the evolution of a population. For example, in the Pingelapese, nonrandom mating resulted in an increased frequency of colorblindness.

No Natural Selection

A population in Hardy-Weinberg equilibrium has phenotypes that are equally likely to survive and reproduce—no one geno-type has an advantage over another. But in nature we know that some phenotypes do have a reproductive advantage. Individuals who have an advantageous phenotype pass on the allele for this trait to their offspring. Over time, selection for this advanta-geous trait increases the frequency of the alleles associated with it, while other alleles decrease.

Natural selection is the foundation of Darwin's theory of evolution. In the next section we discuss in detail how natural selection works within populations.

Check Your Progress   16.1

1. List the five conditions necessary for Hardy-Weinberg equilibrium and describe what happens to allele frequencies in a population if these conditions are not met.
2. Estimate the equilibrium genotype frequencies from a population with allele frequencies $p = 0.10$, $q = 0.90$.

16.2 Natural Selection

Learning Outcomes

Upon completion of this section, you should be able to

1. Differentiate between stabilizing, directional, and disruptive selection.
2. Interpret the type of natural selection operating on a trait by the change in shape of a phenotype distribution.
3. Explain how sexual selection drives adaptation for increased fitness.

In this chapter, we consider natural selection in a genetic context. Many traits are **polygenic** (controlled by many genes). If a particular trait is measured for all individuals in a population, a graph of the measurement of each individual will display a bell-shaped curve. When this range of variation is exposed to the environment, natural selection favors the variant that is adaptive, or provides an advantage, under the present environmental circumstances. Natural selection acts much the same way as a governing board that decides which students will be admitted to a college. Some students will be favored and allowed to enter, while others will be rejected and not allowed to enter. Of course, in the case of natural selection, the chance to reproduce is the prize awarded.

Types of Natural Selection

Investigators have defined three general types of natural selection; these are stabilizing selection, directional selection, and disruptive selection (Fig. 16.8).

Stabilizing selection occurs when an intermediate phenotype can improve the adaptation of the population to those aspects of the environment that remain constant. With stabilizing selection, extreme phenotypes are selected against, and the intermediate phenotype is favored. As an example, consider that when Swiss starlings lay four or five eggs, more young survive than when the female lays more or fewer than this number. Genes determining physiological characteristics such as the production of yolk, and behavioral characteristics such as how long the female mates, are involved in determining clutch size.

Human birth weight is another example of stabilizing selection. Over many years, hospital data have shown that human infants born with an intermediate birth weight (3–4 kg) have a better chance of survival than those at either extreme (either much less or much greater than average). When a baby is small, its systems may not be fully functional, and when a baby is large, it may have experienced a difficult delivery. Stabilizing selection reduces the variability in birth weight in human populations (Fig. 16.9).

Directional selection occurs when an extreme phenotype is favored, and the distribution curve shifts in that direction. Over time, directional selection changes the average phenotype of a population. Such a shift can occur when a population is adapting to a changing environment. For example, the modern horse, *Equus*, shows a gradual increase in body size as the environment changed from forest to grassland (Fig. 16.10). The ancestor of the modern horse, *Hyracotherium*, was around the size of a dog and lived in forested environments of the Eocene. Its smaller body size and low-crowned teeth were well-suited to finding refuge among trees and eating leaves. In the Miocene, grasslands began to

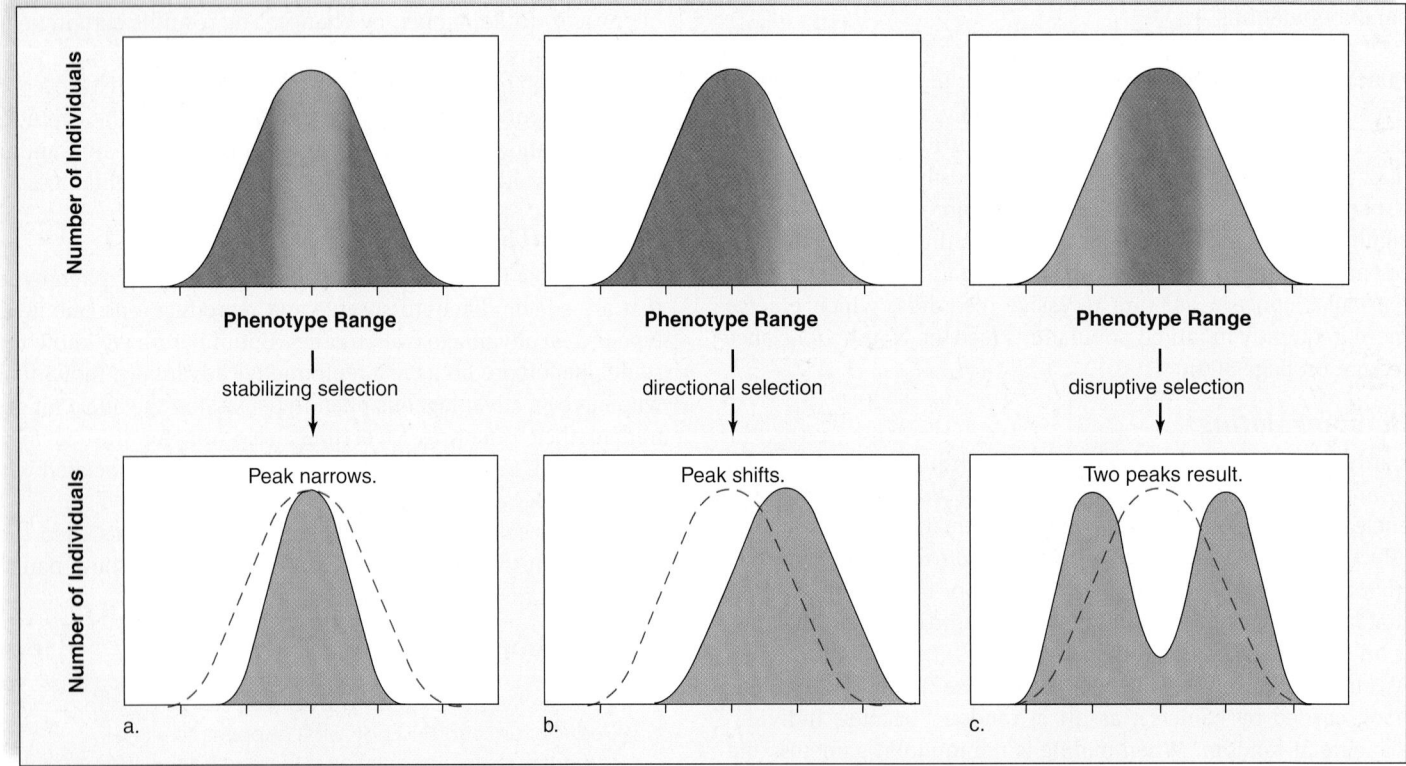

Figure 16.8 Three types of natural selection. Natural selection shifts the average value of a phenotype over time. **a.** During stabilizing selection, the intermediate phenotype increases in frequency; **(b)** during directional selection, an extreme phenotype is favored, which changes the average phenotype value; and **(c)** during disruptive selection, two extreme phenotypes are favored, creating two new average phenotype values, one for each phenotype.

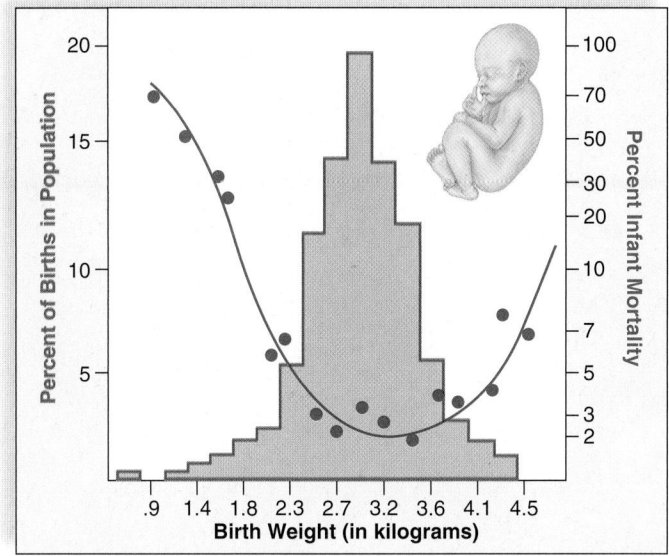

Figure 16.9 Human birth weight.   The birth weight (blue) is influenced by the mortality rate (red).

replace forests. Modern horses are much larger than their ancestors, and are adapted for speed and long-distance movements in a grassland environment. Long legs provide strength and speed for running, and durable teeth are suited for grinding grasses.

Disruptive selection is found when two or more extreme phenotypes are favored over any intermediate phenotype. For example, British land snails have a wide habitat range that includes low-vegetation areas (grass fields and hedgerows) and forests. In forested areas, thrushes feed mainly on light-banded snails, and the snails with dark shells become more prevalent. In low-vegetation areas, thrushes feed mainly on snails with dark shells, and light-banded snails become more prevalent. Therefore, these two distinctly different phenotypes are found in the population (Fig. 16.11).

Sexual Selection

Sexual selection refers to adaptive changes in males and females that lead to an increased ability to secure a mate. Sexual selection in males may result in an increased ability to compete with other males for a mate, while females may select a male with the best **fitness** (the ability to produce surviving offspring). In that way, the female increases her own fitness. Many consider sexual selection a form of natural selection because it affects fitness.

Female Choice

Females produce few eggs compared to a male's production of sperm, so the choice of a mate becomes a serious consideration. In a study of satin bowerbirds, two opposing hypotheses regarding female choice were tested:

1. *Good genes hypothesis:* Females choose mates on the basis of traits that improve the chance of survival.
2. *Runaway hypothesis:* Females choose mates on the basis of traits that improve male appearance. (The term *runaway* pertains to the possibility that over generations, the trait will become exaggerated in the male until its mating benefit is checked by the trait's unfavorable survival cost.)

As investigators observed the behavior of satin bowerbirds, they discovered that aggressive males were usually chosen as mates by females. It could be that inherited aggressiveness does improve the chance of survival, or it could be aggressive males are good at stealing blue feathers from other males—females prefer blue feathers as bower decorations. Therefore, the data did not clearly support either hypothesis.

The Raggiana Bird of Paradise exhibits remarkable **sexual dimorphism**, meaning that males and females differ in size and other traits. The males are larger than the females and have beautiful orange flank plumes. In contrast, the females are drab (Fig. 16.12). Female choice can explain why male birds are more

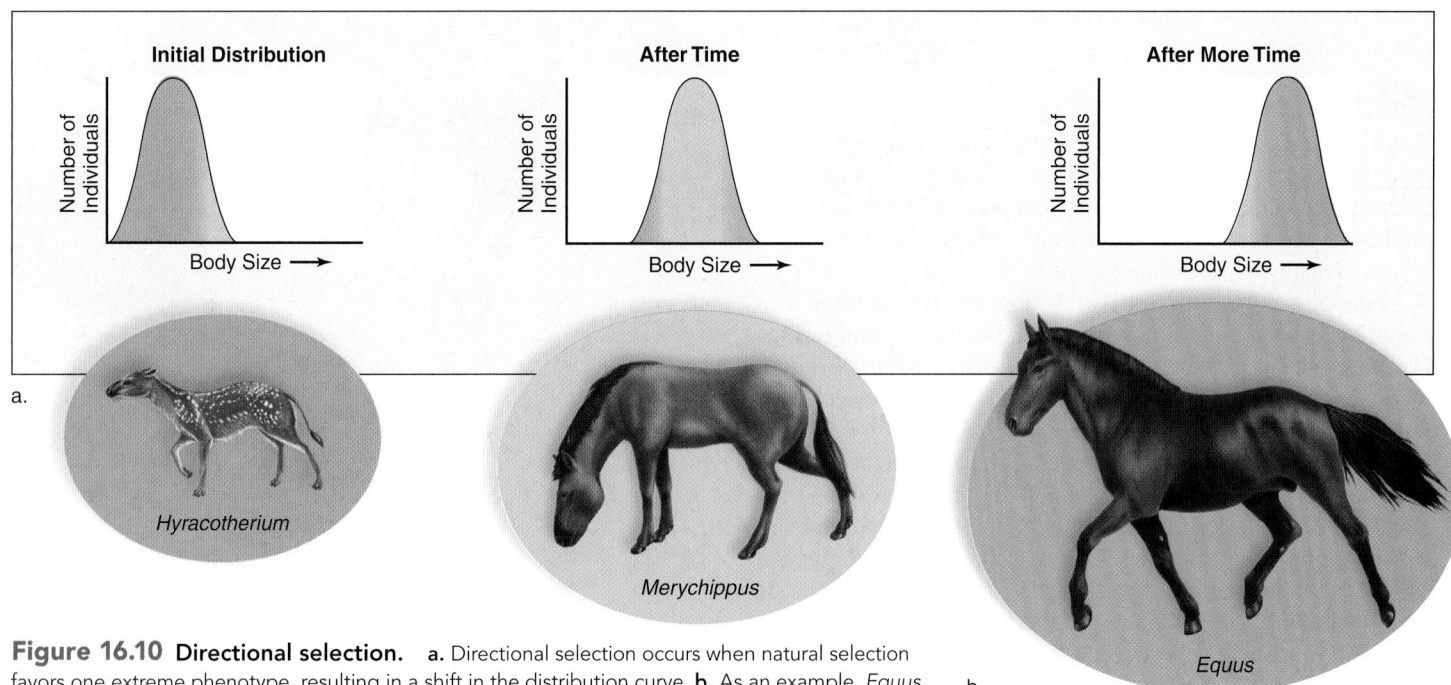

Figure 16.10 Directional selection.   **a.** Directional selection occurs when natural selection favors one extreme phenotype, resulting in a shift in the distribution curve. **b.** As an example, *Equus*, the modern-day horse, is adapted to a grassland habitat with a larger body size than its ancestor, *Hyracotherium*, which was adapted to a forest habitat.

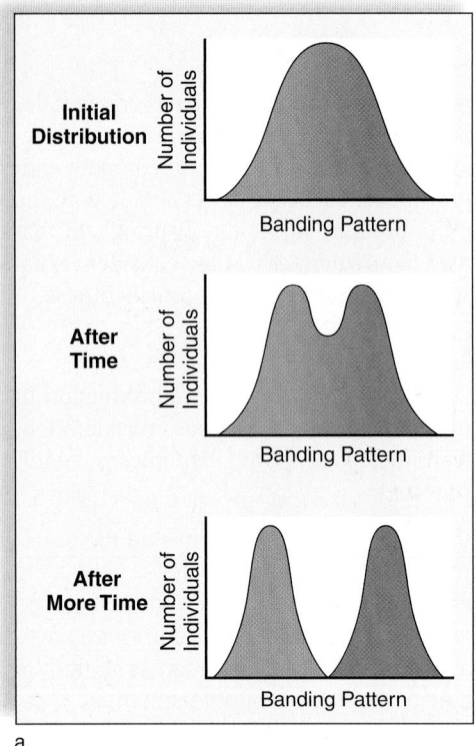

a.

b.

Figure 16.11 Disruptive selection. **a.** Disruptive selection favors two or more extreme phenotypes. **b.** Today, British land snails comprise mainly two different phenotypes, each adapted to a different habitat. Snails with dark shells are more prevalent in forested areas, and light-banded snails in areas with low-lying vegetation.

ornate than females. Consistent with the two hypotheses, it is possible that the remarkable plumes of the male signify health and vigor to the female. Or, it's possible that females' choice of flamboyant males gives their sons an increased chance of being selected by females.

Some investigators have hypothesized that extravagant male features could indicate males that are relatively parasite-free. In barn swallows, females choose males with the longest tails, and investigators have shown that males that are relatively free of parasites have longer tails than those carrying more parasites.

Video
Cichlid Territoriality

Male Competition

Males can father many offspring because they continuously produce sperm in great quantity. We expect males to compete in order to inseminate as many females as possible. **Cost–benefit analyses** have been done to determine whether the *benefit* of access to mating is worth the *cost* of competition among males.

Baboons, a type of Old World monkey, live together in a troop. Males and females have separate **dominance hierarchies** in which a higher-ranking animal has greater access to resources than a lower-ranking animal. Dominance is decided by confrontations, resulting in one animal giving way to the other.

Baboons are dimorphic; the males are larger than the females, and they can threaten other members of the troop with their long, sharp canine teeth. One or more males become dominant by frightening the other males. However, the male baboon pays a cost for his dominant position. Being larger means that he needs more food, and being willing and able to fight predators means that he may get hurt, and so forth. Is there a reproductive benefit to his behavior? Yes, in that dominant males do indeed monopolize females when they are most fertile. Nevertheless, there may

Figure 16.12 Dimorphism. In the Raggiana Bird of Paradise, *Paradisaea raggiana*, males have brilliantly colored plumage brought about by sexual selection. The drab females tend to choose flamboyant males as mates.

be other ways to father offspring. A male may act as a helper to a female and her offspring; then, the next time she is in estrus, she may mate preferentially with him instead of with a dominant male. Or, subordinate males may form a friendship group that opposes a dominant male, making him give up a receptive female.

A **territory** is an area that is defended against competitors. Scientists are able to track an animal in the wild to determine its home range or territory. **Territoriality** includes the type of defensive behavior needed to defend a territory. Baboons travel within a home range, foraging for food each day and sleeping in trees at night. Dominant males decide where and when the troop will move. If the troop is threatened, dominant males protect the troop as it retreats and attack intruders when necessary.

Vocalization and displays, rather than outright fighting, may be sufficient to defend a territory (Fig. 16.13). In songbirds, for example, males use singing to announce their willingness to defend a territory. Other males of the species become reluctant to make use of the same area.

Red deer stags (males) on the Scottish island of Rhum compete to be the harem master of a group of hinds (females) that mate only with them. The reproductive group occupies a territory that the harem master defends against other stags. Harem masters first attempt to repel challengers by roaring. If the challenger remains, the two lock antlers and push against one another (Fig. 16.14). If the challenger then withdraws, the master pursues him for a short distance, roaring the whole time. If the challenger wins, he becomes the harem master.

A harem master can father two dozen offspring at most, because he is at the peak of his fighting ability for only a short time. And there is a cost to being a harem master. Stags must be large and powerful in order to fight; therefore, they grow faster and have less body fat. During bad times, they are more likely to die of starvation, and in general, they have shorter lives.

Figure 16.13 A male olive baboon displaying full threat. In olive baboons, *Papio anubis*, males are larger than females and have enlarged canines. Competition between males establishes a dominance hierarchy for the distribution of resources.

Harem master behavior will persist in the population only if its cost (reduction in the potential number of offspring because of a shorter life) is lower than its benefit (increased number of offspring due to harem access).

Check Your Progress   16.2

1. Evaluate female choice and male competition in relation to the size and number of gametes.
2. Explain why sexual selection is a form of natural selection.
3. Construct an example of a phenotype distribution curve for a population under directional selection that is increasing body size.

a.

Figure 16.14 Competition between male red deer. Male red deer, *Cervus elaphus*, compete for a harem within a particular territory. **a.** Roaring alone may frighten off a challenger, but **(b)** outright fighting may be necessary, and the victor is most likely the stronger of the two animals.

b.

Nature of Science

Inbreeding in Populations

One of the requirements of a population in Hardy-Weinberg equilibrium is that mates are chosen at random, that is, without preference for a particular trait. Most populations, however, do not meet this requirement because some traits are more attractive in a mate than others. Humans, for example, select mates based on a set of traits that we find appealing in a mate. This type of assortative mating based on trait preference is common in many species. Inbreeding is a unique form of nonrandom, or assortative, mating where individuals mate with close relatives, such as cousins.

One consequence of inbreeding is an increase in the frequency of homozygous genotypes in a population. For the most part, an increase in homozygotes does not have a large detrimental effect, especially if the population is very large. But when populations are small, inbreeding can have a large impact on the health of a population. Many human diseases are caused by the inheritance of two recessive alleles, such that the disease appears only in persons who are homozygous recessive for the disease-causing allele. In very small populations, the probability that individuals carrying the recessive allele will mate increases because the mating pool is very small. In this case, inbreeding significantly increases the frequency of homozygous recessive genotypes, and thus those afflicted with a disease.

Human cultures tend to have social rules that discourage inbreeding, but in very small populations, such as those following a bottleneck or founder event, inbreeding is sometimes unavoidable. One example of the effects of inbreeding on human populations is the occurrence of a rare form of non-sex-linked colorblindness, achromatopsia, on the island of Pingelap, a small coral atoll in the Pacific Ocean. People who are achromatic are completely colorblind (Fig. 16A). Normal human color vision is possible because of cells in the back of the eye called cones. Those with achromatopsia do not have any cone cells because they are homozygous for a rare recessive allele that prevents cone cells from developing. Complete colorblindness is so common on Pingelap that it is considered part of everyday life for most families.

Experts propose that the high frequency of achromatopsia appeared following a severe population bottleneck in 1775 when a typhoon killed 90% of the inhabitants. Approximately 20 people survived the typhoon. Four generations after the typhoon struck, achromatopsia began to appear frequently in the population. How could this happen? Figure 16B shows how inbreeding in a population can produce homozygous recessive genotypes. Geneticists explain that the high frequency of this rare genetic disorder on Pingelap is consistent with a large degree of intermarriage among relatives, or inbreeding, following the typhoon.

Mwanenised, a male survivor of the typhoon, was a carrier, or a heterozygote, for the achromatopsia allele. He had 10 children, which was a large proportion of the 1st generation of Pingelapese after the typhoon. Geneticists would predict that on average 50% of his children would have been homozygous for the normal allele, and 50% would have been heterozygous carriers of the colorblind allele. Thus none of Mwanenised's children would have been colorblind (Fig. 16B). However, intermarriage of next generations would unavoidably have brought together men and women who were carriers for the

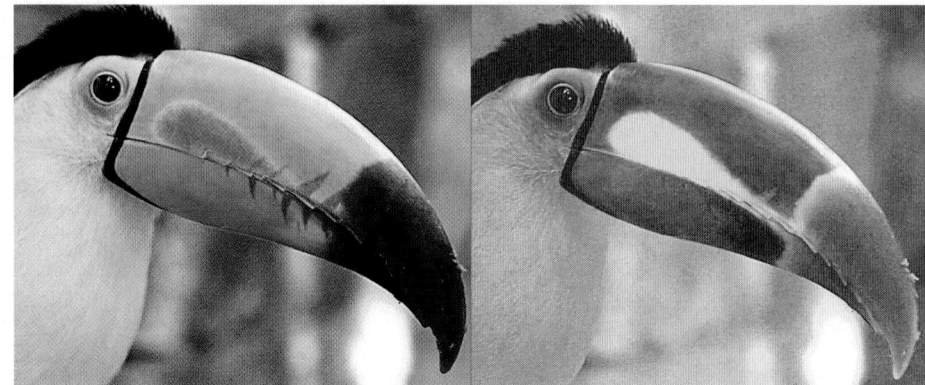

Figure 16A Achromatopsia.   Complete achromatopsia is a recessive genetic disorder that causes complete colorblindness. People with complete achromatopsia see no color, only shades of gray. The image above shows how a person with this form of colorblindness would see the world.

16.3 Maintenance of Diversity

Learning Outcomes

Upon completion of this section, you should be able to

1. List two examples of how diversity is maintained in populations.
2. Describe why heterozygote advantage is a form of stabilizing selection.

Diversity is maintained in a population for any number of reasons. Mutations create new alleles, and sexual reproduction recombines alleles due to meiosis and fertilization. Genetic drift also occurs, particularly in small populations, and the end result may be contrary to adaptation to the environment.

Natural Selection

The process of natural selection itself causes imperfect adaptation to the environment. First, it is important to realize that evolution doesn't start from scratch. Just as you can only bake a cake with the ingredients available to you, evolution is constrained by the available diversity. Lightweight titanium bones might benefit birds, but their bones contain calcium and other minerals the same as other reptiles. When you mix the

achromatopsia allele. Thus, in subsequent generations the homozygous recessive genotype, and complete colorblindness, did appear due to inbreeding within a small population (Fig. 16B). On Pingelap, an increase in colorblindness was observed by the 4th generation.[1]

1 See Sacks, Oliver. 1998. *The Island of the Colorblind*. (Vintage/Anchor Books, New York.) An interesting, nontechnical account of the Pingelapese.

The effect of inbreeding can be long-term, especially if a population remains relatively small and isolated. Today one in 12 Pingelapese suffers from achromatopsia, more than 3,000 times more frequent than in the United States where it occurs in approximately 1/40,000 births.

Questions to Consider

1. What might have happened to the colorblind allele in the Pingelapese population following the typhoon if Mwanenised had had no children? Only 5 children?
2. Would you predict that the Pingelapese population is in Hardy-Weinberg equilibrium? How would you measure this?
3. Has the Pingelapese population evolved? Explain. (Hint: Does inbreeding cause a change in allele frequencies?)

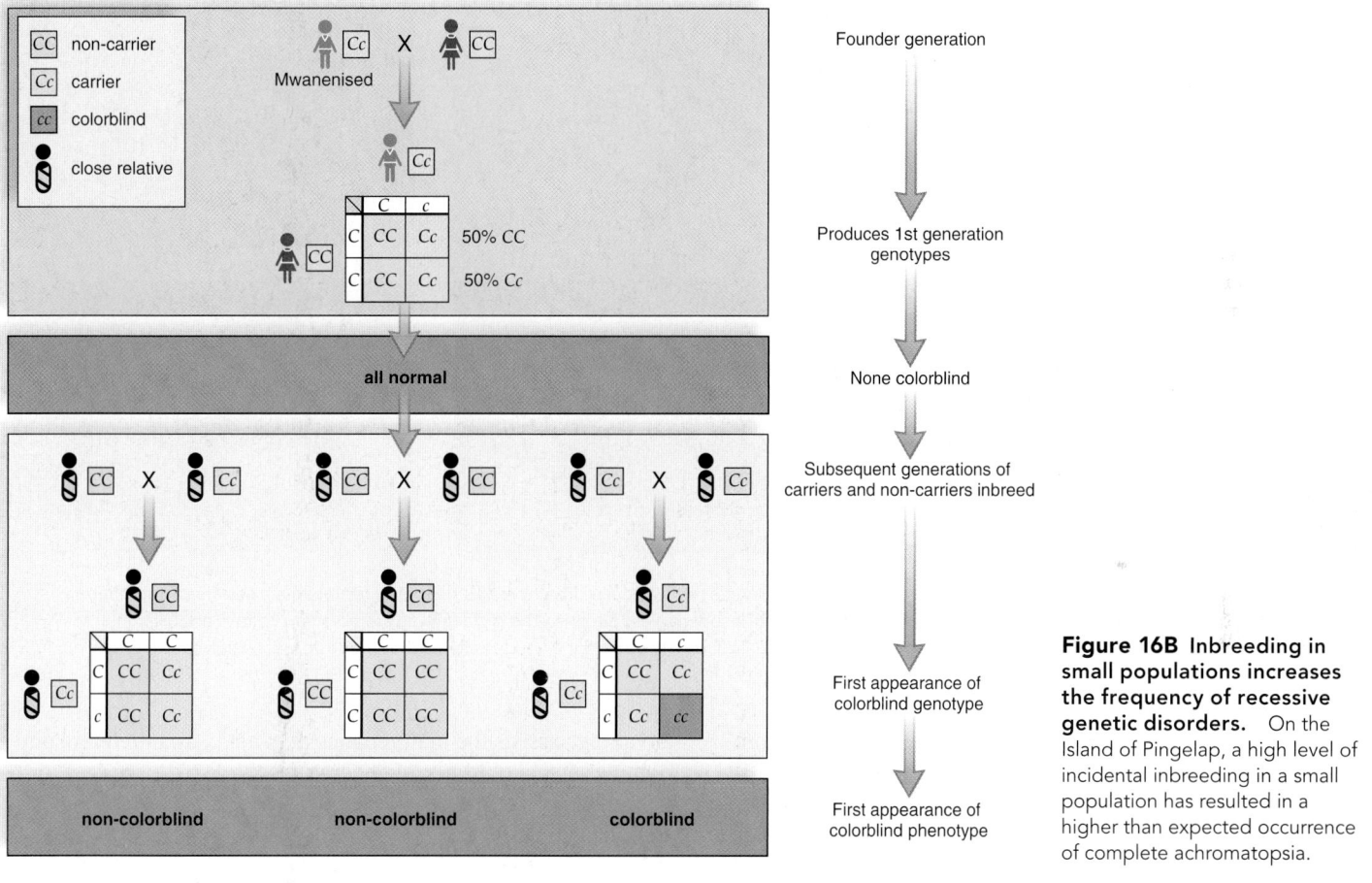

Figure 16B Inbreeding in small populations increases the frequency of recessive genetic disorders.   On the Island of Pingelap, a high level of incidental inbreeding in a small population has resulted in a higher than expected occurrence of complete achromatopsia.

ingredients for a cake, you probably follow the same steps taught to you by your elders. Similarly, the processes of development prevent the emergence of novel features, and therefore the wing of a bird has the same bones as those of other vertebrate forelimbs.

Imperfections are common because of necessary compromises. The success of humans is attributable to their dexterous hands, but the spine is subject to injury because the vertebrate spine did not originally support the body in an erect position in our ancestors. A feature that evolves has a benefit that is worth the cost. For example, the benefit of freeing the hands must have been worth the cost of spinal injuries from assuming an erect posture. We should also consider that sexual selection has a reproductive benefit, but not necessarily an adaptive benefit.

Second, we want to realize that the environment plays a role in maintaining diversity. It's easy to see that disruptive selection, dependent on an environment that differs widely, promotes polymorphisms in a population (see Fig. 16.11). Then, too, if a population occupies a wide range, as shown in Figure 16.15, it may have several subpopulations designated as subspecies because of recognizable differences. (Subspecies are given a third name in addition to the usual binomial name.) Each subspecies is partially adapted to its own environment and can serve

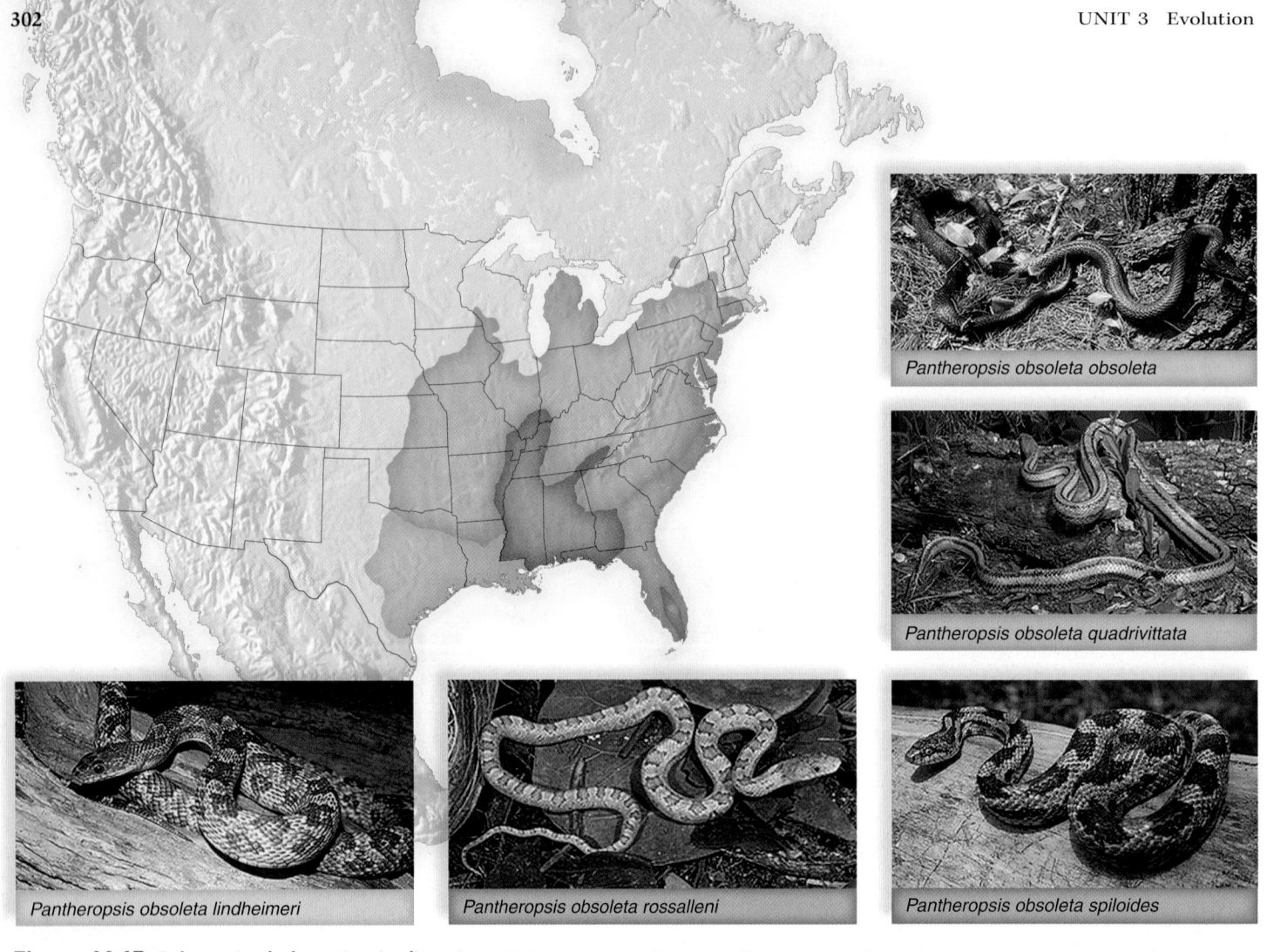

Pantheropsis obsoleta obsoleta

Pantheropsis obsoleta quadrivittata

Pantheropsis obsoleta lindheimeri

Pantheropsis obsoleta rossalleni

Pantheropsis obsoleta spiloides

Figure 16.15 Subspecies help maintain diversity. Each subspecies of rat snake (*Pantheropsis obsoleta*) represents a separate population of snakes. Each subspecies has a reservoir of alleles different from another subspecies. Because the populations are adjacent to one another, they may interbreed, and therefore, gene flow may occur among the populations. This interbreeding introduces alleles that may keep each subspecies from fully adapting to their particular environment.

as a reservoir for a different combination of alleles that flow from one group to the next when adjacent subspecies interbreed.

The environment also includes specific selecting agents that help maintain diversity. We have already seen how insectivorous birds can help maintain the frequencies of both the light-colored and dark-colored moths, depending on the color of background vegetation. Some predators have a search image that causes them to select the most common phenotype among their prey. This promotes the survival of the rare forms and helps maintain variation. Or, a herbivore can oscillate in its preference for food. In Figure 15.11, we observed that the average beak size of the medium ground finch on the Galápagos Islands depended on the available food supply. In times of drought, when only large seeds were available, birds with larger beaks were favored. In this case, we can clearly see that maintenance of variation among a population's members has survival value for the species.

Video
Finches Natural Selection

Heterozygote Advantage

Heterozygote advantage occurs when the heterozygote is favored over the two homozygotes. In this way, heterozygote

advantage assists the maintenance of genetic, and therefore phenotypic, diversity in future generations.

Sickle-Cell Disease

Sickle-cell disease can be a devastating condition. Patients can have severe anemia, physical weakness, poor circulation, impaired mental function, pain and high fever, rheumatism, paralysis, spleen damage, low resistance to disease, and kidney and heart failure. In these individuals, the red blood cells are sickle-shaped and tend to pile up and block flow through tiny capillaries. The condition is due to an abnormal form of hemoglobin (*Hb*), the molecule that carries oxygen in red blood cells. People with sickle-cell disease ($Hb^S Hb^S$) tend to die early and leave few offspring, due to hemorrhaging and organ destruction.

Interestingly, however, geneticists studying the distribution of sickle-cell disease in Africa have found that the recessive allele (Hb^S) has a higher frequency in regions (blue color) where the disease malaria is also prevalent (Fig. 16.16). Malaria is caused by a protozoan parasite that lives in and destroys the red blood cells of the normal homozygote ($Hb^A Hb^A$). Individuals

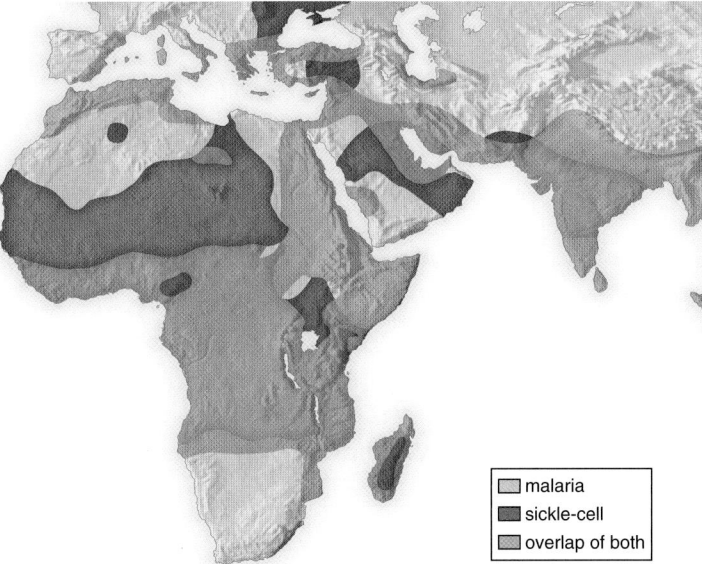

Figure 16.16 Sickle-cell disease. Sickle-cell disease is more prevalent in areas of Africa where malaria is more common.

with this genotype also have fewer offspring, due to an early death or to debilitation caused by malaria.

Heterozygous individuals ($Hb^A Hb^S$) have an advantage because they don't die from sickle-cell disease, and they don't die from malaria. The parasite causes any red blood cell it infects to become sickle-shaped. Sickle-shaped red blood cells lose potassium, and this causes the parasite to die. Heterozygote advantage causes all three alleles to be maintained in the population. As long as the protozoan that causes malaria is present in the environment, it is advantageous to maintain the recessive allele.

Heterozygote advantage is also an example of stabilizing selection because the genotype $Hb^A Hb^S$ is favored over the two extreme genotypes, $Hb^A Hb^A$ and $Hb^S Hb^S$. In the parts of Africa where malaria is common, one in five individuals is heterozygous (has sickle-cell trait) and survives malaria, while only 1 in 100 is homozygous and dies of sickle-cell disease. What happens in the United States where malaria is not prevalent? As you would expect, the frequency of the Hb^S allele is declining among African Americans because the heterozygote has no particular advantage in this country.

Cystic Fibrosis

Stabilizing selection is also thought to have influenced the frequency of other alleles. Cystic fibrosis is a debilitating condition that leads to lung infections and digestive difficulties. In this instance, the recessive allele, common among individuals of northwestern European descent, causes the person to have a defective plasma membrane protein. The agent that causes typhoid fever can use the normal version of this protein, but not the defective one, to enter cells. Here again, heterozygote superiority caused the recessive allele to be maintained in the population.

Check Your Progress   16.3

1. Identify the ways in which diversity is maintained in a population.
2. Demonstrate how sickle-cell disease is an example of stabilizing selection. Do the same for cystic fibrosis.
3. Use the Hardy-Weinberg principle to demonstrate why natural selection cannot eliminate diversity in a population.

CONNECTING *the* CONCEPTS *with the* BIG IDEAS

Evolution

- Natural selection acts on trait variation, and trait variation is determined by genes. Whether or not a trait gives an advantage depends upon the environment. Thus genes, traits, environment, and natural selection are all involved in microevolution. (1A2a)
- Microevolution occurs when allele frequencies in a population change over time; Hardy and Weinberg devised a mathematical method by which geneticists can measure that change. (1A1h, 4C3c)
- If there is no gene flow or genetic drift, random mating, occurrence of mutation, or any type of selection, microevolution should not occur. (1A1g)
- Small populations are especially vulnerable to genetic drift—chance events that in a small population may remove some alleles completely and may cause others to become more frequent. (1A1f, 1A3a)

- Some genes, such as those for sickle cell anemia, are maintained in populations living in particular environments due to stabilizing selection. (1A1e, 1A2cIE, 4C1b1)
- Mutation and genetic variation are the ultimate sources of the raw material for evolution. (1A1c)

*Find the unabridged version of all EK citations at www.glencoe.com/maderAP11.

![Media Study Tools]

www.glencoe.com/maderAP11

Enhance your study of this chapter with study tools and practice tests. Also ask your instructor about the resources available through

ConnectPlus, including the media-rich eBook, interactive learning tools, and animations.

Summarize

16.1 Genes, Populations, and Evolution

Individuals can alter their phenotype during their lifetime, but this is not evolution. Evolution is about change in a trait within populations, not individuals, over many generations. Because genes and traits are linked, evolution is about genetic change—the change in allele frequencies in a population over time. Microevolution pertains to evolutionary change within populations.

Population geneticists study microevolution by measuring allele and genotype frequencies over generations. A population in which allele frequencies do not change over time is said to be in Hardy-Weinberg equilibrium—a stable, non-evolving state. The Hardy-Weinberg equilibrium is a constancy of gene-pool allele frequencies that remains from generation to generation if certain conditions are met. The conditions are: no mutations, no gene flow, random mating, no genetic drift, and no selection. Because these conditions are rarely met, a change in allele frequencies is likely.

When gene-pool frequencies change, microevolution has occurred. Deviations from a Hardy-Weinberg equilibrium allow us to detect microevolutionary shifts. Thus mutation, gene flow, nonrandom mating, genetic drift, and selection cause populations to evolve. The way allele frequencies deviate from Hardy-Weinberg equilibrium tells us which of these processes is causing the population to evolve (see Table 16.1).

16.2 Natural Selection

Most of the traits of evolutionary significance are polygenic; the diversity in a population results in a bell-shaped curve. Three types of selection occur: (1) directional—the average trait value shifts in one direction, as when average body size increases over time; (2) stabilizing—the variation in a trait decreases as the extreme forms reduce in number, and the average trait value becomes more common, as is found in optimum clutch size for survival of Swiss starling young; and (3) disruptive—the trait has two values that are adaptive; the curve forms two peaks, as when banding patterns of *Cepaea* snails have two dominant variants, each in a different environment.

Sexual selection is about reproductive success, or fitness. Males produce many sperm and compete to inseminate females. Females produce few eggs and are selective about their mates. Traits that promote reproductive success, such as dimorphism, are shaped by sexual selection. It is possible that male competition and female choice also occur among humans. Biological differences between the sexes may promote certain mating behaviors because they increase fitness.

16.3 Maintenance of Diversity

Despite constant natural selection, genetic diversity is maintained. Mutations and recombination still occur; gene flow among small populations can introduce new alleles; and natural selection itself sometimes results in variation. In sexually reproducing diploid organisms, the heterozygote acts as a repository for recessive alleles whose frequency is low. In regard to sickle-cell disease, the heterozygote is more fit in areas where malaria occurs, and therefore both homozygotes are maintained in the population.

Key Terms

allele frequency 290	cost–benefit analysis 298
assortative mating 295	directional selection 296
bottleneck effect 295	disruptive selection 297
dominance hierarchy 298	microevolution 290
fitness 297	nonrandom mating 295
founder effect 295	polygenic 296
gene flow 294	population 290
gene pool 290	population genetics 290
genetic drift 294	reproductively isolated 294
Hardy-Weinberg	sexual dimorphism 297
equilibrium 290	sexual selection 297
Hardy-Weinberg principle 291	stabilizing selection 296
heterozygote advantage 302	territoriality 299
inbreeding 295	territory 299

Assess

Reviewing This Chapter

1. What is the Hardy-Weinberg principle?   290–92
2. Name the five conditions of evolutionary change. Predict the outcome of deviation from these conditions on a population gene pool.   292–95
3. Distinguish among directional, stabilizing, and disruptive selection by giving examples.   296–97
4. What is sexual selection, and why does it foster female choice and male competition during mating?   297–99
5. State ways in which diversity is maintained in a population.   301–03

Testing Yourself

Choose the best answer for each question.

1. Assuming a Hardy-Weinberg equilibrium, 21% of a population is homozygous dominant, 50% is heterozygous, and 29% is homozygous recessive. What percentage of the next generation is predicted to be homozygous recessive?
   a. 21%   d. 42%
   b. 50%   e. 58%
   c. 29%

2. A human population has a higher-than-usual percentage of individuals with a genetic disorder. The most likely explanation is
   a. mutations and gene flow.
   b. mutations and natural selection.
   c. nonrandom mating and founder effect.
   d. nonrandom mating and gene flow.
   e. All of these are correct.

3. The offspring of better-adapted individuals are expected to make up a larger proportion of the next generation. The most likely explanation is
   a. mutations and nonrandom mating.
   b. gene flow and genetic drift.
   c. mutations and natural selection.
   d. mutations and genetic drift.

4. The continued occurrence of sickle-cell disease with malaria in parts of Africa is due to
   a. continual mutation.
   b. gene flow between populations.
   c. relative fitness of the heterozygote.
   d. disruptive selection.
   e. protozoan resistance to DDT.

5. Which of these is necessary for natural selection to occur?
   a. diversity   d. differential adaptiveness
   b. differential reproduction   e. All of these are correct.
   c. inheritance of differences

6. When a population is small, there is a greater chance of
   a. gene flow.
   b. genetic drift.
   c. natural selection.
   d. mutations occurring.
   e. sexual selection.

7. Which of these is an example of stabilizing selection?
   a. Over time, *Equus* developed strength, intelligence, speed, and durable grinding teeth.
   b. British land snails mainly have two different phenotypes.
   c. Swiss starlings usually lay four or five eggs, thereby increasing their chances of more offspring.
   d. Drug resistance increases with each generation; the resistant bacteria survive, and the nonresistant bacteria get killed off.
   e. All of these are correct.

8. Which of these cannot occur if a population is to maintain an equilibrium of allele frequencies?
   a. People leave one country and relocate in another.
   b. A disease wipes out the majority of a herd of deer.
   c. Members of an Indian tribe only allow the two tallest people in a tribe to marry each spring.
   d. Large black rats are the preferred males in a population of rats.
   e. All of these are correct.

9. In some bird species, the female chooses a mate that is most similar to her in size. This supports
   a. the good genes hypothesis.
   b. the runaway hypothesis.
   c. Either hypothesis could be true.
   d. Neither hypothesis is true.

10. A red deer harem master typically dies earlier than other males because he is
    a. likely to get expelled from the herd and cannot survive alone.
    b. more prone to disease because he interacts with so many animals.
    c. in need of more food than other males.
    d. apt to place himself between a predator and the herd to protect the herd.

11. Which one of the following statements would *not* pertain to a Punnett square that involves the alleles of a gene pool?
    a. The results tell you the chances that an offspring can have a particular condition.
    b. The results tell you the genotype frequencies of the next generation.
    c. The eggs and sperm are the gamete frequencies of the previous generation.
    d. All of these are correct.

12. Which of the following applies to the Hardy-Weinberg expression: $p^2 + 2pq + q^2$?
    a. Knowing either p^2 or q^2, you can calculate all the other frequencies.
    b. applies to Mendelian traits that are controlled by one pair of alleles
    c. $2pq$ = heterozygous individuals
    d. can be used to determine the genotype and allele frequencies of the previous and the next generations
    e. All of these are correct.

13. Following genetic drift, the
    a. genotype and allele frequencies would not change.
    b. genotype and allele frequencies would change.
    c. adaptation would occur.
    d. population would have more phenotypic variation but less genotypic variation.

14. For disruptive selection to occur,
    a. the population has to contain diversity.
    b. the environment has to contain diversity.
    c. pollution must be present.
    d. natural selection must occur.
    e. All but c are correct.

Engage

Thinking Scientifically

1. A farmer uses a new pesticide. He applies the pesticide as directed by the manufacturer and loses about 15% of his crop to insects. A farmer in the next state learns of these results, uses three times as much pesticide, and loses only 3% of her crop to insects. Each farmer follows this pattern for five years. At the end of five years, the first farmer is still losing about 15% of his crop to insects, but the second farmer is losing 40% of her crop to insects. How could these observations be interpreted on the basis of natural selection?

2. You are observing a grouse population in which two feather phenotypes are present in males. One is relatively dark and blends into shadows well, and the other is relatively bright and so is more obvious to predators. The females are uniformly dark-feathered. Observing the frequency of mating between females and the two types of males, you have recorded the following:

   matings with dark-feathered males: 13

   matings with bright-feathered males: 32

   Propose a hypothesis to explain why females apparently prefer bright-feathered males. What selective advantage might there be in choosing a male with alleles that make it more susceptible to predation? What data would help test your hypothesis?

3. If $p^2 = 0.36$, what percentage of the population has the recessive phenotype, assuming a Hardy-Weinberg equilibrium?

4. If 1% of a human population has the recessive phenotype, what percentage has the dominant phenotype, assuming a Hardy-Weinberg equilibrium?

5. In a population of snails, ten had no antennae (*aa*); 180 were heterozygous with antennae (*Aa*); and 810 were homozygous with antennae (*AA*). What is the frequency of the *a* allele in the population?

17

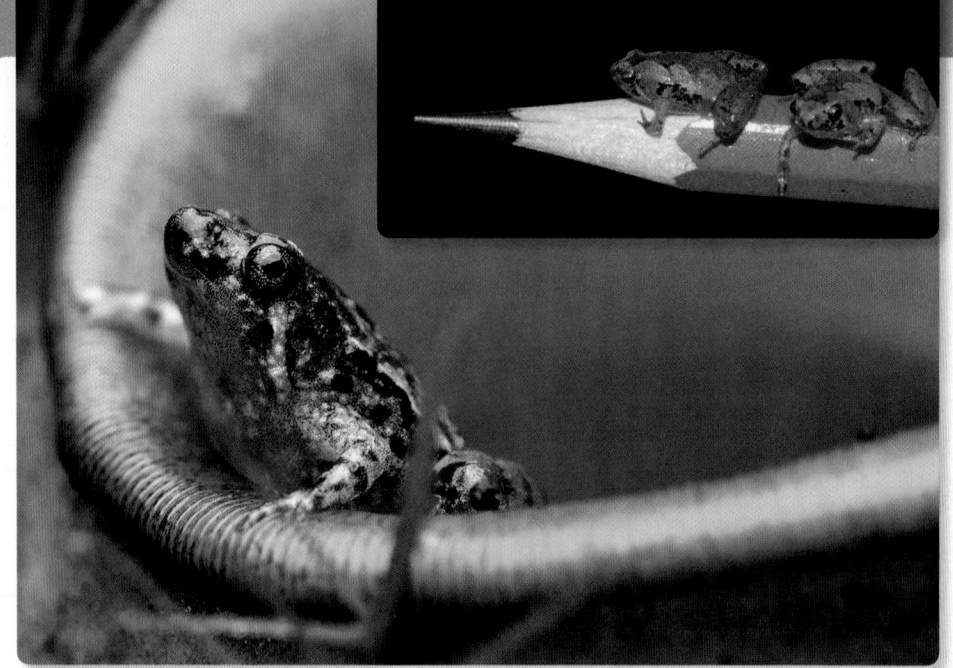

Speciation and Macroevolution

BEFORE YOU BEGIN

Before you begin this chapter, take a few moments to review the following discussions.

Sections 13.1 and 13.2 What is gene regulation and how are genes regulated?

Section 15.3 What is the genetic, fossil, and morphological evidence for evolution?

Section 16.1 What processes drive evolution within populations (microevolution)?

I n 2007, a new species of micro-frog, *Microhyla nepenthicola*, was discovered in the forests of Borneo. Only a fraction of the size of a penny, this 12 mm frog mates and lays eggs in a pitcher plant, *Nepenthes ampullaria*, that lives on the forest floor. The pitcher plant collects water in which the frog tadpoles develop into adult frogs. For a long time, *M. nepenthicola* was considered to be a juvenile of a different species, but a closer look revealed a suite of unique features that allow it to scale the waxy, smooth walls of the pitcher plant. The origin of *M. nepenthicola* has been shaped by a tightly knit relationship with *N. ampullaria*, its nest and nursery.

In this chapter we take microevolution one step further and look at how a population, over time, accumulates differences large enough to become a new species. The origin of species is the key to understanding the origin of the diversity of all life on Earth.

As you read this chapter, think about the following questions:

1. How do scientists determine whether an organism is a new species?

2. What processes drive the evolution of new species? Are they different from those that drive the evolution of populations?

3. What can the fossil record tell us about the origin and extinction of species over time?

FOLLOWING *the* BIG IDEAS

CHAPTER 17   SPECIATION AND MACROEVOLUTION

Evolution   Macroevolution, or the origin of new species, results from the accumulation of microevolutionary change over time.

17.1 How New Species Evolve

Learning Outcomes

Upon completion of this section, you should be able to

1. Compare and contrast the processes of microevolution and macroevolution.
2. Identify and compare features of prezygotic and postzygotic reproductive isolation.
3. Explain three ways that species are defined.

In Chapter 16, we examined microevolution, or small-scale evolution within populations. Microevolution is measured as change in allele frequencies in a population over generations. In this chapter, we turn our attention to evolution on a large scale, that is, **macroevolution.** The history of life on Earth is a part of macroevolution. Macroevolution involves **speciation,** or the splitting of one species into two or more species. The same microevolutionary mechanisms that are at play within populations—genetic drift, natural selection, mutation, and migration—also shape the evolution of new species. Thus microevolution and macroevolution are the result of the same processes, differing only at the scale at which they occur. Macroevolution is the result of the accumulation of microevolutionary change that results in the formation of new species (see the Evolution feature, page 309).

Species originate, adapt to their environment, and then may become extinct. In fact, much of the biodiversity that has existed on Earth is now extinct. For example, mammals experienced many periods of speciation in the past that resulted in high levels of diversity; but the majority of those species are now extinct. Without the continuous origin and extinction of species, life on Earth would not have the ever-changing history that we find in the fossil record.

Darwin devoted his life to understanding the "mystery of mysteries"—how new species originate. We have learned a lot about this mystery since Darwin's time, including some of the processes that cause species to form. In this chapter we take a closer look at what constitutes a species and how new species evolve.

What Is a Species?

When you take a walk in the forest, you see a lot of different "types" of plants and animals. If you've had a biology class, you would probably call these different types "species." Although you may not be able to identify each species, you would intuitively recognize many of these organisms as different because of their appearance. Many differences are obvious, for example, clearly an oak tree is a different species from a squirrel. If you look a little closer, you might recognize two different kinds of oak trees, one with large acorns and one with small acorns, and with leaves of a slightly different shape. Would you recognize these two oaks as different species? Or would you think they were variations of the same species? It depends on how species are defined.

Scientists define species based on many types of evidence. When presented with two organisms, a **taxonomist,** a scientist that classifies organisms into groups, makes a working hypothesis about whether they are different species based on the evidence provided, such as their external features. In this manner, each species that is defined is a hypothesis about how the Earth's tree of life is organized.

As with any hypothesis, the addition of new information can result in the redefining of species. For example, Linnaeus once hypothesized that birds and bats should be together in the same group because both have wings and fly. We now know that birds and bats are very different organisms on separate branches of the tree of life.

How species are defined is an exciting area of study, because all of the diversity of life on Earth has originated from the evolution of new species. Up until now, we have defined a species as a type of living thing, but in this chapter we characterize species in more depth. First, we examine three **species concepts,** or the different ways in which a species is defined, and then we examine some of the mechanisms by which new species originate.

Morphological Species Concept

Linnaeus, the father of taxonomy, identified new species by differences in their appearance, or **morphology.** In the **morphological species concept,** each species is defined by one or more distinct physical characteristics called **diagnostic traits** that distinguish one species from another. It turns out that Linnaeus was very adept at recognizing species, and many of the morphological species he defined have held up to 200 years of scrutiny.

But the morphological species concept has some disadvantages that Linnaeus could not have predicted. Bacteria and other microorganisms do not have many measurable traits. Also, we know that similarities and differences between organisms can be very subtle and sometimes misleading. Some organisms look so similar that they appear to be the same species. **Cryptic species** are species that look almost identical but are very different in other traits, such as habitat use or courtship behaviors. As one example, species of leopard frogs in North America are very difficult to distinguish in appearance, but the males of each species have a unique courtship call (Fig. 17.1).

Fossils do not provide information about color or anatomy of soft tissues, or behavioral traits. Subtle changes in skeletal features, often represented by partial pieces of bone, are used to diagnose new species in the fossil record. In this case, the morphological species concept is useful for paleontologists as a way to define fossil species based only on those traits that are preserved in the fossil record.

Evolutionary Species Concept

The **evolutionary species concept** was proposed to explain speciation in the fossil record. Thus, like the morphological species concept, the evolutionary species concept relies on identification of certain morphological diagnostic traits to distinguish one species from another. In addition, the evolutionary species concept, as its name implies, also requires that the members of a species share the same distinct, evolutionary pathway. That is, small, transitional changes in a trait are not used to define new species, because these transitional forms are part of the same evolutionary pathway. However, abrupt changes in traits indicate the evolution of a new species in the fossil record.

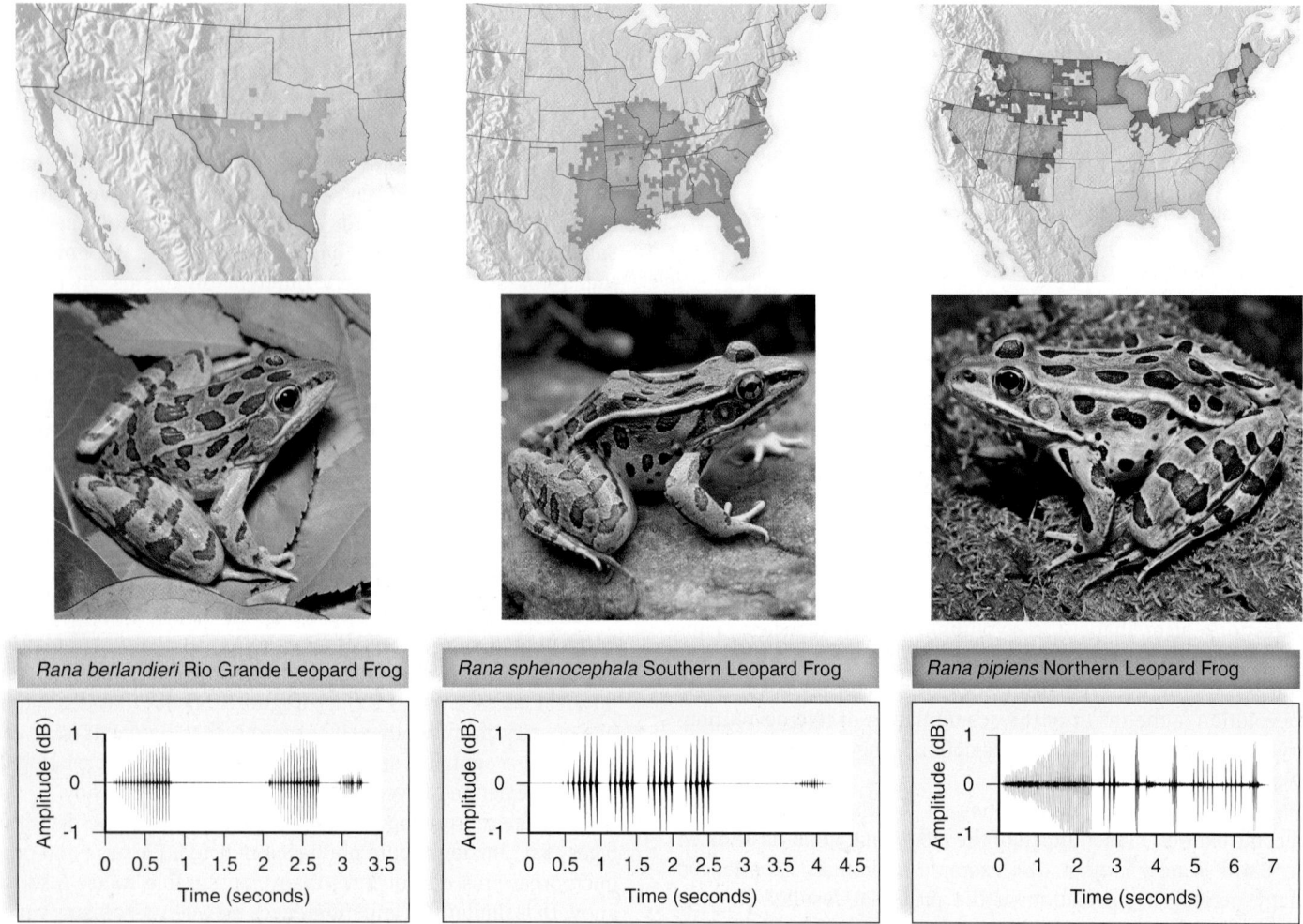

Rana berlandieri Rio Grande Leopard Frog

Rana sphenocephala Southern Leopard Frog

Rana pipiens Northern Leopard Frog

Figure 17.1 Cryptic species of leopard frogs. Leopard frogs are common throughout North America and were once considered to be a single species. Further investigation has revealed that there are at least 3 different species, the Rio Grande, southern, and northern leopard frogs. Although they look similar, they are reproductively isolated because each has a unique mating call.

As an example, consider that the species depicted in Figure 17.2 are part of the evolutionary history of *Orcinus orca*, a toothed whale. These species can be recognized individually by differences in diagnostic traits (hindlimbs), but collectively they share an evolutionary pathway distinct from those of other whale species.

Phylogenetic Species Concept

In the **phylogenetic species concept,** an evolutionary "family tree"—or phylogeny—is used to identify species based on a common ancestor, that is, a single ancestor for two or more different groups (see the Evolution feature). For you and your cousins, your shared grandmother is a common ancestor. Similarly, groups of organisms have a common ancestor.

According to the phylogenetic species concept, a species is the smallest set of interbreeding organisms—usually a population—that shares a common ancestor. In a phylogeny, a branch that contains all of the descendants of a common ancestor is said to be **monophyletic.** Monophyly is the main criterion for defining species in the phylogenetic species concept.

One advantage of the phylogenetic species concept is that it does not rely only on morphological traits to define a species. The nucleotide sequence of a region of an organism's DNA can be compared to identify individual A, C, G, or T nucleotide differences that are diagnostic of a species. Thus species of microorganisms and cryptic species can be identified with the phylogenetic species concept because traits other than morphology can be diagnostic. One example is the giraffe, which

Evolution

The Anatomy of Speciation

The primary goal of evolutionary biology is to infer the processes of evolution that produce the patterns of diversity we see in nature. More simply put, the evolutionary biologist is interested in explaining biodiversity. At the heart of this endeavor is the phylogeny, the most important tool of the evolutionary biologist, because it represents the history of evolution among organisms. Evolution is considered one of the unifying theories of biology; thus understanding how to interpret a phylogeny is fundamental to understanding how evolution works. This guide to the anatomy of a phylogeny should help you with the interpretation of a phylogeny.

Macroevolution is about the origin of new species. Darwin's theory proposes that all life on Earth shares a common ancestor. This means that a species originates from evolutionary changes to preexisting species. In this manner, all life on Earth can be envisioned as a large tree of life, with many branches, or species, radiating from it (see Nature of Science feature Chapter 15).

Microevolution is about populations, while macroevolution is about the origin of new species (Fig. 17Ab). Yet both are governed by the same processes. Different populations of a single species can accumulate genetic differences as microevolution shapes allele frequencies over time. As more and more genetic differences accumulate, a population may no longer be able to recognize members of another population as potential mates. Under the biological species concept, this would be evidence that the populations are different species (Fig. 17Ac). In a phylogeny constructed from DNA nucleotide sequences, for example, these two populations would likely be represented by two different lineages. In this case, both the phylogenetic and biological species concepts would support the origin of a new species.

Phylogenetic Tree Terms

Figure 17Aa shows the different parts of a phylogeny. *Node:* The point at which two branches, or lineages, intersect. The node represents a shared common ancestor for all of the species that branch from it. For example, Node 1 is the common ancestor of "A," "B," and "C," while Node 2 is a common ancestor of "A" and "B."

Root: The point to which all species in the phylogeny can trace their ancestry; the origin of their shared common ancestry.

Extinction: A taxon that is represented in the fossil record, but is now extinct, is represented by a shortened branch correlated with the time at which the extinction occurred.

Monophyletic group (monophyly): A group of species and their common ancestor. For example, all of the taxa that share Node 2 ("A," "B," and one extinct fossil species) are members of a monophyletic group (dark-shaded purple rectangle). Also called a *clade*.

Animation
Phylogenetic Trees

Questions to Consider

1. How could an evolutionary biologist use a phylogeny to find out if two populations have evolved into two different species?
2. Would you expect two different organisms on separate branches of a phylogeny to be able to mate? Why or why not?

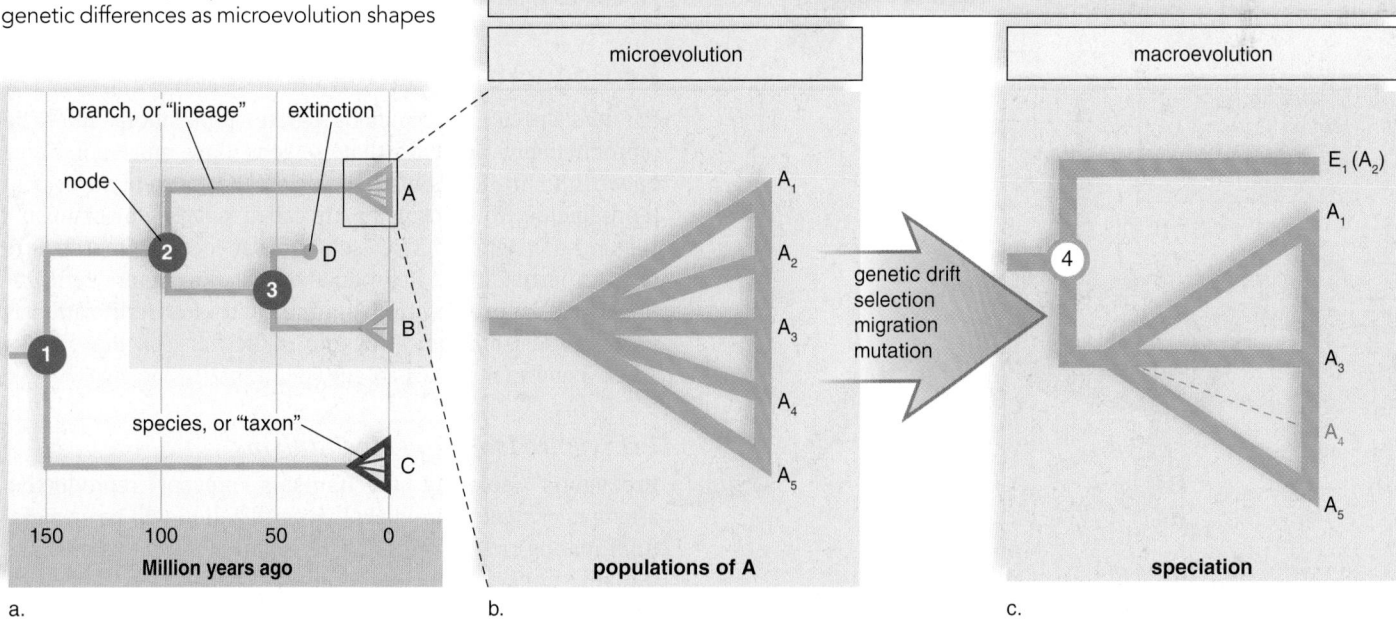

Figure 17A Anatomy of speciation.   A phylogeny, or "family tree," is a hypothesis of evolutionary history of taxa such as species or genera. **a.** A phylogeny has many parts, each of which tells us something about evolutionary relationships among taxa, such as a species or genera. For example, species "A" is more closely related to "B" than to "C," because "A" and "B" share a more recent common ancestor (node 2). **b.** Species "A" is comprised of many populations (A_1–A_5) that are all part of the same branch of the phylogeny. Microevolution occurs at the level of the population. **c.** Microevolution and macroevolution are governed by the same processes, that is genetic drift, selection, migration, and mutation, but at different scales. Speciation is the result of the accumulation of microevolutionary change in a population over time. Eventually, microevolution can result in the divergence of a population (such as A_2) to form a new species (E_1). Another outcome of evolution is the extinction of populations (A_4) and species (D).

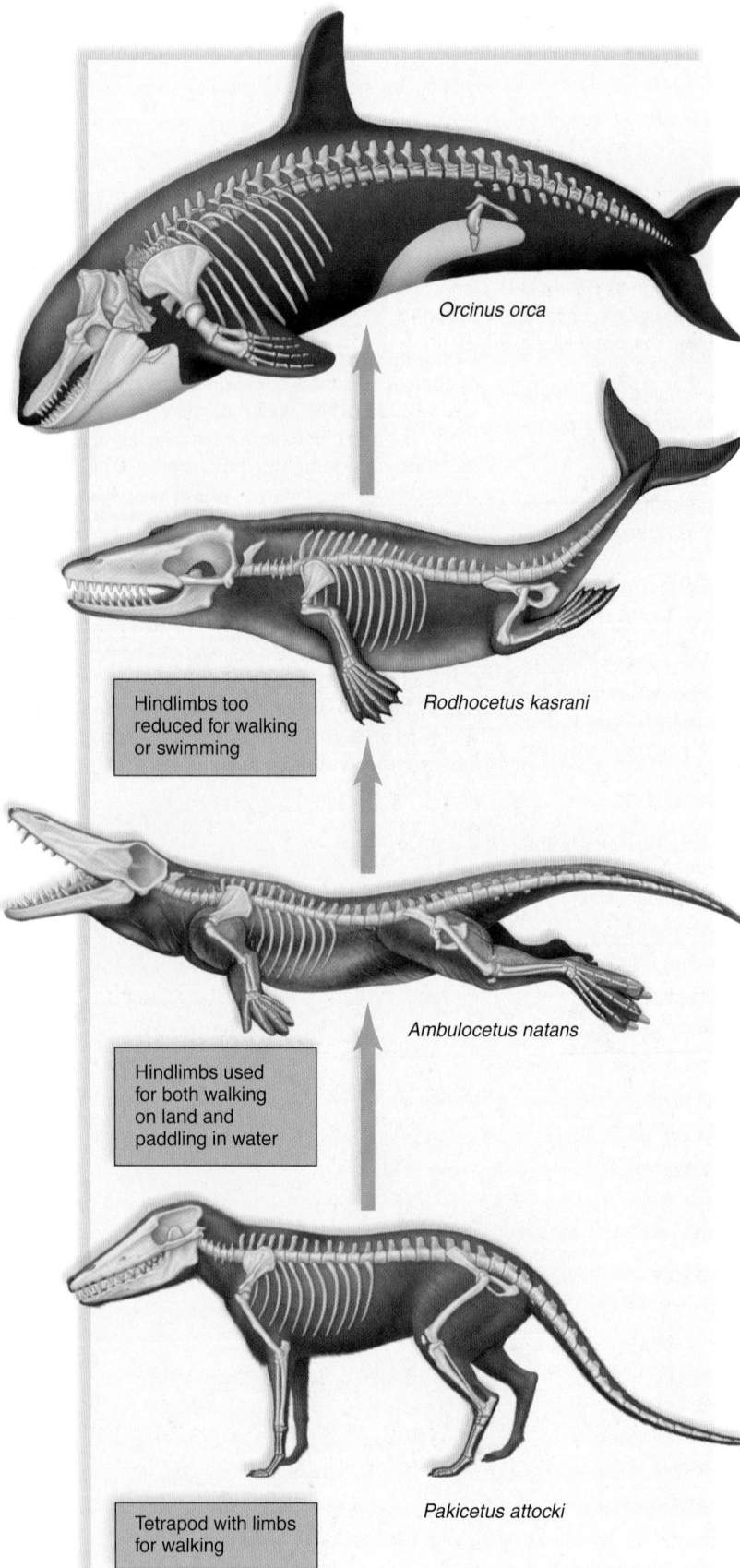

Orcinus orca

Hindlimbs too
reduced for walking
or swimming

Rodhocetus kasrani

Hindlimbs used
for both walking
on land and
paddling in water

Ambulocetus natans

Tetrapod with limbs
for walking

Pakicetus attocki

Figure 17.2 Evolutionary species concept.   Diagnostic traits
can be used to distinguish these species known only from the fossil
record. Such traits no doubt would include the anatomy of the limbs.

has several regional populations distributed around Africa
each distinguishable only by a unique spot shape (Fig. 17.3).
Historically, each population was considered as members of
a single species, *Giraffa camelopardalis*. A recent phylogeny
based on DNA data hypothesizes that each regional population represents a monophyletic group, and thus each population should be recognized as individual species (Fig. 17.3).

Biological Species Concept

The **biological species concept** relies primarily on reproductive isolation rather than trait differences or shared evolutionary history to define a species. In other words, although traits
can help us distinguish species, the most important criterion,
according to the biological species concept, is **reproductive
isolation**—physiological, behavioral, and genetic processes
that inhibit interbreeding. Specifically, if organisms cannot
mate and produce offspring in nature, or if their offspring are
sterile, they are defined as different species.

Although useful, the biological species concept often
cannot be tested in nature because many potential species
do not overlap in their distribution, and thus do not have an
opportunity to determine whether they are reproductively
isolated. Furthermore, the biological species concept cannot
be applied to asexually reproducing organisms or fossils. The
benefit of the concept is that, when applicable, it confirms the
lack of gene flow—the best indicator that two populations are
following independent evolutionary pathways. For example, a
group of birds collectively called the flycatchers all look very
similar, but they do not reproduce with one another; therefore, they are separate species. Like the leopard frogs (see Fig.
17.1), they not only live in different habitats, but also each
group has a unique courtship song.

Reproductive Isolating Mechanisms

For two species to remain separate, populations must be
reproductively isolated—that is, gene flow must not occur
between them. Reproductive barriers that prevent successful
reproduction from occurring are called isolating mechanisms
(Fig. 17.4). Reproductive isolation can occur either before
or after fertilization. Reproductive isolation before fertilization is called **prezygotic isolation**, and after fertilization is
postzygotic isolation. A **zygote** is the first cell that results
when a sperm fertilizes an egg.

Prezygotic Isolating Mechanisms

Prezygotic isolating mechanisms prevent reproductive
attempts or make it unlikely that fertilization will be successful if mating is attempted. These isolating mechanisms make
it highly unlikely **hybridization,** or the mating between two
species, will occur. Several isolating mechanisms can be distinguished, and we describe each of these next.

Habitat isolation When two species occupy different
habitats, even within the same geographic range, they
are less likely to meet and attempt to reproduce. This
is one of the reasons that flycatchers do not mate. In

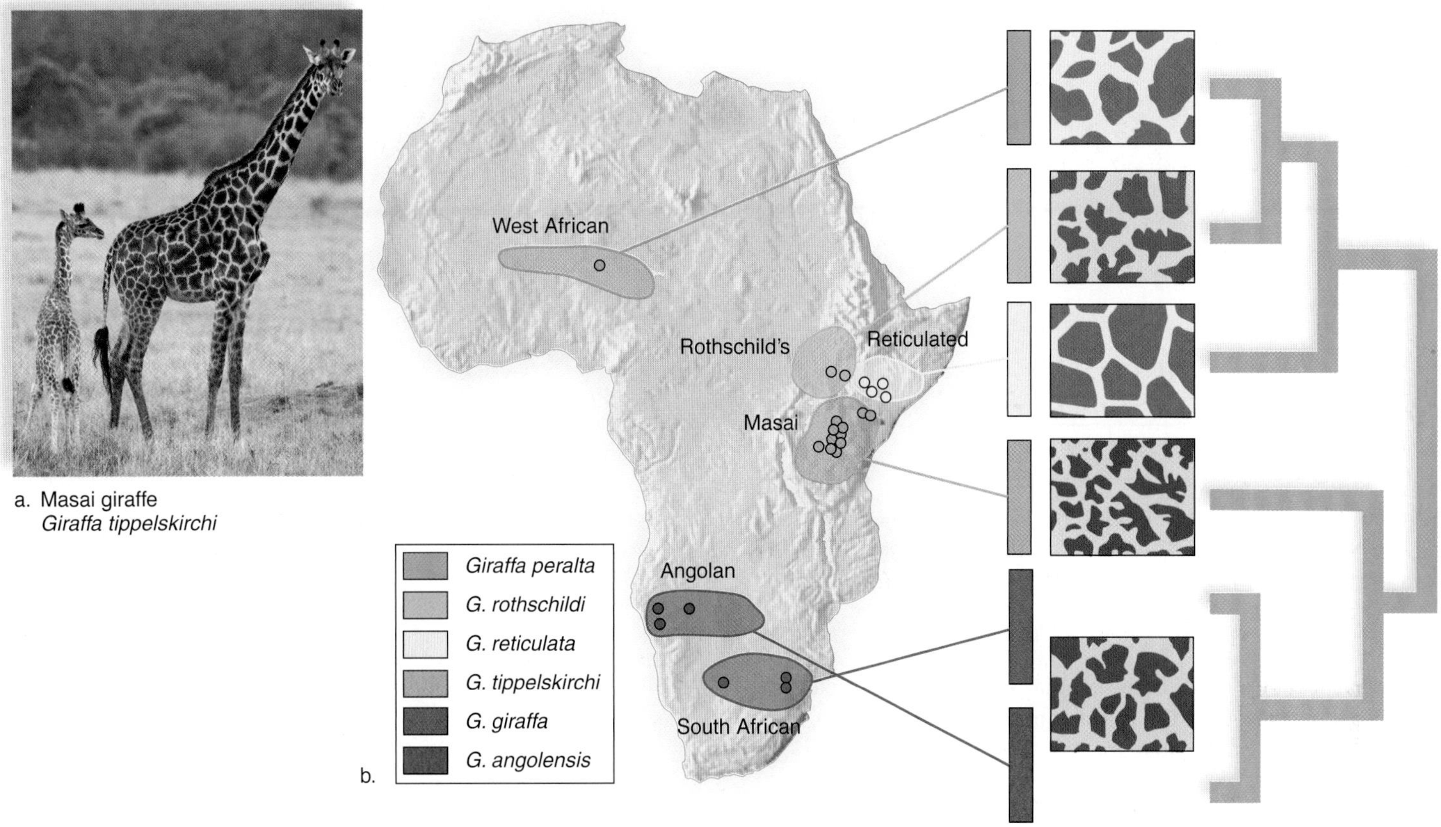

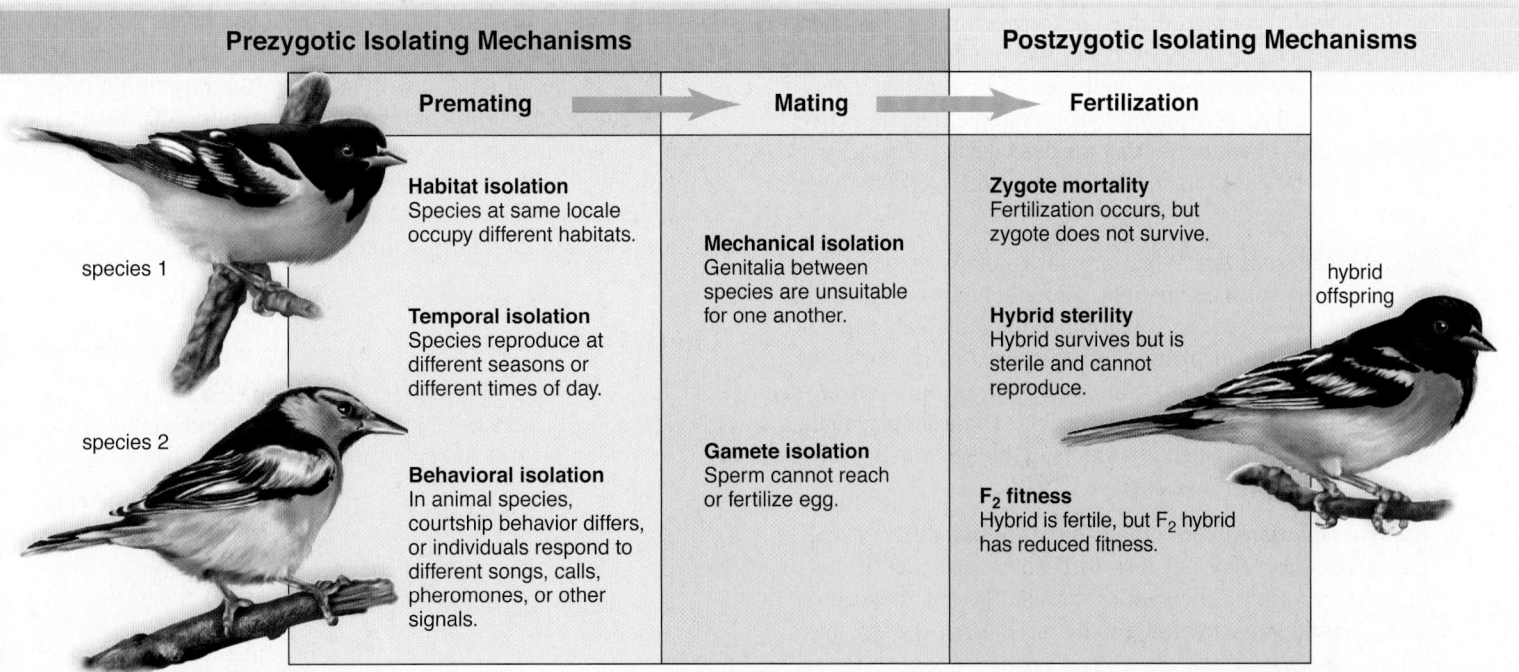

a. Masai giraffe
Giraffa tippelskirchi

Legend:
- *Giraffa peralta*
- *G. rothschildi*
- *G. reticulata*
- *G. tippelskirchi*
- *G. giraffa*
- *G. angolensis*

Map labels: West African, Rothschild's, Reticulated, Masai, Angolan, South African

b.

Figure 17.3 The phylogenetic species concept defines species from an evolutionary tree. **a.** *Giraffa tippelskirchi* is known for its unique ragged-edge spot pattern. Historically, regional variations in spot pattern were used to define what were once considered multiple populations of the same species, *Giraffa camelopardalis*. **b.** Recent studies show that each regional population represents a unique evolutionary branch of the giraffe family tree. According to the phylogenetic species concept, each of these branches should be recognized as one of six unique giraffe species.

Figure 17.4 Reproductive barriers. Prezygotic isolating mechanisms prevent mating attempts or a successful outcome should mating take place. No zygote ever forms. Postzygotic isolating mechanisms prevent the zygote from developing—or should an offspring result, it is not fertile.

Prezygotic Isolating Mechanisms	Postzygotic Isolating Mechanisms

Premating → Mating → Fertilization

species 1

species 2

Habitat isolation
Species at same locale occupy different habitats.

Temporal isolation
Species reproduce at different seasons or different times of day.

Behavioral isolation
In animal species, courtship behavior differs, or individuals respond to different songs, calls, pheromones, or other signals.

Mechanical isolation
Genitalia between species are unsuitable for one another.

Gamete isolation
Sperm cannot reach or fertilize egg.

Zygote mortality
Fertilization occurs, but zygote does not survive.

Hybrid sterility
Hybrid survives but is sterile and cannot reproduce.

F_2 fitness
Hybrid is fertile, but F_2 hybrid has reduced fitness.

hybrid offspring

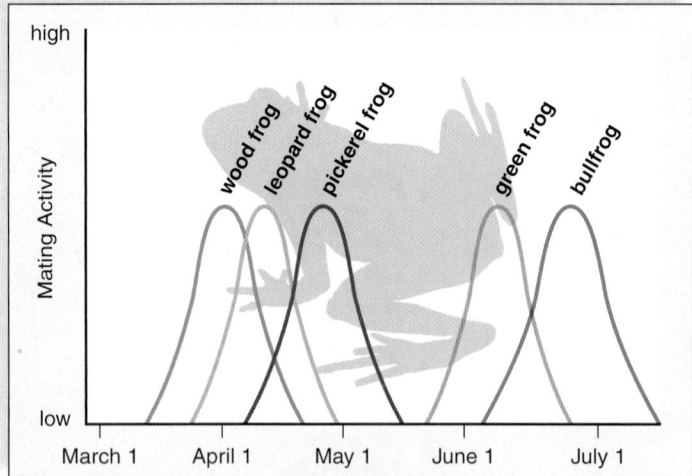

Figure 17.5 Temporal isolation. Five species of frogs of the genus *Rana* are all found at Ithaca, New York. The species remain separate due to breeding peaks at different times of the year, as indicated by this graph.

Figure 17.6 Prezygotic isolating mechanism. An elaborate courtship display allows the blue-footed boobies of the Galápagos Islands to select a mate. The male lifts up his feet in a ritualized manner that shows off their bright blue color.

tropical rain forests, many animal species are restricted to a particular level of the forest canopy, and in this way they are isolated from similar species.

Temporal isolation Several related species can live in the same locale, but if each reproduces at a different time of year, they do not attempt to mate. Five species of frogs of the genus *Rana* are all found at Ithaca, New York (Fig. 17.5). The species remain separate because the period of most active mating differs and so do the breeding sites. For example, wood frogs breed in woodland ponds or shallow water, leopard frogs in lowland swamps, and pickerel frogs in streams and ponds on high ground.

Behavioral isolation Many animal species have courtship patterns that allow males and females to recognize one another. The male blue-footed boobie in Figure 17.6 does a dance. Male fireflies are recognized by females of their species by the pattern of their flashings; similarly, female crickets recognize male crickets by their chirping. Many males recognize females of their species by sensing chemical signals called pheromones. For example, female gypsy moths release pheromones that are detected miles away by receptors on the antennae of males.

Video Flirting Flies

Mechanical isolation When animal genitalia or plant floral structures are incompatible, reproduction cannot occur. Inaccessibility of pollen to certain pollinators can prevent cross-fertilization in plants, and the sexes of many insect species have genitalia that do not match, or other characteristics that make mating impossible. For example, male dragonflies have claspers that are suitable for holding only the females of their own species.

Gamete isolation Even if the gametes of two different species meet, they may not fuse to become a zygote. In animals, the sperm of one species may not be able to survive in the reproductive tract of another species, or the egg may have receptors only for sperm of its species. In plants, only certain types of pollen grains can germinate so that sperm successfully reach the egg.

Postzygotic Isolating Mechanisms

Postzygotic isolating mechanisms, those that operate after formation of the zygote, prevent hybrid offspring from developing, even if reproduction attempts have been successful. Or, if a hybrid is born, it is infertile and cannot reproduce. Either way, the genes of the parents are unable to be passed on.

Hybrid inviability A hybrid zygote may not be viable, and so it dies. A zygote with two different chromosome sets may fail to go through mitosis properly, or the developing embryo may receive incompatible instructions from the maternal and paternal genes so that it cannot continue to exist.

Hybrid sterility The hybrid zygote may develop into a sterile adult. As is well known, a cross between a female horse and a male donkey produces a mule, which is usually sterile—it cannot reproduce (Fig. 17.7). Sterility of hybrids generally results from complications in meiosis that lead to an inability to produce viable gametes. Similarly, a cross between a cabbage and a radish produces offspring that cannot form gametes, most likely because the cabbage chromosomes and the radish chromosomes cannot align during meiosis.

Check Your Progress   17.1

1. Identify the type of reproductive isolation that explains why a female bark beetle that feeds on pine does not recognize the attractant pheromones of a male that feeds on oak.
2. List three different species concepts and explain the main requirements of each.
3. Explain how frogs that have different courtship calls but look similar can be different species.

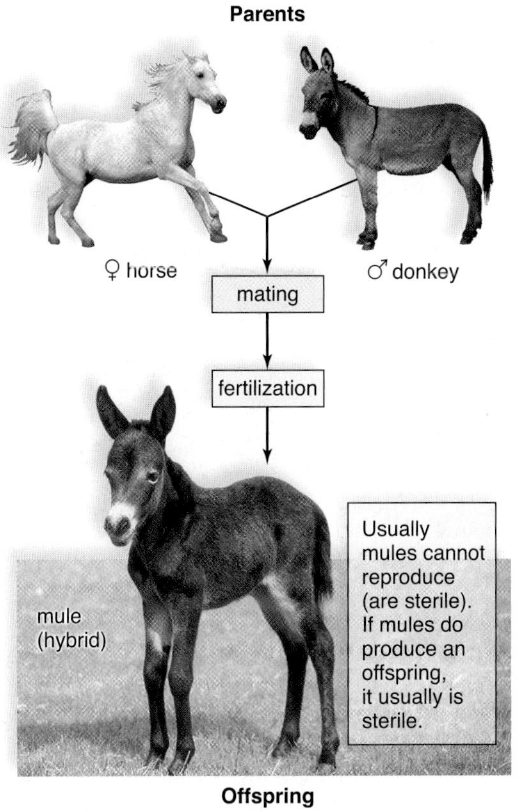

Parents

♀ horse   ♂ donkey

mating

fertilization

mule
(hybrid)

Usually
mules cannot
reproduce
(are sterile).
If mules do
produce an
offspring,
it usually is
sterile.

Offspring

Figure 17.7 Postzygotic isolating mechanism. Mules are infertile. Horse and donkey chromosomes cannot pair to produce gametes.

17.2 Modes of Speciation

Learning Outcomes

Upon completion of this section, you should be able to

1. Define two modes of speciation and give examples of each.
2. Identify an example of adaptive radiation.
3. Distinguish between coevolution and convergent evolution.

Researchers recognize two principal modes of **speciation,** which is the splitting of one species into two or more species, or the transformation of one species into new species over time. One mode requires populations to be physically isolated from one another, and the other mode does not.

Geographic isolation is helpful because it allows populations to continue on their own evolutionary path, which eventually causes them to be reproductively isolated from other species and from one another. Once reproductive isolation has begun, it can be reinforced by the evolution of more traits that prevent breeding with related species. Geographic isolation can repeatedly occur, so one ancestral species can give rise to several other species.

Allopatric Speciation

In 1942, Ernst Mayr, an evolutionary biologist, published the book *Systematics and the Origin of Species,* in which he proposed the biological species concept and a process by which speciation could occur. This process, termed **allopatric speciation**

313

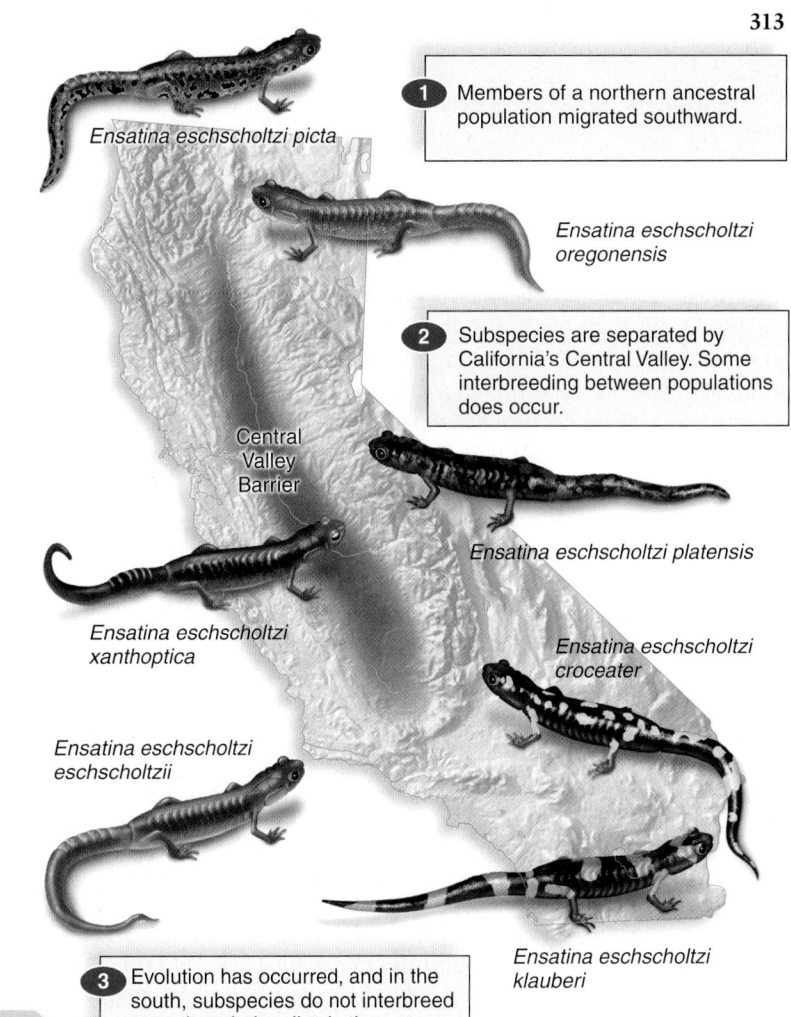

Ensatina eschscholtzi picta

Ensatina eschscholtzi oregonensis

Central Valley Barrier

Ensatina eschscholtzi platensis

Ensatina eschscholtzi xanthoptica

Ensatina eschscholtzi croceater

Ensatina eschscholtzi eschscholtzii

Ensatina eschscholtzi klauberi

1 Members of a northern ancestral population migrated southward.

2 Subspecies are separated by California's Central Valley. Some interbreeding between populations does occur.

3 Evolution has occurred, and in the south, subspecies do not interbreed even though they live in the same environment.

Figure 17.8 Allopatric speciation in progress among *Ensatina* salamanders. The Central Valley of California is reproductively separating a range of populations of *Ensatina eschscholtzi* that are all descended from the same northern ancestral species.

[Gk. *allo*, other, and *patri*, fatherland], is the eventual result of populations separated by a geographic or some other physical barrier. Mayr said that when populations of a species become geographically isolated, microevolutionary processes, such as genetic drift and natural selection, alter the gene pool of each population independently. If the differences between the groups becomes large enough, reproductive isolation may occur, resulting in the formation of new species.

Examples of Allopatric Speciation

Figure 17.8 features an example of allopatric speciation that has been extensively studied in Southern California. An ancestral population of *Ensatina* salamanders lives in the Pacific Northwest. **1** Members of this ancestral population migrated southward, establishing a series of populations. Each population was exposed to unique selective pressures along the coastal mountains and the Sierra Nevada mountains. **2** Due to the presence of the Central Valley of California, gene flow rarely occurs between the eastern populations and the western populations. **3** Genetic differences increased from north to south, resulting in two distinct forms of *Ensatina* salamanders in Southern California that differ dramatically in color and rarely interbreed.

Geographic isolation is even more obvious in other examples. The green iguana of South America is hypothesized to be the common ancestor for both the marine iguana on the Galápagos Islands to the west, and the rhinoceros iguana on Hispaniola, an island to the north. If so, how could it happen? Green iguanas are strong swimmers, so by chance, a few could have migrated to these islands, where they formed populations separate from each other and from the parent population back in South America. Each population continued on its own evolutionary path as new mutations, genetic drift, and different selection pressures occurred. Eventually, reproductive isolation developed, and the result was three species of iguanas that are reproductively isolated from each other.

It is interesting to note that the ability of an organism to move about has a large impact on whether or not allopatric speciation can occur. For example, the oceans of the world are all interconnected, and wide-ranging animals, such as humpback whales, are members of a single species even though they are seasonally thousands of miles apart. Conversely, the scale of distance and the size of the organism is important, too. Often, small organisms such as parasites are tightly linked to their hosts. In fact, a species and its parasites can coevolve, because their evolutionary pathways are interdependent.

Another example of allopatric speciation involves sockeye salmon in Washington State. In the 1930s and 1940s, hundreds of thousands of sockeye salmon were introduced into Lake Washington. Some colonized an area of the lake near Pleasure Point Beach (Fig. 17.9a). Others migrated into the Cedar River (Fig. 17.9b). Andrew Hendry, a biologist at McGill University, is able to tell a Pleasure Point Beach salmon from a Cedar River salmon because they differ in shape and size due to the demands of reproducing. Males in rivers where the waters are fast-moving tend to be more slender than those at the beach. Sockeye salmon turn sideways into the strong current as part of their mating ritual, and a male with a slender body is better able to perform this maneuver. In contrast, the females in rivers tend to be larger than those at the beach. This larger body helps them dig slightly deeper nests in the gravel beds on the river bottom. Deeper nests are not disturbed by river currents and remain warm enough for eggs to survive and hatch. These differences have resulted in reproductive isolation between these two populations. Not all river salmon remain near the beach their whole lives, in fact, a third of the sockeye males in Pleasure Point grew up in the river population. But the two populations are not interbreeding because the size and shape of females and males in both populations remains.

Reinforcement of Reproductive Isolation

As seen in sockeye salmon and other animals, independent evolution of populations can result in reproductive isolation. Another example is seen among *Anolis* lizards in which males court females by extending a colorful flap of skin, called a "dewlap." The dewlap must be seen in order to attract mates. Populations of *Anolis* in a dim forest tend to evolve a light-colored dewlap, while populations in open habitats tend to evolve dark-colored ones. This change in dewlap color causes the populations to be reproductively isolated, because females distinguish males of their species by their dewlaps.

As populations become reproductively isolated, postzygotic isolating mechanisms may arise before prezygotic isolating mechanisms. As we have seen, when a horse and a donkey reproduce, the hybrid is not fertile. Therefore, the process of natural selection would favor any variation that would prevent the production of hybrids that are unable to reproduce. Indeed, natural selection would favor the continual development of prezygotic isolating mechanisms until the two populations are completely reproductively isolated.

The term **reinforcement** is given to the process of natural selection that 'reinforces' reproductive isolation. Reinforcement occurs when two populations, formerly of the same species, come back in contact after being isolated. These two species are not able to reproduce when they come into contact, because they no longer recognize each other as mates. An example of reinforcement has been seen in the pied and collared flycatchers of the Czech Republic and Slovakia, where both species occur in close proximity. Only here have the pied flycatchers evolved a different coat color than the collared flycatchers. The difference in color reinforces the choice to mate with their own species.

Sympatric Speciation

Speciation without the presence of a geographic barrier is termed **sympatric speciation** [Gk. *sym*, together, and *patri*, fatherland]. Sympatric speciation is more difficult to observe in nature because no physical barrier prevents mating between populations, as in allopatric speciation. Some of the best examples of sympatric speciation in nature have involved divergence in diet, microhabitat, or both. In these cases, a new species evolves when a population becomes specialized to live in a different microhabitat.

One example is the midas and arrow cichlid fishes that live in a small lake in Nicaragua. The midas cichlid colonized the lake and occupied its usual rocky, coastal habitat. Over time, a new species of cichlid, the arrow cichlid, evolved from a population of the midas cichlid that adapted to living and feeding in open water habitat. The partitioning of lake habitats and dietary preferences resulted in a shift in body size, jaw morphology, and tooth size and shape. Now the midas and arrow cichlids are two distinct species.

Lake male

Lake female

River male

River female

a. Sockeye salmon at Pleasure Point Beach, Lake Washington

b. Sockeye salmon in Cedar River. The river connects with Lake Washington.

Figure 17.9 Allopatric speciation among sockeye salmon. In Lake Washington, salmon that matured (a) at Pleasure Point Beach do not reproduce with those that matured (b) in the Cedar River. The females from Cedar River are noticeably larger and the males are more slender than those from Pleasure Point Beach, and these shapes help them reproduce in the river.

Sympatric speciation involving **polyploidy** (a chromosome number beyond the diploid [2n] number) is well documented in plants. A polyploid plant can reproduce with itself, but produces only sterile offspring when mated with 2n individuals because not all the chromosomes would be able to pair during meiosis. Two types of polyploidy are known: autoploidy and alloploidy.

Autoploidy occurs when a diploid plant produces diploid gametes due to nondisjunction during meiosis (see Fig. 10.10). If this diploid gamete fuses with a haploid gamete, a triploid plant results. A triploid (3n) plant is sterile and cannot produce offspring because the chromosomes cannot pair during meiosis. Humans have found a use for sterile plants because they produce fruits without seeds. If two diploid gametes fuse, the plant is a tetraploid (4n) and the plant is fertile, so long as it reproduces with another of its own kind. The fruits of polyploid plants are much larger than those of diploid plants. The huge strawberries of today are produced by octaploid (8n) plants.

Alloploidy [Gk. *allo*, other, and *ploidy*, uncountable] requires a more complicated process than autoploidy because it requires that two different but related species of plants hybridize. Hybridization is followed by doubling of the chromosomes. For example, the California wildflower *Clarkia concinna* is a diploid plant with fourteen chromosomes (seven pairs). A related species, *C. virgata*, is a diploid plant with ten chromosomes (five pairs). A hybrid of these two species is not fertile because seven chromosomes from one plant cannot pair evenly with five chromosomes from the other plant (Fig. 17.10). However, if the chromosome number doubles in the hybrid, the chromosomes can pair during meiosis, resulting in a fertile plant. The species *C. pulchella* could have arisen this way. Recent molecular data tell

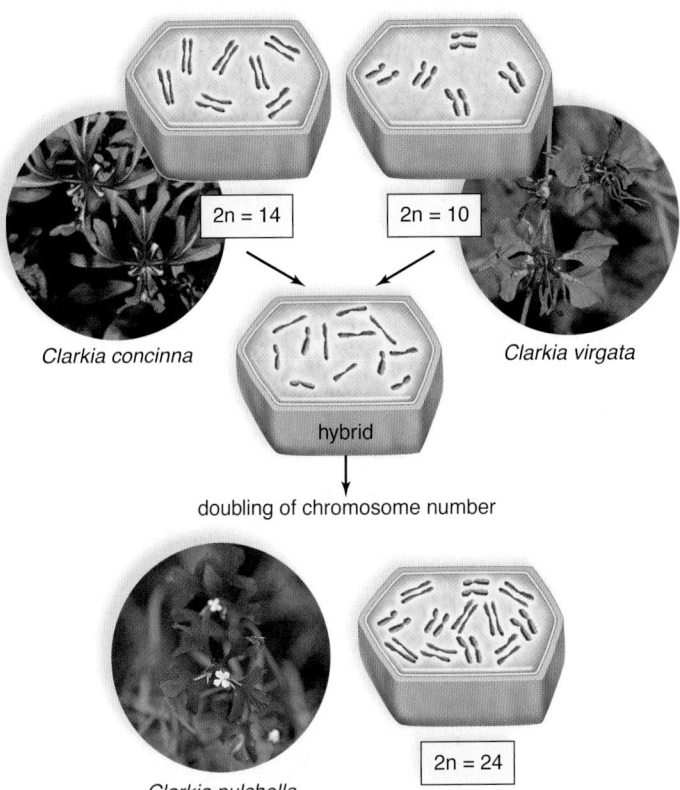

Clarkia concinna

2n = 14

2n = 10

Clarkia virgata

hybrid

doubling of chromosome number

Clarkia pulchella

2n = 24

Figure 17.10 Polyploidy produces a new species.
Reproduction between two species of *Clarkia* results in a sterile hybrid. Doubling of the chromosome number results in a fertile third *Clarkia* species that can reproduce with itself only.

us that polyploidy is common in plants and makes a significant contribution to the evolution of new plant species.

Adaptive Radiation

Adaptive radiation is a type of speciation that occurs when a single ancestral species rapidly gives rise to a radiation of new species as each adapts to a specific environment. Many instances of adaptive radiation involve sympatric speciation following the removal of a competitor, a predator, or a change in the environment. Reduced competition provides **ecological release,** or the freedom for a species to expand its use of resources within habitats where competition has been removed. Ecological release provides opportunity for new species to originate as populations become specialized to newly available microhabitats.

Examples of Adaptive Radiation

Darwin proposed that a breeding pair of ancestral finches colonized the Galápagos Islands and their descendants spread out to occupy various niches. Geographic isolation of the various finch populations caused their gene pools to become isolated. Because of natural selection, each population adapted to a particular habitat on its island. In time, the many populations became so genotypically different that now, when by chance they reside on the same

Figure 17.11 Adaptive radiation in Hawaiian honeycreepers. More than 20 species, now classified in separate genera, evolved from a single species of a goldfinch-like bird that colonized the Hawaiian Islands.

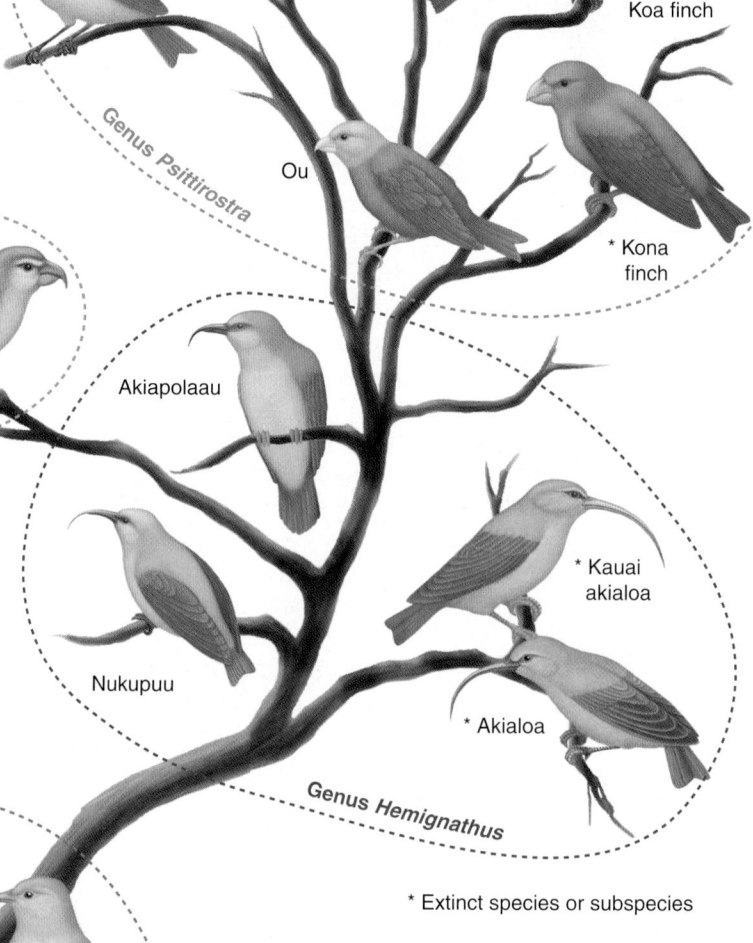

* Extinct species or subspecies

island, they do not interbreed, and are therefore separate species. During mating, female finches use beak shape to recognize members of the same species, and suitors with the wrong type of beak are rejected. (The genetic mechanisms by which beak shape changed are described in the Nature of Science feature on page 318.)

Similarly, on the Hawaiian Islands a wide variety of honeycreepers are descended from a common goldfinchlike ancestor that arrived from Asia or North America about 5 million years ago. Today, honeycreepers have a range of beak sizes and shapes for feeding on various food sources, including seeds, fruits, flowers, and insects (Fig. 17.11). Adaptive radiation also occurs among plants; a good example is the silversword alliance, which includes plants adapted to moist and dry environments and even lava fields.

Adaptive radiation has occurred throughout the history of life on Earth when a group of organisms exploits new environments. For example, with the demise of the dinosaurs about 66 million years ago, mammals underwent adaptive radiation as they exploited environments previously occupied by the dinosaurs. Mammals diversified in just 10 million years to include the early representatives of all the mammalian orders, including hoofed mammals (e.g., horses and pigs), aquatic mammals (e.g., whales and seals), primates (e.g., lemurs and monkeys), flying mammals (e.g., bats), and rodents (e.g., mice and squirrels). A changing world presented new environmental habitats and new food sources also. Insects fed on flowering plants and, in turn, became food for mammals. Primates lived in trees where fruits were available.

Convergent Evolution

Convergent evolution is said to occur when a similar biological trait evolves in two unrelated species as a result of exposure to similar environments. For example, both birds and bats have wings, but not because they share a recent common ancestor that had wings. Flight has evolved in birds and bats independently, and in so doing, has resulted in two different, although similar-looking, solutions to the requirements of flight—an airfoil, or wing; muscles to move the wings; and a lightweight body.

Traits that evolve convergently in two unrelated lineages because of a response to a similar lifestyle or habitat are said to be **analogous**—such as the wings of birds and bats. The opposite of analogous is **homologous**—traits that are similar because they evolved from a common ancestor (see Fig. 15.15). For example, the wings of butterflies are homologous to wings of moths because both are members of the same lineage of insects called the Lepidoptera. All of Lepidoptera evolved from a winged common ancestor.

Recent examples of convergent evolution involve adaptive radiations of species in similar, but unconnected, habitats. Lake Malawi and Lake Tanganyika are two African Rift Valley lakes (Fig. 17.12). Each lake has within it a set of 200–500 species of cichlid fishes, each adapted to feed on prey in a particular microhabitat of the lake. For example, each lake has fish that are adapted to feeding on the sandy bottom, and species that feed along the rocky shore. Diet specialization has produced a suite of different jaw and tooth shapes and sizes, each adapted to a particular food type. The outcome is an amazing example of convergent evolution. Each lake's assemblage of cichlids has evolved independently of the other, yet if you compare the assemblage in each lake, you find amazing similarities in coloration, body shape, the size and shape of jaws and teeth (Fig. 17.12). The convergent evolution is apparent in the pairing of Lake Malawi and Lake Tanganyika species that have the same features—evidence for evolution of the same, but independently derived, set of features, adapted to forage in similar habitats.

Video Cichlid Specialization

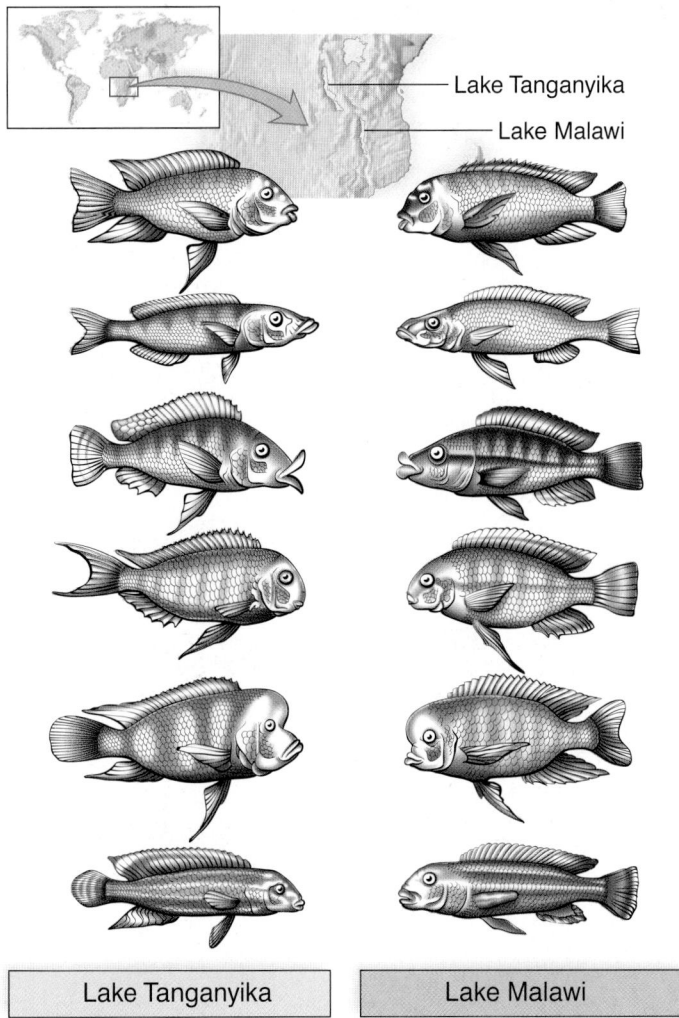

| Lake Tanganyika | Lake Malawi |

Figure 17.12 Convergent evolution of African lake fish. Cichlids exhibit remarkable evolutionary convergence. In Lake Malawi and Lake Tanganyika, a very similar set of body shapes and sizes has evolved independently of each other, with each type adapted to feed on a different type of food source. Although they appear morphologically similar, all of the cichlids from Lake Malawi are more closely related to one another than to any species within Lake Tanganyika.

Check Your Progress 17.2

1. List the evidence you would need to show that the five species of big cats, *Panthera leo* (lion), *P. tigris* (tiger), *P. pardus* (leopard), *P. onca* (jaguar), and *P. uncia* (snow leopard) are an example of an adaptive radiation.
2. Predict the outcome of convergent evolution on the variety of cichlid fish in a newly discovered African Rift Valley lake compared to other lakes with similar microhabitats.
3. During the last Ice Age, deer mice in Michigan became separated by a large glacial lake and are now two different species. Identify the mode of speciation.

Nature of Science

Genetic Basis of Beak Shape in Darwin's Finches

Darwin's finches are a famous example of how many species originate from a common ancestor. Over time, each of the type of finches on the Galápagos Islands adapted to a unique way of life, with beak size and shape related to their diet. Ground finches have thick, short beaks adept at crushing hard seeds. Cactus finches have long, thin beaks well-suited to probing flowers and fruit of cacti. The warbler finch feeds on both seeds and insects, and has a thin, short beak useful for a mixed diet. Multiple sources of evidence in DNA sequences and morphology support the hypothesis that Darwin's finches are closely related to one another (Fig. 17B).

The differences in beak shape have been recorded by decades of research.

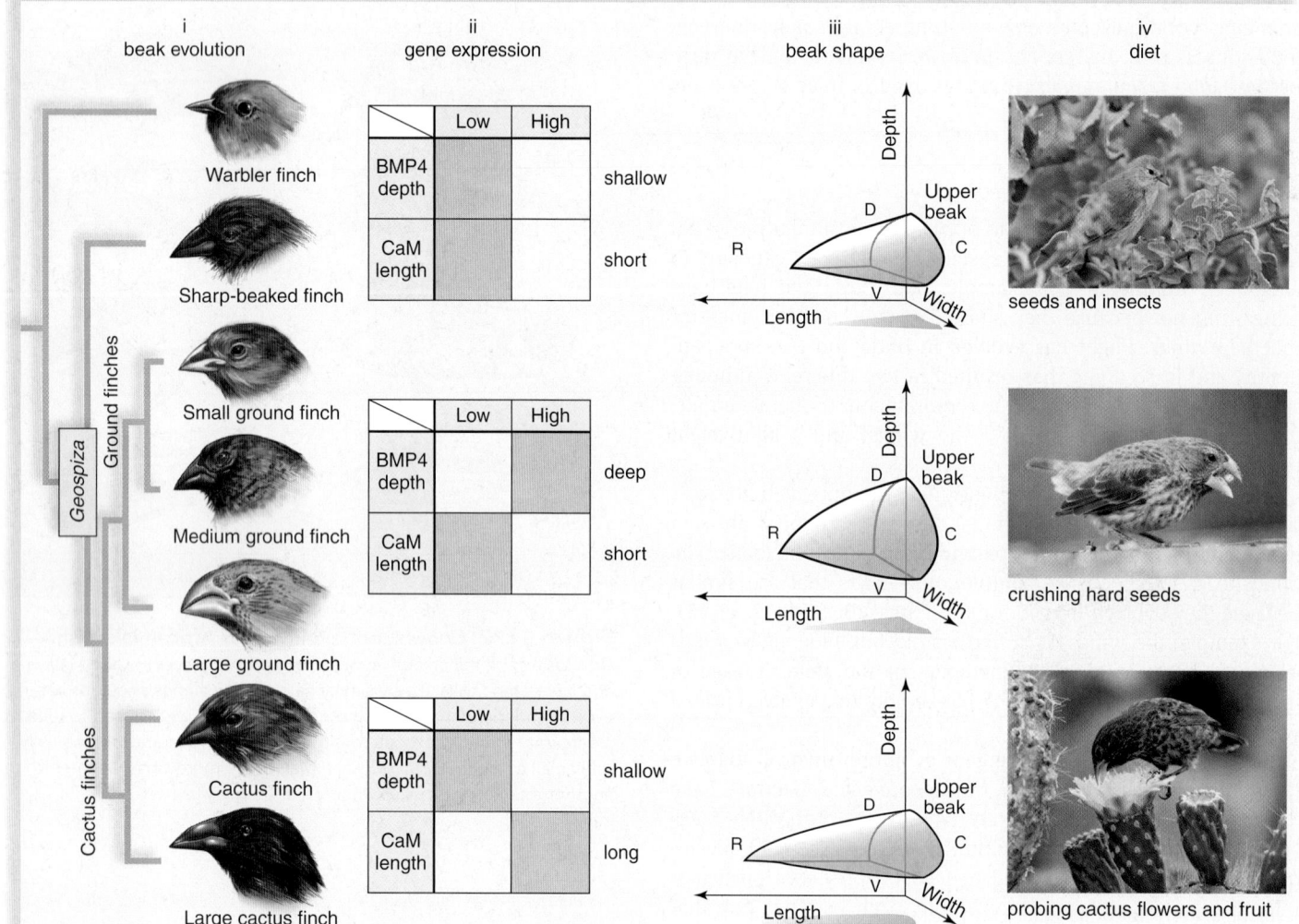

Figure 17B Genetic basis of finch beak size and shape. Bone morphogenetic protein 4 and calmodulin genes regulate the depth and length of the beaks of Darwin's finches. An increase or decrease in gene activity work together to fine tune beak morphology.

17.3 Principles of Macroevolution

Learning Outcomes

Upon completion of this section, you should be able to

1. Discriminate between the gradualistic and punctuated equilibrium models of evolution.
2. Explain how gene expression can influence speciation.
3. Support, by providing an example, the argument that macroevolution is not goal oriented.

Many evolutionary biologists hypothesize, as Darwin did, that macroevolution occurs gradually. After all, natural selection can only do so much to bring about change in each generation. The gradual evolution of new species is the basis of the *gradualistic model* of evolution, which proposes that speciation occurs after populations become isolated, with each group continuing slowly on its own evolutionary pathway. The proponents of the gradual model often show the history of groups of organisms by drawing the type of diagram shown in Figure 17.13*a*. Note that in this diagram, an ancestral species has given rise to two separate

Without any additional information, scientists proposed, as did Darwin, that there must be a genetic explanation for the difference in beak shape among species. In 2006, the genes that are responsible for the variation in finch beak shape were discovered. These findings are direct evidence—the actual mechanism—for how macroevolution occurs.

Two genes control beak depth and shape. The gene for bone morphogenetic protein 4 (*BMP4*) determines how deep, or tall, the beak will be. The gene for calmodulin (*CaM*) regulates how long a beak will grow. For example, a high level of *BMP4* creates a deep, wide beak. A high level of *CaM* produces a long beak. In Darwin's finches, a combination of *BMP4* and *CaM* determines overall beak shape (Fig. 17B-ii). More precisely, the degree of expression of each gene—or how much of each protein is produced—affects how the beak develops in the embryo.

The cactus finch, for example, has a low level of *BMP4* and a high level of *CaM* expression, which produces a shallow, long beak (Fig. 17B-iii, -iv). In contrast, the ground finch has the opposite pattern, with a high level of *BMP4* and low level of *CaM* expression, producing a short, deep beak (Fig. 17B-iii, -iv). One of the most interesting findings of this research is that evolution of beak shape did not require changes to the *BMP4* or the *CaM* genes—an increase or decrease in the expression of these genes during embryo development was enough to change beak shape!

The ability of *BMP4* and *CaM* to affect beak morphology is not limited to Darwin's finches. Variation in beak shape was reproduced in chicken embryos (Figure 17C). Using molecular tools, finch *BMP4* and *CaM* genes were expressed in the beaks of developing chicken embryos. The expression of *BMP4* caused the chick beaks to deepen, and *CaM* expression caused their beaks to get longer (Figure 17C*b, c*). Overall, this is strong evidence for the genetic basis of macroevolution.

Questions to Consider

1. How might have small, or microevolutionary, changes in *BMP4* and *CaM* in finch populations resulted in new species of finch?
2. Use Darwin's theory of evolution by natural selection to explain the match between finch beak shape and diet on the Galápagos Islands.

No CaM expression **CaM expression**

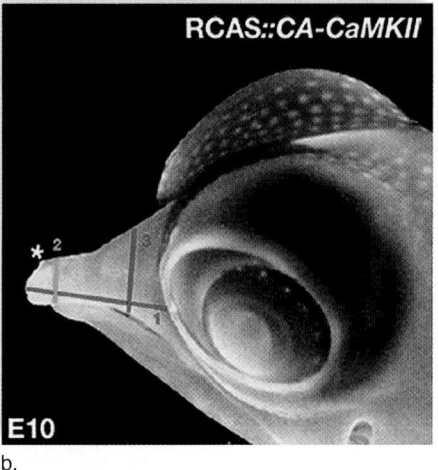

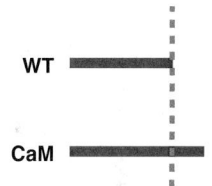

a.   b.   c.

Figure 17C Expression of *BMP4* and *CaM* in chicks. Chicken embryos were genetically modified to express *CaM* in their beaks during development. **a.** The normal, or wild type (WT) chick did not have *CaM* expression in the beak during development and the length of the beak was normal. **b.** The *CaM* chick produced an elongated beak. **c.** A side-by-side comparison of the length of WT and *CaM* beaks. Similarly, a separate study of *BMP4* expression in chick embryos produced deeper beaks compared to the WT embryos.

species, represented by a slow change in plumage color. The gradualistic model suggests that it is difficult to indicate when speciation occurred because there would be so many transitional links.

After studying the fossil record, some paleontologists tell us that species can appear quite suddenly, and then they remain essentially unchanged during a period of stasis (sameness) until they either undergo extinction or evolve in response to change in the environment. Based on these findings, they developed a *punctuated equilibrium model* to explain the fluctuating pace of evolution. This model says that the assembly of species in the fossil record can be explained by periods of equilibrium, or stasis, that are punctuated (interrupted) by periods of rapid, abrupt speciation, or change. Figure 17.13*b* shows this way of representing the history of evolution over time.

A strong argument can be made that it is not necessary to choose between these two models of evolution, and that both could very well assist us in interpreting the fossil record. In other words, some fossil species may fit one model, and some may fit the other model. In a stable environment, a species may

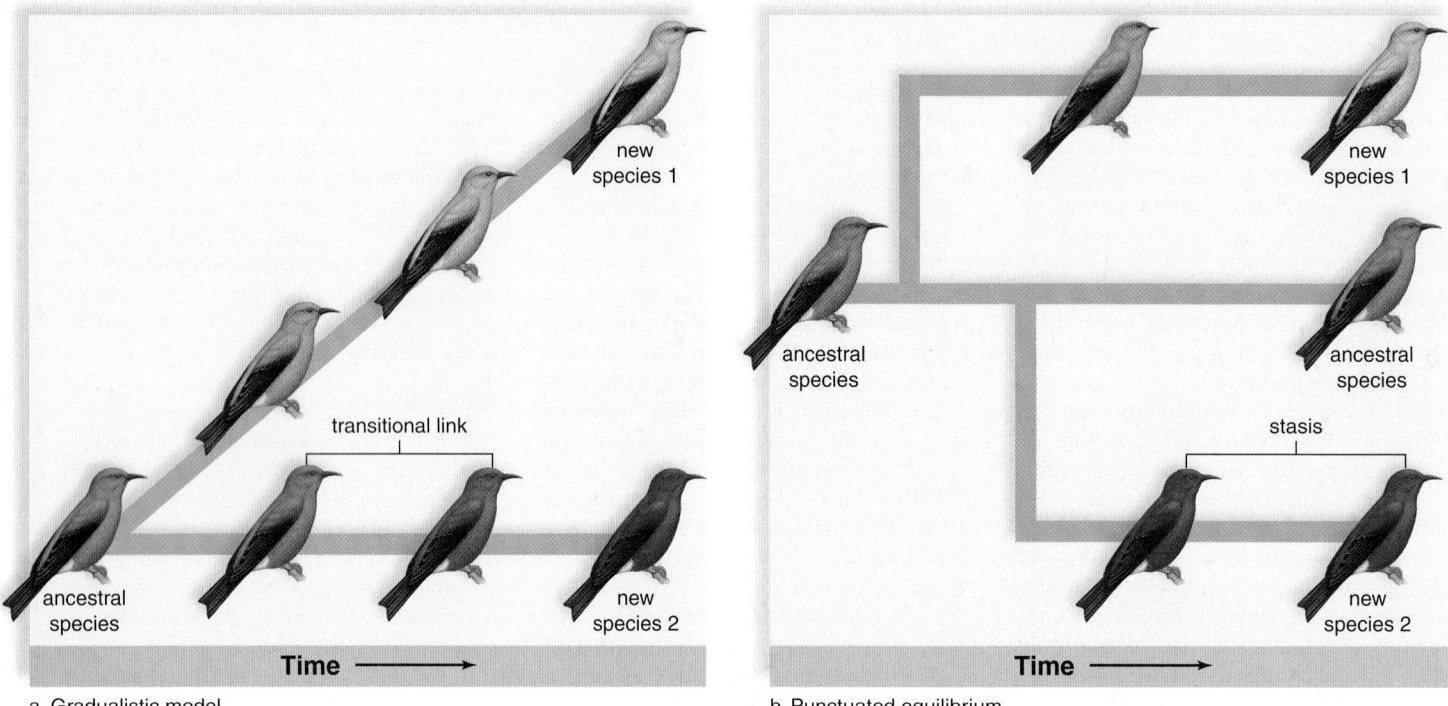

Figure 17.13 Gradualistic and punctuated equilibrium models.   **a.** Under the gradualism model, new species evolve from a series of small changes that occur constantly over time. This process brings about a lot of transitional forms. **b.** Under the punctuated model, new species evolve from a series of abrupt, rapid changes after a period of little or no change. This process would result in different species with few transitional forms.

be kept in equilibrium by stabilizing selection for a long period. If the environment changes slowly, a species may be able to adapt gradually. And, if environmental change is rapid, a new species may arise suddenly before the parent species goes on to extinction. The differences between all possible patterns of evolutionary change are rather subtle, especially when we consider that, because geologic time is measured in millions of years, the "sudden" appearance of a new species in the fossil record could actually represent many thousands of years.

Developmental Genes and Macroevolution

Investigators have discovered genes that can bring about radical changes in body shapes and organs. For example, it is now known that the *Pax6* gene is involved in eye formation in all animals, and that homeotic (*Hox*) genes determine the location of repeated structures in all vertebrates (see Chapter 42).

Whether slow or fast, how could evolution have produced the myriad of animals in the history of life? Or, to ask the question in a genetic context, how can genetic changes bring about such major differences in form? It has been suggested since the time of Darwin that the answer must involve the processes that shape development. In 1917, D'Arcy Thompson asked us to imagine an ancestor in which all parts are developing at a particular rate. A change in gene expression could stop a developmental process or could continue it beyond its normal time. For instance, if the growth of limb bones were stopped early,

the result would be shorter limbs, and if it were extended, the result would be longer limbs compared to those of an ancestor. Or, if the whole period of growth were extended, a larger animal would result, accounting for why some species of horses are so large today.

Using the modern techniques of cloning and manipulating genes, investigators have indeed discovered genes whose differences in expression (the timing and location in the body where proteins they encode are synthesized) can bring about changes in body shapes and organs. This result suggests that these genes must date back to a common ancestor that lived more than 600 MYA, and that despite millions of years of divergent evolution, all animals share the same control switches for development (see Chapter 28).

Development of the Eye

The animal kingdom contains many different types of eyes, and it was long thought that each type would require its own set of genes. Flies, crabs, and other arthropods have compound eyes that have hundreds of individual visual units. Humans and all other vertebrates have a camera-type eye with a single lens. So do squids and octopuses. Humans are not closely related to either flies or squids, so wouldn't it seem as if all three types of animals evolved "eye" genes separately? But this is now known not to be the case.

In 1994, Walter Gehring and his colleagues at the University of Basel, Switzerland, discovered that a gene called *Pax6* is

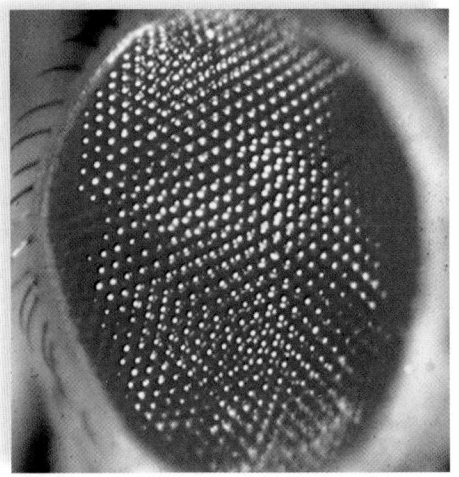 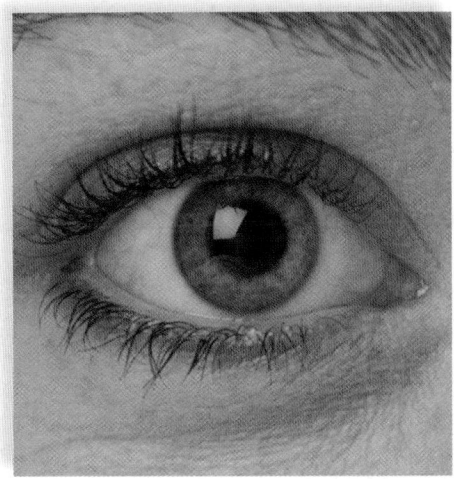 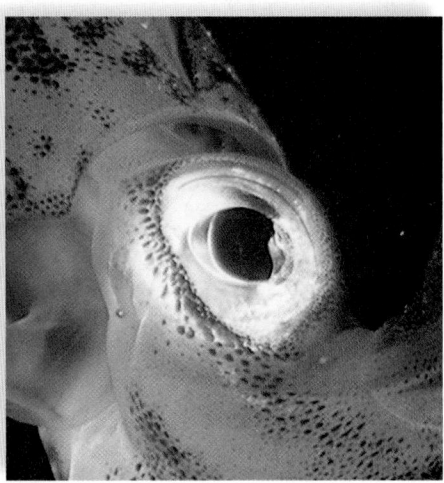

Figure 17.14 *Pax6 gene and eye development.* *Pax6* is involved in eye development in a fly, a human, and a squid.

required for eye formation in all animals (Fig. 17.14). Mutations in the *Pax6* gene lead to failure of eye development in both people and mice, and remarkably, the mouse *Pax6* gene can cause an eye to develop on the leg of a fruit fly (Fig. 17.15).

Development of Limbs

Wings and arms are very different, but both humans and birds express the *Tbx5* gene in developing limb buds. *Tbx5* encodes a protein that is a transcription factor that turns on the genes needed to make a limb during development. Birds and humans both express *Tbx5*, but differ in which genes Tbx5 turns on. Perhaps in an ancestral tetrapod, the Tbx5 protein triggered the transcription of only one gene, and the evolution of limb formation evolved in vertebrates such that changes in the genes regulated by Tbx5 and other transcription factors contributed to the variation we see in tetrapod limb structure. Therefore subtle changes in gene control can have profound effects on the shape of body, which could explain the abundance of variation we see in plant and animal shape and form.

There is also the question of timing. Changing the timing of gene expression, as well as which genes are expressed, can result in dramatic changes in shape.

Development of Overall Shape

Vertebrates have repeating segments, as exemplified by the vertebral column. Changes in the number of segments can lead to changes in overall shape. In general, *Hox* genes control the number and appearance of repeated structures along the main body axes of vertebrates. Shifts in when *Hox* genes are expressed in embryos are responsible for why a snake has hundreds of

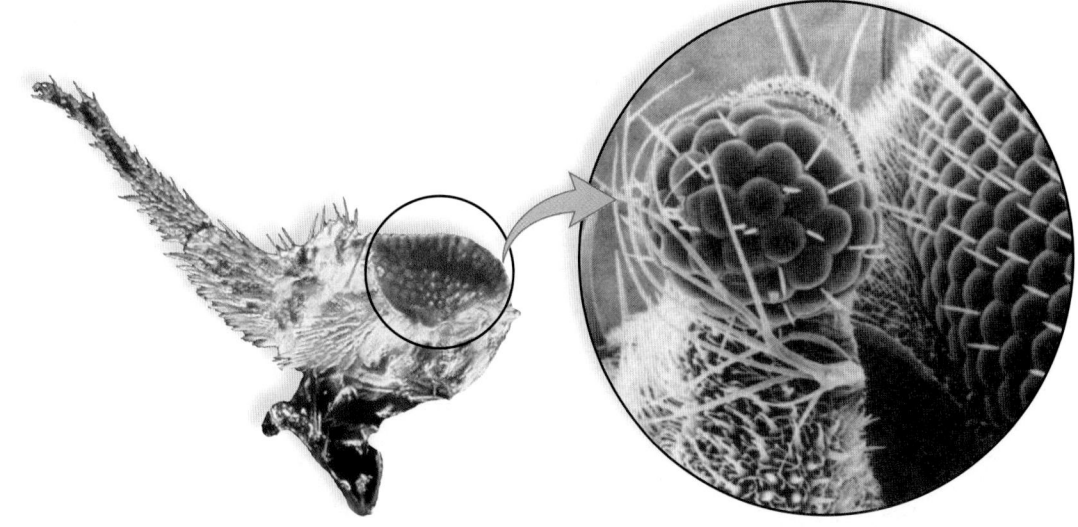

Figure 17.15 Study of Pax6 gene. The mouse *Pax6* gene makes a compound eye on the leg of a fruit fly.

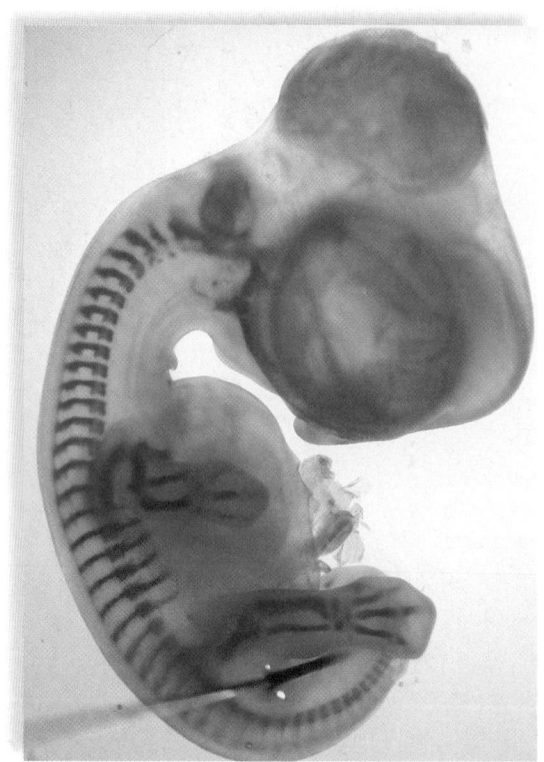

Figure 17.16
Hox6 genes.
Differential expression of *Hox6* genes causes a chick to have seven vertebrae (*purple*) and a snake to have many more vertebrae (*purple*) in corresponding regions. Notice that no vertebrae corresponding to the chick's upper thoracic and neck vertebrae develop in the snake.

Source: Burke, A. C. 2000. Hox genes and the global patterning of the somitic mesoderm. In C. Ordahl (ed.), Somitogenesis. Current Topics in Developmental Biology (series), Vol. 47. (Academic Press [Elsevier], Waltham, Mass.)

rib-bearing vertebrae and essentially no neck, in contrast to other vertebrates, such as a chick (Fig. 17.16).

Hox genes have been found in all animals, and other shifts in the expression of these genes can explain why insects have just six legs and other arthropods, such as crayfish, have ten legs. In general, the study of *Hox* genes has shown how animal diversity is due to variations in the expression of ancient genes, rather than to wholly new and different genes (see the Nature of Science feature in Chapter 28).

Pelvic-Fin Genes

The three-spined stickleback fish occurs in two forms in North American lakes. In the open waters of a lake, long pelvic spines help protect the stickleback from being eaten by large predators. But on the lake bottom, long pelvic spines are a disadvantage because dragonfly larvae seize and feed on young sticklebacks by grabbing them by their spines.

The presence of short spines in bottom-dwelling fish can be traced to a reduction in the development of the pelvic-fin bud in the embryo, and this reduction is due to the altered expression of a particular gene called *Pitx1*.

Hindlimb reduction has occurred during the evolution of other vertebrates. The hindlimbs became greatly reduced in size as whales and manatees evolved from land-dwelling ancestors into fully aquatic forms (see Fig. 15.13). Similarly, legless lizards have evolved many times. The stickleback study has shown how natural selection can lead to major skeletal changes in a relatively short time.

Human Evolution

The sequencing of genomes has shown us that our DNA base sequence is very similar to that of chimpanzees, mice, and, indeed, all vertebrates. The human genome has around 23,000

genes. Based on this knowledge and the work just described, investigators no longer expect to find new genes to account for the evolution of humans. Instead, they predict that differential gene expression, new functions for "old" genes, or both will explain how humans evolved. We discuss the details of human evolution in Chapter 30.

As with all genes, mutations of developmental genes occur by chance, and it is this random process that creates variation. Without variation, evolution cannot occur. Even though mutation is random, natural selection is not a random process. Rather, natural selection acts on the variation that is present, in a way that favors the survival of advantageous traits, in a particular environment at a particular time. This should not be misinterpreted as evidence that evolution "works" toward an end goal or optimum. Evolution is a perpetual process that shapes variation from generation to generation. Where this process leads is unpredictable, and is dependent upon a complicated array of external forces. In the next section, we observe that evolution is not directed toward any particular end.

Macroevolution Is Not Goal-Oriented

The evolution of the horse, *Equus*, has been studied since the 1870s, and at first the ancestry of this genus seemed to represent a model for gradual, directed evolution toward the "goal," of the modern horse. Three trends were particularly evident during the evolution of the horse: increase in overall size, toe reduction, and change in tooth size and shape.

By now, however, many more fossils have been found, making it easier to tell that the evolutionary history of the horse is complicated by the presence of many lineages that evolved, went extinct, and thus were not on the lineage that led to the modern horse. The evolutionary tree of the horse in Figure 17.17 is an

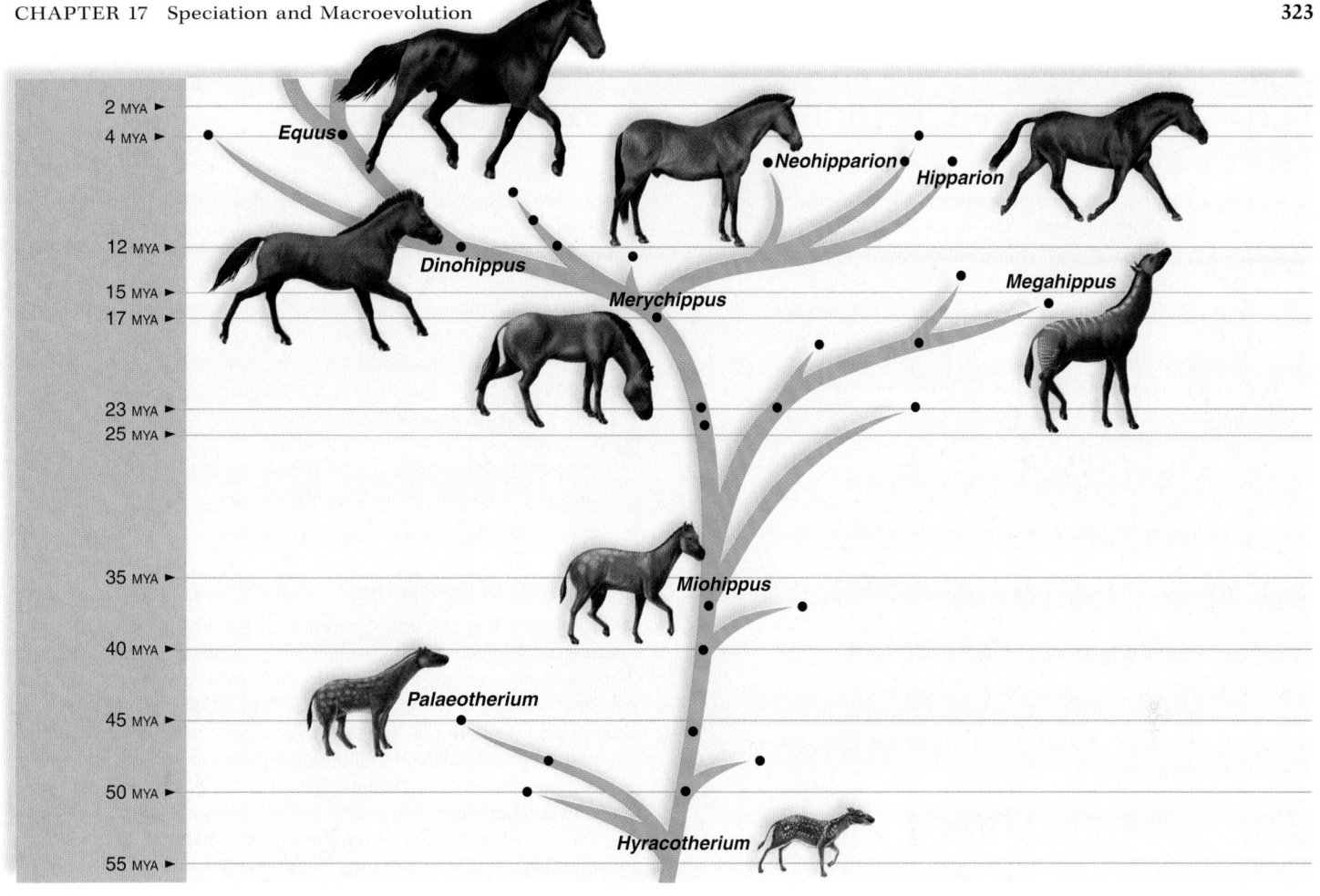

Figure 17.17 Simplified family tree of *Equus*. Every dot represents a genus.

oversimplification because it is based only upon the evidence from the few fossils that we have, and there are likely many more fossils that have yet to be discovered. If the evolution of the horse were directed toward the "goal" of the modern horse, we would expect to see a single branch on this tree with intermediate fossils leading directly from the ancestor to the horse. Yet the actual evolutionary tree of *Equus* has many branches, and will have even more as new fossils are discovered.

Each of the ancestral species of the horse was adapted to its environment. Adaptation occurs only because the members of a population with an advantage are able to have more offspring than other members. Natural selection is opportunistic, not goal-directed.

Fossils named *Hyracotherium* have been designated as the first probable members of the horse family, living about 57 MYA. These animals had a wooded habitat, ate leaves and fruit, and were about the size of a dog. Their short legs and broad feet with several toes would have allowed them to scamper from thicket to thicket to avoid predators. *Hyracotherium* was obviously well adapted to its environment because this genus survived for 20 million years.

The family tree of *Equus* does tell us once more that speciation, diversification, and extinction are common occurrences in the fossil record. The first adaptive radiation of horses occurred about 35 MYA. The weather was becoming drier, and grasses were evolving. Eating grass requires tougher teeth, and an increase in size and longer legs would have permitted greater speed

to escape enemies. The second adaptive radiation of horses occurred about 15 MYA and included *Merychippus* as a representative of those groups that were speedy grazers who lived on the open plain. By 10 MYA, the horse family was quite diversified. Some species were large forest browsers, some were small forest browsers, and others were large plains grazers. Many species had three toes, but some had one strong toe. (The hoof of the modern horse includes only one toe.)

Modern horses evolved about 4 MYA from ancestors who had features that are adaptive for living on an open plain, such as large size, long legs, hoofed feet, and strong teeth. The other groups of horses prevalent at the time became extinct, no doubt for complex reasons. Humans have corralled modern horses for various purposes, and this makes it difficult to realize that the traits of a modern horse are adaptive for living in a grassland environment.

Check Your Progress 17.3

1. Explain how the punctuated equilibrium model provides an alternative explanation to the theory of catastrophism proposed by Cuvier (see Chapter 15).
2. Discuss how the study of developmental genes supports the possibility of rapid speciation in the fossil record.
3. Explain why the development of an eye on the leg of a fruit fly because of *Pax6* gene expression provides support for macroevolution.

CONNECTING *the* CONCEPTS *with the* BIG IDEAS

Evolution

- A group of interbreeding organisms whose offspring are fertile constitute a species. (1C narrative)
- Isolating mechanisms which promote speciation can be of the pre-zygotic or the post-zygotic nature. (1C2a)
- The environment has a significant impact on how species originate. An ecological or physical barrier promotes allopatric speciation while microhabitat specialization drives sympatric speciation. (1C2a, 1C2b)
- Adaptive radiation occurs when multiple species develop in specialized habitat niches from one original ancestral species. (1C1a)

- The rate of speciation can be very slow (gradualism) or can occur in periodic and sometimes rapid spurts (punctuated equilibrium). (1C narrative, 1C1a, 1C2b)
- Evolution does not have direction; variation arises by random mutation, not because a particular solution is "needed." Natural selection "weeds out" variation in a particular environment at a particular time. (1A1e)
- Molecular biologists are now able to document shared ancestry of species at the level of DNA sequences, and all life on Earth shares some common developmental genes. These universal genes support the theory that macroevolution is the source of biodiversity. (1C3a-b)

*Find the unabridged version of all EK citations at www.glencoe.com/maderAP11.

Media Study Tools

www.glencoe.com/maderAP11

Enhance your study of this chapter with study tools and practice tests. Also ask your instructor about the resources available through ConnectPlus, including the media-rich eBook, interactive learning tools, and animations.

Summarize

17.1 How New Species Evolve

The morphological species concept identifies species based on diagnostic traits. Linnaeus, the father of taxonomy, identified species according to the morphological species concept. The morphological species concept has limitations. Not all organisms have measurable diagnostic traits, such as bacteria and other microorganisms. Cryptic species can look very similar but have other life history or behavioral traits that distinguish them as different species.

The evolutionary species concept requires that each species have its own evolutionary pathway and can be recognized by certain diagnostic morphological traits. The phylogenetic species concept recognizes a new species as a set of populations that share a common ancestor, or are monophyletic on an evolutionary tree. Populations that have different variants of a trait, such as spot shape or color, but still meet the criterion of monophyly are considered the same species under this concept.

The biological species concept identifies species based on whether two or more populations are reproductively isolated from each other. Today, DNA sequence data can also be used to distinguish one species from another.

Prezygotic isolating mechanisms (habitat, temporal, behavior, mechanical, and gamete isolation) prevent mating from being attempted or prevent fertilization from being successful if mating is attempted.

Postzygotic isolating mechanisms (hybrid inviability, hybrid sterility, and F_2 fitness) prevent hybrid offspring from surviving or reproducing.

17.2 Modes of Speciation

During allopatric speciation, geographic or physical separation precedes reproductive isolation. Isolation of populations allows genetic changes to accumulate over time via microevolution. Eventually, the ancestral species and the new species no longer breed with one another. As one example, series of salamander subspecies on either side of the Central Valley of California has resulted in two populations of the same species that are unable to successfully reproduce when they come in contact.

During sympatric speciation, a geographic barrier is not required, and speciation is simply a change in genotype that prevents successful reproduction. Sympatric speciation in animals is relatively rare, but can occur when populations of the same species become specialized on a particular subhabitat and/or food item in the same geographic area. For example, a new species of cichlid, the arrow cichlid, diverged from the midas cichlid in a single lake in Nicaragua because it specialized in feeding in open water. Another example of sympatric speciation is occurrence of polyploidy in plants.

Adaptive radiation is a type of speciation that occurs when a single ancestral species rapidly gives rise to a radiation of new species as each adapts to a specific environment. Many instances of adaptive radiation involve sympatric speciation following ecological release. Ecological release provides opportunity for new species to originate as populations become specialized in newly available microhabitats. Finches on the Galápagos Islands, Honeycreepers of the Hawaiian Islands, and the rise of mammals after the extinction of dinosaurs are examples of adaptive radiation events.

Convergent evolution has occurred when the same biological trait has evolved in two unrelated species as a result of adaptation to a similar set of conditions or lifestyle. The wings of birds and bats are examples of convergent evolution.

An adaptive radiation accompanied by convergent evolution can produce similar assemblages of morphological types in geographically isolated, but similar, environments. Lake Tanganyika and Lake Malawi contain assemblages of cichlids that are similar as a result of convergent evolution during two independent adaptive radiations in each lake.

17.3 Principles of Macroevolution

Macroevolution is evolution of new species and higher levels of classification. The fossil record gives us a view of life many millions of years ago. The hypothesis that species evolve gradually is now being challenged by the hypothesis that speciation can also occur rapidly.

In that case, the fossil record could show periods of stasis interrupted by spurts of change, that is, a punctuated equilibrium. Transitional fossils would be expected with gradual change but not with punctuated equilibrium.

It could be that both models are seen in the fossil record, but rapid change can occur by differential expression of regulatory genes. The same regulatory gene (*Pax6*) controls the development of both the camera-type and the compound-type eye. The *Tbx5* gene controls development of limbs, whether the wing of a bird or the leg of a tetrapod. *Hox* genes control the number and appearance of a repeated structure along the main body axes of vertebrates. The same pelvic-fin genes control the development of a pelvic girdle. Variation in the expression of *BMP4* and *CaM* produces different beak shapes in each of Darwin's finches. Changing the timing of gene expression, as well as which genes are expressed, can result in dramatic changes in shape.

Speciation, diversification, and extinction are seen during the evolution of *Equus*. These three processes are commonplace in the fossil record and illustrate that macroevolution is not goal-directed. The life we see about us represents adaptations to particular environments. Such adaptations have changed in the past and will change in the future.

Key Terms

adaptive radiation 316
allopatric speciation 313
alloploidy 315
analogous 317
autoploidy 315
biological species
  concept 310
convergent evolution 317
cryptic species 307
diagnostic traits 307
ecological release 316
evolutionary species
  concept 307
homologous 317
hybridization 310
macroevolution 307
monophyletic 308

morphological species concept
  307
morphology 307
phylogenetic species
  concept 308
polyploidy 315
postzygotic isolating
  mechanism 312
prezygotic isolating
  mechanism 310
reinforcement 314
reproductive isolation 310
speciation 307, 313
species concepts 307
sympatric speciation 314
taxonomist 307
zygote 310

 **Assess**

Reviewing This Chapter

1. Give the pros and cons of the evolutionary species concept and the biological species concept. Give an example to show that DNA sequence data can distinguish species. 307–10
2. List and discuss five prezygotic isolating mechanisms and three postzygotic isolating mechanisms. 310–12
3. Use the *Ensatina* salamander example to explain allopatric speciation. 313
4. How does sympatric speciation differ from allopatric speciation, and why is sympatric speciation common in plants but rare in animals. 313–14
5. Use the honeycreepers of Hawaii and the Galápagos finches to explain adaptive radiation. 316
6. With regard to the speed of speciation, how do the gradualistic model and the punctuated equilibrium model differ? Which model predicts the occurrence of many transitional fossils? Explain. 318–20

7. What types of genes are pertinent to the discussion of the speed of speciation? Explain. 320–22
8. Use the evolution of the horse to show that speciation is not goal-oriented. 322–23

Testing Yourself

Choose the best answer for each question.

1. A biological species
   a. always looks different from other species.
   b. always has a different chromosome number from that of other species.
   c. is reproductively isolated from other species.
   d. never occupies the same niche as other species.

2. Which of these is a prezygotic isolating mechanism?
   a. habitat isolation   d. zygote mortality
   b. temporal isolation   e. Both a and b are correct.
   c. hybrid sterility

3. Male moths recognize females of their species by sensing chemical signals called pheromones. This is an example of
   a. gamete isolation.
   b. habitat isolation.
   c. behaviorial isolation.
   d. mechanical isolation.
   e. temporal isolation.

4. Which of these is mechanical isolation?
   a. Sperm cannot reach or fertilize an egg.
   b. Courtship pattern differs.
   c. The organisms live in different locales.
   d. The organisms reproduce at different times of the year.
   e. Genitalia are unsuitable to each other.

5. Complete the following diagram illustrating allopatric speciation by using these phrases: genetic changes (used twice), geographic barrier, species 1, species 2, species 3.

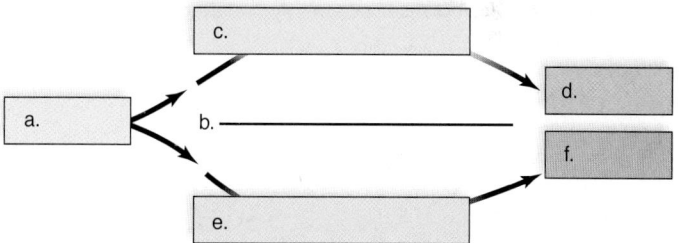

6. The creation of new species due to geographic barriers is called
   a. isolation speciation.
   b. allopatric speciation.
   c. allelomorphic speciation.
   d. sympatric speciation.
   e. symbiotic speciation.

7. The many species of Galápagos finches are each adapted to eating different foods. This is the result of
   a. gene flow.
   b. adaptive radiation.
   c. sympatric speciation.
   d. genetic drift.
   e. All of these are correct.

8. Allopatric, but not sympatric, speciation requires
   a. reproductive isolation.
   b. geographic isolation.
   c. spontaneous differences in males and females.
   d. prior hybridization.
   e. rapid rate of mutation.

9. Which of the following is not a characteristic of plant alloploidy?
   a. hybridization
   b. chromosome doubling
   c. related species mating
   d. All of these are characteristics of plant alloploidy.

10. Corn is an allotetraploid, which means that its
    a. chromosome number is 4n.
    b. occurrence resulted from hybridization.
    c. occurrence required a geographic barrier.
    d. Both a and b are correct.

11. Transitional links are least likely to be found if evolution proceeds according to the
    a. gradualistic model.
    b. punctuated equilibrium model.

12. Adaptive radiation is only possible if evolution is punctuated.
    a. true
    b. false

13. Which of the following can influence the rapid development of new types of animals?
    a. The influence of molecular clocks.
    b. A change in the expression of regulating genes.
    c. The sequential expression of genes.
    d. All of these are correct.

14. Which gene is incorrectly matched to its function?
    a. *Hox*—body shape
    b. *Pax6*—body segmentation
    c. *Tbx5*—limb development
    d. All of these choices are correctly matched.

15. In the evolution of the modern horse, which was the goal of the evolutionary process?
    a. large size
    b. single toe
    c. Both a and b are correct.
    d. Neither a nor b is correct.

16. Which of the following was not a characteristic of *Hyracotherium*, an ancestral horse genus?
    a. small size
    b. single toe
    c. wooded habitat
    d. All of these are characteristics of *Hyracotherium*.

17. Which statement about speciation is not true?
    a. Speciation can occur rapidly or slowly.
    b. Developmental genes can account for rapid speciation.
    c. The fossil record gives no evidence that speciation can occur rapidly.
    d. Speciation always requires genetic changes, such as mutations, genetic drift, and natural selection.

18. Which statement concerning allopatric speciation would come first?
    a. Genetic and phenotypic changes occur.
    b. Subspecies have a three-part name.
    c. Two subpopulations are separated by a barrier.

Engage

Thinking Scientifically

1. You want to decide what definition of a species to use in your study. What are the advantages and disadvantages of one based on DNA sequences as opposed to the evolutionary and biological species concept?

2. You decide to create a hybrid by crossing two species of plants. If the hybrid is a fertile plant that produces normal size fruit, what conclusion is possible?

Many transitional forms, such as *Microraptor* shown here in reconstruction, indicate a link between dinosaurs and birds.

18

Origin and History of Life

Today, paleontologists are setting the record straight about dinosaurs. It now appears that some dinosaurs nested in the same manner as some species of birds! Bowl-shaped nests containing dinosaur eggs have been found in Mongolia, Argentina, and the United States. These nests contain fossilized eggs and bones along with eggshell fragments. From this evidence, it seems that baby dinosaurs stayed in the nest after hatching until they were big enough to walk around and fragment the eggshells. The spacing between the nests of *Maiasaura* (meaning good mother lizard in Greek), suggests that this dinosaur fed its young. The remains of an enormous herd of about 10,000 *Maiasaura* found in Montana is further evidence that this dinosaur was indeed social in its behavior.

Maiasaura and *Microraptor*, the winged gliding dinosaur featured above, provide us with structural and behavioral evidence of the link between dinosaurs and birds. In this chapter, we trace the origin of life before considering the history of life.

As you read through this chapter, think about the following questions:

1. Does the theory of evolution explain the origins of life? Why or why not?
2. How do scientists measure the age of life on Earth from the fossil record?
3. What types of scientific evidence, including the fossil record, tell us something about the history of life on Earth?

CHAPTER OUTLINE

18.1 Origin of Life 328

18.2 History of Life 333

18.3 Geological Factors That Influence Evolution 342

BEFORE YOU BEGIN

Before beginning this chapter, take a few moments to review the following discussions.

Section 1.1 What are the basic characteristics that define life?

Section 2.1 What are chemical isotopes?

Section 15.3 How do the fossil and biogeographical records provide evidence for the evolution of new organisms?

FOLLOWING *the* BIG IDEAS

CHAPTER 18 ORIGIN AND HISTORY OF LIFE

Evolution

The theory of evolution does not explain the origin of life but how life on Earth became diverse after life began, and that history can be summarized by macroevolutionary change witnessed in the fossil record. However, early life scientists have developed several hypotheses on possible life origins.

327

18.1 Origin of Life

Learning Outcomes

Upon completion of this section, you should be able to

1. List and describe the four stages of the origin of life.
2. Differentiate between the stages of chemical and biological evolution.
3. Summarize a protocell membrane structure and its importance to the evolution of the first living cell.
4. Summarize at least one hypothesis that explains each of the four stages of the origin of life.

At the heart of Darwin's theory of evolution by natural selection is the principle of common ancestry, that all life on Earth can be traced back to a single ancestor. This ancestor, also called the **last universal common ancestor (LUCA)**, is common to all organisms that live, and have lived, on Earth since life began (Fig. 18.1). What were the properties of this common ancestor? How did life first begin on Earth? Darwin mused that perhaps the first living organism arose in a "warm little pond with all sorts of ammonia and phosphoric salts, light, heat, electricity, etc." What Darwin unknowingly described were some of the conditions of ancient Earth that gave rise to the first forms of life.

In Chapter 1 we considered the characteristics shared by all living things. Living things acquire energy through metabolism, or the chemical reactions that occur within cells. Living things also respond and interact with their environment, self-replicate, and are subject to the forces of natural selection that drive adaptation to the environment. The molecules of living things, called **biomolecules**, are organic molecules. The first living organisms on Earth would have had all of these characteristics. Yet early Earth was very different from the Earth we know today, and consisted mainly of inorganic substances. How, then, did life get started in this inorganic "warm little pond?"

Advances and discoveries in chemistry, evolutionary biology, paleontology, microbiology and other branches of science have helped to develop new, and test old, hypotheses about the origin of life. These studies contribute to an ever-growing body of scientific evidence that life originated 3.5–4 billion years ago (BYA) from nonliving matter in a series of four stages:

Stage 1. *Organic monomers.* Simple organic molecules, called monomers, evolved from inorganic compounds prior to the existence of cells. Amino acids, the basis of proteins, and nucleotides, the building blocks of DNA and RNA, are examples of organic monomers.

Stage 2. *Organic polymers.* Organic monomers were joined, or polymerized, to form organic polymers, such as DNA, RNA, and proteins.

Stage 3. *Protocells.* Organic polymers became enclosed in a membrane to form the first cell precursors, called *protocells* or *probionts.*

Stage 4. *Living cells.* Probionts acquired the ability to self-replicate as well as other cellular properties.

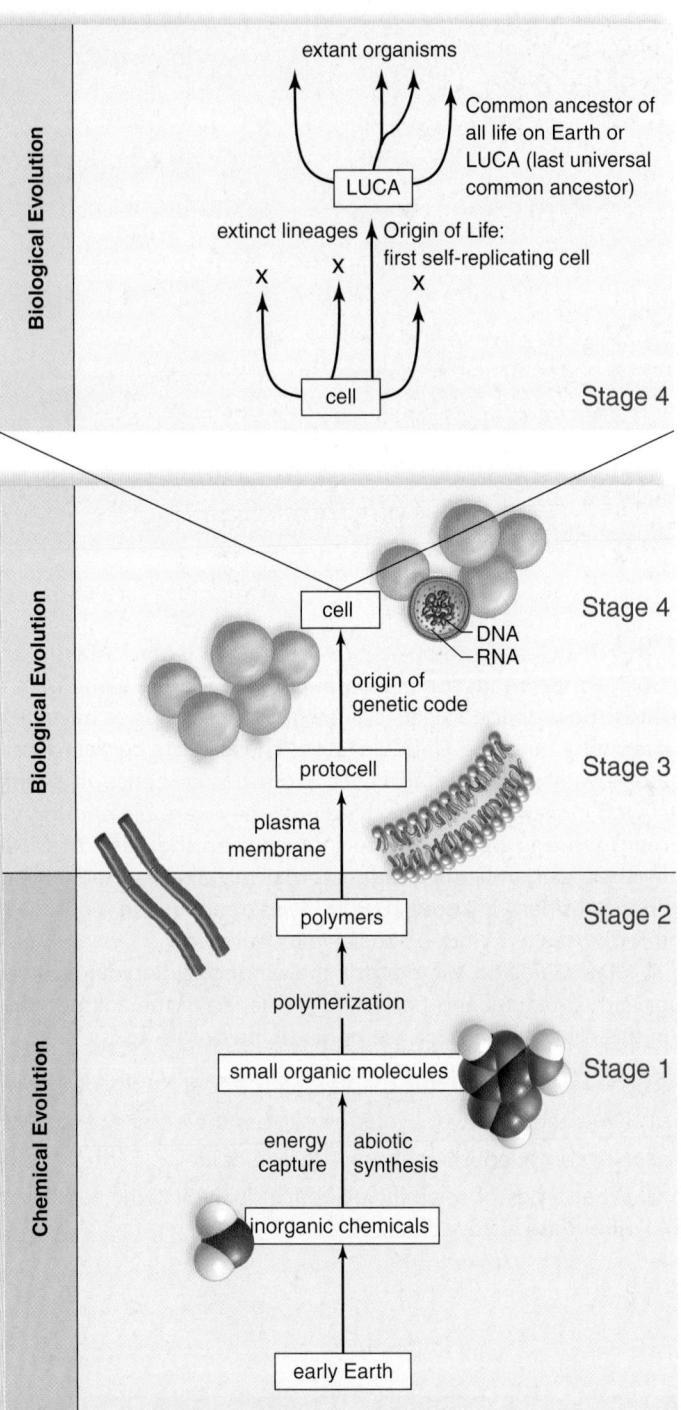

Figure 18.1 Stages of the origin of life. The first organic molecules (bottom) originated from chemically altered inorganic molecules present on early Earth (Stage 1). More complex organic macromolecules were synthesized to create polymers (Stage 2) that were then enclosed in a plasma membrane to form the protocells, or probionts (Stage 3). The protocell underwent biological evolution to produce the first true, self-replicating, living cell (Stage 4). This first living cell underwent continued biological evolution (top), with a single surviving lineage, the LUCA, that became the common ancestor of all life on Earth.

Scientists have performed experiments to test hypotheses at each stage of the origin of life. Stages 1–3 involve the processes of a "chemical evolution," before the origin of life (Fig. 18.1). Stage 4 is when life first evolved through the processes of "biological evolution" (Fig. 18.1). In this chapter we examine the hypotheses and supporting scientific evidence for each stage of chemical and biological evolution, or **abiogenesis,** the origin of life from nonliving matter.

Stage 1. Evolution of Monomers

In the 1920s Alexander Oparin, a notable biochemist, and J. B. S. Haldane, a geneticist and evolutionary biologist, independently proposed a hypothesis stating that the first stage in the origin of life was the evolution of simple organic molecules, or monomers (Fig. 18.1), from the inorganic compounds that were present in the Earth's early atmosphere. The Oparin-Haldane hypothesis, sometimes called the "**primordial soup" hypothesis**, proposes that early Earth had very little oxygen (O_2), but instead was made up of water vapor (H_2O), hydrogen gas (H_2), methane (CH_4), and ammonia (NH_3).

Methane and ammonia are reducing agents because they readily donate their electrons. In the absence of oxygen, their reducing capability is powerful. Thus, the early Earth had a reducing atmosphere in which oxidation-reduction (redox) reactions could have driven the chemical evolution, or **abiotic synthesis,** of organic monomers from inorganic molecules in the presence of strong energy sources. Note that 'oxidation' used in this context refers to a chemical redox reaction (see Section 6.4), and not to 'oxygen' gas.

Primordial Soup Hypothesis and the Miller-Urey Experiment

In 1953 Stanley Miller, under the mentorship of Harold Urey, performed a famous experiment to test Oparin's and Haldane's hypothesis of early chemical evolution (Fig. 18.2). The energy sources on early Earth included heat from volcanoes and meteorites, radioactivity from isotopes, powerful electric discharges in lightning, and solar radiation, especially ultraviolet radiation.

Animation
Miller-Urey

For his experiment, Miller placed a mixture of methane (CH_4), ammonia (NH_3), hydrogen (H_2), and water (H_2O), in a closed system, heated the mixture, and circulated it past an electric spark (simulating lightning). After a week's run, Miller discovered that a variety of amino acids and other organic acids had been produced.

The Miller-Urey experiment has been tested and re-examined over the decades since it was first performed. Other investigators have achieved similar results as Miller by using other, less-reducing combinations of gases dissolved in water. In 2008, a group of scientists examined 11 vials of compounds produced from variations of the Miller-Urey experiment, and found a greater variety of organic molecules than Miller reported, including all 22 amino acids.

Recent evidence suggests that nitrogen gas (N_2), and not ammonia (NH_3), would have been abundant in the primitive

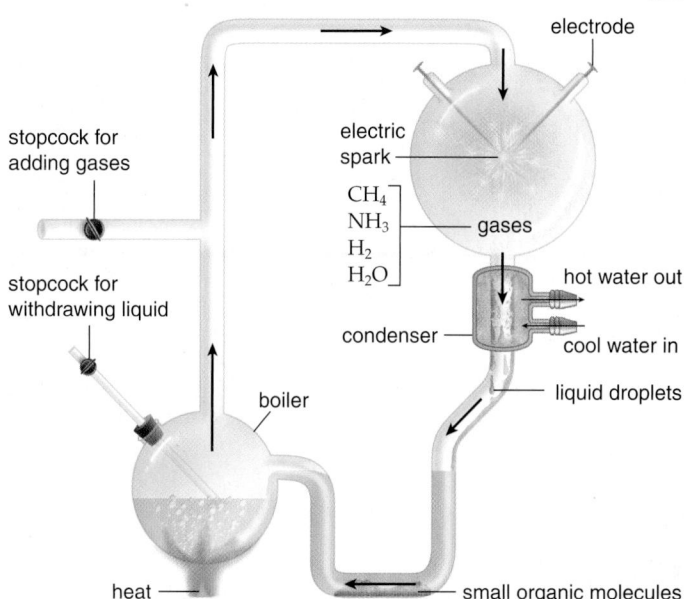

Figure 18.2 **Stanley Miller's experiment.** Gases that were thought to be present in the early Earth's atmosphere were admitted to the apparatus, circulated past an energy source (electric spark), and cooled to produce a liquid that could be withdrawn. Upon chemical analysis, the liquid was found to contain various small organic molecules, which could serve as monomers for large cellular polymers.

atmosphere. The scarcity of ammonia challenged the Miller-Urey experiment as a valid test of early Earth conditions. However, later experiments showed that ammonia could have been produced when various mixes of iron-nickel sulfides catalyzed the change of N_2 to NH_3. A laboratory test of this hypothesis worked perfectly. Under conditions simulating that of deep-ocean thermal vents (Fig. 18.3), 70% of various nitrogen sources were converted to ammonia within 15 minutes. Thus the early formation of organic monomers could have been supported by the production of ammonia from these vents.

If early atmospheric gases did react with one another to produce small organic compounds, neither oxidation (no free oxygen was present) nor decay (no bacteria existed) would have destroyed these molecules, and rainfall would have washed them into the ocean, where they would have accumulated for hundreds of millions of years. Therefore, the oceans would have been a thick, warm organic soup—much like Darwin's "warm little pond."

Iron-Sulfur World Hypothesis

The Oparin-Haldane hypothesis was an important contribution to our understanding of the early stages of life's origins, but other hypotheses have also been proposed and tested. In the late 1980s biochemist Günter Wächtershäuser proposed that thermal vents at the bottom of the Earth's oceans provided all the elements and conditions necessary to synthesize organic monomers. According to his **iron-sulfur world** hypothesis, dissolved gases emitted from thermal vents, such as carbon monoxide (CO), ammonia, and hydrogen sulfide, pass over iron and nickel sulfide minerals, also present at thermal vents. The iron and nickel sulfide molecules act as catalysts that drive the chemical evolution from inorganic to organic molecules.

plume of hot water
rich in iron-nickel sulfides

hydrothermal
vent

Figure 18.3 Chemical evolution at hydrothermal vents.
Minerals that form at deep-sea hydrothermal vents like this one can
catalyze the formation of ammonia and even monomers of larger organic
molecules that occur in cells.

Extraterrestrial Origins Hypothesis

Comets and meteorites have constantly pelted the Earth through-out history. In recent years, scientists have confirmed the presence of organic molecules in some meteorites. Many scientists, perhaps the foremost being Chandra Wickramsinghe, feel that these organic molecules could have seeded the chemical origin of life on early Earth. Others even hypothesize that bacterium-like cells evolved first on another planet and then were carried to Earth. A meteorite from Mars labeled ALH84001 landed on Earth some 13,000 years ago. When examined, experts found tiny rods similar in shape to fossilized bacteria. This hypothesis continues to be investigated.

Stage 2. Evolution of Polymers

Within a cell's cytoplasm, organic monomers join to form poly-mers in the presence of enzymes—such as the synthesis of protein polymers from amino acids by ribosomes. Enzymes themselves are proteins, but proteins were not present on early Earth. How did the first organic polymers form if no enzymes were present?

Iron-Sulfur World Hypothesis

The iron-sulfur world hypothesis of Günter Wächtershaüser and the research of Claudia Huber have shown that organic molecules will react and amino acids will form peptides in the presence of iron-nickel sulfides under conditions found at thermal vents. The inorganic iron-nickel sulfides have a charged surface that attracts amino acids and binds them together to make proteins.

Protein-First Hypothesis

Sidney Fox has shown that amino acids polymerize abiotically when exposed to dry heat. He suggests that once amino acids were present in the oceans, they could have collected in shallow puddles along the rocky shore. Then, the heat of the Sun could have caused them to form **proteinoids,** small polypeptides that have some catalytic properties.

The formation of proteinoids has been simulated in the laboratory. When placed in water, proteinoids form **microspheres** [Gk. *mikros,* small, little, and *sphaera,* ball], structures composed only of protein that have many properties of a cell. It's possible that even newly formed polypeptides had enzymatic properties, and some may have been more enzymatically active than others, which perhaps would be a trait with a selective advantage. If a certain level of enzyme activity provided an advantage over others, this would have set the stage for natural selection to shape the evolution of these first organic polymers. Those that evolved to be part of the first cell or cells would have had a selective advantage over those that did not become part of a cell.

Fox's **protein-first hypothesis** assumes that protein enzymes arose prior to the first DNA molecule. Thus the genes that encode proteins followed the evolution of the first polypeptides.

RNA-First Hypothesis

The **RNA-first hypothesis** suggests that only the macromolecule RNA was needed to progress toward formation of the first cell or cells. Thomas Cech and Sidney Altman shared a Nobel Prize in 1989 for their discovery that RNA can be both a substrate and an enzyme. Some viruses today have RNA genes; therefore, the first genes could have been RNA. It would seem, then, that RNA could have carried out the processes of life commonly associated today with DNA and proteins. Those who support this hypothesis say that it was an "RNA world" some 4 BYA.

Stage 3. Evolution of Protocells

Before the first true cell arose, a **protocell** (or **protobiont**) would have emerged—a structure that is characterized by having an outer membrane. After all, life requires chemical reactions to take place within a boundary, protecting them from disruption of conditions.

The Plasma Membrane

The plasma membrane of a cell provides a boundary between the inside of the cell and its outside world. This membrane is critical to the proper regulation and maintenance of cellular activities, and therefore the evolution of a membrane was a critical step in the origin of life.

The modern cell membrane is made up of phospholipids assembled in a bilayer (Fig. 18.4). The first plasma membranes were likely made up of fatty acids, which are smaller than phospholipids but like phospholipids have a hydrophobic "tail"

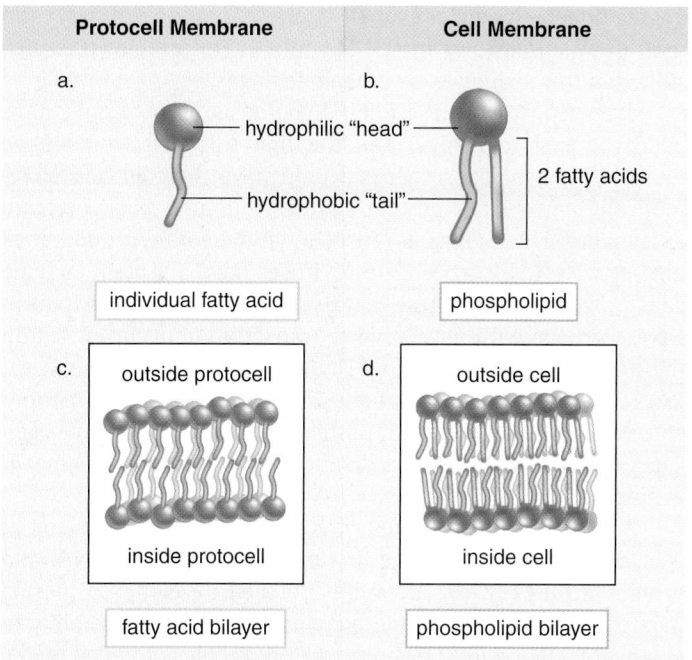

Figure 18.4. A comparison of protocell and modern cell plasma membranes. The first lipid membrane was likely made of a single layer of fatty acids. These first protocell membranes and the modern cell membrane have features in common. **a.** Individual fatty acids have a single fatty acid chain with a hydrophilic head and hydrophobic tail. **b.** Phospholipids of modern cell membranes are made of two hydrophobic fatty acid chains (the "tails") attached to a hydrophilic head. **c.** Protocell membranes were likely made up of a bilayer of fatty acids, with hydrophilic heads pointing outward, and hydrophilic tails pointing inward. **d.** Modern cells are organized in a similar fashion. Both bilayers create a semipermeable barrier between the inside and outside of a cell.

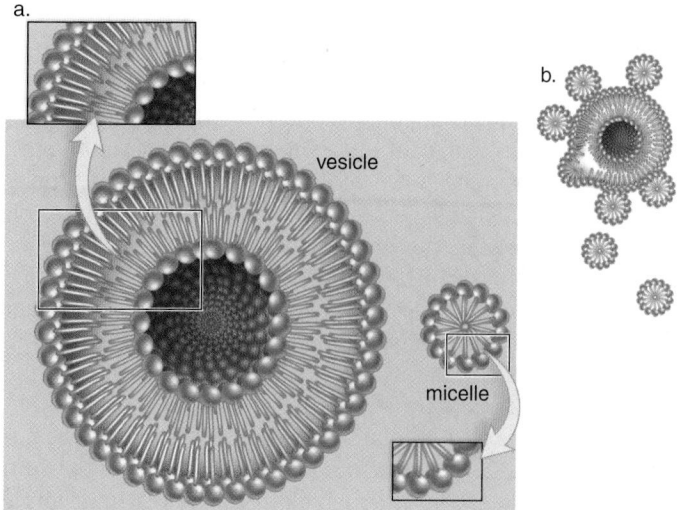

Figure 18.5. Structure and growth of the first plasma membrane. The first plasma membrane was likely made of a fatty acid bilayer, similar to that seen in vesicles. Protocells, the ancestor of modern cells, are thought to have had this type of membrane. **a.** A cross section of a vesicle reveals the fatty acid bilayer of a vesicle membrane. Micelles are spherical droplets formed by a single layer of fatty acids, and are much smaller than vesicles. **b.** Under proper conditions, micelles can merge to form vesicles. As micelles are added to the growing vesicle, individual fatty acids flip their hydrophilic heads toward the inside and outside of the vesicle, and their hydrophilic tails toward each other. This process forms a bilayer of fatty acids.

and hydrophilic "head" (Fig. 18.4). Fatty acids are one of the organic polymers that could have formed from chemical reactions at deep-water thermal vents early in the history of life. In water, fatty acids assemble into small spheres called **micelles** (Fig. 18.5). Micelles are a single layer of fatty acids organized with their heads pointing out and tails pointing toward the center of the sphere.

Under appropriate conditions micelles can merge to form **vesicles.** Vesicles are larger than micelles and are surrounded by a bilayer (two layers) of fatty acids (Fig. 18.5), similar to the *phospholipid bilayer* of modern cell membranes. An important feature of a vesicle lipid bilayer is that the individual fatty acids can flip between the two layers, which helps to move select molecules, such as amino acids, from outside to the inside of the vesicle. The first protocell would likely have been a type of vesicle with this type of fatty acid bilayer membrane (Fig. 18.5). Interestingly, if lipids are made available to protein microspheres, lipids tend to become associated with microspheres, producing a lipid-protein membrane. Lipid-protein microspheres share some interesting properties with modern cells: They resemble bacteria, they have an electrical potential difference, and they divide and perhaps are subject to selection.

In the 1920s, Alexander Oparin demonstrated that under appropriate conditions of temperature, ionic composition, and pH, concentrated mixtures of macromolecules tend to give rise to complex units called **coacervate droplets.** Coacervate droplets have a tendency to absorb and incorporate various substances from the surrounding solution. Eventually, a semipermeable-type boundary, also a trait of the modern cell membrane, may form about the droplet.

In the early 1960s, biophysicist Alec Bangham of the Animal Physiology Institute in Cambridge, England, discovered that when he extracted lipids from egg yolks and placed them in water, the lipids would naturally organize themselves into double-layered bubbles roughly the size of a cell. Bangham's bubbles soon became known as **liposomes** [Gk. *lipos,* fat, and *soma,* body].

Later, biophysicist David Deamer of the University of California and Bangham realized that liposomes might have provided life's first membranous boundary. Perhaps liposomes with a phospholipid membrane engulfed early molecules that had enzymatic, even replicative abilities. The liposomes would have protected the molecules from their surroundings and concentrated them so they could react (and evolve) quickly and efficiently. These investigators called this the **membrane-first hypothesis,** meaning that the first cell had to have a plasma membrane before any of its other parts. Perhaps the first membrane formed in this manner.

Nutrition

A protocell would have had to acquire nutrition, that is, other molecules, so that it could grow. One hypothesis suggests that protocells were heterotrophic [Gk. *hetero*, different, and *trophe*, food], organisms that consume preformed organic molecules. However, if the protocell evolved at hydrothermal vents, it may have carried out chemosynthesis—the synthesis of organic molecules from inorganic molecules and nutrients. Chemoautotrophic bacteria, for example, obtain energy by oxidizing inorganic compounds, such as hydrogen sulfide (H_2S), a molecule that is abundant at thermal vents. When hydrothermal vents in the deep and extremely dark ocean were first discovered in the 1970s, investigators were surprised to discover complex vent ecosystems supported by organic molecules formed by chemosynthesis, a process that does not require the energy of the Sun.

Glycolysis is a critical metabolic pathway that transforms high-energy chemical bonds into energy for a cell to do work. Glycolysis is the first stage of cellular respiration (see Section 8.2), and occurs outside of the mitochondria. ATP (adenosine triphosphate) is the most important energy-carrying molecule in living things. In the early stages of the origin of life, ATP would have been available to protocells in a preformed state, but over time, ATP would be transformed to ADP as it was used up. Natural selection would have favored the evolution of ATP/ADP recycling as a means to provide a renewable energy supply to the first cells. Modern cells have a way to produce ATP from ADP and inorganic phosphate via substrate-level phosphorylation (see Fig. 8.3) during glycolysis.

The greatest amount of ATP is synthesized by oxidative phosphorylation, in particular via the electron transport chain (see Section 8.2). Because all life on Earth uses ATP to fuel cellular metabolism, the evolution of a means to synthesize ATP must have occurred very early in the history of life. The first bacteria evolved in the oxygen-poor environment of early Earth (see Table 18.1). Thus it is probable that ATP was likely synthesized first by fermentation (see Section 8.3). The evolution of oxidative phosphorylation in eukaryotes provided an advantage because it greatly increased the amount of ATP synthesized per unit of energy. Mitochondria share a common ancestor with a group of bacteria that synthesize ATP via an electron transport chain. Oxidative phosphorylation is possible in eukaryotes because mitochondria provide an electron-transport-chain ATP factory.

At first the protocell must have had limited ability to break down organic molecules, and scientists speculate that it took millions of years for glycolysis to evolve completely. Interestingly, some evidence suggests that microspheres from which protocells may have evolved have some catalytic ability. Oparin's coacervates incorporate enzymes if they are available in the medium.

Stage 4. Evolution of a Self-Replication System

Today's cell is able to carry on protein synthesis in order to produce the enzymes that allow DNA to replicate. The central dogma of genetics states that DNA directs protein synthesis and that information flows from DNA to RNA to protein. It is possible that this sequence developed in stages.

According to the RNA-first hypothesis, RNA would have been the first to evolve, and the first true cell would have had RNA genes. These genes would have directed and enzymatically carried out protein synthesis. Today, ribozymes are enzymatic RNA molecules. Also, today we know a number of viruses have RNA genes. These viruses have a protein enzyme called reverse transcriptase that uses RNA as a template to form DNA. Perhaps with time, reverse transcription occurred within the protocell, and this is how DNA-encoded genes arose. If so, RNA was responsible for both DNA and protein formation. Once DNA genes developed, protein synthesis would have been carried out in the manner dictated by the central dogma of genetics.

According to the protein-first hypothesis, proteins, or at least polypeptides, were the first of the three (DNA, RNA, and protein) to arise. Only after the protocell developed a plasma membrane and sophisticated enzymes did it have the ability to synthesize DNA and RNA from small molecules provided by the ocean. Because a nucleic acid is a complicated molecule, the likelihood RNA arose *de novo* is minimal. It seems more likely that enzymes were needed to guide the synthesis of nucleotides and then nucleic acids. Again, once there were DNA genes, protein synthesis would have been carried out in the manner dictated by the central dogma of genetics.

Cairns-Smith proposed that polypeptides and RNA evolved simultaneously. Therefore, the first true cell would have contained RNA genes that could have replicated because of the presence of proteins. This eliminates the baffling chicken-and-egg paradox: Assuming a plasma membrane, which came first, proteins or RNA? It means, however, that two unlikely events would have had to happen at the same time.

After DNA formed, the genetic code had to evolve before DNA could store genetic information. The present genetic code is subject to fewer errors than a million other possible codes. Also, the present code is among the best at minimizing the effect of mutations. A single-base change in a present codon is likely to result in the substitution of a chemically similar amino acid and, therefore, minimal changes in the final protein. This evidence suggests that the genetic code did undergo a natural selection process before finalizing into today's code.

Check Your Progress   18.1

1. Match each step of the Miller-Urey experiment with the part of the Oparin-Haldane hypothesis it was designed to test.
2. List two alternative hypotheses that explain the origin of organic molecules from inorganic matter, and identify the key features of these hypotheses.
3. Compare and contrast the features of the protocell membrane with a modern cell membrane.

a.

Figure 18.6 The history of life. **a.** Strata, layers seen in sedimentary rock as exposed by road cuts, are the source of fossils, one source of information about the history of life. **b.** The blue ring of this diagram shows the history of life as it would be measured on a 24-hour timescale starting at midnight. (The red ring shows the actual years going back in time to 4.6 BYA.) The fossil record suggests that a very large portion of life's history was devoted to the evolution of unicellular organisms. The first multicellular organisms do not appear in the fossil record until just before 8 P.M., and humans are not on the scene until less than a minute before midnight. BYA = billion years ago

b.

18.2 History of Life

Learning Outcomes

Upon completion of this section, you should be able to

1. Explain the processes of relative and absolute dating of fossils.
2. List three sources of evidence that supports the endosymbiotic theory of organelle evolution.
3. Discuss when and where the first multicellular organisms evolved.
4. Describe in chronological order the periods of Earth's history, and identify one major biological event that took place in each.

Macroevolution is the origin of new species and other taxonomic categories. The fossil record documents the history of macroevolution over long periods of time.

Fossils Tell a Story

Fossils [L. *fossilis,* dug up] are the remains and traces of past life or any other direct evidence of past life. Recall from Chapter 15 that fossils consist mainly of hard parts such as shells, bones, or teeth, because these are usually not consumed or destroyed. However, some fossilized traces can be trails, footprints, or the impressions

of soft body parts. **Paleontology** [Gk. *palaios,* ancient, *ontos,* having existed, and *-logy,* study of] is the science of discovering and studying the fossil record and, from it, making decisions about the history of life, ancient climates, and environments.

The great majority of fossils are found embedded in or recently eroded from sedimentary rock. **Sedimentation** [L. *sedimentum,* a settling], a process that has been going on since the Earth was formed, is the gradual settling of particles of eroded and weathered rock and soil, called silt, that are carried by a flow of water. Silt is gradually deposited, forming layers of particles that vary in size and nature called **sediment.** Sediment becomes a **stratum** (pl., strata), a recognizable layer of sediment in a stratigraphic sequence (Fig. 18.6a).

The law of superposition states that a given stratum is considered to be older than the one above it and younger than the one immediately below it. However, the layers of sedimentation can be disturbed by geological forces, and such disturbances can complicate the interpretation of the stratigraphic sequence. Figure 18.6b shows the history of the Earth as if it had occurred during a 24-hour time span that starts at midnight. (The actual years are shown on an inner ring of the diagram.) This figure illustrates dramatically that only unicellular organisms were present during most (about 80%) of the history of the Earth.

If the Earth formed at midnight, prokaryotes do not appear until about 5 A.M., eukaryotes are present at approximately

4 P.M., and multicellular forms do not appear until around 8 P.M. Invasion of the land doesn't occur until about 10 P.M., and humans don't appear until 30 seconds before the end of the day. This timetable has been worked out by studying the fossil record. In addition to sedimentary fossils, more recent fossils can be found in tar, ice, bogs, and amber. Shells, bones, leaves, and even footprints, are commonly found in the fossil record (Fig. 18.7).

Relative Dating of Fossils

In the early nineteenth century, even before the theory of evolution was formulated, geologists sought to correlate the strata worldwide. The problem was that strata change their character over great distances, and therefore a stratum in England might contain different sediments than one of the same age in Russia. Geologists discovered that each stratum of the same age contained certain **index fossils** that serve to identify deposits made at apparently the same time in different parts of the world. These index fossils are used in **relative dating** methods. For example, a particular species of fossil ammonite (an animal related to the chambered nautilus) has been found over a wide range and for a limited time period. Therefore, all strata around the world that contain this fossil must be of the same age.

Absolute Dating of Fossils

Absolute **dating** methods rely on **radiometric** techniques to assign an actual date to a fossil. All radioactive isotopes have a particular half-life, the length of time it takes for half of the radioactive isotope to change into another stable element.

Radiocarbon dating uses the radioactive decay of ^{14}C, a rare carbon isotope. Over time, ^{14}C will stabililize into ^{14}N (see Section 2.1). This process of radioactive decay, that is, when ^{14}C becomes ^{14}N, occurs at a constant rate. Willard Libby won the Nobel Prize in Chemistry for his discovery in 1949 that shows the half-life of ^{14}C is 5,730 years, meaning that half of the original ^{14}C in organic matter decays to ^{14}N in this period of time. All living things have in them about the same relative amount of ^{14}C, so the absolute date of a fossil can be determined by comparing the difference in the relative amount of ^{14}C to other forms of carbon in the fossil to that of a living organism. The amount of ^{14}C that remains can be used to calculate the age of the fossil.

Radiocarbon dating is accurate only for fossils up to approximately 100,000 years old; after this time, the amount of ^{14}C radioactivity is so low that it cannot be used to measure the age of a fossil accurately. ^{14}C is the only radioactive isotope contained within organic matter, but it is possible to use others to date rocks, and from that to infer the age of a fossil contained in the rock. Potassium-argon (K-Ar) dating can be used to date rocks that are 100,000 to 4.5 billion years old, and thus can be used to indirectly estimate the age of fossils well beyond the limits of radiocarbon dating.

The Precambrian Time

Geologists have devised the **geologic timescale,** which divides the history of the Earth into eras and then periods and epochs (Table 18.1). The geologic timescale was derived from

Figure 18.7 Fossils.
Fossils are the remains of living organisms from the past. They can be impressions left in rocks, footprints, mineralized bones, shells, or any other evidences of life-forms that lived in the past.

trilobite

ichthyosaur

fossil fern

placoderm

ammonites

dinosaur footprint

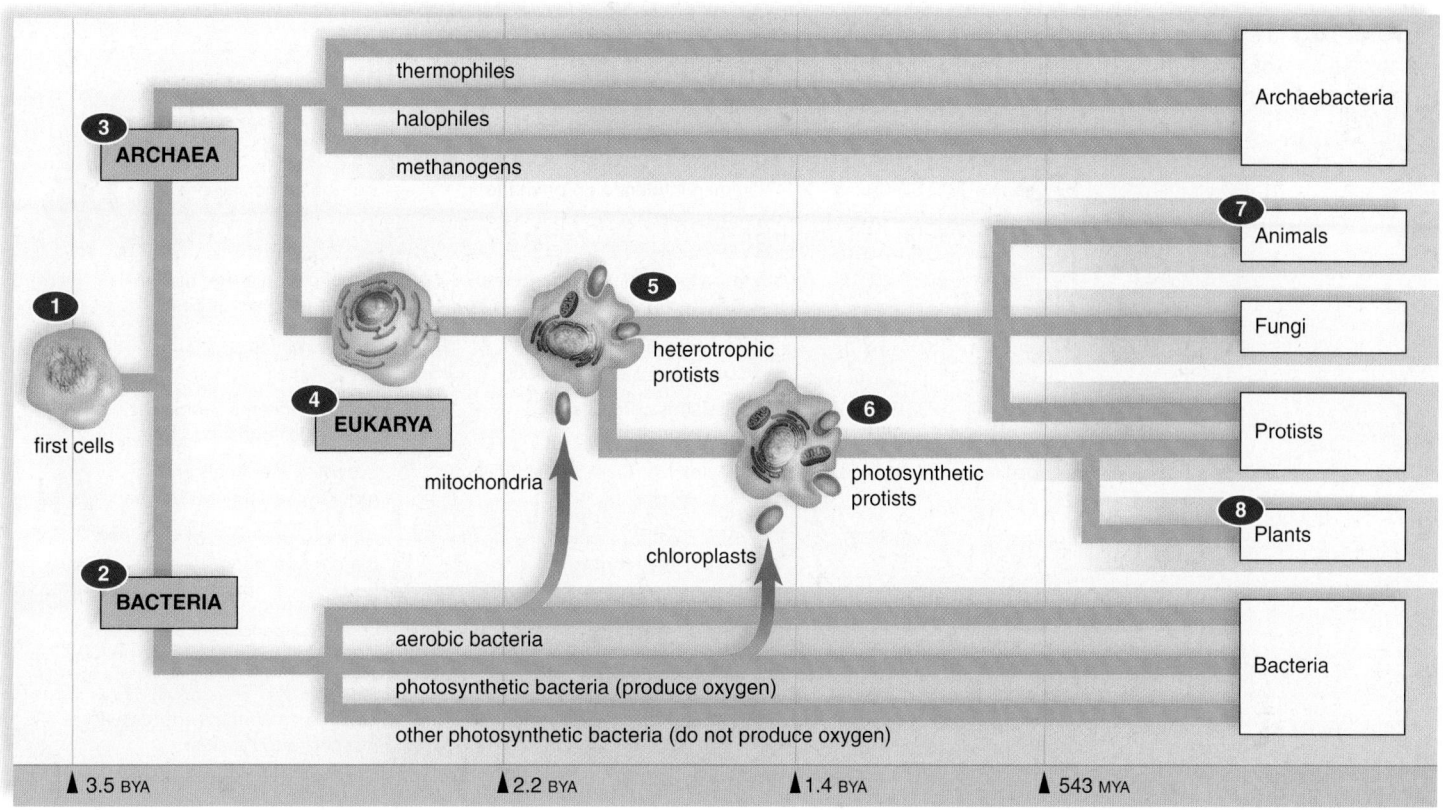

Figure 18.8 The tree of life. During the Precambrian time, ❶ the first cell or cells give rise to ❷ bacteria and ❸ archaea; ❹ the first eukaryotic cell evolves from archaea. ❺ Heterotrophic protists arise when eukaryotic cells gain mitochondria by engulfing aerobic bacteria, and ❻ photosynthetic protists arise when these cells gain chloroplasts by engulfing photosynthetic bacteria. ❼ Animals (and fungi) evolve from heterotrophic protists, and ❽ plants evolve from photosynthetic protists. BYA = billion years ago; MYA = million years ago

accumulation of data from the age of fossils in strata all over the world. Biologists traditionally begin the geologic timescale at the Precambrian time, since this is when life began (Table 18.1). The Precambrian is a very long period of time, comprising about 87% of the geologic timescale. During this time, life arose and the first cells came into existence (Fig. 18.8).

Animation
Geological History of the Earth

The first modern-type cells were probably prokaryotes, which first appeared approximately 3.5 BYA (Table 18.1). Prokaryotes are relatively simple cells—they do not have a nucleus or membrane-bounded organelles. Prokaryotes can live in the most inhospitable of environments, such as hot springs, salty lakes, and oxygen-free swamps—all of which may typify habitats on early Earth. The cell wall, plasma membrane, RNA polymerase, and ribosomes of one lineage of prokaryotes, the archaea, are more like those of eukaryotes than those of bacteria (see Chapter 20).

The first identifiable fossils are those of complex prokaryotes. Chemical fingerprints of complex cells are found in sedimentary rocks from southwestern Greenland, dated at 3.8 BYA. The oldest prokaryotic fossils have been discovered in western Australia. These 3.46-billion-year-old microfossils resemble today's cyanobacteria, prokaryotes that carry on photosynthesis in the same manner as plants. At this time, only volcanic rocks jutted above the waves, and there were as yet no continents. Strange-looking boulders, called **stromatolites,** littered beaches and shallow waters (Fig. 18.9a). Living stromatolites can still be found today along Australia's western coast.

The outer surface of a stromatolite is alive with cyanobacteria. The cyanobacteria in ancient stromatolites added oxygen to the atmosphere (Fig. 18.9b). By 2.0 BYA, the presence of oxygen made most environments unsuitable for anaerobic prokaryotes. Photosynthetic cyanobacteria and aerobic bacteria proliferated as new metabolic pathways evolved. Due to the presence of oxygen, the atmosphere became an oxidizing one instead of a reducing one. Oxygen in the upper atmosphere forms ozone (O_3), which filters out the ultraviolet (UV) rays of the Sun. Before the formation of the **ozone shield,** the amount of ultraviolet radiation reaching the Earth could have helped create organic molecules, but it would have destroyed any land-dwelling organisms. Once the ozone shield was in place, living things were sufficiently protected and able to live on land.

Eukaryotic Cells Arise

The eukaryotic cell, which originated around 2.1 BYA, obtains energy from cellular metabolism in the presence of oxygen, and contains a nucleus as well as other membranous organelles. The eukaryotic cell acquired its organelles gradually. The nucleus may have developed by an invagination of the plasma membrane. The ancestors of the mitochondria in eukaryotic cells were free-living bacteria that synthesized ATP via an electron transport chain, and chloroplasts were free-living photosynthetic prokaryotes. The **endosymbiotic theory** states that a nucleated cell engulfed these prokaryotes, which then became organelles.

Table 18.1 The Geologic Timescale: Major Divisions of Geologic Time and Some of the Major Evolutionary Events of Each Time Period

Era	Period	Epoch	Million Years Ago (MYA)	Plant Life	Animal Life
		Holocene	(0.01–0)	Human influence on plant life	Age of *Homo sapiens*
		Significant Mammalian Extinction			
	Quaternary	Pleistocene	(1.80–0.01)	Herbaceous plants spread and diversify.	Presence of Ice Age mammals. Modern humans appear.
		Pliocene	(5.33–1.80)	Herbaceous angiosperms flourish.	First hominids appear.
		Miocene	(23.03–5.33)	Grasslands spread as forests contract.	Apelike mammals and grazing mammals flourish; insects flourish.
Cenozoic	Tertiary	Oligocene	(33.9–23.03)	Many modern families of flowering plants evolve.	Browsing mammals and monkeylike primates appear.
		Eocene	(55.8–33.9)	Subtropical forests with heavy rainfall thrive.	All modern orders of mammals are represented.
		Paleocene	(65.5–55.8)	Flowering plants continue to diversify.	Primitive primates, herbivores, carnivores, and insectivores appear.
		Mass Extinction: 50% of all Species, Dinosaurs and Most Reptiles			
	Cretaceous		(145.5–65.5)	Flowering plants spread; conifers persist.	Placental mammals appear; modern insect groups appear.
Mesozoic	Jurassic		(199.6–145.5)	Flowering plants appear.	Dinosaurs flourish; birds appear.
		Mass Extinction: 48% of All Species, Including Corals and Ferns			
	Triassic		(251–199.6)	Forests of conifers and cycads dominate.	First mammals appear; first dinosaurs appear; corals and molluscs dominate seas.
		Mass Extinction ("The Great Dying"): 83% of All Species on Land and Sea			
	Permian		(299–251)	Gymnosperms diversify.	Reptiles diversify; amphibians decline.
	Carboniferous		(359.2–299)	Age of great coal-forming forests; ferns, club mosses, and horsetails flourish.	Amphibians diversify; first reptiles appear; first great radiation of insects.
		Mass Extinction: Over 50% of Coastal Marine Species, Corals			
Paleozoic	Devonian		(416–359.2)	First seed plants appear. Seedless vascular plants diversify.	First insects and first amphibians appear on land.
	Silurian		(443.7–416)	Seedless vascular plants appear.	Jawed fishes diversify and dominate the seas.
		Mass Extinction: Over 57% of Marine Species			
	Ordovician		(488.3–443.7)	Nonvascular land plants appear on land.	First jawless and then jawed fishes appear.
	Cambrian		(542–488.3)	Marine algae flourish.	All invertebrate phyla present; first chordates appear.
			630	Soft-bodied invertebrates	
			1,000	Protists diversify.	
Precambrian Time			2,100	First eukaryotic cells	
			2,700	O_2 accumulates in atmosphere.	
			3,500	First prokaryotic cells	
			4,570	Earth forms.	

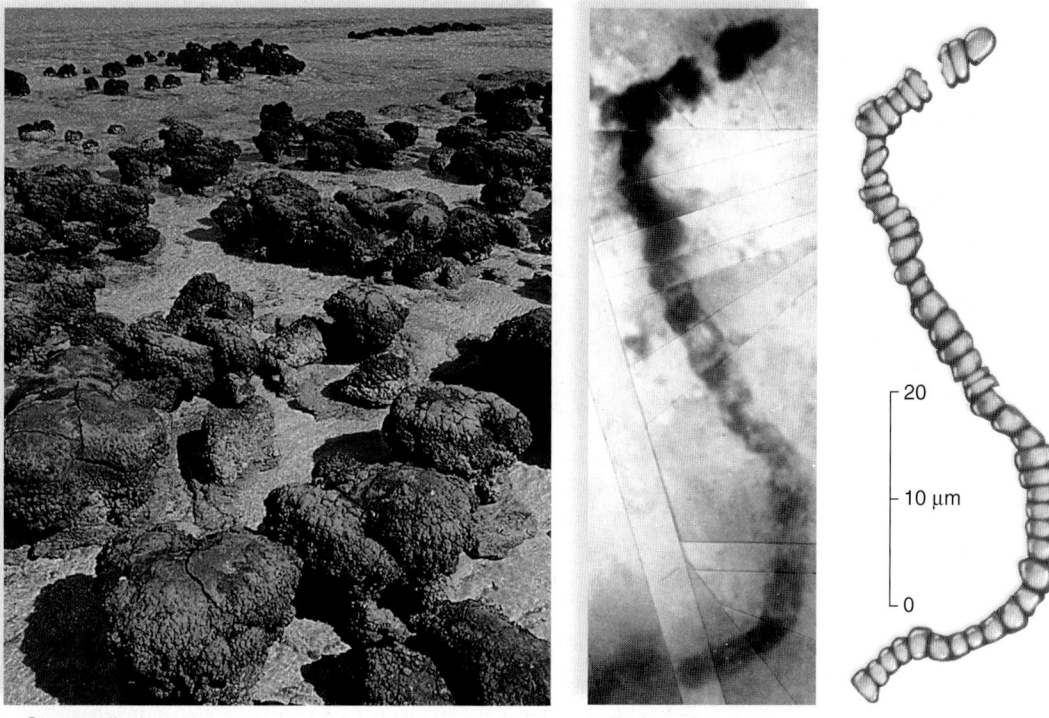

a. Stromatolites   b. *Primaevifilum*

Figure 18.9 Prokaryote fossils of the Precambrian. **a.** Stromatolites date back to 3.46 BYA. Living stromatolites are located in shallow waters off the shores of western Australia and also in other tropical seas. **b.** The prokaryotic microorganism *Primaevifilum* (with interpretive drawing) was found in stromatolites.

The nucleated cell and the engulfed bacteria co-evolved the ability to synthesize ATP via oxidative phosphorylation (see Fig. 18.8). The evidence for the theory is as follows:

Animation
Endosymbiosis

1. The present-day mitochondria and chloroplasts have a size that lies within the range of that for bacteria.
2. Mitochondria and chloroplasts have their own DNA and make some of their own proteins. (The DNA of the nucleus also codes for some of the mitochondrial proteins.)
3. The mitochondria and chloroplasts divide by binary fission, the same as bacteria do.
4. The outer membrane of mitochondria and chloroplasts differ—the outer membrane resembles that of a eukaryotic cell and the inner membrane resembles that of a bacterial cell.

It's been suggested that flagella (and cilia) also arose by endosymbiosis. First, slender undulating prokaryotes could have attached themselves to a host cell to take advantage of food leaking from the host's plasma membrane. Eventually, these prokaryotes adhered to the host cell and became the flagella and cilia we know today. The first eukaryotes were unicellular, as are prokaryotes.

Multicellularity Arises

Fossils of multicellular protists at least as old as 1.4 BYA have been found in arctic Canada. It's possible that the first multicellular organisms practiced sexual reproduction. Among today's protists we find colonial forms in which some cells are specialized to produce gametes needed for sexual reproduction.

Separation of germ cells, which produce gametes, from somatic cells may have been an important first step toward the evolution of the Ediacaran invertebrates, which appeared about 630 MYA (million years ago) and died out about 545 MYA. The first fossils of these organisms were found in the Ediacara Hills of Australia. Since then, similar fossils have been discovered on a number of other continents.

Many of the fossils, dated 630–545 MYA, are thought to be of soft-bodied invertebrates (animals without a vertebral column) that most likely lived on mudflats in shallow marine waters. Some may have been mobile, but others were large, immobile, bizarre creatures resembling spoked wheels, corrugated ribbons, and lettuce-like fronds. All were flat and probably had two tissue layers; few had any type of skeleton (Fig. 18.10). They apparently had no mouths; perhaps they absorbed nutrients from the sea or else had photosynthetic organisms living on their tissues. With few exceptions, this group disappears from the fossil record at 545 MYA, but it may have given rise to modern cnidarians and related animals.

Check Your Progress   18.2A

1. Describe how the half-life of a radioactive isotope can be used to estimate the age of fossils.
2. List the sequence of events in the Precambrian that led to the evolution of heterotrophic and photosynthetic protists.
3. Identify two features of organelles that support the endosymbiotic theory of organelle evolution.

a.   b.

Figure 18.10 Ediacaran fossils. The Ediacaran invertebrates lived from about 600–545 MYA. They were all flat, soft-bodied invertebrates. **a.** Classified as *Spriggina*, this bilateral organism had a crescent-shaped head and numerous segments tapering to a posterior end. **b.** Classified as *Dickinsonia*, these fossils are often interpreted to be segmented worms. However, in the opinion of some, they may be cnidarian polyps.

The Paleozoic Era

The Paleozoic era lasted about 300 million years. Even though the era was quite short compared to the Precambrian, three major mass extinctions occurred during this era (see Table 18.1). An **extinction** is the total disappearance of all the members of a species or higher taxonomic group. **Mass extinctions** are the disappearance of a large number of species or higher taxonomic groups within an interval of just a few million years (see page 343).

Cambrian Animals

The seas of the Cambrian period, which began at about 542 MYA, teemed with invertebrate life (Fig. 18.11). Life became so abundant that scientists refer to this period in Earth's history as the Cambrian explosion. All of today's groups of animals can trace their ancestry to this time, and perhaps earlier, according to new molecular clock data. A *molecular clock* is based on the principle that mutations in certain parts of the genome occur at a fixed rate and are not tied to natural selection. Therefore, the number of DNA base-pair differences tells how long two species have been evolving separately.

Animals that lived during the Cambrian possessed protective outer skeletons known as exoskeletons. These hard-body parts were fossilized more readily than Precambrian soft-bodied organisms. Thus fossils are more abundant during the Cambrian. For example, Cambrian seafloors were dominated by now-extinct trilobites, which had thick, jointed armor covering them from head to tail. Trilobites, a common Cambrian fossil, are classified as arthropods, a major phylum of animals today.

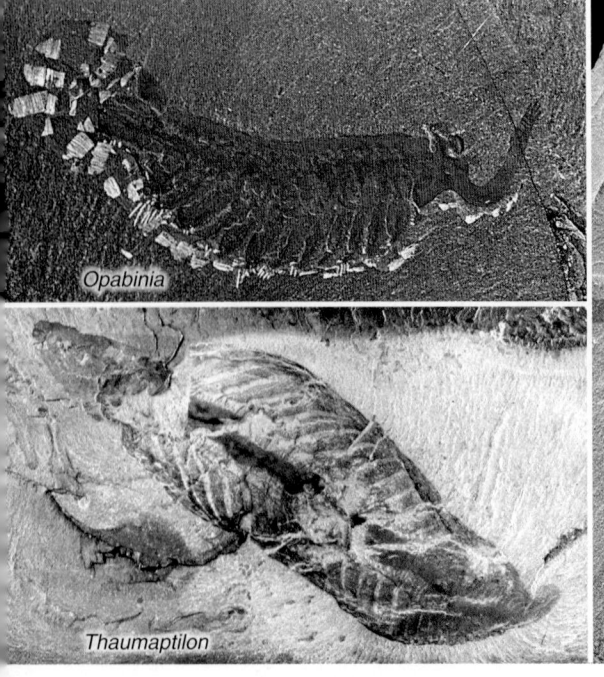

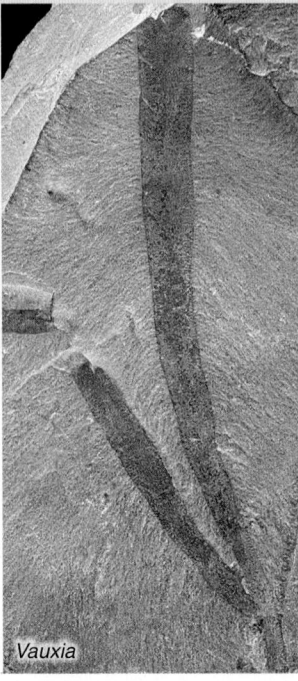

Opabinia

Thaumaptilon

Vauxia

Wiwaxia

Figure 18.11 Sea life of the Cambrian period. The animals depicted here are found as fossils in the Burgess Shale, a formation of the Rocky Mountains of British Columbia, Canada. Some lineages represented by these animals are still evolving today. *Opabinia* has been classified as a crustacean, *Thaumaptilon* a sea pen, *Vauxia* a sponge, and *Wiwaxia* a segmented worm.

b.

a.

c.

Figure 18.12 Swamp forests of the Carboniferous period. **a.** Vast swamp forests of treelike club mosses and horsetails dominated the land during the Carboniferous period (see Table 18.1). The air contained insects with wide wingspans, such as the predecessors to dragonflies shown here, and amphibians lumbered from pool to pool. **b.** Dragonfly fossil from the Carboniferous period. **c.** Modern-day dragonfly.

Invasion of Land

Life first began to move out of the ocean and onto land around 500 MYA. The process of moving from water to land occurred over a long period of time. Different life-forms, such as plants, invertebrates, and vertebrates, had independent histories, and each group colonized the land at different times.

Plants. During the Ordovician period, algae, which were abundant in the seas, most likely began to take up residence in bodies of fresh water. Eventually, algae invaded damp areas on land. The first land plants were nonvascular (did not possess water-conducting tissues), similar to the mosses and liverworts that survive today. The lack of water-conducting tissues limited the height of these plants to a few centimeters. Although the Ordovician evidence is scarce, spore fossils from this time support this hypothesis.

Fossils of seedless vascular plants (those having tissue for water and organic nutrient transport) date back to the Silurian period. They later flourished in the warm swamps of the Carboniferous period. Club mosses, horsetails, and seed ferns were the trees of that time, and they grew to enormous size. A wide variety of smaller ferns and fernlike plants formed an underbrush (Fig. 18.12).

Invertebrates. The jointed appendages and exoskeleton of arthropods are adapted to life on land. Various arthropods—spiders, centipedes, mites, and millipedes—all preceded the appearance of insects on land. Insects, also arthropods, enter the fossil record in the Carboniferous period. One fossil dragonfly from the Carboniferous had a wingspan of nearly a meter. The evolution of wings provided advantages that allowed insects to radiate into the most diverse and abundant group of animals today. Flying provides a way to escape enemies, find food, and disperse to new territories.

Vertebrates. Vertebrates are animals with a vertebral column. The vertebrate line of descent began in the early Ordovician period with the evolution of jawless fishes. Jawed fishes appeared later in the Silurian period. Fishes are ectothermic (see Section 29.3) aquatic vertebrates that have gills, scales, and fins. The cartilaginous and ray-finned fishes made their appearance in the Devonian period, which is called the Age of Fishes.

At this time, the seas contained giant predatory fishes covered with protective armor made of external bone. Sharks cruised up deep, wide rivers, and smaller lobe-finned fishes lived at the river's edge in waters too shallow for large predators. Fleshy fins helped the small fishes push aside debris or hold their place in strong currents, and the fins may also have allowed these fishes to venture onto land and lay their eggs safely in inland pools. Much data tells us that lobe-finned fishes were ancestral to the amphibians and to modern-day lobe-finned fishes.

Amphibians are thin-skinned vertebrates that are not fully adapted to life on land, particularly because they must return to water to reproduce. The Carboniferous swamp forests provided the water they needed, and amphibians adaptively radiated into many different sizes and shapes. Some superficially resembled alligators and were covered with protective scales; others were small and snakelike; and a few were larger plant-eaters. The largest measured 6 m from snout to tail. The Carboniferous period is called the Age of Amphibians.

Video
Amphibian Origin

The process that turned the great Carboniferous forests into the coal we use today to fuel our modern society started during the Carboniferous period. The weather turned cold and dry, and this brought an end to the Age of Amphibians. A major mass extinction event occurred at the end of the Permian period, bringing an end to the Paleozoic era and setting the stage for the Mesozoic era.

Check Your Progress 18.2B

1. List the major events in the history of life that occurred in the Precambrian.
2. Identify the feature that evolved in animals in the Cambrian period that contributed to an abundant fossil record.

Figure 18.13 Dinosaurs of the late Cretaceous period.   *Parasaurolophus walkeri*, although not as large as other dinosaurs, was one of the largest plant-eaters of the late Cretaceous period. Its crest atop the head was about 2 m long, and may have been used as a way to regulate body temperature or as a means to communicate by making booming calls. Also living at this time were the rhinolike dinosaurs represented here by *Triceratops* (left), another herbivore.

The Mesozoic Era

Although a severe mass extinction occurred at the end of the Paleozoic era, the evolution of certain types of plants and animals continued into the Triassic, the first period of the Mesozoic era. Nonflowering seed plants (collectively called gymnosperms), which had evolved and then spread during the Paleozoic, became dominant. Cycads are short and stout with palmlike leaves, and they produce large cones. Cycads and related plants were so prevalent during the Triassic and Jurassic periods that these periods are sometimes called the Age of Cycads. Reptiles can be traced back to the Permian period of the Paleozoic era. Unlike amphibians, reptiles can thrive in a dry climate because they have scaly skin and lay a shelled egg that hatches on land. Reptiles underwent an adaptive radiation during the Mesozoic era to produce forms that lived in the sea and on the land, and that could fly. One group of reptiles, the therapsids, had several mammalian skeletal traits.

During the Jurassic period, large flying reptiles called pterosaurs ruled the air, and giant marine reptiles with paddle-like limbs ate fishes in the sea. But on land, dinosaurs dominated, and the evolving mammals remained small and less conspicuous.

Although the average size of the dinosaurs was about that of a crow, many giant species developed. The gargantuan *Apatosaurus* and the armored, tractor-sized *Stegosaurus* fed on cycad seeds and conifer trees. The size of a dinosaur such as *Apatosaurus* is hard for us to imagine. It was 4.5 m tall at the hips and 27 m in length, and weighed about 40 tons. How might dinosaurs have benefited from being so large? One hypothesis is that, being ectothermic (cold-blooded), the surface-area-to-volume-ratio was favorable for retaining heat. Some data also suggest that dinosaurs were endothermic (warm-blooded).

During the Cretaceous period, great herds of rhinolike dinosaurs, *Triceratops*, roamed the plains, as did *Tyrannosaurus rex*, which was a carnivore, perhaps a part-time scavenger, filling the same ecological role as lions do today. *Parasaurolophus* was a long-crested, duck-billed dinosaur (Fig. 18.13). The long, hollow crest was bigger than the rest of its skull and may have functioned as a thermoregulatory organ, or as a resonating chamber for making booming calls used during mating or to help members of a herd locate each other. In comparison to *Apatosaurus*, *Parasaurolophus* was small. It was less than 3 m tall at the hips and weighed only about 3 tons. Still, it was one of the largest plant-eaters of the late Cretaceous period and fed on pine needles, leaves, and twigs. *Parasaurolophus* was easy prey for large predators; its main defense would have been running away in large herds.

At the end of the Cretaceous period, the dinosaurs became victims of a mass extinction, which is discussed on page 343.

One group of bipedal dinosaurs, called theropods, includes the Tyrannosaurs and various raptors, such as *Velociraptor*. This group most likely gave rise to the birds, whose fossil record includes the famous *Archaeopteryx*.

Up until 1999, Mesozoic mammalian fossils largely consisted of teeth. This changed when a fossil found in China was dated at 120 MYA and named *Jeholodens*. The animal, identified as a mammal, apparently looked like a long-snouted rat. Surprisingly, *Jeholodens* had sprawling hindlimbs as do reptiles, but its forelimbs were under the belly, as in today's mammals.

Small mammals appeared early in the Mesozoic and were very diverse in the Jurassic and early Cretaceous periods. By the end of the Cretaceous, the common ancestors of each of the modern orders of mammals had evolved. However, by the late Cretaceous, the majority of the Jurassic and early Cretaceous mammalian diversity had gone extinct.

Figure 18.14 Mammals of the Oligocene epoch. The artist's representation of these mammals and their habitat vegetation is based on fossil remains.

Figure 18.15 Woolly mammoth of the Pleistocene epoch. Woolly mammoths were animals that lived along the borders of continental glaciers.

The Cenozoic Era

Classically, the Cenozoic era is divided into two periods, the Tertiary period and the Quaternary period. Another scheme, dividing the Cenozoic into the Paleogene and the Neogene periods, is gaining popularity. This new system divides the epochs differently. In any case, we are currently living in the Holocene epoch.

Mammalian Diversification

At the end of the Mesozoic era, mammals began an adaptive radiation into the many habitats now left vacant by the demise of the dinosaurs. Mammals are endotherms, and they have hair, which helps keep body heat from escaping. Their name refers to the presence of mammary glands, which produce milk to feed their young. Mammals were very diverse in the Mesozoic, but by the start of the Paleocene epoch, the majority of Mesozoic mammal diversity had gone extinct.

By the end of the Eocene epoch, all of the modern orders of mammals had evolved—representatives of modern mammal lineages were the only mammals left in existence. Mammals underwent adaptive radiation into a number of environments. Several species of mammals, including the bats, took to the air. Ancestors of modern whales, dolphins, and manatees began their return to an aquatic lifestyle. On land, herbivorous hoofed mammals populated the forests and grasslands and were preyed upon by carnivorous mammals. Many of the types of herbivores and carnivores of the Oligocene epoch are extinct today (Fig. 18.14).

Evolution of Primates

The ancestors of modern primates appeared during the Eocene epoch about 55 MYA. The first primates were small, squirrel-like animals. Flowering plants (collectively called angiosperms) were already diverse and plentiful by the Cenozoic era, and many species of tree-sized angiosperms had evolved by the Eocene. Some primates are adapted to living in trees, where protection from predators and food in the form of fruit is plentiful.

Ancestral apes appeared during the Oligocene epoch. These primates were adapted to living in the open grasslands and savannas. Apes diversified during the Miocene and Pliocene epochs and gave rise to the first hominids, a group that includes humans. Many of the skeletal differences between apes and humans relate to the fact that humans walk upright. Exactly what caused humans to adopt bipedalism is still being debated.

The world's climate became progressively colder during the Tertiary period. The Quaternary period begins with the Pleistocene epoch, which is known for multiple ice ages in the Northern Hemisphere. During periods of glaciation, snow and ice covered about one-third of the land surface of the Earth. The Pleistocene epoch was an age of not only humans, but also mammalian **megafauna**—giant ground sloths, beavers, wolves, bison, woolly rhinoceroses, mastodons, and mammoths (Fig. 18.15). Humans have survived, but what happened to the oversized mammals just mentioned? Mammalian megafauna had died out by the end of the Pleistocene. One hypothesis is that human hunting was at least partially responsible for the extinction of these animals. Climate change is another hypothesis.

Check Your Progress   18.2C

1. Describe the major events in the history of life that occurred in the Mesozoic.
2. List the groups of plants and animals that were abundant during the Mesozoic era.

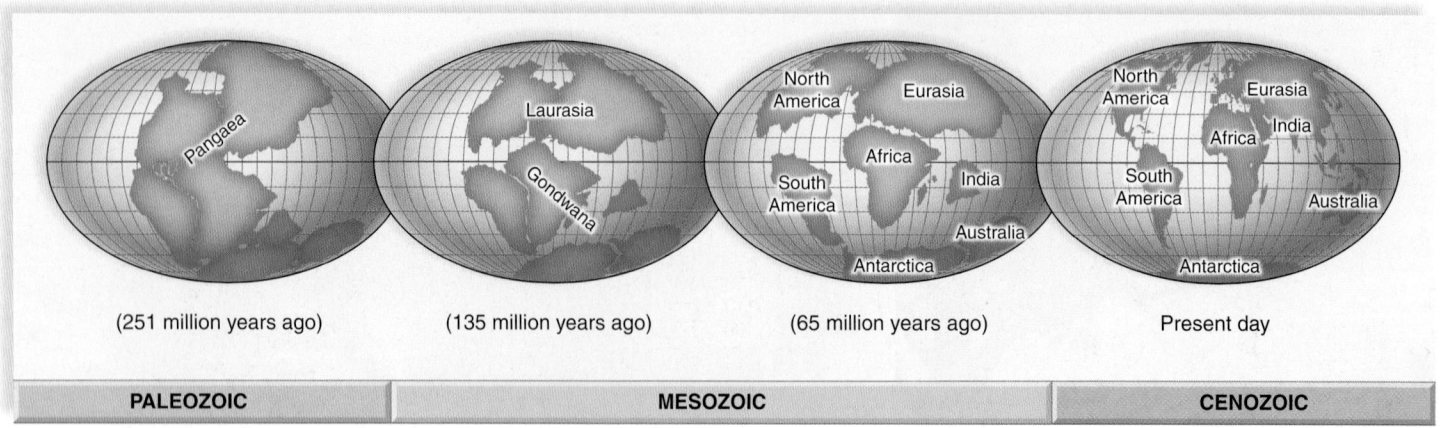

(251 million years ago)   (135 million years ago)   (65 million years ago)   Present day

PALEOZOIC	MESOZOIC	CENOZOIC

Figure 18.16 Continental drift.   About 251 MYA, all the continents were joined into a supercontinent called Pangaea. During the Mesozoic era, the joined continents of Pangaea began moving apart, forming two large continents called Laurasia and Gondwana. Then all the continents began to separate. Presently, North America and Europe are drifting apart at a rate of about 2 cm per year.

18.3 Geological Factors That Influence Evolution

Learning Outcomes

Upon completion of this section, you should be able to

1. Identify who proposed the theory of continental drift, and the evidence provided to support this theory.
2. Describe plate tectonics and how it explains the drifting of continents.
3. Interpret correlated biogeographic and geologic evidence in support of continental drift.
4. Describe the geological points at which mass extinctions occurred and give one proposed cause for each extinction event.

In the past, it was thought that the Earth's crust was immobile, that the continents had always been in their present positions, and that the ocean floors were only a catch basin for the debris that washed off the land. But in 1920, Alfred Wegener, a German meteorologist, presented data from a number of disciplines to support his hypothesis of continental drift.

Continental Drift

Continental drift was finally confirmed in the 1960s, establishing that the continents are not fixed; instead, their positions and the positions of the oceans have changed over time (Fig. 18.16). During the Paleozoic era, the continents joined to form one supercontinent that Wegener called Pangaea [Gk. *pangea*, all lands]. First, Pangaea divided into two large subcontinents, called Gondwana and Laurasia, and then these also split to form the continents of today. Presently, the continents are still drifting in relation to one another.

Continental drift explains why the coastlines of several continents are roughly mirror images of each other—for example, the outline of the west coast of Africa matches that of the east coast of South America. The same geological structures are also found in many of the areas where the continents touched. For example, based on geological data, a single mountain range runs through South America, Antarctica, and Australia.

Continental drift also explains the unique distribution patterns of several fossils. Fossils of the seed ferns (*Glossopteris*) have been found on all the continents, which supports that all continents were once joined (Fig. 18.17). Similarly, the fossil reptile *Cynognathus* is found in Africa and in South America, and an assemblage of early mammal-like reptiles, the *Lystrosaurus* group, was discovered in Antarctica, far from Africa and southeast Asia, where it also occurs.

With mammalian fossils, the situation is different: Australia, South America, and Africa each share lineages of mammals that were widely distributed across Gondwana. For example, each continent has representatives of the rodent lineage. However, each continent also has its own unique mammal lineages that evolved in isolation after the continents separated.

The mammalian biological diversity of today's world is the result of isolated evolution on separate continents. For example, as mentioned in Chapter 15, marsupials are endemic only to South America and Australia, which were connected when the

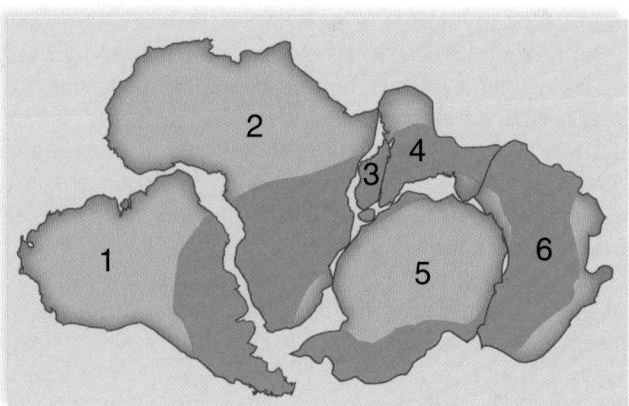

Figure 18.17. Biogeography of *Glossopteris* supports continental drift.   Fossils of the ferns *Glossopteris* are distributed throughout what was Gondwana land.

landmasses were a supercontinent (Fig. 18.16). The fossil record suggests that marsupials were abundant in South America up until about 7 MYA. Why are marsupials now prevalent in the Australian biogeographic region but not as diverse in South America? Most likely, marsupials started evolving in the Americas and were able to reach Australia when the southern continents were still joined. Once Australia separated off, marsupials were able to diversify without competition from placental mammals that evolved elsewhere. Today, Australia has no native placental mammals except for bats, which were introduced to the continent relatively recently. In contrast, placental mammals are prevalent in the Americas, and much of the marsupial diversity has disappeared on this continent.

Plate Tectonics

Why do the continents drift? An answer has been suggested through a branch of geology known as **plate tectonics** [Gk. *tektos*, fluid, molten, able to flow], which says that the Earth's crust is fragmented into slablike plates that float on a lower, hot mantle layer. The continents and the ocean basins are a part of these rigid plates, which move like conveyor belts. At ocean ridges, seafloor spreading occurs as molten mantle rock rises and material is added to the ocean floor. Seafloor spreading causes the continents to move a few centimeters a year on the average. At *subduction zones*, the forward edge of a moving plate sinks into the mantle and is destroyed, forming deep-ocean trenches bordered by volcanoes or volcanic island chains.

The Earth isn't getting bigger or smaller, so the amount of oceanic crust being formed is as much as that being destroyed.

When two continents collide, the result is often a mountain range; for example, the Himalayas resulted when India collided with Eurasia. The place where two plates meet and scrape past one another is called a *transform boundary*. The San Andreas fault in Southern California is at a transform boundary, and the movement of the two plates is responsible for the many earthquakes in that region. Earthquakes leave visible evidence of the movement of plates at transform boundaries. In rare cases the movement of the plates has actually been witnessed by people as it occurs—roads diverge, the ground uplifts, and cracks emerge during seismic activity. Examples of this were seen in the massive earthquake that took place in Japan in 2011.

Animation Plate Tectonics

Mass Extinctions

At least five mass extinctions have occurred throughout history: at the ends of the Ordovician, Devonian, Permian, Triassic, and Cretaceous periods (Fig. 18.18; see Table 18.1). Is a mass extinction due to some cataclysmic event, or is it a more gradual process brought on by environmental changes, including tectonic, oceanic, and climatic fluctuations?

This question was brought to the fore when Walter and Luis Alvarez proposed in 1977 that the Cretaceous extinction, during which the dinosaurs died out, was due to a bolide. A **bolide** is a large, crater-forming projectile that impacts Earth. The Alvarezes found that Cretaceous clay contains an abnormally high level of iridium, an element that is rare in the Earth's crust but more common in asteroids and meteorites. Walter Alvarez stated that the "10^8-megaton impact of the comet (or asteroid)

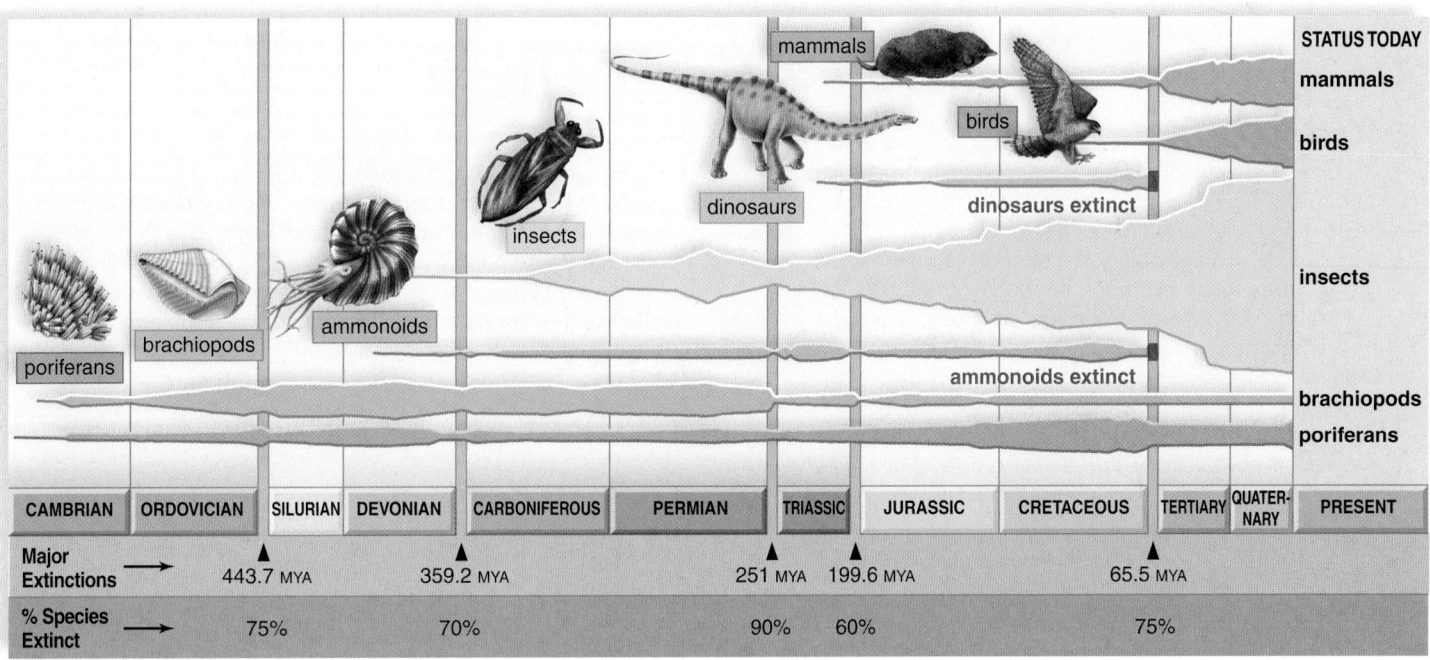

Figure 18.18 Mass extinctions.   Five significant mass extinctions and their effects on the abundance of certain forms of marine and terrestrial life. The thickness of the horizontal colored bars indicates the varying abundance of each life-form considered. MYA=million years ago

which ended the Cretaceous was . . . the equivalent of the explosion of 10,000 times the entire nuclear arsenal of the world."[1]

An impact of this magnitude would have blasted into the atmosphere a layer of ash and soot so expansive it would have blocked the Sun. A layer of soot has been identified in the strata alongside the iridium, and a huge crater, called the Chicxulub crater, that could have been caused by a meteorite impact of this magnitude was found in the Caribbean–Gulf of Mexico region on the Yucatán peninsula. The Chicxulub crater has been precisely dated to when the Cretaceous period ended and the Tertiary period began.

Certainly, continental drift contributed to the Ordovician extinction. This extinction occurred after Gondwana arrived at the South Pole. Immense glaciers, which drew water from the oceans, chilled even once-tropical land. Marine invertebrates and coral reefs, which were especially hard hit, didn't recover until Gondwana drifted away from the pole and warmth returned.

The mass extinction at the end of the Devonian period saw an end to 70% of marine invertebrates. Helmont Geldsetzer of Canada's Geological Survey notes that iridium has also been found in Devonian rocks in Australia, suggesting it's possible that a bolide event was involved. Some scientists believe that this mass extinction could have been due to movement of Gondwana back to the South Pole.

The extinction at the end of the Permian period was quite severe; 90% of species disappeared. The latest hypothesis attributes the Permian extinction to excess carbon dioxide. When

---

1 Alvarez, Walter. 2008. *T. rex and the Crater of Doom.* (Princeton University Press, Princeton, N.J.)

Pangaea formed, there were no polar ice caps to initiate ocean currents. The lack of ocean currents caused organic matter to stagnate at the bottom of the ocean. Then, as the continents drifted into a new configuration, ocean circulation switched back on. Now, the extra carbon on the seafloor was swept up to the surface where it became carbon dioxide, a deadly gas for sea life. The trilobites became extinct, and the crinoids (sea lilies) barely survived. Excess carbon dioxide on land led to climate change that altered the pattern of vegetation. Areas that were wet and rainy became dry and warm, and vice versa. Burrowing animals that could escape land surface changes seemed to have the best chance of survival.

The extinction at the end of the Triassic period is another that has been attributed to the environmental effects of a meteorite collision with Earth. Central Quebec has a crater half the size of Connecticut that some believe is the impact site. The dinosaurs may have benefited from this event because this is when the first of the gigantic dinosaurs took charge of the land. A second wave occurred in the Cretaceous period, but it ended in dinosaur extinction as discussed previously.

Check Your Progress   18.3

1. Explain why continents drift, and summarize the geological evidence to support plate tectonics.
2. Defend the theory of continental drift with evidence from the biogeographic distribution of organisms.
3. List the mass extinction events that have occurred on Earth, and describe the hypotheses that explain the cause of each.

CONNECTING *the* CONCEPTS *with the* BIG IDEAS

Evolution

- "Chemical evolution" involves the formation of organic monomers, and then polymers, from inorganic elements present on early Earth. (ME2, 1D1a1-4, 1D2a2)
- Metabolism and the ability to self-replicate were necessary for biological evolution to begin. Protocells, true cells, and multicellular organisms followed, shaped by natural selection. (1D1a3)
- Only one lineage, the LUCA (last universal common ancestor), gave rise to all life on Earth. (1D2b)
- The ability to measure the rate of radioisotope decay has allowed us to calculate the absolute age of many fossils and set the age of Earth at around 4.6 billion years. (1A4b1, 1D2a1)
- The fossil record provides evidence for life from 3.5 BYA in the Precambrian to current life in the Cenozoic. (1D2a1)

- Biogeography is used to reconstruct the evolutionary history of life; plate tectonics indicate that continents drift over geologic time, and the distribution of organisms correlates with plate movement. (1C2a)
- Extinction, the disappearance of species from Earth, often occurs during times of great ecological change with documentation of at least five mass extinctions being found in the fossil record. (1C1b*IE*)

---

*Find the unabridged version of all EK citations at www.glencoe.com/maderAP11.

Media Study Tools

www.glencoe.com/maderAP11

Enhance your study of this chapter with study tools and practice tests. Also ask your instructor about the resources available through ConnectPlus, including the media-rich eBook, interactive learning tools, and animations.

Summarize

18.1 Origin of Life

The unique conditions of the early Earth allowed a chemical evolution to occur. An abiotic synthesis of small organic molecules such as amino acids and nucleotides occurred, possibly either in the atmosphere or at hydrothermal vents. These monomers joined together to form polymers either on land (warm seaside rocks or clay) or at the vents. The first polymers could have been proteins or RNA, or they could have evolved together.

The aggregation of polymers inside a fatty acid plasma membrane produced a protocell having some enzymatic properties such that it could grow. If the protocell developed in the ocean, it was a heterotroph; if it developed at hydrothermal vents, it was a chemoautotroph. A true cell had evolved once the protocell contained DNA genes. The ability to synthesize ATP, the source of energy for cellular work, likely evolved early in the history of life. The first genes may have been RNA molecules, but later DNA became the information storage molecule of heredity. Biological evolution began with the origin of self-replicating molecules such as DNA and RNA.

18.2 History of Life

The fossil record allows us to trace the history of life. The oldest prokaryotic fossils are cyanobacteria, dated about 3.5 BYA, and they were the first organisms to add oxygen to the atmosphere. The eukaryotic cell evolved about 2.2 BYA, but multicellular animals (the Ediacaran animals) do not occur until 600 MYA.

A rich animal fossil record starts at the Cambrian period of the Paleozoic era. The evolution of the exoskeleton seems to explain the increased number of fossils at this time. Predation pressures may also have selected for the evolution of a hard external enclosure like an exoskeleton. The fishes were the first vertebrates to diversify and become dominant. Amphibians are descended from lobe-finned fishes.

Plants invaded the land from water during the Ordovician period. The swamp forests of the Carboniferous period contained seedless vascular plants, insects, and amphibians. This period is sometimes called the Age of Amphibians.

The Mesozoic era was the Age of Cycads and Reptiles. Mammals and birds evolved from reptilian ancestors. During this era, dinosaurs of enormous size were present. By the end of the Cretaceous period, the dinosaurs were extinct.

The Cenozoic era is divided into the Tertiary period and the Quaternary period. The Tertiary is associated with the adaptive radiation of mammals and flowering plants that formed vast tropical forests. The Quaternary is associated with the evolution of primates; first monkeys appeared, then apes, and then humans. Grasslands

were replacing forests, and this put pressure on primates, who were adapted to living in trees, to respond to an ever-expanding terrestrial habitat. The result may have been the evolution of humans—primates who left the trees.

18.3 Geological Factors That Influence Evolution

The continents are on massive plates that move, carrying the land with them. Plate tectonics is the study of the movement of the plates. Continental drift helps explain the distribution pattern of today's land organisms.

Mass extinctions have played a dramatic role in the history of life. It has been suggested that the extinction at the end of the Cretaceous period was caused by the impact of a large meteorite, and evidence indicates that other extinctions have a similar cause as well. It has also been suggested that tectonic, oceanic, and climatic fluctuations, particularly due to continental drift, can bring about mass extinctions.

Key Terms

abiogenesis 329	microsphere 330
abiotic synthesis 329	ozone shield 335
absolute dating (of fossils) 334	paleontology 333
biomolecules 328	plate tectonics 343
bolide 343	"primordial soup"
coacervate droplet 331	hypothesis 329
continental drift 342	protein-first hypothesis 330
endosymbiotic theory 335	proteinoid 330
extinction 338	protobiont (protocell) 330
fossil 333	radiocarbon dating 334
geologic timescale 334	relative dating (of fossils) 334
index fossil 334	RNA-first hypothesis 330
iron-sulfur world 329	sediment 333
last universal common ancestor	sedimentation 333
(LUCA) 328	stratum 333
liposome 331	stromatolite 335
mass extinction 338	vesicle 331
megafauna 341	
membrane-first	
hypothesis 331	
micelle 331	

Assess

Reviewing This Chapter

1. List and describe the various hypotheses concerning the chemical evolution that produced polymers. 328–30
2. Trace in general the steps by which the protocell may have evolved from polymers. 330–32
3. List and describe the various hypotheses concerning the origin of a self-replication system. 332
4. Explain how the fossil record develops and how fossils are dated relatively and absolutely. 333–34
5. When did prokaryotes arise, and what are stromatolites? 334–35
6. When and how might the eukaryotic cell have arisen? 335–37
7. Describe the first multicellular animals found in the Ediacara Hills in southern Australia. 337
8. Why might there be so many fossils from the Cambrian period? 338
9. Which plants, invertebrates, and vertebrates were present on land during the Carboniferous period? 339

10. Which type vertebrate was dominant during the Mesozoic era? Which types began evolving at this time?   340

11. Which type of vertebrate underwent an adaptive radiation in the Cenozoic era?   341

12. What is continental drift, and how is it related to plate tectonics? Give examples to show how biogeography supports the occurrence of continental drift.   342–43

13. Identify five significant mass extinctions during the history of the Earth. What may have caused mass extinctions?   343–44

Testing Yourself

Choose the best answer for each question.

For questions 1–6, match the statements with events in the key. Answers may be used more than once.

KEY:

a. early Earth   d. protocell evolves
b. monomers evolve   e. self-replication system evolves
c. polymers evolve

1. The heat of the Sun could have caused amino acids to form proteinoids.

2. In a liquid environment, phospholipid molecules automatically form a membrane.

3. As the Earth cooled, water vapor condensed, and subsequent rain produced the oceans.

4. Miller's experiment shows that under the right conditions, inorganic chemicals can react to form small organic molecules.

5. Some investigators believe that RNA was the first nucleic acid to evolve.

6. An abiotic synthesis may have occurred at hydrothermal vents.

7. Which of these did Stanley Miller place in the experimental system to show that organic monomers could have arisen from inorganic molecules on the early Earth?
   a. microspheres   d. only RNA
   b. purines and pyrimidines   e. All of these are correct.
   c. early atmospheric gases

8. Which of these is the chief reason the protocell was probably a fermenter?
   a. The protocell didn't have any enzymes.
   b. The atmosphere didn't have any oxygen.
   c. Fermentation provides the most energy.
   d. There was no ATP yet.
   e. All of these are correct.

9. Evolution of the DNA ⟶ RNA ⟶ protein system was a milestone because the protocell could now
   a. be a heterotrophic fermenter.
   b. pass on genetic information.
   c. use energy to grow.
   d. take in preformed molecules.
   e. All of these are correct.

10. Fossils
   a. are the remains and traces of past life.
   b. can be dated absolutely according to their location in strata.
   c. are usually remains of hard tissues such as bone.
   d. have been found for all types of animals except humans.

11. Which of these events did not occur during the Precambrian?
   a. evolution of the prokaryotic cell
   b. evolution of the eukaryotic cell
   c. evolution of multicellularity
   d. evolution of the first animals
   e. All of these occurred during the Precambrian.

For questions 12–16, match the phrases with divisions of geologic time in the key. Answers may be used more than once.

KEY:

a. Cenozoic era   c. Paleozoic era
b. Mesozoic era   d. Precambrian time

12. dinosaur diversity, evolution of birds and mammals

13. contains the Carboniferous period

14. prokaryotes abound; eukaryotes evolve and become multicellular

15. mammalian diversification

16. invasion of land

17. Which of these occurred during the Carboniferous period?
   a. Dinosaurs evolved twice and became huge.
   b. Human evolution began.
   c. The great swamp forests contained insects and amphibians.
   d. Prokaryotes evolved.
   e. All of these are correct.

18. Continental drift helps explain
   a. mass extinctions.
   b. the distribution of fossils on the Earth.
   c. geological upheavals such as earthquakes.
   d. climatic changes.
   e. All of these are correct.

19. Which of these pairs is mismatched?
   a. Mesozoic—cycads and dinosaurs
   b. Cenozoic—grasses and humans
   c. Paleozoic—rise of prokaryotes and unicellular eukaryotes
   d. Cambrian—marine organisms with external skeletons
   e. Precambrian—origin of the cell at hydrothermal vents

20. Which statement is not correct? The tree of life shows that
   a. endosymbiotic events can account for at least some of the organelles in a eukaryotic cell.
   b. evolution proceeds from the simple to the complex.
   c. both plants and animals can trace their ancestry to the protists.
   d. humans hold a special place in the evolution of animals.
   e. the prokaryotic cell preceded the eukaryotic cell.
   f. All of these are correct.

Engage

Thinking Scientifically

1. You were asked to supply an evolutionary tree of life and decided to use Figure 18.8. How is this tree consistent with evolutionary principles?

2. Explain the occurrence of living fossils, such as horseshoe crabs, that closely resemble their ancestors known from the fossil record.

3. Illustrate the evolution of the first cellular membrane (see http://exploringorigins.org/fattyacids.html).

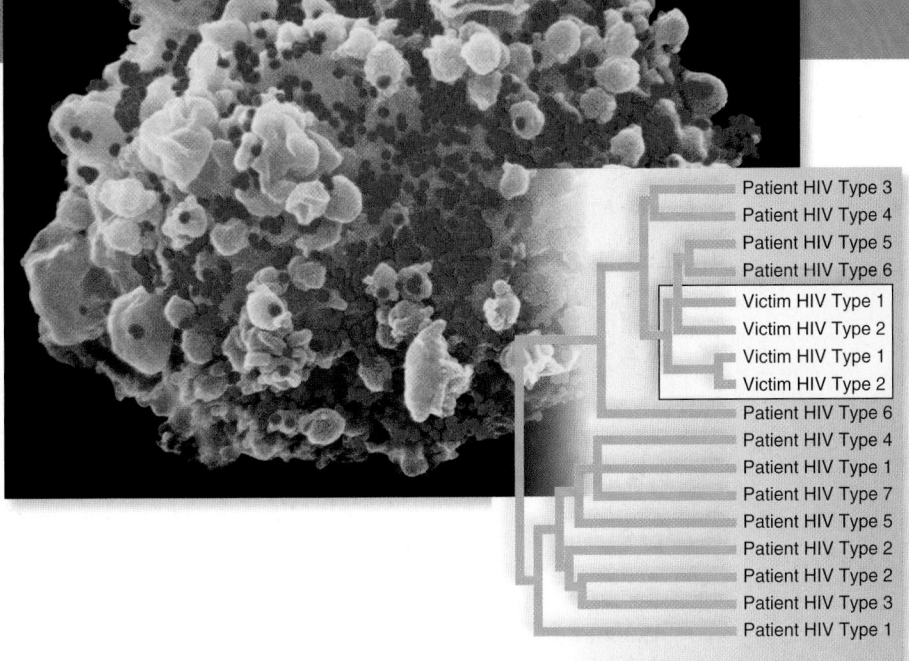

Taxonomy, Systematics, and Phylogeny

Application of phylogeny to forensics in a case of attempted murder by HIV infection. *Left.* T4 cell with HIV (red dots) on the cell surface. *Right.* The HIV lineages from the doctor's victim (small box) are nested within, and thus share a common ancestor with, those of the doctor's patient (large box).

Molecular technology offers a powerful new set of tools to address issues important to human society. The ability to sequence segments of an organism's DNA has helped to address problems in conservation, agriculture, medicine, and forensic science. In the courtroom, a segment of a gene from HIV was used to convict a gastroenterologist for attempted murder of his girlfriend by purposely injecting her with HIV disguised in a vitamin B_{12} shot. A phylogeny, or "evolutionary tree," of the HIV gene traced the girlfriend's HIV to one of the doctor's patients with HIV/AIDS. The doctor was convicted to a 50-year prison sentence for attempted murder. Systematic biology uses phylogenies, in a similar way, to trace the origin of biodiversity back to a common ancestor. In this chapter we are introduced to how systematic biologists use traits to reconstruct evolutionary history and classify organisms according to shared ancestry.

As you read through the chapter, think about the following questions:

1. How are the study of macroevolution and systematic biology interrelated?

2. Why is it important to classify biodiversity, and why is evolutionary history important to this classification?

3. What is a phylogeny? What does it represent, and what can one tell you about the history of life on Earth?

CHAPTER OUTLINE

19.1 Systematic Biology 348

19.2 The Three-Domain System 351

19.3 Phylogeny 354

BEFORE YOU BEGIN

Before beginning this chapter, take a few moments to review the following discussions.

Section 1.3 What is biodiversity?

Section 12.3 Why is DNA called the "genetic code of life?"

Chapter 17 Evolution feature "The Anatomy of Speciation." How do microevolutionary processes lead to macroevolutionary change?

FOLLOWING *the* BIG IDEAS

CHAPTER 19 TAXONOMY, SYSTEMATICS, AND PHYLOGENY

Evolution	Macroevolution explains how new species originate. Systematic biology reconstructs "a tree" of common ancestors and their descendants, investigating the evolutionary relationships of living things.

19.1 Systematic Biology

Learning Outcomes

Upon completion of this section, you should be able to

1. Differentiate between taxonomy, classification, and systematic biology.
2. Reconstruct the levels of the Linnaean classification hierarchy.
3. Identify the genus and species of an organism from its scientific name.

In Chapter 17 we examined macroevolution, or evolutionary change that results in the formation of new species. Macroevolution is the source of past and present biodiversity, and **systematic biology,** or **systematics,** is the study of the history of biodiversity. **Systematic biology** is a quantitative science that uses characteristics of living and fossil organisms, or **traits,** to infer the relationships among organisms over time.

The Field of Taxonomy

Taxonomy [Gk. *tasso*, arrange, classify, and *nomos*, usage, law] is the branch of systematic biology that identifies, names, and organizes biodiversity into related categories (Fig. 19.1). A **taxon** (pl. taxa) is the general name for a group containing an organism or group of organisms that exhibit a set of shared traits. **Classification** is the process of naming and assigning organisms or groups of organisms to a taxon. For example, the taxon Vertebrata contains organisms with a bony spinal column. As another example, the taxon Canidae contains the wolf (*Canis lupus*) and its close relative, the domestic dog (*Canis familiaris*).

Taxonomists, scientists that study taxonomy, strive to classify all living things on Earth. The methods used to classify living things have changed throughout history. The famous Greek philosopher Aristotle (384–322 BCE) was interested in taxonomy, and he sorted organisms into a particular group, such as horses, birds, and oaks, based on a set of shared traits. Similarly, early taxonomists after Aristotle relied on physical traits to classify organisms. This method proved to be problematic because many features of organisms were similar, not because they share a common ancestor, but because of convergent evolution (see Chapter 17). For example, animals that have wings could be grouped into a single taxonomic group, but birds, bats, and beetles, all of which have wings, are profoundly different in many other ways.

Today, taxonomists attempt to classify organisms into **natural groups,** those groupings of organisms that represent a shared evolutionary history. Natural groups are classified by using a set of traits to construct a **phylogeny,** or evolutionary "family tree," that represents the evolutionary history of taxa. This evolutionary history is then used to classify taxa based on shared ancestry.

Figure 19.1 Classifying organisms. How would you name and classify these organisms? After naming them, how would you assign each to a particular group? Based on what criteria? An artificial system would not take into account how they might be related through evolution, as would a natural system.

b. *Lilium canadense*

c. *Lilium bulbiferum*

Figure 19.2 Carolus Linnaeus. **a.** Linnaeus was the father of taxonomy and devised the binomial system of naming and classifying organisms. His original name was Karl von Linne, but he later latinized it because of his fascination with scientific names. Linnaeus was particularly interested in classifying plants. **b, c.** Each of these two lilies are species in the same genus, *Lilium*.

a.

Advances in DNA technology allow modern systematic biologists to compare traits other than external features to classify organisms. For example, a phylogeny of animals constructed from DNA sequences clearly shows that beetles, birds, and bats have wings that evolved at different times in the history of life (see Fig.19.9 for an example of DNA sequence differences). This means that birds, bats, and beetles do not share a common ancestor with wings. Rather, wings originated three times independently, on three different branches of the tree of life, as a result of convergent evolution.

Linnaean Taxonomy

The classification hierarchy that taxonomists use today was created by Carolus Linnaeus (1707–78), the father of modern taxonomy (see Section 15.1). Linnaeus' system was developed as a way to organize biodiversity. In the mid-eighteenth century, Europeans traveled to distant parts of the world and described, collected, and sent back to Europe examples of plants and animals they had not encountered before. During this time of discovery, Linnaeus developed **binomial nomenclature,** part of his classification system in which each species receives a unique two-part Latin name (Fig. 19.2).

As an example, *Lilium bulbiferum* and *Lilium canadense* are two different species of lily. The first word, *Lilium*, is the genus (pl., genera), a classification category that can contain many species. The second word, known as the **specific epithet,** refers to one species within that genus. The specific epithet sometimes tells us something descriptive about the organism. Notice that the scientific name is in italics. The species name is designated by the full binomial name—in this case, either *Lilium bulbiferum* or *Lilium canadense.* The specific epithet without the genus gives no clue as to species—just as a house number alone without the street name is useless for finding an address. The genus name can be used alone, however, to refer to a group of related species.

Scientific names are derived in a number of ways. Some scientific names are descriptive in nature, for example, *Acer rubrum* for the red maple (*Acer* = maple, *rubrum* = red). Other scientific names may include geographic descriptions such as *Alligator mississippiensis* for the American alligator. Scientific names can also include eponyms (the name of a person), such as the owl mite *Strigophilus garylarsonii* (named after the cartoonist Gary Larson). Many scientific names are derived from mythical characters, such as *Iris versicolor*, named for Iris, the goddess of the rainbow.

Why do organisms need scientific names? And why do scientists use Latin, rather than common names, to describe organisms? There are several reasons. First, a common name varies from country to country because of language differences. Second, even people who speak the same language sometimes use different common names to describe the same organism. For example, bowfin, grindle, choupique, and cypress trout all refer to the same common fish, *Amia calva.* Furthermore, between countries, the same common name is sometimes given to different organisms. A "robin" (*Erithacus rubecula*) in England is very different from a "robin" (*Turdus migratorius*) in the United States.

Latin, however, is a universal language that not too long ago was well known by most scholars, many of whom were physicians or clerics. When scientists throughout the world use the scientific binomial name, they know they are speaking of the same organism.

Linnaean Classification Hierarchy

The binomial nomenclature system of Linnaeus is used to classify species. Today, taxonomists use a nested, hierarchical set of categories to classify organisms (Fig. 19.3). Each taxon is given a name and a rank according to which of the following set of major taxonomic groups it belongs: **species, genus, family, order, class, phylum, kingdom,** and more recently, a higher taxonomic category, the **domain.** The organisms that fill a particular classification category are distinguishable from other organisms by sharing a set of traits.

Organisms in the same domain have general traits in common, whereas those in the same species have quite specific traits in common. For example, the kingdom Animalia includes all animals. Within the kingdom Animalia is the phylum Chordata, a taxonomic group that contains only those animals with a spinal chord. Within the phylum Chordata is the class Mammalia

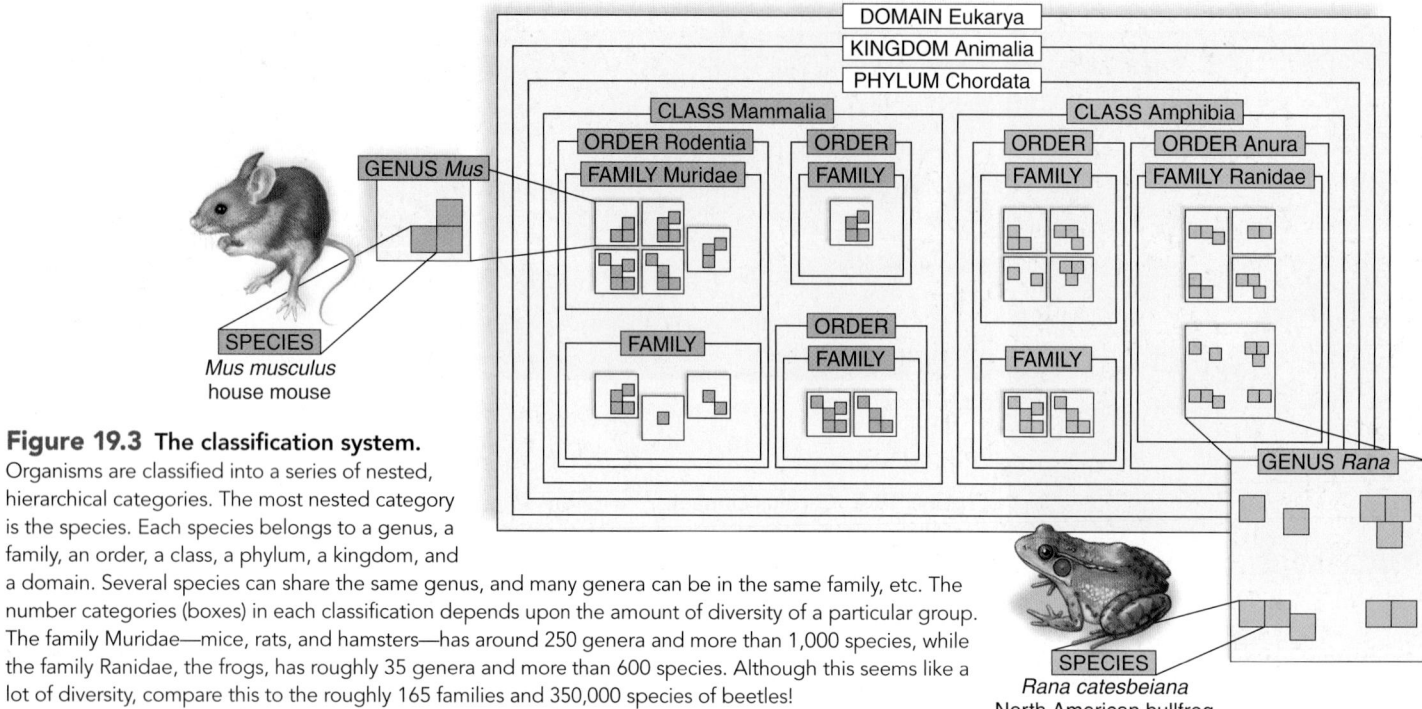

Figure 19.3 The classification system.
Organisms are classified into a series of nested, hierarchical categories. The most nested category is the species. Each species belongs to a genus, a family, an order, a class, a phylum, a kingdom, and a domain. Several species can share the same genus, and many genera can be in the same family, etc. The number categories (boxes) in each classification depends upon the amount of diversity of a particular group. The family Muridae—mice, rats, and hamsters—has around 250 genera and more than 1,000 species, while the family Ranidae, the frogs, has roughly 35 genera and more than 600 species. Although this seems like a lot of diversity, compare this to the roughly 165 families and 350,000 species of beetles!

that contains animals that have spinal chords (phylum Chordata), and among other characteristics, have mammary glands. The species is the most exclusive of all the categories as it contains only a single type of organism. The house mouse, *Mus musculus*, is a single species of mouse in the family Muridae, a family in the order Rodentia that is one of several orders in the class Mammalia (Fig. 19.3).

The classification hierarchy is very useful because it allows scientists to organize the diversity of life, but it is important to remember that the hierarchy was created by scientists, and thus does not represent any special relationship between organisms in nature. For example, grouping monkeys and apes into the order Primates tells scientists something about their evolutionary history, but being in the same order does not mean much to the monkeys or the apes! The classification hierarchy, if based on natural groups, should be viewed as the best working hypothesis of evolutionary relationships.

As with any scientific hypothesis, the hierarchy, and the placement of organisms within it, can be (and is) revised with the addition of new information. This is why you can find the same organism classified differently in older textbooks or even among different taxonomists. This uncertainty is part of the nature of science and is one of its greatest principles.

Linnaeus set the standard of binomial nomenclature in the mid-1700s, but there were no official rules for classifying organisms until the mid-1800s. During this 100-year period, problems with name confusion arose when scientists in different parts of the world, who did not communicate with each other (remember, there was no internet!), began to give the same Latin names to different species, or to develop their own version of Linnaeus' classification system.

The first attempt to make standardized rules of **nomenclature**, the procedure of assigning scientific names to taxonomic groups, occurred in 1842 at a meeting of scientists, among them Charles Darwin. In 1961, after 120 more years of revisions,

the International Code of Zoological Nomenclature (ICZN) was officially accepted as the universal guide for naming taxonomic groups. Today, the International Commission on Zoological Nomenclature is the keeper of the ICZN, and all scientists of the world use the rules of the ICZN when classifying organisms.

Despite a universal set of rules and over 250 years of taxonomy, only a fraction of the estimated 3–30 million species now living on Earth have been classified. Some taxonomic groups are classified more completely than others; we may have finished the birds and mammals, but there are millions of insects and microorganisms that remain to be discovered and classified.

The task of identifying and naming the species of the world is a daunting one. A new fast and efficient way of identifying species that is based on their DNA is described in the Nature of Science feature on page 352. This method, called DNA "barcoding," compares a short fragment of DNA sequence from an unknown organism to a large database of sequences from known organisms. The similarities and differences in the nucleotide sequence of this fragment of DNA can help determine which taxonomic group a new organism likely belongs to.

DNA barcoding does not always get the taxonomy correct, however, and for this reason it has been criticized by some taxonomists as being too simplistic. Nevertheless, it is a potentially powerful way to rapidly and inexpensively catalog at least a portion of the world's biodiversity.

Check Your Progress 19.1

1. Use the Linnaean classification system to fully classify the human species, *Homo sapiens*.
2. Explain why the grouping together of birds and bats based on having wings does not represent a natural group.
3. Describe the relationship between the terms classification, systematic biology, and taxonomy.

19.2 The Three-Domain System

Learning Outcomes

Upon completion of this section, you should be able to

1. List the three domains of life.
2. Summarize two characteristics that define each of the three domains.

From Aristotle's time to the middle of the twentieth century, biologists recognized only two kingdoms: kingdom Plantae (plants) and kingdom Animalia (animals). Plants were literally organisms that were planted and immobile, while animals were animated and moved about. In the 1880s, a German scientist, Ernst Haeckel, proposed adding a third kingdom. The kingdom Protista (protists) included unicellular microscopic organisms but not multicellular, largely macroscopic ones.

In 1969, R. H. Whittaker expanded the classification system to the **five-kingdom system:** Monera, Protista, Fungi, Plantae, and Animalia. Organisms were placed in these kingdoms based on the type of cell (prokaryotic or eukaryotic), complexity (unicellular or multicellular), and type of nutrition. Kingdom Monera contained all the prokaryotes, which are organisms that lack a membrane-bounded nucleus. These unicellular organisms were collectively called the bacteria. The other four kingdoms contain types of eukaryotes, which we describe later.

Defining the Domains

In the late 1970s, Carl Woese and his colleagues at the University of Illinois were studying relationships among the prokaryotes by comparing the nucleotide sequences of ribosomal RNA (rRNA). Woese found that the rRNA sequences of prokaryotes that lived at high temperatures or produced methane were quite different from that of all the other types of prokaryotes and from all eukaryotes. Therefore, he proposed that there are two groups of prokaryotes (rather than one group, the Monera, in the five-kingdom system). Further, Woese found that these two groups of prokaryotes have rRNA sequences so fundamentally different from each other that they should be assigned to separate domains, a category of classification that is higher than the kingdom category. The two designated domains are **domain Bacteria** and **domain Archaea.** He placed the eukaryotes in a third domain, the **domain Eukarya.**

The phylogenetic tree shown in Figure 19.4 is based on Woese's rRNA sequencing data. The data suggested that both bacteria and archaea evolved early in the history of life from the last universal common ancestor

(LUCA) (see Fig. 18.1). Later, the eukarya diverged from the archaea lineage. This means that the bacteria are members of the oldest lineage of living organisms on Earth, and the eukaryotes, the youngest lineage, are more closely related to the archaea than to the bacteria (Fig. 19.4).

Animation
Three Domains

Domain Bacteria

Bacteria are such a diversified group that they are found in large numbers nearly everywhere on Earth. The archaea are structurally similar to bacteria, but are placed in a separate domain, domain Archaea, because of biochemical differences (Table 19.1). The details of these differences are covered in Chapter 20.

The cyanobacteria are large, photosynthetic prokaryotes. They carry on photosynthesis in the same manner as plants in that they use solar energy to convert carbon dioxide and water to a carbohydrate, and in the process give off oxygen. The cyanobacteria belong to a very ancient lineage of bacteria, and may have been the first organisms to contribute oxygen to early

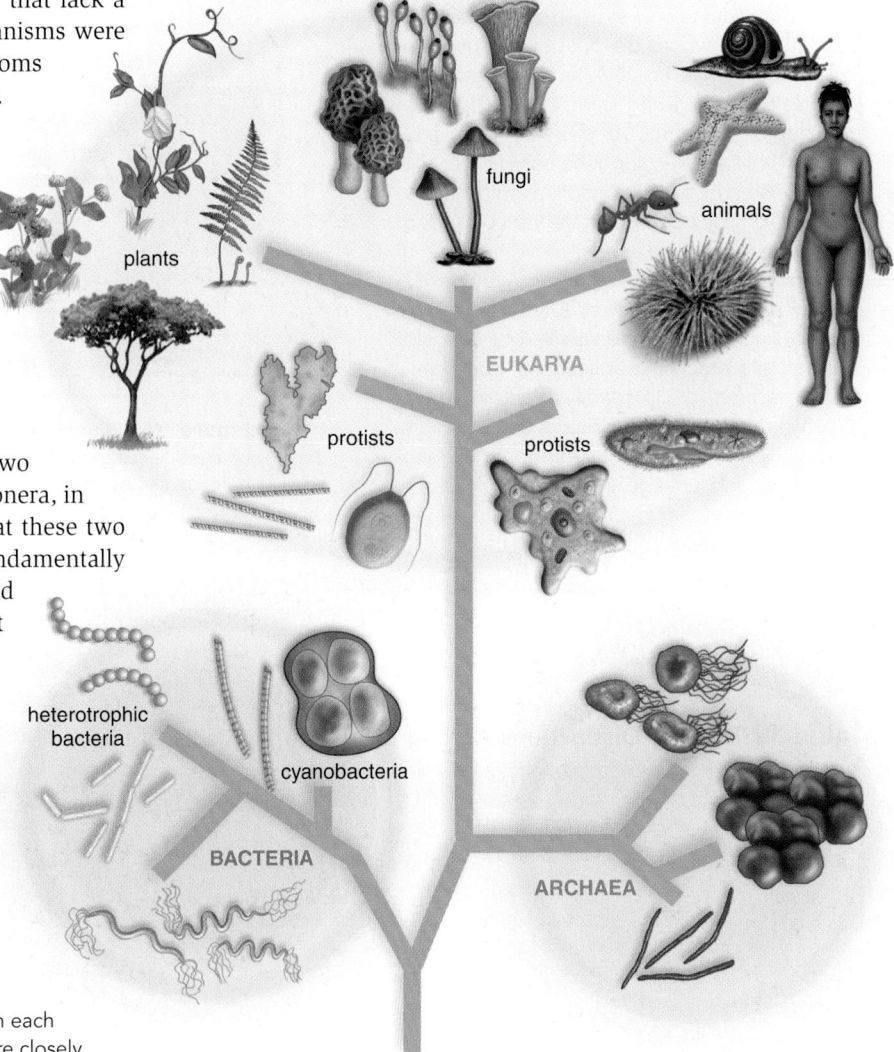

Figure 19.4 A tree of life showing the three domains. Representatives of each domain are depicted in each shaded area, and the tree shows that domain Archaea is more closely related to domain Eukarya than either is to domain Bacteria.

Nature of Science

DNA Barcoding of Life

Traditionally, taxonomists relied on anatomical traits to tell species apart. Although useful, physical features have several limitations when used as the only means to define a species. The identification of anatomical traits often requires the assistance of taxonomic experts that specialize on a specific group of organisms. In recent time, the number of species has far exceeded the number of available experts. This puts the cataloging of the world's biodiversity on a slow path—much slower than the rate at which our natural areas are disappearing. So far, scientists have identified only about 1.5 million species out of a potential 30 million.

The Consortium for the Barcode of Life (CBOL) proposes that any scientist, not just taxonomists, could use a sample of DNA to identify any organism on Earth. Just like a barcode, or UPC code, is a unique identifier of products on a store's shelf, the CBOL suggests it would be possible to use the base sequence in DNA to develop a barcode for each living thing (Fig. 19A). The order of DNA nucleotides—A, T, C, and G—within a particular gene common to the organisms in each kingdom would fill the role taken by numbers in the barcode used in warehouses and stores.

Speedy DNA barcoding would not only be a boon to efforts to catalog a rapidly disappearing biodiversity, but it would also have practical applications. For example, farmers could readily identify a pest attacking their crops, doctors could rapidly identify the correct antivenin for snakebite victims, and college students could identify the plants, animals, and protists on an ecological field trip. Already, the CBOL has accumulated hundreds of thousands of DNA barcodes representing species across the diversity of life.

The CBOL initiative has the potential to be a powerful tool for conservation biologists and wildlife officials worldwide. A DNA barcode can be used to identify illegal trade in endangered species, and for the early detection of invasive species that arrive into other countries as a consequence of global transportation.

In 2008, a pair of New York City high school students found a commercial application for the CBOL database (Fig. 19B). Kate Stoeckle and Louisa Strauss, two Trinity School seniors, did a project on the

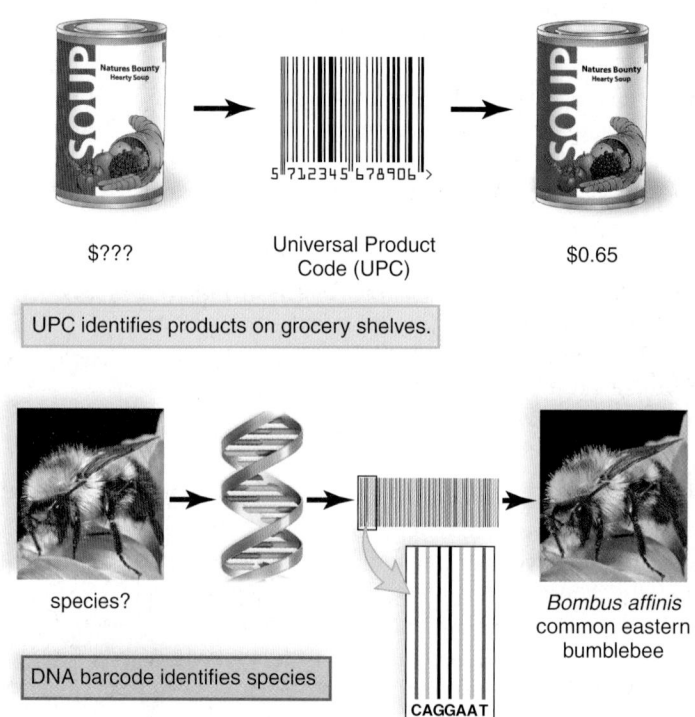

$???   Universal Product Code (UPC) $0.65

UPC identifies products on grocery shelves.

species?

DNA barcode identifies species

CAGGAAT

Bombus affinis common eastern bumblebee

Figure 19A Illustration of the barcoding concept. Much like a barcode (UPC) can identify the type and price of a can of soup, a DNA barcode can be used to identify a species. This is possible because each UPC is unique to a particular product, and each DNA barcode, based on a DNA sequence, is unique to an individual species.

Table 19.1 Major Distinctions Among the Three Domains of Life

	Bacteria	Archaea	Eukarya
Unicellularity	Yes	Yes	Some, many multicellular
Membrane lipids	Phospholipids, unbranched	Varied branched lipids	Phospholipids, unbranched
Cell wall	Yes (contains peptidoglycan)	Yes (no peptidoglycan)	Some yes, some no
Nuclear envelope	No	No	Yes
Membrane-bounded organelles	No	No	Yes
Ribosomes	Yes	Yes	Yes
Introns	No	Some	Yes

identification of fishes sold in markets and sushi restaurants in Manhattan, New York. They collected 60 fish samples from 4 restaurants and 10 grocery stores in Manhattan, which they sent off to have the DNA segment, the barcode, sequenced and compared to a global library of fish barcodes representing nearly 5,500 fish species.

Their results sent a wave of controversy throughout Manhattan and beyond:

2 of the 4 restaurants, and 6 of the 10 grocery stores, sold fish that were mislabeled. Most of the mislabeled fish were being sold as more expensive species. For example, Mozambique tilapia, a commonly farmed fish selling for $1.70 per pound wholesale, was being sold as albacore tuna at $8.50 per pound (Fig. 19C). In one case they found an endangered fish, the Acadian redfish, being sold as red snapper!

Questions to Consider

1. How might DNA barcoding be used by systematic biologists to speed up the classification of biodiversity?
2. Propose additional ways that DNA barcoding could aid in managing modern societal problems, such as the conservation of biodiversity, global warming, crime, and disease.

Figure 19B Katie Stoeckle (left) and Louisa Strauss uncover mislabeled fish in Manhattan.

Some fish sold in New York City are mislabeled as more expensive varieties

Sold as:
White (Albacore) Tuna
$8.50/lb wholesale

DNA ID:
Mozambique Tilapia
$1.70/lb wholesale

Photo Fishbase M Bariche   Photo Fishbase B Gratwicke

Figure 19C Mislabeled fish.

Earth's atmosphere, and in the process making an environment hospitable for the evolution of oxygen-using organisms, including animals.

Bacteria have a wide variety of means to obtain nutrients, but most are heterotrophic. *Escherichia coli,* which lives in the human intestine, is heterotrophic. *Clostridium tetani* (cause of tetanus), *Bacillus anthracis* (cause of anthrax), and *Vibrio cholerae* (cause of cholera) are disease-causing species of bacteria. Heterotrophic bacteria are beneficial in ecosystems because they break down organic remains. Along with fungi, they keep chemical cycling going so that plants always have a source of inorganic nutrients.

Domain Archaea

Like bacteria, archaea are prokaryotic unicellular organisms that reproduce asexually. Archaea don't look that different from bacteria under the microscope, and the extreme conditions under which many species live has made it difficult to grow them in the laboratory. This may have been the reason that their unique place among the living organisms went unrecognized for a long time.

The archaea are distinguishable from bacteria by a difference in their rRNA nucleotide sequences and also by their unique plasma membrane and cell wall chemistry. The chemical nature of the archaeal cell wall is diverse and is never the same as that of a bacterial cell. The unique cell wall structure of archaea could possibly help them to live in extreme conditions in which they are found. In addition, the branched nature of diverse lipids in the archaeal plasma membrane are very different from those of bacteria.

The archaea live in all sorts of environments, but they are known for thriving in those extreme environments thought to be similar to those of the early Earth. For example, the methanogens live in environments without oxygen, such as swamps, marshes and the guts of animals where methane is abundant; the halophiles thrive in salty environments such as the Great Salt Lake in Utah; and the thermoacidophiles are both high temperature and acid loving. These archaea live in extremely hot acidic environments, such as hot springs and geysers.

Domain Eukarya

Eukaryotes are unicellular to multicellular organisms whose cells have a membrane-bounded nucleus. They also have various organelles, some of which arose through endosymbiosis of other unicellular organisms (see page 335). Sexual reproduction is common in eukaryotes, and various types of life cycles are seen.

Later in this text, we study the individual kingdoms that occur within the domain Eukarya. In the meantime, we can note that the protists are a diverse group of single-celled eukaryotes that are hard to classify and define. Some protists have filaments and form colonies or multicellular sheets. Even so, protists do not have true tissues. Nutrition is diverse; some are heterotrophic by ingestion or absorption and some are photosynthetic. Green algae, paramecia, and slime molds are representative protists. There has been considerable debate over the classification of protists, and presently they are placed in six supergroups (see Fig. 21.2) within the domain Eukarya.

Fungi are eukaryotes that form spores, lack flagella, and have cell walls containing chitin. They are multicellular with a few exceptions. Fungi are saprotrophic by absorption—they secrete digestive enzymes and then absorb nutrients from decaying organic matter. Mushrooms, molds, and yeasts are representative fungi. Despite appearances, molecular data suggest that fungi and animals are more closely related to each other than either is to plants.

Plants are photosynthetic, multicellular organisms that have become adapted to a land environment. They share a common ancestor, which is an aquatic photosynthetic protist. Land plants possess true tissues and have the organ-system level of organization. Examples include cacti, ferns, and cypress trees.

Animals are motile, eukaryotic, multicellular organisms that evolved from a heterotrophic protist. Like land plants, animals have true tissues and the organ-system level of organization. Animals are heterotrophs. Examples include the worms, whales, and insects.

Check Your Progress   19.2

1. List traits that separate the Archaea from other unicellular organisms.
2. Explain the evidence indicating that fungi are more closely related to animals than to plants.
3. Identify the domain that includes all multicellular organisms.

19.3 Phylogeny

Learning Outcomes

Upon completion of this section, you should be able to

1. Discriminate between ancestral and derived traits.
2. Interpret the evolutionary relationships depicted in a phylogeny.
3. List the types of traits used to construct a phylogeny.

Systematic biologists use characters from the fossil record, comparative anatomy and development, and the sequence, structure, and function of RNA and DNA molecules to construct a phylogeny. Systematic biologists study the evolutionary history of biodiversity, represented by a phylogeny. In essence, systematic biology is the study of the evolutionary history of biodiversity, and a phylogeny is the visual representation of this history.

Interpreting a Phylogeny

Systematic biologists construct a phylogeny from traits that are unique to, and shared by, a taxon and their **common ancestor** (an ancestor to two or more lines of descent). Each branch, or **lineage,** in a phylogeny represents a descendant of a common ancestor. When a new character evolves, a new evolutionary path can begin, or **diverge,** from the old, a new lineage is formed, and a new branch of the phylogeny arises (Fig. 19.5).

Animation
Phylogenetic Tree

Not all traits have equal value when making a phylogeny. **Ancestral traits,** or those found in the common ancestor, are not useful for determining the evolutionary relationships of an ancestor's descendants. For example, deer, cattle, monkeys, and apes, all examples of mammals, share a common ancestor that had mammary glands (Fig. 19.5). Because all mammals have mammary glands, this trait is an ancestral trait, and is not helpful for understanding how deer, cattle, monkeys, and apes are related to each other. In contrast, **derived traits,** or those not found in the common ancestor of a taxonomic group, are the most important traits for clarifying evolutionary relationships. For example, both monkeys and apes have an opposable thumb capable of grasping, a trait not present in the common ancestor of mammals. This shared derived trait places monkeys and apes on a separate lineage of mammals called "primates" (Fig. 19.5).

Whether a trait is derived or ancestral is dependent upon whether it is present in the common ancestor, and thus relative to its location within a phylogeny. For example, an opposable thumb is a derived trait for all monkeys and apes when compared to the common ancestor of all mammals. But the opposable thumb, while a derived trait when compared to the mammal common ancestor, is nevertheless an ancestral trait of primates (Fig. 19.5). Similarly, deer and cattle, both artiodactyls, have even-toed hooves, a trait not found in the common ancestor of mammals or in primates. Thus even-toed hooves is a derived trait when compared to the common ancestor of mammals, but an ancestral trait of artiodactyls (Fig. 19.5). Both even-toed hooves and the opposable thumb suggest that the evolutionary history of artiodactyls and primates became independent as each group diverged from the mammal common ancestor.

Derived traits provide a more detailed phylogeny as they define closer and closer evolutionary relationships. Within the primates, a fully rotating shoulder joint, which allows apes to swing from limb to limb of a tree, and the prehensile tail of monkeys are derived traits that define separate ape and monkey branches within the primates. Similarly, horns and antlers are derived traits that divide the cattle and deer into two individual artiodactyl branches (Fig. 19.5).

The hierarchical classification system of Linnaeus defines species as closely related to other species within the same genus. A genus is related to other genera in the same family, and so forth, from order to class to phylum to kingdom to domain (see Fig. 19.3). When we say that two species (or genera, families, etc.) are closely related, we mean that they share a recent common ancestor. For example, all the animals in Figure 19.5 are related because we can trace their ancestry back to a common ancestor. Taxonomists use the pattern of branching in a phylogeny constructed from an analysis of derived traits to classify taxa into natural groups.

For example, all mammals with even-toed hooves form a single lineage that is assigned to the order Artiodactyla (Fig. 19.5). Furthermore, artiodactyls that have antlers form a lineage within the order Artiodactyla that is classified as the family Cervidae. Antlers are grown only in males during the breeding season, and can grow quite large and can be highly branched. In contrast, artiodactyls that have horns form a lineage within the order Artiodactyla that is classified as the family Bovidae. Unlike antlers, horns are not shed seasonally and are found on both males and females, although they are smaller in females.

Similarly, the order Primates is a lineage of mammals with opposable thumbs. The rotating shoulder of apes and the prehensile tail of monkeys form two independent lineages in the order Primates that are classified as the family Hominidae and family Cebidae, respectively.

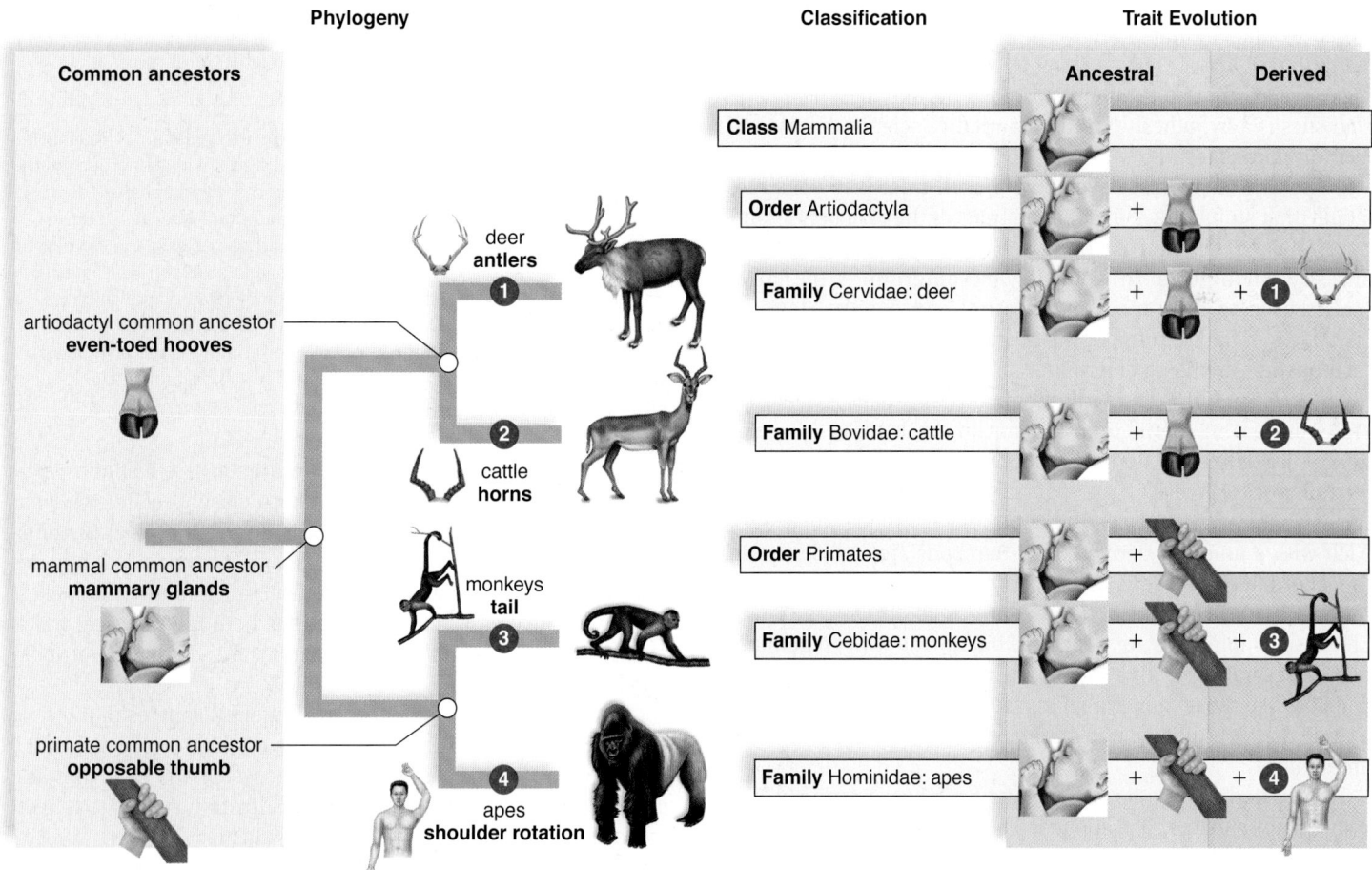

Figure 19.5 The relationship between phylogeny, classification, and traits. The phylogeny is a representation of the evolutionary history of a few members of the class Mammalia. Each common ancestor has a trait that is present in all of its descendants. For example, the red-numbered traits are derived traits present in all descendants of a clade: 1. deer have antlers, 2. cattle have horns, 3. monkeys have tails, 4. apes have full shoulder rotation. As you move toward the tips of the tree, new derived traits provide a greater resolution to the classifications. For example, mammals have mammary glands. Artiodactyls have mammary glands and even-toed hooves. Deer have mammary glands, even-toed hooves, and antlers.

Cladistics

When constructing any phylogeny, many characters from many different organisms are compared at the same time (Fig. 19.6). This approach can produce many possible evolutionary trees, because not all traits are equally useful to the study of evolutionary history. How do we know which phylogeny of the many possible phylogenies is the best hypothesis of evolutionary history? One answer is through the use of cladistics.

Cladistics is a method that uses shared, derived traits to develop a hypothesis of evolutionary history. The evolutionary history of derived traits is interpreted into a type of phylogeny constructed with cladistic methods—called a **cladogram.** In a cladogram, a common ancestor and all of its descendant lineages is called a **clade.**

Cladistics applies the principle of **parsimony** [L. *parsimonia,* frugality, thrift] to a set of traits to construct a cladogram. Parsimony considers the simplest solution to be the "optimal" solution. Thus the cladogram that represents the simplest evolutionary history, that is, the one that requires the fewest number of evolutionary changes, is considered the best hypothesis based on the traits used to construct the cladogram. As with any hypothesis, a cladogram may change, and often does change, when new traits are discovered and included in construction of a cladogram. The important point is that our understanding of evolutionary history is a working hypothesis, constantly changing as we learn more about organisms' traits and lifestyles. Thus cladistics is a hypothesis-based, quantitative science that is subject to rigorous testing.

The first step when developing a cladogram is to construct a table that summarizes the derived traits of the taxa being compared (Fig. 19.6). Derived traits are used to determine shared ancestry among taxa. In cladistics, the **outgroup** is the taxon that is used to determine the ancestral and derived states of characters in the **ingroup,** or the taxa for which the evolutionary relationships are being determined. In Figure 19.6, the outgroup is the lancelet, and the ingroup contains all other vertebrates. Traits present in the ingroup but not in the lancelet (the outgroup) are defined as derived traits. For example, all **chordates,** including the lancelet, have a dorsal or spinal nerve chord, so this is an ancestral trait. Nested within the chordates are clades, each with a uniquely derived trait. Tetrapods are a clade within the chordates that does not include fish, because fish have a dorsal nerve chord, but do not have four limbs (Fig. 19.7). Likewise, amphibians are tetrapods, but do not have an amnion, one of several protective membranes that surround a growing embryo, like those found in an amniotic egg. Thus amniotes are a clade that does not include amphibians, but all organisms with an amnion.

Once derived and ancestral traits have been identified, the principle of parsimony is applied. The cladogram in Figure 19.7 is considered the best hypothesis, or explanation, of the evolutionary history based on the traits used.

	Species							
	ingroup							
Traits	chimpanzee	dog	finch	crocodile	lizard	frog	tuna	lancelet (outgroup)
mammary glands	X	X						
hair	X	X						
gizzard			X	X				
epidermal scales			X	X	X			
amniotic egg	X	X	X	X	X			
four limbs	X	X	X	X	X	X		
vertebrae	X	X	X	X	X	X	X	
notochord in embryo	X	X	X	X	X	X	X	X

Figure 19.6 Constructing a cladogram: the data. This lancelet is in the outgroup, and all the other species listed are in an ingroup (study group). The species in the ingroup have shared derived traits—derived because a lancelet does not have the trait, and shared because certain species in the study group do have them. All the species in the ingroup have vertebrae, all but a fish have four limbs, and so forth. The shared derived traits indicate which species are distantly related and which are closely related. For example, a human is more distantly related to a fish, with which it shares only one trait, namely vertebrae, than to an iguana, with which it shares three traits—vertebrae, four limbs, and an amniotic egg (the amnion layer protects the embryo; see Chapter 29).

Tracing Phylogeny

Traditionally, systematic biologists relied on morphological data to study evolutionary relationships between taxa. However, morphology can be misleading. For example, recent studies suggest that birds and crocodiles are more closely related to each other than crocodiles are to lizards. The morphological traits that Linnaeus used, namely wings and feathers, led him to classify birds as a different lineage from crocodiles. This makes sense when we consider that Linnaeus had only physical traits at his disposal when classifying organisms—birds do not look much like crocodiles!

Today, we have a wide range of different sources of traits to assist with understanding the evolutionary history of organisms. Armed with these new sources of data, systematic biologists are continually revising the historical classification system to reflect our best understanding of evolutionary history.

Fossil Traits

One of the advantages of fossils is that they can be dated (see Chapter 18), but unfortunately it is not always possible to tell to which lineage, living or extinct, a fossil is related. At present, paleontologists are discussing whether fossil turtles indicate

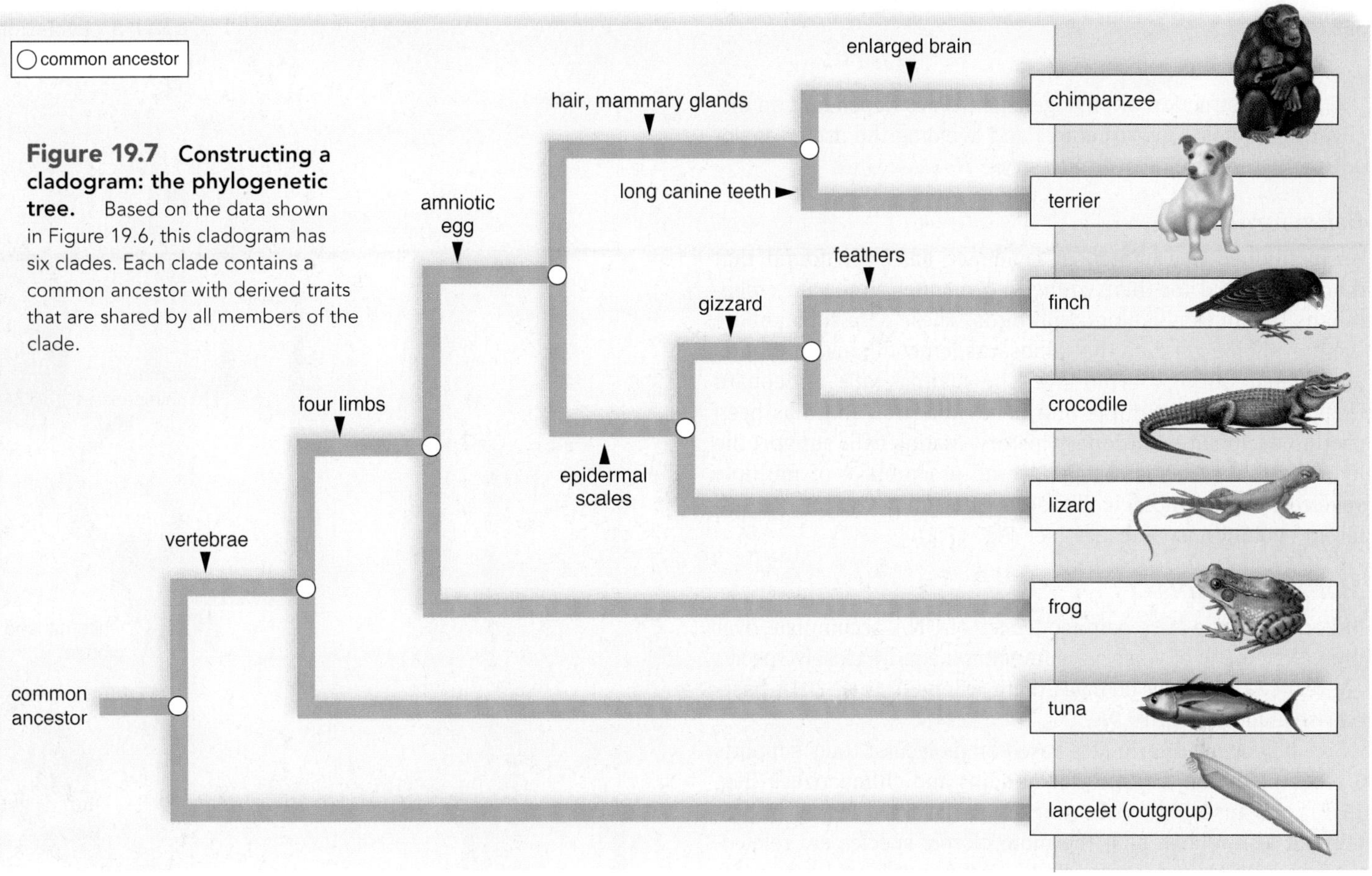

Figure 19.7 Constructing a cladogram: the phylogenetic tree.

Figure 19.7 Constructing a cladogram: the phylogenetic tree. Based on the data shown in Figure 19.6, this cladogram has six clades. Each clade contains a common ancestor with derived traits that are shared by all members of the clade.

○ common ancestor

enlarged brain

hair, mammary glands

long canine teeth ►

amniotic egg

feathers

gizzard

four limbs

epidermal scales

vertebrae

common ancestor

chimpanzee

terrier

finch

crocodile

lizard

frog

tuna

lancelet (outgroup)

that turtles are distantly or closely related to crocodiles. On the basis of his interpretation of fossil turtles, Olivier C. Rieppel of the Field Museum of Natural History in Chicago is challenging the conventional interpretation that turtles are ancestral (have traits seen in a common ancestor to all reptiles) and are not closely related to crocodiles, which evolved later. His interpretation is being supported by molecular data that show turtles and crocodiles are closely related (see Section 29.5 for an overview of reptile evolution).

If the fossil record were more complete, fewer controversies might arise about the interpretation of fossils. One reason the fossil record is incomplete is that most fossils exist as only harder body parts, such as bones and teeth. Soft parts are usually eaten or decayed before they have a chance to be buried and preserved. This may be one reason it has been difficult to discover when angiosperms (flowering plants) first evolved. A Jurassic fossil recently found may help to pinpoint the date of origin (Fig. 19.8). As paleontologists continue to discover new fossils, the fossil record will reveal more traits useful to systematic biologists.

Morphological Traits

Homology [Gk. *homologos*, agreeing, corresponding] is structural similarity that stems from having a common ancestor. Comparative anatomy, including developmental evidence provides information regarding homology (see Figs. 15.15 and 15.16).

Homologous structures are similar to each other because of common descent. The forelimbs of vertebrates contain the same bones organized just as they were in a common ancestor, despite adaptations to different environments. As Figure 15.15 shows, even though a horse has but a single digit and toe (the hoof),

while a bat has four lengthened digits that support its wing, a horse's forelimb and a bat's forelimb contain the same bones.

Deciphering homology is sometimes difficult because of convergent evolution. **Convergent evolution** has occurred when distantly related species have a structure that looks the same only because of adaptation to the same type of environment (see Fig. 17.12 for an example of convergent evolution in fish). Similarity due to convergence is termed **analogy.** The wings of an insect and the wings of a bat are analogous.

Analogous structures have the same function in different groups but do not have a common ancestry. Both cacti and spurges are adapted similarly to a hot, dry environment, and both are succulent (thick, fleshy) with spines that originate from modified leaves. However, the details of their flower structure indicate that these plants are not closely related.

fruits

paired stamens

Figure 19.8 Ancestral angiosperm. The fossil *Archaefructus liaoningensis*, dated from the Jurassic period, may be the earliest angiosperm to be discovered. Without knowing the anatomy of the first flowering plant, it has been difficult to determine the ancestry of angiosperms.

The construction of phylogenetic trees is dependent on discovering homologous structures and avoiding the use of analogous structures to uncover ancestry.

Behavioral Traits

As mentioned in Chapter 18, evidence has been found that dinosaurs cared for their young in a manner similar to crocodilians (includes alligators) and birds. These data substantiate the morphological data that dinosaurs, crocodilians, and birds are related through evolution. The mating calls of leopard frogs are another example of a behavioral trait that has been used to decipher evolutionary history. Mating calls support the hypothesis that leopard frogs are an assemblage of multiple species that morphologically look quite similar but are on different evolutionary lineages (see Fig. 17.2).

Molecular Traits

Mutations in the base-pair sequences of DNA accumulate over time. Systematic biologists assume that the more closely species are related, the fewer changes there will be in their DNA base-pair sequences (Fig. 19.9).

A phylogeny of primates based on molecular traits supports a recent common ancestor for humans and chimpanzees (Fig. 19.10). Because DNA codes for amino acid sequences in proteins, it also follows that the more closely species are related, the fewer differences there will be in the amino acid sequences within their proteins.

Advances in molecular biology have made it very quick, easy, and inexpensive to collect nucleotide sequences for many different taxa. Software breakthroughs have made it possible to analyze nucleotide sequences or amino acid sequences quickly

	1	2	3	4	5	6	7	8	9	10	11	12	13	14	15	16	17	18	19	20	21	22	23
Cow	c	c	c	c	g	t	g	g	a	g	g	t	a	c	g	c	t	t	c	a	c	t	c
Pig	c	c	c	c	g	t	g	g	a	g	g	t	g	c	g	c	t	t	c	a	c	t	c
Horse	t	c	c	g	g	t	g	g	a	g	g	t	g	c	g	c	t	t	c	g	c	c	c
Mouse	c	c	c	c	g	t	g	g	a	g	g	t	g	c	g	c	t	t	c	a	c	c	c
Rat	c	c	c	c	g	t	a	g	a	g	g	t	g	c	g	c	t	t	c	a	c	c	c
Dog	c	c	c	t	g	t	g	g	a	g	g	t	c	c	g	c	t	t	c	a	c	c	c
Guinea Pig	c	c	c	t	g	t	g	g	g	g	g	t	g	c	g	c	t	t	c	a	c	c	c
Chimp	c	c	t	g	g	t	g	g	g	g	c	t	a	c	g	c	t	t	c	a	c	c	t
Human	c	c	t	g	g	t	g	g	g	g	g	t	a	c	g	c	t	t	c	a	c	c	t
Orangutan	c	c	c	g	g	t	g	g	g	g	g	t	g	c	g	c	t	t	c	a	c	c	c
Macca	a	c	c	g	g	t	g	g	g	g	g	t	g	c	g	c	t	t	c	a	c	c	c

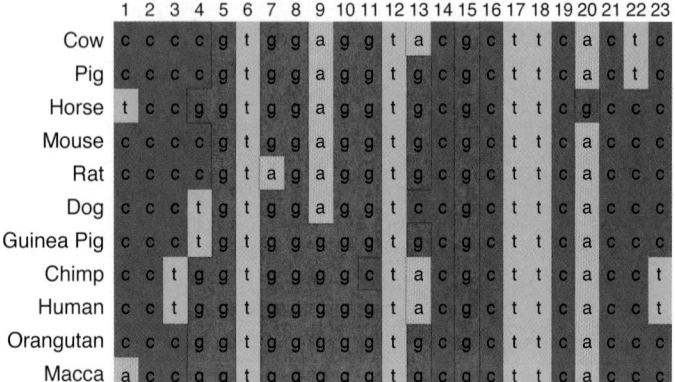

Figure 19.9 DNA sequence alignment. The DNA sequences of a small section of a gene are aligned to determine the evolutionary relationships among some mammals. Each nucleotide is aligned in columns and assigned a number (in this case, 1–23). Just as with physical traits, individual nucleotides provide information about how organisms are related. For example, the 6th nucleotide is a T shared among all mammals. Thus this T would be an ancestral trait of the class Mammalia (see Fig. 19.5). Chimpanzees and humans have in common a T at nucleotides 3 and 23 that are not found in other mammals. Thus these two nucleotides would be shared, derived traits that support a close relationship between chimps and humans. Some nucleotide sequence alignments are thousands of nucleotides long, thus computers are used to perform comparisons.

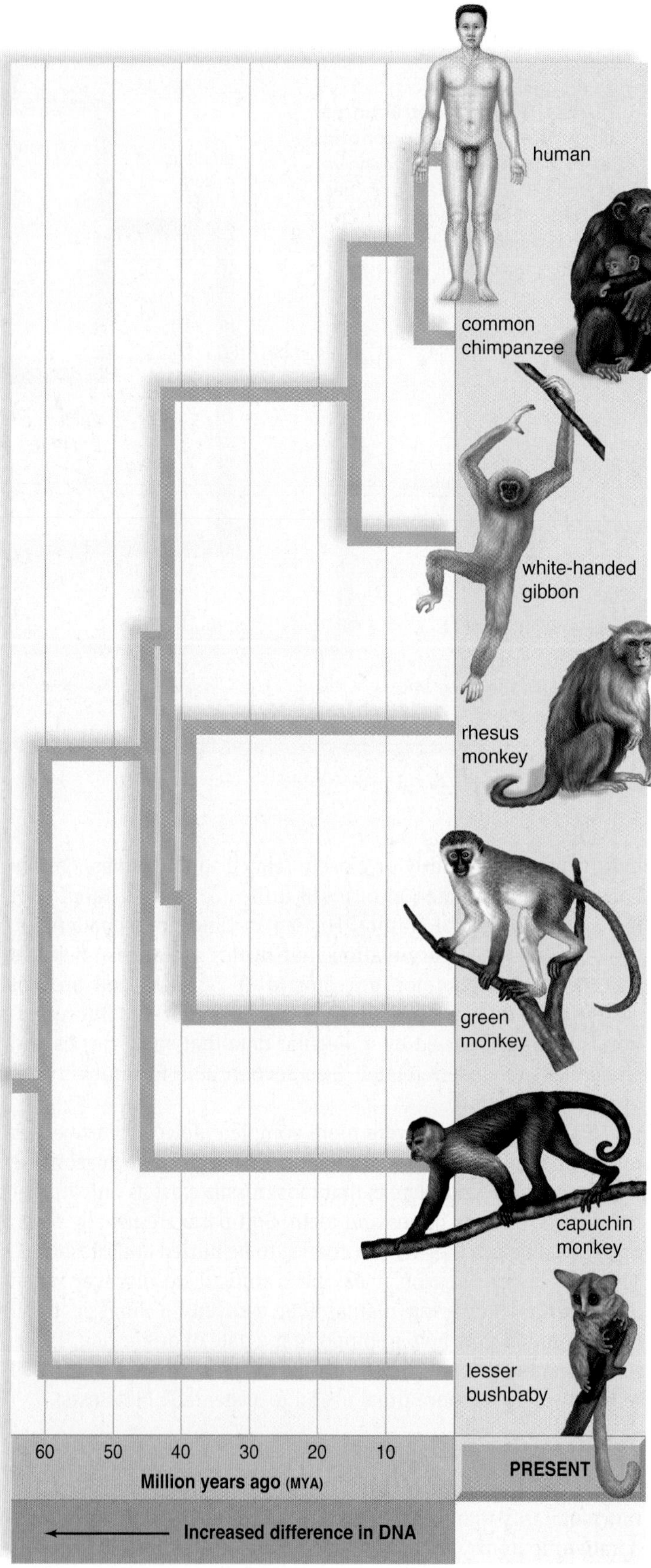

Figure 19.10 A phylogeny determined from molecular data. The relationship of certain primate species based on a study of their genomes. The length of the branches indicates the relative number of nucleotide pair differences that were found between groups. These data, along with knowledge of the fossil record for one divergence, make it possible to suggest a date for the other divergences in the tree.

and accurately using a computer. Archives of DNA sequences from different genes for thousands of organisms are freely available to anyone doing systematic biology research.

Mitochondrial DNA (mtDNA) mutates ten times faster than nuclear DNA. Therefore, when determining the phylogeny of closely related species, investigators often choose to sequence mtDNA instead of nuclear DNA. One such study concerned North American songbirds. Ornithologists believed for a long time that these birds diverged into eastern and western subspecies due to retreating glaciers some 250,000–100,000 years ago. Sequencing of mtDNA allowed investigators to conclude that the two groups of North American songbirds diverged from one another an average of 2.5 million years ago (MYA). Because the old hypothesis based on glaciation is apparently flawed, a new hypothesis is required to explain why eastern and western subspecies arose among these songbirds.

Phylogenetic trees derived from molecular data are routinely used to apply evolutionary theory to all areas of biology, and have helped human society in agriculture, medicine, and forensic science. For example, a phylogeny of the HIV types in an individual guide doctors' decisions about antiviral drug treatment. Likewise phylogenies of insects have assisted agriculturalists to design effective control of crop pests.

Protein Comparisons. Before amino acid sequencing became routine, immunological techniques were used to roughly judge the similarity of plasma membrane proteins. In one procedure, antibodies are produced by transfusing a rabbit with the cells of one species. Cells of the second species are exposed to these antibodies, and the degree of the reaction is observed. The stronger the reaction, the more similar the cells from the two species.

Later, it became customary to use amino acid sequencing to determine the number of amino acid differences in a particular protein. Cytochrome c is a protein found in all aerobic organisms, so its sequence has been used to examine evolutionary relationships among many organisms. There are 3 amino acid differences in the cytochrome c of chickens and ducks, but between chickens and humans there are 13 amino acid differences. From these data we can conclude that, as expected, chickens and ducks are more closely related than are chickens and humans. In addition to amino acids, differences in individual nucleic acids in DNA and RNA have become a powerful tool for examining evolutionary relationships.

Molecular Clocks. Some nucleic acid changes are neutral (not under the influence of natural selection), and thus accumulate at a fairly constant rate. These neutral mutations can be used as a kind of **molecular clock** to construct a timeline of evolutionary history. For example, songbird subspecies have mtDNA with 5.1% nucleic acid differences. Researchers know the average rate at which mtDNA nucleotide changes occur, called the mutation rate, measured in the number of mutations per unit time. The researchers doing comparative mtDNA sequencing used their data as a molecular clock when they equated a 5.1% nucleic acid difference among songbird subspecies to 2.5 MYA.

In Figure 19.10, the researchers used their DNA sequence data to suggest how long the different types of primates have been separate. The fossil record was used to calibrate the clock: When the fossil record for one divergence is known, it indicates how long it probably takes for each nucleotide pair difference to occur. When the fossil record and molecular clock data agree, researchers have more confidence that the proposed phylogenetic tree is correct.

Animation
Molecular Clock

Check Your Progress 19.3

1. Identify a branch, a node, a common ancestor, and a taxon in Figure 19.4.
2. Interpret the ancestral or derived state of traits relative to their position on the phylogeny in Figure 19.4.
3. Compare two phylogenies of the same set of organisms; one requires 10 evolutionary changes, the other 15. Explain which phylogeny would be the best hypothesis for evolutionary history, and why.

CONNECTING *the* CONCEPTS *with the* BIG IDEAS

Evolution

- Systematic biologists have modified Linnaeus' classification system, and use fossil, anatomical, and molecular data to categorize species into taxa that reflect shared evolutionary relationships. (1B2c)
- Organisms can be organized into phylogenetic trees that represent their evolutionary relationships, showing common ancestors and descendents. (1B2)
- Cladistics employs similarities in morphology and in DNA sequencing to create phylogenetic trees called cladograms. (1A4b4, 1B2 a, c)
- Cladograms chart the relatedness of groups and document speciation from ancestral stock. (1B2b)
- Cladograms undergo constant revision as new information about species is uncovered. (1B2d)

- Domains represent the highest taxa to which organisms belong. There are major distinctions between the Bacteria, Archaea, and Eukarya domains, though all share some basic common characteristics which lead scientists to hypothesize they had an original common ancestor. (1D2b)

*Find the unabridged version of all EK citations at www.glencoe.com/maderAP11.

Media Study Tools

www.glencoe.com/maderAP11

Enhance your study of this chapter with study tools and practice tests. Also ask your instructor about the resources available through ConnectPlus, including the media-rich eBook, interactive learning tools, and animations.

Summarize

19.1 Systematic Biology

Systematics is dedicated to understanding the evolutionary history of life on Earth. Taxonomy, a part of systematics, deals with the naming of organisms; each species is given a binomial name consisting of the genus and specific epithet. Taxonomists strive to classify organisms based on natural groups—those that share a common evolutionary ancestry. Traditional Linnaean taxonomy did not consider evolutionary history when classifying organisms; thus, today parts of the historical taxonomy are being revised under the guidance of evolutionary principles.

Classification involves the assignment of species to categories. When an organism is named, a species has been assigned to a particular genus. There are eight main categories of classification: species, genus, family, order, class, phylum, kingdom, and domain. The categories are hierarchical, such that each higher category is more inclusive; species in the same kingdom share general characters, and species in the same genus share quite specific characters.

19.2 The Three-Domain System

On the basis of molecular data, three evolutionary domains have been established: Bacteria, Archaea, and Eukarya. The first two domains contain prokaryotes; the domain Eukarya contains the protists, fungi, plants, and animals.

19.3 Phylogeny

Phylogenetic trees are visual representations of evolutionary history. Each phylogeny is a hypothesis concerning the evolutionary relationships between designated species. Taxonomists use phylogenies to classify organisms into natural groups—those based on shared evolutionary history. Classification is the naming and grouping of organisms into a taxon. A taxon is the general name for a category containing an organism or a group of organisms that exhibit the same traits.

Cladistics uses shared derived traits to distinguish different groups of species from one another. The phylogeny that results from cladistic analysis is called a cladogram, with lineages called clades. The principle of parsimony is used to select the tree with the simplest explanation of evolutionary history given the set of traits used. The most parsimonious cladogram is the one that requires the fewest number of evolutionary steps.

The fossil record, homology, and molecular data, in particular, are used to reconstruct evolutionary history. Because fossils can be dated, available fossils can establish the antiquity of a species. If the fossil record is complete enough, we can sometimes trace a lineage through time.

Homology helps indicate when species share a common ancestor; however, convergent evolution sometimes makes it difficult to distinguish homologous structures from analogous structures. DNA nucleotide sequence data are commonly used to help determine evolutionary relationships.

Key Terms

analogous structure 357	ingroup 356
analogy 357	kingdom 349
ancestral traits 354	lineage 354
binomial nomenclature 349	molecular clock 359
chordates 356	natural group 348
clade 356	nomenclature 350
cladistics 356	order 349
cladogram 356	outgroup 356
class 349	parsimony 356
classification 348	phylogeny (or phylogenetic
common ancestor 354	tree) 348
convergent evolution 357	phylum 349
derived trait 354	species 349
diverge 354	specific epithet 349
domain 349	systematic biology (or
domain Archaea 351	systematics) 348
domain Bacteria 351	systematics 348
domain Eukarya 351	taxon (pl., taxa) 348
family 349	taxonomist 348
five-kingdom system 351	taxonomy 348
genus 349	trait 348
homologous structure 357	
homology 357	

Assess

Reviewing This Chapter

1. Explain the binomial system of naming organisms. Why must species be designated by a complete name?   349
2. Why is it necessary to give organisms scientific names?   349
3. What are the eight obligatory classification categories? In what way are they a hierarchy?   349–50
4. Compare the five-kingdom system of classification to the three-domain system.   351
5. Contrast the characteristics of the bacteria, the archaea, and the eukarya.   351–54
6. Contrast the eukaryotic protists, fungi, plants, and animals.   354
7. Discuss the principles of cladistics, and explain how to construct a cladogram.   356
8. With reference to the phylogenetic tree shown in Figure 19.7, why are birds in a clade with crocodiles? In a clade with other reptiles?   357
9. What types of data help systematists construct phylogenetic trees?   356–59

Testing Yourself

Choose the best answer for each question.

1. Which is the scientific name of an organism?
   a. *Rosa rugosa*
   b. *Rosa*
   c. *rugosa*
   d. *Rugosa rugosa*
   e. Both a and d are correct.

2. Which of these describes systematics?
   a. studies evolutionary relationships
   b. includes taxonomy and classification
   c. includes phylogenetic trees
   d. utilizes fossil, morphological, and molecular data
   e. All of these are correct.

3. The classification category below the level of family is
   a. class.
   b. species.
   c. phylum.
   d. genus.
   e. order.

4. Which of these are domains? Choose more than one answer if correct.
   a. Bacteria
   b. Archaea
   c. Eukarya
   d. Animals
   e. Plants

5. Which of these are eukaryotes? Choose more than one answer if correct.
   a. bacteria
   b. archaea
   c. eukarya
   d. animals
   e. plants

6. Which of these characteristics is shared by bacteria and archaea? Choose more than one answer if correct.
   a. presence of a nucleus
   b. absence of a nucleus
   c. presence of ribosomes
   d. absence of membrane-bounded organelles
   e. presence of a cell wall

7. Which is mismatched?
   a. Fungi—prokaryotic single cells
   b. Plants—nucleated
   c. Plants—flowers and mosses
   d. Animals—arthropods and humans
   e. Protists—unicellular eukaryotes

8. Which is mismatched?
   a. Fungi—heterotrophic by absorption
   b. Plants—usually photosynthetic
   c. Animals—rarely ingestive
   d. Protists—various modes of nutrition
   e. Both c and d are mismatched.

9. Concerning a phylogenetic tree, which is incorrect?
   a. Dates of divergence are always given.
   b. Common ancestors give rise to descendants.
   c. The more recently evolved are always at the top of the tree.
   d. Ancestors have primitive characters.

10. Which pair is mismatched?
    a. homology—character similarity due to a common ancestor
    b. molecular data—DNA strands match
    c. fossil record—bones and teeth
    d. homology—functions always differ
    e. molecular data—molecular clock

11. One benefit of the fossil record is
    a. that hard parts are more likely to fossilize.
    b. fossils can be dated.
    c. its completeness.
    d. fossils congregate in one place.
    e. All of these are correct.

12. The discovery of common ancestors in the fossil record, the presence of homologies, and nucleic acid similarities help scientists decide
    a. how to classify organisms.
    b. the proper cladogram.
    c. how to construct phylogenetic trees.
    d. how evolution occurred.
    e. All of these are correct.

13. Molecular clock data are based on
    a. common adaptations among animals.
    b. DNA dissimilarities in living species.
    c. DNA fingerprinting of fossils.
    d. finding homologies among plants.
    e. All of these are correct.

Engage

Thinking Scientifically

1. Recent DNA evidence suggests to some plant taxonomists that the traditional way of classifying flowering plants is not correct, and that flowering plants need to be completely reclassified. Other botanists disagree, saying it would be chaotic and unwise to disregard the historical classification groups. Argue for and against keeping traditional classification schemes.

2. What data might make you conclude that the eukaryotes should be in more than one domain? What domains would you hypothesize might be required?

Bioethical Issue

Classifying Chimpanzees

Because the genomes of chimpanzees and humans are almost identical, and the differences between them are no greater than between any two human beings, their classification has been changed. Chimpanzees and humans are placed in the same family and subfamily. They are in different "tribes," which is a rarely used classification category between subfamily and genus.

The former classification of chimpanzees and humans placed the two animals in different families. Do you believe the chimpanzees should be classified in the same family and subfamily as humans, or do you prefer the classification used formerly? Which way seems prejudicial? Give your reasons for preferring one method over the other.

UNIT 4

Microbiology and Evolution

Viruses. Every fall, headlines will again warn of the possibility of another flu epidemic; you may have been touched by them as you suffer with a cold, fever blisters, or a wart. Some hypothesize viruses are the remnants of the first life form on Earth; many others see them as degenerate cells that have simply been streamlined for success. Whatever their origin, viruses continue to be of tremendous importance both as pirates who ultimately take over cellular functions and as creative transporters of genetic information both in nature and in bioengineering labs.

Bacteria have been around for eons, at least 3.5 billion years. These prokaryotes were the pioneers in cellular respiration, photosynthesis, and are the only creatures on Earth that can be autotrophic without the sun's energy input. The endosymbiotic theory states that some of these bacteria reside in eukaryotic organisms alive today as the chloroplasts and/or mitochondria that power all of life. Food-spoilers, nitrogen-fixers, gold-depositors, disease-causers, and drug producers...we know them and their activities well.

There are "fungus among us" everywhere, from the yeast that produce our bread and wine to blights that destroy our crops to the drugs that save our lives. They stealthily grow without the need of sunlight, introducing their walled hyphae into their source of nourishment, be it a rotting log or a human foot. With bizarre life cycles and strange folklore tales about their powers, fungi remain among life's oddest living things.

So, not unexpectedly, segments of the Big Ideas ask you to focus on these mostly microscopic entities of paramount importance:

 Simple bacteria-like cells are hypothesized to be the original form of life on Earth and the common ancestor of all other organisms.

 Viruses, bacteria, and fungi engage in symbiotic relationships with many other organisms, influencing the survival of each.

 The genomes of these groups hold many clues about early life and are now harnessed in many biotechnological endeavors.

 Viruses, bacteria, and fungi impact their ecosystems by limiting population growth, recycling nutrients, and/or serving as the basis of so many food chains.

UNIT OUTLINE

Two women in Mexico City wear face masks to protect against the outbreak of H1N1 in 2009.

On April 24, 2009, the World Health Organization announced an outbreak of a new "swine flu" influenza virus in Mexico. This virus, called H1N1, rapidly spread across the world. Within four weeks, 12,000 confirmed cases of H1N1 had been reported in 43 countries, marking the beginning of a pandemic. Certain viruses and other microbes, including some species of bacteria and archaea, are the cause of many serious diseases in plants and animals, including humans. The reason is that microbes such as the H1N1 virus have the ability to evolve, often fast enough to outmaneuver our own immune systems and our ability to develop effective immunizations and medical treatments. Yet not all microbes cause illness. Our skin is home to about 182 different bacterial species, and our guts are full of bacteria that assist with digestion. Bacteria have uses that include cleaning up oil spills, treating sewage, and even producing human proteins through genetic engineering. In addition, microbes are an essential component of ecosystems. Archaea and bacteria are at the base of the Tree of Life, and were the first living organisms on Earth. Surprisingly, molecular biologists tell us we are more closely related to archaea than to bacteria. In this chapter, we examine these amazing microbes—viruses and prokaryotes.

As you read through the chapter, think about the following questions:

1. Based on the characteristics shared by all living things, should viruses be considered living?

2. How might a rapid mutation rate of virus and bacteria genomes affect our ability to treat diseases?

3. Although some cause disease, why are microorganisms essential to life?

BEFORE YOU BEGIN

Before beginning this chapter, take a few moments to review the following discussions.

Section 1.1 What characteristics are necessary for an organism to be considered "living?"

Section 4.2 What is the structure of a prokaryotic cell?

Section 19.2 What characteristics divide microorganisms into the domains Bacteria and Archaea?

FOLLOWING *the* BIG IDEAS

CHAPTER 20   VIRUSES, BACTERIA, AND ARCHAEA

Evolution	Bacteria are the most ancient form of life on Earth; the origin of viruses remains unknown.
Energy and Homeostasis	Prokaryotes include members who use energy in many forms and under diverse conditions.
Information and Signaling	Viral and bacterial genomes benefit from their ability to evolve rapidly.
Interactions and Systems	Prokaryotes can achieve cooperative relationships which benefit all the members.

20.1 Viruses, Viroids, and Prions

Learning Outcomes

Upon completion of this section, you should be able to

1. Identify the basic structures of a virus.
2. Describe the stages of the lysogenic and lytic reproductive cycle of viruses.
3. Reconstruct the reproductive cycle of HIV.

The term **virus** [L. *virus*, poison] is associated with a number of plant, animal, and human diseases (Table 20.1). The mere mention of the term brings to mind serious illnesses such as polio, rabies, and AIDS (acquired immunodeficiency syndrome), as well as formerly common childhood maladies such as measles, chickenpox, and mumps. Viral diseases are of concern to everyone; it is estimated that the average person catches a cold two or three times a year.

Viruses are a biological enigma. They have some characteristics of living organisms, such as a DNA or RNA genome, and the ability to evolve and replicate. Yet viruses are not considered living organisms. They can reproduce only by using the metabolic machinery of a host cell. Furthermore, viruses do not have a metabolism and do not respond to stimuli. In this section we discuss features of this unique and often puzzling entity.

Table 20.1 Viral Diseases in Humans

Category	Disease
Sexually transmitted diseases	AIDS (HIV), genital warts, genital herpes
Childhood diseases	Mumps, measles, chickenpox, German measles
Respiratory diseases	Common cold, influenza, severe acute respiratory syndrome (SARS)
Skin diseases	Warts, fever blisters, shingles
Digestive tract diseases	Gastroenteritis, diarrhea
Nervous system diseases	Poliomyelitis, rabies, encephalitis
Other diseases	Smallpox, hemorrhagic fevers, cancer, hepatitis, mononucleosis, yellow fever, dengue fever, conjunctivitis, hepatitis C

Discovery and Classification of Viruses

A virus is a small packet of genetic instructions for making copies of itself inside living cells. There are thousands of known viruses that invade all types of living cells, including eukaryotic cells, bacteria, and archaea.

Discovery of Viruses

Our knowledge of viruses began in 1884 when the French chemist Louis Pasteur (1822–95) suggested that something smaller

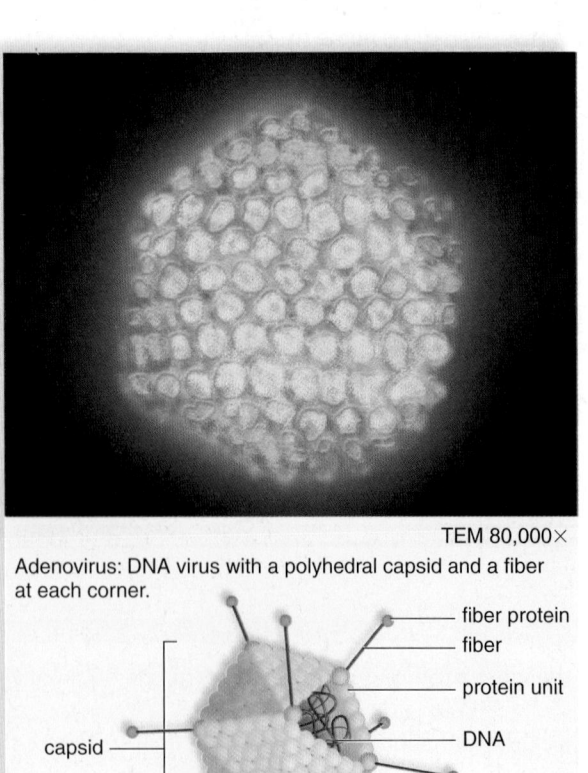

Figure 20.1 Viruses. Despite their diversity, all viruses have an outer capsid composed of protein subunits and a nucleic acid core—composed of either DNA or RNA, but not both. Some types of viruses also have a membranous envelope.

TEM 80,000×

Adenovirus: DNA virus with a polyhedral capsid and a fiber at each corner.

fiber protein
fiber
protein unit
DNA
capsid

a.

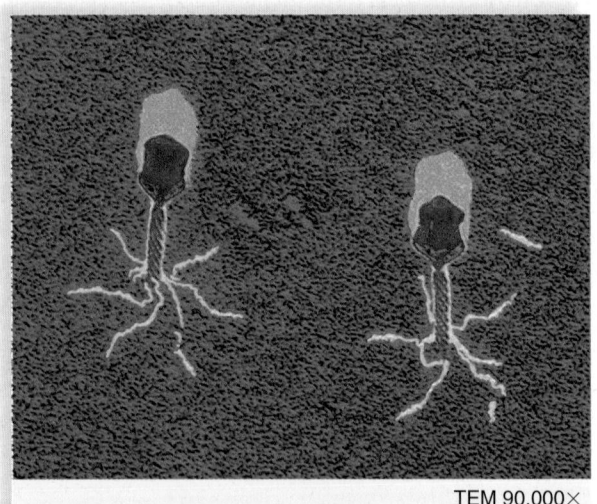

TEM 90,000×

T-even bacteriophage: DNA virus with a polyhedral head and a helical tail.

capsid
DNA
neck
tail sheath
tail fiber
pins
base plate

b.

than a bacterium was the cause of rabies, and he chose the word *virus* from the Latin word meaning poison.

In 1892, Dimitri Ivanowsky (1864–1920), a Russian microbiologist, was studying a disease of tobacco leaves, called tobacco mosaic disease because of the leaves' mottled appearance. He noticed that even when an infective extract was filtered through a fine-pore porcelain filter that retained bacteria, the extract still caused disease. This substantiated Pasteur's belief because it meant that the disease-causing agent was smaller than any known bacterium.

In the twentieth century, electron microscopy was born, and viruses were seen for the first time. By the 1950s, virology was an active field of research; the study of viruses, and now also viroids and prions, has contributed much to our understanding of disease, genetics, and the characteristics of living things.

Classification System of Viruses

Although viruses are not living things, a system for classifying them into natural groups has been devised, based on evolutionary relationships.

The classification system of viruses has two parts. The first part sorts viruses into one of four groups (Groups I–IV) based upon the structure of their genome, for example, whether the genome is DNA or RNA, or whether the DNA or RNA is single- or double-stranded. Group I viruses have double-stranded DNA; the chicken pox and cold sore viruses are Group I viruses.

The second step of virus classification is similar to the Linnaean classification system of organisms. The International Committee on Taxonomy of Viruses (ICTV) has identified taxonomic groups for viruses—order, family, genus, and species. For example, the virus that causes chicken pox is in the genus *Varicellovirus*, and the single species is human herpes virus 3 (HHV-3); the cold sore virus belongs to the genus *Simplexvirus*, and one species is herpes simplex virus 1 (HSV-1).

Because viruses mutate so rapidly, it is difficult to classify them into higher-order taxonomic groups such as phyla, kingdoms, and domains. In addition, a new strain, or genetic type, of virus often emerges within a single species of virus. The seasonal flu is an example of a different strain of virus that emerges each fall and winter.

Structure of Viruses

The size of a virus is comparable to that of a large protein macromolecule, approximately in the range of 10 to 400 nm. Viruses are best studied through electron microscopy (Fig. 20.1). Many viruses can be purified and crystallized, and the crystals can be stored just as chemicals are stored. Still, viral crystals become infectious when the viral particles they contain are given the opportunity to invade a host cell.

Viruses are categorized by (1) their size and shape; (2) their type of nucleic acid, including whether it is single stranded or

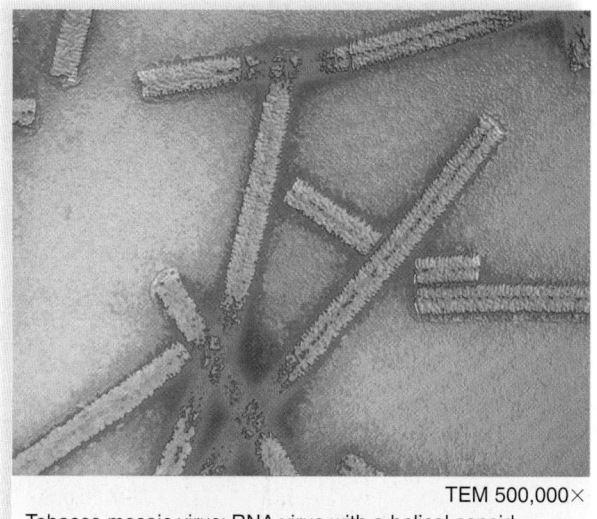

TEM 500,000×

Tobacco mosaic virus: RNA virus with a helical capsid.

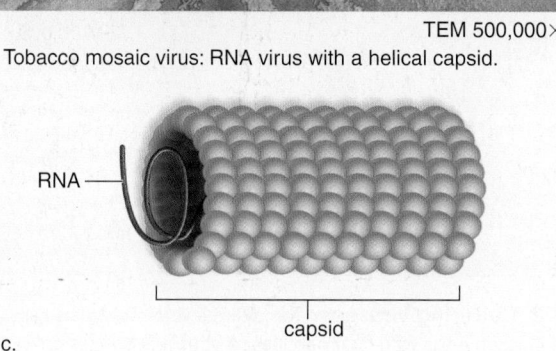

RNA

capsid

c.

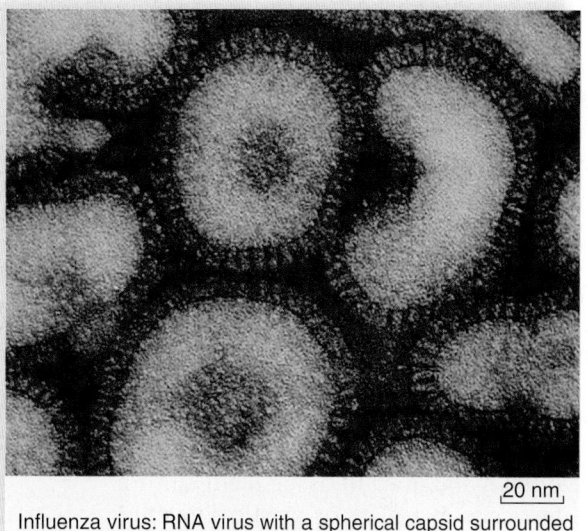

20 nm

Influenza virus: RNA virus with a spherical capsid surrounded by an envelope with spikes.

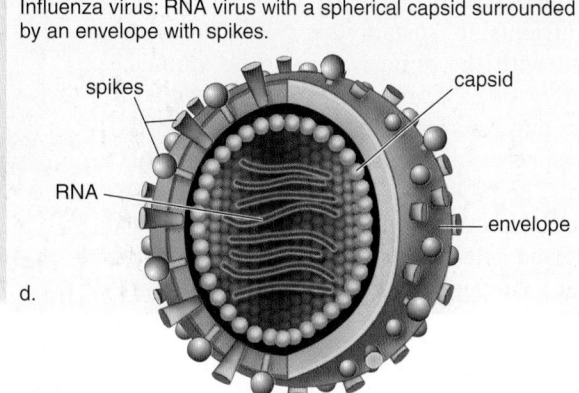

spikes

capsid

RNA

envelope

d.

double stranded; and (3) the presence or absence of an outer envelope. The structure of a virus can be summarized by the following diagram:

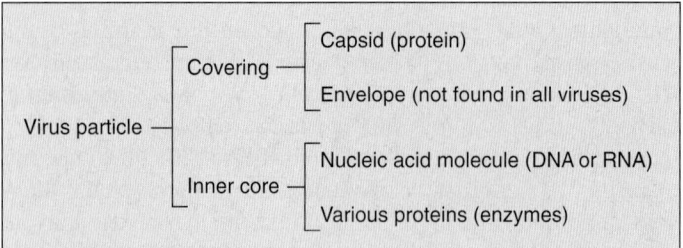

Viruses vary in shape from threadlike to polyhedral. However, all viruses possess the same basic anatomy: an outer **capsid** composed of protein subunits and an inner core of nucleic acid—either DNA or RNA, but not both. A viral genome may have as few as 3 to as many as 100 genes; a human cell, in contrast, contains tens of thousands of genes.

The viral capsid may be surrounded by an outer membranous envelope; if not, the virus is said to be naked. Figure 20.1*a, b, c* gives examples of naked viruses; Figure 20.1*d* is an example of an enveloped virus. The envelope is actually a piece of the host's plasma membrane that also contains viral glycoprotein spikes. Aside from its genome, a viral particle may also contain various proteins, especially enzymes such as the polymerases, needed to produce viral DNA and/or RNA.

Parasitic Nature of Viruses

Viruses are *obligate intracellular parasites,* which means they cannot reproduce outside a living cell. Like prokaryotic and eukaryotic cells, viruses have genetic material. Whereas a cell is capable of copying its own genetic material in order to reproduce, a virus cannot duplicate its genetic material or any of its other components on its own. For a virus to reproduce, it must infect a living cell. Once inside a living cell, the virus "hijacks" the cell's protein synthesis machinery to replicate the nucleic acid and other parts of the virus, including the capsid, viral enzymes, and for some viruses, the envelope.

To maintain animal viruses in the laboratory, technicians sometimes inject them into live chicken embryos (Fig. 20.2). Today, host cells are often maintained in tissue (cell) culture by simply placing a few cells in a glass or plastic container with appropriate nutrients to sustain the cells. The cells can then be infected with the animal virus to be studied. Viruses infect a variety of cells, but they are *host specific.* Bacteriophages infect only bacteria, the tobacco mosaic virus infects only certain species of plants, and the rabies virus infects only mammals. Some human viruses even specialize in a particular tissue. Human immunodeficiency virus (HIV) enters only certain blood cells, the polio virus reproduces in spinal nerve cells, and the hepatitis viruses infect only liver cells.

What could cause this remarkable parasite–host cell correlation? Some scientists hypothesize that viruses are derived from the very cell they infect; in other words, that the nucleic acid of viruses came from their host cell genomes! In that case, viruses evolved after cells came into existence, and new viruses may be evolving even now. In 2000, using a combination of DNA and protein analyses, scientists hypothesized that viruses arose early in the origin of life, predating the three domains.

Viruses can also mutate; therefore, it is correct to say that they evolve. Those that undergo rapid mutation can be quite troublesome because a vaccine that is effective today may not be effective tomorrow. Flu viruses are well known for mutating, and this is why it is necessary to have a flu shot every year—antibodies generated from last year's shot are not expected to be effective this year.

The HIV is a very rapidly evolving virus, which makes it difficult to find a cure. Viruses such as HIV are often engaged in an "arms race" with an animal's immune system. As new vaccines and drugs are discovered for HIV, new forms evolve that evade the new treatments. In the case of HIV, the result of rapid evolution is the presence of many different virus types within each infected individual. Each individual can carry billions of virions. Thus it is very likely that at least one virus type will by chance evolve resistance to treatment. The rapid evolution of resistance in HIV is why there is currently no cure.

Reproduction of Viruses

Viruses are microscopic pirates, commandeering the metabolic machinery of a host cell. Viruses gain entry into a host cell because portions of a naked capsid (or one of the types of envelope spikes) attach in a lock-and-key manner with a receptor on the host cell's outer surface. The attachment of the capsid or spikes of a virus to particular host cell receptors is responsible for the remarkable specificity between viruses and their host cells. A virus cannot infect a host cell to which it is unable to

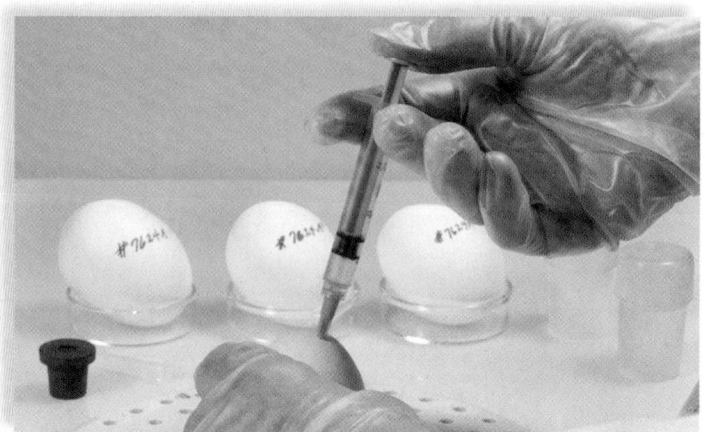

Figure 20.2 Culturing viruses.   To culture a virus, scientists can inoculate live chicken eggs with viral particles. A virus reproduces only inside a living cell because it takes over the machinery of the cell.

attach. For example, the tobacco mosaic virus cannot infect an exposed human because its capsid cannot attach to the receptors on the surfaces of human cells. Many antiviral medications are effective because they interfere with the lock-and-key attachment of viruses to the host cell.

Animation
Antiviral Agents

After a virus has become attached to a suitable host cell, the viral nucleic acid enters the cell. Once inside, the nucleic acid codes for the protein units in the capsid. In addition, the virus may have genes for special enzymes needed for the virus to reproduce and exit from the host cell. In large measure, however, a virus relies on the host's enzymes, ribosomes, transfer RNA (tRNA), and ATP (adenosine triphosphate) for its own reproduction. Because the host cell's metabolism is diverted from meeting the needs of the cell, infected cells may have an abnormal appearance.

Video
How Viruses Attack

Reproduction of Bacteriophages

Bacteriophages [Gk. *bacterion*, rod, and *phagein*, to eat], or simply phages, are viruses that parasitize bacteria. Figure 20.3 shows two alternative life cycles of bacteriophages, called the lytic cycle and the lysogenic cycle. All viruses have a lytic cycle, a lysogenic cycle, or both at different times. In the lytic cycle, viral reproduction occurs right after the virus enters the cell, and the host cell undergoes *lysis,* a breaking open of the cell to release new viral particles. In the lysogenic cycle, viral reproduction does not immediately occur, but reproduction may take place sometime in the future.

Figure 20.3 Lytic and lysogenic cycles in prokaryotes. **a.** In the lytic cycle, viral particles escape when the cell is lysed (broken open). In the lysogenic cycle, viral DNA is integrated into host DNA. At some time in the future, the lysogenic cycle can be followed by the lytic cycle. **b.** Micrograph of bacteriophage viruses attaching to a bacterium, *Escherichia coli.*

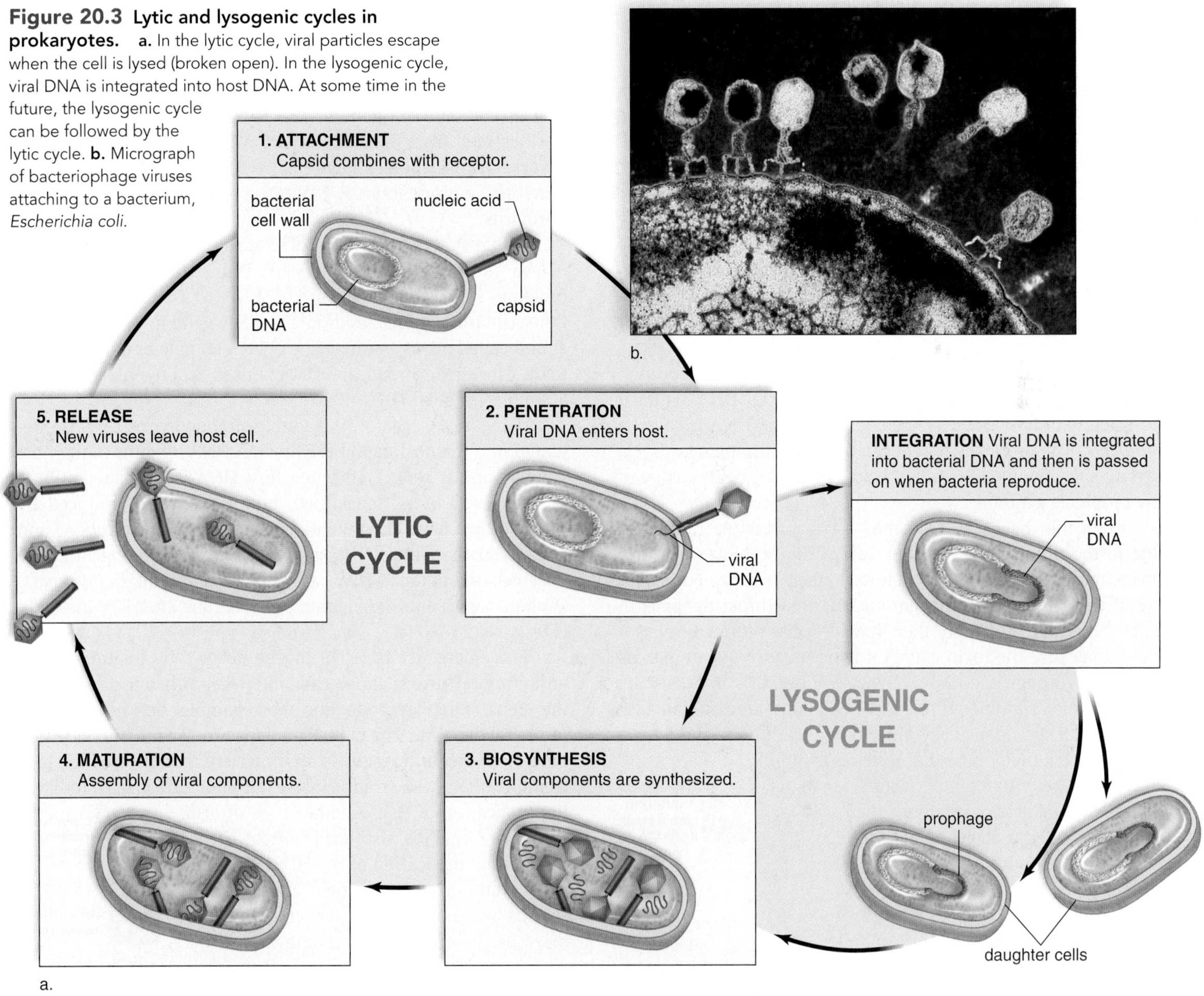

1. ATTACHMENT
Capsid combines with receptor.

bacterial cell wall
nucleic acid
bacterial DNA
capsid

b.

5. RELEASE
New viruses leave host cell.

LYTIC CYCLE

2. PENETRATION
Viral DNA enters host.

viral DNA

INTEGRATION Viral DNA is integrated into bacterial DNA and then is passed on when bacteria reproduce.

viral DNA

LYSOGENIC CYCLE

4. MATURATION
Assembly of viral components.

3. BIOSYNTHESIS
Viral components are synthesized.

prophage

daughter cells

a.

Viruses that are highly virulent enter directly into the lytic cycle, causing rapid and severe destruction of host cells. The Ebola virus, which causes Ebola hemorrhagic fever, is highly virulent. Up to 90% of those infected die from the disease within 2–21 days after infection. In contrast, HIV enters first into the lysogenic cycle, and thus can lay inactive, or latent, for many years before AIDS symptoms emerge. The following discussion is based on the DNA bacteriophage lambda, which undergoes both lytic and lysogenic cycles.

Lytic Cycle.   The **lytic cycle** may be divided into five stages: attachment, penetration, biosynthesis, maturation, and release.

During *attachment,* portions of the capsid combine with a receptor on the rigid bacterial cell wall in a lock-and-key manner. During *penetration,* a viral enzyme digests away part of the cell wall, and viral DNA is injected into the bacterial cell. *Biosynthesis* of viral components begins after the virus brings about inactivation of host genes not necessary to viral replication. The virus takes over the machinery of the cell in order to carry out viral DNA replication and production of multiple copies of the capsid protein subunits. During *maturation,* viral DNA and capsids assemble to produce several hundred viral particles. Lysozyme, an enzyme coded for by a viral gene, is produced; this disrupts, or lyses, the cell wall, and the *release* of new viruses occurs. The bacterial cell dies as a result.

Lysogenic Cycle.   With the **lysogenic cycle**, an infecting phage does not immediately proliferate, but may do so sometime in the future. In the meantime, the phage is *latent*—not actively replicating.

Following attachment and penetration, *integration* occurs: Viral DNA becomes incorporated into bacterial DNA with no destruction of host DNA. While latent, the viral DNA is called a *prophage.* The prophage is replicated along with the host DNA, and all subsequent cells, called **lysogenic cells,** carry a copy of the prophage genome.

Lysogenic bacterial cells may have distinctive properties due to the prophage genes they carry. The presence of a prophage may cause a bacterial cell to produce a toxin. For example, if the same bacterium that causes strep throat happens to carry a certain prophage, then it will cause scarlet fever, so-named because the toxin causes a widespread red skin rash as it spreads through the body. Likewise, diphtheria is caused by a bacterium carrying a prophage. The diphtheria toxin damages the lining of the upper respiratory tract, resulting in the formation of a thick membrane that restricts breathing.

Certain environmental factors, such as ultraviolet radiation, can induce the prophage to enter the lytic stage of biosynthesis, followed by maturation and release.

Reproduction of Animal Viruses

Animal viruses reproduce in a manner similar to that of bacteriophages, but with modifications. Various animal viruses have different ways of introducing their genetic material into their host cells. For some enveloped viruses, the process is as simple as attachment and fusion of the spike-studded envelope with the host cell's plasma membrane. Many naked and some enveloped viruses are taken into host cells by endocytosis. Once inside, the virus is uncoated—that is, the capsid and, if necessary, the envelope are removed. The viral genome, either DNA or RNA, is now free of its covering, and biosynthesis plus the other steps then proceed.

Viral release is just as variable as penetration for animal viruses. Some mature viruses are released by budding. During budding, the virus picks up its envelope, consisting of lipids, proteins, and carbohydrates, from the host cell. Most enveloped animal viruses acquire their envelope from the plasma membrane of the host cell, but some take envelopes from other membranes, such as the nuclear envelope or Golgi apparatus. Envelope markers, such as the glycoprotein spikes that allow the virus to enter a host cell, are coded for by viral genes. Naked animal viruses are usually released by host cell lysis.

Retroviruses.   Retroviruses [L. *retro,* backward] are animal viruses with an RNA genome that is converted into DNA within the host cell by a special enzyme called **reverse transcriptase.** Figure 20.4 illustrates the reproduction of HIV, a type of retrovirus.

Before a retrovirus can integrate into the host's genome, or use the host cell's machinery to transcribe and translate its proteins, it must first convert its RNA to DNA. First, reverse transcriptase synthesizes from its RNA genome a single DNA strand called cDNA because it is a DNA complement of the viral RNA. The single strand of cDNA is used as a template to make a double-stranded DNA.

Using host enzymes, the double-stranded virus DNA is integrated into the host genome. The viral DNA remains in the host genome and is replicated when host DNA is replicated. When and if this DNA is transcribed, new viruses are produced by the steps we have already cited: biosynthesis, maturation, and release; in the case of HIV, it is released by budding from the host cell membrane. HIV can remain latent for many years. Without treatment, the median survival time after HIV infection is 9–11 years.

The emergence of AIDS can be delayed by treatment with antiretroviral drugs, those that interfere with one or more of the steps of HIV reproduction. For example, one type of antiretroviral drug consists of reverse transcriptase inhibitors, which bind to reverse transcriptase and interfere with its function. For example, the reverse transcriptase inhibitor AZT is used to block the replication of HIV. Another type of drug, Acyclovir, which is also used to treat herpes, inhibits the replication of the HIV viral DNA.

Certain viral infections are especially serious because they lead to even more severe diseases. In humans, papillomaviruses, herpesviruses, hepatitis viruses, adenoviruses, and retroviruses are associated with specific types of cancer. Emerging viruses are of recent concern.

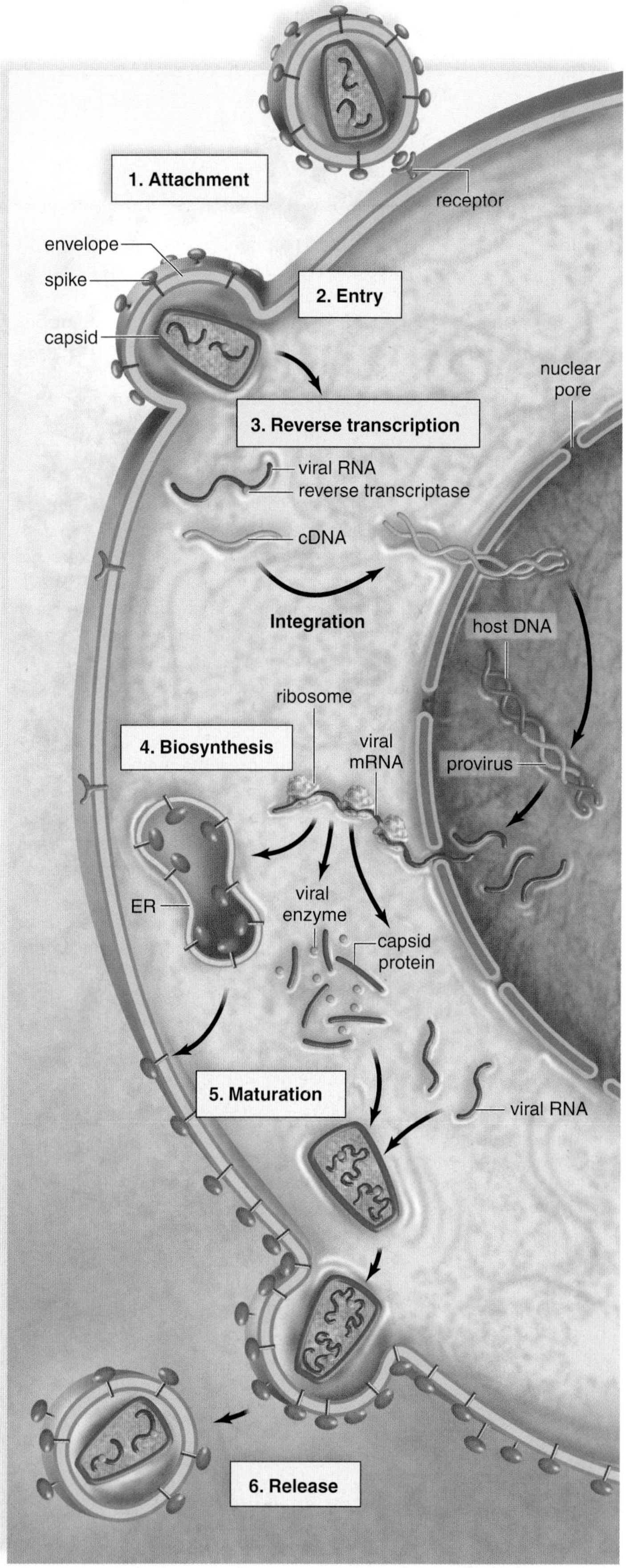

Figure 20.4

1. Attachment

receptor

envelope

spike

capsid

2. Entry

nuclear
pore

3. Reverse transcription

viral RNA
reverse transcriptase

cDNA

Integration

host DNA

ribosome

4. Biosynthesis

viral
mRNA

provirus

ER

viral
enzyme

capsid
protein

5. Maturation

viral RNA

6. Release

Figure 20.4 Reproduction of the retrovirus HIV. HIV uses reverse transcription to produce a DNA copy (cDNA) of RNA genes; double-stranded DNA integrates into the cell's chromosomes before the virus reproduces and buds from the cell. The steps in color are unique to retroviruses.

Emerging Viruses

Some emerging diseases—new or previously uncommon illnesses—are caused by viruses that are now able to infect large numbers of humans. These viruses are known as **emerging viruses.** Examples of emerging viral diseases are AIDS, West Nile encephalitis, hantavirus pulmonary syndrome (HPS), severe acute respiratory syndrome (SARS), Ebola hemorrhagic fever, and avian influenza (bird flu) (Fig. 20.5).

Several different types of events can cause a viral disease to suddenly "emerge" and start causing a widespread human illness. A virus can extend its range when it is transported from one part of the world to another. West Nile encephalitis is a virus that extended its range after being transported into the United States, where it took hold in bird and mosquito populations. SARS was transported from Southeast Asia to Toronto, Canada. In 2009, a new strain of influenza virus, H1N1, became an emerging viral disease. Although sometimes called the "swine flu," H1N1 is actually a form of the human influenza A virus, the same one that causes seasonal flu. However, genetic analysis shows that it is actually a combination of influenza viruses from pigs, birds, and humans.

Two other widespread emergences of H1N1 type viruses have had a large impact on human populations. The largest emergence of a swine flu virus was the flu pandemic in 1918 that killed 50 million people worldwide, approximately one-third of the world's population at the time. Another type of H1N1 virus emerged as another flu pandemic in the 1970s. People exposed to the 1970s H1N1 virus likely retained some immunity to other swine flu viruses. This is the reason that the 2009 H1N1 is thought to have had a greater impact on the young; older people may have had the advantage of earlier exposure, and some immunity, to H1N1.

Viruses are well known for their ability to move between different species. Sometimes viruses that formerly infected animals other than humans can "jump" species and start infecting humans due to a change in their capsids or spikes that enables them to attach to human cell receptors, as discussed in the Biological Systems feature on page 370. The 2009 H1N1 pandemic was a result of the evolution of novel ways of infecting human cells.

Video
Killer Flu Recreated

Video
Virus Crisis

Viroids and Prions

At least a thousand different viruses cause diseases in plants. About a dozen diseases of crops, including potatoes, coconuts, and citrus, have been attributed not to viruses but to **viroids,** which are naked strands of RNA (not covered by a capsid). Like viruses, however, viroids direct the cell to produce more viroids.

A number of fatal brain diseases, known as *transmissible spongiform encephalopathies,* or TSEs, have been attributed to **prions,** a term coined for *proteinaceous infectious* particles. Prions are proteins that normally exist in an animal, but have a different conformation, or structure. Like viruses, prions cannot replicate on their own, but cause infection by interacting with a normal protein and altering its structure. The process that

Biological Systems

Flu Pandemic

If you've ever had seasonal flu (influenza), you know how miserable it can be. The flu is a viral infection that causes runny nose, cough, chills, fever, head and body aches, and nausea. You catch the flu by inhaling virus-laden droplets that have been coughed or sneezed into the air by an infected person, or by contact with contaminated objects, such as door handles or bedding. The viruses then attach to and infect cells of the respiratory tract.

Flu Viruses

A flu virus has an H (hemagglutinin) spike and an N (neuraminidase) spike (Fig. 20A*a*, *left*). Its H spike allows the virus to bind to its receptor, and its N spike attacks host plasma membranes in a way that allows mature viruses to exit the cell.

Just as purses and wallets can each be shaped differently, so can H spikes and N spikes: 16 types of H and 9 types of N spikes are known. Worse yet, just as any shape purse or wallet can be a different color, so each type of spike can occur in different varieties called subtypes. Many of the flu viruses are assigned specific codes based on the type of spike. For example, H5N1 virus gets its name from its variety of H5 spikes and its variety of N1 spikes. Our immune system can recognize only the particular variety of H spikes and N spikes it has been exposed to in the past by infection or immunization. When a new flu virus arises, one for which there is little or no immunity in the human population, a flu pandemic (global outbreak) may occur.

Possible Bird Flu Pandemic of the Future

Currently, the H5N1 subtype of flu virus is of great concern because of its potential to reach pandemic proportions. An H5N1 is common in wild birds such as waterfowl, and can readily infect domestic poultry such as chickens, which is why it is referred to as an avian influenza or a bird flu virus. An H5N1 virus has infected waterfowl for some time without causing serious illness. A more pathogenic version of H5N1 appeared about a decade ago in China, and promptly started to cause widespread and severe illness in domestic chickens. Scientists are still trying to determine what made H5N1 become so lethal, first to chickens, and then to humans.

Why can the bird flu H5N1 infect humans? Because the virus can attach to both a bird flu receptor and to a human flu receptor. Close contact between domestic poultry and humans is necessary for this to happen. At this time, the virus has rarely been transmitted from one human to another, and only among people who have close contact with one another, such as members of the same household. The concern is that with additional mutations, the H5N1 virus could become capable of sustained human-to-human transmission, and then spread around the world.

How could H5N1 become better at spreading within the human population? At this time, bird flu H5N1 infects mostly the lungs. Most human flu viruses infect the upper respiratory tract, trachea, and bronchi and can be spread by coughing. If a spontaneous mutation in the H spike of H5N1 enabled it to attack the upper respiratory tract, then it could be easily spread from human to human by coughing and sneezing (Fig. 20A*a*). Or, another possibility is that a combining of spikes could occur in a person who is infected with both the bird flu and the human flu viruses (Fig. 20A*b*). According to the CDC (Centers for Disease Control and Prevention), over the past decade an increasing number of humans infected with an H5N1 virus have been reported in Asia, the Pacific, the Near East, Africa, and Europe. Over half of these people have died. Currently, there are no available vaccines for an H5N1 virus.

How to Be Prepared

A flu pandemic presents many challenges, including rapid spread of the virus, the overload of our health-care systems, inadequate medical supplies, and economic and social disruption. Vaccines and antiviral medications would become very short in supply. Knowing what a pandemic is, what needs to be done to prepare for one, and what could happen during a pandemic will help us as individuals and citizens to make wise decisions. For more information on flu pandemics, visit www.pandemicflu.gov.

> Video
> Flu Fighter

Questions to Consider

1. Based on how the H5N1 pandemic began, how might have the H1N1 swine flu have been transmitted from animals to humans?

2. Compare the codes of the H5N1 and the H1N1 viruses. What can their codes tell you about differences in their structure?

3. Why do humans need a flu vaccine every year?

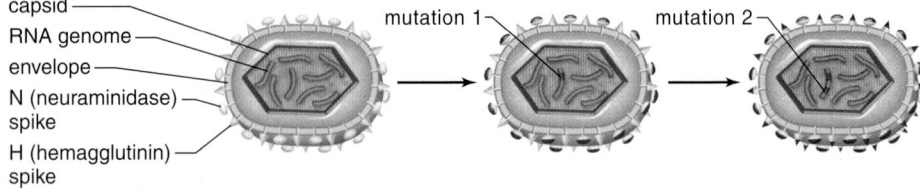

capsid
RNA genome
envelope
N (neuraminidase) spike
H (hemagglutinin) spike

mutation 1   mutation 2

a. Viral genetic mutations occur in a bird host

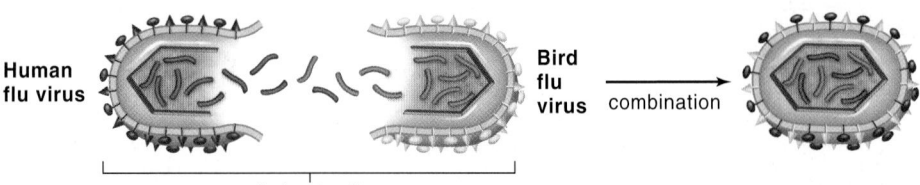

Human flu virus   Bird flu virus   combination

in host cell

b. Combination of viral genes occurs in a human host

Figure 20A Spikes of bird flu virus.
a. Genetic mutations in bird flu viral spikes could allow the virus to infect the human upper respiratory tract. **b.** Alternatively, combination of bird flu and human spikes could allow the virus to infect the human upper respiratory tract.

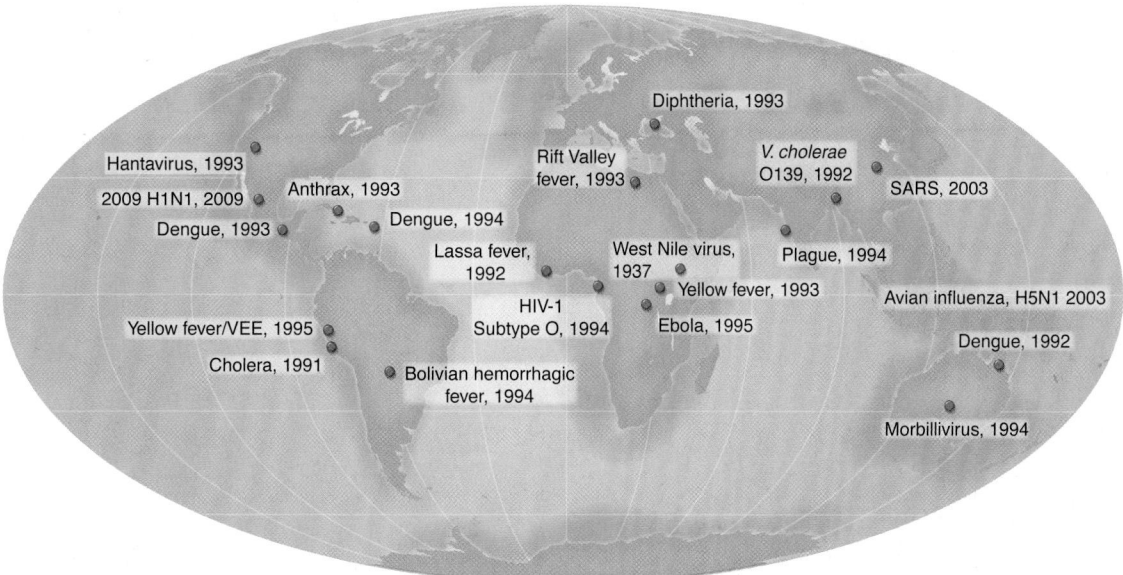

Figure 20.5 Emerging diseases.   Emerging diseases, such as those noted here according to their country of origin, are new or demonstrate increased prevalence. These disease-causing agents may have acquired new virulence factors, or environmental factors may have encouraged their spread to an increased number of hosts.

changes the structure from the normal protein conformation to the prion conformation can be as simple as changing a chemical bond. Once a prion infects tissue, a chain reaction begins that converts normal proteins to prions at an exponential rate.

TSEs are **neurodegenerative diseases**, or those that destroy nerve tissue in the brain. In the brain, prion proteins form clusters that break down normal brain tissue, creating small holes that give the brain a spongy appearance. All TSEs are untreatable and fatal. Mad cow disease, or bovine spongiform encephalopathy (BSE), is a neurodegenerative disease of cattle that is transmissible to humans by eating cattle brains or meat contaminated with prion-infected brain tissue. The discovery of prions began when it was observed that members of a primitive tribe in the highlands of Papua New Guinea died from a disease commonly called kuru (meaning trembling with fear). The disease occurred after the individual had participated in the cannibalistic practice of eating a deceased person's brain; the brain was evidently infected with prions.

 **Animation**
How Prions Arise

Animation
Prion Diseases

Check Your Progress   20.1

1. Describe the two features shared by all viruses.
2. List the features that viruses and prions do not have that are required to be considered living organisms.
3. Distinguish between the structure of a virus and a prion.
4. Explain from an evolutionary standpoint why it is beneficial to a virus if its host lives.

20.2 The Prokaryotes

Learning Outcomes

Upon completion of this section, you should be able to

1. List the two domains of life that contain prokaryotes.
2. Identify structural features of prokaryotes.
3. Describe at least four ways in which the cells of prokaryotes differ from eukaryotic cells.

Prokaryotes include bacteria and archaea, which are fully functioning, living, single-celled organisms. Because they are microscopic, the prokaryotes were not discovered until the Dutch microscopist Antonie van Leeuwenhoek (1632–1723) first described them along with many other microorganisms (see the Nature of Science feature "Microscopy Today," in Chapter 4).

Leeuwenhoek and others after him believed that the "little animals" that he observed could arise spontaneously from inanimate matter. *Spontaneous generation*, the idea that living organisms can emerge from nonliving things, was common at the time. When meat spoiled, for example, it was thought that the maggots arose from the meat spontaneously.

For about 200 years, scientists carried out various experiments to determine the origin of microorganisms in laboratory cultures. Finally, in about 1850, Louis Pasteur devised an experiment for the French Academy of Sciences that is described in Figure 20.6. It showed that a previously sterilized broth cannot become cloudy with microorganism growth unless it is exposed directly to the air where bacteria are abundant.

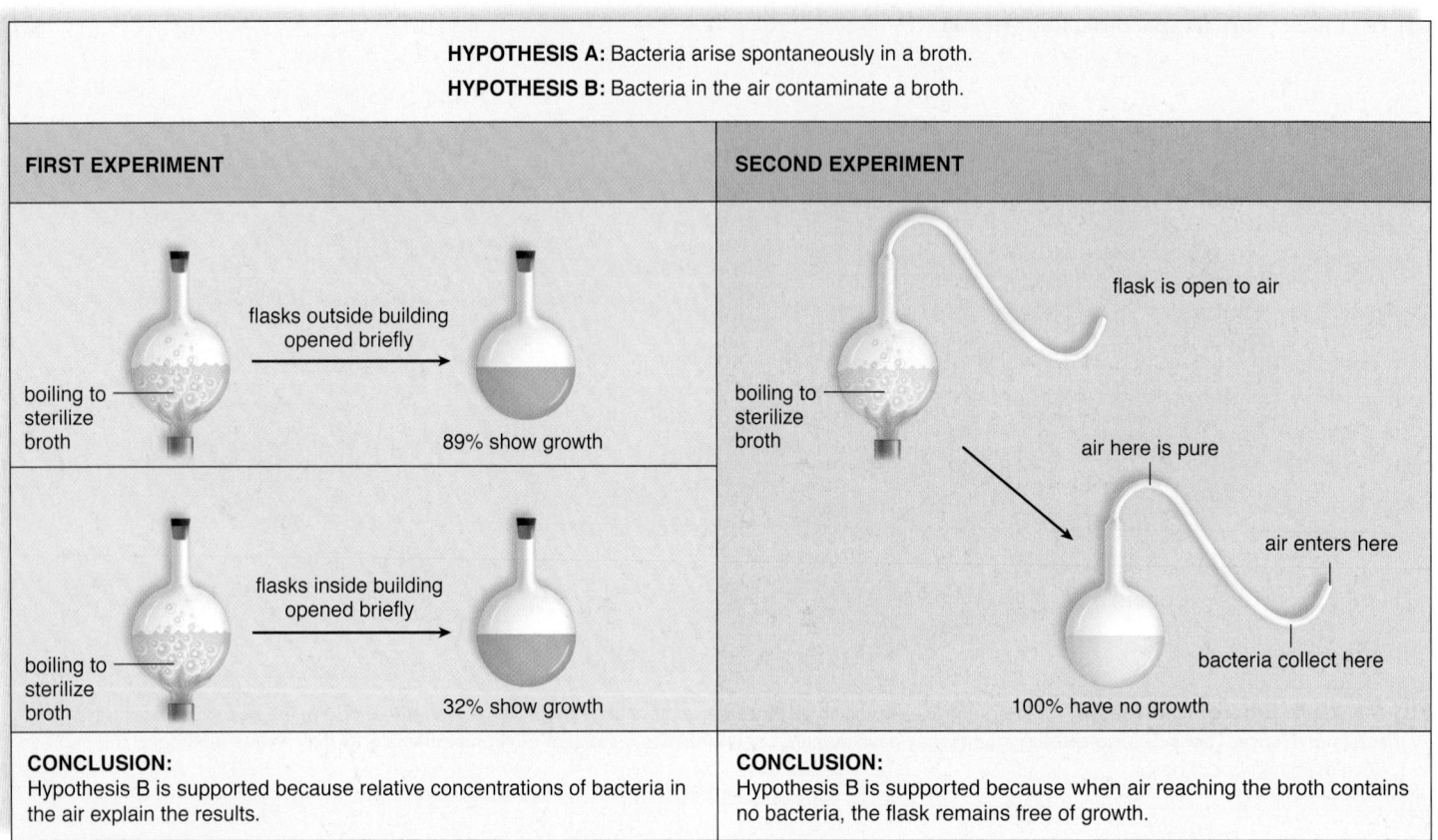

HYPOTHESIS A: Bacteria arise spontaneously in a broth.
HYPOTHESIS B: Bacteria in the air contaminate a broth.

FIRST EXPERIMENT

boiling to sterilize broth → flasks outside building opened briefly → 89% show growth

boiling to sterilize broth → flasks inside building opened briefly → 32% show growth

CONCLUSION:
Hypothesis B is supported because relative concentrations of bacteria in the air explain the results.

SECOND EXPERIMENT

flask is open to air

boiling to sterilize broth

air here is pure

air enters here

bacteria collect here

100% have no growth

CONCLUSION:
Hypothesis B is supported because when air reaching the broth contains no bacteria, the flask remains free of growth.

Figure 20.6 Pasteur's experiments. Pasteur disproved the theory of spontaneous generation of microbes by performing these types of experiments.

Today we know that bacteria are plentiful in air, water, and soil, and that new bacteria arise from the division of pre-existing bacteria—not by spontaneous generation. We also know a lot about the structure of bacteria, their membranes, DNA, and proteins, how and where they live, where they get their nutrition, and how they coordinate replication. In the pages following, the general characteristics of prokaryotes are discussed before those specific to the bacteria (domain Bacteria) and then the archaea (domain Archaea) are considered in more detail.

Structure of Prokaryotes

Prokaryotes generally range in size from 1 to 10 μm in length and from 0.7 to 1.5 μm in width. To put this into perspective, a human is on average 1 m tall, or 1 million times longer than a bacterium (see Fig. 4.2). The term *prokaryote* means "before a nucleus," and these organisms lack a membrane-bound nucleus like that found in eukaryotes. Prokaryotic fossils exist that are dated as long ago as 3.5 billion years, and the fossil record indicates that the prokaryotes were alone on Earth for at least 1.3 billion years. During that time, they became extremely diverse in structure and especially diverse in metabolic capabilities. Prokaryotes are adapted to living in most environments because they have a wide variety of ways that they can acquire and use energy.

A typical prokaryotic cell has a cell wall situated outside the plasma membrane (see Fig. 4.4). The cell wall prevents a prokaryote from bursting or collapsing due to fluctuations in the amount of fluid inside the cell. Yet another layer may exist outside the cell wall; the structure and composition of this layer vary among the different kinds of prokaryotes.

In many bacteria, the cell wall is surrounded by a layer of polysaccharides called a *glycocalyx*. A well-organized glycocalyx is called a capsule (Fig. 20.7a), while a loosely organized one is called a slime layer. Many bacteria and archaea have a layer comprised of protein, or glycoprotein, instead of a glycocalyx; such a layer is called an *S-layer*. In parasitic forms of bacteria, these outer coverings help protect the cell from host defenses.

Some prokaryotes move by means of **flagella** (Fig. 20.7b). A bacterial flagellum has a filament composed of strands of the protein flagellin wound in a helix. The filament is inserted into a hook that is anchored by a basal body. The 360° rotation of the flagellum causes the cell to spin and move forward. The archaeal flagellum is similar, but more slender and apparently lacking a basal body.

Many prokaryotes adhere to surfaces by means of **fimbriae,** short bristlelike fibers extending from the surface (Fig. 20.7a). The fimbriae of the bacterium *Neisseria gonorrhoeae* allow it to attach to host cells and cause gonorrhea, a sexually transmitted disease.

Animation Bacterial Locomotion

A prokaryotic cell lacks the membranous organelles of a eukaryotic cell; instead the various metabolic pathways take place within the plasma membrane. Although prokaryotes do not have a nucleus, they do have a dense area called a **nucleoid** where a single chromosome consisting of a circular strand of DNA is found. Many prokaryotes also have accessory rings of DNA called **plasmids.** Plasmids can be extracted and used to carry foreign DNA into host bacteria during genetic engineering processes.

Protein synthesis in a prokaryotic cell is carried out by thousands of ribosomes, which are smaller than eukaryotic ribosomes. The diagram on the next page (Fig. 20.7) summarizes the structure of a prokaryotic cell.

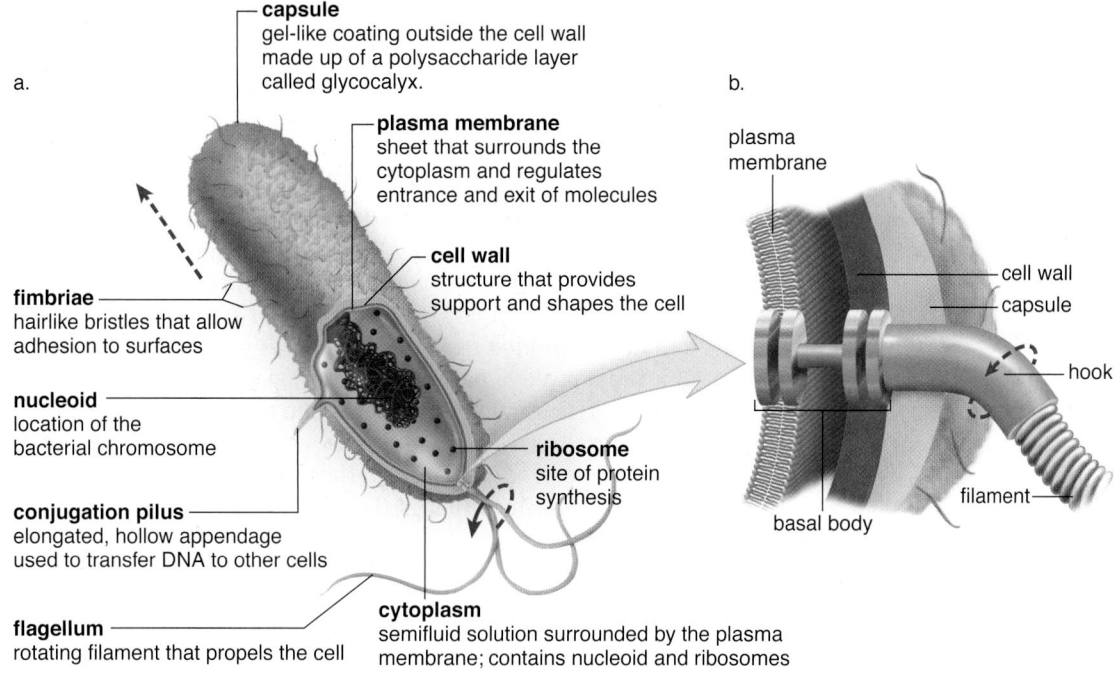

a.

capsule
gel-like coating outside the cell wall
made up of a polysaccharide layer
called glycocalyx.

plasma membrane
sheet that surrounds the
cytoplasm and regulates
entrance and exit of molecules

cell wall
structure that provides
support and shapes the cell

fimbriae
hairlike bristles that allow
adhesion to surfaces

nucleoid
location of the
bacterial chromosome

conjugation pilus
elongated, hollow appendage
used to transfer DNA to other cells

flagellum
rotating filament that propels the cell

ribosome
site of protein
synthesis

cytoplasm
semifluid solution surrounded by the plasma
membrane; contains nucleoid and ribosomes

b.

plasma
membrane

cell wall
capsule

hook

filament

basal body

Figure 20.7 Features of prokaryotic cells.
a. Structural components of a generalized prokaryotic cell. b. Each flagellum of a bacterium contains a basal body, a hook, and a filament. The red-dashed arrows indicate that the hook and filament rotate 360°.

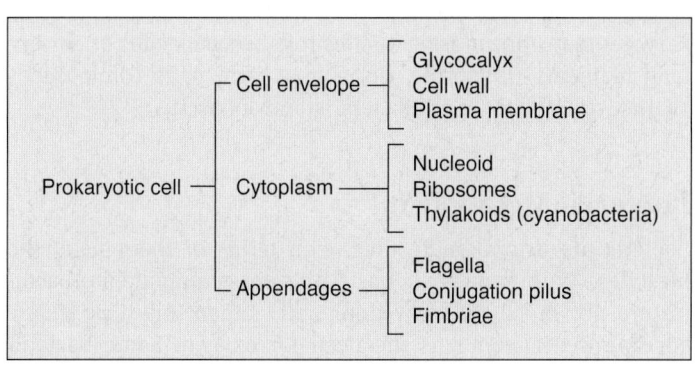

Prokaryotic cell ─ Cell envelope ─ Glycocalyx / Cell wall / Plasma membrane
Cytoplasm ─ Nucleoid / Ribosomes / Thylakoids (cyanobacteria)
Appendages ─ Flagella / Conjugation pilus / Fimbriae

Reproduction in Prokaryotes

Mitosis, which requires the formation of a spindle apparatus, does not occur in prokaryotes. Instead, prokaryotes reproduce asexually by means of **binary fission** (Fig. 20.8).

The single circular chromosome replicates, and then the two copies separate as the cell enlarges. Newly formed plasma membrane and cell wall separate the cell into two cells. Prokaryotes have a generation time as short as 12 minutes under favorable conditions. Mutations are generated and passed on to offspring more quickly than in eukaryotes. Also, prokaryotes are haploid, and so mutations are immediately subjected to natural selection, which determines any possible adaptive benefit in the particular environment.

 **Animation** Binary Fission

In eukaryotes, genetic recombination occurs as a result of sexual reproduction. Sexual reproduction does not occur among prokaryotes, but three means of genetic recombination have been observed in prokaryotes.

- During **conjugation,** two bacteria are temporarily linked together, often by means of a **conjugation pilus** (see Fig. 20.7a). While they are linked, the donor cell passes DNA to a recipient cell.
- **Transformation** occurs when a cell picks up free pieces of DNA from its surrounding medium; this DNA has

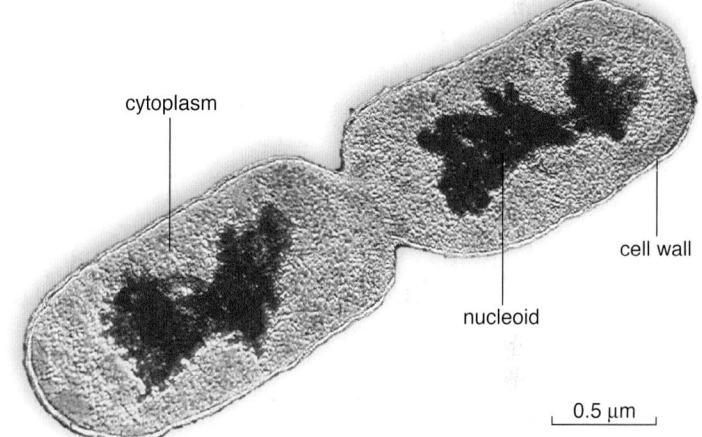

cytoplasm

cell wall

nucleoid

0.5 μm

Figure 20.8 Binary fission. When conditions are favorable for growth, prokaryotes divide to reproduce. This is a form of asexual reproduction because the daughter cells have exactly the same genetic material as the parent cell.

been secreted by live prokaryotes or released by dead prokaryotes.
- During **transduction,** bacteriophages carry portions of DNA from one bacterial cell to another. Viruses have also been found to infect archaeal cells, and so transduction may play an important role in gene transfer for both domains of prokaryotes.

Animation Bacterial Conjugation

Animation Bacterial Transformation

Check Your Progress 20.2

1. Explain the connection between Pasteur's experiment and the sterilization of surgical instruments.
2. Describe the difference between a prokaryote nucleoid and a eukaryote nucleus.
3. Define three ways in which prokaryotes can recombine their genetic material without sexual reproduction.

20.3 The Bacteria

Learning Outcomes

Upon completion of this section, you should be able to

1. Identify the similarities and differences in the cell wall structure of Gram-positive and Gram-negative bacteria.
2. Identify three different metabolic types of bacteria and describe how they obtain nutrients from their environments.
3. Describe the unique properties of cyanobacteria.

Bacteria (domain Bacteria) are the more common type of pro-karyote. The number of species of bacteria is amazing: To date, over 9,000 different bacteria species have been named, but the actual number of species is likely in the tens of millions. They are found in practically every kind of environment on Earth. In this section we consider the bacteria—their characteristics, metabolism, and lifestyle.

Characteristics of Bacterial Cells

Most bacterial cells are protected by a cell wall that contains layers of plasma membrane and the unique molecule **peptidoglycan.** Peptidoglycan is a complex of polysaccharides linked by amino acids. Two types of bacterial cell walls have been distinguished based on staining properties of their peptidoglycan layer.

In the late 1880s, Hans Christian Gram, a Danish bacteriologist, formulated an iodine-dye complex that stains peptidoglycan purple; this dye, or stain, is named after him. When exposed to the Gram stain, some bacteria stain a darker purple than others. *Gram-positive bacteria,* with a thicker peptidoglycan layer, are darker purple; *Gram-negative bacteria,* are lighter purple or lack purple color completely.

Gram-positive bacteria have a single plasma membrane and a thick outer layer of peptidoglycan. Further investigation showed that in the Gram-negative bacteria, the peptidoglycan layer is sandwiched between two plasma membranes. This extra, outer membrane reduces the amount of Gram stain that binds to the peptidoglycan, so during wash steps, the iodine-dye complex is washed away. A second dye, safranin, is needed to stain Gram-negative bacteria pink so that they can be seen under a microscope.

Most of the bacteria that are harmful to humans are Gram-negative, but Gram-positive bacteria also are responsible for human illnesses. *Vibrio cholerae,* which infects the small intestine and causes cholera, is an example of a Gram-negative bacterium. *Clostridium tetani,* which causes tetanus, is an example of a Gram-positive bacterium. Even today, the Gram stain is usually the first test used to identify unknown bacteria.

Animation
Gram Stain

Bacteria (and archaea) can also be described in terms of their three basic cell shapes (Fig. 20.9):

- spirilli (sing., spirillum), spiral-shaped or helical-shaped;
- bacilli (sing., bacillus), rod-shaped; and
- cocci (sing., coccus), round or spherical.

These three basic shapes may be augmented by particular arrangements or shapes of cells. For example, rod-shaped prokaryotes may appear as very short rods (coccobacilli) or as very long filaments (fusiform). Cocci may form pairs (diplococci), chains (streptococci), or clusters (staphylococci).

Bacterial Metabolism

Bacteria are astoundingly diverse in terms of their metabolic lifestyles. With respect to basic nutrient requirements, bacteria are not much different from other organisms. One difference, however, concerns the need for oxygen. Some bacteria are **obligate anaerobes** and are unable to grow in the presence of free oxygen. A few serious illnesses—such as botulism, gas gangrene, and tetanus—are caused by anaerobic bacteria that infect oxygen-free environments in the human body, such as in the intestine or in deep puncture wounds. Other bacteria, called **facultative anaerobes,** are able to grow in either the presence or the absence of gaseous oxygen. Most bacteria, however, are aerobic and, like animals, require a constant supply of oxygen to carry out cellular respiration.

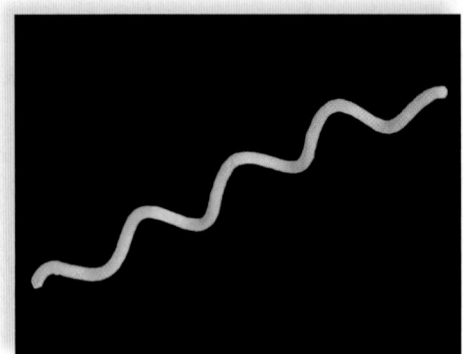

a. Spirillum: SEM 3,520×
   Spirillum volutans

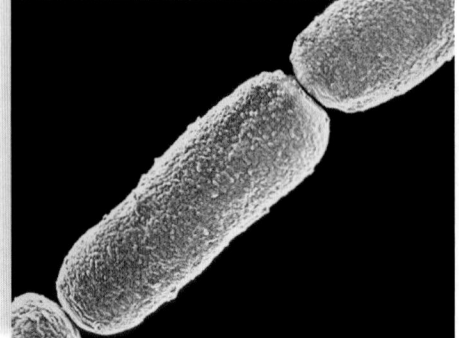

b. Bacilli: SEM 35,000×
   Bacillus anthracis

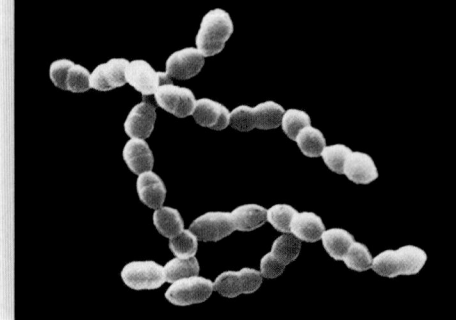

c. Cocci: SEM 6,250×
   Streptococcus thermophilus

Figure 20.9 Diversity of bacteria. **a.** Spirillum, a spiral-shaped bacterium. **b.** Bacilli, rod-shaped bacteria. **c.** Cocci, round bacteria.

Autotrophic Bacteria

Bacteria called **photoautotrophs** [Gk. *photos*, light, *auto*, self, and *trophe*, food] are photosynthetic (for a review of photosynthesis, see Section 7.2). They use solar energy to reduce carbon dioxide to organic compounds. There are two types of photoautotrophic bacteria: those that evolved first and do not give off oxygen (O_2), and those that evolved later and do give off oxygen. Their characteristics are shown here:

Photoautotrophic Bacteria	
Do Not Produce O_2	**Do Produce O_2**
- Photosystem I only	- Photosystems I and II
- Unique type of chlorophyll called bacteriochlorophyll	- Type of chlorophyll *a* found in plants

Green sulfur bacteria and some purple bacteria carry on the first type of photosynthesis. These bacteria usually live in anaerobic (oxygen poor) conditions such as the muddy bottom of a marsh. They cannot photosynthesize in the presence of oxygen, and they do not emit oxygen. In contrast, the cyanobacteria (see Fig. 20.12) contain chlorophyll *a* and carry on photosynthesis in the second way, just as algae and plants do; that is, they reduce carbon dioxide to organic compounds and give off oxygen as a by-product.

Bacteria called **chemoautotrophs** [Gk. *chemo*, pertaining to chemicals, *auto*, self, and *trophe*, food] carry out chemosynthesis. They oxidize inorganic compounds such as hydrogen gas, hydrogen sulfide, and ammonia to obtain the necessary energy to reduce CO_2 to an organic compound. The nitrifying bacteria oxidize ammonia (NH_3) to nitrites (NO_2^-) and nitrites to nitrates (NO_3^-). Their metabolic abilities keep nitrogen cycling through ecosystems. Other bacteria oxidize sulfur compounds. They live in environments such as deep-sea vents 2.5 km below sea level.

The organic compounds produced by such bacteria and also archaea (see page 379) support the growth of a community of organisms found at vents. This discovery lends support to the suggestion that the first cells originated at deep-sea vents.

Heterotrophic Bacteria

Bacteria called **chemoheterotrophs** [Gk. *chemo*, pertaining to chemicals, *hetero*, different, and *trophe*, food] take in organic nutrients. They are aerobic **saprotrophs** that decompose almost any large organic molecule into smaller ones that can be absorbed. Probably no natural organic molecule exists that cannot be digested by at least one prokaryotic species. In ecosystems, saprotrophic bacteria are called *decomposers*. They play a critical role in recycling matter and making inorganic molecules available to photosynthesizers.

The metabolic capabilities of chemoheterotrophic bacteria have long been exploited by human beings. Bacteria are used commercially to produce chemicals, such as ethyl alcohol, acetic acid, butyl alcohol, and acetones. Bacterial action is also involved in the production of butter, cheese, sauerkraut, rubber, silk, coffee, and cocoa. Even antibiotics are produced by some bacteria.

Symbiotic Relationships

Bacteria (and archaea) form **symbiotic relationships** [Gk. *sym*, together, and *bios*, life] in which two different species live together in an intimate way.

- In **mutualism,** both species benefit from the association.
- In **commensalism,** only one species benefits while the other is unaffected.
- In **parasitism,** one species benefits while harming the other.

Mutualistic bacteria live in human intestines, where they release vitamins K and B_{12}, which we can use to help produce blood components. In the stomachs of cows and goats, special mutualistic prokaryotes digest cellulose, enabling these animals to feed on grass. Mutualistic bacteria live in the root nodules of soybean, clover, and alfalfa plants where they reduce atmospheric nitrogen (N_2) to ammonia, a process called nitrogen fixation (Fig. 20.10). Plants are unable to fix atmospheric nitrogen, and those without nodules take up nitrate and ammonia from the soil.

Commensalism often occurs when one population modifies the environment in such a way that a second population benefits. Obligate anaerobes can live in our intestines only because the bacterium *Escherichia coli* uses up the available oxygen.

Parasitic bacteria cause diseases, and therefore are called **pathogens.** Some of the deadliest pathogens form **endospores** [Gk. *endon*, within, and *spora*, seed] when faced with unfavorable environmental conditions. A portion of the cytoplasm and a copy of the chromosome become dehydrated and are then

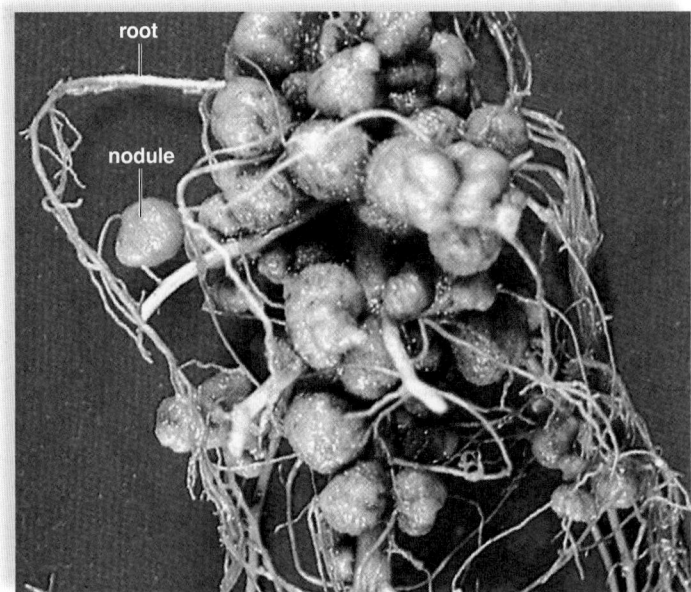

Figure 20.10 Nodules of a legume. Some free-living bacteria carry on nitrogen fixation; however, bacteria of the genus *Rhizobium* invade the roots of legumes, with the resultant formation of nodules. Here the bacteria convert atmospheric nitrogen to an organic nitrogen that the plant can use. These are nodules on the roots of a soybean plant (*Glycine* sp.).

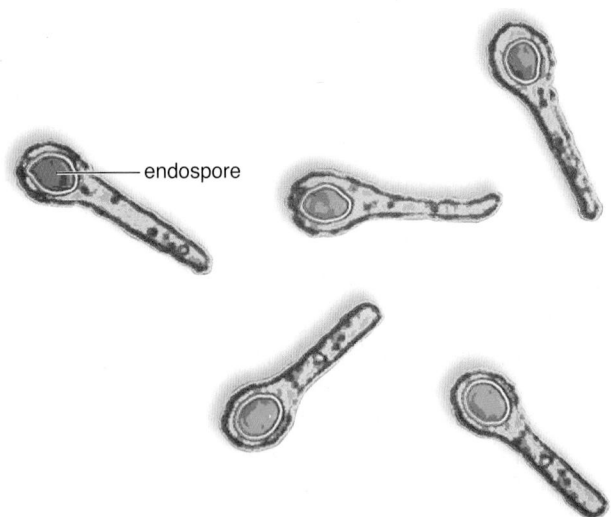

Figure 20.11 The endospore of *Clostridium tetani.*

C. tetani produces a terminal endospore that causes it to have a drumstick appearance. If endospores gain access to a wound, they germinate and release bacteria that produce a neurotoxin. The patient develops tetanus, a progressive rigidity that can result in death; immunization can prevent tetanus.

Table 20.2 Bacterial Diseases in Humans

Category	Disease
Sexually transmitted diseases	Syphilis, gonorrhea, chlamydia
Respiratory diseases	Strep throat, scarlet fever, tuberculosis, pneumonia, Legionnaires disease, whooping cough, inhalation anthrax
Skin diseases	Erysipelas, boils, carbuncles, impetigo, acne, infections of surgical or accidental wounds and burns, leprosy (Hansen disease)
Digestive tract diseases	Gastroenteritis, food poisoning, dysentery, cholera, peptic ulcers, dental caries
Nervous system diseases	Botulism, tetanus, leprosy, spinal meningitis
Systemic diseases	Plague, typhoid fever, diphtheria
Other diseases	Tularemia, Lyme disease

encased by a heavy, protective endospore coat (Fig. 20.11). In some bacteria, the rest of the cell deteriorates, and the endospore is released.

Endospores survive in the harshest of environments—desert heat and dehydration, boiling temperatures, polar ice, and extreme ultraviolet radiation. They also survive for very long periods. When anthrax endospores 1,300 years old germinate, they can still cause a severe infection (usually seen in cattle and sheep). Humans also fear a deadly but uncommon type of food poisoning called botulism that is caused by the germination of endospores inside cans of food. To germinate, the endospore absorbs water and grows out of the endospore coat. In a few hours' time, it becomes a typical bacterial cell, capable of reproducing once again by binary fission. Endospore formation is not a means of reproduction, but it does allow survival and dispersal of bacteria to new places.

Many other bacteria cause diseases in humans; a few are listed in Table 20.2. In almost all cases, the growth of microbes themselves does not cause disease; the poisonous substances they release, called **toxins,** are the pathological portion. When Gram-negative bacteria are killed by an antibiotic, the cell wall releases toxins called lipopolysaccharide fragments, and the result may be a high fever and a drop in blood pressure. Other bacteria secrete toxins while they are living. Some of these bacteria have a needle-shaped secretion apparatus they can use to inject toxins directly into host cells!

When someone steps on a rusty nail, bacteria can be injected deep into damaged tissue. If the deep damaged area does not have a good blood flow, then conditions can become anaerobic. The endospores of *Clostridium tetani* germinate and produce a toxin that causes the disease tetanus. The bacteria never leave the site of the wound, but the tetanus toxin they

produce does move throughout the body. This toxin prevents the relaxation of muscles. In time, the body contorts because all the muscles have contracted. Eventually, suffocation occurs.

Fimbriae allow a pathogen to bind to certain cells, and their specificity determines which organs or cells of the body are its host. Like many bacteria that cause dysentery (severe diarrhea), *Shigella dysenteriae* is able to stick to the intestinal wall. In addition, *S. dysenteriae* produces a toxin called Shiga toxin that increases the potential for fatality. Also, invasive mechanisms that give a pathogen the ability to move through tissues and into the bloodstream result in a more medically significant disease than if it were localized. Usually a person can recover from food poisoning caused by *Salmonella*. But some strains of *Salmonella* have virulence factors—including a needle-shaped toxin secretion apparatus—that allow the bacteria to penetrate the lining of the colon and move beyond this organ. Typhoid fever, a life-threatening disease, can then result.

Antibiotics

A number of antibiotic compounds are active against bacteria and are widely prescribed. Most antibacterial compounds fall within two classes, those that inhibit protein biosynthesis and those that inhibit cell wall biosynthesis. Two types of antibiotics, erythromycin and tetracyclines, inhibit bacterial protein synthesis by affecting bacterial ribosomes because they function somewhat differently than eukaryotic ribosomes. Antibiotics that inhibit cell wall biosynthesis generally block the formation of peptidoglycan, a compound necessary to maintain the integrity of bacterial cell walls. Penicillin, ampicillin, and fluoroquinolone (e.g., Cipro) inhibit bacterial cell wall biosynthesis without harming animal cells.

Animation
Antibiotic Inhibition of Protein Synthesis

Antibiotics are heavily prescribed to treat infection, often when not needed. One outcome of excessive and improper use of antibiotics has been increasing bacterial resistance to antibiotics. Genes conferring resistance to antibiotics can be transferred between infectious bacteria by transformation, conjugation, or

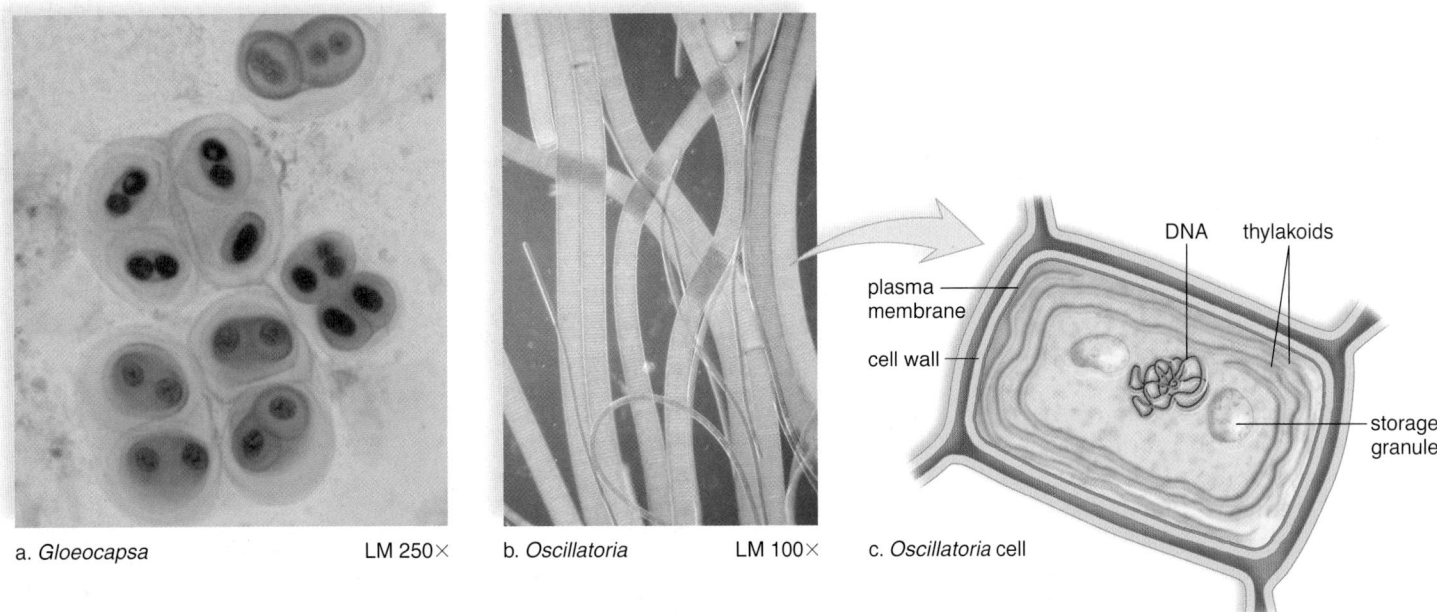

a. *Gloeocapsa*   LM 250×   b. *Oscillatoria*   LM 100×   c. *Oscillatoria* cell

plasma membrane
cell wall
DNA   thylakoids
storage granule

Figure 20.12 Diversity among the cyanobacteria. **a.** In *Gloeocapsa*, single cells are grouped in a common gelatinous sheath. **b.** Filaments of cells occur in *Oscillatoria*. **c.** One cell of *Oscillatoria* as it appears through the electron microscope.

transduction. When penicillin was first introduced, less than 3% of *Staphylococcus aureus* strains were resistant to it. Now, because of selective advantage, 90% or more are resistant to penicillin and, increasingly, to methicillin, an antibiotic developed in 1957. A strain of methicillin resistant *S. aureus* (abbreviated MRSA) is responsible for many difficult to treat infections that have become a threat to human health. MRSA is common in hospitals and nursing facilities where it causes problems for those with a greater risk of infection, especially those with open wounds and weakened immune systems.

Cyanobacteria

Cyanobacteria [Gk. *kyanos,* blue, and *bacterion,* rod] are Gram-negative bacteria with a number of unusual traits. They photosynthesize in the same manner as plants and are believed to be responsible for first introducing oxygen into the primitive atmosphere. Formerly, the cyanobacteria were called blue-green algae and were classified with eukaryotic algae, but now they are classified as prokaryotes. Cyanobacteria can have other pigments that mask the color of chlorophyll so that they appear red, yellow, brown, or black, rather than only blue-green (Fig. 20.12).

Cyanobacterial cells are rather large, ranging from 1 to 50 μm in width. They can be unicellular, colonial, or filamentous. Cyanobacteria lack any visible means of locomotion, although some glide when in contact with a solid surface and others oscillate (sway back and forth). Some cyanobacteria have a special advantage because they possess heterocysts, which are thick-walled cells without nuclei, where nitrogen fixation occurs. The ability to photosynthesize and also to fix atmospheric nitrogen (N$_2$) means that their nutritional requirements are minimal. They can serve as food for heterotrophs in ecosystems.

Cyanobacteria are common in fresh and marine waters, in soil, and on moist surfaces, but they are also found in harsh habitats, such as hot springs. They are symbiotic with a number of organisms, including liverworts, ferns, and even at times invertebrates such as corals. In association with fungi, they form **lichens** that can grow on rocks. In a lichen, the cyanobacterium mutualistically provides organic nutrients to the fungus, while the fungus possibly protects and furnishes inorganic nutrients to the cyanobacterium. It is also possible that the fungus is parasitic on the cyanobacterium. Lichens help transform rocks into soil; other forms of life then may follow. It is hypothesized that cyanobacteria were the first colonizers of land during the course of evolution.

Cyanobacteria are ecologically important in still another way. If care is not taken in disposing of industrial, agricultural, and human wastes, phosphates drain into lakes and ponds, resulting in a "bloom" of these organisms. The surface of the water becomes turbid, and light cannot penetrate to lower levels. When a portion of the cyanobacteria die off, the decomposers feeding on them use up the available oxygen, causing fish to die from lack of oxygen.

Check Your Progress   20.3

1. Describe how the peptidoglycan layer is different in Gram-positive and Gram-negative cells.
2. Discuss two ways that antibiotics kill bacteria.
3. Define the function of bacterial endospores.
4. Construct a hypothesis about how cyanobacteria may have affected the atmosphere of early Earth.

20.4 The Archaea

a.

At one time, **archaea** (domain Archaea) were considered to be a unique group of bacteria. Archaea came to be viewed as a distinct domain of organisms in 1977, when Carl Woese and George Fox discovered that the rRNA of archaea has a different sequence of bases than the rRNA of bacteria. He chose rRNA because of its involvement in protein synthesis—any changes in rRNA sequence probably occur at a slow, steady pace as evolution occurs.

As discussed in Chapter 19, it is proposed that the tree of life contains three domains: Archaea, Bacteria, and Eukarya. Because archaea and some bacteria are found in extreme environments (hot springs, thermal vents, salt basins), they may have diverged from a common ancestor relatively soon after life began. Then later, the eukarya diverged from the archaeal line of descent. In other words, the eukarya are more closely related to the archaea than to the bacteria. Archaea and eukarya share some of the same ribosomal proteins (not found in bacteria), initiate transcription in the same manner, and have similar types of tRNA.

b.

Structure of Archaea

Archaea are prokaryotes with biochemical characteristics that distinguish them from both bacteria and eukaryotes. The plasma membranes of archaea contain unusual lipids that allow many of them to function at high temperatures. The lipids of archaea contain glycerol linked to branched-chain hydrocarbons, in contrast to the lipids of bacteria, which contain glycerol linked to fatty acids.

The archaea also evolved diverse cell wall types, which facilitate their survival under extreme conditions. The cell walls of archaea do not contain peptidoglycan as do the cell walls of bacteria. In some archaea, the cell wall is largely composed of polysaccharides, and in others, the wall is pure protein. A few have no cell wall.

Types of Archaea

Archaea were originally discovered living in extreme environmental conditions. Three main types of archaea are still distinguished based on their unique habitats: methanogens, halophiles, and thermoacidophiles (Fig. 20.13).

Methanogens

The **methanogens** (methane makers) are obligate anaerobes found in environments such as swamps, marshes, and the intestinal tracts of animals. Methanogenesis, the ability to form methane (CH_4), is a type of metabolism performed only by some archaea. Methanogens are chemoautotrophs, using hydrogen

c.

Figure 20.13 Extreme habitats. **a.** Halophilic archaea can live in salt lakes. **b.** Thermoacidophilic archaea can live in the hot springs of Yellowstone National Park. **c.** Methanogens live in swamps and in the guts of animals.

gas (H_2) to reduce carbon dioxide (CO_2) to methane and couple the energy released to ATP production.

Methane, also called biogas, is released into the atmosphere where it contributes to the greenhouse effect and climate change. About 65% of the methane found in our atmosphere is produced by these methanogenic archaea.

Methanogenic archaea may help us anticipate what life may be like on other celestial bodies. Consider, for instance, the unusual microbial community residing in the Lidy Hot Springs of eastern

Idaho. The springs, which originate 200 m (660 feet) beneath the Earth's surface, are lacking in organic nutrients, but rich in H$_2$. Scientists have found the springs to be inhabited by vast numbers of microorganisms; over 90% are archaea, and the overwhelming majority are methanogens. The researchers who first investigated the Lidy Hot Springs microbes point out that similar methanogenic communities may someday be found beneath the surfaces of Mars and Europa (one of Jupiter's moons). Because hydrogen is the most abundant element in the universe, it would be readily available for use by methanogens everywhere.

Halophiles

The **halophiles** require high salt concentrations (usually 12–15%; the ocean, in contrast, is about 3.5%). They have been isolated from highly saline environments in which few organisms are able to survive, such as the Great Salt Lake in Utah, the Dead Sea, solar salt ponds, and hypersaline soils.

These archaea have evolved a number of mechanisms to survive in environments that are high in salt. This survival ability benefits the halophiles, as they do not have to compete with as many microorganisms as they would encounter in a more moderate environment. The proteins of halophiles have unique chloride pumps that use halorhodopsin (related to the rhodopsin pigment found in our eyes) to pump chloride to the inside of the cell, and this prevents water loss.

These organisms are aerobic chemoheterotrophs; however, some species can carry out a unique form of photosynthesis if their oxygen supply becomes scarce, as commonly occurs in highly saline conditions. Instead of chlorophyll, these halophiles use a purple pigment called bacteriorhodopsin to capture solar energy for use in ATP synthesis. Interestingly, most halophiles are so adapted to a high-saline environment that they perish if placed in a solution with a low salt concentration (such as pure water).

Thermoacidophiles

A third major type of archaea are the **thermoacidophiles.** These archaea are isolated from extremely hot, acidic environments such as hot springs, geysers, submarine thermal vents, and around volcanoes. They are chemoautotrophic anaerobes that use hydrogen (H$_2$) as the electron donor, and sulfur (S) or sulfur compounds as terminal electron acceptors, for their electron transport chains. Hydrogen sulfide (H$_2$S) and protons (H$^+$) are common products.

Recall that the greater the concentration of protons, the lower (and more acidic) the pH. Thus, it is not surprising that thermoacidophiles grow best at extremely low pH levels, between pH 1 and 2. Due to the unusual lipid composition of their plasma membranes, thermoacidophiles survive best at temperatures above 80°C; some can even grow at 105°C (remember that water boils at 100°C)!

Archaea in Moderate Habitats

Although archaea are capable of living in extremely stressful conditions, they are found in all moderate environments as well. For example, some archaea have been found living in symbiotic relationships with animals, including sponges and sea cucumbers. Such relationships are sometimes mutualistic or even commensalistic, but there are no parasitic archaea—that is, they are not known to cause infectious diseases.

The roles of archaea in activities such as nutrient cycling are still being explored. For example, a group of nitrifying marine archaea has recently been discovered. Some scientists think that these archaea may be major contributors to the supply of nitrite in the oceans. Nitrite can be converted by certain bacteria to nitrate, a form of nitrogen that can be used by plants and other producers to construct amino acids and nucleic acids. Archaea have also been found inhabiting lake sediments, rice paddies, and soil, where they are likely to be involved in nutrient cycling.

Check Your Progress   **20.4**

1. Identify the differences between archaea and bacteria.
2. List the three types of archaea distinguished by their unique habitats.
3. Archaea are thought to be closely related to eukaryotes. Explain the evidence that supports this possibility.

CONNECTING *the* CONCEPTS *with the* BIG IDEAS

Evolution

- Bacteria and archaea originated at least 3.5 billion years ago. Many are able to live in extreme habitats. (1D2a1)
- The overuse of antibiotics has lead to selection of new variations of bacteria. (1A2d*IE*, 1C3b*IE*, 3C1d*IE*)
- The phenomenon of emergent diseases supports present-day evolution. (1C3b*IE*)

Energy and Homeostasis

- Prokaryotes lack membranous organelles. (2B1c2, 2B3c)
- Some prokaryotes originated photosynthesis; others are chemosynthetic. (2A2a2, 2A2e)
- Bacteria respond to environmental changes and often live in mutualistic relationships. (2C2a, 2D2c*IE*, 2E3b4*IE*)

Information and Signaling

- Viruses evolve rapidly, engaging in lytic and lysogenic cycles as well as transduction. (3C3a-b)
- Retroviruses use reverse transcriptase. (3A1a6)
- Bacteria contain circular chromosomal and plasmid DNA and modify it through conjugation, transduction, and transformation. (3A1a2-3, 3C2b)

Interactions and Systems

- Unicellular populations are capable of interactions which affect their efficiency and productivity. (4B2a3)

*Find the unabridged version of all EK citations at www.glencoe.com/maderAP11.

Media Study Tools

www.glencoe.com/maderAP11

Enhance your study of this chapter with study tools and practice tests. Also ask your instructor about the resources available through ConnectPlus, including the media-rich eBook, interactive learning tools, and animations.

Summarize

20.1 Viruses, Viroids, and Prions

Viruses are nonliving, while prokaryotes are fully functioning organisms. All viruses have at least two parts: an outer capsid composed of protein subunits and an inner core of nucleic acid, either DNA or RNA, but not both. Some also have an outer membranous envelope. The type of genetic material is used to classify viruses into different groups.

Viruses are obligate intracellular parasites that can be maintained only inside living cells, such as those of a chicken egg, or those propagated in cell (tissue) culture.

Bacteriophages are viruses that infect prokaryotes. The lytic cycle of a bacteriophage consists of attachment, penetration, biosynthesis, maturation, and release. In the lysogenic cycle of a bacteriophage, viral DNA is integrated into bacterial DNA for an indefinite period of time, but it can undergo the lytic cycle when stimulated.

The reproductive cycle differs for viruses that infect animal cells. Uncoating is needed to free the genome from the capsid, and either budding or lysis releases the viral particles from the cell. Retroviruses have an enzyme, reverse transcriptase, that carries out reverse transcription. This enzyme produces one strand of DNA (cDNA) using viral RNA as a template, and then another DNA strand that is complementary to the first one. The resulting double-strand DNA becomes integrated into host DNA. The AIDS virus is a retrovirus.

Viruses cause various diseases in plants and animals, including human beings. Viroids are naked strands of RNA (not covered by a capsid) that can cause disease in plants. Prions are protein molecules that have a misshapen tertiary structure. Prions cause neurodegenerative diseases such as TSEs in humans and mad cow disease in cattle when they cause normal proteins of their own kind to become misshapen.

20.2 The Prokaryotes

The bacteria (domain Bacteria) and archaea (domain Archaea) are prokaryotes. Prokaryotic cells lack a nucleus and most of the other cytoplasmic organelles found in eukaryotic cells. Louis Pasteur performed an experiment that proved the existence of microorganisms.

Prokaryotes reproduce asexually by binary fission. Their chief method for achieving genetic variation is mutation, but genetic recombination by means of conjugation, transformation, and transduction has been observed.

Prokaryotes differ in their need (and tolerance) for oxygen. There are obligate anaerobes, facultative anaerobes, and aerobic prokaryotes. Some prokaryotes are autotrophic, and some are heterotrophic.

20.3 The Bacteria

Bacteria (domain Bacteria) are the more prevalent type of prokaryote. The classification of bacteria is still being developed. Of primary importance at this time are the shape of the cell and the structure of the cell wall, which affects Gram staining. Bacteria occur in three basic shapes: spiral-shaped (spirillum), rod-shaped (bacillus), and round (coccus).

Some prokaryotes are autotrophic—either photoautotrophs (photosynthetic) or chemoautotrophs (chemosynthetic). Some photosynthetic bacteria (cyanobacteria) give off oxygen, and some (purple and green sulfur bacteria) do not. Chemoautotrophs oxidize inorganic compounds such as hydrogen gas, hydrogen sulfide, and ammonia to acquire energy to make their own food. Surprisingly, chemoautotrophs support communities at deep-sea vents.

Many bacteria are chemoheterotrophs (aerobic heterotrophs) and are saprotrophic decomposers that are absolutely essential to the cycling of nutrients in ecosystems. Their metabolic capabilities are so vast that they are used by humans both to dispose of and to produce substances.

Many heterotrophic bacteria are symbiotic. The mutualistic nitrogen-fixing bacteria live in nodules on the roots of legumes. Of special interest are the cyanobacteria, which were the first organisms to photosynthesize in the same manner as plants. When cyanobacteria are symbionts with fungi, they form lichens.

Some bacterial symbionts, however, are parasitic and cause plant and animal, including human, diseases. Certain bacteria form endospores, which are extremely resistant to destruction. Their genetic material can thereby survive unfavorable conditions.

20.4 The Archaea

The archaea (domain Archaea) are a second type of prokaryote. On the basis of rRNA sequencing, it is thought that there are three evolutionary domains: Bacteria, Archaea, and Eukarya. In addition, the archaea appear to be more closely related to the eukarya than to the bacteria. Archaea do not have peptidoglycan in their cell walls, as do the bacteria, and they share more biochemical characteristics with the eukarya than do bacteria.

Three types of archaea live under harsh conditions, such as anaerobic marshes (methanogens), salty lakes (halophiles), and hot sulfur springs (thermoacidophiles). Archaea are also found in moderate environments.

Key Terms

archaea 378
bacteria 374
bacteriophage 367
binary fission 373
capsid 366
chemoautotroph 375
chemoheterotroph 375
commensalism 375
conjugation 373
conjugation pilus 373
cyanobacteria 377
emerging virus 369
endospore 375
facultative anaerobe 374
fimbriae 372
flagellum (pl., flagella) 372
halophile 379
lichen 377
lysogenic cell 368
lysogenic cycle 368
lytic cycle 368
methanogen 378

mutualism 375
neurodegenerative disease 371
nucleoid 372
obligate anaerobe 374
parasitism 375
pathogen 375
peptidoglycan 374
photoautotroph 375
plasmid 372
prion 369
prokaryote 371
retrovirus 368
reverse transcriptase 368
saprotroph 375
symbiotic relationship 375
thermoacidophile 379
toxin 376
transduction 373
transformation 373
viroid 369
virus 364

Assess

Reviewing This Chapter

1. Contrast viruses with cells in terms of the characteristics of life. 364
2. Describe the general structure of viruses, and describe both the lytic cycle and the lysogenic cycle of bacteriophages. 365–68
3. How do animal viruses differ in structure and reproductive cycle? 368
4. How do retroviruses differ from other animal viruses? Describe the reproductive cycle of retroviruses in detail. 368
5. Explain Pasteur's experiment, which showed that organisms do not arise spontaneously. 371–72
6. Provide a diagram of prokaryotic cell structure and discuss. 372–73
7. How do prokaryotes introduce variation? How does genetic recombination occur in bacteria? 373
8. How do all prokaryotes differ in their tolerance of and need for oxygen? 374–75
9. Compare photosynthesis between the green sulfur bacteria and the cyanobacteria. 375, 377
10. What are chemoautotrophic prokaryotes, and where have they been found to support whole communities? 375
11. What role do endospores play in disease? 375–76
12. Discuss the importance of cyanobacteria in ecosystems and in the history of the Earth. 377
13. How do archaea differ from bacteria? 378
14. What three different types of archaea may be distinguished based on their different habitats? 378–79

Testing Yourself

Choose the best answer for each question.

1. Label this condensed version of bacteriophage reproductive cycles using these terms: penetration, maturation, release, prophage, attachment, biosynthesis, and integration.

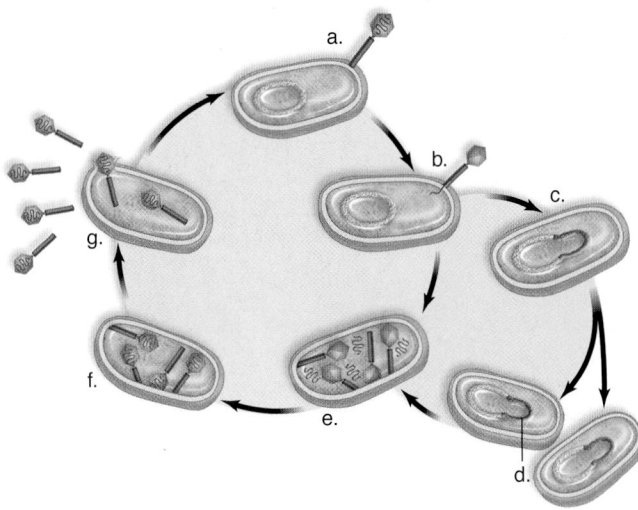

2. Viruses are considered nonliving because
   a. they do not locomote.
   b. they cannot reproduce independently.
   c. their nucleic acid does not code for protein.
   d. they are noncellular.
   e. Both b and d are correct.

3. Which of the following statements about viruses is incorrect?
   a. The nucleic acid may be either DNA or RNA, but not both.
   b. The capsid may be polyhedral or helical.
   c. Viruses do not fit into the current system for naming organisms.
   d. The nucleic acid may be either single stranded or double stranded.
   e. Viruses are rarely, if ever, host specific.

4. The envelope of an animal virus is derived from the _____ of its host cell.
   a. cell wall
   b. membrane
   c. glycocalyx
   d. receptors

5. A prophage occurs during the reproduction cycle of
   a. a lysogenic virus.
   b. a poxvirus.
   c. a lytic virus.
   d. an enveloped virus.

6. Which of these are found in all viruses?
   a. envelope, nucleic acid, capsid
   b. DNA, RNA, and proteins
   c. proteins and a nucleic acid
   d. proteins, nucleic acids, carbohydrates, and lipids
   e. tail fibers, spikes, and rod shape

7. Which would be the worst choice for cultivation of viruses?
   a. tissue culture
   b. bird embryos
   c. live mammals
   d. sterile broth

8. A pathogen would most accurately be described as
   a. a parasite.
   b. a commensal.
   c. a saprobe.
   d. a symbiont.

9. RNA retroviruses have a special enzyme that
   a. disintegrates host DNA.
   b. polymerizes host DNA.
   c. transcribes viral RNA to cDNA.
   d. translates host DNA.
   e. produces capsid proteins.

10. Which is not true of prokaryotes? They
    a. are living cells.
    b. lack a nucleus.
    c. all are parasitic.
    d. are both archaea and bacteria.
    e. evolved early in the history of life.

11. Facultative anaerobes
    a. require a constant supply of oxygen.
    b. are killed in an oxygenated environment.
    c. do not always need oxygen.
    d. are photosynthetic but do not give off oxygen.
    e. All of these are correct.

12. Which of these is most apt to be a prokaryotic cell wall function?
    a. transport
    b. motility
    c. support
    d. adhesion

13. Cyanobacteria, unlike other types of bacteria that photosynthesize, do
    a. give off oxygen.
    b. not have chlorophyll.
    c. not have a cell wall.
    d. need a fungal partner.

14. Chemoautotrophic prokaryotes
    a. are chemosynthetic.
    b. use the rays of the Sun to acquire energy.
    c. oxidize inorganic compounds to acquire energy.
    d. are always bacteria, not archaea.
    e. Both a and c are correct.

15. Archaea differ from bacteria in that
    a. some can form methane.
    b. they have different rRNA sequences.
    c. they do not have peptidoglycan in their cell walls.
    d. they rarely photosynthesize.
    e. All of these are correct.

16. Which of these archaea would live at a deep-sea vent?
    a. thermoacidophile
    b. halophile
    c. methanogen
    d. parasitic forms
    e. All of these are correct.

17. A prokaryote that can synthesize all its required organic components from CO_2 using energy from the Sun is a
    a. photoautotroph.
    b. photoheterotroph.
    c. chemoautotroph.
    d. chemoheterotroph.

18. While testing some samples of marsh mud, you discover a new microbe that has never been described before. You examine it closely and find that, although it is cellular, there is no nucleus. Biochemical analysis reveals that the plasma membrane is made up of glycerol linked to branched-chain hydrocarbons, and the cell wall contains no peptidoglycan. This microbe is most likely a(n)
    a. enveloped virus.
    b. archaean.
    c. prion.
    d. bacterium.
    e. bacteriophage.

19. The Nobel laureate Peter Medawar called a certain type of microbe "a piece of bad news wrapped up in protein." Which of the following was he describing?
    a. archaea
    b. bacteria
    c. prion
    d. virus
    e. viroid

Engage

Thinking Scientifically

1. While a few drugs are effective against some viruses, they often impair the function of body cells and thereby have a number of side effects. Most antibiotics (antibacterial drugs) do not cause side effects. Why would antiviral medications be more likely to produce side effects?

2. Model organisms are those widely used by researchers who wish to understand basic processes that are common to many species. Bacteria such as *Escherichia coli* are model organisms for modern geneticists. Give three reasons why bacteria would be useful in genetic experiments.

Bioethical Issue

Identifying Carriers

Carriers of disease are persons who do not appear to be ill but can nonetheless pass on an infectious disease. The only way society can protect itself is to identify carriers and remove them from areas or activities where transmission of the pathogen is most likely. Sometimes it's difficult to identify all activities that might pass on a pathogen—for example, HIV. A few people believe that they have acquired HIV from their dentists, and while this is generally believed to be unlikely, medical personnel are still required to identify themselves when they are carriers of HIV.

Transmission of HIV is believed to be possible in certain sports. In a statistical study, the Centers for Disease Control and Prevention figured that the odds of acquiring HIV from another football player were 1 in 85 million. But the odds might be higher for boxing, a bloody sport. When two brothers, one of whom had AIDS, got into a vicious fight, the infected brother repeatedly bashed his head against his brother's. Both men bled profusely, and soon after, the previously uninfected brother tested positive for the virus. The possibility of transmission of HIV in the boxing ring has caused several states to require boxers to undergo routine HIV testing. If they are HIV positive, they can't fight.

Should all people who are HIV positive always be required to identify themselves, no matter what the activity? Why or why not? By what method would they identify themselves at school, at work, and in other places?

Patients in South African communities receiving donated medications to fight NTDs.

21

Protist Evolution and Diversity

A s of 2010, the World Health Organization (WHO) has recognized sixteen Neglected Tropical Diseases (NTDs) that persist only in the poorest nations—largely forgotten by the rest of the world. NTDs affect more than 1 billion people worldwide, even though they are in many cases easily preventable and economically treatable. Three of these NTDs are caused by parasitic protists. Protists are the simplest of all eukaryotes—all but a few are microscopic, single-celled organisms. Not all protists cause disease, however. The algae and the protozoans, a group of animal-like protists, are the foundation of the food chain of the world's oceans. Algin, a seaweed product, is used in the production of pharmaceuticals, cosmetics, paper, and textiles—even the tungsten filaments of lightbulbs. Protists were the first eukaryotes to appear on Earth, and without them, life on Earth would not exist or be able to persist as we know it today. In this chapter we tour this widely diverse and interesting group of microscopic living organisms.

As you read through the chapter, think about the following questions:

1. How do microorganisms, such as protists, impact human health and welfare?

2. Why is the theory of endosymbiosis so important to our understanding of the origin of eukaryotes?

3. Why are microorganisms able to exploit extreme environments that more complex multicellular organisms cannot?

BEFORE YOU BEGIN

Before beginning this chapter, take a few moments to review the following discussions.

Section 4.3 How does the endosymbiotic theory explain the origin of energy-producing organelles in the eukaryotic cell?

Section 4.4 What is the basic structure of a eukaryotic cell?

Section 6.4 How are eukaryotic organelles involved in the production and flow of energy in a cell?

FOLLOWING *the* BIG IDEAS

CHAPTER 21   PROTIST EVOLUTION AND DIVERSITY

Evolution	The inferred ancestry and diversification of protists can be charted via cladograms.
Energy and Homeostasis	Both biotic and abiotic factors affect the size and activity levels of members of the kingdom Protista.
Interactions and Systems	Protists may interact with each other or with other species producing beneficial or harmful consequences.

21.1 General Biology of Protists

Protists are the simplest, but most diverse, of the eukaryotes. Most are unicellular, but some exist as colonies of cells or are multicellular. As eukaryotes, protists have membranous organelles, such as mitochondria and plastids, that serve as the energy centers of the cell. (Section 4.7 describes the structure and function of mitochondria and plastids.)

The endosymbiotic theory proposes that eukaryotic cells acquired mitochondria and plastids, including chloroplasts, by engulfing a free-living bacterium that developed a symbiotic relationship within the host cell, a process termed **endosymbiosis** (see Section 18.2). Mitochondria were derived first from the endosymbiosis of an aerobic bacterium, and chloroplasts were derived later from the endosymbiosis of a cyanobacterium (see Fig. 4.5). Much of the endosymbiotic bacteria's genomes have been incorporated into the genome of the host cell and now compliment the life processes of the host.

**Animation
Endosymbiosis**

Characteristics of Protists

Protists vary in size from microscopic algae and protozoans, to kelp that can exceed 200 m in length. Kelp, a brown alga, is multicellular; *Volvox*, a green alga, is colonial; while *Spirogyra*, also a green alga, is filamentous. Most protists are unicellular, but despite their small size they have attained a high level of complexity. The amoeboids and ciliates possess unique organelles—their contractile vacuole is an organelle that assists in water regulation.

Protists are sometimes grouped according to how they acquire organic nutrients. The algae are a diverse group of photoautotrophic protists that synthesize organic compounds via photosynthesis. Protozoans are a group of heterotrophic protists that obtain organic compounds from the environment. Some protozoans, such as *Euglena*, are **mixotrophic,** meaning they are able to combine autotrophic and heterotrophic nutritional modes.

Protists reproduce sexually and asexually. Asexual reproduction by mitosis is the norm in protists. Sexual reproduction generally occurs only when environmental conditions are unfavorable. Protists can form spores or **cysts** that are dormant phases of the protist life cyle that can survive until favorable conditions return. Parasitic protists form cysts for the transfer to a new host.

Many protists cause diseases in humans, but many others have significant ecological importance. Aquatic photoautotrophic protists produce oxygen and are the foundation of the food chain in both freshwater and saltwater ecosystems. They are a part of **plankton** [Gk. *plankt,* wandering], organisms that are suspended in the water and serve as food for heterotrophic protists and animals. Interestingly, whales, the largest animals in the sea, feed on plankton, one of the smallest!

Evolution and Diversity of Protists

Protists were once classified together as a single kingdom. Recently, DNA evidence suggests that protists are not **monophyletic,** that is, they do not all belong to the same evolutionary lineage. In fact, protists and other eukaryotes, including the plants, fungi, and animals, are currently classified into six supergroups (Table 21.1). A **supergroup** is a high-level taxonomic group below domain and above kingdom. Each supergroup represents a separate evolutionary lineage.

The DNA evidence supports multiple protist lineages, but the relationships among the protist lineages are difficult to decipher. Protist lineages are very long and old, dating back to the origin of the first eukaryotes. As lineages stretch back in time, we can be less and less certain about how they are related to each other, just as the history of humans is less complete the further back in time we look.

New research in the evolution of protists has helped to clarify some of the evolutionary relationships among eukaryote lineages (Fig. 21.1), but a lot of research still needs to be done.

Check Your Progress   21.1

1. Explain how mitochondria and chloroplasts originated in eukaryotic cells.
2. Describe how algae and protozoans are nutritionally different from one another.
3. Identify the eukaryote supergroups that (1) include both plants and protists and (2) includes fungi, animals, and protists.

21.2 Diversity of Protists

In this section, we describe the different supergroups into which protists and other eukaryotes have been placed. The supergroups are summarized as follows:

- Archaeplastida—red and green algae (also includes land plants); have plastids.

Table 21.1 Protist Diversity

Supergroup	Members	Distinguishing Features	Protist Example
Archaeplastids	Green algae, red algae, land plants, charophytes	Plastids; unicellular, colonial, and multicellular	red alga; *Chlamydomonas*; *Closterium*, a green alga
Chromalveolates	Stramenopiles: brown algae, diatoms, golden brown algae, water molds Alveolates: ciliates, apicomplexans, dinoflagellates	Most with plastids; unicellular and multicellular Alveoli support plasma membrane; unicellular	*Blepharisma*, a ciliate with visible vacuoles; Assorted fossilized diatoms; *Ceratium*, an armored dinoflagellate
Excavates	Euglenids, kinetoplastids, parabasalids, diplomonads	Feeding groove; unique flagella; unicellular	*Giardia*, a single-celled, flagellated diplomonad; *Euglena*
Amoebozoans	Amoeboids, plasmodial and cellular slime molds	Pseudopods; unicellular	*Dictyostelium*, a slime mold with fruiting bodies; *Amoeba proteus*, a protozoan
Rhizarians	Foraminiferans, radiolarians	Thin pseudopods; some with tests; unicellular	*Nonionina*, a foraminiferan; Radiolarians (assorted), produce a calcium carbonate shell
Opisthokonts	Choanoflagellates, animals, nucleariids, fungi	Some with flagella; unicellular and colonial	Choanoflagellate (unicellular), animal-like protist

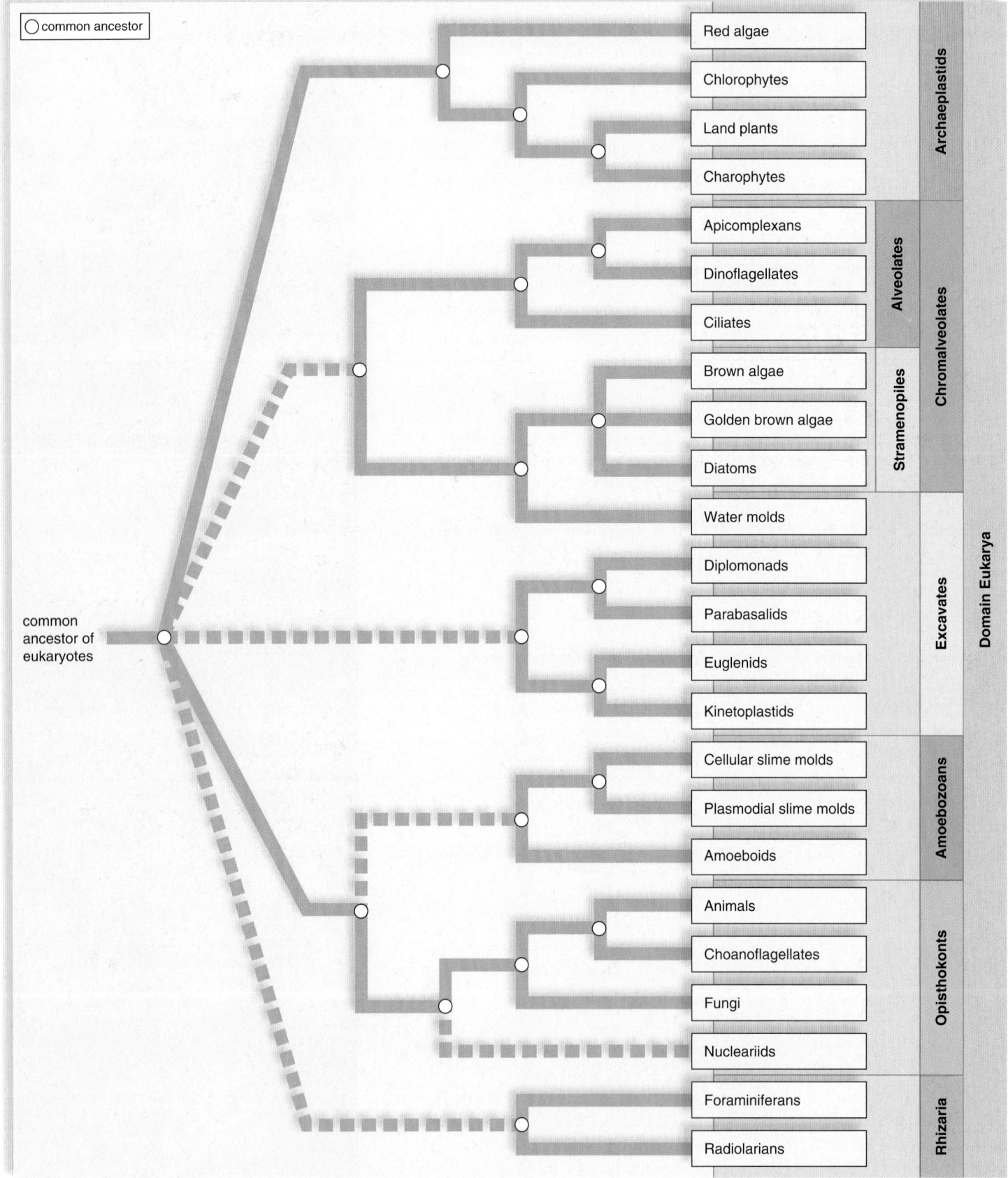

Figure 21.1 Evolutionary relationships between the eukaryotic supergroups. Molecular data are used to determine the relatedness of the supergroups and their constituents. The dashed lines indicate relationships that are not certain at this time. This is a simplified tree that does not include all members of each supergroup.

- Chromalveolata—brown and golden brown algae, water molds; also the alveolates; most have plastids.
- Excavata—zooflagellates, often with distinctive oral grooves.
- Amoebozoa—protozoans that move via pseudopods.
- Opisthokonta—unicellular and multicellular protists, including choanoflagellates (also includes animals and fungi).
- Rhizaria—foraminiferans and radiolarians.

Supergroup Archaeplastida

The **archaeplastids** [Gk. *archeos*, ancient, *plastikos*, moldable] include land plants and other photosynthetic organisms, such as green and red algae that have plastids derived from endosymbiotic cyanobacteria (see Fig. 4.5).

Green Algae

The **green algae** are protists that contain both chlorophylls *a* and *b*. They inhabit a variety of environments including oceans, fresh water, snowbanks, the bark of trees, and the backs of turtles. Some of the 17,000 species of green algae also form symbiotic relationships with plants, animals, and fungi in lichens (see Chapter 22).

Green algae occur in many different forms. The majority are unicellular; however, filamentous and colonial forms exist. Seaweeds are multicellular green algae that resemble lettuce leaves. Despite the name, green algae are not always green; some possess additional pigments that give them an orange, red, or rust color.

Biologists propose that land plants are closely related to the green algae because both land plants and green algae have chlorophylls *a* and *b*, a cell wall that contains cellulose, and food reserves made of starch. Molecular data suggest that the green algae are subdivided into two groups, the **chlorophytes** and the **charophytes.** Charophytes are thought to be the green algae group most closely related to land plants.

Chlorophytes. *Chlamydomonas* is a tiny, photoautotrophic chlorophyte that inhabits still, freshwater pools. Its fossil ancestors date back over a billion years. The anatomy of *Chlamydomonas* is best seen in an electron micrograph because it is less than 25 μm long (Fig. 21.2). It has a defined cell wall and a single, large, cup-shaped chloroplast that contains a *pyrenoid*, a dense body where starch is synthesized. In many species, a bright red light-sensitive eyespot helps guide individuals toward light for photosynthesis.

When conditions are favorable, that is, proper nutrients and sunlight are available, *Chlamydomonas* exists as haploid cells. These haploid cells, called *vegetative cells*, have two long, whiplike flagella projecting from the end. These operate with a breaststroke-like motion. *Chlamydomonas* vegetative cells often reproduce asexually by mitosis. Each mitosis event produces two haploid daughter cells, and as many as 16 daughter cells can form inside the parental cell wall (Fig. 21.3). Each daughter cell then secretes a cell wall and acquires flagella. The fully formed, functional haploid daughter cells emerge from within the parent by secreting an enzyme that digests the parent cell wall.

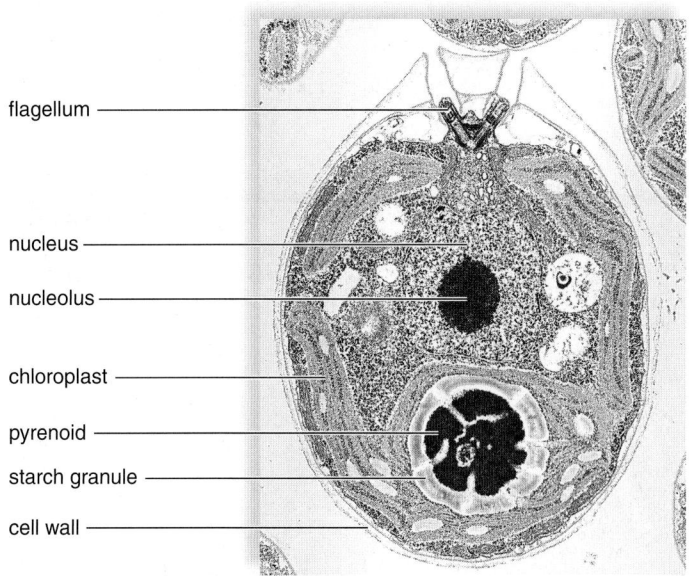

Figure 21.2 Electron micrograph of *Chlamydomonas.*
Chlamydomonas is a microscopic unicellular chlorophyte.

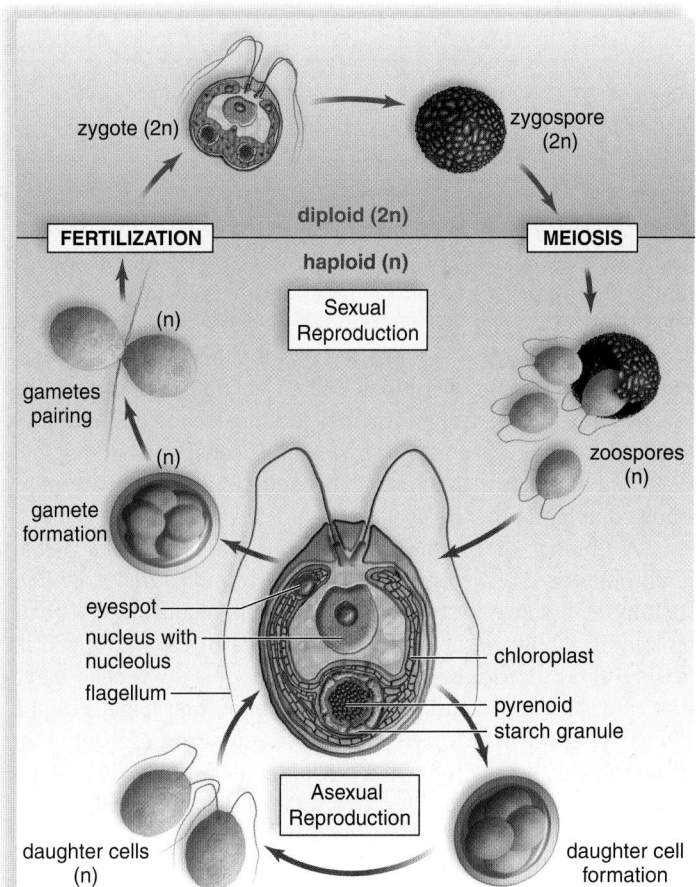

Figure 21.3 Haploid life cycle of *Chlamydomonas.*
Chlamydomonas reproduction is an example of the haploid life cycle common to algae. During asexual reproduction, all structures are haploid; during sexual reproduction, meiosis follows the zygospore stage, which is the only diploid part of the cycle.

40 µm

daughter colony

vegetative cells

Figure 21.4 Volvox. *Volvox* is a colonial chlorophyte. The adult *Volvox* colony often contains daughter colonies, which are asexually produced by special cells.

When growth conditions are unfavorable, *Chlamydomonas* reproduces sexually. Two haploid vegetative cells of two different mating types come into contact and fuse to form a diploid zygote. A heavy wall forms around the zygote, and it becomes a zygospore that undergoes a period of dormancy where it is resistant to unfavorable conditions. When conditions improve, the zygospore emerges from dormancy, undergoes meiosis and produces four haploid zoospores by meiosis. **Zoospores** are haploid flagellated spores that grow to become adult vegetative cells, thus completing the life cycle.

A number of colonial forms occur among the flagellated chlorophytes. *Volvox* is a well-known colonial green alga. A **colony** is a loose association of independent cells. A *Volvox* colony is a hollow sphere with thousands of cells arranged in a single layer surrounding a watery interior. *Volvox* cells move the colony by coordinating the movement of their flagella. Some *Volvox* cells are specialized for reproduction, and each of these can divide asexually to form a new daughter colony (Fig. 21.4). This daughter colony resides for a time within the parent colony, but then it escapes by releasing an enzyme that dissolves away a portion of the parent colony.

Ulva is a multicellular chlorophyte called sea lettuce because it lives in the sea and has a leafy appearance (Fig. 21.5*a*). The body of *Ulva* is two cells thick and can be as much as a meter long. *Ulva* has an alternation-of-generations life cycle (Fig. 21.5*c*) like that of land plants.

Charophytes. The charophytes are filamentous algae. **Filaments** [L. *filum*, thread] are end-to-end chains of cells.

Charophytes have both branched and unbranched filaments. *Spirogyra* is an example of an unbranched charophyte (Fig. 21.6). Charophytes often grow on aquatic flowering plants. Others attach to rocks or other objects under water or are suspended in the water column.

Spirogyra is found in green masses on the surfaces of ponds and streams. It has ribbonlike, spiralled chloroplasts (Fig. 21.6). *Spirogyra* undergoes sexual reproduction via **conjugation** [L. *conjugalis*, pertaining to marriage], a temporary union during which the cells exchange genetic material. Two haploid filaments line up parallel to each other, and the cell contents of one filament move into the cells of the other filament, forming diploid zygospores. Diploid zygospores survive the winter, and in the spring they undergo meiosis to produce new haploid filaments.

Chara (Fig. 21.7) is a charophyte that lives in freshwater lakes and ponds. It is commonly called a stonewort because it is encrusted with calcium carbonate deposits. The main strand of the alga, which can be over a meter long, is a single-file

a. *Ulva*, several individuals   b. One individual

Figure 21.5 Ulva.
a. *Ulva* is a multicellular chlorophyte known as sea lettuce. **b.** A single *Ulva* individual has a flat, leaflike appearance. **c.** Alternation-of-generations life cycle of *Ulva*.

sporophyte (2n)

zygote   sporangium

diploid (2n)

FERTILIZATION   MEIOSIS

haploid (n)

gametes   spore

gametophyte (n)

Alternation of generations
• Sporophyte is 2n generation.
• Meiosis produces spores.
• Gametophyte is n generation.

c. Alternation-of-generations life cycle

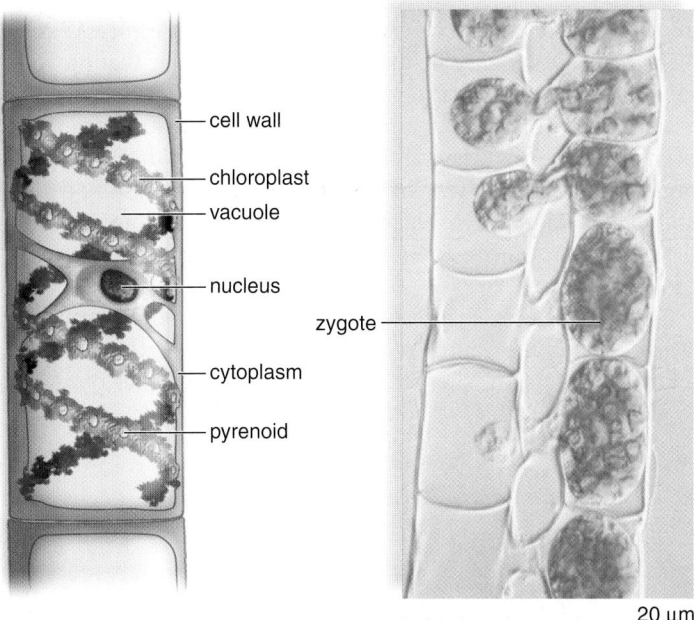

a. Cell anatomy

cell wall
chloroplast
vacuole
nucleus
zygote
cytoplasm
pyrenoid

b. Conjugation

20 µm

Figure 21.6 *Spirogyra*. **a.** *Spirogyra* is an unbranched charophyte in which each cell has a ribbonlike chloroplast. **b.** During conjugation, the cell contents of one filament enter the cells of another filament. Zygote formation follows.

strand of very long cells anchored by rhizoids, which are colorless, hairlike filaments. Only the cell at the upper end of the main strand produces new cells. Whorls of branches occur at multicellular nodes, regions between the giant cells of the main strand. Each of the branches is also a single-file thread of cells (Fig. 21.7*b*).

Male and female multicellular reproductive structures grow at the nodes, and in some species they occur on separate individuals. The male structure produces flagellated sperm, and the female structure produces a single egg. The gametes fuse to

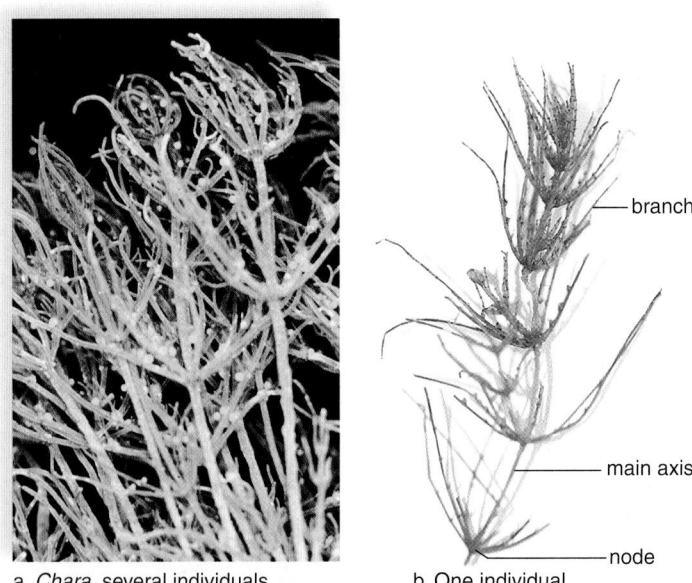

a. *Chara*, several individuals

b. One individual

branch
main axis
node

Figure 21.7 *Chara*. *Chara* is an example of a stonewort, a charophyte that shares a common ancestor with land plants.

Figure 21.8 Red alga. Red algae are multicellular seaweeds, represented by *Rhodoglossum affine*.

produce a diploid zygote that is retained until it is enclosed by tough walls.

DNA sequencing data suggest that among green algae, the stoneworts are most closely related to land plants.

Red Algae

The **red algae** are multicellular seaweeds that possess red and blue **accessory pigments** that transfer energy from absorbed light to the photopigment chlorophyll during photosynthesis (Fig. 21.8). These algae live in warm seawater, some at depths exceeding 70 m. Their accessory pigments allow them to absorb the wavelengths of light that penetrate into deep water.

Most of the more than 5,000 species of red algae are much smaller and more delicate than brown algae, but some species can exceed a meter in length. Red algae can be filamentous but most have feathery, flat, or, ribbonlike branches. Coralline red algae have cell walls that contain calcium carbonate, a mineral that contributes to the growth of coral reefs.

Red algae are economically important. Agar is a gelatin-like product made primarily from the algae *Gelidium* and *Gracilaria*. Agar is used commercially to make capsules for vitamins and drugs, as a material for making dental impressions, and as a base for cosmetics. In the laboratory, agar is a solidifying agent for a bacterial culture medium. When purified, it becomes the gel for electrophoresis, a procedure that separates proteins or nucleotides. Agar is also used in food preparation as an antidrying agent for baked goods and to make jellies and desserts set rapidly.

Carrageenan, extracted from various red algae, is an emulsifying agent for the production of chocolate and cosmetics. *Porphyra*, another red alga, is the basis of a billion-dollar aquaculture industry in Japan. The reddish-black wrappings around sushi rolls consist of processed *Porphyra* blades.

Check Your Progress   21.2A

1. Identify the haploid and diploid stages of the life cycle of *Chlamydomonas*.
2. Describe the function of accessory pigments in red algae.

Supergroup Chromalveolata

The **chromalveolates** [Gk. *chroma*, color, L. *alveolus*, hollow] include two large subgroups: the stramenopiles and the alveolates.

Stramenopiles

The **stramenopiles** include the brown algae, diatoms, golden brown algae, and water molds.

Brown Algae.   The **brown algae** have chlorophylls *a* and *c* in their chloroplasts and an accessory carotenoid pigment that gives them their characteristic brown color. Food reserves are stored as a carbohydrate called *laminarin.* The brown algae range from small forms with simple filaments to large multicellular forms that may reach 100 m in length (Fig. 21.9). The vast majority of the 1,500 species live in cold ocean waters.

The multicellular brown algae are seaweeds that live along the rocky coasts in the north temperate zone. They are pounded by waves as the tide comes in and are exposed to dry air as the tide goes out. They dry out slowly, however, because their cell walls contain a water-retaining material.

Laminaria, commonly called *kelp,* and *Fucus,* known as rockweed, are examples of brown algae that grow along the shoreline. They have a structure called a holdfast that allows them to cling to rocks. The giant kelps *Macrocystis* and *Nereocystis* form dense kelp forests in deeper water that provide food and habitat for marine organisms.

Laminaria is unique among the protists because members of this genus show tissue differentiation—that is, they transport organic nutrients by way of a tissue that resembles phloem in land plants. Most brown algae have an alternation-of-generations life cycle, but some species of *Fucus* have an exclusively sexual life cycle.

Brown algae are harvested for human food and for fertilizer. *Macrocystis* is the source of algin, a pectinlike material that is added to ice cream, sherbet, cream cheese, and other products to give them a stable, smooth consistency.

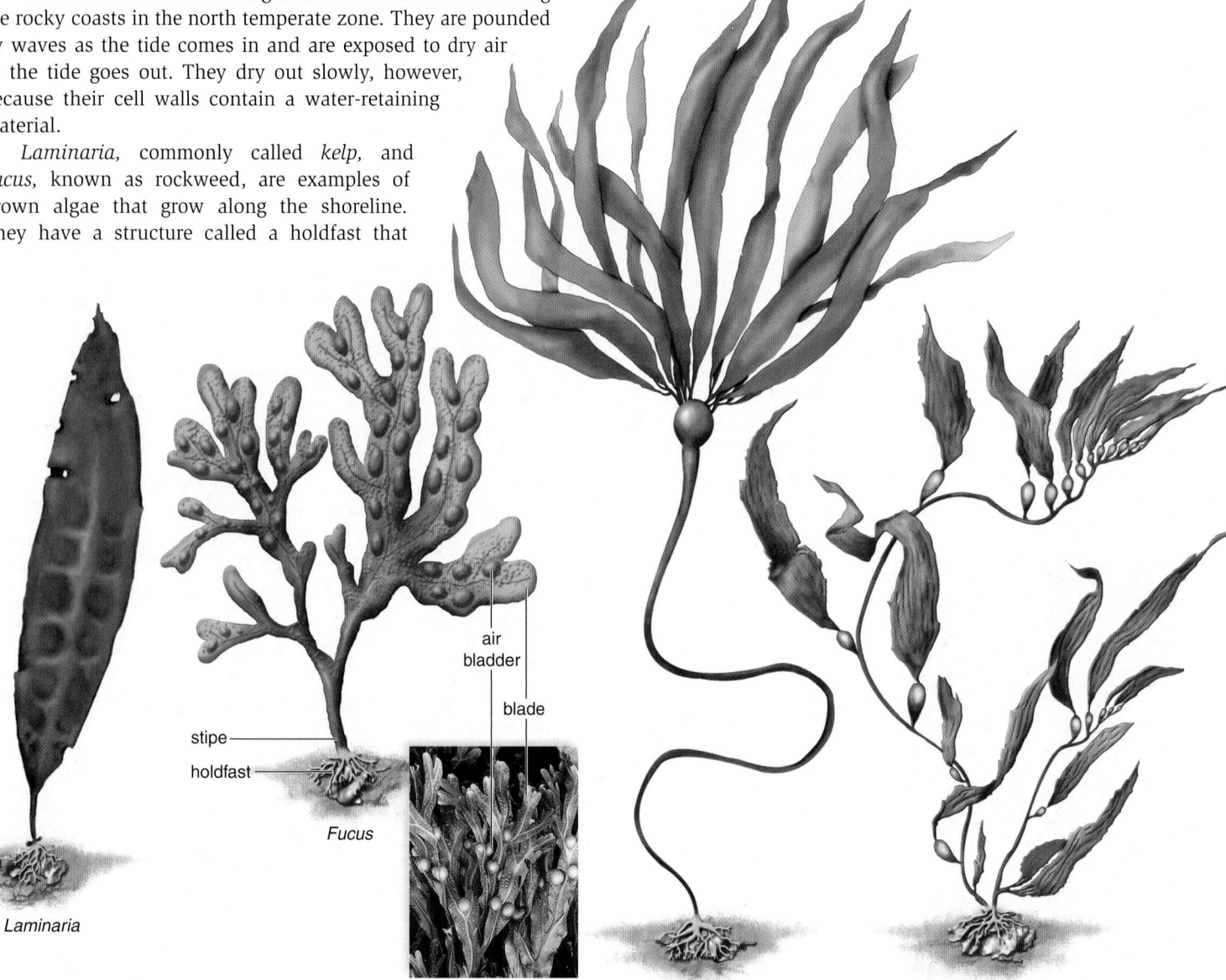

Laminaria

stipe

holdfast

Fucus

air bladder

blade

Rockweed, *Fucus*

Nereocystis

Macrocystis

Figure 21.9 Brown algae.   *Laminaria* and *Fucus* are seaweeds known as kelps. They live along rocky coasts of the north temperate zone. The other brown algae featured, *Nereocystis* and *Macrocystis,* form spectacular underwater "forests" at sea.

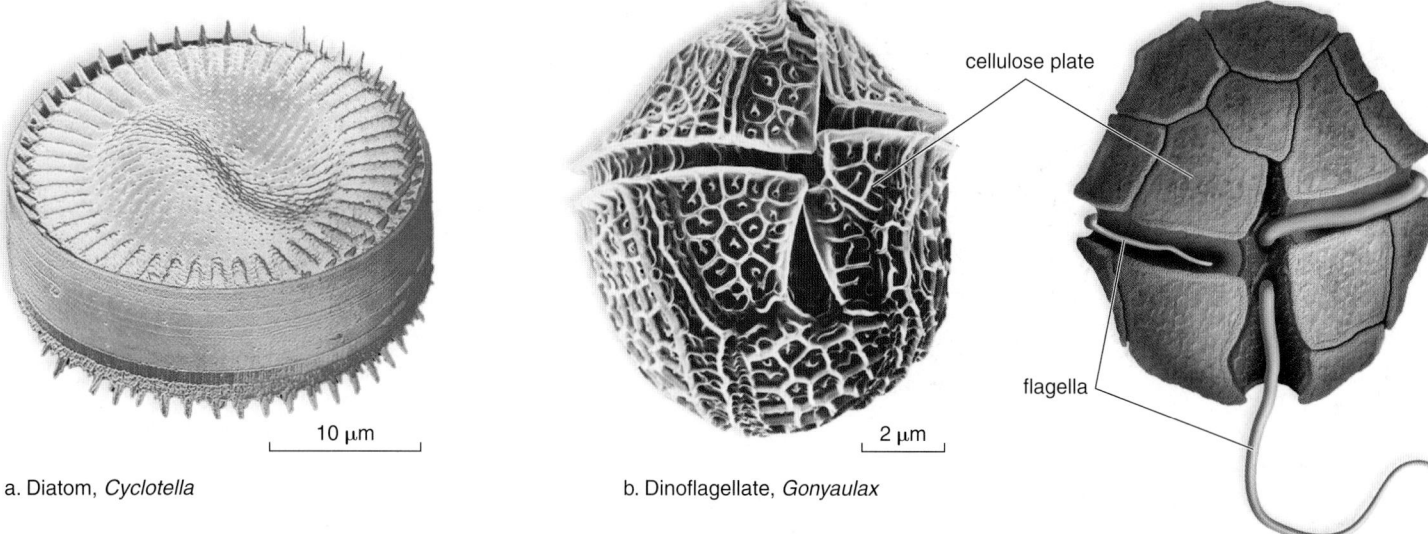

cellulose plate

flagella

a. Diatom, *Cyclotella*   b. Dinoflagellate, *Gonyaulax*

Figure 21.10 Diatoms and dinoflagellates.   a. Diatoms are stramenopiles of various colors, but even so their chloroplasts contain a unique golden brown pigment (*fucoxanthin*), in addition to chlorophylls *a* and *c*. The beautiful pattern results from markings on the silica-embedded wall. **b.** Dinoflagellates are alveolates, such as *Gonyaulax,* with cellulose plates.

Diatoms.   A **diatom** [Gk. *dia*, through, and *temno*, cut] is a tiny, single-celled stramenopile with an ornate silica shell. The shell is made up of an upper and lower shelf, called a valve, that fit together. (Fig. 21.10*a*). Diatoms have a carotenoid accessory pigment that gives them an orange-yellow color. Diatoms make up a significant part of plankton, which serves as a source of oxygen and food for heterotrophs in both freshwater and marine ecosystems.

Diatoms reproduce asexually and sexually. Asexual reproduction occurs by diploid parents undergoing mitosis to produce two diploid daughter cells. Each time a diatom reproduces asexually, the size of the daughter cells decrease until diatoms are about 30% of their original size. At this point, they begin to reproduce sexually. The diploid cell produces gametes by meiosis. Gametes fuse to produce a diploid zygote that grows and then divides via mitosis to produce new diploid diatoms of normal size.

The valves of diatoms are covered with a great variety of striations and markings that form beautiful patterns when observed under the microscope. These are actually depressions or pores through which the organism makes contact with the outside environment. The remains of diatoms, called diatomaceous earth, accumulate on the ocean floor and are mined for use as filtering agents, soundproofing materials, and gentle polishing abrasives, such as those found in silver polish and toothpaste.

Golden Brown Algae.   The **golden brown algae** derive their distinctive color from yellow-brown carotenoid accessory pigments. The cells of these unicellular or colonial protists typically have two flagella with tubular hairs, a characteristic of stramenopiles. Golden brown algae cells may be naked, covered with organic or silica scales, or enclosed in a secreted cagelike structure called a lorica. Many golden brown algae, such as *Ochromonas* (Fig. 21.11), are mixotrophs, capable of photosynthesis as well as phagocytosis. Like diatoms, the golden brown algae contribute to freshwater and marine phytoplankton.

Water Molds.   **Water molds** form furry growths when they parasitize fishes or insects and decompose remains. In spite of their common name, some water molds live on land and parasitize insects and plants. Nearly 700 species of water molds have been described. A water mold, *Phytophthora infestans*, was responsible for the 1840s potato famine in Ireland, and another, *Plasmopara viticola,* for the downy mildew of grapes that ravaged the vineyards of France in the 1870s. However, most water molds are saprotrophic and live off dead organic matter. Another well-known water mold is *Saprolegnia*, which is often seen as a cottonlike white mass on dead organisms (Fig. 21.12).

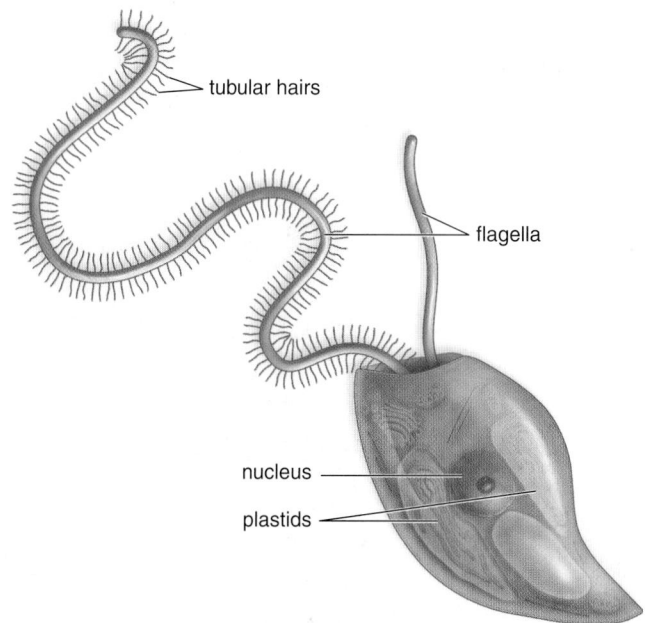

tubular hairs

flagella

nucleus

plastids

Figure 21.11 *Ochromonas,* a golden brown alga.   Golden brown algae have a type of flagella that characterizes the stramenopiles. The longer of the two flagella has rows of tubular hairs.

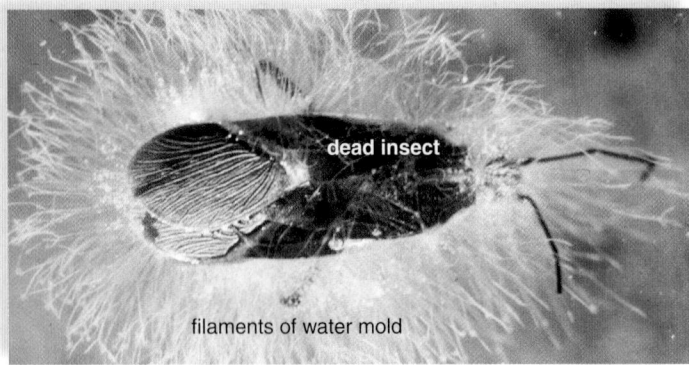

LM 10×

Figure 21.12 Water mold. *Saprolegnia*, a water mold, feeding on a dead insect, is not a fungus.

Water molds, also called oomycetes (or "egg fungi"), used to be grouped with fungi, because they are similar to fungi in many ways. The funguslike water molds have a filamentous body similar to the hyphae of fungi, but their cell walls are composed of cellulose instead of chitin. The life cycle of water molds also differs from that of the fungi.

Water molds are diploid, but can reproduce both asexually and sexually. During asexual reproduction, water molds produce flagellated, motile, diploid zoospores inside structures called *sporangia* (or *zoosporangia*). During sexual reproduction, structures called oogonia, which produce haploid eggs, and antheridia, which produce haploid sperm, form at the tips of filaments. Sperm and eggs come together when the antheridium and oogonium come in contact and sperm is inserted into the oogonium. Eggs and sperm fuse and produce diploid zoospores that emerge from the oogonium.

Check Your Progress   21.2B

1. Identify the characteristic common to all stramenopiles.
2. List the major groups of stramenopiles and give one example from each.

Alveolates

Alveolates have alveoli (small air sacs) lying just beneath their plasma membranes that are thought to lend support to the cell surface. Alveolates are unicellular.

Dinoflagellates.   **Dinoflagellates** are unicellular, photoautotrophic algae encased by protective cellulose and silicate plates (see Fig. 21.10). Dinoflagellates typically have two flagella: one flagellum acts as a rudder, and the other causes the cell to spin as it moves forward. Dinoflagellates have chlorophylls *a* and *c* and carotenoid accessory pigments that give them a yellow-green to brown color.

There are approximately 4,000 species of dinoflagellates. Some species, such as *Noctiluca*, are capable of bioluminescence (producing light). Dinoflagellates are a component of plankton and thus an important source of food for small animals in the ocean.

Zooxanthellae are dinoflagellates that form symbiotic relationships with invertebrates such as corals and other protozoans. Zooxanthellae are endosymbionts, living within the bodies of their hosts. Endosymbiotic dinoflagellates lack cellulose plates and flagella. Corals (see Chapter 28), members of the animal kingdom, usually contain large numbers of zooxanthellae, which provide their animal hosts with organic nutrients. In return, the corals provide the zooxanthellae with shelter, nutrients, and protection.

Dinoflagellates are one of the most important groups of primary producers in the marine ecosystem. Under unusually high nutrient conditions, populations of dinoflagellates and other algae can undergo a population explosion called an algal bloom. During an algal bloom, a single milliliter of water can contain more than 30,000 algae. Some algal blooms caused by dinoflagellates are so large that they turn the water brown or red because of the high density of the algae (Fig. 21.13*a*). These colorful algal blooms, called **red tides,** are sometimes so extensive that they can be seen from space. The dinoflagellate *Alexandrium catanella* can cause a Harmful Algal Bloom (HAB) because it secretes saxitoxin, a neurotoxin that is responsible for neurotoxic shellfish poisoning. Massive fish kills can occur as

Figure 21.13 Dinoflagellate bloom and fish kill. **a.** A dinoflagellate bloom, often called a red tide because of the color of the water, occurring near California's central coast. **b.** Fish kills, such as this one in University Lake in Baton Rouge, Louisiana, can be the result of a dinoflagellate bloom.

a.

b.

the result of saxitoxin produced during a red tide (Fig. 21.13b). Humans who consume shellfish that have fed during an *A. catanella* outbreak can die from paralytic shellfish poisoning, which paralyzes the respiratory organs.

Dinoflagellates usually reproduce asexually. Each daughter cell inherits half of the parent's cellulose plates. During sexual reproduction the daughter cells act as gametes and fuse to form a diploid zygote. The zygote enters a resting stage until signaled to undergo meiosis. The product of meiosis is a single haploid cell because the other cells disintegrate.

Ciliates. The **ciliates** are unicellular protists that move by means of cilia. They are the most structurally complex and specialized of all protozoa. Members of the genus *Paramecium* are classic examples of ciliates (Fig. 21.14a). Hundreds of cilia project through tiny holes in a semirigid outer covering, or pellicle. *Paramecium* beat their cilia in a coordinated and rhythmic manner to "swim" through their environment. Oval capsules that lie just beneath the pellicle contain **trichocysts.** Upon mechanical

or chemical stimulation, trichocysts discharge long, barbed threads that are useful for defense and for capturing prey. Some trichocysts release poisons.

Most ciliates ingest their food. *Paramecium* feed by sweeping food particles down a gullet, below which food vacuoles form. Following digestion, the soluble nutrients are absorbed by the cytoplasm, and the nondigestible residue is eliminated through the anal pore.

The ciliates are a diverse group of protozoans which range in size from 10 to 3,000 μm. The majority of the 8,000 species of ciliates are free-living, mobile, and unicellular; however, several parasitic, sessile, and colonial forms exist. *Suctoria* are sessile ciliates that have specialized microtubules for capturing, paralyzing, and ingesting other ciliates. *Stentor* may be the most elaborate ciliate, resembling a giant blue vase decorated with stripes (Fig. 21.14c). Cilia at the opening of the "vase" sweep in food particles. They are one of the largest unicellular protozoans, reaching lengths of 2 mm. *Ichthyophthirius* is an ectoparasitic protozoan that causes a common disease in fishes called "ick." If left untreated, it can be fatal.

During asexual reproduction, ciliates divide by transverse binary fission. Ciliates have two types of nuclei: a large macronucleus and one or more small micronuclei. The macronucleus controls the normal metabolism of the cell, whereas the micronuclei participate in reproduction. During sexual reproduction, the micronuclei undergo meiosis. Sexual reproduction involves conjugation (Fig. 21.14b), during which two ciliates unite and

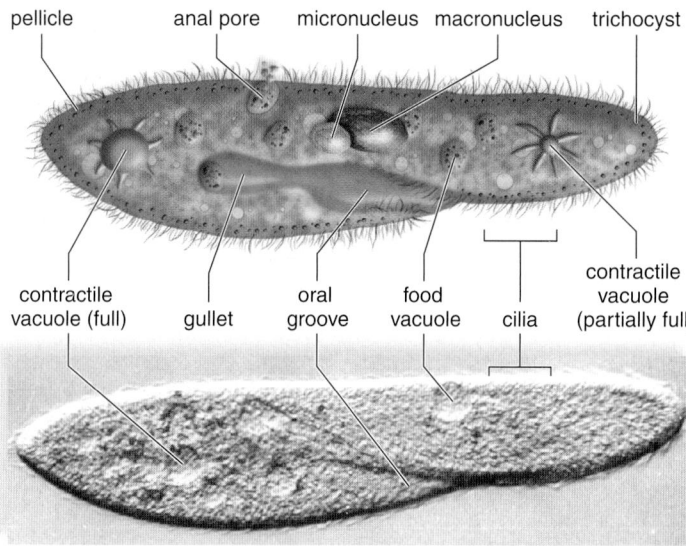

pellicle   anal pore   micronucleus   macronucleus   trichocyst

contractile vacuole (full)   gullet   oral groove   food vacuole   cilia   contractile vacuole (partially full)

a. *Paramecium*

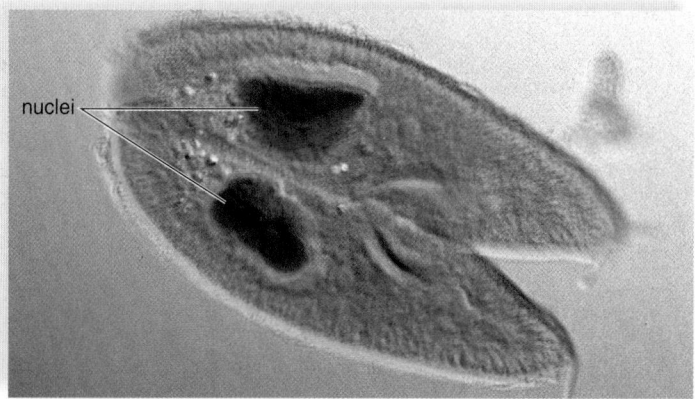

nuclei

b. During conjugation two *paramecia* first unite at oral areas

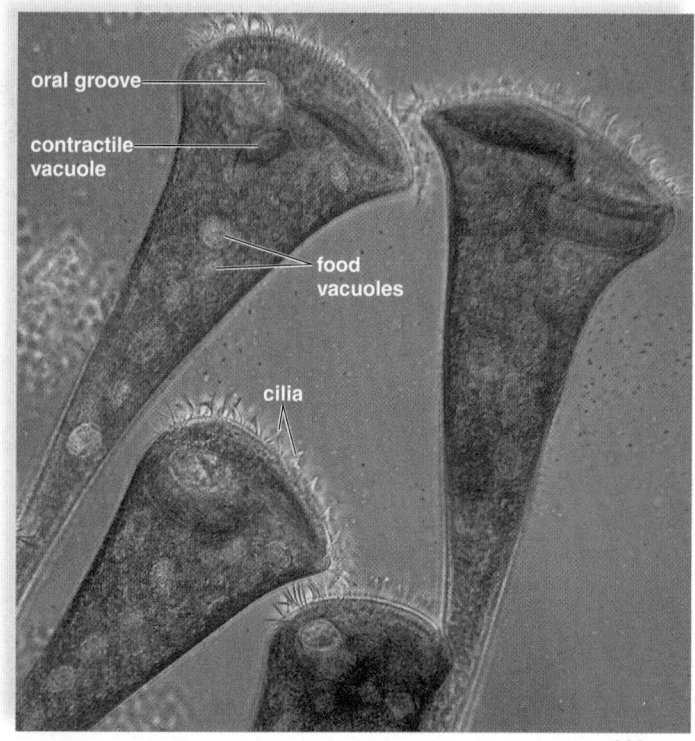

oral groove

contractile vacuole

food vacuoles

cilia

c. *Stentor*   200 μm

Figure 21.14 Ciliates. Ciliates are the most complex of the protists. **a.** Structure of *Paramecium,* adjacent to an electron micrograph. Note the oral groove, the gullet, and anal pore. **b.** A form of sexual reproduction called conjugation occurs periodically. **c.** *Stentor,* a large, vase-shaped, freshwater ciliate.

exchange haploid micronuclei. After conjugation the macronuclei dissolve and new macronuclei are formed from the fusion of the micronuclei.

Apicomplexans.

The **apicomplexans,** also known as *sporozoans,* are nearly 3,900 species of nonmotile, parasitic, sporeforming protozoans. Apicomplexans have a unique organelle called an apicoplast which is used to penetrate a host cell. All apicomplexans are parasites of animals, and some can infect multiple hosts.

Plasmodium, an apicomplexan, is responsible for malaria, a disease that causes more than one million fatalities a year. *Plasmodium* is a parasitic protozoan that infects human red blood cells. The chills and fever of malaria appear when the infected cells burst and release toxic substances into the blood (Fig. 21.15). In humans, malaria is caused by four distinct members of the genus *Plasmodium. Plasmodium vivax,* the cause of one type of malaria, is the most common.

The transmission cycle of *Plasmodium* involves the *Anopheles* mosquito as an intermediate organism, or **vector,** that transmits the disease between the host and other organisms. The life cycle of *Plasmodium* alternates between a sexual and an asexual phase dependent upon whether reproduction takes place inside the mosquito (sexual) or human host (asexual) (Fig. 21.15). Female *Anopheles* mosquitoes acquire protein for production of eggs by biting humans and other animals.

Despite efforts to control malaria, more than 250 million new cases of malaria appear each year. Malaria is most common in densely populated, tropical areas of the globe. Because of its prevalence in these areas, 3.5 billion people, nearly half of the world's population, are at risk from the disease.

Apicomplexans include other human parasites. Cyclosporiasis is an infection of the intestine caused by the parasitic apicomplexan *Cyclospora cayetanensis.* The cyclosporin parasite is transmitted by feces-contaminated fresh produce and water. Outbreaks of cyclosporiasis in the United States have been attributed to contaminated fresh raspberries, basil, snow peas, and mesclun lettuce. *Toxoplasma gondii,* another apicomplexan, causes toxoplasmosis that is transmitted to humans from the feces of infected cats. In pregnant women, the parasite can infect the fetus and cause birth defects and mental retardation; in AIDS patients, it can infect the brain and cause neurological problems.

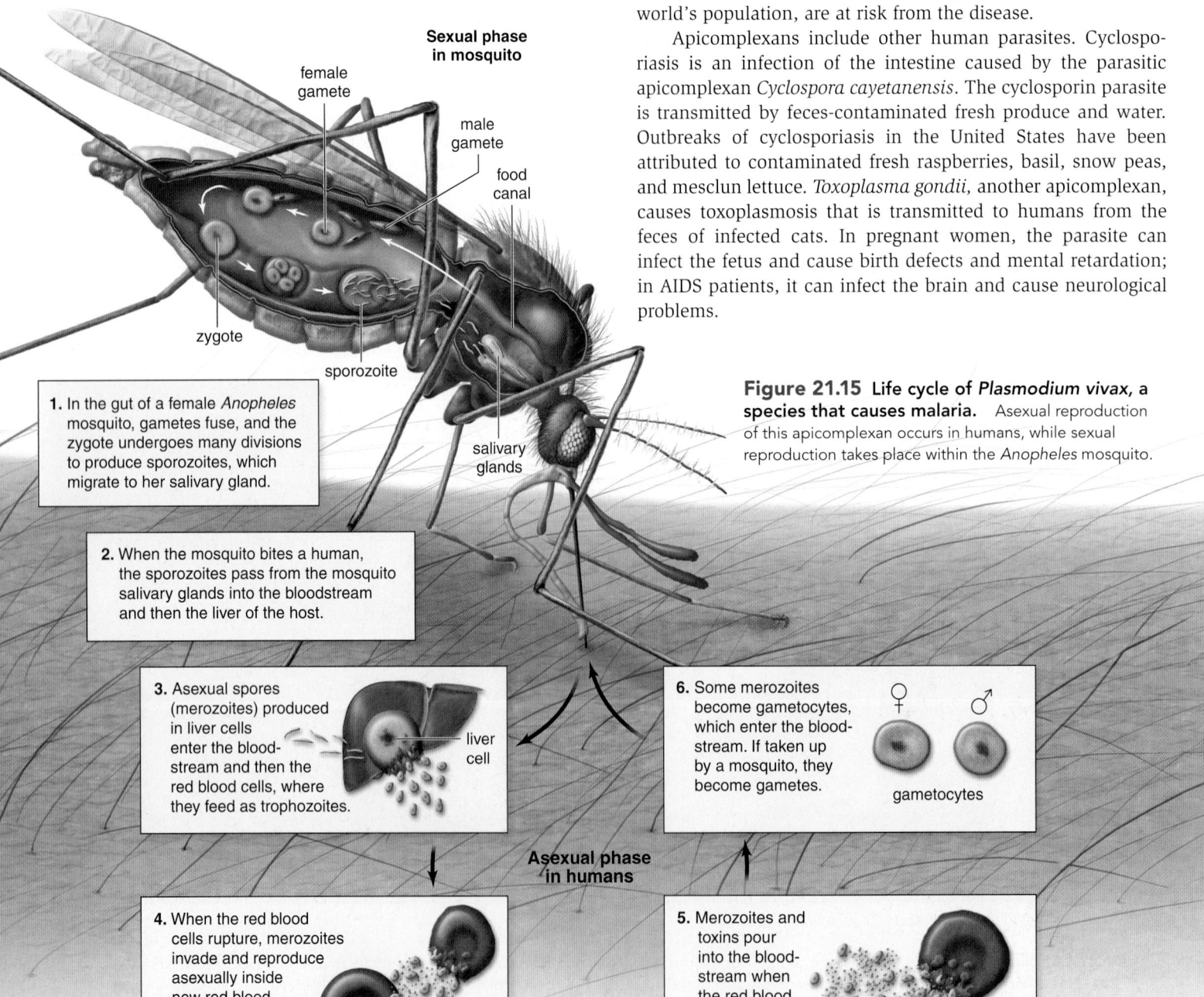

Sexual phase in mosquito

female gamete

male gamete

food canal

zygote

sporozoite

salivary glands

1. In the gut of a female *Anopheles* mosquito, gametes fuse, and the zygote undergoes many divisions to produce sporozoites, which migrate to her salivary gland.

2. When the mosquito bites a human, the sporozoites pass from the mosquito salivary glands into the bloodstream and then the liver of the host.

3. Asexual spores (merozoites) produced in liver cells enter the bloodstream and then the red blood cells, where they feed as trophozoites.

liver cell

6. Some merozoites become gametocytes, which enter the bloodstream. If taken up by a mosquito, they become gametes.

♀ ♂

gametocytes

Asexual phase in humans

4. When the red blood cells rupture, merozoites invade and reproduce asexually inside new red blood cells.

5. Merozoites and toxins pour into the bloodstream when the red blood cells rupture.

Figure 21.15 Life cycle of *Plasmodium vivax,* a species that causes malaria. Asexual reproduction of this apicomplexan occurs in humans, while sexual reproduction takes place within the *Anopheles* mosquito.

Supergroup Excavata

The **excavates** [L. *cavus*, hollow] include **zooflagellates** that have atypical or absent mitochondria and distinctive flagella and/or deep (excavated) oral grooves.

Euglenids

The **euglenids** are small (10–500 µm), freshwater unicellular organisms. *Euglena deses* is a common inhabitant of freshwater ditches and ponds. Classifying the approximately 1,000 species of euglenids is problematic because they are very diverse. Some euglenids are mixotrophic, some are photoautotrophic, and others heterotrophic. One-third of all genera have chloroplasts; the rest do not. Those that lack chloroplasts ingest or absorb their food.

When present, the chloroplasts are surrounded by three rather than two membranes. Carbohydrates are synthesized in a special region of the chloroplast called a pyrenoid. Euglenids produce an unusual type of carbohydrate called paramylon.

Euglenids have two flagella, one much longer than the other (Fig. 21.16). The longer flagellum is called a tinsel flagellum because it has hairs on it. Near the base of the tinsel flagellum is an eyespot that has a photoreceptor capable of detecting light.

Euglenids are bounded by a flexible *pellicle* composed of protein bands lying side by side; this arrangement allows euglenids to assume different shapes. A contractile vacuole rids the body of excess water. Euglenids reproduce by longitudinal cell division, and sexual reproduction is not known to occur.

Parabasalids and Diplomonads

Parabasalids and diplomonads are unicellular, flagellated, excavates that are endosymbionts of animals. They are able to survive in anaerobic, or low oxygen, environments. These protozoans lack mitochondria; instead they rely on fermentation for the production of ATP. A variety of forms are found in the guts of termites where they assist with the breakdown of cellulose.

Parabasalids have a unique, fibrous connection between the Golgi apparatus and flagella. The most common sexually transmitted disease, trichomoniasis, is caused by the parabasalid *Trichomonas vaginalis*. Infection causes vaginitis in women. The parasite may also infect the male genital tract; however, the male may have no symptoms.

A **diplomonad** [Gk. *diplo*, double, and *monas*, unit] cell has two nuclei and two sets of flagella. The diplomonad *Giardia lamblia* forms cysts that are transmitted by contaminated water. *Giardia* attaches to the human intestinal wall, causing severe diarrhea (Fig. 21.17). This protozoan lives in the digestive tracts

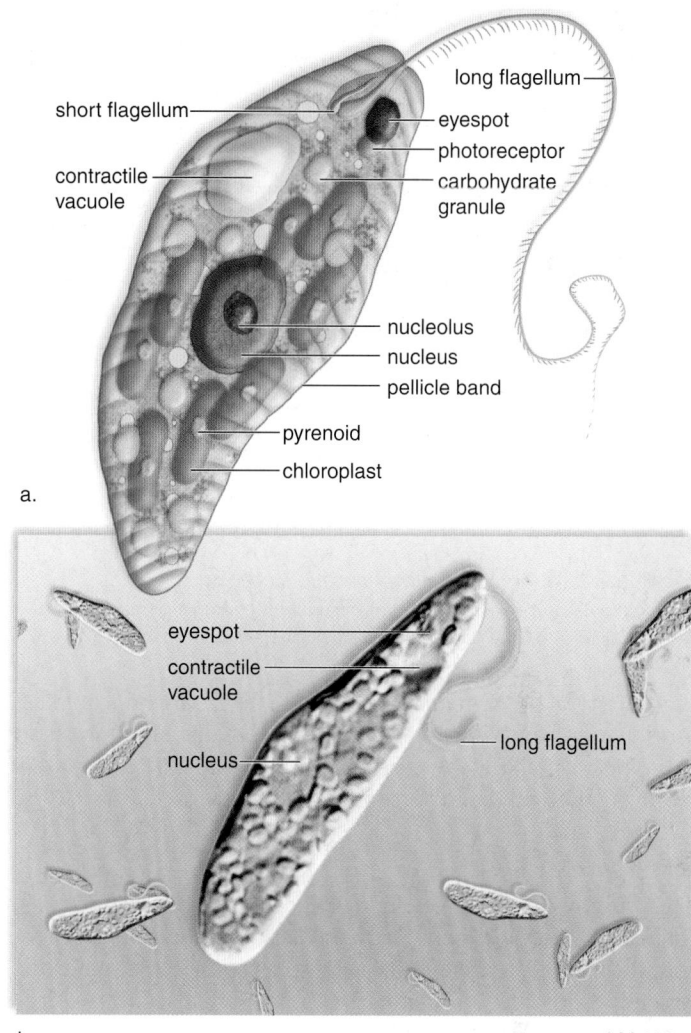

Figure 21.16 *Euglena.* a. In *Euglena*, a very long flagellum propels the body, which is enveloped by a flexible pellicle. **b.** Micrograph of several specimens.

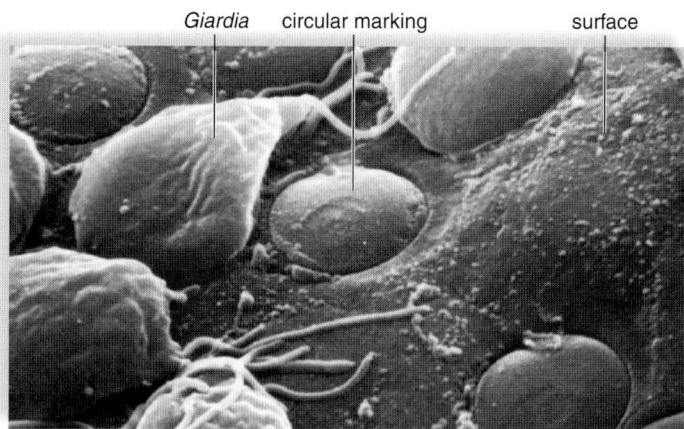

Figure 21.17 *Giardia lamblia.* This diplomonad adheres to any surface, including epithelial cells, by means of a sucking disk. Characteristic markings can be seen after the disk detaches.

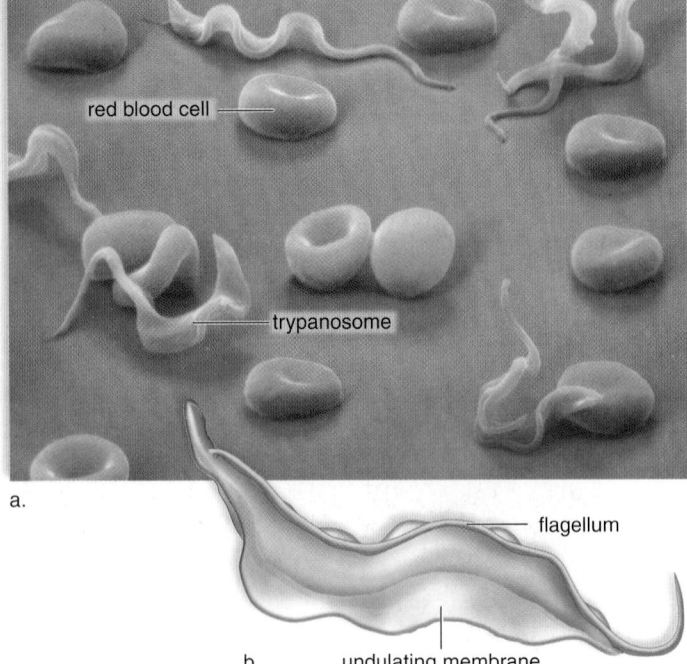

a.

b.

Figure 21.18 *Trypanosoma brucei.* **a.** Micrograph of *Trypanosoma brucei*, a causal agent of African sleeping sickness, among red blood cells. **b.** The drawing shows its general structure.

of a variety of other mammals as well. Beavers are known to be a reservoir of *Giardia* infection in the mountains of the western United States, and many cases of infection have been acquired by hikers who fill their canteens at a beaver pond.

Kinetoplastids

The **kinetoplastids** are single-celled, flagellated protozoans named for their distinctive *kinetoplasts*, large masses of DNA found in their mitochondria. Trypanosomes are parasitic kinetoplastids that are passed to humans by insect bites. *Trypanosoma brucei* (Fig. 21.18) is the cause of African sleeping sickness (see the Biological Systems feature on page 398). It is transmitted by an insect vector, the tsetse fly (*Glossina*). The lethargy characteristic of the disease is caused by an inadequate supply of oxygen to the brain. Many thousands of cases are diagnosed each year. Fatalities or permanent brain damage are common.

Trypanosoma cruzi causes Chagas disease in humans in Central and South America. Approximately 45,000 people die yearly from this parasite.

Check Your Progress   **21.2D**

1. Identify two distinctive features of the excavates.
2. Identify one parasitic and one free-living excavate.
3. Match two excavates with the human disease it causes.

Supergroup Amoebozoa

The **amoebozoans** [Gk. *ameibein*, to change, *zoa*, animal] are protozoans that move by **pseudopods** [Gk. *pseudes*, false, and *podos*, foot]. Pseudopods form when an amoebozoan's microfilaments contract and extend as the cytoplasm streams toward a particular direction. Amoebozoans usually live in aquatic environments where they are often a part of the plankton.

Amoeboids

The **amoeboids** are protists that move and ingest their food with pseudopods. Hundreds of species of amoeboids have been identified. *Amoeba proteus* is a commonly studied freshwater amoeba (Fig. 21.19). Amoeboids feed by **phagocytosis** [Gk. *phagein*, eat, and *kytos*, cell], by which they engulf their prey with a pseudopod; prey organisms may be algae, bacteria, or other protists. Digestion occurs within a *food vacuole*. Freshwater amoeboids, including *Amoeba proteus*, have contractile vacuoles that excrete excess water from the cytoplasm.

Video Amoeba Locomotion

Entamoeba histolytica is a parasitic amoeboid that can live in the human large intestine and cause amoebic dysentery. Infectious *Entamoeba* cysts pass out of the body with infected stool. Others become infected from ingesting water and food contaminated with cysts. Infection of the liver and brain can also be fatal.

Slime Molds

Slime molds are important decomposers that feed on dead plant material. Slime molds were once classified as fungi, but unlike fungi, they lack cell walls, and they have flagellated cells at some time during their life cycle. The vegetative state of the slime molds is mobile and amoeboid. Slime molds produce spores by meiosis; the spores germinate to form gametes.

Video Decomposers

Usually, **plasmodial slime molds** exist as a *plasmodium*, a diploid, multinucleated, cytoplasmic mass enveloped by a slime sheath. (This term should not be confused with the genus *Plasmodium*, which is in the alveolate group.) This sluglike slime mass consumes decaying plant material as it creeps along a forest floor or agricultural field (Fig. 21.20). Approximately 700

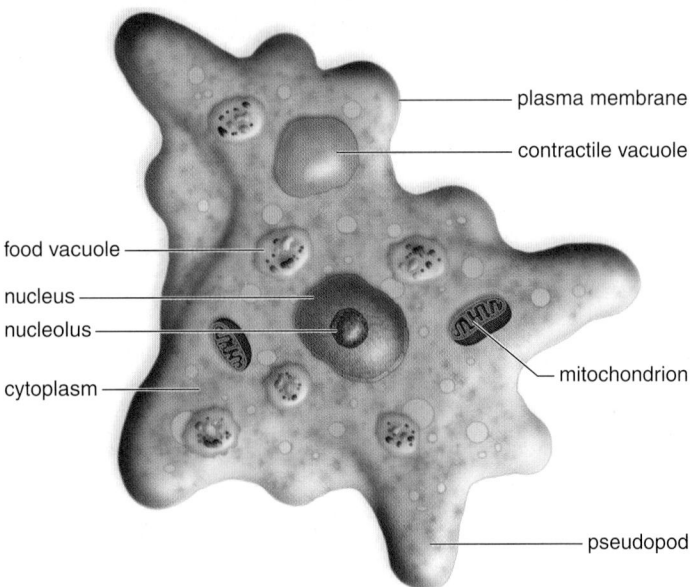

Figure 21.19 *Amoeba proteus.* This amoeboid is common in freshwater ponds. Bacteria and other microorganisms are digested in food vacuoles, and contractile vacuoles rid the body of excess water.

Plasmodium, *Physarum*   Sporangia, *Hemitrichia* |— 1 mm —|

mature
plasmodium

young
plasmodium

sporangia
formation begins

zygote

young
sporangium

FERTILIZATION   diploid (2n)   **MEIOSIS**
haploid (n)

mature
sporangium

fusion

spores

amoeboid cells

germinating
spore

flagellated cells

Figure 21.20 Plasmodial slime molds. The diploid adult forms sporangia during sexual reproduction, when conditions are unfavorable to growth. Haploid spores germinate, releasing haploid amoeboid or flagellated cells that fuse.

species of plasmodial slime molds have been described. Many species are brightly colored.

When conditions are unfavorable, a plasmodium develops many sporangia. A **sporangium** [Gk. *spora*, seed, and *angeion* (dim. of *angos*), vessel] is a reproductive structure that produces spores. An aggregate of sporangia is called a fruiting body. The spores produced by a plasmodial slime-mold sporangium can survive until moisture is sufficient for them to germinate. In plasmodial slime molds, spores release either a haploid flagellated cell or an amoeboid cell. Eventually, two of the haploid cells fuse to form a zygote that feeds and grows, producing a multinucleated plasmodium once again.

Cellular slime molds exist as individual amoeboid cells. They are common in soil, where they feed on bacteria and yeasts. Nearly 70 species of cellular slime molds have been described.

As unfavorable environmental conditions develop, the slime mold cells release a chemical that causes them to aggregate into a pseudoplasmodium. The pseudoplasmodium stage is temporary and eventually gives rise to a fruiting body in which sporangia produce spores. When favorable conditions return, the spores germinate, releasing haploid amoeboid cells, and the asexual cycle begins again.

Supergroup Opisthokonta

Animals and fungi are **opisthokonts** [Gk. *opisthos*, behind, *kontos*, pole] along with several closely related protists. This supergroup includes both unicellular and multicellular protozoans. Among the opisthokonts are the **choanoflagellates,** animal-like protozoans that are closely related to sponges. The choanoflagellates, including unicellular as well as colonial forms, are filter feeders with cells that bear a striking resemblance to the choanocytes that line the inside of sponges (see Section 28.2). Each choanoflagellate has a single posterior flagellum surrounded by a collar of slender microvilli. Beating of the flagellum creates a water current that flows through the collar, where food particles are taken in by phagocytosis. Colonial choanoflagellates such as *Codonosiga* (Fig. 21.21*a*) commonly attach to surfaces with a stalk, but sometimes float freely like *Proterospongia* (Fig. 21.21*b*).

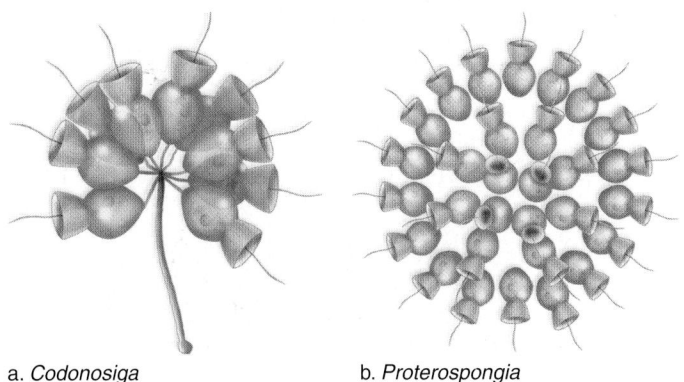

a. *Codonosiga*   b. *Proterospongia*

Figure 21.21 Colonial choanoflagellates. a. A *Codonosiga* colony can anchor itself with a slender stalk. **b.** A *Proterospongia* colony is unattached.

Biological Systems

African Sleeping Sickness

The World Health Organization (WHO) recognizes sixteen Neglected Tropical Diseases (NTDs) that affect more than 1 billion people exclusively in the most impoverished communities. Three of these NTDs—leishmaniasis, African sleeping sickness, and Chagas disease—are caused by parasitic protists that are transmitted to humans from the bite of an infected insect (Table 21.2).

African sleeping sickness, also called African human trypanosomiasis, is the deadliest of all of the NTDs. It is caused by a type of parasitic, single-celled protist,

Trypanosoma brucei, that is transmitted to humans in the bite of the blood sucking tsetse fly (*Glossina*) (Fig. 21A). The tsetse fly is the transmission vector of the disease because it carries the trypanosomes and transmits the long, blade-shaped protists into the human bloodstream while feeding. The flies first become infected with the parasite when they bite either an infected human or other mammal, such as domestic cattle and pigs, that carry *T. brucei*.

Figure 21A illustrates the relationship between humans, the tsetse fly, and domes-

tic livestock. Unlike humans, livestock can carry *T. brucei* without getting sick. Livestock that carry *T. brucei* are called disease reservoirs because they carry a reservoir of *T. brucei* infection that is picked up by the tsetse fly and transmitted to humans. Humans also can serve as a reservoir for new tsetse fly infections because the parasite can be picked up from humans by a fly bite and transmitted to others.

It is estimated that as many as 50,000 people are plagued by this disease. In the later stages of infection, *T. brucei* attacks the brain, causing behavioral changes and a shift in sleeping patterns. If caught early enough, it can be cured with medication, but because it is most common in very poor nations, medication and treatment are difficult to come by. Without treatment, the disease is fatal.

The tsetse fly prefers shaded, cool, moist habitats, such as the sandy edges of pools and rivers in central Africa (Fig. 21B). Regions that are plagued by sleeping sickness have no running water, so daily trips to local rivers and watering holes are necessary to collect water. Local people also use rivers to fish and to water livestock, their sole source of food and income (Fig. 21B). The people who live and work in tsetse habitat are the most vulnerable to the disease. As a result, many communities have abandoned the areas infested with the tsetse fly out of fear of sleeping sickness. River valleys, prime tsetse habitat, are the most fertile agricultural areas in Africa, but many of these rich lands have been abandoned for drier, outlying areas. The abandonment of these fertile valleys has had a huge impact on the ecology of surrounding areas and

Table 21.2 Neglected Tropical Diseases Caused by Parasitic Protists

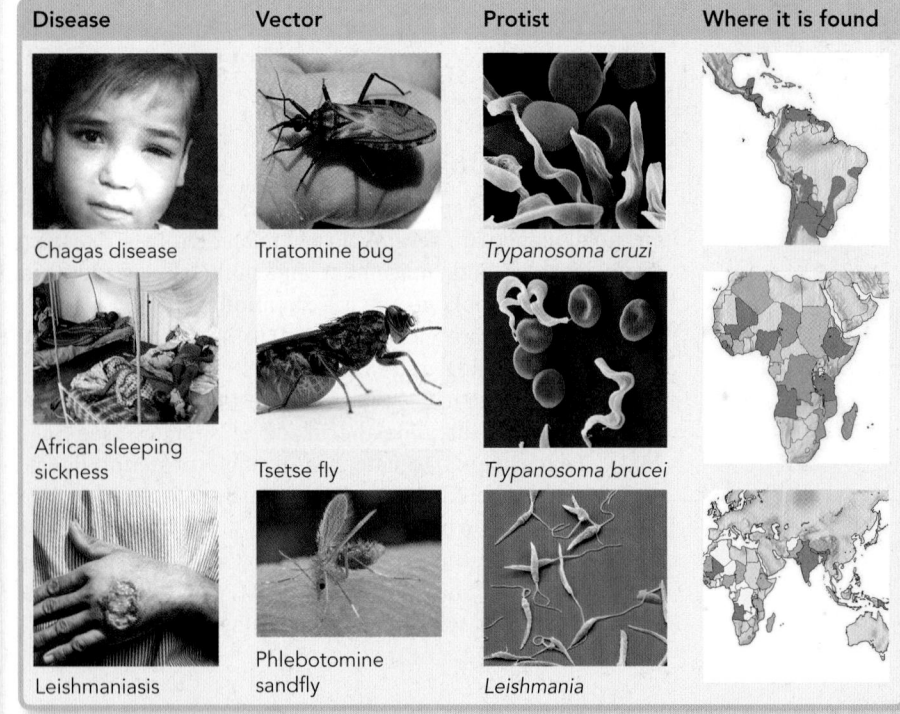

Disease	Vector	Protist	Where it is found
Chagas disease	Triatomine bug	*Trypanosoma cruzi*	
African sleeping sickness	Tsetse fly	*Trypanosoma brucei*	
Leishmaniasis	Phlebotomine sandfly	*Leishmania*	

Nucleariids are opisthokonts with a rounded or slightly flattened cell body and threadlike pseudopods called filopodia.

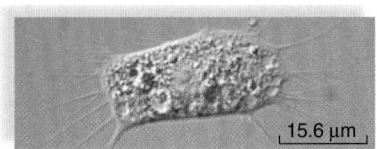

15.6 μm

Most feed on algae or cyanobacteria. Although they lack the characteristic cell walls found in fungi, molecular similarities suggest nucleariids are close relatives of fungi.

Supergroup Rhizaria

The **rhizarians** [Gk. *rhiza*, root] consist of the **foraminiferans** and the **radiolarians,** organisms with fine, threadlike pseudopods. Although rhizarians were once classified with amoebozoans, they are now assigned to a different supergroup. Foraminiferans and radiolarians both have a skeleton called a **test** made of calcium carbonate.

The tests of foraminiferans and radiolarians are intriguing and beautiful. In the foraminiferans, the calcium carbonate test is often multichambered. The pseudopods extend through

the economy of communities and nations on a large scale.

Recently, an effort to eliminate the tsetse fly from these lands has been successful, so much so that the World Health Organization (WHO) reports that the number of new cases of sleeping sickness has dropped to the lowest levels in 50 years. This decline in the incidences of new cases is a direct result of efforts by public health agencies and the WHO to control tsetse populations. The treatment and control of sleeping sickness has allowed the resettlement of more than 25 million hectares of prime agricultural land.

Questions to Consider

1. How does poverty reinforce a high occurrence of African sleeping sickness?
2. Livestock are called "disease reservoirs." Explain why.
3. How are the economy, ecology, and disease biology of African sleeping sickness interdependent?

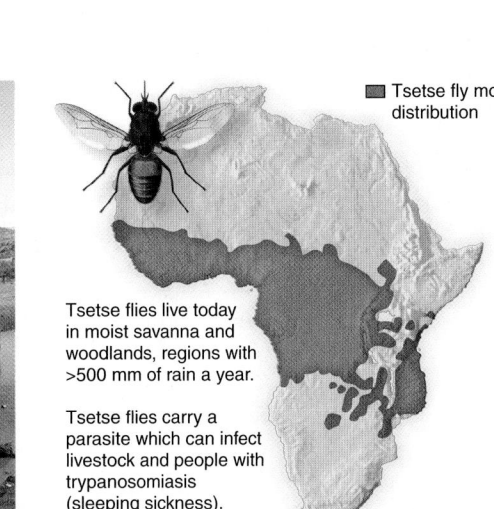

Figure 21A The transmission pattern of *Trypanosoma brucei. T. brucei* is a parasitic protist that causes African sleeping sickness. It can be transmitted from livestock or other humans to new human hosts.

Tsetse flies live today in moist savanna and woodlands, regions with >500 mm of rain a year.

Tsetse flies carry a parasite which can infect livestock and people with trypanosomiasis (sleeping sickness).

■ Tsetse fly modern distribution

Figure 21B Areas affected by sleeping sickness. Sleeping sickness is found in moist tropical regions of Africa. People who live near wet areas or engage in fishing, raising livestock, and farming are at greater risk.

---

openings in the test, which covers the plasma membrane (Fig. 21.22*a*). In the radiolarians, the glassy silicon test is internal and usually has a radial arrangement of spines (Fig. 21.22*b*). The pseudopods are external to the test.

The tests of dead foraminiferans and radiolarians form a layer of sediment 700–4,000 m deep on the ocean floor. The presence of tests is used as an indicator of oil deposits on land and sea. Their fossils date as far back as the Precambrian, and are evidence of the antiquity of the protists. Each geological period has a distinctive form of foraminiferan, and therefore

foraminiferans can be used as index fossils to date sedimentary rock. Millions of years of foraminiferan deposits formed the White Cliffs of Dover along the southern coast of England. Also, the great Egyptian pyramids are built of foraminiferan limestone. One foraminiferan test found in the pyramids is about the size of an old silver dollar—about an inch and a half across! This species, known as *Nummulites,* produced a flattened, coiled test, and its fossils have been found in deposits worldwide, including central-eastern Mississippi.

a. Foraminiferan, *Globigerina*, and the White Cliffs of Dover, England

b. Radiolarian tests   SEM 200×

Figure 21.22 Foraminiferans and radiolarians.   a. Pseudopods of a live foraminiferan project through holes in the calcium carbonate shell. Fossilized shells were so numerous they became a large part of the White Cliffs of Dover when a geological upheaval occurred. **b.** Skeletal test of a radiolarian. In life, pseudopods extend outward through the openings of the glassy silicon shell.

© Dr. Richard Kessel & Dr. Gene Shih/Visuals Unlimited.

Check Your Progress   21.2E

1. Describe how opisthokonts differ from the amoebozoans and the rhizarians.
2. Summarize the evidence that suggests animals are opisthokonts.

CONNECTING *the* CONCEPTS *with the* BIG IDEAS

Evolution

- Phylogenetic trees and cladograms model the possible evolutionary history of groups of living organisms. (1B2a-d)

Energy and Homeostasis

- Protists are often classified based on their energy acquisition, with many autotrophic and heterotrophic members. (2A2a-b)
- Populations' size is affected by both biotic and abiotic factors, as seen in algal blooms. (2D1c*IE*)
- Environmental cycles can help synchronize physiological events, such as fruiting body formation in slime molds. (2E2c*IE*)
- Cooperation that occurs between some protists and other types of organisms improves the survival chances of both, as in lichens. (2E3b4*IE*)

Interactions and Systems

- Unicellular populations are capable of interactions which affect their efficiency and productivity. (4B2a3)
- Introduction of some protists into a new ecosystem or into a population with little diversity may lead to devastation of a native species, as in potato blight. (4B4a*IE*, 4C3a*IE*)

*Find the unabridged version of all EK citations at www.glencoe.com/maderAP11.

Media Study Tools

www.glencoe.com/maderAP11

Enhance your study of this chapter with study tools and practice tests. Also ask your instructor about the resources available through ConnectPlus, including the media-rich eBook, interactive learning tools, and animations.

Summarize

21.1 General Biology of Protists

Protists are in the domain Eukarya. Independent endosymbiotic events may account for the presence of mitochondria and chloroplasts, as well as other plastids, in eukaryotic cells. Protists are generally unicellular, but complex organisms because they (1) are morphologicaly diverse, (2) employ various nutrition modes, and (3) have multi-stage life cycles that include the ability to withstand hostile environments.

Protists are of great ecological importance because they are the the main producers in aquatic ecosystems. Protists also enter into various types of symbiotic relationships. The domain Eukarya can be broken into six supergroups, or major eukaryotic lineages. However, classification of protists within the eukaryote family tree is problematic, because relationships among major eukaryote lineages remain unresolved.

21.2 Diversity of Protists

Supergroup Archaeplastida includes land plants and all green and red algae. Like plants, green algae possess chlorophylls, store energy as starch, and have cell walls made of cellulose. Chlorophytes, a group of green algae, include unicellular (*Chlamydomonas*), colonial (*Volvox*), and multicellular *(Ulva)* forms. Charophytes include *Spirogyra* as well as stoneworts (*Chara*), and are thought to be the closest living relatives of land plants. Green algae zygotes undergo meiosis and the adult is haploid. *Ulva* has an alternation-of-generations life cycle like land plants. Red algae contain plastids with pigments that give them a reddish or reddish-brown appearance. Red algae have notable economic importance.

Supergroup Chromalveolata consists of stramenopiles and alveolates. Stramenopiles include brown algae, diatoms, golden brown algae, and water molds. Brown algae have chlorophylls *a* and *c* plus a brownish carotenoid accessory pigment. Seaweeds are economically important, large, complex, brown algae. Diatoms and golden brown algae have an outer layer of silica. Water molds are similar to fungi except that they produce flagellated, diploid zoospores. Alveolates include dinoflagellates, ciliates, and apicomplexans. Dinoflagellates can form harmful algal blooms called red tides. The ciliates, such as *Paramecium*, move by coordinated movement of their many cilia. The apicomplexans are nonmotile parasites that form spores. *Plasmodium*, an apicomplexan that causes malaria, is transmitted between disease reservoirs and hosts by *Anopheles* mosquitos.

Supergroup Excavata includes single-celled, motile protists, including euglenids, parabasalids, diplomonads, and kinetoplastids. Euglenids are photoautotrophic, flagellated cells with a pellicle instead of a cell wall. Many of the kinetoplastids are parasites, including trypanosomes, which cause Chagas disease and African sleeping sickness when transmitted by the bite of an insect vector. Parabasalids and diplomonads, such as *Trichomonas vaginalis* and *Giardia lamblia*, are common parasites of animal hosts. They thrive in low-oxygen conditions; they lack mitochondria and cannot perform aerobic respiration.

Supergroup Amoebozoa contains amoeboids and slime molds—protists that use pseudopods for motility and feeding. Amoeboids are heterotrophs that move and feed by forming pseudopods. In *Amoeba proteus*, food vacuoles form following phagocytosis of prey. Contractile vacuoles discharge excess water. Slime molds, which produce nonmotile spores, are similar to fungi except they have an amoeboid stage.

Supergroup Opisthokonta includes animals, the animal-like choanoflagellates, the fungi, and the funguslike protists called nucleariids.

Supergroup Rhizaria includes the foraminiferans and radiolarians that have threadlike pseudopods and skeletons called tests. The tests of foraminiferans and radiolarians form a deep layer of sediment on the ocean floor. The tests of foraminiferans can be used as index fossils.

Key Terms

accessory pigment 389
alveolate 392
amoeboid 396
amoebozoan 396
apicomplexan 394
archaeplastid 387
brown algae 390
cellular slime mold 397
charophyte 387
chlorophyte 387
choanoflagellate 397
chromalveolate 390
ciliate 393
colony 388
conjugation 388
cyst 384
diatom 391
dinoflagellate 392
diplomonad 395
endosymbiosis 384
euglenid 395
excavate 395
filament 388
foraminiferan 398
golden brown algae 391

green algae 387
kinetoplastid 396
mixotrophic 384
monophyletic 384
nucleariid 398
opisthokont 397
parabasalid 395
phagocytosis 396
plankton 384
plasmodial slime mold 396
protist 384
pseudopod 396
radiolarian 398
red algae 389
red tide 392
rhizarian 398
sporangium 397
stramenopile 390
supergroup 384
test 398
trichocyst 393
vector 394
water mold 391
zooflagellate 395
zoospore 388

Assess

Reviewing This Chapter

1. List and discuss ways that protists are varied. 384
2. Describe the structures of *Chlamydomonas* and *Volvox*, and contrast how they reproduce. 387–88
3. Describe the structures of *Ulva* and *Spirogyra*, and explain how they reproduce. 388
4. Describe the structure of red algae, and discuss their economic importance. 389
5. Describe the structure of brown algae, and discuss their ecological and economic importance. 390
6. Describe the structures of diatoms and dinoflagellates. What is a red tide? 392–93
7. Describe the life cycle of *Plasmodium vivax*, the most common causative agent of malaria. 394
8. Describe the unique structure of euglenids. 395
9. Summarize the transmission cycle of trypanosomes that cause African sleeping sickness. 398
10. How are ciliates like and different from amoeboids? 393, 396
11. What features distinguish slime molds and water molds from fungi? Describe the life cycle of a plasmodial slime mold. 391–92, 396–97
12. Distinguish between amoeboids, foraminiferans, and radiolarians. 396, 398–99

Testing Yourself

Choose the best answer for each question.
For questions 1–6, match each item to those in the key.

KEY:

a. Amoebozoa   d. Excavata
b. Archaeoplastida e. Opisthokonta
c. Chromalveolata   f. Rhizaria

1. foraminiferans   4. amoeboids

2. ciliates   5. green algae

3. brown algae   6. choanoflagellates

7. Which of these is not a green alga?
   a. *Volvox*   d. *Chlamydomonas*
   b. *Fucus*   e. *Ulva*
   c. *Spirogyra*

8. Which is not a characteristic of brown algae?
   a. multicellular
   b. chlorophylls *a* and *b*
   c. live along rocky coast
   d. harvested for commercial reasons
   e. contain a brown pigment

9. In *Chlamydomonas*,
   a. the adult is haploid.
   b. the zygospore survives times of stress.
   c. sexual reproduction occurs.
   d. asexual reproduction occurs.
   e. All of these are correct.

10. *Ulva*
    a. undergoes alternation of generations.
    b. is sea lettuce.   d. is an archaeplastid.
    c. is multicellular.   e. All of these are correct.

11. Which pair is mismatched?
    a. diatoms—silica shell, resemble a petri dish, free-living
    b. euglenids—flagella, pellicle, eyespot
    c. *Fucus*—adult is diploid, seaweed, chlorophylls *a* and *c*
    d. *Paramecium*—cilia, calcium carbonate shell, gullet
    e. foraminiferan—test, pseudopod, digestive vacuole

12. Which pair is mismatched?
    a. trypanosome—African sleeping sickness
    b. *Plasmodium vivax*—malaria
    c. amoeboid—severe diarrhea
    d. AIDS—*Giardia lamblia*
    e. dinoflagellates—coral

13. Which is found in slime molds but not in fungi?
    a. nonmotile spores   d. photosynthesis
    b. flagellated cells   e. chitin in cell walls
    c. zygote formation

14. Which pair is properly matched?
    a. water mold—flagellate
    b. trypanosome—zooflagellate
    c. *Plasmodium vivax*—mold
    d. amoeboid—algae
    e. golden brown algae—kelp

15. In the haploid life cycle (e.g., *Chlamydomonas*),
    a. meiosis occurs following zygote formation.
    b. the adult is diploid.
    c. fertilization is delayed beyond the diploid stage.
    d. the zygote produces sperm and eggs.

16. Dinoflagellates
    a. usually reproduce sexually.
    b. have protective cellulose plates.
    c. are insignificant producers of food and oxygen.
    d. have cilia instead of flagella.
    e. tend to be larger than brown algae.

17. Ciliates
    a. move by pseudopods.
    b. are not as varied as other protists.
    c. have a gullet for food gathering.
    d. do not divide by binary fission.
    e. are closely related to the radiolarians.

18. A(n) _____ is a collared, flagellated, heterotrophic protist that is closely related to animals such as sponges.
    a. radiolarian   d. amoeboid
    b. kinetoplastid   e. trypanosome
    c. choanoflagellate

19. Label this diagram of the *Chlamydomonas* life cycle.

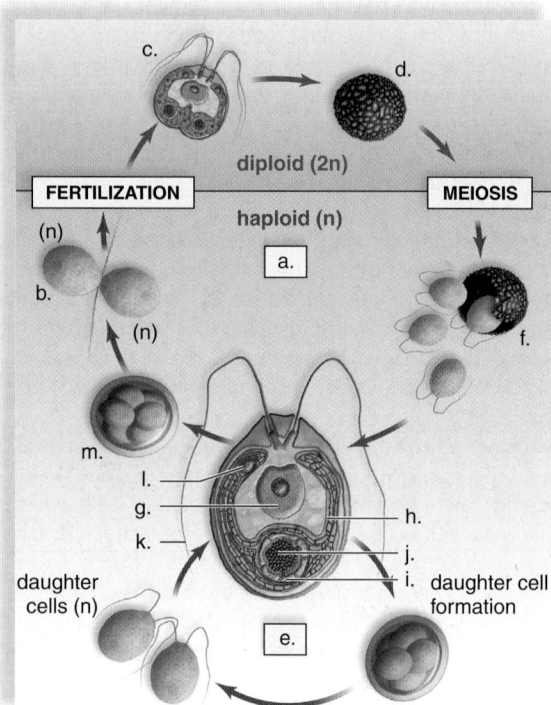

Engage

Thinking Scientifically

1. While studying a unicellular alga, you discover a mutant in which the daughter cells do not separate after mitosis. This gives you an idea about how filamentous algae may have evolved. You hypothesize that the mutant alga is missing a protein or making a new form of a protein. How might each possibly lead to a filamentous appearance?

2. You are trying to develop a new anti-termite chemical that will not harm environmentally beneficial insects. Because termites are adapted to eat only wood, they will starve if they cannot digest this food source. Termites have two symbiotic partners: the protozoan *Trichonympha collaris* and the bacteria it harbors that actually produce the enzyme that digests the wood. Knowing this, how might you prevent termite infestations without targeting the termites directly?

Mycobond™ is an ecologically sustainable insulation and packaging product derived entirely from fungal mycelium and other organic materials.

Fungi Evolution and Diversity

F ungi were the first eukaryotes to invade land. They are ancient organisms, with deep underground networks of mycelium, or "fungal roots" that service the ecosystem by recycling organic debris back into topsoil. The part of the fungi that we see, and sometimes eat, is the fruiting body, or mushroom, that is the center of reproduction for many species.

The medicinal properties of fungi have been known for centuries—distillates from fungi have antimicrobial and antiviral properties that are just now being tapped by pharmaceutical research. Fungi also hold potential to replace the plastics and styrofoam packing materials that we depend on so heavily today. As one example, in 2007 a new eco-minded company, Ecovative Design, developed a compound, called MycoBond™, created entirely from the mycelium of fungi and agricultural waste products. MycoBond™ can be grown into any shape, creating custom-made home insulation and packaging materials for any type of product. MycoBond™ requires much less energy to create than styrofoam or plastic and is completely organic, non-toxic, and one hundred percent compostable.

Products such as antibiotics, antivirals, and MycoBond™, derived from the natural processes of fungi, have the potential to revolutionize human society and our impacts on the environment.

As you read through the chapter, think about the following questions:

1. Why is it important to human health to preserve fungi biodiversity?
2. How might the commercial production of materials from fungi reduce our impact on the environment?
3. How would ecosystems be impacted if fungi were to go extinct?

CHAPTER OUTLINE

BEFORE YOU BEGIN

Before beginning this chapter, take a few moments to review the following discussions.

1. **Section 9.2** What is the difference between a haploid and a diploid eukaryotic cell?
2. **Section 10.1** Why is meiosis essential to sexual reproduction?
3. **Table 21.1** What are the six supergroups of eukaryotes, and which one contains the fungi and animals?

FOLLOWING *the* BIG IDEAS

CHAPTER 22 FUNGI EVOLUTION AND DIVERSITY	
Evolution	The inferred ancestry and diversification of fungi can be charted via cladograms.
Energy and Homeostasis	The environment exerts influence on many aspects of fungal physiology and reproduction.
Interactions and Systems	Fungi engage in both beneficial and harmful relationships with other organisms.

22.1 Evolution and Characteristics of Fungi

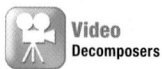

Learning Outcomes

Upon completion of this section, you should be able to

1. Identify two traits that are similar between animals and fungi.
2. Recognize fungi as the closest living relatives of animals.
3. Define and identify the structural features of fungi.

The **fungi** include over 80,000 species of mostly multicellular eukaryotes that share a common mode of nutrition. Mycologists, scientists that study fungi, expect this number of species to increase to over 1.5 million in the future as new species are discovered and described.

Like animals, fungi are heterotrophic and consume preformed organic matter. Animals, however, are heterotrophs that ingest food, while fungi are **saprotrophs** that absorb food. Their cells send out digestive enzymes into the immediate environment that break down dead and decaying organic matter. The resulting nutrient molecules are then absorbed by the fungus cells.

Video Decomposers

Evolution of Fungi

Figure 22.1 illustrates the evolutionary relationships among the five groups of fungi we will be discussing. The evolutionary tree is a hypothesis about how these groups are related. The chytrids are different from all other fungi because they are aquatic and have flagellated spores and gametes. Our description of fungal structure applies best to the zygospore fungi, sac fungi, and club fungi. The AM fungi are notable because they exist only as mycorrhizae in association with plant roots!

Protists evolved some 1.5 BYA (billion years ago). Plants, animals, and fungi can all trace their ancestry to protists, but molecular data tells us that animals and fungi shared a more recent common ancestor than animals and plants. Therefore, animals and fungi, both in the supergroup Opisthokonta, are more closely related to each other than either is to plants (Fig. 22.1). The common ancestor of animals and fungi was most likely an aquatic, flagellated, unicellular protist. Multicellular forms evolved sometime after animals and fungi split into two

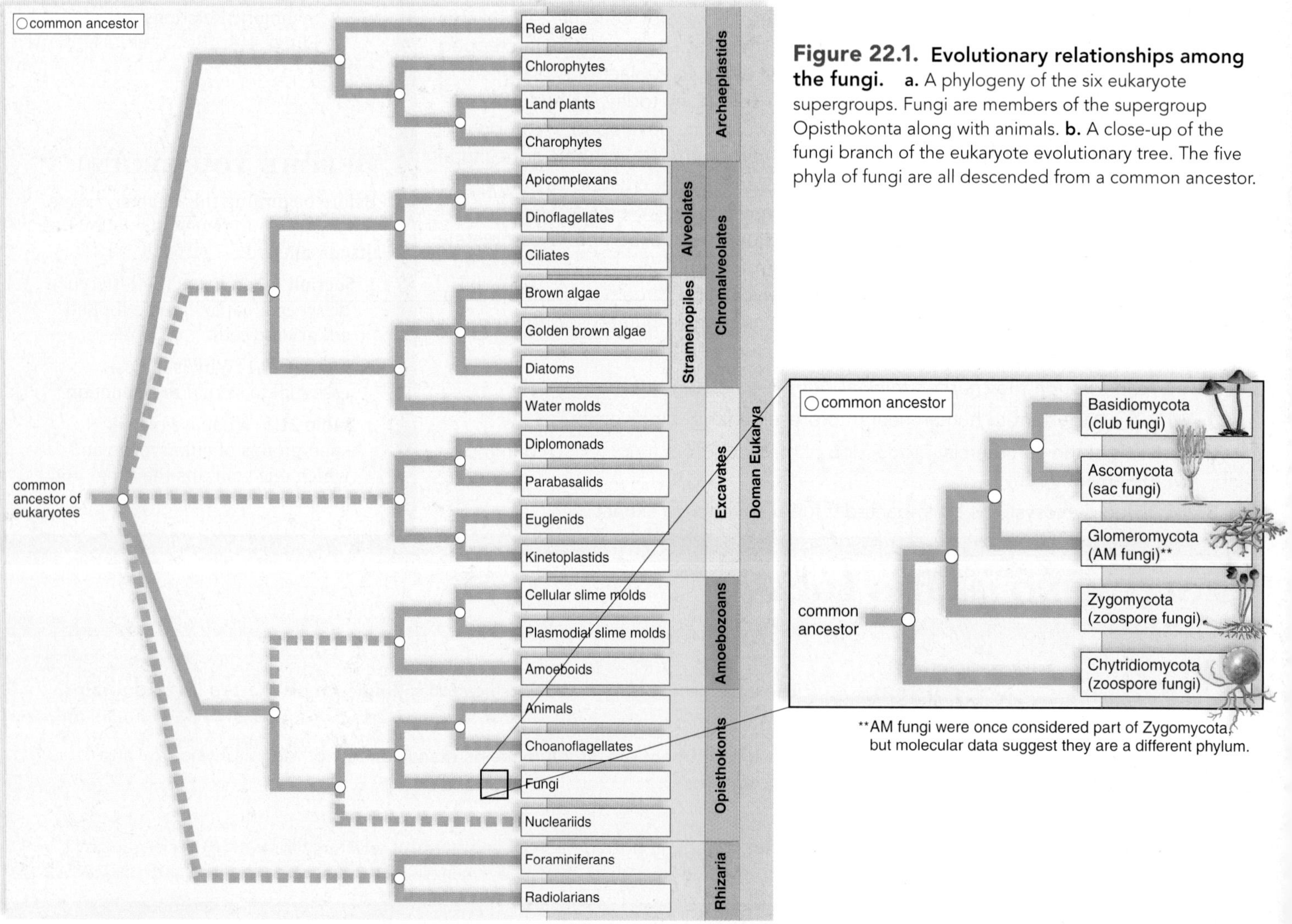
common ancestor
Red algae
Chlorophytes
Land plants
Charophytes
Archaeplastids
Apicomplexans
Dinoflagellates
Ciliates
Alveolates
Brown algae
Golden brown algae
Diatoms
Stramenopiles
Chromalveolates
Water molds
Diplomonads
Parabasalids
Euglenids
Kinetoplastids
Excavates
Doman Eukarya
common ancestor of eukaryotes
Cellular slime molds
Plasmodial slime molds
Amoeboids
Amoebozoans
Animals
Choanoflagellates
Fungi
Nucleariids
Opisthokonts
Foraminiferans
Radiolarians
Rhizaria
common ancestor
Basidiomycota (club fungi)
Ascomycota (sac fungi)
Glomeromycota (AM fungi)**
Zygomycota (zoospore fungi)
Chytridiomycota (zoospore fungi)
common ancestor
**AM fungi were once considered part of Zygomycota, but molecular data suggest they are a different phylum.

Figure 22.1. Evolutionary relationships among the fungi. **a.** A phylogeny of the six eukaryote supergroups. Fungi are members of the supergroup Opisthokonta along with animals. **b.** A close-up of the fungi branch of the eukaryote evolutionary tree. The five phyla of fungi are all descended from a common ancestor.

different lineages. Some animals have retained flagellated cells but most groups of fungi do not have flagella today.

Fungi do not fossilize well, so it is difficult to estimate from the fossil record when they first evolved. The earliest known fossil fungi is dated at 450 MYA (million years ago), but fungi probably evolved a lot earlier. We do know that while animals were still swimming in the seas during the Silurian, plants were beginning to live on the land, and they brought fungi with them. Mycorrhizae are evident in plant fossils, also some 450 MYA. Perhaps fungi were instrumental in the colonization of land by plants. Much of the fungal diversity we observed most likely had its origin in an adaptive radiation when organisms began to colonize land.

Structure of Fungi

Some fungi, including the yeasts, are unicellular; however, the vast majority of species are multicellular. The thallus or body of most fungi is a multicellular structure known as a mycelium (Fig. 22.2a). A **mycelium** [Gk. *mycelium*, fungus filaments] is a network of fungal filaments; these filaments are called **hyphae** [Gk. *hyphe*, web]. Hyphae give the mycelium quite a large surface-to-volume ratio, and this maximizes the absorption of nutrients into the body of a fungus.

Hyphae grow from their tips and in some fungi **septa** (sing. septum), or walls of tissue, are formed behind the growing tip, partitioning the hyphae into individual cells. Fungi that have septa in their hyphae are called **septate** [L. *septum*, fence, wall]. Pores in the septa allow cytoplasm and sometimes even organelles to pass freely from one cell to another. The septa that separate reproductive cells, however, are completely closed in all fungal groups. **Nonseptate** fungi are not divided into cells, and many nuclei are present in the cytoplasm of a single hypha (Fig. 22.2c). Some hyphae are capable of penetrating rigid substances, such as plant tissues. When a fungus reproduces, a specific portion of the mycelium becomes a reproductive structure that is nourished by the rest of the mycelium (Fig. 22.2b).

Fungal cells are quite different from plant cells because they lack chloroplasts and have a cell wall made of chitin and not cellulose. **Chitin,** like cellulose, is a polymer of glucose organized into microfibrils, but each glucose molecule of chitin has a nitrogen-containing amino group attached to it. Chitin is also found in the exoskeleton of arthropods, a major phylum of animals that includes the insects and crustaceans. Unlike plants that store energy as starch, fungi store energy as glycogen, the same molecule that animals use to store energy. Except for the aquatic chytrids, fungi

are not motile. Terrestrial fungi lack basal bodies (see Chapter 4) and do not have flagella at any stage in their life cycle. They move toward a food source by hyphae growing toward it. Growing hyphae can cover as much as a kilometer a day!

Reproduction of Fungi

Both sexual and asexual reproduction occur in fungi. Sexual reproduction of terrestrial fungi occurs in these stages:

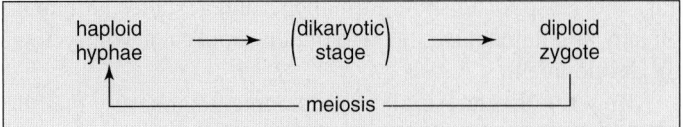

The relative length of time of each phase varies with the species.

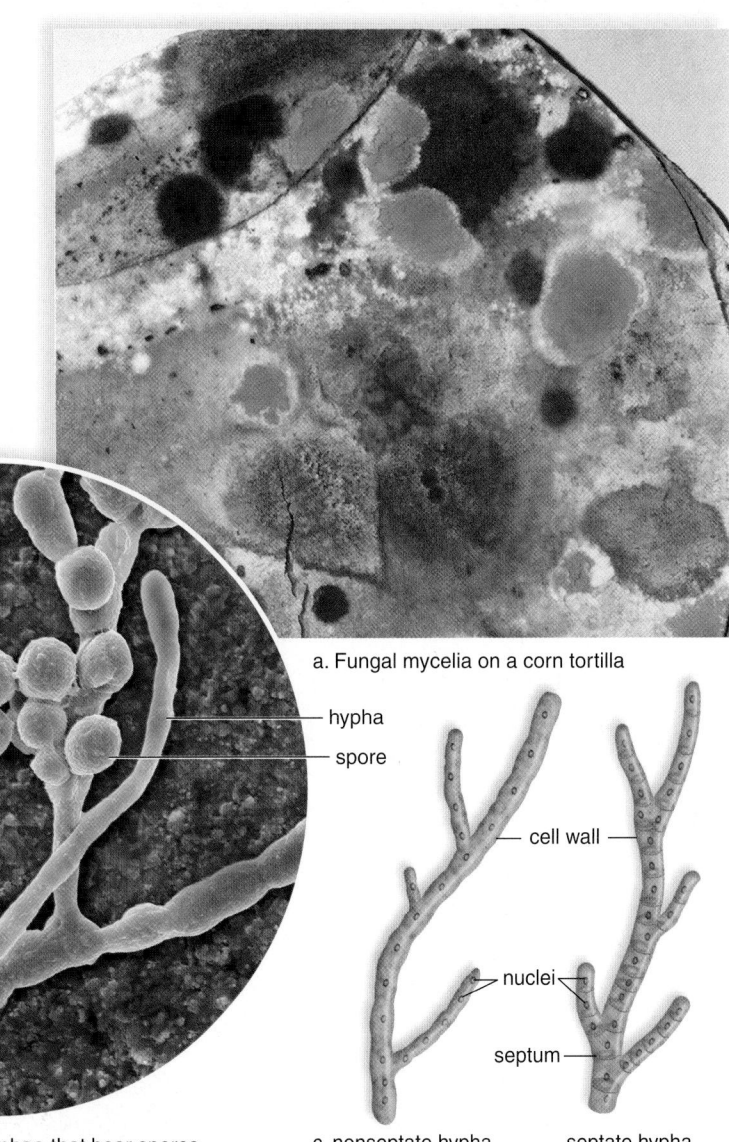

a. Fungal mycelia on a corn tortilla

hypha
spore

cell wall

nuclei

septum

SEM 300×

b. Specialized fungal hyphae that bear spores

c. nonseptate hypha   septate hypha

Figure 22.2 Mycelia and hyphae of fungi. a. Each mycelium grown from a different spore on a corn tortilla is quite symmetrical. **b.** Scanning electron micrograph of specialized aerial fungal hyphae that bear spores. **c.** Hyphae are either nonseptate (do not have cross-walls) or septate (have cross-walls).

During sexual reproduction, hyphae (or a portion thereof) from two different mating types make contact and fuse. In some species, the nuclei from the two mating types fuse immediately. In other species, the nuclei pair but do not fuse for days, months, or even years. The nuclei continue to divide in such a way that every cell (in septate hyphae) has at least one of each nucleus. A hypha that contains paired haploid nuclei is said to be n + n or **dikaryotic** [Gk. *dis*, two, and *karyon*, nucleus, kernel]. When the nuclei do eventually fuse, the zygote undergoes meiosis prior to spore formation. Fungal spores germinate directly into haploid hyphae without any noticeable embryological development.

How can the terrestrial and nonmotile fungi ensure that the offspring will be dispersed to new locations? As an adaptation to life on land, fungi usually produce nonmotile, but normally windblown, spores during both sexual and asexual reproduction. A **spore** is a reproductive cell that develops into a new organism without the need to fuse with another reproductive cell. A large mushroom may produce billions of spores within a few days. When a spore lands upon an appropriate food source, it germinates and begins to grow.

Asexual reproduction usually involves the production of spores by a specialized part of a single mycelium. Alternatively, asexual reproduction can occur by fragmentation—a portion of a mycelium begins a life of its own. Also, unicellular yeasts reproduce asexually by **budding;** a small cell forms and gets pinched off as it grows to full size (see Fig. 22.5).

Check Your Progress　　　　　　　　　　　22.1

1. Describe how animals and fungi differ with respect to nutritional mode.
2. Explain how fungal cell walls differ from plant cell walls.
3. Describe the function of a fungal spore.

22.2 Diversity of Fungi

Learning Outcomes

Upon completion of this section, you should be able to

1. List the major phyla of fungi.
2. Summarize the life cycle typical of fungi in each of the six phyla.
3. Identify one benefit and one disadvantage of human and fungi interactions.

In 1969 R. H. Whittaker classified fungi as a separate group from protists, plants, animals, and prokaryotic organisms. He based his reasoning on the observation that fungi are the only type of multicellular organism to be saprotrophic. However, fungi are now considered to be the closest multicellular relative of animals. Both fungi and animals are placed in the eukaryote supergroup Opisthokonta, which also includes certain heterotrophic protists (see Fig. 22.1).

The fossil record of the fungi is not very good, so mycologists rely on comparative molecular data to decipher evolutionary relationships among fungi groups. Currently five phyla of fungi are recognized: the Chytridiomycota (chytrids), Zygomycota (zygospore fungi), Glomeromycota (AM fungi), Ascomycota (sac fungi), and Basidiomycota (club fungi). In addition to molecular properties, each phylum has independently evolved unique characteristics of their life cycles and reproduction, which support the hypothesis that each phylum evolved from a unique common ancestor (Fig. 22.1, Table 22.1).

Chytrids Are Aquatic Fungi

The **Chytridiomycota,** or **chytrids,** include about 790 species of the simplest fungi, which may resemble the first fungi to have evolved. Some chytrids are single cells; others form branched

Table 22.1 Features of the Fungi Phyla

Phylum (common name)	Reproduction	Key Features	Examples	
Basidiomycota (club fungi)	basidiospores, sexual	basidiocarp fruiting body (mushroom)	most edible mushrooms, common button mushroom (*Agaricus bisporus*)	
Ascomycota (sac fungi)	conidiospores, asexual, ascus with spores, sexual	ascocarp fruiting body (mushroom), yeasts, molds	morel mushroom (*Morchella*), cup fungus (*Sarcoscypha*)	
Glomeromycota (AM fungi)	Spores, asexual	arbuscules, symbiotic with plants	*Glomerales*	
Chytridiomycota (chytrid fungi)	zoospores with flagella; most asexual; some alternation of generations	single celled, simplest fungi, aquatic	*Chytriomyces hyalinus*	
Zygomycota (zygospore fungi)	zygospores, sexual; sporangiospores, asexual	sporangia, gametangia saprotrophic, feed on animal remains or bakery goods	black bread mold (*Rhizopus stolonifer*)	

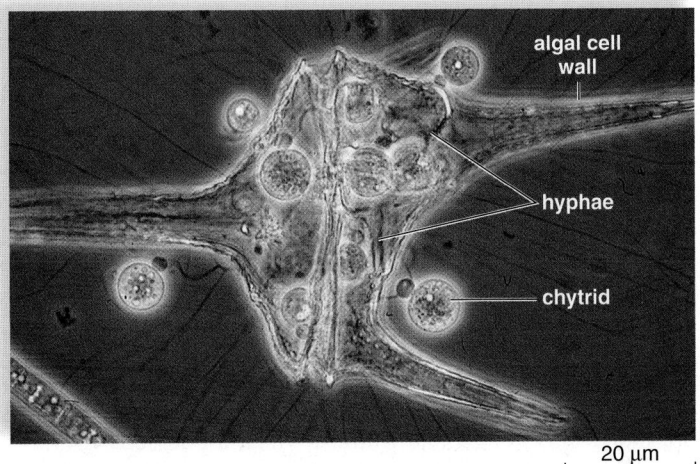

algal cell wall

hyphae

chytrid

20 µm

Figure 22.3 Chytrids parasitizing a protist. These aquatic chytrids *(Chytriomyces hyalinus)* have penetrated the cell walls of this dinoflagellate and are absorbing nutrients meant for their host. They will produce flagellated zoospores that will go on to parasitize other protists.

nonseptate hyphae. Chytrids are unique among fungi because they have flagellated gametes and spores, a feature consistent with their aquatic lifestyle, although some also live in moist soil. The placement of the flagella in their spores, called **zoospores,** suggests a shared ancestry of fungi and choanoflagellates and animals all of which are placed in the supergroup Opisthokonta (see Table 21.1).

Most chytrids reproduce asexually through the production of zoospores within a single cell. The zoospores grow into new chytrids. However, some have an alternation-of-generations life cycle, much like that of green plants and certain algae (see Fig. 21.5), which is quite uncommon among fungi.

Chytrids play a role in the decay and digestion of dead aquatic organisms, but some are parasites of living plants, animals, and protists (Fig. 22.3). They are also known to cause diseases such as the brown spot of corn and the black wart of the potato. The parasitic chytrid *Batrachochytrium dendrobatidis* has recently decimated populations of harlequin frogs *(Atelopus)* in Central and South America. They grow inside skin cells and disrupt the ability of frogs to acquire oxygen through their skin.

Zygomycota Produce Zygospores

The **Zygomycota,** or **zygospore fungi,** include approximately 1,050 species of fungi. Zygospore fungi live off plant and animal remains in the soil or in bakery goods in the pantry. Some are parasites of soil protists, worms, and insects such as the housefly.

The black bread mold, *Rhizopus stolonifer,* is a common example of this phylum. *Rhizopus* has both a sexual and asexual phase in its life cycle (Fig. 22.4). The body of this fungus is composed of mostly nonseptate hyphae that can specialize to perform various tasks. *Rhizopus* exhibits three kinds of specialized hyphae:

- *stolons*—horizontal hyphae that exist on the surface of the bread.

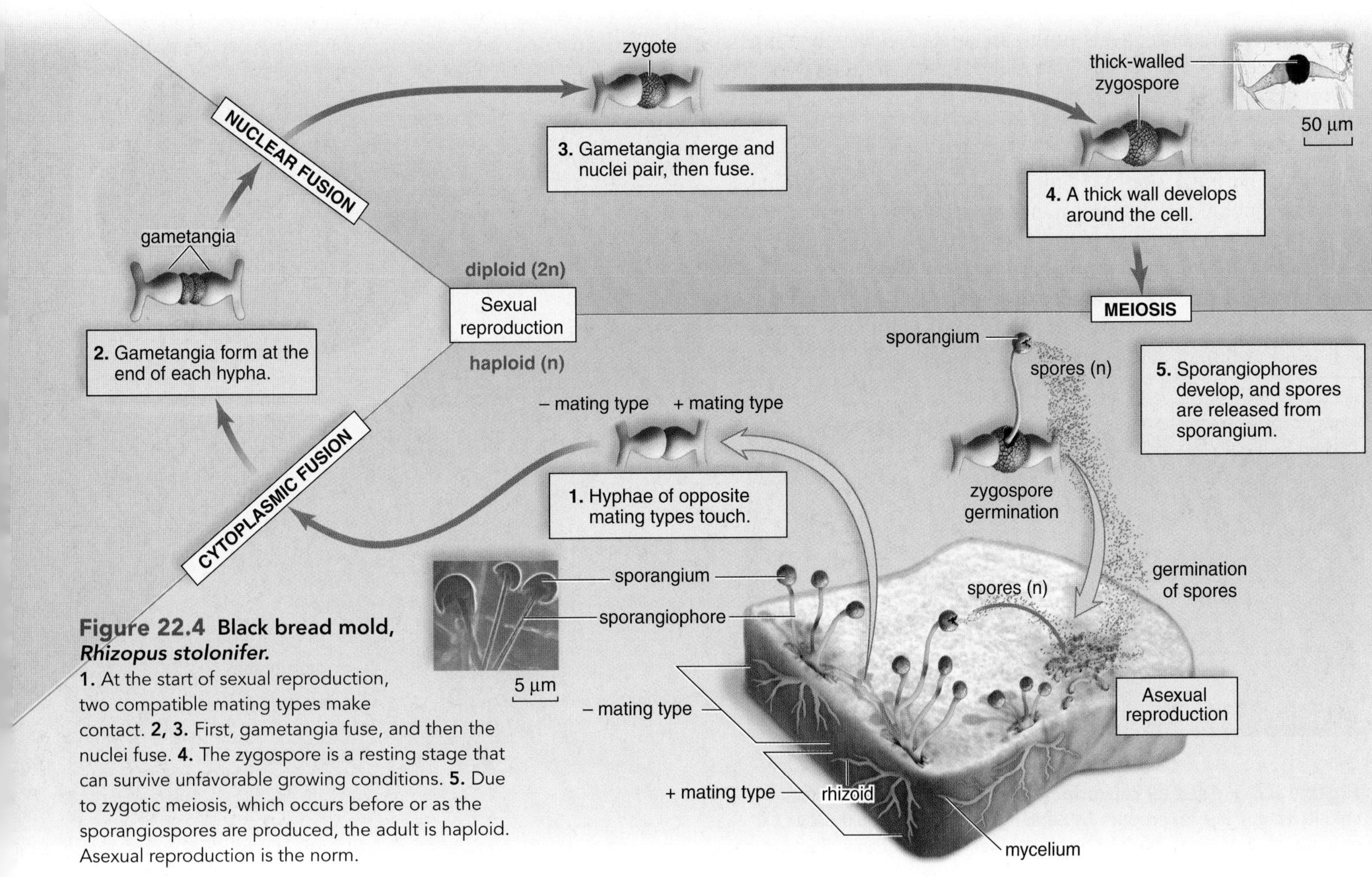

zygote

NUCLEAR FUSION

thick-walled zygospore

50 µm

gametangia

3. Gametangia merge and nuclei pair, then fuse.

4. A thick wall develops around the cell.

diploid (2n)

Sexual reproduction

haploid (n)

2. Gametangia form at the end of each hypha.

MEIOSIS

sporangium

spores (n)

5. Sporangiophores develop, and spores are released from sporangium.

– mating type   + mating type

CYTOPLASMIC FUSION

1. Hyphae of opposite mating types touch.

zygospore germination

Figure 22.4 Black bread mold, *Rhizopus stolonifer.*
1. At the start of sexual reproduction, two compatible mating types make contact. **2, 3.** First, gametangia fuse, and then the nuclei fuse. **4.** The zygospore is a resting stage that can survive unfavorable growing conditions. **5.** Due to zygotic meiosis, which occurs before or as the sporangiospores are produced, the adult is haploid. Asexual reproduction is the norm.

sporangium

sporangiophore

5 µm

– mating type

+ mating type

rhizoid

spores (n)

germination of spores

Asexual reproduction

mycelium

- *rhizoids*—hyphae that grow into bread, anchor the mycelium, and carry out digestion.
- *sporangiophores*—aerial hyphae that bear sporangia.

A **sporangium** (pl., sporangia) is a capsule that produces haploid spores called *sporangiospores* during the asexual phase of reproduction (Fig. 22.4).

Zygospores are diploid spores produced during sexual reproduction. Two mating types of hyphae, termed plus (+) and minus (−), are chemically attracted to one another, and they grow toward each other until they touch (Fig. 22.4). The ends of the touching hyphae swell as nuclei are directed toward, and enter, the tips of the hyphae. Cross-walls then develop a short distance behind the swollen end of each hypha, forming an isolated capsule called a **gametangium** (pl., gametangia). The gametangia of each hypha merge, and the nuclei fuse to form a diploid zygote.

A thick wall develops around the zygote, which is now called a **zygospore**. The zygospore undergoes a period of dormancy before new haploid sporangiospores are formed by meiosis. Sporangiophores with sporangia at their tips germinate from the zygospore, and many sporangiospores are released. The spores, dispersed by air currents, give rise to new haploid mycelia that will continue the sexual phase of the life cycle. Spores from black bread mold have been found in the air above the North Pole, in the jungle, and far out at sea.

Glomeromycota Are Important Symbiotic Fungi

The **Glomeromycota**, or **AM fungi**, are a relatively small group (160 species) of fungi. The name AM stands for *arbuscular mycorrhizal* fungi. Arbuscules are branching invaginations that the fungus makes when it invades plant roots.

Mycorrhizae (see Section 22.3) are a mutualistic association of plants and fungi, and AM fungi are the most common fungi to form symbiotic relationships with plants. The AM fungi were classified with the zygospore fungi for a long time, but are now recognized as a separate group based on molecular data. They play a critical role in the ability of plants to absorb nutrients with their roots. The majority of plants have a mutually beneficial relationship, or symbiosis, with AM fungi.

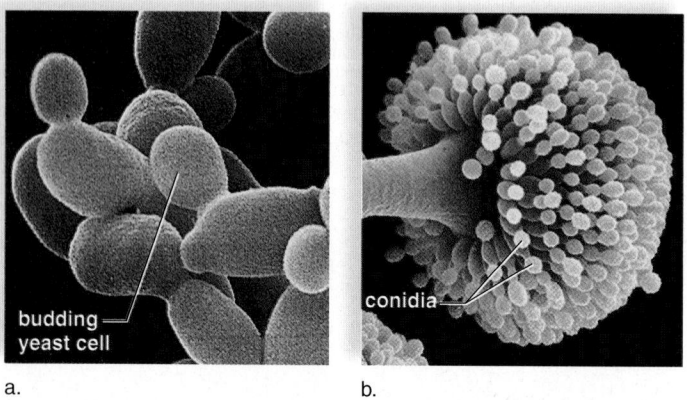

a.   b.

Figure 22.5 Asexual reproduction in sac fungi. **a.** Yeasts, unique among fungi, reproduce by budding. **b.** The sac fungi usually reproduce asexually by producing spores called conidia or conidiospores.

Ascomycota Produce a Fruiting Body Called an Ascocarp

The **Ascomycota**, or **sac fungi**, consist of about 50,000 species of fungi. The sac fungi can be thought of as having two main groups: the sexual sac fungi and the asexual sac fungi.

The sexual sac fungi include such organisms as the **yeast** *Saccharomyces*, the unicellular fungi important in the baking and brewing industries and also in various molecular biological studies. *Neurospora*, another model organism for molecular biology, and the other red bread molds are also sexual sac fungi. So are morel mushrooms and truffles, which are famous gourmet delicacies revered throughout the world.

The asexual sac fungi were formerly in the phylum Deuteromycota, sometimes called the imperfect fungi, because their means of sexual reproduction was unknown. However, on the basis of

Figure 22.6 Sexual reproduction in sac fungi. The sac fungi reproduce sexually by producing asci, in fruiting bodies called ascocarps. **a.** In the ascocarp of cup fungi, dikaryotic hyphae terminate forming the asci, where meiosis follows nuclear fusion and spore formation takes place. **b.** In morels, the asci are borne on the ridges of pits. **c.** Peach leaf curl, a parasite of leaves, forms asci as shown.

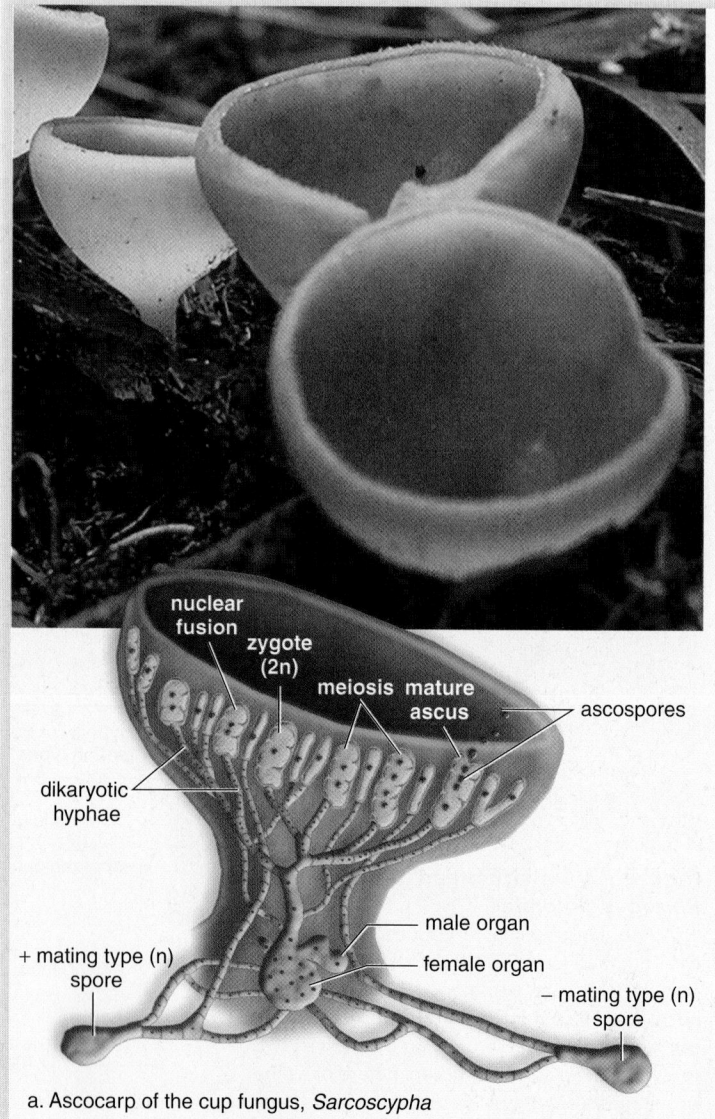

a. Ascocarp of the cup fungus, *Sarcoscypha*

molecular data and structural characteristics, these fungi have now been identified as sac fungi. The asexual sac fungi include the yeast *Candida* and the **molds** *Aspergillus* and *Penicillium*.

Biology of the Sac Fungi

Sac fungi can be unicellular, as in yeasts, but most are multicellular with mycelium composed of septate hyphae. The sac fungi are distinguished by the structures they form when they reproduce asexually and sexually.

Asexual Reproduction. Asexual reproduction is the norm among sac fungi. The yeasts usually reproduce by asexual budding in which a small cell forms and pinches off as it grows to full size (Fig. 22.5a). The other asexual sac fungi produce spores called conidia or conidiospores that vary in size and shape and may be multicellular.

The **conidiospores** usually develop at the tips of specialized aerial hyphae called *conidiophores* (Fig. 22.5b). The structure of conidiophores help mycologists identify a particular sac fungus. When released, the conidiospores are dispersed by wind. The conidiospores of the allergy-causing mold *Cladosporium* are carried easily through the air and can traverse oceans. One researcher found a concentration of more than 35,000 *Cladosporium* conidiospores in a single square meter of air over Leiden (Germany).

Sexual Reproduction. The name ascomycota refers to the **ascus** (pl. asci) [Gk. *askos*, bag, sac], a fingerlike sac that develops during sexual reproduction. On occasion, the asci are surrounded and protected by sterile hyphae within a fruiting body called an *ascocarp* (Fig. 22.6a, b). A **fruiting body** is a reproductive structure where spores are produced and released. Ascocarps can have different shapes; in cup fungi they are cup shaped, and in morels they are stalked and crowned by a pitted, bell-shaped structure.

Ascus-producing hyphae are dikaryotic, with a walled-off portion that becomes the ascus. Within the ascus, the two nuclei fuse to create diploid cells that then undergo meiosis to become new haploid cells. Each ascus in the fruiting body contains eight nuclei that fuse, divide, and become eight haploid spores. In most ascomycetes, the asci become swollen as they mature, and then they burst, expelling the ascospores. If released into the air, the spores are then dispersed by the wind.

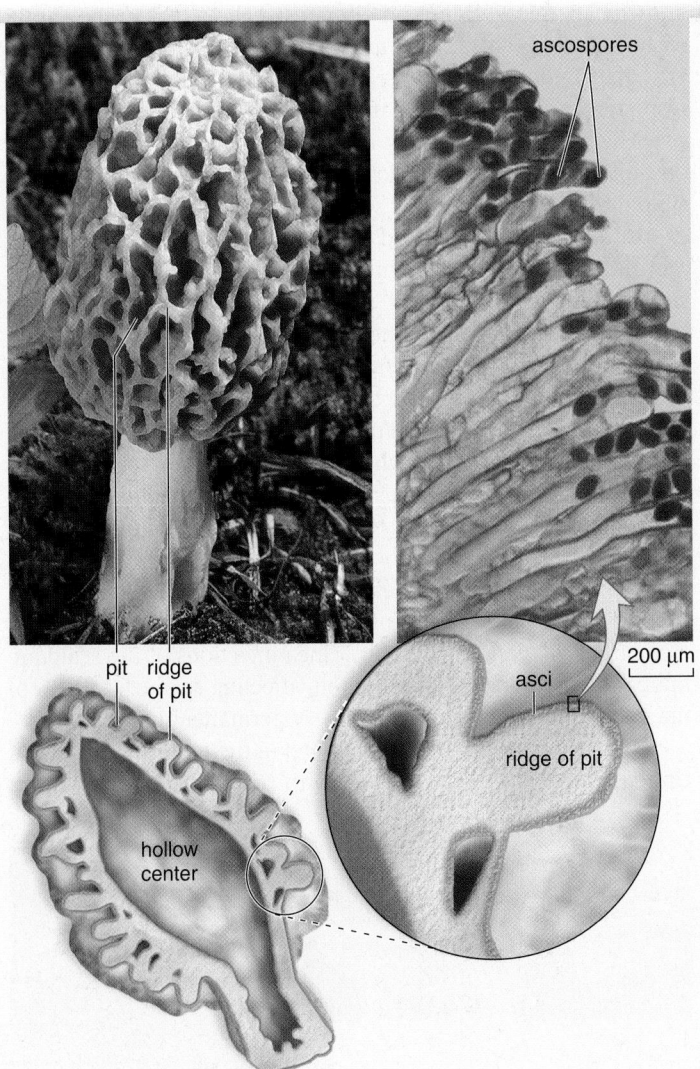

b. Ascocarp of the morel, *Morchella*

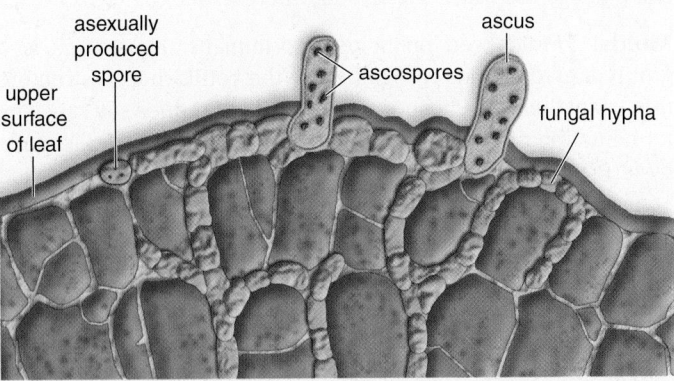

c. Peach leaf curl, *Taphrina*

The Benefits and Drawbacks of Sac Fungi

The sac fungi play an essential role in recycling by digesting resistant (not easily decomposed) materials containing cellulose, lignin, or collagen. Species are also known that can even consume jet fuel and wall paint. Some are symbiotic with algae, forming lichens, and with plant roots, forming mycorrhizae. They also account for most of the known fungal pathogens causing various plant diseases. Powdery mildews grow on leaves, as do leaf curl fungi (Fig. 22.6c); chestnut blight and Dutch elm disease destroy trees. A parasitic sac fungus infects rye and (less commonly) other grains (see the Biological Systems feature).

Many sac fungi have antibiotic and antimicrobial properties. One sac fungus, *Penicillium*, produces the antibiotic penicillin, which is used to treat bacterial infections; another produces cyclosporine, which controls the immune system to prevent rejection of an organ transplant. Although the biosynthesis of steroids in fungi differs somewhat from the same pathways in humans, compounds excreted from sac fungi are used in the synthesis of the steroids in birth control pills. They are also used during the production of various foods such as blue cheese. In contrast, many human diseases such as ringworm and athlete's foot (see Fig. 22.8) are caused by sac fungi.

Yeasts. Yeasts can be both beneficial and harmful to humans. In the wild, yeasts grow on fruits; historically the yeasts already present on grapes contributed to the fermentation of grape juice to make wine. The yeast *Saccharomyces* is added to prepared grains to make beer. When *Saccharomyces* ferments, it produces ethanol and also carbon dioxide. Both the ethanol and the carbon dioxide are retained for beers and sparkling wines; carbon dioxide is released for still wines. In baking, the carbon dioxide given off is the leavening agent that causes bread to rise. *Saccharomyces* is also an important model organism for genetic engineering experiments requiring a eukaryote.

Yeasts can be harmful to humans. *Candida albicans* is a yeast that causes fungal infections called candidiasis. *Candida albicans* is a normal component of the microorganism population in and on our bodies. Candidiasis occurs when the balance between *Candida* and other microorganisms is disturbed. A vaginal infection results from a proliferation of *Candida*, which causes inflammation, itching, and discharge. Oral thrush is a *Candida* infection of the mouth, common in newborns and AIDS patients. In immunocompromised individuals, *Candida* can move through the body, causing a systemic infection that can damage the heart, brain, and other organs.

Molds. Molds can be helpful to humans. *Aspergillus* is a group of green molds recognized by the bottle-shaped structure that bears their conidiospores. It is used to produce soy sauce by fermentation of soybeans. A Japanese food called miso is made by fermenting soybeans and rice with *Aspergillus*. In the United States, *Aspergillus* is used to produce citric and gallic acids, which serve as additives during the manufacture of a wide variety of products, including foods, inks, medicines, dyes, plastics, toothpaste, soap, and chewing gum.

Molds can also be harmful to humans. Commonly isolated from soil, plant debris, and house dust, *Aspergillus* is sometimes pathogenic to humans. *Aspergillus flavus*, which grows on moist seeds, secretes a toxin that is the most potent natural carcinogen

SEM 1,800×

Figure 22.7 Black mold. *Stachybotrys chartarum*, or black mold, grows well in moist areas, including the walls of homes. It represents a potential health risk.

known. Therefore, in humid climates such as that in the southeastern United States, care must be taken to store grains properly. *Aspergillus* also causes a potentially deadly disease of the respiratory tract that arises after spores have been inhaled.

The mold *Stachybotrys chartarum* (Fig. 22.7) grows well on building materials. It is known as black mold and is responsible for the "sick-building" syndrome. Individuals with chronic exposure to toxins produced by this fungus have reported cold and flulike symptoms, fatigue, and dermatitis. The toxins may suppress and could destroy the immune system, affecting the lymphoid tissue and the bone marrow.

Moldlike fungi cause infections of the skin called tineas. Athlete's foot, caused by a species of *Trichophyton*, is a tinea characterized by itching and peeling of the skin between the toes (Fig. 22.8a). In ringworm, which can be caused by several different fungi, the fungus releases enzymes that degrade keratin and collagen in skin. The area of infection becomes red and inflamed. The fungal colony grows outward, forming a ring of inflammation. The center of the lesion begins to heal, thereby giving ringworm its characteristic appearance, a red ring surrounding an area of healed skin (Fig. 22.8b). Tinea infections of the scalp are rampant among school-age children, affecting as much as 30% of the population, with the possibility of permanent hair loss.

The vast majority of people living in the eastern and central United States have been infected with *Histoplasma capsulatum*,

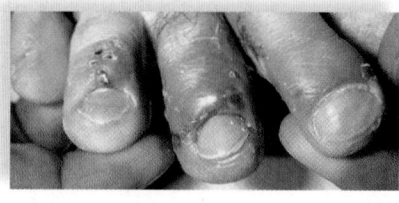

a.

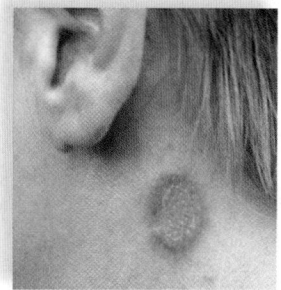

b.

Figure 22.8 Tineas. Athlete's foot (**a**) and ringworm (**b**) are termed tineas.

Biological Systems

Deadly Fungi

It is unwise and potentially fatal for amateurs to collect mushrooms in the wild because certain mushroom species are poisonous. The red and yellow *Amanitas* are one example. These species are also known as fly agaric because they were once thought to kill flies (the mushrooms were gathered, crushed, and then sprinkled into milk to attract flies). Its toxins include muscarine and muscaridine, which produce symptoms similar to those of acute alcoholic intoxication. In one to six hours, the victim staggers, loses consciousness, and becomes delirious, sometimes suffering from hallucinations, manic conditions, and stupor. Luckily, it also causes vomiting, which rids the system of the poison, so death occurs in less than 1% of cases.

The death cap mushroom (*Amanita phalloides*, Fig. 22A) causes 90% of the fatalities attributed to mushroom poisoning. When this mushroom is eaten, symptoms don't begin until 10–12 hours later. Abdominal pain, vomiting, delirium, and hallucinations are not the real problem; rather, a poison interferes with RNA transcription by inhibiting RNA polymerase, and the victim dies from liver and kidney damage.

Some hallucinogenic mushrooms are used in religious ceremonies, particularly among Mexican Indians. *Psilocybe mexicana* contains a chemical called psilocybin that is a structural analogue of LSD and mescaline. It produces a dreamlike state in which visions of colorful patterns and objects seem to fill up space and dance past in endless succession. Other senses are also sharpened to produce a feeling of intense reality.

The only reliable way to tell a nonpoisonous mushroom from a poisonous one is to be able to correctly identify the species. Poisonous mushrooms cannot be identified with simple tests, such as whether they peel easily, have a bad odor, or blacken a silver coin during cooking. Only consume mushrooms that have been identified by a bona fide expert!

Like club fungi, some sac fungi also contain chemicals that can be dangerous to people. *Claviceps purpurea*, the ergot fungus, infects rye and replaces the grain with ergot—hard, purple-black bodies consisting of tightly cemented hyphae (Fig. 22B). When ground with the rye and made into bread, the fungus releases toxic alkaloids that cause the disease ergotism. In humans, vomiting, feelings of intense heat or cold, muscle pain, a yellow face, and lesions on the hands and feet are accompanied by hysteria and hallucinations.

The alkaloids that cause ergotism have medicinal properties. They stimulate smooth muscle and block the sympathetic nervous system, and can be used to cause uterine contractions and to treat migraine headaches. Although the ergot fungus can be cultured in petri dishes, no one has successfully produced ergot in the laboratory. The only way to obtain ergot is to collect it from an infected field of rye.

Ergotism was common in Europe during the Middle Ages. During this period, it was known as St. Anthony's Fire and was responsible for 40,000 deaths in an epidemic in 994 AD. We now know that ergot contains lysergic acid, from which LSD is easily synthesized. Based on recorded symptoms, some historians believe that those individuals who claimed to have been "bewitched" in Salem, Massachusetts, during the seventeenth century were actually suffering from ergotism. The ensuing mass hysteria, however, led to the execution of 20 people for the crime of witchcraft.

Questions to Consider

1. How might a sequence of DNA from an unknown species of fungus help to identify it?
2. Why are those toxins that interfere with RNA polymerase fatal? That is, why is RNA polymerase needed for normal body function?

Figure 22A A poisonous mushroom species, *Amanita phalloides*.

Figure 22B Ergot infection of rye, caused by *Claviceps purpurea*.

a thermally dimorphic fungus that grows in mold form at 25°C and in yeast form at 37°C. This common soil fungus, often associated with bird droppings, leads in most cases to a mild "fungal flu." Less than half of those infected notice any symptoms, with 3,000 showing severe disease and about 50 dying each year from histoplasmosis. The fungal pathogen lives and grows within cells of the immune system and causes systemic illness. Lesions are formed in the lungs that leave calcifications, visible in X-ray images, that resemble those of tuberculosis.

Control of Fungal Infections.   The strong similarities between fungal and human cells make it difficult to design fungal medications that do not also harm humans. Researchers exploit any biochemical differences they can discover. Fungi, like animals, synthesize steroids. A variety of fungicides interfere with steroid biosynthesis, including some that are applied to fields of grain. Fungicides based on heavy metals are applied to seeds of sorghum and other crops. Topical agents are available for the treatment of yeast infections and tineas, and systemic medications are available for systemic sac fungi infections.

Basidiomycota Produce a Fruiting Body Called a Basidium

The **Basidiomycota,** or **club fungi,** consist of over 30,000 species. Mushrooms, toadstools, puffballs, shelf fungi, jelly fungi,

bird's-nest fungi, and stinkhorns are basidiomycetes. In addition, fungi that cause plant diseases such as the smuts and rusts are placed in this phylum. Several mushrooms, such as the portabella and shiitake mushrooms, are savored as foods by humans. Approximately 75 species of basidiomycetes are considered poisonous. The poisonous "death cap" mushroom is discussed in the Biological Systems feature.

Biology of Club Fungi

The body of a basidiomycete is a mycelium composed of septate hyphae. Most members of this phylum are saprotrophs, feeding on dead and decaying organic matter, although several parasitic species exist that obtain nutrition from living hosts.

Reproduction.   Although club fungi occasionally produce conidia asexually, they usually reproduce sexually. Their formal name, Basidiomycota, refers to the **basidium** [L. *basidi*, small pedestal], a club-shaped structure in which spores called basidiospores develop. Basidia are located within a fruiting body called a basidiocarp, which we recognize as a mushroom (Fig. 22.9). Prior to formation of a basidiocarp, haploid hyphae of opposite mating types meet and fuse, producing a dikaryotic mycelium. The dikaryotic mycelium continues its existence year after year, even for hundreds of years on occasion. In many species of mushrooms, the

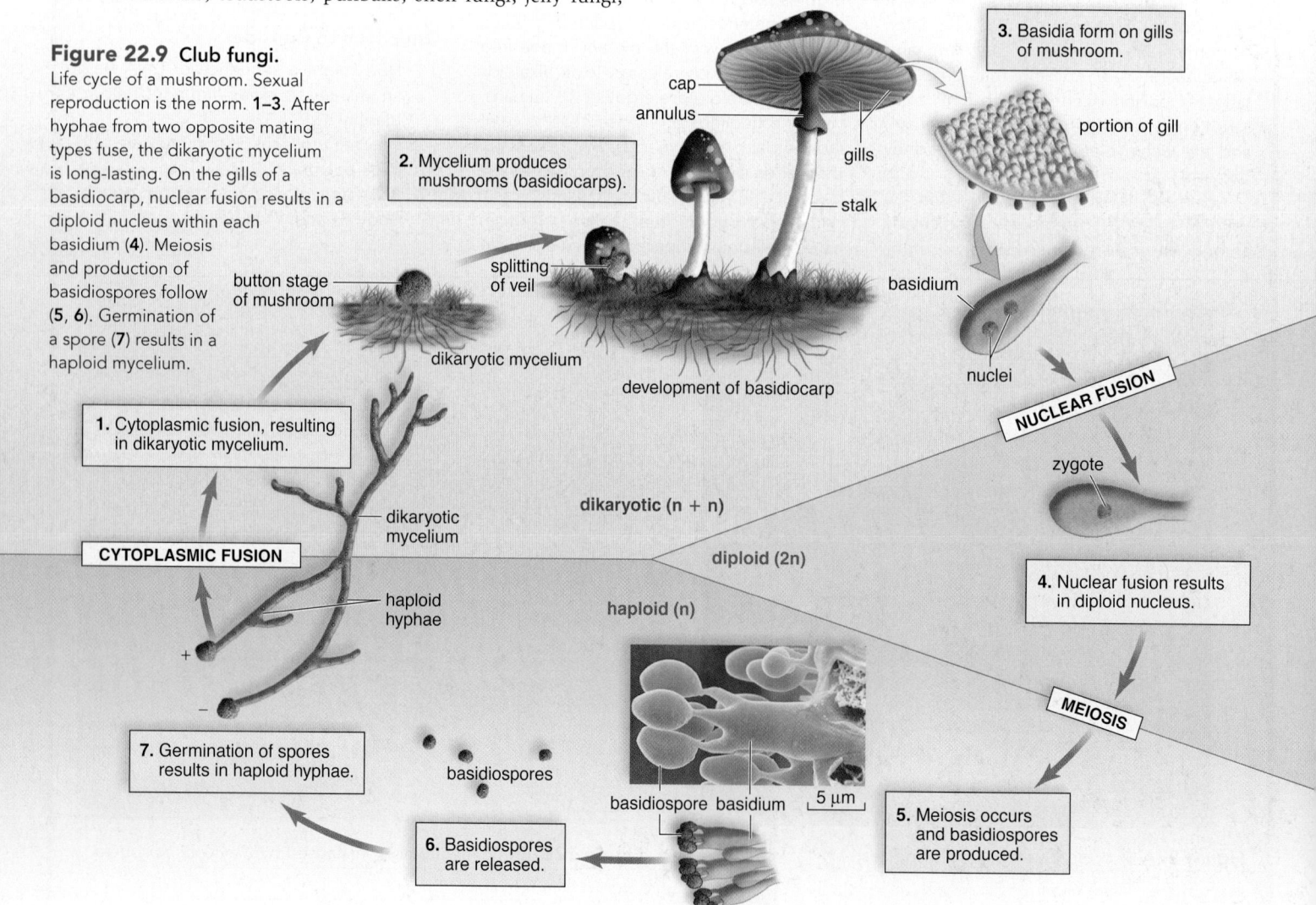

Figure 22.9 Club fungi.
Life cycle of a mushroom. Sexual reproduction is the norm. **1–3.** After hyphae from two opposite mating types fuse, the dikaryotic mycelium is long-lasting. On the gills of a basidiocarp, nuclear fusion results in a diploid nucleus within each basidium (**4**). Meiosis and production of basidiospores follow (**5, 6**). Germination of a spore (**7**) results in a haploid mycelium.

Mature mushroom

3. Basidia form on gills of mushroom.

portion of gill

cap

annulus

gills

stalk

basidium

nuclei

2. Mycelium produces mushrooms (basidiocarps).

splitting of veil

button stage of mushroom

dikaryotic mycelium

development of basidiocarp

NUCLEAR FUSION

zygote

1. Cytoplasmic fusion, resulting in dikaryotic mycelium.

dikaryotic mycelium

dikaryotic (n + n)

diploid (2n)

CYTOPLASMIC FUSION

haploid hyphae

haploid (n)

4. Nuclear fusion results in diploid nucleus.

+

−

MEIOSIS

7. Germination of spores results in haploid hyphae.

basidiospores

basidiospore basidium   5 µm

5. Meiosis occurs and basidiospores are produced.

6. Basidiospores are released.

a. Fairy ring

b. Shelf fungus

c. Pore mushroom, *Boletus*

d. Puffball, *Calvatiga gigantea*

Figure 22.10 Club fungi. **a.** Fairy ring. Mushrooms develop in a ring on the outer living fringes of a dikaryotic mycelium. The center has used up its nutrients and is no longer living. **b.** A shelf fungus. **c.** Fruiting bodies of *Boletus*. This mushroom is not gilled; instead, it has basidia-lined tubes that open on the undersurface of the cap. **d.** In puffballs, the spores develop inside an enclosed fruiting body. Giant puffballs are estimated to contain 7 trillion spores.

dikaryotic mycelium often radiates out and produces mushrooms in an ever larger circle, or "fairy ring" (Fig. 22.10*a*).

Mushrooms are composed of tightly packed hyphae whose walled-off ends become basidia. In gilled mushrooms, the basidia are located on radiating lamellae called gills. In shelf fungi and pore mushrooms (Fig. 22.10*b, c*), the basidia terminate in tubes. The extensive surface area of a basidiocarp is lined by basidia, where nuclear fusion, meiosis, and spore production occur. A basidium has four projections into which cytoplasm and a haploid nucleus enter as the basidiospore forms. Basidiospores are windblown; when they germinate, a new haploid mycelium forms. It is estimated that some large mushrooms can produce up to 40 million spores per hour.

In puffballs, spores are produced inside parchmentlike membranes, and the spores are released through a pore or when the membrane breaks down (Fig. 22.10*d*). In bird's-nest fungi, falling raindrops provide the force that causes the nest's basidiospore-containing "eggs" to fly through the air and land on vegetation. Stinkhorns resemble a mushroom with a spongy stalk and a compact, slimy cap. The long stalk bears the elongated

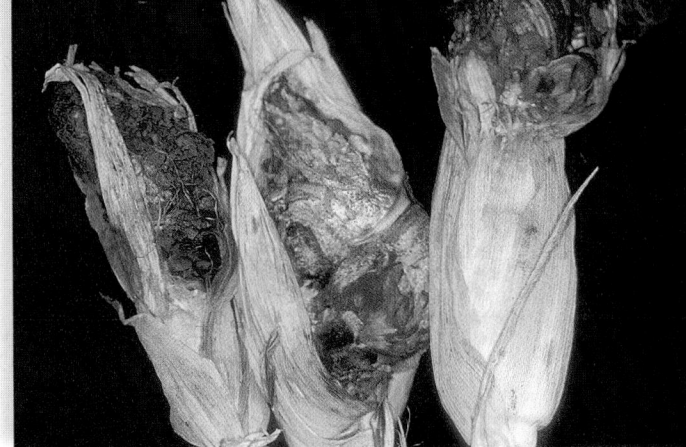

a. Corn smut, *Ustilago*

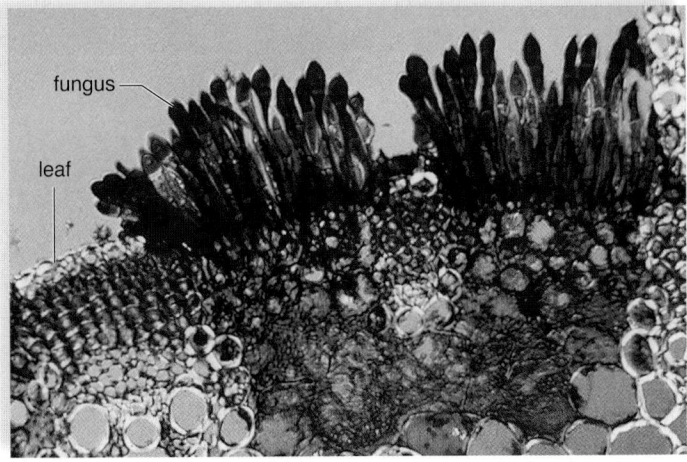

fungus

leaf

b. Wheat rust, *Puccinia*

Figure 22.11 Smuts and rusts. **a.** Corn smut. **b.** Micrograph of wheat rust.

basidiocarp. Stinkhorns emit an incredibly disagreeable odor; flies are attracted by the odor, and when they linger to feed on the sweet jelly, the flies pick up spores that they later distribute.

Smuts and Rusts

Smuts and rusts are club fungi that parasitize cereal crops such as corn, wheat, oats, and rye. They are of great economic importance because of the crop losses they cause every year. Smuts and rusts don't form basidiocarps, and their spores are small and numerous, resembling soot. Some smuts enter seeds and exist inside the plant, becoming visible only near maturity. Other smuts externally infect plants. In corn smut, the mycelium grows between the corn kernels and secretes substances that cause the development of tumors on the ears of corn (Fig. 22.11*a*).

The life cycle of rusts requires alternate hosts, and one way to keep them in check is to eradicate the alternate host. Wheat rust (Fig. 22.11*b*) is also controlled by producing new and resistant strains of wheat. The process is continuous, because rust can mutate to cause infection once again.

Check Your Progress   22.2

1. List the features that make chytrids different from all other fungi.
2. Identify three fungal infections of plants or animals and the fungus responsible for the infections.
3. Name one example of fungi for each of these: puffballs, ergots, athlete's foot, and black bread mold.

22.3 Symbiotic Relationships of Fungi

Learning Outcomes

Upon completion of this section, you should be able to

1. Summarize the association that occurs betwen cyanobacteria and fungi in lichens.
2. Define mycorrhizae.
3. Explain the mutualistic relationship between mycorrhizae and plants.

Several instances in which fungi are parasites of plants and animals have been mentioned. Two other symbiotic associations are of interest: lichen associations and mycorrhizae.

Lichens

Lichens are an association between a fungus, usually a sac fungus, and a cyanobacterium or a green alga. As one example, a crustose lichen has a body consisting of three layers. The fungus forms a thin, tough upper layer and a loosely packed lower layer that shield the photosynthetic cells in the middle layer (Fig. 22.12*a*). Specialized fungal hyphae, which penetrate or envelop the photosynthetic cells, transfer nutrients directly to the rest of the fungus. Lichens can reproduce asexually by releasing fragments that contain hyphae and an algal cell. In fruticose lichens, the sac fungus reproduces sexually (Fig. 22.12*b*).

In the past, lichens were assumed to be mutualistic relationships in which the fungus received nutrients from the algal cells, and the algal cells were protected from desiccation by the fungus. Actually, lichens may involve a controlled form of parasitism of the algal cells by the fungus, with the algae not benefiting at all from the association. This idea is supported by experiments in which the fungal and algal components are removed and grown separately. The algae grow faster when they are alone than when they are part of a lichen. In contrast, it is difficult to cultivate the fungus, which does not naturally grow alone. The different lichen species are identified according to the fungal partner.

Three types of lichens are recognized. Compact crustose lichens are often seen on bare rocks or on tree bark; fruticose lichens are shrublike; and foliose lichens are leaflike (Fig. 22.12*c*). Lichens are efficient at acquiring nutrients and moisture, and therefore they can survive in areas of low moisture and low temperature as well as in areas with poor or no soil. They produce and improve the soil, thus making it suitable for plants to invade the area. Unfortunately, lichens also take up pollutants, and they cannot survive where the air is polluted. Therefore, lichens can serve as air pollution sensors.

Mycorrhizae

Mycorrhizae [Gk. *mykes,* fungus, and *rhizion,* little root] are mutualistic relationships between soil fungi and the roots of most plants. Plants whose roots are invaded by mycorrhizae grow more successfully in poor soils—particularly soils deficient in phosphates—than do plants without mycorrhizae (Fig. 22.13).

Figure 22.12 Lichen morphology. **a.** A section of a compact crustose lichen shows the placement of the algal cells and the fungal hyphae, which encircle and penetrate the algal cells. **b.** Fruticose lichens are shrublike. **c.** Foliose lichens are leaflike.

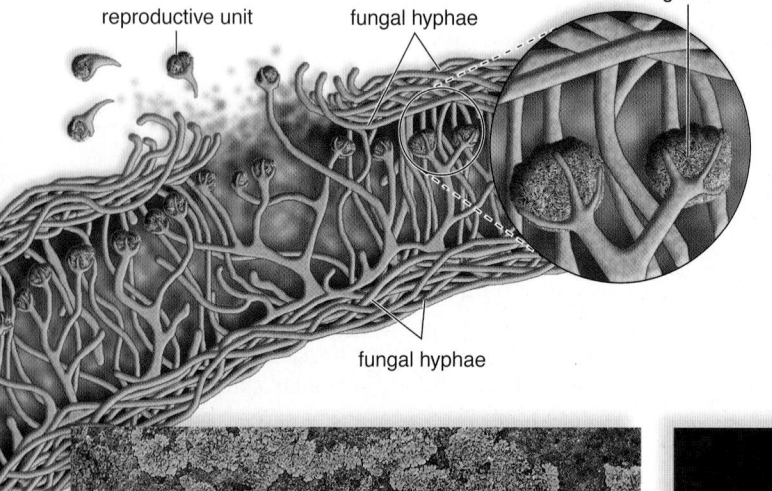

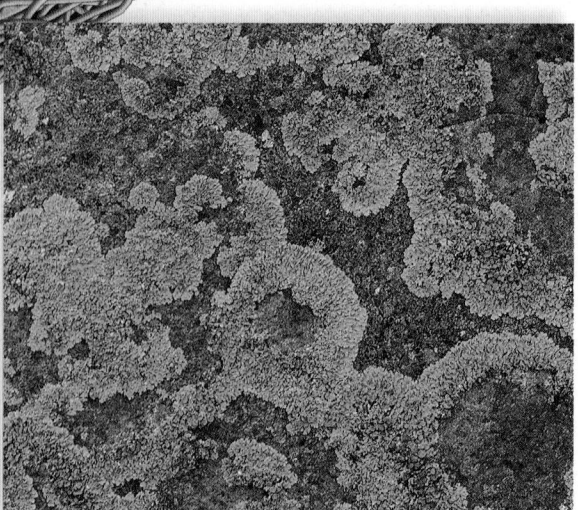
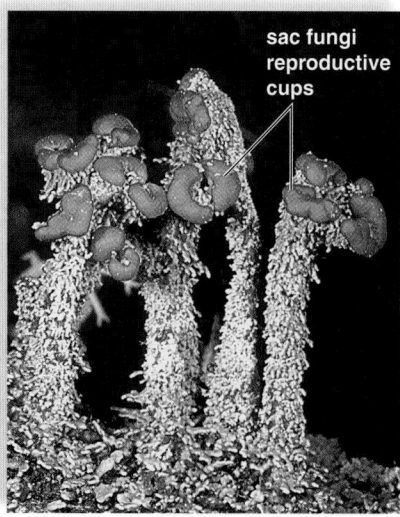

a. Crustose lichen, *Xanthoria*   b. Fruticose lichen, *Lobaria*   c. Foliose lichen, *Xanthoparmelia*

Figure 22.13 Plant growth experiment. A soybean plant without mycorrhizae (*left*) grows poorly compared to two other plants infected with different strains of mycorrhizae (*center, right*).

The fungal partner, either a glomerulomycete or a sac fungus, may enter the cortex of roots but does not enter the cytoplasm of plant cells. Ectomycorrhizae form a mantle that is exterior to the root, and they grow between cell walls.

Endomycorrhizae, such as the AM fungi mentioned earlier, penetrate only the cell walls. The presence of the fungus gives the plant a greater absorptive surface for the intake of minerals. The fungus also benefits from the association by receiving carbohydrates from the plant. As mentioned, even the earliest fossil plants have mycorrhizae associated with them. It would appear, then, that mycorrhizae helped plants adapt to and flourish on land.

The truffle, an underground fungus that is a gourmet delight and that has an ascocarp somewhat prunelike in appearance, is a mycorrhizal sac fungus living in association with oak and beech tree roots. In the past, the French used pigs ("truffle-hounds") to sniff out and dig up truffles, but now they have succeeded in cultivating truffles by inoculating the roots of seedlings with the proper mycelium.

Check Your Progress   22.3

1. Identify the components of a lichen.
2. Explain how a lichen reproduces.
3. Summarize the symbiotic relationship between mycorrhizae and plants.

CONNECTING *the* CONCEPTS *with the* BIG IDEAS

Evolution

- Phylogenetic trees and cladograms model the possible evolutionary history of groups of living organisms. (1B2a-d)

Energy and Homeostasis

- Fungal cell walls provide structural integrity and permeability. (2B1c2)
- Environmental conditions and available energy affect fungal reproduction. (2A1d2, 2C2a*IE*)
- Environmental cycles can help synchronize physiological events, such as fruiting body formation. (2E2c*IE*)
- Cooperation that occurs between some fungi and other organisms improves the survival chances of both, as in lichens and mycorrhizzae. (2E3b4*IE*)

Interactions and Systems

- Specific fungi are often able to attack species with little diversity, impacting species abundance and distribution, such as corn rust and Dutch elm disease. (4B3c*IE*, 4C3a*IE*)
- The environment can influence traits of all types, including specific mating type pheromones in fungi. (4C2a*IE*)

*Find the unabridged version of all EK citations at www.glencoe.com/maderAP11.

Media Study Tools

www.glencoe.com/maderAP11

Enhance your study of this chapter with study tools and practice tests. Also ask your instructor about the resources available through ConnectPlus, including the media-rich eBook, interactive learning tools, and animations.

Summarize

22.1 Evolution and Characteristics of Fungi

Fungi are multicellular, saprotrophic, eukaryotes. After external digestion, they absorb the resulting nutrient molecules. Saprotrophic fungi aid the cycling of organic molecules in ecosystems by decomposing dead remains. Some fungi are parasitic, especially on plants, and others are mutualistic with plant roots and algae.

The body of a fungus is composed of thin filaments called hyphae, which collectively are termed a mycelium. The cell wall contains chitin, and the energy reserve is glycogen. With the notable exception of the chytrids, which have flagellated spores and gametes, fungi do not have flagella at any stage in their life cycle. Nonseptate hyphae have no cross-walls; septate hyphae have cross-walls, but there are pores that allow the cytoplasm and even organelles to pass through.

Fungi produce spores during both asexual and sexual reproduction. During sexual reproduction, hyphae tips fuse so that dikaryotic (n + n) hyphae usually result, depending on the type of fungus. Following nuclear fusion, zygotic meiosis occurs during the production of the sexual spores.

22.2 Diversity of Fungi

The fungi are divided into five phyla, the Chytridiomycota (chytrids), Zygomycota (zygospore fungi), Glomeromycota (AM fungi), Ascomycota (sac fungi), and Basidiomycota (club fungi).

The chytrids are predominately aquatic, single-celled fungi with motile zoospores and gametes. Some chytrids have an alternation-of-generations life cycle similar to that of plants and certain algae. There are also some multicelluar, hyphae-forming chytrids. When hyphae form, they are nonseptate. *Chytriomyces* is an example of a chytrid.

The zygospore fungi are multicellular fungi with nonseptate hyphae. During sexual reproduction they have a dormant stage consisting of a thick-walled zygospore. Meiosis occurs in the diploid zygospore. The zygospore germinates a sporangium with haploid sporangiospores. An example of a zygomycete is the black bread mold *Rhizopus*.

The AM fungi were once classified with the zygospore fungi but are now viewed as a distinct group. AM fungi exist in mutualistic associations with the roots of most land plants.

The sac fungi are septate, and during sexual reproduction saclike cells called asci produce spores. Asci are sometimes located in fruiting bodies called ascocarps. Asexual reproduction, which is dependent on the production of conidiospores, is more common. Sexual reproduction is unknown in some sac fungi. Sac fungi include *Talaromyces*, *Aspergillus*, *Candida*, morels and truffles, and various yeasts and molds, some of which cause disease in plants and animals, including humans.

The club fungi are septate, and during sexual reproduction club-shaped structures called basidia produce spores. Basidia are located in fruiting bodies called basidiocarps. Club fungi have a prolonged dikaryotic stage, and asexual reproduction by conidiospores is rare. A dikaryotic mycelium periodically produces fruiting bodies. Mushrooms and puffballs are examples of club fungi.

22.3 Symbiotic Relationships of Fungi

Lichens are an association between a fungus, usually a sac fungus, and a cyanobacterium or a green alga. Traditionally, this association was considered mutualistic, but experimentation suggests a controlled parasitism by the fungus on the alga. Lichens can live in extreme environments and on bare rocks; they allow other organisms to colonize these harsh environments, and together they eventually form soil.

The term *mycorrhizae* refers to an association between a fungus and the roots of a plant. The fungus helps the plant absorb minerals, and the plant supplies the fungus with carbohydrates.

Key Terms

AM fungi (Glomeromycota) 408
ascus 409
basidium 412
budding 406
chitin 405
chytrids (Chytridiomycota) 406
club fungi (Basidiomycota) 412
conidiospore 409
dikaryotic 406
fruiting body 409
fungus (pl., fungi) 404
gametangium (pl., gametangia) 408
hypha (pl., hyphae) 405
lichen 414
mold 410
mycelium 405
mycorrhizae 414
nonseptate 405
sac fungi (Ascomycota) 408
saprotroph 404
septate 405
septum (pl., septa) 405
sporangium 408
spore 406
yeast 408
zoospore 407
zygospore 408
zygospore fungi (Zygomycota) 407

Assess

Reviewing This Chapter

1. Which characteristics best define fungi? Describe the body of a fungus and how fungi reproduce.   404–5
2. Discuss the evolution and classification of fungi.   404–5
3. Explain how chytrids are different from the other phyla of fungi.   406–7
4. Explain the term *zygospore fungi*. How does black bread mold reproduce asexually? Sexually?   407–8
5. Explain the terms *sexual sac fungi* and *asexual sac fungi*.   408
6. How do sac fungi reproduce asexually? Describe the structure of an ascocarp.   408–9
7. Describe the structure of yeasts, and explain how they reproduce. How are yeasts and molds useful/harmful to humans?   409–10
8. Explain the term *club fungi*. Draw and explain a diagram of the life cycle of a typical mushroom.   412
9. What is the economic importance of smuts and rusts? How can their numbers be controlled?   413
10. Describe the structure of a lichen, and name the three different types. What is the nature of this fungal association?   414
11. Describe the association known as mycorrhizae, and explain how each partner benefits.   414–15

Testing Yourself

Choose the best answer for each question.
For questions 1–3, match the fungi to the phyla in the key.

KEY:
  a. Chytridiomycota
  b. Zygomycota
  c. Ascomycota
  d. Basidiomycota

1. club fungi
2. zygospore fungi
3. sac fungi
4. During sexual reproduction, the zygospore fungi produce
  a. an ascus.   c. a sporangium.
  b. a basidium.   d. a conidiophore.
5. Which of the following represents an organism that uses decomposition as its mode of nutrition?
  a. parasite   c. autotroph
  b. saprotroph   d. All of the above are correct.
6. Hyphae are generally characterized by
  a. strong, impermeable walls. d. pigmented cells.
  b. rapid growth.   e. Both b and c are correct.
  c. large surface area.

7. A fungal spore
   a. contains an embryonic organism.
   b. germinates directly into an organism.
   c. is always windblown, because none are motile.
   d. is most often diploid.
   e. Both b and c are correct.

8. Fungal groups have different
   a. sexual reproductive structures.
   b. sporocarp shapes.
   c. modes of nutrition.
   d. types of cell wall.
   e. levels of organization.

9. In the life cycle of black bread mold, the zygospore
   a. undergoes meiosis and produces zoospores.
   b. produces spores as a part of asexual reproduction.
   c. is a thick-walled dormant stage.
   d. is equivalent to asci and basidia.
   e. All of these are correct.

10. In an ascocarp,
    a. there are fertile and sterile hyphae.
    b. hyphae fuse, forming the dikaryotic stage.
    c. a sperm fertilizes an egg.
    d. hyphae do not have chitinous walls.
    e. conidiospores form.

11. In which fungus is the dikaryotic stage longer lasting?
    a. zygospore fungus   c. club fungus
    b. sac fungus   d. chytrids

12. Conidiospores are formed
    a. asexually at the tips of special hyphae.
    b. during sexual reproduction.
    c. by all types of fungi except water molds.
    d. when it is windy and dry.
    e. as a way to survive a harsh environment.

13. The asexual sac fungi are so called because
    a. they have no zygospore.
    b. they cause diseases.
    c. they form conidiospores.
    d. sexual reproduction has not been observed.
    e. All of these are correct.

14. Label this diagram of black bread mold structure and asexual reproduction.

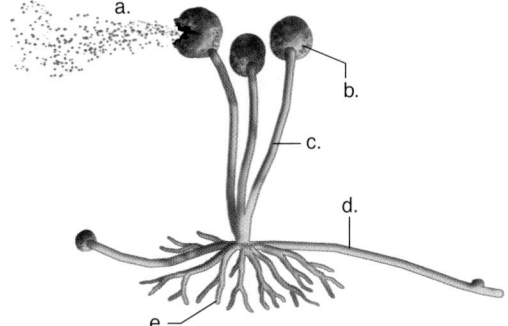

15. Why is it challenging to treat fungal infections of the human body?
    a. Human and fungal cells share common characteristics.
    b. Fungal cells share no features in common with human cells.
    c. All fungal cells are highly resistant to drug treatments.
    d. Both b and c are correct.

16. Mycorrhizae
    a. are a type of lichen.
    b. are mutualistic relationships.
    c. help plants gather solar energy.
    d. help plants gather inorganic nutrients.
    e. Both b and d are correct.

17. Which stage(s) in the chytrid life cycle is/are motile?
    a. diploid zoospores   d. female gametes
    b. haploid zoospores   e. All of these are correct.
    c. male gametes

18. Yeasts are what type of fungi?
    a. zygospores   d. Both b and c are correct.
    b. single cell   e. All of these are correct.
    c. sac fungi

19. Symbiotic relationships of fungi include
    a. athlete's foot.   d. Only b and c are correct.
    b. lichens.   e. All three examples are
    c. mycorrhizae.   correct.

20. Label this diagram of the life cycle of a mushroom.

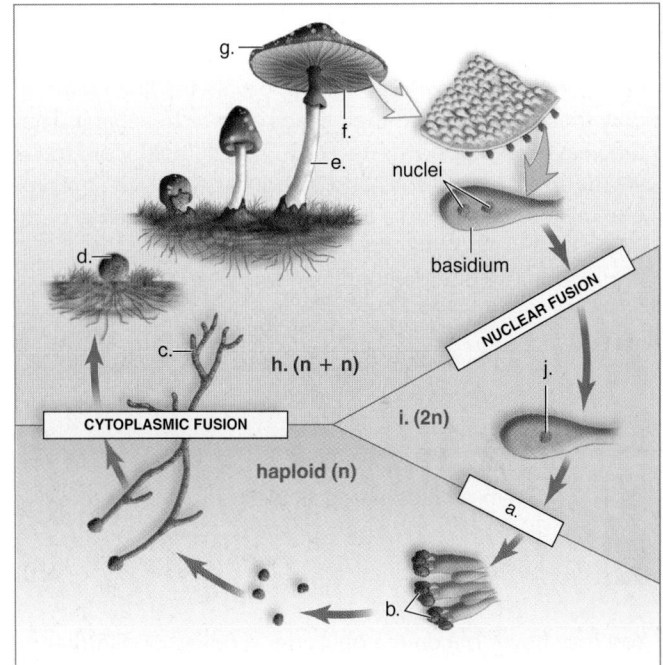

Engage

Thinking Scientifically

1. The very earliest bakers observed that dough left in the air would rise. Unknown to them, yeast from the air "contaminated" the bread, began to grow, and produced carbon dioxide. Carbon dioxide caused the bread to rise. Later, cooks began to save some soft dough (before much flour was added) from the previous loaf to use in the next loaf. The saved portion was called the mother. What is in the mother, and why was it important to save it in a cool place?

2. A fine line seems to exist between symbiosis and parasitism when you examine the relationships between fungi and plants. What hypotheses could explain how different selective pressures may have caused particular fungal species to adopt one or the other relationship? Under what circumstances might a mutualistic relationship evolve between fungi and plants? Under what circumstances might a parasitic relationship evolve?

UNIT 5

Plant Evolution and Biology

Life as we know it on Earth today would not exist without plants. The producers of much of the oxygen we breathe and the food we eat, the suppliers of materials with which we build our homes and clothe ourselves, the sources of many medicines that ease our pains and eradicate our enemies, the creators of breathtaking landscapes, plants are unquestionably linked to all of our lives in countless ways.

What are the secrets behind these green entities? Their life cycles are alien to most of the organisms they support. Not one but two different forms appear as their generations alternate between the gametophytes whose main job is to create eggs and sperm and the sporophytes that create spores via meiosis. They lead mostly sessile "rooted" lives, with only their gametes and their seeds able to travel afar. They have adapted to the arctic and to the tropical rain forests; they have developed strategies to survive heat, cold, drought, and aquatic submersion. They protect themselves with thorns and bark, poisons and camouflage, and their reproductive feats are legendary—a single orchid can produce over one million seeds, and a Judean date palm seed has germinated after 2,000 years of dormancy.

Portions of the Big Idea outlines deal with adaptations and diversification of plants:

 Plant evolution can be traced using phylogenetic trees.

 Plants exhibit many adaptations involving timing and coordination mechanisms.

 Chemical messaging allows plants to regulate gene expression and responses to the environment.

 Roots, stems, and leaves interact to achieve gas exchange, material transport, and other essential functions.

You may never become a gardener, a florist, or a farmer, but every day you come in direct and personal contact with plants, and you should know something about their history, architecture, and adaptations that continue to astound engineers and poets with their simplicity, efficacy, and beauty. We humans should be green with envy at the accomplishments of these masters of photosynthesis...Plants!

UNIT OUTLINE

Produce aisle in a grocery store.

Plant Evolution and Diversity

Aaas you wander down the produce aisle of the grocery store you can't help but notice the large diversity of fruits and vegetables. What is even more amazing is the fact that, despite the diversity of shapes, colors, textures and tastes, all fruits and vegetables have evolved from a common ancestor. In many respects, plants owe their characteristics to their evolutionary past. Plants, like other organisms, have evolved to adapt to a wide variety of environments over millions of years. But plants have also been influenced by human activity. For thousands of years humans have been directing the evolution of plants as a result of artificial selection and genetic engineering.

This chapter traces the evolutionary history of plants from their green algal ancestor to the various groups we depend upon today for our survival. During the course of this evolutionary journey some groups have proven to be more successful than others due to various structural adaptations. In this chapter we discuss the similarities and differences of the major plant groups and the evolutionary adaptations that have contributed to their success. We pay special attention to the reasons behind the tremendous success of the flowering plants.

As you read through the chapter, think about the following questions:

1. What environmental challenges did plants have to overcome in order to survive on land?
2. Why are angiosperms more widespread than all other groups of plants?
3. What characteristics are unique to each of the major groups of plants?

CHAPTER OUTLINE

23.1 The Green Algal Ancestor of Plants 420

23.2 Evolution of Bryophytes: Colonization of Land 423

23.3 Evolution of Lycophytes: Vascular Tissue 426

23.4 Evolution of Pteridophytes: Megaphylls 427

23.5 Evolution of Seed Plants: Full Adaptation to Land 430

BEFORE YOU BEGIN

Before beginning this chapter, take a few moments to review the following discussions.

Figure 4.7 What cellular structures are unique to plants?

Figure 10.6 What is the end result of meiosis?

Section 16.2 What is the role of natural selection in the evolutionary process?

FOLLOWING *the* BIG IDEAS

CHAPTER 23 PLANT EVOLUTION AND DIVERSITY

Evolution — Plants evolved from aquatic ancestors, eventually adapting to land with developments of vascular tissue, fertilization without water, and seeds.

23.1 The Green Algal Ancestor of Plants

Plants are multicellular, photosynthetic eukaryotes whose evolution is marked by adaptations to a land existence. Plants are thought to have evolved from a freshwater green alga. In this section we consider the adaptations necessary for plants to live on land and also their evolutionary origins, life cycle, and other traits.

Adaptation to Land

A land environment does offer certain advantages to plants. Water, even if clear, filters light, so life on land tends to receive a greater amount of sunlight for photosynthesis. Carbon dioxide is also present in higher concentrations and diffuses more readily in air than in water.

The land environment, however, requires adaptations, in particular to deal with the constant threat of desiccation (drying out). The most successful land plants are those that protect all phases of reproduction (sperm, egg, embryo) from drying out and have an efficient means of dispersing offspring on land. Seeds provide the developing embryo a source of food and a protective seed coat, both of which tend to increase the chances of the embryo's survival.

The water environment not only provides plentiful water, it also offers support for the body of a plant. To conserve water, the land plant body, at the very least, is covered by a waxy cuticle that prevents loss of water while still allowing carbon dioxide to enter so that photosynthesis can continue. In many land plants, the roots absorb water from the soil, and a vascular system transports water to other parts of the plant body. The vascular system that evolved in land plants allows them to stand tall and support expansive, broad leaves that efficiently collect sunlight. The flowering plants, the last type to evolve, often employ animals to assist with reproduction and dispersal of seed. The evolutionary history of plants given in Figure 23.1 shows the sequence in which land plants evolved adaptive features for an existence on land.

The Ancestry of Plants

The plants (listed in Table 23.1) evolved from a freshwater green algal species some 450 MYA (million years ago). Research shows that both green algae and plants contain chlorophylls *a* and *b* and various accessory pigments; store excess carbohydrates as starch; and have cellulose in their cell walls. These

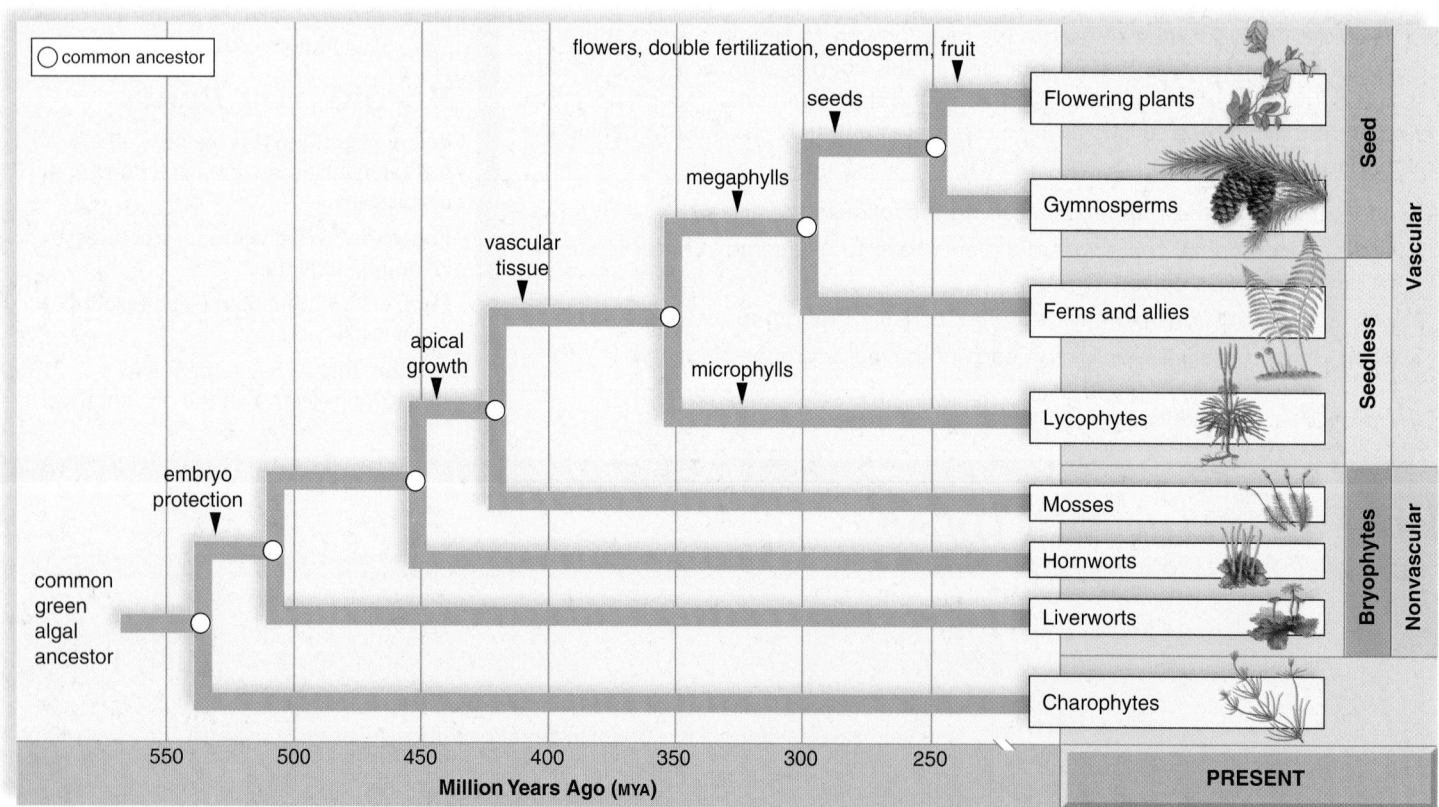

Figure 23.1 Evolutionary history of plants.   The evolution of plants involves these significant innovations. In particular, protection of a multicellular embryo was seen in the first plants to live on land. Vascular tissue permits the transport of water and nutrients. The evolution of the seed increased the chance of survival for the next generation.

characteristics support the hypothesis that the green algae are the ancestral group of the land plants.

In recent years, molecular systematists have compared the sequence of ribosomal RNA bases between organisms. The results suggest that among the green algae, land plants are most closely related to freshwater green algae, known as **charophytes.** Fresh water, of course, exists in bodies of water on land, and natural selection would have favored those specimens best able to make the transition to the land itself.

Charophytes are classified into several different types— *Spirogyra,* for example, is a charophyte. But botanists tell us that among living charophytes, Charales (an order with 300 macroscopic species) and the *Coleochaete* (a genus with 30 microscopic species), featured in Figure 23.2, are most like land

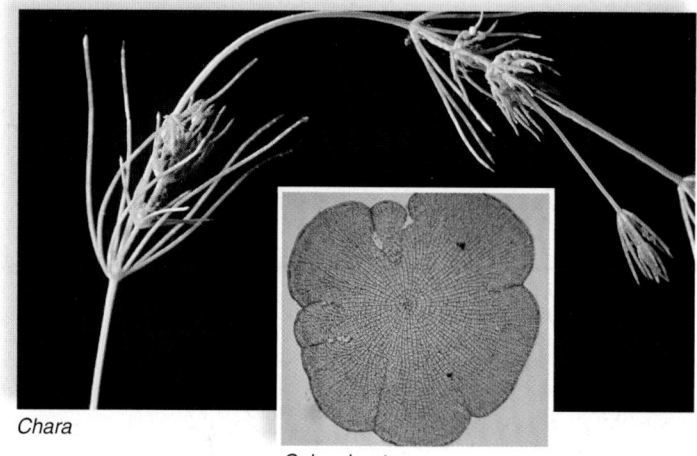

Chara

Coleochaete

Figure 23.2 Charophytes.   The charophytes (represented here by *Chara* and *Coleochaete*) are the green algae most closely related to the land plants.

Table 23.1 Key Groups of Plants and Their Features

DOMAIN: Eukarya
KINGDOM: Plantae

CHARACTERISTICS
Multicellular, usually with specialized tissues; photosynthesizers that became adapted to living on land; most have alternation-of-generations life cycle.

Charophytes
Live in water; haploid life cycle; share certain traits with the land plants

LAND PLANTS (embryophytes)
Alternation-of-generations life cycle; protect a multicellular sporophyte embryo; gametangia produce gametes; apical tissue produces complex tissues; waxy cuticle prevents water loss.

  Bryophytes (liverworts, hornworts, mosses)
  Low-lying, nonvascular plants that prefer moist locations: Dominant gametophyte produces flagellated sperm; unbranched, dependent sporophyte produces windblown spores.

VASCULAR PLANTS (lycophytes, ferns and their allies, seed plants)
Dominant, branched sporophyte has vascular tissue: Lignified xylem transports water, and phloem transports organic nutrients; typically has roots, stems, and leaves; and gametophyte is eventually dependent on sporophyte.

  Lycophytes (club mosses)
  Leaves are microphylls with a single, unbranched vein; sporangia borne on sides of leaves produce windblown spores; independent and separate gametophyte produces flagellated sperm.

  Ferns and Allies (pteridophytes)
  Leaves are megaphylls with branched veins; dominant sporophyte produces windblown spores in sporangia borne on leaves; and independent and separate gametophyte produces flagellated sperm.

SEED PLANTS (gymnosperms and angiosperms)
Leaves are megaphylls; dominant sporophyte produces heterospores that become dependent male and female gametophytes. Male gametophyte is pollen grain and female gametophyte occurs within ovule, which becomes a seed.

  Gymnosperms (cycads, ginkgoes, conifers, gnetophytes)
  Usually large; cone-bearing; existing as trees in forests. Sporophyte bears pollen cones, which produce windblown pollen (male gametophyte), and seed cones, which produce seeds.

  Angiosperms (flowering plants)
  Diverse; live in all habitats. Sporophyte bears flowers, which produce pollen grains, and bear ovules within ovary. Following double fertilization, ovules become seeds that enclose a sporophyte embryo and endosperm (nutrient tissue). Fruit develops from ovary.

plants. Charophytes and land plants are in the same clade and form a monophyletic group (Fig. 23.1). Their common ancestor no longer exists, but if it did, it would have features that resemble those of the Charales and *Coleochaete.*

First, let's take a look at these filamentous green algae (Fig. 23.2). The Charales (e.g., *Chara*) are commonly known as stoneworts because some species are encrusted with calcium carbonate deposits. The body consists of a single file of very long cells anchored in mud by thin filaments. **Whorls,** or clusters of branches, occur at multicellular **nodes,** regions between the enlarged cells of the main axis. Male and female reproductive structures grow at the nodes. The zygote is retained until it is enclosed by tough walls. A *Coleochaete* looks like a flat pancake, but the body is actually composed of elongated branched filaments of cells that spread flat across the substrate or form a three-dimensional cushion. The zygote is also retained in *Coleochaete.*

These two groups of charophytes have several features that would have promoted the evolution of multicellular land plants listed in Table 23.1, which have complex tissues and organs. These features are present in charophytes and land plants, and have been improved upon in modern land plants:

1. The *cellulose cell walls* of charophytes and the land plant lineage are laid down by the same unique type of cellulose-synthesizing complexes. Charophytes also have a mechanism of cell-wall formation during cytokinesis that is nearly identical to that of land plants. In land plants, a strong cell wall supports an upright posture.
2. The *apical cells* of charophytes produce cells that allow their filaments to increase in length. **Apical** [L. *apex,* farthest point] refers to the tip of a filament or branch. At the nodes, other cells can divide asymmetrically to produce reproductive structures. Land plants are noted for their apical tissue, which produces specialized tissues that add to or develop into new organs, such as new branches and leaves.
3. The *plasmodesmata* of charophytes provide a means of communication between neighboring cells, otherwise separated by cell walls (see Section 5.4). Evolutionary researchers believe that the plasmodesmata played a role in the evolution of land plants by allowing for the exchange of nutrients from one cell to the next.

4. The *placenta* (certain designated cells of the ovary) of charophytes transfers nutrients from haploid cells of the previous generation to the diploid zygote. Both charophytes and land plants retain and care for the zygote.

Alternation of Generations

All land plants exhibit **alternation of generations,** meaning that an organism has two alternating forms in the course of its life cycle. Study the alternation-of-generations life cycle in Figure 23.3 to understand that the sporophyte (2n) is so named for its production of spores by meiosis. A **spore** is a haploid reproductive cell that develops into a new organism without the need to fuse with another reproductive cell. In the plant life cycle, a spore undergoes mitosis and becomes a gametophyte.

The gametophyte (n) is so named for its production of gametes. In plants, eggs and sperm are produced by mitotic cell division. A sperm and egg fuse, forming a diploid zygote that undergoes mitosis—becoming the sporophyte embryo, and then the 2n generation.

Two observations can be made. First, meiosis produces haploid spores. This is consistent with the sporophyte being the diploid generation and producing spores, which are haploid reproductive cells. Second, mitosis occurs as a spore becomes a gametophyte, and mitosis occurs again as a zygote becomes a sporophyte. The occurrence of mitosis at these times defines the two generations.

We can contrast the haploid life cycle, such as that of *Chlamydomonas* (see Fig. 21.3), with the alternation-of-generations life cycle. In the haploid life cycle, the zygote undergoes meiosis, and therefore only four zoospores are produced per zygote. In the alternation-of-generations life cycle, the zygote undergoes mitosis to become a multicellular sporophyte with one or more sporangia that produce many windblown spores. The production of an abundance of spores would most likely have assisted land plants in colonizing the land environment.

Dominant Generation

Land plants differ as to which generation is dominant—that is, more conspicuous. In plants like mosses, the gametophyte is dominant, but in plants like ferns, pine trees, and peach trees, the sporophyte is dominant (Fig. 23.4). In the history of land plants, a vascular system evolves only in the sporophyte; therefore, the shift to sporophyte dominance is an adaptation to life on land. Notice that as the sporophyte gains in dominance, the gametophyte becomes microscopic. The gametophyte also becomes dependent on the sporophyte.

Other Derived Traits of Land Plants

The following traits are additional characteristics of reproduction in land plants:

1. Not just the zygote but also the multicellular 2n embryo are retained and protected from drying out. Because land plants protect the embryo, an alternate name for this clade is **embryophyta.**

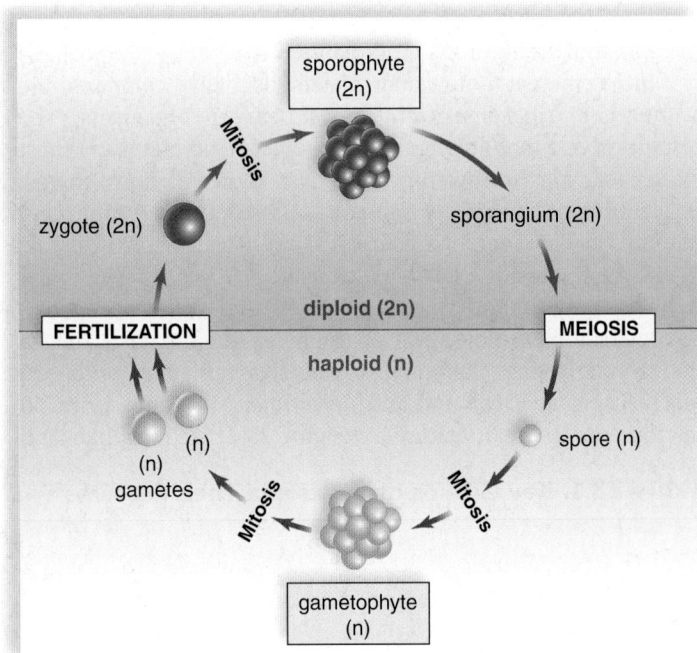

Figure 23.3 Alternation of generations in land plants. The zygote develops into a multicellular 2n generation, and meiosis produces spores in multicellular sporangia. The gametophyte generation produces gametes within multicellular gametangia.

2. This 2n generation, called a **sporophyte,** produces at least one, and perhaps several, multicellular, sporangia.
3. **Sporangia** (sing., **sporangium**) produce spores by meiosis. Spores (and pollen grains, if present) have a wall that contains *sporopollenin,* a molecule that prevents drying out.
4. Spores become an n generation, called a **gametophyte,** that bears multicellular gametangia, which have an outer layer of sterile cells and an inner mass of cells that become the gametes:

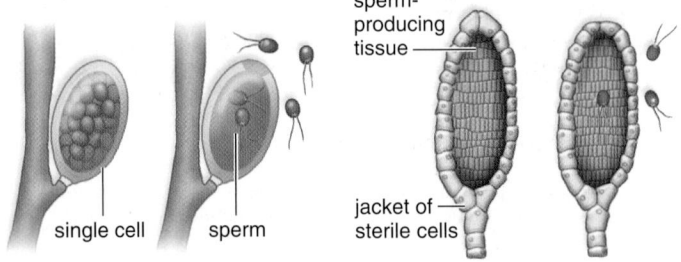

a. sperm production in algae   b. sperm production in bryophytes

A male gametangium is called an **antheridium,** and a female gametangium is called an **archegonium.**

Aside from innovations associated with the life cycle, the exposed parts of land plants are covered by an impervious waxy **cuticle,** which prevents loss of water. Most land plants also have **stomata** (sing., **stoma** [Gk. *stoma,* mouth]), little openings, that allow gas exchange, despite the plant being covered by a cuticle (Fig. 23.5).

Another trait seen in most land plants is the presence of apical tissue, which has the ability to produce complex tissues and organs.

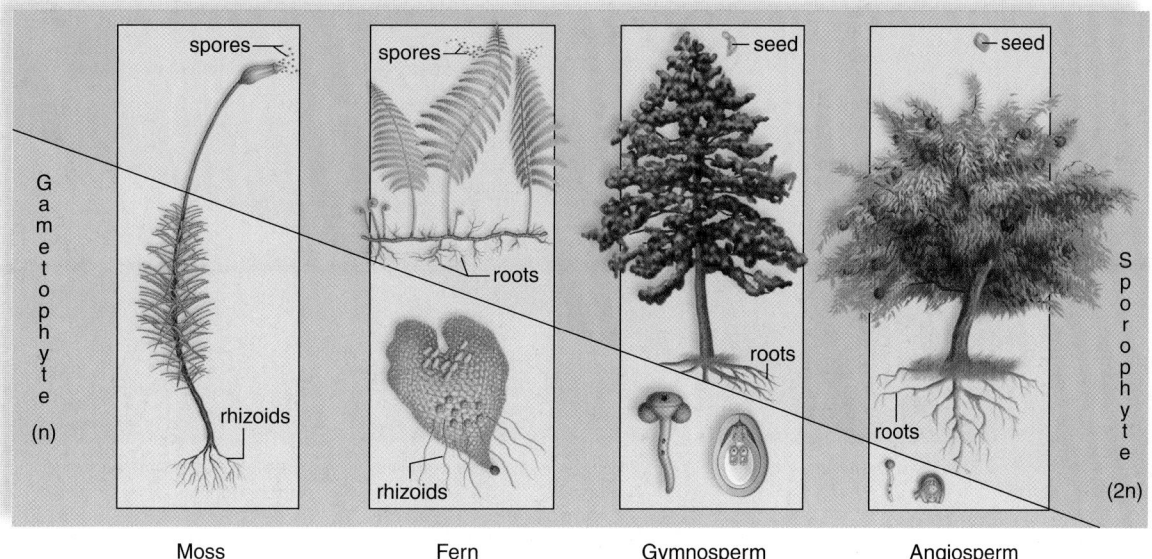

Figure 23.4
Reduction in the size of the gametophyte.
Notice the reduction in the size of the gametophyte and the increase in the size of the sporophyte among these representatives of today's land plants. This trend occurred as these plants became adapted for life on land. In the moss and fern, spores disperse the gametophyte. In gymnosperms and angiosperms, seeds disperse the sporophyte.

Labels in figure: spores, spores, seed, seed; Gametophyte (n); Sporophyte (2n); rhizoids, roots, rhizoids, roots, roots; Moss; Fern; Gymnosperm; Angiosperm

Check Your Progress   **23.1**

1. List the benefits of a land existence for plants.
2. Compare and contrast the traits of charophytes and land plants.
3. Identify the role of each generation in the alternation-of-generations life cycle.

23.2 Evolution of Bryophytes: Colonization of Land

Learning Outcomes

Upon completion of this section, you should be able to

1. List the traits that classify a plant as a bryophyte.
2. Compare the three groups of bryophytes.
3. Identify the key structures and stages in the life cycle of a moss.

The **bryophytes**—the liverworts, hornworts [Anglo-Saxon *wort*, herb], and mosses—were the first plants to colonize land. They only superficially appear to have roots, stems, and leaves because, by definition, true roots, stems, and leaves must contain vascular tissue. **Vascular tissue** is specialized for the transport of water and organic nutrients throughout the body of a plant. Bryophytes, which lack vascular tissue, are often called the **nonvascular plants.** Vascular tissue also provides support to the plant body, so bryophytes typically are low-lying; some mosses reach a maximum height of only about 20 cm.

The fossil record contains some evidence that the various bryophytes evolved during the Ordovician period (488.3–443.7 MYA). An incomplete fossil record makes it difficult to tell how closely related the various bryophytes are. Molecular data, in particular, suggest that these plants have individual lines of descent, as shown in Figure 23.1, and that they do not form a monophyletic group. The observation that today's mosses have a rudimentary form of vascular tissue suggests that they are more closely related to vascular plants than the hornworts and liverworts.

Bryophytes do share other traits with the vascular plants. For example, they have an alternation-of-generations life cycle and they have the traits listed on page 421. Their bodies are covered by a cuticle that is interrupted in hornworts and mosses by stomata, and they have apical tissue that produces complex tissues. However, bryophytes are the only land plants in which the gametophyte is dominant (see Fig. 23.4). Antheridia produce flagellated sperm, which means they need a film of moisture in order to swim to eggs located inside archegonia. The bryophytes' lack of vascular tissue and the presence of flagellated sperm means that you are apt to find bryophytes in moist

cuticle

a. Stained photomicrograph of a leaf cross section

Plant leaves have a cuticle and stomata.

stomata

400 ×

b. Falsely colored scanning electron micrograph of leaf surface

Figure 23.5 Leaf adaptation. **a.** A cuticle keeps the underlying cells and tissues from drying out. **b.** The uptake of carbon dioxide is possible because the cuticle is interrupted by stomata.

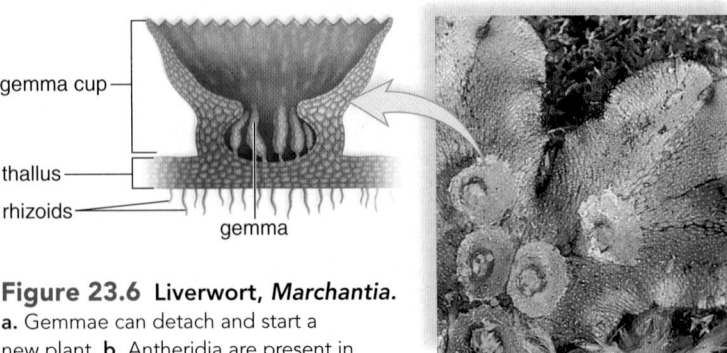

gemma cup

thallus

rhizoids

gemma

Figure 23.6 Liverwort, *Marchantia*.
a. Gemmae can detach and start a new plant. **b.** Antheridia are present in disk-shaped structures. **c.** Archegonia are present in umbrella-shaped structures.

a. Thallus with gemmae cups

male gametophyte

b. Male gametophytes bear antheridia

female gametophyte

c. Female gametophytes bear archegonia

locations. Some bryophytes compete well in harsh environments because they reproduce asexually.

The Biological Systems feature "The Uses of Bryophytes" describes the value of these plants.

Liverworts

Liverworts are divided into two groups—the thallose liverworts with flattened bodies, known as a thallus; and the leafy liverworts, which superficially resemble mosses. The name liverwort refers to the lobes of the thallus, which to some resemble lobes of the liver. The majority of liverwort species are the leafy types.

The liverworts in the genus *Marchantia* have a thin thallus, about 30 cells thick in the center. Each branched lobe of the thallus is approximately a centimeter in length; the upper surface is divided into diamond-shaped segments with a small pore, and the lower surface bears numerous hairlike extensions called **rhizoids** [Gk., *rhizion*, dim. of root] that project into the soil (Fig. 23.6). Rhizoids serve in anchorage and limited absorption.

Marchantia species reproduce both asexually and sexually. Gemmae cups on the upper surface of the thallus contain gemmae, groups of cells that detach from the thallus and can start a new plant. Sexual reproduction depends on disk-headed stalks that bear antheridia, and on umbrella-headed stalks that bear archegonia. Following fertilization, tiny sporophytes composed of a foot, a short stalk, and a capsule begin growing within archegonia. Windblown spores are produced within the capsule.

sporophyte

gametophyte

Figure 23.7 Hornwort, *Anthoceros* sp. The "horns" of a hornwort are sporophytes that grow continuously from a base anchored in gametophyte tissue.

Hornworts

The **hornwort** gametophyte usually grows as a thin rosette or ribbonlike thallus between 1 and 5 cm in diameter. Although some species of hornworts live on trees, the majority of species live in moist, well-shaded areas. They photosynthesize, but they also have a symbiotic relationship with cyanobacteria, which, unlike plants, can fix nitrogen from the air.

The small sporophytes of a hornwort resemble tiny green broom handles rising from a thin gametophyte, usually less than 2 cm in diameter (Fig. 23.7). Like the gametophyte, a sporophyte can photosynthesize, although it has only one chloroplast per cell. A hornwort can bypass alternation of generations by producing asexually through fragmentation.

Mosses

Mosses are the largest phyla of nonvascular plants, with over 15,000 species. There are three distinct groups of mosses: peat mosses, granite mosses, and true mosses. Although most prefer damp, shaded locations in the temperate zone, some survive in deserts, and others inhabit bogs and streams. In forests, they frequently form a mat that covers the ground and rotting logs. In dry environments, they may become shriveled, turn brown, and look completely dead. As soon as it rains, however, the plant becomes green and resumes metabolic activity.

Figure 23.8 describes the life cycle of a typical temperate-zone moss. The gametophyte of mosses begins as an algalike branching filament of cells, the protonema, which precedes and produces upright leafy shoots that sprout rhizoids. The shoots bear either antheridia or archegonia. The dependent sporophyte consists of a foot, which is enclosed in female gametophyte tissue; a stalk; and an upper capsule, the sporangium, where spores are produced. A moss sporophyte is always attached to the gametophyte. At first, the sporophyte is green and photosynthetic; at maturity, it is brown and nonphotosynthetic. In some species, the sporangium can produce as many as 50 million spores. The spores disperse the new gametophyte generation.

Check Your Progress   23.2

1. Explain the various methods of bryophyte reproduction.
2. List the characteristics that enabled the bryophytes to successfully colonize land.
3. Explain which portion of the bryophyte life cycle is the most important.

Biological Systems

The Uses of Bryophytes

Mosses, and also liverworts and hornworts, are of great ecological significance. They contribute to the lush beauty of rain forests and the conversion of mountain rocks to soil because of their ability to hold moisture and metals in the soil, and tolerate desiccation. The genus *Sphagnum* (peat moss) has great commercial and ecological importance to humans.

The cell walls of peat moss have a tremendous ability to absorb water, which is why it is used in gardening to improve the water-holding capacity of the soil. One percent of the Earth's surface is peatlands, where dead *Sphagnum* accumulates and does not decay. It can then be extracted and used for various purposes, such as fuel and building materials. Peat holds much CO_2, and some are concerned that when peat is extracted, we are losing this depository for CO_2, the gas that contributes the most to climate change.

Bryophytes are expected to become valuable to genetic research and applications. Once scientists know which genes give bryophytes the ability to resist chemical reagents and decay, as well as animal attacks, these qualities could be transferred to other plants through genetic engineering.

Questions to Consider

1. What role do the bryophytes play in their environment?
2. What economic value do the bryophytes provide to human society?

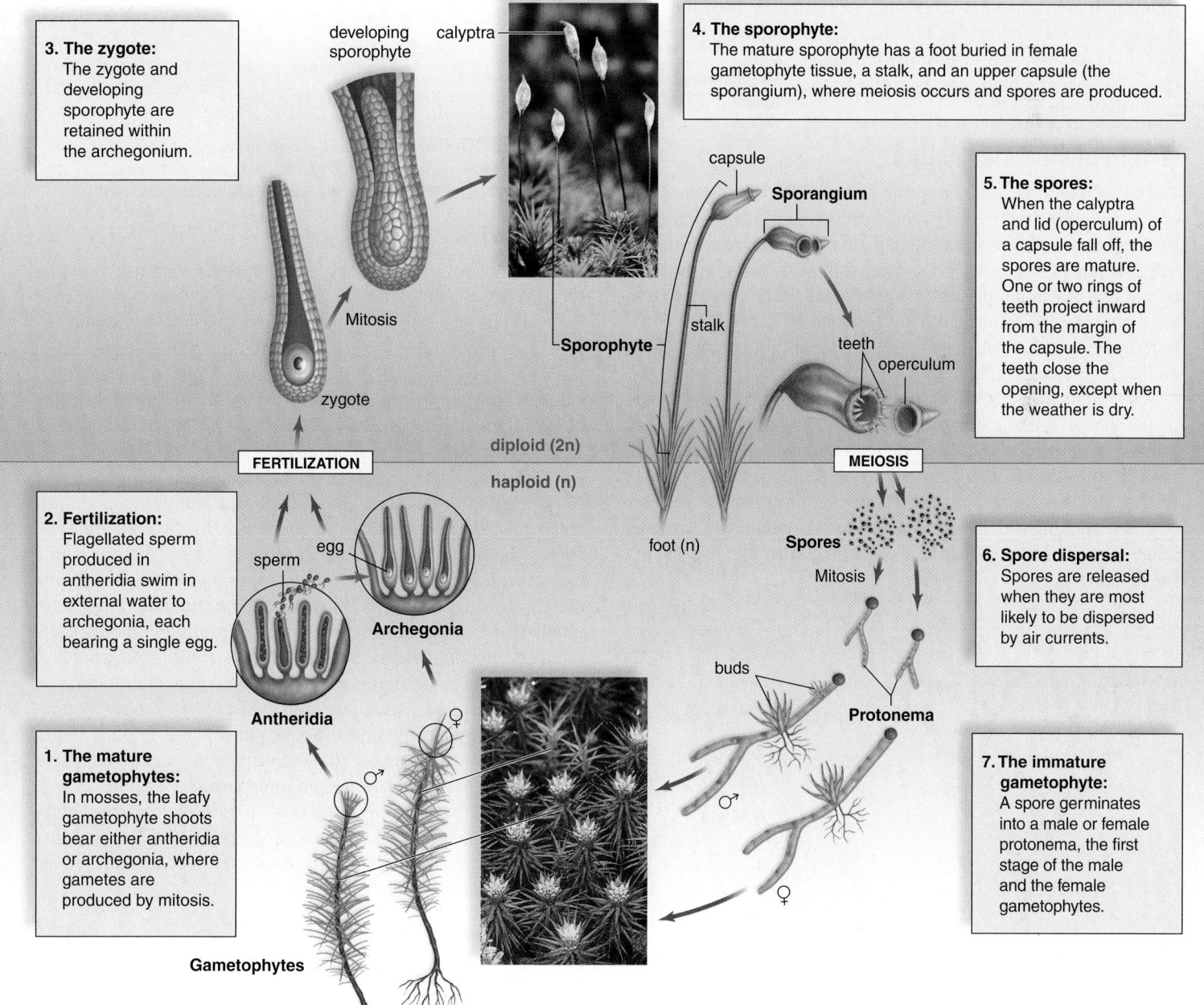

3. The zygote: The zygote and developing sporophyte are retained within the archegonium.

developing sporophyte

calyptra

4. The sporophyte: The mature sporophyte has a foot buried in female gametophyte tissue, a stalk, and an upper capsule (the sporangium), where meiosis occurs and spores are produced.

Mitosis

zygote

capsule

Sporangium

5. The spores: When the calyptra and lid (operculum) of a capsule fall off, the spores are mature. One or two rings of teeth project inward from the margin of the capsule. The teeth close the opening, except when the weather is dry.

stalk

Sporophyte

teeth

operculum

diploid (2n)

haploid (n)

FERTILIZATION

MEIOSIS

2. Fertilization: Flagellated sperm produced in antheridia swim in external water to archegonia, each bearing a single egg.

sperm

egg

Archegonia

Antheridia

foot (n)

Spores

Mitosis

6. Spore dispersal: Spores are released when they are most likely to be dispersed by air currents.

buds

Protonema

1. The mature gametophytes: In mosses, the leafy gametophyte shoots bear either antheridia or archegonia, where gametes are produced by mitosis.

♂

♀

7. The immature gametophyte: A spore germinates into a male or female protonema, the first stage of the male and the female gametophytes.

Gametophytes

rhizoids

23.3 Evolution of Lycophytes: Vascular Tissue

Today, **vascular plants** dominate the natural landscape in nearly all terrestrial habitats. Trees are vascular plants that achieve great height because they have roots that absorb water from the soil and a vascular tissue called **xylem,** which transports water through the stem to the leaves. (Another conducting tissue called **phloem** transports nutrients in a plant.) Further, the cell walls of the conducting cells in xylem contain **lignin,** a material that strengthens plant cell walls; therefore, the evolution of xylem was essential to the evolution of trees.

Animation
Vascular Tissue

Origin of Vascular Plants

The fossil record indicates that the first vascular plants, such as *Cooksonia,* were more likely a bush than a tree. *Cooksonia* is a rhyniophyte, a group of vascular plants that flourished during the Silurian period (443.7–416 MYA), but then became extinct by the mid-Devonian period (416–359.2 MYA). The rhyniophytes were only about 6.5 cm tall and had no roots or leaves. They consisted simply of a stem that forked evenly to produce branches ending in sporangia (Fig. 23.9).

The branching of *Cooksonia* was significant because instead of the single sporangium produced by a bryophyte, the plant produced many sporangia, and therefore many more spores. For branching to occur, meristem has to be positioned at the apex (tip) of stems and branches, as it is in vascular plants today.

The sporangia of *Cooksonia* produced windblown spores, which classifies it as a **seedless vascular plant,** like the rest of the lycophytes.

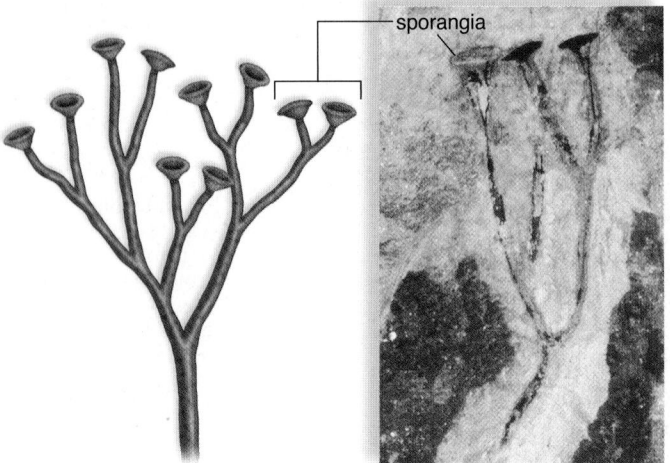

Figure 23.9 A *Cooksonia* fossil. The upright branches of a *Cooksonia* fossil, no more than a few centimeters tall, terminated in sporangia as seen here in the drawing and photo.

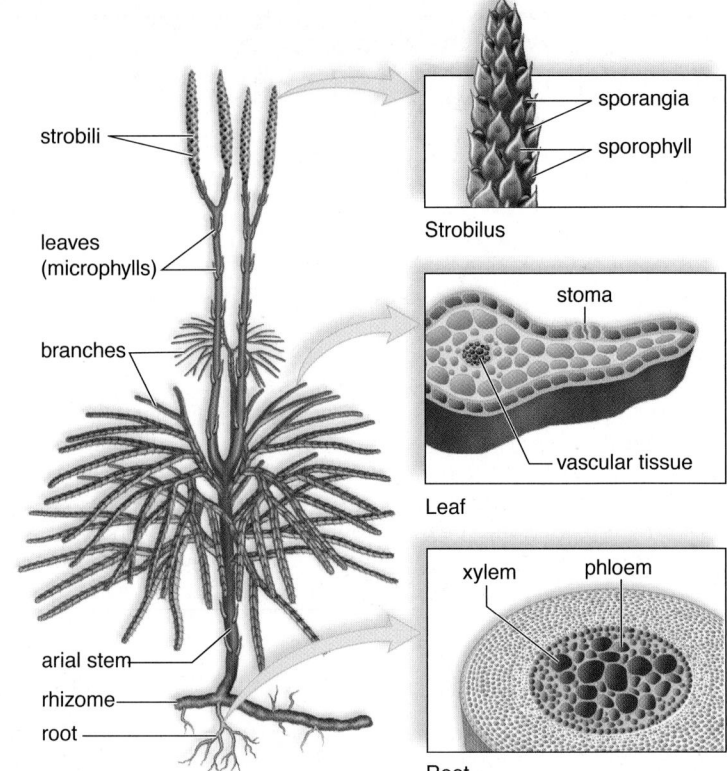

Figure 23.10 Ground pine, *Lycopodium*. The *Lycopodium* sporophyte develops an underground rhizome system. A rhizome is an underground stem. This rhizome produces true roots along its length.

Lycophytes

In addition to the stem of early vascular plants, the first **lycophytes** also had leaves and roots. The leaves are called **microphylls** because they had only one strand of vascular tissue. Microphylls most likely evolved as simple side extensions of the stem (see Fig 23.11*a*). Roots evolved simply as lower extensions of the stem; the organization of vascular tissue in the roots of lycophytes today is much like it was in the stems of fossil vascular plants—the vascular tissue is centrally placed.

Today's lycophytes, also called club mosses, include around 1,150 species in three groups: the ground pines (*Lycopodium*), spike mosses (*Selaginella*), and quillworts (*Isoetes*). Figure 23.10 shows the structure of *Lycopodium;* note the structure of the roots and leaves and the location of sporangia. The roots come off a branching, underground stem called a **rhizome.** The microphylls that bear sporangia are called **sporophylls,** and they are grouped into club-shaped **strobili** (sing., **strobilus** [Gr. *strobilos,* pinecone]), accounting for their common name, club mosses.

Vascular plants, which include the lycophytes, have a dominant sporophyte generation. Some plants, like ground pines, have spores that germinate into inconspicuous and independent gametophytes and are termed **homosporous** (see Fig. 23.15). By contrast, **heterosporous** plants, like spike mosses, quillworts, and the seed plants, have **microspores** that develop into male gametophytes, and **megaspores** that develop into female gametophytes.

Check Your Progress 23.3

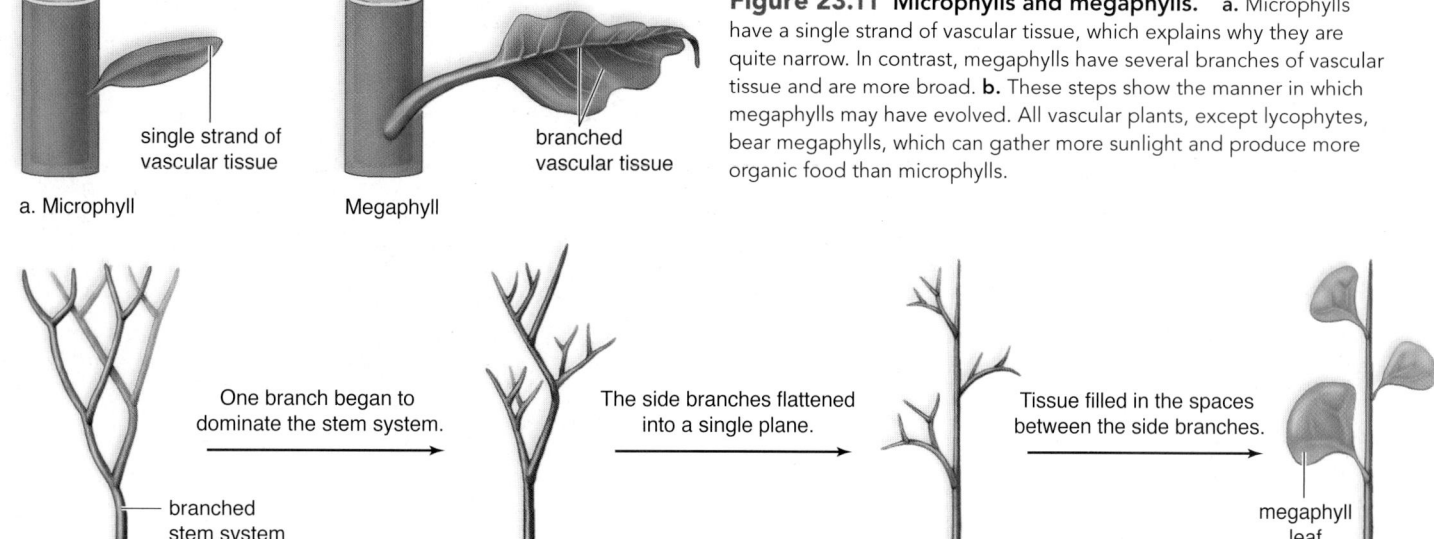

single strand of
vascular tissue

a. Microphyll

Megaphyll

branched
vascular tissue

Figure 23.11 Microphylls and megaphylls. **a.** Microphylls have a single strand of vascular tissue, which explains why they are quite narrow. In contrast, megaphylls have several branches of vascular tissue and are more broad. **b.** These steps show the manner in which megaphylls may have evolved. All vascular plants, except lycophytes, bear megaphylls, which can gather more sunlight and produce more organic food than microphylls.

One branch began to
dominate the stem system.

The side branches flattened
into a single plane.

Tissue filled in the spaces
between the side branches.

branched
stem system

megaphyll
leaf

b. Megaphyll evolution process

23.4 Evolution of Pteridophytes: Megaphylls

Learning Outcomes

Upon completion of this section, you should be able to

1. Identify three types of pteridophytes.
2. Compare and contrast microphylls and megaphylls.
3. Identify the components of the ferns' life cycle.

Pteridophytes, consisting of ferns and their allies, the horsetails and whisk ferns, are seedless vascular plants. However, both pteridophytes and the seed plants, which are studied in Section 23.5, have megaphylls. **Megaphylls** are broad leaves with several strands of vascular tissue. Figure 23.11a shows the difference between microphylls and megaphylls, and Figure 23.11b shows how megaphylls could have evolved.

Megaphylls, which evolved about 370 MYA, allow plants to efficiently collect solar energy, leading to the production of more food and the possibility of producing more offspring than plants without megaphylls. Therefore, the evolution of megaphylls made plants more fit. Recall that fitness, in an evolutionary sense, is judged by the number of living offspring an organism produces relative to others of its own kind.

The pteridophytes, like the lycophytes, were dominant from the late Devonian period through the Carboniferous period. Today, the lycophytes are quite small, but some of the extinct relatives of today's club mosses were 35 m tall and dominated the Carboniferous swamps. The horsetails (around 18 m tall) and ancient tree ferns (8 + m in height) also contributed significantly to the great swamp forests of the time (see the Evolution feature "Carboniferous Forests" on page 435).

Horsetails

Today, **horsetails** consist of one genus, *Equisetum*, and approximately 25 species of distinct seedless vascular plants. Most horsetails inhabit wet, marshy environments around the globe. About 300 MYA, horsetails were dominant plants and grew as large as modern trees. Today, horsetails have a rhizome that produces hollow, ribbed aerial stems and reaches a height of 1.3 m (Fig. 23.12). The leaves may have been megaphylls at one time but now they are reduced and form whorls at the nodes. The whorls of slender, green side branches make the plant bear a resemblance to a horse's tail. Many horsetails have strobili at the tips of all stems; others send up special buff-colored stems that bear the strobili. The spores germinate into inconspicuous and independent gametophytes.

The stems are tough and rigid because of silica deposited in cell walls. Early Americans, in particular, used horsetails for

strobilus

branches

node

leaves

rhizome

root

Figure 23.12 Horsetail, *Equisetum*. Whorls of branches and tiny leaves are at the nodes of the stem. Spore-producing sporangia are borne in strobili.

scouring pots and called them "scouring rushes." Today, they are still used as ingredients in a few abrasive powders.

Whisk Ferns

Whisk ferns are represented by the genera *Psilotum* and *Tmesipteris*. Depending on the climate, *Psilotum* and *Tmesipteris* can be found living as **epiphytes** (plants that live in or on trees), or on the ground. The two *Psilotum* species resemble a whisk broom (Fig. 23.13) because they have no leaves. A horizontal rhizome gives rise to an aerial stem that repeatedly forks. The sporangia are borne on short side branches. The two to three species of *Tmesipteris* have appendages that some maintain as reduced megaphylls.

Ferns

Ferns include approximately 11,000 species. Ferns are most abundant in warm, moist, tropical regions, but they can also be found in temperate regions and as far north as the Arctic Circle. Several species live in dry, rocky places whereas others have adapted to an aquatic life. Ferns range in size from tiny aquatic species less than 1 cm in diameter to modern giant tropical tree ferns that exceed 20 m in height.

The megaphylls of ferns, called **fronds,** are commonly divided into leaflets. The royal fern has fronds that stand about 1.8 m tall; those of the hart's tongue fern are straplike and leathery; and those of the maidenhair fern are broad, with subdivided leaflets (Fig. 23.14). In nearly all ferns, the leaves first appear in a curled-up form called a fiddlehead, which unrolls as it grows.

The life cycle of a typical temperate zone fern, shown in Figure 23.15, applies in general to the other types of vascular seedless plants. The dominant sporophyte produces windblown spores by meiosis within sporangia. In a fern, the sporangia can be located within *sori* (sing., *sorus*) on the underside of the

Cinnamon fern, *Osmunda cinnamomea*

Hart's tongue fern, *Campyloneurum scolopendrium*

Maidenhair fern, *Adiantum pedatum*

Figure 23.14 Diversity of ferns.   Structural adaptations found in a variety of ferns.

leaflets. The windblown spores disperse the gametophyte, the generation that lacks vascular tissue. The separate heart-shaped gametophyte produces flagellated sperm that swim in a film of water from the antheridium to the egg within the archegonium, where fertilization occurs. Eventually, the gametophyte disappears and the sporophyte is independent. Both generations of a fern are considered to be separate and independent of one another. Many ferns can also reproduce asexually by fragmentation of the fern rhizome.

Figure 23.13 Whisk fern, *Psilotum*.   *Psilotum* has no leaves—the branches carry on photosynthesis. The sporangia are yellow.

Check Your Progress   23.4

1. Identify two ways in which the fern life cycle is dependent on external water.
2. Compare the life cycle of a fern to that of a moss.

The Uses of Ferns

Ostrich fern (*Matteuccia struthiopteris*) is the only edible fern to be traded as a food, and it comes to the table in North America as "fiddleheads." In tropical regions of Asia, Africa, and the western Pacific, dozens of types of ferns are taken from the wild and used as food.

The fern *Azolla* harbors *Anabaena*, a nitrogen-fixing cyanobacteria, and *Azolla* is grown in rice paddies, where it fertilizes rice plants. It is estimated that each year, *Azolla* converts more atmospheric nitrogen into a form available for plant growth than all the legumes. And like legumes, its use avoids the problems associated with artificial fertilizer applications.

Several authorities have noted that ferns and their allies are used as medicines in China. Among the conditions being treated are boils and ulcers, whooping cough, and dysentery. Extracts from ferns have also been used to kill insects because they inhibit insect molting.

Ferns beautify gardens, and horticulturists may use them in floral arrangements. Vases, small boxes and baskets, and also jewelry are made from the trunk of tree ferns. Their black and very hard vascular tissue provides many interesting patterns.

Questions to Consider

1. Explain the impact the extinction of ferns would have upon human society.
2. What are the environmental benefits of ferns?

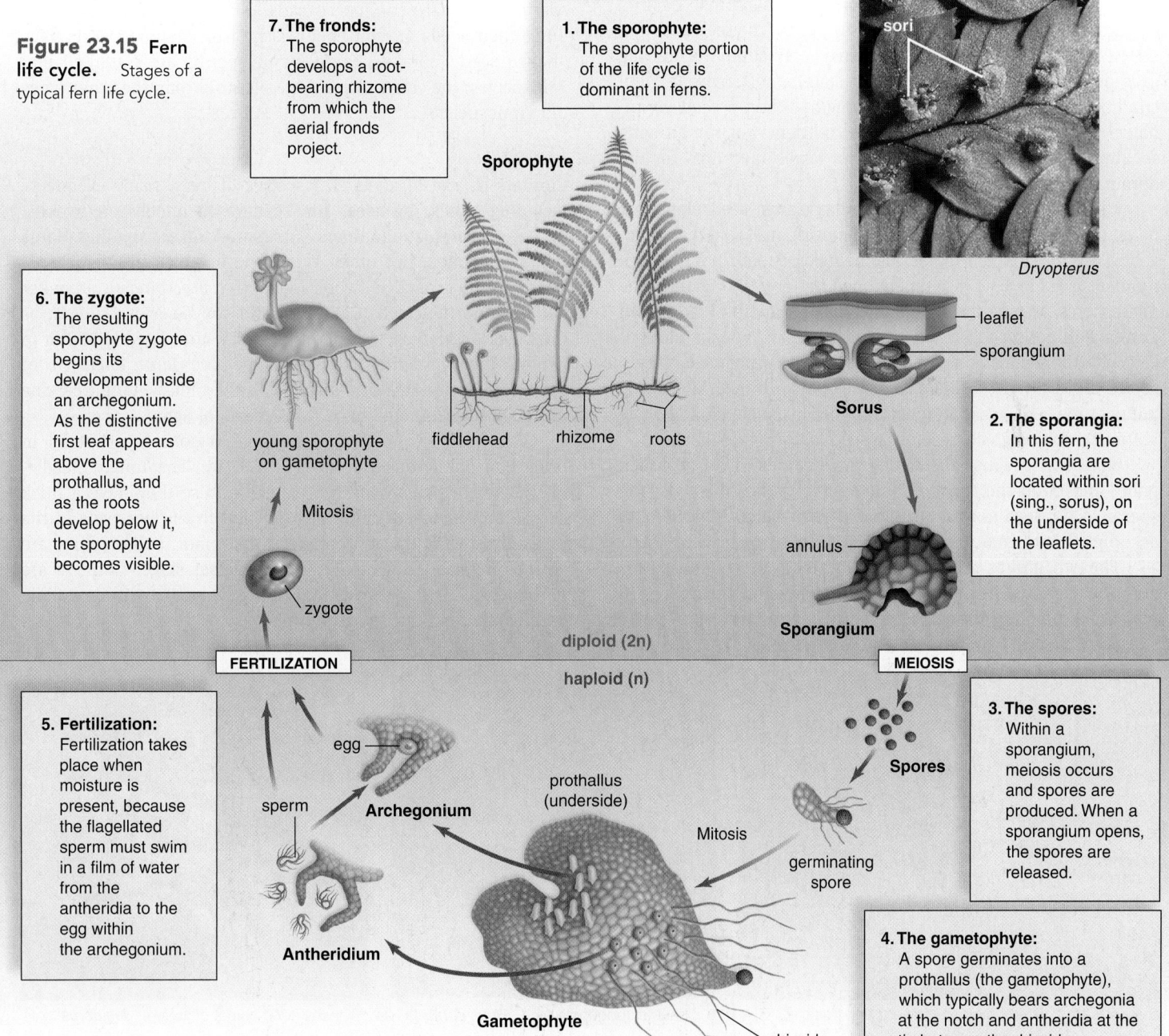

Figure 23.15 Fern life cycle. Stages of a typical fern life cycle.

7. The fronds: The sporophyte develops a root-bearing rhizome from which the aerial fronds project.

1. The sporophyte: The sporophyte portion of the life cycle is dominant in ferns.

sori

Dryopterus

Sporophyte

leaflet
sporangium

Sorus

6. The zygote: The resulting sporophyte zygote begins its development inside an archegonium. As the distinctive first leaf appears above the prothallus, and as the roots develop below it, the sporophyte becomes visible.

young sporophyte on gametophyte

fiddlehead   rhizome   roots

annulus

Sporangium

2. The sporangia: In this fern, the sporangia are located within sori (sing., sorus), on the underside of the leaflets.

Mitosis

zygote

diploid (2n)

MEIOSIS

FERTILIZATION

haploid (n)

3. The spores: Within a sporangium, meiosis occurs and spores are produced. When a sporangium opens, the spores are released.

Spores

5. Fertilization: Fertilization takes place when moisture is present, because the flagellated sperm must swim in a film of water from the antheridia to the egg within the archegonium.

egg

sperm

Archegonium

Antheridium

prothallus (underside)

Mitosis

germinating spore

Gametophyte

rhizoids

4. The gametophyte: A spore germinates into a prothallus (the gametophyte), which typically bears archegonia at the notch and antheridia at the tip between the rhizoids.

23.5 Evolution of Seed Plants: Full Adaptation to Land

Seed plants are vascular plants that use seeds during the dispersal stage of their life cycle. **Seeds** contain a sporophyte embryo and stored food within a protective seed coat. The seed coat and stored food allow an embryo to survive harsh conditions during long periods of dormancy (arrested state) until environmental conditions become favorable for growth. When a seed germinates, the stored food is a source of nutrients for the growing seedling. The survival value of seeds largely accounts for the dominance of seed plants today.

Like a few of the seedless vascular plants, seed plants are heterosporous (have microspores and megaspores), but their innovation was to retain the spores, and not release them into the environment (see Fig. 23.17). As mentioned earlier in this chapter, the microspores become male gametophytes; these are called **pollen grains. Pollination** occurs when a pollen grain is brought into contact with the female gametophyte by wind or a pollinator. Then, sperm move toward the female gametophyte through a growing **pollen tube.** A megaspore develops into a female gametophyte within an **ovule,** which becomes a seed following fertilization.

Note that because the whole male gametophyte (a pollen grain) moves to the female gametophyte, rather than just the sperm as in seedless plants, no external water is needed to accomplish fertilization.

The two groups of seed plants alive today are gymnosperms and angiosperms. In **gymnosperms** (mostly cone-bearing seed plants), the ovules are not completely enclosed by sporophyte tissue at the time of pollination. In **angiosperms** (flowering plants), the ovules are completely enclosed within diploid sporophyte tissue (ovaries), which becomes a fruit.

The first type of seed plant was a woody plant that appeared during the Devonian period; it has been erroneously named a seed fern. The seed ferns of the Devonian were not ferns at all; they were progymnosperms. It's possible that these were the type of progymnosperm that gave rise to today's gymnosperms and angiosperms. All gymnosperms are still woody plants, but whereas the first angiosperms were woody, many today are nonwoody. Progymnosperms, including seed ferns, were part of the Carboniferous swamp forests (see the Evolution feature "Carboniferous Forests" on page 435).

Gymnosperms

The four groups of living gymnosperms [Gk. *gymnos,* naked, and *sperma,* seed] are conifers, cycads, ginkgoes, and gnetophytes. Since their seeds are not enclosed by fruit, gymnosperms have "naked seeds." Today, living gymnosperms are classified into 780 species; the conifers are more plentiful than the other types of gymnosperms.

Conifers

Conifers consist of about 575 species of trees, many evergreen, including pines, spruces, firs, cedars, hemlocks, redwoods, cypresses, yews, and junipers. The name *conifers* signifies plants that bear **cones,** but other gymnosperm phyla are also cone-bearing. The coastal redwood (*Sequoia sempervirens*), a conifer native to northwestern California and southwestern Oregon, is the tallest living vascular plant and may attain nearly 100 m in height. Another conifer, the bristlecone pine (*Pinus longaeva*) of the White Mountains of California, is the oldest living tree; one individual is estimated to be 4,900 years of age.

Vast areas of northern temperate regions are covered in evergreen coniferous forests (Fig. 23.16). The tough, needle-like leaves of pines conserve water because they have a thick cuticle and recessed stomata. Note that in the life cycle of the pine (Fig. 23.17) the sporophyte is dominant, pollen grains are windblown, and the seed is the dispersal stage. Conifers are **monoecious,** which means that a single plant carries both male and female reproductive structures.

a. A northern coniferous forest of evergreen trees

b. Cones of lodgepole pine, *Pinus contorta*

c. Fleshy seed cones of juniper, *Juniperus*

Figure 23.16 Conifers. Various types of conifers and their cones.

Biological Systems

The Uses of Pines

Today, pines are often grown in dense plantations for the purpose of providing wood for construction of all sorts. Although technically a softwood, some pinewoods are actually harder than so-called hardwoods. The foundations of the 125-year-old Brooklyn Bridge are made of Southern yellow pine.

Pines are well known for their beauty and pleasant smell; they make attractive additions to parks and gardens, and a number of dwarf varieties are now available for smaller gardens. Some pines are used as Christmas trees or for Christmas decorations.

Pine needles and the inner bark are rich in vitamins A and C. Pine needles can be boiled to make a tea to ease the symptoms of a cold, and the inner bark of white pines can be used in wound dressings or to provide a medicine for colds and coughs. Large pine seeds, called pine nuts, are sometimes harvested for use in cooking and baking.

Pine oil is distilled from the twigs and needles of Scotch pines and is used to scent a number of household and personal care products, such as room sprays and masculine perfumes. Resin, made by pines as an insect and fungal deterrent, is harvested commercially for a derived product called turpentine.

Question to Consider

1. Explain the impact the extinction of pines would have upon human society.

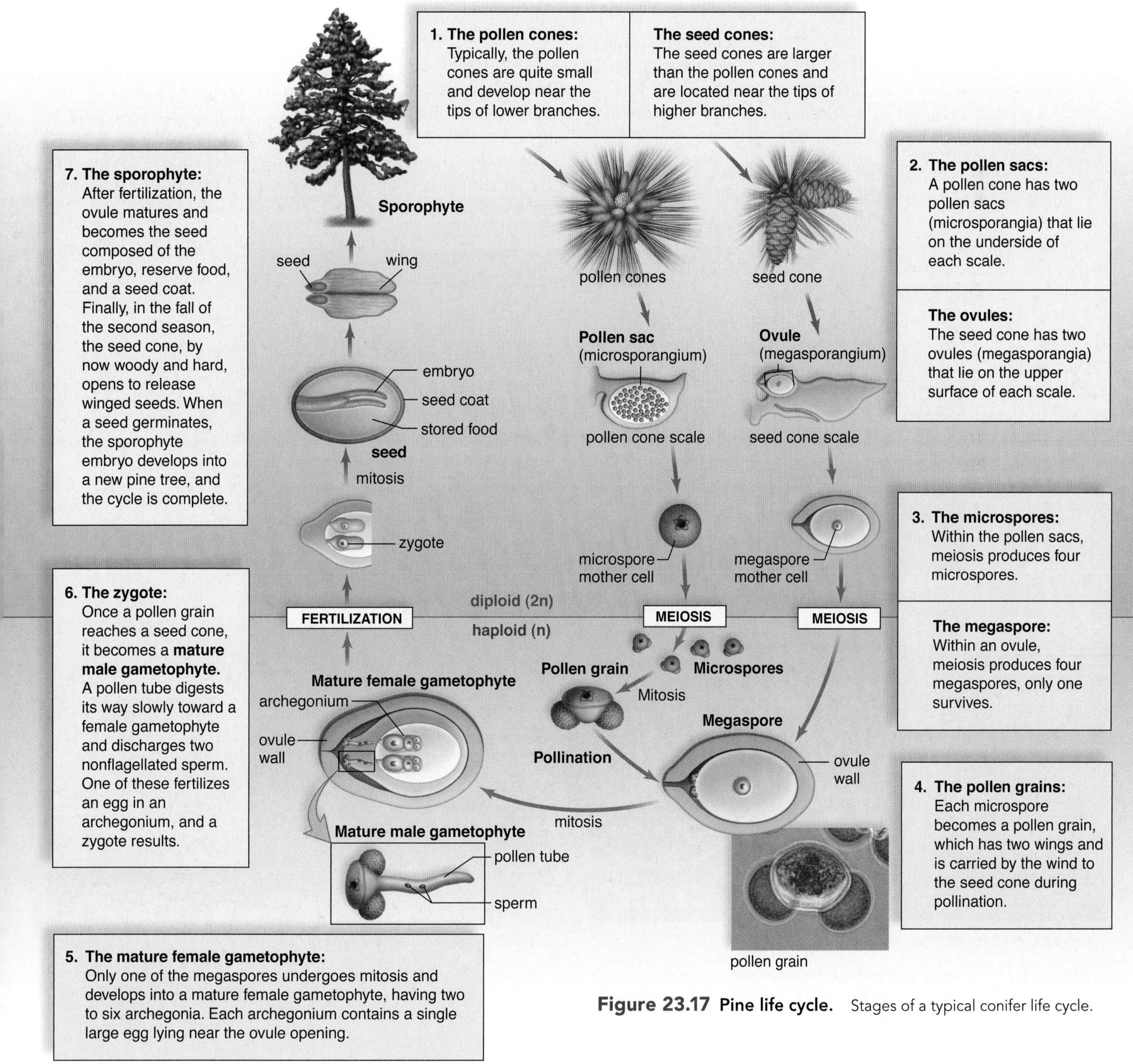

1. The pollen cones: Typically, the pollen cones are quite small and develop near the tips of lower branches.
The seed cones: The seed cones are larger than the pollen cones and are located near the tips of higher branches.
2. The pollen sacs: A pollen cone has two pollen sacs (microsporangia) that lie on the underside of each scale.
The ovules: The seed cone has two ovules (megasporangia) that lie on the upper surface of each scale.
3. The microspores: Within the pollen sacs, meiosis produces four microspores.
The megaspore: Within an ovule, meiosis produces four megaspores, only one survives.
4. The pollen grains: Each microspore becomes a pollen grain, which has two wings and is carried by the wind to the seed cone during pollination.
5. The mature female gametophyte: Only one of the megaspores undergoes mitosis and develops into a mature female gametophyte, having two to six archegonia. Each archegonium contains a single large egg lying near the ovule opening.
6. The zygote: Once a pollen grain reaches a seed cone, it becomes a **mature male gametophyte.** A pollen tube digests its way slowly toward a female gametophyte and discharges two nonflagellated sperm. One of these fertilizes an egg in an archegonium, and a zygote results.
7. The sporophyte: After fertilization, the ovule matures and becomes the seed composed of the embryo, reserve food, and a seed coat. Finally, in the fall of the second season, the seed cone, by now woody and hard, opens to release winged seeds. When a seed germinates, the sporophyte embryo develops into a new pine tree, and the cycle is complete.
Sporophyte
seed
wing
pollen cones
seed cone
Pollen sac (microsporangium)
Ovule (megasporangium)
pollen cone scale
seed cone scale
embryo
seed coat
stored food
seed
mitosis
zygote
microspore mother cell
megaspore mother cell
diploid (2n)
haploid (n)
FERTILIZATION
MEIOSIS
MEIOSIS
Pollen grain
Microspores
Mitosis
Mature female gametophyte
archegonium
ovule wall
Megaspore
Pollination
ovule wall
mitosis
Mature male gametophyte
pollen tube
sperm
pollen grain

Figure 23.17 Pine life cycle. Stages of a typical conifer life cycle.

Cycads

Cycads include 10 genera and 140 species of distinctive gymnosperms. The cycads are native to tropical and subtropical forests. *Zamia pumila,* found in Florida, is the only species of cycad native to North America. Cycads are commonly used in landscaping. One species, *Cycas revoluta,* referred to as the sago palm, is a common landscaping plant. Their large, finely divided leaves grow in clusters at the top of the stem, and therefore they resemble palms or ferns, depending on their height. The trunk of a cycad is unbranched, even if it reaches a height of 15–18 m, as is possible in some species.

Cycads have pollen and seed cones on separate plants. The cones, which grow at the top of the stem surrounded by the leaves, can be huge—more than a meter long with a weight of 40 kg (Fig. 23.18*a*). Cycads exhibit the life cycle of a gymnosperm, except they are pollinated by insects rather than by wind. Also, the pollen tube bursts in the vicinity of the archegonium, and multiflagellated sperm swim to reach an egg.

Cycads were plentiful in the Mesozoic era at the time of the dinosaurs, and it's likely that dinosaurs fed on cycad seeds. Now, cycads are in danger of extinction because they grow very slowly.

Ginkgoes

Although **ginkgoes** are plentiful in the fossil record, they are represented today by only one surviving species, *Ginkgo biloba,* the ginkgo or maidenhair tree. Ginkgoes are **dioecious,** which means that a single plant produces either male or female reproductive structures, but not both (Fig. 23.18*b*). The fleshy seeds, which ripen in the fall, give off such a foul odor that male trees are usually preferred for planting. Ginkgo trees are resistant to pollution and do well along city streets and in city parks. Ginkgo is native to China, and in Asia, ginkgo seeds are considered a delicacy. Extracts from ginkgo trees have been used to improve blood circulation.

Like cycads, the pollen tube of ginkgo bursts to release multiflagellated sperm that swim to the egg produced by the female gametophyte, located within an ovule.

Gnetophytes

Gnetophytes are represented by three living genera and 70 species of plants that are very diverse in appearance. In all gnetophytes, xylem is structured similarly, none have archegonia, and their strobili (cones) have a similar construction. The reproductive structures of some gnetophyte species produce nectar, and insects play a role in the pollination of these species.

a. *Encephalartos transvenosus,* an African cycad

b. *Ginkgo biloba,* a native of China

c. *Ephedra,* a type of gnetophyte

d. *Welwitschia mirabilis,* a type of gnetophyte

Figure 23.18 Three groups of gymnosperms.   **a.** Cycads may resemble ferns or palms but they are cone-producing gymnosperms. **b.** A ginkgo tree has broad leaves and fleshy seeds borne at the end of stalklike megasporophylls. **c.** *Ephedra,* a type of gnetophyte, is a branched shrub. This specimen produces pollen in microsporangia. **d.** *Welwitschia mirabilis,* another type of gnetophyte, produces two straplike leaves that split one to several times.

Gnetum, which occurs in the tropics, consists of trees or climbing vines with broad, leathery leaves arranged in pairs. *Ephedra,* occurring only in southwestern North America and southeast Asia, is a shrub with small, scalelike leaves (Fig. 23.18c). Ephedrine, a medicine with serious side effects, is extracted from *Ephedra. Welwitschia,* living in the deserts of southwestern Africa, has only two enormous, straplike leaves (Fig. 23.18d).

Angiosperms

Angiosperms [Gk. *angion,* dim. of *angos,* vessel, and *sperma,* seed] are the flowering plants. They are an exceptionally large and successful group of plants, with 240,000 known species—six times the number of all other plant groups combined. Angiosperms live in all sorts of habitats, from fresh water to desert, and from the frigid north to the torrid tropics. They range in size from the tiny, almost microscopic duckweed to *Eucalyptus* trees over 100 m tall.

It would be impossible to exaggerate the importance of angiosperms in our everyday lives. Angiosperms include all the hardwood trees of temperate deciduous forests and all the broad-leaved evergreen trees of tropical forests. Also, all herbaceous (nonwoody plants, such as grasses) and most garden plants are flowering plants. This means that all fruits, vegetables, nuts, herbs, and grains, which are the staples of the human diet, are angiosperms. They provide us with clothing, food, medicines, and other commercially valuable products. Over the past 12,000 years, humans have also artificially selected plants to serve us better, resulting in cultivation of plants as food crops and for other uses. The Evolution feature "Evolutionary History of Maize" (page 438) describes corn (*Zea mays*) as an example.

Video Plants

The flowering plants are called angiosperms because their ovules, unlike those of gymnosperms, are always enclosed within diploid tissues. In the Greek derivation of their name, *angio* ("vessel") refers to the ovary, which develops into a fruit, a unique angiosperm feature.

Origin and Radiation of Angiosperms

Although the first fossils of angiosperms are no older than about 135 million years (see Fig. 19.8), the angiosperms probably arose much earlier. Indirect evidence suggests the possible ancestors of angiosperms may have originated as long ago as 200 MYA. But their exact ancestral past has remained a mystery since Charles Darwin pondered it.

To find the angiosperm of today that might be most closely related to the first angiosperms, botanists have turned to DNA comparisons. Gene-sequencing data singled out *Amborella trichopoda* (Fig. 23.19) as having ancestral traits. This small woody shrub, with small cream-colored flowers, lives only on the island of New Caledonia in the South Pacific. Its flowers are only about 4–8 mm wide and the petals and sepals look the same; therefore, they are called tepals. Plants bear either male or female flowers, with a variable number of stamens or carpels.

Although *A. trichopoda* may not be the original angiosperm species, it is sufficiently close that much may be learned from

Figure 23.19 *Amborella trichopoda.* Molecular data suggest this plant is most closely related to the first flowering plants.

studying its reproductive biology. Botanists hope that this knowledge will help them understand the early adaptive radiation of angiosperms during the Tertiary period. The gymnosperms were abundant during the Mesozoic era but declined during the mass extinction that occurred at the end of the Cretaceous Period (145.5–65.5 MYA). Angiosperms survived and went on to become the dominant plants during modern times.

Monocots and Eudicots

Most flowering plants belong to one of two classes. These classes are the **Monocotyledones,** often shortened to simply the **monocots** (about 65,000 species), and the **Eudicotyledones,** shortened to **eudicots** (about 175,000 species). The term *eudicot* (meaning true dicot) is more specific than the term *dicot.* It was discovered that some of the plants formerly classified as dicots diverged before the evolutionary split that gave rise to the two major classes of angiosperms. These earlier-evolving plants are not included in the designation eudicots.

Monocots are so called because their seeds have only one cotyledon. **Cotyledons** [Gk. *kotyledon,* cuplike cavity] are the "seed leaves" that contain nutrients that nourish the plant embryo. Several common monocots include corn, tulips, pineapples, bamboos, and sugarcane. Eudicot seeds possess two cotyledons. Several common eudicots include cacti, strawberries, dandelions, poplars, and beans. Table 23.2 lists several fundamental features found in monocots and eudicots.

Table 23.2 Key Features of Monocots and Eudicots

Monocots	Eudicots
One cotyledon	Two cotyledons
Flower parts in threes or multiples of three	Flower parts in fours or fives or multiples of four or five
Pollen grain with one pore	Pollen grain with three pores
Usually herbaceous	Woody or herbaceous
Usually parallel venation	Usually net venation
Scattered bundles in stem	Vascular bundles in a ring
Fibrous root system	Taproot system

a. Monocot

b. Eudicot

Figure 23.20 Flower diversity.   Regardless of size and shape, flowers, such as this **(a)** monocot, and **(b)** eudicot, share certain features.

The Flower

Although **flowers** vary widely in appearance (Fig. 23.20), most have certain structures in common. The **peduncle,** a flower stalk, expands slightly at the tip into a **receptacle,** which bears the other flower parts. These parts, called sepals, petals, stamens, and carpels, are attached to the receptacle in four whorls (Fig. 23.21).

1. The **sepals,** collectively called the calyx, protect the flower bud before it opens. The sepals may drop off or may be colored like the petals. Usually, however, sepals are green and remain attached to the receptacle.

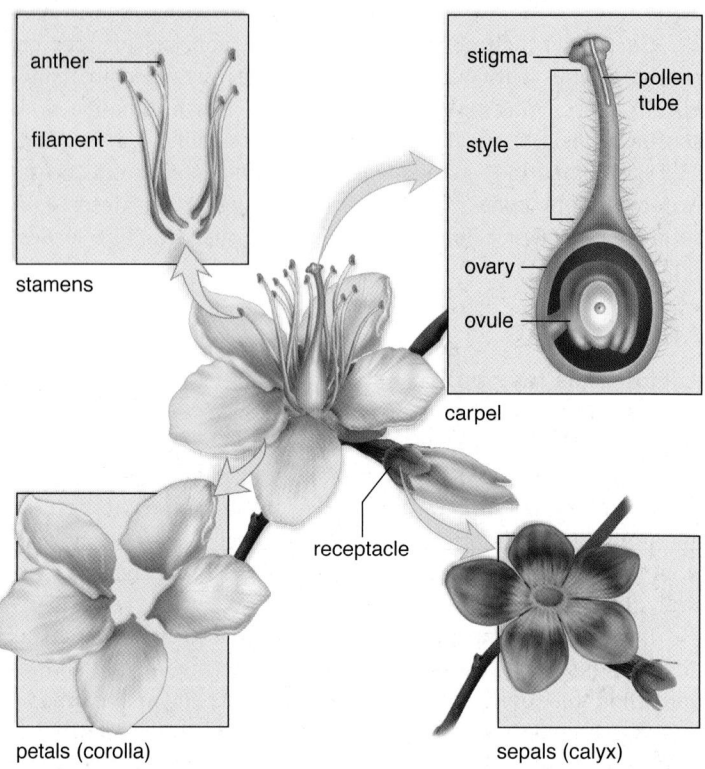

Figure 23.21 Generalized flower.   A flower has four main parts: sepals, petals, stamens, and carpels. A stamen has an anther and filament. A carpel has a stigma, style, and ovary. An ovary contains ovules.

2. The **petals,** collectively called the corolla, are quite diverse in size, shape, and color. The petals are often used to attract a particular pollinator.
3. Next are the **stamens.** Each stamen consists of two parts: a saclike container called the anther, and a slender stalk called the filament. Pollen grains develop from microspores produced in the anther.
4. At the very center of a flower is the **carpel,** a vaselike structure with three major regions: the **stigma,** an enlarged sticky knob; the **style,** a slender stalk; and the **ovary,** an enlarged base that encloses one or more ovules. The ovule becomes the seed, and the ovary becomes the fruit. Fruit is often instrumental in the distribution of seeds.

Note that not all flowers have all these parts (Table 23.3). A flower is said to be *complete* if it has all four parts; otherwise it is termed *incomplete.*

Table 23.3 Other Flower Terminology

Term	Type of Flower
Complete	All four parts (sepals, petals, stamens, and carpels) present
Incomplete	Lacks one or more of the four parts
Perfect	Has both stamens and (a) carpel(s)
Imperfect	Has stamens or (a) carpel(s), but not both
Inflorescence	A cluster of flowers
Composite	Appears to be a single flower but consists of a group of tiny flowers

Evolution

Carboniferous Forests

Our industrial society runs on fossil fuels such as coal. The term *fossil fuel* might seem odd at first until one realizes that it refers to the remains of organic material from ancient times. During the Carboniferous period more than 300 million years ago, a great swamp forest (Fig. 23A) encompassed what is now northern Europe, the Ukraine, and the Appalachian Mountains in the United States. The weather was warm and humid, and the plants grew very tall. These were not plants that would be familiar to us; instead, they are related to various groups of plants that are less well known— the lycophytes, horsetails, and ferns.

Lycophytes today may stand as high as 30 cm, but their ancient relatives were 35 m tall and 1 m wide. The stroboli were up to 30 cm long, and some had leaves more than 1 m long. Horsetails too—at 18 m tall—were giants compared to today's specimens. Tree ferns were also taller than tree ferns found in the tropics today. The progymnosperms, including "seed ferns," were significant plants of a Carboniferous swamp.

The amount of biomass in a Carboniferous swamp forest was enormous, and occasionally the swampy water rose, covering the plants that had died. Dead plant material under water does not decompose well. The partially decayed remains became covered by sediment that changed them into sedimentary rock. Exposed to significant amounts of pressure and over long periods of time, the organic material became coal. This process continued for millions of years, resulting in immense deposits of coal. Geological upheavals raised the deposits to the levels where they can be mined today.

With a change of climate, many of the plants of the Carboniferous period became extinct. Some of their smaller herbaceous relatives have evolved and have survived to our time. We owe the industrialization of today's society to these ancient forests.

Questions to Consider

1. How might a geologist use this information to look for new sources of coal?
2. Why is coal not considered to be a renewable resource?

Fossil seed ferns

Figure 23A Swamp forest of the Carboniferous period.   Nonvascular plants and early gymnosperms dominated the swamp forests of the Carboniferous period. Among the early gymnosperms were the seed ferns, so named because their leaves looked like fronds, as shown in a micrograph of fossil remains in the upper right.

lycophytes

horsetail

seed fern

progymnosperm

fern

Flowering Plant Life Cycle

Figure 23.22 depicts the life cycle of a typical flowering plant. Like the gymnosperms, flowering plants are heterosporous, producing two types of spores. A megaspore located in an ovule within an ovary of a carpel develops into an egg-bearing female gametophyte called the embryo sac. In most angiosperms, the embryo sac has seven cells; one of these is an egg, and another contains two polar nuclei. (These two nuclei are called the polar nuclei because they came from opposite ends of the embryo sac.)

Microspores, produced within anthers, become pollen grains that, when mature, are sperm-bearing male gametophytes. The full-fledged mature male gametophyte consists of only three cells: the tube cell and two sperm cells.

Figure 23.22 Flowering plant life cycle. The parts of the flower involved in reproduction are the stamens and the carpel. Reproduction has been divided into significant stages of female gametophyte development, male gametophyte development, and also significant stages of sporophyte development.

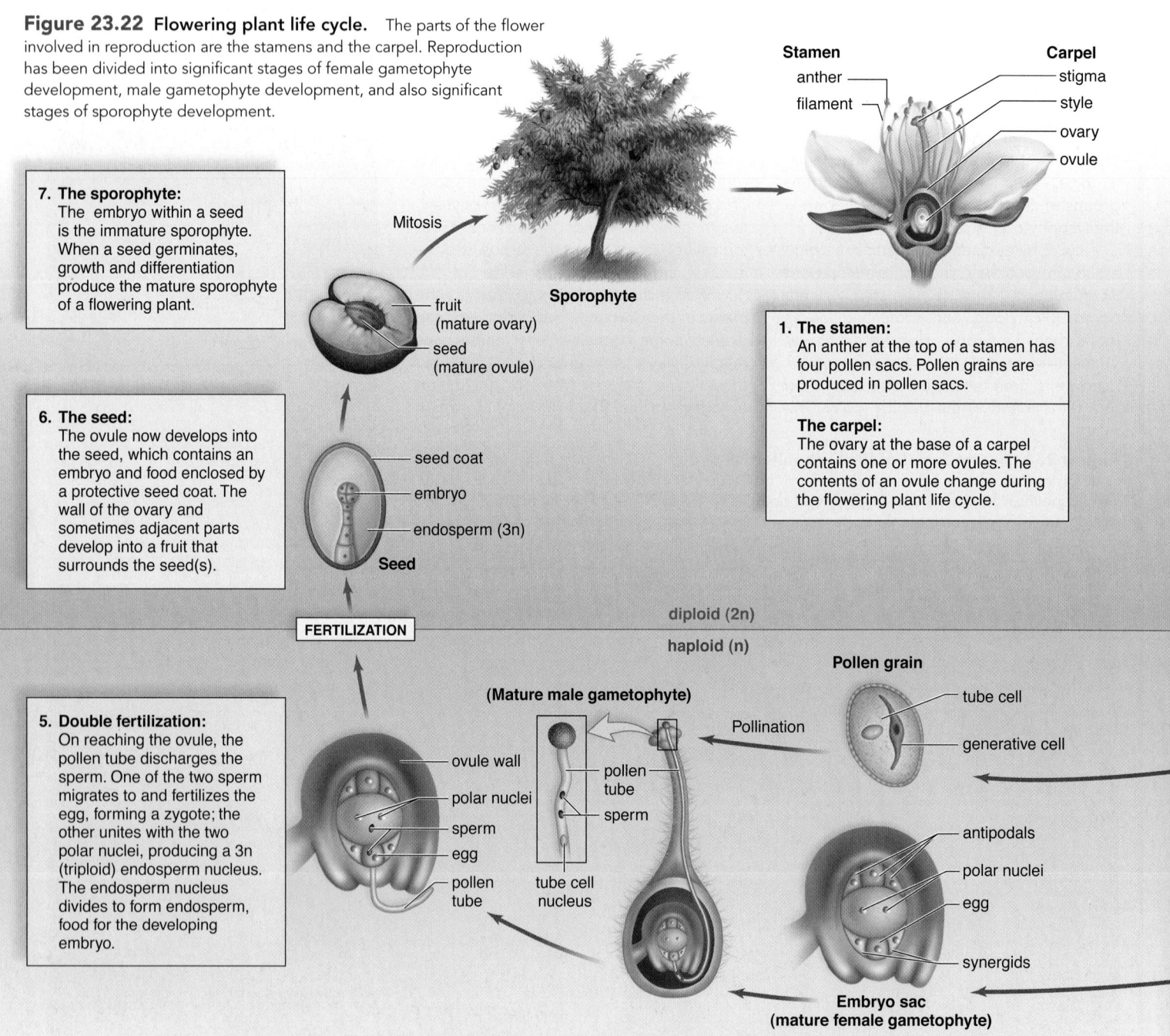

7. The sporophyte:
The embryo within a seed is the immature sporophyte. When a seed germinates, growth and differentiation produce the mature sporophyte of a flowering plant.

6. The seed:
The ovule now develops into the seed, which contains an embryo and food enclosed by a protective seed coat. The wall of the ovary and sometimes adjacent parts develop into a fruit that surrounds the seed(s).

5. Double fertilization:
On reaching the ovule, the pollen tube discharges the sperm. One of the two sperm migrates to and fertilizes the egg, forming a zygote; the other unites with the two polar nuclei, producing a 3n (triploid) endosperm nucleus. The endosperm nucleus divides to form endosperm, food for the developing embryo.

1. The stamen:
An anther at the top of a stamen has four pollen sacs. Pollen grains are produced in pollen sacs.

The carpel:
The ovary at the base of a carpel contains one or more ovules. The contents of an ovule change during the flowering plant life cycle.

Mitosis

Sporophyte

fruit (mature ovary)
seed (mature ovule)

seed coat
embryo
endosperm (3n)
Seed

FERTILIZATION

diploid (2n)
haploid (n)

Stamen
anther
filament

Carpel
stigma
style
ovary
ovule

Pollen grain
tube cell
generative cell

(Mature male gametophyte)
Pollination

ovule wall
polar nuclei
sperm
egg
pollen tube

pollen tube
sperm

tube cell nucleus

antipodals
polar nuclei
egg
synergids

Embryo sac
(mature female gametophyte)

4. The mature male gametophyte:
A pollen grain that lands on the carpel of the same type of plant germinates and produces a pollen tube, which grows within the style until it reaches an ovule in the ovary. Inside the pollen tube, the generative cell nucleus divides and produces two nonflagellated sperm. A fully germinated pollen grain is the mature male gametophyte.

The mature female gametophyte:
The ovule now contains the mature female gametophyte (embryo sac), which typically consists of eight haploid nuclei embedded in a mass of cytoplasm. The cytoplasm differentiates into cells, one of which is an egg and another of which contains two polar nuclei.

During pollination, a pollen grain is transported by various means from the anther to the stigma of a carpel, where it germinates. During germination, the tube cell produces a pollen tube. The pollen tube carries the two sperm to the micropyle (small opening) of an ovule. Flowering plants undergo **double fertilization:** one sperm unites with an egg, forming a diploid zygote, and the other unites with polar nuclei, forming a triploid endosperm nucleus (see Chapter 27).

Ultimately, the ovule becomes a seed that contains the embryo (the sporophyte of the next generation) and stored food enclosed within a seed coat. Endosperm in some seeds is absorbed by the cotyledons, whereas in other seeds endosperm is digested as the seed matures.

A **fruit** is derived from an ovary, and in some instances it is an accessory part of the flower. Some fruits, such as apples and tomatoes, provide a moist, fleshy covering; other fruits, such as pea pods and acorns, provide a dry covering for seeds.

Flowers and Diversification

Flowers are involved in the production and development of spores, gametophytes, gametes, and embryos enclosed within seeds. Successful completion of sexual reproduction in angiosperms requires the effective dispersal of pollen and then seeds. The various ways pollen and seeds can be dispersed have resulted in many different types of flowers (see Chapter 27).

Wind-pollinated flowers are usually not showy, whereas insect-pollinated flowers and bird-pollinated flowers are often colorful. Night-blooming flowers attract nocturnal mammals or insects; these flowers are usually aromatic and white or cream-colored.

Although some flowers disperse their pollen by wind, many are adapted to attract specific pollinators, such as bees, wasps, flies, butterflies, moths, and even bats, that carry pollen from one flower to another flower of the same type. For example, glands located in the region of the ovary produce nectar, a nutrient that is gathered by pollinators as they go from flower to flower. Bee-pollinated flowers are usually blue or yellow and have ultraviolet shadings that lead the pollinator to the location of nectar. The mouthparts of bees are fused into a long tube that is able to obtain nectar from the base of the flower.

Video
Pollinators

The fruits of flowers protect and aid in the dispersal of seeds. Dispersal occurs when seeds are transported by wind, gravity, water, and animals to another location. Fleshy fruits may be eaten by animals, which transport the seeds to a new location and then deposit them when they defecate. Because animals live in particular habitats and/or have particular migration patterns, they are apt to deliver the fruit-enclosed seeds to a suitable location for seed germination (when the embryo begins to grow again) and development of the plant.

Video
Dung Seed Dispersal

Video
Fruit Bat Seed Dispersal

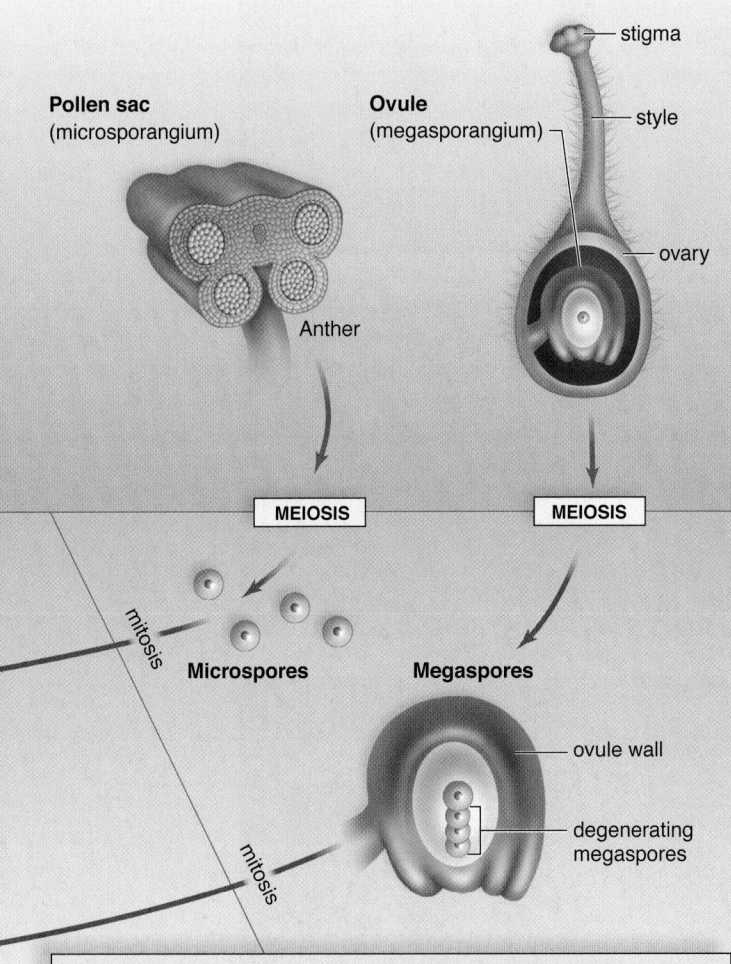

2. The pollen sacs:
In pollen sacs (microsporangia) of the anther, meiosis produces microspores.

The ovules:
In an ovule (megasporangium) within an ovary, meiosis produces four megaspores.

Pollen sac
(microsporangium)

Ovule
(megasporangium)

stigma

style

ovary

Anther

MEIOSIS

MEIOSIS

mitosis

Microspores

Megaspores

mitosis

ovule wall

degenerating megaspores

3. The microspores:
Each microspore in a pollen sac undergoes mitosis to become an immature pollen grain with two cells: the tube cell and the generative cell. The pollen sacs open, and the pollen grains are windblown or transported by an animal carrier, usually to other flowers. This is known as pollination.

The megaspores:
Inside the ovule of an ovary, three megaspores disintegrate, and only the remaining one undergoes mitosis to become a female gametophyte.

Check Your Progress   23.5

1. List the life cycle changes that have enabled pines to better adapt to life on land.
2. Compare and contrast the four types of gymnosperms.
3. List the function of the key structures required for angiosperm reproduction.
4. Identify which groups of plants produce naked seeds by citing specific examples.
5. Compare and contrast animal-pollinated plants to wind-pollinated plants.

Evolution

Evolutionary History of Maize

Approximately 12,000 years ago, humans began shifting away from a hunter-gatherer society and moving toward an agriculturally based one. As early groups began to spend more time in the same location, they would have required a larger and more readily available source of food. This would have prompted them to select for plants that would produce a greater yield of food with less investment of time and energy.

Maize, which we commonly call corn, was first cultivated in Central America about 7,000 years ago in the central highlands of Mexico. By the time Europeans were exploring Central America, over 300 varieties were in existence. We now typically grow six major varieties of corn: sweet, pop, flour, dent, pod, and flint (Fig. 23B). All of these varieties can trace their ancestral lineage back to a wild Mexican grass known as teosinte.

One of the main differences between teosinte and maize is that the kernels of teosinte are enveloped in a hardened protective casing, while the kernels of maize are exposed on the surface of the ear (Fig. 23C). The stony fruitcase of teosinte was one of the biggest obstacles that plant breeders had to overcome in order to provide a seed that would be more readily available for consumption.

sweet

pop

flour

dent

pod

flint

Figure 23B Modern varieties of corn.   These six types of corn are most commonly grown for human purposes.

Molecular evidence indicates that changes in only a few genes can introduce dramatic changes in the teosinte/maize plant's phenotype during the domestication process. Researchers have begun to identify key genes that control some of the critical morphological changes associated with the domestication process. *Teosinte branched1* (*tb1*) was identified as a gene that controls the apical dominance between maize and teosinte, and *teosinte glume architecture1* (*tga1*) controls the formation of the hard casing that surrounds the seeds of teosinte. Small mutations in these genes could have led to traits that were more desirable to the early human groups who were starting to cultivate teosinte.

So how did early hunter-gatherers manage to artificially select for these traits? Hunter-gatherers often followed seasonal migration routes between base camps, returning to the same locations year after year. The disturbance of the natural vegetation at these sites would have provided an ideal site for the colonization of species that were being cultivated by these groups. The seeds of one year's crop would have provided the starting point for the next year's crop. If seeds were collected from the plants with the most desirable traits, then over time the frequency of plants with specific traits would increase.

Because of the selective nature of harvesting the plants with the best traits, however, the genetic diversity within the population would have begun to decline. This decline would have produced a genetic bottleneck effect, in which various traits were lost while others increased in frequency, giving rise to a change in the phenotypic frequency within the population as well (see Chapter 16).

Over the seven thousand years of maize domestication, the plant has become completely dependent on humans for continued successful reproduction. This domesticated crop produces larger grain, generally a more robust plant, an increased amount of apical growth, and a loss of natural seed dispersal so that seeds remain attached to the plant more firmly than they do for its wild ancestor.

As maize has continued to evolve into the more familiar crop of modern society, humans have become more and more dependent upon it for our survival as well. Only in the past several decades have humans reached the point of altering maize through genetic engineering to select for unique traits (Fig. 23D). Genetic engineering has opened up new windows of evolutionary options for the future of maize and with that the future of the human species.

Questions to Consider

1. What potentially beneficial traits may have been lost during artificial selection in maize?
2. What other food plants might have undergone a similar form of natural selection?

Figure 23C Teosinte and maize. Teosinte has a stony fruitcase compared to the softer kernels of modern corn.

teosinte   hybrid of teosinte and maize   maize

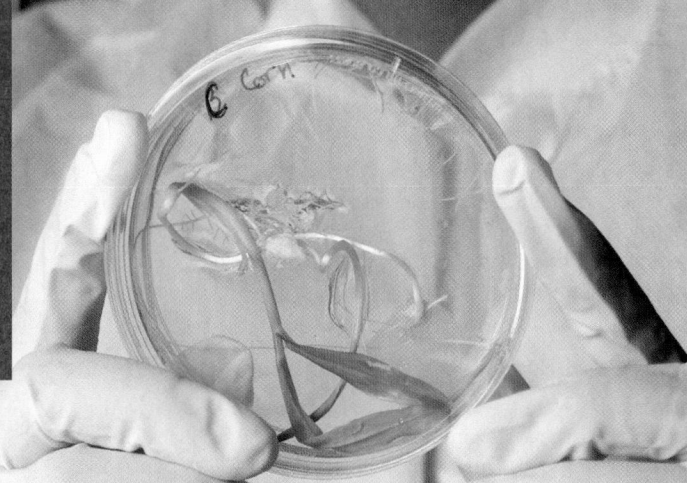

Figure 23D Genetic engineering. Genetic engineering is responsible for the majority of corn grown worldwide today.

CONNECTING *the* CONCEPTS *with the* BIG IDEAS

Evolution

- Phylogenetic trees and cladograms model the possible evolutionary history of groups of living organisms. (1B2a-d)
- The mechanism of polyploidy allows plants to achieve rapid speciation. (1C2b)

*Find the unabridged version of all EK citations at www.glencoe.com/maderAP11.

Media Study Tools

www.glencoe.com/maderAP11

Enhance your study of this chapter with study tools and practice tests. Also ask your instructor about the resources available through ConnectPlus, including the media-rich eBook, interactive learning tools, and animations.

Summarize

23.1 The Green Algal Ancestor of Plants

Land plants evolved from a common ancestor with multicellular, freshwater algae about 450 MYA. During the evolution of plants, adaptations for successful colonization of land included protecting the embryo, apical growth, development of vascular tissue, development of megaphylls, dispersal of offspring by seeds, and the production of flowers.

23.2 Evolution of Bryophytes: Colonization of Land

Ancient bryophytes were the first plants to colonize land. Liverworts, hornworts, and mosses are examples of bryophytes, which lack well-developed vascular tissue. The sporophyte is nutritionally dependent on the gametophyte because the gametophyte is larger and photosynthetic. The life cycle of mosses demonstrates the reproductive strategies of flagellated sperm and dispersal by means of windblown spores.

23.3 Evolution of Lycophytes: Vascular Tissue

Vascular plants, such as rhyniophytes, evolved during the Silurian period. The sporophytes contain two types of conducting tissues. Xylem is specialized to conduct water and dissolved minerals whereas phloem is specialized to conduct organic nutrients. Lycophytes are descended from these first plants and also contain vascular tissue.

Ancient lycophytes were the first plants to have leaves known as microphylls. The lycophyte life cycle is similar to that found in ferns.

23.4 Evolution of Pteridophytes: Megaphylls

In the pteridophytes (ferns and their allies, horsetails and whisk ferns) and the lycophytes, the sporophyte is the dominant stage of the life cycle and is separate from the tiny gametophyte. Windblown spores are the dispersal agents for these plants. The ferns found on Earth today have an obvious megaphyll; horsetails and whisk ferns have reduced megaphylls.

"Seedless vascular" is a description that applies to lycophytes, ferns, and fern allies that grew to enormous sizes during the Carboniferous period. Today, the seedless vascular plants that live in temperate zones use asexual propagation to spread into environments that are not favorable to the water-dependent gametophyte generation.

23.5 Evolution of Seed Plants: Full Adaptation to Land

Seed plants also have an alternation of generations, but they are heterosporous, producing both microspores and megaspores. Microspores become the windblown or animal-transported male gametophytes—the pollen grains. Pollen grains carry sperm to the female gametophyte (megaspore), which is the ovule. Following fertilization, the ovule becomes the seed, which contains a sporophyte embryo. Fertilization no longer requires external water, and sexual reproduction is fully adapted to the terrestrial environment.

The gymnosperms (cone-bearing plants) and also possibly angiosperms (flowering plants) evolved from woody seed ferns during the Devonian period. The conifers, represented by the pine tree, exemplify the traits of these plants. Gymnosperms have "naked seeds" because they are not enclosed by fruit, as are those of flowering plants.

A woody shrub, *Amborella trichopoda*, has been identified as most closely related to the common ancestor for the angiosperms. In angiosperms, the reproductive organs are found in flowers. After fertilization, the ovules become seeds, which are located in the ovary. This ultimately becomes the fruit. Therefore, angiosperms have "covered seeds."

In many angiosperms, pollen is transported from flower to flower by various pollinators. Both flowers and fruits are found only in angiosperms and may account for the extensive colonization of terrestrial environments by the flowering plants.

Key Terms

alternation of generations 422	Monocotyledone 433
angiosperm 430, 433	monoecious 430
antheridium 422	moss 424
apical 421	node 421
archegonium 422	nonvascular plant 423
bryophyte 423	ovary 434
carpel 434	ovule 430
charophyte 421	peduncle 434
cone 430	petal 434
conifer 430	phloem 426
cotyledon 433	plant 420
cuticle 422	pollen grain 430
cycad 432	pollen tube 430
dioecious 432	pollination 430
double fertilization 437	pteridophyte 427
embryophyta 422	receptacle 434
epiphytes 428	rhizoid 424
eudicot 433	rhizome 426
Eudicotyledone 433	seed 430
fern 428	seed plant 430
flower 434	seedless vascular plant 426
frond 428	sepal 434
fruit 437	sporangia (sing., sporangium) 422
gametophyte 422	spore 422
ginkgo 432	sporophyll 426
gnetophyte 432	sporophyte 422
gymnosperm 430	stamen 434
heterosporous 426	stigma 434
homosporous 426	stomata (sing., stoma) 422
hornwort 424	strobilus (pl., strobili) 426
horsetail 427	style 434
lignin 426	vascular plant 426
liverwort 424	vascular tissue 423
lycophyte 426	whisk fern 428
megaphyll 427	whorl 421
megaspore 426	xylem 426
microphyll 426	
microspore 426	
monocot 433	

Assess

Reviewing This Chapter

1. Refer to Figure 23.1, and trace the evolutionary history of land plants. What traits do charophytes have that are shared by land plants? 420
2. What is meant when it is said that a plant alternates generations? Distinguish between a sporophyte and a gametophyte. 422
3. Describe the various types of bryophytes and the life cycle of mosses. Discuss the ecological and commercial importance of mosses. 423–24
4. When do vascular plants appear in the fossil record, and why do lycophytes perhaps resemble the first vascular plants? Mention the importance of branching. 426
5. Draw a diagram to describe the life cycle of a fern, pointing out significant features. What are the human uses of ferns? 428–29

6. What features do all seed plants have in common? When do seed plants appear in the fossil record? 430
7. List and describe the four phyla of gymnosperms. What are the human uses of gymnosperms? 430–33
8. Use a diagram of the pine life cycle to point out significant features, including those that distinguish a seed plant's life cycle from that of a seedless vascular plant. 431
9. What is known about the ancestry of flowering plants? 433
10. How do monocots and eudicots differ? What are the parts of a flower? 433–34
11. Use a diagram to explain and point out significant features of the flowering plant life cycle. 436–37
12. Offer an explanation as to why flowering plants are the dominant plants today. 437

Testing Yourself

Choose the best answer for each question.

1. Which of these are characteristics of land plants?
   a. multicellular with specialized tissues and organs
   b. photosynthetic and contain chlorophylls *a* and *b*
   c. protect the developing embryo from desiccation
   d. have an alternation-of-generations life cycle
   e. All of these are correct.

2. In bryophytes, sperm usually move from the antheridium to the archegonium by
   a. swimming.   d. worm pollination.
   b. flying.   e. bird pollination.
   c. insect pollination.

3. Ferns have
   a. a dominant gametophyte generation.
   b. vascular tissue.
   c. seeds.
   d. Both a and b are correct.
   e. Choices a, b, and c are correct.

4. The spore-bearing structure that gives rise to a female gametophyte in seed plants is called a
   a. microphyll.   d. microsporangium.
   b. spore.   e. sporophyll.
   c. megasporangium.

5. A small, upright plant that resembles a tiny upright pine tree with club-shaped strobili and microphylls is a
   a. whisk fern.   d. horsetail.
   b. lycophyte.   e. fern.
   c. conifer.

6. Trends in the evolution of plants include all of the following except
   a. from homospory to heterospory.
   b. from less to more reliance on water for life cycle.
   c. from nonvascular to vascular.
   d. from nonwoody to woody.

7. Gymnosperms
   a. have flowers.
   b. are eudicots.
   c. are monocots.
   d. do not have spores in their life cycle.
   e. reproduce by seeds.

8. In the moss life cycle, the sporophyte
   a. consists of leafy green shoots.
   b. is the heart-shaped prothallus.
   c. consists of a foot, a stalk, and a capsule.
   d. is the dominant generation.
   e. All of these are correct.

9. Microphylls
   a. have a single strand of vascular tissue.
   b. evolved before megaphylls.
   c. evolved as extensions of the stem.
   d. are found in lycophytes.
   e. All of these are correct.

10. How are ferns different from mosses?
    a. Only ferns produce spores as dispersal agents.
    b. Ferns have vascular tissue.
    c. In the fern life cycle, the gametophyte and sporophyte are both independent.
    d. Ferns do not have flagellated sperm.
    e. Both b and c are correct.

11. Which of these pairs is mismatched?
    a. pollen grain—male gametophyte
    b. ovule—female gametophyte
    c. seed—immature sporophyte
    d. pollen tube—spores
    e. tree—mature sporophyte

12. In the life cycle of the pine tree, the ovules are found on
    a. needlelike leaves.   d. root hairs.
    b. seed cones.   e. All of these are correct.
    c. pollen cones.

13. Monocotyledonous plants often have
    a. parallel leaf venation.
    b. flower parts in units of four or five.
    c. leaves with petioles only.
    d. flowers with stipules.
    e. Choices b, c, and d are correct.

14. Which of these pairs is mismatched?
    a. anther—produces microspores
    b. carpel—produces pollen
    c. ovule—becomes seed
    d. ovary—becomes fruit
    e. flower—reproductive structure

15. Which of these plants contributed the most to our present-day supply of coal?
    a. bryophytes   d. angiosperms
    b. seedless vascular plants   e. Both b and c are correct.
    c. gymnosperms

16. Which of these is found in seed plants?
    a. complex vascular tissue
    b. pollen grains that are not flagellated
    c. retention of female gametophyte within the ovule
    d. roots, stems, and leaves
    e. All of these are correct.

17. Which of these is a seedless vascular plant?
    a. gymnosperm   d. monocot
    b. angiosperm   e. eudicot
    c. fern

18. Label this diagram of alternation-of-generations life cycle.

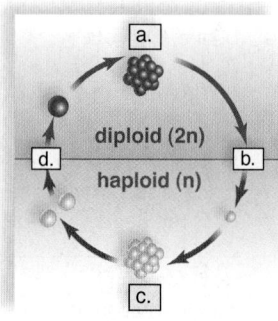

Engage

Virtual Lab
Classifying Using
Biotechnology

The virtual lab "Classifying Using Biotechnology" provides an interactive look at how scientists analyze molecular data to establish evolutionary relationships.

Thinking Scientifically

1. Using as many terms as necessary (from both X and Y axes), fill in the proposed phylogenetic tree for vascular plants.

	ferns	conifers	ginkgoes	monocots	eudicots
vascular tissue	X	X	X	X	X
seed plants		X	X	X	X
naked seeds		X	X		
needlelike leaves		X			
fan-shaped leaves			X		
enclosed seeds				X	X
one embryonic leaf				X	
two embryonic leaves					X

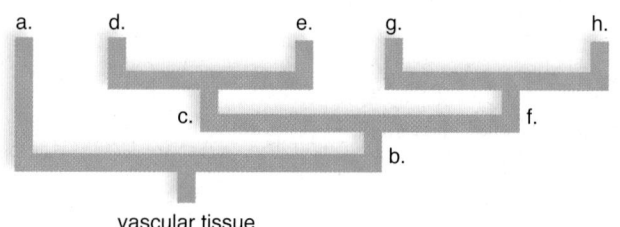

vascular tissue

2. Using Figure 23.1, distinguish between the (a) microphyll and the (b) megaphyll clade.

Bioethical Issue
Saving Plant Species

Pollinator populations have been decimated by pollution, pesticide use, and destruction or fragmentation of natural areas. Belatedly, we have come to realize that many types of bees are responsible for pollinating cash crops such as blueberries, cranberries, and squash, and are partly responsible for pollinating apple, almond, and cherry trees.

Why are we so shortsighted when it comes to protecting the environment and living creatures like pollinators? The reason may be that pollinators are a resource held in common. The term *commons* originally meant a piece of land where all members of a village were allowed to graze their cattle. The farmer who thought only of himself and grazed more cattle than his neighbor was better off. The difficulty is, of course, that eventually the resource is depleted, and everyone loses.

So, a farmer or property owner who uses pesticides is thinking only of his or her field or lawn, and not the good of the whole. The commons can only be protected if citizens have the foresight to enact rules and regulations by which all abide. DDT was outlawed in this country in part because it led to the decline of birds of prey. Similarly, we may need legislation to protect pollinators from factors that kill them off. Legislation to protect pollinators would protect the food supply for all of us.

The various components, or organs, of flowering plants consist of leaves, roots, stems, and flowers.

24

Flowering Plants: Structure and Organization

A stunning array of plant life covers the Earth, and over 80% of all living plants are flowering plants, or angiosperms. The organization of flowering plants allows them to efficiently run photosynthesis in a variety of terrestrial environments. The elevated leaves have a shape that facilitates absorption of solar energy and carbon dioxide. Strong stems conduct water up to the leaves from the roots, which not only anchor the plant but also absorb water and minerals. All the vegetative organs of flowering plants have a role to play in photosynthesis.

When plants photosynthesize, they take in large amounts of CO_2, storing it as carbohydrates. Therefore, keeping our world green by preserving plants, particularly forests, is a way to remove CO_2 from the atmosphere and reduce the dangers of climate change. In this chapter we consider the structure of roots, stems, and leaves. The following two chapters cover physiology and nutrition of plants, and Chapter 27 deals with reproduction.

As you read through the chapter, think about the following questions:

1. What structural features are necessary for the function of flowering plants?
2. How do monocot and eudicot structural differences adapt them to their various roles in the environment?

CHAPTER OUTLINE

BEFORE YOU BEGIN

Before beginning this chapter, take a few moments to review the following discussions.

Figure 4.7 What cellular structures are necessary for plant cells to function?

Section 7.2 What are the reactants, intermediates, and end products of photosynthesis?

Section 23.5 What structural features helped promote angiosperm success?

FOLLOWING *the* BIG IDEAS

CHAPTER 24   FLOWERING PLANTS: STRUCTURE AND ORGANIZATION

Energy and Homeostasis	Important adaptations including increased surface area and specialized exterior surfaces benefit plants.
Interactions and Systems	The environment affects all aspects of a plant's life; adaptations allow them to be successful in dramatically varying climates.

24.1 Organs of Flowering Plants

From cacti living in hot deserts to water lilies growing in a nearby pond, the flowering plants, or angiosperms, are extremely diverse. Despite their great diversity in size and shape, flowering plants share many common structural features. Most flowering plants possess a root system and a shoot system (Fig. 24.1). The **root system** simply consists of the roots, while the **shoot system** consists of the stem and leaves. A typical plant features three vegetative **organs** (structures that contain different tissues and perform one or more specific functions) that allow them to live and grow. The roots, stems, and leaves are the vegetative organs. *Vegetative organs* are concerned with growth and nutrition and not with reproduction. Flowers, seeds, and fruits are structures involved in reproduction (Chapter 27).

Roots

The root system in the majority of plants is located underground. As a rule of thumb, the root system is at least equivalent in size and extent to the shoot system. An apple tree has a much larger root system than a corn plant, for example. A single corn plant may have roots as deep as 2.5 m and spread out over 1.5 m; in contrast, a mesquite tree that lives in the desert may have roots that penetrate to a depth of over 20 m.

The extensive root system of a plant anchors it in the soil and gives it support (Fig. 24.2a). The root system absorbs water and minerals from the soil for the entire plant. The cylindrical shape of a root allows it to penetrate the soil as it grows and permits water to be absorbed from all sides. The absorptive capacity of a root is dependent on its many branches, which all bear root hairs in a special zone near the tip. Root hairs, which are projections from epidermal root-hair cells, are the structures that absorb water and minerals. Root hairs are so numerous that they tremendously increase the absorptive surface of a root. It has been estimated that a single rye plant has about 14 billion hair cells, and if placed end to end, the root hairs would stretch 10,626 km. Root-hair cells are constantly being replaced, so this same rye plant forms about 100 million new root-hair cells every day. A plant roughly pulled out of the soil will not do well when transplanted; the reason is that small lateral roots and root hairs are torn off. Transplantation is more apt to be successful if you take a part of the surrounding soil along with the plant, leaving as much of the lateral roots and the root hairs intact as possible.

Roots have still other functions. Roots produce hormones that stimulate the growth of stems and coordinate their size with the size of the root. It is most efficient for a plant to have root and stem sizes that are appropriate to each other. **Perennial** plants have vegetative structures that live year after

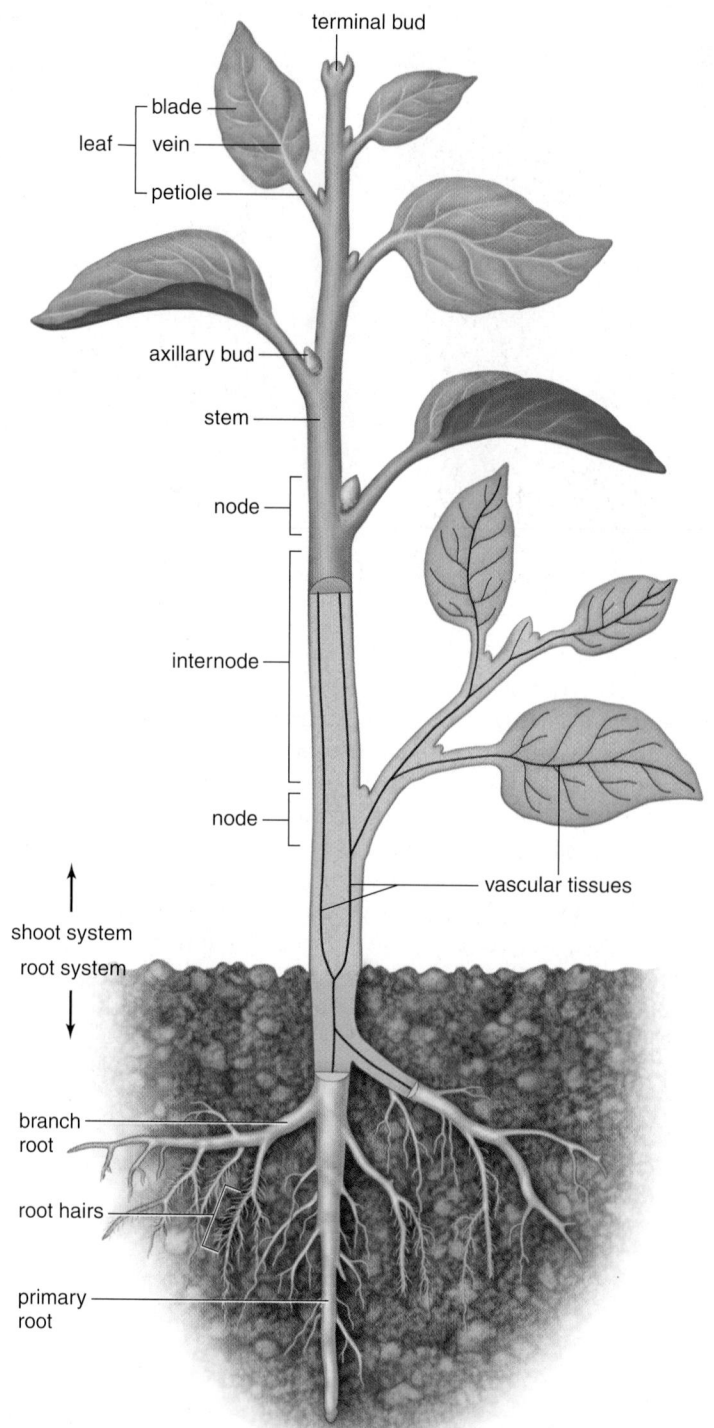

Figure 24.1 Organization of a plant body. The body of a plant consists of a root system and a shoot system. The shoot system contains the stem and leaves, two types of plant vegetative organs. Axillary buds can develop into branches of stems or flowers, the reproductive structures of a plant. The root system is connected to the shoot system by vascular tissue (brown) that extends from the roots to the leaves.

year. Herbaceous perennials generally live in temperate areas and die back during the late fall and winter seasons. They store the products of photosynthesis in their roots. Carrots and sweet potatoes are examples of the roots of such plants.

a. Root system, dandelion

b. Shoot system, bean seedling

c. Leaves, pumpkin seedling

Figure 24.2 Vegetative organs of several eudicots. **a.** The root system anchors the plant and absorbs water and minerals. **b.** The shoot system consists of a stem and its branches, which support the leaves and transport water and organic nutrients. **c.** The leaves, which may be broad and thin, carry on photosynthesis.

Stems

The shoot system of a plant is composed of the stem, the branches, and the leaves. A **stem,** the main axis of a plant, has a terminal bud that allows the stem to elongate and produce new leaves (Fig. 24.2b). If upright, as most are, a stem supports leaves in a way that exposes each one to as much sunlight as possible. A **node** occurs where leaves are attached to the stem; the region between nodes is called an **internode** (see Fig. 24.1). An **axillary bud,** located at a node in the upper angle between the leaf and the stem can produce new branches of the stem (or flowers). The presence of nodes and internodes is used to identify a stem, even if it happens to be an underground stem. A horizontal underground stem, called a rhizome, sends out roots below and shoots above at the nodes as it grows. Therefore, a rhizome allows a plant, such as ginger and bamboo, to increase its territory.

In addition to supporting the leaves, a stem has vascular tissue that transports water and minerals from the roots through the stem to the leaves and transports the products of photosynthesis, usually in the opposite direction. Nonliving cells form a continuous pipeline for water and mineral transport, while living cells join end to end for organic nutrient transport. A cylindrical stem can sometimes expand in girth as well as length. As trees grow taller each year, they accumulate woody tissue that adds to the strength of their stems.

Stems may have functions other than those mentioned: increasing the length of shoot system and transporting water and nutrients. In some plants (e.g., cactus), the stem is the primary photosynthetic organ. The stem is also a water reservoir in succulent plants. Some underground branches of a stem, or a portion of the root called a tuber, store nutrients.

Leaves

Leaves are the part of a plant that generally carries on the majority of photosynthesis, a process that requires water, carbon dioxide, and sunlight. Leaves receive water from the root system by way of the stem.

The size, shape, color, and texture of leaves are highly variable. These characteristics are fundamental in plant identification. The leaves of some aquatic duckweeds may be less than 1 mm in diameter, while some palms may have leaves that exceed 6 m in length. The shape of leaves can vary from cactus spines to deeply lobed white oak leaves. Leaves can exhibit a variety of colors from various shades of green to deep purple. The texture of leaves varies from smooth and waxy like a magnolia to coarse like a sycamore. Plants that bear leaves the entire year are called **evergreens** and those that lose all of their leaves at the end of their growing season are called **deciduous.**

Broad and thin plant leaves have the maximum surface area for the absorption of carbon dioxide and the collection of solar energy needed for photosynthesis. Also unlike stems, leaves are almost never woody. With few exceptions, their cells are living, and the bulk of a leaf contains photosynthetic tissue.

The wide portion of a foliage leaf is called the **blade.** The **petiole** is a stalk that attaches the blade to the stem (Fig. 24.2c). The upper acute angle between the petiole and stem is the leaf axil where the axillary bud is found. Not all leaves are foliage leaves. Some are specialized to protect buds, attach to objects (tendrils), store food (bulbs), or even capture insects. Specialized leaves are discussed in Section 24.5.

Monocot Versus Eudicot Plants

As described in the preceding chapter, flowering plants are divided into two groups, depending on the number of *cotyledons,* or seed leaves, in the embryonic plant (Fig. 24.3). Plants with a seed containing only one cotyledon are referred to as monocotyledons, or **monocots.** Plants with seeds that contain two cotyledons are known as eudicotyledons, or **eudicots.** Cotyledons of eudicots supply nutrients for seedlings, but the cotyledon of monocots acts as a transfer tissue, and the nutrients are derived from the endosperm before the true leaves begin photosynthesizing.

The vascular (transport) tissue is organized differently in monocots and eudicots. In the monocot root, vascular tissue occurs in a ring, and in the monocot stem, the vascular bundles, which contain vascular tissue surrounded by a bundle sheath, are scattered. In the eudicot root, the xylem, which transports water and minerals, is star-shaped; and the phloem, which transports organic nutrients, is located between the points of the star. In a eudicot stem, the vascular bundles occur in a ring.

Leaf veins are vascular bundles within a leaf. Monocots exhibit parallel venation, whereas eudicots exhibit netted venation, which may be either pinnate or palmate. Pinnate venation means that major veins originate from points along the centrally placed main vein, and palmate venation means that the major veins all originate at the point of attachment of the blade to the petiole:

Netted venation: pinnately veined   palmately veined

Adult monocots and eudicots have other structural differences, such as the number of flower parts and the number of pores in the wall of their pollen grains. Monocots have their flower parts arranged in multiples of three, and eudicots have their flower parts arranged in multiples of four or five. Eudicot pollen grains usually have three pores, and monocot pollen grains usually have one pore.

Although the distinctions between monocots and eudicots may seem of limited importance, they do in fact affect many aspects of their structure. The eudicots are the larger group and include some of our most familiar flowering plants—from dandelions to oak trees. The monocots include grasses, lilies, orchids, and palm trees, among others. Some of our most significant food sources are monocots, including rice, wheat, and corn.

Figure 24.3 Flowering plants are either monocots or eudicots.   Five features illustrated here are used to distinguish monocots from eudicots: number of cotyledons; the arrangement of vascular tissue in roots, stems, and leaves; and the number of flower parts.

Check Your Progress   24.1

1. List the three vegetative organs in a plant and state their major functions.
2. Compare and contrast the structural differences between monocots and eudicots.

	Seed	Root	Stem	Leaf	Flower
Monocots	One cotyledon in seed	Root xylem and phloem in a ring	Vascular bundles scattered in stem	Leaf veins form a parallel pattern	Flower parts in threes and multiples of three
Eudicots	Two cotyledons in seed	Root phloem between arms of xylem	Vascular bundles in a distinct ring	Leaf veins form a net pattern	Flower parts in fours or fives and their multiples

24.2 Tissues of Flowering Plants

Learning Outcomes

Upon completion of this section, you should be able to

1. Identify the three types of tissues found in angiosperms.
2. Recognize the differences between the location, structure, and function of various angiosperm tissues.
3. Identify the cell types that comprise the tissue types found in angiosperms.

A flowering plant has the ability to grow throughout its entire life because it possesses meristematic (embryonic) tissue. **Apical meristems** are located at or near the tips of stems and roots, where they increase the length of these structures. This increase in length is called primary growth. In addition to apical meristems, monocots have a type of **meristem** called intercalary [L. *intercalare,* to insert] meristem, which allows them to regrow lost parts. Intercalary meristems occur between mature tissues; they account for why grass can so readily regrow after being grazed by a cow or cut by a lawnmower.

> **Video**
> **Seedling Growth**

Apical meristem continually produces three types of meristems, which develop into the three types of specialized primary tissues in the body of a plant: Protoderm gives rise to epidermis; ground meristem produces ground tissue; and procambium produces vascular tissue. The functions of these three specialized tissues include:

1. **Epidermal tissue** forms the outer protective covering of a plant.
2. **Ground tissue** fills the interior of a plant.
3. **Vascular tissue** transports water and nutrients within the plant as well as providing support.

Epidermal Tissue

The entire body of both nonwoody (herbaceous) and young woody plants contains closely packed epidermal cells called the epidermis [Gk. *epi,* over, and *derma,* skin]. The walls of epidermal cells that are exposed to air are covered with a waxy **cuticle** [L. *cutis,* skin] to minimize water loss. The cuticle also protects against bacteria and other organisms that might cause disease.

In roots, certain epidermal cells have long, slender projections called **root hairs** (Fig. 24.4*a*). The hairs increase the surface area of the root for absorption of water and minerals as well as anchoring the plant to various substrates.

On stems, leaves, and reproductive organs, epidermal cells produce hairs called **trichomes** [Gk. *trichos,* hair] that have two important functions: to protect the plant from too much sun, and to conserve moisture. Sometimes trichomes, particularly glandular ones, help protect a plant from herbivores by producing a toxic substance. Under the slightest pressure the stiff trichomes of the stinging nettle lose their tips, forming "hypodermic needles" that inject an intruder with a stinging secretion. (See the Evolution feature for discussion of some of the defenses that have evolved in angiosperms.)

In leaves, the lower epidermis of eudicots and both surfaces of monocots contain specialized cells called guard cells (Fig. 24.4*b*). Guard cells, which are epidermal cells with chloroplasts, surround microscopic pores called **stomata** (sing., **stoma**). When the stomata are open, gas exchange and water loss occur.

In older woody plants, the epidermis of the stem is replaced by **periderm** [Gk. *peri,* around; *derma,* skin]. The majority component of periderm is boxlike **cork** cells. At maturity, cork cells can be sloughed off (Fig. 24.4*c*). New cork cells are made by a meristem called **cork cambium.** As the new cork cells mature, they increase slightly in volume, and their walls become encrusted with suberin, a lipid material, so that they are waterproof and chemically inert. These nonliving cells protect the plant and help it resist fungal, bacterial, and animal attacks. Some cork tissues, notably from the cork oak (*Quercus suber*), are commercially used for bottle corks and other products.

The cork cambium overproduces cork in certain areas of the stem surface causing ridges and cracks to appear. These features on the surface are called *lenticels.* Lenticels are the site of gas exchange between the interior of a stem and the air.

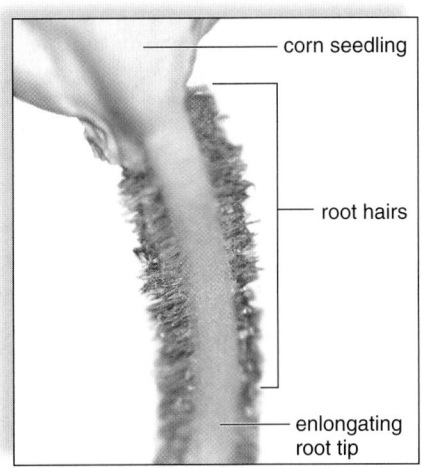

a. Root hairs

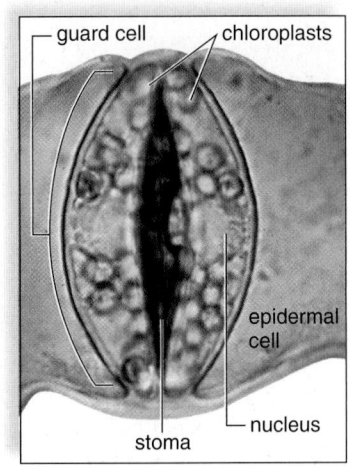

b. Stoma of leaf

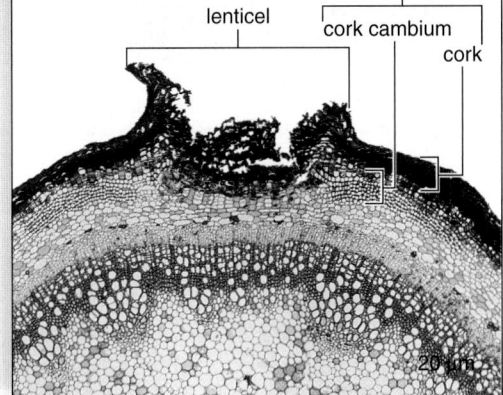

c. Cork of older stem

Figure 24.4 Modifications of epidermal tissue. **a.** Root epidermis has root hairs to absorb water. **b.** Leaf epidermis contains stomata (sing., stoma) for gas exchange. **c.** Periderm includes cork and cork cambium. Lenticels in cork are important in gas exchange.

Evolution

Survival Mechanisms of Plants

Plants first made their appearance on land approximately 450 million years ago. Since then they have evolved a wide variety of mechanisms in order to survive and have established the base of terrestrial ecosystems.

Some groups of plants employ defensive strategies in an attempt to deter predation (Figure 24A). A defensive strategy is a mechanism that has arisen through a process of natural selection in which the members of a group that possess the strategy compete better than those without it. The more successful competitors usually have a greater opportunity to pass on their genes.

Thorns and spines like those found on black locust trees and cacti are often employed to repel large herbivores, but they are generally ineffective against the smaller herbivores. If a tree does become injured, the tracheids and vessel elements of xylem immediately plug up with chemicals that block them off above and below the site of the injury. This response acts to prevent the damage from spreading to other locations on the tree.

Other plants produce toxins or sticky secretions in an attempt to deter predation. Anyone who has come into contact with poison ivy or the sap of a pine tree knows firsthand the effectiveness of this defensive mechanism.

Cellulose is the polysaccharide found in the cell walls of plants. This indigestible substance makes it difficult for many predators to obtain nutrients from eating the leaves, discouraging continued predation.

Dormancy enables plants to survive in environments that have seasonal conditions that do not allow year-round growth. Deciduous trees shed their leaves and transfer their nutrients into their root systems in response to the decrease in light, temperature, and moisture levels that occurs during the fall.

Seed dormancy allows the next generation to wait until growing conditions are optimal before germinating and competing for resources. Many plant life cycles are timed so that the seeds are produced during the summer, sit dormant throughout the winter, and germinate the following spring.

In some plants, the germination of seeds is triggered only after they have undergone some form of physical trauma, or seed scarification. Many species of plants found in chaparral regions, which are hot and dry, germinate only after they have been slightly burned, allowing them to germinate in an environment that has minimal competition for resources.

Evolutionary success is not measured by the survival of a single individual but by the passing of one's genes to the next generation. A number of groups of plants have evolved the ability to reproduce both sexually and asexually. Stolons, rhizomes, and tubers are asexual methods of reproduction. Strawberries, irises, and potatoes are plants that use these methods as well as sexual reproduction.

The wide variety of survival mechanisms has ensured that plants will be present on Earth for a very long time.

Questions to Consider

1. Which plant defense mechanisms would be the most effective against large predators? Small predators?
2. How would the supression of fires in a chaparral region impact the plant diversity?

a.

b.

Figure 24A Survival mechanisms used by plants.   Plants employ a wide variety of mechanisms to ensure their survival. **a.** Thorns of a locust tree **b.** Poison ivy contains a toxin within the leaves.

Ground Tissue

Ground tissue forms the bulk of a flowering plant and contains parenchyma, collenchyma, and sclerenchyma cells (Fig. 24.5). **Parenchyma** [Gk. *para*, beside, and *enchyma*, infusion] cells are the most abundant and correspond best to the typical plant cell. These are the least specialized of the cell types and are found in all the organs of a plant. They may contain chloroplasts and carry on photosynthesis (chlorenchyma), or they may contain colorless plastids that store the products of photosynthesis. A juicy bite from an apple yields mostly storage parenchyma cells. Parenchyma cells line the connected air spaces of a water lily and other aquatic plants. Parenchyma cells can divide and give rise to more specialized cells, such as when roots develop from stem cuttings placed in water.

Collenchyma cells are like parenchyma cells except they have thicker primary walls. The thickness is uneven and usually involves the corners of the cell. Collenchyma cells often form bundles just beneath the epidermis and give flexible support to

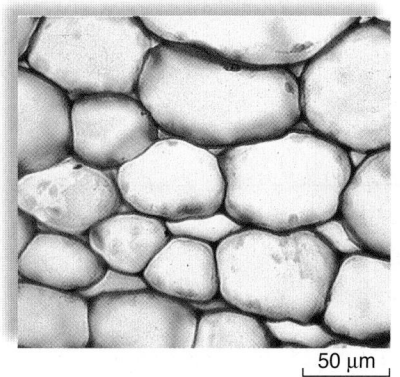

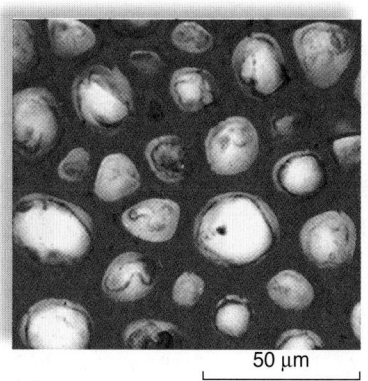

50 μm

50 μm

50 μm

a. Parenchyma cells

b. Collenchyma cells

c. Sclerenchyma cells

Figure 24.5 Ground tissue cells.
a. Parenchyma cells are the least specialized of the plant cells. **b.** Collenchyma cells. Notice how much thicker and irregular the walls are compared to those of parenchyma cells. **c.** Sclerenchyma cells have very thick walls and are nonliving—their only function is to give strong support.

immature regions of a plant body. The familiar strands in celery stalks (leaf petioles) are composed mostly of collenchyma cells.

Sclerenchyma cells have thick secondary cell walls impregnated with **lignin,** which is a highly resistant organic substance that makes the walls tough and hard. Most sclerenchyma cells are nonliving; their primary function is to support the mature regions of a plant. Two types of sclerenchyma cells are *fibers* and *sclereids.* Although fibers are occasionally found in ground tissue, most are in vascular tissue, which is discussed next. Fibers are long and slender and may be grouped in bundles that are sometimes commercially important. Hemp fibers can be used to make rope, and flax fibers can be woven into linen. Flax fibers, however, are not lignified, which is why linen is soft. Sclereids, which are shorter than fibers and more varied in shape, are found in seed coats and nutshells. Sclereids, or "stone cells," are responsible for the gritty texture of pears. The hardness of nuts and peach pits is due to sclereids.

Vascular Tissue

There are two types of vascular (transport) tissue. **Xylem** transports water and minerals from the roots to the leaves, and **phloem** transports sucrose and other organic compounds, usually from the leaves to the roots. Both xylem and phloem are considered **complex tissues** because they are composed of two or more kinds of cells.

Xylem contains two types of conducting cells: tracheids and vessel elements (VEs), which are modified sclerenchyma cells (Fig. 24.6). Both types of conducting cells are hollow and nonliving, but the **vessel elements** are larger, may have perforation plates in their end walls, and are arranged to form a continuous vessel for water and mineral transport.

The elongated **tracheids,** with tapered ends, form a less obvious means of transport, but water can move across the end walls and side walls because there are **pits,** or depressions,

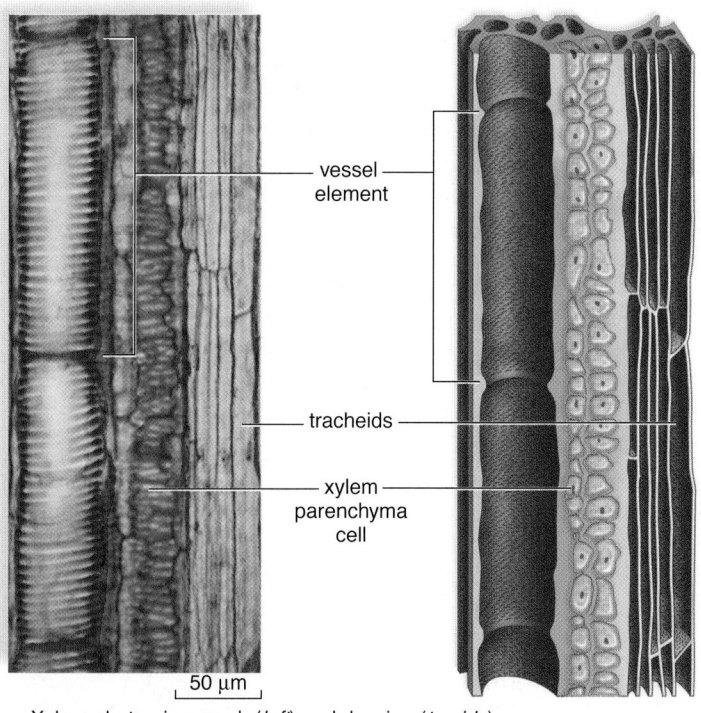

vessel element

tracheids

xylem parenchyma cell

50 μm

a. Xylem photomicrograph (*left*) and drawing (*to side*)

Figure 24.6 Xylem structure. **a.** Photomicrograph of xylem vascular tissue and drawing showing general organization of xylem tissue. **b.** Drawing of two types of vessels (composed of vessel elements)—the perforation plates differ. **c.** Drawing of tracheids.

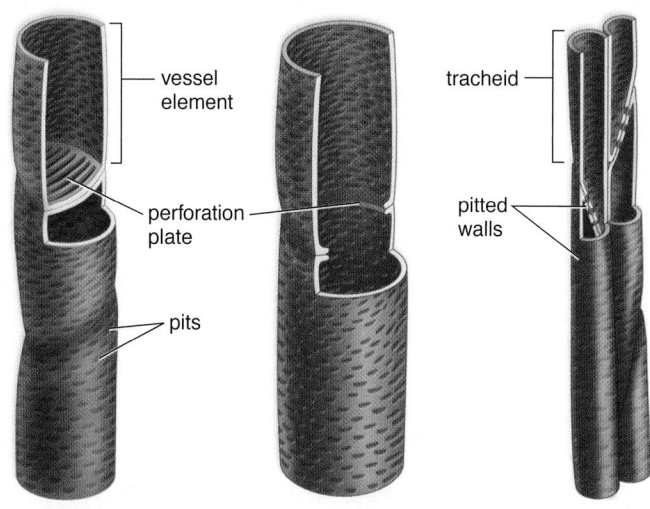

vessel element

perforation plate

pits

tracheid

pitted walls

b. Two types of vessels

c. Tracheids

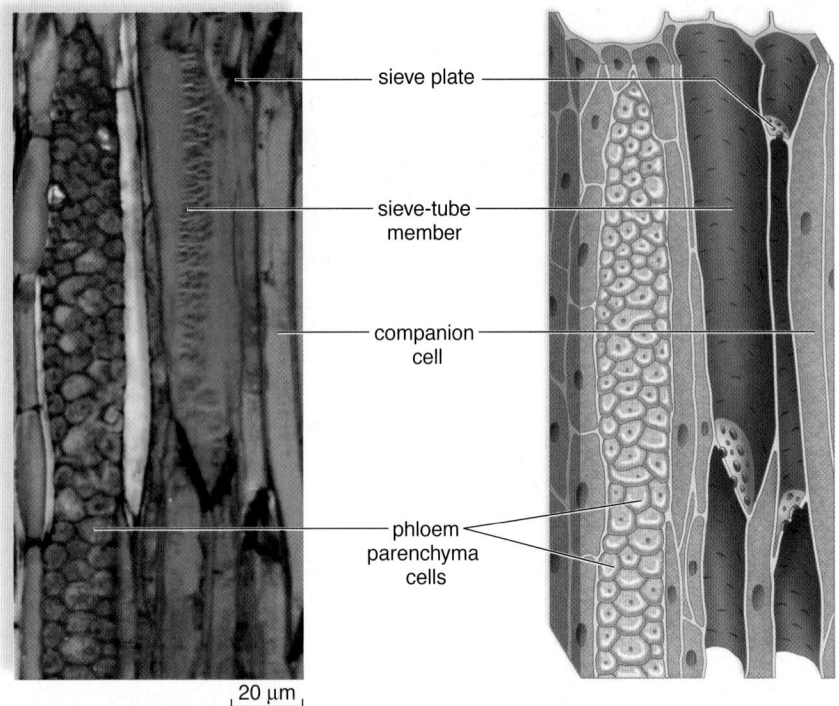

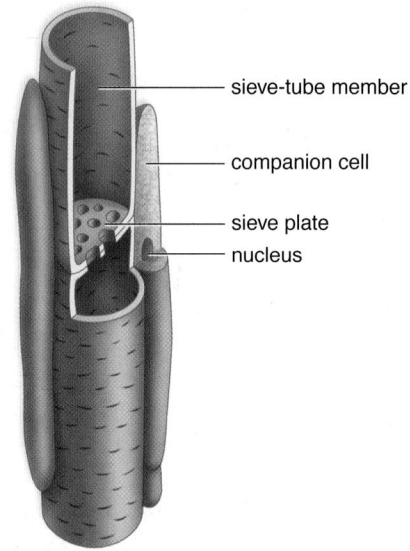

a. Phloem photomicrograph (*left*) and drawing (*to side*)

b. Sieve-tube member and companion cells

Figure 24.7 Phloem structure.   a. Photomicrograph of phloem vascular tissue and drawing showing general organization of phloem tissue. **b.** Drawing of sieve tube (composed of sieve-tube members) and companion cells.

where the secondary wall does not form. In addition to vessel elements and tracheids, xylem can contain sclerenchyma fibers that lend additional support as well as parenchyma cells that store various substances. Vascular rays, which are flat ribbons or sheets of parenchyma cells located between rows of tracheids, conduct water and minerals across the width of a plant.

Animation
Vascular System
of Plants

The conducting cells of phloem are specialized parenchyma cells called **sieve-tube members** that are arranged to form a continuous sieve tube (Fig. 24.7). Sieve-tube members contain cytoplasm but no nuclei. The term *sieve* refers to a cluster of pores in the end walls, which is known as a sieve plate. Each sieve-tube member has a companion cell, which contains a nucleus. The two are connected by numerous plasmodesmata, and the nucleus of the companion cell may control and maintain the life of both cells. The companion cells are also believed to be involved in the transport function of phloem. Sclerenchyma fibers also lend support to phloem.

It is important to realize that vascular tissue (xylem and phloem) extends from the root through the stems to the leaves, and vice versa (see Fig. 24.1). In the roots, the vascular tissue is located in the **vascular cylinder;** in the stem, it forms **vascular bundles;** and in the leaves, it is found in **leaf veins.**

Check Your Progress   24.2

1. List the three specialized tissues in angiosperms and the cells that make up these tissues.
2. Compare the transport function of xylem and phloem.

24.3 Organization and Diversity of Roots

Learning Outcomes

Upon completion of this section, you should be able to

1. Describe the tissue types that are found in each zone of a root.
2. Identify the structural differences between the roots of monocots and eudicots.
3. Describe the various adaptations, associations, and specializations that lead to root diversity.

Figure 24.8*a*, a longitudinal section of a eudicot root, reveals zones where cells are in various stages of differentiation as primary growth occurs. The **root apical meristem** is in the region protected by the **root cap.** Root cap cells have to be replaced constantly because they get ground off by rough soil particles as the root grows. The primary meristems are in the zone of cell division, which continuously provides new cells to the zone of elongation. In the zone of elongation, the cells lengthen as they become specialized. The zone of maturation, which contains fully differentiated cells, is recognizable because here root hairs are found on many of the epidermal cells.

Tissues of a Eudicot Root

Figure 24.8*a* also shows a cross section of a root at the region of maturation. These specialized tissues are identifiable:

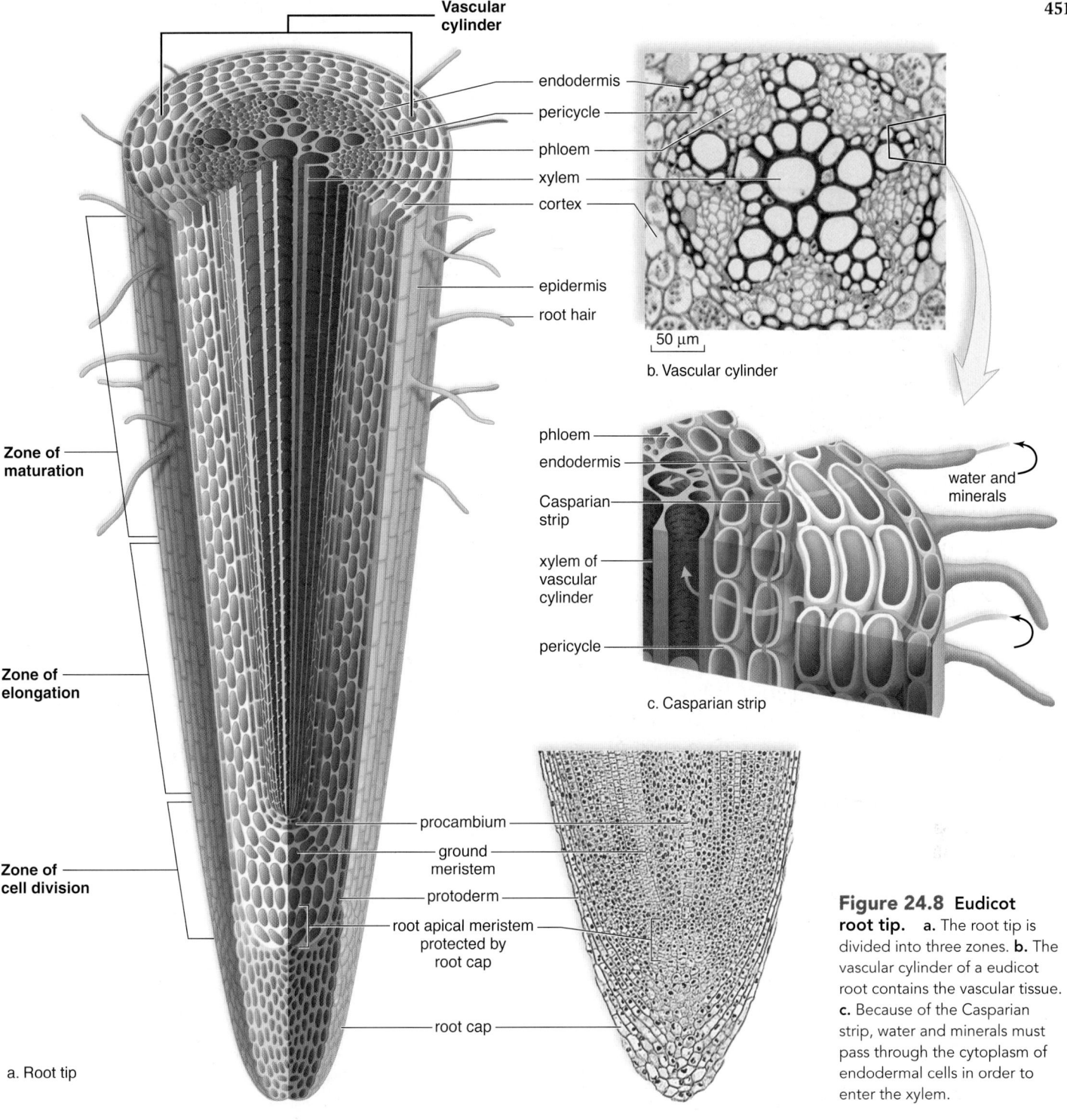

Vascular cylinder

endodermis
pericycle
phloem
xylem
cortex

epidermis
root hair

50 μm

b. Vascular cylinder

Zone of maturation

Zone of elongation

Zone of cell division

phloem
endodermis
Casparian strip
xylem of vascular cylinder
pericycle

water and minerals

c. Casparian strip

procambium
ground meristem
protoderm
root apical meristem protected by root cap
root cap

a. Root tip

Figure 24.8 Eudicot root tip. a. The root tip is divided into three zones. **b.** The vascular cylinder of a eudicot root contains the vascular tissue. **c.** Because of the Casparian strip, water and minerals must pass through the cytoplasm of endodermal cells in order to enter the xylem.

Epidermis   The epidermis forms the outer layer of the root and consists of only a single layer of cells. The majority of epidermal cells are thin-walled and rectangular, but in the zone of maturation, many epidermal cells have root hairs. These can project as far as 5–8 mm into the soil particles.

Cortex   Moving inward, next to the epidermis are the large, thin-walled parenchyma cells that make up the **cortex** of the root. These irregularly shaped cells are loosely packed, making it possible for water and minerals to move through the cortex without entering the cells. The cells contain starch granules, and the cortex functions in food storage.

Endodermis   The **endodermis** [Gk. *endon,* within, and *derma,* skin] is a single layer of rectangular cells that forms a boundary between the cortex and the inner vascular cylinder. The endodermal cells fit snugly together and are bordered on four sides (but not the two sides that contact the cortex and the vascular cylinder) by a layer of impermeable lignin and suberin known as the **Casparian strip** (Fig. 24.8c). This strip prevents the passage of water and mineral ions between adjacent cell walls. Therefore, the only access to the vascular cylinder is through the endodermal cells themselves, as shown by the arrow in Figure 24.8c. This arrangement regulates the entrance of minerals into the vascular cylinder.

Vascular tissue   The **pericycle,** the first layer of cells within the vascular cylinder, has retained its capacity to divide and

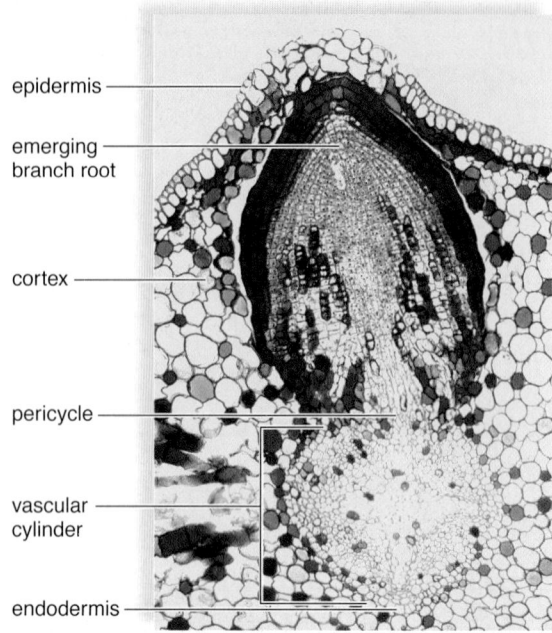

epidermis

emerging
branch root

cortex

pericycle

vascular
cylinder

endodermis

Figure 24.9 Branching of eudicot root. This cross section of a willow, *Salix*, shows the origination and growth of a branch root from the pericycle.

can start the development of branch, or lateral, roots (Fig. 24.9). The main portion of the vascular cylinder contains xylem and phloem. The xylem appears star-shaped in eudicots because several arms of tissue radiate from a common center (see Fig. 24.8*b*). The phloem is found in separate regions between the arms of the xylem.

Organization of Monocot Roots

Monocot roots have the same growth zones as eudicot roots, and undergo the same secondary growth as eudicot roots. Also, the organization of their tissues is slightly different. The ground tissue of a monocot root's **pith** is centrally located and is surrounded by a vascular ring composed of alternating xylem and phloem bundles (Fig. 24.10). Monocot roots also have pericycle, endodermis, cortex, and epidermis.

Root Diversity

Roots possess a variety of adaptations and associations to better perform their functions: anchorage, absorption of water and minerals, and storage of carbohydrates.

In some plants, notably eudicots, the first or **primary root** grows straight down and remains the dominant root of the plant. This so-called **taproot** is often fleshy and stores food (Fig. 24.11*a*). Carrots, beets, turnips, and radishes have taproots that we consume as vegetables. Sweet potato plants don't have taproots, but they do have roots that expand to store starch. These storage roots are the sweet potatoes we eat.

In other plants, notably monocots, there is no single, main root; instead, a large number of slender roots grow from the lower nodes of the stem when the first (primary) root dies. These slender roots and their lateral branches make up a **fibrous root system** (Fig. 24.11*b*). Most grasses have a fibrous root system that helps anchor the plant to the soil.

Root Specializations

When roots develop from organs of the shoot system instead of the root system, they are known as *adventitious roots*. One style of adventitious root is typically found in corn plants. These are called prop roots because they emerge above the soil line and act as anchors for the plant (Fig. 24.11*c*). Other examples of adventitious roots are those found on horizontal stems (see Fig. 24.19*a*) or at the nodes of climbing English ivy (Fig. 24.11*e*). As the vines climb, the rootlets attach the plant to any available vertical structure.

Black mangroves live in marshy environments and have pneumatophores, root projections that rise above the water that allow roots to acquire oxygen for cellular respiration (Fig. 24.11*d*).

Some plants, such as dodders and broomrapes, are parasitic on other plants. Their stems have rootlike projections called haustoria (sing., haustorium) that grow into the host plant and make contact with vascular tissue from which they extract water and nutrients (see Fig. 25.8*a*). As described in Chapter 22, *mycorrhizae* are associations between roots and fungi. Plants that have mycorrhizae are able to extract water and minerals from the soil better than those with roots that lack a fungus partner. This relationship is mutualistic because the fungus receives sugars and amino acids from the plant, while the plant receives increased water and minerals via the fungus.

Peas, beans, and other legumes have **root nodules** where nitrogen-fixing bacteria live. Plants cannot extract nitrogen from the air, but the bacteria within the nodules can take up and reduce atmospheric nitrogen. This means that the plant is no longer dependent on a supply of nitrogen (i.e., nitrate or ammonium) from the soil.

Animation
Root Nodule Formation

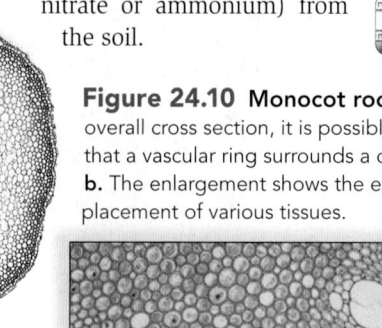

vascular
cylinder

a.

Figure 24.10 Monocot root. a. In this overall cross section, it is possible to observe that a vascular ring surrounds a central pith. **b.** The enlargement shows the exact placement of various tissues.

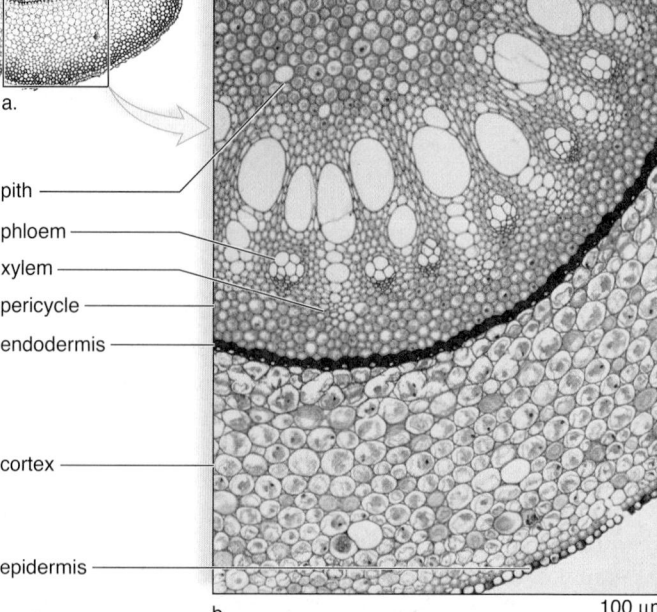

pith

phloem

xylem

pericycle

endodermis

cortex

epidermis

b.   100 µm

a. Taproot

b. Fibrous root system

c. Prop roots, a type of adventitious root

d. Pneumatophores of black mangrove trees

e. Aerial roots of English ivy clinging to tree trunks

Figure 24.11 Root diversity. **a.** A taproot may have branch roots in addition to a main root. **b.** A fibrous root has many slender roots with no main root. **c.** Prop roots are specialized for support. **d.** The pneumatophores of a black mangrove tree allow it to acquire oxygen even though it lives in swampy water. **e.** *Left:* English ivy climbs up the trunk because it has aerial roots (*right*) that cling to tree bark.

Check Your Progress 24.3

1. Explain the relationship between the root apical meristem and the root cap.
2. List the function of the cortex, the endodermis, and the pericycle in a root.

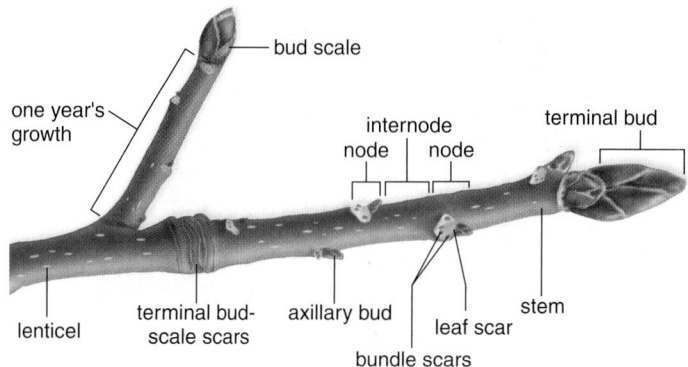

Figure 24.12 Woody twig. The major parts of a stem are illustrated by a woody twig collected in winter.

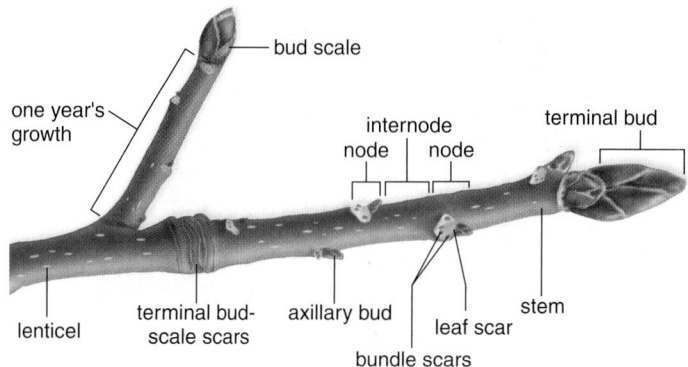

453

24.4 Organization and Diversity of Stems

Learning Outcomes

Upon completion of this section, you should be able to

1. Identify the anatomical structures of a woody twig.
2. Recognize the differences in the arrangement of vascular tissue in herbaceous dicot and monocot stems.
3. Describe how secondary growth of a woody stem results in the various tissues found within it.
4. Identify the various adaptations that lead to stem diversity.

The anatomy of a woody twig helps us review the organization of a stem (Fig. 24.12). The **terminal bud** contains the shoot tip protected by modified leaves called bud scales. Each spring when growth resumes, bud scales fall off and leave a scar. Each bud-scale scar indicates one year of growth. Leaf scars and bundle scars mark the location of leaves that have dropped. Dormant axillary buds that will give rise to branches or flowers are also found here.

As seasonal growth resumes, the apical meristem at the shoot tip produces new cells that increase the height of the stem. The **shoot apical meristem** is protected within the terminal bud, where leaf primordia (immature leaves) envelop it (Fig. 24.13). The leaf primordia mark the location of a node; the portion of stem in between nodes is an internode. As a stem grows, the internodes increase in length.

In addition to leaf primordia, the three specialized types of primary meristem (see Section 24.2) develop from a shoot apical meristem (Fig. 24.13*b*). These primary meristems contribute to the length of a shoot. The *protoderm*, the outermost primary meristem, gives rise to the epidermis. The *ground meristem* produces two tissues composed of parenchyma cells: the pith and the cortex. The *procambium* (Figure 24.13*a*) produces the first xylem cells, called primary xylem, and the first phloem cells, called primary phloem.

Differentiation continues as certain cells become the first tracheids or vessel elements of the xylem within a vascular bundle. The first sieve-tube members of a vascular bundle do not have companion cells and are short-lived (some live only a day before being replaced). Mature vascular bundles contain fully

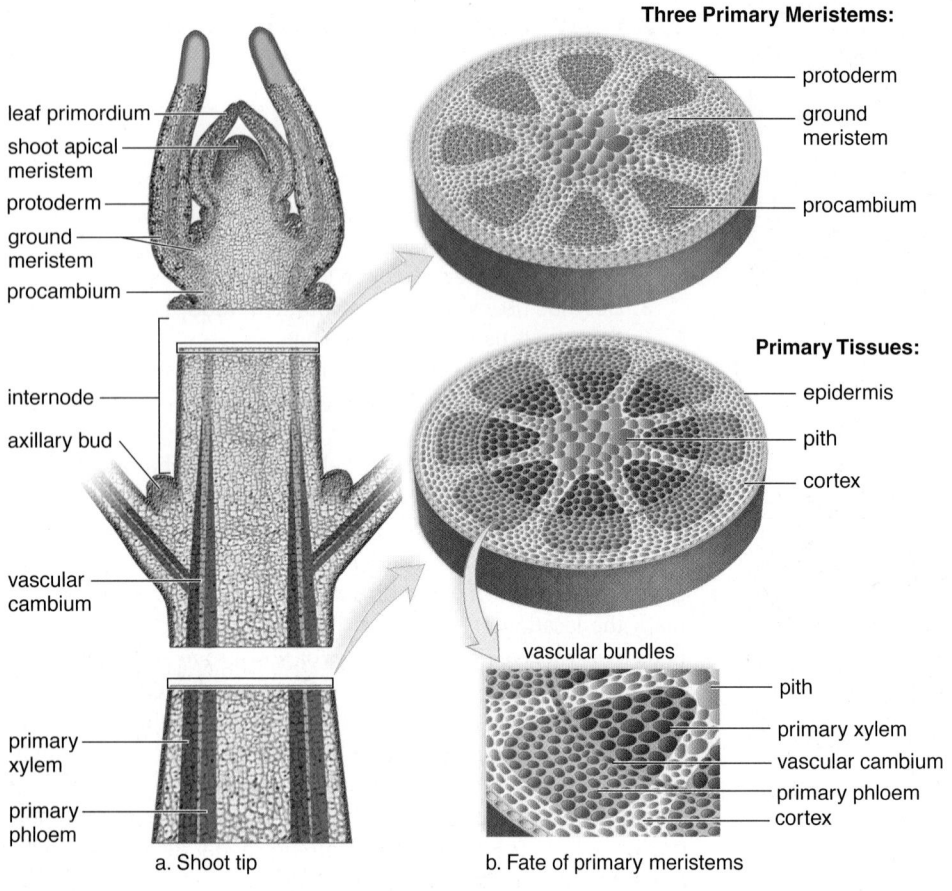

Three Primary Meristems:

- protoderm
- ground meristem
- procambium

Primary Tissues:

- epidermis
- pith
- cortex

vascular bundles

- pith
- primary xylem
- vascular cambium
- primary phloem
- cortex

leaf primordium
shoot apical meristem
protoderm
ground meristem
procambium

internode

axillary bud

vascular cambium

primary xylem

primary phloem

a. Shoot tip

b. Fate of primary meristems

Figure 24.13 Shoot tip and primary meristems. **a.** The shoot apical meristem within a terminal bud is surrounded by leaf primordia. **b.** The shoot apical meristem produces the primary meristems: Protoderm gives rise to epidermis; ground meristem gives rise to pith and cortex; and procambium gives rise to vascular tissue, including primary xylem, primary phloem, and vascular cambium.

differentiated xylem, phloem, and a lateral meristem called **vascular cambium** [L. *vasculum*, dim. of *vas*, vessel, and *cambio*, exchange]. Vascular cambium is discussed more fully in the section on woody stems, a little later.

Herbaceous Stems

Mature nonwoody stems, called **herbaceous stems** [L. *herba*, vegetation, plant], exhibit only primary growth. The outermost tissue of herbaceous stems is the epidermis, which is covered by a waxy cuticle to prevent water loss. These stems have distinctive vascular bundles, where xylem and phloem are found. In each bundle, xylem is typically found toward the inside of the stem, and phloem is found toward the outside.

In the herbaceous eudicot stem such as a sunflower, the vascular bundles are arranged in a distinct ring in which the cortex is separated from the central pith, which stores water and products of photosynthesis (Fig. 24.14). The cortex is sometimes green and carries on photosynthesis.

In a monocot stem, such as a corn stalk, the vascular bundles are scattered throughout the stem, and often the cortex and pith are not clearly distinguishable (Fig. 24.15). The stems of one monocot in the grass family, bamboo, have been of

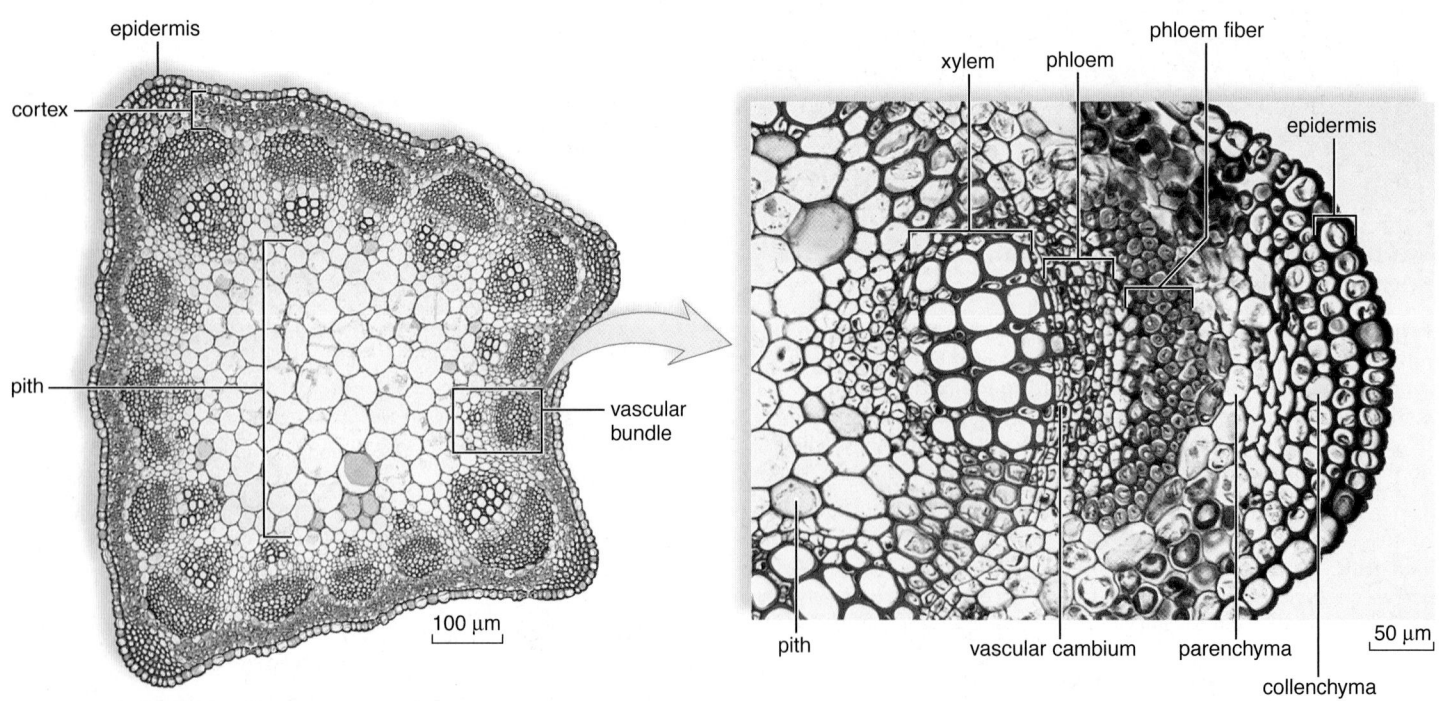

epidermis

cortex

pith

vascular bundle

100 μm

Figure 24.14 Herbaceous eudicot stem.

xylem   phloem   phloem fiber

epidermis

pith   vascular cambium   parenchyma

collenchyma

50 μm

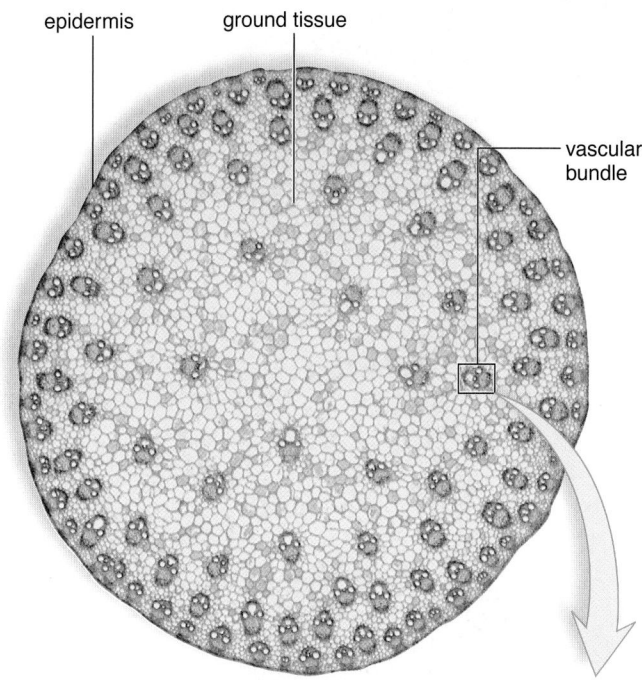

epidermis   ground tissue

vascular bundle

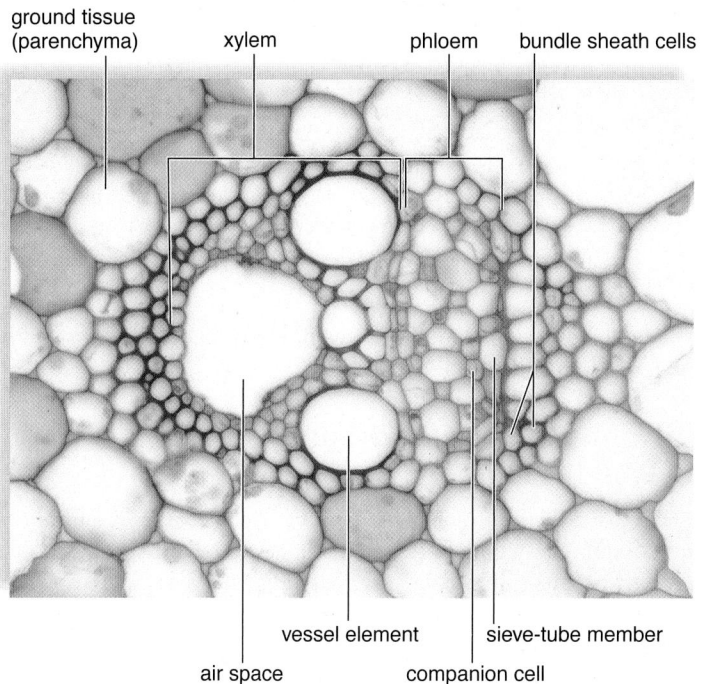

ground tissue (parenchyma)   xylem   phloem   bundle sheath cells

vessel element   sieve-tube member

air space   companion cell

Figure 24.15 Monocot stem.

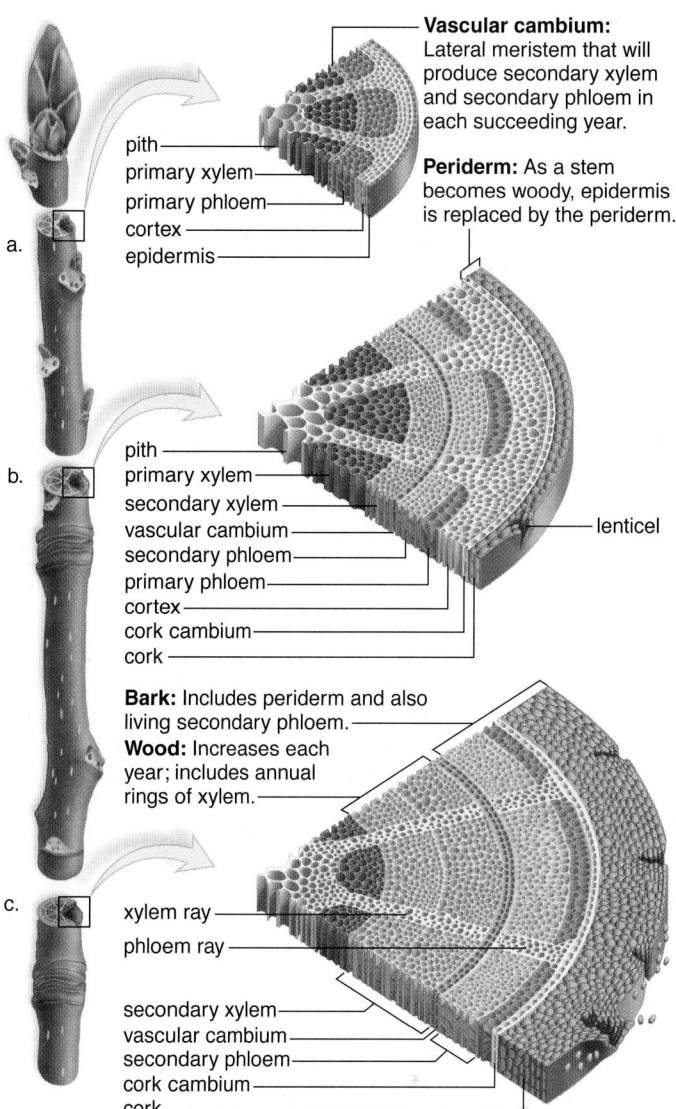

Vascular cambium: Lateral meristem that will produce secondary xylem and secondary phloem in each succeeding year.

pith
primary xylem
primary phloem
cortex
epidermis

a.

Periderm: As a stem becomes woody, epidermis is replaced by the periderm.

pith
primary xylem
secondary xylem
vascular cambium
secondary phloem
primary phloem
cortex
cork cambium
cork

b.

lenticel

Bark: Includes periderm and also living secondary phloem.
Wood: Increases each year; includes annual rings of xylem.

c.

xylem ray
phloem ray

secondary xylem
vascular cambium
secondary phloem
cork cambium
cork

Figure 24.16 Diagrams of secondary growth of stems.
a. Diagram showing eudicot herbaceous stem just before secondary growth begins. **b.** Diagram showing that secondary growth has begun. Periderm has replaced the epidermis. Vascular cambium produces secondary xylem and secondary phloem each year. **c.** Diagram showing a two-year-old stem. The primary phloem and cortex will eventually disappear, and only the secondary phloem (within the bark) produced by vascular cambium will be active that year. Secondary xylem builds up to become the annual rings of a woody stem.

great benefit in human history, and this plant continues to be useful to us today (see the Nature of Science feature).

Woody Stems

A woody plant such as an oak tree has both primary and secondary tissues. Primary tissues are those new tissues formed each year from the primary meristems. Secondary tissues develop during the first and subsequent years of growth from lateral meristems forming the vascular cambium and cork cambium. *Primary growth,* which occurs in all plants, increases the length of plant stems and roots; *secondary growth,* which occurs only in

conifers and woody eudicots, increases the girth of trunks, stems, branches, and roots.

Animation
Woody Dicot

Trees and shrubs undergo secondary growth because of a change in the location and activity of vascular cambium (Fig. 24.16). In herbaceous plants, vascular cambium is present between the xylem and phloem of each vascular bundle. In woody plants, the vascular cambium develops to form a ring of meristem that divides parallel to the surface of the plant, and produces new xylem toward the inside and phloem toward the outside on a yearly basis.

Eventually, a woody eudicot stem has an entirely different organization from that of a herbaceous eudicot stem. A woody stem forms three distinct areas: the bark, the wood, and the pith. Vascular cambium occurs between the bark and the wood, which causes woody plants to increase in girth. Cork cambium, occurring first beneath the epidermis, is instrumental in the production of cork in woody plants.

Also notice in Figure 24.16 the *xylem rays* and *phloem rays* that are visible in the cross section of a woody stem. Rays consist of parenchyma cells that permit lateral conduction of nutrients from the pith to the cortex, as well as some storage of food. A phloem ray can vary in width and is a continuation of a xylem ray.

Bark

The **bark** of a tree contains periderm (cork and cork cambium), and phloem. Although secondary phloem is produced each year by vascular cambium, phloem does not build up from season to season. The bark of a tree can be removed; however, this is very harmful because, without phloem, organic nutrients cannot be transported. Girdling, removing a ring of bark from around a tree, can be lethal to the tree. Overgrazing by some herbivores can result in the girdling of trees.

At first, cork cambium is located beneath the epidermis, and then later, it is found beneath the periderm. When cork cambium first begins to divide, it produces tissue that disrupts the epidermis and replaces it with cork cells. Cork cells are impregnated with suberin, a waxy layer that makes them waterproof but also causes them to die. This development is advantageous because it makes the stem less edible. But an impervious barrier means that gas exchange is impeded except near pockets of loosely arranged cork cells not impregnated with suberin, called lenticels.

Wood

Wood is secondary xylem that builds up year after year, thereby increasing the girth of trees. In trees that have a growing season, vascular cambium is dormant during the winter. In the spring, when moisture is plentiful and leaves require much water for growth, the secondary xylem contains wide vessel elements with thin walls. In this so-called *spring wood,* wide vessels transport sufficient water to the growing leaves. Later in the season, moisture is scarce, and the wood at this time, called *summer wood,* has a lower proportion of vessels (Fig. 24.17). Strength is required because the tree is growing larger, and summer wood contains numerous thick-walled tracheids. At the end of the growing season, just before the cambium becomes dormant again, only heavy fibers with especially thick secondary walls may develop.

When the trunk of a tree has spring wood followed by summer wood, the two together make up one year's growth, or an **annual ring.** You can tell the age of a tree by counting the annual rings (Fig. 24.18a). The outer annual rings, where transport occurs, are called sapwood.

In older trees, the inner annual rings, called heartwood, no longer function in water transport. The cells become plugged with deposits, such as resins, gums, and other substances that inhibit the growth of bacteria and fungi. Heartwood may help support a tree, although some trees stand erect and live for many years after the heartwood has rotted away. Figure 24.18b shows the layers of a woody stem in relation to one another.

The annual rings are used to tell the age of a tree as well as the historical record of tree growth. For example, if rainfall and other conditions have been extremely favorable during a season, the annual ring may be wider than usual. If the tree has been

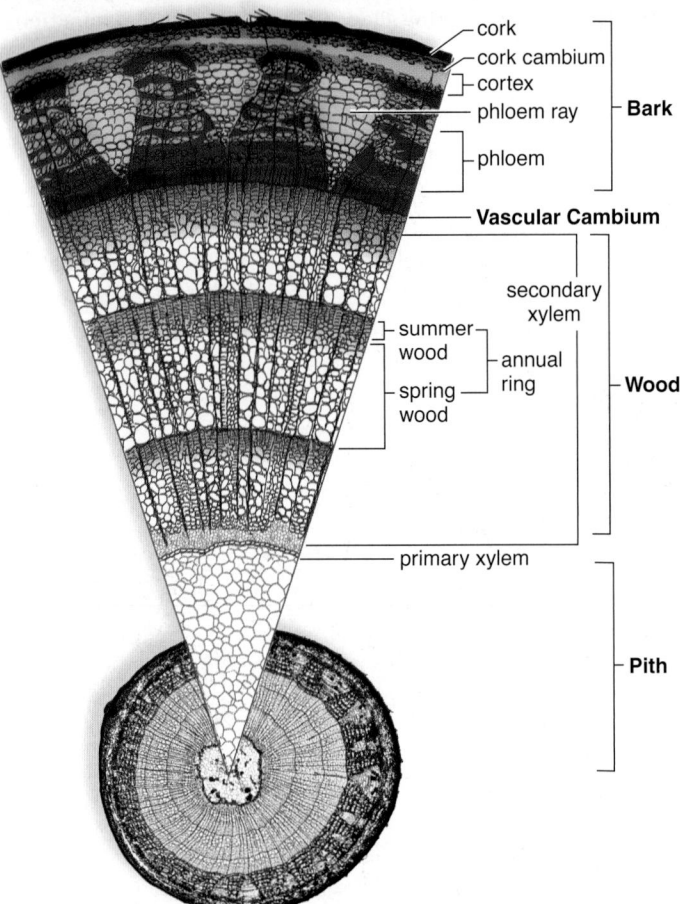

Figure 24.17 Three-year-old woody twig. The buildup of secondary xylem in a woody stem results in annual rings, which tell the age of the stem. The rings can be distinguished because each one begins with spring wood (large vessel elements) and ends with summer wood (smaller and fewer vessel elements).

shaded on one side by another tree or building, the rings may be wider on the sunnier side.

Advantages and Disadvantages of Woody Plants

What are the evolutionary benefits of woody plants? With adequate rainfall, woody plants can grow taller and have more growth because they have adequate vascular tissue to support and service their leaves. Furthermore, a long life may mean more opportunity to reproduce.

However, it takes energy to produce secondary growth and to prepare the body for winter if the plant lives in the temperate zone. Also, woody plants need more defense mechanisms because a long-lived plant is likely to be attacked by herbivores and parasites. Trees usually do not reproduce until after they have grown for several seasons, by which time they may have been attacked by predators or been infected with a disease. In certain habitats, it is more advantageous for a plant to put most of its energy into producing a large number of seeds rather than being woody.

Stem Diversity

Stem diversity is illustrated in Figure 24.19. Aboveground horizontal stems, called **stolons** [L. *stolo,* shoot] or runners, produce new plants where nodes touch the ground. The strawberry plant

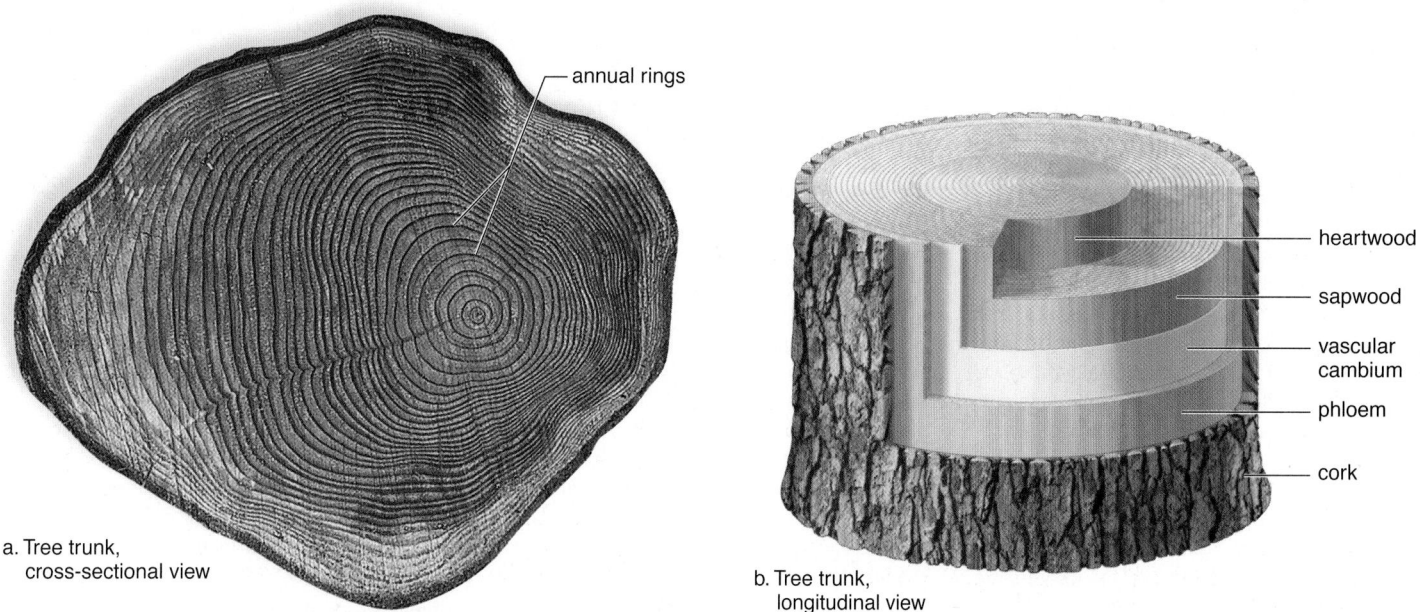

a. Tree trunk,
cross-sectional view

annual rings

heartwood

sapwood

vascular cambium

phloem

cork

b. Tree trunk,
longitudinal view

Figure 24.18 Tree trunk. **a.** A cross section of a 39-year-old larch, *Larix decidua*. The xylem within the darker heartwood is inactive; the xylem within the lighter sapwood is active. **b.** The relationship of bark, vascular cambium, and wood is retained in a mature stem. The pith has been buried by the growth of layer after layer of new secondary xylem.

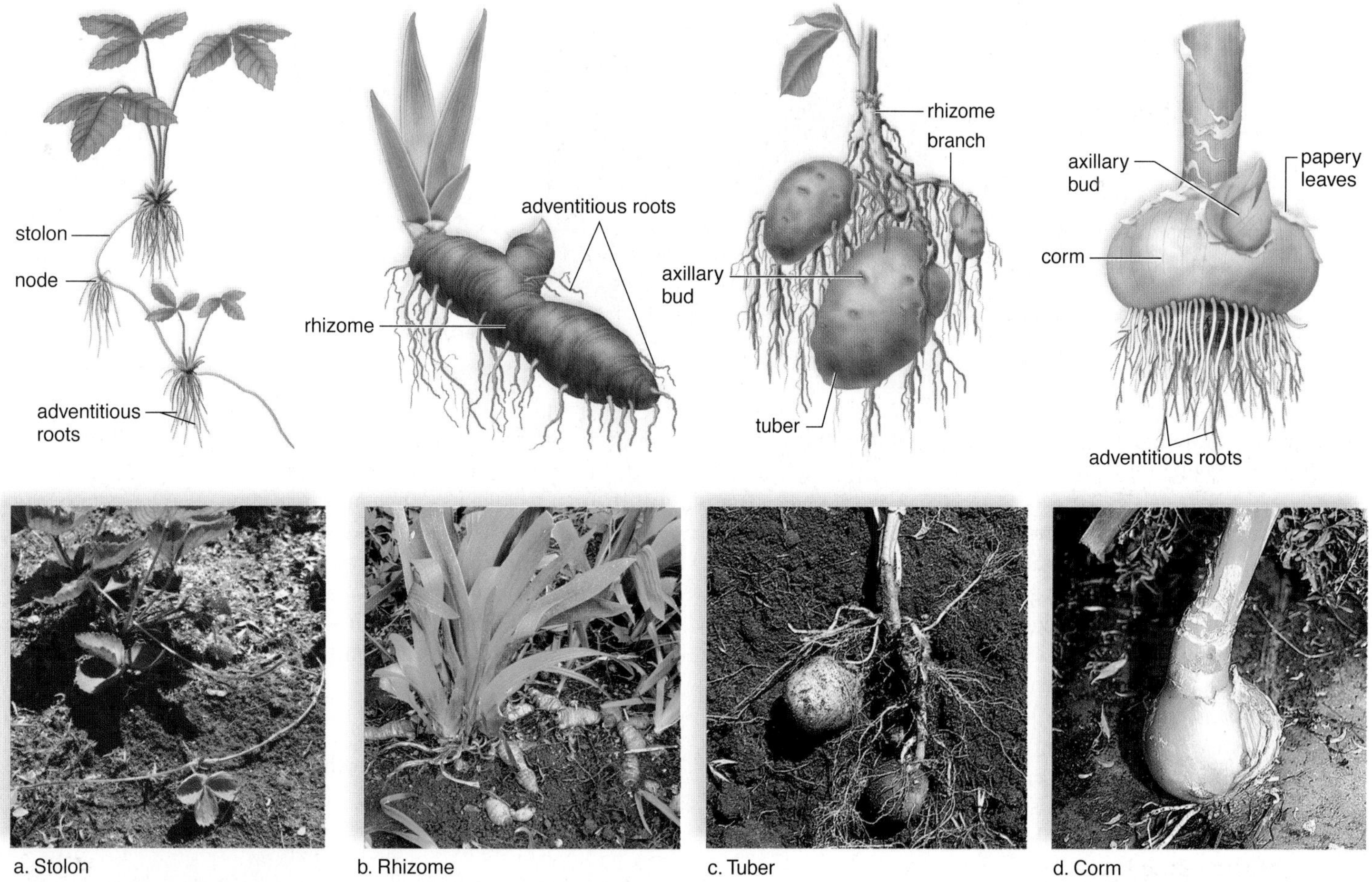

stolon

node

adventitious roots

rhizome

adventitious roots

rhizome branch

axillary bud

tuber

axillary bud

corm

papery leaves

adventitious roots

a. Stolon

b. Rhizome

c. Tuber

d. Corm

Figure 24.19 Stem diversity. **a.** A strawberry plant has aboveground horizontal stems called stolons. Every other node produces a new shoot system. **b.** The underground horizontal stem of an iris is a fleshy rhizome. **c.** The underground stem of a potato plant has enlargements called tubers. We call the tubers potatoes. **d.** The corm of a gladiolus is a stem covered by papery leaves.

The Many Uses of Bamboo

Because of its resilience, amazing rate of growth, and ability to grow in a variety of climates, bamboo is quickly becoming a valuable crop. Certain varieties of bamboo are capable of growing up to a foot per day and can reach their full height within one year in the right conditions.

Bamboo is classified as a grass that contains varieties ranging in height from one foot to over 100 feet. Globally there are over 1,400 species of bamboo. Approximately 900 can be found in tropical climates, and the remaining 500 are found in temperate environments. Several varieties of bamboo are even native to the Eastern and Southeastern United States.

Bamboo is recognized for its versatility as a building material, commercial food product, and clothing material. It can be processed into roofing material, flooring (Fig. 24B), and support beams, as well as a variety of other construction materials. The mature stalks can be used as support columns in "green" construction. It can also be used to replace steel reinforcing rods that are typically used in concrete-style construction. Bamboo products are 3 times harder than oak.

Although not used extensively in the commercial food market, bamboo does have a variety of culinary uses. The shoots are often used in Asian dishes as a type of vegetable (Fig. 24B). Shoots can be boiled and added to a variety of dishes or eaten raw. In China, bamboo is used to make certain alcoholic drinks; other Asian countries make bamboo soups, pancakes, and broths. The hollow bamboo stalk can serve as a container to cook rice and soups, giving the foods a subtle but distinctive taste. Additional uses include modifying bamboo into cooking and eating utensils, most notably chopsticks.

Clothing products are now being made out of bamboo. Clothing made of bamboo fibers is reported as being very light and extremely soft. Bamboo also has the ability to wick moisture away from the skin, making it ideal to wear during exercise. Several lines of baby clothing are being made from bamboo.

Bamboo is being recognized as one of the most eco-friendly crops. It requires lower amounts of chemicals or pesticides to grow. It removes nearly 5 times more greenhouse gases and produces nearly 35% more oxygen than an equivalent stand of trees. Harvesting can be done from the 2nd or 3rd year of growth through the 5th to 7th year of growth. Because bamboo is a perennial, a stand can regrow after yearly harvesting instead of requiring replanting.

The bamboo-goods industry started increasing in popularity in the United States during the mid-1990s and is expected to reach $25 billion by the year 2012. Versatility, hardiness, ease of growing, and a greater awareness of environmental issues is quickly making bamboo an ideal natural product that may someday replace metal, wood, and plastics.

Questions to Consider

1. What might be some drawbacks to planting bamboo for human uses?
2. Why hasn't bamboo become a more popular crop in the United States?

Figure 24B **The many uses of bamboo.**
Bamboo is quickly becoming a multifunctional product in today's society. Uses range from building materials to food to clothing.

is a common example of this type of stem, which functions in vegetative reproduction.

Aboveground vertical stems can also be modified. For example, cacti have succulent stems specialized for water storage, and the tendrils of grape plants (which are stem branches) allow them to climb. Morning glory and relatives have stems that twine around support structures. Such tendrils and twining shoots help plants expose their leaves to the Sun.

Underground horizontal stems, **rhizomes** [Gk. *rhiza*, root], may be long and thin, as in sod-forming grasses, or thick and fleshy, as in irises. Rhizomes survive the winter and contribute to asexual reproduction because each node bears a bud. Some rhizomes have enlarged portions called tubers, which function in food storage. Potatoes are tubers, and the potato "eyes" are buds that mark the nodes.

Corms are bulbous underground stems that lie dormant during the winter, just as rhizomes do. They also produce new plants the next growing season. Gladiolus corms are referred to as bulbs by laypersons, but the botanist reserves the term *bulb* for a structure composed of modified leaves attached to a short vertical stem. An onion is a bulb.

Humans make use of stems in many ways. The stem of the sugarcane plant is a primary source of table sugar; cinnamon and the drug quinine are derived from the bark of *Cinnamomum verum* and various *Cinchona* species, respectively; and wood is necessary for the production of paper, as building materials, and as fuel in many parts of the world.

Check Your Progress   24.4

1. Describe transport tissues that are found in a vascular bundle.
2. Compare the arrangement of the vascular bundles in monocot stems and eudicot stems.
3. Contrast primary growth with secondary growth.
4. List the components of bark.
5. Compare the features found in the annual rings of spring wood and summer wood.

24.5 Organization and Diversity of Leaves

Leaves are the organs of photosynthesis in most vascular plants. As mentioned earlier, a leaf usually consists of a flattened blade and a petiole connecting the blade to the stem. The blade may be single or composed of several leaflets. Externally, it is possible to see the pattern of the leaf veins, which contain vascular tissue. Leaf veins have a net pattern in eudicot leaves and a parallel pattern in monocot leaves (see Fig. 24.3).

Figure 24.20 Leaf structure.   Photosynthesis takes place in mesophyll tissue of leaves. The leaf is enclosed by epidermal cells covered with a waxy layer, the cuticle. Leaf hairs are also protective. The veins contain xylem and phloem for the transport of water and solutes. A stoma is an opening in the epidermis that permits the exchange of gases.

Leaf Morphology

Figure 24.20 shows a cross section of a typical eudicot leaf of a temperate zone plant. At the top and bottom are layers of epidermal tissue that often bear trichomes, protective hairs often modified as glands that secrete irritating substances. These features help deter insects from eating the leaf. The epidermis characteristically has an outer, waxy cuticle that helps keep the leaf from drying out. The cuticle also prevents gas exchange because it is not gas permeable. However, the lower epidermis of eudicot and both surfaces of monocot leaves contain stomata that allow gases to move into and out of the leaf. Water loss also occurs at stomata, but each stoma has two guard cells that regulate its opening and closing, and stomata close when the weather is hot and dry.

The body of a leaf is composed of **mesophyll** [Gk. *mesos,* middle, and *phyllon,* leaf] tissue. Most eudicot leaves have two distinct regions: **palisade mesophyll,** containing elongated cells, and **spongy mesophyll,** containing irregular cells bounded by air spaces. The parenchyma cells of these layers have many chloroplasts and carry on most of the photosynthesis for the plant. The loosely packed arrangement of the cells in the spongy layer increases the amount of surface area for gas exchange.

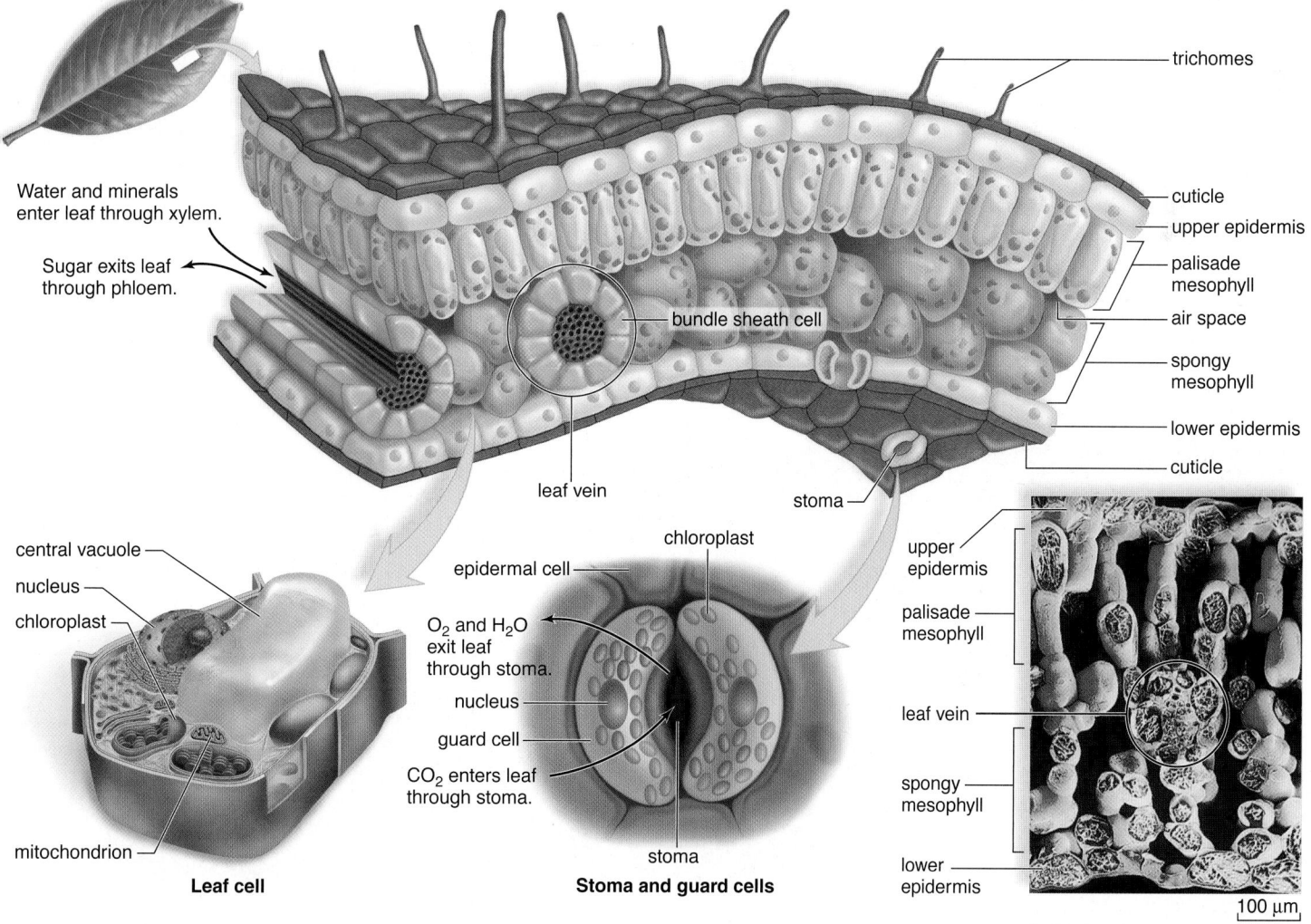

Water and minerals enter leaf through xylem.

Sugar exits leaf through phloem.

bundle sheath cell

leaf vein

trichomes

cuticle
upper epidermis
palisade mesophyll
air space
spongy mesophyll
lower epidermis
cuticle

stoma

central vacuole
nucleus
chloroplast

epidermal cell

chloroplast

O_2 and H_2O exit leaf through stoma.

nucleus

guard cell

CO_2 enters leaf through stoma.

mitochondrion

stoma

Leaf cell

Stoma and guard cells

upper epidermis

palisade mesophyll

leaf vein

spongy mesophyll

lower epidermis

100 μm

SEM of leaf cross section

Leaf Diversity

The blade of a leaf can be simple or compound (Fig. 24.21). A simple leaf has a single blade in contrast to a compound leaf, which is divided in various ways into leaflets. An example of a plant with simple leaves is a magnolia, and a plant with compound leaves is a pecan tree. Pinnately compound leaves have the leaflets occurring in pairs, such as in a black walnut tree, while palmately compound leaves have all of the leaflets attached to a single point, as in a buckeye tree.

Leaves can be arranged on a stem in three ways: alternate, opposite, or whorled. The leaves are alternate in the American beech; in a maple, the leaves are opposite, with two leaves being attached to the same node. Bedstraw has a whorled leaf arrangement with several leaves originating from the same node.

Leaves are adapted to environmental conditions. Plants that grow in shade tend to have broad, wide leaves, and desert plants tend to have reduced leaves with sunken stomata. The spines of a cactus are actually modified leaves attached to the succulent (water-containing) stem (Fig. 24.22*a*).

An onion bulb is made up of leaves surrounding a short stem. In a head of cabbage, large leaves overlap one another. The petiole of a leaf can be thick and fleshy, as in celery and rhubarb. Climbing leaves, such as those of peas and cucumbers, are modified into tendrils that can attach to nearby objects (Fig. 24.22*b*).

The leaves of a few plants are specialized for catching insects. A sundew has sticky trichomes that trap insects and other trichomes that secrete digestive enzymes. The Venus flytrap has hinged leaves that snap shut and interlock when an

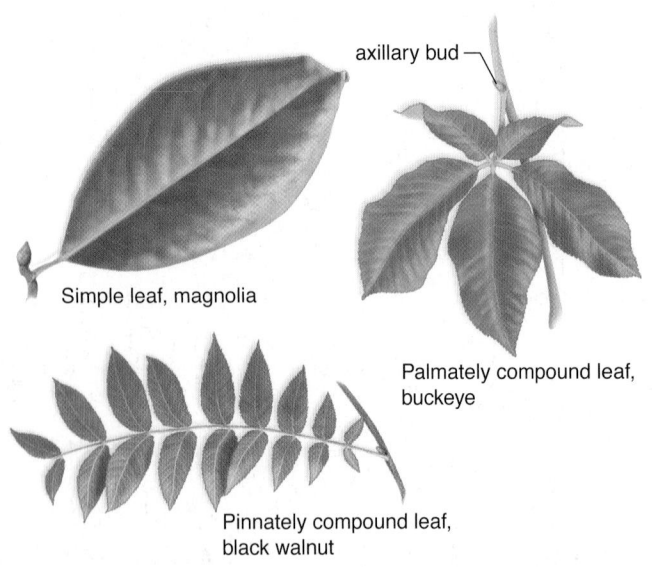

Simple leaf, magnolia

axillary bud

Palmately compound leaf, buckeye

Pinnately compound leaf, black walnut

a. Simple versus compound leaves

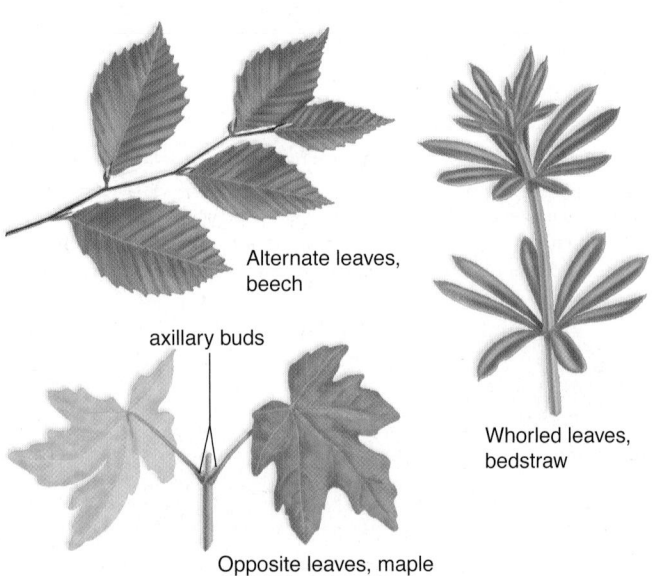

Alternate leaves, beech

axillary buds

Whorled leaves, bedstraw

Opposite leaves, maple

b. Arrangement of leaves on stem

Figure 24.21 Classification of leaves.   **a.** Leaves are either simple or compound, being either pinnately compound or palmately compound. Note the one axillary bud per compound leaf. **b.** Leaf arrangement on stem can be alternate, opposite, or whorled.

stem   spine

a. Cactus, *Opuntia*

tendril

b. Cucumber, *Cucumis*

Figure 24.22 Leaf diversity.   **a.** The spines of a cactus plant are modified leaves that protect the fleshy stem from animal predation. **b.** The tendrils of a cucumber are modified leaves that attach the plant to a physical support. **c.** The modified leaves of the Venus flytrap serve as a trap for insect prey. When triggered by an insect, the leaf snaps shut. Once shut, the leaf secretes digestive juices that break down the soft parts of the insect's body.

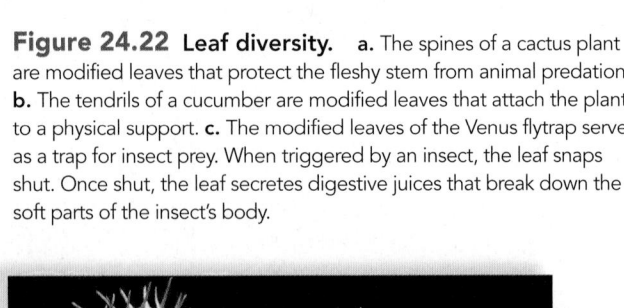

trigger hairs

c. Venus flytrap, *Dionaea*

insect triggers sensitive trichomes that project from inside the leaves (Fig. 24.22c). Certain leaves of a pitcher plant resemble a pitcher and have downward-pointing hairs that lead insects into a pool of digestive enzymes secreted by trichomes. Insectivorous plants commonly grow in marshy regions, where the supply of soil nitrogen is severely limited. The digested insects provide the plants with a source of organic nitrogen.

Video Carnivorous Plant

Check Your Progress 24.5

1. Explain the importance of the leaf tissue called mesophyll.
2. Compare the structural and functional differences between the palisade and spongy mesophyll.
3. Give examples of different types of leaves and their functions.

CONNECTING *the* CONCEPTS *with the* BIG IDEAS

Energy and Homeostasis

- Root hairs allow plants to increase surface area and allow greater exchange of materials. (2A3b1*IE*)

Interactions and Systems

- The roots, stems, and leaves of a plant interact in various ways to accomplish essential life functions such as gas exchange, waste removal, and materials transport. (4A4a*IE*, 4B2a2)
- The vascular system provides structural support and allows the flow of water and nutrients to and from the leaves. (4A4b*IE*)
- Specializations in roots, stems and leaves are often related to acquiring energy and needed materials in particular environments. (4A6g)

*Find the unabridged version of all EK citations at www.glencoe.com/maderAP11.

Media Study Tools

www.glencoe.com/maderAP11

Enhance your study of this chapter with study tools and practice tests. Also ask your instructor about the resources available through ConnectPlus, including the media-rich eBook, interactive learning tools, and animations.

Summarize

24.1 Organs of Flowering Plants

Flowering plants have three vegetative organs. The root anchors the plant and absorbs water and minerals, as well as storing the products of photosynthesis. Stems produce new tissues, support the leaves, transport minerals from the root system to the leaves and back, as well as storing plant products. Leaves are specialized for gas exchange and carry on the majority of photosynthesis within the plant.

Flowering plants are classified into the monocots and the eudicots. This distinction is based upon the number of cotyledons in the seed, the arrangement of the vascular tissue in the roots, stems, and leaves, and the number of flowering parts.

24.2 Tissues of Flowering Plants

Flowering plants contain apical meristems along with three types of primary meristems. The protoderm produces the epidermal tissue. Within the roots, the epidermal cells bear root hairs, and in the leaves, the epidermis contains the guard cells. In a woody stem the epidermis is replaced by the periderm.

The procambium produces vascular tissue. The vascular tissue consists of xylem and phloem. The xylem contains two types of conducting cells: vessel elements and tracheids. Vessel elements are larger and have perforated plates that form a continuous pipeline from the roots to the leaves. The elongated tracheids contain tapered ends that allow water to move through pits located in the end walls and side walls. Xylem transports water and minerals.

The phloem contains sieve tubes that are made up of sieve-tube members, each of which is associated with a companion cell. The phloem transports sucrose and other organic compounds, including plant hormones.

24.3 Organization and Diversity of Roots

The root tip has three main zones: the zone of cell division (contains primary meristems), the zone of elongation, and the zone of maturation.

A cross section of a herbaceous eudicot root reveals the epidermis, which functions in protection; the cortex, which stores food; the endodermis, which regulates the movement of minerals; and the vascular cylinder, which is composed of vascular tissue. Within the vascular cylinder of a eudicot, the xylem appears star-shaped; the phloem is found in separate regions between the points of the star. In contrast, a monocot root has a ring of vascular tissue with alternating bundles of xylem and phloem surrounding the pith.

Roots are highly diverse. Taproots are specialized to store the products of photosynthesis; a fibrous root system anchors the plant to the ground with a wide network of fibers; prop roots are adventitious roots that also increase the anchorage of the plant to the ground.

24.4 Organization and Diversity of Stems

The activity of the shoot apical meristem accounts for the primary growth of a stem. The terminal bud contains internodes and leaf primordia at the nodes. The lengthening of the internodes allows for stem growth.

A cross section of a nonwoody eudicot stem reveals epidermis, cortex, vascular bundles in a ring, and an inner pith. Monocots stems have scattered vascular bundles and a cortex and pith that are not well defined.

Secondary growth of a woody stem is due to the vascular cambium, which produces new xylem and phloem on an annual basis. The cork cambium produces new cork cells when needed. Cork is part of the bark and replaces the epidermis in woody plants. In a cross section of a woody stem all of the tissue outside of the vascular cambium is bark. It consists of secondary phloem, cork cambium, and cork. Wood consists of secondary xylem that builds up year after year and forms annual rings.

24.5 Organization and Diversity of Leaves

The bulk of a leaf is made up of mesophyll tissue that is bordered by an upper and lower layer of epidermis. The epidermis is covered by a cuticle that may contain trichomes. Stomata tend to be located in the lower layer of the leaf; vascular tissue is found in the leaf veins.

Leaves come in a variety of forms. Cacti spines, onion bulbs, and the tendrils of peas are examples of different types of leaves.

Key Terms

annual ring 456	monocot 446
apical meristem 447	node 445
axillary bud 445	organ 444
bark 456	palisade mesophyll 459
blade 445	parenchyma 448
Casparian strip 451	perennial 444
collenchyma 448	pericycle 451
complex tissue 449	periderm 447
cork 447	petiole 445
cork cambium 447	phloem 449
cortex 451	pit 449
cuticle 447	pith 452
deciduous 445	primary root 452
endodermis 451	rhizome 458
epidermal tissue 447	root apical meristem 450
epidermis 447, 451	root cap 450
eudicot 446	root hair 447
evergreen 445	root nodule 452
fibrous root system 452	root system 444
ground tissue 447	sclerenchyma 449
herbaceous stem 454	shoot apical meristem 453
internode 445	shoot system 444
leaf vein 450	sieve-tube member 450
leaves 445	spongy mesophyll 459
lignin 449	stem 445
meristem 447	stolon 456
mesophyll 459	stomata (sing., stoma) 447

taproot 452	vascular cylinder 450
terminal bud 453	vascular tissue 447
tracheid 449	vessel element 449
trichome 447	wood 456
vascular bundle 450	xylem 449
vascular cambium 454	

 **Assess**

Reviewing This Chapter

1. Name and discuss the vegetative organs of a flowering plant. 444–45
2. List five differences between monocots and eudicots.   446
3. Epidermal cells are found in what type of plant tissue? Explain how epidermis is modified in various organs of a plant. Contrast an epidermal cell with a cork cell.   447
4. Contrast the structure and function of parenchyma, collenchyma, and sclerenchyma cells. These cells occur in what type of plant tissue?   448–49
5. Contrast the structure and function of xylem and phloem. Xylem and phloem occur in what type of plant tissue?   449–50
6. Name and discuss the zones of a root tip. Trace the path of water and minerals across a root from the root hairs to xylem. Be sure to mention the Casparian strip.   450–51
7. Contrast a taproot with a fibrous root system. What are adventitious roots?   452
8. Describe the primary growth of a stem.   455–56
9. Describe cross sections of a herbaceous eudicot, a monocot, and a woody stem.   454–56
10. Discuss the diversity of stems by giving examples of several adaptations.   457–58
11. Describe the structure and organization of a typical eudicot leaf.   459
12. Note the diversity of leaves by giving examples of several adaptations.   460–61

Testing Yourself

Choose the best answer for each question.

1. Which of these is an incorrect contrast between monocots (stated first) and eudicots (stated second)?
   a. one cotyledon—two cotyledons
   b. leaf veins parallel—net veined
   c. vascular bundles in a ring—vascular bundles scattered
   d. flower parts in threes—flower parts in fours or fives
   e. All of these are correct contrasts.

2. Which of these types of cells is most likely to divide?
   a. parenchyma   d. xylem
   b. meristem   e. sclerenchyma
   c. epidermis

3. Which of these cells in a flowering plant is apt to be nonliving?
   a. parenchyma   d. epidermal cells
   b. collenchyma   e. guard cells
   c. sclerenchyma

4. Root hairs are found in the zone of
   a. cell division.   d. apical meristem.
   b. elongation.   e. All of these are correct.
   c. maturation.

5. Cortex is found in
   a. roots, stems, and leaves. d. stems and leaves.
   b. roots and stems.   e. roots only.
   c. roots and leaves.

6. Between the bark and the wood in a woody stem, there is a layer of meristem called
   a. cork cambium.   d. the zone of cell division.
   b. vascular cambium.   e. procambium preceding bark.
   c. apical meristem.

7. Which part of a leaf carries on most of the photosynthesis of a plant?
   a. epidermis   d. guard cells
   b. mesophyll   e. Both a and b are correct.
   c. epidermal layer

8. Annual rings are the
   a. internodes in a stem.
   b. rings of vascular bundles in a monocot stem.
   c. layers of xylem in a woody stem.
   d. bark layers in a woody stem.
   e. Both b and c are correct.

9. The Casparian strip is found
   a. between all epidermal cells.
   b. between xylem and phloem cells.
   c. on four sides of endodermal cells.
   d. within the secondary wall of parenchyma cells.
   e. in both endodermis and pericycle.

10. Which of these is a stem?
    a. taproot of carrots   d. prop roots
    b. stolon of strawberry plants e. Both b and c are correct.
    c. spine of cacti

11. Meristem tissue that gives rise to epidermal tissue is called
    a. procambium.   d. protoderm.
    b. ground meristem. e. periderm.
    c. epiderm.

12. New plant cells originate from the
    a. parenchyma.   d. base of the shoot.
    b. collenchyma.   e. apical meristem.
    c. sclerenchyma.

13. Ground tissue does not include
    a. collenchyma cells.   c. parenchyma cells.
    b. sclerenchyma cells.   d. chlorenchyma cells.

14. Evenly thickened cells that function to support mature regions of a flowering plant are called
    a. guard cells.   d. sclerenchyma cells.
    b. aerenchyma cells.   e. xylem cells.
    c. parenchyma cells.

15. Roots
    a. are the primary site of photosynthesis.
    b. give rise to new leaves and flowers.
    c. have a thick cuticle to protect the epidermis.
    d. absorb water and nutrients.
    e. contain spores.

16. Monocot stems have
    a. vascular bundles arranged in a ring.
    b. vascular cambium.
    c. scattered vascular bundles.
    d. a cork cambium.
    e. a distinct pith and cortex.

17. Secondary thickening of stems occurs in
    a. all angiosperms.   c. many eudicots.
    b. most monocots.   d. few eudicots.

18. How are compound leaves distinguished from simple leaves?
    a. Compound leaves do not have axillary buds at the base of leaflets.
    b. Compound leaves are smaller than simple leaves.
    c. Simple leaves are usually deciduous.
    d. Compound leaves are found only in pine trees.
    e. Simple leaves are found only in gymnosperms.

19. Label this root using these terms: endodermis, phloem, xylem, cortex, and epidermis.

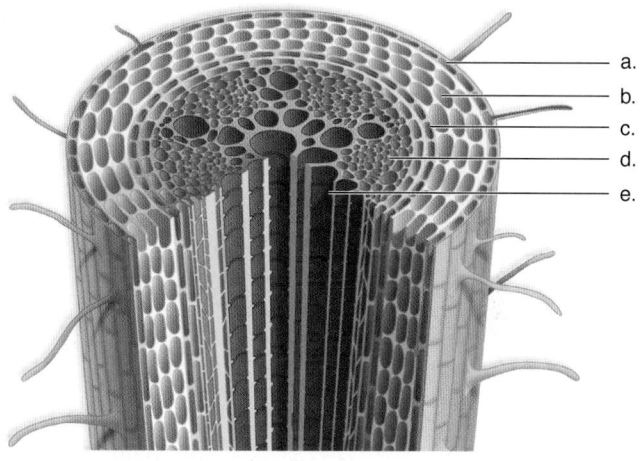

a.
b.
c.
d.
e.

20. Label this leaf using these terms: leaf vein, lower epidermis, palisade mesophyll, spongy mesophyll, and upper epidermis.

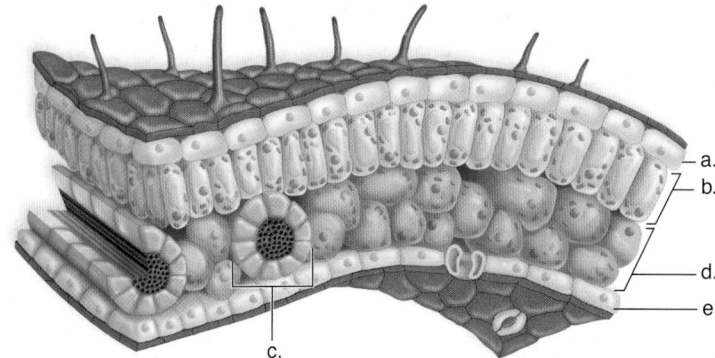

a.
b.
c.
d.
e.

Engage

Thinking Scientifically

1. Utilizing an electron microscope, how might you confirm that a companion cell communicates with its sieve-tube member?

2. Design an experiment that tests the hypothesis that new plants arise at the nodes of a stolon according to environmental conditions (temperature, water, and sunlight).

25

Flowering Plants: Nutrition and Transport

BEFORE YOU BEGIN

Before beginning this chapter, take a few moments to review the following discussions.

Section 2.3   Which properties of water are essential for conduction of water from the root system to the leaves?

Figure 8.12   What are the basic requirements for photosynthesis and cellular respiration?

Section 22.3   Which organisms have evolved symbiotic relationships with flowering plants?

Flowers may be colored artificially by taking advantage of their transport systems.

Every year during homecoming, proms, and the thousands of weddings that occur around the country, we see a dazzling array of floral creations. Blue carnations, green daisies, and purple roses are artificially colored to increase the variations available to consumers. Florists have learned how to alter flower color by using the plants' natural conducting system.

As a flower blossoms, the petals unfold as it matures and enters the blooming stage. The resulting blossom will be the color dictated by the plant's genes. Daisies naturally come in pink, yellow, blue, and white, while carnations are naturally found in white, pink, and red.

To accomplish the artificial color change, the florist needs the flower and its stem. The stem is cut under water to prevent air bubbles from getting trapped within the conduction tubes of the stem. An air bubble will block the transport of fluid up the stem. The flowers are placed in a vase of water containing dye. The dye is transported up the stem and into the flower due to water potential and the cohesion of water, both of which are described in this chapter. A wide variety of floral color can be created by using dye and the natural conducting system within the plant.

In this chapter we explore how plants use water to conduct essential minerals and nutrients throughout their systems—from the highest leaves to the tips of the deepest roots, and vice versa.

As you read through the chapter, think about the following questions:

1. Which nutrients are essential for plant growth?

2. What structures enable plants to absorb water and minerals from the soil?

3. Why does fluid "leak" from a branch when it is cut?

FOLLOWING *the* BIG IDEAS

CHAPTER 25   FLOWERING PLANTS: NUTRITION AND TRANSPORT

Energy and Homeostasis | Acquisition of water and nutrients by plants involve specialized mechanisms and structures.

25.1 Plant Nutrition and Soil

Learning Outcomes

Upon completion of this section, you should be able to

1. Identify the macronutrients and micronutrients that are required by plants.
2. List the ions that are absorbed or leached away by a root system.
3. Explain a simplified soil profile.

The ancient Greeks believed that plants were "soil-eaters" and somehow converted soil into plant material. Apparently to test this hypothesis, a seventeenth-century Dutchman named Jean-Baptiste van Helmont (1579–1644) planted a willow tree weighing 5 lb in a large pot containing 200 lb of soil. He watered the tree regularly for five years and then reweighed both the tree and the soil. The tree weighed 170 lb, and the soil weighed only a few ounces less than the original 200 lb. Van Helmont concluded that the increase in weight of the tree was due primarily to the addition of water.

Water is a vitally important nutrient for a plant, but van Helmont was unaware that water and carbon dioxide (diffusing through the leaves) combine in the presence of sunlight to produce carbohydrates, the chief

**Figure 25.1
Overview of plant nutrition.** Carbon dioxide, which enters leaves, and water, which enters roots, are combined during photosynthesis to form carbohydrates, with the release of oxygen from the leaves. Root cells, and all other plant cells, carry on cellular respiration, which uses oxygen and gives off carbon dioxide. Aside from the elements carbon, hydrogen, and oxygen, plants require nutrients that are absorbed as minerals by the roots.

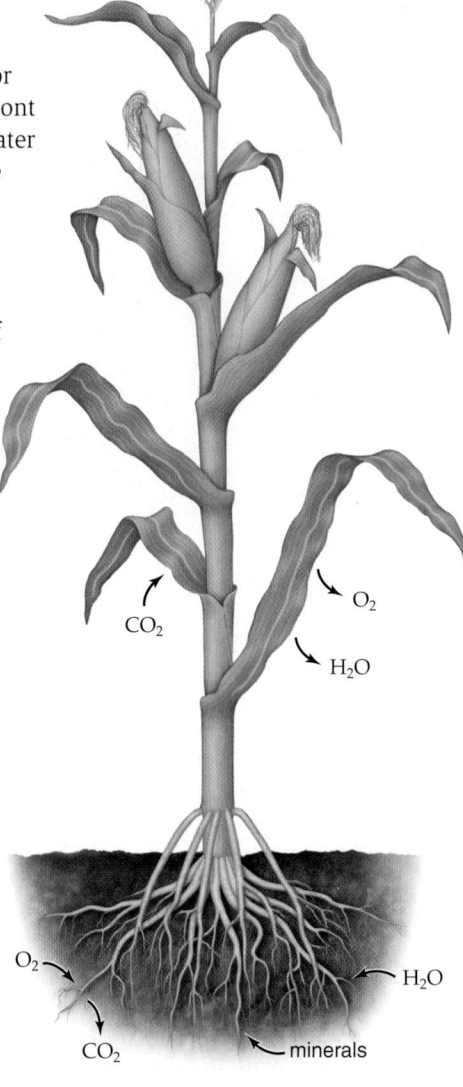

organic matter of plants. Much of the water entering a plant evaporates at the leaves. Roots, like all plant organs, carry on cellular respiration, a process that uses oxygen and gives off carbon dioxide (Fig. 25.1).

Essential Inorganic Nutrients

Approximately 95% of a typical plant's dry weight (weight excluding free water) is carbon, hydrogen, and oxygen. Why? Because these are the elements found in most organic compounds, such as carbohydrates. Carbon dioxide (CO_2) supplies the carbon, and water (H_2O) supplies the hydrogen and oxygen found in the organic compounds of a plant.

In addition to carbon, hydrogen, and oxygen, plants require certain other nutrients that are absorbed as minerals by the roots. A **mineral** is an inorganic substance usually containing two or more elements. Why do plants need minerals from the soil? In plants, nitrogen is a major component of nucleic acids and proteins, magnesium is a component of chlorophyll, and iron is a building block of cytochrome molecules.

The major functions of the **essential nutrients** for plants are listed in Table 25.1. A nutrient is essential if (1) it has an identifiable role, (2) no other nutrient can substitute and fulfill the same role, and (3) a deficiency of this nutrient causes a plant to die without completing its life cycle. Essential nutrients are divided into **macronutrients** and **micronutrients** according to their relative concentrations in plant tissue. The following diagram shows the macronutrients and the micronutrients for plants:

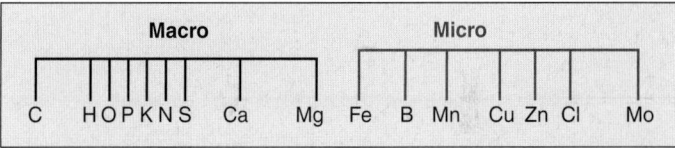

Beneficial nutrients are another category of elements taken up by plants. **Beneficial nutrients** are either required for growth or enhance the growth of a particular plant. Horsetails require silicon as a mineral nutrient, and sugar beets show enhanced growth in the presence of sodium. Nickel is a beneficial mineral nutrient in soybeans when root nodules are present. Aluminum is used by some ferns, and selenium, which is often fatally poisonous to livestock, is utilized by locoweeds.

Determination of Essential Nutrients

When a plant is burned, its nitrogen component is given off as ammonia and other gases, but most other essential minerals remain in the ash. The presence of a mineral in the ash, however, does not necessarily mean that the plant normally requires it. The preferred method for determining the mineral requirements of a plant was developed at the end of the nineteenth century by the German plant physiologists Julius von Sachs (1832–97) and Wilhem Knop (1817–91). This method is called water culture, or **hydroponics** [Gk. *hydrias*, water, and *ponos*, hard work]. Hydroponics allows plants to grow well if they are supplied with all the nutrients they need. The investigator omits

Table 25.1 Some Essential Inorganic Nutrients in Plants

Elements	Symbol	Form	Major Functions
Macronutrients			
Carbon	C	CO_2	Major component of organic molecules
Hydrogen	H	H_2O	
Oxygen	O	O_2	
Phosphorus	P	$H_2PO_4^-$ HPO_4^{2-}	Part of nucleic acids, ATP, and phospholipids
Potassium	K	K^+	Cofactor for enzymes; water balance and opening of stomata
Nitrogen	N	NO_3^- NH_4^+	Part of nucleic acids, proteins, chlorophyll, and coenzymes
Sulphur	S	SO_4^{2-}	Part of amino acids, some coenzymes
Calcium	Ca	Ca^{2+}	Regulates responses to stimuli and movement of substances through plasma membrane; involved in formation and stability of cell walls
Magnesium	Mg	Mg^{2+}	Part of chlorophyll; activates a number of enzymes
Micronutrients			
Iron	Fe	Fe^{2+} Fe^{3+}	Part of cytochrome needed for cellular respiration; activates some enzymes
Boron	B	BO_3^{3-} $B_4O_7^{2-}$	Role in nucleic acid synthesis, hormone responses, and membrane function
Manganese	Mn	Mn^{2+}	Required for photosynthesis; activates some enzymes such as those of the citric acid cycle
Copper	Cu	Cu^{2+}	Part of certain enzymes, such as redox enzymes
Zinc	Zn	Zn^{2+}	Role in chlorophyll formation; activates some enzymes
Chlorine	Cl	Cl^-	Role in water-splitting step of photosynthesis and water balance
Molybdenum	Mo	MoO_4^{2-}	Cofactor for enzyme used in nitrogen metabolism

a. Solution lacks nitrogen   Complete nutrient solution

b. Solution lacks phosphorus   Complete nutrient solution

c. Solution lacks calcium   Complete nutrient solution

Figure 25.2 Nutrient deficiencies. The nutrient cause of poor plant growth is diagnosed when plants are grown in a series of complete nutrient solutions except for the elimination of just one nutrient at a time. These experiments show that sunflower plants respond negatively to a deficiency of **(a)** nitrogen, **(b)** phosphorus, and **(c)** calcium.

a particular mineral and observes the effect on plant growth. If growth suffers, it can be concluded that the omitted mineral is an essential nutrient (Fig. 25.2).

This method has been more successful for macronutrients than for micronutrients. For studies involving the latter, the water and the mineral salts used must be absolutely pure; however, purity is difficult to attain, because even instruments and glassware can introduce micronutrients. Then, too, the element in question may already be present in the seedling used in the experiment. These factors complicate the determination of essential plant micronutrients by means of hydroponics.

Soil

Plants acquire carbon when the diffusion of carbon dioxide occurs at the stomata. Oxygen can also enter from the air, but all of the other essential nutrients are absorbed by roots from the soil. It would not be an exaggeration to say that terrestrial life is dependent on the quality of the soil and the ability of soil to provide plants with the nutrients they need.

Soil Formation

Soil formation begins with the weathering of rock. Weathering first gradually breaks down rock to rubble and then to soil particles. Some weathering mechanisms, such as the freeze-thaw cycle of ice or the grinding of rock on rock by the action of glaciers or river flow, are purely mechanical. Other forces include a chemical effect, as when acidic rain leaches (washes away) soluble components of rock or when oxygen combines with the iron of rocks.

In addition to these forces, organisms also play a role in the formation of soil. Lichens and mosses grow on pure rock and trap particles that later allow grasses, herbs, and soil animals to follow. When these die, their remains are decomposed, notably by bacteria and fungi. Decaying organic matter, called **humus,** begins to accumulate. Humus supplies nutrients to plants, and its acidity also leaches minerals from rock.

Building soil takes a long time. Under ideal conditions, depending on the type of parent material (the original rock) and the various processes at work, a centimeter of soil may take 15 years to develop.

The Nutritional Function of Soil

Soil is defined as a mixture of mineral particles, decaying organic material, living organisms, air, and water, which together support the growth of plants. In a good agricultural soil, the first three components come together in such a way that there are spaces for air and water (Fig. 25.3). It's best if the soil contains particles of different sizes because only then can spaces for air be present. Roots take up oxygen from air spaces. Ideally, water clings to particles by capillary action and does not fill the spaces. Overwatering can block the airspaces and result in killing the plant.

Mineral Particles.   Mineral particles vary in size: Sand particles are the largest (0.05–2.0 mm in diameter); silt particles have an intermediate size (0.002–0.05 mm); and clay particles are the smallest (less than 0.002 mm). Soils are a mixture of these three types of particles. Because sandy soils have many large particles, they have large spaces, and the water drains readily between the particles. In contrast to sandy soils, a soil composed mostly of clay particles has small spaces that fill completely with water. Most likely, you have experienced the feel of sand and clay: Sand, having no moisture, flows right through your fingers, while clay clumps together in one large mass because of its water content.

Clay particles have another benefit that sand particles do not have. As Table 25.1 indicates, some minerals are negatively charged and others are positively charged. Clay particles are negative, and they can retain positively charged minerals such as calcium (Ca^{2+}) and potassium (K^+), preventing these minerals from being washed away by leaching. Plants exchange hydrogen ions for these minerals when they take them up (Fig. 25.3). If rain is acidic, its hydrogen ions displace positive mineral ions and cause them to drain away; this is one reason acid rain kills trees.

Because clay particles are unable to retain negatively charged NO_3^-, the nitrogen content of soil is apt to be low. Legumes (see Fig. 1.13) are sometimes planted to replenish the nitrogen in the soil in preference to relying solely on the addition of fertilizer.

The type of soil called loam is composed of roughly one-third each of sand, silt, and clay particles. This combination sufficiently retains water and nutrients while still allowing the drainage necessary to provide air spaces. Some of the most productive soils are loam.

Humus.   Humus, which mixes with the top layer of soil particles, increases the benefits of soil. Plants do well in soils that contain 10–20% humus.

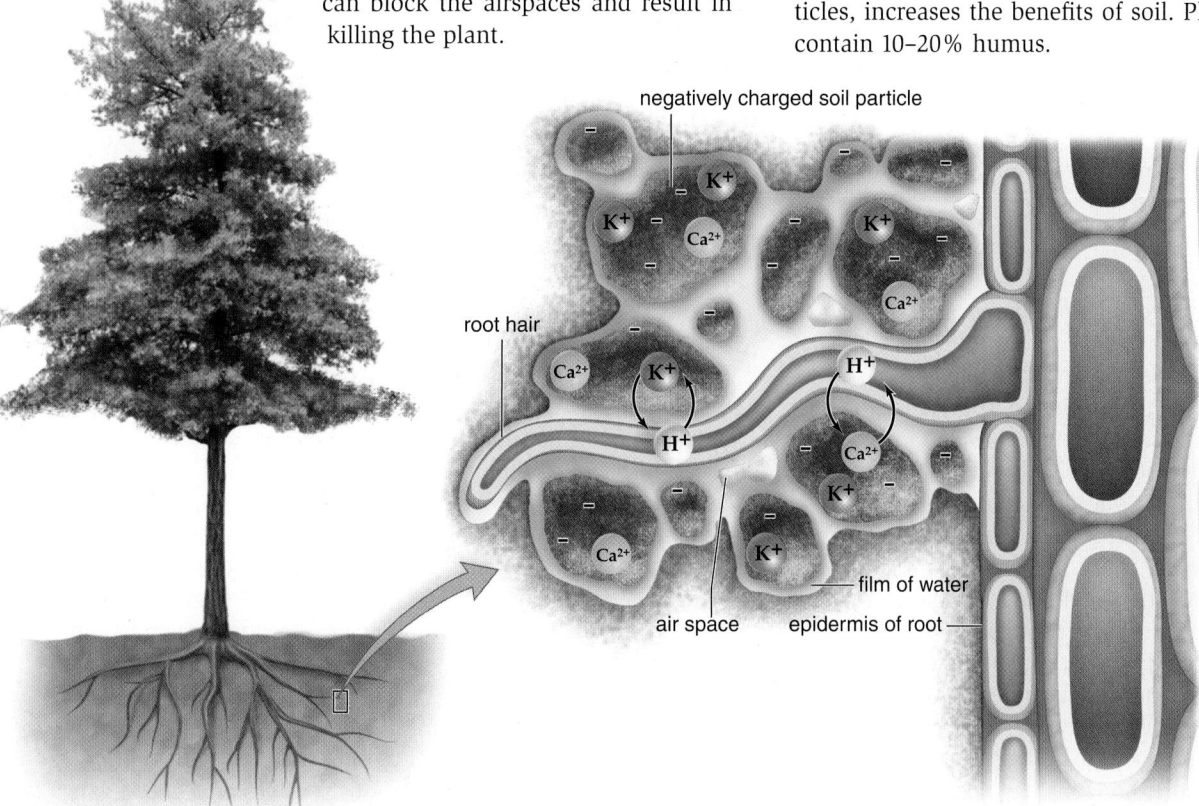

negatively charged soil particle

root hair

air space

film of water

epidermis of root

Figure 25.3
Absorbing minerals.
Negatively charged clay particles bind positively charged minerals such as Ca^{2+} and K^+. Plants extract these minerals by exchanging H^+ for them (see also Fig. 25.5).

Humus causes soil to have a loose, crumbly texture that allows water to soak in without doing away with air spaces. After a rain, the presence of humus decreases the chances of runoff. Humus swells when it absorbs water and shrinks as it dries. This action helps aerate soil.

Soil that contains humus is nutritious for plants. Humus is acidic; therefore, it retains positively charged minerals until plants take them up. When the organic matter in humus is broken down by bacteria and fungi, inorganic nutrients are returned to plants. Although soil particles are the original source of minerals in soil, recycling of nutrients, as you know, is a major characteristic of ecosystems.

Living Organisms.   Small plants play a major role in the formation of soil from bare rock. Due to the process of succession (see Fig. 45.13), larger plants eventually become dominant in certain ecosystems. The roots of larger plants penetrate soil even to the bedrock layer. This action slowly opens up soil layers, allowing water, air, and animals to follow.

A wide variety of animals dwell in the soil, at least part of the time. The largest of them, such as toads, snakes, moles, badgers, and rabbits, disturb and mix soil by burrowing. Smaller animals like earthworms ingest fine soil particles and deposit them on the surface as worm casts. Earthworms also loosen and aerate the soil. A range of small soil animals, including mites, springtails, and millipedes, help break down leaves and other plant remains by eating them. Soil-dwelling ants construct tremendous colonies with massive chambers and tunnels. These ants also loosen and aerate the soil.

The microorganisms in soil, such as protozoans, fungi, algae, and bacteria, are responsible for the final decomposition of organic remains in humus to inorganic nutrients. As mentioned, plants are unable to make use of atmospheric nitrogen (N_2), and soil bacteria play an important nutrient role because they make nitrate available to plants.

Insects may improve the properties of soil, but they can be a major crop pest when they feed on plant roots. Certain soil organisms, such as grubs, can severely impact our lawns by feeding on the roots of the grass.

Soil Profiles

A **soil profile** is a vertical section from the ground surface to the unaltered rock below. Usually, a soil profile has parallel layers known as **soil horizons.** Mature soil generally has three horizons (Fig. 25.4). The *A horizon* is the uppermost (or topsoil) layer that contains litter and humus, although most of the soluble chemicals may have been leached away. The *B horizon* has little or no organic matter but does contain the inorganic nutrients leached from the A horizon. The *C horizon* is a layer of weathered and shattered rock.

Because the parent material (rock) and climate (e.g., temperature and rainfall) differ in various parts of the biosphere, the soil profile varies according to the particular ecosystem. Soils formed in grasslands tend to have a deep A horizon built up from decaying grasses over many years, but because of limited rain, little leaching into the B horizon has occurred. In forest soils, both the A and B horizons have enough inorganic

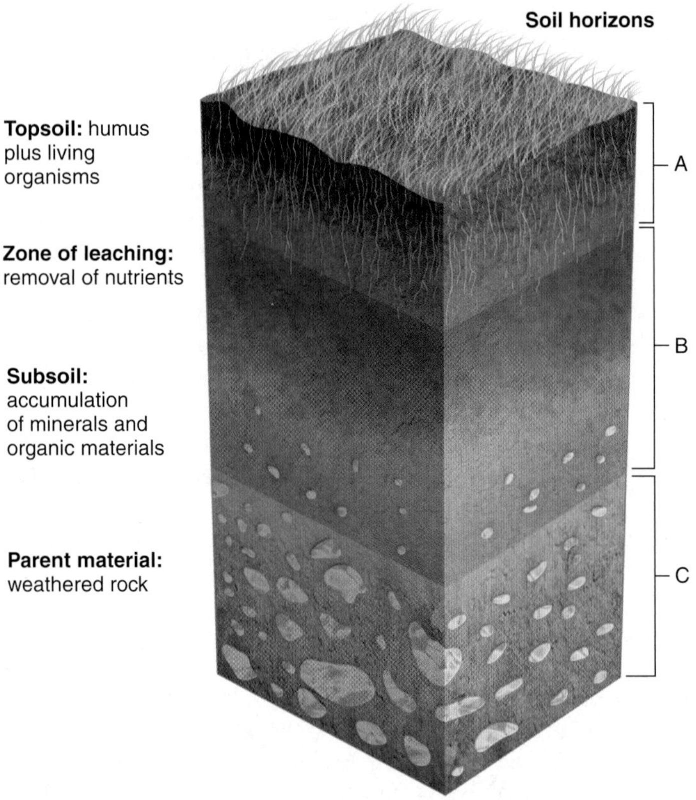

Figure 25.4 Simplified soil profile.   The top layer (A horizon) contains most of the humus; the next layer (B horizon) accumulates materials leached from the A horizon; and the lowest layer (C horizon) is composed of weathered parent material. Erosion removes the A horizon, a primary source of humus and minerals in soil.

nutrients to allow for root growth. In tropical rain forests, the A horizon is more shallow than the generalized profile, and the B horizon is deeper, signifying that leaching is more extensive. Since the topsoil of a rain forest lacks nutrients, it can only support crops for a few years before it is depleted.

Soil Erosion

Soil erosion occurs when water or wind carries soil away to a new location. Erosion removes about 25 billion tons of topsoil yearly, worldwide. If this rate of loss continues, some scientists predict that the Earth will lose practically all of its topsoil by the middle of the next century. Deforestation (removal of trees) and desertification (increase in deserts due to overgrazing and over-farming marginal lands) contribute to the occurrence of erosion, and so do poor farming practices in general.

In the United States, soil is eroding faster than it is being formed on about one-third of all cropland. Fertilizers and pesticides, carried by eroding soil into groundwater and rivers, are threatening human health. To make up for the loss of soil due to erosion, more energy is used to apply more fertilizers and pesticides to crops. Instead, it would be best to stop erosion before it occurs by following sound, environmentally sustainable agricultural practices.

The coastal wetlands are losing soil at a tremendous rate. These wetlands are important as nurseries for many species of organisms, such as shrimp and redfish, and as protection against storm surge from hurricanes. In Louisiana, 24 square miles of wetlands are lost each year. This equates to one football field being lost every 38 minutes. The storm surge produced by Hurricane Katrina in 2005 was exceptionally destructive due to the long-term loss of the wetlands protecting the shoreline. These wetlands were further impacted by the Deep Water Horizon oil spill of 2010.

Check Your Progress   25.1

1. Name the element(s), aside from C, H, and O, that are required to form proteins and nucleic acids. How does a plant acquire these elements?
2. Identify the benefits of leaving the remains of the previous year's crops in the field to overwinter.
3. List the benefits of humus in the soil.

Figure 25.5 Water and mineral uptake. **a.** Pathways of water and minerals. Water and minerals can travel via porous cell walls but then must enter endodermal cells because of the Casparian strip (pathway A). Alternatively, water and minerals can enter root hairs and move from cell to cell (pathway B). **b.** Transport of minerals across an endodermal plasma membrane. ❶ An ATP-driven pump removes hydrogen ions (H⁺) from the cell. ❷ This establishes an electrochemical gradient that allows potassium (K⁺) and other positively charged ions to cross the membrane via a channel protein. ❸ Negatively charged mineral ions (I⁻) can cross the membrane by way of a carrier when they "hitch a ride" with hydrogen ions (H⁺), which are diffusing down their concentration gradient.

25.2 Water and Mineral Uptake

Learning Outcomes

Upon completion of this section, you should be able to

1. Choose the correct order of mineral uptake across the plasma membrane within a plant cell.
2. Describe the mutualistic relationships that assist plants in acquiring nutrients from the soil.

Water and mineral uptake and transport within a plant occurs via a single pathway. As Figure 25.5a shows, water along with minerals can enter the root of a flowering plant from the soil simply by passing between the porous cell walls. Eventually, however, the *Casparian strip*, a band of suberin and lignin bordering four sides of root endodermal cells (see Fig. 24.8), forces water to enter endodermal cells. Alternatively, water can enter epidermal cells at their root hairs and then progress through cells across the cortex and endodermis of a root by means of cytoplasmic strands within plasmodesmata (see Fig. 5.15). Regardless of the pathway, water enters root cells when osmotic pressure in the root tissues is lower than that of the soil solution.

Mineral Uptake

In contrast to water, minerals can be actively taken up by plant cells. Plants possess an astonishing ability to concentrate minerals—that is, to take up minerals until they are many times more concentrated in the plant than in the surrounding medium. The concentration of certain minerals in roots is as much as 10,000 times greater than in the surrounding soil. The Nature of Science feature explains how this ability of plants can be exploited for environmental cleanup.

Following their uptake by root cells, minerals move into xylem and are transported into leaves by the upward movement

endodermis
pericycle
phloem
xylem
cortex

50 μm

vascular cylinder

pericycle

endodermis and Casparian strip

cortex

epidermis

pathway A of water and minerals

root hair

pathway B of water and minerals

a.

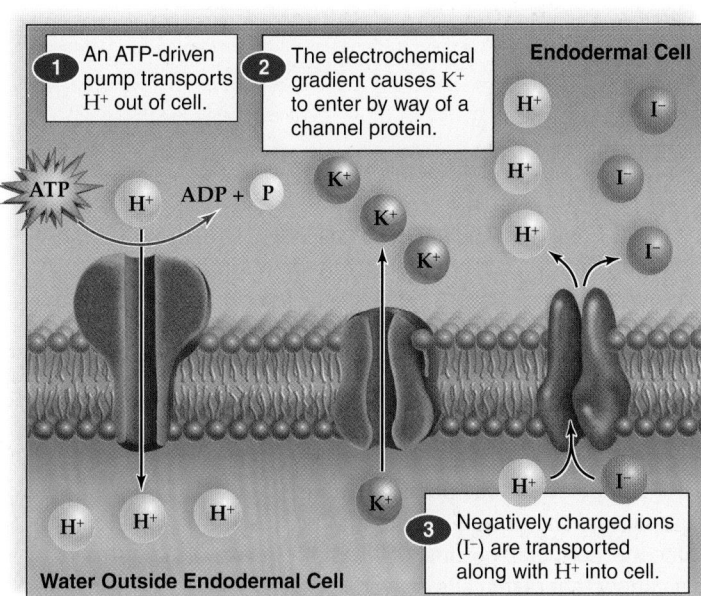

b.

Nature of Science

Plants Can Clean Up Toxic Messes

Phytoremediation uses plants like mulberry, poplar, and canola to clean up environmental pollutants. The genetic makeup of these plants allows them to absorb, store, degrade, or transform substances that normally kill or harm other plants and animals. "It's an elegantly simple solution to pollution problems," says Louis Licht, who runs Ecolotree, an Iowa City phytoremediation company.

The idea behind phytoremediation is not new; scientists have long recognized certain plants' abilities to absorb and tolerate toxic substances. But the idea of using these plants on contaminated sites has just gained support in the last twenty years. Different plants work on different contaminants. The mulberry bush, for instance, is effective on industrial sludge; some grasses attack petroleum wastes; and sunflowers (together with soil additives) remove lead.

The plants clean up sites in different ways depending on the substance involved. If it is an organic contaminant, such as spilled oil, the plants, or microbes around their roots, break down the substance. The remnants can either be absorbed by the plant or left in the soil or water. For an inorganic contaminant such as cadmium or zinc, the plants absorb the substance and trap it. The plants must then be harvested and disposed of, or processed to reclaim the trapped contaminant.

Figure 25A Poplar trees cleaning up nitrogen. Poplars are able to remove large amounts of nitrogen from runoff.

Poplars Take Up Excess Nitrates

Most trees planted along the edges of farms are intended to break the wind, but another use of poplars is to remove excess minerals from runoff. The poplars act like vacuum cleaners, sucking up nitrate-laden runoff from a fertilized cornfield before this run-off reaches a nearby brook—and perhaps other waters (Fig. 25A). Nitrate runoff into the Mississippi River from Midwest farms is a major cause of the large "dead zone" of oxygen-depleted water that develops each summer in the Gulf of Mexico.

In one case, a brook's nitrate levels were as much as ten times the amount considered safe. But then Licht, a University of Iowa graduate student, had the idea that poplars, which absorb lots of water and tolerate pollutants, could help. In 1991, Licht tested his hunch by planting poplars along a field owned by a corporate farm. The brook's nitrate levels subsequently dropped more than 90%, and the trees have thrived.

Canola Plants Take Up Selenium

Canola plants (*Brassica rapa* and *B. napa*) are grown in California's San Joaquin Valley to soak up excess selenium in the soil to help prevent an environmental catastrophe like the one that occurred there in the 1980s. Back then, irrigated farming caused naturally occurring selenium to rise to the soil surface. When excess water was pumped onto the fields, some selenium would flow

of water. Along the way, minerals can exit xylem and enter those cells that require them. Some eventually reach leaf cells. In any case, minerals must again cross a selectively permeable plasma membrane when they exit xylem and enter living cells.

By what mechanism do minerals cross plasma membranes? As it turns out, the energy of ATP is involved, but only indirectly. Recall that plant cells absorb minerals in the ionic form: Nitrogen is absorbed as nitrate (NO_3^-), phosphorus as phosphate (HPO_4^{2-}), potassium as potassium ions (K^+), and so forth. Ions cannot cross the plasma membrane because they are unable to enter the nonpolar portion of the lipid bilayer. Plant physiologists know that plant cells expend energy to actively take up and concentrate mineral ions. If roots are deprived of oxygen or are poisoned so that cellular respiration cannot occur, mineral ion uptake is diminished.

As shown in Figure 25.5*b*, a plasma-membrane pump, called a proton pump, hydrolyzes ATP and uses the energy released to transport hydrogen ions (H^+) out of the cell. The result is an electrochemical gradient that drives positively charged ions such as K^+ through a channel protein into the cell. Negatively charged mineral ions are transported, along with H^+, by carrier proteins. Because H^+ is moving down its concentration gradient, no energy is required. Notice that this model of mineral ion transport in plant cells is based on *chemiosmosis*, the establishment of an electrochemical gradient to perform work.

Animation
Proton Pump

off into drainage ditches, eventually ending up in Kesterson National Wildlife Refuge.

The selenium in ponds at the refuge accumulated in plants and fish and subsequently deformed and killed waterfowl, says Gary Bañuelos, a plant scientist with the U.S. Department of Agriculture who helped remedy the problem. He recommended that farmers add selenium-accumulating canola plants to their crop rotations. As a result, selenium levels in runoff are being managed. Although the underlying problem of excessive selenium in soils has not been solved, "this is a tool to manage mobile selenium and prevent another unlikely selenium-induced disaster," Bañuelos says.

Limitations of Phytoremediation

Phytoremediation does have its limitations, however. One of them is its slow pace.

Depending on the contaminant, it can take several growing seasons to clean a site—much longer than conventional methods. "We normally look at phytoremediation as a target of one to three years to clean a site," notes Edenspace's Mike Blaylock. "People won't want to wait much longer than that."

Phytoremediation is also only effective at depths that plant roots can reach, making it useless against deep-lying contamination unless the contaminated soils are excavated. Phytoremediation does not work on lead and other metals unless chemicals are added to the soil. Plants also vary in the amount of pollutant that they can phytoremediate (Fig. 25B).

Despite its shortcomings, experts see a bright future for this technology. For one thing, the costs are relatively small

compared to those of traditional remediation technologies. Traditional methods of cleanup require much energy input and therefore have higher cost. In general, phytoremediation is a low-cost alternative to traditional methods because less energy is required for operation and maintenance.

Phytoremediation is a promising solution to pollution problems, but according to the EPA's Walter W. Kovalick, "it's not a panacea. It's another arrow in the quiver. It takes more than one arrow to solve most problems."

Questions to Consider

1. What happens to the pollutants when the plant dies?
2. Why would one plant be more adapted at absorbing a particular pollutant than another plant?

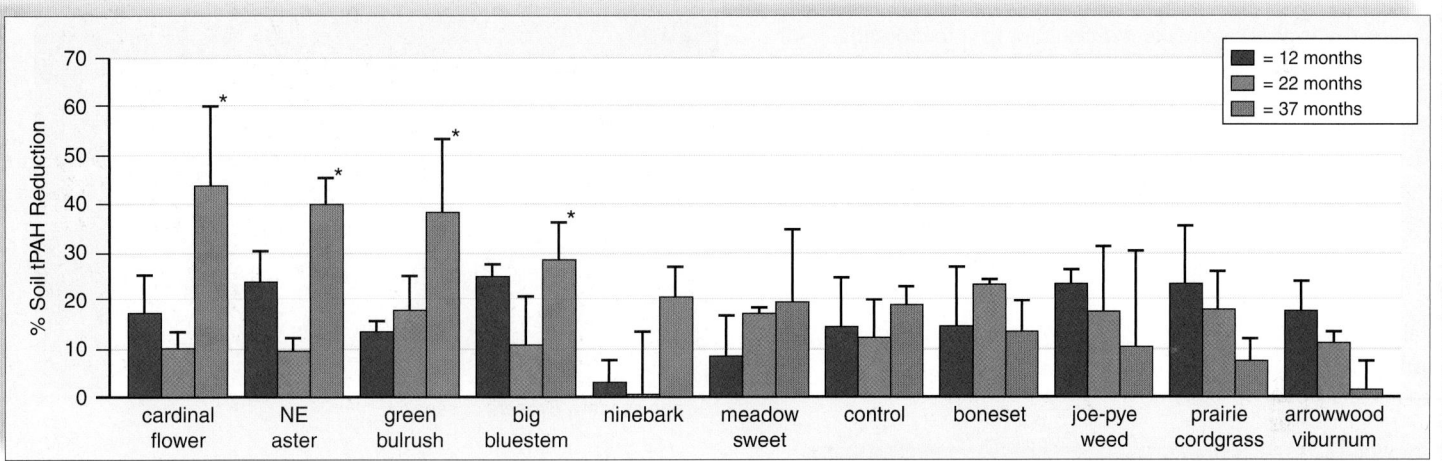

Figure 25B Phytoremediation ability of various plants. Plants vary in their ability to remove pollutants from the environment.

Source: Modified from Thomas, J. C., and Rugh, C. 2007. Final Report: Root Exudate Biostimulation for Polyaromatic Hydrocarbon Phytoremediation. Figure 6. EPA Grant R829479C020. Available at http://cfpub.epa.gov/ncer_abstracts/INDEX.cfm/fuseaction/display.abstractDetail/abstract/7556/report/F.

Adaptations of Roots for Mineral Uptake

Two mutualistic relationships assist roots in obtaining mineral nutrients. Root nodules involve a mutualistic relationship with bacteria, and mycorrhizae are a mutualistic relationship with fungi.

As described earlier, some plants, such as members of the legume family (e.g., beans, clover, and alfalfa), have roots colonized by *Rhizobium* bacteria, which can fix atmospheric nitrogen (N_2). They break the $N \equiv N$ bond and reduce nitrogen to NH_4^+ for incorporation into organic compounds. The bacteria live in **root nodules** [L. *nodulus*, dim. of *nodus*, knot] and are supplied with carbohydrates by the host plant (Fig. 25.6). The bacteria, in turn, furnish their host with nitrogen compounds.

Animation Root Nodule Formation

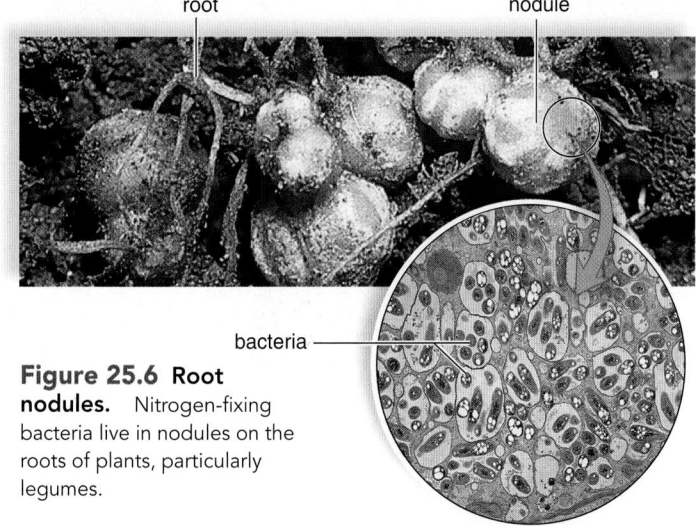

Figure 25.6 Root nodules. Nitrogen-fixing bacteria live in nodules on the roots of plants, particularly legumes.

Portion of infected cell

The second type of mutualistic relationship, called **mycorrhizae,** involves fungi and almost all plant roots (Fig. 25.7; see Section 22.3). Only a small minority of plants do not have mycorrhizae, and these plants are most often limited as to the environment in which they can grow. The fungus increases the surface area available for mineral and water uptake and breaks down organic matter in the soil, releasing nutrients that the plant can use. In return, the root furnishes the fungus with sugars and amino acids. Plants are extremely dependent on mycorrhizae. Orchid seeds, which are quite small and contain limited nutrients, do not germinate until a mycorrhizal fungus has invaded their cells.

Other means of acquiring nutrients also occur. Parasitic plants such as dodders, broomrapes, and pinedrops send out rootlike projections called haustoria that tap into the xylem and phloem of the host stem (Fig. 25.8*a*). Carnivorous plants such as the Venus flytrap and sundews obtain some nitrogen and minerals when their leaves capture and digest insects (Fig. 25.8*b*).

Check Your Progress   25.2

1. List the factors that make it difficult for ions to cross the plasma membrane.
2. Describe the significance to plants of nitrogen-fixing bacteria in the soil.
3. Explain how both partners benefit from a mycorrhizal association.

Figure 25.7 Mycorrhizae.   Plant growth is better when mycorrhizae are present.

Mycorrhizae present

Mycorrhizae not present

mycorrhizae

dodder (brown)

a. Dodder, *Cuscuta* sp.

Figure 25.8 Other ways to acquire nutrients.   **a.** Some plants, such as the dodder, are parasitic. **b.** Other plants, such as the sundew, are carnivorous.

bulbs release digestive enzymes

Sundew leaf enfolds prey

sticky hairs

narrow leaf form

b. Cape sundew, *Drosera capensis*

25.3 Transport Mechanisms in Plants

Learning Outcomes

Upon completion of this section, you should be able to

1. Describe the structure and function of xylem and phloem.
2. Identify the properties of water that influence water transportation with flowering plants.
3. Explain how environmental factors influence the opening and closing of stomata.
4. List the correct sequence of events during xylem and phloem transport.

Flowering plants are well adapted to living in a terrestrial environment. Their leaves, which carry on photosynthesis, are positioned to catch the rays of the Sun because they are held aloft by the stem (Fig. 25.9). Carbon dioxide enters leaves at the stomata, but water, the other main requirement for photosynthesis, is absorbed by the roots. Water must be transported from the roots through the stem to the leaves.

Reviewing Xylem and Phloem Structure

Vascular plants have a transport tissue, called *xylem,* that moves water and minerals from the roots to the leaves. Xylem, with its strong-walled, nonliving cells, gives trees much-needed internal support. Xylem contains two types of conducting cells: tracheids and vessel elements (see Fig. 24.6).

- *Tracheids* are tapered at both ends. The ends overlap with those of adjacent tracheids, and pits allow water to pass from one tracheid to the next.
- *Vessel elements* are long and tubular with perforation plates at each end. Vessel elements placed end to end form a completely hollow pipeline from the roots to the leaves.

The process of photosynthesis results in sugars, which are used as a source of energy and building blocks for other organic molecules throughout a plant. *Phloem* is the type of vascular tissue that transports organic nutrients to all parts of the plant. Roots buried in the soil cannot possibly carry on photosynthesis, but they still require a source of energy so that they can carry on cellular metabolism. In flowering plants, phloem consists of two types of cells: sieve-tube members and companion cells (see Fig. 24.7).

- *Sieve-tube members* are the conducting cells of phloem. The end walls are called sieve plates and have numerous pores; strands of cytoplasm extend from one sieve-tube member to another through the pores. Sieve-tube members lack nuclei.
- *Companion cells,* which do have nuclei, provide proteins to sieve-tube members.

In this way, sieve-tube members form a continuous *sieve tube* for organic nutrient transport throughout the plant.

Determining Xylem and Phloem Function

Knowing that vascular plants are structured in a way that allows materials to move from one part to another does not tell us the mechanisms by which these materials move. Plant physiologists have performed numerous experiments to determine how water and minerals rise to the tops of very tall trees in xylem, and how organic nutrients move in the opposite direction in phloem. We might expect that these processes are mechanical in nature and are based on the properties of water, because water is a large part of both *xylem sap* and *phloem sap.*

In living systems, water molecules diffuse freely across plasma membranes from the area of higher concentration to the area of lower concentration. Botanists favor describing the movement of water in terms of **water potential:** Water always flows passively from the area of higher water potential to the area of lower water potential. As can be seen in the Biological Systems feature on page 474, the concept of water potential has the benefit of considering water pressure in addition to osmotic pressure.

Chemical properties of water are also important in movement of xylem sap. The polarity of water molecules and the hydrogen bonding between water molecules allow water to fill xylem cells.

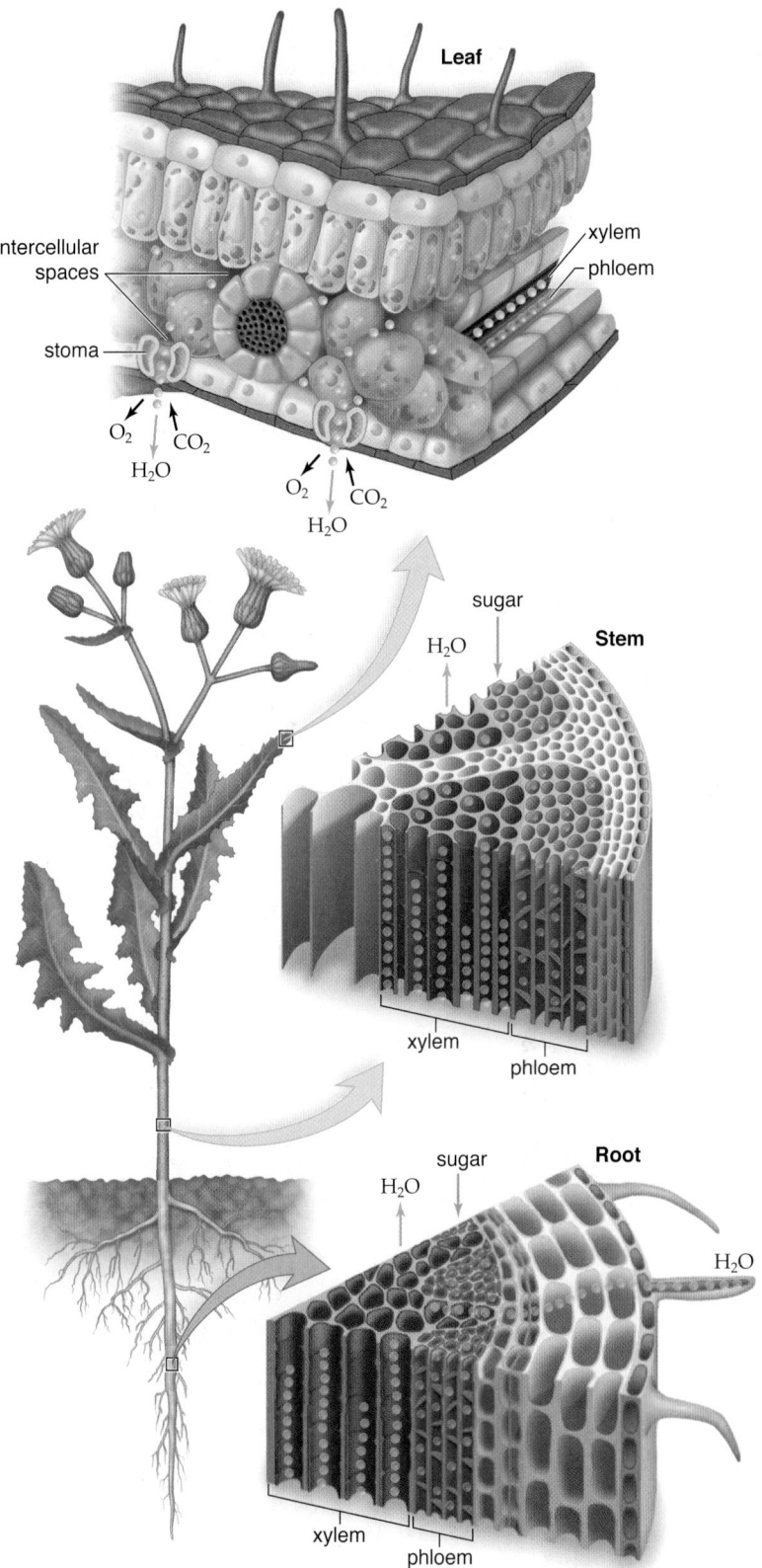

Figure 25.9 Plant transport system. Vascular tissue in plants includes xylem, which transports water and minerals from the roots to the leaves, and phloem, which transports organic nutrients oftentimes in the opposite direction. Notice that xylem and phloem are continuous from the roots through the stem to the leaves, which are the vegetative organs of a plant.

Biological Systems

The Concept of Water Potential

As you learned in Chapter 6, potential energy is stored energy. Potential energy can exist in an object's position, such as that of a boulder at the top of a hill (mechanical energy), or in chemical bonds, such as the bonds between phosphate groups in ATP (chemical energy). Kinetic energy is the energy of motion; it is energy actively engaged in doing work. A boulder rolling down a hillside is exhibiting kinetic energy, as is an enzyme reaction that breaks a bond, converting ATP to ADP and releasing energy in the process.

Water potential is defined as the mechanical energy of water. Just like the boulder, water at the top of a waterfall has a higher water potential than water at the bottom of the waterfall. As illustrated by this example, water moves from a region of higher water potential to a region of lower water potential.

In terms of cells, two factors usually determine water potential, which in turn determines the direction in which water moves across a plasma membrane. These factors concern differences in:

1. Water pressure across a membrane
2. Solute concentration across a membrane

Pressure potential is the effect that pressure has on water potential. Water moves across a membrane from an area of higher pressure to an area of lower pressure. The higher the water pressure, the higher the water potential, and vice versa. Pressure potential is the concept that best explains the movement of sap in xylem and phloem.

Osmotic potential, in contrast, takes into account the effects of solutes on the movement of water. The presence of solutes restricts the movement of water because water tends to engage in molecular interactions with solutes, such as hydrogen bonding. Water therefore tends to move across a membrane from an area of lower solute concentration to an area of higher solute concentration. The lower the concentration of solutes, the higher the water potential, and vice versa.

It is not surprising that increasing the water pressure can counter the effects of solute concentration. This situation is common in plant cells. As water enters a plant cell by osmosis because of the higher solute concentration, water pressure increases inside the cell—a plant cell has a strong cell wall that allows water pressure to build up. When the pressure potential inside the cell balances the osmotic potential outside the cell, the flow of water in and out becomes the same.

Pressure potential that increases due to osmosis is often called *turgor pressure.* Turgor pressure is critical because plants depend on it to maintain the turgidity of their bodies (Fig. 25C). The cells of a wilted plant have insufficient turgor pressure, and the plant droops as a result.

Questions to Consider

1. What variables will restrict the movement of water across the cell membrane?
2. What structures are necessary for a plant to maintain turgidity?
3. What environmental conditions might cause a plant to lose its turgidity?

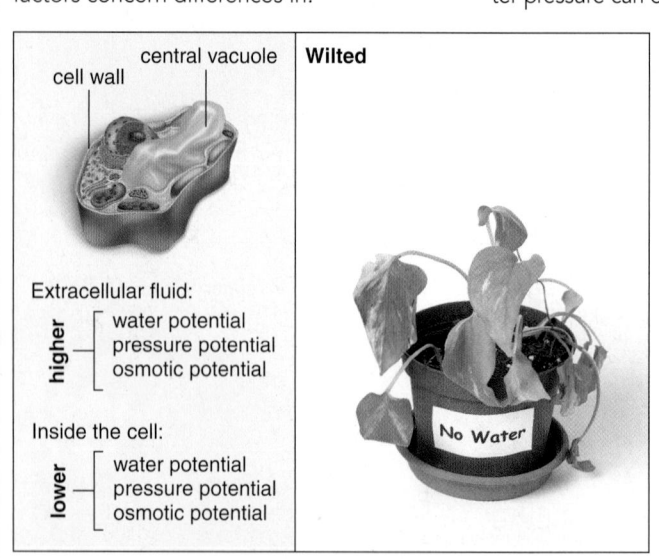

a. Plant cells need water.

b. Plant cells are turgid.

Figure 25C Water potential and turgor pressure.   Water flows from an area of higher water potential to an area of lower water potential. **a.** The cells of a wilted plant have a lower water potential; therefore, water enters the cells. **b.** Equilibrium is achieved when the water potential is equal inside and outside the cell. Cells are now turgid, and the plant is no longer wilted.

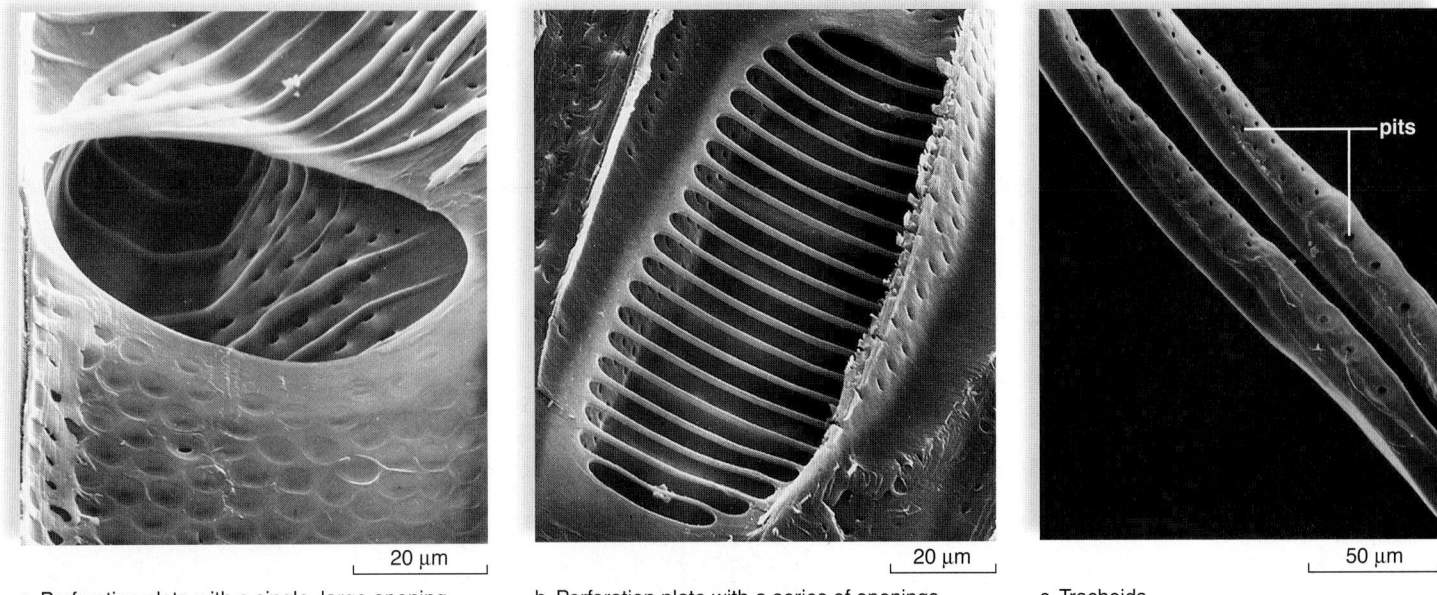

a. Perforation plate with a single, large opening   b. Perforation plate with a series of openings   c. Tracheids

Figure 25.10 Conducting cells of xylem.   Water can move from vessel element to vessel element through perforation plates (**a** and **b**). Vessel elements can also exchange water with tracheids through pits. **c.** Tracheids are long, hollow cells with tapered ends. Water can move into and out of tracheids through pits only.

Water Transport

Figure 25.5 traces the path of water from the root hairs to the xylem. As you know, xylem vessels constitute an open pipeline because the vessel elements have perforation plates separating one from the other (Fig. 25.10*a, b*). The tracheids, which are elongated with tapered ends, form a less obvious means of transport, but water can move across the end and side walls of tracheids because of pits where the secondary wall does not form (Fig. 25.10*c*).

Water entering root cells creates a positive pressure called **root pressure.** Root pressure, which primarily occurs at night, tends to push xylem sap upward. Root pressure may be responsible for **guttation** [L. *gutta,* drops, spots] which occurs when drops of water are forced out of vein endings along the edges of leaves (Fig. 25.11). Although root pressure may contribute to the upward movement of water in some instances, it is not responsible for the movement of water rising to the tops of trees. After an injury or pruning, especially in spring, some plants appear to "bleed" as water exudes from the site. This phenomenon is the result of root pressure.

Figure 25.11 Guttation.   Drops of guttation water on the edges of a strawberry leaf. Guttation, which occurs at night, may be due to root pressure. Root pressure is a positive pressure potential caused by the entrance of water into root cells. Often guttation is mistaken for early morning dew.

Cohesion-Tension Model of Xylem Transport

Water that enters xylem must be transported to all parts of the plant. Transporting water can appear to be a daunting task, especially for plants such as redwood trees, which can exceed 90 m (almost 300 ft) in height.

The **cohesion-tension model** of xylem transport, outlined in Figure 25.12, describes a mechanism for xylem transport that requires no expenditure of energy by the plant and is dependent on the properties of water. The term *cohesion* refers to the tendency of water molecules to cling together. Because of hydrogen bonding, water molecules interact with one another and form a continuous **water column** in xylem, from the leaves to the roots, that is not easily broken.

In addition to cohesion, another property of water called *adhesion* plays a role in xylem transport. Adhesion refers to the ability of water, a polar molecule, to interact with the molecules making up the walls of the vessels in xylem. Adhesion gives the water column extra strength and prevents it from slipping back.

3D Animation
Water Transport in Xylem

The Leaves. When the stomata of a leaf are open, the cells of the spongy layer are exposed to the air, which can be quite dry. Water then evaporates as a gas or vapor from the spongy layer into the intercellular spaces. Evaporation of water through leaf stomata is called **transpiration.** At least 90% of the water taken up by the roots is eventually lost by transpiration. This means that the total amount of water lost by a plant over a long period of time is surprisingly large. A single *Zea mays* (corn) plant loses somewhere between 135 and 200 liters of water through transpiration during a growing season. An average-sized birch tree with over 200,000 leaves will transpire up to 3,700 liters of water per day during the growing season.

The water molecules that evaporate from cells into the intercellular spaces are replaced by other water molecules from the leaf veins. Because the water molecules are cohesive, transpiration exerts a *pulling force*, or *tension*, that draws the water column through the xylem to replace the water lost by leaf cells.

Note that the loss of water by transpiration is the mechanism by which minerals are transported throughout the plant body. Also, evaporation of water moderates the temperature of leaf tissues.

The way water is transported in plants has an important consequence. When a plant is under water stress, the stomata close. Now the plant loses little water because the leaves are protected against water loss by the waxy *cuticle* of the upper and lower epidermis. When stomata are closed, however, carbon dioxide cannot enter the leaves, and many plants are unable to photosynthesize efficiently. Photosynthesis, therefore, requires an abundant supply of water so that stomata remain open, allowing carbon dioxide to enter.

The Stem. The tension in xylem created by evaporation of water at the leaves pulls the water column in the stem upward. Usually, the water column in the stem is continuous because of the cohesive property of water molecules. The water molecules

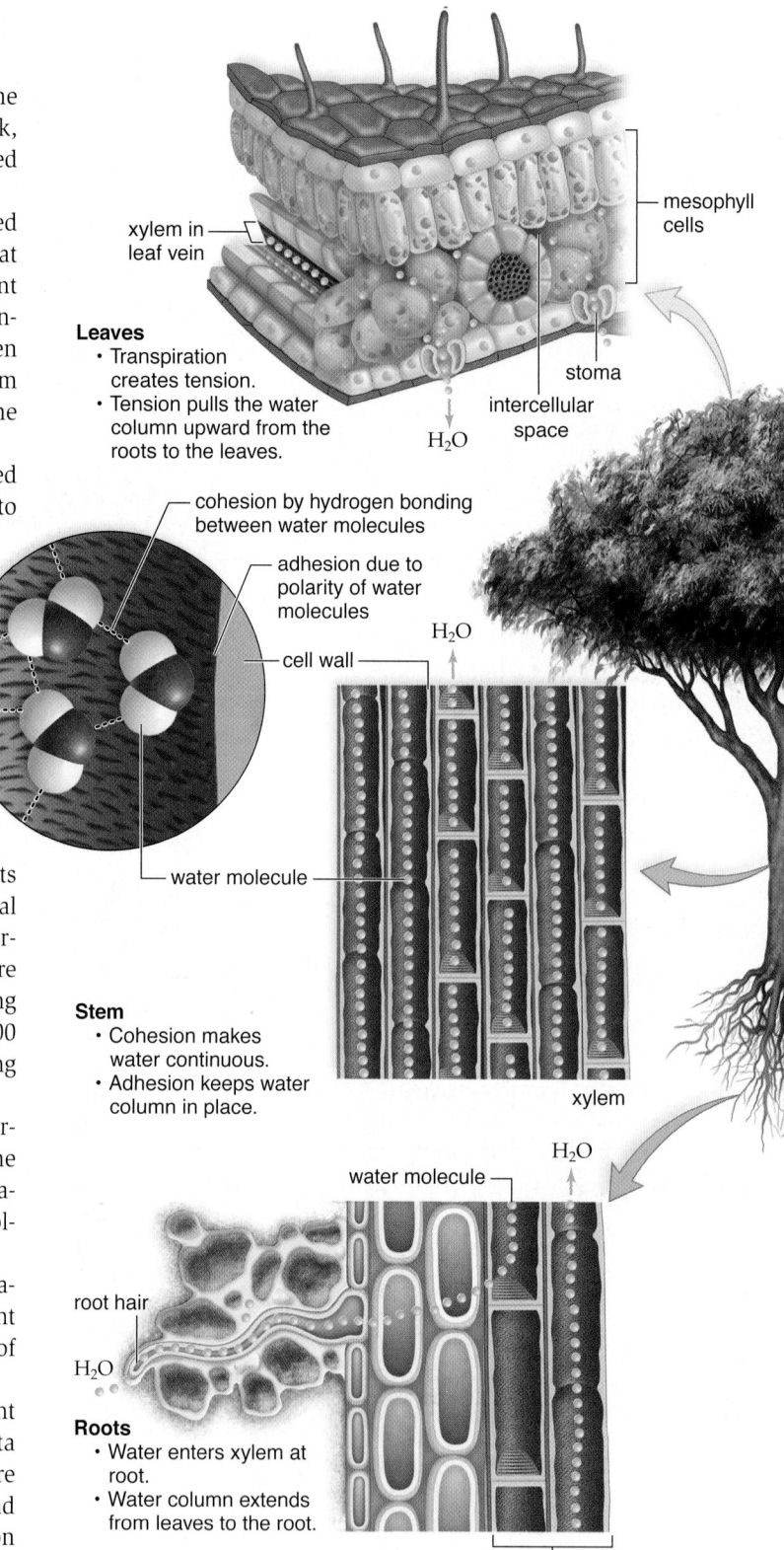

Leaves
- Transpiration creates tension.
- Tension pulls the water column upward from the roots to the leaves.

xylem in leaf vein

mesophyll cells

stoma

intercellular space

H_2O

cohesion by hydrogen bonding between water molecules

adhesion due to polarity of water molecules

cell wall

water molecule

H_2O

Stem
- Cohesion makes water continuous.
- Adhesion keeps water column in place.

xylem

water molecule

H_2O

root hair

H_2O

Roots
- Water enters xylem at root.
- Water column extends from leaves to the root.

xylem

Figure 25.12 Cohesion-tension model of xylem transport. Tension created by evaporation (transpiration) at the leaves pulls water along the length of the xylem—from the roots to the leaves.

also adhere to the sides of the vessels. If the water column within xylem breaks, then the water column "snaps back" down the xylem vessel away from the site of breakage, making it more difficult for conduction to occur.

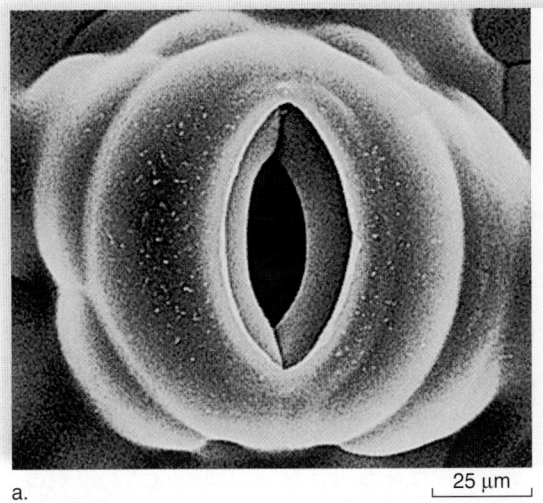

a.

25 μm

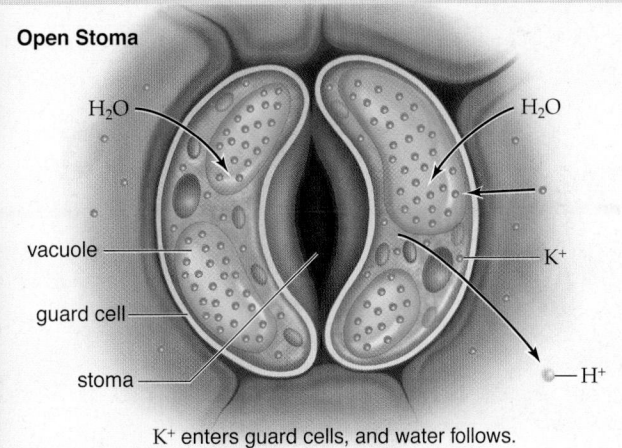

Open Stoma

H$_2$O

H$_2$O

vacuole

guard cell

K$^+$

stoma

H$^+$

K$^+$ enters guard cells, and water follows.

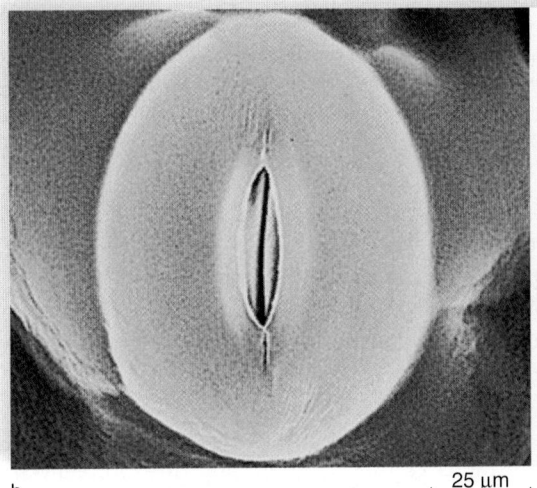

b.

25 μm

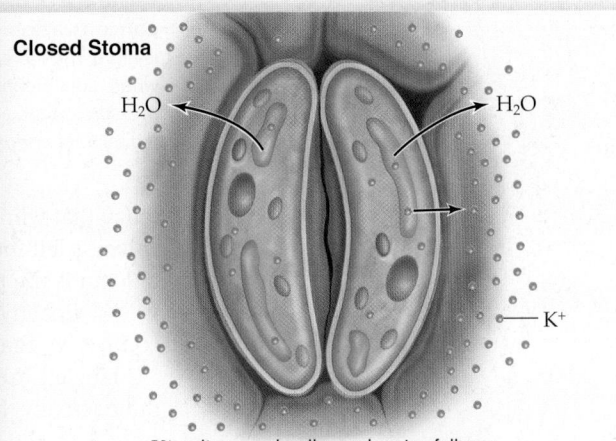

Closed Stoma

H$_2$O

H$_2$O

K$^+$

K$^+$ exits guard cells, and water follows.

**Figure 25.13
Opening and
closing of stomata.**
a. A stoma opens
when turgor pressure
increases in guard cells
due to the entrance
of K$^+$ followed by the
entrance of water.
b. A stoma closes
when turgor pressure
decreases due to the
exit of K$^+$ followed by
the exit of water.

Next time you use a straw to drink a soda, notice that pulling the liquid upward is fairly easy, as long as there is liquid at the end of the straw. When the soda runs low and you begin to get air, it takes considerably more suction to pull up the remaining liquid. This property also explains why flower stems are best cut when underwater to preserve an unbroken water column and the life of the flowers.

The Roots. In the root, water enters xylem passively by osmosis because xylem sap always has a greater concentration of solutes than do the root cells. The water column in xylem extends from the leaves down to the root. Water is pulled upward from the roots due to the tension in xylem created by the evaporation of water at the leaves.

Opening and Closing of Stomata

Each *stoma,* a small pore in leaf epidermis, is bordered by **guard cells.** When water enters the guard cells and turgor pressure increases, the stoma opens; when water exits the guard cells and turgor pressure decreases, the stoma closes. Notice in Figure 25.13 that the guard cells are attached to each other at their ends and that the inner walls are thicker than the outer walls. When water enters, a guard cell's radial expansion is restricted because of cellulose microfibrils in the walls, but lengthwise expansion

of the outer walls is possible. When the outer walls expand lengthwise, they buckle out from the region of their attachment, and the stoma opens.

Since about 1968, plant physiologists have known that potassium ions (K$^+$) accumulate within guard cells when stomata open. In other words, active transport of K$^+$ into guard cells causes water to follow by osmosis and stomata to open. Another interesting observation is that hydrogen ions (H$^+$) accumulate outside guard cells as K$^+$ moves into them. A proton pump run by the hydrolysis of ATP transports H$^+$ to the outside of the cell. This establishes an electrochemical gradient that allows K$^+$ to enter by way of a channel protein (see Fig. 25.5b).

The blue-light component of sunlight has been found to regulate the opening and closing of stomata. Evidence suggests that a flavin pigment absorbs blue light, and then this pigment sets in motion the cytoplasmic response that leads to activation of the proton pump. In a similar way, a receptor in the plasma membrane of guard cells could bring about inactivation of the pump when carbon dioxide (CO$_2$) concentration rises, as might happen when photosynthesis ceases. Abscisic acid (ABA), which is produced by cells in wilting leaves, can also cause stomata to close (see Chapter 26). Although photosynthesis cannot occur, water is conserved.

If plants are kept in the dark, stomata open and close just about every 24 hours, as though they were responding to the

a. An aphid feeding on a plant stem

Figure 25.14 Acquiring phloem sap. Aphids are small insects that remove nutrients from phloem by means of a needlelike mouthpart called a stylet. **a.** Excess phloem sap appears as a droplet after passing through the aphid's body. **b.** Micrograph of stylet in plant tissue. When an aphid is cut away from its stylet, phloem sap becomes available for collection and analysis.

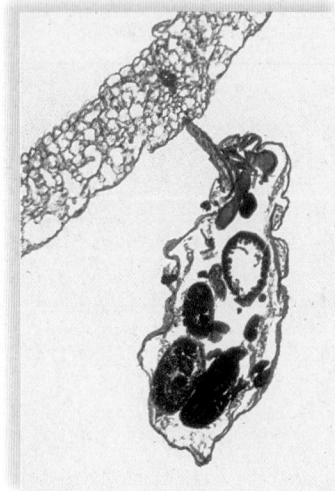

b. Aphid stylet in place

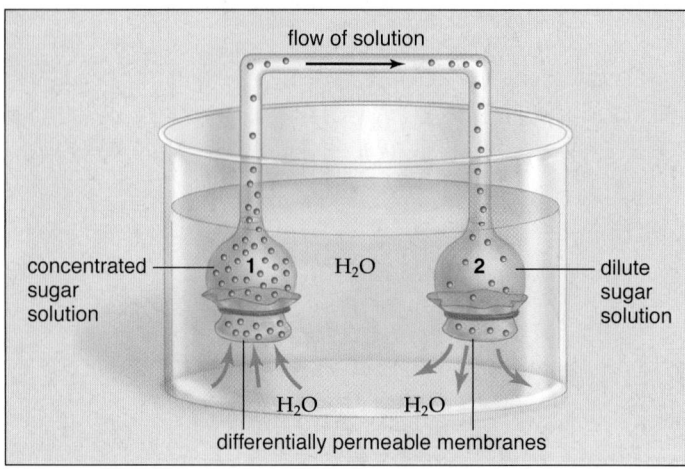

Figure 25.15 Flow of solution due to solute concentration. Water enters bulb 1, where the solute concentration is high. The increased water pressure causes water to flow into bulb 2, where solute concentration is lower. Water is forced out of bulb 2 by this pressure. In the same way, the sieve tube of a plant conducts a solution of high solute concentration to an area of low solute concentration.

presence of sunlight in the daytime and the absence of sunlight at night. The implication is that some sort of internal biological clock must be keeping time. Circadian rhythms (a behavior that occurs nearly every 24 hours) and biological clocks are areas of intense investigation at this time. Other factors that influence the opening and closing of stomata include temperature, humidity, and stress.

Organic Nutrient Transport

In addition to moving water and minerals, plants also transport organic nutrients to the parts of plants that need them. These regions include young leaves that have not yet reached their full photosynthetic potential; flowers that are in the process of making seeds and fruits; and the roots, whose location in the soil prohibits them from carrying on photosynthesis.

Role of Phloem

As long ago as 1679, Marcello Malpighi (1628–94) suggested that bark is involved in translocating sugars from leaves to roots. He observed the results of removing a strip of bark from around a tree, a procedure called **girdling.** If a tree is girdled below the level of the majority of leaves, the bark swells just above the cut, and sugar accumulates in the swollen tissue. We know

today that when a tree is girdled, the phloem is removed, but the xylem is left intact. Therefore, the results of girdling suggest that phloem is the tissue that transports sugars.

Radioactive tracer studies with carbon 14 (^{14}C) have confirmed that phloem transports organic nutrients. When ^{14}C-labeled carbon dioxide (CO_2) is supplied to mature leaves, radioactively labeled sugar is soon found moving down the stem into the roots. It's difficult to get samples of sap from phloem without injuring the phloem, but this problem is solved by using aphids, small insects that are phloem feeders. The aphid inserts its stylet, which is a sharp mouthpart that functions like a hypodermic needle, between the epidermal cells, and sap enters its body from a sieve-tube member (Fig. 25.14). If the aphid is anesthetized with ether, its body can be carefully cut away, leaving the stylet in place. Phloem can then be collected and analyzed by a researcher. By the use of radioactive tracers and aphids, it is known that the movement through phloem can be as fast as 60–100 cm per hour and possibly up to 300 cm per hour.

Pressure-Flow Model of Phloem Transport

The **pressure-flow model** is a current explanation for the movement of organic materials in phloem (see Fig. 25.16). Consider the experiment shown in Figure 25.15 in which two bulbs are connected by a glass tube. The first bulb (bulb 1) contains solute at a higher concentration than the second bulb (bulb 2). Each bulb is bounded by a differentially permeable membrane, and the entire apparatus is submerged in distilled water.

Distilled water flows into the first bulb because it has the higher solute concentration. The entrance of water creates a positive *pressure*, and water *flows* toward the second bulb. This flow not only drives water toward the second bulb, but it also provides enough force for water to move out through

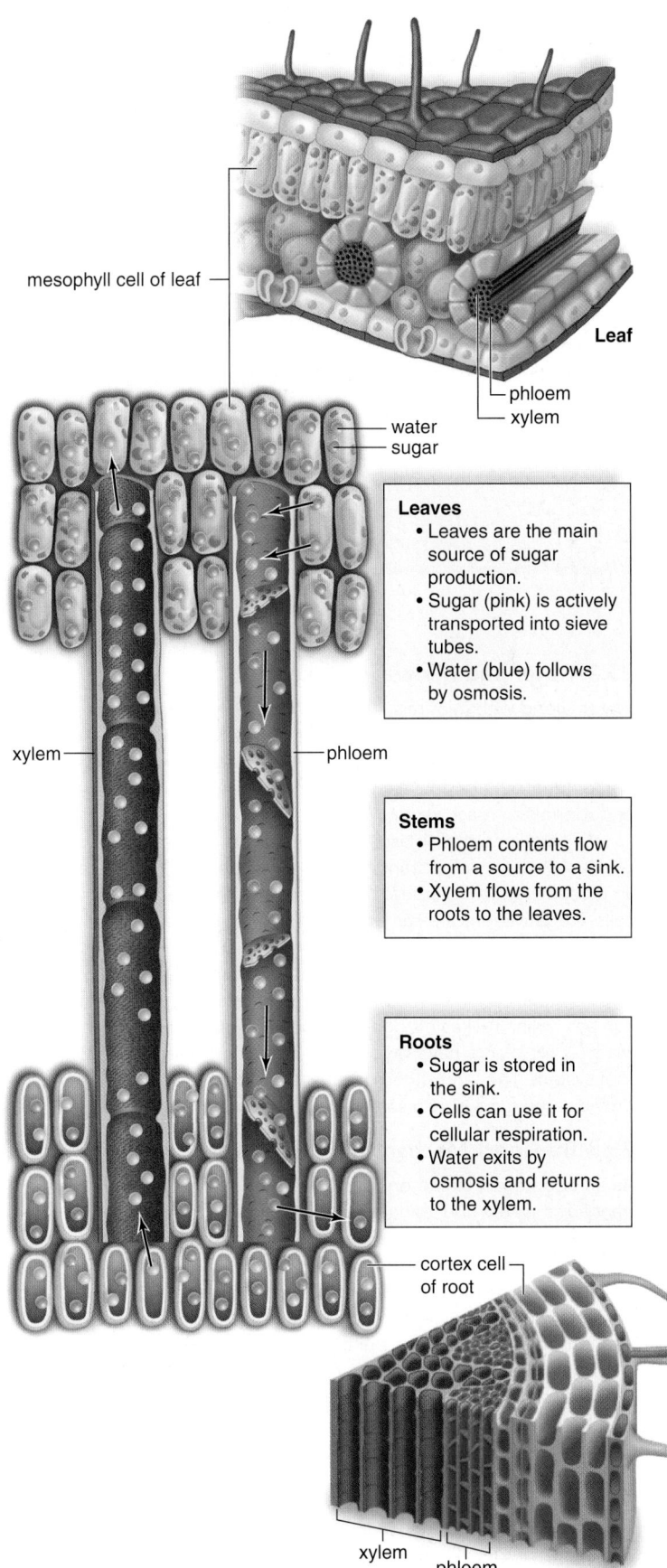

mesophyll cell of leaf

Leaf

phloem
xylem

water
sugar

Leaves
- Leaves are the main source of sugar production.
- Sugar (pink) is actively transported into sieve tubes.
- Water (blue) follows by osmosis.

xylem

phloem

Stems
- Phloem contents flow from a source to a sink.
- Xylem flows from the roots to the leaves.

Roots
- Sugar is stored in the sink.
- Cells can use it for cellular respiration.
- Water exits by osmosis and returns to the xylem.

cortex cell of root

xylem   phloem

Root

Figure 25.16 Pressure-flow model of phloem transport.
Sugars are produced at the source (leaves) and dissolve in water to form phloem. In the sieve tubes, water is pulled in by osmosis. The phloem follows positive pressure and moves toward the sink (root system).

the membrane of the second bulb—even though the second bulb contains a higher concentration of solute than the distilled water.

In plants, sieve tubes are analogous to the glass tube that connects the two bulbs. The sieve-tube members are aligned end to end, and strands of cytoplasm extend through sieve plates from one sieve-tube member to the other through plasmodesmata. Sieve tubes, therefore, form a continuous pathway for organic nutrient transport throughout a plant (Fig. 25.16).

3D Animation Translocation in Phloem

At the Source (e.g., Leaves). The **source** refers to the site of photosynthesis that ultimately results in the production of sugar. Generally the leaves and any portion of the stem that can photosynthesize are considered sources. The sugar produced is actively transported into phloem. Again, transport is dependent on an electrochemical gradient established by a proton pump, a form of active transport. Sugar is carried across the membrane in conjunction with hydrogen ions (H^+), which are moving down their concentration gradient (see Fig. 25.5). After sugar enters sieve tubes, water follows passively by osmosis.

In the Stem. The buildup of water within sieve tubes creates the positive pressure that accounts for the flow of phloem contents.

At the Sink (e.g., Roots). The **sink** refers to the site of sugar accumulation within the plant. At the sink, the sugar may be used for cellular respiration or stored for future use. The roots (and other growth areas) are a sink for sugar. After sugar is actively transported out of sieve tubes, water exits phloem passively by osmosis and is taken up by xylem. Xylem transports this water to leaves, where it is used for photosynthesis. Phloem contents continue to flow from the leaves (source) to the roots (sink).

The pressure-flow model of phloem transport can account for any direction of flow in sieve tubes if we consider that the direction of flow is always from source to sink. For example, recently formed leaves can be a sink, and they will receive sucrose until they begin to fully photosynthesize.

Check Your Progress 25.3

1. Describe why water is under tension in stems.
2. Identify the cohesion and adhesion properties of water that pertain to water transport.
3. Define the process in which sugars move from source to sink in a plant.

CONNECTING *the* CONCEPTS *with the* BIG IDEAS

Energy and Homeostasis

- Plants depend on water's special properties of adhesion and cohesion to accomplish movement of water via transpiration pull when stomata are open. (2A3a3*IE*)
- Plants employ stomata and root hairs to facilitate gas exchange. (2D2b*IE*)
- Negative feedback allows plants to regulate their transpiration based on water availability. (2C1a*IE*)
- Related methods of osmoregulation are used throughout the plant kingdom, showing their common ancestry. (2D2c*IE*)

- Important cooperative relationships that assist plants in their acquisition of nutrients include mycorrhizae and bacterial root nodules. (2E3b4*IE*)

*Find the unabridged version of all EK citations at www.glencoe.com/maderAP11.

Media Study Tools

www.glencoe.com/maderAP11

Enhance your study of this chapter with study tools and practice tests. Also ask your instructor about the resources available through ConnectPlus, including the media-rich eBook, interactive learning tools, and animations.

3D Animation
Plant Transport   For an interactive exploration of the processes of plant transport, take a moment to watch McGraw-Hill's new 3D animation "Plant Transport."

Summarize

25.1 Plant Nutrition and Soil

Plants need various inorganic nutrients. Carbon, hydrogen, and oxygen make up 95% of a plant's dry weight. The other necessary nutrients are taken up by the roots as mineral ions. Even nitrogen (N), which is present in the atmosphere, is most often taken up as NO_3^-.

Plant life is dependent on soil, which is formed by the weathering of rock. Soil is a mixture of mineral particles, humus, living organisms, air, and water. Soil particles are of three types from the largest to the smallest: sand, silt, and clay. Loam, which contains about equal proportions of all three types, retains water but still has air spaces. Humus contributes to the texture of soil and its ability to provide inorganic nutrients to plants. Topsoil (a horizon of a soil profile) contains humus, and this is the layer that is lost by erosion.

25.2 Water and Mineral Uptake

Water, along with minerals, can enter a root by passing between the porous cell walls, until it reaches the Casparian strip, after which it passes through an endodermal cell before entering xylem. Water can also enter root hairs and then pass through the cells of the cortex and endodermis to reach xylem.

Mineral ions cross plasma membranes by a chemiosmotic mechanism. A proton pump transports H^+ out of the cell. This establishes an electrochemical gradient that causes positive ions to flow into the cells. Negative ions are carried across the membrane when H^+ moves along its concentration gradient.

Plants have a number of adaptations that assist them in acquiring nutrients. Legumes have nodules infected with the bacterium *Rhizobium*, which makes nitrogen compounds available to these plants. Many other plants have mycorrhizae, or fungus associated with the root system. The fungus gathers nutrients from the soil, while the root provides the fungus with sugars and amino acids.

25.3 Transport Mechanisms in Plants

As an adaptation to life on land, plants have a vascular system that transports water and minerals from the roots to the leaves. Sugars are then transported in the opposite direction. Vascular tissue includes xylem and phloem.

In xylem, vessels composed of vessel elements aligned end to end form an open pipeline from the roots to the leaves. Particularly at night, root pressure can build in the root. However, this does not contribute significantly to xylem transport.

The cohesion-tension model of xylem transport states that transpiration creates a tension that pulls water upward in xylem from the roots to the leaves. This means of transport works only because water molecules are cohesive with one another and adhesive with xylem walls.

Most of the water taken in by a plant is lost through stomata by transpiration. Only when there is plenty of water do stomata remain open, allowing carbon dioxide to enter the leaf and photosynthesis to occur.

Stomata open when guard cells take up water. The guard cells stretch lengthwise because the microfibrils in their walls prevent lateral expansion. Water enters the guard cells after potassium ions (K^+) have entered, causing these cells to buckle outward. Light signals stomata to open, and a high carbon dioxide (CO_2) level may signal stomata to close. Abscisic acid produced by wilting leaves also signals for closure.

In phloem, sieve tubes composed of sieve-tube members aligned end to end form a continuous pipeline from the leaves to the roots. Sieve-tube members have sieve plates through which strands of cytoplasm extend through plasmodesmata from one member to the other.

The pressure-flow model of phloem transport proposes that a positive pressure drives phloem contents in sieve tubes. Sucrose is actively transported into sieve tubes—by a chemiosmotic mechanism—at a source, and water follows by osmosis. The resulting increase in pressure creates a flow that moves water and sucrose to a sink. A sink can be at the roots or at any other part of the plant that requires organic nutrients.

Key Terms

beneficial nutrient 465	pressure-flow model 478
cohesion-tension model 476	root nodule 471
essential nutrient 465	root pressure 475
girdling 478	sink 479
guard cell 477	soil 467
guttation 475	soil erosion 468
humus 467	soil horizon 468
hydroponics 465	soil profile 468
macronutrient 465	source 479
micronutrient 465	transpiration 476
mineral 465	water column 476
mycorrhizae 472	water potential 473
phytoremediation 470	

Assess

Reviewing This Chapter

1. Name the elements that make up most of a plant's body. What are essential mineral nutrients and beneficial mineral nutrients? 465–66
2. Briefly describe the use of hydroponics to determine the mineral nutrients of a plant. 465–66
3. How is soil formed, and how does humus provide nutrients to plants? Describe a generalized soil profile and how a profile is affected by erosion. 467–68
4. Give two pathways by which water and minerals can cross the epidermis and cortex of a root. What feature allows endodermal cells to regulate the entrance of molecules into the vascular cylinder? 469
5. Describe the chemiosmotic mechanism by which mineral ions cross plasma membranes. 470
6. Name two symbiotic relationships that assist plants in taking up minerals and two types of plants that have other means of acquiring nutrients. 471–72
7. A vascular system is adaptive for a land existence. Explain. Describe the composition of a plant's vascular system. 472–73
8. What is root pressure, and why can't it account for the transport of water in xylem? 475–76

9. Describe and give evidence for the cohesion-tension model of water transport. 476
10. Describe the structure of stomata and explain how they can open and close. By what mechanism do guard cells take up potassium (K^+) ions? 477
11. What data are available to show that phloem transports organic compounds? Explain the pressure-flow model of phloem transport. 478–79

Testing Yourself

Choose the best answer for each question.

1. Which of these molecules is not a nutrient for plants?
   a. water
   b. carbon dioxide gas
   c. mineral ions
   d. nitrogen gas
   e. None of these are nutrients.

2. Which is a component of soil?
   a. mineral particles   d. air and water
   b. humus   e. All of these are correct.
   c. organisms

3. The Casparian strip affects
   a. how water and minerals move into the vascular cylinder.
   b. vascular tissue composition.
   c. how soil particles function.
   d. how organic nutrients move into the vascular cylinder.
   e. Both a and d are correct.

4. Which of these is not a mineral ion?
   a. NO_3^- d. Al^{3+}
   b. MgI_2 e. All of these are correct.
   c. CO_2

5. What role do cohesion and adhesion play in xylem transport?
   a. Like transpiration, they create a tension.
   b. Like root pressure, they create a positive pressure.
   c. Like sugars, they cause water to enter xylem.
   d. They create a continuous water column in xylem.
   e. All of these are correct.

6. The pressure-flow model of phloem transport states that
   a. phloem content always flows from the leaves to the root.
   b. phloem content always flows from the root to the leaves.
   c. water flow brings sucrose from a source to a sink.
   d. water pressure creates a flow of water toward the source.
   e. Both c and d are correct.

7. Root hairs do not play a role in
   a. oxygen uptake.   d. carbon dioxide uptake.
   b. mineral uptake.   e. the uptake of any of these.
   c. water uptake.

8. Xylem includes all of these except
   a. companion cells.   c. tracheids.
   b. vessels.   d. dead tissue.

9. After sucrose enters sieve tubes,
   a. it is removed by the source.
   b. water follows passively by osmosis.
   c. it is driven by active transport to the source, which is usually the roots.
   d. stomata open so that water flows to the leaves.
   e. All of these are correct.

10. An opening in the leaf that allows gas and water exchange is called
    a. the lenticel.   d. the guard cell.
    b. the hole.   e. the accessory cell.
    c. the stoma.

11. What main force drives absorption of water, creates tension, and draws water through the plant?
    a. adhesion   d. transpiration
    b. cohesion   e. absorption
    c. tension

12. A nutrient element is considered essential if
    a. plant growth increases with a reduction in the concentration of the element.
    b. plants die in the absence of the element.
    c. plants can substitute a similar element for the missing element with no ill effects.
    d. the element is a positive ion.

13. Humus
    a. supplies nutrients to plants.
    b. is basic in its pH.
    c. is found in the deepest soil horizons.
    d. is inorganic in origin.

14. Soils rich in which type of mineral particle will have a high water-holding capacity?
    a. sand   c. clay
    b. silt   d. All soil particles hold water equally well.

15. Stomata are usually open
    a. at night, when the plant requires a supply of oxygen.
    b. during the day, when the plant requires a supply of carbon dioxide.
    c. day or night if there is excess water in the soil.
    d. during the day, when transpiration occurs.
    e. Both b and d are correct.

16. Why might the water column in tracheids be less susceptible to breakage than in vessels?
    a. Tracheids are more narrow, giving more opportunity for adhesion to play a role in maintaining the water column.
    b. The end walls of tracheids are more slanted than the end walls of vessel elements.
    c. Tracheids receive support from vessel elements, but not vice versa.
    d. All of these are correct.

17. Explain why this experiment supports the hypothesis that transpiration can cause water to rise to the tops of tall trees.

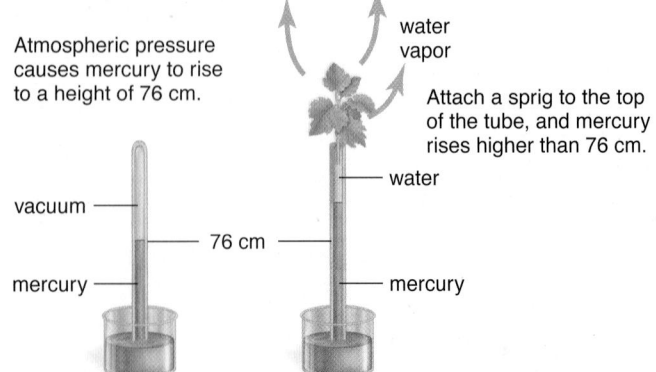

Atmospheric pressure causes mercury to rise to a height of 76 cm.

water vapor

Attach a sprig to the top of the tube, and mercury rises higher than 76 cm.

vacuum

water

76 cm

mercury

mercury

18. Negatively charged clay particles attract
    a. K^+.
    b. NO_3^-.
    c. Ca^{2+}.
    d. Both a and b are correct.
    e. Both a and c are correct.

19. a. Label water (H_2O) and potassium ions (K^+) appropriately in these diagrams. b. What is the role of K^+ in the opening and closing of stomata?

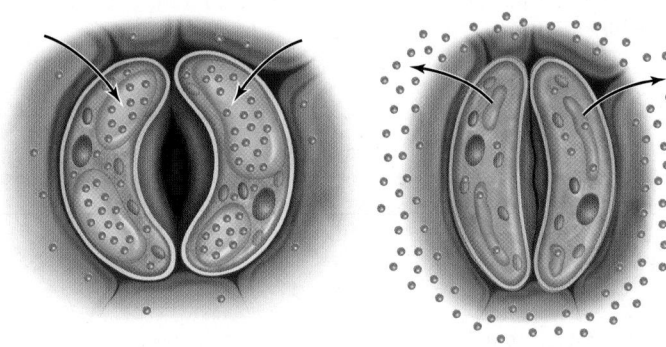

20. Explain why solution flows from the left bulb to the right bulb.

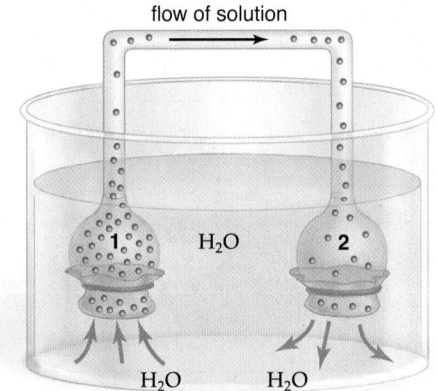

flow of solution

1   H_2O 2

H_2O H_2O

Engage

Virtual Lab
Plant Transpiration

The virtual lab "Plant Transpiration" provides a more detailed look at the environmental conditions associated with the transpiration of water from the leaves of a plant.

Thinking Scientifically

1. Design an experiment to determine whether calcium, nitrogen, or phosphate is an essential plant nutrient. State the possible results.

2. *Welwitschia* is a genus of plant that lives in the Namib and Mossamedes deserts in Africa. Annual rainfall averages only 2.5 cm (1 inch) per year. *Welwitschia* plants contain a large number of stomata (22,000 per cm²), which remain closed most of the time. Can you suggest how a large number of stomata would be beneficial to these desert plants?

Time-lapse photograph of sunflowers tracking the movement of the Sun.

Flowering Plants: Control of Growth Responses

The observation that sunflowers track the Sun as it moves through the sky is a striking example of a flowering plant's ability to respond to environmental stimuli. Other responses to light can take longer than sun-tracking because they involve hormones and an alteration in growth. For example, flowering plants will exhibit a bend toward the light within a few hours because a hormone produced by the growing tip has moved from the sunny side to the shady side of the stem. Hormones also help flowering plants respond to stimuli in a coordinated manner. In the spring, seeds germinate and growth begins if the soil is warm enough to contain liquid water. In the fall, when temperatures drop, shoot and root apical growth ceases. Some plants also flower according to the season. The pigment phytochrome is instrumental in detecting the photoperiod and bringing about changes in gene expression, which determine whether a plant flowers or does not flower.

Plant defenses include physical barriers, chemical toxins, and even mutualistic animals. This chapter discusses the variety of ways flowering plants can respond to their environment, including other organisms.

As you read through the chapter, think about the following questions:

1. Which hormones are essential for plant growth?
2. What environmental stimuli trigger the various plant responses?

CHAPTER OUTLINE

26.1 Plant Hormones 484
26.2 Plant Responses 490

BEFORE YOU BEGIN

Before beginning this chapter, take a few moments to review the following discussions.

Figure 5.3 Which membrane proteins are essential for the control of growth responses?

Figure 5.8 What role does turgor pressure play in the plant response to stimuli?

Section 13.2 How do eukaryotic organisms regulate the expression of their genes?

FOLLOWING *the* BIG IDEAS

CHAPTER 26 FLOWERING PLANTS: CONTROL OF GROWTH RESPONSES

Energy and Homeostasis	Growth and timing responses are essential to plant energy acquisition and survival.
Information and Signaling	Plant hormones typically work by affecting gene expression in some manner.

26.1 Plant Hormones

Learning Outcomes

Upon completion of this section, you should be able to

1. Explain the role of hormones when plant cells utilize signal transduction to respond to stimuli.
2. Compare and contrast the effects of auxins, gibberellins, cytokinins, abscisic acid, and ethylene on plant growth and development.

All organisms are capable of responding to environmental stimuli. Being able to respond to stimuli is a beneficial adaptation because it leads to organisms' longevity and ultimately to the survival of the species. Flowering plants perceive and react to a variety of environmental stimuli. Some examples include light, gravity, carbon dioxide levels, pathogen infection, drought, and touch. Their responses can be short term, as when stomata open and close in response to light levels, or long term, as when plants respond to gravity with the downward growth of the root and the upward growth of the stem.

Although we think of responses in terms of a plant structure, the mechanism that brings about a response occurs at the cellular level. Research has shown that plant cells respond to stimuli by utilizing **signal transduction,** the binding of a molecular

Figure 26.1 Signal transduction in plants. ❶ The hormone auxin enters the cell and is received by a receptor in the nucleus. This complex alters gene expression. ❷ A light receptor in the plasma membrane is sensitive to and activated by blue light. Activation leads to stimulation of a transduction pathway that ends with gene expression changes. ❸ When attacked by a herbivore, the flowering plant produces defense hormones that bind to a plasma membrane receptor. Again, the transduction pathway results in a change in gene expression.

"signal" that initiates and amplifies a cellular response. You first encountered the concept of signal transducers in Chapter 13 in the description of tumor suppressor genes and oncogenes (see Figs. 13.13 and 13.14).

Notice in Figure 26.1 that signal transduction involves the following:

Receptors—proteins activated by a specific signal. Receptors can be located in the plasma membrane, the cytoplasm, the nucleus, or even the endoplasmic reticulum. A receptor that responds to light has a pigment component. For example, phytochrome has a region that is sensitive to red light, and phototropin has a region that is sensitive to blue light.

Transduction pathway—a series of relay proteins or enzymes that amplify and transform the signal to one understood by the machinery of the cell. In some instances, a stimulated receptor immediately communicates with the transduction pathway, and in other instances, a second messenger, such as Ca^{2+}, initiates the response.

Animation
Second
Messengers

Cellular response—the result of the transduction pathway. Very often, the response is either the transcription of particular genes or the end product of an activated metabolic pathway. The cellular response brings about the observed macroscopic response, such as stomata closing or a stem that turns toward the light.

How do plant **hormones** [Gk. *hormao,* instigate] fit into this model for the ability of flowering plants to respond to both abiotic and biotic stimuli? The answer is that they serve as chemical signals to coordinate cell responses. These molecules are

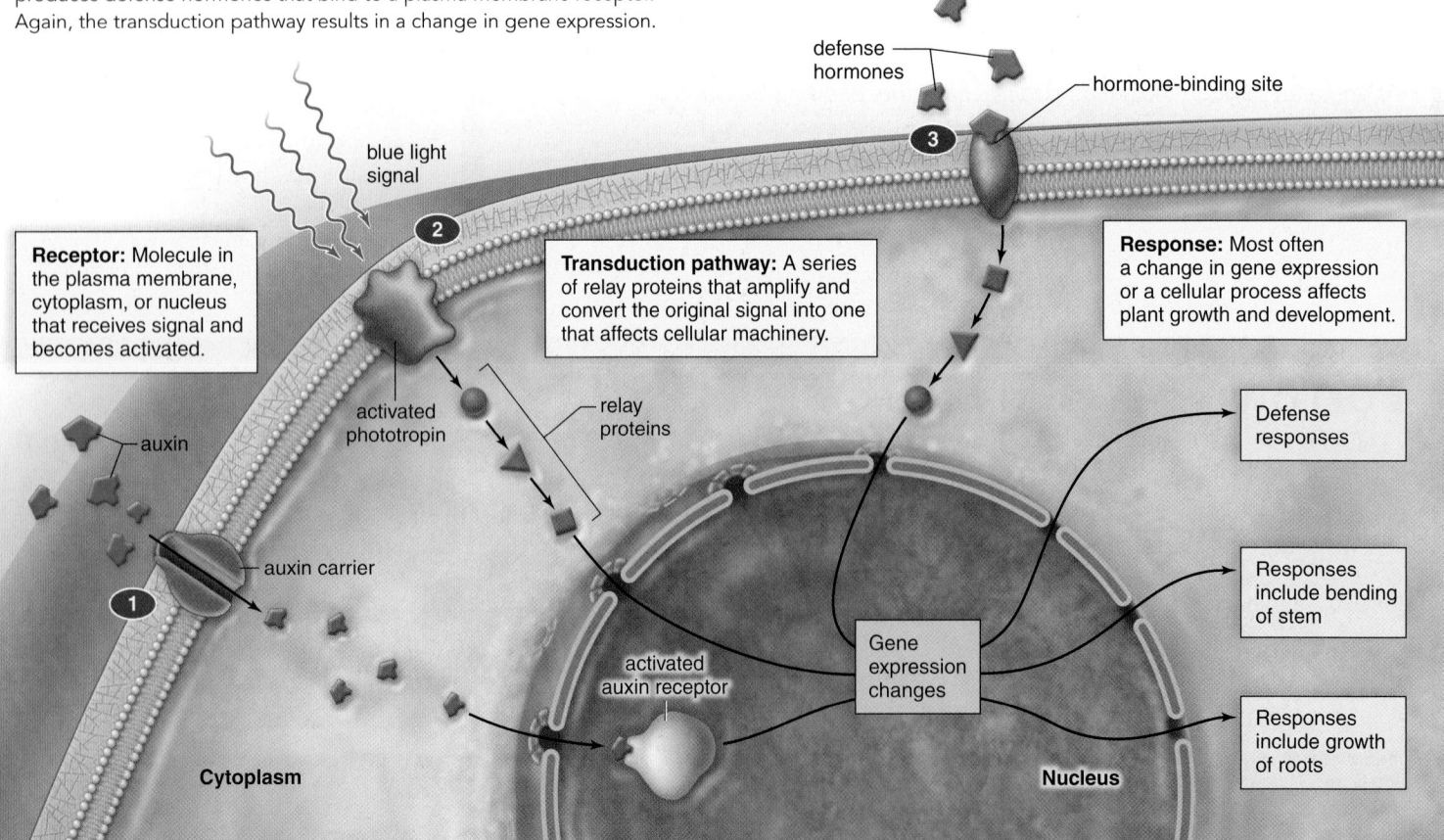

defense hormones

hormone-binding site

blue light signal

Receptor: Molecule in the plasma membrane, cytoplasm, or nucleus that receives signal and becomes activated.

Transduction pathway: A series of relay proteins that amplify and convert the original signal into one that affects cellular machinery.

Response: Most often a change in gene expression or a cellular process affects plant growth and development.

activated phototropin

relay proteins

Defense responses

auxin

auxin carrier

activated auxin receptor

Gene expression changes

Responses include bending of stem

Responses include growth of roots

Cytoplasm

Nucleus

produced in very low concentrations and are active in another part of the organism. Hormones such as auxin, for example, are synthesized or stored in one part of the plant, but they travel within phloem or from cell to cell in response to the appropriate stimulus.

In this section we present descriptions of five major types of plant hormones: auxins, gibberellins, cytokinins, abscisic acid, and ethylene. Each of these affects different aspects of plant responses to stimuli.

Auxins

Auxins [Gk. *auximos,* promoting growth] are produced in shoot apical meristem and are found in young leaves and in flowers and fruits. The most common naturally occurring auxin is indoleacetic acid (IAA):

$$CH_2\!-\!COOH$$

Structure of
indoleacetic acid (IAA)

Auxins Affect Growth and Development

Auxins affect many aspects of plant growth and development. Auxins, or more simply, auxin, is responsible for **apical dominance,** which occurs when the terminal bud produces new growth instead of the axillary buds. When a terminal bud is removed deliberately or accidentally, the nearest axillary buds begin to grow, and the plant branches. Therefore, pruning the top of a flowering plant generally achieves a fuller look. This removes apical dominance and causes more branching of the main body of the plant.

Auxin causes the growth of roots and fruits and prevents the loss of leaves and fruit. The application of an auxin paste to a

stem cutting causes adventitious roots to develop more quickly than they would otherwise. Auxin production by seeds promotes the growth of fruit. As long as auxin is concentrated in leaves or fruits rather than in the stem, leaves and fruits do not drop off. Therefore, trees can be sprayed with auxin to keep mature fruit from falling to the ground.

Synthetic auxins are used today in a number of applications. These auxins are sprayed on plants such as tomatoes to induce the development of fruit without pollination. Thus, seedless tomatoes can be commercially developed. Synthetic auxins such as 2,4-D and 2,4,5-T have been used as herbicides to control broadleaf weeds, such as dandelions and other plants. These substances have little effect on grasses. 2,4-D is still used, but 2,4,5-T was banned in 1979 because of its detrimental effects on human and animal life. A mixture of 2,4-D and 2,4,5-T is best known as the defoliant Agent Orange, used in the Vietnam War.

Gravitropism and Phototropism. After the direction of gravity has been detected by a flowering plant, auxin moves to the lower surface of roots and stems. Thereafter, roots curve downward and stems curve upward, the effect is termed gravitropism. (Gravitropism is discussed at more length on page 491.)

The role of auxin in the positive phototropism of stems has been studied for quite some time. The experimental material of choice has been oat seedlings with coleoptiles intact. A **coleoptile** is a protective sheath for the young leaves of the seedling. In 1881, Charles Darwin and his son Francis found that phototropism does not occur if the tip of the seedling is cut off or covered by a black cap. They concluded that some influence that causes curvature is transmitted from the coleoptile tip to the rest of the shoot.

In 1926, Frits W. Went cut off the tips of coleoptiles and placed them on agar blocks (agar is a gelatin-like material). Then he placed an agar block to one side of a tipless coleoptile and found that the shoot would curve away from that side. The bending occurred even though the seedlings were not exposed to light (Fig. 26.2). Went concluded that the agar block contained a chemical that had been produced by the coleoptile tips.

1. Coleoptile tip is intact.

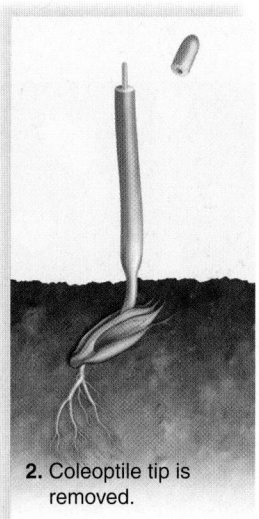

2. Coleoptile tip is removed.

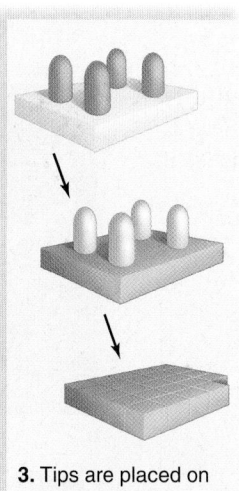

3. Tips are placed on agar, and auxin diffuses into the agar.

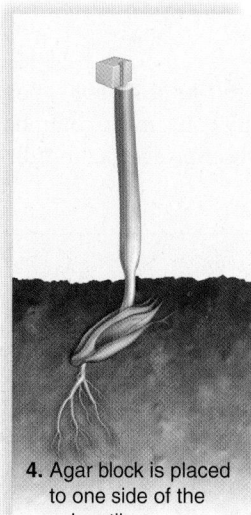

4. Agar block is placed to one side of the coleoptile.

5. Curvature occurs beneath the block.

Figure 26.2 Auxin and phototropism.   Oat seedlings are protected by a hollow sheath called a coleoptile. After coleoptile tips are removed and placed on agar, a block of the agar to one side of the cut coleoptile can cause it to curve due to the presence of auxin (pink) in the agar. This shows that auxin causes the coleoptile to bend, as it does when exposed to a light source.

Figure 26.3 Expansion of the cell wall. ❶ Auxin leads to activation of a proton pump and entrance of hydrogen ions in the cell wall. ❷ As the pH decreases, enzymes are activated and break down cellulose fibers in the cell wall. ❸ Cellulose fibers burst and the cell expands as turgor pressure inside the cell increases.

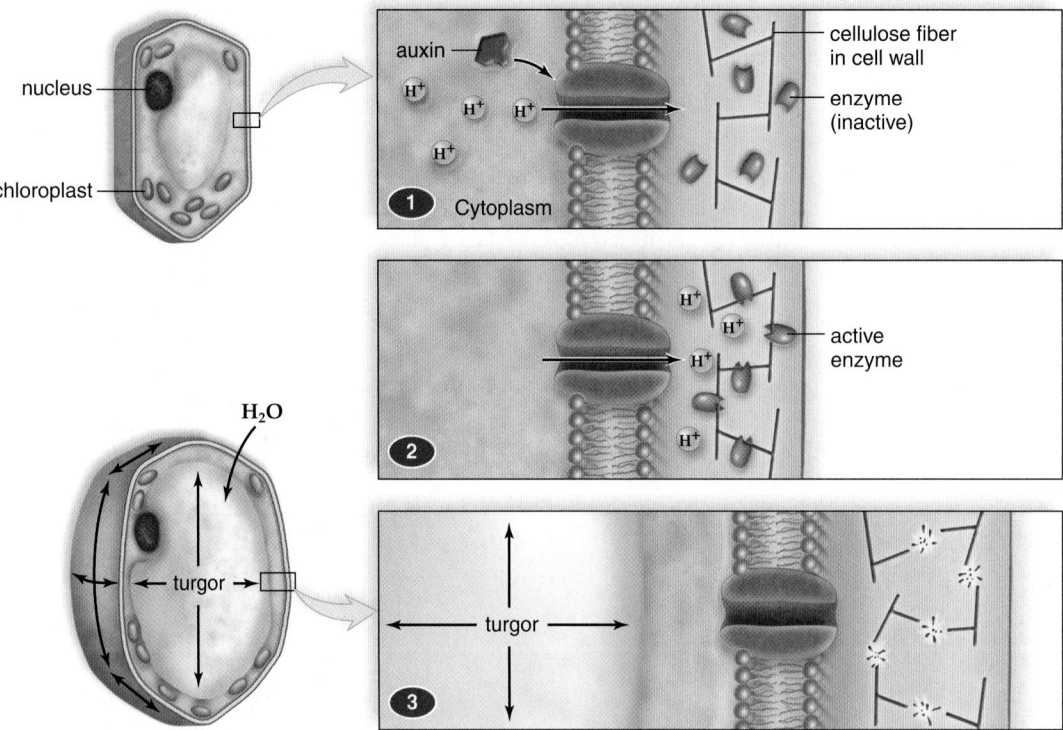

This chemical, he decided, had caused the shoots to bend. He named the chemical substance auxin after the Greek word *auximos,* which means promoting growth.

How Auxins Cause Stems to Bend

When a stem is exposed to unidirectional light, auxin moves to the shady side, where it enters the nuclei of cells and attaches to a receptor. The complex leads to the activation of a proton (H⁺) pump, and the resulting acidic conditions loosen the cell wall: hydrogen bonds are broken and cellulose fibrils are weakened by enzymatic action. The end result of these activities is elongation of the stem on the shady side so that the stem tip bends toward the light (Fig. 26.3).

Gibberellins

We know of about 70 **gibberellins** [L. *gibbus,* bent], and they differ chemically only slightly. The most common of these is gibberellic acid, GA_3 (the subscript designation distinguishes it from other gibberellins):

Structure of gibberellic acid (GA_3)

Gibberellins Promote Stem Elongation

When gibberellins are applied externally to plants, the most obvious effect is stem elongation (Fig. 26.4a). Gibberellins can cause dwarf plants to grow, cabbage plants to become 2 m tall, and bush beans to become pole beans.

Gibberellins were discovered in 1926, the same year that Went performed his classic experiments with auxin. Eiichi Kurosawa, a Japanese scientist, was investigating a fungal disease of rice plants called "foolish seedling disease." The plants elongated too quickly, causing the stem to weaken and the plant to collapse. Kurosawa found that the fungus infecting the plants produced an excess of a chemical he called gibberellin, named after the fungus *Gibberella fujikuroi.* It wasn't until 1956 that gibberellic acid was isolated from a flowering plant rather than from a fungus. Sources of gibberellin in flowering plant parts are young leaves, roots, embryos, seeds, and fruits.

a.   b.

Figure 26.4 Gibberellins cause stem elongation. **a.** The *Cyclamen* plant on the right was treated with gibberellins; the plant on the left was not treated. **b.** The grapes are larger on the right because gibberellins caused an increase in the space between the grapes, allowing them to grow larger.

Gibberellins Have Commercial Uses

Commercially, gibberellins are helpful in a number of ways. Gibberellins induce the growth of plants and increase the size of flowers. Gibberellins have also been successfully used to produce larger seedless grapes. In Figure 26.4b, gibberellins caused an increase in the space between the grapes, allowing them to grow larger.

Dormancy is a period of time when plant growth is suspended. Gibberellins can break the dormancy of buds and seeds. Therefore, application of gibberellins is one way to hasten the development of a flower bud. When gibberellins break the dormancy of barley seeds, a large, starchy endosperm is broken down into sugars to provide energy for growth. This occurs because amylase, an enzyme that breaks down starch, makes its appearance. It would seem, then, that gibberellins are involved in a transduction pathway that leads to the production of amylase.

Cytokinins

The **cytokinins** [Gk. *kytos*, cell, and *kineo*, move] are derivatives of adenine, one of the purine bases in DNA and RNA. A naturally occurring cytokinin was not isolated until 1967. Because it came from the kernels of maize (*Zea*), it was called zeatin:

$$N-CH_2-C=C \begin{matrix} CH_3 \\ CH_2OH \end{matrix}$$

Structure of zeatin

Cytokinins Promote Cell Division

Cytokinins were discovered as a result of attempts to grow plant tissue and organs in culture vessels in the 1940s. It was found that cell division occurred when coconut milk (a liquid endosperm) and yeast extract were added to the culture medium. The effective components were collectively called cytokinins because cytokinesis means cell division. Since then, cytokinins have been isolated from various seed plants, where they occur in the actively dividing tissues of roots and also in seeds and fruits. Cytokinins have been used to prolong the life of flower cuttings as well as vegetables in storage.

Plant tissue culturing is now common practice. Researchers are well aware that the ratio of auxin to cytokinin and the acidity of the culture medium determine whether the plant tissue forms an undifferentiated mass, called a callus, or differentiates to form roots, vegetative shoots, leaves, or floral shoots (Fig. 26.5). These effects illustrate that a plant hormone rarely acts alone; it is the relative concentrations of hormones and their interactions that produce an effect.

Researchers have reported that chemical fragments released from the cell wall, termed oligosaccharins, are also effective in directing differentiation. They hypothesize that auxin and cytokinins are a part of a signal transduction pathway that leads to the activation of enzymes that release these fragments from the cell wall.

Cytokinins Prevent Senescence

When a plant organ, such as a leaf, loses its natural color, it is most likely undergoing an aging process called **senescence.** During senescence, large molecules within the leaf are broken down and transported to other parts of the plant. Senescence does not always affect the entire plant at once; for example, as some plants grow taller, they naturally lose their lower leaves. It

 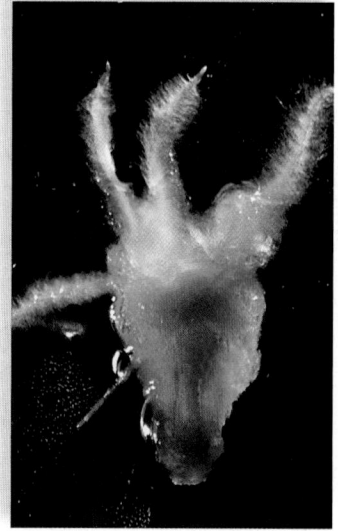  

a.   b.   c.   d.

Figure 26.5 Interaction of hormones. Tissue culture experiments have revealed that auxin and cytokinin interact to affect differentiation during development. **a.** In tissue culture that has the usual amounts of these two hormones, tobacco strips develop into a callus of undifferentiated tissue. **b.** If the ratio of auxin to cytokinin is appropriate, the callus produces roots. **c.** Change the ratio, and vegetative shoots and leaves are produced. **d.** Yet another ratio causes floral shoots. It is now clear that each plant hormone rarely acts alone; it is the relative concentrations of hormones that produce an effect. The modern emphasis is to look for an interplay of hormones when a growth response is studied.

has been found that senescence of leaves can be prevented by the application of cytokinins. Also, axillary buds begin to grow, despite apical dominance, when cytokinin is applied to them.

Abscisic Acid

Abscisic acid (ABA) is produced by any "green tissue" (that contains chloroplasts). ABA is also produced in monocot endosperm and roots where it is derived from carotenoid pigments:

Structure of
abscisic acid (ABA)

Abscisic acid is sometimes called the stress hormone because it initiates and maintains seed and bud dormancy and brings about the closure of stomata. It was once believed that ABA functioned in **abscission,** the dropping of leaves, fruits, and flowers from a plant. But although the external application of ABA promotes abscission, this hormone is no longer believed to function naturally in this process. Instead, the hormone ethylene seems to bring about abscission.

ABA Promotes Dormancy

Recall that dormancy is a period of low metabolic activity and arrested growth. Dormancy occurs when a plant organ readies itself for adverse conditions by ceasing to grow (even though conditions at the time may be favorable for growth). For example, it is believed that ABA moves from leaves to vegetative buds in the fall, and thereafter these buds are converted to winter buds. A winter bud is covered by thick, hardened scales (Fig. 26.6). A reduction in the level of ABA and an increase

Figure 26.6 Dormancy and winter buds. Abscisic acid promotes the formation of winter buds.

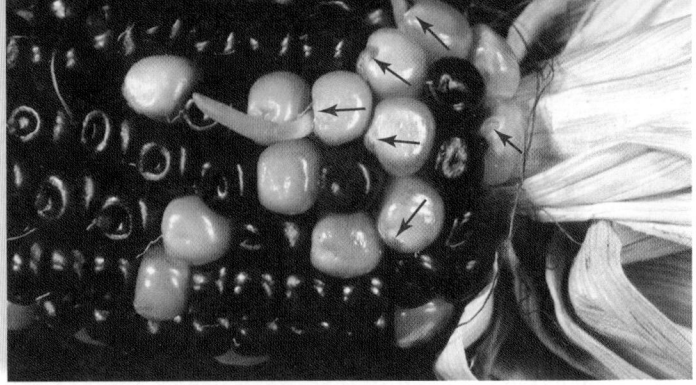

Figure 26.7 Dormancy and germination. Image of viviparous mutant of maize (Indian corn) showing germination on the cob due to reduced sensitivity to abscisic acid.

in the level of gibberellins are believed to break seed and bud dormancy. Then seeds germinate, and buds send forth leaves. In Figure 26.7, corn kernels have begun to germinate on the developing cob because this maize mutant is deficient in ABA. Abscisic acid is needed to maintain the dormancy of seeds.

ABA Closes Stomata

The reception of abscisic acid brings about the closing of stomata when a plant is under water stress, as described in Figure 26.8. Investigators have also found that ABA induces rapid depolymerization of actin filaments and formation of a new type of actin that is randomly oriented throughout the cell. This change in actin organization may also be part of the transduction pathways involved in stomata closure.

Ethylene

Ethylene ($H_2C = CH_2$) is a gas formed from the amino acid methionine. This hormone is involved in abscission and the ripening of fruits.

Ethylene Causes Abscission

The absence of auxin, and perhaps gibberellin, probably initiates abscission. But once abscission has begun, ethylene stimulates certain enzymes, such as cellulase, which helps cause leaf, fruit, or flower drop. In Figure 26.9, a ripe apple, which gives off ethylene, is under the bell jar on the right, but not under the bell jar on the left. As a result, only the holly plant on the right loses its leaves.

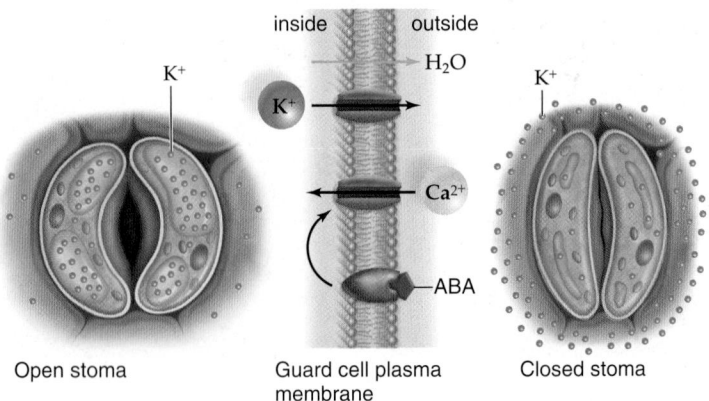

Open stoma   Guard cell plasma membrane   Closed stoma

Figure 26.8 Abscisic acid promotes closure of stomata. The stoma is open (*left*). When ABA (the first messenger) binds to its receptor in the guard cell plasma membrane, the second messenger (Ca^{2+}) enters (*middle*). Now, K^+ channels open, and K^+ exits the guard cells. After K^+ exits, so does water. The stoma closes (*right*).

Ethylene Ripens Fruit

In the early 1900s, it was common practice to prepare citrus fruits for market by placing them in a room with a kerosene stove. Only later did researchers realize that an incomplete combustion product of kerosene, namely ethylene, ripens fruit. It does so by increasing the activity of enzymes that soften fruits. For example, in addition to stimulating the production of cellulase, it promotes the activity of enzymes that produce the flavor and smell of ripened fruits and breaks down chlorophyll, inducing the color changes associated with fruit ripening.

Ethylene moves freely through a plant by diffusion, and because it is a gas, ethylene also moves freely through the air. That is why a basket of ripening apples can induce ripening of a bunch of bananas some distance away. Ethylene is released at the site of a plant wound due to physical damage or infection (which is why one rotten apple spoils the whole bushel).

The use of ethylene in agriculture is extensive. It is used to hasten the ripening of green fruits, such as melons and honeydews, and is also applied to citrus fruits to attain pleasing colors before being put out for sale. Normally, tomatoes ripen on the vine because the plants produce ethylene. Today, tomato plants can be genetically modified to not produce ethylene. This facilitates shipping because green tomatoes are not subject to as much damage (Fig. 26.10). Once the tomatoes have arrived at their destination, they can be exposed to ethylene so that they ripen.

Other Effects of Ethylene

Ethylene is involved in axillary bud inhibition. Auxin, transported down from the apical meristem of the stem, stimulates the production of ethylene, which suppresses axillary bud development. Ethylene also suppresses stem and root elongation, even in the presence of other hormones.

This completes our discussion of plant hormones. Table 26.1 summarizes the five hormones, their actions, and their commercial uses. The next section of the chapter explores plant responses to environmental stimuli.

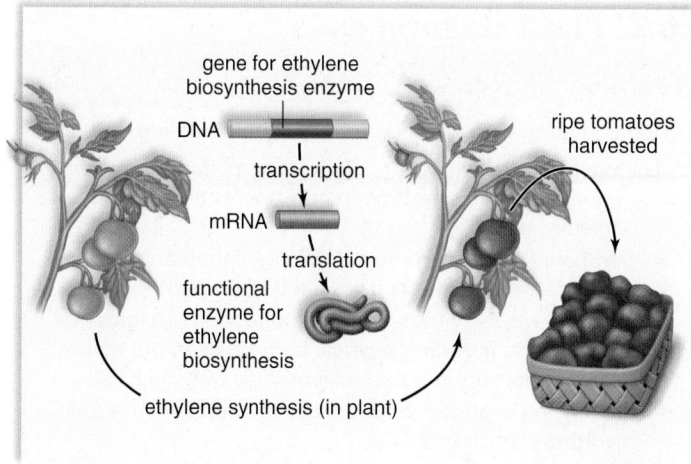

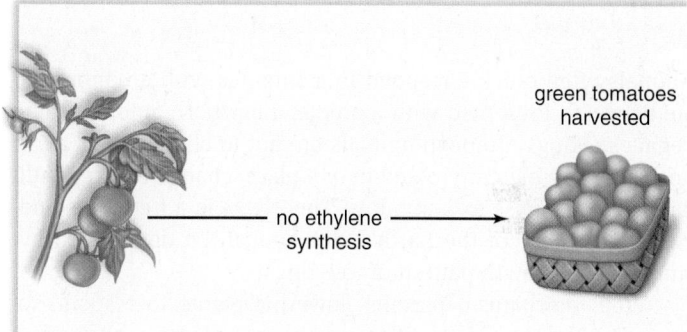

Figure 26.10 Ethylene and fruit ripening. Wild-type tomatoes (*top*) ripen on the vine after producing ethylene. Tomatoes (*bottom*) are genetically modified to produce no ethylene and stay green for shipping.

Table 26.1 Plant Hormones

Hormone	Plant use	Commercial use
auxin	growth of roots and fruits, prevents the loss of leaves	induce fruit production without pollination, used in commercial herbicides (2,4-D)
gibberellins	stem elongation	increase growth and size of the plants, breaks the dormancy cycle
cytokinins	cell division, prevents senescence	prolongs the shelf life of flowers and vegetables
abscisic acid	initiates and maintains seed and bud dormancy	commercial thinning of fruits to promote growth in the remaining fruit
ethylene	abscission and ripening of fruit	ripening of fruits and vegetables for market

No abscission   Abscission

Figure 26.9 Ethylene and abscission. Normally, there is no abscission when a holly twig is placed under a glass jar for a week. When an ethylene-producing ripe apple is also under the jar, abscission of the holly leaves occurs.

Check Your Progress   26.1

1. Explain how hormones assist in bringing about responses to stimuli.
2. Determine which hormones would produce an increase in the size of a plant's organs.
3. Explain why abscisic acid is sometimes referred to as an inhibitory hormone.

26.2 Plant Responses

a.

Animals often quickly respond to a stimulus with an appropriate behavior. Presented with a nipple, a newborn automatically begins sucking. Although animals are apt to change their location, plants, which are rooted in one place, change their growth pattern in response to a stimulus. The events in a tree's life, and even the history of the Earth's climate, can be determined by studying the growth pattern of tree rings!

What mechanism permits flowering plants to respond to stimuli? Although details differ, animals and plants go through a similar sequence of events. When humans respond to light, the stimulus is first received by a pigment in the retina at the back of the eyes, and then nerve impulses are generated that go to the brain. Only then do humans perform an appropriate behavior.

As shown in Figure 26.1, the first step toward a response is *reception* of the stimulus. The next step is *transduction*, meaning that the stimulus has been changed into a form that is meaningful to the organism. (In the case of human vision, the light stimulus is changed to nerve impulses.) Finally, a *response* is made by the organism. Animals and plants go through this same sequence of events when they respond to a stimulus; however, in the case of plants, no nerves are present—instead, chemical signals are released, and binding of these signals brings about transduction and response.

In this section, we consider four plant responses: tropisms, nastic movements, photoperiodism, and responses to the biotic environment—the other organisms present.

Tropisms

Growth toward or away from a unidirectional stimulus is called a **tropism** [Gk. *tropos*, turning]. Unidirectional means that the stimulus is coming from only one direction instead of multiple directions. Growth toward a stimulus is called a positive tropism, and growth away from a stimulus is called a negative tropism. Tropisms are due to differential growth—one side of an organ elongates faster than the other, and the result is a curving toward or away from the stimulus.

A number of tropisms have been observed in plants. The three best-known tropisms are gravitropism (gravity), phototropism (light), and thigmotropism (touch).

b.

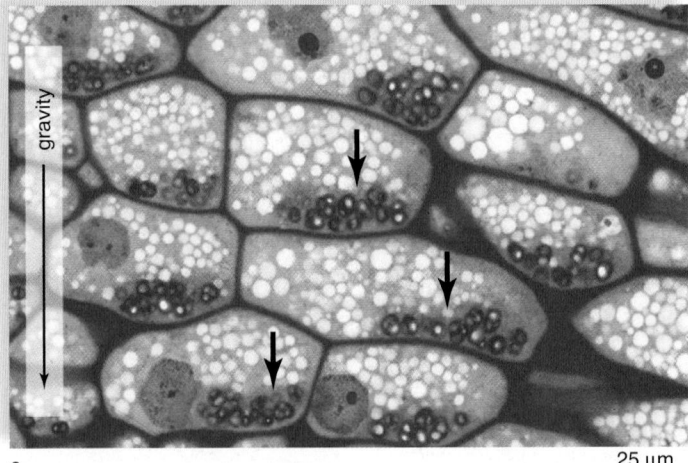

gravity

c.   25 µm

Figure 26.11 Gravitropism. **a.** Negative gravitropism of the stem of a *Coleus* plant 24 hours after the plant was placed on its side. **b.** Positive gravitropism of a root emerging from a corn kernel. **c.** Sedimentation of statoliths (see arrows), which are amyloplasts containing starch granules, is thought to explain how roots perceive gravity.

Gravitropism: a movement in response to gravity

Phototropism: a movement in response to a light stimulus

Thigmotropism: a movement in response to touch

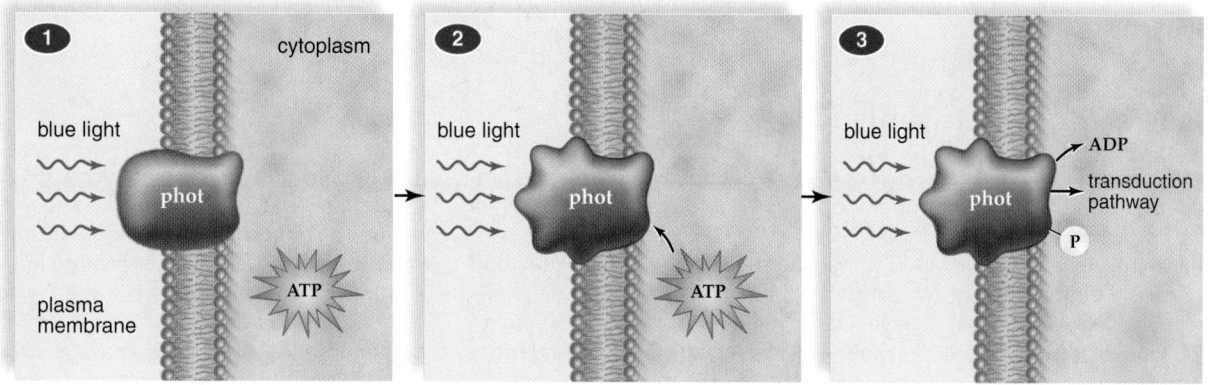

Figure 26.12 Phototropin. In the presence of blue light, a photoreceptor called phototropin (phot) is activated and becomes phosphorylated. A transduction pathway begins.

Other tropisms include chemotropism (chemicals), traumotropism (trauma), skototropism (darkness), and aerotropism (oxygen).

Gravitropism

As you might have expected from the previous discussion on page 485, when an upright plant is placed on its side, the stem displays negative **gravitropism** [L. *gravis*, heavy; Gk. *tropos*, turning] because it grows upward, opposite the pull of gravity (Fig. 26.11*a*). Charles and Francis Darwin were among the first to say that roots, in contrast to stems, show positive gravitropism (Fig. 26.11*b*). Further, they discovered that if the root cap is removed, roots no longer respond to gravity.

Later, investigators came up with an explanation. Root cap cells contain sensors called **statoliths,** which are starch grains located within amyloplasts, a type of plastid (Fig. 26.11*c*). Under the influence of gravity, the amyloplasts settle to a lower part of the cell where they come in contact with cytoskeletal elements. How amyloplasts affect auxin levels is uncertain, but some change in auxin distribution appears to take place. Stems and roots may respond differently to auxin, such that the cells on the upper surface of a root elongate so that the root curves downward, and cells on the lower surface of a stem elongate, so that the stem curves upwards.

Phototropism

As also discussed previously, positive **phototropism** of stems occurs because the cells on the shady side of the stem elongate due to the presence of auxin. Plant growth that curves away from light is called negative phototropism. Roots, depending on the species examined, are either insensitive to light or exhibit negative phototropism.

Through the study of mutant *Arabidopsis* plants (see the Nature of Science feature on page 492), botanists now know that phototropism occurs because plants respond to blue light (Fig. 26.12). ❶ When blue light is absorbed, the pigment portion of a photoreceptor, called phototropin (phot), undergoes a conformation change. ❷ This change results in the transfer of a phosphate group from ATP (adenosine triphosphate) to a protein portion of the photoreceptor. ❸ The phosphorylated photoreceptor triggers a transduction pathway that, in some unknown way, leads to the entry of auxin into the cell.

Thigmotropism

Unequal growth due to contact with solid objects is called **thigmotropism** [Gk. *thigma*, touch, and *tropos*, turning]. An example of this response is the coiling of tendrils or the stems of plants, such as the stems of pea and morning glory plants (Fig. 26.13).

A flowering plant grows straight until it touches something. Then the cells in contact with an object, such as a pole, grow less while those on the opposite side elongate. Thigmotropism can be quite rapid; tendrils have been observed to encircle an object within 10 minutes. Several minutes of touching can bring about a response that lasts for several days. The response isn't always immediate—tendrils touched in the dark will respond once they are illuminated. ATP rather than light initiates the response. It is possible that auxin and ethylene play a role in

Figure 26.13 Coiling response. The stem of a morning glory plant, *Ipomoea*, coiling around a pole illustrates thigmotropism.

Nature of Science

Arabidopsis Is a Model Organism

Arabidopsis thaliana is a small flowering plant related to cabbage and mustard plants (Fig. 26A). *Arabidopsis* has no commercial value—in fact, it is a weed! However, it has become a model organism for the study of plant molecular genetics, including signal transduction. Unlike crop plants used formerly, *Arabidopsis* has characteristics that make it an ideal model organism.

- It is small, so many hundreds of plants can be grown in a small amount of space. *Arabidopsis* consists of a flat rosette of leaves from which grows a short flower stalk.
- Generation time is short. It takes only 5–6 weeks for plants to mature, and each one produces about 10,000 seeds!
- It normally self-pollinates, but it can easily be cross-pollinated. This feature facilitates gene mapping and the production of strains with multiple mutations.
- The number of base pairs in its DNA is relatively small: 125 million base pairs are distributed in 5 chromosomes (2n = 10) and 25,500 genes.

In contrast to *Arabidopsis*, crop plants, such as corn, have generation times of at least several months, and they require a great deal of field space for a large number to grow. Crop plants also have much larger genomes than *Arabidopsis*. For comparison, the genome sizes for rice (*Oryza sativa*), wheat (*Triticum aestivum*), and corn (*Zea mays*) are 420 million, 16 billion, and 2.5 billion base pairs, respectively. However, crop plants have about the same number of functional genes as *Arabidopsis*, and they occur in the same sequence. Therefore, knowledge of the *Arabidopsis* genome can be used to help locate specific genes in the genomes of other plants.

Now that the *Arabidopsis* genome has been sequenced, genes of interest can be cloned from the *Arabidopsis* genome and then used as probes for the isolation of homologous genes from plants of economic value. Also, cellular processes controlled by a family of genes in other plants require only a single gene or fewer genes in *Arabidopsis*. This, too, facilitates molecular biological studies of the plant.

The creation of *Arabidopsis* mutants plays a significant role in discovering what each of its genes do. For example, if a mutant plant lacks stomata (openings in leaves), then we know that the affected gene influences the formation of stomata. Transformation has emerged as a powerful way to create *Arabidopsis* mutants. The transforming DNA often gets inserted directly within a particular gene sequence.

This usually destroys the function of the disrupted gene, resulting in a "knockout mutant." Furthermore, the piece of transformed DNA (T-DNA) that is inserted in the disrupted plant gene can serve as a flag for tracking down the gene by molecular biology methods. Large-scale projects using this T-DNA insertion technique are under way to mutate, identify, and characterize every gene in the *Arabidopsis* genome.

Researchers have discovered three classes of genes that are essential to normal floral pattern formation. These are identified as *homeotic genes* because they cause sepals, petals, stamens, or carpels to appear in place of one another. Triple mutants that lack all three types of genetic activities have flowers that consist entirely of leaves arranged in whorls. And a mutation of a regulatory gene results in flowers that have three whorls of petals. These floral-organ-identity genes appear to be regulated by transcription factors that are expressed and required for extended periods.

The application of *Arabidopsis* genetics to other plants has been demonstrated. For example, one of the mutant genes that alters the development of flowers has been cloned and introduced into tobacco plants, where, as expected, it causes sepals and stamens to appear where normally petals would develop. The investigators commented that knowledge about the development of flowers in *Arabidopsis* can have far-ranging applications. It will undoubtedly lead someday to more productive crops.

A study of the *Arabidopsis* genome will undoubtedly promote plant molecular genetics in general. And because *Arabidopsis* is a model organism, genetic findings from this plant may have applications to humans just as Mendel's work with pea plants led to formuation of genetic laws. It's far easier to study signal transduction in *Arabidopsis* cells than in human cells.

Questions to Consider

1. Why is *Arabidopsis* a better study organism than cabbage or mustard?
2. What are the potential applications of *Arabidopsis* research?
3. Explain how transformation is used for genetic studies.

Arabidopsis thaliana

Figure 26A Photograph showing overall appearance of *Arabidopsis thaliana*. Many investigators have turned to this weed as an experimental organism to study the actions of genes, including those that control growth and development.

the process since they are capable of inducing the curvature of tendrils even in the absence of touch.

Thigmomorphogenesis is a touch response related to thigmotropism. In thigmomorphogenesis, the entire plant responds to the presence of environmental stimuli, such as wind or rain. A tree growing in a windy location often has a shorter, thicker trunk than the same type of tree growing in a more protected location. Even simple mechanical stimulation, such as rubbing a plant with a stick, can inhibit cellular elongation and produce a sturdier plant with increased amounts of support tissue.

Video
Plant Tactile Response

Nastic Movements

Recall that a plant cell exhibits turgor when it fills with water:

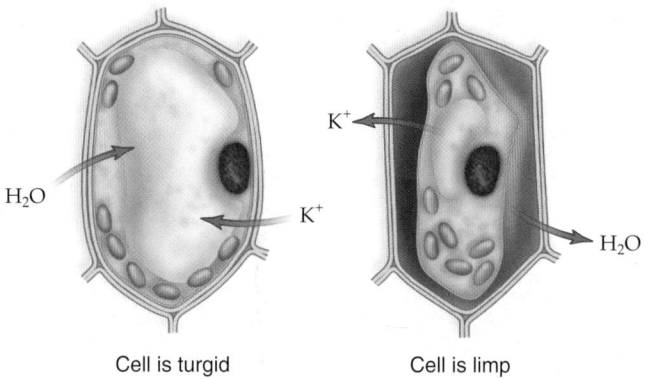

Cell is turgid   Cell is limp

In general, if water exits the many cells of a leaf, the leaf goes limp. Conversely, if water enters a limp leaf, and cells exhibit turgor, the leaf moves as it regains its former position. Nastic movements, also known as **turgor movements,** are dependent on turgor pressure changes in plant cells. In contrast to tropisms, turgor movements do not involve growth and are not directly related to the source of the stimulus.

Turgor Responses to Touch

Turgor movements can result from touch, shaking, or thermal stimulation. The sensitive plant, *Mimosa pudica*, has compound leaves, meaning that each leaf contains many leaflets. Touching one leaflet collapses the whole leaf (Fig. 26.14). *Mimosa* is remarkable because the progressive response to the stimulus takes only a second or two.

The portion of a flowering plant involved in controlling turgor movement is a thickening called a pulvinus at the base of each leaflet. A leaf folds when the cells in the lower half of the pulvinus, called the motor cells, lose potassium ions (K^+), and then water. When the pulvinus cells lose turgor, the leaflets of the leaf collapse. An electrical mechanism may cause the response to move from one leaflet to another. The speed of an electrical charge transmission has been measured at about 1 cm/sec.

A Venus flytrap closes its trap in less than 1 second when three hairs at the base of the trap, called the trigger hairs, are touched by an insect. When the trigger hairs are stimulated by the insect, an electrical charge is transmitted throughout the

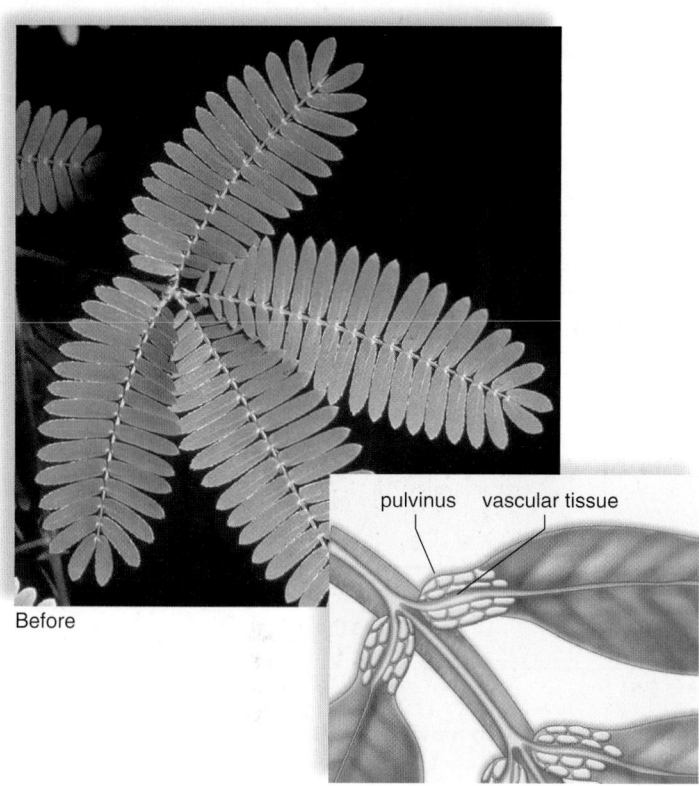

Before

pulvinus   vascular tissue

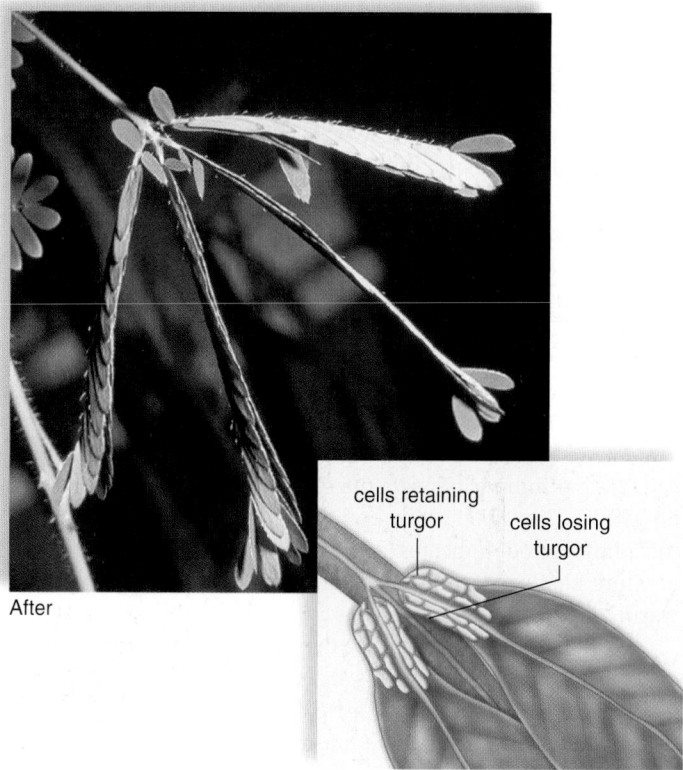

After

cells retaining turgor   cells losing turgor

Figure 26.14 Turgor movement.   A leaf of the sensitive plant, *Mimosa pudica,* before and after it is touched.

lobes of a leaf. Exactly what causes this electrical charge is being studied. Two possible explanations have been suggested: (1) Perhaps the cells located near the outer region of the lobes rapidly secrete hydrogen ions into their cell walls, loosening them, and allowing the walls to swell rapidly by osmosis; or (2) Perhaps the cells in the inner portion of the lobes and the midrib rapidly lose ions, leading to a loss of water by osmosis and collapse of these cells. In any case, it appears that turgor movements are involved.

Sleep Movements and Circadian Rhythms

Leaves that close at night are said to exhibit sleep movements. Activities that occur regularly in a 24-hour cycle, such as sleep movements, are called **circadian rhythms.** One of the most common examples occurs in a houseplant called the prayer plant (*Maranta leuconeura*) because at night the leaves fold upward into a shape resembling hands at prayer (Fig. 26.15, *top*). This movement is due to changes in the turgor pressure of motor cells in a pulvinus located at the base of each leaf.

Morning glory (*Ipomoea leptophylla*) is a plant that opens its flowers in the early part of the day and closes them at night (Fig. 26.15, *bottom*). In most plants, stomata open in the morning and close at night, and some plants secrete nectar at the same time of the day or night.

To qualify as a circadian rhythm, the activity must (1) occur every 24 hours; (2) take place in the absence of external stimuli, such as in dim light; and (3) be able to be reset if external cues are provided. For example, if you take a transcontinental flight, you will likely suffer jet lag at the destination because your body will still be attuned to the day/night pattern of its previous environment. But after several days, you probably will adjust and will be able to go to sleep and wake up according to your new time.

Biological Clock

The internal mechanism by which a circadian rhythm is maintained in the absence of appropriate environmental stimuli is termed a **biological clock.** If organisms are sheltered from environmental stimuli, their biological clock keeps the circadian rhythms going, but the cycle extends. In prayer plants, for example, the sleep cycle changes to 26 hours when the plant is kept in constant dim light, as opposed to 24 hours when in traditional day/night conditions. Therefore, it is suggested that biological clocks are synchronized by external stimuli to 24-hour rhythms.

The length of daylight compared to the length of darkness, called the **photoperiod,** sets the clock. Temperature has little or no effect. This synchronization with light is adaptive because the photoperiod indicates seasonal changes better than temperature changes. Spring and fall, in particular, can have both warm and cold days.

Work with *Arabidopsis* (see the Nature of Science feature on page 492) and other organisms suggests that the biological clock involves the transcription of a small number of "clock genes." One model proposes that the information-transfer system from DNA to RNA to enzyme to metabolite, with all its feedback controls, is intrinsically cyclical and could be the basis for biological

Prayer plant (morning) Prayer plant (night)
a.

Morning glory (morning) Morning glory (night)
b.

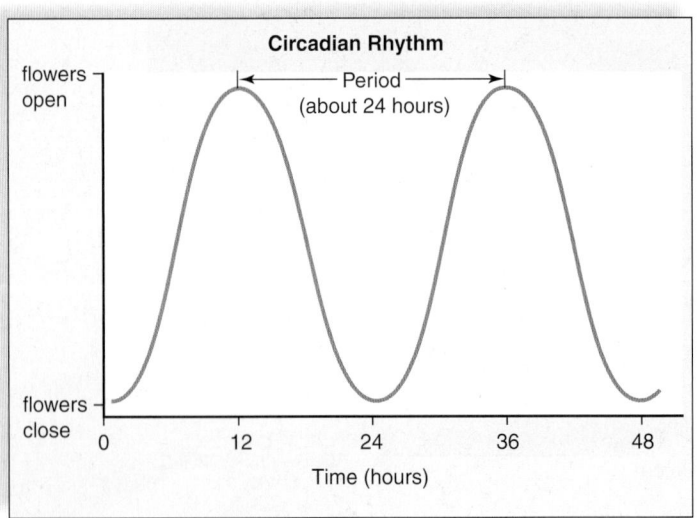

c.

Figure 26.15 Sleep movements and circadian rhythms.
a. The leaves of a prayer plant, *Maranta leuconeura,* fold every 24 hours at night. **b.** The flowers of the morning glory, *Ipomoea leptophylla,* close at night. **c.** Graph of circadian rhythm exhibited by morning glory plant.

clocks. In *Arabidopsis,* the biological clock involves about 5% of the genome. These genes control sleep movements, the opening and closing of stomata, the discharge of floral fragrances, and the metabolic activities associated with photosynthesis. The biological clock also influences seasonal cycles that depend on day/night lengths, including the regulation of flowering.

Although circadian rhythms are outwardly very similar in all species, the clock genes that have been identified are not the same in all species. It would seem, then, that biological clocks have evolved several times to perform similar tasks.

Photoperiodism

As just noted, many physiological changes in flowering plants are related to a seasonal change in day length. Such changes include seed germination, the breaking of bud dormancy, and the onset of senescence. A physiological response prompted by changes in the length of day or night in a 24-hour daily cycle is called **photoperiodism** [Gk. *photos,* light, and *periodus,* completed course]. In some plants, photoperiodism influences flowering; for example, violets and tulips flower in the spring, and asters and goldenrod flower in the fall. Photoperiodism requires the participation of a biological clock and the activity of a plant photoreceptor called phytochrome.

Phytochrome

Phytochrome [Gk. *phyton,* plant, and *chroma,* color] is a blue-green leaf pigment that is present in the cytoplasm of plant cells. A phytochrome molecule is composed of two identical proteins (Fig. 26.16). Each protein has a larger portion in which a light-sensitive region is located. The smaller portion is a kinase that can link light absorption with a transduction pathway within the cytoplasm. Phytochrome can be said to act like a light switch because, like a light switch, it can be in the down (inactive) position or in the up (active) position.

Red light prevalent in daylight activates phytochrome, and it assumes its active conformation known as P_{fr}. When P_{fr} moves into the nucleus, it interacts with specific proteins, such as a transcription factor. The complex activates certain genes and inactivates others. The "fr" subscript in P_{fr} indicates that it absorbs *far-red* light. Far-red light is prevalent in the evening, and it serves to change P_{fr} to P_r (for *red* light), which is the inactive form of phytochrome.

Animation
Phytochrome Signaling

Functions of Phytochrome

The $P_r \rightarrow P_{fr}$ conversion cycle is now known to control various growth functions in plants. P_{fr} promotes seed germination and inhibits shoot elongation, for example. The presence of P_{fr} indicates to some seeds that sunlight is present and conditions are favorable for germination. This explains why some seeds must be only partly covered with soil when planted. Germination of other seeds, such as those of *Arabidopsis,* is inhibited by light, so they must be planted deeper.

Following germination, the presence of P_{fr} indicates that sunlight is available, and the seedlings begin to grow normally—the leaves expand and become green and the stem begins branching. Seedlings that are grown in the dark **etiolate**—that is, the shoot increases in length, and the leaves remain small (Fig. 26.17). Only when P_r is converted to P_{fr} does the seedling grow normally.

Flowering and Photoperiodism

Flowering plants can be divided into three groups on the basis of their flowering status:

1. **Short-day plants** flower when the day length is shorter than a *critical length.* (Examples are cocklebur, goldenrod, poinsettia, and chrysanthemum.)

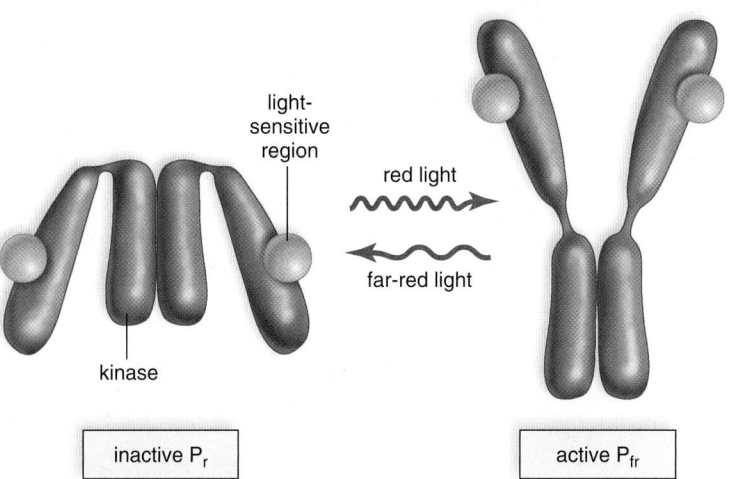

Figure 26.16 Phytochrome conversion cycle. The inactive form of phytochrome (P_r) is converted to the active form P_{fr} in the presence of red light, which is prevalent in daylight. P_{fr}, the active form of phytochrome, is involved in various plant responses such as seed germination, shoot elongation, and flowering. P_{fr} is converted to P_r whenever light is limited, such as in the shade or during the night.

light-sensitive region

red light

far-red light

kinase

inactive P_r

active P_{fr}

a. Normal growth   b. Etiolation

Figure 26.17 Phytochrome control of shoot elongation.
a. If red light is prevalent, as it is in bright sunlight, normal growth occurs.
b. If far-red light is prevalent, as it is in the shade, etiolation occurs. These effects are due to phytochrome.

2. **Long-day plants** flower when the day length is longer than a critical length. (Examples are wheat, barley, rose, iris, clover, and spinach.)

3. **Day-neutral plants** are not dependent on day length for flowering. (Examples are tomato and cucumber.)

The criterion for designating plants as short-day or long-day is not an absolute number of hours of light, but a critical number that either must be or cannot be exceeded. Spinach is a long-day plant that has a critical day length of 14 hours; ragweed is a short-day plant with the same critical length. Spinach, however, flowers in the summer when the day length increases to 14 hours or more, and ragweed flowers in the fall, when the day length shortens to 14 hours or fewer. In addition, we now know that some plants require a specific sequence of day lengths in order to flower.

Soon after the three groups of flowering plants were distinguished, researchers began to experiment with artificial lengths of light and dark that did not necessarily correspond to a normal 24-hour day. These investigators discovered that the cocklebur, a short-day plant, does not flower if a required long dark period is interrupted by a brief flash of white light. (Interrupting the light period with darkness has no effect.) In contrast, a long-day plant does flower if an overly long dark period is interrupted by a brief flash of white light. They concluded that the length of the dark period, not the length of the light period, controls flowering. Of course, in nature, short days always go with long nights, and vice versa.

To recap, let's consider Figure 26.18:

• Cocklebur is a short-day plant (Fig. 26.18, *left*). ❶ When the night is longer than a critical length, cocklebur flowers. ❷ The plant does not flower when the night is shorter than the critical length. ❸ Cocklebur also

does not flower if the longer-than-critical-length night is interrupted by a flash of light.

• Clover is a long-day plant (Fig. 26.18, *right*). ❹ When the night is shorter than a critical length, clover flowers. ❺ The plant does not flower when the night is longer than a critical length. ❻ Clover does flower when a slightly longer-than-critical-length night is interrupted by a flash of light.

Responses to the Biotic Environment

Plants are always under attack by herbivores as well as parasites. Fortunately, they have an arsenal of defense mechanisms to help them deal with their predators (Fig. 26.19).

Physical and Chemical Defenses

A plant's epidermis and bark do a good job of discouraging attackers. But, unfortunately, herbivores have ways around a plant's first line of defense. A fungus can invade a leaf via the stomata and set up shop inside the leaf, where it feeds on nutrients meant for the plant. Underground nematodes have sharp mouthparts to break through the epidermis of a root and establish a parasitic relationship. Tiny insects called aphids have styletlike mouthparts that allow them to tap into the phloem of a nonwoody stem. These examples illustrate why plants need a variety of defenses that are not dependent on its outer surface.

The primary metabolites of plants, such as sugars and amino acids, are necessary to the normal workings of a cell. Plants also produce molecules termed **secondary metabolites** as a defense mechanism. Secondary metabolites were once thought to be waste products, but now we know that they are part of a plant's arsenal to prevent predation.

Tannins, present in or on the epidermis of leaves, are defensive compounds that interfere with the outer proteins of bacteria

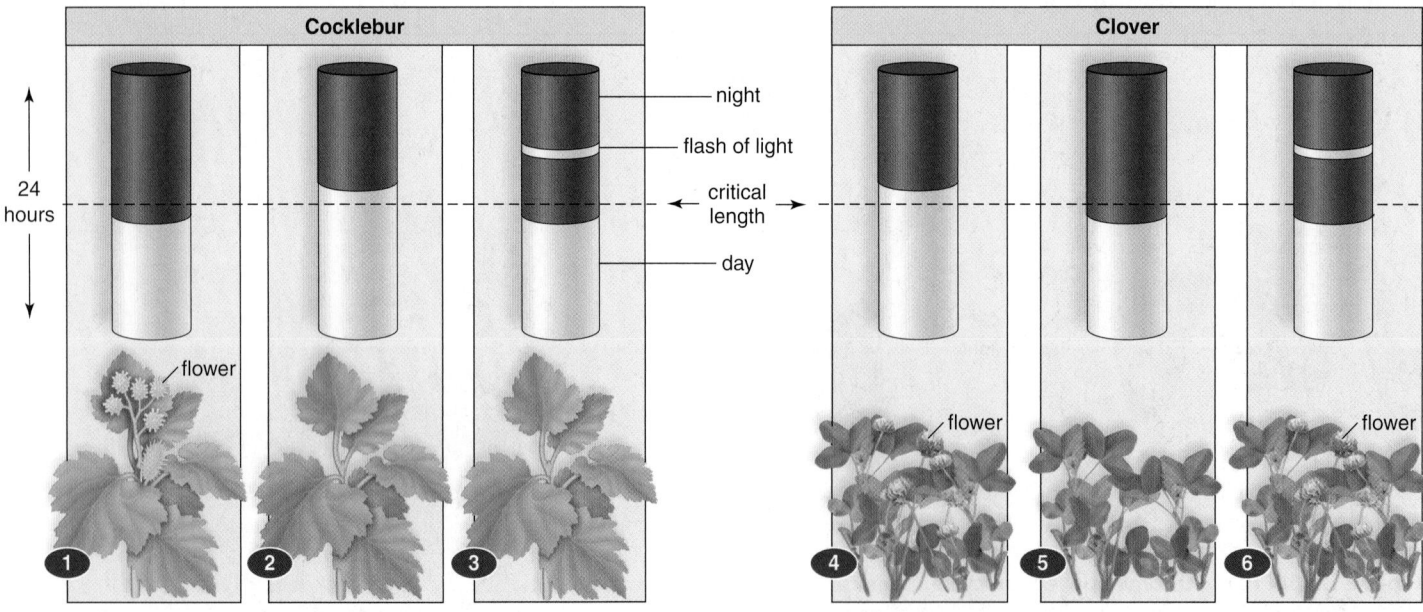

Figure 26.18 Photoperiodism and flowering.   a. Short-day plant. When the day is shorter than a critical length, this type of plant flowers. The plant does not flower when the day is longer than the critical length. It also does not flower if the longer-than-critical-length night is interrupted by a flash of light. **b.** Long-day plant. The plant flowers when the day is longer than a critical length. When the day is shorter than a critical length, this type of plant does not flower. However, it does flower if the slightly longer-than-critical-length night is interrupted by a flash of light.

Alfalfa plant bug

Fungus infection

Monarch caterpillar and butterfly

Figure 26.19 Plant predators and parasites. Insects are predators and fungi are parasites of flowering plants.

and fungi. They also deter herbivores because of their astringent effect on the mouth and their interference with digestion. Some secondary metabolites, such as bitter nitrogenous substances called **alkaloids** (e.g., morphine, nicotine, and caffeine), are well-known to humans because we use them for our own purposes. The seedlings of coffee plants contain caffeine at a concentration high enough to kill insects and fungi by blocking DNA and RNA synthesis. Other secondary metabolites include the **cyanogenic glycosides** (molecules containing a sugar group) that break down to cyanide and inhibit cellular respiration. Foxglove (*Digitalis purpurea*) produces deadly cardiac and steroid glycosides, which cause nausea, hallucinations, convulsions, and death in animals that ingest them. Taxol, an unsaturated hydrocarbon from the Pacific yew (*Taxus brevifolia*), is now a well-known cancer-fighting drug.

Predators can be one step ahead of a plant's secondary metabolites. Monarch caterpillars, for example, are able to feed on milkweed plants despite the presence of a poisonous glycoside, and they even store the chemical in their bodies. In this way, the Monarch caterpillar and the butterfly become poisonous to their own predators (Fig. 26.19). Birds that become sick after eating a monarch butterfly leave them alone thereafter.

Video
Cranberries versus Bacteria

Wound Responses

Wound responses illustrate that plants can make use of transduction pathways to produce chemical defenses only when they are needed. After a leaf is chewed or injured, a plant produces *proteinase inhibitors*, chemicals that destroy the digestive enzymes of a predator feeding on them. The proteinase inhibitors are produced throughout the plant, not just at the wound site. The defense hormone that brings about this effect is a small peptide called **systemin** (Fig. 26.20). Systemin is produced in the wound area in response to the predator's saliva, but then it travels between cells to reach phloem, which distributes it about the plant. A transduction pathway is activated in cells with systemin

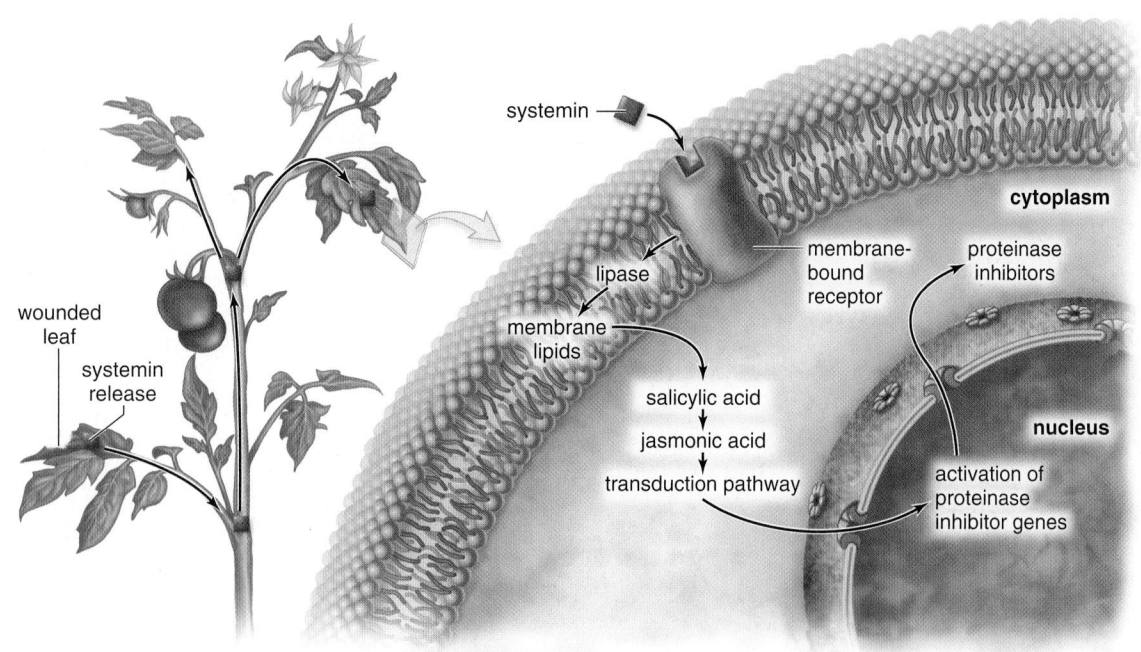

Figure 26.20 Wound response in tomato. Wounded leaves produce systemin, which travels in phloem to all parts of a plant where it binds to cells that have a systemin receptor. These cells then produce jasmonic acid, a molecule that initiates a transduction pathway that leads to production of proteinase inhibitors, which limit insect feeding.

receptors, and the cells produce proteinase inhibitors. A chemical called jasmonic acid, and also possibly a chemical called salicylic acid, are part of this transduction pathway. Salicylic acid (the active chemical in aspirin) has been known since the 1930s to bring about a phenomenon called systemic acquired resistance (SAR), the production of antiherbivore chemicals by defense genes. Recently, companies have begun marketing salicylic acid and other similar compounds as a way to activate SAR in crops, including tomato, spinach, lettuce, and tobacco.

Hypersensitive Response

On occasion, plants produce a gene product that binds to a specific viral, bacterial, or fungal gene product made within the cell. This combination offers a way for the plant to "recognize" a particular pathogen. A transduction pathway now ensues, and the final result is a **hypersensitive response (HR)** that seals off the infected area and also initiates the wound response just discussed.

Indirect Defenses

Some defenses do not kill or discourage herbivores outright. For example, female butterflies are less likely to lay their eggs on plants that already have butterfly eggs. The leaves of some passion flowers display physical structures resembling the yellow eggs of *Heliconius* butterflies, and as a result, these butterflies do not lay eggs on this plant. Other plants produce hormones that prevent caterpillars from metamorphosing into adults and laying more eggs.

Certain plants attract the natural enemies of the caterpillars that are feeding upon them. The plants produce volatile molecules that diffuse into the air and advertise that food is available for the predator of the caterpillar. For example, lima beans produce volatiles that attract carnivore mites only when the beans are being damaged by a spider mite. Corn and cotton plants release volatiles that attract wasps, which then inject their eggs into the caterpillars munching on the plants' leaves. The wasp eggs develop into larvae that eat the caterpillars, preventing them from eating the leaves and becoming reproductive adults.

Relationships with Animals

Mutualism is a relationship between two species in which both species benefit. As evidence that a mutualistic relationship can help protect a plant from predators, consider the bullhorn acacia tree, and the acacia ant, *Pseudomyrmex ferruginea*. Unlike other acacias, this tree has swollen thorns with a hollow interior where ant larvae can grow and develop.

In addition to housing the ants, acacias provide them with food. The ants feed from nectaries at the base of leaves and eat fat- and protein-containing nodules called Beltian bodies found at the tips of the leaves. In return, the ants constantly protect the plant by attacking and stinging any would-be herbivores. In fact, when the ants on experimental trees were removed, the acacia trees died.

Check Your Progress　　　　　　　　26.2

1. Explain why it is adaptive for roots to grow toward water.
2. Distinguish between a positive and a negative geotropism.
3. Describe the structure of phytochrome and how it functions in plant cells.
4. Explain why a long-day plant still flowers if the long day is interrupted by a period of darkness.
5. Describe several methods that flowering plants use to protect themselves from insect predators.

CONNECTING *the* CONCEPTS *with the* BIG IDEAS

Energy and Homeostasis

- Phototropism facilitates plant response to light changes using molecular signals which maximize photosynthetic surface area. (2C2a*IE*, 2E2a1, 2E3b1)
- Photoperiodism involves phytochrome's control of flowering and seasonal changes. (2C2a*IE*, 2E2a2, 2E3b2, 2A1f*IE*)
- Circadian rhythms, including stomata openings, allow plants to adjust to environmental conditions. (2C2a*IE*)
- Defensive chemical signals and deterrents allow plants to recognize and destroy infected cells. (2D4a*IE*, 3D2b*IE*)

Information and Signaling

- Cytokines trigger mitosis and cytokinesis by regulating gene expression. (3B2a*IE*)
- Increase in ethylene levels induce enzyme production that promotes fruit ripening. (3B2a*IE*)
- Gibberellins promote signal transmissions that affect specific genes, triggering seed germination. (3B2a*IE*)

*Find the unabridged version of all EK citations at www.glencoe.com/maderAP11.

▓▓▓ Media Study Tools ▓▓▓

www.glencoe.com/maderAP11

Enhance your study of this chapter with study tools and practice tests. Also ask your instructor about the resources available through

ConnectPlus, including the media-rich eBook, interactive learning tools, and animations.

Summarize

26.1 Plant Hormones

Like animals, flowering plants use a signal transduction pathway when they respond to a stimulus. The process involves receptor activation, transduction of the signal by relay proteins, and a cellular response, which can consist of the turning on of a gene or an enzymatic pathway.

Auxin-controlled cell elongation is involved in phototropism and gravitropism. When a plant is exposed to light, auxin moves laterally from the bright to the shady side of a stem.

Gibberellin causes stem elongation between nodes and also breaks bud and seed dormancy. After this hormone binds to a plasma membrane receptor, a DNA-binding protein activates a gene leading to the production of amylase. Amylase is an enzyme that speeds the breakdown of amylose.

Cytokinins cause cell division, the effects of which are especially obvious when plant tissues are grown in culture.

Abscisic acid (ABA) and ethylene are two plant growth inhibitors. ABA is well known for causing stomata to close, and ethylene is known for causing fruits to ripen.

26.2 Plant Responses

When flowering plants respond to stimuli, growth, movement, or both occur. Tropisms are growth responses toward or away from unidirectional stimuli. The positive phototropism of stems results in a bending toward light, and the negative gravitropism of stems results in a bending away from the direction of gravity. Roots that bend toward the direction of gravity show positive gravitropism. Thigmotropism occurs when a plant part makes contact with an object, as when tendrils coil about a pole.

Nastic movements, or turgor movements, are not directional. Due to turgor pressure changes, some plants respond to touch and some perform sleep movements. Plants exhibit circadian rhythms, which are believed to be controlled by a biological clock. The sleep movements of prayer plants, the closing of stomata, and the daily opening of certain flowers have a 24-hour cycle.

Phytochrome is a pigment that is involved in photoperiodism, the ability of plants to sense the length of the day and night during a 24-hour period. This sense leads to seed germination, shoot elongation, and flowering during favorable times of the year. The conformation, and activity, of phytochrome is influenced by daylight. Phytochrome in the P_{fr} form leads to a biological response such as flowering.

Short-day plants flower only when the days are shorter than a critical length, and long-day plants flower only when the days are longer than a critical length. Research has shown that actually it is the length of darkness that is critical. Interrupting the dark period with a flash of white light prevents flowering in a short-day plant and induces flowering in a long-day plant.

Flowering plants have defenses against predators and parasites. The first line of defense is their outer covering. They also routinely produce secondary metabolites that protect them from herbivores, particularly insects. Wounding causes plants to produce systemin, which travels about the plant and causes cells to produce proteinase inhibitors that destroy an insect's digestive enzymes. During a hypersensitive response, an infected area is sealed off. As an indirect response, plants temporarily attract animals that destroy predators, and going one step further, plants have permanent relationships with animals, such as ants, that attack predators.

Key Terms

abscisic acid (ABA) 488	hypersensitive response
abscission 488	(HR) 498
alkaloid 497	long-day plant 496
apical dominance 485	photoperiod 494
auxin 485	photoperiodism 495
biological clock 494	phototropism 491
cellular response 484	phytochrome 495
circadian rhythm 494	receptor 484
coleoptile 485	secondary metabolite 496
cyanogenic glycoside 497	senescence 487
cytokinin 487	short-day plant 495
day-neutral plant 496	signal transduction 484
dormancy 487	statolith 491
etiolate 495	systemin 497
ethylene 488	thigmotropism 491
gibberellin 486	transduction pathway 484
gravitropism 491	tropism 490
hormone 484	turgor movement 493

Assess

Reviewing This Chapter

1. Name and describe the three stages of signal transduction in plant cells.   484
2. Why does removing a terminal bud cause a plant to get bushier?   485
3. What experiments led to knowledge that a hormone is involved in phototropism? Explain the mechanism by which auxin brings about elongation of cells.   485–86
4. Gibberellin research supports the hypothesis that plant hormones initiate a reception-transduction-response pathway. Explain.   486–87
5. What is the function of cytokinins? Discuss experimental evidence to suggest that hormones interact when they bring about an effect.   487
6. What are some of the primary effects of abscisic acid and how does it bring about these effects?   488
7. What hormones are involved in abscission? How does ethylene bring about ripening of fruits?   488–89
8. Tropisms are responses to stimuli. Why are stems said to exhibit positive phototropism but negative gravitropism?   491
9. What are nastic movements, and how do turgor pressure changes bring about movements of plants?   493–94
10. What is a biological clock, how does it function, and what is its primary usefulness in plants?   494–95
11. Define photoperiodism, and discuss its relationship to flowering in certain plants.   495–96
12. Describe the structure of phytochrome and its response to red and far-red light.   495
13. What mechanisms allow a plant to defend itself?   496–98

Testing Yourself

Choose the best answer for each question.

1. During which step of signal transduction is a second messenger released into the cytoplasm?
   a. reception   c. transduction
   b. response   d. final step

2. Which of the following plant hormones causes apical dominance?
    a. auxin
    b. gibberellins
    c. cytokinins
    d. abscisic acid
    e. ethylene

3. Internode elongation is stimulated by
    a. abscisic acid.
    b. ethylene.
    c. cytokinin.
    d. gibberellin.
    e. auxin.

4. Which of the following plant hormones is responsible for a plant losing its leaves?
    a. auxin
    b. gibberellins
    c. cytokinins
    d. abscisic acid
    e. ethylene

For questions 5-9, match each statement with a hormone in the key.

KEY:
    a. auxin
    b. gibberellin
    c. cytokinin
    d. ethylene
    e. abscisic acid

5. One rotten apple can spoil the bushel.

6. Cabbage plants bolt (grow tall).

7. Stomata close when a plant is water-stressed.

8. Stems bend toward the Sun.

9. Coconut milk causes plant tissues to undergo cell division.

10. The sensors in the cells of the root cap are called
    a. mitochondria.
    b. central vacuoles.
    c. statoliths.
    d. chloroplasts.
    e. intermediate filaments.

11. A student places 25 morning glory (see Fig. 26.13) seeds in a large pot and allows the seeds to germinate in total darkness. Which of the following growth or movement activities would the seedlings exhibit?
    a. gravitropism, as the roots grow down and the shoots grow up
    b. phototropism, as the shoots search for light
    c. thigmotropism, as the tendrils coil around other seedlings
    d. Both a and c are correct.

12. A plant requiring a dark period of at least 14 hours will
    a. flower if a 14-hour night is interrupted by a flash of light.
    b. not flower if a 14-hour night is interrupted by a flash of light.
    c. not flower if the days are 14 hours long.
    d. not flower if the nights are longer than 14 hours.
    e. Both b and c are correct.

13. Primary metabolites are needed for _____ while secondary metabolites are produced for _____.
    a. growth, signal transduction
    b. normal cell functioning, defense
    c. defense, growth
    d. signal transduction, normal cell functioning

14. Which of the following is a plant secondary metabolite used by humans to treat disease?
    a. morphine
    b. codeine
    c. quinine
    d. penicillin
    e. All but d are correct.

15. Which of these is an indirect defense?
    a. secondary chemical
    b. making a predator think that butterfly eggs are already on the leaves
    c. inviting ants to live on a plant
    d. All of these are correct.

Engage

Thinking Scientifically

1. You hypothesize that abscisic acid (ABA) is responsible for the turgor pressure changes that permit a plant to track the Sun (see the chapter opening photograph). What observations could you make to support your hypothesis?

2. You formulate the hypothesis that the negative gravitropic response of stems is greater than the positive phototropism of stems. How would you test your hypothesis?

Honeybee feeding on a yellow aster; the reproductive parts are actually in the center.

Flowering Plants: Reproduction

The flowering plants, or angiosperms, are the most diverse and widespread of all the plants, and their means of sexual reproduction, centered in the flower, is well adapted to life on land. The flower shields the female gametophyte and produces pollen grains that protect the male gametophyte until fertilization takes place. The presence of an ovary allows angiosperms to produce seeds within fruits. Seeds guard the embryo until conditions are favorable for germination.

The evolution of the flower has permitted pollination not only by wind but also by animals. Flowering plants that rely on animals for pollination have a mutualistic relationship with them. The flower provides nutrients for a pollinator such as a bee, a fly, a beetle, a bird, or even a bat. The animals, in turn, inadvertently carry pollen from one flower to another, allowing pollination to occur. Similarly, animals help flowering plants disperse their fruits, and therefore, their seeds. As you'll learn in this chapter, the diversity of flowering plants can, in part, be attributed to their relationships with a great variety of animals.

As you read through the chapter, think about the following questions:

1. How are flowering plants adapted to a terrestrial lifestyle?
2. What would happen to the diversity of flowering plants if a large percentage of insect species were to disappear in a given community?
3. What are the advantages of sexual reproduction versus asexual reproduction in flowering plants?

BEFORE YOU BEGIN

Before beginning this chapter, take a few moments to review the following discussions.

Section 10.3 What role does meiosis play in sexual reproduction?

Section 17.1 What types of isolating mechanisms prevent two different species of flowers from reproducing sexually?

Table 18.1 How long have angiosperms and modern insects been coevolving?

FOLLOWING *the* BIG IDEAS

CHAPTER 27   FLOWERING PLANTS: REPRODUCTION

Energy and Homeostasis — Angiosperm reproductive efforts involve specialized organismal responses and cooperation with other species.

Interactions and Systems — Plant reproduction is directly affected by environmental factors.

27.1 Sexual Reproductive Strategies

Sexual reproduction in plants is advantageous because it generates variation among the offspring through the process of meiosis and fertilization. In a changing environment, a new variation may be better adapted for survival and reproduction than either parent.

Life Cycle Overview

When plants reproduce sexually, they alternate between two multicellular stages, one diploid and one haploid (Fig. 27.1).

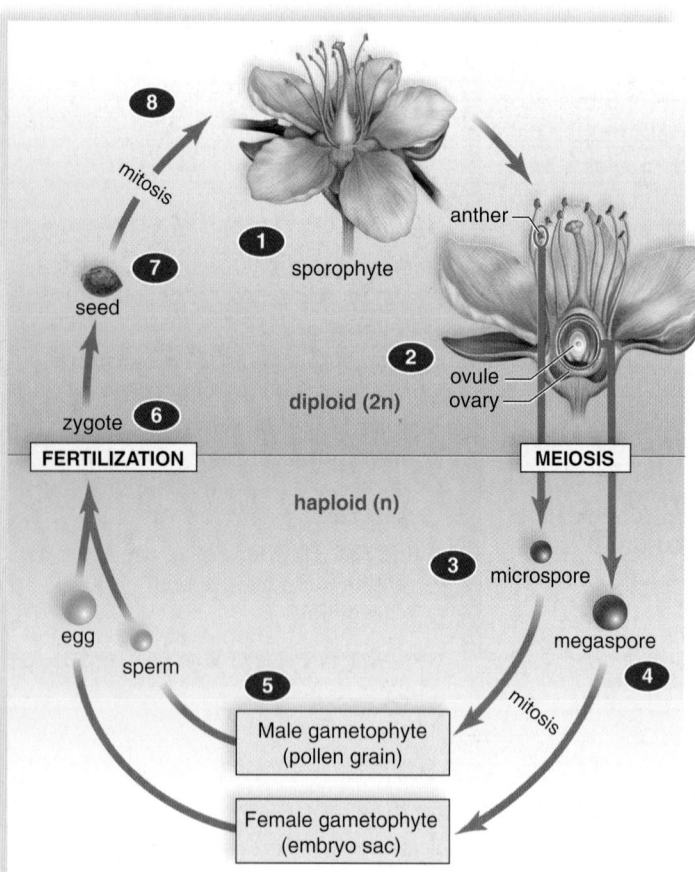

Figure 27.1 Sexual reproduction in flowering plants.   The sporophyte bears flowers. The flower produces microspores within anthers and megaspores within ovules by meiosis. A megaspore becomes a female gametophyte, which produces an egg within an embryo sac, and a microspore becomes a male gametophyte (pollen grain), which produces sperm. Fertilization results in a zygote and in sustenance for the embryo. A seed contains an embryo and stored food within a seed coat. After dispersal, a seed becomes a new sporophyte plant.

① In flowering plants, the diploid sporophyte is dominant, and it is the portion of the life cycle that bears flowers. **②** A **flower,** which is the reproductive structure of angiosperms, produces two types of spores by meiosis, microspores and megaspores. **③** A **microspore** [Gk. *mikros,* small, little] undergoes mitosis and becomes a pollen grain, which is either windblown or carried by an animal to the vicinity of the female gametophyte. **④** In the meantime, the **megaspore** [Gk. *megas,* great, large] undergoes mitosis and becomes the female gametophyte. The female gametophyte is an embryo sac located within the ovule that is found within an ovary. **⑤** At maturity, a pollen grain contains nonflagellated sperm, which travel by way of a pollen tube to the embryo sac. **⑥** Once a sperm fertilizes an egg, the zygote becomes an embryo, still within an ovule. **⑦** The ovule develops into a **seed,** which contains the embryo and stored food surrounded by a seed coat. The ovary becomes a fruit, which aids in dispersing the seeds. **⑧** When a seed germinates, a new sporophyte emerges and through mitosis and growth becomes a mature organism.

Notice that the sexual life cycle of flowering plants is adapted to a land-based existence. The microscopic female gametophytes develop completely within the sporophyte and are thereby protected from desiccation. Pollen grains (male gametophytes) are not released until they develop a thick wall. No external water is needed to bring about fertilization in flowering plants. Instead, the pollen tube provides passage for a sperm to reach an egg. Following fertilization, the embryo and its stored food are enclosed within a protective seed coat until external conditions are favorable for germination.

Flowers

The flower is unique to angiosperms (Fig. 27.2). Aside from producing the spores and protecting the gametophytes, flowers often attract pollinators, which aid in transporting pollen from plant to plant. Flowers also produce the fruits that enclose the seeds.

The evolution of the flower was a major factor leading to the success of angiosperms, with over 240,000 species. Flowering is often a response to environmental signals such as the

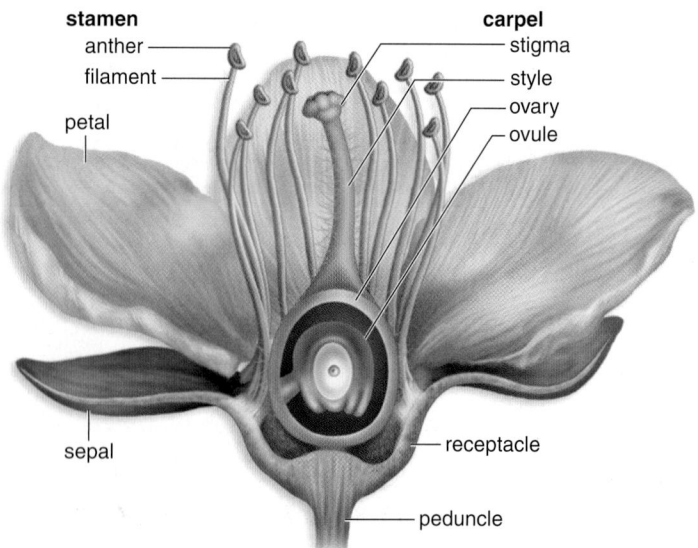

Figure 27.2 Anatomy of a flower.   A complete flower has all flower parts: sepals, petals, stamens, and at least one carpel.

a. Daylily, *Hemerocallis* sp.

b. Festive azalea, *Rhododendron* sp.

Figure 27.3 Monocot versus eudicot flowers. **a.** Monocots, such as daylilies, have flower parts usually in threes. In particular, note the three petals and three sepals, both of which are colored in this flower. **b.** Azaleas are eudicots. They have flower parts in fours or fives; note the five petals of this flower. p = petal, s = sepal

length of the day (see Chapter 26). In many plants, a flower develops when shoot apical meristem that previously formed leaves suddenly stops producing leaves and starts producing a flower enclosed within a bud. In other plants, axillary buds develop directly into flowers. In monocots, flower parts occur in multiples of three; in eudicots, flower parts occur in multiples of four or five (Fig. 27.3).

Flower Structure

The typical eudicot flower has four whorls of modified leaves attached to a receptacle at the end of a flower stalk called a peduncle.

1. The **sepals** are the most leaflike of all the flower parts and are usually green. Sepals protect the bud as the flower develops within. Collectively, the sepals are called the **calyx.**
2. An open flower next has a whorl of **petals,** whose color accounts for the attractiveness of many flowers. The size, the shape, and the color of petals are attractive to a specific pollinator. Wind-pollinated flowers may have no petals at all. Collectively, the petals are called the **corolla.**
3. **Stamens** are the "male" portion of the flower. Each stamen contains two parts: the slender stalk called the *filament* [L. *filum*, thread], and the saclike *anther* the filament

supports. Pollen grains develop from the microspores produced in the anther.

4. At the very center of a flower is the **carpel,** a vaselike structure that represents the "female" portion of the flower. A carpel usually has three parts: the **style** is a slender stalk that supports the **stigma,** an enlarged sticky knob, and the **ovary,** an enlarged base that encloses one or more *ovules* (see Fig. 27.2).

Ovules [L. *ovulum,* little egg] play a significant role in the production of megaspores, and therefore, female gametophytes, as described shortly.

A flower can have a single carpel or multiple carpels. Sometimes several carpels are fused into a single structure, in which case the ovary is termed compound; it has several chambers, each of which contains ovules. For example, an orange develops from a compound ovary, and every section of the orange is a chamber.

Variations in Flower Structure

We have space to mention only a few variations in flower structure. Not all flowers have sepals, petals, stamens, or carpels. Those that do are said to be *complete* and those that do not are said to be *incomplete.* Flowers that have both stamens and carpels are called *perfect* (bisexual) flowers; flowers with only stamens and those with only carpels are *imperfect* (unisexual) flowers.

If both staminate flowers and carpellate flowers occur on a single plant, the plant is *monoecious* [Gk. *monos,* one, and *oikos,* home, house] (Fig. 27.4*a*). Corn is an example of a plant that is monoecious. If staminate and carpellate flowers occur on separate plants, the plant is *dioecious.* Holly trees are dioecious; if you hope to have holly berries, you need one plant with staminate flowers and another plant with carpellate flowers (Fig. 27.4*b*).

a.   b.

Figure 27.4 Monoecious and dioecious plants. **a.** Corn plants are monoecious, having both male and female flowers on the same plant. **b.** Holly trees are dioecious; berries are produced only by female plants, and pollen only by male plants.

Life Cycle in Detail

In all land plants, the sporophyte produces haploid spores by meiosis. The haploid spores grow and develop into haploid **gametophytes,** which produce gametes by mitotic division. Flowering plants, however, are *heterosporous*—they produce microspores and megaspores. Microspores become mature male gametophytes (sperm-bearing pollen grains), and megaspores become mature female gametophytes (egg-bearing embryo sacs).

Development of Male Gametophyte

Microspores are produced in the anthers of flowers (Fig. 27.5). An anther has four pollen sacs, each containing many microspore mother cells. A microspore mother cell undergoes meiosis to produce four haploid microspores. In each, the haploid nucleus divides mitotically, followed by unequal cytokinesis, and the result is two cells enclosed by a finely sculptured wall. This structure, called the **pollen grain,** is at first an immature male gametophyte that consists of a tube cell and a generative cell. The larger tube cell will eventually produce a *pollen tube.* The smaller generative cell divides mitotically either now or later to produce two sperm. Once these events have taken place, the pollen grain has become the mature male gametophyte.

Development of Female Gametophyte

The ovary contains one or more ovules. An ovule has a central mass of parenchyma cells almost completely covered by layers of tissue called integuments, except where there is an opening called the micropyle. One parenchyma cell enlarges to become a megaspore mother cell, which undergoes meiosis, producing four haploid megaspores (Fig. 27.5). Three of these megaspores are nonfunctional, and one is functional. In a typical pattern, the nucleus of the functional megaspore divides mitotically until there are eight nuclei in the *female gametophyte.* When cell walls form later, there are seven cells, one of which is binucleate.

Figure 27.5 Life cycle of flowering plants. *Development of gametophytes (far page):* A pollen sac in the anther contains microspore mother cells, which produce microspores by meiosis. A microspore develops into a pollen grain, which germinates and has two sperm. An ovule in an ovary contains a megaspore mother cell, which produces a megaspore by meiosis. A megaspore develops into an embryo sac containing seven cells, one of which is an egg. *Development of sporophyte (this page):* A pollen grain contains two sperm by the time it germinates and forms a pollen tube. During double fertilization, one sperm fertilizes the egg to form a diploid zygote, and the other fuses with the polar nuclei to form a triploid (3n) endosperm cell. A seed contains the developing sporophyte embryo plus stored food.

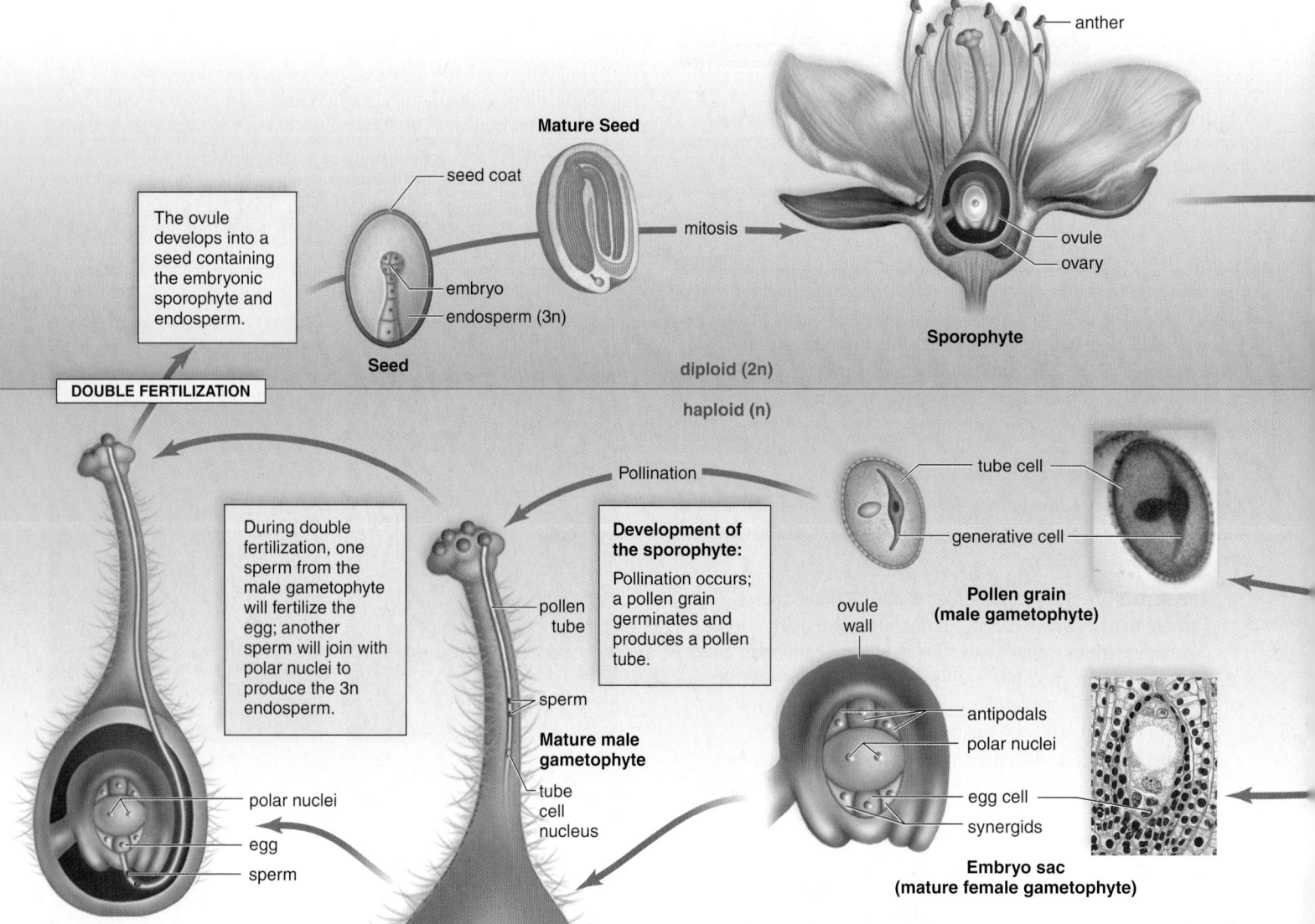

The female gametophyte, also called the **embryo sac,** consists of these seven cells:

- one egg cell, associated with two synergid cells;
- one central cell, with two polar nuclei; and
- three antipodal cells

Development of New Sporophyte

The walls separating the pollen sacs in the anther break down when the pollen grains are ready to be released. **Pollination** is simply the transfer of pollen from an anther to the stigma of a carpel. Self-pollination occurs if the pollen is from the same plant, and cross-pollination occurs if the pollen is from a different plant of the same species.

When a pollen grain lands on the stigma of the same species, it germinates, forming a pollen tube (see Fig. 27.5). The germinated pollen grain, containing a tube cell and two sperm, is the mature male gametophyte. As it grows, the pollen tube passes between the cells of the stigma and the style to reach the micropyle, a pore of the ovule. When the pollen tube reaches the micropyle, **double fertilization** occurs: One of the sperm unites with the egg to form a 2n zygote; however, the second sperm unites with the two polar nuclei centrally placed in the embryo

Figure 27.6 Pollination. **a.** Cocksfoot grass, *Dactylus glomerata,* releasing pollen. **b.** Pollen grains of Canadian goldenrod, *Solidago canadensis.* **c.** Pollen grains of pussy willow, *Salix discolor.* The shape and pattern of pollen grain walls are quite distinctive, and experts can use them to identify the genus, and even sometimes the species, that produced a particular pollen grain. Pollen grains have strong walls resistant to chemical and mechanical damage; therefore, they frequently become fossils.

sac to form a 3n endosperm nucleus. This latter fertilization is unique to angiosperms. The endosperm nucleus eventually develops into the **endosperm** [Gk. *endon,* within, and *sperma,* seed], a nutritive tissue that the developing embryonic sporophyte will use as an energy source.

Now the ovule begins to develop into a seed. One important aspect of seed development is formation of the seed coat from the ovule wall. A mature seed contains the embryo, stored food, and the seed coat (see Fig. 27.8).

Cross Pollination

Some species of flowering plants, such as grasses and grains, rely on wind pollination (Fig. 27.6), as do the gymnosperms, the other type of seed plant. Much of the plant's energy goes into making pollen to ensure that some pollen grains actually reach a stigma. The amount of pollen successfully transferred is staggering: A single corn plant may produce from 20 to 50 million pollen grains a season. In corn, the flowers tend to be monoecious, and clusters of tiny male flowers move in the wind, freely releasing pollen into the air.

Most angiosperms rely on animals—insects (e.g., bumble-bees, flies, butterflies, and moths), birds (e.g., hummingbirds), or mammals (e.g., bats)—to carry out pollination. The use of animal pollinators is unique to flowering plants, and it helps account for why these plants are so successful on land.

By the time flowering plants appear in the fossil record some 135 MYA, insects had long been present. For millions of years,

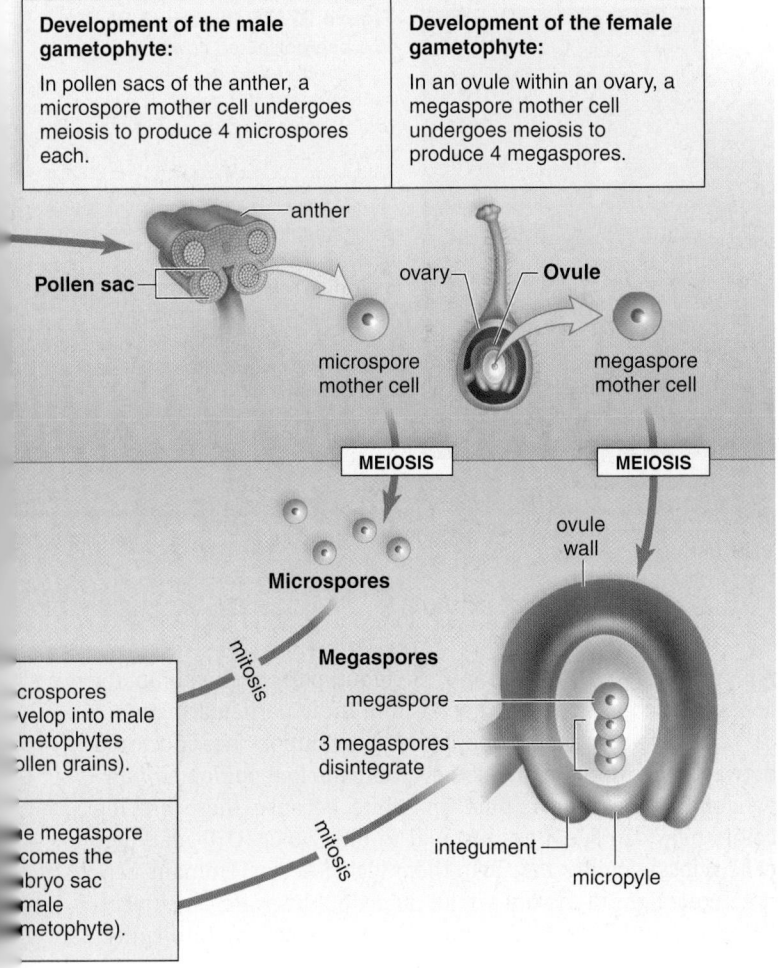

Development of the male gametophyte:	Development of the female gametophyte:
In pollen sacs of the anther, a microspore mother cell undergoes meiosis to produce 4 microspores each.	In an ovule within an ovary, a megaspore mother cell undergoes meiosis to produce 4 megaspores.

anther

Pollen sac

ovary   **Ovule**

microspore mother cell

megaspore mother cell

ovule wall

MEIOSIS   **MEIOSIS**

Microspores

mitosis

Megaspores

megaspore

3 megaspores disintegrate

crospores velop into male ametophytes ollen grains).

e megaspore comes the bryo sac male ametophyte).

mitosis

integument

micropyle

Evolution

Plants and Their Pollinators

Plants and their pollinators have adapted to one another. They have a mutualistic relationship in which each benefits—the plant uses its pollinator to ensure that cross-pollination takes place, and the pollinator uses the plant as a source of food. This mutualistic relationship came about through the process of coevolution—that is, the interdependency of the plant and the pollinator is the result of suitable changes in the structure and function of each.

The evidence for coevolution is observational. For example, floral coloring and odor are suited to the sense perceptions of the pollinator; the mouthparts of the pollinator are suited to the structure of the flower; the type of food provided is suited to the nutritional needs of the pollinator; and the pollinator forages at the time of day that specific flowers are open. Here we present some examples of coevolution.

Bee-Pollinated Flowers

There are 20,000 known species of bees that pollinate flowers. The best-known pollinators are the honeybees (Fig. 27Aa). As noted in the text, bee eyes see ultraviolet (UV) wavelengths. Therefore, bee-pollinated flowers are usually brightly colored and are predominantly blue or yellow; they are not entirely red. They may also have ultraviolet shadings called nectar guides, which highlight the portion of the flower that contains the reproductive structures.

The mouthparts of bees are fused into a long tube that contains a tongue. This tube is an adaptation for sucking up nectar provided by the plant, usually at the base of the flower. Bees also collect pollen as a food.

Bee flowers are delicately sweet and fragrant to advertise that nectar is present. The nectar guides often point to a narrow floral tube large enough for the bee's feeding apparatus but too small for other insects to reach the nectar. Bee-pollinated flowers are sturdy and may be irregular in shape because they often have a landing platform where the bee can alight. The flower structure requires the bee to brush up against the anther and stigma as it moves toward the floral tube to feed.

One type of orchid, *Ophrys*, has evolved a unique adaptation. The flower resembles a female wasp, and when the male of that species attempts to copulate with the flower, the wasp receives or transfers pollen.

Moth- and Butterfly-Pollinated Flowers

Contrasting moth- and butterfly-pollinated flowers emphasizes the close adaptation between pollinator and flower. Both moths and butterflies have a long, thin, hollow proboscis, but they differ in other character

Figure 27A Pollinators.
a. A bee-pollinated flower is a color other than red (bees cannot detect this color). The reproductive structures of the flower brush up against the bee's body, ensuring that pollen is transferred. **b.** A butterfly-pollinated flower is often a composite, containing many individual flowers. The broad expanse provides room for the butterfly to land, after which it lowers its proboscis into each flower in turn.

a.   b.

plants and their animal pollinators have coevolved. **Coevolution** means that as one species changes, the other species undergoes adaptation in response, so that in the end, the two species are suited to one another. Plants with flowers that attracted a pollinator had an advantage because, in the end, they produced more seeds. Similarly, pollinators that were able to find and remove food from the flower were more successful. Today, we see that the reproductive parts of the flower are positioned so that the pollinator picks up pollen from one flower and delivers it to another. And concurrently, the mouthparts of the pollinator are suited to gathering the nectar from these particular plants.

One well-studied example of coevolution has occurred between bees and the plants they pollinate. Bee-pollinated flowers tend to be yellow, blue, or white because these are the colors bees can see. Bees respond to ultraviolet (UV) markings called nectar guides that help them locate nectar. Humans cannot detect light in the ultraviolet range, but bees are sensitive to UV. A bee has a feeding proboscis of the right length to collect

istics. Moths usually feed at night and have a well-developed sense of smell. The flowers they visit are visible at night because they are lightly shaded (white, pale yellow, or pink), and they have strong, sweet perfume, which helps attract moths. Moths hover when they feed, and their flowers have deep tubes with open margins that allow the hovering moths to reach the nectar with their long proboscis.

Butterflies, in comparison, are active in the daytime and have good vision but a weak sense of smell. Their flowers have bright colors—even red because butterflies can see the color red—but the flowers tend to be odorless. Unable to hover, butterflies need a place to land. Flowers that are visited by butterflies often have flat landing platforms (Fig. 27A*b*). Composite flowers (composed of a compact head of numerous individual flowers) are especially favored by butterflies. Each flower has a long, slender floral tube, accessible to the long, thin butterfly proboscis.

Bird- and Bat-Pollinated Flowers

In North America, the most well-known bird pollinators are the hummingbirds. These small animals have good eyesight but do not have a well-developed sense of smell. Like moths, they hover when they feed. Typical flowers pollinated by hummingbirds are red, with a slender floral tube and margins that are curved back and out of the

way. And although they produce copious amounts of nectar, the flowers have little odor. As a hummingbird feeds on nectar with its long, thin beak, its head comes into contact with the stamens and pistil (Fig. 27B*a*).

Bats are adapted to gathering food in various ways, including feeding on the nectar and pollen of plants. Bats are nocturnal and have an acute sense of smell. Those that are pollinators also have keen vision and a long, extensible, bristly tongue. Typically, bat-pollinated flowers open only at night and are light-colored or white. They have a strong, musky smell similar to the odor that bats produce to attract one another. The flowers are generally large and sturdy and are able to hold up when a bat inserts part of its head to reach the nectar. While the bat is at the flower, its head becomes dusted with pollen (Fig. 27B*b*).

Coevolution

How did this coevolution of plants and pollinators come about? Some 200 million years ago, when seed plants were just be-

ginning to evolve and insects were not as diverse as they are today, wind alone carried pollen. Wind pollination, however, is a hit-or-miss affair. Perhaps beetles feeding on vegetative leaves were the first insects to carry pollen directly from plant to plant by chance. Because flowers undergoing direct cross-fertilization would likely produce more fruit, natural selection favored flowers with features that would attract pollinators.

As cross-fertilization continued, more and more flower variations likely developed, and pollinators became increasingly adapted to specific angiosperm species. Today, there are some 240,000 species of flowering plants and over 900,000 species of insects. This diversity suggests that the success of angiosperms has contributed to the success of insects, and vice versa.

Questions to Consider

1. What are the potential consequences if honeybees were to go extinct?
2. How does coevolution cause two species to change over time?

Figure 27B More pollinators.
a. Hummingbird-pollinated flowers are curved back, allowing the bird to insert its beak to reach the rich supply of nectar. While doing this, the bird's forehead and other body parts touch the reproductive structures. **b.** Bat-pollinated flowers are large, sturdy flowers that can take rough treatment. Here the head of the bat is positioned so that its bristly tongue can lap up nectar.

a.

b.

nectar from certain flowers and a pollen basket on its hind legs that allows it to carry pollen back to the hive.

Because many fruits and vegetables are dependent on bee pollination, many people have great concern today that the number of bees is declining due to disease and the use of pesticides. More examples of the coevolution between plants and their pollinators are given in the Evolution feature "Plants and Their Pollinators."

**Video
Pollinators**

Check Your Progress   27.1

1. Compare the development and structure of the male gametophyte and the female gametophyte.
2. Describe the products of double fertilization in angiosperms.
3. Explain how a flowering plant may coevolve in response to an increase in the body size of its pollinator.

27.2 Seed Development

Development of the embryo within the seed is the next event in the life cycle of the angiosperm. Plant growth and development involves cell division, cell elongation, and differentiation of cells into tissues that then develop into organs. **Development** is a programmed series of stages from a simple to a more complex form. Cellular **differentiation,** or specialization of structure and function, occurs as development proceeds.

Stages of Eudicot Development

Figure 27.7 shows the stages of development for a eudicot embryo.

Zygote and Proembryo Stages

1 Immediately after double fertilization the zygote and the endosperm become visible. The zygote is small with dense cytoplasm. **2** The zygote divides repeatedly in different planes, forming several cells called a proembryo. Also present is an

Figure 27.7 Development of a eudicot embryo.   The embryo undergoes six developmental stages: Zygote, proembryo, globular stage, heart stage, torpedo stage, and mature embryo.

elongated structure called a suspensor. The suspensor transfers and produces nutrients from the endosperm, which allows the embryo to grow.

Globular Stage

3 During the globular stage, the proembryo is largely a ball of cells. The root-shoot axis of the embryo is already established at this stage. The embryonic cells near the suspensor will go on to become a root, while those at the other end will ultimately become a shoot.

   The outermost cells of the plant embryo will become the dermal tissue. These cells divide with their cell plate perpendicular to the surface; therefore, they produce a single outer layer of cells. Recall that dermal tissue protects the plant from desiccation and includes the stomata, which open and close to facilitate gas exchange and minimize water loss.

The Heart Stage and Torpedo Stage Embryos

4 The embryo has a heart shape when the **cotyledons,** or seed leaves, appear because of local, rapid cell division. **5** As the embryo continues to enlarge and elongate, it takes on a torpedo shape. Now the root and shoot apical meristems are distinguishable. The shoot apical meristem is responsible for aboveground growth while the root apical meristem is responsible for underground growth. Ground meristem gives rise to the bulk of the embryonic interior, which is now present.

The Mature Embryo

6 In the mature embryo, the *epicotyl* is the portion between the cotyledon(s) that contributes to shoot development. The *plumule* is found at the tip of the epicotyl and consists of the

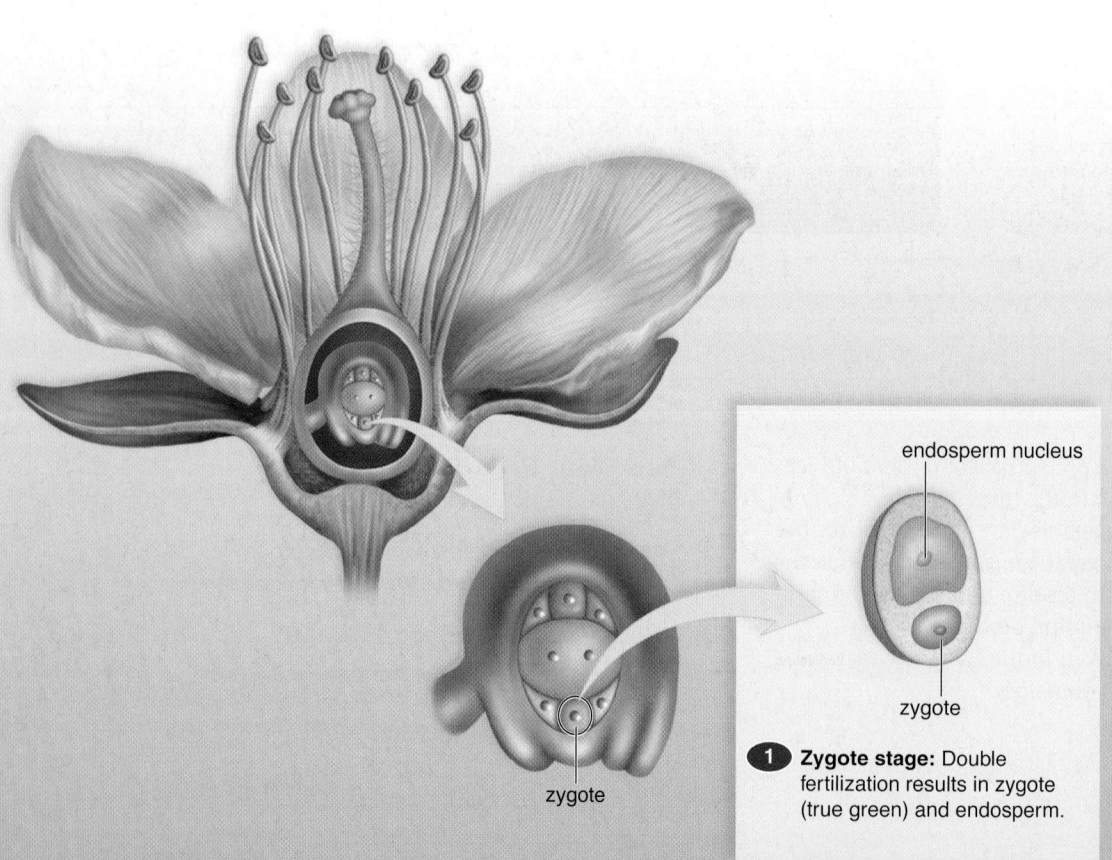

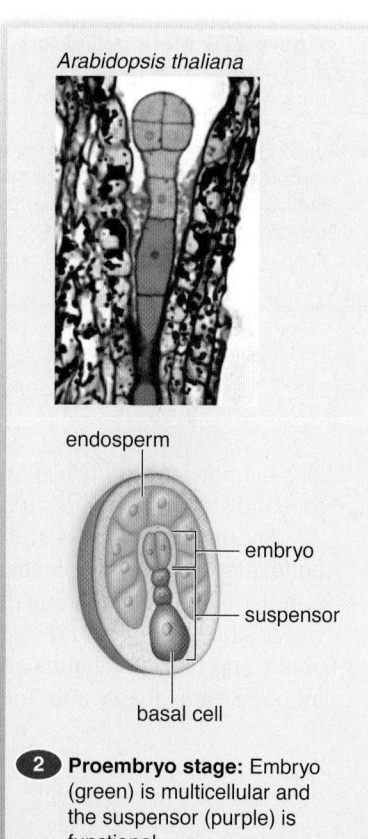

Arabidopsis thaliana

endosperm

embryo

suspensor

basal cell

zygote

endosperm nucleus

zygote

1 **Zygote stage:** Double fertilization results in zygote (true green) and endosperm.

2 **Proembryo stage:** Embryo (green) is multicellular and the suspensor (purple) is functional.

Figure 27.8 top diagrams

seed coat
plumule
hypocotyl
radicle
cotyledon
embryo

a.

pericarp
endosperm
coleoptile
cotyledon
plumule
radicle
coleorhiza
embryo

b.

Figure 27.8 Monocot versus eudicot. **a.** In a bean seed (eudicot), the endosperm has disappeared; the bean embryo's cotyledons take over food storage functions. The hypocotyl becomes the shoot system, which will include the plumule (first leaves). The radicle becomes the root system. **b.** The corn kernel (monocot) has endosperm that is still present at maturity. The coleoptile is a protective sheath for the shoot system; the coleorhiza similarly protects the future root system. The pericarp of the fruit develops from the ovary wall.

shoot tip and a pair of small leaves. The *hypocotyl* is the portion below the cotyledon(s). It contributes to stem development and terminates in the radicle or embryonic root.

The cotyledons are quite noticeable in a eudicot embryo and may fold over. Procambium is located at the core of the embryo and is destined to form the future vascular tissue responsible for water and nutrient transport.

As the embryo develops, the integuments of the ovule become the seed coat. The seed coat encloses and protects the embryo and its food supply.

Monocot Versus Eudicot Seeds

Monocots, unlike eudicots, have only one cotyledon. Another important difference between monocots and eudicots is the manner in which nutrient molecules are stored in the seed. In monocots, the cotyledon, in addition to storing certain nutrients,

absorbs other nutrient molecules from the endosperm and passes them to the embryo. In eudicots, the cotyledons usually store all the nutrient molecules that the embryo uses. Therefore, in Figure 27.7 you can see that the endosperm seemingly disappears. Actually, it has been taken up by the two cotyledons.

Figure 27.8 contrasts the structure of a bean seed (eudicot) and a corn kernel (monocot). The size of seeds may vary from the dust-sized seeds of orchids to the 27-kg seed of the double coconut.

Check Your Progress   27.2

1. Identify the origin of each of the three parts of a seed.
2. Explain why the seed coat and the embryo are both 2n.
3. Describe the structure and function of the cotyledon.

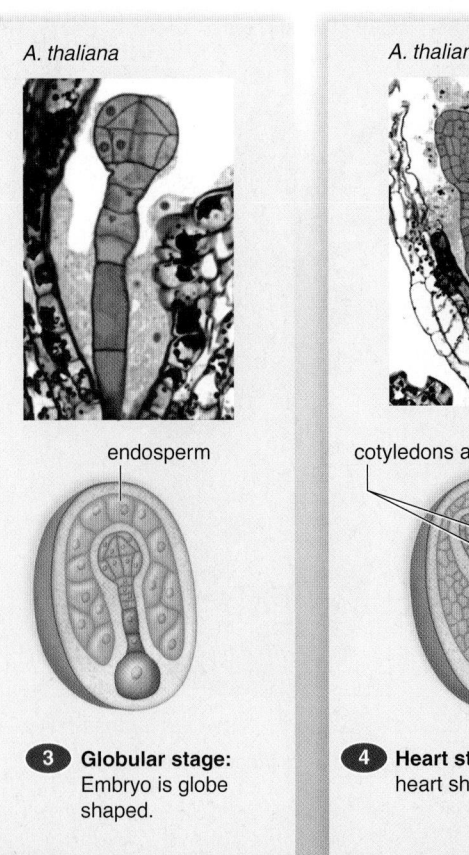

A. thaliana

endosperm

3 Globular stage: Embryo is globe shaped.

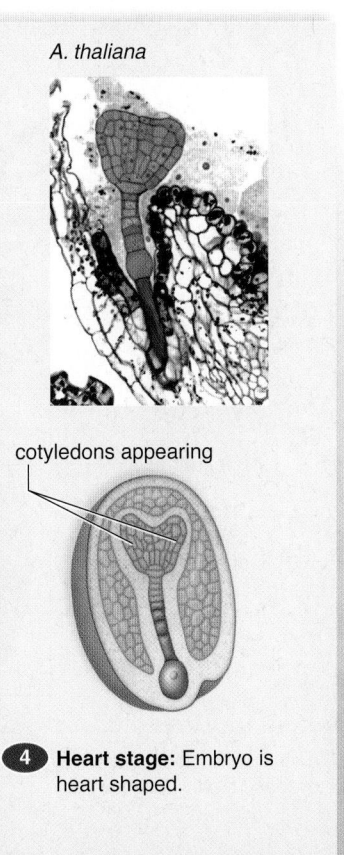

A. thaliana

cotyledons appearing

4 Heart stage: Embryo is heart shaped.

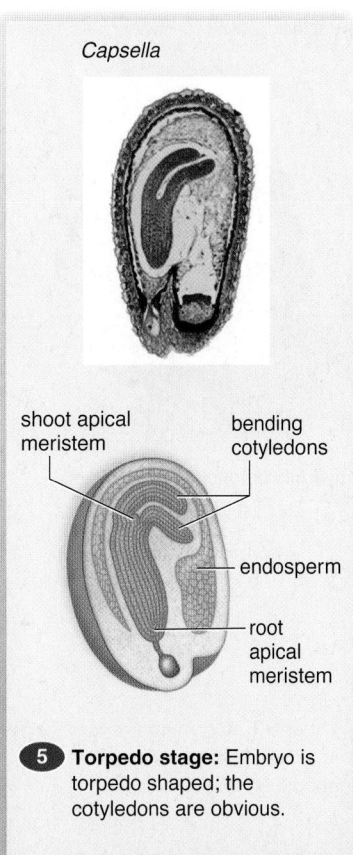

Capsella

shoot apical meristem

bending cotyledons

endosperm

root apical meristem

5 Torpedo stage: Embryo is torpedo shaped; the cotyledons are obvious.

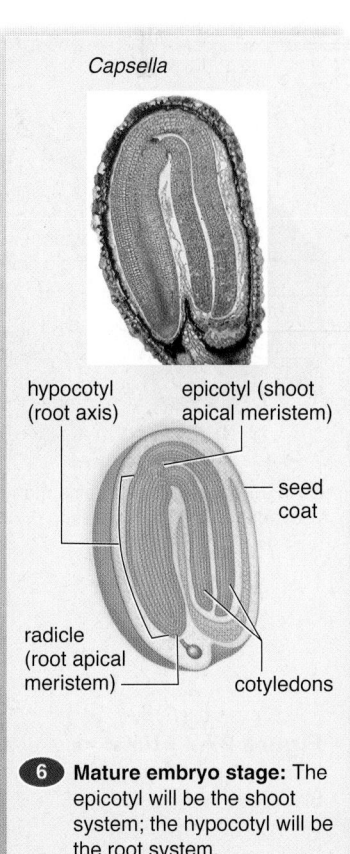

Capsella

hypocotyl (root axis)

epicotyl (shoot apical meristem)

seed coat

radicle (root apical meristem)

cotyledons

6 Mature embryo stage: The epicotyl will be the shoot system; the hypocotyl will be the root system.

Drupe

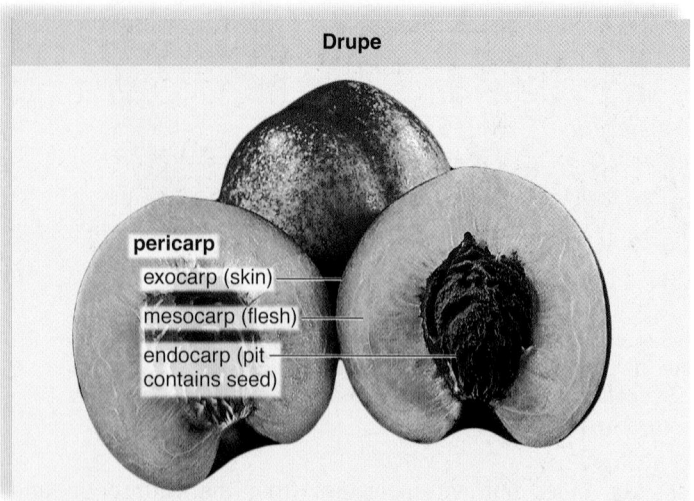

a. A drupe is a fleshy fruit with a pit containing a single seed produced from a simple ovary.

True Berry

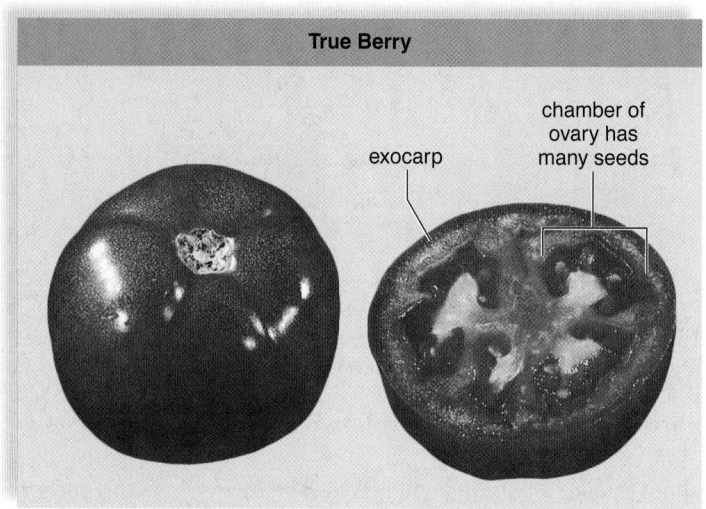

b. A berry is a fleshy fruit having seeds and pulp produced from a compound ovary.

Legume

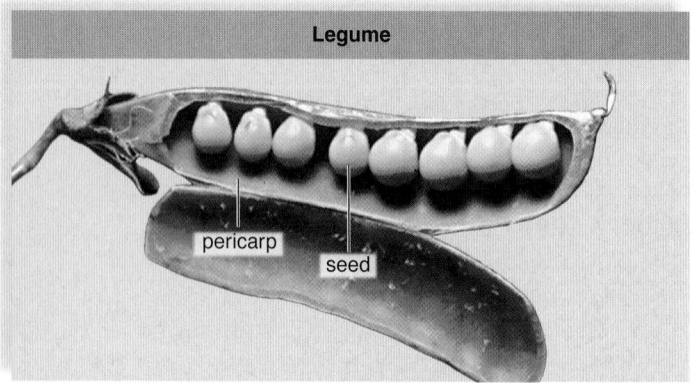

c. A legume is a dry dehiscent fruit produced from a simple ovary.

Samara

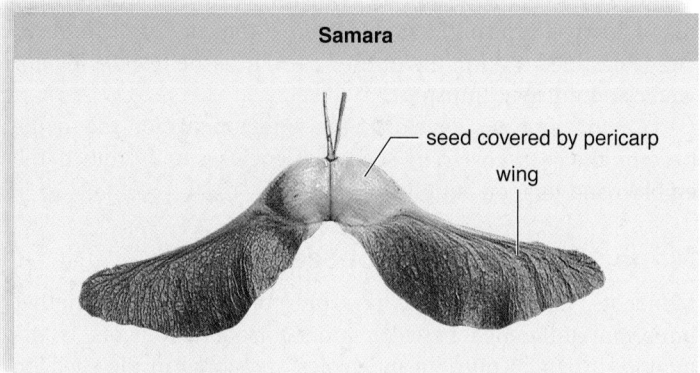

d. A samara is a dry indehiscent fruit produced from a simple ovary.

Aggregate Fruit

e. An aggregate fruit contains many fleshy fruits produced from simple ovaries of the same flower.

Multiple Fruit

f. A multiple fruit contains many fused fruits produced from simple ovaries of individual flowers.

Figure 27.9 Fruits.   a. A peach is a drupe. **b.** A tomato is a true berry. **c.** A pea is a legume. **d.** The fruit of a maple tree is a samara. **e.** A blackberry is an aggregate fruit. **f.** A pineapple is a multiple fruit.

a.

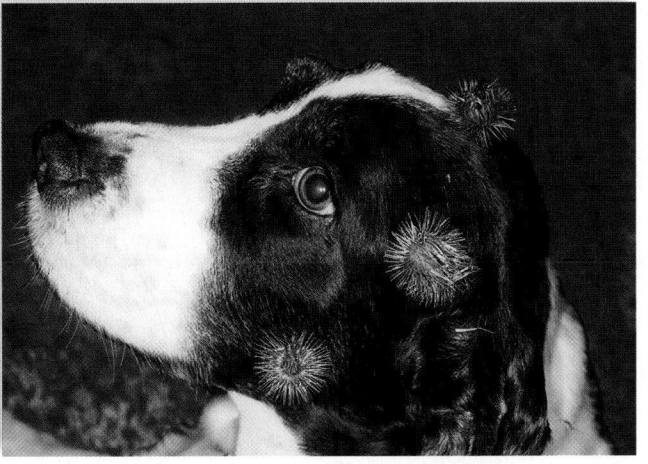

b.

Figure 27.10 Fruit dispersal by animals. **a.** When birds eat fleshy fruits, seeds pass through their digestive system. **b.** Burdock, a dry fruit, clings to the fur of animals.

27.3 Fruit Types and Seed Dispersal

Learning Outcomes

Upon completion of this section, you should be able to

1. Identify examples of fleshy and dry fruits that are simple, compound, aggregate, or accessory fruits.
2. Explain the sequence of events during seed germination.

Most people are unfamiliar with the botanical definition of a fruit (Table 27.1). A **fruit** in botany is a mature ovary that can also contain other flower parts, such as the receptacle. This means that pea pods, tomatoes, and what are usually called winged maple seeds are actually fruits (Fig. 27.9). Fruits protect and help disperse seeds. Some fruits are better at one function than the others. The fruit of a peach protects the seed well, but the pit may make it difficult for germination to occur. Peas easily escape from pea pods, but once they are free, they are protected only by the seed coat.

Kinds of Fruits

Fruits can be simple or compound (Fig. 27.9). A *simple fruit* is derived from a single ovary that can have one or several chambers (Fig. 27.9a–d). A *compound fruit* is derived from several groups of ovaries (Fig. 27.9e–f). If a single flower has multiple ovaries, as in a blackberry, then it produces an *aggregate fruit* (Fig. 27.9e). In contrast, a pineapple comes from many individual ovaries. Because the flowers had only one receptacle, the ovaries fused to form a large, *multiple fruit* (Fig. 27.9f).

As a fruit develops, the ovary wall thickens to become the pericarp, which can have as many as three layers: exocarp, mesocarp, and endocarp.

- The *exocarp* forms the outermost skin of a fruit.
- The *mesocarp* is often the fleshy tissue between the exocarp and endocarp of the fruit.
- The *endocarp* serves as the boundary around the seed(s). The endocarp may be hard, as in peach pits, or papery, as in apples.

Some fruits, such as legumes and cereal grains of wheat, rice, and corn, are *dry fruits.* The fruits of grains can be mistaken for seeds because a dry pericarp adheres to the seed within. Legume fruits such as the pea pod (Fig. 27.9c) are dehiscent because they split open when ripe. Grains are indehiscent—they

Table 27.1 Fruit Classification Based on Composition and Texture

Composition (based on type and arrangement of ovaries and flowers)
Simple: develops from a simple ovary or compound ovary
Compound: develops from a group of ovaries
Aggregate: ovaries are from a single flower on one receptacle
Multiple: ovaries are from separate flowers on a common receptacle
Texture (based on mature pericarp)
Fleshy: the entire pericarp or portions of it are soft and fleshy at maturity
Dry: the pericarp is papery, leathery, or woody when the fruit is mature
Dehiscent: the fruit splits open when ripe
Indehiscent: the fruit does not split open when ripe

don't split open. Humans gather grains before they are released from the plant and then process them to acquire their nutrients.

You are probably more familiar with fleshy fruits, such as the peach and tomato. In these fruits, the mesocarp is well developed.

Dispersal of Fruits

Generally it is beneficial for plants to diperse their fruits away from the parent plant so that seedlings do not have to compete with the parent for nutrients. Fruits may drift or be blown to new locations by wind, or be carried away by animals.

Dispersal by Air

Many dry fruits are dispersed by wind. Woolly hairs, plumes, and wings are all adaptations for this type of dispersal. The somewhat heavier dandelion fruit uses a tiny "parachute" for dispersal. Milkweed pods split open to release seeds that float away on puffy white threads. The winged fruit of a maple tree has been known to travel up to 10 km from its parent. Other fruits depend on animals for dispersal.

Dispersal by Animals

Ripe, fleshy, colorful fruits, such as peaches and cherries, often attract animals and provide them with food (Fig. 27.10a). Their hard endocarp protects the seed so it can pass through the digestive system of an animal and remain unharmed. As the flesh of

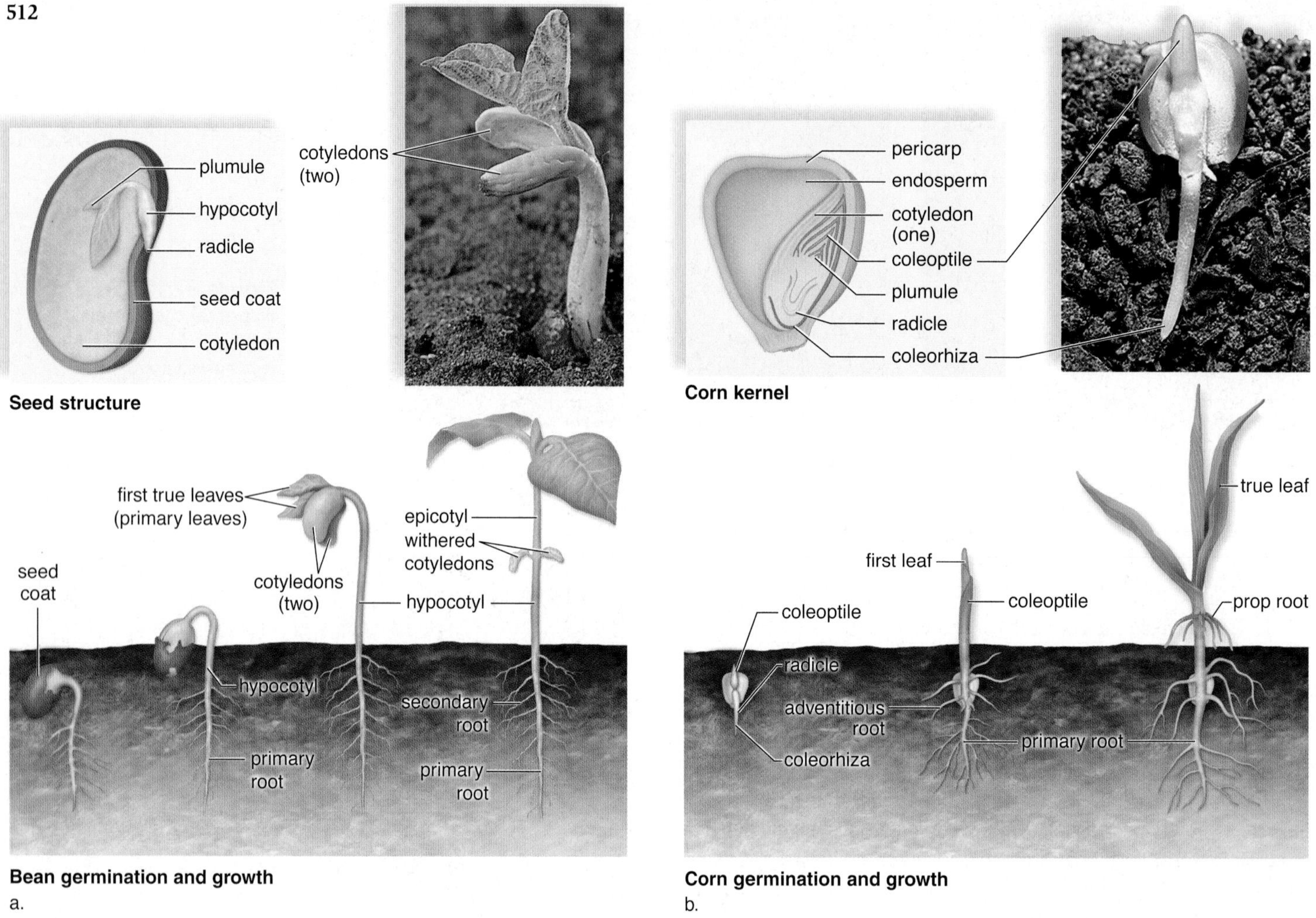

Seed structure

- plumule
- hypocotyl
- radicle
- seed coat
- cotyledon

cotyledons (two)

Corn kernel

- pericarp
- endosperm
- cotyledon (one)
- coleoptile
- plumule
- radicle
- coleorhiza

Bean germination and growth

first true leaves (primary leaves)

seed coat

cotyledons (two)

epicotyl
withered cotyledons
hypocotyl

hypocotyl

secondary root

primary root

primary root

a.

Corn germination and growth

true leaf

first leaf

coleoptile

coleoptile

prop root

radicle

adventitious root

coleorhiza

primary root

b.

Figure 27.11 **Eudicot and monocot seed structure and germination.** **a.** Bean (eudicot) seed structure and germination. **b.** Corn (monocot) kernel structure and germination.

a tomato is eaten, the small size of the seeds and the slippery seed coat means that tomato seeds rarely get crushed by the teeth of animals. The seeds swallowed by birds and mammals are defecated (passed out of the digestive tract with the feces) some distance from the parent plant. Squirrels and other animals that gather seeds and fruits bury them some distance away and may even forget where they have been stored. The hooks and spines of clover, bur, and cocklebur attach a dry fruit to the fur of animals and the clothing of humans (Fig. 27.10b).

 **Video** Fruit Bat Seed Dispersal

Seed Germination

Following dispersal, if conditions are right, seeds may **germinate** to form a seedling. Germination doesn't usually take place until there is sufficient water, warmth, and oxygen to sustain growth. These requirements help ensure that seeds do not germinate until the most favorable growing season has arrived.

Some seeds do not germinate until they have been dormant for a period of time. For seeds, *dormancy* is the time during which no growth occurs, even though conditions may be favorable for growth. In the temperate zone, seeds often have to be exposed to a period of cold weather before dormancy is broken. Fleshy fruits (e.g., apples, pears, oranges, and tomatoes) contain inhibitors so that germination does not occur while the fruit is

still on the plant. For seeds to take up water, bacterial action and even fire may be needed. Once water enters, the seed coat bursts and the seed germinates.

If the two cotyledons of a bean seed are parted, the rudimentary plant with immature leaves is exposed (Fig. 27.11a). As the eudicot seedling starts to form, the root emerges first. The shoot is hook-shaped to protect the immature leaves as they emerge from the soil. The cotyledons provide the new seedlings with enough energy for the stem to straighten and the leaves to grow. As the mature leaves of the plant begin photosynthesizing, the cotyledons shrivel up.

A corn kernel is actually a fruit, and therefore its outer covering is the pericarp and seed coat combined (Fig. 27.11b). Inside is the single cotyledon. Also, the immature leaves and the radicle are covered, respectively, by a coleoptile and a coleorhiza. These sheaths are discarded as the root grows directly downward into the soil and the shoot of the seedling begins to grow directly upward.

Check Your Progress 27.3

1. Compare the structure and dispersal methods of dry and fleshy fruits.
2. Compare the protective methods of monocot and eudicot seedlings used to protect their first true leaves.

27.4 Asexual Reproductive Strategies

Learning Outcomes

Upon completion of this section, you should be able to

1. Identify the asexual methods of reproduction in plants.
2. Describe how tissue culture can be used to clone plants with desirable traits.

Asexual reproduction is the production of an offspring identical to a single parent. Asexual reproduction is less complicated in plants because pollination and seed production is not required. Therefore, it can be advantageous when the parent is already well adapted to a particular environment, and the production of genetic variations is not an apparent necessity.

Asexual Reproduction from Stems, Roots, and Cuttings

Figure 27.12 features a strawberry plant that has produced a stolon. **Stolons** are horizontal stems that can be seen because they run aboveground. As you know, the nodes of stems are regions where new growth can occur. In the case of stolons, a new shoot system appears above the node, and a new root system appears below the node. The larger plant on the left is the parent plant, and the smaller plant on the right is the asexual offspring that has arisen at a node. The characteristics of new offspring produced by stolons are identical to those of the parent plant.

Rhizomes are underground stems that produce new plants asexually. Irises are examples of plants that have no aboveground stem because their main stem is a rhizome that grows horizontally underground. As with stolons, new plants arise at the nodes of a rhizome. White potatoes are expanded portions of a rhizome branch called tubers, and each "eye" is a bud that can produce a new potato plant if it is planted with a portion of the swollen tuber.

Sweet potatoes, by contrast, are modified roots; they can be propagated by planting sections of the root, which are called slips.

You may have noticed that the roots of some fruit trees, such as cherry and apple trees, produce "suckers," small plants that can be used to grow new trees. In addition, pineapple, sugarcane, azalea, gardenia, and many other food and ornamental plants have been propagated from stem cuttings. In these plants, a cut stem automatically produces roots. The discovery that the plant hormone auxin can cause roots to develop has expanded the list of plants that can be propagated from stem cuttings.

Tissue Culture of Plants

Tissue culture is the growth of a tissue in an artificial liquid or solid culture medium. Somatic embryogenesis, meristem tissue culture, and anther tissue culture are three methods of cloning plants due to the ability of plants to grow from single cells. Many

Figure 27.12 Asexual reproduction in plants. Meristem tissues at nodes can generate new plants, as when the stolons of strawberry plants, *Fragaria*, give rise to new plants.

Parent plant

stolon

Asexually produced offspring

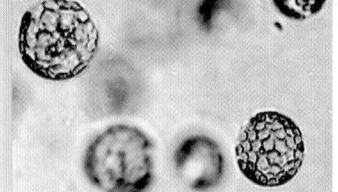

a. Protoplasts, naked cells

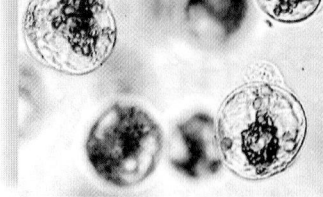

b. Cell wall regeneration

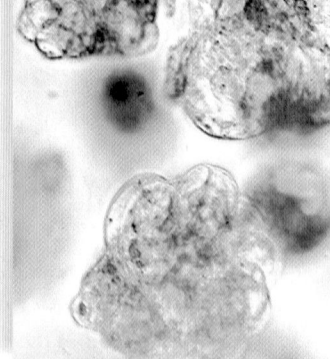

c. Aggregates of cells

d. Callus, undifferentiated mass

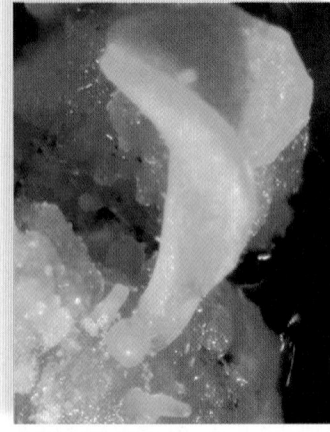

e. Somatic embryo

f. Plantlet

Figure 27.13 Asexual reproduction through tissue culture. **a.** When plant cell walls are removed by digestive enzyme action, the result is naked cells, or protoplasts. **b.** Regeneration of cell walls and the beginning of cell division. **c.** Cell division produces aggregates of cells. **d.** An undifferentiated mass, called a callus. **e.** Somatic cell embryos such as this one appear. **f.** The embryos develop into plantlets that can be transferred to soil for growth into adult plants.

plant cells are **totipotent,** which means that each one has the genetic capability of becoming an entire plant.

During *somatic embryogenesis,* hormones are added to the medium, and they cause leaf or other tissue cells to generate small masses of cells, which can be genetically engineered before being allowed to become many new identical plants. Thousands of little "plantlets" can be produced by using this method of plant tissue culture (Fig. 27.13). Many important crop plants, such as tomato, rice, celery, and asparagus, as well as ornamental plants such as lilies, begonias, and African violets, have been produced using somatic embryogenesis.

Plants generated from somatic embryos are not always genetically identical clones. They can vary because of mutations that arise spontaneously during the production process. These mutations, called *somaclonal variations*, are another way to produce new plants with desirable traits. Somatic embryos can be

Figure 27.14 Producing whole plants from meristem tissue. Each flask on the vertically rotating tables contains a growing meristem clone.

encapsulated in hydrated gel, creating artificial "seeds" that can be shipped anywhere.

Meristem tissue can also be used as a source of plant cells. In this case, the resulting products are *clonal plants* that always have the same traits. In Figure 27.14, culture flasks containing meristematic orchid tissue are rotated under lights. If the correct proportions of hormones are added to the liquid medium, many new shoots develop from a single shoot tip. When these new shoots are removed, more shoots form. Another advantage to producing identical plants from meristem tissue is that the plants are virus free. (The presence of plant viruses weakens plants and makes them less productive.)

Anther tissue culture is a technique in which the haploid cells within pollen grains are cultured in order to produce haploid plantlets. Conversely, a diploid (2n) plantlet can be produced if chemical agents that encourage chromosomal doubling are added to the anther culture. Anther tissue culture is a direct way to produce plants that are certain to have the same characteristics.

Cell Suspension Culture

A technique called **cell suspension culture** allows scientists to extract chemicals (i.e., secondary metabolites) from plant cells, which may have been genetically modified. This technique allows scientists to avoid over-collection of wild plants from their natural environments. These cells produce the same chemicals that the plant produces. For example, cell suspension cultures of *Cinchona ledgeriana* produce quinine, which is used to treat leg cramping, a major symptom of malaria. Several *Digitalis* species produce digitalis, digitoxin, and digoxin, which are useful in the treatment of heart disease.

Check Your Progress   27.4

1. Identify the possible benefits of asexual reproduction.
2. Describe methods of asexual reproduction in wild plants.
3. Identify methods by which new plants are produced asexually in the laboratory.

CONNECTING *the* CONCEPTS *with the* BIG IDEAS

Energy and Homeostasis

- Organisms regulate their metabolism and internal temperatures via specialized strategies. (2A1d1*IE*)
- Plant reproductive strategies coincide with favorable energy availability. (2A1d2*IE*)
- Apoptosis is vital to normal flower development. (2E1c*IE*)
- Flower colorations and patterns attract specific pollinators, highlighting their interdependence. (2E3b4*IE*, 3E1b*IE*)
- Positive feedback impacts fruit's ripening process. (2C1b*IE*)
- Seed germination is triggered by adequate moisture and temperatures. (2E1b3)

Interactions and Systems

- Flowers may display differing colors due to changes in soil pH. (4C2a*IE*)
- Climate change may alter flowering season due to adaptations of a flexible genome. (4C2b*IE*, 1A2a*IE*)

*Find the unabridged version of all EK citations at www.glencoe.com/maderAP11.

Media Study Tools

www.glencoe.com/maderAP11

Enhance your study of this chapter with study tools and practice tests. Also ask your instructor about the resources available through ConnectPlus, including the media-rich eBook, interactive learning tools, and animations.

Summarize

27.1 Sexual Reproductive Strategies

Flowering plants exhibit an alternation-of-generations life cycle. Flowers borne by the sporophyte produce microspores and megaspores by meiosis. Microspores develop into a male gametophyte, and megaspores develop into a female gametophyte. Following fertilization, the sporophyte is enclosed within a seed covered by the fruit.

The flowering plant life cycle is adapted to a land existence. The microscopic gametophytes are protected from desiccation by the sporophyte; the pollen grain has a protective wall; and fertilization does not require external water. The seed has a protective seed coat, and seed germination does not occur until conditions are favorable.

A typical flower has several parts: Sepals, which are usually green in color, form an outer whorl; petals, often colored, are the next whorl; and stamens, each having a filament and anther, form a whorl around the base of at least one carpel. The carpel, in the center of a flower, consists of a stigma, style, and ovary. The ovary contains ovules.

The anthers contain microspore mother cells which divide meiotically to produce four haploid microspores. These divide into a two-celled pollen grain, which is the male gametophyte. One of these cells will divide later to become two sperm cells. After pollination, the pollen grain germinates, and as the pollen tube grows, the sperm cells travel to the embryo sac. Pollination is simply the transfer of pollen from anther to stigma.

Each ovule contains a megaspore mother cell, which divides meiotically to produce an egg cell. The embryo sac contains seven cells that include the egg cell and a central cell with two nuclei.

Flowering plants undergo double fertilization. One sperm nucleus unites with the egg nucleus, forming a 2n zygote, and the other unites with the polar nuclei of the central cell, forming a 3n endosperm cell.

After fertilization, the endosperm cell divides to form multicellular endosperm. The zygote becomes the sporophyte embryo. The ovule matures into the seed (its integuments become the seed coat). The ovary becomes the fruit.

Coevolution occurs when specific traits or behaviors evolve in one species in response to changes that have occurred in another species. The coevolution of flowering plants with animal pollinators may help explain the success of flowering plants on land. Flowering plants would have gained an increased chance of seed production. The pollinators would have gained more feeding opportunities. In the end, both species became dependent upon each other for success.

27.2 Seed Development

As the ovule is becoming a seed, the zygote is becoming an embryo. After the first several divisions, it is possible to discern the embryo and the suspensor. The suspensor attaches the embryo to the ovule and supplies it with nutrients.

The eudicot embryo becomes first heart-shaped and then torpedo-shaped. Once you can see the two cotyledons, it is possible to distinguish the shoot tip and the root tip, which contain the apical meristems. In eudicot seeds, the cotyledons frequently absorb the endosperm.

27.3 Fruit Types and Seed Dispersal

The seeds of flowering plants are enclosed by fruits. Simple fruits are derived from a single ovary (which can be simple or compound). Some simple fruits are fleshy, such as a peach or an apple. Others are dry, such as peas, nuts, and grains. Compound fruits consist of aggregate fruits, which develop from a number of ovaries of a single flower, and multiple fruits, which develop from a number of ovaries of separate flowers.

Flowering plants have several ways to disperse seeds. Seeds may be blown by the wind, be attached to animals that carry them away, or be eaten by animals that defecate them some distance away.

Prior to germination, you can distinguish the two cotyledons and plumule of a bean seed (eudicot). The plumule is the shoot that bears leaves. Also present are the epicotyl, the hypocotyl, and the radicle. In a corn kernel (monocot), the endosperm, the cotyledon, the plumule, and the radicle are visible.

27.4 Asexual Reproductive Strategies

Many flowering plants reproduce asexually, as when the nodes of stems (either aboveground or underground) give rise to entire plants, or when roots produce new shoots.

Somatic embryogenesis is the development of adult plants from single cells in tissue culture. Micropropagation, the production of clonal plants as a result of meristem culture in particular, is now a commercial venture. Flower meristem culture results in somatic embryos that can be packaged in gel for worldwide distribution. Anther culture results in homozygous plants that express recessive genes. Leaf, stem, and root culture can result in cell suspensions that allow plant chemicals to be produced in large tanks.

Key Terms

asexual reproduction 513	megaspore 502
calyx 503	microspore 502
carpel 503	ovary 503
cell suspension culture 514	petal 503
coevolution 506	pollen grain 504
corolla 503	pollination 505
cotyledon 508	rhizome 513
development 508	seed 502
differentiation 508	sepal 503
double fertilization 505	stamen 503
embryo sac 504	stigma 503
endosperm 505	stolon 513
flower 502	style 503
fruit 511	tissue culture 513
gametophyte 504	totipotent 514
germinate 512	

▊ Assess

Reviewing This Chapter

1. Draw a diagram of alternation of generations in flowering plants, and indicate which structures are protected by the sporophyte. Explain.   502
2. Draw a diagram of a flower, and name the parts.   502–3
3. Describe the development of a male gametophyte, from the microsporocyte to the production of sperm.   504
4. Describe the development of a female gametophyte, from the megasporocyte to the production of an egg.   504
5. What is the difference between pollination and fertilization? Why doesn't fertilization require any external water? What is double fertilization?   505
6. Describe methods of cross-pollination, including the use of animal pollinators.   505–7
7. Describe the sequence of events as a eudicot zygote becomes an embryo enclosed within a seed.   508–9
8. Distinguish between simple dry fruits and simple fleshy fruits. Give an example of each type. What is an aggregate fruit? A multiple fruit?   511
9. Name several mechanisms of seed and/or fruit dispersal.   511–12
10. What are the requirements for seed germination? Contrast the germination of a bean seed with that of a corn kernel.   512
11. In what ways do plants ordinarily reproduce asexually? What is the importance of totipotency with regard to tissue culture?   513–14

Testing Yourself

Choose the best answer for each question.

1. In plants,
   a. a gamete becomes a gametophyte.
   b. a spore becomes a sporophyte.
   c. both sporophyte and gametophyte produce spores.
   d. only a sporophyte produces spores.
   e. Both a and b are correct.
2. The flower part that contains ovules is the
   a. carpel.   d. petal.
   b. stamen.   e. seed.
   c. sepal.
3. The megaspore and the microspore mother cells
   a. both produce pollen grains.
   b. both divide meiotically.
   c. both divide mitotically.
   d. produce pollen grains and embryo sacs, respectively.
   e. All of these are correct.
4. A pollen grain is
   a. a haploid structure.
   b. a diploid structure.
   c. first a diploid and then a haploid structure.
   d. first a haploid and then a diploid structure.
   e. the mature gametophyte.
5. Which of these pairs is incorrectly matched?
   a. polar nuclei—plumule
   b. egg and sperm—zygote
   c. ovule—seed
   d. ovary—fruit
   e. stigma—carpel
6. Animals assist with
   a. pollination and seed dispersal.
   b. control of plant growth and response.
   c. translocation of organic nutrients.
   d. asexual propagation of plants.
   e. germination of seeds.
7. A seed contains
   a. a seed coat.   d. cotyledon(s).
   b. an embryo.   e. All of these are correct.
   c. stored food.
8. Which of these is not a common procedure in the tissue culture of plants?
   a. shoot tip culture for the purpose of micropropagation
   b. meristem culture for the purpose of somatic embryos
   c. leaf, stem, and root culture for the purpose of cell suspension cultures
   d. culture of hybridized mature plant cells
9. In the life cycle of flowering plants, a microspore develops into
   a. a megaspore.   d. an ovule.
   b. a male gametophyte.   e. an embryo.
   c. a female gametophyte.
10. Carpels
   a. are the female part of a flower.
   b. contain ovules.
   c. are the innermost part of a flower.
   d. may be absent in a flower.
   e. All of these are correct.

11. Bat-pollinated flowers
    a. are colorful.
    b. are open throughout the day.
    c. are strongly scented.
    d. have little scent.
    e. Both b and c are correct.

12. Heart, torpedo, and globular refer to
    a. embryo development.
    b. sperm development.
    c. female gametophyte development.
    d. seed development.
    e. Both b and d are correct.

13. Fruits
    a. nourish embryo development.
    b. help with seed dispersal.
    c. signal gametophyte maturity.
    d. attract pollinators.
    e. signal when they are ripe.

14. Plant tissue culture takes advantage of
    a. a difference in flower structure.
    b. sexual reproduction.
    c. gravitropism.
    d. phototropism.
    e. totipotency.

15. Label this diagram of alternation of generations in flowering plants.

Engage

Thinking Scientifically

1. You notice that a type of wasp has been visiting a flower type in your garden. What web/library research would allow you to hypothesize that this wasp is a pollinator for this flower type?

2. The pollinator for a very rare plant has become extinct. **a.** What laboratory technique would you use to prevent the plant from also becoming extinct? **b.** How might you improve the hardiness of the plant?

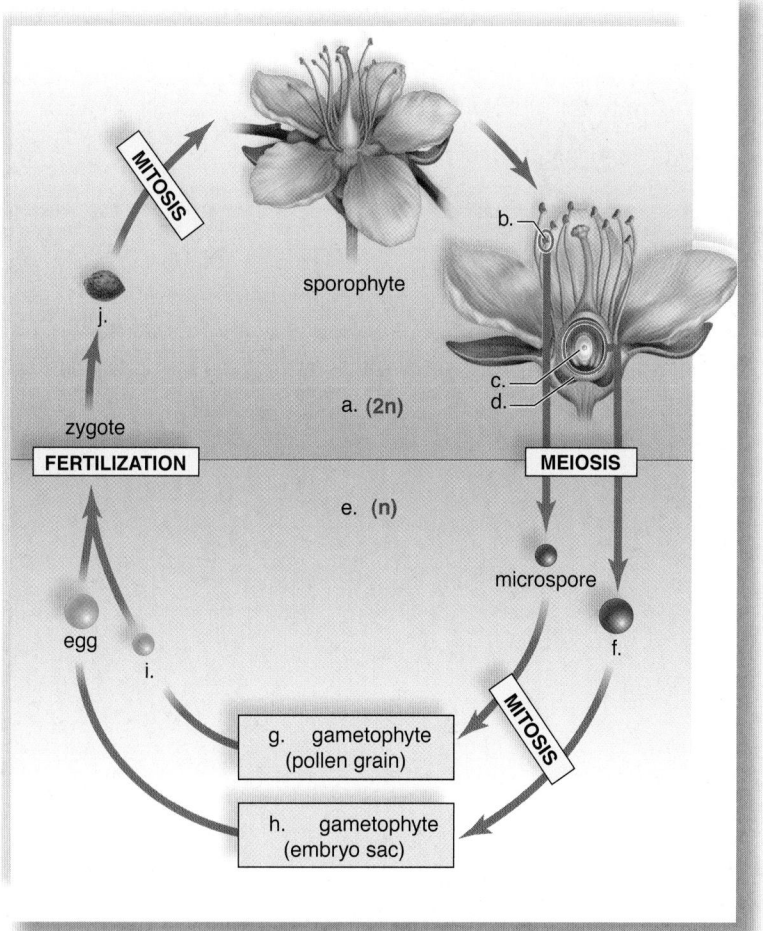

UNIT **6**

Animal Evolution and Diversity

Imagine starting from scratch to develop an animal; in a world filled with only microscopic organisms, fungi, and plants, an animal would be a very alien creature! The earliest "Animalia" were almost indistinguishable from their distant plant cousins; immobile, nondescript sponges and tiny flower-like corals were all in fact classified as plants through the nineteenth century. But with the flatworms and their novel bilateral symmetry and cephalization, the basic idea of an animal was advancing. Moving, responsive heterotrophs with heads and tails, worms enlarged, developed one-way digestive and circulatory systems, and their simple body plans began embellishment. Mollusks often protected their epidermis with shells while arthropods developed chitinous exoskeletons with flexible joints, bizarre and utilitarian in their form. Some even "slipped the surly bonds" of Earth's surface with the first wings. Jaws and antennae and appendages continued to evolve. Internal organs became specialized for living in water or on land, for eating vegetation or dining carnivorously, and for surviving the pressure of the abyssal ocean or the scorching temperatures of the barren desert. Vertebrates exploded into streamlined snakes and cheetahs, gargantuan elephants and blue whales, armored armadillos and turtles, aerial bats and sugar gliders. Obviously, there is no lack of diversity among animals: insects alone sport over a million species!

Portions of the Big Ideas will help you understand the history of animals on Earth:

 The fossil record documents the advance of the size, complexity and variety of animal life.

 Specialized mechanisms for such jobs as respiration and circulation help animals maintain homeostasis and adapt to their environment.

 Animal systems cooperate to accomplish efficient body function.

So after surveying the animal kingdom from jellyfish to lobster to human, you can perhaps understand why, on the final page of his *Origin of Species*, Darwin penned this eloquent sentiment: "There is grandeur in this view of life… (that) from so simple a beginning endless forms most beautiful and most wonderful have been and are being evolved."

UNIT OUTLINE

The tapeworm *Dipylidium*, commonly found in cats, dogs, and humans.

28

Invertebrate Evolution

BEFORE YOU BEGIN

Before beginning this chapter, take a few moments to review the following discussions.

Figure 18.7 When did the lineage of animals first appear in the tree of life?

Figure 21.3 Where do animals fit into the evolutionary history of eukaryote supergroups?

Section 21.2 Which single-celled eukaryotes are thought to be the common ancestor of all animals?

A fad diet, called the "tapeworm diet," proposes that people intentionally ingest human tapeworms, a parasitic invertebrate, to decrease their body's absorption of nutrients, thus promoting weight loss. The side effects of tapeworm infection, however, are potentially very severe. In impoverished nations the World Health Organization (WHO) lists cysticercosis, an advanced tapeworm infection, as a serious Neglected Tropical Diseases (NTD) that threatens more than 50 million people worldwide. Cysticercosis causes severe malnutrition in growing children, as well as anemia and organ damage in children and adults. This diet fad is now banned by the FDA in the United States. Public education about the risks of parasite infection is our greatest defense against such a potentially life-threatening pursuit of an ideal body type.

The invertebrates—animals that lack a backbone—are far more diverse and numerous than the vertebrates. Many invertebrates enhance the quality of our lives. For example, insects are essential pollinators of food crops and other plants. However, not all invertebrates are beneficial. Mosquitoes are vectors of malaria, and parasitic worms are responsible for other NTDs such as elephantiasis and schistosomiasis.

In this chapter, you will read about the evolution of the major groups of invertebrates and the suite of traits that have allowed them to successfully adapt to a wide range of habitats.

As you read through the chapter, think about the following questions:

1. Where do invertebrates fit into the evolutionary history of eukaryotes? Of animals?
2. What evolutionary advantage might there be to invertebrates having multiple developmental stages, each with a different body form, habitat, and lifestyle?

FOLLOWING *the* BIG IDEAS

CHAPTER 28   INVERTEBRATE EVOLUTION

Evolution	Invertebrate evolutionary trends include bilateral symmetry, cephalization, and segmentation.
Energy and Homeostasis	Animal body systems help maintain homeostasis but are tailored to specific needs and environments.

28.1 Evolution of Animals

The traditional five-kingdom classification placed animals in the kingdom Animalia. The modern three domain system places animals in the domain Eukarya. Within the Eukarya, they are placed in the supergroup Opisthokonta (see Chapter 21), along with fungi and certain protozoans (notably the choanoflagellates). In this section, we consider what characteristics distinguish animals from other eukaryotes and look at how these traits evolved.

Characteristics of Animals

Animals, fungi, and plants are all multicellular eukaryotes, but unlike plants, which make their food through photosynthesis, fungi and animals are heterotrophs and must acquire nutrients from an external source. Animals differ from fungi, which are saprotrophs, because animals ingest (eat) food and digest it internally, while fungi dissolve food externally and absorb nutrients. Animal cells also lack a cell wall, which is common to both the plants and fungi, although the cell walls in fungi and plants are made of different organic molecules. In general, animals are mobile, have nerves and muscles, and reproduce sexually. However, the tremendous diversity of animals makes assigning specific characteristics to all animals difficult.

Animals have a variety of life cycles. Many reproduce sexually, some reproduce asexually, and some combine both life cycles. While many animals have a diploid life cycle, other life cycles, such as haploid or haplodiploid, are also represented. Animals undergo a series of developmental stages to produce an organism that has specialized tissues that carry on specific functions.

Muscle and nerve tissues, both characteristic of animals, allow motility and a variety of flexible movements. In turn, motility enables animals to search actively for food and to prey on other organisms. Coordinated movements also allow animals to seek mates, shelter, and a suitable climate—behaviors that have allowed animals to live in all habitats and to become vastly diverse.

Animals are descended from a single common ancestor, and thus form a single lineage on the tree of life. Within the animal lineage are two main branches, the vertebrates and invertebrates. **Vertebrates** are animals that at some stage of their lives

Figure 28.1 Animals—multicellular, heterotrophic eukaryotes. Animals begin life as a 2n zygote that undergoes development to produce a multicellular organism that has specialized tissues. Animals depend on a source of external food to carry on life's processes. This series of images shows the development and metamorphosis of the frog, a complex animal.

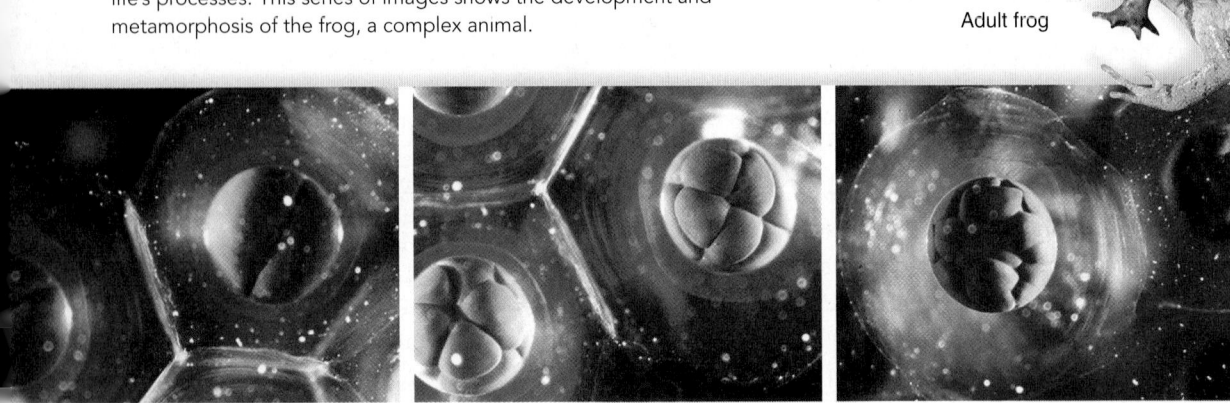

Adult frog

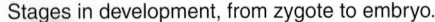

Stages in development, from zygote to embryo.

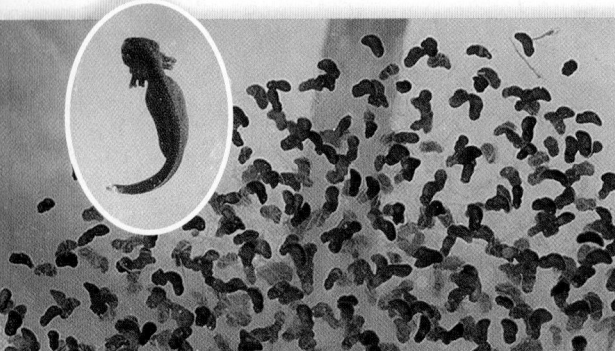

Stages in metamorphosis, from hatching to tadpole.

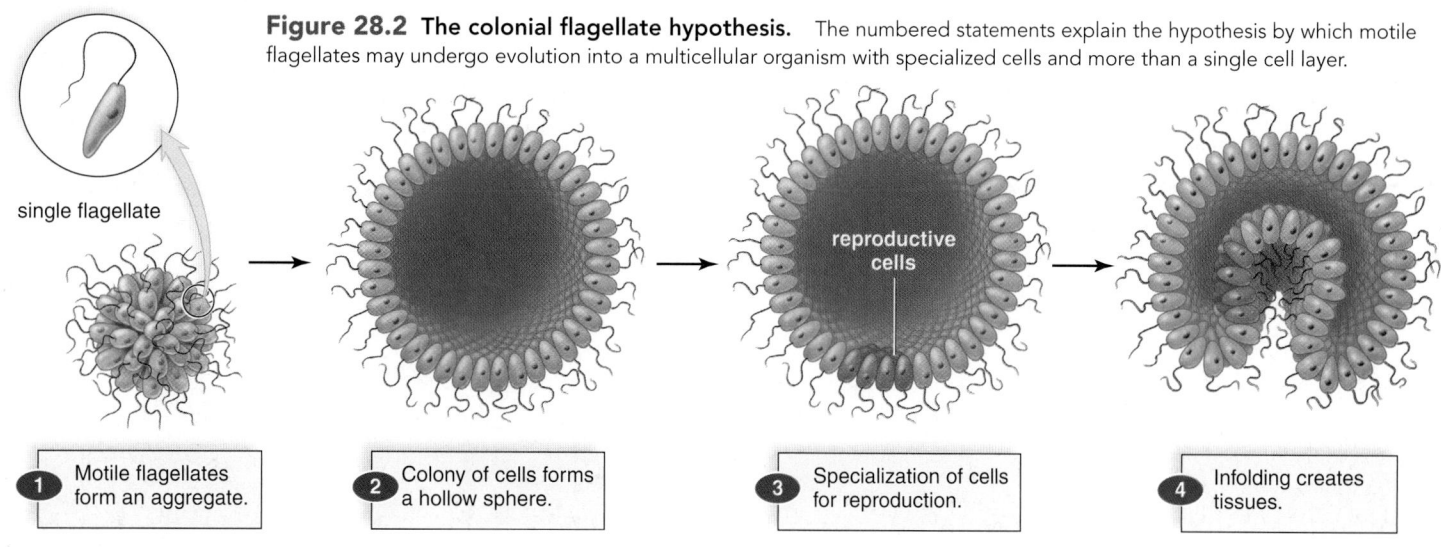

Figure 28.2 The colonial flagellate hypothesis. The numbered statements explain the hypothesis by which motile flagellates may undergo evolution into a multicellular organism with specialized cells and more than a single cell layer.

single flagellate

reproductive cells

1 Motile flagellates form an aggregate.

2 Colony of cells forms a hollow sphere.

3 Specialization of cells for reproduction.

4 Infolding creates tissues.

have a spinal cord (or backbone) running down the center of the back, whereas **invertebrates,** the topic of this chapter, do not have a backbone. The frog in Figure 28.1 is a vertebrate, while the damselfly it is devouring is an invertebrate.

Figure 28.1 illustrates the characteristics of a complex animal, using the frog as an example. A frog goes through a number of embryonic stages to become a larval form (the tadpole) with many specialized tissues. A *larva* is an immature stage that typically lives in a different habitat and feeds on different foods than the adult. By means of a change in body form called **metamorphosis,** the larva, which typically swims, turns into a sexually mature adult frog that swims and hops. The tadpole feeds on tiny aquatic organisms, while the terrestrial adult typically feeds on insects and worms. A large African bullfrog will try to eat just about anything, including other frogs, small fish, reptiles, and mammals.

Video Tadpole Development

Ancestry of Animals

In Chapter 23, we discussed evidence that plants most likely share a protist (green algae) ancestor with the charophytes. What about animals? Did they also evolve from a protist, most likely a particular motile protozoan? The **colonial flagellate hypothesis** states that animals are descended from an ancestor that resembled a hollow spherical colony of flagellated cells.

Among the protists, the choanoflagellates most likely resemble the last unicellular ancestor of animals, and molecular data tells us that they are the closest living relatives of animals! A choanoflagellate is a single cell, 3–10 μm in diameter, with a flagellum surrounded by a collar of 30–40 microvilli. Movement of the flagellum creates water currents that pull the protist along. As the water moves through the microvilli, they engulf bacteria and debris from the water. Choanoflagellates also exist as a colony of cells (see Fig. 21.23). Several can be found together at the end of a stalk or simply clumped together like a bunch of grapes.

Figure 28.2 shows how the process of transition from colonial flagellates to multicellular animals might have begun with an aggregate of a few flagellated cells. From there, a larger number of cells could have formed a hollow sphere. Individual cells within the colony would have become specialized for particular functions, such as reproduction. Two tissue layers could have arisen by an infolding of certain cells into a hollow sphere. Tissue layers arise in this manner during the development of animals today. The colonial flagellate hypothesis is also attractive because of its implications regarding animal symmetry, which is discussed shortly.

Evolution of Body Plans

All of the various animal body plans we see today were present by the Cambrian period (see Table 18.1). How could such diversity have arisen within a relatively short period of geological time? As an animal develops, there are many possible outcomes regarding the number, position, size, and patterns of its body parts. Different combinations could have led to the great variety of animal forms in the past and present. We now think that slight shifts in the DNA code and expression of genes called *Hox* (homeotic) genes are responsible for the major differences between animals that arise during development, as described in the Evolution feature on pages 524–25.

The Phylogenetic Tree of Animals

The fossil record of the early evolution animals is very sparse. Therefore, the best hypothesis of the evolution of animals, represented by the phylogenetic tree of animals shown in Figure 28.3, is based on a combination of morphological characters of living and fossil organisms, developmental homologies, and molecular (DNA) characters. The more closely related two organisms are, the more similar their DNA sequences. The addition of molecular data has resulted in an updated phylogenetic tree that is quite different from hypotheses that were historically based solely upon morphological characteristics.

Refer to the phylogeny in Figure 28.3 as we discuss the anatomical characteristics that are shared among animals of each branch of the animal family tree.

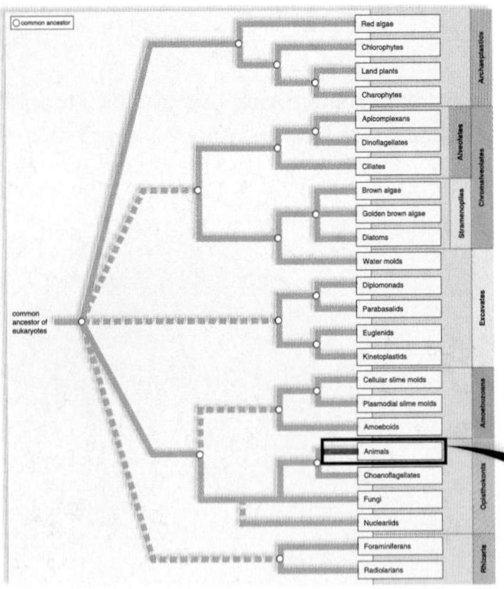

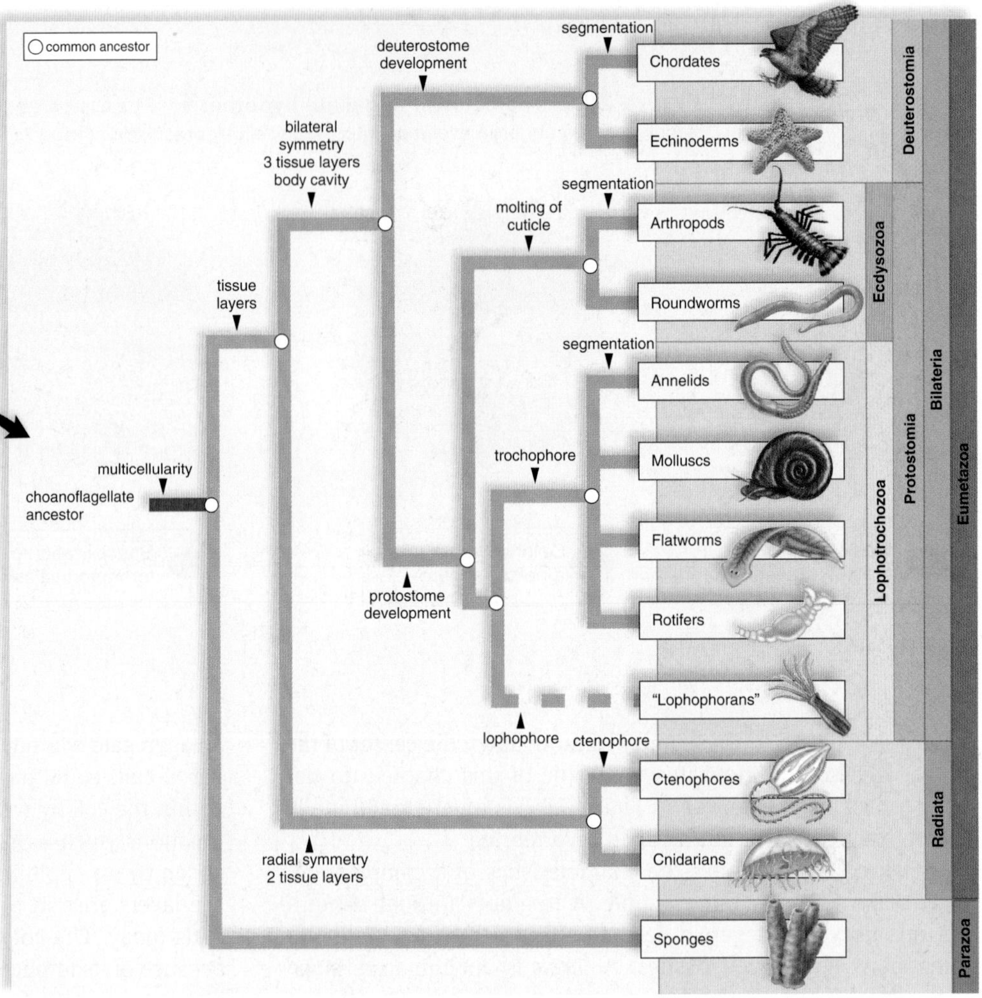

Type of Symmetry

Symmetry [Gk. *syn,* together, and *metron,* measure] refers to a pattern of similarity that is observed in objects. Three types of symmetry exist in the animal kingdom. **Asymmetry,** or lack of symmetry, is seen in sponges that have no particular pattern to body shape (see Fig. 28.6).

The cnidarians and comb jellies exhibit **radial symmetry**—their bodies are organized circularly, similar to a wheel, such that any longitudinal cut through the central point produces two identical halves (Fig. 28.4a). Many adult and immature or larval forms of animals are radially symmetrical. Some radially symmetrical animals are mobile and others are immobile, or sessile, when attached to a substrate. Radial symmetry allows an organism to extend out in all directions from one center. Floating animals with radial symmetry, such as jellyfish, also have this feature.

The rest of the animals exhibit **bilateral symmetry** as adults—they have a definite left and right half, and only a single longitudinal cut down the centerline of the animal produces two equal halves (Fig. 28.4b). Bilaterally symmetrical animals have defined anterior and posterior ends, and forward movement is guided with the anterior end. The colonial flagellate hypothesis, mentioned earlier, is attractive because it implies that radial symmetry preceded bilateral symmetry in animal history, and a phylogeny supports this hypothesis.

During the evolution of animals, bilateral symmetry was accompanied by **cephalization,** localization of a brain and specialized sensory organs at the anterior end of an animal. This development was critical to an animal's ability to engage in directed movement—toward food or mates, and away from danger.

Embryonic Development

The simplest animals are the sponges. Like all animals, sponges are multicellular but they do not have true specialized tissues. True tissues develop in the more complex animals as they undergo embryological development. The first tissue layers that appear are called **germ layers,** and they give rise to the organs and organ systems of complex animals.

Animals such as the cnidarians, which as embryos have only two tissue layers—the ectoderm and endoderm—are termed

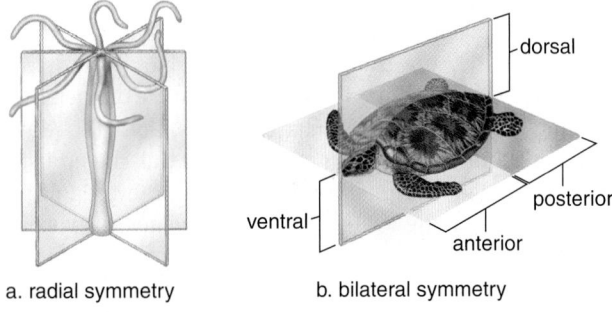

a. radial symmetry b. bilateral symmetry

Figure 28.4 Symmetry in animals The body plans of animals can be radially symmetrical **(a)** or bilaterally symmetrical **(b)**.

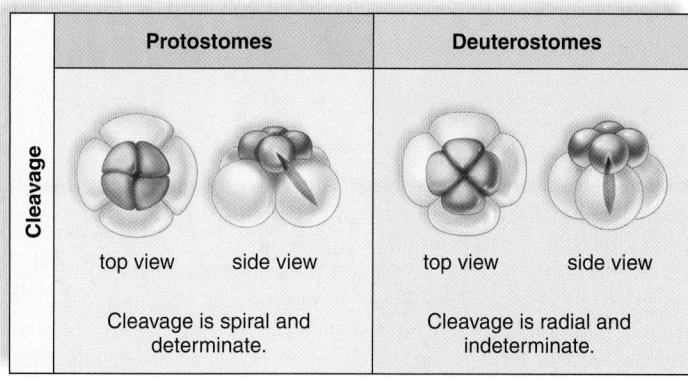

Protostomes	Deuterostomes

Cleavage

top view side view

Cleavage is spiral and determinate.

top view side view

Cleavage is radial and indeterminate.

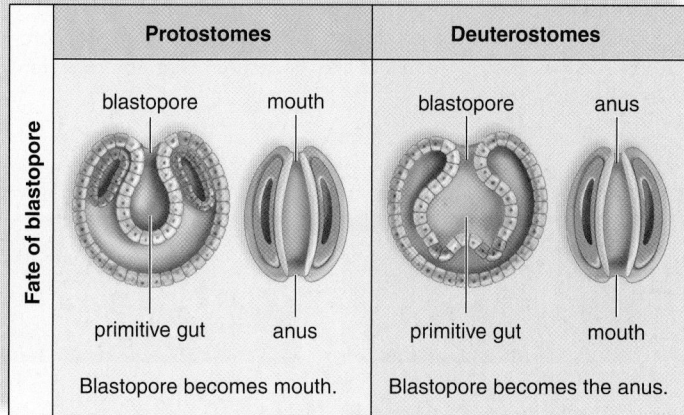

Protostomes	Deuterostomes

Fate of blastopore

blastopore mouth

primitive gut anus

Blastopore becomes mouth.

blastopore anus

primitive gut mouth

Blastopore becomes the anus.

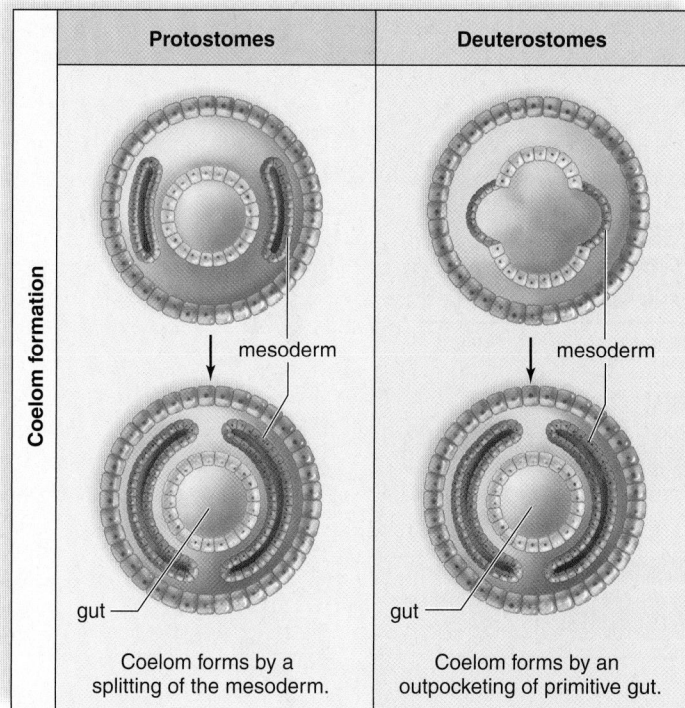

Protostomes	Deuterostomes

Coelom formation

mesoderm

gut

Coelom forms by a splitting of the mesoderm.

mesoderm

gut

Coelom forms by an outpocketing of primitive gut.

Figure 28.5 Protostomes compared to deuterostomes.
Left: In the embryo of protostomes, cleavage is spiral—new cells are at an angle to old cells—and each cell has limited potential and cannot develop into a complete embryo; the blastopore is associated with the mouth; and the coelom, if present, develops by a splitting of the mesoderm. *Right:* In deuterostomes, cleavage is radial—new cells sit on top of old cells—and each one can develop into a complete embryo; the blastopore is associated with the anus; and the coelom, if present, develops by an outpocketing of the primitive gut.

diploblastic. Diploblastic animals develop tissues, but no specialized organs. The animals that develop specialized organs are termed *triploblastic* because as embryos they have three tissue layers—the ectoderm, mesoderm, and endoderm. All animals except the sponges (Parazoa) and cnidarians and comb jellies (Radiata) are triploblastic.

Notice in the phylogenetic tree (see Fig. 28.3) the triploblastic animals are either **protostomes** [Gk. *proto,* first, and *stoma,* mouth] or **deuterostomes** [Gk. *deuter,* second]. These terms have to do with whether the mouth or the anus develops first in the embryo. In protostomes, the mouth develops prior to the anus, whereas in deuterostomes, the anus develops prior to the mouth. Figure 28.5 shows that protostome and deuterostome development are differentiated by three major events: cleavage, blastula formation, and coelom development.

Cleavage. The first developmental event after fertilization is **cleavage,** cell division without cell growth. In protostomes, spiral cleavage occurs, and daughter cells sit in grooves formed by the previous cleavages. The fate of these cells is fixed and determinate in protostomes; each can contribute to development in only one particular way.

In deuterostomes, radial cleavage occurs, and the daughter cells sit right on top of the previous cells. The fate of these cells is indeterminate—that is, if they are separated from one another, each cell can go on to become a complete organism.

Blastula Formation. As development proceeds, a hollow sphere of cells, or **blastula,** forms, and the indentation that follows produces an opening called the blastopore. In protostomes, the mouth appears at or near the blastopore.

In deuterostomes, the anus appears at or near the blastopore, and only later does a second opening form the mouth.

 **Video Blastocyst Formation**

Coelom Development. Certain protostomes and all deuterostomes have a body cavity lined by mesoderm called a **coelom** [Gk. koiloma, cavity]. More specifically, the coelom in these groups is a **true coelom** because the mesoderm cells line the cavity completely. However, the coelom develops differently in the two groups. In protostomes, the mesoderm arises from cells located near the embryonic blastopore, and a splitting occurs that produces the coelom.

In deuterostomes, the coelom arises as a pair of mesodermal pouches from the wall of the primitive gut. The pouches enlarge until they meet and fuse.

Table 28.1 summarizes the classification of organisms found in kingdom Animalia. In the sections that follow, we describe and compare the major groups of invertebrate animals. All of these except the echinoderms are protostomes. We begin with the simplest multicellular animals, the sponges.

Check Your Progress 28.1

1. State three characteristics that all animals have in common.
2. Explain the colonial flagellate hypothesis about the origin of animals.
3. List two differences between deuterostomes and protostomes.

Evolution

Evolution of the Animal Body Plan

The animal body plan can be divided into three categories based upon symmetry (see Fig. 28.3). The general trend seems to be for body plans to become increasingly complex, from a lack of symmetry in the sponges, to radial symmetry in the ctenophores, to bilateral symmetry in more recently evolved groups such as the arthropods and chordates that have multiple tissue types and organ systems.

The body plan of an animal is the result of a carefully orchestrated pattern of genes being expressed (or not expressed) at the right time and in the correct region of the developing embryo. In the first stage of development, the anterior (front) and posterior (rear) ends of the embryo are determined (Fig. 28A). In bilaterally symmetrical animals with segments, such as insects and chordates, the next step in development is to divide the embryo into segments, each of which will become a different part of the body. In fruit flies, genes such as the *gap* and the *pair-rule* genes determine the number of segments. In vertebrates, *FGF-8* is one of many genes that determine segmentation pattern.

Once the segmentation pattern is established, homeotic, or *Hox*, genes determine the ultimate developmental fate of each segment. *Hox* genes encode homeotic proteins that bind to the regulatory region of genes that determine the body plan during development. Homeotic proteins act like "switches" that control when,

Ph se   **Ph se**

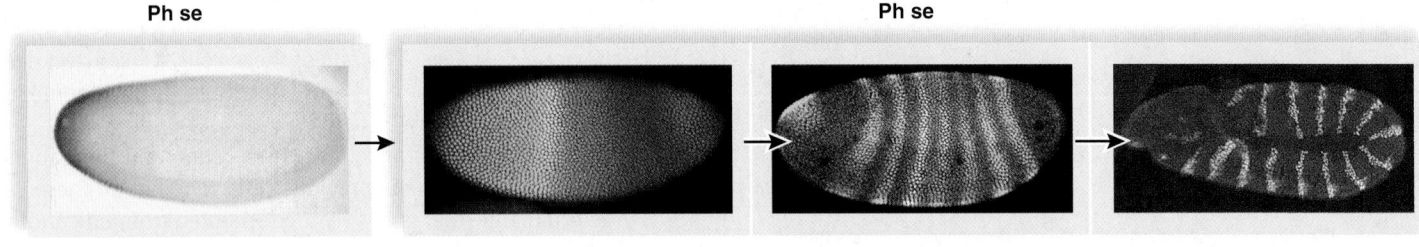

he a ter ior a d o ter ior
re io are deter mi ed

he umber a d atter o e me t are deter   mi ed

Figure 28A Development stages in the fruit fly embryo. Fruit fly embryos are sectioned and stained during different stages of development. The anterior and posterior regions of the embryo are determined in Phase 1 by genes such as *bicoid*. Genes such as *gap* and *pair-rule* determine the segmentation pattern in Phase 2.

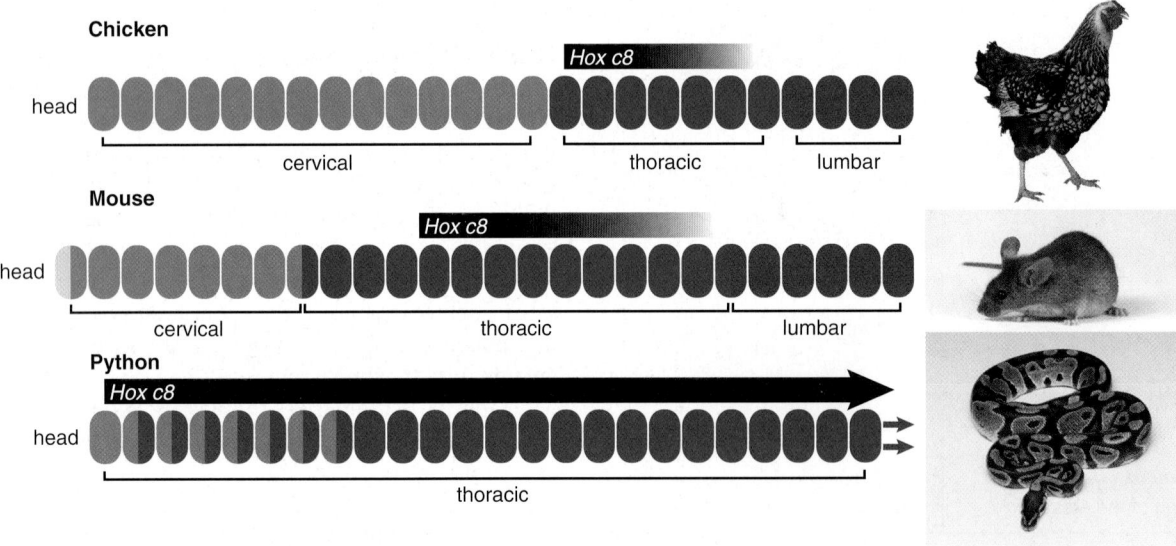

Figure 28B *Hox* **genes orchestrate the development of the body plan of animals.** In animal development, the number of segments, or vertebrae in vertebrates, is controlled by *Hox* genes. One *Hox* gene, *Hox c8*, determines the number of thoracic vertebrae. Variation in the number of thoracic vertebrae in the spine of a chicken, mouse, and snake is caused by a shift in when, and for how long, *Hox c8* is active during embryonic development. In the chicken, 7 thoracic vertebrae are formed, in the mouse, 12. A large increase in the number of thoracic vertebrae results in the long and thin body plan of a snake.

where, and for how long a particular developmental gene is active.

Each *Hox* gene orchestrates the developmental fate of a particular region of the body. In mice, *Hox C8* sets the fate of 12 segments to become thoracic vertebrae, while in snakes *Hox C8* orchestrates the development of hundreds of thoracic vertebrae (Fig. 28B).

Hox genes have played a role in the development of the body plan since the early stages of animal evolution. We know this because *Hox* genes are found in all animals with multiple types of tissues, and there is a shared similarity of *Hox* genes across animal groups (Fig. 28C). For example, the *Hox* genes that determine the fate of the head region have the same evolutionary origin in flies, worms, and mice. Even cnidarians, with a simple body plan, have some *Hox* genes shared in common with more complex animals. This implies that all *Hox* genes evolved in animals from a common *Hox* gene ancestor. *Hox* genes are found in linear clusters on the same chromosome. However, not all *Hox* genes are found in all animals. Some groups of animals have more than one cluster of genes with duplicate copies of genes within each cluster.

Questions to Consider

1. Why are *Hox* genes evidence for a common ancestor of all life?
2. What would you expect to happen if a particular *Hox* gene, such as *Hox C8*, was not functioning properly?
3. What experiments could you plan to test what a particular *Hox* gene controlled at a particular time of development?

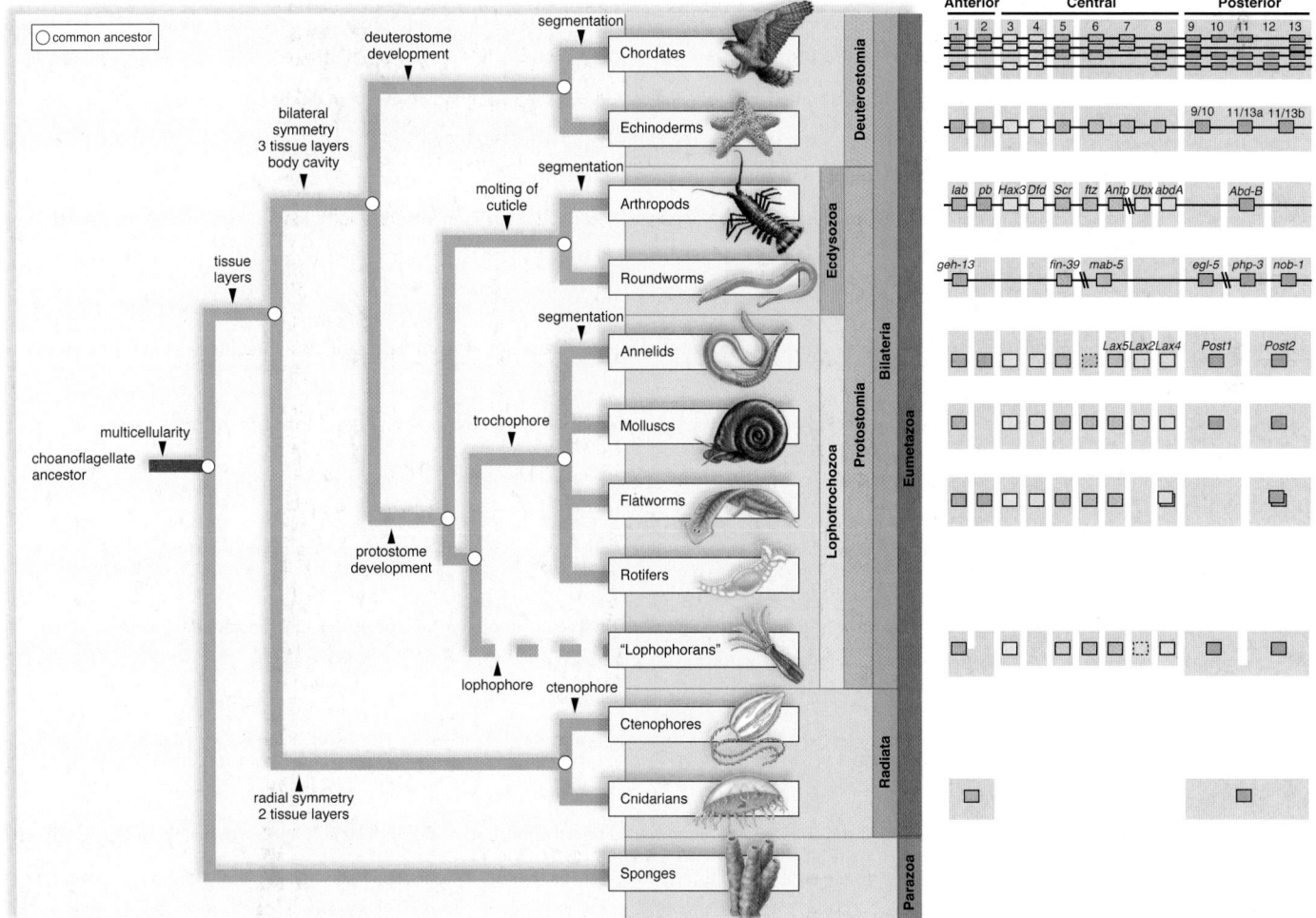

Figure 28C Shared evolutionary history of *Hox* genes in animals. *Hox* genes are found in all groups of animals with multiple tissue layers. A general trend has been observed for an increase in the number of *Hox* genes and *Hox* gene clusters with an increase in body plan complexity. Although the number of *Hox* genes varies among animals, there is strong evidence for common ancestry of each *Hox* gene. The color coding indicates a particular *Hox* gene is shared between lineages of animals.

Table 28.1 The Animal Kingdom

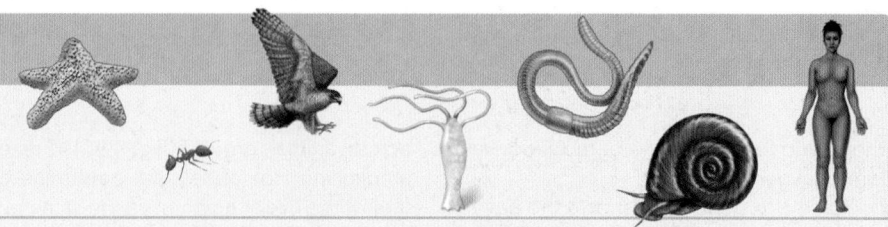

DOMAIN: Eukarya
KINGDOM: Animalia

CHARACTERISTICS

Multicellular, usually with specialized tissues; ingest or absorb food; diploid life cycle

INVERTEBRATES

Sponges (bony, glass, spongin):* Asymmetrical, saclike body perforated by pores; internal cavity lined by choanocytes; spicules serve as internal skeleton. 5,150+

Radiata

Cnidarians (hydra, jellyfish, corals, sea anemones): Radially symmetrical with two tissue layers; sac body plan; tentacles with nematocysts. 10,000+

Comb jellies: Have the appearance of jellyfish; the "combs" are eight visible longitudinal rows of cilia that can assist locomotion; lack the nematocysts of cnidarians but some have two tentacles. 150+

Protostomia (Lophotrochozoa)

Lophophorates (lampshells, bryozoa): Filter feeders with a circular or horseshoe-shaped ridge around the mouth that bears feeding tentacles. 5,935+

Flatworms (planarians, tapeworms, flukes): Bilateral symmetry with cephalization; three tissue layers and organ systems; acoelomate with incomplete digestive tract that can be lost in parasites; hermaphroditic. 20,000+

Rotifers (wheel animals): Microscopic animals with a corona (crown of cilia) that looks like a spinning wheel when in motion. 2,000+

Molluscs (chitons, clams, snails, squids): Coelom, all have a foot, mantle, and visceral mass; foot is variously modified; in many, the mantle secretes a calcium carbonate shell as an exoskeleton; true coelom and all organ systems. 110,000+

Annelids (polychaetes, earthworms, leeches): Segmented with body rings and setae; cephalization in some polychaetes; hydroskeleton; closed circulatory system. 16,000+

Protostomia (Ecdysozoa)

Roundworms (*Ascaris*, pinworms, hookworms, filarial worms): Pseudocoelom and hydroskeleton; complete digestive tract; free-living forms in soil and water; parasites common. 25,000+

Arthropods (crustaceans, spiders, scorpions, centipedes, millipedes, insects): Chitinous exoskeleton with jointed appendages undergoes molting; insects—most have wings—are most numerous of all animals. 1,000,000+

Deuterostomia

Echinoderms (sea stars, sea urchins, sand dollars, sea cucumbers): Radial symmetry as adults; unique water-vascular system and tube feet; endoskeleton of calcium plates. 7,000+

Chordates (tunicates, lancelets, vertebrates): All have notochord, dorsal tubular nerve cord, pharyngeal pouches, and postanal tail at some time; contains mostly vertebrates in which notochord is replaced by vertebral column. 56,000+

VERTEBRATES

Fishes (jawless, cartilaginous, bony): Endoskeleton, jaws, and paired appendages in most; internal gills; single-loop circulation; usually scales. 28,000+

Amphibians (frogs, toads, salamanders): Jointed limbs; lungs; three-chambered heart with double-loop circulation; moist, thin skin. 5,383+

Reptiles (snakes, turtles, crocodiles): Amniotic egg; rib cage in addition to lungs; three- or four-chambered heart typical; scaly, dry skin; copulatory organ in males and internal fertilization. 8,000+

Birds (songbirds, waterfowl, parrots, ostriches): Endothermy, feathers, and skeletal modifications for flying; lungs with air sacs; four-chambered heart. 10,000+

Mammals (monotremes, marsupials, eutherians): Hair and mammary glands. 4,800+

*After a character is listed, it is present in the rest, unless stated otherwise.
+Number of species.

28.2 The Simplest Invertebrates

Sponges, cnidarians, and comb jellies represent the most ancient and the simplest animals. In this section, we discuss their distinguishing characteristics.

Sponges

All animals are multicellular; **sponges** (phylum Porifera [L. *porus*, pore, and *ferre*, to bear]) are the only animals to lack true tissues and to have only a cellular level of organization. They have only a few cell types, and lack nerve and muscle cells like more complex animals. Molecular data place them at the base of the evolutionary tree of animals (see Fig. 28.3).

The sac-like body of a sponge is perforated by many pores (Fig. 28.6). Sponges are aquatic, largely marine animals that vary greatly in size, shape, and color. But, they all have a canal system of pores of varying complexity that allows water to move through their bodies.

The interior of the canals is lined with flagellated cells that resemble choanoflagellates. In a sponge, these cells are called collar cells, or choanocytes. The beating of the flagella produces water currents that flow through the pores into the central cavity and out through the osculum, the upper opening of the body. Even a simple sponge only 10 cm tall is estimated to filter as much as 100 liters of water each day. It takes this much water to supply the needs of the sponge.

A sponge is a sessile filter feeder, also called a suspension feeder, because it filters suspended particles from the water by means of a straining device—in this case, the pores of the walls and the microvilli making up the collar of collar cells. Microscopic food particles that pass between the microvilli are engulfed by the collar cells and digested by them in food vacuoles.

The skeleton of a sponge prevents the body from collapsing. All sponges have fibers of spongin, a modified form of collagen; a natural bath sponge is the dried spongin skeleton from which all living tissue has been removed. Today, however, commercial "sponges" are usually synthetic.

Typically, the endoskeleton of sponges also contains **spicules**—small, needle-shaped structures with one to six rays. Traditionally, the type of spicule has been used to classify sponges; there are bony, glass, and spongin sponges. The success of sponges—they have existed longer than any other animal group—can be attributed to their spicules. They have few predators because a mouth full of spicules is an unpleasant experience. Also, they produce a number of foul-smelling and toxic substances that discourage predators.

Sponges can reproduce both asexually and sexually. They reproduce asexually by fragmentation or by budding. During budding, a small protuberance appears and gradually increases in size until a complete organism forms. Budding produces colonies of sponges that can become quite large. During sexual reproduction, eggs and sperm are released into the central cavity, and the zygote develops into a flagellated larva that may swim to a new location.

osculum → H₂O out — sponge wall
— spicule
— pore
— amoebocyte

H₂O in through pores

central cavity

epidermal cell — amoebocyte
collar — nucleus

flagellum — collar cell (choanocyte)

a. Yellow tube sponge, *Aplysina fistularis*   b. Sponge organization

Figure 28.6 Simple sponge anatomy.

rows of cilia

tentacle

a.

tentacles

b.

Figure 28.7 Comb jelly compared to cnidarian.
a. *Pleurobrachia pileus,* a comb jelly.
b. *Polyorchis penicillatus,* medusan form of a cnidarian. Both animals have similar symmetry, diploblastic organization, and gastrovascular cavities.

inner tissue layer, which is derived from endoderm, secretes digestive juices into the internal cavity, called the **gastrovascular cavity** [Gk. *gastros,* stomach; L. *vasculum,* dim. of *vas,* vessel] because it serves for digestion of food and circulation of nutrients. The fluid-filled gastrovascular cavity also serves as a supportive **hydrostatic skeleton,** so called because it offers some resistance to the contraction of muscle but permits flexibility. The two tissue layers are separated by mesoglea.

Two basic body forms are seen among cnidarians. The mouth of a **polyp** is directed upward, while the mouth of a jellyfish, or **medusa,** is directed downward. The bell-shaped medusa has more mesoglea than a polyp, and the tentacles are concentrated on the margin of the bell.

At one time, both body forms may have been a part of the life cycle of all cnidarians. When both are present, the animal is dimorphic: The sessile polyp stage produces medusae by asexual budding, and the motile medusan stage produces egg and sperm. In some cnidarians, one stage is dominant and the other is reduced; in other species, one form is absent altogether.

Cnidarian Diversity

Sea anemones (Fig. 28.8*a*) are sessile polyps that live attached to submerged rocks, timbers, or other substrate. Most sea anemones range in size from 0.5–20 cm in length and 0.5–10 cm in diameter and are often colorful. Their upward-turned oral disk contains the mouth and is surrounded by a large number of hollow tentacles containing nematocysts.

Corals (Fig. 28.8*b*) resemble sea anemones encased in a calcium carbonate (limestone) house. The coral polyp can extend into the water to feed on microorganisms and retreat into the house for safety. Some corals are solitary, but the vast majority live in colonies that vary in shape from rounded to branching. Corals are the animals that build coral reefs. They exhibit elaborate geometric designs and stunning colors.

Coral reefs are built from the slow accumulation of limestone produced by corals. Over hundreds of years, this accumulation can result in massive structures such as the Great Barrier Reef along the eastern coast of Australia. Coral reef ecosystems are very productive, and a diverse group of marine life call the reef home.

Video Coral Reef Spawning

The hydrozoans have a dominant polyp stage. *Hydra* (see Fig. 28.9) is a hydrozoan, and so is a Portuguese man-of-war. You might think the Portuguese man-of-war is an odd-shaped medusa, but actually it is a colony of polyps (Fig. 28.8*c*). The original polyp becomes a gas-filled float that provides buoyancy, keeping the colony afloat. Other polyps, which bud from this one, are specialized for feeding or for reproduction. A long, single tentacle armed with numerous nematocysts arises from the base of each feeding polyp. Swimmers who accidentally come upon a Portuguese man-of-war can receive painful, even serious, injuries from these stinging tentacles.

Video Portuguese Man-of-War

In true jellyfish, also known as sea jellies (Fig. 28.8*d*), the medusa is the primary stage, and the polyp remains small. Jellyfish are zooplankton and depend on tides and currents for their primary means of movement. They feed on a variety of invertebrates and fishes and are themselves food for marine animals.

If the cells of a sponge are mechanically separated, they will reassemble into a complete and functioning organism! Like many less specialized organisms, sponges are also capable of regeneration, or growth of a whole from a small part.

Comb Jellies and Cnidarians

These two groups of animals (Fig. 28.7) have true tissues, and as embryos, they have two germ layers, ectoderm and endoderm. They are radially symmetrical as adults.

Comb Jellies

Comb jellies (phylum Ctenophora) are solitary, mostly free-swimming marine invertebrates that are usually found in warm waters. Ctenophores represent the largest of these animals; they are propelled by beating cilia and range in size from a few centimeters to 1.5 m in length. Their body is made up of a transparent jellylike substance called **mesoglea.** Most ctenophores do not have stinging cells and capture their prey by using sticky adhesive cells called colloblasts. Some ctenophores are *bioluminescent,* meaning that they are capable of producing their own light.

Video Ctenophores

Cnidarians

Cnidarians (phylum Cnidaria) are tubular or bell-shaped animals that reside mainly in shallow coastal waters; however, some freshwater, brackish, and oceanic forms are known. The term *cnidaria* is derived from the presence of specialized stinging cells called cnidocytes. Each cnidocyte has a fluid-filled capsule called a **nematocyst** [Gk. *nema,* thread, and *kystis,* bladder] that contains a long, spirally-coiled hollow thread. When the trigger of the cnidocyte is touched, the nematocyst is discharged. Some threads merely trap a prey, and others have spines that penetrate and inject paralyzing toxins.

The body of a cnidarian is a two-layered sac. The outer tissue layer is a protective epidermis derived from ectoderm. The

a. Sea anemone, *Corynactis*

b. Cup coral, *Tubastrea*

Figure 28.8 Cnidarian diversity. **a.** The anemone, which is sometimes called the flower of the sea, is a solitary polyp. **b.** Corals are colonial polyps residing in a calcium carbonate or proteinaceous skeleton. **c.** The Portuguese man-of-war is a colony of modified polyps and medusae. **d.** Jellyfish, *Aurelia*.

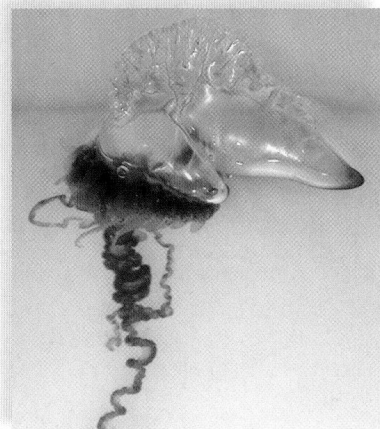

c. Portuguese man-of-war, *Physalia*

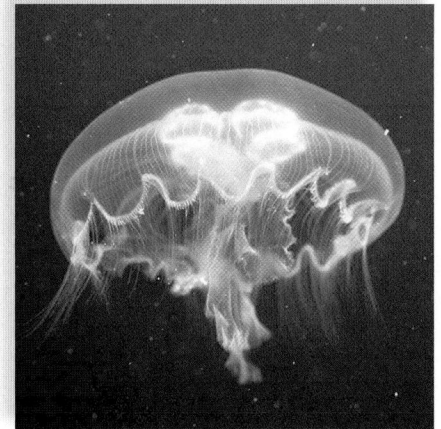

d. Jellyfish, *Aurelia*

A Typical Cnidarian: Hydra

Hydras are often studied as an example of a cnidarian. Hydras are likely to be found attached to underwater plants or rocks in most lakes and ponds. The body of a hydra is a small tubular polyp about one-quarter inch in length. The only opening (the mouth) is in a raised area surrounded by four to six tentacles that contain a large number of nematocysts.

Figure 28.9 shows the microscopic anatomy of *Hydra*. The cells of the epidermis are termed epitheliomuscular cells because they contain muscle fibers. Also present in the epidermis are nematocyst-containing cnidocytes and sensory cells that make contact with the nerve cells within a **nerve net.** These interconnected nerve cells allow transmission of impulses in several directions at once. The body of a hydra can contract or extend, and the tentacles that ring the mouth can reach out and grasp prey and discharge nematocysts.

Hydras reproduce asexually by forming buds, small outgrowths that develop into a complete animal and then detach. Interstitial cells of the epidermis are capable of becoming other types of cells such as an ovary and/or a testis. When hydras reproduce sexually, sperm from a testis swim to an egg within an ovary. The embryo is encased within a hard, protective shell that allows it to survive until conditions are optimum for it to emerge and develop into a new polyp.

Like the sponges, cnidarians have great regenerative powers, and hydras can grow an entire organism from a small piece.

Check Your Progress 28.2

1. List three ways in which cnidarians are more complex than the sponges.
2. Summarize how a sponge obtains nutrients.
3. Describe the medusa and polyp body forms of a cnidarian.
4. Explain how a cnidarian, such a jellyfish, stings its prey.

Figure 28.9 Anatomy of *Hydra*. *Top:* The body of *Hydra* is a small tubular polyp that reproduces asexually by forming outgrowths called buds. The buds develop into a complete animal. *Bottom:* The body wall contains two tissue layers separated by mesoglea. Cnidocytes are cells that contain nematocysts.

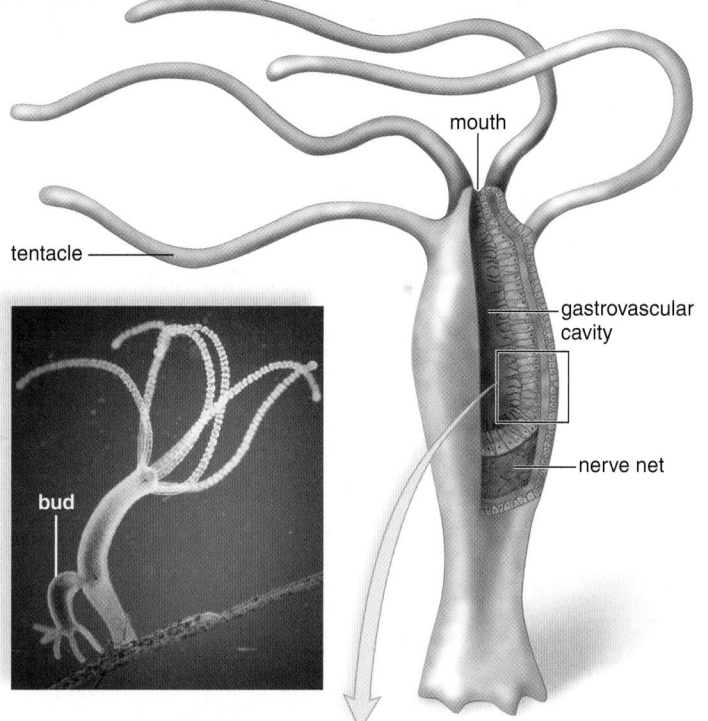

mouth

tentacle

gastrovascular cavity

nerve net

bud

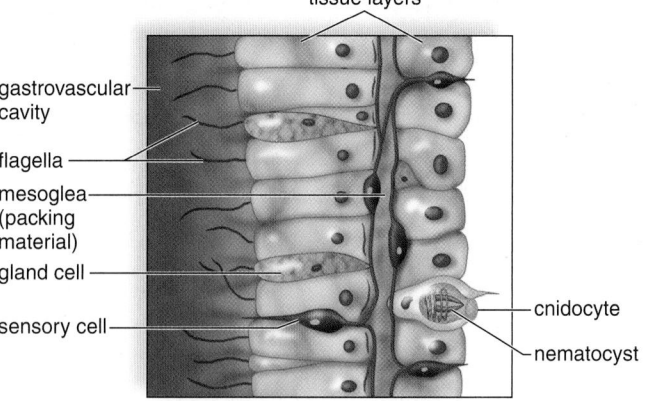

tissue layers

gastrovascular cavity

flagella

mesoglea (packing material)

gland cell

sensory cell

cnidocyte

nematocyst

28.3 Diversity Among the Lophotrochozoans

Learning Outcomes

Upon completion of this section, you should be able to

1. List the basic features of lophotrochozoans.
2. Summarize the steps in the life cycles of *Schistosoma* and *Taenia*.
3. Identify morphological features of molluscs, bivalves, rotifers, and annelids.

The **lophotrochozoa** are the most diverse of the protostomes. These animals are bilaterally symmetrical during at least one stage of their development. As embryos, they have three germ layers, and as adults, they have the organ level of organization. Lophotrochozoans are protostomes. Some have a true coelom (see Section 28.1), as exemplified best in the annelid worms.

The lophotrochozoans can be divided into two groups: the **lophophorans** (e.g., bryozoans, phoronids, and brachiopods) [Gk. *lophos,* crest; *phoros,* bearing] and the **trochozoans** (e.g, flatworms, rotifers, molluscs, and annelids) [Gk. *trochos,* wheel] (see Fig. 28.4).

All lophophorans are aquatic and have a feeding apparatus called the lophophore, which is a mouth surrounded by ciliated tentacle-like structures (Fig. 28.10*a*).

A trochophore is a free-swimming, marine larva with bands of cilia that control the direction of movement (Fig. 28.10*b*). The trochozoans either have a trochophore stage of development today (e.g., molluscs and annelids), or had an ancestor that had a trochophore stage at some point in the past (e.g., flatworms and rotifers).

Figure 28.11 Lophophorans. A bryozoan (*Canda* sp.) with a horseshoe-shaped tentacle crown.

Lophophorans

Traditionally, the "lophophorans" were considered a lineage of protostomes that included the bryozoans and brachiopods, but recent evidence from analysis of the small subunit of ribosomal RNA (rRNA) (see Fig. 12.16) and other genes suggest the evolutionary relationships among the lophophorans within the Protostomia are more complex. However, for our purposes, we represent the "lophophorans" in the traditional sense as a single lineage (see Fig. 28.3), containing three closely related groups, the bryozoans, the phoronids, and the brachiopods (Fig. 28.11).

Bryozoans (phylum Bryozoa) are aquatic, colonial lophophorans. Colonies are made up of individuals called zooids. Zooids are not independent animals, but are single members of a colony that cooperate together as a single organism. Some zooids specialize in feeding and filter particles from the water with the lophophore; some specialize in reproduction; and some can perform both functions. Zooids coordinate functions within a colony by communicating with chemical signals. Zooids have protective exoskeletons that they use to attach to substrates, including the bottom of ships where they cause a nuisance by increasing drag and impeding maneuverability.

Brachiopods (phylum Brachiopoda) are a small group of lophophorans that have two hinged shells, much like a mollusc—but instead of having a left and right shell, have a top and bottom shell. Brachiopods affix themselves to hard surfaces with a muscular pedicle. Like other lophophorans, brachiopods use their lophophore to feed by filtering particles from the water.

Phoronids (phylum Phoronida) live inside a long tube that is formed from their own chitinous secretions. The tube is buried in the ground and their lophophore extends from it, but can retract very quickly when needed. Only about 15 species of phoronids exist worldwide.

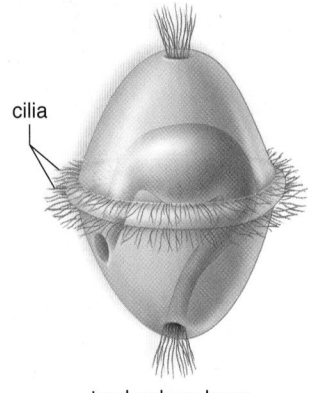

lophophore

cilia

trochophore larva

a.

b.

Figure 28.10 Lophotrochozoan characteristics. a. The lophophore feeding apparatus. **b.** A trochophore larvae.

Trochozoans

Flatworms (phylum Platyhelminthes) are trochozoans with an extremely flat body. Like the cnidarians, flatworms have a sac body plan with only one opening, the mouth. Organisms with a single opening are said to have an *incomplete digestive tract,* whereas a *complete digestive tract* has two openings. Flatworms have no body cavity, and instead the third germ layer, mesoderm, fills the space between their organs.

Among flatworms, planarians are free-living; flukes and tapeworms are parasitic.

Free-living Flatworms

Planarians are a group of non-parasitic, free-living flatworms. *Dugesia* is a planarian that lives in freshwater lakes, streams, and ponds, where it feeds on small living or dead organisms. A planarian captures food by wrapping itself around the prey, entangling it in slime, and pinning it down. Then a muscular pharynx is extended through the mouth and a sucking motion takes pieces of the prey into the pharynx. The pharynx leads into a three-branched gastrovascular cavity in which digestion is both extracellular and intracellular (Fig. 28.12*a*). The digestive system delivers nutrients and oxygen to the cells; the animal has no circulatory system or respiratory system. Waste molecules exit through the mouth.

Planarians have a well-developed excretory system (Fig. 28.12*b*). The excretory organ functions in osmotic regulation, as well as in water excretion. The organ consists of a series of interconnecting canals that run the length of the body on each side. Bulblike structures containing cilia are at the ends of the side branches of the canals. The cilia move back and forth, bringing water into the canals that empty at pores. The excretory system often functions as an osmotic-regulating system. The beating of the cilia reminded an early investigator of the flickering of a flame, and so the excretory organ of the flatworm is called a flame cell.

Planarians usually reproduce sexually. They are **hermaphroditic** or monoecious, which means that they possess both male and female sex organs and gametes in a single individual (Fig. 28.12*c*). The worms cross-fertilize when the penis of one is inserted into the genital pore of the other. During this process each planarian gives and receives sperm. The fertilized eggs are enclosed in a cocoon and hatch in two or three weeks as tiny worms.

Planarians also can reproduce asexually via regeneration. The tail portion of the planaria breaks off, and each part grows into a new worm. Because planarians have the ability to regenerate, they have been the subject of a field of research called regenerative medicine. Many animals, including humans, do not have the ability to regenerate parts of the body after amputation. The study of how planarians are able to regenerate may lead to advances in regenerative medicine for humans and other animals.

The nervous system of planarians is called a *ladder-type* because the two lateral nerve cords plus transverse nerves look like a ladder (Fig. 28.12*d*). Paired **ganglia,** or collections of nerve cells, function as a brain.

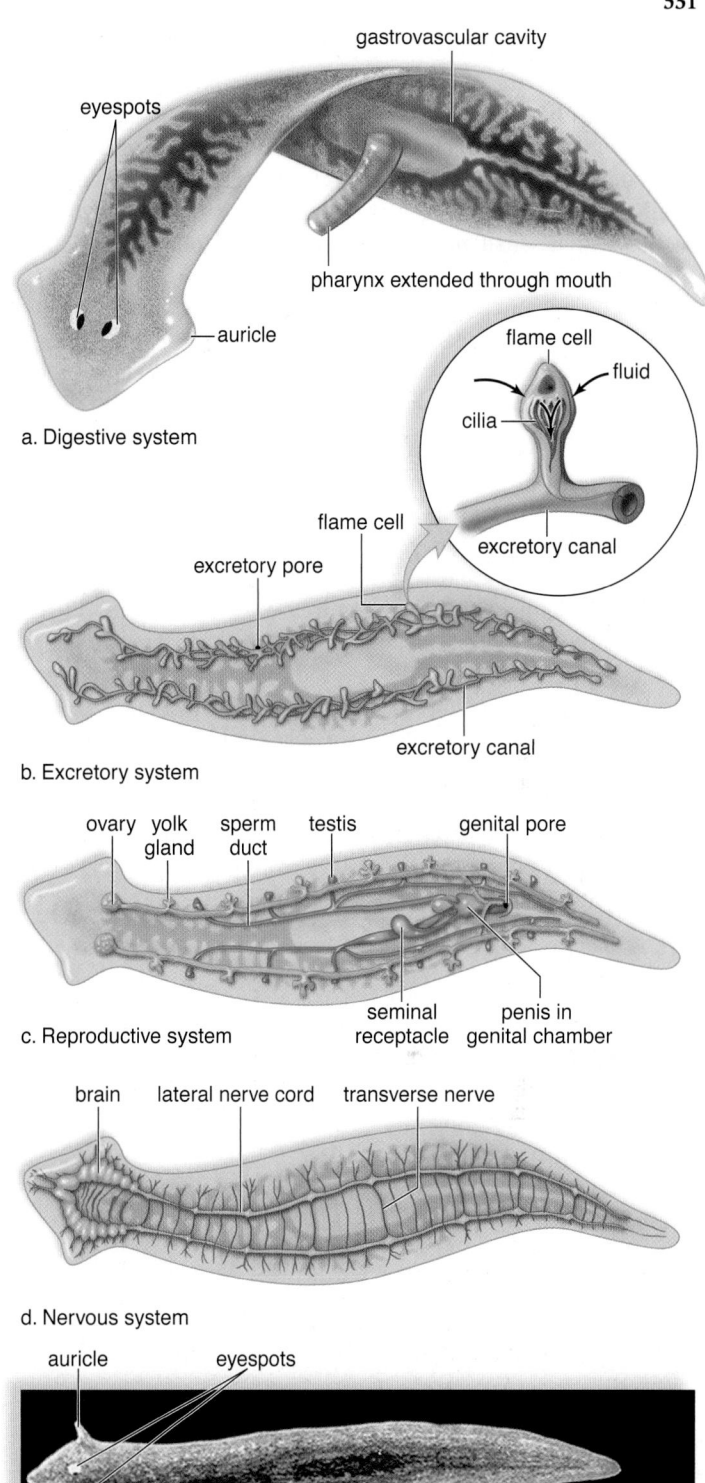

a. Digestive system

b. Excretory system

c. Reproductive system

d. Nervous system

e. Photomicrograph

5 mm

Figure 28.12 Planarian anatomy. **a.** When a planarian extends the pharynx, food is sucked up into a gastrovascular cavity that branches throughout the body. **b.** The excretory system with flame cells is shown in detail. **c.** The reproductive system (shown in pink and blue) has both male and female organs. **d.** The nervous system has a ladderlike appearance. **e.** The photomicrograph shows that a planarian, *Dugesia,* is bilaterally symmetrical and has a head region with eyespots.

Planarians are bilaterally symmetrical and exhibit cephalization. The head of a planarian is bluntly arrow shaped, with lateral extensions called auricles that contain chemosensory cells and tactile cells used to detect potential food sources and enemies. They do not have complex eyes, but rather two pigmented, light-sensitive eyespots on the top of the head that make the worm look "cross-eyed" (Fig. 28.12e).

Planarians have three kinds of muscle layers that allow for quite varied movement: an outer circular layer, an inner longitudinal layer, and a diagonal layer. In larger forms, locomotion is accomplished by the movement of cilia on the ventral and lateral surfaces. Numerous gland cells secrete a mucus upon which the animal glides.

Parasitic Flatworms

Flukes (trematodes) and tapeworms (cestodes) are parasitic flatworms. The bodies of both groups are highly modified for a parasitic mode of life. Flukes and tapeworms feed on nutrients provided by the host, and are covered by a protective tegument, which is a specialized body covering resistant to host digestive juices.

The parasitic flatworms have lost cephalization, and the head with sensory structures has been replaced by an anterior end with hooks and/or suckers for attachment to the host. They no longer hunt for prey, and the nervous system is not well developed. In contrast, a well-developed reproductive system helps ensure transmission to a new host.

Both flukes and tapeworms utilize a secondary, or intermediate, host to transmit offspring from primary host to primary host. The primary host is infected with the sexually mature adult; the secondary host contains the larval stage or stages. Several human diseases are caused by fluke and tapeworm infections.

Flukes. Flukes are named for the organ they inhabit; for example, there are liver, lung, and blood flukes (Fig. 28.13). The almost 11,000 species have an oval to more elongated flattened body about 2.5 cm long. At the anterior end is an oral sucker surrounded by sensory papilla and at least one other sucker used for attachment to a host.

Schistosomiasis is a serious disease caused by a genus of blood fluke, *Schistosoma,* which occurs predominantly in the Middle East, Asia, and Africa. The World Health Organization also lists schistosomiasis as one of several Neglected Tropical Diseases (NTDs) that afflict the poor people in impoverished nations (see the Biological Systems feature, "African Sleeping Sickness," in Chapter 21). In schistosomiasis, female flukes deposit their eggs in small blood vessels close to the lumen of the intestine, and the eggs make their way into the digestive tract by a slow migratory process (Fig. 28.13). After the eggs pass out with the feces, they hatch into tiny larvae that swim about in rice paddies and elsewhere until they enter a particular species of snail. Within the snail, asexual reproduction occurs; sporocysts, which are spore-containing sacs, eventually produce new larval forms that leave the snail. If the larvae penetrate the

Figure 28.13 Life cycle of a blood fluke, *Schistosoma.*
a. Micrograph of *Schistosoma.*
b. *Schistosomiasis,* an infection of humans caused by the blood fluke *Schistosoma,* is an extremely prevalent disease in Egypt—especially since the building of the Aswan High Dam. Standing water in irrigation ditches, combined with unsanitary practices, has created the conditions for widespread infection.

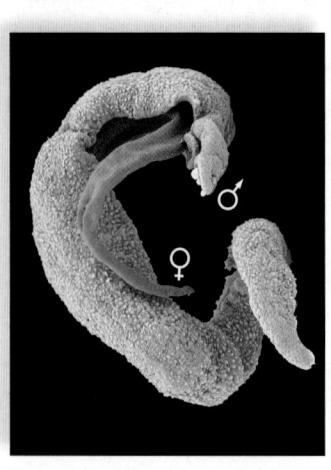

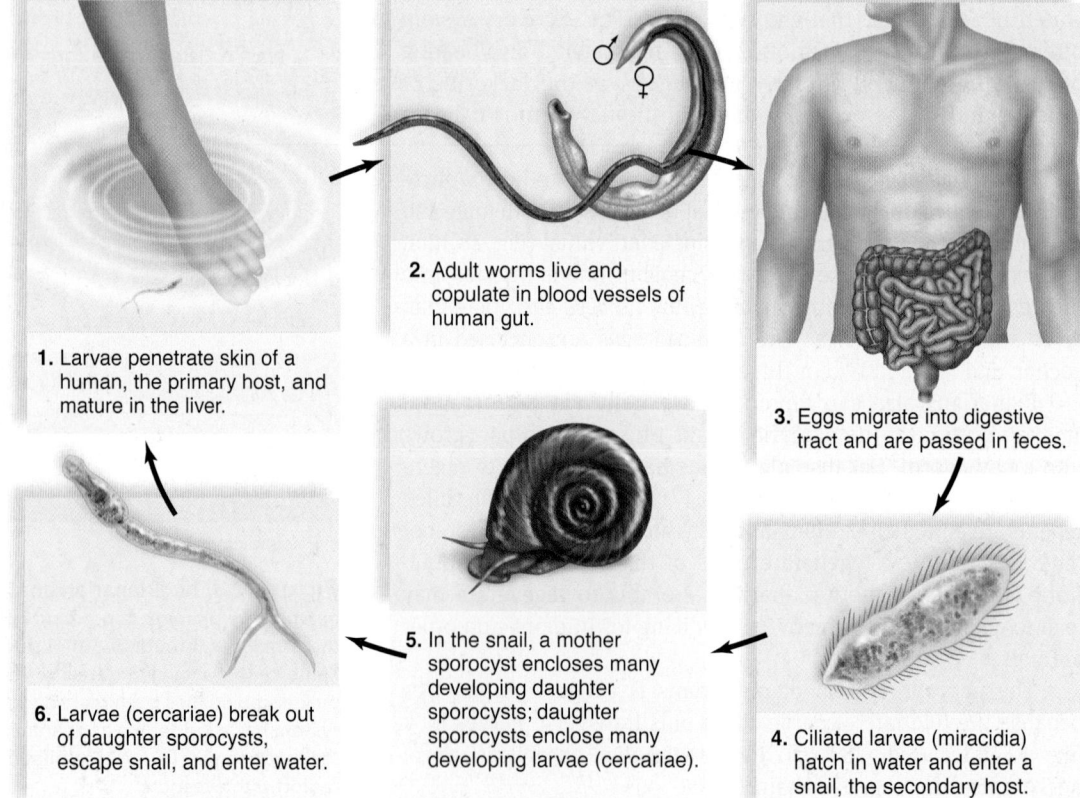

2. Adult worms live and copulate in blood vessels of human gut.

1. Larvae penetrate skin of a human, the primary host, and mature in the liver.

3. Eggs migrate into digestive tract and are passed in feces.

6. Larvae (cercariae) break out of daughter sporocysts, escape snail, and enter water.

5. In the snail, a mother sporocyst encloses many developing daughter sporocysts; daughter sporocysts enclose many developing larvae (cercariae).

4. Ciliated larvae (miracidia) hatch in water and enter a snail, the secondary host.

a.   b.

skin of a human, they begin to mature in the liver and implant themselves in the blood vessels of the small intestine.

The flukes and their eggs can cause dysentery, anemia, bladder inflammation, brain damage, and severe liver complications. Infected persons usually die of secondary diseases brought on by their weakened condition. It is estimated that over 200 million people worldwide are afflicted with this disease.

The Chinese liver fluke, *Clonorchis sinensis*, is a parasite of cats, dogs, pigs, and humans and requires two secondary hosts: a snail and a fish. The adults reside in the liver and deposit their eggs in bile ducts, which carry them to the intestines for elimination in feces. Nonhuman species generally become infected through the fecal route, but humans usually become infected by eating raw fish. A heavy *Clonorchis* infection can cause severe cirrhosis of the liver and death.

Tapeworms.   Tapeworms vary in length from a few millimeters to nearly 20 m. They have a highly modified head region called the **scolex** that contains hooks for attachment to the intestinal wall of the host and suckers for feeding. Tapeworms are hermaphrodites; behind the scolex is a series of reproductive units called **proglottids** that contain a full set of female and male sex organs. The number of proglottids may vary depending on the species.

After fertilization, the organs within a proglottid disintegrate and the proglottids become gravid, or full mature eggs. Gravid proglottids may contain 100,000 eggs. Once mature, the eggs of some species may be released through a pore in the proglottid into the host's intestine where they exit with feces. In other species, the gravid proglottids break off and are eliminated with the feces.

Most tapeworms have complicated life cycles that usually involve several hosts. Figure 28.14 illustrates the life cycle of the pork tapeworm, *Taenia solium*, which has the human as the primary host and the pig as the secondary host. After a pig feeds on feces-contaminated food, the larvae are released. They burrow through the intestinal wall and travel in the bloodstream to finally lodge and encyst in muscle. This **cyst** is a small, hard-walled structure that contains a larva called a bladder worm. Humans become infected with tapeworms when they eat infected meat that has not been thoroughly cooked.

In impoverished nations of the world, more than 50 million people suffer from cysticercosis, an advanced tapeworm infection that interferes with the uptake of nutrients and is therefore particularly serious for growing children. After tainted meat is eaten, the bladder worms break out of the cysts, attach themselves to the intestinal wall, and grow to adulthood. Then the cycle begins again. Generally, tapeworm infections cause diarrhea, weight loss, and fatigue in the primary host.

Rotifers

Rotifers (phylum Rotifera) are trochozoans related to the flatworms. Antonie van Leeuwenhoek (1632–1723), the inventor of the microscope, viewed microscopic rotifers and called them the "wheel animalcules." Rotifers have a crown of cilia, known as the corona, on their heads (Fig. 28.15). When in motion, the corona, which looks like a spinning wheel, serves as an organ of locomotion and also directs food into the mouth.

The approximately 2,000 species primarily live in fresh water; however, some marine and terrestrial forms exist. The

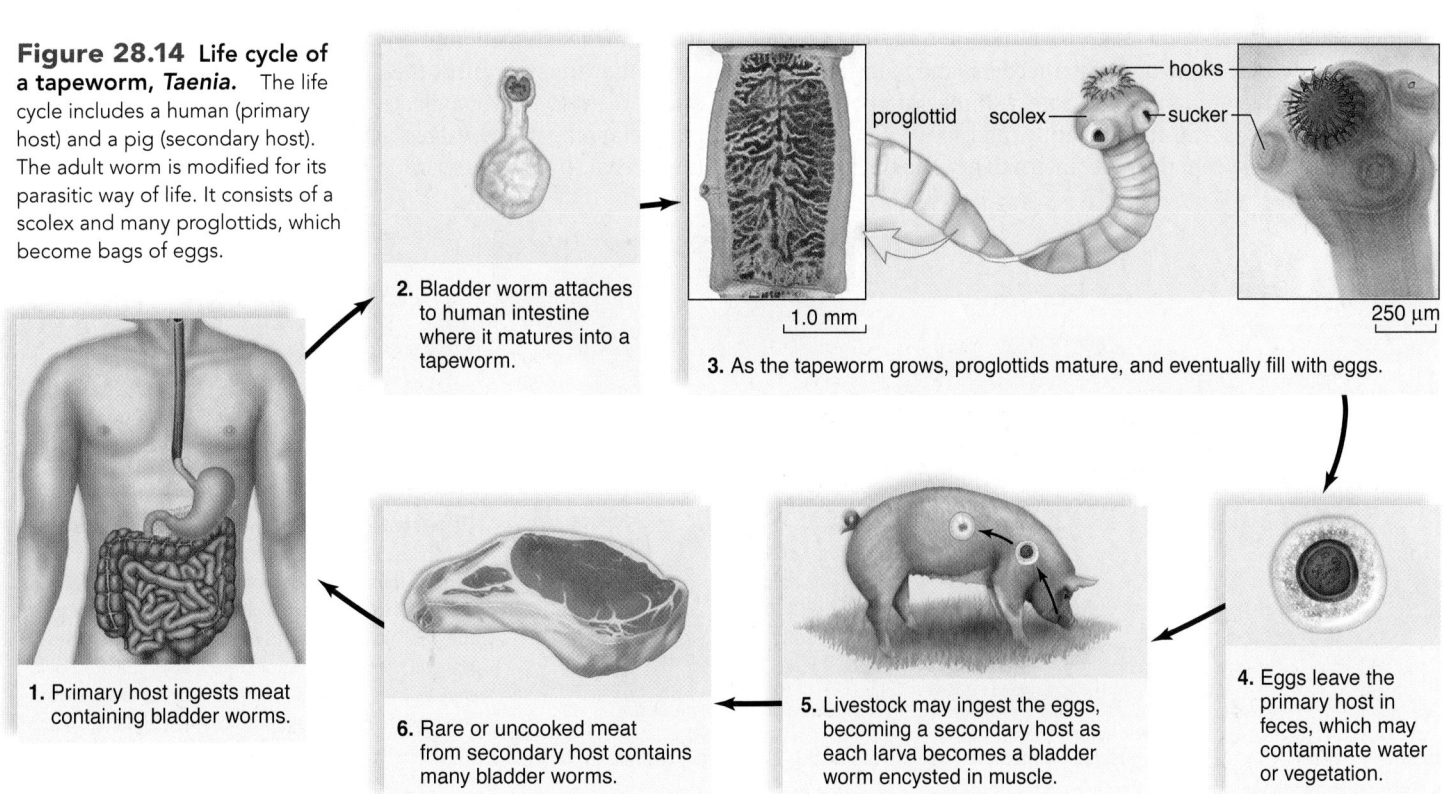

Figure 28.14 Life cycle of a tapeworm, *Taenia*. The life cycle includes a human (primary host) and a pig (secondary host). The adult worm is modified for its parasitic way of life. It consists of a scolex and many proglottids, which become bags of eggs.

2. Bladder worm attaches to human intestine where it matures into a tapeworm.

proglottid   scolex   hooks   sucker

1.0 mm   250 µm

3. As the tapeworm grows, proglottids mature, and eventually fill with eggs.

1. Primary host ingests meat containing bladder worms.

6. Rare or uncooked meat from secondary host contains many bladder worms.

5. Livestock may ingest the eggs, becoming a secondary host as each larva becomes a bladder worm encysted in muscle.

4. Eggs leave the primary host in feces, which may contaminate water or vegetation.

Figure 28.15
Rotifer. Rotifers are microscopic animals only 0.1–3 mm in length. The beating of cilia on two lobes at the anterior end of the animal gives the impression of a pair of spinning wheels.

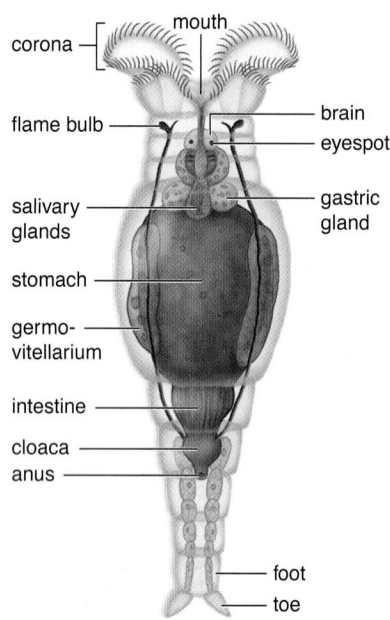

corona — mouth
flame bulb — brain
— eyespot
salivary glands — gastric gland
stomach
germo-vitellarium
intestine
cloaca
anus
— foot
— toe

majority of rotifers are transparent, but some are very colorful. Many species of rotifers can desiccate during harsh conditions and remain dormant for lengthy periods of time.

Molluscs

The **molluscs** (phylum Mollusca) are the second most numerous group of animals. They inhabit a variety of environments, including marine, freshwater, and terrestrial habitats. This diverse phylum includes chitons, limpets, slugs, snails, abalones, conchs, nudibranchs, clams, oysters, scallops, squid, and octopuses. Molluscs vary in size from microscopic to the giant squid, which can attain lengths of over 20 m and weigh over 450 kg. The group includes herbivores, carnivores, filter feeders, and parasites.

Although diverse, molluscs share a three-part body plan consisting of the visceral mass, mantle, and foot (Fig. 28.16a). The visceral mass contains the internal organs, including a highly specialized digestive tract, paired kidneys, and reproductive organs. The **mantle** is a covering that lies to either side of, but does not completely enclose, the visceral mass. It may secrete a shell and/or contribute to the development of gills or lungs. The space between the folds of the mantle is called the mantle cavity. The foot is a muscular organ that may be adapted for locomotion, attachment, food capture, or a combination of functions. Another feature often present in molluscs is a rasping, tonguelike *radula,* an organ that bears many rows of teeth and is used to obtain food (Fig. 28.16b).

The true coelom is reduced in molluscs and largely limited to the region around the heart. Most molluscs have an open circulatory system. The heart pumps blood, more properly called hemolymph, through vessels into sinuses (cavities) collectively called a **hemocoel** [Gk. *haima,* blood, and *koiloma,* cavity]. Blue hemocyanin, rather than red hemoglobin, is the oxygen-carrying pigment. Nutrients and oxygen diffuse into the tissues from these sinuses instead of being carried into the tissues by capillaries, the microscopic blood vessels present in animals with closed circulatory systems.

The nervous system of molluscs consists of several ganglia connected by nerve cords. The amount of cephalization and sensory organs varies from nonexistent in clams to complex in squid and octopuses. The molluscs also exhibit variation in mobility. Oysters are sessile, snails are extremely slow moving, and squid are fast-moving, active predators.

Bivalves

Clams, oysters, shipworms, mussels, and scallops are all **bivalves** (class Bivalvia) with a two-part shell that is hinged and closed by powerful muscles (Fig. 28.17). They have no head, no radula, and very little cephalization. Clams use their hatchet-shaped foot for burrowing in sandy or muddy soil, and mussels use their foot to produce threads that attach them to nearby objects. Scallops both burrow and swim; rapid clapping of the valves releases water in spurts and causes the animal to move forward in a jerky fashion for a few feet.

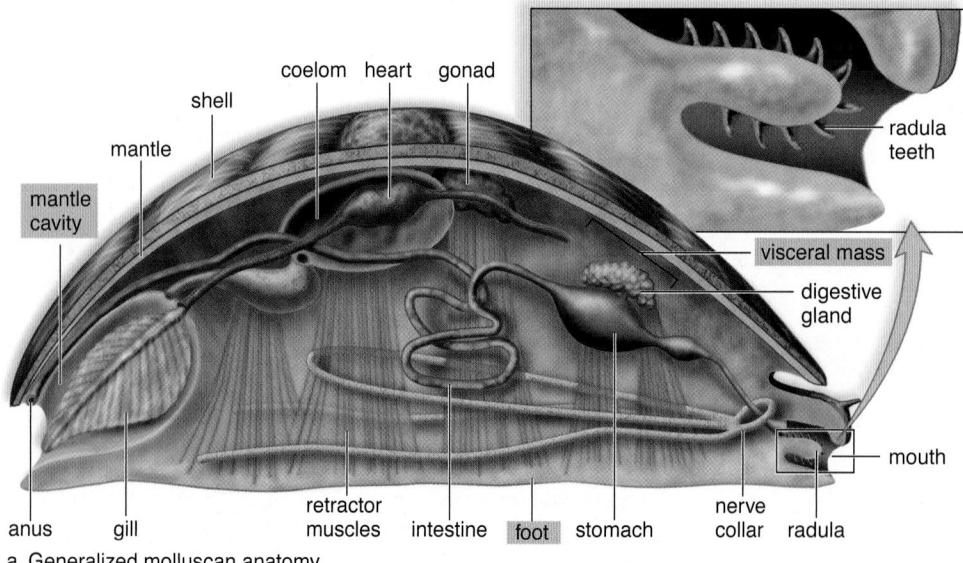

coelom heart gonad
shell
mantle
mantle cavity
radula teeth
visceral mass
digestive gland
mouth
anus   gill   retractor muscles   intestine   foot   stomach   nerve collar   radula

a. Generalized molluscan anatomy

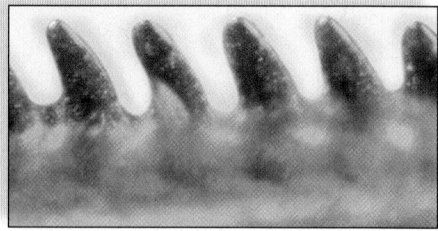

b. Radula

Figure 28.16 Body plan of molluscs.
a. Molluscs have a three-part body consisting of a ventral, muscular foot that is specialized for various means of locomotion; a visceral mass that includes the internal organs; and a mantle that covers the visceral mass and may secrete a shell. Ciliated gills may lie in the mantle cavity and direct food toward the mouth. **b.** In the mouth of many molluscs, such as snails, the radula is a tonguelike organ that bears rows of tiny teeth that point backward, shown here in drawing (a) and a micrograph.

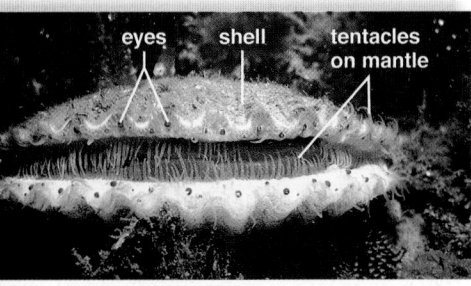

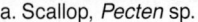

a. Scallop, *Pecten* sp.

b. Mussels, *Mytilus edulis*

Figure 28.17 Bivalve diversity. Bivalves have a two-part shell. **a.** Scallops clap their valves and swim by jet propulsion. This scallop has sensory organs consisting of blue eyes and tentacles along the mantle edges. **b.** Mussels form dense beds in the intertidal zone of northern shores. **c.** In this drawing of a clam, the mantle has been removed from one side. Follow the path of food from the incurrent siphon to the gills, the mouth, the stomach, the intestine, the anus, and the excurrent siphon. Locate the three ganglia: anterior, foot, and posterior. The heart lies in the reduced coelom.

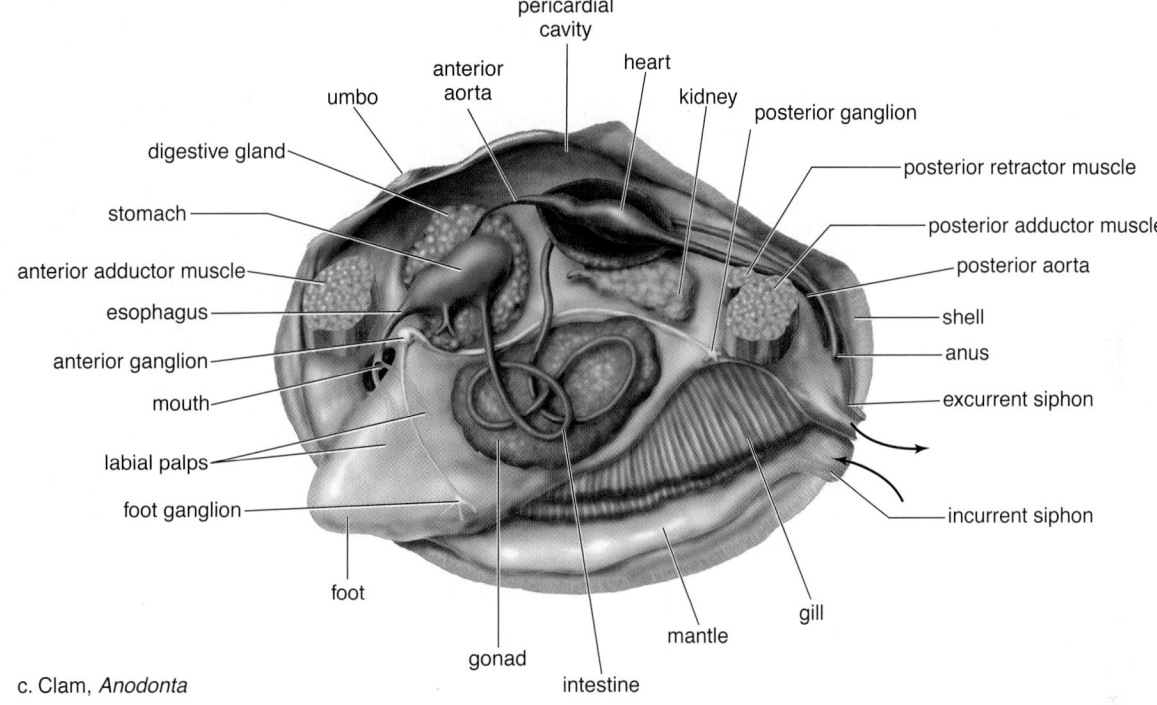

c. Clam, *Anodonta*

In freshwater clams such as *Anodonta* (Fig. 28.17c), the shell, secreted by the mantle, is composed of protein and calcium carbonate with an inner layer, called mother of pearl. If a foreign body is placed between the mantle and the shell, pearls form as concentric layers of shell are deposited about the particle. The compressed muscular foot of a clam projects ventrally from the shell; by expanding the tip of the foot and pulling the body after it, the clam moves forward.

Video Clam Locomotion

Within the mantle cavity, the ciliated gills hang down on either side of the visceral mass. The beating of the cilia causes water to enter the mantle cavity by way of the incurrent siphon and to exit by way of the excurrent siphon. The clam is a filter feeder; small particles in this constant stream of water adhere to the gills, and ciliary action sweeps them toward the mouth.

The mouth leads to a stomach and then to an intestine, which coils about in the visceral mass before going right through the heart and ending in an anus. The anus empties at the excurrent siphon. An accessory organ of digestion called a digestive gland is also present. The heart lies just below the hump of the shell within the pericardial cavity, the only remains of the coelom. The circulatory system is open; the heart pumps hemolymph into vessels that open into the *hemocoel*. The nervous system is composed of three pairs of ganglia (located anteriorly, posteriorly, and in the foot), which are connected by nerves.

Two excretory kidneys lie just below the heart and remove waste from the pericardial cavity for excretion into the mantle cavity. The clam excretes ammonia (NH_3), a toxic substance that requires the excretion of water at the same time.

In freshwater clams, the sexes are separate, and fertilization is internal. Fertilized eggs develop into specialized larvae and are released from the clam. Some larvae attach to the gills of a fish and become a parasite before they sink to the bottom and develop into a clam. Certain clams and annelids have the same type of larva, namely the trocophore larva, and this reinforces the evolutionary relationship between molluscs and annelids.

Other Molluscs

The **gastropods** [Gk. *gastros*, stomach, and *podos*, foot], the largest class of molluscs, include slugs, snails, whelks, conchs, limpets, and nudibranchs (Fig. 28.18a–c). Most are marine; however, slugs and garden snails are adapted to terrestrial environments (Fig. 28.18c).

Video Snail's Pace

Gastropods have an elongated, flattened foot, and most, except for slugs and nudibranchs, have a one-piece coiled shell that protects the visceral mass. The anterior end bears

Figure 28.18 Gastropod and cephalopod diversity.
a, b. Gastropods have the three parts of a mollusc, and the foot is muscular, elongated, and flattened. **c, d.** Cephalopods have tentacles and/or arms in place of a head. Speed suits their predatory lifestyle, and only the chambered nautilus has a shell.

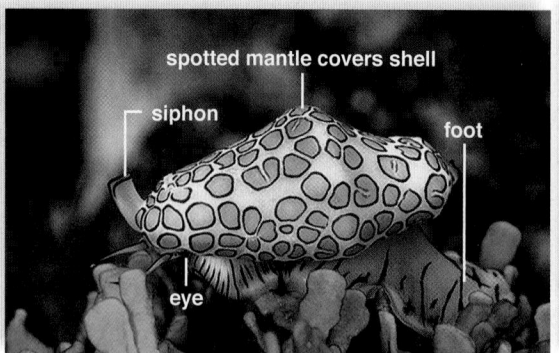

a. Flamingo tongue shell, *Cyphoma gibbosum*

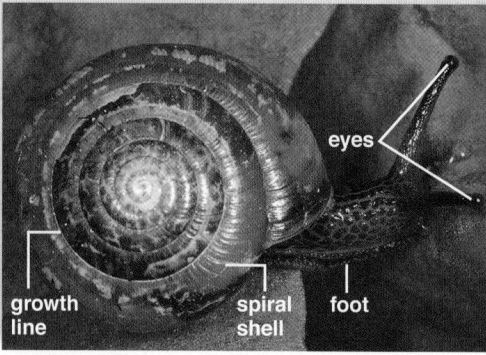

b. Land snail, *Helix aspersa*

c. Chambered nautilus, *Nautilus belauensis*

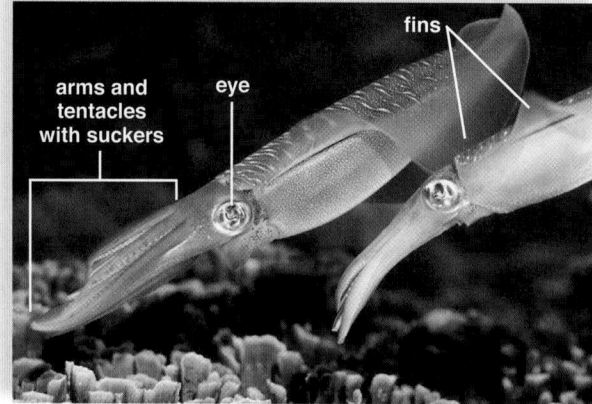

d. Bigfin reef squid, *Sepioteuthis lessoniana*

a well-developed head region with a cerebral ganglion and eyes on the ends of tentacles. Land snails are hermaphroditic; when two snails meet, they shoot calcareous darts into each other's body wall as a part of premating behavior. Then each inserts a penis into the vagina of the other to provide sperm for the future fertilization of eggs, which are deposited in the soil. Development proceeds directly without the formation of larvae.

Cephalopods [Gk. *kaphale*, head, and *podos*, foot] range in length from 2 cm to 20 m as in the giant squid, *Architeuthis*. Cephalopod means head-footed; both squids and octopi can squeeze their mantle cavity so that water is forced out through a funnel, propelling them by jet propulsion (Fig. 28.18*d*). Also, the tentacles and arms that circle the head capture prey by adhesive secretions or by suckers. A powerful, parrotlike beak is used to tear prey apart. They have well-developed sense organs, including eyes that are similar to those of vertebrates and focus like a camera. Cephalopods, particularly octopuses, have well-developed brains and show a remarkable capacity for learning.

Annelids

Annelids (phylum Annelida [L. *anellus*, little ring]), which are sometimes called the segmented worms, vary in size from microscopic to tropical earthworms that can be over 4 m long. The most familiar members of this group are earthworms, marine worms, and leeches.

Annelids are the only trochozoan with segmentation and a well-developed coelom. **Segmentation** is the repetition of body parts along the length of the body. The well-developed coelom is fluid-filled and serves as a supportive *hydrostatic skeleton*. A hydrostatic skeleton, along with partitioning of the coelom, permits independent movement of each body segment. Instead of just burrowing in the mud, an annelid can crawl on a surface.

Setae [L. *seta*, bristle] are bristles that protrude from the body wall, can anchor the worm, and help it move. The *oligochaetes* are annelids with few setae, and the *polychaetes* are annelids with many setae.

Earthworms

The common earthworm, *Lumbricus terrestris,* is an oligochaete (Fig. 28.19). Earthworm setae protrude in pairs directly from the surface of the body. Locomotion, which is accomplished section by section, uses muscle contraction and the setae. When longitudinal muscles contract, segments bulge and their setae protrude into the soil; then, when circular muscles contract, the setae are withdrawn, and these segments move forward.

Earthworms reside in soil where there is adequate moisture to keep the body wall moist for gas exchange. They are scavengers that feed on leaves or any other organic matter conveniently taken into the mouth along with dirt. Segmentation and a complete digestive tract have led to increased specialization of digestive system components. Food drawn into the mouth by the action of the muscular pharynx is stored in a crop and ground up in a thick, muscular gizzard (Fig. 28.19*a*). Digestion and absorption occur in a long intestine whose dorsal surface has an expanded region called a typhlosole that increases the surface for absorption. Waste is eliminated through the anus.

Earthworm segmentation, which is obvious externally, is also internally evidenced by septa that occur between segments. The long, ventral nerve cord leading from the brain has ganglionic swellings and lateral nerves in each segment. The excretory system consists of paired **nephridia** [Gk. *nephros*, kidney], which are coiled tubules in each segment (Fig. 28.19*b*). A nephridium has two openings: One is a ciliated funnel that collects coelomic fluid, and the other is an exit in the body wall. Between the two openings is a convoluted region where waste material is removed from the blood vessels about the tubule.

Annelids have a *closed circulatory system,* which means that the blood is always contained in blood vessels that run the length of the body. Oxygenated blood moves anteriorly in the dorsal

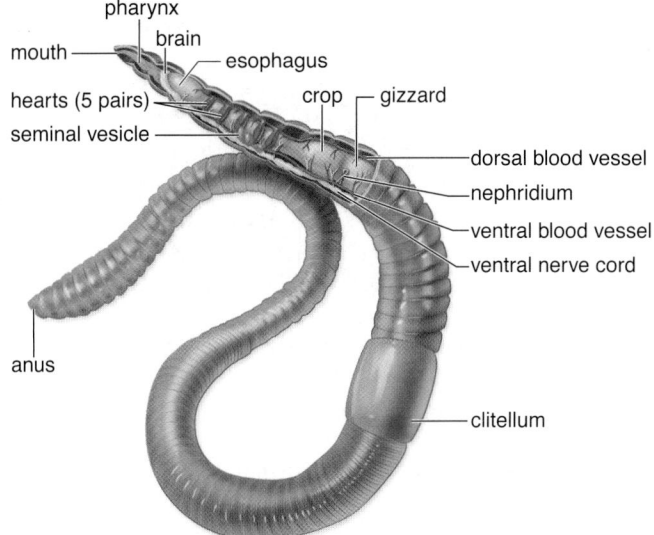

pharynx
brain
mouth
esophagus
hearts (5 pairs)
crop — gizzard
seminal vesicle
dorsal blood vessel
nephridium
ventral blood vessel
ventral nerve cord
anus
clitellum

a.

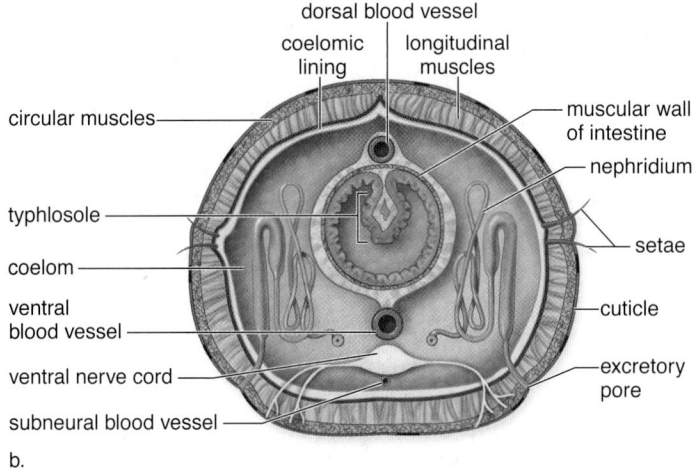

dorsal blood vessel
coelomic lining
longitudinal muscles
circular muscles
muscular wall of intestine
nephridium
typhlosole
setae
coelom
ventral blood vessel
cuticle
ventral nerve cord
excretory pore
subneural blood vessel

b.

anterior end
clitellum
clitellum
anterior end

c.

Figure 28.19 Earthworm, *Lumbricus terrestris*.
a. In the longitudinal section, note the specialized parts of the digestive tract. **b.** In cross section, note the spacious coelom, the paired setae and nephridia, and a ventral nerve cord that has branches in each segment. **c.** When earthworms mate, they are held in place by a mucus secreted by the clitellum. The worms are hermaphroditic, and when mating, sperm pass from the seminal vesicles of each to the seminal receptacles of the other.

blood vessel, which connects to the ventral blood vessel by five pairs of muscular vessels called "hearts." Pulsations of the dorsal blood vessel and the five pairs of hearts are responsible for blood flow. As the ventral vessel takes the blood toward the posterior regions of the worm's body, it gives off branches in every segment.

Earthworms are *hermaphroditic;* the male organs are the testes, the seminal vesicles, and the sperm ducts, and the female organs are the ovaries, the oviducts, and the seminal receptacles. During mating, two worms lie parallel to each other facing in opposite directions (Fig. 28.19c). The fused midbody segment, called a *clitellum,* secretes mucus, protecting the sperm from drying out as they pass between the worms. After the worms separate, the clitellum of each produces a slime tube, which is moved along over the anterior end by muscular contractions. As it passes, eggs and the sperm received earlier are deposited, and fertilization occurs. The slime tube then forms a cocoon to protect the hatched worms as they develop. There is no larval stage in earthworms.

Other Annelids

Approximately two-thirds of annelids are marine polychaetes. Polychaetes have setae in bundles on parapodia [Gk. *para,* beside, and *podos,* foot], which are paddlelike appendages found on most segments. These are used not only in swimming but also as respiratory organs, where the expanded surface area allows for exchange of dissolved gases. Some polychaetes are free-swimming, but the majority live in crevices or burrow into the ocean bottom. Clam worms, such as *Nereis* (Fig. 28.20a), are predators. They prey on crustaceans and other small animals, which are captured by a pair of strong chitinous jaws that

Figure 28.20 Polychaete diversity. **a.** Clam worms are predaceous polychaetes that undergo cephalization. Note also the parapodia, which are used for swimming and as respiratory organs. **b.** Christmas tree worms (a type of tube worm) are sessile feeders whose radioles (ciliated mouth appendages) spiral as shown here. **c.** Medical leech is a blood sucker.

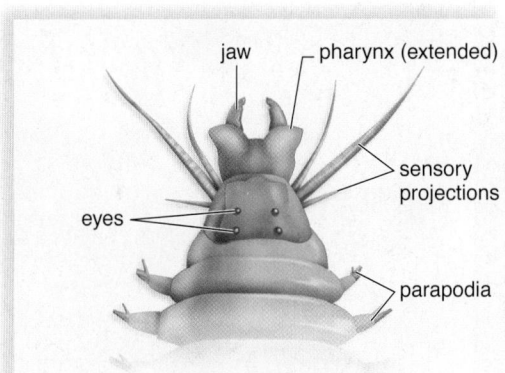

jaw
pharynx (extended)
sensory projections
eyes
parapodia

a. Clam worm, *Nereis succinea*

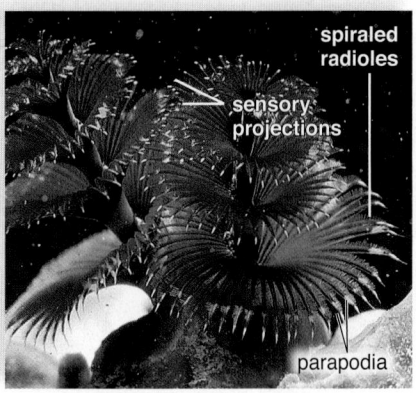

spiraled radioles
sensory projections
parapodia

b. Christmas tree worm, *Spirobranchus giganteus*

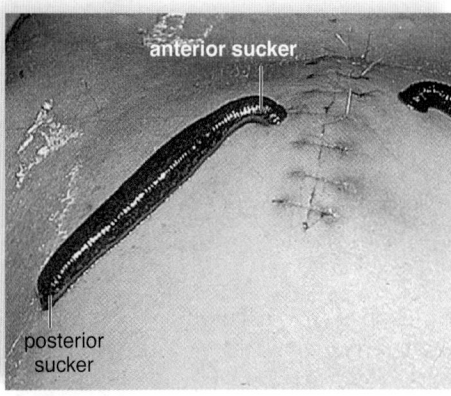

anterior sucker
posterior sucker

c. Medicinal leech, *Hirudo medicinalis*

extend with a part of the pharynx when the animal is feeding. Associated with its way of life, *Nereis* is cephalized, having a head region with eyes and other sense organs.

Another group of marine polychaetes are sessile tube worms, with *radioles* (ciliated mouth appendages) used to gather food (Fig 28.20b). Christmas tree worms, fan worms, and featherduster worms all have radioles. In featherduster worms, the beautiful radioles cause the animal to look like an old fashioned feather duster.

Polychaetes have breeding seasons, and only during these times do the worms have sex organs. In *Nereis,* many worms simultaneously shed a portion of their bodies containing either eggs or sperm, and these float to the surface where fertilization takes place. The zygote rapidly develops into a trochophore larva.

Leeches are annelids that normally live in freshwater habitats. They range in size from less than 2 cm to the medicinal leech, which can be 20 cm in length. They exhibit a variety of patterns and colors but most are brown or olive green. The body of a leech is flattened dorsoventrally. They have the same body plan as other annelids, but they have no setae and each body ring has several transverse grooves.

While some are free-living, most leeches are fluid feeders that attach themselves to open wounds. Among their modifications are two suckers, a small oral one around the mouth and a posterior one. Some bloodsuckers, such as the medicinal leech, can cut through tissue. Leeches are able to keep blood flowing by means of hirudin, a powerful anti-coagulant in their saliva. Medicinal leeches have been used for centuries in blood-letting and other procedures. Today, they are used in reconstructive surgery for severed digits and in plastic surgery (Fig. 28.20c).

Check Your Progress 28.3

1. List the characteristics that unite the flatworms, molluscs, and annelids.
2. Compare features of the flatworm, mollusc, and annelid body cavity, digestive tract, and circulatory system.
3. Describe the life cycle of two lophotrochozoan parasites.
4. List three phyla that feed with lophophores.

28.4 Diversity of the Ecdysozoans

Learning Outcomes

Upon completion of this section, you should be able to
1. Identify the characteristics unique to ecdysozoans.
2. Compare three key characteristics of roundworms to that of arthropods.
3. Describe the five characteristics responsible for the success of the arthropods.

The **ecdysozoans** [Gk. *ecdysis,* stripping off] are another large group of protostomes. Both roundworms and arthropods, which belong to this group, periodically shed their outer covering.

Roundworms

Roundworms (phylum Nematoda) are nonsegmented worms that are prevalent in almost any environment. Generally, nematodes are colorless and range in size from microscopic to exceeding 1 m in length. Internal organs, including the tubular reproductive organs, lie within the pseudocoelom. A **pseudocoelom** [Gk. *Pseudes,* false, and *koiloma,* cavity] is a body cavity that is incompletely lined by mesoderm. In other words, mesoderm occurs inside the body wall but not around the digestive cavity (gut).

Nematodes have developed a variety of lifestyles from free-living to parasitic. One species, *Caenorhabditis elegans,* a free-living nematode, is a model animal used in genetics and developmental biology.

Parasitic Roundworms

An *Ascaris* infection typically occurs in humans, cats, dogs, pigs, and a number of other vertebrates. As in other nematodes, *Ascaris* males tend to be smaller (15–31 cm long) than females (20–49 cm long) (Fig. 28.21a). In males, the posterior end is curved and comes to a point. Both sexes move by means of a characteristic whiplike motion using longitudinal muscles that lie next to the body wall.

Figure 28.21 Roundworm diversity.
a. *Ascaris,* a common cause of a roundworm infection in humans. **b.** Encysted *Trichinella* larva in muscle. **c.** A filarial worm infection causes elephantiasis, which is characterized by a swollen body part when the worms block lymphatic vessels.

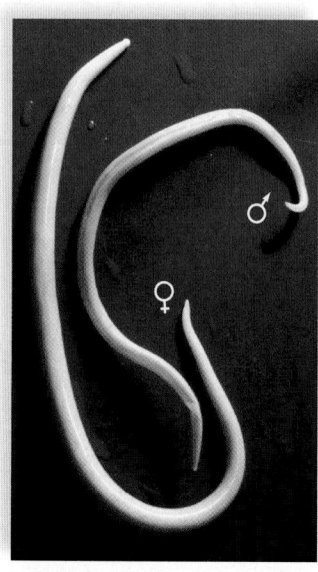 a. 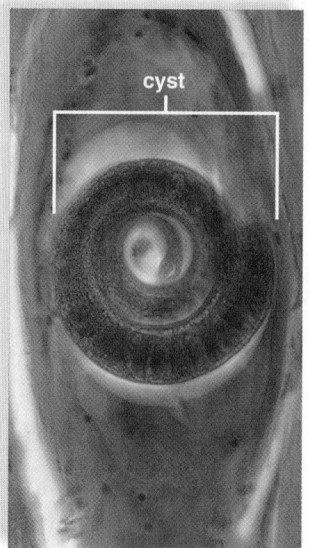 b. SEM 400× 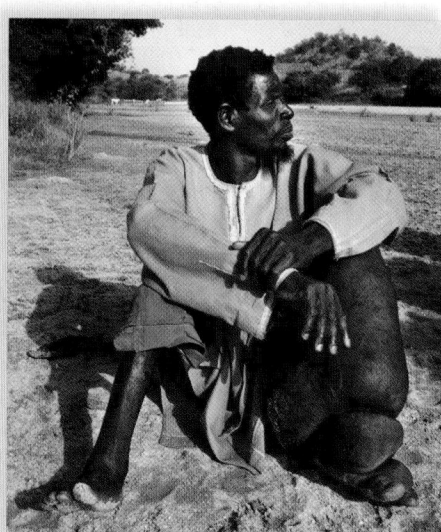 c.

A typical female *Ascaris* produces over 200,000 eggs daily. The eggs are eliminated with the host's feces and can remain viable in the soil for many months. Eggs enter the body via uncooked vegetables, soiled fingers, or ingested fecal material and hatch in the intestines. The juveniles make their way into the veins and lymphatic vessels and are carried to the heart and lungs. From the lungs, the larvae travel up the trachea, where they are swallowed and eventually reach the intestines. There, the larvae mature and begin feeding on intestinal contents.

The symptoms of an *Ascaris* infection depend on the stage of the infection. Larval *Ascaris* in the lungs can cause pneumonia-like symptoms. In the intestines, *Ascaris* can cause malnutrition; blockage of the bile duct, pancreatic duct, and appendix; and poor health.

Trichinosis, caused by *Trichinella spiralis,* is a serious infection that humans can contract when they eat rare pork containing encysted *Trichinella* larvae. After maturation, the female adult burrows into the wall of the small intestine and produces living offspring that are carried by the bloodstream to the skeletal muscles, where they encyst (Fig. 28.21*b*). Heavy infections can be painful and lethal.

Filarial worms, a type of roundworm, cause various diseases. In the United States, mosquitoes transmit the larvae of a parasitic filarial worm to dogs. Because the worms live in the heart and the arteries that serve the lungs, the infection is called *heartworm disease.* The condition can be fatal; therefore, heartworm medicine is recommended as a preventive measure for all dogs.

Elephantiasis is a disease of humans caused by *filiariasis,* or infection by the filarial worm, *Wuchereria bancrofti.* Restricted to tropical areas of Africa, this parasite not only infects mosquitos, but also uses the mosquito as a vector to transmit the disease to humans. Because the adult filarial worms reside in lymphatic vessels, the removal of lymph fluid is impeded, and the limbs of an infected person may swell to a monstrous size

resulting in elephantiasis (Fig. 28.21*c*). Elephantiasis is treatable in its early stages but usually not after scar tissue has blocked lymphatic vessels. The World Health Organization lists elephantiasis as one of the 16 Neglected Tropical Diseases (NTDs), which affect impoverished nations. It is estimated that 120 million people in subtropical and tropical areas of the world have filariasis.

Pinworms are the most common nematode parasite in the United States. The adult parasites live in the cecum and large intestine. Females migrate to the anal region at night and lay their eggs. Scratching the resultant itch can contaminate hands, clothes, and bedding. The eggs are swallowed, and the life cycle begins again.

Arthropods

The **arthropods** (phylum Arthropoda [Gk. *arthron,* joint, and *podos,* foot]) are a very large group of protostomes that have exoskeletons and jointed appendages. The phylum Arthropoda includes insects, crustaceans, millipedes, centipedes, spiders, and scorpions. The parasitic mite measures less than 0.1 mm in length, while the Japanese crab measures up to 4 m in length. Arthropods, which also occupy every type of habitat, are considered the most successful group of all the animals.

Characteristics of Arthropods

The remarkable success of arthropods is dependent on five characteristics: an exoskeleton, segmentation, a well-developed nervous system, a variety of respiratory systems, and a life cycle that includes metamorphosis.

Exoskeleton. Arthropods feature a rigid but jointed exoskeleton (Fig. 28.22*a, b*). The exoskeleton is composed primarily of chitin, a strong, flexible, nitrogenous polysaccharide. The exoskeleton serves many functions, including protection,

Figure 28.22 Arthropod skeleton and eye. a. The joint in an arthropod skeleton is a region where the cuticle is thinner and not as hard as the rest of the cuticle. The direction of movement is toward the flexor muscle or the extensor muscle, whichever one has contracted. **b.** The exoskeleton is secreted by the epidermis and consists of the endocuticle; the exocuticle, hardened by the deposition of calcium carbonate; and the epicuticle, a waxy layer. Chitin makes up the bulk of the exo- and endocuticles. **c.** Because the exoskeleton is nonliving, it must be shed through a process called molting for the arthropod to grow. **d.** Arthropods have a compound eye that contains many individual units, each with its own lens and photoreceptors.

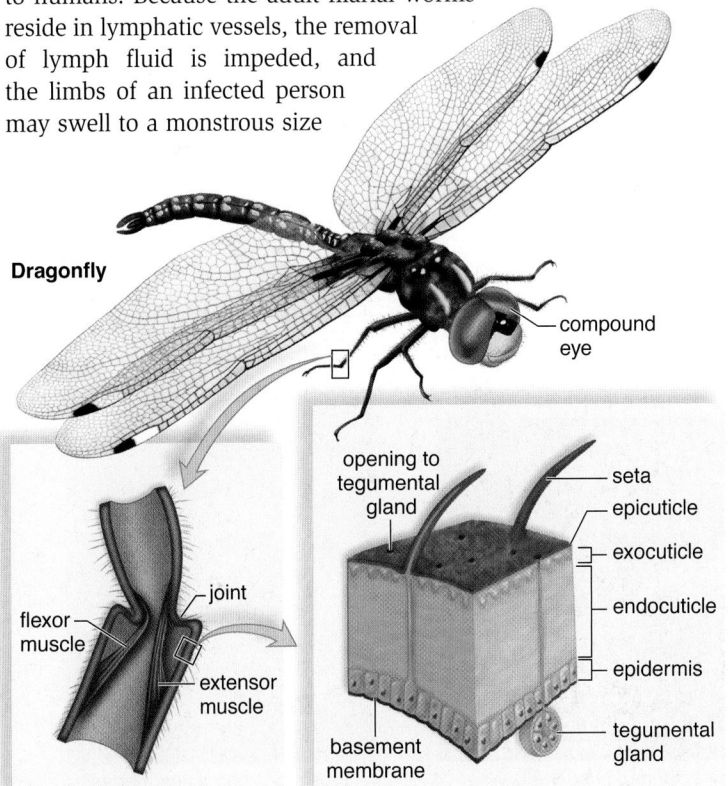

Dragonfly

compound eye

opening to tegumental gland

seta
epicuticle
exocuticle
endocuticle
epidermis
tegumental gland

flexor muscle
joint
extensor muscle

basement membrane

a. Joint movement

b. Exoskeleton composition

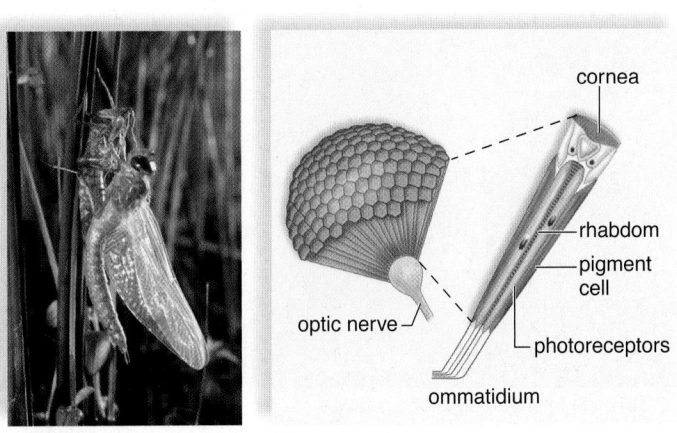

cornea
rhabdom
pigment cell
photoreceptors
optic nerve
ommatidium

c. Molting

d. Compound eye

attachment for muscles, locomotion, and prevention of desic-cation. However, because it is hard and nonexpandable, arthro-pods must molt, or shed, the exoskeleton as they grow larger. Arthropods have this in common with other ecdysozoans.

Before molting, the body secretes a new, larger exoskel-eton, which is soft and wrinkled, underneath the old one. After enzymes partially dissolve and weaken the old exoskeleton, the animal breaks it open and wriggles out. The new exoskeleton then quickly expands and hardens (Fig. 28.22c).

Segmentation.   Segmentation is readily apparent because each segment has a pair of jointed appendages, even though certain segments are fused into a head, thorax, and abdomen. The jointed appendages of arthropods are basically hollow tubes moved by muscles. Typically, the appendages are highly adapted for a particular function, such as food gathering, repro-duction, and locomotion. In addition, many appendages are associated with sensory structures and used for tactile purposes.

Nervous system.   Arthropods have a brain and a ventral nerve cord. The head bears various types of sense organs, including eyes of two types—simple and compound. The com-pound eye is composed of many complete visual units, each of which operates independently (Fig. 28.22d). The lens of each visual unit focuses an image on the light-sensitive membranes of a small number of photoreceptors within that unit. The sim-ple eye, like that of vertebrates, has a single lens that brings the image to focus onto many receptors, each of which receives only a portion of the image.

In addition to sight, many arthropods have well-developed touch, smell, taste, balance, and hearing. Arthropods display many complex behaviors and methods of communication.

Variety of respiratory organs.   Marine forms use gills, which are vascularized, highly convoluted, thin-walled tissue special-ized for gas exchange. Terrestrial forms have book lungs (e.g., spi-ders) or air tubes called tracheae [L. *trachea*, windpipe]. Tracheae serve as a rapid way to transport oxygen directly to the cells.

Metamorphosis.   Many arthropods undergo a drastic change in form and physiology that occurs as an immature stage, or larva, becomes an adult. This change is termed **metamorphosis**

[Gk. *meta*, implying change, and *morphe*, shape, form]. Among arthropods, the larva eats different food and lives in a differ-ent environment than does the adult. For example, larval crabs live among and feed on plankton, while adult crabs are bottom dwellers that catch live prey or scavenge dead organic mat-ter. Among insects, such as butterflies, the caterpillar feeds on leafy vegetation, while the adult feeds on nectar. The result is reduced competition for resources, increasing the animal's over-all fitness.

While keeping these five arthropod features in mind, let's look at representatives of this large and varied phylum.

Crustaceans

The name **crustacean** is derived from their hard, crusty exoskel-eton, which contains calcium carbonate in addition to the typi-cal chitin. Although crustaceans are extremely diverse, the head usually bears a pair of compound eyes and five pairs of append-ages. The first two pairs, called antennae and antennules, lie in front of the mouth and have sensory functions. The other three pairs (mandibles and first and second maxillae) lie behind the mouth and are mouthparts used in feeding. Biramous [Gk. *bi-*, two, and *ramus*, a branch] appendages on the thorax and abdo-men are segmentally arranged; one branch is the gill branch, and the other is the leg branch.

The majority of crustaceans live in marine and aquatic environments (Fig. 28.23). Decapods, which are the most famil-iar and numerous crustaceans, include lobsters, crabs, crayfish, hermit crabs, and shrimp. These animals have a thorax that bears five pairs of walking appendages. Typically, the gills are positioned above the walking legs. The first pair of walking legs may be modified as claws.

Video
Voice of the Lobster

Copepods and krill are small, free-swimming crustaceans that live in the water where they feed on algae. In the marine environment, they serve as food for fishes, sharks, and whales. They are so numerous that, despite their small size, some believe they are harvestable as food. Barnacles are sessile crus-taceans with a thick, heavy, protective shell. Barnacles can live on wharf pilings, ship hulls, seaside rocks, and even the bodies of whales. They begin life as free-swimming larvae, but they

Figure 28.23 Crustacean diversity.   Crabs **(a)** and shrimp **(b)** are decapods—they have five pairs of walking legs. Shrimp resemble crayfish more closely than crabs, which have a reduced abdomen. The gooseneck barnacle **(c)** is attached to an object by a long stalk.

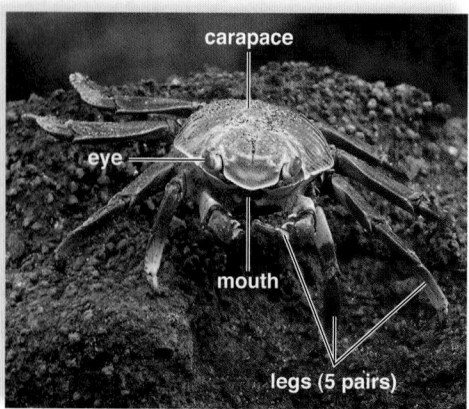

a. Sally lightfoot crab, *Grapsus grapsus*

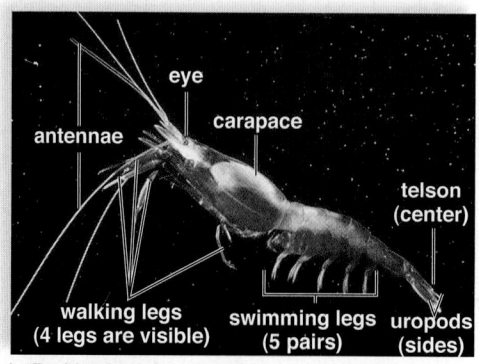

b. Red-backed cleaning shrimp, *Lysmata grasbhami*

c. Gooseneck barnacles, *Lepas anatifera*

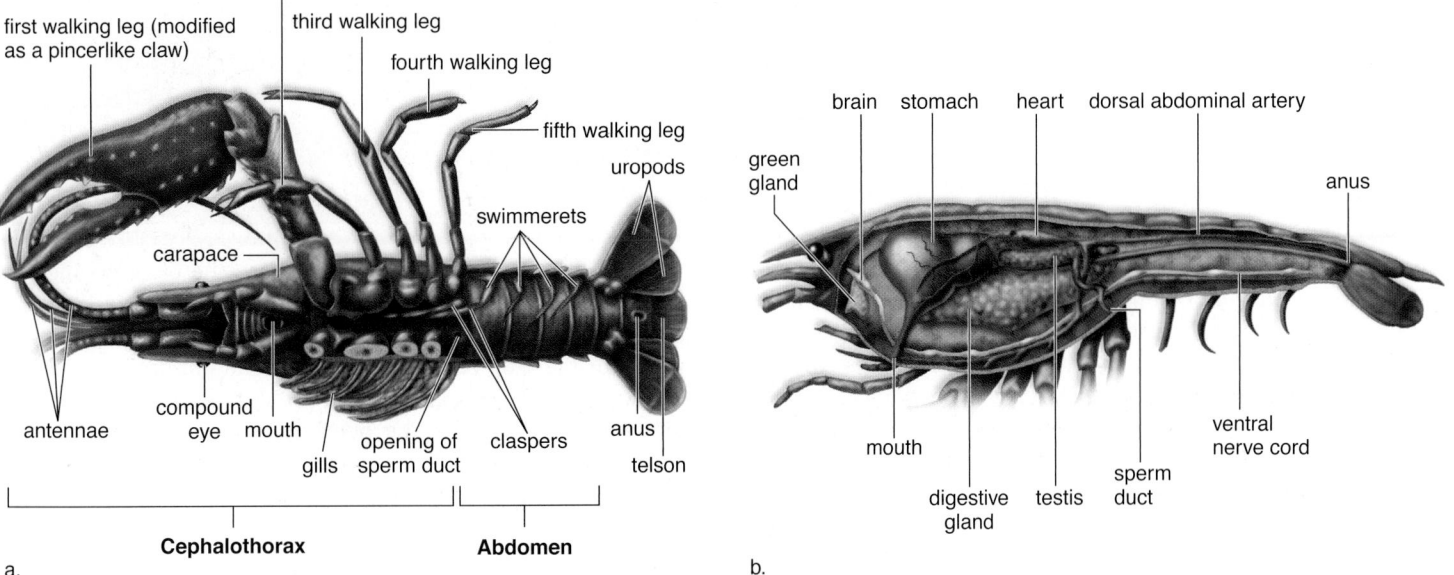

a.

b.

Figure 28.24 Male crayfish, *Cambarus*. **a.** Externally, it is possible to observe the jointed appendages, including the swimmerets, and the walking legs, which include the claws. These appendages, plus a portion of the carapace, have been removed from the right side so that the gills are visible. **b.** Internally, the parts of the digestive system are particularly visible. The circulatory system can also be clearly seen. Note the ventral nerve cord.

undergo a metamorphosis that transforms their swimming appendages to cirri, feathery structures that are extended and allow them to filter feed when they are submerged.

Video Barnacle-Free Boats

A Typical Crustacean: Crayfish. Figure 28.24*a* gives a view of the external anatomy of the crayfish. The head and thorax are fused into a *cephalothorax* [Gk. *kephale*, head, and *thorax*, breastplate], which is covered on the top and sides by a nonsegmented carapace. The abdominal segments are equipped with swimmerets, small paddlelike structures. The first two pairs of swimmerets in the male, known as claspers, are quite strong and are used to pass sperm to the female. The last two segments bear the uropods and the telson, which make up a fan-shaped tail.

Ordinarily, a crayfish lies in wait for prey. It faces out from an enclosed spot with the claws extended, and the antennae moving about. The claws seize any small animal, either dead or living, that happens by, and carry it to the mouth. When a crayfish moves about, it generally crawls slowly, but may swim rapidly by using its heavy abdominal muscles and tail.

The respiratory system consists of gills that lie above the walking legs protected by the carapace. As shown in Figure 28.24*b*, the digestive system includes a stomach, which is divided into two main regions: an anterior portion called the gastric mill, equipped with chitinous teeth to grind coarse food; and a posterior region, which acts as a filter to prevent coarse particles from entering the digestive glands, where absorption takes place.

Green glands lying in the head region, anterior to the esophagus, excrete metabolic wastes through a duct that opens externally at the base of the antennae. The coelom, which is well developed in the annelids, is reduced in the arthropods and is composed chiefly of the space about the reproductive system. A heart pumps hemolymph containing the blue respiratory pigment hemocyanin into a *hemocoel* consisting of sinuses (open spaces), where the hemolymph flows about the organs. As in the molluscs, this is an open circulatory system because blood is not contained within blood vessels.

The crayfish nervous system is well developed. Crayfish have a brain and a ventral nerve cord that passes posteriorly. Along the length of the nerve cord, periodic ganglia give off lateral nerves. Sensory organs are well developed. The compound eyes are found on the ends of movable eyestalks. These eyes are accurate and can detect motion and respond to polarized light. Other sensory organs include tactile antennae and chemosensitive setae. Crayfish also have statocysts that serve as organs of equilibrium.

The sexes are separate in the crayfish, and the gonads are located just ventral to the pericardial cavity. In the male, a coiled sperm duct opens to the outside at the base of the fifth walking leg. Sperm transfer is accomplished by the modified first two swimmerets of the abdomen. In the female, the ovaries open at the bases of the third walking legs. A stiff fold between the bases of the fourth and fifth pairs serves as a seminal receptacle. Following fertilization, the eggs are attached to the swimmerets of the female. Young hatchlings are miniature adults, and no metamorphosis occurs.

Centipedes and Millipedes

The centipedes and millipedes are known for their many legs (Fig. 28.25*a*). In **centipedes** ("hundred-leggers"), each of their many body segments has a pair of walking legs. The approximately 3,000 species prefer to live in moist environments such as under logs, in crevices, and in leaf litter, where they are active predators on worms, small crustaceans, and insects. The head of a centipede includes paired antennae and jawlike mandibles. Appendages on the first trunk segment are clawlike venomous jaws that kill or immobilize prey, while mandibles chew.

In **millipedes** ("thousand-leggers"), each of four thoracic segments bears one pair of legs (Fig. 28.25*b*), while abdominal segments have two pairs of legs. Millipedes live under stones or burrow in the soil as they feed on leaf litter. Their cylindrical bodies have a tough chitinous exoskeleton. Some secrete hydrogen cyanide, a poisonous substance.

Insects

Insects are adapted for an active life on land, although some have secondarily invaded aquatic habitats. The body of an insect

Figure 28.25 Centipede and millipede.
a. A centipede has a pair of appendages on almost every segment. **b.** A millipede has two pairs of legs on most segments.

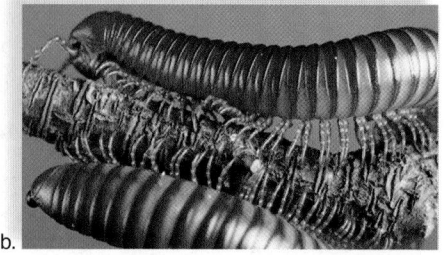

a.   b.

is divided into a head, a thorax, and an abdomen. The head bears the sense organs and mouthparts (Fig. 28.26). The thorax bears three pairs of legs and possibly one or two pairs of wings; and the abdomen contains most of the internal organs. Wings enhance an insect's ability to survive by providing a way of escaping enemies, finding food, facilitating mating, and dispersing offspring.

Many insects, such as butterflies, undergo *complete metamorphosis,* involving drastic changes in form. At first, the animal is a wormlike larva (caterpillar) with chewing mouthparts. It then forms a case, or cocoon, about itself and becomes a pupa. During this stage, the body parts are completely reorganized; the adult then emerges from the cocoon. As mentioned earlier, this life cycle allows the larvae and adults to use different food sources.

Insects show remarkable behavioral adaptations. Bees, wasps, ants, termites, and other colonial insects have complex societies.

A Typical Insect: Grasshopper   In the grasshopper (Fig. 28.27), the third pair of legs is suited to jumping. This insect has two pairs of wings. The forewings are tough and leathery, and when folded back at rest, they protect the broad, thin hindwings.

On each lateral surface, the first abdominal segment bears a large tympanum for the reception of sound waves. The posterior region of the exoskeleton in the female has an ovipositor, used to dig a hole in which eggs are laid.

The digestive system is suitable for a herbivorous diet. In the mouth, food is broken down mechanically by mouthparts and enzymatically by salivary secretions. Food is temporarily stored in the crop before passing into the gizzard, where it is finely ground. Digestion is completed in the stomach, and nutrients are absorbed into the hemocoel from outpockets called gastric ceca (*cecum,* a cavity open at one end only).

The excretory system consists of *Malpighian tubules,* which extend into a hemocoel and collect nitrogenous wastes that are concentrated and excreted into the digestive tract. The formation of a solid nitrogenous waste, namely uric acid, conserves water.

The respiratory system begins with openings in the exoskeleton called spiracles. From here, air enters small tubules called **tracheae** (Fig. 28.27*a*). The tracheae branch and rebranch, finally ending in moist areas where the actual exchange of gases takes place. No individual cell is very far from a site of gas exchange. The movement of air through this complex of tubules is not a passive process; air is pumped through by a series of bladderlike structures (air sacs) attached to the tracheae near the spiracles. Air enters the anterior four spiracles and exits by the posterior six spiracles. Breathing by tracheae may account for the small size of insects (most are less than 6 cm in length), because the tracheae are so tiny and fragile that they would be crushed by any amount of weight.

Figure 28.26 Insect diversity.

white, granular secretion

piercing-sucking mouthparts

Mealybug, order Homoptera

antenna

Hard forewings cover membranous hindwings and abdomen.

chewing mouthparts

Beetle, order Coleoptera

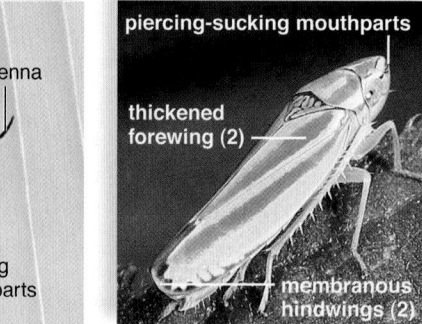

piercing-sucking mouthparts

thickened forewing (2)

membranous hindwings (2)

Leafhopper, order Homoptera

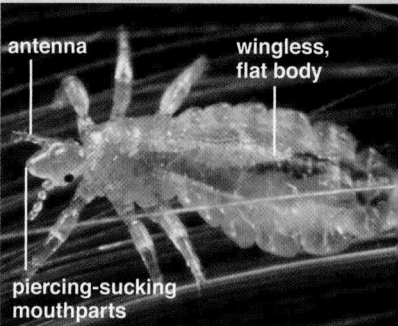

antenna

wingless, flat body

piercing-sucking mouthparts

Head louse, order Anoplura

narrow, membranous forewing

constricted waist

chewing mouthparts

ovipositor
stinger

Wasp, order Hymenoptera

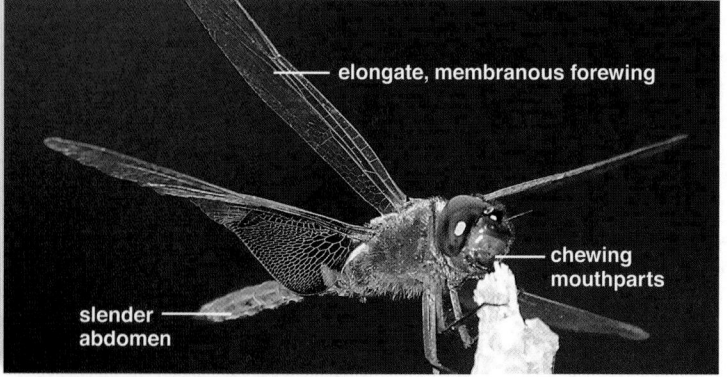

elongate, membranous forewing

chewing mouthparts

slender abdomen

Dragonfly, order Odonata

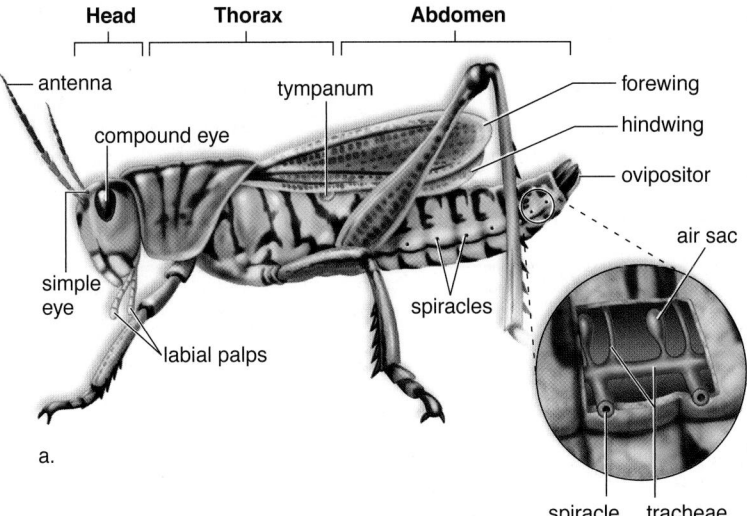

Head   Thorax   Abdomen

antenna
compound eye
tympanum
forewing
hindwing
ovipositor
air sac
simple eye
spiracles
labial palps
spiracle   tracheae

a.

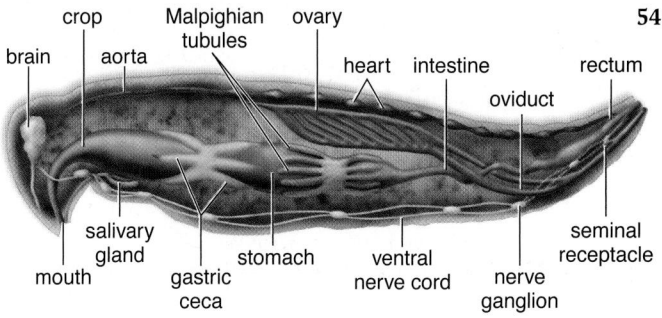

crop   Malpighian   ovary
brain   aorta   tubules   heart   intestine   rectum
oviduct
salivary   seminal   receptacle
gland   stomach   ventral   nerve
mouth   gastric   nerve cord   ganglion
ceca

b.

543

Figure 28.27 Female grasshopper, *Romalea*. **a.** Externally, the body of a grasshopper is divided into three sections and has three pairs of legs. The tympanum receives sound waves, and the jumping legs and the wings are for locomotion. **b.** Internally, the digestive system is specialized. The Malpighian tubules excrete a solid nitrogenous waste (uric acid). A seminal receptacle receives sperm from the male, which has a penis.

The circulatory system contains a slender, tubular heart that lies against the dorsal wall of the abdominal exoskeleton and pumps hemolymph into the hemocoel, where it circulates before returning to the heart again. The hemolymph is colorless and lacks a respiratory pigment, and so transports nutrients and wastes. The highly efficient tracheal system transports respiratory gases.

Grasshoppers undergo *incomplete metamorphosis,* a gradual change in form as the animal matures. The immature grasshopper, called a nymph, is recognizable as a grasshopper, even though it differs in body proportions from the adult.

Chelicerates

The **chelicerates** live in terrestrial, aquatic, and marine environments. The first pair of appendages is the pincerlike chelicerae, used in feeding and defense. The second pair is the pedipalps, which can have various functions. A cephalothorax (fused head and thorax) is followed by an abdomen that contains internal organs.

Horseshoe crabs of the genus *Limulus* are familiar along the east coast of North America (Fig. 28.28*a*). The body is covered by exoskeletal shields. The anterior shield is a horseshoe-shaped carapace, which bears two prominent compound eyes. Ticks, mites, scorpions, spiders, and harvestmen are all arachnids. Over 25,000 species of mites and ticks have been classified, some of which are parasitic on a variety of other animals. Ticks

are ectoparasites of various vertebrates, and they are carriers for such diseases as Rocky Mountain spotted fever and Lyme disease. When not attached to a host, ticks hide on plants and in the soil.

Video
Lyme Disease

Scorpions can be found on all continents except Antarctica (Fig. 28.28*b*). North America is home to approximately 1,500 species. Scorpions are nocturnal and spend most of the day hidden under a log or a rock. Their pedipalps are large pincers, and their long abdomen ends with a stinger that contains venom.

Presently, over 35,000 species of spiders have been classified (Fig. 28.28*c*). Spiders, the most familiar chelicerates, have a narrow waist that separates the cephalothorax from the abdomen. Spiders do not have compound eyes; instead, they have numerous simple eyes that perform a similar function. The chelicerae are modified as fangs, with ducts from poison glands, and the pedipalps are used to hold, taste, and chew food. The abdomen often contains silk glands, and spiders spin a web in which to trap their prey.

Check Your Progress   28.4

1. Name two ways in which the roundworms are anatomically similar to the arthropods.
2. List two ways that crustaceans are adapted to an aquatic life and insects are adapted to living on land.
3. Describe the features chelicerates have in common.

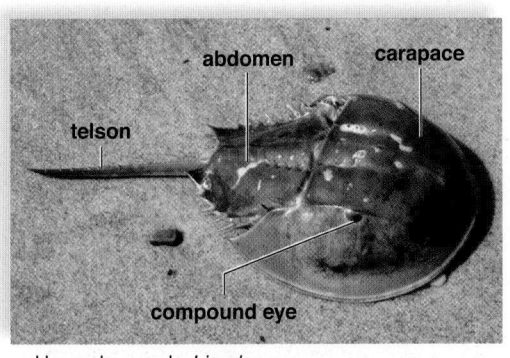

telson
abdomen   carapace
compound eye

a. Horseshoe crab, *Limulus*

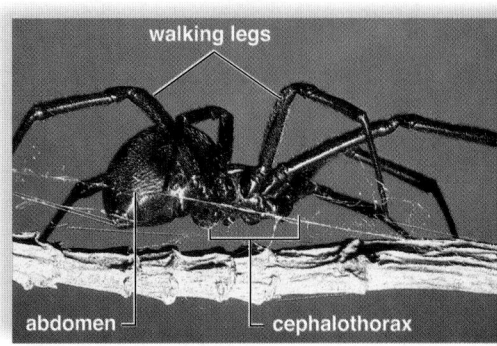

stinger   pedipalp
chelicera
walking legs
abdomen   cephalothorax

b. Kenyan giant scorpion, *Pandinus*   c. Black widow spider, *Latrodectus*

Figure 28.28 Chelicerate diversity. **a.** Horseshoe crabs are common along the east coast. **b.** Scorpions are more common in tropical areas. **c.** The black widow spider is a poisonous spider that spins a web.

28.5 Invertebrate Deuterostomes

Deuterostomes include invertebrate echinoderms, such as sea stars (also known as starfish) and chordates. Molecular data tell us that echinoderms and chordates are closely related. Morphological data indicate that these two groups share the deuterostome pattern of development (see Fig. 28.5). The echinoderms and a few chordates are invertebrates; most of the chordates are vertebrates.

Echinoderms

Echinoderms (phylum Echinodermata [Gk. *echinos,* spiny, and *derma,* skin]) are primarily bottom-dwelling marine animals. They range in size from brittle stars less than 1 cm in length to giant sea cucumbers over 2 m long (Fig. 28.29*c*).

The most striking feature of echinoderms is their 5-pointed radial symmetry, as illustrated by a sea star (Fig. 28.29*b*). Although echinoderms are radially symmetrical as adults, their larvae are free-swimming filter feeders with bilateral symmetry. Echinoderms have an endoskeleton of spiny calcium-rich plates called ossicles (Fig. 28.29*a*). Another innovation of echinoderms is their unique **water vascular system** consisting of canals and appendages that function in locomotion, feeding, gas exchange, and sensory reception.

The more familiar of the echinoderms are the Asteroidea, containing the sea stars (Fig. 28.29*a, b*); the Holothuroidea, including the sea cucumbers, which have long leathery bodies in the shape of a cucumber (Fig. 28.29*c*); and the Echinoidea, including the sea urchin and sand dollar, both of which use their spines for locomotion, defense, and burrowing (Fig. 28.29*d*). Less familiar are the Ophiuroidea, which includes the brittle stars, with a central disk surrounded by radially flexible arms; and the Crinoidea, the oldest group, which includes the stalked feather stars and the motile feather stars.

Video
Sea Urchin Reproduction

A Typical Echinoderm: Sea Star

Sea stars number about 1,600 species that are commonly found along rocky coasts, where they feed on clams, oysters, and other bivalve molluscs. Various structures project through the body wall: (1) spines from the endoskeletal plates offer some protection; (2) pincerlike structures around the bases of spines keep the surface free of small particles; and (3) skin gills, tiny fingerlike extensions of the skin, are used for respiration. On the oral surface, each arm has a groove lined by little *tube feet* (Fig. 28.29*a*).

To feed, a sea star positions itself over a bivalve and attaches some of its tube feet to each side of the shell. By working its tube feet in alternation, it gradually pulls the shell open. A very small crack is enough for the sea star to evert its cardiac stomach and push it through the crack, so that it contacts the soft parts of the bivalve. The stomach secretes enzymes, and digestion begins, even while the bivalve is attempting to close its shell. Later, partly digested food is taken into the sea star's body, where digestion continues in the pyloric stomach using enzymes from the digestive glands found in each arm. A short intestine opens at the anus on the aboral side (side opposite the mouth).

In each arm, the well-developed coelom contains a pair of digestive glands and gonads (either male or female) that open on the aboral surface by very small pores. The nervous system consists of a central nerve ring that gives off radial nerves into each arm. A light-sensitive eyespot is at the tip of each arm.

Figure 28.29 Echinoderms. **a.** Sea star (starfish) anatomy. Like other echinoderms, sea stars have a water vascular system that begins with the sieve plate and ends with expandable tube feet. **b.** The red sea star *Mediastar* uses the suction of its tube feet to open a clam, a primary source of food. **c.** Sea cucumber (*Pseudocolochirus*). **d.** Sea urchin (*Strongylocentrotus*).

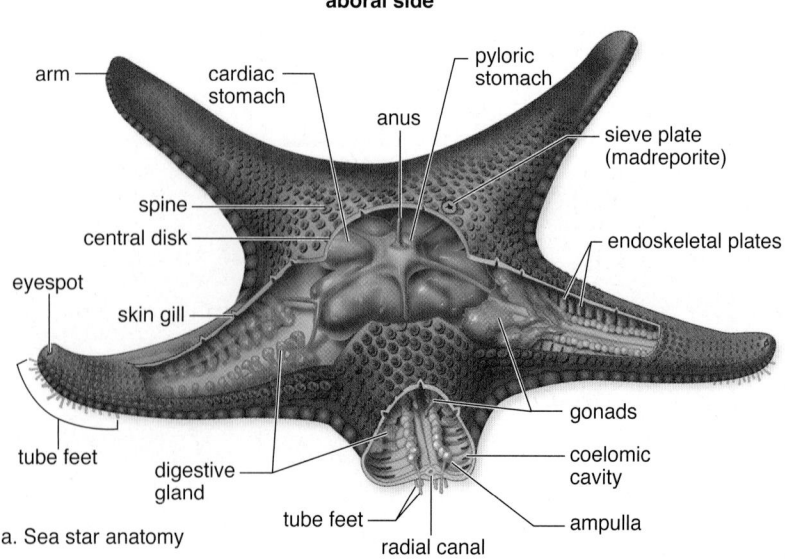

a. Sea star anatomy

b. Red sea star, *Mediastar*

Locomotion depends on the water vascular system. Water enters this system through a structure on the aboral side called the *sieve plate,* or *madreporite* (Fig. 28.29*a*). From there it passes through a stone canal to a ring canal, which circles around the central disc, and then to a radial canal in each arm. From the radial canals, many lateral canals extend into the tube feet, each of which has an ampulla. Contraction of an ampulla forces water into the tube foot, expanding it. When the foot touches a surface, the center is withdrawn, giving it suction so that it can adhere to the surface. By alternating the expansion and contraction of the tube feet, a sea star moves slowly along.

Echinoderms do not have a respiratory, excretory, or circulatory system. Fluids within the coelomic cavity and the water vascular system carry out many of these functions. For example, gas exchange occurs across the skin gills and the tube feet. Nitrogenous wastes diffuse through the coelomic fluid and the body wall. Cilia on the coelom and other structures keep the coelomic fluid moving.

Sea stars reproduce asexually and sexually. If the body is fragmented, each fragment can regenerate a whole animal as long as a portion of the central disc is present. Sea stars spawn and release either eggs or sperm at the same time. The bilaterally symmetrical larva undergo metamorphosis to become a radially symmetrical adult.

Check Your Progress 28.5

1. Describe the evidence that supports the evolution of echinoderms from bilaterally symmetrical animals.
2. Describe the functions of the water vascular system in sea stars.

CONNECTING *the* CONCEPTS *with the* BIG IDEAS

Evolution

- Phylogenetic trees and cladograms model the possible evolutionary history of groups of living organisms. (1B2a-d)

Energy and Homeostasis

- Animals display a variety of mechanisms for food digestion, including specialized vacuoles, gastrovascular cavities, and one-way flow digestive tracts. (2D2b*IE*)
- Animals display a variety of mechanisms for nitrogen waste removal, including flame cells, nephridia, and Malphigian tubules. (2D2b*IE*)
- Animals display a variety of mechanisms for gas exchange, including simple diffusion, gills, trachea/spiracles, and lungs. (2D2b*IE*)
- Excretory and osmoregulatory systems have similar characteristics across the phyla, supporting the idea of common ancestry. (2D2c*IE*)

*Find the unabridged version of all EK citations at www.glencoe.com/maderAP11.

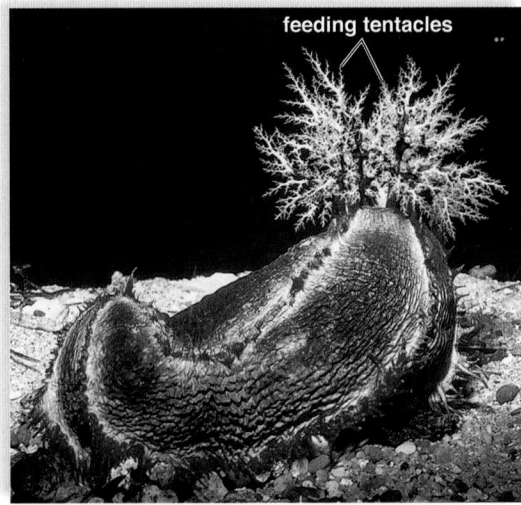

c. Sea cucumber, *Pseudocolochirus*

d. Purple sea urchin, *Strongylocentrotus*

Media Study Tools

Summarize

28.1 Evolution of Animals

Animals are multicellular organisms that are heterotrophic and ingest their food. They commonly have a diploid life cycle, but other life cycles are represented. Typically, they have the power to move by means of contracting fibers. The *colonial flagellate hypothesis* proposes that animals evolved from a protist that resembles the choanoflagellates of today.

The traditional evolutionary tree of invertebrates was based only on morphological characters. A modern phylogeny constructed with molecular and morphological data presents an alternative evolutionary history, mainly that the lophophorans are a set of many different lineages, and not a monophyletic group.

All animals except sponges, cnidarians, and comb jellies are triploblastic. Triploblastic animals are either deuterostomes or protostomes, which differ in the development of the coelom and whether the blastophore gives rise to the mouth (protostomes) or the anus (deuterostomes). The evolution of the body plan of animals is controlled by *Hox* genes, a set of regulatory genes that are present in the genome of all animals.

28.2 The Simplest Invertebrates

Sponges are the oldest lineage of animals. Sponges resemble colonial protozoans. They have a cellular level of organization, lack tissues, and are asymmetrical. Sponges are sessile filter feeders that depend on a flow of water through the body to acquire food. Food is digested in vacuoles within collar cells that line a central cavity.

Comb jellies and cnidarians are diploblastic. They are aquatic and radially symmetrical, with bands of cilia that propel them through the water.

Cnidarians have a sac body plan. They exist as either polyps or medusae, or they can alternate between the two. Hydras and their relatives—sea anemones and corals—are polyps; in jellyfishes, the medusan stage is most common. In *Hydra* and other cnidarians, an outer epidermis is separated from an inner gastrodermis by mesoglea. They possess tentacles to capture prey and cnidocytes armed with nematocysts to stun it. A nerve net coordinates movements.

28.3 Diversity Among the Lophotrochozoans

Lophotrochozoans are the most diverse group of protostomes. "Lophophorans" are a group of lophotrochozoans that feed with a feathery lophophore that filters microorganisms from water. Trocophorans have a trochophore larva either in living forms or in their ancestors.

Free-living flatworms (planarians) have three tissue layers and no coelom. Planarians have muscles and a ladder-type nervous system with cephalization. They take in food through an extended pharynx leading to a gastrovascular cavity that extends through the body. An osmotic-regulating organ contains flame cells.

Flukes and tapeworms are parasitic. Flukes have two suckers by which they attach to and feed from their hosts. Tapeworms have a scolex with hooks and suckers for attaching to the host's intestinal wall. The body of a tapeworm is made up of proglottids, which, when mature, contain thousands of eggs. If these eggs are taken up by pigs or cattle, larvae become encysted in their muscles. If humans eat this meat, they too may become infected with a tapeworm. Tapeworm infection is a serious concern in impoverished nations.

Rotifers are microscopic and aquatic and have a corona that resembles a spinning wheel when in motion.

The body of a mollusc typically contains a visceral mass, a mantle, and a foot. Many also have a head and a radula for feeding. The nervous system consists of several ganglia connected by nerve cords. They have a reduced coelom and an open circulatory system. Clams (bivalves) are adapted to a sedentary coastal life, squids (cephalopods) to an active life in the sea, and snails (gastropods) to life on land.

Annelids are segmented worms. They have a well-developed coelom divided by septa, a closed circulatory system, a ventral solid nerve cord, and paired nephridia. Earthworms are oligochaetes ("few bristles") that use the body wall for gas exchange. Polychaetes ("many bristles") are marine worms that have parapodia. They may be predators, with a definite head region, or they may be filter feeders, with ciliated tentacles to filter food from the water. Leeches also belong to this phylum.

28.4 Diversity of the Ecdysozoans

Ecdysozoans are protostomes. Roundworms have a pseudocoelom and are usually small and very diverse; they are present almost everywhere in great numbers. Many are significant parasites of humans. The parasite *Ascaris* is representative of the group. Infections can also be caused by *Trichinella,* whose larval stage encysts in the muscles of humans. Elephantiasis is caused by a filarial worm that blocks lymphatic vessels. This disease is of concern to poor tropical and subtropical regions where millions of people have limited access to clean food, water, and sanitation.

Arthropods are the most varied and numerous of animals. Their success is largely attributable to a flexible exoskeleton, specialized body regions, and jointed appendages. Also important are a high degree of cephalization, a variety of respiratory organs, and many undergo some form of metamorphosis. The hard exoskeleton requires a molt at different stages of growth. Crustaceans, insects, and chelicerates are representative groups of arthropods.

Crustaceans (crayfish, lobsters, shrimps, copepods, krill, and barnacles) have a head that bears compound eyes, antennae, antennules, and mouthparts. Crayfish have other features such as an open circulatory system, respiration by gills, and a ventral solid nerve cord.

Insects include butterflies, grasshoppers, bees, and beetles. The anatomy of the grasshopper is adapted to life on land. Like other insects, grasshoppers have wings and three pairs of legs attached to the thorax. Grasshoppers have a tympanum for sound reception, and a digestive system specialized for a grass diet, Malpighian tubules for excretion of solid nitrogenous waste, tracheae for respiration, internal fertilization, and incomplete metamorphosis.

Chelicerates (horseshoe crabs, spiders, scorpions, ticks, and mites) have chelicerae, pedipalps, and four pairs of walking legs attached to a cephalothorax.

28.5 Invertebrate Deuterostomes

Echinoderms (sea stars, sea urchins, sea cucumbers, and sea lilies) are radially symmetrical as adults but not as larvae, and have internal calcium-rich plates with spines. Typical of echinoderms, sea stars have tiny skin gills, a central nerve ring with branches, and a water vascular system for locomotion. Each arm of a sea star contains branches from the nervous, digestive, and reproductive systems.

Key Terms

![Assess] Assess

Reviewing This Chapter

1. What does the phylogenetic tree (see Fig. 28.3) tell you about the evolution of the animals studied in this chapter? 522
2. What features make sponges different from the other organisms placed in the animal kingdom? 527–28
3. What are the two body forms found in cnidarians? Explain how they function in the life cycle of various types of cnidarians. 528–29
4. Describe the anatomy of *Hydra*, pointing out those features that typify cnidarians. 529
5. Describe the anatomy of a free-living planarian, and how it differs from the parasitic flatworms. 531–33
6. What are the general characteristics of molluscs and the specific features of bivalves, cephalopods, and gastropods? 534–36
7. What are the general characteristics of annelids and the specific features of earthworms? 536–37
8. Describe the anatomy of *Ascaris*, pointing out those features that typify roundworms. 538
9. What are the general characteristics of arthropods, specifically crustaceans and insects? 540–42
10. What other types of arthropods were discussed in the chapter? 541–42
11. What are the general characteristics of echinoderms? Explain how the water vascular system works in sea stars. 544–45

Testing Yourself

Choose the best answer for each question.

1. Which of these is not a characteristic of animals?
   a. heterotrophic
   b. diploid life cycle
   c. have contracting fibers
   d. single cells or colonial
   e. lack of chlorophyll

2. The phylogenetic tree of animals shows that
   a. three germ layers evolved before a coelom.
   b. both molluscs and annelids are protostomes.
   c. some animals have radial symmetry.
   d. sponges were the first to evolve from an ancestral protist.
   e. All of these are correct.

3. Which of these descriptions does not pertain to both protostomes and deuterostomes?
   a. three germ layers, bilateral symmetry, first opening is mouth
   b. bilateral symmetry, first opening is mouth, all have a true coelom
   c. spiral cleavage, first opening is anus, true coelom develops by a splitting of mesoderm
   d. bilateral symmetry, three germ layers, second opening is mouth
   e. None pertain to both protostomes and deuterostomes.

4. Which of these pairs is mismatched?
   a. sponges—spicules
   b. tapeworms—proglottids
   c. cnidarians—nematocysts
   d. roundworms—cilia
   e. cnidarians—polyp and medusa

5. Flukes and tapeworms
   a. show cephalization.
   b. have well-developed reproductive systems.
   c. have well-developed nervous systems.
   d. are free-living.

6. *Ascaris* is a parasitic
   a. roundworm.
   b. flatworm.
   c. hydra.
   d. sponge.
   e. comb jelly.

7. The phylogenetic tree of animals shows that
   a. cnidarians evolved directly from sponges.
   b. flatworms evolved directly from roundworms.
   c. rotifers are closely related to flatworms.
   d. coelomates gave rise to the acoelomates.
   e. All of these are correct.

8. Comb jellies are most closely related to
   a. cnidarians.
   b. sponges.
   c. flatworms.
   d. roundworms.
   e. Both a and b are correct.

9. Write the correct type of animal beside each of the following terms.
   a. proglottids:
   b. mantle cavity:
   c. collar cells:
   d. cnidocytes:
   e. crown of cilia:
   f. branched gastrovascular cavity:
   g. exoskeleton contains chitin:

10. Which of these does not pertain to a protostome?
    a. spiral cleavage
    b. blastopore is associated with the anus
    c. coelom, splitting of mesoderm
    d. annelids, arthropods, and molluscs
    e. mouth is associated with first opening

11. Which of these best shows that snails are not closely related to crayfish?
    a. Snails are terrestrial, and crayfish are aquatic.
    b. Snails have a broad foot, and crayfish have jointed appendages.
    c. Snails are hermaphroditic, and crayfish have separate sexes.
    d. Snails are insects, but crayfish are fishes.
    e. Snails are bivalves, and crayfish are chelicerates.

12. A radula is a unique organ for feeding found in
    a. molluscs.
    b. annelids.
    c. arthropods.
    d. only insects.
    e. All of these are correct.

13. Label this diagram.

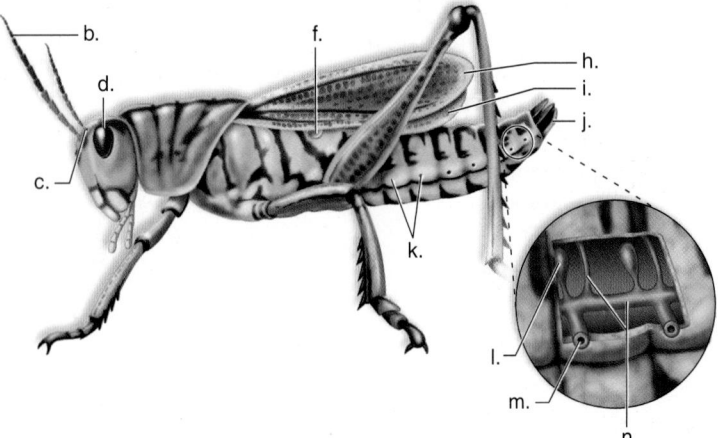

14. Segmentation in the earthworm is not exemplified by
    a. body rings.
    b. coelom divided by septa.
    c. setae on most segments.
    d. nephridia interior in most segments.
    e. tympanum exterior to segments.

15. Which characteristic accounts for the success of arthropods?
    a. jointed exoskeleton
    b. well-developed nervous system
    c. segmentation
    d. respiration adapted to environment
    e. All of these are correct.

Engage

 Virtual Labs
Earthworm Dissection
Classifying Arthropods

The virtual lab "Earthworm Dissection" provides an interactive look at the anatomy of the earthworm. In addition, the virtual lab "Classifying Arthropods" provides an exploration of the key characteristics used to identify groups of arthropods.

Thinking Scientifically

1. How is the lifestyle of radially symmetrical animals different from that of bilaterally symmetrical animals? Explain.
2. Animals, like other life-forms, evolved in the ocean. What made water a better nursery than land for the evolution of animals?

New Guinea singing dog, *Canis familiaris hallstromi.*

29

Vertebrate Evolution

T he New Guinea singing dog lives on the island of New Guinea, but it doesn't actually sing. It yelps, whines, and howls at different pitches unlike other dogs. The howl is eerie to the point of causing goose bumps when one tone blends with the next. The evolutionary history of the New Guinea singing dog is a matter of debate. Is it a unique species, a type of gray wolf, or is it related to the dingo, the nonbarking canine that lives in Australia? Some suggest that the New Guinea singing dog might be the most primitive known "breed" of domestic dog because Stone Age people brought it to New Guinea 6,000 years ago! Living on an island, it has remained geographically and reproductively isolated ever since. Therefore, it could be a living fossil, an organism with the same genes and characteristics as its original ancestor. If so, the New Guinea singing dog offers an opportunity to study what the first domesticated dog was like.

However, like many other vertebrates, which are the topic of this chapter, the New Guinea singing dog is threatened with extinction. Expeditions into the highlands of New Guinea have yielded only a few droppings, tracks, and haunting howls in the distance.

As you read through the chapter, think about the following questions:

1. What are the traits that set vertebrates apart from other animals?
2. How are vertebrates important to the day-to-day life of humans?
3. When in the history of life did vertebrates evolve?

BEFORE YOU BEGIN

Before beginning this chapter, take a few moments to review the following discussions.

Section 15.3 How does *Archaeopteryx*, and similar transitional fossils, support the close relationship between birds and dinosaurs?

Figure 18.8 When in the history of life on Earth did the vertebrates first appear?

Figure 21.1 Which eukaryotes are the closest relatives of vertebrates? In which eukaryotic supergroups are vertebrates found?

FOLLOWING *the* BIG IDEAS

CHAPTER 29   VERTEBRATE EVOLUTION

Evolution	Evolutionary trends in vertebrates move toward endothermy, a four-chambered heart, air breathing, and internal fertilization and fetal development.
Energy and Homeostasis	Animal body systems help maintain homeostasis but are tailored to specific needs and environments.
Interactions and Systems	Organ systems in animals promote efficient work.

29.1 The Chordates

Chordates (phylum Chordata), like echinoderms, are euterostomes; however, the chordates do not have an exoskeleton like the invertebrates, but have an internal skeleton made of bone and cartilage to which the muscles are attached. This arrangement allows the chordates to enjoy freedom of movement and attainment of a larger body size than invertebrates.

Characteristics of Chordates

All chordates have four basic characteristics that appear at some point during development (Figure 29.1). Although all of these characteristics were likely present in the adult stage of the common ancestor of chordates, not all of the traits are present in the adults of modern chordates. However, the presence of these characteristics during the development of all chordates indicates their evolutionary ties to a single common ancestor. The four characteristics are as follows:

Notochord. [Gk. *notos,* back, and *chorde,* string]. A dorsal supporting rod; the **notochord** is located just below the nerve cord. The majority of vertebrates have an embryonic notochord that is replaced by the vertebral column during embryonic development.

Dorsal tubular nerve cord. In contrast to the arthropods, which have a ventral nerve cord, chordates have a tubular cord situated dorsally. The anterior portion becomes the brain in most chordates. In vertebrates, the nerve cord, often called the spinal cord, is protected by vertebrae.

Pharyngeal pouches. These are seen only during embryonic development in most vertebrates. In the nonvertebrate chordates, the fishes, and amphibian larvae, the pharyngeal pouches become functioning **gills** (respiratory organs of aquatic vertebrates). Water passing into the mouth and the

Figure 29.2 Lancelet, *Branchiostoma.* Lancelets are filter feeders. Water enters the mouth and exits at the atriopore after passing through the gill slits.

pharynx goes through the gill slits, which are supported by gill arches. In terrestrial vertebrates, the pouches are modified for various purposes. For example, in humans the first pair of pouches become the auditory tubes.

Postanal tail. A tail—in the embryo, if not in the adult—extends beyond the anus. In other groups of animals, the anus is terminal.

Nonvertebrate Chordates

The nonvertebrate chordates do not have a spine made of bony vertebrae. They are divided into two groups: the cephalochordates and the urochordates.

Lancelets (genus *Branchiostoma,* previously called *Amphioxus*) are the **cephalochordates** [Gk. *kephalo,* head]. These marine chordates, which are only a few centimeters long, are named for their resemblance to a lancet—a small, two-edged surgical knife (Fig. 29.2). Lancelets are found in the shallow water along most coasts, where they usually lie partly buried in the sandy or muddy bottom with only their mouth and gill apparatus exposed. They feed on microscopic particles that they filter out of a constant stream of water that enters the mouth and passes through the gill slits into a chamber, or *atrium,* before exiting through an opening called an *atriopore.*

Many chordates lose one or more of the defining characteristics as adults, but lancelet adults possess all four general chordate characteristics. For that reason lancelets are important

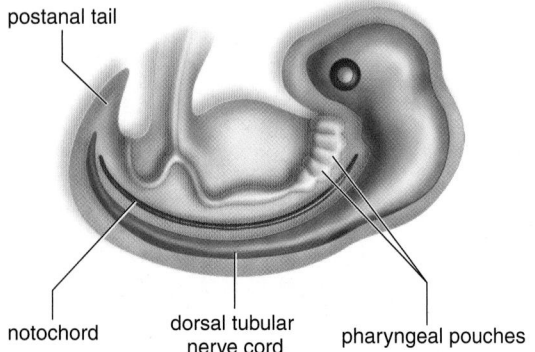

Figure 29.1 Characteristics of the chordates. Chordates have four distinctive traits. The notochord is present in all chordates.

in comparative anatomy and evolutionary studies. In lancelets, the notochord extends from the head to the tail. In addition to the four features of all chordates, segmentation is present, as witnessed by the fact that the muscles are segmentally arranged, and the dorsal tubular nerve cord has periodic branches. Segmentation may not be an important feature in lancelets, but it is in the vertebrates where it leads to specialization of parts, as we also witnessed in the annelids and arthropods in the preceding chapter.

Sea squirts, or **urochordates,** are also called tunicates because adults have a tunic (outer covering) that makes them look like thick-walled, squat sacs. They live on the ocean floor, where they squirt out water from their excurrent siphon when disturbed. The sea squirt larva is bilaterally symmetrical and has the four chordate characteristics. Metamorphosis produces the sessile adult with an incurrent and excurrent siphon (Fig. 29.3). The pharynx is lined by numerous cilia whose beating creates a current of water that moves into the pharynx and out the numerous gill slits, the only chordate characteristic that remains in the adult.

Many evolutionary biologists hypothesize that the sea squirts are directly related to the vertebrates. It has been suggested that a larva with the four chordate characteristics may have become sexually mature without developing the other adult sea squirt characteristics. Then it may have evolved over

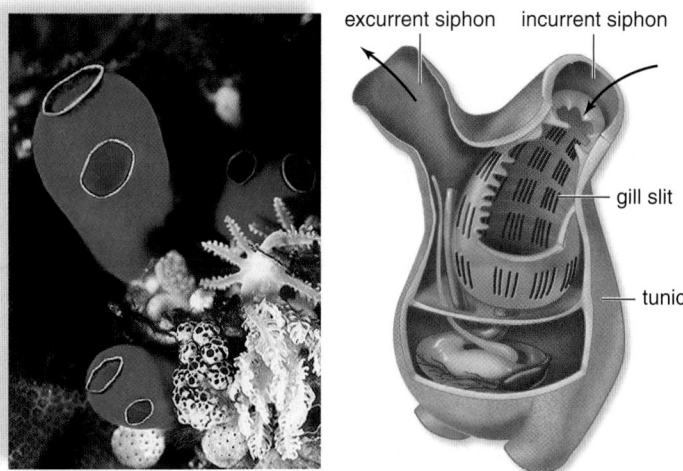

Figure 29.3 Sea squirt, *Halocynthia*.   Note that the only chordate characteristic remaining in the adult is gill slits.

time into a fishlike vertebrate. Figure 29.4 shows how the main groups of chordates may have evolved.

Check Your Progress 29.1

1. Discuss how humans, as chordates, possess all four characteristics either as embryos or adults.
2. Explain why adult sea squirts are classified as chordates although they look like thick-walled, squat sacs.

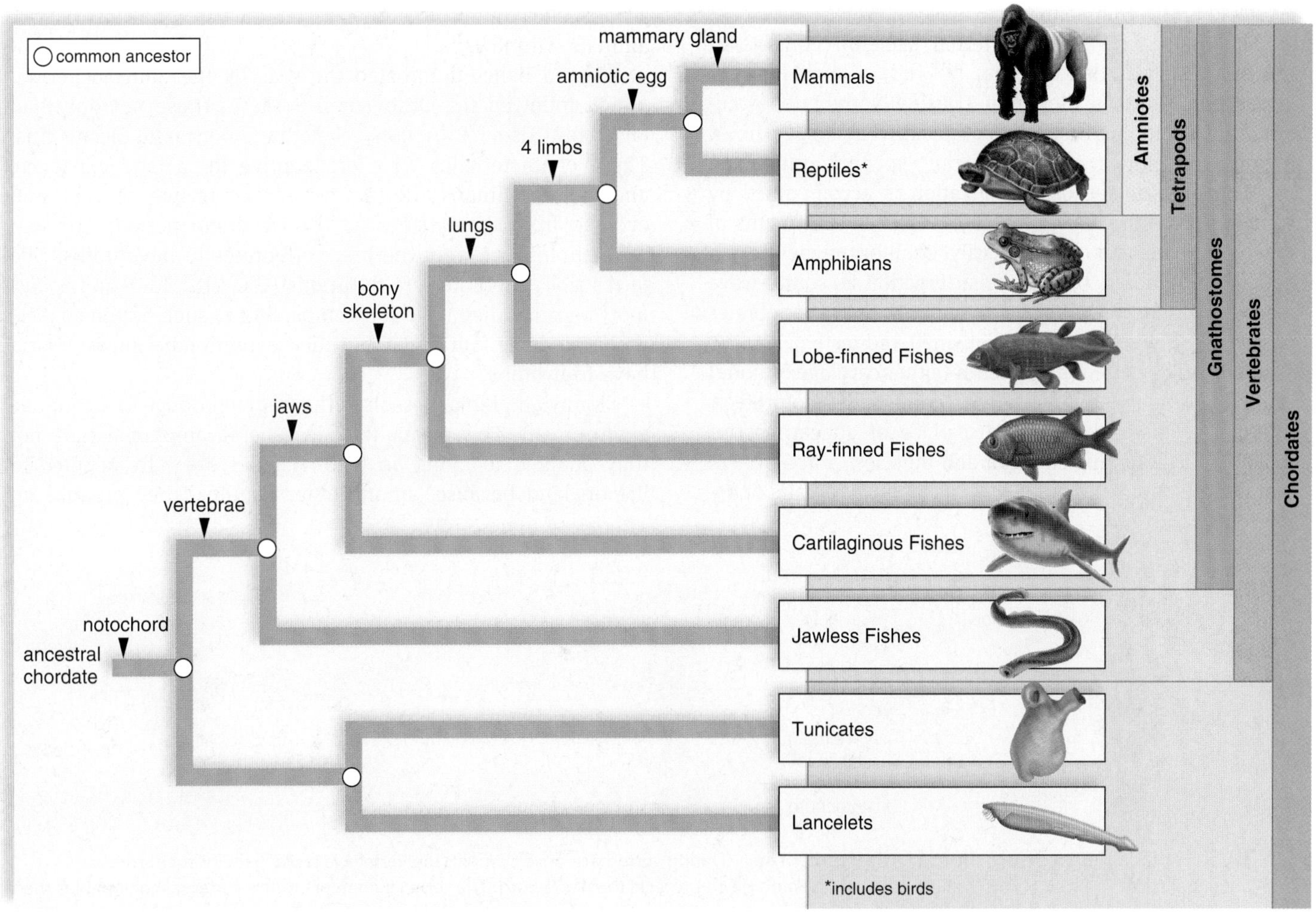

Figure 29.4 Phylogenetic tree of the chordates.   The evolution of vertebrates is marked by at least seven derived characteristics.

29.2 The Vertebrates

Vertebrates are chordates with vertebral columns and certain other features that distinguish them from the nonvertebrate chordates.

Characteristics of Vertebrates

As embryos, vertebrates have the four chordate characteristics. In addition, vertebrates have these features:

Vertebral column. The embryonic notochord is generally replaced by a vertebral column composed of individual vertebrae (Fig. 29.5). Remnants of the notochord are seen in the intervertebral discs, which are compressible cartilaginous pads found between the vertebrae. The vertebral column, which is a part of the flexible but strong endoskeleton, gives evidence that vertebrates are segmented.

Skull. The main axis of the internal skeleton consists of not only the vertebral column, but also a skull that encloses and protects the brain. During vertebrate evolution, the brain increases in complexity, and specialized regions developed to carry out specific functions.

The high degree of cephalization is accompanied by complex sense organs. The eyes develop as outgrowths of the brain. The ears are primarily equilibrium devices in aquatic vertebrates, but they also function as sound-wave receivers in land vertebrates. In addition, many vertebrates possess well-developed senses of smell and taste.

Endoskeleton. The vertebrate skeleton (either cartilage or bone) is a living tissue that grows with the animal. It also protects internal organs and serves as a place of attachment for muscles. Together, the skeleton and muscles form a system that permits rapid and efficient movement. Two pairs of appendages are characteristic. Fishes typically have pectoral and pelvic fins, while terrestrial tetrapods have four limbs.

Internal organs. Vertebrates have a large coelom and a complete digestive tract. The blood in vertebrates is contained entirely within blood vessels, and is therefore a closed circulatory system. The respiratory system consists of gills or lungs, which obtain oxygen from the environment. The kidneys are important excretory and water-regulating organs that conserve or rid the body of water as necessary. The sexes are generally separate, and reproduction is usually sexual.

Vertebrate Evolution

The Paleozoic era is distinguished by the arising of the chordates and first vertebrates. Chordates appear on the scene suddenly at the start of the Cambrian period, 542 MYA. By the end of the Ordovician period that followed, both jawless and jawed fishes had appeared; during this period, nonvascular plants had moved onto the land.

Even though we do not know the precise origin of vertebrates, we can trace their evolutionionary history, as shown in Figure 29.4. (See also Table 18.1 to review evolutionary events on the geologic time scale.) The earliest vertebrates were fishes, organisms that are abundant today both in marine and freshwater habitats. A few of today's fishes lack jaws and have to suck and otherwise engulf their prey. Most fishes have jaws, which are a more efficient means of grasping and eating prey. The jawed fishes and all the other vertebrates are **gnathostomes**—animals with jaws.

Jawed fishes dominated the seas by the Silurian period, which followed the Ordovician. Some of these not only had jaws, but also had a bony skeleton, lungs, and fleshy fins. These characteristics were preadaptive for a land existence, and the amphibians, the first vertebrates to live on land, had evolved from these fishes by the Devonian period, 416 MYA. The amphibians were the first vertebrates to have limbs. The terrestrial vertebrates are **tetrapods** [Gk. *tetra*, four, and *podos*, foot] because they have four limbs. Some, such as the snakes, no longer have four limbs, but their evolutionary ancestors did have four limbs.

Many amphibians, such as the frog, reproduce in an aquatic environment. This means that, in general, amphibians are not fully adapted to living on land. Reptiles are fully adapted to life on land because, among other features, they produce an

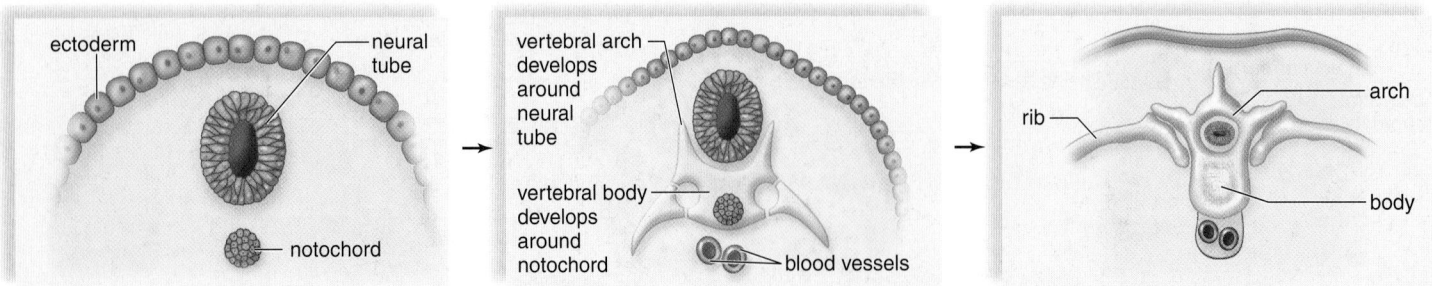

Figure 29.5 Replacement of notochord by the vertebrae. During vertebrate development, the vertebrae replace the notochord and surround the neural tube. The result is the flexible vertebral column, which protects the nerve cord. The term spine refers to the vertebral column plus the nerve cord.

amniotic egg. The amniotic egg is so named because the embryo is surrounded by an amniotic membrane that encloses the amniotic fluid. Therefore, **amniotes,** animals that exhibit an amniotic membrane, develop within an aquatic environment but one of their own making. In placental mammals, such as ourselves, the fertilized egg develops inside the female, where it is surrounded by an amniotic membrane.

A watertight skin, which is seen among the reptiles and mammals, is also a good feature to have when living on land.

Check Your Progress 29.2

1. Describe the advantages of an endoskeleton.
2. Explain how four legs would be useful in terrestrial environments.

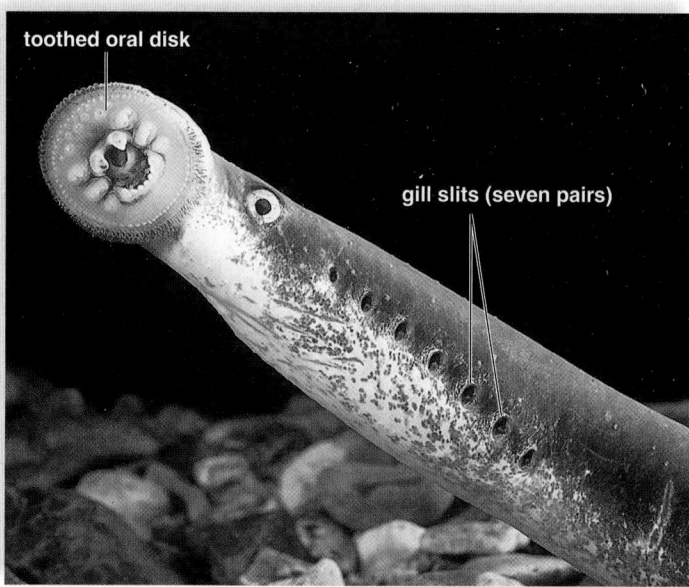

Figure 29.6 Lamprey, _Petromyzon._ Lampreys, which are agnathans, have an elongated, rounded body and nonscaly skin. Note the lamprey's toothed oral disk, which is attached to the aquarium glass.

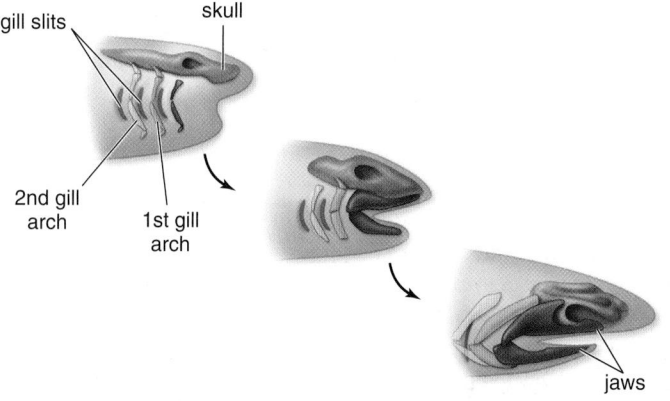

Figure 29.7 Evolution of the jaw. The first jaw evolved from the first and second gill arches of fishes.

29.3 The Fishes

Learning Outcomes

Upon completion of this section, you should be able to

1. List two features of jawless fishes.
2. Describe the three characteristics shared by all jawed fishes.
3. List features that distinguish cartilaginous fishes from bony fishes.

Fishes are the largest group of vertebrates with nearly 28,000 recognized species. They range in size from a few millimeters in length to the whale shark, which may reach lengths of 12 m. The fossil record of the fishes is extensive.

Jawless Fishes

The earliest fossils of Cambrian origin were the small, filter-feeding, jawless and finless **ostracoderms.** Several groups of ostracoderms developed heavy dermal armor for protection.

Today's **jawless fishes,** or **agnathans,** have a cartilaginous skeleton and persistent notochord. They are cylindrical, up to a meter long, and have smooth, nonscaly skin (Fig. 29.6). The hagfishes are exclusively marine scavengers that feed on soft-bodied invertebrates and dead fishes. Many species of lampreys are filter feeders like their ancestors. Parasitic lampreys have a round, muscular mouth used to attach themselves to another fish and suck nutrients from the host's cardiovascular system. The parasitic sea lamprey gained access to the Great Lakes in 1829, and by 1950 it had almost demolished the resident trout and whitefish populations.

Fishes with Jaws

Fishes with jaws have these characteristics:

Ectothermy. Like all fishes, jawed fishes are **ectotherms** [Gk. _ekto,_ outer, and _therme,_ heat], which means that they depend on the environment to regulate their temperature.

Gills. Like all fishes, jawed fishes breathe with gills and have a single-looped, closed circulatory system with a heart that pumps the blood first to the gills. Then, oxygenated blood passes to the rest of the body.

Cartilaginous or bony endoskeleton. The endoskeleton of jawed fishes includes the vertebral column, a skull with jaws, and paired pectoral and pelvic **fins,** projections that are controlled by muscles. The large muscles of the body actually do most of the work of locomotion, but the fins help with balance and turning.

Jaws evolved from the first pair of gill arches present in ancestral agnathans. The second pair of gill arches became support structures for the jaws (Fig. 29.7).

Scales. The skin of the jawed fishes is covered by scales, and therefore, it is not exposed directly to the environment. A scientist can tell the age of a fish from examining the growth of the scales.

The **placoderms,** extinct jawed fishes of the Devonian period, are probably the ancestors of early sharks and bony fishes.

a. Sand tiger shark, *Carcharias taurus*

b. Blue-spotted stingray, *Taeniura lymma*

Figure 29.8 Cartilaginous fishes. **a.** Sharks are predators or scavengers that move gracefully through open ocean waters. **b.** Most stingrays grovel in the sand, feeding on bottom-dwelling invertebrates. This blue-spotted stingray is protected by the spine on its whiplike tail.

Placoderms were armored with heavy, bony plates and had strong jaws. Like modern-day fishes, they also had paired pectoral and pelvic fins (Fig. 29.8).

Cartilaginous Fishes

Sharks, rays, skates, and chimaeras are marine **cartilaginous fishes** (Chondrichthyes). The cartilaginous fishes have a skeleton composed of cartilage instead of bone; have five to seven gill slits on both sides of the pharynx; and lack the gill cover of bony fishes. In addition, many have openings to the gill chambers located behind the eyes called spiracles. Their body is covered with dermal denticles, tiny teethlike scales that project posteriorly, which is why a shark's skin feels like sandpaper. The menacing teeth of sharks and their relatives are simply larger, specialized versions of these scales. At any one time, a shark such as the great white shark may have up to 3,000 teeth in its mouth, arranged in 6 to 20 rows. Only the first row or two are actively used for feeding; the other rows are replacement teeth.

Three well-developed senses enable sharks and rays to detect their prey. They have the ability to sense electric currents in water—even those generated by the muscle movements of animals. They have a lateral line system, a series of pressure-sensitive cells that lie within canals along both sides of the body, which can sense pressure caused by a fish or other animal

swimming nearby. They also have a very keen sense of smell; the part of the brain associated with this sense is very well developed. Sharks can detect about one drop of blood in 115 liters of water.

The largest sharks are filter feeders, not predators. The basking sharks and whale sharks ingest tons of small crustaceans, collectively called krill. Many sharks are fast-swimming predators in the open sea (Fig. 29.8a). The great white shark, about 7 m in length, feeds regularly on dolphins, sea lions, and seals. Humans are normally not attacked except when mistaken for sharks' usual prey. Tiger sharks, so named because the young have dark bands, reach 6 m in length and are unquestionably one of the most predaceous sharks. As it swims through the water, a tiger shark will swallow anything, including rolls of tar paper, shoes, gasoline cans, paint cans, and even human parts. The number of bull shark attacks has increased in the Gulf of Mexico in recent years. This increase is due to more swimmers in the shallow waters and loss of the sharks' natural food supply.

In rays and skates (Fig. 29.8b), the pectoral fins are greatly enlarged into a pair of large, winglike fins, and the bodies are dorsoventrally flattened. The spiracles are enlarged and allow them to move water over the gills while resting on the bottom, where they feed on organisms such as crustaceans, small fishes, and molluscs.

Stingrays have a whiplike tail that has serrated spines with venom glands at the base. This tail spike is a defensive weapon, but it can deliver a harmful or even fatal wound. Manta rays are harmless oceanic filter-feeding giants with a fin span of up to 6 m and a weight of 2,000 kg. Sawfish rays are named for their large, protruding anterior "saw." Some species of stingrays can deliver an electric shock. Their large electric organs, located at the base of their pectoral fins, can discharge over 300 volts. Skates resemble stingrays but possess two dorsal fins and a caudal fin.

The chimaeras, or ratfishes, are a group of cartilaginous fishes that live in cold marine waters. They are known for their unusual shape and iridescent colors.

Bony Fishes

The majority of living vertebrates, approximately 25,000 species, are **bony fishes** (Osteichthyes). The bony fishes range in size from gobies, which are less than 7.5 mm long, to the giant sturgeons, which can obtain a length of 4 m.

The majority of fish species are **ray-finned bony fishes** with fan-shaped fins supported by a thin, bony ray. These fishes are the most successful and diverse of all the vertebrates (Fig. 29.9). Some, such as herrings, are filter feeders; others, such as trout, are opportunists; and still others are predaceous carnivores, such as piranhas and barracudas.

Video
Cichlid Specialization

Despite their diversity, bony fishes have many features in common. They lack external gill slits, and instead, their gills are covered by an *operculum*. Many bony fishes have a **swim bladder,** a gas-filled sac into which they can secrete gases or from which they can absorb gases, altering its pressure. This results in a change in the fishes' buoyancy and, therefore, their depth in the water. Bony fishes have a single-loop, closed cardiovascular system (see Fig. 29.11a).

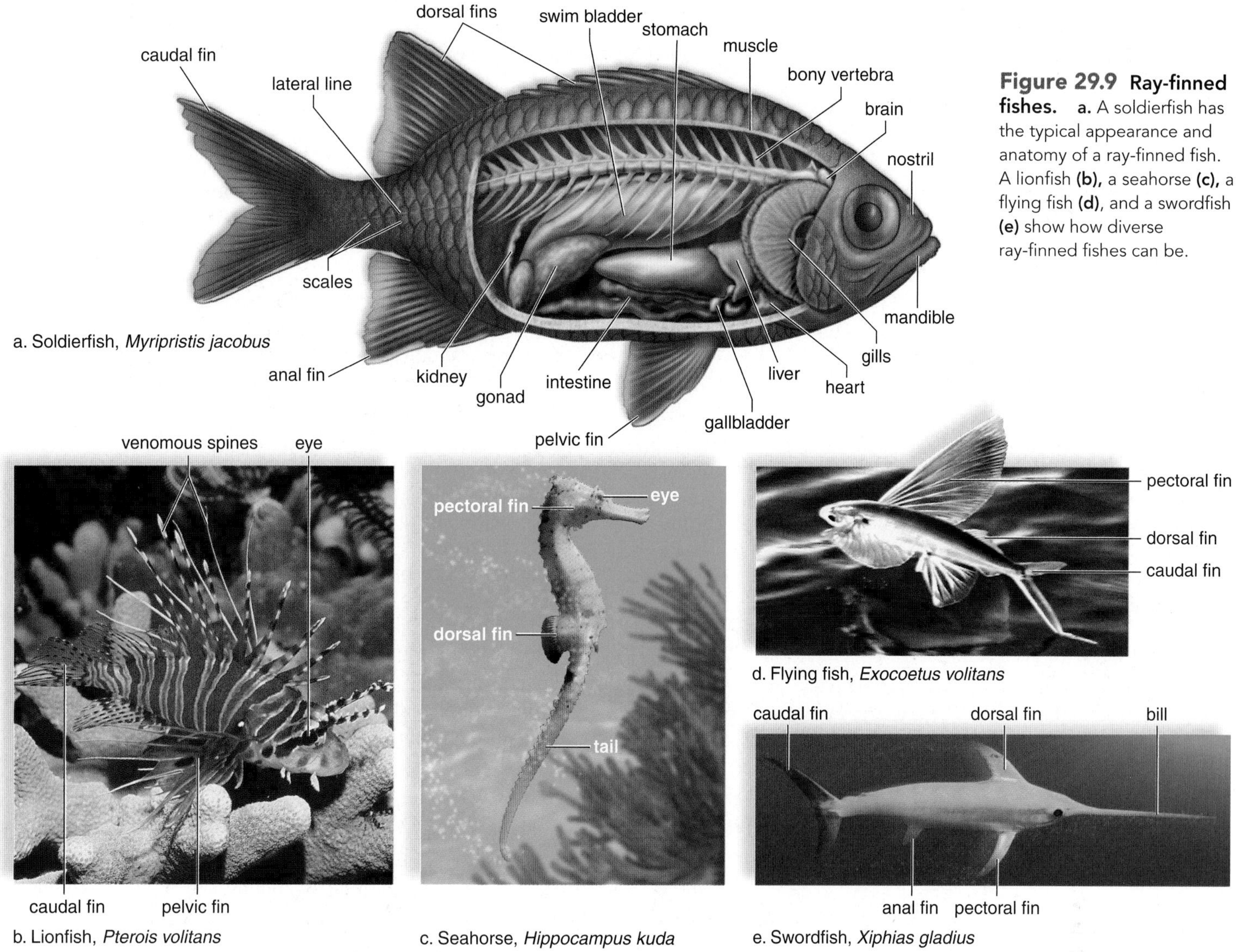

caudal fin
lateral line
dorsal fins
swim bladder
stomach
muscle
bony vertebra
brain
nostril
scales
a. Soldierfish, *Myripristis jacobus*
anal fin
kidney
gonad
intestine
pelvic fin
gallbladder
liver
heart
gills
mandible

Figure 29.9 Ray-finned fishes. a. A soldierfish has the typical appearance and anatomy of a ray-finned fish. A lionfish (**b**), a seahorse (**c**), a flying fish (**d**), and a swordfish (**e**) show how diverse ray-finned fishes can be.

venomous spines
eye
caudal fin
pelvic fin
b. Lionfish, *Pterois volitans*

pectoral fin
eye
dorsal fin
tail
c. Seahorse, *Hippocampus kuda*

pectoral fin
dorsal fin
caudal fin
d. Flying fish, *Exocoetus volitans*

caudal fin
dorsal fin
bill
anal fin
pectoral fin
e. Swordfish, *Xiphias gladius*

The nervous system and brain in bony fishes are well developed, and complex behaviors are common. Bony fishes have separate sexes, and the majority of species undergo external fertilization after females deposit eggs and males deposit sperm into the water.

Lobe-Finned Fishes. **Lobe-finned fishes** possess fleshy fins supported by bones. Ancestral lobe-finned fishes gave rise to modern-day lobe-finned fishes, the lungfishes, and to the amphibians. **Lungfishes** have lungs and also gills for gas exchange. The lobe-finned fishes and lungfishes are grouped together as the Sarcopterygii. Today, only two species of lobe-finned fishes are known (the coelacanths), and six species of lungfishes. Lungfishes live in Africa, South America, and Australia, either in stagnant fresh water or in ponds that dry up annually.

In 1938, a coelacanth was caught from the deep waters of the Indian Ocean off the eastern coast of South Africa. It took the scientific world by surprise because these animals were thought to be extinct for 70 million years. Approximately 200 coelacanths have been captured since that time (Fig. 29.10).

 **Animation** Early Vertebrates

 **Animation** Bony Fish

Check Your Progress   **29.3**

1. List and describe the characteristics that fishes have in common.
2. Distinguish between lobe-finned and ray-finned bony fishes.

Figure 29.10 Coelacanth, *Latimeria chalumnae*. A coelacanth is a lobe-finned fish once thought to be extinct.

lobed fins

29.4 The Amphibians

Learning Outcomes

Upon completion of this section, you should be able to

1. List the seven characteristics that define the amphibians.
2. Describe features of the three groups of living amphibians.
3. Summarize the two hypotheses that explain the evolution of amphibians from lobe-finned fishes.

Amphibians (class Amphibia [Gk. *amphibios*, living both on land and in water]) were abundant during the Carboniferous period and exhibit these characteristics:

Limbs. Typically, amphibians are tetrapods, as mentioned earlier. The skeleton, particularly the pelvic and pectoral girdles, is well developed to promote locomotion.

Smooth and nonscaly skin. The skin, which is kept moist by mucous glands, plays an active role in water balance and respiration and can also help in temperature regulation when on land through evaporative cooling. A thin, moist skin does mean, however, that most amphibians stay close to water, or else risk drying out.

Lungs. If lungs are present, they are relatively small, and respiration is supplemented by exchange of gases across the porous skin (called cutaneous respiration).

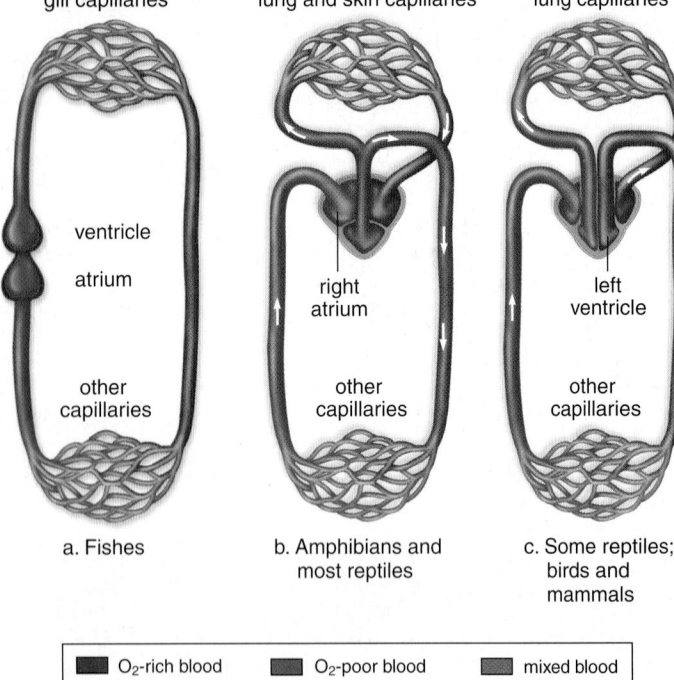

Figure 29.11 Vertebrate circulatory pathways. a. The single-loop pathway of fishes has a two-chambered heart. **b.** The double-loop pathway of other vertebrates sends blood to the lungs and to the body. In amphibians and most reptiles, limited mixing of oxygen-rich and oxygen-poor blood takes place in the single ventricle of their three-chambered heart. **c.** The four-chambered heart of some reptiles (crocodilians and birds) and mammals sends only oxygen-poor blood to the lungs and oxygen-rich blood to the body.

Double-loop circulatory pathway. (Fig. 29.11*b, c*). A three-chambered heart with a single ventricle and two atria pumps blood to both the lungs and to the body.

Sense organs. Special senses, such as sight, hearing, and smell, are fine-tuned for life on land. Amphibian brains are larger than those of fish, and the cerebral cortex is more developed. These animals have a specialized tongue for catching prey, eyelids for keeping their eyes moist, and a sound-producing larynx.

Ectothermy. Like fishes, amphibians are ectotherms, but they are able to live in environments where the temperature fluctuates greatly. During winters in the temperate zone, they become inactive and enter torpor. The European common frog can survive in temperatures dropping to as low as −6°C.

Aquatic reproduction. Their name, amphibians, is appropriate because many return to water for the purpose of reproduction. They deposit their eggs and sperm into the water, where external fertilization takes place. Generally, the eggs are protected only by a jelly coat and not by a shell. When the young hatch, they are tadpoles (aquatic larvae with gills) that feed and grow in the water. After amphibians undergo a *metamorphosis* (change in form), they emerge from the water as adults that breathe air. Some amphibians, however, have evolved mechanisms that allow them to bypass this aquatic larval stage and reproduce on land.

Evolution of Amphibians

Amphibians evolved from the lobe-finned fishes with lungs by way of transitional forms. Two hypotheses have been suggested to account for the evolution of amphibians from lobe-finned fishes. Perhaps lobe-finned fishes had an advantage over others because they could use their lobed fins to move from pond to pond. Or, perhaps the supply of food on land in the form of plants and insects—and the absence of predators—promoted further adaptations to the land environment. Paleontologists have recently found a well-preserved transitional fossil from the late Devonian period in Arctic Canada that represents an intermediate between lobe-finned fishes and tetrapods with limbs. This fossil, named *Tiktaalik roseae* (see Chapter 15), provides unique insights into how the legs of tetrapods arose (Fig. 29.12).

Diversity of Living Amphibians

The amphibians of today occur in three groups: salamanders and newts; frogs and toads; and caecilians. Salamanders and newts have elongated bodies, long tails, and usually two pairs of limbs (Fig. 29.13*a*). Salamanders and newts range in size from less than 15 cm to the giant Japanese salamander, which exceeds 1.5 m in length. Most have limbs that are set at right angles to the body and resemble the earliest fossil amphibians. They move like a fish, with side-to-side, sinusoidal (S-shaped) movements.

Both salamanders and newts are carnivorous, feeding on small invertebrates such as insects, slugs, snails, and worms. Salamanders practice internal fertilization; in most, males produce a sperm-containing spermatophore that females pick up

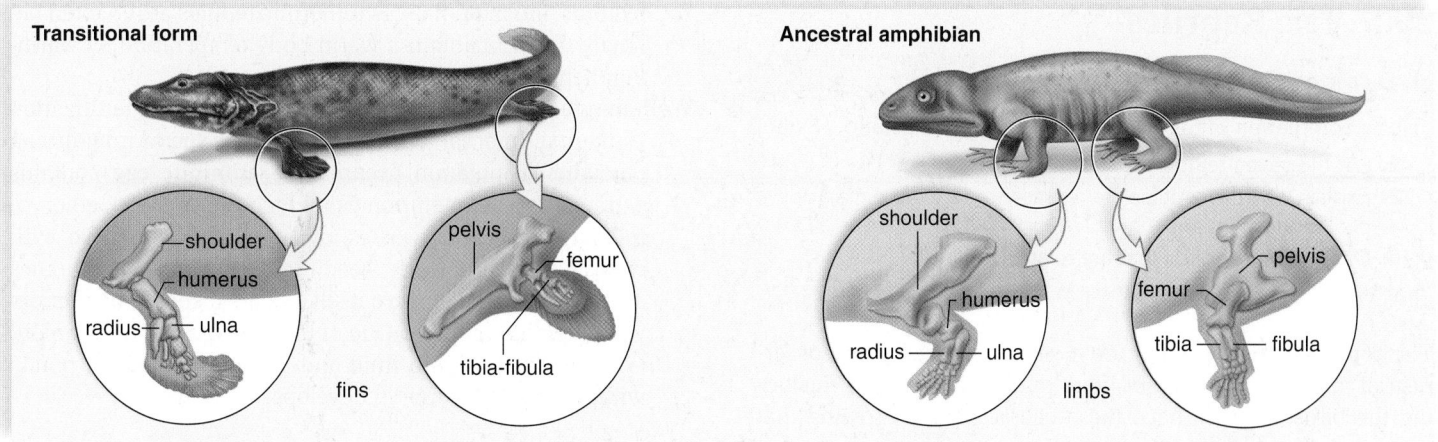

Figure 29.12 Lobe-finned fishes to amphibians.   This transitional form links the lobes of lobe-finned fishes to the limbs of ancestral amphibians. Compare the fins of the transitional form (*left*) to the limbs of the ancestral amphibian (*right*).

with their *cloaca* (the terminal chamber common to the urinary, digestive, and genital tracts). Then the fertilized eggs are laid in water or on land, depending on the species. Some amphibians, such as the mudpuppy of eastern North America, remain in the water and retain the gills of the larva.

Frogs and toads, which range in length from less than 1 cm to 30 cm, are common in subtropical to temperate to desert climates around the world. In these animals, which lack tails as adults, the head and trunk are fused, and the long hindlimbs are specialized for jumping (Fig. 29.13*b*). All species are carnivorous and have a tremendous array of specializations depending on their habitats. Glands in the skin secrete poisons that make the animal distasteful to eat and protect them from microbial infections. Some tropical species with brilliant fluorescent green and red coloration are particularly poisonous (see Fig. 29A*a*). Colombian Indians dip their

darts in the deadly secretions of these frogs, aptly called poison-dart frogs. The tree frogs have adhesive toepads that allow them to climb trees, while others, the spadefoots, have hardened spades that act as shovels enabling them to dig into the soil.

Video
Frog Reproduction

Caecilians are legless, often sightless, worm-shaped amphibians that range in length from about 10 cm to more than 1 m (Fig. 29.13*c*). Most burrow in moist soil, feeding on worms and other soil invertebrates. Some species have folds of skin that make them look like a segmented earthworm.

Check Your Progress   29.4

1. List the characteristics that amphibians have in common.
2. Describe the usual life cycle of amphibians.

a. Barred tiger salamander,
   Ambystoma tigrinum

b. Tree frog,
   Hyla andersoni

c. Caecilian,
   Caecilia nigricans

Figure 29.13 Amphibians.   Living amphibians are divided into three orders: **a.** Salamanders and newts. Members of this order have a tail throughout their lives and, if present, unspecialized limbs. **b.** Frogs and toads. Like this frog, members of this order are tailless and have limbs specialized for jumping. **c.** The caecilians are wormlike burrowers.

29.5 The Reptiles

The **reptiles** (class Reptilia) are a very successful group of terrestrial animals consisting of more than 17,000 species, including the birds. Reptiles have these characteristics showing that they are fully adapted to life on land:

Paired limbs. Two pairs of limbs, usually with five toes each. Reptiles are adapted for climbing, running, paddling, or flying.

Skin. A thick and dry skin is impermeable to water. Therefore, the skin prevents water loss. In reptiles, the skin is wholly or in part scaly (Fig. 29.14). Many reptiles (e.g., snakes and lizards) molt several times a year.

Efficient breathing. The lungs are more developed than in amphibians. Also in many reptiles, an expandable rib cage assists breathing.

Efficient circulation. The heart prevents mixing of blood. A septum divides the ventricle either partially or completely. If it partially divides the ventricle, the mixing of oxygen (O_2)-poor blood and oxygen-rich blood is reduced. If the septum is complete, oxygen-poor blood is completely separated from oxygen-rich blood (see Fig. 29.11c).

Efficient excretion. The kidneys are well developed. The kidneys excrete uric acid, and therefore less water is required to rid the body of nitrogenous wastes.

Ectothermy. Most reptiles are ectotherms, and this allows them to survive on a fraction of the food per body weight required by birds and mammals. Ectothermic reptiles are adapted behaviorally to maintain a warm body temperature by warming themselves in the sun.

Well-adapted reproduction. Sexes are separate and fertilization is internal. Internal fertilization prevents sperm from drying out when copulation occurs. The **amniotic egg** contains extraembryonic membranes, which protect the embryo, remove nitrogenous wastes, and provide the embryo with oxygen, food, and water (see Fig. 29.16). These membranes are not part of the embryo itself and are disposed of after development is complete. One of the membranes, the amnion, is a sac that fills with fluid and provides a "private pond" within which the embryo develops.

Evolution of Amniotes

An ancestral amphibian gave rise to the amniotes at some point in the Carboniferous period, beginning some 359 MYA. The amniotes include animals now classified as the reptiles (including birds) and the mammals. The embryo of an amniote has extracellular membranes, including an amnion (see Fig. 29.16).

Figure 29.15 shows that the amniotes consist of three lineages: (1) the turtles, in which the skull is **anapsid**, that is, it has no openings behind the orbit—eye socket; (2) all the other reptiles including the birds, in which the skull is **diapsid**, or has two openings behind the orbit; and (3) the mammals, in which the skull is **synapsid** and has one opening behind the orbit.

The reptile group is an artificial grouping because it has no common ancestor. In other words, reptiles are a paraphyletic group and not a monophyletic group. Therefore, authorities are in the process of dividing the reptiles into a number of monophyletic groups. For example, the classical view of reptile evolution considers the turtles, or anapsids, as an independent lineage, separate from that of the rest of the reptiles, which are diapsids. The modern view of turtle evolution places turtles within the archosaurs with birds and crocodiles. This would mean that the anapsids are really just highly specialized diapsids. The

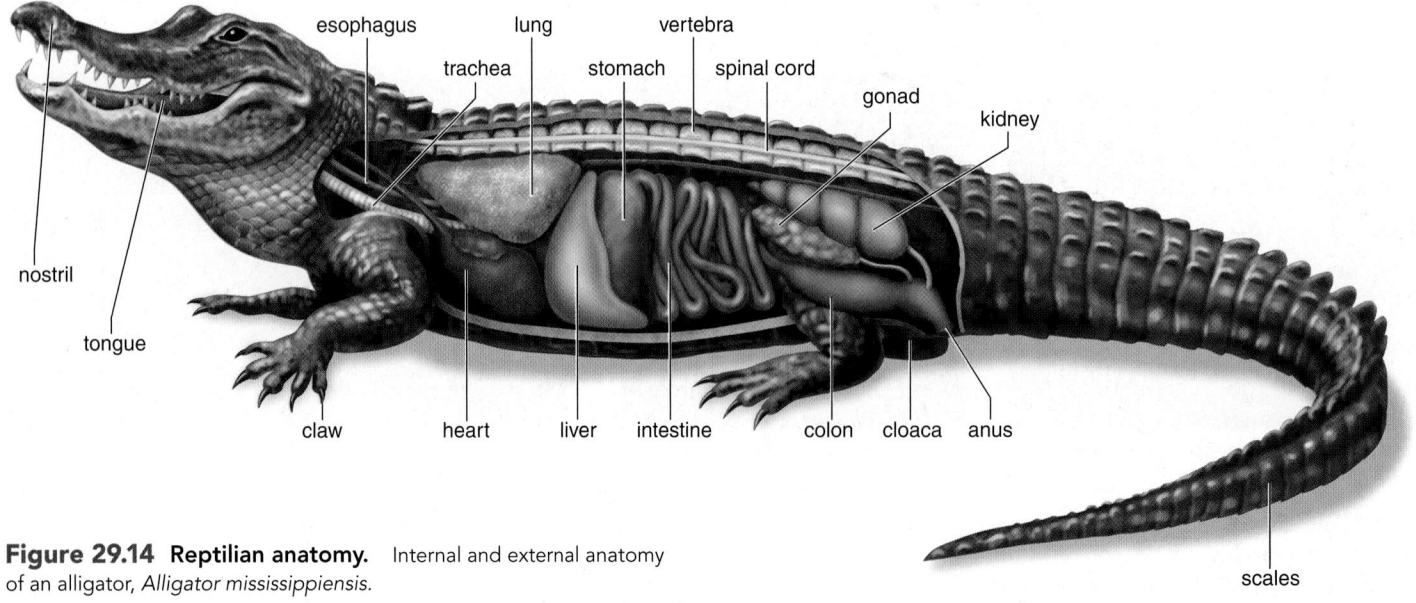

Figure 29.14 Reptilian anatomy. Internal and external anatomy of an alligator, *Alligator mississippiensis*.

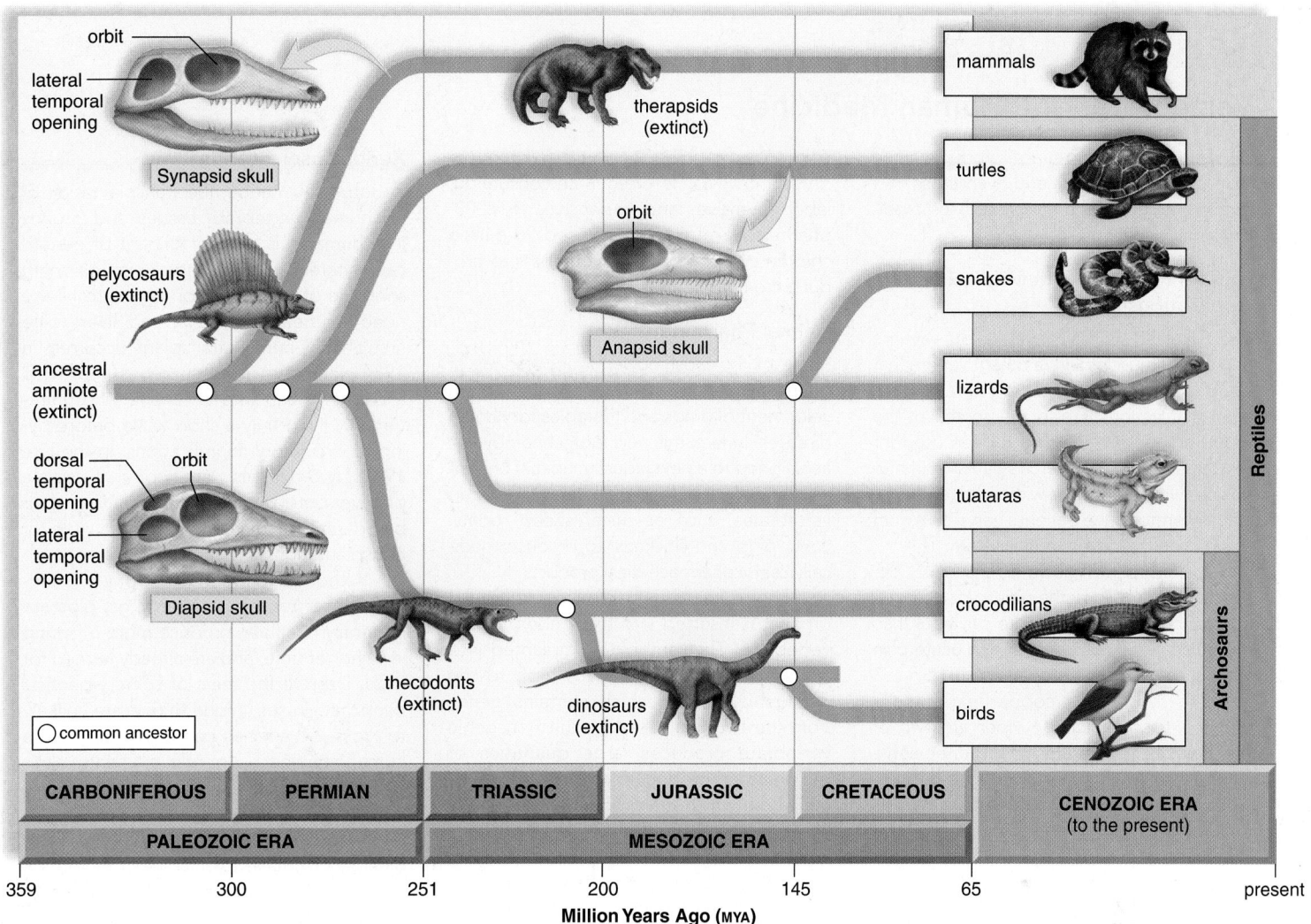

Figure 29.15 Timeline of the evolution of the amniotes. This diagram shows an overview of the presumed evolutionary relationships among amniotes, including the major groups of reptiles. The amniote ancestor evolved in the Paleozoic era. Historically, the openings in the skull were considered evidence that there are two major groups of reptiles, with turtles separate from the other reptile groups. This historical understanding of turtle evolution is presented in this figure. A modern view, not shown, proposes that the turtles are archosaurs along with crocodiles and birds.

evidence in support of this modern view is still being debated, and thus the classical view of reptile evolution is presented here (Figure 29.15).

According to the classical view, all other reptiles except the turtles are diapsids because they have a skull with two openings behind the eyes. The thecodonts are diapsids that gave rise to the ichthyosaurs, which returned to the aquatic environment, and the pterosaurs of the Jurassic period, which had a keel for the attachment of large flight muscles and air spaces in their bones to reduce weight. Their wings were membranous and supported by elongated bones of the fourth finger. *Quetzalcoatlus,* the largest flying animal ever to live, had an estimated wingspan of nearly 13.5 m.

Of interest to us, the thecodonts gave rise to the crocodiles and dinosaurs. A sequence of now known transitional forms occurs between the dinosaurs and the birds. The crocodilians and birds share derived features, such as skull openings in front of the eyes and clawed feet. It is customary now to use the designation archosaurs for the crocodilians, dinosaurs, and birds.

This means that these animals are more closely related to each other than they are to snakes and lizards.

The **dinosaurs** varied greatly in size and behavior. The average size of a dinosaur was about the size of a chicken. Some of the dinosaurs, however, were the largest land animals ever to live. *Brachiosaurus,* a herbivore, was about 23 m long and about 17 m tall. *Tyrannosaurus rex,* a carnivore, was 5 m tall when standing on its hind legs. A bipedal stance freed the forelimbs and allowed them to be used for purposes other than walking, such as manipulating prey. It was also preadaptive for the evolution of wings in the birds.

Dinosaurs dominated the Earth for about 170 million years before they died out at the end of the Cretaceous period, 65 MYA. One hypothesis for this mass extinction is that a massive meteorite struck the Earth near the Yucatán Peninsula (see Section 18.3). The resultant cataclysmic events disrupted existing ecosystems, destroying many living things. This hypothesis is supported by the presence of a layer of the mineral iridium, which is rare on Earth but common in meteorites in the late Cretaceous strata.

Nature of Science

Vertebrates and Human Medicine

Hundreds of pharmaceutical products come from other vertebrates, and even those that produce poisons and toxins give us medicines that benefit us.

Natural Products with Medical Applications

The Thailand cobra paralyzes its victim's nerves and muscles with a potent venom that eventually leads to respiratory arrest. However, that venom is also the source of the drug Immunokine, which has been used for ten years in multiple sclerosis patients. Immunokine, which is almost without side effects, actually protects the patient's nerve cells from destruction by their immune system.

A compound known as ABT-594, derived from the skin of the poison-dart frog, is approximately 50 times more powerful than morphine in relieving chronic and acute pain without the addictive properties.

The southern copperhead snake and the fer-de-lance pit viper are two of the unlikely vertebrates that either serve as the source of pharmaceuticals or provide a chemical model for the synthesis of effective drugs in the laboratory. These drugs include anticoagulants ("clot busters"), painkillers, antibiotics, and anticancer drugs.

A variety of friendlier vertebrates produce proteins that are similar enough to human proteins to be used for medical treatment. Until 1978, when recombinant DNA human insulin was produced, diabetics injected insulin purified from pigs. Currently, the flu vaccine is produced in fertilized chicken eggs. The production of these drugs, however, is often time-consuming, labor intensive, and expensive. In 2003, pharmaceutical companies used 90 million chicken eggs and took nine months to produce the flu vaccine.

Animal Pharming

Some of the most powerful applications of genetic engineering can be found in the development of drugs and therapies for human diseases. In fact, this new biotechnology has actually led to a new industry: animal pharming. Animal pharming uses genetically altered vertebrates, such as mice, sheep, goats, cows, pigs, and chickens, to produce medically useful pharmaceutical products.

The human gene for some useful product is inserted into the embryo of the vertebrate. That embryo is implanted into a foster mother, which gives birth to the transgenic animal, which contains genes from the two sources. An adult transgenic vertebrate produces large quantities of the pharmed product in its blood, eggs, or milk, from which the product can be easily harvested and purified.

An example of a pharmed product advanced in development and in the FDA approval process is ATIII, a bioengineered form of human antithrombin. This medication is important in the treatment of individuals who have a hereditary deficiency of this protein and so are at high risk for life-threatening blood clots, especially during such events as surgery or childbirth procedures.

Xenotransplantation

Xenotransplantation, the transplantation of nonhuman vertebrate tissues and organs into humans, is another benefit of genetically altered animals. There is an alarming shortage of human donor organs to fill the need for hearts, kidneys, and livers. The first animal–human transplant occurred in 1984 when a team of surgeons implanted a baboon heart into an infant, who unfortunately lived only a short while before dying of circulatory complications. In the late 1990s, two patients were kept alive using a pig liver outside of their body to filter their blood until a human organ was available for transplantation.

Although baboons are phylogenetically closer to humans than pigs, pigs are generally healthier, produce more offspring in a shorter time, and are already farmed for food. Despite the fears of some, scientists think that viruses unique to pigs are unlikely to cross the species barrier and infect the human recipient. Currently, pig heart valves and skin are routinely used for treatment of humans. Miniature pigs, whose heart size is similar to humans, are being genetically engineered to make their tissues less foreign to the human immune system, in order to avoid rejection.

Questions to Consider

1. Is it ethical to change the genetic makeup of vertebrates in order to use them as drug or organ factories?
2. What are some of the health concerns that may arise due to xenotransplantation?

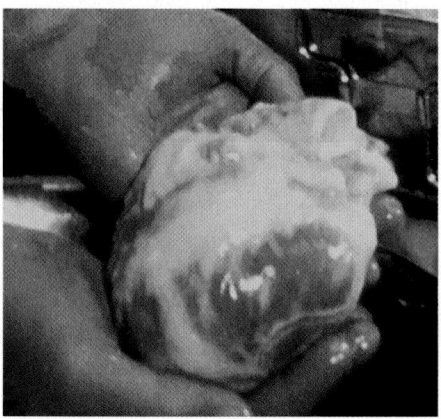

a. Poison-dart frogs, source of a medicine   b. Pigs, source of organs   c. Heart for transplantation

Figure 29A Use of other vertebrates for medical purposes. **a.** The poison-dart frog is the source of a pain medication. **b.** Pigs are now being genetically altered to provide a supply of **(c)** hearts for heart transplant operations.

Diversity of Living Reptiles

Living reptiles are represented by turtles, lizards, snakes, tuataras, crocodilians, and birds. Figure 29.16 shows representatives of all but the birds.

Along with tortoises, turtles can be found in marine, freshwater, and terrestrial environments. Most turtles have ribs and thoracic vertebrae that are fused into a heavy shell. They lack teeth but have a sharp beak. The legs of sea turtles are flattened and paddlelike (Fig. 29.16a), while terrestrial tortoises have strong limbs for walking.

Lizards have four clawed feet and resemble their prehistoric ancestors in appearance (Fig. 29.16b), although some species have lost their limbs and superficially resemble snakes. Typically, they are carnivorous and feed on insects and small animals, including other lizards. Marine iguanas of the Galápagos Islands are adapted to spending time each day at sea, where they feed on sea lettuce and other algae. Chameleons are adapted to live in trees and have long, sticky tongues for catching insects some distance away. They can change color to blend in with their background. Geckos are primarily nocturnal lizards with adhesive pads on their toes. Skinks are common elongated lizards with reduced limbs and shiny scales. Monitor lizards and Gila monsters, despite their names, are generally not a dangerous threat to humans.

Video Basilisk Lizards

Video Leaf-Tailed Gecko

Although most snakes (Fig. 29.16c) are harmless, several venomous species, including rattlesnakes, cobras, mambas, and copperheads, have given the whole group a reputation of being dangerous. Snakes evolved from lizards and have lost their limbs as an adaptation to burrowing. A few species such as pythons and boas still possess the vestiges of pelvic girdles. Snakes are carnivorous and have a jaw that is loosely attached to the skull; therefore, they can eat prey that is much larger than their head size. When snakes and lizards flick out their tongues, they are collecting airborne molecules and transferring them to a *Jacobson's organ* at the roof of the mouth and sensory cells on the floor of the mouth. The Jacobson's organ is an olfactory organ for the analysis of airborne chemicals. Snakes possess internal ears that are capable of detecting low-frequency sounds and vibrations. Their ears lack external ear openings.

Video Two-Headed Snake

Video Snake Eating

Two species of tuataras are found in New Zealand (Fig. 29.16d). They are lizardlike animals that can attain a length of 66 cm and can live for nearly 80 years. These animals possess a well-developed "third" eye, known as a pineal eye, which is light sensitive and buried beneath the skin in the upper part of the head. The tuataras are the only member of an ancient group of reptiles that included the common ancestor of modern lizards and snakes.

The majority of crocodilians (including alligators and crocodiles) live in fresh water feeding on fishes, turtles, and terrestrial animals that venture too close to the water. They have long, powerful jaws (Fig. 29.16e) with numerous teeth and a muscular tail that

a. Green sea turtle, *Chelonia mydas*

b. Gila monster, *Heloderma suspectum*

c. Diamondback rattlesnake, *Crotalus atrox*

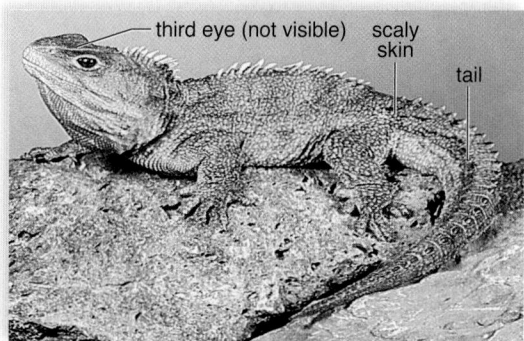

d. Tuatara, *Sphenodon punctatus*

e. American crocodile, *Crocodylus acutus*

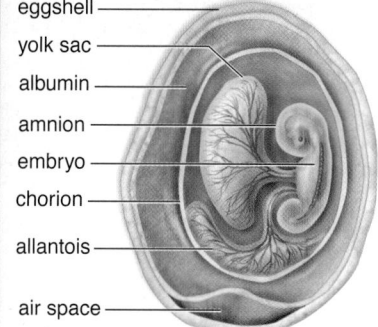

Figure 29.16 Reptilian diversity other than birds. Representative living reptiles include **(a)** green sea turtles, **(b)** the venomous Gila monster, **(c)** the diamondback rattlesnake, and **(d)** the tuatara. **e.** A young crocodile hatches from an egg. The eggshell is leathery and flexible, not brittle like birds' eggs. Inside the egg, the embryo is surrounded by three membranes. The chorion aids in gas exchange, the allantois stores waste, and the amnion encloses a fluid that prevents drying out and provides protection. The yolk sac provides nutrients for the embryo.

serves as both a weapon and a paddle. Male crocodiles and alligators bellow to attract mates. In some species, the male protects the
eggs and cares for the young.

Birds

Birds share a common ancestor with crocodilians and have traits
such as the presence of scales (feathers are modified scales), a
tail with vertebrae, and clawed feet that show they are indeed
reptiles.

To many people, birds are the most conspicuous, melodic,
beautiful, and fascinating group of vertebrates. Birds range in
size from the tiny "bee" hummingbird at 1.8 g (less than a
penny) and 5 cm long to the ostrich at a maximum weight of
160 kg and a height of 2.7 m.

Nearly every anatomical feature of a bird can be related to
its ability to fly (Fig. 29.17). These features are involved in the
action of flight, providing energy for flight or the reduction of
the bird's body weight, making flight less energetically costly:

Feathers. Soft down keeps birds warm, wing feathers allow
flight, and tail feathers are used for steering. A feather is a
modified reptilian scale with the complex structure shown
in Figure 29.17a. Nearly all birds molt (lose their feathers)
and replace their feathers about once a year.

Modified skeleton. Unique to birds, the collarbone is fused (the
wishbone), and the sternum has a keel (Fig. 29.17b). Many
other bones are fused, making the skeleton more rigid than
the reptilian skeleton. The breast muscles are attached to
the keel, and their action accounts for a bird's ability to fly.

A horny beak has replaced jaws equipped with teeth,
and a slender neck connects the head to a rounded, compact torso.

Modified respiration. In birds, unlike other reptiles, the lobular
lungs connect to anterior and posterior air sacs. The presence of these sacs means the air circulates one way through
the lungs, and gases are continuously exchanged across
respiratory tissues. Another benefit of air sacs is that they

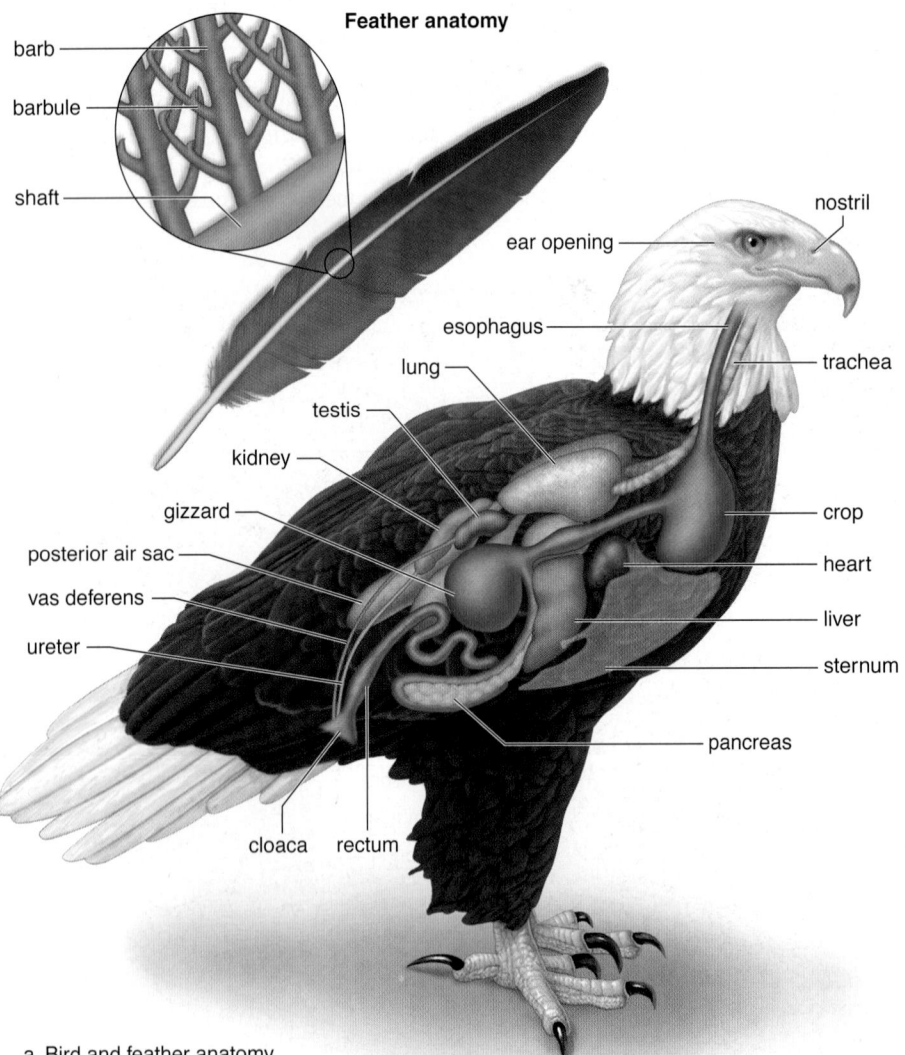

a. Bird and feather anatomy

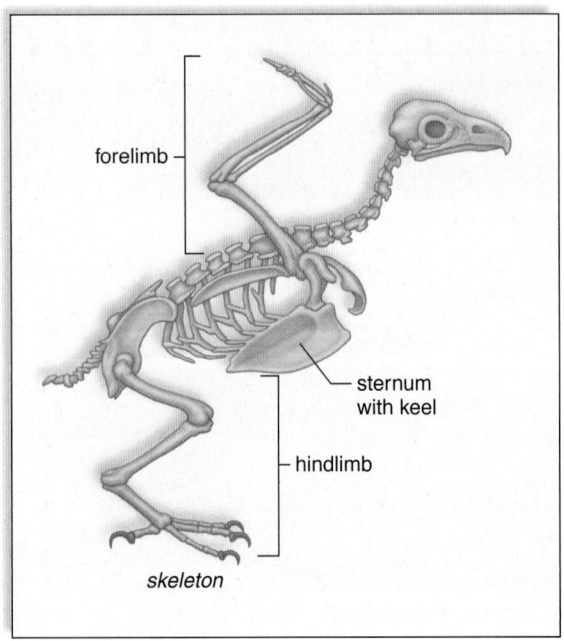

Figure 29.17   Bird anatomy and flight.
a. Bird anatomy. *Top:* In feathers, a hollow central
shaft gives off barbs and barbules, which interlock in
a latticelike array. *Bottom:* The anatomy of an eagle
is representative of bird anatomy. **b.** Bird flight. The
skeleton of an eagle shows that birds have a large,
keeled sternum to which flight muscles attach. The
bones of the forelimb help support the wings.

b. Bald eagle, *Haliaetus*

lighten the body and bones for flying. Some of the air sacs are present in cavities within the bones.

Endothermy. Birds, unlike other reptiles, generate internal heat. Many **endotherms** can use metabolic heat to maintain a constant internal temperature. Endothermy may be associated with their efficient nervous, respiratory, and circulatory systems.

Well-developed sense organs and nervous system. Birds have particularly acute vision and well-developed brains. Their muscle reflexes are excellent. An enlarged portion of the brain seems to be the area responsible for instinctive behavior.

A ritualized courtship often precedes mating. Many newly hatched birds require parental care before they are able to fly away and seek food for themselves.

A remarkable aspect of bird behavior is the seasonal migration of many species over long distances. Birds navigate by day and night, whether it is sunny or cloudy, by using the Sun and stars and even the Earth's magnetic field to guide them.

Birds are very vocal animals. Their vocalizations are distinctive and so convey an abundance of information.

Diversity of Living Birds

The majority of birds, including eagles, geese, and mockingbirds, have the ability to fly. However, some birds, such as emus, penguins, kiwis, and ostriches, are flightless. Traditionally, birds have been classified according to beak and foot type (Fig. 29.18) and, to some extent, on their habitat and behavior. The birds of prey have notched beaks and sharp talons; shorebirds have long, slender, probing beaks and long, stiltlike legs; woodpeckers have sharp, chisel-like beaks and grasping feet; waterfowl have broad beaks and webbed toes; penguins have wings modified as paddles; songbirds have perching feet; and parrots have short, strong pliers-like beaks and grasping feet.

 Video Finches Adaptive Radiation

 Video Harris Hawk

Check Your Progress   29.5

1. Contrast the characteristics of crocodilians to those of snakes.
2. Explain what features indicate that birds are reptiles.

a. Bald eagle, *Haliaetus leucocephalus*

b. Pileated woodpecker, *Dryocopus pileatus*

c. Blue-and-yellow macaw, *Ara ararauna*

d. Cardinal, *Cardinalis cardinalis*

Figure 29.18 Bird beaks. a. A bald eagle's beak allows it to tear apart prey. **b.** A woodpecker's beak is used to chisel in wood. **c.** A parrot's beak is modified to pry open nuts. **d.** A cardinal's beak allows it to crack tough seeds.

29.6 The Mammals

Learning Outcomes

Upon completion of this section, you should be able to

1. Describe five features of mammals.
2. Identify unique features that define each of the three living lineages of mammals.
3. Explain why mammals underwent adaptive radiation after the demise of dinosaurs.

The **mammals** (class Mammalia) include the largest animal ever to live, the blue whale (130 metric tons); the smallest mammal, the Kitti's bat (1.5 g); and the fastest land animal, the cheetah (110 km/hr). These characteristics distinguish mammals:

Hair. The most distinguishing characteristics of mammals are the presence of hair and milk-producing mammary glands. Hair provides insulation against heat loss, and being endothermic allows mammals to be active even in cold weather. The color of hair can camouflage a mammal and help the animal blend into its surroundings. In addition, hair can be ornamental and can serve sensory functions.

Mammary glands. These glands enable females to feed (nurse) their young without having to leave them to collect food, as birds do. Nursing also creates a bond between mother and offspring that helps ensure parental care while the young are helpless, and provides antibodies to the young from the mother through the milk.

Skeleton. The mammalian skull accommodates a larger brain relative to body size than does the reptilian skull. Also, mammalian cheek teeth are differentiated as premolars and molars. The vertebrae of mammals are highly differentiated; typically the middle region of the vertebral column is arched, and the limbs are under the body rather than out to the sides.

Internal organs. Efficient respiratory and circulatory systems ensure a ready oxygen supply to muscles whose contraction produces body heat. Like birds, mammals have a double-loop circulatory pathway and a four-chambered heart. The kidneys are adapted to conserving water in terrestrial mammals. The nervous system of mammals is highly developed. Special senses in mammals are well developed, and mammals exhibit complex behavior.

Internal development. In most mammals, the young are born alive after a period of development in the uterus, a part of the female reproductive tract. Internal development shelters the young and allows the female to move actively about while the young are maturing.

Evolution of Mammals

Mammals share an amniote ancestor with reptiles (see Fig. 29.15). Their more immediate ancestors in the Mesozoic era had a synapsid skull. The first true mammals appeared during the Triassic period, about the same time as the first dinosaurs, and were similar in size to mice. During the reign of the dinosaurs (170 million years), mammals were a minor group that changed little.

The common ancestor of all three mammal groups appeared in the late Triassic–early Jurassic period, about 200 MYA (Fig. 29.19).

The earliest mammalian group is the monotremes. The marsupials probably originated in the Americas and then spread through South America and Antarctica to Australia before these continents separated. Placental mammals, the third branch of the mammalian lineage (Fig. 29.19), originated in Eurasia and spread to the Americas also by land connections that existed between the continents during the Mesozoic. The placental mammals underwent an adaptive radiation into the habitats previously occupied by the dinosaurs.

Monotremes

Monotremes [Gk. *monos*, one, and *trema*, hole] are egg-laying mammals that include only the duckbill platypus (Fig. 29.20a) and two species of echidna, or spiny anteaters. The term monotreme refers to the presence of a single urogenital opening, the cloaca, which is a shared extretory and reproductive canal. Monotremes, unlike other mammals, lay hard-shelled amniotic eggs (Table 29.1). No embryonic development occurs inside the female's body. The female duckbill platypus lays her eggs in a burrow in the ground. She incubates the eggs, and after hatching, the young lick up milk that seeps from modified sweat glands on the mother's abdomen.

Echidnas, which actually feed mainly on termites, have pores that seep milk in a shallow belly pouch formed by skin folds on each side. The egg moves from the cloaca to this pouch, where hatching takes place and the young remain for about 53 days. Then they stay in a burrow, where the mother periodically visits and nurses them.

Marsupials

The **marsupials** [Gk. *marsupium*, pouch] are also known as the pouched mammals. Marsupials include kangaroos, koalas, Tasmanian devils, wombats, sugar gliders, and opossums. Marsupials have a true uterus (Table 29.1). The embryos of marsupials are nourished by a yolk-based type of placenta inside the female's body, but they are born in a very immature condition. Newborns are typically hairless and have yet to open their eyes, but they crawl up into a pouch on their mother's abdomen. Inside the pouch, they attach to nipples of mammary glands and continue to develop. Frequently, more are born than can be accommodated by the number of nipples, and it's "first come, first served."

Today, marsupial mammals are most abundant in Australia and New Guinea, filling all the typical roles of placental mammals on other continents. For example, among herbivorous marsupials in Australia today, koalas are tree-climbing browsers (Fig. 29.20b), and kangaroos are grazers. A significant number of marsupial species are also found in South and Central America. The opossum is the only North American marsupial (Fig. 29.20c).

Placental Mammals

The **placental mammals,** also known as the eutherians, are the dominant group of mammals on Earth. Developing placental mammals are dependent on the placenta (Table 29.1). The **placenta** in placental mammals is a very specialized organ for the exchange of substances between maternal blood and fetal blood. Nutrients are supplied to the growing offspring, and wastes are passed to the mother for excretion. Although the fetus is clearly parasitic on the female, she has the advantage of being able to freely move about while the fetus develops.

Placental mammals lead an active life. The senses are acute, and the brain is enlarged due to the convolution and expansion of the foremost part—the cerebral hemispheres. The brain is not

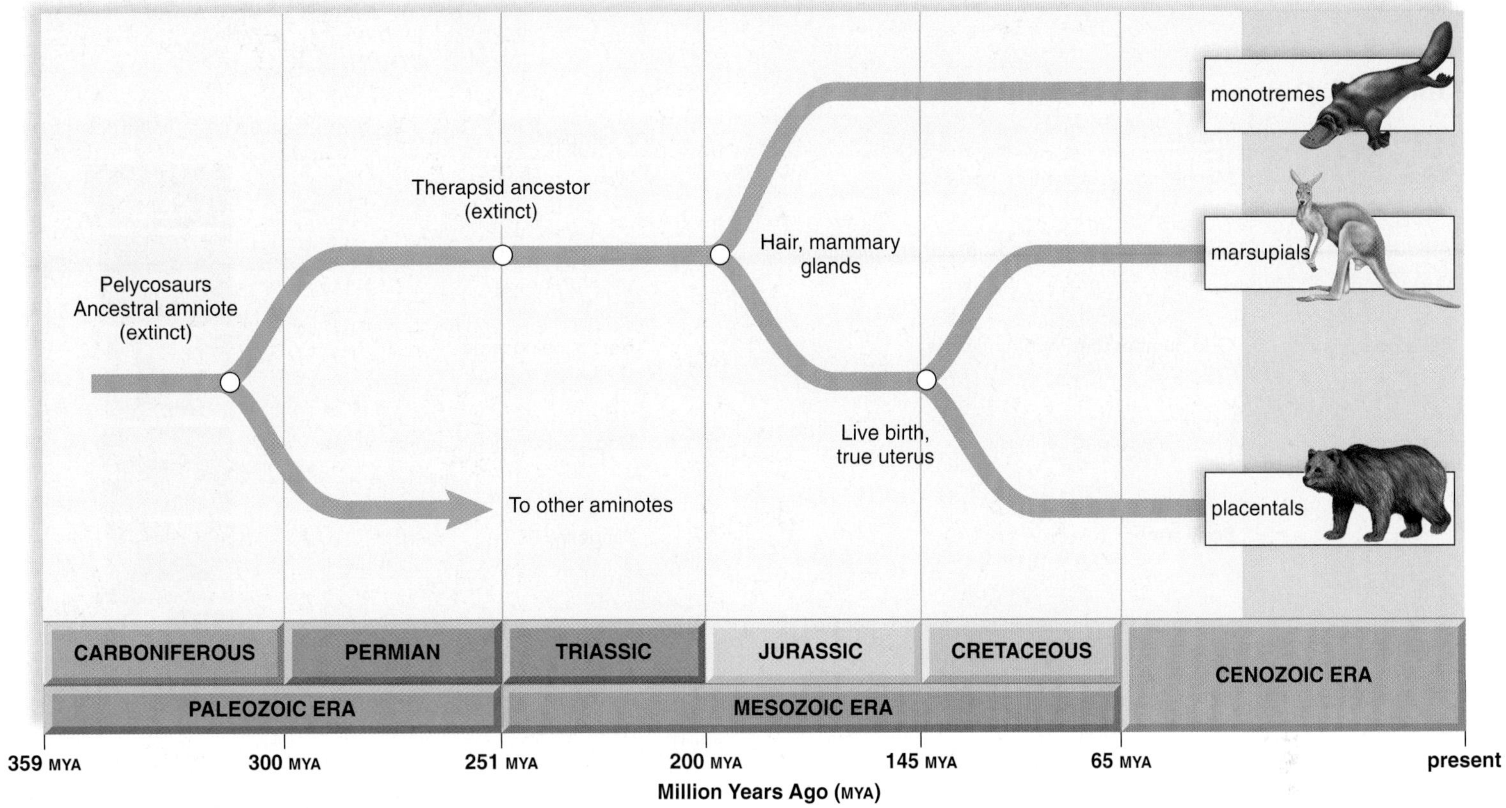

Figure 29.19 Timeline of the evolution of mammal groups. Only three lineages of mammals exist on Earth today, namely the monotremes, marsupials, and placentals (eutherians). Monotremes are the oldest group of mammals, have hair and mammary glands, but do not give live birth—they lay eggs. The marsupials and placental lineages evolved in the early Cretaceous. Both marsupials and placentals give live birth. In the placental mammals, however, the placenta is more complex and specialized than in the marsupials. Both marsupials and placentals have a true uterus but it is structurally different.

a. Duckbill platypus, *Ornithorhynchus anatinus*

b. Koala, *Phascolarctos cinereus*

c. Virginia opossum, *Didelphis virginianus*

Figure 29.20 Monotremes and marsupials. **a.** The duckbill platypus is a monotreme that inhabits Australian streams. **b.** The koala is an Australian marsupial that lives in trees. **c.** The opossum is the only marsupial in North America. The Virginia opossum is found in a variety of habitats.

Table 29.1 Female reproductive traits of mammal lineages.

	Milk	Nipples	Birth	Uterus	Placenta	Birth canal
monotremes	Yes	No	Eggs	None	None	Cloaca
marsupials	Yes	Yes	Live	True (>1)	Simple (yolk)	Vagina (>1)
placentals	Yes	Yes	Live	True (1)	Complex (tissue)	Vagina (1)

fully developed for some time after birth, and there is a long period of dependency on the parents, during which the young learn to take care of themselves.

Most mammals live on land, but some (e.g., whales, dolphins, seals, sea lions, and manatees) are secondarily adapted to live in water, and bats are able to fly. Although bats are the only mammal that can actually fly, three types of placentals can glide: the flying squirrels, scaly-tailed squirrels, and flying lemurs. There are 20 different orders of placental mammals (Table 29.2).

Humans are mammals, and in the following chapter we consider the evolutionary history of humans, an enormously successful mammalian group.

Video Mom Grizzly Teaches Her Cubs

Video Bat Echolocation

Check Your Progress 29.6

1. Identify two traits that are unique to mammals.
2. Describe features that distinguish the three groups of mammals.

Table 29.2 Orders of placental mammals

Order	Traits	Examples	
Cetacea	Marine, no fur, streamline bodies	Whales, dolphins	
Artiodactyla	Even-toed ungulates, horns or antlers, ruminants	Cattle, deer, antelope	
Perissodactyla	Odd number toes, non-ruminants	Horses, rhinoceros	
Carnivora	Meat eaters, long canines	Dogs, cats, weasels, minks, stoats	
Pholidota	Scaly armor, no teeth	Pangolin	
Chiroptera	Fly with membranous wings, some echolocate	Bats	
Erinaceomorpha**	Spines, insectivores	Hedgehogs	
Soricomorpha**	Small, high metabolism, insectivorous	Shrews, moles	
Rodentia	One pair of specialized incisors that grow continuously	Mice, rats, voles, beavers, squirrels	
Lagomorpha	Two pair of incisors that grow continuously, long hindlimbs	Rabbits, hares	
Dermoptera	Membrane between hands and feet	Colugo, or flying lemur	
Scandentia	Large forward-facing eyes, grasping hand	Tree shrew	
Primates	Opposable thumb, developed brains, social	Monkeys, apes	
Xenarthra	Special projections on the spine	Sloths, armadillos	
Afrocoricida**	Insectivores	Golden moles and tenrecs	
Macroscelidia	Small, long snout, long hindlimbs	Elephant shrews	
Tubulidentata	Insectivore, long snout, coarse fur	Aardvark	
Sirenia	Marine, long whiskers	Manatee, dugong	
Hyracoidea	Small mammals	Hyraxes	
Proboscidea	Largest land mammals, trunk, tusks	Elephants	

CONNECTING *the* CONCEPTS *with the* BIG IDEAS

Evolution

- Phylogenetic trees demonstrate how traits, such as opposable thumbs, the number of chambers in the heart, or legs in oceanic mammals, develop or are lost as evolution occurs. (1B2a*IE*)

Energy and Homeostasis

- Natural selection for different environments has shaped the gas exchange and excretory systems of animals. (2D2b*IE*)
- Fish, amphibian, and mammal circulatory systems differ in number of heart chambers and circuits, but retain the closed system of their common ancestor. (2D2c*IE*)
- Endothermy and ectothermy provide different strategies for regulation of energy use and internal temperature. (2A1d1*IE*)

Interactions and Systems

- Organ specialization (gills/lungs, hearts, digestive tract members, kidneys) promotes energy efficiency in body systems. (4B2*IE*)

*Find the unabridged version of all EK citations at www.glencoe.com/maderAP11.

Media Study Tools

www.glencoe.com/maderAP11

Enhance your study of this chapter with study tools and practice tests. Also ask your instructor about the resources available through ConnectPlus, including the media-rich eBook, interactive learning tools, and animations.

Summarize

29.1 The Chordates

Chordates (sea squirts, lancelets, and vertebrates) have a notochord, a dorsal tubular nerve cord, pharyngeal pouches, and a postanal tail at some time in their life history.

Lancelets and sea squirts are the nonvertebrate chordates. Lancelets are the only chordate to have the four characteristics in the adult stage. Sea squirts lack chordate characteristics (except gill slits) as adults, but they have a larva that could be ancestral to the vertebrates.

29.2 The Vertebrates

Vertebrates have the four chordate characteristics as embryos. As adults, the notochord is replaced by the vertebral column. Vertebrates undergo cephalization, and have an endoskeleton, paired appendages, and well-developed internal organs.

Vertebrate evolution is marked by the evolution of vertebrae, jaws, a bony skeleton, lungs, limbs, and the amniotic egg.

29.3 The Fishes

The first vertebrates lacked jaws and paired appendages. They are represented today by the hagfishes and lampreys. Ancestral bony fishes, which had jaws and paired appendages, gave rise during the Devonian period to two groups: today's cartilaginous fishes (skates, rays, and sharks) ven the bony fishes, including the ray-finned fishes

and the lobe-finned fishes. The ray-finned fishes (Actinopterygii) became the most diverse group among the vertebrates. Ancient lobe-finned fishes (Sarcopterygii) gave rise to the coelacanths and amphibians.

29.4 The Amphibians

Amphibians are tetrapods represented primarily today by frogs and salamanders. Most frogs and some salamanders return to the water to reproduce and then metamorphose into terrestrial adults.

29.5 The Reptiles

Reptiles (today's alligators and crocodiles, birds, turtles, tuataras, lizards, and snakes) lay a shelled amniotic egg, which allows them to reproduce on land. There are two views of reptile evolution. The classical view considers turtles, with an anapsid skull, as having a separate ancestry from the other reptiles. The other reptiles, with a diapsid skull, include the crocodilians, dinosaurs, and the birds. A more recent view of reptile evolution proposes that turtles are actually very specialized diapsid archosaurs along with birds and crocodiles. Birds have reptilian features, including scales (feathers are modified scales), tail with vertebrae, and clawed feet.

The feathers of birds help them maintain a constant body temperature. Birds are adapted for flight: Their bones are hollow, their shape is compact, their breastbone is keeled, and they have well-developed sense organs.

29.6 The Mammals

Mammals share an amniote ancestor with reptiles but they have a synapsid skull. Mammals remained small and insignificant while the dinosaurs existed, but when dinosaurs became extinct at the end of the Cretaceous period, mammals became the dominant land organisms.

Mammals are vertebrates with hair and mammary glands. Hair helps them maintain a constant body temperature, and the mammary glands allow them to feed and establish an immune system in their young. Monotremes lay eggs, while marsupials have a pouch in which the newborn crawls and continues to develop. The placental mammals, or eutherians, are the most varied and numerous, and retain offspring inside the female until birth.

Key Terms

agnathan 553	jawless fishes 553
amniote 553	lobe-finned fishes 555
amniotic egg 558	lungfishes 555
amphibian 556	mammal 564
anapsid 558	marsupial 564
bird 562	monotreme 564
bony fish 554	notochord 550
cartilaginous fish 554	ostracoderm 553
cephalochordate 550	placenta 564
chordate 550	placental mammal 564
diapsid 558	placoderm 553
dinosaur 559	ray-finned bony fishes 554
ectotherm 553	reptile 558
endotherm 563	swim bladder 554
fin 553	synapsid 558
fishes 553	tetrapod 552
gills 550	urochordate 551
gnathostome 552	

 **Assess**

Reviewing This Chapter

1. What four characteristics do all chordates have at some time in their development?   550
2. Describe the two groups of nonvertebrate chordates, and explain how the sea squirts might be ancestral to vertebrates.   550–51
3. Discuss the distinguishing characteristics and the evolution of vertebrates.   552–53
4. Describe the jawless fishes, including ancient ostracoderms.   553
5. Describe the characteristics of fishes with jaws. What is the significance of having jaws? Describe today's cartilaginous and bony fishes. The amphibians evolved from what type of ancestral fish?   553–55
6. Discuss the characteristics of amphibians, stating which ones are especially adaptive to a land existence. Explain how their name (amphibians) characterizes these animals.   556–57
7. What is the significance of the amniotic egg? What other characteristics make reptiles less dependent on a source of external water?   558
8. Draw a simplified phylogenetic tree that includes the anapsid, diapsid, and synapsid skulls.   558–59
9. What is the significance of wings? In what other ways are birds adapted to flying?   562–63
10. What are the three major groups of mammals, and what are their primary characteristics?   564–65

Testing Yourself

Choose the best answer for each question.

1. Which of these is not a chordate characteristic?
   a. dorsal supporting rod, the notochord
   b. dorsal tubular nerve cord
   c. pharyngeal pouches
   d. postanal tail
   e. vertebral column

2. Adult sea squirts
   a. do not have all five chordate characteristics.
   b. are also called tunicates.
   c. are fishlike in appearance.
   d. are the first chordates to be terrestrial.
   e. All of these are correct.

3. Cartilaginous fishes and bony fishes are different in that only
   a. bony fishes have paired fins.
   b. bony fishes have a keen sense of smell.
   c. bony fishes have an operculum.
   d. cartilaginous fishes have a complete skeleton.
   e. cartilaginous fishes are predaceous.

4. Amphibians evolved from what type of ancestral fish?
   a. sea squirts and lancelets
   b. cartilaginous fishes
   c. jawless fishes
   d. ray-finned fishes
   e. lobe-finned fishes

5. Which of these is not a feature of amphibians?
   a. dry skin that resists desiccation
   b. metamorphosis from a swimming form to a land form
   c. small lungs and a supplemental means of gas exchange
   d. reproduction in the water
   e. a single ventricle

6. Reptiles
   a. were dominant during the Mesozoic era.
   b. include the birds.
   c. lay shelled eggs.
   d. are ectotherms, except for birds.
   e. All of these are correct.

7. Which of these is a true statement?
   a. In all mammals, offspring develop completely within the female.
   b. All mammals have hair and mammary glands.
   c. All mammals have one birth at a time.
   d. All mammals are land-dwelling forms.
   e. All of these are true.

8. Which of these is not an invertebrate? Choose more than one answer if correct.
   a. tunicate
   b. frog
   c. lancelet
   d. squid
   e. roundworm

9. Which of these is not a characteristic of vertebrates? Choose more than one answer if correct.
   a. All vertebrates have a complete digestive system.
   b. Vertebrates have a closed circulatory system.
   c. Sexes are usually separate in vertebrates.
   d. Vertebrates have a jointed endoskeleton.
   e. Most vertebrates never have a notochord.

10. Bony fishes are divided into which two groups?
    a. hagfishes and lampreys
    b. sharks and ray-finned fishes
    c. ray-finned fishes and lobe-finned fishes
    d. jawless fishes and cartilaginous fishes

11. Which of these is an incorrect difference between reptiles and birds?

Reptiles	**Birds**
a. shelled egg	partial internal development
b. scales	feathers
c. tetrapods	wings
d. ectothermy	endothermy
e. no air sacs	air sacs

12. Which of these does not produce an amniotic egg? Choose more than one answer if correct.
    a. bony fishes
    b. duckbill platypus
    c. snake
    d. robin
    e. frog

13. Label the following diagram of a chordate embryo.

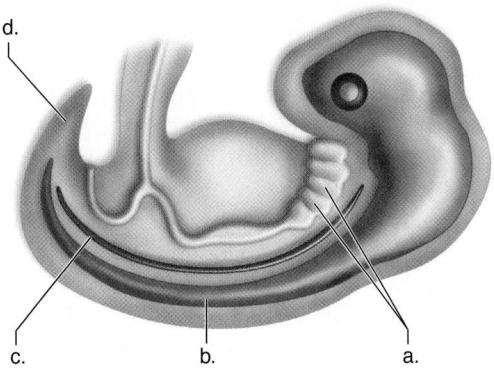

14. The amniotes include all but the
    a. birds.
    b. mammals.
    c. reptiles.
    d. amphibians.

15. Which of the following groups has a three-chambered heart?
    a. all birds
    b. all reptiles
    c. all mammals
    d. all amphibians

Engage

Thinking Scientifically

1. *Archaeopteryx* was a birdlike reptile that had a toothed beak. Give an evolutionary explanation for the elimination of teeth in a bird's beak.

2. While amphibians have rudimentary lungs, skin is also a respiratory organ. Why would a thin skin be more sensitive to pollution than lungs?

30

Human Evolution

A comparison of the Neandertal and *Homo sapiens* genomes is yielding insight into human evolution.

BEFORE YOU BEGIN

Before beginning this chapter, take a few moments to review the following discussions.

Section 15.3 What is a transitional fossil?

Section 17.1 What is the difference between the evolutionary species concept and the biological species concept?

Table 29.2 To which order of mammals do the hominins belong?

For years, Neandertals have been perceived as an offshoot of human evolution, a branch of the evolutionary tree that was only distantly connected with the evolution of our species, *Homo sapiens*. This changed in 2010 when researchers in Germany obtained DNA samples from Neandertal skeletons that were between 38,000 and 45,000 years old. Upon comparing this DNA to five populations of modern humans, the research team uncovered two surprises. First, even though Neandertals and *Homo sapiens* were believed to have diverged from a common ancestor a brief 500,000 years ago, there were already 15 areas of the *Homo sapiens* genome that showed evidence of selection following the split. These included areas of the genome associated with cranial development, metabolic functions, and cognitive (thinking) abilities.

The second discovery reported by the research team was even more interesting—that Neandertals and *Homo sapiens* may have interbred, perhaps as little as 50,000 years ago. In fact, the research suggests that Neandertals are more closely related to populations of *Homo sapiens* from Europe and Asia than to African populations, meaning that Neandertals and humans interacted after *Homo sapiens* migrated from Africa. These discoveries have renewed an interest in discovering the genes that distinguish us from our closest ancestors, and in the process, develop a greater understanding of the evolutionary history of our species.

As you read through this chapter, think about the following questions:

1. What was the last common ancestor of both Neandertals and *Homo sapiens*?
2. What does the replacement model tell us about the evolution of our species?
3. How might an understanding of evolutionary patterns in other animals help us understand our evolutionary history?

FOLLOWING *the* BIG IDEAS

Evolution

Fossil evidence suggests humans have evolved over millions of years, and that apes and humans had a common ancestor.

30.1 Evolution of Primates

Learning Outcomes

Upon completion of this section, you should be able to

1. Identify the major groups of primates.
2. Discuss the four traits common to the primates.
3. Arrange the groups of primates in an evolutionary tree that shows their relationships.

The order **primates** [L. *primus,* first] include prosimians, monkeys, apes, and humans (Fig. 30.1). In contrast to other types of mammals, primates are adapted for an **arboreal** life—that is, for living in trees. The evolution of primates is characterized by trends toward mobile limbs, grasping hands, a flattened face and stereoscopic vision, a large and complex brain, and a reduced reproductive rate. These traits are particularly useful for living in trees.

Mobile Forelimbs and Hindlimbs

Primates tend to have prehensile hands and feet, meaning that they are adapted for grasping and holding. In most primates,

flat nails have replaced the claws of ancestral primates, and sensitive pads on the undersides of fingers and toes assist the grasping of objects. All primates have a thumb, but it is only truly opposable in Old World monkeys, great apes, and humans. Because an **opposable thumb** can touch each of the other fingers, the grip is both powerful and precise (Fig. 30.2). In all but humans, primates with an opposable thumb also have an opposable toe.

The evolution of the primate limb was a very important adaptation for their life in trees. Mobile limbs with clawless opposable digits allow primates to freely grasp and release tree limbs. They also allow primates to easily reach out and bring food, such as fruit, to the mouth.

Stereoscopic Vision

A foreshortened snout and a relatively flat face are also evolutionary trends in primates. These may be associated with a general decline in the importance of smell and an increased reliance on vision. In most primates, the eyes are located in the front, where they can focus on the same object from slightly

PROSIMIANS

Tarsier, *Tarsius bancanus*

NEW WORLD MONKEY

White-faced monkey, *Cebus capucinus*

OLD WORLD MONKEY

Anubis baboon, *Papio anubis*

ASIAN APES

Orangutan, *Pongo pygmaeus*

AFRICAN APES

Chimpanzee, *Pan troglodytes*

HOMININS

Humans, *Homo sapiens*

Figure 30.1 Primate diversity. Today's prosimians may resemble the first group of primates to evolve. Modern monkeys are divided into the New World monkeys and the Old World monkeys. The apes can be divided into the Asian apes (orangutans and gibbons) and the African apes (chimpanzees and gorillas). Humans (hominins) are also primates.

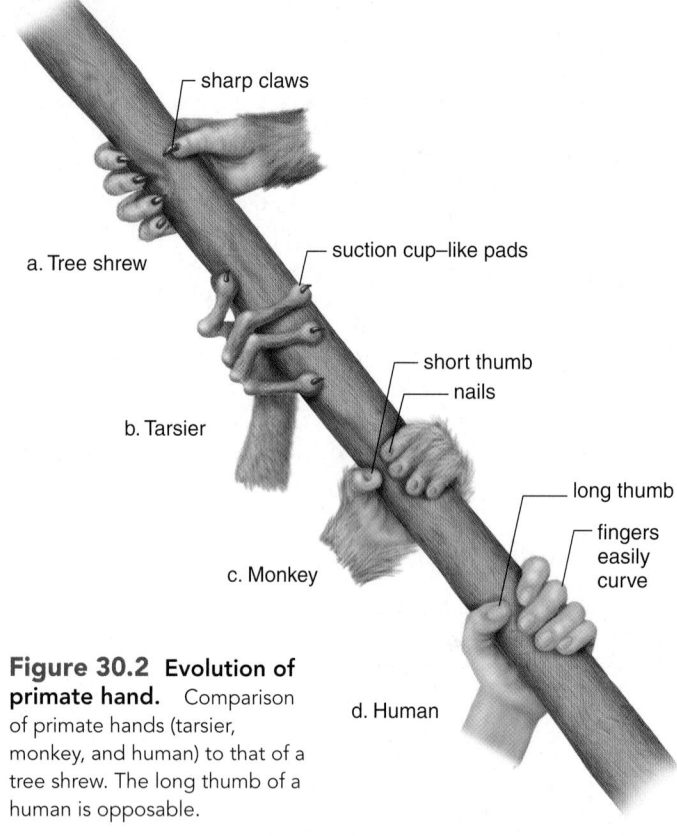

Figure 30.2 Evolution of primate hand. Comparison of primate hands (tarsier, monkey, and human) to that of a tree shrew. The long thumb of a human is opposable.

a. Tree shrew

sharp claws

b. Tarsier

suction cup–like pads

c. Monkey

short thumb

nails

d. Human

long thumb

fingers easily curve

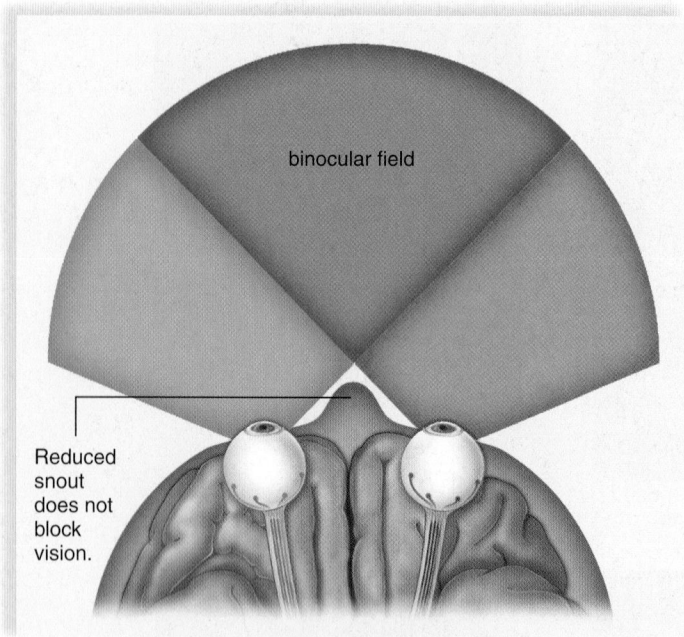

binocular field

Reduced snout does not block vision.

Figure 30.3 Stereoscopic vision. In primates, the snout is reduced, and the eyes are at the front of the head. The result is a binocular field that aids depth perception and provides stereoscopic vision.

different angles (Fig. 30.3). The result is **stereoscopic vision** (three-dimensional) with good depth perception that permits primates to make accurate judgments about the distance and position of adjacent tree limbs.

Some primates, humans in particular, have color vision and greater visual acuity because the retina contains cone cells in addition to rod cells. Rod cells are activated in dim light, but the blurry image is in shades of gray. Cone cells require bright light,

but the image is sharp and in color. The lens of the eye focuses light directly on the fovea, a region of the retina where cone cells are concentrated (see Chapter 38).

Large, Complex Brain

Sense organs are only as beneficial as the brain that processes their input. The evolutionary trend among primates is toward a larger and more complex brain. This is evident when comparing the brains of prosimians, such as lemurs and tarsiers, with that of apes and humans. In apes and humans, the portion of the brain devoted to smell is smaller, and the portions devoted to sight have increased in size and complexity. Also, more of the brain is devoted to controlling and processing information received from the hands and the thumb. The result is good hand–eye coordination. A larger portion of the brain is devoted to communication skills, which supports primates' tendency to live in social groups.

Reduced Reproductive Rate

One other trend in primate evolution is a general reduction in the rate of reproduction, associated with increased age of sexual maturity and extended life spans. Gestation is lengthy, allowing time for forebrain development. One birth at a time is the norm in primates; it is difficult to care for several offspring in the trees while moving from limb to limb. The juvenile period of dependency is extended, and there is an emphasis on learned behavior and complex social interactions.

Sequence of Primate Evolution

Figure 30.4 traces the evolution of primates during the Cenozoic era. **Hominins** (the designation that includes humans, and species very closely related to humans) first evolved about 5 MYA. Molecular data shows that hominins and gorillas are closely related and that these two groups must have shared a common ancestor sometime during the Miocene. Hominins, chimpanzees, and gorillas are now grouped together as **hominines.** The Nature of Science feature on page 574 presents DNA comparisons between modern humans and chimpanzees.

The **hominids** [L. *homo*, man; Gk. *eides*, like] include the hominines and the orangutan. The **hominoids** include the gibbon and the hominids. The hominoid common ancestor first evolved at the beginning of the Miocene about 23 MYA.

The **anthropoids** [Gk. *anthropos*, man, and *eides*, like] include the hominoids and the Old World monkeys and New World monkeys. Old World monkeys lack tails and have protruding noses. Some of the better-known Old World monkeys are the baboon, a ground dweller, and the rhesus monkey, which has been used in medical research. The New World monkeys often have long prehensile tails and flat noses. Two of the well-known New World monkeys are the spider monkey and the capuchin, the "organ grinder's monkey."

Primate fossils similar to monkeys are first found in Africa, dated about 45 MYA. At that time, the Atlantic Ocean would have been too expansive for some of them to have easily made their way to South America, where the New World monkeys live today. It is hypothesized that a common ancestor to both the New World and Old World monkeys arose much earlier when a narrower Atlantic would have made crossing much more

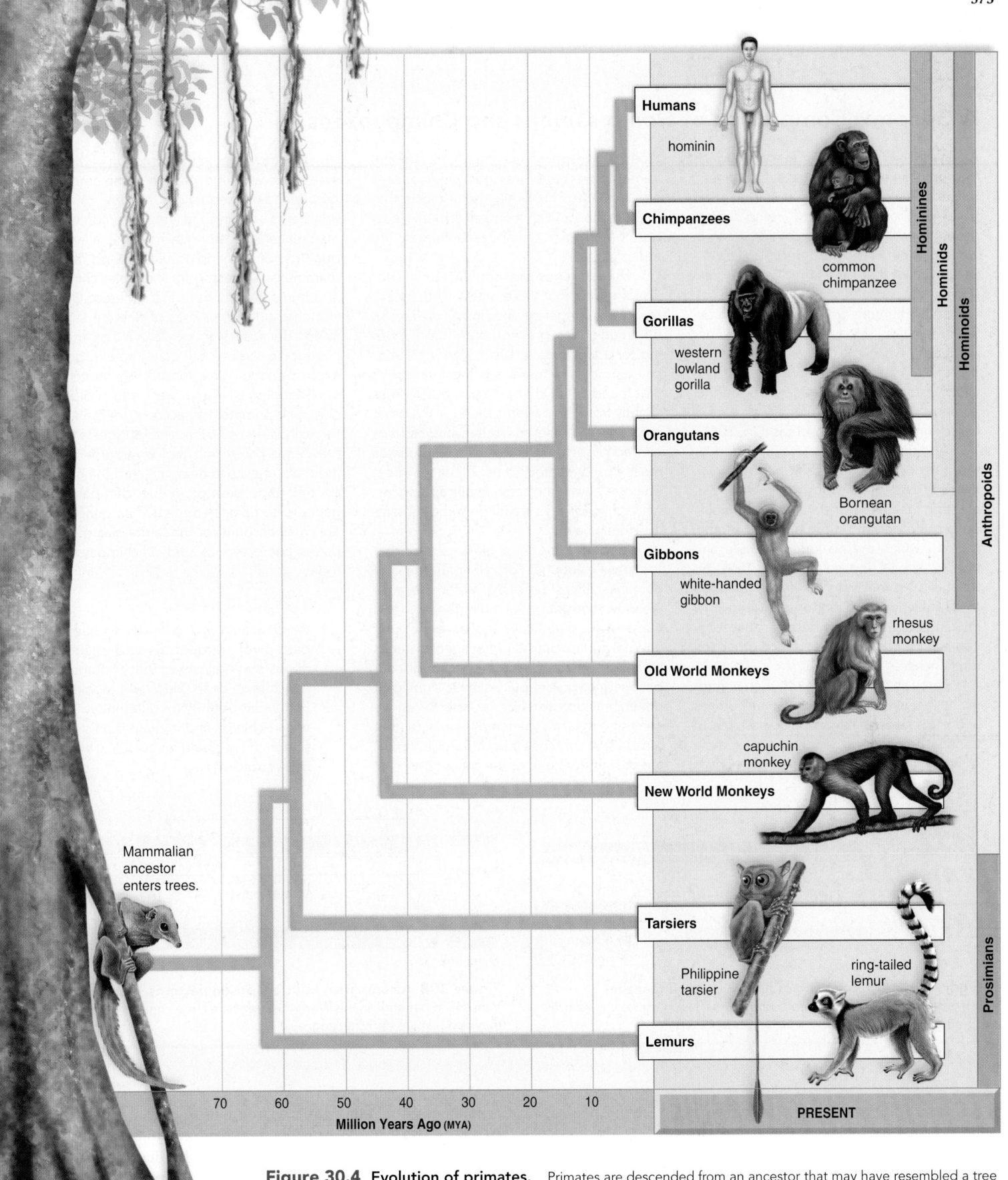

Humans

hominin

Chimpanzees

common chimpanzee

Gorillas

western lowland gorilla

Orangutans

Bornean orangutan

Gibbons

white-handed gibbon

rhesus monkey

Old World Monkeys

capuchin monkey

New World Monkeys

Tarsiers

ring-tailed lemur

Philippine tarsier

Lemurs

Mammalian ancestor enters trees.

Homininnes

Hominids

Hominoids

Anthropoids

Prosimians

70 60 50 40 30 20 10

Million Years Ago (MYA)

PRESENT

Figure 30.4 Evolution of primates. Primates are descended from an ancestor that may have resembled a tree shrew. The time when each type of primate diverged from the main line of descent is known from the fossil record. A common ancestor was living at each point of divergence; for example, a common ancestor for hominines was present about 7 MYA, for the hominoids about 15 MYA, and for anthropoids about 45 MYA.

Nature of Science

A Genomic Comparison of *Homo sapiens* and Chimpanzees

A wealth of genetic evidence suggests that humans and chimpanzees are closely related, despite the fact that chimpanzees have 48 chromosomes and *Homo sapiens* have 46. At first, the difference in chromosome number was considered significant—so significant that for a long time the great apes (chimpanzees, gorillas, and orangutans) and humans were classified into different families. The apes were in the family Pongidae, while the humans were in the the family Hominidae. However, in 1991, investigators at Yale University showed that human chromosome 2 is actually a fusion of two chimpanzee chromosomes (Fig. 30A).

Additional evidence was provided by the study of transposons. In Chapter 14 you learned that transposons are mobile elements in the genome. Many copies of transposons are present in the human genome, most of which are no longer mobile and instead remain in one location. They are relics of past retrovirus infections. Because these infections are random, any similarity in transposon patterns between two species can be considered evidence of a common ancestor.

Through the course of many studies, investigators have found that humans and chimpanzees have similar patterns of transposons in their genome. An example is shown in Fig. 30B of an *Alu* element (a

type of transposon) in the vicinity of the hemoglobin gene in both chimpanzees and humans. This similarity suggests that the transposon inserted itself into this location before the chimpanzee–human lineages split.

Pseudogenes are nonfunctional copies of genes that were active in the past. Most pseudogenes are inactive due to mutations that prevent them from coding for a functional protein. The pattern of pseudogenes in humans is most similar to that found in the chimpanzees, but it varies slightly from the patterns found in the other great apes. These and other studies have caused a reclassification of the primates most closely related to us. All of the great apes are now in the same family as humans, while chimpanzees are in the same subfamily (Homininae).

Modern genomic data show that the base sequence of chimpanzees and humans differs only by 1.5%. Because we now consider that chimpanzees and humans are genetically similar, geneticists have begun to focus on what specific genes make us different—for even though our base sequences are quite similar, significant differences do exist. For example, when we compare the two genomes, we find that many DNA stretches (about 5 million in all) are absent from one or the other genomes.

We know that in the evolution of mammals there was an explosion in the amount of noncoding sequences relative to the number of coding genes. So a common mammalian ancestor may have had a larger quantity of noncoding DNA, and each mammalian ancestor (for example, chimps and humans) may have lost different DNA stretches since their lines of descent separated. Because we now know that these noncoding regions still play a role in gene expression (see Chapter 14), the stretches of noncoding DNA that were retained by a particular primate may account for the anatomical differences. Evidence suggests that the genes controlling the development of the brain may have been affected the most by the gaps present in the chimpanzee genome compared to the human genome. This may account for the larger size of our brains compared to that of chimpanzees today.

Questions to Consider

1. Which would you consider to have a pattern of transposons and pseudogenes that is closer to that of humans, a prosimian or an Old World Monkey?
2. For what additional differences between chimps and humans would you screen the genome for evidence of lost DNA stretches?

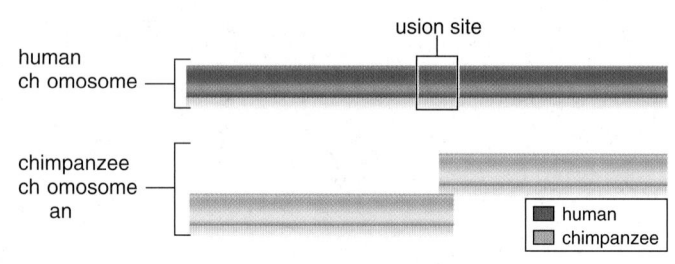

Figure 30A A comparison of human and chimpanzee chromosomes.   Human chromosome 2 is a fusion of two chimpanzee chromosomes.

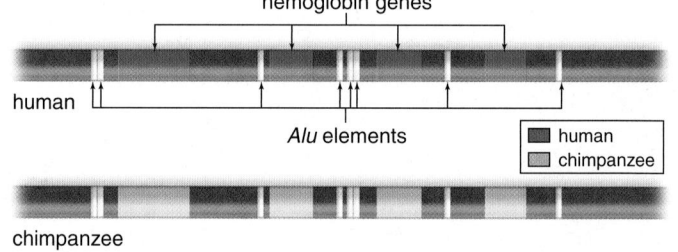

Figure 30B A comparison of transposon patterns.   Similarities in transposon patterns, in this case *Alu*, suggest a close common ancestor between humans and chimpanzees.

reasonable. The New World monkeys evolved in South America, and the Old World monkeys evolved in Africa.

The transitional link between the monkeys and the hominoids (around 35 MYA) is best represented by a fossil classified as *Proconsul*. *Proconsul* was about the size of a baboon, and the size of its brain (165 cc) was also comparable. This fossil species didn't have the tail of a monkey (Fig. 30.5), but its limb proportions suggest that it walked as a quadruped on top of tree limbs

as monkeys do. Although primarily a tree dweller, *Proconsul* may have also spent time exploring nearby environs for food.

Proconsul was probably ancestral to the **dryopithecines,** from which the hominoids arose. About 10 MYA, Africarabia (Africa plus the Arabian Peninsula) joined with Asia, and the apes migrated into Europe and Asia. In 1966, Spanish paleontologists announced the discovery of a specimen of *Dryopithecus* dated at 9.5 MYA near Barcelona. The anatomy of these bones

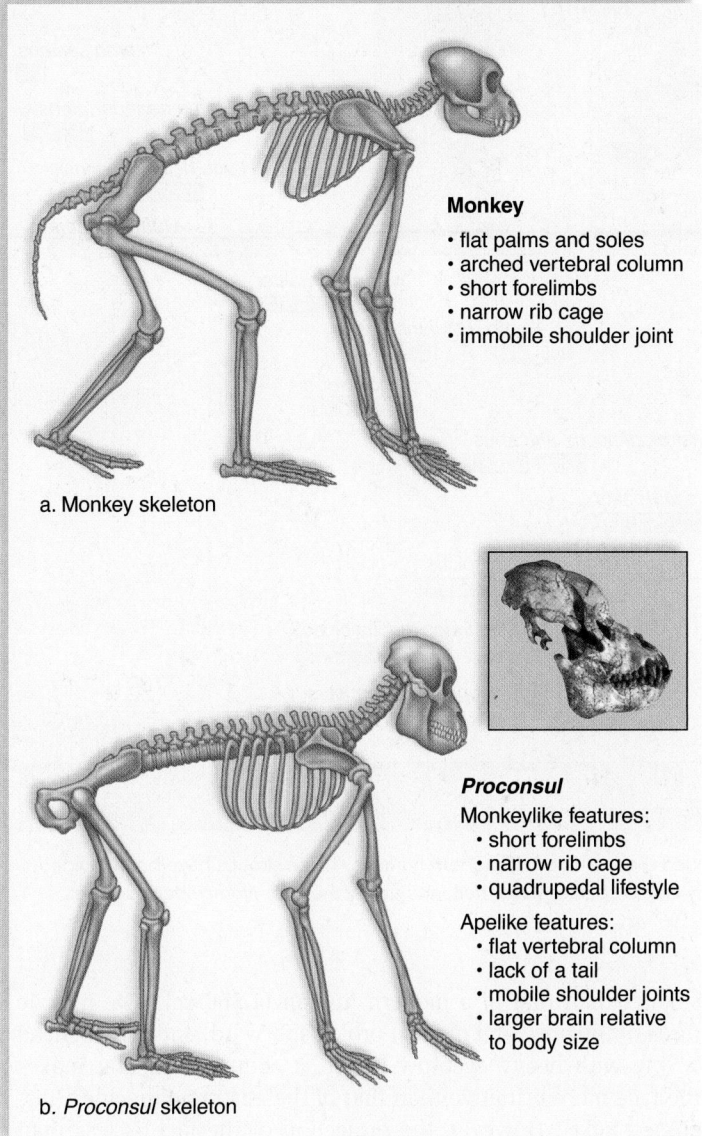

Monkey
• flat palms and soles
• arched vertebral column
• short forelimbs
• narrow rib cage
• immobile shoulder joint

a. Monkey skeleton

Proconsul
Monkeylike features:
• short forelimbs
• narrow rib cage
• quadrupedal lifestyle

Apelike features:
• flat vertebral column
• lack of a tail
• mobile shoulder joints
• larger brain relative to body size

b. _Proconsul_ skeleton

Figure 30.5 Monkey skeleton compared to _Proconsul_ skeleton. Comparison of a monkey skeleton **(a)** with that of _Proconsul_ **(b)** shows various dissimilarities, indicating that _Proconsul_ is more related to today's apes than to today's monkeys.

clearly indicates that _Dryopithecus_ was a tree dweller and locomoted by swinging from branch to branch as gibbons do today. They did not walk along the top of tree limbs as _Proconsul_ did.

Note that **prosimians** [L. _pro,_ before, and _simia,_ ape, monkey], represented by lemurs and tarsiers, were the first type of primate to diverge from the common ancestor for all the primates. All primates share one common mammalian ancestor, which lived about 55 MYA. This ancestor may have resembled today's tree shrews.

Check Your Progress 30.1

1. Identify the location of hominins, hominines, hominoids, hominids, anthropoids, and prosimians in the evolutionary tree of the primates.
2. Describe the characteristics that humans share with the chimpanzees.

30.2 Evolution of Humanlike Hominins

Learning Outcomes

Upon completion of this section, you should be able to
1. Explain the significance of bipedalism in hominin evolution.
2. Summarize the importance of both ardipithecines and australopithecines in hominin evolution.

The relationship of hominins to the other primates is shown in the classification box below. Molecular data have been used to determine when hominin evolution began. When two lines of descent first diverge from a common ancestor, the genes of the two lineages are nearly identical. But as time goes by, each lineage accumulates genetic changes. Genetic changes compared to the other hominines suggest that hominin evolution began about 5 MYA.

Evolution of Bipedalism

The anatomy of humans is suitable for standing erect and walking on two feet, a characteristic called _bipedalism._ Humans are bipedal, while apes are quadrupedal (walk on all fours). Early humanlike hominins are not in the genus _Homo,_ but they are considered closely related to humans because they exhibit bipedalism. Although bipedalism places stress on the spinal column, the upright posture frees the hands for tool use.

Figure 30.6 provides a timeline of hominin evolution. The orange and green bars signify the early humanlike hominins, a lavender bar signifies an early species of the genus _Homo,_ while the later members of genus _Homo_ are in blue. The length of each bar indicates when evidence of the species first appears in the fossil record until the estimated time of its extinction.

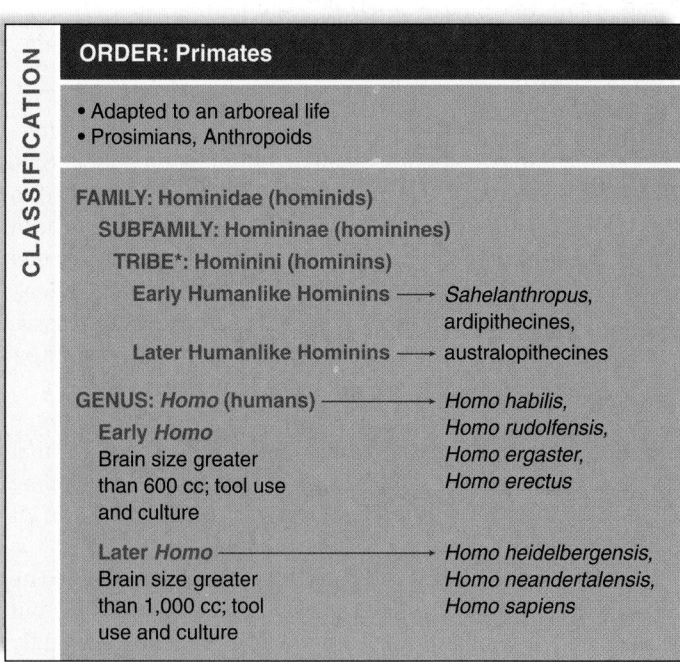

CLASSIFICATION

ORDER: Primates
• Adapted to an arboreal life
• Prosimians, Anthropoids

FAMILY: Hominidae (hominids)
SUBFAMILY: Homininae (hominines)
TRIBE*: Hominini (hominins)
Early Humanlike Hominins ⟶ _Sahelanthropus,_ ardipithecines,
Later Humanlike Hominins ⟶ australopithecines

GENUS: _Homo_ (humans) ⟶ _Homo habilis,_
Early _Homo_ _Homo rudolfensis,_
Brain size greater _Homo ergaster,_
than 600 cc; tool use _Homo erectus_
and culture

Later _Homo_ ⟶ _Homo heidelbergensis,_
Brain size greater _Homo neandertalensis,_
than 1,000 cc; tool _Homo sapiens_
use and culture

* A new taxonomic level that lies between subfamily and genus.

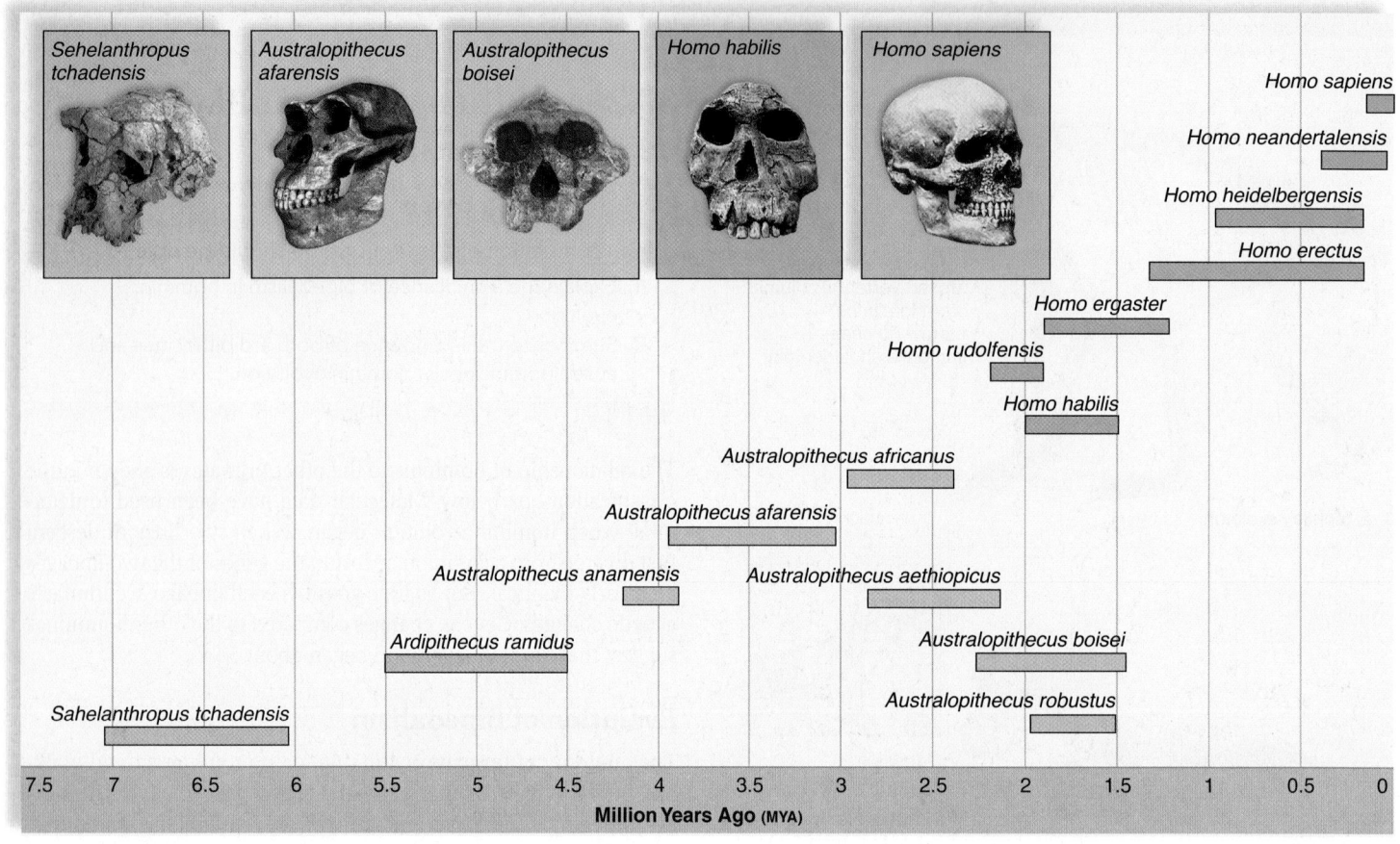

Figure 30.6 Human Evolution. Several groups of extinct hominins preceded the evolution of modern humans. These groups have been divided into the early humanlike hominins (*orange*), later humanlike hominins (*green*), early *Homo* species (*lavender*), and finally the later *Homo* species (*blue*). Only modern humans are classified as *Homo sapiens*.

Paleontologists have identified several fossils dated around the time of the split between the human and ape lineages. One of these is *Sahelanthropus tchadensis*, a species that lived in West Central Africa between 6 and 7 MYA. Only cranial fragments of this species have been uncovered to date, but the point on the back of the skull where the neck muscles would have attached suggests bipedalism. The skull of this fossil is very similar to that of the ardipithecines, discussed next.

Ardipithecines

Two species of ardipithicines have been uncovered, *Ardipithecus kadabba* and *A. ramidus*. Only teeth and a few bone bits have been found for *A. kadabba,* and these have been dated to around 5.6 MYA. A more extensive collection of fossils has been collected for *A. ramidus.* To date, over 100 skeletons, all dated to 4.4 MYA, have been identified from this species; all were collected near a small town in Ethiopia, East Africa. These fossils have been reconstructed to form a female fossil specimen affectionately called Ardi.

Some of Ardi's features are primitive, like that of an ape such as *Dryopithecus* (see Section 30.1), but others are like that of a human. Ardi was about the size of a chimpanzee, standing about 120 cm (4 ft) tall and weighing about 55 kg (110 lb). It appears that males and females were about the same size.

Ardi had a small head compared to the size of her body. The skull has the same features as *Sahelanthropus tchadensis,* but was smaller. Ardi's brain size was around 300 to 350 cc, slightly less than that of a chimpanzee brain (around 400 cc), and much

smaller than that of a modern human (1,360 cc). The muzzle (area of the nose and mouth) projects forward, and the forehead is low with heavy eyebrow ridges, a combination that makes the face more primitive than that of the australopithecines (discussed next). However, the projection of the face is less than that of a chimpanzee because Ardi's teeth were small and like those of an omnivore. She lacked the strong sharp canines of a chimpanzee, and her diet probably consisted mostly of soft, rather than tough, plant material.

Ardi could walk erect, but she spent a lot of time in trees. Notice in Figure 30.7 that in both the human and Ardi skeletons, the spine exits from the center of the skull rather than toward the rear. Also, the femurs angle inward toward the knees (red arrows). These skeletal features assist walking erect by placing the trunk's center of gravity squarely over the feet. Also, Ardi's pelvis and hip joint are broad enough to keep her from swaying from side to side (as chimps do) as she walked. The knee joint in both humans and Ardi is modified to support the body's weight because the bones broaden at this joint.

Ardi's feet have a bone, missing in apes, that would keep her feet squarely on the ground, a sure sign that she was bipedal and not a quadruped like the apes. Nevertheless, like the apes, she has an opposable big toe. Opposable toes allow an animal's feet to grab hold of a tree limb.

The wrists of Ardi's hands were flexible, and most likely she moved along tree limbs on all fours, as ancient apes did. Modern apes *brachiate*—use their arms to swing from limb to limb. Ardi did not do this, but her shoulders were flexible enough to allow

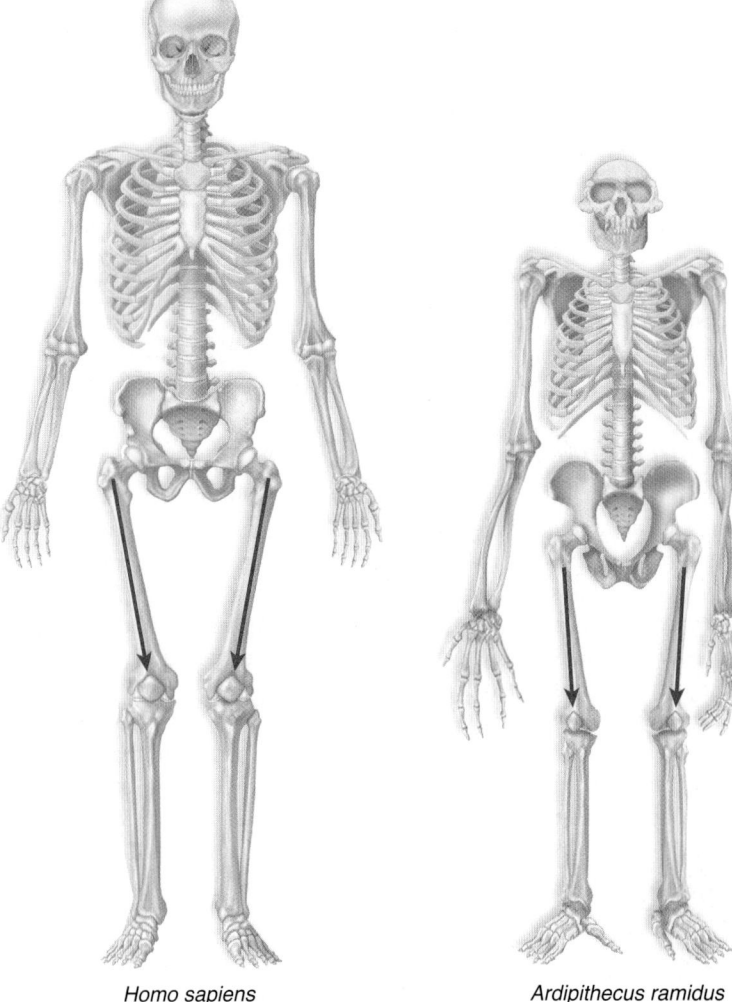

Homo sapiens

Ardipithecus ramidus

Figure 30.7 Adaptations for Walking Erect. A human skeleton compared to an ardipithecine (Ardi). In both skeletons, the spine exits from the center of the skull. A broad pelvis (green) causes the femurs to angle (red arrows) toward the broadened knee joints. However, the skeleton of Ardi has an opposable toe, indicating that this species still lived in trees.

her to reach for limbs to the side or over her head. The general conclusion is that Ardi moved carefully in trees. Although the top of her pelvis is like that of a human, and probably served for the attachment of muscles needed for walking, the bottom of the pelvis served as an attachment for the strong muscles needed for climbing trees.

Until recently, it's been suggested that bipedalism evolved when a dramatic change in climate caused the forests of East Africa to be replaced by grassland. However, evidence suggests that Ardi lived in the woods, which questions the advantage that walking erect would have afforded her. Bipedalism does provide an advantage in caring for a helpless infant by allowing it to be carried by hand from one location to another. It is also possible that bipedalism may have benefited the males of the species as they foraged for food on the floor of the forests. More evidence is needed to better understand this mystery, but one thing is clear— the ardipithecines represent a link between our quadruped ancestors and the bipedal hominins.

Australopithecines

The **australopithecines** (called australopiths for short) are a group of hominins that evolved and diversified in Africa from 4 MYA until about 1.5 MYA. In Figure 30.6, the australopiths are represented by green-colored bars. The australopiths had a small brain (an apelike characteristic) and walked erect (a humanlike characteristic). Therefore, it seems that human characteristics did not evolve all together at the same time. The ardipithecines and australopiths give evidence of **mosaic evolution,** meaning that different body parts change at different rates and, therefore, at different times.

Australopiths stood about 100–115 cm in height and had relatively small brains averaging from about 370–515 cc—slightly larger than that of a chimpanzee. Males were distinctly larger than females. Some australopiths were slight of frame and termed *gracile* (slender). Others were *robust* (powerful) and tended to have massive jaws because of their large grinding teeth. The larger species, now placed in the genus *Paranthropus*, had well-developed chewing muscles that were anchored to a prominent bony crest along the top of the skull. Their diet included seeds and roots. The gracile types of genus *Australopithecus* most likely fed on soft fruits and leaves. Therefore, the australopiths show adaptations to different ways of life. Fossil remains of australopiths have been found in both southern Africa and in eastern Africa. The exact relationship between these two groups is still uncertain.

East African Australopiths

The most significant fossil from East Africa is from a species of australopiths called *Australopithecus afarensis*. The female specimen of this species, known as Lucy, (Fig. 30.8) had a low forehead and a face that projected forward with large canine teeth. The body of Lucy was broader than that of an ardipithecine. Although the brain size was small (around 400 cc), Lucy's skeleton indicates that she was a biped that stood upright. She stooped a bit like a chimpanzee, and the arms were somewhat proportionally longer than the legs. This suggests brachiation as a possible mode of locomotion in trees. Otherwise, the skeleton was humanlike, even though the pelvis lacked refinements that would have allowed Lucy to walk with a striding gait in a manner similar to modern humans.

Even better evidence of bipedal locomotion comes from a trait of fossilized footprints in Laetoli (Tanzania) dated to about 3.7 MYA (Fig. 30.8). The larger footprints are double, as though a smaller-sized being was stepping in the footprints of another, and there are additional footprints off to one side, within hand-holding distance.

Some 30 years after Lucy was discovered, paleontologists discovered a skeleton of a child that has been dated to be 0.1 million years older than Lucy. The face of this skeleton, named Selam but sometimes referred to as "Lucy's baby," looks more like that of an ardipithecine, and the structure of the bones suggests that Selam was not as agile a walker as Lucy.

A. afarensis is a gracile form of australopith, and is believed to be ancestral to the robust types found in eastern Africa, *A. aethiopicus* and *A. boisei*. *A. boisei* had a powerful upper body and the largest molars of any hominin. *A. afarensis* is generally considered more directly related to the early members of the genus *Homo* than are the South African species.

30.3 Evolution of Early Genus *Homo*

Learning Outcomes

Upon completion of this section, you should be able to

1. Arrange the early species of *Homo* in evolutionary order.
2. Explain the signficance of *Homo habilis*, *H. ergaster*, and *H. erectus* in the study of human evolution.

Early *Homo* species (lavender bars in Fig. 30.6) appear in the fossil record somewhat earlier or later than 2 MYA. They all have a brain size that is 600 cc or greater, their jaws and teeth resemble those of modern humans, and tool use is in evidence.

Homo habilis and *Homo rudolfensis*

Homo habilis and *Homo rudolfensis* are closely related and are considered together here. *Homo habilis* means handyman, and these two species are credited by some as being the first hominins to use stone tools, as discussed in the Biological Systems feature on page 582. Most believe that although they appear to have been socially organized, they were probably scavengers rather than hunters. The cheek teeth of these hominins tend to be smaller than even those of the gracile australopiths. This is also evidence that they were omnivorous and ate meat, in addition to plant material.

Compared to australopiths, the face protruded less, and the brain was larger. Although the height of *H. rudolfensis* did not exceed that of the australopiths, some of this species' fossils have a brain size as large as 800 cc, which is considerably larger than that of *A. afarensis*.

Homo ergaster and *Homo erectus*

Homo ergaster evolved in Africa, perhaps from *H. rudolfensis*. Similar fossils found in Asia are different enough to be classified as *Homo erectus* [L. *homo*, man, and *erectus*, upright]. These fossils span the dates between 1.9 and 0.3 MYA, and many other fossils belonging to both species have been found in Africa and Asia.

Compared to *H. rudolfensis*, *H. ergaster* had a larger brain (about 1,000 cc), a rounder jaw, prominent brow ridges, and a projecting nose. This type of nose is adaptive for a hot, dry climate because it permits water to be removed before air leaves the body. The recovery of an almost complete skeleton of a 10-year-old boy indicates that *H. ergaster* was much taller than the hominins discussed thus far (Fig. 30.9). Males were 1.8 m tall, and females were 1.55 m tall. Indeed, these hominins stood erect and most likely had a striding gait like that of modern humans. The robust and probably heavily muscled skeleton still retained some australopithecine features. Even so, the size of the birth canal in female specimens indicates that infants were born in an immature state that required an extended period of care.

H. ergaster first appeared in Africa but then migrated into Europe and Asia sometime between 2 MYA and 1 MYA. Most likely, *H. erectus* evolved from *H. ergaster* after *H. ergaster* arrived in Asia. In any case, such an extensive population movement is a first in the history of humankind and a tribute to the intellectual and physical skills of these hominins. They also had a knowledge of fire and may have been the first to cook meat.

b.

Figure 30.8 *Australopithecus afarensis.* **a.** A reconstruction of Lucy on display at the St. Louis Zoo. **b.** These fossilized footprints occur in ash from a volcanic eruption some 3.7 MYA. The larger footprints are double, and a third, smaller individual was walking to the side. The footprints suggest that *A. afarensis* walked bipedally.

a.

South African Australopiths

The first australopith to be discovered was unearthed in southern Africa in the 1920s. This hominin, named *Australopithecus africanus,* is a gracile type. A second south African specimen called *A. robustus,* discovered in the 1930s, is a robust type that is believed to have had a brain size of around 530 cc.

In 2008, the American anthropologist Lee Berger discovered the bones of an australopithecine he named *A. sediba* (or "wellspring"). The small brain (500 cc) and long arms suggests that this species is an australopith that still climbed trees. However, the humanlike pelvis and striding gait suggests that this species may actually be a member of the genus *Homo*. The skull suggests a projecting nose and dentition (teeth) similar to that of either *Homo habilis* or *Homo erectus*. This fossil, dated to around 2.0 MYA, supports the hypothesis that *A. sediba* may be a transitional link between the early hominins and later *Homo*.

Check Your Progress   30.2

1. Compare the characteristics of an australopith with an ardipithecine.
2. Explain why an understanding of bipedalism and brain size is important in understanding the evolution of the hominins.

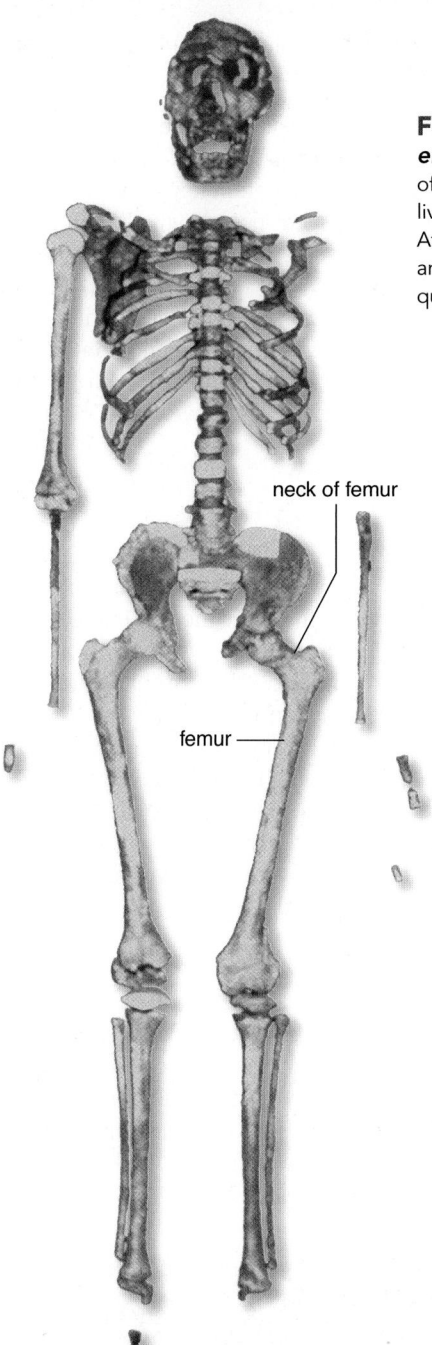

Figure 30.9 *Homo ergaster.* This skeleton of a 10-year-old boy who lived 1.6 MYA in eastern Africa shows femurs that are angled because the neck is quite long.

neck of femur

femur

Homo floresiensis

In 2004, scientists announced the discovery of the fossil remains of *Homo floresiensis.* The 18,000-year-old fossil of a 1 m tall, 25 kg adult female was discovered on the island of Flores in the South Pacific. The specimen was the size of a 3-year-old *Homo sapiens* but possessed a braincase only one-third the size of a modern human. A 2007 study supports the hypothesis that this diminutive hominin and her peers evolved from normal-sized, island hopping *Homo erectus* populations that reached Flores about 840,000 years ago. Apparently, *H. floresiensis* used tools and fire.

Check Your Progress 30.3

1. Discuss the general evolutionary trends in the early species of *Homo.*
2. Discuss how the evolution of bipedalism and increased brain size probably contributed to *H. ergaster's* migration from Africa.

30.4 Evolution of Later Genus *Homo*

Learning Outcomes

Upon completion of this section, you should be able to

1. Explain how the replacement model explains the evolutionary trends of later members of the genus *Homo.*
2. Discuss the significance of increased tool use in Cro-Magnons.
3. Summarize how the replacement model explains the major human ethnic groups.

Later *Homo* species are represented by blue-colored bars in Figure 30.6. The evolution of these species from older *Homo* species has been the subject of much debate. Most researchers believe that modern humans (*Homo sapiens*) evolved from *H. ergaster,* but they differ as to the details.

Evolutionary Hypotheses

Many disparate early *Homo* species in Europe are now classified as *Homo heidelbergensis.* Just as *H. erectus* is believed to have evolved from *H. ergaster* in Asia, so *H. heidelbergensis* is believed to have evolved from *H. ergaster* in Europe. Further, for the sake of discussion, *H. ergaster* in Africa, *H. erectus* in Asia, and *H. heidelbergensis* (and *H. neandertalensis*) in Europe can be grouped together as archaic humans who lived between 1.5 and 0.25 MYA.

The most widely accepted hypothesis for the evolution of modern humans from archaic humans is referred to as the *replacement model* or *out-of-Africa hypothesis,* which proposes that modern humans evolved from archaic humans only in Africa, and then modern humans migrated to Asia and Europe, where they replaced the archaic species about 100,000 years BP (before the present) (Fig. 30.10).

The replacement model is supported by the fossil record. The earliest remains of modern humans (the Cro-Magnon), dating at least 130,000 years BP, have been found only in Africa. Modern humans are not found in Asia until 100,000 years BP and not in Europe until 60,000 years BP. Until earlier modern human fossils are found in Asia and Europe, the replacement model is supported.

The replacement model is also supported by DNA data. Several years ago, a study showed that the mitochondrial DNA of Africans is more diverse than the DNA of the people in Europe (and the world). This is significant because if mitochondrial DNA has a constant rate of mutation, Africans should show the greatest diversity, since modern humans have existed the longest in Africa. Called the "mitochondrial Eve" hypothesis by the press (note that this is a misnomer because no single ancestor is proposed), the statistics that calculated the date of the African migration were found to be flawed. Still, the raw data—which indicate a close genetic relationship among all Europeans—support the replacement model.

An opposing hypothesis to the out-of-Africa hypothesis does exist. This hypothesis, called the *multiregional continuity hypothesis,* proposes that modern humans arose from archaic humans in essentially the same manner in Africa, Asia, and Europe. The hypothesis is multiregional because it applies

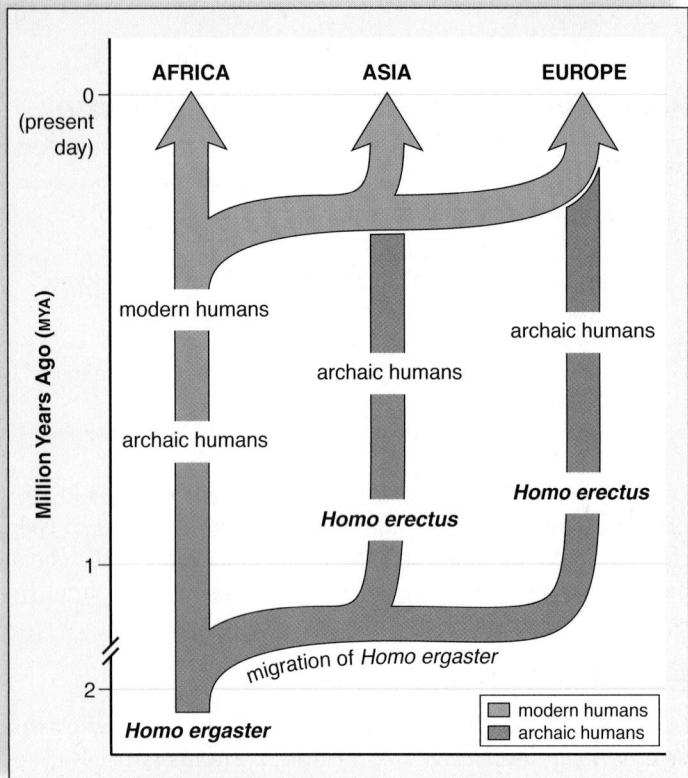

Figure 30.10 Replacement model. Modern humans evolved in Africa and then replaced archaic humans in Asia and Europe.

equally to Africa, Asia, and Europe, and it proposes that in these regions, genetic continuity will be found between modern populations and archaic populations. This hypothesis has sparked many innovative studies to test which hypothesis is correct.

Neandertals

The **Neandertals,** *Homo neandertalensis,* are an intriguing species of archaic humans that lived between 200,000 and 30,000 years ago. Neandertal fossils are known from the Middle East and throughout Europe. Neandertals take their name from Germany's Neander Valley, where one of the first Neandertal skeletons, dated some 200,000 years ago, was discovered.

According to the replacement model, the Neandertals were also supplanted by modern humans. Surprisingly, however, the Neandertal brain was, on the average, slightly larger than that of *Homo sapiens* (1,400 cc, compared with 1,360 cc in most modern humans). The Neandertals had massive brow ridges and wide, flat noses. They also had a forward-sloping forehead and a receding lower jaw. Their nose, jaws, and teeth protruded far forward. Physically, the Neandertals were powerful and heavily muscled, especially in the shoulders and neck. The bones of Neandertals were shorter and thicker than those of modern humans. New fossils show that the pubic bone was long compared to that of modern humans. The Neandertals lived in Europe and Asia during the last Ice Age, and their sturdy build could have helped conserve heat.

Archaeological evidence suggests that Neandertals were culturally advanced. Some Neandertals lived in caves; however, others probably constructed shelters. They manufactured a variety of stone tools, including spear points, which could have been used for hunting, and scrapers and knives, which would have helped in food preparation. They most likely successfully hunted bears, woolly mammoths, rhinoceroses, reindeer, and other contemporary animals. They used and could control fire, which probably helped in cooking frozen meat and in keeping warm. They even buried their dead with flowers and tools and may have had a religion.

Cro-Magnons

The **Cro-Magnons** are the oldest fossils to be designated *Homo sapiens.* In keeping with the replacement model, the Cro-Magnons, who are named after a fossil location in France, were the modern humans who entered Asia from Africa about 100,000 years BP and then spread to Europe. They probably reached western Europe about 40,000 years ago.

Cro-Magnons had a thoroughly modern appearance (Fig. 30.11). They had lighter bones, flat high foreheads, domed skulls housing brains of 1,590 cc, small teeth, and a distinct chin. They were **hunter-gatherers** who collected food from the environment rather than domesticating animals and growing food plants. *H. erectus* were also hunter-gatherers, but Cro-Magnons hunted more efficiently.

Tool Use in Cro-Magnons

Cro-Magnons designed and manipulated tools and weapons of increasing sophistication. They made advanced stone tools, including compound tools, as when stone flakes were fitted to a wooden handle. They may have been the first to make knife-like blades and to throw spears, enabling them to kill animals

Figure 30.11 Cro-Magnons. Cro-Magnon people are the first to be designated *Homo sapiens.* Their tool-making ability and other cultural attributes, such as their artistic talents, are legendary.

from a distance. They were such accomplished hunters that some researchers believe they may have been responsible for the extinction of many larger mammals, such as the giant sloth, the mammoth, the saber-toothed tiger, and the giant ox, during the late Pleistocene epoch. This event is known as the Pleistocene overkill.

Language and Cro-Magnons

A more highly developed brain may have also allowed Cro-Magnons to perfect a language composed of patterned sounds. Language greatly enhanced the possibilities for cooperation and a sense of cohesion within the small bands that were the predominant form of human social organization, even for the Cro-Magnons. They combined hunting and fishing with the gathering of fruits, berries, grains, and root crops that grew in the wild.

The Cro-Magnons were highly creative. They sculpted small figurines and jewelry out of reindeer bones and antlers. These sculptures could have had religious significance or possibly have been seen as a way to increase fertility. The most impressive artistic achievements of the Cro-Magnons were cave paintings, realistic and colorful depictions of a variety of animals, from woolly mammoths to horses, that have been discovered deep in caverns in southern France and Spain. These paintings suggest that Cro-Magnons had the ability to think symbolically, as would be needed in order to speak.

Rise of Agriculture

The Cro-Magnons combined hunting and fishing with gathering fruits, berries, grains, and root crops that grew in the wild. With the rise of agriculture about 10,000 BP, modern humans are no longer considered Cro-Magnon. However, full dependency on domestic crops and animals did not occur until humans started making tools of bronze (instead of stone), about 4,500 BP.

Anthropologists previously thought that early humans turned to agriculture because life as a hunter-gatherer had its drawbacks. However, skeletal evidence suggests that early agricultural societies experienced an increase in infectious diseases, malnutrition, and anemia compared to earlier hunter-gatherer groups. Many anthroplogists now think that the change in lifestyle was dictated by extinctions of the large game animals, and a general warming of the climate. As the glaciers retreated, fertile soil was deposited into rivers and streams full of fish. In suitable locations, such as the fertile crescent of Mesopotamia, fishing villages may have developed, causing populations to settle in one location. Combined with a beneficial climate, the increase in food supplies would have resulted in an increase in the population, reducing its ability to migrate easily. An increase in agriculture would also have supported the specialization of tasks in the population and enhanced the process of **biocultural evolution**, in which cultural achievements, and not individual phenotypes, are influenced by natural selection (see the Biological Systems feature on page 582).

Human Variation

Human beings have been widely distributed about the globe ever since they evolved. As with any other species that has a wide geographic distribution, phenotypic and genotypic variations are noticeable between populations. Today, we say that people have different ethnicities (Fig. 30.12a).

Evolutionists have hypothesized that human variations evolved as adaptations to local environmental conditions. One obvious difference among people is skin color. A darker skin is protective against the high UV intensity of bright sunlight. On the other hand, a white skin ensures vitamin D production in the skin when the UV intensity is low. Harvard University geneticist Richard Lewontin points out, however, that this hypothesis concerning the survival value of dark and light skin has never been tested.

Two correlations between body shape and environmental conditions have been noted since the nineteenth century. The first, known as Bergmann's rule, states that animals in colder

a.

b.

c.

Figure 30.12 Ethnic groups. **a.** Some of the differences between the various prevalent ethnic groups in the United States may be due to adaptations to their original environment. **b.** The Maasai live in East Africa. **c.** Eskimos live near the Arctic Circle.

Biological Systems

Biocultural Evolution Began with *Homo*

The term *culture* encompasses human activities and products that are passed on from one generation to another outside of direct biological inheritance. *Homo habilis* (and *Homo rudolfensis*) could make the simplest of stone tools, called Oldowan tools after a location in Africa where the tools were first found. The main tool could have been used for hammering, chopping, and digging. A flake tool was a type of knife sharp enough to scrape away hide and remove meat from bones. The diet of *H. habilis* most likely consisted of collected plants, but they probably had the opportunity to eat meat scavenged from kills abandoned by lions, leopards, and other large predators in Africa.

Homo erectus, who lived in Eurasia, also made stone tools, but the flakes were sharper and had straighter edges. They are called Acheulian tools for a location in France where they were first found. Their "multipurpose" hand axes were large flakes with an elongated oval shape, a pointed end, and sharp edges on the sides. Supposedly they were handheld, but no one knows for sure. *H. erectus* also made the same core and flake tools as *H. habilis*. In addition, *H. erectus* could have also made many other implements out of wood or bone, and even grass, which can be twisted together to make string and rope. Excavation of *H. erectus* campsites dated 400,000 years ago have uncovered literally tens of thousands of tools.

H. erectus, like *H. habilis*, also gathered plants as food. However, *H. erectus* may have also harvested large fields of wild plants. The members of this species were not master hunters, but they gained some meat through scavenging and hunting. The bones of all sorts of animals litter the areas where they lived. Apparently, they ate pigs, sheep, rhinoceroses, buffalo, deer, and many other smaller animals. *H. erectus* lived during the last Ice Age, but even so, moved northward. No wonder *H. erectus* is believed to have used fire. A campfire would have protected them from wild beasts and kept them warm at night. And the ability to cook would have made meat easier to eat.

For early humans to survive during the winter in northern climates, meat must have become a substantial part of the diet because plant sources are not available in the dead of winter. It is even possible that the campsites of *H. erectus* were "home bases" to which the group of individuals returned after the day's search for food. If so, these people may have been the first hunter-gatherers (Fig. 30C)—that is, they hunted animals and gathered plant products. This was a successful way of life that allowed the hominin populations to increase from a few thousand australopiths in Africa 2 MYA to hundreds of thousands of *H. erectus* by 300,000 years ago.

The hunter-gatherer lifestyle most likely encourages the development and spread of culture between individuals and generations. Those who could speak a language would have been able to cooperate better as they hunted and sought places to gather food. Among animals, only humans have a complex language that allows them to communicate their experiences symbolically. Words stand for objects and events that can be pictured in the mind.

The cultural achievements of *H. erectus* essentially began a new phase of human evolution, called *biocultural evolution*, in which natural selection is influenced by cultural achievements rather than by anatomic phenotype. *H. erectus* succeeded in new, colder environments because these individuals occupied caves, used fire, and became more capable of obtaining and eating meat as a substantial part of their diet.

Questions to Consider

1. Explain why the development of culture plays such an important role in human evolution.
2. Give an example of how natural selection may interact with culture.

Figure 30C *Homo erectus*. The *Homo erectus* people may have been hunter-gatherers.

regions of their range have a bulkier body build. The second, known as Allen's rule, states that animals in colder regions of their range have shorter limbs, digits, and ears. Both of these effects help regulate body temperature by increasing the surface-area-to-volume ratio in hot climates and decreasing the ratio in cold climates. For example, Figure 30.12*b, c* shows that the Maasai of East Africa tend to be slightly built with elongated limbs, while the Eskimos, who live in northern regions, are bulky and have short limbs.

Other anatomic differences among ethnic groups, such as hair texture, a fold on the upper eyelid (common in Asian peoples), or the shape of lips, cannot be explained as adaptations to the environment. Perhaps these features became fixed in different populations due simply to genetic drift. As far as intelligence is concerned, no significant disparities have been found among different ethnic groups.

Genetic Evidence for a Common Ancestry

The replacement model for the evolution of humans, discussed earlier in this section, pertains to the origin of ethnic groups. This hypothesis proposes that all modern humans have a relatively recent common ancestor, that is, Cro-Magnon, who evolved in Africa and then spread into other regions. Paleontologists tell us that the variation among modern populations is considerably

less than among archaic human populations some 250,000 years ago. If so, all ethnic groups evolved from the same single, ancestral population.

A comparative study of mitochondrial DNA shows that the differences among human populations are consistent with their having a common ancestor no more than a million years ago. Lewontin has also found that the genotypes of different modern populations are extremely similar. He examined variations in 17 genes, including blood groups and various enzymes, among seven major geographic groups: Europeans (caucasians), black Africans, mongoloids, south Asian Aborigines, Amerinds, Oceanians, and Australian Aborigines. He found that the great majority of genetic variation—85%—occurs within ethnic groups, not between them. In other words, the amount of genetic variation between individuals of the same ethnic group is greater than the variation between any two ethnic groups.

Check Your Progress   30.4

1. Explain how the replacement model explains both the dominance of Cro-Magnon and the formation of human ethnic groups.
2. Discuss what factors led to the development of biocultural evolution as a factor in human evolution.

CONNECTING *the* CONCEPTS *with the* BIG IDEAS

Evolution

- The discovery, dating, and study of different fossils have been a major source of evidence for evolution. (1A4b1)

*Find the unabridged version of all EK citations at www.glencoe.com/maderAP11.

Media Study Tools

www.glencoe.com/maderAP11

Enhance your study of this chapter with study tools and practice tests. Also ask your instructor about the resources available through ConnectPlus, including the media-rich eBook, interactive learning tools, and animations.

Summarize

30.1 Evolution of Primates

Primates, in contrast to other types of mammals, are adapted for an arboreal life. The evolution of primates is characterized by trends

toward mobile limbs; grasping hands; a flattened face; stereoscopic vision; a large, complex brain; and birth of one offspring at a time. These traits are particularly useful for living in trees.

The term *hominin* is used for humans and their closely related, but extinct, relatives. A hominin is a member of a larger group, hominines, which also includes the chimpanzees and gorillas. Hominids, hominoids, and anthropoids are groupings that contain additional, more distant relatives.

Proconsul is a transitional link between monkeys and the hominoids, which include the gibbons, orangutans, and hominines.

30.2 Evolution of Humanlike Hominins

Fossil and molecular data tell us humanlike hominins shared a common ancestor with chimpanzees until about 5 MYA, and the split between their lineage and the human lineage occurred around this time.

Adaptations that allow humans to walk erect have resulted in our anatomy differing from that of the apes. In humans, the spinal cord curves and exits from the center of the skull, rather than from the rear of the skull. The human pelvis is broader and more bowl-shaped to place the weight of the body over the legs. Humans use only the longer, heavier lower limbs for walking upright bipedally; in apes, all four limbs are used for walking, and the upper limbs are longer than the lower limbs.

Several early humanlike hominin fossils, such as *Sahelanthropus tchadensis*, have been dated around the time of a shared ancestor for apes and humans (7 MYA). The ardipithecines appeared about 4.5 MYA. Ardi (*Ardipithecus ramidus*) is an example of an ardipithecine. All the early humanlike hominins have a chimp-sized braincase but are believed to have walked erect.

It is possible that an australopith (4 MYA–1 MYA) is a direct ancestor for humans. These hominins walked upright and had a brain size of 370–515 cc. In southern Africa, hominins classified as australopiths include *Australopithecus africanus*, a gracile form, and *Paranthropus robustus*, a robust form. In eastern Africa, hominins classified as australopiths include *A. afarensis* (Lucy), a gracile form, and also robust forms. Many of the australopiths coexisted, and one of these species is the probable ancestor to the genus *Homo*.

30.3 Evolution of Early Genus *Homo*

Early *Homo*, such as *Homo habilis* and *Homo rudolfensis*, dated around 2 MYA, is characterized by a brain size of at least 600 cc, a jaw with teeth that resembled those of modern humans, and the use of tools.

Homo ergaster and *Homo erectus* (1.9–0.3 MYA) had a striding gait, made well-fashioned tools, and could control fire. *Homo ergaster* migrated into Asia and Europe from Africa between 2 and 1 MYA. *Homo erectus* evolved in Asia and gave rise to *H. floresiensis*.

30.4 Evolution of Later Genus *Homo*

The replacement model of human evolution says that modern humans originated only in Africa and, after migrating into Europe and Asia, replaced the archaic *Homo* species found there. The multiregional continuity model suggests that modern humans arose in several regions.

The Neandertals, a group of archaic humans, lived in Europe and Asia. Their sloping chins, squat frames, and heavy muscles are apparently adaptations to the cold. Cro-Magnon is a name often given to modern humans. Their tools were sophisticated, and they definitely had a culture, as witnessed by the paintings on the walls of caves.

The human ethnic groups of today differ in ways that can be explained in part by adaptation to the environment. Genetic studies tell us that there are more genetic differences between people of the same ethnic group than between ethnic groups. We are one species.

Key Terms

anthropoid 572	hunter-gatherer 580
arboreal 571	mosaic evolution 577
australopithecine 577	Neandertal 580
biocultural evolution 581	opposable thumb 571
Cro-Magnon 580	primate 571
dryopithecine 574	prosimian 575
hominid 572	stereoscopic vision 572
hominin 572	
hominine 572	
hominoid 572	

Assess

Reviewing This Chapter

1. List and discuss various evolutionary trends among primates, and state how they would be beneficial to animals with an arboreal life.   571–72

2. What is the significance of the fossils known as *Proconsul*?   574–75
3. How does an upright stance cause human anatomy to differ from that of chimpanzees?   574
4. Discuss the possible benefits of bipedalism in early hominins.   575
5. What is the evolutionary significance of the ardipithecines? 576
6. How does the term *mosaic evolution* apply to both the ardipithecines and the the australopiths?   577
7. Why are the early *Homo* species classified as humans? If these hominins did make tools, what does this say about their probable way of life?   578–79, 582
8. What role might *H. ergaster* have played in the evolution of modern humans according to the replacement model?   579
9. Who were the Neandertals and the Cro-Magnons, and what is their place in the evolution of humans according to the replacement model mentioned in question 8?   580–81

Testing Yourself

Choose the best answer for each question.

1. Which of the following terms would include only humans, and their immediate, bipedal ancestors?
   a. hominoid   c. hominin
   b. prosimian   d. anthropoid

2. Lucy is a(n)
   a. early *Homo*.   c. ardipithecine.
   b. australopith.   d. modern human.

3. What possibly influenced the evolution of bipedalism?
   a. Humans wanted to stand erect in order to use tools.
   b. With bipedalism, it's possible to reach food overhead.
   c. With bipedalism, sexual intercourse is facilitated.
   d. An upright stance exposes more of the body to the Sun, and vitamin D production requires sunlight.
   e. All of these are correct.

4. *H. ergaster* could have been the first to
   a. use and control fire.   d. have a brain of about 1,000 cc.
   b. migrate out of Africa.   e. All of these are correct.
   c. make axes and cleavers.

5. Which of these statements is correct? The last common ancestor for chimpanzees and hominins
   a. has been found, and it resembles a gibbon.
   b. was probably alive around 5 MYA.
   c. has been found, and it has been dated at 30 MYA.
   d. is not expected to be found because there was no such common ancestor.
   e. is now believed to have lived in Asia, not Africa.

6. Which of these pairs is incorrectly matched?
   a. gibbon—hominoid   d. *H. erectus—H. ergaster*
   b. *A. africanus*—hominin   e. early *Homo—H. habilis*
   c. tarsier—anthropoid

7. If the replacment (out-of-Africa) model is correct, then
   a. human fossils in China after 100,000 years BP would not be expected to resemble earlier fossils.
   b. human fossils in China after 100,000 years BP would be expected to resemble earlier fossils.
   c. humans did not migrate out of Africa.
   d. Both b and c are correct.
   e. Both a and c are correct.

8. Which of these pairs is incorrectly matched?
   a. *H. erectus*—made tools   d. Cro-Magnon—good artist
   b. Neandertal—good hunter   e. *A. robustus*—fibrous diet
   c. *H. habilis*—controlled fire

9. Which hominins could have inhabited the Earth at the same time?
   a. australopiths and Cro-Magnons
   b. *Paranthropus robustus* and *Homo habilis*
   c. *Homo neandertalensis* and *Homo sapiens*
   d. gibbons and humans

10. Which of these is an incorrect statement?
    a. *H. habilis* and *H. rudolfensis* were omnivores with a brain size of about 800 cc.
    b. *H. ergaster* had a brain size larger than that of *H. erectus*.
    c. *H. floresiensis*, discovered in 2004, used tools and fire.
    d. All of these are correct.

For questions 11–13, indicate whether the statement is true (T) or false (F).

11. Australopiths were adapted to different diets. _____

12. *Homo habilis* made stone tools. _____

13. Mitochondrial DNA differences are inconsistent with the existence of a recent human common ancestor for all ethnic groups. _____

For questions 14–17, fill in the blanks.

14. The replacement model proposes that modern humans evolved in _____ only.

15. The australopiths could probably walk _____, but they had a _____ brain.

16. The only fossil rightly called *Homo sapiens* is that of _____.

17. Modern humans evolved _____ (choose billions, millions, thousands) of years ago.

18. Complete this diagram of the replacement model by filling in the blanks.

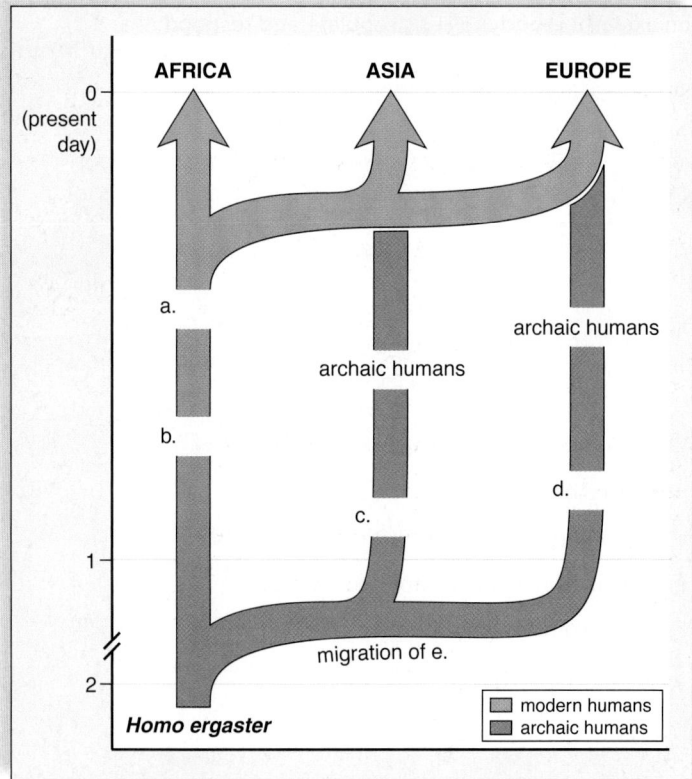

Replacement Model

Engage

Thinking Scientifically

1. Bipedalism has many selective advantages. However, there is one particular disadvantage to walking on two feet: Giving birth to an offspring with a large head through the smaller pelvic opening that is necessitated by upright posture is very difficult. This situation results in a high percentage of deaths (of both mother and child) during birth compared to other primates. How do you explain the selection of a trait that is both positive and negative?

2. How might you use biotechnology to show that humans today have Neandertal genes, and therefore, Cro-Magnons and Neandertals interbred with one another?

Bioethical Issue

Manipulation of Evolution

Since the dawn of civilization, humans have carried out cross-breeding programs to develop plants and animals of use to them. With the advent of DNA technology, we have entered a new era in which even greater control can be exerted over the evolutionary process. We can manipulate genes and give organisms traits that they would not ordinarily possess. Some plants today produce human proteins that can be extracted from their seeds, and some animals grow larger because we have supplied them with an extra gene for growth hormone. Does this type of manipulation seem justifiable?

What about the possibility that we are manipulating our own evolution? Should doctors increase the fitness of certain couples by providing them with a means to reproduce that they cannot achieve on their own? Is the use of alternative means of reproduction bioethically justifiable? In the near future, it may be possible for parents to choose the phenotypic traits of their offspring; in effect, this might enable humans to ensure that their offspring are stronger and brighter than their parents. Does this choosing of "designer babies" seem ethical to you? Explain your thinking.

UNIT 7

Comparative Animal Biology

 AP Biology's Big Idea 2 reads "Biological systems utilize free energy and molecular building blocks to grow, to reproduce, and to maintain dynamic homeostasis." Though every organism does these things, it is the personal interest in and familiar nature of animals (after all, you are one!) that makes this study so intriguing. How did your brain arrive at such a complex and convoluted form? Why is the cardiovascular and respiratory system in one body cavity, the digestive and excretory in another? How is your body temperature kept at such a constant level and what happens when it isn't? Why are their two lungs, two kidneys but only one pancreas? How is it possible for the immune system to protect you against new enemies that have never been seen on Earth before? Why do females have monthly reproductive cycles and males don't? Why is the liver the only internal organ that can "regenerate" itself? How does the digestive system avoid being digested? What about the structure of bones makes them stronger than concrete? How does the heart manage to work without rest for a lifetime? You will get a glimpse of answers to all of these questions as you study the long evolution and the delicately intertwined anatomy and physiology of animal body systems when the Big Ideas focus on these structures, mechanisms, and relationships:

 Evolutionary history can be gleaned not only from external characteristics but also from internal structures.

 All body systems exhibit regulatory and feedback mechanisms that maintain homeostasis.

 Communication via chemical signals or direct cellular action allows the body to both regulate and respond.

 Internal systems do not operate independently but integrate with other body components.

Enjoy and learn from these chapters, serving in part as an owner's manual for your own "machine!" As an ancient Greek physician once penned, "There is nothing on Earth that can surpass the miraculous and intricate workings of the body human."

UNIT OUTLINE

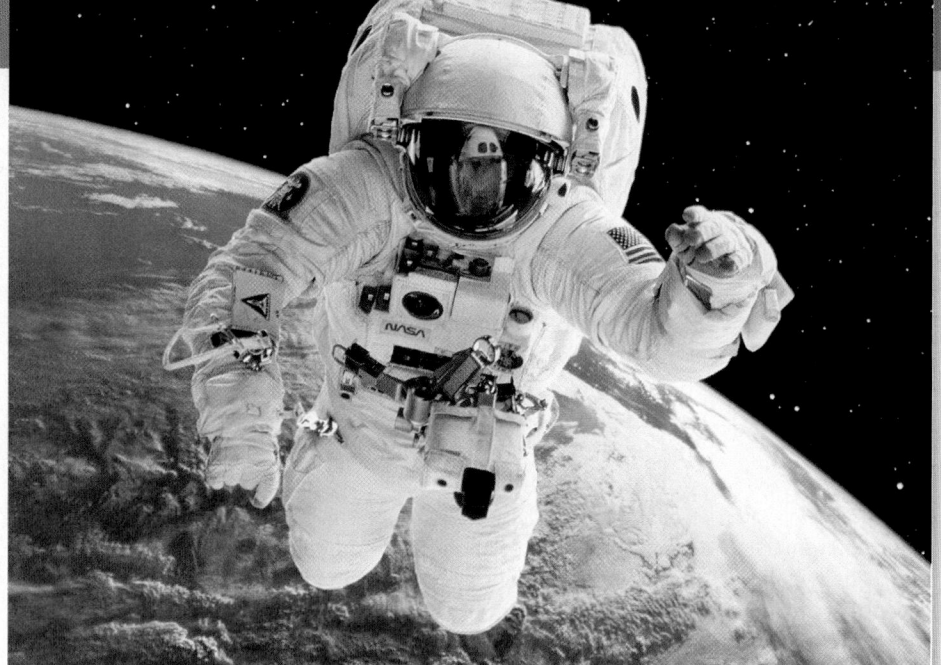

Astronauts need a special suit to take a walk outside their spacecraft.

31

Animal Organization and Homeostasis

The fact that an astronaut needs to wear a special suit to take a space walk reminds us that organ systems function best if the internal environment stays within normal limits. For example, a warm temperature speeds enzymatic reactions, a moderate blood pressure helps circulate blood, and a sufficient oxygen concentration facilitates ATP production. Working in harmony and under the coordination of the nervous and endocrine systems, healthy organ systems are capable of maintaining homeostasis, a dynamic equilibrium of the internal environment. Swim the English Channel, cross the Sahara Desert by camel, visit the South Pole, or take a space walk—your body temperature will stay at just about 37°C as long as you take proper precautions. An astronaut depends on artificial systems in addition to natural systems to maintain homeostasis.

This chapter discusses homeostasis after a look at the body's organization. Just as in other complex animals, each organ system of the human body contains a particular set of organs. The circulatory system contains the heart and blood vessels and the nervous system contains the brain and nerves. Organs are composed of tissues, and each type of tissue has like cells that perform specific functions. We begin the chapter by examining several of the major types of tissues.

As you read through the chapter, think about the following questions:

1. How did the evolution of specialized tissues, organs, and organ systems allow animals to better adapt to their environment?

2. What are some of the most important functions of animal skin?

3. How does the disruption of homeostasis lead to disease?

BEFORE YOU BEGIN

Before beginning this chapter, take a few moments to review the following discussions.

Figure 1.2 What levels of biological organization are found in animals?

Section 24.2 What types of tissues are found in flowering plants?

Section 28.1 Which types of tissues are most characteristic of animals?

FOLLOWING *the* BIG IDEAS

Energy and Homeostasis

Homeostasis is essential for survival, with regulation and feedback mechanisms operating constantly to achieve it.

31.1 Types of Tissues

Learning Outcomes

Upon completion of this section, you should be able to

1. List and describe the four major types of tissues found in animals.
2. Identify the common locations of the various types of animal tissues.
3. Explain how specialization of cells in tissues enhances tissue function.

Like all living things, animals are highly organized. Animals begin life as a single cell, the fertilized egg or *zygote*. The zygote undergoes cell division, producing cells that will eventually form the variety of tissues that make up organs and organ systems. Although all of these cells carry out a number of common functions—such as obtaining nutrients, synthesizing basic cellular constituents, and in most cases, reproducing themselves—the cells of multicellular organisms further differentiate so that they can perform additional, unique functions.

A **tissue** is composed of specialized cells of the same or similar type that perform a common function in the body. The tissues of most complex animals can be categorized into four major types:

1. *Epithelial tissue* covers body surfaces, lines body cavities, and forms glands.
2. *Connective tissue* binds and supports body parts.
3. *Muscular tissue* moves the body and its parts.
4. *Nervous tissue* receives stimuli and transmits nerve impulses.

 **MP3** Overview of Tissues

Epithelial Tissue

Epithelial tissue, also called *epithelium* (pl., epithelia), consists of tightly packed cells that form a continuous layer. Epithelial tissue covers surfaces and lines body cavities. Usually, it has a protective function, but it can also be modified to carry out secretion, absorption, excretion, and filtration.

Epithelial cells may be connected to one another by three types of junctions composed of proteins (see Fig. 5.14). Regions where proteins join them together are called tight junctions. In the intestine, the gastric juices stay out of the body, and in the kidneys, the urine stays within kidney tubules because epithelial cells are joined by tight junctions. In the skin, adhesion junctions add strength and allow epithelial cells to stretch and bend, whereas gap junctions are protein channels that permit the passage of molecules between two adjacent cells. (These junctions are described in more detail in Section 5.4.)

Epithelial tissues are often exposed to the environment on one side, but on the other side they are attached to a *basement membrane.* The basement membrane is simply a thin layer of various types of proteins that anchors the epithelium to the extracellular matrix, which is often a type of connective tissue. The basement membrane should not be confused with the plasma membrane or with the body membranes we will be discussing.

Simple Epithelia

Epithelial tissue is either simple or complex. Simple epithelia have only a single layer of cells (Fig. 31.1) and are classified according to cell type. **Squamous epithelium,** which is composed of flattened cells, is found lining blood vessels and the air sacs of lungs. **Cuboidal epithelium** contains cube-shaped

Figure 31.1 Types of epithelial tissues in vertebrates. Basic epithelial tissues found in vertebrates are shown, along with locations of the tissue and the primary function of the tissue at these locations.

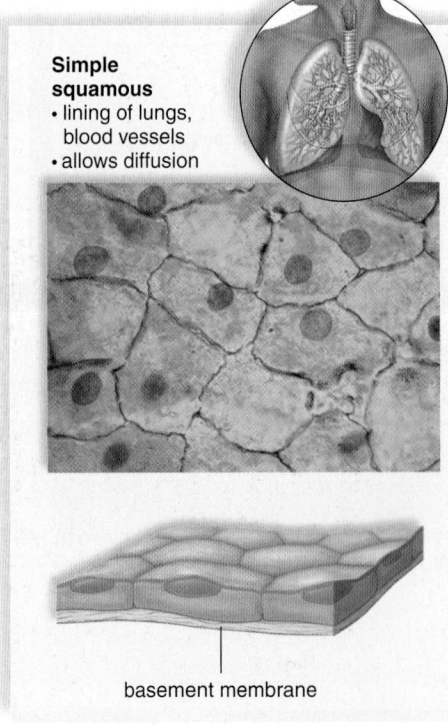

Simple squamous
• lining of lungs, blood vessels
• allows diffusion

basement membrane

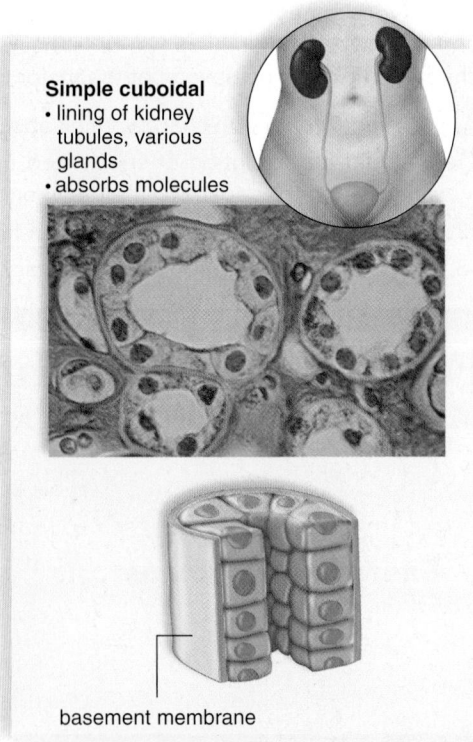

Simple cuboidal
• lining of kidney tubules, various glands
• absorbs molecules

basement membrane

cells and is found lining the kidney tubules and various glands. **Columnar epithelium** has cells resembling rectangular pillars or columns, with nuclei usually located near the bottom of each cell. This epithelium is found lining the digestive tract, where it efficiently absorbs nutrients from the small intestine because of minute cellular extensions called microvilli. Ciliated columnar epithelium is found lining the oviducts, where it propels the egg toward the uterus.

When an epithelium is pseudostratified, it appears to be layered, but true layers do not exist because each cell touches the basement membrane. The lining of the windpipe, or trachea, is pseudostratified ciliated columnar epithelium. A secreted covering of mucus traps foreign particles, and the upward motion of the cilia carries the mucus to the back of the throat, where it may be either swallowed or expectorated. Smoking can cause a change in mucus secretion and inhibit ciliary action, resulting in a chronic inflammatory condition called bronchitis.

Stratified Epithelia

Stratified epithelia have layers of cells piled one on top of the other. Only the bottom layer touches the basement membrane. The nose, mouth, esophagus, anal canal, and vagina are all lined with stratified squamous epithelium. As you'll see, the outer layer of skin is also stratified squamous epithelium, but the cells have been reinforced by keratin, a protein that provides strength. Stratified cuboidal and stratified columnar epithelia also occur in the body.

Glandular Epithelia

When an epithelium secretes a product, it is said to be glandular. A **gland** can be a single epithelial cell, as in the case of mucus-secreting goblet cells within the columnar epithelium lining the digestive tract, or a gland can contain many cells. Glands that secrete their product into ducts are called **exocrine glands.**

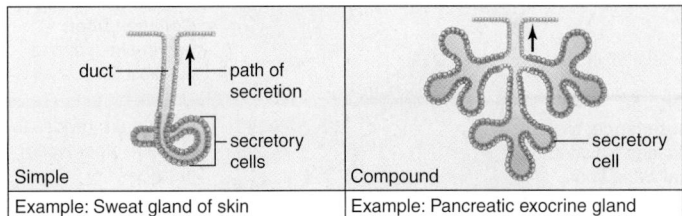

duct ——— path of secretion	
——— secretory cells	——— secretory cell
Simple	Compound
Example: Sweat gland of skin	Example: Pancreatic exocrine gland

Glands that have no duct are known as **endocrine glands.** Endocrine glands (e.g., pituitary gland and thyroid) secrete hormones internally, so they are transported by the bloodstream (see Chapter 40).

MP3
Epithelial Tissue

Connective Tissue

Connective tissue is the most abundant and widely distributed tissue in complex animals. It is quite diverse in structure and function, but, even so, all types have three components: specialized cells, ground substance, and protein fibers (Fig. 31.2).

The ground substance is a noncellular material that separates the cells and varies in consistency from solid to semifluid to fluid. The fibers[1] are of three possible types. White **collagen fibers** contain collagen, a protein that gives them flexibility and strength. **Reticular fibers** are very thin collagen fibers that are highly branched and form delicate supporting networks. Yellow **elastic fibers** contain elastin, a protein that is not as strong

---

1 In connective tissue, a fiber is a component of the matrix; in muscular tissue, a fiber is a muscle cell; in nervous tissue, a nerve fiber is an axon and its myelin sheath.

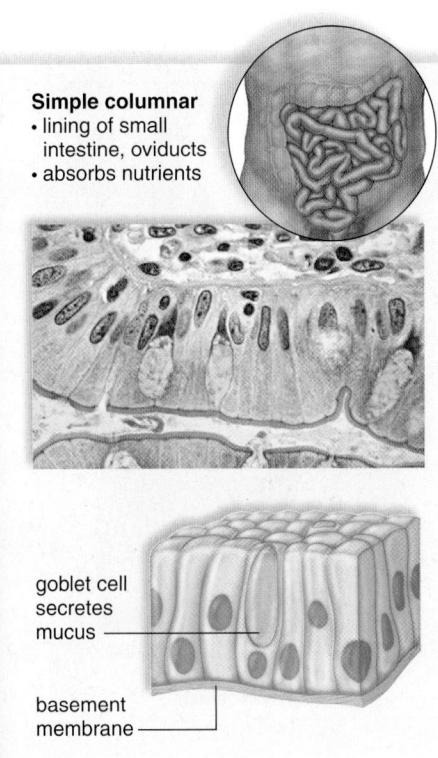

Simple columnar
- lining of small intestine, oviducts
- absorbs nutrients

goblet cell secretes mucus

basement membrane

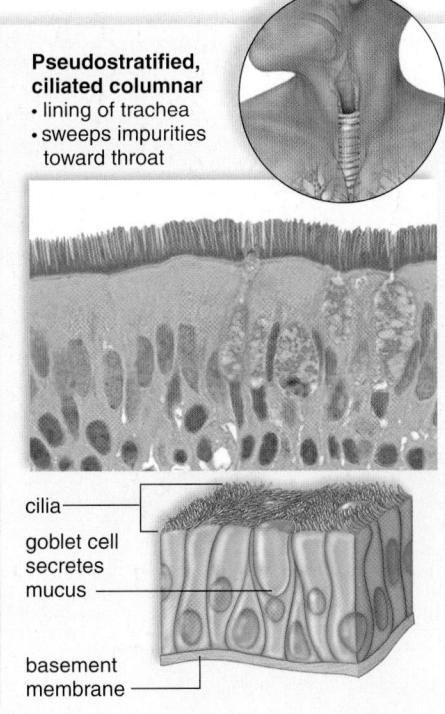

Pseudostratified, ciliated columnar
- lining of trachea
- sweeps impurities toward throat

cilia

goblet cell secretes mucus

basement membrane

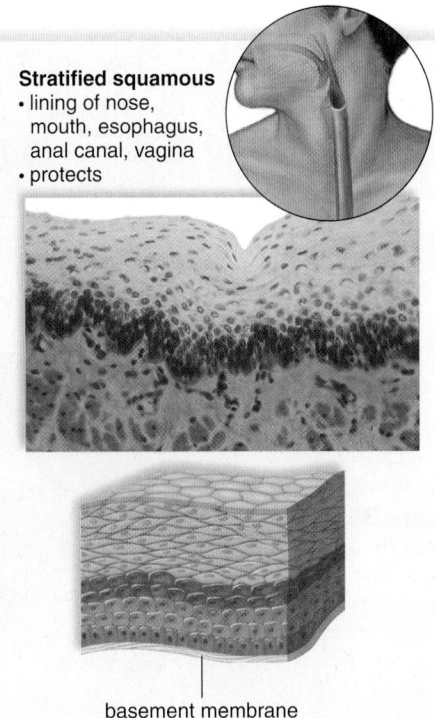

Stratified squamous
- lining of nose, mouth, esophagus, anal canal, vagina
- protects

basement membrane

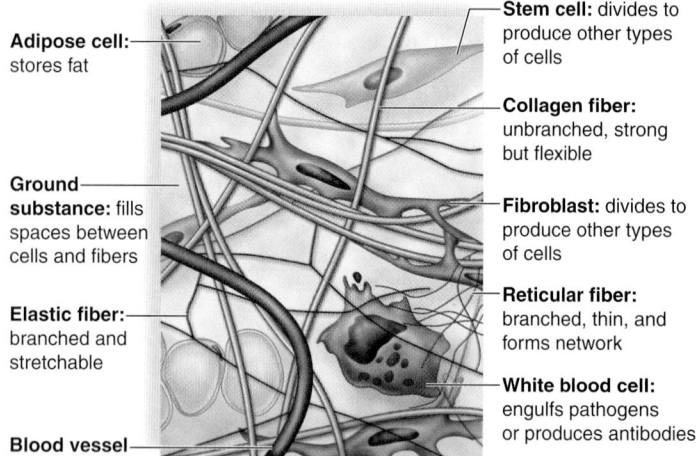

Adipose cell: stores fat

Ground substance: fills spaces between cells and fibers

Elastic fiber: branched and stretchable

Blood vessel

Stem cell: divides to produce other types of cells

Collagen fiber: unbranched, strong but flexible

Fibroblast: divides to produce other types of cells

Reticular fiber: branched, thin, and forms network

White blood cell: engulfs pathogens or produces antibodies

Figure 31.2 Diagram of fibrous connective tissue.

as collagen but is more elastic. The ground substance plus the fibers together are referred to as the connective tissue *matrix.*

Connective tissue is classified into three major categories: fibrous, supportive, and fluid. Each category includes a number of different types of tissues.

Fibrous Connective Tissue

Both loose fibrous and dense fibrous connective tissues have cells called **fibroblasts** [L. *fibra,* thread, and Gk. *blastos,* bud] that are located some distance from one another and are separated by a jellylike matrix containing white collagen fibers and yellow elastic fibers.

Loose fibrous connective tissue supports epithelium and also many internal organs (Fig. 31.3a). Its presence in lungs, arteries, and the urinary bladder allows these organs to expand.

It forms a protective covering enclosing many internal organs, such as muscles, blood vessels, and nerves.

Adipose tissue [L. *adipalis,* fatty] serves as the body's primary energy reservoir (Fig. 31.3b). It is loose fibrous connective tissue composed mostly of enlarged fibroblasts that store fat. These specialized fibroblasts are called *adipocytes.* Adipose tissue also insulates the body, contributes to body contours, and provides cushioning. In mammals, adipose tissue is found particularly beneath the skin, around the kidneys, and on the surface of the heart.

The number of adipocytes in an individual is fixed. When a person gains weight, the cells become larger, and when weight is lost, the cells shrink. In obese people, the individual cells may be up to five times larger than normal. Most adipose tissue is white, but in newborns and hibernating mammals, some is brown due to an increased number of mitochondria that can produce heat.

Dense fibrous connective tissue contains many collagen fibers that are packed together (Fig. 31.3c). This type of tissue has more specific functions than does loose connective tissue. For example, dense fibrous connective tissue is found in **tendons** [L. *tendo,* stretch], which connect muscles to bones, and in **ligaments** [L. *ligamentum,* band], which connect bones to other bones at joints.

Supportive Connective Tissue

Cartilage and bone are the two main supportive connective tissues that provide structure, shape, protection, and leverage for movement. Generally cartilage is more flexible than bone because it lacks mineralization of the matrix.

Cartilage.   In **cartilage,** the cells lie in small chambers called lacunae (sing., **lacuna**), separated by a matrix that is solid yet flexible. Unfortunately, because this tissue lacks a direct blood

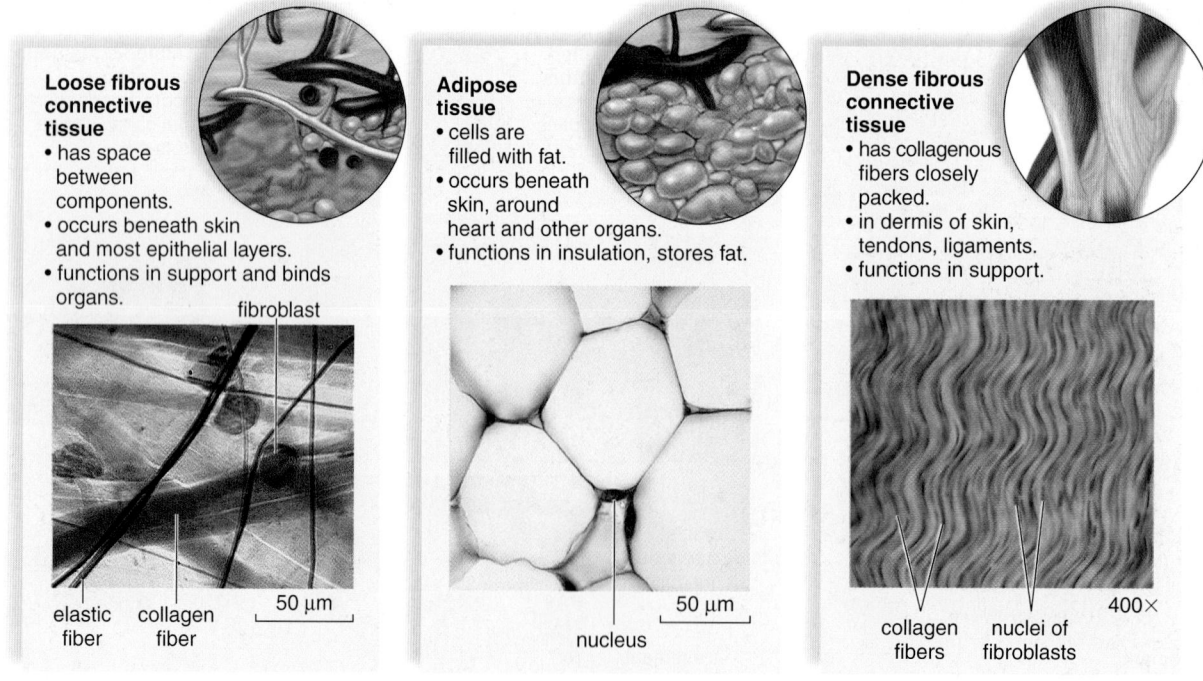

Figure 31.3 Types of connective tissue in vertebrates. Pertinent information about each type of connective tissue is given.

Loose fibrous connective tissue
- has space between components.
- occurs beneath skin and most epithelial layers.
- functions in support and binds organs.

fibroblast

elastic fiber   collagen fiber   50 µm

a.

Adipose tissue
- cells are filled with fat.
- occurs beneath skin, around heart and other organs.
- functions in insulation, stores fat.

50 µm

nucleus

b.

Dense fibrous connective tissue
- has collagenous fibers closely packed.
- in dermis of skin, tendons, ligaments.
- functions in support.

400×

collagen fibers   nuclei of fibroblasts

c.

supply, it heals very slowly. There are three types of cartilage, distinguished by the type of fiber in the matrix.

Hyaline cartilage (Fig. 31.3*d*), the most common type of cartilage, contains only very fine collagen fibers. The matrix has a white, translucent appearance. Hyaline cartilage is found in the nose and at the ends of the long bones and the ribs, and it forms rings in the walls of respiratory passages. The fetal skeleton also is made of this type of cartilage. Later, the cartilaginous fetal skeleton is replaced by bone.

Elastic cartilage has more elastic fibers than hyaline cartilage. For this reason, it is more flexible and is found, for example, in the framework of the outer ear.

Fibrocartilage has a matrix containing strong collagen fibers. Fibrocartilage is found in structures that withstand tension and pressure, such as the pads between the vertebrae in the backbone and the wedges in the knee joint.

Bone. Of all the connective tissues, **bone** is the most rigid. It consists of an extremely hard matrix of inorganic salts, notably calcium salts, deposited around protein fibers, especially collagen fibers. The inorganic salts give bones rigidity, and the protein fibers provide elasticity and strength, much as steel rods do in reinforced concrete.

Compact bone makes up the shaft of a long bone (Fig. 31.3*e*). It consists of cylindrical structural units called osteons (Haversian systems). The central canal of each osteon is surrounded by rings of hard matrix. Bone cells are located in spaces called lacunae between the rings of matrix. Blood vessels in the central canal carry nutrients that allow bone to renew itself. Thin extensions of bone cells within canaliculi (minute canals) connect the cells to each other and to the central canal. The hollow shaft of long bones such as the femur (thigh bone) is filled with yellow bone marrow (see Chapter 39).

Figure 31.4 Blood, a liquid connective tissue. a. Blood is classified as connective tissue because the cells are separated by a matrix—plasma. Plasma, the liquid portion of blood, usually contains several types of cells. **b.** Drawing of the components seen in a stained blood smear: red blood cells, white blood cells, and platelets (which are actually fragments of a larger cell).

plasma

white blood cells (leukocytes)

red blood cells (erythrocytes)

a. Blood sample after centrifugation

white blood cell

platelets

red blood cell

plasma

b. Blood smear

The ends of a long bone contain spongy bone, which has an entirely different structure. **Spongy bone** contains numerous bony bars and plates, separated by irregular spaces. Although lighter than compact bone, spongy bone is still designed for strength. Just as braces are used for support in buildings, the solid portions of spongy bone follow lines of stress. Spongy bone is also the site of red bone marrow, which is critical to production of blood cells (see Chapters 32 and 33).

Fluid Connective Tissues

Blood, which consists of formed elements and plasma, is a fluid connective tissue located in blood vessels (Fig. 31.4). Formed elements in the blood consist of the many kinds of blood cells and the platelets.

The internal environment of the body consists of blood and **tissue fluid.** The systems of the body help keep blood composition and chemistry within normal limits, and blood in turn creates tissue fluid. Blood transports nutrients and oxygen to tissue fluid and removes carbon dioxide and other wastes. It helps distribute heat and also plays a role in fluid, ion, and pH balance. The formed elements, discussed following, each have specific functions.

The **red blood cells** are small, disk-shaped cells without nuclei. The absence of a nucleus makes the cells biconcave. The presence of the red pigment hemoglobin

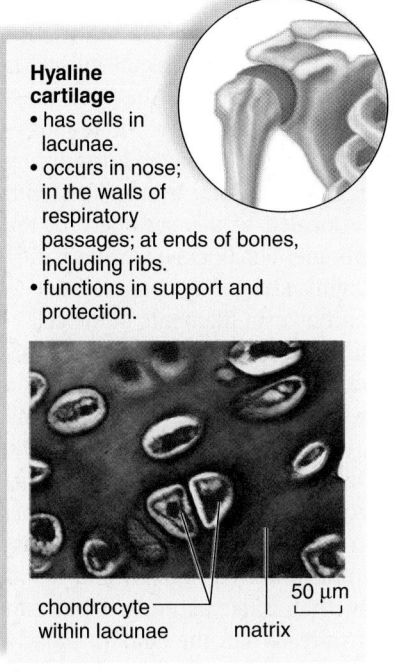

Hyaline cartilage
• has cells in lacunae.
• occurs in nose; in the walls of respiratory passages; at ends of bones, including ribs.
• functions in support and protection.

chondrocyte within lacunae

50 μm

matrix

d.

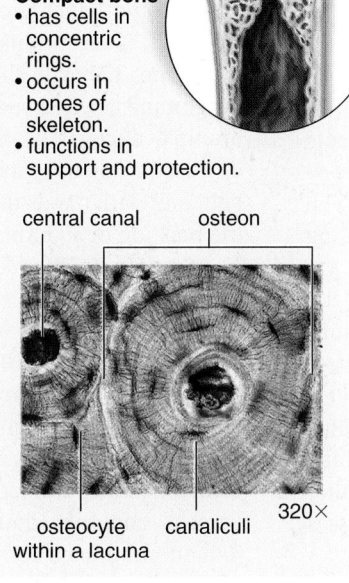

Compact bone
• has cells in concentric rings.
• occurs in bones of skeleton.
• functions in support and protection.

central canal

osteon

osteocyte within a lacuna

canaliculi

320×

e.

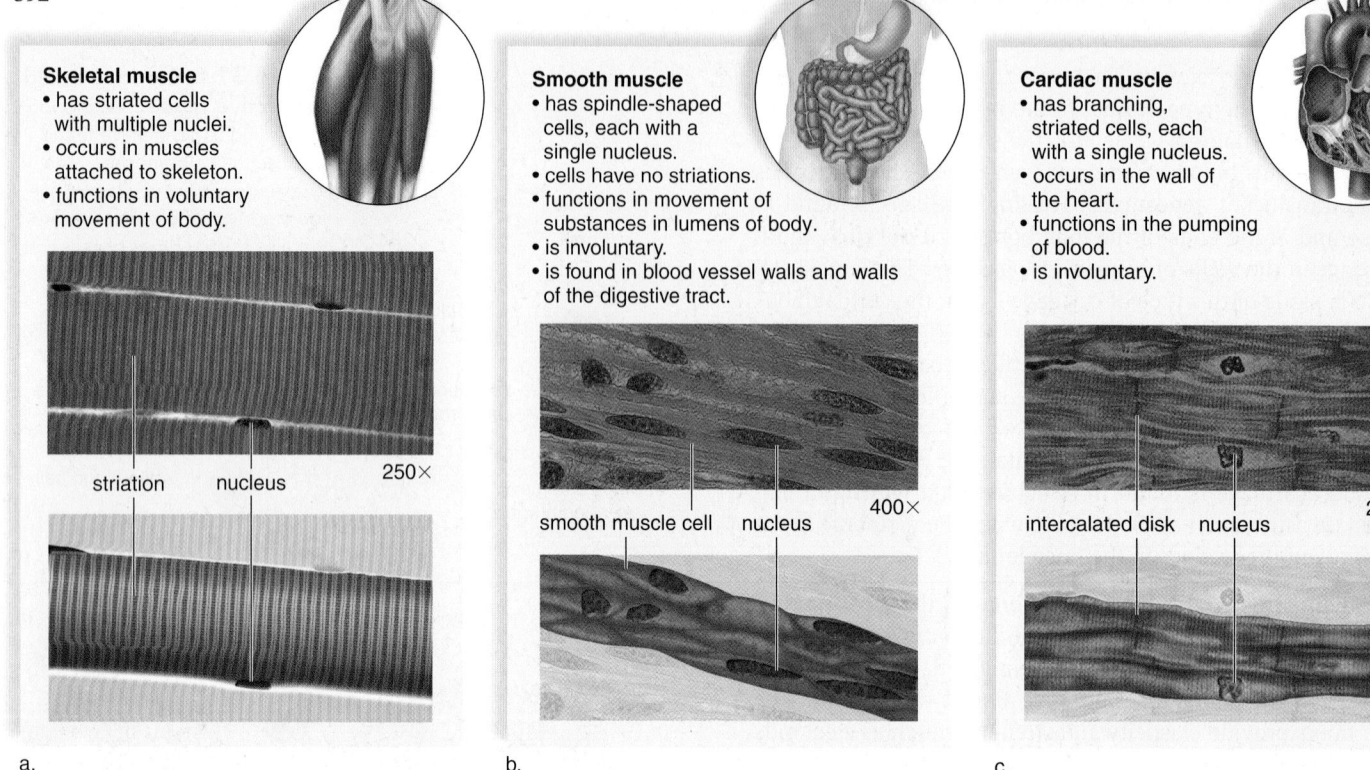

Skeletal muscle
- has striated cells with multiple nuclei.
- occurs in muscles attached to skeleton.
- functions in voluntary movement of body.

striation   nucleus   250×

Smooth muscle
- has spindle-shaped cells, each with a single nucleus.
- cells have no striations.
- functions in movement of substances in lumens of body.
- is involuntary.
- is found in blood vessel walls and walls of the digestive tract.

smooth muscle cell   nucleus   400×

Cardiac muscle
- has branching, striated cells, each with a single nucleus.
- occurs in the wall of the heart.
- functions in the pumping of blood.
- is involuntary.

intercalated disk   nucleus   250×

a.   b.   c.

Figure 31.5 Muscular tissue. **a.** Skeletal muscle is voluntary and striated. **b.** Smooth muscle is involuntary and nonstriated. **c.** Cardiac muscle is involuntary and striated. Cardiac muscle cells branch and fit together at intercalated disks.

makes them red, and in turn, makes the blood red. Hemoglobin is composed of four units; each unit is composed of the protein globin and a complex iron-containing structure called heme. The iron forms a loose association with oxygen, and in this way red blood cells transport oxygen and readily give it up in the tissues.

White blood cells may be distinguished from red blood cells by the fact that they are usually larger, have a nucleus, and without staining would appear translucent. When blood is smeared onto a microscope slide and stained, the nucleus of a white blood cell typically looks blue or purple. White blood cells fight infection, primarily in two ways. Some white blood cells are phagocytic and engulf infectious **pathogens,** while other white blood cells either produce antibodies, molecules that combine with foreign substances to inactivate them, or they kill cells outright.

Platelets are not complete cells; rather, they are fragments of large cells present only in bone marrow. When a blood vessel is damaged, platelets help to form a plug that seals the vessel, and injured tissues release molecules that help the clotting process.

Lymph is a fluid connective tissue located in lymphatic vessels. Lymphatic vessels absorb excess tissue fluid and return it to the cardiovascular system. Special lymphatic capillaries, called lacteals, also absorb fat molecules from the small intestine. Lymph nodes, composed of fibrous connective tissue plus specialized white blood cells called lymphocytes, occur along the length of lymphatic vessels. These lymphocytes and other cells remove any foreign material from the lymph as it passes through lymph nodes. Lymph nodes may enlarge when these cells respond to an infection.

MP3
Connective Tissue

Muscular Tissue

Muscular (contractile) tissue is composed of cells called muscle fibers. Muscle fibers contain actin filaments and myosin filaments, whose interaction accounts for movement. The muscles are also important in the generation of body heat. There are three distinct types of muscle tissue: skeletal, smooth, and cardiac. Each type differs in appearance, physiology, and function.

Skeletal muscle, also called voluntary muscle (Fig. 31.5a), is attached by tendons to the bones of the skeleton, and when it contracts, body parts move. Contraction of skeletal muscle is under voluntary control and occurs faster than in the other muscle types. Skeletal muscle fibers are cylindrical and quite long—sometimes they run the length of the muscle. They arise during development when several cells fuse, resulting in one fiber with multiple nuclei. The nuclei are located at the periphery of the cell, just inside the plasma membrane. The fibers have alternating light and dark bands that give them a **striated** appearance, due to the position of actin filaments and myosin filaments in the cell.

Smooth (visceral) muscle is so named because the cells lack striations. The spindle-shaped cells, each with a single nucleus, form layers in which the thick middle portion of one cell is opposite the thin ends of adjacent cells. Consequently, the nuclei form an irregular pattern in the tissue (Fig. 31.5b). Smooth muscle is not under voluntary control and therefore is said to be involuntary. Smooth muscle, found in the walls of viscera (intestine, stomach, and other internal organs) and blood vessels, contracts more slowly than skeletal muscle but can remain contracted for a longer time. When the smooth muscle of the intestine contracts, food moves along its lumen (central cavity). When the smooth muscle of the blood vessels contracts, blood vessels constrict, helping to raise blood pressure. Small amounts of smooth muscle are also found in the iris of the eye and in the skin.

Cardiac muscle (Fig. 31.5c) is found only in the walls of the heart. Its contraction pumps blood and accounts for the heart-beat. Cardiac muscle combines features of both smooth muscle and skeletal muscle. Like skeletal muscle, it has striations, but the contraction of the heart is involuntary for the most part. Cardiac muscle cells also differ from skeletal muscle cells in that they usually have a single, centrally placed nucleus. The cells are branched and seemingly fused one with the other, and the heart appears to be composed of one large interconnecting mass of muscle cells. Actually, cardiac muscle cells are separate and individual, but they are bound end to end at *intercalated disks,* areas where folded plasma membranes between two cells contain adhesion junctions and gap junctions.

MP3
Muscle Tissue

Nervous Tissue

Nervous tissue contains nerve cells called neurons and supporting cells called neuroglia. An average human body has about 1 trillion neurons. The nervous system conveys signals termed nerve impulses throughout the body.

Neurons

A **neuron** is a specialized cell that has three parts: dendrites, a cell body, and an axon (Fig. 31.6a). A dendrite is a process that conducts signals toward the cell body. The cell body contains the major portion of the cytoplasm and the nucleus of the neuron. An axon is a process that typically conducts nerve impulses away from the cell body. Long axons are covered by myelin, a white, fatty substance. The term *fiber* is used here to refer to an axon along with its myelin sheath if it has one. Outside the brain and spinal cord, fibers bound by connective tissue form **nerves.**

The nervous system has just three functions: sensory input, integration of data, and motor output. Nerves conduct impulses from sensory receptors to the spinal cord and the brain, where integration occurs. The phenomenon called sensation occurs only in the brain, however. Nerves also conduct nerve impulses away from the spinal cord and brain to the muscles and glands, causing them to contract and secrete, respectively. In this way, a coordinated response to the stimulus is achieved.

Neuroglia

In addition to neurons, nervous tissue contains cells called **neuroglia**. In the human brain, these cells outnumber neurons as much as ten to one, and make up approximately half the volume of the organ. Although the primary function of neuroglia is to support and nourish neurons, recent research has shown that some neuroglia directly contribute to brain function.

Several types of neuroglia are found in the brain. Microglia, astrocytes, and oligodendrocytes are shown in Figure 31.6a. Microglia, in addition to supporting neurons, engulf bacterial and cellular debris. Astrocytes provide nutrients to neurons and produce a hormone known as glial cell-derived growth factor, which is being studied as a possible treatment for Parkinson disease and other diseases caused by neuron degeneration. Oligodendrocytes form myelin in the brain.

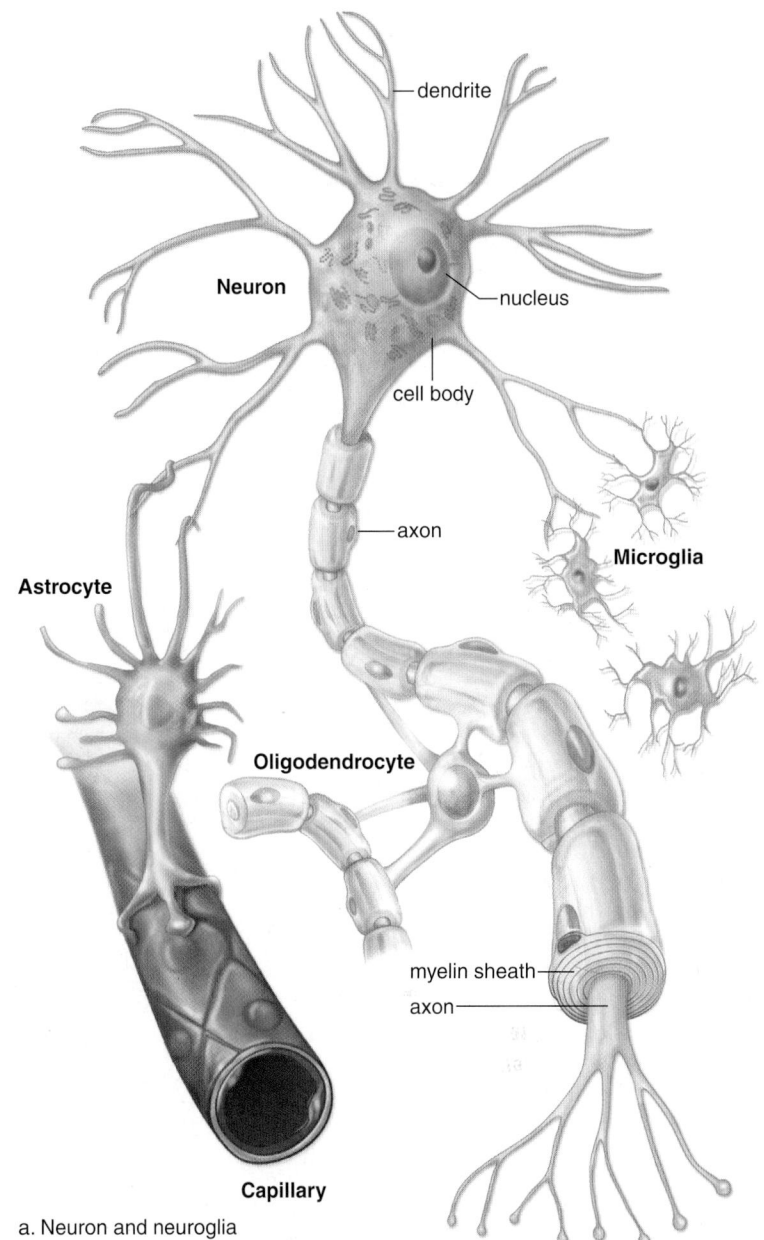

a. Neuron and neuroglia

b. Micrograph of a neuron 200×

Figure 31.6 Neurons and neuroglia. Neurons conduct nerve impulses. Neuroglia consist of cells that support and service neurons and have various functions: Microglia become mobile in response to inflammation and phagocytize debris. Astrocytes lie between neurons and a capillary; therefore, nutrients must first pass through astrocytes before entering neurons. Oligodendrocytes form the myelin sheaths around fibers in the brain and spinal cord.

Nature of Science

Regenerative Medicine

The ability of salamanders and a few other animals to regenerate amputated tails, limbs, and other organs has fascinated scientists for centuries. Although many invertebrate species can regenerate missing parts (some decapitated snails can regrow a new head!), most vertebrates can regenerate body parts only during embryonic stages of development. The salamander is an exception to this rule, however, with many species retaining the ability to replace lost limbs, tails, and even jaws throughout their lives. Unlike salamanders, humans who lose limbs or organs must either get by without them, or rely on artificial limbs, machines, or transplanted organs.

Much like human arms or legs, salamander limbs contain many different tissue types: bone, muscle, nerve, connective tissue, blood vessels, and skin. Until quite recently, most scientists believed that when a salamander limb is severed, the cells that migrate to the site of the injury revert to being truly undifferentiated, pluripotent stem cells capable of becoming any kind of tissue. That hypothesis leads to a number of questions: How did an injury cause the cells to become "reprogrammed"? And how do the cells know which types of tissues to become?

Some recent research has cast doubt on the idea that the cells that regrow a salamander's limb are true stem cells. In a 2009 study, scientists in a German laboratory devised a new method to track the fate of cells rebuilding the severed limb. The researchers first inserted a gene coding for a jellyfish protein called green fluorescent protein (GFP) into the genome of embryonic axolotl salamanders. The result was transgenic salamanders, in which every cell of the animal gave off a detectable green glow when illuminated with ultraviolet light (Figure 31A).

Different tissues from these transgenic salamanders could then be transplanted into nontransgenic salamanders of the same species. When the researchers were certain that the transplanted tissues had survived, different limb amputations were performed. As the limb regenerated, the fate of the transplanted cells could be easily followed by examining the regenerating limb for GFP+ (green) cells. This tracking allowed the researchers to answer the question of whether the GFP+ donor cells would be found throughout the regenerating limb, or only in the same type of tissue that was originally transplanted (Figure 31B).

Somewhat surprisingly, the results showed that most GFP+ tissue types gave rise only to the same tissue type, or to a very limited set of tissues (Figure 31B). For example, when GFP+ skin was transplanted, GFP+ cells were found only in skin or cartilage, but never in muscle. Similarly, when GFP+ cartilage was transplanted, GFP+ cells were found in cartilage but never in muscle, and GFP+ muscle tissue never produced cartilage. Additionally, although previous research had suggested that neuroglia-type cells present in the regenerating limb might revert to pluripotent cells capable of forming many types of tissues, in these experiments GFP+ neuroglia-type cells gave rise only to nerve tissue.

Although the ability to regenerate human limbs and organs is still far in the future, the key to teaching human cells how to regenerate may lie in experiments like these. Beyond regenerating limbs and organs, scientists working in the field of regenerative medicine are closing in on ways to replace insulin-producing cells in type 1 diabetes, to grow new blood vessels to replace blocked ones in failing hearts, and even to replace damaged nerves as well as bone, muscle, and cartilage. Although a

Figure 31A Creating a new type of salamander for limb regeneration research. *Left:* Wild type axolotl salamander showing normal coloration. *Right:* Researchers used genetic engineering techniques to produce an axolotl whose cells would fluoresce green under UV illumination.

One fundamental difference between neurons and neuroglia is that the neurons of adult animals usually cannot undergo cell division, but neuroglial cells retain this capacity. As a result, the majority of brain tumors in adults involve actively dividing neuroglial cells. Most of these tumors have to be treated with surgery or radiation therapy because a large number of tight junctions in the epithelial cells of brain capillaries prevent many substances (including anti-cancer drugs) from entering the brain tissue.

MP3
Nervous Tissue

Check Your Progress   31.1

1. List five types of epithelium and identify an example of where each type could be found.
2. Compare and contrast the three major types of connective tissue.
3. Describe the structure and function of skeletal, smooth, and cardiac muscle.
4. Recall the three parts of a neuron, and explain the function of each.

large amount of financial support for this research is needed initially, regenerative medicine offers the potential to cure many conditions that currently can only be managed over the lifetime of a patient.

Questions to Consider

1. Certain primitive invertebrate animals, such as sponges and hydras, can re-generate their entire body from one or a few cells. How does this phenomenon complicate the definition of what constitutes an "animal" versus a single-celled life-form?

2. Even in humans, certain cell types (e.g., neurons) are relatively less likely to regenerate after an injury than other types (e.g., skin). What are some rel-evant differences between these two types of cells that might explain this observation?

3. Beyond regenerating limbs, what are some other potential human diseases that could be treated or cured if we completely understood regeneration?

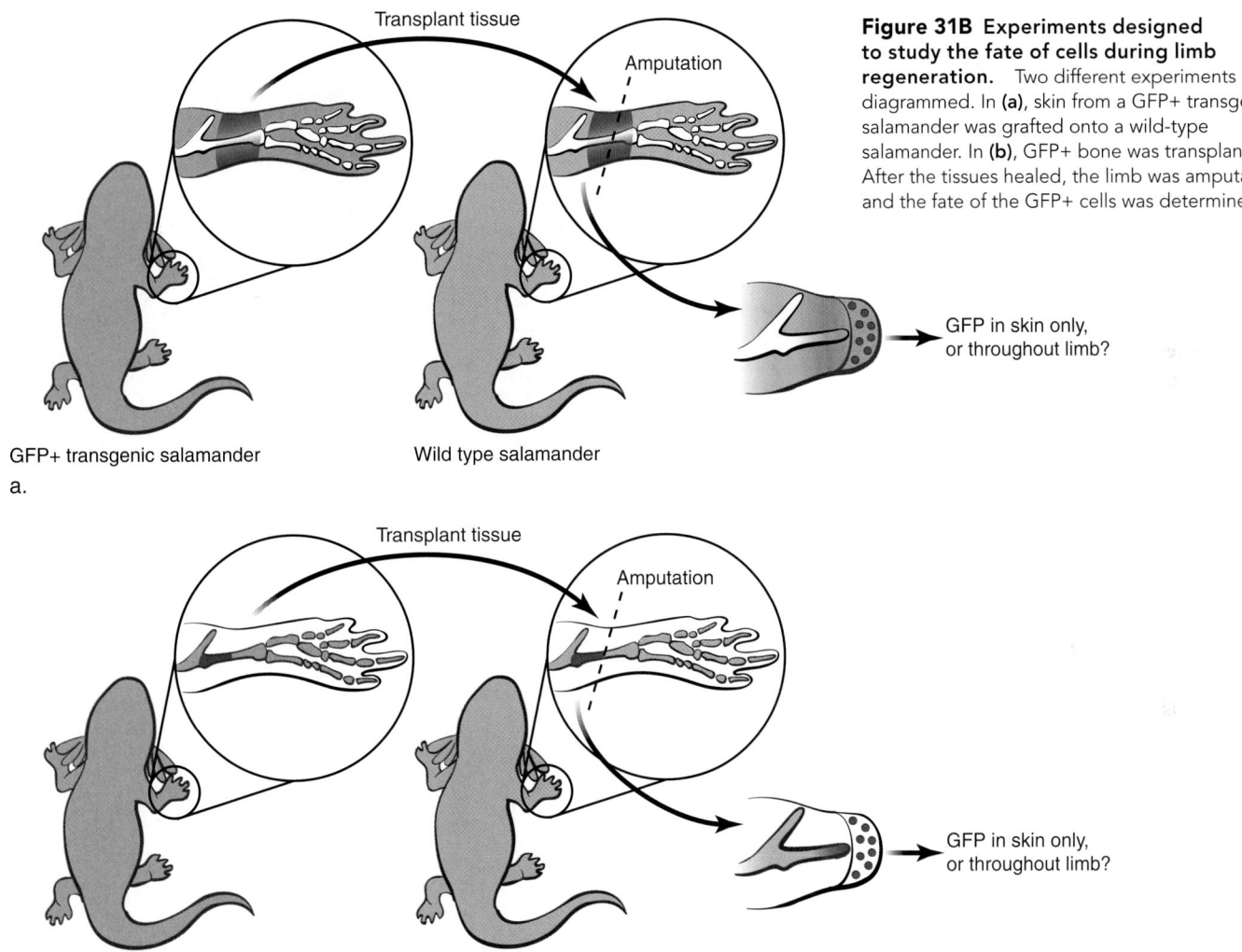

Figure 31B Experiments designed to study the fate of cells during limb regeneration. Two different experiments are diagrammed. In **(a)**, skin from a GFP+ transgenic salamander was grafted onto a wild-type salamander. In **(b)**, GFP+ bone was transplanted. After the tissues healed, the limb was amputated, and the fate of the GFP+ cells was determined.

a.

b.

31.2 Organs, Organ Systems, and Body Cavities

Learning Outcomes

Upon completion of this section, you should be able to

1. Distinguish between tissues, organs, and organ systems.
2. List the major life processes carried out by each organ system in vertebrate animals.
3. Describe the two main cavities of the human body, and the major organs found in each.

Tissues have four main types; however, two or more of these types may be arranged together in a structure termed an organ that has a specific function; in turn, several organs may work together in an organ system to accomplish a general process. The evolution of organs and organ systems allowed animals to localize these processes to certain areas of the body, and to accomplish them more efficiently.

Organs

We first described the concept of organs in Chapter 24 when we discussed the vegetative and reproductive organs of flowering plants. An **organ** is composed of two or more types of tissues working together to perform a particular function. For example, a kidney is an organ that contains a variety of epithelial and connective tissues, and these tissues are specialized for the function of eliminating waste products from the blood.

Organ Systems

In most animals, individual organs function as part of an organ system. An **organ system** contains many different organs that cooperate to carry out a general process, such as the digestion of food. Similar types of organ systems are found in most invertebrates, and in all vertebrate animals. These organ systems carry out the life processes that all of these animals, including humans, must carry out.

Life Processes	Organ Systems
Coordinate body activities	Nervous system Endocrine system
Acquire materials and energy (food)	Skeletal system Muscular system Digestive system
Maintain body shape	Skeletal system Muscular system
Exchange gases	Respiratory system
Transport materials	Cardiovascular system
Eliminate wastes	Urinary system Digestive system
Protect the body from pathogens	Lymphatic system Immune system
Produce offspring	Reproductive system

Body Cavities

Each organ system has a particular distribution within the body. Vertebrates have two main **body cavities:** the smaller dorsal cavity and the larger ventral cavity (Fig. 31.7a). The brain and the spinal cord are in the dorsal cavity.

During development, the ventral cavity develops from the coelom. In humans and other mammals, the coelom is divided by a muscular diaphragm that assists breathing. The heart and the lungs are located in the upper (thoracic or chest) cavity (Fig. 31.7b). The major portions of the digestive system, including the accessory organs (e.g., the liver and pancreas) are located in the abdominal cavity, as are the kidneys of the urinary system. The urinary bladder, the female reproductive organs, or certain of the male reproductive organs, are located in the pelvic cavity.

MP3
Body Cavities

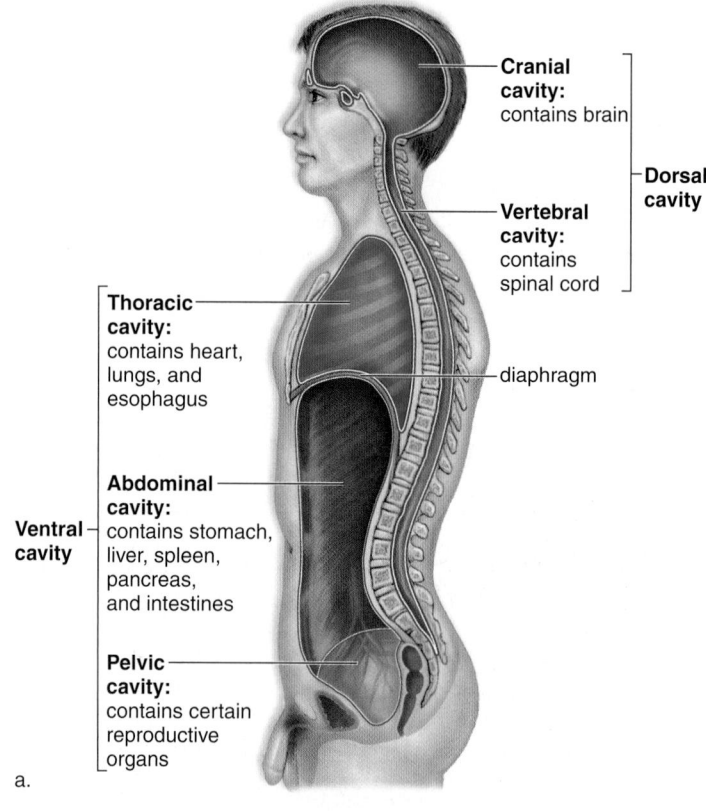

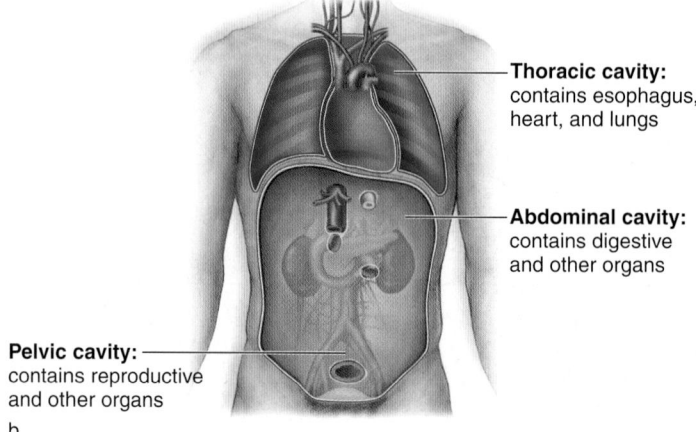

Figure 31.7 Mammalian body cavities. **a.** Side view. The dorsal (toward the back) cavity contains the cranial cavity and the vertebral canal. The brain is in the cranial cavity, and the spinal cord is in the vertebral canal. The well-developed ventral (toward the front) cavity is divided by the diaphragm into the thoracic cavity and the abdominopelvic cavity (abdominal cavity and pelvic cavity). The heart and lungs are in the thoracic cavity, and most other internal organs are in the abdominal cavity. **b.** Frontal view of the thoracic cavity.

Check Your Progress   31.2

1. Explain the difference between an organ and an organ system.
2. Identify two organ systems that protect the body from disease.
3. List and locate the two major body cavities in humans, as well as the two cavities found in each of these.

31.3 The Integumetary System

Learning Outcomes

Upon completion of this section, you should be able to

1. Distinguish between the functions of skin that are common to all animals versus those that are unique to specific groups.
2. Identify the two main regions of skin, and how these differ from the subcutaneous layer.
3. Explain the function of melanocytes in the skin and the effects of UV radiation.
4. Describe the makeup and function of the accessory structures of human skin.

The integumentary system, consisting of the skin, its derivatives, and its accessory organs, is the largest and most conspicuous organ system in the body. The skin of an average human covers an area of 21 square feet and accounts for nearly 15% of the body weight.

Derivatives of the skin differ throughout the vertebrate world. Most fishes have protective outgrowths of the skin called scales. Amphibian skin is usually covered with mucous glands. Scales are characteristic of reptiles, and are also found on the legs and feet of birds, but only birds have feathers, which grow from specialized follicles in the skin. Hair is found only on the skin of mammals, and all mammals (even whales) have hair at some stage of their life.

Functions of Skin

Skin covers the body, protecting underlying parts from dessication, physical trauma, and pathogen invasion. It is also important in regulating body temperature. The skin of small aquatic or semiaquatic animals is often involved in the exchange of gases with the environment (see Chapter 35). In contrast, the dry, scaly skin of reptiles is very poor at gas exchange, but prevents water loss and thus was probably an important evolutionary adaptation to life on land. Feathers are unique appendages of bird skin that function in insulation, waterproofing, and of course, flight. Skin is also equipped with a variety of sensory structures that monitor touch, pressure, temperature, and pain. In addition, skin cells manufacture precursor molecules that are converted to vitamin D after exposure to UV (ultraviolet) light.

Regions of Skin

The **skin** has two main regions: the epidermis and the dermis (Fig. 31.8). A *subcutaneous layer*, also known as the hypodermis, is found between the skin and any underlying structures, such as muscle or bone.

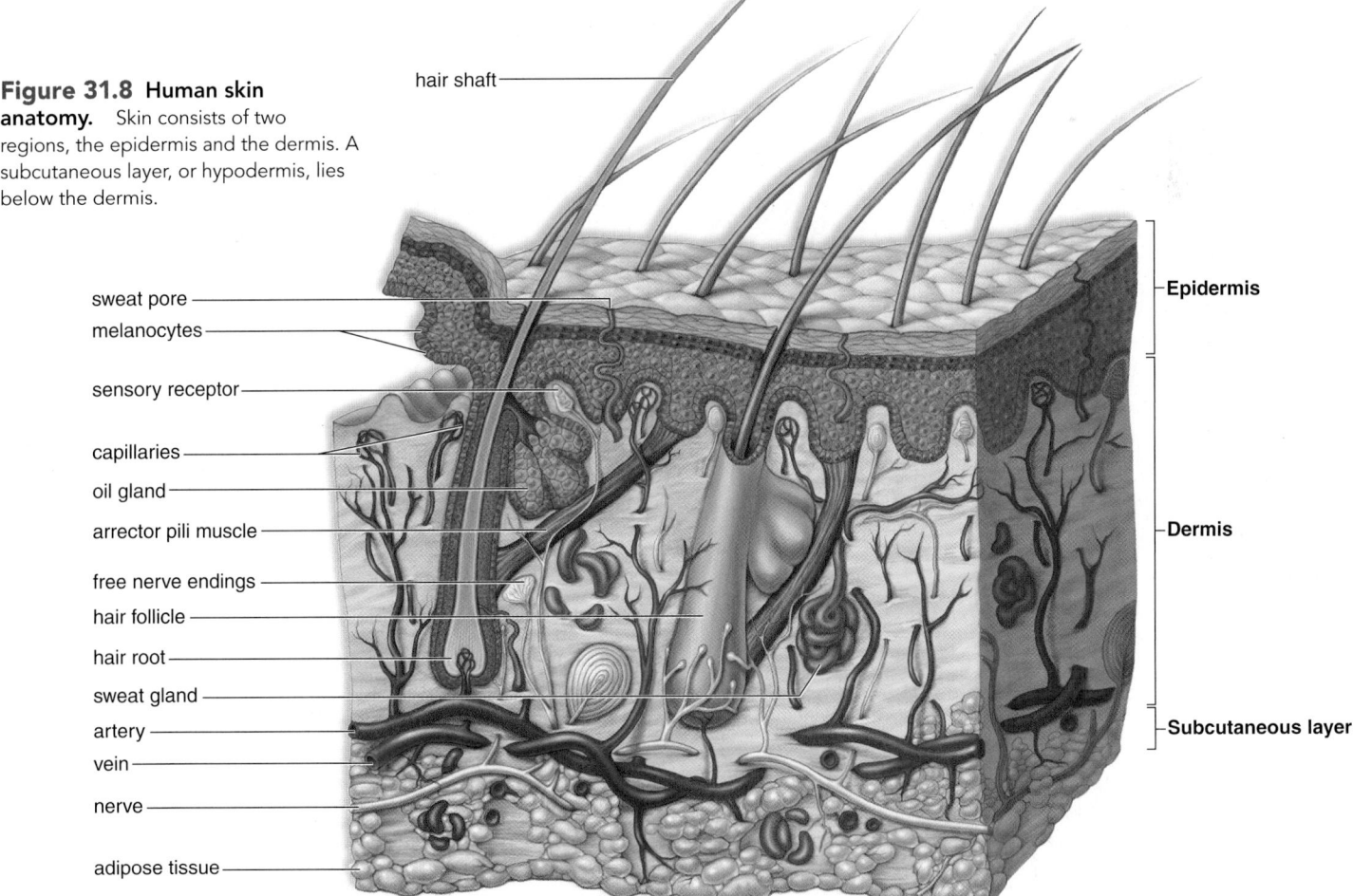

Figure 31.8 Human skin anatomy. Skin consists of two regions, the epidermis and the dermis. A subcutaneous layer, or hypodermis, lies below the dermis.

hair shaft

sweat pore
melanocytes
sensory receptor
capillaries
oil gland
arrector pili muscle
free nerve endings
hair follicle
hair root
sweat gland
artery
vein
nerve
adipose tissue

Epidermis

Dermis

Subcutaneous layer

The Epidermis

The **epidermis** [Gk. *epi,* over, and *derma,* skin] is made up of stratified squamous epithelium. Human skin can be described as thin skin or thick skin based on the thickness of the epidermis. Thin skin covers most of the body and is associated with hair follicles, sebaceous (oil) glands, and sweat glands. Thick skin appears in regions of wear and tear, such as the palms of the hands and soles of the feet. Thick skin has sweat glands but no sebaceous glands or hair follicles.

In both types of skin, new cells derived from stem (basal) cells become flattened and hardened as they push to the surface (Fig. 31.9*a*). Hardening takes place because the cells produce keratin, a waterproof protein. It is estimated that 1.5 million of these cells are shed from the human body every day! A thick layer of dead keratinized cells, arranged in spiral and concentric patterns, forms fingerprints (and toe prints, too), which are thought to increase friction and aid in gripping objects.

Specialized cells in the epidermis called **melanocytes** produce melanin, the pigment responsible for skin color. The amount of melanin varies throughout the body. It is concentrated in freckles and moles. Tanning occurs after a lighter-skinned person is exposed to sunlight because melanocytes produce more melanin, which is distributed to epidermal cells before they rise to the surface. Although we tend to associate a tan with health, in reality it signifies that the body is trying to protect itself from the dangerous rays of the Sun. Some ultraviolet (UV) radiation can benefit health, however. As mentioned, certain cells in the epidermis convert a steroid related to

cholesterol into *vitamin D* only with the aid of ultraviolet radiation. Vitamin D is required for proper bone growth, and perhaps for many other bodily functions.

In contrast to its beneficial effects, UV radiation can cause mutations in the DNA of skin cells, leading to skin cancer. Basal cell carcinoma (Fig. 31.9*b*) derived from stem cells gone awry is the more common type of skin cancer and the most curable. Melanoma (Fig. 31.9*c*), the type of skin cancer derived from melanocytes, is the most deadly form. New melanoma treatments are being developed that activate the immune system to fight the tumors more effectively.

The Dermis

The **dermis** [Gk. *derma,* skin] is a region of dense fibrous connective tissue beneath the epidermis. As seen in Figure 31.9*a*, the deeper epidermis forms ridges that interact with projections of the dermis. The dermis contains collagen and elastic fibers. The collagen fibers are flexible but offer great resistance to overstretching; they prevent the skin from being torn. Stretching of the dermis, as occurs in obesity and pregnancy, can produce stretch marks, or striae.

The elastic fibers maintain normal skin tension but also stretch to allow movement of underlying muscles and joints. (The number of collagen and elastic fibers decreases with age and with exposure to the Sun, causing the skin to become less supple and more prone to wrinkling.) The dermis also contains blood vessels that nourish the skin. When blood rushes into these vessels, a person blushes, and when blood is minimal in them, a person turns "blue."

Sensory receptors are specialized nerve endings in the dermis that respond to external stimuli. There are sensory receptors for touch, pressure, pain, and temperature. The fingertips contain the most touch receptors, and these add to our ability to use our fingers for delicate tasks.

The Subcutaneous Layer

Technically speaking, the subcutaneous layer (the hypodermis) beneath the dermis is not a part of skin. It is composed of loose connective tissue and adipose tissue. Subcutaneous adipose tissue helps to thermally insulate the body from either gaining heat from the outside or losing heat from the inside. Excessive development of the subcutaneous layer accompanies obesity.

Accessory Structures of Human Skin

Nails, hair, and glands are of epidermal origin, even though some parts of hair and glands are largely found in the dermis.

Nails are a protective covering of the distal part of fingers and toes, collectively called digits. Nails grow from special epithelial cells at the base of the nail in the portion called the nail root. The cuticle is a fold of skin that hides the nail root. The whitish color of the half-moon-shaped base, or lunula, results

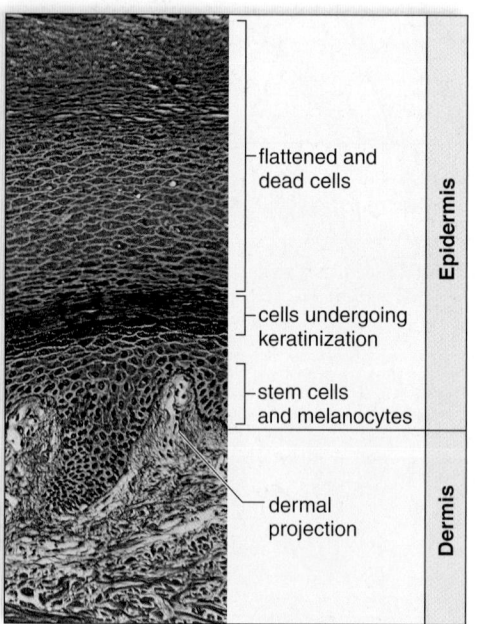

flattened and dead cells

Epidermis

cells undergoing keratinization

stem cells and melanocytes

dermal projection

Dermis

a. Photomicrograph of skin

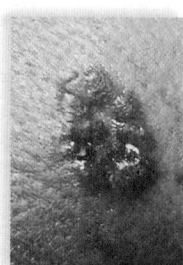

b. Basal cell carcinoma

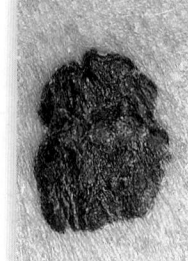

c. Melanoma

Figure 31.9 The epidermis.   a. Epidermal ridges following dermal projections are clearly visible. Stem cells and melanocytes are in this region. Types of skin cancers include **(b)** basal cell carcinoma derived from stem cells, and **(c)** melanoma derived from melanocytes.

from the thick layer of cells in this area. The cells of a nail become keratinized as they grow out over the nail bed. The appearance of nails can reveal clues about a person's health. For example, in clubbing of the nails, the nails turn down instead of lying flat. This condition is associated with a deficiency of oxygen in the blood.

Hair follicles begin in the dermis and continue through the epidermis, where the hair shaft extends beyond the skin. Contraction of the arrector pili muscles attached to hair follicles causes the hairs to "stand on end" and goose bumps to develop. Epidermal cells form the root of hair, and their division causes a hair to grow. The cells become keratinized and die as they are pushed farther from the root.

Hair, except for the root, is formed of dead, hardened epidermal cells; the root is alive and resides at the base of a follicle in the dermis. A person's scalp has about 100,000 hair follicles on average. The number of follicles varies from one body region to another. The texture of hair is dependent on the shape of the hair shaft. In wavy hair, the shaft is oval, and in straight hair, the shaft is round. Hair color is determined by pigmentation. Dark hair is due to melanin concentration, and blond hair has scanty amounts of melanin. Red hair is caused mainly by the presence of an iron-containing pigment called pheomelanin. Gray or white hair results from a lack of pigment. A hair on the scalp grows about 1 mm every three days.

Each hair follicle has one or more **oil glands,** also called sebaceous glands, which secrete sebum, an oily substance that lubricates the hair within the follicle and the skin itself. If the sebaceous glands fail to discharge, the secretions collect and form "whiteheads" or "blackheads." The color of blackheads is due to oxidized sebum. Acne is an inflammation of the sebaceous glands that most often occurs during adolescence due to hormonal changes.

On average, a person's skin has about 250,000 **sweat glands**, which are present in all regions of skin. A sweat gland is a tubule that begins in the dermis and either opens into a hair follicle, or more often opens onto the surface of the skin. Sweat glands located all over the body play a role in modifying body temperature. When the body temperature starts to rise, sweat glands become active. Sweat absorbs body heat as it evaporates. Once the body temperature lowers, sweat glands are no longer active. Other sweat glands occur in the groin and axillary regions and are associated with distinct scents.

MP3 Human Skin

Check Your Progress 31.3

1. List one function of skin that is common to all animals, and one that is unique to a single group.
2. Compare the structure and function of the epidermal and dermal layers of the skin.
3. Discuss why a dark-skinned individual living in northern Canada might develop bone problems.
4. Describe the structure of nails, hair, sweat glands, and oil glands.

31.4 Homeostasis

Learning Outcomes

Upon completion of this section, you should be able to

1. Define homeostasis, and explain why it is an essential feature of all living organisms.
2. Evaluate the evolutionary benefits of regulating an internal variable, such as body temperature, versus the cost.
3. Differentiate between positive and negative feedback mechanisms, and list one specific example of each in animals.

All organ systems of animals contribute to **homeostasis**, the ability of an organism to maintain a relatively constant internal environment. Although all organisms must carry out some degree of homeostasis, animals vary in the degree to which they regulate these internal variables.

Examples of Homeostatic Regulation

With respect to body temperature, all invertebrates, as well as fish, amphibians, and reptiles, are "cold-blooded," or **poikilothermic,** meaning their body temperature fluctuates depending on their environmental temperature. This approach saves energy, but it also may restrict the ability of these species to live in extremely cold or hot environments.

Birds and mammals tend to be "warm-blooded," or **homeothermic,** and they have mechanisms for regulating their body temperature toward an optimum. This approach is energetically expensive, but provides the evolutionary advantage of being able to adapt to many different environments.

Homeostasis does not mean a rigid or unvarying stability, but rather a dynamic fluctuation around a set point. Besides temperature, animal systems regulate pH, salt balance, and the concentrations of many other body constituents such as glucose, oxygen, CO_2, and various minerals. All organ systems participate in this regulation:

- The digestive system takes in and digests food, providing nutrient molecules to replace those constantly being consumed by the body cells.
- The respiratory system adds oxygen to the blood and removes carbon dioxide, to meet body needs.
- The liver removes glucose from the blood and stores it as glycogen; later, glycogen is broken down to supply the needs of body cells. Blood glucose levels remain fairly constant.
- The pancreas secretes insulin in response to elevated blood glucose; insulin helps regulate glycogen storage.
- Under hormonal control, the kidneys excrete wastes and salts, substances that can affect the pH of the blood.

When homeostasis fails, disease or death often results.

MP3 Homeostasis

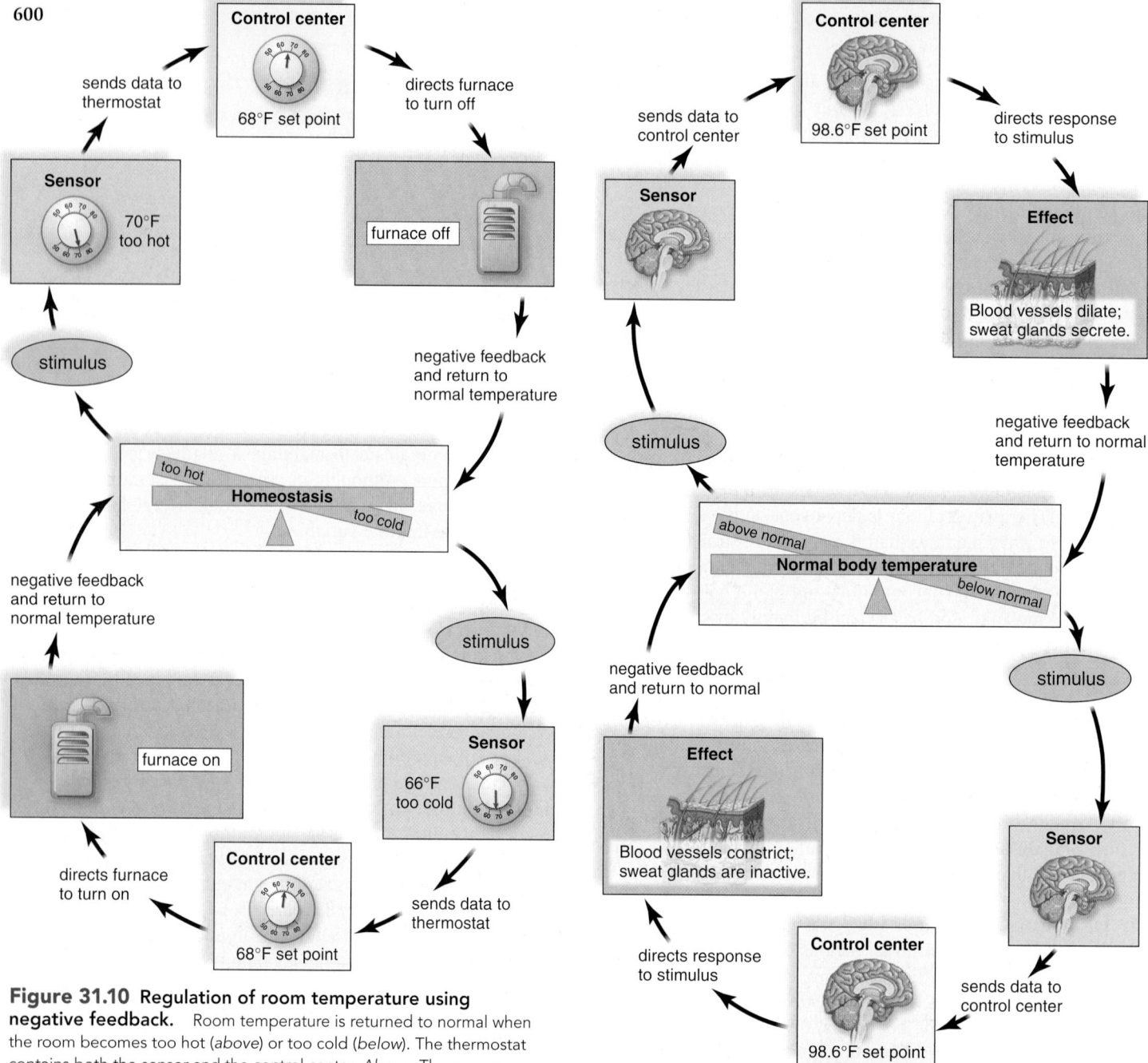

Figure 31.10 Regulation of room temperature using negative feedback. Room temperature is returned to normal when the room becomes too hot (*above*) or too cold (*below*). The thermostat contains both the sensor and the control center. *Above:* The sensor detects that the room is too hot, and the control center turns the furnace off. The room cools, removing the stimulus (negative feedback). *Below:* The sensor detects that the room is too cold, and the control center turns the furnace on. When the temperature returns to normal, the stimulus is no longer present.

Figure 31.11 Regulation of body temperature by negative feedback. *Above:* When body temperature rises above normal, the hypothalamus senses the change and causes blood vessels to dilate and sweat glands to secrete so that body temperature returns to normal. *Below:* When body temperature falls below normal, the hypothalamus senses the change and causes blood vessels to constrict. In addition, shivering may occur to bring body temperature back to normal. In this way, the original stimulus was removed (negative feedback).

Although homeostasis is, to a degree, controlled by hormones, it is ultimately controlled by the nervous system. In humans, the brain contains regulatory centers that control the function of other organs, maintaining homeostasis. These regulatory centers are often a part of negative feedback systems.

Negative Feedback

Negative feedback is the primary homeostatic mechanism that keeps a variable, such as the blood glucose level, close to a particular value, or set point.

A homeostatic mechanism has at least two components: a sensor and a control center. The sensor detects a change in the internal environment; the control center then initiates an action to bring conditions back to normal again. When normal conditions are reached, the sensor is no longer activated. In other words, a negative feedback mechanism is present when the output of the system dampens the original stimulus. As an example, when blood pressure rises, sensory receptors signal a control center in the brain. The center stops sending nerve impulses to the arterial walls, and they relax. Once the blood pressure drops, signals no longer go to the control center.

A home heating system is often used to illustrate how a more complicated negative feedback mechanism works (Fig. 31.10).

You set the thermostat at, say, 68°F. This is the *set point*. The thermostat contains a thermometer, a sensor that detects when the room temperature is above or below the set point. The thermostat also contains a control center; it turns the furnace off when the room is warm and turns it on when the room is cool. When the furnace is off, the room cools a bit, and when the furnace is on, the room warms a bit. In other words, typical of negative feedback mechanisms, there is a fluctuation above and below normal.

Human Example: Regulation of Body Temperature

The sensor and control center for body temperature are located in a part of the brain called the hypothalamus.

Above Normal Temperature. When the body temperature is above normal, the control center directs the blood vessels of the skin to dilate. The result is that more blood flows near the surface of the body, where heat can be lost to the environment. In addition, the nervous system activates the sweat glands, and the evaporation of sweat helps lower body temperature. Gradually, body temperature decreases to 37.0 degrees C (98.6°F). Body temperature does not get colder and colder because a body temperature below normal brings about a change toward a warmer body temperature.

Below Normal Temperature. When the body temperature falls below normal, the control center directs (via nerve impulses) the blood vessels of the skin to constrict (Fig. 31.11). This action conserves heat. If body temperature falls even lower, the control center sends nerve impulses to the skeletal muscles, and shivering occurs. Shivering generates heat, and gradually body temperature rises to 37.0°C. When the temperature rises to normal, the control center is inactivated.

MP3 Temperature Regulation

Positive Feedback

Positive feedback is a mechanism that brings about a continually greater change in the same direction (Fig. 31.12).

When a woman is giving birth, the head of the baby begins to press against the cervix (opening to the birth canal) stimulating sensory receptors there. When nerve impulses reach the brain, the brain causes the pituitary gland to secrete the hormone

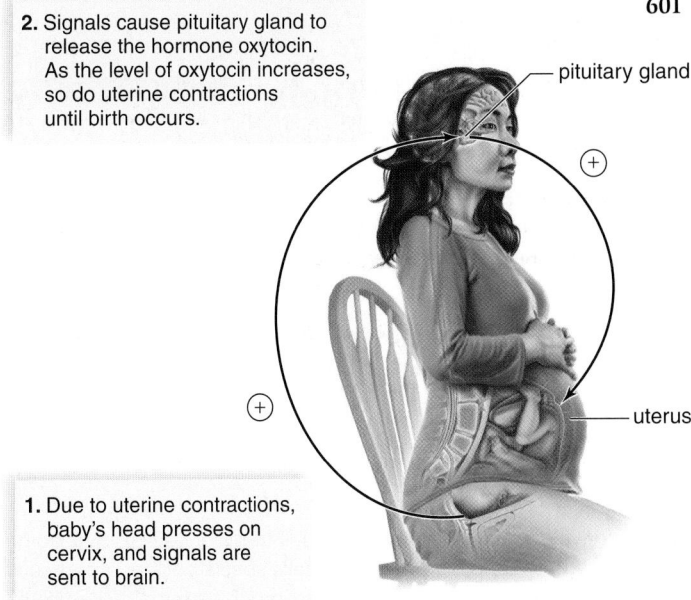

2. Signals cause pituitary gland to release the hormone oxytocin. As the level of oxytocin increases, so do uterine contractions until birth occurs.

pituitary gland

(+)

(+)

uterus

1. Due to uterine contractions, baby's head presses on cervix, and signals are sent to brain.

Figure 31.12 Positive feedback. This diagram shows how positive feedback works. The signal causes a change in the same direction until there is a definite cutoff point, such as birth of a child.

oxytocin. Oxytocin travels in the blood and causes the uterus to contract. As labor continues, the cervix is increasingly more stimulated and uterine contractions become stronger until birth occurs.

A positive feedback mechanism can be harmful, as when a fever causes metabolic changes that push the fever still higher. Death occurs at a body temperature of 113°F because cellular proteins denature at this temperature and metabolism stops. Still, positive feedback loops such as those involved in childbirth, blood clotting, and the stomach's digestion of protein assist the body in completing a process that has a definite cutoff point.

Animation Feedback Mechanisms

Check Your Progress 31.4

1. Define homeostasis and discuss why is it important to body function.
2. Explain how the circulatory, respiratory, and urinary systems specifically contribute to homeostasis.
3. Discuss how negative feedback is similar to the way a thermostat works.

CONNECTING *the* CONCEPTS *with the* BIG IDEAS

Energy and Homeostasis

- Regulation of body temperature by negative feedback is essential in normal function for many organisms. (2C1a*IE*, 2C2a*IE*)
- The positive feedback regulation of labor and lactation insure changes continue when required. (2C1b*IE*)
- Cellular respiration can aid in thermoregulation by uncoupling electron transport from ATP production to produce heat. (2A2g5)
- There is typically an inverse relationship between body mass and metabolic rate: the larger the organism, the lower its metabolism. (2A1d3)

- Common ancestry can be inferred from similar homeostatic mechanisms though adaptations may have occurred due to differing environmental pressures. (2D2a)

*Find the unabridged version of all EK citations at www.glencoe.com/maderAP11.

Media Study Tools

Summarize

31.1 Types of Tissues

During development, the zygote divides to produce cells that go on to form tissues composed of similar cells specialized for a particular function. Tissues make up organs, and organ systems make up the organism. This sequence describes the levels of organization within an organism.

Tissues are categorized into four groups. Epithelial tissue, which covers the body and lines cavities, is of three types: squamous, cuboidal, and columnar epithelium. Each type can be simple or stratified; it can also be glandular or have modifications, such as cilia. In all cases, epithelial tissues are attached to a basement membrane. Epithelial tissue protects, absorbs, secretes, and excretes.

Connective tissue contains cells and a matrix containing fibers and ground substance. The three major categories of connective tissue are fibrous, supportive, and fluid. Loose fibrous and dense fibrous connective tissues contain cells called fibroblasts. Loose fibrous connective tissue has both collagen and elastic fibers. Dense fibrous connective tissue, like that of tendons and ligaments, contains closely packed collagen fibers. Adipose tissue contains cells called adipocytes that enlarge and store fat.

Supportive connective tissue includes cartilage and bone. Both have cells within lacunae, but the matrix for cartilage is more flexible than that for bone, which contains calcium salts. In bone, the lacunae lie in concentric circles within an osteon (or Haversian system) about a central canal. Blood is a fluid connective tissue in which the matrix is a liquid called plasma.

Muscular (contractile) tissue can be smooth or striated, and involuntary (smooth and cardiac) or voluntary (skeletal). In humans, skeletal muscle is attached to bone, smooth muscle is in the wall of internal organs, and cardiac muscle makes up the heart.

Nervous tissue has one main type of conducting cell, the neuron, and several types of neuroglia. The majority of neurons have dendrites, a cell body, and an axon. The brain and spinal cord contain complete neurons, while nerves contain only axons. Axons are specialized to conduct nerve impulses.

31.2 Organs, Organ Systems, and Body Cavities

Organs contain two or more tissues. Organ systems are made up of several organs that function together in a particular life process.

The human body has two main cavities. The dorsal cavity contains the brain and spinal cord. The ventral cavity is divided into the thoracic cavity (containing the heart and lungs) and the abdominal cavity (containing most other internal organs).

31.3 The Integumentary System

The integumentary system includes the skin and its derivatives, such as scales (in fish, reptiles, and birds), feathers (birds only), and hair (mammals only). This organ system protects underlying tissues against dessication, trauma, and pathogens, plus functions in thermoregulation, sensory perception, and vitamin D synthesis.

Skin itself has two regions. Epidermis (stratified squamous epithelium) overlies the dermis (fibrous connective tissue containing sensory receptors, hair follicles, blood vessels, and nerves). The epidermis contains melanocytes, cells that produce a pigment called melanin that helps protect against the harmful effects of UV light. A subcutaneous layer is composed of loose connective tissue and adipose tissue.

31.4 Homeostasis

Homeostasis is the ability of an organism to maintain a relatively constant internal environment. Animals differ in their approach to regulating certain variables like body temperature, but all must be capable of maintaining a certain degree of homeostasis.

All organ systems contribute to homeostasis, but special contributions are made by the liver, which keeps the blood glucose constant, and the kidneys, which regulate the pH. The nervous and hormonal systems regulate the other body systems. Both of these are controlled by negative feedback mechanisms, which result in slight fluctuations above and below desired levels. Less commonly, positive feedback mechanisms can bring about an increasingly greater change it the same direction.

Key Terms

adipose tissue 590	lymph 592
blood 591	melanocyte 598
body cavity 596	muscular (contractile)
bone 591	tissue 592
cardiac muscle 593	nail 598
cartilage 590	negative feedback 600
collagen fiber 589	nerve 593
columnar epithelium 589	nervous tissue 593
compact bone 591	neuroglia 593
connective tissue 589	neuron 593
cuboidal epithelium 588	oil gland 599
dense fibrous connective	organ 596
tissue 590	organ system 596
dermis 598	pathogen 592
elastic cartilage 591	platelet 592
elastic fiber 589	poikilothermic 599
endocrine gland 589	positive feedback 601
epidermis 598	red blood cell 591
epithelial tissue 588	reticular fiber 589
exocrine gland 589	skeletal muscle 592
fibroblast 590	skin 597
fibrocartilage 591	smooth (visceral) muscle 592
gland 589	spongy bone 591
hair follicle 599	squamous epithelium 588
homeostasis 599	striated 592
homeothermic 599	sweat gland 599
hyaline cartilage 591	tendon 590
lacuna 590	tissue 588
ligament 590	tissue fluid 591
loose fibrous connective	white blood cell 592
tissue 590	

Assess

Reviewing This Chapter

1. Name the four major types of tissues.   588
2. Describe the structure and the functions of three types of simple epithelial tissue.   588–89
3. Describe the structure and the functions of six major types of connective tissue.   590–92
4. Describe the structure and the functions of three types of muscular tissue.   592–93
5. Nervous tissue contains what types of cells?   593
6. In general terms, describe the locations of the human organ systems.   596
7. Describe the structure of skin, and state five functions of the integumentary system.   597–99
8. Tell how the various systems of the body contribute to homeostasis.   599
9. What are the functions of sensors, the regulatory center, and effectors in a negative feedback mechanism? Why is it called negative feedback?   600–1

Testing Yourself

Choose the best answer for each question.

1. Which of these is not a type of epithelial tissue?
   a. simple cuboidal
   b. cartilage
   c. stratified squamous
   d. striated
   e. All of these are epithelial tissue.

2. Tendons and ligaments
   a. are dense fibrous connective tissue.
   b. are associated with the bones.
   c. are found in vertebrates.
   d. contain collagen.
   e. All of these are correct.

3. Which tissue has cells in lacunae?
   a. epithelial tissue
   b. cartilage
   c. bone
   d. smooth muscle
   e. Both b and c are correct.

4. Cardiac muscle is
   a. striated.
   b. involuntary.
   c. smooth.
   d. many fibers fused together.
   e. Both a and b are correct.

5. Which of these components of blood fights infection?
   a. red blood cells
   b. white blood cells
   c. platelets
   d. hydrogen ions
   e. All of these are correct.

6. Which of these body systems contribute to homeostasis?
   a. digestive and urinary systems
   b. respiratory and nervous systems
   c. nervous and endocrine systems
   d. immune and cardiovascular systems
   e. All of these are correct.

7. In a negative feedback mechanism,
   a. the output cancels the input.
   b. there is a fluctuation above and below the average.
   c. there is self-regulation.
   d. a regulatory center communicates with other body parts.
   e. All of these are correct.

8. When a human being is cold, the superficial blood vessels
   a. dilate, and the sweat glands are inactive.
   b. dilate, and the sweat glands are active.
   c. constrict, and the sweat glands are inactive.
   d. constrict, and the sweat glands are active.
   e. contract so that shivering occurs.

9. Give the name, the location, and the function for each of the illustrated tissues in the human body.
   a. type of epithelial tissue _____
   b. type of muscular tissue _____
   c. type of connective tissue _____

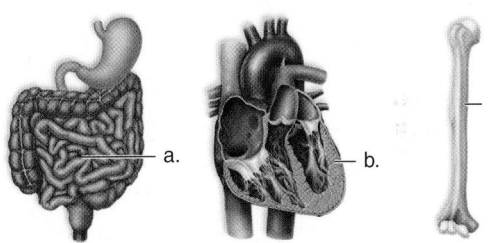

10. Which of these is a function of skin?
    a. temperature regulation
    b. manufacture of vitamin D
    c. collection of sensory input
    d. protection from invading pathogens
    e. All of these are correct.

11. Which of these is an example of negative feedback?
    a. Air conditioning switches off when room temperature lowers.
    b. Insulin decreases blood sugar levels after eating a meal.
    c. Heart rate increases when blood pressure drops.
    d. All of these are examples of negative feedback.

12. Which of these correctly describes a layer of the skin?
    a. The epidermis is simple squamous epithelium in which hair follicles develop and blood vessels expand when we are hot.
    b. The subcutaneous layer lies between the epidermis and the dermis. It contains adipose tissue, which keeps us warm.
    c. The dermis is a region of connective tissue that contains sensory receptors, nerve endings, and blood vessels.
    d. The skin has a special layer, still unnamed, in which there are all the accessory structures such as nails, hair, and various glands.

13. The _____ separates the thoracic cavity from the abdominal cavity.
    a. liver
    b. pancreas
    c. diaphragm
    d. pleural membrane
    e. intestines

In questions 14–16, match each type of muscle tissue to as many terms in the key as possible.

KEY:

   a. voluntary
   b. involuntary
   c. striated
   d. nonstriated
   e. spindle-shaped cells
   f. branched cells
   g. long, cylindrical cells

14. skeletal muscle

15. smooth muscle

16. cardiac muscle

In questions 17–20, match each description to the tissues in the key.

KEY:

   a. loose fibrous connective tissue
   b. hyaline cartilage
   c. adipose tissue
   d. compact bone

17. occurs in nose and walls of respiratory passages

18. occurs only within bones of skeleton

19. occurs beneath most epithelial layers

20. occurs beneath skin, and around organs, including the heart

Engage

Virtual Lab
Virtual Frog Dissection

The Virtual Lab "Virtual Frog Dissection" provides an interactive look at the major organ systems found in vertebrate animals.

Thinking Scientifically

1. Many cancers develop from epithelial tissue. These include lung, colon, and skin cancers. What are two attributes of this tissue type that make cancer more likely to develop?

2. Bacterial or viral infections can cause a fever. Fevers occur when the hypothalamus changes its temperature set point. Signaling of the hypothalamus could be direct (from the infectious agent itself) or indirect (from the immune system). Which of these would enable the hypothalamus to respond to the greatest variety of infectious agents? Is there any disadvantage to such a signaling system?

Bioethical Issue

Organ Transplants

Despite widespread efforts to convince people of the need for organ donors, supply continues to lag far behind demand. One proposed strategy to bring supply and demand into better balance is to develop an "insurance" program for organs. In this program, participants would pay the "premium" by promising to donate their organs at death. They, in turn, would receive priority for transplants as their "benefit." To avoid the problem of too many high-risk people applying for this type of insurance, a medical exam would be required so that only people with a normal risk of requiring a transplant would be accepted. Do you suppose this system would be an improvement over the current system in which organ donation is voluntary, with no tangible benefit? Can you think of any other strategy that would be more effective for increasing organ donations?

Movie and television star John Ritter died unexpectedly at a relatively young age due to a sudden failure of his circulatory system.

32

Circulation and Cardiovascular Systems

J ohn Ritter was an actor best known for his roles on the TV sitcoms *Three's Company* and *8 Simple Rules for Dating My Teenage Daughter*. In September of 2003, Ritter, then age 54, complained of nausea and chest pains while rehearsing for the latter show. He was taken to a hospital across the street, where he died later that evening.

On autopsy, Ritter's death was confirmed to have been due to an aortic dissection, a condition in which the walls of the body's largest blood vessel, the aorta, separate and fill with blood. This can affect blood flow through the aorta, which may also rupture, filling the chest cavity with blood. Deaths due to aortic dissection are uncommon, with an estimated 5 to 30 such cases per million people per year. Interestingly, evidence of some aortic dissection is found in 1% to 3% of all autopsies performed in the United States; clearly many people are unaware that they have this defect.

After her husband's death, John Ritter's widow filed a $67 million wrongful death lawsuit against the doctors who she believed failed to diagnose and treat his condition properly. In 2008, a Los Angeles jury split 9 to 3 in favor of the doctors, but the plaintiff felt that she had helped to educate people about the condition. Indeed, after Ritter died, his brother was diagnosed with a similar condition, which was then surgically corrected. This tragic situation demonstrates the critical importance of the circulatory system, the topic of this chapter.

As you read through the chapter, think about the following questions:

1. What are the essential components of any circulatory system, and their functions?

2. Through what major blood vessels does a human red blood cell normally pass as it travels from the aorta to a capillary in the lower leg, and back to the heart?

3. Why are the processes that occur in capillaries essential to life?

CHAPTER OUTLINE

BEFORE YOU BEGIN

Before beginning this chapter, take a few moments to review the following discussions.

Figure 8.1 Animal cells use glucose and oxygen for what specific purpose(s)? Where is carbon dioxide generated?

Figure 29.9 How do the circulatory pathways of fishes, amphibians, reptiles, birds, and mammals resemble each other? How are they different?

Section 31.3 What types of tissues comprise the various parts of the cardiovascular system?

FOLLOWING *the* BIG IDEAS

CHAPTER 32 CIRCULATION AND CARDIOVASCULAR SYSTEMS

Energy and Homeostasis	Feedback mechanisms throughout a cardiovascular system allow extreme responses with an eventual return to homeostasis.
Interactions and Systems	The circulatory system is inextricably tied to every other system in the body.

32.1 Transport in Invertebrates

All animal cells require a steady supply of oxygen and nutrients, and their waste products must be removed. In most animals these tasks are facilitated by a **circulatory system**, which moves fluid between various parts of the body. However, some invertebrates, such as sponges, cnidarians (e.g., hydras, sea anemones), and flatworms (e.g., planarians) lack a circulatory system (Fig. 32.1a, b). Their thin body wall makes a circulatory system unnecessary.

In hydras, cells are either part of an external layer, or they line the gastrovascular cavity. Each cell is exposed to water and can independently exchange gases and rid itself of wastes. The cells that line the gastrovascular cavity are specialized to complete the digestive process. They pass nutrient molecules to other cells by diffusion. In planarians, a trilobed gastrovascular cavity branches throughout the small, flattened body. No cell is very far from one of the three digestive branches, so nutrient molecules can diffuse from cell to cell. Similarly, diffusion meets the respiration and elimination needs of the cells.

Pseudocoelomate invertebrates, such as nematodes, use the coelomic fluid of their body cavity for transport purposes. The coelomate echinoderms also rely on movement of coelomic fluid within a body cavity as a circulatory system (Fig. 32.1c).

Invertebrates with a Circulatory System

Most invertebrates have a circulatory system that transports oxygen and nutrients, such as glucose and amino acids, to their cells. There it picks up wastes, which are later excreted from the body by the lungs or kidneys. There are two types of circulatory fluids: **blood,** which is always contained within blood vessels, and **hemolymph,** a mixture of blood and tissue fluid, which fills the body cavity and surrounds the internal organs.

Open Circulatory Systems

Hemolymph is seen in animals that have an **open circulatory system** that consists of blood vessels plus open spaces. Open circulatory systems were likely the first to evolve, as they are present in simpler and evolutionarily older animals. For example, in most molluscs and arthropods, the heart pumps hemolymph via vessels into tissue spaces that are sometimes enlarged into saclike sinuses (Fig. 32.2a). Eventually, hemolymph drains back to the heart. In the grasshopper, an arthropod, the dorsal tubular heart pumps hemolymph into a dorsal aorta, which empties into the hemocoel. When the heart contracts, openings called ostia (sing., ostium) are closed; when the heart relaxes, the hemolymph is sucked back into the heart by way of the ostia.

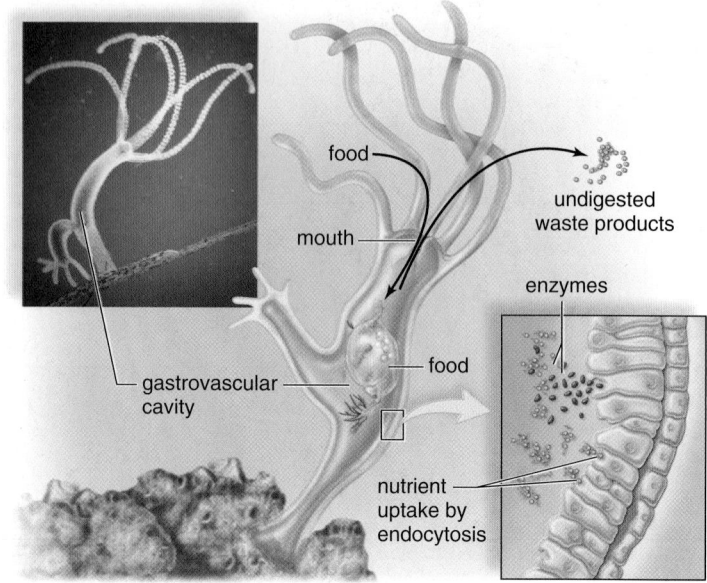

a. Hydra

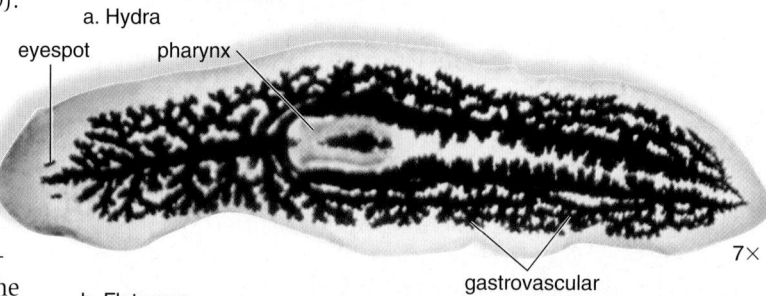

b. Flatworm

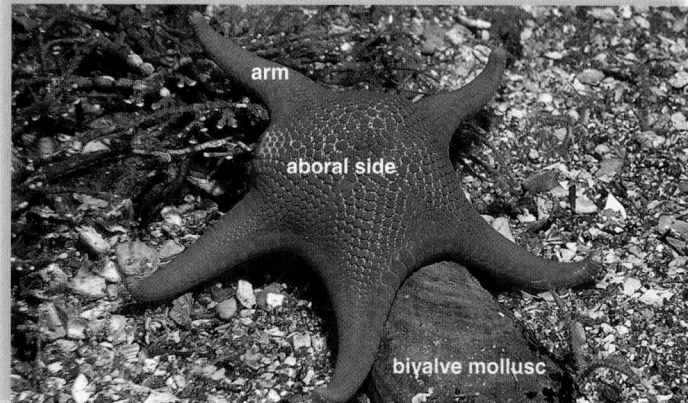

c. Red sea star, *Mediastar*

Figure 32.1 Aquatic animals without a circulatory system.
a. In a hydra, a cnidarian, the gastrovascular cavity makes digested material available to the cells that line the cavity. These cells can also acquire oxygen from the watery contents of the cavity and discharge their wastes there. **b.** In a planarian, a flatworm, the gastrovascular cavity branches throughout the body, bringing nutrients to body cells. **c.** In a sea star, the coelomic fluid distributes oxygen and picks up wastes.

The hemolymph of a grasshopper is colorless because it does not contain hemoglobin or any other respiratory pigment. It carries nutrients but no oxygen. Oxygen is taken to cells, and carbon dioxide is removed from them, by way of air tubes called tracheae, which are found throughout the body. The tracheae provide efficient transport and delivery of respiratory gases, while at the same time restricting water loss.

Closed Circulatory Systems

Blood is seen in animals that have a **closed circulatory system,** in which blood does not leave the vessels. For example, in annelids, such as earthworms, and in some molluscs, such as squid and octopuses, blood consisting of cells and plasma (a liquid) is pumped by the heart into a system of blood vessels (Fig. 32.2*b*). Valves prevent the backward flow of blood.

In the segmented earthworm, five pairs of anterior hearts (aortic arches) pump blood into the ventral blood vessel (an artery), which has a branch called a lateral vessel in every segment of the worm's body. Blood moves through these branches into capillaries, the thinnest of the blood vessels, where exchanges with tissue fluid take place. Both gas exchange and nutrient-for-waste exchange occur across the capillary walls. (Most cells in the body of an animal with a closed circulatory system are not far from a capillary.) In an earthworm, after leaving a capillary, blood moves from small veins into the dorsal blood vessel (a vein). This dorsal blood vessel returns blood to the heart for repumping.

The earthworm has red blood that contains the respiratory pigment hemoglobin. Hemoglobin is dissolved in the blood and is not contained within blood cells. The earthworm has no specialized organ, such as lungs, for gas exchange with the external environment. Gas exchange takes place across the body wall, which must always remain moist for this purpose.

Check Your Progress   32.1

1. List the general functions of all circulatory systems.
2. Explain how blood differs from hemolymph.
3. Regarding oxygen transport, deduce the specific additional step that must occur in animals with a closed circulatory system, compared to those with an open system.

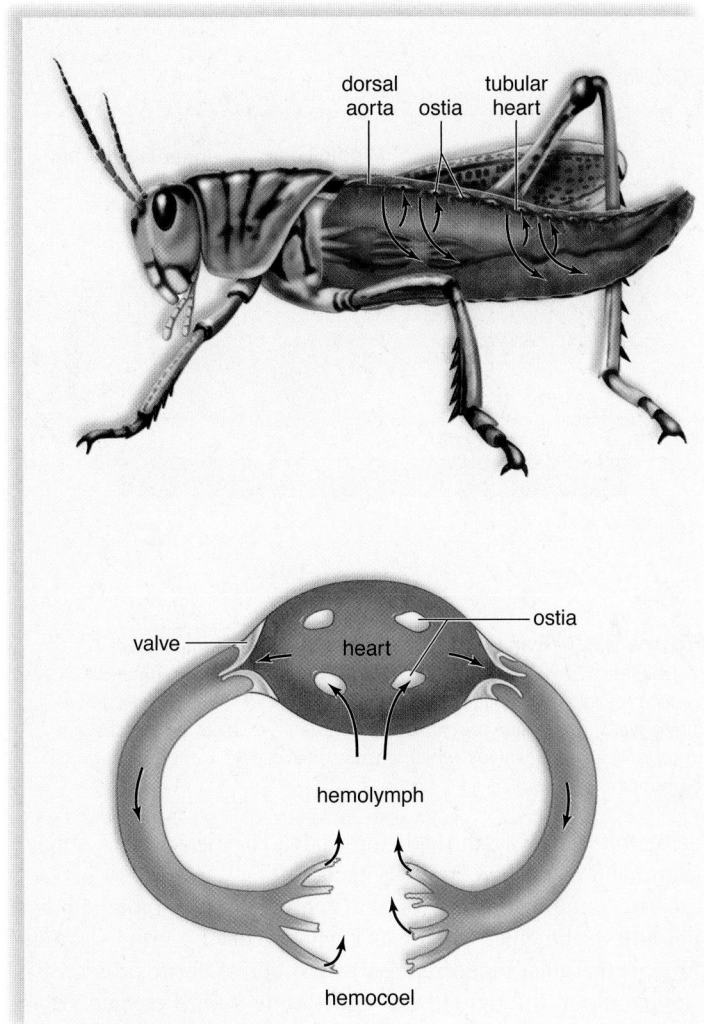

a. Open circulatory system

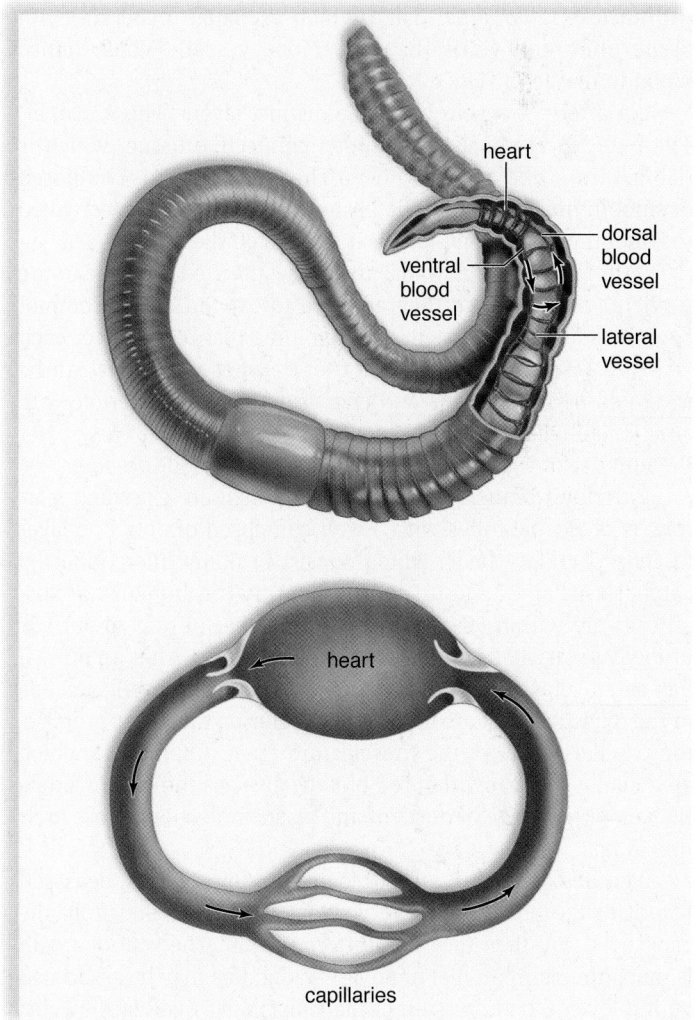

b. Closed circulatory system

Figure 32.2 Open versus closed circulatory systems. **a.** *Top:* The grasshopper, an arthropod, has an open circulatory system. *Bottom:* A hemocoel is a body cavity filled with hemolymph, which freely bathes the internal organs. The heart, a pump, sends hemolymph out through vessels and collects it through ostia (openings). This open system probably could not supply oxygen to wing muscles rapidly enough. These muscles receive oxygen directly from tracheae (air tubes). **b.** *Top:* The earthworm, an annelid, has a closed circulatory system. The dorsal and ventral blood vessels are joined by five pairs of anterior hearts, which pump blood. *Bottom:* The lateral vessels distribute blood to the rest of the worm.

32.2 Transport in Vertebrates

Learning Outcomes

Upon completion of this section, you should be able to

1. Distinguish the structure and functions of arteries, veins, and capillaries.
2. Compare the path of blood in animals with a one-circuit circulatory pathway vs. a two-circuit pathway.
3. Identify the number of atria and ventricles in each type of vertebrate animal: fish, amphibians, most reptiles, crocodilians, birds, and mammals.

All vertebrate animals have a closed circulatory system, which is called a **cardiovascular system** [Gk. *kardia*, heart; L. *vascular*, vessel]. It consists of a strong, muscular heart in which the atria (sing., atrium) receive blood and the muscular ventricles pump blood through the blood vessels. There are three kinds of blood vessels: **arteries,** which carry blood away from the heart; **capillaries** [L. *capillus*, hair], which exchange materials with tissue fluid; and **veins** [L. *vena*, blood vessel], which return blood to the heart (Fig. 32.3).

An artery or a vein has three distinct layers (Fig. 32.3*a, c*). The outer layer consists of fibrous connective tissue, which is rich in elastic and collagen fibers. The middle layer is composed of smooth muscle and elastic tissue. The innermost layer, called the endothelium, is similar to squamous epithelium.

Arteries have thick walls, and those attached to the heart are resilient, meaning that they are able to expand and accommodate the sudden increase in blood volume that results after each heartbeat. **Arterioles** are small arteries whose diameter can be regulated by the nervous and endocrine systems. Arteriole constriction and dilation affect blood pressure in general. The greater the number of vessels dilated, the lower the blood pressure.

Arterioles branch into capillaries, which are extremely narrow, microscopic tubes with a wall composed of only one layer of cells. Capillary beds, which consist of many interconnected capillaries (Fig. 32.4), are so prevalent that in humans, almost all cells are within 60–80 μm of a capillary. But only about 5% of the capillary beds are open at the same time. After an animal has eaten, precapillary sphincters relax, and the capillary beds in the digestive tract are usually open. During muscular exercise, the capillary beds of the muscles are open. Capillaries, which are usually so narrow that red blood cells pass through in single file, allow exchange of nutrient and waste molecules across their thin walls.

Venules and veins collect blood from the capillary beds and take it to the heart. First, the venules drain the blood from the capillaries, and then they join to form a vein. The wall of a vein is much thinner than that of an artery, and this may be associated with a lower blood pressure in the veins. Valves within the veins point, or open, toward the heart, preventing a backflow of blood when they close (Fig. 32.3*c*).

MP3
Classification of Blood Vessels

Comparison of Circulatory Pathways

Two different types of circulatory pathways are seen among vertebrate animals. In fishes, blood follows a one-circuit (single-loop)

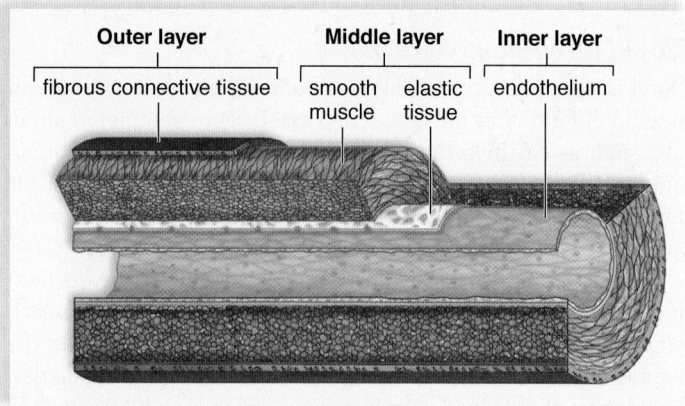

a. Artery

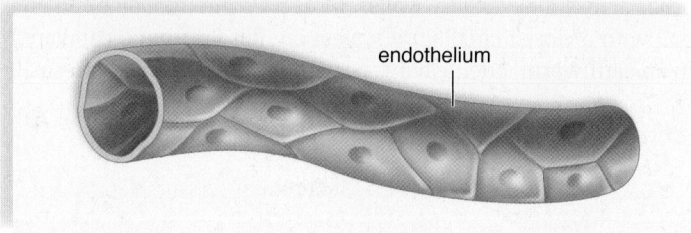

b. Capillary

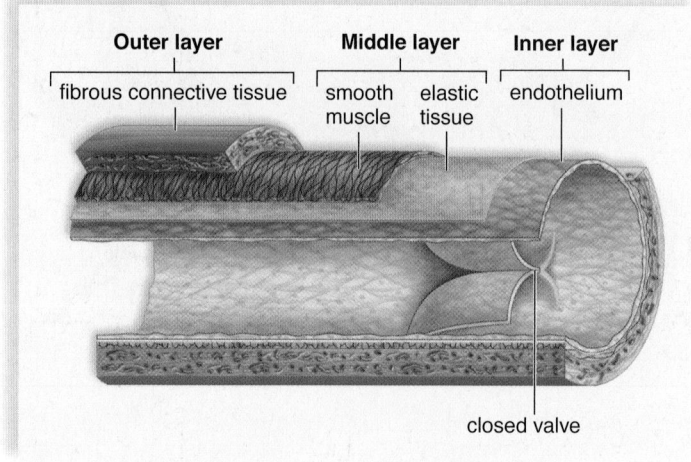

c. Vein

Figure 32.3 Transport in vertebrates. **a.** Arteries have well-developed walls with a thick middle layer of elastic tissue and smooth muscle. **b.** Capillary walls are only one cell thick. **c.** Veins have flabby walls, particularly because the middle layer is not as thick as in arteries. Veins have valves, which ensure one-way flow of blood back to the heart.

circulatory pathway through the body. The heart has a single atrium and a single ventricle (Fig. 32.5*a*).

The pumping action of the ventricle sends blood under pressure to the gills, where gas exchange occurs. After passing through the gills, blood returns to the dorsal aorta, which distributes blood throughout the body. Veins return oxygen (O_2)-poor blood to an enlarged chamber called the sinus venosus that leads to the atrium. The atrium pumps blood back to the ventricle. This single circulatory loop has an advantage in that the gill capillaries receive oxygen-poor blood and the capillaries of the body, called systemic capillaries, receive fully oxygen-rich blood. It is disadvantageous in that after leaving the gills, the blood is under reduced pressure.

As a result of evolutionary changes, other vertebrates have a two-circuit (double-loop) circulatory pathway. The heart pumps blood to the tissues through a **systemic circuit,** and also pumps blood to the lungs through a **pulmonary circuit** [L. *pulmonarius*, of the lungs]. This double-pumping action is an adaptation to breathing air on land.

In amphibians, the heart has two *atria* and a single *ventricle* (Fig. 32.5*b*). Oxygen-poor blood from the systemic veins returns to the right atrium. Oxygen-rich blood returning from the lungs passes to the left atrium. Both of the atria empty into the single ventricle. Oxygen-rich and oxygen-poor blood are kept somewhat separate because oxygen-poor blood is pumped out of the ventricle before the oxygen-rich blood enters. When the ventricle contracts, the division of the main artery also helps keep the blood somewhat separated. More oxygen-rich blood is distributed to the body, and more oxygen-poor blood is delivered to the lungs, and perhaps to the skin, for recharging with oxygen.

In most reptiles, a septum partially divides the ventricle. In these animals, mixing of oxygen-rich and oxygen-poor blood is kept to a minimum. In crocodilians (alligators and crocodiles), the septum completely separates the ventricle. These reptiles have a four-chambered heart. The heart of birds and mammals is also divided into left and right halves (Fig. 32.5*c*). The right ventricle pumps blood to the lungs, and the larger left ventricle pumps blood to the rest of the body. This arrangement provides adequate blood pressure for both the pulmonary and systemic circuits.

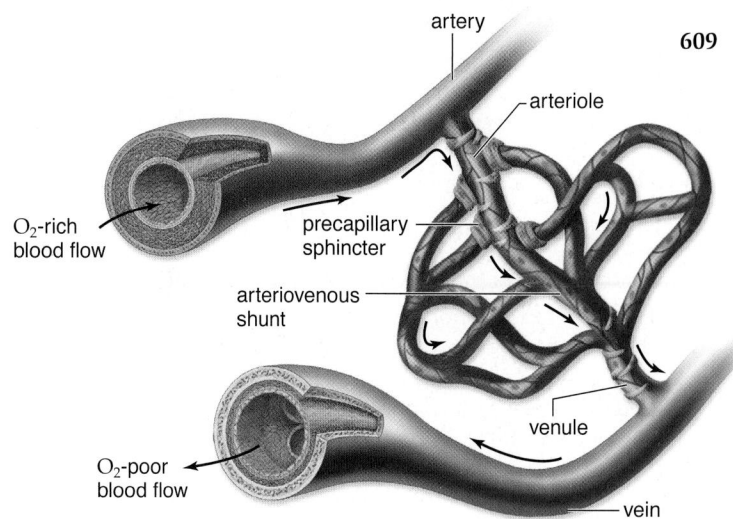

Figure 32.4 Anatomy of a capillary bed. When a capillary bed is open, sphincter muscles are relaxed and blood flows through the capillaries. When precapillary sphincter muscles are contracted, the bed is closed and blood flows through an arteriovenous shunt that carries blood directly from an arteriole to a venule.

Check Your Progress 32.2

1. List and describe the functions of three types of vessels in a cardiovascular system.
2. Identify which of the three types of vessels listed in Question 1 contain valves, and analyze why.
3. Examine the evolutionary benefits of a two-circuit circulatory pathway compared to a one-circuit pathway, especially for animals that breathe air on land.

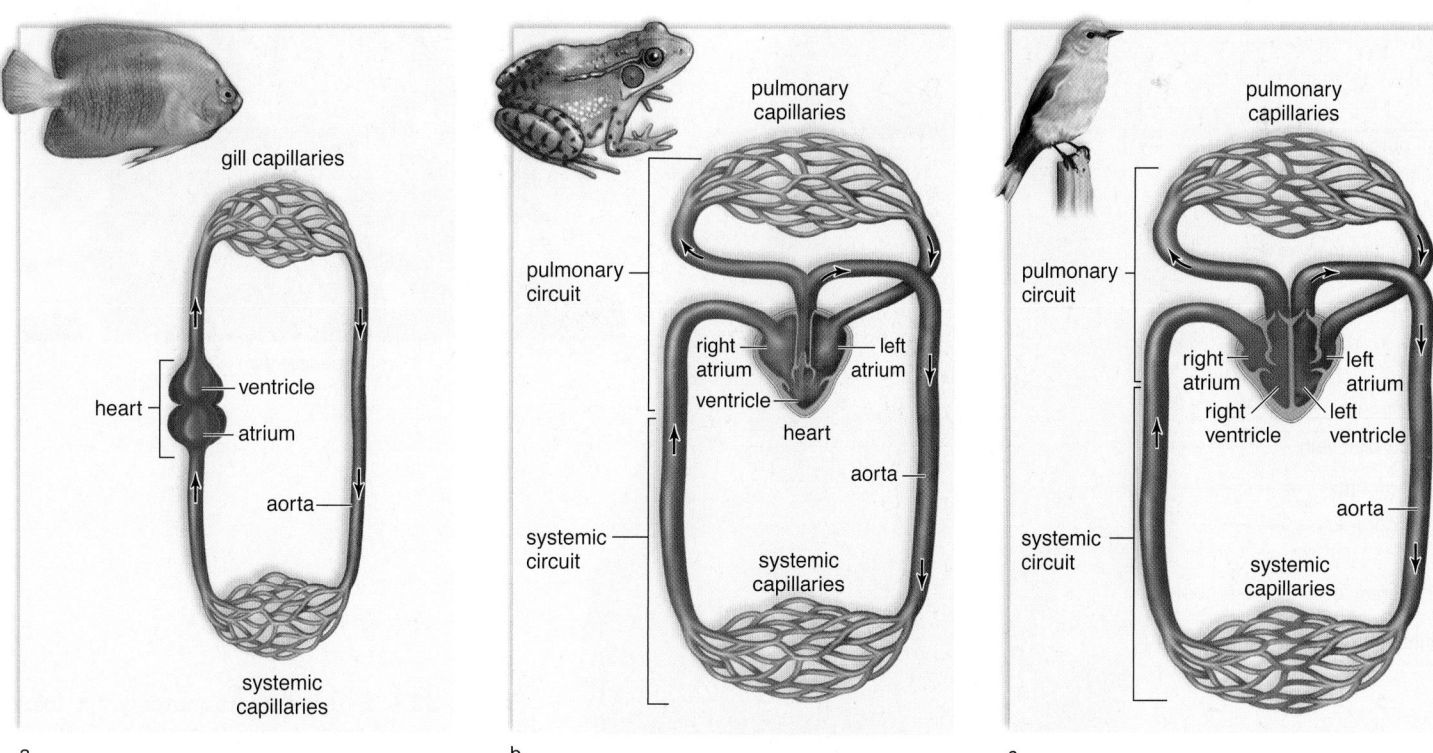

a. b. c.

Figure 32.5 Comparison of circulatory circuits in vertebrates. **a.** In fishes, the blood moves in a single circuit. Blood pressure created by the pumping of the heart is dissipated after the blood passes through the gill capillaries. This is a disadvantage of this one-circuit system. **b.** Amphibians and most reptiles have a two-circuit system in which the heart pumps blood to both the pulmonary capillaries in the lungs and the systemic capillaries in the body itself. Although there is a single ventricle, little mixing of oxygen-rich and oxygen-poor blood takes place. **c.** The pulmonary and systemic circuits are completely separate in crocodiles (a reptile) and in birds and mammals, because the heart is divided by a septum into right and left halves. The right side pumps blood to the lungs, and the left side pumps blood to the rest of the body.

32.3 The Human Cardiovascular System

In the cardiovascular system of humans, the pumping of the heart keeps blood moving primarily in the arteries. Skeletal muscle contraction pressing against veins is the main force responsible for the movement of blood in the veins.

The Human Heart

The **heart** is a cone-shaped, muscular organ about the size of a fist (Fig. 32.6). It is located between the lungs directly behind the sternum (breastbone) and is tilted so that the apex (the pointed end) is oriented to the left.

Structure of the Heart

The major portion of the heart, called the myocardium, consists largely of cardiac muscle tissue. The myocardium receives oxygen and nutrients from the coronary arteries, not from the blood it pumps. The muscle fibers of the myocardium are branched and tightly joined to one another at intercalated disks.

The heart lies within the pericardium, a thick, membranous sac that secretes a small quantity of lubricating liquid. The inner surface of the heart is lined with endocardium, a membrane composed of connective tissue and endothelial tissue. The lining is continuous with the endothelium lining of the blood vessels.

Internally, a wall called the **septum** separates the heart into a right side and a left side (Fig. 32.7). The heart has four chambers. The two upper, thin-walled atria (sing., **atrium**) have wrinkled, protruding appendages called auricles. The two lower

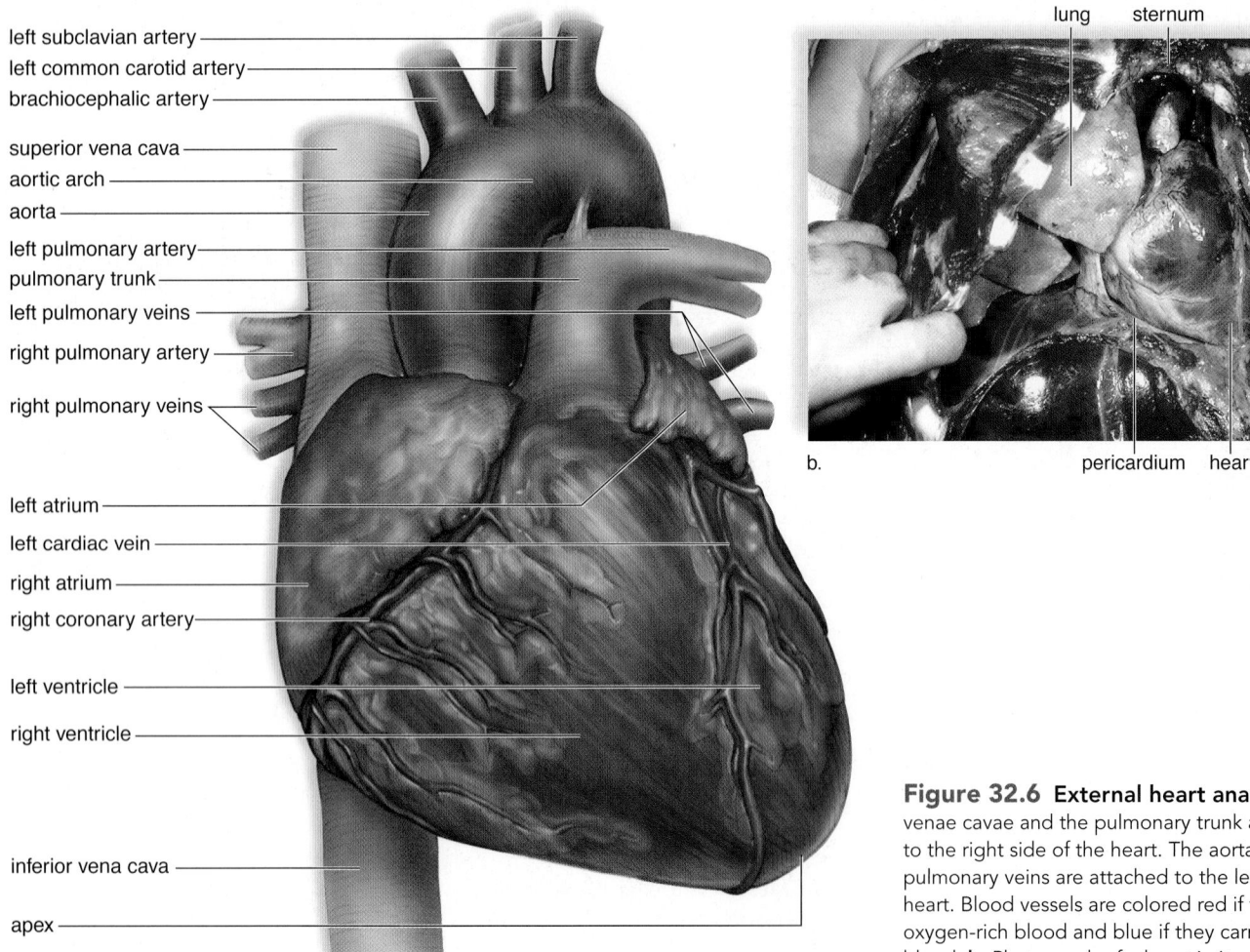

Figure 32.6 External heart anatomy. **a.** The venae cavae and the pulmonary trunk are attached to the right side of the heart. The aorta and the pulmonary veins are attached to the left side of the heart. Blood vessels are colored red if they carry oxygen-rich blood and blue if they carry oxygen-poor blood. **b.** Photograph of a heart in its natural position in the chest.

chambers are the thick-walled **ventricles,** which pump the blood away from the heart.

The heart also has four valves, which direct the flow of blood and prevent its backward movement. The two valves that lie between the atria and the ventricles are called the **atrioventricular valves.** These valves are supported by strong fibrous strings called chordae tendineae. The chordae, which are attached to muscular projections of the ventricular walls, support the valves and prevent them from inverting when the heart contracts. The atrioventricular valve on the right side is called the tricuspid valve because it has three flaps, or cusps. The valve on the left side is called the bicuspid (or the mitral) because it has two flaps.

The remaining two valves are the **semilunar valves,** whose flaps resemble half-moons, between the ventricles and their attached vessels. The pulmonary semilunar valve lies between the right ventricle and the pulmonary trunk. The aortic semilunar valve lies between the left ventricle and the aorta.

Path of Blood Through the Heart

Even though both atria and then both ventricles contract simultaneously due to the presence of intercalated disks (Fig. 32.7b),

we can trace the path of blood through the heart in the following manner:

- The superior vena cava and the inferior vena cava, which carry oxygen (O_2)-poor blood that is relatively high in carbon dioxide, empty into the right atrium.
- The right atrium sends blood through an atrioventricular valve (the tricuspid valve) to the right ventricle.
- The right ventricle sends blood through the pulmonary semilunar valve into the pulmonary trunk and the two pulmonary arteries to the lungs.
- Four pulmonary veins, which carry oxygen-rich blood, empty into the left atrium.
- The left atrium sends blood through an atrioventricular valve (the bicuspid or mitral valve) to the left ventricle.
- The left ventricle sends blood through the aortic semilunar valve into the aorta and to the rest of the body.

From this description, it is obvious that oxygen-poor blood never mixes with oxygen-rich blood, and that blood must go through the lungs in order to pass from the right side to the left side of the heart, as is typical in a double-loop circulatory system. Because

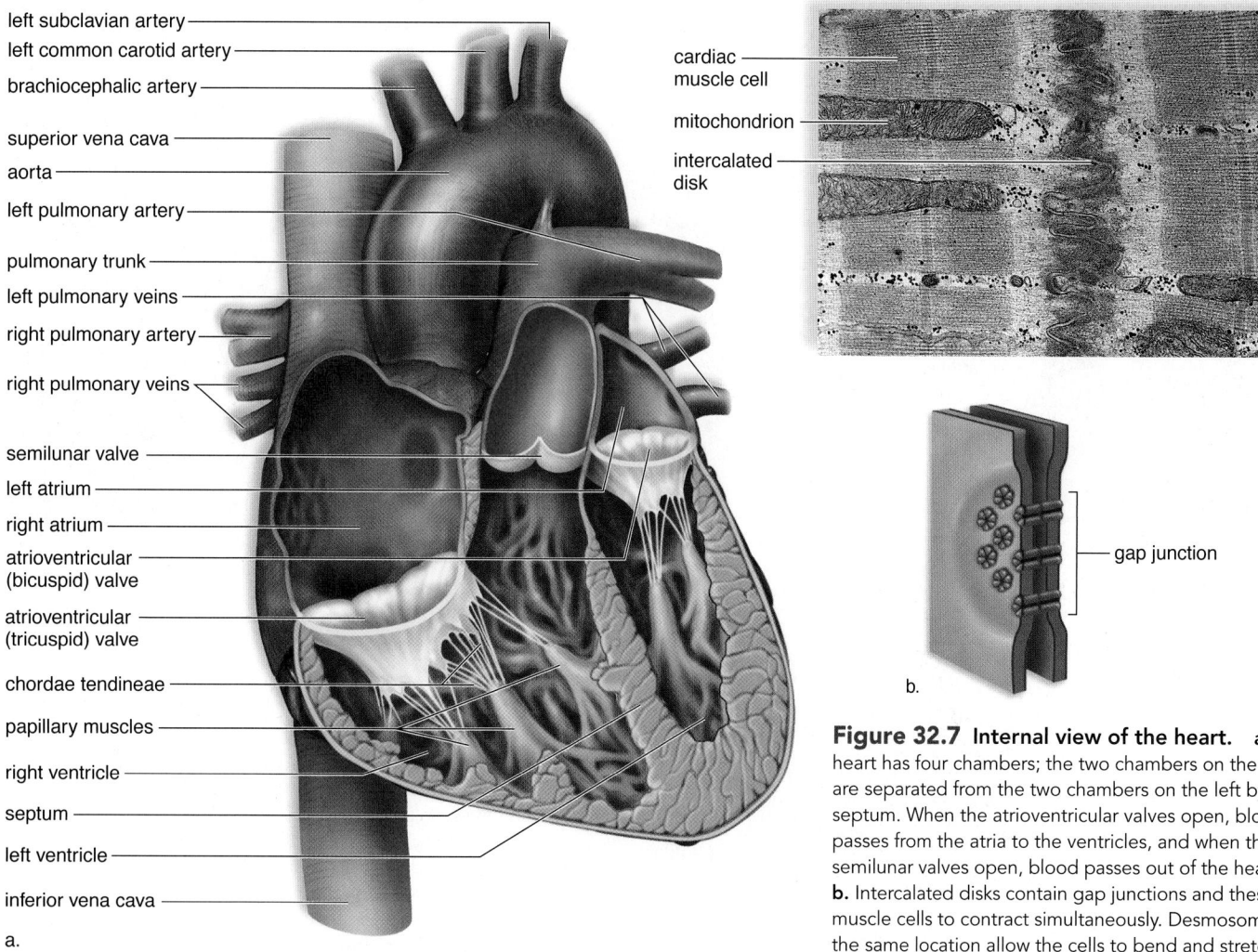

Figure 32.7 Internal view of the heart.   a. The heart has four chambers; the two chambers on the right are separated from the two chambers on the left by a septum. When the atrioventricular valves open, blood passes from the atria to the ventricles, and when the semilunar valves open, blood passes out of the heart. **b.** Intercalated disks contain gap junctions and these allow muscle cells to contract simultaneously. Desmosomes at the same location allow the cells to bend and stretch.

the left ventricle has the harder job of pumping blood to the entire body, its walls are thicker than those of the right ventricle, which pumps blood a relatively short distance to the lungs.

People often associate oxygen-rich blood with all arteries and oxygen-poor blood with all veins, but this idea is incorrect: Pulmonary arteries and pulmonary veins are just the reverse. That is why pulmonary arteries are colored blue and pulmonary veins are colored red in Figures 32.6 and 32.7.

The pumping of the heart sends blood out under pressure into the arteries. Because the left side of the heart is the stronger pump, blood pressure is greatest in the aorta. Blood pressure then decreases as the cross-sectional area of arteries and then arterioles increases. Therefore, a different mechanism is needed to move blood in the veins, as we shall discuss later.

The Heartbeat

The average human heart contracts, or beats, about 70 times a minute, so each heartbeat lasts about 0.85 second. This adds up to about 100,000 beats per day, and over a 70-year lifespan, the average human heart will have contracted about 2.5 billion times! The term **systole** [Gk. *systole,* contraction] refers to contraction of the heart chambers, and the word **diastole** [Gk. *diastole,* dilation, spreading] refers to relaxation of these chambers. Each heartbeat, or **cardiac cycle,** consists of the following phases, which are also depicted in Figure 32.8:

Cardiac Cycle		
Time	**Atria**	**Ventricles**
0.15 sec	Systole	Diastole
0.30 sec	Diastole	Systole
0.40 sec	Diastole	Diastole

First the atria contract (while the ventricles relax), then the ventricles contract (while the atria relax), and then all chambers rest. Note that the heart is in diastole about 50% of the time. The short systole of the atria is appropriate because the atria send blood only into the ventricles. It is the muscular ventricles that actually pump blood out into the cardiovascular system proper.

The volume of blood that the left ventricle pumps per minute into the systemic circuit is called the **cardiac output.** A person with a heartbeat of 70 beats per minute has a cardiac output of 5.25 liters a minute. This is almost equivalent to the amount of blood in the body, and adds up to about 2,000 gallons a day. During heavy exercise, the cardiac output can increase manyfold.

When the heart beats, the familiar lub-dub sound is heard as the valves of the heart close. The longer and lower-pitched

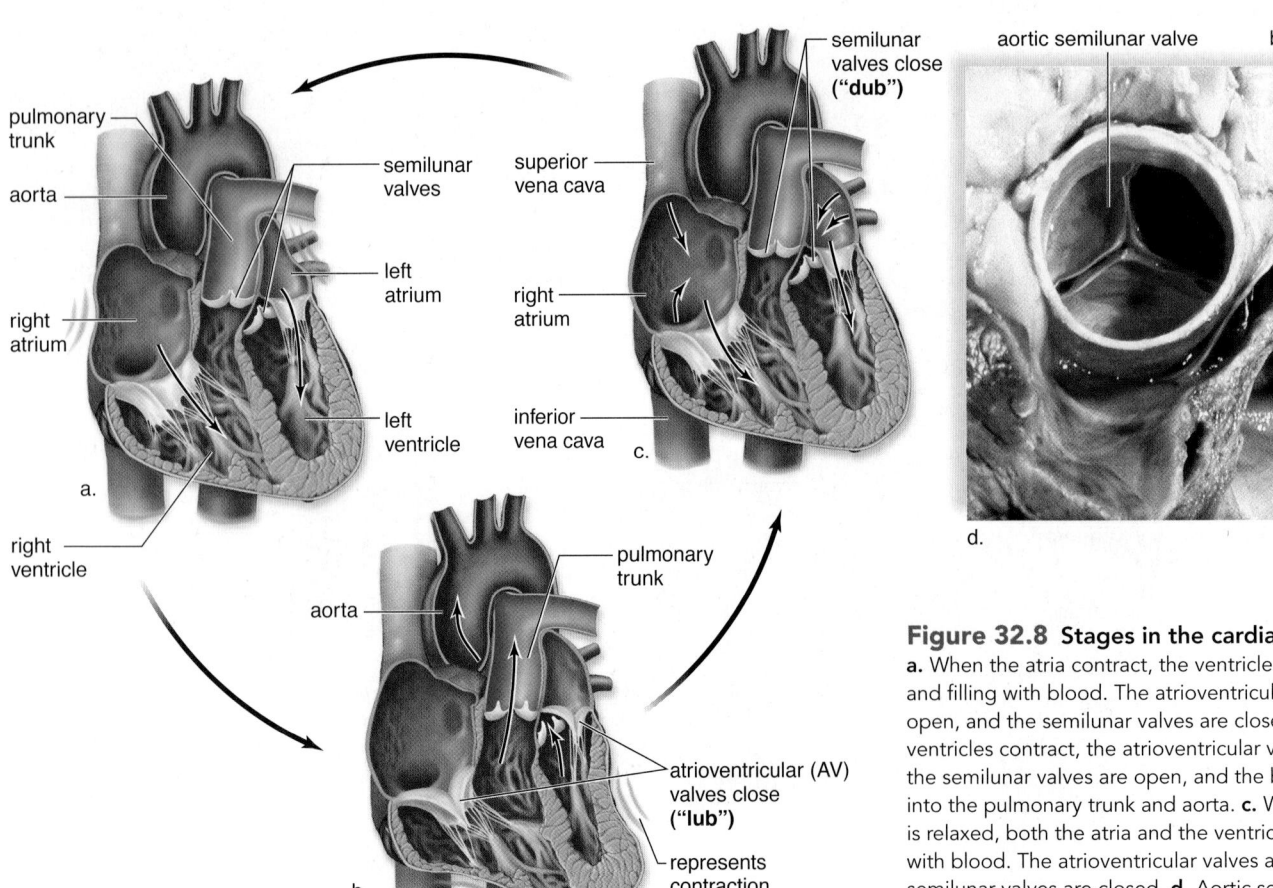

Figure 32.8 Stages in the cardiac cycle.
a. When the atria contract, the ventricles are relaxed and filling with blood. The atrioventricular valves are open, and the semilunar valves are closed. **b.** When the ventricles contract, the atrioventricular valves are closed, the semilunar valves are open, and the blood is pumped into the pulmonary trunk and aorta. **c.** When the heart is relaxed, both the atria and the ventricles are filling with blood. The atrioventricular valves are open, and the semilunar valves are closed. **d.** Aortic semilunar valve and bicuspid valve, an atrioventricular valve on left.

lub is caused by vibrations of the heart when the atrioventricular valves close due to ventricular contraction. The shorter and sharper *dub* is heard when the semilunar valves close due to back pressure of blood in the arteries. A heart murmur, a slight slush sound after the lub, is often due to ineffective valves, which allow blood to pass back into the atria after the atrioventricular valves have closed.

The **pulse** is a wave effect that passes down the walls of the arterial blood vessels when the aorta expands and then recoils following ventricular systole. Because there is one arterial pulse per ventricular systole, the arterial pulse rate can be used to determine the heart rate.

The rhythmic contraction of the atria and ventricles is due to the internal (intrinsic) conduction system of the heart. Nodal tissue, which has both muscular and nervous characteristics, is a unique type of cardiac muscle located in two regions of the heart. The *SA (sinoatrial) node* is found in the upper dorsal wall of the right atrium; the *AV (atrioventricular) node* is found in the base of the right atrium very near the septum (Fig. 32.9*a*). The SA node initiates the heartbeat about every 0.85 second by automatically sending out an excitation impulse, which causes the atria to contract. Therefore, the SA node is called the **pacemaker** because it usually keeps the heartbeat regular. When the impulse reaches the AV node, the AV node signals the ventricles to contract by way of large fibers terminating in the more numerous and smaller Purkinje fibers.

Although the heart muscle will contract without any external nervous stimulation, input from the brain can increase or decrease the rate and strength of heart contractions. In addition, the hormones epinephrine and norepinephrine, secreted into the blood by the adrenal glands, also stimulate the heart. When a person is frightened, for example, the heart pumps faster and stronger due to both nervous and hormonal stimulation.

MP3 Cardiac Cycle

The Electrocardiogram. An **electrocardiogram (ECG)** is a recording of the electrical changes that occur in the myocardium during a cardiac cycle. Body fluids contain ions that conduct electrical currents, and therefore these electrical changes can be detected on the body surface. During an ECG procedure, these changes pass from electrodes placed on the skin through wires to an instrument, generating "waves" that can be traced onto paper. Figure 32.9*b* depicts the pattern that results from a normal cardiac cycle.

When the SA node triggers an impulse, the atrial fibers produce an electrical change called the P wave. The P wave indicates that the atria are about to contract. After that, the QRS complex signals that the ventricles are about to contract and the atria are relaxing. The electrical changes that occur as the ventricular muscle fibers recover produce the T wave.

Animation Cardiac Cycle

Various types of abnormalities can be detected by an electrocardiogram. One of these, called ventricular fibrillation, is

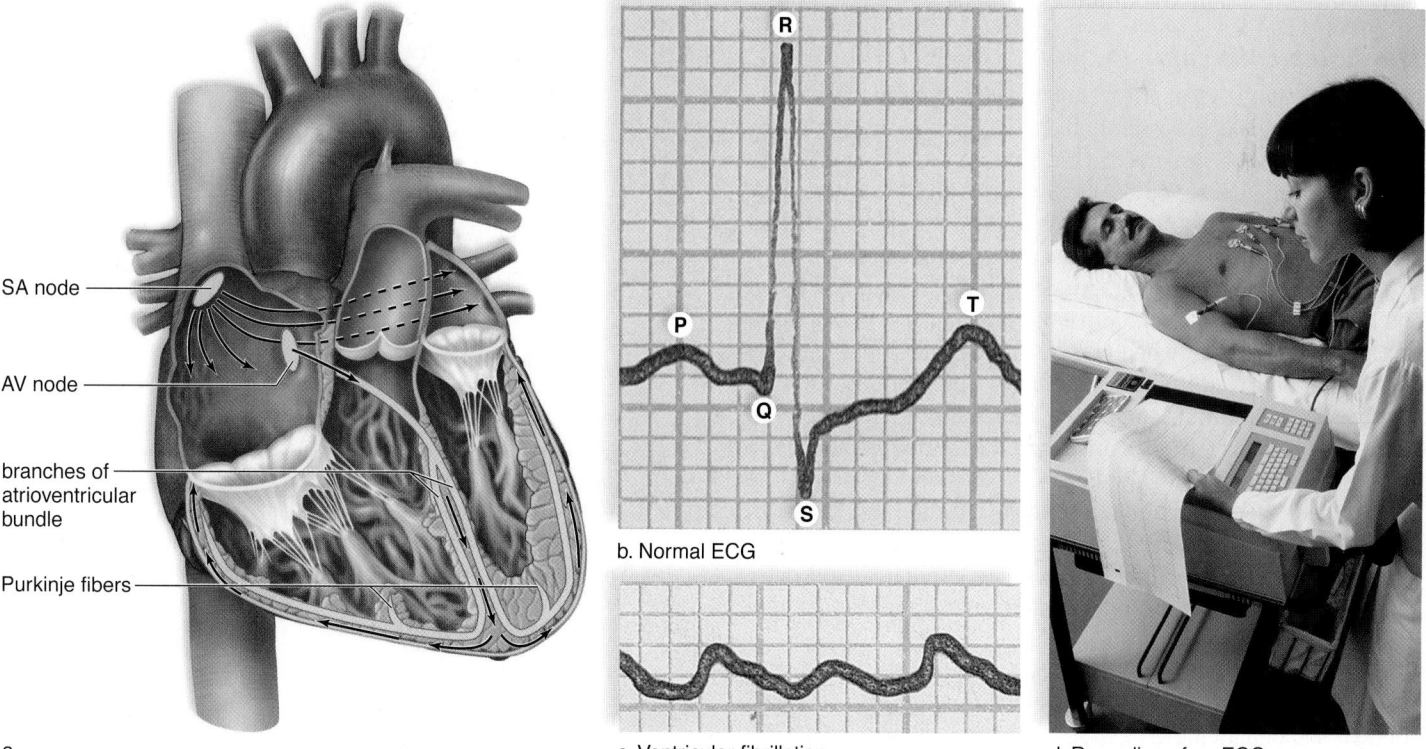

SA node

AV node

branches of atrioventricular bundle

Purkinje fibers

a.

R

P

Q

S

T

b. Normal ECG

c. Ventricular fibrillation

d. Recording of an ECG

Figure 32.9 Conduction system of the heart.   **a.** The SA node sends out a stimulus (black arrows), which causes the atria to contract. When this stimulus reaches the AV node, it signals the ventricles to contract. Impulses pass down the two branches of the atrioventricular bundle to the Purkinje fibers, and thereafter the ventricles contract. **b.** A normal ECG usually indicates that the heart is functioning properly. The P wave occurs just prior to atrial contraction; the QRS complex occurs just prior to ventricular contraction; and the T wave occurs when the ventricles are recovering from contraction. **c.** Ventricular fibrillation produces an irregular electrocardiogram due to irregular stimulation of the ventricles. **d.** The recording of an ECG.

caused by uncoordinated contraction of the ventricles (Fig. 32.9c). Ventricular fibrillation is of special interest because it can be caused by an injury or drug overdose. It is the most common cause of sudden cardiac death in a seemingly healthy person. When the ventricles are fibrillating, they can be defibrillated by applying a strong electric current for a short period of time. Then the SA node may be able to reestablish a coordinated beat. Many public places, and even private homes, have automatic external defibrillators. These are small devices that can be used to determine whether a person is suffering from ventricular fibrillation. If so, the AED administers an appropriate electrical shock to the chest.

Comparison of Circulatory Circuits

As mentioned, the human cardiovascular system includes two major circular pathways, the pulmonary circuit and the systemic circuit (Fig. 32.10).

The Pulmonary Circuit

In the pulmonary circuit, the path of blood can be traced as follows: oxygen-poor blood from all regions of the body collects in the right atrium and then passes into the right ventricle, which pumps it into the pulmonary trunk. The pulmonary trunk divides into the right and left pulmonary arteries, which carry blood to the lungs. As blood passes through pulmonary capillaries, carbon dioxide is given off and oxygen is picked up. Oxygen-rich blood returns to the left atrium of the heart, through pulmonary venules that join to form pulmonary veins.

The Systemic Circuit

The **aorta** [L. *aorte*, great artery] and the **venae cavae** (sing., vena cava) [L. *vena*, blood vessel, and *cavus*, hollow] are the major blood vessels in the systemic circuit. To trace the path of blood to any organ in the body, you need only start with the left ventricle, mention the aorta, the proper branch of the aorta, the organ, and the vein returning blood to the vena cava, which enters the right atrium. In the systemic circuit, arteries contain oxygen-rich blood and have a bright red color, but veins contain oxygen-poor blood and appear dull red or, when viewed through the skin, blue.

The coronary arteries are extremely important because they supply oxygen and nutrients to the heart muscle itself (see Fig. 32.6). The coronary arteries arise from the aorta just above the aortic semilunar valve. They lie on the exterior surface of the heart, where they branch into arterioles and then capillaries. In the capillary beds, nutrients, wastes, and gases are exchanged between the blood and the tissues. The capillary beds enter venules, which join to form the cardiac veins, and these empty into the right atrium.

A **portal system** [L. *porto*, carry, transport] is a structure in which blood from capillaries travels through veins to reach another set of capillaries, without first traveling through the heart. The hepatic portal system takes blood from the intestines directly to the liver. The liver then performs such functions as metabolizing nutrients and removing toxins (liver functions are explored further in Chapter 34). Blood leaves the liver by way of the hepatic vein, which enters the inferior vena cava.

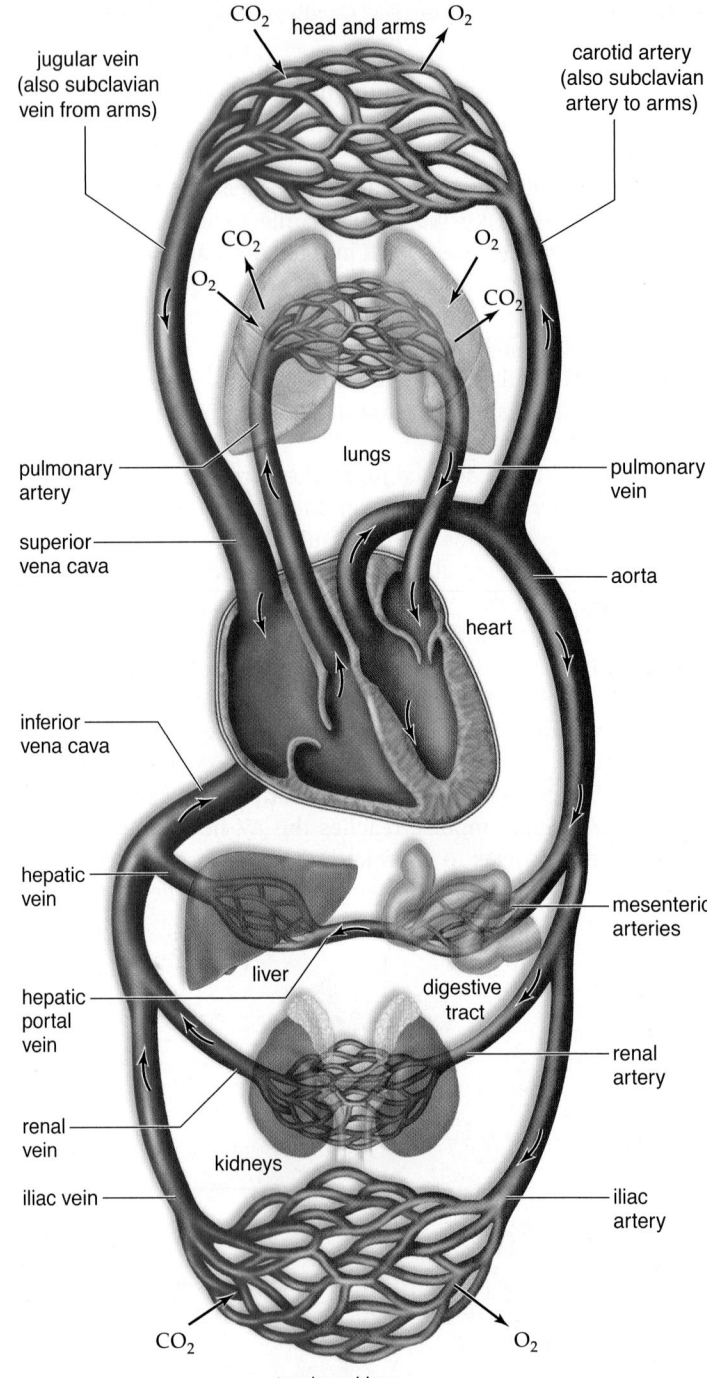

Figure 32.10 Path of blood. When tracing blood from the right to the left side of the heart in the pulmonary circuit, you must mention the pulmonary vessels. When tracing blood from the digestive tract to the right atrium in the systemic circuit, you must mention the hepatic portal vein, the hepatic vein, and the inferior vena cava. The blue-colored vessels carry oxygen-poor blood, and the red-colored vessels carry oxygen-rich blood; the arrows indicate the flow of blood.

Tracing the Path of Blood. Branches from the aorta go to the organs and major body regions. For example, this is the path of blood to and from the lower legs:

left ventricle—aorta—common iliac artery—femoral artery—lower leg capillaries—femoral vein—common iliac vein—inferior vena cava—right atrium

In most instances, the artery and the vein that serve the same region are given the same name. For example, iliac and femoral are names applied to both arteries and veins. What happens in between the artery and the vein? Arterioles from the artery branch into capillaries, where exchange takes place, and then venules join to form the vein that enters a vena cava. An exception occurs between the digestive tract and the liver, where blood must pass through two sets of capillaries because of the hepatic portal system.

Blood Pressure

When the left ventricle contracts, blood is forced into the aorta and then other systemic arteries under pressure. Systolic pressure results from blood being forced into the arteries during ventricular systole, and diastolic pressure is the pressure in the arteries during ventricular diastole. Human **blood pressure** can be measured with a *sphygmomanometer*, which has a pressure cuff that determines the amount of pressure required to stop the flow of blood through an artery.

Blood pressure is normally measured on the brachial artery, an artery in the upper arm. Today, digital manometers are often used to take one's blood pressure instead of the older type with a dial. Blood pressure is given in millimeters of mercury (mm Hg). A blood pressure reading consists of two numbers—for example, 120/80—that represent systolic and diastolic pressures, respectively.

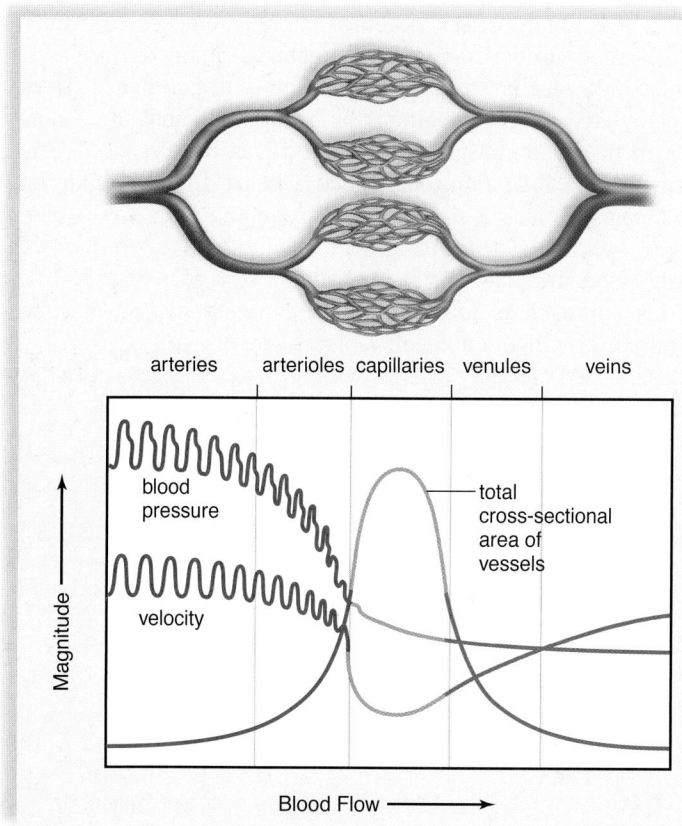

Figure 32.11 Velocity and blood pressure related to vascular cross-sectional area. In capillaries, blood is under minimal pressure and has the least velocity. Blood pressure and velocity drop off because capillaries have a greater total cross-sectional area than arterioles.

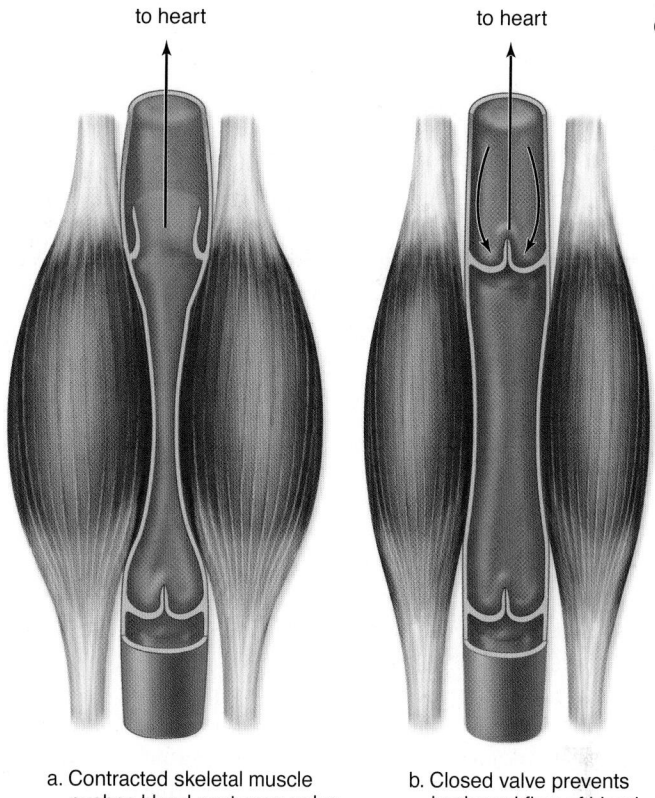

to heart   to heart

a. Contracted skeletal muscle pushes blood past open valve.

b. Closed valve prevents backward flow of blood.

Figure 32.12 Cross section of a valve in a vein. **a.** Pressure on the walls of a vein, exerted by skeletal muscles, increases blood pressure within the vein and forces a valve open. **b.** When external pressure is no longer applied to the vein, blood pressure decreases, and back pressure forces the valve closed. Closure of the valves prevents the blood from flowing in the opposite direction.

As blood flows from the aorta into the various arteries and arterioles, blood pressure falls. Also, the difference between systolic and diastolic pressure gradually diminishes. In the capillaries, there is a slow, fairly even flow of blood. This may be related to the very high total cross-sectional area of the capillaries (Fig. 32.11). It has been calculated that if all the blood vessels in a human body were connected end to end, the total distance would reach around the Earth at the equator two times! Most of this distance would be due to the large number of capillaries.

Blood pressure in the veins is low and is insufficient for moving blood back to the heart, especially from the limbs of the body. Venous return is dependent on three factors:

- Skeletal muscles near veins put pressure on the collapsible walls of the veins, and therefore on the blood contained in these vessels, when they contract.
- Valves in the veins prevent the backward flow of blood, and therefore pressure from muscle contraction moves blood toward the heart (Fig. 32.12). Varicose veins, abnormal dilations in superficial veins, develop when the valves of the veins become weak and ineffective due to a backward pressure of the blood.
- Variations in pressure in the chest cavity during breathing, also known as the *respiratory pump*, cause blood to flow from areas of higher pressure (such as the abdominal cavity) to lower pressure (in the thoracic cavity) during each inhalation.

MP3
Blood Flow and
Blood Pressure

Cardiovascular Disease

Cardiovascular disease (CVD) is the leading cause of death in most Western countries. According to the American Heart Association, CVD has been the most common cause of death in the United States every year since 1900. The only exception to this statistic was 1918, the worst year of a global influenza pandemic. In 2010, nearly 2,300 Americans died of CVD each day, an average of 1 death every 38 seconds. The Nature of Science feature "New Information About Preventing Cardiovascular Disease" emphasizes the possible prevention of CVD.

Hypertension

It is estimated that about 30% of Americans suffer from **hypertension**, which is high blood pressure. Another 30% are thought to have a condition called prehypertension, which can lead to hypertension. Under age 45, a reading above 130/90 is hypertensive, and beyond age 45, a reading above 140/95 is hypertensive.

Hypertension is most often caused by a narrowing of arteries due to atherosclerosis (described next). This narrowing causes the heart to work harder to supply the required amount of blood. The resulting increase in blood pressure can damage the heart, arteries, and other organs. Other risk factors that can contribute to hypertension include obesity, smoking, chronic stress, and a high dietary salt intake (which causes retention of fluid). Only about two-thirds of people with hypertension seek medical help for their condition, and it is likely that many people with high blood pressure are unaware of it.

Atherosclerosis

Atherosclerosis is an accumulation of soft masses of fatty materials, particularly cholesterol, beneath the inner linings of arteries (see Fig. 32A). Such deposits are called plaque. As deposits occur, plaque tends to protrude into the lumen of the vessel, interfering with the flow of blood. Plaque can also cause a clot to form on the irregular arterial wall. As long as the clot remains stationary, it is called a *thrombus*, but when and if it dislodges and moves along with the blood, it is called an *embolus*. If thromboembolism is not treated, complications can arise (see the following section).

Atherosclerosis often begins in early adulthood and develops progressively through middle age, but symptoms may not appear until an individual is 50 or older. See the Nature of Science feature "New Information About Preventing Cardiovascular Disease" for some recent recommendations on reducing risk.

Stroke and Heart Attack

Strokes and heart attacks are associated with hypertension and atherosclerosis. A **stroke,** or disruption of blood supply to the brain, often results when a small cranial arteriole bursts or is blocked by an embolus. A lack of oxygen causes a portion of the brain to die, and paralysis or death can result. A person is sometimes forewarned of a stroke by a feeling of numbness in

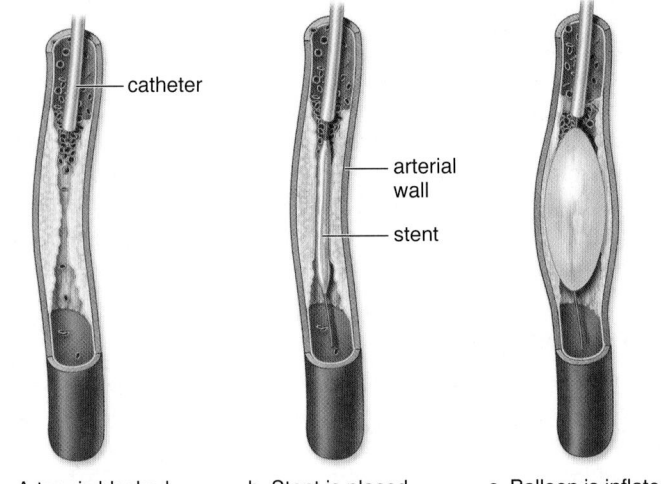

a. Artery is blocked.   b. Stent is placed.   c. Balloon is inflated.

Figure 32.13 Angioplasty with stent placement.   **a.** A plastic tube (catheter) is inserted into the coronary artery until it reaches the clogged area. **b.** A metal stent with a balloon inside it is pushed out the end of the plastic tube into the clogged area. **c.** When the balloon is inflated, the vessel opens, and the stent is left in place to keep the vessel open.

the hands or the face, difficulty in speaking, or temporary blindness in one eye.

If a coronary artery becomes partially blocked, the individual may suffer from **angina pectoris,** characterized as a squeezing sensation or a flash of burning. If a coronary artery is completely blocked, perhaps by a thromboembolism, a portion of the heart muscle dies due to a lack of oxygen. This is a myocardial infarction, also called a **heart attack.** It may be necessary to place a stent, or self-expanding wire mesh tube, inside a blocked artery to keep it open. About 1.3 million of these stents are placed in U.S. patients every year (Fig. 32.13). If this approach is unsuccessful, a coronary bypass may be required, in which a surgeon replaces one or more blocked coronary arteries with an artery taken from elsewhere in the patient's body. In 2009, about 450,000 of these procedures were performed in the United States.

Video
Cardiac Repair

Video
Heart Stem Cells

Check Your Progress   32.3

1. Name each blood vessel and heart chamber that blood passes through on its way from the venae cavae to the aorta, and identify which artery carries oxygen-poor blood.
2. Explain what specifically causes the sounds of the heartbeat.
3. Discuss why systolic blood pressure is higher than diastolic.
4. Predict what type of conditions might occur as a result of chronic hypertension and plaque.

Nature of Science

New Information About Preventing Cardiovascular Disease

For decades several factors have been associated with an increased risk of cardiovascular disease (CVD), especially atherosclerosis (Fig. 32A). Some of these cannot be avoided, such as increasing age, male gender, family history of heart disease, and belonging to certain races, including African American, Mexican American, and American Indian. Other risk factors—smoking, obesity, high cholesterol, hypertension, diabetes, physical inactivity—can be avoided or at least affected by changing one's behavior or taking medications. In recent years however, other factors have been under consideration; we discuss a few of them below.

- **Alcohol.** Alcohol abuse can destroy just about every organ in the body, the heart included. But recent research suggests that a moderate level of alcohol intake can improve cardiovascular health by improving the blood cholesterol profile, decreasing unwanted clot formation, increasing blood flow in the heart, and reducing blood pressure. According to the American Heart Association, people who consume one or two drinks per day have a 30% to 50% reduction in cardiovascular disease compared to nondrinkers.

  Please note, however, that the maximum protective effect is achieved with only one or two drinks per day—con-suming more than that increases the risk of many alcohol-related problems. The American Heart Assocation does *not* recommend that nondrinkers start using alcohol, or that drinkers increase their consumption, based on these findings.

- **Resvertrol.** The "red wine effect" or "French paradox" refers to the observation that levels of CVD in France are relatively low, despite the consumption of a high fat diet. One possible explanation is that wine is frequently consumed with meals.

  In addition to its alcohol content, red wine contains especially high levels of antioxidants, including resvertrol. Resvertrol is mainly produced in the skin of grapes, so it is also found in grape juice. Resvertrol supplements are also available at health food stores.

  The benefits of resvertrol alone are questionable, however, and most controlled studies to date have demonstrated no beneficial effects. The lower incidence of CVD in the French may be due to multiple factors, including lifestyle and genetic differences.

- **Omega-3 Fatty Acids.** The influence that diet has on blood cholesterol levels has been well studied. It is generally beneficial to minimize our intake of foods high in saturated fat (red meat, cream, and butter) and trans-fats (most margarines, commercially baked goods, and deep-fried foods). Replacing these harmful fats with healthier ones, such as monounsaturated fats (olive and canola oils) and polyunsaturated fats (corn, safflower, and soybean oils), is beneficial.

  In addition, the American Heart Association now recommends eating at least two servings of fish a week, especially fish like salmon, mackerel, herring, lake trout, sardines, and albacore tuna, which are are high in omega-3 fatty acids. These essential fatty acids can decrease triglyceride levels, slow the growth rate of atherosclerotic plaque, and lower blood pressure.

  However, children and pregnant women are advised to limit their fish consumption because of the high levels of mercury contamination in some fishes. For middle-aged and older men and postmenopausal women, the benefits of fish consumption far outweigh the potential risks.

Questions to Consider

1. Would you be helping your health if you decided to eat mackerel every day and drink two glasses of red wine with it? Why or why not?
2. What what would be some difficulties in trying to determine the true cause of the "French paradox"?

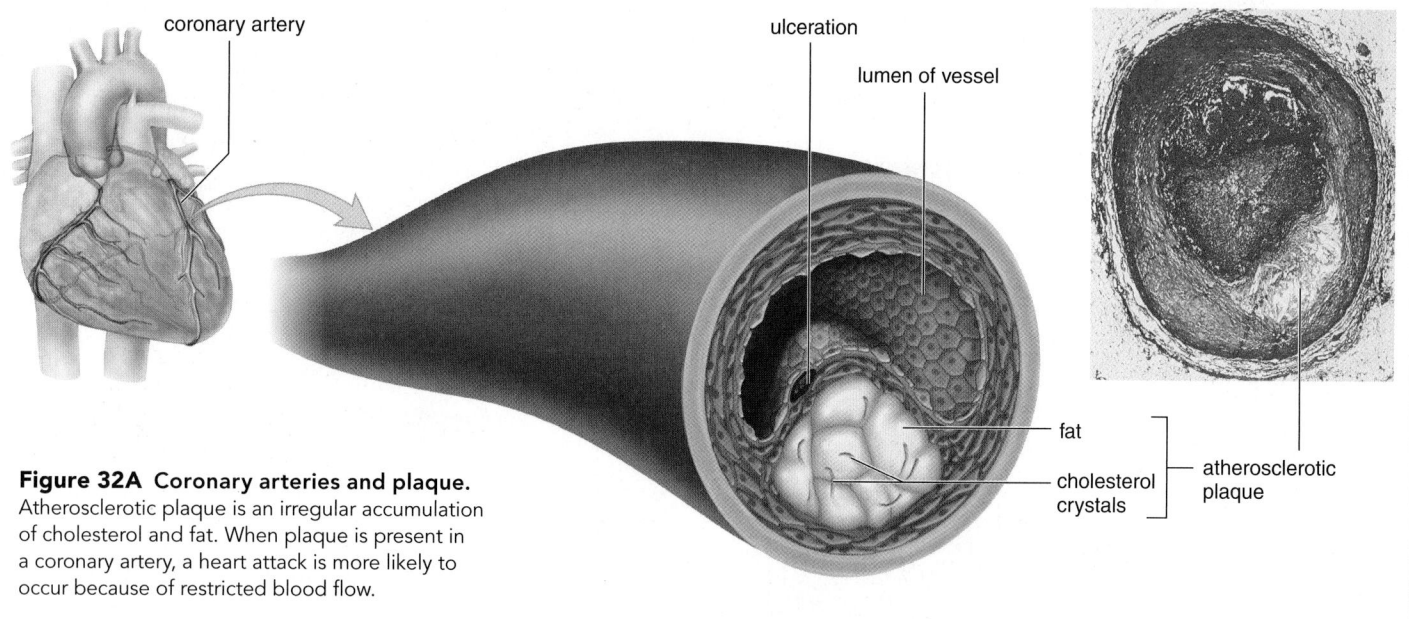

Figure 32A Coronary arteries and plaque. Atherosclerotic plaque is an irregular accumulation of cholesterol and fat. When plaque is present in a coronary artery, a heart attack is more likely to occur because of restricted blood flow.

coronary artery

ulceration

lumen of vessel

fat

cholesterol crystals

atherosclerotic plaque

32.4 Blood

As discussed in Chapter 31, blood is considered to be a connective tissue with a fluid matrix. In contrast to the hemolymph found in open circulatory systems, blood is normally contained within blood vessels. The blood of mammals has a number of functions that help maintain homeostasis, including:

- transporting gases, nutrients, waste products, and hormones throughout the body;
- combating pathogenic microorganisms;
- helping to maintain water balance and pH;
- regulating body temperature; and
- carrying platelets and factors that ensure clotting to prevent blood loss.

Blood has two main portions: a liquid portion, called plasma, and the formed elements, consisting of cells and platelets (Fig. 32.14).

Plasma

Plasma [Gk. *plasma*, something molded] contains many types of molecules, including nutrients, wastes, salts, and hundreds of different types of proteins. Some of these proteins are involved in buffering the blood, effectively keeping the pH near 7.4. They also maintain the blood's osmotic pressure so that water has an automatic tendency to enter blood capillaries. Several plasma proteins are involved in blood clotting, and others transport large organic molecules in the blood.

Albumin, the most plentiful of the plasma proteins, transports bilirubin, a breakdown product of hemoglobin, and various types of lipoproteins transport cholesterol. Another very significant group of plasma proteins are the **antibodies,** which are proteins produced by the immune system in response to specific pathogens and other foreign materials (see Chapter 33).

Formed Elements

The **formed elements** are of three types: red blood cells, or erythrocytes [Gk. *erythros*, red, and *kytos,* cell]; white blood cells, or leukocytes [Gk. *leukos*, white]; and platelets, or thrombocytes [Gk. *thrombos*, blood clot].

Red Blood Cells

Red blood cells (RBCs) are small, biconcave disks that at maturity lack a nucleus and contain the respiratory pigment hemoglobin. The average adult human has five to six million RBCs per cubic millimeter (mm³) of whole blood, and each one of these cells contains about 250 million hemoglobin molecules. **Hemoglobin** [Gk. *haima*, blood; L. *globus*, ball] contains four globin protein chains, each associated with heme, an iron-containing group.

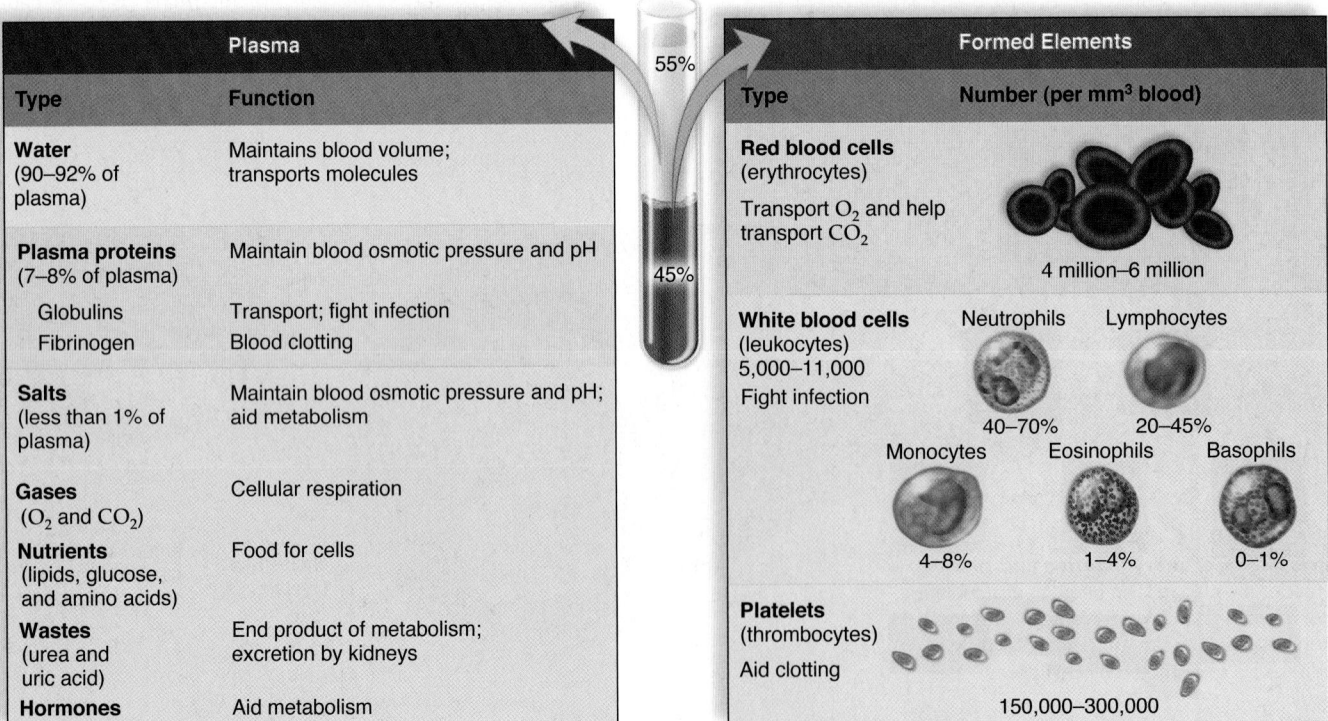

Figure 32.14 Composition of blood.

Iron combines loosely with oxygen, and in this way oxygen is carried in the blood. If the number of RBCs is insufficient, or if the cells do not have enough hemoglobin, the individual suffers from anemia and has a tired, run-down feeling.

In adults, RBCs are manufactured in the red bone marrow of the skull, the ribs, the vertebrae, and the ends of the long bones. The hormone erythropoietin, produced by the kidneys, stimulates RBC production. Now available as a drug, erythropoietin is helpful to persons with anemia and is also sometimes abused by athletes who want to enhance the oxygen-carrying capacity of their blood.

Before they are released from the bone marrow into blood, RBCs synthesize hemoglobin and lose their nuclei. After living about 120 days, they are destroyed chiefly in the liver and the spleen, where they are engulfed by large phagocytic cells. When RBCs are destroyed, hemoglobin is released. The iron is recovered and returned to the red bone marrow for reuse. The heme portions of the molecules undergo chemical degradation and are excreted by the liver as bile pigments in the bile. The bile pigments are primarily responsible for the color of feces.

MP3
General Functions and
Composition of Blood

Blood Types

The earliest attempts at blood transfusions resulted in illness and even death of some recipients. Eventually, it was discovered that only certain transfusion donors and recipients are compatible because red blood cell membranes carry specific proteins or carbohydrates that are antigens to blood recipients. An **antigen** [Gk. *anti*, against; L. *genitus*, forming, causing] is a molecule, usually a protein or carbohydrate, that can trigger a specific immune response. Several groups of RBC antigens exist, the most significant being the ABO and Rh systems. Clinically, it is very important that the blood groups be properly cross-matched to avoid a potentially deadly transfusion reaction.

ABO System

In the ABO system, the presence or absence of type A and type B antigens on RBCs determines a person's blood type. For example, if a person has type A blood, the A antigen is on his or her RBCs. Because it is considered by the immune system to be "self," this molecule is not recognized as an antigen by this individual, although it can be an antigen to a recipient who does not have type A blood.

In the ABO system, there are four blood types: A, B, AB, and O. Because the A and B antigens are also commonly found on microorganisms present in and on our bodies, a person's plasma usually contains antibodies to the A or B antigens not present on his or her RBCs. These antibodies are called anti-A and anti-B. This chart explains what antibodies are present in the plasma of each blood type:

Blood Type	Antigen on Red Blood Cells	Antibody in Plasma
A	A	Anti-B
B	B	Anti-A
AB	A, B	None
O	None	Anti-A and anti-B

Because type A blood has anti-B and not anti-A antibodies in the plasma, a donor with type A blood can give blood to a recipient with type A blood (Fig. 32.15). However, if type A blood is given to a type B recipient, **agglutination** (Fig. 32.16), the clumping of RBCs, can cause blood to stop circulating in small blood vessels, leading to organ damage.

Theoretically, a person with which blood type could donate to all recipients? The answer is that type O RBCs have no A or B antigens, and this is sometimes called the universal donor type. A person with which blood type could receive blood from any donor? Type AB blood has no anti-A or anti-B antibodies, and thus it is sometimes called the universal recipient. In practice, however, it is not safe to rely solely on the ABO system when matching blood. Instead, samples of the two types of blood are physically mixed, and the result is microscopically examined for agglutination before blood transfusions are done.

An equally important concern when transfusing blood is to make sure that the donor is free from transmissible infectious agents, such as the microbes that cause AIDS, hepatitis, and syphilis.

Rh System

Another important antigen on RBCs is the Rh factor. Eighty-five percent of the U.S. population have this particular antigen on their RBCs and are Rh positive. Fifteen percent do not have the

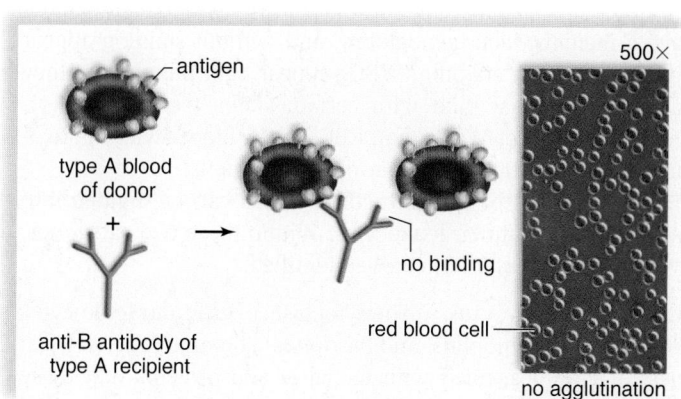

Figure 32.15 Matched blood transfusion. No agglutination occurs when the donor and recipient have the same type blood.

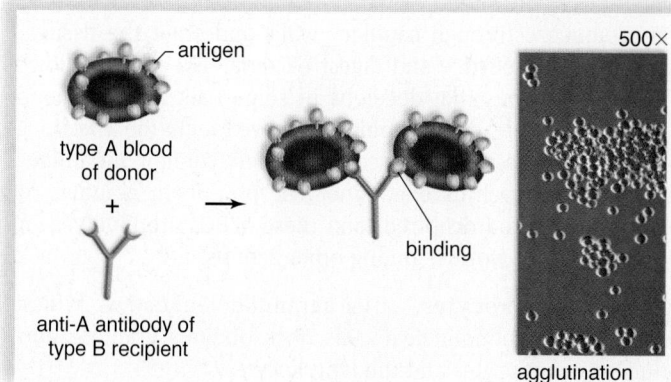

Figure 32.16 Mismatched blood transfusion. Agglutination occurs because blood type B has anti-A antibodies in the plasma.

antigen and are Rh negative. The designation of blood type usually includes whether the person has or does not have the Rh factor on the RBCs, e.g., type A-positive (A +). Unlike the case with the A and B antigens, Rh-negative individuals normally do not have antibodies to the Rh factor, but they may make them when exposed to the Rh factor.

During pregnancy, if the mother is Rh negative and the father is Rh positive, the child may be Rh positive. If the Rh-positive fetal RBCs leak across the placenta, the mother may produce anti-Rh antibodies. In this or a subsequent pregnancy with another Rh-positive baby, these antibodies may cross the placenta and destroy the child's RBCs. This condition, called hemolytic disease of the newborn (HDN), can be fatal without an immediate blood transfusion after birth.

The problem of Rh incompatibility can be prevented by giving Rh-negative women an Rh immunoglobulin injection toward the end of pregnancy and within 72 hours of giving birth to an Rh-positive child. This treatment contains a relatively low level of anti-Rh antibodies that help destroy any Rh-positive blood cells in the mother's blood before her immune system produces high levels of anti-Rh antibodies.

MP3
Blood Groupings

White Blood Cells

Because they are a critical component of the immune system, the functions of white blood cells are discussed in detail in Chapter 33 and only briefly here. **White blood cells (WBCs)**, or **leukocytes**, differ from RBCs in that they are usually larger, have a nucleus, lack hemoglobin, and without staining appear translucent. With staining, WBCs appear light blue unless they have granules that bind with certain stains (see Fig. 32.14). There are far fewer WBCs than RBCs in the blood, with approximately 5,000–11,000 WBCs per mm^3 in humans.

On the basis of their structure, WBCs can be divided into granular and agranular leukocytes. Within these two categories, five main types of WBCs can be identified.

Granular Leukocytes.   The cytoplasm of **granular leukocytes** (neutrophils, eosinophils, and basophils) contains spherical vesicles, or granules, filled with enzymes and proteins that these cells use to help defend the body against invading microbes and other parasites.

Neutrophils [Gk. *neuter*, neither, and *phileo*, love] have a multilobed nucleus, resulting in their other name, polymorphonuclear cells. They are the most abundant of the WBCs and are able to squeeze through capillary walls and enter the tissues, where they phagocytize and digest bacteria. The thick, yellowish fluid called *pus* that develops in some bacterial infections contains mainly dead neutrophils that have fought the infection. **Basophil** granules stain a deep blue and contain inflammatory chemicals such as histamine. The prominent granules of **eosinophils** stain a deep red, and these WBCs are involved in fighting parasitic worms, among other actions.

Agranular Leukocytes.   The **agranular leukocytes**, which are also called mononuclear cells, lack obvious granules and include the monocytes and the lymphocytes.

Monocytes are the largest of the WBCs, and they tend to migrate into tissues in response to chronic, ongoing infections, where they differentiate into large phagocytic **macrophages** [Gk. *makros*, long, and *phagein*, to eat]. These long-lived cells not only fight infections directly, but they also release growth factors that increase the production of different types of WBCs by the bone marrow. Some of these factors are available for medicinal use and may be helpful to people with low immunity, such as AIDS patients or people on chemotherapy for cancer. A third function of macrophages is to interact with lymphocytes to help initiate the adaptive immune response (see Chapter 33).

Lymphocytes [L. *lympha*, clear water; Gk. *kytos*, cell] are the second most common type of WBC in the blood. The two major types of lymphocytes, T cells and B cells, each play a distinct role in adaptive immune responses to specific antigens. One type of T cell, the helper T cell, initiates and influences most of the other cell types involved in adaptive immunity. The other type, the cytotoxic T cell, attacks infected cells that contain viruses. In contrast, the main function of B cells is to produce antibodies. Each B cell produces just one type of antibody, which is specific for one type of antigen. As mentioned earlier in this section, an antigen is a molecule that causes a specific immune response because the immune system recognizes it as "foreign." When antibodies combine with antigens, the complex is often phagocytized by a macrophage. The activities of lymphocytes, along with other aspects of animal immune systems, are discussed in more detail in Chapter 33.

Platelets and Blood Clotting

Platelets (thrombocytes) result from fragmentation of large cells, called *megakaryocytes*, in the red bone marrow. Platelets are produced at a rate of 200 billion a day, and the blood contains 150,000–300,000 per mm^3. These formed elements are involved in blood **clotting**, or coagulation.

When a blood vessel in the body is damaged, platelets clump at the site of the puncture and partially seal the leak. Platelets and the injured tissues release a clotting factor called prothrombin activator that converts prothrombin in the plasma to thrombin. This reaction requires calcium ions (Ca^{2+}). *Thrombin*, in turn, acts as an enzyme that severs two short amino acid chains from a *fibrinogen* molecule, one of the proteins in plasma. These activated fragments then join end to end, forming long threads of *fibrin*.

Fibrin threads wind around the platelet plug in the damaged area of the blood vessel and provide the framework for the clot. Red blood cells also are trapped within the fibrin threads; these cells make a clot appear red (Fig. 32.17). A fibrin clot is present only temporarily. As soon as blood vessel repair is initiated, an enzyme called *plasmin* destroys the fibrin network and restores the fluidity of plasma.

The Nature of Science feature "How Horseshoe Crabs Save Human Lives" on page 622 describes how a clotting reaction in these arthropods can help identify bacterial contamination.

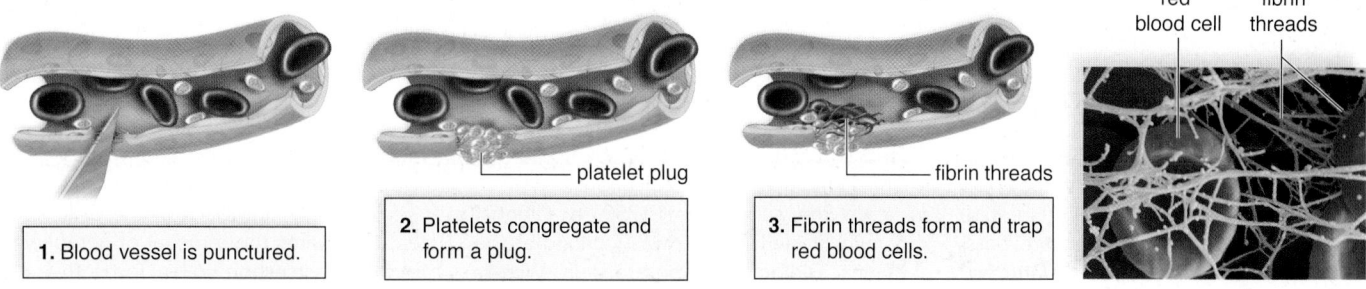

red
blood cell   fibrin
threads

1. Blood vessel is punctured.

platelet plug

2. Platelets congregate and
form a plug.

fibrin threads

3. Fibrin threads form and trap
red blood cells.

Figure 32.17 Blood clotting.   A number of plasma proteins participate in a series of enzymatic reactions that lead to the formation of fibrin threads.

Capillary Exchange

Figure 32.18 illustrates capillary exchange between a systemic capillary and tissue fluid, the fluid between the body's cells. Blood that enters a capillary at the arterial end is rich in oxygen and nutrients, and it is under pressure created by the pumping of the heart. Two forces primarily control movement of fluid through the capillary wall: (1) osmotic pressure, which tends to cause water to move from tissue fluid into blood, and (2) blood pressure, which tends to cause water to move in the opposite direction. At the arterial end of a capillary, the osmotic pressure of blood (21 mm Hg) is lower than the blood pressure (30 mm Hg). Osmotic pressure is created by the presence of salts and the plasma proteins. Because osmotic pressure is lower than blood pressure at the arterial end of a capillary, water exits a capillary at this end.

Midway along the capillary, where blood pressure is lower, the two forces essentially cancel each other, and there is no net movement of water. Solutes now diffuse according to their concentration gradient: Oxygen and nutrients (glucose and amino acids) diffuse out of the capillary; carbon dioxide and wastes diffuse into the capillary. Red blood cells and almost all plasma proteins remain in the capillaries.

The substances that leave a capillary contribute to **tissue fluid,** the fluid between the body's cells. Because plasma proteins are too large to readily pass out of the capillary, tissue fluid tends to contain all components of plasma but has much lower amounts of protein.

At the venule end of a capillary, where blood pressure has fallen even more, osmotic pressure is greater than blood pressure, and water tends to move into the capillary. Almost the same amount of fluid that left the capillary returns to it, although some

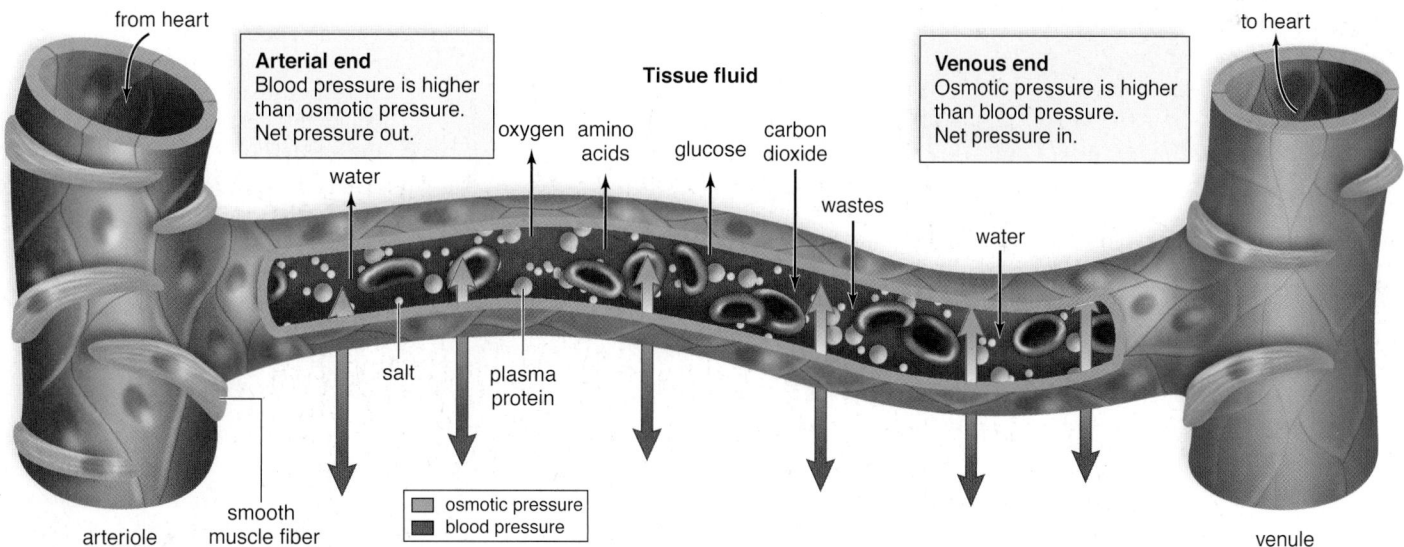

from heart

Arterial end
Blood pressure is higher
than osmotic pressure.
Net pressure out.

water

Tissue fluid

oxygen amino
acids

glucose

carbon
dioxide

to heart

Venous end
Osmotic pressure is higher
than blood pressure.
Net pressure in.

wastes

water

salt

plasma
protein

osmotic pressure
blood pressure

arteriole

smooth
muscle fiber

venule

Figure 32.18 Capillary exchange.   A capillary, illustrating the exchanges that take place and the forces that aid the process. At the arterial end of a capillary, the blood pressure is higher than the osmotic pressure; therefore, water (H_2O) tends to leave the bloodstream. In the midsection, molecules, including oxygen (O_2) and carbon dioxide (CO_2), follow their concentration gradients. At the venous end of a capillary, the osmotic pressure is higher than the blood pressure; therefore, water tends to enter the bloodstream. Notice that the red blood cells and the plasma proteins are too large to exit a capillary.

Nature of Science

How Horseshoe Crabs Save Human Lives

Take a walk along a beach on the northeastern U.S. coast, and you are likely to see horseshoe crabs (*Limulus polyphemus*) (Fig. 32B). Although not truly "crabs," they are classified in the phylum Arthropoda, along with insects, arachnids, and crustaceans. Because of their prehistoric appearance, horseshoe crabs are sometimes called "living fossils," and in fact they were living on Earth before the dinosaurs. They have an open circulatory system, with an elongated heart that pumps hemolymph between the gills and the body, without returning to the heart in between.

Instead of hemoglobin, the hemolymph of horseshoe crabs contains hemocyanin, which binds to oxygen using copper instead of iron, giving the blood a light blue color. The blood also contains amebocytes, cells that are analogous to the neutrophils or macrophages of higher animals, which serve a similar role in protecting against bacterial infections. Oddly enough, these amebocytes turned out to be the key to developing a method for detecting potentially fatal bacterial contamination of

medical products such as IV solutions, vaccines, and injectable medications.

In the summer of 1950, a scientist named Frederick Bang was working at the Marine Biological Laboratory in Woods Hole, Massachusetts. Bang was interested in the immune system of primitive organisms, and so he chose to inject various types of bacteria into horseshoe crabs to study their immune response. What he found was that injection of any bacteria of the Gram-negative type, or an extract of their cell walls, caused the horseshoe crabs to quickly die, not from the infection, but from a massive coagulation of their circulatory fluid.

After many experiments and collaborations with other scientists, Bang developed a test using an extract of the amebocytes, which could be mixed with any sample to determine if that sample contained any contamination by Gram-negative bacteria. If so, the material would clot within 45 minutes. This LAL test, as it is now called, replaced the existing pyrogen test, which required that materials be injected into

rabbits, took more time, and was less sensitive to low levels of contamination.

The only drawback of the LAL test is that it requires the removal of hemolymph from the horseshoe crabs. To do this, biomedical companies hire trawlers to catch adult horseshoe crabs. These are brought to a laboratory, washed, and about 30% of the animal's hemolymph removed from the animal's heart with a large gauge needle. The hemolymph is then centrifuged to separate the amebocytes, distilled water is added to lyse the cells, and the proteins responsible for the clotting reaction are separated and processed into the product used for the LAL test.

The horseshoe crabs are usually returned to the ocean within 72 hours of bleeding, and studies suggest that most of them survive, perhaps to be caught and bled again. Because one quart of hemolymph is worth about $15,000, the companies have good reason to preserve this ancient, fascinating species.

Questions to Consider

1. Compared to hemoglobin, hemocyanins are much larger, free-floating molecules. Why might hemocyanins work better with an open circulatory system, compared to hemoglobin?
2. The amebocytes of horseshoe crabs cause the animal's hemolymph to clot in response to certain bacteria. How could this response be beneficial to the animal?

Figure 32B Horseshoe crabs. Horseshoe crabs have lived on the Earth for an estimated 450 million years.

excess tissue fluid is always collected by the lymphatic capillaries (Fig. 32.19). Tissue fluid contained within lymphatic vessels is called **lymph.** Lymph is returned to the systemic venous blood when the major lymphatic vessels enter the subclavian veins in the shoulder region. See Chapter 33 for more information about the lymphatic system.

Figure 32.19 Capillary bed.   A lymphatic capillary bed lies near a blood capillary bed. When lymphatic capillaries take up excess tissue fluid, it becomes lymph. Precapillary sphincters can shut down a blood capillary, and blood then flows through the shunt.

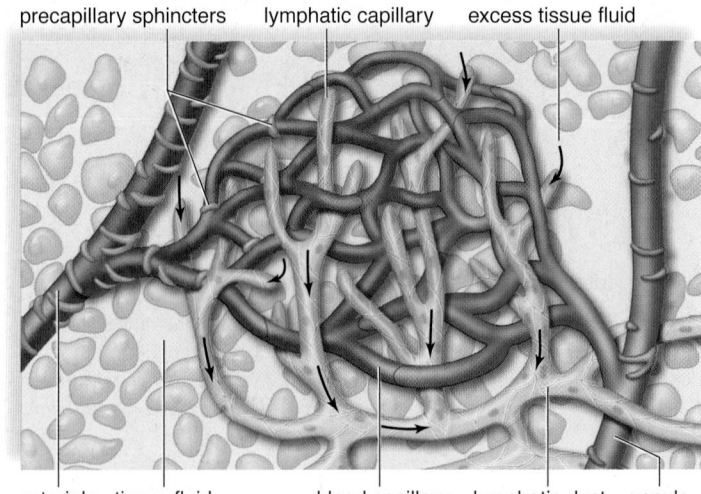

precapillary sphincters   lymphatic capillary   excess tissue fluid

arteriole   tissue fluid   blood capillary   lymphatic duct   venule

Not all capillary beds are open at the same time. When the precapillary sphincters (circular muscles) shown in Figure 32.4 are relaxed, the capillary bed is open and blood flows through the capillaries. When precapillary sphincters are contracted, blood flows through a shunt that carries blood directly from an arteriole to a venule.

In addition to nutrients and wastes, the blood distributes heat to body parts. When you are warm, many capillaries that serve the skin are open, and your face is flushed. This helps rid the body of excess heat. When you are cold, skin capillaries close, conserving heat, and your skin takes on a bluish tinge.

 **MP3**
Capillary Exchange and Bulk Flow

 **Animation**
Fluid Exchange Across the Walls of Capillaries

Check Your Progress 32.4

1. List the major components of blood and the functions of each.
2. Name the major events, in chronological order, that result in a blood clot.
3. Explain why Rh incompatibility is a problem only when a fetus is Rh-positive and the mother is Rh-negative, but not vice-versa.
4. Describe the major factors that affect the rate of capillary exchange.

Connecting *the* Concepts *with the* Big Ideas

Energy and Homeostasis

- A disturbance in the feedback mechanism that controls blood clotting can cause serious consequences. (2C1c*IE*)

Interactions and Systems

- The circulatory and respiratory systems work together to insure the well being of the body. (4A4b*IE*)
- Organ specialization within the circulatory system increases efficiency for the organism. (4B2a2*IE*)
- Molecular variations that produce different varieties of hemoglobin help organisms adapt to different conditions. (4C1a*IE*)

*Find the unabridged version of all EK citations at www.glencoe.com/maderAP11.

Media Study Tools

www.glencoe.com/maderAP11

Enhance your study of this chapter with study tools and practice tests. Also ask your instructor about the resources available through ConnectPlus, including the media-rich eBook, interactive learning tools, and animations.

Summarize

32.1 Transport in Invertebrates

Some invertebrates do not have a transport system. The presence of a gastrovascular cavity allows diffusion alone to supply the needs of cells in cnidarians and flatworms. Roundworms make use of their pseudocoelom in the same way that echinoderms use their coelom to circulate materials.

Other invertebrates do have a transport system. Insects have an open circulatory system, and earthworms have a closed one.

32.2 Transport in Vertebrates

Vertebrates have a closed system in which arteries carry blood away from the heart to capillaries, where exchange takes place, and veins carry blood to the heart.

Fishes have a one-circuit circulatory pathway because the heart, with the single atrium and ventricle, pumps blood to the gills and then to the body, without a second pass through the heart. The other vertebrates have both pulmonary and systemic circuits. Amphibians have two atria but a single ventricle. Crocodilians, birds, and mammals, including humans, have a heart with two atria and two ventricles, in which oxygen-rich blood is always separate from oxygen-poor blood.

32.3 The Human Cardiovascular System

The heartbeat in humans begins when the SA (sinoatrial) node (pacemaker) causes the two atria to contract, and blood moves through the atrioventricular valves to the two ventricles. The SA node also stimulates the AV (atrioventricular) node, which in turn causes the two ventricles to contract. Ventricular contraction sends blood through the semilunar valves to the pulmonary trunk and the aorta. Now all chambers rest. The heart sounds, lub-dub, are caused by the closing of the valves.

In the pulmonary circuit, blood travels to and from the lungs. In the systemic circuit, the aorta divides into blood vessels that serve the body's cells. The venae cavae return oxygen-poor blood to the heart.

Blood pressure created by the pumping of the heart accounts for the flow of blood in the arteries, but skeletal muscle contraction is largely responsible for the flow of blood in the veins, which have valves preventing a backward flow.

Hypertension and atherosclerosis are two circulatory disorders that can lead to heart attack and to stroke. Following a heart-healthy diet, getting regular exercise, maintaining a proper weight, and not smoking cigarettes can help protect against the development of these conditions.

32.4 Blood

Blood has two main parts: plasma and formed elements. Plasma is mostly water (90–92%), but also contains proteins (7–8%), nutrients, and wastes.

The formed elements include red blood cells, white blood cells, and platelets. Red blood cells contain hemoglobin, which functions in oxygen transport.

White blood cells, which include granulocytes and agranulocytes, defend the body against infections. Three types of granulocytes are the neutrophils, which are phagocytes; basophils, which are involved in inflammation; and eosinophils, which are important in parasitic infections.

The two types of agranulocytes are the lymphocytes, which carry out adaptive (specific) immunity to disease; and monocytes, which enter tissues to become phagocytic macrophages.

The platelets and two plasma proteins, prothrombin and fibrinogen, function in blood clotting, an enzymatic process that results in fibrin threads. Blood clotting includes three major events: (1) Platelets and injured tissue release prothrombin activator, which (2) enzymatically changes prothrombin to thrombin, which is an enzyme that (3) causes fibrinogen to be converted to fibrin threads.

The ABO blood typing system is based on the presence or absence of A and B antigens on the red blood cells (A and B). If a mismatched transfusion is given, antibodies in the recipient's blood may react to these antigens. A second type of red blood cell antigen is the Rh factor. If an Rh-negative woman becomes pregnant with an Rh-positive fetus, she may produce anti-Rh antibodies that could damage any Rh-positive fetus she carries.

When blood reaches a capillary, water moves out at the arterial end due to blood pressure. At the venous end, water moves in due to osmotic pressure. In between, nutrients diffuse out of, and wastes diffuse into, the capillary according to concentration gradients.

Key Terms

agglutination 619
agranular leukocyte 620
angina pectoris 616
antibody 618
antigen 619
aorta 614
arteriole 608
artery 608
atherosclerosis 616
atrioventricular valve 611
atrium 610
basophil 620
blood 606
blood pressure 615
capillary 608
cardiac cycle 612
cardiac output 612
circulatory (or cardiovascular)
    system 606, 608

closed circulatory system 607
clotting 620
diastole 612
electrocardiogram (ECG) 613
eosinophil 620
formed elements 618
granular leukocyte 620
heart 610
heart attack 616
hemoglobin 618
hemolymph 606
hypertension 616
leukocyte 620
lymph 622
lymphocyte 620
macrophage 620
monocyte 620
neutrophil 620

open circulatory system 606
pacemaker 613
plasma 618
platelet 620
portal system 614
pulmonary circuit 609
pulse 613
red blood cell 618
semilunar valve 611
septum 610

stroke 616
systemic circuit 609
systole 612
tissue fluid 621
vein 608
vena cava 614
ventricle 611
venule 608
white blood cell 620

 Assess

Reviewing This Chapter

1. Describe transport in invertebrates that have no circulatory system; in those that have an open circulatory system; and in those that have a closed circulatory system. 606–7
2. Compare the circulatory systems of a fish, an amphibian, and a mammal. 608–9
3. Trace the path of blood in humans from the right ventricle to the left atrium; from the left ventricle to the kidneys and to the right atrium; from the left ventricle to the small intestine and to the right atrium. 611
4. Describe the mechanism of a heartbeat, mentioning all the factors that account for this repetitive process. Describe how the heartbeat affects blood flow. What other factors are involved in blood flow? 612–13
5. Define these terms: pulmonary circuit, systemic circuit, and portal system. 614
6. Discuss the life cycle and function of red blood cells. 618–19
7. Explain the ABO and Rh systems of typing blood. 619–20
8. How are white blood cells classified? What are the functions of neutrophils, monocytes, and lymphocytes? 620
9. Name the steps that take place when blood clots. Which substances are present in blood at all times, and which appear during the clotting process? 620
10. What forces facilitate exchange of molecules across the capillary wall? 621–23

Testing Yourself

Choose the best answer for each question.

1. Which one of these would you expect to be part of a closed, but not an open, circulatory system?
   a. ostia
   b. capillary beds
   c. hemocoel
   d. heart
   e. All of these are correct.

2. In which animal does aortic blood have less oxygen than blood in the pulmonary vein?
   a. frog
   b. chicken
   c. monkey
   d. fish
   e. All of these are correct.

3. In humans, blood returning to the heart from the lungs returns to
   a. the right ventricle.
   b. the right atrium.
   c. the left ventricle.
   d. the left atrium.
   e. both the right and left sides of the heart.

4. Systole refers to the contraction of the
   a. major arteries.
   b. SA node.
   c. atria and ventricles.
   d. major veins.
   e. All of these are correct.

5. Which of these is an incorrect association?
   a. white blood cells—infection fighting
   b. red blood cells—blood clotting
   c. plasma—water, nutrients, and wastes
   d. red blood cells—hemoglobin
   e. platelets—blood clotting

6. Water enters capillaries on the venous end as a result of
   a. active transport from tissue fluid.
   b. an osmotic pressure gradient.
   c. higher blood pressure on the venous end.
   d. higher blood pressure on the arterial side.
   e. higher red blood cell concentration on the venous end.

7. Macrophages are derived from
   a. basophils.   d. lymphocytes.
   b. eosinophils.   e. monocytes.
   c. neutrophils.

8. Which of the following is not a formed element of blood?
   a. leukocyte   c. fibrinogen
   b. eosinophil   d. platelet

9. Which of these is an incorrect statement concerning the heartbeat?
   a. The atria contract at the same time.
   b. The ventricles relax at the same time.
   c. The atrioventricular valves open at the same time.
   d. The semilunar valves open at the same time.
   e. First, the right side contracts, and then the left side contracts.

10. All arteries in the body contain O_2-rich blood, with the exception of the
    a. aorta.   c. renal arteries.
    b. pulmonary arteries.   d. coronary arteries.

11. The "lub," the first heart sound, is produced by the closing of
    a. the aortic semilunar valve.
    b. the pulmonary semilunar valve.
    c. the right (atrioventricular) tricuspid valve.
    d. the left (atrioventricular) bicuspid valve, or mitral valve.
    e. both atrioventricular valves.

12. Label this diagram of the heart.

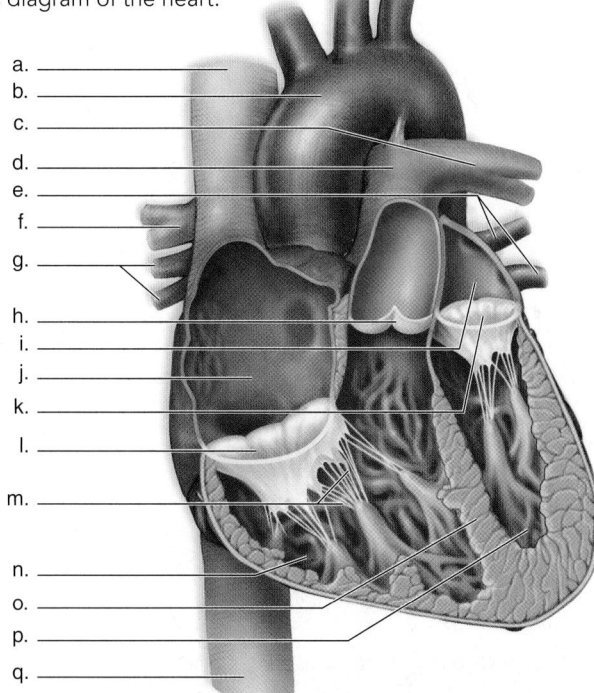

a. _____
b. _____
c. _____
d. _____
e. _____
f. _____
g. _____
h. _____
i. _____
j. _____
k. _____
l. _____
m. _____
n. _____
o. _____
p. _____
q. _____

Engage

Virtual Labs
Virtual Frog Dissection
Earthworm Dissection

The Virtual Labs "Virtual Frog Dissection" and "Earthworm Dissection" provide an interactive look at the circulatory systems of vertebrate and invertebrate animals.

Thinking Scientifically

1. For several years, researchers have attempted to produce an artificial blood for transfusions. Artificial blood would most likely be safer and more readily available than human blood. While artificial blood might not have all the characteristics of human blood, it would be useful on the battlefield and in emergency situations. Which characteristics of normal blood must artificial blood have to be useful, and which would probably be too difficult to reproduce?

2. You have to stand in front of the class to give a report. You are nervous, and your heart is pounding. What is the specific mechanism behind this reaction? How would your ECG appear?

Bioethical Issue

A Healthy Lifestyle

Many deaths a year could be prevented if people adopted the healthy lifestyle described in the Nature of Science feature "New Information About Preventing Cardiovascular Disease." Tobacco, lack of exercise, and a high-fat diet probably cost the nation about $200 billion per year in health-care costs. To what lengths should we go to prevent these deaths and reduce health-care costs?

E. A. Miller, a meat-packing subsidiary of ConAgra in Hyrum, Utah, charges extra for medical coverage of employees who smoke. Eric Falk, Miller's director of human resources, says, "We want to teach employees to be responsible for their behavior." Anthem Blue Cross–Blue Shield of Cincinnati, Ohio, takes a more positive approach. They give insurance plan participants $240 a year in extra benefits, such as additional vacation days, if they get good scores in five out of seven health-related categories. The University of Alabama, Birmingham, School of Nursing has a health-and-wellness program that counsels employees about how to get into shape in order to keep their insurance coverage. Audrey Brantley, who is in the program, has mixed feelings. She says, "It seems like they are trying to control us, but then, on the other hand, I know of folks who found out they had high blood pressure or were borderline diabetics and didn't know it."

Does it really work? Turner Broadcasting System in Atlanta has a policy that affects all employees hired after 1986. They will be fired if caught smoking—whether at work or at home—but some admit they still manage to sneak a smoke. What steps do you think are ethical to encourage people to adopt a healthy lifestyle?

33

The Lymphatic and Immune Systems

David Vetter, who suffered from severe combined immune deficiency and had to be kept in an isolated environment.

BEFORE YOU BEGIN

Before beginning this chapter, take a few moments to review the following discussions.

Figure 5.3 Which of the major protein functions are most important in the immune system?

Figure 21.21 In what ways do the amoeboid protists resemble macrophages of the mammalian immune system?

Section 31.2 What life processes are carried out by the lymphatic and immune systems?

The opening story for Chapter 31 noted how an astronaut walking in space needs a protective suit to provide his or her body with the oxygen, warmth, and other factors needed for internal homeostasis. Equally important is our ability to keep a variety of microbial and other threats from damaging our body. The many and varied molecules, cells, and tissues that carry out this task are collectively known as the immune system.

Unfortunately not everyone inherits a healthy immune system. Without treatment, most children born with a severe immunodeficiency disorder die at a few months of age. In the 1970s, almost everyone in the U.S. was familiar with the case of David Vetter, better known as "The Boy in the Plastic Bubble." Born with a faulty immune system, David spent most of his life inside a series of sterile plastic enclosures. Because of the high risk of infection, he was never allowed to hug his family or walk barefoot in the grass (as was his dream), although at age six he was able to briefly explore the outside world in a special suit produced for him by NASA. David died at age 12 after a failed bone marrow transplant. His sad and controversial story has inspired many articles, songs, TV episodes, and at least two movies. It can also serve to remind us of the critical role of our immune system in protecting us against a vast array of viruses, bacteria, fungi, parasites, and toxins in our environment. In this chapter we explore that system, along with the lymphatic system that helps to produce and distribute the cells of the immune system.

As you read through the chapter, think about the following questions:

1. What are the most essential components of the immune system?

2. How do the lymphatic and immune systems work together?

3. We no longer hear much about keeping severely immunodeficient patients in plastic bubbles today, so how do you suppose they are treated?

FOLLOWING *the* BIG IDEAS

CHAPTER 33   THE LYMPHATIC AND IMMUNE SYSTEMS

Energy and Homeostasis	Rapid recognition and specialized destruction of foreign antigens are hallmarks of an advanced immune system.
Information and Signaling	Communication between immune system cells ensures appropriate defensive action.
Interactions and Systems	The extraordinary variety of antigen destructors makes the immune system so successful.

33.1 Evolution of Immune Systems

Learning Outcomes

Upon completion of this section, you should be able to

1. Summarize the evidence suggesting that cellular slime molds can form a rudimentary "immune system."
2. Define PAMPs and explain how they enable many animals to identify the presence of harmful microbes.
3. Compare the types of antigens recognized by the innate versus the adaptive immune system.

Our **immune system** protects us from all sorts of harmful invaders, including bacterial and viral pathogens, various toxins, and perhaps even cancerous cells that occasionally arise. It is a very intricate system made up of many components that work together, perhaps rivaling the nervous system in its complexity. But how did such a complicated system first evolve? And which components developed first? Scientists are beginning to answer those questions.

Immunity in Cellular Slime Molds

In August 2007, researchers at Baylor College of Medicine in Houston, Texas, discovered that very simple creatures called cellular slime molds can exhibit signs of a rudimentary immune system. These organisms are commonly known as "social amoebas" because of their unique life cycle. When food is plentiful, these protists live as separate amoeboid cells (Figure 33.1a), ingesting bacteria through phagocytosis and reproducing by binary fission (dividing into identical copies). As the food supply dwindles, many thousands of these amoebas can join together to form a slug two to four millimeters long (Figure 33.1b). The slug migrates toward light as a single, multicellular body, then differentiates into a reproductive structure that releases spores, many of which disperse to become new amoebas.

For several years biologists have known that individual cells within the slug can become specialized to perform various functions, but the Baylor group discovered a new type of cell within the slug that they named *sentinel cells*. These cells circulate throughout the slug, engulfing bacteria and toxins. Eventually the sentinel cells are sloughed from the body of the slug, thereby "sacrificing themselves" for the good of the organism. Some scientists believe that phagocytic white blood cells in the human body, such as neutrophils and macrophages, may have evolved from these types of cells.

Immunity in *Drosophila*

Although social amoebas may provide an example of how phagocytic cells first became specialized to protect multicellular organisms, the immune system is far more developed in invertebrates such as the fruit fly, *Drosophila melanogaster*. It was in this well-studied insect that scientists discovered the existence of a group of cellular receptors that could recognize common components found in many pathogenic microbes, but not in the insect's own cells. When these receptors bind to these

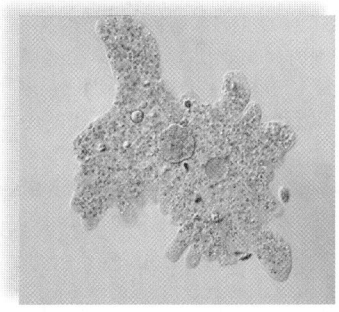

a.   b.

Figure 33.1 Social amoebas (*Dictyostelium discoideum*) **a.** The single-celled form, shown here at a high magnification. **b.** Under certain conditions thousands of amoebas can form a multicellular slug, in which some cells develop protective functions.

pathogen-associated molecular patterns, or *PAMPs*, they trigger an immune reaction, increasing the odds that the pathogen can be eliminated from the fly.

Examples of PAMPs found on pathogenic microbes include the double-stranded RNA that is produced during the replication cycle of many viruses, and certain arrangements of carbohydrates, lipids, or proteins found only on bacterial cell walls. Receptors for PAMPs have been found in organisms as diverse as fruit flies, plants, and humans, suggesting that they were one of the earliest, and most successful, types of cellular receptors that evolved for the recognition of pathogens.

The Rise of Adaptive Immunity

As you will see later in Section 33.3, these first two examples illustrate a type of host defense known as **innate immunity,** which can recognize common microbial invaders very quickly, but shows no signs of an increased response upon repeated exposure to the same invader. Most vertebrate animals also exhibit **adaptive immunity,** characterized by the production of a very large number of diverse receptors that are found on the surface of specialized white blood cells (such as B and T lymphocytes in humans). These receptors bind very specifically to molecules called **antigens**, much as a key fits a lock. This binding stimulates lymphocytes to divide and become much more numerous, resulting in characteristic features of adaptive immunity, such as greatly increased responses to specific antigens and immunological memory after the initial exposure to an antigen.

The generation of such a diverse array of antigen receptors depends on a rearrangement of the DNA that codes for these receptors, somewhat like choosing different combinations of cards from a deck. Scientists have now discovered that this process developed quite suddenly in an ancestor that gave rise to the jawed vertebrates—including the cartilaginous fishes (sharks and rays), bony fishes, amphibians, reptiles, birds, and mammals.

The precise mechanism by which this "explosion" of adaptive immunity occurred is still incompletely understood, although it now seems quite likely that it involved the insertion of a small piece of DNA (a transposon, or "jumping gene"; see Section 14.4) into a gene coding for a more primitive, less variable antigen receptor—perhaps similar to the receptors for

PAMPs mentioned earlier here. Note that in contrast to the relatively "fixed" receptors for PAMPs recognized by the innate immune system, the generation of antigen receptors by gene rearrangement allows the adaptive immune system to be able to respond to new antigens that evolve, for example, in emerging infectious agents. In other words, the vertebrate immune system has evolved an ability to respond to the continuing evolution of pathogenic microbes and other threats to our health.

Check Your Progress 33.1

1. Describe the function of sentinel cells found in cellular slime molds.
2. List three specific types of PAMPs found on microbes.
3. Describe three ways that the evolution of receptors for specific antigens increased the effectiveness of the immune system.

33.2 The Lymphatic System

Learning Outcomes

Upon completion of this section, you should be able to

1. Describe three major functions of the lymphatic system.
2. Distinguish between the roles of primary and secondary lymphoid tissues, and list examples of each.

The **lymphatic system,** which is closely associated with the cardiovascular and immune systems, includes the lymphatic vessels and the lymphoid organs. It has three main functions that contribute to homeostasis: 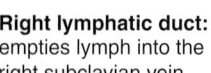 **MP3** Lymphatic System

- Lymphatic capillaries absorb excess tissue fluid and return it to the bloodstream.
- In the small intestine, lymphatic capillaries called lacteals absorb fats in the form of lipoproteins and transport them to the bloodstream.
- The lymphoid organs and lymphatic vessels are sites of production and distribution of lymphocytes, which help defend the body against pathogens.

Lymphatic Vessels

Lymphatic vessels form a one-way system that begins with lymphatic capillaries (Fig. 33.2). Most regions of the body are

Figure 33.2 Lymphatic system. Lymphatic vessels drain excess fluid from the tissues and return it to the cardiovascular system. The enlargement shows that lymphatic vessels, like cardiovascular veins, have valves to prevent backward flow. The lymph nodes, spleen, thymus, and red bone marrow are the main lymphoid organs that assist immunity.

Right lymphatic duct: empties lymph into the right subclavian vein

Right subclavian vein: transports blood away from the right arm and the right ventral chest wall toward the heart

Axillary lymph nodes: located in the underarm region

Thoracic duct: empties lymph into the left subclavian vein

Inguinal lymph nodes: located in the groin region

Tonsils: aggregates of lymphoid tissue that respond to pathogens in the pharynx

Left subclavian vein: transports blood away from the left arm and the left ventral chest wall toward the heart

Red bone marrow: site for the origin of all types of blood cells

Thymus: lymphoid organ where T cells mature

Spleen: resident T cells and B cells respond to the presence of antigen in blood

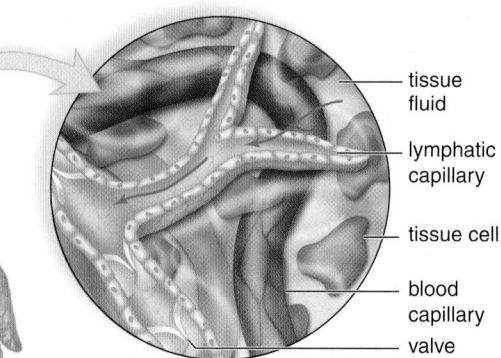

tissue fluid

lymphatic capillary

tissue cell

blood capillary

valve

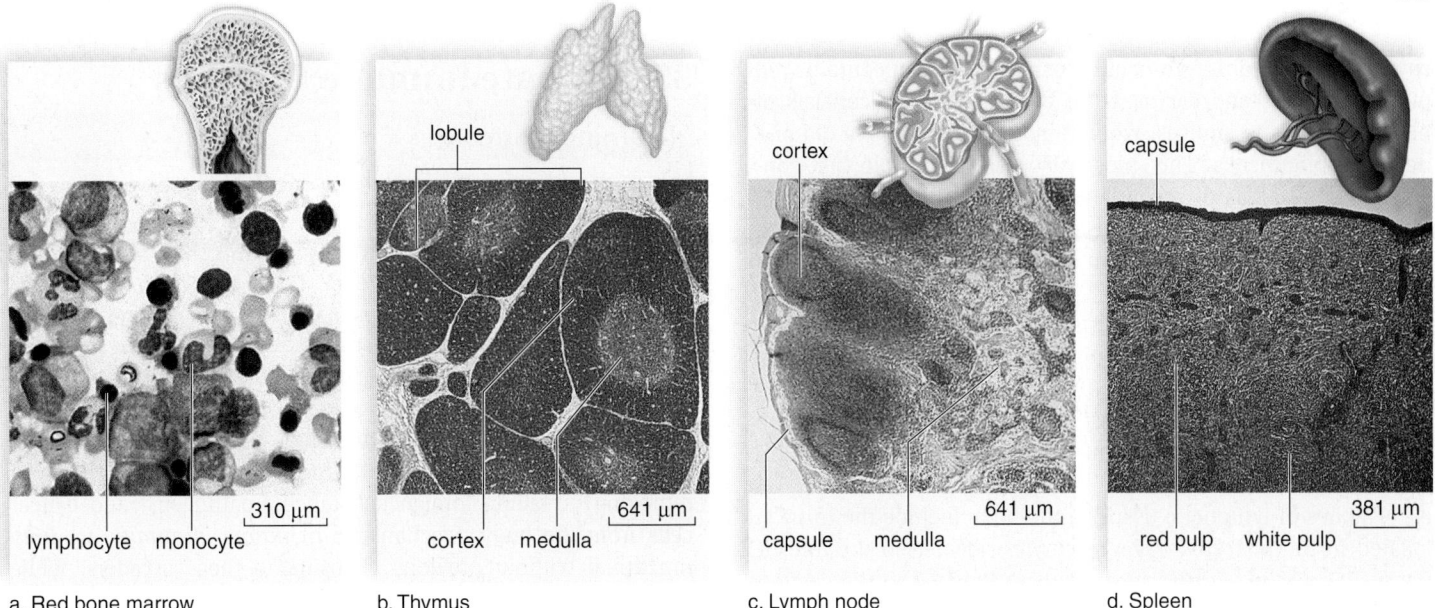

a. Red bone marrow

310 µm

lymphocyte monocyte

b. Thymus

lobule

cortex medulla

641 µm

c. Lymph node

cortex

capsule medulla

641 µm

d. Spleen

capsule

red pulp white pulp

381 µm

Figure 33.3 The lymphatic organs. **a.** Blood cells, including lymphocytes, are produced in red bone marrow. B cells mature in the bone marrow, but (**b**) T cells mature in the thymus. **c.** Lymph is cleansed in lymph nodes, while (**d**) blood is cleansed in the spleen.

richly supplied with **lymphatic capillaries**—tiny, closed-ended vessels. Lymphatic capillaries take up excess tissue fluid. The fluid inside lymphatic capillaries is called **lymph.**

The lymphatic capillaries join to form lymphatic vessels that merge before entering either the thoracic duct or the right lymphatic duct. The larger thoracic duct returns lymph to the left subclavian vein. The right lymphatic duct returns lymph to the right subclavian vein.

The construction of the larger lymphatic vessels is similar to that of cardiovascular veins. Skeletal muscle contraction forces lymph through lymphatic vessels, and it is prevented from flowing backward by one-way valves.

A number of diseases may result in an increased amount of fluid leaving the blood capillaries, or an insufficient return of fluid to the blood via the lymphatic vessels. In either case, a localized accumulation of tissue fluid called *edema* may result, illustrating the importance of this aspect of lymphatic system function.

Animation
Lymphatic System

Lymphoid Organs

The **lymphoid (lymphatic) organs** are reviewed in Figures 33.2 and 33.3. Lymphocytes develop and mature in **primary lymphoid organs** such as bone marrow and the thymus; **secondary lymphoid organs** are sites where some lymphocytes are activated by antigens.

A major primary lymphoid organ is the **red bone marrow**, a spongy, semisolid red tissue where hematopoietic stem cells divide and produce all the types of blood cells, including lymphocytes (Fig. 33.3*a*). In a child, most of the bones have red bone marrow, but in an adult, it is present only in the bones of the skull, the sternum (breastbone), the ribs, the clavicle (collarbone), the pelvic bones, the vertebral column, and the proximal heads of the femur and humerus.

There are two main types of lymphocytes: B lymphocytes (B cells) and T lymphocytes (T cells). Although both types begin their development in the red bone marrow, **B cells** remain there until they are mature. In contrast, immature **T cells** migrate from

the bone marrow via the bloodstream to the thymus, where they mature.

The soft, bilobed **thymus** is a primary lymphoid organ located in the thoracic cavity between the trachea and the sternum ventral to the heart (see Fig. 33.2). It is in the thymus that T cells learn to recognize the combinations of self-molecules and foreign molecules; this recognition characterizes mature T-cell responses (see Section 33.4). The thymus varies in size, but it is largest in children and shrinks as we get older. When well developed, it contains many lobules (Fig. 33.3*b*).

Once lymphocytes are mature, they enter the bloodstream. From there they frequently migrate into secondary lymphoid organs, such as the lymph nodes and spleen. Here lymphocytes may encounter foreign molecules or cells, and in response they proliferate and become activated. Activated lymphocytes then reenter the bloodstream, searching for signs of infection or inflammation, like a squadron of highly trained military personnel seeking to destroy a specific enemy.

Lymph nodes are small (about 1–25 mm in diameter) ovoid structures occurring along lymphatic vessels. They are a major type of secondary lymphoid organ. As lymph percolates through the cortex and medulla of a lymph node (Fig. 33.3*c*), resident phagocytic cells engulf any foreign debris and pathogens. These phagocytes can then "present" these foreign materials to T cells in the lymph node (see Section 33.4).

Sometimes incorrectly called "lymph glands," lymph nodes are named for their location. For example, inguinal lymph nodes are in the groin, and axillary lymph nodes are in the armpits. Physicians often feel for the presence of swollen, tender lymph nodes as evidence that the body is fighting an infection. Unfortunately, cancer cells sometimes enter lymphatic vessels and congregate in lymph nodes. Therefore, when a person undergoes surgery for cancer, it is a common procedure to remove some lymph nodes and examine them to determine whether the cancer has spread to other regions of the body.

The **spleen,** an oval secondary lymphoid organ with a dull purplish color, is located in the upper left side of the abdominal cavity posterior to the stomach. Most of the spleen contains red pulp that filters and cleanses the blood. Red pulp consists of blood vessels and sinuses, where macrophages remove old and defective blood cells. The spleen also has white pulp that consists of small areas of secondary lymphoid tissue (Fig. 33.3d). Much as the lymph nodes serve as sites for lymphocytes to respond to foreign material from the tissues, the spleen serves a similar role for the blood.

The spleen's outer capsule is relatively thin, and an infection or trauma can cause the spleen to burst, necessitating surgical removal. Although some of the spleen's functions can be largely replaced by other organs, an asplenic individual is more susceptible to certain types of infections and may require antibiotic therapy indefinitely.

Patches of lymphatic tissue in the body include the *tonsils*, located in the pharynx; *Peyer's patches*, located in the intestinal wall; and the *vermiform appendix*, attached to the cecum. These structures encounter pathogens and antigens that enter the body by way of the mouth.

Check Your Progress   33.2

1. Write a brief description of the lymphatic system.
2. Summarize the functions of the lymphatic system.
3. Describe the general appearance, location, and function of the lymphatic organs.

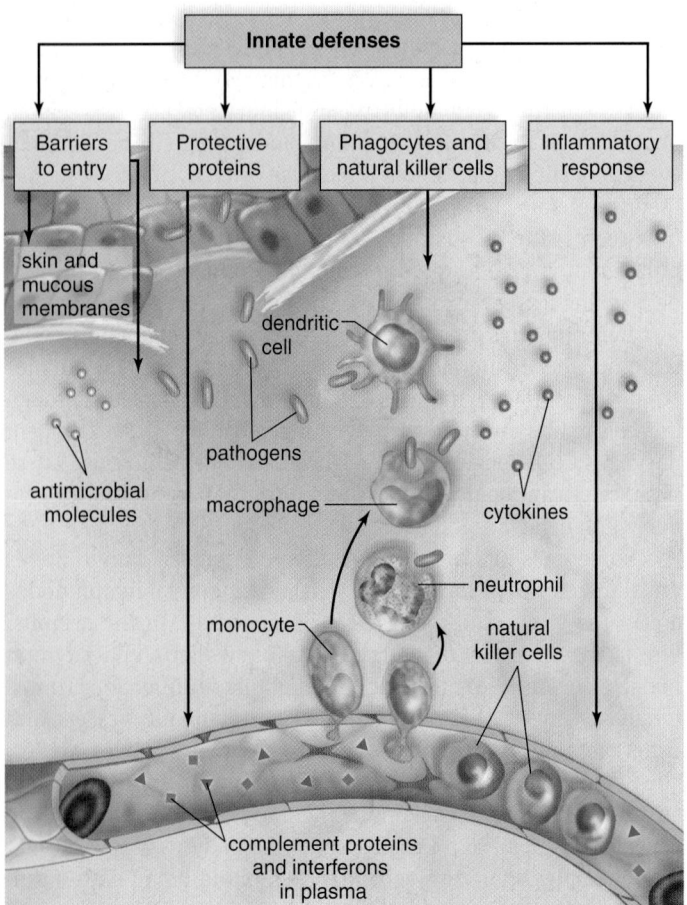

33.3 Innate Immune Defenses

Learning Outcomes

Upon completion of this section, you should be able to

1. Define innate immunity.
2. Describe four mechanisms of innate immunity, and the major tissues, molecules, and/or cells involved.
3. Explain some specific ways that the innate immune system interacts with and influences the adaptive immune system.

We are constantly exposed to microbes such as viruses, bacteria, and fungi in our environment. **Immunity** is the capability of removing or killing foreign substances, pathogens, and cancer cells from the body. Mechanisms of *innate immunity* are fully functional without previous exposure to these invaders, while *adaptive immunity* (see Section 33.4) is initiated and amplified by exposure.

As summarized in Figure 33.4, innate immune defenses include:

- physical and chemical barriers,
- the inflammatory response,
- phagocytes and natural killer cells, and
- protective proteins such as complement and interferons.

Innate defenses occur immediately or very shortly after an infection occurs. With innate immunity, there is no recognition that an intruder has attacked before, and therefore no immunological "memory" is present for the attacker.

Physical and Chemical Barriers

Physical barriers to various types of invaders include the skin as well as the mucous membranes lining the respiratory, digestive, and urinary tracts. As you saw in Chapter 31, the outer layers of our skin are composed of dead, keratinized cells that form a relatively impermeable barrier. But when the skin has been injured, one of the first concerns is the possibility of an infection.

The mucus produced by mucous membranes physically ensnares microbes. The upper respiratory tract is lined by ciliated cells that sweep mucus and trapped particles up into the throat, where they can be swallowed or expectorated (coughed out). In addition, various bacteria that normally reside in the intestine and in other areas, such as the vagina, take up nutrients and block binding sites that potentially could be exploited by pathogens.

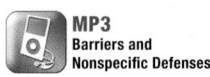

The secretions of oil glands found in human skin also contain chemicals that weaken or kill certain bacteria; saliva, tears, milk, and mucus contain lysozyme, an enzyme that can lyse bacteria; and the stomach has an acidic pH, which inhibits or kills many microbes.

Figure 33.4 Overview of innate immune defenses.   Most innate defenses act rapidly to detect and respond to various, highly conserved molecules expressed by pathogens.

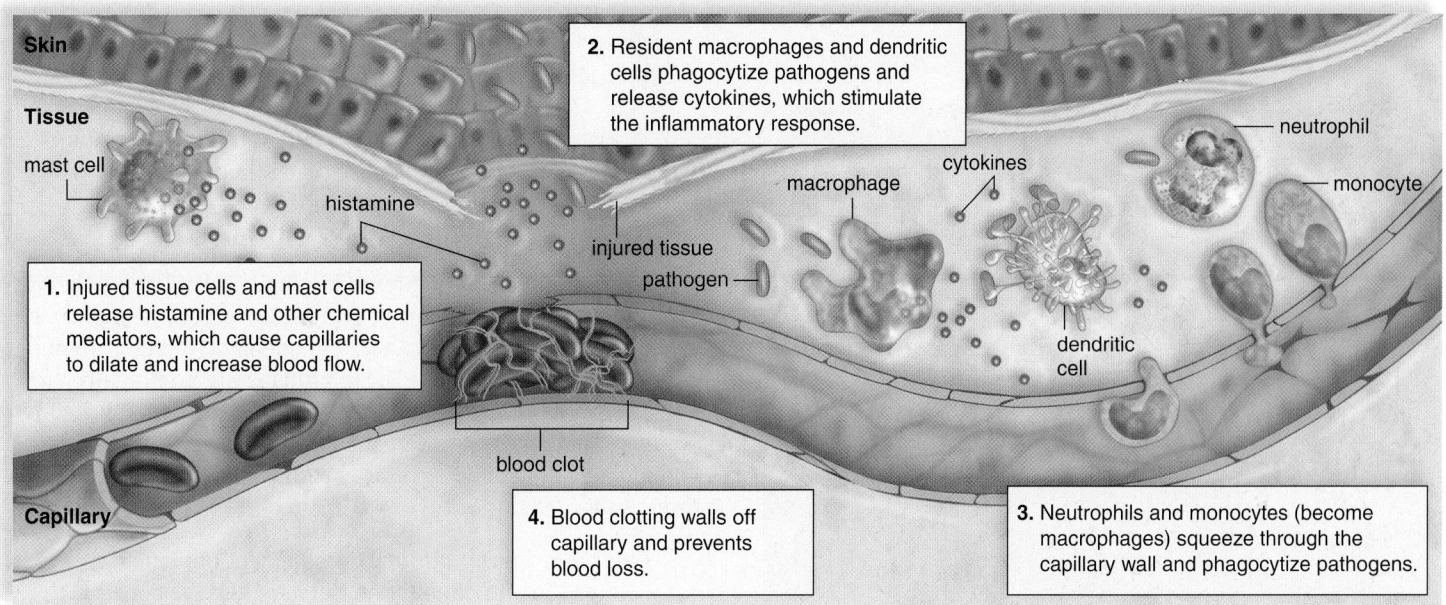

Figure 33.5 Inflammatory response. Due to capillary changes in a damaged area and the release of chemical mediators, such as histamine by mast cells, an inflamed area exhibits redness, heat, swelling, and pain. The inflammatory response can be accompanied by other reactions to the injury. Macrophages and dendritic cells, present in the tissues, phagocytize pathogens, as do neutrophils, which squeeze through capillary walls from the blood. Macrophages and dendritic cells release cytokines, which stimulate the inflammatory and other immune responses. A blood clot can form to seal a break in a blood vessel.

Inflammatory Response

When tissues are damaged by a variety of causes, including pathogens, a series of events known as the **inflammatory response** occurs. An inflamed area has at least four common signs: redness, heat, swelling, and pain. Most of these signs are due to capillary changes in the damaged area, as illustrated in Figure 33.5. Chemical mediators released by damaged cells, including **histamine** that is mainly secreted by tissue-dwelling cells of the innate immune system called **mast cells,** cause capillaries to dilate and become more permeable. Increased blood flow to the area causes the skin to redden and become warm. Increased permeability of the capillaries allows proteins and fluids to escape into the tissues, resulting in swelling. Various chemicals released by damaged cells stimulate free nerve endings, causing the sensation of pain.

Inflammation also causes various types of white blood cells to migrate from the bloodstream into damaged tissues. Once in the tissues, monocytes can differentiate into dendritic cells and macrophages, both of which are able to devour many pathogens and still survive (Fig. 33.6). Macrophages also release colony-stimulating factors, namely cytokines, which pass by way of the blood to the red bone marrow where they stimulate the production and release of white blood cells.

Animation Inflammatory Response

Sometimes an inflammation persists, and the result is chronic inflammation that can itself become damaging to tissues. Examples include the chronic responses to the bacterium that causes tuberculosis, or to asbestos fibers that, once inhaled into the lungs, cannot be removed. Some cases of chronic inflammation are treated by administering anti-inflammatory drugs such as aspirin, ibuprofen, or cortisone. These medications inhibit the responses to inflammatory chemicals being released in the damaged area.

From the site of their production in damaged tissues, various inflammatory mediators are absorbed into the bloodstream, where they can affect several other organs. Although it is not normally thought of as a part of the immune system, the liver responds to these chemicals by increasing production of various *acute phase proteins,* some of which can coat microbial invaders, making them easier for phagocytes to engulf. One type of acute phase protein, called C-reactive protein, is frequently measured to assess levels of inflammation in patients suffering from certain diseases.

Inflammatory chemicals in the blood may also act on the brain to initiate an elevated body temperature or *fever.* Although the exact function of the body's fever response is unknown, many speculate that certain bacteria or viruses may not survive as well at higher temperatures, or that certain immune mechanisms work better at higher body temperatures. Experimental data have been collected that support both hypotheses, and in fact, both may be true. In either case, because mild to moderate fever appears to help the body fight off invaders more effectively, the wisdom of using drugs to treat mild fevers can be questioned. However, body temperatures above 103°F can be fatal, especially in children, so obviously this situation must be treated as an emergency.

Inflammatory responses are accompanied by other responses to the injury. The clotting system can be activated to seal a break in a blood vessel. Antigens, chemical mediators, dendritic cells, and macrophages move from the damaged tissue via the lymph to the lymph nodes. There, these phagocytes interact with T cells and B cells to activate a specific, adaptive response to the infection (see Section 33.4). Finally, inflammation also initiates the healing response, in which macrophages play an essential role.

Phagocytes and Natural Killer Cells

Several types of white blood cells are phagocytic, meaning that they can engulf and digest relatively large particles such as viruses and bacteria. **Neutrophils** are able to leave the bloodstream and phagocytize bacteria in tissues. They have multiple

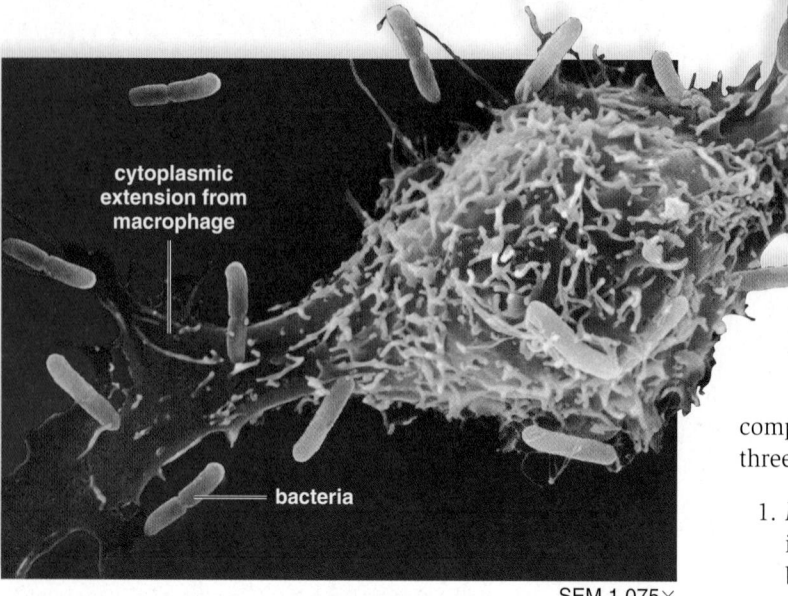

cytoplasmic
extension from
macrophage

bacteria

SEM 1,075×

Figure 33.6 Macrophage engulfing bacteria. Monocyte-derived macrophages are the body's scavengers. They engulf microbes and debris in the body's fluids and tissues, as illustrated in this colorized scanning electron micrograph.

ways of killing bacteria. The cytoplasm of a neutrophil is packed with granules that contain antimicrobial peptides, as well as enzymes that can digest bacteria. Other enzymes inside neutrophil granules generate highly reactive free radicals such as superoxide and hydrogen peroxide, all of which participate in killing engulfed bacteria.

As the infection is being overcome, some neutrophils die. These—along with dead tissue cells, dead bacteria, and living white blood cells—may form pus, a whitish material. The presence of pus usually indicates that the body is trying to overcome a bacterial infection.

Video Neutrophils

Eosinophils can be phagocytic, but they are better known for mounting an attack against animal parasites such as tapeworms that are too large to be phagocytized.

As mentioned, the two longer-lived types of phagocytic white blood cells are **macrophages** (see Fig. 33.6) and **dendritic cells.** Macrophages are found in all sorts of tissues, whereas dendritic cells are especially prevalent in the skin. Both cell types engulf pathogens, which are then digested and broken down into smaller molecular components. They then travel to lymph nodes, where they stimulate T cells, which are responsible for initiating adaptive immune responses.

Natural killer (NK) cells are large, granular lymphocytes that kill virus-infected cells and cancer cells by cell-to-cell contact. NK cells do their work while adaptive defenses are still mobilizing, and they produce cytokines that promote adaptive immunity.

What makes NK cells attack and kill a cell? NK cells seek out cells that lack a particular type of "self" molecule, called MHC-I (major histocompatibility complex I), on their surface. Because some virus-infected and cancer cells may lack these MHC-I molecules, they may be recognized by NK cells, which kill these cells by inducing them to undergo cellular suicide (apoptosis). Because NK cells do not recognize specific viral

or tumor antigens, and do not proliferate when exposed to a particular antigen, their numbers do not increase after stimulation.

Protective Proteins

Complement is composed of a number of blood plasma proteins, produced mainly by the liver, that "complement" certain immune responses. These proteins are continually present in the blood plasma but must be activated by pathogens to exert their effects. The complement system helps destroy pathogens in three ways:

Animation Activation of Complement

1. *Enhanced inflammation.* Complement proteins are involved in and amplify the inflammatory response because certain ones can bind to mast cells and trigger histamine release, and others can attract phagocytes to the scene.
2. *Increased phagocytosis.* Some complement proteins bind to the surface of pathogens, increasing the odds that the pathogens will be phagocytized by a neutrophil or macrophage.
3. *Membrane attack complexes.* Certain other complement proteins join to form a membrane attack complex that produces holes in the surface of some bacteria and viruses. Fluids and salts then enter the bacterial cell or virus to the point that it bursts (Fig. 33.7).

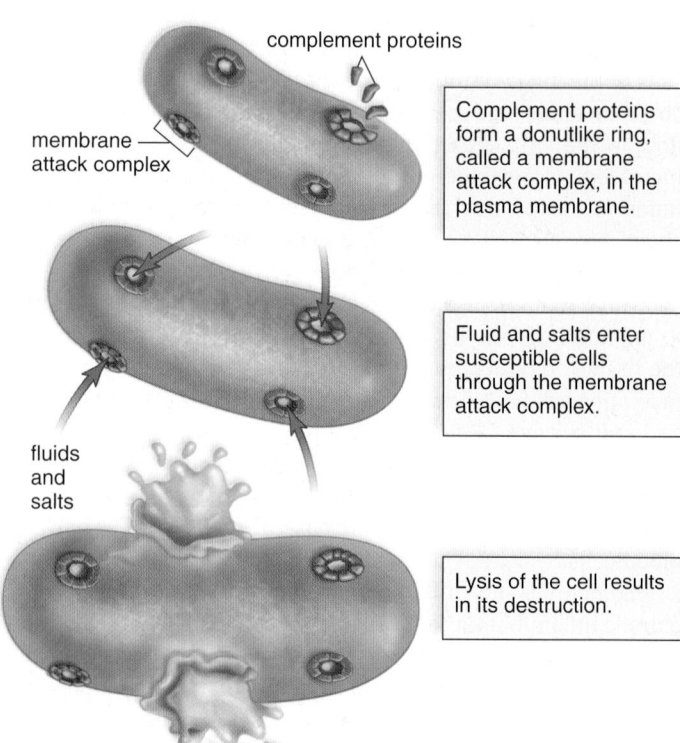

complement proteins

membrane attack complex

fluids and salts

Complement proteins form a donutlike ring, called a membrane attack complex, in the plasma membrane.

Fluid and salts enter susceptible cells through the membrane attack complex.

Lysis of the cell results in its destruction.

Figure 33.7 Action of the complement system against a bacterium. When complement proteins in the blood plasma are activated by an immune response, they form a membrane attack complex that makes holes in bacterial cell walls and plasma membranes, allowing fluids and salts to enter until the cell eventually bursts.

Interferons come in several different types, but all are **cytokines,** soluble proteins that affect the behavior of other cells. Most interferons are made by virus-infected cells. They bind to the receptors of noninfected cells, causing them to produce substances that slow cellular metabolism and interfere with viral replication. Interferons are used to treat certain cancers and viral infections such as hepatitis C.

 **Animation**
Antiviral Activity
of Interferon

Check Your Progress   33.3

1. List three physical and three chemical barriers.
2. Describe the four cardinal signs associated with an inflammatory response. How is this response beneficial?
3. Name five cell types involved in innate immunity, and the major functions of each.
4. Summarize three specific functions of the complement system.

33.4 Adaptive Immune Defenses

Learning Outcomes

Upon completion of this section, you should be able to

1. Compare and contrast the activities of B cells and T cells.
2. Describe the basic structure of an antibody molecule, explain the different functions of IgG, IgA, IgM, and IgE.
3. Define monoclonal antibodies, and list some specific applications of this technology.
4. Discuss active and passive immune responses, giving specific examples of each.

Even while innate defenses are trying to fight an infection, adaptive defenses also begin to respond. Because these defenses do not normally react to our own cells or molecules, it is said that the adaptive immune system can distinguish "self" from "nonself." Adaptive defenses usually take from five to seven days to become activated, but they may last for years. This explains why, once we recover from some infectious diseases, we usually do not get the same disease a second time. Because we are not born with these defenses, some prefer the term *acquired immunity* to describe this type of immunity.

Adaptive defenses depend primarily on the activities of B cells and T cells (Fig. 33.8). Both B cells and T cells are manufactured in the red bone marrow. As mentioned earlier, B cells mature there, but T cells mature in the thymus. Both cell types are capable of binding to and thus "recognizing" specific antigens because they have **antigen receptors** on their plasma membrane. Pathogens, cancer cells, and transplanted tissues and organs bear antigens the immune system usually recognizes as nonself.

During our lifetime, we need a diversity of B cells and T cells to recognize these antigens and protect us against them. Remarkably, diversification occurs during the lymphocyte

maturation process to so great an extent that there are specific B cells and/or T cells for almost every possible antigen. Because B and T cells defend us from disease by specifically reacting to antigens, they can be likened to special forces that can attack selected targets without harming nearby residents (uninfected cells).

B Cells and Antibody-Mediated Immunity

The receptor for antigen on the surface of a B cell is called a **B-cell receptor (BCR)**. B cells are usually activated in a lymph node or the spleen, after their BCRs bind to a specific antigen. Subsequently, the B cell divides by mitosis many times, making many copies (clones) of itself. The **clonal selection theory** states that the antigen receptor of each B cell or T cell binds to only a single type of antigen.

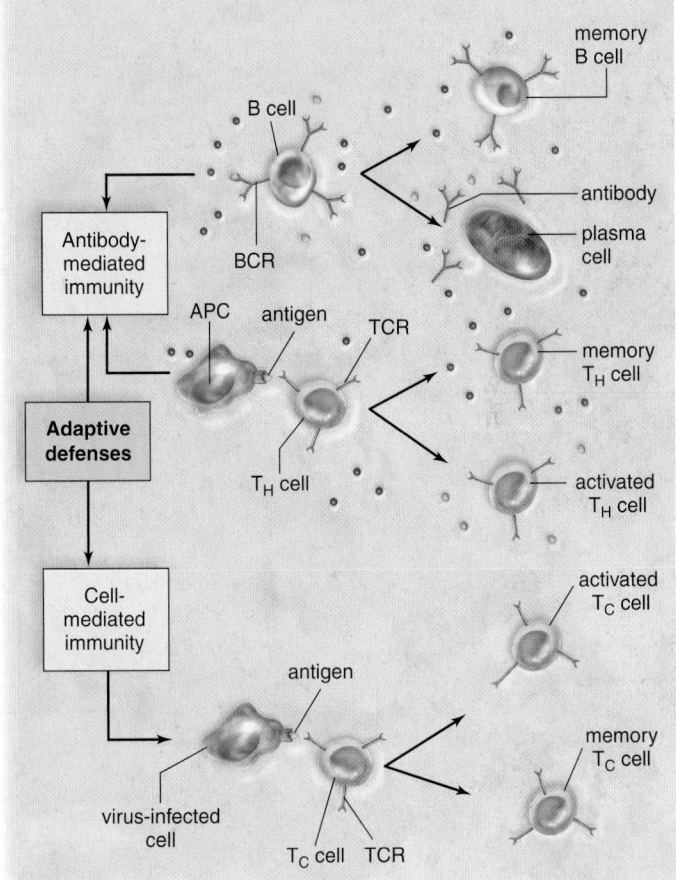

Figure 33.8 Overview of adaptive immune defenses. B cells, helper T (T$_H$) cells, and cytotoxic T (T$_C$) cells respond to specific antigens by dividing and differentiating. The BCRs of B cells bind to whole, intact antigens, while the TCRs of T$_H$ and T$_C$ cells only bind to antigens that are processed and presented by MHC proteins on the surface of other cells. When activated by antigens, B cells differentiate into antibody-secreting plasma cells, T$_H$ cells become cytokine-secreting cells, and T$_C$ cells are able to destroy virus-infected or cancer cells. Each cell type also produces memory cells that can respond more quickly to a subsequent exposure to the same antigen.

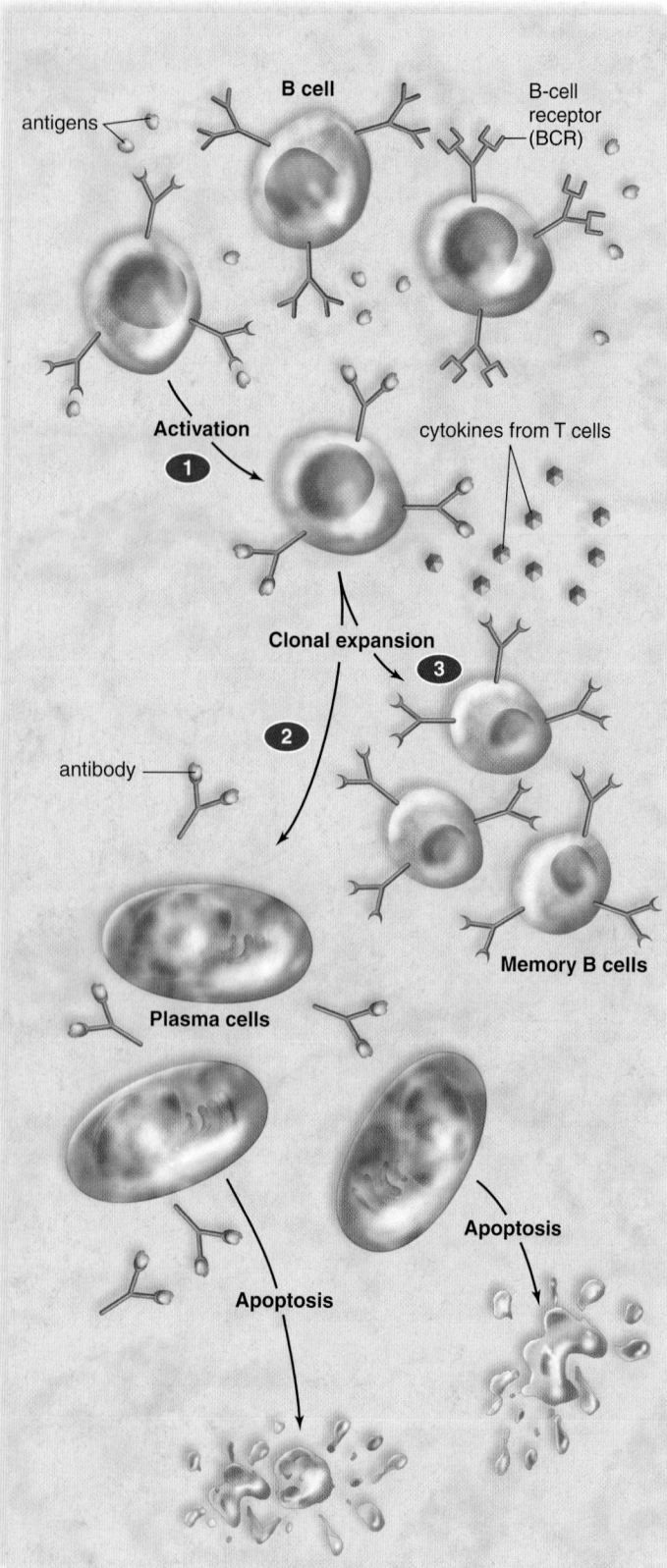

Figure 33.9 Clonal selection theory as it applies to B cells.
Each B cell has a B-cell receptor (BCR) designated by shape that will combine with a specific antigen. Activation of a B cell occurs when its BCR can combine with an antigen (colored green). In the presence of cytokines, the B cell undergoes clonal expansion, producing many plasma cells and memory B cells. These plasma cells secrete antibodies specific to the antigen, and memory B cells immediately recognize the antigen in the future. After the infection passes, plasma cells undergo apoptosis, also called programmed cell death.

As illustrated in Fig. 33.9, ❶ many B cells are present, but only those that have BCRs that can combine with the specific antigen(s) present go on to divide and produce many new cells. Therefore, the antigen is said to "select" the B cells that will begin dividing. At the same time, cytokines secreted by helper T cells stimulate B cells to differentiate. ❷ Many of these B cells become **plasma cells**, which are specialized for secretion of antibodies. Plasma cells are larger than regular B cells because they have extensive rough endoplasmic reticulum for the mass production and secretion of antibodies that bind to a specific antigen. Antibodies are the secreted form of the BCR of an activated B cell, and these antibodies react to the same antigen as the original B cell. Once the threat of an infection has passed, the development of new plasma cells ceases, and those present undergo apoptosis. ❸ Other progeny of the dividing B cells become **memory B cells**, so named because these cells always "remember" a particular antigen and make us immune to a particular illness, but not to any other illness.

Defense of the body by B cells is known as **antibody-mediated immunity** (see Fig. 33.8, *top*). It is also called humoral immunity because these antibodies are present in blood and lymph. (Historically, the term *humor* referred to any fluid normally occurring in the body.)

Structure of Antibodies

The basic unit of antibody structure is a Y-shaped protein molecule with two arms. Each arm has a "heavy" (long) polypeptide chain and a "light" (short) polypeptide chain (Fig. 33.10). These chains have constant (C) regions, located at the trunk of the Y, where the sequence of amino acids is set. The variable (V) regions at the tips of the Y form two antigen-binding sites, and their shape is specific to a particular antigen. The antigen combines with the antibody at the antigen-binding site in a lock-and-key manner.

The binding of antibodies to an antigen can have several outcomes. Often, the reaction produces a clump of antigens combined with antibodies, termed an *immune complex*. The antibodies in an immune complex are like a beacon that attracts white blood cells that move in for the kill. For example, immune complexes may be engulfed by neutrophils or macrophages, or they may activate NK cells to destroy a cell coated with antibodies. Immune complexes may also activate the complement system. Antibodies may also "neutralize" viruses or toxins by preventing them from binding to specific receptors on cells.

There are several types, or classes, of antibodies, also called **immunoglobulins (Ig)**. The class of antibody is determined by the structure of the antibody's constant region. The major class of antibody found in the blood, called *IgG*, is a single Y-shaped molecule (see Fig. 33.10). IgG antibodies are the major type that can cross the placenta from a mother to her fetus, to provide some temporary protection to the newborn. IgG is also found in breast milk, along wth a type called IgA. *IgA* is also the main class secreted in milk, tears, saliva, and at mucous membranes. Antibodies of the *IgM* class are pentamers—that is, clusters of

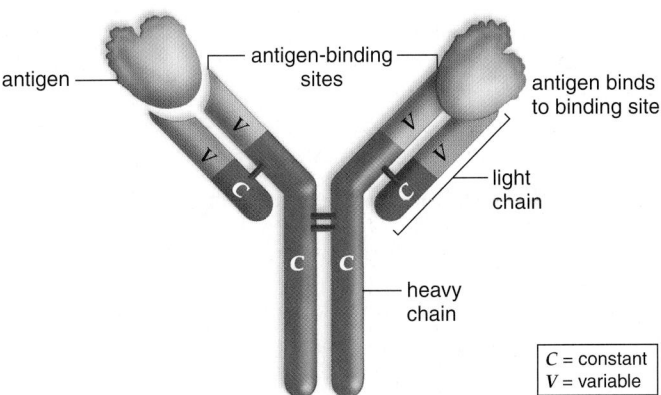

Figure 33.10 Structure of antibodies. An antibody contains two heavy (long) polypeptide chains and two light (short) chains arranged so there are two variable regions, where a particular antigen is capable of binding with the antibody. The shape of the antigen fits the shape of the binding site.

five Y-shaped molecules linked together. IgM antibodies are the first antibodies produced during most B-cell responses. As such, their presence is often interpreted as indicating a recent infection. The other major type of antibody is *IgE*, which is mainly found bound to receptors on eosinophils and on mast cells in the tissues.

Monoclonal Antibodies

As has been noted, every plasma cell derived from a single B cell secretes antibodies that bind to a single antigen. This discovery, along with the development of techniques for growing cells in the laboratory, led to the production of **monoclonal antibodies**. Niels Jerne, George Köhler, and César Milstein were awarded the Nobel Prize in 1984 for their development of this technology.

Monoclonal antibodies are produced by cells derived from a single plasma cell, thus all these antibodies have identical specificity for one antigen. Monoclonal antibodies are typically produced by first immunizing an animal (usually a mouse) with the antigen of interest. The animal is then killed, its spleen removed, and the spleen cells (including a large number of B cells) are fused with mouse myeloma cells (malignant plasma cells that live and divide indefinitely). The fused cells are called *hybridomas*: *hybrid-* because they result from the fusion of two different cells, and *-oma* because one of the cells is a cancer cell. The hybridomas are then isolated as individual cells and screened to select only those that are producing the desired monoclonal antibody.

Research Uses for Monoclonal Antibodies. The ability to quickly produce monoclonal antibodies in the laboratory has made them an important tool for academic research. Monoclonal antibodies are very useful because of their extreme specificity for only a particular molecule. A monoclonal antibody can be used to select out a specific molecule among many others, much like

finding needles in a haystack. Now the target molecule can be purified from all the others that are also present in a sample. In this way, monoclonal antibodies have simplified formerly tedious laboratory tasks.

Animation
Monoclonal Antibody
Production

Medical Uses for Monoclonal Antibodies. Monoclonal antibodies also have many applications in medicine. In one application, they can be used to make quick and certain diagnoses of infections and other conditions. For example, a particular hormone called hCG is present in the urine of a woman only if she is pregnant. An anti-hCG monoclonal antibody can be used to detect this hormone. Thanks to this technology, pregnancy tests that once required a visit to a doctor's office and the use of expensive laboratory equipment can now be performed at home, and at minimal expense.

Monoclonal antibodies can be used not only to diagnose infections and illnesses but also to fight them. RSV, a common virus that causes serious respiratory tract infections in very young children, is now being successfully treated with a monoclonal antibody drug. The antibody recognizes a protein on the viral surface, and when it binds very tightly to the surface of the virus, the patient's own immune system can easily recognize the virus and destroy it before it has a chance to cause serious illness.

Because monoclonal antibodies can distinguish some cancer cells from normal tissue cells, they may also be used to identify cancers at very early stages when treatment can be most effective. Herceptin (trastuzumab) is a monoclonal antibody used to treat breast cancer. Given intravenously, it binds to a protein receptor found on some breast cancer cells and prevents them from dividing so quickly.

Since the first therapeutic monoclonal antibody was approved by the FDA in 1986, over 20 are now available, and hundreds more are currently being tested.

T Cells and Cell-Mediated Immunity

After a T cell completes its development in the thymus, it has a unique **T-cell receptor (TCR)** similar to the BCR on B cells. Unlike B cells however, T cells are unable to recognize an antigen without help. The antigen must be displayed, or "presented," to the TCR by an **MHC (major histocompatibility complex) protein** on the surface of another cell.

There are two major types of T cells: **helper T cells,** (T_H cells) which regulate adaptive immunity, and **cytotoxic T cells** (T_C cells, or CTLs), which attack and kill virus-infected cells and cancer cells. Each type of T cell has a TCR that can recognize an antigen fragment in combination with an MHC molecule. A major difference, however, is that the T_H cells only recognize and respond to antigens presented by specialized **antigen-presenting cells (APCs)** with MHC class II proteins on their surface, while T_C cells only recognize and respond to antigens presented by various cell types with MHC class I proteins on their surface.

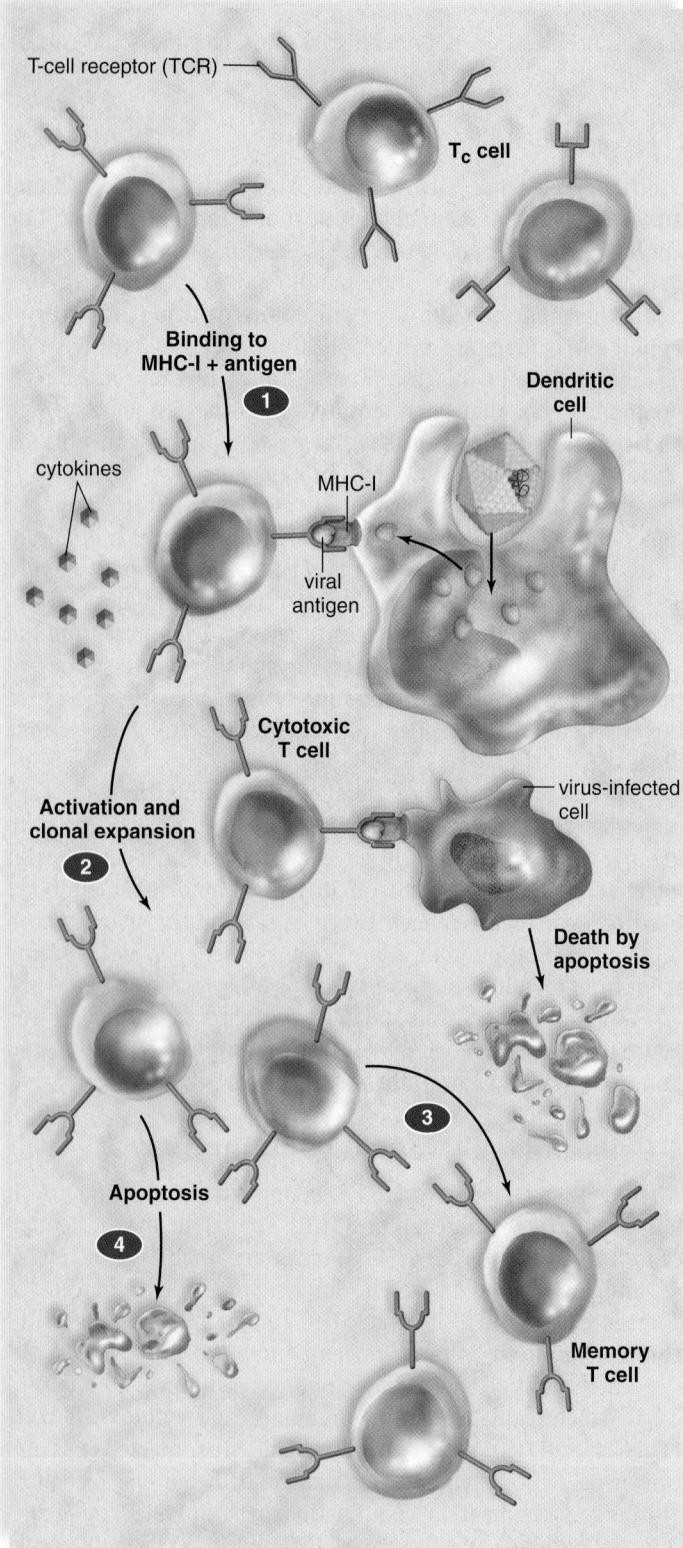

Figure 33.11 Clonal selection theory as it applies to T_C cells. Each T cell has a T-cell receptor (TCR) designated by a shape that will combine only with a specific antigen. Activation of a T cell occurs when its TCR can combine with an antigen. A dendritic cell presents the antigen (colored green) in the groove of an MHC class I (MHC-I) molecule. Thereafter, the T_C cell undergoes clonal expansion, and many copies of the same type of T cell are produced. After the immune response has been successful, the majority of T cells undergo apoptosis, but a small number are memory T cells. Memory T cells provide protection should the same antigen enter the body again at a future time.

Figure 33.11 illustrates the important role that APCs such as macrophages and dendritic cells play in stimulating T$_H$ cells. After phagocytizing a pathogen, APCs travel to a lymph node or the spleen, where T$_H$ cells, T$_C$ cells, and B cells also congregate. In the meantime, the APC has broken the pathogen apart in a lysosome. A fragment of the pathogen is then displayed in association with an MHC class II protein on the cell's surface ❶ which can bind to and select any T$_H$ cell that has a TCR capable of combining with a particular antigen/MHC combination. This interaction stimulates the T$_H$ cell to divide, thereby cloning itself. ❷ Some of these proliferating T$_H$ cells secrete various cytokines that influence B cells, T$_C$ cells, and other cell types. ❸ Many other cloned T$_H$ cells become **memory T cells.** Like memory B cells, memory T cells are long-lived, and their number is far greater than the original number of T cells that could recognize a specific antigen. Therefore, when the same antigen enters the body later on, the immune response may occur so rapidly that no detectable illness occurs. As the illness disappears, the immune response wanes, and activated T cells become susceptible to apoptosis, ❹ just as plasma cells do during a B-cell response.

By releasing a variety of cytokines, activated helper T cells are largely responsible for **cell-mediated immunity** (see Fig. 33.8, *middle*), which refers to the destruction or elimination of pathogens and other threats by T$_C$ cells, macrophages, natural killer cells, or other cells. For example, certain cytokines cause T$_C$ cells to proliferate (see Fig. 33.8, *bottom*), while others activate macrophages to more actively seek, engulf, and destroy pathogens. Other cytokines released by T$_H$ cells influence B-cell activities, even though B cells are responsible for antibody-mediated immunity. Cytokines are also used for immunotherapy purposes, as discussed in the next section.

Functions of Cytotoxic T Cells

A major difference in recognition of an antigen by helper T cells and cytotoxic T cells is that helper T cells recognize an antigen only in combination with MHC class II proteins, while T$_C$ cells recognize an antigen only in combination with MHC class I proteins, which are found on almost all types of cells. The outcome of this recognition is also very different—activated T$_H$ cells secrete a variety of cytokines that "help" other cells, but activated T$_C$ cells specialize in killing other cells.

The cytoplasm of a T$_C$ cell contains storage vacuoles that are filled with a chemical called *perforin*, as well as enzymes called *granzymes* (Fig. 33.12). After a T$_C$ cell is activated, it travels via the bloodstream to areas of inflammation in the body. Upon migrating into the tissues, if the T$_C$ cell recognizes its unique target combination of antigen plus MHC-I protein on a virus-infected cell or a cancer cell, it releases perforin molecules, which perforate the plasma membrane, forming a pore. Granzymes then use the pore to enter the abnormal cell, causing it to undergo apoptosis and die. Once T$_C$ cells have released the perforins and granzymes, they move on to the next target cell. Because of this ability, some have referred to T$_C$ cells as the "serial killers" of the immune system. As with B cells and T$_H$ cells, most activated T$_C$ cells are short-lived, while others become long-lived memory T$_C$ cells, ever ready to defend against the same virus or kill the same type of cancer cell again.

Video
T Lymphocyte

Animation
Cytotoxic T-Cell Activity
Against Target Cells

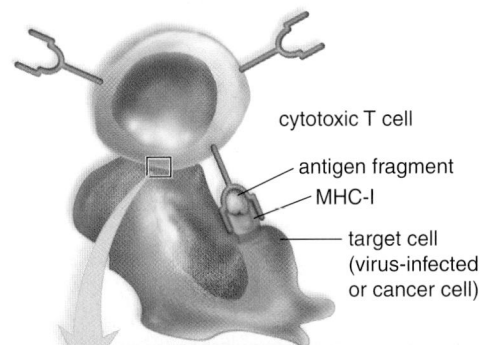

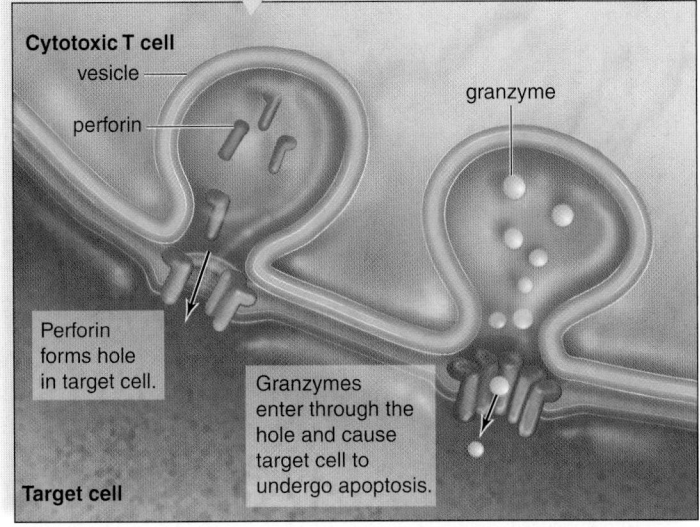

a.

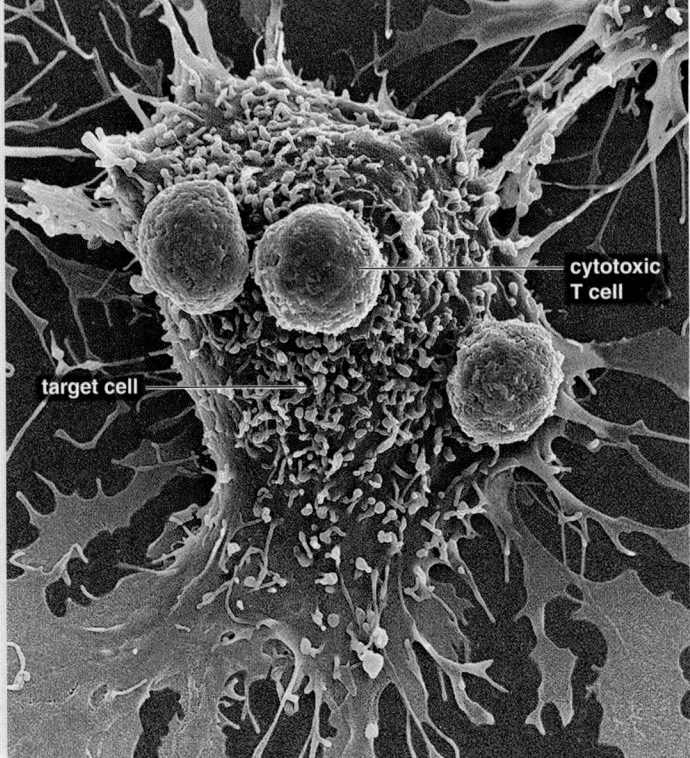

b.   SEM 1,250×

Figure 33.12 Cell-mediated immunity. a. How a T cell destroys a virus-infected cell or cancer cell. **b.** The scanning electron micrograph shows cytotoxic T cells attacking and destroying a cancer cell (target cell).

HIV Infections

By the end of 2009, an estimated 33.3 million people worldwide were infected with **HIV (human immunodeficiency virus),** and more than 25 million people have died of AIDS since 1981. Helper T cells are the major host cell for HIV, but macrophages and dendritic cells can also be infected.

After HIV enters a host cell, it reproduces as was illustrated in Figure 20.4, inserting itself into the host cell DNA, where it will remain for the life of the cell. Perhaps years later, this hidden viral DNA will begin producing more HIV particles, which go on to infect and destroy more and more helper T cells. (Figure 33A in the Biological Systems feature, on page 638, details some key features of the progression of an HIV infection over time.)

At first the immune system is able to control the virus, probably through a combination of antibodies and cell-mediated immunity. But gradually the virus mutates so that it is no longer recognized by the immune system, and as the HIV count rises, the helper T-cell count drops to way below normal. As noted in the Biological Systems feature, the HIV-infected person begins to develop opportunistic infections—infections that would be unable to take hold in a person with a healthy immune system. Now the individual has AIDS (acquired immunodeficiency syndrome).

Animation
How the HIV Infection Cycle Works

Fortunately, antiretroviral drugs have been developed that have made HIV a manageable infection in most treated individuals, although the drugs have many side effects. Unfortunately, there is still a shortage of these expensive drugs in many developing countries.

Animation
Treatment of HIV

Cytokines as Therapeutic Agents

Cytokines are produced by T cells, macrophages, and many other cells. Because they affect white blood cell formation and/or function, some cytokines are approved for use as therapy for cancer and certain other conditions. As mentioned, interferon-alpha is used to treat hepatitis C and certain cancers, and it may also slow the progression of multiple sclerosis. The T-cell activating cytokine interleukin-2 is used to treat some forms of melanoma and kidney cancer.

In some cases it may be desirable to inhibit the activity of certain cytokines, especially those involved in promoting chronic inflammatory diseases. For example, tumor necrosis factor (TNF) is a major cytokine produced by macrophages that has the ability to promote the inflammatory response. Several treatments that inhibit the TNF response, including anti-TNF monoclonal antibodies, are currently being tested as potential treatments for inflammatory diseases such as rheumatoid arthritis, Crohn's disease, and asthma. As would be expected however, these medications may also inhibit some of the beneficial effects of inflammation, and patients taking them may be more prone to infections.

Active Versus Passive Immunity

In general, adaptive immune responses can be induced actively or passively. **Active immunity** occurs when an individual produces their own immune response against an antigen. For example, when you catch a cold, you recover because your body

Biological Systems

AIDS and Opportunistic Infections

AIDS (acquired immunodeficiency syndrome) is caused by HIV, the human immunodeficiency virus. An HIV infection leads to the eventual destruction of helper T (T_H) cells. HIV kills T_H cells by directly infecting them, and it also causes many uninfected T_H cells to die by a variety of mechanisms.

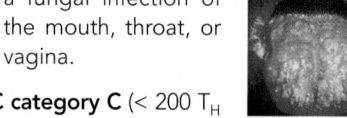

AIDS victim: Kaposi sarcoma is evident

Many HIV-infected T_H cells are also killed by the person's own immune system.

A healthy individual typically has 800–1,000 T_H cells per mm³ of blood (Fig. 33A). After an initial HIV infection, it may take several years for an individual's T_H-cell numbers to drop below 500 cells per mm³ of blood, at which point the HIV-infected individual usually begins to suffer from many unusual types of infections that would not cause disease in a person with a healthy immune system. Such infections are known as "opportunistic infections" (OIs). The U.S. Centers for Disease Control and Prevention (CDC) have defined three categories of HIV infection, based mainly on the types of OIs seen in the patient. These OIs also tend to be associated with a decreasing number of T_H cells in the blood, as indicated below:

CDC category A (>500 T_H cells/mm³ blood)

- typically no OIs seen; may have persistently enlarged lymph nodes

CDC category B (200-499 T_H cells/mm³ blood)

- Shingles, a painful infection with varicella zoster (chickenpox) virus.
- Candidiasis, or thrush, a fungal infection of the mouth, throat, or vagina.

Shingles

CDC category C (< 200 T_H cells/mm³ blood)

- Pneumocystis pneumonia, a fungal infection causing the lungs to become useless as they fill with fluid and debris.

Candidiasis

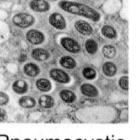

- Kaposi's sarcoma, a cancer of blood vessels due to human herpesvirus 8 gives rise to reddish purple, coin-sized spots and lesions on the skin.

Pneumocystis pneumonia

- Toxoplasmic encephalitis, a protozoan infection characterized by severe headaches, fever, seizures, and coma.

- *Mycobacterium avium* complex (MAC), a bacterial infection resulting in persistent fever, night sweats, fatigue, weight loss, and anemia.
- Cytomegalovirus, a viral infection that leads to blindness, inflammation of the brain, and throat ulcerations.

Thanks to development of powerful drug therapies that inhibit the life cycle of HIV, people infected with HIV in the United States are suffering lower incidence of OIs than in the 1980s and 1990s.

Questions to Consider

1. Why has it been so difficult to develop an effective vaccine for HIV?
2. What are some possible differences between the types of OIs typically seen in category B and those seen in category C (i.e., why do shingles and candidiasis occur in B, but others more commonly in C)?
3. What, if any, obligation do relatively wealthy countries like the United States have in providing anti-HIV drugs to poorer countries?

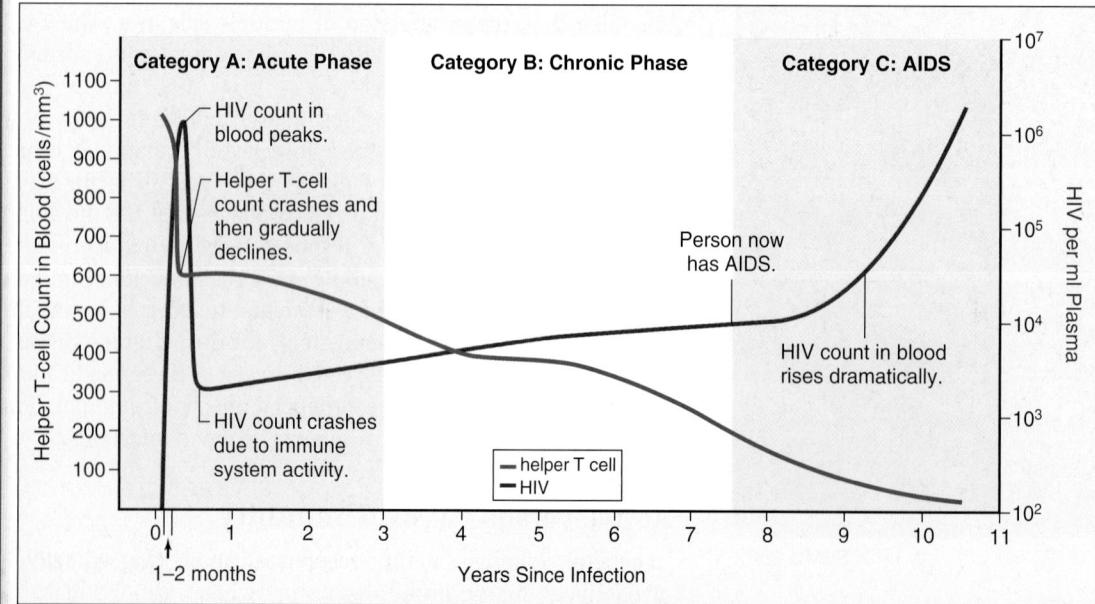

Figure 33A Progression of HIV infection during its three stages, called categories A, B, and C. In category A, the individual may have no symptoms or very mild symptoms associated with the infection. By category B, opportunistic infections have begun to occur, such as candidiasis, shingles, and diarrhea. Category C is characterized by more severe opportunistic infections and is clinically described as AIDS.

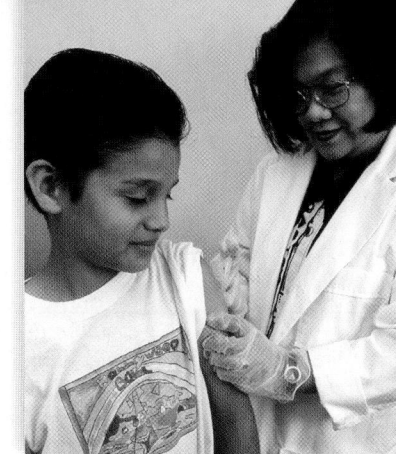

Figure 33.13 Antibody titers. During immunization, the primary response after the first injection of a vaccine is minimal, but the secondary response, which occurs after the second injection, shows a dramatic rise in the amount of antibody present in plasma.

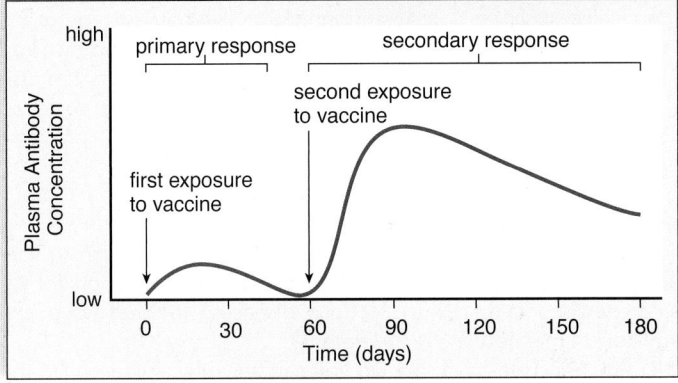

Figure 33.14 Passive immunity. Breast-feeding is believed to prolong the passive immunity an infant receives from the mother during pregnancy because antibodies are present in the mother's milk.

produces the T-cell and B-cell responses that eventually clear the offending viruses from your body. Active immunity can also be induced artificially when a person is well, to prevent infection in the future. **Immunization** involves the use of vaccines, substances that contain an antigen to which the immune system responds (Fig. 33.13). Traditionally, vaccines are the pathogens themselves, or their products, that have been treated so they are no longer virulent (able to cause disease). The use of vaccines has been effective in reducing the rates of bacterial diseases such as diphtheria, tetanus, and whooping cough, as well as viral diseases such as measles, mumps, and rubella. In fact, vaccination against smallpox was so successful that the disease was eradicated from the planet in 1977.

Today, it is possible to genetically engineer bacteria to mass-produce a protein from pathogens, and this protein can be used as a vaccine. This method was used to produce a vaccine against hepatitis B, a viral disease, and is being used to prepare a potential vaccine against malaria. The Nature of Science feature describes a recently FDA-approved type of cancer vaccine that boosts the immunity of prostate cancer patients by stimulating their own antigen-presenting cells.

Video Potato Vaccine

Animation Constructing Vaccines

After most vaccines are given, it is possible to determine the antigen-specific antibody titer (the amount of antibody present in a sample of plasma). After the first vaccination, a primary response occurs. For a period of several days, no antibodies are present; then the titer rises slowly, followed by first a plateau and then a gradual decline as the antibodies bind to the antigen or simply break down (see Fig. 33.13). After a second exposure, a secondary response occurs. The second exposure is called a "booster" because it boosts the immune response to a high level. The high levels of antigen-specific T cells and antibodies are

expected to prevent disease symptoms if the individual is later exposed to the disease-causing agent. Even years later, if the antigen enters the body, memory B cells can quickly give rise to more plasma cells capable of producing the correct type of antibody.

Passive immunity occurs when an individual receives another person's antibodies or immune cells. The passive transfer of antibodies is a common natural process. For example, newborn infants are passively immune to some diseases because antibodies have crossed the placenta from the mother's blood. These antibodies soon disappear, however, so that within a few months, infants become more susceptible to infections as their own immune system must now protect them. Breast-feeding may prolong the natural passive immunity an infant receives from its mother because antibodies are present in the mother's milk (Fig. 33.14).

Even though passively administered antibodies last only a few weeks, they can sometimes be used to prevent illness in a patient who has been unexpectedly exposed to certain infectious agents or toxins. Examples of diseases that may be prevented or treated in this manner include rabies, tetanus, botulism, and snakebites. Individuals with certain types of genetic immunodeficiencies may also benefit greatly from regular, intravenous injections of human IgG, extracted from a large, diverse adult population.

Instead of antibodies, cells of the immune system can also be transferred into a patient. The best example is a bone marrow transplant, in which stem cells that produce blood cells are replenished after a cancer patients's own marrow has been intentionally destroyed by radiation or chemotherapy.

MP3 Specific Immunity

Check Your Progress 33.4

1. Describe the three types of blood cells that are mainly responsible for adaptive defenses, and explain the major function of each type.
2. List three possible outcomes of antigen complex formation in the body.
3. Explain the difference between active and passive immunity, and list three examples of each.

Nature of Science

Cancer Vaccines: Becoming a Reality

Approximately 200 different types of human cancer have been identified, all of which involve an uncontrolled proliferation of cells. Some of these cancer cells express unusual molecules on their surface, which may allow the immune system to identify and destroy them. Unfortunately, many cancers cells grow unchecked in the body, apparently avoiding the immune system's many weapons.

Vaccines against infectious disease are an essential component of modern medicine. Most vaccines are administered before a person is exposed to an infection, but in a few cases (e.g., rabies, tetanus) vaccines can be used therapeutically, after exposure to a pathogen. Medical researchers are now developing vaccines to treat certain types of cancer, and in April, 2010, the U.S. Food and Drug Administration approved the first therapeutic cancer vaccine for use in humans.

Prostate cancer is the second most common type of cancer among men in the United States (behind skin cancer); about 27,000 men died from the disease in 2009. Provenge (sipuleucel-T) can now be used to treat certain types of advanced prostate cancer.

Provenge is prepared in a very different manner than a typical vaccine (Fig. 33B). First, the patient undergoes a procedure called leukapheresis, in which a catheter is inserted into a large vein, and blood flows into a machine that removes some of the white blood cells. Another catheter returns the plasma and red blood cells to the patient. Within the white blood cells are the antigen-presenting cells, such as dendritic cells. These cells are exposed to a protein called PAP that is commonly expressed on prostate cancer cells. The APCs can take up the PAP molecule, and present it on MHC molecules. These APCs are also activated using a cytokine called granulocyte-macrophage-CSF. Three days later, the cells are shipped back and infused into the patient. Typically three doses of Provenge are administered, each containing a minimum of 50 million stimulated cells, at a cost of about $31,000 per dose.

Once inside the patient, the APCs are able to activate a cytotoxic T-cell response that targets prostate cancer cells for destruction. In a 2010 trial of 512 men with advanced prostate cancer, the median survival for patients receiving Provenge was 25.8 months, compared to 21.7 months for those who received a placebo.

Although this approach to cancer therapy is a significant advance, there are also drawbacks, including the fact that a separate vaccine must be prepared for each patient. Other cancer vaccines are currently in clinical testing. Another 2010 study showed that prostate cancer patients who received Prostvac, a vaccine containing harmless viruses that express prostate specific antigen (PSA), lived an average of 8.5 months longer than patients who received a placebo injection.[1] Researchers are also making progress on therapeutic vaccines for cancers of the lung, breast, prostate, colon, and skin.

Questions to Consider

1. In what ways is this approach to vaccination similar to vaccination for infectious diseases? In what ways is it different?

2. Do you see any ethical issues with the cost of this cancer vaccine? Even if a prostate cancer patient lives only an additional four months, is it ever reasonable to place a monetary value on human life?

---

1 Kantoff, Philip W., Schuetz, Thomas J., et al. 2010. Overall survival analysis of a Phase II randomized controlled trial of a poxviral-based PSA-targeted immunotherapy in metastatic castration-resistant prostate cancer. *Journal of Clinical Oncology* 28: 1099–1105.

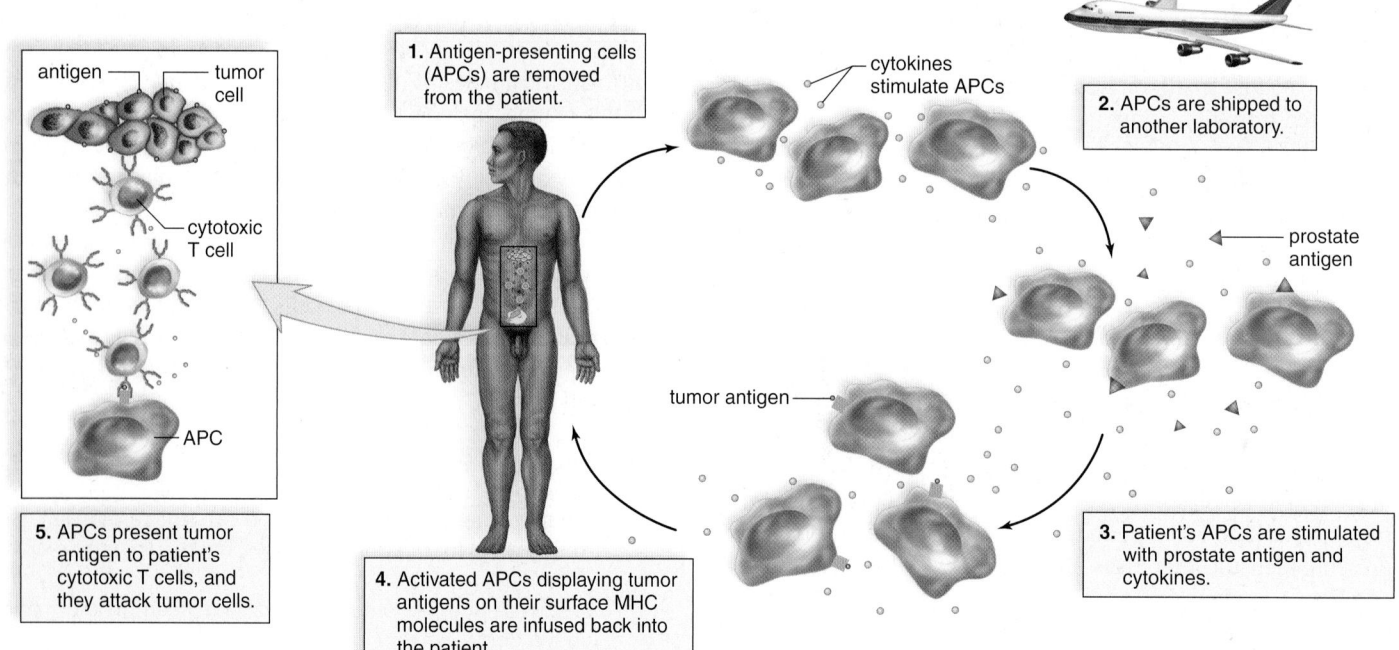

Figure 33B Prostate cancer vaccine. The first therapeutic vaccine for a human cancer involves a complex procedure in which the patient's own antigen-presenting cells (APCs) are removed, shipped to another laboratory, stimulated with cytokines and prostate antigen, then returned to be injected back into the patient.

33.5 Immune System Disorders and Adverse Reactions

Learning Outcomes

Upon completion of this section, you should be able to

1. Describe the two main types of immunodeficiency disorders, and provide examples of each.
2. Discuss the most common immunological mechanisms responsible for allergies, and how these may be treated.
3. Define autoimmune disease, and list several specific examples of these diseases.
4. Explain the types of precautions that must be taken when transplanting organs.

The immune system can be thought of as a "double-edged sword." It is essential for our health and survival, as demonstrated by the diseases, some of them fatal, that occur in people who are immunodeficient. In other instances, the immune system may work against the best interests of the body, as occurs in allergies, autoimmune disorders, and rejection of transplanted organs.

Immunodeficiencies

A number of immunodeficiency disorders are known, but all result in some degree of increased susceptibility to infections. Primary immunodeficiencies are genetic in nature, meaning they are passed from parents to offspring. In **severe combined immunodeficiency (SCID),** both T cells and B cells are either lacking completely or not functioning well enough to protect the body from a variety of infections that aren't a problem for most people. SCID only occurs in about one in 500,000 births.

A variety of faulty genes can cause SCID, but in most cases, by about three months of age (when most of the antibodies obtained from the mother have been degraded), untreated infants usually die. Possible treatments include a bone marrow transplant to replace the stem cells that form all of our white blood cells, or gene therapy to replace the faulty DNA. As was seen in the chapter-opening story of David Vetter, if these treatments are unsuccessful, the outcome is usually poor.

Another primary immunodeficiency is X-linked agammaglobulinemia (XLA), which is due to a mutated gene on the X-chromosome that is needed for proper development of B cells. XLA affects only males, because their cells only have one X chromosome. A female with a normal gene on at least one of her two X chromosomes does not develop the disease. Because their T cells are unaffected, boys with XLA can live relatively normal lives as long as they receive regular injections of human IgG. About 1 in 50,000 males has XLA.

We have already seen that HIV infection causes AIDS, which is an example of a secondary immunodeficiency. These disorders are not genetic, but instead are acquired after birth. Besides infections, other potential causes of secondary immunodeficiencies include malnutrition, irradiation, certain drugs and toxins, and certain cancers. Some of these can be cured by addressing the initiating cause.

Allergies

Allergies are hypersensitivities to substances such as pollen, food, or animal hair, that ordinarily would do no harm to the body. The response to these antigens, called allergens, usually includes some degree of tissue damage.

An **immediate allergic response** can occur within seconds of contact with an allergen. The response is caused by antibodies of the IgE class (Table 33.1). IgE antibodies are attached to receptors on the plasma membrane of mast cells in the tissues, and also to basophils and eosinophils in the blood. When an allergen attaches to these IgE antibodies, the cells release histamine and other substances that bring about the symptoms of an allergy (Fig. 33.15). When an allergen such as pollen is inhaled, histamine stimulates the inflammatory response in the mucous membranes of the nose and eyes typical of *hay fever.* If a person has **asthma,** the airways leading to the lungs constrict, resulting in difficult breathing accompanied by wheezing. When an allergen is in food, nausea, vomiting, and diarrhea typically occur.

Anaphylactic shock is an immediate allergic response that occurs after an allergen has entered the bloodstream. Bee stings, foods, various medications, and latex rubber are all known to cause this reaction in some individuals. Anaphylactic shock is characterized by a sudden and life-threatening drop in blood

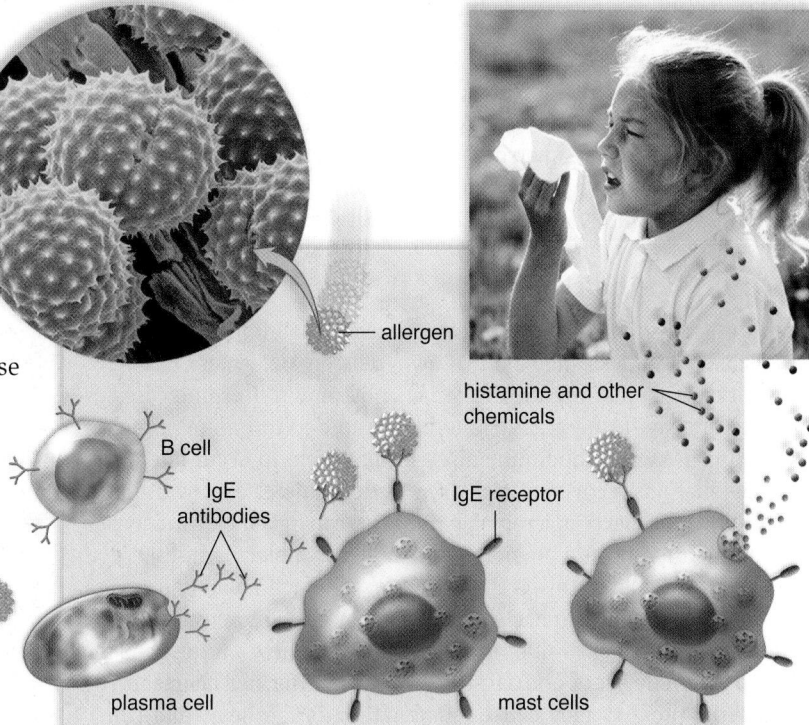

Figure 33.15 An allergic reaction. An allergen attaches to IgE antibodies, which then cause mast cells to release histamine and other chemicals that are responsible for the allergic reaction.

Table 33.1 Comparison of Immediate and Delayed Allergic Responses

	Immediate Response	Delayed Response
Onset of Symptoms	Takes several minutes	Takes 1 to 3 days
Lymphocytes Involved	B cells	T cells
Immune Reaction	IgE antibodies	Cell-mediated immunity
Type of Symptoms	Hay fever, asthma, and many other allergic responses	Contact dermatitis (e.g., poison ivy)
Therapy	Antihistamine and epinephrine	Cortisone

pressure, due to an increased dilation of the capillaries by histamine throughout the body. The smooth muscle lining the bronchi may also be strongly stimulated to constrict, resulting in an inability to breathe. Injecting epinephrine can counteract this reaction until medical help is available, and some people carry an epinephrine-containing spring-loaded syringe (sometimes called an EpiPen) for this purpose.

Mild to moderate allergies are usually treated with antihistamines, which compete with histamine for binding to histamine receptors. In more serious cases, injections of the allergen can be given in an effort to stimulate the immune system to produce high quantities of IgG against the allergen. The hope is that these IgG antibodies will combine with the allergen molecules before they have a chance to reach the IgE antibodies. A monoclonal antibody called Xolair is also available, which blocks the binding of IgE to its receptor on inflammatory cells.

A **delayed allergic response** is initiated by memory T cells at the site of allergen contact in the body. The allergic response is regulated by the cytokines secreted by these "sensitized" T cells at the site. A classic example of a delayed allergic response is the skin test for tuberculosis (TB). When the test result is positive, the tissue where the antigen was injected becomes red and hardened. This indicates prior exposure to *Mycobacterium tuberculosis*, the bacterium that causes TB. Contact dermatitis, which occurs when a person's skin reacts to poison ivy, jewelry, cosmetics, or many other substances that touch the skin, is another example of a delayed allergic response.

Autoimmune Diseases

When a person has an **autoimmune disease,** their immune system mistakenly attacks their body's own cells or molecules. Exactly what causes autoimmune diseases is not known. In some cases, there appears to be a genetic tendency to develop autoimmune diseases. Many autoimmune diseases also seem to occur after an individual has recovered from an infection. It is also known that certain antigens of microbial pathogens can resemble antigens found in their host, a phenomenon known as molecular mimicry. A good example of this is rheumatic fever, which sometimes follows an infection with bacteria of the genus *Streptococcus*. Certain proteins in the cell wall of these bacteria are known to resemble proteins in the heart, and as a result antibodies formed against the bacterial proteins may cause inflammation of the heart that can continue even after the bacteria are cleared from the body.

Some autoimmune diseases affect only specific tissues. Rheumatoid arthritis is a common autoimmune disorder that causes recurring inflammation in synovial joints (Fig. 33.16). Complement proteins, T cells, and B cells all participate in the destruction of the joints, which eventually become immobile. In myasthenia gravis, antibodies interfere with the functioning of neuromuscular junctions, causing muscular weakness. In multiple sclerosis, T cells attack the myelin sheath of nerve fibers, causing a variety of symptoms related to the defective transmission of messages by nerves.

Systemic lupus erythematosus (lupus) is a chronic autoimmune disorder that affects multiple tissues and organs. It is characterized by the production of antibodies that react with the DNA contained in almost every cell of the body. The symptoms vary somewhat, but most patients experience a characteristic skin rash (Fig. 33.17), joint pain, and kidney damage, which may be life-threatening. About 500,000 to 1.5 million people in the United States have lupus, 90 percent of whom are women of childbearing age. Because little is known about the origin of autoimmune disorders, no cures are currently available. The symptoms can sometimes be controlled using immunosuppressive drugs such as cortisone, but these drugs can have serious side effects.

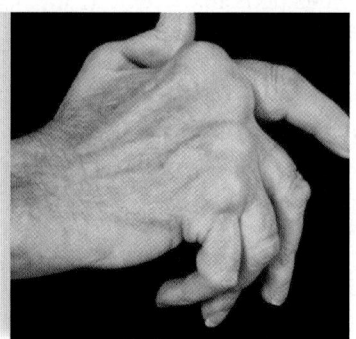

**Figure 33.16
Rheumatoid arthritis.** Rheumatoid arthritis is due to recurring inflammation in skeletal joints, due to immune system attack.

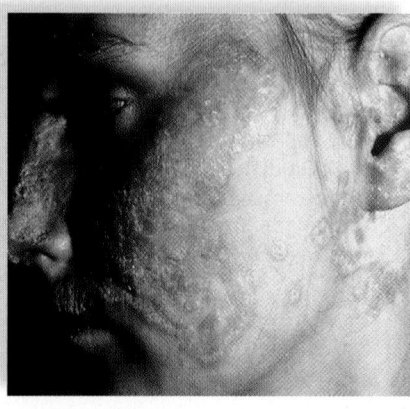

**Figure 33.17
Systemic lupus.** A butterfly-shaped rash appears on the face.

Transplant Rejection

Certain organs, such as skin, the heart, and the kidneys, could be transplanted relatively easily from one person to another if the body did not attempt to reject them. Unfortunately, like the immune system itself, MHC proteins are a double-edged sword. In addition to their beneficial role in presenting antigens to T cells, MHC proteins are major targets of the immune response during rejection of a transplanted organ.

The MHC proteins of different individuals differ by the sequence of their amino acids, and the immune system will attack any foreign tissue that bears MHC antigens different from those of the individual. Ideally, a transplant donor would have exactly the same type of MHC proteins as those of the recipient, but because it is difficult to find a perfect MHC match, the odds of organ rejection can be reduced by administering immunosuppressive drugs. Two commonly used drugs, cyclosporine and tacrolimus, act by inhibiting the production of certain cytokines by T cells.

Xenotransplantation, the transplantation of animal tissues and organs into human beings, is a potential way to solve the shortage of organs from human donors. The pig is the most popular choice as an organ source, because pig organs are generally the right size, and pigs are a common meat source and are thus readily available. Genetic engineering can make pig organs less antigenic by removing the MHC antigens. The ultimate goal is to make pig organs as widely accepted as type O blood cells.

Other researchers hope that tissue engineering, including the production of human organs from stem cells, will one day do away with the problem of rejection. Scientists have recently grown new heart valves in the laboratory using stem cells gathered from amniotic fluid following amniocentesis, and surgeons have sucessfully used lab-grown urinary bladder tissue to rebuild defective bladders in human patients.

Check Your Progress 33.5

1. Explain the general type of abnormality that causes most primary immunodeficiencies.
2. Describe the treatments for autoimmune diseases.
3. Define xenotransplantation.

CONNECTING *the* CONCEPTS *with the* BIG IDEAS

Energy and Homeostasis

- Immune responses may cause disruptions to normal function and health. (2D3a*IE*, 2D4a*IE*)
- Cytotoxic T cells attack surface antigens while B cells produce antibodies that recognize specific foreign markers. (2D4b1-5)
- "Immunity" to recurring pathogen attack is achieved because memory cells allow rapid and enlarged response to its antigens. (2D4b6)
- Apoptosis pares down and personalizes a specific organism's immune system. (2E1c*IE*)

Information and Signaling

- Antigen presenting cells must make cell-to-cell contact to communicate with helper and killer T-cells. (3D2a*IE*)

Interactions and Systems

- Molecular variations that produce different MHC and antibody proteins give organisms more varieties of cellular markers and defense molecules. (4C1a*IE*)

*Find the unabridged version of all EK citations at www.glencoe.com/maderAP11.

Media Study Tools

www.glencoe.com/maderAP11

Enhance your study of this chapter with study tools and practice tests. Also ask your instructor about the resources available through ConnectPlus, including the media-rich eBook, interactive learning tools, and animations.

Summarize

33.1 Evolution of Immune Systems

From the available evidence, it appears that innate immunity is much older than adaptive immunity, because aspects of innate immunity are present in relatively simple multicellular animals. Innate immunity is quite well developed in insects and often involves recognition of pathogen-associated molecular patterns (PAMPs). Available evidence indicates that the ability to generate a diverse array of antigen receptors first appeared in an ancestor of the jawed vertebrates.

33.2 The Lymphatic System

The lymphatic system consists of lymphatic vessels and lymphoid organs. The lymphatic system (1) removes excess tissue fluid collected by lymphatic capillaries; (2) absorbs fats from the small intestine; and (3) produces and distributes lymphocytes.

Lymphocytes are produced in the primary lymphoid organs (red bone marrow and thymus) and lymphocytes respond to antigens in secondary lymphoid organs (lymph nodes, spleen, etc.) T lymphocytes mature in the thymus, while B lymphocytes mature in the red bone marrow.

33.3 Innate Immune Defenses

Innate defenses are always present, or occur very soon after exposure to an infection. These include barriers to entry, the inflammatory response, phagocytes and natural killer cells, and protective proteins.

33.4 Adaptive Immune Defenses

Adaptive defenses (also called acquired immunity) involve B lymphocytes and T lymphocytes, also called B cells and T cells. After exposure to specific antigen, both types of lymphocytes undergo clonal selection. Upon binding of their B-cell receptor with a specific antigen, B cells divide and differentiate into plasma cells and memory B cells. Plasma cells secrete antibodies and are responsible for antibody-mediated immunity.

Antibodies are Y-shaped molecules that have at least two binding sites for a specific antigen. Monoclonal antibodies, which are produced by cells derived from a single plasma cell fused to a myeloma cancer cell, have various uses, from detecting infections to treating cancer.

The two main types of T cells are helper T (T_H) cells and cytotoxic T (T_C) cells. For a T_H cell to recognize an antigen, the antigen must be presented by MHC (major histocompatibility complex) class II proteins on the surface of an antigen-presenting cell (APC). Types of APCs include dendritic cells and macrophages. Thereafter, the activated T cell undergoes clonal expansion forming activated T_H cells and memory T_H cells. Activated T_H cells produce cytokines that affect many other immune cells. HIV infection induces an immunodeficiency by infecting and reducing the numbers of T_H cells.

T_C cells recognize antigens presented by MHC class I proteins on the surface of virus-infected or cancer cells. They then kill these cells by releasing perforin and granzymes, inducing apoptosis.

Active immunity occurs as a response to an illness or the administration of vaccines. Passive immunity is needed when an individual is in immediate danger of succumbing to an infectious disease. Passive immunity can occur naturally (as in the case of transfer of antibodies from mother to infant) or may be used as a medical treatment.

33.5 Immune System Disorders and Adverse Reactions

Immunodeficiencies can be primary (genetic) or secondary (due to some other cause). Allergic responses occur when the immune system reacts to substances not normally recognized as dangerous. Autoimmune diseases occur when the immune system attacks the body's own cells or tissues. In a transplant rejection, the immune system is usually responding to unmatched MHC proteins found on the cells of a donated organ.

Key Terms

active immunity   637
adaptive immunity   627
allergy   641
anaphylactic shock   641
antibody-mediated
    immunity   634
antigen   627
antigen-presenting cell
    (APC)   635
antigen receptor   633
asthma   641
autoimmune disease   642
B cell   629
B cell receptor (BCR)   633
cell-mediated immunity   636
clonal selection theory   633
complement   632
cytokine   633
cytotoxic T cell   635
delayed allergic response   642
dendritic cell   632
eosinophil   632
helper T cell   635
histamine   631
human immunodeficiency virus
    (HIV)   637
immediate allergic
    response   641
immune system   627
immunity   630
immunization   639
immunoglobulin (Ig)   634
inflammatory response   631
innate immunity   627
interferon   633
lymph   629
lymphatic capillary   629
lymphatic system   628
lymphatic vessel   628
lymph node   629
lymphoid (lymphatic) organ 629
macrophage   632
mast cell   631
memory B cell   634
memory T cell   636
MHC (major histocompatibility
    complex) protein   635
monoclonal antibody   635
natural killer (NK) cell   632
neutrophil   631
passive immunity   639
plasma cell   634
primary lymphoid organ   629
red bone marrow   629
secondary lymphoid organ   629
severe combined
    immunodeficiency (SCID)
    641
spleen   629
T cell   629
T-cell receptor (TCR)   635
thymus   629

![Assess banner] **Assess**

Reviewing This Chapter

1. Which functions of the lymphatic system are not assisted by another system? Explain.   628
2. Describe the microscopic structure and the function of lymph nodes, the spleen, the thymus, and red bone marrow.   629–30
3. Discuss the body's innate defense mechanisms.   630–33
4. Describe the inflammatory response, and give a role for each type of cell and molecule that participates in the response.   630–33
5. Describe the clonal selection theory as it applies to B cells. B cells are responsible for which type of immunity?   633–34
6. How is active immunity artificially achieved? How is passive immunity achieved?   637, 639
7. Describe the structure of an antibody, and define the terms variable regions and constant regions.   634
8. How are monoclonal antibodies produced, and what are their applications?   635
9. Discuss the clonal selection theory as it applies to T cells.   635–36
10. Name the two main types of T cells and state their functions.   635–36
11. What are cytokines, and how are they used in immunotherapy?   636–37
12. Discuss autoimmune diseases and allergies as they relate to the immune system.   641–42

Testing Yourself

Choose the best answer for each question.

1. Both veins and lymphatic vessels
   a. have thick walls of smooth muscle.
   b. contain valves for one-way flow of fluids.
   c. empty directly into the heart.
   d. are fed fluids from arterioles.

2. Complement
   a. is an innate defense mechanism.
   b. is involved in the inflammatory response.
   c. is a series of proteins present in the plasma.
   d. plays a role in destroying bacteria.
   e. All of these are correct.

3. Which of these pertain(s) to T cells?
   a. have specific receptors
   b. recognize antigen presented by MHC proteins
   c. are responsible for cell-mediated immunity
   d. stimulate antibody production by B cells
   e. All of these are correct.

4. Which one of these does not pertain to B cells?
   a. have passed through the thymus
   b. have specific receptors
   c. are responsible for antibody-mediated immunity
   d. become plasma cells that synthesize and release antibodies

5. The clonal selection model says that
   a. an antigen selects certain B cells and suppresses them.
   b. an antigen stimulates the multiplication of B cells that produce antibodies against it.
   c. T cells select those B cells that should produce antibodies, regardless of antigens present.
   d. T cells suppress all B cells except the ones that should multiply and divide.
   e. Both b and c are correct.

6. Plasma cells are
   a. the same as memory cells.
   b. formed from blood plasma.
   c. B cells that are actively secreting antibody.
   d. inactive T cells carried in the plasma.
   e. a type of red blood cell.

7. Which of these pairs is incorrectly matched?
   a. cytotoxic T cells—help complement react
   b. cytotoxic T cells—active in tissue rejection
   c. macrophages—activate T cells
   d. memory T cells—long-living T cells
   e. T cells—mature in thymus

8. Vaccines are
   a. the same as monoclonal antibodies.
   b. treated bacteria or viruses, or one of their proteins.
   c. short-lived.
   d. MHC proteins.
   e. All of these are correct.

9. Which is an innate defense against pathogens?
   a. skin
   b. gastric juice
   c. complement
   d. interferons
   e. All of these are correct.

10. Which cell is not a phagocyte?
    a. neutrophil
    b. lymphocyte
    c. dendritic cell
    d. macrophage

11. B cells mature within
    a. the lymph nodes.
    b. the spleen.
    c. the thymus.
    d. the bone marrow.

12. Plasma cells secrete
    a. antibodies.
    b. perforins.
    c. lysosomal enzymes.
    d. histamine.
    e. lymphokines.

13. Immediate hypersensitivity occurs after an allergen combines with
    a. IgG antibodies.
    b. IgE antibodies.
    c. IgM antibodies.
    d. IgA antibodies.

14. After a second exposure to a vaccine,
    a. antibodies are made quickly and in greater amounts.
    b. immunity lasts longer than after the first exposure.
    c. antibodies of the IgG class are produced.
    d. plasma cells are active.
    e. All of these are correct.

15. Active immunity may be produced by
    a. having a disease.
    b. receiving a vaccine.
    c. receiving gamma globulin injections.
    d. Both a and b are correct.
    e. Both b and c are correct.

Engage

Thinking Scientifically

1. The transplantation of organs from one person to another was impossible until the discovery of immunosuppressant drugs. Now, with the use of drugs such as cyclosporine, organs can be transplanted without rejection. Transplant patients must take immunosuppressant drugs for the remainder of their lives. What is a major expected side effect of this lifelong therapy?

2. Laboratory mice are immunized with a measles vaccine. When the mice are challenged with measles virus to test the strength of their immunity, the memory cells do not completely prevent replication of the measles virus. The virus undergoes a few rounds of replication before the immune response is observed. You have developed a strain of mice with a much faster response to a viral challenge, but these mice often develop an autoimmune disease. Speculate on the connection between speed of response and an autoimmune disease.

Bioethical Issue

Cost of Drugs to Treat AIDS

As noted in this chapter, an estimated 33.3 million people worldwide were living with HIV infection at the end of 2009. Most of these people will develop AIDS if their HIV infection is not treated. Drug companies typically charge a high price for anti-HIV medications because people in developed countries and their insurance companies can afford to pay for them. However, these prices are out of reach in many countries, such as those in Africa, where AIDS is a widespread problem.

Some people argue that drug companies should use the profits from other drugs (such as those for heart disease, depression, and impotence) to make AIDS drugs affordable to those who need them. This has not happened yet. In some countries, governments have allowed companies to infringe on foreign patents held by major drug companies so that affordable AIDS drugs can be produced. Do drug companies have a moral obligation to provide low-cost AIDS drugs, even if they have to do so at a loss of revenue? Is it right for governments to ignore patent laws in order to provide their citizens with affordable drugs?

34

Digestive Systems and Nutrition

Cattle, like sheep, goats, and other ruminants, are able to digest the cellulose found in grasses because of their highly specialized digestive system.

BEFORE YOU BEGIN

Before beginning this chapter, take a few moments to review the following discussions.

Chapter 3 What are some structural differences between carbohydrates, lipids, proteins, and nucleic acids?

Figure 6.1 How does energy flow from the Sun, into chemical energy, to be ultimately dissipated as heat?

Figure 8.11 How do components of the human diet enter common metabolic pathways?

Humans first domesticated cattle around 8,000 years ago. Cattle are part of a large group of mammals called ruminants, which use a process of digestion that begins when plant material is swallowed and enters a large chamber called the rumen. Here, a rich population of bacteria and other microbes break down the cellulose present in plant material. During this process some solid material is also regurgitated as the cud, which is chewed slowly to break down the plant fibers into a more digestible size.

This ability of ruminants to utilize the cellulose present in grasses and other plants is the main advantage of using these animals as a source of meat for human consumption. Rangeland that is not suitable for growing other kinds of crops can be used to raise cattle (although most beef cattle in the United States are fed grain). However, due largely to the growing human population and high demand for meat in some countries, the total number of domesticated cattle on Earth has more than doubled in the last 40 years to an estimated 1.5 billion in 2009. Estimates vary, but ruminants account for about 15–20% of the global production of methane, an important contributor to climate change. Most medical experts also believe a diet containing too much red meat is an important factor in major diseases like atherosclerosis, diabetes, and many cancers. As you will see in this chapter, eating a well-balanced diet is one of the most important things we can do to maintain good health.

As you read through the chapter, think about the following questions:

1. What different types of strategies have animals evolved to efficiently obtain nutrients?
2. In what ways do the types of diets that humans choose to consume play a role in our health, as well as in the quality of our environment?

FOLLOWING *the* BIG IDEAS

CHAPTER 34 DIGESTIVE SYSTEMS AND NUTRITION

Energy and Homeostasis	Increased surface area for digestion and absorption allows animals to obtain adequate nutrients.
Interactions and Systems	Compartmentalization and specialization of organs promotes a more efficient disassembly line for digestion.

34.1 Digestive Tracts

Learning Outcomes

Upon completion of this section, you should be able to

1. Compare the structural features of incomplete versus complete digestive tracts.
2. Describe several examples of animals that are either continuous or discontinuous feeders.
3. Discuss some specific adaptations that are seen in omnivores, herbivores, and carnivores.

A digestive system includes all the organs, tissues, and cells involved in ingesting food and breaking it down into smaller components. Digestion contributes to homeostasis by providing the body with the nutrients needed to sustain the life of cells. A digestive system:

1. ingests food,
2. breaks food down into small molecules that can cross plasma membranes,
3. absorbs these nutrient molecules, and
4. eliminates undigestible remains.

A digestive tract, or *gut*, is typically defined as a long tube through which food passes as it is being digested. The majority of animals have some sort of digestive tract, but some, like sponges, have no digestive tract at all. Instead, as water from the aqueous environment flows through the sponge (see Fig. 28.6), food particles are removed by cells that make up the inner lining of the organism. Cells in the sponge called *archaeocytes* may also ingest and distribute food to the rest of the organism.

Incomplete Versus Complete Tracts

An **incomplete digestive tract** has a single opening, usually called a mouth; however, the single opening is used both as an entrance for food and an exit for wastes. Planarians, which are flatworms, have an incomplete tract (Fig. 34.1). It begins with a mouth and muscular pharynx, and then the tract, a gastrovascular cavity, branches throughout the body.

Planarians are primarily carnivorous and feed largely on smaller aquatic animals, as well as bits of organic debris. When a planarian is feeding, the pharynx actually extends beyond the mouth. The body is wrapped about the prey and the pharynx sucks up small quantities at a time. Digestive enzymes present in the tract allow some extracellular digestion to occur. Digestion is finished intracellularly by the cells that line the tract. No cell in the body is far from the digestive tract; therefore, diffusion alone is sufficient to distribute nutrient molecules.

The digestive tract of a planarian is notable for its lack of specialized parts. It is saclike because the pharynx serves not only as an entrance for food but also as an exit for undigestible material. This use of the same body parts for more than one function tends to minimize the evolution of more specialized parts, such as those seen in complete tracts.

Planarians have some modified parasitic relatives. Tapeworms, which are parasitic flatworms, lack a digestive system.

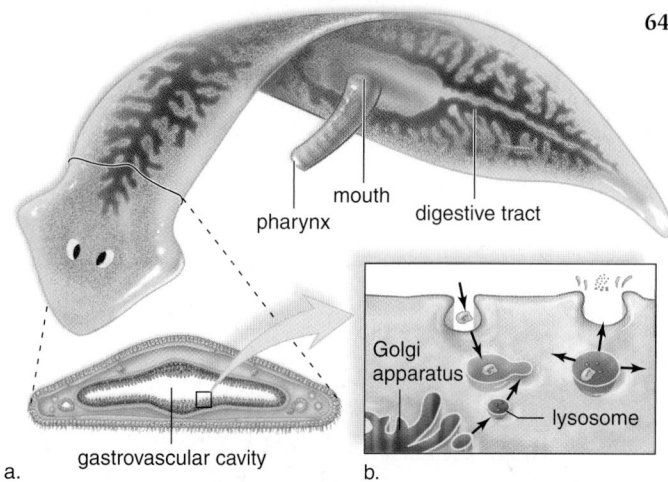

Figure 34.1 Incomplete digestive tract of a planarian.
a. Planarians, which are flatworms, have a gastrovascular cavity with a single opening that acts as both an entrance and an exit. Like hydras, planarians rely on intracellular digestion to complete the digestive process. **b.** Phagocytosis produces a vacuole, which joins with an enzyme-containing lysosome. The digested products pass from the vacuole into the cytoplasm before any undigestible material is eliminated at the plasma membrane.

Nutrient molecules are absorbed by the tapeworm from the intestinal juices of the host, which surround the tapeworm's body. The integument and body wall of the tapeworm are highly modified for this purpose. They have millions of microscopic, fingerlike projections that increase the surface area for absorption.

In contrast to planarians, earthworms, which are annelids, have a **complete digestive tract,** meaning that the tract has a mouth and an anus (Fig. 34.2). Earthworms feed mainly on the decayed organic matter found in soil. The muscular pharynx draws in a large amount of soil with a sucking action. Soil then enters the crop, which is a storage area with thin, expansive walls. From there, it goes to the gizzard, where thick, muscular walls crush the food and ingested sand grinds it. Digestion is extracellular within the intestine. The surface area of digestive tracts is often increased for absorption of nutrient molecules, and in earthworms, this is accomplished by an intestinal fold called the *typhlosole*. Undigested remains pass out of the body at the

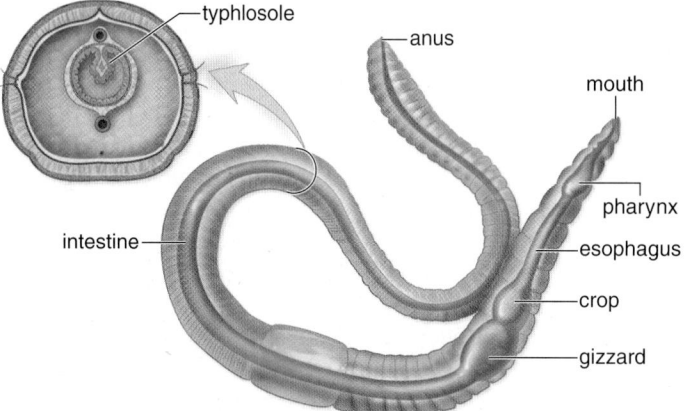

Figure 34.2 Complete digestive tract of an earthworm.
Complete digestive tracts have both a mouth and an anus and can have many specialized parts, such as those labeled in this drawing. Also in earthworms, which are annelids, the absorptive surface of the intestine is increased by an internal fold called the typhlosole.

Figure 34.3 Nutritional mode of a clam compared to a squid. Clams and squids are molluscs. A clam burrows in the sand or mud, where it filter feeds, whereas a squid swims freely in open waters and captures prey. In keeping with their lifestyles, a clam **(a)** is a continuous feeder and a squid **(b)** is a discontinuous feeder. Digestive system labels are shaded green.

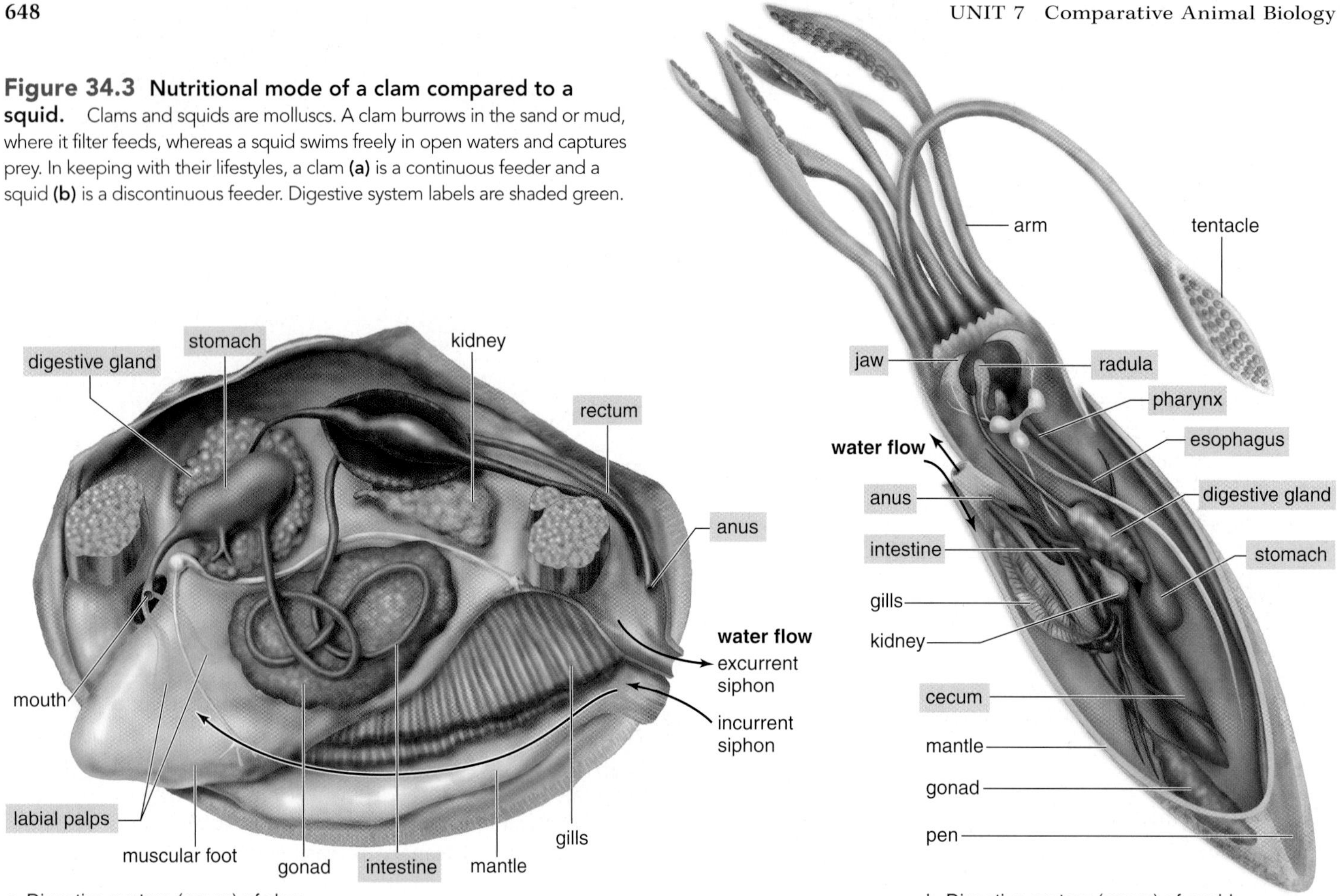

a. Digestive system (green) of clam

b. Digestive system (green) of squid

anus. Specialization of parts is obvious in the earthworm because the pharynx, the crop, the gizzard, and the intestine have particular functions as food passes through the digestive tract.

Continuous Versus Discontinuous Feeders

Some aquatic animals acquire their nutrients by continuously passing water through some type of apparatus that captures food. Clams, which are molluscs, are filter feeders (Fig. 34.3*a*). Water is always moving into the mantle cavity by way of the incurrent siphon (slitlike opening) and depositing particles, including algae, protozoans, and minute invertebrates, on the gills. The size of the incurrent siphon permits the entrance of only small particles, which adhere to the gills. Ciliary action moves suitably sized particles to the labial palps, which force them through the mouth into the stomach. Digestive enzymes are secreted by a large digestive gland, but amoeboid cells present throughout the tract are believed to complete the digestive process by intracellular digestion.

Not all filter feeders are relatively small invertebrates. A baleen whale, such as the blue whale, is an active filter feeder. Baleen, a keratinized curtainlike fringe, hangs from the roof of the mouth and filters small shrimp called krill from the water. A baleen whale filters up to a ton of krill every few minutes.

Discontinuous feeders have evolved the ability to store food temporarily while it is being digested, enabling them to spend

less time feeding and more time engaging in other activites. Discontinuous feeding requires a storage area for food, which can be a crop, where no digestion occurs, or a stomach, where digestion begins.

Squids, which are molluscs, are discontinuous feeders (Fig. 34.3*b*). The body of a squid is streamlined, and the animal moves rapidly through the water using jet propulsion (forceful expulsion of water from a tubular funnel). The head of a squid is surrounded by ten arms, two of which have developed into long, slender tentacles whose suckers have toothed, horny rings. These tentacles seize prey (fishes, shrimps, and worms) and bring it to the squid's beaklike jaws, which bite off pieces pulled into the mouth by the action of a *radula,* a tonguelike structure. An esophagus leads to a stomach and a cecum (blind sac), where digestion occurs. The stomach, supplemented by the cecum, retains food until digestion is complete.

Adaptations to Diet

Beyond the general categories of continuous versus discontinuous feeders, some animals have further adapted to more specialized diets. Some animals are omnivores; they eat both plants and animals. Others are strict herbivores; they feed only on plants. Still others are strict carnivores; they eat only other animals. Among invertebrates, filter feeders such as clams and tube worms are omnivores. Land snails, which are terrestrial

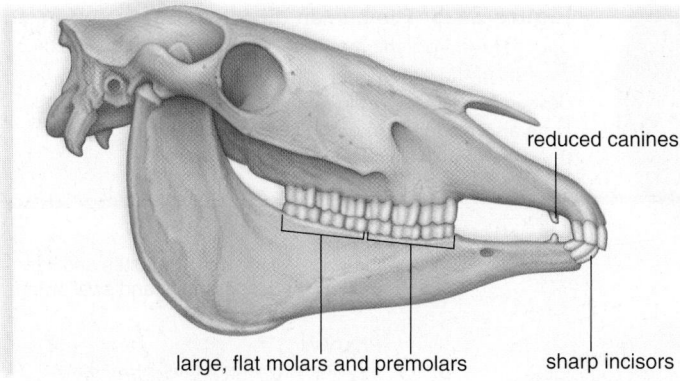

a. Horses are herbivores.

reduced canines

large, flat molars and premolars

sharp incisors

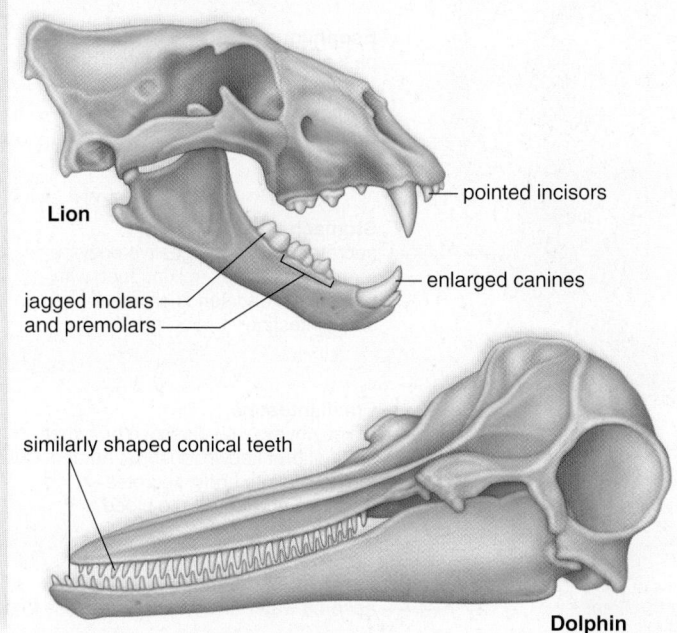

Lion

pointed incisors

enlarged canines

jagged molars
and premolars

similarly shaped conical teeth

Dolphin

b. Lions and dolphins are carnivores.

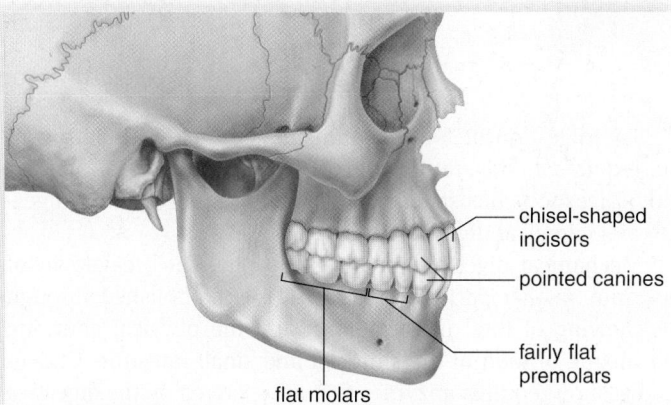

chisel-shaped
incisors

pointed canines

fairly flat
premolars

flat molars

c. Humans are omnivores.

Figure 34.4 Dentition among mammals. **a.** Horses are herbivores and have teeth suitable to clipping and chewing grass. **b.** Lions and dolphins are carnivores. Dentition in a lion is suitable for killing large animals such as zebras and wildebeests and tearing apart their flesh. Dentition in a dolphin is suitable to grasping small animals like fishes, which are swallowed whole. **c.** Humans are omnivores and have teeth suitable to a mixed diet of vegetables and meat.

molluscs, and some insects, such as grasshoppers and locusts, are herbivores. Spiders (arthropods) are carnivores, as are sea stars (echinoderms), which feed on clams. A sea star positions itself above a clam and uses its tube feet to pull the valves of the shell apart (see Fig. 28.29). Then, it everts a part of its stomach to start the digestive process, even while the clam is trying to close its shell. Some invertebrates are cannibalistic. A female praying mantis (an insect), if starved, will feed on her mate as the reproductive act is taking place!

Mammals have also adapted to consume a variety of different food sources. Among herbivores, the koala of Australia is famous for its diet of only eucalyptus leaves, and likewise many other mammals are browsers, feeding on bushes and trees. Grazers, like the horse, feed off grasses. The horse has sharp, even incisors for neatly clipping off blades of grass and large, flat premolars and molars for grinding and crushing the grass (Fig. 34.4a). Extensive grinding and crushing disrupts plant cell walls, allowing bacteria located in a part of the digestive tract called the cecum to digest cellulose.

As mentioned in the chapter-opening story, ruminants such as cattle, sheep, and deer have a large, four-chambered stomach. In contrast to horses, they graze quickly and swallow partially chewed grasses into the **rumen,** which is the first chamber. The rumen serves as a fermentation vat where microorganisms break down material such as cellulose that the animal could not otherwise digest. Later on, when the ruminant is no longer feeding, undigested solid material called cud is regurgitated and chewed again to facilitate more complete digestion.

Many mammals, including dogs, lions, toothed whales, and dolphins, are carnivores. Lions use pointed canine teeth for killing, short incisors for scraping bones, and pointed molars for slicing flesh (Fig. 34.4b, top). Dolphins and toothed whales swallow food whole without chewing it first; they are equipped with many identical, conical teeth that are used to catch and grasp their slippery prey before swallowing (Fig. 34.4b, bottom). Meat is rich in protein and fat and is easier to digest than plant material. The intestine of a rabbit, a herbivore, is much longer than that of a similarly sized cat, a carnivore.

Humans, like pigs, raccoons, mice, and most bears, are omnivores. Therefore, the dentition has a variety of specializations to accommodate both a vegetable diet and a meat diet. An adult human has 32 teeth. One-half of each jaw has teeth of four different types: two chisel-shaped incisors for shearing; one pointed canine (cuspid) for tearing; two fairly flat premolars (bicuspids) for grinding; and three molars, well flattened for crushing (Fig. 34.4c). Omnivores are generally better able to adapt to different food sources, which can vary in different locations and seasons.

Check Your Progress 34.1

1. Compare the digestive tract of a planarian with that of an earthworm.
2. Describe some of the limitations of an incomplete digestive tract.
3. Compare the teeth of carnivores to those of herbivores.

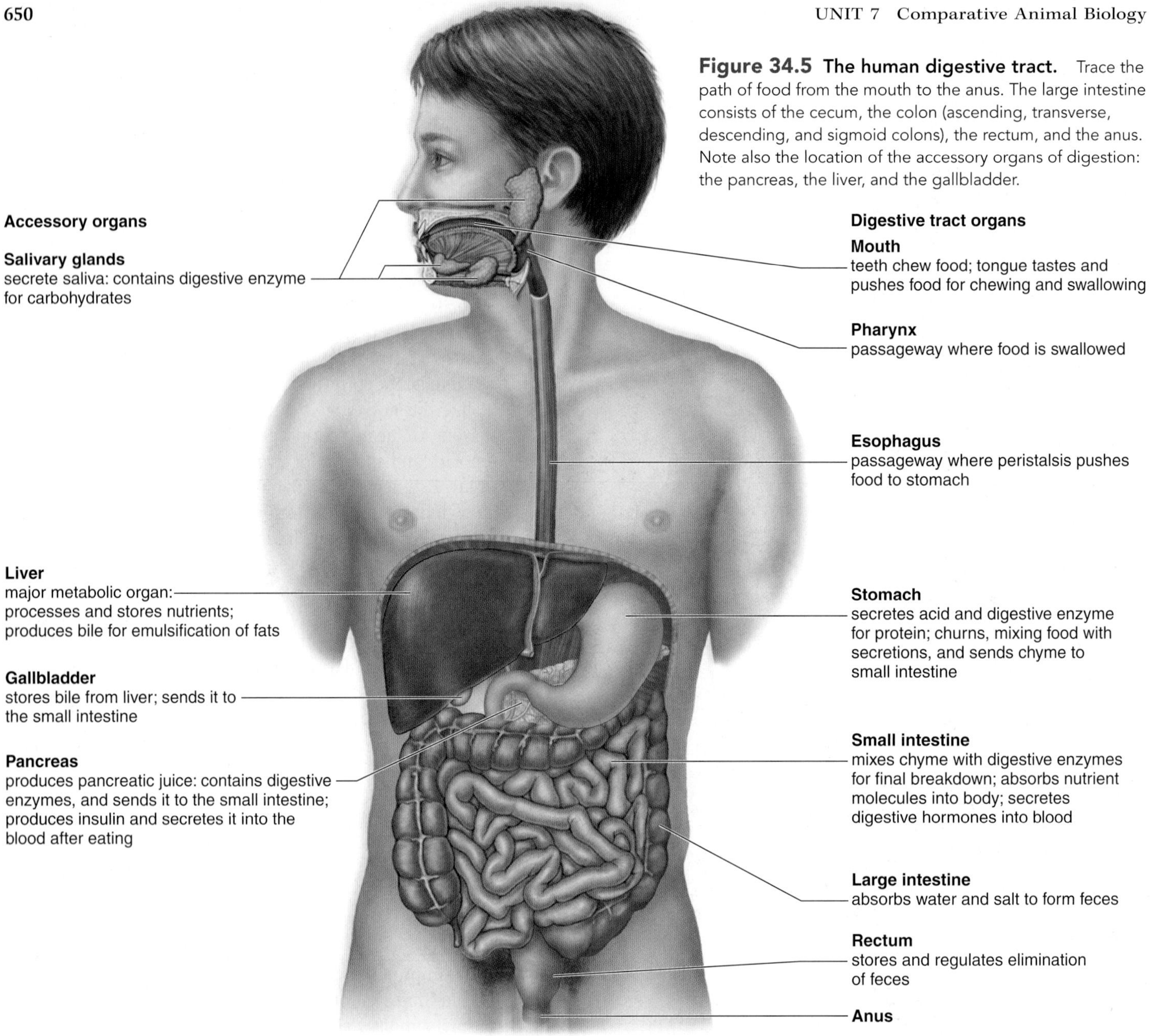

Figure 34.5 The human digestive tract. Trace the path of food from the mouth to the anus. The large intestine consists of the cecum, the colon (ascending, transverse, descending, and sigmoid colons), the rectum, and the anus. Note also the location of the accessory organs of digestion: the pancreas, the liver, and the gallbladder.

Accessory organs

Salivary glands
secrete saliva: contains digestive enzyme for carbohydrates

Liver
major metabolic organ: processes and stores nutrients; produces bile for emulsification of fats

Gallbladder
stores bile from liver; sends it to the small intestine

Pancreas
produces pancreatic juice: contains digestive enzymes, and sends it to the small intestine; produces insulin and secretes it into the blood after eating

Digestive tract organs

Mouth
teeth chew food; tongue tastes and pushes food for chewing and swallowing

Pharynx
passageway where food is swallowed

Esophagus
passageway where peristalsis pushes food to stomach

Stomach
secretes acid and digestive enzyme for protein; churns, mixing food with secretions, and sends chyme to small intestine

Small intestine
mixes chyme with digestive enzymes for final breakdown; absorbs nutrient molecules into body; secretes digestive hormones into blood

Large intestine
absorbs water and salt to form feces

Rectum
stores and regulates elimination of feces

Anus

34.2 The Human Digestive System

Learning Outcomes

Upon completion of this section, you should be able to

1. List all the major components of the human digestive tract, from the mouth to the anus.
2. Compare and contrast the structural features of the small intestine vs. large intestine.
3. Discuss the major functions of the pancreas, liver, and gallbladder.

Humans have a complete digestive tract, which begins with a mouth and ends in an anus. The major structures of the human digestive tract are illustrated in Figure 34.5. The pancreas, liver, and gallbladder are accessory organs that aid digestion.

MP3
An Overview of the Digestive System

Animation
Organs of Digestion

The digestion of food in humans is an extracellular event and requires a cooperative effort between different parts of the body. Digestion consists of two major stages: mechanical digestion and chemical digestion.

Mechanical digestion involves the physical breakdown of food into smaller particles. This task is accomplished through the chewing of food in the mouth and the physical churning and mixing of food in the stomach and small intestine. Chemical digestion requires enzymes that are secreted by the digestive tract or by accessory glands that lie nearby. Specific enzymes break down particular macromolecules into smaller molecules that can be absorbed.

Mouth

The **mouth,** or oral cavity, serves as the beginning of the digestive tract. The *palate,* or roof of the mouth, separates the oral cavity from the nasal cavity. It consists of the anterior hard palate and the posterior soft palate. The fleshy *uvula* is the posterior

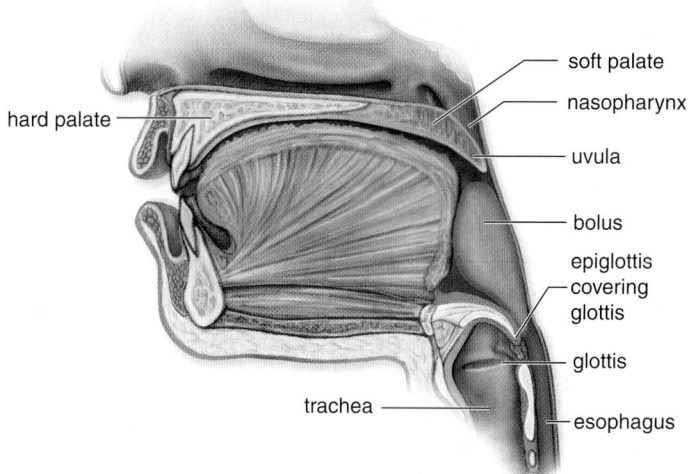

Figure 34.6 Swallowing. Respiratory and digestive passages converge and diverge in the pharynx. When food is swallowed, the soft palate closes off the nasopharynx, and the epiglottis covers the glottis, forcing the bolus to pass down the esophagus. Therefore, a person does not breathe when swallowing.

extension of the soft palate (Fig. 34.6). The cheeks and lips retain food while it is chewed by the teeth and mixed with saliva.

Three major pairs of **salivary glands** send their juices by way of ducts into the mouth. Saliva contains the enzyme **salivary amylase,** which begins to digest the starch that is present in many foods of plant origin. The disaccharide maltose is a typical end product of salivary amylase digestion (see Section 34.3).

While in the mouth, food is manipulated by a muscular tongue, which has touch and pressure receptors similar to those in the skin. Taste buds, sensory receptors that are stimulated by the chemical composition of food, are also found primarily on the tongue as well as on the surface of the mouth. The tongue, which is composed of striated muscle and an outer layer of mucous membrane, mixes the chewed food with saliva. It then forms this mixture into a mass called a *bolus* in preparation for swallowing.

MP3
Oral Cavity, Esophagus, and the Swallowing Reflex

The Pharynx and the Esophagus

The digestive and respiratory passages come together in the **pharynx** and then separate. The **esophagus** [Gk. *eso,* within, and *phagein,* eat] is a tubular structure, of about 25 cm in length, that takes food to the stomach. *Sphincters* are muscles that encircle tubes and act as valves; tubes close when sphincters contract, and they open when sphincters relax. The lower gastroesophageal sphincter is located where the esophagus enters the stomach. When food enters the stomach, the sphincter relaxes for a few seconds and then it closes again. Heartburn occurs due to acid reflux, when some of the stomach's contents escape into the esophagus. When vomiting occurs, the abdominal muscles and the diaphragm, a muscle that separates the thoracic and abdominal cavities, contract.

When food is swallowed, the soft palate, the rear portion of the mouth's roof, moves back to close off the nasopharynx. A flap of tissue called the epiglottis covers the glottis, or opening into the trachea. Now the bolus must move through the pharynx into the esophagus because the air passages are blocked (see Fig. 34.6).

The central space of the digestive tract, through which food passes as it is digested, is called the **lumen** (Fig. 34.7). From the esophagus to the large intestine, the wall of the digestive tract is composed of four layers. The innermost layer next to the lumen is called the **mucosa**. The mucosa is a type of mucous membrane and therefore it produces mucus, which protects the wall from the digestive enzymes inside the lumen.

The second layer in the digestive wall is called the **submucosa**. The submucosal layer is a broad band of loose connective tissue that contains blood vessels, lymphatic vessels, and nerves. Lymph nodules, called Peyer's patches, are also in the submucosa. Like other secondary lymphoid tissues, they are sites of lymphocyte responses to antigens (see Chapter 33).

The third layer is termed the **muscularis**, and it contains two layers of smooth muscle. The inner, circular layer encircles the tract; the outer, longitudinal layer lies in the same direction as the tract. The contraction of these muscles, which are under involuntary nervous control, accounts for the movement of the gut contents from the esophagus to the rectum by **peristalsis** [Gk. *peri,* around, and *stalsis,* compression], a rhythmic contraction that serves to move the contents along in various tubular organs. (Fig. 34.8).

The fourth layer of the wall is the **serosa**, which secretes a watery fluid that lubricates the outer surfaces of the digestive tract and reduces friction as various parts rub against each

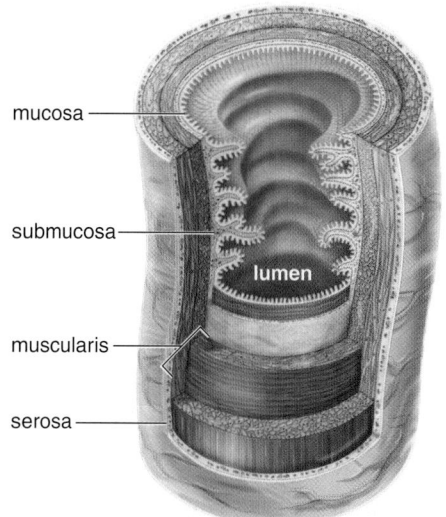

Figure 34.7 Wall of the digestive tract. The esophagus, stomach, small intestine, and large intestine all have a lumen and walls composed of similar layers.

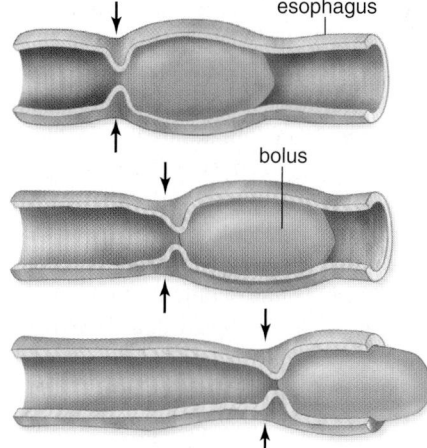

Figure 34.8 Peristalsis in the digestive tract. These three drawings show how a peristaltic wave moves through a single section of the esophagus over time. The arrows point to areas of contraction.

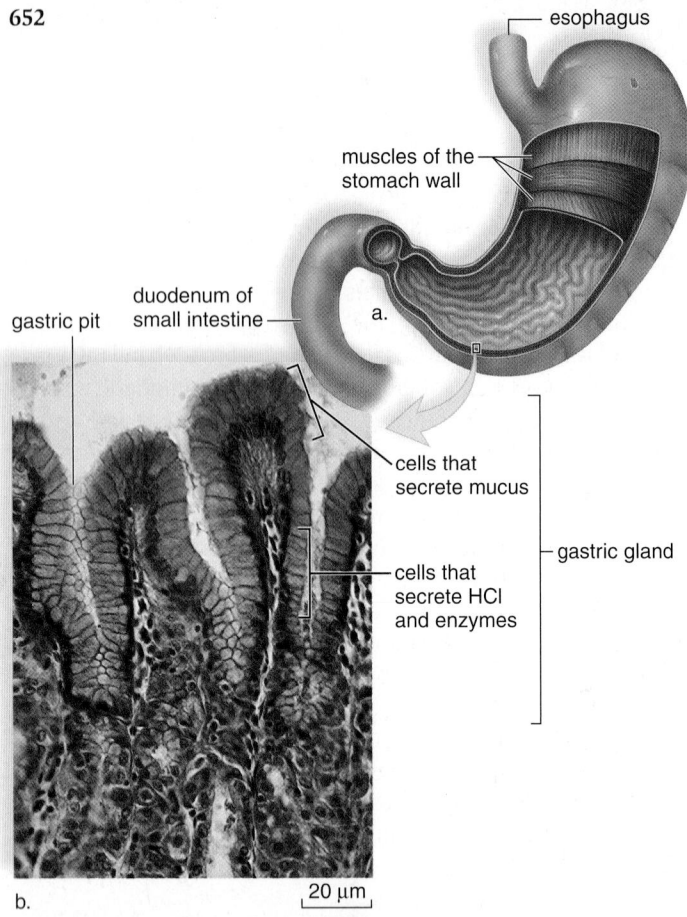

Figure 34.9 Anatomy of the stomach. **a.** The stomach, which has thick walls, expands as it fills with food. **b.** The mucous membrane layer of its walls secretes mucus and contains gastric glands, which secrete a gastric juice active in the digestion of protein.

The epithelial lining of the stomach has millions of *gastric pits,* which lead into *gastric glands* (see Fig. 34.9). The gastric glands produce gastric juice. So much hydrochloric acid is secreted by the gastric glands that the stomach routinely has a pH of about 2.0. Such a high acidity usually is sufficient to kill bacteria and other microorganisms that might be in food. This low pH also stops the activity of salivary amylase, which functions optimally at the near-neutral pH of saliva.

As with the rest of the digestive tract, a thick layer of mucus protects the wall of the stomach from enzymatic action. Still, some individuals develop ulcers, which are areas where the protective epithelial layer has been damaged. Gastric ulcers can be caused by an acid-resistant bacterium, *Helicobacter pylori,* which is able to attach to the epithelial lining. Wherever the bacterium attaches, the lining stops producing mucus, and the area becomes damaged by acid and digestive enzymes. If promptly diagnosed, antibiotic treatment is usually curative.

Eventually, food mixing with gastric juice in the stomach contents becomes **chyme,** which has a thick, creamy consistency. At the base of the stomach is a narrow opening controlled by a sphincter. Whenever the sphincter relaxes, a small quantity of chyme passes through the opening into the small intestine. When chyme enters the small intestine, it sets off a neural reflex that causes the muscles of the sphincter to contract vigorously and to close the opening temporarily. Then the sphincter relaxes again and allows more chyme to enter. The slow manner in which chyme enters the small intestine allows for thorough digestion.

MP3
Stomach

Animation
Three Phases of
Gastric Secretion

The Small Intestine

The **small intestine** is named for its small diameter (compared to that of the large intestine), but perhaps it should be called the long intestine. The small intestine averages about 6 m in length, compared to the large intestine, which is about 1.5 m in length.

The first 25 cm of the small intestine is called the **duodenum.** A duct brings bile from the liver and gallbladder, and pancreatic juice from the pancreas, into the small intestine (see Fig. 34.12*a*). **Bile** emulsifies fat—emulsification causes fat droplets to disperse in water. The intestine has a slightly basic pH because pancreatic juice contains sodium bicarbonate ($NaHCO_3$), which neutralizes chyme. The enzymes in pancreatic juice and enzymes produced by the intestinal wall complete the process of food digestion.

It has been estimated that the surface area of the small intestine is approximately that of a tennis court. What factors contribute to increasing its surface area? First, the wall of the small intestine contains fingerlike projections called villi (sing., **villus**), which give the intestinal wall a soft, velvety appearance (Fig. 34.10). Second, a villus has an outer layer of columnar epithelial cells, and each of these cells has thousands of microscopic extensions called microvilli. Collectively, in electron micrographs, microvilli give the villi a fuzzy border, known as a "brush border." Because the microvilli bear the intestinal enzymes, these enzymes are called brush-border enzymes. The

other and other organs. The serosa is actually a part of the *peritoneum,* the internal lining of the abdominal cavity.

Stomach

The **stomach** (Fig. 34.9) is a thick-walled, J-shaped organ that lies on the left side of the body beneath the diaphragm. The wall of the stomach has deep folds (rugae) that disappear as the stomach fills to its capacity (approximately 1 liter in humans). Therefore, many animals can periodically eat relatively large meals and spend the rest of their time at other activities.

The stomach is much more than a storage organ, as was discovered by William Beaumont (1785–1853) in the mid-nineteenth century. Beaumont, an American doctor, had a French Canadian patient who had been shot in the stomach, and when the wound healed, he was left with a fistula, or opening, that allowed Beaumont to look inside the stomach and to collect gastric (stomach) juices produced by gastric glands. Beaumont was able to determine that the muscular walls of the stomach contract vigorously and mix food with juices that are secreted whenever food enters the stomach. He found that gastric juice contains hydrochloric acid (HCl) and a substance, now called pepsin, that is active in digestion. Beaumont's work, which was carefully and painstakingly done, pioneered the study of digestive physiology. For similar reasons, modern animal scientists can surgically create fistulas into the rumen of cattle in order to study ruminant nutrition.

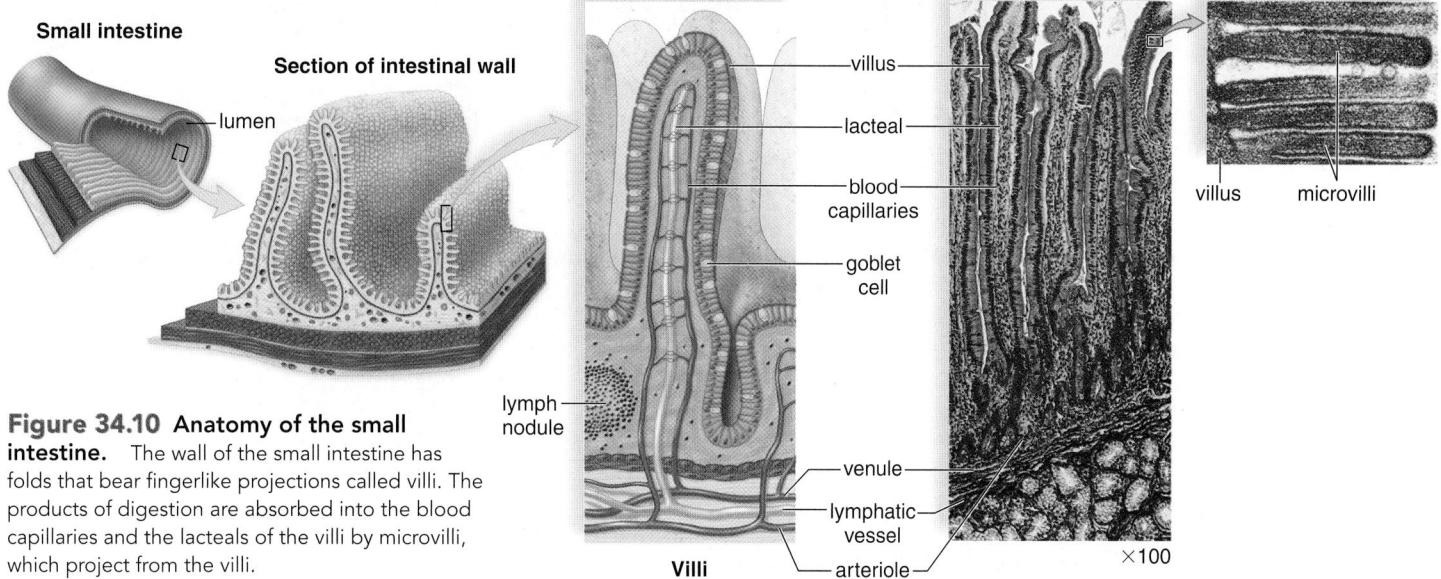

Figure 34.10 Anatomy of the small intestine. The wall of the small intestine has folds that bear fingerlike projections called villi. The products of digestion are absorbed into the blood capillaries and the lacteals of the villi by microvilli, which project from the villi.

microvilli greatly increase the surface area of the villus for the absorption of nutrients.

Nutrients are absorbed into the vessels of a villus, which contains blood capillaries and a lymphatic capillary, called a **lacteal.** Sugars (digested from carbohydrates) and amino acids (digested from proteins) enter the blood capillaries of a villus. Glycerol and fatty acids (digested from fats) enter the epithelial cells of the villi, and within these cells they are joined and packaged as lipoprotein droplets, which enter a lacteal. After nutrients are absorbed, they are eventually carried to all the cells of the body by the bloodstream.

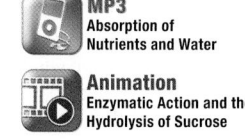

MP3 Absorption of Nutrients and Water

Animation Enzymatic Action and the Hydrolysis of Sucrose

Large Intestine

The **large intestine,** which includes the cecum, the colon, the rectum, and the anus, is larger in diameter (6.5 cm) but shorter in length (1.5 m) than the small intestine. The large intestine absorbs water, salts, and some vitamins. It also stores undigestible material until it is eliminated as feces.

The cecum, which lies below the junction with the small intestine, is the blind end of the large intestine. The cecum has a small projection called the vermiform **appendix** [L. *verm,* worm, and *form,* shape; and *append,* an addition] (Fig. 34.11). The function of the human appendix is unclear, although many experts suggest it may serve as a reservoir for the "good bacteria" that help to maintain our intestinal health. In the case of appendicitis, the appendix becomes infected and so filled with fluid that it may burst. If an infected appendix bursts before it can be removed, it can lead to a serious, generalized infection of the abdominal lining called peritonitis.

The colon joins the rectum, the last 20 cm of the large intestine. About 1.5 liters of water enters the digestive tract daily as a result of eating and drinking. An additional 8.5 liters enter the digestive tract each day carrying the various substances secreted by the digestive glands. About 95% of this water is absorbed by the small intestine, and much of the remaining portion is absorbed by the colon. If this water is not reabsorbed, **diarrhea,** the passing of watery feces, can lead to serious dehydration and ion loss, especially in children.

The large intestine has a large population of bacteria, including *Escherichia coli* and perhaps 400 other species. By taking up space and nutrients, these bacteria provide protection against more pathogenic species. They also produce some vitamins, such as vitamin K, which is necessary to blood clotting. Digestive wastes, or feces, eventually leave the body through the **anus,** the opening of the anal canal.

Feces are normally about 75% water and 25% solid matter. Almost one-third of this solid matter is made up of intestinal bacteria. In fact, there are about 100 billion bacteria per gram of feces! The rest of the solids are undigested plant material, fats, waste products (such as bile pigments), inorganic material, mucus, and dead cells from the intestinal lining. The color of feces is the result of bilirubin breakdown and the presence of oxidized iron. The foul odor is the result of bacterial action.

Video Fat Microbes

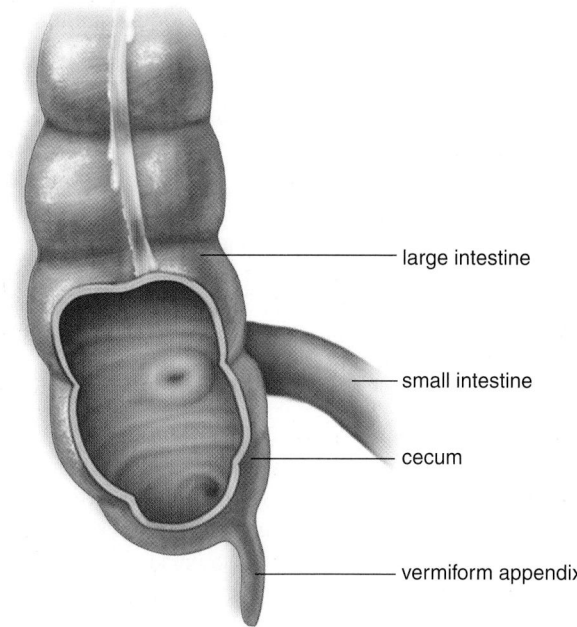

Figure 34.11 Junction of the small intestine and the large intestine.

The colon is subject to the development of **polyps,** which are small growths arising from the mucosa. Polyps, whether they are benign or cancerous, can be removed surgically. Some investigators believe that dietary fat increases the likelihood of colon cancer. Dietary fat causes an increase in bile secretion, and it could be that intestinal bacteria convert bile salts to substances that promote the development of colon cancer. Dietary fiber absorbs water and adds bulk, thereby diluting the concentration of bile salts and facilitating the movement of substances through the intestine. Regular elimination reduces the time that the colon wall is exposed to any cancer-promoting agents in feces.

Three Accessory Organs

The pancreas, liver, and gallbladder are accessory digestive organs. Figure 34.12a shows how the pancreatic duct from the pancreas and the common bile duct from the liver and gallbladder enter the duodenum.

The Pancreas

The **pancreas** lies deep in the abdominal cavity, resting on the posterior abdominal wall. It is an elongated and somewhat flattened organ that has both an endocrine and an exocrine function. As an endocrine gland, it secretes insulin and glucagon, hormones that help keep the blood glucose level within normal limits (see Chapter 40). In this chapter, however, we are interested in its exocrine function. Most pancreatic cells produce pancreatic juice, which contains sodium bicarbonate ($NaHCO_3$) and digestive enzymes for all types of food. Sodium bicarbonate neutralizes acid chyme from the stomach. Pancreatic amylase digests starch, trypsin digests protein, and lipase digests fat.

The Liver

The **liver,** which is the largest gland in the body, lies mainly in the upper right section of the abdominal cavity, under the diaphragm (see Fig. 34.5). The liver contains approximately 100,000 lobules that serve as its structural and functional units (Fig. 34.12b). Triads, located between the lobules, consist of a bile duct, which takes bile away from the liver; a branch of the hepatic artery, which brings oxygen-rich blood to the liver; and a branch of the hepatic portal vein, which transports nutrients to the liver from the intestines (see Fig. 34.10). The central veins of lobules enter a hepatic vein. Blood moves from the intestines to the liver via the hepatic portal vein and from the liver to the inferior vena cava via the hepatic veins.

In some ways, the liver acts as the gatekeeper to the blood. As blood in the hepatic portal vein passes through the liver, it removes many toxic substances and metabolizes them. The liver also removes and stores iron and the vitamins A, B_{12}, D, E, and K. The liver makes many of the proteins found in blood plasma and helps regulate the quantity of cholesterol in the blood.

The liver maintains the blood glucose level at about 100 mg/100 mL (0.1%), even though a person eats intermittently. When insulin is present, any excess glucose present in blood is removed and stored by the liver as glycogen. Between meals, glycogen is broken down to glucose, which enters the hepatic veins, and in this way, the blood glucose level remains constant.

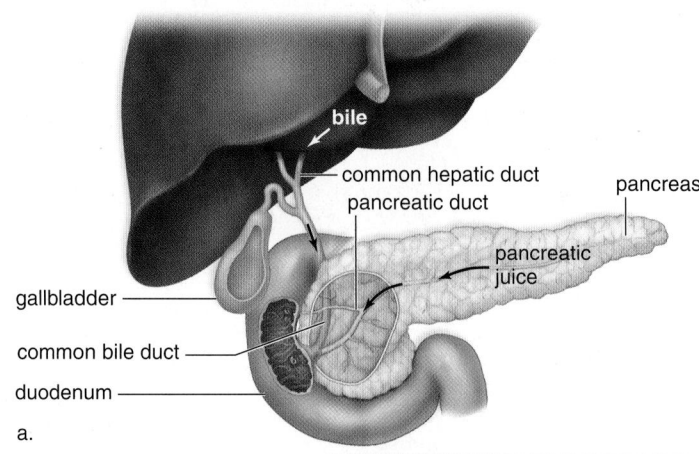

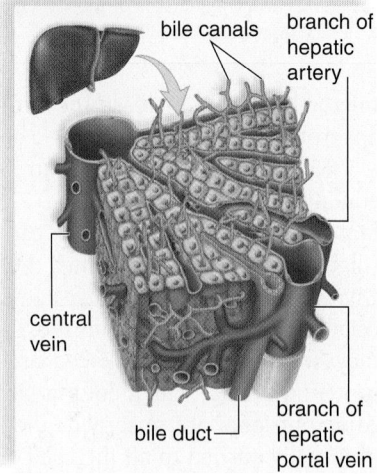

Figure 34.12 Liver, gallbladder, and pancreas.   a. The liver makes bile, which is stored in the gallbladder and sent (black arrow) to the small intestine by way of the common bile duct. The pancreas produces digestive enzymes that are sent (black arrows) to the small intestine by way of the pancreatic duct. **b.** The liver contains over 100,000 lobules. Each lobule contains many cells that perform the various functions of the liver. They remove and add materials to the blood and deposit bile in a duct.

If the supply of glycogen is depleted, the liver converts glycerol (from fats) and amino acids to glucose molecules. The conversion of amino acids to glucose necessitates deamination, the removal of amino groups. By a complex metabolic pathway, the liver then combines ammonia with carbon dioxide to form urea. Urea is the usual nitrogenous waste product from amino acid breakdown in humans.

The liver produces bile, which is stored in the gallbladder. Bile has a yellowish green color because it contains the bile pigment *bilirubin,* derived from the breakdown of hemoglobin, the red pigment of red blood cells. Bile also contains bile salts. Bile salts are derived from cholesterol, and they emulsify fat in the small intestine. When fat is emulsified, it breaks up into droplets, providing a much larger surface area, which can be acted upon by a digestive enzyme from the pancreas.

Liver Disorders.   Because the liver performs so many vital functions, serious disorders of the liver can be life-threatening. When a person has a liver ailment, a yellowing of the skin and the sclera of the eyes called **jaundice** may occur. Jaundice results when the liver is not helping the body excrete excess bilirubin, which is then deposited in the tissues.

Regardless of the cause, inflammation of the liver is called **hepatitis.** The most common causes of hepatitis are viruses. Hepatitis A virus is usually acquired from food or water that has been contaminated with feces. Hepatitis B, which is usually spread by

Nature of Science

New Approaches to Treating Obesity

Diet and exercise. We all know that the most critical factor in weight gain is consuming more calories than we need for our level of physical activity. But considering the rising rates of obesity in developed countries, and the many associated health risks, researchers are investigating factors and approaches that might help some people to reduce their weight:

Adequate Sleep

According to the National Sleep Foundation's annual poll in 2009, Americans got an average of 6.7 hours of sleep on weekdays, down from 7 hours in 2001. Whatever the reasons for this decline, an increasing amount of scientific evidence points to a link between declining sleep and increasing obesity rates.

Even a modest amount of sleep deprivation can cause alterations in hormones that control appetite and regulate metabolism. In a 2004 study, 12 healthy college-age males were divided into two groups: one that slept ten hours a night, and a second that slept only four hours. After two days, the sleep-deprived subjects averaged an 18% decrease in serum leptin (a hormone that normally suppresses appetite) and a 28% increase in ghrelin (an appetite stimulator). Those with the greatest hormonal differences also reported greater increases in hunger, especially for high-carb foods.[1] Other, larger studies have mostly confirmed these findings, suggesting that adequate sleep may be a significant factor in avoiding obesity.

Anti-Obesity Drugs

In general, two types of drugs are available for the treatment of obesity: fat absorption inhibitors and appetite suppressants. Xenical (Orlistat), taken three times a day with meals, interferes with fat digestion by inhibiting pancreatic lipase. Various studies have shown that Xenical is more effective than dietary management alone in promoting weight loss, although the typical patient may lose only a few pounds. Because an increased amount of fat is passing undigested through the digestive tract, some people experience side effects such as abdominal

pain, increased frequency of bowel movements, or even fecal incontinence (inability to control fecal release).

Although several appetite suppressor drugs are available, some have been removed from the market due to safety concerns. Research is continuing on identifying strategies to control hunger using drugs that effect various appetite control mechanisms.

Surgical Procedures

Despite advances in understanding the factors behind obesity, many individuals still struggle with this problem. The number of bariatric (weight loss) surgeries performed per year in the United States has increased from about 16,000 in the early 1990s to an estimated 220,000 in 2008. Many different types of procedures are performed, but all are intended to reduce the size of the stomach, to decrease the absorption of nutrients, or both. When successful, any procedure that reduces the weight of an obese person to a value closer to normal can lead to a significant reduction in the risk of other health problems. However, as with any major surgery, there is also a substantial risk of harm, including death.

Traditionally, bariatric surgery has been recommended only for people who are morbidly obese (BMI greater than 40), or a BMI greater than 35 but with weight-related med-

ical problems. Many of these procedures require removal of stomach tissue, or bypassing the stomach altogether. However, less invasive procedures are being developed, and in December 2010 a panel of FDA advisors recommended that a product called Lap-Band be approved for use in patients whose BMI is as low as 30. As shown in Figure 34A, the Lap-Band is an adjustable plastic strip that is inserted into the abdomen and over the top of the stomach via a small incision. Advantages of this procedure over more traditional surgeries are that the tension can be adjusted for greater or less restriction of stomach volume, and it can be removed. Many experts seem to believe that gastric banding procedures are not as effective as other types of bariatric surgery, however.

Questions to Consider

1. From an evolutionary perspective, why are humans predisposed to gain weight when food supplies are readily available?
2. Besides diet, exercise, and lack of sleep, what are some some other factors that could cause some people to be more prone to obesity?
3. The average cost for various bariatric surgeries is $18,000 to $35,000. With limited health care dollars to go around, do you think this is a justifiable expense, either practically or ethically?

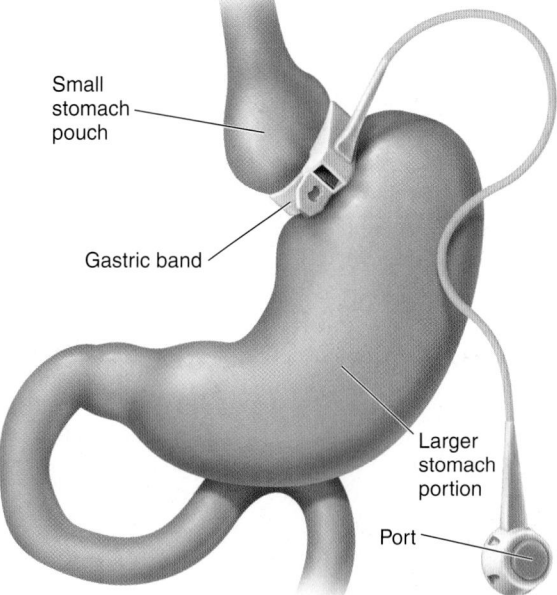

Small stomach pouch

Gastric band

Larger stomach portion

Port

Figure 34A The Lap-Band. The Lap-Band is an inflatable loop that is surgically placed around the upper part of the stomach to reduce the amount of food that can enter. It is approved for use only in obese people for whom other weight loss therapies have failed.

1 Spiegel, K., et al. 2004. Sleep Curtailment in Healthy Young Men. . . . *Ann. Internal Medicine* 141: 846–50.

sexual contact, can also be spread by blood transfusions or contaminated needles. The hepatitis B virus is more contagious than the AIDS virus, which is spread in the same way. A vaccine is now available for hepatitis B, however. Hepatitis C, which is usually acquired by contact with infected blood and for which no vaccine is available, can lead to chronic hepatitis, liver cancer, and death.

Video
Halting Hepatitis

Cirrhosis is another chronic disease of the liver. First, the organ becomes fatty, and then liver tissue is replaced by inactive fibrous scar tissue. Cirrhosis of the liver is often seen in alcoholics, due to malnutrition and to the excessive amounts of alcohol (a toxin) the liver is forced to break down.

The liver has amazing regenerative powers and can recover if the rate of regeneration exceeds the rate of damage. During liver failure, however, there may not be enough time to let the liver heal itself. Liver transplantation is usually the preferred treatment for liver failure, but currently an estimated 4,000 people in the United States alone are waiting for a liver transplant.

Because the liver serves so many functions, artificial livers have been difficult to develop. One type is a cartridge that contains cultured liver cells (either human or pig). Like kidney dialysis, the patient's blood is passed outside of the body and through an apparatus containing the liver cells, which perform their normal functions, and the blood is returned to the patient.

Progress is also being made in the area of transplanting a smaller number of liver cells, as opposed to the entire organ. These cells can either be grown from stem cells, or may be derived from the livers of donors who have died, but whose livers as entire organs are not suitable for transplantation.

The Gallbladder

The **gallbladder** is a pear-shaped, muscular sac attached to the surface of the liver (see Fig. 34.5). About 1,000 mL of bile is produced by the liver each day, and any excess is stored in the gallbladder. Water is reabsorbed by the gallbladder so that bile becomes a thick, mucuslike material. When bile is needed, the gallbladder contracts, releasing bile into the duodenum via the common bile duct (see Fig. 34.12).

The cholesterol content of bile can come out of solution and form crystals called gallstones. These stones can be as small as a grain of sand or as large as a golf ball. The passage of the stones from the gallbladder may block the common bile duct, causing pain as well as possible damage to the liver or pancreas. Then, the gallbladder must be removed.

Check Your Progress   34.2

1. Trace the path of food from the mouth to the large intestine.
2. Describe the likely selective pressures that resulted in the evolution of taste buds.
3. Explain how the stomach, small intestine, and large intestine are each adapted to perform their particular functions.
4. Discuss how each accessory organ contributes to the digestion of food.

34.3 Digestive Enzymes

Learning Outcomes

Upon completion of this section, you should be able to

1. Describe the overall characteristics and functions of digestive enzymes.
2. Compare the specific types of nutrients that are digested in the mouth, stomach, and small intestine.

The various digestive enzymes present in the digestive juices, mentioned earlier, help break down carbohydrates, proteins, nucleic acids, and fats, the major nutritional components of food. Starch is a polysaccharide, and its digestion begins in the mouth. Saliva from the salivary glands has a neutral pH and contains **salivary amylase,** the first enzyme to act on starch.

$$\text{starch} + H_2O \xrightarrow{\text{salivary amylase}} \text{maltose}$$

Maltose molecules cannot be absorbed by the intestine; additional digestive action in the small intestine converts maltose to glucose, which can be absorbed.

Protein digestion begins in the stomach. Gastric juice secreted by gastric glands has a very low pH—about 2.0—because it contains hydrochloric acid (HCl). Pepsinogen, a precursor that is converted to **pepsin** when exposed to HCl, is also present in gastric juice. Pepsin acts on protein to produce peptides.

$$\text{protein} + H_2O \xrightarrow{\text{pepsin}} \text{peptides}$$

Peptides are usually too large to be absorbed by the intestinal lining, but later they are broken down to amino acids in the small intestine.

Starch, proteins, nucleic acids, and fats are all enzymatically broken down in the small intestine. Pancreatic juice, which enters the duodenum, has a basic pH because it contains sodium bicarbonate ($NaHCO_3$). One pancreatic enzyme, **pancreatic amylase,** digests starch (Fig. 34.13a).

$$\text{starch} + H_2O \xrightarrow{\text{pancreatic amylase}} \text{maltose}$$

Another pancreatic enzyme, **trypsin,** digests protein (Fig. 34.13b).

$$\text{protein} + H_2O \xrightarrow{\text{trypsin}} \text{peptides}$$

Trypsin is secreted as trypsinogen, which is converted to trypsin in the duodenum.

Maltase and peptidases, enzymes produced by the small intestine, complete the digestion of starch to glucose and protein to amino acids, respectively. Glucose and amino acids are small molecules that cross into the cells of the villi and enter the blood (Fig. 34.13a, b).

Maltose, a disaccharide that results from the first step in starch digestion, is digested to glucose by **maltase.**

$$\text{maltose} + H_2O \xrightarrow{\text{maltase}} \text{glucose} + \text{glucose}$$

The brush border of the small intestine produces other enzymes for digestion of specific disaccharides. The absence of any one of these enzymes can cause illness. For example, approximately 75% of the world's adult human population is estimated to be lactose intolerant, because of a decreased expression of the enzyme lactase beyond the age of childhood. When such a person ingests milk or other products containing lactose, the undigested sugar is fermented by intestinal bacteria, resulting in a variety of unpleasant intestinal symptoms.

Peptides, which result from the first step in protein digestion, are digested to amino acids by **peptidases.**

$$\text{peptides} + \text{H}_2\text{O} \xrightarrow{\text{peptidases}} \text{amino acids}$$

Lipase, a third pancreatic enzyme, digests fat molecules in fat droplets after they have been emulsified by bile salts.

$$\text{fat} \xrightarrow{\text{bile salts}} \text{fat droplets}$$

$$\text{fat droplets} + \text{H}_2\text{O} \xrightarrow{\text{lipase}} \text{glycerol} + 3 \text{ fatty acids}$$

Specifically, the end products of lipase digestion are monoglycerides (glycerol + one fatty acid) and fatty acids. These enter the cells of the villi, and within these cells, they are rejoined and packaged as lipoprotein droplets, called chylomicrons. Chylomicrons enter the lacteals (Fig. 34.13c).

MP3
Chemical Digestion in the Small Intestine

Check Your Progress 34.3

1. Describe the location(s) in the digestive tract where each of the major types of nutrients is broken down.
2. Explain what final molecule (monomer) results from the digestion of carbohydrates, proteins, and fats.

34.4 Nutrition and Human Health

Learning Outcomes

Upon completion of this section, you should be able to

1. List the major types of nutrients, and provide examples of foods that are a good source of each.
2. Describe the connection between a person's diet and the likely development of obesity, type 2 diabetes, and cardiovascular disease.
3. Distinguish between vitamins, coenzymes, and minerals.

This section of the chapter discusses the components of a balanced human diet, as well as some problems that may arise from consuming a poor diet.

Carbohydrates

Carbohydrates are present in food in the form of sugars, starch, and fiber. Fruits, vegetables, milk, and honey are natural sources of sugars. Glucose and fructose are monosaccharide sugars, and lactose (milk sugar) and sucrose (table sugar) are disaccharides. Disaccharides are broken down in the small intestine, and monosaccharides are absorbed into the bloodstream and delivered to cells. Once inside animal cells, monosaccharides are converted to glucose, much of which is used for the production of ATP by cellular respiration (see Section 8.1).

Plants store glucose as starch, and animals store glucose as glycogen. Good sources of starch are beans, peas, cereal grains, and potatoes. Starch is digested to glucose in the digestive tract, and excess glucose is stored as glycogen. The human liver and muscles can only store a total of about 600 g of glucose in the form of glycogen; excess glucose is converted into fat and stored in adipose tissues.

Although other animals likewise store glucose as glycogen in liver or muscle tissue (meat), little is left by the time an animal is eaten for food. Except for honey and milk, which contain sugars, animal foods do not contain high levels of carbohydrates.

Fiber includes various undigestible carbohydrates derived from plants. Food sources rich in fiber include beans, peas,

Figure 34.13 Digestion and absorption of nutrients. **a.** Starch is digested to glucose, which is actively transported into the epithelial cells of intestinal villi. From there, glucose moves into the bloodstream. **b.** Proteins are digested to amino acids, which are actively transported into the epithelial cells of intestinal villi. From there, amino acids move into the bloodstream. **c.** Fats are emulsified by bile and digested to monoglycerides and fatty acids. These diffuse into epithelial cells, where they recombine and join with proteins to form lipoproteins, called chylomicrons. Chylomicrons enter a lacteal.

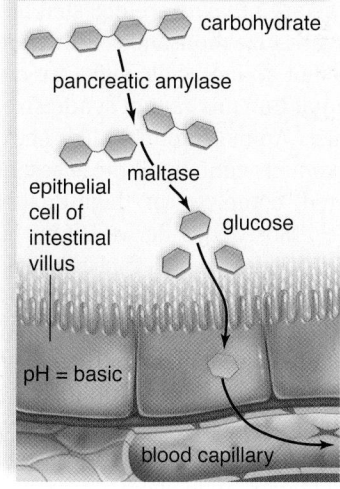

a. Carbohydrate digestion

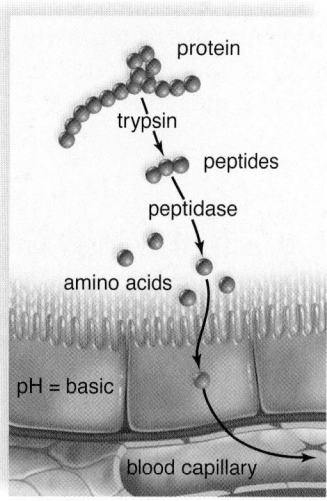

b. Protein digestion

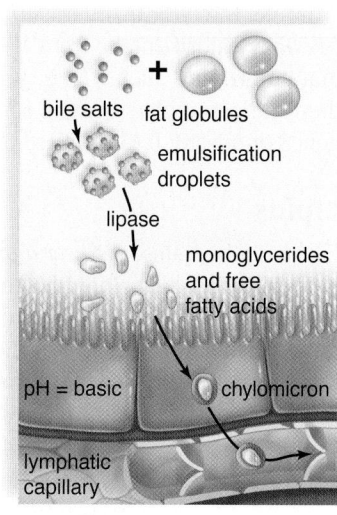

c. Fat digestion

nuts, fruits, and vegetables. Whole-grain products are also a good source of fiber, and are therefore more nutritious than food products made from refined grains. During *refinement*, fiber and also vitamins and minerals are removed from grains, so that primarily starch remains. For example, a slice of bread made from whole-wheat flour contains 3 g of fiber; a slice of bread made from refined wheat flour contains less than 1 g of fiber.

Technically, fiber is not a nutrient for humans because it cannot be digested to small molecules that enter the bloodstream. Insoluble fiber, however, adds bulk to fecal material, which stimulates movement in the large intestine, preventing constipation. Soluble fiber combines with bile acids and cholesterol in the small intestine and prevents them from being absorbed. In this way, high-fiber diets may protect against heart disease. The typical American consumes only about 15 g of fiber each day; the recommended daily intake of fiber is 25 g for women and 38 g for men. To increase your fiber intake, eat whole-grain foods, snack on fresh fruits and raw vegetables, and include nuts and beans in your diet (Fig. 34.14).

If you, or someone you know, has lost weight by following low carbohydrate diets, you may think "carbs" are unhealthy and should be avoided. According to the American Dietetic Association, however, some low-carbohydrate, high-fat diets are potentially hazardous and have no benefits over well-balanced diets that include the same number of calories. In fact, a recent study of over 4,400 Canadian adults found the lowest risk of obesity in people who consumed about half of their calories from carbohydrates.[1] Evidence also suggests that many Americans are not eating the right kind of carbohydrates. In some countries, the traditional diet is 60–70% high-fiber carbohydrates, and these people have a low incidence of the diseases that plague Americans.

A current controversy in human nutrition is the relative risk of consuming high levels of high fructose corn syrup (HFCS), compared to other sweeteners. HFCS, or corn sugar, is now the most commonly used sweetening agent, found in soft drinks and a huge variety of foods that end up on our plates. Many websites and a few research studies have suggested that HFCS is a major factor responsible for the rising epidemic of obesity and related diseases, but many nutritionists contend that the *type* of sugar consumed is not as important as the *amount*. As an example, the typical American obtains about one-sixth of his or her daily caloric intake from HFCS and other sugars. It is likely that consuming such a high percentage of "empty calories" in the form of simple sugars is contributing to the increasing incidence of obesity in the United States.

Lipids

Like carbohydrates, *triglycerides* (fats and oils) supply energy for cells, but *fat* is also stored for the long term in the body. Dietary experts generally recommend that people include unsaturated, rather than saturated, fats in their diets (see Fig. 3.10 to review these structures). Two unsaturated fatty acids, alpha-linolenic and linoleic acids (also called omega-3 fatty acids), are *essential* in the diet, meaning that we cannot synthesize them. Delayed

1 Merchant, A. T., et al. 2009. Carbohydrate intake and overweight-obesity among healthy adults. *J. Am. Dietetic Assn.* 109: 1165–72.

Figure 34.14 Fiber-rich foods. Plants provide a good source of carbohydrates. They also provide a good source of vitamins, minerals, and fiber when they are not processed (refined).

growth and skin problems can develop in people who consume an insufficient amount of these essential unsaturated fatty acids, which are found in high amounts in certain fish and in plant oils like canola and soybean oils.

Another type of lipid, *cholesterol*, is a necessary component of the plasma membrane of all animal cells. It is also a precursor for synthesis of various compounds, including bile, steroid hormones, and vitamin D. Plant foods do not contain cholesterol, but animal foods such as cheese, egg yolks, liver, and certain shellfish (shrimp and lobster) are rich in cholesterol. Elevated blood cholesterol levels are associated with an increased risk of cardiovascular disease, the number one cause of disease-related death in the U.S. (described later on).

Animal-derived foods, such as butter, red meat, whole milk, and cheeses, contain saturated fats, which are also associated with cardiovascular disease. Statistical studies suggest that trans-fatty acids (trans-fats) are even more harmful than saturated fatty acids. Trans-fatty acids arise when unsaturated oils are hydrogenated to produce a solid fat, as in shortening and some margarines. Trans-fats may reduce the function of the plasma membrane receptors that clear cholesterol from the bloodstream. Trans-fats are found in commercially packaged foods, such as cookies and crackers; in commercially fried foods, such as french fries; and in packaged snacks.

Proteins

Dietary *proteins* are digested to amino acids, which cells use to synthesize thousands of different cellular proteins. Of the 20 different amino acids, eight are *essential amino acids* that normal adult humans cannot synthesize and thus must be present in the diet. Animal products like beef, pork, poultry, eggs, and dairy products contain all these essential amino acids and are considered "complete" or "high-quality" protein sources.

Most foods derived from plants do not have as much protein per serving as those derived from animals, and some types of plant foods lack one or more of the essential amino acids. For example, the proteins found in corn have a low content of the essential amino acid lysine (although high-lysine corn has been produced through genetic engineering technology). Approximately 3% of Americans (and millions of people in other countries) are either vegetarians, who avoid eating animal flesh, or vegans, who avoid consuming any products derived from animals. Neither group needs to rely on animal sources of protein.

To meet their protein needs, vegetarians and vegans can eat grains, beans, and nuts in various combinations. Also, tofu, soymilk, and other foods made from processed soybeans are complete protein sources. A 2009 report from the American Dietetic Association states that "Well-planned vegetarian diets are appropriate for individuals during all stages of the life cycle, including pregnancy, lactation, infancy, childhood, and adolescence, and for athletes."[2]

Although a severe deficiency in dietary protein intake can be life-threatening, most Americans probably consume too much protein. Even further, some health food stores are full of protein supplements, aimed mainly at athletes who are trying to build muscle mass. However, both the American and Canadian Dietetic Associations recommend that even athletes should consume only 1 to 1.5 grams of protein per kilogram of body weight per day, which is just slightly higher than the 0.8 grams per kilogram recommended for sedentary people. This means an inactive 150 pound person would need to consume only about 60 grams of protein per day, which is about the amount contained in two cheeseburgers.

When amino acids are broken down, the liver removes the nitrogen portion (*deamination*) and uses it to form urea, which is excreted in urine. The water needed for excretion of urea can cause dehydration when a person is exercising and losing water by sweating. High-protein diets can also increase calcium loss in the urine and encourage the formation of kidney stones. Furthermore, high-protein foods derived from animals often contain a high amount of fat.

Diet and Obesity

As mentioned, the consumption of an excess amount of calories (relative to calories expended) from any source causes storage of these calories in the form of body fat. Obesity can be defined in several ways, including: 1) a condition in which excess body fat has an adverse effect on normal activity and health; 2) weighing over 20% more than the ideal for your height and body build, and 3) having a body mass index (BMI) over 30. A person's BMI can be calculated by dividing weight in kilograms by height in meters squared, or by using an online BMI calculator. As of 2009, most estimates indicate that about 30% of Americans are obese. Obesity raises the risk of many medical conditions, including type 2 diabetes and cardiovascular disease. The seriousness of obesity as a health care problem is evidenced by increasing popularity of surgical procedures designed to reduce food consumption (see the Nature of Science feature).

Type 2 Diabetes

Diabetes mellitus occurs when the hormone insulin is not functioning properly, resulting in abnormally high levels of glucose in the blood. This may occur due to a deficiency of insulin secretion by the pancreas, as in type 1 diabetes, or to an inability of cells to respond to insulin (also called insulin resistance), defined as type 2 diabetes. In both types, the excess blood glucose spills over into the urine, leading to increased urination, thirst, and weight loss. Over time, the high levels of blood glucose, and

2 Craig, W. J., and Mangels, A. R., 2009. Position of the American Dietetic Association: Vegetarian diets. *J. Am. Dietetic Assn.* 109: 1266–82.

Figure 34.15 Exercising for good health. Regular exercise helps prevent and control type 2 diabetes.

lack of other insulin functions, can lead to damage to blood vessels, nerves, eyes, kidneys, and even to death. Type 1 diabetes can usually be successfully managed with insulin injections, but type 2 diabetes can be much more resistant to treatment.

In a 2010 report published in the Journal of the American Medical Association, 4,193 adults were studied for an average of 12.4 years, during which 339, or 8.1%, developed diabetes. Among the key findings, people who gained 20 pounds or more after age 50 had three times the risk of developing diabetes, and the risk was four times greater for those with the biggest waist circumferences and highest BMIs. Because we tend to lose muscle and gain fat as we age, the study authors noted the importance of exercise, as opposed to weight loss alone, in maintaining good health (Fig. 34.15).

Animation
Blood Sugar Regulation in Diabetics

Cardiovascular Disease

Cardiovascular disease is the leading cause of death in the United States. Heart attacks and strokes often occur when arteries become blocked by plaque, which contains saturated fats and cholesterol. Cholesterol is carried in the blood by two types of lipoproteins: low-density lipoprotein (LDL) and high-density lipoprotein (HDL). LDL molecules are considered "bad" because they are like delivery trucks that carry cholesterol from the liver to the cells and to the arterial walls. HDL molecules are considered "good" because they are like garbage trucks that dispose of cholesterol. HDL transports cholesterol from the cells to the liver, which converts it to bile salts that enter the small intestine.

According to the American Heart Association, diets high in saturated fats, trans-fats, and/or cholesterol tend to raise LDL cholesterol levels, while eating unsaturated fats may actually lower LDL cholesterol levels. Furthermore, coldwater fish (e.g., herring, sardines, tuna, and salmon) contain polyunsaturated fatty acids and especially omega-3 fatty acids, which are believed to reduce the risk of cardiovascular disease. However, taking fish oil supplements to obtain omega-3s is not recommended without a physician's approval, because too much of these fatty acids can interfere with normal blood clotting.

The American Heart Association also recommends limiting total cholesterol intake to 300 mg per day. This requires careful selection of the foods we include in our daily diets. For example, an egg yolk contains about 210 mg of cholesterol, which would be two-thirds of the recommended daily intake. Still, this doesn't mean eggs should be eliminated from a healthy diet, because the proteins in them are very nutritious; in fact, most healthy people can eat a couple of whole eggs each week without experiencing an increase in their blood cholesterol levels.

A physician can determine whether blood lipid levels are normal. If a person's cholesterol and triglyceride levels are elevated, modifying the fat content of the diet, losing excess body fat, and exercising regularly can reduce them. If lifestyle changes do not lower blood lipid levels enough to reduce the risk of cardiovascular disease, a physician may prescribe cholesterol-lowering medications.

Video
A Burger a Day

Vitamins and Minerals

Vitamins are organic compounds other than carbohydrates, fats, and proteins that regulate various metabolic activities and must be present in the diet. Many vitamins are part of coenzymes; for example, niacin is the name for a portion of the coenzyme NAD^+, and riboflavin is a part of FAD. Coenzymes are needed in small amounts because they are used over and over again in cells. Not all vitamins are coenzymes, however; vitamin A, for example, is a precursor for the pigment that prevents night blindness.

It has been known for some time that the absence of a vitamin can be associated with a particular disorder. Vitamins are especially abundant in fruits and vegetables, and so it is suggested that we eat about $4^1/_2$ cups of fruits and vegetables per day. Although many foods are now enriched or fortified with vitamins, some individuals are still at risk for vitamin deficiencies, generally as a result of poor food choices.

Animation
B Vitamins

The body also needs about 20 elements called *minerals* for various physiological functions, including regulation of biochemical reactions, maintenance of fluid balance, and incorporation into certain structures and compounds. Occasionally individuals (especially women) do not receive enough iron, calcium, magnesium, or zinc in their diets. Adult females need more iron in the diet than males (18 mg compared to 10 mg) if they are menstruating each month. Many people take calcium supplements, as directed by a physician, to counteract osteoporosis, a degenerative bone disease that especially affects older men and women. Many people consume too much sodium, even double the amount needed. Excess sodium can cause water retention and contribute to hypertension.

Animation
Osteoporosis

Check Your Progress   34.4

1. Review several reasons why a diet that includes plenty of vegetables is generally better for you than a diet that includes excess protein.
2. Discuss the relationship between blood cholesterol, saturated fat intake, and cardiovascular disease.
3. Define vitamin.

CONNECTING *the* CONCEPTS *with the* BIG IDEAS

Energy and Homeostasis

- The villi and microvilli of the intestine help increase surface area, allowing more efficient absorption of materials. (2A3b1*IE*)

Interactions and Systems

- Cooperation between the stomach and small intestine assures an effective digestive process. (4A4a*IE*)
- Organ specialization within the digestive system increases efficiency for the organism. (4B2a2*IE*)

*Find the unabridged version of all EK citations at www.glencoe.com/maderAP11.

▓▓▓ Media Study Tools ▓▓▓

www.glencoe.com/maderAP11

Enhance your study of this chapter with study tools and practice tests. Also ask your instructor about the resources available through ConnectPlus, including the media-rich eBook, interactive learning tools, and animations.

▓▓▓ Summarize ▓▓▓

34.1 Digestive Tracts

A few animals (e.g., sponges) lack a digestive tract; others, such as planarians, have an incomplete digestive tract that has only one opening. An incomplete tract has little specialization. Many other animals, such as earthworms, have a complete digestive tract that has both a mouth and an anus. A complete tract tends to have specialized regions.

Some animals are continuous feeders (e.g., clams, which are filter feeders); others are discontinuous feeders (e.g., squid). Discontinuous feeders need a storage organ for food.

Most mammals have teeth. Herbivores need teeth that can clip off plant material and grind it up. Also, the herbivore's stomach contains bacteria that can digest cellulose. Carnivores need teeth that can tear and rip animal flesh into pieces. Meat is easier to digest than plant material, so the digestive system of carnivores has fewer specialized regions and the intestine is shorter than that of herbivores.

34.2 The Human Digestive System

In the human digestive system, both mechanical and chemical digestion begins in the mouth, where food is chewed and mixed with saliva. Saliva contains salivary amylase, which begins carbohydrate digestion.

Food then passes to the pharynx and down the esophagus by peristalsis to the stomach. The stomach stores and mixes food with mucus and gastric juices to produce chyme. Pepsin begins protein digestion in the stomach.

Chyme passes into the small intestine. The duodenum of the small intestine receives bile from the liver and pancreatic juice from the pancreas. Bile emulsifies fat and readies it for digestion by pancreatic lipase. The pancreas also produces amylase and proteases. These and other intestinal enzymes finish the process of chemical digestion.

The walls of the small intestine have fingerlike projections called villi where small nutrient molecules are absorbed. Amino acids and glucose enter the blood vessels of a villus. Glycerol and fatty acids are joined and packaged as lipoproteins before entering lymphatic vessels called lacteals in a villus.

The large intestine consists of the cecum, the colon, and the rectum, which ends at the anus. The cecum, a blind pouch at the junction of the small and large intestine, has a small projection called the appendix, which sometimes becomes infected and inflamed, necessitating its removal. The large intestine does not produce digestive enzymes; it does absorb water, salts, and some vitamins. Reduced water absorption results in diarrhea. The intake of water and fiber help prevent constipation.

Three accessory organs of digestion—the pancreas, liver, and gallbladder—send secretions to the duodenum via ducts. The pancreas produces pancreatic juice, which contains digestive enzymes for carbohydrates, protein, and fat.

The liver produces bile, which is stored in the gallbladder. The liver receives blood from the small intestine by way of the hepatic portal vein. It has numerous important functions, and any malfunction of the liver is a matter of considerable concern.

34.3 Digestive Enzymes

Digestive enzymes are present in digestive juices and break down food into the nutrient molecules glucose, amino acids, fatty acids, and glycerol. Salivary amylase and pancreatic amylase begin the digestion of starch. Pepsin and trypsin digest protein to peptides. Following emulsification by bile, lipase digests fat to glycerol and fatty acids. Intestinal enzymes finish the digestion of starch and protein.

Each digestive enzyme is present in a particular part of the digestive tract. Salivary amylase functions in the mouth; pepsin functions in the stomach; trypsin, lipase, and pancreatic amylase occur in the intestine along with the various enzymes that digest disaccharides and peptides.

34.4 Nutrition and Human Health

The nutrients released by the digestive process should provide us with an adequate amount of major nutrients, essential amino acids and fatty acids, and all necessary vitamins and minerals.

Carbohydrates are necessary in the diet, but simple sugars and refined starches are not as healthy because they provide calories but little or no fiber, vitamins, or minerals. Proteins supply us with essential amino acids, but many Americans may consume more protein than is healthy. It is also wise to restrict one's intake of meats that are fatty because animal fats are saturated fats. Unsaturated fatty acids, particularly the omega-3 fatty acids, are protective against cardio-

vascular disease, whereas saturated fatty acids may lead to plaque formation, which blocks blood vessels.

Obesity is becoming an increasingly serious problem, especially because it is associated with the development of type 2 diabetes and cardiovascular disease.

Key Terms

anus 653	mouth 650
appendix (vermiform appendix) 653	mucosa 651
	muscularis 651
bile 652	pancreas 654
chyme 652	pancreatic amylase 656
cirrhosis 656	pepsin 656
complete digestive tract 647	peptidase 657
diarrhea 653	peristalsis 651
duodenum 652	pharynx 651
esophagus 651	polyp 654
gallbladder 656	rumen 649
hepatitis 654	salivary amylase 651, 656
incomplete digestive tract 647	salivary gland 651
jaundice 654	serosa 651
lacteal 653	small intestine 652
large intestine 653	stomach 652
lipase 657	submucosa 651
liver 654	trypsin 656
lumen 651	villus 652
maltase 656	

Assess

Reviewing This Chapter

1. Contrast the incomplete digestive tract with the complete digestive tract, using the planarian and earthworm as examples.   647–48
2. Contrast a continuous feeder with a discontinuous feeder, using the clam and squid as examples.   648
3. Contrast the dentition of the mammalian herbivore with that of the mammalian carnivore, using the horse and lion as examples.   649
4. List the parts of the human digestive tract, anatomically describe them, and state the contribution of each to the digestive process.   650–54, 656
5. Discuss the absorption of the products of digestion into the lymphatic and cardiovascular systems.   652–53
6. Describe the structure and function of the large intestine. Name several medical conditions associated with the large intestine.   653–54
7. State the location and describe the functions of the pancreas, the liver, and the gallbladder.   654, 656
8. Name and discuss two serious illnesses of the liver.   654, 656
9. Assume that you have just eaten a ham sandwich. Discuss the digestion of the contents of the sandwich. Mention all the necessary enzymes.   656–57
10. Explain why good nutrition is important to human health. Explain why obesity can lead to type 2 diabetes and cardiovascular disease.   657–60

Testing Yourself

Choose the best answer for each question.

1. Animals that feed discontinuously
   a. have digestive tracts that permit storage.
   b. are always filter feeders.
   c. exhibit extremely rapid digestion.
   d. have a nonspecialized digestive tract.
   e. usually eat only meat.

2. Archaeocytes are associated with
   a. digestion in bacteria.
   b. the digestive tracts of sponges.
   c. the filter organs of continuous feeders.
   d. the human appendix.
   e. ingestion and distribution of food in sponges.

3. The typhlosole within the gut of an earthworm compares best to which of these organs in humans?
   a. teeth in the mouth
   b. esophagus in the thoracic cavity
   c. folds in the stomach
   d. villi in the small intestine
   e. the large intestine because it absorbs water

4. Which of these animals is a continuous feeder with a complete digestive tract?
   a. planarian   d. lion
   b. clam   e. human
   c. squid

5. Tracing the path of food in humans, which step is out of order first?
   a. mouth   c. esophagus   e. stomach
   b. pharynx   d. small intestine   f. large intestine

6. The products of digestion are
   a. large macromolecules needed by the body.
   b. enzymes needed to digest food.
   c. small nutrient molecules that can be absorbed.
   d. regulatory hormones of various kinds.
   e. the foods we eat.

7. Why can a person not swallow food and talk at the same time?
   a. To swallow, the epiglottis must close off the trachea.
   b. The brain cannot control two activities at once.
   c. To speak, air must come through the larynx to form sounds.
   d. A swallowing reflex is only initiated when the mouth is closed.
   e. Both a and c are correct.

8. Which of these could be absorbed directly without need of digestion?
   a. glucose   d. nucleic acid
   b. fat   e. All of these are correct.
   c. protein

9. Which association is incorrect?
   a. protein—trypsin   d. starch—amylase
   b. fat—lipase   e. protein—pepsin
   c. maltose—pepsin

10. Most of the absorption of the products of digestion takes place in humans across
    a. the squamous epithelium of the esophagus.
    b. the convoluted walls of the stomach.
    c. the fingerlike villi of the small intestine.
    d. the smooth wall of the large intestine.
    e. the lacteals of the lymphatic system.

11. The hepatic portal vein is located between
    a. the hepatic vein and the vena cava.
    b. two capillary beds.
    c. the pancreas and the small intestine.
    d. the small intestine and the liver.
    e. Both b and d are correct.

12. Bile in humans
    a. is an important enzyme for the digestion of fats.
    b. is made by the gallbladder.
    c. emulsifies fat.
    d. must be activated during the first few weeks of life.
    e. All of these are correct.

13. Which of these is not a function of the liver in adults?
    a. produces bile
    b. stores glucose
    c. produces urea
    d. makes red blood cells
    e. produces proteins needed for blood clotting

14. The large intestine in humans
    a. digests all types of food.
    b. is the longest part of the intestinal tract.
    c. absorbs water.
    d. is connected to the stomach.
    e. All of these are correct.

15. The appendix connects to the
    a. cecum.   d. large intestine.
    b. small intestine.   e. liver.
    c. esophagus.   f. All of these are correct.

16. Which association is incorrect?
    a. mouth—starch digestion
    b. esophagus—protein digestion
    c. small intestine—starch, lipid, protein digestion
    d. stomach—food storage
    e. liver—production of bile

17. Predict and explain the expected digestive results per test tube for this experiment.

Incubator

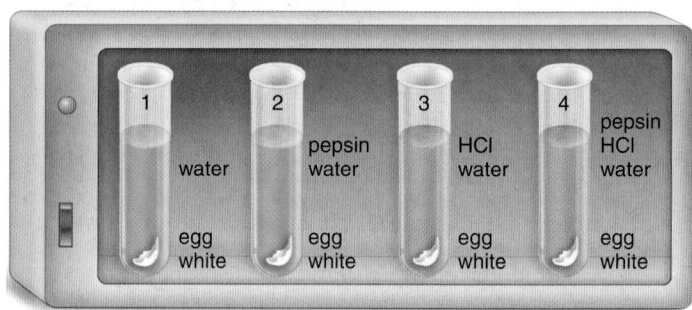

Engage

Virtual Labs
Earthworm Dissection
Virtual Frog Dissection
Nutrition

The virtual labs "Earthworm Dissection" and "Virtual Frog Dissection" provide you with a more detailed look at digestive systems. The virtual lab "Nutrition" allows you to input your personal information to assess the health of your diet.

Thinking Scientifically

1. A drug for leukemia is not broken down in the stomach and is well absorbed by the intestine. However, the molecular form of the drug collected from the blood is not the same as the form that was swallowed by the patient. What explanation is most likely?

2. Snakes often swallow whole animals, a process that takes a long time. Then snakes spend some time digesting their food. What structural modifications would allow slow swallowing and storage of a whole animal to occur? What chemical modifications would be necessary to digest a whole animal?

David Blaine displays an amazing ability to hold his breath while under water, which he has learned through training. Mammals living in aquatic environments have gained this ability through adaptation.

35

Respiratory Systems

O n April 30, 2008, magician David Blaine set a world record by holding his breath for 17 minutes, 4 seconds, while submerged in a glass globe filled with cold water. While this was an amazing feat, Blaine may have been aided by an ancient evolutionary adaptation called the "diving response"—simply immerse your face in cold water, and your heart rate decreases, your spleen may contract (to release stored red blood cells), and blood vessels in your extremities constrict. Taken to the extreme, however, this response—along with decreasing oxygen levels—can lead to painful muscle cramping, and even tissue damage.

Despite efforts to push the limits of human physiology, the true breath-holding champions are aquatic mammals such as the elephant seal, which can dive almost a mile deep and hold its breath for up to two hours. These animals benefit from various evolutionary adaptations: They have more red blood cells per body weight than we do; their muscles contain more oxygen-storing proteins; and they have a particularly effective diving response. Research indicates that elephant seals also tolerate exceptionally low levels of oxygen in their blood. Wherever they live, animals have evolved an amazing variety of strategies for delivering oxygen to their cells and removing carbon dioxide.

As you read through the chapter, think about the following questions:

1. What are some possible evolutionary pressures that might explain why a strictly terrestrial species, such as humans, would have a diving response?

2. Considering the adaptations that are required, what kinds of physiological limitations prevent elephant seals from being able to hold their breath for even longer?

BEFORE YOU BEGIN

Before beginning this chapter, take a few moments to review the following discussions.

Figure 7.4 During which specific parts of photosynthesis do plants produce oxygen, and use carbon dioxide?

Section 8.4 At what point is carbon dioxide produced inside mitochondria, and why is oxygen required?

Figure 32.5 What path does blood travel from the heart to the site of gas exchange in fish, amphibians, and birds?

FOLLOWING *the* BIG IDEAS

CHAPTER 35   RESPIRATORY SYSTEMS

Energy and Homeostasis	The design of animal respiratory systems achieves maximum surface area for gas exchange.
Interactions and Systems	Cooperation between respiratory and circulatory systems is a mark of advanced animals.

35.1 Gas Exchange Surfaces

Learning Outcomes

Upon completion of this section, you should be able to

1. Distinguish between ventilation, external respiration, and internal respiration.
2. Compare and contrast the gas-exchange mechanisms of hydras, earthworms, insects, aquatic vertebrates, and terrestrial vertebrates.
3. Trace the path of a molecule of oxygen as it passes from the human nose to an alveolus.

Respiration is the sequence of events that results in gas exchange between the body's cells and the environment. In terrestrial vertebrates, respiration includes these steps:

- **Ventilation** (i.e., breathing) includes inspiration (entrance of air into the lungs) and expiration (exit of air from the lungs).
- **External respiration** is gas exchange between the air and the blood within the lungs. Blood then transports oxygen from the lungs to the tissues.
- **Internal respiration** is gas exchange between the blood and the tissue fluid. (The body's cells exchange gases with the tissue fluid.) The blood then transports carbon dioxide to the lungs.

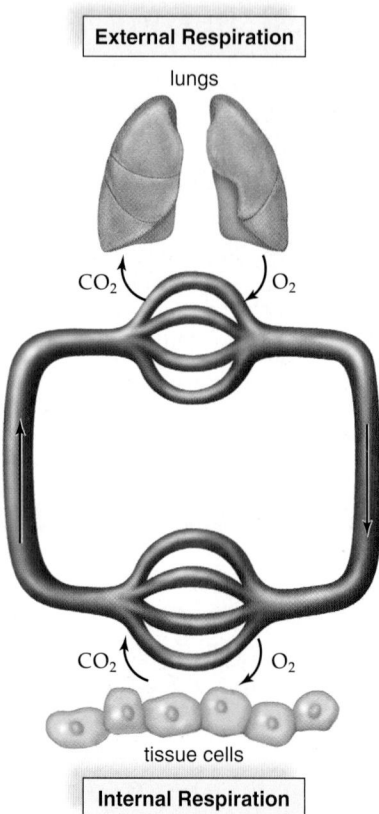

Gas exchange takes place by the physical process of diffusion (see Section 5.2). For external respiration to be effective, the gas-exchange region must be (1) moist, (2) thin, and (3) large in relation to the size of the body.

Some animals are small and shaped in a way that allows the surface of the animal to be the gas-exchange surface. Most complex animals have evolved specialized tissues for external respiration, such as gills in aquatic animals and lungs in terrestrial animals. The effectiveness of diffusion is enhanced by vascularization (the presence of many capillaries), and delivery of oxygen to the cells is promoted when the blood contains a respiratory pigment, such as hemoglobin.

Regardless of the particular external respiration surface and the manner in which gases are delivered to the cells, in the end, oxygen enters mitochondria, where cellular respiration takes place (see Section 8.1). A rare exception to this was discovered in April 2010, when a team of Italian and Danish deep-sea divers discovered a new species of tiny jellyfish-like animals called loriciferans living in sediment more than 10,000 feet below the surface of the Mediterranean Sea that contains almost no oxygen. This discovery represents the first known multicellular animals that do not appear to require oxygen! Subsequent studies have indicated that the cells of these animals may lack mitochondria, but instead contain structures that resemble those used by anaerobic bacteria to undergo cellular respiration in the absence of oxygen. For most animals however, if internal respiration does not occur, ATP production declines dramatically, and life ceases.

Overview of Gas-Exchange Surfaces

It is more difficult for animals to obtain oxygen from water than from air. Water fully saturated with air contains only a fraction of the amount of oxygen that is present in the same volume of air. Also, water is more dense than air. Therefore, aquatic animals expend more energy carrying out gas exchange than do terrestrial animals. Fishes use as much as 25% of their energy output to respire, while terrestrial mammals use only 1–2% of their energy output for that purpose.

Hydras, which are cnidarians, and planarians, which are flatworms, have a large surface area in comparison to their size. This makes it possible for most of their cells to exchange gases directly with the environment. In hydras, the outer layer of cells is in contact with the external environment, and the inner layer can exchange gases with the water in the gastrovascular cavity (Fig. 35.1).

The earthworm is an example of a terrestrial invertebrate that is able to use its body surface for respiration because the capillaries come close to the surface (Fig. 35.2). An earthworm keeps its body surface moist by secreting mucus and by releasing fluids from excretory pores. Further, the worm is behaviorally adapted to remain in damp soil during the day, when the air is driest.

Aquatic invertebrates (e.g., clams and crayfish) and aquatic vertebrates (e.g., fish and tadpoles) have gills that extract oxygen from a watery environment. **Gills** are finely divided, vascularized outgrowths of the body surface or the pharynx (Fig. 35.3a). Various mechanisms are used to pump water across the gills, depending on the organism.

Insects have a system of air tubes called **tracheae** through which oxygen is delivered directly to the cells without entering the blood (Fig. 35.3b). Air sacs located near the wings, legs,

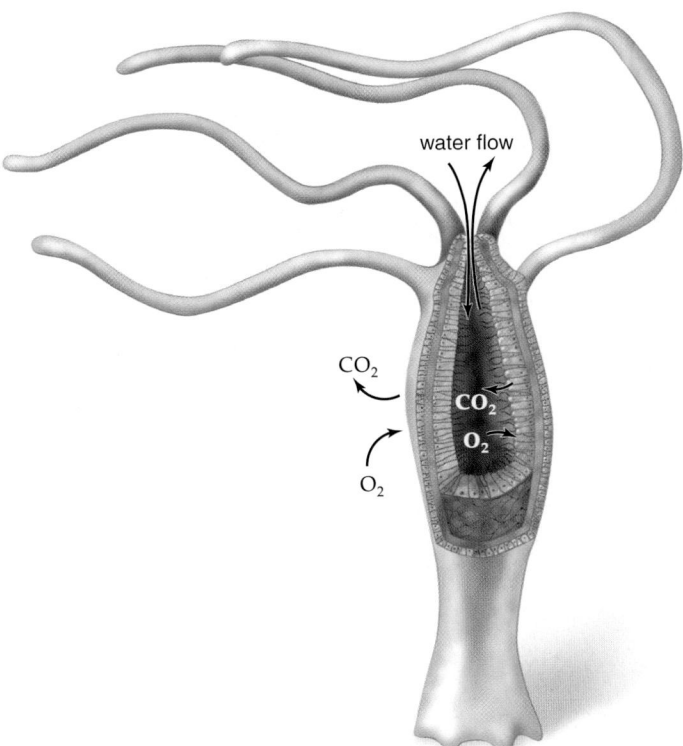

Figure 35.1 Hydra. Some small aquatic animals, such as a hydra, use their body surface for gas exchange. This works because the body surface is large compared to the size of the animal.

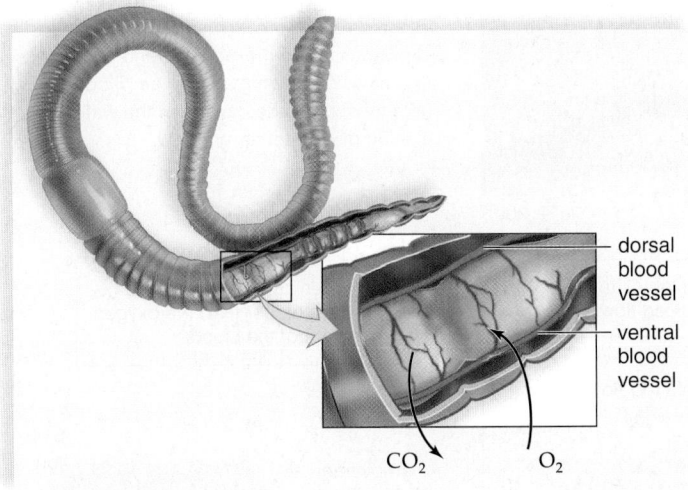

Figure 35.2 Earthworm. An earthworm's entire external surface functions in external respiration.

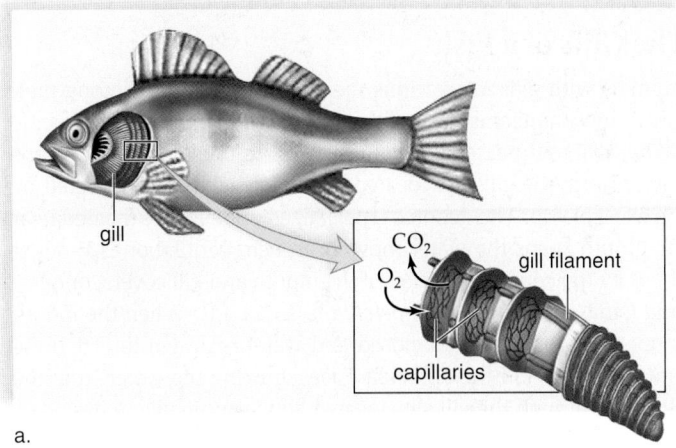

a.

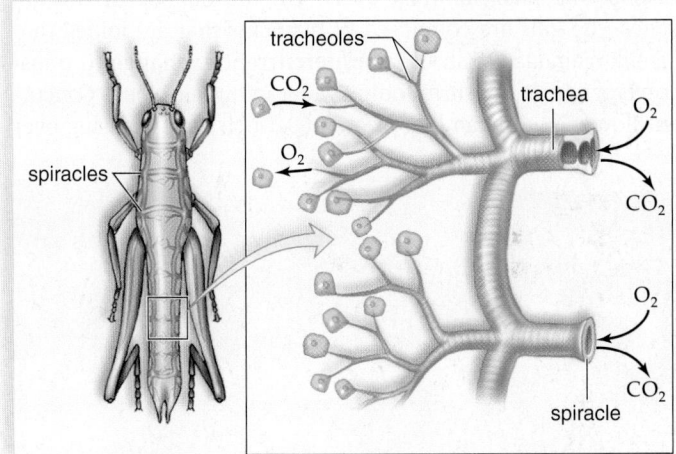

b.

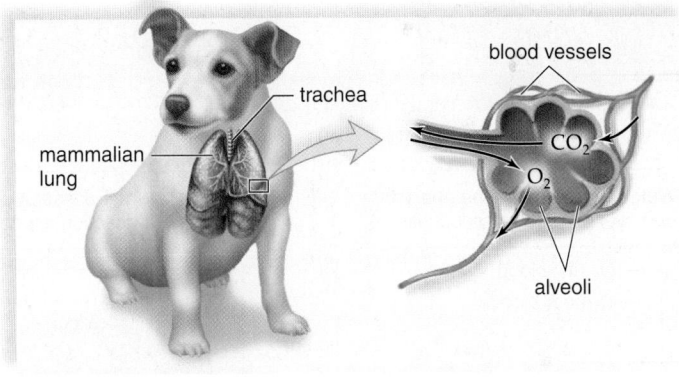

c.

Figure 35.3 Respiratory organs. **a.** Fish have gills to assist external respiration. **b.** Insects have a tracheal system that delivers oxygen directly to their cells. **c.** Vertebrates have lungs with a large total external respiration surface.

and abdomen act as bellows to help move the air into the tubes through external openings.

Terrestrial vertebrates usually have **lungs,** which are vascularized outgrowths from the lower pharyngeal region. The tadpoles of frogs live in the water and have gills as external respiratory organs, but adult amphibians possess simple, saclike lungs. Most amphibians respire to some extent through the skin, and some salamanders depend entirely on the skin, which is kept moist by mucus produced by numerous glands on the surface of the body.

The lungs of birds and mammals are elaborately subdivided into small passageways and spaces (Fig. 35.3c). It has been estimated that human lungs have a total surface area of about 70 square meters, which is about 50 times the skin's surface area. Air is a rich source of oxygen compared to water; however, it does have a drying effect on external respiratory surfaces. A human loses about 350 mL of water per day through respiration when the air has a relative humidity of only 50%. To keep the lungs from drying out, air is moistened as it moves in through the passageways leading to the lungs.

The Gills of a Fish

Animals with gills use various means of ventilation. Among molluscs, such as clams or squids, water is drawn into the mantle cavity, where it passes through the gills. In crustaceans like crabs and shrimp, the gills are located in thoracic chambers covered by the exoskeleton. The action of specialized appendages located near the mouth keeps the water moving. In fish, ventilation is brought about by the combined action of the mouth and gill covers, or opercula (sing., operculum; L. *operculum,* small lid). When the mouth is open, the opercula are closed and water is drawn in. Then the mouth closes, and the opercula open, drawing the water from the pharynx through the gill slits located between the gill arches.

As mentioned, the gills of bony fishes are outward extensions of the pharynx (Fig. 35.4). On the outside of the gill arches, the gills are composed of filaments that are folded into platelike lamellae. Fish use **countercurrent exchange** to transfer oxygen from the surrounding water into their blood. *Concurrent flow would mean that oxygen (O_2)-rich water passing over

the gills would flow in the same direction as oxygen-poor blood in the blood vessels. This arrangement results in an equilibrium point, at which only half the oxygen in the water would be captured. *Counter*current flow, in contrast, means that the two fluids flow in opposite directions. With countercurrent flow, as blood gains oxygen, it always encounters water having an even higher oxygen content. A countercurrent mechanism prevents an equilibrium point from being reached, and about 80–90% of the initial dissolved oxygen in water is extracted.

The Tracheal System of Insects

Arthropods are coelomate animals, but the coelom is reduced and the internal organs lie within a cavity called the hemocoel because it contains hemolymph, a mixture of blood and lymph (see Chapter 32). Hemolymph flows freely through the hemocoel, making circulation in arthropods inefficient. Many insects are adapted for flight, and their flight muscles require a steady supply of oxygen.

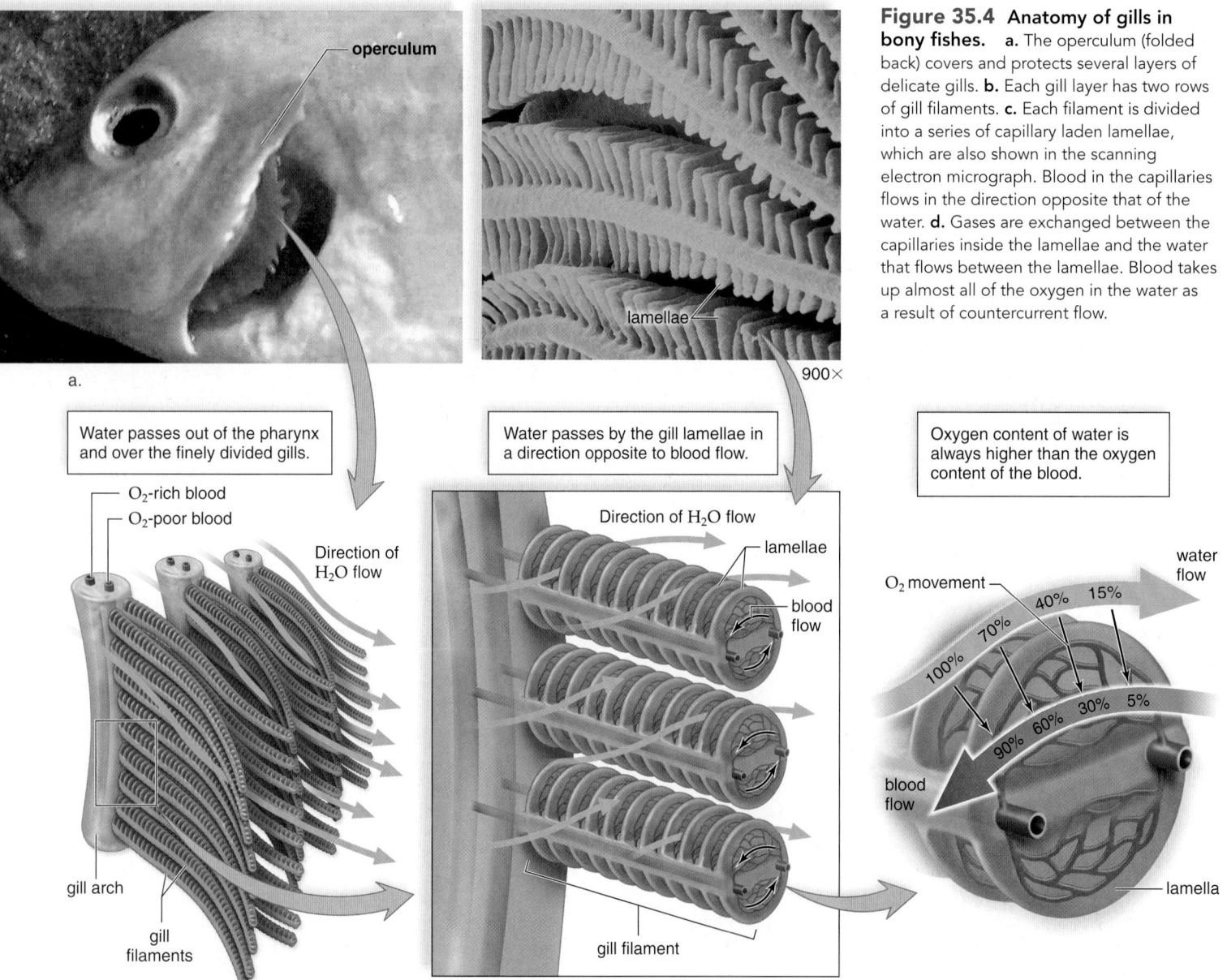

Figure 35.4 Anatomy of gills in bony fishes. a. The operculum (folded back) covers and protects several layers of delicate gills. **b.** Each gill layer has two rows of gill filaments. **c.** Each filament is divided into a series of capillary laden lamellae, which are also shown in the scanning electron micrograph. Blood in the capillaries flows in the direction opposite that of the water. **d.** Gases are exchanged between the capillaries inside the lamellae and the water that flows between the lamellae. Blood takes up almost all of the oxygen in the water as a result of countercurrent flow.

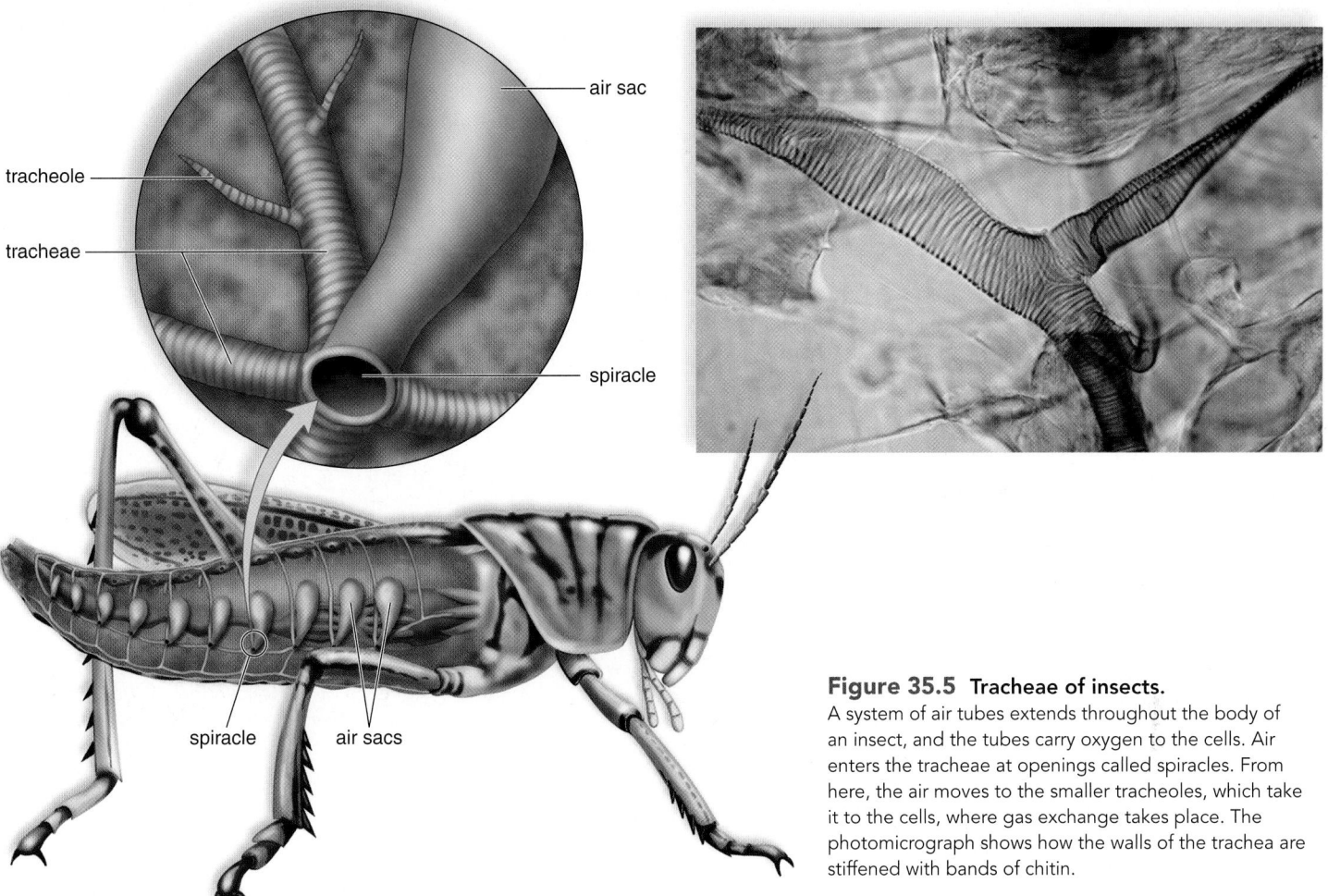

tracheole

tracheae

air sac

spiracle

spiracle   air sacs

Figure 35.5 Tracheae of insects.
A system of air tubes extends throughout the body of an insect, and the tubes carry oxygen to the cells. Air enters the tracheae at openings called spiracles. From here, the air moves to the smaller tracheoles, which take it to the cells, where gas exchange takes place. The photomicrograph shows how the walls of the trachea are stiffened with bands of chitin.

Insects overcome the inefficiency of their blood flow by having a respiratory system that consists of *tracheae,* tiny air tubes that take oxygen directly to the cells (Fig. 35.5). The tracheae branch into even smaller tubules called tracheoles, which also branch and rebranch until finally the air tubes are only about 0.1 μm in diameter. There are so many fine tracheoles that almost every cell is near one. Also, the tracheoles indent the plasma membrane so that they terminate close to mitochondria. Therefore, O_2 can flow more directly from a tracheole to mitochondria, where cellular respiration occurs. The tracheae also dispose of CO_2.

The tracheoles are fluid-filled, but the larger tracheae contain air and open to the outside by way of spiracles (Fig. 35.5). Usually, the spiracle has some sort of closing device that reduces water loss, and this may be why insects have no trouble inhabiting drier climates. Scientists have determined that the tracheae can actually expand and contract, thereby drawing air into and out of the system. To improve the efficiency of the tracheal system, many larger insects also have air sacs, which are thin-walled and flexible, located near major muscles. Contraction of these muscles causes the air sacs to empty, and relaxation causes the air sacs to expand and draw in air. This method is comparable to the way that human lungs expand to draw air into them.

Even with all these adaptations, insects still lack the efficient circulatory system of birds and mammals that is able to pump oxygen-rich blood through arteries to all the cells of the body. This may be why insects have remained relatively small, despite the attempts of science-fiction movies to make us think otherwise.

A tracheal system is an adaptation to breathing air, and yet some insect larval stages, and even some adult insects, live in the water. In these instances, the tracheae do not receive air by way of spiracles. Instead, diffusion of oxygen across the body wall supplies the tracheae with oxygen. Mayfly and stonefly nymphs have thin extensions of the body wall called tracheal gills—the tracheae are particularly numerous in this area. This is an interesting adaptation because it dramatizes that tracheae function to deliver oxygen in the same manner as vertebrate blood vessels.

Some aquatic insects have developed a different strategy. Like most insects, water beetles breathe through spiracles. Because they live in water however, they capture a bubble of air from the surface, and carry it with them, exchanging the oxygen inside for CO_2. Water spiders even spin an underwater web, which they fill with air bubbles, forming what some scientists have called an "external lung."

The Lungs of Humans

The human respiratory system includes all of the structures that conduct air in a continuous pathway to and from the lungs (Fig. 35.6*a*). The lungs lie deep in the body, within the thoracic cavity, where they are protected from drying out. As air moves through the nose, the pharynx, the trachea, and the bronchi to the lungs, it is filtered so that it is free of debris, warmed, and humidified. By the time the air reaches the lungs, it is at body temperature and saturated with water.

In the nose, hairs and cilia act as a screening device. In the trachea and the bronchi, cilia beat upward, carrying mucus, dust, and occasional small bits of food that "went down the wrong way" back into the throat, where the accumulation may be swallowed or expectorated (Fig. 35.6b).

The hard and soft palates separate the nasal cavities from the mouth, but the air and food passages cross in the **pharynx.** This arrangement may seem inefficient, and there is danger of choking if food accidentally enters the trachea; however, it does have the advantage of letting you breathe through your mouth in case your nose is plugged up. In addition, it permits greater intake of air during heavy exercise, when greater gas exchange is required.

Air passes from the pharynx through the **glottis,** which is an opening into the **larynx,** or voice box. At the edges of the glottis are two folds of connective tissue covered by mucous membrane, called the **vocal cords.** These flexible and pliable bands vibrate against each other, producing sound when air is expelled past them through the glottis from the larynx.

The larynx and the trachea remain open to receive air at all times. The larynx is held open by a complex of nine cartilages, among them the Adam's apple. Easily seen in many men, the Adam's apple resembles a small, rounded apple just under the skin in the front of the neck. The **trachea** (windpipe) is held open by a series of C-shaped, cartilaginous rings that do not completely meet in the rear. When food is being swallowed, the larynx rises, and the glottis is covered by a flap of tissue called the **epiglottis.** A backward movement of the soft palate covers

the entrance of the nasal passages into the pharynx. The food then enters the esophagus, which lies behind the larynx.

The trachea divides into two primary **bronchi,** which enter the right and left lungs. Branching continues, eventually forming a great number of smaller passages called **bronchioles.** The two bronchi resemble the trachea in structure, but as the bronchial tubes divide and subdivide, their walls become thinner, and rings of cartilage are absent. Each bronchiole terminates in an elongated space enclosed by a multitude of air pockets, or sacs, called **alveoli,** which fill the internal region of the lungs (Fig. 35.6c). Internal gas exchange occurs between the air in the alveoli and the blood in the capillaries.

Check Your Progress   35.1

1. List some features common to animals like hydras, earthworms, and salamanders, which are able to exchange gases directly with their environment.
2. Explain why the countercurrent flow that occurs in the gills of fish is much more efficient than concurrent flow would be.
3. Describe the role of each of the following in insect respiration: hemolymph, tracheae, tracheoles, spiracles, air sacs, tracheal gills.

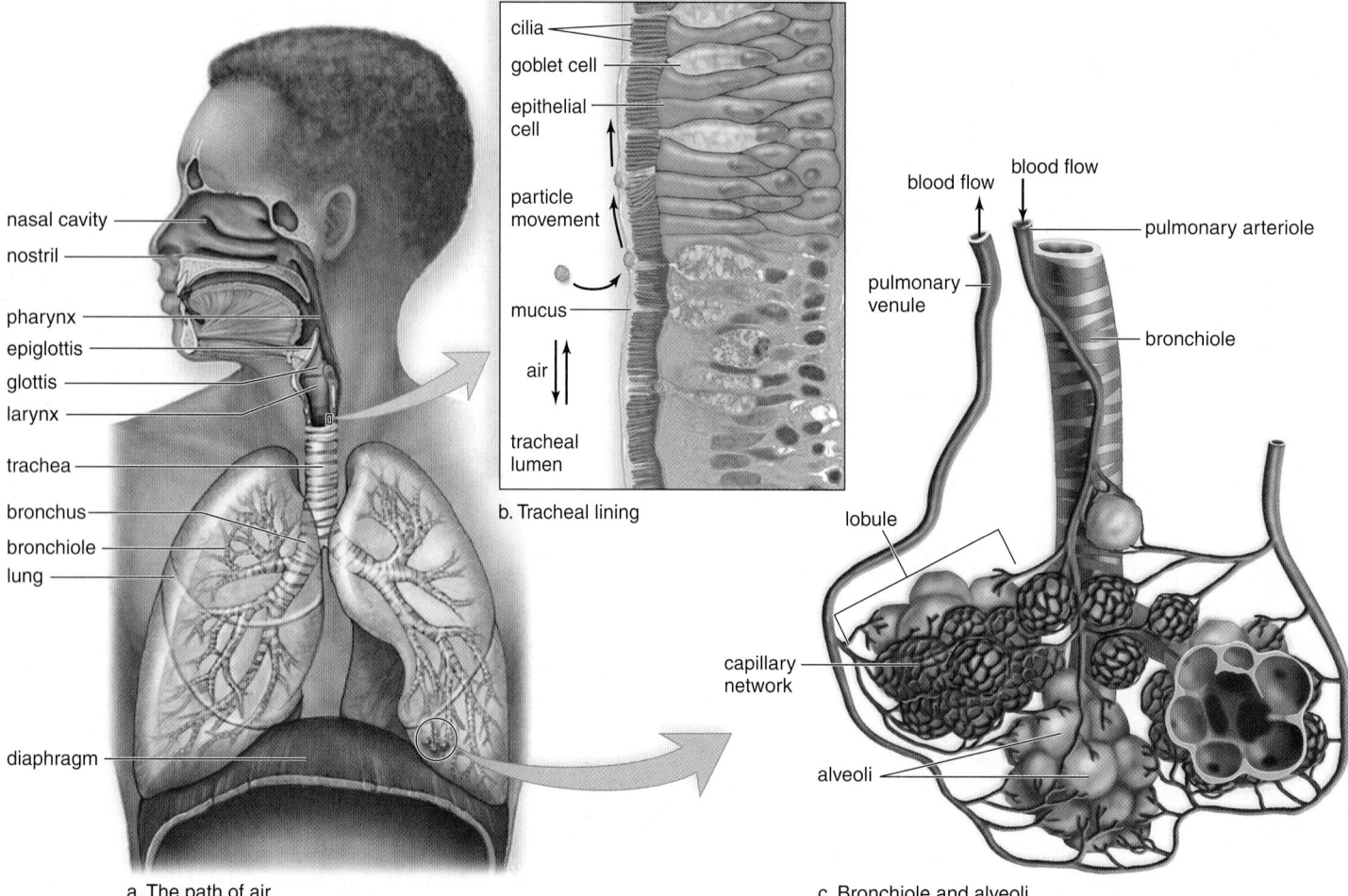

Figure 35.6 The human respiratory tract.   a. The respiratory tract extends from the nose to the lungs, which are composed of air sacs called alveoli. **b.** The lining of the trachea is a ciliated epithelium with mucus-producing goblet cells. The lining prevents inhaled particles from reaching the lungs: the mucus traps the particles, and the cilia help move the mucus toward the throat, where it can be swallowed or expectorated. **c.** Gas exchange occurs between air in the alveoli and blood within a capillary network that surrounds the alveoli.

35.2 Breathing and Transport of Gases

Learning Outcomes

Upon completion of this section, you should be able to

1. Compare the mechanisms used by amphibians, mammals, and birds to inflate their lungs.
2. Explain how the breathing rate in humans is influenced by both physical and chemical factors.
3. Describe how carbon dioxide (CO_2) is carried in the blood, and the effect that blood P_{CO_2} has on blood pH.

During breathing, the lungs are ventilated. Oxygen (O_2) moves into the blood, and carbon dioxide (CO_2) moves out of the blood into the lungs. Blood transports O_2 to the body's cells and CO_2 from the cells to the lungs.

Breathing

Terrestrial vertebrates ventilate their lungs by moving air into and out of the respiratory tract. Amphibians use positive pressure to force air into the respiratory tract. With the mouth and nostrils firmly shut, the floor of the mouth rises and pushes the air into the lungs. Reptiles, birds, and mammals use negative pressure to move air into the lungs and positive pressure to move it out. **Inspiration** (or inhalation) is the act of moving air into the lungs, and **expiration** (or exhalation) is the act of moving air out of the lungs.

Reptiles have jointed ribs that can be raised to expand the lungs, but mammals have both a rib cage and a diaphragm. The

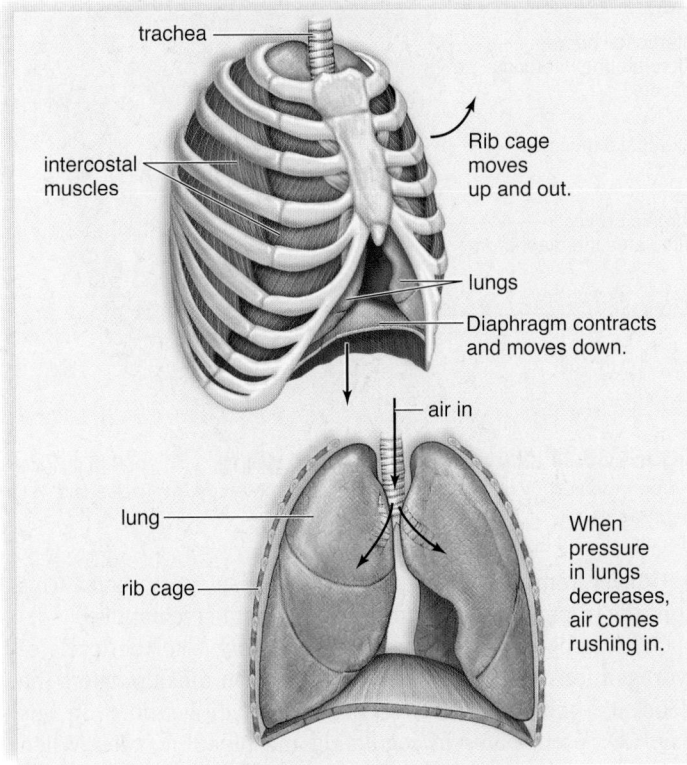

Figure 35.7 Inspiration. During inspiration, the thoracic cavity and lungs expand so that air is drawn in.

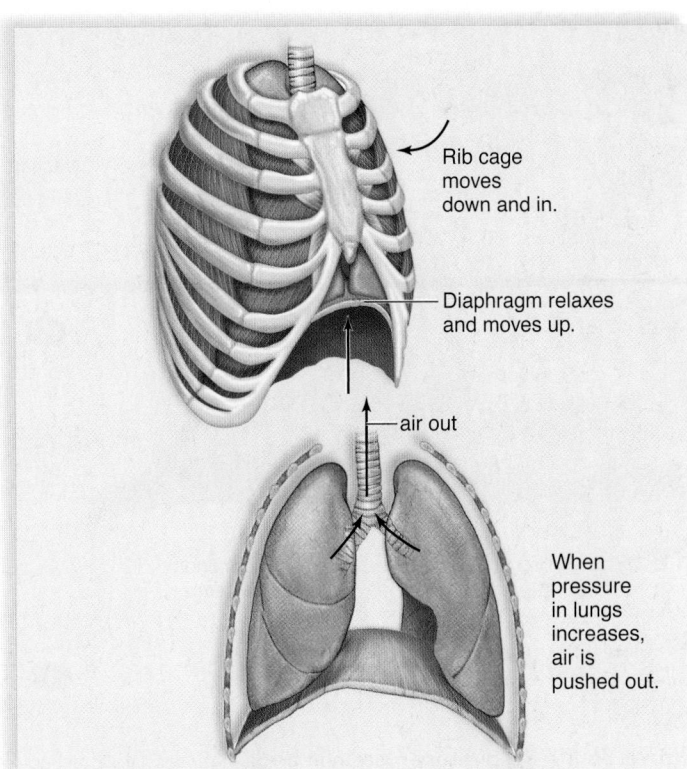

Figure 35.8 Expiration. During expiration, the thoracic cavity and lungs resume their original positions and pressures. Now, air is forced out.

diaphragm is a horizontal muscle that divides the thoracic cavity (above) from the abdominal cavity (below). During inspiration in mammals, the rib cage moves up and out, and the diaphragm contracts and moves down (Fig. 35.7). As the thoracic (chest) cavity expands and lung volume increases, air flows into the lungs due to decreased air pressure in the thoracic cavity and lungs. Also, inspiration is the active phase of breathing in reptiles and mammals.

During expiration in mammals, the rib cage moves down, and the diaphragm relaxes and moves up to its former position (Fig. 35.8). No muscle contraction is required, thus expiration is the passive phase of breathing in reptiles and mammals. During expiration, air flows out as a result of increased pressure in the thoracic cavity and lungs.

We can compare ventilation in reptiles and mammals to the way a bellows, used to fan a fire, functions (Fig. 35.9). First, the handles of the bellows are pulled apart, decreasing the air pressure inside the bellows. This causes air to automatically flow into the bellows, just as air automatically enters the lungs

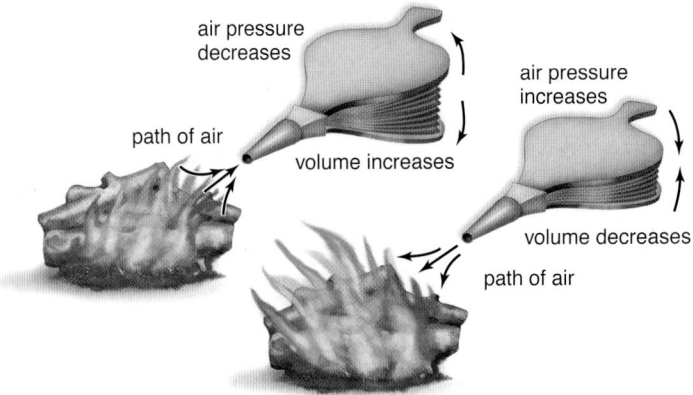

Figure 35.9 Bellows. Lungs function much as bellows do.

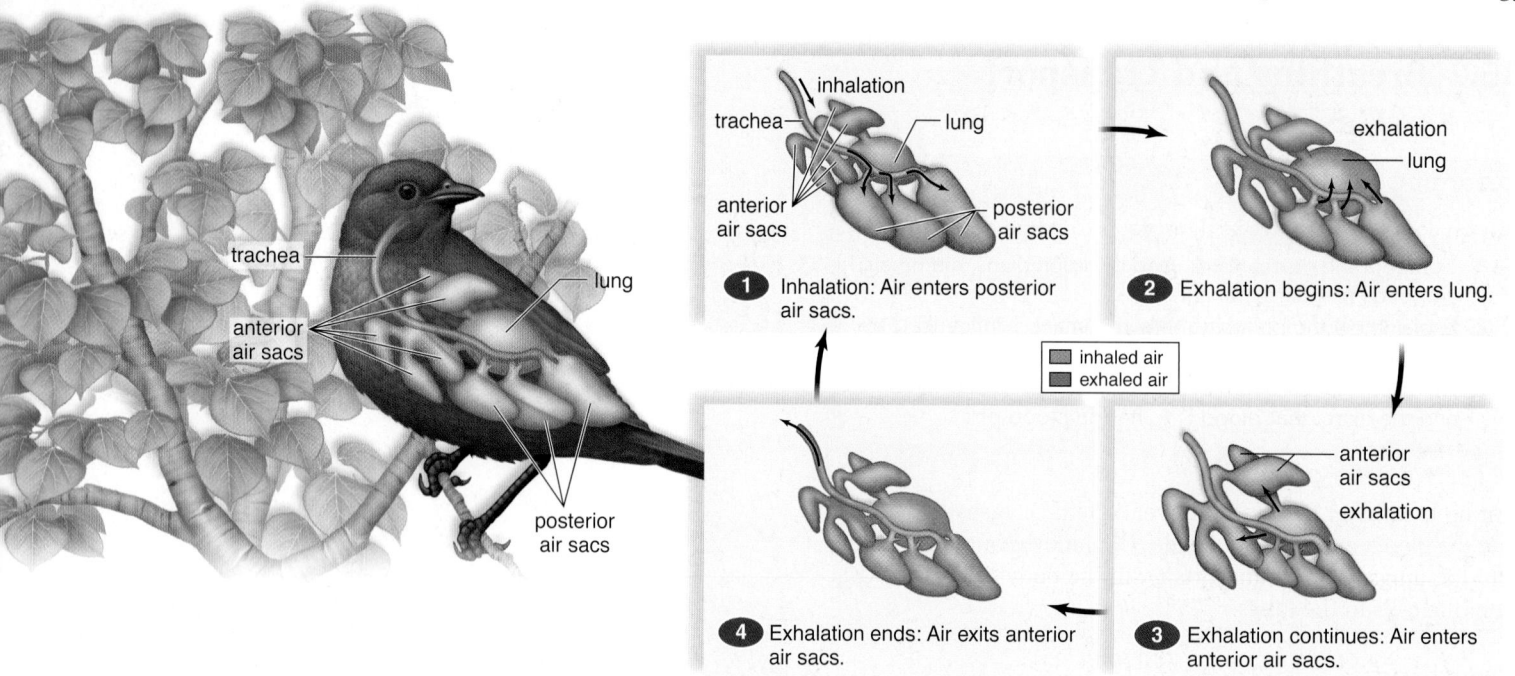

Figure 35.10 Respiratory system in birds. Air sacs are attached to the lungs of birds. These allow birds to have a one-way mechanism of ventilating their lungs.

because the rib cage moves up and out during inspiration. Then, when the handles of the bellows are pushed together, air automatically flows out because the air pressure increases inside the bellows. Similarly, air automatically exits the lungs when the rib cage moves down and in during expiration. The analogy is not exact, however, because no force is required for the rib cage to move down, and inspiration is the only active phase of breathing. Forced expiration can occur if we so desire, however.

All terrestrial vertebrates, except birds, use a *tidal ventilation mechanism,* so called because the air moves in and out by the same route. This means that the lungs of amphibians, reptiles, and mammals are not completely emptied and refilled during each breathing cycle. Because of this, the air entering mixes with used air remaining in the lungs. Although this does help conserve water, it also decreases gas-exchange efficiency. In contrast, birds use a *one-way ventilation mechanism* (Fig. 35.10). Incoming air is carried past the lungs by a trachea, which takes it to a set of posterior air sacs. The air then passes forward through the lungs into a set of anterior air sacs. From here, it is finally expelled. Notice that fresh air never mixes with used air in the lungs of birds, thereby greatly improving gas-exchange efficiency.

Modifications of Breathing in Humans

Normally, adults have a breathing rate of 12 to 20 ventilations per minute. The rhythm of ventilation is controlled by a **respiratory center** in the medulla oblongata of the brain. The respiratory center automatically sends out impulses by way of a spinal nerve to the diaphragm (phrenic nerve) and intercostal nerves to the intercostal muscles of the rib cage (Fig. 35.11). Now inspiration occurs. Then, when the respiratory center stops sending neuronal signals to the diaphragm and the rib cage, expiration occurs.

Although the respiratory center automatically controls the rate and depth of breathing, its activity can also be influenced by nervous input and chemical input. Following forced inhalation, stretch receptors in the alveolar walls initiate inhibitory nerve impulses

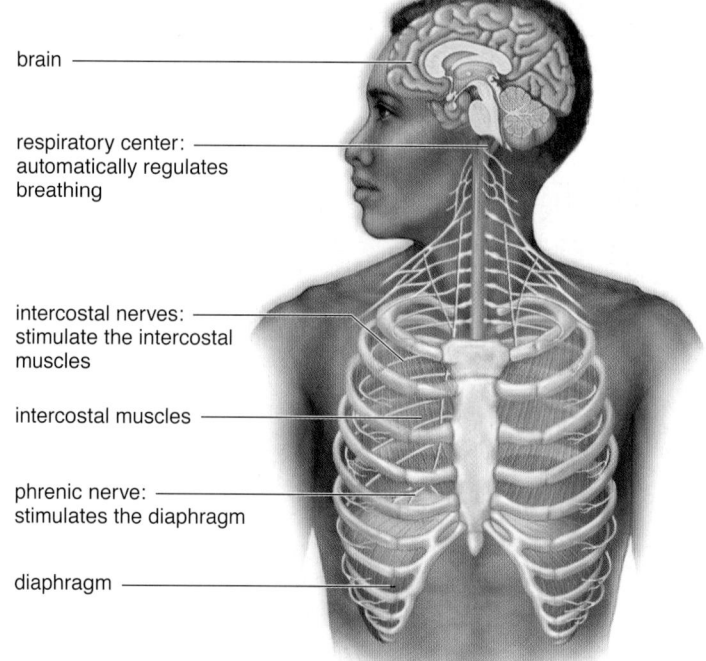

Figure 35.11 Nervous control of breathing. The breathing rate can be modified by nervous stimulation of the intercostal muscles and diaphragm.

that travel from the inflated lungs to the respiratory center. This stops the respiratory center from sending out nerve impulses.

The respiratory center is directly sensitive to the levels of hydrogen ions (H^+). However, when carbon dioxide enters the blood, it reacts with water and releases hydrogen ions. In this way, CO_2 participates in regulating the breathing rate. When hydrogen ions rise in the blood and the pH decreases, the respiratory center increases the rate and depth of breathing. The

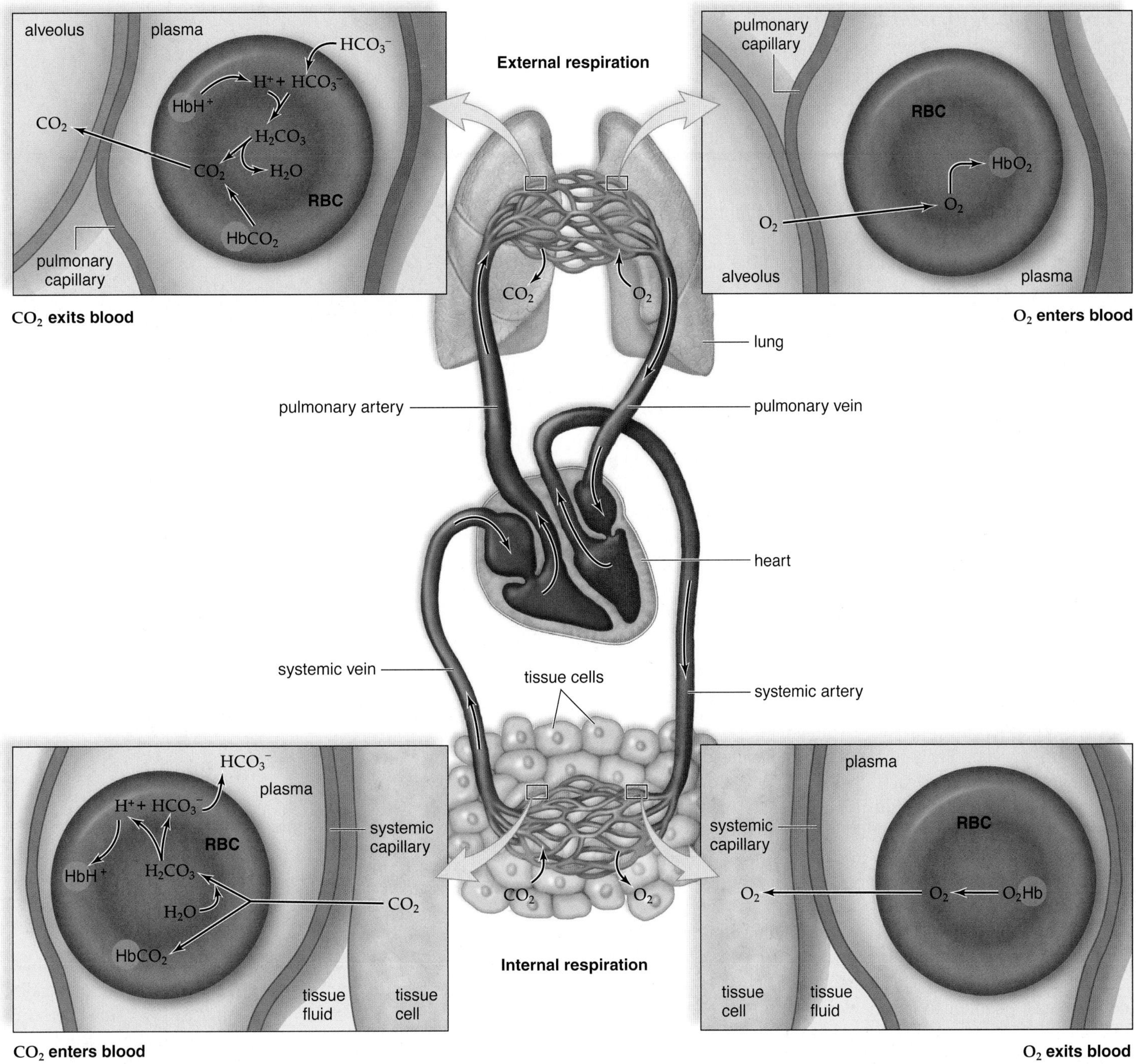

CO₂ exits blood

alveolus plasma

HCO_3^-

$H^+ + HCO_3^-$

HbH^+

CO_2

H_2CO_3

CO_2 H_2O

RBC

$HbCO_2$

pulmonary capillary

External respiration

O₂ enters blood

pulmonary capillary

RBC

HbO_2

O_2

O_2

alveolus   plasma

lung

pulmonary artery

pulmonary vein

CO_2 O_2

heart

systemic vein

tissue cells

systemic artery

Internal respiration

CO₂ enters blood

HCO_3^-

plasma

$H^+ + HCO_3^-$

RBC

HbH^+ H_2CO_3

H_2O CO_2

$HbCO_2$

systemic capillary

tissue fluid   tissue cell

O₂ exits blood

plasma

systemic capillary

RBC

O_2 O_2 O_2Hb

tissue cell   tissue fluid

CO_2 O_2

Figure 35.12 External and internal respiration. During external respiration (*top*) in the lungs, carbon dioxide (CO₂) leaves blood, and oxygen (O₂) enters blood. During internal respiration (*bottom*) in the tissues, oxygen leaves blood, and carbon dioxide enters blood.

chemoreceptors in the **carotid bodies,** located in the carotid arteries, and in the **aortic bodies,** located in the aorta, stimulate the respiratory center during intense exercise due to a reduction in pH, and also if and when arterial oxygen decreases to 50% of normal.

MP3 Control of Respiration

Gas Exchange and Transport

Respiration includes the exchange of gases in our lungs, called external respiration, as well as the exchange of gases in the tissues, called internal respiration (Fig. 35.12). The principles of diffusion largely govern the movement of gases into and out of blood vessels in the lungs and in the tissues. Gases exert pressure, and the amount of pressure each gas exerts is called the **partial pressure,** symbolized as P_{O_2} and P_{CO_2}. If the partial pressure of oxygen differs across a membrane, oxygen will diffuse from the higher to the lower pressure. Similarly, carbon dioxide diffuses from the higher to the lower partial pressure.

Ventilation causes the alveoli of the lungs to have a higher P_{O_2} and a lower P_{CO_2} than the blood in pulmonary capillaries, and this accounts for the exchange of gases in the lungs. When blood reaches the tissues, cellular respiration in cells causes the tissue fluid to have a lower P_{O_2} and a higher P_{CO_2} than the blood in the systemic capillaries, and this accounts for the exchange of gases in the tissues.

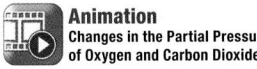

MP3 Gas Exchange

Animation Changes in the Partial Pressure of Oxygen and Carbon Dioxide

Transport of Oxygen and Carbon Dioxide

The transport of O_2 and CO_2 is somewhat different in external respiration than in internal inspiration, although the driving forces of diffusion are the same.

External Respiration. As blood enters the lungs, a small amount of CO_2 is being carried by hemoglobin with the formula $HbCO_2$. Also, some hemoglobin is carrying hydrogen ions with the formula HbH^+. Most of the CO_2 in the pulmonary capillaries is carried as bicarbonate ions (HCO_3^-) in the plasma. As the free CO_2 from the following equation begins to diffuse out, this reaction is driven to the right:

$$H^+ + HCO_3^- \rightarrow H_2CO_3 \rightarrow H_2O + CO_2$$

$$\text{hydrogen} \quad \text{bicarbonate} \quad \text{carbonic} \quad \text{water} \quad \text{carbon}$$
$$\text{ion} \qquad \text{ion} \qquad \text{acid} \qquad\qquad \text{dioxide}$$

The reaction occurs in red blood cells, where the enzyme **carbonic anhydrase** speeds the breakdown of carbonic acid (see Fig. 35.12, *top left*). Pushing this equation to the far right by breathing fast can cause you to stop breathing for a time; pushing this equation to the left by not breathing is even more temporary because breathing will soon resume due to the rise in H^+.

Most oxygen entering the pulmonary capillaries from the alveoli of the lungs combines with **hemoglobin (Hb)** in red blood cells (RBCs) to form **oxyhemoglobin** (Fig. 35.12, *top right*):

$$\underset{\text{deoxyhemoglobin}}{Hb} + \underset{\text{oxygen}}{O_2} \rightarrow \underset{\text{oxyhemoglobin}}{HbO_2}$$

At the normal P_{O_2} in the lungs, hemoglobin is practically saturated with oxygen. Each hemoglobin molecule contains four polypeptide chains, and each chain is folded around an iron-containing group called **heme** (Fig. 35.13). The iron forms a loose association with oxygen. Because there are about 250 million hemoglobin molecules in each red blood cell, each red blood cell is capable of carrying at least one billion molecules of oxygen.

Carbon monoxide (CO) is an air pollutant that is produced by the incomplete combustion of natural gas, gasoline, kerosene, and even wood and charcoal. Because CO is a colorless, odorless gas, people can be unaware that they are breathing it. But once CO is in the bloodstream, it combines with the iron of hemoglobin 200 times more tightly than oxygen, and the result can be death. This is the reason that homes are equipped with CO detectors.

Internal Respiration. Blood entering the systemic capillaries is a bright red color because RBCs contain oxyhemoglobin. Because the temperature in the tissues is higher and the pH is lower than in the lungs, oxyhemoglobin has a tendency to give up oxygen:

$$HbO_2 \rightarrow Hb + O_2$$

Oxygen diffuses out of the blood into the tissues because the P_{O_2} of tissue fluid is lower than that of blood (see Fig. 35.12, *bottom right*). The lower P_{O_2} is due to cells continuously using up oxygen in cellular respiration. After oxyhemoglobin gives up O_2, this oxygen leaves the blood and enters tissue fluid, where it is taken up by cells.

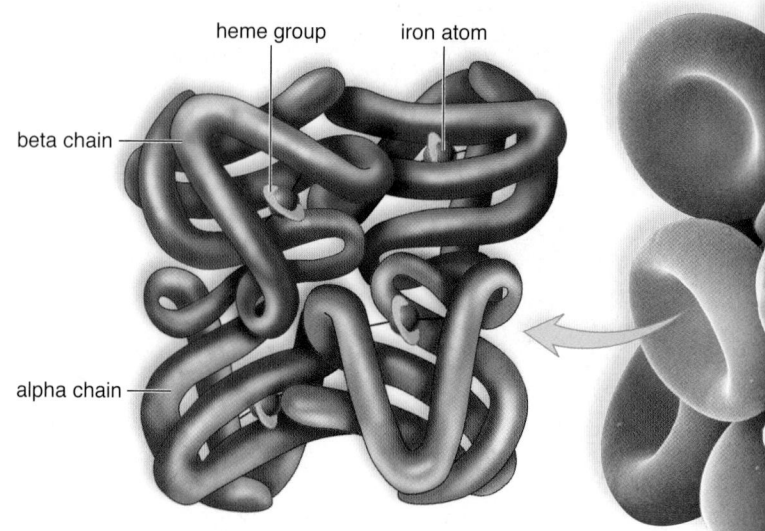

Figure 35.13 Hemoglobin.   Hemoglobin consists of four polypeptide chains, two alpha (*red*) and two beta (*purple*), each associated with a heme group. Each heme group contains an iron atom, which can bind to O_2.

Carbon dioxide, in contrast, enters blood from the tissues because the P_{CO_2} of tissue fluid is higher than that of blood. Carbon dioxide, produced continuously by cells, collects in tissue fluid. After CO_2 diffuses into the blood, it enters the red blood cells, where a small amount combines with the protein portion of hemoglobin to form **carbaminohemoglobin** ($HbCO_2$). Most of the CO_2, however, is transported in the form of the **bicarbonate ion** (HCO_3^-). First, CO_2 combines with water, forming carbonic acid, and then this dissociates to a hydrogen ion (H^+) and HCO_3^-:

$$CO_2 + H_2O \rightarrow H_2CO_3 \rightarrow H^+ + HCO_3^-$$

$$\text{carbon} \quad \text{water} \qquad \text{carbonic} \qquad \text{hydrogen} \quad \text{bicarbonate}$$
$$\text{dioxide} \qquad\qquad \text{acid} \qquad\qquad \text{ion} \qquad \text{ion}$$

Carbonic anhydrase also speeds this reaction. The HCO_3^- diffuses out of the red blood cells to be carried in the plasma (see Fig. 35.12, *bottom left*).

The release of H^+ from this reaction could drastically change the pH of the blood, which is highly undesirable because cells require a normal pH in order to remain healthy. However, the H^+ is absorbed by the globin portions of hemoglobin. Hemoglobin that has combined with H^+ is called reduced hemoglobin and has the formula HbH^+. HbH^+ plays a vital role in maintaining the normal pH of the blood. Blood that leaves the systemic capillaries is a dark maroon color because red blood cells contain reduced hemoglobin.

  Video Gas Transport

  Animation Gas Exchange During Respiration

Check Your Progress   35.2

1. Compare the mechanism of ventilation in reptiles and mammals as being similar to the operation of a bellows.
2. Explain how the carotid bodies and aortic bodies affect the rate of respiration.
3. Define the role of oxyhemoglobin, reduced hemoglobin, and carbaminohemoglobin in homeostasis.

35.3 Respiration and Human Health

Learning Outcomes

Upon completion of this section, you should be able to

1. Describe several common disorders that mainly affect the upper respiratory tract, and several that affect the lower respiratory tract.
2. Classify several common respiratory disorders according to whether they are mainly caused by allergies, infections, a genetic defect, or toxin exposure.

The human respiratory tract is constantly exposed to environmental air that may contain infectious agents, allergens, tobacco smoke, or other toxins. This results in the respiratory tract being susceptible to a number of diseases. Some of the most important of these are summarized here.

Disorders of the Upper Respiratory Tract

The upper respiratory tract consists of the nasal cavities, sinuses, pharynx, and larynx. Because the upper part of the respiratory tract filters out many pathogens and other materials that may be present in the air, it is commonly affected by a variety of infections, which may also spread to the middle ear or the sinuses.

The Common Cold

Most "colds" are relatively mild viral infections of the upper respiratory tract characterized by sneezing, rhinitis (runny nose), and perhaps a mild fever. Most colds last a few days, after which the immune response is able to eliminate the inciting virus. However, since colds are caused by several different viruses, and by several hundred strains of these viruses, we usually have no immunity to the next strain that "goes around," and vaccines are very difficult to develop. As with all viral infections, antibiotics like penicillin are useless in treating colds.

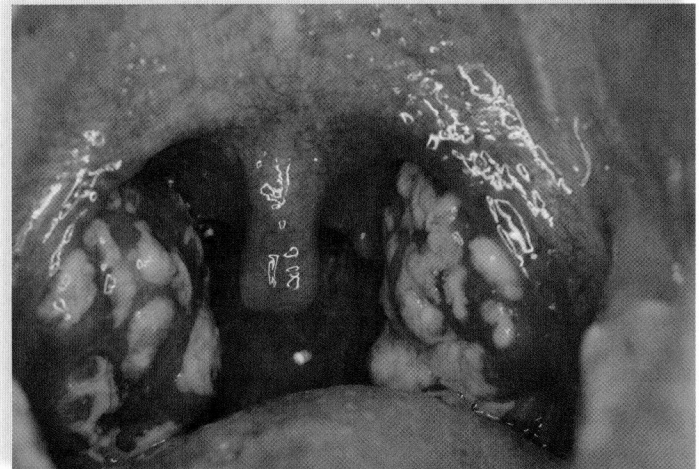

Figure 35.14 Strep throat. Pharyngitis caused by the bacterium *Streptococcus pyogenes* can cause swollen tonsils, as shown here. The whitish patches are areas of pus formation, indicating that white blood cells are fighting the infection.

Strep Throat

Most cases of **pharyngitis**, or inflammation of the pharynx, are caused by viruses, but strep throat is an acute pharyngitis caused by the bacterium *Streptococcus pyogenes*. Typical symptoms include severe sore throat, high fever, and white patches in the tonsillar area (Fig. 35.14). Adults experience about half as many sore throats as do children, who average about five upper respiratory infections per year, and about one strep throat infection every four years. Many untreated strep infections probably resolve on their own, but some can lead to more serious conditions like scarlet fever or rheumatic fever. Fortunately, infection with *S. pyogenes* can be easily and quickly diagnosed with specific laboratory tests, and is usually curable with antibiotics.

Disorders of the Lower Respiratory Tract

Several common disorders affecting the lower respiratory tract are summarized in Figure 35.15.

Disorders Affecting the Trachea and Bronchi

One of the most obvious and life-threatening disorders that can affect the trachea is choking. The best way for a person without extensive medical training to help someone who is choking is to perform the Heimlich maneuver, which involves grabbing the choking person around the waist from behind, and forcefully pulling both hands into their upper abdomen to expel whatever is lodged. If this fails, trained medical personnel may be able to quickly insert a breathing tube through an incision made in the trachea. This procedure is called a tracheotomy, and the opening is a *tracheostomy*.

If infections of the upper respiratory tract spread into the lower respiratory tract, *acute bronchitis*, or inflammation of the bronchi, often results. Other causes of acute bronchitis include allergic reactions and damage from environmental toxins, such as those present in cigarette smoke. It is estimated that approximately 5% of the U.S. population suffers from a bout of acute bronchitis in any given year. Symptoms include fever and a cough that produces phlem or pus, and chest pain. Depending on the cause, acute bronchitis maybe be treatable with antibiotics, or may resolve with time or progress to more serious conditions.

If the inciting cause (such as smoking) persists, acute bronchitis can develop into *chronic bronchitis* (Fig. 35.15), in which the airways are inflamed and filled with mucus. Over time, the bronchi undergo degenerative changes, including the loss of cilia and their normal cleansing action. Under these conditions, infections are more likely to occur. Smoking and exposure to other airborne toxins are the most frequent causes of chronic bronchitis. Along with emphysema, chronic bronchitis is a major component of *chronic obstructive pulmonary disease*, the fourth leading cause of death in the U.S.

Asthma is a disease of the bronchi and bronchioles marked by coughing, wheezing, and breathlessness. The airways are unusually sensitive to various irritants, which can include allergens such as pollen, animal dander, dust, and/or cigarette smoke. Even cold air or exercise can be an irritant.

An asthmatic attack results from inflammation in the airways and contraction of smooth muscle lining their walls,

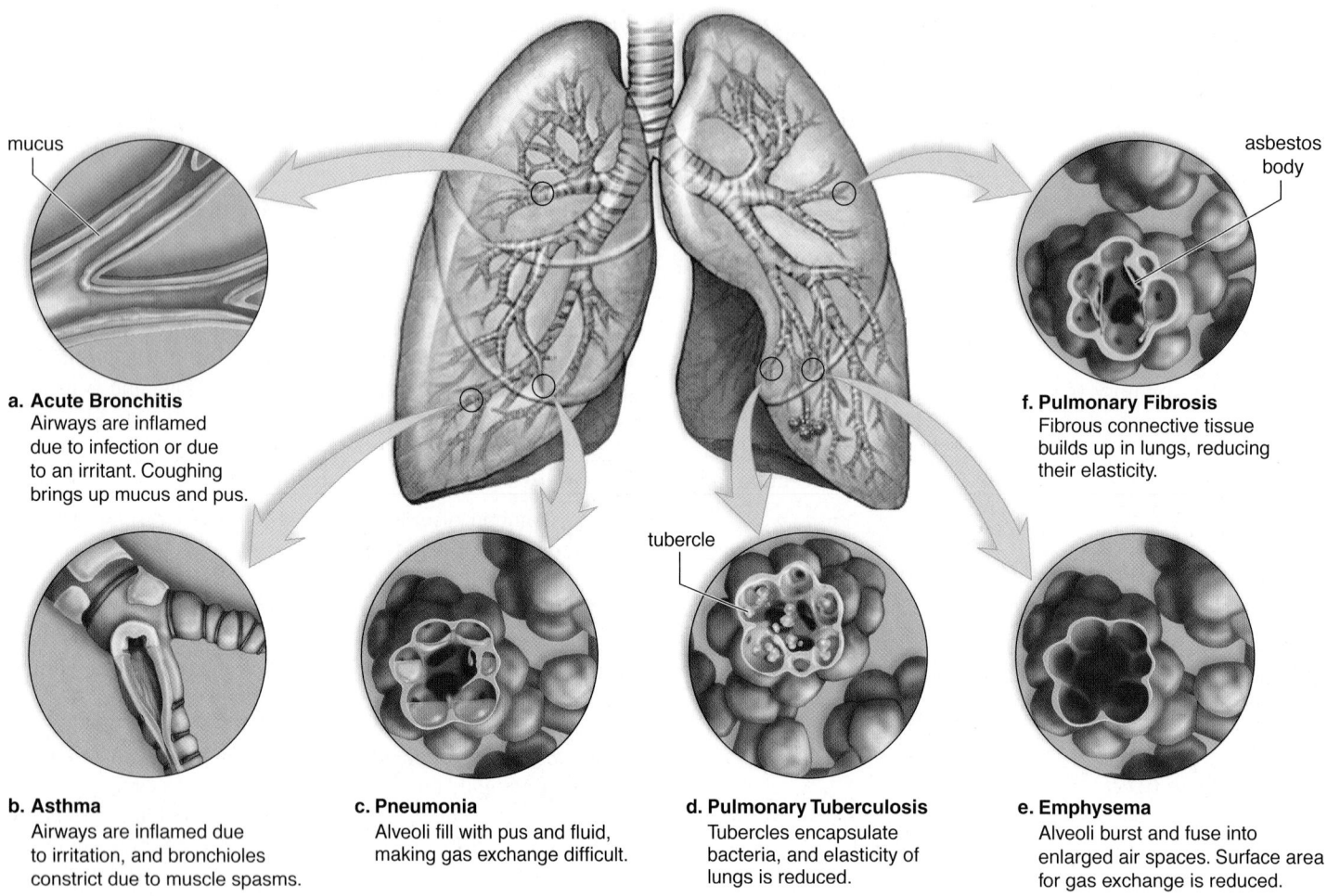

mucus

a. Acute Bronchitis
Airways are inflamed
due to infection or due
to an irritant. Coughing
brings up mucus and pus.

b. Asthma
Airways are inflamed due
to irritation, and bronchioles
constrict due to muscle spasms.

c. Pneumonia
Alveoli fill with pus and fluid,
making gas exchange difficult.

tubercle

d. Pulmonary Tuberculosis
Tubercles encapsulate
bacteria, and elasticity of
lungs is reduced.

asbestos
body

f. Pulmonary Fibrosis
Fibrous connective tissue
builds up in lungs, reducing
their elasticity.

e. Emphysema
Alveoli burst and fuse into
enlarged air spaces. Surface area
for gas exchange is reduced.

Figure 35.15 Common bronchial and pulmonary diseases.   Exposure to infectious pathogens and/or polluted air, including tobacco smoke, causes the diseases and disorders shown here.

resulting in a narrowing of the diameter of the airways (Fig. 35.15). All estimates indicate that the incidence of asthma in American children has been increasing steadily since the early 1980s. Possible explanations for this include a more sedentary lifestyle, with more exposure to indoor toxins, and less frequent exposure to beneficial microbes. Asthma is not curable, but it is treatable. Drugs administered by inhalers can help to prevent the inflammation and dilate the bronchi.

Disorders Affecting the Lungs

Combined together, various diseases of the lung cause about 400,000 deaths per year in the United States, and affect hundreds of millions of people worldwide. Depending on the cause, treatments for lung disease may include antibiotics, supplemental oxygen, and administration of anti-inflammatory drugs. For patients with serious lung conditions who have exhausted all other treatment options, a lung transplant may be the best option, but there aren't enough donor lungs to meet the need. The Nature of Science feature on page 676 describes some recent advances in artificial-lung technology.

Pneumonia is a viral, bacterial, or fungal infection of the lungs in which bronchi and alveoli fill with a discharge, such as pus and fluid (Fig. 35.15). Along with coughing and difficulty breathing, people suffering from pneumonia often have a high

fever, sharp chest pain, and a cough that produces thick phlem or even pus. Several bacteria can cause pneumonia, as can the influenza virus, especially in the very young, very old, or people with a suppressed immune system. AIDS patients are subject to a particularly rare form of pneumonia caused by a fungus of the genus *Pneumocystis*, but they suffer from many other types of pneumonias as well.

Pulmonary tuberculosis is caused by the bacterium *Mycobacterium tuberculosis*. Tuberculosis (TB) was a major killer in the United States before the middle of the twentieth century, after which antibiotic therapy brought it largely under control. However, the incidence of TB is rising in certain areas of the world, especially where HIV infection (which reduces immunity to *M. tuberculosis*) is common, and treatments are not widely available. According to the Centers for Disease Control, in 2007 approximately one-third of the world's population were infected with *M. tuberculosis*, and TB was the actual cause of death for as many as half of all persons with AIDS.

When tubercle bacilli invade the lung tissue, the cells build a protective capsule about the organisms, isolating them from the rest of the body. This tiny capsule is called a tubercle (Fig. 35.15). If the resistance of the body is high, the imprisoned organisms die, but if the resistance is low, the organisms can escape and spread.

It is possible to tell if a person has ever been exposed to *M. tuberculosis* with a TB skin test, in which a highly diluted extract of the bacteria is injected into the skin of the patient. A person who has never been exposed to the bacterium shows no reaction, but one who has previously been infected develops an area of inflammation that peaks in about 48 hours.

Inhaling particles such as silica (sand), coal dust, or asbestos can lead to **pulmonary fibrosis**, a condition in which fibrous connective tissue builds up in the lungs. The lungs cannot inflate properly and are always tending toward deflation (Fig. 35.15). Breathing asbestos is also associated with the development of cancer, including a type called mesothelioma. In the U.S., the use of asbestos as a fireproofing and insulating agent has been limited since the 1970s; however, many thousands of lawsuits are filed each year by patients suffering from asbestos-related illnesses.

Emphysema is a chronic and incurable lung disorder in which the alveoli are distended and their walls damaged so that the surface area available for gas exchange is reduced (Fig. 35.15). Emphysema is often preceded by chronic bronchitis. Air trapped in the lungs leads to alveolar damage and a noticeable ballooning of the chest. The elastic recoil of the lungs is reduced, so not only are the airways narrowed, but the driving force behind expiration is also reduced. The patient is breathless and may have a cough. Because the surface area for gas exchange is reduced, less oxygen reaches the heart and brain, leaving the person feeling depressed, sluggish, and irritable. Exercise, drug therapy, and supplemental oxygen, along with giving up smoking, may relieve the symptoms and possibly slow the progression of emphysema. Upon death, the lungs are decidedly abnormal (Fig. 35.16*b*).

Lung cancer is the leading cause of cancer-related death in the U.S. and worldwide. It is slightly more common in men than women, but that gap has been narrowing in recent years as lung cancer rates in women have increased, due to an increasing number of women who smoke. Lung cancer rates remain low until about age 40, when they gradually start to rise, peaking at around age 70. Symptoms may include coughing, shortness of breath, blood in the sputum, and chest pain. Many other symptoms can occur if the cancer spreads to other parts of the body, which is common.

The only treatment that offers a possibility of cure is to remove a lobe or the whole lung before metastasis has had time to occur. Chemotherapy and radiation may also be used. Even with treatment, lung cancer is highly lethal—five-year survival rates range from 15% in the United States to 8% in less developed countries. About 150,000 people in the United States die of lung cancer each year (Fig. 35.16*c*). The American Cancer Society links over 85% of these deaths to smoking. Smoking is also associated with bronchitis, emphysema, heart disease, and other types of cancer. Considering that the nicotine in cigarette smoke is addictive, it is better never to start smoking than to try quitting later on.

Cystic fibrosis (CF) is an example of a lung disease that is genetic rather than infectious, although infections also play a role in the disease. One in 31 Americans carries the defective gene, but a child must inherit two copies of the gene to have the disease. Still, CF is the most common genetic disease in the U.S. white population.

The gene that is defective in CF codes for cystic fibrosis transmembrane regulator (CFTR), a protein needed for proper transport of chloride (Cl^-) ions out of the epithelial cells of the lung. Because this also reduces the amount of water transported out of lung cells, the mucus secretions become very sticky and can form plugs that interfere with breathing.

Symptoms of CF include coughing and shortness of breath; part of the treatment involves clearing mucus from the airways by vigorously slapping the patient on the back as well as by administering mucus-thinning drugs. None of these treatments is curative, however, and because the lungs can be severely affected, the median survival age for people with CF is only 30 years. Researchers are attempting to develop gene therapy strategies to replace the faulty CFTR gene.

Video
Good Poison

Check Your Progress 35.3

1. Explain why antibiotic drugs such as penicillin are ineffective at treating the common cold.
2. Name two disorders of the lower respiratory tract that mainly cause a narrowing of the airways, and two that restrict the lungs' ability to expand.
3. List six illnesses associated with smoking cigarettes.

Figure 35.16 Smoking and lung disorders.
Smoking causes almost 90% of all lung cancers and is also a major cause of emphysema. **a.** Normal lungs. **b.** The lungs of a person who died from emphysema are shrunken and blackened from trapped smoke. **c.** The lungs of a person who died from lung cancer are blackened from smoke except for the presence of the tumor, which is a mass of malformed soft tissue.

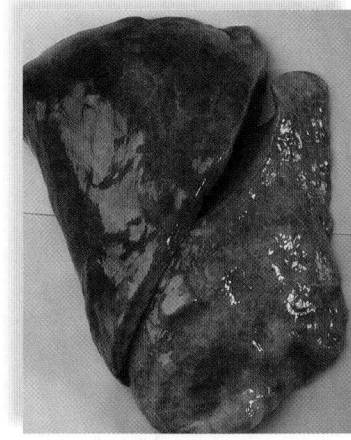

a. Normal lungs

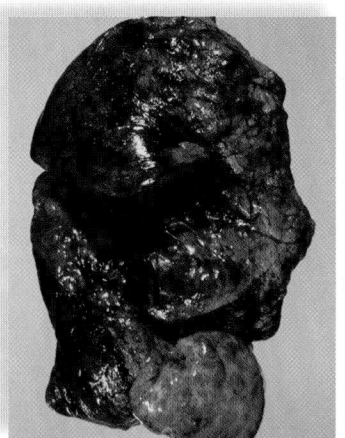

b. Emphysema

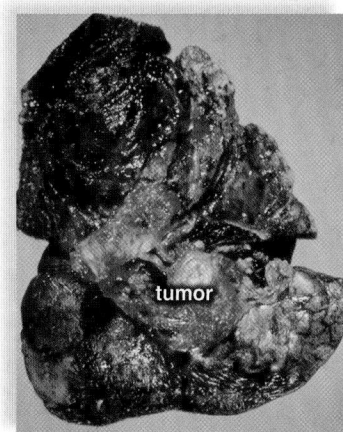

c. Lung cancer

Nature of Science

Artificial Lung Technology

Some organs, such as the kidneys, are relatively easy to transplant from one well-matched individual to another, with a high rate of success. Lungs are more difficult to transplant, however, with a ten-year survival rate of only 10 to 20 percent. Some very recent research is showing how one day it may be possible to replace diseased lungs with laboratory-grown versions.

In early 2010, a group of scientists at Yale University anesthetized a group of adult rats, surgically removed their left lungs, and replaced them with tissue-engineered lungs that had been produced in the laboratory. For periods of up to two hours, the implanted lung tissue exchanged oxygen and carbon dioxide at similar rates as the natural lungs.[1]

In order to build the artificial rat lungs, the Yale researchers began by removing lungs from adult rats, and treating the organs to remove most of the cellular components, while preserving the airways and supporting connective tissue matrix. Next, they placed these decellularized structures, along with various types of cultured cells, into a sterile container designed to mimic some aspects of the fetal environment in which the lungs normally first develop. The scientists found that the cells were able to form much of the lung tissues as well as the blood vessels needed to supply the tissue with blood and transport gases.

Although these results are an important early step toward growing replacement lungs in the lab, there is a long way to go before anyone will contemplate implanting engineered lungs into humans.

In a separate study published on the same day as the Yale study, a Harvard group announced that they had developed a pea-sized device that mimics human lung tissues. Made of human lung cells, a permeable membrane, plus blood capillary cells, all mounted on a microchip (Fig. 35A), the device is able to mimic the function of alveoli. When the researchers placed bacteria on the alveolar side of the device, and white blood cells on the capillary side, the blood cells crossed the membrane, mimicking an immune response.[2] At a minimum, the researchers hope that their "lung-on-a-chip" device can be used for testing certain drugs or the effects of various toxins on the lungs, which might replace much of the animal testing that is currently performed.

Questions to Consider

1. With regard to producing tissue-engineered human lungs, what is the major drawback in the procedure developed by the Yale group?
2. What are some of the aspects of lung structure that make it a more difficult organ to grow in the lab than, for example, a urinary bladder?
3. Besides increased availability, what are two other potential advantages of laboratory-grown lungs (or other tissues) compared to regular donor tissues?

1 Petersen, T. H., et al. 2010. Tissue-engineered lungs for in vivo implantation. *Science* 329: 538–41.

2 Huh, D., et al. 2010. Reconstituting organ-level lung functions on a chip. *Science* 328: 1662–68.

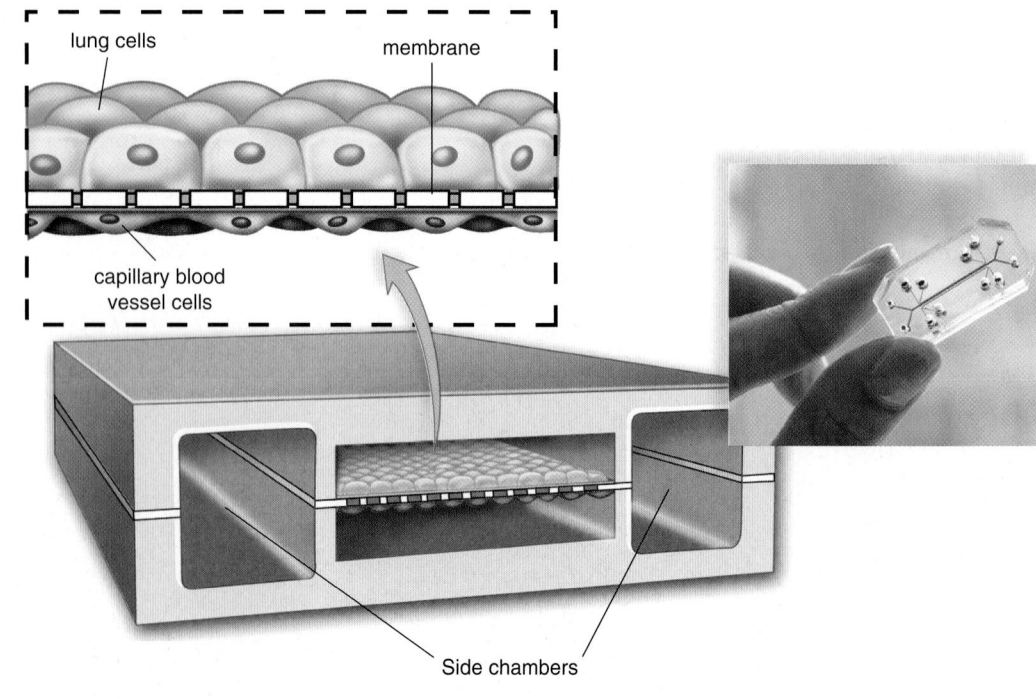

Figure 35A Lung on a chip developed by Harvard researchers. About the size of a credit card, the device consists of a semipermeable membrane with lung cells on one side and blood vessel cells on the other. Liquid medium flows in a channel on the blood cell side, while air flows in and out on the lung cell side.

CONNECTING *the* CONCEPTS *with the* BIG IDEAS

Energy and Homeostasis

- The structure of the alveoli allows a tremendous increase in surface area for exchange of gases. (2A3b1*IE*)

Interactions and Systems

- The respiratory and circulatory systems work together to ensure the well being of the body. (4A4b*IE*)
- Organ specialization within the respiratory system increases efficiency for the organism. (4B2a2*IE*)

*Find the unabridged version of all EK citations at www.glencoe.com/maderAP11.

Media Study Tools

www.glencoe.com/maderAP11

Enhance your study of this chapter with study tools and practice tests. Also ask your instructor about the resources available through ConnectPlus, including the media-rich eBook, interactive learning tools, and animations.

Summarize

35.1 Gas Exchange Surfaces

Some aquatic animals, such as hydras, earthworms, and some amphibians, use their entire body surface for gas exchange.

Most animals have a specialized gas-exchange area. Large aquatic animals usually pass water through gills. In bony fishes, blood in the capillaries flows in the direction opposite that of the water. Blood takes up almost all of the oxygen in the water as a result of this countercurrent flow.

On land, insects use tracheal systems, and vertebrates have lungs. In insects, air enters the tracheae at openings called spiracles. From there, the air moves to ever smaller tracheoles until gas exchange takes place at the cells themselves. Lungs are found inside the body, where water loss is reduced. To ventilate the lungs, some vertebrates use positive pressure, but most inhale, using muscular contraction to produce a negative pressure that causes air to rush into the lungs. When the breathing muscles relax, air is exhaled.

Birds have a series of air sacs attached to the lungs. When a bird inhales, air enters the posterior air sacs, and when a bird exhales, air moves through the lungs to the anterior air sacs before exiting the respiratory tract. The one-way flow of air through the lungs allows more fresh air to be present in the lungs with each breath, and this leads to greater uptake of oxygen from one breath of air. Mammals have two-way or tidal airflow in and out of the lungs, and as a result some mixing is always occurring between fresh air and previously inhaled air, with less uptake of oxygen than in birds.

35.2 Breathing and Transport of Gases

During inspiration, air enters the body at nasal cavities and then passes from the pharynx through the glottis, larynx, trachea, bronchi, and bronchioles to the alveoli of the lungs, where exchange occurs; during expiration, air passes in the opposite direction. Humans breathe by negative pressure, as do other mammals. During inspiration, the rib cage goes up and out, and the diaphragm lowers. The lungs expand and air comes rushing in. During expiration, the rib cage goes down and in, and the diaphragm rises. Therefore, air rushes out.

The rate of breathing increases when the amount of H^+ and carbon dioxide in the blood rises, as detected by chemoreceptors such as the aortic and carotid bodies.

Gas exchange in the lungs and tissues is brought about by diffusion. Hemoglobin transports oxygen in the blood; carbon dioxide is mainly transported in plasma as the bicarbonate ion. Excess hydrogen ions are transported by hemoglobin. The enzyme carbonic anhydrase found in red blood cells speeds the formation of the bicarbonate ion.

35.3 Respiration and Human Health

The respiratory tract is susceptible to a wide variety of infections and other disease conditions. Infections of the upper respiratory tract include the common cold, caused by several different viruses, and strep throat, caused by a bacterium. In the lower tract, bronchitis is an inflammation of the bronchi, usually due to infections, whereas in asthma the inflammation and smooth muscle contraction in the bronchi and bronchioles is due to sensitivity to various irritants. Many important diseases affect the lungs, including pneumonia, tuberculosis, pulmonary fibrosis, emphysema, lung cancer, and cystic fibrosis.

Key Terms

alveolus (pl., alveoli) 668	carotid body 671
aortic body 671	countercurrent exchange 666
asthma 673	cystic fibrosis (CF) 675
bicarbonate ion 672	diaphragm 669
bronchiole 668	emphysema 675
bronchus (pl., bronchi) 668	epiglottis 668
carbaminohemoglobin 672	expiration 669
carbonic anhydrase 672	external respiration 664

▅▅ Assess

Reviewing This Chapter

1. Compare the respiratory organs of aquatic animals to those of terrestrial animals. 664–67
2. Why is it beneficial for the body wall of earthworms to be moist? Why don't insects require circulatory system involvement in air transport? 664
3. How does the countercurrent flow of blood within gill capillaries and water passing across the gills assist respiration in fishes? 666
4. Name the parts of the human respiratory system, and list a function for each part. How is air reaching the lungs cleansed? 669–70
5. Explain the phrase "breathing by using negative pressure." 669
6. Contrast the tidal ventilation mechanism in humans with the one-way ventilation mechanism in birds, and explain the benefits of the ventilation mechanism in birds. 670
7. The concentration of what substances in blood controls the breathing rate in humans? Explain. 670–71
8. How are oxygen and carbon dioxide transported in blood? What does carbonic anhydrase do? 671–72
9. Which conditions depicted in Figure 35.15 are due to infection? Which are due to behavioral or environmental factors? Explain. 673–75

Testing Yourself

Choose the best answer for each question.

1. Label the following diagram depicting respiration.

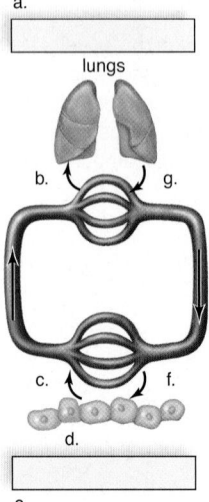

a.

lungs

b.   g.

c.   f.

d.

e.

2. Birds have more efficient lungs than humans because the flow of air in birds
   a. is the same during both inspiration and expiration.
   b. travels in only one direction through the lungs.
   c. follows a tidal ventilation pattern.
   d. is not hindered by a larynx.
   e. enters their bones.

3. Which animal breathes by positive pressure?
   a. fish
   b. human
   c. bird
   d. frog
   e. planarian

4. Which of these is a true statement?
   a. In lung capillaries, carbon dioxide combines with water to produce carbonic acid.
   b. In tissue capillaries, carbonic acid breaks down to carbon dioxide and water.
   c. In lung capillaries, carbonic acid breaks down to carbon dioxide and water.
   d. In tissue capillaries, carbonic acid combines with hydrogen ions to form the carbonate ion.
   e. All of these statements are true.

5. Air enters the human lungs because
   a. atmospheric pressure is less than the pressure inside the lungs.
   b. atmospheric pressure is greater than the pressure inside the lungs.
   c. although the pressures are the same inside and outside, the partial pressure of oxygen is lower within the lungs.
   d. the residual air in the lungs causes the partial pressure of oxygen to be less than it is outside.
   e. the process of breathing pushes air into the lungs.

6. If the digestive and respiratory tracts were completely separate in humans, there would be no need for
   a. swallowing.
   b. a nose.
   c. an epiglottis.
   d. a diaphragm.
   e. All of these are correct.

7. In humans, the respiratory control center
   a. is stimulated by carbon dioxide.
   b. is located in the medulla oblongata.
   c. controls the rate of breathing.
   d. is stimulated by hydrogen ion concentration.
   e. All of these are correct.

8. Carbon dioxide is carried in the plasma
   a. in combination with hemoglobin.
   b. as the bicarbonate ion.
   c. combined with carbonic anhydrase.
   d. only as a part of tissue fluid.
   e. All of these are correct.

9. Which of these is anatomically incorrect?
   a. The nose has two nasal cavities.
   b. The pharynx connects the nasal and oral cavities to the larynx.
   c. The larynx contains the vocal cords.
   d. The trachea enters the lungs.
   e. The lungs contain many alveoli.

10. The chemical reaction that converts carbon dioxide to a bicarbonate ion takes place in
    a. the blood plasma.
    b. red blood cells.
    c. the alveolus.
    d. the hemoglobin molecule.

11. Which of these is incorrect concerning inspiration?
    a. Rib cage moves up and out.
    b. Diaphragm contracts and moves down.
    c. Pressure in lungs decreases, and air comes rushing in.
    d. The lungs expand because air comes rushing in.

12. Asthma
    a. mainly affects the upper respiratory tract.
    b. is usually caused by an infection.
    c. is considered to be a genetic disorder.
    d. is usually curable.
    e. None of these statements are true.

13. Label this diagram of the human respiratory system.

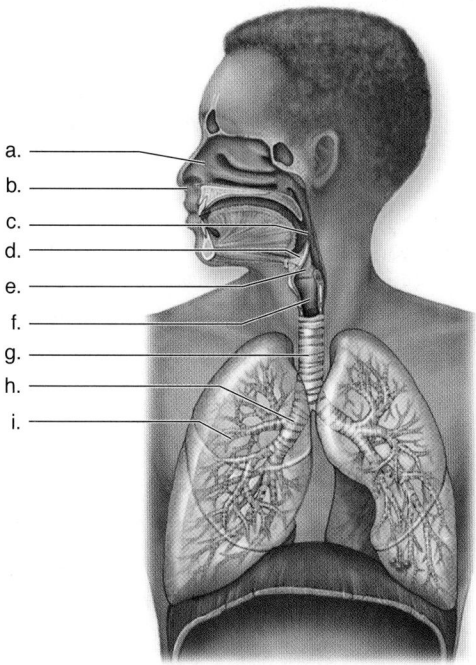

a.
b.
c.
d.
e.
f.
g.
h.
i.

Engage

Thinking Scientifically

1. You are a physician who witnessed Christopher Reeve's riding accident. Why might you immediately use mouth to mouth resuscitation until mechanical ventilation becomes available?

2. Fetal hemoglobin picks up oxygen from the maternal blood. If the oxygen-binding characteristics of hemoglobin in the fetus were identical to the hemoglobin of the mother, oxygen could never be transferred at the placenta to fetal circulation. What hypothesis about the oxygen-binding characteristics of fetal hemoglobin would explain how fetuses get the oxygen they need?

Bioethical Issue
Antibiotic Therapy

Antibiotics can cure many respiratory infections, but problems are associated with antibiotic therapy. Aside from a possible allergic reaction, antibiotics not only kill off disease-causing bacteria, but they also reduce the number of beneficial bacteria in the intestinal tract and other locations. These beneficial bacteria hold in check the growth of other pathogens that now begin to flourish. Diarrhea can result, as can yeast infections.

Especially alarming is the occurrence of bacterial resistance. Resistance takes place when vulnerable bacteria are killed off by an antibiotic, while resistant bacteria continue to live and thus become more prevalent. Many bacteria that cause ear, nose, and throat infections as well as scarlet fever and pneumonia have developed resistant populations because we have not been using antibiotics properly. Moreover, new strains of the bacterium that causes tuberculosis are resistant to the usual combined antibiotic therapy. As mentioned in Chapter 20, methicillin resistant *Staphylococcus aureus* (MRSA) infection has become a major concern in hospitals and other institutions; infection with this bacterium can be deadly.

Every citizen needs to be aware of the serious problem of antibiotic resistance. Stuart Levy, a Tufts University School of Medicine microbiologist, says that we should do what is ethical for society and ourselves. Antibiotics kill bacteria, not viruses—therefore, we shouldn't take antibiotics unless we know for sure we have a bacterial infection. And we shouldn't take them prophylactically—that is, just in case we might need one. If antibiotics are taken in low dosages and intermittently, resistant strains are more likely to arise. Many experts advocate that the use of antibiotics in animals produced for food should be pared down, and household disinfectants should no longer be spiked with antibacterial agents such as triclosan, often found in hand soaps. Perhaps then, Levy says, vulnerable bacteria will begin to supplant the resistant ones in the population.[1]

1 Levy, Stuart B., and Marshall, B. 2004. Antibacterial resistance worldwide: causes, challenges and responses. *Natural Medicine* 10: S122–29.

Marine organisms rid the body of excess salt; fishes extrude salt at their gills, and turtles do so near their eyes.

36

Body Fluid Regulation and Excretory Systems

CHAPTER OUTLINE

BEFORE YOU BEGIN

Before beginning this chapter, take a few moments to review the following discussions.

Figure 5.8 What occurs when a cell is surrounded by a solution having a greater or lesser solute concentration than that inside the cell?

Section 8.5 What process results in the production of urea in humans?

Section 34.4 What happens to excess nutrients and minerals that cannot be stored?

I f the salt concentration in body fluids is too high, cells shrivel and die. If it is too low, cells swell and rupture. Yet animals are found in all sorts of environments, including marine environments that are too salty, freshwater environments that don't have enough salt, and even terrestrial environments that are simply too dry. Animals clearly spend a lot of energy regulating the composition of their body fluids, and chief among the organs that help are the kidneys of the urinary system. Sometimes animals such as marine birds and reptiles get some assistance from accessory glands. Sea turtles have salt glands located above their eye that, true to their name, rid the body of salt. When the glands excrete a salty solution collected from body fluids, sea turtles appear to cry. Humans lack salt glands and cannot survive after drinking too much salt water because the kidneys alone can't handle all the salt.

In this chapter, you'll learn how animals maintain their normal water-salt balance while excreting various metabolic wastes and regulating their pH. All these functions are of primary importance to homeostasis and continued good health.

As you read through the chapter, think about the following questions:

1. Besides salt glands, what other types of strategies have animals developed to either conserve or excrete excess salt?

2. If you were stranded on a desert island with plenty of food but only seawater to drink, how long do you think you could survive?

3. By what mechanisms is the human kidney able to regulate the salt concentration of urine it produces?

FOLLOWING *the* BIG IDEAS

CHAPTER 36 BODY FLUID REGULATION AND EXCRETORY SYSTEMS

Energy and Homeostasis	Differing environments will shape osmoregulation systems handed down from a common ancestor.
Interactions and Systems	Excretory organs must both extract wastes and maintain homeostasis by interactions inside and outside the system.

36.1 Animal Excretory Systems

Learning Outcomes

Upon completion of this section, you should be able to

1. Describe the overall, specific functions of animal excretion systems.
2. List the costs and benefits of the excretion of ammonia, urea, or uric acid as nitrogenous waste products.
3. Compare and contrast the excretory organs of earthworms, arthropods, aquatic vertebrates, and terrestrial vertebrates.

An important part of maintaining homeostasis in animals involves **osmoregulation**, or balancing the levels of water and salts in the body. Often the osmoregulatory system of an animal also removes metabolic wastes from the body, a process called **excretion**.

Nitrogenous Waste Products

The breakdown of nitrogen-containing molecules, such as amino acids and nucleic acids, results in excess nitrogen that must be excreted. When amino acids are broken down by the body to generate energy, or are converted to fats or carbohydrates, the amino ($-NH_2$) groups must be removed because they are not needed, and they may be toxic at high levels. Depending on the species, this excess nitrogen may be excreted in the form of ammonia, urea, or uric acid. Removal of amino groups from amino acids requires a fairly constant amount of energy; however, the amount of energy required to convert amino groups to ammonia, urea, or uric acid differs, as indicated in Figure 36.1.

Ammonia

Amino groups removed from amino acids immediately form **ammonia** (NH_3) by the addition of a third hydrogen ion (H^+). This reaction requires little or no energy. Ammonia is quite toxic, but it can be a nitrogenous excretory product if sufficient water is available to wash it from the body. Ammonia is excreted by most fishes and other aquatic animals whose gills and/or body surfaces are in direct contact with the water of the environment.

Urea

Sharks, adult amphibians, and mammals usually excrete **urea** as their main nitrogenous waste. Urea is much less toxic than ammonia and can be excreted in a moderately concentrated solution. This elimination strategy allows body water to be conserved, an important advantage for terrestrial animals with limited access to water. Production of urea requires the expenditure of energy, however, because it is produced in the liver by a set of energy-requiring enzymatic reactions, known as the urea cycle. In this cycle, carrier molecules take up carbon dioxide and two molecules of ammonia, finally releasing urea.

Uric Acid

Uric acid is synthesized by a long, complex series of enzymatic reactions that requires expenditure of even more energy than does urea synthesis. Uric acid is not very toxic, and it is poorly soluble in water. Poor solubility is an advantage if water conser-

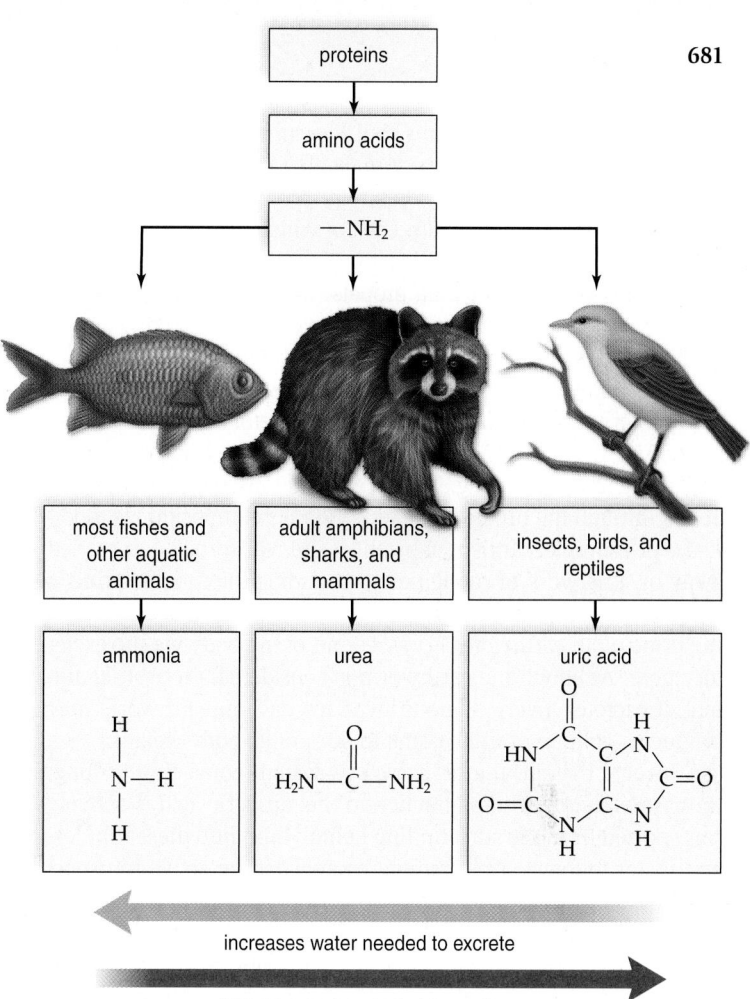

Figure 36.1 Nitrogenous wastes. Proteins are hydrolyzed to amino acids, whose breakdown results in carbon chains and amino groups ($-NH_2$). The carbon chains can be used as an energy source, but the amino groups must be excreted as ammonia, urea, or uric acid.

vation is needed, because uric acid can be concentrated even more readily than can urea.

Uric acid is routinely excreted by insects, reptiles, and birds. In reptiles and birds, a dilute solution of uric acid passes from the kidneys to the *cloaca*, a common reservoir for the products of the digestive, urinary, and reproductive systems. The cloacal contents are refluxed into the large intestine, where water is reabsorbed. The white substance in bird feces is uric acid.

Embryos of reptiles and birds develop inside completely enclosed shelled eggs. The production of insoluble, relatively nontoxic uric acid is advantageous for shelled embryos because all nitrogenous wastes are stored inside the shell until hatching takes place. For all these reasons, the evolutionary advantages of uric acid production have outweighed the disadvantage of energy expenditure needed for its synthesis.

Humans have retained the ability to produce uric acid, mainly from the breakdown of excess purine and pyrimidine nucleic acids in the diet. Although the exact causes are unknown, in some individuals uric acid builds up in the blood and can precipitate in and around the joints, producing a painful ailment called *gout*.

Excretory Organs Among Invertebrates

Most invertebrates have tubular excretory organs that regulate the water-salt balance of the body and excrete metabolic wastes into the environment.

The planarians, flatworms that live in fresh water, have two strands of branching excretory tubules that open to the outside of the body through excretory pores (Fig. 36.2a). Located along the tubules are bulblike *flame cells,* each of which contains a cluster of beating cilia that looks like a flickering flame under the microscope. The beating of flame-cell cilia propels fluid through the excretory tubules and out of the body. The system is believed to function in ridding the body of excess water and in excreting wastes.

The body of an earthworm is divided into segments, and nearly every body segment has a pair of excretory structures called *nephridia.* Each nephridium is a tubule with a ciliated opening and an excretory pore (Fig. 36.2b). As fluid from the coelom is propelled through the tubule by beating cilia, its composition is modified. For example, nutrient substances are reabsorbed and carried away by a network of capillaries surrounding the tubule. **Urine** is a liquid that contains metabolic wastes, excreted salts, and water; the urine of an earthworm is passed out of the body via the excretory pore. Although the earthworm is considered a terrestrial animal, it excretes a very dilute urine. Each day, an earthworm may produce a volume of urine equal to 60% of its body weight.

Insects have a unique excretory system consisting of long, thin *Malpighian tubules* attached to the gut. Uric acid is actively transported from the surrounding hemolymph into these tubules, and water follows a salt gradient established by active transport of K^+. Water and other useful substances are reabsorbed at the rectum, but the uric acid leaves the body through the anus. Insects that live in water, or eat large quantities of moist food, reabsorb little water. But insects in dry environments reabsorb most of the water and excrete a dry, semisolid mass of uric acid.

The excretory organs of other arthropods are given different names, although they function similarly. In aquatic crustaceans (e.g., crabs, crayfish), nitrogenous wastes are generally removed by diffusion across the gills. Some crustaceans also possess excretory organs called *green glands* located in the ventral portion of the head region. Fluid collects within the tubules from the surrounding blood of the hemocoel, but this fluid is modified by the time it leaves the tubules. The secretion of salts into the tubule regulates the amount of urine excreted.

In shrimp and pillbugs, the excretory organs are located in the maxillary segments and are called *maxillary glands.* Spiders, scorpions, and other arachnids possess *coxal glands,* which are located near one or more appendages and used for excretion. Coxal glands are spherical sacs resembling annelid nephridia. Wastes are collected from the surrounding blood of the hemocoel and discharged through pores at one to several pairs of appendages.

Osmoregulation by Aquatic Vertebrates

In most vertebrates the kidneys are the most important organs involved in osmoregulation. As described later in this chapter, the kidneys perform several functions critical to homeostasis, including maintaining the balance between water and several types of salts. This is a necessity because ions such as Na^+, Ca^{2+}, K^+, and PO_4^- greatly affect the workings of the body systems, such as the skeletal, nervous, and muscular systems.

The kidneys produce urine, a liquid that contains a number of different metabolic wastes. The concentration of urine produced by an animal varies depending on its environment, as well as on factors like water and salt intake.

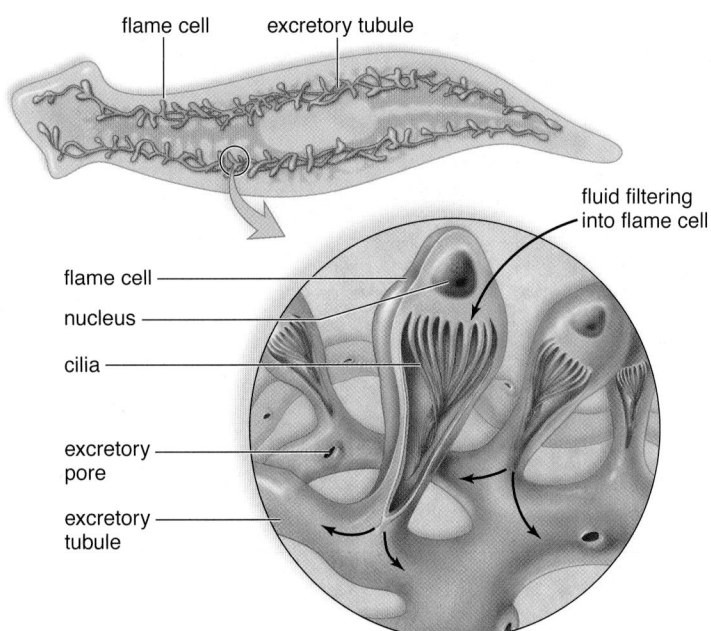

a. **Flame-cell excretory system in planarians**

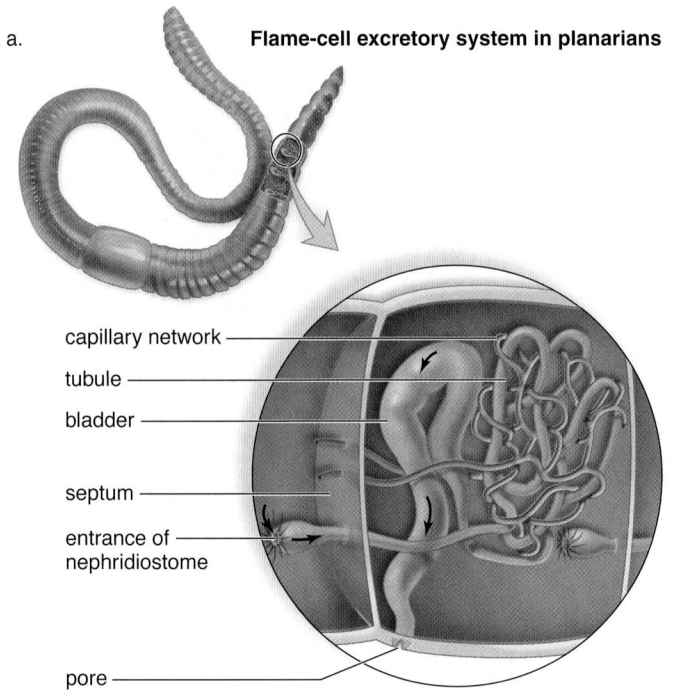

b. **Earthworm nephridium**

Figure 36.2 Excretory organs in animals. **a.** Two or more tracts of branching tubules run the length of the body and open to the outside by pores. At the ends of side branches are small bulblike cells called flame cells. **b.** The nephridium has a ciliated opening, the nephridiostome, that leads to a coiled tubule surrounded by a capillary network. Urine can be temporarily stored in the bladder before being released to the outside via a pore termed a nephridiopore.

Cartilaginous Fishes

The total concentration of the various ions in the blood of sharks, rays, and skates is less than that in seawater. Their blood plasma is nearly isotonic to seawater because they pump it full of urea, and this molecule gives their blood the same tonicity as seawater. Excess salts are secreted by the kidneys and by a special excretory organ, the rectal gland.

Marine Bony Fishes

The marine environment, which is high in salts, is hypertonic to the blood plasma of bony fishes. Apparently, the common ancestor of marine fishes evolved in fresh water, and only later did some groups invade the sea. Therefore, marine bony fishes must avoid the tendency to become dehydrated (Fig. 36.3a).

As the sea washes over their gills, marine bony fishes lose water by osmosis. To counteract this, they drink seawater almost constantly. On the average, marine bony fishes swallow an amount of water equal to 1% of their body weight every hour. This is equivalent to a human drinking about 700 mL of water every hour around the clock. But while they get water by drinking, this habit also causes these fishes to acquire salt. To rid the body of excess salt, they actively transport it into the surrounding seawater at the gills. The kidneys conserve water, and marine bony fishes produce a scant amount of isotonic urine.

Freshwater Bony Fishes

The osmotic problems of freshwater bony fishes and the response to their environment are exactly opposite those of marine bony fishes (Fig. 36.3b). Freshwater fishes tend to gain water by osmosis across the gills and the body surface. As a consequence, these fishes never drink water. They actively transport salts into the blood across the membranes of their gills. They eliminate excess water by producing large quantities of dilute (hypotonic) urine. They discharge a quantity of urine equal to one-third their body weight each day.

Osmoregulation by Terrestrial Vertebrates

An important evolutionary adaptation that allowed animals to survive on land was the development of a kidney that could produce a concentrated (hypertonic) urine. The need for water conservation is particularly well illustrated in desert mammals such as the kangaroo rat, as well as in animals that drink seawater.

Kangaroo Rat

Dehydration threatens all terrestrial animals, especially those that live in a desert, like the kangaroo rat. During daylight hours, these animals remain in a cool burrow, a behavioral adaptation to conserve water. In addition, the kangaroo rat's nasal passages have a highly convoluted mucous membrane surface that captures condensed water from exhaled air. Exhaled air is usually full of moisture, which is why you can see it on cold winter mornings—the moisture in exhaled air is condensing.

A major adaptation that allows the kangaroo rat to conserve water is the ability to form a very hypertonic urine—20 times more concentrated than its blood plasma. The kidneys of a kangaroo rat are able to accomplish this feat because the structure in their kidneys that is largely responsible for producing concentrated urine, called the loop of the nephron (see Fig. 36.8) is much longer and more efficient than that in most other animals. Also, kangaroo rats produce fecal material that is almost completely dry.

Most terrestrial animals need to drink water at least occasionally to make up for the water lost from the skin and respiratory passages and through urination. However, the kangaroo rat is so adapted to conserving water that it can survive by using metabolic water derived from cellular respiration, and it never drinks water (Fig. 36.4).

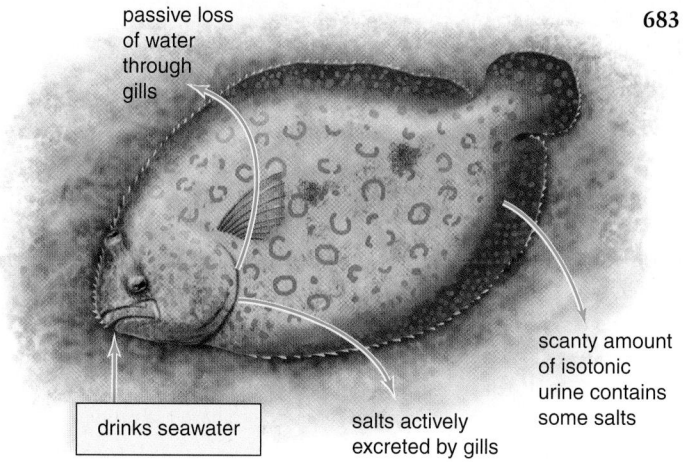

a. Marine bony fish (a flounder)

passive loss of water through gills

drinks seawater

salts actively excreted by gills

scanty amount of isotonic urine contains some salts

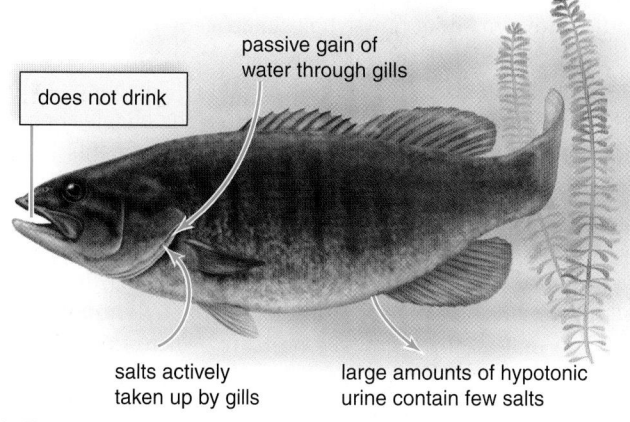

does not drink

passive gain of water through gills

salts actively taken up by gills

large amounts of hypotonic urine contain few salts

b. Freshwater bony fish (a trout)

Figure 36.3 Body fluid regulation in bony fishes. Marine bony fishes **(a)** employ different mechanisms compared to freshwater fishes **(b)** in order to osmoregulate their body fluids.

Exhaled air is cooled and dried in long convoluted air passages.

Animal fur prevents evaporative loss of water at skin.

Urine is the most hypertonic known among animals.

Oxidation of food results in metabolic water.

Fecal pellets are dry.

Figure 36.4 Adaptations of a kangaroo rat to a dry environment. The ways that a kangaroo rat minimizes water loss.

Seagulls, Reptiles, and Mammals

Birds, reptiles, and mammals evolved on land, and their kidneys are especially good at conserving water. However, some of these vertebrates have become secondarily adapted to living in or near the sea. They can drink seawater and still manage to survive. If

Figure 36.5 Adaptations of marine birds to a high salt environment. Many marine birds and reptiles have glands that pump salt out of the body.

salt solution exits here

salt solution runs down beak here

humans drink seawater, we lose more water than we take in just ridding the body of all that salt!

Little is known about how whales manage to get rid of extra salt, but we know that their kidneys are enormous. In some marine animals, however, the kidneys are not efficient enough to secrete all of the excess salt. As mentioned in the chapter opening story, some animals living in high salt environments have developed specialized glands for excreting these salts. These glands work by actively transporting salt from the blood into the gland, where it can be excreted as a concentrated solution.

In sea birds, salt-excreting glands are located near the eyes. The glands produce a salty solution that is excreted through the nostrils and moves down grooves on their beaks until it drips off (Fig. 36.5). In marine turtles, the salt gland is a modified tear (lacrimal) gland, and in sea snakes, a salivary sublingual gland beneath the tongue gets rid of excess salt. The work of these glands is regulated by the nervous system. Osmoreceptors, perhaps located near the heart, are thought to stimulate the brain, which then directs the gland to excrete salt until the salt concentration in the blood decreases to a tolerable level.

Check Your Progress   36.1

1. Distinguish between osmoregulation and excretion.
2. Describe two advantages of excreting urea instead of ammonia or uric acid.
3. List five strategies used by kangaroo rats to conserve water.

36.2 The Human Urinary System

Learning Outcomes

Upon completion of this section, you should be able to

1. Trace the anatomical path taken by urine from the glomeruli to its exit from the body.
2. Discuss the contributions of glomerular filtration, tubular reabsorption, and tubular secretion to the formation of urine.
3. Summarize the four major functions of human kidneys in maintaining homeostasis.

The major excretory organs of humans, as with most other vertebrates, are the kidneys (Fig. 36.6). The kidneys are the ultimate regulators of blood composition because they can remove various unwanted products from the body.

Human **kidneys** are bean-shaped, reddish-brown organs, each about the size of a fist. They are located on each side of the vertebral column just below the diaphragm, in the lower back, where they are partially protected by the lower rib cage. The right kidney is slightly lower than the left kidney.

Urine made by the kidneys is conducted from the body by the other organs in the urinary system. Each kidney is connected to a **ureter,** a duct that takes urine from the kidney to the **urinary bladder,** where it is stored until it is voided from the body through the single **urethra.** In males, the urethra passes through the penis, and in females, the opening of the urethra is ventral to that of the vagina. No connection exists between the genital (reproductive) and urinary systems in females, but in males, the urethra also carries sperm during ejaculation.

Figure 36.6 The human urinary system. a. The kidneys are well supplied with blood, as shown in the angiogram. b. Urine is found only within the kidneys, the ureters, the urinary bladder, and the urethra.

renal artery

renal vein

aorta

inferior vena cava

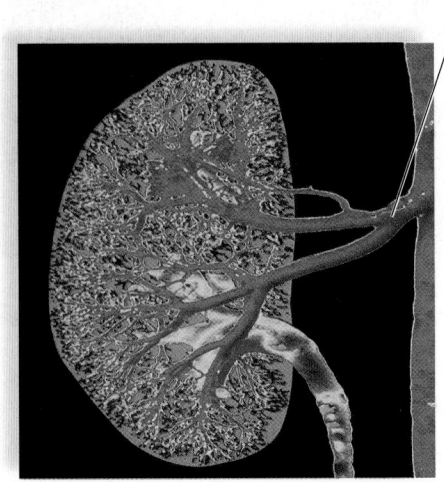

a.

b.

1. Kidneys produce urine.

2. Ureters transport urine.

3. Urinary bladder stores urine.

4. Urethra passes urine to outside.

Kidneys

If a kidney is sectioned longitudinally, three major parts can be distinguished (Fig. 36.7). The *renal cortex*, which is the outer region of a kidney, has a somewhat granular appearance. The *renal medulla* consists of six to ten cone-shaped renal pyramids that lie on the inner side of the renal cortex. The innermost part of the kidney is a hollow chamber called the *renal pelvis*. Urine collects in the renal pelvis and then is carried to the bladder by a ureter.

Nephrons

Microscopically, each kidney is composed of over 1 million tiny tubules called **nephrons** [Gk. *nephros*, kidney]. The nephrons of a kidney produce urine. Some nephrons are located primarily in the renal cortex, but others dip down into the renal medulla, as shown in Figure 36.7b. Each nephron is made of several parts (Fig. 36.8). The blind end of a nephron is pushed in on itself to form a cuplike structure called the **glomerular capsule** [L. *glomeris*, ball] (also known as Bowman's capsule). The outer layer of the glomerular capsule is composed of squamous epithelial

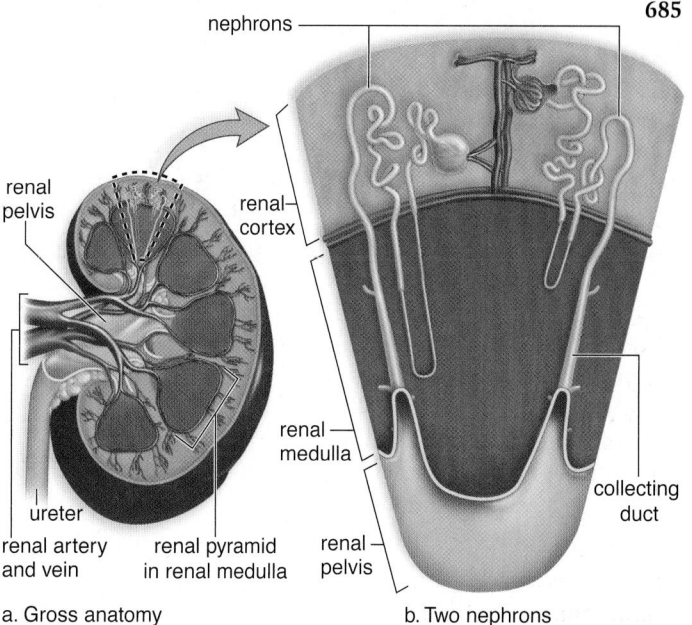

a. Gross anatomy

b. Two nephrons

Figure 36.7 Macroscopic and microscopic anatomy of the kidney. a. Longitudinal section of a kidney, showing the location of the renal cortex, the renal medulla, and the renal pelvis. **b.** An enlargement of one renal lobe, showing the placement of nephrons.

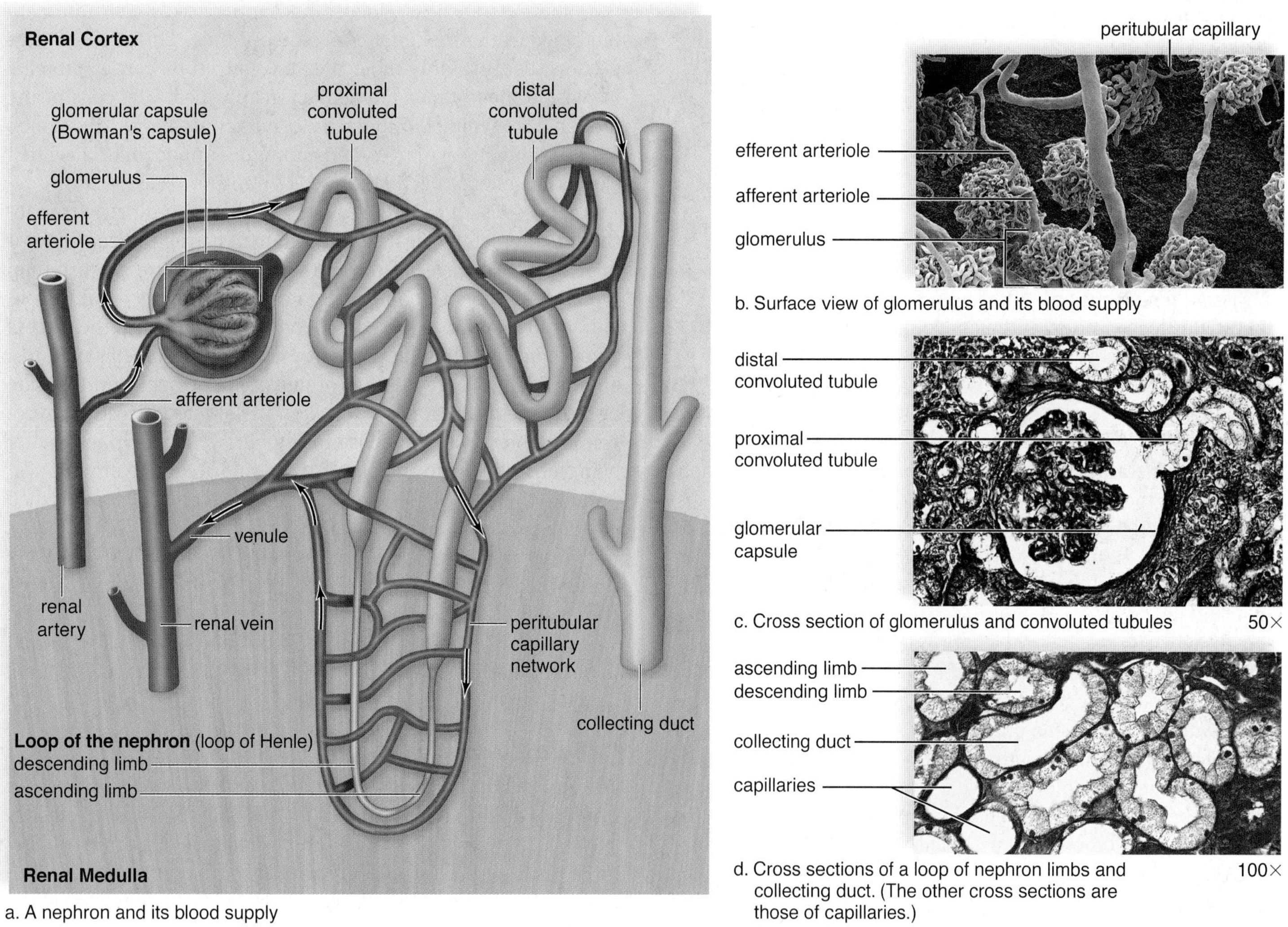

Renal Cortex

glomerular capsule (Bowman's capsule)

glomerulus

efferent arteriole

proximal convoluted tubule

distal convoluted tubule

afferent arteriole

venule

renal artery

renal vein

peritubular capillary network

collecting duct

Loop of the nephron (loop of Henle)
descending limb
ascending limb

Renal Medulla

a. A nephron and its blood supply

peritubular capillary

efferent arteriole

afferent arteriole

glomerulus

b. Surface view of glomerulus and its blood supply

distal convoluted tubule

proximal convoluted tubule

glomerular capsule

c. Cross section of glomerulus and convoluted tubules   50×

ascending limb
descending limb

collecting duct

capillaries

d. Cross sections of a loop of nephron limbs and collecting duct. (The other cross sections are those of capillaries.)   100×

Figure 36.8 Nephron anatomy.   a. You can trace the path of blood about a nephron by following the arrows. A nephron is made up of a glomerular capsule, the proximal convoluted tubule, the loop of the nephron, the distal convoluted tubule, and the collecting duct. The micrographs in **(b)**, **(c)**, and **(d)** show these structures.

cells; the inner layer is composed of specialized cells that allow easy passage of molecules.

Leading from the glomerular capsule is a portion of the nephron known as the **proximal convoluted tubule** [L. *proximus,* nearest], which is lined by cells with many mitochondria and tightly packed microvilli. Then, simple squamous epithelium appears in the **loop of the nephron** (loop of Henle), which has a descending limb and an ascending limb. This is followed by the **distal convoluted tubule** [L. *distantia,* far]. Several distal convoluted tubules enter one **collecting duct.** The collecting duct transports urine down through the renal medulla and deliver it to the renal pelvis.

Each nephron has its own blood supply (Fig. 36.8). The renal artery branches into numerous small arteries, which branch into arterioles, one for each nephron. Each arteriole, called an afferent arteriole, divides to form a capillary bed, the **glomerulus** [L. *glomeris,* ball], which is surrounded by the glomerular capsule. The glomerulus drains into an efferent arteriole, which subsequently branches into a second capillary bed around the tubular parts of the nephron. These capillaries, called peritubular capillaries, lead to venules that join to form veins leading to the renal vein, a vessel that enters the inferior vena cava.

Urine Formation

An average human produces between 1 and 2 liters of urine daily. The fundamental process of urine formation involves initially filtering a large amount of water and a collection of solutes out of the blood, then reabsorbing much of the water, along with other material the body needs to conserve.

Urine production requires three distinct processes (see Fig. 36.9*a*), and, as you can see, the entire tubule portion of a nephron participates in the last two steps in urine formation:

1. glomerular filtration at the glomerular capsule;
2. tubular reabsorption at the convoluted tubules; and
3. tubular secretion at the convoluted tubules.

 **MP3**
An Overview of
Urine Formation

Glomerular Filtration

Glomerular filtration (Fig. 36.9*a*) is the movement of small molecules across the glomerular wall into the glomerular capsule as a result of blood pressure. When blood enters the glomerulus, blood pressure is sufficient to cause small molecules, such as water, nutrients, salts, and wastes, to move from the glomerulus to the inside of the glomerular capsule, especially since the glomerular walls are 100 times more permeable than the walls of most capillaries elsewhere in the body. The molecules that leave the blood and enter the glomerular capsule are called the *glomerular filtrate.* Plasma proteins and blood cells are too large to be part of this filtrate, so they remain in the blood as it flows into the efferent arteriole.

Glomerular filtrate is essentially protein free, but otherwise it has the same composition as blood plasma. If this composition were not altered in other parts of the nephron, death from starvation (loss of nutrients) and dehydration (loss of water) would quickly follow. The total blood volume averages about 5 liters, and this amount of fluid is filtered every 40 minutes.

Thus 180 liters of filtrate is produced daily, some 60 times the amount of blood plasma in the body. Most of the filtered water is obviously quickly returned to the blood, or a person would actually die from urination. Tubular reabsorption prevents this from happening.

Tubular Reabsorption

Tubular reabsorption (Fig. 36.9*a*) takes place when substances move across the walls of the tubules into the associated peritubular capillary network (Fig. 36.9*a, b*). Here, osmosis comes into play. You may remember that *osmosis* is the diffusion of water down its concentration gradient across a membrane (see Section 5.2). *Osmolarity* is a measure of the potential for osmosis; water tends to move from a solution with low osmolarity into a solution with high osmolarity.

The osmolarity of the blood is essentially the same as that of the filtrate within the glomerular capsule, and therefore osmosis of water from the filtrate into the blood cannot yet occur. However, sodium ions (Na^+) are actively pumped into the peritubular capillary, and then chloride ions (Cl^-) follow passively. Now the osmolarity of the blood is such that water moves passively from the tubule into the blood. About 60–70% of salt and water are reabsorbed at the proximal convoluted tubule.

Nutrients such as glucose and amino acids also return to the blood at the proximal convoluted tubule. This is a selective process, because only molecules recognized by carrier proteins in plasma membranes are actively reabsorbed. The cells of the proximal convoluted tubule have numerous microvilli, which increase the surface area, and numerous mitochondria, which supply the energy needed for active transport (Fig. 36.9*b*).

Glucose is an example of a molecule that ordinarily is reabsorbed completely because the supply of carrier molecules for it is plentiful. However, if the filtrate contains more glucose than there are carriers to handle it, glucose exceeds its renal threshold, or transport maximum. When this happens, the excess glucose in the filtrate appears in the urine. In diabetes mellitus, an abnormally large amount of glucose is present in the filtrate because the liver cannot store all of the excess glucose as glycogen. The presence of glucose in the filtrate results in less water being absorbed; the increased thirst and frequent urination in untreated diabetics are a result of less water being reabsorbed into the peritubular capillary network.

Urea is an example of a substance that is passively reabsorbed from the filtrate. At first, the concentration of urea within the filtrate is the same as that in blood plasma. But after water is reabsorbed, the urea concentration is greater than that of peritubular plasma. In the end, about 50% of the filtered urea is reabsorbed.

Tubular Secretion

Tubular secretion is the second way substances are removed from blood and added to tubular fluid (Fig. 36.9*a*). Substances such as uric acid, hydrogen ions, ammonia, creatinine, histamine, and penicillin are eliminated by tubular secretion. The process of tubular secretion may be viewed as helping to rid the body of potentially harmful compounds that were not filtered into the glomerulus. ▶ **Animation**
Kidney Function

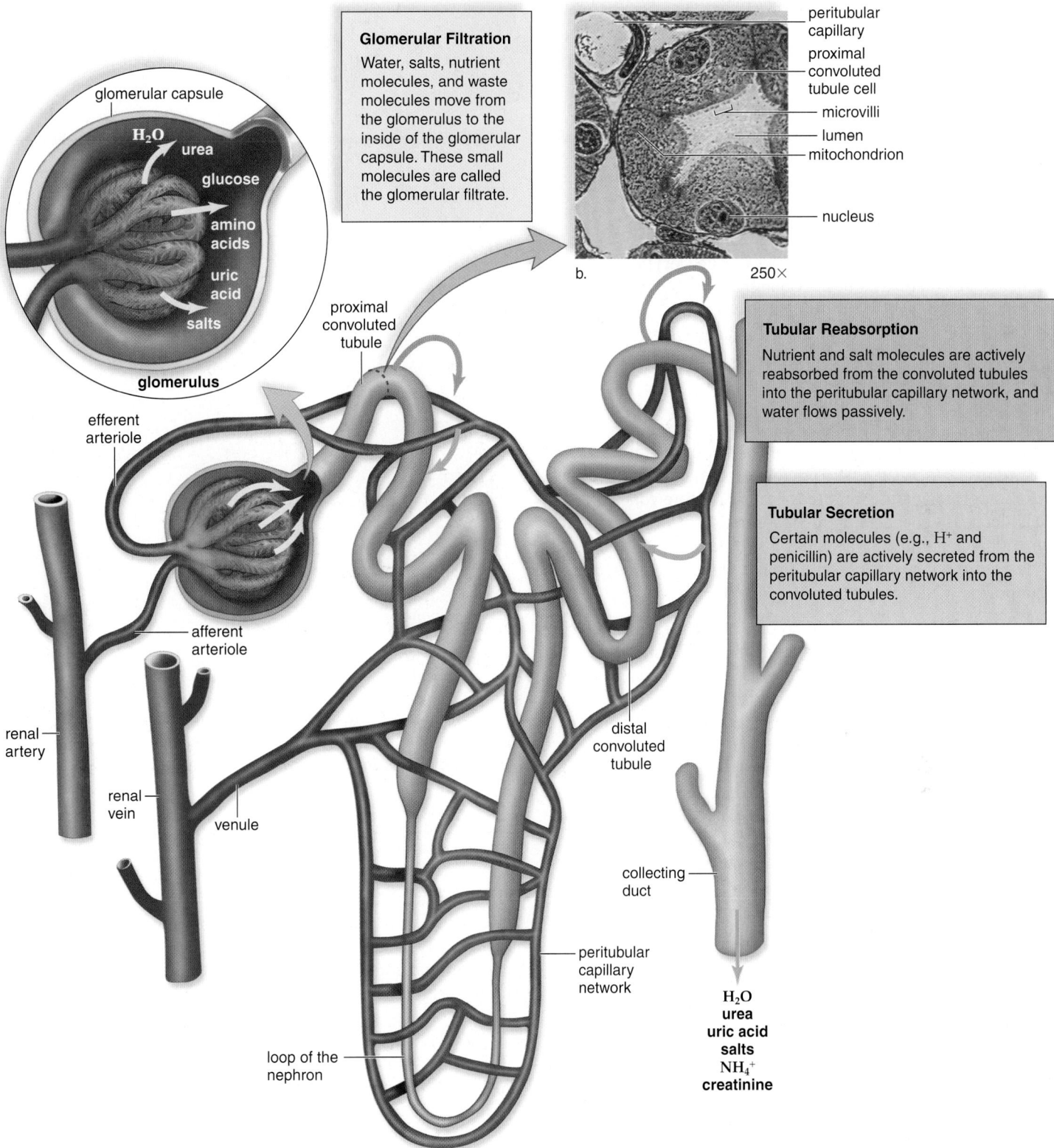

Glomerular Filtration

Water, salts, nutrient molecules, and waste molecules move from the glomerulus to the inside of the glomerular capsule. These small molecules are called the glomerular filtrate.

glomerular capsule

H₂O
urea
glucose
amino acids
uric acid
salts

glomerulus

peritubular capillary
proximal convoluted tubule cell
microvilli
lumen
mitochondrion
nucleus

b.   250×

efferent arteriole

proximal convoluted tubule

Tubular Reabsorption

Nutrient and salt molecules are actively reabsorbed from the convoluted tubules into the peritubular capillary network, and water flows passively.

Tubular Secretion

Certain molecules (e.g., H⁺ and penicillin) are actively secreted from the peritubular capillary network into the convoluted tubules.

afferent arteriole

renal artery

renal vein

venule

distal convoluted tubule

collecting duct

peritubular capillary network

loop of the nephron

H₂O
urea
uric acid
salts
NH₄⁺
creatinine

Figure 36.9 Processes in urine formation. **a.** The three main processes in urine formation are described in boxes and color coded to arrows that show the movement of molecules into or out of the nephron at specific locations. In the end, urine is composed of the substances within the collecting duct (see blue arrow). **b.** This photomicrograph shows that the cells lining the proximal convoluted tubule have a brush border composed of microvilli, which greatly increases the surface area exposed to the lumen. The peritubular capillary adjoins the cells.

The Kidneys and Homeostasis

The kidneys are organs of homeostasis for four main reasons:

1. The kidneys *excrete metabolic wastes* such as urea, which is the primary nitrogenous waste of humans.
2. They *maintain the water-salt balance,* which in turn affects blood volume and blood pressure.
3. Kidneys *maintain the acid-base balance,* and therefore, the pH balance.
4. They *secrete hormones.*

One hormone secreted by the kidneys, called **erythropoietin,** stimulates the stem cells in the bone marrow to produce more red blood cells. The Nature of Science feature examines the potential abuse of this hormone by endurance athletes. Another substance produced by the kidneys, called renin, is discussed later in this section.

Maintaining the Water-Salt Balance

Most of the water and salt (NaCl) present in the filtrate is reabsorbed across the wall of the proximal convoluted tubule. The excretion of a hypertonic urine (one that is more concentrated than blood) is dependent on the reabsorption of water from the loop of the nephron and the collecting duct. During the process of reabsorption, water passes through water channels called **aquaporins,** which were discovered in 1992.

Loop of the Nephron.   A long loop of the nephron, which typically penetrates deep into the renal medulla, is made up of a descending (downward) limb and an ascending (upward) limb. Salt (NaCl) passively diffuses out of the lower portion of the ascending limb, but the upper, thick portion of the limb actively extrudes salt out into the tissue of the outer renal medulla (Fig. 36.10). Less and less salt is available for transport as fluid moves up the thick portion of the ascending limb. Because of these circumstances, an osmotic gradient is created within the tissues of the renal medulla: The concentration of salt is greater in the direction of the inner medulla. (Note that water cannot leave the ascending limb because this limb is impermeable to water.)

The innermost portion of the inner medulla has the highest concentration of solutes. This cannot be due to salt because active transport of salt does not start until fluid reaches the thick portion of the ascending limb. Urea is believed to leak from the lower portion of the collecting duct, and it is this molecule that contributes to the high solute concentration of the inner medulla.

Because of the osmotic gradient within the renal medulla, water leaves the descending limb along its entire length. This is a countercurrent mechanism: As water diffuses out of the descending limb, the remaining fluid within the limb encounters an even greater osmotic concentration of solute; therefore, water continues to leave the descending limb from the top to the bottom. Filtrate within the collecting duct also encounters the same osmotic gradient mentioned earlier (Fig. 36.10). Therefore, water diffuses out of the collecting duct into the renal medulla, and the urine within the collecting duct becomes hypertonic to blood plasma.

Antidiuretic hormone (ADH) [Gk. *anti,* against; L. *ouresis,* urination] is released by the posterior lobe of the pituitary in response to an increased concentration of salts in the blood. To

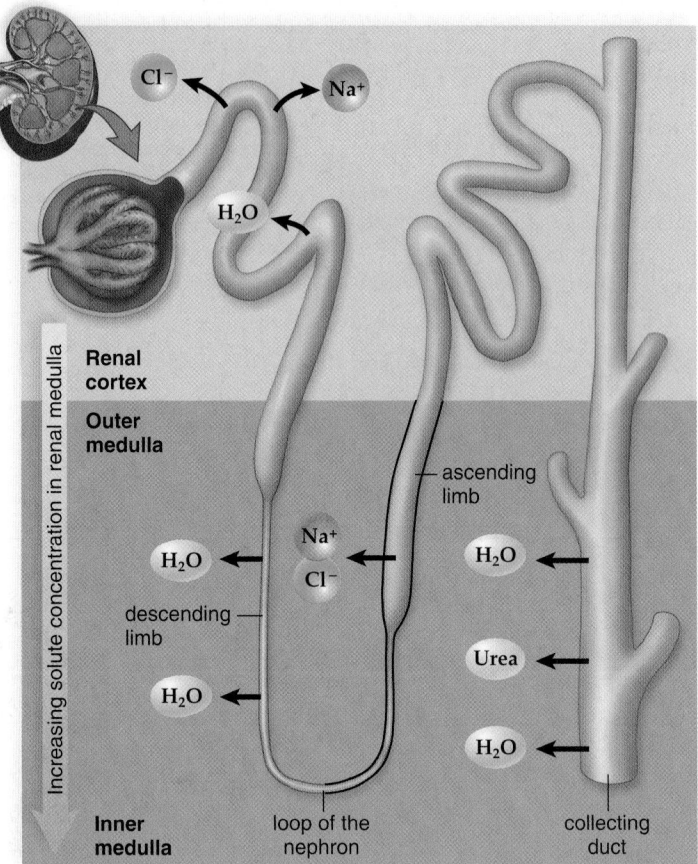

Figure 36.10   Reabsorption of salt and water.   Salt (NaCl) diffuses and is actively transported out of the ascending limb of the loop of the nephron into the renal medulla; also, urea leaks from the collecting duct and enters the tissues of the renal medulla. These actions create a hypertonic environment, which draws water out of the descending limb and the collecting duct. This water is returned to the cardiovascular system.

understand the action of this hormone, consider its name. *Diuresis* means increased amount of urine, and *antidiuresis* means decreased amount of urine. When ADH is present, more water is reabsorbed (blood volume and pressure rise), and a decreased amount of more concentrated urine is produced. One way by which ADH accomplishes this change is by causing the insertion of additional aquaporin water channels into the epithelial cells of the distal tubule and collecting duct, allowing more water to be reabsorbed.

In practical terms, if an individual does not drink much water on a certain day, the posterior lobe of the pituitary releases ADH, causing more water to be reabsorbed and less urine to form. On the other hand, if an individual drinks a large amount of water and does not perspire much, ADH is not released. Now more water is excreted, and more urine forms. Diuretics, such as caffeine and alcohol, increase the flow of urine by interfering with the action of ADH. ADH production also decreases at night, an adaptation that allows longer periods of sleep without the need to wake up to urinate.

Hormones Control the Reabsorption of Salt.   Usually, more than 99% of Na^+ filtered at the glomerulus is returned to the blood. Most sodium (67%) is reabsorbed at the proximal

Misuse of Erythropoietin in Sports

Lance Armstrong is a hero to many sports fans (Fig. 36A). He is a seven-time winner of the Tour de France cycling event, a survivor of testicular cancer, and a sponsor of many charitable causes. On May 22, 2011, the TV news program 60 Minutes aired a segment in which Tyler Hamilton, a former teammate of Armstrong's, said he witnessed Armstrong using performance enhancing drugs on several occasions. Armstrong has denied these charges, noting that he tested negative nearly 500 times during his 20-year career.

The substance that Armstrong is most often accused of abusing is erythropoietin (EPO), a hormone secreted by the kidney in response to low blood oxygen levels. EPO binds to a specific cellular receptor, found mainly on bone marrow cells that produce red blood cells (RBCs). The gene coding for EPO was isolated in 1985, and since 1989, the use of recombinant human EPO (rHuEPO) has been approved for medical purposes, such as treating the anemia that is often associated with kidney failure. However, rHuEPO use can have serious side effects, including increased blood clotting, high blood pressure, and even sudden death.

In addition to its legitimate medical uses, rHuEPO has been misused by athletes in several endurance sports. Research has generally verified that EPO is effective at increasing athletic performance. A small study published in 2007 showed a 13% increase in peak power output, and a 54% increase in time to exhaustion in cyclists

who used rHuEPO (Fig. 36B). Because this is an unfair advantage, rHuEPO use has been banned by the Tour de France, the Olympics, and other sports organizations.

To determine whether an athlete is using rHuEPO, an accurate test is needed. One method of testing athletes for any treatment designed to increase the RBC count (including collecting one's own blood, storing it, and transfusing it back into the athlete soon before an event) is to monitor the *hematocrit;* the percentage of the blood that is comprised of cells. Since 1997, sports cycling governing bodies have decreed that any athlete whose hematocrit is over 50% may be suspended from competition. However, normal hematocrit values for men can vary widely, from 40% to 54%. This means an athlete with a naturally high hematocrit could be unfairly accused of cheating.

A better approach would be to directly measure levels of rHuEPO in an athlete's body. Unfortunately, these tests have limitations as well. Part of the problem is that when rHuEPO is injected, it may persist in the body for as little as 24 hours, but its effect continues for as long as two weeks. Unless a test is conducted during the short time that rHuEPO is present, even the most accurate test may miss it. A second issue is that rHuEPO must be distinguished from the EPO produced by an athlete's own kidneys. rHuEPO is produced in cultured hamster ovary cells, which process the protein differently, attaching different sugar

molecules to the amino acids. This should result in slight differences in the movement of rHuEPO versus EPO through a gel material that is subjected to an electrical field (i.e., electrophoresis).

Unfortunately, direct testing for rHuEPO may also be unreliable. In a 2008 study, eight adult male volunteers received weekly rHuEPO injections, and identical samples of their urine were submitted to two labs that were approved for rHuEPO testing. "Lab A" found that 6 of 16 samples were positive, but "Lab B" concluded that all samples were negative.[1]

Because of these issues, the question of whether Armstrong abused rHuEPO may never be resolved. It will also be difficult to eliminate the misuse of rHuEPO in endurance sports, especially in participants whose motivation to win overpowers their concern about the risks of abusing rHuEPO.

Questions to Consider

1. Certain kidney tumors secrete large amounts of EPO. What types of symptoms might this cause?
2. Some medical conditions, e.g., bacterial infections, cause the numbers of WBCs to increase, perhaps even to twice the normal levels. Why does this have very little effect on the hematocrit?

---

1 Lundby, C., et al. 2008. Testing for recombinant human erythropoietin in urine: problems associated with current anti-doping testing. *J. Applied Physiology* 105: 417–19.

Figure 36A Lance Armstrong and Tyler Hamilton. Hamilton (*green and white helmet*) says he observed Armstrong (*red, white, and blue helmet*) using performance enhancing drugs, including rHuEPO.

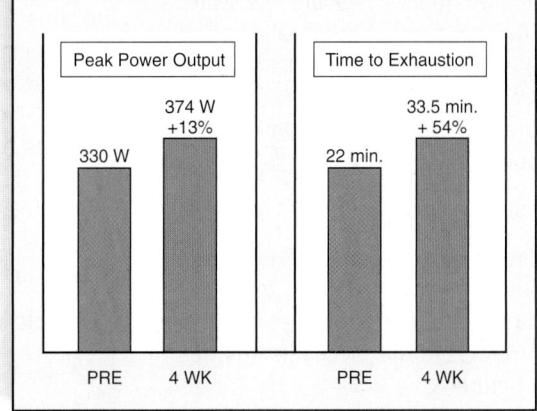

Peak Power Output

374 W
+13%

330 W

PRE   4 WK

Time to Exhaustion

33.5 min.
+ 54%

22 min.

PRE   4 WK

Figure 36B Use of rHuEPO improves athletic performance. Four weeks of rHuEPO treatments increased the peak power output and prolonged the time that relatively fit cyclists were able to ride (at 80% of their maximum exertion level).

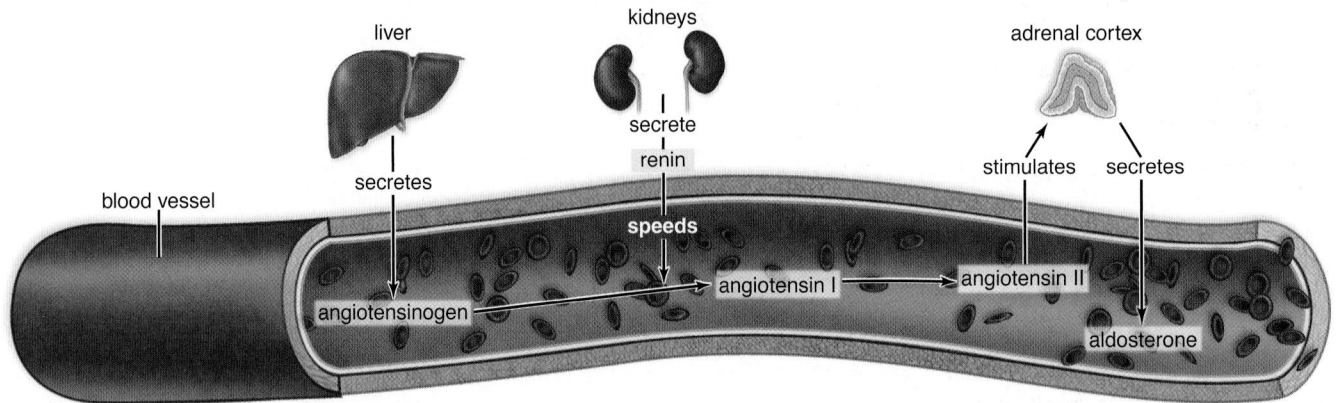

Figure 36.11 The renin-angiotensin-aldosterone system. The liver secretes angiotensinogen into the bloodstream. Renin from the kidneys initiates the chain of events that results in angiotensin II. Angiotensin II acts on the adrenal cortex to secrete aldosterone, which causes reabsorption of sodium ions by the kidneys and a subsequent rise in blood pressure.

convoluted tubule, and a sizable amount (25%) is extruded by the ascending limb of the loop of the nephron. The rest is reabsorbed from the distal convoluted tubule and collecting duct.

Blood volume and pressure is, in part, regulated by salt reabsorption. When blood volume, and therefore blood pressure, is not sufficient to promote glomerular filtration, a cluster of cells near the glomerulus called the *juxtaglomerular apparatus* secretes renin. **Renin** is an enzyme that changes angiotensinogen (a large plasma protein produced by the liver) into angiotensin I. Later, angiotensin I is converted to **angiotensin II**, a powerful vasoconstrictor that also stimulates the adrenal glands, which lie on top of the kidneys, to release aldosterone (Fig. 36.11). **Aldosterone** is a hormone that promotes the excretion of potassium ions (K^+) and the reabsorption of sodium ions (Na^+) at the distal convoluted tubule. The reabsorption of sodium ions is followed by the reabsorption of water. Therefore, blood volume and blood pressure increase.

Atrial natriuretic hormone (ANH) is a hormone secreted by the atria of the heart when cardiac cells are stretched due to increased blood volume. ANH inhibits the secretion of renin by the juxtaglomerular apparatus and the secretion of aldosterone by the adrenal cortex. Its effect, therefore, is to promote the excretion of Na^+—that is, natriuresis. When Na^+ is excreted, so is water, and therefore blood volume and blood pressure decrease.

These examples show that the kidneys regulate the water balance in blood by controlling the excretion and the reabsorption of ions. Sodium is an important ion in plasma that must be regulated, but the kidneys also excrete or reabsorb other ions, such as potassium ions (K^+), bicarbonate ions (HCO_3^-), and magnesium ions (Mg^{2+}), as needed.

MP3
Water Conservation

Maintaining the Acid-Base Balance

The functions of cells are influenced by pH. Therefore regulation of pH is extremely important to good health. The *bicarbonate (HCO_3^-) buffer system* and breathing work together to help maintain the pH of the blood. Central to the mechanism is this reaction, which you have seen before:

$$H^+ + HCO_3^- \rightleftharpoons H_2CO_3 \rightleftharpoons H_2O + CO_2$$

The *excretion of carbon dioxide (CO_2) by the lungs* helps keep the pH within normal limits, because when carbon dioxide is exhaled, this reaction is pushed to the right, and hydrogen ions are tied up in water. As you learned in the preceding chapter, when blood pH

decreases, chemoreceptors in the carotid bodies (in the carotid arteries) and in aortic bodies (in the aorta) stimulate the respiratory control center, and the rate and depth of breathing increase. And, when blood pH begins to rise, the respiratory control center is depressed, and the amount of bicarbonate ion increases in the blood.

As powerful as this system is, only the kidneys can rid the body of a wide range of acidic and basic substances. The kidneys are slower acting than the buffer/breathing mechanism, but they have a more powerful effect on pH. For the sake of simplicity, we can think of the kidneys as reabsorbing bicarbonate ions and excreting hydrogen ions as needed to maintain the normal pH of the blood:

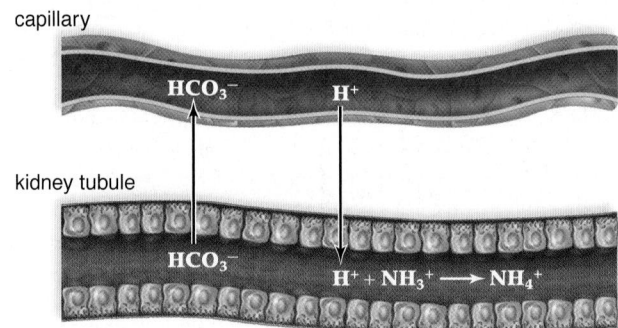

If the blood is acidic, hydrogen ions are excreted and bicarbonate ions are reabsorbed. If the blood is basic, hydrogen ions are not excreted and bicarbonate ions are not reabsorbed. The fact that urine is typically acidic (pH about 6) shows that usually an excess of hydrogen ions are excreted. Ammonia (NH_3) provides a means for buffering these hydrogen ions in urine: ($NH_3 + H^+ \longrightarrow NH_4^+$). Ammonia is produced in tubule cells by the deamination of amino acids. Phosphate provides another means of buffering hydrogen ions in urine.

MP3
Acid-Base Balance

Check Your Progress   36.2

1. Describe which of the four major functions of the human urinary system are accomplished solely by the kidneys, and which are shared with other body systems.
2. Explain the effect of the renin-angiotensin-aldosterone system on water-salt balance.
3. Describe how the kidneys contribute to the maintenance of normal blood pH.

CONNECTING *the* CONCEPTS *with the* BIG IDEAS

Energy and Homeostasis

- Organisms of all types, from bacteria to protists to fish, employ similar methods of osmoregulation to achieve homeostasis. (2D2c*IE*)

Interactions and Systems

- The kidney and bladder interact to ensure an efficient waste disposal system. (4A4a*IE*)
- Organ specialization within the excretory system increases efficiency for the organism. (4B2a2*IE*)

*Find the unabridged version of all EK citations at www.glencoe.com/maderAP11.

Media Study Tools

www.glencoe.com/maderAP11

Enhance your study of this chapter with study tools and practice tests. Also ask your instructor about the resources available through ConnectPlus, including the media-rich eBook, interactive learning tools, and animations.

Summarize

36.1 Animal Excretory Systems

Animals excrete nitrogenous wastes. Aquatic animals usually excrete ammonia (requires little energy but much water to excrete), and land animals excrete either urea or uric acid (requires much energy to produce).

Most animals have specialized excretory organs. The flame cells of planarians rid the body of excess water. Earthworm nephridia exchange molecules with the blood in a manner similar to that of vertebrate kidneys. Malpighian tubules in insects take up metabolic wastes and water from the hemolymph. Later, the water is absorbed by the gut.

Osmotic regulation is important to animals. Most must balance their water and salt intake and excretion to maintain normal solute and water concentration in body fluids. Marine fishes constantly drink water, excrete salts at the gills, and pass an isotonic urine. Freshwater fishes never drink water; they take in salts at the gills and excrete a hypotonic urine.

Some terrestrial animals have adapted to extreme environments. For example, the desert kangaroo rat can survive on metabolic water; marine birds and reptiles have glands that extrude salt.

36.2 The Human Urinary System

The human urinary system includes the kidneys, ureters, urinary bladder, and urethra. The kidneys serve four basic homeostatic functions: excretion of metabolic waste; maintenance of water-salt balance; maintenance of pH balance; and production of hormones, such as erythropoietin.

Kidneys are made up of nephrons, each of which has several parts. Urine formation by a nephron requires three steps: glomerular filtration, during which nutrients, water, and wastes enter the nephron's glomerular capsule; tubular reabsorption, when nutrients and most water are reabsorbed into the peritubular capillary network; and tubular secretion, during which additional wastes are added to the convoluted tubules.

In order to excrete a hypertonic urine, the ascending limb of the loop of the nephron extrudes salt so that the renal medulla is increasingly hypertonic. Since urea leaks from the lower end of the collecting duct, the inner renal medulla has the highest concentration of solute. Therefore, a countercurrent mechanism ensures that water diffuses out of the descending limb and the collecting duct.

Three hormones are involved in maintaining the water-salt balance of the blood. ADH (antidiuretic hormone), which makes the collecting duct more permeable to water, is secreted by the posterior pituitary in response to an increase in the osmotic pressure of the blood. Aldosterone is secreted by the adrenal cortex after low blood pressure has caused the kidneys to release renin. The presence of renin leads to the formation of angiotensin II, which causes the adrenal cortex to release aldosterone. Aldosterone causes the kidneys to retain Na^+; therefore, water is reabsorbed and blood pressure rises. Atrial natriuretic hormone, in contrast, prevents the secretion of renin and aldosterone.

The kidneys also keep blood pH within normal limits. They reabsorb HCO_3^- and excrete H^+ as needed to maintain the pH at about 7.4.

Key Terms

aldosterone 690	kidneys 684
ammonia 681	loop of the nephron 686
angiotensin II 690	nephron 685
antidiuretic hormone	osmoregulation 681
(ADH) 688	proximal convoluted
aquaporin 688	tubule 686
atrial natriuretic hormone	renin 690
(ANH) 690	tubular reabsorption 686
collecting duct 686	tubular secretion 686
distal convoluted tubule 686	urea 681
erythropoietin 688	ureter 684
excretion 681	urethra 684
glomerular capsule 685	uric acid 681
glomerular filtration 686	urinary bladder 684
glomerulus 686	urine 682

Assess

Reviewing This Chapter

1. Relate the three primary nitrogenous wastes to the habitat of animals.   681
2. Describe how the excretory organs of the earthworm and the insect function.   682
3. Give examples of how other types of animals regulate their water and salt balance.   682–84
4. Contrast the osmotic regulation of a marine bony fish with that of a freshwater bony fish.   683
5. Describe the path of urine in humans, and give a function for each structure mentioned.   684
6. Describe the macroscopic anatomy of a human kidney, and relate it to the placement of nephrons.   685
7. List the parts of a nephron, and give a function for each structure mentioned.   685–86
8. Describe how urine is made by outlining what happens at each part of the nephron.   686–87
9. Describe the reabsorption of water and salt along the length of the nephron. Include the contribution of the loop of the nephron.   688, 690
10. Name and describe the action of antidiuretic hormone (ADH), the renin-aldosterone connection, and the atrial natriuretic hormone (ANH).   688, 690
11. How does the nephron regulate the pH of the blood?   690

Testing Yourself

Choose the best answer for each question.

1. Which of these pairs is mismatched?
   a. insects—excrete uric acid
   b. humans—excrete urea
   c. fishes—excrete ammonia
   d. birds—excrete ammonia
   e. All of these are correct.

2. One advantage of excretion of urea instead of uric acid is that urea
   a. requires less energy than uric acid to produce.
   b. can be concentrated to a greater extent.
   c. is not a toxic substance.
   d. requires no water to excrete.
   e. is a larger molecule.

3. Freshwater bony fishes maintain water balance by
   a. excreting salt across their gills.
   b. periodically drinking small amounts of water.
   c. excreting a hypotonic urine.
   d. excreting wastes in the form of uric acid.
   e. Both a and c are correct.

4. Animals with which of these are most likely to excrete a semisolid nitrogenous waste?
   a. nephridia
   b. Malpighian tubules
   c. flame cells
   d. All of these are correct.

5. Excretion of a hypertonic urine in humans is associated best with
   a. the glomerular capsule.
   b. the proximal convoluted tubule.
   c. the loop of the nephron.
   d. the collecting duct.
   e. Both c and d are correct.

6. ADH (antidiuretic hormone) causes an individual to excrete
   a. less salt.
   b. less water.
   c. more water.
   d. Both a and c are correct.

7. Which of these causes blood pressure to decrease?
   a. aldosterone
   b. antidiuretic hormone (ADH)
   c. renin
   d. atrial natriuretic hormone (ANH)

8. Which of these materials is not filtered from the blood at the glomerulus?
   a. water
   b. urea
   c. protein
   d. glucose

9. By what process are most molecules secreted from the blood into the convoluted tubules?
   a. osmosis
   b. diffusion
   c. active transport
   d. facilitated diffusion

10. Which of these is not correct?
   a. Uric acid is produced from the breakdown of nucleic acids.
   b. Urea is produced from the breakdown of proteins.
   c. Ammonia results from the deamination of amino acids.
   d. All of these are correct.

11. When tracing the path of blood, the blood vessel that follows the renal artery is the
   a. peritubular capillary.
   b. efferent arteriole.
   c. afferent arteriole.
   d. renal vein.

12. Label this diagram of a nephron.

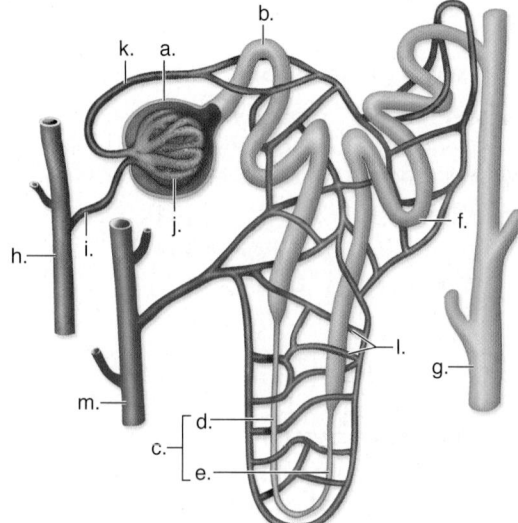

Engage

Virtual Labs
Earthworm Dissection
Virtual Frog Dissection

The Virtual Labs "Earthworm Dissection" and "Virtual Frog Dissection" allow you to explore the excretory systems of invertebrate and vertebrate animals.

Thinking Scientifically

1. High blood pressure often is accompanied by kidney damage. In some people, the kidney damage is subsequent to the high blood pressure, but in others the kidney damage is what caused the high blood pressure. Explain how a low-salt diet would enable you to determine whether the high blood pressure or the kidney damage came first?

2. The renin-angiotensin-aldosterone system can be inhibited in order to reduce high blood pressure. Usually, the angiotensin-converting enzyme, which converts angiotensin I to angiotensin II, is inhibited by drug therapy. Why would this enzyme be an effective point to disrupt the system?

Muhammad Ali, nicknamed "The Greatest," in the prime of his boxing career.

37

Neurons and Nervous Systems

BEFORE YOU BEGIN

Before beginning this chapter, take a few moments to review the following discussions.

Figure 5.10 How does the sodium-potassium pump function?

Section 29.1 From what embryonic structures are the brain, spinal cord, and (in vertebrates) vertebral column derived?

Section 30.1 How has the evolution of a large, complex brain allowed humans to become the most dominant species on the planet?

Muhammad Ali was perhaps the greatest heavyweight boxer of all time. A three-time world champion, Ali ruled boxing during the early to mid-1970s. Although he was 6 feet 3 inches tall and fought at a weight of about 220 pounds, Ali was incredibly fast, often pummeling opponents with flurries of punches while gliding around the ring with a style that became known as "float like a butterfly, sting like a bee."

To watch a professional athlete like Ali display his skills is to watch the nervous system performing at a high level. No computer has yet been invented that could coordinate the complex muscular movements involved not only in delivering punches but also avoiding the aggression of one's opponent. Sadly, in 1984 Ali was diagnosed with a form of Parkinson disease (PD), a nervous system disorder that affects his ability to speak or control his muscles. Although his thought processes remain intact, Ali's speech is slow and slurred, and his muscles tremble when he attempts to move. It is unknown whether the years of trauma to Ali's brain caused his condition, especially since the great majority of PD patients have suffered no apparent brain trauma. Later on in the chapter you can learn more about one method scientists are using to investigate the cause of PD.

As you read through the chapter, think about the following questions:

1. How did the evolution of the nervous system provide advantages to animals?
2. What specific types of processes occur uniquely in nervous tissues?
3. Why are diseases of the human nervous system generally difficult to treat?

FOLLOWING *the* BIG IDEAS

CHAPTER 37   NEURONS AND NERVOUS SYSTEMS

Information and Signaling	Neurons conduct electrochemical messaging throughout the animal body.
Interactions and Systems	The nervous system provides information input and output upon which other systems rely for their operation.

37.1 Evolution of the Nervous System

The nervous system is vitally important in complex animals, enabling them to seek food and mates and avoid danger. It ceaselessly monitors internal and external conditions and makes appropriate changes to maintain homeostasis. A comparative study of animal nervous systems shows the evolutionary trends that led to the nervous system of mammals.

Invertebrate Nervous-System Organization

The simplest multicellular animals, such as sponges, lack neurons (nerve cells), and therefore have no nervous system. However, their cells can respond to their environment and can communicate with each other, perhaps by releasing calcium or other ions; the most common example is closure of the osculum (central opening) in response to various stimuli.

Hydras, which are cnidarians with the tissue level of organization and radial symmetry, have a **nerve net** that is composed of neurons in contact with one another and with contractile cells in the body wall (Fig. 37.1*a*). They can contract and extend their bodies, move their tentacles to capture prey, and even turn somersaults. Sea anemones and jellyfish, which are also cnidarians, seem to have two nerve nets: A fast-acting one allows major responses, particularly in times of danger; and the slower one coordinates slower and more delicate movements.

Planarians (flatworms) have a nervous organization that reflects their bilateral symmetry. They have a ladderlike nervous system, with two ventrally located lateral or longitudinal nerve cords (bundles of nerves) that extend from the cerebral ganglia to the posterior end of their body. Transverse nerves connect the nerve cords, as well as the cerebral ganglia, to the eyespots. **Cephalization**, or concentration of nervous tissue in the anterior or head region, has occurred. A cluster of neuron cell bodies is called a **ganglion** (pl., ganglia), and the anterior cerebral ganglia of flatworms receive sensory information from photoreceptors in the eyespots and sensory cells in the auricles (Fig. 37.1*b*). The two lateral nerve cords allow a rapid transfer of information from the cerebral ganglia to the posterior end, and the transverse nerves between the nerve cords keep the movement of the two sides coordinated. Bilateral symmetry plus cephalization are two significant trends in the development of a nervous organization that is adaptive for an active way of life.

Annelids (e.g., earthworm, Fig. 37.1*c*) and arthropods (e.g., crab, Fig. 37.1*d*), are complex animals with the typical invertebrate nervous system. A brain is present, and a ventral nerve cord has a ganglion in each segment. The brain, which normally receives sensory information, controls the activity of the ganglia

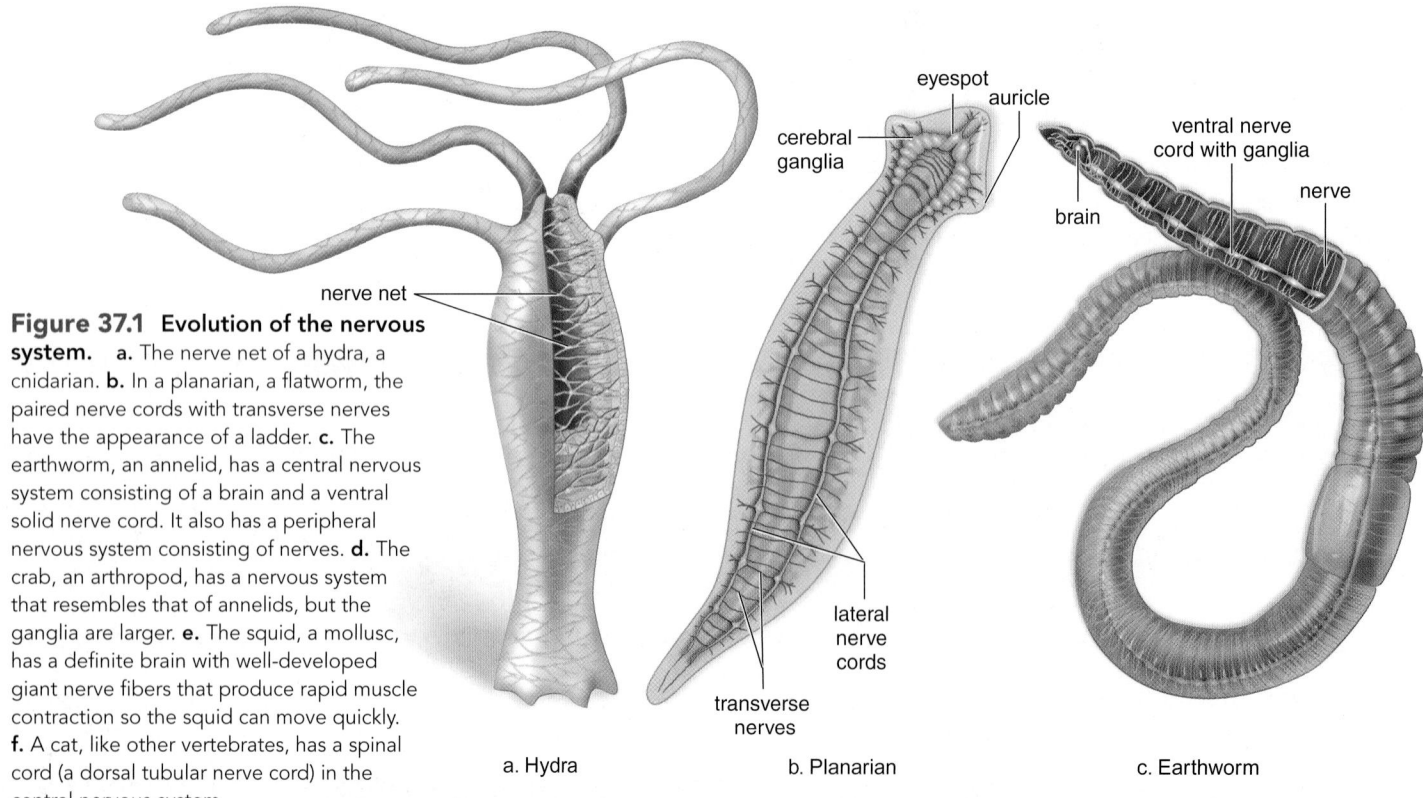

Figure 37.1 Evolution of the nervous system. a. The nerve net of a hydra, a cnidarian. **b.** In a planarian, a flatworm, the paired nerve cords with transverse nerves have the appearance of a ladder. **c.** The earthworm, an annelid, has a central nervous system consisting of a brain and a ventral solid nerve cord. It also has a peripheral nervous system consisting of nerves. **d.** The crab, an arthropod, has a nervous system that resembles that of annelids, but the ganglia are larger. **e.** The squid, a mollusc, has a definite brain with well-developed giant nerve fibers that produce rapid muscle contraction so the squid can move quickly. **f.** A cat, like other vertebrates, has a spinal cord (a dorsal tubular nerve cord) in the central nervous system.

a. Hydra

b. Planarian

c. Earthworm

and assorted nerves so that the muscle activity of the entire animal is coordinated.

A group of molluscs called cephalopods (e.g. squid, Fig. 37.1*e*) show marked cephalization—the anterior end has a well-defined brain, and well-developed sense organs, such as eyes, are present. The cephalopods are widely regarded as the most intelligent invertebrates; many are highly social creatures and some, such as the octopus, have been observed to collect, transport, and assemble coconut shells for later use as a shelter.

Vertebrate Nervous-System Organization

Vertebrates have many more neurons than do invertebrates. For example, an insect's entire nervous system contains a total of about 1 million neurons, while a cat's nervous system may contain many thousand times that number (Fig. 37.1*f*). The human cerebral cortex alone contains an estimated 11 billion neurons; some whales have even more.

All vertebrates have a brain that controls the nervous system. It is customary to divide the vertebrate brain into the hindbrain, midbrain, and forebrain (Fig. 37.2), although the relative sizes of the parts vary greatly among species. The hindbrain is the most ancient part of the brain, which regulates motor activity below the level of consciousness. For example, the lungs and heart function even when an animal is sleeping. The medulla oblongata contains control centers for breathing and heart rate. Coordination of motor activity associated with limb movement, posture, and balance eventually became centered in the cerebellum.

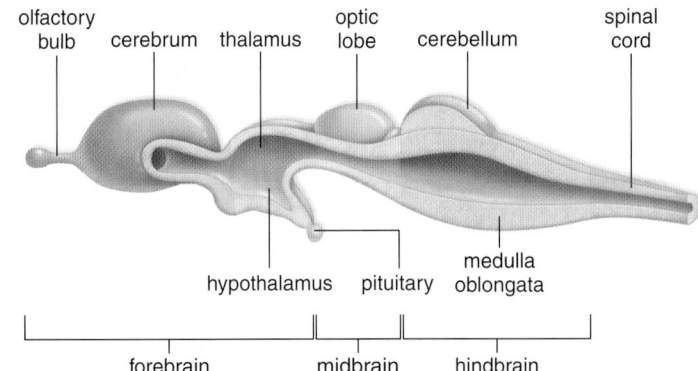

Figure 37.2 Organization of the vertebrate brain. The vertebrate brain is divided into the forebrain, the midbrain, and the hindbrain.

Several types of paired sensory receptors, including the eyes, ears, and olfactory structures, allow the animal to gather information from the environment. These sense organs are generally located at the anterior end of the animal, because this end is usually the first to enter new environments. The optic lobes are part of the midbrain, which was originally a center for coordinating reflexes involving the eyes and ears. In early vertebrate evolution, the forebrain was concerned mainly with the sense of smell. Beginning with the amphibians and continuing in the other vertebrates, the forebrain processes sensory information. Later, the thalamus evolved to receive sensory input from the midbrain and the hindbrain and to pass it on to the cerebrum,

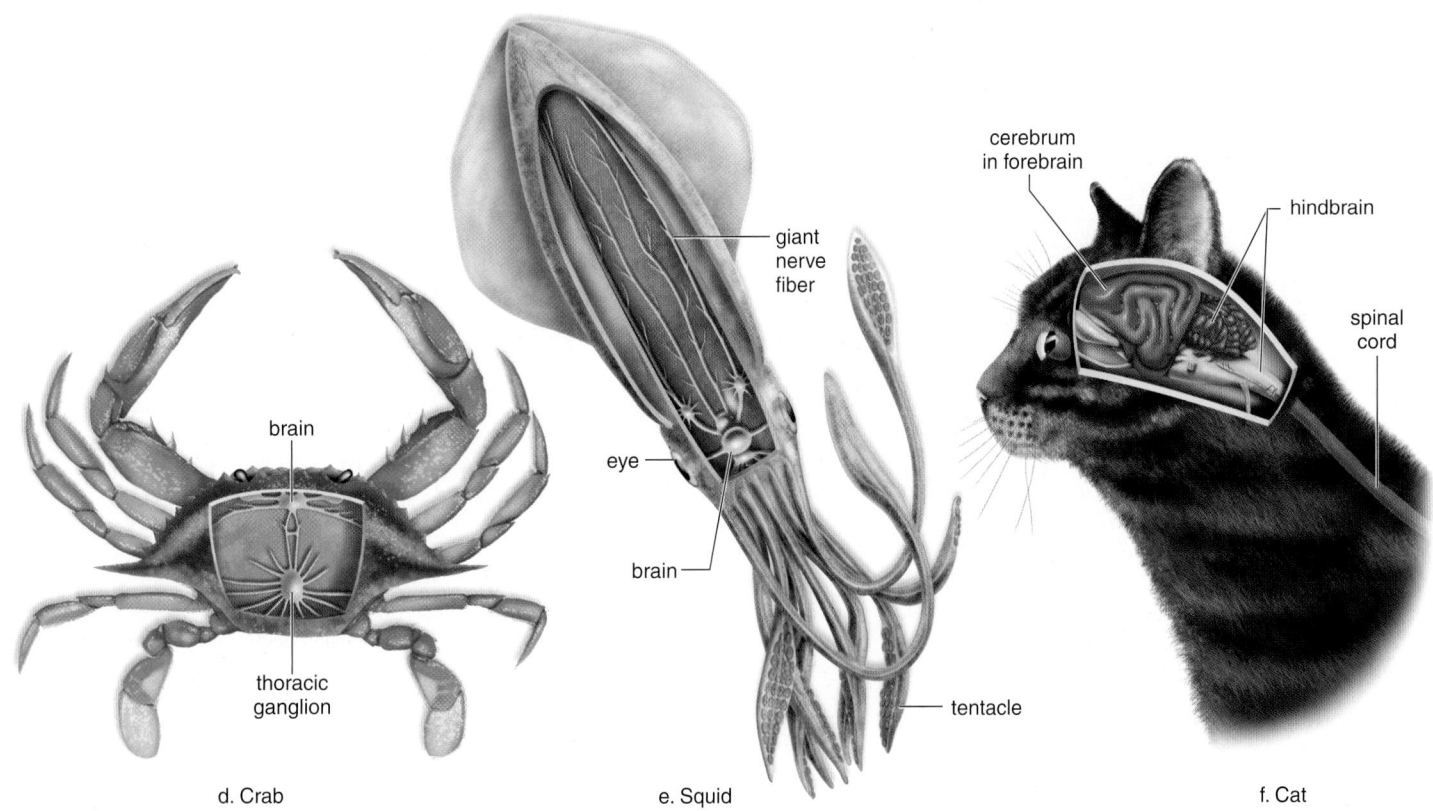

d. Crab

e. Squid

f. Cat

the anterior part of the forebrain in vertebrates. In the forebrain, the hypothalamus is particularly concerned with homeostasis, and in this capacity, the hypothalamus communicates with the medulla oblongata and the pituitary gland.

The **central nervous system (CNS)** consists of the brain and spinal cord (Fig. 37.3). The **peripheral nervous system (PNS)** [Gk. *periphereia,* circumference] consists of all the nerves and ganglia that lie outside the central nervous system. The CNS and PNS are considered in more detail in sections 37.3 and 37.4.

The Mammalian Nervous System

The hindbrain and midbrain of mammals are similar to those of other vertebrates. However, the forebrain of mammals is greatly enlarged, due to the addition of an outermost layer called the *neocortex*, which is seen only in mammals. It functions in higher mental processes such as spatial reasoning, conscious thought, and language.

Although all mammals have a neocortex, all neocortexes are not the same. Large variation has been observed in the number of crevices and folds, which can greatly increase the surface area and numbers of connections between regions. The frontal lobes (see Section 37.3) are especially large and complex in primates, and in humans other parts of the cortex are also enlarged and form very complex connections with other parts of the brain. It is likely that this greatly increased brain capacity allowed mammals, and especially humans, to become increasingly adept at higher mental activities such as manipulating the environment, complex learning, and anticipating the future, all of which have provided tremendous evolutionary advantages.

Check Your Progress   37.1

1. Define nerve net, ganglion, and brain.
2. Describe the major functions of the hindbrain, midbrain, and forebrain.
3. Identify the specific location of the more recently evolved parts of the brain, compared to the older parts.

Figure 37.3 Organization of the nervous system in humans.
a. The central nervous system (CNS) is composed of brain and spinal cord; the peripheral nervous system (PNS) consists of nerves. **b.** In the somatic system of the PNS, nerves conduct impulses from sensory receptors located in the skin and internal organs to the CNS and motor impulses from the CNS to the skeletal muscles. In the autonomic system, consisting of the sympathetic and parasympathetic divisions, motor impulses travel to smooth muscle, cardiac muscle, and glands.

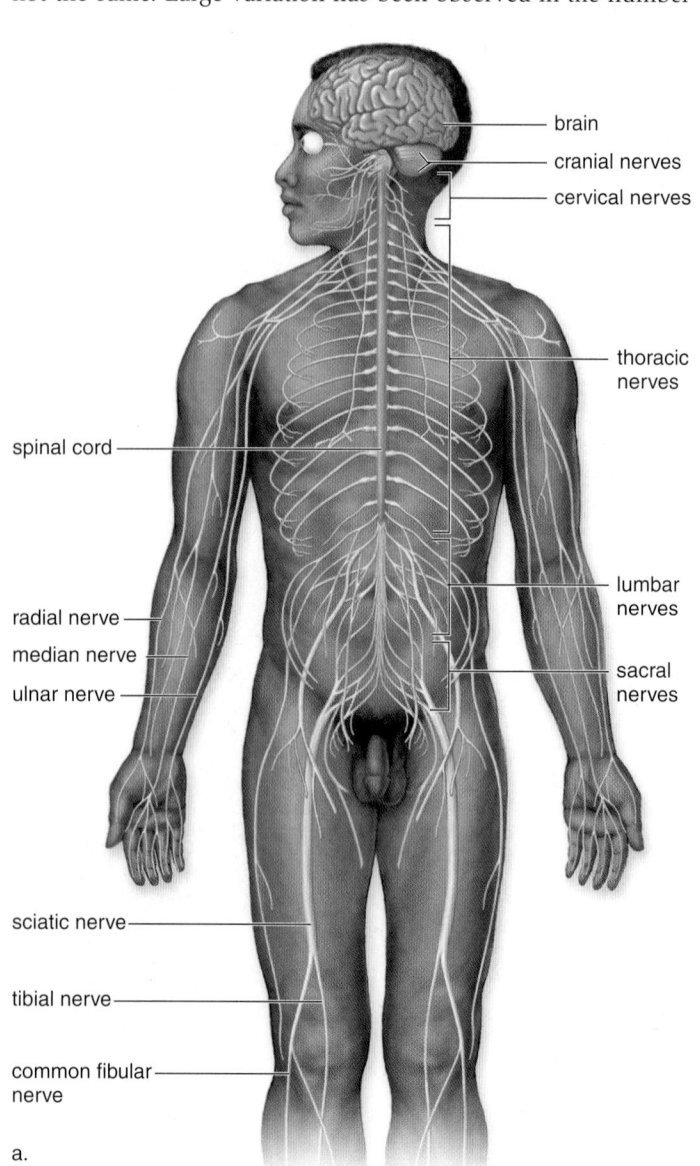

- brain
- cranial nerves
- cervical nerves
- thoracic nerves
- spinal cord
- radial nerve
- median nerve
- ulnar nerve
- lumbar nerves
- sacral nerves
- sciatic nerve
- tibial nerve
- common fibular nerve

a.

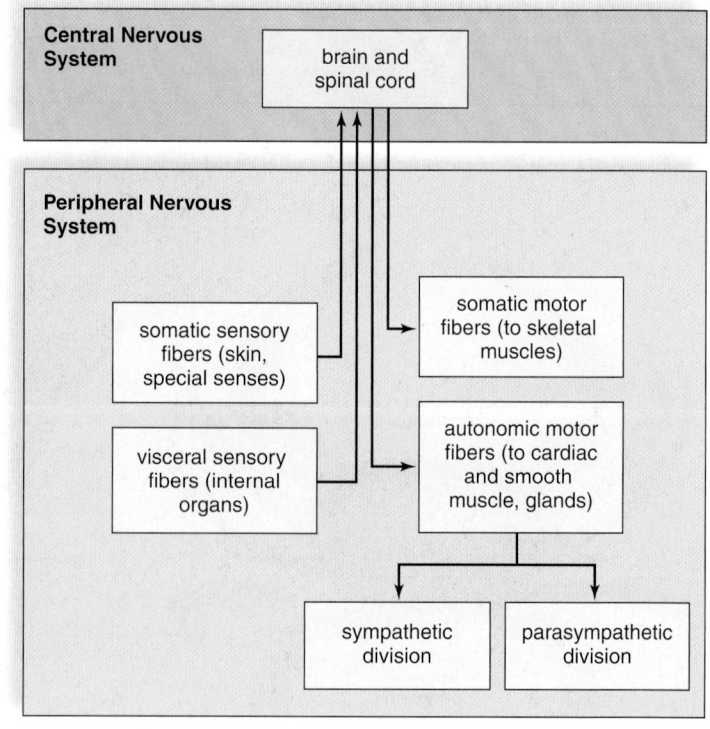

b.

37.2 Nervous Tissue

Although complex, nervous tissue is composed of just two principal types of cells. **Neurons,** also known as nerve cells, are the functional units of the nervous system. They receive sensory information, convey the information to an integration center such as the brain, and conduct signals from the integration center to effector structures such as the glands and muscles. **Neuroglia** serve as supporting cells, providing support and nourishment to the neurons.

Neurons and Neuroglia

Neurons vary in appearance depending on their function and location. They consist of three major parts: a cell body, dendrites, and an axon (Fig. 37.4). The **cell body** contains a nucleus and a variety of organelles. The **dendrites** [Gk. *dendron*, tree] are short, highly branched processes that receive signals from the sensory receptors or other neurons and transmit them to the cell body. The **axon** [Gk. *axon*, axis] is the portion of the neuron that conveys information to another neuron or to other cells. Axons can be bundled together to form nerves. For this reason, axons are often called **nerve fibers.** Many axons are covered by a white insulating layer called the **myelin sheath** [Gk. *myelos*, spinal cord].

Neuroglia, or glial cells (see Section 3.1), greatly outnumber neurons in the brain. There are several different types of neuroglia in the CNS, each with specific functions. **Microglia** are phagocytic cells that help remove bacteria and debris, while **astrocytes** (astrocytes) provide metabolic and structural support directly to the neurons. The myelin sheath is formed from the membranes of tightly spiraled neuroglia. In the PNS, **Schwann cells** perform this function, leaving gaps called **nodes of Ranvier.** In the CNS, neuroglial cells called **oligodendrocytes** form the myelin sheath.

MP3
Cells of the
Nervous System

Types of Neurons

Neurons can be described in terms of their function and shape. **Motor (efferent) neurons** take nerve impulses from the CNS to muscles or glands. Motor neurons are said to have a *multipolar* shape because they have many dendrites and a single axon (Fig. 37.4*a*). Motor neurons cause muscle fibers to contract or glands to secrete, and therefore they are said to innervate these structures.

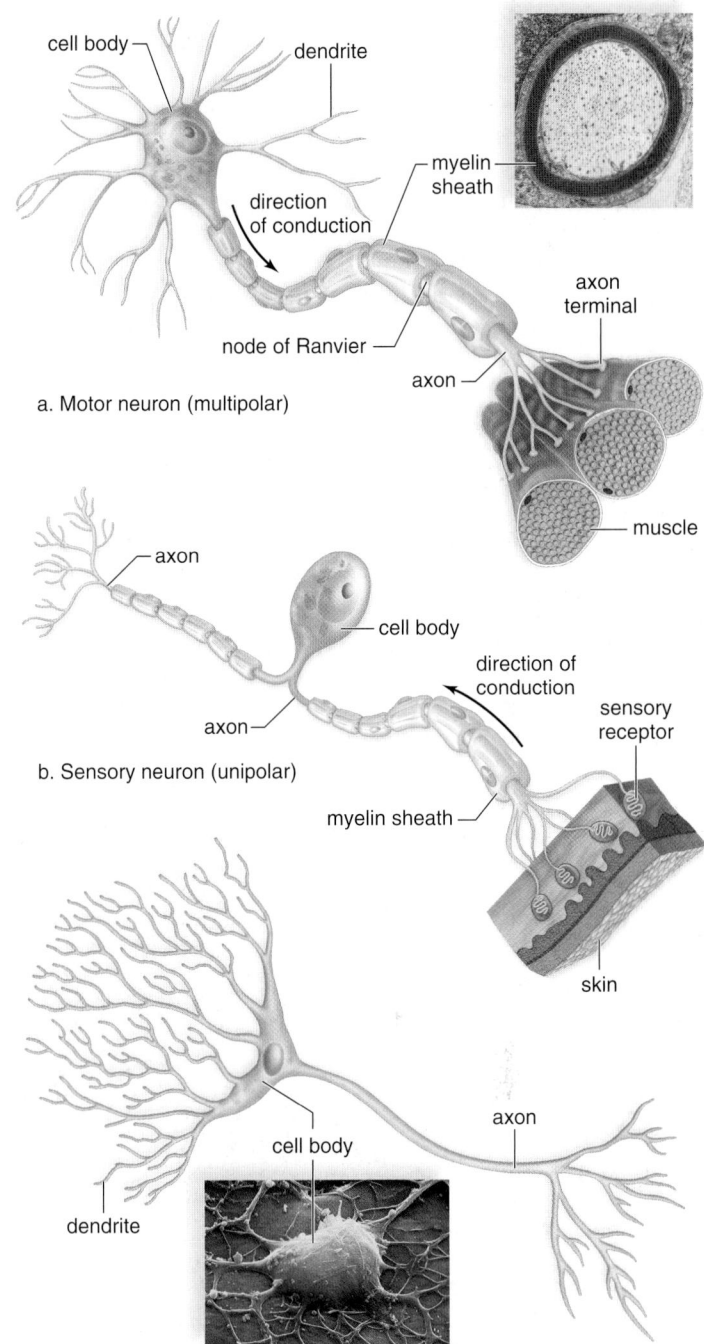

Figure 37.4 Neuron anatomy. **a.** Motor neuron. Note the branched dendrites and the single, long axon, which branches only near its tip. **b.** Sensory neuron with dendritelike structures projecting from the peripheral end of the axon. **c.** Interneuron (from the cortex of the cerebellum) with very highly branched dendrites.

Sensory (afferent) neurons take nerve impulses from sensory receptors to the CNS. The sensory receptor, which is the distal end of the long axon of a sensory neuron, may be as simple as a naked nerve ending (a pain receptor), or may be built into a highly complex organ, such as the eye or ear. Almost all sensory neurons have a structure that is termed *unipolar* (Fig. 37.4*b*). In unipolar neurons, the process that extends from the cell body divides into a branch that extends to the periphery and another that extends to the CNS.

Interneurons [L. *inter*, between] occur entirely within the CNS. Interneurons, which are typically multipolar in shape (Fig. 37.4*c*), convey nerve impulses between various parts of the CNS. Some lie between sensory neurons and motor neurons; some take messages from one side of the spinal cord to the other or from the brain to the cord, and vice versa. They also form complex pathways in the brain, leading to higher mental functions like thinking, memory, and language.

Video
Making Brain Cells

Transmission of Nerve Impulses

In the early 1900s, scientists first hypothesized that the nerve impulse is an electrochemical phenomenon involving the movement of unequally distributed ions on either side of an axonal membrane, the plasma membrane of an axon. It was not until the 1960s, however, that experimental techniques were developed to test this hypothesis. Investigators were able to insert a tiny electrode into the giant axon of the squid *Loligo*. This internal electrode was then connected to a voltmeter, an instrument with a screen that shows voltage differences over time

(Fig. 37.5). Voltage is a measure of the electrical potential difference between two points, which in this case is the difference between the electrode placed inside and another placed outside the axon. An electrical potential difference across a membrane is called the *membrane potential*.

Resting Potential

When the axon is not conducting an impulse, the voltmeter records a membrane potential equal to about −70 mV (millivolts), indicating that the inside of the neuron is more negative than the outside (Fig. 37.5*a*). This is called the **resting potential** because the axon is not conducting an impulse.

The existence of this polarity can be correlated with a difference in ion distribution on either side of the axonal membrane. As Figure 37.5*a* shows, there is a higher concentration of sodium ions (Na^+) outside the axon and a higher concentration of potassium ions (K^+) inside the axon.

The unequal distribution of these ions is in part due to the activity of the *sodium-potassium pump* (described in Section 5.3; see Figure 5.10). This pump is an active transport system in the plasma membrane that pumps three sodium ions out of the axon and two potassium ions into the axon. The pump is always working because the membrane is somewhat permeable

Figure 37.5 Resting and action potential of the axonal membrane.   **a.** Resting potential. A voltmeter that records voltage changes indicates the axonal membrane has a resting potential of −70 mV. There is a preponderance of Na^+ outside the axon and a preponderance of K^+ inside the axon. The permeability of the membrane to K^+ compared to Na^+, and the presence of large, negatively charged proteins (not shown) within the axon, causes the inside to be negative compared to the outside. **b.** Action potential. Depolarization occurs when Na^+ gates open and Na^+ moves inside the axon, and (**c**) repolarization occurs when K^+ gates open and K^+ moves outside the axon. **d.** Graph of the action potential.

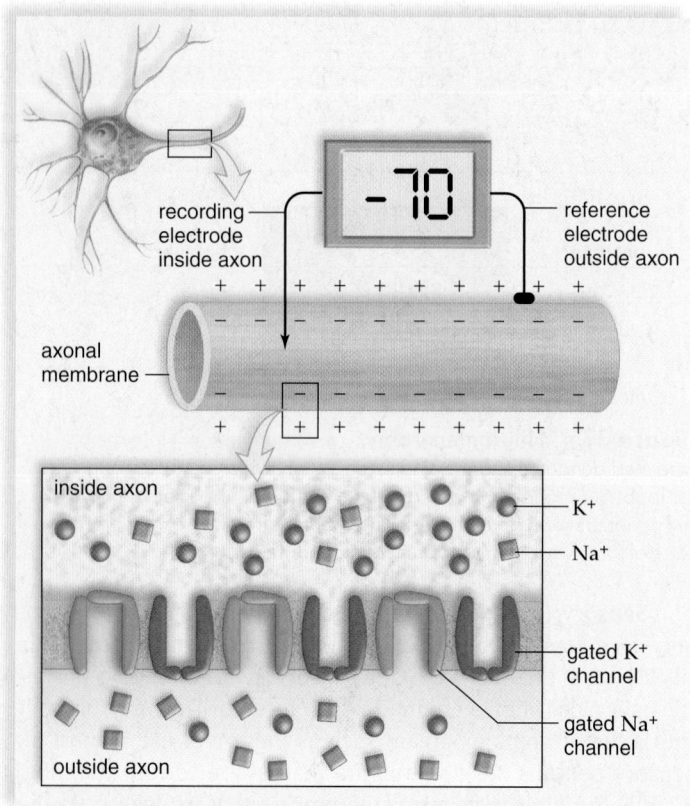

a. Resting potential: more Na^+ outside the axon and more K^+ inside the axon causes polarization.

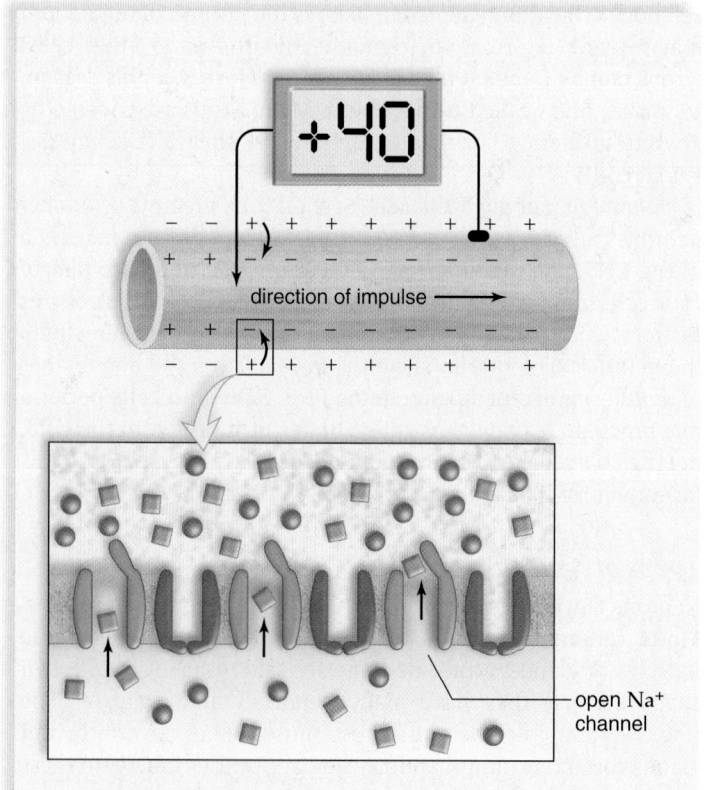

b. Action potential begins: depolarization occurs when Na^+ gates open and Na^+ moves to inside the axon.

to these ions, and they tend to diffuse toward areas of lesser concentration. Because the membrane is more permeable to potassium than to sodium, there are always more positive ions outside the membrane than inside; this accounts for some of the membrane potential recorded by the voltmeter. The axon cytoplasm also contains large, negatively charged proteins. Altogether, then, the voltmeter records that the resting potential is −70 mV inside the cell.

Animation
How the Sodium-Potassium Pump Works

Action Potential

An **action potential** is a rapid change in polarity across a portion of an axonal membrane as the nerve impulse occurs. An action potential involves two types of gated ion channels in the axonal membrane, one that allows passage of Na⁺, and another that allows passage of K⁺. In contrast to ungated ion channels, which constantly allow ions across the membrane, gated ion channels open and close in response to a stimulus, such as a signal from another neuron.

The *threshold* is the minimum change in polarity across the axonal membrane that is required to generate an action potential. Therefore, the action potential is an all-or-none event. During *depolarization,* the inside of a neuron becomes positive because of the sudden entrance of sodium ions. If threshold is reached, many more sodium channels open, and the action potential begins. As sodium ions rapidly move across

the membrane to the inside of the axon, the action potential swings up from −70 mV to +40 mV (Fig. 37.5b). This reversal in polarity causes the sodium channels to close and the potassium channels to open. As potassium ions leave the axon, the membrane potential swings down from +40 mV to −70 mV. In other words, a *repolarization* occurs (Fig. 37.5c). An action potential only takes 2 msec (milliseconds). To visualize such rapid fluctuations in voltage across the axonal membrane, researchers generally find it useful to plot the voltage changes over time (Fig. 37.5d).

Animation
Nerve Impulse

Propagation of Action Potentials

In nonmyelinated axons (such as sensory receptors in the skin), the action potential travels down an axon one small section at a time, at a speed of about 1 m/sec (meter per second). In myelinated axons, the gated ion channels that produce an action potential are concentrated at the nodes of Ranvier. *Saltar* in Spanish means "to jump," and so this mode of conduction, called **saltatory conduction,** means that the action potential "jumps" from node to node:

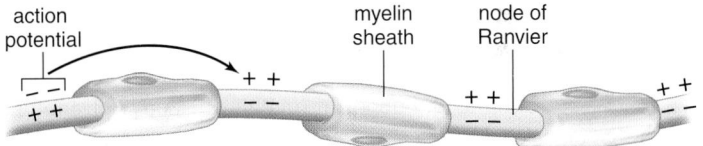

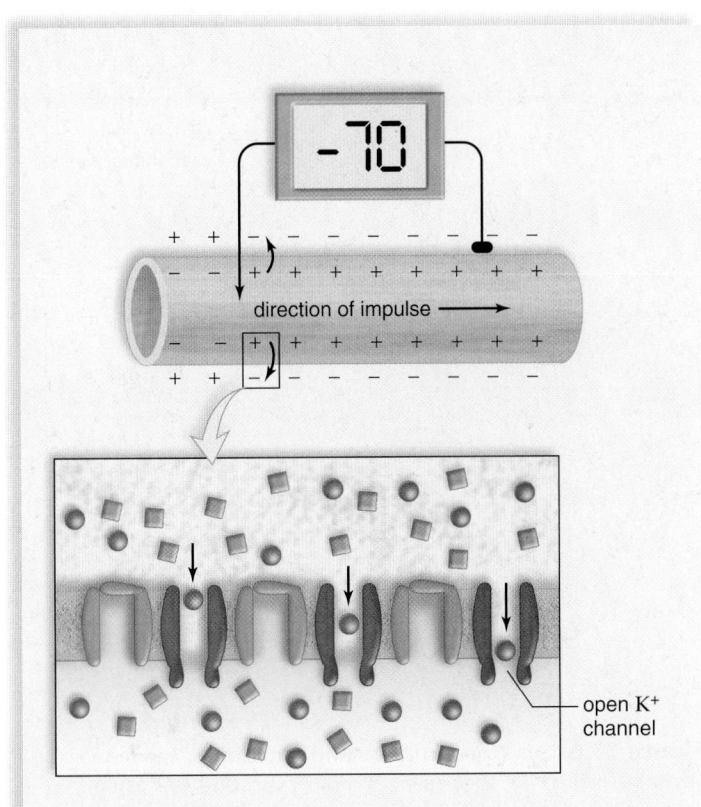

c. Action potential ends: repolarization occurs when K⁺ gates open and K⁺ moves to outside the axon.

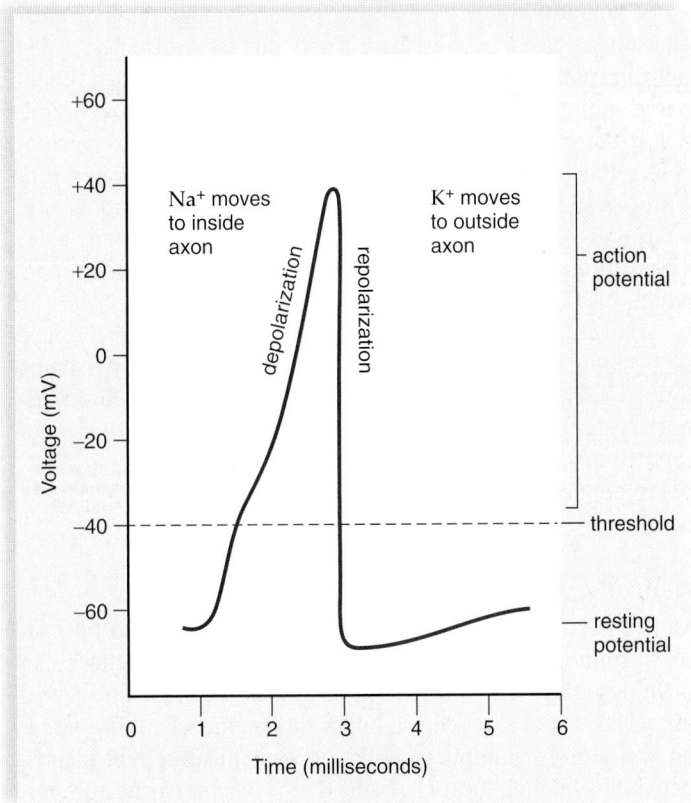

d. An action potential can be visualized if voltage changes are graphed over time.

Speeds of 200 m/sec (about 450 miles per hour) have been recorded. As you can see, this speed is considerably greater than the rate of travel in nonmyelinated axons and allows what seems to be an instantaneous response.

As soon as an action potential has moved on, the previous section undergoes a **refractory period,** during which the Na$^+$ gates are unable to open. Notice, therefore, that the action potential cannot move backward and instead always moves down an axon toward its terminals. The intensity of a signal traveling down a nerve fiber is determined by how many nerve impulses are generated within a given time span.

Transmission Across a Synapse

Every axon branches into many fine endings, each tipped by a small swelling, called an axon terminal (Fig. 37.6). Each terminal lies very close to the dendrite (or the cell body) of another neuron. This region of close proximity is called a **synapse.** At a synapse, the membrane of the first neuron is called the *presynaptic* membrane, and the membrane of the next neuron is called the *post*synaptic membrane. The small gap between the neurons is called the **synaptic cleft.**

A nerve impulse cannot cross a synaptic cleft. Transmission across a synapse is carried out by molecules called **neurotransmitters,** which are stored in synaptic vesicles. When nerve impulses traveling along an axon reach an axon terminal, gated channels for calcium ions (Ca^{2+}) open, and calcium enters the terminal. This sudden rise in Ca^{2+} stimulates synaptic vesicles to merge with the presynaptic membrane, and neurotransmitter molecules are released into the synaptic cleft. They diffuse across the cleft to the postsynaptic membrane, where they bind with specific receptor proteins.

Depending on the type of neurotransmitter and/or the type of receptor, the response of the postsynaptic neuron can be toward excitation or toward inhibition. Excitatory neurotransmitters that use gated ion channels are fast acting. Other neurotransmitters affect the metabolism of the postsynaptic cell and therefore are slower acting.

Animation
Action Potential Propagation

Neurotransmitters

More than 100 substances are known or suspected to be neurotransmitters in both the CNS and PNS. Many of these can have opposing effects on different tissues. *Acetylcholine (ACh)* excites skeletal muscle but inhibits cardiac muscle. It has either an excitatory or inhibitory effect on smooth muscle or glands, depending on their location. In the CNS, *norepinephrine* is important to dreaming, waking, and mood. *Dopamine* is involved in emotions, learning, and attention, and *serotonin* is involved in

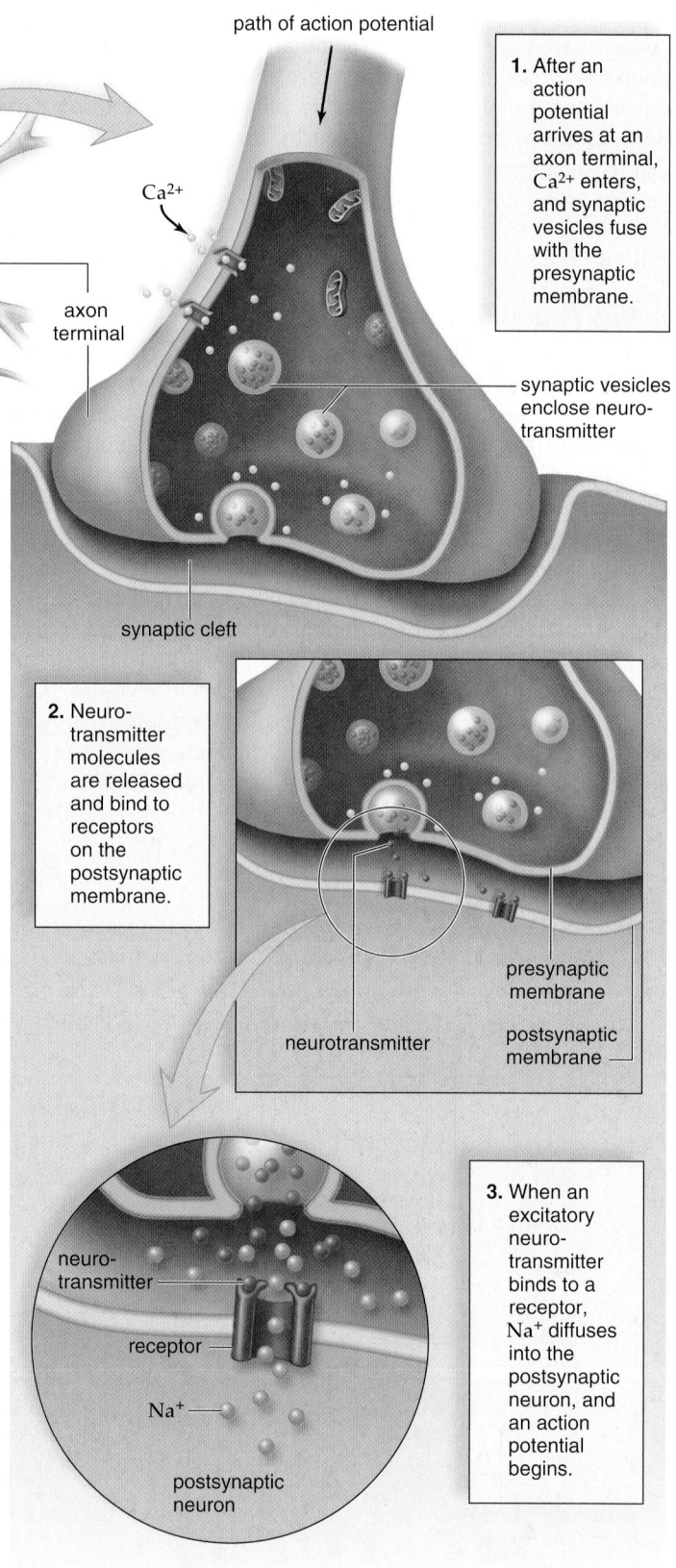

1. After an action potential arrives at an axon terminal, Ca^{2+} enters, and synaptic vesicles fuse with the presynaptic membrane.

path of action potential

Ca^{2+}

axon terminal

cell body of postsynaptic neuron

synaptic vesicles enclose neurotransmitter

synaptic cleft

2. Neurotransmitter molecules are released and bind to receptors on the postsynaptic membrane.

presynaptic membrane

neurotransmitter

postsynaptic membrane

3. When an excitatory neurotransmitter binds to a receptor, Na$^+$ diffuses into the postsynaptic neuron, and an action potential begins.

neurotransmitter

receptor

Na$^+$

postsynaptic neuron

Figure 37.6 Synapse structure and function. Transmission across a synapse from one neuron to another occurs when a neurotransmitter is released at the presynaptic membrane, diffuses across a synaptic cleft, and binds to a receptor in the postsynaptic membrane. An action potential may begin.

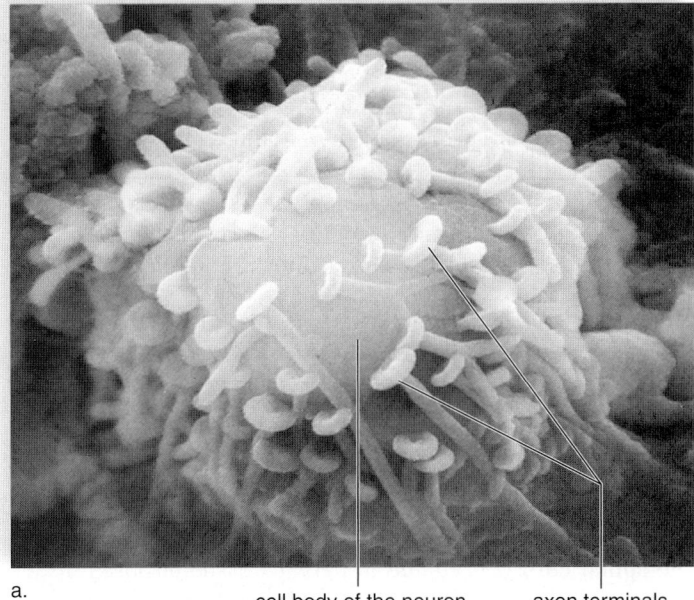

a.

cell body of the neuron   axon terminals

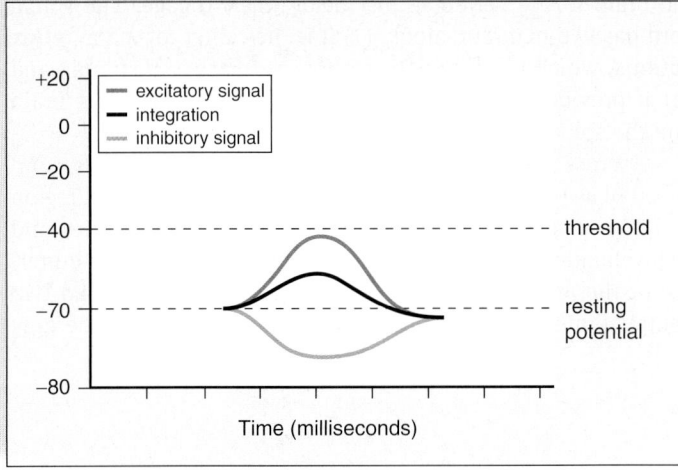

b.

Figure 37.7 Synaptic integration.   a. Many neurons synapse with a cell body. **b.** Both inhibitory signals (blue) and excitatory signals (red) are summed up in the dendrite and cell body of the postsynaptic neuron. Only if the combined signals cause the membrane potential to rise above threshold does an action potential occur. In this example, threshold was not reached.

thermoregulation, sleeping, emotions, and perception. *Endorphins* are neurotransmitters that bind to natural opioid receptors in the brain. They are associated with the "runner's high" of exercisers because they also produce a feeling of tranquility. Endorphins are produced by the brain not only when there is physical stress but also when emotional stress is present.

After a neurotransmitter has been released into a synaptic cleft and has initiated a response, it is removed from the cleft. In some synapses, the postsynaptic membrane contains enzymes that rapidly inactivate the neurotransmitter. For example, the enzyme *acetylcholinesterase (AChE)* breaks down acetylcholine. In other synapses, the presynaptic cell is responsible for reuptake, a process in which it rapidly reabsorbs the

neurotransmitter, possibly for repackaging in synaptic vesicles or for molecular breakdown. The short existence of neurotransmitters at a synapse prevents continuous stimulation (or inhibition) of postsynaptic membranes.

Animation
Chemical
Synapse

It is of interest to note here that many drugs affecting the nervous system act by interfering with or potentiating the action of neurotransmitters. Such drugs can enhance or block the release of a neurotransmitter, mimic the action of a neurotransmitter or block the receptor, or interfere with the removal of a neurotransmitter from a synaptic cleft. Depression, a common mood disorder, appears to involve imbalances in norepinephrine and serotonin. Some antidepressant drugs, such as fluoxetine (Prozac®), prevent the reuptake of serotonin, and others, including bupropion hydrochloride (Wellbutrin®), prevent the reuptake of both serotonin and norepinephrine. Blocking reuptake prolongs the effects of these two neurotransmitters in networks of neurons within the brain that are involved in the emotional state.

Drugs that affect neurotransmitter activity are often abused for "recreational" purposes, with often unfortunate and sometimes deadly results. The Biological Systems feature on page 710 describes a number of these dangerous drugs.

Synaptic Integration

A single neuron has many dendrites plus the cell body, and both can have synapses with many other neurons. One thousand to 10,000 synapses per single neuron is not uncommon. Therefore, a neuron is on the receiving end of many excitatory and inhibitory signals. An excitatory signal produces a potential change that causes the neuron to become less polarized, or closer to triggering an action potential. An inhibitory signal causes the neuron to become hyperpolarized, or farther from an action potential.

Neurons integrate these incoming signals, and do so specifically at the area of the neuron cell body where the axon emerges, called the *axon hillock*. **Integration** is the summing up of excitatory and inhibitory signals (Fig. 37.7). If a neuron receives many excitatory signals (either from different synapses or at a rapid rate from one synapse), chances are the axon will transmit a nerve impulse. In Fig. 37.7, the inhibitory signals (shown in blue) are cancelling out the excitatory signals, resulting in no nerve impulse.

MP3
Synapses

Check Your Progress   37.2

1. Explain why a nerve impulse travels more quickly down a myelinated axon as compared with an unmyelinated axon.
2. Describe the movement of specific ions during the generation of a nerve impulse.
3. Analyze how the bite of a black widow spider, which contains a powerful AChE inhibitor, might cause each of these common symptoms: muscle cramps, salivation, fast heart rate, high blood pressure.

37.3 The Central Nervous System

Learning Outcomes

Upon completion of this section, you should be able to

1. Describe the anatomy of the spinal cord and spinal nerves.
2. List the major regions of the human brain and describe some major functions of each.
3. Compare the causes and types of symptoms seen in some common CNS disorders.

The central nervous system (CNS) consists of the spinal cord and the brain. It has three specific functions:

1. *Receives sensory input*—sensory receptors in skin and other organs respond to external and internal stimuli by generating nerve impulses that travel to the CNS.
2. *Performs integration*—the CNS sums up the input it receives from all over the body.
3. *Generates motor output*—nerve impulses from the CNS go to the muscles and glands. Muscle contractions and gland secretions are responses to stimuli received by sensory receptors.

As an example of the operation of the CNS, consider the events that occur as a person raises a glass to the lips. Continuous sensory input to the CNS from the eyes and hand informs the CNS of the position of the glass, and the CNS continually sums up the incoming data before commanding the hand to proceed. At any time, integration with other sensory data might cause the CNS to command a different motion instead. The lips detect the

MP3
Organization of the
Nervous System

arrival of the glass, passing this information to the CNS, which then directs the actions of drinking.

The spinal cord and the brain are both protected by bone; the spinal cord is surrounded by vertebrae, and the brain is enclosed by the skull. Both the spinal cord and the brain are wrapped in three protective membranes known as **meninges** (see Fig. 37.8). The spaces between the meninges are filled with **cerebrospinal fluid,** which cushions and protects the CNS. Cerebrospinal fluid, produced by a type of glial cell, is contained in the central canal of the spinal cord and within the **ventricles** of the brain, which are interconnecting spaces that produce and serve as reservoirs for cerebrospinal fluid. **Meningitis** (inflammation of the meninges) is a serious disorder caused by a number of bacteria or viruses that can invade the meninges.

The Spinal Cord

The **spinal cord** is a bundle of nervous tissue enclosed in the vertebral column (see Fig. 37.12); it extends from the base of the brain to the vertebrae just below the rib cage. The spinal cord has two main functions: (1) it is the center for many **reflex actions,** which are automatic responses to external stimuli, and (2) it provides a means of communication between the brain and the spinal nerves, which leave the spinal cord.

A cross section of the spinal cord reveals that it is composed of a central portion of gray matter and a peripheral region of white matter. The **gray matter** consists of cell bodies and unmyelinated fibers. In cross section it is shaped like a butterfly, or the letter H, with two dorsal (posterior) horns and two ventral (anterior) horns surrounding a central canal. The gray

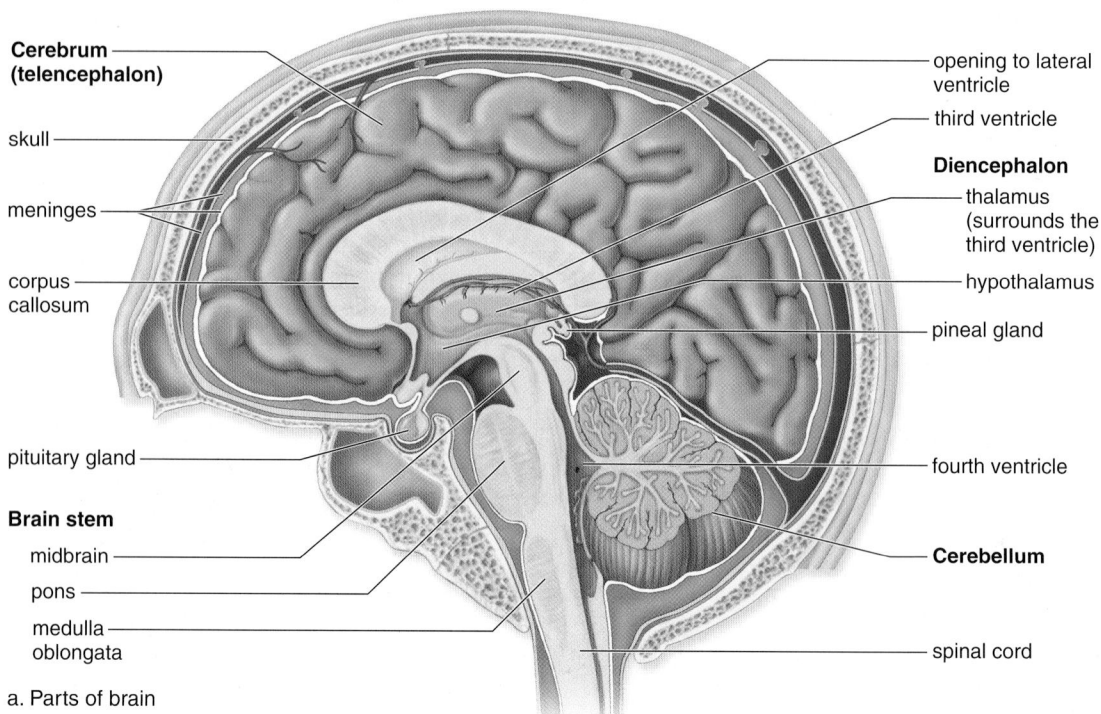

Cerebrum (telencephalon)

skull

meninges

corpus callosum

pituitary gland

Brain stem
   midbrain
   pons
   medulla oblongata

a. Parts of brain

opening to lateral ventricle

third ventricle

Diencephalon
thalamus (surrounds the third ventricle)

hypothalamus

pineal gland

fourth ventricle

Cerebellum

spinal cord

Figure 37.8 The human brain. a. The right cerebral hemisphere is shown here, along with other, closely associated structures. The hemispheres are connected by the corpus callosum. **b.** The cerebrum is divided into the right and left cerebral hemispheres.

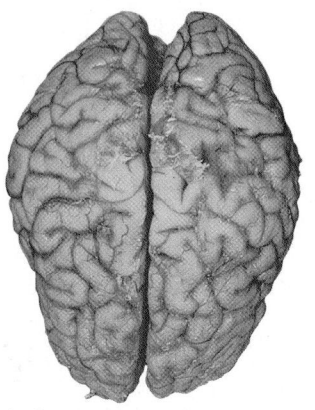

b. Cerebral hemispheres

matter contains portions of sensory neurons and motor neurons, as well as short interneurons that connect sensory and motor neurons.

Myelinated long fibers of interneurons that run together in bundles called **tracts** give **white matter** its color. These tracts connect the spinal cord to the brain. These tracts are like a busy superhighway, by which information continuously passes between the brain and the rest of the body. In the dorsal part of the cord the tracts are primarily ascending, taking information *to* the brain. Ventrally, the tracts are primarily descending, carrying information *from* the brain. Because the tracts cross over at one point, the left side of the brain controls the right side of the body, and the right side of the brain controls the left side of the body. Researchers estimate that there are about 100,000 miles of myelinated nerve fibers in the adult human brain.

If the spinal cord is damaged as the result of an injury, paralysis may result. If the injury occurs in the cervical (neck) region, all four limbs are usually paralyzed, a condition known as quadriplegia. If the injury occurs in the thoracic region, the lower body may be paralyzed, a condition called paraplegia.

Other disease processes can cause paralysis. In **amyotrophic lateral sclerosis (ALS)**, or Lou Gehrig's disease, motor neurons in the brain and spinal cord degenerate and die, leaving patients weakened, then paralyzed, then unable to breathe properly. Although there is no cure, some drugs slow the disease, and others are currently in clinical trials.

The Brain

Nerve impulses are the same in all neurons, so how is it that stimulation of our eyes causes us to see, and stimulation of our ears causes us to hear? Essentially, the central nervous system carries out the function of integrating incoming data. The **brain** allows us to perceive our environment, reason, and remember. The exact processes by which the brain generates these higher functions remain largely mysterious.

Video Brain Bank

The Cerebrum

The **cerebrum** is the largest, outermost portion of the brain in humans (Fig. 37.8*a*). The cerebrum is the last center to receive sensory input and carry out integration before commanding voluntary motor responses. It communicates with and coordinates the activities of the other parts of the brain. The cerebrum also contains the two lateral ventricles; the third ventricle is surrounded by the diencephalon, and the fourth ventricle lies between the cerebellum and the pons.

Cerebral Hemispheres. The cerebrum is divided into two halves, called **cerebral hemispheres** (see Fig. 37.8*b*). A deep groove called the *longitudinal fissure* divides the cerebrum into the right and left hemispheres. Each hemisphere receives information from and controls the opposite side of the body. Although the hemispheres appear the same, the right hemisphere is associated with artistic and musical ability, emotion, spatial relationships, and pattern recognition. The left hemisphere is more adept at mathematics, language, and analytical reasoning. The two cerebral hemispheres are connected by a bridge of tracts within the *corpus callosum*.

Shallow grooves called *sulci* (sing., *sulcus*) divide each hemisphere into paired lobes (Fig. 37.9). The *frontal lobes* lie toward the front of the hemispheres and are associated with

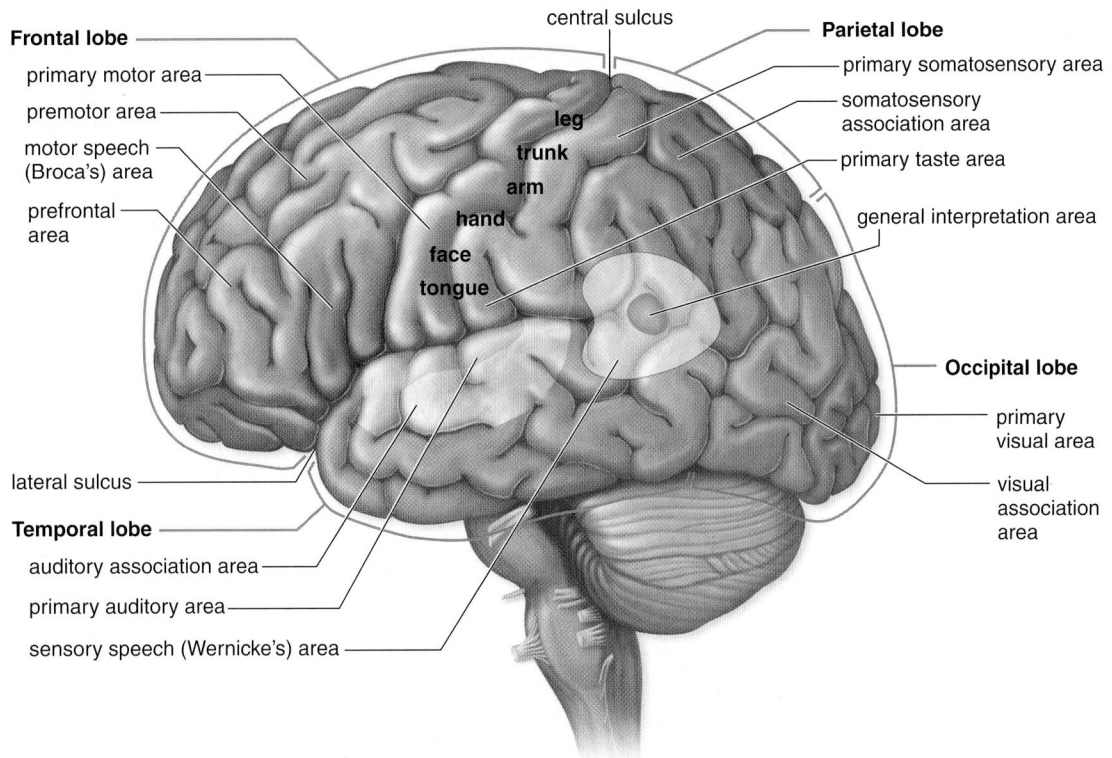

Frontal lobe
primary motor area
premotor area
motor speech (Broca's) area
prefrontal area
lateral sulcus
Temporal lobe
auditory association area
primary auditory area
sensory speech (Wernicke's) area

central sulcus
leg
trunk
arm
hand
face
tongue

Parietal lobe
primary somatosensory area
somatosensory association area
primary taste area
general interpretation area

Occipital lobe
primary visual area
visual association area

Figure 37.9 The lobes of a cerebral hemisphere. Each cerebral hemisphere is divided into four lobes: frontal, parietal, temporal, and occipital. These lobes contain centers for reasoning and movement, somatic sensing, hearing, and vision, respectively.

motor control, memory, reasoning, and judgment. For example, if a fire occurs, the frontal lobes enable you to decide whether to exit via the stairs or the window, or if it is winter, how to dress if the temperature plummets to subzero. The left frontal lobe contains *Broca's area,* which organizes motor commands to produce speech.

The *parietal lobes* lie posterior to the frontal lobe and are concerned with sensory reception and integration, as well as taste. A *primary taste area* in the parietal lobe accounts for taste sensations.

The *temporal lobes* are located laterally. A primary auditory area in each temporal lobe receives information from our ears. The *occipital lobes* are the most posterior lobes. A *primary visual area* in each occipital lobe receives information from the eyes.

MP3
The Cerebrum

The Cerebral Cortex. The **cerebral cortex** is a thin (less than 5 mm thick), but highly convoluted, outer layer of gray matter that covers the cerebral hemispheres. The convolutions increase the surface area of the cerebral cortex. The cerebral cortex contains tens of billions of neurons and is the region of the brain that accounts for sensation, voluntary movement, and all the thought processes required for learning, memory, language, and speech.

Two regions of the cerebral cortex are of particular interest. The *primary motor area* is in the frontal lobe just ventral to (before) the central sulcus. Voluntary commands to skeletal muscles begin in the primary motor area, and each part of the body is controlled by a certain section. The size of the section indicates the precision of motor control. For example, controlling the muscles of the face and hands takes up a much larger portion of the primary motor area than controlling the entire trunk. The *primary somatosensory area* is just dorsal to the central sulcus in the parietal lobe. Sensory information from the skin and skeletal muscles arrives here.

Video
Brain Surgery

When the blood supply to any area of the brain is disrupted, a **stroke** results. Stroke is the third leading cause of death in the United States. The most common type is *ischemic* stroke, in which there is a sudden loss of blood supply to an area of the brain, usually due to arterial blockage or clot formation. The area(s) of the brain affected by a stroke will determine what type of symptoms arise. For example, a stroke that affects only the motor areas of the cerebral cortex might paralyze one side of the body, while a stroke involving Broca's area might render a stroke victim unable to speak.

Although strokes are most common in older people, a study released in 2011 noted a 51 percent increase in strokes in young men aged 15 through 34, and a 17 percent increase in women the same age. Because many of the risk factors for stroke are similar to those for cardiovascular disease, see the Nature of Science feature, "New Information About Preventing Cardiovascular Disease," in Chapter 32 to learn how to reduce your risk.

Video
Brain Healing

Basal Nuclei. Although the bulk of the cerebrum is composed of white matter (i.e., tracts), masses of gray matter are located deep within the white matter. These so-called **basal nuclei** (basal ganglia) integrate motor commands, ensuring that proper muscle groups are activated or inhibited. As mentioned in the chapter opening essay about Muhammad Ali, **Parkinson disease (PD),** is a brain disorder characterized by tremors, speech difficulties, and difficulty standing and walking. PD results from a loss of cells in the basal nuclei that normally produce the neurotransmitter dopamine. See the Nature of Science feature on page 706 to learn about the somewhat strange history of research into PD, as well as some treatments for this disease.

MP3
The Brain

Other Parts of the Brain

The hypothalamus and the thalamus are in the *diencephalon,* a region that encircles the third ventricle. The **hypothalamus** forms the floor of the third ventricle. It is an integrating center that helps maintain homeostasis by regulating hunger, sleep, thirst, body temperature, and water balance. The hypothalamus controls the pituitary gland and, thereby, serves as a link between the nervous and endocrine systems (see Chapter 40).

The **thalamus** consists of two masses of gray matter located in the sides and roof of the third ventricle. It receives all sensory input except smell. The thalamus integrates this information and sends it on to the appropriate portions of the cerebrum. For this reason, the thalamus is often referred to as the "gatekeeper" for sensory information en route to the cerebral cortex. The thalamus also participates in higher mental functions such as memory and emotions.

The **pineal gland,** which secretes the hormone melatonin, is also located in the diencephalon. *Melatonin* is a hormone that is involved in maintaining our normal sleep-wake cycle. It is sometimes recommended for people suffering from insomnia, but many side effects can occur. Relatively high levels of melatonin are a key ingredient in "relaxation brownies" that are sold at convenience stores as well as online; many states are banning their sale, however.

Video
Winter Mood

The **cerebellum** lies under the occipital lobe of the cerebrum and is separated from the brain stem by the fourth ventricle. It is the largest part of the hindbrain. The cerebellum receives sensory input from the eyes, ears, joints, and muscles about the present position of body parts, and it also receives motor output from the cerebral cortex about where these parts should be located. After integrating this information, the cerebellum sends motor impulses by way of the brain stem to the skeletal muscles. In this way, the cerebellum maintains posture and balance. It also ensures that all of the muscles work together to produce smooth, coordinated voluntary movements, such as playing the piano or hitting a baseball.

The **brain stem** contains the midbrain, the pons, and the medulla oblongata (see Fig. 37.8). The **midbrain** acts as a relay station for tracts passing between the cerebrum and the spinal cord or cerebellum. The tracts cross in the brain stem so that the right side of the body is controlled by the left portion of the brain, and the left side of the body is controlled by the right portion of the brain.

The **pons** [L. *pons,* bridge] contains bundles of axons that form a "bridge," traveling between the cerebellum and the rest of the CNS. The pons also works with the medulla oblongata to regulate many basic body functions.

The **medulla oblongata** lies just superior to the spinal cord, and it contains tracts that ascend or descend between the spinal cord and higher brain centers. It regulates heartbeat, breathing, swallowing, and blood pressure. It also contains reflex centers for vomiting, coughing, sneezing, hiccuping, and swallowing.

The most common neurological disease of young adults is **multiple sclerosis (MS)**. This disease typically affects myelinated nerves in the cerebellum, brain stem, basal ganglia, and optic nerve. MS is considered an autoimmune disease, in which the patient's own white blood cells attack the myelin, oligodendrocytes, and eventually, neurons in the CNS. The word sclerosis refers to the multiple scars, or plaques, that can be seen using various types of scans. The damage to myelin affects transmission of nerve impulses, resulting in the most common symptoms like fatigue, vision problems, weakness, numbness, and tingling. Nearly 350,000 people in the United States have MS, and about 10,000 new cases are diagnosed each year, mainly in young adults.

The Reticular Activating System. The **reticular activating system (RAS)** contains the reticular formation, a complex network of nuclei and nerve fibers that extend the length of the brain stem (Fig. 37.10). The reticular formation receives sensory signals that it sends up to higher centers, and motor signals that it sends to the spinal cord.

The RAS arouses the cerebrum via the thalamus and causes a person to be alert. Apparently, the RAS can filter out unnecessary sensory stimuli, explaining why you can study with the TV on. If you want to awaken the RAS, surprise it with a sudden stimulus, like splashing your face with cold water; if you want to deactivate it, remove visual and auditory stimuli. A severe injury to the RAS can cause a person to be comatose, from which recovery may be impossible.

The Limbic System

The **limbic system** is a complex network of tracts and nuclei that incorporates medial portions of the cerebral lobes, the basal

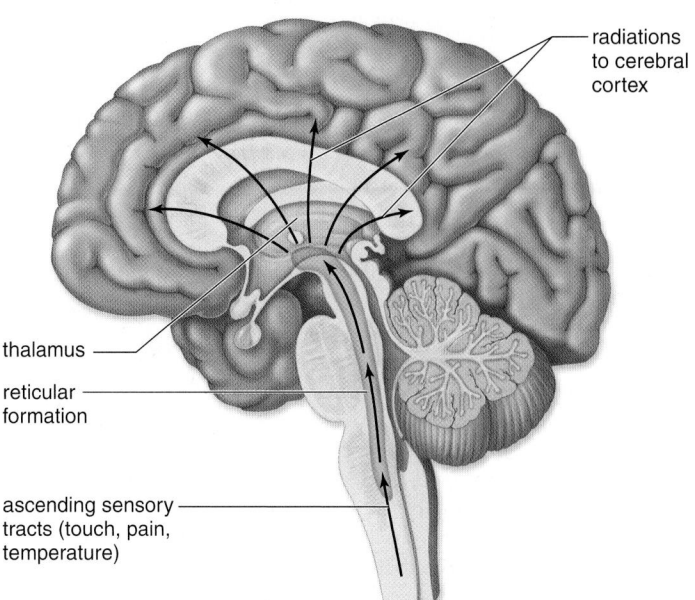

Figure 37.10 The reticular activating system. The reticular formation receives and sends on motor and sensory information to various parts of the CNS. One portion, the reticular activating system (RAS; see arrows), arouses the cerebrum and, in this way, controls alertness versus sleep.

nuclei, and the diencephalon (Fig. 37.11). The limbic system blends higher mental functions and primitive emotions into a united whole. It accounts for why activities like sexual behavior and eating seem pleasurable, and also why, as an example, mental stress can cause high blood pressure.

Two significant structures within the limbic system are the hippocampus and the amygdala, which are essential for learning and memory. The **hippocampus**, a seahorse-shaped structure that lies deep in the temporal lobe, is well situated in the brain to make the prefrontal area aware of past experiences

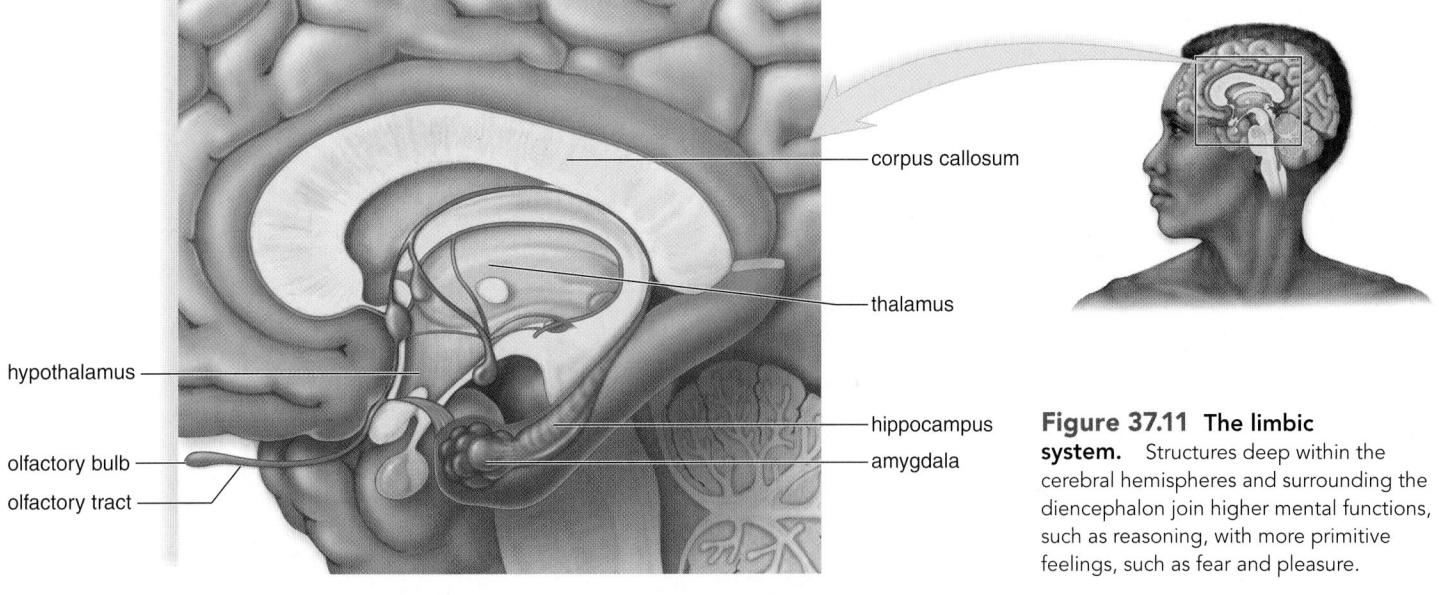

Figure 37.11 The limbic system. Structures deep within the cerebral hemispheres and surrounding the diencephalon join higher mental functions, such as reasoning, with more primitive feelings, such as fear and pleasure.

Nature of Science

An Accidental Experimental Model for Parkinson Disease

Parkinson disease (PD) was first described as the "Shaking Palsy" in 1817 by English surgeon James Parkinson, for whom the disease was later named. It affects about 1.5 million people in the United States, and over six million worldwide. PD is most common in people over 60 and rarely affects those under 40. The symptoms of PD include bradykinesia (slow movements), tremors, rigidity of the limbs and trunk, and impaired speech, balance, and coordination. PD is caused by injury to or death of dopamine-producing neurons in the basal ganglia (specifically the substantia nigra), deep in the forebrain, that normally help control voluntary movement. The initial event(s) that cause the neuron damage aren't well understood, and there is no cure.

Frozen Junkies

In 1976, Barry Kidston, a graduate student in chemistry at the University of Maryland, wanted to experiment with hard drugs and decided to synthesize a narcotic he had read about in a 1947 scientific paper. The drug, called MPPP, was said to be less addictive than morphine, and was technically not illegal (although it is now). Kidston was successful at first, and was apparently able to achieve a satisfactory "high" by intravenously injecting the compound, which he synthesized in a makeshift lab he set up in his parent's basement. His luck ran out, however, when he accidently produced a related compound called MPTP instead of MPPP. Soon after injecting the MPTP, Kidston's speech became slurred, he had trouble walking, and within three days he could hardly move.

Kidston's doctors were baffled, but after a neurologist noted that his symptoms resembled Parkinson disease, he was treated with anti-PD drugs, and he dramatically improved. For two years he was able to function well on medication, but then he died, somewhat ironically, from a cocaine overdose. His autopsy revealed a substantial loss of dopamine-producing cells in the substantia nigra, which is a hallmark of PD (Fig. 37A).

Kidston's case was published in a medical journal, but it was barely noticed until 1981, when six IV drug users turned up at various emergency rooms in the San Francisco area, all showing very similar symptoms, as if they had "turned to stone." Some quick investigative work turned up the fact that all had tried some new heroin that had become available on the street.

When some of this material was sent in for analysis, one of the lab toxicologists remembered seeing the article about Kidston a few years earlier, and quickly determined that the drug all the new patients had injected indeed contained MPTP.[1]

New Treatments

As news that PD could be induced by MPTP in humans reached the biomedical research community, scientists were quickly able to demonstrate that MPTP could also induce PD in animals such as monkeys and mice. Since the mid-1980s, animals with MPTP-induced PD have been used in hundreds of studies that have advanced our understanding of how PD develops, and have been instrumental in the development of new therapies, such as:

- Drugs that increase dopamine. These vary from L-Dopa, which is converted into dopamine in the brain, and is still the most effective drug therapy, to newer drugs like Entacapone that inhibit the normal breakdown of dopamine.

- Surgical treatments. The most common of these is deep brain stimulation, in which an electrode is inserted into an area of the brain called the subthalamic nucleus, which seems to be overactive in PD. Coincidentally, one of the original "frozen addicts," still serving time in prison, had this procedure done and has had very significant improvement in his condition.

- Stem cell transplants. The use of fetal stem cells to replace diseased or dying neurons in PD remains controversial in many countries. However, results from animal models continue to be encouraging, and private groups such as the Michael J. Fox Foundation for Parkinson Disease Research continue to support stem cells as the most likely approach that may result in a cure for PD.

Questions to Consider

1. What are some other instances where accidental findings have resulted in scientific discoveries?
2. Should the families of people like Kidston, who accidentally benefit medicine, be compensated financially?
3. Why is it so difficult to determine the specific causes of PD and other brain diseases?

---

1 Further details about these cases can be read in Langston, J. W., and Palfreman, J. 1995. *The Case of The Frozen Addicts* (Pantheon, NY). A PBS video of the same name is also available.

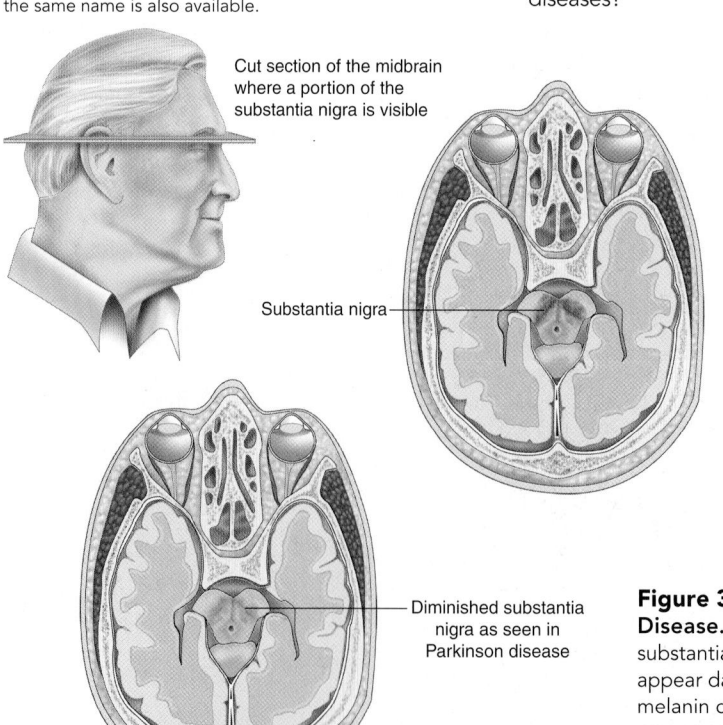

Cut section of the midbrain where a portion of the substantia nigra is visible

Substantia nigra

Diminished substantia nigra as seen in Parkinson disease

Figure 37A Parkinson Disease. Neurons in the substantia nigra of the brain appear dark due to a high melanin content. In PD, this area appears lighter due to a loss of these neurons.

stored in sensory association areas. The **amygdala**, in particular, can cause these experiences to have emotional overtones. For example, the smell of smoke may serve as an alarm to search for fire in the house. The inclusion of the frontal lobe in the limbic system gives us the capability of restraining ourselves from acting out on strong feelings by using reason.

Learning and Memory. **Memory** is the ability to hold a thought in mind or recall events from the past, ranging from a word we learned only yesterday to an early emotional experience that has shaped our lives. Learning takes place when we retain and use past memories.

The prefrontal area in the frontal lobe is active during short-term memory as when we temporarily recall a telephone number. Some telephone numbers go into long-term memory. Think of a telephone number you know by heart, and see if you can bring it to mind without also thinking about the place or person associated with that number. Most likely you cannot, because typically long-term memory is a mixture of what is called semantic memory (numbers, words, etc.) and episodic memory (persons, events, etc.). Skill memory is a type of memory that can exist independently of episodic memory. Skill memory is being able to perform motor activities like riding a bike or playing ice hockey.

What parts of the brain are functioning when you remember something from long ago? The hippocampus gathers our long-term memories, which are stored in bits and pieces throughout the sensory association areas, and makes them available to the frontal lobe. Why are some memories so emotionally charged? The amygdala is responsible for fear conditioning and associating danger with sensory information received from the thalamus and the cortical sensory areas.

Several diseases of the brain can affect memory. **Alzheimer disease (AD)** is the most common cause of dementia, or a loss of reasoning, memory, and other higher brain functions, especially in people over age 65. AD patients have abnormal neurons throughout the brain, but especially in the hippocampus and amygdala. These neurons have two abnormalities: 1) plaques, containing a protein called beta amyloid, accumulate around the axons, and 2) neurofibrillary tangles (bundles of fibrous protein) surround the nucleus. The cause of these protein abnormalites is unknown, although several genes that predispose a person to develop AD have been identified. Although no cure is available, most of the drugs that are currently approved to treat symptoms of AD are cholinesterase inhibitors, which effectively increase the levels of acetylcholine in the AD patient's brain. This in turn can improve learning and memory, at least temporarily.

Check Your Progress   37.3

1. Trace the path of a nerve impulse from a stimulus in an internal organ (such as food in the intestine stimulating peristalsis) to the brain and back.
2. Name the four major lobes of the human brain.
3. List at least two common diseases of the CNS and describe their symptoms and causes.

37.4 The Peripheral Nervous System

Learning Outcomes

Upon completion of this section, you should be able to

1. Describe the overall anatomy of the PNS, including the cranial nerves and spinal nerves.
2. Explain how the somatic system differs from the autonomic system.
3. Contrast the functions of the sympathetic and parasympathetic divisions of the autonomic nervous system.

The peripheral nervous system (PNS) lies outside the central nervous system and contains **nerves,** which are bundles of axons. Axons that occur in nerves are also called *nerve fibers*. The cell bodies of neurons are found in the CNS and in ganglia, collections of cell bodies outside the CNS.

MP3
Organization of the Nervous System

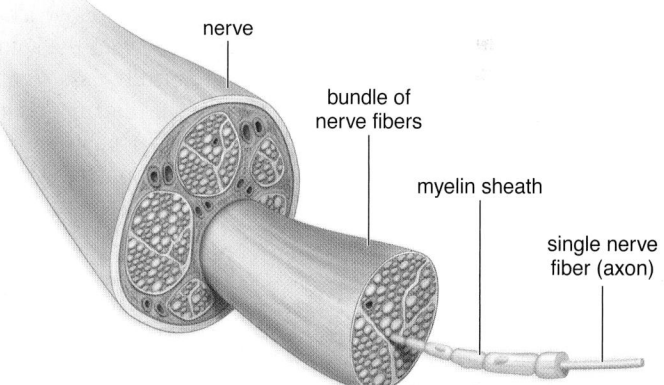

nerve

bundle of nerve fibers

myelin sheath

single nerve fiber (axon)

The paired cranial and spinal nerves are part of the PNS. In the PNS, the somatic nervous system has sensory and motor functions that control the skeletal muscles. Ascending tracts carry sensory information to the brain, and descending tracts carry motor commands to the neurons in the spinal cord that control the muscles. The autonomic nervous system controls smooth muscle, cardiac muscle, and the glands. It is further divided into the sympathetic and parasympathetic divisions.

Humans have 12 pairs of **cranial nerves** attached to the brain (Fig. 37.12*a*). Some of these are sensory nerves; that is, they contain only sensory nerve fibers. Some are motor nerves that contain only motor fibers, and others are mixed nerves that contain both sensory and motor fibers. Cranial nerves are largely concerned with the head, neck, and facial regions of the body. However, the vagus nerve has branches not only to the pharynx and larynx but also to most of the internal organs.

Humans also have 31 pairs of **spinal nerves** (Figs. 37.12*b* and 37.13) that emerge from the spinal cord via two short branches, or roots. The dorsal roots contain axons of sensory neurons, which conduct impulses to the spinal cord from sensory receptors. The cell body of a sensory neuron is in the **dorsal root ganglion.** The ventral roots contain axons of motor neurons, which conduct impulses away from the spinal cord to effectors. These two roots join to form a spinal nerve. All spinal nerves are mixed nerves that contain many sensory and motor fibers.

Somatic System

The PNS has two divisions—somatic and autonomic. The nerves in the **somatic system** serve the skin, joints, and skeletal muscles. Therefore, the somatic system includes nerves that

- take sensory information from external sensory receptors in the skin and joints to the CNS, and
- carry motor commands away from the CNS to the skeletal muscles.

The neurotransmitter acetylcholine (ACh) is active in the somatic system.

Voluntary control of skeletal muscles always originates in the brain. Involuntary responses to stimuli, called reflex actions, can involve only the spinal cord. Reflexes enable the body to react swiftly to stimuli that could disrupt homeostasis. Flying objects cause our eyes to blink, and sharp pins cause our hands to jerk away, even without our having to think about it.

The Reflex Arc.   Figure 37.13 illustrates the path of a reflex that involves only the spinal cord. If your hand touches a sharp pin, **sensory receptors** generate nerve impulses that move along sensory axons through a dorsal root ganglion toward the spinal cord. Sensory neurons that enter the cord dorsally pass signals on to many interneurons in the gray matter of the spinal cord. Some of these interneurons synapse with motor neurons. The short dendrites and the cell bodies of motor neurons are also in the spinal cord, but their axons leave the cord ventrally. Nerve impulses travel along motor axons to an **effector,** which brings about a response to the stimulus. In this case, a muscle contracts so that you withdraw your hand from the pin. (Sometimes an effector is a gland.)

Various other reactions are possible—you will most likely look at the pin, wince, and cry out in pain. This whole series of responses is explained by the fact that some of the interneurons in the white matter of the cord carry nerve impulses in tracts to the brain. The brain makes you aware of the stimulus and directs subsequent reactions to the situation. You don't feel pain until the brain receives the information and interprets it! Visual information received directly by way of a cranial nerve may make you aware that your finger is bleeding. Then you might decide to look for a Band-Aid.

Autonomic System

The **autonomic system** of the PNS regulates the activity of cardiac and smooth muscle and glands. It carries out its duties without our awareness or intent. The system is divided into the sympathetic and parasympathetic divisions (see Table 37.1 and Fig. 37.14). Both of these divisions

- function automatically and usually in an involuntary manner,
- innervate all internal organs, and
- use two neurons and one ganglion for each impulse.

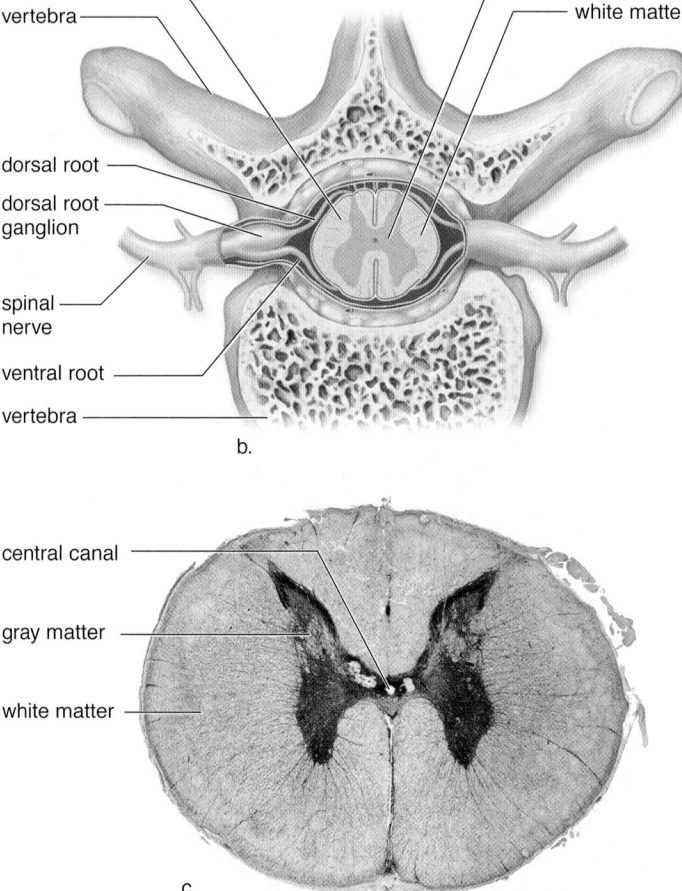

Figure 37.12 Cranial and spinal nerves.   a. Ventral surface of the brain, showing the attachment of the cranial nerves. **b.** Cross section of the vertebral column and spinal cord, showing a spinal nerve. Each spinal nerve has a dorsal root and a ventral root attached to the spinal cord. **c.** Photomicrograph of spinal cord cross section.

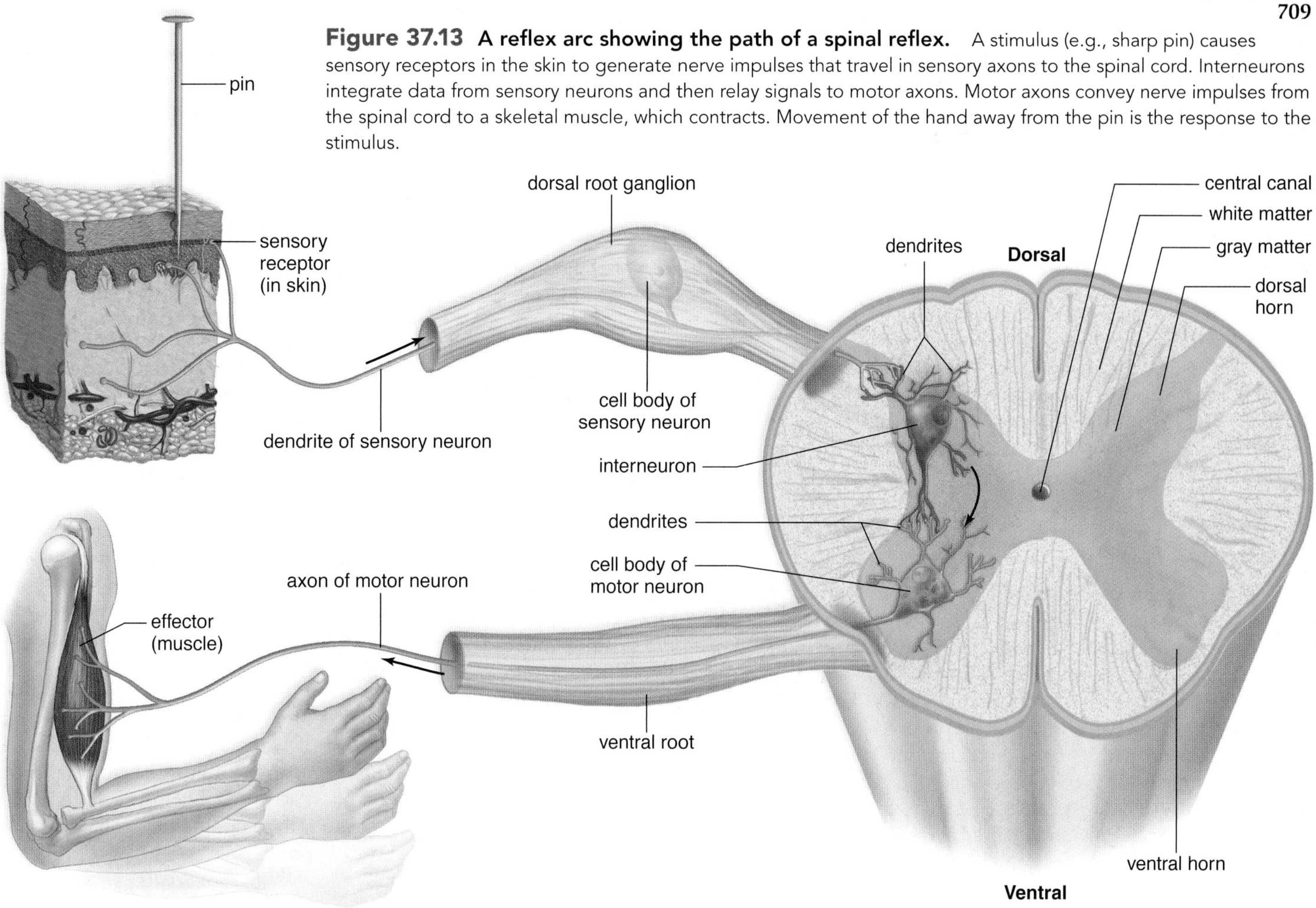

Figure 37.13 A reflex arc showing the path of a spinal reflex. A stimulus (e.g., sharp pin) causes sensory receptors in the skin to generate nerve impulses that travel in sensory axons to the spinal cord. Interneurons integrate data from sensory neurons and then relay signals to motor axons. Motor axons convey nerve impulses from the spinal cord to a skeletal muscle, which contracts. Movement of the hand away from the pin is the response to the stimulus.

pin

sensory receptor (in skin)

dorsal root ganglion

dendrites

Dorsal

central canal

white matter

gray matter

dorsal horn

cell body of sensory neuron

dendrite of sensory neuron

interneuron

dendrites

cell body of motor neuron

axon of motor neuron

effector (muscle)

ventral root

ventral horn

Ventral

The first neuron has a cell body within the CNS and a preganglionic fiber. The second neuron has a cell body within the ganglion and a postganglionic fiber.

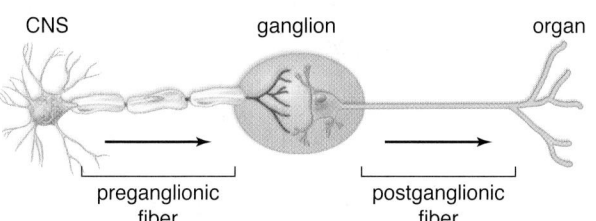

CNS   ganglion   organ

preganglionic fiber   postganglionic fiber

Reflex actions, such as those that regulate blood pressure and breathing rate, are especially important to the maintenance of homeostasis. These reflexes begin when the sensory neurons in contact with internal organs send information to the CNS. They are completed by motor neurons within the autonomic system.

Sympathetic Division

Most preganglionic fibers of the **sympathetic division** arise from the middle, or thoracolumbar, portion of the spinal cord and almost immediately terminate in ganglia that lie near the cord. Therefore, in this division, the preganglionic fiber is short, but the postganglionic fiber that makes contact with an organ is long.

Table 37.1 Comparison of Somatic Motor and Autonomic Motor Pathways

	Somatic Motor Pathway	Autonomic Motor Pathways	
		Sympathetic	Parasympathetic
Type of control	Voluntary/involuntary	Involuntary	Involuntary
Number of neurons per message	One	Two (preganglionic shorter than postganglionic)	Two (preganglionic longer than postganglionic)
Location of motor fiber	Most cranial nerves and all spinal nerves	Thoracolumbar spinal nerves	Cranial (e.g., vagus) and sacral spinal nerves
Neurotransmitter	Acetylcholine	Norepinephrine	Acetylcholine
Effectors	Skeletal muscles	Smooth and cardiac muscle, glands	Smooth and cardiac muscle, glands

Biological Systems

Drugs of Abuse

Drug abuse is apparent when a person takes a drug at a dose level and under circumstances that increase the potential for a harmful outcome. Addiction is present when more of the drug is needed to get the same effect, and withdrawal symptoms occur when the user stops taking the drug. This is true not only for teenagers and adults, but also for newborn babies of mothers who abuse and are addicted to drugs.

Alcohol

With the exception of caffeine, alcohol (ethanol) consumption is the most socially accepted form of drug use in the United States. According to a 2006 survey, nearly one-third of all U.S. high school students reported hazardous drinking (five or more drinks in one setting) during the 30 days preceding the survey. Notably, 80% of college-age young adults drink. According to a U.S. government study, drinking in college contributes to an estimated 1,400 student deaths, 500,000 injuries, and 70,000 cases of sexual assault or date rape each year.

Alcohol acts as a *depressant* on many parts of the brain by increasing the action of GABA, an inhibitory neurotransmitter. Depending on the amount consumed, the effects of alcohol on the brain can lead to a feeling of relaxation, lowered inhibitions, impaired concentration and coordination, slurred speech, and vomiting. If the blood level of alcohol becomes too high, coma or death can occur.

Beginning in about 2005, several manufacturers began selling alcoholic energy drinks. With names like Four Loco, Joose, and Sparks, these drinks combine fairly high levels of alcohol with caffeine and other ingredients. Although interactions between drugs can be complex, the stimulant effects of caffeine can counteract some of the depressant effects of alcohol, so that users feel able to drink more. Because caffeine does not reduce the intoxicating effects of alcohol, many state legislatures are banning these products, and in November 2010, the U.S. Food and Drug Administration warned several manufacturers that they would no longer be allowed to mix caffeine with alcohol in their products.

Nicotine

About 23% of U.S. high school students, and 8% of middle school students, reported smoking cigarettes in 2006. Young adults between age 18 and 25 reported the highest tobacco usage of any age group, at 45%. When tobacco is smoked or chewed, nicotine is rapidly delivered throughout the body. It causes a release of epinephrine from the adrenal glands, increasing blood sugar and causing the initial feeling of stimulation. As blood sugar falls, depression and fatigue set in, causing the user to seek more nicotine. In the CNS, nicotine stimulates neurons to release dopamine, a neurotransmitter that promotes a temporary sense of pleasure, and reinforces dependence on the drug. About 70% of people who try smoking become addicted.

As mentioned in earlier chapters, smoking is strongly associated with serious diseases of the cardiovascular and respiratory systems. Once addicted, however, only 10% to 20% of smokers are able to quit. Most medical approaches to quitting smoking involve the administration of nicotine in safer forms, like skin patches, gum, or a newly developed nicotine inhaler, so that withdrawal symptoms can be minimized while dependence is gradually reduced. As of 2010, an anti-nicotine vaccine called NicVAX was showing promise in early clinical trials. The vaccine stimulates the production of antibodies that prevent nicotine from entering the brain.

Club and Date Rape Drugs

Methamphetamine and Ecstasy are considered club or party drugs. Methamphetamine (commonly called meth or crank) is a powerful CNS stimulant. Meth is often produced in makeshift home laboratories, usually starting with ephedrine or pseudoephedrine, common ingredients in many cold and asthma medicines. As a result, many states have passed laws making these medications more difficult to purchase. The number of toxic chemicals used to prepare the drug makes a former meth lab site hazardous to humans and to the environment. Over 9 million people in the United States have used methamphetamine at least once in their lifetime. It is available as a powder that can be snorted, or as crystals (crystal meth or ice) that can be smoked.

The structure of methamphetamine is similar to that of dopamine, and the most immediate effect of taking meth is a rush of euphoria, energy, alertness, and elevated mood. However, this is typically followed by a state of agitation that, in some individuals, leads to violent behavior. Chronic use can result in what is called an amphetamine psychosis, characterized by paranoia, hallucinations, irritability, and aggressive, erratic behavior.

Video
Meth and the Brain

Ecstasy is the street name for MDMA (methylenedioxymethamphetamine), which is chemically similar to methamphetamine. Many users say that "X," taken as a pill that looks like an aspirin or candy, increases their feelings of well-being and love for other people. However, it has many of the same side effects as other stimulants, plus it can interfere with temperature regulation, leading to hyperthermia, high blood pressure, and seizures. In June 2010, a 15-year-old girl died from Ecstasy-related causes after attending a large rave party in Los Angeles.

Drugs with sedative effects, known as date rape or predatory drugs, include Rohypnol (roofies), gamma-hydroxybutyric acid (GHB), and ketamine (special K). Ketamine is actually a drug that veterinarians sometimes use to perform surgery on animals. Any of these drugs can be given to an unsuspecting person, who may fall into a dreamlike state where they are unable to move, and thus are vulnerable to sexual assault.

Cocaine and Crack

Cocaine is an alkaloid derived from the shrub *Erythroxylon coca*. Approximately 35 million Americans have used cocaine by sniffing/snorting, injecting, or smoking. Cocaine is a powerful *stimulant* in the CNS that interferes with the reuptake of dopamine at synapses, increasing overall brain activity (Fig. 37B). The result is a rush of well-being that lasts from 5 to 30 minutes.

"Crack" is the street name given to cocaine that is processed to a free base form for smoking. The term *crack* refers to the crackling sound heard when the drug is smoked. Smoking allows high doses of the drug to reach the brain rapidly, providing an intense and immediate high, or "rush." Approximately 8 million Americans use crack.

A cocaine binge is a period in which a user takes the drug at ever-higher doses. The user is hyperactive, with little desire for food or sleep but an increased sex drive. This is followed by a crash period, characterized by fatigue, depression, irritability, and lack of interest in sex. In fact, men who use cocaine often become impotent.

Cocaine is highly addictive; with continued use, the brain makes less dopamine to compensate for a seemingly endless supply. The user experiences withdrawal symptoms and an intense craving for cocaine. Overdosing on cocaine can cause cardiac and/or respiratory arrest.

Heroin

Heroin is derived from the resin or sap of the opium poppy plant, which is widely grown—from Turkey to Southeast Asia and in parts of Latin America. Drugs derived from opium are called opiates, a class that also includes morphine and codeine. After heroin is injected, snorted, or smoked, a feeling of euphoria, along with relief of any pain, occurs within a few minutes. It is estimated that 4 million Americans have used heroin some time in their lives, and over 300,000 people use heroin annually.

As with other drugs of abuse, addiction is common. Heroin binds to receptors meant for the endorphins, naturally occurring neurotransmitters that kill pain and produce feelings of tranquility. With repeated heroin use, the body's production of endorphins decreases. Tolerance develops so that the user needs to take more of the drug just to prevent withdrawal symptoms (tremors, restlessness, cramps, vomiting), and the original euphoria is no longer felt. Long-term users commonly acquire hepatitis, HIV/AIDS, and various bacterial infections due to the use of shared needles, and heavy users may experience convulsions and death by respiratory arrest.

Heroin addiction can be treated with synthetic opiate compounds, such as methadone or suboxone, that decrease withdrawal symptoms and block heroin's effects. However, methadone itself can be addictive, and methadone-related deaths are on the rise.

Marijuana and K2

The dried flowering tops, leaves, and stems of the marijuana plant, *Cannabis sativa*, contain and are covered by a resin that is rich in THC (tetrahydrocannabinol). The names *cannabis* and *marijuana* apply to either the plant or THC. Marijuana can be ingested, but usually it is smoked in a cigarette called a "joint." Although the drug was banned in the United States in 1937, an estimated 22 million Americans use marijuana, making it the most commonly used illegal drug in the U.S. Beginning with California in 1996, several states have legalized its use for medical purposes, such as treating cancer, AIDS, or glaucoma. In 2005 however, the Supreme Court ruled that patients prescribed medical marijuana can still be prosecuted by federal agencies.

Researchers have found that THC binds to a receptor for anandamide, a naturally occurring neurotransmitter that is important for short-term memory processing, and perhaps for feelings of contentment. The occasional marijuana user experiences mild euphoria, along with alterations in vision and judgment. Heavy use can cause hallucinations, anxiety, depression, paranoia, and psychotic symptoms. Some researchers believe that long-term marijuana use leads to brain impairment.

In recent years, awareness is increasing about a synthetic compound called K2 or Spice. Originally synthesized by an organic chemist at Clemson University, K2 is about ten times as potent as THC. The chemical is typically sprayed onto a mixture of other herbal products, and smoked. However, because there is no regulation of how it is produced, the amount of K2 itself, or contaminants, can vary greatly. This may account for the several reports of serious medical problems and even deaths in K2 users.

Questions to Consider

1. Suppose a form of heroin could be synthesized that had only the desired effects (euphoria and pain relief) with no side effects. Should such a drug be legal?
2. Should medical marijuana be legal for use in all states? If so, how should it be regulated?
3. In November 2010, the U.S. Drug Enforcement Agency banned the sale of five chemicals used to make K2. Is this an overreaction?

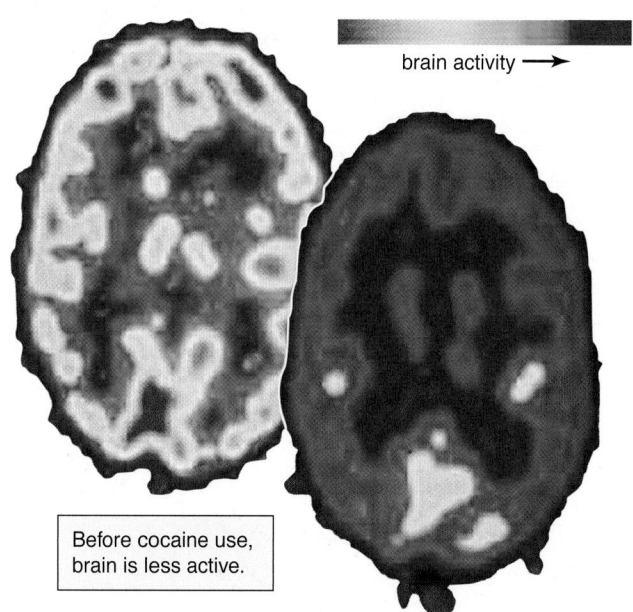

brain activity ⟶

Before cocaine use, brain is less active.

After cocaine use, brain is more active.

Figure 37B Drug use. Brain activity before and after the use of cocaine.

Figure 37.14 Autonomic system structure and function. Sympathetic preganglionic fibers (*left*) arise from the cervical, thoracic, and lumbar portions of the spinal cord; parasympathetic preganglionic fibers (*right*) arise from the cranial and sacral portions of the spinal cord. Each system innervates the same organs but has contrary effects.

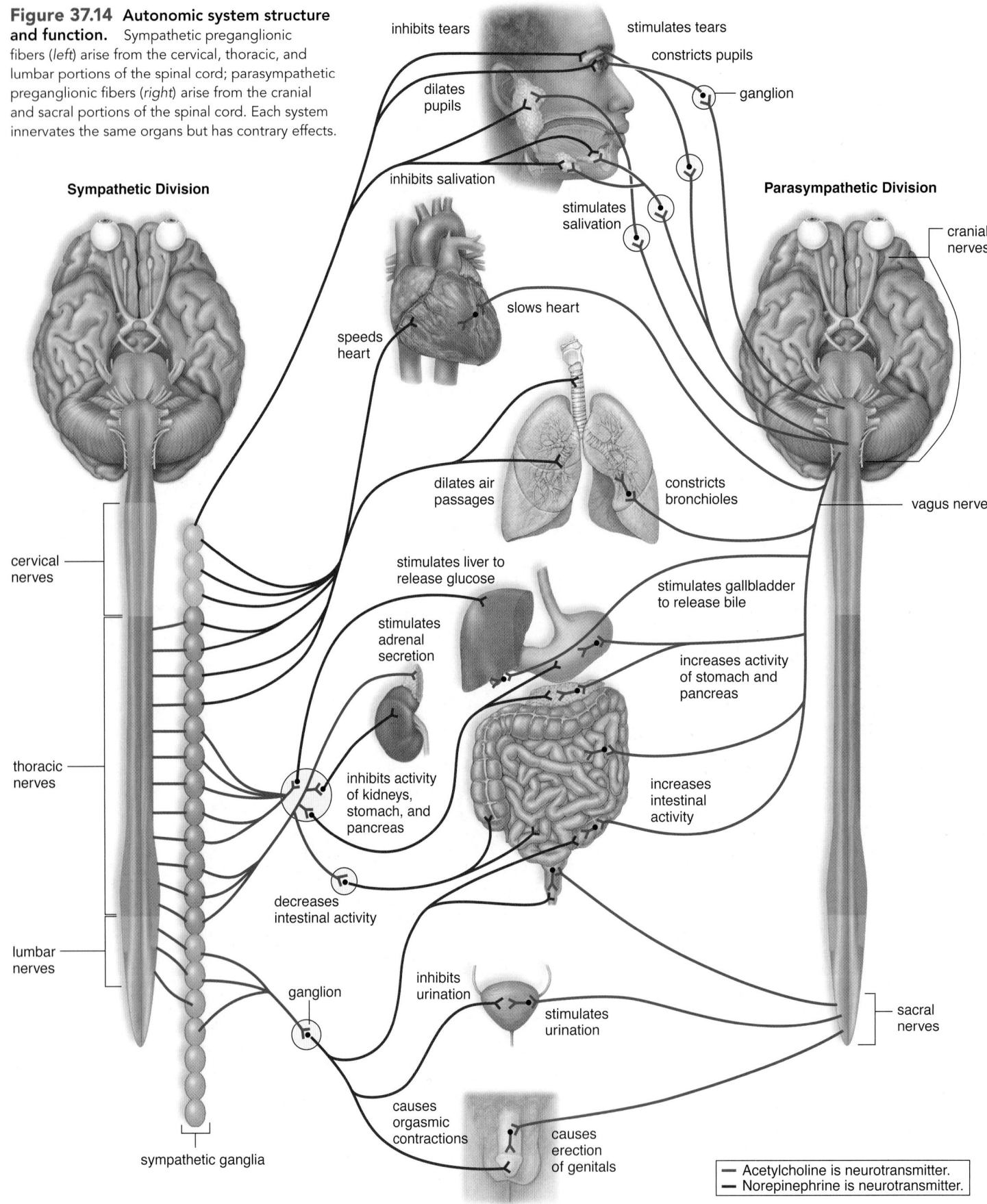

Sympathetic Division

inhibits tears

dilates pupils

inhibits salivation

speeds heart

dilates air passages

stimulates liver to release glucose

stimulates adrenal secretion

inhibits activity of kidneys, stomach, and pancreas

decreases intestinal activity

cervical nerves

thoracic nerves

lumbar nerves

ganglion

sympathetic ganglia

Parasympathetic Division

stimulates tears

constricts pupils

ganglion

stimulates salivation

slows heart

constricts bronchioles

stimulates gallbladder to release bile

increases activity of stomach and pancreas

increases intestinal activity

cranial nerves

vagus nerve

sacral nerves

inhibits urination

stimulates urination

causes orgasmic contractions

causes erection of genitals

— Acetylcholine is neurotransmitter.
— Norepinephrine is neurotransmitter.

The sympathetic division is especially important during emergency situations and is associated with "fight or flight" (Fig. 37.14). If you need to fend off a foe or flee from danger, active muscles require a ready supply of glucose and oxygen. The sympathetic division accelerates the heartbeat and dilates the bronchi. At the same time, the sympathetic division inhibits the digestive tract, because digestion is not an immediate necessity if you are under attack. The neurotransmitter released by the postganglionic axon is primarily norepinephrine (NE). Structurally, NE resembles epinephrine (adrenaline), an adrenal medulla hormone that usually increases heart rate and contraction (see Chapter 40).

Parasympathetic Division

The **parasympathetic division** includes a few cranial nerves (e.g., the vagus nerve) and also fibers that arise from the sacral (bottom) portion of the spinal cord. Therefore, this division often is referred to as the craniosacral portion of the autonomic system. In the parasympathetic division, the preganglionic fiber is long, and the postganglionic fiber is short because the ganglia lie near or within the organ.

The parasympathetic division, sometimes called the "housekeeper" or "rest and digest division," promotes all the internal responses we associate with a relaxed state; for example, it causes the pupil of the eye to contract, promotes digestion of food, and slows the heartbeat. The neurotransmitter used by the parasympathetic division is acetylcholine (ACh).

Check Your Progress   37.4

1. Review the neurological explanation for the observation that, after you touch a hot stove, you withdraw your hand before you feel any pain.
2. Apply your knowledge of the autonomic nervous system to explain why, if you eat a big lunch, then go out for a jog, your stomach may ache.
3. Describe the shift in autonomic system activity that occurs when you go from quietly reading to having ice dumped down the back of your shirt.

CONNECTING *the* CONCEPTS *with the* BIG IDEAS

Information and Signaling

- Neurons exhibit specialized dendrites and axons (some with myelin sheaths) which receive, transmit, and integrate information. (3E2a1-3)
- Polarized neuronal membranes depolarize when stimulation causes the ATP-powered Na+ and K+ gated channels to open. (3E2b1-3)
- Neurotransmitters (acetylcholine, GABA, etc.) are used for short-distance cell communication. (3D2b*IE*, 3E2c1*IE*)
- Synapse transmission stimulates or inhibits responses in receiving neurons. (3E2c2-3)
- The compartmentalized vertebrate brain (left/right cerebral hemispheres, cerebellum, brainstem) oversees many functions: vision, hearing, movement, emotion and thought, production of neuro-hormones, etc. (3E2d*IE*)

Interactions and Systems

- The nervous and muscular systems work together to ensure normal body function. (4A4b*IE*)

*Find the unabridged version of all EK citations at www.glencoe.com/maderAP11.

Media Study Tools

www.glencoe.com/maderAP11

Enhance your study of this chapter with study tools and practice tests. Also ask your instructor about the resources available through ConnectPlus, including the media-rich eBook, interactive learning tools, and animations.

McGraw Hill **connect**™ plus+
|BIOLOGY

Summarize

37.1 Evolution of the Nervous System

A comparative study of the invertebrates shows a gradual increase in the complexity of the nervous system. Vertebrate nervous systems are much more complex, and the mammalian nervous system is the most complex of all.

37.2 Nervous Tissue

The anatomical unit of the nervous system is the neuron, of which there are three types: sensory neuron, motor neuron, and interneuron. Each of these is made up of a cell body, an axon, and dendrites.

When an axon is not conducting an action potential (nerve impulse), the resting potential indicates that the inside of the fiber is negative compared to the outside. The sodium-potassium pump helps maintain this resting potential. When the axon is conducting a nerve impulse, an action potential (i.e., a change in membrane potential) travels along the fiber. Depolarization occurs (inside becomes positive) due to the movement of Na+ to the inside, and then repolarization occurs (inside becomes negative again) due to the movement of K+ to the outside of the fiber.

Transmission of the nerve impulse from one neuron to another takes place across a synapse. Synaptic vesicles usually release a chemical, known as a neurotransmitter, into the synaptic cleft. The binding of neurotransmitters to receptors in the postsynaptic membrane can either increase the chance of an action potential (stimulation) or decrease the chance of an action potential (inhibition) in the next neuron. A neuron may transmit several nerve impulses, one after the other.

37.3 The Central Nervous System

The CNS consists of the spinal cord and brain, which are both protected by bone. The CNS receives and integrates sensory input and formulates motor output. The gray matter of the spinal cord contains neuron cell bodies; the white matter consists of myelinated axons that occur in bundles called tracts. The spinal cord sends sensory information to the brain, receives motor output from the brain, and carries out reflexes.

In the brain, the cerebrum has two cerebral hemispheres connected by the corpus callosum. Sensation, reasoning, learning and memory, and language and speech take place in the cerebrum. The cerebral cortex is a thin layer of gray matter covering the cerebrum.

The cerebral cortex of each cerebral hemisphere has four lobes: a frontal, parietal, occipital, and temporal lobe. The primary motor area in the frontal lobe sends out motor commands to lower brain centers, which pass them on to motor neurons. The primary somatosensory area in the parietal lobe receives sensory information from lower brain centers in communication with sensory neurons. Association areas for vision are in the occipital lobe, and those for hearing are in the temporal lobe.

The brain has a number of other regions. The hypothalamus controls homeostasis, and the thalamus specializes in sending sensory input on to the cerebrum. The cerebellum primarily coordinates skeletal muscle contractions. The medulla oblongata and the pons have centers for vital functions such as breathing and the heartbeat.

A number of important diseases affect the human nervous system. In amyotrophic lateral sclerosis (ALS), motor neurons in the brain and spinal cord degenerate and die. Strokes can affect any area of the brain, often due to a blockage in the blood supply. Multiple sclerosis (MS) is an autoimmune disease that affects the myelin sheaths, disrupting nerve transmission. Parkinson disease and Alzheimer disease are brain disorders that mainly affect older individuals.

37.4 The Peripheral Nervous System

The PNS contains the somatic system and the autonomic system. Reflexes are automatic, and some do not require involvement of the brain. A simple reflex utilizes neurons that make up a reflex arc. In the somatic system, a sensory neuron conducts nerve impulses from a sensory receptor to an interneuron, which in turn transmits impulses to a motor neuron, which stimulates an effector to react.

The motor portion of the somatic system of the PNS controls skeletal muscle; in contrast, the motor portion of the autonomic system controls smooth muscle of the internal organs and glands. The sympathetic division, which is often associated with reactions that occur during times of stress, and the parasympathetic division, which is often associated with activities that occur during times of relaxation, are both parts of the autonomic system.

Key Terms

action potential   699
Alzheimer disease   707
amygdala   707
amyotrophic lateral sclerosis
    (ALS)   703
astrocyte   697
autonomic system   708
axon   697

basal nuclei   704
brain   703
brain stem   704
cell body   697
central nervous system
    (CNS)   696
cephalization   694
cerebellum   704

cerebral cortex   704
cerebral hemisphere   703
cerebrospinal fluid   702
cerebrum   703
cranial nerve   707
dendrite   697
dorsal root ganglion   707
effector   708
ganglion   694
gray matter   702
hippocampus   705
hypothalamus   704
integration   701
interneuron   698
limbic system   705
medulla oblongata   705
memory   707
meninges   702
meningitis   702
microglia   697
midbrain   704
motor (efferent) neuron   697
multiple sclerosis   705
myelin sheath   697
nerve   707
nerve fiber   697
nerve net   694
neuroglia   697
neuron   697

neurotransmitter   700
nodes of Ranvier   697
oligodendrocyte   697
parasympathetic division   713
Parkinson disease (PD)   704
peripheral nervous system
    (PNS)   696
pineal gland   704
pons   704
reflex action   702
refractory period   700
resting potential   698
reticular activating system
    (RAS)   705
saltatory conduction   699
Schwann cell   697
sensory (afferent) neuron   697
sensory receptor   708
somatic system   708
spinal cord   702
spinal nerve   707
stroke   704
sympathetic division   709
synapse   700
synaptic cleft   700
thalamus   704
tract   703
ventricle   702
white matter   703

▦ Assess

Reviewing This Chapter

1. Trace the evolution of the nervous system by contrasting its organization in hydras, planarians, earthworms, and mammals.   694–96
2. Describe the structure of a neuron, and give a function for each part mentioned. Name three types of neurons, and give a function for each.   697–98
3. What are the major events of an action potential, and what ion changes are associated with each event?   699–700
4. Describe the mode of action of a neurotransmitter at a synapse, including how it is stored and how it is destroyed.   700–1
5. Name the major parts of the human brain, and give a principal function for each part.   702–4
6. Describe the limbic system, and discuss its possible involvement in learning and memory.   705, 707
7. Discuss the structure and function of the peripheral nervous system.   707–9, 712–13
8. Trace the path of a spinal reflex.   708–9
9. Contrast the sympathetic and parasympathetic divisions of the autonomic system.   708–9, 712–13

Testing Yourself

Choose the best answer for each question.

1. Which of these are the first and last elements in a spinal reflex?
   a. axon and dendrite
   b. sense organ and muscle effector
   c. ventral horn and dorsal horn
   d. motor neuron and sensory neuron
   e. sensory receptor and the brain

2. A spinal nerve takes nerve impulses
   a. to the CNS.
   b. away from the CNS.
   c. both to and away from the CNS.
   d. only inside the CNS.
   e. only from the cerebrum.

3. Which of these correctly describes the distribution of ions on either side of an axon when it is not conducting a nerve impulse?
   a. more sodium ions (Na$^+$) outside and fewer potassium ions (K$^+$) inside
   b. K$^+$ outside and Na$^+$ inside
   c. charged proteins outside; Na$^+$ and K$^+$ inside
   d. charged proteins inside
   e. Both a and d are correct.

4. When the action potential begins, sodium gates open, allowing Na$^+$ to cross the membrane. Now the polarity changes to
   a. negative outside and positive inside.
   b. positive outside and negative inside.
   c. There is no difference in charge between outside and inside.
   d. Any one of these could be correct.

5. Transmission of the nerve impulse across a synapse is accomplished by
   a. the release of Na$^+$ at the presynaptic membrane.
   b. the release of neurotransmitters at the postsynaptic membrane.
   c. the reception of neurotransmitters at the postsynaptic membrane.
   d. Only a and c are correct.

6. The autonomic system has two divisions, called the
   a. CNS and PNS.
   b. somatic and skeletal systems.
   c. efferent and afferent systems.
   d. sympathetic and parasympathetic divisions.

7. Synaptic vesicles are
   a. at the ends of dendrites and axons.
   b. at the ends of axons only.
   c. along the length of all long fibers.
   d. at the ends of interneurons only.
   e. Both b and d are correct.

8. Which of these pairs is mismatched?
   a. cerebrum—thinking and memory
   b. thalamus—motor and sensory centers
   c. hypothalamus—internal environment regulator
   d. cerebellum—motor coordination
   e. medulla oblongata—fourth ventricle

9. Repolarization of an axon during an action potential is produced by
   a. inward diffusion of Na$^+$.
   b. active extrusion of K$^+$.
   c. outward diffusion of K$^+$.
   d. inward active transport of Na$^+$.

10. Which two parts of the brain are least likely to work directly together?
    a. thalamus and cerebrum
    b. cerebrum and cerebellum
    c. hypothalamus and medulla oblongata
    d. cerebellum and medulla oblongata

11. Which of the following is not part of the spinal cord?
    a. central canal   d. tracts
    b. dorsal horn   e. ventral horn
    c. association areas

12. A drug that inactivates acetylcholinesterase
    a. stops the release of ACh from presynaptic endings.
    b. prevents the attachment of ACh to its receptor.
    c. increases the ability of ACh to stimulate postsynaptic cells.
    d. All of these are correct.

13. Which of these statements about autonomic neurons is correct?
    a. They are motor neurons.
    b. Preganglionic neurons have cell bodies in the CNS.
    c. Postganglionic neurons innervate smooth muscles, cardiac muscle, and glands.
    d. All of these are correct.

14. Which of these fibers release norepinephrine?
    a. preganglionic sympathetic axons
    b. postganglionic sympathetic axons
    c. preganglionic parasympathetic axons
    d. postganglionic parasympathetic axons

15. Sympathetic nerve stimulation does not cause
    a. the liver to release glycogen.
    b. dilation of bronchioles.
    c. the gastrointestinal tract to digest food.
    d. an increase in the heart rate.

16. The limbic system
    a. involves portions of the cerebral lobes, basal nuclei, and the diencephalon.
    b. is responsible for our deepest emotions, including pleasure, rage, and fear.
    c. is not responsible for reason and self-control.
    d. All of these are correct.

17. Label this diagram of a reflex arc.

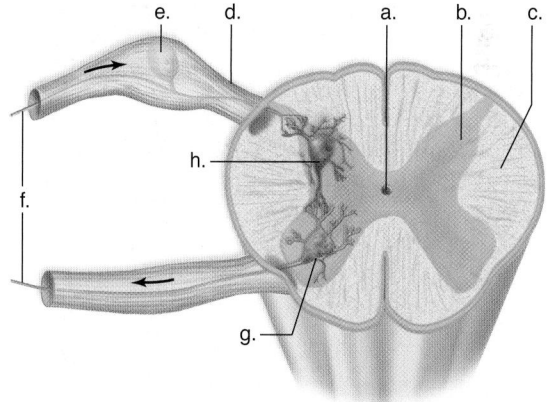

Engage

Thinking Scientifically

1. In individuals with panic disorder, the fight-or-flight response is activated by inappropriate stimuli. How might it be possible to directly control this response in order to treat panic disorder? Why is such control often impractical?

2. A man who lost his leg several years ago continues to experience pain as though it were coming from the missing limb. What hypothesis could explain the neurological basis of this pain?

38

Sense Organs

Before You Begin

Before beginning this chapter, take a few moments to review the following discussions.

Sections 37.3 and 37.4 What are the roles of the central and peripheral nervous systems in an animal's responses to its environment?

Figure 37.9 How is the cerebral cortex involved in the processing of sensory information?

Figure 37.13 How do sensory receptors in the skin stimulate a spinal reflex?

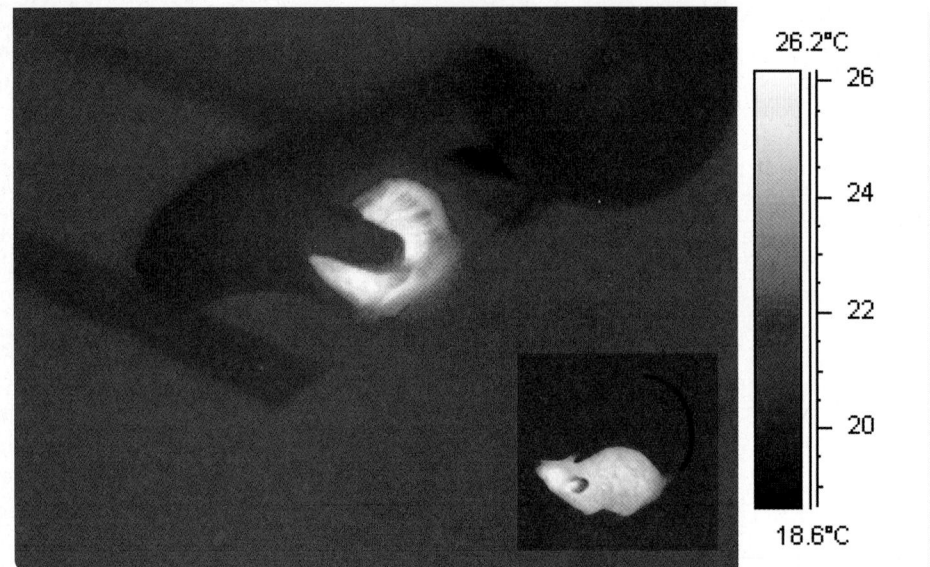

Certain snakes can detect infrared energy emitted by their prey. In this infrared image, the red-orange areas are the warmest, and the blue-black areas are the coldest.

Visible light is made up of waves of electromagnetic energy with different wavelengths. As you'll see in this chapter, sensory receptors in the human eye contain pigments that undergo a chemical change when exposed to electromagnetic energy at these various wavelengths, which is then perceived by the brain as vision. Electromagnetic energy with a wavelength at the shortest end of the visible spectrum is perceived as violet; that with a wavelength at the highest end is perceived as red, with other colors in between.

A few types of animals have evolved the ability to detect electromagnetic energy with longer wavelengths, known as the infrared spectrum. Just about any source of heat—the Sun, a fire, or a warm body—emits energy in the infrared spectrum. Certain kinds of snakes, such as the pit vipers, have evolved specialized infrared sensory organs. Located in a pit below each eye, these organs are very sensitive to infrared waves emitted by their warm-blooded prey; the photo above, taken using an infrared-sensitive camera, shows how a mouse might "appear" to a snake. Even when placed in total darkness, snakes with this ability can track and find such prey quickly. Although the ability to detect infrared energy with such precision is unusual, all animals rely upon many types of sensory systems to maintain homeostasis.

As you read through the chapter, think about the following questions:

1. How does the ability to detect infrared energy provide snakes with a competitive advantage over predators lacking this ability?

2. Of the types of sensory receptors described in this chapter—chemoreceptors, photoreceptors, mechanoreceptors, or thermoreceptors—which is the most necessary for an animal to survive?

Following *the* Big Ideas

CHAPTER 38 SENSE ORGANS	
Evolution	Basic abilities to perceive environmental information in primitive animals have evolved into complex and discriminating sense organs.

38.1 Sensory Receptors

Learning Outcomes

Upon completion of this section, you should be able to

1. Explain the differences between sensory receptors, sensory transduction, and perception.
2. Describe four types of sensory receptors, and list examples of each.

In order to survive, animals must be able to maintain homeostasis as well as locate required nutrients, avoid dangers, and learn from previous experiences. These kinds of selective pressures have resulted in the evolution of the many different types of **sensory receptors,** which are specialized cells capable of detecting changes in internal or external conditions, and of communicating that information to the central nervous system.

A sensory receptor is able to convert some type of event, or *stimulus*, occurring in the environment into a nerve impulse. This process is known as **sensory transduction.** Some sensory receptors are modified neurons, and others are specialized cells closely associated with neurons.

The plasma membrane of a sensory receptor contains proteins that react to a stimulus. For example, these membrane proteins might be sensitive to temperature, or react with a certain chemical. When this happens, ion channels open, and ions flow across the plasma membrane. If the stimulus is sufficient, nerve impulses begin and are carried by a sensory nerve fiber within the PNS to the CNS.

Note that there is no difference between the nerve impulses carried by the different types of sensory nerves. All these impulses are simply the action potentials discussed in the preceding chapter, regardless of whether they arise in the eyes, ears, nose, mouth, skin, or internal organs. The interpretation of

these nerve impulses by the brain brings about a response that is appropriate for the particular type of stimulus. That is why artificial stimulation of the nerves that normally carry impulses generated in the ear or the eye are interpreted by the brain as sound or light, respectively (see the Nature of Science feature on page 725).

What's more, not all of these sensory impulses are received at the conscious levels of the brain—for example, we are not aware of the constant adjustments that are occurring in response to various internal stimuli. Any sensory stimuli of which humans, and perhaps other animals, become conscious are known as *perceptions*.

Although the extent to which nonhuman animals have perceptions is largely unknown, it is likely that some of them perceive their world in very different ways. As noted in the opening essay, some snakes can detect infrared energy that is completely invisible to humans. Bats, dolphins, and whales are capable of *echolocation*, meaning they can produce very high frequency sounds, and then learn about objects in their environment by listening for echoes. Some whales can also hear very low frequency sounds emitted by other whales hundreds of miles away. Dogs have a sense of smell that is many times more sensitive than that of humans, and they can be trained to detect drugs, human remains, blood, and even bedbugs. Several scientific studies have confirmed that dogs can detect some types of human cancer just by sniffing the appropriate samples. Clearly the sensory systems are a subject of much fascination to biologists, and they are the systems that literally determine how we experience our world.

An animal's ability to detect information in its environment is dependent on just a few types of sensory receptors. **Chemoreceptors** [Gk. *chemo,* pertaining to chemicals; L. *receptor,* receiver] can respond to a diverse range of chemical substances, from oxygen levels in the blood, to molecules of food in the mouth or nasal passages. **Photoreceptors** [Gk. *photos,* light], such as those found in the human retina, respond to light energy. **Mechanoreceptors** are stimulated by mechanical forces, usually pressure of some sort. Mechanoreceptors are responsible for detecting changes that are perceived as sound or touch, as well as for maintaining our equilibrium, balance, and proper tone in muscles and joints. **Thermoreceptors,** located in the hypothalamus and skin, are stimulated by changes in temperature.

Check Your Progress   38.1

1. Define sensory transduction.
2. List three examples of sensory capabilities found in animals that are lacking in humans.

38.2 Chemical Senses

Learning Outcomes

Upon completion of this section, you should be able to

1. Discuss the locations of chemoreceptors in arthropods, crustaceans, and vertebrates.
2. Describe the types and locations of taste receptors in humans.
3. Compare and contrast how the brain receives information about taste versus smell.

Chemoreception is found almost universally in animals and is therefore believed to be the most primitive sense. Chemoreceptors sensitive to certain chemical substances can be important in locating food, finding a mate, and detecting potentially dangerous chemicals in the environment.

The location and sensitivity of chemoreceptors vary throughout the animal kingdom. Although chemoreceptors are present throughout the body of planarians, they are concentrated in the auricles located on the sides of the head. Many insects have taste receptors on their mouthparts, but in the housefly, chemoreceptors are located primarily on the feet. Insects also detect airborne *pheromones*, which are chemical messages passed between individuals.

Video Sexy Bugs

In crustaceans such as lobsters and crabs, chemoreceptors are widely distributed on their appendages and antennae. Many fish have chemoreceptors scattered over the surface of their skin. Snakes possess *Jacobson's organs*, a pair of sensory pitlike organs located in the roof of the mouth. When a snake flicks its forked tongue, scent molecules are carried to the Jacobson's organs and sensory information is transmitted to the brain for interpretation.

Sense of Taste in Humans

In adult humans, approximately 3,000 **taste buds** are located primarily on the tongue (Fig. 38.1). Many taste buds lie along the walls of the papillae, the small elevations on the tongue that are visible to the unaided eye. Isolated taste buds are also present on the hard palate, the pharynx, and the epiglottis.

Taste buds on the tongue open at a taste pore. Taste buds have supporting cells and a number of elongated taste cells that end in microvilli. The microvilli, which project into the taste pore, bear receptor proteins for certain molecules. When molecules bind to receptor proteins, nerve impulses are generated in associated sensory nerve fibers. These nerve impulses travel to the brain, where they are interpreted as tastes.

 **Animation Taste**

Humans have five main types of taste receptors: sweet, sour, salty, bitter, and umami (Japanese: savory, delicious). Foods rich in certain amino acids, such as the common seasoning monosodium glutamate (MSG), as well as certain flavors of cheese, beef broth, and some seafood, produce the taste of umami. Taste buds for each of these tastes are located throughout the tongue, although certain regions may be slightly more sensitive to particular tastes. A food can stimulate more than one of these types of taste buds. The brain appears to survey the overall pattern of incoming sensory impulses and to take a "weighted average" of their taste messages as the perceived taste.

In 2010, researchers at the University of Maryland found chemoreceptors in the human lung that are sensitive only to chemicals that normally taste bitter. These receptors are not clustered in buds, and they do not send taste signals to the brain. Stimulation of these receptors causes the airways to dilate, leading the researchers to speculate about implications for new medications to treat diseases like asthma.

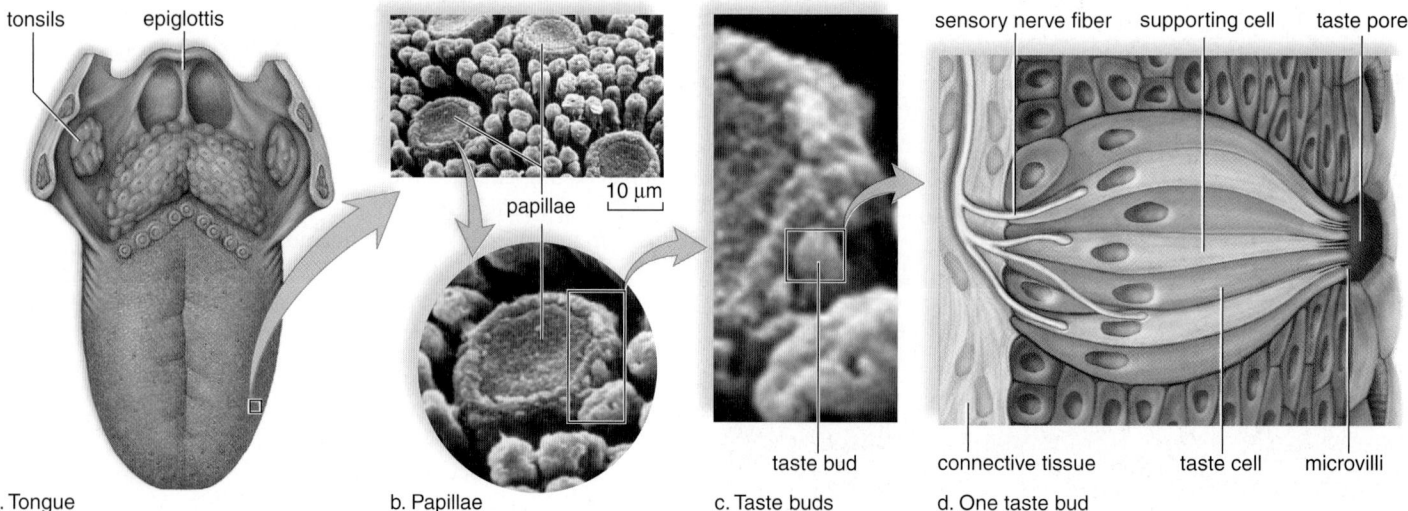

a. Tongue   b. Papillae   c. Taste buds   d. One taste bud

Figure 38.1 Taste buds in humans. **a.** Papillae on the tongue contain taste buds that are sensitive to sweet, sour, salty, bitter, and umami. **b.** Photomicrograph and enlargement of papillae. **c.** Taste buds occur along the walls of the papillae. **d.** Taste cells end in microvilli that bear receptor proteins for certain molecules. When molecules bind to the receptor proteins, nerve impulses are generated and go to the brain, where the sensation of taste occurs.

Sense of Smell in Humans

In humans, the sense of smell, or olfaction, is dependent on between 10 and 20 million **olfactory cells.** These structures are located within olfactory epithelium high in the roof of the nasal cavity (Fig. 38.2). Olfactory cells are modified neurons. Each cell ends in a tuft of about five olfactory cilia that bear receptor proteins for odor molecules. Each olfactory cell has only 1 out of 1,000 different types of receptor proteins. Nerve fibers from similar olfactory cells lead to the same neuron in the olfactory bulb, an extension of the brain.

An odor contains many odor molecules that activate a characteristic combination of receptor proteins. A rose might stimulate certain olfactory cells, designated by blue and green in Figure 38.2, whereas a gardenia might stimulate a different combination. When the neurons communicate this information via the olfactory tract to the olfactory areas of the cerebral cortex, we perceive that we have smelled a rose or a gardenia.

Have you ever noticed that a certain aroma vividly brings to mind a certain person or place? A whiff of perfume may remind you of someone you knew, or the smell of boxwood may remind you of your grandfather's farm. The olfactory bulbs have direct connections with the limbic system and its centers for emotions and memory. One study found that participants with previous negative experiences of visiting the dentist rated the smell of a chemical often encountered in dentist's offices as unpleasant, while those lacking such negative experiences rated it as pleasant.

The number of olfactory cells declines with age, and the remaining population of receptors becomes less sensitive. Thus, older people may tend to apply excessive amounts of perfume or aftershave. The ability to smell can also be lost as the result of head trauma, respiratory infection, or brain disease. This condition can become dangerous if these individuals cannot smell spoiled food, smoke, or a gas leak.

Usually, the sense of taste and the sense of smell work together to create a combined effect when interpreted by the cerebral cortex. For example, when you have a cold, you may think food has lost its taste, but most likely you have lost the ability to detect its smell. This method works in reverse also. When you smell something, some of the molecules move from the nose down into the mouth region and stimulate the taste buds there. Therefore, part of what we refer to as smell may in fact be taste.

MP3
Taste and Smell

Video
Mosquitoes and Sweat

Check Your Progress   38.2

1. Compare and contrast the senses of smell and taste.
2. List the five types of taste receptors in humans.
3. Discuss what could account for nerve impulses, which are all action potentials, being interpreted as arising from different sense organs.

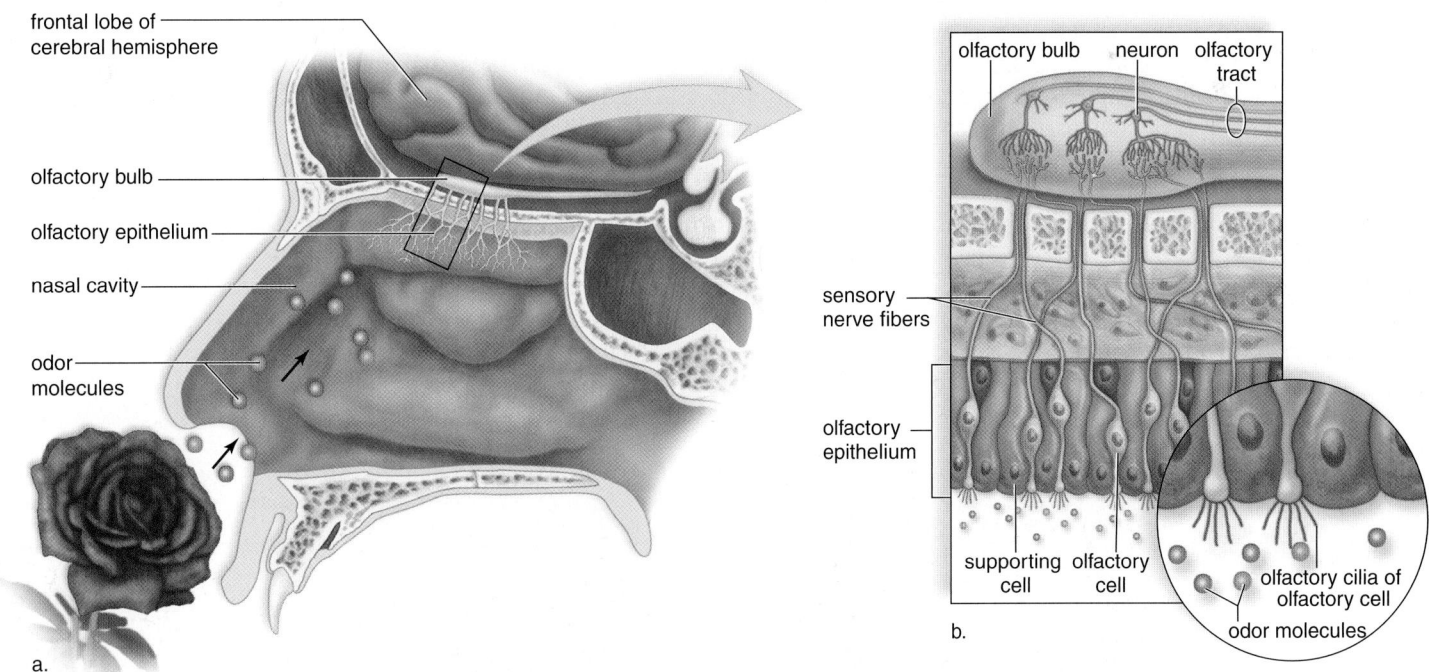

Figure 38.2 Olfactory cell location and anatomy.   **a.** The olfactory epithelium in humans is located high in the nasal cavity. **b.** Olfactory cells end in cilia that bear receptor proteins for specific odor molecules. The cilia of each olfactory cell can bind to only one type of odor molecule (signified here by color). For example, if a rose causes olfactory cells sensitive to "blue" and "green" odor molecules to be stimulated, then neurons designated by blue and green in the olfactory bulb are activated. The primary olfactory area of the cerebral cortex interprets the pattern of stimulation as the scent of a rose.

38.3 Sense of Vision

Vision is an important capability for many, but not all, animals. Like the senses of smell and hearing, vision allows us to perceive the environment at a distance, which can have survival value. In this section we review how animals detect light and how the human eye accomplishes vision.

How Animals Detect Light

Photoreceptors are sensory receptors that are sensitive to light. Some animals lack photoreceptors and depend on senses such as smell and hearing instead; other animals have photoreceptors but live in environments that do not require them. For example, moles live underground and use their senses of smell and touch rather than eyesight.

Not all photoreceptors form images. The "eyespots" of planarians allow these animals to sense and move away from light. Image-forming eyes are found among four invertebrate groups: cnidarians, annelids, molluscs, and arthropods. Arthropods have **compound eyes** composed of many independent visual units called ommatidia [Gk. *ommation*, dim. of *omma*, eye], each possessing all the elements needed for light reception (Fig. 38.3). Both the cornea and crystalline cone function as lenses to direct light rays toward the photoreceptors. The photoreceptors generate nerve impulses, which pass to the brain by way of optic nerve fibers. The outer pigment cells absorb stray light rays so that the rays do not pass from one visual unit to the other.

nectar guides

Figure 38.4 Nectar guides. Evening primrose, *Oenothera*, as seen by humans (*left*) and insects (*right*). Humans see no markings, but insects see distinct lines and central blotches because their eyes respond to ultraviolet rays. These types of markings, known as nectar guides, often highlight the reproductive parts of flowers, where insects feed on nectar and pick up pollen at the same time.

Flies and mosquitoes can only see a few millimeters in front of them, but dragonflies can see small prey insects several meters away. Research has shown that foraging bees use their sense of vision as a sort of "odometer" to estimate how far they have flown from their hive.

Most insects have color vision, but they see a limited number of colors compared to humans. However, many insects can also see some ultraviolet rays, and this enables them to locate the particular parts of flowers, such as nectar guides, that have ultraviolet patterns (Fig. 38.4). Some fishes, all reptiles, and most birds are believed to have color vision, but among mammals, only humans and other primates have color vision. It would seem, then, that this trait was adaptive for a diurnal habit (active during the day), which accounts for its retention in only a few mammals.

Vertebrates (including humans) and certain molluscs, such as the squid and the octopus, have a **camera-type eye.** Because molluscs and vertebrates are not closely related, this similarity is an example of convergent evolution. A single lens focuses an image of the visual field on photoreceptors, which are closely packed together. In vertebrates, the lens changes shape to aid focusing, but in molluscs the lens moves back and forth. All of

Figure 38.3 Compound eye. Each visual unit of a compound eye has a cornea and a lens that focus light onto photoreceptors. The photoreceptors generate nerve impulses that are transmitted to the brain, where interpretation produces a mosaic image.

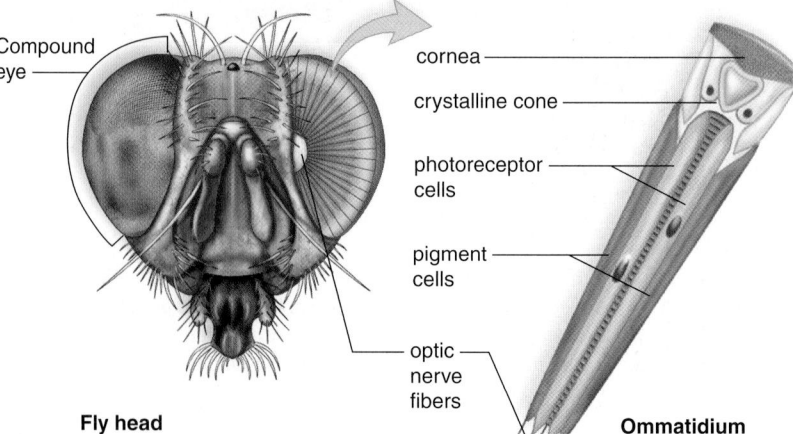

Compound eye

cornea

crystalline cone

photoreceptor cells

pigment cells

optic nerve fibers

Fly head

Ommatidium

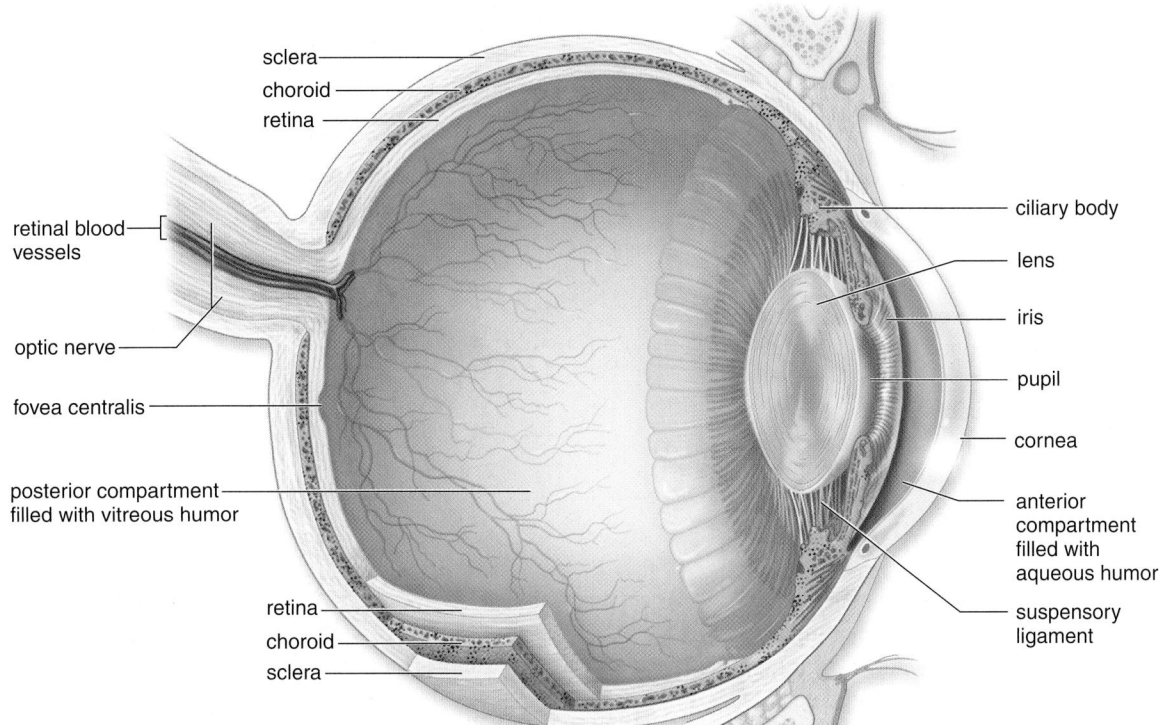

sclera
choroid
retina
retinal blood vessels
optic nerve
fovea centralis
posterior compartment filled with vitreous humor
retina
choroid
sclera

ciliary body
lens
iris
pupil
cornea
anterior compartment filled with aqueous humor
suspensory ligament

Figure 38.5 Anatomy of the human eye. Notice that the sclera, the outer layer of the eye, becomes the cornea and that the choroid, the middle layer, is continuous with the ciliary body and the iris. The retina, the inner layer, contains the photoreceptors for vision. The fovea centralis is the region where vision is most acute.

the photoreceptors taken together can be compared to a piece of film in a camera. The human eye is more complex than a camera, however, as you will see.

Animals with two eyes facing forward have three-dimensional vision, or **stereoscopic vision.** The visual fields overlap, and each eye is able to view an object from a different angle. Predators tend to have stereoscopic vision and so do humans. Animals with eyes facing sideways, such as rabbits, don't have stereoscopic vision, but they do have **panoramic vision,** meaning that the visual field is very wide. Panoramic vision is useful to prey animals because it makes it more difficult for a predator to sneak up on them.

Many vertebrates have a membrane in the back of their eye called a *tapetum lucidum,* which reflects light back into the photoreceptor cells of the retina to increase sensitivity to light. This explains the eerie glowing appearance of some animals' eyes at night.

The Human Eye

The human eye, which is an elongated sphere about 2.5 cm in diameter, has three layers: the sclera, the choroid, and the retina (Fig. 38.5). The outer **sclera** [Gk. *skleros,* hard], is an opaque, white, fibrous layer that covers most of the eye; in front of the eye, the sclera becomes the transparent **cornea,** the window of the eye. A thin layer of epithelial cells forms a mucous membrane called the **conjunctiva** that covers the surface of the sclera and keeps the eyes moist.

The middle, thin, dark-brown **choroid** [Gk. *chorion,* membrane] layer contains many blood vessels and a brown pigment that absorbs stray light rays. Toward the front of the eye, the

choroid thickens and forms the ring-shaped ciliary body and a thin, circular, muscular diaphragm, the iris. The **iris** is the colored portion of the eye and regulates the size of an opening called the pupil. The **pupil,** like the aperture of a camera lens, regulates light entering the eye. The **lens,** which is attached to the ciliary body by ligaments, divides the cavity of the eye into two portions and helps form images. A basic, watery solution called aqueous humor fills the anterior compartment between the cornea and the lens. The aqueous humor provides a fluid cushion and nutrient and waste transport for the eye.

The inner layer of the eye, the **retina** [L. *rete,* net], is located in the posterior compartment. The retina contains photoreceptors called **rod cells** and **cone cells.** The rods are very sensitive to light, but they do not respond to colors; therefore, at night or in a darkened room, we see only shades of gray. Rods are distributed in the peripheral regions of the retina. The cones, which require bright light, are sensitive to different wavelengths of light, and therefore, humans have the ability to distinguish colors. The retina has a central region called the **fovea centralis,** where cone cells are densely packed. Light is normally focused on the fovea when we look directly at an object. This is helpful because vision is most acute in the fovea centralis.

Sensory fibers form the optic nerve, which takes nerve impulses to the brain. No rods or cones are present where the optic nerve exits the retina (see Fig. 38.9). Therefore, no vision is possible in this area, and it is termed the **blind spot.** You can detect your own blind spot by putting a dot to the right of center on a piece of paper. Use your right hand to move the paper slowly toward your right eye while you look straight ahead. The dot will disappear at one point—this is your blind spot.

Video Artificial Eye

Focusing of the Eye

When we look directly at something, such as the printed letters on this page, light rays pass through the pupil and are focused on the retina. The image produced is much smaller than the object because light rays are bent (refracted) when they are brought into focus. The image on the retina is also upside down and is reversed from left to right. When information from the retina reaches the brain, it is processed so that we perceive our surroundings in the correct orientation.

Focusing starts at the cornea and continues as the rays pass through the lens. The lens provides additional focusing power as **visual accommodation** occurs for close vision. The shape of the lens is controlled by the **ciliary muscle** within the ciliary body. When we view a distant object, the ciliary muscle is relaxed, causing the suspensory ligaments attached to the ciliary body to be taut; therefore, the lens remains relatively flat (Fig. 38.6*a*). When we view a near object, the ciliary muscle contracts, releasing the tension on the suspensory ligaments, and the lens becomes more round due to its natural elasticity (Fig. 38.6*b*). Because close work requires contraction of the ciliary muscle, it very often causes muscle fatigue known as eyestrain. With normal aging, the lens loses its ability to accommodate for near objects (Fig. 38.6*b*); therefore, people frequently need reading glasses once they reach middle age.

Photoreceptors of the Eye

Sensory transduction occurs once light has been focused on the photoreceptors in the retina. Figure 38.7 illustrates the structure of these rod cells and cone cells. Both rods and cones have an outer segment joined to an inner segment by a stalk. Pigment molecules that react to light are embedded in the membrane of the many disks present in the outer segment. Synaptic vesicles are located at the synaptic endings of the inner segment.

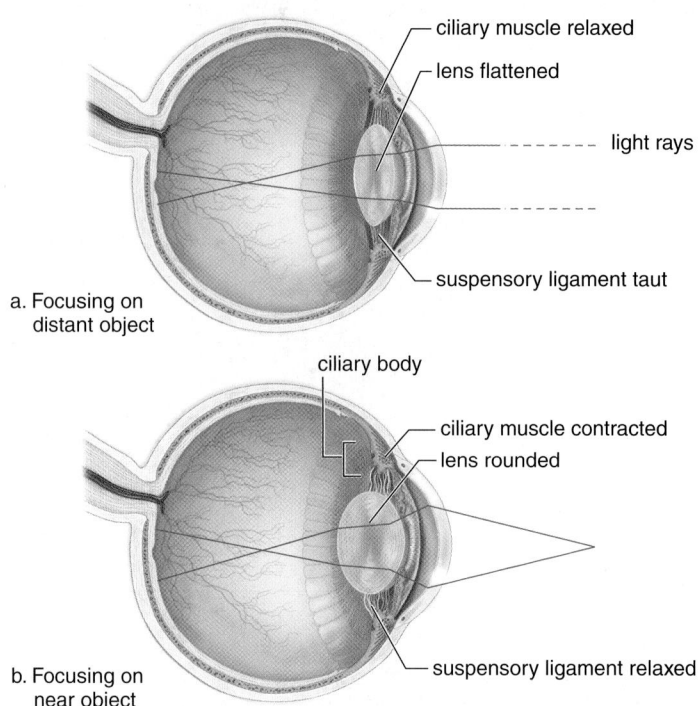

a. Focusing on distant object

b. Focusing on near object

Figure 38.6 Focusing of the human eye. Light rays from each point on an object are bent by the cornea and the lens in such a way that an inverted and reversed image of the object forms on the retina. **a.** When focusing on a distant object, the lens is flat because the ciliary muscle is relaxed and the suspensory ligament is taut. **b.** When focusing on a near object, the lens accommodates; that is, it becomes rounded because the ciliary muscle contracts, causing the suspensory ligament to relax.

The visual pigment in rods is called rhodopsin. **Rhodopsin** is a complex molecule made up of the protein opsin and a light-absorbing molecule called *retinal*, which is a derivative of vitamin A. When a rod absorbs light, rhodopsin splits into opsin and retinal, leading to a cascade of reactions and the closure of ion channels in the rod cell's plasma membrane. The release of

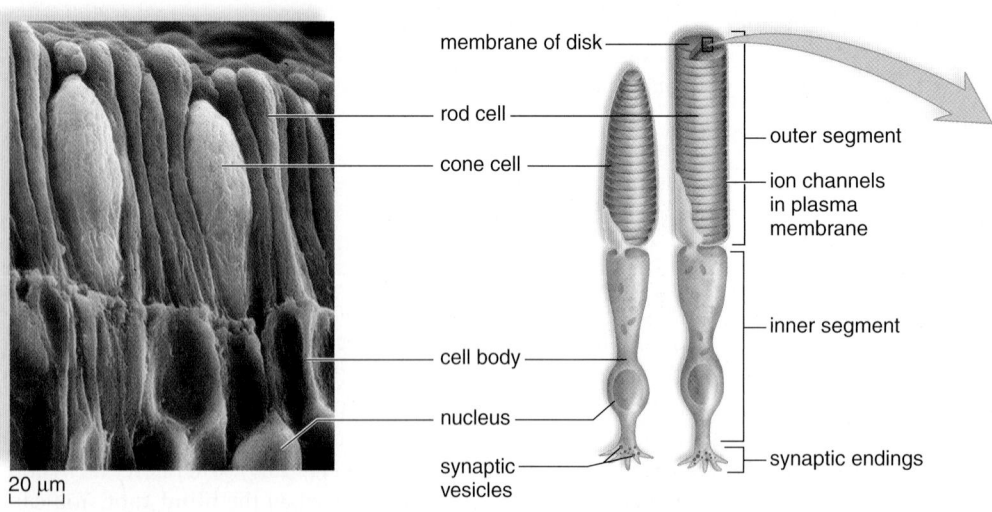

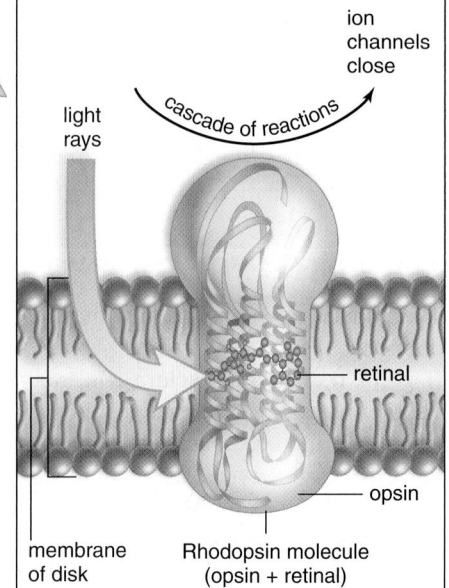

Figure 38.7 Photoreceptors in the eye. The outer segment of rods and cones contains stacks of membranous disks, which contain visual pigments. In rods, the membrane of each disk contains rhodopsin, a complex molecule containing the protein opsin and the pigment retinal. When rhodopsin absorbs light energy, it splits, releasing opsin, which sets in motion a cascade of reactions that cause ion channels in the plasma membrane to close. Thereafter, nerve impulses go to the brain.

inhibitory transmitter molecules from the rod's synaptic vesicles ceases. Thereafter, nerve impulses go to the visual areas of the cerebral cortex. Rods are very sensitive to light and, therefore, are suited to night vision. Carrots and other brightly-colored vegetables are rich in *carotenoids*, some of which can be easily converted to vitamin A in the body, so eating these foods may improve your night vision. Rod cells are plentiful in the peripheral region of the retina; therefore, they also provide us with peripheral vision and perception of motion.

The cones, by contrast, are located primarily in the fovea centralis and are activated by bright light. They allow us to detect the fine detail and the color of an object. Color vision depends on three different kinds of cones, which contain pigments called the B (blue), G (green), and R (red) pigments. Each pigment is made up of retinal and opsin, but a slight difference is present in the opsin structure of each, which accounts for their individual absorption patterns. Various combinations of cones are believed to be stimulated by in-between shades of color. For example, the color yellow is perceived when green cones are highly stimulated, red cones are partially stimulated, and blue cones are not stimulated. In color blindness, an individual lacks certain visual pigments. As indicated in Chapter 11, color blindness is often a sex-linked hereditary disorder; however, the Nature of Science feature, "Inbreeding in Populations," in Chapter 16 describes a case of non-sex-linked color blindness.

Integration of Visual Signals in the Retina

The retina has three layers of neurons (Fig. 38.8). The layer closest to the choroid contains the rod cells and cone cells; the middle layer contains bipolar cells; and the innermost layer contains ganglion cells, whose sensory fibers become the optic nerve. Only the rod cells and the cone cells are sensitive to light, and therefore light must penetrate to the back of the retina before they are stimulated.

The rod cells and the cone cells synapse with the bipolar cells, which in turn synapse with ganglion cells that initiate nerve impulses. Notice in Figure 38.8 that there are many more rod cells and cone cells than ganglion cells. In fact, the human retina has about 150 million rod cells and 6 million cone cells, but only 1 million ganglion cells.

The sensitivity of cones versus rods is mirrored by how directly they connect to ganglion cells. As many as 150 rods may excite the same ganglion cell, meaning that each ganglion cell receives signals from rod cells covering about 1 mm² of retina (about the size of a thumbtack hole). Therefore, stimulation of rods results in vision that is blurred and indistinct. In contrast, some cone cells in the fovea centralis excite only one ganglion cell. This explains why cones provide us with a sharper, more detailed image of an object.

Animation Vision

MP3 Sense of Vision

Figure 38.8 Structure and function of the retina.
a. The retina is the inner layer of the eyeball. Rod cells and cone cells, located at the back of the retina nearest the choroid, synapse with bipolar cells, which synapse with ganglion cells. Further, notice that many rod cells share one bipolar cell, but cone cells do not. Certain cone cells synapse with only one ganglion cell. Cone cells, in general, distinguish more detail than do rod cells. **b.** Micrograph of retina.

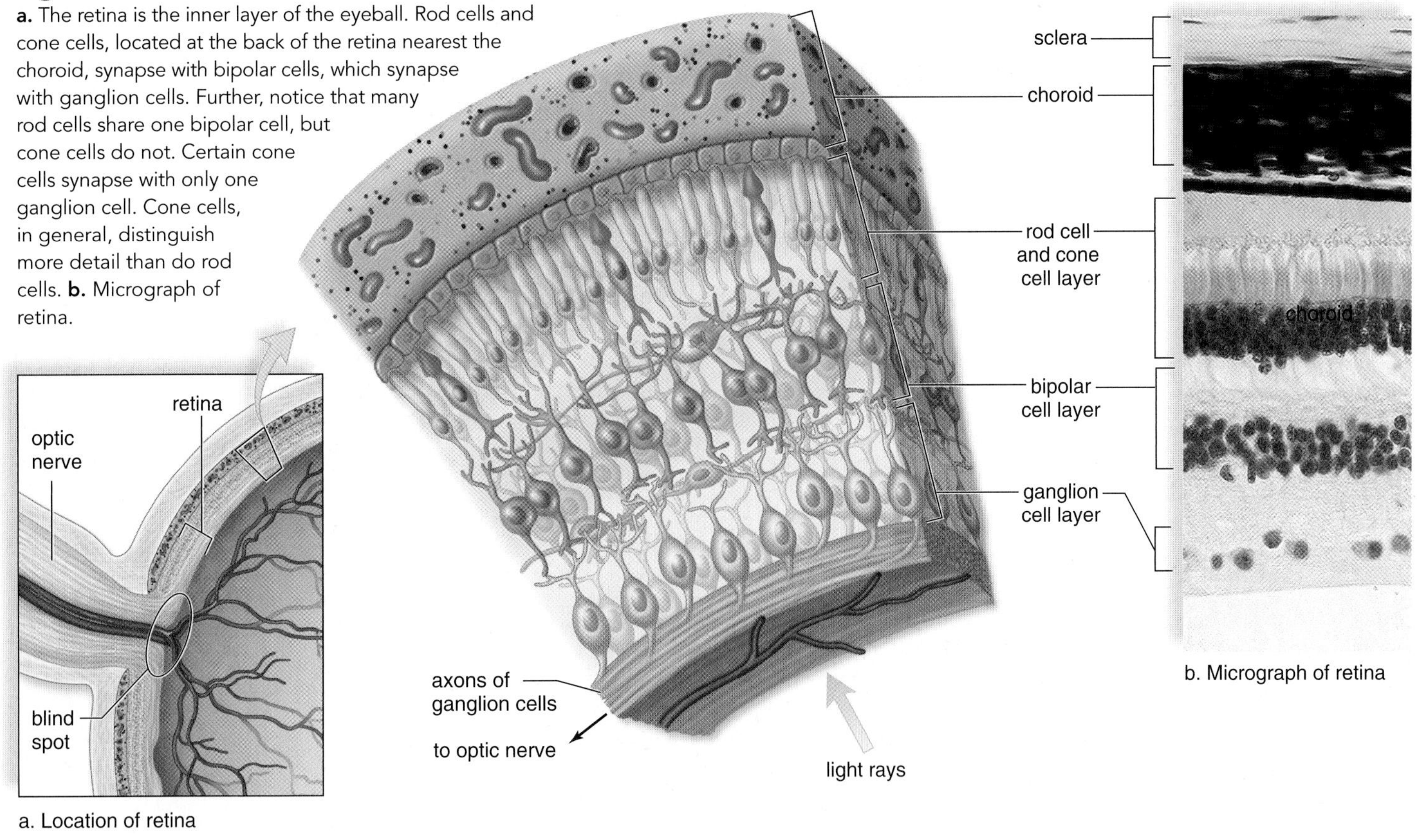

optic nerve

retina

blind spot

a. Location of retina

axons of ganglion cells

to optic nerve

light rays

sclera

choroid

rod cell and cone cell layer

bipolar cell layer

ganglion cell layer

choroid

b. Micrograph of retina

Disorders of Vision

Disorders of vision include diseases of the retina, glaucoma and cataracts, and problems with visual focus.

Diseases of the Retina

Diseases of the retina are the most common cause of blindness in adults. These include **diabetic retinopathy,** in which capillaries to the retina become damaged secondary to diabetes. In **macular degeneration,** the leading cause of blindness for people in the U.S. under the age of 65, capillaries supplying the retinas become damaged, and hemorrhages and blocked vessels can occur. Smokers are 20 times more likely to acquire this disorder than are nonsmokers. If diagnosed early, both of these retinal diseases can be treated, using drugs that are injected directly into the vitreous humor. With **retinal detachment,** the retina peels away from the supportive choroid layer, due to eye trauma or other diseases. A detached retina can often be reattached using a laser. As noted in the Nature of Science feature, artificial retina technology is becoming an option for some blind people whose visual pathways remain intact.

Glaucoma and Cataracts

Glaucoma is the second most common cause of blindness in the United States. Glaucoma occurs when the drainage system of the eyes fails, so that aqueous humor builds up and increases intraocular pressure. In the early stages, this pressure tends to destroy nerve fibers responsible for peripheral vision, but untreated glaucoma can result in total blindness. Eye doctors always check intraocular pressure, but the disorder can come on quickly, and any nerve damage is permanent. Treatment usually involves drugs that increase the outflow of aqueous humor, but surgery may be required.

With aging, the eye is increasingly subject to **cataracts,** in which the lens can become opaque and therefore incapable of transmitting light rays. Cataracts occur in 50% of people between the ages of 65 and 74, and in 70% of those ages 75 or older. In most cases, vision can be restored by surgical removal of the unhealthy lens and replacement with a clear plastic artificial lens. Cataract surgery is the most frequently performed surgery in the United States. Surgeons may be able to replace cataracts with multifocal lenses that allow the eye to focus, eliminating the need for glasses in some patients.

Visual Focus Disorders

People who can read what are designated as size 20 letters on an optometrist's chart 20 ft away are said to have 20/20 vision. Those who cannot read these letters but can focus on close objects are said to **nearsighted (myopic).** These individuals often have an elongated eyeball, and when they attempt to look at a distant object, the image is brought to focus in front of the retina. Concave lenses, which diverge the light rays so that the image can be focused on the retina, usually correct this problem (Fig. 38.9a).

Those who can easily read the optometrist's chart but cannot easily focus on near objects are said to be **farsighted (hyperopic).** They often have a shortened eyeball, and when they try to view near objects, the image is focused behind the

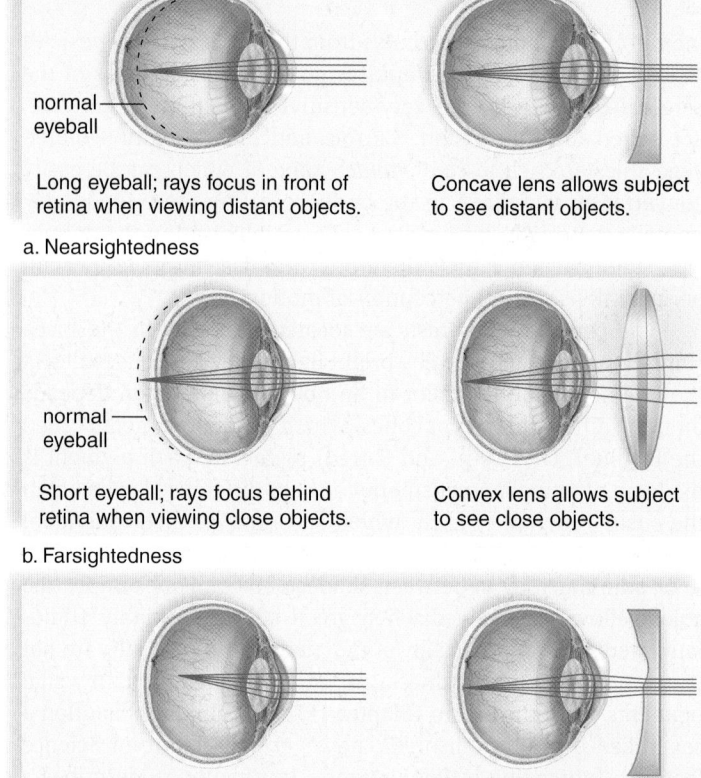

a. Nearsightedness

Long eyeball; rays focus in front of retina when viewing distant objects.

Concave lens allows subject to see distant objects.

b. Farsightedness

Short eyeball; rays focus behind retina when viewing close objects.

Convex lens allows subject to see close objects.

c. Astigmatism

Uneven cornea; rays do not focus evenly.

Uneven lens allows subject to see objects clearly.

Figure 38.9 Common abnormalities of the eye with possible corrective lenses. **a.** A concave lens in nearsighted persons focuses light rays on the retina. **b.** A convex lens in farsighted persons focuses light rays on the retina. **c.** An uneven lens in persons with astigmatism focuses light rays on the retina.

retina. Convex lenses that increase the bending of light rays allow the image to be focused on the retina (Fig. 38.9b).

When the cornea or lens is uneven, the image appears fuzzy. This condition, called **astigmatism,** can sometimes be corrected by an unevenly ground lens to compensate for the uneven cornea (Fig. 38.9c).

Rather than wear glasses or contact lenses, many people with visual focus problems are now choosing to undergo laser-assisted in situ keratomileusis, or LASIK. First, specialists determine how much the cornea needs to be flattened to achieve optimum vision. Controlled by a computer, the laser then removes this amount of the cornea. Various surveys indicate that about 95% of people who have the LASIK procedure are satisfied with the results.

Check Your Progress 38.3

1. Compare rods and cones in terms of main functions, light sensitivity, and excitation of ganglion cells.
2. List the three layers of cells in the human retina. Why do you think this arrangement is sometimes described as "backwards"?
3. Summarize how the shape of the eyeball can result in nearsightedness or farsightedness. Also, describe the cause of astigmatism.

Nature of Science

Artificial Retinas Come into Focus

Over 25 million people worldwide are blind due to diseases of the retinas. In most cases, there are no cures and few effective treatments. However, after years of work, medical researchers have now developed artificial retina technology that is restoring some vision for people who were completely blind.

Retinal diseases like macular degeneration cause the rods and cones to die but leave the ganglion cells and neurons intact. In 1988, researchers first demonstrated that a blind person could see light if the nerve cells behind the retina were stimulated with an electrical current. Since then, scientists have been working on developing technological approaches that could take the place of the diseased photoreceptors. In March 2011, a California company received approval in Europe to sell the Argus II, a retinal prosthesis designed for implantation in blind patients who still have intact connections between their retinas and brain.

Priced at about $115,000, the device includes a tiny digital camera embedded in a pair of glasses that are worn by the patient. Images from the camera are translated into electrical signals representing patterns of light and dark. These signals are transmitted wirelessly to a receiver implanted above the ear or near the eye. This receiver in turn sends signals via a tiny wire that attaches to a 1 mm × 1 mm microchip that has been surgically implanted under the retina (Fig. 38A). The chip contains 60 electrodes that can stimulate the ganglion cells, producing electrical impulses that travel via the optic nerve to the brain, where they are interpreted as vision.

Although it takes time for patients to learn to interpret the patterns of light and dark transmitted by the device, some formerly blind patients have been able to recognize simple objects, see people in front of them and follow their movement, and even read large print slowly. The Argus II device may be approved for use in patients in the United States by 2012. Moreover, researchers are already working on upgraded models that incorporate up to 1,500 electrodes, to increase the resolution of the image produced. This is anticipated to be the threshold of information that will be required for formerly blind patients to recognize faces. Other groups are working on artificial retinas that can be implanted inside the eye itself.

Although the amount of information that is transmitted by an artificial retina device is important, other scientists are focusing on the type of information that is transmitted. A 2010 study conducted at Weill Cornell Medical College in New York by investigators Shiela Nirenberg and Chethan Pandarinath demonstrated that a different approach could restore normal vision in blind mice. First, these scientists focused on deciphering the patterns by which the ganglion cells normally transmit information from the photoreceptors to the brain. Next, they tested these codes in blind mice that had been genetically altered so that their ganglion cells express a protein called channelrhodopsin, which initiates a nerve impulse when it is exposed to light. That way the investigators didn't have to implant artificial retinas in the mice, but instead exposed the mice to different patterns of flashing light. The results of these experiments showed that the mice receiving light information using the proper code appeared to be able to see at nearly normal levels (Fig. 38A).

While eliminating the need for surgery is a major advantage of this approach, it is impossible (or at least unethical) to genetically engineer humans to express certain genes. The solution to this problem may lie in gene therapy, in which the ganglion cells of blind patients would be treated so that they express the *channelrhodopsin* gene. If that approach is successful, patients would again wear a pair of glasses with an embedded camera, only this time the camera would emit light that had been translated into the patterns that the brain can understand.

Questions to Consider

1. Although this technology is very exciting, it is also expensive. Do you believe that health insurance companies should pay the $115,000 cost of an artificial retina?
2. Suppose this technology advances to the point where vision can not only be restored, but also greatly enhanced. Should it then be legally available to anyone who can afford it, for example, highly paid professional baseball players? What about for children who want to be professional athletes?

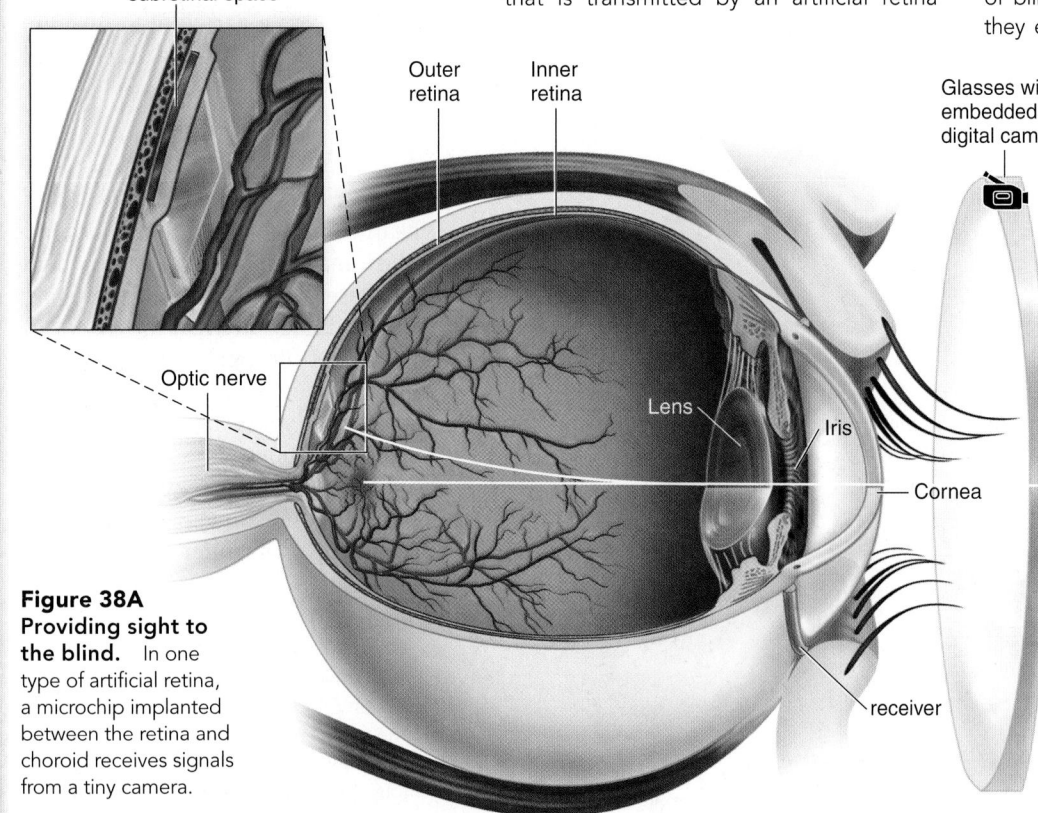

Figure 38A Providing sight to the blind. In one type of artificial retina, a microchip implanted between the retina and choroid receives signals from a tiny camera.

38.4 Senses of Hearing and Balance

Learning Outcomes

Upon completion of this section, you should be able to

1. Compare several strategies that different types of animals use to hear sounds.
2. Distinguish the parts of the human ear that make up the outer, middle, and inner ear.
3. Review the differences between rotational and gravitational equilibrium, and how each is accomplished.
4. Describe several common disorders affecting hearing and/ or equilibrium.

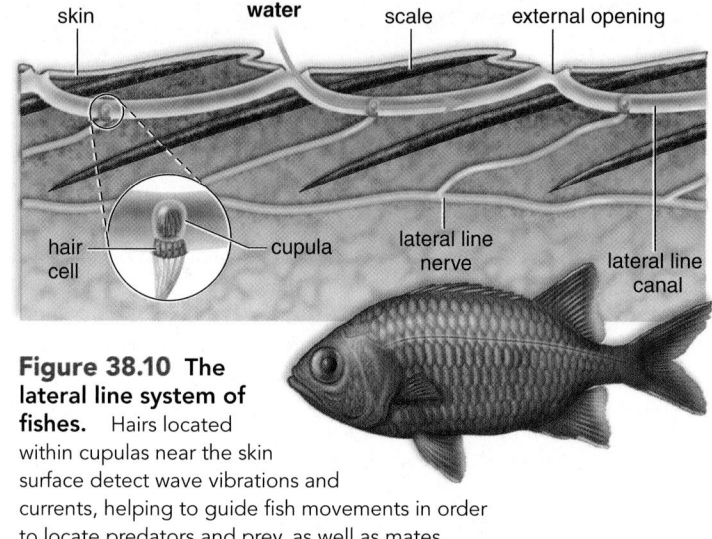

Figure 38.10 The lateral line system of fishes. Hairs located within cupulas near the skin surface detect wave vibrations and currents, helping to guide fish movements in order to locate predators and prey, as well as mates.

Mechanoreception is sensing physical contact on the surface of the skin or movement of the surrounding environment (such as sound waves in air or water). The simplest mechanoreceptors are free nerve endings found in the skin. At the other end of the spectrum, the most complex mechanoreception occurs in the middle and inner ear of vertebrates.

The evolutionary advantage of hearing is that it allows animals to receive information at a distance, and from any direction. Hearing plays an important role in avoiding danger, detecting prey, finding mates, and communication. In the most basic sense, hearing is caused by the vibration in a surrounding medium to resonate some part of an animal's body. This resonance is converted into electrical signals through some means that can then be interpreted by the animal's brain.

How Animals Detect Sound Waves

Many insects can detect sounds. A common structure involved in insect hearing is a thin membrane, or tympanum, that stretches across an air space of some sort, such as the tracheae that also function in insect respiration (see Chapter 35). Tympanal organs are also located on the thorax (chest) of grasshoppers, and on the front

legs of crickets. Similar to a mammalian eardrum, the membrane is stimulated to vibrate by sound waves, but in insects, this directly activates nerve impulses in attached receptor cells.

The **lateral line** system of fishes (Fig. 38.10) guides them in their movements and in locating other fishes, including predators, prey, and mates. Usually running along both sides of a fish from the gills to the tail, the system detects water currents and pressure waves from nearby objects in a similar manner as the sensory receptors in the human ear. Water from the environment enters tiny canals that contain hair cells with cilia embedded in a gelatinous cupula. When the cupula bends due to pressure waves, the hair cells initiate nerve impulses.

Most terrestrial vertebrates can hear sound traveling in air, but some, like amphibians and snakes, are also sensitive to vibrations from the ground, which travel to their inner ear via various parts of their skeleton. A middle ear with an eardrum that transmits sound waves to the inner ear via three small bones is unique to mammals.

The Human Ear

The ear has two sensory functions: hearing and balance (equilibrium). The mechanoreceptors for both of these are located in the inner ear, and each consists of *hair cells* with stereocilia (long microvilli) that are sensitive to mechanical stimulation.

MP3 The Sense of Hearing and Equilibrium

The ear has three distinct divisions: the outer, inner, and middle ear (Fig. 38.11). The **outer ear** consists of

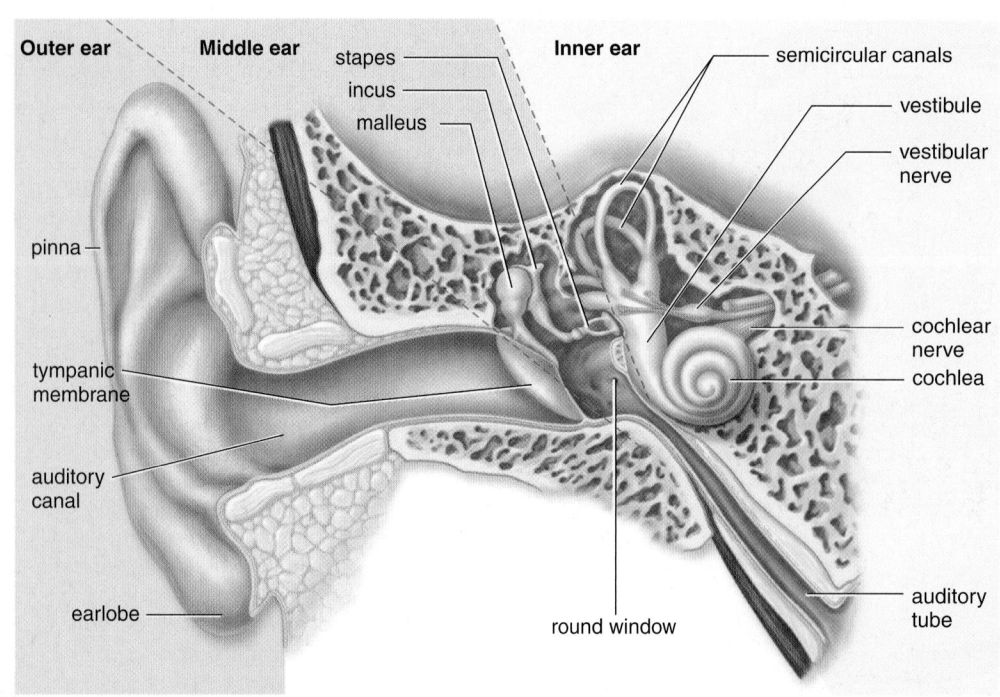

Figure 38.11 Anatomy of the human ear. In the middle ear, the malleus (hammer), the incus (anvil), and the stapes (stirrup) amplify sound waves. In the inner ear, the mechanoreceptors for equilibrium are in the semicircular canals and the vestibule, and the mechanoreceptors for hearing are in the cochlea.

the pinna (external "ear") and the auditory canal. The opening of the auditory canal is lined with fine hairs and glands. Glands that secrete earwax, or cerumen, are located in the upper wall of the auditory canal. Earwax helps guard the ear against the entrance of foreign materials, such as air pollutants and microorganisms.

The **middle ear** begins at the **tympanic membrane** (eardrum) and ends at a bony wall containing two small openings covered by membranes. These openings are called the *oval window* and the *round window*. Three small bones are found between the tympanic membrane and the oval window. Collectively called the **ossicles,** individually they are the *malleus* (hammer), the *incus* (anvil), and the *stapes* (stirrup) because their shapes resemble these objects. The malleus adheres to the tympanic membrane, and the stapes touches the oval window.

An **auditory tube** (eustachian tube), which extends from each middle ear to the nasopharynx, permits equalization of air pressure. Chewing gum, yawning, and swallowing in elevators and airplanes help move air through the auditory tubes upon ascent and descent. As this occurs, we often hear the ears "pop."

Whereas the outer ear and the middle ear contain air, the inner ear is filled with fluids. Anatomically speaking, the **inner ear** has three areas: the **semicircular canals** and the **vestibule** are both concerned with equilibrium; the **cochlea** is concerned with hearing. The cochlea resembles the shell of a snail because it spirals.

The Auditory Canal and Middle Ear

The process of hearing begins when sound waves enter the auditory canal. Just as ripples travel across the surface of a pond, sound waves travel by the successive vibrations of molecules. Ordinarily, sound waves do not carry much energy, but when a large number of waves strike the tympanic membrane, it moves back and forth (vibrates) ever so slightly. The malleus then takes the pressure from the inner surface of the tympanic membrane and passes it by means of the incus to the stapes in such a way that the pressure is multiplied about 20 times as it moves. The stapes strikes the membrane of the oval window, causing it to vibrate, and in this way, the pressure is passed to the fluid within the cochlea.

Inner Ear

When the snail-shaped cochlea is examined in cross section (Fig. 38.12), the vestibular canal, the cochlear canal, and the tympanic canal become apparent. The cochlear canal contains endolymph, which is similar in composition to tissue fluid. The vestibular and tympanic canals are filled with perilymph, which is continuous with the cerebrospinal fluid. Along the length of the basilar membrane, which forms the lower wall of the *cochlear canal*, are little hair cells whose stereocilia are embedded within a gelatinous material called the *tectorial membrane.* The hair cells of the cochlear canal, called the **organ of Corti,** or spiral organ, synapse with nerve fibers of the *cochlear nerve* (auditory nerve).

When the stapes strikes the membrane of the oval window, pressure waves move from the vestibular canal to the tympanic canal across the basilar membrane, and the round window membrane bulges. The basilar membrane moves up and down,

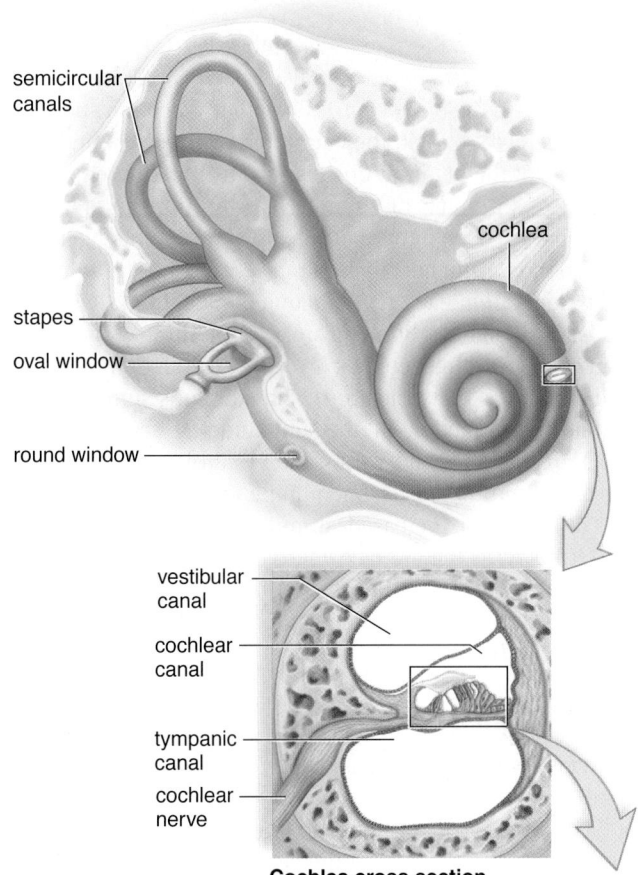

Cochlea cross section

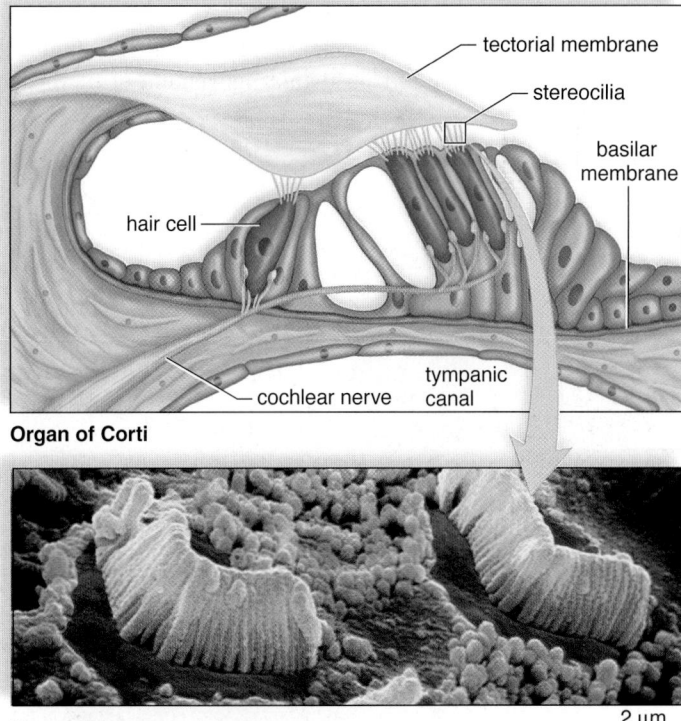

Organ of Corti

Stereocilia   2 μm

Figure 38.12 Mechanoreceptors for hearing. The organ of Corti is located within the cochlea. In the uncoiled cochlea, note that the organ consists of hair cells resting on the basilar membrane, with the tectorial membrane above. Pressure waves move from the vestibular canal to the tympanic canal, causing the basilar membrane to vibrate. This causes the stereocilia (or at least a portion of the more than 20,000 hair cells) embedded in the tectorial membrane to bend. Nerve impulses traveling in the cochlear nerve result in hearing.

and the stereocilia of the hair cells embedded in the tectorial membrane bend. Then nerve impulses begin in the cochlear nerve and travel to the brain stem. When they reach the auditory areas of the cerebral cortex, they are interpreted as a sound.

Each part of the organ of Corti is sensitive to different wave frequencies, or the pitch of sound. Near the tip, the organ of Corti responds to low pitches, such as a tuba, and near the base, it responds to higher pitches, such as a bell or a whistle. The nerve fibers from each region along the length of the organ of Corti lead to slightly different areas in the brain. The pitch sensation we experience depends on which region of the basilar membrane vibrates and which area of the brain is stimulated.

Volume is a function of the amplitude of sound waves. Loud noises cause the fluid within the vestibular canal to exert more pressure and the basilar membrane to vibrate to a greater extent. The resulting increased stimulation is interpreted by the brain as volume. It is believed that the brain interprets the tone of a sound based on the distribution of the hair cells stimulated.

Animation
Hearing

Animation
Effect of Sound Waves
on Cochlear Structures

Sense of Balance

Gravitational equilibrium organs, called statocysts (Fig. 38.13), are found in cnidarians, molluscs, and crustaceans, which are arthropods. When the head stops moving, a small particle called a statolith stimulates the cilia of the closest hair cells, and these cilia generate impulses that are interpreted as the position of the head.

In the human ear, mechanoreceptors in the semicircular canals detect rotational and/or angular movement of the head

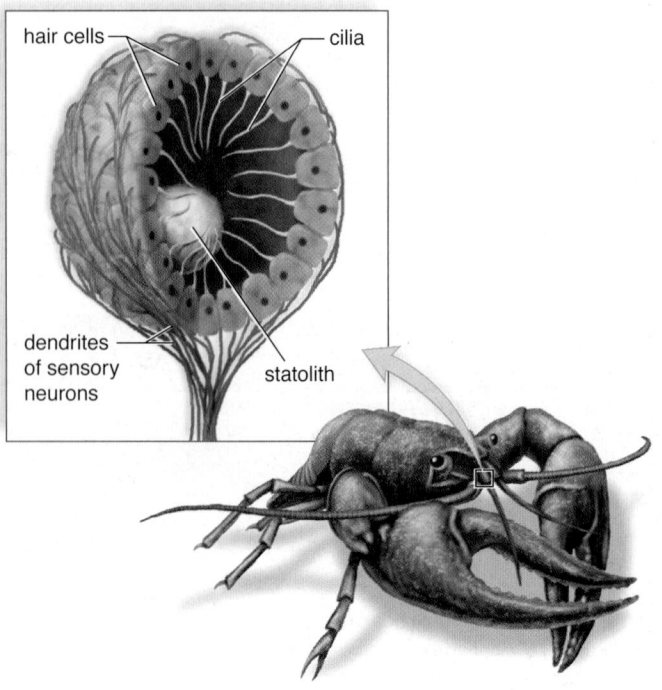

Figure 38.13 Sense of balance in an invertebrate. Within a statocyst, a small particle (the statolith) comes to rest on hair cells and allows a crustacean to sense the position of its head.

(i.e., **rotational equilibrium**), while mechanoreceptors in the utricle and saccule detect straight-line movement of the head in any direction (i.e., **gravitational equilibrium**).

Rotational Equilibrium

Rotational equilibrium (Fig. 38.14a) involves the semicircular canals, which are arranged so that there is one in each dimension of space. The base of each of the three canals, called the ampulla, is slightly enlarged. Little hair cells, whose stereocilia are embedded within a gelatinous material called a cupula, are found within the ampullae. Because there are three semicircular canals, each ampulla responds to head movement in a different plane of space. As fluid (endolymph) within a semicircular canal flows over and displaces a cupula, the stereocilia of the hair cells bend, and the pattern of impulses carried by the vestibular nerve to the brain changes. The brain uses information from the semicircular canals to maintain equilibrium through appropriate motor output to various skeletal muscles that can right our position in space as need be.

Gravitational Equilibrium

Gravitational equilibrium (Fig. 38.14b) depends on the **utricle** and **saccule,** two membranous sacs located in the vestibule. Both of these sacs contain little hair cells, whose stereocilia are embedded within a gelatinous material called an otolithic membrane. Calcium carbonate ($CaCO_3$) granules, or **otoliths,** rest on this membrane. The utricle is especially sensitive to horizontal (back and forth) movements of the head, while the saccule responds best to vertical (up and down) movements.

When the head is still, the otoliths in the utricle and the saccule rest on the otolithic membrane above the hair cells. When the head moves in a straight line, the otoliths are displaced and the otolithic membrane sags, bending the stereocilia of the hair cells beneath. If the stereocilia move toward the largest stereocilium, called the kinocilium, nerve impulses increase in the vestibular nerve. If the stereocilia move away from the kinocilium, nerve impulses decrease in the vestibular nerve. If you are upside down, nerve impulses in the vestibular nerve cease. These data tell the brain the direction of the movement of the head.

Disorders of Hearing and Equilibrium

Hearing loss can develop gradually or suddenly and has many potential causes. Especially in children, the middle ear is subject to infections that, in severe cases, can lead to hearing impairment. Age-associated hearing loss usually develops gradually, beginning at around age 20. By age 60, about one in three people report significant hearing loss.

Most cases of hearing loss can be attributed to the effect of years of frequent (and preventable) exposure to loud noise that can damage the stereocilia of the spiral organ. Prolonged exposure to noise above a level of 80 decibels can damage the hair cells of the organ of Corti (Fig. 38.15). The first signs of noise-induced hearing loss are usually muffled hearing, pain or a "full" feeling in the ears, or tinnitus (ringing in the ears). If you have any of these symptoms, take steps immediately to prevent

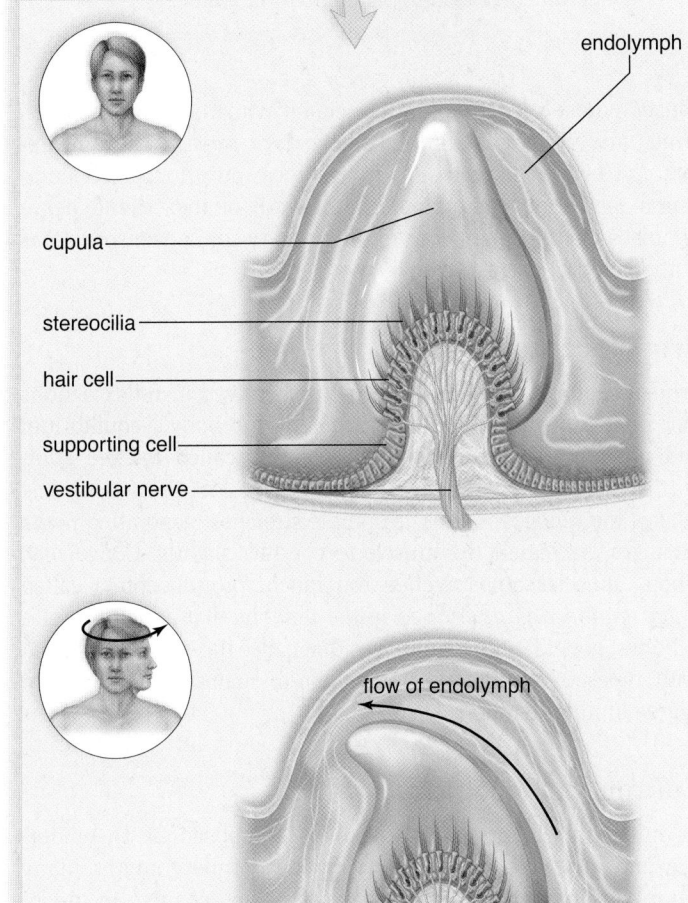

semicircular canals

ampullae

receptor in ampulla

vestibular nerve

cochlea

endolymph

cupula

stereocilia

hair cell

supporting cell

vestibular nerve

flow of endolymph

a. Rotational equilibrium: receptors in ampullae of semicircular canal

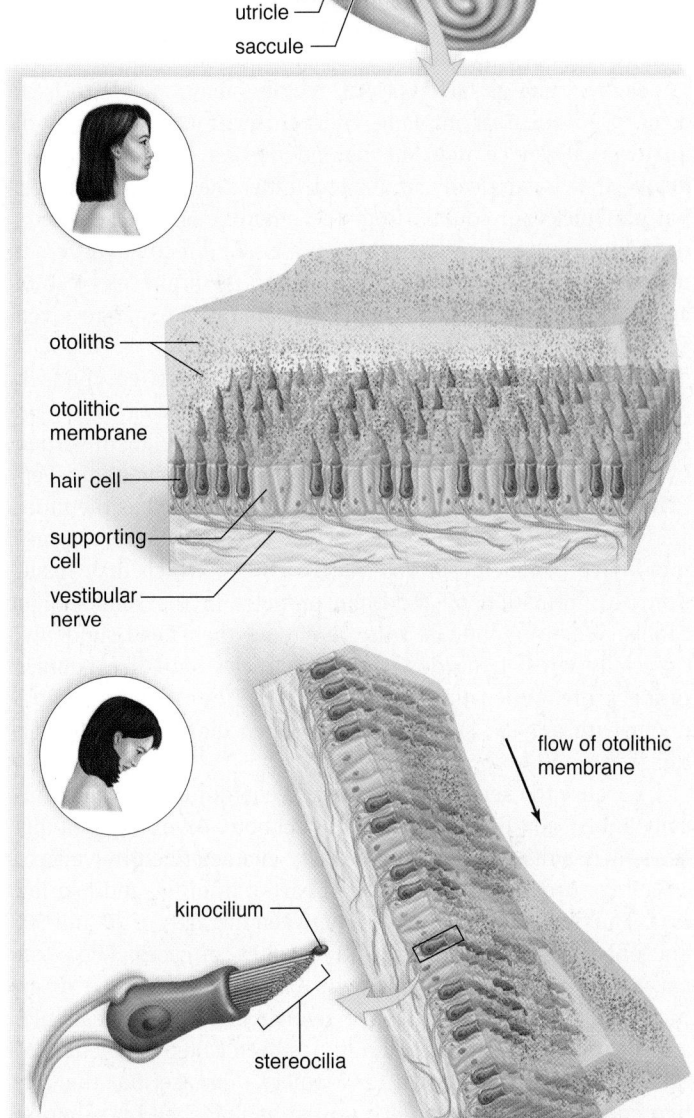

endolymph

utricle

saccule

otoliths

otolithic membrane

hair cell

supporting cell

vestibular nerve

flow of otolithic membrane

kinocilium

stereocilia

b. Gravitational equilibrium: receptors in utricle and saccule of vestibule

Figure 38.14 Mechanoreceptors for equilibrium.
a. Rotational equilibrium. The ampullae of the semicircular canals contain hair cells with stereocilia embedded in a cupula. When the head rotates, the cupula is displaced, bending the stereocilia. Thereafter, nerve impulses travel in the vestibular nerve to the brain. **b.** Gravitational equilibrium. The utricle and the saccule contain hair cells with stereocilia embedded in an otolithic membrane. When the head bends, otoliths are displaced, causing the membrane to sag and the stereocilia to bend. If the stereocilia bend toward the kinocilium, the longest of the stereocilia, nerve impulses increase in the vestibular nerve. If the stereocilia bend away from the kinocilium, nerve impulses decrease in the vestibular nerve. This difference tells the brain in which direction the head moved.

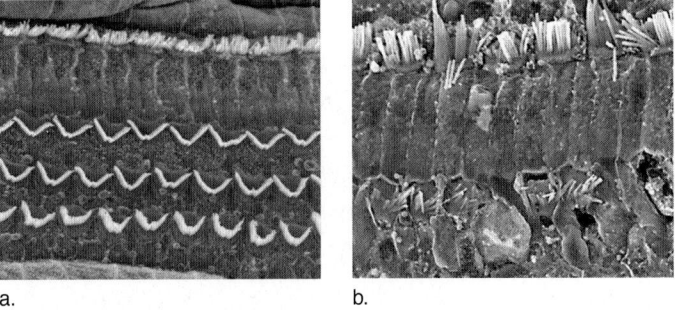

a.

b.

Figure 38.15 Studies of hearing loss. **a.** Microscopic view of normal hair cells in the organ of Corti of a guinea pig. **b.** Damage to these hair cells occurred after 24 hours of exposure to a noise level typical of a rock concert.

further damage. If exposure to loud noise is unavoidable, noise-reducing earmuffs or earplugs are available.

Some types of deafness can be present at birth. Several genetic disorders can interfere with the ability to hear, as can infections with German measles (rubella) or mumps virus when contracted by a woman during her pregnancy. For this reason, every girl should be vaccinated against these viruses before she reaches childbearing age.

People with certain types of deafness may be able to hear again with cochlear implants. A cochlear implant consists of an external device that sits behind the ear, and an internal device that is surgically implanted under the skin. The external part picks up sounds from the environment and converts them into electrical impulses, which are sent directly to different regions of the auditory nerve and then to the brain. As of 2010, about 22,000 adults and 15,000 children in the U.S. had received cochlear implants.

Disorders of equilibrium often manifest as **vertigo**, the feeling that a person or their environment is moving when no motion is occurring. It is possible to simulate a feeling of vertigo by spinning your body rapidly and then stopping suddenly. Vertigo can be caused by problems in the brain as well as the inner ear. An estimated 20% of those who experience these symptoms have benign positional vertigo (BPV), which may result from the formation of abnormal particles in the semicircular canals. When individuals with BPV move their head suddenly, especially when lying down, these particles shift like pebbles inside a tire. When the movement stops, the particles tumble down with gravity, stimulating the stereocilia and resulting in the sensation of movement.

Because the senses of hearing and equilibrium are anatomically linked, certain disorders can affect both. An example of this is Meniere's disease, which is usually characterized by vertigo, a feeling of fullness in the affected ear(s), tinnitus, and hearing loss. The disease usually strikes between the ages of 20 and 50, and only one ear is affected in about 80% of cases. The exact cause of this disease is unknown, but it seems related to an increased volume of fluid in the semicircular canals, vestibule, and/or cochlea. For this reason, it has been called "glaucoma of the ear." No cure is known for Meniere's disease, but the vertigo and feeling of pressure can often be managed by adhering to a low-salt diet, which is thought to decrease the amount of fluid present. Any hearing loss that occurs, however, is usually permanent.

Check Your Progress   38.4

1. Determine whether each of the following belongs to the outer, middle, or inner ear: **a.** ossicles, **b.** pinna, **c.** semicircular canals, **d.** cochlea, **e.** vestibule, **f.** auditory canal.
2. List, in order, the structures that must conduct a sound wave from the time it enters the auditory canal until it reaches the cochlea.
3. Identify which structures of the inner ear are responsible for gravitational equilibrium, and for rotational equilibrium.

38.5 Somatic Senses

Learning Outcomes

Upon completion of this section, you should be able to

1. Compare and contrast the functions of proprioceptors, cutaneous receptors, and pain receptors.
2. List specific types of cutaneous receptors that are sensitive to fine touch, pressure, pain, and temperature.

Senses whose receptors are associated with the skin, muscles, joints, and viscera are termed the *somatic senses*. These receptors can be categorized into three types: proprioceptors, cutaneous receptors, and pain receptors. All of these send nerve impulses via the spinal cord to the primary somatosensory areas of the cerebral cortex (see Fig. 37.9).

Proprioceptors

Proprioceptors are mechanoreceptors involved in reflex actions that maintain muscle tone, and thereby the body's equilibrium and posture. For example, proprioceptors called *muscle spindles* are embedded in muscle fibers (Fig. 38.16). If a muscle relaxes too much, the muscle spindle stretches, generating nerve impulses that cause the muscle to contract slightly. Conversely, when muscles are stretched too much, proprioceptors called *Golgi tendon organs*, buried in the tendons that attach muscles to bones, generate nerve impulses that cause the muscles to relax. Both types of receptors act together to maintain a functional degree of muscle tone.

Cutaneous Receptors

As noted in Chapter 31, the skin is composed of an epidermis and a dermis (see Fig. 31.8). The dermis contains many **cutaneous receptors**, which make the skin sensitive to touch, pressure, pain, and temperature.

Four types of cutaneous receptors are sensitive to fine touch. *Meissner corpuscles* and *Krause end bulbs* are concentrated in the fingertips, palms, lips, tongue, nipples, penis, and clitoris. *Merkel disks* are found where the epidermis meets the dermis. A free nerve ending called a *root hair plexus* winds around the base of a hair follicle and fires if a hair is touched.

Two types of cutaneous receptors are sensitive to pressure. *Pacinian corpuscles* are onion-shaped sensory receptors that lie deep inside the dermis. *Ruffini endings* are encapsulated by sheaths of connective tissue and contain lacy networks of nerve fibers.

At least two types of free nerve endings in the epidermis are thermoreceptors. Both cold and warm receptors contain ion channels with activities that are affected by temperature. Cold receptors generate nerve impulses at an increased frequency as the temperature drops; warm receptors increase activity as temperature rises. Some chemicals (e.g., menthol) can stimulate cold receptors.

Figure 38.16 Muscle spindles and Golgi tendon organs. **①** When a muscle is stretched, muscle spindles send sensory nerve impulses to the spinal cord. **②** Motor nerve impulses from the spinal cord cause slight muscle contraction. **③** When tendons are stretched excessively, Golgi tendon organs cause muscle relaxation.

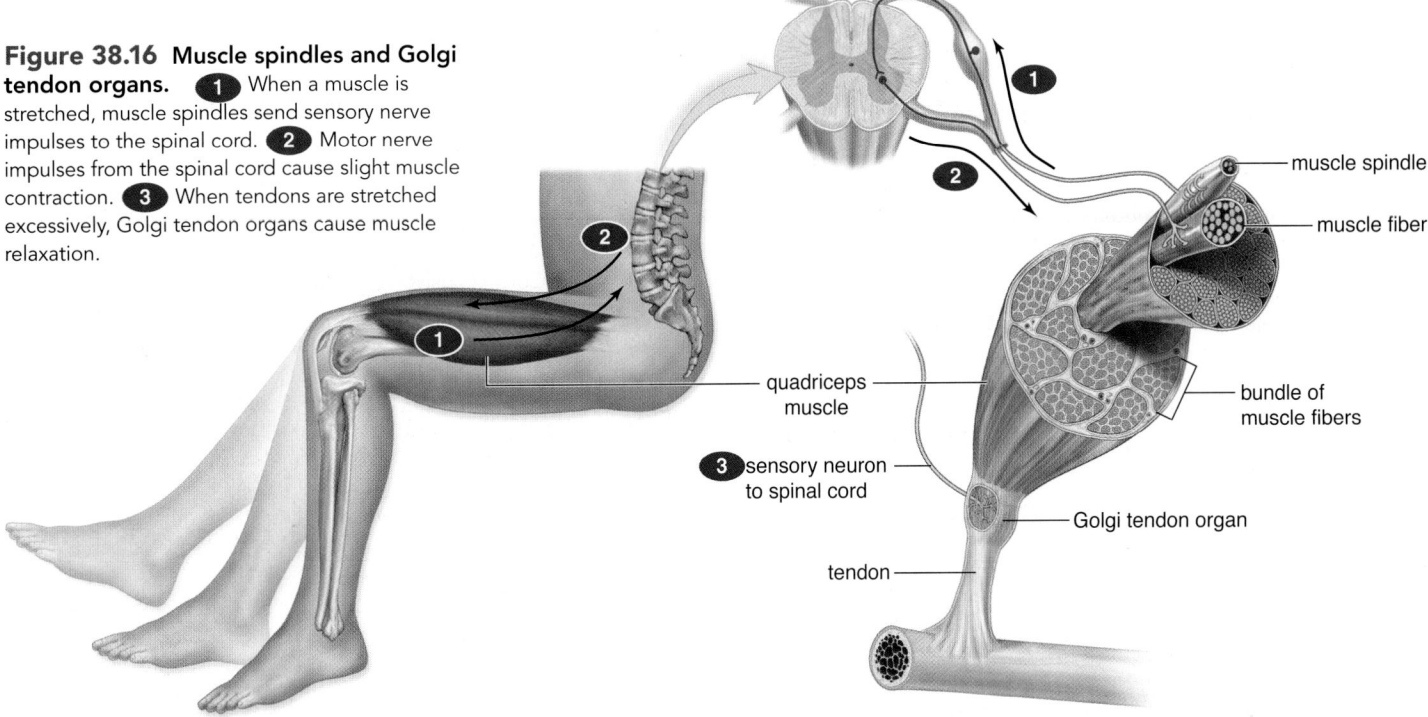

muscle spindle

muscle fiber

quadriceps muscle

bundle of muscle fibers

③ sensory neuron to spinal cord

Golgi tendon organ

tendon

Pain Receptors

The skin and many internal organs and tissues have pain receptors, also called free nerve endings or *nociceptors*. Regardless of the cause, damaged cells release chemicals that cause nociceptors to generate nerve impulses that are interpreted by the brain as pain. Other types of nociceptors are sensitive to extreme temperatures, or excessive pressure.

Pain receptors have arisen in evolution because they alert us to potential danger. If you accidentally reach too close to a fire, for example, the reflex action of withdrawing your hand will help protect you from further tissue damage, while the unpleasant sensation perceived by your brain will help you to remember not to reach too close to a fire again.

Check Your Progress 38.5

1. Identify the likely problems that would occur if a person lacked muscle spindles, or nociceptors.
2. In evolutionary terms, assess why cutaneous receptors quickly become adapted to stimuli (i.e., we don't continue to feel a chair once we settle in), while the sense of pain seems to be much less adaptable (i.e., many people suffer from chronic pain).

CONNECTING *the* CONCEPTS *with the* BIG IDEAS

Evolution

- Documented evolution of various sensory structures are seen in the animal kingdom. (1C3b*IE*)

*Find the unabridged version of all EK citations at www.glencoe.com/maderAP11.

Media Study Tools

www.glencoe.com/maderAP11

Enhance your study of this chapter with study tools and practice tests. Also ask your instructor about the resources available through ConnectPlus, including the media-rich eBook, interactive learning tools, and animations.

Summarize

38.1 Sensory Receptors

Sensory receptors evolved to enable animals to receive and respond to information about their internal and external environments. Sensory receptors are specialized cells capable of detecting various stimuli, then performing sensory transduction, or conversion of those events into nerve impulses. Four types of sensory receptors are chemoreceptors, photoreceptors, proprioceptors, and thermoreceptors.

38.2 Chemical Senses

Chemoreception is found universally in animals and is, therefore, believed to be the most primitive sense. Chemoreceptors are present in a wide variety of locations in animals. Examples of chemoreceptors in humans include taste buds and olfactory cells.

38.3 Sense of Vision

Most animals have photoreceptors that are sensitive to light. Arthropods have compound eyes; vertebrates and some molluscs have a camera-type eye. In humans, vision is dependent on the eye, the optic nerves, and the visual areas of the cerebral cortex. The human eye has three layers. The sclera can be seen as the white of the eye; it also becomes the transparent bulge in the front of the eye called the cornea. The middle choroid layer contains blood vessels and pigment that absorbs stray light rays. The rod cells (sensory receptors for dim light) and the cone cells (sensory receptors for bright light and color) are located in the retina, the inner layer.

The cornea and lens focus light rays on the retina. The retina is composed of three layers of cells: the rod and cone layer, the bipolar cell layer, and the ganglion cell layer. When light strikes rhodopsin within the membranous disks of rod cells, rhodopsin splits into opsin and retinal, generating nerve impulses that are transmitted from the ganglion cells, via the optic nerve, to the brain. To see a close object, accommodation occurs as the lens rounds up.

Common causes of blindness include retinal disorders, glaucoma, and cataracts. Problems with visual focus are usually managed with corrective lenses.

38.4 Senses of Hearing and Balance

Animals use a variety of strategies to detect sounds or vibrations in their environment. Hearing in humans is dependent on the ear, the cochlear nerve, and the auditory areas of the cerebral cortex. The ear is divided into three parts. The outer ear consists of the pinna and the auditory canal, which direct sound waves to the middle ear. The

middle ear begins with the tympanic membrane and contains the ossicles (malleus, incus, and stapes). The malleus is attached to the tympanic membrane, and the stapes is attached to the oval window, which is covered by membrane. The inner ear contains the cochlea and the semicircular canals, plus the utricle and saccule.

Hearing begins when the outer and middle portions of the ear convey and amplify the sound waves that strike the oval window. Its vibrations set up pressure waves within the cochlea, which contains the organ of Corti, consisting of hair cells whose stereocilia are embedded within the tectorial membrane. When the stereocilia of the hair cells bend, nerve impulses begin in the cochlear nerve and are carried to the brain.

The ear also contains receptors for our sense of equilibrium. Rotational equilibrium is dependent on the stimulation of hair cells within the ampullae of the semicircular canals. Gravitational equilibrium relies on the stimulation of hair cells within the utricle and the saccule. Disorders affecting hearing and balance include age-associated hearing loss, vertigo, and Meniere's disease.

38.5 Somatic Senses

The somatic sensory receptors include proprioceptors, sensory receptors, and pain receptors. Proprioceptors, such as muscle spindles and Golgi tendon organs, help maintain equilibrium and posture. Cutaneous receptors in the skin detect touch, pressure, pain, and temperature. Pain receptors are free nerve endings that respond to chemicals released by damaged cells.

Key Terms

astigmatism 724	olfactory cell 719
auditory tube 727	organ of Corti 727
blind spot 721	ossicle 727
camera-type eye 720	otolith 728
cataracts 724	outer ear 726
chemoreceptor 717	panoramic vision 721
choroid 721	photoreceptor 717, 720
ciliary muscle 722	proprioceptor 730
cochlea 727	pupil 721
compound eye 720	retina 721
cone cell 721	retinal detachment 724
conjunctiva 721	rhodopsin 722
cornea 721	rod cell 721
cutaneous receptor 730	rotational equilibrium 728
diabetic retinopathy 724	saccule 728
fovea centralis 721	sclera 721
glaucoma 724	semicircular canal 727
gravitational	sensory receptors 717
equilibrium 728	sensory transduction 717
farsighted (hyperopic) 724	stereoscopic vision 721
inner ear 727	taste bud 718
iris 721	thermoreceptor 717
lateral line 726	tympanic membrane 727
lens 721	utricle 728
macular degeneration 724	vertigo 730
mechanoreceptor 717	vestibule 727
middle ear 727	visual accommodation 722
nearsighted (myopic) 724	

Assess

Reviewing This Chapter

1. Discuss the structure and the function of human chemoreceptors.   718–19
2. In general, how does the eye in arthropods differ from that in humans?   720–21
3. Name the parts of the human eye, and give a function for each part.   721–23
4. Explain focusing and accommodation in terms of the anatomy of the human eye.   722
5. Contrast the location and the function of rod cells to those of cone cells.   722–23
6. Explain the process of integration in the retina and the brain.   723
7. Review some common disorders of the eye.   724
8. Describe some of the ways that animals detect sound waves.   726
9. Describe the structure of the human ear.   726–27
10. Describe how we hear.   727–28
11. Describe the role of the semicircular canals, utricle, and saccule in balance.   728
12. Discuss some common problems with hearing and equilibrium.   728–30
13. Distinguish between the functions of proprioceptors, cutaneous receptors, and pain receptors.   730–31

Testing Yourself

Choose the best answer for each question.

1. Which type of sensory receptor would be involved in detecting blood pressure?
   a. chemoreceptors
   b. mechanoreceptors
   c. photoreceptors
   d. thermoreceptors
2. Which of these gives the correct path for light rays entering the human eye?
   a. sclera, retina, choroid, lens, cornea
   b. fovea centralis, pupil, aqueous humor, lens
   c. cornea, pupil, lens, vitreous humor, retina
   d. optic nerve, sclera, choroid, retina, humors
   e. All of these are correct.
3. Which gives an incorrect function for the structure?
   a. lens—focusing
   b. iris—regulation of amount of light
   c. choroid—location of cones
   d. sclera—protection
   e. fovea centralis—acute vision
4. Retinal is
   a. a derivative of vitamin A.
   b. sensitive to light energy.
   c. a part of rhodopsin.
   d. found in rod cells.
   e. All of these are correct.
5. Which association is incorrect?
   a. taste buds—humans
   b. compound eye—arthropods
   c. camera-type eye—squid
   d. statocysts—sea stars
   e. chemoreceptors—planarians
6. Which one of these wouldn't you mention if you were tracing the path of sound vibrations?
   a. auditory canal
   b. tympanic membrane
   c. semicircular canals
   d. cochlea
   e. ossicles
7. Which one of these correctly describes the location of the organ of Corti?
   a. between the tympanic membrane and the oval window in the inner ear
   b. in the utricle and saccule within the vestibule
   c. between the tectorial membrane and the basilar membrane in the cochlear canal
   d. between the outer and inner ear within the semicircular canals
8. Which of these pairs is mismatched?
   a. semicircular canals—inner ear
   b. utricle and saccule—outer ear
   c. auditory canal—outer ear
   d. ossicles—middle ear
   e. cochlear nerve—inner ear
9. Both olfactory receptors and sound receptors have cilia, and they both
   a. are chemoreceptors.
   b. are a part of the brain.
   c. are mechanoreceptors.
   d. initiate nerve impulses.
   e. All of these are correct.
10. Stimulation of hair cells in the semicircular canals results from the movement of
    a. endolymph.
    b. aqueous humor.
    c. basilar membrane.
    d. otoliths.
11. To focus on objects that are close to the viewer,
    a. the suspensory ligaments must be pulled tight.
    b. the lens needs to become more rounded.
    c. the ciliary muscle will be relaxed.
    d. the image must focus on the area of the optic nerve.
12. Which abnormality of the eye is incorrectly matched?
    a. astigmatism—either the lens or cornea is not even
    b. farsightedness—eyeball is shorter than usual
    c. nearsightedness—image focuses behind the retina
    d. color blindness—genetic disorder in which certain types of cones may be missing
13. Which of these would allow you to know that you were upside down, even if you were in total darkness?
    a. utricle and saccule
    b. cochlea
    c. semicircular canals
    d. tectorial membrane

14. The thin, darkly pigmented layer that underlies most of the sclera is the
    a. conjunctiva.
    b. cornea.
    c. retina.
    d. choroid.

15. Adjustment of the lens to focus on objects close to the viewer is called
    a. convergence.
    b. accommodation.
    c. focusing.
    d. constriction.

16. Which of the following is/are not involved with detection of sound waves?
    a. tympanum of insects
    b. statocysts of crustaceans
    c. lateral line of fishes
    d. skeleton of amphibians and snakes

17. Label this diagram of the human eye. State a function for each structure labeled.

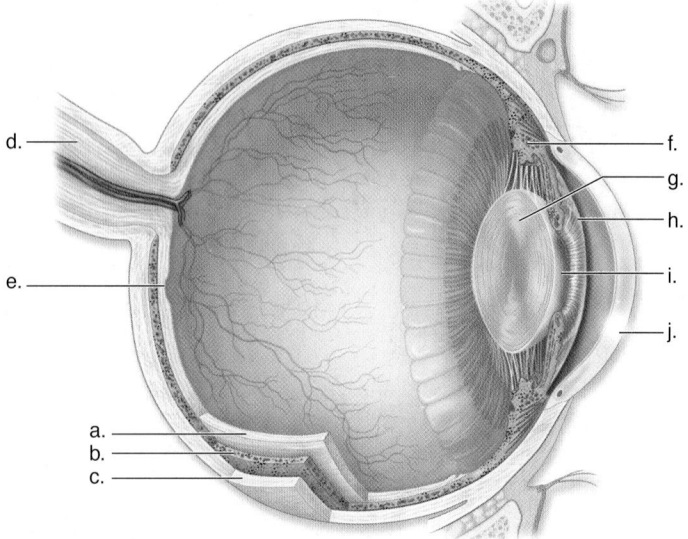

Engage

Thinking Scientifically

1. The density of taste buds on the tongue can vary. Some obese individuals have a lower density of taste buds than usual. Assume that taste perception is related to taste bud density. If so, what hypothesis would you test to see if there is a relationship between taste bud density and obesity?

2. A man who has spent many years serving on submarines complains of hearing loss, particularly the inability to hear high tones. When a submarine submerges, the inside air pressure intensifies. What hypothesis or hypotheses might explain hearing loss in this individual?

Bioethical Issue

Preventable Hearing Loss

The National Institute on Deafness and Other Communication Disorders (NIDCD) tells us that, of the approximately 28 million people in the United States with hearing loss, over one-third have been affected to some degree by noise exposure. Simple steps can be taken to prevent noise-induced hearing loss, such as wearing ear protection when engaged in loud activities such as mowing the lawn. Portable music devices are common culprits in hearing loss: If you can hear sound coming from someone's earbuds or headphones from a distance of 3 feet, that person is inflicting a damaging level of noise on him- or herself.

It's clear that for many of us, our own behavior contributes to the occurrence of hearing loss. Those with hearing loss may cope by using hearing aids, which can be costly and do not restore one's hearing to "normal." A device called a cochlear implant can compensate for damaged hair cells by directly stimulating the auditory nerve. However, the device must be surgically implanted, requires extensive postsurgical rehabilitation, does not replicate unassisted hearing, and is extremely expensive. Today, most health insurance provides at least partial assistance for hearing aids and cochlear implants. However, it is estimated that over the past 30 years, the number of people in the United States with hearing loss has doubled. Therefore, health insurance costs must rise in order to assist those with hearing loss.

What is your responsibility as an individual when it comes to protecting your own hearing? Does this responsibility change if you are in a group, say, taking a road trip in a car with the stereo turned all the way up? If you are aware that a friend is damaging his or her hearing in preventable ways, should you say anything? What is society's responsibility for educating its members about the importance of hearing protection? And does society have a responsibility to restrict sound levels at certain events, or to require that hearing protection be made available?

Gymnastics requires coordination between the nervous and support systems.

39

Locomotion and Support Systems

BEFORE YOU BEGIN

Before beginning this chapter, take a few moments to review the following discussions.

Figure 37.6 What is the function of acetylcholine in the transmission of a nerve impulse to skeletal muscle?

Sections 37.3 and 37.4 How does the primary motor area of the cerebral cortex generate commands to skeletal muscle, and how does the somatic division of the PNS control the muscles?

Chapter 38 How do the various types of sensory receptors provide information and feedback to the brain?

When Paul Hamm, a three-time Olympic medalist in gymnastics, does routines on rings, his muscular and skeletal systems are working together under the control of his nervous system. The same is true when eagles fly, fish swim, or when animals feed, escape prey, reproduce, or simply play. Although some animals lack muscles and bones, they all use contractile fibers to move about at some stage of their life. In many invertebrates, muscles push against body fluids located inside either a gastrovascular cavity or a coelom.

Only in vertebrates are muscles attached to a bony endoskeleton. Both the skeletal system and the muscular system contribute to homeostasis. Aside from giving the body shape and protecting internal organs, the skeleton serves as a storage area for inorganic calcium and produces blood cells. The skeleton also protects internal organs while supporting the body against the pull of gravity. While contributing to body movement, the skeletal muscles give off heat, which warms the body. This chapter compares locomotion in animals and reviews the musculoskeletal system of vertebrates.

As you read through the chapter, think about the following questions:

1. What advantages do animals with a skeletal system have over animals that completely lack such a system?

2. How does the nervous system specifically control the skeletal system?

3. How do the sensory systems exert influence over the nervous system, and therefore over the muscular and skeletal systems?

FOLLOWING *the* BIG IDEAS

CHAPTER 39   LOCOMOTION AND SUPPORT SYSTEMS

Interactions and Systems — The contraction of muscles is dependent upon their close relationship with the nervous system.

39.1 Diversity of Skeletons

Learning Outcomes

Upon completion of this section, you should be able to

1. Describe a typical hydrostatic skeleton, and list some examples of animals that possess one.
2. Discuss some advantages of having an endoskeleton versus an exoskeleton.
3. Provide several examples of how mammalian skeletons are adapted to particular forms of locomotion.

Skeletons serve as support systems for animals, providing rigidity, protection, and surfaces for muscle attachment. Several different kinds of skeletons occur in the animal kingdom. Cnidarians, flatworms, roundworms, and annelids have a hydrostatic skeleton. Typically, molluscs and arthropods have an exoskeleton (external skeleton) composed of calcium carbonate or chitin, respectively. Sponges, echinoderms, and vertebrates possess an endoskeleton (internal skeleton). In echinoderms, the endoskeleton is composed of calcareous plates and in vertebrates, the endoskeleton is comprised of cartilage, bone, or both.

Hydrostatic Skeleton

In animals that lack a hard skeleton, a fluid-filled gastrovascular cavity or a fluid-filled coelom can act as a hydrostatic skeleton. A **hydrostatic skeleton** utilizes fluid pressure to offer support and resistance to the contraction of muscles so that mobility results. As analogies, consider that a garden hose stiffens when filled with water, and that a water-filled balloon changes shape when squeezed at one end. Similarly, an animal with a hydrostatic skeleton can change shape and perform a variety of movements.

Hydras and planarians use their fluid-filled gastrovascular cavity as a hydrostatic skeleton. The tentacles of a hydra also have hydrostatic skeletons, allowing them to be extended to capture food. Roundworms have a fluid-filled pseudocoelom and move in a whiplike manner when their longitudinal muscles contract.

The coelom of annelids, such as earthworms, is segmented and has septa that divide it into compartments (Fig. 39.1). Each segment has its own set of longitudinal and circular muscles and its own nerve supply, so each segment or group of segments may function independently. When circular muscles contract, the segments become thinner and elongate. When longitudinal muscles contract, the segments become thicker and shorten. By alternating circular muscle contraction and longitudinal muscle contraction and by using its setae to hold its position during contractions, the animal moves forward.

Use of Muscular Hydrostats

Even animals that have an exoskeleton or an endoskeleton move selected body parts by means of *muscular hydrostats*, meaning that fluid contained within the individual cells that comprise a muscle assists with movement of that part. Muscular hydrostats are used by clams to extend their muscular foot and by sea stars to extend their tube feet. Spiders depend on them to move their legs, and moths depend on them to extend their proboscis. In vertebrates, movement of an elephant's trunk involves a muscular hydrostat that allows the animal to reach high into trees, pick up a morsel of food off the ground, or manipulate other objects.

Exoskeletons and Endoskeletons

Molluscs and arthropods have a rigid **exoskeleton,** an external covering composed of a stiff material. The strength of an exoskeleton can be improved by increasing its thickness and weight, but this leaves less room for internal organs.

In molluscs, such as snails and clams, a thick and nonmobile calcium carbonate shell is primarily used for protection against the environment and predators. A mollusc's shell can grow as the animal grows.

The exoskeleton of arthropods, such as insects and crustaceans, is composed of chitin, a strong, flexible nitrogenous polysaccharide. Their exoskeleton protects them against wear and tear, predators, and desiccation (drying out)—an important feature for arthropods that live on land. Working together with muscles, the jointed and movable appendages of arthropods allow them to crawl, fly, and/or swim. Because their exoskeleton is of fixed size, however, arthropods must *molt*, or shed their skeleton, in order to grow (Fig. 39.2).

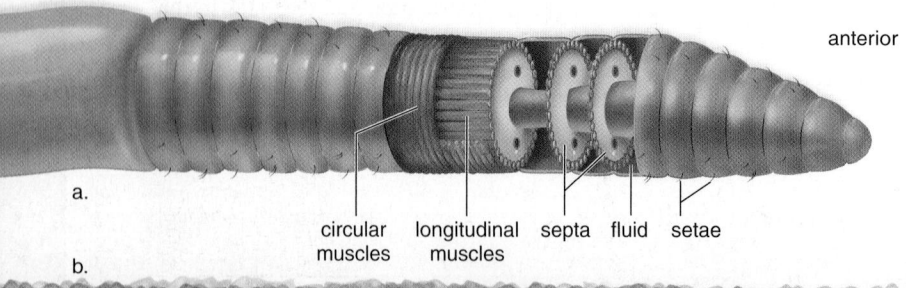

anterior

Figure 39.1 Locomotion in an earthworm.
a. The coelom is divided by septa, and each body segment is a separate locomotor unit. Both circular and longitudinal muscles are present. **b.** As circular muscles contract, a few segments extend. The worm is held in place by setae, needlelike chitinous structures on each segment of the body. Then, as longitudinal muscles contract, a portion of the body is brought forward. This series of events occurs down the length of the worm.

a.

circular　longitudinal　septa　fluid　setae
muscles　　muscles

b.

circular　longitudinal
muscles　　muscles
contracted　contracted

circular muscles
contract, and anterior
end moves forward

longitudinal muscles
contract, and segments
catch up

circular muscles
contract, and anterior
end moves forward

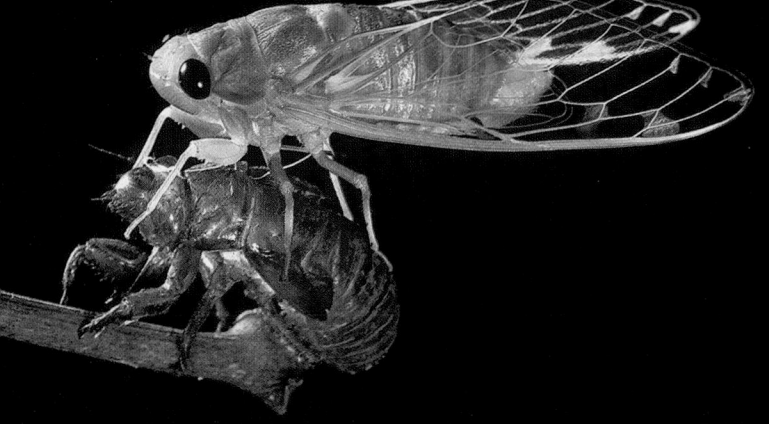

Figure 39.2 Exoskeleton. Exoskeletons support muscle contraction and prevent drying out. The chitinous exoskeleton of an arthropod is shed as the animal molts; until the new skeleton dries and hardens, the animal is more vulnerable to predators, and muscle contractions may not translate into body movements. In this photo, a dog-day cicada, *Tibicen*, has just finished molting.

Both echinoderms and vertebrates have an **endoskeleton,** which is made up of rigid internal structures. The skeleton of echinoderms consists of spicules and plates of calcium carbonate embedded in the living tissue of the body wall. In contrast, the vertebrate endoskeleton is living tissue. Sharks and rays have skeletons composed only of cartilage. Other vertebrates, such as bony fishes, amphibians, reptiles, birds, and mammals, have endoskeletons composed of bone and cartilage.

The advantages of the jointed vertebrate endoskeleton are listed in Figure 39.3. An endoskeleton grows with the animal, thus molting is not required. It supports the weight of a large animal without limiting the space for internal organs. An endoskeleton also offers protection to vital internal organs, but it is protected by the soft tissues around it. Injuries to soft tissue are usually easier to repair than injuries to a hard skeleton. Compared to the relatively limited mobility of arthropod appendages,

vertebrate limbs are generally more flexible and have different types of joints, allowing for even more complex movements.

The skeletons of mammals come in many sizes and shapes, which are often adapted to a particular mode of locomotion. Aquatic animals like seals, sea lions, whales, and dolphins have a streamlined, torpedo-shaped skeleton that facilitates movement through water. Many animals that jump, such as kangaroos and rabbits, have a compact skeleton with elongated hindlimbs that propel them forward. Carnivores, such as members of the cat family, walk on their toes, which is an adaptation to running and chasing prey. (Note that when humans run, we push off with our toes to move faster.) Hoofed mammals, such as horses or deer, have evolved long legs and run on the tips of elongated phalanges. The hoof of a horse is its third phalange only.

Humans are bipedal and walk on the soles of feet formed by the tarsal and metatarsal bones. This form of locomotion allows the hands to be free and may have evolved from the monkeys' and apes' habit of using only forelimbs as they swing through the branches of trees. Dexterity of hands and feet is actually the ancestral mammalian condition. In humans and apes, the bones of the hands and feet are not fused, and the wrist and ankle can rotate in three dimensions.

Check Your Progress   39.1

1. Identify the type of skeleton of each of the following animals: butterfly, crow, shrimp, oyster, and sea urchin.
2. Describe the type of support system that makes it possible to stick out your tongue.
3. Explain why an earthworm loses its cylindrical shape when it dies.

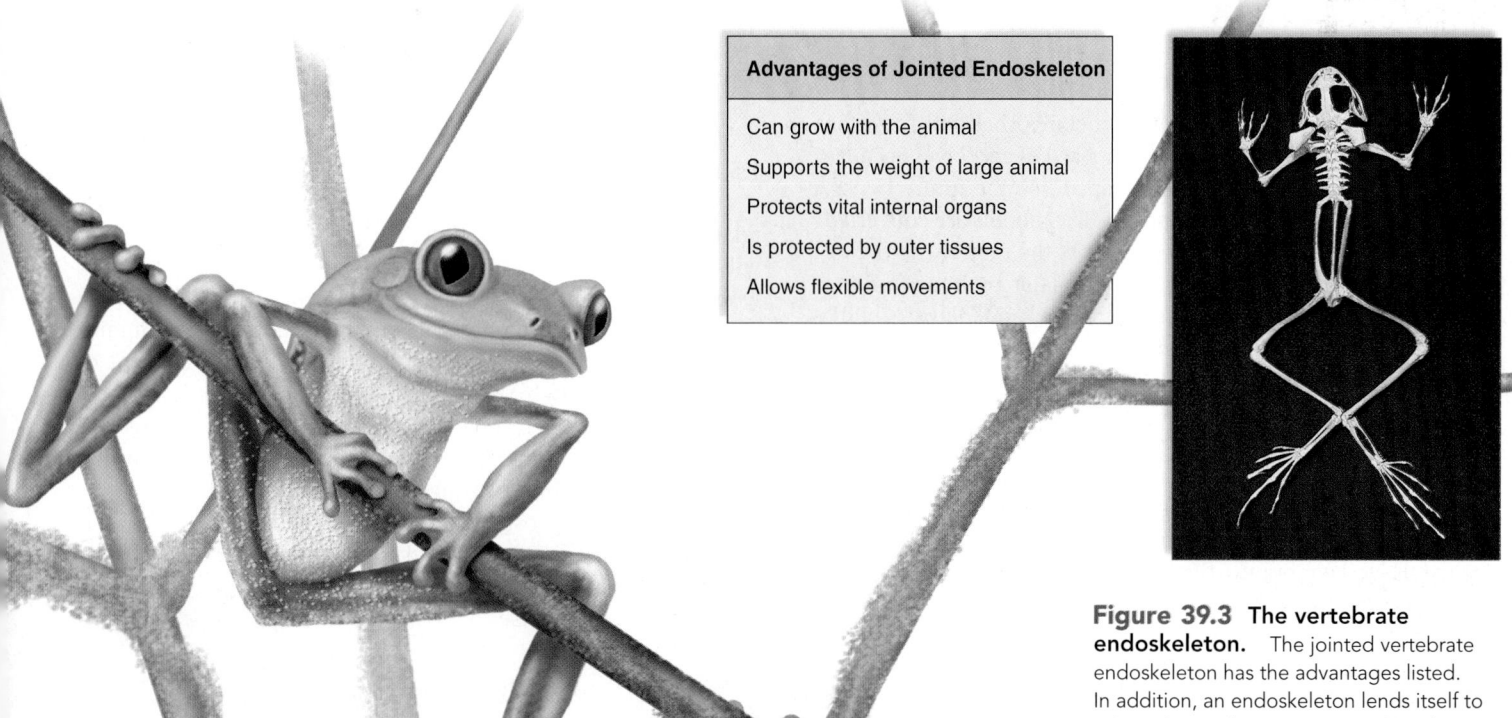

Advantages of Jointed Endoskeleton
Can grow with the animal
Supports the weight of large animal
Protects vital internal organs
Is protected by outer tissues
Allows flexible movements

Figure 39.3 The vertebrate endoskeleton. The jointed vertebrate endoskeleton has the advantages listed. In addition, an endoskeleton lends itself to adaptation to the environment. Vertebrates move in various ways (e.g., jumping, flying, swimming, running).

39.2 The Human Skeletal System

Learning Outcomes

Upon completion of this section, you should be able to

1. Review the five major functions of the skeletal system.
2. Describe the macroscopic and microscopic structure of bone.
3. List the major bones that comprise the human axial and appendicular skeletons.

The skeletal system has many functions that contribute to homeostasis:

- *Support of the body.* The rigid skeleton provides an internal framework that largely determines the body's shape.
- *Protection of vital internal organs,* such as the brain, heart, and lungs. The bones of the skull protect the brain; the rib cage protects the heart and lungs. The vertebrae protect the spinal cord.
- *Sites for muscle attachment.* The pull of muscles on the bones makes movement possible. Articulations (joints) occur between all the bones, but we associate body movement particularly with jointed appendages.
- *Storage reservoir for ions.* All bones have a matrix that contains calcium phosphate, a source of calcium ions and phosphate ions in the blood.
- *Production of blood cells.* Blood cells and other blood elements are produced within the red bone marrow of the skull, ribs, sternum, pelvis, and long bones.

In this section, we describe the characteristics of human bones, the components of various regions of the skeleton, and the different types and functions of joints.

Bone Growth and Renewal

During prenatal development, the structures that will form the bones of the human skeleton are composed of cartilage. Because these cartilaginous structures are shaped like the future bones, they provide "models" of these bones. The models are converted to bones as calcium salts are deposited in the matrix (nonliving material), first by the cartilage cells and later by bone-forming cells called **osteoblasts** [Gk. *osteon,* bone, and *blastos,* bud]. The conversion of cartilaginous models to bones is called endochondral ossification.

In some cases, ossification occurs without any previous cartilaginous model. This type of ossification occurs in the dermis and forms bones called dermal bones. Examples include the mandible (lower jaw), certain bones of the skull, and the clavicle (collarbone). During intramembranous ossification, fibrous connective tissue membranes give support as ossification begins.

Endochondral ossification of a long bone begins in a region called a primary ossification center, located in the middle of the cartilaginous model. In the primary ossification center, the cartilage is broken down and invaded by blood vessels, and cells in the area mature into bone-forming osteoblasts. Later, secondary ossification centers form at the ends of the model. A cartilaginous *growth plate* remains between the primary ossification center and each secondary center. As long as these plates remain, growth is possible. The rate of growth is controlled by hormones, particularly growth hormone (GH) and the sex hormones. Eventually, the plates become ossified, causing the primary and secondary centers of ossification to fuse, and the bone stops growing.

 **Animation** Bone Growth

 Video Nano Bones

In the adult, bone is continually being broken down and built up again. Bone-absorbing cells called **osteoclasts** [Gk. *osteon,* bone, and *klastos,* broken in pieces] break down bone, remove worn cells, and deposit calcium in the blood. In this way, osteoclasts help maintain the blood calcium level and contribute to homeostasis.

Among many other functions, calcium ions play a major role in muscle contraction and nerve conduction. The blood calcium level is closely regulated by the antagonistic hormones parathyroid hormone (PTH) and calcitonin. PTH promotes the activity of osteoclasts, and calcitonin inhibits their activity to keep the blood calcium level within normal limits.

MP3 Calcium Homeostasis

Assuming that the blood calcium level is normal, bone destruction caused by the work of osteoclasts is repaired by osteoblasts. As they form bone, some of these osteoblasts get caught in the matrix (nonliving material) they secrete and are converted to **osteocytes** [Gk. *osteon,* bone, and *kytos,* cell]. These cells live within the lacunae of osteons, where they continue to affect the timing and location of bone remodeling.

While a child is growing, the rate of bone formation is greater than the rate of bone breakdown. The skeletal mass continues to increase until ages 20 to 30. After that, the rate of formation and rate of breakdown of bone mass are equal, until ages 40 to 50. Then, reabsorption begins to exceed formation, and the total bone mass slowly decreases.

As people age, an abnormal thinning of the bones called **osteoporosis** can lead to an increased risk of fractures, especially of the wrist, vertebrae, and pelvis. About ten million people in the United States have osteoporosis, which results in 1.5 million fractures each year.

Women are twice as likely as men to have an osteoporosis-related fracture in their lifetime, and about one in four women will experience such a fracture. This is partly due to the fact that women have about 30% less bone mass than men to begin with. Also, because sex hormones play an important role in maintaining bone strength, women typically lose about two percent of their bone mass each year after menopause. Estrogen replacement therapy has been shown to increase bone mass and reduce fractures, but it can also increase the risk of cardiovascular disease and certain types of cancer. Other strategies for decreasing osteoporosis risk include consuming 1,000-1,500 mg of calcium per day, obtaining sufficient vitamin D, and engaging in regular physical exercise.

 **Animation** Osteoporosis

Video Bear Bones

Anatomy of a Long Bone

A long bone, such as the humerus, illustrates principles of bone anatomy. When the bone is split open, as in Figure 39.4, the longitudinal section shows that it is not solid but has a

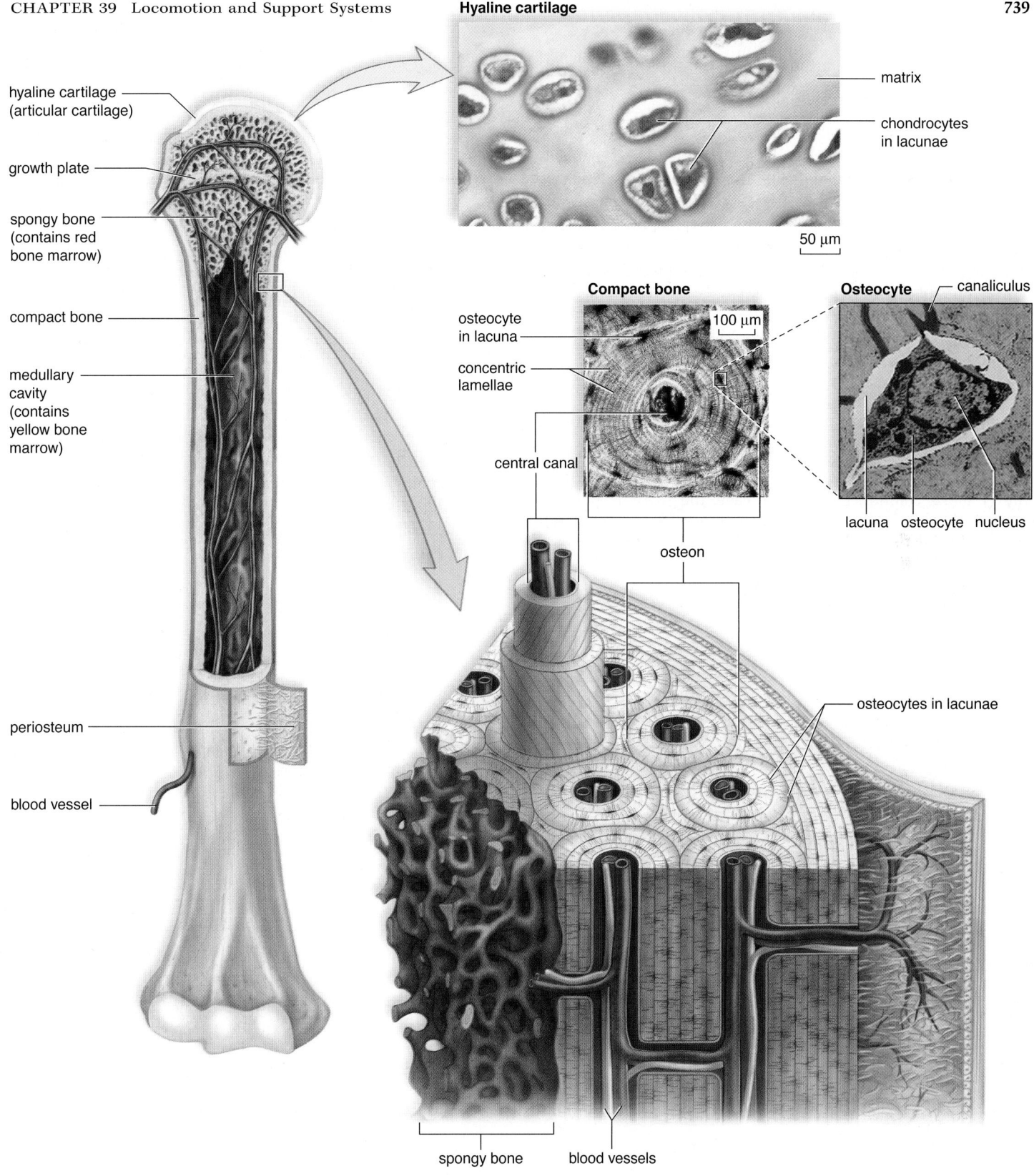

Figure 39.4   Anatomy of a long bone.   *Left:* A long bone is encased by fibrous membrane (periosteum) except where it is covered at the ends by hyaline cartilage (see micrograph). Spongy bone located beneath the cartilage may contain red bone marrow. The central shaft contains yellow bone marrow and is bordered by compact bone, which is shown in the enlargement and micrograph (*right*).

cavity called the medullary cavity bounded at the sides by compact bone and at the ends by spongy bone. Beyond the spongy bone, there is a thin shell of compact bone and finally a layer of hyaline cartilage. The cavity of an adult long bone usually contains yellow bone marrow, which is a fat-storage tissue.

Compact bone contains many osteons (also called Haversian systems), where osteocytes lie in tiny chambers called lacunae. The lacunae are arranged in concentric circles around central canals that contain blood vessels and nerves. The lacunae are separated by a matrix of collagen fibers and mineral deposits, primarily calcium and phosphorus salts.

Spongy bone has numerous bony bars and plates separated by irregular spaces. Although lighter than compact bone, spongy bone is still designed for strength. Just as braces are used for support in buildings, the solid portions of spongy bone follow lines of stress. The spaces in spongy bone are often filled with **red bone marrow,** a specialized tissue that produces blood cells. This is an additional way the skeletal system assists homeostasis. As you know, red blood cells transport oxygen, and white blood cells are a part of the immune system, which fights infection.

MP3
Bone Structure

The Axial Skeleton

The **axial skeleton** [L. *axis,* axis, hinge; Gk. *skeleton,* dried body] lies in the midline of the body and consists of the skull, the vertebral column, the thoracic cage, the sacrum, and the coccyx (blue labels in Fig. 39.5). A total of 80 bones make up the axial skeleton.

The Skull

The skull, which protects the brain, is formed by the cranium and the facial bones (Fig. 39.6). In newborns, certain bones of the cranium are joined by membranous regions called *fontanels* (or "soft spots"), all of which usually close and become **sutures** by the age of two years. The bones of the cranium contain the sinuses [L. *sinus,* hollow], air spaces lined by mucous membrane that reduce the weight of the skull and give a resonant sound to the voice. Two sinuses, called the mastoid sinuses, drain into the middle ear. Mastoiditis, a condition that can lead to deafness, is an inflammation of these sinuses.

The major bones of the cranium have the same names as the lobes of the brain. On the top of the cranium, the frontal bone forms the forehead, and the parietal bones extend to the sides. Below the much larger parietal bones, each temporal bone has an opening that leads to the middle ear. In the rear of the skull, the occipital bone curves to form the base of the skull. At the base of the skull, the spinal cord passes upward through a large opening called the *foramen magnum* [L. *foramen,* hole, and *magnus,* great, large] and becomes the brain stem.

The temporal and frontal bones are cranial bones that contribute to the face. The sphenoid bones account for the flattened

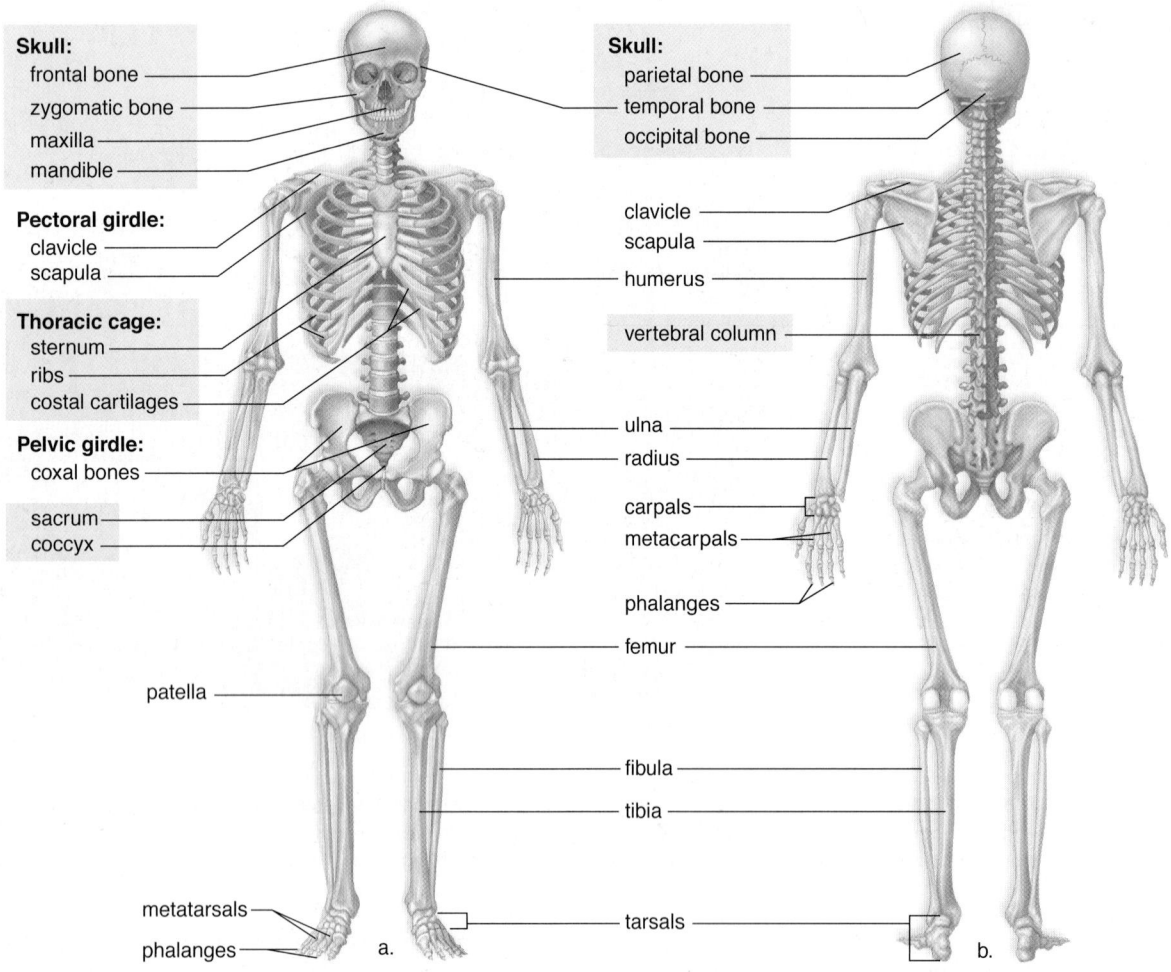

Skull:
 frontal bone
 zygomatic bone
 maxilla
 mandible

Pectoral girdle:
 clavicle
 scapula

Thoracic cage:
 sternum
 ribs
 costal cartilages

Pelvic girdle:
 coxal bones

 sacrum
 coccyx

patella

metatarsals
phalanges

a.

Skull:
 parietal bone
 temporal bone
 occipital bone

clavicle
scapula
humerus

vertebral column

ulna
radius

carpals
metacarpals

phalanges

femur

fibula
tibia

tarsals

b.

Figure 39.5 The human skeleton.   a. Anterior view. **b.** Posterior view. The bones of the axial skeleton are in blue and the rest is the appendicular skeleton.

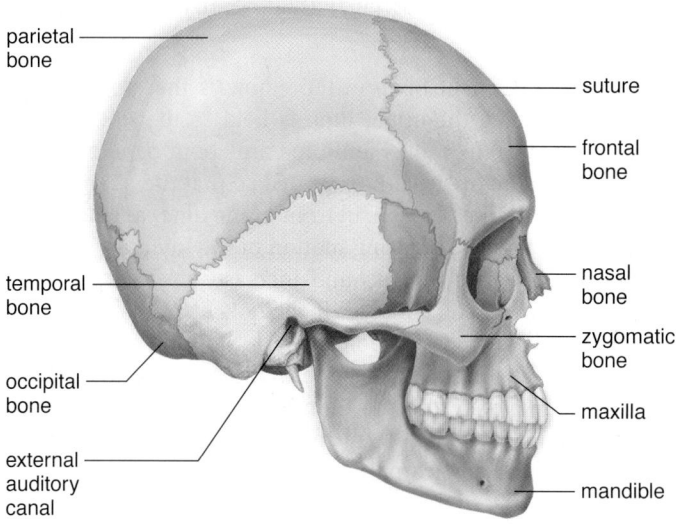

Lateral view

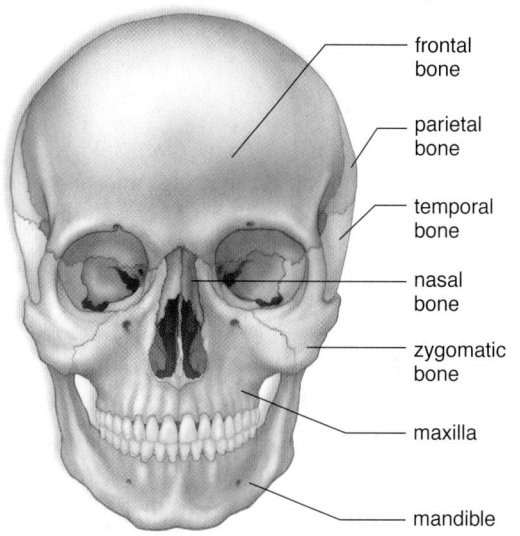

Frontal view

parietal bone
suture
frontal bone
nasal bone
zygomatic bone
maxilla
temporal bone
occipital bone
external auditory canal
mandible

frontal bone
parietal bone
temporal bone
nasal bone
zygomatic bone
maxilla
mandible

Figure 39.6
The skull.
The skull consists of the cranium and the facial bones. The frontal bone is the forehead; the zygomatic arches form the cheekbones, and the maxillae form the upper jaw. The mandible has a projection we call the chin.

areas on each side of the forehead, which we call the temples. The frontal bone not only forms the forehead, but it also has supraorbital ridges where the eyebrows are located. Glasses sit where the frontal bone joins the nasal bones.

The most prominent of the facial bones are the mandible [L. *mandibula*, jaw], the maxillae, the zygomatic bones, and the nasal bones. The mandible, or lower jaw, is the only freely movable portion of the skull (Fig. 39.6), and its action permits us to chew our food. It also forms the "chin." Tooth sockets are located in the mandible and on the maxillae, which form the upper jaw and a portion of the hard palate. The zygomatic bones are the cheekbone prominences, and the nasal bones form the bridge of the nose. Other bones make up the nasal septum, which divides the nose cavity into two regions.

Whereas the ears are formed only by cartilage and not by bone, the nose is a mixture of bones, cartilage, and connective tissues. The lips and cheeks have a core of skeletal muscle.

MP3
The Skull

The Vertebral Column and Rib Cage

The **vertebral column** [L. *vertebra*, bones of backbone] supports the head and trunk and protects the spinal cord and the roots of the spinal nerves. It is a longitudinal axis that serves either directly or indirectly as an anchor for all the other bones of the skeleton.

Twenty-four vertebrae make up the vertebral column. Seven cervical vertebrae are located in the neck; 12 thoracic vertebrae are in the thorax; five lumbar vertebrae are in the lower back; five fused sacral vertebrae form a single sacrum; and several fused vertebrae are in the coccyx, or tailbone. Normally, the vertebral column has four curvatures that provide more resilience and strength for an upright posture than could a straight column.

Intervertebral disks, composed of fibrocartilage between the vertebrae, provide padding. They prevent the vertebrae from grinding against one another and absorb shock caused by movements such as running, jumping, and even walking. The presence of the disks allows the vertebrae to move as we bend forward, backward, and from side to side. Unfortunately, these disks become weakened with age and can herniate and rupture. Pain results if a disk presses against the spinal cord

and/or spinal nerves. The body may heal itself, or the disk can be removed surgically. If the latter, the vertebrae can be fused together, but this limits the flexibility of the body.

The thoracic vertebrae are a part of the *rib cage*, sometimes also called the thoracic cage. The rib cage also contains the ribs, the costal cartilages, and the sternum, or breastbone (Fig. 39.7).

There are 12 pairs of ribs. The upper seven pairs are "true ribs" because they attach directly to the sternum. The lower five pairs do not connect directly to the sternum and are called the "false ribs." Three pairs of false ribs attach by means of a common cartilage, and two pairs are "floating ribs" because they do not attach to the sternum at all.

The rib cage demonstrates how the skeleton is protective but also flexible. The rib cage protects the heart and lungs; yet it swings outward and upward upon inspiration and then downward and inward upon expiration.

MP3
The Vertebral Column and Thoracic Cage

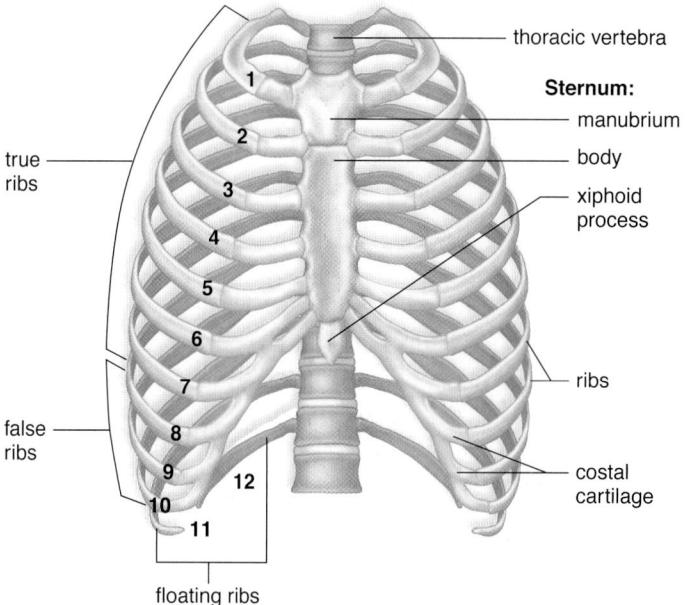

thoracic vertebra
Sternum:
manubrium
body
xiphoid process
ribs
costal cartilage

true ribs
false ribs
floating ribs

1 2 3 4 5 6 7 8 9 10 11 12

Figure 39.7 The rib cage. The rib cage consists of the thoracic vertebrae, the 12 pairs of ribs, the costal cartilages, and the sternum, or breastbone.

The Appendicular Skeleton

The **appendicular skeleton** [L. *appendicula,* dim. of *appendix,* appendage] consists of the bones within the pectoral and pelvic girdles and the attached limbs (see Fig. 39.5). The pectoral (shoulder) girdle and upper limbs are specialized for flexibility, but the pelvic girdle (hipbones) and lower limbs are specialized for strength. A total of 126 bones make up the appendicular skeleton.

MP3
The Appendicular Skeleton

The Pectoral Girdle and Upper Limb

The components of the **pectoral girdle** [Gk. *pechys,* forearm] are only loosely linked together by ligaments (Fig. 39.8). Each clavicle (collarbone) connects with the sternum and the scapula (shoulder blade), but the scapula is held in place only by muscles. This allows it to glide and rotate on the clavicle.

The single long bone in the arm, the humerus, has a smoothly rounded head that fits into a socket of the scapula. The socket, however, is very shallow and much smaller than the head. Although this means that the arm can move in almost any direction, the joint lacks stability. Therefore, this is the joint that is most apt to dislocate. The opposite end of the humerus meets the two bones of the lower arm, the ulna and the radius, at the elbow. (The prominent bone in the elbow is the topmost part of the ulna.) When the upper limb is held so that the palm is turned frontward, the radius and ulna are about parallel to one another. When the upper limb is turned so that the palm is next to the body, the radius crosses in front of the ulna, a feature that contributes to the easy twisting motion of the forearm.

The many bones of the hand increase its flexibility. The wrist has eight carpal bones, which look like small pebbles. From these, five metacarpal bones fan out to form a framework for the palm. The metacarpal bone that leads to the thumb is placed in such a way that the thumb can reach out and touch the other digits (digit is a term that refers to either fingers or toes). Beyond the metacarpals are the phalanges, the bones of the fingers and the thumb. The phalanges of the hand are long, slender, and lightweight.

The Pelvic Girdle and Lower Limb

The **pelvic girdle** [L. *pelvis,* basin] (Fig. 39.9) consists of two heavy, large coxal bones (hip bones). The coxal bones are anchored to the sacrum, and together these bones form a hollow cavity called the pelvic cavity. The wider pelvic cavity in females compared to that of males accommodates pregnancy and childbirth. The weight of the body is transmitted through the pelvis to the lower limbs and then onto the ground. The largest bone in the body is the femur, or thighbone.

In the leg, the larger of the two bones, the tibia, has a ridge we call the shin. Both of the bones of the leg have a prominence that contributes to the ankle—the tibia on the inside of the ankle and the fibula on the outside of the ankle.

Although there are seven tarsal bones in the ankle, only one receives the weight and passes it on to the heel and the ball of the foot. If you wear high-heeled shoes, the weight is thrown toward the front of your foot. The metatarsal bones participate in forming the arches of the foot. There is a longitudinal arch from the heel to the toes and a transverse arch across the foot. These provide a stable, springy base for the body. If the tissues that bind the metatarsals together become weakened, "flat feet" are apt to result. The bones of the toes are called phalanges, just as are those of the fingers, but in the foot the phalanges are stout and extremely sturdy.

Classification of Joints

Bones are connected at the **joints,** which are classified as fibrous, cartilaginous, and synovial. Most fibrous joints, such as the sutures that exist between the cranial bones, are immovable. Cartilaginous joints, such as those between the vertebrae, are slightly movable. The vertebrae are also separated by disks, which increase their flexibility. The two hipbones are slightly movable because they are ventrally joined by cartilage at the pubic symphysis. Owing to hormonal changes, this joint becomes more flexible during late pregnancy, allowing the pelvis to expand during childbirth.

In freely movable **synovial joints,** the two bones are separated by a cavity. **Ligaments,** composed of fibrous connective tissue, bind the two bones to each other, holding them in place

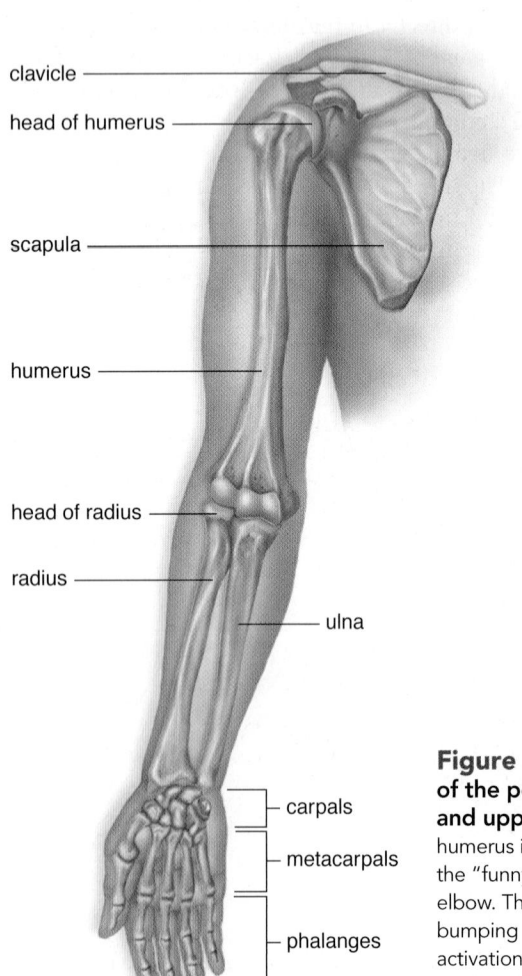

clavicle
head of humerus
scapula
humerus
head of radius
radius
ulna
carpals
metacarpals
phalanges

Figure 39.8 Bones of the pectoral girdle and upper limb. The humerus is known as the "funny bone" of the elbow. The sensation upon bumping it is due to the activation of a nerve that passes across its end.

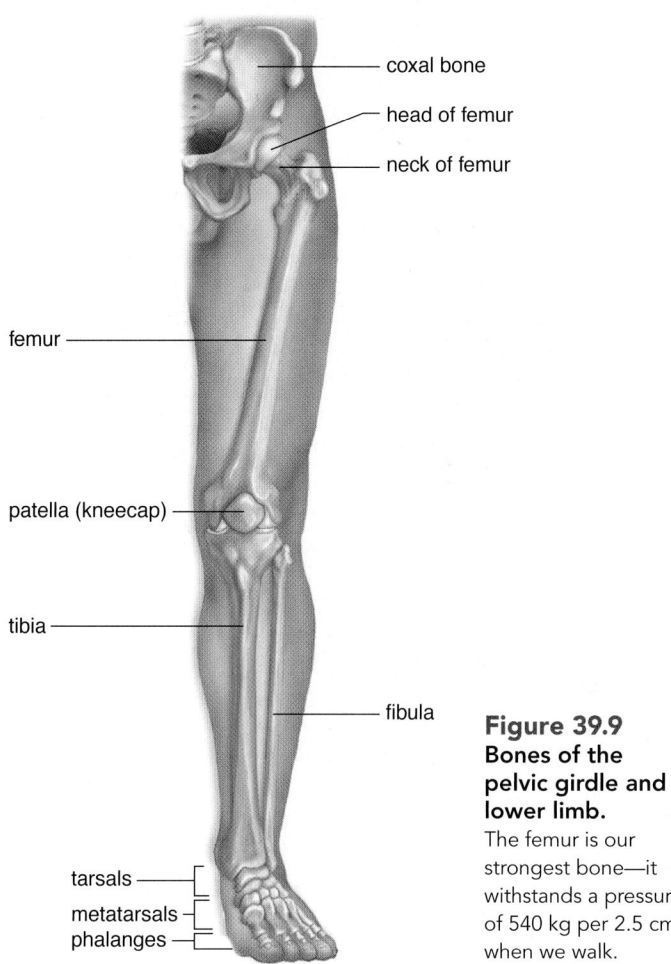

**Figure 39.9
Bones of the
pelvic girdle and
lower limb.**
The femur is our
strongest bone—it
withstands a pressure
of 540 kg per 2.5 cm^3
when we walk.

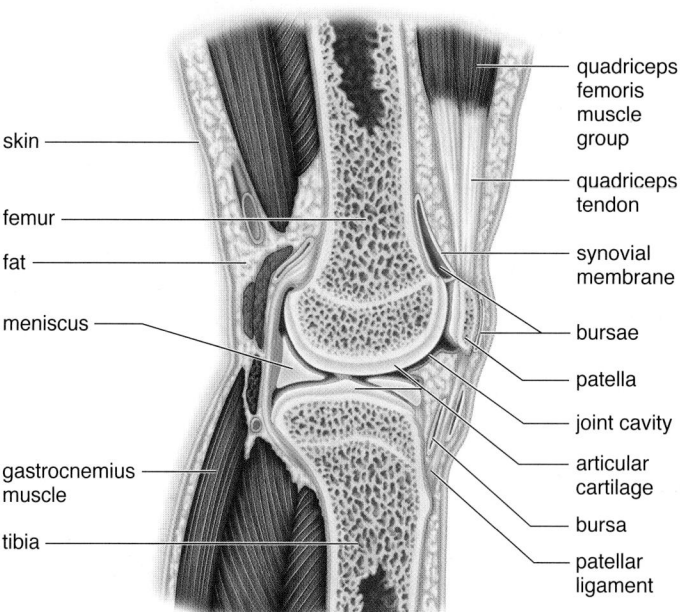

Figure 39.10 Knee joint.   The knee joint is an example of a synovial joint. The cavity between the bones is encased by ligaments and lined by synovial membrane. The patella (kneecap) serves to guide the quadriceps tendon over the joint when flexion or extension occurs.

as they form a capsule. In a "double-jointed" individual, the ligaments are unusually loose. The joint capsule is lined by synovial membrane, which produces synovial fluid, a lubricant for the joint.

The shoulders, elbows, hips, and knees are examples of synovial joints (Fig. 39.10). In the knee, as in other freely movable joints, the bones are capped by a layer of articular cartilage. In addition, crescent-shaped pieces of cartilage called menisci [Gk. *meniscus,* crescent] lie between the bones. These give added stability, helping to support the weight placed on the knee joint. Unfortunately, athletes often suffer injury of the meniscus, known as torn cartilage. Thirteen fluid-filled sacs called bursae (sing., bursa; L. *bursa,* purse) occur around the knee joint. Bursae ease friction between tendons and ligaments and between tendons and bones. Inflammation of the bursae is called bursitis. Tennis elbow is a form of bursitis.

Different types of synovial joints can be distinguished. The knee and elbow joints are *hinge joints* because, like a hinged door, they largely permit movement in one direction only. The joint between the first two cervical vertebrae, which permits side-to-side movement of the head, is an example of a *pivot joint,* which only allows rotation. More movable are the *ball-and-socket joints;* for example, the ball of the femur fits into a

socket on the hipbone. Ball-and-socket joints allow movement in all planes and even rotational movement.

All types of joints are subject to **arthritis,** or inflammation of the joints. The Arthritis Foundation estimates that nearly one in three adult Americans has some degree of chronic joint pain, and arthritis is secondary only to heart disease as a cause of work disability in the United States. The most common type of arthritis is osteoarthritis, which results from the deterioration of the cartilage in one or more synovial joints. The hands, hips, knees, lower back, and neck are most commonly affected. Rheumatoid arthritis is considered to be an autoimmune disease, in which the immune system attacks the joints for reasons that are not well understood.

Regardless of the cause, arthritis is most often treated by over-the-counter anti-inflammatory drugs such as aspirin and ibuprofen, or more powerful prescription drugs such as the corticosteroids. In severe cases, certain joints can be replaced with artificial versions made of ceramic, metal, and/or plastic. About 500,000 artificial knees and 200,000 artificial hips are installed in U.S. patients each year.

Check Your Progress   39.2

1. Describe the function of osteoblasts, osteoclasts, and osteocytes.
2. Distinguish between the structure and functions of spongy bone and compact bone.
3. Determine whether each of the following bones belongs to the axial or appendicular skeleton: sacrum, frontal bone, humerus, tibia, vertebra, coxal bone, temporal bone, scapula, and sternum.

39.3 The Muscular System

Learning Outcomes

Upon completion of this section, you should be able to

1. Describe the macroscopic and microscopic structure of a muscle fiber.
2. Explain the molecular mechanism of muscle contraction.
3. Indicate three ways that muscle cells can generate ATP.
4. Explain the specific role of acetylcholine (ACh) in stimulating a muscle fiber to contract.

Muscles are composed of contractile tissue that is capable of changing its length by contracting and relaxing. Most animals rely on muscle tissue to produce movement—to swim, crawl, walk, run, jump, or fly. Clearly there are strong evolutionary advantages to be able to move into new environments, to flee from danger, to seek and/or capture food, and to find new mates. Only a few animals are nonmotile (also called sessile), and most of these live in water, where currents can bring a supply of food to them.

As discussed in Chapter 31, humans and other vertebrates have three distinct types of muscle tissue: smooth, cardiac, and skeletal. Most of the focus in this chapter is on skeletal muscle, or striated voluntary muscle, which is important in maintaining posture, providing support, and allowing for movement. The processes responsible for skeletal muscle contraction also release heat that is distributed throughout the body, helping to maintain a constant body temperature.

MP3 Muscle Tissue

Macroscopic Anatomy and Physiology

The nearly 700 skeletal muscles and their associated tissues make up approximately 40% of the weight of an average human. Muscle tissue is approximately 15% more dense than fat tissue, so that a pound of muscle takes up less space than does a pound of fat. However, even at rest, muscle tissue consumes about three times more energy than adipose tissue.

Several of the major human superficial muscles are illustrated in Figure 39.11. Skeletal muscles are attached to the skeleton by bands of fibrous connective tissue called **tendons** [L. *tendo*, stretch]. When muscles contract, they shorten. Therefore, muscles can only pull; they cannot push. Because of this, skeletal muscles must work in antagonistic pairs. One muscle of an antagonistic pair flexes the joint and bends the limb; the other one extends the joint and straightens the limb. Figure 39.12 illustrates this principle.

In the laboratory, if a muscle is given a rapid series of threshold stimuli (i.e., stimuli strong enough to bring about action potentials, as described on page 699), it can respond to the next stimulus without relaxing completely. In this way, muscle contraction builds, or summates, until maximal sustained contraction, called **tetany,** is achieved. Tetanic contractions ordinarily occur in the body's muscles whenever skeletal muscles are actively used.

Even when muscles appear to be at rest, they exhibit **tone,** in which some of their fibers are contracting. As you saw in the preceding chapter, sensory receptors called muscle spindles and

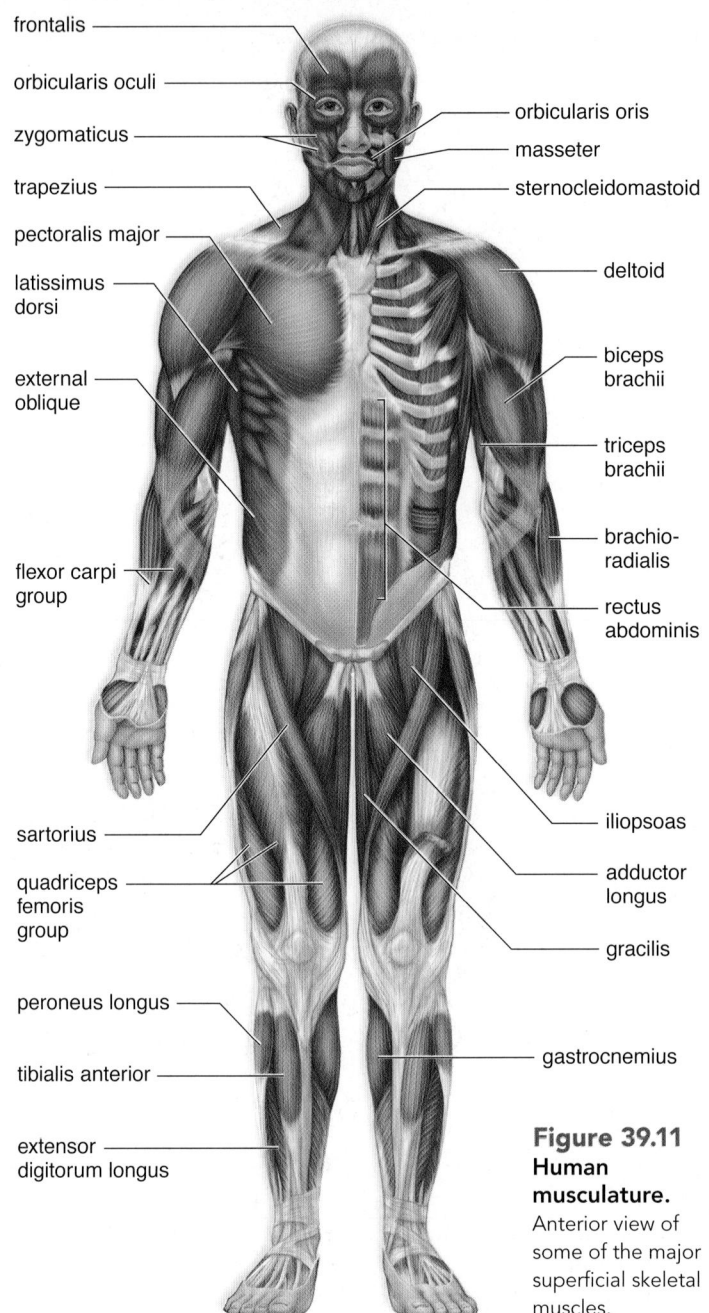

Figure 39.11 Human musculature. Anterior view of some of the major superficial skeletal muscles.

Labels: frontalis, orbicularis oculi, zygomaticus, trapezius, pectoralis major, latissimus dorsi, external oblique, flexor carpi group, sartorius, quadriceps femoris group, peroneus longus, tibialis anterior, extensor digitorum longus, orbicularis oris, masseter, sternocleidomastoid, deltoid, biceps brachii, triceps brachii, brachioradialis, rectus abdominis, iliopsoas, adductor longus, gracilis, gastrocnemius

Golgi tendon organs are partly responsible for maintaining tone. Muscle tone is particularly important in maintaining posture. If all the fibers within the muscles of the neck, trunk, and legs were to suddenly relax, the body would collapse.

Muscle tone has also been implicated in the formation of facial wrinkles. As described the Nature of Science feature, medical injections of Botox® interfere with muscle contraction, smoothing wrinkles.

Microscopic Anatomy and Physiology

A vertebrate skeletal muscle is composed of a number of muscle fibers in bundles. Each muscle fiber is a cell containing the usual cellular components, but some components have special features (Fig. 39.13).

The **sarcolemma,** or plasma membrane [Gk. *sarkos,* flesh, and *lemma,* sheath], forms a transverse system, or T system. The T tubules penetrate, or dip down, into the cell so that they come into contact—but do not fuse—with the **sarcoplasmic reticulum,** which consists of expanded portions of modified endoplasmic

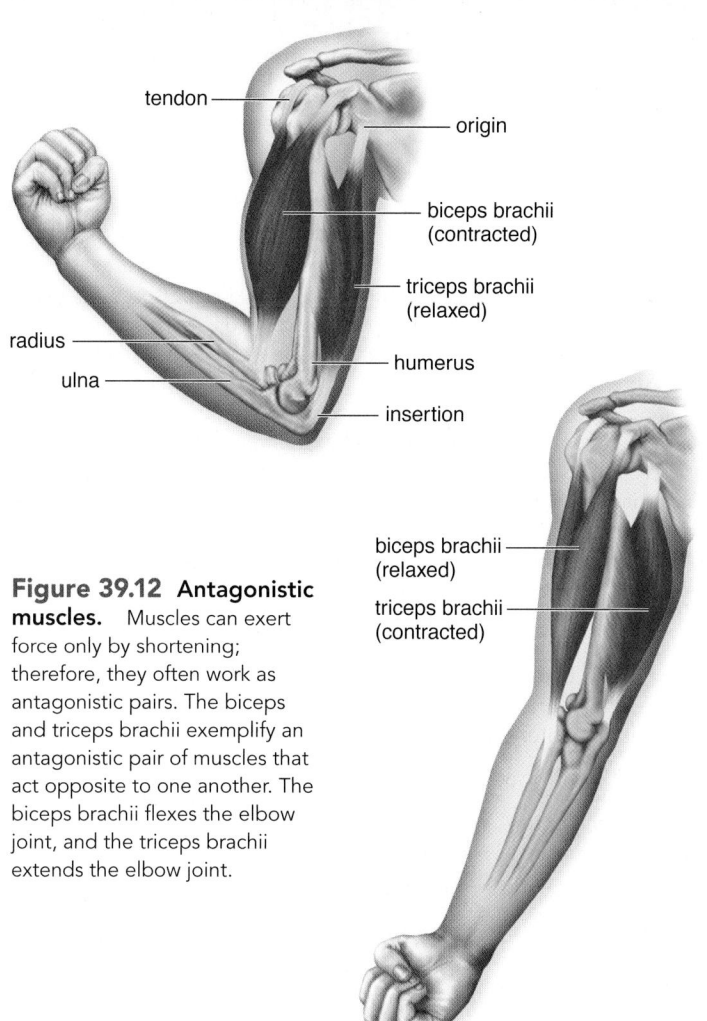

Figure 39.12 Antagonistic muscles. Muscles can exert force only by shortening; therefore, they often work as antagonistic pairs. The biceps and triceps brachii exemplify an antagonistic pair of muscles that act opposite to one another. The biceps brachii flexes the elbow joint, and the triceps brachii extends the elbow joint.

Labels (top arm): tendon, origin, biceps brachii (contracted), triceps brachii (relaxed), radius, humerus, ulna, insertion

Labels (bottom arm): biceps brachii (relaxed), triceps brachii (contracted)

reticulum (ER). These expanded portions serve as storage sites for calcium ions (Ca²⁺), which are essential for muscle contraction. The sarcoplasmic reticulum encases hundreds and sometimes even thousands of **myofibrils** [Gk. *myos*, muscle; L. *fibra*, thread], which are the contractile portions of a muscle fiber.

Myofibrils are cylindrical in shape and run the length of the muscle fiber. The light microscope shows that a myofibril has light and dark bands termed striations. These bands are responsible for skeletal muscle's striated appearance. The electron microscope reveals that the striations of myofibrils are formed by the placement of protein filaments within contractile units called **sarcomeres** [Gk. *sarkos*, flesh, and *meros*, part].

Examining sarcomeres when they are relaxed shows that a sarcomere extends between two dark lines called the Z lines (Fig. 39.13). There are two types of protein filaments: thick filaments made up of **myosin,** and thin filaments made up of **actin.** The I band is light colored because it contains only actin filaments attached to a Z line. The dark regions of the A band contain overlapping actin and myosin filaments, and its H zone has only myosin filaments.

MP3 Muscle Structure

Sliding Filament Model

Examining muscle fibers when they are contracted reveals that the sarcomeres within the myofibrils have shortened. When a sarcomere shortens, the actin (thin) filaments slide past the myosin (thick) filaments and approach one another. This causes the I band to shorten and the H zone to nearly or completely disappear.

The movement of actin filaments in relation to myosin filaments is called the **sliding filament model** of muscle

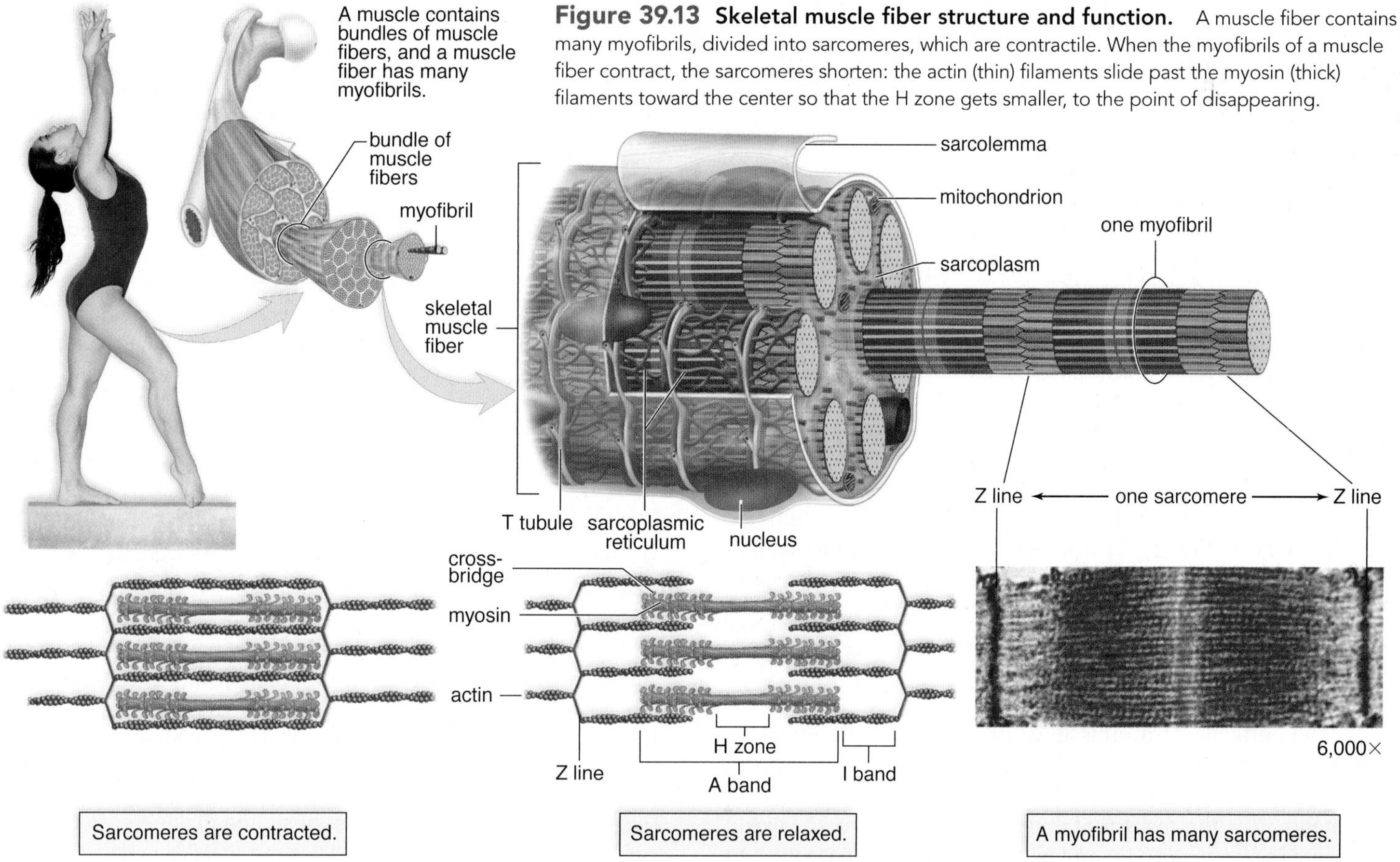

A muscle contains bundles of muscle fibers, and a muscle fiber has many myofibrils.

Figure 39.13 Skeletal muscle fiber structure and function. A muscle fiber contains many myofibrils, divided into sarcomeres, which are contractile. When the myofibrils of a muscle fiber contract, the sarcomeres shorten: the actin (thin) filaments slide past the myosin (thick) filaments toward the center so that the H zone gets smaller, to the point of disappearing.

Labels: bundle of muscle fibers, myofibril, skeletal muscle fiber, sarcolemma, mitochondrion, sarcoplasm, one myofibril, T tubule, sarcoplasmic reticulum, nucleus, Z line, one sarcomere, Z line, cross-bridge, myosin, actin, Z line, H zone, A band, I band

6,000×

Sarcomeres are contracted.

Sarcomeres are relaxed.

A myofibril has many sarcomeres.

Nature of Science

The Accidental Discovery of Botox®

Several of the most important bacterial pathogens that cause human diseases—including cholera, diphtheria, tetanus, and botulism—do so by secreting potent toxins capable of sickening or killing their victims. The botulinum toxin, produced by the bacterium *Clostridium botulinum*, is one of the most lethal substances known. Less than a microgram of the purified toxin can kill an average size person, and four kilograms (8.8 pounds) would be enough to kill all humans on Earth! Given this scary fact, it seems that the scientists who discovered the lethal activity of this bacterial toxin nearly 200 years ago could never have anticipated that the intentional injection of a very dilute form of botulinum toxin (now known as Botox) would become the most common non-surgical cosmetic procedure performed by many physicians.

As with many breakthroughs in science and medicine, the pathway from thinking about botulism as a deadly disease to using botulinum toxin as a beneficial treatment involved the hard work of many scientists, mixed with a considerable amount of luck.

In the 1820s, a German scientist, Justinus Kerner (1786–1862), was able to prove that the deaths of several people had been caused by their consumption of spoiled sausage (in fact, botulism is named for the Latin word for sausage, *botulus*). A few decades later a Belgian researcher named Emile Pierre van Ermengem (1851–1932) identified the specific bacterium responsible for producing the botulinum toxin, which could cause symptoms ranging from droopy eyelids to paralysis and respiratory failure.

By the 1920's, medical scientists at the University of California had obtained the toxin in pure form, which allowed them to determine that it acted by preventing nerves from communicating with muscles, specifically by interfering with the release of acetylcholine from the axon terminals of motor nerves (see Section 39.3).

Scientists soon began testing very dilute concentrations of the toxin as a treatment for conditions in which the muscles contract too much, such as crossed eyes or spasms of the facial muscles or vocal cords, and in 1989 the FDA first approved diluted botulinum toxin (Botox) for treating specific eye conditions called blepharospasm (eyelid spasms) and strabismus (crossing of the eyes).

Right around this time, a lucky break occurred that eventually would open the medical community's eyes to the greater potential of the diluted toxin. A Canadian ophthalmologist, Jean Carruthers, had been using it to treat her patients' eye conditions when she noticed that some of their wrinkles had also subsided. One night at a family dinner, Dr. Carruthers shared this information with her husband, a dermatologist, who decided to investigate whether he could

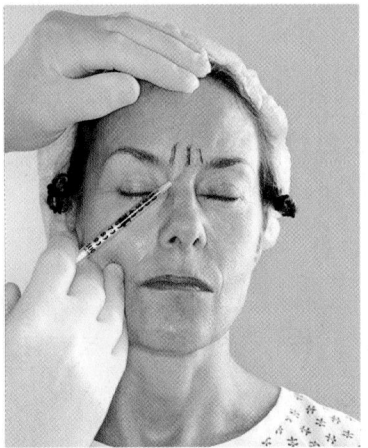

Figure 39A Treating wrinkles with diluted botulinum toxin.

reduce the deep wrinkles of some of his patients by injecting the dilute toxin into their skin. The treatment worked well, and after trying it on several more patients, as well as on themselves (!), the Canadian doctors spent several years presenting their findings at scientific meetings and in research journals. Although they were initially considered "crazy," the Carruthers eventually were able to convince the scientific community that diluted botulinum toxin was effective in treating wrinkles; however, they never patented it for that use, so they missed out on much of the $1.3 billion in annual sales the drug now earns for the company that did patent it.

The uses of diluted botulinum toxin seem to be growing since it was FDA-approved for the treatment of frown lines in 2002 (Fig. 39A). In March 2010, it was approved for the treatment of muscle stiffness in people with upper limb spasticity, and the company is currently seeking approval for as many as 90 uses of diluted botulinum toxin, including treatment of migraine headaches.

Perhaps all of this would have eventually happened even without the observations of an alert eye doctor, but progress would have very likely been slower. As the French microbiologist Louis Pasteur observed in 1854, "Chance favors the prepared mind," meaning that many scientific discoveries involve many investigators, and years of work, mixed with a flash of inspiration.

Questions to Consider

1. Considering that botulism is caused by a preformed toxin, how do you suppose it can be treated?
2. Do you think companies should be allowed to patent a naturally occurring molecule like botulinum toxin? Why or why not?

---

contraction. During the sliding process, the sarcomere shortens, even though the filaments themselves remain the same length. When you play "tug of war," your hands grasp the rope, pull, let go, attach farther down the rope, and pull again. The myosin heads are like your hands—grasping, pulling, letting go, and then repeating the process.

Animation
Sarcomere Contraction

The participants in muscle contraction have the functions listed in Table 39.1. ATP supplies the energy for muscle contraction. Although the actin filaments slide past the myosin filaments, it is the myosin filaments that do the work. Myosin filaments break down ATP and form cross-bridges that attach to and pull the actin filaments toward the center of the sarcomere.

Use of ATP in Contraction

ATP provides the energy for muscle contraction. Although muscle cells contain *myoglobin*, a molecule that stores oxygen, cellular respiration does not immediately supply all the ATP that is needed. In the meantime, muscle fibers rely on *creatine phosphate* (phosphocreatine), a storage form of high-energy phosphate. Creatine phosphate cannot directly participate in muscle contraction. Instead, it anaerobically regenerates ATP by the following reaction:

$$\text{creatine—P} + \text{ADP} \longrightarrow \text{ATP} + \text{creatine}$$

This reaction occurs in the midst of sliding filaments, and therefore, this method of supplying ATP is the speediest energy source available to muscles.

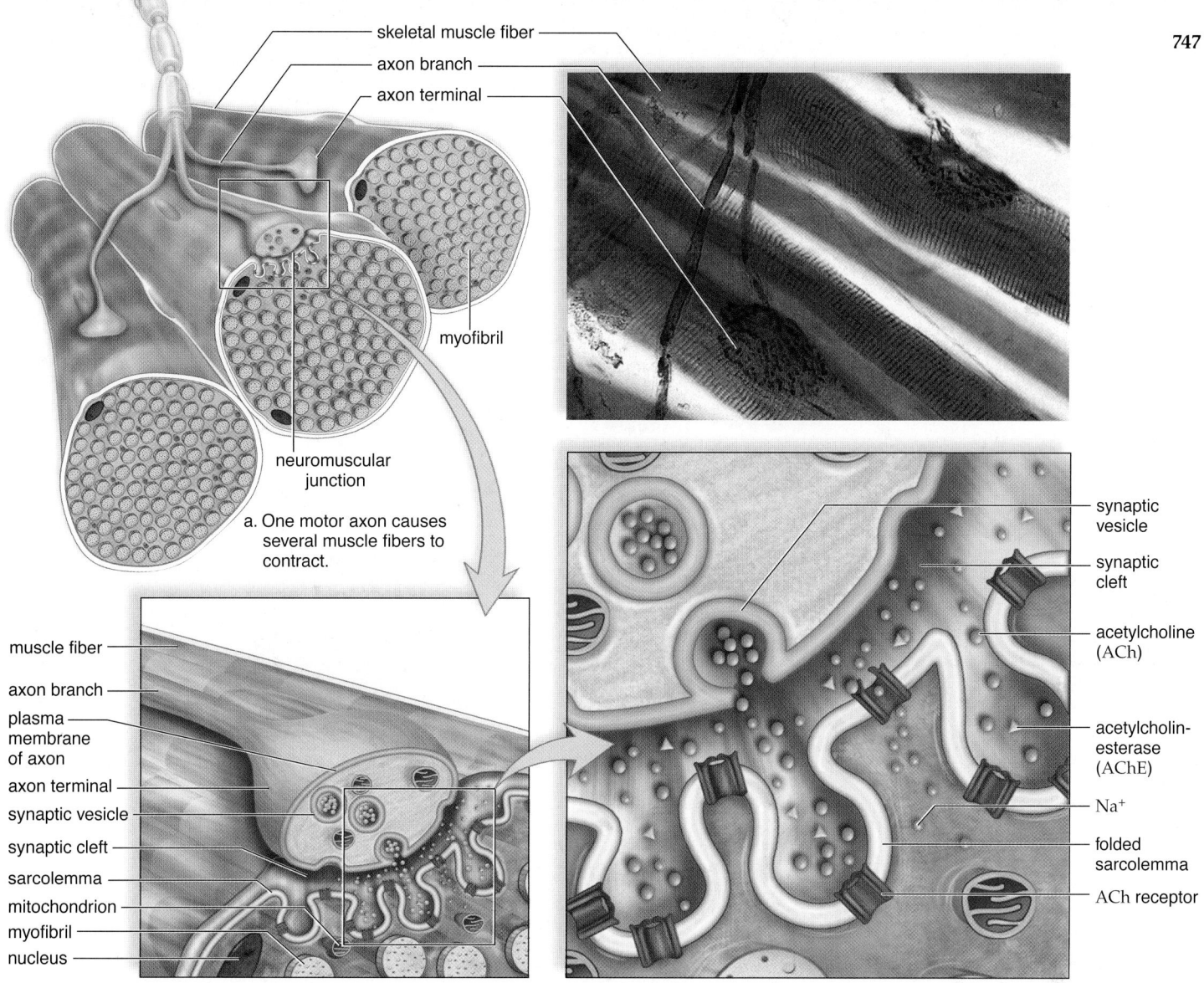

skeletal muscle fiber
axon branch
axon terminal

myofibril

neuromuscular junction

a. One motor axon causes several muscle fibers to contract.

muscle fiber
axon branch
plasma membrane of axon
axon terminal
synaptic vesicle
synaptic cleft
sarcolemma
mitochondrion
myofibril
nucleus

synaptic vesicle
synaptic cleft
acetylcholine (ACh)
acetylcholinesterase (AChE)
Na+
folded sarcolemma
ACh receptor

b. A neuromuscular junction is the juxtaposition of an axon terminal and the sarcolemma of a muscle fiber.

c. The release of a neurotransmitter (ACh) causes receptors to open and Na+ to enter a muscle fiber.

Figure 39.14 Neuromuscular junction. The branch of a motor nerve fiber **(a)** ends in an axon terminal **(b)** that meets but does not touch a muscle fiber. A synaptic cleft separates the axon terminal from the sarcolemma of the muscle fiber. Nerve impulses traveling down a motor fiber cause synaptic vesicles **(c)** to discharge a neurotransmitter that diffuses across the synaptic cleft. When the neurotransmitter is received by the sarcolemma of a muscle fiber, impulses begin that lead to muscle fiber contraction.

Table 39.1 Muscle Contraction

Name	Function
Actin filaments	Slide past myosin, causing contraction
Ca^{2+}	Needed for myosin to bind to actin
Myosin filaments	Pull actin filaments by means of cross-bridges; are enzymatic and split ATP
ATP	Supplies energy for muscle contraction

When all of the creatine phosphate is depleted, mitochondria may by then be producing enough ATP for muscle contraction to continue. If not, fermentation is a second way for muscles to supply ATP without consuming oxygen. Fermentation, which is apt to occur when strenuous exercise first begins, supplies ATP for only a short time, and lactate builds up. Whether lactate causes muscle aches and fatigue upon exercising is now being questioned.

We all have had the experience of needing to continue deep breathing following strenuous exercise. This continued intake of oxygen, which is required to complete the metabolism of lactate and restore cells to their original energy state, offsets what is known as **oxygen debt.** The lactate is transported to the liver, where 20% of it is completely broken down to carbon dioxide (CO_2) and water (H_2O). The ATP gained by this respiration is then used to reconvert 80% of the lactate to glucose.

In persons who regularly exercise, such as athletes in training, the number of mitochondria increases, and muscles rely on them rather than on fermentation to produce ATP. Less lactate is produced, and there is less oxygen debt.

Muscle Innervation

Muscles are stimulated to contract by motor nerve fibers. Nerve fibers have several branches, each of which ends at an axon terminal that lies in close proximity to the sarcolemma of a muscle fiber. A small gap, called a synaptic cleft, separates the axon terminal from the sarcolemma. This entire region is called a **neuromuscular junction** (Fig. 39.14).

Axon terminals contain synaptic vesicles that are filled with the neurotransmitter acetylcholine (ACh). When nerve

actin filament   troponin   myosin binding sites   Ca²⁺

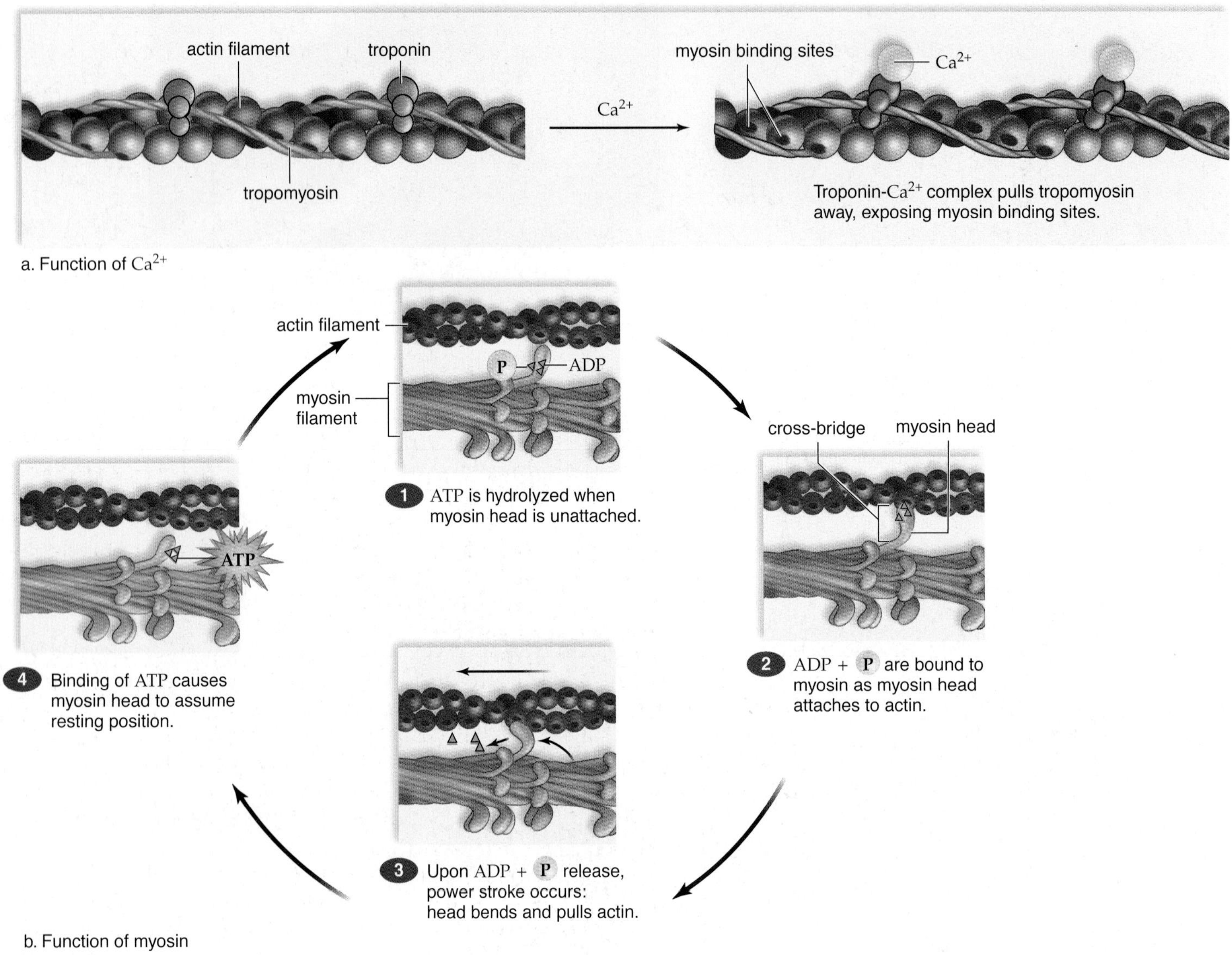

Ca²⁺

tropomyosin

Troponin-Ca²⁺ complex pulls tropomyosin
away, exposing myosin binding sites.

a. Function of Ca²⁺

actin filament

myosin
filament

P ── ADP

1 ATP is hydrolyzed when
myosin head is unattached.

cross-bridge   myosin head

2 ADP + **P** are bound to
myosin as myosin head
attaches to actin.

ATP

4 Binding of ATP causes
myosin head to assume
resting position.

3 Upon ADP + **P** release,
power stroke occurs:
head bends and pulls actin.

b. Function of myosin

Figure 39.15 The role of calcium and myosin in muscle contraction.   a. Upon release, calcium binds to troponin, exposing myosin binding sites.
b. After breaking down ATP **1** , myosin heads bind to an actin filament **2** , and later, a power stroke causes the actin filament to move **3** . When
another ATP binds to myosin, the head detaches from actin **4** , and the cycle begins again. Although only one myosin head is featured, many heads are
active at the same time.

impulses traveling down a motor neuron arrive at an axon terminal, the synaptic vesicles release ACh into the synaptic cleft.
ACh quickly diffuses across the cleft and binds to receptors in
the sarcolemma. Now the sarcolemma generates impulses that
spread over the sarcolemma and down T tubules to the sarcoplasmic reticulum. The release of calcium from the sarcoplasmic
reticulum causes the filaments within sarcomeres to slide
past one another. Sarcomere contraction results in myofibril contraction, which in turn results in muscle fiber, and
finally muscle, contraction.

Once a neurotransmitter has been released into a neuromuscular junction and has initiated a response, it is removed
from the junction. When the enzyme acetylcholinesterase (AChE) breaks down acetylcholine, muscle contraction ceases due to
reasons we discuss next.

**Animation
Function of the
Neuromuscular
Junction**

Role of Calcium in Muscle Contraction

Figure 39.15 illustrates the placement of two other proteins associated with a thin filament, which is composed of a double row
of twisted actin molecules. Threads of tropomyosin wind about
an actin filament, and troponin occurs at intervals along the
threads. Calcium ions (Ca²⁺) that have been released from the
sarcoplasmic reticulum combine with troponin. After binding
occurs, the tropomyosin threads shift their position, and myosin
binding sites are exposed.

Thick filaments are bundles of myosin molecules with double globular heads. Myosin heads function as ATPase enzymes,
splitting ATP into ADP and Ⓟ. This reaction activates the heads
so that they can bind to actin. The ADP and Ⓟ remain on the
myosin heads until the heads attach to actin, forming cross-
bridges. Now, ADP and Ⓟ are released, and this causes the
cross-bridges to change their positions. This is the power stroke

that pulls the thin filaments toward the middle of the sarcomere. When more ATP molecules bind to myosin heads, the cross-bridges are broken as the heads detach from actin. The cycle begins again; the actin filaments move nearer the center of the sarcomere each time the cycle is repeated.

Contraction continues until nerve impulses cease and calcium ions are returned to their storage sites. The membranes of the sarcoplasmic reticulum contain active transport proteins that pump calcium ions back into the calcium storage sites, and muscle relaxation occurs. When a person or animal dies, ATP production ceases. Without ATP, the myosin heads cannot detach

from actin, nor can calcium be pumped back into the sarcoplasmic reticulum. As a result, the muscles remain contracted, a phenomenon called *rigor mortis*.

MP3
Sliding Filament Theory

Check Your Progress 39.3

1. Define an "antagonistic pair" of muscles.
2. Describe the microscopic levels of structure in a skeletal muscle.
3. Discuss the specific role of ATP in muscle contraction.

CONNECTING *the* CONCEPTS *with the* BIG IDEAS

Interactions and Systems

- The nervous and muscular systems work together to assure normal body function. (4A4b*IE*)

*Find the unabridged version of all EK citations at www.glencoe.com/maderAP11.

Media Study Tools

www.glencoe.com/maderAP11

Enhance your study of this chapter with study tools and practice tests. Also ask your instructor about the resources available through ConnectPlus, including the media-rich eBook, interactive learning tools, and animations.

Summarize

39.1 Diversity of Skeletons

Three types of skeletons are found in the animal kingdom: hydrostatic skeleton (cnidarians, flatworms, and segmented worms); exoskeleton (certain molluscs and arthropods); and endoskeleton (sponges, echinoderms, and vertebrates). The rigid but jointed skeleton of arthropods and vertebrates helped them colonize the terrestrial envi-

ronment. The overall shape of an animal's skeleton is adapted to its environment and the type(s) of locomotion it uses.

39.2 The Human Skeletal System

The human skeleton gives support to the body, helps protect internal organs, provides sites for muscle attachment, and is a storage area for calcium and phosphorus salts, as well as a site for blood cell formation.

Most bones are cartilaginous in the fetus but are converted to bone during development. A long bone undergoes endochondral ossification in which a cartilaginous growth plate remains between the primary ossification center in the middle and the secondary centers at the ends of the bones. Growth of the bone is possible as long as the growth plates are present, but eventually they too are converted to bone.

Bone is constantly being renewed; osteoclasts break down bone, and osteoblasts build new bone. Osteocytes are in the lacunae of osteons; a long bone has a shaft of compact bone and two ends that contain spongy bone. The shaft contains a medullary cavity with yellow marrow, and the ends contain red marrow. Osteoporosis, or loss of bone density, is a common disease in older adults.

The human skeleton is divided into two parts: (1) the axial skeleton, which is made up of the skull, the vertebral column, the sternum, and the ribs; and (2) the appendicular skeleton, which is composed of the girdles and their appendages.

Joints are classified as immovable, like those of the cranium; slightly movable, like those between the vertebrae; and freely movable (synovial joints), like those in the knee and hip. In synovial joints, ligaments bind the two bones together, forming a capsule containing synovial fluid.

39.3 The Muscular System

Whole skeletal muscles can only shorten when they contract; therefore, they work in antagonistic pairs. For example, if one muscle flexes the joint and brings the limb toward the body, the other muscle of the antagonistic pair extends the joint and straightens the limb. A muscle at rest exhibits tone, which is dependent on tetanic contractions.

A whole skeletal muscle is composed of muscle fibers. Each muscle fiber is a cell that contains myofibrils in addition to the usual cellular components. Longitudinally, myofibrils are divided into sarcomeres, which display the arrangement of actin and myosin filaments.

The sliding filament model of muscle contraction states that myosin filaments have cross-bridges, which attach to and detach from actin filaments, causing actin filaments to slide and the sarcomere to shorten. (The H zone disappears as actin filaments approach one another.) Myosin breaks down ATP, and this supplies the energy for muscle contraction. Anaerobic creatine phosphate breakdown and fermentation quickly generate ATP. Sustained exercise requires cellular respiration for the generation of ATP.

Nerves innervate muscles. Nerve impulses traveling down motor neurons to neuromuscular junctions cause the release of ACh, which binds to receptors on the sarcolemma (plasma membrane of a muscle fiber). Impulses begin and move down T tubules that approach the sarcoplasmic reticulum (endoplasmic reticulum of muscle fibers), where calcium is stored. Thereafter, calcium ions are released and bind to troponin. The troponin-Ca^{2+} complex causes tropomyosin threads winding around actin filaments to shift their position, revealing myosin binding sites. Myosin filaments are composed of many myosin molecules with double globular heads. When myosin heads break down ATP, they are ready to attach to actin. The release of ADP + Ⓟ causes myosin heads to change their position. This is the power stroke that causes the actin filament to slide toward the center of a sarcomere. When more ATP molecules bind to myosin, the heads detach from actin, and the cycle begins again.

Key Terms

actin   745
appendicular skeleton   742
arthritis   743
axial skeleton   740
compact bone   739
endoskeleton   737
exoskeleton   736
hydrostatic skeleton   736
joint   742
ligament   742
myofibril   745
myosin   745
neuromuscular junction   747
osteoblast   738
osteoclast   738
osteocyte   738

osteoporosis   738
oxygen debt   747
pectoral girdle   742
pelvic girdle   742
red bone marrow   740
sarcolemma   744
sarcomere   745
sarcoplasmic reticulum   744
sliding filament model   745
spongy bone   740
suture   740
synovial joint   742
tendon   744
tetany   744
tone   744
vertebral column   741

Assess

Reviewing This Chapter

1. What are the three types of skeletons found in the animal kingdom and how do they differ? Cite animals that have these types of skeletons.   736–37
2. Give several functions of the skeletal system in humans. How does the skeletal system contribute to homeostasis?   738
3. Contrast compact bone with spongy bone. Explain how bone grows and is renewed.   738–40
4. Distinguish between the axial and appendicular skeletons.   740–42
5. List the bones that form the pectoral girdle and upper limb; the pelvic girdle and lower limb.   742
6. How are joints classified? Describe the anatomy of a freely movable joint.   742–43
7. Give several functions of the muscular system in humans. How does the muscular system contribute to homeostasis?   744
8. Describe how muscles are attached to bones. What is accomplished by muscles acting in antagonistic pairs?   744
9. Discuss the microscopic structural features of a muscle fiber and a sarcomere. What is the sliding filament model?   744–46
10. Discuss the availability and the specific role of ATP during muscle contraction. What is oxygen debt, and how is it repaid?   746–47
11. Describe the structure and function of a neuromuscular junction.   747–48
12. Describe the cyclical events as myosin pulls actin toward the center of a sarcomere.   748–49

Testing Yourself

Choose the best answer for each question.
For questions 1–4, match each bone to the location in the key.

KEY:

a. arm
b. forearm
c. pectoral girdle
d. pelvic girdle
e. thigh
f. leg

1. ulna
2. tibia
3. clavicle
4. femur

5. Spongy bone
   a. is a storage area for fat.
   b. contains red bone marrow, where blood cells are formed.
   c. lends strength to bones.
   d. Both b and c are correct.

6. The human skeletal system does not
   a. produce blood cells.
   b. store minerals.
   c. help produce movement.
   d. store fat.
   e. produce body heat.

7. All blood cells—red, white, and platelets—are produced by which of the following?
   a. yellow bone marrow
   b. red bone marrow
   c. periosteum
   d. medullary cavity

8. Which of the following is not a bone of the appendicular skeleton?
   a. the scapula
   b. a rib
   c. a metatarsal bone
   d. the patella
   e. the ulna

9. The vertebrae that articulate with the ribs are the
   a. lumbar vertebrae.
   b. sacral vertebrae.
   c. thoracic vertebrae.
   d. cervical vertebrae.
   e. coccyx.

10. In a muscle fiber,
    a. the sarcolemma is connective tissue holding the myofibrils together.
    b. the sarcoplasmic reticulum stores calcium.
    c. both myosin and actin filaments have cross-bridges.
    d. there is a T system but no endoplasmic reticulum.
    e. All of these are correct.

11. When muscles contract,
    a. sarcomeres shorten.
    b. myosin heads break down ATP.
    c. actin slides past myosin.
    d. the H zone disappears.
    e. All of these are correct.

12. Nervous stimulation of muscles
    a. occurs at a neuromuscular junction.
    b. involves the release of ACh.
    c. results in impulses that travel down the T system.
    d. causes calcium to be released from the sarcoplasmic reticulum.
    e. All of these are correct.

13. When calcium is released from the sarcoplasmic reticulum, it binds to
    a. myosin.
    b. actin.
    c. troponin.
    d. sarcomeres.
    e. Both b and d are correct.

14. Acetylcholine
    a. is active at somatic synapses but not at neuromuscular junctions.
    b. binds to receptors in the sarcolemma.
    c. precedes the buildup of ATP in mitochondria.
    d. is stored in the sarcoplasmic reticulum.
    e. Both b and d are correct.

15. Label this diagram of a muscle fiber, using these terms: myofibril, Z line, T tubule, sarcomere, sarcolemma, sarcoplasmic reticulum.

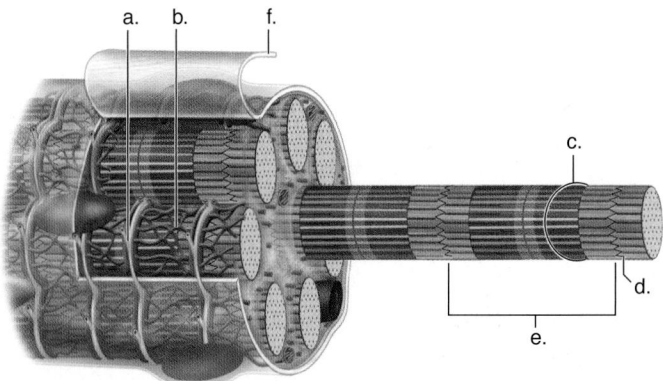

Engage

Virtual Lab
Muscle Stimulation

The virtual lab "Muscle Stimulation" allows you to visualize how the load on a muscle influences the structures of the muscle at the microscopic level.

Thinking Scientifically

1. It is observed that some motor neurons innervate only a few muscle fibers in the biceps brachii. Other motor neurons each innervate many muscle fibers. How might this observation correlate with our ability to pick up a pencil or a 2-liter soda bottle? On what basis would the brain bring about the correct level of contraction?

2. Some athletes believe that taking oral creatine will increase their endurance because it will increase the amount of phosphate available to their muscles for ATP synthesis. This statement can be regarded as two hypotheses: (a) oral creatine increases endurance, and (b) oral creatine increases the amount of creatine available in muscles for ATP synthesis. How could these two hypotheses be tested?

Bioethical Issue

Anabolic Steroids and Muscle Enhancement

A natural advantage does not bar an athlete from participating in and winning a medal in a particular sport at the Olympic Games. Nor are athletes restricted to a certain amount of practice or required to eliminate certain foods from their diets.

Athletes are, however, prevented from participating in the Olympic Games if they have taken certain performance-enhancing drugs. There is no doubt that regular use of drugs such as anabolic steroids leads to kidney disease, liver dysfunction, hypertension, increased aggression, and a myriad of other undesirable side effects (see Fig. 40.16). Even so, shouldn't the individual be allowed to take these drugs if he or she wants to? Anabolic steroids are synthetic forms of the male sex hormone testosterone. Taking large doses, along with strength training, leads to much larger muscles than otherwise. Extra strength and endurance can give an athlete an advantage in certain sports, such as racing, swimming, and weight lifting.

On what basis have anabolic steroids been banned by the Olympic Committee and other sports organizations? The basis can't be an unfair advantage, because some athletes naturally have an unfair advantage over other athletes. Should these drugs be outlawed because of long-term health effects? Excessive practice or a purposeful decrease or increase in weight to better perform in a sport can also be injurious to one's health. In other words, how can we justify allowing some behaviors that enhance performance and not others? (See also the Nature of Science feature, "Misuse of Erythropoietin in Sports," in Chapter 36, which described the controversy over use of erythropoietin in the Tour de France.)

40

Hormones and Endocrine Systems

BEFORE YOU BEGIN

Before beginning this chapter, take a few moments to review the following discussions.

Section 3.3 What is the general structure and function of steroids?

Chapter 5 Nature of Science What are the general classifications of chemical signaling molecules?

Section 37.3 What is the location and function of the hypothalamus?

The sphinx moth (*Manduca sexta*) begins life as a caterpillar. The caterpillar molts and undergoes metamorphosis, as orchestrated by hormones.

Hormones, chemical messengers of the endocrine system, regulate the metamorphosis of many insects from wormlike larval stages to their adult forms. One hormone, ecdysone, initiates shedding of the exoskeleton as the larva passes through a series of growth stages. A decline in the production of another hormone triggers the final metamorphosis into an adult, as shown in the inset above for the sphinx moth (adult form), also referred to as the tobacco hornworm (caterpillar form).

Along with the nervous system, the endocrine system coordinates the activities of the body's other organ systems and helps maintain homeostasis. In contrast to the nervous system, the endocrine system is not centralized, but consists of several organs scattered throughout the body. The hormones secreted by endocrine glands travel through the bloodstream and tissue fluid to reach their target tissues. The metabolism of a cell changes when it has a plasma membrane or nuclear receptor for that hormone. In this chapter, you'll learn how hormones exert their slow but powerful influences on the body. You'll see how the endocrine system maintains homeostasis when working properly, as well as some consequences of endocrine malfunction.

As you read through the chapter, think about the following questions:

1. Why do more complex animals, such as mammals, tend to use some of the same hormones that are present in more primitive invertebrates, instead of evolving completely new hormones?

2. What are some specific examples where the nervous system works with the endocrine system to control body functions?

3. What are some specific examples where the endocrine system works independently?

FOLLOWING *the* BIG IDEAS

CHAPTER 40 HORMONES AND ENDOCRINE SYSTEMS	
Energy and Homeostasis	Endocrine imbalance leads to major physiological problems.
Information and Signaling	Hormones are chemical signals whose messages are received only by their target cells.

40.1 Animal Hormones

Learning Outcomes

Upon completion of this section, you should be able to

1. Distinguish between the mode of action of a neurotransmitter and that of a hormone.
2. Identify the major endocrine glands of the human body.
3. Compare the mechanisms of action of peptide and steroid hormones.

The nervous and endocrine systems work together to regulate the activities of the other organs. Both systems use chemical signals when they respond to changes that might threaten homeostasis. However, they have different means of delivering these signals (Fig. 40.1). As discussed in Chapters 37 and 38, sensory receptors detect changes in the internal and external environment and transmit that information to the CNS, which responds by stimulating muscles and glands. Communication depends on nerve signals, conducted in axons, and neurotransmitters, which cross synapses. Axon conduction occurs rapidly and so does diffusion of a neurotransmitter across the short distance of a synapse. In other words, the nervous system is organized to respond rapidly to stimuli. This is particularly useful if the stimulus is an external event that endangers our safety—we can move quickly to avoid being hurt.

The **endocrine system** functions differently. The endocrine system is largely composed of glands (Fig. 40.2). These glands secrete **hormones,** such as insulin, which are carried by the bloodstream to target cells throughout the body. It takes time to deliver hormones, and it takes time for cells to respond, but the effect is longer lasting. In other words, the endocrine system is organized for a slower but prolonged response.

Endocrine glands can be contrasted with exocrine glands. **Exocrine glands** secrete their products into ducts, which take them to the lumens of other organs or outside the body. For example, the salivary glands send saliva into the mouth by way of the salivary ducts. **Endocrine glands,** as stated, secrete their products into the bloodstream, which delivers them throughout the body.

Hormones influence almost every basic homeostatic function of an organism, including metabolism, growth, reproduction, osmoregulation, and digestion. Therefore, it is not surprising that hormones are produced by invertebrates as well as vertebrates. For example, the hormone insulin is a key regulator of metabolism in vertebrates, and insulin-related peptides have been identified in insects and molluscs, suggesting an early evolutionary origin of this hormone.

Hormones also control some processes that are unique to invertebrates. As mentioned in the chapter-opening story, hormones control insect *metamorphosis*, the dramatic transformation that some insects undergo between hatching from an egg as a wormlike larva, going through several molts where the

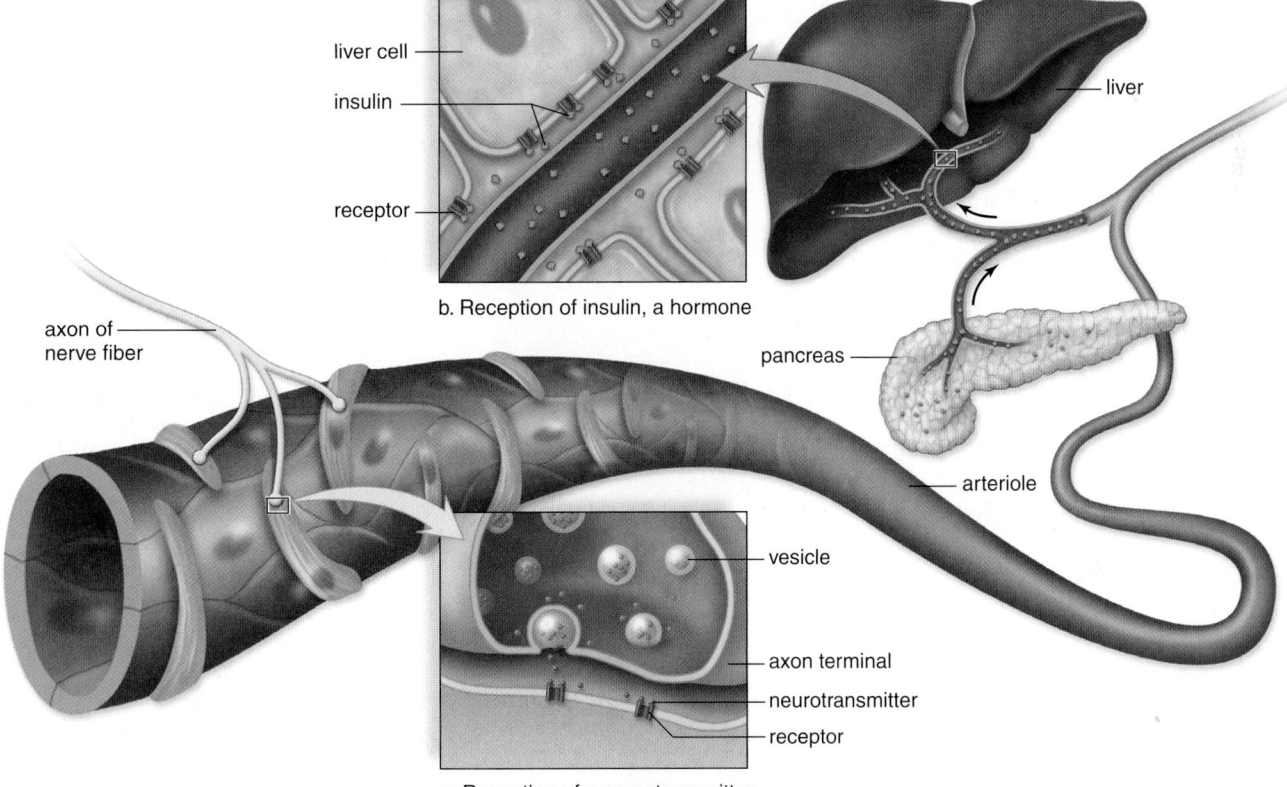

b. Reception of insulin, a hormone

a. Reception of a neurotransmitter

Figure 40.1 Modes of action of the nervous and endocrine systems. **a.** Nerve impulses passing along an axon cause the release of a neurotransmitter. The neurotransmitter, a chemical signal, binds to a receptor and causes the wall of an arteriole to constrict. **b.** The hormone insulin, a chemical signal, travels in the cardiovascular system from the pancreas to the liver, where it binds to a receptor and causes liver cells to store glucose as glycogen.

HYPOTHALAMUS
Releasing and inhibiting hormones: regulate the anterior pituitary

PITUITARY GLAND
Posterior Pituitary

Antidiuretic (ADH): water reabsorption by kidneys

Oxytocin: stimulates uterine contraction and milk letdown

Anterior Pituitary

Thyroid stimulating (TSH): stimulates thyroid

Adrenocorticotropic (ACTH): stimulates adrenal cortex

Gonadotropic (FSH, LH): egg and sperm production; sex hormone production

Prolactin (PL): milk production

Growth (GH): bone growth, protein synthesis, and cell division

THYROID
Thyroxine (T_4) and triiodothyronine (T_3): increase metabolic rate; regulates growth and development

Calcitonin: lowers blood calcium level

ADRENAL GLAND
Adrenal cortex

Glucocorticoids (cortisol): raises blood glucose level; stimulates breakdown of protein

Mineralocorticoids (aldosterone): reabsorption of sodium and excretion of potassium

Sex hormones: reproductive organs and bring about sex characteristics

Adrenal medulla

Epinephrine and norepinephrine: active in emergency situations; raise blood glucose level

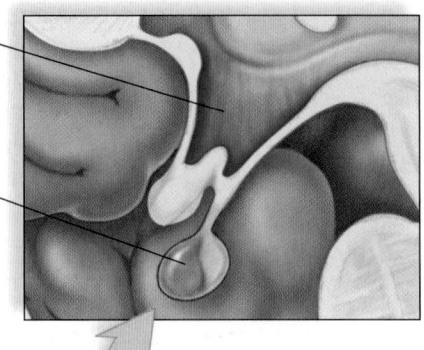

Figure 40.2 Major glands of the human endocrine system.   Major glands and the hormones they produce are depicted. Also, the endocrine system includes other organs such as the kidneys, gastrointestinal tract, and the heart, which also produce hormones but not as a primary function of these organs.

PINEAL GLAND
Melatonin: controls circadian and circannual rhythms

PARATHYROIDS
Parathyroid hormone (PTH): raises blood calcium level

parathyroid glands (posterior surface of thyroid)

THYMUS
Thymosins: production and maturation of T lymphocytes

PANCREAS
Insulin: lowers blood glucose level and promotes glycogen buildup

Glucagon: raises blood glucose level and promotes glycogen breakdown

testis (male)

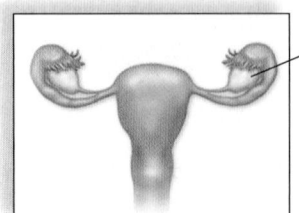

ovary (female)

GONADS
Testes

Androgens (testosterone): male sex characteristics

Ovaries

Estrogens and progesterone: female sex characteristics

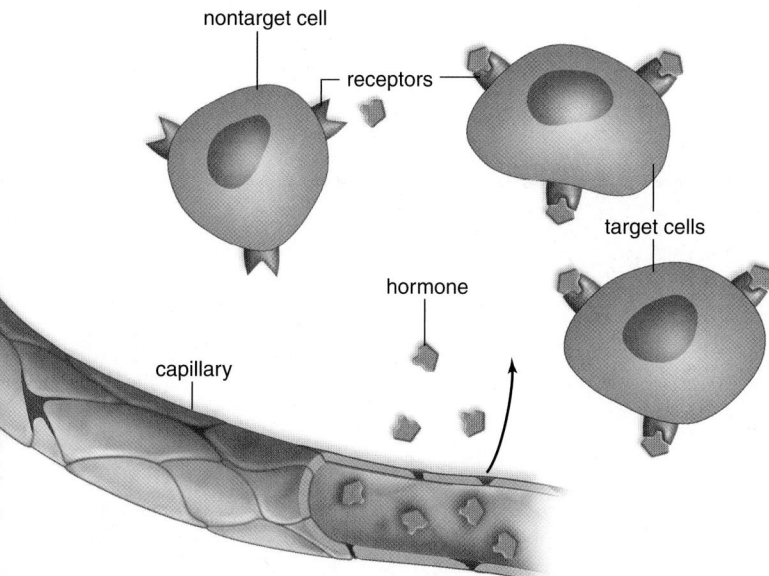

Figure 40.3 Target cell concept. Most hormones are distributed by the bloodstream to target cells. Target cells have receptors for the hormone, and the hormone combines with the receptor as a key fits a lock.

exoskeleton is shed, and maturing into adults. Several hormones have been identified that control this process.

Sometimes evolution produces new uses for the same hormones. In the freshwater snail *Lymnaea*, a peptide related to insulin is involved in body and shell growth as well as in energy metabolism. Variable hormone functions are seen in vertebrates as well. All vertebrates synthesize thyroid hormones, which generally increase metabolism, as you'll see later in this chapter. In amphibians, a surge of thyroid hormones also seems to promote metamorphosis from a tadpole into an adult. In contrast, the hormone prolactin inhibits metamorphosis in amphibians, stimulates skin pigmentation in reptiles, initiates incubation of eggs in birds, and stimulates milk production in mammals.

Hormones Are Chemical Signals

Like other **chemical signals,** hormones are a means of communication between cells, between body parts, and even between individuals. It must be stressed that only certain cells, called target cells, can respond to a specific hormone. A target cell for a particular hormone carries a receptor protein for that hormone (Fig. 40.3). The hormone and receptor protein bind together like a key fits a lock. The target cell then responds to that hormone. In a condition called *androgen insensitivity*, an individual has X and Y sex chromosomes, and the testes, which remain in the abdominal cavity, produce the sex hormone testosterone. However, the body cells lack receptors that are able to combine with testosterone, and the individual appears to be a normal female.

Chemical signals that influence the behavior of other individuals are called **pheromones.** Pheromones have been well documented in several animal species, although their influence

has been more difficult to prove in humans. Women who live in the same household tend to have synchronized menstrual cycles, perhaps because pheromones released by a woman who is menstruating affect the menstrual cycle of other women in the household. Studies also suggest that women prefer the smell of tee shirts worn by men who have a different MHC type from themselves. As noted in Chapter 33, MHC molecules are involved in immunity, and choosing a mate of a different MHC type could conceivably improve the health of offspring.

In a small study reported in 2011, men had lower testosterone levels in their saliva after they smelled a jar containing tears from women who were sad, as compared to saline droplets that were trickled down the women's cheeks. While the significance of these studies is unclear, they do suggest that humans may in fact be influenced by pheromones.

The Action of Hormones

Hormones exert a wide range of effects on cells. Some hormones induce target cells to increase their uptake of particular molecules, such as glucose, or ions, such as calcium. Others bring about an alteration of the target cell's structure in some way.

Most endocrine glands secrete **peptide hormones.** These hormones are peptides, proteins, glycoproteins, and modified amino acids. **Steroid hormones,** in contrast, all have the same molecular complex of four carbon rings because they are all derived from cholesterol (see Fig. 3.12).

The Action of Peptide Hormones. The actions of peptide hormones can vary, and as an example here, we concentrate on what happens in muscle cells after the hormone epinephrine binds to a receptor in the plasma membrane (Fig. 40.4). In muscle cells, the reception of epinephrine leads to the breakdown of glycogen to glucose, which provides energy for ATP production.

The immediate result of epinephrine binding is the formation of **cyclic adenosine monophosphate (cAMP).** Cyclic AMP contains one phosphate group attached to adenosine at two locations. Therefore, the molecule is cyclic. Cyclic AMP activates a protein kinase enzyme in the cell, and this enzyme, in turn, activates another enzyme, and so forth. The series of enzymatic reactions that follows cAMP formation is called an enzyme cascade (or signaling cascade). Because each enzyme can be used over and over again, more enzymes become involved at every step of the cascade. Finally, many molecules of glycogen are broken down to glucose, which enters the bloodstream.

Animation
Peptide Hormone Action

Typical of a peptide hormone, epinephrine never enters the cell. Therefore, the hormone is called the **first messenger,** whereas cAMP, which sets the metabolic machinery in motion, is called the **second messenger.** To explain this terminology, let's imagine that the adrenal medulla, which produces epinephrine, is like the home office that sends out a courier (i.e., the hormone epinephrine—the first messenger) to a factory (the cell). The courier doesn't have a pass to enter the factory, but tells a supervisor through the screen door that the home office wants the factory to produce a particular product. The supervisor (i.e., cAMP—the second messenger) walks over and flips a

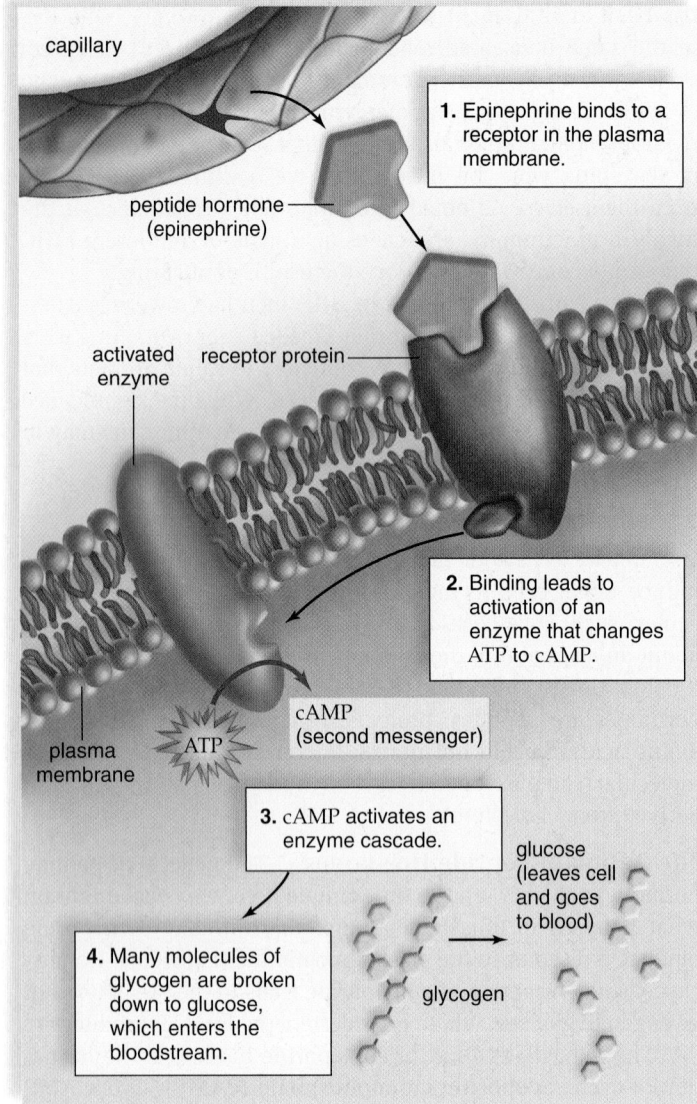

Figure 40.4 Epinephrine, a peptide hormone. Peptide hormones (epinephrine in this example) act as first messengers, binding to specific receptors in the plasma membrane. First messengers activate second messengers (cAMP in this case) that influence various cellular processes.

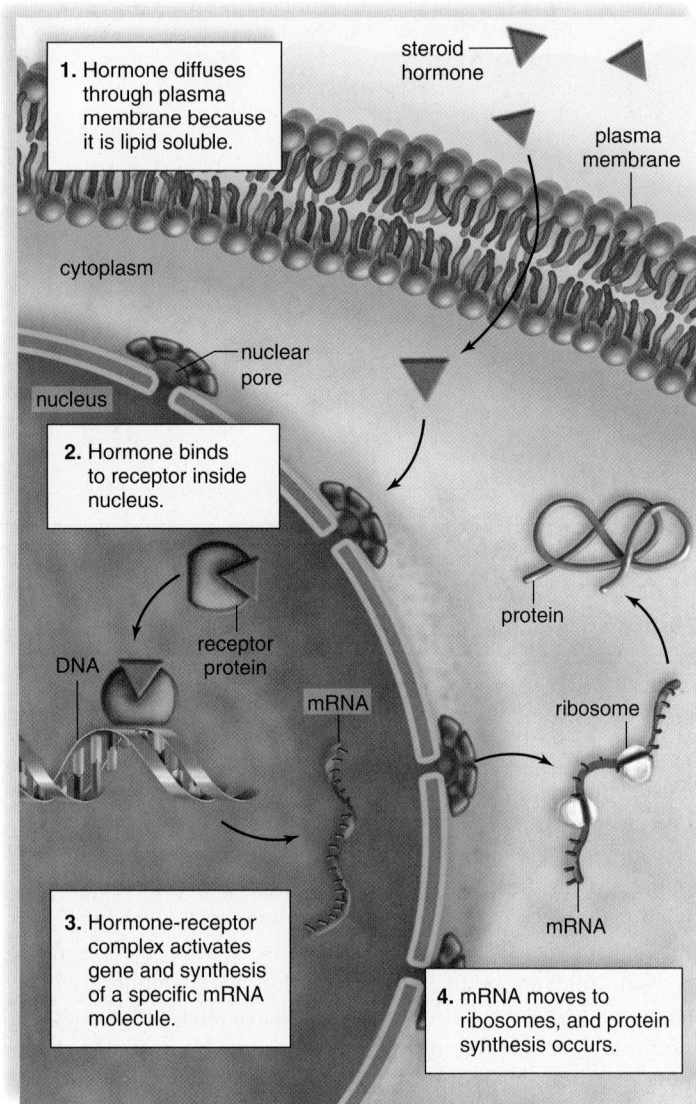

Figure 40.5 Steroid hormone. A steroid hormone passes directly through the target cell's plasma membrane before binding to a receptor in the nucleus or cytoplasm. The hormone-receptor complex binds to DNA and gene expression follows.

switch that starts the machinery (the enzymatic pathway), and a product is made.

Animation
Second Messengers

The Action of Steroid Hormones. Only the adrenal cortex, the ovaries, and the testes produce steroid hormones. Steroid hormones do not bind to plasma membrane receptors; instead they are able to enter the cell because they are lipids (Fig. 40.5). Once inside, a steroid hormone binds to an internal receptor, usually in the nucleus but sometimes in the cytoplasm. Inside the nucleus, the hormone-receptor complex binds with DNA and activates certain genes. Messenger RNA (mRNA) moves to the ribosomes in the cytoplasm and protein synthesis (e.g., an enzyme) follows. To continue the analogy, a steroid hormone is like a courier that has a pass to enter the factory (the cell). Once inside, the courier makes contact with the plant manager

(DNA), who sees to it that the factory (cell) is ready to produce a product.

Steroids act more slowly than peptides because it takes more time to synthesize new proteins than to activate enzymes already present in cells. Their action lasts longer, however.

Animation
Mechanism of Steroid Hormone Action

Check Your Progress   40.1

1. Compare and contrast the nervous and endocrine systems with regard to function and the types of signals used.
2. Compare the location of the receptors for peptide and steroid hormones.
3. Explain why second messengers are needed for most peptide hormones.

40.2 Hypothalamus and Pituitary Gland

Learning Outcomes

Upon completion of this section, you should be able to

1. Describe the relationship between the hypothalamus and the pituitary gland.
2. List and describe the functions of the hormones released by the anterior and posterior pituitary gland.
3. Explain how some hormones are regulated by negative feedback, and some by positive feedback; give an example of each.

The **hypothalamus** helps to regulate the internal environment of the body in two ways. Through the autonomic nervous system, it influences the heartbeat, blood pressure, appetite, body temperature, and water balance. It also controls the glandular secretions of the **pituitary gland** (hypophysis), a small gland about 1 cm in diameter that is connected to the hypothalamus by a stalklike structure. The pituitary has two portions: the posterior pituitary and the anterior pituitary.

MP3
Endocrine System

Posterior Pituitary

Neurons in the hypothalamus called neurosecretory cells produce the hormones **antidiuretic hormone (ADH)** [Gk. *anti*, against, and *ouresis*, urination] and oxytocin (Fig. 40.6, *left*). These hormones pass through axons into the **posterior pituitary,** where they are stored in axon terminals. Certain neurons in the hypothalamus are sensitive to the water-salt balance of the blood. When these cells determine that the blood is too concentrated, ADH is released from the posterior pituitary. Upon reaching the kidneys, ADH causes water to be reabsorbed. As the blood becomes dilute, ADH is no longer released. This is an example of control by **negative feedback** because the *effect* of the hormone (to dilute blood) acts to shut down the *release* of the hormone. Negative feedback maintains stable conditions and homeostasis.

Animation
Hormonal Communication

If too little ADH is secreted, or if the kidneys become unresponsive to ADH, a condition known as *diabetes insipidus* results. Patients with this condition are usually very thirsty; they produce copious amounts of urine and can become severely dehydrated if the condition is untreated.

The consumption of alcohol inhibits ADH release. This effect helps to explain the frequent urination associated with drinking alcohol.

Oxytocin [Gk. *oxys*, quick, and *tokos*, birth], the other hormone made in the hypothalamus, causes uterine contractions during childbirth and milk letdown when a baby is nursing. The more the uterus contracts during labor, the more nerve impulses reach the hypothalamus, causing oxytocin to be released. Similarly, the more a baby suckles, the more oxytocin is released. In both instances, the release of oxytocin from the posterior pituitary is controlled by **positive**

feedback—that is, the stimulus continues to bring about an effect that ever increases in intensity. Oxytocin may also play a role in the propulsion of semen through the male reproductive tract and may affect feelings of sexual satisfaction and emotional bonding.

Anterior Pituitary

A portal system, which consists of two capillary networks connected by a vein, lies between the hypothalamus and the anterior pituitary (Fig. 40.6, *right*). The hypothalamus controls the anterior pituitary by producing **hypothalamic-releasing hormones** and in some instances **hypothalamic-inhibiting hormones.** For example, one hypothalamic-releasing hormone stimulates the anterior pituitary to secrete a thyroid-stimulating hormone, and a particular hypothalamic-inhibiting hormone prevents the anterior pituitary from secreting prolactin.

Anterior Pituitary Hormones Affecting Other Glands

Some of the hormones produced by the **anterior pituitary** affect other glands. **Gonadotropic hormones** stimulate the gonads—the testes in males and the ovaries in females—to produce gametes and sex hormones. **Adrenocorticotropic hormone (ACTH)** stimulates the adrenal cortex to produce cortisol. **Thyroid-stimulating hormone (TSH)** stimulates the thyroid to produce thyroxine (T_4) and triiodothyronine (T_3). In each instance, the blood level of the last hormone in the sequence exerts negative feedback control over the secretion of the first two hormones. This is how it works for TSH:

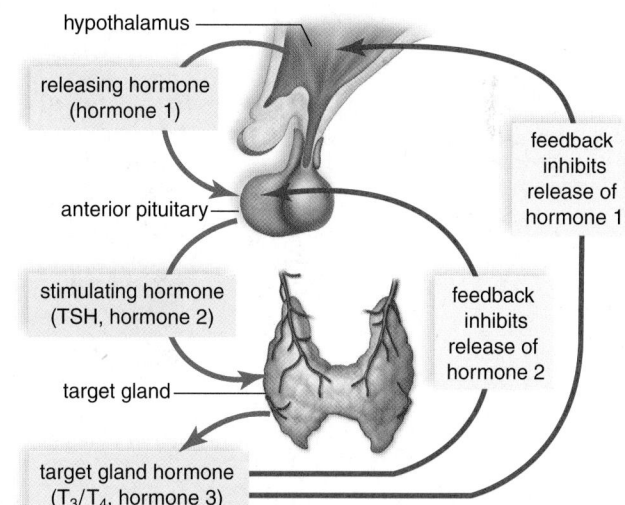

hypothalamus

releasing hormone (hormone 1)

anterior pituitary

feedback inhibits release of hormone 1

stimulating hormone (TSH, hormone 2)

feedback inhibits release of hormone 2

target gland

target gland hormone (T_3/T_4, hormone 3)

Anterior Pituitary Hormones Not Affecting Other Glands

Three hormones produced by the anterior pituitary do not affect other endocrine glands. **Prolactin (PRL)** [L. *pro*, before, and *lactis*, milk] is produced in quantity only after childbirth. It causes the mammary glands in the breasts to develop and produce milk. It also plays a role in carbohydrate and fat metabolism.

Melanocyte-stimulating hormone (MSH) [Gk. *melanos*, black, and *kytos*, cell] causes skin-color changes in many fishes, amphibians, and reptiles having melanophores, special skin cells that produce color variations. The concentration of this hormone in humans is very low.

Growth hormone (GH), or somatotropic hormone, promotes skeletal and muscular growth (Fig. 40.6, *right*). It increases the rate at which amino acids enter cells and protein synthesis occurs. It also promotes fat metabolism as opposed to glucose metabolism. The amount of GH produced is greatest during childhood and adolescence.

If too little GH is produced during childhood, the individual has *pituitary dwarfism,* characterized by normal proportions but small stature. Such children also have problems with low blood sugar (hypoglycemia), because GH normally helps oppose the effect of insulin on glucose uptake. Through the administration of GH, growth patterns can be restored and blood sugar problems alleviated. If too much GH is secreted

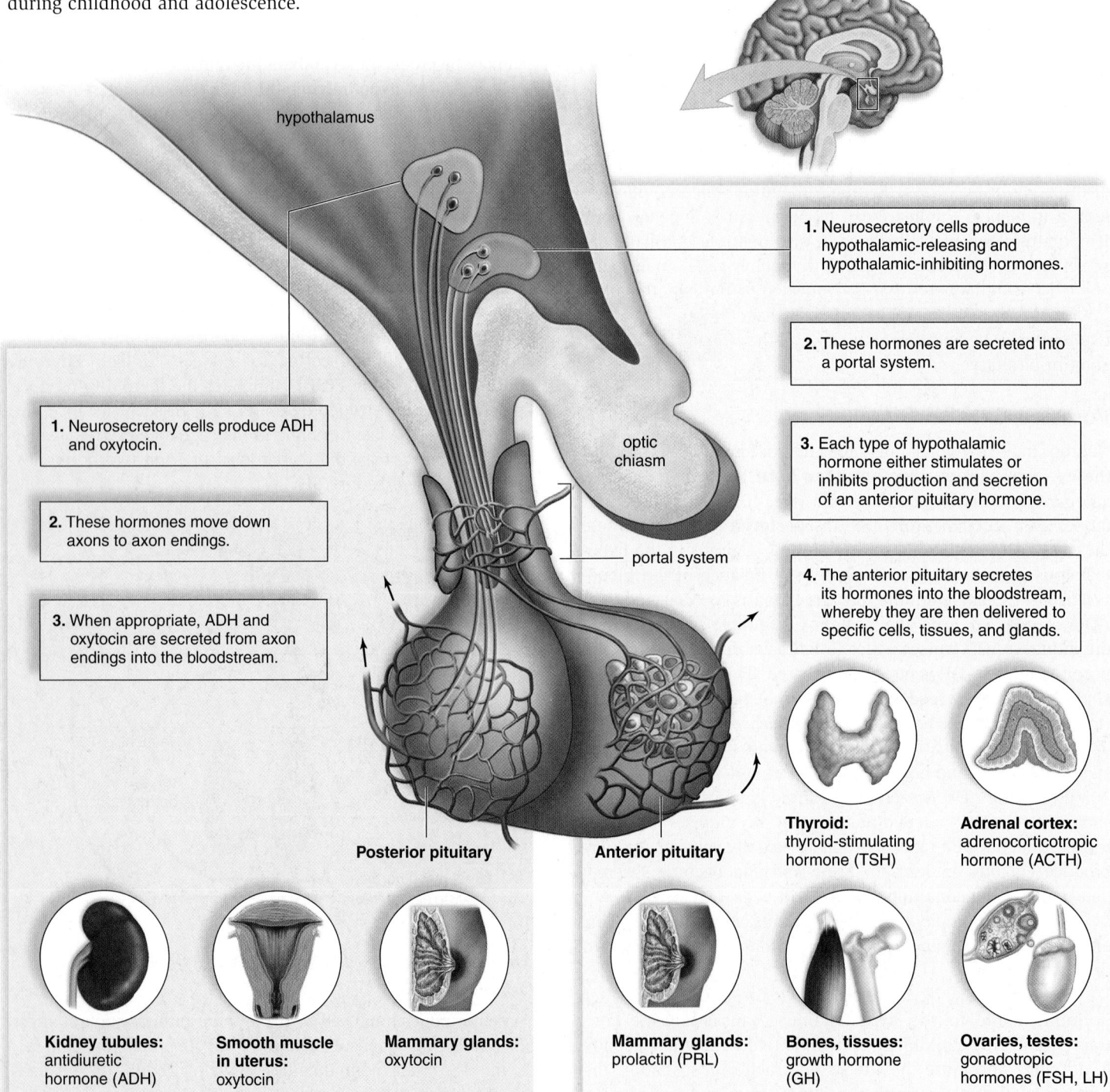

hypothalamus

1. Neurosecretory cells produce hypothalamic-releasing and hypothalamic-inhibiting hormones.

2. These hormones are secreted into a portal system.

optic chiasm

1. Neurosecretory cells produce ADH and oxytocin.

2. These hormones move down axons to axon endings.

3. When appropriate, ADH and oxytocin are secreted from axon endings into the bloodstream.

3. Each type of hypothalamic hormone either stimulates or inhibits production and secretion of an anterior pituitary hormone.

portal system

4. The anterior pituitary secretes its hormones into the bloodstream, whereby they are then delivered to specific cells, tissues, and glands.

Posterior pituitary

Anterior pituitary

Thyroid: thyroid-stimulating hormone (TSH)

Adrenal cortex: adrenocorticotropic hormone (ACTH)

Kidney tubules: antidiuretic hormone (ADH)

Smooth muscle in uterus: oxytocin

Mammary glands: oxytocin

Mammary glands: prolactin (PRL)

Bones, tissues: growth hormone (GH)

Ovaries, testes: gonadotropic hormones (FSH, LH)

Figure 40.6 Hypothalamus and the pituitary.   In this diagram, the name of the hormone is given below its target organ, which is depicted in the circle. *Left:* The hypothalamus produces two hormones, ADH and oxytocin, which are stored and secreted by the posterior pituitary. *Right:* The hypothalamus controls the secretions of the anterior pituitary, and the anterior pituitary controls the secretions of the thyroid, adrenal cortex, and gonads, which are also endocrine glands.

a.

b.

Figure 40.7
Effect of growth hormone.
a. The amount of growth hormone produced by the anterior pituitary during childhood affects the height of an individual. Plentiful growth hormone produces very tall basketball players. **b.** Too much growth hormone can lead to giantism, while an insufficient amount results in limited stature and even pituitary dwarfism.

during childhood, the person may become a giant (Fig. 40.7b). Giants usually have poor health, primarily because elevated GH cancels out the effects of insulin, promoting diabetes mellitus (see the discussion in Section 40.3).

On occasion, GH is overproduced in the adult, and a condition called *acromegaly* results. Because long bone growth is no longer possible in adults, only the feet, hands, and face (particularly the chin, nose, and eyebrow ridges) can respond, and these portions of the body become overly large (Fig. 40.8).

A quick internet search reveals that many websites offer human growth hormone for sale as a "fountain of youth" that can help adults lose weight, add muscle, and reduce the effects of aging. Indeed, several professional actors and athletes have admitted to using human GH to build muscle and reduce body

fat. (Note that GH is not a "steroid," which refers to testosterone or related hormones.) However, using human GH in this manner can have many undesired side effects, such as joint and muscle pain, high blood pressure, and diabetes mellitus.

MP3
Hormonal Secretion Action

Check Your Progress 40.2

1. Explain how the hypothalamus communicates with the endocrine system.
2. List the hormones produced by the posterior pituitary and provide a function for each.
3. List the hormones produced by the anterior pituitary and provide a function for each.

Age 9   Age 16   Age 33   Age 52

Figure 40.8
Acromegaly.
Acromegaly is caused by overproduction of GH in the adult. It is characterized by enlargement of the bones in the face, the fingers, and the toes as a person ages.

40.3 Other Endocrine Glands and Hormones

Learning Outcomes

Upon completion of this section, you should be able to

1. Distinguish between the functions of T_3, T_4, calcitonin, and parathyroid hormone.
2. Compare and contrast the mineralocorticoids and glucocorticoids.
3. Identify the cause and major symptoms of the major conditions associated with the endocrine system.

In this section we discuss the thyroid and parathyroid glands, adrenal glands, pancreas, pineal gland, thymus, and other tissues that produce hormones secondarily. All the hormone products of these glands and tissues play a role in health and homeostasis.

Thyroid and Parathyroid Glands

The **thyroid gland** [Gk. *thyreos,* large, door-shaped shield] is attached to the trachea just below the larynx (see Fig. 40.2). Weighing approximately 20 grams, the thyroid gland is composed of a large number of follicles, each a small spherical structure made of thyroid cells that produce the hormones triiodothyronine (T_3), which contains three iodine atoms, and **thyroxine (T_4),** which contains four iodine atoms. Cells of a different type in the thyroid gland produce the hormone calcitonin. The parathyroid glands, which produce parathyroid hormone, are embedded in the posterior surface of the thyroid gland.

Effects of T_3 and T_4

Both of these thyroid hormones increase the overall metabolic rate. They do not have a single target organ; instead, they stimulate most the cells of the body to metabolize at a faster rate. More glucose is broken down, and more energy is used. Interestingly, even though T_3 and T_4 are peptide hormones because they are derived from the amino acid tyrosine, their receptor is actually located inside cells, more like a steroid hormone receptor.

Animation Mechanism of Thyroxine Action

To produce T_3 and T_4, the thyroid gland actively acquires iodine. The concentration of iodine in the thyroid gland is approximately 25 times that found in the blood. If a person consumes an insufficient amount of iodine, the thyroid gland is unable to produce the required amount of T_3 and T_4. This results in constant stimulation of the thyroid by TSH released by the anterior pituitary. The thyroid gland enlarges, resulting in a *simple goiter* (Fig. 40.9*a*). In the 1920s, scientists discovered that the use of iodized salt allows the thyroid to produce thyroid hormones and therefore helps prevent simple goiter. However, iodine deficiency is still extremely common in some parts of the world, with an estimated two billion people (one-third of the world's population) still suffering from some degree of iodine deficiency.

An insufficiency of T_3 and T_4 in the newborn is called *congenital hypothyroidism* (cretinism) (Fig. 40.9*b*). Babies with this condition are short and stocky, and are often mentally retarded. The causes vary from iodine deficiency in the mother during her pregnancy, to genetic defects affecting the production of TSH, T_3, T_4, or the receptor for any of these hormones. Once detected, iodine deficiency is easily treated by insuring appropriate levels of iodine consumption. However, according to the American Thyroid Association, congenital hypothyroidism due to iodine deficiency remains the most common preventable cause of mental retardation in the world.

Even mild iodine deficiency during pregnancy, which may be present in some women in the United States, may be associated with low intelligence in children. In cases due to problems with the thyroid gland itself, thyroxine therapy is curative, but must begin as early as possible to avoid permanently stunted growth and mental retardation.

a. Simple goiter

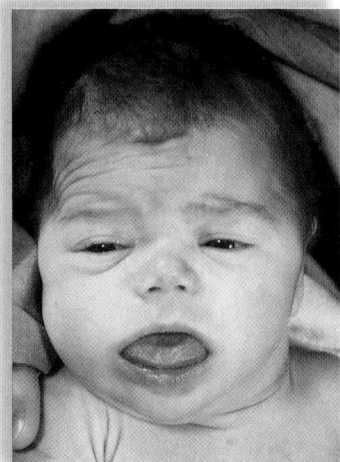
b. Congenital hypothyroidism

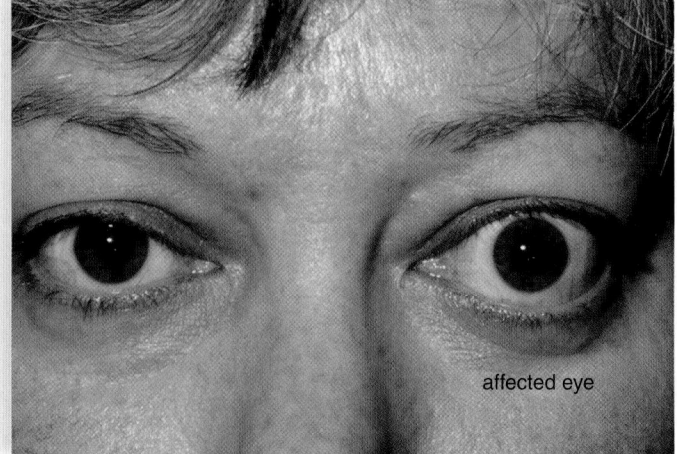
affected eye

c. Exophthalmic goiter

Figure 40.9 Abnormalities of the thyroid. **a.** An enlarged thyroid gland is often caused by a lack of iodine in the diet. Without iodine, the thyroid is unable to produce its hormones and continued anterior pituitary stimulation causes the gland to enlarge. **b.** Individuals who develop hypothyroidism during infancy or childhood do not grow and develop as others do. Unless medical treatment is begun, the body is short and stocky; mental retardation is also likely. **c.** In exophthalmic goiter, a goiter is due to an overactive thyroid, and the eye protrudes because of edema in eye socket tissue.

Hypothyroidism can also occur in adults, most often when the immune system produces antibodies that destroy the thyroid gland. Untreated hypothyroidism in adults results in *myxedema,* which is characterized by lethargy, weight gain, loss of hair, slower heart rate, lowered body temperature, and thickness and puffiness of the skin. The administration of thyroxine usually restores normal body functions and appearance.

Hyperthyroidism results from the oversecretion of T_3 or T_4. In Graves disease, antibodies are produced that react with the TSH receptor on thyroid follicular cells, mimicking the effect of TSH and causing the overproduction of T_3 and T_4. One typical sign of Graves disease is **exophthalmos** (exophthalmia) or excessive protrusion of the eyes due to edema in eye socket tissues and inflammation of the muscles that move the eyes (Fig. 40.9c). The patient usually becomes hyperactive, nervous, and irritable and suffers from insomnia. Graves disease is the most common cause of hyperthyroidism in children and adolescents, and is five to ten times more common in females. Available treatments include drugs that block iodine uptake by the thyroid gland, surgical removal of part or all of the gland, or administration of radioactive iodine to destroy the overactive tissue.

Hyperthyroidism can also be caused by thyroid cancer, the most common cancer of the endocrine system, which is usually detected as a lump during physical examination. Again, the treatment is surgery in combination with administration of radioactive iodine. The prognosis for most patients is excellent.

Effects of Calcitonin

Calcium (Ca^{2+}) plays a significant role in both nervous conduction and muscle contraction. It is also necessary for blood clotting and the maintenance of healthy bones and teeth. The blood calcium level is regulated in part by **calcitonin,** a hormone secreted by the thyroid gland when the blood calcium level rises. The primary effect of calcitonin is to bring about the deposit of calcium in the bones (Fig. 40.10, *top*). It does this by temporarily reducing the activity and number of osteoclasts. When the blood calcium lowers to normal, the release of calcitonin by the thyroid is inhibited.

While calcitonin appears to play a very important role in regulating calcium homeostasis in fish and a few other animals, it appears to be less significant in humans. As evidence for this, a deficiency of calcitonin (as occurs when the thyroid glands are removed) is not linked with any specific disorder. However, because of its bone-building effects, calcitonin is an FDA-approved drug for reducing bone loss in osteoporosis.

Parathyroid Glands

Parathyroid hormone (PTH), produced by the **parathyroid glands,** causes the blood calcium level to increase, and the blood phosphate (HPO_4^{2-}) level to decrease. Low blood calcium stimulates the release of PTH, which promotes the activity of osteoclasts, releasing calcium from the bones. PTH also promotes the reabsorption of calcium by the kidneys, lessening its excretion. In the kidneys, PTH also brings about activation of vitamin D. Vitamin D, in turn, stimulates the absorption of calcium from the intestine (Fig. 40.10, *bottom*). These effects bring the blood calcium level back to the normal range, and PTH secretion stops.

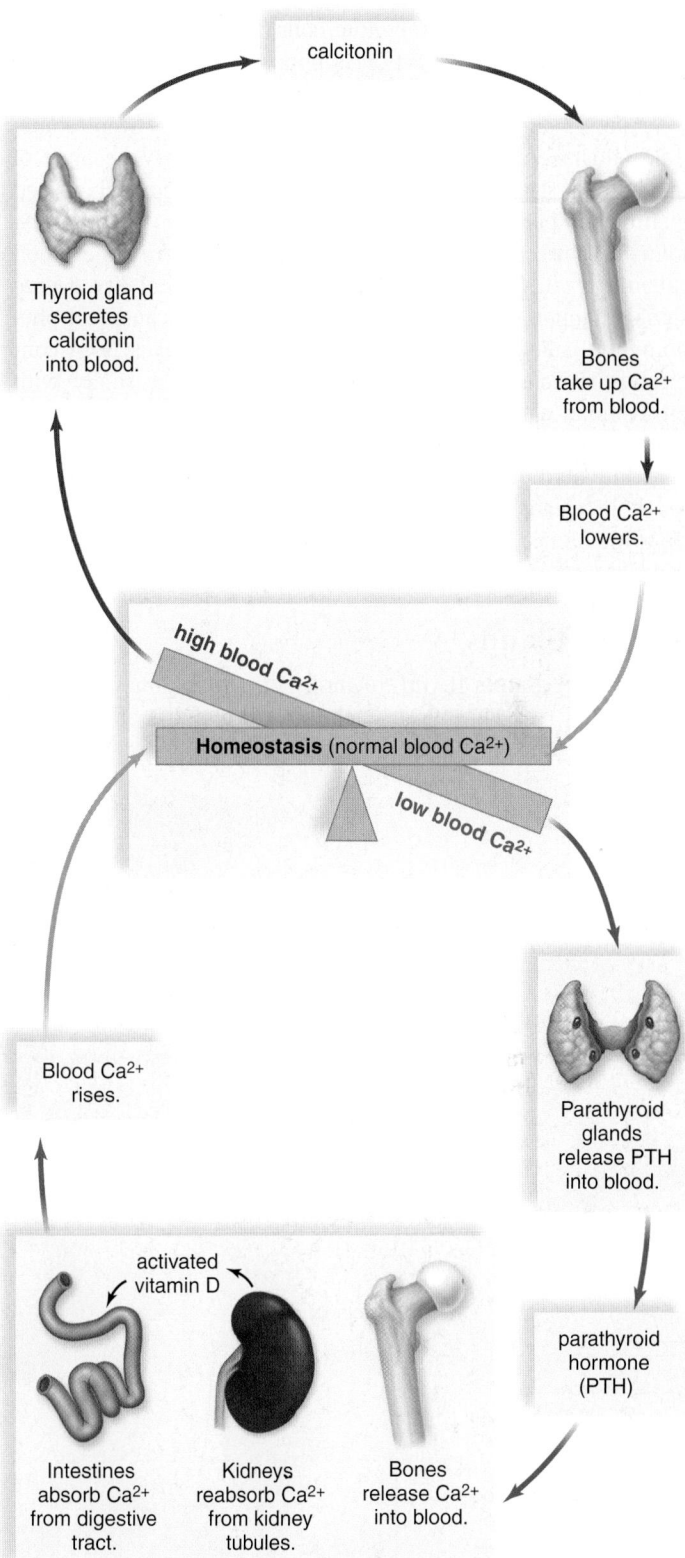

Figure 40.10 Regulation of blood calcium level. *Top:* When the blood calcium (Ca^{2+}) level is high, the thyroid gland secretes calcitonin. Calcitonin promotes the uptake of Ca^{2+} by the bones, and therefore the blood Ca^{2+} level returns to normal. *Bottom:* When the blood Ca^{2+} level is low, the parathyroid glands release parathyroid hormone (PTH). PTH causes the bones to release Ca^{2+} and the kidneys to reabsorb Ca^{2+} and activate vitamin D; thereafter, the intestines absorb Ca^{2+}. Therefore, the blood Ca^{2+} level returns to normal.

Calcitonin and PTH are therefore considered to be *antagonistic hormones* because their action is opposite to one another, and both hormones work together to regulate the blood calcium level.

Many years ago, the four parathyroid glands were sometimes mistakenly removed during thyroid surgery because of their small size and hidden location. Gland removal caused insufficient parathyroid hormone production or *hypoparathyroidism*. This condition leads to a dramatic drop in the blood calcium level, followed by excessive nerve excitability. Nerve signals happen spontaneously and without rest, causing a phenomenon called tetany. In *tetany*, the body shakes from continuous muscle contraction. Without treatment (usually with intravenous calcium), severe hypoparathyroidism causes seizures, heart failure, and death.

Untreated *hyperparathyroidism*, (oversecretion of PTH) can result in osteoporosis because of continuous calcium release from the bones. Hyperparathyroidism can also cause the formation of calcium kidney stones.

MP3 Calcium Homeostasis

Adrenal Glands

The **adrenal glands** [L. *ad*, toward, and *renis*, kidney] sit atop the kidneys (see Fig. 40.2). Each adrenal gland is about 5 cm long and 3 cm wide, weighs about 5 g, and consists of an inner portion called the **adrenal medulla** and an outer portion called the **adrenal cortex.** These portions, like the anterior and posterior pituitary, are two functionally distinct endocrine glands. Stress of all types, including both emotional and physical trauma, prompts the hypothalamus to stimulate both portions of the adrenal glands. The adrenal cortex is also involved in regulating the salt and water balance, and secreting a small amount of male and female sex hormones.

Adrenal Medulla

As noted in Chapter 37, during emergency situations that call for a "fight or flight" reaction, the hypothalamus sends nerve impulses by way of sympathetic nerve fibers to many organs, including the adrenal medulla (see Fig. 37.14). This neurological response to danger quickly dilates the pupils, speeds the heart, dilates the air passages, and inhibits many nonessential bodily functions. Meanwhile, the adrenal medulla secretes **epinephrine** (adrenaline) and **norepinephrine** (noradrenaline) into the bloodstream (Fig. 40.11). These hormones continue the response to stress throughout the body, for example by accelerating the breakdown of glucose to form ATP, triggering

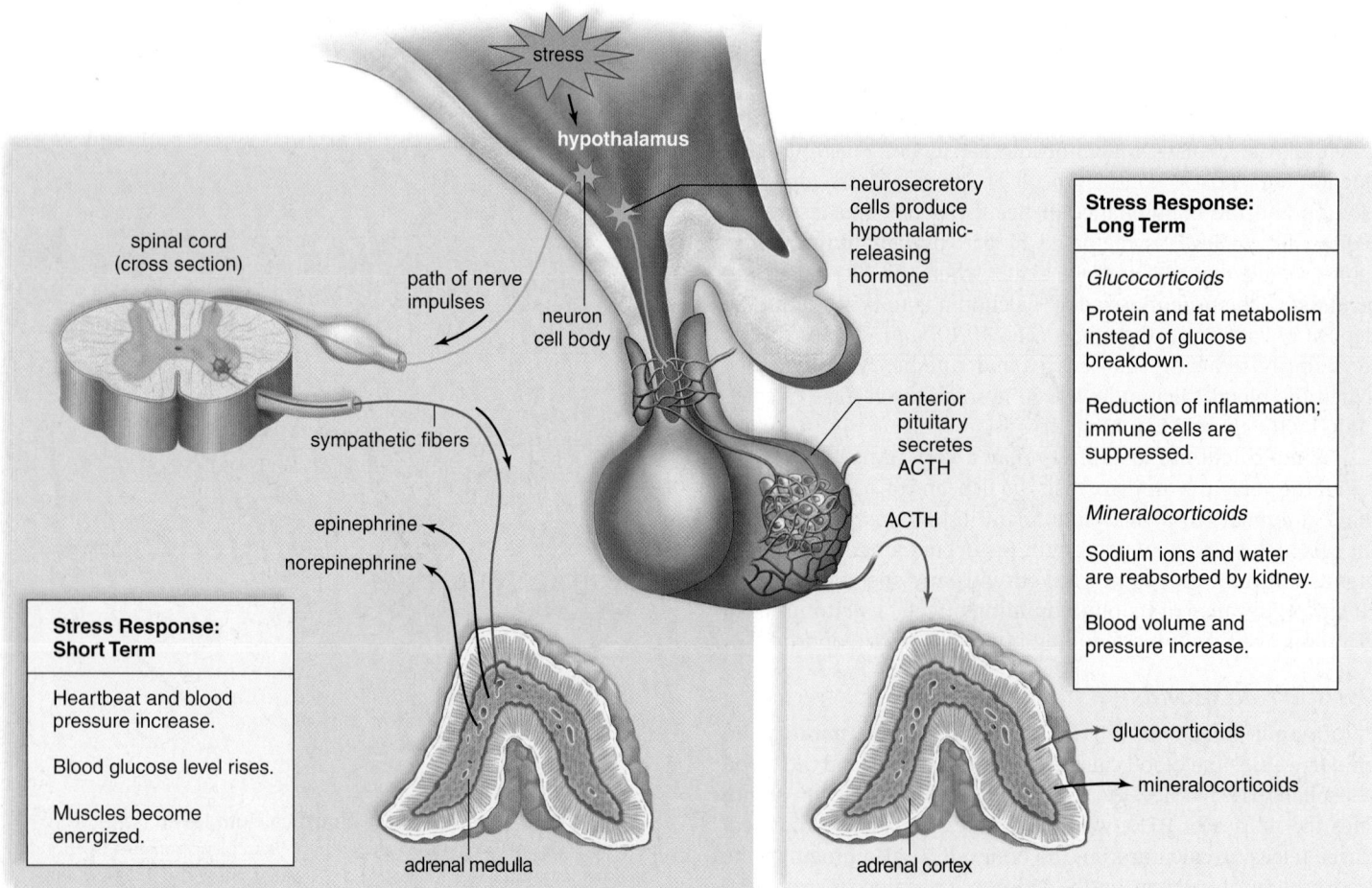

Figure 40.11 Adrenal glands. Both the adrenal cortex and the adrenal medulla are under the control of the hypothalamus when they help us respond to stress. *Left:* Nervous stimulation causes the adrenal medulla to provide a rapid, but short-term, stress response. *Right:* ACTH from the anterior pituitary causes the adrenal cortex to release glucocorticoids. Independently, the adrenal cortex releases mineralocorticoids. The adrenal cortex provides a slower, but long-term, stress response.

the mobilization of glycogen reserves in skeletal muscle, and increasing the cardiac rate and force of contraction. These effects are usually short-lived however, as these two hormones are rapidly metabolized by the liver.

Adrenal Cortex

In contrast to the rapid response of the sympathetic nervous system and adrenal medulla, the hypothalamus produces a longer-term response to stress by stimulating the anterior pituitary to produce ACTH, which in turn causes the adrenal cortex to secrete glucocorticoids.

Glucocorticoids.   **Cortisol** is the most important **glucocorticoid** produced by the human adrenal cortex. Cortisol raises the blood glucose level in at least two ways:

1. It promotes the breakdown of muscle proteins to amino acids, which are taken up by the liver from the bloodstream and converted into glucose.
2. It promotes the catabolism of fatty acids rather than carbohydrates, and this spares glucose. The rise in blood glucose is beneficial to an animal under stress because glucose is the preferred energy source for neurons.

Glucocorticoids also counteract the inflammatory response, including the type of reaction that leads to the pain and swelling of joints in arthritis and bursitis. Cortisone, a glucocorticoid, is often used to treat these conditions because it reduces inflammation. However, very high levels of glucocorticoids in the blood can suppress the body's defense system, rendering an individual more susceptible to injury and infection.

Animation
Action of Glucocorticoid

Mineralocorticoids.   **Mineralocorticoids** produced by the adrenal cortex regulate salt and water balance, leading to increases in blood volume and blood pressure. **Aldosterone** is the most important of the mineralocorticoids. Aldosterone primarily targets the kidneys, where it promotes renal absorption of sodium (Na^+) and renal excretion of potassium (K^+), and thereby helps regulate blood volume and blood pressure.

The secretion of mineralocorticoids is not controlled by the anterior pituitary. In Chapter 36 we noted that when the atria of the heart are stretched due to increased blood volume, cardiac cells release a hormone called **atrial natriuretic hormone (ANH),** which inhibits the secretion of aldosterone from the adrenal cortex. (Note that the heart is one of several organs in the body that release a hormone but are not considered among the major endocrine organs.) The effect of ANH is to cause *natriuresis,* the excretion of sodium ions (Na^+). When sodium is excreted, so is water, and therefore blood pressure lowers to normal (Fig. 40.12, *top*).

We also noted in Chapter 36 that when the blood sodium (Na^+) level, and therefore blood pressure, is low, the kidneys secrete **renin** (Fig. 40.12, *bottom*). Renin is an enzyme that converts the plasma protein angiotensinogen to angiotensin I, which is changed to angiotensin II by an enzyme in lung capillaries. Angiotensin II stimulates the adrenal cortex to release aldosterone. The effect of this process, called the renin-angiotensin-aldosterone system, is to raise blood pressure in two ways:

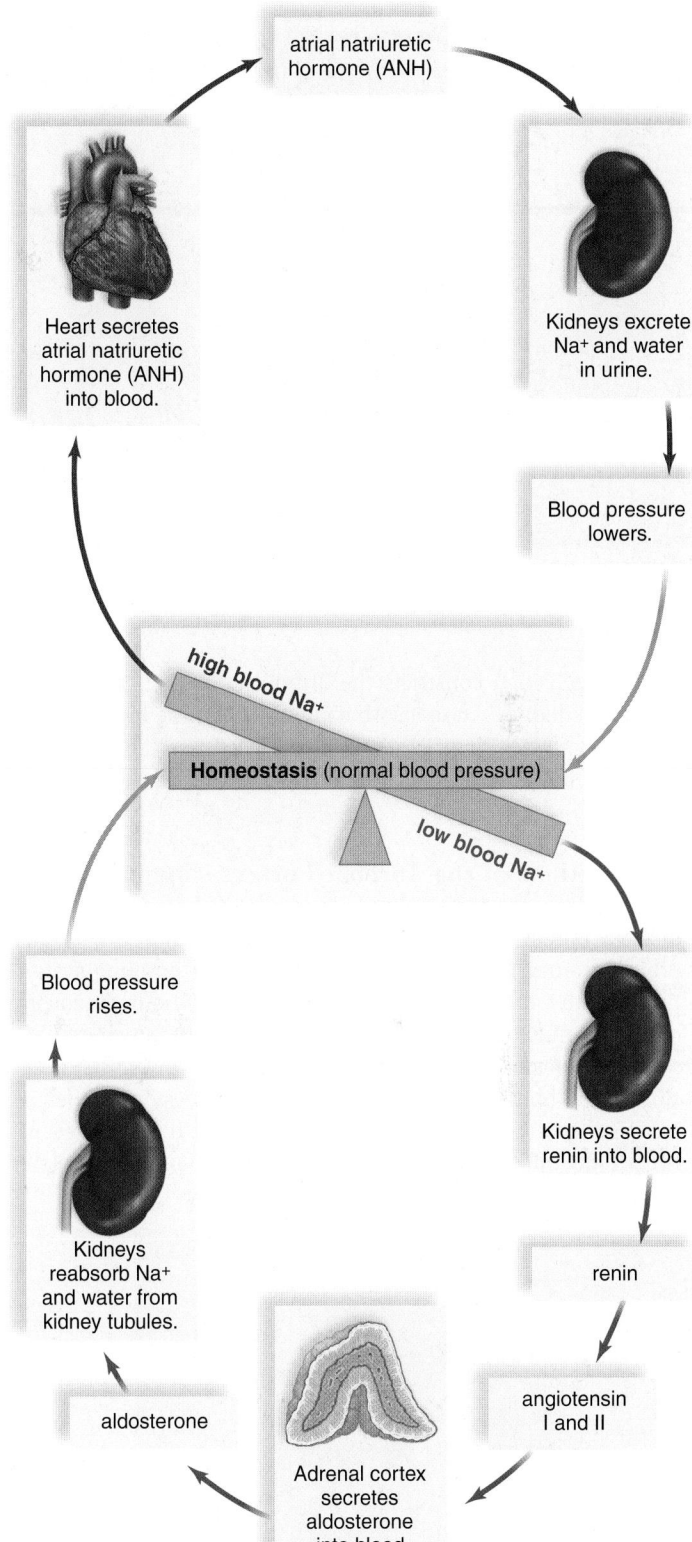

atrial natriuretic hormone (ANH)

Heart secretes atrial natriuretic hormone (ANH) into blood.

Kidneys excrete Na^+ and water in urine.

Blood pressure lowers.

high blood Na^+

Homeostasis (normal blood pressure)

low blood Na^+

Blood pressure rises.

Kidneys reabsorb Na^+ and water from kidney tubules.

Kidneys secrete renin into blood.

aldosterone

renin

Adrenal cortex secretes aldosterone into blood.

angiotensin I and II

Figure 40.12 Regulation of blood pressure and volume. *Top:* When the blood Na^+ is high, a high blood volume causes the heart to secrete atrial natriuretic hormone (ANH). ANH causes the kidneys to excrete Na^+, and water follows. The blood volume and pressure return to normal. *Bottom:* When the blood sodium (Na^+) level is low, a low blood pressure causes the kidneys to secrete renin. Renin leads to the secretion of aldosterone from the adrenal cortex. Aldosterone causes the kidneys to reabsorb Na^+, and water follows, so that blood volume and pressure return to normal.

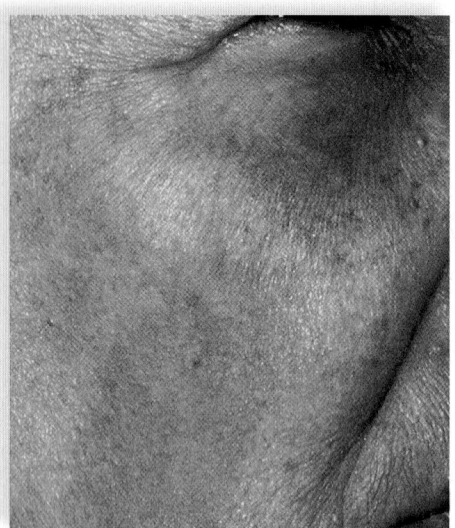

a.

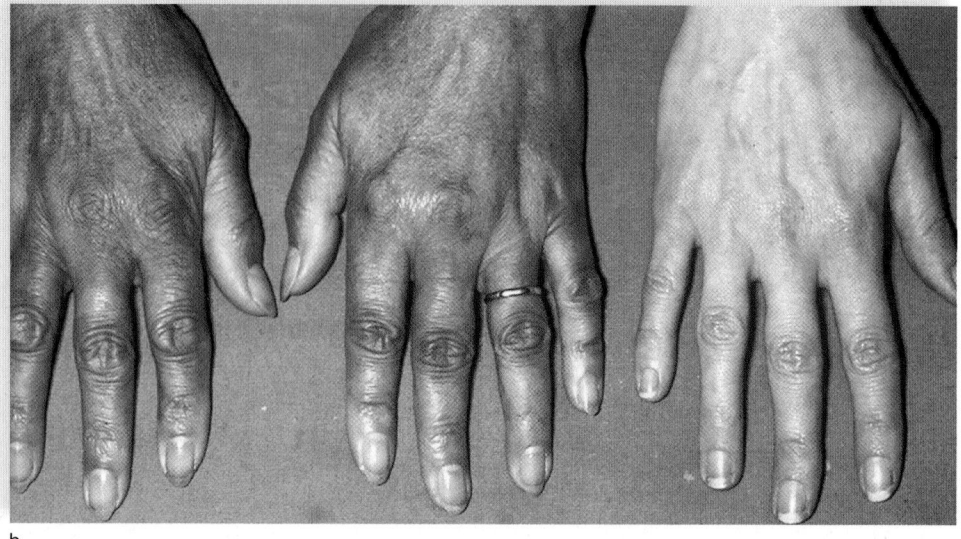

b.

Figure 40.13 Addison disease. Addison disease is characterized by a peculiar bronzing of the skin, particularly noticeable in light-skinned individuals. Note the color of the face **(a)** and the hands **(b)** compared to the hand of an individual without the disease.

(1) angiotensin II constricts the arterioles, and (2) aldosterone causes the kidneys to reabsorb sodium. When the blood sodium level rises, water is reabsorbed, in part because the hypothalamus secretes ADH (see Section 40.2). Then blood pressure rises to normal.

Malfunctions of the Adrenal Cortex. Insufficient secretion of hormones by the adrenal cortex, also known as *Addison disease*, is relatively rare. The most common cause is an inappropriate attack on the adrenal cortex by the immune system. Because the disease affects the secretion of both glucocorticoids and mineralocorticoids, a variety of symptoms may occur, such as dehydration, weakness, weight loss, and hypotension (low blood pressure). The presence of excessive but ineffective ACTH often causes increased pigmentation of the skin because ACTH, like MSH, can stimulate melanocytes to produce melanin

(Fig. 40.13). Treatment involves replacement of the missing hormones. Left untreated, Addison disease can be fatal.

Excessive levels of glucocorticoids results in *Cushing syndrome*. This disorder can be caused by tumors that affect either the pituitary gland, resulting in excess ACTH secretion, or the adrenal cortex itself. The most common cause, however, is the administration of glucocorticoids to treat other conditions (e.g., to suppress chronic inflammation). Regardless of the source, excess glucocorticoids cause muscle protein to be metabolized and subcutaneous fat to be deposited in the midsection (Fig. 40.14). Excess production of adrenal male sex hormones in women may result in masculinization, including an increase in body hair, deepening of the voice, and beard growth. Depending on the cause, treatment of Cushing syndrome may involve a careful reduction in the amount of cortisone being taken, the use of cortisol-inhibiting drugs, or surgery to remove any existing pituitary or adrenal tumor.

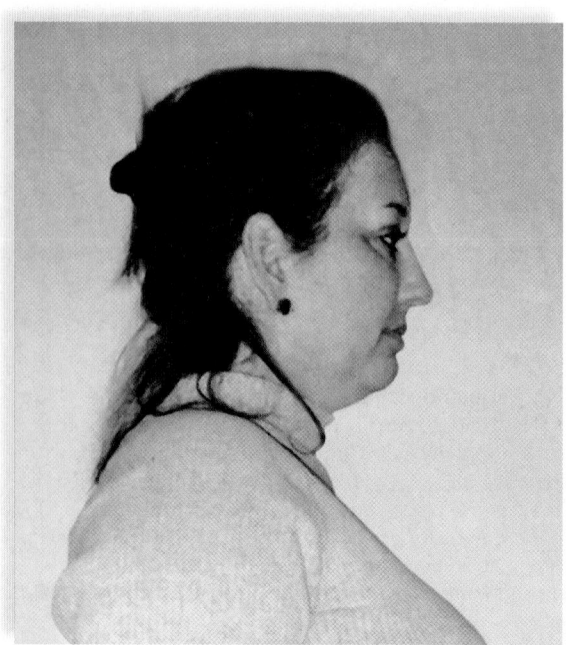

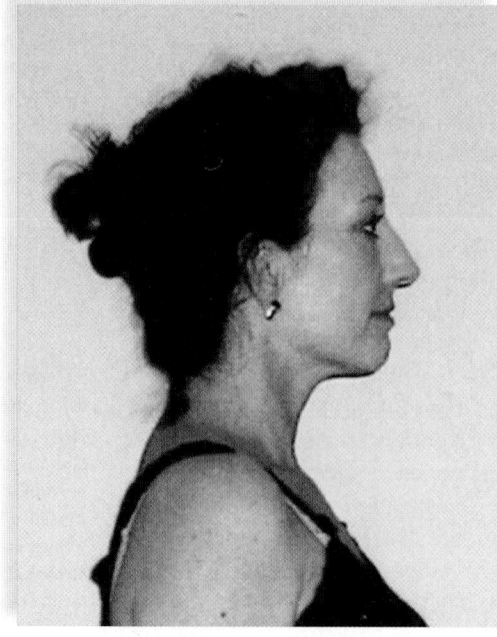

Figure 40.14 Cushing syndrome. This 40-year-old woman was diagnosed with a small tumor in her pituitary gland, which secreted large amounts of ACTH. The high ACTH levels stimulated the adrenal glands to produce excessive amounts of cortisol. *Left:* Patient at the time of surgery to remove her pituitary tumor. *Right:* Patient's appearance one year later.

Pancreas

The **pancreas** (see Fig. 40.2) is a slender, fish-shaped organ that stretches across the abdomen behind the stomach and near the duodenum of the small intestine. Approximately six inches long and weighing about 80 grams, the pancreas is composed of two types of tissue. Exocrine tissue produces and secretes digestive juices that go by way of ducts to the small intestine. Endocrine tissue, called the **pancreatic islets** (islets of Langerhans), produces and secretes the hormones insulin and glucagon directly into the blood. The Nature of Science feature on page 766 describes the discovery of insulin.

The pancreas is not under pituitary control. **Insulin** is secreted when there is a high blood glucose level, which usually occurs just after eating (Fig. 40.15, *top*). Insulin stimulates the uptake of glucose by cells, especially liver cells, muscle cells, and adipose tissue cells. In liver and muscle cells, glucose is then stored as glycogen, and in fat cells the breakdown of glucose supplies glycerol for the formation of fat. In these ways, insulin lowers the blood glucose level.

Glucagon is secreted from the pancreas, usually in between meals, when blood glucose is low (Fig. 40.15, *bottom*). The major target tissues of glucagon are the liver and adipose tissue. Glucagon stimulates the liver to break down glycogen to glucose and to use fat and protein in preference to glucose as energy sources. Adipose tissue cells break down fat to glycerol and fatty acids. The liver takes these up and uses them as substrates for glucose formation. In these ways, glucagon raises the blood glucose level. Insulin and glucagon are another example of antagonistic hormones, which work together to maintain the blood glucose level.

Diabetes Mellitus

In 2011, the National Diabetes Information Clearinghouse estimated that 25.8 million Americans (8.3% of the population) have *diabetes mellitus*, often referred to simply as diabetes. Of these, an estimated 7.0 million are undiagnosed. Diabetes is characterized by an inability of the body's cells, especially liver and muscle cells, to take up glucose as they should. This causes blood glucose to be higher than normal, and cells rely on other "fuels" like fatty acids for energy. Therefore, cellular famine exists in the midst of plenty. As the blood glucose level rises, glucose, along with water, is excreted in the urine (*mellitus*, from Greek, refers to "honey" or "sweetness"). This results in frequent urination and causes the diabetic to be extremely thirsty.

Other symptoms of diabetes include fatigue, constant hunger, and weight loss. Diabetics often experience vision problems due to diabetic retinopathy (see Section 38.2) and swelling in the lens of the eye due to the high blood sugar levels. If untreated, diabetics often develop serious and even fatal complications. Sores that don't heal develop into severe infections. Blood vessel damage causes kidney failure, nerve destruction, heart attack, or stroke.

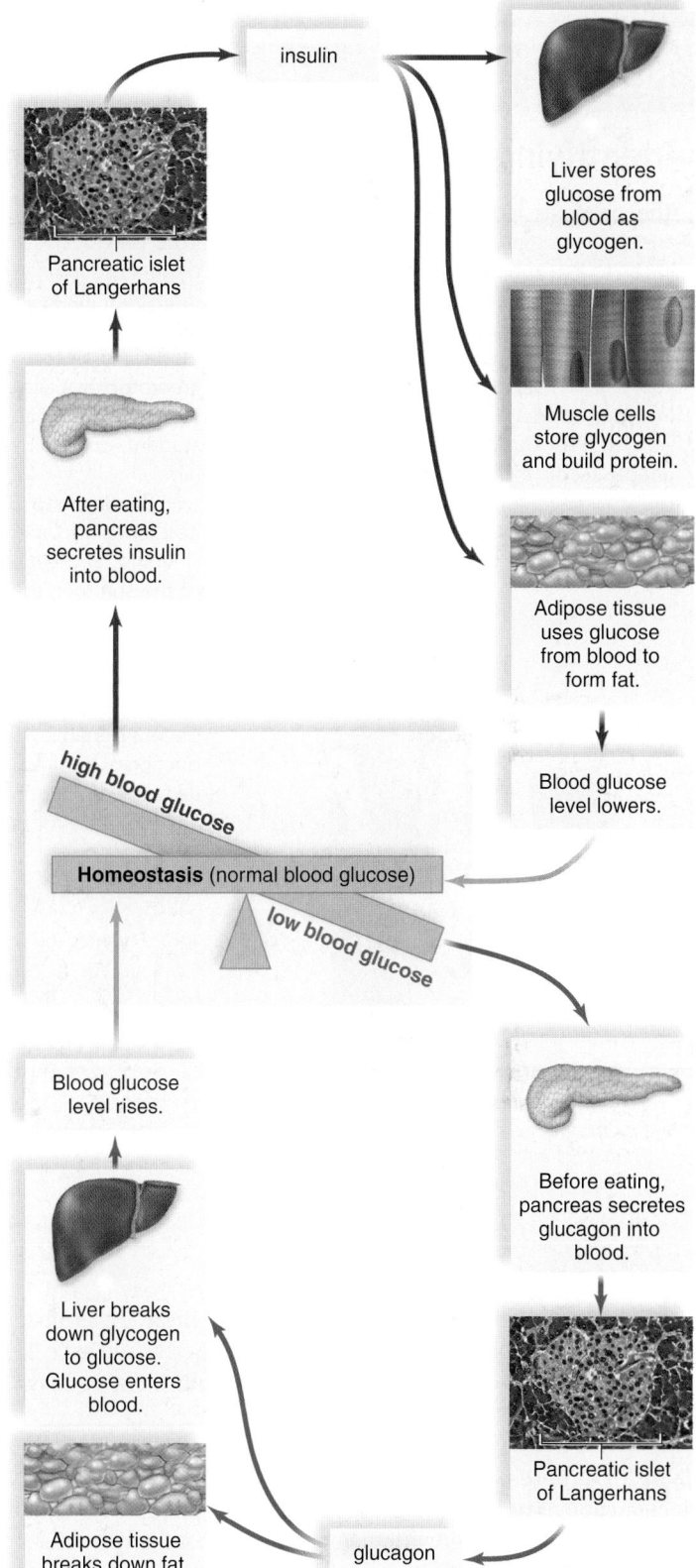

Figure 40.15 Regulation of blood glucose level.   *Top:* When the blood glucose level is high, the pancreas secretes insulin. Insulin promotes the storage of glucose as glycogen and the synthesis of proteins and fats (as opposed to their use as energy sources). Therefore, insulin lowers the blood glucose level to normal. *Bottom:* When the blood glucose level is low, the pancreas secretes glucagon. Glucagon acts opposite to insulin; therefore, glucagon raises the blood glucose level to normal.

Nature of Science

Identifying Insulin as a Chemical Messenger

The pancreas is both an exocrine gland and an endocrine gland. It sends digestive juices to the duodenum by way of the pancreatic duct, and it secretes the hormones insulin and glucagon into the bloodstream.

In 1920, physician Frederick Banting (1891–1941) decided to try to isolate insulin in order to identify it as a chemical messenger. Previous investigators had been unable to do this because the enzymes in the digestive juices destroyed insulin (a protein) during the isolation procedure. Banting hit upon the idea of ligating (tying

Figure 40A   **Early insulin experiments.**   Charles H. Best and Sir Frederick Banting in 1921 with the first dog to be kept alive by insulin.

off) the pancreatic duct, which he knew from previous research would lead to the degeneration only of the cells that produce digestive juices and not of the pancreatic islets (of Langerhans), where insulin is made. His professor, J. J. Macleod, made a laboratory available to him at the University of Toronto and also assigned a graduate student, Charles Best (1899–1978), to assist him.

Banting and Best (Fig. 40A) had limited funds and spent that summer working, sleeping, and eating in the lab. By the end of the summer, they had obtained pancreatic extracts that did lower the blood glucose level in diabetic dogs. Macleod then brought in biochemists, who purified the extract. Insulin therapy for the first human patient began in 1922, and large-scale production of purified insulin from pigs and cattle followed. Banting and Macleod received a Nobel Prize for their work in 1923.

The amino acid sequence of insulin was determined in 1953. Insulin is now synthesized using recombinant DNA technology, using the bacterium *E. coli* to produce the hormone. Banting and Best performed the required steps (given in the chart below) to identify a chemical messenger.

Questions to Consider

1. What type of disease or symptoms would you expect to occur after ligating the pancreatic ducts of dogs?
2. What are some advantages, and potential disadvantages, of producing a medicine destined to be injected in humans (such as insulin) in a bacterium like *E. coli*?
3. Some people oppose the use of animals for medical research. Do you think that insulin would have eventually been discovered without animal experimentation? Why or why not?

Steps to Identify a Chemical Messenger as Used by Banting and Best	
1. Identify the source of the chemical	Pancreatic islets are source
2. Identify the effect to be studied	Presence of pancreas in body lowers blood glucose
3. Isolate the chemical	Insulin isolated from pancreatic secretions
4. Show that the chemical has the desired effect	Insulin lowers blood glucose

Two types of diabetes have been identified; these are termed type 1 diabetes, sometimes called juvenile diabetes, and type 2 diabetes, or adult-onset diabetes; however, both diseases may occur in children or adults.

Type 1 Diabetes. In *type 1 diabetes,* the pancreas is not producing enough insulin. This condition is believed to be brought on by exposure to an environmental agent, most likely a virus, whose presence causes cytotoxic T cells to destroy the pancreatic islets. The body turns to the metabolism of fat, which leads to the buildup of ketones in the blood, called ketoacidosis, which increases acidity of the blood and can lead to coma and death.

Individuals with type 1 diabetes must have daily insulin injections. These injections control the diabetic symptoms but still can cause inconveniences because the blood sugar level may swing between hypoglycemia (low blood glucose) and

hyperglycemia (high blood glucose). Without testing the blood glucose level, it is difficult to be certain which of these is present because the symptoms can be similar. These symptoms include perspiration, pale skin, shallow breathing, and anxiety. Whenever these symptoms appear, immediate attention is required to bring the blood glucose back within the normal range. If the problem is hypoglycemia, the treatment is one or two glucose tablets, hard candy, or orange juice. If the problem is hyperglycemia, the treatment is insulin. Better control of blood glucose levels can often be achieved with an insulin pump, a small device worn outside the body that is connected to a plastic catheter inserted under the skin.

Because diabetes is such a common problem, many researchers are working to develop more effective methods for treating diabetes. The most desirable would be an artificial pancreas, defined as an automated system that would provide insulin based on real-time changes in blood sugar levels. It is also possible to transplant a working pancreas, or even fetal pancreatic islet cells, into patients with type 1 diabetes. Another possibility is *xenotransplantation,* in which insulin-producing islet cells of another species, such as pigs, are placed inside a capsule that allows insulin to exit but prevents the immune system from attacking the foreign cells. Finally, researchers are close to testing a vaccine that could block the immune system's attack on the islet cells, perhaps by inducing T cells capable of suppressing these responses.

Animation
Blood Sugar Regulation
in Diabetics

Type 2 Diabetes.   Most adult diabetics have *type 2 diabetes.* Often, the patient is overweight or obese, and adipose tissue produces a substance that impairs insulin receptor function. However, complex genetic factors can be involved, as shown by the tendency for type 2 diabetes to occur more often in certain families, or even ethnic groups. For example, the condition is 77% more common in African Americans than in non-Hispanic whites.

Normally, the binding of insulin to its plasma membrane receptor causes the number of protein carriers for glucose to increase, causing more glucose to enter the cell. In the type 2 diabetic, insulin still binds to its receptor, but the number of glucose carriers does not increase. Therefore, the cell is said to be insulin resistant.

It is possible to prevent or at least control type 2 diabetes by adhering to a low-fat, low-sugar diet and exercising regularly. If this fails, oral drugs are available that stimulate the pancreas to secrete more insulin and enhance the metabolism of glucose in the liver and muscle cells. Millions of Americans may have type 2 diabetes without being aware of it; yet, the effects of untreated type 2 diabetes are as serious as those of type 1 diabetes.

Testes and Ovaries

The activity of the testes and ovaries is controlled by the hypothalamus and pituitary. The **testes** are located in the scrotum, and the **ovaries** are located in the pelvic cavity. The testes produce **androgens** (e.g., **testosterone**), the male sex hormones. The ovaries produce **estrogens** and **progesterone,** the female sex hormones. These hormones provide feedback that controls the hypothalamic secretion of gonadotropin-releasing hormone (GnRH). The pituitary gland secretion of follicle-stimulating hormone (FSH) and luteinizing hormone (LH), the gonadotropic hormones, is controlled by feedback from the sex hormones, too. The activities of FSH and LH are discussed in Chapter 41.

Under the influence of the gonadotropic hormones, the testes release an increased amount of testosterone at the time of puberty, which stimulates growth of the penis and the testes. Testosterone also brings about and maintains the male secondary sex characteristics that develop during puberty. These include the growth of facial, axillary (underarm), and pubic hair. It prompts the larynx and the vocal cords to enlarge, causing the voice to lower. Testosterone also stimulates the activity of oil and sweat glands in the skin. Another side effect of testosterone is baldness. Genes for baldness are inherited by both sexes, but baldness is seen more often in males because of the presence of testosterone.

Testosterone is partially responsible for the muscular strength of males, and this is the reason some athletes take supplemental amounts of **anabolic steroids,** which are either testosterone or related chemicals. The dangerous side effects of taking anabolic steroids are listed in Figure 40.16.

The female sex hormones, estrogens (often referred to in the singular) and progesterone, have many effects on the body. In

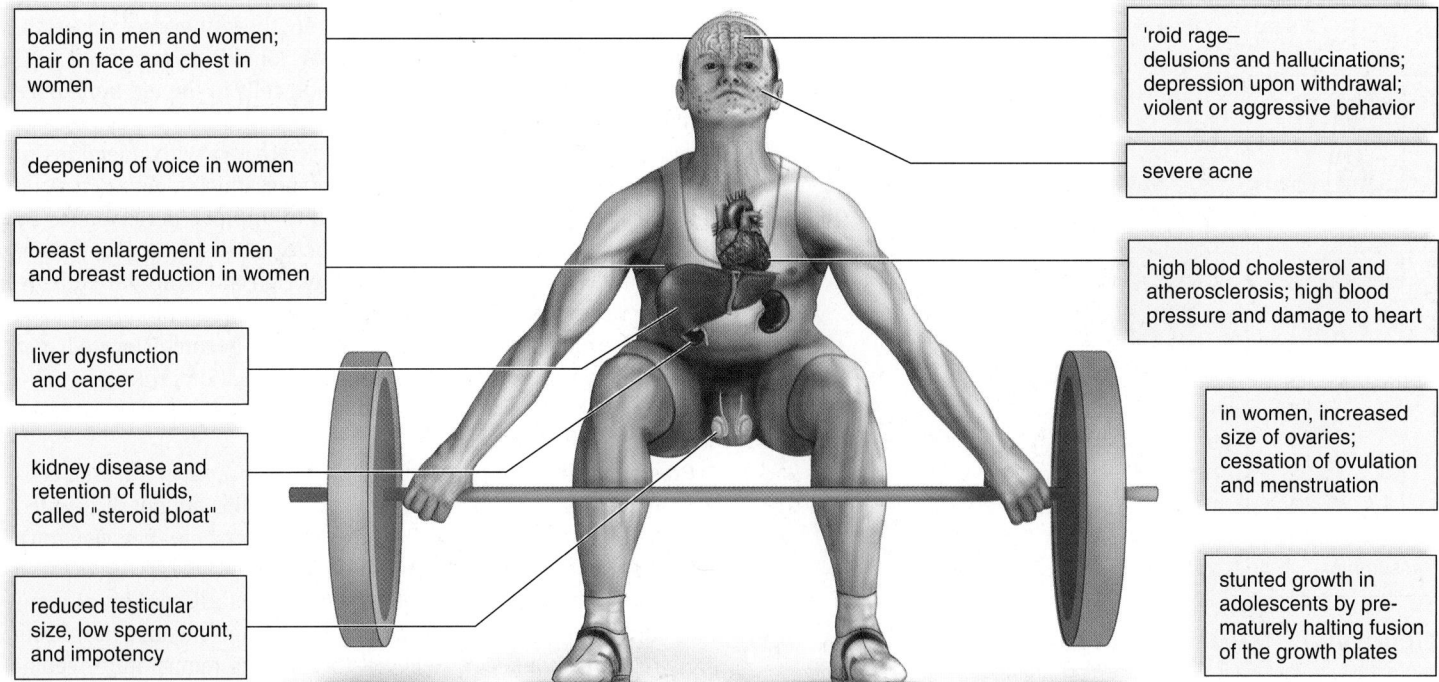

balding in men and women; hair on face and chest in women

deepening of voice in women

breast enlargement in men and breast reduction in women

liver dysfunction and cancer

kidney disease and retention of fluids, called "steroid bloat"

reduced testicular size, low sperm count, and impotency

'roid rage— delusions and hallucinations; depression upon withdrawal; violent or aggressive behavior

severe acne

high blood cholesterol and atherosclerosis; high blood pressure and damage to heart

in women, increased size of ovaries; cessation of ovulation and menstruation

stunted growth in adolescents by prematurely halting fusion of the growth plates

Figure 40.16 The effects of anabolic steroid use.

particular, estrogen [Gk. *oistros,* sexual heat; L. *genitus,* producing] secreted at the time of puberty stimulates the growth of the uterus and the vagina. Estrogen is necessary for egg maturation and is largely responsible for the secondary sex characteristics in females, including female body hair and fat distribution. In general, females have a more rounded appearance than males because of a greater accumulation of fat beneath the skin. Also, the pelvic girdle is wider in females than in males, resulting in a larger pelvic cavity. Both estrogen and progesterone are required for breast development and regulation of the uterine cycle. This includes monthly menstruation (discharge of blood and mucosal tissues from the uterus).

Pineal Gland

The **pineal gland** (epiphysis), located deep in the human brain (see Fig. 40.2), produces the hormone **melatonin,** primarily at night. Melatonin is involved in our daily sleep-wake cycle; normally we grow sleepy at night when melatonin levels increase and awaken once daylight returns and melatonin levels are low (Fig. 40.17). Daily 24-hour cycles such as this are called **circadian rhythms** [L. *circum,* around, and *dies,* day], and circadian rhythms are controlled by an internal timing mechanism called a biological clock.

Instead of being buried deep in the brain, the pineal gland of some vertebrates is located on top of the brain, and in certain fossilized reptiles and even some primitive extant reptiles and amphibians, an additional opening in the skull is present, covered only by a thin layer of skin. This, along with the presence of light-sensing cells in the pineal gland, has led some investigators to conclude that this gland functioned as a "third eye" at some point in evolution. The exact function(s) of this structure are not completely understood, although it may have aided in determining the position of the Sun, or in establishing circadian rhythms.

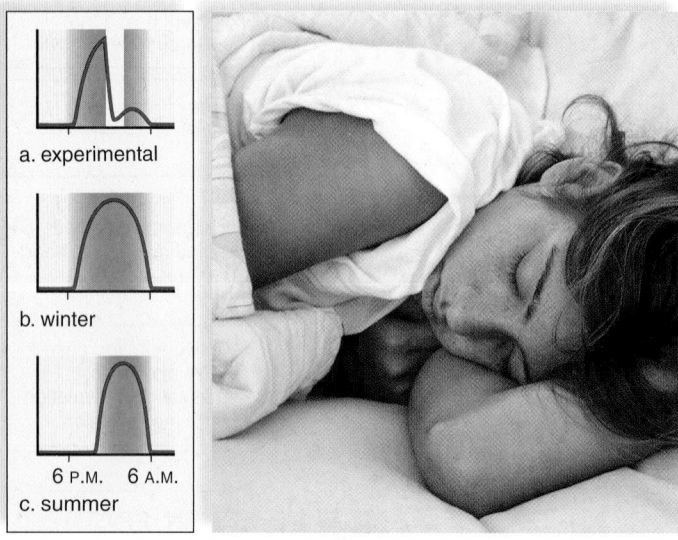

Figure 40.17 Melatonin production. Melatonin production is greatest at night when we are sleeping. Light suppresses melatonin production (**a**), so its duration is longer in the winter (**b**) than in the summer (**c**).

Animal research suggests that melatonin also regulates sexual development. In keeping with these findings, it has been noted that children whose pineal glands have been destroyed due to a brain tumor experience early puberty.

Thymus Gland

The lobular **thymus gland** lies just beneath the sternum (see Fig. 40.2). This organ reaches its largest size and is most active during childhood. With aging, the thymus gets smaller and becomes fatty. Lymphocytes that originate in the bone marrow and then pass through the thymus become T lymphocytes. The lobules of the thymus are lined by epithelial cells that secrete hormones called thymosins. These hormones aid in the differentiation of T lymphocytes packed inside the lobules.

Hormones from Other Organs or Tissues

Some organs not usually considered as endocrine glands do secrete hormones. Two examples already mentioned in this chapter are the excretion of renin by the kidneys, and atrial natriuretic hormone by the heart. A number of other tissues produce hormones.

Leptin

Leptin is a peptide hormone produced by adipose tissue throughout the body. Leptin acts on the hypothalamus, where it signals satiety or fullness. After leptin was discovered in the 1990s, researchers hoped that it could be used to control obesity in humans. Unfortunately, the trials have not yielded satisfactory results. In fact, the blood of obese individuals may be rich in leptin. It is possible that the leptin they produce is ineffective because of a genetic mutation or because their hypothalamic cells lack a suitable number of receptors for leptin.

Erythropoietin

As previously mentioned in Chapter 36, the kidneys secrete erythropoietin (EPO) in response to a low blood oxygen level. EPO stimulates the production of red blood cells in the red bone marrow. People with anemia, which is common in kidney disease, cancer, and AIDS, may be effectively treated with injections of recombinant EPO. In recent years, some athletes have practiced blood doping, in which EPO is used to improve performance by increasing the oxygen-carrying capacity of the blood. The potential dangers of blood doping far outweigh the temporary advantages. Because EPO increases the number of red blood cells, the blood becomes thicker, blood pressure can become elevated, and the athlete is at increased risk of heart attack or stroke.

Prostaglandins

Prostaglandins are potent chemical signals produced within cells from arachidonate, a fatty acid. Prostaglandins are not distributed in the blood. They act locally, quite close to where they were produced. In the uterus, prostaglandins cause muscles to contract; therefore, they are implicated in the pain and discomfort of menstruation in some women. Also, prostaglandins mediate the effects of *pyrogens,* chemicals believed to reset the temperature regulatory center in the brain. Aspirin reduces body temperature and controls pain because it prevents the synthesis of prostaglandins.

Certain prostaglandins increase the secretion of protective mucus in the stomach and thus are used to prevent gastric ulcers. Others lower blood pressure and have been used to treat hypertension. Still others inhibit platelet aggregation and have been used to prevent thrombosis. Because prostaglandins can affect various tissues however, unwanted side effects can be a problem. For example, Misoprostol, a prostaglandin commonly used to prevent stomach ulcers, should not be taken by pregnant women, as it may also cause uterine contractions, resulting in miscarriage or premature labor.

Check Your Progress 40.3

1. Explain how the renin-angiotensin-aldosterone system raises blood pressure.
2. List the endocrine gland that secretes each of the following hormones: aldosterone, melatonin, epinephrine, EPO, leptin, glucagon, ANH, cortisol, and calcitonin.
3. Name one hormone that stimulates the activity of osteoclasts, and one that inhibits them.

CONNECTING the CONCEPTS with the BIG IDEAS

Energy and Homeostasis

- Problems in a normal hormonal feedback system can cause major physiological problems, as in diabetes mellitus, Grave's disease, and dehydration due to ADH shortage. (2C1c*IE*)

Information and Signaling

- Signals in the form of hormones are sent by endocrine cells through the blood to reach their targets. (3D2c1)
- Insulin, human growth hormone, thyroid hormones, testosterone and estrogen illustrate the many different release mechanisms and types of target cell effects of endocrine hormones. (3D2c1*IE*)
- Signals via epinephrine can lead to glycogen breakdown and sugar release in mammals. (3D1d*IE*)

*Find the unabridged version of all EK citations at www.glencoe.com/maderAP11.

Media Study Tools

www.glencoe.com/maderAP11

Enhance your study of this chapter with study tools and practice tests. Also ask your instructor about the resources available through ConnectPlus, including the media-rich eBook, interactive learning tools, and animations.

Summarize

40.1 Animal Hormones

The nervous and endocrine systems both use chemical signals. Endocrine glands secrete hormones into the bloodstream, and from there they are distributed to target organs or tissues.

Hormones are a type of chemical signal that usually act at a distance between body parts. Hormones are either peptides or steroids. Reception of a peptide hormone at the plasma membrane activates an enzyme cascade inside the cell. Steroid hormones combine with a receptor inside the cell, and the complex attaches to and activates DNA. Protein synthesis follows.

40.2 Hypothalamus and Pituitary Gland

Neurosecretory cells in the hypothalamus produce antidiuretic hormone (ADH) and oxytocin, which are stored in axon endings in the posterior pituitary until they are released.

The hypothalamus produces hypothalamic-releasing and hypothalamic-inhibiting hormones, which pass to the anterior pituitary by way of a portal system. The anterior pituitary produces several types of hormones, and some of these stimulate other endocrine glands to secrete hormones.

40.3 Other Endocrine Glands and Hormones

The thyroid gland, controlled by TSH, requires iodine to produce thyroxine (T_4) and triiodothyronine (T_3), which increase the metabolic rate. Depending on the age of an individual, a deficiency of T_3 and T_4 may result in congenital hypothyroidism, simple goiter, or myxedema. Hyperthyroidism may be the result of Graves disease or thyroid cancer. The thyroid gland also produces calcitonin, which helps lower the blood calcium level.

The parathyroid glands secrete parathyroid hormone, which raises the blood calcium and decreases the blood phosphate levels.

The adrenal glands respond to stress: Immediately, the adrenal medulla secretes epinephrine and norepinephrine, which bring about responses we associate with emergency situations. On a long-term basis, the adrenal cortex, controlled by ACTH, produces the glucocorticoids (e.g., cortisol) and the mineralocorticoids (e.g., aldosterone). Cortisol stimulates hydrolysis of proteins to amino acids that are converted to glucose; in this way, it raises the blood glucose level. Aldosterone causes the kidneys to reabsorb sodium ions (Na^+) and to excrete potassium ions (K^+). Addison disease develops when the adrenal cortex is underactive, and Cushing syndrome develops when the adrenal cortex is overactive.

The pancreatic islets secrete insulin, which lowers the blood glucose level, and glucagon, which has the opposite effect. The most common illness caused by hormonal imbalance is diabetes mellitus, which is due to the failure of the pancreas to produce insulin or the failure of the cells to take it up.

The gonads, controlled by gonadotropic hormones, produce the sex hormones; the pineal gland produces melatonin, which may be involved in circadian rhythms and the development of the reproductive organs; and the thymus secretes thymosins, which stimulate T-lymphocyte production and maturation.

Tissue and organs having other functions also produce hormones. Leptin is a newly described hormone that regulates appetite, and erythropoietin stimulates the production of red blood cells. Prostaglandins are produced and act locally, with a variety of effects on different tissues.

Key Terms

adrenal cortex   762
adrenal gland   762
adrenal medulla   762
adrenocorticotropic hormone
   (ACTH)   757
aldosterone   763
anabolic steroid   767
androgen   767
anterior pituitary   757
antidiuretic hormone
   (ADH)   757
atrial natriuretic hormone
   (ANH)   763
calcitonin   761
chemical signal   755
circadian rhythm   768
cortisol   763
cyclic adenosine
   monophosphate
   (cAMP)   755
endocrine gland   753
endocrine system   753
epinephrine   762
estrogen   767
exocrine gland   753
exophthalmos   761
first messenger   755
glucagon   765
glucocorticoid   763
gonadotropic hormone   757
growth hormone (GH)   758
hormone   753
hyperthyroidism   761

hypothalamic-inhibiting
   hormone   757
hypothalamic-releasing
   hormone   757
hypothalamus   757
hypothyroidism   761
insulin   765
leptin   768
melanocyte-stimulating
   hormone (MSH)   757
melatonin   768
mineralocorticoid   763
negative feedback   757
norepinephrine   762
ovary   767
oxytocin   757
pancreas   765
pancreatic islet   765
parathyroid gland   761
parathyroid hormone
   (PTH)   761
peptide hormone   755
pheromone   755
pineal gland   768
pituitary gland   757
positive feedback   757
posterior pituitary   757
progesterone   767
prolactin (PRL)   757
prostaglandin   768
renin   763
second messenger   755
steroid hormone   755

testes   767
testosterone   767
thymus gland   768
thyroid gland   760

thyroid-stimulating hormone
   (TSH)   757
thyroxine (T_4)   760

Assess

Reviewing This Chapter

1. In what ways are the nervous and endocrine systems alike, and how are they different?   753
2. Explain how steroid hormones and peptide hormones affect the metabolism of the cell.   755–56
3. Explain the relationship of the hypothalamus to the posterior pituitary gland and to the anterior pituitary gland. List the hormones secreted by the posterior and anterior pituitary.   757–59
4. Give an example of the negative feedback relationship involving the hypothalamus, the anterior pituitary, and other endocrine glands.   757
5. Discuss the effect of too much or too little growth hormone when a young person is growing. What is the result if the anterior pituitary overproduces growth hormone in an adult?   758–59
6. What types of diseases are associated with a malfunctioning thyroid? Explain each type.   760–61
7. How do the thyroid and the parathyroid work together to control the blood calcium level?   761–62
8. How do the adrenal glands respond to stress? What hormones are secreted by the adrenal medulla, and what effects do these hormones have?   762–64
9. Name the most significant glucocorticoid and mineralocorticoid, and discuss the function of each. Explain the symptoms of Addison disease and Cushing syndrome.   763–64
10. Draw a diagram to explain how insulin and glucagon maintain the blood glucose level. Use your diagram to explain three major symptoms of type 1 diabetes mellitus.   765–67
11. Name the other endocrine glands cited in this chapter, and discuss the functions of the hormones they secrete. Also, discuss the hormones not produced by endocrine glands.   767–69

Testing Yourself

Choose the best answer for each question.
For questions 1–5, match each hormone to a gland in the key.

KEY:
   a. pancreas
   b. anterior pituitary
   c. posterior pituitary
   d. thyroid
   e. adrenal medulla
   f. adrenal cortex

1. cortisol

2. growth hormone (GH)

3. oxytocin storage

4. insulin

5. epinephrine

6. The anterior pituitary controls the secretion(s) of both
   a. the adrenal medulla and the adrenal cortex.
   b. the thyroid and the adrenal cortex.
   c. the ovaries and the testes.
   d. Both b and c are correct.

7. Diabetes mellitus is associated with
   a. too much insulin in the blood.
   b. too much glucose in the blood.
   c. blood that is too dilute.
   d. Both b and c are correct.

8. Which of these is not a pair of antagonistic hormones?
   a. insulin—glucagon
   b. calcitonin—parathyroid hormone
   c. aldosterone—atrial natriuretic hormone (ANH)
   d. thyroxine—growth hormone

9. Which hormone and condition are mismatched?
   a. cortisol—myxedema
   b. growth hormone—acromegaly
   c. thyroxine—goiter
   d. parathyroid hormone—tetany
   e. insulin—diabetes

10. The difference between type 1 and type 2 diabetes is that
    a. for type 2 diabetes, insulin is produced but not used; type 1 results from lack of insulin production.
    b. treatment for type 2 involves insulin injections, while type 1 can be controlled, usually by diet.
    c. only type 1 can result in complications such as kidney disease, reduced circulation, or stroke.
    d. type 1 can be a result of lifestyle, and type 2 is thought to be caused by a virus or other agent.

11. Which of the following hormones is/are found in females?
    a. estrogen
    b. testosterone
    c. follicle-stimulating hormone
    d. Both a and c are correct.
    e. All of these are correct.

12. Parathyroid hormone causes
    a. the kidneys to excrete more calcium ions.
    b. bone tissue to break down and release calcium into the bloodstream.
    c. fewer calcium ions to be absorbed by the intestines.
    d. more calcium ions to be deposited in bone tissue.

For questions 13–18, match the function to a hormone in the key. Choose more than one answer if correct. Answers may be used more than once.

KEY:
   a. antidiuretic hormone
   b. oxytocin
   c. glucocorticoids
   d. glucagon
   e. parathyroid hormone

13. raises blood glucose level

14. stimulates uterine muscle contraction

15. stimulates water reabsorption by kidneys

16. stimulates release of milk by mammary glands

17. raises blood glucose and stimulates breakdown of protein

18. raises blood calcium level

19. Steroid hormones are secreted by
    a. the adrenal cortex.
    b. the gonads.
    c. the thyroid.
    d. Both a and b are correct.
    e. Both b and c are correct.

20. Which one of the following statements about the pituitary gland is incorrect?
    a. The pituitary lies inferior to the hypothalamus.
    b. Growth hormone and prolactin are secreted by the anterior pituitary.
    c. The anterior pituitary and posterior pituitary communicate with each other.
    d. Axons run between the hypothalamus and the posterior pituitary.

Engage

Thinking Scientifically

1. Caffeine inhibits the breakdown of cAMP in the cell. Referring to Figure 40.4, how would this influence a stress response brought about by epinephrine?

2. Both males and females can develop secondary sex characteristics of the opposite sex if they take enough of the appropriate sex hormone. Hypothesize the mechanism that would make this possible after reviewing Figure 40.4.

Bioethical Issue
Growth Hormone

Untreated GH deficiency in childhood results in pituitary dwarfism. If young children with GH deficiency receive daily GH injections through adolescence, most will attain normal stature. Few would argue that treatment of pituitary dwarfism is not justified. But is it justified to use GH therapy to make a child of normal height taller?

GH levels naturally decline with age, and there is evidence that treating older adults with GH boosts muscle mass and reduces body fat. Pills do not work, and GH injections are by prescription only and very expensive, being around $1,000 a month. Furthermore, GH therapy can result in undesirable side effects, such as elevated blood sugar, fluid retention, and joint pain. Nevertheless, there is keen interest in GH therapy—not only from those who wish to delay or avoid the effects of aging, but also from athletes in search of performance-enhancing substances. Thus far, like anabolic steroids and erythropoietin (EPO), GH is banned by most competitive sports authorities, including the U.S. and International Olympic Committees.

The existence of GH therapy raises questions about medical treatment for disease versus enhancement of an already healthy body. Under what circumstances is it suitable to increase a child's growth? Should the physical decline of aging be accepted, or should we try to maintain a youthful build using GH? If an athlete wants to make himself or herself more competitive using GH, and is willing to accept the risks, should therapy be permitted? Who decides?

41

Reproductive Systems

Male potbelly seahorses have a brood pouch that can hold 150–300 developing embryos.

CHAPTER OUTLINE

BEFORE YOU BEGIN

Before beginning this chapter, take a few moments to review the following discussions.

Chapter 10 How does meiosis halve the chromosome number to produce haploid cells?

Section 16.2 What influences does sexual selection have on mating patterns in animals?

Figure 40.6 What is the relationship between the hypothalamus/pituitary and the reproductive systems of males and females?

Seahorses are fishes with an unusual style of sexual reproduction: The males become pregnant and give birth to the young. When seahorses mate, the female uses a tiny structure called an ovipositor to insert her unfertilized eggs into a brood pouch on the male's abdomen, where they are fertilized by his sperm. Tissues of the pouch then grow around the eggs, forming a placenta-like structure that provides oxygen and nutrition to the developing offspring. After about a month, the pouch opens and he gives birth.

Although the evolutionary benefit conferred by male pregnancy in seahorses is unclear, researchers are learning how this phenomenon can affect sexual selection. One finding is that the typical sex roles are often reversed—for example, because the males invest more resources into producing the offspring, they tend to be choosier about selecting mates, whereas females sometimes mate with multiple males during a breeding season.

In this chapter we focus on the more common strategies of asexual and sexual reproduction. Most of the chapter is devoted to the human male and female reproductive systems, methods that are available to help prevent unwanted pregnancies, and sexually transmitted infections that can have a detrimental effect on human reproduction.

As you read through the chapter, think about the following questions:

1. Considering this description of seahorse reproduction, define the terms "male," "female," "sex," and "pregnant."

2. List some common examples of secondary sex characteristics that develop in male animals in order to attract female mates.

FOLLOWING *the* BIG IDEAS

Energy and Homeostasis

Reproductive systems must respond to feedback from both internal and external signals.

41.1 How Animals Reproduce

Learning Outcomes

Upon completion of this section, you should be able to

1. Compare and contrast asexual and sexual reproduction, and list several animals that undergo each.
2. Describe some advantages of each of the following life histories: oviparous, ovoviviparous, viviparous.

For an animal species to survive, individuals must reproduce. As was discussed in Chapter 10, sexual reproduction involves a reshuffling of genetic material during *meiosis*. In animals, this process forms the sex cells, or gametes, which subsequently unite to form genetically unique offspring. The tremendous amount of genetic variation that is generated during meiosis can be advantageous to the survival of a species, especially during times when the environment is changing.

In asexual reproduction, a single parent gives rise to offspring that are identical to the parent, unless mutations have occurred. The adaptive advantage of asexual reproduction is that organisms can reproduce rapidly and colonize favorable environments quickly. Although the majority of animals reproduce sexually, a few groups of animals are also capable of asexual reproduction.

Asexual Reproduction

Several types of invertebrates, such as sponges, cnidarians, flatworms, annelids, and echinoderms, can reproduce asexually. In cnidarians, such as hydras, new individuals may arise asexually as an outgrowth (bud) of the parent (Fig. 41.1). Some species of hydras can also reproduce sexually.

Many flatworms can reproduce asexually by splitting in half, generating two identical individuals. In the laboratory, a planarian can be cut into as many as ten pieces, each of which will grow into a new planarian. To varying degrees, many other animals, including sponges, annelids, and echinoderms, also have the ability to regenerate from fragments. Scientists are studying the ability of some animals to regenerate, with the hope of learning how to regenerate certain human tissues (see the Nature of Science feature, "Regenerative Medicine," in Chapter 31).

Sexual Reproduction

The majority of animals are *dioecious*, which means having separate sexes. Usually during sexual reproduction, the egg of one parent is fertilized by the sperm of another. *Monoecious*, or **hermaphroditic,** animals have both male and female sex organs in a single body. Some hermaphroditic organisms, like tapeworms, are capable of self-fertilization, but the majority, such as earthworms, practice cross-fertilization with other individuals. Sequential hermaphroditism, or sex reversal, also occurs. In coral reef fishes called wrasses, a male has a harem of several females. If the male dies, the largest female becomes a male.

Animals usually produce gametes in specialized organs called **gonads.** Sponges are an exception to this rule because the collar cells lining the central cavity of a sponge give rise to sperm and eggs. Hydras and other cnidarians produce only temporary gonads in the fall, when sexual reproduction occurs. Animals in other phyla have permanent reproductive organs.

The gonads are **testes,** which produce sperm, and **ovaries,** which produce eggs. Eggs or sperm are derived from **germ cells,** which become specialized for this purpose during early development. Other cells present in a gonad support and nourish the developing gametes or produce hormones necessary to the reproductive process. The reproductive system also usually has a number of accessory structures, such as ducts and storage areas, that aid in bringing the gametes together.

Several types of flatworms, roundworms, crustaceans, annelids, insects, fishes, lizards, and even some turkeys have the ability to reproduce parthenogenetically. **Parthenogenesis** is a modification of sexual reproduction in which an unfertilized egg develops into a complete individual. In honeybees, the queen bee makes and stores sperm she uses to selectively fertilize eggs. Any unfertilized eggs become haploid males.

Many aquatic animals practice *external fertilization*, in which the gametes are released and unite outside the bodies of the reproducing animals. Among fishes, methods of fertilization and care of the developing embryos vary greatly. Some species simply scatter large numbers of eggs and sperm in the environment, and may even eat some of their eggs or offspring. Mouthbrooders scoop up their eggs or newly hatched young into their mouths, protecting them from predators. Other fish species build elaborate nests for their eggs, made from plant debris and saliva-coated bubbles.

Video Sea Urchin Reproduction

Video Coral Reef Spawning

An invertebrate aquatic species called palolo worms, which inhabit coral reefs in the South Pacific, exhibit a unique fertilization strategy. Always at the same time of year and during a particular phase of the moon, the worms break in half. During

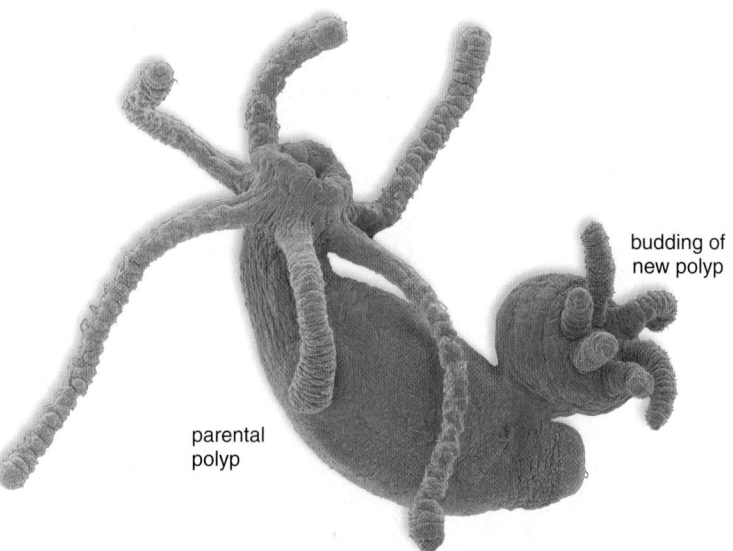

budding of
new polyp

parental
polyp

Figure 41.1 Reproduction in *Hydra*.   Hydras reproduce asexually and sexually. During asexual reproduction, a new polyp buds from the parental polyp. During sexual reproduction, temporary gonads develop in the body wall.

a period of just a few hours, hundreds of thousands of tail sections, containing the gonads, swim to the surface, releasing eggs and sperm. Meanwhile, local human populations take to the sea, using nets to scoop up large numbers of the half-worms, which are eaten raw or saved in buckets to be baked or fried, and consumed at an annual feast.

Copulation is sexual union to facilitate the reception of sperm, resulting in *internal fertilization*. In terrestrial vertebrates, males typically have a penis for depositing sperm into the vagina of females. But not so in birds, which lack a penis and vagina. They have a cloaca, a chamber that receives products from the digestive, urinary, and reproductive tracts. A male transfers sperm to a female after placing his cloacal opening against hers. In damselflies, the female curls her abdomen forward to receive sperm previously deposited in a pouch by the male, and copulation in the usual sense doesn't occur (Fig. 41.2).

Video
Frog Reproduction

Life History Strategies

Any animal that deposits an egg in the external environment is **oviparous.** Many oviparous animals have a larval stage, an immature form capable of feeding. Because the larva has a different lifestyle, it is able to use a different food source than the adult. Other animals, like crayfish, do not have a larval stage; the egg hatches into a tiny juvenile with the same form as the adult.

Some aquatic animals are **ovoviviparous,** meaning they retain their eggs in some way and release young able to fend for themselves. For example, oysters, which are molluscs, retain their eggs in the mantle cavity. As mentioned in the chapter-opening story, male seahorses have a brood pouch in which fertilized eggs develop.

Usually, reptiles (e.g., turtles and crocodiles) lay a leathery-shelled egg that contains **extraembryonic membranes** to serve

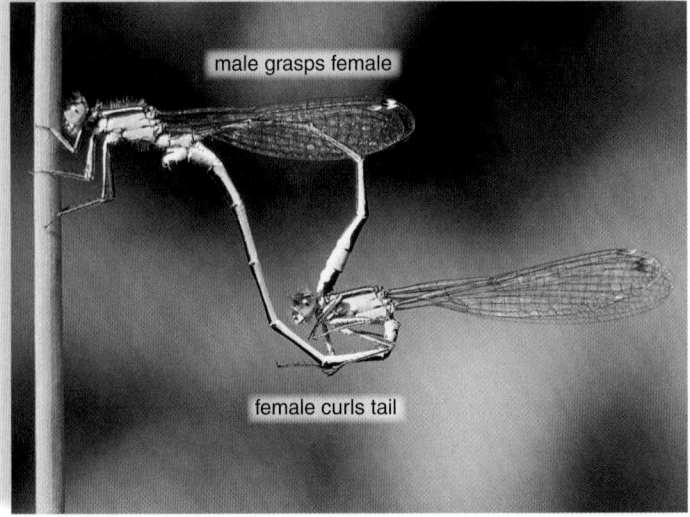

Figure 41.2 Damselflies mating on land.   On land, animals have various means to ensure that the gametes do not dry out. Here, a female damselfly receives sperm from the male.

male grasps female

female curls tail

Figure 41.3 Parenting in birds.   Birds, such as the American goldfinch, *Carduelis tristis,* are oviparous and lay hard-shelled eggs. They are well known for incubating their eggs and caring for their offspring after they hatch.

the needs of the embryo and prevent drying out. One membrane surrounds an abundant supply of **yolk,** which is a nutrient-rich material. The shelled egg with its internal extraembryonic membranes is a significant adaptation to the terrestrial environment. Birds lay and care for hard-shelled eggs; the newly hatched birds usually have to be fed before they are able to fly away and seek food for themselves (Fig. 41.3). Complex hormonal and neural regulation are involved in the reproductive behavior of parental birds.

Among mammals, the duckbill platypus and spiny anteater lay shelled eggs. Marsupials and placental mammals do not lay eggs. In marsupials, the yolk sac membrane functions briefly to supply the unborn embryo with nutrients acquired internally from the mother. Immature young finish their development within a pouch where they are nourished on milk.

The placental mammals are termed **viviparous,** because they do not lay eggs and development occurs inside the female's body until offspring can live independently. Their **placenta** is a complex structure derived, in part, from the chorion, another of the reptilian extraembryonic membranes. The evolution of this type of placenta allowed the developing young to internally exchange materials with the mother until they are able to function on their own. Viviparity represents the ultimate in caring for the unborn, and in placental mammals, the mother continues to supply milk to nourish her offspring after birth.

In the sections that follow, we take an in-depth look at the reproductive systems of humans and the issues of birth control and health.

Check Your Progress   41.1

1. List one advantage of asexual reproduction, and one advantage of sexual reproduction.
2. Distinguish between oviparous, ovoviviparous, and viviparous.
3. Explain how a shelled egg allows reproduction on land.

41.2 Human Male Reproductive System

Learning Outcomes

Upon completion of this section, you should be able to

1. Identify the structures of the human male reproductive system and provide a function for each.
2. Describe the location and stages of spermatogenesis.
3. Summarize how hormones regulate the male reproductive system.

The human male reproductive system includes the organs pictured in Figure 41.4 and listed in Table 41.1. The primary sex organs, or gonads, of males are the paired **testes** (sing., testis), which are suspended within the sacs of the scrotum.

The Testes

The testes produce sperm and the male sex hormones. They begin their development inside the abdominal cavity, but descend into the scrotal sacs during the last two months of fetal development. If the testes fail to descend—and the male does not receive hormone therapy or undergo surgery to place the testes in the scrotum—male infertility may result. The reason is that undescended testes developing in the body cavity are subject to

Table 41.1 Male Reproductive System

Organ	Function
Testes	Produce sperm and sex hormones
Epididymides	Sites of maturation and some storage of sperm
Vasa deferentia	Conduct and store sperm
Seminal vesicles	Contribute fluid to semen
Prostate gland	Contributes fluid to semen
Urethra	Conducts sperm (and urine)
Bulbourethral glands	Contribute fluid to semen
Penis	Organ of copulation

higher body temperatures than testes that have descended into the scrotum. The scrotum helps regulate testicular temperature by holding them closer or farther away from the body. In fact, any activity that increases testicular temperature, such as taking hot baths, can decrease sperm production. However, despite the frequently held belief that wearing boxer-style underwear favors higher sperm production compared to briefs, this assumption is unsupported by research.

Sperm produced in the testes mature within the **epididymis** (pl., epididymides), a tightly coiled duct lying just outside each testis. Maturation seems to be required for sperm to swim to the egg. Once the sperm have matured, they enter the **vas deferens** (pl., vasa deferentia), also called the ductus deferens. Sperm may be stored for a time in the vasa deferentia. Each

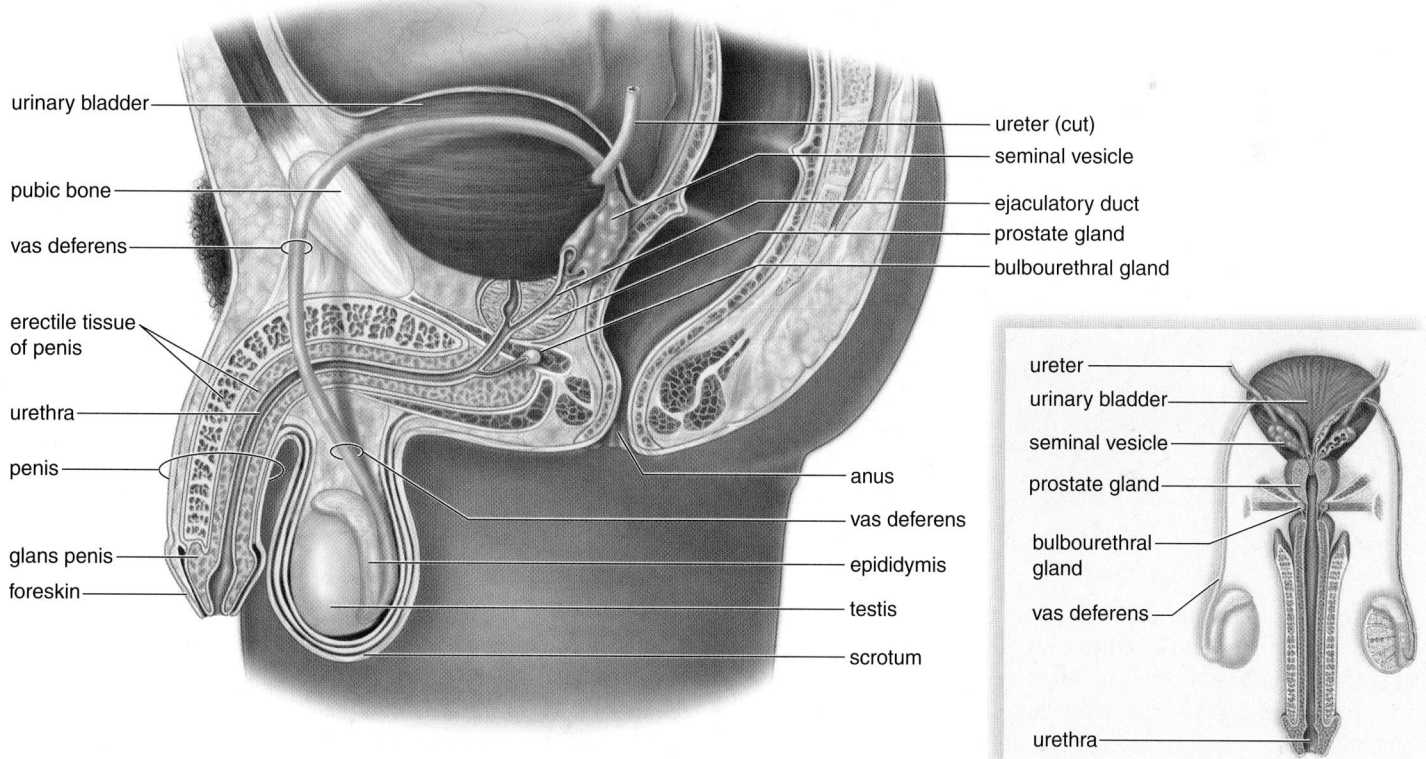

Figure 41.4 The male reproductive system.   The testes produce sperm. The seminal vesicles, the prostate gland, and the bulbourethral glands provide a fluid medium for the sperm, which move from the vas deferens through the ejaculatory duct to the urethra in the penis. The foreskin (prepuce) is removed when a penis is circumcised.

Figure 41.5 Testis and sperm. **a.** The lobules of a testis contain seminiferous tubules. **b.** Light micrograph of a cross section of the seminiferous tubules, where spermatogenesis occurs. Note the location of interstitial cells in clumps among the seminiferous tubules. **c.** Diagrammatic representation of spermatogenesis, which occurs in wall of tubules. **d.** A sperm has a head, a middle piece, and a tail. The nucleus is in the head, which is capped by the enzyme-containing acrosome.

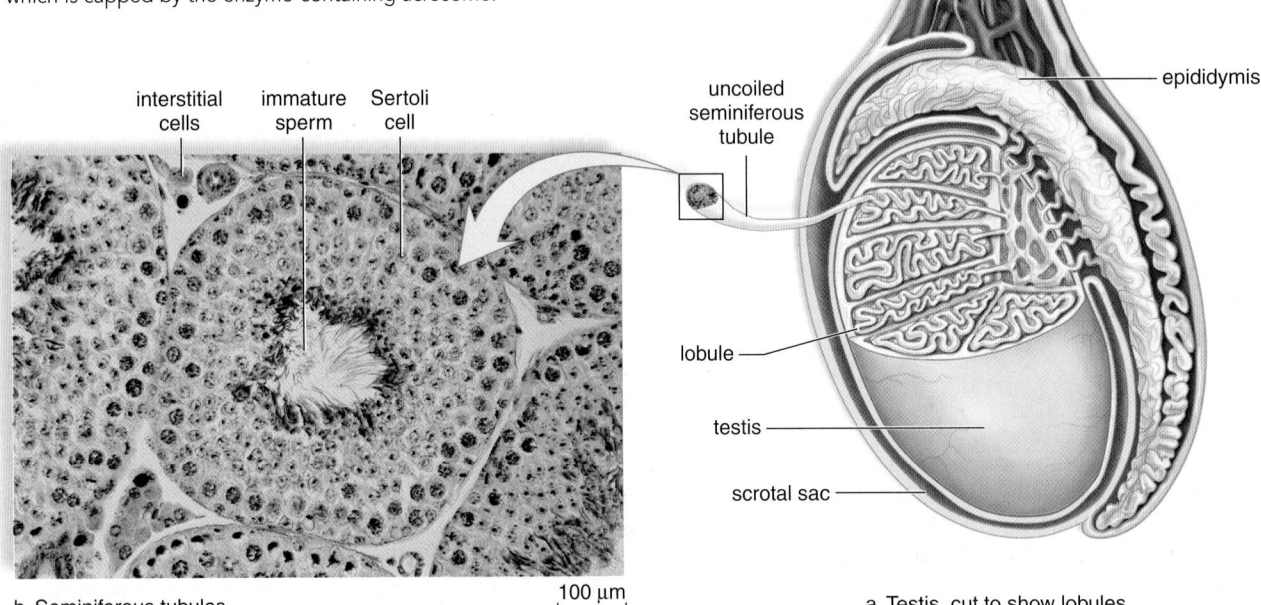

b. Seminiferous tubules

100 μm

a. Testis, cut to show lobules

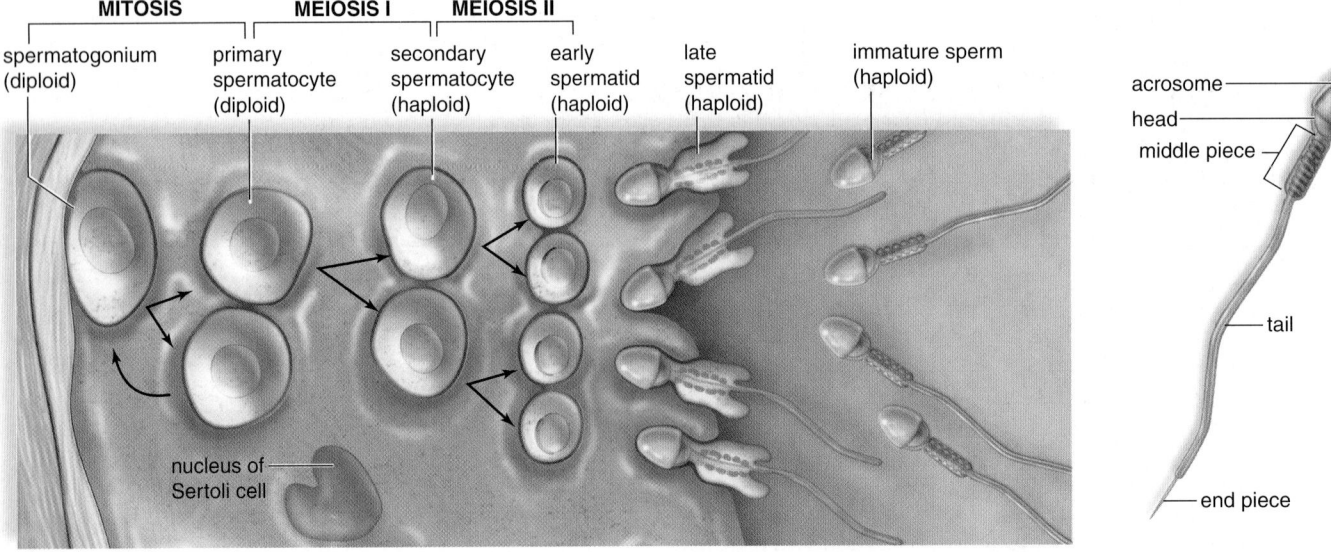

c. Spermatogenesis

d. Sperm

vas deferens passes into the abdominal cavity, where it curves around the bladder and empties into an ejaculatory duct. The vasa deferentia are severed or blocked in a surgical form of birth control called a *vasectomy*. Sperm are then unable to complete their journey through the male reproductive tract, and are phagocytized and destroyed by macrophages.

Testicular cancer is the most common type of cancer in young men between the ages of 15 and 34. The cure rate is high if early treatment is received. All men should perform regular self-exams to detect unusual lumps or swellings of the testes, and report any changes to their doctors.

Production of Sperm

A longitudinal section of a testis shows that it is composed of compartments called lobules, each of which contains one to three tightly coiled **seminiferous tubules** (Fig. 41.5*a*). Altogether, these tubules have a combined length of approximately 250 meters (over two and a half football fields). This provides a large amount of surface area for sperm development. A microscopic cross section of a seminiferous tubule shows that it is packed with cells undergoing *spermatogenesis* (Fig. 41.5*b*), the production of sperm.

During spermatogenesis, spermatogonia divide to produce primary spermatocytes (which have the 2n number of chromosomes).

Primary spermatocytes move away from the outer wall, increase in size, and undergo meiosis I to produce secondary spermatocytes. Each secondary spermatocyte has only 23 chromosomes (Fig. 41.5c). Secondary spermatocytes (n) undergo meiosis II to produce four spermatids, each of which also has 23 chromosomes. Spermatids then differentiate into sperm (Fig. 41.5d). Note the presence of *Sertoli cells* (purple in Fig. 41.5c), which support and nourish the developing sperm, and regulate spermatogenesis. It takes approximately 74 days for a spermatogonium to develop into sperm.

Animation
Spermatogenesis

Mature **sperm,** or *spermatozoa,* have three distinct parts: a head, a middle piece, and a tail (Fig. 41.5d). Mitochondria in the middle piece provide energy for the movement of the tail, which is a flagellum. Like other eukaryotic flagella and cilia, microtubules in the sperm tail form the characteristic 9 + 2 pattern (see Fig. 4.20). The head contains a nucleus covered by a cap called the *acrosome* [Gk. *akros,* at the tip, and *soma,* body], which stores enzymes needed to penetrate the thick membrane surrounding the egg.

Video
Human Sperm

The Penis and Male Orgasm

The **penis** (Fig. 41.6) is the male organ of sexual intercourse. The penis has a long shaft and an enlarged tip called the glans penis. The glans penis is normally covered by a layer of skin called the *foreskin*. Male *circumcision*, the surgical removal of the foreskin, is performed on more than 50% of newborn males in the United States, often for cultural, religious, or health reasons. Research shows that circumcised males are 25–35% less likely to acquire certain STDs, including AIDS.

Three cylindrical columns of spongy, erectile tissue containing distensible blood spaces extend through the shaft of the penis. During sexual arousal, autonomic nerves release nitric oxide (NO). This stimulus leads to the production of cGMP (cycline guanosine monophosphate), a type of second messenger similar to cAMP (see Section 40.1). The cGMP causes the smooth muscle of the incoming arterial walls to relax and the erectile tissue to fill with blood. The veins that normally take blood away from the penis are compressed, and the penis becomes erect.

Semen (seminal fluid) [L. *semen,* seed] is a thick, whitish fluid that contains sperm and secretions from three glands (Fig. 41.4 and Table 41.1). The paired **seminal vesicles** lie at the base of the bladder, and each has a duct that joins with a vas deferens. As sperm pass from the vasa deferentia into the ejaculatory ducts, the seminal vesicles secrete a thick, viscous fluid containing nutrients (fructose) for possible use by the sperm. The fluid also contains prostaglandins that stimulate smooth muscle contraction along the male and female reproductive tracts.

The **prostate gland** is a single, doughnut-shaped gland that surrounds the upper portion of the urethra just below the bladder. The prostate gland secretes a milky alkaline fluid believed to activate or increase the motility of sperm. By age 50, over half of all men experience a benign enlargement (hypertrophy) of the prostate gland (BPH). Because of its location, BPH can result in an increased frequency of urination, and a weak urinary stream. Also, after skin cancer, prostate cancer is the most common type of cancer among men in the United States. Based on current rates, 13–16% of men born today can expect to be diagnosed with prostate cancer during their lifetime.

Bulbourethral glands are pea-sized organs that lie posterior to the prostate on either side of the urethra. They produce a clear, viscous secretion known as *pre-ejaculate*. Secretions from the bulbourethral glands are the first to enter the urethra, where they may function to cleanse the urethra of acidic residue from urine. Since this fluid can also pick up sperm that remain in the urethra from previous ejaculations, it is possible for sperm to enter the genital tract of a female during sexual intercourse, even when ejaculation has not yet occurred.

As sexual stimulation intensifies, *ejaculation* may occur. During the first phase of ejaculation, called emission, the spinal cord sends sympathetic nerve impulses to the epididymides and vasa deferentia. Their muscular walls contract, causing sperm to enter the ejaculatory duct, whereupon the seminal vesicles, prostate gland, and bulbourethral glands release their secretions.

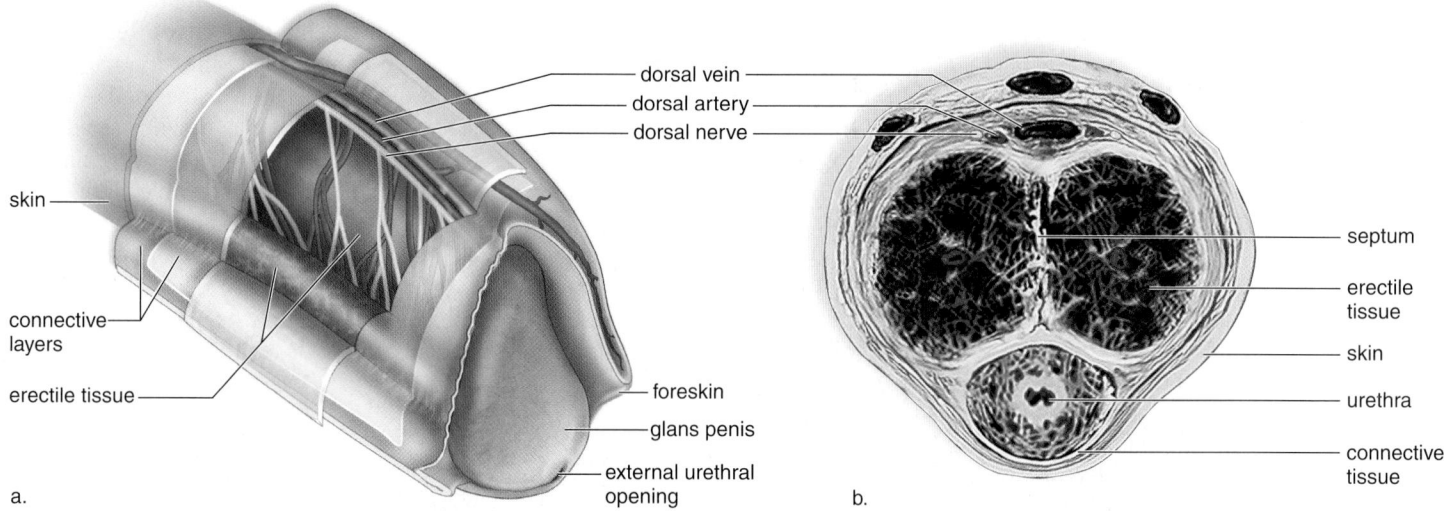

Figure 41.6 Penis anatomy. **a.** Beneath the skin and the connective tissue lies the urethra, surrounded by erectile tissue. This tissue expands to form the glans penis, which in uncircumcised males is partially covered by the foreskin (prepuce). **b.** Micrograph of shaft in cross section showing location of erectile tissue. One column surrounds the urethra.

During the second phase of ejaculation, called expulsion, rhythmical contractions of muscles at the base of the penis and within the urethral wall expel semen in spurts from the opening of the urethra. During ejaculation, the urinary bladder sphincter normally contracts so that no semen can enter the bladder. (Note that the urethra carries either urine or semen at different times.)

The contractions that expel semen from the penis are a part of male *orgasm* [Gk. *orgasmos*, sexual excitement], the physiological and psychological sensations that occur at the climax of sexual stimulation. The physiological reactions of both male and female orgasm involve the genital (reproductive) organs and assocated muscles, as well as the entire body. Marked muscular tension is followed by contraction and relaxation. In the male, following ejaculation and/or loss of sexual arousal, the penis returns to its flaccid state. After ejaculating, males typically experience a period of time, called the *refractory period*, during which stimulation does not bring about an erection. The length of the refractory period increases with age.

An average male releases 2–5 mL of semen during orgasm. The ejaculated semen of a normal human male contains between 50 and 150 million sperm per mL, meaning there may be more than 400 million sperm released in one ejaculation. According to the World Health Organization, a low sperm concentration, or *oligozoospermia*, is defined as less than 15 million sperm per mL of ejaculate. Because only one sperm typically enters an egg, however, men with low sperm counts can still be fertile.

MP3
Male Reproductive Anatomy and Physiology

Erectile dysfunction (ED), formerly called impotency, is the inability to achieve or maintain an erection suitable for sexual intercourse. ED may have a number of causes, including any condition that affects blood flow, certain medications or illicit drugs, and psychological factors. Since 1998, medications have been available for the treatment of ED. All of these drugs—such as Viagra, Levitra, and Cialis—work by inhibiting an enzyme called PDE-5, found mainly in the penis, which normally breaks down cGMP. This extends the activity of cGMP, increasing the likelihood of achieving a full erection. Because the PDE-5 enzyme is also found in the retinas and certain other tissues, ED medications can cause vision problems and other undesirable side effects.

Hormonal Regulation in Males

The hypothalamus has ultimate control of the testes' sexual function because it secretes a hormone called gonadotropin-releasing hormone (GnRH). GnRH stimulates the anterior pituitary to produce the gonadotropic hormones, follicle-stimulating hormone (FSH) and luteinizing hormone (LH), which are present in both males and females. In males, FSH promotes spermatogenesis in the seminiferous tubules.

LH in males controls the production of the androgen testosterone by the interstitial cells (Leydig cells), which are scattered in the spaces between the seminiferous tubules (Fig. 41.5*b*). All these hormones, including inhibin, a hormone released by the seminiferous tubules, are involved in a negative feedback relationship that maintains the fairly constant production of sperm and testosterone (Fig. 41.7).

Testosterone is the main sex hormone in males. It is essential for the normal development and functioning of the organs

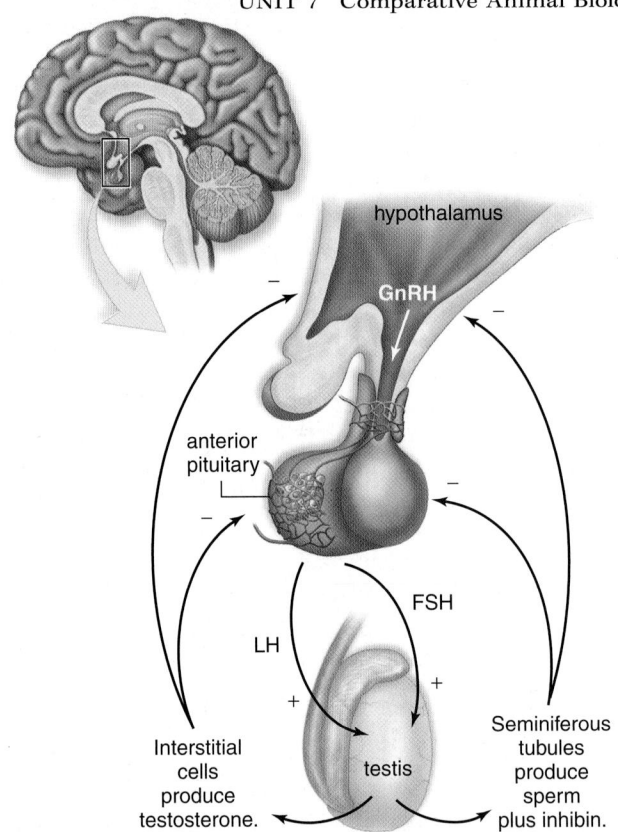

Figure 41.7 Hormonal control of the testes. GnRH (gonadotropin-releasing hormone) stimulates the anterior pituitary to produce FSH and LH. FSH stimulates the testes to produce sperm, and LH stimulates the testes to produce testosterone. Testosterone from interstitial cells and inhibin from the seminiferous tubules exert negative feedback control over the hypothalamus and the anterior pituitary, and this ultimately regulates the level of testosterone in the blood.

listed in Table 41.1. Testosterone also brings about and maintains the male *secondary sex characteristics* that develop at the time of puberty. Males are generally taller than females and have broader shoulders and longer legs relative to trunk length. The deeper voice of males compared to females is due to the development in males of a larger larynx with longer vocal cords. The "Adam's apple," which is a protrusion at the front of the larynx, is usually more prominent in males than in females. Testosterone causes males to develop noticeable hair on the face, chest, and occasionally other regions of the body, such as the back. A chemical called dihydrotestosterone (DHT) which is synthesized from testosterone in the hair follicles, leads to the receding hairline and male-pattern baldness in genetically susceptible individuals.

Testosterone is responsible for the greater muscular development in males. Knowing this, both males and females sometimes take anabolic steroids, either testosterone or related steroid hormones resembling testosterone, to build up their muscles. For more information about the risks of anabolic steroids, review Figure 40.16.

Check Your Progress   41.2

1. Trace the pathway a sperm must follow from its origin to its exit from the male reproductive tract.
2. List the glands that contribute fluids to the semen.
3. Describe the effects of FSH and LH in males.

41.3 Human Female Reproductive System

The human female reproductive system includes organs depicted in Figure 41.8 and listed in Table 41.2. The female gonads, or primary sex organs, are paired **ovaries** (sing., ovary), located on each side of the upper pelvic cavity.

The Ovaries, Uterus, and Vagina

The ovaries produce the female sex hormones and a secondary **oocyte** each month. The oviducts [L. *ovum*, egg, and *duco*, lead out], also called uterine or fallopian tubes, extend from the ovaries to the uterus; however, the oviducts are not attached to the ovaries. Instead, they have fingerlike projections called fimbriae (sing., fimbria) that sweep over the ovaries. When an oocyte bursts from an ovary during ovulation, it usually is swept into an oviduct by the combined action of the fimbriae and the beating of cilia that line the oviducts. Fertilization, if it occurs, normally takes place in an oviduct. The developing embryo is propelled slowly by ciliary movement and tubular muscle contraction to the uterus. Some women opt to permanently prevent pregnancy by having a surgery called *tubal ligation*, in which the oviducts are closed off so that sperm cannot reach the oocyte.

Normally, the ovaries alternate in producing one oocyte each month. For the sake of convenience, we will refer to the released oocyte as an **egg.** The ovaries also produce the female sex hormones; the **estrogens,** collectively called estrogen, and **progesterone,** during the ovarian cycle.

The **uterus** [L. *uterus*, womb] is a thick-walled muscular organ about the size and shape of an inverted pear. The narrow end of the uterus is called the **cervix.** The embryo completes its development after embedding itself in the uterine lining, called the **endometrium** [Gk. *endon*, within, and *metra*, womb]. If, by chance, the embryo should embed itself in another location, such as an oviduct, an *ectopic pregnancy* results.

A small opening in the cervix leads to the vaginal canal. The **vagina** [L. *vagina*, sheath] is a tube at a 45-degree angle to the small of the back. The mucosal lining of the vagina lies in folds and the vagina can be distended. This is especially important when the vagina serves as the birth canal, and it also can facilitate intercourse, when the vagina receives the penis during copulation. Several different types of bacteria normally reside in the vagina and create an acidic environment. This environment is protective against the possible growth of pathogenic bacteria, but sperm prefer the basic environment provided by semen.

Table 41.2 Female Reproductive Organs

Organ	Function
Ovaries	Produce egg and sex hormones
Oviducts (fallopian tubes)	Conduct egg; location of fertilization
Uterus (womb)	Houses developing embryo and fetus
Vagina	Receives penis during copulation and serves as birth canal

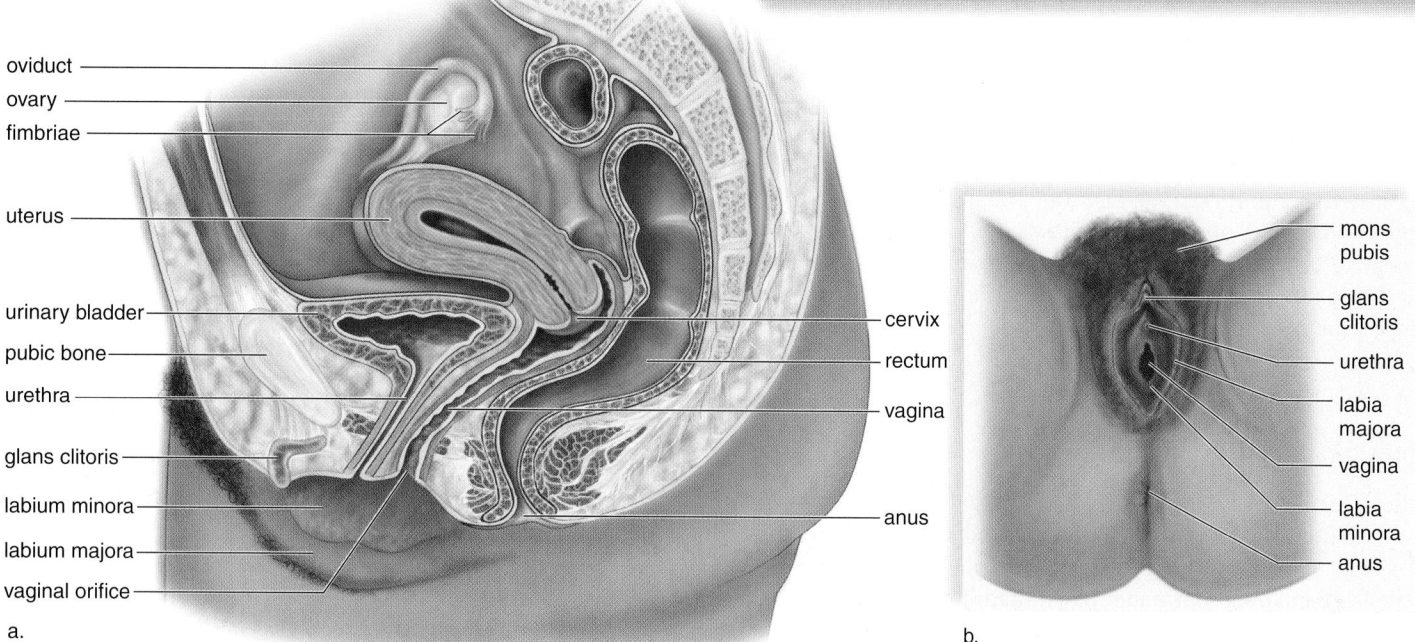

a.   b.

Figure 41.8 Female reproductive system.   a. The ovaries produce one oocyte per month. Fertilization occurs in the oviduct, and development occurs in the uterus. The vagina is the birth canal and organ of sexual intercourse. **b.** Vulva. At birth, the opening of the vagina is partially occluded by a membrane called the hymen. Physical activities and sexual intercourse disrupt the hymen.

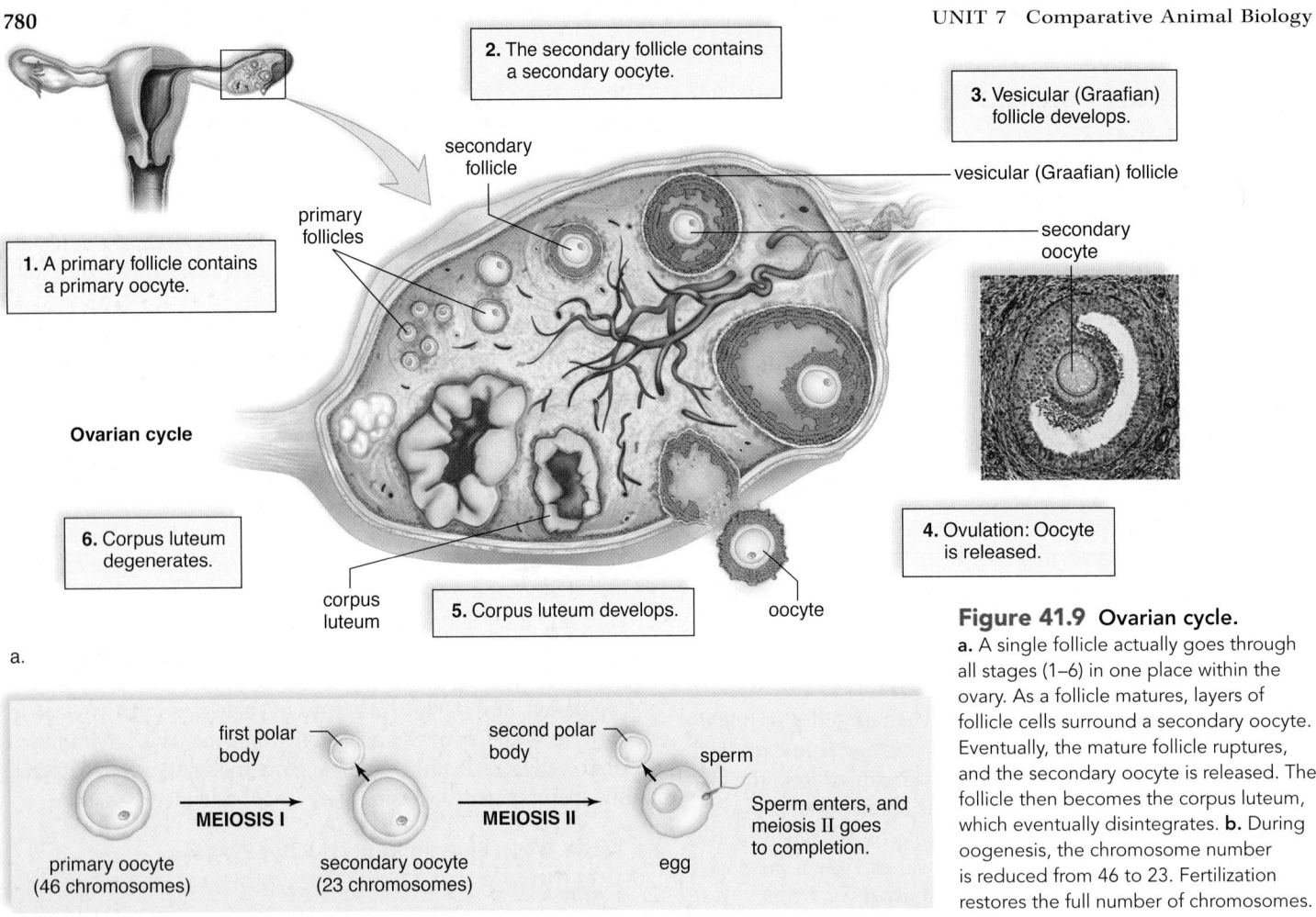

2. The secondary follicle contains a secondary oocyte.

3. Vesicular (Graafian) follicle develops.

secondary follicle

vesicular (Graafian) follicle

primary follicles

secondary oocyte

1. A primary follicle contains a primary oocyte.

Ovarian cycle

6. Corpus luteum degenerates.

corpus luteum

5. Corpus luteum develops.

4. Ovulation: Oocyte is released.

oocyte

a.

first polar body

second polar body

sperm

MEIOSIS I

MEIOSIS II

Sperm enters, and meiosis II goes to completion.

primary oocyte (46 chromosomes)

secondary oocyte (23 chromosomes)

egg

b.

Figure 41.9 Ovarian cycle.
a. A single follicle actually goes through all stages (1–6) in one place within the ovary. As a follicle matures, layers of follicle cells surround a secondary oocyte. Eventually, the mature follicle ruptures, and the secondary oocyte is released. The follicle then becomes the corpus luteum, which eventually disintegrates. **b.** During oogenesis, the chromosome number is reduced from 46 to 23. Fertilization restores the full number of chromosomes.

The Female Orgasm

The external genital organs of a female are known collectively as the vulva (Fig. 41.8b). The mons pubis and two folds of skin called labia minora and labia majora are on either side of the urethral and vaginal openings. Beneath the labia majora, pea-sized greater vestibular glands (Bartholin glands) open on either side of the vagina. They keep the vulva moist and lubricated during intercourse.

At the juncture of the labia minora is the *clitoris*, which is homologous to the penis in males. The clitoris has a shaft of erectile tissue and is capped by a pea-shaped glans. The many sensory receptors of the clitoris allow it to function as a sexually sensitive organ. The clitoris has twice as many nerve endings as the penis. As in males, female orgasm involves both psychological and physiological reactions. In the female there is also a release of neuromuscular tension in the muscles of the genital area, vagina, and uterus.

The Ovarian Cycle

The **ovarian cycle** occurs as a **follicle** [L. *folliculus*, little bag] changes from a primary, to a secondary, and finally to a vesicular (Graafian) follicle (Fig. 41.9a) under the influence of follicle-stimulating hormone (FSH) and luteinizing hormone (LH) from the anterior pituitary (Fig. 41.10). Epithelial cells of a primary follicle surround a primary oocyte. Pools of follicular fluid surround the oocyte in a secondary follicle. In a vesicular follicle, a fluid-filled cavity increases to the point that the follicle wall balloons out on the surface of the ovary.

Oogenesis and Ovulation

As a follicle matures, *oogenesis*, a form of meiosis depicted in Figure 41.9b, is initiated and continues. The primary oocyte divides, producing two haploid cells. One cell is a secondary oocyte, and the other is a polar body. The vesicular follicle bursts, releasing the secondary oocyte (egg) surrounded by a clear membrane. This process is referred to as **ovulation.** Once a vesicular follicle has lost the secondary oocyte, it develops into a **corpus luteum** [L. *corpus*, body, and *luteus*, yellow], a glandlike structure.

The secondary oocyte enters an oviduct. If fertilization occurs, a sperm enters the secondary oocyte and then the oocyte completes meiosis and becomes an egg. An egg with 23 chromosomes and a second polar body results. When the sperm nucleus unites with the egg nucleus, a zygote with 46 chromosomes is produced. If zygote formation and pregnancy do not occur, the corpus luteum begins to degenerate after about ten days.

Phases of the Ovarian Cycle

The ovarian cycle is controlled by the gonadotropic hormones, FSH and LH (Fig. 41.11). The gonadotropic hormones are not present in constant amounts and instead are secreted at different rates during the cycle. For simplicity's sake, it is convenient to emphasize that during the first half, or **follicular phase,** of the cycle, FSH promotes the development of a follicle that primarily secretes estrogen. As the estrogen level in the blood rises, it exerts negative feedback control over the anterior pituitary secretion of FSH so that the follicular phase comes to an end. At the same time that FSH along with LH release is being dampened by

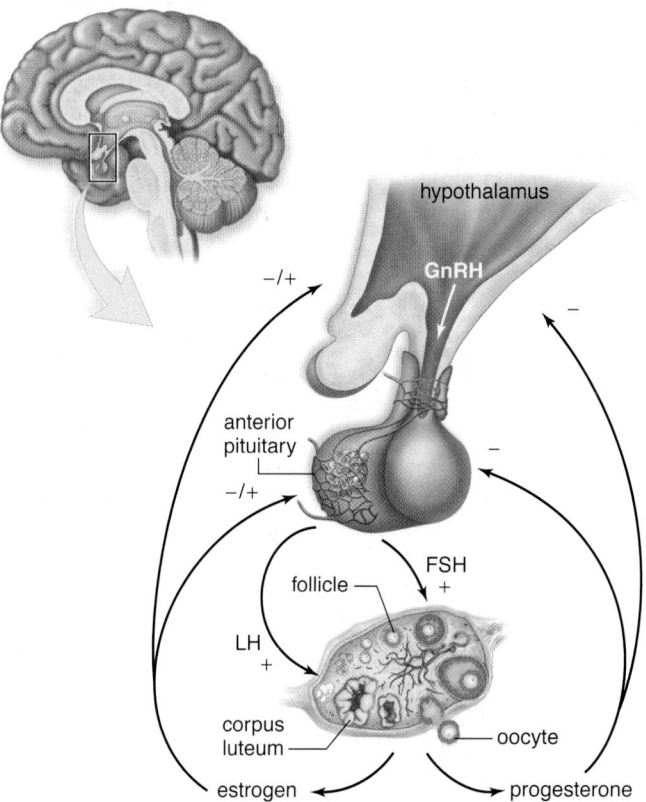

Figure 41.10 Hormonal control of ovaries. The hypothalamus produces GnRH (gonadotropic-releasing hormone). GnRH stimulates the anterior pituitary to produce FSH (follicle-stimulating hormone) and LH (luteinizing hormone). FSH stimulates the follicle to produce primarily estrogen, and LH stimulates the corpus luteum to produce primarily progesterone. Estrogen and progesterone maintain the sex organs (e.g., uterus) and the secondary sex characteristics, and they exert feedback control over the hypothalamus and the anterior pituitary. Feedback control regulates the relative amounts of estrogen and progesterone in the blood.

moderate amounts of estrogen, the synthesis of the gonadotropic hormones continues, and they build up in the anterior pituitary.

When the level of estrogen in the blood becomes very high, it exerts positive feedback on the hypothalamus and anterior pituitary. The hypothalamus is stimulated to suddenly secrete a large amount of GnRH. This leads to a surge of LH (and to a lesser degree, FSH) by the anterior pituitary and to ovulation at about the fourteenth day of a 28-day cycle (Fig. 41.11, *top*).

During the second half, or **luteal phase,** of the ovarian cycle, it is convenient to emphasize that LH promotes the development of the corpus luteum, which primarily secretes progesterone. As the blood level of progesterone rises, it exerts negative feedback control over anterior pituitary secretion of LH so that the corpus luteum begins to degenerate. As the luteal phase comes to an end, the low levels of progesterone and estrogen in the body cause menstruation to begin, as discussed next.

Animation
Maturation of the Follicle and Oocyte

The Uterine Cycle

The female sex hormones produced in the ovarian cycle (estrogen and progesterone) affect the endometrium of the uterus, causing the cyclical series of events known as the **uterine cycle** (Fig. 41.11, *bottom*). The twenty-eight-day cycle is divided into the three phases described next.

Phases of the Uterine Cycle

During *days 1–5*, there is a low level of female sex hormones in the body, causing the endometrium to disintegrate and its blood vessels to rupture. A flow of blood passes out of the vagina during **menstruation** [L. *menstrualis*, happening monthly], also known as the menstrual period. This is the *menstrual phase* of the uterine cycle.

During *days 6–13*, increased production of estrogen by an ovarian follicle causes the endometrium to thicken and to become vascular and glandular. This is called the *proliferative phase* of the uterine cycle.

Ovulation usually occurs on *day 14* of the 28-day cycle. During *days 15–28*, increased production of

Figure 41.11 Female hormone levels during the ovarian and uterine cycles. During the follicular phase of the ovarian cycle (*top*), FSH released by the anterior pituitary promotes the maturation of a follicle in the ovary. The ovarian follicle produces increasing levels of estrogen, which causes the endometrium to thicken during the proliferative phase of the uterine cycle (*bottom*). After ovulation and during the luteal phase of the ovarian cycle, LH promotes the development of the corpus luteum. This structure produces increasing levels of progesterone, which causes the endometrium to become secretory. Menstruation and the proliferative phase begin when progesterone production declines to a low level.

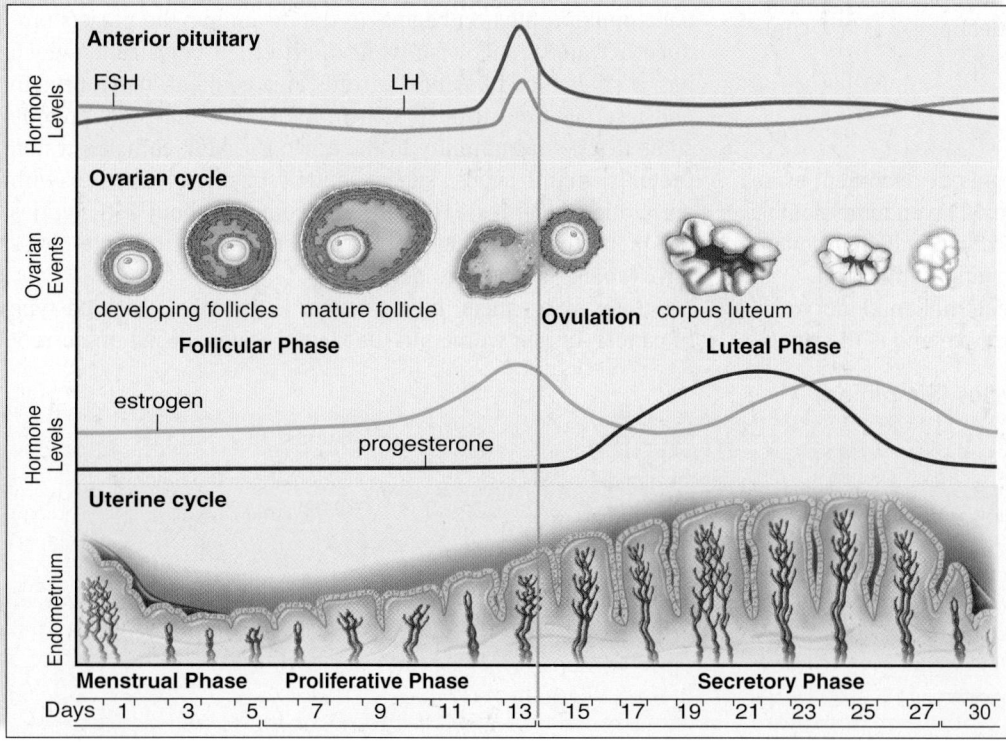

progesterone by the corpus luteum causes the endometrium to double in thickness and the uterine glands to mature, producing a thick mucoid secretion. This is called the *secretory phase* of the uterine cycle. The endometrium now is prepared to receive the developing embryo. If pregnancy does not occur, the corpus luteum degenerates, and the low level of sex hormones in the female body causes the endometrium to break down as menstruation occurs. Table 41.3 compares the stages of the uterine cycle with those of the ovarian cycle.

Menstruation

Seven to ten days before the start of menstruation, some women suffer from premenstrual syndrome (PMS). During this time a woman may exhibit breast enlargement and tenderness, achiness, headache, and irritability. The exact cause for PMS has yet to be discovered.

Animation
Female Reproductive System

During menstruation, arteries that supply the lining constrict and the capillaries weaken. Blood spilling from the damaged vessels detaches layers of the lining, not all at once but in random patches. Mucus, blood, and degenerating endometrium descend from the uterus, through the vagina, creating menstrual flow. Fibrinolysin, an enzyme released by dying cells, prevents the blood from clotting. Menstruation lasts from three to five days, as the uterus sloughs off the thick lining that was three weeks in the making.

The first menstrual period, called *menarche*, typically occurs between the ages of 11 and 13. Menarche signifies that the ovarian and uterine cycles have begun. If menarche does not occur by age 16, or if normal uterine cycles are interrupted for six months or more without pregnancy being the cause, the condition is termed amenorrhea. Primary amenorrhea is usually caused by nonfunctional ovaries or developmental abnormalities. Secondary amenorrhea may be caused by weight loss and/or excessive exercise.

Menopause, which usually occurs between ages 45 and 55, is the time in a woman's life when menstruation ceases because the ovaries are no longer functioning. Menopause is not complete until menstruation is absent for a year.

Fertilization and Pregnancy

If fertilization does occur, an embryo begins development even as it travels down the oviduct to the uterus. The endometrium is now prepared to receive the developing embryo, which becomes embedded in the lining several days following fertilization.

The placenta originates from both maternal and embryonic tissues. It is shaped like a large, thick pancake and is the site of the exchange of gases and nutrients between fetal and maternal blood, although the two rarely mix. At first, the placenta produces **human chorionic gonadotropin (HCG),** which maintains the corpus luteum until the placenta begins its own production of progesterone and estrogen. HCG is the hormone detected in pregnancy tests; it is present in the mother's blood and urine as soon as ten days after conception.

Progesterone and estrogen have two effects. They shut down the anterior pituitary so that no new follicles mature, and they maintain the lining of the uterus so that the corpus luteum is not needed. No menstruation occurs during pregnancy.

Estrogen and Progesterone

Estrogen in particular is essential for the normal development and functioning of the female reproductive organs listed in Table 41.2. Estrogen is also largely responsible for the secondary sex characteristics in females, including body hair and fat distribution. In general, females have a more rounded appearance than males because of a greater accumulation of fat beneath their skin. Also, the pelvic girdle enlarges so that females have wider hips than males, and the thighs converge at a greater angle toward the knees. Both estrogen and progesterone are required for breast development as well.

The Female Breast

A female breast contains between 15 and 24 lobules, each with its own mammary duct (Fig. 41.12). A duct begins at the nipple and divides into numerous other ducts that end in blind sacs called alveoli.

Lactation, the production of milk by the cells of the alveoli, is stimulated by the hormone prolactin. Milk is not produced during pregnancy because production of prolactin is suppressed by the feedback inhibition effect of estrogen and progesterone on the anterior pituitary. A couple of days after delivery of a baby, milk production begins. In the meantime, the breasts produce a watery, yellowish-white fluid called colostrum, which has a similar composition to milk but contains more protein and less fat. Colostrum is rich in antibodies that may provide some degree of immunity to the newborn. Milk contains water, proteins, amino acids, sugars, and lysozymes (enzymes with antibiotic properties). Human milk contains about 750 calories per liter, which is considerably higher than even whole cow's milk (about 500 calories per liter).

After skin cancer, breast cancer is the most common type of cancer among women in the United States. Based on current

Table 41.3 Ovarian and Uterine Cycles (Simplified)

Ovarian Cycle	Events	Uterine Cycle	Events
Follicular phase—Days 1–13	FSH	Menstrual phase—Days 1–5	Endometrium breaks down
	Follicle maturation	Proliferative phase—Days 6–13	Endometrium rebuilds
	Estrogen		
Ovulation—Day 14*	LH spike		
Luteal phase—Days 15–28	LH	Secretory phase—Days 15–28	Endometrium thickens and
	Corpus luteum		glands are secretory
	Progesterone		

*Assuming a 28-day cycle

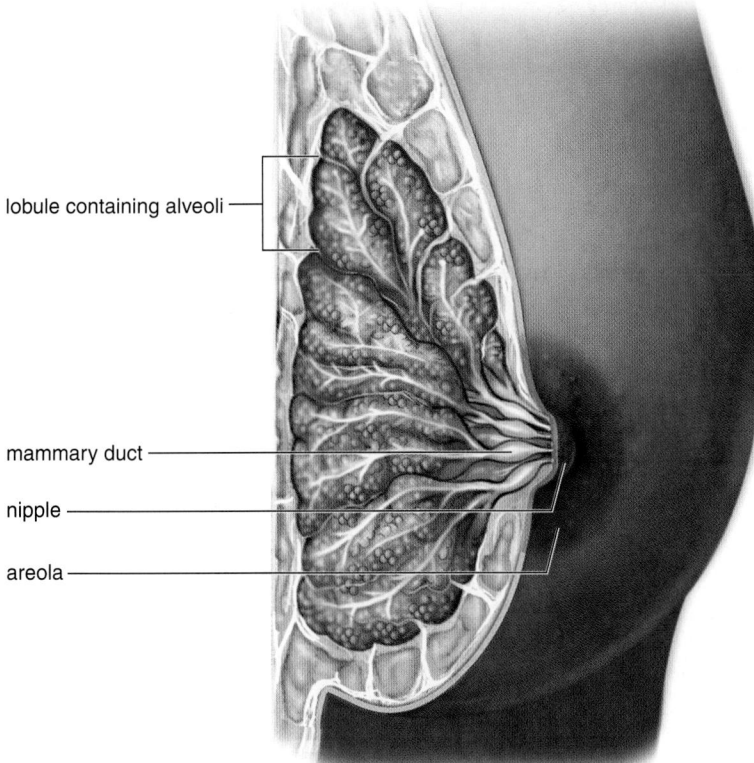

lobule containing alveoli

mammary duct

nipple

areola

Figure 41.12 Anatomy of the breast. The female breast contains lobules consisting of ducts and alveoli. In the lactating breast, cells lining the alveoli have been stimulated to produce milk by the hormone prolactin.

rates, about 12% of women born today can expect to be diagnosed with cancer of the breast during their lifetime. Women should regularly check their breasts for lumps and other irregularities and have mammograms (X-ray photographs) taken as recommended by their physician. Although breast cancer genes have been described, most forms of breast cancer are nonhereditary. The number of treatment options for various types of breast cancer has expanded in recent years, resulting in increased survival rates. Currently, about 86% of women survive at least five years after their initial diagnosis.

Besides more traditional surgery, radiation, and chemotherapy, two newer types of breast cancer treatments are hormonal therapy and targeted therapy. Some breast cancers have receptors for estrogen, and they grow in response to increased estrogen levels. To treat these tumors, various drugs are available to either decrease estrogen synthesis, or to block the interaction of estrogen with its receptor. Targeted therapies usually involve the administration of a monoclonal antibody (see Chapter 33) that can react very specifically with antigens found on tumor cells. The best known example, called Herceptin, blocks a protein called HER2, expressed on 20–30% of breast cancer cells, where it is involved in stimulating uncontrolled cell growth. The routine use of this drug is somewhat controversial, however, partly because of its estimated cost of $100,000 per year.

Check Your Progress   **41.3**

1. Name the structures of the female reproductive system that (a) produce the egg, (b) transport the egg, (c) house a developing embryo, and (d) serve as the birth canal.
2. Explain the effects of FSH and LH on the ovarian cycle.
3. Summarize the roles of estrogen and progesterone in the ovarian and uterine cycles.

41.4 Control of Human Reproduction

Learning Outcomes

Upon completion of this section, you should be able to

1. List five commonly available methods of birth control that are designed to prevent conception.
2. Note which birth control methods are effective at preventing the spread of sexually transmitted diseases.
3. Describe several reproductive technologies that can help infertile couples to have children.

Several means are available to dampen or enhance the human reproductive potential. Contraceptives are medications and devices that reduce the chance of pregnancy.

Traditional Birth Control Methods

The most reliable method of birth control is abstinence—that is, not engaging in sexual intercourse. This form of birth control has the added advantage of preventing transmission of a sexually transmitted disease. The male and female condoms also offer some protection against sexually transmitted diseases in addition to helping prevent pregnancy. Some of the common means of birth control used in the United States are shown in Figure 41.13.

Morning-After Pills

The term "morning-after pill," or emergency contraception, refers to medications that can prevent pregnancy after unprotected intercourse. The expression "morning-after" is a misnomer, in that some treatments can be started up to five days after unprotected intercourse.

The first FDA-approved medication produced for emergency contraception was a kit called Preven. Preven includes four synthetic progesterone pills; two are taken up to 72 hours after unprotected intercourse, and two more are taken 12 hours later. The hormone upsets the normal uterine cycle, making it difficult for an embryo to implant in the endometrium. A recent study estimated that Preven was 85% effective in preventing unintended pregnancies. The Preven kit also includes a pregnancy test; women are instructed to take the test first before using the hormone, because the medication is not effective on an established pregnancy.

In 2006, the FDA approved another drug called Plan B, which is up to 89% effective in preventing pregnancy if taken within 72 hours after unprotected sex. It is available without a prescription to women age 17 and older. In August 2010, ulipristal acetate (also known as ella) was also approved for emergency contraception. It can be taken up to five days after unprotected sex, and studies indicate it is somewhat more effective than Plan B. Unlike Plan B, however, a prescription is required.

Mifepristone, also known as RU-486 or the "abortion pill," can cause the loss of an implanted embryo by blocking the progesterone receptors of endometrial cells. This causes the endometrium to slough off, carrying the embryo with it. When taken in conjunction with a prostaglandin to induce uterine

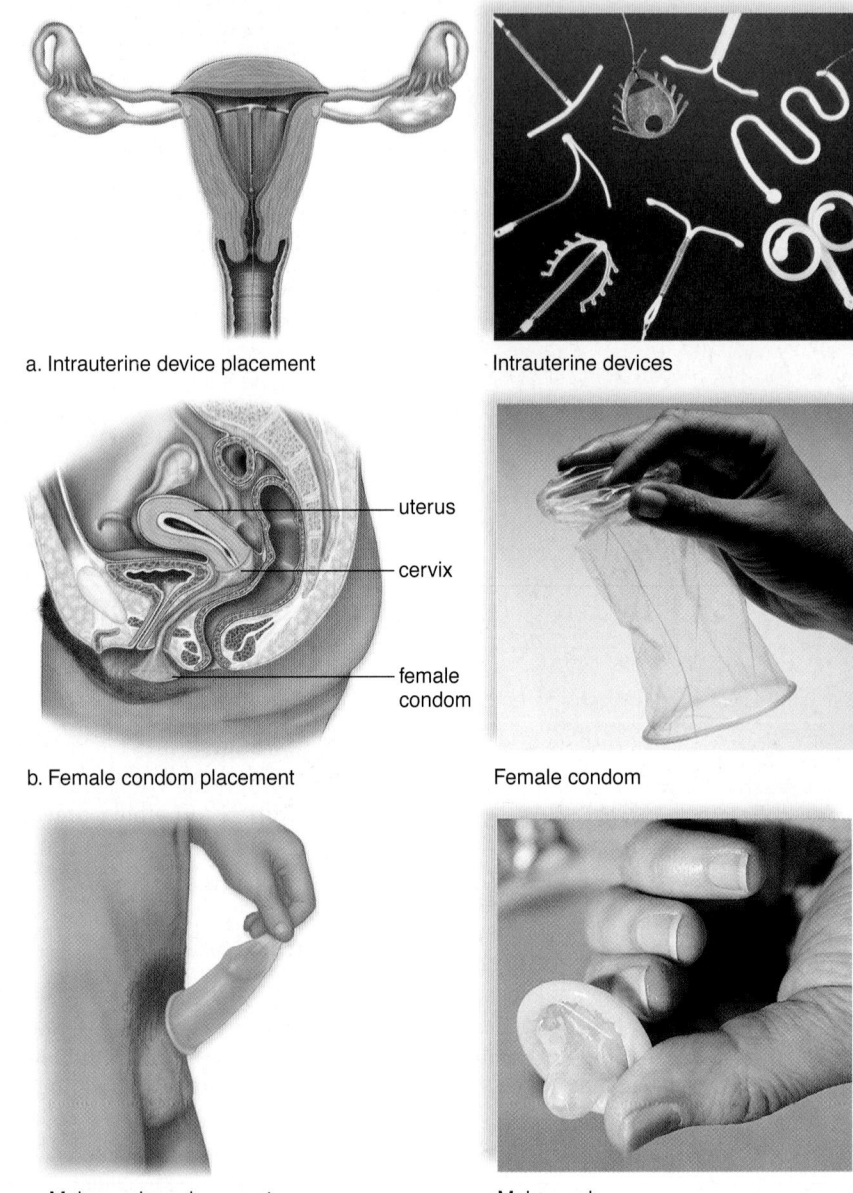

a. Intrauterine device placement

Intrauterine devices

uterus

cervix

female condom

b. Female condom placement

Female condom

c. Male condom placement

Male condom

Figure 41.13 Various birth control methods. **a.** Intrauterine devices mechanically prevent implantation and can contain progesterone to prevent ovulation, prevent implantation, and thicken cervical mucus. **b.** Female condom that is fitted inside vagina prevents sperm entry and protects against STDs. **c.** Male condom that fits over penis prevents sperm from entering the vagina and protects against STDs.

contractions, RU-486 is 95% effective at inducing an abortion up to the 49th day of gestation. Because of its mechanism of action, the use of RU-486 is more controversial compared to other medications, and while it is currently available in the U.S. for early medical abortion, it is not approved for emergency contraception.

Vaccines

Contraceptive vaccines are now being developed. For example, a vaccine intended to immunize women against HCG, the hormone so necessary to maintaining the implantation of the embryo, was successful in a limited clinical trial. Because HCG is not normally present in the body, no autoimmune reaction is expected, but the immunization does wear off with time. Others believe that it would also be possible to develop a safe antisperm vaccine that could be used in women.

Male Hormonal Birth Control

A great deal of research is being devoted to developing safe and effective hormonal birth control for men. Implants, pills, patches, and injections are being explored as ways to deliver testosterone and/or progesterone at adequate levels to suppress sperm production. Even the most successful formulations are still in the experimental stage and are unlikely to be available outside of clinical trials for at least a few more years.

Reproductive Technologies

Infertility is the inability of a couple to achieve pregnancy after one year of regular, unprotected intercourse. The American Medical Association estimates that 15% of all couples are infertile. The cause of infertility can be attributed to the man (40%), the woman (40%), or both (20%).

Sometimes the causes of infertility may be corrected by medical intervention so that couples can have children. If no obstruction is apparent and body weight is normal, it is possible for women to take fertility drugs, which are gonadotropic hormones that stimulate the ovaries and bring about ovulation. Such hormone treatments may cause multiple ovulations and multiple births.

When reproduction does not occur in the usual manner, many couples adopt a child. Others sometimes try one of the assisted reproductive technologies (ARTs) developed to increase the chances of pregnancy. In these cases, sperm and/or oocytes are often retrieved from the testes and ovaries, and fertilization takes place in a clinical or laboratory setting.

Artificial Insemination by Donor (AID)

During artificial insemination, sperm from a donor are introduced into the woman's vagina by a physician. This technique is especially helpful if the male partner has a low sperm count, because the sperm can be collected over a period of time and concentrated so that the sperm count is sufficient to result in fertilization. Often, however, a woman is inseminated by sperm acquired from a different donor. At times, a combination of partner and donor sperm is used.

A variation of AID is *intrauterine insemination (IUI)*. In IUI, fertility drugs are given to stimulate the ovaries, and then the donor's sperm is placed into the uterus, rather than into the vagina.

If the prospective parents wish, sperm can be sorted into those believed to be X-bearing or Y-bearing, to increase the chances of having a child of the desired sex. First, the sperm are treated with a DNA-staining chemical. Because the X chromosome has slightly more DNA than the Y chromosome, it takes up more dye. When a laser beam shines on the sperm, the X-bearing sperm shine a little more brightly than the Y-bearing sperm. A machine sorts the sperm into two groups on this basis. Parents can expect about a 65% success rate when selecting for males and about 85% when selecting for females.

In Vitro Fertilization (IVF)

During IVF, fertilization occurs in laboratory glassware. Ultrasound machines can now spot follicles in the ovaries that hold immature oocytes; therefore, the latest method is to forgo the administration of fertility drugs and retrieve immature oocytes by using a needle. The immature oocytes are then brought to maturity in glassware before concentrated sperm are added. After about two to four days, the embryos are ready to be transferred to the uterus of the woman, who is now in the secretory phase of her uterine cycle. If desired, the embryos can be tested for a genetic disease, as discussed in the Nature of Science feature on page 786. If implantation is successful, development continues to term.

 **Video** One Healthy Baby

 **Video** In Vitro Fertilization

Gamete Intrafallopian Transfer (GIFT)

Recall that the term **gamete** refers to a sex cell, either a sperm or an oocyte. Gamete intrafallopian transfer was devised to overcome the low success rate (15–20%) of in vitro fertilization. The

Figure 41.14   Intracytoplasmic sperm injection (ICSI).
A microscope connected to a television screen is used to carry out in vitro fertilization. A pipette holds the oocyte steady while a needle (not visible) introduces the sperm into the oocyte.

method is exactly the same as for in vitro fertilization, except the oocytes and the sperm are placed in the oviducts immediately after they have been brought together. GIFT has the advantage of being a one-step procedure for the woman—the oocytes are removed and reintroduced all in the same time period. A variation on this procedure is to fertilize the eggs in the laboratory and then place the zygotes in the oviducts.

Surrogate Mothers

In some instances, women are contracted and paid to carry babies as *surrogate mothers*. The sperm and even the oocyte can be contributed by the contracting parents.

Intracytoplasmic Sperm Injection (ICSI)

In this highly sophisticated procedure, a single sperm is injected into an oocyte (Fig. 41.14). It is used effectively when a man has severe infertility problems.

If all the assisted reproductive technologies discussed were employed simultaneously, it would be possible for a baby to have five parents: (1) sperm donor, (2) oocyte donor, (3) surrogate mother, (4) contracting mother, and (5) contracting father. The legal definitions of "parent" can become a matter of contention in some cases.

Check Your Progress   41.4

1. Identify which of the birth control methods discussed in this section physically block sperm from entering the uterus.
2. Rank the following birth control methods regarding their effectiveness at preventing pregnancy: abstinence, female birth control pill, condoms, natural family planning, and vasectomy.
3. Briefly describe the following processes: artificial insemination by donor (AID), in vitro fertilization (IVF), gamete intrafallopian transfer (GIFT), and intracytoplasmic sperm injection (ICSI).

Preimplantation Genetic Diagnosis

If prospective parents are heterozygous for a genetic disorder, they may want the assurance that their offspring will be free of the disorder. Determining the genotype of an embryo can provide this assurance. For example, if both parents are *Aa* for a recessive disorder, the embryo will develop normally if it has the genotype *AA* or *Aa*. In contrast, if one of the parents is *Aa* for a dominant disorder, the embryo will develop normally only if it has the genotype *aa*.

Following in vitro fertilization (IVF), the zygote (fertilized egg) divides. When the embryo has six to eight cells (Fig. 41Aa), removal of one of these cells for testing purposes has no effect on normal development. Only embryos with a cell that tests negative for the genetic disorder of interest are placed in the uterus to continue developing.

So far, about 1,000 children with normal genotypes for genetic disorders that run in their families have been born worldwide following embryo testing. In the future, it's possible that embryos who test positive for a disorder could be treated by gene therapy, so that they, too, would be allowed to continue to term.

Testing the oocyte is possible if the condition of concern is recessive. Recall that meiosis in females results in a single egg and at least two polar bodies.[1] Polar bodies later disintegrate, and they receive very little cytoplasm, but they do receive a haploid number

1 Once meiosis is complete, the oocyte is termed an egg.

of chromosomes. When a woman is heterozygous for a recessive genetic disorder, about half the first polar bodies have received the genetic defect, and in these instances, the egg has received the normal allele. Therefore, if a polar body tests positive for a recessive defect, then the egg has received the normal dominant allele. Only normal eggs are then used for IVF. Even if the sperm should happen to carry the same mutant allele, the zygote will, at worst, be heterozygous. But the phenotype will appear normal (Fig. 41A*b*).

If, in the future, gene therapy becomes routine, it's possible that an egg could be given genes that control traits desired by the parents, such as musical or athletic ability, prior to IVF.

Questions to Consider

1. Some biomedical ethicists are concerned that embryo testing could lead to a form of eugenics, in which wealthier parents are able to afford improving their children's physical traits. Others argue that this technology should be developed as much as possible to reduce the occurrence of terrible diseases. Which argument is more compelling to you, and why?
2. Some experts feel that pre-implantation screening should be required for all embryos that are conceived by in vitro fertilization, to reduce the occurrence of disease. Do you agree or disagree?

Figure 41A Preimplantation genetic diagnosis.   a. Following IVF and cleavage, genetic analysis is performed on one cell removed from an eight-celled embryo. If it is found to be free of the genetic defect of concern, the seven-celled embryo is implanted in the uterus and develops into a fetus. **b.** Chromosome and genetic analysis is performed on a polar body attached to an oocyte. If the oocyte is free of a genetic defect, the egg is used for IVF, and the embryo is implanted in the uterus for further development.

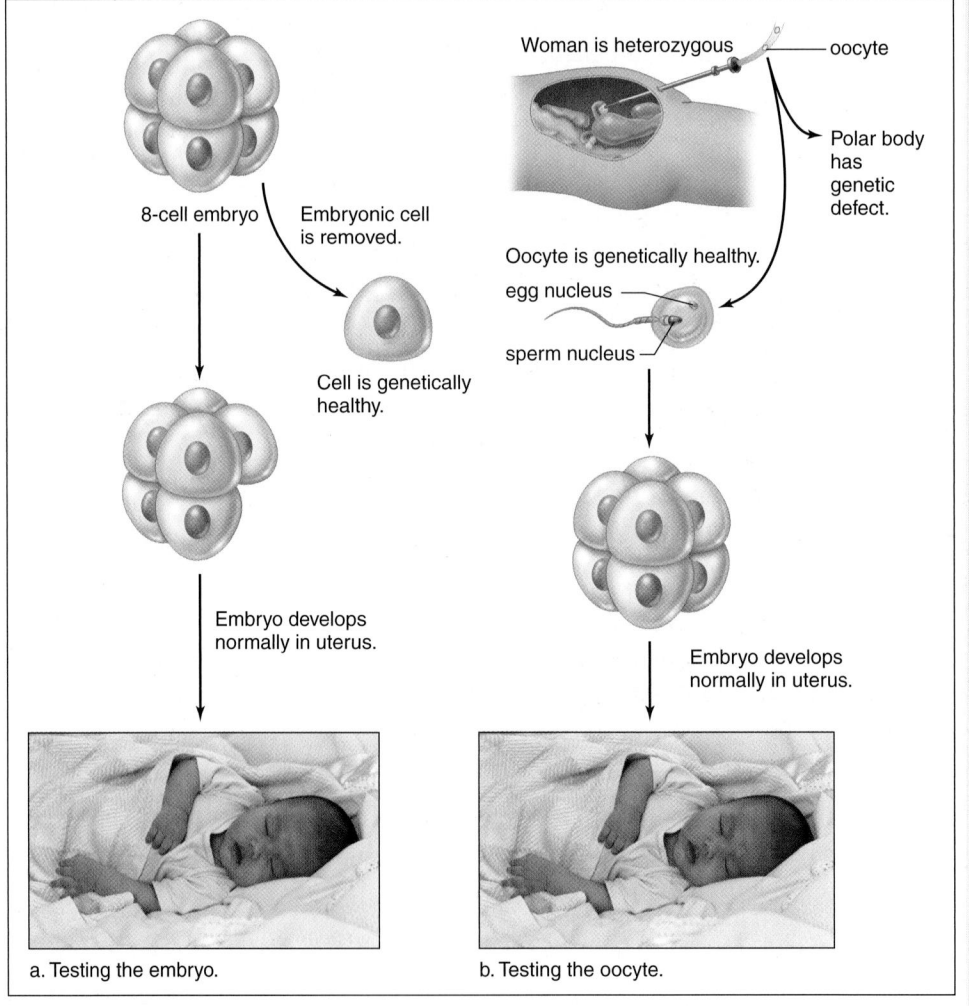

8-cell embryo   Embryonic cell is removed.

Cell is genetically healthy.

Embryo develops normally in uterus.

a. Testing the embryo.

Woman is heterozygous — oocyte

Polar body has genetic defect.

Oocyte is genetically healthy.

egg nucleus

sperm nucleus

Embryo develops normally in uterus.

b. Testing the oocyte.

41.5 Sexually Transmitted Diseases

Learning Outcomes

Upon completion of this section, you should be able to

1. List three STDs that are caused by viruses and three that are caused by bacteria.
2. Discuss how HIV affects the immune system and how this is related to the three categories of HIV infection.
3. Describe one STD that is currently preventable by vaccination.

According to a 2010 report, about 19 million new cases of sexually transmitted diseases (STDs) occur each year in the United States. Approximately 30 different STDs are caused by organisms ranging from viruses to arthropods; however, this discussion centers on the most prevalent and serious ones. In addition, the Biological Systems feature on page 789 describes methods to avoid transmission of STDs.

Viral STDs

AIDS, genital herpes, genital warts, and hepatitis A and B are STDs caused by viruses. Therefore, they do not respond to treatment with traditional antibiotics. Antiviral drugs have been developed to treat some of these viruses, but in many cases the infection persists for life.

AIDS

Acquired immunodeficiency syndrome (AIDS) is caused by the **human immunodeficiency virus (HIV).** An introduction to HIV and AIDS was provided in Chapter 33. As noted there, by the end of 2008, about 33.4 million people worldwide were infected with HIV, and more than 25 million people have died of AIDS since 1981. The great majority of these people live in underdeveloped countries; an estimated 68% of HIV-infected individuals in the world live in sub-Saharan Africa.

On a somewhat more positive note, total AIDS-related deaths have been decreasing worldwide since about 2004, mainly due to increased availability of antiretroviral therapy. In the United States, an estimated one million people are infected, and over 500,000 have died. Homosexual men make up the largest proportion of people living with HIV in the U.S., but the fastest rate of increase is now seen in heterosexuals. Over half of new HIV infections are now occurring in people under the age of 25.

Transmission. HIV is transmitted by sexual contact with an infected person, including vaginal or anal intercourse and oral-genital contact. Also, needle-sharing among intravenous drug users is a very efficient way to transmit HIV. Some of the earliest cases of AIDS resulted from transfusion of HIV-infected blood, but this is now extremely rare thanks to effective screening methods. Babies born to HIV-infected mothers may become infected before or during birth or through breast-feeding. Transmission through tears, saliva, or sweat is thought to be highly unlikely.

Cultural factors may also play an important role in transmission of HIV. For example, polygamy is common in many African countries, as are misconceptions about the effectiveness of condoms in preventing HIV transmission, and even about the fact that HIV causes AIDS. Such practices and beliefs may significantly increase the risk of HIV infection.

Stages and Symptoms of an HIV Infection. HIV infects helper T cells, macrophages, and other cells that have a molecule called CD4 on their surface (Fig. 41.15). As you may recall from Chapter 33, helper T cells control the activities of many other cells in the immune system. After HIV enters a host cell, it converts its viral RNA to DNA, which is then integrated into the host cell DNA.

In the early stages after an HIV infection, the person's blood contains a high amount of virus, even though the individual may experience only mild flulike symptoms, which soon disappear. Even if an HIV-infected person undergoes an HIV test at this time, it would probably be negative because this tests for the presence of anti-HIV antibodies, which are usually not detectable for an average of 25 days. However, the person can still be highly infectious at this stage.

Animation
How the HIV Infection Cycle Works

After the initial surge of HIV replication, the immune system temporarily reduces the virus levels in the blood. Over an average period of eight to ten years, the HIV that has been established in cells gradually produces increasing amounts of HIV, which infects and destroys more and more helper T cells. As the virus reproduces, it also mutates so that it is no longer recognized by the immune system.

Three categories of HIV infection are distinguished, based mainly on the type of disease seen in the patient. Within each category, the number of CD4 + T cells is also monitored. Patients in category A usually have greater than 500 CD4 + T cells per microliter of blood, and either lack symptoms completely or have persistently enlarged lymph nodes. Patients in category B have a CD4 + T-cell count between 200 and 499 per microliter, and begin to suffer from opportunistic infections (OIs)—infections that would be unable to take hold in a person with a healthy immune system. Once the CD4 + T-cell count falls below 200 per microliter, the patient is in category C, or true AIDS. The deficiency of helper T-cell functions in these patients results in their contracting one or more AIDS-defining OIs (see Chapter 33).

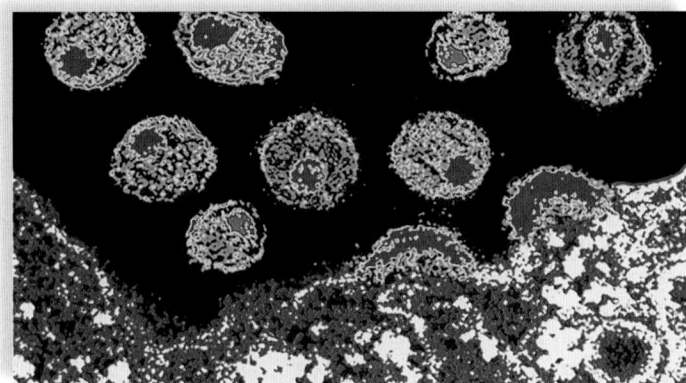

Figure 41.15 HIV, the AIDS virus. False-colored micrograph showing HIV particles budding from an infected helper T cell. These viruses can infect other helper T cells and also macrophages, which work with helper T cells to stem infection.

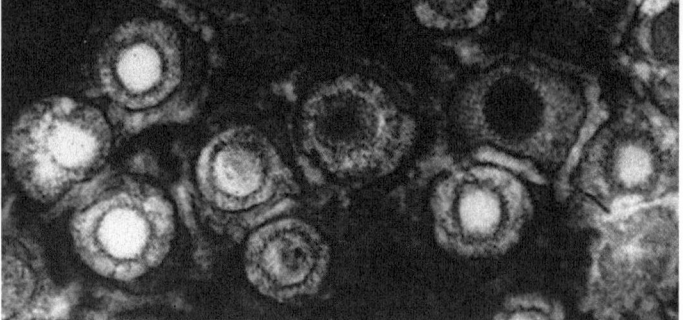

Figure 41.16 Genital warts. A photomicrograph of human papillomaviruses.

Treating or Preventing HIV Infection.

Before the development of antiretroviral drugs, most HIV-infected individuals eventually progressed to category C AIDS, which was almost invariably fatal. There is still no cure for HIV infection, but a treatment called highly active antiretroviral therapy (HAART) is often able to stop HIV replication to such an extent that the viral load becomes undetectable.

HAART uses a combination of drugs that interfere with the life cycle of HIV. Entry inhibitors stop HIV from entering a cell by, for example, preventing the virus from binding to CD4. Reverse transcriptase inhibitors, such as AZT, interfere with the conversion of viral RNA into DNA. Integrase inhibitors prevent HIV from inserting its DNA into that of the host cells. Protease inhibitors prevent a viral enzyme from processing newly created polypeptides. By giving patients a combination of these drugs, the virus is less likely to become resistant to drug therapy, because several mutations would need to occur at once.

Throughout the world, more than 5 million people are now receiving HIV treatment. In 2009 alone, 1.2 million people received HIV antiretroviral therapy for the first time—a 30% increase in the number of people receiving treatment in a single year. This is largely due to the President's Emergency Plan for AIDS Relief, or PEPFAR program, which began as a commitment from the United States to provide $15 billion in antiretroviral drugs to resource-poor countries from 2003 to 2008. Since then, the program has been continued and expanded. A study published in 2009 showed the PEPFAR program had cut AIDS death rates by more than 10% in targeted countries in Africa, though it had no appreciable effect on prevalence of the disease in those nations.

An HIV-positive pregnant woman who takes reverse transcriptase inhibitors during her pregnancy reduces the chances of HIV transmission to her newborn. If possible, drug therapy should be delayed until the tenth to twelfth week of pregnancy to minimize any adverse effects of AZT on fetal development. If treatment begins at this time and delivery is by cesarean section, the chance of transmission from mother to infant is very slim (about 1%).

The general consensus is that control of the AIDS epidemic will not occur until a vaccine is available. Many different approaches have thus far been tried with limited success. The most encouraging news has been a large study in Thailand that concluded in 2009, showing that a combination of two different types of HIV vaccines reduced the risk of con- tracting the virus by nearly a third (31.2%).

Animation
Treatment of HIV Infection

Genital Warts

An estimated 20 million Americans are currently infected with the human papillomavirus (HPV) (Fig. 41.16). There are over

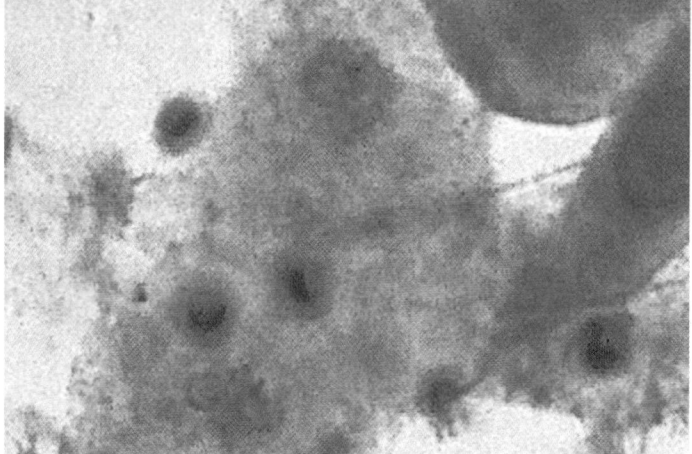

Figure 41.17 Genital herpes. A photomicrograph of cells infected with the herpes simplex virus.

100 types of HPV. Most cause warts, and some specifically cause genital warts, which are an STD. Genital warts may appear as flat or raised lesions on the penis and foreskin of males, and on the vuvla, vagina, and cervix in females. Note that if warts are present only on the cervix, there may be no outward signs of the disease. Newborns can also be infected with HPV during passage through the birth canal.

Currently, there is no cure for an HPV infection, but the warts can be treated effectively by surgery, freezing, application of an acid, or laser burning. Even after treatment, however, the virus can sometimes be transmitted. Therefore, once someone has been diagnosed with genital warts, abstinence or the use of a condom is recommended.

About ten types of HPV can cause cancer of the cervix, as well as tumors of the vulva, vagina, anus, and penis. Cervical cancer is the second leading cause of cancer death in women in the United States (approximately 500,000 per year). Early detection of cervical cancer is possible by means of a *Pap smear*, in which a few cells are removed from the cervix for microscopic examination. If the cells are cancerous, a hysterectomy (removal of the uterus) may be recommended. According to a 2008 study, HPV now causes as many cancers of the mouth and throat in U.S. males as does tobacco, perhaps due to an increase in oral sex in combination with a decline in tobacco use.

A vaccine called Gardasil is available to prevent infection with the four most common types of HPV found in the U.S., including two types that cause the majority of cervical cancers. The vaccine was initially approved only for use in girls, and it is recommended that girls be vaccinated before they become sexually active. In late 2009, the Food and Drug Administration also approved Gardasil for use in boys ages 9 through 26.

Genital Herpes

Genital herpes is caused by the herpes simplex virus (Fig. 41.17), of which there are two types. Type 1 usually causes cold sores and fever blisters, while type 2 more often causes genital herpes. In the United States, it is estmated that approximately one in four women, and one in five men, have had a genital herpes infection, but about 20% of these never have symptoms. Even without symptoms, however, the virus can be transmitted.

After becoming infected, most people experience a tingling or itching sensation before blisters appear on the genitals within 2–20 days. Once the blisters rupture, they leave painful ulcers that take from five days to three weeks to heal. These symptoms may be accompanied by fever, pain on urination, swollen

Biological Systems

Preventing Transmission of STDs

Being aware of how STDs are spread and then observing the following guidelines will greatly help prevent the transmission of STDs.

Sexual Activities Can Transmit STDs

Abstain from sexual intercourse or develop a long-term monogamous relationship (i.e., having only one sexual partner) with someone who is free of STDs (Fig. 41B).

Refrain from multiple sex partners or having relations with someone who has multiple sex partners. If you have sex with two other people and each of these has sex with two people and so forth, the number of people who can potentially be exposed to disease is quite large.

Remember that anyone can be at risk for HIV. The highest rate of increase in HIV infection is now occurring among heterosexuals.

Be aware that having sex with an intravenous drug user is risky because needle-sharing can lead to infection with hepatitis and HIV. Also, anyone who already has another STD is more susceptible to becoming infected with HIV.

Uncircumcised males are more likely to become infected with an STD than circumcised males, at least partly because viruses and bacteria can remain under the foreskin for a long period of time.

Avoid anal-rectal intercourse (in which the penis is inserted into the rectum) because the lining of the rectum is thin and HIV can easily enter the body there.

Safer Sex Practices Can Help Prevent STD Transmission

Always use a latex condom during sexual intercourse if you do not know for certain that your partner is free of STDs. Be sure to follow the directions supplied by the manufacturer.

Avoid fellatio (kissing and insertion of the penis into a partner's mouth) *and cunnilingus* (kissing and insertion of the tongue into the vagina) because they may be a means of transmission. The mouth and gums often have cuts and sores that facilitate the entrance of infectious agents.

Be cautious about the use of alcohol or any drug that may prevent you from being able to control your behavior. Females have to be particularly aware of "date-rape" drugs like GHB (gamma-hydroxy-butyramine), that puts a person in an uninhibited state with no memory of what transpired. These drugs can easily be slipped into a beverage without the victim's knowledge.

Drug Use Transmits Hepatitis and HIV

Do not inject drugs. Be aware that hepatitis and HIV can be spread by blood-to-blood contact. If you are currently an IV drug user, stop immediately.

Always use a new, sterile needle for injection or one that has been cleaned in a bleach solution and then well rinsed if you are a drug user and cannot stop your behavior (Fig. 41C).

Questions to Consider

1. How can you be certain that a potential sexual partner doesn't have any STDs?
2. What does it mean if a person tells you he or she just had an HIV test, and it was negative?
3. Which methods of birth control are most effective at preventing transmission of STDs? Which are least effective or provide no protection?

Figure 41B Sexual activities can transmit STDs. People of all sexual orientations need to be aware of STD dangers and use precautions.

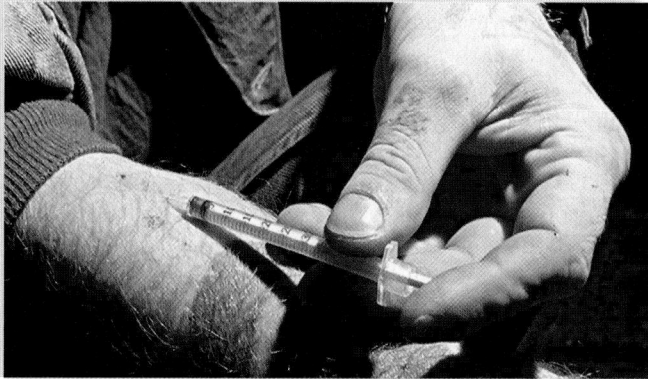

Figure 41C Sharing needles transmits STDs. Intravenous drug use with unclean, shared needles can transmit viral STDs.

lymph nodes in the groin, and in women, a watery vaginal discharge. At this time, the individual also has an increased risk of acquiring an HIV infection. Exposure to herpes during a vaginal delivery can cause an infection in the newborn, which can lead to neurological disorders and even death. Birth by cesarean section prevents this possibility.

After the blisters heal, the virus becomes latent (dormant). Blisters can recur repeatedly, at variable intervals and with milder symptoms. Sunlight, sexual intercourse, menstruation, and stress are associated with these recurrences. While the virus is latent, it resides in the ganglia of the sensory nerves associated with the infected skin. The ability of the virus to "hide" in this manner has made it difficult to develop an effective vaccine for herpes simplex viruses.

Viral Hepatitis

Several different viruses can cause *hepatitis*, or inflammation of the liver. Of these, hepatitis A and B viruses are the most important. Hepatitis A is usually acquired from sewage-contaminated drinking water, but this infection can also be sexually transmitted through oral-anal contact. Hepatitis B is spread through sexual contact and by blood-borne transmission (accidental needle-stick on the job, receiving a contaminated blood transfusion, sharing needles while injecting drugs, from mother to fetus, etc.). Simultaneous infection with hepatitis B virus and HIV is common, because both share the same routes of transmission. Fortunately, a combined vaccine is available for hepatitis A and B. It is recommended that all children receive the vaccine to prevent these infections.

Video Halting Hepatitis

Bacterial STDs

Chlamydia, gonorrhea, and syphilis are major bacterial STDs that are usually curable with antibiotic therapy, if diagnosed early enough. If properly used, latex condoms can help prevent the spread of most STDs.

Chlamydia

Infection with the tiny bacterium *Chlamydia trachomatis*, simply known as chlamydia, is the most common STD in the United States today, with the highest rates of infection occurring in 15- to 19-year-olds. The rate of new chlamydial infections has increased steadily since 1987 (Fig. 41.18).

Chlamydial infections of the lower reproductive tract are usually mild or asymptomatic, especially in women. About 8–21 days after infection, men may experience a mild burning sensation on urination and a mucoid discharge. Women may have a vaginal discharge along with symptoms resembling a urinary tract infection. Chlamydia also causes cervical ulcerations, which increase the risk of aquiring HIV. If the infection is misdiagnosed or if a woman does not seek medical help, the infection may spread to the oviducts so that pelvic inflammatory disease (PID) results.

PID is characterized by inflamed oviducts that become partially or completely blocked by scar tissue. As a result, the woman can become infertile or at increased risk of ectopic pregnancy, which can be a medical emergency. If a baby comes in contact with chlamydia during delivery, pneumonia or eye infections can result.

Gonorrhea

Gonorrhea is caused by the bacterium *Neisseria gonorrhoeae* (Fig. 41.19). About 75% of reported cases occur in people between 15 and 29 years of age, and rates among African Americans are about 20 times greater than among Caucasians. Although as many as 20% of infected males are asymptomatic, usually they experience pain upon urination and a thick, greenish yellow urethral discharge.

Untreated gonorrhea can cause infertility in both males and females, by causing inflammation and scarring in either the vas deferens or the oviducts, respectively. If a baby is exposed

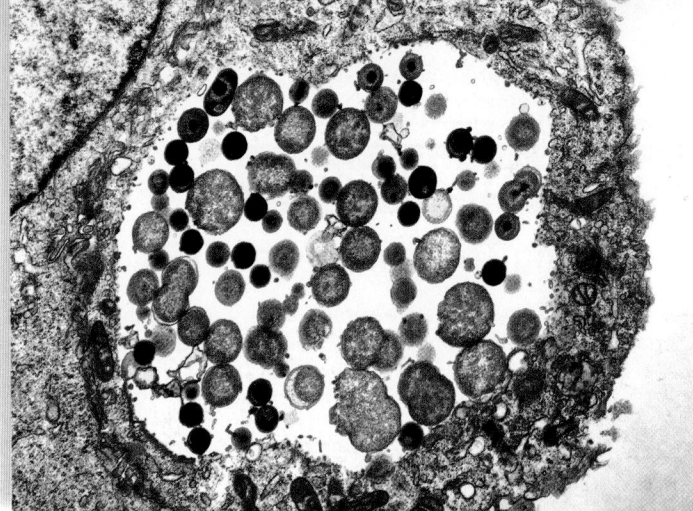

5000×

Figure 41.18 Chlamydial infection. Different stages of *Chlamydia trachomatis* are stained red, brown, and black, inside an infected cell.

during birth, an eye infection leading to blindness can result. All vaginally delivered newborns are treated with antibiotic eye-drops to prevent this possibility.

N. gonorrhoeae is also the most common cause of proctitis, an inflammation of the anus characterized by anal pain and blood or pus in the feces. Oral-genital contact can also cause infection of the mouth, throat, and tonsils. Gonorrhea can spread to internal parts of the body, causing heart damage or arthritis. If, by chance, the person touches infected genitals and then touches his or her eyes, a severe eye infection can result.

Gonorrhea is still considered to be curable by antibiotics. However, antibiotic-resistant strains of *N. gonorrhoeae* are becoming more common each year.

Syphilis

Syphilis is caused by the bacterium *Treponema pallidum* (Fig. 41.20). It is an STD of considerable historical importance that was first described hundreds of years ago. Untreated syphilis progresses through three stages, separated by latent periods during which the bacteria are not multiplying. In the primary stage, a hard chancre (ulcerated sore with hard edges) indicates the site of infection. During the secondary stage, the individual breaks out in a rash that does not itch, coinciding with the replication and spread of the bacteria all over the body.

Not all cases of syphilis progress to the tertiary stage, during which syphilis may affect the cardiovascular system, nervous system, or both. An infected person may become mentally disabled, become blind, walk with a shuffle, or show signs of insanity. Gummas, which are large, destructive ulcers, may develop on the skin or within the internal organs.

T. pallidum bacteria can cross the placenta, causing birth defects or a stillbirth. If diagnosed and treated in its early stages, syphilis is easy to cure with antibiotics.

Vaginal Infections

Vaginitis, meaning vaginal inflammation from any cause, is the most commonly diagnosed gynecologic condition. Bacterial vaginosis (BV), or overgrowth of certain bacteria inhabiting the vagina, accounts for 40–50% of vaginitis cases in American women. A common culprit is the bacterium *Gardnerella*

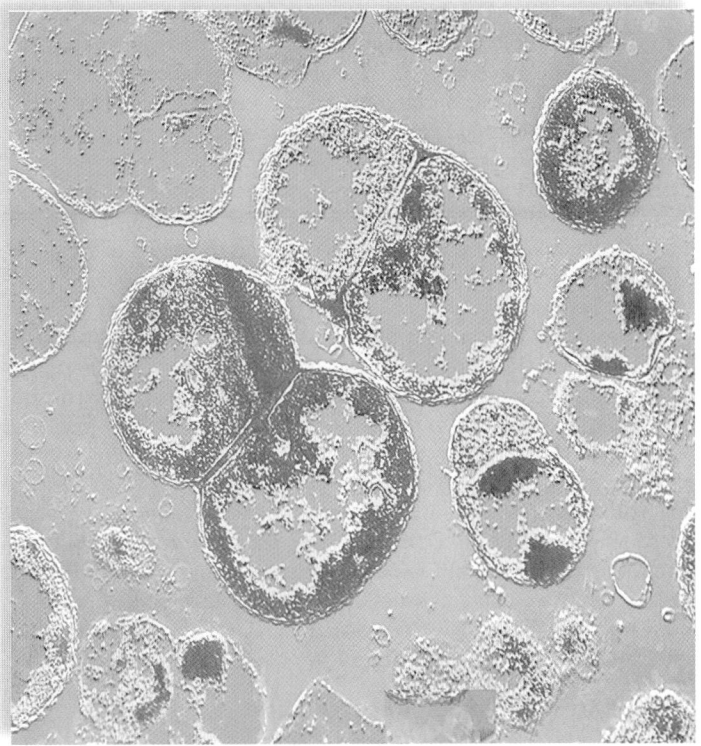

Figure 41.19 Gonorrhea.   A photomicrograph of a urethral discharge from an infected male. Gonorrheal bacteria (*Neisseria gonorrhoeae*) occur in pairs; for this reason, they are called diplococci.

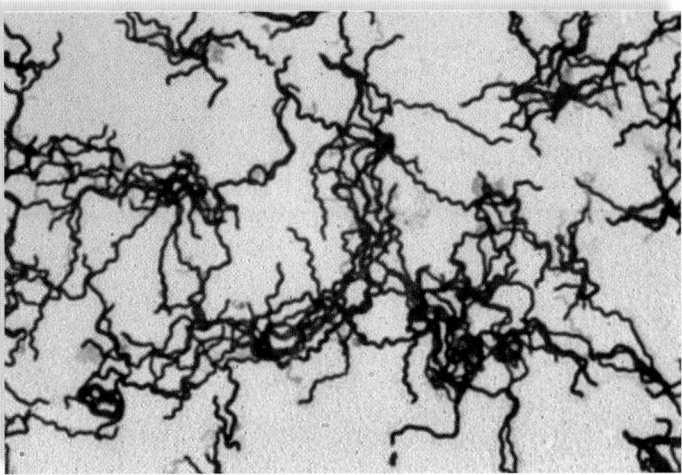

Figure 41.20 Syphilis.   The syphilis-causing bacterium (*Treponema pallidum*) is a spirochete.

vaginalis. How women acquire this infection is not well understood. According to the Centers for Disease Control, BV is not considered an STD, but it is more common in women who are sexually active.

In addition to BV, the yeast *Candida albicans* can also cause vaginitis. Yeast is normally found living in the vagina; under certain circumstances, its growth increases above normal, causing vaginitis. For exmple, women taking birth control pills or antibiotics may be prone to yeast infections. Both can alter the normal balance of vaginal organisms, causing a yeast infection. Typical signs of a yeast infection include a thickish, white vaginal discharge and itching of the vulva and/or vagina. Antifungal medications inserted into the vagina are used to treat yeast infections.

The protozoan *Trichomonas vaginalis* can also infect the urethra of males and the vagina of females. Infected males are usually asymptomatic, but can pass the parasite to their partner through sexual intercourse. Symptoms of trichomoniasis in females are a foul-smelling, yellow-green exudate and itching of the vulva/vagina. Effective antiprotozoal drugs are available by prescription, but if one partner remains infected, reinfection will occur. For this reason, both people should be treated simultaneously.

Check Your Progress   41.5

1. Describe four classes of antiretroviral drugs, and the specific stage of the HIV reproduction cycle targeted by each class.
2. Describe the medical complications in women that are associated with HPV infection (genital warts).
3. Identify a serious condition that can occur in women due to both chlamydia and gonorrhea.

CONNECTING *the* CONCEPTS *with the* BIG IDEAS

Energy and Homeostasis
- Feedback mechanisms of many types work in the reproductive systems of animals to achieve gamete production. (2C1)

Media Study Tools

www.glencoe.com/maderAP11

Enhance your study of this chapter with study tools and practice tests. Also ask your instructor about the resources available through ConnectPlus, including the media-rich eBook, interactive learning tools, and animations.

Summarize

41.1 How Animals Reproduce

Ordinarily, asexual reproduction may quickly produce a large number of offspring genetically identical to the parent. Sexual reproduction involves gametes and produces offspring that are genetically slightly different from the parents. The gonads are the primary sex organs. The accessory sex organs consist of storage areas for sperm and ducts that conduct the gametes. In males, they also contribute to formation of the semen.

The egg of oviparous animals contains yolk, and in terrestrial animals a shelled egg prevents drying out. The amount of yolk is dependent on whether there is a larval stage.

Eggs of reptiles have extraembryonic membranes that allow them to develop on land; these same membranes are modified for internal development in mammals. Ovoviviparous animals retain their eggs until the offspring have hatched, and viviparous animals retain the embryo and give birth. Placental mammals exemplify viviparous animals.

41.2 Human Male Reproductive System

In human males, sperm are produced in the testes, mature in the epididymides, and may be stored in the vasa deferentia before entering the urethra, along with seminal fluid (produced by seminal vesicles, the prostate gland, and bulbourethral glands). Sperm are ejaculated during male orgasm, when the penis is erect.

Spermatogenesis occurs in the seminiferous tubules of the testes, which also produce testosterone in interstitial cells. Testosterone brings about the maturation of the primary sex organs during puberty and promotes the secondary sex characteristics of males, such as low voice, facial hair, and increased muscle strength.

Follicle-stimulating hormone (FSH) from the anterior pituitary stimulates spermatogenesis, and luteinizing hormone (LH) stimulates testosterone production. A hypothalamic-releasing hormone, gonadotropin-releasing hormone (GnRH), controls anterior pituitary production and FSH and LH release. The level of testosterone in the blood controls the secretion of GnRH and the anterior pituitary hormones by a negative feedback system.

41.3 Human Female Reproductive System

In females, an oocyte produced by an ovary enters an oviduct, which leads to the uterus. The uterus opens into the vagina. The external genital area of women includes the vaginal opening, the clitoris, the labia minora, and the labia majora.

In either ovary, one follicle a month matures, produces a secondary oocyte, and becomes a corpus luteum. This process is called the ovarian cycle. The follicle and the corpus luteum produce estrogens, collectively called estrogen, and progesterone, the female sex hormones.

The uterine cycle occurs concurrently with the ovarian cycle. In the first half of these cycles (days 1–13, before ovulation), the anterior pituitary produces FSH and the follicle produces estrogen. Estrogen causes the endometrium to increase in thickness. In the second half of these cycles (days 15–28, after ovulation), the anterior pituitary produces LH and the follicle produces progesterone. Progesterone causes the endometrium to become secretory. Feedback control of the hypothalamus and anterior pituitary causes the levels of estrogen and progesterone to fluctuate. When they are at a low level, menstruation begins.

If fertilization occurs, a zygote is formed, and development begins. The resulting embryo travels down the oviduct and implants itself in the endometrium. A placenta, which is the region of exchange between the fetal blood and the mother's blood, forms. At first, the placenta produces HCG, which maintains the corpus luteum; later, it produces progesterone and estrogen.

The female sex hormones, estrogen and progesterone, also affect other traits of the body. Primarily, estrogen brings about the maturation of the primary sex organs during puberty and promotes the secondary sex characteristics of females, including body hair (usually less than in males), a wider pelvic girdle, a more rounded appearance, and development of breasts.

41.4 Control of Human Reproduction

Numerous birth control methods and devices, such as the birth control pill, diaphragm, and condom, are available for those who wish to prevent pregnancy. A "morning-after pill," RU-486, is now available.

Some couples are infertile, and if so, they may use assisted reproductive technologies to have a child. Artificial insemination and in vitro fertilization have been followed by more sophisticated techniques, such as intracytoplasmic sperm injection.

41.5 Sexually Transmitted Diseases

Important sexually transmitted diseases caused by viruses include AIDS, genital warts and cervical cancer, genital herpes, and hepatitis A and B. Major bacterial STDs include chlamydia and gonorrhea, which may cause pelvic inflammatory disease (PID), and syphilis, which has cardiovascular and neurological complications if untreated. Vaginal infections may have many causes; the most common are bacterial vaginosis (BV), yeast infection, and *Trichomonas* infection.

Key Terms

acquired immunodeficiency syndrome (AIDS) 787	human chorionic gonadotropin (HCG) 782
bulbourethral glands 777	human immunodeficiency virus (HIV) 787
cervix 779	
contraceptive vaccine 784	infertility 784
copulation 774	lactation 782
corpus luteum 780	luteal phase 781
egg 779	menopause 782
endometrium 779	menstruation 781
epididymis 775	oocyte 779
estrogen 779	ovarian cycle 780
extraembryonic membrane 774	ovary 773, 779
follicle 780	oviparous 774
follicular phase 780	ovoviviparous 774
gamete 785	ovulation 780
germ cell 773	parthenogenesis 773
gonad 773	penis 777
hermaphroditic 773	placenta 774
	progesterone 779

▨ Assess

Reviewing This Chapter

1. Contrast asexual reproduction with sexual reproduction, reproduction in water with reproduction on land, and the life history of an insect with that of a bird. 773–74
2. Discuss the anatomy and physiology of the testes. Describe the structure of sperm. 775–77
3. Trace the path of sperm in a human male. What glands contribute fluids to semen? 775–78
4. Name the endocrine glands involved in maintaining the sex characteristics of males and the hormones produced by each. 778
5. Trace the path of an oocyte in a human female. Where do fertilization and implantation occur? When does the oocyte become an egg? 779
6. Describe the external genital organs in females. Name two functions of the vagina. 779
7. Discuss the anatomy and physiology of the ovaries. Describe the ovarian cycle and ovulation. 780–81
8. Describe the uterine cycle, and relate it to the ovarian cycle. In what way is menstruation prevented if pregnancy occurs? 781–82
9. What events occur at fertilization? Name three functions of the female sex hormones, aside from their involvement in the uterine cycle. 782
10. Describe the anatomy and physiology of the breast. 782–83
11. Which means of birth control require surgery, use hormones, use barrier methods, or are dependent on none of these? 783–84
12. If couples are infertile, what assisted reproductive technologies are available? 784–85
13. List the cause, symptoms, and treatment for the most common types of sexually transmitted diseases. 787–91

Testing Yourself

Choose the best answer for each question.

1. Which of these is a requirement for sexual reproduction?
   a. male and female parents
   b. production of gametes
   c. optimal environmental conditions
   d. aquatic habitat
   e. All of these are correct.

2. Internal fertilization
   a. can prevent the drying out of gametes and zygotes.
   b. must take place on land.
   c. is practiced by humans.
   d. requires that males have a penis.
   e. Both a and c are correct.

3. Which of these pairs is mismatched?
   a. interstitial cells—testosterone
   b. seminiferous tubules—sperm production
   c. vasa deferentia—seminal fluid production
   d. urethra—conducts sperm
   e. Both c and d are mismatched.

4. Follicle-stimulating hormone (FSH)
   a. is secreted by females but not males.
   b. stimulates the seminiferous tubules to produce sperm.
   c. secretion is controlled by gonadotropin-releasing hormone (GnRH).
   d. is the same as luteinizing hormone.
   e. Both b and c are correct.

5. Which of these combinations is most likely to be present before ovulation occurs?
   a. FSH, corpus luteum, estrogen, secretory uterine lining
   b. luteinizing hormone (LH), follicle, progesterone, thick endometrium
   c. FSH, follicle, estrogen, endometrium becoming thick
   d. LH, corpus luteum, progesterone, secretory endometrium
   e. Both c and d are correct.

6. In tracing the path of sperm, you would mention vasa deferentia before
   a. testes.
   b. epididymides.
   c. urethra.
   d. seminiferous tubules.
   e. All of these are correct.

7. An oocyte is fertilized in
   a. the vagina.
   b. the uterus.
   c. the oviduct.
   d. the ovary.
   e. All of these are correct.

8. During pregnancy,
   a. the ovarian and uterine cycles occur more quickly than before.
   b. GnRH is produced at a higher level than before.
   c. the ovarian and uterine cycles do not occur.
   d. the female secondary sex characteristics are not maintained.
   e. Both b and c are correct.

9. Which means of birth control is most effective in preventing sexually transmitted diseases?
   a. condom
   b. pill
   c. diaphragm
   d. spermicidal jelly
   e. vasectomy

10. Which of these is a sexually transmitted disease caused by a bacterium?
    a. gonorrhea
    b. hepatitis B
    c. genital warts
    d. genital herpes
    e. HIV

11. The HIV virus has a preference for binding to
    a. B lymphocytes.
    b. cytotoxic T lymphocytes.
    c. helper T lymphocytes.
    d. All of these are correct.

For questions 12–14, match the descriptions with the sexually transmitted diseases in the key.

KEY:
   a. AIDS
   b. hepatitis B
   c. genital herpes
   d. genital warts
   e. gonorrhea
   f. chlamydia
   g. syphilis

12. blisters, ulcers, pain on urination, swollen lymph nodes

13. flulike symptoms, jaundice; eventual liver failure possible

14. males have a thick, greenish-yellow discharge; no symptoms in female; can lead to PID

15. Which of these is the primary sex organ of the male?
   a. penis
   b. scrotum
   c. testis
   d. prostate
   e. vasectomy

16. The luteal phase of the uterine cycle is characterized by
   a. high levels of LH and progesterone.
   b. low levels of estrogen and progesterone.
   c. increasing estrogen and little or no progesterone.
   d. high levels of LH only.

17. The secretory phase of the uterine cycle occurs during which ovarian phase?
   a. follicular phase
   b. ovulation
   c. luteal phase
   d. menstrual phase

18. Which of these is the correct path of an oocyte?
   a. oviduct, fimbriae, uterus, vagina
   b. ovary, fimbriae, oviduct, uterine cavity
   c. oviduct, fimbriae, abdominal cavity
   d. fimbriae, uterine tube, uterine cavity, ovary

19. Following ovulation, the corpus luteum develops under the influence of
   a. progesterone.
   b. FSH.
   c. LH.
   d. estradiol.

Engage
Virtual Lab
Virtual Frog Dissection

The Virtual Lab "Virtual Frog Dissection" provides an examination of the reproductive system of a vertebrate.

Thinking Scientifically

1. Female athletes who train intensively often stop menstruating. The important factor appears to be the reduction of body fat below a certain level. Give a possible evolutionary explanation for a relationship between body fat in females and reproductive cycles.

2. The average sperm count in males is now lower than it was several decades ago. The reasons for the lower sperm count usually seen today are not known. What data might be helpful in order to formulate a testable hypothesis?

Bioethical Issue
Fertility Drugs

Higher-order multiple births (triplets or more) in the United States increased 19% between 1980 and 1994. During these years, it became customary to use fertility drugs (gonadotropic hormones) to stimulate the ovaries. Although a variety of assisted reproductive technologies have become commonplace in recent years, fertility drugs are still routinely prescribed for women who are infertile due to ovulation disorders. However, fertility drugs frequently result in the release and subsequent fertilization of multiple oocytes.

The risks for premature delivery, low birth weight, and developmental abnormalities rise sharply for higher-order multiple births. In addition, the physical and emotional burden placed on the parents is extraordinary. They face endless everyday chores and find it difficult to maintain normal social relationships, if only because they get insufficient sleep. Finances are strained to provide for the children's needs, including housing and child-care assistance. About one-third report that they received no help from relatives, friends, or neighbors in the first year after the birth. Trips to the hospital for accidental injury are more frequent because parents often have difficulties keeping so many children safe at one time.

A higher-order multiple pregnancy can be terminated, or selective reduction can be done. During selective reduction, one or more of the fetuses is killed by an injection of potassium chloride. Selective reduction, however, could very well result in psychological and social complications for the mother and surviving children.

Many clinicians are now urging that all possible steps be taken to ensure that the chance of higher-order multiple births be reduced. For example, pre-ovulation oocytes may be obtained and subjected to in vitro fertilization (in which the eggs are fertilized in the lab), as shown in Figure 41A on page 786. Even this option poses an ethical dilemma: In vitro fertilization may be carried out with the intent that only one or two zygotes will be placed in the woman's womb. But then any "leftover" zygotes may never have an opportunity to continue development.

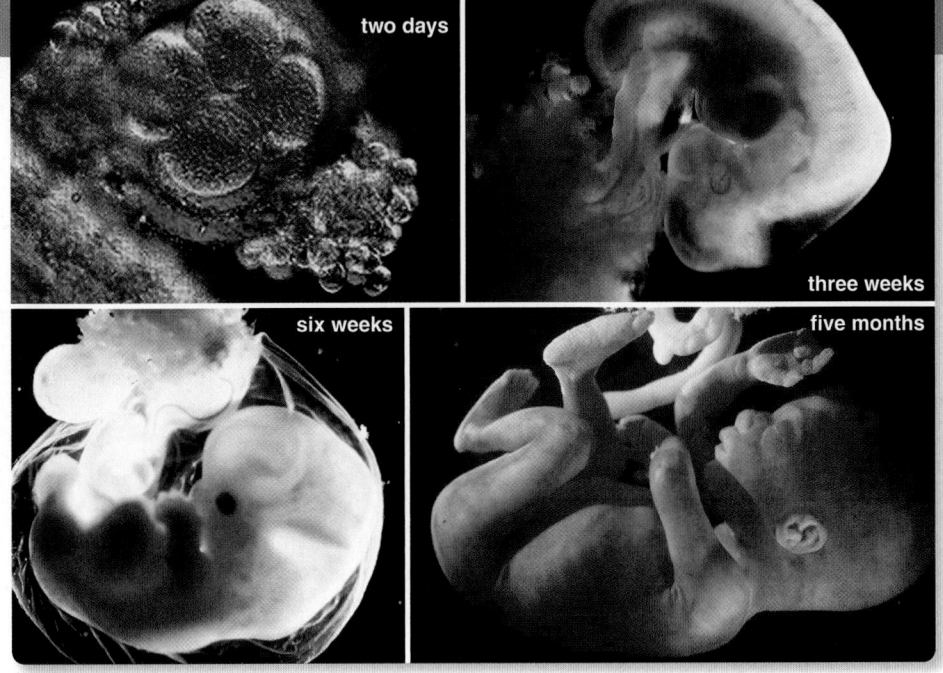

two days

three weeks

six weeks

five months

As development occurs, complexity increases.

42

Animal Development and Aging

BEFORE YOU BEGIN

Before beginning this chapter, take a few moments to review the following discussions.

Figure 13.7 How do transcription factors control the expression of eukaryotic genes?

Figure 41.8 In the human female reproductive system, where are the oviducts located relative to the ovaries? To the uterus?

Section 41.3 What are the roles of estrogen and progesterone in the ovarian and uterine cycles?

M any of the fundamental processes that guide development, from fertilization onward, are shared throughout most of the animal kingdom. Whether the embryo will develop into a worm, insect, amphibian, bird, or mammal, it must first establish an anterior and a posterior end. As the embryo continues to grow, the body segments form and give rise to structures such as appendages. The coordination of these developmental events is quite intricate and is easily disrupted by harmful environmental factors such as chemicals and radiation. Interestingly, some of these same environmental factors are thought to cause cellular damage that, over time, contributes to the aging process.

Although the main focus of this chapter is on events that occur either at the beginning, or toward the end, of an animal's life span, we should note that once an animal has reached maturity, development does not cease. If an animal's body is injured, the wound must be repaired. Not surprisingly, many of the same genes involved in wound repair are those which were active during earlier stages in development. In this chapter, we examine the basic processes of early animal development, as well as human fetal development, and the consequences of aging.

As you read through the chapter, think about the following questions:

1. Why are so many similarities present in the early embryonic structures of animals as diverse as the lancelet, the frog, the chicken, and the human?

2. Why is the developing embryo of any species more susceptible to harmful environmental factors, compared to the mature animal?

3. What are some of the fundamental mechanisms by which cells can become specialized to form organs and tissues?

FOLLOWING *the* BIG IDEAS

CHAPTER 42   ANIMAL DEVELOPMENT AND AGING

Energy and Homeostasis	Control and timing of developmental events are achieved by differential gene expression.
Information and Signaling	Transmission of various signals control gene expression and cell differentiation.

42.1 Early Developmental Stages

Learning Outcomes

Upon completion of this section, you should be able to

1. Identify the structures of an egg and sperm that are directly involved in fertilization.
2. Compare and contrast the cellular, tissue, and organ stages of embryonic development.
3. Describe the major steps in the development of the central nervous system of vertebrates.

As noted in the preceding chapter, animals that reproduce sexually employ a wide variety of strategies to achieve **fertilization,** the union of a sperm and an egg to form a zygote. Figure 42.1 illustrates how fertilization occurs in mammals, including humans.

Details of Fertilization in Humans

Human sperm have three distinct parts: a tail, a middle piece, and a head. The head contains the sperm nucleus and is capped by a membrane-bounded *acrosome*. The egg of a mammal (actually the secondary oocyte) is surrounded by a few layers of adhering follicular cells, collectively called the *corona radiata*. These cells nourished the oocyte when it was in a follicle of the ovary. The oocyte also has an extracellular matrix termed the *zona pellucida* just outside the plasma membrane, but beneath the corona radiata.

Video Human Sperm

Fertilization requires a series of events that result in the diploid zygote. Although only one sperm actually fertilizes the oocyte, millions of sperm begin this journey, and perhaps a few hundred succeed in reaching the oocyte. The sperm cover the surface of the oocyte and secrete enzymes that help weaken the corona radiata. They squeeze through the corona radiata and bind to the zona pellucida. After a sperm head binds tightly to

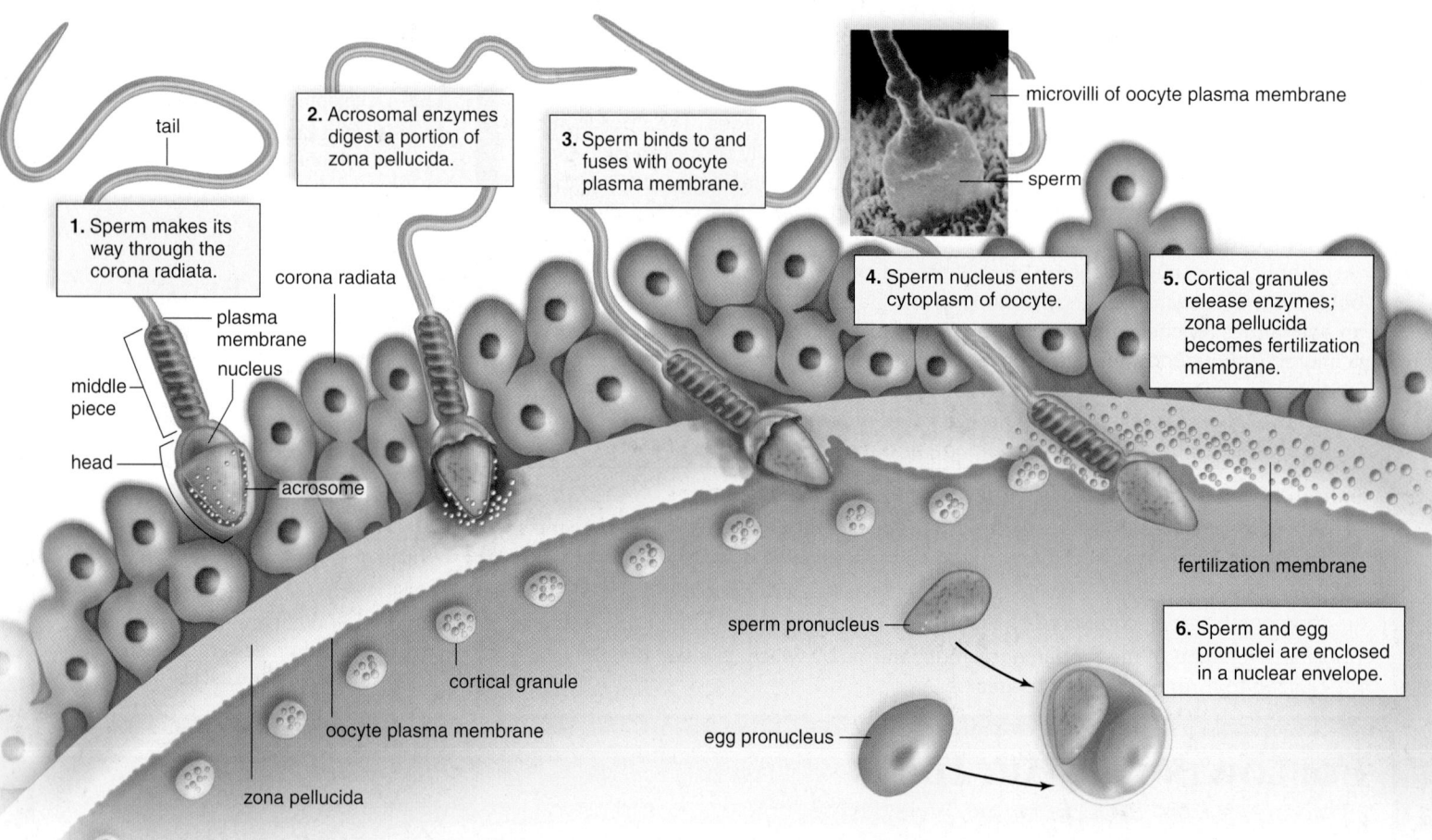

Figure 42.1 Fertilization.   During fertilization, a single sperm is drawn into the oocyte by microvilli of its plasma membrane (micrograph). The head of a sperm has a membrane-bounded acrosome filled with enzymes. When released, these enzymes digest a pathway for the sperm through the zona pellucida. After a sperm binds to the plasma membrane of the oocyte, changes occur that prevent other sperm from entering the oocyte. The oocyte finishes the second meiotic division and is now an egg. Fertilization is complete when the sperm pronucleus and the egg pronucleus contribute chromosomes to the zygote.

the zona pellucida, the acrosome releases digestive enzymes that forge a pathway for the sperm through the zona pellucida to the oocyte plasma membrane.

When the first sperm binds to the oocyte plasma membrane, the next few events prevent *polyspermy* (entrance of more than one sperm). As soon as a sperm touches the plasma membrane of an oocyte, the oocyte's plasma membrane depolarizes, and this change in charge, known as the "fast block," serves to repel sperm only for a few seconds. Then vesicles in the oocyte called *cortical granules* secrete enzymes that turn the zona pellucida into an impenetrable *fertilization membrane.* The longer-lasting cortical reaction is known as the "slow block."

The last of the events includes formation of the diploid zygote. Microvilli extending from the plasma membrane of the oocyte (Fig. 42.1) bring the entire sperm into the oocyte. The sperm nucleus releases its chromatin, which re-forms into chromosomes enclosed within the sperm *pronucleus.* In the meantime, the secondary oocyte completes meiosis, becoming an egg whose chromosomes are also enclosed in a pronucleus. A single nuclear envelope soon surrounds both sperm and egg pronuclei. Cell division is imminent, and the centrosomes that give rise to a spindle apparatus are derived from the basal body of the sperm's flagellum. The two haploid sets of chromosomes share the first spindle apparatus of the newly formed zygote.

Embryonic Development

Development includes all the changes that occur during the life cycle of an organism. You first encountered the basics of animal development in Chapter 28. Although development is a continuous, complex process, it can be divided into three major stages: (1) cellular, (2) tissue, and (3) organ.

Cellular Stages of Development

During the early stages of development, an organism is called an **embryo.** The cellular stages of development are

1. cleavage, resulting in a multicellular embryo, and
2. formation of the blastula.

Cleavage is cell division without growth; DNA replication and mitotic cell division occur repeatedly, and the cells get smaller with each division.

As shown in Figure 42.2, cleavage in a primitive animal called a lancelet results in uniform cells that form a **morula** [L. *morula,* little mulberry], which is a solid ball of cells. The morula continues to divide, forming a **blastula** [Gk. *blastos,* bud; L. *-ula,* small], a hollow ball of cells having a fluid-filled cavity called a **blastocoel.** The blastocoel forms when the cells of the morula extrude Na^+ into extracellular spaces, and water follows by osmosis, collecting in the center.

All vertebrates have a blastula stage, but the appearance of the blastula can be different from that of a lancelet. Chickens lay a hard-shelled egg containing plentiful **yolk,** a dense nutrient material. Because yolk-filled cells do not participate in cleavage,

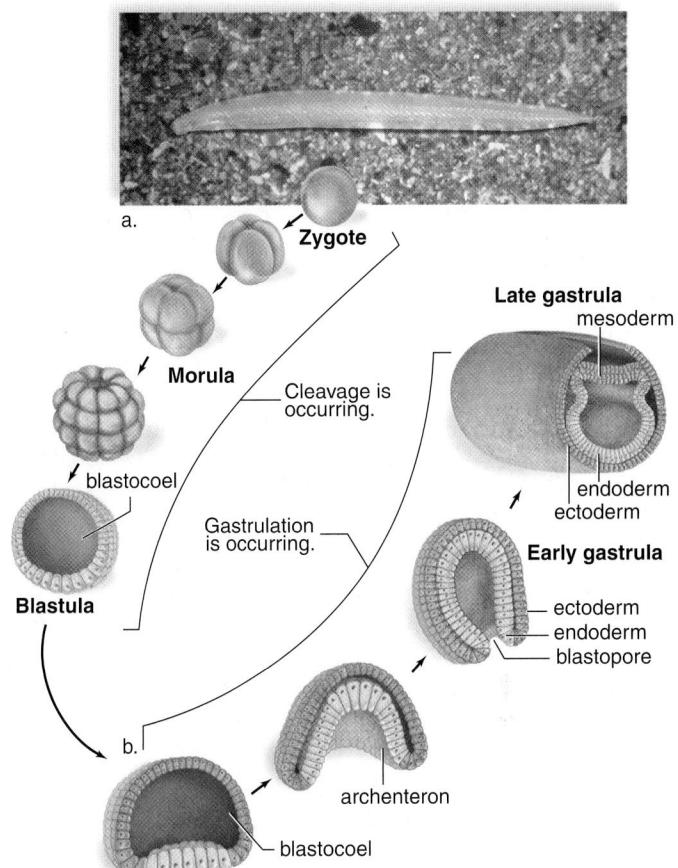

Figure 42.2 Lancelet early development. a. A lancelet. **b.** The early stages of development are exemplified in the lancelet. Cleavage produces a number of cells that form a cavity, the blastocoel. Invagination during gastrulation produces the germ layers ectoderm and endoderm. Then the mesoderm arises.

the blastula is a layer of cells that spreads out over the yolk. The blastocoel then forms as a space that separates these cells from the yolk:

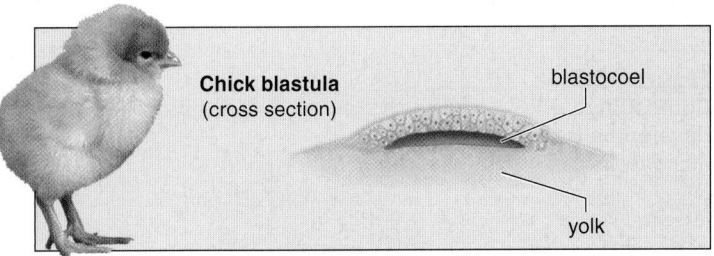

The zygotes of vertebrates, such as frogs, chickens, and humans, also undergo cleavage and form a morula. In frogs, cleavage is not equal because of the presence of the yolk. When yolk is present, the zygote and embryo exhibit polarity, and the embryo has an *animal pole* and a *vegetal pole.* The animal pole contains faster-growing, smaller cells that will eventually develop into the ectoderm and endoderm layers; the vegetal pole contains slower-growing, larger cells that develop into endoderm (see Fig. 42.3).

Tissue Stages of Development

The tissue stages of development involve **gastrulation,** the formation of a **gastrula** [Gk. *gaster*, belly; L. *-ula*, small] from the blastula. The two tissue stages are

(1) the early gastrula, and
(2) the late gastrula.

The early gastrula stage begins when certain cells begin to push, or invaginate, into the blastocoel, creating a double layer of cells (see Fig. 42.2). Cells migrate during this and other stages of development, sometimes traveling quite a distance before reaching a destination, where they continue developing. As cells migrate, they "feel their way" by changing their pattern of adhering to extracellular proteins.

An early gastrula has two layers of cells. The outer layer of cells is called the **ectoderm,** and the inner layer is called the **endoderm.** The endoderm borders the gut, but at this point, it is termed either the archenteron or the primitive gut. The pore, or hole, created by invagination (inward folding) is the **blastopore,** and in a lancelet, the blastopore eventually becomes the anus.

Gastrulation is not complete until three layers of cells that will develop into adult organs are produced. In addition to ectoderm and endoderm, the late gastrula has a middle layer of cells called the **mesoderm.**

Figure 42.2 illustrates gastrulation in a lancelet, and Figure 42.3 compares the lancelet, frog, and chicken late gastrula stages. In the lancelet, mesoderm formation begins as outpocketings from the primitive gut (Fig. 42.3). These outpocketings grow in size until they meet and fuse, forming two layers of mesoderm. The space between them is the coelom (see Section 28.1). The coelom is a body cavity lined by mesoderm that contains internal organs. In humans, the coelom becomes the thoracic and abdominal cavities of the body.

In the frog, the cells containing yolk do not participate in gastrulation and, therefore, they do not invaginate. Instead, a

Table 42.1 Embryonic Germ Layers

Embryonic Germ Layer	Vertebrate Adult Structures
Ectoderm (outer layer)	Nervous system; epidermis of skin and derivatives of the epidermis (hair, nails, glands); tooth enamel, dentin, and pulp; epithelial lining of oral cavity and rectum
Mesoderm (middle layer)	Musculoskeletal system; dermis of skin; cardiovascular system; urinary system; lymphatic system; reproductive system—including most epithelial linings; outer layers of respiratory and digestive systems
Endoderm (inner layer)	Epithelial lining of digestive tract and respiratory tract, associated glands of these systems; epithelial lining of urinary bladder; thyroid and parathyroid glands

slitlike blastopore is formed when the animal pole cells begin to invaginate from above, forming endoderm. Animal pole cells also move down over the yolk, to invaginate from below. Some yolk cells, which remain temporarily in the region of the blastopore, are called the yolk plug. Mesoderm forms when cells migrate between the ectoderm and endoderm. Later, a splitting of the mesoderm creates the coelom.

A chicken egg contains so much yolk that endoderm formation does not occur by invagination. Instead, an upper layer of cells becomes ectoderm, and a lower layer becomes endoderm. Mesoderm arises by an invagination of cells along the edges of a longitudinal furrow in the midline of the embryo. Because of its appearance, this furrow is called the *primitive streak*. Later, the newly formed mesoderm splits to produce a coelomic cavity.

Ectoderm, mesoderm, and endoderm are called the embryonic **germ layers.** No matter how gastrulation takes place, the result is the same: three germ layers are formed. It is possible to relate the development of future organs to these germ layers (Table 42.1).

**Figure 42.3
Comparative development of mesoderm.
a.** In the lancelet, mesoderm forms by an outpocketing of the archenteron. **b.** In the frog, mesoderm forms by migration of cells between the ectoderm and endoderm. **c.** In the chick, mesoderm also forms by invagination of cells.

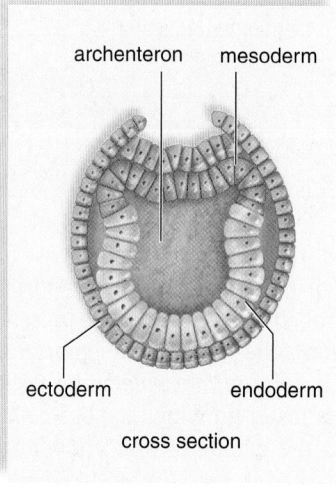

a. Lancelet late gastrula

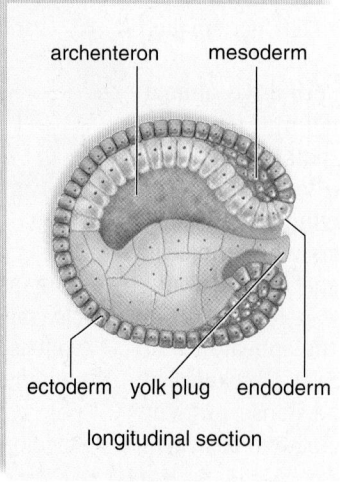

b. Frog late gastrula

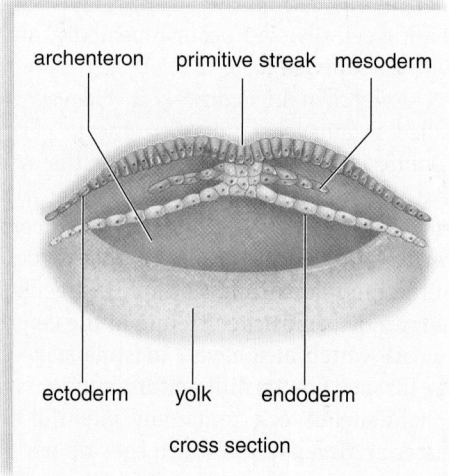

c. Chick late gastrula

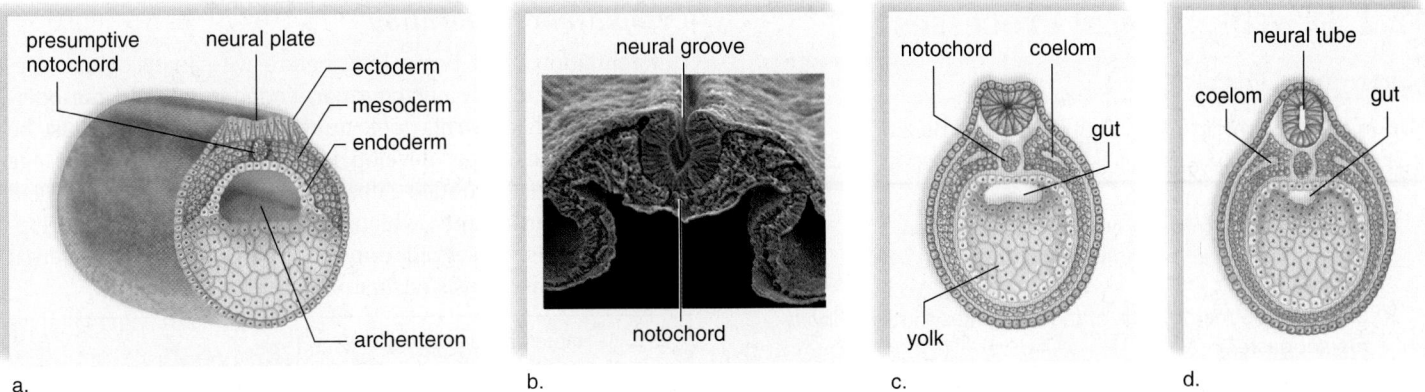

presumptive notochord
neural plate
ectoderm
mesoderm
endoderm
archenteron

a.

neural groove
notochord

b.

notochord coelom
gut
yolk

c.

neural tube
coelom gut

d.

Figure 42.4 Development of neural tube and coelom in a frog embryo. **a.** Ectodermal cells that lie above the future notochord (called presumptive notochord) thicken to form a neural plate. **b.** The neural groove and folds are noticeable as the neural tube begins to form. **c.** A splitting of the mesoderm produces a coelom, which is completely lined by mesoderm. **d.** A neural tube and a coelom have now developed.

Organ Stages of Development

The organs of an animal's body develop from the three embryonic germ layers. Much study has been devoted to how the nervous system develops in chordates, so we summarize that process here.

The newly formed mesoderm cells lie along the main longitudinal axis of the animal and coalesce to form a dorsal supporting rod called the **notochord.** The notochord persists in lancelets, but in frogs, chickens, and humans, it is replaced later in development by the vertebral column, giving these animals the name vertebrates.

The nervous system develops from midline ectoderm located just above the notochord. At first, a thickening of cells, called the **neural plate,** is seen along the dorsal surface of the embryo. Then, neural folds develop on either side of a neural groove, which becomes the **neural tube** when these folds fuse. Figure 42.4 shows cross sections of frog development to illustrate the formation of the neural tube. At this point, the embryo is called a *neurula.* Later, the anterior end of the neural tube develops into the brain, and the rest becomes the spinal cord.

In addition, the *neural crest* is a band of cells that develops where the neural tube pinches off from the ectoderm. Neural crest cells migrate to various locations, where they contribute to formation of skin and muscles, in addition to the adrenal medulla and the ganglia of the peripheral nervous system.

Midline mesoderm cells that did not contribute to the formation of the notochord now become two longitudinal masses of tissue. These two masses become blocked off into *somites,* which are serially arranged along both sides along the length of the notochord. Somites give rise to muscles associated with the axial skeleton and to the vertebrae. The serial origin of axial muscles and the vertebrae illustrates that vertebrates (including humans) are segmented animals. Lateral to the somites, the mesoderm splits, forming the mesodermal lining of the coelom.

In development of other organs, a primitive gut tube is formed by endoderm as the body itself folds into a tube. The heart, too, begins as a simple tubular pump. Organ formation continues until the germ layers have given rise to the specific organs listed in Table 42.1. Figure 42.5 helps you relate the formation of vertebrate structures and organs to the three embryonic layers of cells: the ectoderm, the mesoderm, and the endoderm.

Check Your Progress 42.1

1. Describe the fast and slow blocks to polyspermy.
2. Name the germ layer of the gastrula that gives rise to each of the following: notochord, thyroid and parathyroid glands, nervous system, epidermis, skeletal muscles, kidneys, bones, and pancreas.
3. Identify the stage(s) of embryonic development in which cross sections of all chordate embryos closely resemble one another.

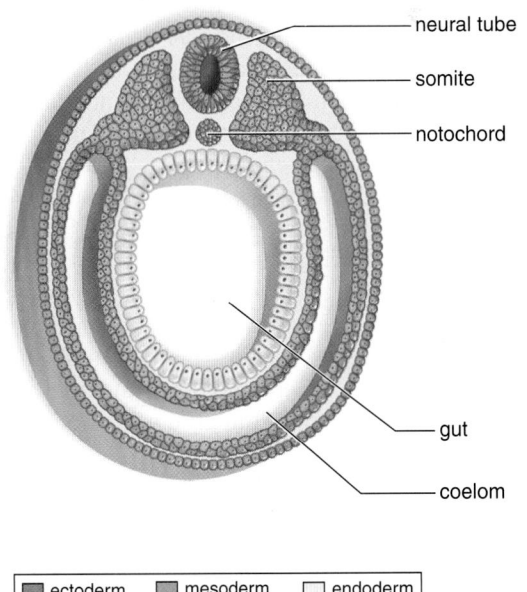

neural tube
somite
notochord
gut
coelom

ectoderm   mesoderm   endoderm

Figure 42.5 Vertebrate embryo, cross section. At the neurula stage, each of the germ layers, indicated by color (see key), can be associated with the later development of particular parts. The somites give rise to the muscles of each segment and to the vertebrae, which replace the notochord in vertebrates.

42.2 Developmental Processes

Learning Outcomes

Upon completion of this section, you should be able to

1. Distinguish between cellular differentiation and morphogenesis.
2. Define cytoplasmic segregation and induction, and explain how these phenomena contribute to cellular differentiation.
3. Describe the major steps of morphogenesis in *Drosophila melanogaster*.

Development requires three interconnected processes: growth, cellular differentiation, and morphogenesis. **Cellular differentiation** occurs when cells become specialized in structure and function; that is, a muscle cell looks different and acts differently than a nerve cell. **Morphogenesis** produces the shape and form of the body. One of the earliest indications of morphogenesis is cell movement. Later, morphogenesis includes **pattern formation,** which means how tissues and organs are arranged in the body. **Apoptosis,** or programmed cell death (see Fig. 9.2), is an important part of pattern formation.

Developmental genetics has benefited from research using the roundworm *Caenorhabditis elegans* and the fruit fly *Drosophila melanogaster.* These organisms are referred to as model organisms because the study of their development produced concepts that help us understand development in general.

Cellular Differentiation

At one time, investigators mistakenly believed that different cell types, e.g., human liver cells and brain cells, must inherit different DNA sequences from the original single-celled zygote. Perhaps, they speculated, the genes are parceled out as development occurs, and that is why cells of the body have a different structure and function. We now know that is not the case; rather, every cell in the body (except for the sperm and unfertilized egg) has a full complement of genes.

The zygote is **totipotent;** it has the ability to generate the entire organism and, therefore, must contain all the instructions needed by any other specialized cell in the body. For the first few days of cell division, all the embryonic cells are totipotent. When the embryonic cells begin to specialize and lose their totipotency, they do not lose genetic information. In fact, our ability today to clone mammals such as sheep, mice, and cats from specialized adult cells shows that every cell in an organism's body has the same collection of genes.

The answer to this puzzle becomes clear when we consider that only muscle cells produce the proteins myosin and actin; only red blood cells produce hemoglobin; and only skin cells produce keratin. In other words, we now know that specialization is not due to a parceling out of genes; rather, it is due to differential gene expression. Certain genes and not others are turned on (transcribed) in differentiated cells. In recent years, investigators have turned their attention to discovering the mechanisms that lead to differential gene expression. Two mechanisms—cytoplasmic segregation and induction—seem to be especially important.

Cytoplasmic Segregation

Differentiation must begin long before we can recognize specialized types of cells. Ectodermal, endodermal, and mesodermal cells in the gastrula look quite similar, but they must be different because they develop into different organs. The egg is now known to contain substances in the cytoplasm called **maternal determinants,** which influence the course of development. **Cytoplasmic segregation** is the parceling out of maternal determinants as mitosis occurs:

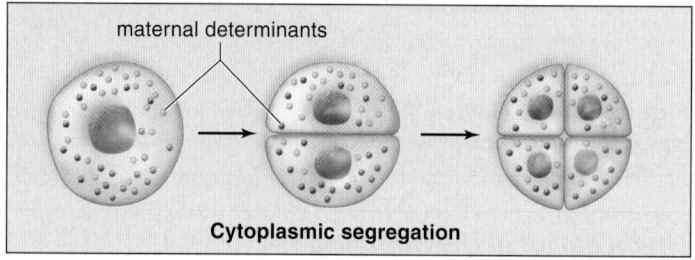

maternal determinants

Cytoplasmic segregation

An experiment conducted in 1935 showed that the cytoplasm of a frog's egg is not uniform. It is polar, having both an anterior/posterior axis and a dorsal/ventral axis, which can be correlated with the *gray crescent,* a gray area that appears after the sperm fertilizes the egg (Fig. 42.6*a*). If the gray crescent is divided equally by the first cleavage, each experimentally separated daughter cell develops into a complete embryo (Fig. 42.6*b*). However, if the zygote divides so that only one daughter cell receives the gray crescent, only that cell becomes a complete embryo (Fig. 42.6*c*). This experiment allowed scientists to speculate that the gray crescent must contain particular chemical signals that are needed for development to proceed normally.

Induction and Frog Experiments

As development proceeds, specialization of cells and formation of organs are influenced not only by maternal determinants but also by signals given off by neighboring cells. **Induction** [L. *in,* into, and *duco,* lead] is the ability of one embryonic tissue to influence the development of another tissue.

A frog embryo's gray crescent becomes the dorsal lip of the blastopore, where gastrulation begins. Because this region is necessary for complete development, the dorsal lip of the blastopore was termed the *primary organizer.* The cells closest to the primary organizer become endoderm, those farther away become mesoderm, and those farthest away become ectoderm. This suggests that a molecular concentration gradient may act as a chemical signal to induce germ layer differentiation.

The gray crescent in the zygote of a frog marks the dorsal side of the embryo where the mesoderm becomes notochord and ectoderm becomes nervous system. In a classic experiment researchers showed that presumptive (potential) notochord tissue induces the formation of the nervous system (Fig. 42.7). If presumptive nervous system tissue, located just above the presumptive notochord, is cut out and transplanted to the belly region of the embryo, it does not form a neural tube. But in contrast, if presumptive notochord tissue is cut out and transplanted beneath what would be belly ectoderm, this ectoderm does differentiate into neural tissue. Many other examples of induction are now known.

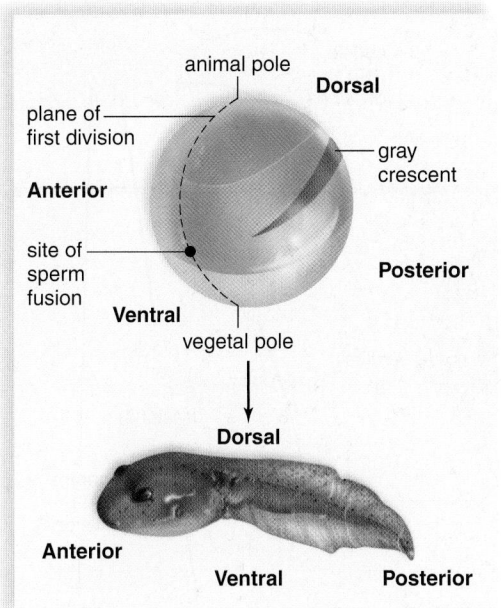

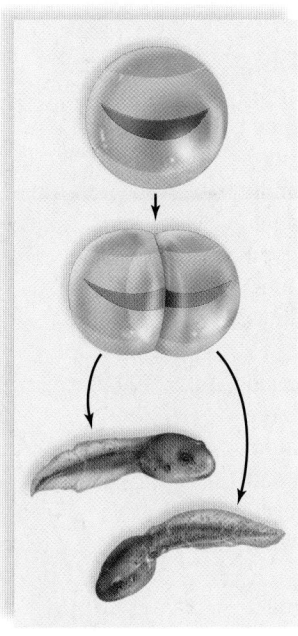

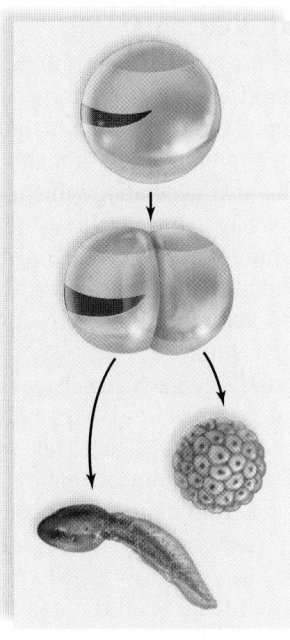

Figure 42.6
Cytoplasmic influence on development. **a.** The zygote of a frog has anterior/posterior and dorsal/ventral axes that correlate with the position of the gray crescent. **b.** The first cleavage normally divides the gray crescent in half, and each daughter cell is capable of developing into a complete tadpole. **c.** But if only one daughter cell receives the gray crescent, then only that cell can become a complete embryo. This shows that maternal determinants are present in the cytoplasm of a frog's egg.

a. Zygote of a frog is polar and has axes.

b. Each cell receives a part of the gray crescent.

c. Only the cell on the left receives the gray crescent.

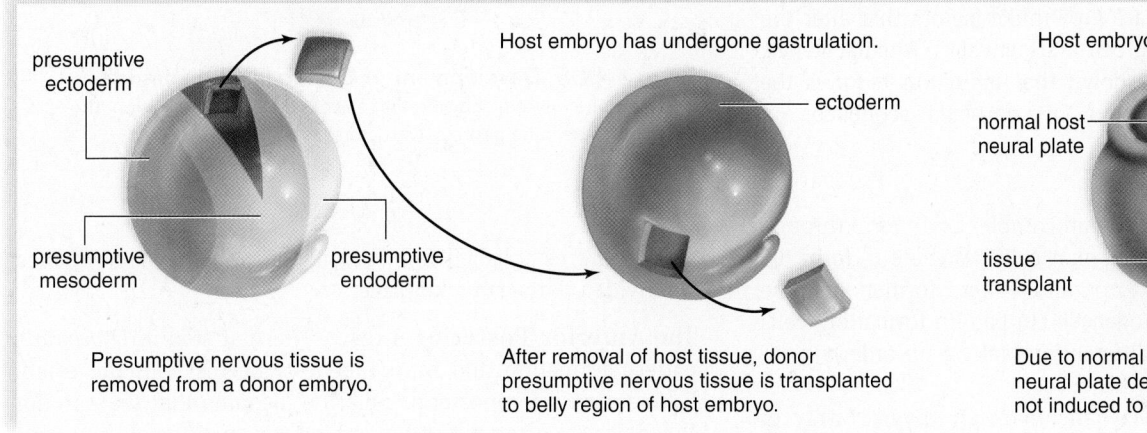

Host embryo has undergone gastrulation.

Host embryo undergoes neurulation.

Presumptive nervous tissue is removed from a donor embryo.

After removal of host tissue, donor presumptive nervous tissue is transplanted to belly region of host embryo.

Due to normal induction process, a host neural plate develops. But donated tissue is not induced to develop into a neural plate.

a.

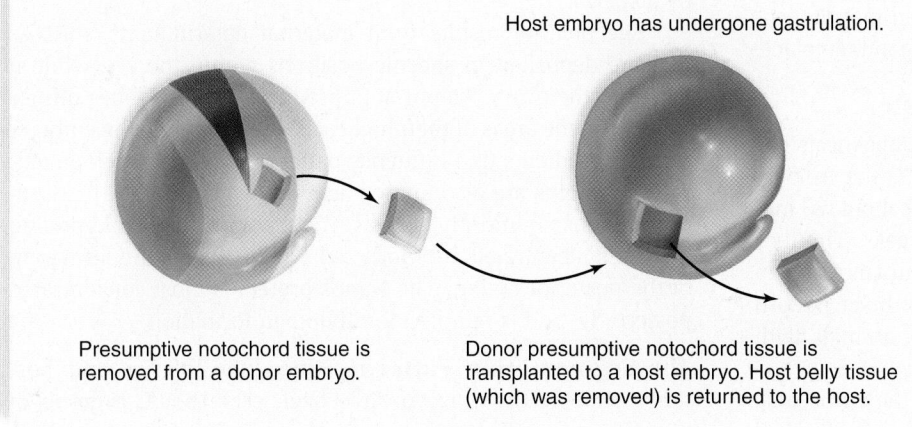

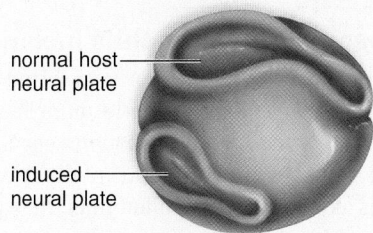

Host embryo has undergone gastrulation.

Host embryo undergoes neurulation.

Presumptive notochord tissue is removed from a donor embryo.

Donor presumptive notochord tissue is transplanted to a host embryo. Host belly tissue (which was removed) is returned to the host.

Host develops two neural plates—one induced by host notochord tissue, the second induced by transplanted notochord tissue.

b.

Figure 42.7 Control of nervous system development. **a.** In this experiment, the presumptive nervous system (*blue*) does not develop into the neural plate if moved from its normal location. **b.** In this experiment, the presumptive notochord (*pink*) can cause even belly ectoderm to develop into the neural plate (*blue*). This shows that the notochord induces ectoderm to become a neural plate, most likely by sending out chemical signals.

Induction in Caenorhabditis elegans

The tiny nematode *Caenorhabditis elegans* is only 1 mm long, and vast numbers can be raised in the laboratory either in petri dishes or a liquid medium. The worm is hermaphroditic, and self-fertilization is the rule. Therefore, even though induced mutations may be recessive, the next generation will yield individuals that are homozygous recessive and show the mutation. Many modern genetic studies have been performed on *C. elegans,* and the entire genome has been sequenced. Individual genes have been altered and cloned and their products injected into cells or extracellular fluid.

As the result of genetic studies, much has been learned about *C. elegans.* Development of *C. elegans* takes only three days, and the adult worm contains only 959 cells. Investigators have been able to watch the process from beginning to end because the worm is transparent. *Fate maps* have been developed that show the destiny of each cell as it arises following successive cell divisions (Fig. 42.8).

Some investigators have studied in detail the development of the worm's vulva, a pore through which eggs are laid. A cell called the anchor cell induces the vulva to form. The cell closest to the anchor cell receives the most inducer and becomes the inner vulva. This cell in turn produces another inducer, which acts on its two neighboring cells, and they become the outer vulva. The inducers are growthlike factors that alter the metabolism of the receiving cell and activate particular genes. Work with *C. elegans* has shown that induction requires the transcriptional regulation of genes in a particular sequence.

Morphogenesis

An animal achieves its ordered and complex body form through morphogenesis, which requires that cells associate to form tissues, and tissues give rise to organs. Pattern formation is the process that enables morphogenesis. In pattern formation, cells of the embryo divide and differentiate, taking up orderly positions in tissues and organs.

Although animals display an amazingly diverse array of morphologies, or body forms, most share common sets of genes that direct pattern formation. When pattern formation has ensured that key cells are properly arranged, then morphogenesis, the construction of the ultimate body form, can take place.

Morphogenesis in Drosophila melanogaster

Much of the study of pattern formation and morphogenesis has been done using relatively simple models such as the fruit fly *Drosophila melanogaster.* A *Drosophila* egg is only about 0.5 mm long, and develops into an adult in about two weeks. The fly's genome is considerably smaller than that of humans or even mice (see Table 14.1). However, the genes that direct pattern formation appear to be highly conserved among animals with segmented body morphologies, including humans.

As pattern formation occurs in *Drosophila,* the embryonic cells begin to express genes differently in graded, periodic, and eventually striped arrangements. Boundaries between large body

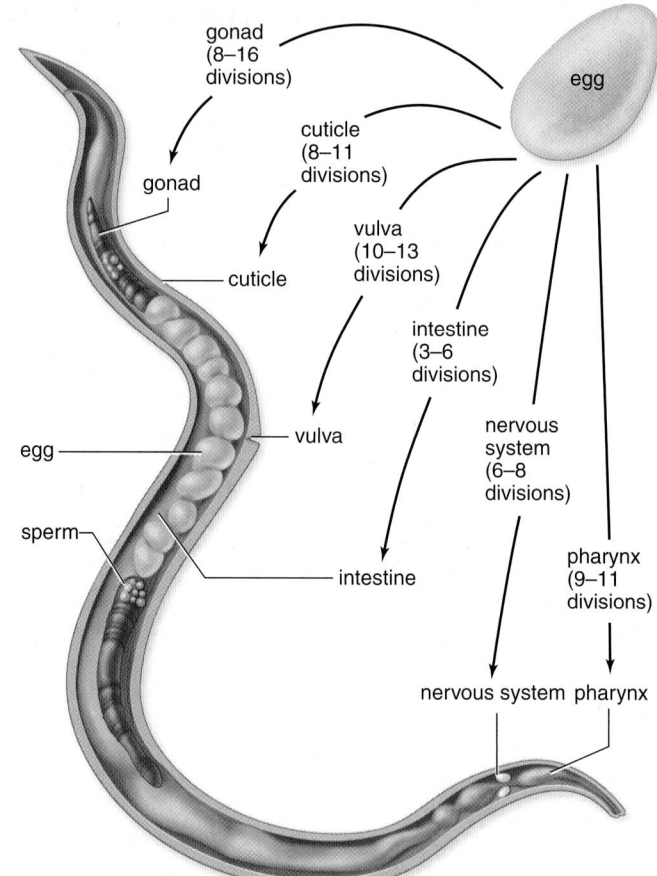

Figure 42.8 Development of *C. elegans,* a nematode. A fate map of the worm showing that as cells arise by cell division, they are destined to become particular structures.

regions are established first, before the refinement of smaller, subdivided parts can take place.

The Anterior/Posterior Axes. The first step in *Drosophila* pattern formation and morphogenesis begins with the establishment of anteroposterior polarity, meaning that the anterior (head) and posterior (abdomen) ends are different from one another. Such polarity is present in the egg before it is fertilized by a sperm.

Egg polarity results from maternal determinants, mRNAs that are deposited in specific positions within the egg while it is still in the ovary. The protein products of these genes diffuse away from the areas of their highest concentration in the embryo, forming gradients that influence patterns of tissue development. These proteins are also known as *morphogens* due to their crucial influence in morphogenesis. For example, the Bicoid protein is most concentrated anteriorly, where it prevents the formation of the posterior region. The Nanos protein is most concentrated posteriorly, and is required for abdomen formation.

The Segmentation Pattern. Once the anterior and posterior ends of the embryo have been established, a group of zygotic genes called *gap* genes come into play. The task of gap genes is to divide the anteroposterior axis into broad regions

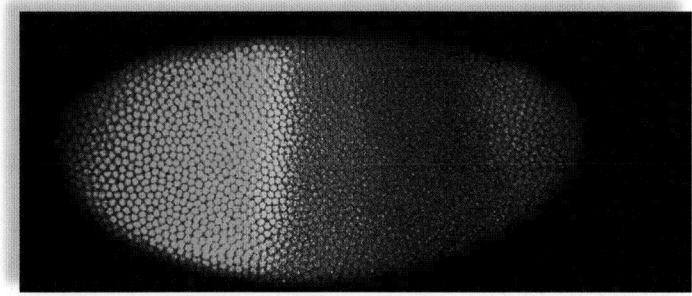

a. Protein products of gap genes

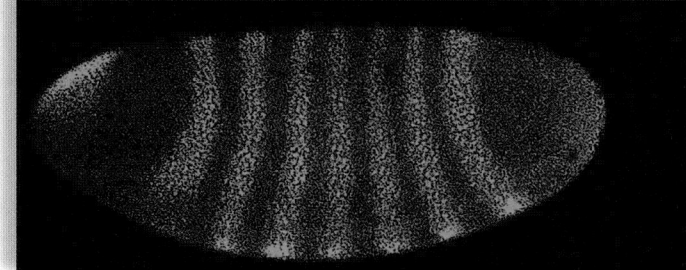

b. Protein products of pair-rule genes

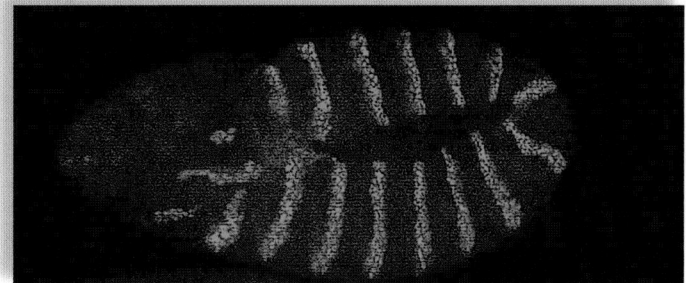

c. Protein products of segment-polarity genes

Figure 42.9 **Development in *Drosophila*, a fruit fly.** **a.** The different colors show that two different gap gene proteins are present from the anterior to the posterior end of an embryo. **b.** The green stripes show that a pair-rule gene is being expressed as segmentation of the fly occurs. **c.** Now segment-polarity genes help bring about division of each segment into an anterior and posterior end.

(Fig. 42.9*a*). They are called gap genes because mutations in these genes result in gaps in the embryo, where large blocks of segments are missing. Gap genes are temporarily activated by the gradients of anterior and posterior morphogens, and in turn activate the *pair-rule* genes.

The pair-rule genes are expressed periodically, in alternating stripes (Fig. 42.9*b*). They serve to "rough out" a preliminary segmentation pattern along the anteroposterior axis. The products of pair-rule genes may stimulate or suppress the expression of other genes, particularly the *segment-polarity* genes. These genes ensure that each segment has boundaries, with distinct anterior and posterior halves. The segment-polarity genes are also expressed in a striped fashion, but with twice as many stripes as the pair-rule genes (Fig. 42.9*c*). Mutations in segment-polarity genes result in the loss of one part of each segment, and the duplication of another portion of the same segment.

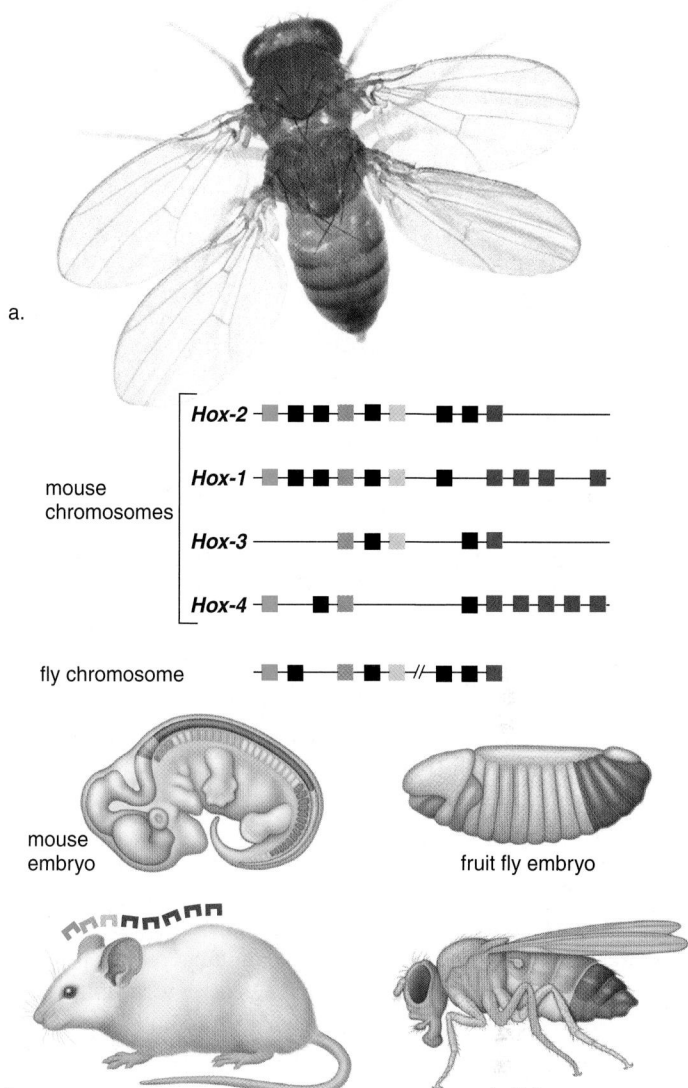

Figure 42.10 **Pattern formation in *Drosophila*.** Homeotic genes control pattern formation, an aspect of morphogenesis. **a.** If homeotic genes are activated at inappropriate times, abnormalities such as a fly with four wings occur. **b.** The green, blue, yellow, and red colors show that homologous homeotic genes occur on four mouse chromosomes and on a fly chromosome in the same order. These genes are color coded to the region of the embryo, and therefore the adult, where they regulate pattern formation. The black boxes are homeotic genes that are not identical between the two animals. In mammals, homeotic genes are called *Hox* genes.

Homeotic Genes.   The **homeotic genes** are often referred to as *selector genes* because they select for segmental identity—in other words, they dictate which body parts arise from the segments. Mutations in homeotic genes may result in the development of body parts in inappropriate areas, such as legs instead of antennae, or wings instead of tiny balancing organs called halteres (Fig. 42.10*a*). Such alterations in morphology are known as *homeotic transformations*.

Interestingly, homeotic genes in *Drosophila* and other organisms have all been found to share a structural feature called a **homeobox**. (*Hox,* the term used for mammalian homeotic genes, is a shortened form of homeobox.) A homeobox is a sequence of nucleotides that encodes a 60-amino-acid sequence called a **homeodomain**:

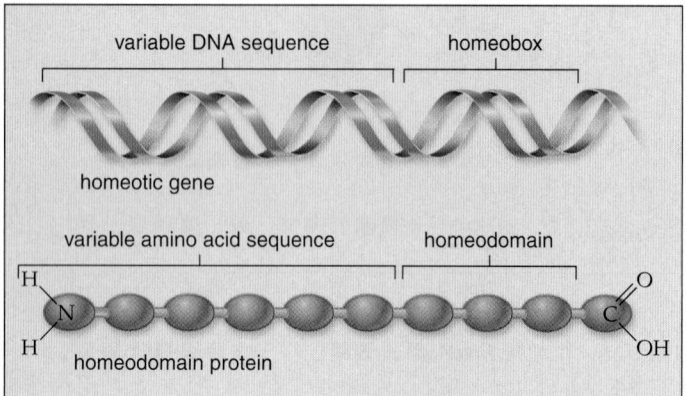

The homeodomain is a functionally important part of the protein encoded by a homeotic gene. Homeotic genes code for transcription factors, proteins that bind to regulatory regions of DNA and determine whether or not specific target genes are turned on. The homeodomain is the DNA-binding portion of the transcription factor, although the other, more variable sequences of a transcription factor determine which target genes are turned on. It is thought that the homeodomain proteins encoded by homeotic genes direct the activities of various target genes involved in morphogenesis, such as those involved in cell-to-cell adhesion. In the end, this orderly process determines the morphology of particular segments.

The importance of homeotic genes is underscored by the finding that the homeotic genes are highly conserved, being present in the genomes of many organisms, including mammals such as mice and even humans. Most homeotic genes have their loci on the same chromosome in *Drosophila,* while in mammals there are four clusters that reside on different chromosomes.

Notice that, in both flies and mammals, the position of the homeotic genes on the chromosome matches their anterior-to-posterior expression pattern in the body (Fig. 42.10*b*). The first gene clusters determine the final development of anterior segments, whereas those later in the sequence determine the final development of posterior segments of the animal's body.

Mutations in homeotic genes have similar effects in the mammalian body to the homeotic transformations observed in *Drosophila.* For instance, mutations in two adjacent *Hox* genes in the mouse result in shortened forelimbs that are missing the radius and ulna bones. In humans, mutations in a different *Hox* gene cause synpolydactyly, a rare condition in which there are extra digits (fingers and toes), some of which are fused to their neighbors.

Apoptosis. We have already discussed the importance of apoptosis (programmed cell death) in the normal day-to-day operation of the immune system and in preventing the occurrence of cancer. Apoptosis is also an important part of morphogenesis. During development of humans, we know that apoptosis is necessary to the shaping of the hands and feet; if it does not occur, the child is born with webbing between its fingers and toes.

The fate maps of *C. elegans* (see Fig. 42.8) indicate that apoptosis occurs in 131 cells as development takes place. When a cell-death signal is received, an inhibiting protein becomes inactive, allowing a cell-death cascade to proceed that ends in enzymes destroying the cell.

Check Your Progress   **42.2**

1. Name and define two mechanisms of cellular differentiation.
2. Define the term "morphogen."
3. Describe the function of the homeobox sequence in a homeotic gene.

42.3 Human Embryonic and Fetal Development

Learning Outcomes

Upon completion of this section, you should be able to

1. Name the membranes surrounding the human embryo, and list their functions.
2. Chronologically list the major events that occur during embryonic and fetal development.
3. Describe the structure and functions of the placenta.

In humans, the length of time from conception (fertilization followed by implantation in the endometrium) to birth (parturition) is approximately nine months. It is customary to calculate the time of birth by adding 280 days to the start of the last menstruation, because this date is usually known, whereas the day of fertilization is usually unknown. Because the time of birth is influenced by so many variables, only about 5% of babies actually arrive on the forecasted date.

Human development is often divided into embryonic development (months 1 and 2) and fetal development (months 3 through 9). During **embryonic development,** the major organs are formed, and during fetal development, these structures become larger and are refined.

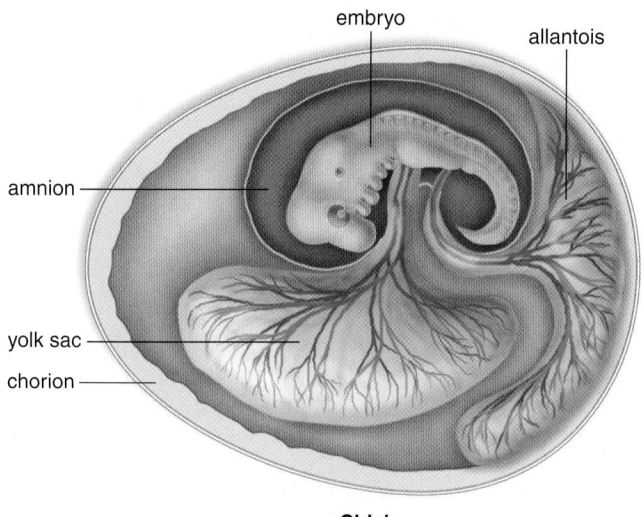

Chick

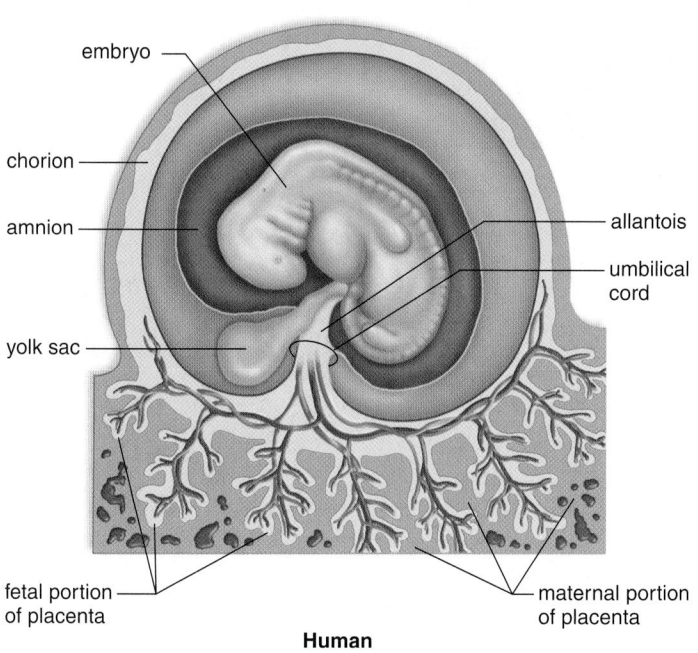

Human

Figure 42.11 Extraembryonic membranes. Extraembryonic membranes, which are not part of the embryo, are found during the development of chicks and humans. Each has a specific function.

Development can also be divided into *trimesters*. Each trimester can be characterized by specific developmental accomplishments. During the first trimester, embryonic and early fetal development occur. The second trimester is characterized by the development of organs and organ systems. By the end of the second trimester, the fetus appears distinctly human. In the third trimester, the fetus grows rapidly and the major organ systems become functional.

Extraembryonic Membranes

Before we consider human development chronologically, we must understand the placement of **extraembryonic membranes** [L. *extra,* on the outside]. Extraembryonic membranes are best understood by considering their function in reptiles and birds. In reptiles, these membranes made development on land first possible. If an embryo develops in the water, the water supplies oxygen for the embryo and takes away waste products. The surrounding water prevents desiccation, or drying out, and provides a protective cushion. For an embryo that develops on land, all these functions are performed by the extraembryonic membranes.

In the chick, the extraembryonic membranes develop from extensions of the germ layers, which spread out over the yolk. Figure 42.11 shows the chick surrounded by the membranes. The **chorion** [Gk. *chorion,* membrane] lies next to the shell and carries on gas exchange. The **amnion** [Gk. *amnion,* membrane around fetus] contains the protective amniotic fluid, which bathes the developing embryo. The **allantois** [Gk. *allantos,* sausage] collects nitrogenous wastes, and the **yolk sac** surrounds the remaining yolk, which provides nourishment.

The function of the extraembryonic membranes in humans has been modified to suit internal development. Their presence, however, shows that we are related to the reptiles.

The chorion develops into the fetal half of the placenta, the organ that provides the embryo/fetus with nourishment and oxygen, and takes away its wastes. Blood vessels within the chorionic villi are continuous with the umbilical blood vessels. The blood vessels of the allantois become the umbilical blood vessels, and the allantois accumulates the small amount of urine produced by the fetal kidneys and later gives rise to the urinary bladder. The yolk sac, which lacks yolk, is the first site of blood cell formation. The amnion contains fluid to cushion and protect the embryo, which develops into a fetus.

It is interesting to note that all chordate animals develop in water—either in bodies of water or surrounded by amniotic fluid within a shell or uterus.

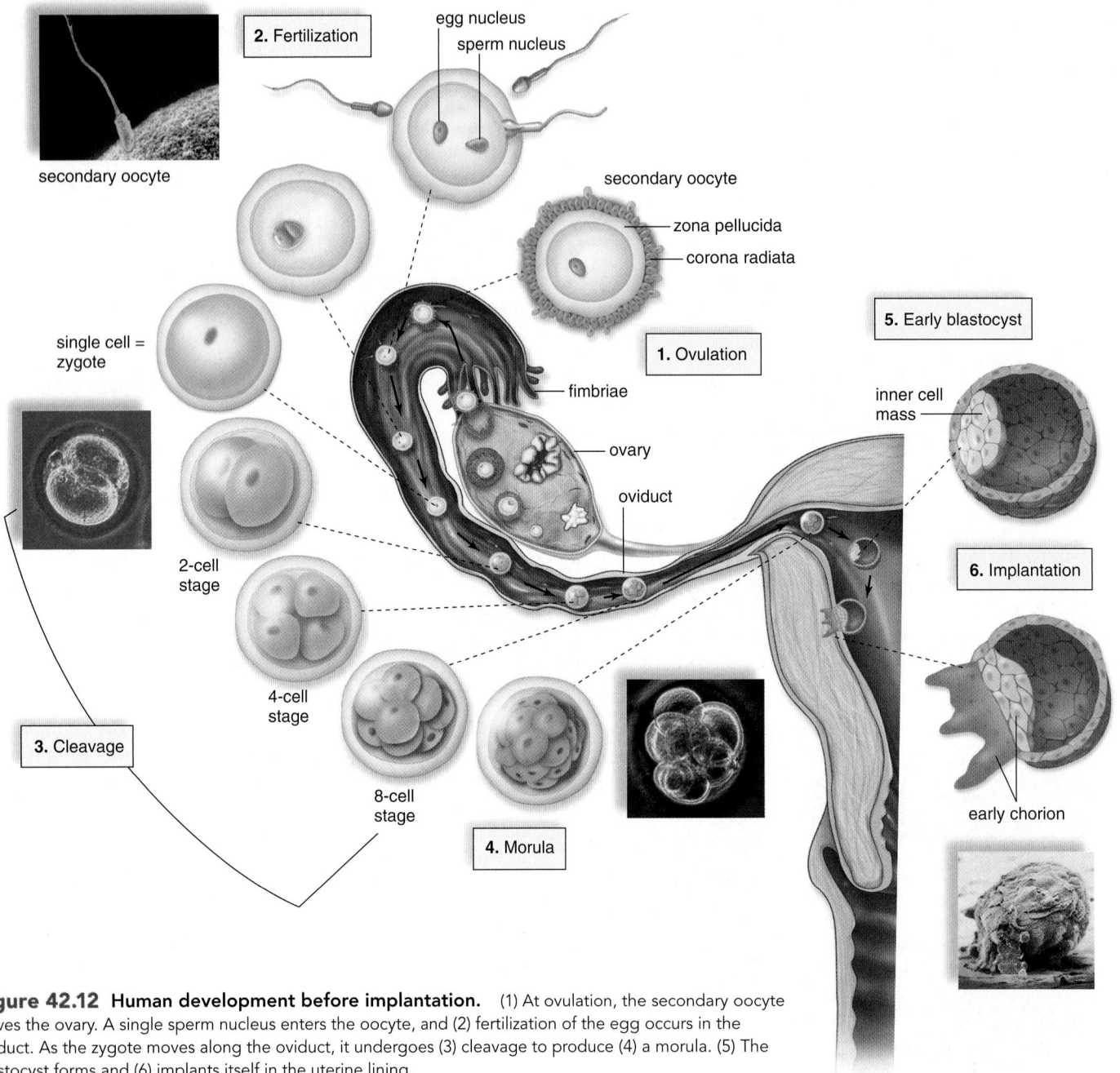

Figure 42.12 Human development before implantation. (1) At ovulation, the secondary oocyte leaves the ovary. A single sperm nucleus enters the oocyte, and (2) fertilization of the egg occurs in the oviduct. As the zygote moves along the oviduct, it undergoes (3) cleavage to produce (4) a morula. (5) The blastocyst forms and (6) implants itself in the uterine lining.

Embryonic Development

Embryonic development includes the first two months of development.

The First Week

Fertilization usually occurs in the upper third of the oviduct (Fig. 42.12). Cleavage begins 30 hours after fertilization and continues as the embryo passes through the oviduct to the uterus. By the time the embryo reaches the uterus on the third day, it is a morula. By about the fifth day, the morula has transformed into a blastula, and in mammals, a blastocyst then forms.

Video Blastocyst Formation

The **blastocyst** has a fluid-filled cavity, a single layer of outer cells called the **trophoblast** [Gk. *trophe*, food, and *blastos*,

bud], and an inner cell mass. The early function of the trophoblast is to provide nourishment for the embryo. Later, the trophoblast, reinforced by a layer of mesoderm, gives rise to the chorion, one of the extraembryonic membranes (see Fig. 42.11). The inner cell mass eventually becomes the embryo, which develops into a fetus.

The Second Week

At the end of the first week, the embryo begins the process of **implantation** in the wall of the uterus. The trophoblast secretes enzymes to digest away some of the tissue and blood vessels of the endometrium of the uterus (Fig. 42.12). The embryo is now about the size of the period at the end of this sentence. The trophoblast begins to secrete **human chorionic gonadotropin (HCG),** the hormone that is the basis for the pregnancy test and

Biological Systems

Preventing and Testing for Birth Defects

Birth defects, or congenital disorders, are abnormal conditions that are present at birth. According to the Centers for Disease Control (CDC), about 1 in 33 babies born in the United States has a birth defect. Those due to genetics can sometimes be detected before birth by one of the methods shown in Figure 42A; however, these methods are not without risk.

Some birth defects are not serious, and not all can be prevented. But women can take steps to increase their chances of delivering a healthy baby.

Eat a Healthy Diet

Certain birth defects occur because the developing embryo does not receive sufficient nutrition. For example, women of childbearing age are urged to make sure they consume ad-

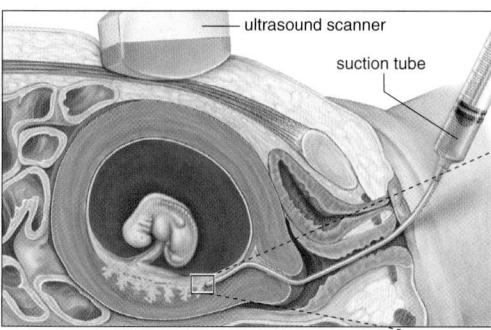

a. Amniocentesis

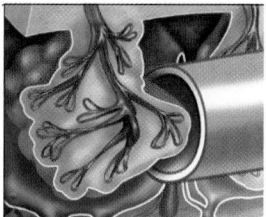

b. Chorionic villi sampling

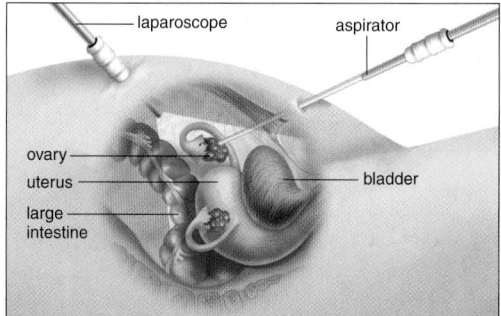

c. Preimplantation genetic diagnosis

equate amounts of the vitamin folic acid in order to prevent neural tube defects such as spina bifida and anencephaly. In spina bifida, part of the vertebral column fails to develop properly and cannot adequately protect the delicate spinal cord. With anencephaly, most of the fetal brain fails to develop. Anencephalic infants are stillborn or survive for only a few days after birth.

Fortunately, folic acid is plentiful in leafy green vegetables, nuts, and citrus fruits. It is also present in multivitamins, and many breads and cereals are fortified with it. The CDC recommends that all women of childbearing age get at least 400 micrograms of folic acid every day through supplements, in addition to eating a healthy, folic acid–rich diet. Unfortunately, neural tube birth defects can occur just a few weeks after conception, when many women are still unaware that they are pregnant—especially if the pregnancy is unplanned.

Avoid Alcohol, Smoking, and Drug Abuse

Alcohol consumption during pregnancy is a leading cause of birth defects. In severe instances, the baby is born with fetal alcohol syndrome (FAS), estimated to occur in 0.2 to 1.5 of every 1,000 live births in the United States. Children with FAS are frequently underweight, have an abnormally small head, abnormal facial development, and mental retardation. As they mature, children with FAS often exhibit short attention span, impulsiveness, and

Figure 42A Three methods for genetic defect testing before birth. About 20% of all birth defects are due to genetic or chromosomal abnormalities, which may be detected before birth. **a.** Amniocentesis is usually performed from the fifteenth to the seventeenth week of pregnancy. **b.** Chorionic villi sampling is usually performed from the eighth to the twelfth week of pregnancy. **c.** Preimplantation genetic diagnosis can be performed prior to in vitro fertilization, either on oocytes that have been collected from the woman, or on the early embryo.

poor judgment, as well as serious difficulties with learning and memory. Heavy alcohol use also reduces a woman's folic acid level, increasing the risk of the neural tube defects just described.

Cigarette smoking causes many birth defects. Babies born to smoking mothers typically have low birth weight, and are more likely to have defects of the face, heart, and brain than those born to nonsmokers.

Illegal drugs should also be avoided. For example, cocaine causes blood pressure fluctuations that deprive the fetus of oxygen. Cocaine-exposed babies may have problems with vision and coordination, and may be mentally retarded.

Alert Medical Personnel If You Are or May Be Pregnant

Several medications that are safe for healthy adults may pose a risk to a developing fetus. If pregnant women need to be immunized, they are usually given killed or inactivated forms of the vaccine, because live forms often present a danger to the fetus.

Because the rapidly dividing cells of a developing embryo or fetus are very susceptible to damage from radiation, pregnant women should avoid unnecessary X-rays. If a medical X-ray is unavoidable, the woman should notify the X-ray technician so that her fetus can be protected as much as possible, for example, by draping a lead apron over her abdomen if this area is not targeted for X-ray examination.

Avoid Infections That Cause Birth Defects

Certain pathogens such as the virus that causes rubella (German measles) can result in birth defects. In the past, this virus caused many birth defects, specifically mental retardation, deafness, blindness, and heart defects. Rubella is much less of a danger in developed countries today because most women have been vaccinated against rubella as children. Other infections that can cause birth defects include toxoplasmosis, herpes simplex, and cytomegalovirus.

Questions to Consider

1. Why is it likely that the actual rate of birth defects is higher than what is reported?
2. Besides tobacco, alcohol, and illegal drugs, what other potential risks should a pregnant woman avoid? Why?

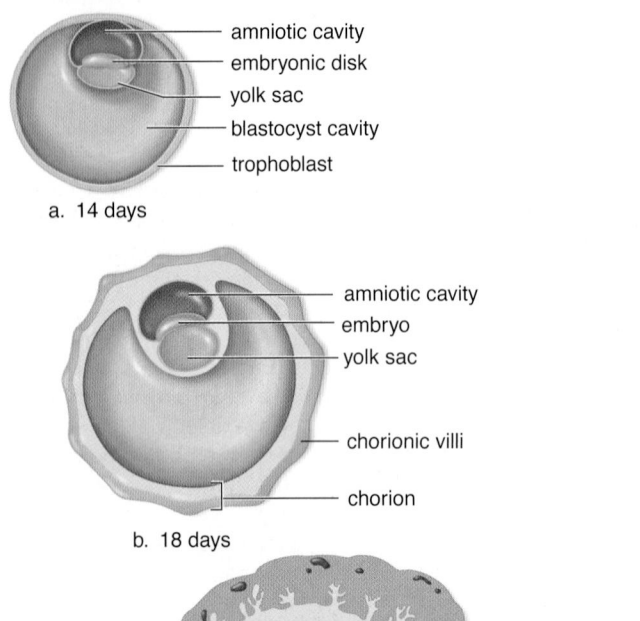

a. 14 days
- amniotic cavity
- embryonic disk
- yolk sac
- blastocyst cavity
- trophoblast

b. 18 days
- amniotic cavity
- embryo
- yolk sac
- chorionic villi
- chorion

c. 21 days
- amniotic cavity
- embryo
- yolk sac
- chorionic villi
- body stalk
- allantois

d. 25 days
- chorion
- amniotic cavity
- amnion
- chorionic villi
- allantois
- yolk sac

that serves to maintain the corpus luteum past the time it normally disintegrates. (Recall that the corpus luteum is a yellow body formed in the ovary from a follicle that has discharged its secondary oocyte.) Because of this, the endometrium is maintained, and menstruation does not occur.

As the week progresses, the inner cell mass detaches itself from the trophoblast, and two more extraembryonic membranes form (Fig. 42.13a). The yolk sac, which forms below the embryonic disk, has no nutritive function as it does in chickens, but it is the first site of blood cell formation. However, the amnion and its cavity are where the embryo (and then the fetus) develops. In humans, amniotic fluid acts as an insulator against cold and heat and also absorbs shock, such as that caused by the mother exercising.

Gastrulation occurs during the second week. The inner cell mass now has flattened into the **embryonic disk,** composed of two layers of cells: ectoderm above and endoderm below. Once the embryonic disk elongates to form the primitive streak, the third germ layer, mesoderm, forms by invagination of cells along the streak. The trophoblast is reinforced by mesoderm and becomes the chorion (Fig. 42.13b). It is possible to relate the development of future organs to these germ layers (see Table 42.1).

The Third Week

Two important organ systems make their appearance during the third week. The nervous system is the first organ system to be visually evident. At first, a thickening appears along the entire

Figure 42.13 Human embryonic development. **a.** At first, the embryo contains no organs, only tissues. The amniotic cavity is above the embryo, and the yolk sac is below. **b.** The chorion develops villi, the structures so important to the exchange between mother and child. **c., d.** The allantois and yolk sac, two more extraembryonic membranes, are positioned inside the body stalk as it becomes the umbilical cord. **e.** At 35+ days, the embryo has a head region and a tail region. The umbilical cord takes blood vessels between the embryo and the chorion (placenta).

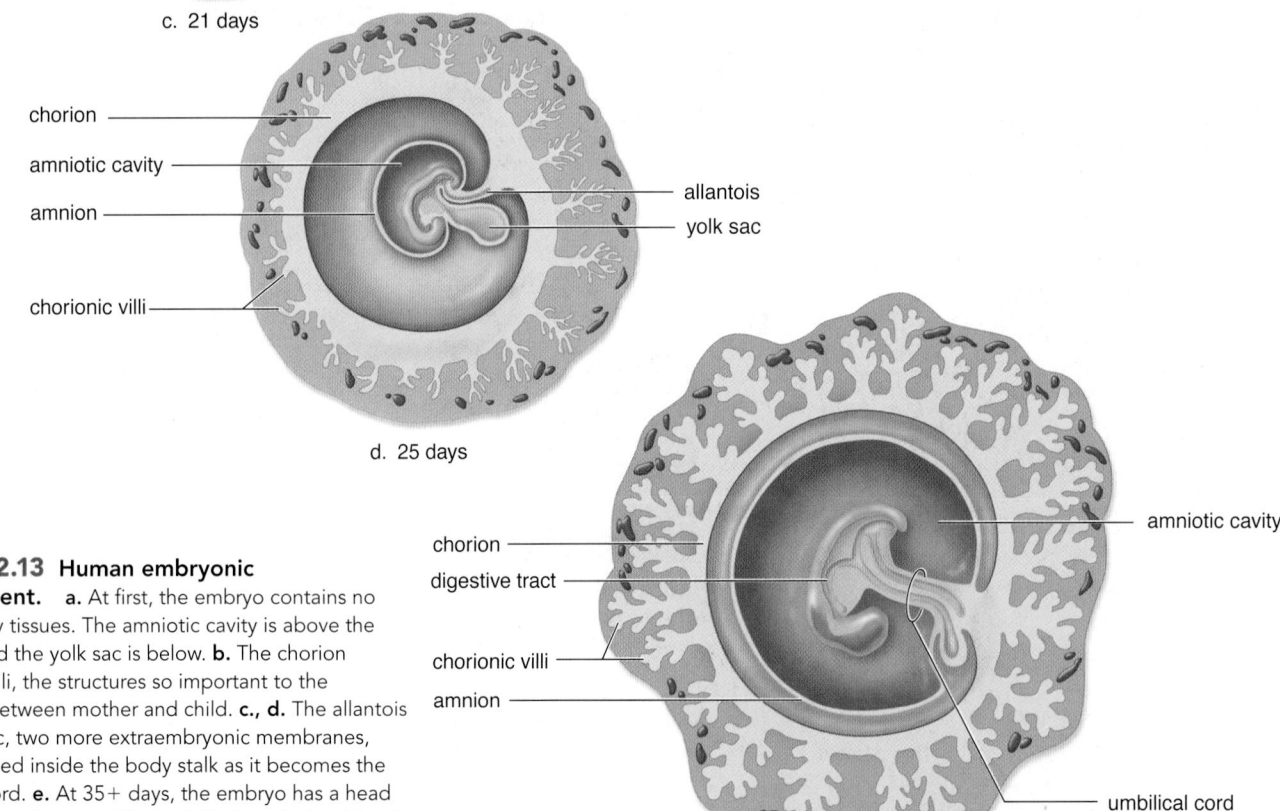

e. 35+ days
- chorion
- digestive tract
- chorionic villi
- amnion
- amniotic cavity
- umbilical cord

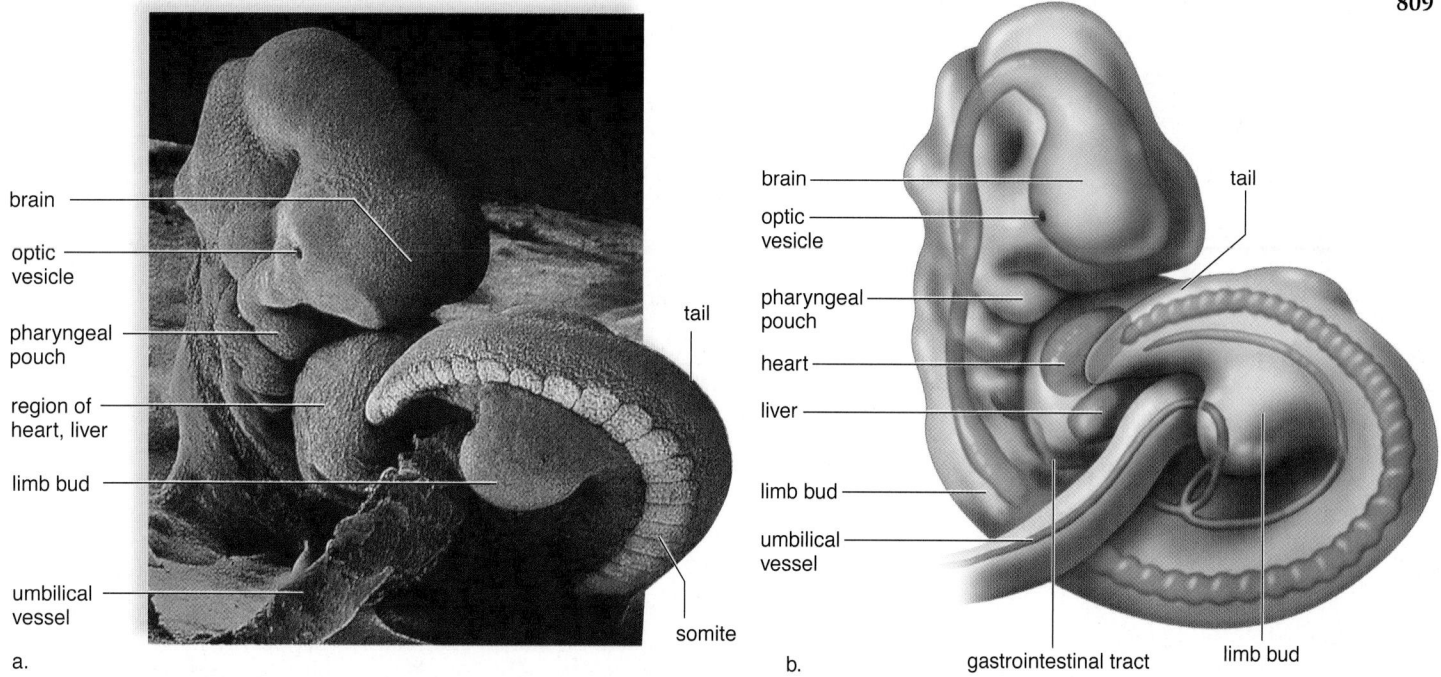

Figure 42.14 Human embryo at beginning of fifth week. **a.** Scanning electron micrograph. **b.** The embryo is curled so that the head touches the heart, the two organs whose development is farther along than the rest of the body. The organs of the gastrointestinal tract are forming, and the arms and the legs develop from the bulges that are called limb buds. The tail is an evolutionary remnant; its bones regress and become those of the coccyx (tailbone). The pharyngeal arches become functioning gills only in fishes and amphibian larvae; in humans, the first pair of pharyngeal pouches becomes the auditory tubes. The second pair becomes the tonsils, while the third and fourth become the thymus gland and the parathyroid glands.

dorsal length of the embryo; then the neural folds appear. When the neural folds meet at the midline, the neural tube, which later develops into the brain and the nerve cord, is formed (see Fig. 42.4). After the notochord is replaced by the vertebral column, the nerve cord is called the spinal cord.

Development of the heart begins in the third week and continues into the fourth week. At first, there are right and left heart tubes; when these fuse, the heart begins pumping blood, even though the chambers of the heart are not fully formed. The veins enter posteriorly, and the arteries exit anteriorly from this largely tubular heart, but later the heart twists so that all major blood vessels are located anteriorly.

The Fourth and Fifth Weeks

At four weeks, the embryo is barely larger than the height of this print. A bridge of mesoderm called the body stalk connects the caudal (tail) end of the embryo with the chorion, which has tree-like projections called **chorionic villi** [Gk. *chorion,* membrane; L. *villus,* shaggy hair] (Fig. 42.13*c, d*). The fourth extraembryonic membrane, the allantois, is contained within this stalk, and its blood vessels become the umbilical blood vessels. These structures form the **umbilical cord** [L. *umbilicus,* navel], which connects the developing embryo to the placenta (Fig. 42.13*e*).

Little protrusions called *limb buds* appear (Fig. 42.14); later, the arms and the legs develop from the limb buds, and even the hands and the feet become apparent. At the same time—during the fifth week—the head enlarges, and the developing eyes, ears, and nose are discernable.

The Sixth through Eighth Weeks

During the sixth to eighth weeks of development, the embryo becomes easily recognizable as human. Concurrent with brain

development, the head achieves its normal relationship with the body as a neck region develops. The nervous system is developed well enough to permit reflex actions, such as a startle response to touch. At the end of this period, the embryo is about 38 mm (1.5 in) long and weighs less than 1 g, even though all organ systems are established.

The Structure and Function of the Placenta

The **placenta** is a mammalian structure that functions in gas, nutrient, and waste exchange between embryonic (later fetal) and maternal cardiovascular systems. The placenta begins formation once the embryo is fully implanted. At first, the entire chorion has chorionic villi that project into endometrium. Later, these disappear in all areas except where the placenta develops. By the tenth week, the placenta (Fig. 42.15) is fully formed and is producing progesterone and estrogen. These hormones have two effects:

- They prevent any new follicles from maturing because of negative feedback control of the hypothalamus and anterior pituitary.
- They maintain the lining of the uterus, so now the corpus luteum is not needed. No menstruation occurs during pregnancy.

The placenta has a fetal side contributed by the chorion, and a maternal side consisting of uterine tissues. Notice in Figure 42.15 how the chorionic villi are surrounded by maternal blood; yet maternal and fetal blood do not mix under normal conditions because exchange always takes place across plasma membranes.

The umbilical cord is the lifeline of the fetus because it contains the umbilical arteries and vein, which transport waste

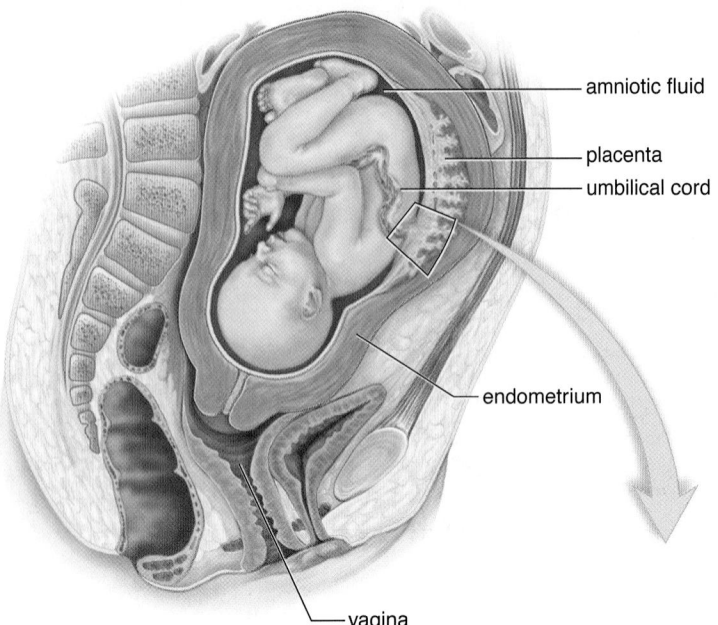

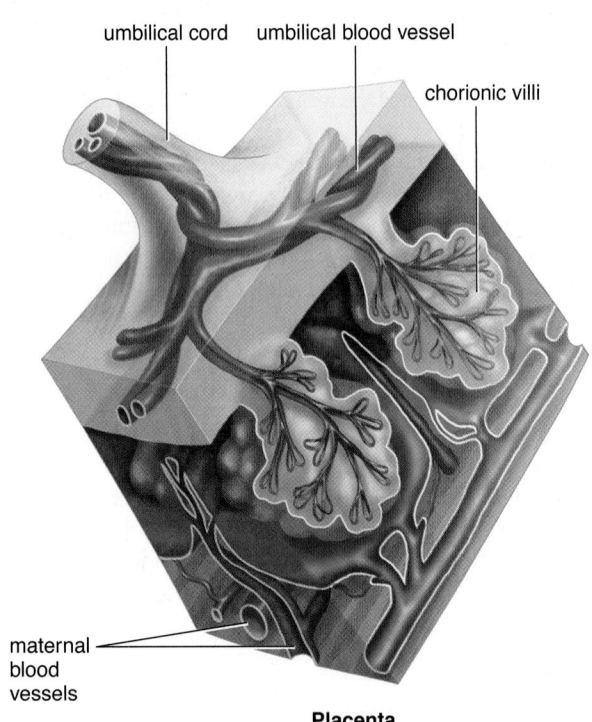

Placenta

Figure 42.15 Anatomy of the placenta in a fetus at six to seven months. The placenta is composed of both fetal and maternal tissues. Chorionic villi penetrate the uterine lining and are surrounded by maternal blood. Exchange of molecules between fetal and maternal blood takes place across the walls of the chorionic villi.

molecules (carbon dioxide and urea) to the placenta for disposal into the maternal blood and take oxygen and nutrient molecules from the placenta to the rest of the fetal circulatory system. If the placenta prematurely tears from the uterine wall, the lives of the fetus and the mother are endangered.

Late in fetal development, and peaking just before birth, maternal antibodies are transported across the placenta, into the fetal circulation. This helps to ensure that the newborn will be protected against common pathogens and other microbes to which the mother has been exposed, until the immune system of the newborn can mature and respond on its own. Harmful chemicals can also cross the placenta, as discussed in the Biological Systems feature on page 807.

Fetal Development and Birth

Fetal development (months 3–9) is marked by an extreme increase in size. Weight multiplies 600 times, going from less than 28 g to 3 kg. During this time, too, the fetus grows to about 50 cm in length. The genitalia appear in the third month, so it is possible to tell if the fetus is male or female.

Soon, hair, eyebrows, and eyelashes add finishing touches to the face and head. In the same way, fingernails and toenails complete the hands and feet. A fine, downy hair (lanugo) covers the limbs and trunk, only to later disappear. The fetus looks very old because the skin is growing so fast that it wrinkles. A waxy, almost cheeselike substance (vernix caseosa) [L. *vernix*, varnish, and *caseus*, cheese] protects the wrinkly skin from the watery amniotic fluid.

From about the fourth month on, the mother can feel movements of the fetal limbs. At about the same time, the fetal heartbeat can be heard through a stethoscope. Conversely, the fetus is able to hear and respond to sounds by about 18 weeks. A fetus born at 24 weeks has a chance of surviving, although the lungs are still immature and often cannot capture oxygen adequately. As a fetus rapidly grows during the third trimester, its chances of surviving being born a month or two prematurely increase dramatically.

The Stages of Birth

When the fetal brain is sufficiently mature, the fetal hypothalamus stimulates the pituitary to release ACTH, causing the adrenal cortex to release androgens into the bloodstream. These diffuse into the placenta, which uses androgens as a precursor for estrogens, hormones that stimulate the production of prostaglandins and oxytocin. All three of these molecules cause the uterus to contract and expel the fetus.

The process of birth (parturition) includes three stages. During the first stage, the cervix dilates to allow passage of the baby's head and body. The amnion usually bursts about this time, an event termed the mother's water breaking. During the second stage, the baby is born and the umbilical cord is cut. During the third stage, the placenta is delivered (Fig. 42.16).

Check Your Progress   42.3

1. Identify the location where fertilization usually occurs in the human female.
2. Name the extraembryonic membrane that gives rise to each of the following: umbilical blood vessels, the first blood cells, and the fetal half of the placenta.
3. Describe the major changes that occur in the human embryo between the third and fifth weeks of development.

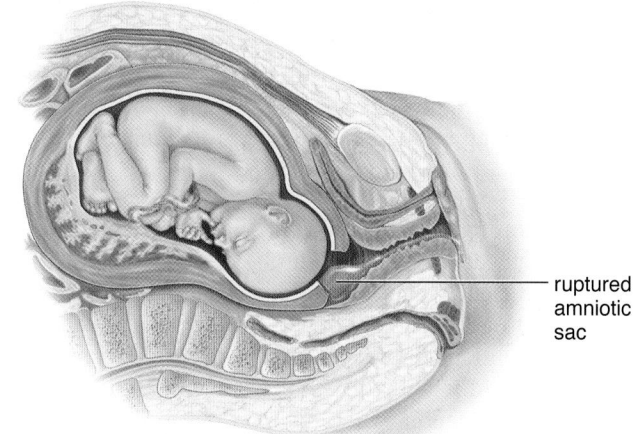

a. First stage of birth: cervix dilates

ruptured amniotic sac

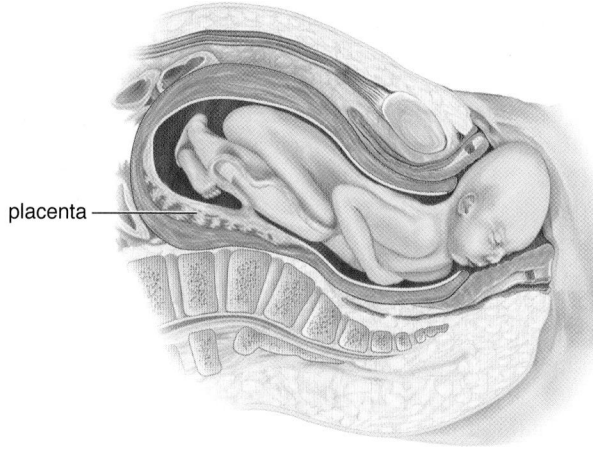

b. Second stage of birth: baby emerges

placenta

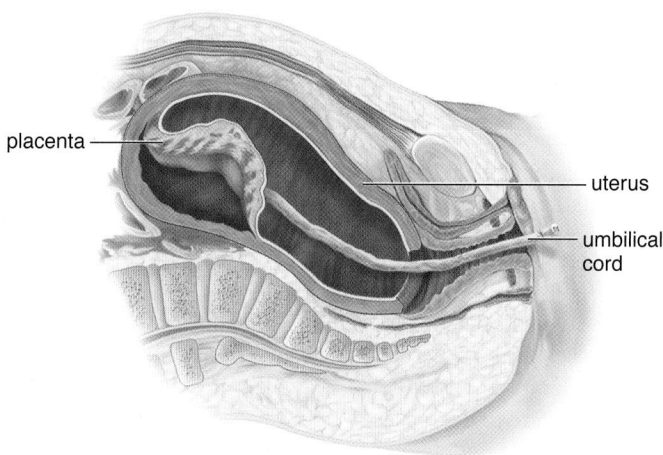

c. Third stage of birth: expelling afterbirth

placenta

uterus

umbilical cord

Figure 42.16 Three stages of parturition. **a.** Position of fetus just before birth begins. **b.** Dilation of cervix and birth of baby. **c.** Expulsion of afterbirth.

42.4 The Aging Process

Learning Outcomes

Upon completion of this section, you should be able to

1. Discuss the effects of aging on various body systems.
2. Compare and contrast the preprogrammed theories to the damage accumulation theories of aging.

Aging and death are as much a part of biology as are development and birth. If animals—and indeed, all forms of life—did not age and eventually die, their exponentially increasing numbers would presumably overwhelm the ability of the Earth to support life. Moreover, aging and death are a critical part of the evolutionary process, because animals that age and die are continuously replaced by individuals with new genetic combinations that can adapt to environmental changes that occur.

The Effects of Aging on Organ Systems

Today's human population is older on average than ever before. In the next half-century, the number of people over age 65 will increase by 147%. Before discussing some possible mechanisms behind the aging process, it seems appropriate to first consider the effects of aging on the various body systems.

Integumentary System

As aging occurs, the skin becomes thinner and less elastic because the number of elastic fibers decreases and the collagen fibers become increasingly cross-linked to each other, reducing their flexibility. There is also less adipose tissue in the subcutaneous layer; therefore, older people are more likely to feel cold. Together these changes typically result in sagging and wrinkling of the skin.

As people age, the sweat glands also become less active, resulting in decreased tolerance to high temperatures. There are fewer hair follicles, so the hair on the scalp and the limbs thins out. Older people also experience a decrease in the number of melanocytes, making their hair gray and their skin pale. In contrast, some of the remaining pigment cells are larger, and pigmented blotches ("age spots") appear on the skin.

Cardiovascular System

Common problems with cardiovascular function are usually related to diseases, especially atherosclerosis (see Chapter 32). However, even with normal aging, the heart muscle does weaken somewhat, and may increase slightly in size as it compensates for its decreasing strength. The maximum heart rate decreases even among the most fit older athlete, and it takes longer for the heart rate and blood pressure to return to normal resting levels following stress. Some part of this decrease in heart function may also be due to blood leaking back through heart valves that have become less flexible.

Aging also affects the blood vessels. The middle layer of arteries contains elastic fibers, which, like collagen fibers in the skin, become more cross-linked and rigid with time. These changes, plus a frequent decrease in the internal diameter of arteries due to atherosclerosis, contribute to a gradual increase in blood pressure with age. Indeed, nearly 50% of older adults have chronic hypertension. Such changes are common in individuals living in Western industrialized countries but not in agricultural societies. This indicates that a diet low in cholesterol and saturated fatty acids, along with a sensible exercise program, may help to prevent age-related cardiovascular disease.

Immune System

As people age, many of their immune system functions become compromised. Because a healthy immune system normally protects the entire body from infections, toxins, and at least some types of cancer, some investigators believe that losses in immune function can play a major role in the aging process.

As noted in Chapter 33, the thymus is an important site for T-cell maturation. Beginning in adolescence, the thymus begins to involute, gradually decreasing in size, and eventually becoming replaced by fat and connective tissue. The thymus of a 60-year-old adult is about 5% of the size of the thymus of a newborn, resulting in a decrease in the ability of older people to generate T-cell responses to new antigens.

The evolutionary rationale for this may be that the thymus is energetically expensive for an organism to maintain, and that compared to younger animals that must respond to a high number of new infections and other antigens, older animals have already responded to most of the antigens to which they will be exposed in their life.

Aging also affects other immune functions. Because most B-cell responses are dependent on T cells, antibody responses also begin to decline. This, in turn, may explain why the elderly do not respond as well to vaccinations as young people do. This presents challenges in protecting older people against diseases like influenza and pneumonia, which can otherwise be prevented by annual vaccination.

Not all immune functions decrease in older animals. The activity of natural killer cells, which are a part of the innate immune system, seems to change very little with age. Perhaps by investigating how these cells remain active throughout a normal human life span, researchers can learn to preserve other aspects of immunity in the elderly.

Digestive System

The digestive system is perhaps less affected by the aging process than other systems. Because secretion of saliva decreases, more bacteria tend to adhere to the teeth, causing more decay and periodontal disease. Blood flow to the liver is reduced, resulting in less efficient metabolism of drugs or toxins. This means that, as a person gets older, less medication is needed to maintain the same level in the bloodstream.

Respiratory System

Cardiovascular problems are often accompanied by respiratory disorders, and vice versa. Decreasing elasticity of lung tissues means that ventilation is reduced. Because we rarely use the entire vital capacity, these effects may not be noticed unless the demand for oxygen increases, such as during exercise.

Excretory System

Blood supply to the kidneys is also reduced. The kidneys become smaller and less efficient at filtering wastes. Salt and water balance are difficult to maintain, and the elderly dehydrate faster than young people. Urinary incontinence (lack of bladder control) increases with age, especially in women. In men, an enlarged prostate gland may reduce the diameter of the urethra, causing frequent or difficult urination.

Nervous System

Between the ages of 20 and 90, the brain loses about 20% of its weight and volume. Neurons are extremely sensitive to oxygen deficiency, and neuron death may be due not to aging itself but to reduced blood flow in narrowed blood vessels. However, contrary to previous scientific opinion, recent studies using advanced imaging techniques show that most age-related loss in brain function is not due to whatever loss of neurons is occurring. Instead, decreased function may occur due to alterations in complex chemical reactions, or increased inflammation in the brain. For example, an age-associated decline in levels of dopamine can affect brain regions involved in complex thinking.

Perhaps more important than the molecular details, recent studies have confirmed that lifestyle factors can affect the aging brain. For example, animals on restricted calorie diets developed fewer Alzheimer-like changes in their brains. Other positive factors that may help maintain a healthy brain include attending college (the "use it or lose it" idea), regular exercise, and sufficient sleep.

Sensory Systems

In general, with aging more stimulation is needed for taste, smell, and hearing receptors to function as before. A majority of people over age 80 have a significant decline in their sense of smell, and about 15% suffer from anosmia, or a total inability to smell. The latter condition can be a serious health hazard, due to the inability to detect smoke, gas leaks, or spoiled food. After age 50, most people gradually begin to lose the ability to hear tones at higher frequencies, and this can make it difficult to identify individual voices and to understand conversation in a group.

Starting at about age 40, the lens of the eye does not accommodate as well, resulting in presbyopia, or difficulty focusing on near objects, which causes many people to require reading glasses as they reach middle age. Finally, as noted in Chapter 38, cataracts and other eye disorders become much more common in the aged.

Musculoskeletal System

For the average person, muscle mass peaks between age 16–19 for females and 18–24 for males. Beginning in the 20s or 30s, but accelerating with increasing age, muscle mass generally decreases, due to decreases in both the size and number of muscle fibers. Most people who reach age 90 have 50% less muscle mass compared to when they were 20. Although some of this loss may be inevitable, regular exercise can slow this decline.

Like muscles, bones tend to shrink in size and density with age. Due to compression of the vertebrae, along with changes in posture, most of us lose height as we age. Those who reach age 80 will be about two inches shorter than they were in their 20s. Women lose bone mass more rapidly than men do, especially after menopause. As noted in Chapter 39, osteoporosis is a common disease in the elderly.

Although some decline in bone mass is a normal result of aging, certain extrinsic factors are also important. A proper diet and moderate exercise program has been found to slow the progressive loss of bone mass.

Endocrine System

As with the immune system, aging of the hormonal system can affect many different organs of the body. These changes are complex, however, with some hormone levels tending to decrease with age, while others increase. The activity of the thyroid gland generally declines, resulting in a lower basal metabolic rate. The production of insulin by the pancreas may remain stable, but cells become less sensitive to its effects, resulting in a rise in fasting glucose levels of about 10 mg/dL each decade after age 50.

Human growth hormone (HGH) levels also decline with age, but it is very unlikely that taking HGH injections will "cure" aging, despite internet claims. In fact, one study found that people with lower levels of HGH actually lived longer than those with higher levels.

Reproductive System

Testosterone levels are highest in men in their 20s. After age 30, testosterone levels decrease by about 1% per year. Extremely low testosterone levels have been linked to a decreased sex drive, excessive weight gain, loss of muscle mass, osteoporosis, general fatigue, and depression. However, the levels below which testosterone treatment should be initiated remain controversial. Testosterone replacement therapy, whether through injection, patches, or gels, is associated with side effects like enlargement of the prostate, acne or other skin reactions, and the production of too many red blood cells.

Menopause, the period in a woman's life during which the ovarian and uterine cycles cease, usually occurs between ages 45 and 55. The ovaries become unresponsive to the gonadotropic hormones produced by the anterior pituitary, and they no longer secrete estrogen or progesterone. At the onset of menopause, the uterine cycle becomes irregular, but as long as menstruation occurs, it is still possible for a woman to conceive. Therefore, a woman is usually not considered to have completed menopause (and thus be infertile) until menstruation has been absent for a year.

The hormonal changes during menopause often produce physical symptoms such as "hot flashes" (caused by circulatory irregularities), dizziness, headaches, insomnia, sleepiness, and depression. To ease these symptoms, female hormone replacement therapy (HRT) was routinely prescribed until 2002, when a large clinical study showed that in the long term, HRT caused more health problems than it prevented. For this reason, most doctors no longer recommend long-term HRT for the prevention of postmenopausal conditions.

It is also of interest that, as a group, females live longer than males. It is likely that estrogen offers women some protection against cardiovascular disorders when they are younger. Males suffer a marked increase in heart disease in their 40s, but an increase is not noted in females until after menopause, when women lead men in the incidence of stroke. Men remain more likely than women to have a heart attack at any age, however.

Hypotheses About Why We Age

Aging is a complex process, and multiple factors can affect the aging process. The many hypotheses about why aging occurs can be grouped into two major categories: (1) theories suggesting that aging is mainly due to preprogrammed genetic events and (2) theories that aging is mainly due to the accumulation of cellular damage.

Preprogrammed Theories

Most scientists who study *gerontology*, or the science of aging, believe that aging is partly genetically preprogrammed. This idea is supported by the observation that longevity runs in families, that is, the children of long-lived parents tend to live longer than those of short-lived parents. As would also be expected, studies show that identical twins have a more similar life span than nonidentical twins.

Many laboratory studies of aging have been performed in the nematode *C. elegans,* in which single-gene mutations have been shown to influence the life span. For example, mutations that decrease the activity of a hormone receptor similar to the insulin receptor more than double the life span of the worms, which also behave and look like much younger worms throughout their prolonged lives. Interestingly, small breed dogs like poodles or terriers, which may live 15–20 years, have lower levels of an analogous receptor, compared to large dog breeds that live 6–8 years.

Studies of the behavior of cells grown in the lab also suggest a genetic influence on aging. Most types of differentiated cells can only divide a limited number of times. One factor that may control the number of cell divisions is the length of the *telomeres,* sequences of DNA at the ends of chromosomes. As you learned in Section 12.2, telomeres protect the ends of chromosomes from deteriorating or fusing with other chromosomes. Each time a cell divides, the telomeres normally shorten, and cells with shorter telomeres tend to undergo fewer divisions. Perhaps as we grow older, more and more cells are unable to divide, and instead, they undergo degenerative changes and die.

However, if aging were mainly a function of genes, we would expect much less variation in life span to be observed among individuals of a given species than is actually seen. For this reason, experts have estimated that in most cases, genes account only for about 25% of what determines the length of life.

Damage Accumulation Theories

A second group of hypotheses postulate that aging involves the accumulation of damage over time. In 1900, the average human life expectancy was around 45 years. A baby born in the United States in 2009 is expected to live an average of about 78 years. Since human genes have presumably not changed much in such a short time, most of this gain in life span is due to better medical care, along with the application of scientific knowledge about how to prolong our lives.

There are two basic types of cellular damage that can accumulate over time. The first type can be thought of as agents that are unavoidable; for example, the accumulation of harmful DNA mutations, or the buildup of harmful metabolites. For some time medical researchers have known that proteins—such as the collagen fibers present in many support tissues—become increasingly cross-linked as people age. This cross-linking may account for the inability of such organs as the blood vessels, the heart, and the lungs to function as they once did. Some researchers have now found that glucose has the tendency to attach to any type of protein, which is the first step in a cross-linking process. They are currently experimenting with drugs that can prevent cross-linking.

Figure 42.17 Supercentenarians.   The oldest verified human life span was that of Jeanne Calment (1875–1997), who lived 122 years and was documented by the Guinness Book of Records as the "World's Oldest Living Person."

Other sources of cellular damage, however, may be avoidable, such as a poor diet or exposure to the Sun. Recent work suggests that when an animal produces fewer free radicals, it lives longer. Free radicals are unstable molecules that have an unpaired electron. In order to stabilize themselves, free radicals react with another molecule, such as DNA or proteins (e.g., enzymes) or lipids found in plasma membranes. Eventually, these molecules are unable to function, and the cell is destroyed. By increasing one's consumption of natural antioxidants, such as those present in brightly colored and dark green vegetables, citrus fruits, nuts, fish, shellfish, and red wine, we can reduce our exposure to free radicals, and slow the aging process.

The longest-lived human whose age has been verified by modern methods was Jeanne Calment (Fig. 42.17), who lived in France her entire life, and died at the age of 122 years, 164 days (she had lived on her own until age 110, when she entered a nursing home). This is currently considered to be the longest potential human life span. However, our understanding of the aging process could be considered, ironically, to be in its infancy, and animal models and other studies suggest that it may be possible to extend human life far beyond that.

Check Your Progress   42.4

1. Describe how involution of the thymus can lead to decreased responses to vaccines in the elderly.
2. Define menopause, and explain why it occurs.
3. Distinguish between the two hypotheses regarding aging, and give examples of each.

CONNECTING *the* CONCEPTS *with the* BIG IDEAS

Energy and Homeostasis

- Normal cell differentiation is a result of specific gene expression, and transplantation experiments confirm that link. (2E1a, 2E1b5, 4A3b)
- The pattern and timing of developmental stages is controlled by homeotic genes. (2E1b1)
- Induction is essential in the timing and sequence of embryonic events. (2E1b2)
- Abnormal development is often a sign of the mutation of developmental genes. (2E1b4)
- New studies show microRNA's importance in both control of normal cell function and in embryonic development. (2E1b6)
- Apoptosis plays crucial roles in all animal development, from *C. elegans* to mammalian digit morphology. (2E1c*IE*)

Information and Signaling

- The SRY gene controls the sexual developmental sequence for "maleness". (3B2a*IE*)
- Morphogens are involved in signal transmission over long or short distances which activate cell differentiation. (3B2b*IE*, 3D2b*IE*)
- Hox or homeotic genes control basic development and placement of body parts in embryos. (3B2b*IE*)

*Find the unabridged version of all EK citations at www.glencoe.com/maderAP11.

Media Study Tools

www.glencoe.com/maderAP11

Enhance your study of this chapter with study tools and practice tests. Also ask your instructor about the resources available through ConnectPlus, including the media-rich eBook, interactive learning tools, and animations.

Summarize

42.1 Early Developmental Stages

Development begins at fertilization. Only one sperm actually enters the oocyte. Both sperm and egg contribute chromosomes to the diploid zygote.

The early developmental stages in animals proceed from cellular stages to tissue stages to organ stages. During the cellular stage, cleavage (cell division) occurs, but there is no overall growth. The result is a morula, which becomes the blastula when an internal cavity (the blastocoel) appears.

During the tissue stage, gastrulation (invagination of cells into the blastocoel) results in formation of the germ layers: ectoderm, mesoderm, and endoderm. Both the cellular and tissue stages can be affected by the amount of yolk.

Organ formation can be related to germ layers. For example, during neurulation, the nervous system develops from midline ectoderm, just above the notochord. At this point, it is possible to draw a typical cross section of a vertebrate embryo in which the notochord has not been replaced by the vertebral column.

42.2 Developmental Processes

Cellular differentiation begins with cytoplasmic segregation in the egg. After the first cleavage of a frog embryo, only a daughter cell that receives a portion of the gray crescent is able to develop into a complete embryo.

Induction is also part of cellular differentiation. For example, the notochord induces the formation of the neural tube in frog embryos. In *C. elegans*, investigators have shown that induction is an ongoing process in which one type of tissue sequentially regulates the development of other types, through chemical signals coded for by particular genes.

Some morphogen genes determine the axes of the body, and others regulate the development of segments. An important concept has emerged: During development, sequential sets of master genes code for morphogen gradients that activate the next set of master genes in turn.

Homeotic genes control pattern formation such as the presence of antennae, wings, and limbs on the segments of *Drosophila*. Homeotic genes code for proteins that contain a homeodomain, a particular sequence of 60 amino acids. These proteins are also transcription factors, and the homeodomain is the portion of the protein that binds to DNA. Homologous homeotic genes have been found in a wide variety of organisms, and therefore they must have arisen early in the history of life and been conserved.

42.3 Human Embryonic and Fetal Development

Human development can be divided into embryonic development (months 1 and 2) and fetal development (months 3 through 9). Fertiliza-

tion occurs in the oviduct, and cleavage occurs as the embryo moves toward the uterus. The morula becomes the blastocyst before implanting in the endometrium of the uterus. The extraembryonic membranes appear early in human development. The trophoblast of the blastocyst is the first sign of the chorion, which goes on to become the fetal part of the placenta.

Exchange occurs between fetal and maternal blood at the placenta. The amnion contains amniotic fluid, which cushions and protects the embryo. The yolk sac and allantois form the umbilical cord, which is a lifeline between the mother and the developing embryo and fetus.

Organ development begins with neural tube and heart formation. There follows a steady progression of organ formation during embryonic development. During fetal development, refinement of organ systems occurs, and the fetus adds weight.

42.4 The Aging Process

The aging process is an essential part of the biology and evolution of animals. Aging affects every organ system of the human body, in mostly predictable ways. Various hypotheses have been advanced to explain the aging process, which can be grouped into those suggesting that aging is genetically preprogrammed, and those emphasizing the accumulation of various types of cellular damage over time.

Key Terms

allantois 805	germ layer 798
amnion 805	homeobox 804
apoptosis 800	homeodomain 804
blastocoel 797	homeotic genes 803
blastocyst 806	human chorionic gonadotropin
blastopore 798	(HCG) 806
blastula 797	implantation 806
cellular differentiation 800	induction 800
chorion 805	maternal determinant 800
chorionic villus 809	menopause 813
cleavage 797	mesoderm 798
cytoplasmic segregation 800	morphogenesis 800
development 797	morula 797
ectoderm 798	neural plate 799
embryo 797	neural tube 799
embryonic development 805	notochord 799
embryonic disk 808	pattern formation 800
endoderm 798	placenta 809
extraembryonic	totipotent 800
membrane 805	trophoblast 806
fertilization 796	umbilical cord 809
gastrula 798	yolk 797
gastrulation 798	yolk sac 805

Assess

Reviewing This Chapter

1. Describe the events of fertilization that (a) allow the sperm to reach the plasma membrane of the secondary oocyte, (b) prevent polyspermy, and (c) result in a diploid zygote. 796–97
2. What happens during the cellular stages, the tissue stages, and the organ stages of early development in animals? 797–99
3. What are the germ layers, and which organs are derived from each of the germ layers? 798–99
4. Draw a cross section of a typical vertebrate embryo at the neurula stage, and label your drawing. 799

5. Describe two mechanisms that are known to be involved in the processes of cellular differentiation.   800–02
6. Describe the classic experiment suggesting that the notochord induces formation of the neural tube.   800–01
7. With regard to *C. elegans*, what is a fate map? How does induction occur?   802
8. With regard to *Drosophila*, what is a morphogen gradient, and what does such a gradient do to bring about morphogenesis?   802
9. What is the function of homeotic genes, and what is the significance of the homeobox within these genes?   803–04
10. List the human extraembryonic membranes, give a function for each, and compare their functions to those in the chick.   805
11. Explain where fertilization, cleavage, the morula stage, and the blastocyst stage occur in humans. What happens to the embryo in the third through fifth weeks?   806, 808–09
12. Describe the structure and the function of the placenta in humans.   809–10
13. List and describe the stages of birth.   810
14. Discuss the effects of aging on the organ systems.   811–13
15. Describe the two major categories of ideas about why we age.   813–14

Testing Yourself

Choose the best answer for each question.

1. Which of these stages is the first one out of sequence?
   a. cleavage
   b. blastula
   c. morula
   d. gastrula
   e. neurula
2. Which of these stages is mismatched?
   a. cleavage—cell division
   b. blastula—gut formation
   c. gastrula—three germ layers
   d. neurula—nervous system
   e. Both b and c are mismatched.
3. A cell that is capable of giving rise to a complete organism is said to be
   a. totipotent.
   b. homeotic.
   c. ectodermal.
   d. differentiated.
   e. induced.
4. Morphogenesis is associated with
   a. protein gradients.
   b. induction.
   c. transcription factors.
   d. homeotic genes.
   e. All of these are correct.
5. In humans, the placenta develops from the chorion. This indicates that human development
   a. resembles that of the chick.
   b. is dependent on extraembryonic membranes.
   c. cannot be compared to that of lower animals.
   d. begins only upon implantation.
   e. Both a and b are correct.

6. In humans, the fetus
   a. is surrounded by four extraembryonic membranes.
   b. has developed organs and is recognizably human.
   c. is dependent on the placenta for excretion of wastes and acquisition of nutrients.
   d. is embedded in the endometrium of the uterus.
   e. Both b and c are correct.
7. Developmental changes
   a. require growth, differentiation, and morphogenesis.
   b. stop occurring when one is grown.
   c. are dependent on a parceling out of genes into daughter cells.
   d. are dependent on activation of master genes in an orderly sequence.
   e. Both a and d are correct.
8. Which of these pairs is mismatched?
   a. brain—ectoderm
   b. gut—endoderm
   c. bone—mesoderm
   d. lens—endoderm
   e. heart—mesoderm
9. Label this diagram illustrating the placement of the extraembryonic membranes, and give a function for each membrane in humans.

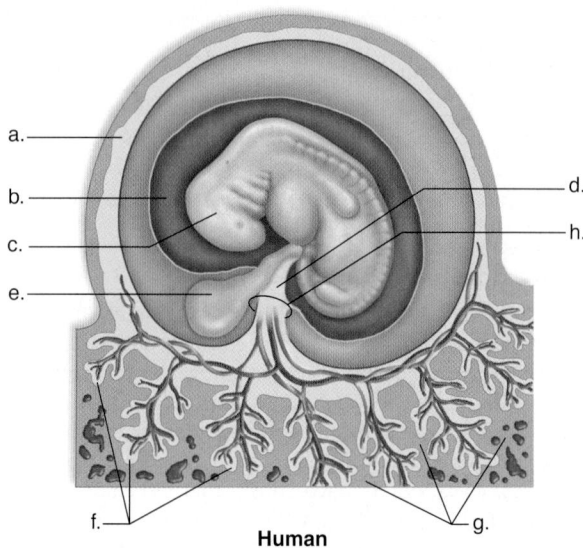

Human

For questions 10–13, match the statement with the terms in the key.

KEY:

   a. apoptosis
   b. homeotic genes
   c. fate maps
   d. morphogen gradients
   e. segment-polarity genes

10. have been developed that show the destiny of each cell that arises through cell division during the development of *C. elegans*
11. occurs when a cell-death cascade is activated
12. can have a range of effects, depending on its concentration in a particular portion of the embryo

13. genes that control pattern formation, the organization of differentiated cells into specific three-dimensional structures

14. The ability of one embryonic tissue to influence the growth and development of another tissue is termed
    a. morphogenesis.
    b. pattern formation.
    c. apoptosis.
    d. cellular differentiation.
    e. induction.

15. Which hormone is sometimes administered to stimulate parturition?
    a. estrogen
    b. oxytocin
    c. prolactin
    d. testosterone
    e. Both b and d are correct.

16. Only one sperm enters and fertilizes a human ovum because
    a. sperm have an acrosome.
    b. the corona radiata gets larger.
    c. the zona pellucida lifts up.
    d. the plasma membrane hardens.
    e. All of these are correct.

17. Which is a correct sequence in humans that ends with the stage that implants?
    a. morula, blastocyst, embryonic disk, gastrula
    b. ovulation, fertilization, cleavage, morula, early blastocyst
    c. embryonic disk, gastrula, primitive streak, neurula
    d. primitive streak, neurula, extraembryonic membranes, chorion
    e. cleavage, neurula, early blastocyst, morula

18. Which of the following is *not* an explanation for why the skin of older people tends to become wrinkled?
    a. increased subcutaneous fat deposition
    b. increased cross-linking of collagen fibers
    c. decreased sebaceous gland activity
    d. decreased number of elastic fibers

19. The onset of menopause is mainly due to
    a. a lack of ovarian follicles that are able to ovulate
    b. an involution of the uterus that begins around age 45
    c. an unresponsiveness of the ovaries to FSH and LH from the pituitary
    d. an increased production of male hormones

20. Possible causes of the cellular damage that can accumulate and contribute to aging include
    a. free radicals
    b. spontaneous DNA mutations
    c. cross-linked proteins
    d. uv radiation
    e. all of the above

Engage

Thinking Scientifically

1. *Drosophila* belongs to the insect order Diptera, in which one pair of wings is the norm. In *Drosophila, Ubx* mutation converts the halteres into an extra pair of wings located posterior to the normal pair (see Fig. 42.10a). Butterflies belong to the order Lepidoptera in which two pairs of wings

is normal. The ancestor of both flies and butterflies is thought to have possessed two pairs of wings. However, in butterflies, mutations in *Ubx* have been found to alter pigmentation, spot pattern, and scale morphology in the hindwings, but not the number of wing pairs. What does this tell us about the role of *Ubx* in insect wing development?

Bioethical Issue

Human Embryonic Stem Cell Research

Human embryonic stem cell lines (also called ES cell lines) are obtained by removing the inner cell mass from a blastocyst and growing the cells in a petri dish. The cells are valued by researchers because they are pluripotent, meaning they have the potential to differentiate into a wide range of different types of cells if properly stimulated. At this time, the federal government funds research on ES cell lines that were started before 2001, obtained from extra embryos left over from in vitro fertilization (IVF), and voluntarily donated for scientific study. However, funding for more new cell lines may be obtained from certain state or private sources, and researchers overseas are not subject to U.S. regulations.

Advocates of ES cell research say that such cells could be used to develop cures for conditions such as Parkinson disease, diabetes, heart disease, Alzheimer disease, cystic fibrosis, and spinal cord injuries. In addition, ES cells could be studied to help scientists understand the basic processes of human development and used to test new drugs.

ES cell research opponents say that it should be restricted because it requires the destruction of human life. Many believe that embryos, even those so early in development, should be considered human beings. If so, then producing an excess of them for IVF and then discarding them would be wrong. It might also be wrong to benefit from their destruction.

Stem cells may be obtained from other sources besides embryos. Some, known as embryonic germ cells, are harvested from aborted or miscarried fetuses, but this source is subject to the same sort of controversy as ES cells. Research using stem cells taken from the umbilical cord and placenta has yielded some very promising results; adult tissues such as bone marrow and parts of the brain are other stem cell sources. And as discussed in the Nature of Science feature, "Reproductive and Therapeutic Cloning," in Chapter 9, genetically modified fibroblasts can serve as stem cells. In fact, some of these nonembryonic cells have already been used to treat medical conditions, including blood disorders, spinal cord injury, and heart attack damage. Such stem cells are obtained without the use of embryos or fetuses. However, they appear to be more limited in their ability to differentiate than ES cells.

Should ES cell research continue to be permitted? If so, should it be supported by government funding? Do the origins of the ES cell lines make a difference? These are all decisions that have been faced by voters in recent years and will certainly continue to generate vigorous debate.

UNIT 8

Behavior and Ecology

The fourth and last of the AP Biology Big Ideas **4** declares "Biological systems interact, and these systems and their interactions possess complex properties." You have seen this principle apply to the cells, tissues, and organs that work together to make efficient and vital systems, and with the body systems that interact to create a successfully functioning organism. But one of the most beautiful and intricate applications of this idea is seen in the workings within the vibrant communities of Earth's ecosystems, as chunks of the Big Ideas Framework suggest:

 The stability of an environment effects the rate of evolution within it.

 Matter cycles while energy flows in all the ecosystems of the world.

 Communication of many types assists organisms in their quest for survival.

 Limiting factors and environmental changes affect the living communities of every ecosystem.

Plants and animals are locked in an eternal partnership of survival. As autotrophs, the plants capture the free energy of sunlight to power the production of food molecules. As heterotrophs, the animals must depend upon those food stores for their survival as they lead herbivorous lives or take advantage of other animals by carnivory. Bacteria, protists, and fungi all intertwine themselves into the life history of their larger cousins, serving as symbiotes, parasites, composters, nitrogen-fixers, and companions in the struggle for existence. When one part of this diverse and intricate web is changed—a tree falls, a species dies out, a grassland is burned—new associations must be made and life styles may change. "Adapt, migrate, or die" is the rule. With luck, the change simply shifts elements and the ecosystem continues on; without luck, the change may so damage the living community that the entire structure collapses.

Populations are the basis of the ecological systems of the world. These groups with their specific niches provide the puzzle pieces that create the big picture called the community. There are producers, consumers, and recyclers. Prey are important for the nutrition of their predators; predators are important for maintaining the health and well-being of their prey population. Photosynthesizers produce food and release oxygen gas. But oxygen is essential to almost every living organism for its cellular respiration, and by using it they produce the carbon dioxide upon which the photosynthesizers build their food molecules. In the background you might be hearing Elton John singing the "Circle of Life," but the truth remains; all life is connected and interactive in untold ways, and those organisms and their relationships have evolved over millennia to produce the beautiful and resilient tapestry that we call the biosphere.

UNIT OUTLINE

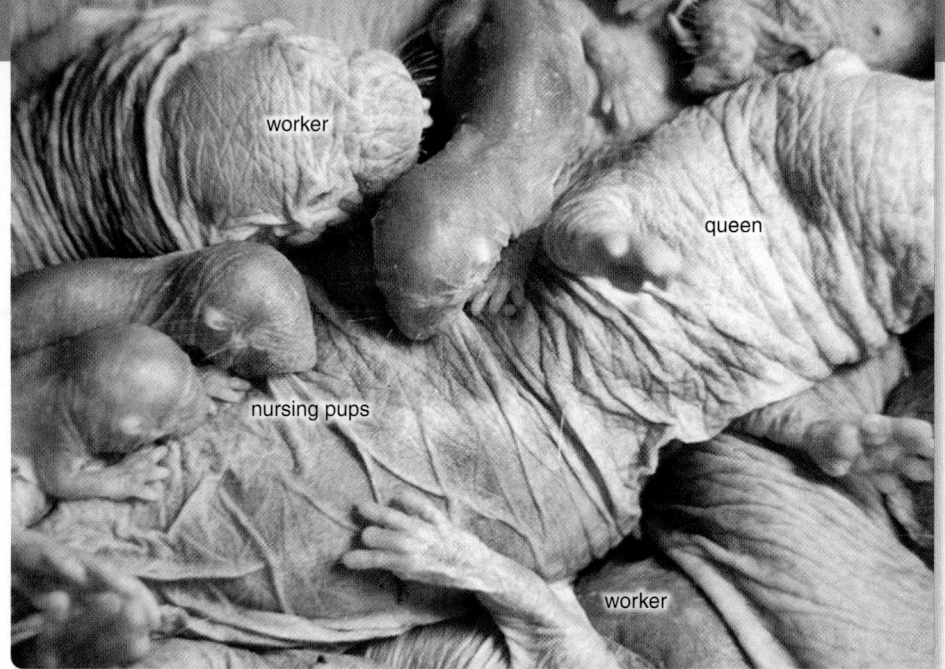

worker

queen

nursing pups

worker

Among naked mole rats, *Heterocephalus glaber*, workers help the queen reproduce.

Behavioral Ecology

Naked mole rats spend almost their entire lives underground and are extremely social in the same manner as ants, which also have underground chambers. Like the ants, mole rats have a rigid social hierarchy. At the top is the queen, the only female who reproduces. At the bottom are workers of both sexes, who do not reproduce, but instead work tirelessly to excavate long intricate tunnels and to help the queen reproduce.

Behaviorists ask: Why are mole rats social? And they can give a two-part answer. Mole rats have to dig underground tunnels in order to obtain the roots and tubers that they rely on for food and moisture. Without efficient tunnel digging, a mole rat can't survive, and it takes several mole rats working together to efficiently dig a tunnel. The second part of the answer has to do with their common genes. The queen keeps a harem of very closely related males with whom she mates. The result of this inbreeding is that all members of the colony share about 80% of the same genes. Members of the group largely see to the propagation of their own genes when they help the queen reproduce.

Behavioral ecology, as discussed in this chapter, is dedicated to the principle that natural selection shapes behavior just as it does the anatomy and physiology of an animal.

As you read through the chapter, think about the following questions:

1. What are the benefits and drawbacks of living in a social group?
2. How do genetics and the environment work together to influence behavior?
3. Why are there so few mammal species that exhibit altruism?

CHAPTER OUTLINE

43.1 Inheritance Influences Behavior 820

43.2 The Environment Influences Behavior 822

43.3 Animal Communication 826

43.4 Behaviors That Increase Fitness 830

BEFORE YOU BEGIN

Before beginning this chapter, take a few moments to review the following discussions.

Section 13.3 How can the environment influence gene expression?

Section 16.2 What role does sexual selection and male competition play in the evolution of the species?

Section 40.2 What role does the endocrine system play in controlling behavior?

FOLLOWING *the* BIG IDEAS

CHAPTER 43   BEHAVIORAL ECOLOGY

Energy and Homeostasis	Both innate and learned behaviors allow all organisms to adjust to individual and environmental changes.
Information and Signaling	External communication of all types allows organisms to function successfully in their community.

43.1 Inheritance Influences Behavior

Learning Outcomes

Upon completion of this section, you should be able to

1. Explain the key aspects of studies that suggest behavior has a genetic basis.
2. Describe the body systems that play a role in influencing behavior.

Behavior encompasses any action that can be observed and described. For example, in Chapter 1, Figure 1.15 described the aggressive behavior of male mountain bluebirds toward other males during mating. An investigator was able to center in on this behavior, observe it, and record it objectively. In the same manner, we wish to pose the question of whether genetics can determine the behavior an animal is capable of performing.

The "nature versus nurture" question asks to what extent our genes (nature) and the environment (nurture) influence behavior. We would expect that genes, which control the development of neural and hormonal mechanisms, also influence the behavior of an animal. The results of experiments done to discover the degree to which genetics controls behavior support the hypothesis that most behaviors have, at least in part, a genetic basis.

Experiments That Suggest Behavior Has a Genetic Basis

Among the many animal behavior studies, we summarize two that suggest that behavior has a genetic basis as well as a type of study in humans that attempts to evaluate nature versus nurture.

Nest-Building Behavior in Lovebirds

Lovebirds are small, green-and-pink African parrots that nest in tree hollows. Several closely related species of lovebirds in the genus *Agapornis* build their nests differently. Fischer's lovebirds cut large leaves (or in the laboratory, pieces of paper) into long strips with their bills. They use their bills to carry the strips (Fig. 43.1a) to the nest, where they weave them in with others to make a deep cup. In contrast, peach-faced lovebirds cut somewhat shorter strips and they carry them to the nest in a very unusual manner. They pick up the strips in their bills and then insert them into their feathers (Fig. 43.1b). In this way, they can carry several of these short strips with each trip to the nest, while Fischer lovebirds can carry only one of the longer strips at a time.

Researchers hypothesized that if the behavior for obtaining and carrying nesting material is inherited, then hybrids might show intermediate behavior. When the two species of birds were mated, the hybrid birds were observed to have difficulty carrying nesting materials. They cut strips of intermediate length and then attempted to tuck the strips into their rump feathers. They did not push the strips far enough into the feathers, however, and when they walked or flew, the strips always came out.

a. Fischer lovebird with nesting material in its beak.

b. Peach-faced lovebird with nesting material in its rump feathers.

Figure 43.1 Nest building behavior in lovebirds. **a.** Fischer lovebirds, *Agapornis fischeri*, carry strips of nesting material in the bill, as do most other birds. **b.** Peach-faced lovebirds, *Agapornis roseicollis*, tuck strips of nesting material into their rump feathers before flying back to the nest.

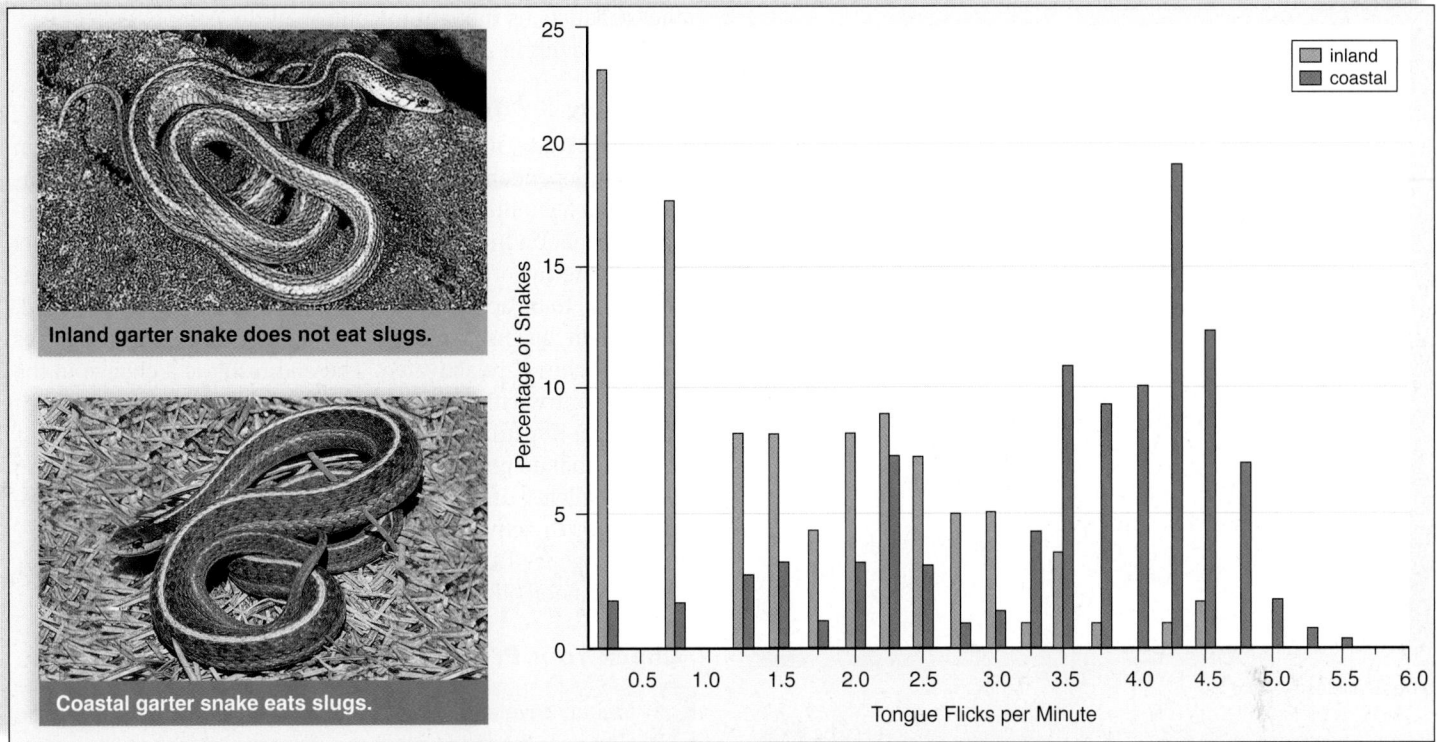

Figure 43.2 Feeding behavior in garter snakes. The number of tongue flicks by inland and coastal garter snakes, *Thamnophis elegans,* is measured in terms of their response to slug extract on cotton swabs. Coastal snakes tongue-flicked more than inland snakes.

Hybrid birds eventually learned, after about three years, to carry the cut strips in their beak, but they still briefly turned their head toward their rump before flying off. These studies support the hypothesis that behavior has a genetic basis.[1]

Food Choice in Garter Snakes

A variety of experiments have been conducted to determine if food preference in garter snakes has a genetic basis. There are two different types of garter snake populations in California. Inland populations are aquatic and commonly feed underwater on frogs and fish. Coastal populations are terrestrial and feed mainly on slugs. In the laboratory, inland adult snakes refused to eat slugs, while coastal snakes readily did so. The experimental results of matings between snakes from the two populations (inland and coastal) show their newborns have an overall intermediate incidence of slug acceptance.

Differences between slug acceptors and slug rejecters appear to be inherited, but what physiological difference is there between the two populations? A clever experiment answered this question. When snakes eat, their tongues carry chemicals to an odor receptor in the roof of the mouth. They use tongue flicks to recognize their prey. Even newborns will flick their tongues at cotton swabs dipped in fluids of their prey. Swabs were dipped in slug extract, and the number of tongue flicks were counted for newborn inland snakes and newborn coastal snakes. Coastal snakes had a higher number of tongue flicks than did inland snakes (Fig. 43.2).

Apparently, inland snakes do not eat slugs because they cannot detect their smell as easily as the coastal snakes. A genetic difference between the two populations of snakes results in a physiological difference in their nervous systems. Although hybrids showed a great deal of variation in the number of tongue flicks, they were generally intermediate, as predicted by the genetic hypothesis.[2]

Twin Studies in Humans

On occasion, human twins have been separated at birth and raised under different environmental conditions. Studies of these twins have shown that they have similar food preferences, activity patterns, and even select mates with similar characteristics. This type of study lends support to the hypothesis that at least certain behaviors are primarily influenced by nature (i.e., the genes).

Experiments That Demonstrate Behavior Has a Genetic Basis

The nervous and endocrine systems are both responsible for the overall coordination of body systems. Studies have shown that the endocrine system plays a role in determining behavior.

Egg-Laying Behavior in Marine Snails

The egg-laying behavior in the marine snail *Aplysia* involves a specific sequence of movements. Following copulation, the

1 Dilger, W. 1962. The behavior of lovebirds. *Scientific American* 206: 89–98.

2 Arnold, S. J. 1981. Behavioral variation in natural populations. I. *Evolution* 35: 489–509.

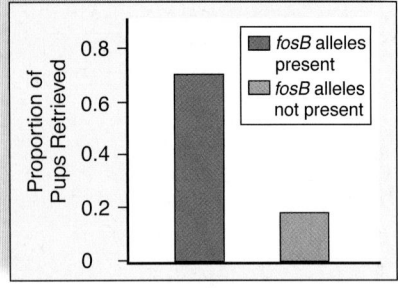

a.

b. *fosB* alleles present.

c. *fosB* alleles not present.

Figure 43.3 Maternal care in mice. **a.** A mother mouse with *fosB* alleles spends time retrieving and crouching over its young, whereas mice that lack these alleles do not display these maternal behaviors. **b.** This typical mother mouse retrieves her young and crouches over them. **c.** This mouse, lacking *fosB*, does not retrieve her young and does not crouch over them.

animal extrudes long strings of more than a million egg cases. The animal takes the egg case string in its mouth, covers it with mucus, waves its head back and forth to wind the string into an irregular mass, and attaches the mass to a solid object, such as a rock.

Several years ago, scientists isolated and analyzed an egg-laying hormone (ELH) that causes the snail to lay eggs, even if it has not mated. ELH was found to be a small protein of 36 amino acids that diffuses into the circulatory system and excites the smooth muscle cells of the reproductive duct, causing them to contract and expel the egg string. Using recombinant DNA techniques, the investigators isolated the ELH gene. The gene's product turned out to be a protein with 271 amino acids. The protein can be cleaved into as many as 11 possible products, and ELH is one of these. ELH alone, or in conjunction with these

other products, is thought to control all the components of egg-laying behavior in *Aplysia*.[3]

Nurturing Behavior in Mice

In another study, investigators found that maternal behavior in mice was dependent on the presence of a gene called *fosB*. Normally, when mothers first inspect their newborn, various sensory information from their eyes, ears, nose, and touch receptors travel to the hypothalamus. This incoming information causes *fosB* alleles to be activated, and a particular protein is produced. The protein begins a process during which cellular enzymes and other genes are activated. The end result is a change in the neural circuitry within the hypothalamus, which manifests itself in maternal nurturing behavior toward the young.

Mice that do not engage in nurturing behavior were found to lack *fosB* alleles, and the hypothalamus fails to make any of the products or to activate any of the enzymes and other genes that lead to maternal nurturing behavior (Fig. 43.3).[4]

Video **Sex and the Senses**

Check Your Progress 43.1

1. Identify several genes that have been linked to behavior.
2. Compare the studies that show how behavior has a genetic basis.

43.2 The Environment Influences Behavior

Learning Outcomes

Upon completion of this section, you should be able to

1. Choose the correct sequence of events required for classical and operant conditioning.
2. Identify examples of how the environment influences behavior.

Even though genetic inheritance serves as a basis for behavior, it is possible that environmental influences (nurture) also affect behavior. For example, behaviorists originally believed that some behaviors were **fixed action patterns (FAP),** that is, an unchanging behavioral response. An FAP is elicited by a **sign stimulus**—a particular trigger in the environment. For example, male stickleback fish aggressively defend a territory against other males. In laboratory studies, the male reacts more aggressively to any model that has a red belly like he has, rather than to a model that otherwise looks like a stickleback fish. In

3 Scheller, R. H., Jackson, J. F., et al. 1982. A family of genes that codes for ELH, a neuropeptide eliciting a stereotyped pattern of behavior in *Aplysia. Cell* 28: 707–19.

4 Brown, J. R., Ye, H., et al. 1996. A defect in nurturing in mice lacking the immediate early gene *fosB. Cell* 86: 297–309.

this instance, the color red is a sign stimulus that triggers the aggressive behavior.

male stickleback

Investigators discovered that many behaviors that were originally thought to be FAPs could improve with practice due to the organism's learning. In this context, **learning** is defined as a durable change in behavior brought about by experience. Habituation is a form of learning because, with experience, an animal no longer responds to a particular stimulus. Deer grazing on the side of a busy highway, ignoring traffic, is an example of habituation.

Instinct and Learning

Laughing gull chicks' begging behavior appears to be a FAP, because it is always performed the same way in response to the parent's red bill (the sign stimulus). A chick directs a pecking motion toward the parent's bill, grasps it, and strokes it downward (Fig. 43.4). Parents can bring about the begging behavior by swinging their bill gently from side to side. After the chick responds, the parent regurgitates food onto the floor of the nest. If need be, the parent then encourages the chick to eat. This interaction between the chicks and their parents suggests that the begging behavior involves learning.

To test this hypothesis, diagrammatic pictures of gull heads were painted on small cards, and then eggs were collected from nests in the field. The eggs were hatched in a dark incubator to eliminate visual stimuli before the test. On the day of hatching, each chick was allowed to make about a dozen pecks at the model. The chicks were returned to the nest, and then each was retested.

The tests showed that on average, only one-third of the pecks by a newly hatched chick strike the model. But one day after hatching, more than half of the pecks are accurate, and two days after hatching, the accuracy reaches a level of more than 75%. Investigators concluded that improvement in motor skills and learning can help improve the development of instinctive behaviors.[5]

Imprinting

Imprinting is considered a form of learning. Konrad Lorenz was one of the first individuals to study imprinting in birds. He observed chicks, ducklings, and goslings following the first moving object they saw after hatching. This object is ordinarily their mother. Imprinting in the wild has survival value and leads to reproductive success. This behavior enables an individual to recognize their own species and, therefore, eventually find an appropriate mate.

But in the laboratory, investigators found that birds can seemingly be imprinted on any object—a human or a red ball—if it is the first moving object they see during a sensitive period of two

5 Hailman, Jack P. 1967. The ontogeny of an instinct. *Behavior* (Suppl. 13). (E. J. Brill Academic Publishers, Leiden, Germany.)

Figure 43.4 Pecking behavior in laughing gulls. At about three days, a laughing gull chick grasps the red bill of a parent, stroking it downward, and the parent then regurgitates food. *Right:* The top diagrams show the accuracy of a chick when striking a test probe, painted red. The bottom diagram shows chick-pecking accuracy graphically. Note from these diagrams that a chick markedly improves its ability (within only two days) to peck a bill, a behavior that normally causes a parent to regurgitate food.

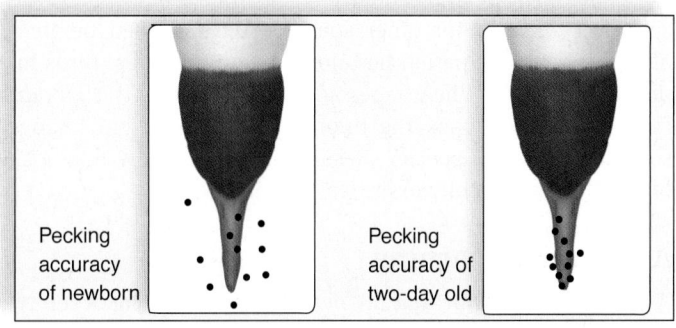

Pecking accuracy of newborn

Pecking accuracy of two-day old

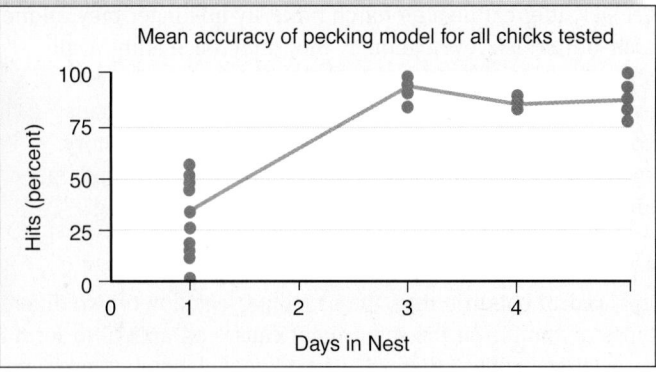

Mean accuracy of pecking model for all chicks tested

to three days after hatching. The term *sensitive period* means the period of time in which a particular behavior develops. A chick imprinted on a red ball follows it around and chirps whenever the ball is moved out of sight.

Social interactions between parent and offspring during the sensitive period seem key to normal imprinting. For example, female mallards cluck during the entire time imprinting is occurring. It could be that vocalization before and after hatching is necessary to normal imprinting.

Goslings imprinted on an investigator

Social Interactions and Learning

White-crowned sparrows sing a species-specific song, but males of a particular region have their own dialect. Birds were caged in order to test the hypothesis that young white-crowned sparrows learn how to sing from older members of their species.

Three groups of birds were tested. Birds in the first group *heard no songs at all.* When grown, these birds sang a song that had a slight resemblance to the adult song. Birds in the second group *heard tapes of white-crowns singing.* When grown, they sang in that dialect, as long as the tapes had been played during a sensitive period from about age 10–50 days. White-crowned sparrows' dialects (or other species' songs) played before or after this sensitive period had no effect on the birds. Birds in a third group did not hear tapes and instead were *given an adult tutor.* No matter when the tutoring began these birds sang a song of a different species. Results like these show that social interactions apparently assist learning in birds.

Associative Learning

A change in behavior that involves an association between two events is termed **associative learning.** For example, birds that get sick after eating a monarch butterfly no longer prey on monarch butterflies, even though they may be readily available. Or the smell of fresh baked bread may entice you, even though you may have just eaten. If so, perhaps you associate the taste of bread with a pleasant memory, such as being at home. Both classical conditioning and operant conditioning are examples of associative learning.

Classical Conditioning

In **classical conditioning,** the paired presentation of two different types of stimuli (at the same time) causes an animal to form an association between them. The best-known laboratory example of classical conditioning is that of an experiment done by the Russian

psychologist Ivan Pavlov. First, Pavlov observed that dogs salivate when presented with food. Then, he rang a bell whenever the dogs were fed. Eventually, the dogs would salivate whenever the bell was rung, regardless of whether food was present (Fig. 43.5).

Classical conditioning suggests that an organism can be trained—that is, conditioned—to associate any response to any stimulus. Unconditioned responses are those that occur naturally, as when salivation follows the presentation of food. Conditioned responses are those that are learned, as when a dog learns to salivate when it hears a bell.

Advertisements attempt to use classical conditioning to sell products. Why do commercials pair attractive people with a product being advertised? The hope is that viewers will associate attractiveness with the product. This pleasant association may cause them to buy the product.

Some types of classical conditioning can be helpful. For example, it's been suggested that you hold a child on your lap when reading to them. Why? Because the child will associate a pleasant feeling with reading.

Operant Conditioning

During **operant conditioning,** a stimulus-response connection is strengthened. Most people know that it is helpful to give an animal a reward, such as food or affection, when teaching it a trick. When we go to an animal show, it is quite obvious that trainers use operant conditioning. They present a stimulus, say, a hoop, and then give a reward (food) for the proper response (jumping through the hoop). Sometimes the reward does not

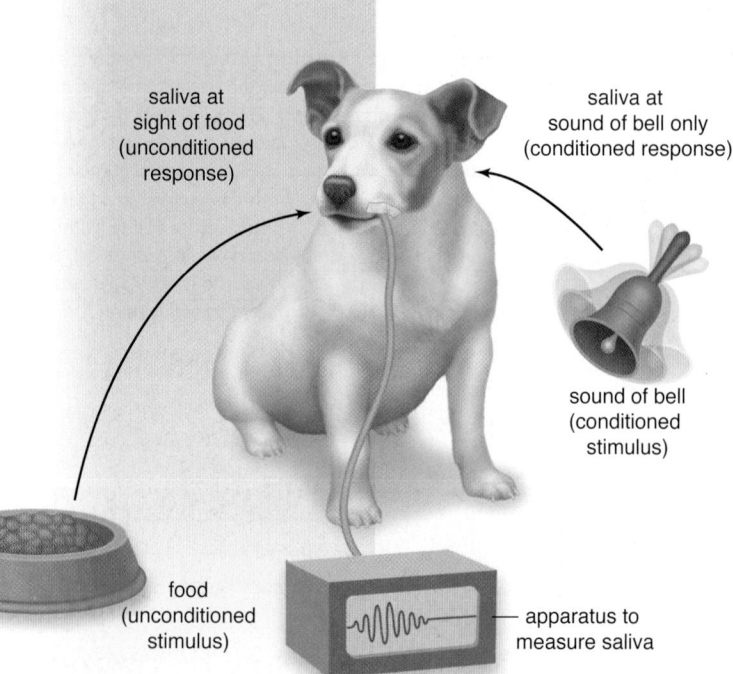
saliva at sight of food (unconditioned response)

saliva at sound of bell only (conditioned response)

sound of bell (conditioned stimulus)

food (unconditioned stimulus)

apparatus to measure saliva

Figure 43.5 Classical conditioning.   Ivan Pavlov discovered classical conditioning by performing this experiment with dogs. A bell is rung when a dog is given food. Salivation is measured. Eventually, the dog salivates when the bell is rung, even though no food is presented. Food is an unconditioned stimulus, and the sound of the bell is a conditioned stimulus that brings about the response—that is, salivation.

need to be immediate. In latent operant conditioning, an animal makes an association without the immediate reward, as when squirrels make a mental map of where they have hidden nuts.

B. F. Skinner was well known for studying this type of learning in the laboratory. In the simplest type of experiment performed by Skinner, a caged rat happens to press a lever and is rewarded with sugar pellets, which it avidly consumes. Thereafter, the rat regularly presses the lever whenever it wants a sugar pellet. In more sophisticated experiments, Skinner even taught pigeons to play Ping-Pong by reinforcing desired responses to stimuli.

In child rearing, it's been suggested that parents who give a positive reinforcement for good behavior will be more successful than parents who punish behaviors they believe are undesirable.

Orientation and Migratory Behavior

Migration is long-distance travel from one location to another. Loggerhead sea turtles hatch on a Florida beach and then travel across the Atlantic Ocean to the Mediterranean Sea, which offers an abundance of food. After several years, pregnant females return to the same beaches to lay their eggs. Every year, monarch butterflies fly from North America to Mexico where they overwinter, and then return in the spring to breed.

At the very least, migration requires **orientation,** the ability to travel in a particular direction, such as south in the winter and north in the spring. Most of the work regarding orientation has been done in birds. Many birds can use the Sun during the day or the stars at night to orient themselves. The Sun moves across the sky during the day, but the birds are able to compensate for this because they have a sense of time. They are presumed to have a biological clock that allows them to know where the Sun will be in relation to the direction they should be going any time of the day.

Experienced birds can also **navigate,** which is the ability to change direction in response to environmental clues. These clues are apt to come from the Earth's magnetic field. Figure 43.6 shows a study that was done with starlings, which typically migrate from the Baltics to Great Britain and return. Test starlings were captured in Holland and transported to Switzerland. Experienced birds corrected their flight pattern and still got to Great Britain. Young, inexperienced birds ended up in Spain instead of Great Britain!

Migratory behavior has a proximate cause and an ultimate cause. The proximate cause is due to environmental stimuli that tell the birds it is time to travel. The ultimate cause is due to the possibility of reaching a more favorable environment for survival and reproduction. Is the benefit worth the cost—the dangers of the journey? It must be, or else the behavior would not persist.

Cognitive Learning

In addition to the modes of learning already discussed, animals may learn through observation, imitation, and insight. For example,

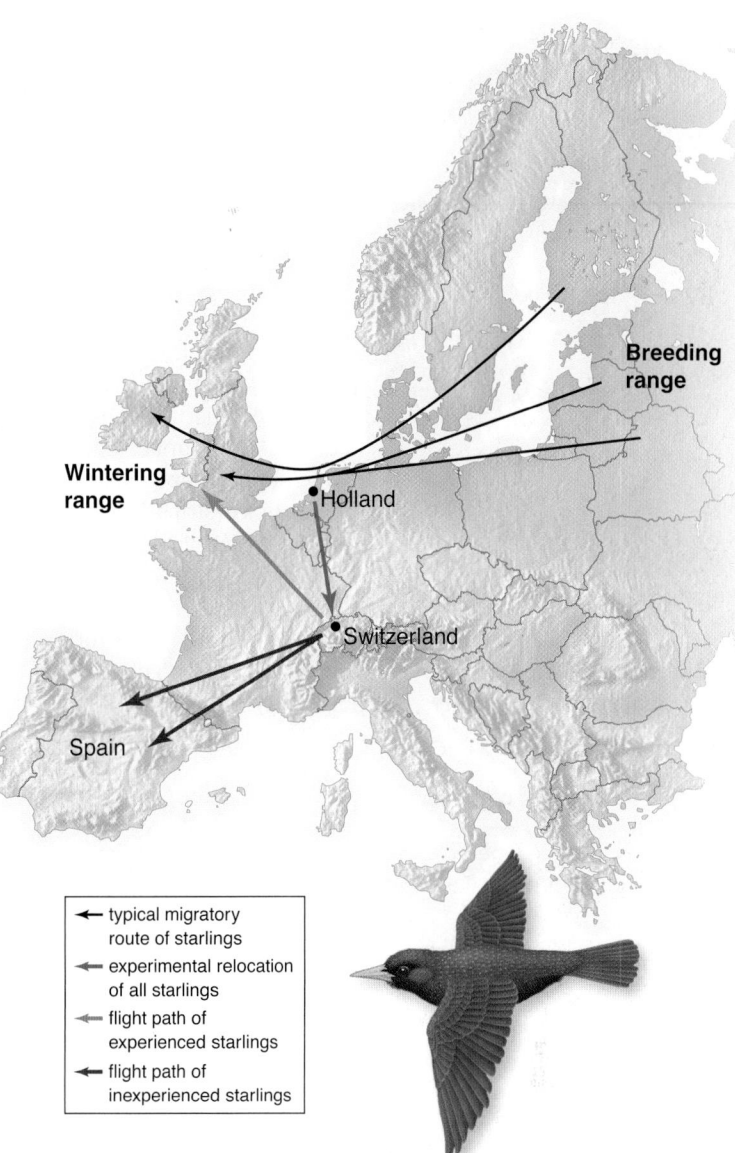

Figure 43.6 Starling migratory experiment. Starlings, *Sturnus vulgaris,* on their way from the Baltics to Great Britain were captured and released in Switzerland. Inexperienced birds kept flying in the same direction and ended up in Spain. Experienced birds had learned to navigate, as witnessed by the fact that they still arrived in Great Britain.

Japanese macaques learn to wash sweet potatoes before eating them by imitating others.

Insight learning occurs when an animal suddenly solves a problem without any prior experience with the situation. The animal appears to call upon prior experience with other circumstances to solve the problem. For example, chimpanzees have been observed stacking boxes to reach bananas in laboratory settings.

Other animals too, aside from primates, seem to be able to reason things out. In one experiment, ravens were offered meat that was attached to string hanging from a branch in a confined aviary. The ravens were accustomed to eating meat, but had no knowledge of how strings work. It took several hours, but eventually one raven flew to the branch, reached down, grabbed

Ravens learn to retrieve food.

the string with its beak, and pulled the string up over and over again, each time securing the string with its foot. Eventually, the meat was within reach, and the raven was able to grab the meat with its beak. Other ravens were then also able to perform this behavior.

Can animals also plan ahead? It seems so. A sea otter saves a particular rock to act as a hard surface against which to bash open clams. A chimpanzee strips leaves off a twig, which it then uses to secure termites from a termite nest.

If animals can think, do they have emotions? This too is an unexplored area that is now an area of interest. The Nature of Science feature explores the possibility that animals have emotions.

Check Your Progress 43.2

1. Describe the type of learning that occurs when an animal no longer tries to eat bumblebees after being stung by one.
2. Give an example that shows how instinct and learning may interact as behavior develops.
3. Discuss evidence that shows that animals have cognitive abilities.

Figure 43.7 Use of a pheromone. This male cheetah is spraying urine onto a tree to mark its territory.

43.3 Animal Communication

Learning Outcomes

Upon completion of this section, you should be able to

1. Recognize the various ways that animals can communicate.
2. Describe the advantages and disadvantages of chemical, auditory, visual, and tactile communication.

Animals exhibit a wide diversity of social behaviors. Some animals are largely solitary and join with a member of the opposite sex only for the purpose of reproduction. Others will find a mate and cooperate in raising offspring. Some form **societies** in which members of a species are organized in a cooperative manner, extending beyond sexual and parental behavior. We have already mentioned the social groups of naked mole rats, baboons, and red deer (see Chapter 16). Social behavior in these and other animals requires that they communicate with one another.

Communicative Behavior

Communication is an action by a sender that may influence the behavior of a receiver. The communication can be purposeful, but it does not have to be. Bats send out a series of sound pulses and listen for the corresponding echoes to find their way through dark caves and locate food at night. Some moths have an ability to hear these sound pulses, and they begin evasive tactics when they sense that a bat is near. Are the bats purposefully communicating with the moths? No, bat sounds are simply a cue to the moths that danger is near.

Chemical Communication

Chemical signals have the advantage of being effective both night and day. The term **pheromone** [Gk. *phero*, bear, carry, and *monos*, alone] designates a chemical signal in low concentration that is passed between members of the same species. Some animals are capable of secreting different pheromones, each with a different meaning. Female moths secrete chemicals from special abdominal glands, which are detected downwind by receptors on male antennae. The antennae are especially sensitive, and this ensures that only male moths of the correct species (and not predators) will be able to detect them.

Ants and termites mark their trails with pheromones. Cheetahs and other cats mark their territories by depositing urine, feces, and anal gland secretions at the boundaries (Fig. 43.7). Klipspringers (small antelope) use secretions from a gland below the eye to mark twigs and grasses of their territory. Pheromones are known to control the behavior of social insects, as when workers slavishly care for the offspring produced by a queen.

To what degree do pheromones, along with hormones, affect the behavior of animals? Are they responsible for determining whether the animal will carry out parental care, become aggressive, or engage in courtship behavior? Some researchers maintain that human behavior is influenced by pheromones

Nature of Science

Do Animals Have Emotions?

In recent years, investigators have become interested in determining whether animals have emotions. The body language of animals can be interpreted to suggest that they have feelings. When wolves reunite, they wag their tails to and fro, whine, and jump up and down; elephants vocalize—emit their "greeting rumble"—flay their ears, and spin about. Many young animals play with one another or even by themselves, as when dogs chase their own tails. On the other end of the spectrum, upon the death of a friend or parent, chimps are apt to sulk, stop eating, and even die.

It seems reasonable to hypothesize that animals are "happy" when they reunite, "enjoy" themselves when they play, and are "depressed" over the loss of a close friend or relative. Even people who rarely observe animals usually agree about what an animal must be feeling when it exhibits certain behaviors (Fig. 43A).

In the past, scientists found it expedient to collect data only about observable behavior and to ignore the possible mental state of the animal. Why? Because emotions are personal, and no one can ever know exactly how another animal is feeling. B. F. Skinner, whose research method is described earlier in this chapter, regarded animals as robots that become conditioned to respond automatically to a particular stimulus. He and others never considered that animals might have feelings. But now, some scientists believe they have sufficient data to suggest that at least other vertebrates and/or mammals do have feelings, including fear, joy, embarrassment, jealousy, anger, love, sadness, and grief. And they believe that those who hypothesize otherwise should have to present the opposing data.

Perhaps it would be reasonable to consider the suggestion of Charles Darwin, who said that animals are different in degree rather than in kind. This means that animals can feel love but perhaps not to the degree that humans can.

Researcher Bendt Würsig watched the courtship of two baleen whales. They touched, caressed, rolled side-by-side, and eventually swam off together. He wondered if their behavior indicated they felt love for one another. When you think about it, it is unlikely that emotions first appeared in humans with no evolutionary homologies in animals.

Iguanas, but not fish and frogs, tend to stay where it is warm. Canadian biologist Michel Cabanac has found that warmth makes iguanas experience a rise in body temperature and an increase in heart rate. These are biological responses associated with emotions in humans. Perhaps the ability of animals to feel pleasure and displeasure is a mental state that rises to the level of consciousness.

Neurobiological data support the hypothesis that other animals, aside from humans, are capable of enjoying themselves when they perform an activity such as play. Researchers have found a high level of dopamine in the brain of rats when they play, and the dopamine level increases simply when rats anticipate the opportunity to play. Certainly even the staunchest critic is aware that many different species of animals have limbic systems and are capable of fight-or-flight responses to dangerous situations. Can we go further and suggest that animals feel fear even when no observable response has yet occurred?

Laboratory animals may be too stressed to provide convincing data on emotions; we have to consider that emotions evolved under an animal's normal environmental conditions. This makes field research more useful. It is possible to fit animals with devices that transmit information on heart rate, body temperature, and eye movements as they go about their daily routine. Such information would help researchers learn how animal emotions might correlate with their behavior, just as emotions influence human behavior. One possible definition describes emotion as a psychological phenomenon that helps animals direct and manage their behavior.

Marc Bekoff, who is prominent in the field of animal behavior, encourages us to be open to the possibility that animals have emotions. He states:

> By remaining open to the idea that many animals have rich emotional lives, even if we are wrong in some cases, little truly is lost. By closing the door on the possibility that many animals have rich emotional lives, even if they are very different from our own or from those of animals with whom we are most familiar, we will lose great opportunities to learn about the lives of animals with whom we share this wondrous planet.[1]

Questions to Consider

1. If it can be shown that animals have emotions, should animal testing be allowed to continue?
2. Do you think invertebrate animals have emotions? If so, what emotions might they experience?

1 Bekoff, M. 2000. Animal emotions: Exploring passionate natures. *Bioscience* 50: 869.

Is the snowshoe hare afraid?

Is the young chimp comforted?

Figure 43A Emotions in animals.

a.

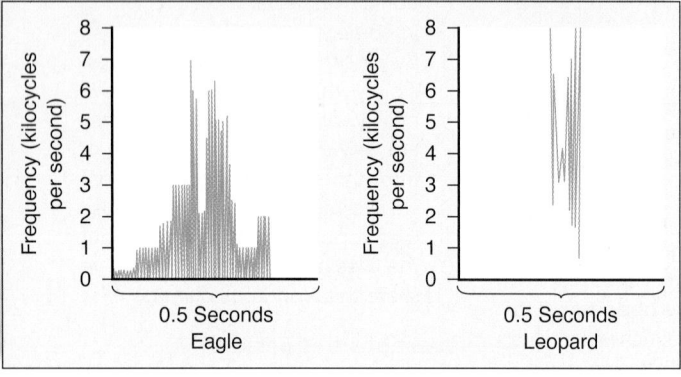

b.

Figure 43.8 Auditory communication. **a.** Vervet monkeys, *Cercopithecus aethiops*, are responding to an alarm call. Vervet monkeys can give different alarm calls according to whether a troop member sights an eagle or a leopard, for example. **b.** The frequency per second of the sound differs for each type call.

Figure 43.9 Male baboon displaying full threat.

wafting through the air—pheromones that are not perceived consciously. They have discovered that like the mouse, humans have an organ in the nose, called the vomeronasal organ (VNO), that can detect not only odors, but also pheromones. The neurons from this organ lead to the hypothalamus, the part of the brain that controls the release of many hormones in the body.

Video Bug Speak

Auditory Communication

Auditory communication (i.e., via sound) has some advantages over other kinds of communication. It is faster than chemical communication, and unlike visual communication, it is effective both night and day. Further, auditory communication can be modified not only by loudness but also by pattern, duration, and repetition. In an experiment with rats, a researcher discovered that an intruder can avoid attack by increasing the frequency with which it makes an appeasement sound.

Male crickets have calls, and male birds have songs for a number of different occasions. For example, birds may have one song for distress, another for courting, and still another for marking territories. Sailors have long heard the songs of humpback whales transmitted through the hull of a ship. But only recently has it been shown that the song has six basic themes, each with its own phrases, that can vary in length and be interspersed with sundry cries and chirps. The purpose of the song is probably sexual, serving to advertise the availability of the singer. Bottlenose dolphins are known to have one of the most complex languages in the animal kingdom.

Language is the ultimate auditory communication. Humans can produce a large number of different sounds and put them together in many different ways. Nonhuman primates have different vocalizations, each having a definite meaning, such as when vervet monkeys give alarm calls (Fig. 43.8). Although chimpanzees can be taught to use an artificial language, they do not progress beyond the capability level of a two-year-old human child. It has also been difficult to prove that chimps understand the concept of grammar or can use their language to reason. It still seems as though humans possess a communication ability greater than that of other animals.

Video Meerkat Warning Calls

Visual Communication

Visual signals are most often used by species that are active during the day. Contests between males often make use of threat postures that possibly prevent outright fighting. A male baboon displaying full threat is an awesome sight that establishes his dominance and keeps peace within the baboon troop (Fig. 43.9). Hippopotamuses perform territorial displays that include mouth opening.

Many animals use complex courtship behaviors and displays. The plumage of a male Raggiana Bird of Paradise allows him to put on a spectacular courtship dance to attract a female, giving her a basis on which to select a mate. Defense and courtship displays are exaggerated and always performed in the same way so that their meaning is clear. Fireflies use a flash pattern to signal females of the same species (Fig. 43.10).

Video Flirting Flies

Visual communication allows animals to signal others of their intentions without the need to provide any auditory or chemical messages. The body language of students during a lecture provides an example. Students who are leaning forward in their seats and making eye contact with the instructor appear interested and engaged with the material. Students who are leaning back in their chairs and gazing around the room or doodling are indicating that they have lost interest. Instructors can use students' body language to determine whether they are effectively presenting the material and make changes accordingly.

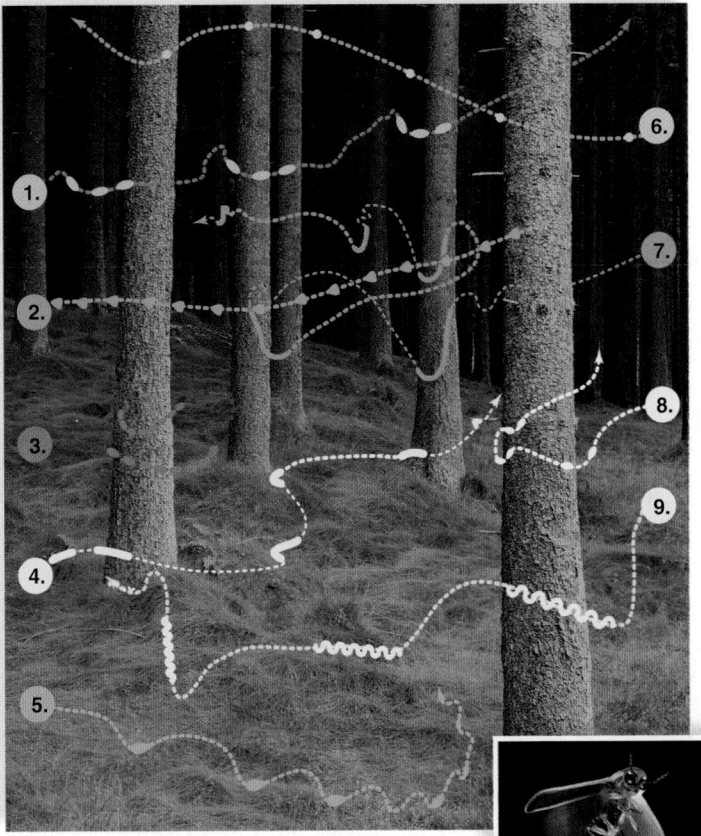

Figure 43.10 Fireflies use visual communication. Each number represents the male flash pattern of a different species. The patterns are a behavorial reproductive isolation mechanism.

The hairstyle and dress of a person, or the way he or she walks and talks, are also ways to send messages. Some studies have suggested that women tend to dress in a more sexually appealing manner when they are ovulating.[6] People who dress in black, move slowly, fail to make eye contact, and sit alone may be telling others that they are unhappy, or simply that they do not want to be socially engaged. Psychologists have long tried to understand how visual clues can be used to better understand human emotions and behavior. Similarly, body language in animals is being evaluated by researchers to determine whether they, as well as humans, have emotions.

Tactile Communication

Tactile communication occurs when one animal touches another. For example, laughing gull chicks peck at the parent's bill to induce the parent to feed them (see Fig. 43.4). A male leopard nuzzles the female's neck to calm her and to stimulate her willingness to mate. In primates, grooming—one animal cleaning the coat and skin of another—helps cement social bonds within a group.

Honeybees use a combination of communication methods, but especially tactile ones, to impart information about the environment. When a foraging bee returns to the hive, it performs a "waggle dance" that indicates the distance and the direction of a food source (Fig. 43.11). As the bee moves between the two

6 Haselton, M. G., et al. 2007. Ovulation and human female ornamentation: Near ovulation, women dress to impress. *Hormones and Behavior* 51: 41–5.

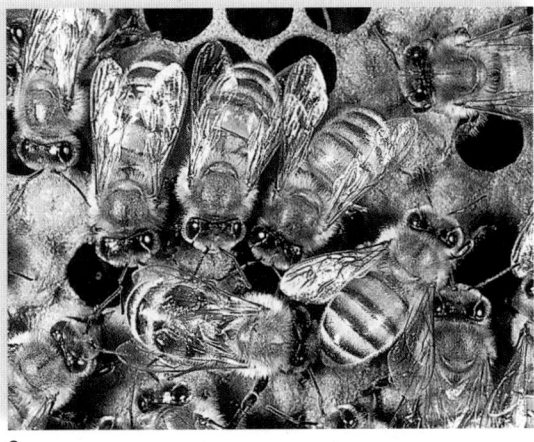

a.

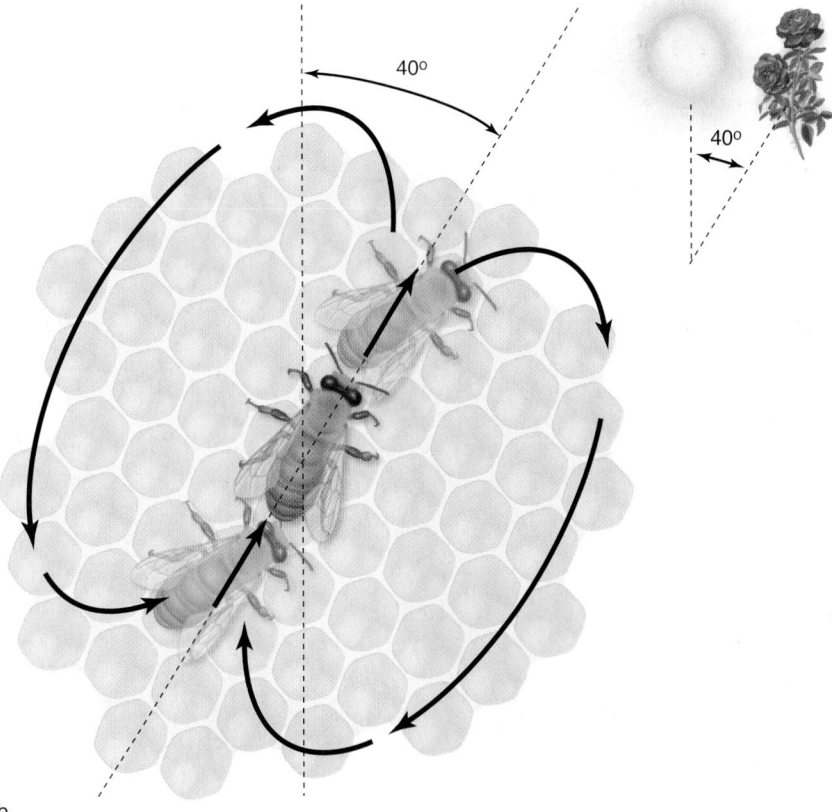

40°

40°

b.

Figure 43.11 Communication among bees. **a.** Honeybees do a waggle dance to indicate the direction of food. **b.** If the dance is done outside the hive on a horizontal surface, the straight run of the dance will point to the food source. If the dance is done inside the hive on a vertical surface, the angle of the straightaway to that of the direction of gravity is the same as the angle of the food source to the Sun.

loops of a figure 8, it buzzes noisily and shakes its entire body in so-called waggles. Outside the hive, the dance is done on a horizontal surface, and the straight run indicates the direction of the food. Inside the hive, the angle of the straight run to that of the direction of gravity is the same as the angle of the food source with the Sun. In other words, a 40° angle to the left of vertical means that food is 40° to the left of the Sun.

Bees can use the Sun as a compass to locate food because they have a biological clock, which allows them to compensate for the movement of the Sun in the sky. A biological clock is an internal means of telling time. Today, we know that the timing of the clock, in both insects and mammals (including humans), requires alterations in the expression of a gene called *period*.

Check Your Progress   43.3

1. Describe examples of how communication is meant to affect the behavior of the receiver.
2. Give an advantage and disadvantage to each type of communication discussed.
3. Identify the human receptor for each type of communication discussed.

Figure 43.12 Male and female gibbons. Siamang gibbons, *Hylobates syndactylus*, are monogamous, and they both share the task of raising offspring. They also share the task of marking their territory by singing. As is often the case in monogamous relationships, the sexes are similar in appearance. Male is above and female is below.

43.4 Behaviors That Increase Fitness

Learning Outcomes

Upon completion of this section, you should be able to

1. Describe how territoriality and the foraging techniques associated with it will increase fitness.
2. Recognize various strategies that will increase an individual's fitness.

Behavioral ecology begins with the assumption that behavior is subject to natural selection. It has been shown that behavior has a genetic basis, and investigators believe that certain behaviors lead to increased survival and production of offspring. Therefore, the behavior we can observe today in a species must have adaptive value. These types of behaviors in particular have been studied for their adaptive value: territoriality, reproductive strategies, social behaviors, and altruistic behaviors.

Territoriality and Fitness

In order to gather food, animals often have a particular home range where they carry out daily activities. The portion of the range that is defended for an animal's exclusive use, in which competing members of its species are not welcome, is called its **territory.** The behavior of defending one's territory is called **territoriality.** An animal's territory may have a good food source, and it may be the area in which the animal will reproduce.

As an example, gibbons live in the tropical rain forest of South and Southeast Asia. Normally, their home range can be covered in about 3–4 days, and they are also monogamous and territorial. Territories are maintained by loud singing (Fig. 43.12). Males sing just before sunrise, and mated pairs sing duets during the morning. Males, but not females, show evidence of fighting to defend their territory in the form of broken teeth and scars. Defense of a territory has a certain cost; it takes energy to sing and fight off others.

What is the adaptive value of being territorial? Chief among the benefits of territoriality are to ensure a source of food, to have breeding opportunities, and to have a place to rear young. The territory has to be the right size for the animal. Too large a territory cannot be defended, and too small a territory may not contain enough resources.

Cheetahs require a large territory in order to hunt for their prey. As a result, they need to mark their territory in a fashion that will last for awhile. As shown in Figure 43.7, cheetahs, like many dogs, use urine to mark their territory. Hummingbirds are known to defend a very small territory because they depend on only a small patch of flowers as their food source.

Territoriality is more likely to occur during times of reproduction. Seabirds have very large home ranges consisting of hundreds of kilometers of open ocean, but when they reproduce they become fiercely territorial. Each bird has a very small territory consisting of only a small patch of beach where they place their nest.

Video
Cichlid Territoriality

Foraging for Food

Food gathering is technically called foraging for food. An animal needs to acquire a food source that will provide more energy than the effort of acquiring the food. In one study, it was shown that shore crabs prefer to eat intermediate-sized mussels if they are given equal numbers of a variety of sizes. The net energy gain is greater than if the crabs eat larger-sized mussels; the large mussels take too much energy to open per the amount of energy they provide (Fig. 43.13). The **optimal foraging model** states that it is adaptive for foraging behavior to be as energetically efficient as possible.

Even though it can be demonstrated that animals that take in more energy during foraging are more likely to produce a greater number of offspring, other factors also come into play in survival. If an animal is killed and eaten, it has no chance of producing offspring—for example, if it forages in a way that puts it into danger. Animals often face trade-offs that lead to modification of their behavior toward maximizing their success.

Reproductive Strategies and Fitness

Many primate species are **polygamous,** meaning that a single male mates with multiple females. Because of gestation and lactation, females tend to invest more energy in their offspring than do males. Under these circumstances, it is adaptive for females to be concerned with a good food source. When food sources are clumped, females congregate in small groups around the food. Because only a few females are expected to be receptive at a time, males will likely be able to defend these few from other males. Males are expected to compete with other males for the limited number of receptive females available (Fig. 43.14).

A limited number of primates are **polyandrous,** meaning that a female mates with multiple males. Tamarins are squirrel-sized New World monkeys that live in Central or South America. Tamarins live together in groups of one or more families in which one female mates with more than one male. The female normally gives birth to twins of such a large size that the fathers, and not the mother, carry it about. This may be the reason these animals are polyandrous. Polyandry also occurs when the

Figure 43.14 Hamadryas baboons.   Among Hamadryas baboons, *Papio hamadryas*, a male, which is silver-white and twice the size of a female, keeps and guards a harem of females with whom he mates exclusively.

environment does not have sufficient resources to support several young at a time.

We have already mentioned the reproductive strategy of gibbons. They are **monogamous,** which means that they pair bond, and both male and female help with the rearing of the young. Males are active fathers, frequently grooming and handling infants. Monogamy is relatively rare in primates, which includes prosimians, monkeys, and apes (only about 18% are monogamous). In primates, monogamy occurs when males have limited mating opportunities, territoriality exists, and the male is fairly certain the offspring are his. In gibbons, females are evenly distributed in the environment, which can lead to an increase in aggression between females. Investigators note that females will attack an electronic speaker when it plays female sounds in their territory.

Sexual Selection

Sexual selection is a form of natural selection that favors features that increase an animal's chances of mating. In other words, these features are adaptive in the sense that they lead to increased fitness.

Sexual selection most often occurs due to female choice, which leads to male competition. Because most females produce a limited number of eggs relative to the number of sperm produced by the male, it is adaptive for females to be choosy about their mate. If they choose a mate that passes on features to a male offspring that will cause him to be chosen by females, the parents' fitness has increased.

Whether females actually choose features that are adaptive to the environment is in question. For example, peahens are likely to choose peacocks that have the most elaborate tails. Such a fancy tail could otherwise be detrimental to the male and make him more likely to be captured by a predator. In one study, an extra ornament was attached to a father zebra finch, and the daughters of this bird underwent the process of imprinting (see Section 43.2). Now, these females were more likely to choose a mate that also had the same artificial ornament.

While females can always be sure an offspring is theirs, males do not have this certainty. However, males produce a plentiful supply of sperm. The best strategy for males to increase their fitness, therefore, is to have as many offspring as possible. Competition may be required for them to gain access to females, and ornaments such as antlers can enhance a male's

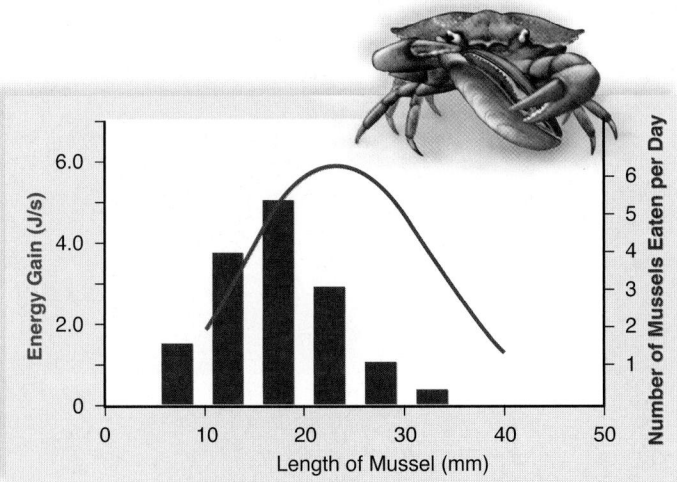

Figure 43.13 Foraging for food.   When offered a choice of an equal number of each size of mussel, the shore crab, *Carcinus maenas,* prefers the intermediate size. This size provides the highest rate of net energy return. Net energy is determined by the yield per time used in breaking open the shell.

Evolution

Sexual Selection in Male Bowerbirds

At the start of the breeding season, male bowerbirds use small sticks and twigs to build elaborate display areas called bowers (Fig. 43B). They clear the space around the bower and decorate the area with items used to attract females. After the bower is complete, the male spends most of his time near his bower, calling to females, renewing his decorations, and guarding his work against possible raids by other males.

A female may approach, and then the male begins his display. He faces her, fluffs up his feathers, and flaps his wings to the beat of a call. If the female enters the bower, the two will mate. Some males mate with up to 25 females per year. Biologists discovered that females often chose males that had well-built bowers with well-decorated platforms.

Male bowerbirds are not gaudy in appearance, but their displays are highly intense and aggressive. Their courtship displays are similar to those used by males during aggressive encounters with other males. Males must display intensely to be attractive, but males that are too intense too soon can startle females. Females may benefit from mating with the most intensely displaying males, but if they are startled, they may not be able to efficiently assess male traits.

Communication between the two sexes might maximize the potential benefits of intense male courtship displays while minimizing the potential costs. Females make crouching motions, and the degree of crouching reflects the level of display intensity she will tolerate without being startled. By giving higher-intensity displays when females increase their crouching, males could increase their courtship success by displaying intensely enough to be attractive without threatening females.

The hypothesis tested was that males respond to female crouching signals by adjusting their intensity, and that a particular male's ability to respond to female signals is related to his success in courtship. A male's ability to modify his courtship display was difficult to measure in natural courtships, because it was not clear whether males were responding to females, or vice versa. To solve this problem, a robotic female bowerbird was used.

Using these "fembots," researchers were able to control female signals and measure male response in experimental courtships. Experiments showed that in general, male Satin Bowerbirds modulate their displays in response to robotic female crouching (Fig 43Ba). Video cameras were used to monitor behaviors at bowers, allowing each male's courtship/mating success to be measured. It was found that males who modulate their displays more effectively in response to robotic female signals startle real females less often in natural courtships, and are thus more successful in courting females (Fig. 43Bb).

The results suggest that females prefer intensely displaying males as mates, and that successful males modulate their intensity in response to female signals, thus producing displays attractive to females without threatening them.[1]

Male responsiveness to female signals may be an important part of successful courtship in many species. Sexual selection may favor the ability of males to read female signals and adjust courtship displays accordingly.

Questions to Consider

1. Why would it be more adaptable for a male bowerbird to be aggressive and have an intense courtship display instead of being gaudy?
2. How would an extremely intense male bowerbird mate?

---

1 Patricelli, G. L., Uy, J. A. C., Walsh, G., and Borgia, G. 2002. Sexual selection: Male displays adjusted to female's response. *Nature* 415: 279–80.

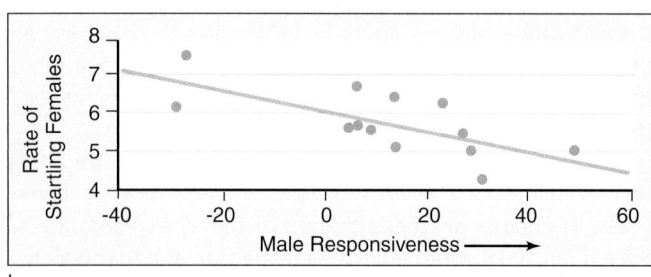

b.

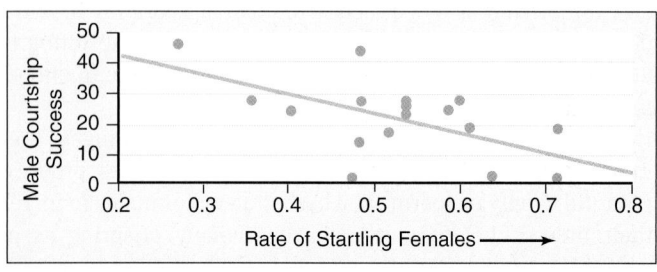

c.

a.

Figure 43B   Courtship success of males.   a. Among Satin Bowerbirds, *Ptilonorhynchus violaceus*, the male bowerbird prepares the bower and typically decorates its platform with blue objects. **b.** Some males are better able to vary the intensity of their courtship display depending on the crouch rate of a robotic female, and these males startle real females less often. **c.** Experimenters found that males who respond best under experimental conditions do startle live females less and do have better courtship success.

Figure 43.15 Competition. During the mating season, bull elk, *Cervus elaphus,* males may find it necessary to engage in antler wrestling in order to have sole access to females in a territory.

ability to fight (Fig. 43.15). When bull elk compete, they issue a number of loud screams that give way to a series of grunts. If a clear winner hasn't emerged, the two bulls walk in parallel to show each other their physique. If this doesn't convince one or the other to back off, the pair resorts to ramming each other with their antlers. The intent of the fight is for one male to gain dominance over the other, not to kill his rival—although these encounters may on occasion have fatal consequences due to injury or mishap.

Societies and Fitness

The principles of evolutionary biology can be applied to the study of social behavior in animals. Sociobiologists hypothesize that living in a society has a greater reproductive benefit than reproductive cost. A cost-benefit analysis can help determine whether this hypothesis is supported.

Group living does have its benefits. It can help an animal avoid predators, rear offspring, and find food. A group of impalas is more likely to hear an approaching predator than a solitary one. Many fish moving rapidly in many directions might distract a would-be predator. Weaver birds form giant colonies that help protect them from predators. The birds may also share information about food sources. Members of the same baboon troop signal to one another when they have found an especially bountiful fruit tree. Lions working together are able to capture large prey, such as zebra and buffalo.

Figure 43.16 Queen ant. A queen ant, *Solenopsis geminata,* has a large abdomen for egg production and is cared for by small ants, called nurses. The idea of inclusive fitness suggests that relatives, in addition to offspring, increase an individual's reproductive success. Therefore, sterile nurses are being altruistic when they help the queen produce offspring to whom they are closely related.

Group living also has its disadvantages. When animals are crowded together into a small area, disputes can arise over access to the best feeding places and sleeping sites. Dominance hierarchies are one way to apportion resources, but this puts subordinates at a disadvantage. Among red deer, sons are preferable because, as a harem master, sons will result in a greater number of grandchildren. However, sons, being larger than daughters, need to be nursed more frequently and for a longer period of time. Subordinate females do not have access to enough food resources to adequately nurse sons, and therefore tend to rear only daughters. Still, like the subordinate males in a baboon troop, subordinate females in a red deer harem may be better off in terms of fitness if they stay with a group, despite the cost involved.

Video
Harris Hawks

Living in close quarters exposes individuals to illness and parasites that can easily pass from one animal to another. Social behavior helps to offset some of the proximity disadvantages. For example, baboons and other types of social primates invest a significant amount of time grooming one another. This most likely decreases their chances of contracting parasites. Humans use extensive medical care to help offset the health problems that arise from living in the densely populated cities of the world.

Altruism Versus Self-Interest

Altruism [L. *alter*, the other] is a behavior that has the potential to decrease the lifetime reproductive success of the altruist, while benefiting the reproductive success of another member of the society. In some insect societies, reproduction is limited to only one pair, the queen and her mate. For example, among army ants, the queen is inseminated only during her nuptial flight. The rest of her life is spent producing her offspring (Fig. 43.16).

The army-ant society has three different sizes of sterile female workers. The smallest workers (3 mm), called the nurses, take care of the queen and larvae, feeding them and keeping them clean. The intermediate-sized workers, constituting most

of the population, go out on raids to collect food. The soldiers (14 mm), with huge heads and powerful jaws, run along the sides and rear of raiding parties and protect the column of ants from attack by intruders.

Although this type of society is rare in mammals, the introduction to this chapter describes how mole rats have a similar type of societal structure.

Can the altruistic behavior of sterile workers be explained in terms of fitness, which is judged by reproductive success? The answer is that genes are passed from one generation to the next in two quite different ways. The first way is direct: A parent can pass a gene directly to an offspring. The second way is indirect: A relative that reproduces can pass a shared gene to the next generation. *Direct selection* is adaptation to the environment due to the reproductive success of an individual. *Indirect selection,* called **kin selection,** is adaptation to the environment due to the reproductive success of the individual's relatives. The **inclusive fitness** of an individual includes personal reproductive success as well as the reproductive success of relatives.

Examples of Inclusive Fitness

Among social bees, social wasps, and ants, the queen is diploid (2n), but her mate is haploid (n). If the queen has had only one mate, sister workers are more closely related to each other. They share, on average, 75% of their genes because they inherit 100% of their father's alleles. Their potential offspring would share, on average, only 50% of their genes with the queen. Therefore, a worker can achieve a greater inclusive fitness by helping her mother (the queen) produce additional sisters than by directly reproducing. Under these circumstances a behavior that appears altruistic is more likely to evolve.

Indirect selection can also occur among animals whose offspring receive only a half set of genes from both parents. Consider that your brother or sister shares 50% of your genes, your niece or nephew shares 25%, and so on. Therefore, the survival of two nieces or nephews is worth the survival of one sibling, assuming they both go on to reproduce.

Among chimpanzees in Africa, a female in estrus frequently copulates with several members of the same group, and the males make no attempt to interfere with each other's matings. How can they be acting in their own self-interest? Genetic relatedness appears to underlie their apparent altruism. Members of a group share more than 50% of their genes in common because members never leave the territory in which they are born.

Reciprocal Altruism

In some bird species, offspring from a previous clutch of eggs may stay at the nest to help parents rear the next batch of offspring. In a study of Florida scrub jays, the number of fledglings produced by an adult pair doubled when they had helpers. Mammalian offspring are also observed to help their parents (Fig. 43.17). Among jackals in Africa, solitary pairs managed to rear an average of 1.4 pups, whereas pairs with helpers reared 3.6 pups.

What are the benefits of staying behind to help? First, a helper is contributing to the survival of its own kin. Therefore, the helper actually gains a fitness benefit. Second, a helper is more likely than a nonhelper to inherit a parental

Figure 43.17 Inclusive fitness. A meerkat is acting as a babysitter for its young sisters and brothers while their mother is away. Researchers point out that the helpful behavior of the older meerkat can lead to increased inclusive fitness.

territory—including other helpers. Helping, then, involves making a short-term reproductive sacrifice in order to increase future reproductive potential. Therefore, helpers at the nest are also practicing a form of **reciprocal altruism.**

Reciprocal altruism also occurs in animals that are not necessarily closely related. In this event, an animal helps or cooperates with another animal with no immediate benefit. However, the animal that was helped repays the debt at some later time. Reciprocal altruism usually occurs in groups of animals that are mutually dependent. Cheaters in reciprocal altruism are recognized and are not reciprocated in future events.

An example of this type of reciprocal altruism occurs in vampire bats that live in the tropics. Bats returning to the roost after a feeding activity share their blood meal with other bats in the roost. If a bat fails to share blood with one that had previously shared blood with it, the cheater bat will be excluded from future blood sharing.

Check Your Progress   43.4

1. Explain how territoriality is related to foraging for food.
2. Compare and contrast reproductive strategies and forms of sexual selection.
3. Describe examples of how altruistic behavior is in the self-interest of the animal.

CONNECTING *the* CONCEPTS *with the* BIG IDEAS

Energy and Homeostasis

- Changes in the environment result in behavioral responses such as migration, hibernation, taxis, and kinesis. (2C2a*IE*)
- Behaviors including circadian rhythms, migration, hibernation, and courtship signals are often synchronized with environmental clues and are shaped by natural selection. (2E2b*IE*, 2E3b3*IE*)
- All innate behavior has a genetic basis. (2E3a1)
- Learning is achieved by interactions with the organism's biotic and abiotic environment. (2E3a2)

Information and Signaling

- Communication signals can change behavior to the advantage of the signal sender and/or receiver. (3E1a-b*IE*)
- Increased reproductive and survival success based on innate and learned behaviors (such as avoidance, foraging, courtship, mating, migration and parent-child behaviors) are favored by natural selection. (3E1c1*IE*)
- Cooperative behaviors of packs, herds, flocks, colonies, and schools benefit both the individual and its population. (3E1c2*IE*)

*Find the unabridged version of all EK citations at www.glencoe.com/maderAP11.

Media Study Tools

www.glencoe.com/maderAP11

Enhance your study of this chapter with study tools and practice tests. Also ask your instructor about the resources available through ConnectPlus, including the media-rich eBook, interactive learning tools, and animations.

Summarize

43.1 Inheritance Influences Behavior

Investigators have long been interested in the degree to which nature (genetics) or nurture (environment) influences behavior. Hybrid studies with lovebirds produce results consistent with the hypothesis that behavior has a genetic basis. Garter snake experiments indicate that the nervous system controls behavior. Studies of human twins have shown that certain types of behavior are apparently inherited. DNA studies of marine snails indicate that the endocrine system also controls behavior.

43.2 The Environment Influences Behavior

Even behaviors formerly thought to be fixed action patterns (FAPs), or otherwise inflexible, sometimes can be modified by learning. The red bill of laughing gulls initiates chick begging behavior. However, with experience, chick begging behavior improves and the chicks demonstrate an increased ability to recognize parents.

Other studies suggest that learning is involved in behaviors. Imprinting in birds, which occurs during a sensitive period, causes them to follow the first moving object they see. Song learning in birds involves various elements—including the existence of a sensitive period during which an animal is primed to learn—and the positive benefit of social interactions.

Associative learning includes classical conditioning and operant conditioning. In classical conditioning, the pairing of two different types of stimuli causes an animal to form an association between them. In this way, dogs will salivate at the sound of a bell. In operant conditioning, animals learn behaviors because they are rewarded when they perform them.

Orientation and migratory behavior occur in several groups of animals. Orientation is the ability to move in a certain direction, but migration can require navigation, and is a learned ability to change direction if need be. Animals use the Sun, stars, and the Earth's magnetic field in order to migrate.

Imitation and insight learning does occur in animals. Insight learning has occurred when an animal can solve a new and different problem without prior experience.

43.3 Animal Communication

Communication is an action by a sender that affects the behavior of a receiver. Chemical, auditory, visual, and tactile signals are forms of communication that foster cooperation that benefits both the sender and the receiver. Pheromones are chemical signals that are passed between members of the same species. Auditory communication includes language, which may occur between other types of animals and not just humans. Visual communication allows animals to signal others without the need of auditory or chemical messages. Tactile communication is often associated with sexual behavior.

43.4 Behaviors That Increase Fitness

Traits that promote reproductive success are expected to be advantageous overall, despite any possible disadvantage. Some animals are territorial and defend a territory where they have food resources and can reproduce. When animals choose those foods that return the most net energy, they have more energy left over for reproduction.

Reproductive strategies include monogamy, polygamy, and polyandry. Which strategy is employed depends on the animal and its environment. Sexual selection is a form of natural selection that selects for traits that increase an animal's fitness. Males produce many sperm and are expected to compete to inseminate females. Females produce few eggs and are expected to be selective about their mates.

Living in a social group can have its advantages (e.g., ability to avoid predators, raise young, and find food). It also has disadvantages (e.g., tension between members, spread of illness and parasites, and reduced reproductive potential). When animals live in groups, the benefits must outweigh the costs or the behavior would not exist.

In most instances, the individuals of a society act to increase their own fitness (ability to produce surviving offspring). In this context, it is necessary to consider inclusive fitness, which includes personal reproductive success and also the reproductive success of relatives. Sometimes animals perform altruistic acts, as when individuals help their parents rear siblings. Social insects help their mother reproduce, but this behavior seems reasonable when we consider that siblings share 75% of their genes. Among mammals, a parental helper may be likely to inherit the parent's territory. In reciprocal altruism, animals aid one another for future benefits.

Key Terms

altruism 833
associative learning 824
auditory communication 828
behavior 820
behavioral ecology 830
classical conditioning 824
communication 826
fixed action pattern
  (FAP) 822
imprinting 823
inclusive fitness 834
insight learning 825
kin selection 834
learning 823
migration 825
monogamous 831

navigate 825
operant conditioning 824
optimal foraging model 831
orientation 825
pheromone 826
polyandrous 831
polygamous 831
reciprocal altruism 834
sexual selection 831
sign stimulus 822
society 826
tactile communication 829
territoriality 830
territory 830
visual communication 828

 **Assess**

Reviewing This Chapter

1. Describe two studies that suggest behavior has a genetic basis. 820–21
2. Describe two studies that show that behavior has a genetic basis. 821–22
3. How does an experiment with laughing gull chicks support the hypothesis that environment (nurture) influences behavior? 823
4. Describe two types of associative learning. 824–25
5. Give examples of the different types of communication among members of a social group. 826, 828–30
6. What is territoriality, and how might it increase fitness? 830
7. Name three types of reproductive strategies, and describe how they can be related to the environment. 831
8. What is sexual selection, and why does it foster female choice and male competition during mating? 831
9. Give examples of behaviors that appear to be altruistic but actually increase the inclusive fitness of an individual. 833–34

Testing Yourself

Choose the best answer for each question.
For questions 1–4, match the type of learning in the key with its description. Answers may be used more than once.

Key:

  a. classical conditioning
  b. insight learning
  c. imprinting
  d. migration

1. Ducks follow the first moving object they see after hatching.
2. A dog salivates when a bell is rung.
3. Starlings fly between the Balkins and Great Britain in the fall and spring.
4. Chimpanzees pile up boxes to reach bananas.
5. Behavior is
  a. any action that is learned.
  b. all responses to the environment.
  c. any action that can be observed and described.

  d. all activity that is controlled by hormones.
  e. unique to birds and humans.

6. Which one of the following is considered, by behaviorists, to control (in part or in whole) animal behavior?
  a. circulatory and respiratory systems
  b. respiratory and digestive systems
  c. digestive and nervous systems
  d. nervous and endocrine systems
  e. All systems of the body control behavior.

7. Which of the following is not an example of a genetically based behavior?
  a. Inland garter snakes do not eat slugs, while coastal populations do.
  b. One species of lovebird carries nesting strips one at a time, while another carries several.
  c. One species of warbler migrates, while another one does not.
  d. Snails lay eggs in response to egg-laying hormone.
  e. Wild foxes raised in captivity are not capable of hunting for food.

8. How would the following graph differ if pecking behavior in laughing gulls was a fixed action pattern?
  a. It would be a diagonal line with an upward incline.
  b. It would be a diagonal line with a downward incline.
  c. It would be a horizontal line.
  d. It would be a vertical line.
  e. None of these is correct.

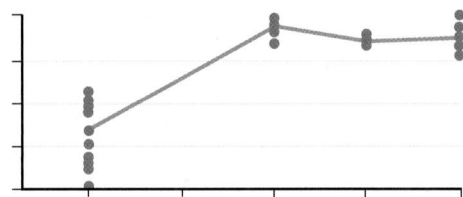

9. Using treats to train a dog to do a trick is an example of
  a. imprinting.
  b. tutoring.
  c. vocalization.
  d. operant conditioning.

10. The benefits of imprinting generally outweigh the costs because
  a. an animal that has been imprinted on the wrong object can be reimprinted on the mother.
  b. imprinting behavior never lasts more than a few months.
  c. animals in the wild rarely imprint on anything other than their mother.
  d. animals that imprint on the wrong object generally die before they pass their genes on.

11. In white-crowned sparrows, social experience exhibits a very strong influence over the development of singing patterns. What observation led to this conclusion?
  a. Birds only learned to sing when they were trained by other birds.
  b. The window in which birds learn from other birds is wider than that when birds learn from tape recordings.
  c. Birds can learn different dialects only from other birds.
  d. Birds that learned to sing from a tape recorder could change their song when they listened to another bird.

12. Which of the following best describes classical conditioning?
    a. The gradual strengthening of stimulus-response connections that seemingly are unrelated.
    b. A type of associative learning in which there is no contingency between response and reinforcer.
    c. The learning behavior in which an organism follows the first moving object it encounters.
    d. The learning behavior in which an organism exhibits a fixed action pattern from the time of birth.

13. The observation that male bowerbirds decorate their nests with blue objects favored by females can best be associated with
    a. insight learning.
    b. imprinting.
    c. sexual selection.
    d. altruism.

14. Migratory behavior
    a. has a higher cost than a benefit.
    b. affects the reproductive strategy.
    c. pertains only to birds.
    d. involves the ability to navigate.

For problems 15–19, match the type of communication in the key with its description. Answers may be used more than once.

Key:

    a. chemical communication
    b. auditory communication
    c. visual communication
    d. tactile communication

15. Aphids (insects) release an alarm pheromone when they sense they are in danger.

16. Male peacocks exhibit an elaborate display of feathers to attract females.

17. Ground squirrels give an alarm call to warn others of the approach of a predator.

18. Male silk moths are attracted to females by a sex attractant released by the female moth.

19. Sage grouses perform an elaborate courtship dance.

20. Bees that do a waggle dance are teaching other bees
    a. how to dance.
    b. where to find food.
    c. how to find and use the Sun for navigation.
    d. how to use auditory communication.

Engage

Virtual Lab
Mealworm Behavior

The virtual lab "Mealworm Behavior" provides an interactive method of studying the factors that influence behavior in animals.

Thinking Scientifically

1. Meerkats are said to exhibit altruistic behavior because certain members of a population act as sentries. How would you test the hypothesis that sentries are engaged in altruistic behavior?

2. You are testing the hypothesis that human infants instinctively respond to higher-pitched voices. Your design is to record head turns toward speakers when they play voices in different pitches. When you do the experiment using several different infants, your data support your hypothesis. However, prior learning by infants is still a serious criticism. What is the basis of this criticism?

Bioethical Issue

Putting Animals in Zoos

In light of the 2010 death of a Sea World trainer by a 5-ton orca whale, many people have questioned the ethical nature of keeping animals in captivity.

If we keep animals in zoos, are we depriving them of their freedom? Some point out that freedom is never absolute. Even an animal in the wild is restricted in various ways by its abiotic and biotic environment. Many modern zoos keep animals in habitats that nearly match their natural ones so that they have some freedom to roam and behave naturally—but is this really possible with an animal such as an orca?

Today, reputable zoos rarely go out and capture animals in the wild—they usually get their animals from other zoos. Most people feel it is not a good idea to take animals from the wild. Certainly, zoos should not be involved in the commercial and often illegal trade of wild animals that still goes on today. Many zoos today are involved in the conservation of animals, rather than just exploitation of animals for exhibit. They provide the best home possible while animals are recovering from injury or increasing their numbers until they can be released to the wild. Can we perhaps look at zoos favorably if they show that they are keeping animals under good conditions and are also involved in preserving animals? In your opinion, which animals are suited to a life in captivity, and which are not?

44

Population Ecology

Asian carp in the Mississippi River system.

BEFORE YOU BEGIN

Before beginning this chapter, take a few moments to review the following discussions.

Section 6.1 How does energy flow from one organism to another impact population growth?

Figure 10.8 Explain what role the human life cycle plays in determining population growth.

Section 16.1 Describe how microevolution is measured within a population.

In the early 1970s several species of Asian carp were imported into Arkansas for the biocontrol of algal blooms in aquaculture facilities. Unfortunately they escaped into the middle and lower Mississippi drainage. Over time they have spread throughout the majority of the Mississippi River system.

Asian carp mainly eat the microscopic algae and zooplankton in freshwater ecosystems. They can achieve weights up to a hundred pounds and grow to a length of more than 4 feet. Because of their voracious appetite they have the potential to reduce the populations of native fish and mussels due to competition for the same food sources.

The main fear of biologists is that the Asian carp will put extreme pressure on the zooplankton populations, which can then result in a dense planktonic algal bloom. This has the potential to completely disrupt the ecology of the ecosystems that they invade. These fish also pose an economic threat by fouling the nets of commercial fishermen. The other major concern is the impact these fish may have upon the Great Lakes seven-billion-dollar fishing industry.

This chapter previews the principles of population ecology, a field that is critical to the preservation and management of species as well as the maintenance of the diversity of life on Earth.

As you read through the chapter think about the following questions:

1. Does the old saying "90% of the fish are found in 10% of the lake" have any truth to it?

2. Do population growth models apply to humans the same way they apply to other species?

FOLLOWING *the* BIG IDEAS

CHAPTER 44   POPULATION ECOLOGY

Energy and Homeostasis	Populations are impacted by both biotic and abiotic factors.
Interactions and Systems	Though populations could grow exponentially, various limiting factors typically keep size at a sustainable level.

44.1 Scope of Ecology

Learning Outcomes

Upon completion of this section, you should be able to

1. Identify what aspects of biology the study of ecology encompasses.
2. Identify the ecological levels that exist within the field of ecology.

In 1866, the German zoologist Ernst Haeckel (1834–1919) coined the word **ecology** [Gk. *oikos,* home, and *-logy,* study of], which he defined as the study of the interactions among all organisms and with their physical environment. Haeckel also pointed out that ecology and evolution are intertwined because ecological interactions act as selection pressures that result in evolutionary change, which in turn affects ecological interactions.

Ecology, like so many biological disciplines, is wide-ranging. At one of its lowest levels, ecologists study how the individual organism is adapted to its environment. For example, they study how a fish is adapted to and survives in its **habitat** (the place where the organism lives) (Fig. 44.1). Most organisms do not exist singly; rather, they are part of a population, the functional unit that interacts with the environment and on which natural selection operates. A **population** is defined as all the organisms belonging to the same species within an area at the same time. At this level of study, ecologists are interested in factors that affect the growth and regulation of population size.

A **community** consists of all the populations of multiple species interacting at a locale. In a coral reef, there are numerous populations of algae, corals, crustaceans, fishes, and so forth. At this level, ecologists want to know how various populations interact with each other. An **ecosystem** is composed of the community of populations along with the abiotic environment (e.g., the availability of sunlight for plants). Energy flow and chemical cycling are significant aspects of understanding how an ecosystem functions. Ecosystems rarely have rigid boundaries. Usually, a transition zone called an ecotone, which has a mixture of organisms from adjacent ecosystems, exists between ecosystems. The **biosphere** encompasses the zones of the Earth's soil, water, and air where living organisms are found.

Video
Coral Reef
Ecosystems

Modern ecology is not just descriptive, it is predictive. It analyzes levels of organization and develops models and hypotheses that can be tested. A central goal of modern ecology is to develop models that explain and predict the distribution and abundance of organisms. Ultimately, ecology considers not one particular area, but the distribution and abundance of populations in the biosphere. For example, what factors have selected for the mix of plants and animals in a tropical rain forest at one latitude and in a desert at another?

Although modern ecology is useful in and of itself, it also has unlimited application possibilities, including the proper management of plants and wildlife, the identification of and efficient use of renewable and nonrenewable resources, the preservation of habitats and natural cycles, the maintenance of food resources, and the ability to predict the impact and course of a disease such as malaria or AIDS.

Check Your Progress   44.1

1. Distinguish between a population and a community.
2. Describe the central goal of modern ecological studies.
3. Explain what is meant by the "abiotic environment."

Figure 44.1 Ecological levels.   The study of ecology encompasses levels of organization that proceed from the individual organism to the population, to the community, and finally to an ecosystem.

Organism ⟶ Population ⟶ Community ⟶ Ecosystem

44.2 Demographics of Populations

Learning Outcomes

Upon completion of this section, you should be able to

1. Recognize how environmental conditions affect the density and distribution patterns of a population.
2. Interpret survivorship curves and life tables.
3. Recognize how the proportion of individuals at varying reproductive stages determines a population's age distribution.

Demography is the statistical study of a population, such as its density, its distribution, and its rate of growth, which is dependent on such factors as its mortality pattern and age distribution.

Density and Distribution

Population density is the number of individuals per unit area; for example, there are 86.8 persons per square mile in the United States. Population density figures make it seem as though individuals are uniformly distributed, but this often is not the case. For example, we know that most people in the United States live in cities, where the number of people per unit area is dramatically higher than in the country. And even within a city, more people live in particular neighborhoods than in others, and such distributions can change over time. Therefore, basing ecological models solely on population density can lead to inaccurate results.

Population distribution is the pattern of dispersal of individuals across an area of interest. The availability of resources can affect where populations live. **Resources** are nonliving (abiotic) and living (biotic) components of an environment that support living organisms. Light, water, space, mates, and food are some important resources for populations. **Limiting factors** are those environmental aspects that determine where an organism lives. For example, trout live only in cool mountain streams, where the oxygen content is high, but carp and catfish are found in rivers near the coast because they can tolerate warm waters with a lower concentration of oxygen. The timberline is the limit of tree growth in mountainous regions or in high latitudes. Trees cannot grow above the high timberline because of low temperatures and the fact that water remains frozen most of the year. Biotic factors can also play a role in the distribution of organisms. In Australia, the red kangaroo does not live outside arid inland areas because it is adapted to feeding on the grasses that grow there.

Three descriptions—*clumped, random,* and *uniform*—are often used to characterize patterns of distribution. Suppose you considered the distribution of a species across its full range. A range is that portion of the globe where the species can be found; red kangaroos live in Australia. On that scale, you would expect to find a clumped distribution. However, organisms are located in areas suitable to their adaptations; as mentioned, red kangaroos live in grasslands, and catfish live in warm river water near the coast.

Within a smaller area, such as a single body of water or a single forest, the availability of resources influences which pattern of distribution is exhibited for a particular population. For example, a study of the distribution of hard clams in a bay on the south shore of Long Island, New York, showed that clam abundance is associated with sediment shell content. Investigators hope to use this information to transform areas that have few clams into high-abundance areas.

Distribution patterns need not be constant. In a study of desert shrubs, investigators found that the distribution changed from clumped to random to a uniform distribution pattern as the plants matured. As time passed, it was found that competition for belowground resources caused the distribution pattern to become uniform (Fig. 44.2).

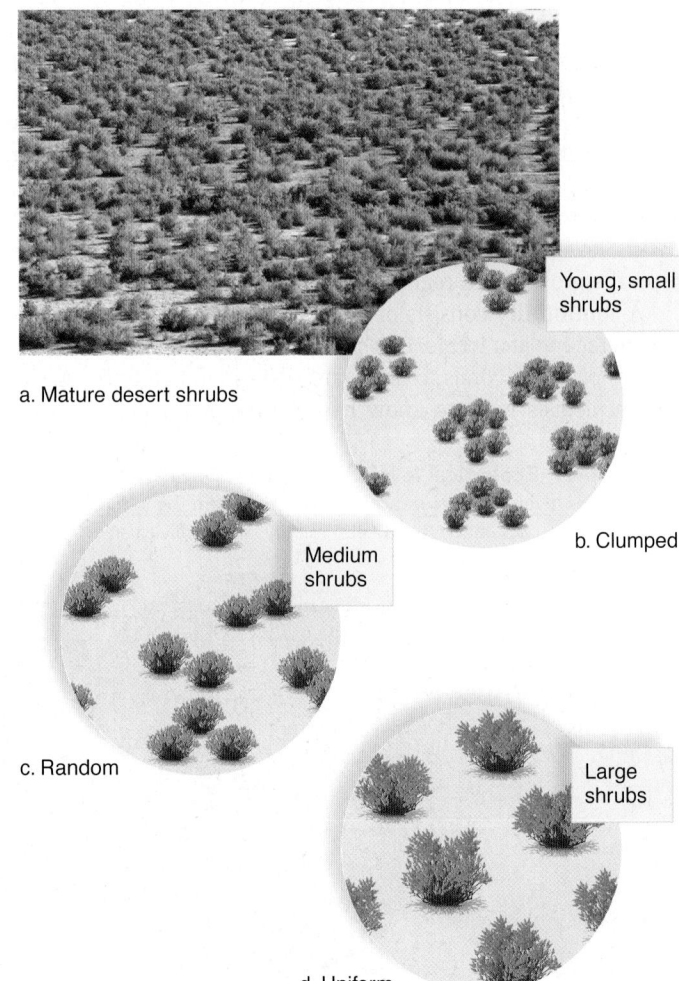

Figure 44.2 Distribution patterns of the creosote bush.
a. Young, small desert shrubs are clumped. **b.** Medium shrubs are randomly distributed. **c.** Mature shrubs are uniformly distributed. **d.** Photograph of mature shrub distribution.

a.

b.

Figure 44.3 Biotic potential. A population's maximum growth rate under ideal conditions—that is, its biotic potential—is greatly influenced by the number of offspring produced in each reproductive event. **a.** Mice, which produce many offspring that quickly mature to produce more offspring, have a much higher biotic potential than the rhinoceros **(b)**, which produces only one or two offspring per infrequent reproductive event.

Other factors besides resource availability can influence distribution patterns. Breeding golden eagles, like many other birds, exhibit territoriality, and this behavioral characteristic discourages a clumped distribution. In contrast, cedar trees tend to be clumped near the parent plant because seeds are not widely dispersed.

Population Growth

The **rate of natural increase (r),** or growth rate, is dependent on the number of individuals born each year and the number of individuals that die each year. It is assumed that immigration and emigration are equal and need not be considered in the calculation of the growth rate. Populations grow when the number of births exceeds the number of deaths. If the number of births is 30 per year and the number of deaths is 10 per year per 1,000 individuals, the growth rate would be:

$$(30 - 10)/1,000 = 0.02 = 2.0\%$$

The highest possible rate of natural increase for a population is called **biotic potential** (Fig. 44.3). Whether the biotic potential is high or low depends on the number of limiting factors that reduce or slow the population's potential reproduction, such as the following:

- Number of offspring per reproductive event that survive until the age of reproduction
- Amount of competition within the population
- Age of and number of reproductive opportunities
- Presence of disease and predators

Mortality Patterns

Population growth patterns assume that populations are made up of identical individuals. Actually, the individuals of a population are in different stages of their life span. A **cohort** is all the members of a population born at the same time. Some investigators study population dynamics and construct life tables that show how many members of a cohort are still alive after certain intervals of time.

For example, Table 44.1 is a life table for a bluegrass cohort. The cohort contains 843 individuals. The table tells us that after three months, 121 individuals have died, and therefore the mortality rate is 0.143 per capita. Another way to express this same statistic, however, is to consider that 722 individuals are still alive—have survived—after three months. **Survivorship** is the probability of newborn individuals of a cohort surviving to particular ages. If we plot the number surviving at each age, a survivorship curve is produced (see Fig. 44.4*b*).

Table 44.1 A Life Table for a Bluegrass Cohort

Age (months)	Number Observed Alive	Number Dying	Mortality Rate per Capita	Avg. Number of Seeds per Individual
0–3	843	121	0.143	0
3–6	722	195	0.271	300
6–9	527	211	0.400	620
9–12	316	172	0.544	430
12–15	144	95	0.626	210
15–18	54	39	0.722	60
18–21	15	12	0.800	30
21–24	3	3	1.000	10
24	0	—	—	—

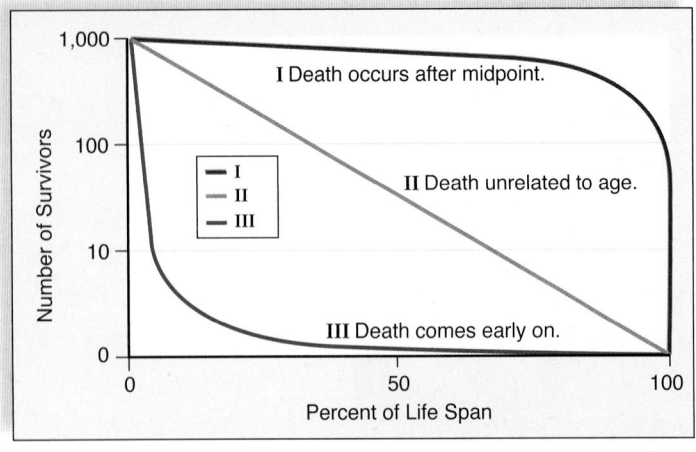

a.

b. Bluegrasses

c. Lizards

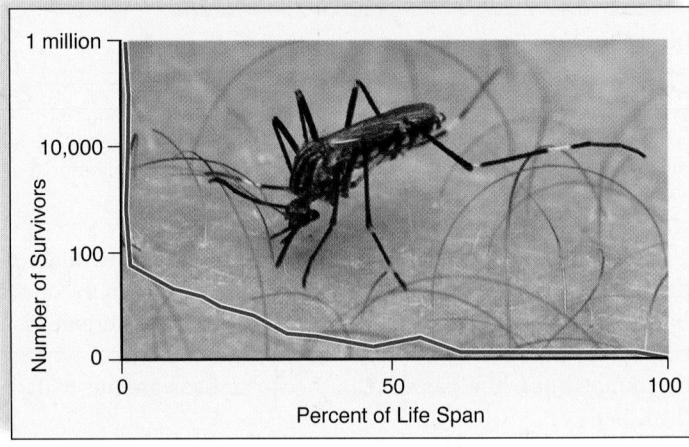

d. Mosquitoes

Figure 44.4 Survivorship curves. Survivorship curves show the number of individuals of a cohort that are still living over time. **a.** Three generalized survivorship curves. **b.** The survivorship curve for bluegrass represents a combination of the type I and type II curves. **c.** The survivorship curve for lizards generally fits a type II curve. **d.** The survivorship curve for mosquitoes is a type III curve.

The results of investigations like this have revealed that each species tends to have one of the typical survivorship curves. Three types of idealized survivorship curves, numbered I, II, and III, are usually recognized (Fig. 44.4*a*). The type I curve is characteristic of a population in which most individuals survive well past the midpoint of the life span, and death does not come until near the end of the life span. Animals that have this type of survivorship curve include large mammals and humans in more-developed countries. In contrast, the type III curve is typical of a population in which most individuals die very young. This type of survivorship curve occurs in many invertebrates, fishes, and humans in less-developed countries. In the type II curve, survivorship decreases at a constant rate throughout the life span. In many songbirds and small mammals, death is usually unrelated to age; thus they represent a type II survivorship curve.

The survivorship curves of natural populations do not fit these three idealized curves exactly. In a bluegrass cohort, as shown in Table 44.1 for example, most individuals survive until six to nine months, and then the chances of survivorship diminish at an increasing rate (Fig. 44.4*b*). Statistics for a lizard cohort are close enough to classify the survivorship curve in the type II category (Fig. 44.4*c*), while a mosquito cohort has a type III curve (Fig. 44.4*d*).

Much can be learned about the life history of a species by studying its life table and the survivorship curve that can be constructed based on this table. In a population with a type III survivorship curve, would you predict that natural selection would favor those who have more offspring, or those who have fewer offspring? Obviously, because death comes early for most members, only a few live long enough to reproduce, and therefore natural selection would favor those who produce more offspring. What about the other two types of survivorship curves?

Other types of information are also available from studying life tables. Looking again at Table 44.1, we can see that per-capita seed production increases as plants mature, and then seed production drops off. How do you predict this would compare to a cohort of human beings?

Age Distribution

When the individuals in a population reproduce repeatedly, several generations may be alive at any given time. From the perspective of population growth, a population contains three major age groups: prereproductive, reproductive, and postreproductive. Populations differ according to what proportion of the population falls in each age group. At least three **age structure diagrams** are possible (Fig. 44.5).

When the prereproductive group is the largest of the three groups, the birthrate is higher than the death rate, and a pyramid-shaped diagram is expected. Under such conditions, even if the growth for that year were matched by the deaths for that year, the population would continue to grow in the following years. Why? Because more individuals are entering than leaving the reproductive group. Eventually, as the size of the reproductive group equals the size of the prereproductive group,

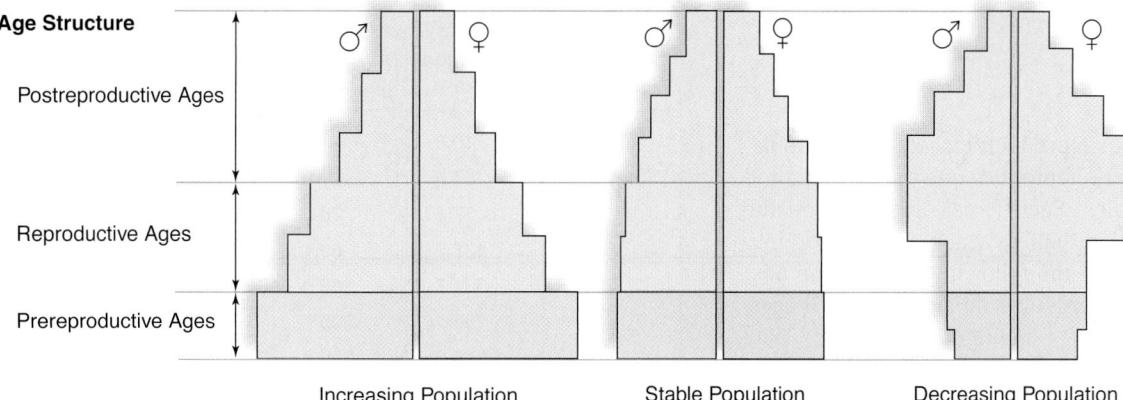

Figure 44.5 **Age structure diagrams.** Typical age structure diagrams for hypothetical populations that are increasing, stable, or decreasing. Different numbers of individuals in each age class create these distinctive shapes. In each diagram, the left half represents males while the right half represents females.

Age Structure

Postreproductive Ages

Reproductive Ages

Prereproductive Ages

Increasing Population   Stable Population   Decreasing Population

a bell-shaped diagram would result. The postreproductive group is still the smallest, however, because of mortality. If the birth-rate falls below the death rate, the prereproductive group would become smaller than the reproductive group. The age structure diagram would then be urn-shaped, because the postreproductive group is now the largest.

The age distribution reflects the past and future history of a population. Because a postwar baby boom occurred in the United States between 1946 and 1964, the postreproductive group will soon be the largest group.

Check Your Progress   44.2

1. Distinguish between population density and population distribution.
2. Describe the differences among type I, II, and III survivorship curves.
3. Explain why a bell-shaped age pyramid indicates a growing population.

44.3 Population Growth Models

Learning Outcomes

Upon completion of this section, you should be able to

1. Describe exponential population growth and the circumstances that encourage it.
2. Identify the features of logistic growth and the carrying capacity of a population.

Based on observation and natural selection principles, ecologists have developed two working models for population growth. In the pattern called **semelparity** [Gk. *seme*, once; L. *parous*, to bear or bring forth], the members of the population have only a single reproductive event in their lifetime. When the time for reproduction draws near, the mature adults cease to grow and expend all their energy in reproduction, and then die. Many insects, such as winter moths, and annual plants, such as zinnias, follow this pattern of reproduction growth. They produce a resting stage of development, such as eggs or seeds that survive unfavorable conditions and resume growth the next favorable season. In other words, semelparity is an adaptation to an unstable environment.

In the pattern called **iteroparity** [Gk. *itero*, repeat], members of the population experience many reproductive events throughout their lifetime. They continue to invest energy in their future survival, which increases their chances of reproducing again. Iteroparity is an adaptation to a stable environment in

which chances of survival for offspring are relatively high. Most vertebrates, shrubs, and trees have this pattern of reproduction.

Figure 44.6 shows that reproduction does not always fit these two patterns. However, ecologists have found it useful to develop mathematical models of population growth based on these two very different patterns of reproduction. Although the mathematical models we describe here are simplifications, they still may be used to predict the distribution and abundance of organisms, or the responses of populations when their environment is altered in some way. Testing predictions permits the development of new hypotheses that can then be evaluated.

a.

b.

Figure 44.6 **Patterns of reproduction.** In general organisms tend to reproduce continuously or have a single reproductive event. **a.** Aphids use asexual reproduction during the summer months and then shift to sexual reproduction right before the onset of winter. **b.** Annual plants tend to produce a large number of seeds per reproductive event. The number that germinates is often dependent upon environmental factors.

Exponential Growth

As an example of semelparous reproduction, consider a population of insects in which females reproduce only once a year and then the adult population dies. Each female produces on the average 4.8 eggs per generation, half of which will develop into female offspring that reproduce the following year. In the next generation, each female again produces an average of 4.8 eggs. In this case of discrete breeding, R = net reproductive rate.[1] Why net reproductive rate? Because it is the observed rate of natural increase after deaths have occurred.

Figure 44.7a shows how the population would grow year after year for ten years, assuming that R stays constant from generation to generation. This growth is equal to the size of the population because all members of the previous generation have died. Mayflies, featured in Figure 44.7, have one reproductive event usually in the spring—hence their name—and then development of the next generation requires as many as 50 molts during the winter. Figure 44.7b shows the growth curve for such a population. This growth curve, which is roughly J shaped, depicts exponential growth. With **exponential growth,** the number of individuals added each generation increases as the total number of females increases.

Notice that the curve has these phases:

Lag phase During this phase, growth is slow because the population is small.
Exponential growth phase During this phase, growth is accelerating.

Figure 44.7c gives the mathematical equation that allows you to calculate growth and size for any population that has discrete (nonoverlapping) generations. In other words, all members of the previous generation die off before the new generation appears. To use this equation to determine future population size, it is necessary to know R, which is the net reproductive rate determined after gathering mathematical data regarding past population increases. Notice that even though R remains constant, growth is exponential because the number of individuals added each year is increasing. Therefore, the growth of the population is accelerating.

For exponential growth to continue unchecked, plenty of habitat, food, shelter, and any other requirements necessary to sustain growth must be available. But in reality, environmental conditions prevent exponential growth. Eventually, any further growth is impossible because of limiting factors—the food supply runs out, and waste products begin to accumulate. Also, as the population increases in size so do the effects of competition between members, predation, parasites, and disease.

Video Exponential Growth

Logistic Growth

What type of growth curve results when limiting environmental factors that oppose growth come into play? In 1930, Raymond Pearl developed a method for estimating the number of yeast cells accruing every two hours in a laboratory culture vessel. His data are shown in Figure 44.8a. When the data are plotted, the

---

1 The change of r to R is simply customary in discrete breeding calculations; both coefficients deal with the same thing (birth minus death).

Generation	Population Size	Number of Females
0	10.0	5
1	24.0	12
2	57.6	28.8
3	138.2	69.1
4	331.7	165.9
5	796.1	398.1
6	1,910.6	955.3
7	4,585.4	2292.7
8	11,005.0	5502.5
9	26,412.0	13206.1
10	63,388.8	31694.5

a.

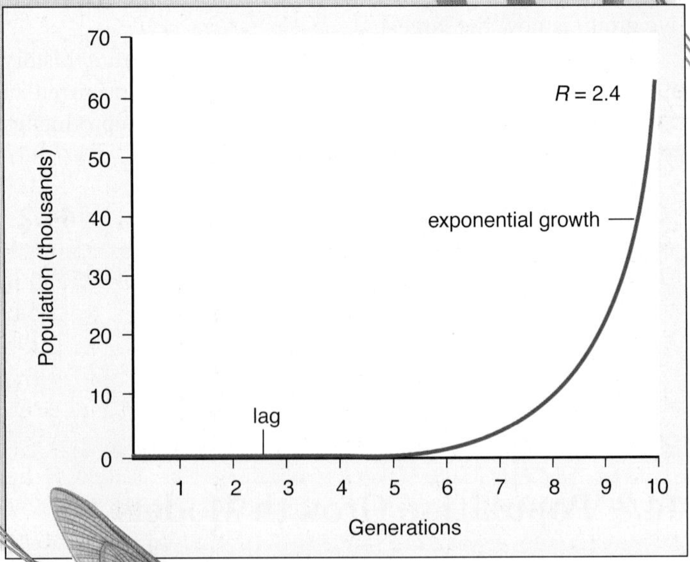

b.

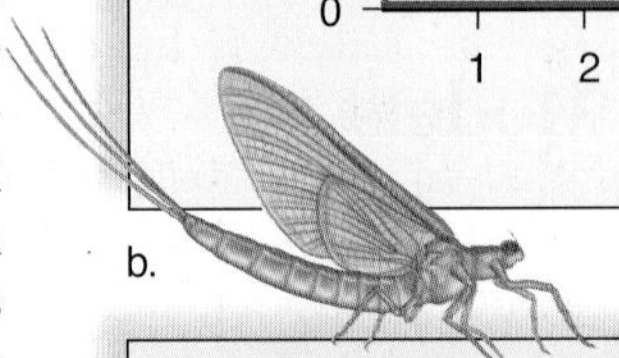

To calculate population size from year to year, use this formula:

$$N_{t+1} = RN_t$$

N_t = number of females already present
R = net reproductive rate
N_{t+1} = population size the following year

c.

Figure 44.7 Model for exponential growth. When the data for discrete reproduction in **(a)** are plotted, the exponential growth curve in **(b)** results. **c.** This formula produces the same results as **(a)** and generates the same graph.

growth curve has the appearance shown in Figure 44.8b. This type of growth curve is a sigmoidal (S) or S-shaped curve.

Notice that this **logistic growth** has these phases:

Lag phase During this phase, growth is slow because the population is small.
Exponential growth phase During this phase, growth is accelerating.

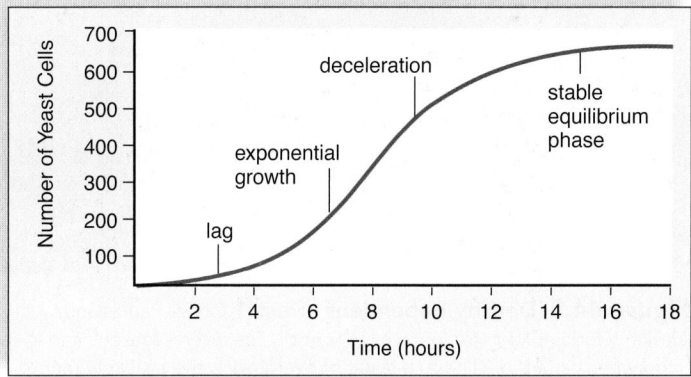

Growth of Yeast Cells in Laboratory Culture

Time (t) (hours)	Number of individuals (N)	Number of individuals added per 2-hour period $\left(\dfrac{\Delta N}{\Delta t}\right)$
0	9.6	0
2	29.0	19.4
4	71.1	42.1
6	174.6	103.5
8	350.7	176.1
10	513.3	162.6
12	594.4	81.1
14	640.8	46.4
16	655.9	15.1
18	661.8	5.9

a.

b.

To calculate population growth as time passes, use this formula:

$$\frac{N}{t} = rN\left(\frac{K-N}{K}\right)$$

N = population size

N/t = change in population size

r = rate of natural increase

K = carrying capacity

$\dfrac{K-N}{K}$ = effect of carrying capacity on population growth

c.

Figure 44.8 Model for logistic growth. When the data for repeated reproduction (a) are plotted, the logistic growth curve in (b) results. c. This formula produces the same results as (a) and generates the same graph.

Deceleration phase During this phase, growth slows down.
Stable equilibrium phase During this phase, there is little if any growth because births and deaths are about equal.

Figure 44.8c gives the mathematical equation that allows us to calculate logistic growth. (The curve is termed "logistic" because the exponential portion of the curve produces a straight line when the log of N is plotted.) The entire equation for logistic growth is:

$$\frac{N}{t} = rN\,\frac{(K-N)}{K}$$

but let's consider each portion of the equation separately.

Because the population has repeated reproductive events, we need to consider growth as a function of change in time (Δ):

$$\frac{\Delta N}{\Delta t} = rN$$

If the change in time is very small, then we can use differential calculus, and the instantaneous population growth (d) is given by:

$$\frac{dN}{dt} = rN$$

This portion of the equation applies to the first two phases of growth—the lag phase and the exponential growth phase. Here, also, we do not expect exponential growth to continue. Charles Darwin calculated that a single pair of elephants could have over 19 million live descendants after 750 years. Others have calculated that a single female housefly could produce over 5 trillion flies in one year! Such explosive growth does not occur because environmental conditions, both abiotic and biotic, cause population growth to slow. The yeast population mentioned earlier was grown in a vessel in which food could run short and waste products could accumulate. These environmental conditions prevent exponential growth from continuing.

Look again at Figure 44.8. Following exponential growth, a population is expected to enter a deceleration phase and then a stable equilibrium phase of the logistic growth curve. Now the population is at the carrying capacity of the environment.

Carrying Capacity

The environmental **carrying capacity (K)** is the maximum number of individuals of a given species the community can support. The closer population size nears the carrying capacity of the community, the more likely it is that resources will become scarce and that biotic effects such as competition and predation will become evident. The birthrate is expected to decline and the death rate is expected to increase. This results in a decrease in population growth; eventually, the population stops growing and its size remains stable. Carrying capacity in any given community can vary throughout time depending on fluctuating conditions, for example, amount of rainfall from one year to the next.

How does the mathematical model for logistic growth take this process into account? To our equation for growth under conditions of exponential growth we add the term:

$$\frac{(K-N)}{K}$$

In this expression, K is the carrying capacity of the environment. The easiest way to understand the effects of this term is to consider two extreme possibilities. First, consider a time at which the population size is well below carrying capacity. Resources are relatively unlimited, and we expect rapid, nearly exponential growth to take place. Does the model predict this? Yes, it does. When N is very small relative to K, the term $(K-N)/K$ is very nearly $(K-0)/K$, or approximately 1. Therefore, $(dN)/(dt)$ is approximately equal to rN.

Similarly, consider what happens when the population reaches carrying capacity. Here, we predict that growth will stop and the population will stabilize. What happens in the model? When N is equal to K, the term $(K-N)/K$ declines from nearly 1 to 0, and the population growth slows to zero.

As mentioned, the model predicts that exponential growth occurs only when population size is much lower than the carrying capacity. So, as a practical matter, if humans are using a fish population as a continuous food source, it would be best to maintain the fish population size in the exponential phase of the logistic growth curve. Biotic potential can have its full effect, and the birthrate is the highest it can be during this phase. If we overfish, the population will sink into the lag phase, and it will be years before exponential growth recurs.

In contrast, if we are trying to limit the growth of a pest, it is best, if possible, to reduce the carrying capacity rather than reduce the population size. Reducing the population size only encourages exponential growth to begin once again. Farmers can reduce the carrying capacity for a pest by alternating rows of different crops rather than growing one type of crop throughout an entire field.

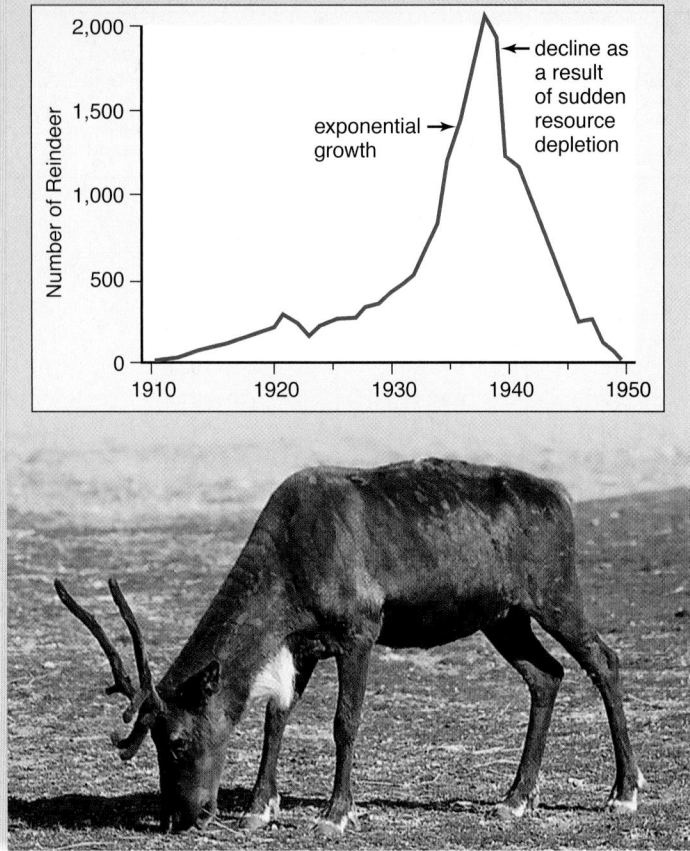

Figure 44.9 Density-dependent effect. On St. Paul Island, Alaska, reindeer, *Rangifer*, grew exponentially for several seasons and then underwent a sharp decline as a result of overgrazing the available range.

Check Your Progress 44.3

1. Describe ecological factors that might result in iteroparity rather than semelparity.
2. Explain how carrying capacity (*K*) limits exponential growth.

44.4 Regulation of Population Size

Learning Outcomes

Upon completion of this section, you should be able to

1. Compare density-independent and density-dependent factors that affect population size.
2. Describe the intrinsic factors that can impact population size and growth.

In a study of winter moth population dynamics, researchers discovered that a large proportion of eggs did not survive the winter and exponential growth never occurred. Perhaps the low number of individuals at the start of each season helps prevent the occurrence of exponential growth.

It is possible that exponential growth may cause population size to rise above the carrying capacity of the environment, and as a consequence a population crash may occur. As one example, in 1911, 4 male and 21 female reindeer were released on St. Paul Island in the Bering Sea off Alaska. St. Paul Island had an undisturbed environment; and there was little hunting pressure and no predators. The herd grew exponentially to about 2,000 reindeer by 1938, overgrazed the habitat, and then abruptly declined to only 8 animals by 1950 (Fig. 44.9).

This pattern of a population explosion eventually followed by a population crash is called irruptive, or Malthusian, growth. It is named in honor of the eighteenth-century economist Thomas Robert Malthus, who had a great influence on Charles Darwin. Populations do not ordinarily undergo Malthusian growth because of factors that regulate population growth.

Video Hungry Reindeer

Ecologists have long recognized that both biotic and abiotic conditions play an important role in regulating population size in natural environments.

a. Low density of mice

b. High density of mice

Figure 44.10 Density-independent effects. The impact of a density-independent factor, such as weather or natural disasters, is not influenced by population density. The impact of a flash flood on **(a)** a low-density population (mortality rate of 3/5, or 60%) is similar to the impact on **(b)** a high-density population (mortality rate of 12/20, also 60%).

a. Low density of birds

b. High density of birds

Figure 44.11 Density-dependent effects—competition. The impact of competition on a population is directly proportional to the density of the population. When density is low (**a**), every member of the population has access to the resource. But when the density is high (**b**), there is competition between members of the population to gain access to available resources.

Density-Independent Factors

Abiotic factors include droughts, freezes, hurricanes, floods, and forest fires. Any one of these natural disasters can kill individuals and lead to a sudden and catastrophic reduction in population size. However, such an event does not necessarily happen to a dense population more often than it happens to a less dense population. Therefore, an abiotic factor is usually a **density-independent factor,** meaning that the intensity of the effect does not increase with increased population density.

For example, the percentage of individuals killed in a flash flood event is independent of density—floods do not necessarily kill a larger percentage of a dense population than of a less dense population. Nevertheless, the larger the population, the greater the number of individuals probably affected. In Figure 44.10, the impact of a flash flood on a low-density population of mice living in a field was 3 out of 5 (60% mortality), whereas the impact on a high-density population was 12 out of 20 (also 60% mortality).

Density-Dependent Factors

Biotic factors are considered **density-dependent factors** because the percentage of the population affected does increase as the density of the population increases. Competition, predation, disease,

and parasitism are all biotic factors that increase in intensity as the density increases.

Competition can occur when members of a species attempt to use the same resources (such as light, food, space) that are in limited supply. As a result, not all members of the population can have access to the resource to the degree necessary to ensure survival or reproduction. As an example, let's consider a western bluebird (*Sialia mexicana*) population in which members have to compete for nesting sites. Each pair of birds requires a tree hole to raise offspring. If there are more holes than breeding pairs, each pair can have a hole in which to lay eggs and rear young birds (Fig. 44.11*a*). But if there are fewer holes than there are breeding pairs, then each pair must compete to acquire a nesting site (Fig. 44.11*b*). Pairs that fail to gain access to holes will be unable to contribute new members to the population.

Video
Booby Chick Competition

Competition for food also controls population growth. However, resource partitioning among different age groups is a way to reduce competition for food. As mentioned earlier in this text, the life cycle of butterflies includes caterpillars, which require a different food from the adults. The caterpillars graze on leaves, while the adults feed on nectar produced by flowers. Therefore, parents do not compete with their offspring for food.

Predation occurs when one living organism, the predator, eats another, the prey. In the broadest sense, predation impacts every species within a community at some level. The effect of predation on a prey population generally increases as the prey population grows denser, because prey is easier to find. Consider a field inhabited by a population of mice (Fig. 44.12). Each mouse must have

a. Low density of mice

b. High density of mice

Figure 44.12 Density-dependent effects—predation. The impact of predation on a population is directly proportional to the density of the population. In a low-density population (**a**), the chances of a predator finding the prey are low, resulting in little predation. But in the higher density population (**b**), there is a greater likelihood of the predator locating potential prey, resulting in a greater predation rate.

a hole in which to hide to avoid being eaten by a hawk. If there are 100 holes, and a low density of 102 mice, then only 2 mice will be left out in the open. It might be hard for the hawk to find only 2 mice in the field. If neither mouse is caught, then the predation rate is 0/2 = 0%. However, if there are 100 holes, and a high density of 200 mice, then the chance is greater that the hawk will be able to find some of these 100 mice without holes. If half of the exposed mice are caught, the predation rate is 50/100 = 50%. Therefore, increasing the density of the available prey has increased the proportion of the population preyed upon.

Video
Harris Hawk

Parasites, such as blood-sucking ticks, are generally much smaller than their host. Although parasites do not always kill their hosts, they do usually weaken them over time. A highly parasitized individual is less apt to produce as many offspring than it would if it were healthy. In this way, parasitism also plays a role in regulating population size.

Disease is much more likely to spread in a dense population due to the amount of contact between healthy individuals and infected individuals. Consider the spread of the *Yersinia pestis* plague, known as the Black Death, in Europe during the 14th century; this outbreak of bacterial disease reduced the population of Europe by 30–60%.

Other Considerations

Density-independent and density-dependent factors are extrinsic to the organism. Intrinsic factors—those based on the anatomy, physiology, or behavior of the organism—can also affect population size and growth rates. Territoriality and dominance hierarchies are behaviors that affect population size and growth rates. Recruitment and migration are other intrinsic social means by which the population sizes of more complex organisms are regulated.

Outside of any regulating factors, it could be that some populations have an innate instability. Ecologists have developed models that predict complex, erratic changes in even simple systems. For example, a computer model of Dungeness crab populations assumed that adults produce many larvae and then die. Most of the larvae do not survive, and those that do stay close to home. Under these circumstances, the model predicted wild fluctuations in population size without a recurring pattern. This type of complex, nonrandom generation is called *deterministic chaos,* or simply *chaos.*

Population growth–regulating factors can serve as selective agents. Some members of a population may possess traits that make it more likely that they, rather than other members of the population, will survive and reproduce when these particular density-independent or density-dependent factors are present in the environment. Therefore, these traits will be more prevalent in the next generation whenever these factors are a part of the environment (see Fig. 15.11).

Check Your Progress 44.4

1. Describe the effect that population density can have on competition and predation.
2. Provide examples that show how a density-independent factor can act as a selective agent.

44.5 Life History Patterns

Learning Outcomes

Upon completion of this section, you should be able to

1. Compare the two life history patterns of species.
2. List organisms that exemplify *r*-selection and those that exhibit *K*-selection.

Populations vary on such particulars as the number of births per reproduction, the age of reproduction, the life span, and the probability of living the entire life span. These particulars are part of a species' life history. Life histories contain characteristics that can be thought of as trade-offs. Each population is able to capture only so much of the available energy, and how this energy is distributed between its life span (short versus long), reproduction events (few versus many), care of offspring (little versus much), and so forth has evolved over time. Natural selection shapes the final life history of individual species, and therefore it is not surprising that related species, such as frogs and toads, may have different life history patterns, if they occupy different environmental roles (Fig. 44.13).

The logistic population growth model has been used to suggest that members of some populations are subject to *r*-selection and members of other populations are subject to *K*-selection.

a. Mouth-brooding frog, *Rhinoderma darwinii*

b. Strawberry poison arrow frog, *Dendrobates pumilio*

c. Midwife toad, *Alyces obstetricans*

Figure 44.13 Parental care among frogs and toads.
a. In mouth-brooding frogs of South America, the male carries the larvae in a vocal pouch (*brown area*), which elongates the full length of his body, before the froglets are released. **b.** In poison arrow frogs of Costa Rica, after the eggs hatch, the tadpoles wiggle onto the parent's back (*at white arrow*) and are then carried to water. **c.** The midwife toad of Europe carries strings of eggs entwined around his hind legs and takes them to water when they are ready to hatch.

**Opportunistic Species
(*r*-strategist)**

- Small individuals
- Short life span
- Fast to mature
- Many offspring
- Little or no care of
  offspring
- Many offspring die
  before reproducing
- Early reproductive age

**Equilibrium Species
(*K*-strategist)**

- Large individuals
- Long life span
- Slow to mature
- Few and large offspring
- Much care of offspring
- Most young survive to
  reproductive age
- Adapted to stable
  environment

Figure 44.14 Life history strategies. Are dandelions *r*-strategists with the characteristics noted, and are bears *K*-strategists with the characteristics noted? Most often the distinctions between these two possible life strategies are not clear-cut.

r-Selected Populations

In fluctuating or unpredictable environments, density-independent factors keep populations in the lag or exponential phase of population growth. Population size is low relative to *K*, and **r-selection** favors *r*-strategists, which tend to be small individuals that mature early and have a short life span. Most energy goes into producing many relatively small offspring, and minimal energy goes into parental care. The more offspring, the more likely it is that some of them will survive any future population crash.

Because of low population densities, density-dependent mechanisms, such as predation and intraspecific competition, are unlikely to play a major role in regulating population size and growth rates most of the time. Such organisms are often very good dispersers and colonizers of new habitats. Classic examples of such *opportunistic species* are bacteria, some fungi, many insects, rodents, and annual plants (Fig. 44.14, *left*).

K-Selected Populations

Some environments are relatively stable and predictable, and in these environments populations tend to be near *K*, with minimal fluctuations in size. Resources such as food and shelter are relatively scarce for these individuals, and those that are best able to compete have the largest number of offspring. **K-selection** favors *K*-strategists, species that allocate energy to their own growth and survival as well as to the growth and survival of their offspring. Therefore, they are usually fairly large, late to mature, have low fecundity and parity, and have a fairly long life span.

Because these organisms, termed *equilibrium species,* are strong competitors, they can become established and exclude opportunistic species. They are specialists rather than colonizers and tend to become extinct when their normal way of life is destroyed. The best possible examples of *K*-strategists include long-lived plants (saguaro cacti, oaks, cypresses, and pines), birds of prey (hawks and eagles), and large mammals (whales, elephants, bears, and humans) (Fig. 44.14, *right*). Another example of a *K*-strategist is the Florida panther, the largest mammal in the Florida Everglades. It requires a very large range and produces few offspring, which require parental care. Currently, the Florida panther is unable to compensate for a reduction in its range and is therefore on the verge of extinction.

The Two Strategies in Nature

Nature is actually more complex than these two possible life history patterns suggest. It now appears that our descriptions of *r*-strategist and *K*-strategist populations are at the ends of a continuum, and most populations lie somewhere in between these two extremes. For example, recall that plants have an alternation-of-generations life cycle, namely the sporophyte generation and the gametophyte generation. Ferns, which could be classified as *r*-strategists, distribute many spores and leave the gametophyte to fend for itself. In contrast, gymnosperms (e.g., pine trees) and angiosperms (e.g., oak trees), which could be classified as *K*-strategists, retain and protect the gametophyte. They produce seeds that contain the next sporophyte generation plus stored food. The added investment is significant, but these plants still release large numbers of seeds.

Also, adult size is not always a determining factor for the life history pattern. For example, a cod is a rather large fish weighing up to 12 kg and measuring nearly 2 m in length—but cod release gametes in vast numbers, the zygotes form in the sea, and the parents make no further investment in developing offspring. Of the 6–7 million eggs released by a single female cod, only a few will become adult fish. Cod are generally considered *r*-strategists.

 **Animation** *r* and *K* Strategies

The Biological Systems feature on page 850 describes the consequences of unchecked population growth in white-tailed deer, a *K*-selected species for which a limiting factor, namely predation, has been reduced.

Check Your Progress   44.5

1. Compare and contrast the general characteristics of a *K*-strategist and an *r*-strategist.
2. Explain why a population may vary between *K* and *r* strategies.

Biological Systems

When a Population Grows Too Large

White-tailed deer (*Odocoileus virginianus;* Fig. 44A*a*), which live from southern Canada to below the equator in South America, are prolific breeders. In one study, investigators found that two male and four female deer were able to produce 160 offspring in six years. Theoretically, the number of offspring could have reached 300 because the majority of does (female deer) tend to produce two young each time they reproduce.

A century ago, the white-tailed deer population across the eastern United States was less than half a million. Today, it is well over 200 million—even more than existed when Europeans first arrived to colonize America. This dramatic increase in population size can be attributed to a reduction of natural predators and a readily available supply of food. The natural predators of deer, such as wolves and mountain lions, are now absent from most regions mostly due to humans' killing of these animals as well as a significant loss of available habitat. In addition, deer hunting is tightly controlled by government agencies and even banned in some urban areas.

The sad reality is that in areas where deer populations have become too large, the deer suffer from starvation as they deplete their own food supply. For example, after deer hunting was banned on Long Island, New York, the deer population quickly outgrew available food resources. The animals became sickly and weak and weighed so little that their ribs, vertebrae, and pelvic bones were visible through their skin.

A very large deer population can also cause humans many problems. A homeowner is dismayed to see new plants decimated and evergreen trees damaged because of munching deer. The economic damage that large deer populations cause to agriculture, landscaping, and forestry exceeds a billion dollars per year. More alarming, over a million deer–vehicle collisions take place in the U.S. each year (Fig. 44A*b*), resulting in over a billion dollars in insurance claims, thousands of human injuries, and hundreds of human deaths—not to mention the injuries and deaths of deer. Lyme disease, transmitted by deer ticks to humans, infects over 3,000 people annually and can lead to debilitating arthritic symptoms.

Deer overpopulation hurts not only deer and humans, but other species as well. The forested areas that are overpopulated by deer have fewer understory plants. Furthermore, the deer selectively eat certain species of plants, while leaving others alone. This selection can cause long-lasting changes in the number and diversity of trees in forests, leading to a negative economic impact on logging and forestry. The number of songbirds, insects, squirrels, mice, and other animals declines with an increasing deer population. It is important, therefore, that we learn to properly manage deer populations.

One example of a successful strategy is that in some states, such as Texas, large landowners now set aside a portion of their property for a deer herd. They improve the nutrition of the herd and restrict the hunting of young bucks, but allow hunting of does. The result is a self-sustaining herd that brings the landowner economic benefits through outfitting services.

Questions to Consider

1. If the deer herd becomes too large should the government pay hunters to reduce the herd size?
2. What is the ideal size of a deer herd for a given area?
3. Is trophy deer hunting an ideal way to control the herd size?

a.

b.

Figure 44A White-tailed deer.   **a.** Buck (male deer) in rut (during mating season). **b.** Deer and car accident.

44.6 Human Population Growth

Learning Outcomes

Upon completion of this section, you should be able to

1. Describe the past and present growth patterns of the human population.
2. Identify the pressures the human population places upon the Earth's resources.

The world's population has risen steadily to a present size of about 6.7 billion people (Fig. 44.15). Prior to 1750, the growth of the human population was relatively slow, but as more reproducing individuals were added, growth increased, until the curve began to slope steeply upward, indicating that the population was undergoing exponential growth. The number of people added annually to the world population peaked at about 87 million around 1990, and currently it is a little over 79 million per year. This is roughly equal to the current populations of Argentina, Ecuador, and Peru combined.

The potential for future population growth can be appreciated by considering the **doubling time,** the length of time it takes for the population size to double. Currently, the doubling time is estimated to be 35 years. Such an increase in population size would put extreme demands on our ability to produce and distribute resources. In 52 years, the world would need double the amount of food, jobs, water, energy, and so on just to maintain the present standard of living.

Many people are gravely concerned that the amount of time needed to add each additional billion persons to the world population has become shorter and shorter. The first billion didn't occur until 1800; the second billion was attained in 1930; the third billion in 1960; and today there are over 6 billion. The world's population may level off at 8, 10.5, or 14.2 billion, depending on the speed at which the growth rate declines. Zero population growth cannot be achieved until on the average each couple has only two children and the number of women entering their reproductive years is the same as those leaving them behind.

More-Developed Versus Less-Developed Countries

The countries of the world can be divided into two general groups. In the **more-developed countries (MDCs),** typified by countries in North America, Europe, Japan, and Australia, population growth is low, and the people enjoy a high standard of living. In the **less-developed countries (LDCs),** such as certain countries in Latin America, Africa, and Asia, population growth is expanding rapidly, and the majority of people live in poverty.

The MDCs doubled their populations between 1850 and 1950. This change was largely due to a decline in the death rate, the development of modern medicine, and improved socioeconomic conditions. The decline in the death rate was followed shortly thereafter by a decline in the birthrate, so that populations in the MDCs experienced only modest growth between 1950 and 1975. This sequence of events (i.e., decreased death rate followed by decreased birthrate) is termed a **demographic transition.** Yearly growth of the MDCs has now stabilized at about 0.2%.

In contrast to the other MDCs, there is no end in sight for the U.S. population growth. Although yearly growth of the U.S. population is only 0.96%, many people immigrate to the United

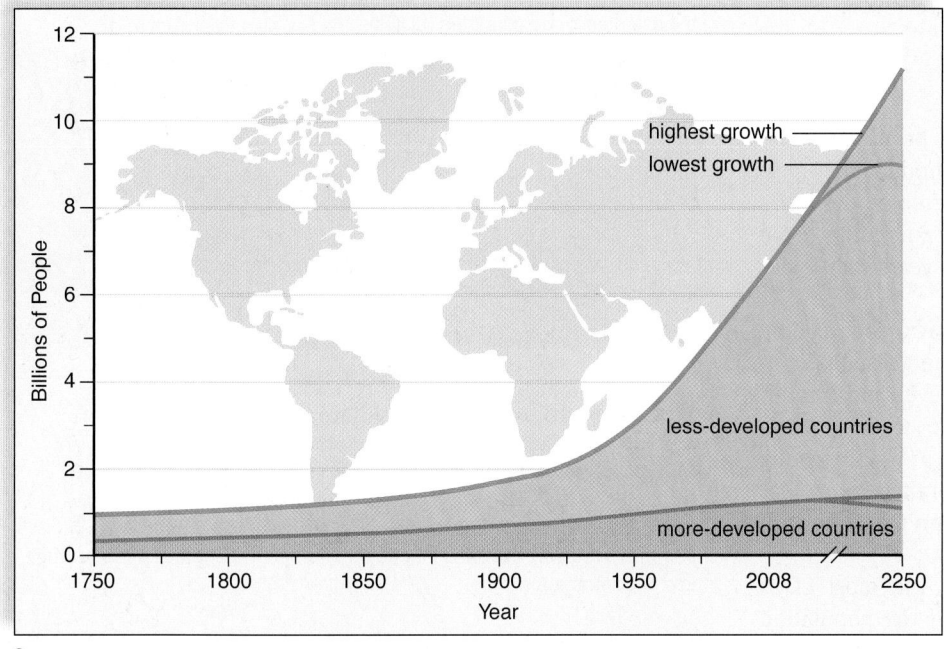

a.

b.

Figure 44.15 World population growth. **a.** The world's' population size is now about 7 billion. It is predicted that the world's population size may level off at 9 billion or increase to more than 11 billion by 2250, depending on the speed with which the growth rate declines. **b.** Lifestyle of most individuals in the MDCs (*above*) is contrasted with most individuals in the LDCs (*below*).

States each year. In addition, an unusually large number of births occurred between 1946 and 1964 (called a baby boom), resulting in a large number of women still of reproductive age.

Most of the world's population (approximately 80%) now lives in LDCs. Although the death rate began to decline steeply in the LDCs following World War II because of importation of modern medicine from the MDCs, the birthrate remained high. The yearly growth of the LDCs peaked at 2.5% between 1960 and 1965. Since that time, a demographic transition has occurred, and the death rate and the birthrate have fallen. The collective growth rate for the LDCs is now 1.5%, but many countries in sub-Saharan Africa have not participated in this decline. In some of these, women average more than five children each.

The population of the LDCs may explode from 5.5 billion today to 8 billion by 2050. Some of this increase will occur in Africa, but most will occur in Asia because the Asian population is now about 4 times the size of the African population. Asia already has 59% of the world's population living on 31% of the world's arable (farmable) land. Therefore, Asia is expected to experience acute water scarcity, a significant loss of biodiversity, and increased urban pollution. Twelve of the world's most polluted cities are in Asia.

These are suggestions about how to reduce the expected population increase in the LDCs:

1. Establish or strengthen family planning programs. A decline in growth is seen in countries with good family planning programs supported by community leaders.
2. Use social progress to reduce the desire for large families. Providing education, raising the status of women, and reducing child mortality are social improvements that could reduce this desire.
3. Delay the onset of childbearing. This, along with wider spacing of births, could help birthrate decline.

Age Distributions

The populations of the MDCs and LDCs can be divided into the three age groups introduced earlier: prereproductive, reproductive, and postreproductive (Fig. 44.16). Currently, the LDCs are experiencing a population momentum because they have more young women entering the reproductive years than older women leaving them.

Most people are under the impression that if each couple has two children, **zero population growth** (no increase in population size) will take place immediately. However, most countries today will continue growing due to the age structure of the population and the human life span. If there are more young women entering the reproductive years than there are older women leaving them, **replacement reproduction** will still result in population growth.

Many MDCs have a stable age structure, but most LDCs have a youthful profile—a large proportion of the population is younger than 15 years old. For example, in Nigeria, 41% of the population is under age 15. On average, Nigerian women have 4.73 children. As a result of these rates, the population of Nigeria is expected to increase from 149 million presently to

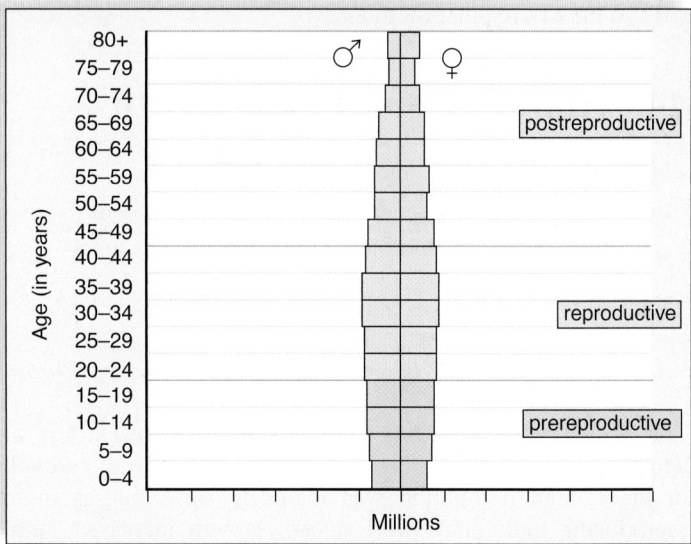

a. More-developed countries (MDCs)

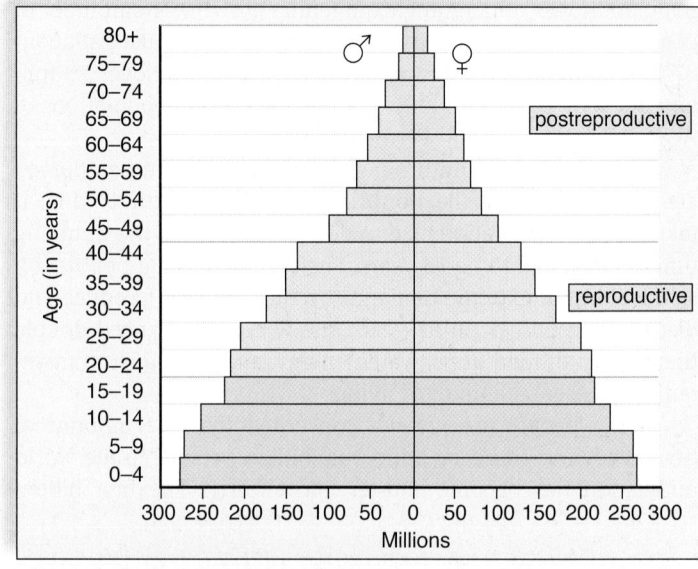

b. Less-developed countries (LDCs)

c.

Figure 44.16 Age-structure diagrams. The shape of these age-structure diagrams allows us to predict that **(a)** the populations of MDCs are approaching stabilization, and **(b)** the populations of LDCs will continue to increase for some time. **c.** Improved women's rights and increasing contraceptive use could change this scenario. Here a community health worker is instructing women in Bangladesh about the use of contraceptives.

289 million by 2050. The population of the continent of Africa is projected to increase from 957 million to 2 billion between 2008 and 2050. This means that the LDC populations will still expand, even after replacement reproduction is attained. The

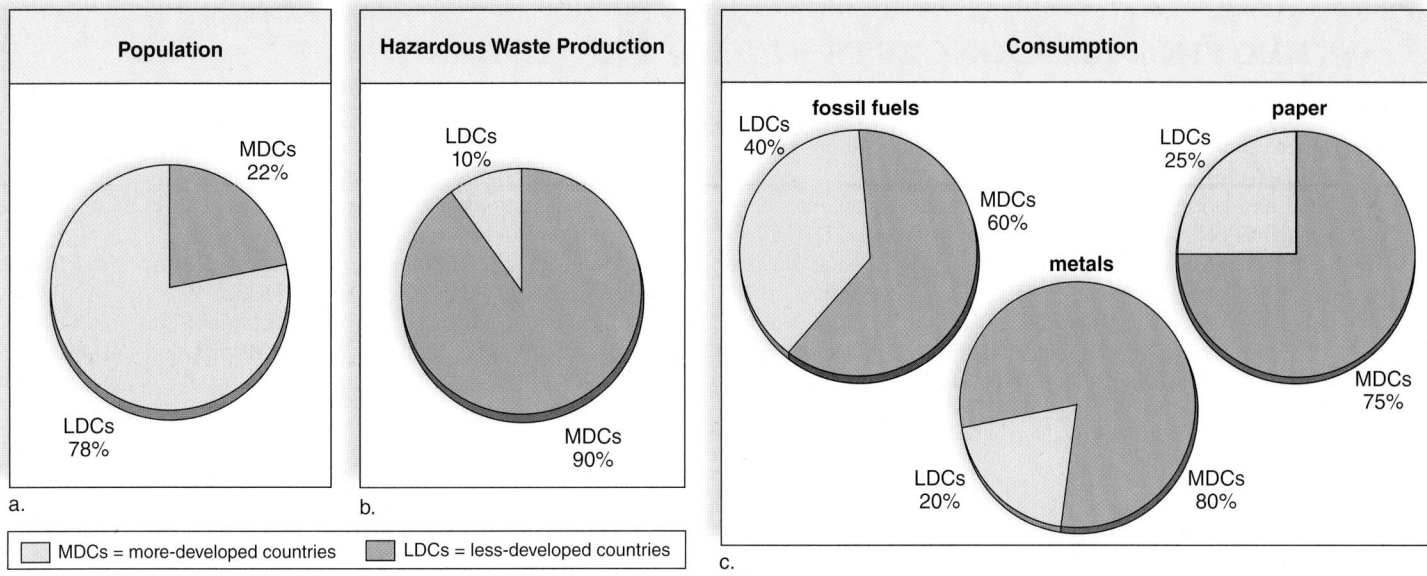

Figure 44.17 Environmental impact caused by MDCs and LDCs. **a.** The combined population of MDCs is smaller than that of LDCs: The MDCs account for 22% of the world's population, and the LDCs account for 78%. **b.** MDCs produce most of the world's hazardous wastes—90% for MDCs compared with 10% for LDCs. The production of hazardous waste is tied to (**c**), the consumption of fossil fuels, metals, and paper, among other resources. The MDCs consume 60% of fossil fuels compared with 40% for the LDCs, 80% of metals compared with 20% for the LDCs, and 75% of paper compared with 25% for the LDCs.

more quickly replacement reproduction is achieved, however, the sooner zero population growth will result.

Population Growth and Environmental Impact

Population growth is putting extreme pressure on each country's social organization, the Earth's resources, and the biosphere. Because the population of the LDCs is still growing at a significant rate, it might seem that their population increase will be the greater cause of future environmental degradation. But this is not necessarily the case because the MDCs consume a much larger proportion of the Earth's resources than do the LDCs. This consumption leads to environmental degradation, which is a major concern.

Environmental Impact

The environmental impact (EI) of a population is measured not only in terms of population size, but also in terms of resource consumption per capita and the pollution that results because of this consumption. In other words:

EI = population size × resource consumption per capita
   = pollution per unit of resource used

Therefore, there are two possible types of overpopulation: The first is simply due to population growth, and the second is due to increased resource consumption caused by population growth. The first type of overpopulation is more obvious in LDCs, and the second type is more obvious in MDCs because the per-capita consumption is so much higher in the MDCs. For example, an average family in the United States, in terms of per-capita resource consumption and waste production, is the equivalent of 30 people in India. We need to realize, therefore, that only a limited number of people can be sustained at anywhere near the standard of living in the MDCs.

The current comparative environmental impact of MDCs and LDCs is shown in Figure 44.17. The MDCs account for only about one-fourth of the world population (Fig. 44.17a). But the MDCs account for 90% of the hazardous waste production (pollution) (Fig. 44.17b). Why should that be? MDCs account for much more pollution than LDCs because they consume much greater amounts of fossil fuel, metal, and paper than do LDCs, as shown in Figure 44.17c.

As the LDCs become more industrialized, their per-capita consumption will also rise and, in some LDCs, it may nearly equal that of a more-developed country. For example, China's economy is growing rapidly and, because China has such a large population (1.3 billion), it is already competing with the United States for oil and metals on the world markets. It could be that as developing countries consume more, people in the United States will have no choice but to consume less.

Results of Environmental Impact

Consumption of resources has a negative effect on the environment. In Chapter 45, you'll learn how resource consumption leads to various pollution problems, and in Chapter 47, you'll see how our increasing environmental impact may cause a mass extinction of wildlife that will significantly impact our entire planet.

Check Your Progress   44.6

1. Compare the population growth of the LDCs with that of the MDCs.
2. Describe why replacement reproduction still leads to continued population growth.
3. Explain how an increase in the consumption of resources by LDCs would affect the consumption by MDCs.

CONNECTING *the* CONCEPTS *with the* BIG IDEAS

Energy and Homeostasis

- When free energy availability is changed, population size fluctuates. (2A1e)
- Individual and population activities are impacted by abiotic factors such as pH, salinity, water, and temperature. (2D1b*IE*)
- The cooperative partitioning of resources and niches as well as mutualistic symbiotic relationships allow populations to achieve greater survival success. (2E3b4*IE*)
- The properties of a population are different than the properties of a single individual of that population. (4B3b)

Interactions and Systems

- Computer and math models elucidate the multiple effects of populations on their communities. (4A5b*IE*)
- Species interactions affect population size and distribution and can be modeled mathematically. (4B3a1-3)
- Behavioral genetic diversity produces unique responses to the environment within a population. (4C3b)
- Environmental changes and density dependent/independent factors affect species abundance and distribution. (4B3c*IE*, 4A5c1-4)

*Find the unabridged version of all EK citations at www.glencoe.com/maderAP11.

Media Study Tools

www.glencoe.com/maderAP11

Enhance your study of this chapter with study tools and practice tests. Also ask your instructor about the resources available through ConnectPlus, including the media-rich eBook, interactive learning tools, and animations.

Summarize

44.1 Scope of Ecology

Ecology is the study of the interactions of organisms with other organisms and with the physical environment. Ecology encompasses several levels of study: organism, population, community, ecosystem, and finally the biosphere. Ecologists are particularly interested in how interactions affect the distribution, abundance, and life history strategies of organisms.

44.2 Demographics of Populations

Demographics include statistical data about a population. For example, population density is the number of individuals per unit area or volume. Distribution of individuals can be uniform, random, or clumped. A population's distribution is often determined by abiotic factors, that is, resources such as water and sunlight, temperature, and availability of nutrients.

Population growth is dependent on number of births and number of deaths, as well as immigration and emigration. The number of births minus the number of deaths results in the rate of natural increase (growth rate). Mortality (deaths per capita) within a population can be recorded in a life table and illustrated by a survivorship curve. The pattern of population growth is reflected in the age distribution of a population, which consists of prereproductive, reproductive, and postreproductive segments. Populations that are growing exponentially have a pyramid-shaped age distribution pattern.

44.3 Population Growth Models

One model for population growth assumes that the environment offers unlimited resources. In the example given, the members of the population have discrete reproductive events, and therefore the size of next year's population is given by the equation $N_{t+1} = RN_t$. Under these conditions, exponential growth results in a J-shaped curve.

Most environments restrict growth, and exponential growth cannot continue indefinitely. Under these circumstances, an S-shaped, or logistic, growth curve results. The growth of the population is given by the equation $N/t = rN (K-N)/K$ for populations in which individuals have repeated reproductive events. The term $(K-N)/K$ represents the unused portion of the carrying capacity (K). When the population reaches carrying capacity, the population stops growing because environmental conditions oppose biotic potential, the maximum rate of natural increase (growth rate) for a population.

44.4 Regulation of Population Size

Population growth is limited by density-independent factors (e.g., abiotic factors such as weather events) and density-dependent factors (biotic factors such as predation and competition). Other means of regulating population growth exist in some populations. For example, territoriality is possibly a means of population regulation. Other populations seem to not be regulated, and their population size fluctuates widely.

44.5 Life History Patterns

The logistic growth model has been used to suggest that the environment promotes either *r*-selection or *K*-selection. In unpredictable environments, *r*-selection occurs due to density-independent factors. Energy is allocated to producing as many small offspring as possible. Adults remain small and do not invest in parental care of offspring. In relatively stable environments, *K*-selection occurs and density-dependent factors affect population size. Energy is allocated to survival and repeated reproductive events. The adults are large and invest in parental care of offspring. Actual life histories contain trade-offs between these two patterns, and such trade-offs are subject to natural selection.

44.6 Human Population Growth

The human population is still expanding, but deceleration has begun. It is unknown when the population size will level off, but it may occur by 2050. Substantial increases are expected in certain LDCs (less-developed countries) of Asia and also Africa. Support for family planning, human development, and delayed childbearing could help prevent the increase.

Key Terms

▨ Assess

Reviewing This Chapter

1. What are the various levels of ecological study?   839
2. What largely determines a population's density, distribution, and growth rate?   840–41
3. How does the mortality pattern and age distribution affect the growth rate?   841–43
4. What type of growth curve indicates that exponential growth is occurring? What are the environmental conditions for exponential growth?   844
5. What type of growth curve levels off because of environmental conditions? What environmental conditions are involved?   844–45
6. What is the carrying capacity of an area?   845
7. Are density-independent or density-dependent factors more likely to regulate population size? Why?   847–48
8. What other types of factors are involved in regulating population size?   848
9. Why would you expect the life histories of natural populations to vary and contain some characteristics that are so-called *r*-selected and some that are so-called *K*-selected?   848–49
10. What type of growth curve presently describes the growth of the human population? In what types of countries is most of this growth occurring, and how might it be curtailed?   851–52

Testing Yourself

Choose the best answer for each question.

1. Which of these levels of ecological study involves an interaction between abiotic and biotic components?
   a. organisms
   b. populations
   c. communities
   d. ecosystem
   e. All of these are correct.

2. When phosphorus is made available to an aquatic community, the algal populations suddenly bloom. This indicates that phosphorus is
   a. a density-dependent regulating factor.
   b. a reproductive factor.
   c. a limiting factor.
   d. an *r*-selection factor.
   e. All of these are correct.

3. A J-shaped growth curve can be associated with
   a. exponential growth.
   b. biotic potential.
   c. unlimited resources.
   d. rapid population growth.
   e. All of these are correct.

4. An S-shaped growth curve
   a. occurs when resources become limited.
   b. includes an exponential growth phase.
   c. occurs in both natural and laboratory populations.
   d. is a way to control overpopulation.
   e. All of these are correct.

5. If a population has a type I survivorship curve (most have a long life span), which of these would you also expect?
   a. a single reproductive event per adult
   b. overlapping generations
   c. reproduction occurring near the end of the life span
   d. a very low birthrate
   e. None of these are correct.

6. A pyramid-shaped age distribution means that
   a. the prereproductive group is the largest group.
   b. the population will grow for some time in the future.
   c. the country is more likely an LDC than an MDC.
   d. fewer women are leaving the reproductive years than entering them.
   e. All of these are correct.

7. Which of these is a density-independent regulating factor?
   a. competition
   b. predation
   c. weather
   d. resource availability
   e. the average age when childbearing begins

8. Fluctuations in population growth can correlate to changes in
   a. predation.
   b. weather.
   c. resource availability.
   d. parasitism.
   e. All of these are correct.

9. A species that has repeated reproductive events, lives a long time, but suffers a crash due to the weather is exemplifying
   a. *r*-selection.
   b. *K*-selection.
   c. a mixture of both *r*-selection and *K*-selection.
   d. density-dependent and density-independent regulation.
   e. a pyramid-shaped age structure diagram.

10. In which pair is the first included in the second?
    a. habitat—population
    b. population—community
    c. community—ecosystem
    d. ecosystem—biosphere
    e. All of these except a are correct.

11. How does population density differ from population distribution?
    a. Actually, population density is the same as population distribution.
    b. The greater the population density, the more likely that the population distribution will be clumped.
    c. Population density has nothing to do with population distribution.
    d. The less dense the population, the more likely that the population distribution will be equally spaced.
    e. Both b and c are correct.

12. Which one of these has nothing to do with a population's growth curve?
    a. exponential growth
    b. biotic potential
    c. environmental conditions
    d. rate of natural increase
    e. All of these pertain to a population's growth curve.

13. When a population is undergoing logistic growth, environmental conditions determine
    a. whether to expect exponential growth.
    b. the carrying capacity.
    c. the length of the lag phase.
    d. whether the population is subject to density-independent regulation.
    e. All of these are correct.

14. Which of these statements is correct?
    a. A life table is based on whether a population has cohorts or not.
    b. The life table supplies the information for constructing the survivorship curve.
    c. A life table, like an ideal weight table, supplies information that can keep members of a population healthy.
    d. A life table tells which ideal survivorship curve is most appropriate to a particular population.
    e. Both b and d are correct.

15. Because of the postwar baby boom, the U.S. population
    a. can never level off.
    b. presently has a pyramid-type age distribution.
    c. is subject to regulation by density-dependent factors.
    d. is more conservation-minded than before.
    e. None of these is correct.

Engage

Virtual Lab
Population Biology

The virtual lab "Population Biology" provides an interactive look at the factors that influence population size.

Thinking Scientifically

1. In the winter moth life cycle, parasites were found to be a less important cause of mortality than cold winter weather and predators. Give an evolutionary explanation for the inefficiency of parasites in controlling population size.

2. You are a river manager charged with maintaining the flow through the use of dams so that trees, which have equilibrium life histories, can continue to grow along the river. What would you do?

Bioethical Issue

Population Control

The answer to how to curb the expected increase in the world's population lies in discovering how to curb the rapid population growth of the less-developed countries. In these countries, population experts have discovered what they call the "virtuous cycle." Family planning leads to healthier women, and healthier women have healthier children, and the cycle continues. Women no longer need to have many babies for only a few to survive. More education is also helpful because better-educated people are more interested in postponing childbearing and promoting women's rights. Women who have equal rights with men tend to have fewer children.

"There isn't any place where women have had the choice that they haven't chosen to have fewer children," says Beverly Winikoff at the Population Council in New York City. "Governments don't need to resort to force." Bangladesh is a case in point. Bangladesh is one of the densest and poorest countries in the world. In 1990, the birthrate was 4.9 children per woman, and now, in 2011, it is 2.6. This achievement was due in part to the Dhaka-based Grameen Bank, which lends small amounts of money mostly to destitute women to start a business. The bank discovered that when women start making decisions about their lives, they also start making decisions about the size of their families. Family planning within Grameen families is twice as common as the national average; in fact, those women who get a loan promise to keep their families small! Also helpful has been the network of village clinics that counsel women who want to use contraceptives. The expression "contraceptives are the best contraceptives" refers to the fact that you don't have to wait for social changes to get people to use contraceptives—the two feed back on each other.

Recently, some of the less-developed countries, faced with economic crises, have cut back on their family planning programs, and the more-developed countries have not taken up the slack. Indeed, some foreign donors have also cut back on aid—the United States by one-third. Are you in favor of foreign aid to help countries develop family planning programs? Why or why not?

A reintroduced wolf in Yellowstone National Park.

45

Community and Ecosystem Ecology

By 1926 the last known wolf pack had been eliminated from Yellowstone National Park. The extinction of the wolves produced a multitude of ecological effects that would ripple through the entire park for the next 70 years. Fewer wolves meant less predation upon the elk and other large herbivores. An increase in herbivores means more pressure upon their food sources. Cottonwoods, willows, and streamside vegetation soon became depleted due to the increasing elk population.

Overbrowsing of the streamside vegetation produced a ripple effect throughout the entire stream ecosystem. This decreased the stability of the stream bank, which in turn caused increased erosion and sediment load in the streams. Fish, insects, birds, and all of the wildlife associated with the streams were negatively impacted.

The reintroduction of wolves into Yellowstone began in 1995. Since then, biologists have seen signs of the stream bank stabilizing due to the decreased browsing by elk and other herbivores. This decrease in predation is due to the wolves' preying upon the elk and large herbivores within the park. Many of the problems that were caused by the eradication of wolves from Yellowstone have begun to be repaired as a direct result of their reintroduction. This chapter examines various types of interactions that occur among the populations of a community. It also looks at the interactions with the nonliving component of an ecosystem.

As you read through the chapter, think about the following questions:

1. In a community, are some species interactions more common than others?
2. Do communities ever truly reach a climax status?
3. What would happen to the trophic levels in the community if the producer base is changed?

BEFORE YOU BEGIN

Before beginning this chapter, take a few moments to review the following discussions.

Section 6.1 What role does thermodynamics play in the flow of energy through an ecosystem?

Section 43.2 Does a species' behavior determine the type of relationship it has with other species?

Section 44.4 How do density-dependent and density-independent factors influence community dynamics?

FOLLOWING *the* BIG IDEAS

CHAPTER 45 COMMUNITY AND ECOSYSTEM ECOLOGY

Evolution	The pace of evolution is tied to the stability of an organism's environment.
Energy and Homeostasis	Food chains and webs assure that biomolecules cycle through the ecosystem.
Interactions and Systems	Communities differ in their species makeup and in their abiotic conditions.

45.1 Ecology of Communities

Populations do not occur as single entities; they are part of a community. A **community** is an assemblage of populations of different species interacting with one another in the same environment. Communities come in different sizes, and it is sometimes difficult to decide where one community ends and another begins.

A fallen log can be considered a community because the many different populations living on and within it interact with one another. The fungi break down the log and provide food for the earthworms and insects living in and on the log. Those insects may feed on one another, too. If birds flying throughout the forest feed on the insects and worms living in and on the log, then they are also part of the larger forest community. In this section we consider characteristics of communities and the interactions that take place in them.

Community Structure

Two characteristics—species composition and diversity—allow us to compare communities. The species composition of a community, also known as **species richness,** is simply a listing of the various species found in that community. **Species diversity** includes both species richness and species evenness, or the relative abundance of the different species.

Species Composition

It is apparent, by comparing the photographs in Figure 45.1, that a coniferous forest has a composition different from that of a tropical rain forest. The narrow-leaved evergreen trees are prominent in a coniferous forest, and broad-leaved evergreen trees are numerous in a tropical rain forest. As the list of mammals demonstrates, a coniferous community and a tropical rain forest community contain different types of mammals. Ecologists comparing these two communities would also find differences in the other plants and animals as well. In the end, ecologists would conclude that not only are the species compositions of these two communities different, but the tropical rain forest has more species, and therefore, higher species richness.

a.

b.

Figure 45.1 Community structure. Communities differ in their composition, as witnessed by their predominant plants and animals. The diversity of communities is described by the richness of species and their relative abundance. **a.** These are mammals commonly found in a coniferous forest. **b.** These are mammals commonly found in a tropical rain forest.

Species Diversity

The *diversity* of a community goes beyond composition because it includes not only a listing of all the species in a community but also the relative abundance of each species. To take an extreme example: A forest sampled in West Virginia has 76 yellow poplar trees but only one American elm (among other species). If we were simply walking through this forest, we might miss seeing the American elm. If, instead, the forest had 36 poplar trees and 41 American elms, the forest would seem more diverse to us, and indeed it would be more diverse. The greater the species richness and the more even the distribution of the species in a community, the greater the species diversity.

Video
Tallgrass Prairie Ecology

The Nature of Science feature "Island Biogeography Pertains to Biodiversity", on page 862, discusses one of the landmark studies in community structure, which led to the **island biogeography model** of species diversity. This model postulates that species diversity on an island depends on the distance from the mainland (closer islands have more diversity than islands farther away) and the total area of the island (large islands have more diversity than small islands). Island biogeography studies play a role in determining the ideal size for conservation areas, among other questions.

Community Interactions

This chapter examines the various types of community interactions and their importance to the structure of a community. Such interactions illustrate some of the most important evolutionary selection pressures acting on individuals. They also help us develop an understanding of how biodiversity can be preserved.

Habitat and Ecological Niche

Each species occupies a particular position in the community, both in a spatial sense (where it lives) and in a functional sense (what role it plays). A particular place where a species lives and reproduces is its **habitat.** The habitat might be the forest floor, a swift stream, or the ocean's edge. The **ecological niche** is the role a species plays in its community. The niche includes the resources used to meet energy, nutrient, and survival demands.

For a dragonfly larva, its habitat is a pond or lake where it eats other insects. The pond must contain vegetation where the dragonfly larva can hide from its predators, such as fish and birds. In addition, the water must be clear enough for it to see its prey and warm enough for it to be in active pursuit.

Because it is difficult to study all aspects of niche, some observations focus only on one aspect, such as study of the Galápagos finches (see Section 15.2). Each of these bird species is a specialist species because each has a limited diet, tolerates only small changes in environmental conditions, and lives in a specific habitat. Some species, such as raccoons, roaches, and house sparrows, are generalist with a broad range of niches. These organisms have a diversified diet, tolerate a wide range of environmental conditions, and can live in a variety of places.

Video
Finches Natural Selection

Video
Cichlid Specialization

Because a species' niche is affected by both abiotic factors (such as climate and habitat) and biotic factors (such as competitors, parasites, and predators), ecologists like to distinguish between the fundamental and the realized niche. The *fundamental niche* comprises all the abiotic conditions under which a species could survive when adverse biotic conditions are absent. The *realized niche* comprises those conditions under which a species does survive when adverse biotic interactions, such as competition and predation, are present. Therefore, a species' fundamental niche tends to be larger than its realized niche.

Competition Between Populations

Competition occurs when members of different species try to use a resource (such as light, space, or nutrients) that is in limited supply. If the resource is not in limited supply, there is no competition. In the 1930s, Russian ecologist G. F. Gause grew two species of *Paramecium* in one test tube containing a fixed amount of food. Although each population survived when grown separately, only one survived when they were grown together (Fig. 45.2). The successful *Paramecium* population had a higher biotic potential than the unsuccessful population. After observing the outcome of other, similar experiments in the laboratory, ecologists formulated the

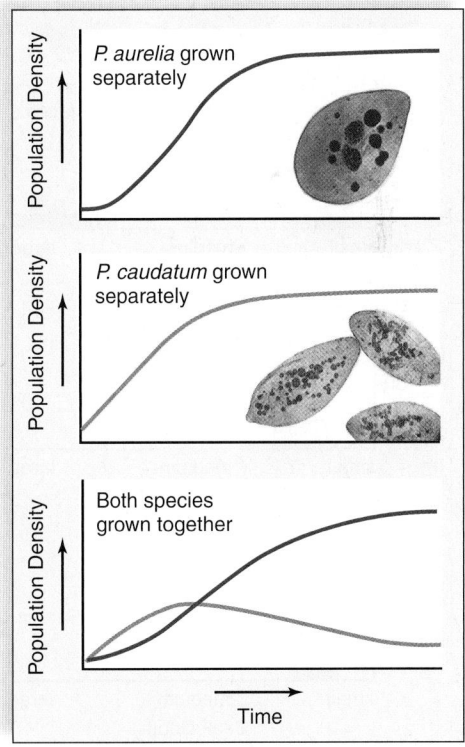

Figure 45.2 Competition between two laboratory populations of *Paramecium*. When grown alone in pure culture, *Paramecium caudatum* and *Paramecium aurelia* exhibit sigmoidal growth. When the two species are grown together in mixed culture, *P. aurelia* is the better competitor, and *P. caudatum* dies out. Both attempted to exploit the same resources, which led to competitive exclusion.
Source: Data from G. F. Gause, The Struggle for Existence, 1934, Williams & Wilkins Company, Baltimore, MD.

competitive exclusion principle, which states that no two species can indefinitely occupy the same niche at the same time.

What does it take to have different ecological niches so that extinction of one species is avoided? In another laboratory experiment, Gause found that two species of *Paramecium* did continue to occupy the same tube when one species fed on food at the bottom of the tube and the other fed on food suspended in solution. Individuals of each species that can avoid competition have a reproductive advantage. **Resource partitioning** decreases competition between two species, leading to increased niche specialization and less niche overlap. An example of resource partitioning involves owl and hawk populations. Owls and hawks feed on similar prey (small rodents), but owls are nocturnal hunters and hawks are diurnal hunters. What could have been one niche became two more specialized niches because of a divergence of behavior.

It is possible to observe the process of niche specialization in nature. When three species of ground finches of the Galápagos Islands occur on separate islands, their beaks tend to be the same intermediate size, enabling each to feed on a wider range of seeds (Fig. 45.3; see also Fig. 15.9). Where the three species co-occur, selection has favored divergence in beak size because the size of the beak affects the kinds of seeds that can be eaten.

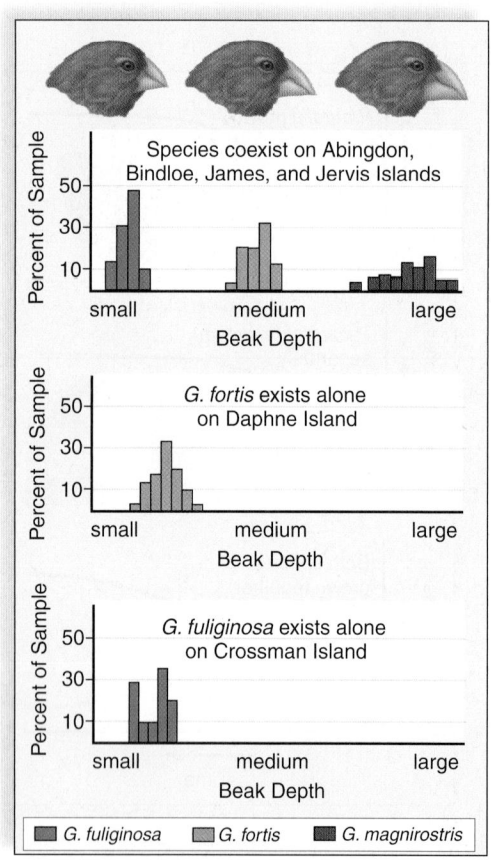

Figure 45.3 Character displacement in finches on the Galápagos Islands. When *Geospiza fuliginosa, G. fortis,* and *G. magnirostris* are on the same island, their beak sizes are appropriate to eating small-, medium-, and large-sized seeds respectively. When *G. fortis* and *G. fuliginosa* are on separate islands, their beaks have the same intermediate size, which allows them to eat seeds that vary in size.

In other words, competition has led to resource partitioning and, therefore, niche specialization.

The tendency for characteristics to be more divergent when populations belong to the same community than when they are isolated is termed **character displacement.** Character displacement is often used as evidence that competition and resource partitioning have taken place.

Video
Finches Adaptive Radiation

Niche specialization can be subtle. Five different species of warblers that occur in North American forests are all nearly the same size, and all feed on a type of spruce tree caterpillar. For all these species to exist in the same community, they must be avoiding direct competition. In a famous study, ecologist Robert MacArthur recorded the amount of time each of five warbler species spent in different regions of spruce canopies to determine where each species did most of its feeding (Fig. 45.4). He discovered that each species primarily used different parts of the tree canopy and, in that way, had a more specialized niche.

As another example, consider that swallows, swifts, and martins all eat flying insects and parachuting spiders. These birds even frequently fly together in mixed flocks. But each type of bird has a different nesting site and migrates at a slightly different time of year. In doing so, they are not competing for the same food source at the same time.

In all these cases of niche partitioning, we have merely assumed that what we observe today is due to competition in the past. Joseph Connell has studied the distribution of barnacles on the Scottish coast, where a small barnacle lives on the high part of the intertidal zone, and a large barnacle lives on the lower part (Fig. 45.5). Free-swimming larvae of both species attach themselves to rocks randomly throughout the intertidal zone; however, in the lower zone, the large *Balanus* barnacles seem to either force the smaller *Chthamalus* individuals off the rocks or grow over them.

To test his observation, Connell removed the larger barnacles and found that the smaller barnacles grew equally well on all parts of the rock. The entire intertidal zone is the fundamental niche for *Chthamalus,* but competition is restricting the range of *Chthamalus* on the rocks. *Chthamalus* is more resistant to drying out than is *Balanus;* therefore, it has an advantage that permits it to grow in the upper intertidal zone where it is exposed more often to the air. In other words, the upper intertidal zone becomes the realized niche for *Chthamalus.*[1]

Predator-Prey Interactions

Predation occurs when one living organism, called the **predator,** feeds on another, called the **prey.** In its broadest sense, predaceous consumers include not only animals such as lions, which kill zebras, but also filter-feeding blue whales, which strain krill from ocean waters, and herbivorous deer, which browse on trees and shrubs.

Video
Harris Hawk

1 Connell, J. H. 1961. The influence of interspecific competition. . . . *Ecology* 42: 710–23.

Cape May
warbler
Black-throated
green warbler
Bay-breasted
warbler
Blackburnian
warbler
Yellow-rumped
warbler

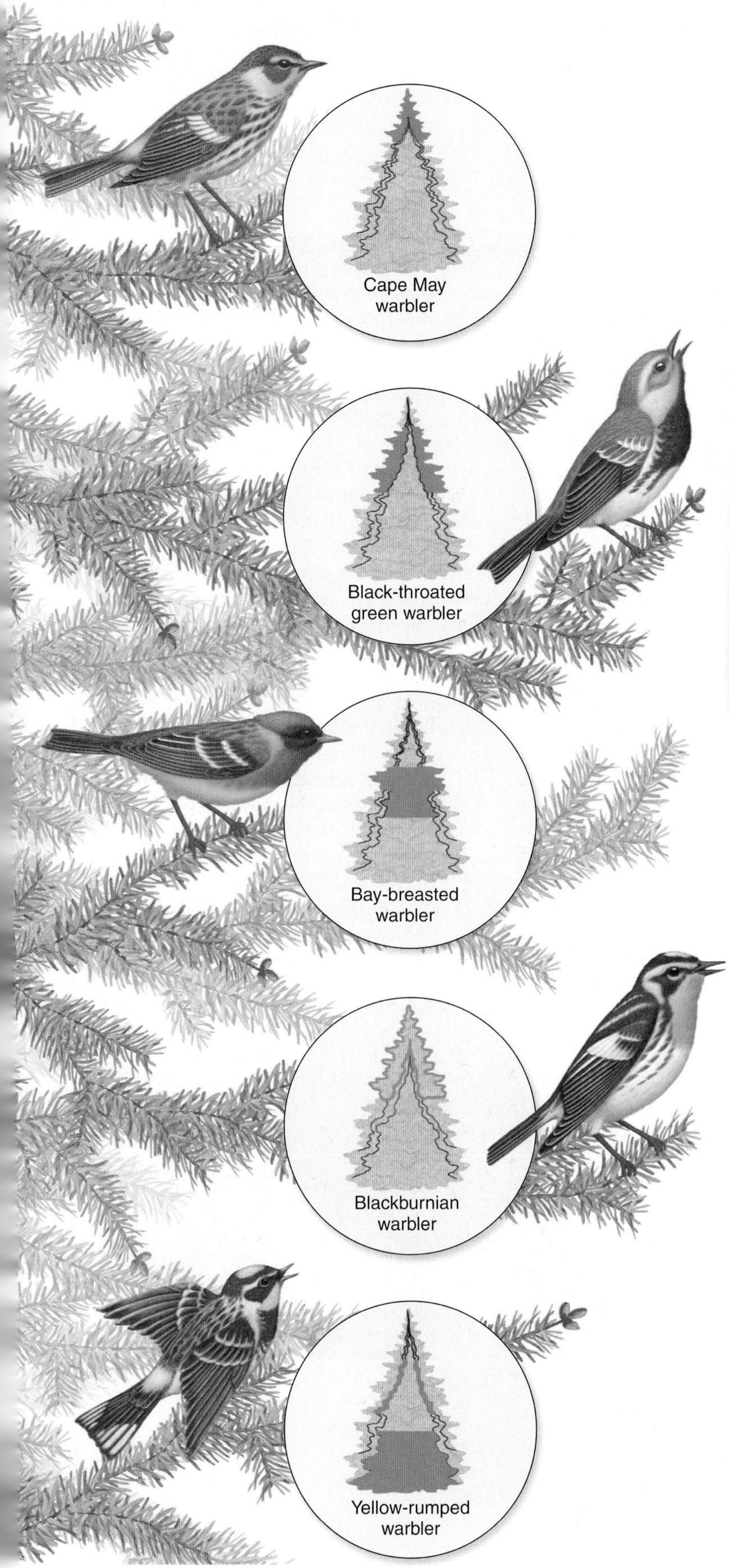

Figure 45.4 Niche specialization among five species of coexisting warblers. The diagrams represent spruce trees. The time each species spent in various portions of the trees was determined; each species spent more than half its time in the blue regions.

high tide
Chthamalus
area of
competition
Balanus
low tide

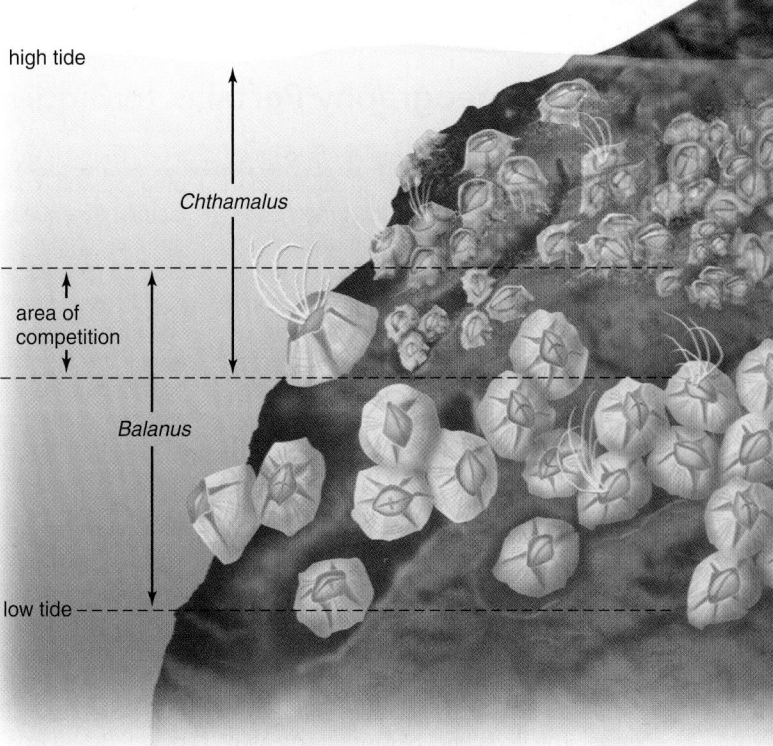

Figure 45.5 Competition between two species of barnacles. Competition prevents two species of barnacles from occupying as much of the intertidal zone as possible. Both exist in the area of competition between *Chthamalus* and *Balanus*. Above this area only *Chthamalus*, a small barnacle, survives, and below it only *Balanus*, a large barnacle, survives.

Parasitism can be considered a type of predation because one individual obtains nutrients from another (see page 865). Parasitoids are organisms that lay their eggs inside of a host. The developing larvae feed on the host and sometimes cause the host to die.

Predation and parasitism tend to increase the abundance of the predator and parasite, but at the expense of the prey or host population.

Predator-Prey Population Dynamics

Do predators reduce the population density of prey? In another classic experiment, G. F. Gause reared the ciliated protozoans *Paramecium caudatum* (prey) and *Didinium nasutum* (predator) together in a culture medium. He observed that *Didinium* ate all the *Paramecium* and then died of starvation.

In nature, we can find a similar example. When a gardener brought prickly-pear cactus to Australia from South America, the cactus spread out of control until millions of acres were covered with nothing but cacti. The cacti were brought under control when a moth from South America, whose caterpillar feeds only on the cactus, was introduced. The caterpillar was a voracious predator on the cactus, efficiently reducing the cactus population. Now both cactus and moth are found at greatly reduced densities in Australia.

Nature of Science

Island Biogeography Pertains to Biodiversity

Would you expect larger coral reefs to have a greater number of species, called species richness, than smaller coral reefs? The area (space) occupied by a community can have a profound effect on its biodiversity. American ecologists Robert MacArthur and E. O. Wilson developed a general model of island biogeography to explain and predict the effects of (1) distance from the mainland and (2) size of an island on community diversity.

Imagine two new islands that contain no species at all. One of these islands is near the mainland, and one is far from the mainland. Which island will receive more immigrants from the mainland? Most likely, the near one, because it's easier for immigrants to get there. Similarly, imagine two islands that differ in size. Which island will be able to support a greater number of species? The large one, because its greater amount of resources can support more populations, while species on the smaller island may eventually face extinction due to scarce resources.

MacArthur and Wilson studied the biodiversity on many island chains, including the West Indies, and discovered that species richness does correlate positively with island distance from mainland and island size. They developed a model of island biogeography that takes into account both factors. An equilibrium is reached when the rate of species immigration matches the rate of species extinction due to limited resources (Fig. 45Aa). Notice that the equilibrium point is highest for a large island that is near the mainland. The equilibrium could be dynamic (new species keep on arriving, and new extinctions keep on occurring), or the composition of the community could remain steady unless disturbed.

Biodiversity

Conservationists note that the trends graphed in Figure 45Ac apply to their studies. Humans often create preserved islands of habitat that are surrounded by farms, towns, and cities, or even water. For example, in Panama, Barro Colorado Island (BCI) (Fig. 45Ab) was created in the 1920s when a river was dammed to form a lake. As predicted by the model of island biogeography, BCI lost species because it was a small island that had been cut off from the mainland. Among the species that became extinct were the top predators on the island, namely the jaguar, puma, and ocelot. Thereafter, medium-sized terrestrial mammals, such as the coatimundi, increased in number. Because the coatimundi is an avid predator of bird eggs and nestlings, soon there were fewer bird species on BCI, even though the island is large enough to support them.

The model of island biogeography suggests that the larger the conserved area, the better the chance of preserving more species. Is it possible to increase the amount of habitat without using more area? If the environment has patches, it has a greater number of habitats—and thus greater diversity. As gardeners, we are urged to create patches in our yards if we wish to attract more butterflies and birds! One way to introduce patchiness is through stratification, the use of layers. Just as a high-rise apartment building allows more human families to live in an area, so can stratification within a community provide more and different types of living space for different species. Stratification could be accomplished by having low plants, medium shrubs, and taller trees.

Questions to Consider

1. Why is the mammalian and reptilian diversity so low on islands?
2. How can city planners apply island biogeography to urban development?

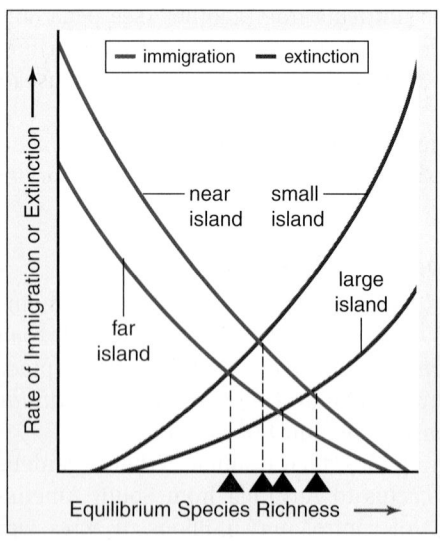

a.

b.

c.

Figure 45A Species richness.
a. The balance between immigration and extinction for four possible types of islands. A large island that is near the mainland will have a higher equilibrium species richness than the other types of islands. **b.** Barro Colorado Island **c.** Central Park in New York City.

This raises an interesting point: The population density of the predator can be affected by the prevalence of the prey. In other words, the predator-prey relationship is actually a two-way street. In that context, consider that at first the biotic potential (maximum reproductive rate) of the prickly-pear cactus was maximized, but factors that oppose biotic potential came into play after the moth was introduced. And the biotic potential of the moth was maximized when it was first introduced, but the carrying capacity decreased after its food supply was diminished.

Sometimes, instead of remaining in a steady state, predator and prey populations increase and then decrease in size. We can appreciate that an increase in predator population size is dependent on an increase in prey population size. But what causes a decrease in population size instead of the establishment of a steady population size? At least two possibilities account for the reduction:

1. Perhaps the biotic potential (reproductive rate) of the predator is so great that its increased numbers overconsume the prey, and then as the prey population declines, so does the predator population; or
2. Perhaps the biotic potential of the predator is unable to keep pace with the prey, and the prey population overshoots the carrying capacity and suffers a crash. Now the predator population follows suit because of a lack of food.

In either case, the result will be a series of peaks and valleys, with the predator population size lagging slightly behind that of the prey population.

A famous example of predator-prey cycles occurs between the snowshoe hare (*Lepus americanus*) and the Canada lynx (*Lynx canadensis*), a type of small predatory cat (Fig. 45.6). The snowshoe hare is a common herbivore in the coniferous forests of North America, where it feeds on terminal twigs of various shrubs and small trees. The Canada lynx feeds on snowshoe hares but also on ruffed grouse and spruce grouse, two types of birds. Studies have revealed that the hare and lynx populations cycle regularly, as graphed in Figure 45.6.

Investigators at first assumed that the lynx brings about a decline in the hare population, and that this accounts for the cycling. But others have noted that the decline in snowshoe hare abundance was accompanied by low growth and reproductive rates, which could be signs of a food shortage. Experiments were done to test whether predation or lack of food caused the decline in the hare population. The results suggest that both factors combine to produce a low hare population and the cycling effect.

Prey Defenses

Prey defenses are mechanisms that thwart the possibility of being eaten by a predator. Prey species have evolved a variety of mechanisms that enable them to avoid predators, including heightened senses, speed, protective armor, protective spines or thorns, tails and appendages that break off, and chemical defenses.

One common strategy to avoid capture by a predator is **camouflage,** or the ability to blend into the background. Some animals have cryptic coloration that blends them into their surroundings. For example, flounders can take on the same

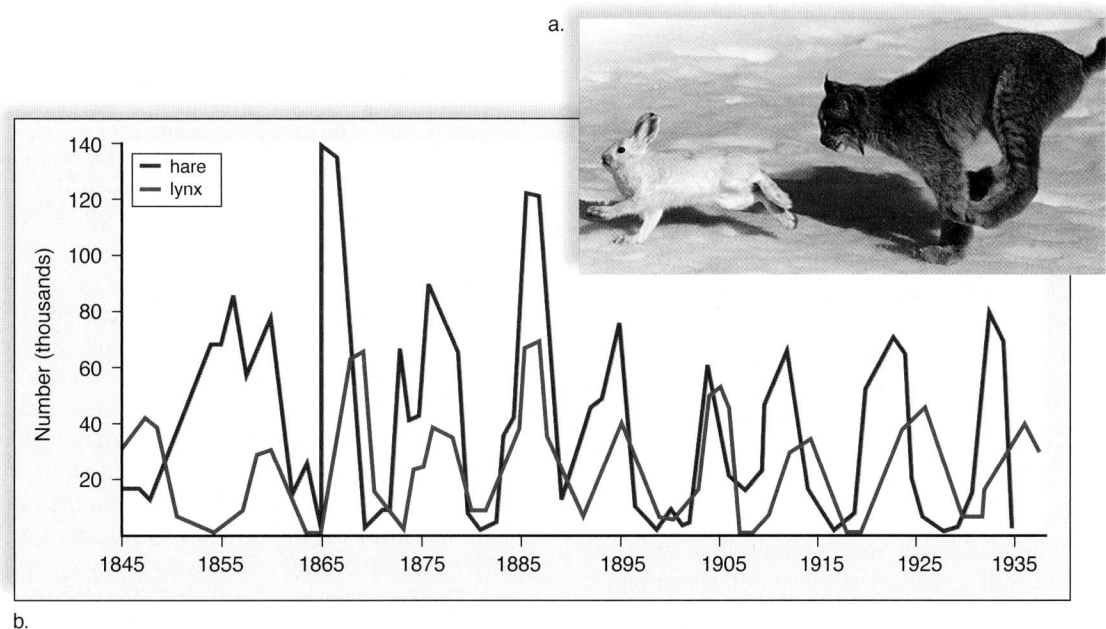

Figure 45.6 Predator-prey interaction between a lynx and a snowshoe hare. **a.** A Canada lynx, *Lynx canadensis*, is a solitary predator. A long, strong forelimb with sharp claws grabs its main prey, the snowshoe hare, *Lepus americanus*. The lynx lives in northern forests. Its brownish gray coat blends in well against tree trunks, and its long legs enable it to walk through deep snow. **b.** The number of pelts received yearly by the Hudson Bay Company for almost 100 years shows a pattern of ten-year cycles in population densities. The snowshoe hare population reaches a peak abundance before that of the lynx by a year or more.

a. Camouflage   b. Warning colorization   c. Fright

Figure 45.7 Antipredator defenses. **a.** Flounder is a fish that blends in with its background. **b.** The skin secretions of poison arrow frogs make them dangerous to touch. **c.** The large false head of the South American lantern fly may startle a potential predator.

coloration as their background (Fig. 45.7*a*). Many examples of protective camouflage are known: Walking sticks look like twigs; katydids look like sprouting green leaves; some caterpillars resemble bird droppings; and some insects, moths, and lizards blend into the bark of trees.

> **Video**
> Flounder
> Camouflage

> **Video**
> Leaf-Tailed Gecko

Another common antipredator defense among animals is *warning coloration*, which tells the predator that the prey is potentially dangerous. As a warning to possible predators, poison arrow frogs are brightly colored (Fig. 45.7*b*). Also, many animals, including caterpillars, moths, and fishes, possess false eyespots that confuse or startle another animal. Other animals have elaborate structures that cause the *startle response*. The South American lantern fly has a large false head with false eyes, making it resemble the head of an alligator (Fig. 45.7*c*). However, antipredator defenses are not always false. A porcupine certainly looks formidable, and for good reason. Its arrowlike quills have barbs that dig into the predator's flesh and penetrate even deeper as the enemy struggles after being impaled. In the meantime, the porcupine runs away.

Association with other prey is another common strategy that may help avoid capture. Flocks of birds, schools of fish, and herds of mammals stick together as protection against predators. Baboons detect predators visually, and antelopes detect predators by smell and sometimes forage together, gaining double protection against stealthy predators. The gazellelike springboks of southern Africa jump stiff-legged 2–4 m into the air when alarmed. Such a jumble of shapes and motions might confuse an attacking lion, allowing the herd to escape.

Mimicry

Mimicry occurs when one species resembles another that possesses an overt antipredator defense. A mimic that lacks the defense of the organism it resembles is called a Batesian mimic (named for Henry Bates [1825–92], who described the phenomenon). Once an animal experiences the defense of the model, it remembers the coloration and avoids all animals that look similar. Figure 45.8*a, b* shows two insects (flower fly and longhorn beetle) that resemble a yellow jacket wasp but lack the wasp's ability to sting. Classic examples of Batesian mimicry include

a. Flower fly

b. Longhorn beetle

c. Bumblebee

d. Yellow jacket

Figure 45.8 Mimicry among insects. **a.** A flower fly, *Chrysotoxum*, and (**b**) a longhorn beetle, *Strophiona*, are Batesian mimics because they are incapable of stinging another animal, yet they have the same appearance as the yellow jacket wasp (**d**). **c.** The bumblebee, *Bombus*, and (**d**) the yellow jacket, *Vespula*, are Müllerian mimics because they both use stinging as a defense.

the scarlet kingsnake mimicking the venomous coral snake and the *Papilio memnon* butterfly mimicking foul tasting butterflies.

Some species that actually have the same defense resemble each other. Many stinging insects—bees, wasps, hornets, and bumblebees—have the familiar alternating black and yellow bands. Once a predator has been stung by a black and yellow insect, it is wary of that color pattern in the future. Mimics that share the same protective defense are called Müllerian mimics, after Fritz Müller (1821–97), who discovered this type of mimicry. The bumblebee in Figure 45.8*c* is a Müllerian mimic of the yellow jacket wasp because both of them can sting.

Symbiotic Relationships

Symbiotic relations are of three types, as shown in Table 45.1. Some biologists argue that the amount of harm or good two

Table 45.1 Symbiotic Relationships

Interaction	Expected Outcome
Parasitism	Abundance of parasite increases, and the abundance of the host decreases.
Commensalism	Abundance of one species increases, and the other is not affected.
Mutualism	Abundance of both species increases.

species do to one another is dependent on which variable the investigator is measuring.

Parasitism is similar to predation in that an organism, called the **parasite,** derives nourishment from another, called the **host.** Parasitism is a type of **symbiosis** in which one of the species causes some harm to the other but tends to not kill it (Table 45.1). Viruses (such as HIV, which reproduce inside human lymphocytes) are always parasitic, and parasites occur in all of the kingdoms of life. Bacteria (e.g., strep infection), protists (e.g., malaria), fungi (e.g., rusts and smuts), plants (e.g., indian pipe), and animals (e.g., tapeworms and fleas) all have parasitic members. Although small parasites can be endoparasites (heartworms) (Fig. 45.9), larger ones are more likely to be ectoparasites (leeches).

The effects of parasites on the health of the host can range from slightly weakening them to actually killing them over time. When host populations are at a high density, parasites readily spread from one host to the next, causing intense infestations and a subsequent decline in host density. Parasites that do not kill their host can still play a role in reducing the host's population density because an infected host is less fertile and becomes more susceptible to other causes of death.

In addition to nourishment, host organisms also provide their parasites with a place to live and reproduce, as well as a mechanism for dispersing offspring to new hosts. Many parasites have both a primary and a secondary host. The secondary host may be a vector that transmits the parasite to the next primary host. Usually both hosts are required in order to complete the life cycle, such as in malaria (see Fig. 21.15). The association between parasite and host is so intimate that parasites will often only parasitize a few select species.

Commensalism

Commensalism is a symbiotic relationship between two species in which one species is benefited, and the other is neither benefited nor harmed.

Instances are known in which one species provides a home and/or transportation for the other species. Barnacles that attach themselves to the backs of whales or the shells of horseshoe crabs are provided with both a home and transportation. It is possible though that the movement of the host is impeded by the presence of the attached animals, and therefore some are reluctant to use these as instances of commensalism.

Epiphytes, such as Spanish moss and some species of orchids and ferns, grow in the branches of trees, where they receive light, but they take no nourishment from the trees. Instead, their roots obtain nutrients and water from the air. Clownfishes live within the waving mass of tentacles of sea anemones (Fig. 45.10). Because most fishes avoid the stinging tentacles of the anemones, clownfishes are protected from predators. Perhaps this relationship borders on mutualism, because the clownfishes actually may attract other fishes on which the anemones can feed, or may provide some cleaning services for the anemones.

Commensalism often turns out, on closer examination, to be an instance of either mutualism or parasitism. Cattle egrets are so named because these birds stay near cattle, which flush out their prey—insects and other animals—from vegetation. The relationship becomes mutualistic when egrets remove ectoparasites from the cattle. Remoras are fishes that attach themselves to the bellies of sharks by means of a modified dorsal fin acting as a suction cup. Remoras benefit by getting a free ride and feeding on a shark's leftovers. However, the shark benefits when remoras remove its ectoparasites.

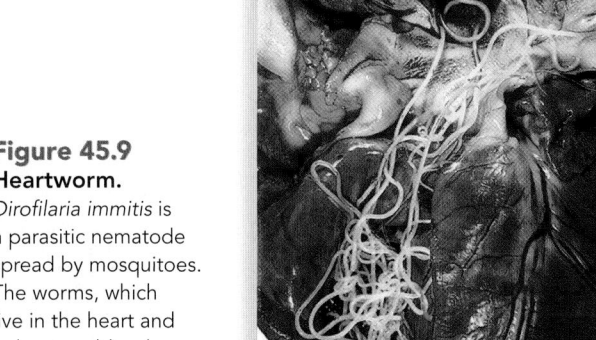

Figure 45.9 Heartworm. *Dirofilaria immitis* is a parasitic nematode spread by mosquitoes. The worms, which live in the heart and pulmonary blood artery, can cause death of the host.

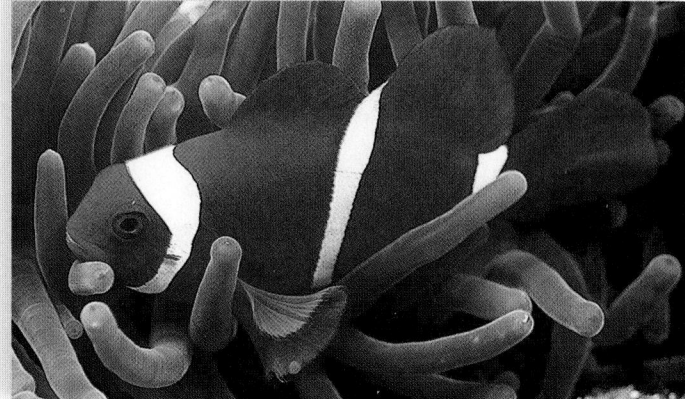

Figure 45.10 Clownfish among sea anemone's tentacles. If the clownfish, *Premnasa biaculeatus*, performs no service for the sea anemone, this association is a case of commensalism. If the clownfish lures other fish to be eaten by the sea anemone, this is a case of mutualism.

Mutualism

Mutualism is a symbiotic relationship in which both members benefit. As with other symbiotic relationships, it is possible to find numerous examples among all organisms. Bacteria that reside in the human intestinal tract acquire food, but they also provide us with vitamins that we cannot synthesize for ourselves. Termites would not be able to digest wood if not for the protozoans that inhabit their intestinal tracts and digest cellulose. Mycorrhizae are mutualistic associations between the roots of plants and fungal hyphae. The hyphae improve the uptake of nutrients for the plant, protect the plant's roots against pathogens, and produce plant growth hormones. In return, the plant provides the fungus with carbohydrates.

Some sea anemones make their home on the backs of crabs. The crab uses the stinging tentacles of the sea anemone to gather food and to protect itself, while the sea anemone gets a free ride that allows it greater access to food. Lichens can grow on rocks because their fungal member conserves water and leaches minerals that are provided to the algal partner. The algae photosynthesize and provide organic food for both populations. However, it's been suggested that the fungus is parasitic, at least to a degree, on the algae.

In Mexico and Central America, the bullhorn acacia tree is adapted to provide a home for ants of the species *Pseudomyrmex ferruginea.* Unlike other acacias, this species has swollen thorns with a hollow interior, where ant larvae can grow and develop. In addition to housing the ants, acacias provide them with food. The ants feed from nectaries at the base of the leaves and eat fat- and protein-containing nodules called Beltian bodies, found at the tips of the leaves. The ants constantly protect the plant from herbivores and other plants that might shade it.

Video Thorn Tree Ants

The relationship between plants and their pollinators, mentioned previously, is a good example of mutualism. Perhaps the relationship began when herbivores feasted on pollen. The provision of nectar by the plant may have spared the pollen

Figure 45.11
Clark's nutcrackers.
Mutualism can take many forms, such as when birds like Clark's nutcrackers, *Nucifraga columbiana,* feed on but also disperse the seeds of whitebark pine trees.

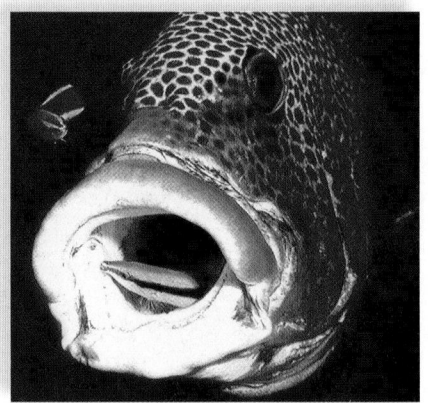

Figure 45.12
Cleaning symbiosis.
A cleaner wrasse, *Labroides dimidiatus,* in the mouth of a spotted sweetlip, *Plectorhincus chaetodontoides,* is feeding off parasites. Does this association improve the health of the sweetlip, or is the sweetlip being exploited? Investigation is under way.

and, at the same time, allowed the animal to become an agent of pollination. By now, pollinator mouthparts are adapted to gathering the nectar of a particular plant species, and this species is dependent on the pollinator for dispersing pollen. The mutualistic relationships between flowers and their pollinators are examples of **coevolution.** The Evolution feature, "Interactions and Coevolution," discusses some examples of coevolution.

Video Pollinators

The outcome of mutualism is an intricate web of species interdependencies critical to the community. For example, in areas of the western United States, the branches and cones of whitebark pine are turned upward, meaning that the seeds do not fall to the ground when the cones open. Birds called Clark's nutcrackers eat the seeds of whitebark pine trees and store them in the ground (Fig. 45.11). Therefore, Clark's nutcrackers are critical seed dispersers for the trees. Also, grizzly bears find the stored seeds and consume them.

Whitebark pine seeds do not germinate unless their seed coats are exposed to fire. When natural forest fires in the area are suppressed, whitebark pine trees decline in number, and so do Clark's nutcrackers and grizzly bears. When lightning-ignited fires are allowed to burn, or prescribed burning is used in the area, the whitebark pine populations increase, as do the populations of Clark's nutcrackers and grizzly bears.

Video Fruit Bat Seed Dispersal

Cleaning symbiosis is a symbiotic relationship in which crustaceans, fish, and birds act as cleaners for a variety of vertebrates. Large fish in coral reefs line up at cleaning stations and wait their turn to be cleaned by small fish that even enter the mouths of the large fish (Fig. 45.12). Whether cleaning symbiosis is an example of mutualism is still being questioned.

Check Your Progress   45.1

1. Identify the difference between an organism's habitat and its niche.
2. Describe the two factors that can cause predator and prey populations to cycle in a predictable manner.
3. Give examples of mutualism and explain how each species benefits.

Evolution

Interactions and Coevolution

Coevolution is present when two species adapt in response to selective pressure imposed by the other. Symbiosis (close association between two species), which includes parasitism, commensalism, and mutualism, is especially prone to the process of coevolution.

Flowering plants often employ brightly colored petals or enticing scents to attract animal pollinators (see the Evolution feature, "Plants and Their Pollinators," in Chapter 27). Butterfly-pollinated flowers often contain a platform for the butterfly to land on while the butterfly has a proboscis that allows it to feed on the flower.

Coevolution also occurs between predators and prey. For example, a cheetah sprints forward to catch its prey, which selects for the gazelles that are fast enough to avoid capture. Over generations, the adaptation of the prey may put selective pressure on the predator for an adaptation to the prey's defense mechanism. In this way, an evolutionary "arms race" can develop.

The process of coevolution has been studied in the cuckoo, a social parasite that reproduces at the expense of other birds by laying its eggs in their nests. It is a strange sight to see a small bird feeding a cuckoo nestling several times its size. Investigators discovered that in order to "trick" a host bird, the adult cuckoo has to quickly lay an egg that mimics the host's egg while the host is away from the nest. The cuckoo will leave most of the host's eggs in the nest to prevent the host from deserting a nest with only one egg. (The cuckoo chick hatches first and is behaviorally adapted to pushing other eggs out of the nest [Fig. 45Ba].)

At this stage in the "arms race," the cuckoo appears to have the upper hand; however, selection may favor host birds that are able to distinguish the cuckoo from their own young, and in this way, the host will evolve.

Coevolution can take many forms. The sexual portion of the life cycle of Plasmodium, the cause of malaria, occurs within mosquitoes (the vector), while the asexual portion occurs in humans. The human immune system uses surface proteins to detect pathogens. Plasmodium stays one step ahead of the host's immune system by changing the surface proteins of its cells to avoid detection. A similar capability of HIV has added to the difficulty of producing an AIDS vaccine.

The relationship between parasite and host can even include the ability of parasites to seemingly manipulate the behavior of their hosts in self-serving ways. Ants infected with the lance fluke mysteriously cling to blades of grass with their mouthparts. There, the infected ants are eaten by grazing sheep, transmitting the flukes to the next host in their life cycle. Similarly, snails of the genus *Succinea* are parasitized by worms of the genus *Leucochloridium*. As the worms mature, they invade the snail's eyestalks, making them resemble edible caterpillars. Birds eat the infected snails allowing the parasites to complete development inside the urinary tracts of birds.

The traditional view was that as host and parasite coevolved, each would become more tolerant of the other. Parasites could start as commensal, or harmless to the host. Then over time the parasite and host might even become mutualists. In fact, the evolution of the eukaryotic cell by endosymbiosis is based on the supposition that some early bacteria took up residence inside a larger cell, and then the parasite and cell became mutualists.

However, this argument is too teleological for some; after all, no organism is capable of "looking ahead" at its evolutionary fate. Rather, if an aggressive parasite could transmit more of itself in less time than a benign one, aggressiveness would be favored by natural selection because the most aggressive would reproduce. Other factors, such as the life cycle of the host, can determine whether aggressiveness is beneficial or not. For example, a benign parasite of newts would do better than an aggressive one. Why? Because newts take up solitary residence in the forest, outside of ponds, for six years, and parasites have to wait that long before they are likely to meet up with another potential host. If a parasite kills its host before it can reach another, it not only has lost its food source, but also its home.

Questions to Consider

1. What happens to a species if it cannot coevolve along with the species it is interacting with?
2. Why is it necessary for a parasite to avoid killing its host?

b.

Figure 45B Social parasitism. a. The cuckoo, *Cuculus*, is a social parasite of the reed warbler, *Acrocephalus*. A cuckoo chick is heaving the eggs of its host out of the nest. **b.** Its own egg (see inset) mimics and is accepted by the host as its own.

a.

Figure 45.13 Secondary succession. This example of secondary succession occurred in a former cornfield in New Jersey on the east coast of the United States. **a.** During the first year, only the remains of corn plants are seen. **b.** During the second year, wild grasses have invaded the area. **c.** By the fifth year, the grasses look more mature, and sedges have joined them. **d.** During the tenth year, there are goldenrod plants, shrubs (blackberry), and juniper trees. **e.** After 20 years, the juniper trees are mature, and there are also birch and maple trees in addition to the blackberry shrubs.

a. First year

b. Second year

45.2 Community Development

Learning Outcomes

Upon completion of this section, you should be able to

1. Choose the correct sequence of events that occur during ecological succession.
2. Compare the two types of ecological succession.

Each community has a history that can be surveyed over a time. We know that the distribution of life has been influenced by dynamic changes occurring during the history of Earth. We have previously discussed how continental drift contributed to various mass extinctions during the past. For example, as the continents slowly came together, forming the supercontinent Pangaea, many forms of marine life became extinct. Or, when the continents drifted toward the poles, immense glaciers drew water from the oceans and even chilled once-tropical lands. During ice ages, glaciers moved southward, and then in between ice ages, glaciers retreated, changing the environment and allowing life to colonize the land once again. Over time, complex communities evolved. Many ecologists, however, try to observe changes as they occur during much shorter timescales.

Ecological Succession

Communities are subject to disturbances that can range in severity from a storm blowing down a patch of trees, to a beaver damming a pond, to a volcanic eruption. We know from observation that after disturbances, changes occur in the plant and animal communities over time.

Ecological succession is a change within a community involving a series of species replacements. *Primary succession* occurs in areas where there is no soil formation, such as following a volcanic eruption or a glacial retreat. Wind, water, and other abiotic factors start the formation of soil from exposed rock. *Secondary succession* occurs in areas where soil is present, as when a cultivated field, such as a cornfield in New Jersey, returns to a natural state (Fig. 45.13). This is disturbance-based

succession. Notice that the progression changes from grasses to shrubs to a mixture of shrubs and trees.

The first species to begin secondary succession are called **pioneer species.** These are the first producers to inhabit a community after a disturbance. The area then progresses through the series of stages over time (Figure 45.14). The initial stage begins with small, short-lived species and proceeds through stages of species of mixed sizes and life spans, until finally large, long-lived species of trees predominate. Ecologists have tried to determine the processes and mechanisms by which the changes described in Figures 45.13 and 45.14 take place—and whether these processes always have the same "end point" of community composition and diversity.

Models About Succession

In 1916, F. E. Clements (1874–1945) proposed that succession in a particular area will always lead to the same type of community, which he called a **climax community.** He hypothesized that climate, in particular, determined whether succession resulted in a desert, a type of grassland, or a particular type of forest. This is the reason, he said, that coniferous forests occur in northern latitudes, deciduous forests in temperate zones, and tropical rain forests in the tropics. Secondarily, he hypothesized that soil conditions might also affect the results. Shallow, dry soil might produce a grassland where otherwise a forest might be expected, or the rich soil of a riverbank might produce a woodland where a prairie would be expected.

Further, Clements hypothesized that each stage facilitated the invasion and replacement by organisms of the next stage. Shrubs can't grow on dunes until dune grass has caused soil to develop. Similarly, in the example given in Figure 45.14, shrubs can't arrive until grasses have made the soil suitable for them. Each group of species prepares the way for the next, so that grass–shrub–forest development occurs in a sequential way. Therefore, in what is sometimes called "climax theory," this is known as the *facilitation model* of succession.

Aside from this facilitation model, an *inhibition model* also has been proposed to help explain succession. That model predicts that colonists hold onto their space and inhibit the growth

c. Fifth year

d. Tenth year

e. Twentieth year

of other plants until the colonists die or are damaged. Still another model, called the *tolerance model*, predicts that different types of plants can colonize an area at the same time. Sheer chance determines which seeds arrive first, and successional stages may simply reflect the length of time it takes species to mature. This alone could account for the grass–shrub–forest development that is often seen (Fig. 45.14).

The length of time it takes for trees to develop might give the impression that there is a recognizable series of plant communities, from the simple to the complex. In reality, succession occurs gradually, and the models mentioned here are probably not mutually exclusive, but a mixture of multiple, complex processes.

Although the dynamic nature of natural communities may not have been apparent to early ecologists, we now recognize it as a community's most outstanding characteristic. Also, it seems obvious to us now that the most complex communities are ones that contain various stages of succession. Each successional stage has its own mix of plants and animals, and if a sample of all stages is present, community diversity is greatest. Furthermore, we do not know whether succession continues to specific end points, because the process may not be complete anywhere on the face of the Earth.

Check Your Progress 45.2

1. Identify the events that would lead to succession.
2. Compare and contrast the various models used to explain succession.

Figure 45.14 Secondary succession in a forest. In secondary succession in a large conifer plantation in central New York State, certain species are common to particular stages. However, the process of regrowth shows approximately the same stages as secondary succession from a cornfield.

grass → low shrub → high shrub → shrub-tree → low tree → high tree

45.3 Dynamics of an Ecosystem

In an **ecosystem,** populations interact with each other and with the physical environment. The abiotic (nonliving) components of an ecosystem include the atmosphere, water, and soil. The biotic (living) components can be categorized according to their food source as autotrophs or heterotrophs (Fig. 45.15).

Autotrophs

Autotrophs require only inorganic nutrients and an outside energy source to produce organic nutrients for their own use and for all the other members of a community. They are called **producers** because they produce food. Autotrophs include photosynthetic organisms such as land plants and algae (photoautotrophs). They possess chlorophyll and carry on photosynthesis in freshwater and marine habitats. Algae make up the phytoplankton, which are photosynthesizing organisms suspended in water. Green plants are the dominant photosynthesizers on land.

> **Video**
> **Plants**

Some autotrophic bacteria are chemosynthetic. They obtain energy by oxidizing inorganic compounds such as ammonia, nitrites, and sulfides, and they use this energy to synthesize organic compounds. Chemoautotrophs have been found to support communities in some caves and also at hydrothermal vents along deep-sea oceanic ridges where sunlight is unavailable.

Heterotrophs

Heterotrophs need a preformed source of organic nutrients as they acquire food from a different (*hetero*) source. They are called **consumers** because they consume food that was generated by a producer.

Herbivores are animals that graze directly on plants or algae. In terrestrial habitats, insects are small herbivores; antelopes and bison are large herbivores. In aquatic habitats, zooplankton are small herbivores; fishes and manatees are large herbivores. **Carnivores** feed on other animals; birds that feed on insects are carnivores, and so are hawks that feed on birds and small mammals. **Omnivores** are animals that feed on both plants and animals. Chickens, raccoons, and humans are omnivores. Some animals are scavengers, such as vultures and jackals, which eat the carcasses of dead animals.

> **Video**
> **Snake Eating**

Detritivores are organisms that feed on detritus, which consists of decomposing particles of organic matter. Marine fan

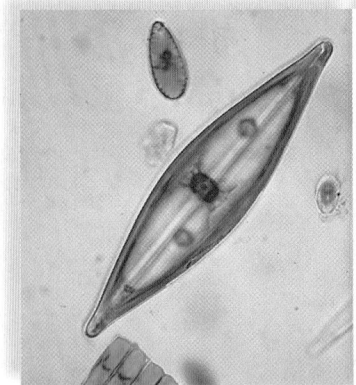

a. Producers

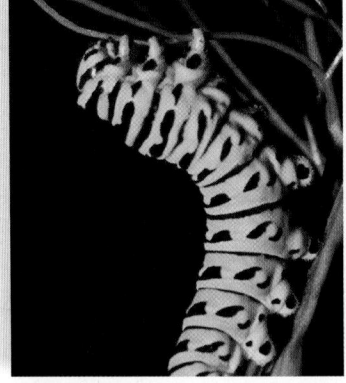

b. Herbivores

c. Carnivores

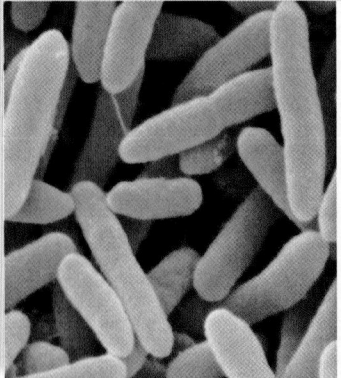

d. Decomposers

Figure 45.15 Biotic components. a. Diatoms and green plants are producers (photoautotrophs). **b.** Caterpillars and rabbits are herbivores. **c.** Spiders and osprey are carnivores. **d.** Bacteria and mushrooms are decomposers.

worms filter detritus from the water, whereas burrowing clams take it from the substratum. Earthworms, some beetles, termites, and ants are all terrestrial detritivores.

Bacteria and fungi, including mushrooms, are **decomposers;** they acquire nutrients by breaking down dead organic matter, including animal wastes. Decomposers perform a valuable service because they release inorganic substances that are taken up by plants once more. Otherwise, plants would be completely dependent only on physical processes, such as the release of minerals from rocks, to supply them with inorganic nutrients.

A diagram of all the biotic components of an ecosystem illustrates that every ecosystem is characterized by two fundamental phenomena: energy flow and chemical cycling (Fig. 45.16). Energy flow begins when producers absorb solar energy, and chemical cycling begins when producers take in inorganic nutrients from the physical environment. Thereafter, via photosynthesis, producers make organic nutrients (food) directly for themselves and indirectly for the other populations within the ecosystem.

Energy Flow

Energy flows through an ecosystem via photosynthesis. As organic nutrients pass from one component of the ecosystem to another, such as when an herbivore eats a plant or a carnivore eats an herbivore, only a portion of the original amount of

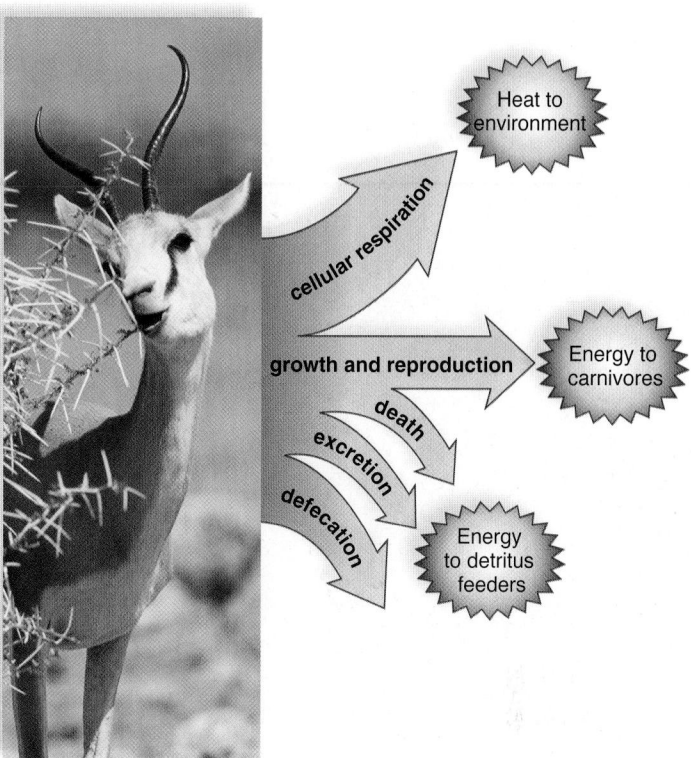

Figure 45.17 Energy balances. Only about 10% of the food energy taken in by an herbivore is passed on to carnivores. A large portion goes to detritus feeders via defecation, excretion, and death, and another large portion is used for cellular respiration.

energy is transferred. Eventually, the energy dissipates into the environment as heat. Therefore, the vast majority of ecosystems cannot exist without a continual supply of solar energy.

Only a portion of the organic nutrients made by producers is passed on to consumers because plants use organic molecules to fuel their own cellular respiration. Similarly, only a small percentage of nutrients consumed by lower-level consumers, such as herbivores, is available to higher-level consumers, or carnivores. As Figure 45.17 demonstrates, a certain amount of the food eaten by an herbivore is not digested and is eliminated as feces. Metabolic nitrogenous wastes are excreted as urine. Of the assimilated energy, a large portion is used during cellular respiration for the production of ATP and thereafter becomes heat. Only the remaining energy, which is converted into increased body weight or additional offspring, becomes available to carnivores.

The elimination of feces and urine by a heterotroph, and the death of an organism, does not mean that organic nutrients are lost to an ecosystem. Instead, they represent the organic nutrients made available to decomposers. Decomposers convert the organic nutrients, such as glucose, back into inorganic chemicals, such as carbon dioxide and water, and release them to the soil or atmosphere. When these inorganic chemicals are absorbed by producers they have completed their cycle within an ecosystem.

The laws of thermodynamics, which were first described in Section 6.1, support the concept that energy flows through an ecosystem. The first law states that energy can neither be created

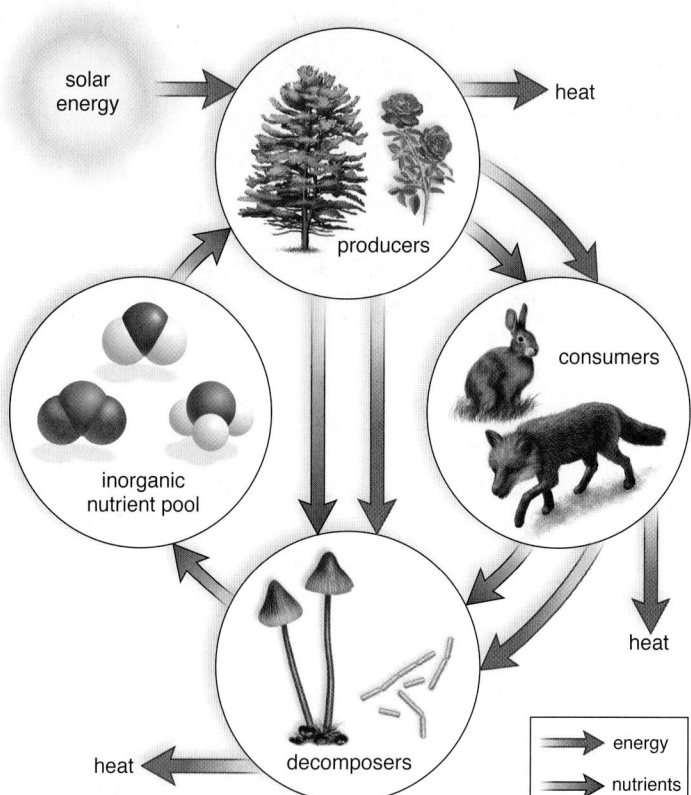

Figure 45.16 Nature of an ecosystem. Chemicals cycle, but energy flows through an ecosystem. As energy transformations repeatedly occur, all the energy derived from the Sun eventually dissipates as heat.

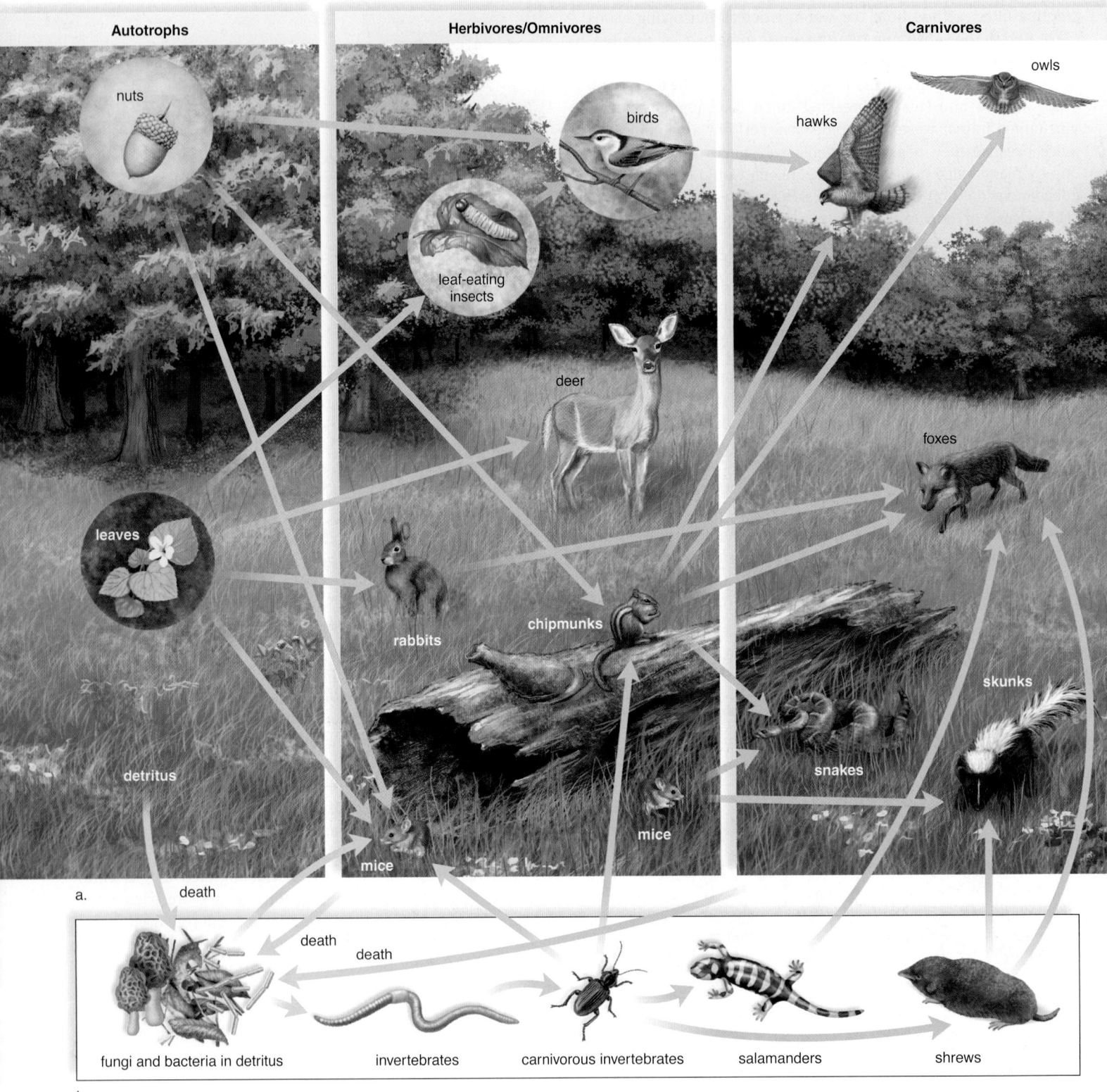

Autotrophs

nuts

leaves

detritus

a.

death

Herbivores/Omnivores

leaf-eating
insects

birds

deer

rabbits

chipmunks

mice

mice

death

Carnivores

owls

hawks

foxes

skunks

snakes

b.

death

fungi and bacteria in detritus   invertebrates   carnivorous invertebrates   salamanders   shrews

Figure 45.18 Grazing and detrital food web.   Food webs are descriptions of who eats whom. **a.** Tan arrows illustrate possible grazing food webs. For example, birds, which feed on nuts, may be eaten by a hawk. Autotrophs such as the tree are producers (first trophic, or feeding, level), the first series of animals are primary consumers (second trophic level), and the next group of animals are secondary consumers (third trophic level). **b.** Green arrows illustrate possible detrital food webs, which begin with detritus—the bacteria and fungi of decay and the remains of dead organisms. A large portion of these remains are from the grazing food web illustrated in (**a**). The organisms in the detrital food web are sometimes fed on by animals in the grazing food web, as when robins feed on earthworms. Thus, the grazing food web and the detrital food web are connected to one another.

nor destroyed. This explains why ecosystems are dependent on a continual outside source of energy, usually solar energy, which is used by photosynthesizers to produce organic nutrients. The second law states that, with every transformation, some energy is degraded into a less available form such as heat. Because plants carry on cellular respiration, for example, only about 55% of the original energy absorbed by plants is available to an ecosystem.

An Example of Energy Flow

Let's apply the principles presented so far to an actual ecosystem—a forest of 132,000 m² in New Hampshire. The various interconnecting paths of energy flow are represented by a **food web,** a diagram that describes trophic (feeding) relationships. Figure 45.18a is a grazing food web that begins with a producer, specifically the oak tree depicted. Insects in the form of caterpillars, mice, rabbits, and deer all feed on leaf tissue. Birds, chipmunks, and mice feed on fruits and nuts, but they are in fact omnivores because they also feed on caterpillars. These herbivores and omnivores all provide food for a number of different carnivores.

Figure 45.18b is a detrital food web, which begins with detritus. Detritus is food for soil organisms such as earthworms. Earthworms are in turn fed on by carnivorous invertebrates, which then may be eaten by shrews or salamanders. Because the members of a detrital food web may become food for aboveground carnivores, the detrital and grazing food webs are connected.

We naturally tend to think that aboveground plants such as trees are the largest storage form of organic matter and energy, but this is not necessarily the case. In this particular forest, the organic matter lying on the forest floor and mixed into the soil contains over twice as much energy as the leaf matter of living trees. Therefore, more energy in a forest may be funneling through the detrital food web than through the grazing food web.

Trophic Levels

The arrangement of the species in Figure 45.18 suggests that organisms are linked to one another in a straight line, according to feeding relationships, or who eats whom. Diagrams that show a single path of energy flow in an ecosystem are called **food chains.** For example, in the grazing food web, we could find this grazing food chain:

leaves ⟶ caterpillars ⟶ birds ⟶ hawks

And in the detrital food web, we could find this detrital food chain:

detritus ⟶ earthworms ⟶ salamanders

A **trophic level** is a level of nutrients within a food web or chain. In the grazing food web in Figure 45.18a the green plants are the producers in the first trophic level, the animals in the center are primary consumers of the second trophic level, and the last group of animals are secondary consumers of the third trophic level.

Ecological Pyramids

The shortness of food chains can be attributed to the loss of energy between trophic levels. In general, only about 10% of the energy of one trophic level is available to the next trophic level. Therefore, if an herbivore population consumes 1,000 kg of plant material, only about 100 kg is converted into the body tissue of an herbivore, 10 kg is converted into the body tissue of the first-level carnivores, and 1 kg into the body tissue of the second-level carnivores. This "10% rule" explains why few carnivores can be supported in a food web. The flow of energy with large losses between successive trophic levels is sometimes depicted as an **ecological pyramid** (Fig. 45.19).

Energy flow from one trophic level to the next produces a pyramid based on the number of organisms or the amount of biomass at each trophic level. When constructing such pyramids, problems arise, however. For example, in Figure 45.18, each tree would contain numerous caterpillars; therefore, there would be more herbivores than autotrophs! The explanation, of course, has to do with size. An autotroph can be as tiny as a microscopic alga or as big as a beech tree; similarly, an herbivore can be as small as a caterpillar or as large as an elephant.

Pyramids of biomass eliminate size as a factor because **biomass** is the number of organisms multiplied by the dry weight of the organic matter within one organism. The biomass of the producers is expected to be greater than the biomass of

Figure 45.19 Ecological pyramid.   The biomass, or dry weight (g/m²), for trophic levels in a grazing food web in a bog at Silver Springs, Florida. There is a sharp drop in biomass between the producer level and herbivore level, which is consistent with the common knowledge that the detrital food web plays a significant role in bogs.

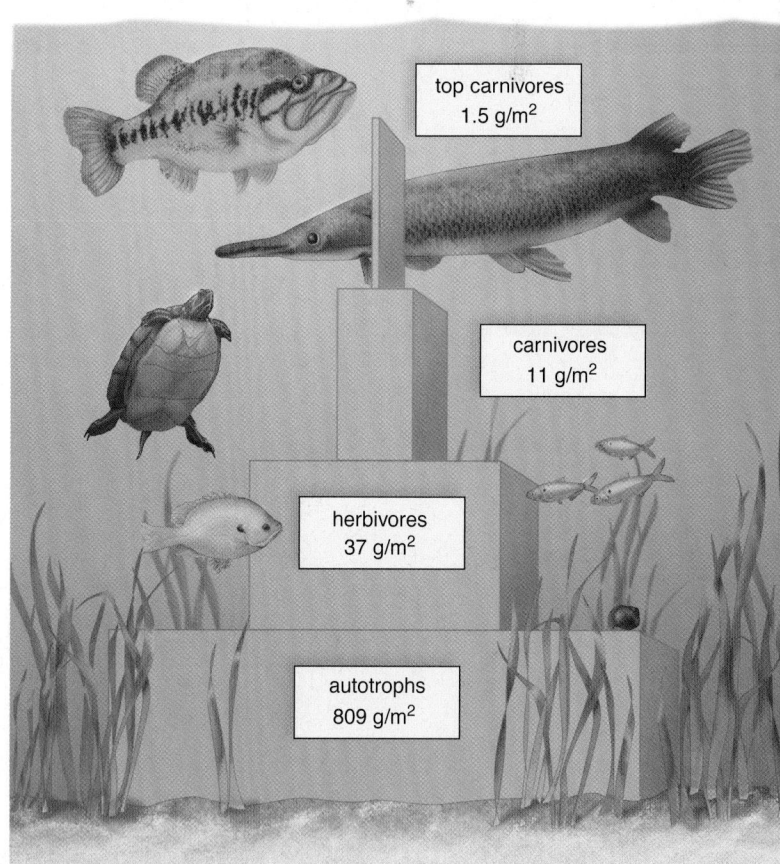

top carnivores
1.5 g/m²

carnivores
11 g/m²

herbivores
37 g/m²

autotrophs
809 g/m²

the herbivores, and that of the herbivores is expected to be greater than that of the carnivores. In aquatic ecosystems, such as some lakes and open seas where algae are the only producers, the herbivores may have a greater biomass than the producers because the algae are consumed at such a high rate. Such pyramids, which have more herbivores than producers, are called inverted pyramids:

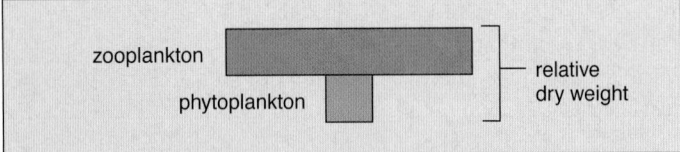

These kinds of drawbacks are making some ecologists hesitant about using pyramids to describe ecological relationships. One more problem is what to do with the decomposers, which are rarely included in pyramids, even though a large portion of energy becomes detritus in many ecosystems.

Chemical Cycling

The pathways by which chemicals circulate through ecosystems involve both living (biotic) and nonliving (geologic) components known as the **biogeochemical cycles.** Here we focus on the four main biogeochemical cycles:

- water cycle
- carbon cycle
- phosphorus cycle
- nitrogen cycle

A biogeochemical cycle may be sedimentary or gaseous. The phosphorus cycle is a sedimentary cycle; the chemical is absorbed from the soil by plant roots, passed to heterotrophs, and eventually returned to the soil by decomposers. The carbon and nitrogen cycles are gaseous, meaning that the chemical returns to and is withdrawn from the atmosphere as a gas.

Chemical cycling involves the components of ecosystems shown in Figure 45.20. A *reservoir* is a source normally unavailable to producers, such as the carbon present in calcium carbonate shells on ocean bottoms. An *exchange pool* is a source from which organisms can take chemicals, such as the atmosphere or

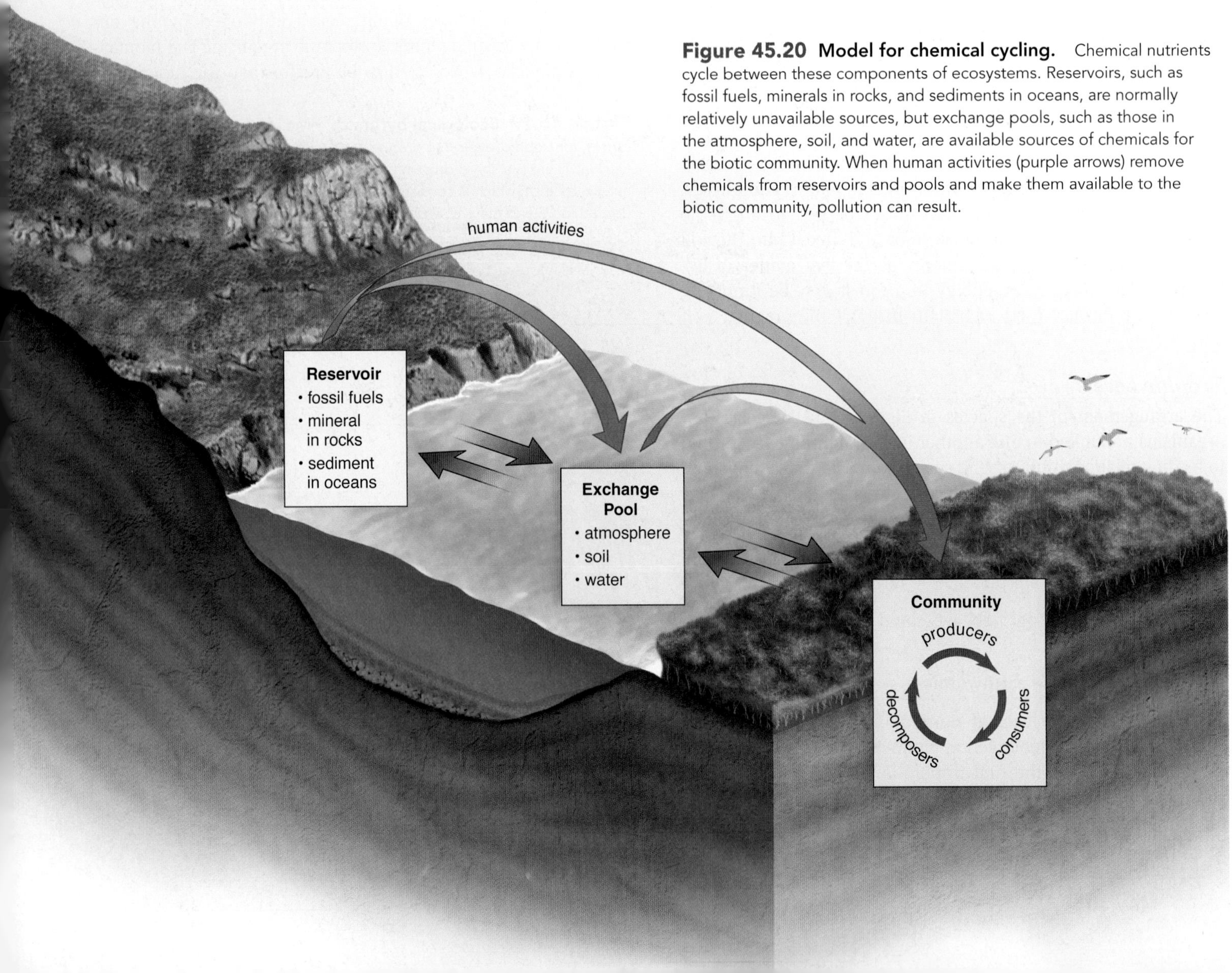

Figure 45.20 Model for chemical cycling.   Chemical nutrients cycle between these components of ecosystems. Reservoirs, such as fossil fuels, minerals in rocks, and sediments in oceans, are normally relatively unavailable sources, but exchange pools, such as those in the atmosphere, soil, and water, are available sources of chemicals for the biotic community. When human activities (purple arrows) remove chemicals from reservoirs and pools and make them available to the biotic community, pollution can result.

soil. Certain chemicals can move along food chains in a *biotic community* and never enter an exchange pool.

Human activities (purple arrows) remove chemicals from reservoirs and exchange pools and make them available to the biotic community. In this way, human activities result in pollution because they upset the normal balance of nutrients for producers in the environment.

The Water Cycle

The **water (hydrologic) cycle** is described in Figure 45.21. A **transfer rate** is defined by the amount of a substance that moves from one component of the environment to another within a specified period of time. The width of the arrows in Figure 45.21 indicates the transfer rate of water.

During the water cycle, fresh water is first distilled from salt water through evaporation. During evaporation, a liquid, in this case water, changes to a gaseous state. The Sun's rays cause fresh water to evaporate from the seawater, and the salts are left behind. Next, condensation occurs. During condensation, a gas is converted into a liquid. For example, vaporized fresh water rises into the atmosphere, collects in the form of a cloud, cools, and falls as rain over the oceans and the land.

Water evaporates from land and from plants (evaporation from plants is called transpiration) and also from bodies of fresh water. Because land lies above sea level, gravity eventually returns all fresh water to the sea. In the meantime, water is contained within standing bodies (lakes and ponds), flowing bodies (streams and rivers), and groundwater.

Some of the water from precipitation (e.g., rain, snow, sleet, hail, and fog) makes its way into the ground and saturates the earth to a certain level. The top of the saturation zone is called the groundwater table, or simply, the water table. Because water infiltrates through the soil and rock layers, groundwater can be located in rock layers called aquifers. Water is usually released in appreciable quantities to wells or springs. Aquifers are recharged when rainfall and melted snow percolate into the soil.

Human Activities and the Water Cycle   In some parts of the United States, especially the arid West and southern Florida, withdrawals from aquifers exceed the level of recharge. This is called "groundwater mining." In these locations, the groundwater level is dropping, and residents may run out of groundwater, at least for irrigation purposes, within a few years.

Fresh water makes up about 3% of the world's supply of water. Water is considered a renewable resource because a new supply is always being produced as a result of the water cycle. It is possible to run out of fresh water, however, when the available supply is not adequate or is so polluted that it is not usable.

Video
Thames River

Figure 45.21 The hydrologic (water) cycle. Evaporation from the ocean exceeds precipitation, so there is a net movement of water vapor onto land, where precipitation results in surface water and groundwater that flow back to the sea. On land, transpiration by plants contributes to evaporation. The numbers in this diagram indicate water flow in cubic kilometers per year.

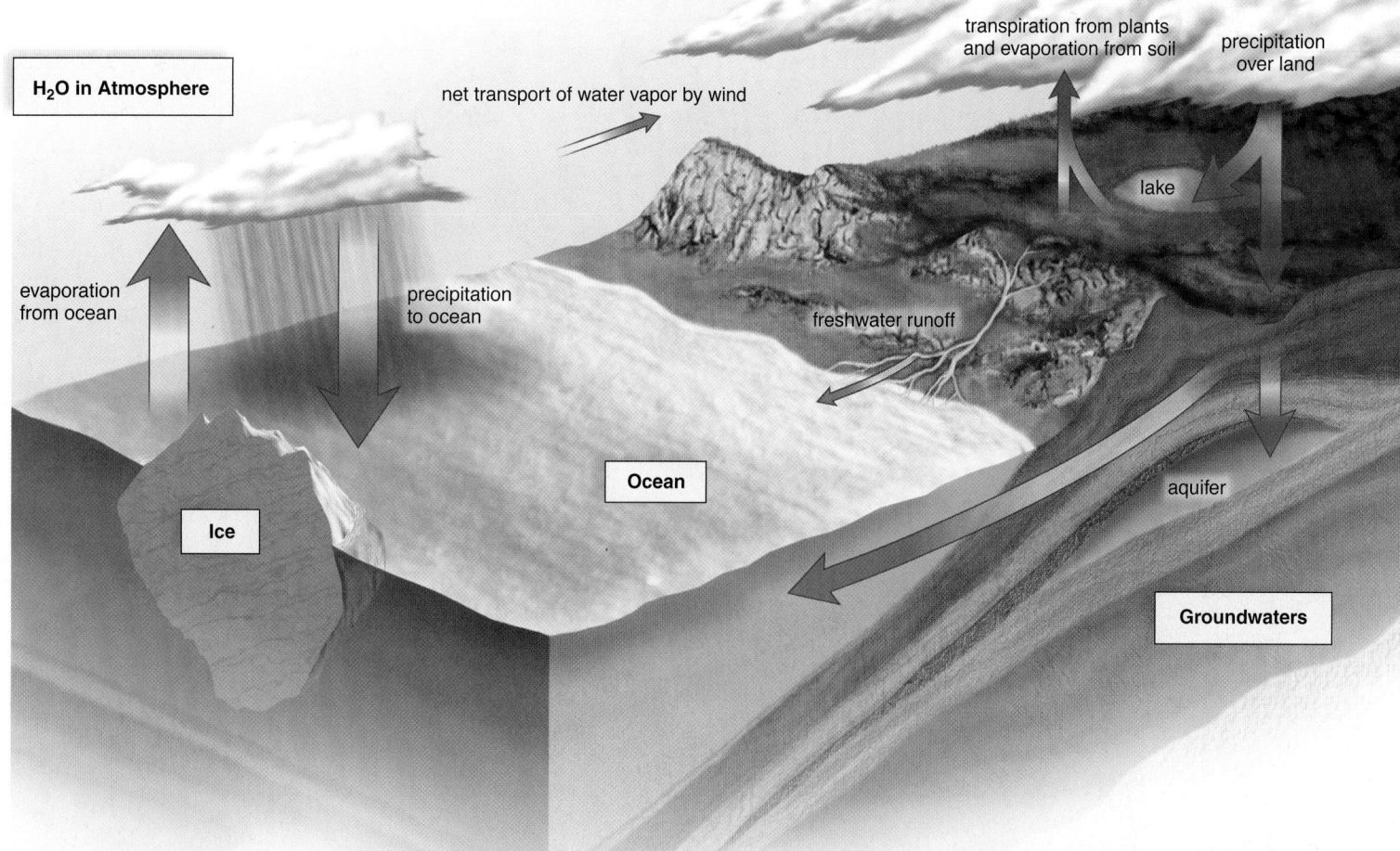

The Carbon Cycle

In the carbon cycle, organisms in both terrestrial and aquatic eco-systems exchange carbon dioxide (CO_2) with the atmosphere (Fig. 45.22). Therefore, the CO_2 in the atmosphere is the exchange pool for the carbon cycle. On land plants take up CO_2 from the air through photosynthesis. The CO_2 is incorporated into nutrients that are used by both autotrophs and heterotrophs alike. **1** When organisms, including plants, respire, carbon is returned to the atmosphere as CO_2. **2** Carbon dioxide then recycles to plants by way of the atmosphere.

In aquatic ecosystems, the exchange of CO_2 with the atmo-sphere is indirect. **3** Carbon dioxide from the air combines with water to produce bicarbonate ion (HCO_3^-). This is the main source of carbon for algae. Similarly, when aquatic organisms respire, the CO_2 they give off becomes HCO_3^-. **4** The amount of bicarbonate in the water is in equilibrium with the amount of CO_2 in the air.

Reservoirs Hold Carbon   Living and dead organisms con-tain organic carbon and serve as a reservoir for the carbon cycle. The world's biotic components, particularly trees, contain 800 billion tons of organic carbon. An additional 1,000–3,000 billion metric tons are estimated to be held in the remains of plants and animals in the soil. **5** Ordinarily, decomposition of organisms returns CO_2 to the atmosphere.

Some 300 MYA, plant and animal remains were transformed into coal, oil, and natural gas, the materials we call fossil fuels. Another reservoir for carbon is the inorganic carbonate that accumulates in limestone and in calcium carbonate shells. Many marine organisms have calcium carbonate shells that remain in sediments long after the organisms have died. Geologic forces change these sediments into limestone.

Human Activities and the Carbon Cycle **6**   More CO_2 is being deposited in the atmosphere than is being removed, largely due to the burning of fossil fuels and the destruction of forests. When we humans do away with forests, we reduce a reservoir and also the very organisms that take up excess carbon dioxide. Today, the amount of CO_2 released into the atmosphere is about twice the amount that remains in the atmosphere. Much of the CO_2 dissolves into the ocean.

Several other gases are emitted into the atmosphere due to human activities. These include nitrous oxide (N_2O) from fertiliz-ers and animal wastes and methane (CH_4) from bacterial decom-position that occurs in anaerobic environments. These gases are known as **greenhouse gases** because, just like the panes of

Figure 45.22 The carbon cycle.   The transfer rate of carbon into the atmosphere due to respiration approximately matches the rate due to withdrawal by plants for photosynthesis. However, due to the burning of fossil fuels and destruction of vegetation by human activities (purple arrows), more carbon dioxide is added to the atmosphere than is withdrawn.

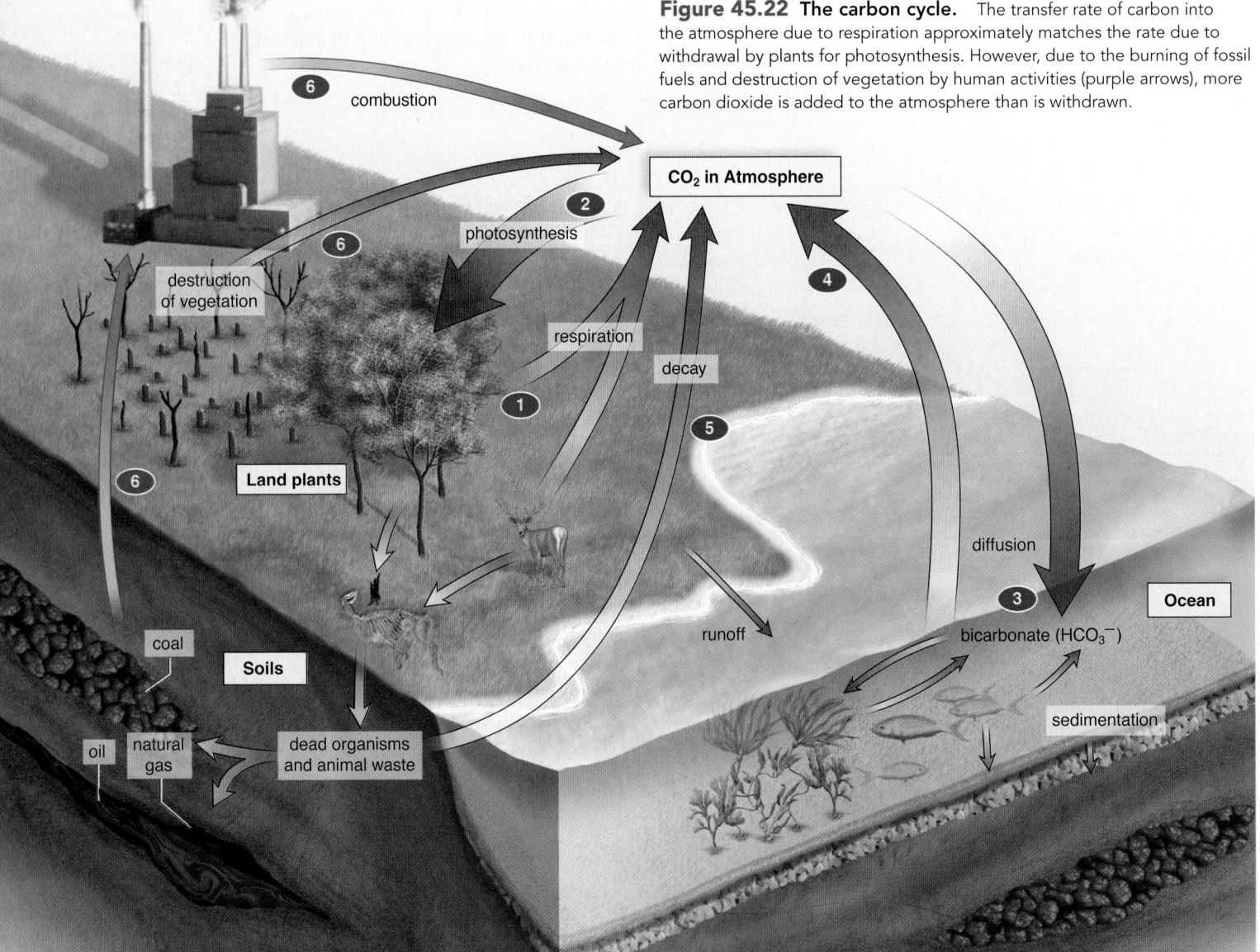

a greenhouse, they allow solar radiation to pass through, but hinder the escape of infrared rays (heat) back into space. This phenomenon has come to be known as the **greenhouse effect.**

The greenhouse gases are contributing significantly to an overall rise in the Earth's ambient temperature, a trend called **global warming** that is leading to climate change. The expression **climate change** refers to the recent changes in the Earth's climate. The global climate has already warmed about 0.6°C since the Industrial Revolution. Computer models are unable to consider all possible variables, but the Earth's temperature may rise 1.5–4.5°C by 2100 if greenhouse emissions continue at the current rates. The Nature of Science feature, on page 879, "Global Climate Change" discusses these effects further.

It is predicted that, as the oceans warm, temperatures in the polar regions will rise to a greater degree than in other regions. If so, glaciers will melt, and sea level will rise, not only due to this melting but also because water expands as it warms. Increased rainfall is likely along the coasts, while drier conditions are expected inland. Coastal agricultural lands, such as the deltas of Bangladesh and China, will be inundated with seawater, and billions of dollars will have to be spent to keep coastal cities such as New Orleans, New York, Boston, Miami, and Galveston from disappearing into the sea.

Video Karoo Global Warming

Video Warming Hurts Rice

Video Global Warming

The Phosphorus Cycle

Figure 45.23 depicts the phosphorus cycle. **1** Phosphorus from oceanic sediments moves onto land due to a geologic uplift. **2** On land, the very slow weathering of rocks places **3** phosphate ions (PO_3^- and HPO_4^+) in the soil. **4** Some of these become available to plants, which use phosphate in a variety of molecules, including phospholipids, ATP, and the nucleotides that become a part of DNA and RNA. **5** Animals eat producers and incorporate some of the phosphate into their teeth, bones, and shells. **6** However, eventually the death and decay of all organisms and the decomposition of animal wastes make phosphate ions available to producers once again. Because the available amount of phosphate is already being used within food chains, phosphate is usually a limiting inorganic nutrient for plants. Phosphate levels will influence the size of populations within an ecosystem.

7 Some phosphate naturally runs off into aquatic ecosystems, where algae acquire it from the water before it can become trapped in sediments. Phosphate in marine sediments does not become available to producers on land until a geologic upheaval exposes sedimentary rocks on land. The cycle then begins again.

Human Activities and the Phosphorus Cycle **8**

Human beings boost the supply of phosphate by mining phosphate ores for producing fertilizer and detergents. Runoff of phosphate and nitrogen from fertilizer use, animal wastes from livestock feedlots, and discharge from sewage treatment plants results in **eutrophication** (over-enrichment) of waterways.

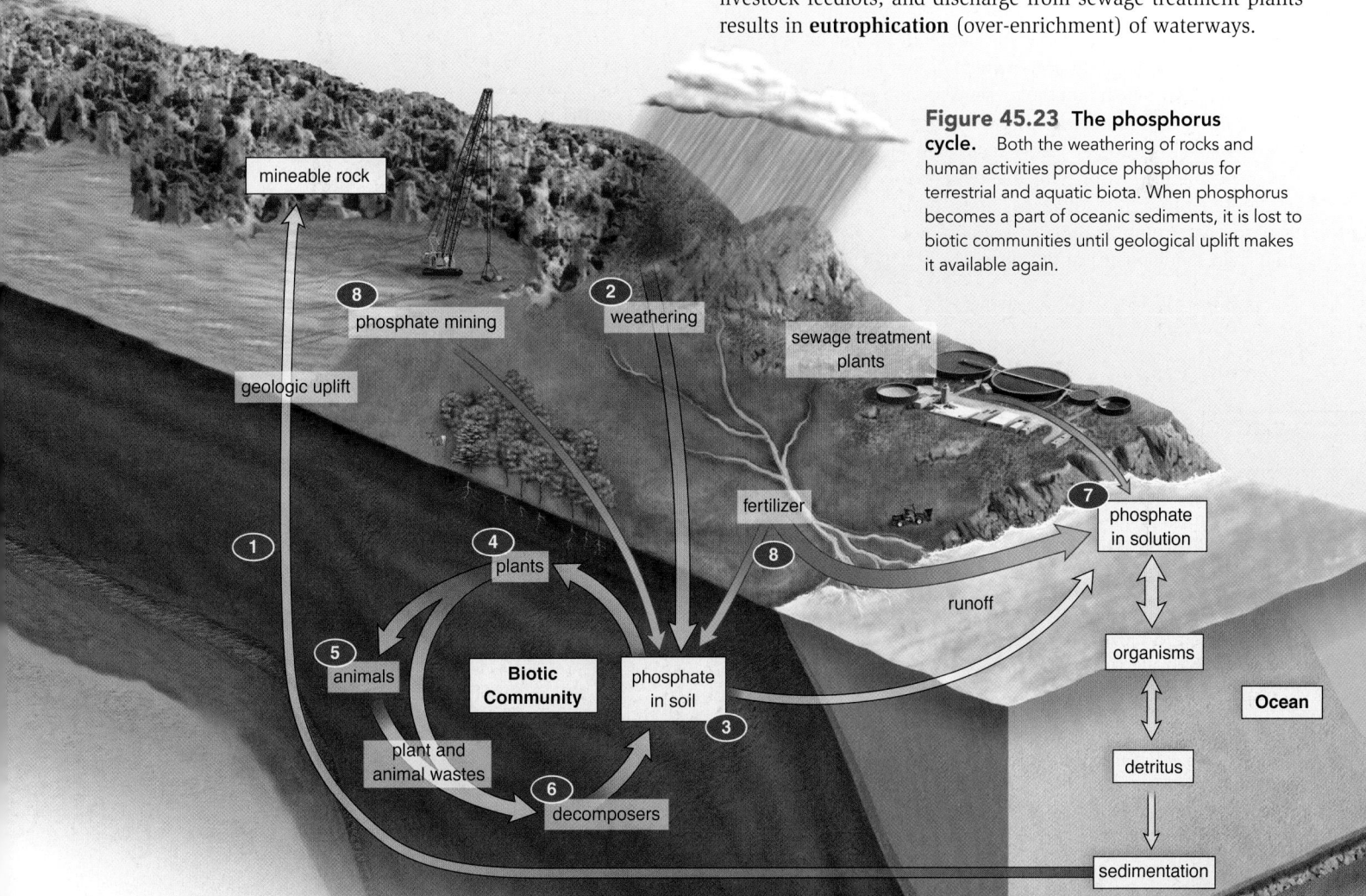

Figure 45.23 The phosphorus cycle. Both the weathering of rocks and human activities produce phosphorus for terrestrial and aquatic biota. When phosphorus becomes a part of oceanic sediments, it is lost to biotic communities until geological uplift makes it available again.

mineable rock

8 phosphate mining

2 weathering

sewage treatment plants

geologic uplift

1

4 plants

5 animals

Biotic Community

phosphate in soil 3

plant and animal wastes

6 decomposers

fertilizer

8

runoff

7 phosphate in solution

organisms

Ocean

detritus

sedimentation

The Nitrogen Cycle

Nitrogen gas (N₂) makes up about 78% of the gases in the atmosphere, but plants cannot make use of nitrogen in its gaseous form. The availability of nitrogen in an ecosystem can limit the size of the producer populations. First, let's consider that **① N₂ (nitrogen) fixation** occurs when nitrogen gas (N₂) is converted to ammonium (NH_4^+), a form plants can use (Fig. 45.24). Some cyanobacteria in aquatic ecosystems and some free-living bacteria in soil are able to fix atmospheric nitrogen in this way. Other nitrogen-fixing bacteria live in nodules on the roots of legumes, such as beans, peas, and clover. They make organic compounds containing nitrogen available to the host plants so that the plant can form proteins and nucleic acids.

② Plants can also use nitrates (NO_3^-) as a source of nitrogen. The production of nitrates during the nitrogen cycle is called **nitrification.** Nitrification can occur in two ways:

1. Nitrogen gas (N₂) is converted to NO_3^- in the atmosphere when cosmic radiation, meteor trails, and lightning provides the energy needed for nitrogen to react with oxygen.
2. Ammonium (NH_4^+) in the soil from various sources, including decomposition of organisms and animal wastes, is converted to NO_3^- by nitrifying bacteria in soil. Specifically, NH_4^+ (ammonium) is converted to NO_2^- (nitrite), and then NO_2^- is converted to NO_3^- (nitrate).

③ During the process of assimilation, plants take up NH_4^+ and NO_3^- from the soil and use these ions to produce proteins and nucleic acids. Notice in Figure 45.24 that the subcycle involving the biotic community, which occurs on land and in the ocean, need not depend on the presence of nitrogen gas at all.

Finally, **④ denitrification** is the conversion of nitrate back to nitrogen gas, which then enters the atmosphere. Denitrifying bacteria living in the anaerobic mud of lakes, bogs, and estuaries carry out this process as a part of their own metabolism. In the nitrogen cycle, denitrification would counterbalance nitrogen fixation if not for human activities.

 **Video** Dung Beetles

Human Activities and the Nitrogen Cycle ⑤
Humans significantly alter the transfer rates in the nitrogen cycle by producing fertilizers from N₂. Fertilizer runs off into lakes and rivers and results in an overgrowth of algae and rooted aquatic plants. When the algae die off, enlarged populations of decomposers use up all the oxygen in the water, and the result is a massive fish kill.

Acid deposition occurs because nitrogen oxides (NO_x) and sulfur dioxide (SO_2) enter the atmosphere from the burning of fossil fuels. Both these gases combine with water vapor to form acids that eventually return to the earth. Acid deposition has drastically affected forests and lakes in northern Europe, Canada, and the northeastern United States by altering the pH of their soils and surface waters. Acid deposition also reduces agricultural yields and corrodes marble, metal, and stonework.

Check Your Progress   45.3

1. Identify the various types of populations that are at the base of an ecological pyramid and the start of a food chain.
2. Compare the flow of energy to the flow of chemicals through an ecosystem.
3. Provide examples of how human activities can alter the biogeochemical cycles.

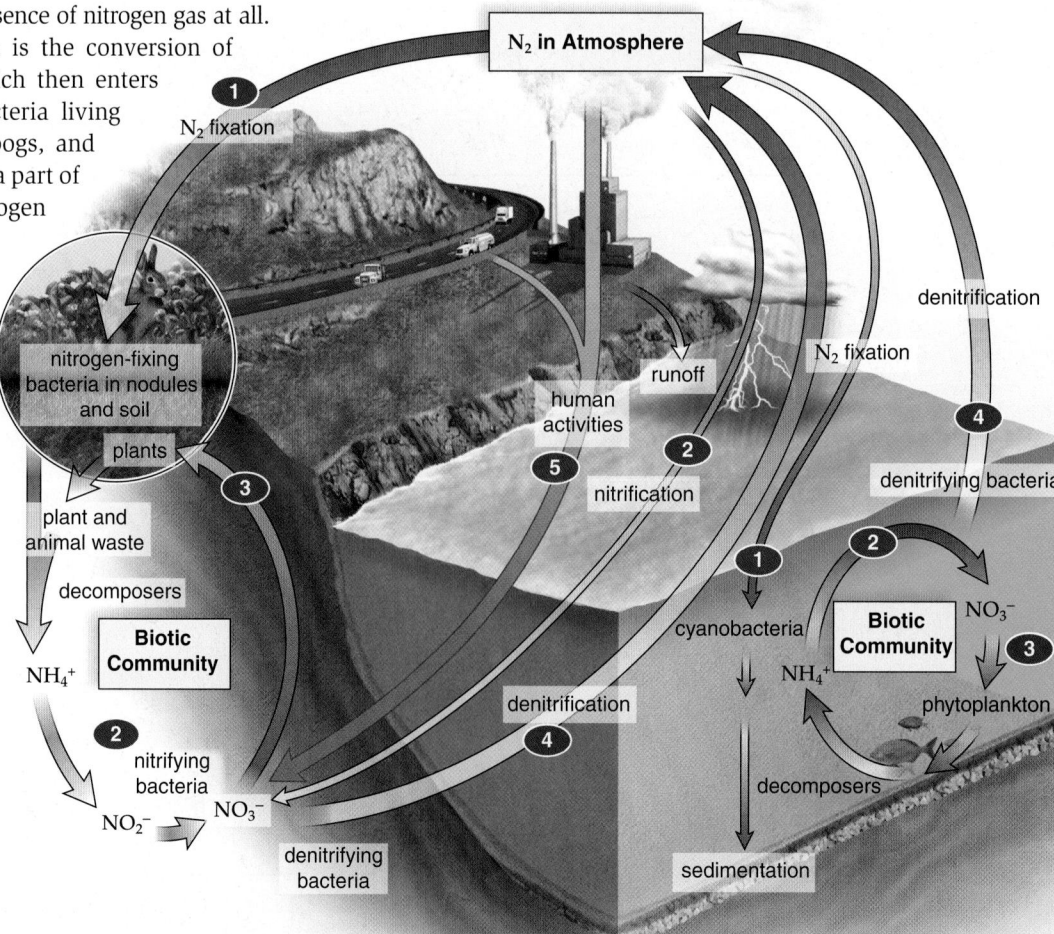

Figure 45.24 The nitrogen cycle. Nitrogen is primarily made available to biotic communities by internal cycling of the element. Without human activities, the amount of nitrogen returned to the atmosphere (denitrification in terrestrial and aquatic communities) exceeds withdrawal from the atmosphere (N₂ fixation and nitrification). Human activities (purple arrows) result in an increased amount of NO_3^- in terrestrial communities with resultant runoff to aquatic biotic communities.

Nature of Science

Global Climate Change

Scientists around the world are working on collecting and interpreting environmental indicators that will help us understand how and why the Earth's climate is changing. Changes in the average temperature of a region, precipitation patterns, sea levels and greenhouse gas concentrations are all indicators that our climate is changing.

Since 1901 the average temperature across the United States has risen, with 2000–2009 being the warmest decade on record worldwide (Fig. 45Ca) with 30 to 60 percent of the U.S. experiencing drought conditions.

Average precipitation rates have also increased by 6 percent over the past century. Since 1990 the U.S. has experienced eight of the top ten years of extreme precipitation events. Increases in sea surface temperatures produce a more active hurricane season. Since the mid-1990s the Atlantic Ocean, Caribbean, and Gulf of Mexico have seen six of the ten most active hurricane seasons.

Sea levels worldwide have risen an average of one inch per decade due to the overall increase in the surface temperature of the world's oceans. Over half of the human population lives within 60 miles of the coast. Climate models suggest that we will see a rise in sea levels of 3 to 4 feet over

the next century. New York City ranges from 5 feet to 16 feet above sea level, while the Florida Keys are an average of 3 to 4 feet above sea level. Even if the rising waters don't produce flooding, many coastal areas will be exposed to increasingly severe storms and storm surges that could lead to significant economic losses.

Between 1990 and 2008, the U.S.-produced greenhouse gas emissions increased by 14 percent. Production of electricity is the largest producer of greenhouse gas emissions in the United States, followed by transportation. We are facing a global problem that will require global cooperation in order to be solved.

The Kyoto Protocol was initially adopted on December 11, 1997, and entered into force on February 16, 2005. The goal of the protocol was to achieve stabilization and reduction of the greenhouse gas concentrations in the atmosphere. As of November 2009, 187 countries had ratified the protocol with the goal of reducing emissions by an average of 5.2% by the year 2012. Even though the United States was responsible for approximately 36.1% of the 1990 baseline emission levels and is the largest per capita emission producer in the world, the U.S. still has not ratified the agreement.

The Copenhagen conference in 2009 ended without any type of binding agreement for long-term action against climate change. It did produce a collective commitment by many developed nations to raise 30 billion dollars to be used to help poor nations cope with the effects of and combat climate change.

The 2010 Cancun summit on climate change helped to solidify this agreement. Because deforestation produces about 15% of the global carbon emissions, many developing countries will be able to receive incentives to prevent the destruction of their rain forests.

The main concern of the conferences is that the hard decisions of making significant changes to our greenhouse emissions continues to be pushed off into the future. The longer we delay in making changes the higher the risks become. (Fig. 45Cb)

Questions to Consider

1. Should the U.S. and other developed nations pay developing nations to preserve their forests?
2. Do individuals have a personal responsibility to help prevent climate change, or is it a governmental responsibility?

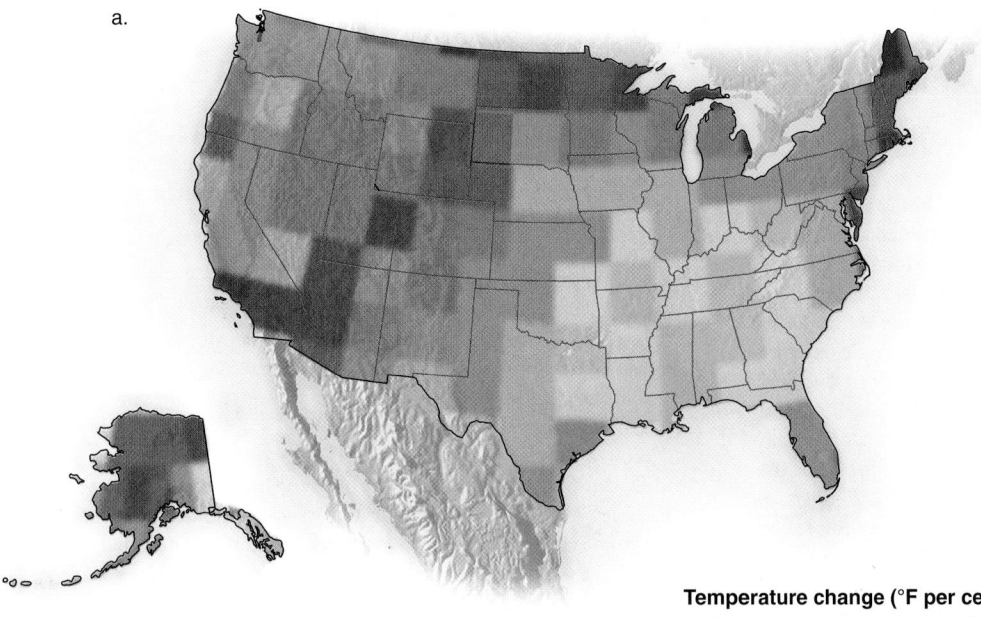

a.

b.

Figure 45C Global Warming and Climate Change. (a) The average temperature in the U.S. has steadily increased over the past 2 centuries, leading to more severe droughts and more erratic periods of precipitation that are altering the composition of many communities. (b) Increased extinction rates are just one of the consequences of climate change.

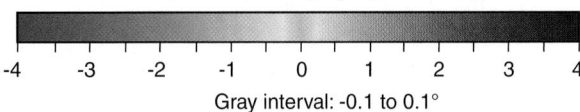

Temperature change (°F per century):

-4   -3   -2   -1   0   1   2   3   4

Gray interval: -0.1 to 0.1°

CONNECTING *the* CONCEPTS *with the* BIG IDEAS

Evolution

- The stability or instability of an environment affects the speed and direction of evolution. (1A1d)

Energy and Homeostasis

- Free energy passes from autotrophs to heterotrophs; its loss limits trophic levels. (2A1f*IE*, 2A2a-b)
- Environmental carbon, phosphorus, and potassium are incorporated into biomolecules by different pathways. (2A3a1-2)
- Symbioses and food chains/webs all demonstrate interactions between community members. (2D1b*IE*, 4A6c)

Interactions and Systems

- Primary productivity levels are subject to atmospheric composition and climate changes. (4A6b)
- Communities are described by their species composition and diversity. (4A5a)

*Find the unabridged version of all EK citations at www.glencoe.com/maderAP11.

Media Study Tools

www.glencoe.com/maderAP11

Enhance your study of this chapter with study tools and practice tests. Also ask your instructor about the resources available through ConnectPlus, including the media-rich eBook, interactive learning tools, and animations.

Summarize

45.1 Ecology of Communities

A community is an assemblage of populations interacting with one another within the same environment. Communities differ in their composition (species found there) and their diversity (species richness and relative abundance).

An organism's habitat is where it lives in the community. An ecological niche is defined by the role an organism plays in its community, including its habitat and how it interacts with other species in the community. Competition, predator-prey, parasite-host, commensalistic, and mutualistic relationships help organize populations into an intricate dynamic system.

The competitive exclusion principle states that no two species can indefinitely occupy the same niche at the same time. Character displacement is a structural change that gives evidence of resource partitioning and niche specialization. When resources are partitioned between two or more species, increased niche specialization occurs. But the difference between species can be more subtle, as when warblers feed at different parts of the tree canopy. Barnacles competing on the Scottish coast may be an example of present ongoing competition.

Predator-prey interactions between two species are especially influenced by amount of predation and the amount of food for the prey. A cycling of population densities may occur. Prey defenses take many forms: Camouflage, use of fright, and warning coloration are three possible mechanisms. Batesian mimicry occurs when one species has the warning coloration but lacks the defense. Müllerian mimicry occurs when two species with the same warning coloration have the same defense.

We would expect coevolution to occur within a community. For example, the better the predator becomes at catching prey, the better the prey becomes at escaping the predator.

Like predators, parasites take nourishment from their host. Whether parasites are aggressive (kill their host) or benign probably depends on which results in the highest fitness.

Symbiotic relationships are classified as commensalistic, parasitic, or mutualistic. Mutualistic relationships, as when Clark's nutcrackers feed on but also disperse whitebark pine seeds, are critical to the cohesiveness of a community.

45.2 Community Development

Ecological succession involves a series of species replacements in a community. Primary succession occurs where there is no soil present. Secondary succession occurs where soil is present and certain plant species can begin to grow. A climax community forms when stages of succession lead to a particular type of community.

45.3 Dynamics of an Ecosystem

Ecosystems have biotic and abiotic components. The biotic components are autotrophs, heterotrophs, detritus feeders, and decomposers. Abiotic components are resources such as nutrients and conditions such as type of soil and temperature.

Ecosystems are characterized by energy flow and chemical cycling. Energy flows in one direction because as food passes from one population to the next, each population makes energy conversions that result in a loss of usable energy. Chemicals cycle because they pass from one population to the next until decomposers return them once more to the producers.

Ecosystems contain food webs in which the various organisms are connected by trophic relationships. In grazing food webs, food chains begin with a producer. In a detrital food web, food chains begin with detritus. Ecological pyramids are graphic representations of the number of organisms, biomass, or energy content of trophic levels.

Biogeochemical cycles may be sedimentary (phosphorus cycle) or gaseous (carbon and nitrogen cycles). Chemical cycling involves a reservoir, an exchange pool, and a biotic community.

In the water cycle, evaporation over the ocean is not compensated for by precipitation. Precipitation over land results in bodies of fresh water plus groundwater, including aquifers. Eventually, all water returns to the oceans.

In the carbon cycle, carbon dioxide in the atmosphere is an exchange pool; both terrestrial and aquatic plants and animals exchange carbon dioxide with the atmosphere. Living and dead organisms serve as reservoirs for the carbon cycle because they contain organic carbon. Human activities increase the level of CO_2 and other greenhouse gases, contributing to global warming and climate change.

In the phosphorus cycle, geological upheavals move phosphorus from the ocean to land. Slow weathering of rocks returns phosphorus to the soil. Most phosphorus is recycled within a community, and phosphorus is a limiting nutrient.

In the nitrogen cycle, plants cannot use nitrogen gas from the atmosphere. During nitrogen fixation, N_2 converts to ammonium, making nitrogen available to plants. Nitrification is the production of nitrates while denitrification is the conversion of nitrate back to N_2, which enters the atmosphere. Human activities increase transfer rates in the nitrogen cycle. Acid deposition occurs when nitrogen oxides enter the atmosphere, combine with water vapor, and return to Earth in precipitation.

Key Terms

acid deposition 878	habitat 859
autotroph 870	herbivore 870
biogeochemical cycle 874	heterotroph 870
biomass 873	host 865
camouflage 863	island biogeography
carnivore 870	model 859
character displacement 860	mimicry 864
climate change 877	mutualism 866
climax community 868	N_2 (nitrogen) fixation 878
coevolution 866	nitrification 878
commensalism 865	omnivore 870
community 858	parasite 865
competitive exclusion	parasitism 865
principle 860	pioneer species 868
consumer 870	predation 860
decomposer 871	predator 860
denitrification 878	prey 860
detritivore 870	producer 870
ecological niche 859	resource partitioning 860
ecological pyramid 873	species diversity 858
ecological succession 868	species richness 858
ecosystem 870	symbiosis 865
eutrophication 877	transfer rate 875
food chain 873	trophic level 873
food web 873	water (hydrologic) cycle 875
global warming 877	
greenhouse effect 877	
greenhouse gas 876	

■ Assess

Reviewing This Chapter

1. What data do you need to describe a community's composition and diversity?   858–59
2. Describe the habitat and ecological niche of a particular species.   859
3. What is the competitive exclusion principle? How does the principle relate to character displacement and niche specialization?   859–60
4. Explain the observation that some predator-prey population densities cycle. Give examples of prey defenses. What is mimicry, and why does it work as a prey defense?   861, 863–64

5. Give examples of parasitism, commensalism, and mutualism, and examples of coevolution that occur within a community.   865–66
6. What are the two types of ecological succession? What is the present controversy surrounding the concept?   868–69
7. Give examples of autotrophs and heterotrophs in an ecosystem.   870–71
8. Distinguish between energy flow and chemical cycling in an ecosystem. Describe two types of food webs and give examples of food chains.   871–73
9. Explain the appearance of an ecological pyramid and the expression trophic level.   873–74
10. Draw diagrams to illustrate the water, carbon, phosphorus, and nitrogen cycles.   875–78
11. List and explain human activities that affect each of these cycles and contribute to a degraded environment for all organisms.   875–78

Testing Yourself

Choose the best answer for each question.

1. According to the competitive exclusion principle,
   a. one species is always more competitive than another for a particular food source.
   b. competition excludes multiple species from using the same food source.
   c. no two species can occupy the same niche at the same time.
   d. competition limits the reproductive capacity of species.

2. Resource partitioning pertains to
   a. niche specialization.
   b. character displacement.
   c. increased species diversity.
   d. the development of mutualism.
   e. All but d are correct.

For statements 3–7, indicate the type of interaction in the key that is described in each scenario.

KEY:

   a. competition   d. commensalism
   b. predation   e. mutualism
   c. parasitism

3. An alfalfa plant gains fixed nitrogen from the bacterial species *Rhizobium* in its root system, while *Rhizobium* gains carbohydrates from the plant.
4. Both foxes and coyotes in an area feed primarily on a limited supply of rabbits.
5. Roundworms establish a colony inside a cat's digestive tract.
6. A fungus captures nematodes as a food source.
7. An orchid plant lives in the treetops, gaining access to sun and pollinators, but not harming the trees.
8. A bullhorn acacia provides a home and nutrients for ants. Which statement is likely?
   a. The plant is under the control of pheromones produced by the ants.
   b. The ants protect the plant.
   c. The plant and the ants compete with each other.
   d. The plant and the ants have coevolved to occupy different ecological niches.
   e. All of these are correct.

9. The frilled lizard of Australia suddenly opens its mouth wide and unfurls folds of skin around its neck. Most likely, this was a way to
   a. conceal itself.
   b. warn that it was noxious to eat.
   c. scare a predator.
   d. scare its prey.
   e. All of these are correct.
10. When one species mimics the appearance of another species, the mimic sometimes
    a. lacks the defense of the model.
    b. possesses the defense of the model.
    c. is brightly colored.
    d. All of these are correct.
11. The species within a community are
    a. used to compare communities.
    b. present due to their abiotic requirements.
    c. more diverse as the size of the area increases.
    d. present due to their biotic interactions.
    e. All of these are correct.
12. Label this diagram of an ecosystem:

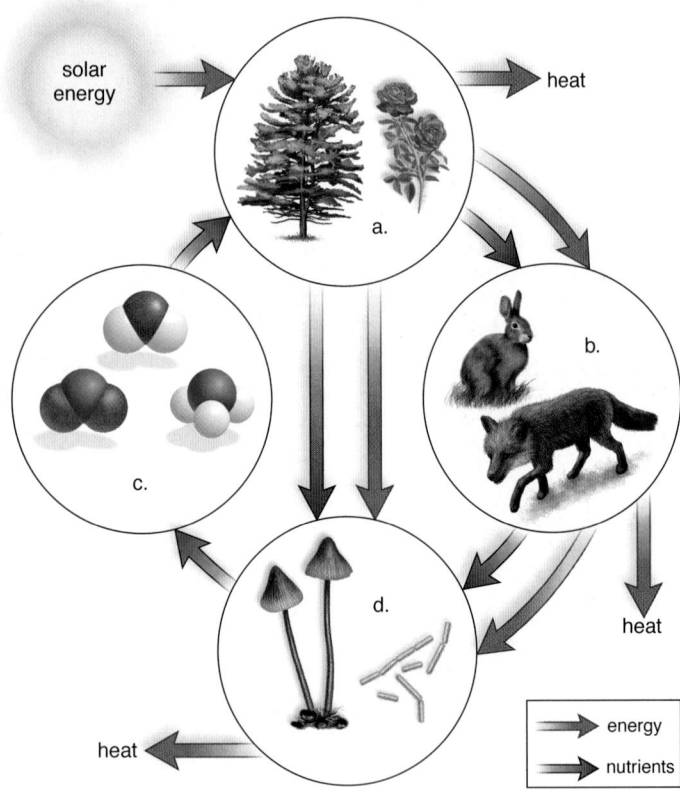

solar energy

heat

a.

b.

c.

d.

heat

heat

energy

nutrients

13. Mosses growing on bare rock will eventually help to create soil. These mosses are involved in _____ succession.
    a. primary
    b. secondary
    c. tertiary
14. The ecological niche of an organism
    a. is the same as its habitat.
    b. includes how it competes and acquires food.
    c. is specific to the organism.
    d. is usually occupied by another species.
    e. Both b and c are correct.

15. In what way are decomposers like producers?
    a. Either may be the first member of a grazing or a detrital food chain.
    b. Both produce oxygen for other forms of life.
    c. Both require nutrient molecules and energy.
    d. Both are present only on land.
    e. Both produce organic nutrients for other members of ecosystems.
16. When a heterotroph takes in food, only a small percentage of the energy in that food is used for growth because
    a. some food is not digested and is eliminated as feces.
    b. some metabolites are excreted as urine.
    c. some energy is given off as heat.
    d. All of these are correct.
    e. None of these are correct.
17. During chemical cycling, inorganic nutrients are typically returned to the soil by
    a. autotrophs.
    b. detritivores.
    c. decomposers.
    d. tertiary consumers.
18. Choose the statement that is true concerning this food chain: grass → rabbits → snakes → hawks
    a. Each predator population has a greater biomass than its prey population.
    b. Each prey population has a greater biomass than its predator population.
    c. Each population is omnivorous.
    d. Each population returns inorganic nutrients and energy to the producer.
    e. Both a and c are correct.
    f. Both a and b are correct.
19. Which of the following could not be a component of the nitrogen cycle?
    a. proteins
    b. ammonium
    c. decomposers
    d. photosynthesis
    e. bacteria in root nodules
20. How do plants contribute to the carbon cycle?
    a. When plants respire, they release CO_2 into the atmosphere.
    b. When plants photosynthesize, they consume CO_2 from the atmosphere.
    c. When plants photosynthesize, they provide oxygen to heterotrophs.
    d. Both a and b are correct.

Engage

Virtual Lab
Model Ecosystems

The virtual lab "Model Ecosystems" allows you to explore an ecological pyramid and the movement of energy through an ecosystem.

Thinking Scientifically

1. As per Figure 17.1, if you observe three species of frogs in the same general area, how would you test a hypothesis that they occupy different niches?

2. In order to improve species richness, you decide to add phosphate to a pond. How might you determine how much phosphate to add in order to avoid eutrophication?

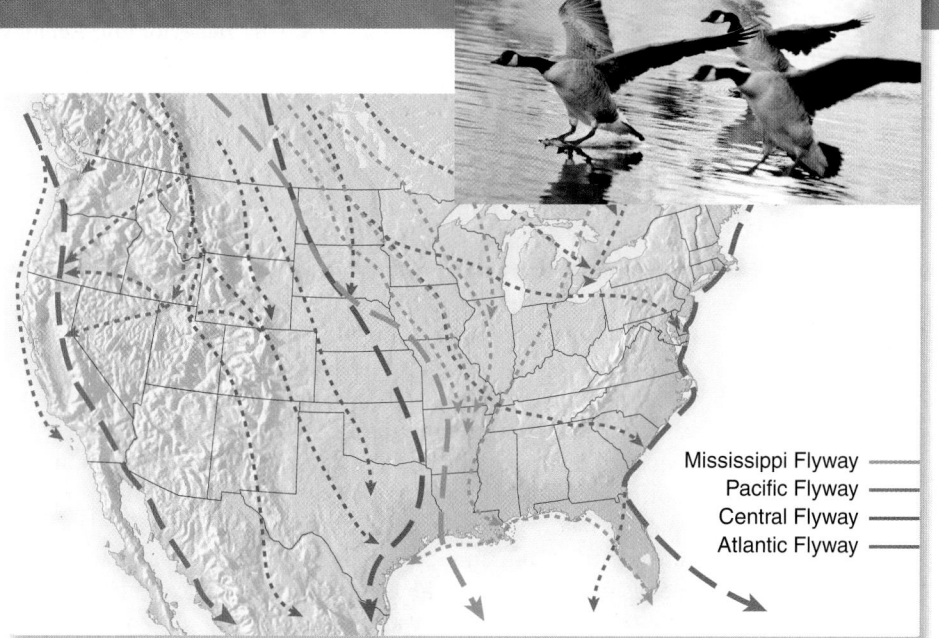

Mississippi Flyway ———
Pacific Flyway ———
Central Flyway ———
Atlantic Flyway ———

Migration route of Canada geese.

46

Major Ecosystems of the Biosphere

Toward fall, as temperatures drop and food becomes scarce, Canada geese (*Branta canadensis*) begin their annual migration. Their primary migration corridor extends from north of the Arctic Circle to south of the Tropic of Cancer. Millions of geese will follow the Atlantic, Mississippi, or Pacific flyways to reach specific wintering grounds. Birds can travel at speeds of 30 to 55 miles per hour and have been known to travel up to 650 miles in a single day. Some migrations will cover 2,000–3,000 miles.

During the winter migration they stop at key staging points to rest and feed. Wetlands, marshes, grasslands and chaparral regions serve as vital habitat for their migration. The high concentration of birds in these areas makes them more susceptible to disease outbreaks and predation. Natural predators range from foxes and owls to bald eagles and humans. Waterfowl hunting in Illinois contributes over 82 million dollars a year to the state's economy.

After the birds have wintered in the southern portion of their range they begin the spring migration toward their breeding grounds. This migration will take them through the same flyways and staging points that the winter migration does. Nesting occurs from the middle of March to the middle of May. By late August the new birds have grown enough to begin the cycle over again. A variety of biomes will play critical roles in the survival of this species.

As you read through the chapter, think about the following questions:

1. What advantage do migratory birds have over their counterparts who do not migrate?

2. How would changes in one land biome impact the other land biomes?

CHAPTER OUTLINE

46.1 Climate and the Biosphere 884
46.2 Terrestrial Ecosystems 887
46.3 Aquatic Ecosystems 897

BEFORE YOU BEGIN

Before beginning this chapter, take a few moments to review the following discussions.

Section 7.5 How does the climate determine the producer base of a biome?

Section 45.1 How do symbiotic relationships enable species to survive in harsh environments?

Figure 45.21 Will pollution in one biome impact other biomes?

FOLLOWING *the* BIG IDEAS

CHAPTER 46   MAJOR ECOSYSTEMS OF THE BIOSPHERE

Interactions and Systems

Ecosystems and their keystone species are often adversely affected by formidable environmental changes.

46.1 Climate and the Biosphere

Learning Outcomes

Upon completion of this section, you should be able to

1. Describe how solar radiation produces variations in Earth's climate.
2. Explain how global air circulation patterns and physical geographic features are associated with the Earth's temperature and rainfall patterns.

Climate refers to the prevailing weather conditions in a particular region. Climate is dictated by temperature and rainfall, which are influenced by the following factors:

- variations in solar radiation distribution due to the tilt of the Earth as it orbits about the Sun; and
- other effects caused by topography (surface features) and whether a body of water is nearby.

Effect of Solar Radiation

Because the Earth is a sphere, it receives direct sunlight at the equator but indirect sunlight at the poles (Fig. 46.1*a*). The region between latitudes approximately 26.5° north and south of the equator is considered the tropics. The tropics are warmer than the areas north of 23.5°N and south of 23.5°S, known as the temperate regions.

The tilt of the Earth as it orbits around the Sun causes one pole or the other to be angled toward the Sun (except at the spring and fall equinoxes, when sunlight aims directly at the equator). This accounts for the seasons that occur in all parts of the Earth except at the equator (Fig. 46.1*b*). When the Northern Hemisphere is having winter, the Southern Hemisphere is having summer, and vice versa.

If the Earth were standing still and were a solid, uniform ball, all air movements would be in two directions. Air at the equator warmed by the Sun would rise and move toward the poles where it would cool and sink. Rising air creates zones of lower air pressure. However, because the Earth rotates on its axis daily and its surface consists of continents and oceans, the flow of warm and cold air is modified into three large circulation cells in each hemisphere (Fig. 46.2).

At the equator, the Sun heats the air and evaporates water. Warm, moist air rises, and as it cools it loses most of its moisture as rain. The greatest amounts of rainfall on Earth are near the equator. The rising air flows toward the poles, but at about 30° north and south latitude, it cools and sinks toward the Earth's surface before reheating. As the dry air descends and warms, areas of high pressure are generated. High-pressure regions are zones of low rainfall. The great deserts of Africa, Australia, and the Americas occur at these latitudes.

At the Earth's surface, the air flows both toward the poles and the equator. As dry air moves across the Earth, moisture from both land and water gets absorbed. At about 60° north and south latitude, the warmed air rises and cools, producing another low-pressure area with high rainfall. This moisture supports the great forests of the temperate zone. Part of this rising air flows toward the equator, and part continues toward the poles, where it descends. The poles are high pressure areas and have low amounts of precipitation.

Besides affecting precipitation, the spinning of the Earth also affects the winds (Fig. 46.2). In the Northern Hemisphere, large-scale winds generally bend clockwise, and in the Southern Hemisphere, they bend counterclockwise. The curving pattern of the winds, ocean currents, and cyclones is the result of the fact that the Earth rotates in an eastward direction. At about 30° north latitude and 30° south latitude, the winds blow from the east-southeast in the Southern Hemisphere and from the

Figure 46.1 Distribution of solar energy.
a. Because the Earth is a sphere, beams of solar energy striking the Earth near one of the poles are spread over a wider area than similar beams striking the Earth at the equator.
b. The seasons of the Northern and Southern Hemispheres are due to the tilt of the Earth on its axis as it rotates about the Sun.

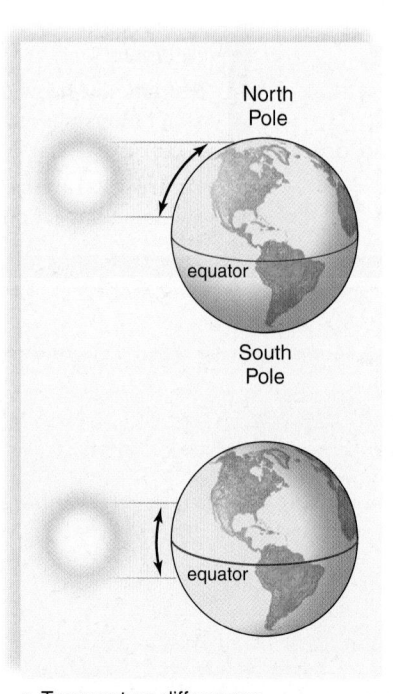

a. Temperature differences

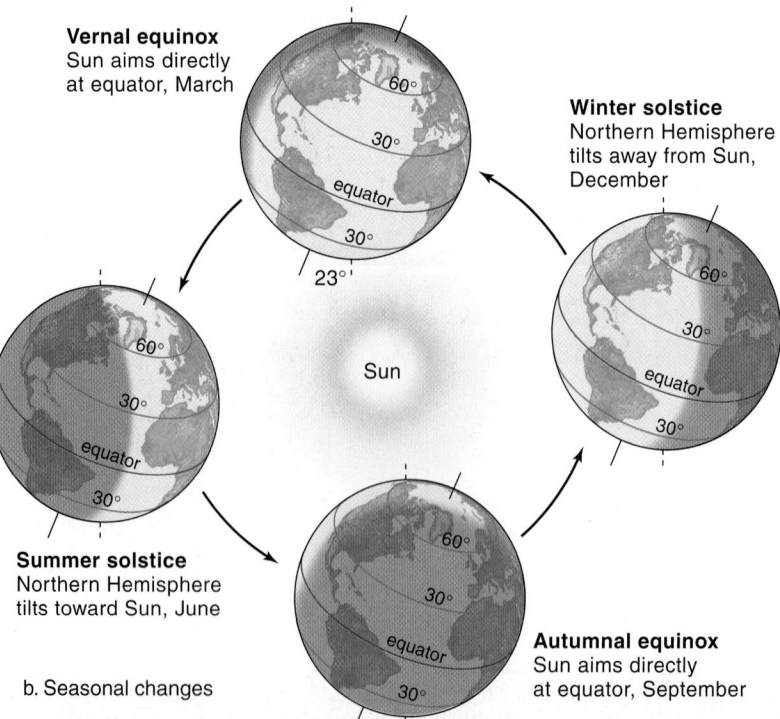

b. Seasonal changes

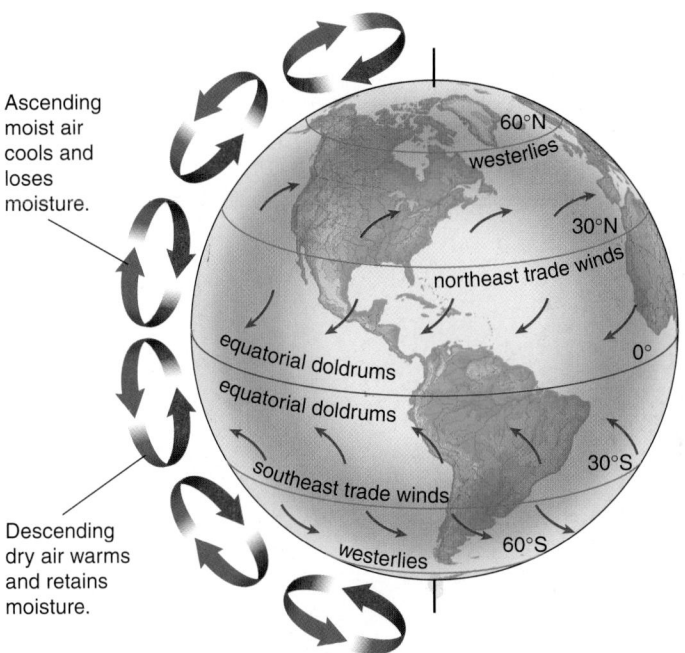

Figure 46.2 Global wind circulation. Air ascends and descends as shown because the Earth rotates on its axis. Also, the trade winds move from the northeast to the west in the Northern Hemisphere, and from the southeast to the west in the Southern Hemisphere. The westerlies move toward the east.

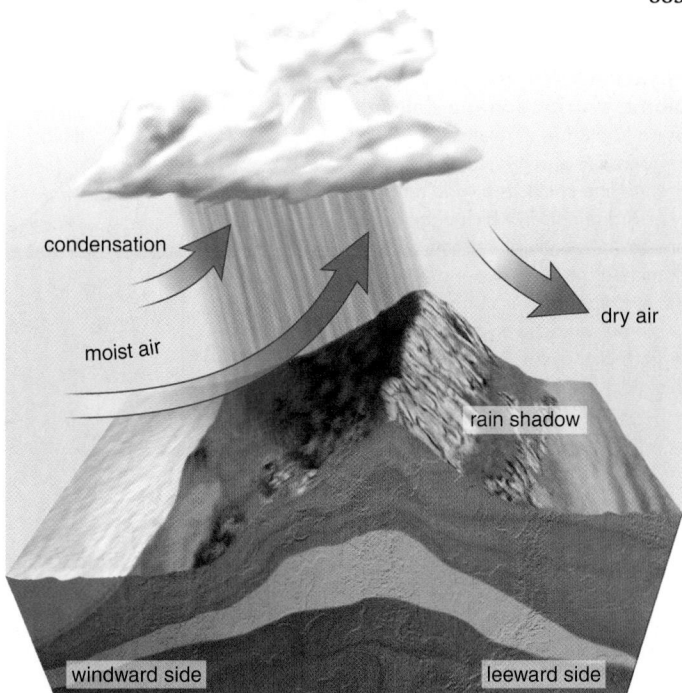

Figure 46.3 Formation of a rain shadow. When winds from the sea cross a coastal mountain range, they rise and release their moisture as they cool this side of a mountain, called the windward side. The leeward side of a mountain receives relatively little rain and is therefore said to lie in a "rain shadow."

east-northeast in the Northern Hemisphere (the east coasts of continents at these latitudes are wet). The doldrums, regions of calm, occur at the equator. The winds blowing from the doldrums toward the poles are called trade winds because sailors depended on them to fill the sails of their trading ships.

Between 30° and 60° north and south latitudes, strong winds called the prevailing westerlies blow from west to east. The west coasts of the continents at these latitudes are wet, as is the Pacific Northwest, where a massive evergreen forest is located. Weaker winds, called the polar easterlies, blow from east to west at still higher latitudes of their respective hemispheres.

Other Effects

Topography means the physical features of the land. One physical feature that affects climate is the presence of mountains. As air blows up and over a coastal mountain range, it rises and cools. One side of the mountain, called the windward side, receives more rainfall than the other side, called the leeward side. On the leeward side, the air descends, absorbs moisture from the ground, and produces clear weather (Fig. 46.3).

The difference between the windward side and the leeward side can be quite dramatic. In the Hawaiian Islands, for example, the windward side of the mountains receives more than 750 cm of rain a year, while the leeward side, which is in a **rain shadow,** gets on the average only 50 cm of rain and is generally sunny. In the United States, the western side of the Sierra Nevada Mountains is lush, while the eastern side is a semidesert.

The temperature of the oceans is more stable than that of landmasses. Oceanic water gains or loses heat more slowly than do terrestrial environments. This difference causes coasts to have a unique weather pattern that is not observed inland.

During the day, the land warms more quickly than the ocean, and the air above the land rises, pulling a cool sea breeze in from the ocean. At night, the reverse happens; the breeze blows from the land toward the sea.

India and some other countries in southern Asia have a **monsoon** climate, in which wet ocean winds blow onshore for almost half the year. The land heats more rapidly than the waters of the Indian Ocean during spring. The difference in temperature between the land and the ocean causes an enormous circulation of air: Warm air rises over the land, and cooler air comes in off the ocean to replace it. As the warm air rises, it loses its moisture, and the monsoon season begins. As just discussed, rainfall is particularly heavy on the windward side of hills. Cherrapunji, a city in northern India, receives an annual average of 1,090 cm of rain a year because of its high altitude. This weather pattern has reversed by November. The land is now cooler than the ocean; therefore, dry winds blow from the Asian continent across the Indian Ocean. In the winter, the air over the land is dry, the skies cloudless, and temperatures pleasant. The chief crop of India is rice, which starts to grow when the monsoon rains begin.

In the United States, people often speak of the "lake effect," meaning that in the winter, arctic winds blowing over the Great Lakes become warm and moisture-laden. When these winds rise and lose their moisture, snow begins to fall. Places such as Buffalo, New York, receive heavy snowfalls due to the lake effect, and snow is on the ground there for an average of 90–140 days every year.

Check Your Progress   46.1

1. Identify the conditions that account for a warm climate at the equator.
2. Name two physical features that can affect rainfall.

Figure 46.4 Pattern of biome distribution. a. Pattern of world biomes in relation to temperature and moisture. The dashed line encloses a wide range of environments in which either grasses or woody plants can dominate the area, depending on the soil type. **b.** The same type of biome can occur in different regions of the world, as shown on this global map.

Arctic-alpine
tundra
cold temperate
taiga
temperate deciduous forest
woodland
shrubland
semidesert
grassland
temperate rain forest
warm temperate
thorn scrub
savanna
desert
tropical seasonal forest
tropical rain forest
tropical
Mean Annual Temperature (°C)
−15
−10
−5
0
5
10
15
20
25
30
50
100
150
200
250
300
350
400
450
Mean Annual Precipitation (cm)

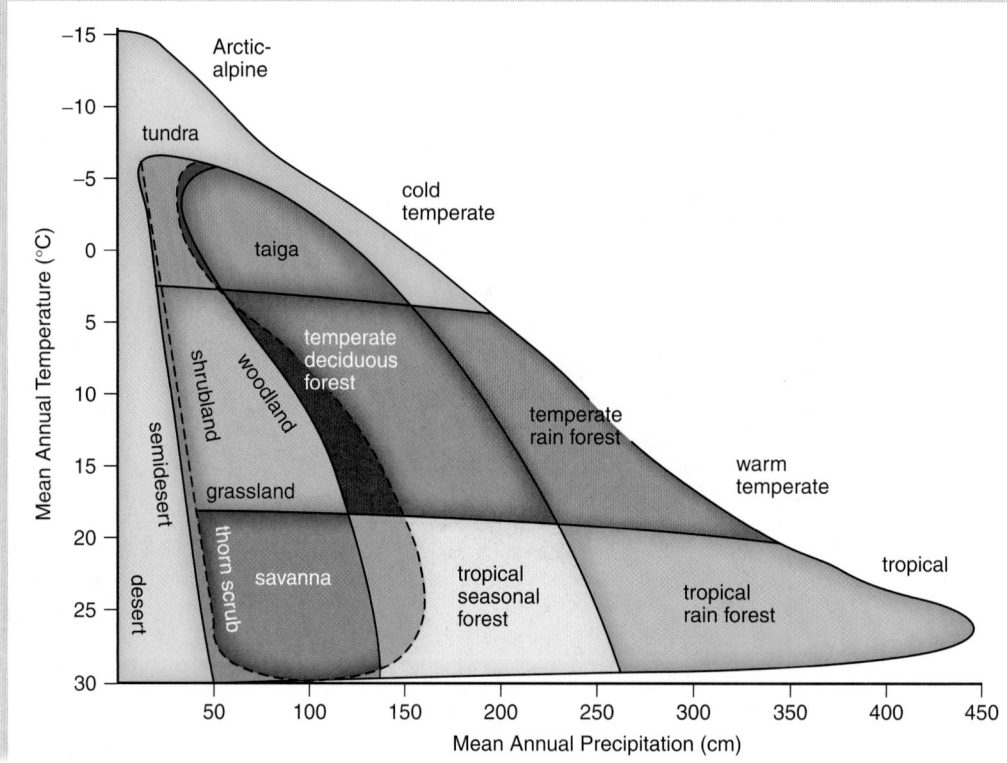

a. Biome pattern of temperature and precipitation

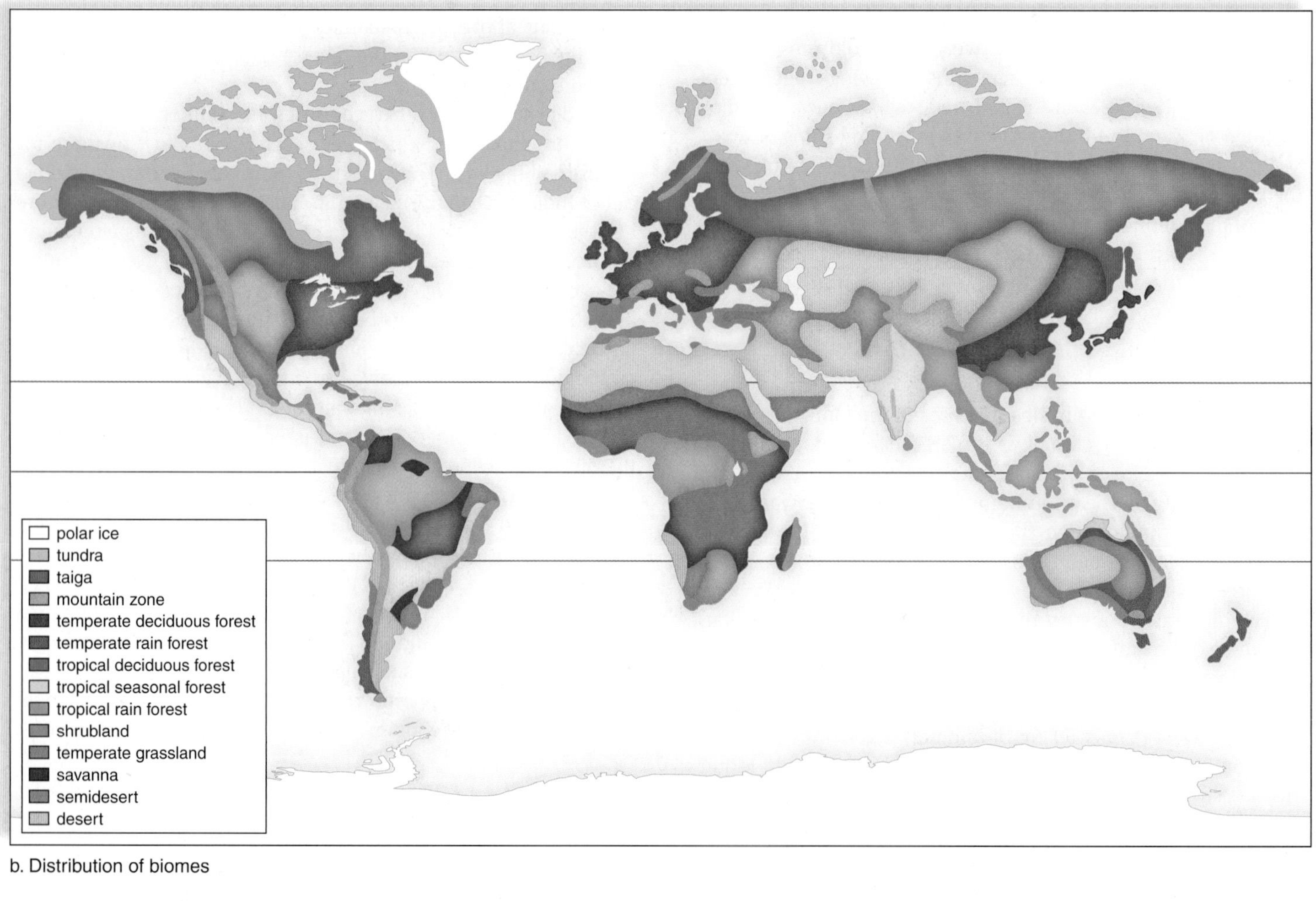

polar ice
tundra
taiga
mountain zone
temperate deciduous forest
temperate rain forest
tropical deciduous forest
tropical seasonal forest
tropical rain forest
shrubland
temperate grassland
savanna
semidesert
desert

b. Distribution of biomes

46.2 Terrestrial Ecosystems

Learning Outcomes

Upon completion of this section, you should be able to

1. Recognize the geographical distribution of the major terrestrial biomes.
2. Identify the key characteristics of the major terrestrial biomes.

Major terrestrial ecosystems, called biomes, are characterized by their climate and geography. A **biome** has a particular mix of plants and animals that are adapted to living under certain environmental conditions, of which climate is an overriding influence.

When terrestrial biomes are plotted according to their mean annual temperature and mean annual precipitation, particular patterns result (Fig. 46.4a). The distribution of biomes is shown in Figure 46.4b. Even though Figure 46.4 shows definite dividing lines, keep in mind that the change from one biome to another is gradual. Each biome has a connection to all the other terrestrial and aquatic ecosystems of the biosphere.

Animation
Biomes

As mentioned, the distribution of the biomes and their corresponding populations are determined principally by differences in climate. Latitude and altitude both play a role in determining temperature gradients. If you travel from the equator to the North Pole, you observe tropical rain forests, followed by a temperate deciduous forest, a coniferous forest, and tundra. A similar sequence is seen when ascending a mountain (Fig. 46.5). The coniferous forest of a mountain is called a **montane coniferous forest,** and the tundra near the peak of a mountain is called an **alpine tundra.** When going from the equator to the South Pole, you would not reach a region corresponding to a coniferous forest or tundra because of the lack of landmasses in the Southern Hemisphere.

Each biome supports characteristic types of animals; however, as described in the opening essay, many animals are able to migrate from one region to another. Animals may breed in one biome, for example, but spend the nonbreeding period in another biome. The Nature of Science feature, on page 889, describes how DNA analysis is helping to clarify not only species relationships but also migration patterns.

We now describe the features of the major biomes, beginning with the tundra.

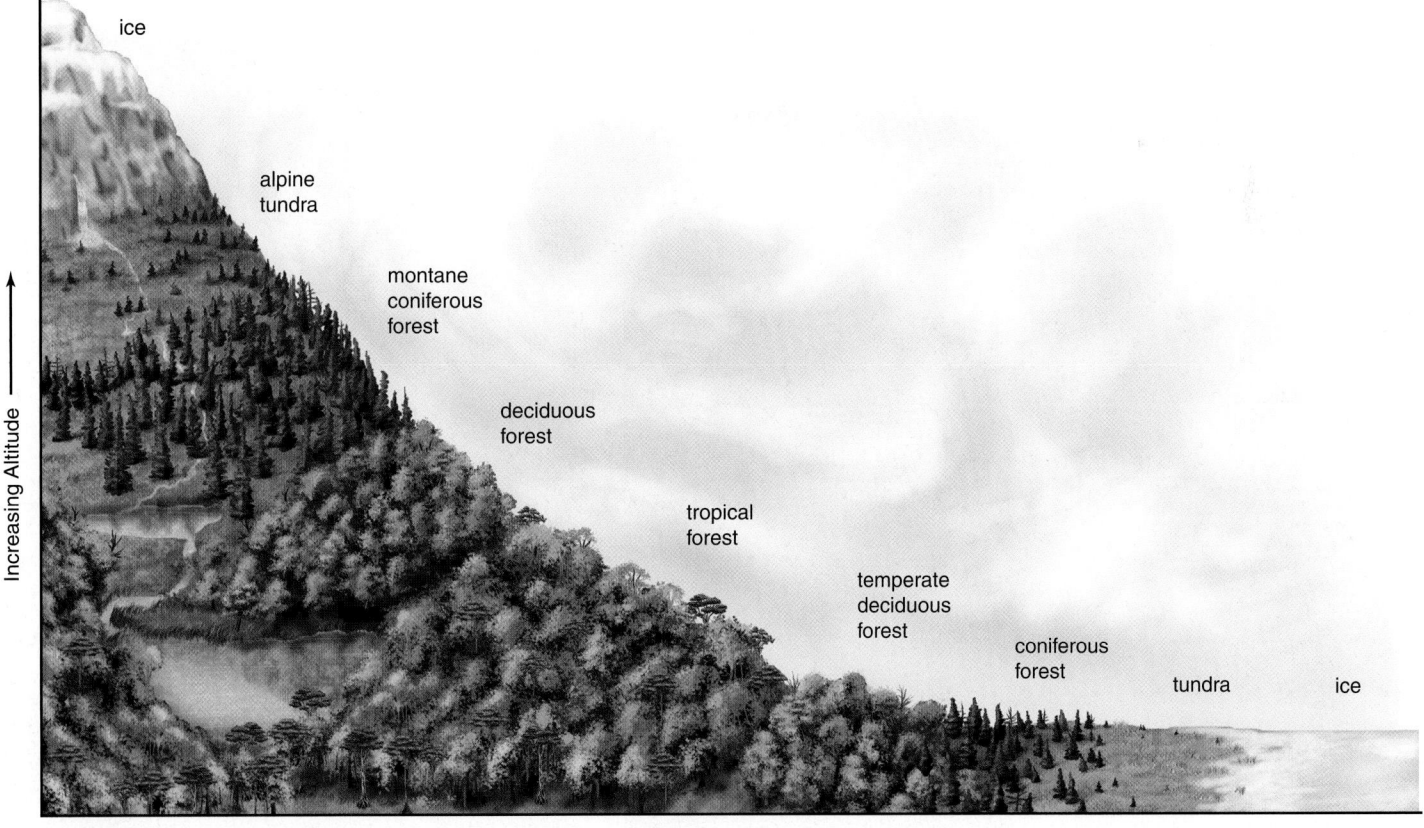

Figure 46.5 Climate and biomes. Biomes change with altitude just as they do with latitude because vegetation is partly determined by temperature. Precipitation also plays a significant role, which is one reason grasslands, instead of tropical or deciduous forests, are sometimes found at the base of mountains.

Tundra

The **Arctic tundra** biome, which encircles the Earth just south of ice-covered polar seas in the Northern Hemisphere, covers about 20% of the Earth's land surface (Fig. 46.6). (As mentioned, a similar ecosystem, the alpine tundra, occurs above the timberline on mountain ranges.) The Arctic tundra is cold and dark much of the year. Arctic tundra has extremely long, cold, harsh winters and short summers (6–8 weeks). Because rainfall amounts to only about 20 cm a year, the tundra could possibly be considered a desert, but melting snow creates a landscape of pools and bogs in the summer. Only the topmost layer of soil thaws; the **permafrost** beneath this layer is always frozen, and therefore, drainage is minimal. The available soil in the tundra is nutrient-poor.

Trees are not found in the tundra because the growing season is too short. The roots cannot penetrate the permafrost and they cannot become anchored in the shallow boggy soil of summer. In the summer, the ground is covered with short grasses and sedges, as well as numerous patches of lichens and mosses. Dwarf woody shrubs, such as dwarf birch, flower and seed during the short growing season.

A few animals live in the tundra year-round. For example, the mouselike lemming stays beneath the snow; the ptarmigan, a grouse, burrows in the snow during storms; and the musk ox conserves heat because of its thick coat and short, squat body. Other animals that live in the tundra include snowy owls, lynxes, voles, Arctic foxes, and snowshoe hares. In the summer, the tundra is alive with numerous insects and birds, particularly shorebirds and waterfowl that migrate inland. Caribou in North America and reindeer in Asia and Europe also migrate to and from the tundra, as do the wolves that prey upon them. Polar bears are common near the coastal regions. All species have adaptations for living in extreme cold with a short growing season.

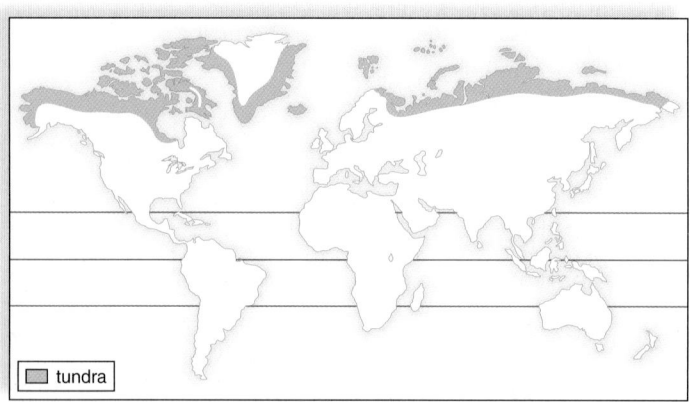

Figure 46.6 The tundra. **a.** In this biome, which is nearest the polar regions, the vegetation consists principally of lichens, mosses, grasses, and low-growing shrubs. Pools of water that do not evaporate or drain into the permanently frozen ground attract many birds. **b.** Caribou, more plentiful in the summer than in the winter, feed on lichens, grasses, and shrubs.

a. Tundra vegetation

b. Tundra wildlife

Nature of Science

Wildlife Conservation and DNA

After DNA analysis, scientists were amazed to find that some 60% of loggerhead turtles drowning in the nets and hooks of fisheries in the Mediterranean Sea were from beaches in the southeastern United States. Because the unlucky creatures were a good representative sample of the turtles in the area, that meant more than half of the young turtles living in the Mediterranean Sea had hatched from nests on beaches in Florida, Georgia, and South Carolina (Fig. 46Aa). Some 20,000–50,000 loggerheads die each year due to the Mediterranean fisheries, which may partly explain the decline in loggerheads nesting on southeastern U.S. beaches for the last 25 years.

The sequencing of DNA from Alaskan brown bears allowed wildlife geneticists Sandra Talbot and Gerald Shields, then at the University of Alaska's Institute of Arctic Biology, to conclude that there are two types of brown bears in Alaska. One type resides only on southeastern Alaska's Admiralty, Baranof, and Chichagof Islands, known as the ABC Islands. The other brown bear in Alaska is found throughout the rest of the state, as well as in Siberia and western Asia (Fig. 46Ab).

A third distinct type of brown bear, known as the Montana grizzly, resides in other parts of North America. These three types comprise all of the known brown bears in the New World.

The ABC bears' uniqueness may be bad news for the timber industry, which has expressed interest in logging parts of the ABC Islands. Says Shields, "Studies show that when roads are built and the habitat is fragmented, the population of brown bears declines. Our genetic observations suggest [these ABC bears] are truly unique, and we should consider their heritage. They could never be replaced by transplants."

In a classic example of how DNA analysis might be used to protect endangered species from future ruin, scientists from the United States and New Zealand carried out discreet experiments in a Japanese hotel room on whale sushi bought in local markets. Sushi, a staple of the Japanese diet, is a rice and meat mixture wrapped in seaweed. Armed with a miniature DNA sampling machine, the scientists found that, of the 16 pieces of whale sushi they examined, many were from whales that are endangered or protected under an international moratorium on whaling. "Their findings demonstrated the true power of DNA

studies," says David Woodruff, a conservation biologist at the University of California, San Diego.

One sample was from an endangered humpback, four were from fin whales, one was from a northern minke, and another from a beaked whale. Stephen Palumbi of the University of Hawaii says the technique could be used for monitoring and verifying catches. Until then, he says, "no species of whale can be considered safe."

Meanwhile, Ken Goddard, director of the unique U.S. Fish and Wildlife Service Forensics Laboratory in Ashland, Oregon, is already on the watch for wildlife crimes in the United States and 122 other countries that send samples to him for analysis. "DNA is one of the most powerful tools we've got," says Goddard, a former California police crime-lab director.

The lab has blood samples, for example, for all of the wolves being released into Yellowstone National Park—"for the obvious reason that we can match those samples to a crime scene," says Goddard. The lab has many cases currently pending in court that he cannot discuss. But

he likes to tell the story of the lab's first DNA-matching case. Shortly after the lab opened in 1989, California wildlife authorities contacted Goddard. They had seized the carcass of a trophy-sized deer from a hunter. They believed the deer had been shot illegally on a 3,000-acre preserve owned by actor Clint Eastwood. The agents found a gut pile on the property but had no way to match it to the carcass. The hunter had two witnesses to deny the deer had been shot on the preserve.

Goddard's lab analysis made a perfect match between tissue from the gut pile and tissue from the carcass. Says Goddard: "We now have a cardboard cutout of Clint Eastwood at the lab saying 'Go ahead: Make my DNA.'"

Questions to Consider

1. Would DNA evidence from a crime scene be enough to convict someone of a crime involving a protected species?
2. What degree of DNA difference is necessary to qualify two organisms as different species?

Figure 46A DNA studies. **a.** Many loggerhead turtles found in the Mediterranean Sea are from the southeastern United States. **b.** These two brown bears appear similar, but one type, known as an ABC bear, resides only on southeastern Alaska's Admiralty, Baranof, and Chichagof islands.

a. Loggerhead turtle, *Caretta caretta*

b. Brown bears, *Ursus arctos*

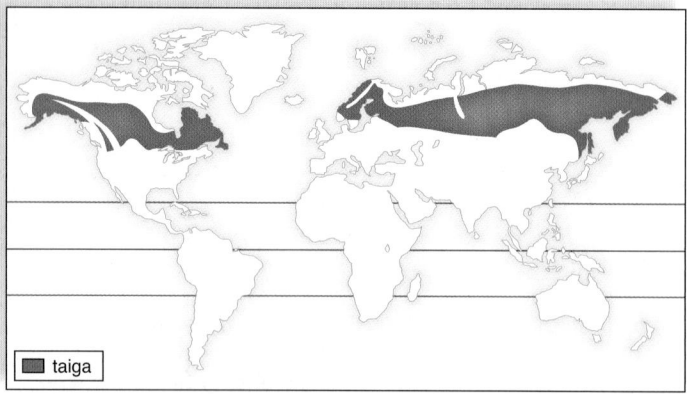

b. Bull moose, *Alces americanus*, a large mammal

Figure 46.7 The taiga.   The taiga, which means swampland, spans northern Europe, Asia, and North America. The appellation "spruce-moose" refers to the **(a)** dominant presence of spruce trees and **(b)** moose, which frequent the ponds.

a. Spruce trees in the taiga biome

Coniferous Forests

Coniferous forests are found in three locations: in the **taiga,** which extends around the world in the northern part of North America and Eurasia; near mountaintops (where it is called a montane coniferous forest); and along the Pacific coast of North America, as far south as northern California.

The taiga, or boreal forest, exists south of the tundra and covers approximately 11% of the Earth's landmasses (Fig. 46.7). There are no comparable biomes in the Southern Hemisphere

because no large landmasses exist at that latitude. The taiga typifies the coniferous forest with its cone-bearing trees, such as spruce, fir, and pine. These trees are well adapted to the cold because both the leaves and bark have thick coverings. Also, the needlelike leaves can withstand the weight of heavy snow. There is a limited understory of plants, but the forest floor is covered by low-lying mosses and lichens beneath a layer of needles. Birds harvest the seeds of the conifers, while bears, deer, moose, beavers, and muskrats live around the lakes and streams. Wolves prey on these larger mammals. A montane coniferous forest also harbors the wolverine and the mountain lion.

The coniferous forest that runs along the west coast of Canada and the United States is sometimes called a **temperate rain forest.** As the prevailing winds move in off the Pacific Ocean, they drop their moisture when they meet the coastal mountain range. The plentiful rainfall and rich soil have produced some of the tallest conifer trees ever in existence, including the coastal redwoods. This forest is also called an old-growth forest because some trees are as old as 800 years. It truly is an evergreen forest because of the abundance of green plants year round.

Squirrels, lynxes, and numerous species of amphibians, reptiles, and birds inhabit the temperate rain forest. The northern spotted owl, *Strix occidentalis caurina,* is an endangered species found in this ecosystem that has received recent conservation attention and efforts.

Temperate Deciduous Forests

Temperate deciduous forests are found south of the taiga in eastern North America, eastern Asia, and much of Europe (Fig. 46.8). The climate in these areas is moderate, with relatively high rainfall (75–150 cm per year). The seasons are well defined, and the growing season ranges between 140 and 300 days. The trees, such as oak, beech, sycamore, and maple, have broad leaves. These are termed deciduous trees because they lose their leaves in the fall and regrow them in the spring.

The tallest trees form a canopy, an upper layer of leaves that are the first to receive sunlight. Enough sunlight penetrates to provide energy for another layer of trees, called understory trees. Beneath these trees are shrubs and herbaceous plants that may flower in the spring before the trees have put forth their leaves. Still another layer of plant growth—mosses, lichens,

and ferns—resides beneath the shrub layer. This stratification provides a variety of habitats for insects and birds. Ground animals are also plentiful. Squirrels, rabbits, woodchucks, and chipmunks are small herbivores. These and ground birds such as turkeys, pheasants, and grouse are preyed on by various carnivores. In contrast to the taiga, amphibians and reptiles occur in this biome because the winters are not as cold. Frogs and turtles prefer an aquatic existence, as do the beaver and muskrat. Autumn fruits, nuts, and berries provide a supply of food for the winter, and the leaves, after turning brilliant colors and falling to the ground, contribute to the rich layer of humus. The minerals within the rich soil are washed far into the ground by spring rains, but the deep tree roots capture these and bring them back up into the forest system again.

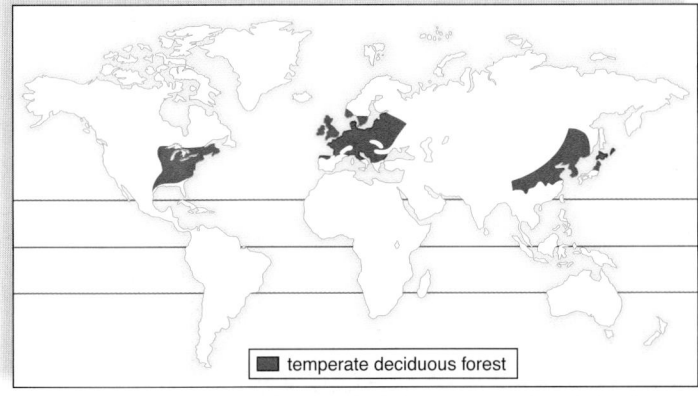

temperate deciduous forest

Figure 46.8 Temperate deciduous forest. **a.** A temperate deciduous forest is home to many varied plants and animals. **b.** Various flowering plants reside here along with chipmunks that feed on acorns, and bobcats prey on these and other small mammals.

Eastern chipmunk, *Tamias striatus*

Marsh marigolds, *Caltha howellii*

Bobcat, *Felis rufus*
b. Animal life of forest biome

a. Temperate deciduous vegetation

Tropical Forests

In the **tropical rain forests** of South America, Africa, and the Indo-Malayan region near the equator, the weather is always warm (between 20° and 25°C), and rainfall is plentiful (with a minimum of 190 cm per year). These are the richest land biomes on Earth. It is estimated that approximately 50% of the species found on Earth live in the tropical rain forests.

A tropical rain forest has a complex structure, with many levels of life, including the forest floor, understory, and canopy. The vegetation of the forest floor tends to be very sparse because most of the sunlight is filtered out by the canopy. The understory consists of smaller plants that are specialized for life in the shade. The canopy, topped by the crowns of tall trees, is the most productive level of the tropical rain forest (Fig. 46.9).

Some of the broadleaf evergreen trees can grow from 15–50 m or more in height. These tall trees often have trunks buttressed at ground level to prevent their toppling over. Lianas, or woody vines, encircle the tree as it grows and help strengthen the trunk. The diversity of species is enormous—a 10-km² area of tropical rain forest may contain 750 species of trees and 1,500 species of flowering plants.

Although some animals live on the forest floor (e.g., pacas, agoutis, peccaries, and armadillos), most live in the trees (Fig. 46.10). Insect life is so abundant that the majority of species have not been identified yet. Termites play a vital role in the decomposition of woody plant material, and ants are found everywhere, particularly in the trees. The various birds, such as hummingbirds, parakeets, parrots, and toucans, are often beautifully colored. Amphibians and reptiles are well represented by many types of frogs, snakes, and lizards. Lemurs, sloths, and monkeys are well-known primates that feed on the fruits of the trees. The largest carnivores are the big cats—the jaguars in South America and the leopards in Africa and Asia.

Many animals spend their entire life in the canopy, as do some plants. **Epiphytes** are plants that grow on other plants but usually have roots of their own that absorb moisture and minerals leached from the canopy. Epiphytes like bromeliads catch rain and debris by forming vases of overlapping leaves. The most common epiphytes are related to pineapples, orchids, and ferns.

Figure 46.9 Levels of life in a tropical rain forest.   The primary levels within a tropical rain forest are the canopy, the understory, and the forest floor. But the canopy (solid layer of leaves) contains levels as well, and some organisms spend their entire life in one particular level. Long lianas (hanging vines) climb into the canopy, where they produce leaves. Epiphytes are air plants that grow on the trees but do not parasitize them.

We usually think of tropical forests as being nonseasonal rain forests, but tropical forests that have wet and dry seasons are found in India, Southeast Asia, West Africa, South and Central America, the West Indies, and northern Australia. Here, there are deciduous trees, with many layers of growth beneath them.

Contrary to popular belief, the soil of a tropical rain forest biome is nutrient poor. Due to the high amount of herbivory, only a small amount of leaf litter makes it to the forest floor. Productivity is high because of high temperatures, a year-long growing season, and nearly twelve hours of sunlight year round. Slash-and-burn agriculture can be successful if it is done on a small scale, but becomes destructive and unsustainable on a large scale. Trees are felled and burned, and the ashes provide enough nutrients for several harvests. Thereafter, the forest must be allowed to regrow, and a new section must be cut and burned. In addition, in humid tropical forests, iron and aluminum oxides occur at the surface, causing a reddish residue known as laterite. When the trees are cleared, laterite bakes in the hot sunlight to a bricklike consistency that will not support crops.

It is estimated that 2.4 acres of rain forest are destroyed per second. This rate equates to nearly 78 million acres annually. Unless conservation strategies are employed soon, rain forests will be destroyed beyond recovery, taking with them unique and interesting life-forms. Ecologists estimate that an average of 137 species are driven to extinction every day in rain forests.

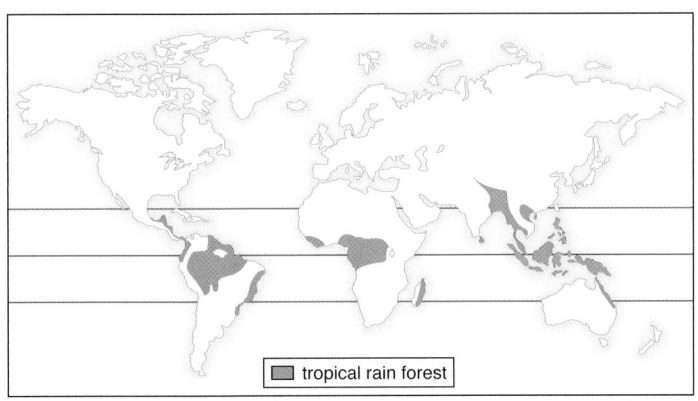

tropical rain forest

Poison arrow frog,
Dendrobates azureus

Cone-headed katydid,
Panacanthus cuspidatus

Panther,
Panthera onca

Scarlet macaw,
Ara macao

Brush-footed butterfly,
Anartia amalthea linnaeus

Black howler monkey,
Alouatta pigra

Arboreal lizard,
Calotes calotes

Figure 46.10 Representative animals of the tropical rain forests of the world.

Shrublands

It is difficult to define a shrub, but in general, shrubs are shorter than trees (4.5–6 m) with woody, persistent stems but no large, central trunk. Shrubs have small but thick evergreen leaves, which are often coated with a waxy material that prevents loss of moisture. Their thick underground roots can survive dry summers and frequent fires and take moisture from the deep soil. Shrubs are adapted to withstand arid conditions and can quickly sprout new growth after a fire. As a point of interest, you may recall from Chapter 45 that a shrub stage is part of the process of both primary and secondary succession.

Shrublands tend to occur along coasts that have dry summers and receive most of their rainfall in the winter. Shrublands are found along the cape of South Africa, the western coast of North America, and the southwestern and southern shores of Australia, as well as around the Mediterranean Sea and in central Chile. The dense shrubland that occurs in California is known as **chaparral** (Fig. 46.11). This type of shrubland lacks an understory and ground litter and is highly flammable. The seeds of many species require the heat and scarring action of fire to induce germination. Other shrubs sprout from the roots after a fire. Typical animals of the chaparral include mule deer, rodents, lizards, and scrub jays.

A northern shrub area lies to the west of the Rocky Mountains. This area is sometimes classified as a cold desert, but the region is dominated by sagebrush. Some of the birds found there are dependent on sagebrush for their existence.

Grasslands

Grasslands occur where annual rainfall is greater than 25 cm but generally insufficient to support trees. For example, in temperate areas, where rainfall is between 25 and 75 cm, it is too dry for forests and too wet for deserts to form.

Grasses are well adapted to a changing environment and can tolerate a high degree of grazing, flooding, drought, and sometimes fire. Where rainfall is high, tall grasses that reach more than 2 m in height (e.g., pampas grass) can flourish. In drier areas, shorter grasses (between 5 and 10 cm) are dominant. Low-growing bunch grasses (e.g., grama grass) grow in the United States near deserts.

The growth of grasses is seasonal. As a result, grassland animals such as bison migrate, and others such as ground squirrels hibernate, when there is little grass for them to eat.

Temperate Grasslands

The **temperate grasslands** include the Russian steppes, the South American pampas, and the North American prairies (Fig. 46.12). In these grasslands, winters are bitterly cold and summers are hot and dry. When traveling across the United States from east to west, the line between the temperate deciduous forest and a tall-grass prairie is roughly along the border between Illinois and Indiana. The tall-grass prairie receives more rainfall than does the short-grass prairie, which occurs near deserts.

Large herds of bison—estimated at hundreds of thousands—once roamed the prairies, as did herds of pronghorn antelope. Now, small mammals, such as mice, prairie dogs, and rabbits, typically live belowground, but usually feed aboveground. Hawks, snakes, badgers, coyotes, and foxes feed on these mammals. Virtually all of these grasslands, however, have been converted to agricultural lands because of their fertile soils.

Video Tallgrass Prairie Ecology

Video Dung Beetles

a. Shrubland overview

b. Wildlife of the chaparral

Figure 46.11 Shrubland.   a. Shrublands, such as chaparral in California, are subject to raging fires, but the shrubs are adapted to quickly regrow. **b.** Greater roadrunners find a home in the chaparral.

Savannas

Savannas occur in regions where a relatively cool dry season is followed by a hot rainy season (Fig. 46.13). The largest savannas are in central and southern Africa. Other savannas exist in Australia, southeast Asia, and South America. The savanna is characterized by large expanses of grasses with relatively few trees. The plants of the savanna have extensive and deep root systems that enable them to survive drought and fire. One tree that can survive the severe dry season is the thorny flat-topped *Acacia*, which sheds its leaves during a drought.

The African savanna supports the greatest variety and number of large herbivores of all the biomes. Elephants and giraffes are browsers that feed on tree vegetation. Antelopes, zebras, wildebeests, water buffalo, and rhinoceroses are grazers that feed on the grasses. Any plant litter that is not consumed by grazers is attacked by a variety of small organisms, among them termites. Termites build towering nests in which they tend fungal gardens, their source of food. The herbivores support a large population of carnivores. Lions, hyenas, cheetahs, and leopards all prey upon the abundant herbivore populations.

Video Thorn Tree Ant

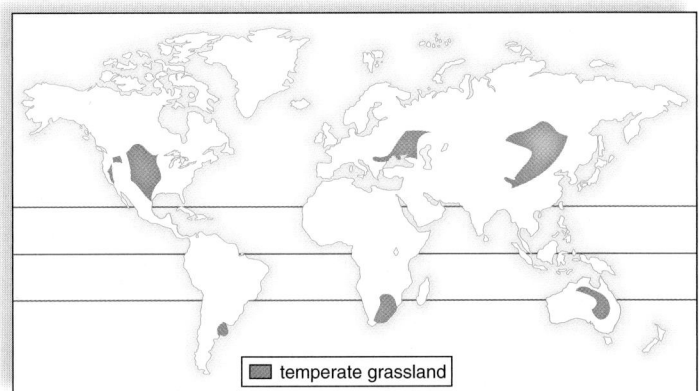

a. Vegetation of the temperate grasslands

b. Wildlife of the temperate grasslands

Figure 46.12 Temperate grassland. **a.** Tall-grass prairies are seas of grasses dotted by pines and junipers. **b.** Bison, once abundant, are now being reintroduced into certain areas.

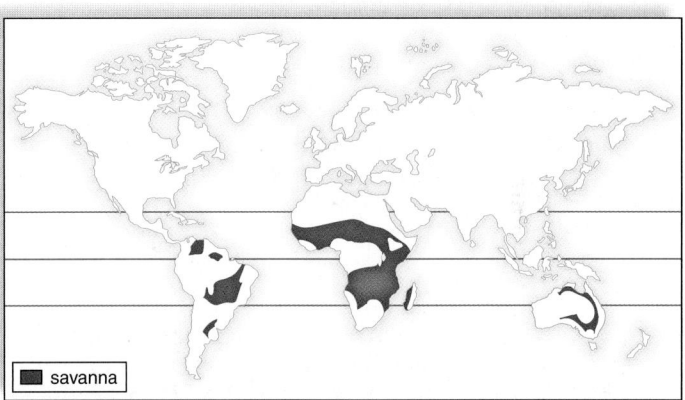

a. Herbivores of the savanna biome

b. A carnivore of the savanna biome

Figure 46.13 The savanna. The African savanna varies from grassland to widely spaced shrubs and trees. **a.** This biome supports a large assemblage of herbivores (e.g., zebras, wildebeests, and giraffes). **b.** Carnivores (e.g., cheetahs) prey on these.

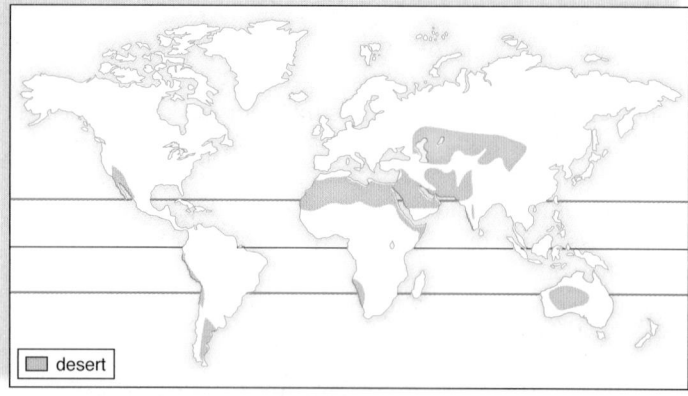

a. Desert vegetation

Bannertail kangaroo rat,
Dipodomys spectabilis

Greater roadrunner,
Geococcyx californianus

Kit fox, *Vulpes velox*
b. Animal life of desert biome

Deserts

Deserts are usually found at latitudes of about 30°, in both the Northern and Southern Hemispheres. Deserts cover nearly 30% of the Earth's land surface. The winds that descend in these regions lack moisture, resulting in an annual rainfall of less than 25 cm. Days are hot because a lack of cloud cover allows the Sun's rays to penetrate easily, but nights are cold because heat escapes easily into the atmosphere.

The Sahara, which stretches all the way from the Atlantic coast of Africa to the Arabian peninsula, along with a few other deserts, have little or no vegetation. However, most deserts contain a variety of plants that are highly adapted to survive long droughts, extreme heat, and extreme cold (Fig. 46.14a). Adaptations to these conditions include thick epidermal layers, water-storing succulent stems and leaves, and the ability to set seeds quickly in the spring. The best-known desert perennials in North America are the spiny cacti, which have stems that store water and carry on photosynthesis. Also common are nonsucculent shrubs, such as the many-branched sagebrush with silvery gray leaves and the spiny-branched ocotillo, which produces leaves during wet periods and sheds them during dry periods.

Some animals are adapted to the desert environment. To conserve water, many desert animals such as reptiles and insects are nocturnal or burrowing and have a protective outer body covering. A desert has numerous insects, which pass through the stages of development in synchrony with the periods of rain. Reptiles, especially lizards and snakes, are perhaps the most characteristic group of vertebrates found in deserts, but running birds (e.g., the roadrunner) and rodents (e.g., the kangaroo rat) are also well known (Fig. 46.14b). Larger mammals, such as the kit fox, prey on the rodents, as do hawks.

Check Your Progress   46.2

1. Contrast the vegetation of the tropical rain forest with that of a temperate deciduous forest.
2. Explain why there are more carnivores on the African savanna than in the tundra.

Figure 46.14 The desert. Plants and animals that live in a desert are adapted to arid conditions. **a.** The plants are either succulents, which retain moisture, or shrubs with woody stems and small leaves, which lose little moisture. **b.** Among the animal life, the kangaroo rat feeds on seeds and other vegetation; the roadrunner preys on insects, lizards, and snakes. The kit fox is a desert carnivore.

46.3 Aquatic Ecosystems

Learning Outcomes

Upon completion of this section, you should be able to

1. Compare the characteristics of freshwater and saltwater ecosystems.
2. Interpret the manner in which ocean currents will affect the climate and weather over the continents.

Aquatic ecosystems are classified in two categories: freshwater (inland) systems or saltwater systems. Brackish water, however, is a mixture of fresh and salt water. Figure 46.15 shows how these ecosystems are joined physically and discusses some of the organisms that are adapted to live in them.

In the water cycle, the Sun's rays cause seawater to evaporate, and the salts are left behind. As discussed in Chapter 45, the evaporated fresh water rises into the atmosphere, cools, and falls as rain. When rain falls, some of the water sinks, or percolates, into the ground and saturates it. The layer of water in the ground is called the groundwater table, or simply the water table.

Because land lies above sea level, gravity eventually returns all fresh water to the sea. Along the route to the sea it is contained in lakes and ponds, or streams and rivers. Sometimes groundwater is also located in underground rivers called aquifers. Whenever the Earth contains basins or channels, water will rise to the level of the water table.

Wetlands

Wetlands are areas that are wet for at least part of the year. Generally, wetlands are classified by their vegetation. **Marshes** are wetlands that are frequently or continually inundated by water. They are characterized by the presence of rushes, reeds, and other grasses, which provide excellent habitat for waterfowl and small mammals. Marshes are one of the most productive ecosystems on Earth. **Swamps** are wetlands that are dominated by either woody plants or shrubs. Common swamp trees include cypress, red maple, and tupelo. The American alligator is a top predator in many swamp ecosystems. **Bogs** are wetlands that are characterized by acidic waters, peat deposits, and sphagnum moss. Bogs receive most of their water from precipitation and are nutrient-poor. Several species of plants thrive in bogs, including cranberries, orchids, and insectivorous plants such as Venus flytraps and pitcher plants. Moose and a number of other animals are inhabitants of bogs in the northern United States and Canada.

Humans have historically channeled and diverted rivers and filled in wetlands with the idea that "useless land" was being improved. However, these activities degrade ecosystems, can cause seasonal flooding, and eliminate food and habitats for many unique fishes, waterfowl, and other wildlife. Wetlands also purify waters by filtering them and by diluting and breaking down toxic wastes and excess nutrients. Wetlands directly absorb storm waters and overflow from lakes and rivers. In this way, they protect farms, cities, and towns from the devastating effects of floods. Federal and local laws have been enacted for the protection of wetlands as more people recognize their value.

Video
Thames River

Figure 46.15 Freshwater and saltwater ecosystems. *Center:* Mountain streams have cold, clear water that flows over waterfalls and rapids. As streams merge, a river forms and gets increasingly wider and deeper until it meanders across broad, flat valleys. At its mouth, a river may divide into many channels, where wetlands and estuaries are located, before flowing into the sea. *To Sides:* Mayfly larvae are found in clean water with a high oxygen content. Trout are a major predator of mayflies. Carp are adapted to water that contains little oxygen and much sediment. Blue crabs are found in estuary regions.

Carp, *Cyprinus carpio*

Blue crab, *Callinectes sapidus*

stream

river

lake

delta

salt marsh

Mayfly larva, *Ephemeroptera sp.*

Rainbow trout, *Salmo gairdneri*

Lakes

Lakes are bodies of fresh water that can be classified by their nutrient status. Oligotrophic (nutrient-poor) lakes are characterized by a small amount of organic matter and low productivity (Fig. 46.16a). Eutrophic (nutrient-rich) lakes are characterized by plentiful organic matter and high productivity (Fig. 46.16b). These latter lakes are usually situated in naturally nutrient-rich regions or are enriched by agricultural or urban and suburban runoff. Oligotrophic lakes can become eutrophic through large inputs of nutrients. This process is called **eutrophication.**

Lake Overturn

In the temperate zone, deep lakes are stratified in the summer and winter and have distinct vertical zones. In summer, lakes in the temperate zone have three layers of water that differ in temperature (Fig. 46.17). The surface layer, the epilimnion, is warm from solar radiation; the middle layer, the thermocline, experiences an abrupt drop in temperature; and the lowest layer, the hypolimnion, is cold. These differences in temperature prevent mixing. The warmer, less dense water of the epilimnion "floats" on top of the colder, more dense water of the thermocline, which floats on top of the hypolimnion.

Phytoplankton found in the sunlit epilimnion use up various nutrients as they photosynthesize. Photosynthesis releases oxygen, giving this layer a ready supply. Detritus accumulates at the bottom of the lake, and there oxygen is used up as decomposition occurs. Decomposition releases nutrients, however. As the season progresses, the epilimnion becomes nutrient-poor, while the hypolimnion begins to be depleted of oxygen.

In the fall, as the epilimnion cools, and in the spring, as it warms, an overturn occurs. In the fall, the upper epilimnion waters become cooler than the hypolimnion waters. This causes the surface water to sink and the deep water to rise. This **fall overturn** continues until the temperature is uniform throughout the lake. At this point, wind aids in the circulation of water so that mixing occurs. Eventually, oxygen and nutrients become evenly distributed.

As winter approaches, the water cools and ice forms. Ice has an insulating effect, preventing further cooling of the water

a. Oligotrophic lake

b. Eutrophic lake

Figure 46.16 Types of lakes. Lakes can be classified according to whether they are **(a)** oligotrophic (nutrient-poor) or **(b)** eutrophic (nutrient-rich). Eutrophic lakes tend to have large populations of algae and rooted plants, resulting in a large population of decomposers that use up much of the oxygen and leave little oxygen for fishes.

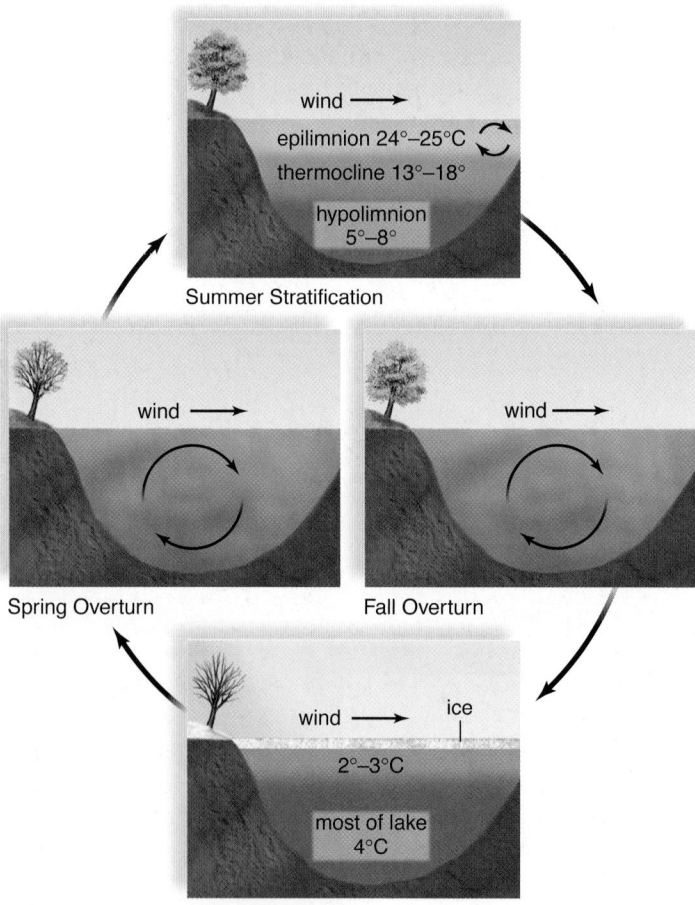

Figure 46.17 Lake stratification in a temperate region. Temperature profiles of a large oligotrophic lake in a temperate region vary with the season. During the spring and fall overturns, the deep waters receive oxygen from surface waters, and surface waters receive inorganic nutrients from deep waters.

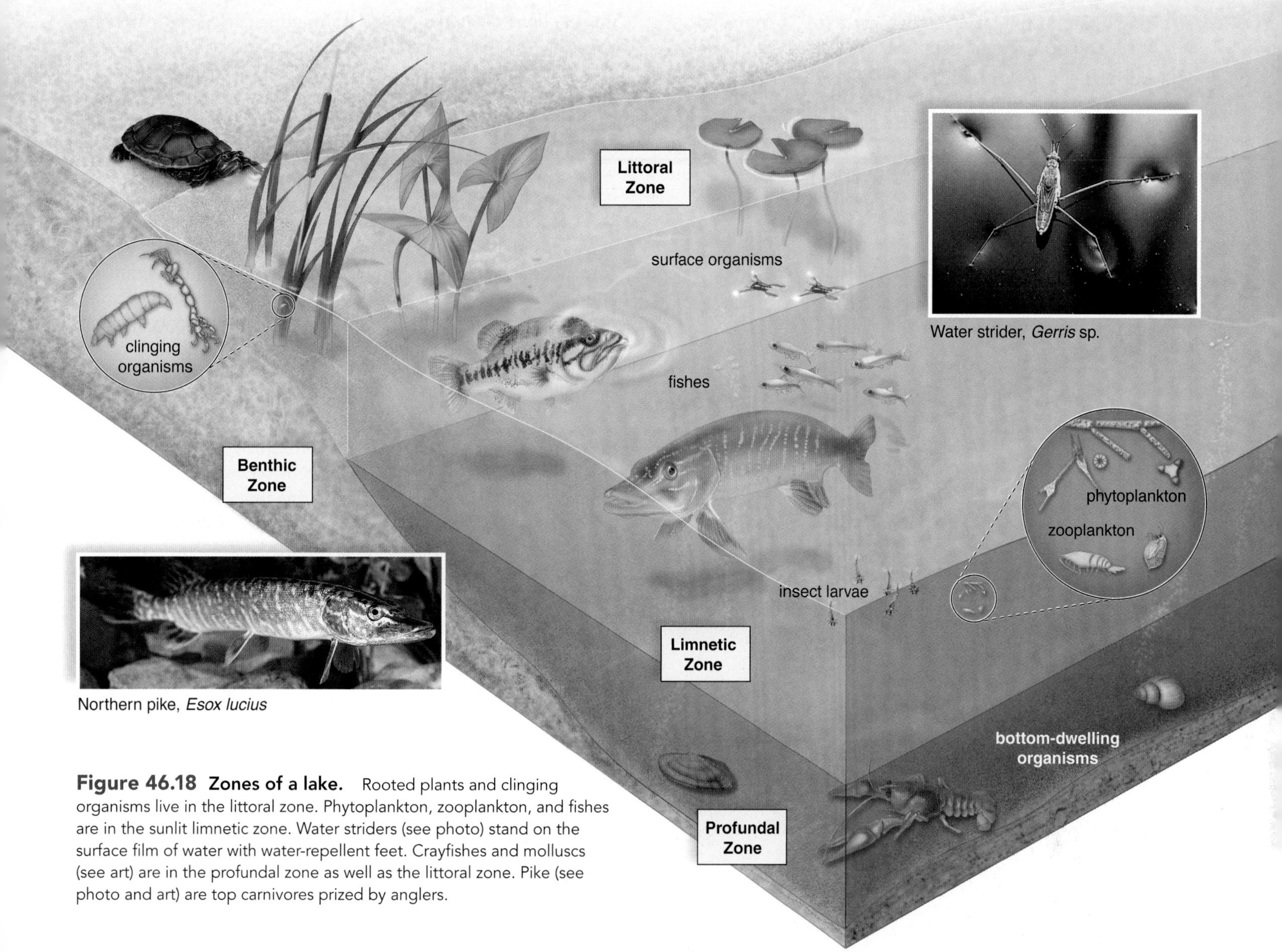

Littoral
Zone

surface organisms

Water strider, *Gerris* sp.

clinging
organisms

fishes

phytoplankton

zooplankton

Benthic
Zone

insect larvae

bottom-dwelling
organisms

Northern pike, *Esox lucius*

Limnetic
Zone

Profundal
Zone

Figure 46.18 Zones of a lake.   Rooted plants and clinging
organisms live in the littoral zone. Phytoplankton, zooplankton, and fishes
are in the sunlit limnetic zone. Water striders (see photo) stand on the
surface film of water with water-repellent feet. Crayfishes and molluscs
(see art) are in the profundal zone as well as the littoral zone. Pike (see
photo and art) are top carnivores prized by anglers.

below. This permits aquatic organisms to survive the winter in
the water beneath the ice.

In the spring, as the ice melts, the cooler water on top sinks
below the warmer water on the bottom. This **spring overturn**
continues until the temperature is uniform throughout the lake.
At this point, wind again aids in the circulation of the water.
When the surface waters absorb solar radiation, thermal stratification occurs once more.

The vertical stratification and seasonal change of temperatures in a lake influence the seasonal distribution of fish
and other aquatic life. For example, coldwater fish move to the
deeper water in summer and inhabit the upper water in winter.
In the fall and spring just after mixing occurs, phytoplankton
growth at the surface is most abundant.

Life Zones

In both fresh and salt water, free-drifting microscopic organisms
called *plankton* [Gk. *planktos,* wandering] are important components of the ecosystem. **Phytoplankton** [Gk. *phyton,* plant, and
planktos, wandering] are photosynthesizing algae that become
noticeable when a green scum or red tide appears on the water.

Zooplankton [Gk. *zoon,* animal, and *planktos,* wandering] are tiny animals that feed on
the phytoplankton.

Video
Plankton Diversity

Lakes and ponds can be divided into several life zones (Fig.
46.18). The *littoral zone* is closest to the shore, the *limnetic zone*
forms the sunlit body of the lake, and the *profundal zone* is
below the level of light penetration. The *benthic zone* includes
the sediment at the soil-water interface.

Aquatic plants are rooted in the shallow littoral zone of
a lake, providing habitat for numerous protozoans, invertebrates, fishes, and some reptiles. Largemouth bass are a type
of ambush predator that waits among vegetation around the
margins of lakes and surges out to capture passing prey. Wad-
ing birds are commonly seen feeding in the littoral zone. Some
organisms, such as the water strider, live at the water-air
interface and can literally walk on water. In the limnetic zone,
small fishes, such as minnows and killifish, feed on plankton
and also serve as food for larger fish, such as bass. In the
profundal zone, zooplankton, invertebrates, and fishes such
as catfish and whitefish feed on debris that falls from higher
zones.

The bottom of the lake, the benthic zone, is composed of mostly silt, sand, inorganic sediment, and dead organic material (detritus). Bottom-dwelling organisms are known as benthic species and include worms, snails, clams, crayfishes, and some insect larvae. Decomposers, such as bacteria, are also found in the benthic zone and serve to break down wastes and dead organisms into nutrients that are eventually used by the producers.

Coastal Ecosystems

Mudflats, mangrove swamps, and rocky shores, featured in Figure 46.19, are ecosystems that occur where fresh water and salt water meet. Mangrove swamps develop in subtropical and

a. Mudflat

b. Mangrove swamp

c. Rocky shore

Figure 46.19 Coastal ecosystems.   a. Mudflats are frequented by migrant birds. **b.** Mangrove swamps skirt the coastlines of many tropical and subtropical lands. **c.** Some organisms of a rocky coast live in tidal pools.

tropical zones, while marshes and mudflats occur in temperate zones. These ecosystems are often designated as an estuary. So are coastal bays, fjords (an inlet of water between high cliffs), and some lagoons (a body of water separated from the sea by a narrow strip of land). Therefore, the term estuary has a very broad definition. An **estuary** is a partially enclosed body of water where fresh water and sea water meet and mix as a river enters the ocean.

Organisms living in an estuary must be able to withstand constant mixing of waters and rapid changes in salinity. Organisms adapted to the estuarine environment benefit from the abundance of nutrients. An estuary acts as a nutrient trap because the sea prevents the rapid escape of nutrients brought by a river. Estuaries are biologically diverse and highly productive communities.

Phytoplankton and shore plants thrive in the nutrient-rich estuaries, providing an abundance of food and habitat for animals. It is estimated that nearly two-thirds of marine fishes and shellfish spawn and develop in the protective and rich environment of estuaries, making the estuarine environment the nursery of the sea. An abundance of larval, juvenile, and mature fish and shellfish attract a wide variety of predators.

Rocky shores (Fig. 46.19c) and sandy shores are constantly bombarded by the sea as the tides roll in and out. The **intertidal zone** lies between the high- and low-tide marks (Fig. 46.20). In the upper portion of the intertidal zone, barnacles are glued so tightly to the stone by their own secretions that their calcareous outer plates remain in place, even after the animal dies. In the midportion of the intertidal zone, brown algae, known as rockweed, may overlie the barnacles. Below the intertidal zone, macroscopic seaweeds, which are the main photosynthesizers, anchor themselves to the rocks by holdfasts.

Figure 46.20 Ocean ecosystems.   Organisms live in the well-lit waters of the euphotic zone and in the increasing darkness of the deep-sea waters of the pelagic zones (see Fig. 46.21).

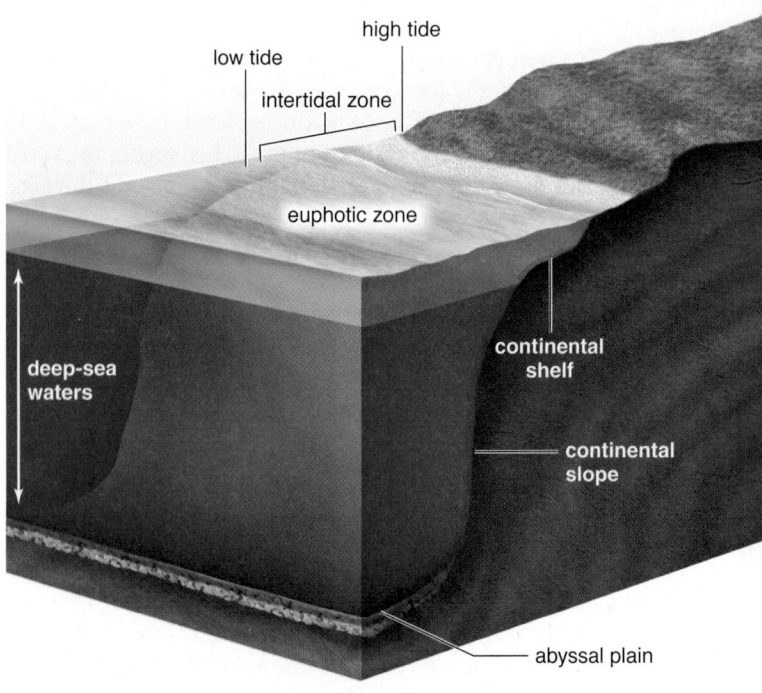

Organisms cannot attach themselves to shifting, unstable sands on a sandy beach; therefore, nearly all the permanent residents dwell underground. They either burrow during the day and surface to feed at night, or they remain permanently within their burrows and tubes. Ghost crabs and sandhoppers (amphipods) burrow themselves above the high-tide mark and feed at night when the tide is out. Sandworms and sand (ghost) shrimp remain within their burrows in the intertidal zone and feed on detritus whenever possible. Still lower in the sand, clams, cockles, and sand dollars are found. A variety of shorebirds visit the beaches and feed on various invertebrates and fishes.

Oceans

Shallow ocean waters (called the *euphotic zone*) contain a greater concentration of organisms than the rest of the sea (see Fig. 46.20). Here, phytoplankton is food not only for zooplankton but also for small fishes. These attract a number of predatory and commercially valuable fishes. On the continental shelf, seaweed can be found growing, even on outcroppings as the water gets deeper. Clams, worms, and sea urchins are preyed upon by sea stars, lobsters, crabs, and brittle stars.

Coral reefs are areas of biological abundance found in shallow, warm, tropical waters. Their chief constituents are stony corals, animals that have a calcium carbonate (limestone) exoskeleton, and calcareous red and green algae. Corals provide a home for microscopic algae called *zooxanthellae*. The corals feed at night, and the algae photosynthesize during the day forming a mutualistic relationship. The algae need sunlight for photosynthesis, which is why coral reefs typically develop in shallow, sunlit waters.

> **Video**
> **Coral Reef Ecosystems**

A reef is densely populated with life. The large number of crevices and caves provide shelter for filter feeders (sponges, sea squirts, and fanworms) and for scavengers (crabs and sea urchins). The barracuda, moray eel, and shark are top predators in coral reefs. Many types of small, beautifully colored fishes live here. These become food for larger fishes, including snappers that are caught for human consumption.

> **Video**
> **Reef Balls**

Most of the ocean's volume lies within the **pelagic zones,** as noted in Figure 46.21. The *epipelagic zone* lacks the inorganic nutrients of shallow waters, which means it does not have as high a concentration of phytoplankton as the shallow areas. Still, the photosynthesizers are food for a large assembly of zooplankton, which then become food for schools of various fishes. A number of porpoise and dolphin species visit and feed in the epipelagic zone. A variety of whales can be found in this zone. Baleen whales strain krill (small crustaceans) from the water, and toothed sperm whales feed primarily on the common squid.

> **Video**
> **Ocean Fishing Ban**

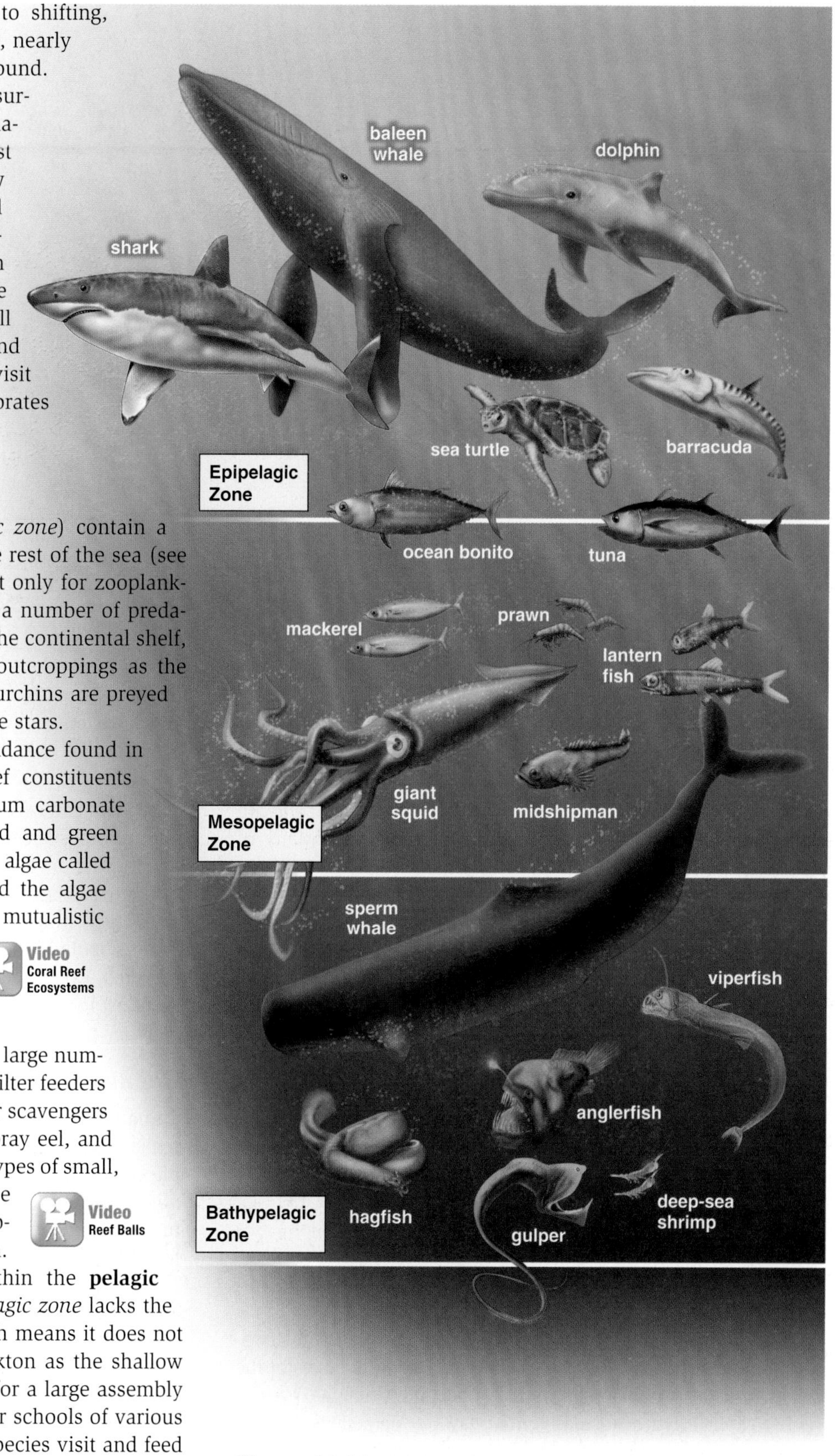

Figure 46.21 Ocean inhabitants of pelagic zones. Different organisms are characteristic of the epipelagic, mesopelagic, and bathypelagic zones.

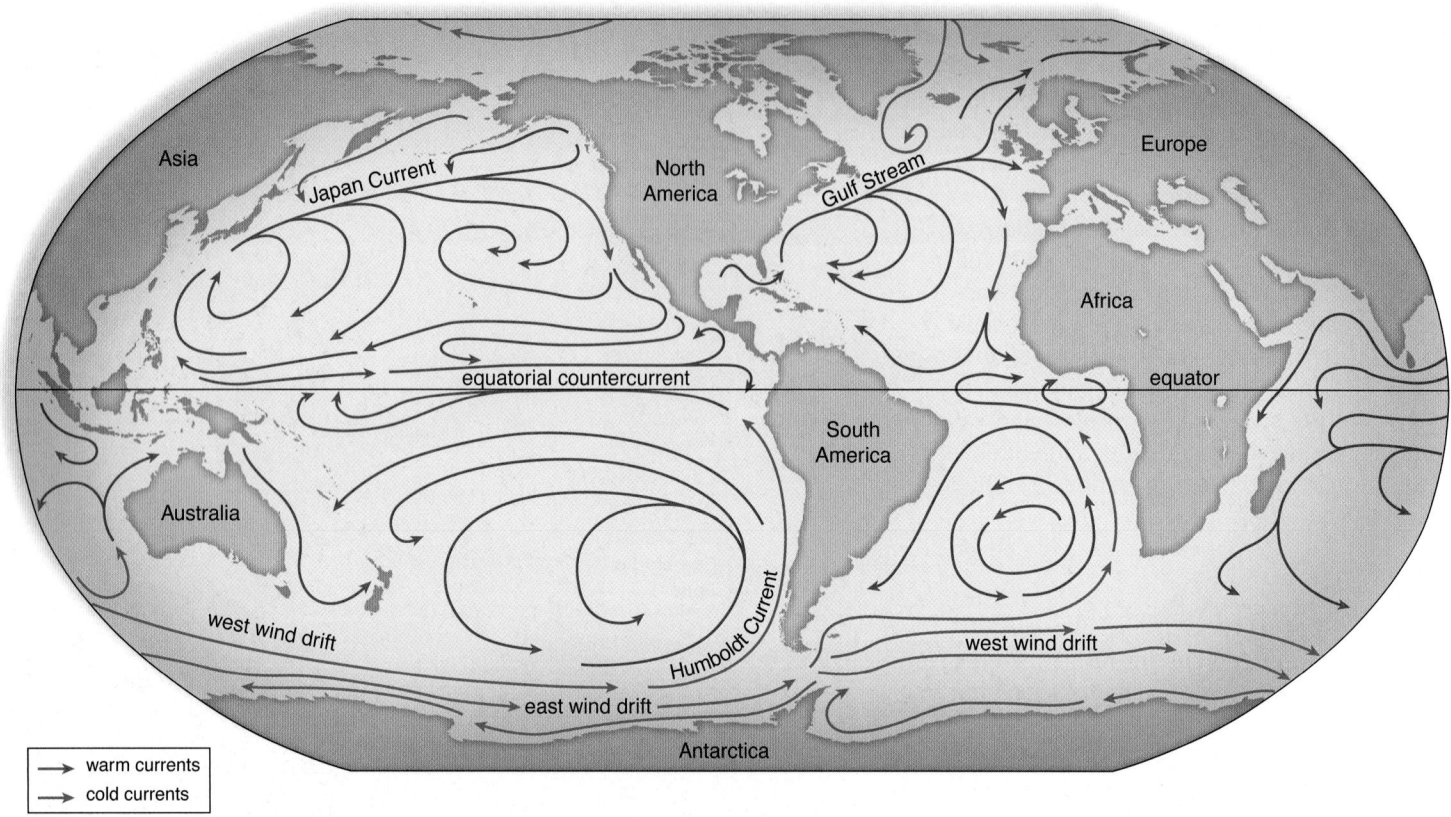

Figure 46.22 Ocean currents. The arrows on this map indicate the locations and directions of the major ocean currents set in motion by the global wind circulation. By carrying warm water to cool latitudes (e.g., the Gulf Stream) and cool water to warm latitudes (e.g., the Humboldt Current), these currents have a major effect on the world's climates.

Animals in the deeper waters of the *mesopelagic zone* are carnivores and are adapted to the absence of light. Many of these organisms tend to be translucent, red colored, or even luminescent; among these are some species of shrimps, squids, and fishes, including lantern and hatchet fishes. Various species of zooplankton, invertebrates, and fishes migrate from the mesopelagic zone to the surface to feed at night.

The deepest waters of the *bathypelagic zone* are in complete darkness except for an occasional flash of bioluminescent light. Carnivores and scavengers are found in this zone. Strange-looking fishes with distensible mouths and abdomens and small, tubular eyes feed on infrequent prey.

It once was thought that minimal life existed on the *abyssal plain* beneath the bathypelagic zone due to the intense pressure and the extreme cold. Yet many invertebrates survive there by feeding on debris floating down from the mesopelagic zone. Sea lilies (crinoids) rise above the seafloor; sea cucumbers and sea urchins crawl around on the sea bottom; and tube worms burrow in the mud.

The flat abyssal plain is interrupted by enormous underwater mountain chains called oceanic ridges. Along the axes of the ridges, crustal plates spread apart, and molten magma rises to fill the gap. At **hydrothermal vents,** seawater percolates through cracks and is heated to about 350°C, causing sulfate to react with water and form hydrogen sulfide (H_2S). Chemoautotrophic bacteria that obtain energy from oxidizing hydrogen sulfide exist freely or live mutualistically within the tissues of organisms. They are the start of food chains for an ecosystem that includes huge tube worms, clams, crustaceans, echinoderms, and fishes. This

ecosystem can exist where light never penetrates because, unlike photosynthesis, chemosynthesis does not require light energy.

In Section 45.3, we described food webs and energy flow through ecosystems. Although energy is lost with transfer to each level, from producers to primary and then secondary consumers, pollutants can become concentrated as they move up the web—an effect termed **biomagnification.** The Biological Systems feature discusses biomagnification as it occurs with mercury pollution, particularly in aquatic ecosystems.

Ocean Currents

Climate is driven by the Sun, but the oceans play a major role in redistributing heat in the biosphere. Water tends to be warm at the equator and much cooler at the poles because of the distribution of the Sun's rays (see Fig. 46.1a). Air takes on the temperature of the water below, and warm air moves from the equator to the poles. In other words, the oceans make the winds blow. (Landmasses also play a role, but the oceans hold heat longer and remain cool longer during periods of changing temperature than do continents.)

When wind blows strongly and steadily across a great expanse of ocean for a long time, friction from the moving air begins to drag the water along with it. Once the water has been set in motion, its momentum, aided by the wind, keeps it moving in a steady flow called a current. Because the ocean currents eventually strike land, they move in a circular path—clockwise in the Northern Hemisphere and counterclockwise in the Southern Hemisphere (Fig. 46.22). As the currents flow, they take warm water from the equator to the poles. One such current,

Biological Systems

Biomagnification of Mercury

Scientists have known since the 1950s that the emissions of mercury into the environment can lead to serious health effects for human beings. Studies show that fish and wildlife exposed to mercury emissions are negatively impacted. Humans are impacted if they come into contact with affected fish and wildlife. Recent fish studies have shown a widespread contamination of mercury in streams, wetlands, reservoirs and lakes throughout the majority of the U.S.

Mercury becomes a serious environmental risk when it undergoes bioaccumulation in an organism's body. Bioaccumulation occurs when an organism accumulates a contaminant faster than it can eliminate it. Most organisms can eliminate about half of the mercury in their bodies every 70 days if they can avoid ingesting any additional mercury during this time. Problems arise when organisms cannot eliminate the mercury before they ingest more.

Mercury tends to enter ecosystems at the base of the food chain and increase in concentration as it moves up each successive trophic level. Top-level predators and organisms that are long lived are the most susceptible to high levels of mercury accumulation in their body tissues.

Mercury exposure for humans generally occurs due to eating contaminated fish or breathing mercury vapor. Methylmercury is the form that leads to health problems such as sterility in men, damage to the central nervous system, and in severe cases, birth defects in infants. Developing fetuses and children can have health consequences from intake levels 5–10 times lower than adults.

Studies have shown such elevated levels of mercury in sharks, tuna (Fig. 46Ba), and swordfish that the EPA has advisories against eating these fish for women who may become pregnant, are pregnant, or are nursing mothers, and for young children. Every state in the U.S., in conjunction with federal agencies, has developed fish advisories for certain bodies of water within that state. Currently, 45 states warn pregnant women to limit their fish consumption from their waters.

Mercury poisoning isn't limited to just aquatic species. Research conducted in the northeastern United States and Canada showed the presence of mercury in a variety of birds ranging from thrushes to loons to bald eagles. It is no surprise that loons and bald eagles can have high levels of mercury accumulation due to their consumption of contaminated fish. It was the presence of mercury in the songbirds that raised serious concerns among ecologists. Some speculate that songbirds in the northeast are ingesting mercury when they feed upon insects that have picked up the toxin from eating smaller insects that ingested it from vegetation. This raises concerns about mercury's ability to enter food webs and bioaccumulate in previously unknown ways.

Ultimately the blame for mercury pollution falls squarely on the shoulders of human beings. Every ecosystem on the planet has some degree of exposure to this pollutant, and with this exposure comes the risk of mercury contamination. Studies on species ranging from polar bears and bald eagles to whales and sharks show us that there are no limits to where mercury can be found. With coal-burning power plants being the largest human-caused source of mercury emissions (Fig. 46Bb), it will be up to us to find a solution to this global problem.

Questions to Consider

1. Would you support higher regulations on coal-fired power plants to reduce their mercury emissions even if it meant an increase in energy costs?
2. What is the easiest way to prevent the bioaccumulation of mercury in an organism?
3. Are there any species in an ecosystem that are not impacted by exposure to mercury?
4. Is there a way to limit the movement of mercury from one ecosystem to the next?

a.

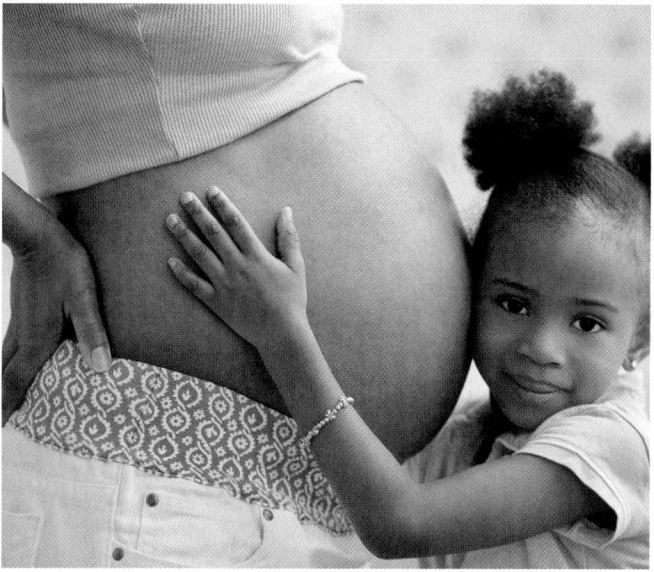

b.

Figure 46B Biomagnification of mercury. **a.** Tuna may contain high levels of mercury due to biomagnification. **b.** Eating tuna or other fish that contain high levels of mercury can lead to problems with fetal development.

called the Gulf Stream, brings tropical Caribbean water to the east coast of North America and the higher latitudes of western Europe. Without the Gulf Stream, Great Britain, which has a relatively warm temperature, would be as cold as Greenland. In the Southern Hemisphere, another major ocean current warms the eastern coast of South America.

Also in the Southern Hemisphere, a current called the Humboldt Current flows toward the equator. The Humboldt Current carries phosphorus-rich cold water northward along the west coast of South America. During a process called **upwelling,** cold offshore winds cause cold nutrient-rich waters to rise and take the place of warm nutrient-poor waters. In South America, the enriched waters cause an abundance of marine life that supports the fisheries of Peru and northern Chile. Birds feeding on these organisms deposit their droppings on land, where they are mined as guano, a commercial source of phosphorus. When the Humboldt Current is not as cool as usual, upwelling does not occur, stagnation results, the fisheries decline, and climate patterns change globally. This phenomenon is called an **El Niño–Southern Oscillation (ENSO).**

Check Your Progress   46.3

1. Identify the steps associated with an overturn.
2. Describe the zones of the open ocean.

CONNECTING *the* CONCEPTS *with the* BIG IDEAS

Interactions and Systems

- Historically, the location of ecosystems and populations can be affected by meteorological and geological occurrences, as in changes due to El Nino, meteor impacts, and continental drift. (4B4b1*IE*)
- Ecosystems with few constituents and meager diversity have difficulty rebounding from environmental changes. (4C4a)

- Keystone species are inordinately important in maintaining the health and diversity of their ecosystem; loss of the keystones often leads to collapse of the community. (4C4b)

*Find the unabridged version of all EK citations at www.glencoe.com/maderAP11.

Media Study Tools

www.glencoe.com/maderAP11

Enhance your study of this chapter with study tools and practice tests. Also ask your instructor about the resources available through ConnectPlus, including the media-rich eBook, interactive learning tools, and animations.

Summarize

46.1 Climate and the Biosphere

Because the Earth is a sphere, the Sun's rays at the poles are distributed out over a larger area than the direct rays at the equator. The temperature at the surface of the Earth therefore decreases from the equator to each pole. The Earth is tilted on its axis, and the seasons change as the Earth revolves annually around the Sun.

Warm air rises near the equator, loses its moisture, and then descends at about 30° north and south latitude to the poles. When the air descends, it absorbs moisture from the land, and therefore the great deserts of the world are formed at 30° latitudes. Because the Earth rotates on its axis daily, the winds blow in opposite directions above and below the equator. Topography also plays a role in the distribution of moisture. Air rising over coastal ranges loses its moisture on the windward side, making the leeward side arid.

46.2 Terrestrial Ecosystems

A biome is a major type of terrestrial community. Biomes are distributed according to climate—that is, temperature and rainfall influence the pattern of biomes about the world. The effect of temperature causes the same sequence of biomes when traveling to northern latitudes as when traveling up a mountain.

The Arctic tundra is the northernmost biome and consists largely of short grasses, sedges, and dwarf woody plants. Because of cold winters and short summers, most of the water in the soil is frozen year-round. This is called the permafrost.

The taiga, a coniferous forest, has less rainfall than other types of forests. The temperate deciduous forest has trees that gain and lose their leaves because of the alternating seasons of summer and winter. Tropical rain forests are continually warm and wet. These are the most complex and productive of all biomes.

Shrublands usually occur along coasts that have dry summers and receive most of their rainfall in the winter. Among grasslands, the savanna, a tropical grassland, supports the greatest number of different types of large herbivores. Temperate grasslands, such as that found in the central United States, have a limited variety of vegetation and animal life.

Deserts are characterized by a lack of water—they are usually found in places with less than 25 cm of precipitation per year. Some desert plants, such as cacti, are succulents with thick stems and leaves, and others are shrubs that are deciduous during dry periods.

46.3 Aquatic Ecosystems

Streams, rivers, lakes, and wetlands are different freshwater ecosystems. In deep lakes of the temperate zone, the temperature and the concentration of nutrients and gases in the water vary with depth. The entire body of water is cycled twice a year, distributing nutrients from the bottom layers. Lakes and ponds have three life zones. Rooted plants and clinging organisms live in the littoral zone, plankton and fishes live in the sunlit limnetic zone, and bottom-dwelling organisms such as crayfishes and molluscs live in the profundal zone.

Marine ecosystems are divided into coastal ecosystems and the oceans. The coastal ecosystems, especially estuaries, are more productive than the oceans. Estuaries (and associated salt marshes, mudflats, and mangrove forests) are near the mouth of a river. Estuaries are considered the nurseries of the sea.

An ocean is divided into the pelagic zone and the ocean floor. The pelagic zone (open waters) has three zones. The epipelagic zone receives adequate sunlight and supports the most life. The mesopelagic zone contains organisms adapted to minimum or no light. The bathypelagic zone is in complete darkness. The ocean floor includes the continental shelf, the continental slope, and the abyssal plain.

Key Terms

alpine tundra 887	monsoon 885
Arctic tundra 888	montane coniferous forest 887
biomagnification 902	pelagic zone 901
biome 887	permafrost 888
bog 897	phytoplankton 899
chaparral 894	rain shadow 885
climate 884	savanna 895
coral reef 901	shrubland 894
desert 896	spring overturn 899
El Niño–Southern	swamp 897
Oscillation 904	taiga 890
epiphyte 892	temperate deciduous
estuary 900	forest 891
eutrophication 898	temperate grassland 894
fall overturn 898	temperate rain forest 890
grassland 894	topography 885
hydrothermal vent 902	tropical rain forest 892
intertidal zone 900	upwelling 904
lake 898	wetland 897
marsh 897	zooplankton 899

Assess

Reviewing This Chapter

1. Describe how a spherical Earth and the path of the Earth about the Sun affect climate.   884
2. Describe the air circulation about the Earth, and tell why deserts are apt to occur at 30° north and south of the equator.   884
3. How does a coastal mountain range affect climate? What causes a monsoon climate?   885
4. Name the terrestrial biomes you would expect to find when going from the base to the top of a mountain.   887
5. Describe the location, the climate, and the populations of the Arctic tundra, coniferous forests (both taiga and temperate rain forest), temperate deciduous forests, tropical rain forests, shrublands, grasslands (both temperate grasslands and savanna), and deserts.   888–96
6. Describe the importance of wetlands and the major types of wetlands.   897
7. Describe the overturn of a temperate lake, the life zones of a lake, and the organisms you would expect to find in each life zone.   898–900
8. Describe the coastal communities, and discuss the importance of estuaries to the productivity of the ocean.   900–01
9. Describe the zones of the open ocean and the organisms you would expect to find in each zone.   901–02
10. Describe the ocean currents and how the Gulf Stream accounts for Great Britain having a mild temperature.   902, 904

Testing Yourself

Choose the best answer for each question.

1. The seasons are best explained by
   a. the distribution of temperature and rainfall in biomes.
   b. the tilt of the Earth as it orbits about the Sun.
   c. the daily rotation of the Earth on its axis.
   d. the fact that the equator is warm and the poles are cold.

2. The mild climate of Great Britain is best explained by
   a. the winds called the westerlies.
   b. the spinning of the Earth on its axis.
   c. Great Britain being a mountainous country.
   d. the flow of ocean currents.

3. Which of these pairs is mismatched?
   a. tundra—permafrost
   b. savanna—*Acacia* trees
   c. prairie—epiphytes
   d. coniferous forest—evergreen trees

4. All of these phrases describe the tundra except
   a. low-lying vegetation.
   b. northernmost biome.
   c. short growing season.
   d. many different types of species.

5. The forest with a multilevel canopy is the
   a. tropical rain forest.
   b. coniferous forest.
   c. tundra.
   d. temperate deciduous forest.

6. Why are lush evergreen forests present in the Pacific Northwest of the United States?
   a. The rotation of the Earth causes this.
   b. Winds blow from the ocean, bringing moisture.
   c. They are located on the leeward side of a mountain range.
   d. Both b and c are correct.
   e. All of these are correct.

7. Which of these influences the location of a particular biome?
   a. latitude
   b. average annual rainfall
   c. average annual temperature
   d. altitude
   e. All of these are correct.

8. Which of these is a function of a wetland?
   a. purifies water
   b. is an area where toxic wastes can be broken down
   c. helps absorb overflow and prevents flooding
   d. is a home for organisms that are links in food chains
   e. All of these are correct.

9. The area of a lake closest to shore and where rooted plants are found is the
   a. littoral zone.
   b. limnetic zone.
   c. profundal zone.
   d. benthic zone.

10. An estuary acts as a nutrient trap because of the
    a. action of rivers and tides.
    b. depth at which photosynthesis can occur.
    c. amount of rainfall received.
    d. height of the water table.

11. Which area of an ocean has the most light and nutrients?
    a. epipelagic zone and abyssal plain
    b. epipelagic zone only
    c. benthic zone only
    d. neritic province

12. Which area of the pelagic zone is completely dark?
    a. epipelagic zone
    b. mesopelagic zone
    c. bathypelagic zone

13. Runoff of fertilizer and animal wastes from a large farm that drains into a lake would be an example of which process?
    a. fall overturn
    b. eutrophication
    c. spring overturn
    d. upwelling

14. Phytoplankton are more likely to be found in which life zone of a lake?
    a. limnetic zone
    b. profundal zone
    c. benthic zone
    d. All of these are correct.

15. Which area of the seashore is only exposed during low tide?
    a. upper littoral zone
    b. mid-littoral zone
    c. lower littoral zone
    d. None of these are correct.

16. Which of these would normally be found in the benthic zone of a lake?
    a. minnows
    b. clams
    c. zooplankton
    d. plants
    e. large fish

17. All of the following phrases describe a tropical rain forest, except
    a. nutrient-rich soil.
    b. many arboreal plant and animals.
    c. canopy composed of many layers.
    d. broad-leaved evergreen trees.

18. An oligotrophic lake
    a. is nutrient-rich.
    b. is cold.
    c. is likely to be found in an agricultural or urban area.
    d. has poor productivity.

19. Energy for the food chain near hydrothermal vents comes from
    a. dead organisms that fall down from above.
    b. highly efficient photosynthetic phytoplankton.
    c. chemoautotrophic bacteria.
    d. heat given off by the vents.

Engage

Thinking Scientifically

1. Pharmaceutical companies are interested in "bioprospecting" in tropical rain forests. These companies are looking for naturally occurring compounds in plants or animals that can be used as drugs for a variety of diseases. The most promising compounds act as antibacterial or antifungal agents. Even discounting the fact that the higher density of species in tropical rain forests would produce a wider array of compounds than another biome, why would antibacterial and antifungal compounds be more likely to evolve in this particular biome?

2. When hurricanes come ashore, the ocean tide may surge dozens of feet above normal into coastal communities. Recently, hurricane Katrina had devastating effects on New Orleans, Louisiana, a city actually built below sea level following the draining of the natural wetlands. How would the presence of wetlands possibly decrease the damage of sea water brought in by an incoming hurricane?

Bioethical Issue
Water Pollution

Agricultural fertilizers are the chief cause of nitrate contamination of drinking-water wells. Excessive nitrates in a baby's bloodstream can lead to slow suffocation, known as blue-baby syndrome. Agricultural herbicides are suspected carcinogens in the tap water of scattered ecosystems coast to coast. What can be done?

Some farmers are already using irrigation methods that deliver water directly to plant roots, no-till agriculture that reduces the loss of topsoil and cuts back on herbicide use, and integrated pest management, which relies heavily on good bugs to kill bad bugs. Perhaps more should do so. Encouraged—in some cases, compelled—by state and federal agents, dairy farmers have built sheds, concrete containments, and underground liquid storage tanks to hold wastes when it rains. Later, the manure is trucked to fields and spread as fertilizer.

Homeowners, like golf clubs and ski resorts, also contribute to the problem. The care of lawns, the use of motor vehicles, and the construction and use of roads and buildings all add contaminants to streams, lakes, and aquifers. Citizens around Grand Traverse Bay on the eastern shore of Lake Michigan have also gotten the message, especially because they want to keep on enjoying water-dependent activities such as boating, swimming, and fishing. James Haverman, a concerned member of the Traverse Bay Watershed Initiative says, "If we can't change the way people live their everyday lives, we are not going to be able to make a difference." Builders in Traverse County are already required to control soil erosion with filter fences, steer rainwater away from exposed soil, build sediment basins, and plant protective buffers. Presently, homeowners must have a 25-foot setback from wetlands and a 50-foot setback from lakes and creeks. They are also encouraged to pump out their septic systems every two years. Do you approve of legislation that requires farmers and homeowners to protect freshwater supplies? Why or why not?

Deformed frogs are undoubtedly due to environmental effects.

47

Conservation of Biodiversity

I t is increasingly more common to find deformed frogs in both urban and rural environments. These frogs often have extra limbs arising from their midsection. Scientists are working to discover why this problem is so widespread, and so far, two hypotheses have been suggested. One hypothesis is that a small trematode (a fluke) burrows into a tadpole and prevents the normal development of legs. The other hypothesis is that a chemical called methoprene, which many farmers use as a pesticide, is affecting development and causing these abnormalities. Once the cause is known, a plan can be developed to counteract the problem.

In this chapter, we discuss the emergence of conservation biology as a branch of environmental science directed at actively improving environmental conditions for wildlife and ourselves. Conservation biology examines the natural and evolutionary origins of biodiversity. As an applied science, it studies the human activities that are presently impacting biodiversity and what steps can be taken to reverse the situation. Conservation biology has as an ultimate goal the preservation and management of ecosystems.

As you read through the chapter, think about the following questions:

1. What is the value of biodiversity to humans?
2. Is the health of an ecosystem more important than human needs?
3. Who is responsible for maintaining the Earth's biodiversity?

BEFORE YOU BEGIN

Before beginning this chapter, take a few moments to review the following discussions.

Section 18.3 Why are the current extinctions of such a great concern to scientists?

Section 44.6 How is human population growth contributing to extinction rates?

Section 45.3 What role does biodiversity play in ecosystem health?

FOLLOWING *the* BIG IDEAS

CHAPTER 47   CONSERVATION OF BIODIVERSITY

Evolution	Humans impact evolution when they cause ecological stress.
Interactions and Systems	The activities of human populations often unbalance their ecosystems.

47.1 Conservation Biology and Biodiversity

Learning Outcomes

Upon completion of this section, you should be able to

1. Identify the role of conservation biology with regard to biodiversity.
2. Describe how conservation biology is an applied, goal-oriented, multidisciplinary field.
3. Recognize the spectrum of biodiversity.

Conservation biology [L. *conservatio*, keep, save] is a relatively new discipline of biology that studies all aspects of biodiversity with the goal of conserving natural resources for this generation and all future generations. Conservation biology is unique in that it is concerned with both the development of scientific concepts and the application of these concepts to the everyday world. The primary goal of conservation biology is the management of **biodiversity,** the variety of life on the Earth. To achieve this goal, conservation biologists come from many subfields of biology that only recently have been brought together into a cohesive whole.

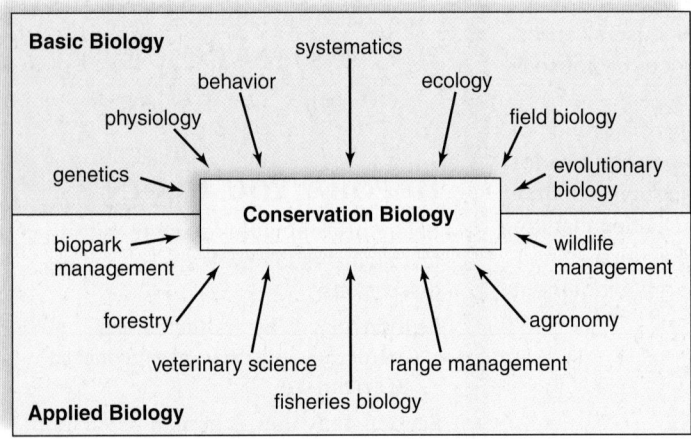

Conservation biologists must be aware of the latest theoretical and practical findings. They need to use this knowledge to identify the source of problems and suggest suitable courses of action. Often conservation biologists work with government officials from the local to the federal levels. Public education is another important component of this profession.

For conservation biology to be an effective field of study, scientists must look at the larger connections within the biosphere. Overall, a high level of biodiversity is desirable because it positively influences the stability of ecosystems. In turn, stable ecosystems contribute to the overall health of the human population. The reverse also tends to be true—if biodiversity decreases, then ecosystem stability decreases, often resulting in negative consequences for the human population.

Conservation biology has emerged in response to the extinction crisis the Earth is currently experiencing. Estimates vary, but at least 10–20% of all species now living most likely will become extinct in the next 20 to 50 years unless planned, coordinated actions are taken. It is urgently important that all citizens understand the

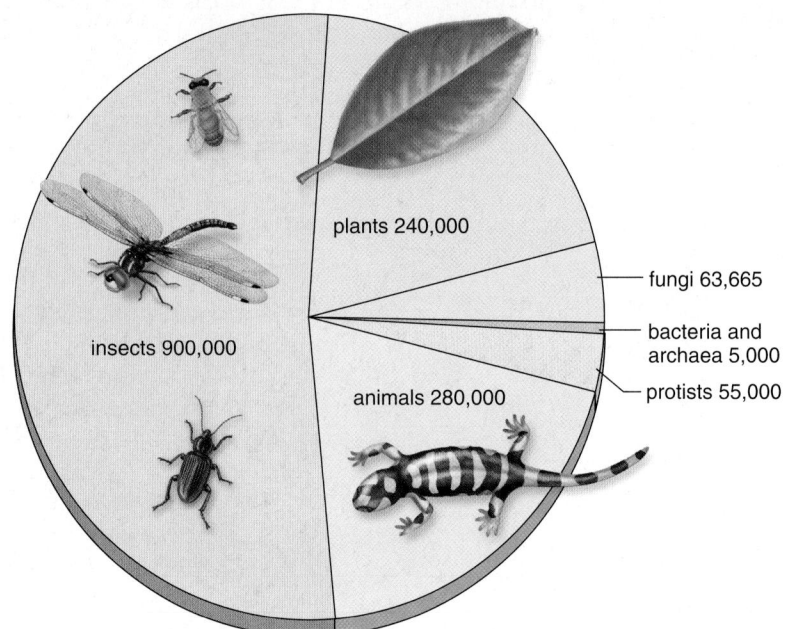

Figure 47.1 Number of described species. There are only about 1.75 million described species; insects are far more prevalent than organisms in other groups. Undescribed species probably number far more than those species that have been described.

concept and importance of biodiversity, the causes of present-day extinctions, how to prevent future extinctions from occurring, and the potential consequences of decreased biodiversity.

To protect biodiversity, scientists can apply the science of **bioinformatics:** the collecting of, analyzing, and making readily available biological information, using modern computer technology. Throughout the world, molecular, descriptive, and biogeographical information on organisms is being collected for study purposes.

Biodiversity

It is common practice to describe biodiversity in terms of the number of species present among various groups of organisms. Figure 47.1 accounts for the species that have so far been described. It has been estimated that there may be between 10 and 50 million species living on Earth.

Of these, nearly 1,200 species in the United States and 30,000 species worldwide are in danger of extinction. An **endangered species** is one that is in peril of immediate extinction throughout all or most of its range. Examples of endangered species include the giant panda, hawksbill turtle, California condor, and snow leopard. **Threatened species** are organisms that are likely to become endangered in the near future. Examples of threatened species include the Navaho sedge, northern spotted owl, and coho salmon.

To develop a meaningful understanding of life on Earth, we need to know more about species than their total number. Ecologists describe biodiversity as a combination of three levels of biological organization:

- genetic diversity
- community diversity
- landscape diversity

Genetic Diversity

Genetic diversity refers to genetic variations among the members of a population. Populations with high genetic diversity are more likely to have some individuals that can survive a change in the structure of their ecosystem. The 1846 potato blight in Ireland, the 1922 wheat failure in the Soviet Union, and the 1984 outbreak of citrus canker in Florida were all made worse by limited genetic variation among these crops. If a species' population is quite small and isolated, it is more likely to become extinct because of a limited genetic diversity. As populations become endangered and threatened, their genetic diversity decreases.

Video
Cloning Endangered Species

Ecosystem Diversity

Ecosystem diversity is dependent on the interactions between species in a given community. One community's species composition can be completely different from that of other communities. A diverse community composition will increase the levels of biodiversity in the biosphere.

Past conservation efforts frequently concentrated on saving a single species with human appeal, termed a charismatic species—such as the the black-footed ferret or spotted owl. This approach, however, is short sighted. A more effective approach is to conserve species that play a critical role in an ecosystem. Saving an entire community can save many species, while disrupting a community threatens the existence of all species.

One example of a short-sighted approach was the introduction of opossum shrimp, *Mysis relicta,* into Flathead Lake in Montana and its tributaries in the 1960s and '70s as food for salmon. The shrimp ate so much zooplankton that in the end far less food was available for the fish. This then resulted in a decrease in the available food for the grizzly bears and bald eagles as well (Fig. 47.2).

Landscape Diversity

Landscape diversity involves a group of interacting ecosystems within one **landscape,** such as mountains, rivers and grasslands. Sometimes ecosystems can be so fragmented that they are connected only by small patches or strips of undeveloped land that allow organisms to move from one ecosystem to the next. Fragmentation of the landscape reduces reproductive capacity and food availability and can disrupt seasonal behaviors.

Distribution of Biodiversity

Biodiversity is not evenly distributed throughout the biosphere. If we are forced to protect only small areas, we should focus our efforts on regions of the greatest biodiversity. Biodiversity is typically higher in the tropics, and declines toward the poles. The coral reefs of the Indonesian archipelago contain the greatest biodiversity of all aquatic ecosystems.

Some regions of the world are called **biodiversity hotspots** because they contain a large concentration of species. Biodiversity in hotspots accounts for about 44% of all known higher plant species and 35% of all terrestrial vertebrate species, but covers only about 1.4% of the Earth's land area. The island of Madagascar, the Cape region of South Africa, Indonesia, the coast of California, and the Great Barrier Reef of Australia are examples of biodiversity hotspots.

Video
Coral Reef Ecosystems

Check Your Progress   47.1

1. Describe how conservation biology is supported by a variety of disciplines.
2. Define biodiversity and explain what is meant by a biodiversity hotspot.

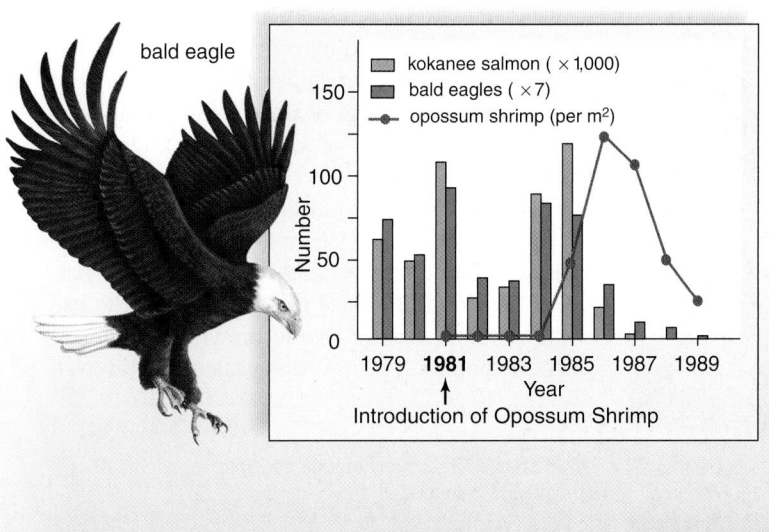

Figure 47.2 Eagles and bears feed on spawning salmon. Humans introduced the opossum shrimp as prey for salmon. Instead, the shrimp competed with salmon for zooplankton as a food source. The salmon, eagle, and bear populations subsequently declined.

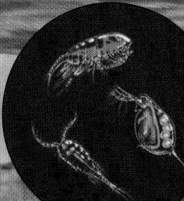

47.2 Value of Biodiversity

Conservation biology strives to reverse the trend toward possible species extinction. Educating people about the value of biodiversity is a key step in making conservation possible.

Direct Value

A great number of species perform services from which humans can derive an economic value. Here we discuss a few of the key economic values that species provide (illustrated in Fig. 47.3).

Medicinal Value

Humans often associate the worth of a resource with its value expressed in monetary terms. Medicines derived from living organisms are a good example.

Most of the prescription drugs used in the United States, valued at over $200 billion, were originally derived from living organisms. The rosy periwinkle from Madagascar is an excellent example of a tropical plant that has provided us with useful medicines. Potent chemicals from this plant are now used to treat two forms of cancer: leukemia and Hodgkin disease.

Because of these drugs, the survival rate for childhood leukemia has gone from 10% to 90%, and Hodgkin disease is now usually curable. Researchers estimate that hundreds of new drugs are yet to be found in tropical rain forests, and that they may have an economic value of over $100 billion.

Video Indigenous Medicine

The antibiotic penicillin is derived from a fungus, and certain species of bacteria produce the antibiotics tetracycline and streptomycin. These drugs have proven to be indispensable in the treatment of bacterial infections.

Leprosy is among those diseases for which no cure is currently available. The bacterium that causes leprosy will not grow in the laboratory, but scientists discovered that it grows naturally in the nine-banded armadillo. Having a source for the bacterium may make it possible to find a cure for leprosy. The blood of horseshoe crabs, *Limulus*, contains a substance called limulus amoebocyte lysate, which is used to ensure that medical devices such as pacemakers, surgical implants, and prosthetic devices are free of bacteria (see the Nature of Science feature, "How Horseshoe Crabs Save Human Lives," in Chapter 32).

Video Good Poison

Agricultural Value

Crops such as wheat, corn, and rice have been derived from wild plants that were modified to increase their yields. The same high-yield, genetically similar strains tend to be grown worldwide. When rice crops in Africa were being devastated by a virus, researchers grew wild rice plants from thousands of seed samples until they found one that contained a gene for resistance to the virus. These wild plants were then used in a breeding program to transfer the gene into high-yield rice plants. If this variety of wild rice had become extinct before it could be discovered, rice cultivation in Africa might have collapsed.

Using natural pest controls is preferable to using chemical pesticides. When a rice pest called the brown planthopper became resistant through natural selection to pesticides, farmers began to use the natural predator of the brown planthopper. The economic savings were calculated at well

Figure 47.3 Direct value of wildlife.
Many wild species have a significant monetary value to humans. The challenge is to manage these species correctly to ensure their continued survival.

Wild species, like the rosy periwinkle, *Catharanthus roseus,* are sources of many medicines.

Wild species, like the nine-banded armadillo, *Dasypus novemcinctus,* play a role in medical research.

Wild species, like many marine species, provide us with food.

over a billion dollars. Similarly, cotton growers in Cañete Valley, Peru, found that pesticides were no longer working against the cotton aphid because of the resistance that had evolved in the aphid population. Research have identified the aphids' natural predators and are using them with great success.

Animals are the main pollinators for flowering plants. The domesticated honeybee, *Apis mellifera,* pollinates almost $10 billion worth of food crops annually in the United States. But dependency on a single species has inherent dangers: Mites that infest bees have wiped out more than 20% of the commercial honeybee population in the United States. In order to find honeybees resistant to the mites, researchers must look to the wild populations. In addition, the value of wild pollinators to the U.S. agricultural economy has been calculated at $4.1–$6.7 billion a year.

Consumptive Use Value

Humans have had much success cultivating crops, domesticating animals, growing trees in plantations, and so forth. But so far, aquaculture, the growing of fish and shellfish for human consumption, has contributed only minimally to human welfare. Most freshwater and marine harvests depend on catching wild animals, such as fishes (e.g., trout, cod, tuna, and flounder), crustaceans (e.g., lobsters, shrimps, and crabs), and mammals (e.g., whales). These aquatic organisms are an invaluable biodiversity resource.  **Video Salmon Farming**

The environment provides a variety of other products that are sold commercially worldwide, including wild fruits and vegetables, skins, fibers, beeswax, and seaweed. Hunting and fishing is the primary method of obtaining meat for many human populations. In one study, researchers calculated that the economic value of wild pigs in the diet of native hunters in Sarawak, East Malaysia, was approximately $40 million per year.

Similarly, many trees are still felled in the natural environment for their wood. Researchers have calculated, however, that a species-rich forest in the Peruvian Amazon is worth far more if the forest is used for fruit and rubber production than for timber production. Fruit and the latex needed to produce rubber can be brought to market for an unlimited number of years. Unfortunately, once the trees are gone, none of these products, including timber, can be harvested.

Indirect Value

The wild species we have been discussing all play a role in their respective ecosystems. If we want to preserve them, it is more beneficial to save large portions of their ecosystems. The indirect value of these ecosystems is based on services they provide simply by existing, such as support of biogeochemical cycles, waste recycling, and provision of fresh water. In addition, these natural regions have intangible value; a walk in the woods or time spent at the local park can have a calming and restorative effect on many people. Reconnecting with nature has proven

Wild species, like the lesser long-nosed bat, *Leptonycteris curasoae*, are pollinators of agricultural and other plants.

Wild species, like ladybugs, *Coccinella*, play a role in biological control of agricultural pests.

Wild species, like rubber trees, *Hevea*, can provide a product indefinitely if the forest is not destroyed.

Figure 47.4 Indirect value of ecosystems. Natural ecosystems provide many physiological and psychological benefits to humans.

physiological and psychological benefits to human health (Fig. 47.4). Our very survival depends on the functions that ecosystems perform for us.

Biogeochemical Cycles

Ecosystems are characterized by energy flow and chemical cycling (see Chapter 45). The biodiversity within ecosystems contributes to the workings of the water, carbon, nitrogen, phosphorus, and other biogeochemical cycles. We are dependent on these cycles for fresh water, removal of carbon dioxide from the atmosphere, uptake of excess soil nitrogen, and provision of phosphate. When human activities upset the natural workings of biogeochemical cycles, the environmental consequences often produce negative consequences to humans. Technology is limited in its ability to mimic the biogeochemical cycles.

Waste Recycling

Decomposers break down dead organic matter and other types of wastes to inorganic nutrients that are used by the producers within ecosystems. This function aids humans immensely because we dump millions of tons of waste material into natural ecosystems each year. If it were not for decomposition, waste would soon cover the entire surface of our planet. We can build sewage treatment plants, but they are expensive, and few of them break down solid wastes completely to inorganic nutrients. It is less expensive and more efficient to provide plants and trees with partially treated wastewater and let soil bacteria cleanse it completely.

Biological communities are also capable of breaking down and immobilizing pollutants, such as heavy metals and pesticides. A review of wetland functions in Canada assigned a value of $50,000 per hectare (100 acres, or 10,000 m^2) per year to the ability of natural areas to purify water and take up pollutants.

Provision of Fresh Water

Few terrestrial organisms are adapted to living in a salty environment—they need fresh water. The water cycle continually supplies fresh water to terrestrial ecosystems. Humans use fresh water in innumerable ways, from drinking water to irrigating crops. Freshwater ecosystems, such as rivers and lakes, provide us with a large diversity of species we can use as a source of food.

Unlike other commodities, there is no substitute for fresh water. We can remove salt from sea water to obtain fresh water, but the cost of desalination is about four to eight times the average cost of fresh water acquired via the water cycle.

Forests and other natural ecosystems exert a "sponge effect." They soak up water and then release it at a regular rate. When rain falls in a natural area, plant foliage and dead leaves lessen its impact, and the soil slowly absorbs it, especially if it has been aerated by soil organisms. The water-holding capacity of forests reduces the possibility of flooding. The value of a marshland outside Boston, Massachusetts, has been estimated at $72,000 per hectare per year, solely for its ability to reduce floods. Forests release water slowly for days or weeks after the rains have ceased. Rivers flowing through forests in West Africa release twice as much water halfway through the dry season, and between three and five times as much at the end of the dry season, as do rivers from agricultural lands.

Prevention of Soil Erosion

Intact ecosystems naturally retain soil and prevent soil erosion. The importance of this ecosystem attribute is especially observed following deforestation. In Pakistan, the world's largest dam, the Tarbela Dam, is losing its storage capacity of 12 billion m^3 many years sooner than expected because sediment is building up behind the dam due to deforestation. At one time, the Philippines exported $100 million worth of oysters, mussels, clams, and cockles each year. Now, silt carried down rivers following deforestation is smothering the mangrove ecosystem that serves as a nursery for the sea. Most coastal ecosystems are not as bountiful as they once were because of deforestation and a myriad of other problems.

Regulation of Climate

At the local level, trees provide shade and reduce the need for fans and air conditioners during the summer. Proper placement of shade trees near a home can reduce energy bills 10–20%.

Globally, forests restore the climate because they take up carbon dioxide. The leaves of trees use carbon dioxide when they photosynthesize, the bodies of the trees store carbon, and oxygen is released as a by-product. When trees are cut and burned, carbon dioxide is released into the atmosphere. The reduction in forests reduces the carbon dioxide uptake and the oxygen output. This change in the atmospheric gases, especially greenhouse gases such as CO_2, affects the amount of solar radiation retained on the Earth's surface. Large-scale deforestation is affecting the global atmosphere, and in turn, changing the Earth's climate.

Ecotourism

In the United States, nearly 100 million people enjoy vacationing in a natural setting. To do so, they spend $4 billion each year on fees, travel, lodging, and food. Many tourists want to go sport fishing, whale watching, boating, hiking, birdwatching, and the

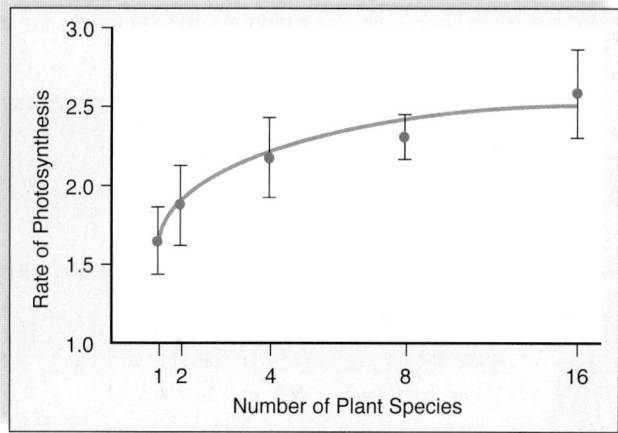

Figure 47.5 The influence of biodiversity on community productivity. Research results show that the higher the biodiversity (measured by the number of plant species), the greater the rate of photosynthesis in an experimental community.

like. Others simply want to immerse themselves in the beauty and serenity of a natural environment. Many underdeveloped countries in tropical regions are taking advantage of this by offering "ecotours" of the local biodiversity. Providing guided tours of forests can often be more profitable than destroying them.

Biodiversity and Natural Ecosystems

Massive changes in biodiversity, such as deforestation, have a significant impact on ecosystems. Researchers are interested in determining whether a high degree of biodiversity also helps ecosystems function more efficiently. To test the benefits of biodiversity in a Minnesota grassland habitat, researchers sowed plots with seven levels of plant diversity. Their study found that ecosystem performance improves with increasing diversity. A similar study in California also showed greater overall resource use in more diverse plots because of resource partitioning among the plants.

Another group of experimenters tested the effects of an increase in diversity at four levels: producers, herbivores, parasites, and decomposers. They found that the rate of photosynthesis increased as diversity increased (Fig. 47.5). A computer simulation has shown that the response of a deciduous forest to elevated carbon dioxide is a function of species diversity. The more complex community, composed of nine tree species, exhibited a 30% greater amount of photosynthesis than a community composed of a single species.

More studies are needed to test whether biodiversity maximizes resource acquisition and retention within an ecosystem. Also, are more diverse ecosystems better able to withstand environmental changes and invasions by other species, including pathogens? Then, too, how does fragmentation affect the distribution of organisms within an ecosystem and the functioning of an ecosystem? These questions and many more are the focus of current conservation biology research.

Check Your Progress   47.2

1. Explain the difference between a direct value of biodiversity and an indirect value of biodiversity.
2. Cite examples of direct and indirect values of biodiversity.

47.3 Causes of Extinction

Learning Outcomes

Upon completion of this section, you should be able to

1. Classify the causes of extinction.
2. Compare natural and human influenced causes of extinction.

To stem the tide of extinction due to human activities, it is first necessary to identify its causes. Researchers examined the records of 1,880 threatened and endangered wild species in the United States and found that habitat loss was involved in 85% of the cases (Fig. 47.6a). Exotic species had a hand in nearly 50%, pollution was a factor in 24%, overexploitation in 17%, and disease in 3%. (The percentages add up to more than 100% because most of these species are imperiled for more than one reason.) Macaws are a good example of how a combination of factors can lead to a species decline (Fig. 47.6b). Not only has their habitat been reduced by encroaching timber and mining companies, but macaws are also hunted for food and collected for the pet trade.

Habitat Loss

Habitat loss has occurred in all ecosystems, but concern has now centered on tropical rain forests and coral reefs because they are particularly rich in species.

A sequence of events in Brazil offers a fairly typical example of the manner in which rain forest is converted to land uninhabitable for wildlife. The construction of a major highway into the forest first provided a way to reach the interior of the forest (see Fig. 47.6c). Small towns and industries sprang up along the highway, and roads branching off the main highway gave rise to even more roads. The result was fragmentation of the once immense forest.

The government offered subsidies to anyone willing to take up residence in the forest, and the people who came cut and burned trees in patches (see Fig. 47.6c). Tropical soils contain limited nutrients, but when the trees are burned, nutrients are released that support a lush growth for the grazing of cattle for about three years. However, once the land was degraded (see Fig. 47.6c), the farmers moved on to another portion of the forest to start over again.

Loss of habitat also affects freshwater and marine biodiversity. Coastal degradation is mainly due to the large concentration of people living on or near the coast. Already, 60% of coral reefs have been destroyed or are on the verge of destruction; it is possible that all coral reefs may disappear during the next 40 years unless our behaviors drastically change. Mangrove forest destruction is also a problem; Indonesia, with the most mangrove acreage, has lost 45% of its mangroves, and the percentage is even higher for other tropical countries. Wetland areas, estuaries, and seagrass beds are also being rapidly destroyed by human actions.

Exotic Species

Exotic species, sometimes called alien species, are nonnative members of an ecosystem. Ecosystems around the globe are

Roads cut through forest

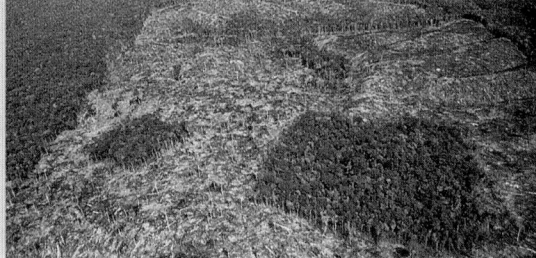
Forest occurs in patches

b. Macaws

Figure 47.6 Habitat loss. **a.** In a study that examined records of imperiled U.S. plants and animals, habitat loss emerged as the greatest threat to wildlife. **b.** Macaws that reside in South American tropical rain forests are endangered for some of the reasons listed in the graph in (**a**). **c.** Habitat loss due to road construction in Brazil. *Top:* Road construction opened up the rain forest and subjected it to fragmentation. *Middle:* The result was patches of forest and degraded land. *Bottom:* Wildlife could not live in destroyed portions of the forest.

Destroyed areas
c. Wildlife habitat is reduced.

characterized by unique assemblages of organisms that have evolved together in one location. Migration of some species to a new location is not usually possible because of barriers such as oceans, deserts, mountains, and rivers. Humans, however, have introduced exotic species into new ecosystems in a variety of ways.

Colonization

Europeans, in particular, brought a number of familiar species with them when they colonized new places. For example, the pilgrims brought the dandelion to the United States as a familiar salad green. In addition, they introduced pigs to North America that have since become feral, reverting to their wild state. In some parts of the United States, feral pigs have become quite destructive.

Horticulture and Agriculture

Some exotics now taking over vast tracts of land have escaped from cultivated areas. Kudzu is a vine from Japan that the U.S. Department of Agriculture thought would help prevent soil erosion. The plant now covers much landscape in the South,

including smothering even walnut, magnolia, and sweet gum trees (Fig. 47.7a). The water hyacinth was introduced to the United States from South America because of its beautiful flowers. Today, it clogs up waterways and diminishes natural diversity.

Accidental Transport

Global trade and travel accidentally bring many new species from one country to another. Researchers found that the ballast water released from ships into Coos Bay, Oregon, contained 367 marine species from Japan. The zebra mussel from the Caspian Sea was accidentally introduced into the Great Lakes in 1988. It now forms dense beds that squeeze out native mussels. Other organisms accidentally introduced into the United States include the Formosan termite, the Argentine fire ant, and the nutria, a type of large rodent.

Exotic species can disrupt food webs. As mentioned earlier, opossum shrimp introduced into a lake in Montana added a trophic level that in the end meant less food for bald eagles and grizzly bears (see Fig. 47.2).

Video Alien Invasion

The Impact of Exotics on Islands

Islands are particularly susceptible to environmental discord caused by the introduction of exotic species. Islands have unique assemblages of native species that are closely adapted to one another and cannot compete well against exotics. Myrtle trees, *Myrica faya,* introduced into the Hawaiian Islands from the Canary Islands, are symbiotic with a type of bacterium that is capable of nitrogen fixation. This feature allows the species to establish itself on nutrient-poor volcanic soil, a distinct advantage in Hawaii. Once established, myrtle trees disrupt the normal succession of native plants on volcanic soil.

The brown tree snake has been introduced onto a number of islands in the Pacific Ocean. The snake eats adult birds, their eggs, and nestlings. On Guam, it has reduced ten native bird species to the point of extinction. On the Galápagos Islands, black rats have reduced populations of giant tortoise, while goats and feral pigs have changed the vegetation from highland forest to pampaslike grasslands and destroyed stands of cacti. In Australia, mice and rabbits have stressed native marsupial populations. Mongooses introduced into the Hawaiian Islands to control rats also prey on native birds (Fig. 47.7*b*).

Pollution

In the present context, **pollution** can be defined as any environmental change that adversely affects the lives and health of living things. Pollution has been identified as the third main cause of extinction. Pollution can also weaken organisms and lead to disease. Biodiversity is particularly threatened by acid deposition, eutrophication, ozone depletion, and synthetic organic chemicals.

Video Ocean Garbage

Acid Deposition

Both sulfur dioxide from power plants and nitrogen oxides in automobile exhaust are converted to acids when they combine with water vapor in the atmosphere. These acids return to Earth as either wet deposition (acid rain or snow) or dry deposition (sulfate and nitrate salts). Sulfur dioxide and nitrogen oxides are not always deposited in the same location where they are emitted, but may be carried far downwind. Acid deposition causes trees to weaken and increases their susceptibility to disease and insects. It also kills small invertebrates and decomposers, so that entire ecosystems are disrupted. Many lakes in the northern United States are now lifeless because of the effects of acid deposition.

Eutrophication

Lakes are also under stress due to over-enrichment. When lakes receive excess nutrients due to runoff from agricultural fields and fertilized lawns, and wastewater from sewage treatment, algae begin to grow in abundance. An algal bloom is apparent as a thick green layer of algae on the surface of the water. Upon death, the decomposers break down the algae, but in so doing, they use up oxygen. A decreased amount of oxygen is available to fish, leading sometimes to a massive fish kill.

Ozone Depletion

The ozone shield is a layer of ozone (O_3) in the stratosphere, some 50 km above the Earth. The ozone shield absorbs most of the wavelengths of harmful ultraviolet (UV) radiation so that they do not strike the Earth. The cause of ozone depletion can be traced to chlorine atoms (Cl^-) that come from the breakdown of chlorofluorocarbons (CFCs). The best-known CFC is Freon, a heat-transfer agent still found in older refrigerators and air conditioners today. Severe ozone shield depletion can impair crop and tree growth and also kill plankton (microscopic plant and animal life) that sustain oceanic life. The immune system and the ability of all organisms to resist infectious diseases can also become weakened.

Organic Chemicals

Our modern society uses synthetic organic chemicals in all sorts of ways. Organic chemicals called nonylphenols are used in products ranging from pesticides to dishwashing detergents, cosmetics, plastics, and spermicides. These chemicals mimic the effects of hormones, which can harm wildlife.

As one example, salmon are born in fresh water but mature in salt water. After investigators exposed young fish to nonylphenol, they found that 20–30% were unable to make the transition between fresh and salt water. Nonylphenols cause the pituitary to produce prolactin, a hormone that may prevent saltwater adaptation in these fish.

Climate Change

As mentioned in Chapter 45, **climate change** refers to recent changes in the Earth's climate. Our planet is experiencing erratic temperature patterns, more severe storms, and melting glaciers as a result of an increase in the Earth's temperature. You may also recall from Chapter 45 that the majority of carbon dioxide and methane produced today is the result of human-based activities. These gases are known as greenhouse gases because they allow solar radiation to reach the Earth but hinder the escape of its heat back into space. Data collected around the world show a steady rise in CO_2 concentration. This data is used to generate

a.   b.

Figure 47.7 Exotic species.   a. Kudzu, a vine from Japan, was introduced in several southern states to control erosion. Today, kudzu has taken over and displaced many native plants. Here it has engulfed an abandoned house. **b.** Mongooses were introduced into Hawaii to control rats, but they also prey on native birds.

computer models that predict how the Earth will warm in the near future. (Fig. 47.8*a*). An upward shift in temperatures could influence everything from growing seasons in plants to migratory patterns in animals.

Video Global Warming

As temperatures rise, regions of suitable climate for human needs may shift toward the poles and higher elevations. Extinctions are expected to increase as species attempt to migrate to more suitable climates. (Plants migrate when seeds disperse and growth occurs in a new locale.) For example, for beech trees to remain in a favorable habitat, it's been calculated that the rate of beech migration would have to be 40 times faster than has ever been observed. It seems unlikely that beech or any other type of tree would be able to meet the pace required.

Many species are confined to relatively small patches of habitat surrounded by agricultural or urban areas that prevent natural migrations. Even if species have the capacity to disperse to new sites, suitable habitats may not be available. If the global climate changes faster than organisms can migrate, extinction of many species is likely to occur, while other species may experience population increase. For example, parasites and pests that are usually killed by cold winters would be able to survive in greater numbers. The tropics may very well expand, and whether present-day temperate-zone agriculture will survive is questionable.

Video Karoo Global Warming

Overexploitation

Overexploitation occurs when the number of individuals taken from a wild population is so great that the population becomes severely reduced in numbers. A positive feedback cycle explains overexploitation: The smaller the population, the more valuable its members, and the greater the incentive to capture the few remaining organisms. Poachers are very active in the collecting and sale of endangered and threatened species because it has become so lucrative. The overall international value of trading wildlife species is $20 billion, of which $8 billion is attributed to the illegal sale of rare species. The Nature of Science feature "Overexploitation of Asian Turtles" provides a case in point.

Markets for rare plants and exotic pets support both legal and illegal trade in wild species. Rustlers dig up rare cacti, such as the crested saguaros, and sell them to gardeners for as much as $15,000 each. Parrots are among birds taken from the wild for sale to pet owners. For every bird delivered alive, many more have died in the process. The same holds true for tropical fish, which often come from the coral reefs of Indonesia and the Philippines. Divers dynamite reefs or use plastic squeeze-bottles of cyanide to stun the fish; in the process, many fish and valuable corals are killed.

The Convention of International Trade of Endangered Species (CITES) was an agreement established in 1973 to ensure that international trade of species does not threaten their survival. Today, over 30,000 species of plants and animals receive some level of protection from over 172 countries worldwide.

Poachers still hunt for hides, claws, tusks, horns, or bones of many endangered mammals. Because of its rarity, a single Siberian tiger is now worth more than $500,000—its bones are pulverized and used as a medicinal powder. The horns of rhinoceroses become ornate carved daggers, and their bones are ground up to sell as a medicine. The ivory of an elephant's tusk is used to make art objects, jewelry, or piano keys. The fur of a Bengal tiger sells for as much as $100,000 in Tokyo.

The U.N. Food and Agricultural organization tells us that we have now overexploited 11 of 15 major oceanic fishing areas. Fish are a renewable resource if harvesting does not exceed the ability of the fish to reproduce. Modern society uses larger and more efficient fishing fleets to decimate fishing stocks. Pelagic species such as tuna are captured by purse-seine fishing, in which a very large net surrounds a school of fish. The net is then closed in the same manner as a drawstring purse. Thousands of dolphins that swim above schools of tuna are often captured and then die in this type of net. Many tuna suppliers now advertise their product as "dolphin safe"; however, this label's meaning varies, depending on the agency or organization that licenses the label.

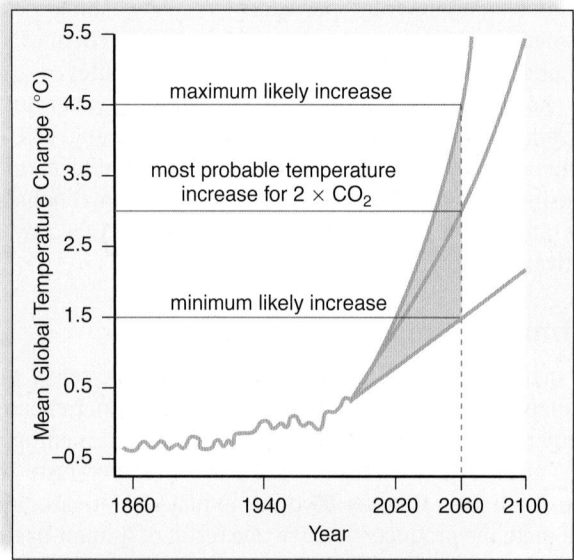

a.

b.

Figure 47.8 Climate Change.   a. Mean global temperature is expected to rise due to the introduction of greenhouse gases into the atmosphere. **b.** The resulting changes in climate have the potential to significantly affect the world's biodiversity. A temperature rise of only a few degrees causes coral reefs to "bleach" and become lifeless. If, in the meantime, migration occurs, coral reefs could move northward.

Nature of Science

Overexploitation of Asian Turtles

Conservation Alert

Collection and trade of terrestrial tortoises and freshwater turtles for human consumption and other uses has surged in Asia over the past two decades and is now spreading to areas around the globe (Fig. 47A). With 40% to 60% of all types of turtles already endangered, these practices have virtually wiped out many tortoise and freshwater turtle populations from wide areas of Asia and have brought others to the brink of extinction in a matter of years.

The wild-collection trade started in Bangladesh in order to supply consumption demands in South China, and then it quickly spread across tropical Asia as one area after another became depleted. Currently, the practice has spread to the United States.

Tortoises and turtles mature late (on the order of 10 to 20+ years), experience great longevity (measured in decades), and have low annual reproductive rates. Traders prefer to collect larger-sized, and therefore breeding-age, individuals. This means that wild populations are not likely to recover after they have been plundered. Presently, the stocking of thousands of new turtle farms (ranging from backyard to industrial-scale operations) is also causing a further run on the last remaining wild stocks of many endangered species. Then, too, there is a very active illegal pet trade in rare tortoises and freshwater turtles for wealthy patrons. This illegal pet trade is now vexing enforcement authorities in Asia, Europe, and the United States.

Major Challenges Today and in the Future

Basic scientific knowledge about the range, natural history, and conservation needs of individual species of tortoises and turtles is lacking. In fact, some traded species are so poorly known, and so endangered, that specimens have sometimes only been found by researchers in wildlife markets. Wildlife inspectors and enforcement officials are often unaware of conservation concerns and are unable to identify turtle species and to associate them with pertinent regulations.

Because the wild-collection trade is now worldwide, it is essential that all nations and states with (remaining) wild turtle populations ensure that their domestic legislation is adequate to secure the future of their turtle populations. Too often, trade starts up and occurs faster than regulatory measures can be put in place to prevent local populations from being decimated.

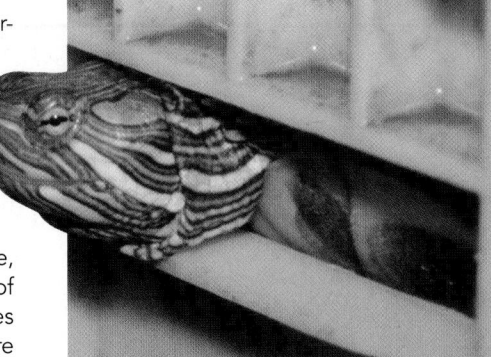

Figure 47B Aquaculture farms. Red-eared slider turtles (*Trachemys scripta elegans*) are now raised in aquaculture farms. Although this takes pressure off wild turtle populations, such turtles reinforce the habit of turtle consumption and can become invasive species when they escape into the wild.

Another issue that needs attention is the threat of invasive species and the spread of diseases from aquaculture facilities to wild populations. It is known that Chinese soft-shell turtles and red-eared slider turtles are highly adaptable to new environments (Fig. 47B). Reports abound from around the globe that turtles have either escaped or have been deliberately released from aquaculture facilities. Ecologists can only speculate about how they will impact wild populations when they become competitors, predators, hybridizers, and disease vectors. Impact studies are desirable, as is a regulatory framework governing the transport and handling of nonlocal turtles.

The public needs to become aware and concerned about the plight of tortoises and freshwater turtle populations because serious conservation research and legislative action is immediately required in order to save local populations from extinction.

Questions to Consider

1. Does the buyer or seller hold more responsibility for the overexploitation of the Asian turtles?
2. What type of ecological impact could an invasive turtle population have upon a native turtle population?

Figure 47A Turtles for sale. Since the 1990s, great numbers of wild-caught tortoises and freshwater turtles can be seen offered for sale at markets in East Asia. Shown here are steppe tortoises (*Testudo horsefieldii*) (*back*) and elongated tortoises (*Indotestudo elongata*) (*front*).

Written by Peter Paul van Dijk with the assistance of Bruce J. Weissgold.

a. Fishing by use of a drag net

b. Result of drag net fishing

Figure 47.9 Trawling.   a. These Alaskan pollock were caught by dragging a net along the seafloor. **b.** Appearance of the seafloor after the net passed.

Other fishing boats drag huge trawling nets, large enough to accommodate 12 jumbo jets, along the seafloor to capture bottom-dwelling fish (Fig. 47.9*a*). Only large fish are kept while the undesirable small fish and sea turtles are discarded back into the ocean to die. Trawling has been called the marine equivalent of clear-cutting forest, because after the net goes by, the sea bottom is devastated (Fig. 47.9*b*).

Today's fishing practices don't allow fisheries to recover. Cod and haddock, once the most abundant bottom-dwelling fish along the northeast coast of the United States, are now often outnumbered by dog-fish and skate.

Video
Big Fish Small Fish

A marine ecosystem can be disrupted by overfishing, as exemplified on the U.S. west coast. When sea otters began to decline in numbers, investigators found that they were being eaten by orcas (killer whales). Usually orcas prefer seals and sea lions over sea otters. The orcas began eating sea otters when the seal and sea lion populations decreased due to the over fishing of the perch and herring. Ordinarily, sea otters keep the population of sea urchins under control. With fewer sea otters around, the sea urchin population exploded and decimated the kelp beds. Overfishing set in motion a chain of events that detrimentally altered the food web of this ecosystem.

Check Your Progress   47.3

1. Identify the five main causes of extinction.
2. Explain why the introduction of exotic species can be detrimental to biodiversity.

47.4 Conservation Techniques

Learning Outcomes

Upon completion of this section, you should be able to

1. Describe the value of preserving biodiversity hotspots.
2. Distinguish between keystone species and flagship species.
3. Identify and explain the most useful procedures for habitat restoration.

Despite the value of biodiversity to our very survival, human activities are causing the extinction of thousands of species a year. Clearly, we need to reverse this trend and preserve as many species as possible. Habitat preservation and restoration are important in preserving biodiversity.

Habitat Preservation

Preservation of a species' habitat is of primary concern, but first we must prioritize which species to preserve. As mentioned previously, the biosphere contains biodiversity hotspots, relatively small areas having a concentration of endemic (native) species not found anyplace else. In the tropical rain forests of Madagascar, 93% of the primate species, 99% of the frog species, and over 80% of the plant species are endemic, found only in Madagascar. Preserving these forests and other hotspots would save a wide variety of organisms.

Keystone Species

Keystone species are species that influence the viability of a community, although their numbers may not be excessively high. The extinction of a keystone species can lead to other extinctions and a loss of biodiversity. For example, bats are designated a keystone species in tropical forests of the Old World. They are pollinators that also disperse the seeds of trees. When bats are killed off and their roosts destroyed, the trees fail to reproduce. The grizzly bear is a keystone species in the northwestern United States and Canada (Fig. 47.10*a*). Bears disperse the seeds of berries; as many as 7,000 seeds may be in one dung pile. Grizzly bears kill and eat the young of many hoofed animals and thereby keep their populations under control. Grizzly bears are also a principal mover of soil when they dig up roots and prey upon hibernating ground squirrels and marmots. Other keystone species are beavers in wetlands, bison in grasslands, alligators in swamps, and elephants in grasslands and forests.

Keystone species should not be confused with **flagship species,** which evoke a strong emotional response in humans. Flagship species are considered charismatic and are treasured for their beauty, "cuteness," strength, and/or regal nature. These species can help motivate the public to preserve biodiversity. Flagship species include Monarch butterflies, lions, tigers, dolphins, and the giant panda. In some cases, a keystone species may also be a flagship species—for example, the grizzly bear—but a keystone species is valued because of its role in an ecosystem.

a. Grizzly bear, *Ursus arctos horribilis*

b. Old-growth forest; northern spotted owl, *Strix occidentalis caurina* (inset)

Figure 47.10 Habitat preservation. When particular species are protected, other wildlife benefit. **a.** The Greater Yellowstone Ecosystem has been delineated in an effort to save grizzly bears, which need a very large habitat. **b.** Currently, the remaining portions of old-growth forests in the Pacific Northwest are not being logged in order to save the northern spotted owl (*inset*).

Metapopulations

The grizzly bear population is actually a **metapopulation** [Gk. *meta,* higher-order], a population subdivided into several small, isolated populations due to habitat fragmentation. Originally there were probably 50,000–100,000 grizzly bears south of Canada, but this number has been reduced because human housing communities have encroached on their home range, and bears are killed by frightened homeowners. Now there are six virtually

isolated subpopulations totaling about 1,000 individuals. The Yellowstone National Park population numbers 200, but the others are even smaller.

Saving metapopulations sometimes requires determining which of the populations is a source and which are sinks. A **source population** is one that most likely lives in a favorable area, and its birthrate is most likely higher than its death rate. Individuals from source populations move into **sink populations,** where the environment is not as favorable and where the birthrate equals the death rate at best. When trying to save the northern spotted owl, conservationists determined that it was best to avoid having owls move into sink habitats. The northern spotted owl reproduces successfully in old-growth rain forests of the Pacific Northwest (Fig. 47.10*b*) but not in nearby immature forests that are in the process of recovering from logging. Scientists have also recognized that competition from the barred owl is also a major threat to the recovery of the northern spotted owl.

Landscape Preservation

Grizzly bears inhabit a number of different types of ecosystems, including plains, mountains, and rivers. Saving any one of these types of ecosystems alone would not be sufficient to preserve the grizzly bears, however. Instead, it is necessary to save diverse ecosystems that are at least connected. As mentioned earlier, a landscape encompasses different types of ecosystems. An area called the Greater Yellowstone Ecosystem, where bears are free to roam, has now been defined. It contains millions of acres in Yellowstone National Park; state lands in Montana, Idaho, and Wyoming; five different national forests; various wildlife refuges; and even private lands.

Landscape protection for one species is often beneficial for other wildlife that share the same space. The last of the contiguous 48 states' harlequin ducks, bull trout, westslope cutthroat trout, lynx, pine martens, wolverines, mountain caribou, and great gray owls are found in areas occupied by grizzly bears. Gray wolves have also recently returned to this territory. Then, too, grizzly bear range overlaps with 40% of Montana's vascular plants of special conservation concern.

The Edge Effect. An *edge* is a transition from one habitat to another; in the case of land development, an edge may be very sharp, as when a forest abuts farm fields. The edge reduces the amount of habitat typical of an ecosystem because the edge habitat is slightly different from that of the interior of the patch—this is termed the **edge effect.** For example, forest edges are brighter, warmer, drier, and windier, with more vines, shrubs, and weeds than the forest interior. Also, Figure 47.11*a* shows that a small and a large patch of habitat both have the same depth of edge; therefore, the effective habitat shrinks as a patch gets smaller.

Many popular game animals, such as turkeys and white-tailed deer, are more plentiful in the edge region of a particular area. However, today it is known that creating edges can be detrimental to wildlife because of habitat fragmentation.

The edge effect can also have a serious impact on population size. Songbird populations west of the Mississippi have been declining, and ornithologists have noticed that the nesting success

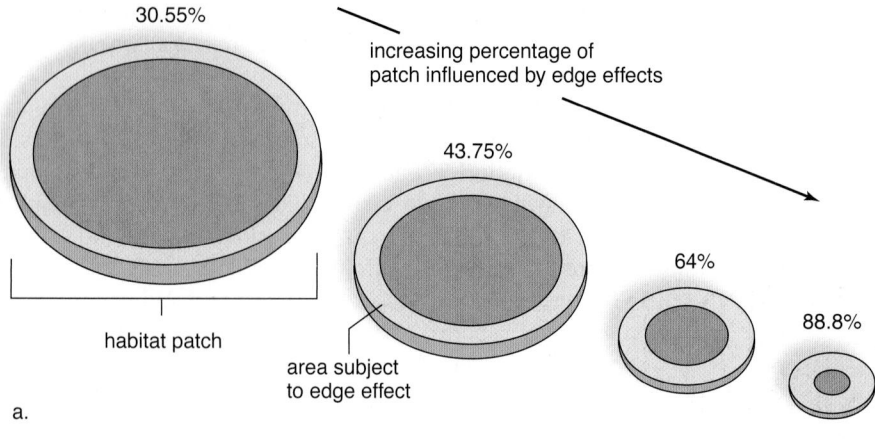

30.55%

increasing percentage of
patch influenced by edge effects

43.75%

64%

88.8%

habitat patch

area subject
to edge effect

a.

brown-headed
cowbird chick

yellow warbler
chick

b.

Figure 47.11 Edge effect. **a.** The smaller the patch, the greater the proportion that is subject to the edge effect. **b.** Cowbirds lay their eggs in the nests of songbirds (yellow warblers). A cowbird is bigger than a warbler nestling and is able to acquire most of the food brought by the warbler parent.

of songbirds is quite low at the edge of a forest. The cause turns out to be the brown-headed cowbird, a social parasite of songbirds. Adult cowbirds prefer to feed in open agricultural areas, and they only briefly enter the forest when searching for a host nest in which to lay their eggs (Fig. 47.11*b*). Cowbirds are therefore benefited, while songbirds are disadvantaged, by the edge effect.

Habitat Restoration

Restoration ecology is a subdiscipline of conservation biology that looks for scientific ways to return ecosystems to their natural state. Restoration ecology has three key principles. First, it should start as quickly as possible to avoid losing the remaining fragments of the original habitat.

Second, once the natural histories of the species in the habitat are understood, it is best to mimic natural processes to bring about restoration. Natural processes may include using controlled burns to bring back grassland habitats, biological pest controls to rid the area of exotic species, or bioremediation techniques to clean up pollutants.

Third, the goal is sustainable development, the ability of an ecosystem to be self-sustaining while still providing services to human beings. The Nature of Science feature "Emiquon Floodplain Restoration" presents an example of a successful project in west-central Illinois.

Check Your Progress   47.4

1. Identify ways in which landscape preservation is more valuable than ecosystem preservation.
2. List and explain the three principles of habitat restoration.

CONNECTING *the* CONCEPTS *with the* BIG IDEAS

Evolution

- Humans influence ecological stress and rates of species extinction. (1C1b*IE*)

Interactions and Systems

- As the human population increases, its effect on ecosystems are magnified; species may be diminished or driven to extinction, and habitat may be damaged or destroyed. (4A6f1-2)
- Human activities from agriculture to logging to urbanization seriously impact ecosystems. (4B4a*IE*)
- Introduced species often experience a population explosion in their new community because it lacks competitors or natural predators. (4B4a*IE*)
- Native species may be devastated by introduction of new pathogens, as seen in smallpox, potato blight, and Dutch elm disease. (4B4a*IE*)

Nature of Science

Emiquon Floodplain Restoration

Emiquon was once a vital component of the Illinois River system, which is a major tributary of the Mississippi River. It included a complex system of backwater wetlands and lakes that originally covered an area of approximately 14,000 acres (Fig. 47C). The vast diversity of native plants and animals found within Emiquon have supported human civilizations for over 10,000 years. Early cultures accessed the abundant inland fisheries and a variety of mussel species that could be found there. River otters, minks, beavers, and migrating birds as well as a variety of prairie grasses on the floodplains would have been used by native peoples (Fig. 47D).

In the early 1900s, the Illinois River was one of the most economically significant river ecosystems in the United States. It boasted the highest mussel abundance and productive commercial fisheries per mile of any stream in the United States. Each year during the seasonal flooding, the river would deposit nutrient-laden sediment onto the floodplain. This created a highly productive floodplain ecosystem. With the abundance of nutrients, a wide diversity of plant life would have been supported. This broad producer base would have been able to support a number of trophic levels as well as a wide diversity of food webs.

Throughout the 20th century, modern cultures altered the natural ebb and flow of the Illinois River to better suit their needs. Levees were constructed, resulting in the isolation of nearly half of the original flood plain from the river. Native habitats were converted to agricultural land. Corn and soy beans soon replaced the variety of prairie and wetland vegetation that were once a hallmark of the Emiquon floodplain. The river itself was also altered to better facilitate the export and import of a variety of commercial goods by boat. With the isolation of these areas from the seasonal flooding, a decrease in the amount of nutrients being deposited also occurred. This resulted in a continuous decline in the natural productivity of the floodplain.

Restoration Plan

The National Research Council has identified Emiquon as one of three large-floodplain river ecosystems in the United States that can be restored to some semblance of its original diversity and function. The floodplain restoration work at Emiquon is a key part of The Nature Conservancy's efforts to conserve the Illinois River.

The Emiquon Science Advisory Council is composed of over 40 scientists and managers who provide guidance for the restoration and managed connection between Emiquon's restored floodplain and the Illinois River. Once established, this connection would allow for the seasonal flooding that it previously experienced. This seasonal flooding would help naturalize the flow of the river and restore the cyclical process of flooding and drying out. Flooding would provide more nutrients for wetland plants as well as helping to improve the water quality. Access between the river and floodplain would once again be restored for various aquatic species, such as paddlefish and gar. Both species need a variety of habitats for reproduction and survival. The wetlands will in turn provide vital nutrients to the species that reside in the river. The Conservancy is working with the Illinois Natural History Survey, the University of Illinois, Springfield, and others to determine the baseline community diversity. These groups will monitor the community diversity to provide vital information about the progress of the restoration plan.

Local communities are investigating the potential economic benefits that can be gained as a result of the restoration. Several communities are exploring the potential for increasing the amount of nature-based

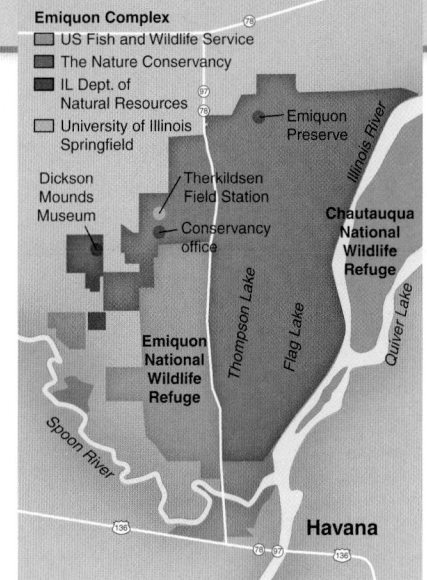

Figure 47C The Emiquon Restoration Project. Restoration plans call for allowing the river to return to some of its former range.

tourism that can be offered at Emiquon and along the Illinois River.

The knowledge gained from the research at Emiquon and The Nature Conservancy's work on the Illinois River and within the Upper Mississippi River system will potentially guide large floodplain-river restoration efforts around the world.

Questions to Consider

1. In what ways might scientists assess whether the restoration plan was successful in its goal?
2. How might the Emiquon Floodplain Restoration plan be used as a model for other restoration projects?

Figure 47D Wildlife of the Illinois River ecosystem. Many species in a number of animal phyla live in this ecosystem.

Media Study Tools

www.glencoe.com/maderAP11

Enhance your study of this chapter with study tools and practice tests. Also ask your instructor about the resources available through ConnectPlus, including the media-rich eBook, interactive learning tools, and animations.

Summarize

47.1 Conservation Biology and Biodiversity

Conservation biology is the scientific study of biodiversity and its management for sustainable human welfare. The unequaled present rate of extinctions has drawn together scientists and environmentalists in basic and applied fields to address the problem.

Biodiversity is the variety of life on Earth; the exact number of species is not known, but there are assuredly many species yet to be discovered and recognized. Biodiversity must also be preserved at the genetic, community (ecosystem), and landscape levels of organization.

Conservationists have discovered that biodiversity is not evenly distributed in the biosphere, and therefore saving particular areas may protect more species than saving other areas.

47.2 Value of Biodiversity

The direct value of biodiversity is seen in the observable services of individual wild species. Wild species are our best source of new medicines to treat human ills, and they help meet other medical needs as well.

Wild species have agricultural value. Domesticated plants and animals are derived from wild species, and wild species can be used as a source of genes for the improvement of their phenotypes. Instead of synthetic pesticides, wild species can be used as biological controls, and most flowering plants benefit from animal pollinators. Much of our food, particularly fish and shellfish, is still caught in the wild. Hardwood trees from natural forests supply us with lumber for many purposes.

The indirect services provided by ecosystems are largely unseen and difficult to quantify, but are absolutely necessary to our well-being. These services include the workings of biogeochemical cycles, waste recycling, provision of fresh water, prevention of soil erosion, and regulation of climate. Many people enjoy living and vacationing in natural settings. Studies show that more diverse ecosystems function better than less diverse systems.

47.3 Causes of Extinction

Researchers have identified the major causes of extinction. Habitat loss is the most frequent cause, followed by introduction of exotic species, pollution and resulting diseases, overexploitation, and climate change.

Habitat loss has occurred in all parts of the biosphere, but concern has now centered on tropical rain forests and coral reefs, where biodiversity is especially high. Exotic species have been introduced into foreign ecosystems through colonization, horticulture and agriculture, and accidental transport. Various causes of pollution include fertilizer runoff, industrial emissions, and improper disposal of wastes, among others. Overexploitation is exemplified by commercial fishing, which is so efficient that fisheries of the world are collapsing. Climate change is contributing to species extinction but at this point the full impact is not yet known.

47.4 Conservation Techniques

To preserve species, it is necessary to preserve their habitat. Some emphasize the need to preserve biodiversity hotspots because of their richness. Often today it is necessary to save metapopulations because of past habitat fragmentation. If so, it is best to determine the source populations and save those instead of the sink populations.

A keystone species such as the grizzly bear requires the preservation of a landscape consisting of several types of ecosystems over millions of acres of territory. Obviously, in the process, many other species will also be preserved.

Because many ecosystems have been degraded, habitat restoration may be necessary before sustainable development is possible. Three principles of restoration are (1) start before sources of wildlife and seeds are lost; (2) use simple biological techniques that mimic natural processes; and (3) aim for sustainable development so that the ecosystem can persist and at the same time fulfill the needs of humans.

Key Terms

biodiversity 908	genetic diversity 908
biodiversity hotspot 909	keystone species 918
bioinformatics 908	landscape 909
climate change 915	landscape diversity 909
conservation biology 908	metapopulation 919
ecosystem diversity 909	overexploitation 916
edge effect 919	pollution 915
endangered species 908	sink population 919
exotic species 913	source population 919
flagship species 918	threatened species 908

Assess

Reviewing This Chapter

1. Explain these attributes of conservation biology: (a) both academic and applied; (b) supports ethical principles; and (c) is responding to a biodiversity crisis.   908
2. Discuss the conservation of biodiversity at the species, genetic, community, and landscape levels.   908–09
3. Describe the uneven distribution of diversity in the biosphere. What is the implication of uneven distribution for conservation biologists?   909
4. List various ways in which individual wild species provide us with valuable services.   910–11
5. List various ways in which ecosystems provide us with indispensable services.   911–13
6. List and discuss the five major causes of extinction, starting with the most frequent cause.   913–18
7. Introduction of exotic species is usually due to what events?   913–14
8. List and discuss five major types of pollution that particularly affect biodiversity.   915

9. Identify the variables that are contributing to climate change and the role it is playing in species extinction. 915–16
10. Use the positive feedback cycle to explain why overexploitation occurs. 916
11. Using the grizzly bear population as an example, explain keystone species, metapopulations, and landscape preservation. 918–19
12. Explain the three principles of habitat restoration with reference to the Emiquon Floodplain restoration. 920–21

Testing Yourself

Choose the best answer for each question.

1. Which of these would not be within the realm of conservation biology?
   a. helping to manage a national park
   b. a government board charged with restoring an ecosystem
   c. writing textbooks and/or popular books on the value of biodiversity
   d. introducing endangered species back into the wild
   e. All of these are concerns of conservation biology.

2. The value of wild pollinators to the U.S. agricultural economy has been calculated to be $4.1–$6.7 billion a year. What is the implication?
   a. Society could easily replace wild pollinators by domesticating various types of pollinators.
   b. Pollinators may be valuable, but that doesn't mean any other species also provide us with valuable services.
   c. If we did away with all natural ecosystems, we wouldn't be dependent on wild pollinators.
   d. Society doesn't always appreciate the services that wild species provide naturally and without any fanfare.
   e. All of these statements are correct.

3. The services provided to us by ecosystems are unseen. This means
   a. they are not valuable.
   b. they are noticed particularly when the service is disrupted.
   c. biodiversity is not needed for ecosystems to keep functioning as before.
   d. we should be knowledgeable about them and protect them.
   e. Both b and d are correct.

4. Which of these is a true statement?
   a. Habitat loss is the most frequent cause of extinctions today.
   b. Exotic species are often introduced into ecosystems by accidental transport.
   c. Climate change may cause many extinctions but also expand the ranges of other species.
   d. Overexploitation of fisheries could very well lead to a complete collapse of the fishing industry.
   e. All of these statements are true.

5. Which of these is expected if the average global temperature increases?
   a. the inability of some species to migrate to cooler climates as environmental temperatures rise
   b. the possible destruction of coral reefs
   c. an increase in the number of parasites in the temperate zone
   d. some species will experience a population decline and others will experience an increase
   e. All of these are expected.

6. Why is the grizzly bear considered a keystone species that exists within a metapopulation?
   a. Grizzly bears require many thousands of miles of preserved land because they are large animals.
   b. Grizzly bears have functions that increase biodiversity, but presently the population is subdivided into isolated subpopulations.
   c. When grizzly bears are present, so are many other types of species within a diverse landscape.
   d. Grizzly bears are a source population for many other types of organisms across several population types.
   e. All of these statements are correct.

7. Sustainable development of the Everglades will mean that
   a. the various populations that make up the Everglades will continue to exist indefinitely.
   b. human needs will also be met while successfully managing the ecosystem.
   c. the means used to maintain the Everglades will mimic the processes that naturally maintain the Everglades.
   d. the restoration plan is a workable plan.
   e. All of these statements are correct.

8. Which statement accepted by conservation biologists best shows that they support ethical principles?
   a. Biodiversity is the variety of life observed at various levels of biological organization.
   b. Wild species directly provide us with all sorts of goods and services.
   c. New technologies can help determine conservation plans.
   d. There are three principles of restoration biology that need to be adhered to in order to restore ecosystems.
   e. Biodiversity is desirable and has value in and of itself, regardless of any practical benefit.

9. A population in an unfavorable area with a high infant mortality rate would be
   a. a metapopulation.
   b. a source population.
   c. a sink population.
   d. a new population.

10. What is the edge effect?
    a. More species live near the edge of an ecosystem, where more resources are available to them.
    b. New species originate at the edge of ecosystems due to interactions with other species.
    c. The edge of an ecosystem is not a typical habitat and may be an area where survival is more difficult.
    d. More species are found at the edge of a rain forest due to deforestation of the forest interior.

11. Consumptive value
    a. means we should think of conservation in terms of the long run.
    b. means we are placing too much emphasis on living things that are useful to us.
    c. means some organisms, other than domestic crops and farm animals, are valuable as sources of food.
    d. is a type of direct value.
    e. Both c and d are correct.

12. Which of these is not an indirect value of species?
    a. participate in biogeochemical cycles
    b. participate in waste disposal
    c. help provide fresh water
    d. prevent soil erosion
    e. All of these are indirect values.

13. Most likely, ecosystem performance improves
    a. the more diverse the ecosystem.
    b. as long as selected species are maintained.
    c. as long as species have both direct and indirect value.
    d. if extinctions are diverse.
    e. Both a and b are correct.

14. Sea urchins feed on kelp beds, and sea otters feed on sea
    urchins. If sea otters are killed off, which of these statement(s)
    is true? Choose more than one answer if correct.
    a. Kelp beds will increase.
    b. Kelp beds will decrease.
    c. Sea urchins will increase.
    d. Sea urchins will decrease.

15. Complete the following graph by labeling each bar (a–e) with
    a cause of extinction, from the most influential to the least
    influential.

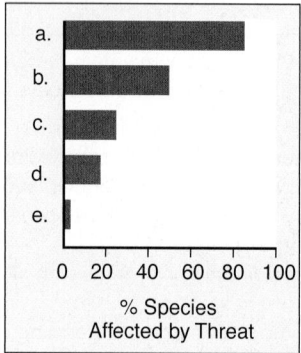

Engage

Thinking Scientifically

1. The scale at which conservation biologists work often makes
   direct experimentation difficult. But computer models can
   assist in predicting the future of populations or ecosystems.
   Some scientists feel that these models are inadequate because
   they cannot reproduce all the variables found in the real world.
   If you were trying to predict the impact on songbirds of clear-
   cutting a portion of a forest, what information would you need
   to develop a good model?

2. Bioprospecting is the search for medically useful molecules
   derived from living things. The desire for monetary
   and medicinal gains from such discoveries can lead to
   the overexploitation of endangered species. However,
   bioprospecting can protect ecosystems, and in this way, saves
   many species rather than individual species. What types of
   living things would bioprospectors be most interested in?

Bioethical Issue

Protecting Bighorn Sheep

To protect the state's declining bighorn sheep populations, the New
Mexico Game Commission approved a plan that calls for killing
scores of mountain lions over several years. The commission pointed
out that the lions have killed the majority of the radio-collared bighorn sheep released into the wild since 1996 and have increasingly
turned to killing sheep as the state's deer herd has declined.

Mountain lions and bighorns have long coexisted as part of a
common ecosystem. Rather than killing mountain lions to increase
the bighorn sheep population, it would be better to consider a restoration management plan that would improve the ecosystem. The
state recently engaged the Hornocker Wildlife Institute of Idaho to
study the situation. The institute's report concluded that the number
of lions does not necessarily affect the size of the sheep population.
Rather, diseases from domestic livestock grazing in the area are
responsible for more than 50% of the deaths in some bighorn sheep
populations.

Opposition to the restoration plan claims that the plan to kill
mountain lions was solidly based in science. The report said that killing lions in selected areas will give the sheep population a chance to
rebound because there are 2,000 lions and only 760 bighorn sheep
in two separate populations. Do you approve of taking steps to protect bighorn sheep from mountain lions? Why or why not? If you do
approve, how would you proceed?

Appendix A

Answer Key

Chapter 1

Check Your Progress

1.1: 1. Populations are all of the organisms of a species in a given area, while an ecosystem contains groups of populations (a community) plus the physical environment. 2. Organization, acquisition and use of energy, homeostasis, response to the environment, reproduction and development, and adaptation. 3. Adaptations allow organisms to survive in changing environments; the changes as a result of adaptations represent evolution. **1.2:** 1. Domain, Kingdom, Phylum, Class, Order, Family, Genus, Species. 2. Protists are generally unicellular organisms; fungi are multicellular filamentous organisms that absorb food; plants are multicellular organisms that are usually photosynthetic; animals are multicellular organisms that must ingest and process their food. 3. Members of a population that have new adaptive traits reproduce more than other members, and therefore pass on these traits to the next generation in a greater proportion. **1.3:** 1. Individuals interact with members of the same species (a population), with other populations (communities), and with the physical environment (ecosystem). 2. By altering the physical environment of the ecosystem such as with pollution, increased urbanization, agriculture, etc. 3. They are more vulnerable, because high biodiversity ecosystems often depend on species with very specific roles that are more vulnerable to changes in the environment. **1.4:** 1. Experiments test the contribution of the experimental variable to a given observation. 2. Test groups are exposed to the experimental variables, while control groups are not. 3. Hypotheses are tested by designing experiments that examine the contribution of a single experimental variable to the observation, and by statistically comparing the response of a test group to that of a control group in an experiment.

Testing Yourself

1. c; **2.** The brain thinks but nerve cells do not think; **3.** b; **4.** b; **5.** d; **6.** d; **7.** b; **8.** a; **9.** c; **10.** e; **11.** b; **12.** b; **13.** Energy is brought into an ecosystem for the first time through photosynthesis. The Sun provides energy for photosynthesis. **14.** Excess carbon dioxide emitted alters its level in the atmosphere. This excess carbon dioxide is more readily available to photosynthetic plants and protists, possibly causing them to overpopulate. **15.** Not necessarily, if the new organism does not serve as prey for one species, allowing it to overpopulate, or if it does not prey on or compete with existing species in the ecosystem.

Chapter 2

Check Your Progress

2.1: 1. Atomic number is the number of protons in an atom. Atomic mass is the approximate sum of the masses of the protons and neutrons. 2. Groups indicate the number of electrons in the outer, or valence, shell of an atom. In Group VII , all the atoms including chlorine have seven electrons in the valence shell. Periods indicate the number of shells present in an atom. All atoms in Period III, including chlorine, have 3 shells. 3. Radiation can be used to kill diseased cells, such as those present in cancer. The radiation is used to damage cancer cell DNA, leading to cell death. **2.2:** 1. Ionic and covalent bonds are similar in that both involve electrons, but differ as to whether the electrons are shared or not. Ionic bonds are created when one atom gives up electron(s) and another gains electron(s) so that both atoms have outer shells filled. Covalent bonds are formed when two atoms share electrons to fill their outer shells. 2. Calcium gives away its two outer electrons in order to have a complete outer shell. Calcium then has two more protons than electrons, creating a net charge of +2. 3. By sharing with four hydrogen atoms, carbon acquires the four electrons it needs to fill its outer shell. Hydrogen has only one shell, which is complete with two electrons; therefore, each of four hydrogens can acquire a needed electron by sharing with carbon. **2.3:** 1. Water can absorb a lot of heat energy without large changes in temperature. Because of water's heat-absorbing properties, coastal cities can maintain relatively constant temperatures, even when the Sun's heat energy fluctuates. 2. Much heat energy can be put into water. As water evaporates, it takes heat energy with it, allowing a child playing in a sprinkler to dissipate body heat during a hot summer day. 3. If frozen water were heavier than liquid water, ponds and other bodies of water would freeze from the bottom up. The insulation provided by a layer of floating ice would be lost, and organisms living in aquatic environments would likely freeze to death. Water freezes from the top down because ice is less dense than water; the reason is that hydrogen bonds become more open when water freezes. **2.4:** 1. An acid dissociates in water to release hydrogen ions. A base either takes up hydrogen ions or releases hydroxide ions. 2. Living organisms require relatively narrow pH conditions to function. Both trees and fish die when their environment becomes too acidic. 3. Decreasing the blood's ability to resist large changes in pH would be catastrophic and likely cause biological malfunction or death.

Testing Yourself

1. d; **2.** e; **3.** c; **4.** c; **5.** a; **6.** d; **7.** e; **8.** b; **9.** b; **10.** Red circle is oxygen (partial negative charge). White circles are hydrogen (partial positive charge); **11.** c; **12.** a; **13.** a; **14.** e; **15.** b; **16.** d; **17.** a; **18.** c; **19.** b.

Chapter 3

Check Your Progress

3.1: 1. A carbon atom bonds with up to four different elements. Carbon–carbon bonds are stable, so long chains can be built; chains can have branching patterns, and chains can form rings. Isomers are possible. 2. 2-carbon alcohols and 2-carbon carboxylic acids are both soluble in water; however, the acid is more soluble because it easily donates its hydrogen. Both alcohols and acids become less soluble as the number of carbons increases. 3. The degradation, or hydrolysis, of a biomolecule requires water. If no water is present the molecule won't be cut into pieces without some other factor, such as heat. **3.2:** 1. Humans have no digestive enzymes capable of breaking the bonds in cellulose. 2. The monomer in cellulose is glucose; in chitin, an amino group is attached to each glucose. Cellulose is found in plant cell walls, while chitin is in the exoskeleton of some animals. **3.3:** 1. In water, oxygen and hydrogen atoms share their electrons unequally in their covalent bonds, resulting in a charge across the bonds, giving the molecule an overall polarity. In lipids, the carbon–carbon covalent bonds tend to share their electrons equally, resulting in no polarity. Because like dissolves like, these physical differences between water and lipids prevent their mixing. 2. A saturated fatty acid contains no double bonds between carbon atoms, while an unsaturated fatty acid contains one or more double bonds. Saturated fatty acids are flat and tend to stack densely in specific locations in blood vessels, reducing blood flow. Unsaturated fatty acids are kinked, and therefore do not block blood vessels to the same degree as saturated fats. 3. Because like dissolves like, phospholipids, which have hydrophilic and hydrophobic domains, arrange themselves so that their hydrophilic heads are adjacent to water, while the hydrophobic tails point inward toward each other. **3.4:** 1. A polypeptide is composed of a sequence of amino acids that are specified by the sequence of bases within a gene in the DNA. 2. Because like dissolves like, hydrophobic amino acids aggregate away from water in an aqueous environment, whereas hydrophilic amino acids associate with water. 3. Hydrophobic interactions play a major role in protein folding, as do pH, temperature, salt concentration, and other factors. **3.5:** 1. A nucleic acid stores information in the sequences of bases within a gene. Gene sequences specify amino acid sequences, thus assembling the proteins that carry out the functions of life. 2. Phosphate, 5-carbon sugar, and nitrogen-containing base. 3. ATP contains 3 phosphate molecules that are strongly negative and repel each other. The covalent bonds attaching them contain considerable potential energy, which can be captured when the bonds are broken and used to do cellular work.

Testing Yourself

1. c; **2.** e; **3.** e; **4.** b; **5.** b; **6.** c; **7.** d; **8.** d; **9.** c; **10. a.** monomer; **b.** monomer; **c.** dehydration reaction; **d.** H_2O; **e.** monomer; **f.** monomer; **g.** polymer; **11.** c; **12.** d; **13.** d; **14.** a; **15.** a; **16.** b; **17.** b; **18.** d; **19.** b; **20.** d.

Chapter 4

Check Your Progress

4.1: 1. Living organisms are made up of one or more cells, which form the basic unit of life. New cells are produced from old cells. Energy flows within cells. 2. Organisms are composed of cells; both cells and organisms are systems in that they balance input and output and use energy to metabolize. The components of each provide structure and function that is greater than the sum of its parts. 3. The movement of materials into and out of a cell requires a sufficiently large surface area relative to cell volume. Cells that are overly large have an insufficient surface-area-to-volume ratio, thus limiting their overall size. **4.2:** 1. Prokaryotic cells lack a membrane-bounded nucleus, which is present in a eukaryotic cell. 2. Cell envelope:

mesosome, plasma membrane, cell wall, glycocalyx. Cytoplasm: inclusion bodies, nucleoid, ribosomes. Appendages: conjugation pilus, fimbriae, flagellum. **4.3:** 1. Separates various metabolic processes; localizes enzymes, substrates, and products; and allows cells to become specialized. 2. Organelles effectively divide cellular work, so that incompatible processes can occur simultaneously. This division of labor enables the cell to more quickly respond to changing environments. 3. The proportion of a given organelle can change based on cellular function. For example, a muscle cell may have a more extensive endoplasmic reticulum due to its need to regulate ions; a nerve cell has more plasma membrane to help conduct electrical signals over long distances. **4.4:** 1. Cells tightly regulate what enters and exits the nucleus by the use of nuclear pores. 2. Hereditary information encoded in genes located in the nucleus is transcribed into RNA molecules, some of which specify amino acid sequence in proteins. Amino acid sequence is determined by RNA sequence through translation. Polypeptides produced from RNA are then folded into a three-dimensional structure that has biological function. **4.5:** 1. Rough ER contains ribosomes, while smooth ER does not. Rough ER synthesizes proteins and modifies them, while smooth ER synthesizes lipids, among other activities. 2. Membrane-enclosed organelles work together as part of a larger endomembrane system within the cell. Transport vesicles from the ER proceed to the Golgi apparatus, which modifies their contents and repackages them in new vesicles to be delivered to specific locations within the cell based on a molecular address. Incoming food and other particles are packaged inside vesicles and delivered to lysosomes, which help to break down the food for energy or for building cellular structures. 3. Golgi malfunction would prevent materials from reaching their necessary destinations, disrupting cellular function and likely leading to cell death. **4.6:** 1. Both peroxisomes and lysosomes enclose enzymes; a peroxisome is a metabolic assistant to other organelles; a lysosome digests molecules and can break down cell parts and their component molecules. 2. Peroxisomal proteins are generally made in the cytoplasm and imported into the peroxisome, whereas lysosomal proteins are made in the ER and sent through the Golgi apparatus. Lysosomal proteins are made in the cytoplasm and the ER. Lysosomal proteins are sent to the lysosome via a molecular address. **4.7:** 1. Chloroplasts and mitochondria are similar in size to bacteria, also have multiple membranes, replicate via binary fission, and have circular genomes. 2. ATP is produced during photosynthesis and cellular respiration via proton gradients across

membranes. In photosynthesis, ATP provides the energy necessary to produce sugar; in respiration, ATP is the primary product and is subsequently used in many metabolic processes. **4.8:** 1. Microtubules help maintain cellular shape and provide an internal road system used to transport intracellular materials. Intermediate filaments anchor cells to each other and components in the external environment. Actin filaments are used in some forms of cellular movement. 2. Cilia and flagella are both composed of microtubules arranged in a particular pattern and enclosed by the plasma membrane. Cilia are shorter than flagella. ATP is used to produce cellular movement via microtubules. 3. Cellular movement from microtubules is accomplished via motor molecules that use ATP.

Testing Yourself

1. c; **2.** c; **3.** d; **4.** a; **5.** c; **6.** c; **7.** a; **8.** c; **9.** b. **10.** b; **11.** e; **12.** a OR d; **13.** a; **14.** b, mitochondria, because both have double membranes and metabolize energy; c, flagellum, because both require the use of microtubules; d, smooth and rough, because both use membrane-embedded proteins to properly function.

Chapter 5

Check Your Progress

5.1: 1. The plasma membrane is a phospholipid bilayer with cholesterol and embedded proteins. Cholesterol helps maintain membrane fluidity across a range of temperatures. Without cholesterol, membranes would become too fluid at high temperatures and too solid at low temperatures, thus preventing necessary life processes. 2. Proteins are embedded or associated with membranes, creating a mosaic. These proteins influence membrane function. Membranes are considered a fluid mosaic because the phospholipids and membrane proteins can freely move within the membrane. 3. Membrane cores are hydrophobic due to the fatty acid tails of phospholipids. Hydrophobic substances can freely move across the membrane. In contrast, hydrophilic substances require transport mediated by membrane proteins or other mechanisms. **5.2:** 1. Hydrophilic water is able to move across hydrophobic membranes by means of an aquaporin channel protein in membranes. 2. Both move molecules from high to low concentration. Diffusion does not require a membrane or transport proteins, while facilitated transport does. **5.3:** 1. Both require a carrier protein, but active transport also requires energy because unlike facilitated diffusion it moves material against its concentration gradient. 2. Both use vesicles to transport materials across the plasma membrane. Vesicles created as materials enter the cell and reduce membrane surface area, whereas vesicles from the Golgi apparatus increase surface area. The net result of this vesicle traffic is little or no net change in the plasma membrane surface area. **5.4:** 1. The extracellular matrix is composed of polysaccharides and proteins that have a multitude of functions. In addition, the matrix in bone contains mineral salts. 2. A primary cell wall is composed of cellulose and other molecules, such as pectin. Secondary cell walls contain cellulose fibrils and lignin that adds rigidity and strength. 3. Both gap junctions in animals and plasmodesmata in plants allow water and other small molecules to freely pass between adjacent cells.

Testing Yourself

1. a. hypertonic—cell shrinks due to loss of water; **b.** hypotonic—central vacuole expands due to gain of water; **2.** b; **3.** b; **4.** b; **5.** c; **6.** c; **7.** e; **8.** e; **9.** b; **10.** b; **11.** See Figure 5.1, page 86; **12.** b; **13.** e; **14.** d; **15.** a; **16.** e.

Chapter 6

Check Your Progress

6.1: 1. Potential energy is stored in the bonds present within the glucose molecule. 2. It takes energy to clean a messy room. The input of energy helps to create the structure associated with a clean room. However, it takes little time and energy for the room to become disorganized and messy again. This is consistent with the second energy law, which states that every energy transformation increases disorder. Entropy is the tendency toward disorder. **6.2:** 1. ATP holds energy but gives it up because it is unstable and the last phosphate group can be easily removed, releasing energy. 2. ATP can donate a phosphate to energize a compound for a reaction. The addition of a high-energy negatively charged phosphate can change the molecule's shape and in that way bring about a structural change that in turn changes its function. **6.3:** 1. They bring reactants together at their active site or position a substrate so it is ready to react. Because enzymes interact very specifically with their reactants, the products produced from enzymes can be regulated, such as is seen in metabolic pathways. 2. Cofactors provide another level of control that cells use to regulate enzyme products. This control enables the cell to maximize its efficiency and ability to survive. **6.4:** 1. Plants fix carbon dioxide from the air to create carbohydrate molecules used for energy and plant structure. Animals consume plants and break these carbohydrates down to derive energy for life processes, and in so

doing produce carbon dioxide as a by-product. This cyclical relationship between plants and animals is essential to life on Earth. 2. Energized electrons are used to create ATP via proton gradients in plants and animals.

Testing Yourself

1. e; **2.** e; **3.** e; **4.** e; **5.** d; **6.** b; **7. a.** active site; **b.** substrates; **c.** product; **d.** enzyme; **e.** enzyme-substrate complex; **f.** enzyme. The shape of an enzyme is important to its activity because it allows an enzyme-substrate complex to form. **8.** a; **9.** a; **10.** c; **11.** b; **12.** c; **13.** c; **14.** c; **15.** e; **16.** c; **17.** e; **18.** c; **19.** e; **20.** See Figure 6.13, page 115.

Chapter 7

Check Your Progress

Plants, algae, and cyanobacteria. 2. Thylakoid membrane absorbs solar energy, and carbohydrate forms in stroma. **7.2:** 1. O_2 is reduced to CH_2O and when the oxygen in water gives its H_2 ($2 H^+ + 2 e^-$), it is oxidized. 2. Enzymes are used to regulate chemical reactions during photosynthesis. **7.3:** 1. Visible light is a small part of the overall electromagnetic spectrum. It ranges from purple to red light. 2. Low energy electrons within chlorophyll in Photosystem II are excited to a higher energy state by sunlight. The high potential energy of these excited electrons is siphoned off during redox reactions in the electron transport chain, creating proton gradients across the thylakoid membrane. These electrons are excited again to a higher energy state by Photosystem I, which harnesses the energy to create NADPH. **7.4:** 1. A 3-carbon G3P molecule would require 3 turns of the Calvin cycle because each turn adds a carbon molecule. RuBP must also be regenerated in order for the cycle to continue. **7.5:** 1. C_4 plants include many grasses, sugarcane, and corn; CAM plants include cacti, stonecrops, orchids, and bromeliads. 2. C_4 plants prevent oxygen from competing with carbon dioxide for an active site on the enzyme rubisco.

Testing Yourself

1. e; **2.** d; **3.** a; **4.** a, b; **5.** c; **6.** a, b, c; **7.** b; **8.** d; **9.** T; **10.** F; **11.** F; **12.** e; **13.** e; **14.** c; **15.** See Figure 7.5, page 123; **f.** thylakoid membranes; **g.** stroma; **16.** a; **17.** e; **18.** e; **19.** c; **20.** d.

Chapter 8

Check Your Progress

8.1: 1. Step-by-step breakdown allows the released energy to be captured and utilized by

the cell. Rapid breakdown would be destructive to the cell. 2. Glycolysis, the preparatory reaction, the citric acid cycle, and the electron transport chain. The prep reaction and the citric acid cycle release CO_2. ETC produces H_2O. **8.2:** 1. During the energy-investment steps, ATP breakdown provides the phosphate groups to activate substrates. During the energy-harvesting steps, NADH and ATP are produced. 2. ADP and phosphate from the cytoplasm are connected together during substrate-level ATP synthesis via enzymes from glycolysis. **8.3:** 1. In the absence of oxygen, cells derive energy through fermentation by continually regenerating NAD^+ so it can be reduced to NADH during glycolysis. **8.4:** 1. Per glucose molecule, the first 2 CO_2 molecules are produced during the preparatory reaction, and the remaining 4 CO_2 molecules are produced during the citric acid cycle. 2. The electron transport chain produces the most ATP during glycolysis breakdown. 3. A dam holds back water, just as the inner membrane holds back hydrogen ions. As water flows over a dam, electricity is produced. As hydrogen ions flow down their concentration gradient through an ATP synthase complex, ATP is produced. **8.5:** 1. Catabolic and anabolic processes are regulated by enzymes, which allow a cell to respond to environmental changes. 2. Both have an inner membrane (thylakoids in chloroplasts; cristae in mitochondria) where complexes form an ETC, and ATP is produced by chemiosmosis. Both have a fluid-filled interior (in the stroma of chloroplasts, NADPH helps reduce CO_2 to a carbohydrate; and in the matrix of mitochondria, NAD helps oxidize glucose products with the release of CO_2).

Testing Yourself

1. b; **2.** c; **3.** a; **4.** c; **5.** c; **6.** c; **7.** a; **8.** b; **9.** e; **10.** c; **11.** a; **12.** c; **13.** b; **14.** b; **15.** c; **16.** a; **17.** c; **18.** a; **19.** b; **20.** b.

Chapter 9

Check Your Progress

9.1: 1. G_1, S, G_2, and M stage. DNA is replicated during S stage, and cell division occurs during M stage. 2. DNA damage, failure to properly replicate the DNA, and failure of chromosomes to attach properly to spindle. 3. Apoptosis controls the overall density of cells by targeting excess cells, or cells with damage, for cell death. **9.2:** 1. Prophase: the nuclear membrane fragments, the nucleus disappears, the chromosomes condense, and the spindle begins to form. Metaphase: the chromosomes align along a central line in the cell. Anaphase: the sister chromatids separate and head to opposite poles. Telophase: the nuclear envelope re-forms, chromosomes decondense; the spindle disassociates. 2. Animal cells furrow and plant cells have a cell plate. Because plant cells have rigid cell walls, they cannot furrow. 3. Because the number of times most cells can divide is controlled, stem cells provide source material for tissues that need to continuously replace cells. **9.3:** 1. Cancer cells lack differentiation, have abnormal nuclei, fail to undergo apoptosis, may form tumors that may undergo angiogenesis, and may also metastasize throughout the body. 2. Malignant tumor cells enter circulation and lodge in other tissues of the body. Benign tumors remain in the tissue in which they form 3. a. Cell commits to cell division even in the absence of proper stimuli. b. Cell fails to stop dividing because the proper stimuli to stop are absent. **9.4:** 1. Binary fission involves inward growth of plasma membrane and cell wall concomitant with the separation of the duplicated chromosome attached to the plasma membrane. Mitosis always involves a mitotic spindle to distribute the daughter chromosomes. 2. Prokaryotes usually have a single, small circular chromosome with a few genes and only a few associated proteins. Eukaryotes have many long, linear chromosomes, which contain many thousands of genes and many more proteins.

Testing Yourself

1. b; **2.** e; **3.** c; **4.** a; **5.** b; **6.** b; **7.** b; **8.** e; **9.** b; **10.** c; **11.** e; **12.** b; **13.** c; **14.** c; **15.** b; **16.** d; **17.** a; **18.** c; **19.** d; **20. a.** chromatid of chromosome; **b.** centriole; **c.** spindle fiber or aster; **d.** nuclear envelope (fragment); early prophase.

Chapter 10

Check Your Progress

10.1: 1. Homologous chromosomes represent two copies of the same kind of chromosome, each containing the same genes for traits in the same order. They are also similar in length and location of the centromere. 2. Homologous chromosomes replicate prior to meiosis I, producing sister chromatids. When they separate each daughter cell receives one pair of sister chromatids, resulting in a reduction from 2n to n chromosomes. Meiosis II splits the sister chromatids, producing n chromosomes in each daughter cell. 3. A bivalent is formed between homologous chromosomes to enable crossing over and strand exchange. This increases the potential genetic variability of the offspring. **10.2:** 1. Independent assortment of chromosomes increases the number of possible combinations of chromosomes in each gamete. Crossing-over shuffles the alleles between homologous

chromosomes to create even more variation. 2. 2^4 or 16 possible combinations. 3. Genetic variability ensures that at least some individuals have traits that may allow a species to survive adverse conditions. **10.3:** 1. Two daughter cells that share the same parent cell from meiosis I are identical unless crossing-over has occurred. 2. Pairs of homologous chromosomes would not split, resulting in cells with incorrect chromosome combinations. Without proper reduction in chromosome number, offspring would have incorrect chromosome combinations and would likely not survive. **10.4:** 1. In metaphase I of meiosis, homologous chromosomes are paired at the metaphase plate with each homologue facing opposite spindle poles. In metaphase II and mitotic metaphase, homologous chromosomes are not paired, and sister chromatids are attached to spindle fibers from opposite spindle poles. 2. Meiosis II resembles mitosis because sister chromatids are separated during both processes. Meiosis II differs from mitosis because the cells are haploid and not diploid. **10.5:** 1. In males, the primary spermatocytes are located within the testes; in females, the primary oocytes are located within the ovaries. 2. Oogenesis produces a single cell, containing the bulk of the cytoplasm and other cellular contents in the one cell that will undergo embryonic development. Spermatogenesis produces four haploid cells, each of which could potentially merge with the haploid egg to create a diploid organism. **10.6:** 1. Nondisjunction in meiosis may cause aneuploidy, an extra or missing chromosome. 2. Sex chromosome aneuploidy is more common because only one of the X chromosomes is active. Any extra X chromosomes become Barr bodies. 3. An inversion involves the reversal of a piece of a chromosome from within, and normally does not cause symptoms. A translocation is swapping of two chromosome fragments from one to the other, and while not usually troublesome, may cause severe problems in offspring if the two chromosomes go into separate cells.

Testing Yourself

1. b; **2.** d; **3.** e; **4.** b; **5.** d; **6.** c; **7.** e; **8.** b; **9.** d; **10.** 24, 12; **11.** spermatogenesis, oogenesis; **12.** fertilization; **13.** diploid, haploid; **14.** b; **15.** b; **16.** a; **17.** d; **18.** a; **19.** a.

Chapter 11

Check Your Progress

11.1: 1. Mendel chose a good model organism, always followed the same well-planned procedure, kept careful records, and analyzed his data mathematically. 2. The garden pea has easily observed traits, a relatively short generation time, each plant

produces many offspring (peas), and cross pollination is only possible by hand. **11.2A:** 1.a. all W; b. 1/2 W, 1/2 w; c. 1/2 T, 1/2 t; d. all T. 2. bb. 3. 3:1, 40. **11.2B:** 1. LG, Lg, lG, lg. 2. 9:3:3:1. **11.2C:** 1. 75% yellow; 25% green. 2. 25%. 3. Tt x tt; **11.2D:** 1. cc, Cc. 2. 25%. 3. woman: Hh; husband: hh. 4. 50%. **11.3A:** 1. A_1A_2. 2. child: ii; mother: I^Ai; father: I^Ai, I^Bi, or i. **11.3B:** 1. Mother X^BX^b; father, X^BY; female offspring are X^BX^b, and males are X^bY. 2. a. 100%; b. none; c. 100%. 3. X^bX^b mother and X^BY father. Female offspring will be X^BX^b and have black hair and be carriers; male children will be X^bY and have brown hair.

Testing Yourself

1. b; **2.** a; **3.** c; **4.** d; **5.** c; **6.** d; **7.** a; **8.** b; **9.** b; **10.** b; **11.** a; **12.** d; **13.** d; **14.** c; **15.** c; **16.** b; **17.** a; **18.** c; **19.** autosomal dominant.

Answers to Figures

11.10. Individual III-1 must be heterozygous since their offspring IV-1 and IV-2 are homozygous recessive (aa).

11.11. Individual III-1 must be heterozygous since their offspring III-3 is homozygous recessive (aa).

Chapter 12

Check Your Progress

12.1: 1. All life on Earth is based on the same four bases in DNA: A, G, C, and T. Because the genetic code is universal, expressing a foreign gene in an organism is possible. 2. DNA is a right-handed double helix with two strands that run in opposite directions. The backbone is composed of alternating sugar-phosphate groups, and the molecule is held together in the center by hydrogen bonds between interacting bases. A always hydrogen-bonds to T, and G to C. **12.2:** 1. (1) The DNA strands are separated by DNA helicase, (2) new nucleotides are positioned by complementary base pairing, and (3) new nucleotides are joined together by DNA polymerase to form a new DNA strand. 2. Prokaryotic DNA replication begins at a single origin of replication and usually proceeds in both directions toward a termination region on the opposite side of the chromosome. Eukaryotic DNA replication begins at multiple origins of replication and continues until the replication forks meet. 3. Telomerase enables eukaryotic chromosomes to fully replicate by using an RNA template in the enzyme to create repeat sequences on the end of linear chromosomes. **12.3:** 1. Information from a gene coded within the DNA is copied into mRNA. mRNA sequence dictates the sequence of amino acids in a protein, which produces the observable traits for an organism. The flow of genetic information is DNA to RNA to protein to trait. 2. mRNA carries information from DNA to direct the synthesis of a protein. rRNA makes up part of the ribosomes that are used to translate messenger RNAs. tRNA transfers amino acids to the ribosome during protein synthesis. 3. Several different codons may specify the same amino acid.

12.4: 1. Transcription reads the template strand in the 3′ to 5′ direction and produces a complementary mRNA strand that is 5′ to 3′. The mRNA has the same orientation and sequence as the gene strand in the DNA and uses uracil instead of thymine. 2. During mRNA processing introns are spliced out and the exons joined together, and a 5′ guanosine cap and a 3′ poly-A tail are added. 3. A pre-mRNA transcript contains noncoding sequences called introns that are spliced out leaving coding sequences, called exons, in the mature mRNA. 4. If a gene strand in the DNA has the sequence 5′ ATGGGCATT 3′, and a template strand sequence of 5′ AATGCCCAT 3′, the mRNA sequence would be 5′ AUGGGCAUU 3′. 5. The ability to create multiple mRNA molecules from a single gene through alternative splicing enables a cell to create a wider variety of proteins whose diversity may positively affect organism survival.

12.5: 1. Transfer RNA delivers amino acids to the ribosome by binding to the appropriate codon on the mRNA being translated. 2. A ribosome consists of a small and large subunit. Each subunit is composed of a mixture of protein and rRNA. Large and small ribosomal subunits work to connect amino acids together in a sequence dictated by the mRNA. 3. Initiation: All components of the translational complex, including the first tRNA carrying methionine, are assembled. Elongation: Amino acids are delivered one by one as tRNA molecules pair with the codons on the mRNA. Termination: A stop codon is reached, a release factor binds to it, and the completed protein is cleaved from the last tRNA as the ribosomal subunits dissociate. **12.6:** 1. Euchromatin may exist in a compact or open configuration. A compact chromatin structure prevents regulatory proteins needed for gene expression from accessing gene control elements. An open configuration allows gene expression to occur.

Testing Yourself

1. c; **2.** a; **3.** c; **4.** b; **5.** d; **6.** e; **7.** b; **8. a.** GGA GGA CUU ACG UUU; **b.** CCU CCU GAA UGC AAA; **c.** glycine-glycine-leucine-threonine-phenylalanine; **9.** d; **10.** a; **11.** a; **12.** a; **13.** c; **14.** d; **15.** d.

Chapter 13

Check Your Progress

13.1: 1. An operon is a group of genes that are regulated in a coordinated manner; a

promotor is a regulatory region that controls the initiation of transcription. 2. In a repressible operon, the operon is normally on and is turned off by the action of a repressor. In an inducible operon, the operon is normally off and is turned on by an environmental condition. 3. A gene under positive control is transcribed when it is regulated by a protein that is an activator and not a repressor, whereas one under negative control is not transcribed when it is regulated by a protein that is a repressor. **13.2:** 1. Chromatin, transcriptional, posttranscriptional, translational, and posttranslational. 2. Packing genes into heterochromatin inactivates a gene, whereas it is genetically active when held in loosely packed euchromatin. 3. siRNA molecules participate in posttranscriptional regulation while proteosomes are active in posttranslational regulation.

13.3: 1. Spontaneous: errors in DNA replication and natural chemical changes in the bases in DNA; Induced: organic chemicals and physical mutagens like X-rays and UV radiation. 2. Mutations in these genes disrupt the normal operation of the checkpoints in the cell cycle, allowing cell division to proceed without regulation.

Testing Yourself

1. a; **2.** e; **3.** a; **4.** b; **5.** d; **6.** e; **7.** b; **8.** b; **9.** b; **10.** b; **11.** b; **12.** e; **13.** e; **14.** e; **15. a.** DNA; **b.** regulator gene; **c.** promoter; **d.** operator; **e.** active repressor; **16.** c; **17.** d; **18.** d; **19.** d.

Chapter 14

Check Your Progress

14.1: 1. To create an rDNA molecule, a piece of foreign DNA is cut with restriction enzymes and mixed with a plasmid vector cut with the same restriction enzyme. The DNAs are mixed together and DNA ligase is added to seal the molecule. Then, it can be inserted into bacteria. 2. DNA molecules amplified by PCR can be used to create DNA profiles, or fingerprints, enabling paternity testing and forensic DNA analysis, among other applications. 3. Patterns in DNA fingerprinting can be analyzed to identify the source of a DNA sample. **14.2:** 1. Bacteria: production of medicine and drugs, bioengineering, production of organic chemicals; Animals: gene pharming, studying human disease; Plants: increased food production, production of human medicines. 2. A transgenic animal contains recombinant DNA molecules in addition to its genome, whereas a cloned animal is genetically identical to the one from which it was created, but does not contain rDNA. **14.3:** 1. Liposomes, nasal sprays, and adenoviruses are currently being used to deliver genes to cells for gene therapy. 2. Ex vivo gene therapy

is being used to treat SCID and familial hypercholesterolemia. In vivo gene therapy is being used to treat cystic fibrosis. **14.4:** 1. The genome is the sum of all of the genes in a cell whereas the proteome represents the sum of all of the proteins encoded by the genome. The proteome is typically bigger than the genome. 2. A tandem repeat consists of repeated sequences that are one next to the other, whereas transposons are short sequences of DNA that may move within the genome. 3. Microarrays allow scientists to analyze the expression of genes under different environmental conditions. Bioinformatics allows scientists to rapidly analyze experimental data and generate large databases of information.

Testing Yourself

1. c; **2.** c; **3.** e; **4.** d; **5.** e; **6.** c; **7.** e; **8.** a; **9.** c; **10.** a; **11.** b; **12.** d; **13.** e; **14.** d; **15.** e; **16.** a; **17.** d.

Chapter 15

Check Your Progress

15.1: 1. Catastrophism (Cuvier) explains a change in the fossil record by mass extinctions in an area replaced by a new set of species after a catastrophe. 2. Lamarck's inheritance of acquired characteristics is flawed as an explanation of biological diversity because features acquired by an individual during his/her lifetime are not heritable to offspring. Evolution occurs only in traits that are heritable. 3. 1707-1778 Linnaeus (classification); 1731-1802 E. Darwin (first theory of evolution); 1769-1832 Cuvier (catastrophism); 1744-1829 Lamarck (acquired characteristics); 1797-1875 Lyell (uniformitarianism); 1831-1836 voyage of H.M.S *Beagle*; 1859 Darwin publishes *Origin of Species*. **15.2:** 1. Fossils, biogeography, change over time (geology). 2. Individuals have variation that is heritable; organisms compete for resources; differential reproductive success; adaptation to environmental change. 3. Evolution can occur over the span of a human lifetime. One example is artificial selection, which can modify plants and animals through selective breeding. Another are beaks of the Galápagos Islands finches, which were observed to evolve between wet and dry seasons over decades. **15.3:** 1. Biomolecules may change over time as a result of natural selection. Examples are the cytochrome *c* protein in cellular metabolism and the homeobox, or *Hox*, genes that have been shaped by natural selection to orchestrate the development of the body plan in animals. 2. Homologous structures are those that are shared because of common ancestry. Analogous structures are shared because of convergent evolution and do not tell us anything about common ancestry. 3. Transitional fossils represent intermediate evolutionary forms of life in transition from one type to another, or a common ancestor of these types. *Tiktaalik* and *Basilosaurus* are examples of intermediate fossils. 4. (One of the following) a. *Evolution is a theory about how life originated.* Evolution is concerned with how life evolved AFTER the origin of life. b. *There are no transitional fossils.* Not all fossils will ever be found because of the nature of how organisms are preserved and discovered. c. *Evolution proposes that life-forms changed as a result of random events.* New variation arises due to random processes, but natural selection is not a random process; it can only shape pre-existing variation. d. *Evolution is not observable or testable.* Evolution is both observable and testable. Many studies show how traits change in response to environmental pressures. There are examples of evolution over long periods of time, and over short periods of time.

Testing Yourself

1. d; **2.** e; **3.** b; **4.** e; **5.** e; **6.** e; **7.** b, c, d, e; **8.** b, d, e; **9.** c; **10.** d, e; **11.** b; **12.** Life has a history, and it's possible to trace the history of individual organisms. **13.** Two different continents can have similar environments, and therefore unrelated organisms that are similarly adapted. **14.** All vertebrates share a common ancestor that had pharyngeal pouches during development. **15.** Similarities are expected because all species share recent and distant common ancestors. DNA base difference through the occurrence of mutations accounts for the diversity of life.

Chapter 16

Check Your Progress

16.1: 1. a. Random mating, allele frequencies do not change. b. No selection, allele frequencies change. c. No mutation, allele frequencies change. d. No migration, allele frequencies change. e. Large population (no genetic drift), allele frequencies change. 2. AA = p^2 = 0.01, Aa = 2pq = 0.18, aa = q^2 = 0.81. **16.2:** 1. Females produce large gametes and have a large investment in eggs so are choosy about their mates. Males produce many small, inexpensive gametes, and value females as a resource for reproduction, resulting in competition for access to females (mates). 2. Sexual selection is a form of natural selection because traits associated with mating determine whether or not traits are passed on to the next generation. 3. See Figure 16.10 on page 297. **16.3:** 1. Mutation, sexual reproduction, genetic drift. 2. In Africa where malaria is

common, being a heterozygote for the sickle-cell trait is advantageous because it provides some resistance to malaria. Those who are homozygous for normal trait get malaria without natural resistance; those homozygous for the sickle-cell trait die young from sickle-cell disease. Heterozygotes for the cystic fibrosis trait have resistance to typhoid fever. As with sickle-cell trait, heterozygotes are maintained because of stabilizing selection that selects against homozygotes for the cystic fibrosis trait (who have the disease) and homozygous normal trait (who get typhoid fever). 3. The dominant allele (p) and the recessive allele (q) must be present in the previous generation in order for the heterozygote (pq) to appear in the next generation.

Testing Yourself

1. c; **2.** c; **3.** c; **4.** c; **5.** e; **6.** b; **7.** c; **8.** e; **9.** d; **10.** c; **11.** a; **12.** e; **13.** b; **14.** e.

Chapter 17

Check Your Progress

17.1: 1. Pre-zygotic reproductive isolation. 2. (Three of the following) Biological species concept: Individuals of a population cannot interbreed. Phylogenetic species concept: Two populations are on different evolutionary lineages in a phylogeny that are monophyletic. Morphological species concept: A species is defined by a diagnostic trait (or traits). Evolutionary species concept: Relies on identification of certain morphological diagnostic traits that define a different evolutionary pathway. 3. If the frogs cannot recognize each other as mates, they will be reproductively isolated. **17.2:** 1. Show that each of the cats is adapted to a different environment. 2. In a newly discovered African Rift Valley lake, you would expect to see in the fishes the same shape of head and teeth for feeding in similar habitats as those observed in other lakes. Similar forms would evolve independently in this new lake—each of which would match the form of a fish that feeds in the same habitat in a different lake. 3. Allopatric speciation. **17.3:** 1. Punctuated equilibrium model proposes periods of rapid evolution and periods of little evolution (stasis). Cuvier proposed that this pattern in the fossil record was because new groups of organisms moved into an area after a mass extinction (catastrophe). 2. *Hox* genes have a powerful effect on development and offer a mechanism by which evolution could occur rapidly. 3. A change in *Pax6* gene expression shows that changes in the genes that regulate body organization can produce rapid change in an organism, leading to speciation.

Testing Yourself

1. c; **2.** e; **3.** c; **4.** e; **5. a.** species 1; **b.** geographic barrier; **c.** genetic changes; **d.** species 2; **e.** genetic changes; **f.** species 3; **6.** b; **7.** b; **8.** b; **9.** d; **10.** d; **11.** b; **12.** b; **13.** b; **14.** b; **15.** d; **16.** b; **17.** c; **18.** c.

Chapter 18

Check Your Progress

18.1: 1. a. Methane, ammonia, hydrogen, water environment: Recreated by Miller-Urey; b. Reducing atmosphere: Electric spark and chemical reactions. 2. a. Iron-sulfur world hypothesis: Iron-nickel sulfides at thermal vents act as catalysts to join organic molecules. b. Protein-first hypothesis: Amino acids collected in small pools and the heat of the Sun formed protenoids, small polypeptides with catalytic properties. c. RNA-first hypothesis: RNA carried out the processes commonly associated with DNA and proteins before either of these evolved. 3. Protocell: single fatty acid hydrophilic tail; modern cell: double fatty acid hydrophilic tail; Protocell: membrane of a single layer of fatty acids; modern cell: bilayer of phospholipids. **18.2A:** 1. Elements that are radioactive decay at a constant rate of time. The half-life is how long it takes for 1/2 of the number of molecules of an element to decay. If we know how much of the radioactive and decayed isotopes of an element we have left, we can estimate the number of years it took for the element to decay since the organism died. 2. Chemical evolution, evolution of first cells, evolution of eukaryotic cells by endosymbiosis, and first heterotrophic protists before photosynthetic protists. 3. DNA content and double-membrane structure of the mitochondria and chloroplasts. **18.2B:** 1. a. Eukaryote cells arise; b. Multicellular organisms arise. 2. The presence of an exoskeleton. **18.2C:** 1. a. Evolution of flowering plants; b. Reptiles undergo an adaptive radiation; c. Dinosaurs go extinct. 2. Flowering plants, conifers, birds, dinosaurs, placental mammals. **18.3:** 1. The theory of plate tectonics says that the Earth's crust is fragmented into slab-like plates that float on a lower hot mantle layer. Growth of mountains, earthquakes and the layout of the continents are all evidence. 2. Distribution of animals would be similar on continents that were once joined together. Marsupials in South America and Australia are an example. 3. Ordovician: continental drift and cooling of the Earth; Devonian: bolide, or impact of a meteorite with the Earth; Permian: excess carbon dioxide; Triassic: meteorite collision with Earth; Cretaceous: bolide, or impact of an asteroid with the Earth.

Testing Yourself

1. c; **2.** d; **3.** a; **4.** b; **5.** e; **6.** b; **7.** c; **8.** b; **9.** b; **10.** e; **11.** e; **12.** b; **13.** c; **14.** d; **15.** a; **16.** c; **17.** c; **18.** e; **19.** c; **20.** d.

Chapter 19

Check Your Progress

19.1: 1. Domain: Eukarya; Kingdom: Animalia; Phylum: Chordata; Class: Mammalia; Order: Primates; Family: Hominidae; Genus: *Homo*; Species: *Homo sapiens*. 2. Wings in birds and bats are analogous because they have independent evolutionary origins. Thus, birds and bats are not a natural group. 3. Classification is the naming of organisms; taxonomy is the assignment of groups to a taxon; systematic biology is the study of classification and taxonomy, or more generally, the study of biodiversity. **19.2:** 1. Archaea have differences in their rRNA, in their cell wall structure, and in the lipids in their plasma membrane. 2. Molecular data (RNA/DNA sequence data) indicates that fungi are related to animals. 3. Eukarya. **19.3:** 1. Answers may vary. Refer to the material on cladistics (page 356) and Figure 19.7 for examples. 2. Ancestral traits are found in the common ancestor of a lineage and may be present in the descendants. Derived traits are found in the descendants but not in the common ancestor. 3. The phylogeny that requires the fewest number of evolutionary steps is the most parsimonious explanation of evolutionary history, and is thus the best hypothesis.

Testing Yourself

1. e; **2.** e; **3.** d; **4.** a, b, c; **5.** c, d, e; **6.** b, c, d, e; **7.** a; **8.** c; **9.** a; **10.** d; **11.** b; **12.** e; **13.** b.

Chapter 20

Check Your Progress

20.1: 1. All viruses have a nucleic acid and a capsid. 2. Viroids and prions are nonliving because, like viruses, they are noncellular and unable to reproduce without a host. 3. A virus consists of both nucleic acids and proteins, whereas a prion is composed of just protein. 4. If a virus's host survives, many more copies of the virus will be produced and spread to other hosts than if the host dies. **20.2:** 1. Bacteria exist in the air and sterilization kills bacteria. 2. The bacterial nucleoid is not surrounded by a membrane; the eukaryote nucleus has its own membrane 3. Transduction: movement of genetic material from one bacterium to another via a virus; transformation: pick up and incorporate genetic material from the environment; conjugation: exchange genetic material between bacteria. **20.3:** 1. The peptidoglycan layer is much thicker in Gram-positive cells. In Gram-negative cells this layer is thinner and is located between layers of the plasma membrane. 2. By inhibiting protein synthesis or construction of the cell wall. 3. Endospores permit survival when environmental conditions are harsh. 4. Cyanobacteria are considered responsible for the production of an oxygen-rich atmosphere. **20.4:** 1. Archaea and bacteria differ in rRNA base sequences, and their plasma membranes and cell walls are biochemically distinct. 2. Methanogens, halophiles, and thermoacidophiles. 3. Archaea and eukaryotes share some of the same ribosomal proteins, initiate transcription in the same way, and have similar tRNA.

Testing Yourself

1. a. attachment; **b.** penetration; **c.** integration; **d.** prophage; **e.** biosynthesis; **f.** maturation; **g.** release; **2.** e; **3.** e; **4.** b; **5.** a; **6.** c; **7.** d; **8.** a; **9.** c; **10.** c; **11.** c; **12.** c; **13.** a; **14.** e; **15.** e; **16.** a; **17.** a; **18.** b; **19.** d.

Chapter 21

Check Your Progress

21.1: 1. Endosymbiosis; mitochondria and chloroplasts were free-living bacteria that were engulfed and incorporated into a eukaryotic protocell. 2. Algae are photoautotrophs; they get their food from photosynthesis; protozoans are heterotrophs and ingest their food from the environment. 3. a. Archaeplastids; b. Opisthokonts. **21.2A:** 1. Haploid stage is the asexual stage of reproduction; diploid is the sexual phase. 2. Accessory pigments allow for the absorption of energy from light of different wavelengths. **21.2B:** 1. Their DNA base sequences are similar, and their ancestor had a flagellum. 2. Brown algae: *Macrocystis, Laminaria, Nereocystis;* Diatoms: *Cyclotella;* Golden brown algae: *Ochromonas*. **21.2C:** 1. Refer to the summary boxes in Fig. 21.15 on page 394. 2. Brown comes from a carotenoid pigment. 3. Aveolates possess small air sacs (alveoli) beneath their plasma membrane; all are unicellular. **21.2D:** 1. Excavates have atypical or absent mitochondria and distinctive flagella and/or deep oral grooves. 2. Parasitic: parabasalids, diplomads, kinetoplastids; free-living: euglenids. 3. Trypanosome: African sleeping sickness and Chagas disease; *Giardia*: giardiasis; *Trichomonas*: vaginitis. **21.2E:** 1. Opisthokonts move using flagella; rhizarians and amoebozoans move using pseudopods. 2. Opisthokonts are multicellular protozoans, animal-like with a flagellum.

Testing Yourself

1. f; **2.** c; **3.** c; **4.** a; **5.** b; **6.** e; **7.** b; **8.** b; **9.** e; **10.** e; **11.** d; **12.** d; **13.** b; **14.** b; **15.** a; **16.** b; **17.** c; **18.** c; **19. a.** sexual reproduction; **b.** gametes pairing; **c.** zygote (2n); **d.** zygospore (2n); **e.** asexual reproduction; **f.** zoospores (n); **g.** nucleus with nucleolus; **h.** chloroplast; **i.** starch granule; **j.** pyrenoid; **k.** flagellum; **l.** eyespot; **m.** gamete formation. See also Figure 21.5, page 388.

Chapter 22

Check Your Progress

22.1: 1. Animals are heterotrophs: they get their nutrition by consuming outside nutrients. Fungi are saprotrophs: they externally digest organic material and absorb nutrients. 2. Plant cell walls are made of cellulose; fungal cell walls are made of chitin. 3. Reproductive structure that can form a new organism without fusing with another reproductive cell. **22.2:** 1. Chitrids are the only fungi to have flagellated cells. 2. Peach leaf curl: *Taphrina;* ergot: *Claviceps;* athletes' foot: *Trichophyton* (tinea); candidiasis: *Candida;* fungal flu: *Histoplasma*. 3. Black mold: *Rhizopus;* ergot: *Claviceps;* athlete's foot: *Trichophyton* (tinea). **22.3:** 1. Lichen has a sac fungus and either a cyanobacterium or green alga. 2. Lichens reproduce asexually by releasing fragments that contain hyphae and an algal cell. In fruticose lichens, the sac fungus reproduces asexually. 3. Fungus enters the cortex of the roots of plants, giving the plants a greater absorptive surface.

Testing Yourself

1. d; **2.** b; **3.** c; **4.** c; **5.** b; **6.** e; **7.** e; **8.** a; **9.** c; **10.** a; **11.** c; **12.** a; **13.** d; **14. a.** spores; **b.** sporangium; **c.** sporangiophore; **d.** stolon; **e.** rhizoid; **15.** a; **16.** e; **17.** e; **18.** d; **19.** e; **20. a.** meiosis; **b.** basidiospores; **c.** dikaryotic mycelium; **d.** button stage of the mushroom (basidiocarp); **e.** stalk; **f.** gill; **g.** cap; **h.** dikaryotic; **i.** diploid; **j.** zygote. See also Figure 22.9, page 412.

Chapter 23

Check Your Progress

23.1: 1. Plentiful light and CO_2. 2. A cellulose cell wall produced in same way; apical cells that produce new tissue; plasmodesmata between cells; transfer of nutrients from haploid cells of previous generation to zygote of new generation. 3. The diploid sporophyte produces haploid spores by meiosis. The haploid gametophyte produces gametes. **23.2:** 1. Bryophytes have a dominant gametophyte generation, produce flagellated sperm that swim to the egg, and can also reproduce asexually. 2. Asexual and sexual reproduction, ability to survive in harsh environments, sporophyte is protected from drying out. 3. The gametophyte portion is most important. The sporophyte is dependent upon the gametophyte. **23.3:** 1. In lycophytes, the dominant sporophyte has vascular tissue, and therefore roots, stems, and leaves. 2. The walls of xylem contains lignin, a strengthening agent. **23.4:** 1. The independent gametophyte generation lacks vascular tissue, and it produces flagellated sperm. 2. In ferns, but not mosses, the sporophyte is dominant and separate from the gametophyte. **23.5:** 1. (1) Water is not required for fertilization because pollen grains (male gametophytes) are windblown, and (2) ovules protect female gametophytes and become seeds that disperse the sporophyte, the generation that has vascular tissue. 2. Conifers: cone bearing and evergreen; cycads: cone bearing, evergreen and wind pollinated; ginkgoes: cone bearing, deciduous; gnetophytes: cone bearing and insect pollinated. 3. The stamen contains the anther and the filament. Pollen forms in the pollen sac of the anther. The carpel contains the stigma, style, and ovary. An ovule in the ovary becomes a seed, and the ovary becomes the fruit. 4. Gymnosperms: cone-bearing, such as cycads, ginkoes, gnetophytes and pine trees. 5. Animal-pollinated flowers are showy in a variety of ways, such as color and fragrance to attract their particular pollinators. Wind-pollinated plants are not very showy.

Testing Yourself

1. e; **2.** a; **3.** b; **4.** c; **5.** b; **6.** b; **7.** e; **8.** c; **9.** c; **10.** e; **11.** d; **12.** b; **13.** a; **14.** b; **15.** b; **16.** e; **17.** c; **18. a.** sporophyte (2n); **b.** meiosis; **c.** gametophyte (n); **d.** fertilization. See also Figure 23.3, page 422.

Chapter 24

Check Your Progress

24.1: 1. Vegetative organs are the leaves (photosynthesis), the stem (support, new growth, transport), and the root (absorb water and minerals). 2. Monocots: embryo with single cotyledon; xylem and phloem in a ring in the root; scattered vascular bundles in the stem; parallel leaf veins; flower parts in multiples of three. Eudicots: embryo with two cotyledons; phloem located between arms of xylem in the root; vascular bundles in a ring in the stem; netted leaf veins; flower parts in multiples of fours or fives. **24.2:** 1. Epidermal tissue: epidermal cells; ground tissue: parenchyma, collenchyma, and sclerenchyma cells; vascular tissue: xylem (vessel elements and tracheids) and phloem (sieve-tube members). 2. Xylem transports water and

minerals usually from roots to leaves. Phloem transports organic compounds throughout the plant. **24.3:** 1. The root apical meristem is located at the tip of the root and is covered by the root cap. 2. Cortex: food storage; endodermis: control of mineral uptake; pericycle: formation of branch roots. **24.4:** 1. A vascular bundle contains xylem and phloem. 2. Vascular bundles are scattered in monocot stems and form a ring in eudicot stems. 3. Primary growth is growth in length and is nonwoody; secondary growth is growth in girth and is woody. 4. Bark is composed of cork, cork cambium, cortex, and phloem. 5. An annual ring is composed of one year's growth of wood—one layer of spring wood followed by one layer of summer wood. **24.5:** 1. Photosynthesis, which produces organic food for a plant, occurs in the mesophyll. 2. The palisade mesophyll contains elongated cells that run photosynthesis; the spongy mesophyll contains irregular cells that increase the surface area for gas exchange. 3. Leaves can be simple with a single blade or compound, which is divided into many leaflets. Pinnate leaves have leaflets occurring in pairs while palmately compound leaves have all the leaflets attached at a single point. Leaves are adapted for environmental conditions.

Testing Yourself

1. c; **2.** b; **3.** c; **4.** c; **5.** b; **6.** b; **7.** b; **8.** c; **9.** c; **10.** b; **11.** d; **12.** e; **13.** d; **14.** d; **15.** d; **16.** c; **17.** c; **18.** a; **19. a.** epidermis; **b.** cortex; **c.** endodermis; **d.** phloem; **e.** xylem. See also Figure 24.8, page 451; **20. a.** upper epidermis; **b.** palisade mesophyll; **c.** leaf vein; **d.** spongy mesophyll; **e.** lower epidermis. See also Figure 24.20, page 459.

Chapter 25

Check Your Progress

25.1: 1. a. Nitrogen and sulfur are needed to form protein. All plant roots take up nitrate (NO_3^-) and sulfate (SO_4^{2-}) from the soil. b. Nitrogen and phosphate (HPO_4^{2-}) are needed to make nucleic acids. Plant roots also take up phosphate from the soil. 2. (1) Helps prevent soil erosion; (2) helps retain moisture; and (3) as the remains decompose, nutrients are returned to the soil. 3. Humus improves soil aeration, soil texture, increases water-holding capacity, decomposes to release nutrients for plant growth, and helps retain positively charged minerals and make them available for plant uptake. **25.2:** 1. The nonpolar tails of phospholipid molecules make the center of the plasma membrane nonpolar. 2. The bacteria convert atmospheric nitrogen to nitrate or ammonium, which can be taken up by plant roots. 3. The fungus

obtains sugars and amino acids from the plant. The plant obtains inorganic nutrients and water from the fungus. **25.3:** 1. Evaporation of water from leaf surfaces causes water to be under tension in stems due to the cohesion of the water molecules. 2. When water molecules are pulled upward during transpiration, their cohesiveness creates a continuous water column. Adhesion allows water molecules to cling to the sides of xylem vessels, so the column of water does not slip down. 3. Sugars enter sieve tubes at sources, creating pressure as water flows in as well. The pressure is relieved at the other end when sugars and water are removed at the sink.

Testing Yourself

1. d; **2.** e; **3.** a; **4.** c; **5.** d; **6.** c; **7.** d; **8.** a; **9.** b; **10.** c; **11.** d; **12.** b; **13.** a; **14.** c; **15.** e; **16.** a; **17.** The diagram shows that air pressure pushing down on mercury in the pan can raise a column of mercury only to 76 cm. When water above the column is transpired, it pulls on the mercury and raises it higher than 76 cm. This suggests that transpiration would be able to raise water to the tops of trees. **18.** e; **19. a.** See Figure 25.13, page 477. **b.** After K^+ enters guard cells, water follows by osmosis and the stoma opens. **20.** There is more solute in bulb 1 than in bulb 2; therefore water enters bulb 1. This creates a positive pressure that causes water, along with solute, to flow toward bulb 2.

Chapter 26

Check Your Progress

26.1: 1. Hormones coordinate the responses of plants to stimuli. 2. You could apply gibberellins to induce growth and cytokinins to increase the number of cells. 3. ABA maintains dormancy and closes stomata. **26.2:** 1. It is adaptive for roots to grow toward water because it enhances their ability to extract water and dissolved minerals from the soil for plant tissues. 2. Positive geotropism is when part of a plant grows downward in response to gravity. Negative geotropism is when a plant grows upward against gravity. 3. Phytochrome is composed of two proteins that contain light-sensitive regions. When triggered by red light phytochrome causes various genes to become active or inactive leading to seed germination, shoot elongation and flowering responses. 4. The plant is responding to a short night, not to the length of the day. 5. Plants have (1) physical and chemical defenses (e.g., secondary metabolites); (2) wound responses (e.g., proteinase inhibitors); (3) hypersensitive responses (e.g., sealing off of infected areas);

and (4) relationships with animals (e.g., acacia with ants).

Testing Yourself

1. c; **2.** a; **3.** d; **4.** e; **5.** d; **6.** b; **7.** e; **8.** a; **9.** c; **10.** c; **11.** d; **12.** e; **13.** b **14.** e; **15.** d.

Chapter 27

Check Your Progress

27.1: 1. Male gametophytes are produced in the anther of the stamen. The female gametophyte is produced in an ovule within the ovary of the carpel. 2. Each microspore produces a two-celled pollen grain. The generative cell produces two sperm, and the tube cell produces a pollen tube. One of the four megaspores produces a seven-celled female gametophyte, called the embryo sac, within the ovule. 3. In order to accommodate the larger size of the pollinator, natural selection in the flowering plant may lead to an increase in the structural support of the part that the pollinator lands upon. It may also lead to an increase in the size of the flower to accommodate a larger pollinator. **27.2:** 1. The embryo is derived from the zygote; the stored food is derived from the endosperm; and the seed coat is derived from the ovule wall. 2. The ovule is a sporophyte structure produced by the female parent. Therefore, the wall (which becomes seed coat) is 2n. The embryo inside the ovule is the product of fertilization and is, therefore, 2n. 3. Cotyledons are embryonic leaves that are present in seeds. Cotyledons store nutrients derived from endosperm (in eudicots). **27.3:** 1. Dry fruits, with a dull, thin, and dry covering derived from the ovary, are more apt to be windblown. Fleshy fruits, with a juicy covering derived from the ovary and possibly other parts of the flower, are more apt to be eaten by animals. 2. Eudicot seedlings have a hook shape, and monocot seedlings have a sheath to protect the first true leaves. **27.4:** 1. Advantages to asexual reproduction include: (1) the newly formed plant is often supported nutritionally by the parent plant until it is established; (2) if the parent is ideally suited for the environment, the offspring will be as well; and (3) if distance between individuals make cross-pollination unlikely, asexual reproduction is a good alternative. 2. For example, stolons and rhizomes produce new shoots and roots; fruit trees produce suckers; and stem cuttings grow new roots and become a shoot system. 3. Tissue from leaves, meristem, and anthers can become whole plants in tissue culture.

Testing Yourself

1. d; **2.** a; **3.** b; **4.** a; **5.** a; **6.** a; **7.** e; **8.** d; **9.** b; **10.** e; **11.** c; **12.** a; **13.** b; **14.** e; **15. a.** diploid; **b.** anther; **c.** ovule; **d.** ovary; **e.** haploid;

f. megaspore; **g.** male; **h.** female; **i.** sperm; **j.** seed. See also Figure 27.1, page 502.

Chapter 28

Check Your Progress

28.1: 1. Multicellular, usually with specialized tissues, ingest food, diploid life cycle. 2. Animals are descended from an ancestor that resembled a hollow spherical colony of flagellated cells. Individual cells became specialized for reproduction. Two tissue layers arose by invagination. 3. Deuterostomes: blastopore becomes anus, radial cleavage, coelum forms from gut; Protostomes: blastopore becomes mouth, spiral cleavage, coelom forms from mesoderm. **28.2:** 1. Sponges are multicellular; no symmetry; no digestive cavity. Cnidarians have true tissues; radial symmetry; have a gastrovascular cavity. 2. Interior of sponges has canals lined with flagellated cells called choanocytes. Flagella produce water current that carries food particles that are filtered out. 3. Polyps have mouths directed upward, medusae are bell-shaped with tentacles around the opening of the bell and mouth downward. 4. Stinging cells called cnidocytes have a fluid-filled capsule called a nematocyst in which a hollow threadlike structure is coiled and is discharged when stimulated. Some trap prey, others contain paralyzing toxins. **28.3:** 1. They all have bilateral symmetry, three tissue layers, and protostome development. They have no body plan, coelom, or any sort of nervous tissue. 2. Annelids and molluscs have a complete digestive tract, a true coelom, and a circulatory system (closed in annelids and open in molluscs). Flatworms have a gastrovascular cavity with only one opening, no coelom, and no circulatory system. 3. Tapeworm, *Taenia*. Host ingests meat with bladder worms; worms attach to intestine and mature; proglottids mature and fill with eggs; eggs leave host in feces; eggs ingested by livestock; worms encysted in muscle of livestock and eaten by host. Blood fluke: *Schistosoma*. Larvae penetrate skin of human, mature in liver; adult worms live and mate in blood vessels of gut; eggs migrate into digestive tract, passed in feces; ciliated larvae hatch in water, enter the snail; larvae emerge from snails, enter skin of humans. 4. Bryozoa, Phoronida, Brachopoda. **28.4:** 1. Roundworms and arthropods are the molting protostomes. They both have a true coelom. 2. Crustaceans breathe by gills and have swimmerets. Insects breathe by tracheae, and they may have wings. 3. The first pair of appendages is the chelicerae (modified fangs), and the second pair is the pedipalps (hold, taste, chew food). **28.5:** 1. The larval stage is bilaterally

symmetrical. 2. The water vascular system functions in locomotion, feeding, gas exchange, and sensory reception.

Testing Yourself

1. d; **2** e; **3.** e; **4.** e; **5.** d; **6.** b; **7.** a; **8.** c; **9.** a; **10. a.** tapeworm; **b.** mollusc; **c.** sponge; **d.** cnidaria; **e.** rotifer; **f.** flatworm; **g.** arthropod; **11.** b; **12.** b; **13.** a; **14.** e; **15.** e.

Chapter 29

Check Your Progress

29.1: 1. Humans are chordates, and they have the four chordate characteristics during the embryonic period of their life cycle. Notochord: replaced by vertebral column during development. Dorsal tubular nerve cord becomes the spinal cord. Pharyngeal pouches: the first pair of pouches develops into auditory tubes. Post-anal tail: is present in developing embryo, but lost during development. 2. A sea squirt larva has the four characteristics as a larva, then undergoes metamorphosis to become an adult, which has gill slits but none of the other characteristics. **29.2:** 1. Endoskeleton protects internal organs, provides a place of attachment for muscles, permits rapid, efficient movement. 2. Four legs are useful for locomotion on land where the body is not supported by water. **29.3:** 1. All fishes are aquatic vertebrates and ectothermic. They all live in water, breathe by gills, and have a single circulatory loop (Fig. 29.11*a*, page 556). 2. Ray-finned bony fishes have fan-shaped fins supported by thin, bony rays. Lobe-finned bony fishes have fleshy fins supported by bones. **29.4:** 1. a. Paired limbs; smooth, nonscaly skin that stays moist; lungs; a three-chambered heart with a double-loop circulatory pathway; sense organs adapted for a land environment; ectothermic; and have aquatic reproduction. b. Lobe-finned fishes and amphibians both have lungs and internal nares that allow them to breathe air. The same bones are present in the front fins of the lobe-finned fishes as in the forelimbs of early amphibians. 2. Usually, amphibians carry out external fertilization in the water. The embryos develop in the eggs until the tadpoles emerge. They then undergo metamorphosis, growing legs and reabsorbing the tail, and become adults. **29.5:** 1. Alligators live in fresh water and have a thick skin, two pairs of legs, powerful jaws, and a long muscular tail that allow them to capture and eat other animals that are in the water or come to the water's edge. Snakes have no limbs and have relatively thin skin. They live close to or in the ground and can escape detection. They use smell (Jacobson's organ) and vibrations to detect prey. Some use venom to subdue prey, which they eat whole because their jaws are distensible. 2. Birds are reptiles: feathers are modified scales; birds have clawed feet and a tail that contains vertebrae. **29.6:** 1. Mammals have hair or fur and mammary glands, endothermy, limbs under body, differentiated teeth, and an enlarged brain. 2. The three groups of mammals are monotremes (have a cloaca and lay eggs), marsupials (young are born immature and finish development in a pouch), and placental (eutherians) mammals (development occurs internally and the fetus is nourished by placenta).

Testing Yourself

1. e; **2.** a; **3.** c; **4.** e; **5.** a; **6.** e; **7.** b; **8.** b; **9.** e; **10.** c; **11.** a; **12.** a; **13. a.** pharyngeal pouches; **b.** dorsal tubular nerve cord; **c.** notochord; **d.** postanal tail; **14.** d; **15.** d.

Chapter 30

Check Your Progress

30.1: 1. Refer to Figure 30.4 on page 573, focusing on the bars on the right of the figure. 2. Both humans and chimpanzees are hominines and they share similar structure of their genomes, plus the general characteristics of all primates. **30.2:** 1. Ardipithecines lived primarily in trees, whereas the australopiths lived both in and out of trees. The brain size of the australopiths was larger, and this group was better adapted for bipedalism. **30.3:** 1. Increasing size of brain cavity, increased height, eventual use of tools and fire. 2. Bipedalism allowed for organisms to move young more easily, increased brain size allowed for higher intellect and thus adaptation to non-forest environments. **30.4:** 1. The replacement model suggests that humans evolved from one group in Africa, and then migrated to other locations. This means that different groups of Cro-Magnon humans could adapt to different locations, eventually forming the major human ethnic groups. 2. Increased brain size and reliance on tool use and a hunter-gatherer lifestyle required development of communication.

Testing Yourself

1. c; **2.** b; **3.** b; **4.** e; **5.** c; **6.** d; **7.** b; **8.** c; **9.** c; **10.** c; **11.** T; **12.** T; **13.** F; **14.** Africa; **15.** erect, small; **16.** Cro-Magnon; **17.** thousands; **18. a.** modern humans; **b.** archaic humans; **c.** *Homo erectus;* **d.** *Homo erectus.* See also Figure 30.10, page 580.

Chapter 31

Check Your Progress

31.1: 1. Squamous epithelium: flat cells that line the blood vessels and air sacs of lungs;

cuboidal epithelium: cube-shaped cells that line the kidney tubules and various glands; columnar epithelium: rectangular cells that line the digestive tract; stratified epithelium: layers of cells that protect various body surfaces; glandular epithelium: modified to secrete products, like the goblet cells of the digestive tract. 2. Fibrous connective tissue has collagen and elastic fibers in a jellylike matrix between fibroblasts; supportive connective tissue has protein fibers in a solid matrix between collagen or bone cells; fluid connective tissue lacks fibers and has a fluid matrix that contains blood cells or lymphatic cells. 3. Skeletal muscle, which is striated with multiple nuclei, causes bones to move when it contracts. Smooth muscle, which is spindle-shaped with a single nucleus, causes the walls of internal organs to constrict. Cardiac muscle, which has branching, striated cells each with a single nucleus, causes the heart to beat. 4. Dendrites conduct signals toward the cell body; the cell body contains most of the cytoplasm and the nucleus; it carries on the usual functions of the cell; the axon conducts nerve impulses. **31.2:** 1. An organ is two or more types of tissues working together to perform a function; an organ system is many organs working to perform a process. 2. The lymphatic and immune systems work together to protect the body from disease (other answers are possible). 3. The dorsal cavity contains the cranial cavity and the vertebral cavity; the ventral cavity contains the thoracic cavity and the abdominopelvic cavity. **31.3:** 1. The skin protects the deeper tissues from pathogens, trauma, and dehydration in all animals. Only birds have feathers, a derivative of skin that are involved in flying. 2. The epidermis is stratified squamous epithelium, and it protects and prevents water loss. The dermis is dense fibrous connective tissue, and it helps regulate body temperature and provides sensory reception. 3. A dark-skinned person in a low sunlight area might develop a deficiency of vitamin D, which is required for normal bone growth. 4. Nails are keratinized cells that grow from the nail root; hair is made of dead epithelial cells; sweat glands are tubules originating from the dermis; oil glands are usually associated with hair follicles. **31.4:** 1. Homeostasis, the dynamic equilibrium of the internal environment, maintains body conditions within a range appropriate for cells to continue living. 2. The circulatory system brings nutrients and removes waste from tissue fluid; the respiratory system carries out gas exchange; the urinary system excretes metabolic wastes and maintains salt-water balance and pH of blood. 3. When conditions go beyond or below a set point, a correction is made to bring conditions back to normality again.

Testing Yourself

1. b; **2.** e; **3.** e; **4.** e; **5.** b; **6.** e; **7.** e; **8.** c; **9. a.** columnar epithelium, lining of intestine (digestive tract), protection and absorption; **b.** cardiac muscle, wall of heart, pumps blood; **c.** compact bone, skeleton, support and protection. **10.** e; **11.** d; **12.** c; **13.** c; **14.** a, c, g; **15.** b, d, e; **16.** b, c, f; **17.** b; **18.** d; **19.** a; **20.** c.

Chapter 32

Check Your Progress

32.1: 1. Circulatory systems carry nutrients and oxygen to cells and to remove their wastes. 2. Blood is always contained within blood vessels, but hemolymph (a mixture of blood and tissue fluid) flows freely in body cavities. 3. With a closed circulatory system, oxygen must diffuse across the capillary wall. **32.2:** 1. Arteries carry blood away from the heart, capillaries exchange their contents with tissue fluid, and veins return blood back to the heart. 2. Veins contain valves, because they are thin-walled and there is insufficient blood pressure in veins to return blood to the heart, so valves plus skeletal muscle contractions are needed. 3. One-circuit pathways are less efficient, because blood supplying the tissues (after leaving the gills) is under low pressure. Two-circuit pathways supply oxygen to tissues more efficiently, which helps land animals meet the increased demands of locomotion on land vs. in water. **32.3:** 1. From the body: venae cavae, right atrium, tricuspid valve, right ventricle, pulmonary semilunar valve, pulmonary trunk and arteries (carrying oxygen-poor blood). From the lungs: pulmonary veins, left atrium, bicuspid valve, left ventricle, aortic semilunar valve, aorta. 2. First the atria contract, then the ventricles contract, and then they both rest. The *lub* sound occurs when the atrioventricular valves close, and the *dub* sound occurs when the semilunar valves close. 3. The wall of the left ventricle is thicker than the wall of the right ventricle, and it generates a greater pressure than the right ventricle. The right ventricle pumps blood into the pulmonary circuit, which takes blood only to the lungs for gas exchange, while the left ventricle pumps blood into the systemic circuit, which takes blood to all the cells of the body. 4. Thromboembolism, stroke, heart attack. **32.4:** 1. Blood contains plasma and formed elements. Plasma transports many substances to and from the capillaries, helps defend against pathogen invasion, helps regulate body temperature, and helps with clotting to prevent excessive blood loss. The formed elements include red blood cells that carry oxygen to tissues and return some carbon dioxide; white blood cells that help protect the body from infections; and platelets that are involved in blood clotting. 2. Platelets accumulate at the site of injury and release a clotting factor that results in the synthesis of thrombin. Thrombin synthesizes fibrin threads that provide a framework for the clot. 3. Any fetal Rh-positive blood cells that enter the circulation of an Rh-negative mother are recognized as foreign antigens, provoking her immune system. An Rh-positive woman does not react to the Rh antigen, because her immune system does not react with "self." 4. Capillary exchange is affected by osmotic pressure (tends to cause water to enter capillaries) and blood pressure (tends to cause water to leave capillaries).

Testing Yourself

1. b; **2.** a; **3.** d; **4.** c; **5.** b; **6.** b; **7.** e; **8.** c; **9.** e; **10.** b; **11.** e; **12. a.** superior vena cava; **b.** aorta; **c.** left pulmonary artery; **d.** pulmonary trunk; **e.** left pulmonary veins; **f.** right pulmonary artery; **g.** right pulmonary veins; **h.** semilunar valve; **i.** left atrium; **j.** right atrium; **k.** left atrioventricular (bicuspid) valve; **l.** right atrioventricular (tricuspid) valve; **m.** right ventricle; **n.** chordae tendinae; **o.** septum; **p.** left ventricle; **q.** superior vena cava (see Figure 32.7, page 611).

Chapter 33

Check Your Progress

33.1: 1. Under certain conditions, slime molds develop specialized sentinel cells that engulf pathogens and toxins. 2. PAMPs are highly conserved microbial molecules such as double-stranded viral RNA as well as carbohydrates and proteins found only in bacteria. 3. Immune cells expressing receptors for specific antigens can multiply and differentiate, resulting in immune memory. **33.2:** 1. The lymphatic system consists of the lymphatic vessels, which have the same structure as cardiovascular veins, and the lymphatic organs: red bone marrow, lymph nodes, and spleen. 2. The lymphatic system absorbs fats, returns excess tissue fluid to the bloodstream, produces lymphocytes, and helps defend the body against pathogens. 3. Red bone marrow is a spongy, semisolid red tissue located in certain bones (e.g., ribs, clavicle, vertebral column, heads of femur, and humerus), which produces all the blood cells of the body. The thymus is a soft, bilobed gland located in the thoracic cavity between the trachea and the sternum where T lymphocytes mature. Lymph nodes are small ovoid structures located along lymphatic vessels through which the lymph travels. The spleen is an oval organ with a dull purplish

color that has both immune as well as red blood cell maintenance functions. **33.3:** 1. Physical barriers include the skin, mucus membranes, and ciliated epithelia; chemical barriers include lysozyme, stomach acid, and protective proteins (e.g., complement). 2. Four cardinal signs of inflammation are heat, swelling, redness, and pain. Inflammation generally directs other components of the immune system (molecules and cells) to the inflamed area. 3. Mast cells initiate inflammation; phagocytes (dendritic cells, macrophages, neutrophils) devour pathogens; natural killer cells kill virus-infected cells and cancer cells by cell-to-cell contact. 4. Complement proteins assist other immune defenses by initiating inflammation, enhancing phagocytosis of pathogens, and forming a membrane attack complex. **33.4:** 1. B cells produce antibodies; cytotoxic T cells attack viral-infected or cancer cells, and helper T cells produce cytokines that stimulate the immune response. 2. Antigen complexes can activate the complement system and natural killer cells, or be engulfed by phagocytes. 3. Active immunity is induced in an individual by natural infection, vaccination, or exposure to a toxin; passive immunity is produced by one individual and transferred into someone else, such as maternal antibodies, administration of antibodies to treat diseases, and bone marrow transplants. **33.5:** 1. Most primary immunodeficiencies are due to a genetic defect in a component of the immune system. 2. Autoimmune diseases (e.g., rheumatoid arthritis) are generally treated with drugs that generally suppress the immune system. 3. The transplanting of animal tissues or organs into humans.

Testing Yourself

1. b; **2.** e; **3.** e; **4.** a; **5.** b; **6.** c; **7.** a; **8.** b; **9.** e; **10.** b; **11.** d; **12.** a; **13.** b; **14.** e; **15.** d.

Chapter 34

Check Your Progress

34.1: 1. Planarians have an incomplete digestive tract, called the gastrovascular cavity; the complete tract of earthworms has a pharynx, crop, gizzard, and intestine. 2. An incomplete tract has only one opening, with parts that cannot become very specialized, because they must serve multiple functions. 3. Carnivores tend to have pointed incisors and enlarged canine teeth to tear off pieces small enough to quickly swallow. The molars are jagged for efficient chewing of meat. Herbivores have reduced canines but sharp even incisors to clip grasses. The large flat molars grind and crush tough grasses. **34.2:** 1. Mouth, pharynx, esophagus, stomach,

duodenum, jejunum, ileum, large intestine. 2. Taste buds enable an animal to discern the difference between nutritious foods, and non-nutritious, potentially dangerous substances that should not be consumed. 3. The stomach mechanically churns food, while acid and pepsin begin protein digestion. The small intestine finishes the digestion of proteins, fats, carbohydrates, and nucleic acids. The villi and microvilli of the intestinal wall greatly enhance the surface area, thereby assisting absorption of the final products of digestion. The large intestine absorbs excess water, and its diameter is larger to permit storage of undigested food prior to defecation. 4. Bile from the liver (stored in the gallbladder) emulsifies fat, while pancreatic amylase, trypsin, and lipase digest carbohydrates, proteins, and lipids. The liver has many other functions, such as the storage of glucose as glycogen. **34.3:** 1. Starch digestion begins in the mouth where salivary amylase digests starch to maltose and pancreatic amylase continues this same process in the small intestine. Maltase and brush-border enzymes digest maltose to glucose, which enters a blood capillary. Protein digestion starts in the stomach where pepsin digests protein to peptides and continues in the small intestine in which trypsin carries out this same process. 2. Carbohydrates are digested to simple sugars (monosaccharides), proteins are digested to amino acids, and fats are digested to glycerol plus fatty acids. **34.4:** 1. Vegetables, if properly chosen, can supply limited calories but all necessary amino acids and vitamins. Much urea results when excess amino acids from proteins are metabolized. The loss of water needed to excrete urea can result in dehydration and loss of calcium ions. 2. A diet high in saturated fats tends to raise LDL cholesterol, which is associated with atherosclerosis. 3. A vitamin is an organic molecule that is required in the diet because it cannot be synthesized.

Testing Yourself

1. a; **2.** e; **3.** d; **4.** b; **5.** d; **6.** c; **7.** e; **8.** a; **9.** c; **10.** c; **11.** e; **12.** c; **13.** d; **14.** c; **15.** a; **16.** b; **17.** Test tube 1: no digestion—no enzyme and no HCl; Test tube 2: some digestion—no HCl; Test tube 3: no digestion—no enzyme; Test tube 4: digestion—both enzyme and HCl are present.

Chapter 35

Check Your Progress

35.1: 1. Air has a drying effect, and respiratory surfaces have to be moist. Hydras are aquatic, while earthworms live in moist earth, and salamanders have skin glands that provide this moisture. These animals also have a large surface area compared to their size, or they have many capillaries close to the skin to facilitate gas exchange. 2. When blood containing a low O_2 level flows in an opposite direction as the O_2-rich water passing over the gills, a higher percentage of oxygen is transferred than if both flowed in the same direction (the highest a concurrent mechanism could achieve would be 50% of the O_2 content in the water, because equilibration would occur). 3. Hemolymph distributes some O_2 in the hemocoel, but it is inefficient. Tracheae are air tubes that branch into ever-smaller tracheoles, which deliver oxygen to most cells. Tracheae open at spiracles, and some larger insects have air sacs that can expand and contract. Some aquatic insects have tracheal gills, expansions of the body wall to provide more oxygen-absorbing surface area. **35.2:** 1. During inspiration, the rib cage moves up and out, and the diaphragm contracts and moves down. As the thoracic cavity expands, air flows into the lungs due to decreased air pressure in the lungs. During expiration, the rib cage moves down and the diaphragm relaxes and moves up to its former position. Air flows out as a result of increased pressure in the lungs. 2. The carotid bodies and aortic bodies contain chemoreceptors that send stimulatory messages to the respiratory center if the pH or O_2 levels are too low. 3. In the lungs, oxygen entering pulmonary capillaries combines with hemoglobin (Hb) in red blood cells to form oxyhemoglobin (HbO$_2$). In the tissues, Hb gives up O_2, while CO_2 enters the blood and the red blood cells. Some CO_2 combines with Hb to form carbaminohemoglobin (HbCO$_2$). Most CO_2 combines with water to form carbonic acid, which dissociates into H^+ and HCO_3^-. The H^+ is absorbed by the globin portions of hemoglobin to form reduced hemoglobin HbH$^+$. This helps stabilize the pH of the blood. The HCO_3^- is carried in the plasma. **35.3:** 1. Penicillin is an antibiotic that kills bacteria, but most colds are caused by viruses. 2. Narrowing of the airways is seen in bronchitis and asthma; reduced lung expansion is seen in pulmonary fibrosis and emphysema. 3. Smoking causes or contributes to acute and chronic bronchitis, asthma, pulmonary fibrosis, emphysema, and lung cancer (along with many cardiovascular disorders).

Testing Yourself

1. a. external respiration; **b.** CO_2; **c.** CO_2; **d.** tissue cells; **e.** internal respiration; **f.** O_2; **g.** O_2; **2.** b; **3.** d; **4.** c; **5.** b; **6.** c; **7.** e; **8.** b; **9.** d; **10.** b; **11.** d; **12.** e; **13. a.** nasal cavity; **b.** nostril; **c.** pharynx; **d.** epiglottis; **e.** glottis;

f. larynx; **g.** trachea; **h.** bronchus; **i.** bronchiole (see also Figure 35.6*a*, page 668).

Chapter 36

Check Your Progress

36.1: 1. Osmoregulation involves the balance between salts and water; metabolic wastes are removed by excretion. 2. Urea is not as toxic as ammonia, and it does not require as much water to excrete; uric acid takes more energy to prepare than urea. 3. Kangaroo rats are active at night, have convoluted nasal passages to capture moisture in exhaled air, fur to prevent water loss from skin, secrete a hypertonic urine, and eliminate very dry feces. **36.2:** 1. The kidneys alone secrete nitrogenous wastes, but share responsibility for regulating blood pressure and water-salt balance. Certain hormones secreted by the kidneys, such as erythropoietin and ADH, have unique functions. 2. In response to low blood pressure, the kidneys secrete renin, which converts angiotensinogen to angiotensin I, which is converted to angiotensin II. Angiotensin II causes blood vessel constriction and stimulates the adrenal glands to release aldosterone, which acts on kidney tubules to increase salt reabsorption. 3. The kidneys maintain blood pH by secreting H^+ ions and reabsorbing bicarbonate ions, as needed.

Testing Yourself

1. d; **2.** a; **3.** c; **4.** b; **5.** e; **6.** b; **7.** d; **8.** c; **9.** c; **10.** d; **11.** c; **12. a.** glomerular capsule; **b.** proximal convoluted tubule; **c.** loop of the nephron; **d.** descending limb; **e.** ascending limb; **f.** distal convoluted tubule; **g.** collecting duct; **h.** renal artery; **i.** afferent arteriole; **j.** glomerulus; **k.** efferent arteriole; **l.** peritubular capillary network; **m.** renal vein (see Figure 36.8, page 685).

Chapter 37

Check Your Progress

37.1: 1. A nerve net is a simple nervous system consisting of interconnected neurons, with no CNS. A ganglion is a cluster of neuron (nerve cell) bodies. In animals with a CNS and a PNS, it is a cluster of neurons located outside the CNS. A centrally located brain controls the ganglia and associated nerves. 2. The hindbrain controls essential functions like breathing, heart functions, and basic motor activity; the midbrain is a relay station connecting the hindbrain with the forebrain; the forebrain, which receives sensory input, includes the hypothalamus (involved with homeostasis) as well as the

cerebrum (involved with higher functions). 3. The more recently evolved parts of the brain are the outer portions, such as the cerebral cortex, and (in mammals) the neocortex. **37.2:** 1. Nerve impulses travel more quickly down myelinated axons due to saltatory conduction ("jumping"). 2. Na^+ moves from the outside of the axon membrane to the inside; K^+ moves from the inside of the axon membrane to the outside. 3. Inhibition of AChE, the enzyme that normally breaks down acetylcholine (ACh), would result in increased activity of nerves that use ACh as a neurotransmitter. **37.3:** 1. Sensory information from internal organs travels via spinal nerves, through the dorsal root ganglia to synapses on spinal cord interneurons whose axons travel in tracts to the brain. These then send motor impulses through tracts, to the ventral root ganglia, and then to the effector organ (smooth muscle in the intestine). 2. The four major lobes of the human brain are the frontal, parietal, temporal, and occipital. 3. Parkinson disease, with symptoms such as tremors, difficult speech, and trouble walking or standing, is associated with a loss of dopamine-producing cells in the basal nuclei of the forebrain (though the inciting cause is unknown); multiple sclerosis is an autoimmune disease that damages the myelin sheaths of neurons in the CNS, resulting in fatigue and in problems with visual and muscular function. **37.4:** 1. A reflex arc can travel from the sensory receptor, synapse on interneurons in the spinal cord, then travel back to the effector (muscle) without being perceived by the brain first. 2. Eating a big lunch mainly stimulates the parasympathetic branch of the autonomic nervous system, diverting blood supply to the digestive tract and away from the muscles. Jogging will engage the sympathetic system, reducing blood flow to the stomach and inhibiting its function. 3. The parasympathetic ("rest and digest") division dominates as you enjoy your meal, but your friend's "surprise" causes a sudden increase in sympathetic ("fight or flight") activity.

Testing Yourself

1. b; **2.** c; **3.** a; **4.** a; **5.** c; **6.** d; **7.** b; **8.** b; **9.** c; **10.** d; **11.** c; **12.** c; **13.** d; **14.** b; **15.** c; **16.** b; **17. a.** central canal; **b.** gray matter; **c.** white matter; **d.** dorsal root; **e.** cell body of sensory neuron in dorsal root ganglion; **f.** spinal nerve; **g.** cell body of motor neuron; **h.** interneuron (see Figure 37.13, page 709).

Chapter 38

Check Your Progress

38.1: 1. Sensory transduction is the conversion of some type of environmental stimulus into a nerve impulse. 2. Some snakes can perceive infrared radiation; bats, dolphins, and whales perceive very high or low frequency sound waves (echolocation); a dog's sense of smell is far more sensitive than a human's. **38.2:** 1. Both are chemical senses that use chemoreceptors to detect molecules in the environment. 2. Sweet, sour, salty, bitter, and umami. 3. Neurons transmit signals from different sensory organs to different areas of the brain, where they are interpreted as different types of information. **38.3:** 1. Rods are for peripheral vision and motion detection; they are well-suited for dim light. Cones are for color perception and fine detail, and are best-suited for bright light. Many rods may excite a single ganglion cell, but much smaller numbers of cones excite individual ganglion cells. 2. Sclera, choroid, retina. Light must pass through the ganglion and bipolar cell layers before reaching the photoreceptor cells. 3. An eyeball that is too long results in nearsightedness; an eyeball that is too short results in farsightedness; an uneven cornea results in astigmatism. **38.4:** 1. a. middle; b. outer; c. inner; d. inner; e. inner; f. outer. 2. Auditory canal, tympanic membrane (eardrum), ossicles (malleus, incus, and stapes), oval window, cochlea. 3. The utricle and saccule are responsible for gravitational equilibrium; the semicircular canals for rotational equilibrium. **38.5:** 1. A person lacking muscle spindles would have trouble walking, sitting, or doing other activities due to a lack of muscle tone. A person lacking nociceptors would be prone to injury due to an absence of warning signs associated with pain. 2. Pain is generally associated with potential harm: i.e., something to be avoided. An animal that quickly adapted to pain would have an increased chance of being injured or killed by a potentially dangerous stimulus.

Testing Yourself

1. b; **2.** c; **3.** c; **4.** e; **5.** d; **6.** c; **7.** c; **8.** b; **9.** d; **10.** a; **11.** b; **12.** c; **13.** a; **14.** d; **15.** b; **16.** d; **17. a.** retina—contains sensory receptors; **b.** choroid—absorbs stray light; **c.** sclera—protects and supports eyeball; **d.** optic nerve—transmits impulses to brain; **e.** fovea centralis—makes acute vision possible; **f.** muscle in ciliary body—holds lens in place, accommodation; **g.** lens—refracts and focuses light rays; **h.** iris—regulates light entrance; **i.** pupil—admits light; **j.** cornea—refracts light rays (see Figure 38.5, page 721).

Chapter 39

Check Your Progress

39.1: 1. Exoskeleton, endoskeleton, exoskeleton, exoskeleton, endoskeleton.

2. The tongue is a muscular hydrostat. 3. Because the muscle layers surrounding the coelom no longer contract, the hydrostatic skeleton cannot provide support for the body. **39.2:** 1. Osteoblasts build bone and osteoclasts break it down. Osteocytes occupy lacunae. 2. Compact bone, which serves mainly to support the body, contains many osteons, in which central canals are surrounded by a hard matrix with lacunae. Spongy bone is lighter, with numerous bars and plates as well as spaces filled with red bone marrow, which produces the blood cells. 3. Axial, axial, appendicular, appendicular, axial, appendicular, axial, appendicular, axial. **39.3:** 1. A pair of muscles that work opposite to one another; for example, if one muscle flexes (bends) the joint the other extends (straightens) it. 2. Myofibrils are tubular contractile units that are divided into sarcomeres. Each sarcomere contains actin (thin filaments) and myosin (thick filaments). 3. Cleavage of ATP allows myosin heads to bind to actin filaments, pulling them toward the center of the sarcomere.

Testing Yourself

1. b; **2.** f; **3.** c; **4.** e; **5.** e; **6.** e; **7.** b; **8.** b; **9.** c; **10.** b; **11.** e; **12.** e; **13.** c; **14.** b; **15. a.** T tubule; **b.** sarcoplasmic reticulum; **c.** myofibril; **d.** Z line; **e.** sarcomere; **f.** sarcolemma of muscle fiber (see Figure 39.13, page 745).

Chapter 40

Check Your Progress

40.1: 1. The nervous system tends to respond rapidly to both external and internal stimuli, while the endocrine system responds more slowly, but often has longer-lasting effects. Both systems use chemicals to communicate with other body systems; the nervous system at synapses, the endocrine system via hormones secreted mainly into the bloodstream. 2. Peptide hormones have receptors in the plasma membrane. Steroid hormones have receptors that are generally in the nucleus, sometimes in the cytoplasm. 3. Since peptide hormones usually bind to receptors on the outside of the cell; they must communicate with the inside of the cell via second messengers. **40.2:** 1. The hypothalamus communicates with the endocrine system via the pituitary gland: two hormones produced by the hypothalamus are stored in the posterior pituitary; several others are produced by the anterior pituitary in response to hypothalamic releasing factors that reach the anterior pituitary via a portal system. 2. ADH conserves body water by causing reabsorption of water by the kidneys; oxytocin causes uterine contractions during

labor and milk letdown during nursing. 3. Thyroid stimulating hormone (TSH) stimulates release of thyroid hormones; adrenocorticotropic hormone (ACTH) stimulates the adrenal glands to produce glucocorticoids, prolactin (PRL) causes breast development and milk production; growth hormone (GH) promotes bone and muscle growth; the gonadotropic hormones FSH and LH stimulate the testes or ovaries to produce gametes and sex hormones; and melanocyte-stimulating hormone (MSH) causes skin color changes in some animals. **40.3:** 1. Angiotensin II causes arterioles to constrict; aldosterone causes reabsorption of Na$^+$, accompanied by water, in the kidneys. 2. Adrenal cortex, pineal gland, adrenal medulla, kidneys, adipose tissue, pancreas, heart, adrenal cortex, thyroid gland. 3. PTH stimulates osteoclasts and calcitonin inhibits them.

Testing Yourself

1. f; **2.** b; **3.** c; **4.** a; **5.** e; **6.** d; **7.** b; **8.** d; **9.** a; **10.** a; **11.** e; **12.** b; **13.** d; **14.** b; **15.** a; **16.** b; **17.** c; **18.** e; **19.** d; **20.** c.

Chapter 41

Check Your Progress

41.1: 1. Asexual reproduction allows organisms to reproduce rapidly and colonize favorable environments quickly. Sexual reproduction produces offspring with a new combination of genes that may be more adaptive to a changed environment. 2. An oviparous animal lays eggs that hatch outside the body. A viviparous animal gives birth after the offspring have developed within the mother's body. Ovoviviparous animals retain fertilized eggs within a parent's body until they hatch; the parent then gives birth to the young. 3. A shelled egg contains extraembryonic membranes which keep the embryo moist, carry out gas exchange, collect wastes, and provide yolk as food. **41.2:** 1. Seminiferous tubule, epididymis, vas deferens, ejaculatory duct, urethra. 2. Seminal vesicles, prostate gland, and bulbourethral glands. 3. In males, FSH stimulates spermatogenesis and LH stimulates testosterone production. **41.3:** 1. Ovary, oviduct, uterus, cervix, vagina. 2. During the follicular phase, FSH stimulates ovarian follicles to produce primarily estrogen. A surge of LH (and FSH) also triggers ovulation. During the luteal phase, LH stimulates the corpus luteum to produce primarily progesterone. 3. Estrogen secreted by the developing follicle inhibits FSH secretion by the anterior pituitary, ending the follicular phase. Progesterone secreted by the developing corpus luteum inhibits LH secretion by the anterior pituitary, ending the luteal phase. Rising estrogen levels cause the endometrium to thicken (proliferative phase), and progesterone causes uterine glands to mature (secretory phase). If no pregnancy occurs, low levels of estrogen and progesterone initiate menstruation. **41.4:** 1. Male and female condoms and the diaphragm prevent sperm from coming in contact with the egg. 2. Abstinence (100%) and vasectomy (nearly 100%) are the most effective, the birth control pill and condoms are the next most effective, and natural family planning is least effective. 3. In AID, sperm are placed in the vagina or sometimes the uterus. In IVF, conception takes place in laboratory glassware and embryos are transferred to the woman's uterus. In GIFT, eggs and sperm are brought together in laboratory glassware, and placed in the oviducts immediately afterward. In ICSI, one sperm is injected directly into an egg. **41.5:** 1. Antiretroviral drug categories include entry inhibitors (viral attachment to host receptor), reverse transcriptase inhibitors (production of DNA from viral RNA), integrase inhibitors (insertion of viral DNA into host DNA), and protease inhibitors (processing of viral proteins). 2. HPV-induced genital warts can be painful and disfiguring, and some strains cause cancer of the cervix, vagina, vulva, penis, and anus. 3. Chlamydia and gonorrhea are associated with pelvic inflammatory disease and infertility.

Testing Yourself

1. b; **2.** e; **3.** c; **4.** e; **5.** c; **6.** c; **7.** c; **8.** c; **9.** a; **10.** a; **11.** c; **12.** c; **13.** b; **14.** e; **15.** c; **16.** a; **17.** c; **18.** b; **19.** c.

Chapter 42

Check Your Progress

42.1: 1. The fast block is the depolarization of the egg's plasma membrane that occurs upon initial contact with a sperm. The slow block occurs when the secretion of cortical granules converts the zona pellucida into the fertilization membrane. 2. Mesoderm, endoderm, ectoderm, ectoderm, mesoderm, mesoderm, mesoderm, endoderm. 3. Neurula stage. **42.2:** 1. Cytoplasmic segregation is the parceling out of maternal determinants as mitosis occurs. Induction is the influence of one embryonic tissue on the development of another. 2. A morphogen is a transcription factor that is distributed along a concentration gradient in the embryo and helps direct morphogenesis. 3. The homeobox encodes the homeodomain region of the protein product of the gene. The homeodomain is the DNA-binding region of the protein, which is a transcription factor. **42.3:** 1. oviduct.

2. Allantois, yolk sac, and chorion. 3. Structures that begin to develop between the third and fifth weeks include the nervous system, heart, chorionic villi, umbilical cord, and limb buds. **42.3:** 1. The thymus is the site where T cells finish their development; T cells are needed to stimulate other types of immune cells, including B cells that produce antibodies. 2. Menopause is the time when ovarian and uterine cycles cease. 3. Preprogrammed theories suggest aging is partly genetically programmed; single gene mutations influence aging in *C. elegans*. Damage accumulation theories suggest aging is due to an accumulation of cellular damage: e.g., DNA mutations, cross-linking of proteins, or oxidation by free radicals.

Testing Yourself

1. b; **2.** b; **3.** a; **4.** e; **5.** b; **6.** e; **7.** e; **8.** d; **9. a.** chorion: contributes to forming placenta where wastes are exchanged for nutrients and oxygen; **b.** amnion: protects and prevents desiccation; **c.** embryo; **d.** allantois: blood vessels become umbilical blood vessels; **e.** yolk sac: first site of blood cell formation; **f.** chorionic villi: embryonic portion of placenta; **g.** maternal portion of placenta; **h.** umbilical cord: connects developing embryo to the placenta (see also Figure 42.11, page 805); **10.** c; **11.** a; **12.** d; **13.** b; **14.** e; **15.** b; **16.** c; **17.** b; **18.** a; **19.** c; **20.** e.

Chapter 43

Check Your Progress

43.1: 1. Fisher lovebird (carry nesting material in beak) mated to Peach-faced lovebird (carry nesting material in rump feathers) result in offspring with intermediate behavior. Offspring of inland garter snakes (do not eat slugs) and coastal garter snakes (eat slugs) show an intermediate liking for slugs. 2. Gene for egg-laying hormone in *Aplysia* was isolated and its protein product controls egg-laying behavior. The gene *fosB* has been found to control maternal behavior in mice. **43.2:** 1. Associative learning. 2. Just hatched, laughing gull chicks instinctively peck at parents' bill to be fed but their accuracy improves after a few days. 3. Chimpanzees pile up boxes to reach food and ravens use their beaks and feet to bring up food attached to a string. **43.3:** 1. Pheromones are used to mark a territory so other individuals of that species will stay away; honeybees do a waggle dance to guide other bees to a food source; vervet monkeys have calls that make other vervets run away. 2. Chemical (effective all the time, not as fast as auditory); auditory (can be modified but the recipient has to be present when message is sent); visual (need not be

accompanied by chemical or auditory, needs light in order to receive); tactile (permits bonding; recipient must be close). 3. chemical: taste buds and olfactory receptors; auditory: ears; visual: eyes; tactile: touch receptors in skin. **43.4:** 1. One benefit of territoriality is to ensure a source of food. 2. Both an animal's reproductive strategy and sexual selection favors features that increase an animal's chance of leaving offspring. 3. Altruistic behavior is supposed to be selfless but when, for example, an offspring helps its parents raise siblings, the helper may be increasing some of its own genes in the next generation.

Testing Yourself

1. c; **2.** a; **3.** d; **4.** b; **5.** c; **6.** d; **7.** e; **8.** c; **9.** d; **10.** c; **11.** c; **12.** a; **13.** c; **14.** d; **15.** a; **16.** c; **17.** b; **18.** a; **19.** c; **20.** b.

Chapter 44

Check Your Progress

44.1: 1. A population is all the members of one species that inhabit a particular area and a community is all the populations that interact within that area. 2. To develop models that explain and predict the distribution and abundance of organisms. 3. Abiotic means the nonliving aspects of an environment such as rainfall and temperature. **44.2:** 1. Population density is the number of individuals per unit area and population distribution is the pattern of dispersal of individuals across an area of interest. 2. In a type I survivorship curve, most individuals survive well past the midpoint of the life span and death does not come until near the end of the life span. In type II, survivorship decreases at a constant rate throughout the life span. In type III, most individuals die young. 3. In a bell-shaped age pyramid, the pre-reproductive members represent the largest portion of the population. **44.3:** 1. An environment in which the weather, food supply, etc., remains stable favors iteroparity. 2. Exponential growth ceases when the environment cannot support a larger population size; that is, the point at which the size of the population has reached the environment's carrying capacity. **44.4:** 1. As population density increases, competition and predation become more intense. 2. If a flash flood occurs, mice that can stay afloat will survive and reproduce whereas those that quickly sink will not survive and will not reproduce. In this way the ability to stay afloat will be more prevalent in the next generation. **44.5:** 1. A *K*-strategist species: allocate energy to their own growth and survival and to the growth and survival of their limited number of offspring. *r*-strategist species: allocate energy to producing a large number of offspring and little or no energy goes into parental care. 2. Populations may vary between *K* and *r* strategies based upon environmental conditions. **44.6:** 1. Less-developed countries have a high population growth while more-developed countries have a low rate of population growth. 2. The parents don't immediately die after they have had their child which causes the population to increase. 3. Since resources are in limited supply an increase in consumption by the LDC will cause more competition for resources.

Testing Yourself

1. d; **2.** c; **3.** e; **4.** e; **5.** b; **6.** e; **7.** c; **8.** e; **9.** c; **10.** e; **11.** c; **12.** e; **13.** b; **14.** e; **15.** e.

Chapter 45

Check Your Progress

45.1: 1. An organism's habitat is the place where it lives and reproduces. The niche is the role it plays in its community such as whether it is a producer or consumer. 2. The two factors are (1) the predator causes the prey population to decline which leads to a decline in the predator population; later when the prey population recovers so does the predator population; (2) lack of food causes the prey population to decline followed by the predator population; later when food is available to the prey population they both recover. 3. Acacias feed the ants that protect them from herbivores; Clark's nutcrackers feed on the seeds of whitebark pine trees but also disperse the seeds; pollinators take nectar from flowers and carry their pollen to other flowers of the same species. **45.2:** 1. Environmental disturbances, fire, flood, etc., lead to succession. 2. The facilitation model predicts that a community will grow toward becoming a climax community. The inhibition model predicts that colonists will inhibit the growth of a community. **45.3:** 1. A producer of food (photosynthesizer) is at the base of an ecological pyramid. 2. Energy passes from one population to the next, and at each step more is converted to heat until all of the original input has become heat. Therefore energy flows though an ecosystem.

Chemicals pass from one population to the next and then recycle back to the producer populations again. 3. Return of CO_2 to the atmosphere because humans burn fossil fuels and destroy forests that take up CO_2.

Testing Yourself

1. c; **2.** e; **3.** e; **4.** a; **5.** c; **6.** b; **7.** d; **8.** b; **9.** c; **10.** d; **11.** e; **12. a.** producers; **b.** consumers; **c.** inorganic nutrient pool; **d.** decomposers; **13.** a; **14.** e; **15.** c; **16.** d; **17.** c; **18.** b; **19.** c; **20.** d.

Chapter 46

Check Your Progress

46.1: 1. Because the Earth is a sphere, the Sun's rays hit the equator straight on but are angled to reach the poles. 2. The windward side of the mountain receives more rainfall than the other side. Winds blowing over bodies of water collect moisture that they lose when they reach land. **46.2:** 1. A tropical rain forest has a canopy (tops of great variety of tall evergreen hardwood trees) with buttressed trunks at ground level. Long lianas (hanging vines) climb into the canopy. Epiphytes grow on the trees. The understory consists of smaller plants and the forest floor is very sparse. A temperate deciduous forest contains trees (oak, beech, sycamore, and maple) that lose their leaves in the fall. Enough light penetrates the canopy to allow a layer of understory trees. Shrubs, mosses, and ferns grow at ground level. 2. The savanna is an expansive grassland that has a moderate climate. Therefore, the grasses keep producing throughout the year and provide plentiful food for a great variety of and number of herbivores that provide food for carnivores. The tundra is cold much of the year and has a limited growing season; therefore, its productivity is low and it supports only small populations of a few types of herbivores. **46.3:** 1. The cooling and warming of the epilimnion causes the layers of water to change position. 2. Most of the open ocean is the pelagic zone (open waters) divided into the epipelagic zone (contains phytoplankton, zooplankton, many types of fishes and also dolphins and whales); the mesopelagic zone (contains only carnivores adapted to the

absence of light); and the bathypelagic zone (in complete darkness that contains strange-looking fishes and invertebrates). Few vertebrates but many invertebrates (echinoderms, tube worms) exist on the abyssal plain and feed on debris that floats down from above.

Testing Yourself

1. b; **2.** d; **3.** c; **4.** d; **5.** a; **6.** b; **7.** e; **8.** e; **9.** a; **10.** a; **11.** b; **12.** d; **13.** b; **14.** a; **15.** c; **16.** b; **17.** a; **18.** d; **19.** c.

Chapter 47

Check Your Progress

47.1: 1. Many disciplines provide data that support conservation biology. 2. Biodiversity includes the number of species on Earth; genetic diversity (variations in a species); ecosystem diversity (interactions of species); and landscape diversity (interactions of ecosystems). A hotspot is a specific region with a large amount of biodiversity. **47.2:** 1. Direct value is a service that is immediately recognizable, while indirect values may not be as noticeable. 2. Direct values include medicine, food, or commercial products. Indirect values include assisting biogeochemical cycles, waste disposal, providing fresh water, preventing soil erosion, regulating climate, or providing a place for recreation. **47.3:** 1. Habitat loss, exotic species, pollution, overexploitation, disease. 2. Exotic plants displace native plants; predators introduced to kill pests also kill native animals; escaped animals may compete with, prey on, hybridize with, or introduce diseases into native populations. **47.4:** 1. A landscape involves more than one ecosystem and sometimes keystone species move between ecosystems. 2. Begin as soon as possible, mimic natural processes, and strive for sustainable development while providing services to humans.

Testing Yourself

1. e; **2.** b; **3.** e; **4.** e; **5.** e; **6.** b; **7.** e; **8.** e; **9.** c; **10.** c; **11.** e; **12.** e; **13.** e; **14.** b, c; **15. a.** habitat loss; **b.** introduction of exotic species; **c.** pollution; **d.** overexploitation; **e.** disease.

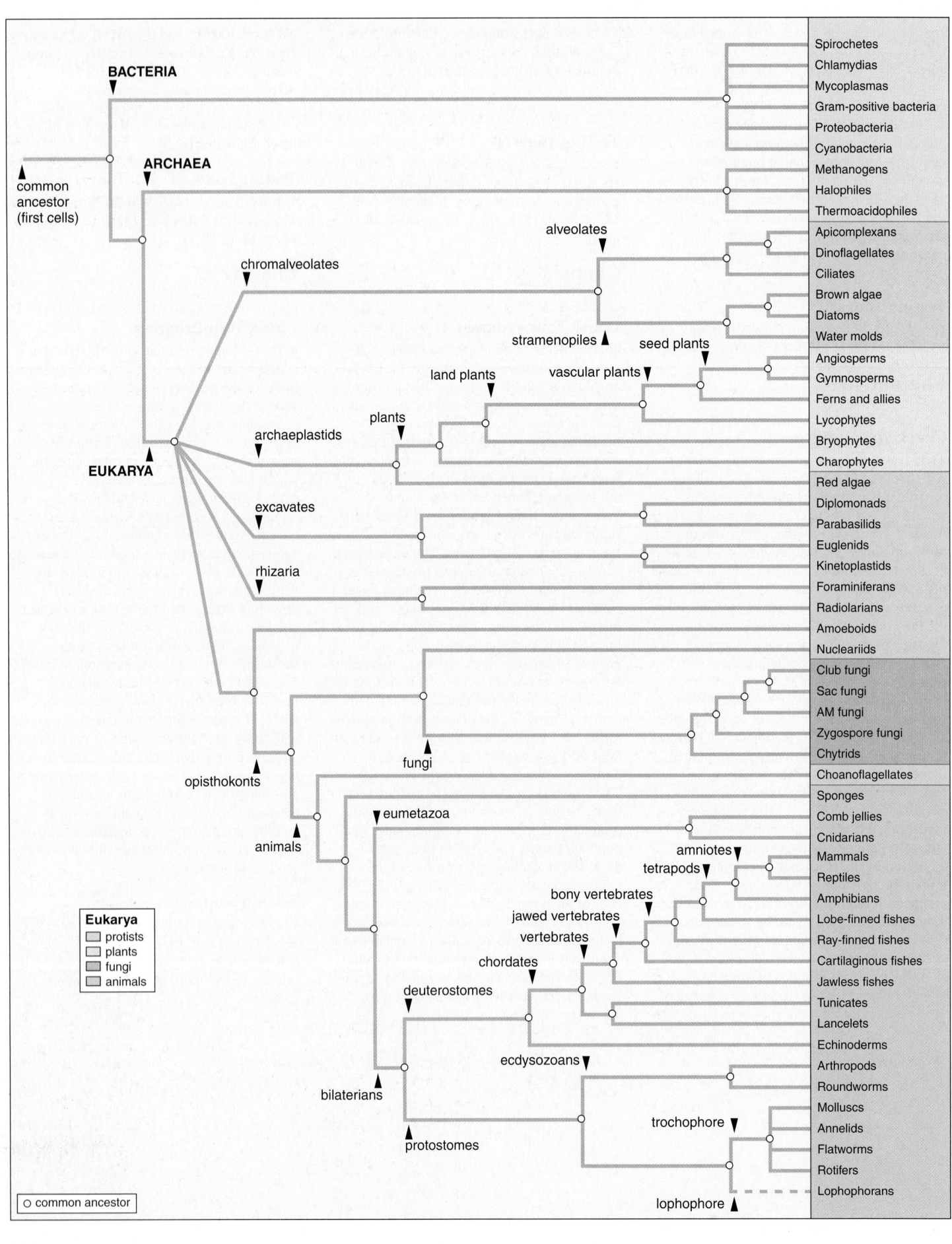

Tree of Life

The tree of life depicted in this appendix is based on the phylogenetic (evolutionary) trees presented in the text. Figure 1.5 showed how the three domains of life—Bacteria, Archaea, and Eukarya—are related. This relationship is also apparent in the tree of life, which combines the individual trees given in the text for eukaryotic protists, plants, fungi, and animals. In combining these trees, we show how all organisms may be related to one another through the evolutionary process.

Prokaryotes

Domains Bacteria and Archaea (Chapter 20) constitute the prokaryotic organisms that are characterized by their simple structure but a complex metabolism. The chromosome of a prokaryote is not bounded by a nuclear envelope, and therefore, these organisms do not have a nucleus. Prokaryotes carry out all the metabolic processes performed by eukaryotes and many others besides. However, they do not have organelles, except for multiple ribosomes.

Domain Bacteria

Bacteria are the most plentiful of all organisms, capable of living in most habitats, and they can carry out many different metabolic processes. Most bacteria are aerobic heterotrophs, but some are photosynthetic, and some are chemosynthetic. Motile forms move by flagella consisting of a single filament. Their cell wall contains peptidoglycan, and they have distinctive RNA sequences.

Domain Archaea

Archaean cell walls lack peptidoglycan, their lipids have a unique branched structure, and their ribosomal RNA sequences are distinctive. Examples are methanogens and the extremophiles.

Domain Eukarya

Eukarya have a complex cell structure with a nucleus and several types of organelles that compartmentalize the cell. Mitochondria that produce ATP and chloroplasts that produce carbohydrates are derived from prokaryotes that took up residence in a larger, nucleated cell. They may be unicellular (the majority of the protists), or multicellular (the plants, fungi, animals). Each multicellular group is characterized by a particular mode of nutrition. Flagella, if present, have a 9 + 2 organization.

Protists

The protists (Chapter 21) represent a group containing those eukaryotes that are not classified as a plant, fungus, or animal. The protists are members of six supergroups whose evolutionary relationships are being actively investigated. A supergroup is a major eukaryotic group, and six supergroups encompass all members of the domain Eukarya, including all protists, plants, fungi, and animals. Examples of protists include amoebas, the brown algae, and the paramecia.

Plants

Plants (Chapter 23) are multicellular photosynthetic eukaryotes that became adapted to living on land. This group includes aquatic green algae called charophytes, which have a haploid life cycle and share certain significant traits with the land plants.

The land plants exhibit an alternation-of-generations life cycle; protect a multicellular sporophyte embryo; produce gametes in gametangia; have apical meristem that produces differentiated tissues and have a waxy cuticle that prevents water loss. Examples include mosses, ferns, conifers, and flowering plants.

Fungi

Fungi (Chapter 22) have multicellular bodies composed of hyphae that absorb nutrients from the environment. They have a haploid life cycle and usually produce nonmotile spores during both asexual and sexual reproduction. Chytrids are aquatic fungi with flagellated spores and gametes. Examples include the mushrooms, cup fungi, and molds.

Animals

Animals (Chapters 28 and 29) are multicellular and have a diploid life cycle. Most have specialized tissues and ingest their food in a digestive cavity. Examples include invertebrate organisms such as sponges, earthworms, molluscs, insects, and vertebrates such as fishes, reptiles, and humans.

Glossary

A

abiogenesis Origin of life from nonliving matter, such as occurred on the early Earth. 329

abiotic synthesis Process of chemical evolution that resulted in the formation of organic molecules (amino acids, monosaccharides, etc.) from inorganic material. 329

abscisic acid (ABA) Plant hormone that causes stomata to close and initiates and maintains dormancy. 488

abscission Dropping of leaves, fruits, or flowers from a land plant. 488

absolute dating (of fossils) Determining the age of a fossil by direct measurement, usually involving radioisotope decay. 334

absorption spectrum For photosynthetic pigments, a graph of how much solar radiation is absorbed versus the wavelength of light. 124

accessory pigment Proteins that assist in the photosynthetic process by transferring energy from the photons to the central chlorophyll molecules. 389

acid Molecules tending to raise the hydrogen ion concentration in a solution and thus to lower its pH numerically. 32

acid deposition The return to Earth in rain or snow of sulfate or nitrate salts of acids produced by commercial and industrial activities. 878

acquired immunodeficiency syndrome (AIDS) Disease caused by the HIV virus that destroys helper T cells and macrophages of the immune system, thus preventing an immune response to pathogens; caused by sexual contact with an infected person, intravenous drug use, and transfusions of contaminated blood (rare). 787

actin One of two major proteins of muscle; makes up thin filaments in myofibrils of muscle fibers. *See also* myosin. 745

actin filament Component of the cytoskeleton; plays a role in the movement of the cell and its organelles; a protein filament in a sarcomere of a muscle, its movement shortens the sarcomere, yielding muscle contraction. 78

action potential Electrochemical changes that take place across the axon membrane; the nerve impulse. 699

active immunity Ability to produce antibodies due to the immune system's response to a microorganism or a vaccine. 637

active site Region of an enzyme where the substrate binds and where the chemical reaction occurs. 110

active transport Use of a plasma membrane carrier protein to move a molecule or ion from a region of lower concentration to one of higher concentration; it opposes equilibrium and requires energy. 95

adaptation Species's modification in structure, function, or behavior that makes a species more suitable to its environment. 278

adaptive immunity Type of immunity that is characterized by the action of lymphocytes to specific antigens. 627

adaptive radiation Rapid evolution of several species from a common ancestor into new ecological or geographical zones. 316

adenine (A) One of four nitrogen-containing bases in nucleotides composing the structure of DNA and RNA. Pairs with uracil (U) and thymine (T). 217

adenosine Portion of ATP and ADP that is composed of the base adenine and the sugar ribose. 54

adhesion junction Junction between cells in which the adjacent plasma membranes do not touch but are held together by intercellular filaments attached to buttonlike thickenings. 99

adipose tissue Connective tissue in which fat is stored. 590

ADP (adenosine diphosphate) Nucleotide with two phosphate groups that can accept another phosphate group and become ATP. 56, 107

adrenal cortex Outer portion of the adrenal gland; secretes mineralocorticoids, such as aldosterone, and glucocorticoids, such as cortisol. 762

adrenal gland Gland that lies atop a kidney; the *adrenal medulla* produces the hormones epinephrine and norepinephrine, and the *adrenal cortex* produces the glucocorticoid and mineralocorticoid hormones. 762

adrenal medulla Inner portion of the adrenal gland; secretes the hormones epinephrine and norepinephrine. 762

adrenocorticotropic hormone (ACTH) Hormone secreted by the anterior lobe of the pituitary gland that stimulates activity in the adrenal cortex. 757

aerobic A chemical process that requires air (oxygen); phase of cellular respiration that requires oxygen. 137

age structure diagram In demographics, a display of the age groups of a population; a growing population has a pyramid-shaped diagram. 842

agglutination Clumping of red blood cells due to a reaction between antigens on red blood cell plasma membranes and antibodies in the plasma. 619

agnathan Fishes that lack jaws; namely, the lampreys and hagfishes. 553

agranular leukocyte Form of white blood cell that lacks spherical vesicles (granules) in the cytoplasm. 620

aldosterone Hormone secreted by the adrenal cortex that regulates the sodium and potassium ion balance of the blood. 690, 763

alkaloid Bitter-tasting nitrogenous compounds that have a basic pH (e.g., caffeine). 497

allantois Extraembryonic membrane that accumulates nitrogenous wastes in the eggs of reptiles, including birds; contributes to the formation of umbilical blood vessels in mammals. 805

allele Alternative form of a gene; alleles occur at the same locus on homologous chromosomes. 196

allele frequency Relative proportion of each allele for a gene in the gene pool of a population. 290

allergy Immune response to substances that usually are not recognized as foreign. 641

allopatric speciation Model that proposes that new species arise due to an interruption of gene flow between populations that are separated geographically. 313

alloploidy Polyploid organism that contains the genomes of two or more different species. 315

allosteric site Site on an allosteric enzyme that binds an effector molecule; binding alters the activity of the enzyme. 112

alpine tundra Tundra near the peak of a mountain. 887

alternation of generations Life cycle, typical of land plants, in which a diploid sporophyte alternates with a haploid gametophyte. 176, 422

altruism Social interaction that has the potential to decrease the lifetime reproductive success of the member exhibiting the behavior. 833

alveolates A group of protists that includes unicellular dinoflagellates, apicomplexans, and ciliates; alveoli support plasma membrane. 392

alveolus (pl., alveoli) In humans, terminal, microscopic, grapelike air sac found in lungs. 668

Alzheimer disease Disease of the central nervous system (brain) that is characterized by an accumulation of beta amyloid protein and neurofibrillary tangles in the hippocampus and amygdala. 707

AM fungi (Glomeromycota) Fungi with branching invaginations (arbuscular mycorrhizae or AM) used to invade plant roots. 408

amino acid Organic molecule composed of an amino group and an acid group; covalently bonds to produce peptide molecules. 50

ammonia Nitrogenous end product that takes a limited amount of energy to produce but requires much water to excrete because it is toxic. 681

amnion Extraembryonic membrane of birds, reptiles, and mammals that forms an enclosing, fluid-filled sac. 805

amniote Vertebrate that produces an egg surrounded by four membranes, one of which is the amnion; amniote groups are the reptiles, (including birds), and mammals. 553

amniotic egg Egg that has an amnion, as seen during the development of reptiles, (including birds), and mammals. 558

amoeboid Cell that moves and engulfs debris with pseudopods. 396

amoebozoan Supergroup of eukaryotes that includes amoebas and slime molds and is characterized by lobe-shaped pseudopodia. 396

amphibian Member of vertebrate class Amphibia that includes frogs, toads, and salamanders; they are tied to a watery environment for reproduction. 556

amygdala Part of the limbic system of the brain; associated with emotional experiences. 707

amyotrophic lateral sclerosis (ALS) Neurodegenerative disease affecting the motor neurons in the brain and spinal cord resulting in paralysis and death; also known as Lou Gehrig's disease. 703

anabolic steroid Synthetic steroid that mimics the effect of testosterone. 767

anabolism Chemical reaction in which smaller molecules (monomers) are combined to form larger molecules (polymers); anabolic metabolism. 147

anaerobic A chemical reaction that occurs in the absence of oxygen; an example is the fermentation reactions. 137

analogous, analogous structure Structure that has a similar function in separate lineages but differs in anatomy and ancestry. 284, 317, 357

analogy Similarity of function but not of origin. 357

anaphase Fourth phase of mitosis; chromosomes move toward the poles of the spindle. 160

anaphylactic shock Severe systemic form of anaphylaxis involving bronchiolar constriction, impaired breathing, vasodilation, and a rapid drop in blood pressure with a threat of circulatory failure. 641

anapsid Characteristic of a vertebrate skull in which there is no opening in the skull behind the eye socket (orbit). 558

ancestral traits Traits that are found in a common ancestor and its descendants. 354

androgen Male sex hormone (e.g., testosterone). 767

aneuploidy Condition in which a cell does not contain the correct number, or combinations, of chromosomes. 183

angina pectoris Condition characterized by thoracic pain resulting from occluded coronary arteries; may precede a heart attack. 616

angiogenesis Formation of new blood vessels; rapid angiogenesis is a characteristic of cancer cells. 164

angiosperm Flowering land plant; the seeds are borne within a fruit. 430, 433

angiotensin II Hormone produced from angiotensinogen (a plasma protein) by the kidneys and lungs; raises blood pressure. 690

animals Multicellular, heterotrophic eukaryote that undergoes development to achieve its final form. In general, animals are mobile organisms, characterized by the presence of muscular and nervous tissue. 8

annelid The segmented worms, such as the earthworm and the clam worm. 536

annual ring Layer of wood (secondary xylem) usually produced during one growing season. 456

anterior pituitary Portion of the pituitary gland that is controlled by the hypothalamus and produces six types of hormones, some of which control other endocrine glands. 757

antheridium (pl., antheridia) Sperm-producing structures, as in the moss life cycle. 422

anthropoid Group of primates that includes monkeys, apes, and humans. 572

antibody Protein produced in response to the presence of an antigen; each antibody combines with a specific antigen. 618

antibody-mediated immunity Specific mechanism of defense in which plasma cells derived from B cells produce antibodies that combine with antigens. 634

anticodon Three-base sequence in a transfer RNA molecule base that pairs with a complementary codon in mRNA. 228

antidiuretic hormone (ADH) Hormone secreted by the posterior pituitary that increases the permeability of the collecting ducts in a kidney. 688, 757

antigen Foreign substance, usually a protein or a polysaccharide, that stimulates the immune system to react, such as to produce antibodies. 627

antigen-presenting cell (APC) Cell that displays an antigen to certain cells of the immune system so they can defend the body against that particular antigen. 635

antigen receptor Receptor proteins in the plasma membrane of immune system cells whose shape allows them to combine with a specific antigen. 633

anus Outlet of the digestive tube. 653

aorta In humans, the major systemic artery that takes blood from the heart to the tissues. 614

aortic body Sensory receptor in the aortic arch sensitive to the O_2, CO_2, and H^+ content of the blood. 671

apical Directional term that refers to the tip of a filament or branch of a plant. 421

apical dominance Influence of a terminal bud in suppressing the growth of axillary buds. 485

apical meristem In vascular land plants, masses of cells in the root and shoot that reproduce and elongate as primary growth occurs. 447

apicomplexan Parasitic protozoans, formerly called sporozoans that lack mobility and form spores; now named for a unique collection of organelles. 394

apoptosis Programmed cell death; involves a cascade of specific cellular events leading to death and destruction of the cell. 155, 800

appendicular skeleton Part of the vertebrate skeleton comprising the appendages, shoulder girdle, and hip girdle. 742

appendix (vermiform appendix) In humans, small, tubular appendage that extends outward from the cecum of the large intestine. 653

aquaporin Channel protein through which water can diffuse across a membrane. 91, 688

arboreal Living in trees. 571

archaeans (archaea) Prokaryotic organisms that are members of the domain Archaea. 65, 378

archaeplastid Supergroup of eukaryotes that includes land plants and red and green algae. Developed from endosymbiotic cyanobacteria. 387

archegonium Egg-producing structures, as in the moss life cycle. 422

Arctic tundra Biome that encircles the Earth just south of ice-covered polar seas in the Northern Hemisphere. 888

arteriole Vessel that takes blood from an artery to capillaries. 608

artery Blood vessel that transports blood away from the heart. 608

arthritis Condition characterized by an inflammation of the joints; two common forms are osteoarthritis and rheumatoid arthritis. 743

arthropod Invertebrates, with an exoskeleton and jointed appendages, such as crustaceans and insects. 539

artificial selection Intentional breeding of certain traits, or combinations of traits, over others to produce a desirable outcome. 278

ascus Fingerlike sac in which nuclear fusion, meiosis, and ascospore production occur during sexual reproduction of sac fungi. 409

asexual reproduction Reproduction that requires only one parent and does not involve gametes. 166, 513

associative learning Acquired ability to associate two stimuli or between a stimulus and a response. 824

assortative mating Mating of individuals with similar phenotypes. 295

aster Short, radiating fibers produced by the centrosomes in animal cells. 159

asthma Condition in which bronchioles constrict and cause difficulty in breathing. 673

astigmatism Uneven shape of the cornea or lens of the eye; causes a distortion in the light reaching the retina of the eye. 724

astrocyte Nervous system cell that provides metabolic and structural support to the neurons. 697

asymmetry Lack of any symmetrical relationship in the morphology of an organism. 522

atherosclerosis Form of cardiovascular disease characterized by the accumulation of fatty materials (usually cholesterol) in the arteries. 616

atom Smallest particle of an element that displays the properties of the element. 22

atomic mass Average of atom mass units for all the isotopes of an atom. 23

atomic number Number of protons within the nucleus of an atom. 23

atomic symbol One or two letters that represent the name of an element—e.g., H stands for a hydrogen atom, and Na stands for a sodium atom. 22

ATP (adenosine triphosphate) Nucleotide with three phosphate groups. The breakdown of ATP into ADP + P makes energy available for energy-requiring processes in cells. 54, 107

ATP synthase Complex of proteins in the cristae of mitochondria and thylakoid membrane of chloroplast that produces ATP from the diffusion of hydrogen ions across a membrane. 114, 126

atrial natriuretic hormone (ANH) Hormone secreted by the heart that increases sodium excretion. 690, 763

atrioventricular valve Heart valve located between an atrium and a ventricle. 611

atrium Chamber; particularly an upper chamber of the heart lying above a ventricle. 610

auditory communication Sound that an animal makes for the purpose of sending a message to another individual. 828

auditory tube Also called the eustachian tube; connects the middle ear to the nasopharynx for the equalization of pressure. 727

australopithecine (australopith) One of several species of *Australopithecus*, a genus that contains the first generally recognized humanlike hominins. 577

autoimmune disease Disease that results when the immune system mistakenly attacks the body's own tissues. 642

autonomic system Portion of the peripheral nervous system that regulates internal organs. 708

autoploidy Polyploid organism with multiple chromosome sets all from the same species. 315

autosome Chromosome pairs that are the same between the sexes; in humans, all but the X and Y chromosomes. 201

autotroph Organism that can capture energy and synthesize organic molecules from inorganic nutrients. 120, 870

auxin Plant hormone regulating growth, particularly cell elongation; also called indoleacetic acid (IAA). 485

axial skeleton Part of the vertebrate skeleton forming the vertical support or axis, including the skull, the rib cage, and the vertebral column. 740

axillary bud Bud located in the axil of a leaf. 445

axon Elongated portion of a neuron that conducts nerve impulses, typically from the cell body to the synapse. 697

B

Bacillus A rod-shaped bacterium; also a genus of bacteria, *Bacillus*. 65

bacteriophage Virus that infects bacteria. 367

bacterium (pl., bacteria) Member of the domain Bacteria. 374

bark External part of a tree, containing cork, cork cambium, and phloem. 456

Barr body Dark-staining body in the cell nuclei of female mammals that contains a condensed, inactive X chromosome; named after its discoverer, Murray Barr. 185, 243

basal body A cytoplasmic structure that is located at the base of—and may organize—cilia or flagella. 80

basal nuclei Subcortical nuclei deep within the white matter that serve as relay stations for motor impulses and produce dopamine to help control skeletal muscle activities. 704

base Molecules tending to lower the hydrogen ion concentration in a solution and thus raise the pH numerically. 32

basidium Clublike structure in which nuclear fusion, meiosis, and basidiospore production occur during sexual reproduction of club fungi. 412

basophil White blood cell with a granular cytoplasm; able to be stained with a basic dye. 620

B cell Lymphocyte that matures in the bone marrow and, when stimulated by the presence of a specific antigen, gives rise to antibody-producing plasma cells. 629

B-cell receptor (BCR) Molecule on the surface of a B cell that binds to a specific antigen. 633

behavior Observable, coordinated responses to environmental stimuli. 820

behavioral ecology Study of how natural selection shapes behavior. 830

beneficial nutrient In plants, a molecule or element that is either required or enhances the growth and production of a plant. 465

benign Mass of cells derived from a single mutated cell that has repeatedly undergone cell division but has remained at the site of origin. 163

bicarbonate ion Ion that participates in buffering the blood, and the form in which carbon dioxide is transported in the bloodstream. 672

bilateral symmetry Body plan having two corresponding or complementary halves. 522

bile Secretion of the liver that is temporarily stored and concentrated in the gallbladder before being released into the small intestine, where it emulsifies fat. 652

binary fission Splitting of a parent cell into two daughter cells; serves as an asexual form of reproduction in bacteria. 166, 373

binomial nomenclature Scientific name of an organism, the first part of which designates the genus and the second part of which designates the specific epithet. 8, 349

biocultural evolution Phase of human evolution in which cultural events affect natural selection. 581

biodiversity Total number of species, the variability of their genes, and the communities in which they live. 10, 908

biodiversity hotspot Region of the world that contains unusually large concentrations of species. 909

biogeochemical cycle Circulating pathway of elements such as carbon and nitrogen involving exchange pools, storage areas, and biotic communities. 874

biogeography Study of the geographical distribution of organisms. 275

bioinformatics Area of scientific study that utilizes computer technologies to analyze large sets of data, typically in the study of genomics and proteomics. 265, 908

biological clock Internal mechanism that maintains a biological rhythm in the absence of environmental stimuli. 494

biological species concept The concept that defines species as groups of populations that have the potential to interbreed and that are reproductively isolated from other groups. 310

biology The branch of science that is concerned with the study of life and living organisms. 2

biomagnification The accumulation of pollutants as they move up the food web. 902

biomass The number of organisms multiplied by their weight. 873

biome One of the biosphere's major communities, characterized in particular by certain climatic conditions and particular types of plants. 887

biomolecule Organic molecules such as proteins, nucleic acids, carbohydrates, and fats. 38, 328

biosphere Zone of air, land, and water at the surface of the Earth in which living organisms are found. 9, 839

biotechnology products Commercial or agricultural products that are made with or derived from transgenic organisms. 258

biotic potential Maximum population growth rate under ideal conditions. 841

bird Endothermic reptile that has feathers and wings, is often adapted for flight, and lays hard-shelled eggs. 562

bivalent Homologous chromosomes, each having sister chromatids that are joined by a nucleoprotein lattice during meiosis; also called a tetrad. 173

bivalve Type of mollusc with a shell composed of two valves; includes clams, oysters, and scallops. 534

blade Broad, expanded portion of a land plant leaf that may be single or compound leaflets. 445

blastocoel Fluid-filled cavity of a blastula. 797

blastocyst Early stage of human embryonic development that consists of a hollow, fluid-filled ball of cells. 806

blastopore Opening into the primitive gut formed at gastrulation. 798

blastula Hollow, fluid-filled ball of cells occurring during animal development prior to gastrula formation. 523, 797

blind spot Region of the retina, lacking rods or cones, where the optic nerve leaves the eye. 721

blood Fluid circulated by the heart through a closed system of vessels; type of connective tissue. 591, 606

blood pressure Force of blood pushing against the inside wall of blood vessels. 615

body cavity In vertebrates, defined regions of the body in which organs reside. 596

bog Wet, spongy ground in a low-lying area, usually acidic and low in organic nutrients. 897

bolide Term for a large, crater-forming meteorite that strikes the Earth's surface. 343

bone Connective tissue having protein fibers and a hard matrix of inorganic salts, notably calcium salts. 591

bony fishes Vertebrates belonging to the class of fish called Osteichthyes that have a bony, rather than cartilaginous, skeleton. 554

bottleneck effect Type of genetic drift; occurs when a majority of genotypes are prevented from participating in the production of the next generation as a result of a natural disaster or human interference. 295

brachiopod Lophophorans in the phyla Brachiopoda that are characterized by two shells, one on the top and one on the bottom. 530

brain Ganglionic mass at the anterior end of the nerve cord; in vertebrates, the brain is located in the cranial cavity of the skull. 703

brain stem In mammals; portion of the brain consisting of the medulla oblongata, pons, and midbrain. 704

bronchiole In terrestrial vertebrates, small tube that conducts air from a bronchus to the alveoli. 668

bronchus (pl., bronchi) In terrestrial vertebrates, branch of the trachea that leads to the lungs. 668

brown algae Marine photosynthetic protists with a notable abundance of xanthophyll pigments; this group includes well-known seaweeds of northern rocky shores. 390

bryophyte A nonvascular land plant—including the mosses, liverworts, and hornworts—in which the gametophyte is dominant. 423

bryozoan Lophophorans of the phylum Bryozoa that are characterized as being aquatic and colonial that form colonies called zooids. 53

budding Asexual form of reproduction whereby a new organism develops as an outgrowth of the body of the parent. 406

buffer Substance or group of substances that tend to resist pH changes of a solution, thus stabilizing its relative acidity and basicity. 33

bulbourethral glands Male sex glands that produce pre-ejaculate fluid that neutralizes acid in the urethra. 777

bulk transport Movement of substances, usually large particles, across the plasma membrane using vesicles. 91

C

C₃ plant Plant that fixes carbon dioxide via the Calvin cycle; the first stable product of C_3 photosynthesis is a 3-carbon compound. 130

C₄ plant Plant that fixes carbon dioxide to produce a C_4 molecule that releases carbon dioxide to the Calvin cycle. 130

calcitonin Hormone secreted by the thyroid gland that increases the blood calcium level. 761

calorie Amount of heat energy required to raise the temperature of one gram of water 1°C. 29

Calvin cycle reaction Portion of photosynthesis that takes place in the stroma of chloroplasts and can occur in the dark; it uses the products of the light reactions to reduce CO_2 to a carbohydrate. 123

calyx The sepals collectively; the outermost flower whorl. 503

CAM Crassulacean-acid metabolism; a form of photosynthesis in succulent plants that separates the light-dependent and Calvin reactions by time. 130

camera-type eye Type of eye found in vertebrates and certain molluscs; a single lens focuses an image on closely packed photoreceptors. 720

camouflage Process of hiding from predators in which an organism's behavior, form, and pattern of coloration allow it to blend into the background and prevent detection. 863

cancer Malignant tumor whose nondifferentiated cells exhibit loss of contact inhibition, uncontrolled growth, and the ability to invade tissue and metastasize. 163

capillary Microscopic blood vessel; gases and other substances are exchanged across the walls of a capillary between blood and tissue fluid. 608

capsid Protective protein containing the genetic material of a virus. 366

capsule A form of glycocalyx that consists of a gelatinous layer; found in blue-green algae and certain bacteria. 65

carbaminohemoglobin Hemoglobin carrying carbon dioxide. 672

carbohydrate Class of organic compounds that typically contain carbon, hydrogen, and oxygen in a 1:2:1 ratio; includes the monosaccharides, disaccharides, and polysaccharides. 41

carbon dioxide fixation Process by which carbon dioxide gas is attached to an organic compound; in photosynthesis, this occurs in the Calvin cycle reactions. 128

carbonic anhydrase Enzyme in red blood cells that speeds the formation of carbonic acid from water and carbon dioxide. 672

carcinogen Environmental agent that causes mutations leading to the development of cancer. 248

cardiac cycle One complete cycle of systole and diastole for all heart chambers. 612

cardiac muscle Striated, involuntary muscle tissue found only in the heart. 593

cardiac output Blood volume pumped by each ventricle per minute (not total output pumped by both ventricles). 612

carnivore Consumer in a food chain that eats other animals. 870

carotenoid An accessory photosynthetic pigment of plants and algae that are often yellow or orange in color; consist of two classes—the xanthophylls and the carotenes. 124

carotid body Structure located at the branching of the carotid arteries; contains chemoreceptors sensitive to the O_2, CO_2, and H^+ content in blood. 671

carpel Ovule-bearing unit that is a part of a pistil. 503

carrier Heterozygous individual who has no apparent abnormality but can pass on an allele for a recessively inherited genetic disorder. 202

carrier protein Protein in the plasma membrane that combines with and transports a molecule or ion across the plasma membrane. 88

carrying capacity (K) Largest number of organisms of a particular species that can be maintained indefinitely by a given environment. 845

cartilage Connective tissue in which the cells lie within lacunae embedded in a flexible, proteinaceous matrix. 590

cartilaginous fish Vertebrates that belong to the class of fish called the Chondrichthyes: possess a cartilaginous, rather than bony, skeleton; includes sharks, rays, and skates. 554

Casparian strip Layer of impermeable lignin and suberin bordering four sides of root endodermal cells; prevents water and solute transport between adjacent cells. 451

catabolism Metabolic process that breaks down large molecules into smaller ones; catabolic metabolism. 147

cataracts Condition where the lens of the eye becomes opaque, preventing the transmission of light to the retina. 724

catastrophism Belief, proposed by Georges Cuvier, that periods of catastrophic extinctions occurred, after which repopulation of surviving species took place, giving the appearance of change through time. 273

cell The smallest unit of life that displays all the properties of life; composed of cytoplasm surrounded by a plasma membrane. 2, 61

cell body Portion of a neuron that contains a nucleus and from which dendrites and an axon extend. 697

cell cycle An ordered sequence of events in eukaryotes that involves cell growth and nuclear division; consists of the stages G_1, S, G_2, and M. 154

cell envelope In a prokaryotic cell, the portion composed of the plasma membrane, the cell wall, and the glycocalyx. 65

cell-mediated immunity Specific mechanism of defense in which T cells destroy antigen-bearing cells. 636

cell plate Structure across a dividing plant cell that signals the location of new plasma membranes and cell walls. 161

cell recognition protein Glycoproteins in the plasma membrane that identify self and help the body defend itself against pathogens. 89

cell suspension culture Small clumps of naked plant cells grown in tissue culture that produce drugs, cosmetics, or agricultural chemicals, among others. 514

cell theory One of the major theories of biology, which states that all organisms are made up of cells; cells are capable of self-reproduction and come only from preexisting cells. 61

cell wall Cellular structure that surrounds a plant, protistan, fungal, or bacterial cell and maintains the cell's shape and rigidity; composed of polysaccharides. 65, 100

cellular differentiation Process and developmental stages by which a cell becomes specialized for a particular function. 800

cellular respiration Metabolic reactions that use the energy from carbohydrate, fatty acid, or amino acid breakdown to produce ATP molecules. 136

cellular response Response to the transduction pathway in which proteins or enzymes change a signal to a format that the cell can understand, resulting in the appropriate response. 484

cellular slime mold Free-living amoeboid cells that feed on bacteria and yeasts by phagocytosis and aggregate to form a plasmodium that produces spores. 397

cellulose Polysaccharide that is the major complex carbohydrate in plant cell walls. 44

centipede Elongated arthropod characterized by having one pair of legs to each body segment; they may have 15 to 173 pairs of legs. 541

central dogma Processes that dictate the flow of information from the DNA to RNA to protein in a cell. 224

central nervous system (CNS) Portion of the nervous system consisting of the brain and spinal cord. 696

central vacuole In a plant cell, a large, fluid-filled sac that stores metabolites. During growth, it enlarges, forcing the primary cell wall to expand and the cell surface-area-to-volume ratio to increase. 75

centriole Cell structure, existing in pairs, that occurs in the centrosome and may help organize a mitotic spindle for chromosome movement during animal cell division. 79, 158

centromere Constriction where sister chromatids of a chromosome are held together. 157

centrosome Central microtubule organizing center of cells. In animal cells, it contains two centrioles. 78, 157

cephalization Having a well-recognized anterior head with a brain and sensory receptors. 522, 694

cephalochordate Small, fishlike invertebrate that is a member of the phylum Chordata. Probably the closest living relative to vertebrates. 550

cephalopod Type of mollusc in which the head is prominent and the foot is modified to form two arms and several tentacles; includes squids, cuttlefish, octopuses, and nautiluses. 536

cerebellum In terrestrial vertebrates, portion of the brain that coordinates skeletal muscles to produce smooth, graceful motions. 704

cerebral cortex Outer layer of cerebral hemispheres; receives sensory information and controls motor activities. 704

cerebral hemisphere Either of the two lobes of the cerebrum in vertebrates. 703

cerebrospinal fluid Fluid found in the ventricles of the brain, in the central canal of the spinal cord, and in association with the meninges. 702

cerebrum Largest part of the brain in mammals. 703

cervix Narrow end of the uterus, which leads into the vagina. 779

channel protein Protein that forms a channel to allow a particular molecule or ion to cross the plasma membrane. 88

chaparral Biome characterized by broad-leafed evergreen shrubs forming dense thickets. 894

chaperone protein Molecule that directs the proper folding of polypeptides. 53

character displacement Tendency for characteristics to be more divergent when similar species belong to the same community than when they are isolated from one another. 860

charophyte Type of living green algae that on the basis of nucleotide sequencing and cellular features is most closely related to land plants. 387, 421

chelicerate Arthropods (e.g., horseshoe crabs, sea spiders, arachnids), that exhibit a pair of pointed appendages used to manipulate food. 543

chemical energy Energy associated with the interaction of atoms in a molecule. 105

chemical signal Molecule that brings about a change in a cell, tissue, organ, or individual when it binds to a specific receptor. 755

chemiosmosis Process by which mitochondria and chloroplasts use the energy of an electron transport chain to create a hydrogen ion gradient that drives ATP formation. 126, 145

chemoautotroph Organism able to synthesize organic molecules by using carbon dioxide as the carbon source and the oxidation of an inorganic substance (such as hydrogen sulfide) as the energy source. 375

chemoheterotroph Organism that is unable to produce its own organic molecules, and therefore requires organic nutrients in its diet. 375

chemoreceptor Sensory receptor that is sensitive to chemical stimulation—for example, receptors for taste and smell. 717

chitin Strong but flexible nitrogenous polysaccharide found in the exoskeleton of arthropods and in the cell walls of fungi. 44, 405

chlorophyll Green photosynthetic pigment of algae and plants that absorbs solar energy; occurs as chlorophyll *a* and chlorophyll *b*. 121

chlorophyte Most abundant and diverse group of green algae, including freshwater, marine, and terrestrial forms that synthesize. Chlorophytes share chemical and anatomical characteristics with land plants. 387

chloroplast Membrane-bounded organelle in algae and plants with chlorophyll-containing membranous thylakoids; where photosynthesis takes place. 76, 121

choanoflagellate Unicellular choanoflagellates have one and colonial forms have many collar cells like those of sponges; choanoflagellates are the protists most closely related to animals. 397

cholesterol A steroid found in the plasma membrane of animal cells and from which other types of steroids are derived. 86

chordate Animals that have a dorsal tubular nerve cord, a notochord, pharyngeal gill pouches, and a postanal tail at some point in their life cycle; includes a few types of invertebrates (e.g., sea squirts and lancelets) and the vertebrates. 356, 550

chorion Extraembryonic membrane functioning for respiratory exchange in birds and reptiles; contributes to placenta formation in mammals. 805

chorionic villus In placental mammals treelike extension of the chorion, projecting into the maternal tissues at the placenta. 809

choroid Vascular, pigmented middle layer of the eyeball. 721

chromalveolate Supergroup of eukaryotes that includes alveolates and stramenopiles. 390

chromatid Following replication, a chromosome consists of a pair of sister chromatids, held together at the centromere; each chromatid is comprised of a single DNA helix. 154

chromatin Network of DNA strands and associated proteins observed within a nucleus of a cell. 70, 157, 241

chromoplast Plastid in land plants responsible for orange, yellow, and red color of plants, including the autumn colors in leaves. 77

chromosome The structure that transmits the genetic material from one generation to the next; composed of condensed chromatin; each species has a particular number of chromosomes that is passed on to the next generation. 70

chyme Thick, semiliquid food material that passes from the stomach to the small intestine. 652

chytrid (Chytridiomycota) Mostly aquatic fungi with flagellated spores that may represent the most ancestral fungal lineage. 406

cilia (sing., cilium) Short, hairlike projections from the plasma membrane, occurring usually in larger numbers. 80

ciliary muscle Within the ciliary body of the vertebrate eye, the ciliary muscle that controls the shape of the lens. 722

ciliate Complex unicellular protist that moves by means of cilia and digests food in food vacuoles. 393

circadian rhythm Biological rhythm with a 24-hour cycle. 494, 768

circulatory system In animals, an organ system that moves substances to and from cells, usually via a heart, blood, and blood vessels. 606, 608

cirrhosis Chronic, irreversible injury to liver tissue; commonly caused by frequent alcohol consumption. 656

citric acid cycle Cycle of reactions in mitochondria that begins with citric acid. This cycle breaks down an acetyl group and produces CO_2, ATP, NADH, and $FADH_2$; also called the Krebs cycle. 137, 143

clade Evolutionary lineage consisting of an ancestral species and all of its descendants, forming a distinct branch on a cladogram. 356

cladistics Method of systematics that uses derived characters to determine monophyletic groups and construct cladograms. 356

cladogram In cladistics, a branching diagram that shows the relationship among species in regard to their shared derived characters. 356

class One of the categories, or taxa, used by taxonomists to group species; the taxon above the order level. 6, 349

classification Process of naming organisms and assigning them to taxonomic groups (taxa). 348

classical conditioning Type of learning whereby an unconditioned stimulus that elicits a specific response is paired with a neutral stimulus so that the response becomes conditioned. 824

cleavage Cell division without cytoplasmic addition or enlargement; occurs during the first stage of animal development. 523, 797

cleavage furrow Indentation in the plasma membrane of animal cells during cell division; formation marks the start of cytokinesis. 160

climate Generalized weather patterns of an area, primarily determined by temperature and average rainfall. 884

climate change Recent changes in the Earth's climate; evidence suggests that this is primarily due to human influence, including the increased release of greenhouse gases. 127, 877, 915

climax community In ecology, community that results when succession has come to an end. 868

clonal selection theory States that the antigen selects which lymphocyte will undergo clonal expansion and produce more lymphocytes bearing the same type of receptor. 633

cloning Production of identical copies. In organisms, the production of organisms with the same genes; in genetic engineering, the production of many identical copies of a gene. 255

closed circulatory system A type of circulatory system where blood is confined to vessels and is kept separate from the interstitial fluid. 607

clotting Also called coagulation, the response of the body to an injury in the vessels of the circulatory system; involves platelets and clotting proteins. 620

club fungi (Basidiomycota) Fungi that produce spores in club-shaped basidia within a fruiting body; includes mushrooms, shelf fungi and puffballs. 412

cnidarian Invertebrates existing as either a polyp or medusa with two tissue layers and radial symmetry. 528

coacervate droplet An aggregate of colloidal droplets held together by electrostatic forces. 331

coccus A spherical-shaped bacterium. 65

cochlea Spiral-shaped structure of the vertebrate inner ear containing the sensory receptors for hearing. 727

codominance Inheritance pattern in which both alleles of a gene are equally expressed in a heterozygote. 205

codon Three-base sequence in messenger RNA that during translation directs the addition of a particular amino acid into a protein or directs termination of the process. 224

coelom Body cavity of an animal; the method by which the coelom is formed (or lack of formation) is an identifying characteristic in animal classification. 523

coenzyme Nonprotein organic molecule that aids the action of the enzyme to which it is loosely bound. 112

coevolution Mutual evolution in which two species exert selective pressures on the other species. 506, 866

cofactor Nonprotein assistant required by an enzyme in order to function; many cofactors are metal ions, others are coenzymes. 112

cohesion-tension model Explanation for upward transport of water in xylem based upon transpiration-created tension and the cohesive properties of water molecules. 476

cohort Group of individuals having a statistical factor in common, such as year of birth, in a population study. 841

coleoptile Protective sheath that covers the young leaves of a seedling. 485

collagen fiber White fiber in the matrix of connective tissue giving flexibility and strength. 589

collecting duct Duct within the kidney that receives fluid from several nephrons; the reabsorption of water occurs here. 686

collenchyma Plant tissue composed of cells with unevenly thickened walls; supports growth of stems and petioles. 448

colonial flagellate hypothesis Hypothesis that all animals are descended from an ancestor that resembled a hollow colony of flagellated cells, most likely resembling modern choanoflagellates. 521

colony Loose association of cells each remaining independent for most functions. 388

columnar epithelium Type of epithelial tissue with cylindrical cells. 589

comb jelly Invertebrates that resemble jelly fishes and are the largest animals to be propelled by beating cilia. 528

commensalism Symbiotic relationship in which one species is benefited, and the other is neither harmed nor benefited. 375, 865

common ancestor Ancestor common to at least two lines of descent. 354

communication Signal by a sender that influences the behavior of a receiver. 826

community Assemblage of species interacting with one another within the same environment. 9, 839, 858

compact bone Type of bone that contains osteons consisting of concentric layers of matrix and osteocytes in lacunae. 591, 739

comparative genomics Study of genomes through the direct comparison of their genes and DNA sequences from multiple species. 263

competition Results when members of a species attempt to use a resource that is in limited supply. 847

competitive exclusion principle Theory that no two species can occupy the same niche in the same place and at the same time. 860

competitive inhibition Form of enzyme inhibition where the substrate and inhibitor are both able to bind to the enzyme's active site. Only when the substrate is at the active site will product form. 113

complement Collective name for a series of enzymes and activators in the blood, some of which may bind to antibody and may lead to rupture of a foreign cell. 632

complementary base pairing Hydrogen bonding between particular purines and pyrimidines; responsible for the structure of DNA, and some RNA, molecules. 55, 219

complementary DNA (cDNA) DNA that has been synthesized from mRNA by the action of reverse transcriptase. 256

complete digestive tract Digestive tract that has both a mouth and an anus. 647

complex tissue In plants, tissue composed of two or more kinds of cells (e.g., xylem, containing tracheids and vessel elements; phloem, containing sieve-tube members and companion cells). 449

compound Substance having two or more different elements in a fixed ratio. 26

compound eye Type of eye found in arthropods; it is composed of many independent visual units. 720

concentration gradient Gradual change in chemical concentration between two areas of differing concentrations. 89

conclusion Statement made following an experiment as to whether or not the results support the hypothesis. 12

cone Reproductive structure in conifers made up of scales bearing sporangia; pollen cones bear microsporangia, and seed cones bear megasporangia. 430

cone cell Photoreceptor in vertebrate eyes that responds to bright light and makes color vision possible. 721

conidiospore Spore produced by sac and club fungi during asexual reproduction. 409

conifer Member of a group of cone-bearing gymnosperm land plants that includes pine, cedar, and spruce trees. 430

conjugation Transfer of genetic material from one cell to another. 388

conjugation pilus (pl., conjugation pili) In a bacterium, elongated, hollow appendage used to transfer DNA to other cells. 66, 373

conjunctiva Delicate membrane that lines the eyelid protecting the sclera. 721

connective tissue Type of animal tissue that binds structures together, provides support and protection, fills spaces, stores fat, and forms blood cells; adipose tissue, cartilage, bone, and blood are types of connective tissue; living cells in a nonliving matrix. 589

conservation biology Discipline that seeks to understand the effects of human activities on species, communities, and ecosystems and to develop practical approaches to preventing the extinction of species and the destruction of ecosystems. 908

consumer Organism that feeds on another organism in a food chain generally; primary consumers eat plants, and secondary consumers eat animals. 870

continental drift The movement of the Earth's crust by plate tectonics resulting in the movement of continents with respect to one another. 342

contraceptive vaccine Under development, this birth control method immunizes against the hormone HCG, crucial to maintaining implantation of the embryo. 784

control Sample that goes through all the steps of an experiment but does not contain the variable being tested; a standard against which the results of an experiment are checked. 12

convergent evolution Similarity in structure in distantly related groups generally due to similiar selective pressures in like environments. 357

copulation Sexual union between a male and a female. 774

coral reef Coral formations in shallow tropical waters that support an abundance of diversity. 901

corepressor Molecule that binds to a repressor, allowing the repressor to bind to an operator in a repressible operon. 239

cork Outer covering of the bark of trees; made of dead cells that may be sloughed off. 447

cork cambium Lateral meristem that produces cork. 447

cornea Transparent, anterior portion of the outer layer of the eyeball. 721

corolla The petals, collectively; usually the conspicuously colored flower whorl. 503

corpus luteum Follicle that has released an egg and increases its secretion of progesterone. 780

cortex In plants, ground tissue bounded by the epidermis and vascular tissue in stems and roots; in animals, outer layer of an organ, such as the cortex of the kidney or adrenal gland. 451

cortisol Glucocorticoid secreted by the adrenal cortex that responds to stress on a long-term basis; reduces inflammation and promotes protein and fat metabolism. 763

cost-benefit analysis A weighing-out of the costs and benefits (in terms of contributions to reproductive success) of a particular strategy or behavior. 298

cotyledon Seed leaf for embryo of a flowering plant; provides nutrient molecules for the developing plant before photosynthesis begins. 433, 508

countercurrent exchange Fluids flow side-by-side in opposite directions, as in the exchange of fluids in the kidneys. 666

coupled reactions Reactions that occur simultaneously; one is an exergonic reaction that releases energy, and the other is an endergonic reaction that requires an input of energy in order to occur. 108

covalent bond Chemical bond in which atoms share one pair of electrons. 27

cranial nerve Nerve that arises from the brain. 707

crenation In animal cells, shriveling of the cell due to water leaving the cell when the environment is hypertonic. 94

cristae (sing., crista) Short, fingerlike projections formed by the folding of the inner membrane of mitochondria. 77

Cro-Magnon Common name for the first fossils to be designated *Homo sapiens*. 580

crossing-over Exchange of segments between nonsister chromatids of a bivalent during meiosis. 174

crustacean Member of a group of aquatic arthropods that contains, among others, shrimps, crabs, crayfish, and lobsters. 540

cryptic species Species that are very similar in appearance but are considered separate species based on other characteristics, such as behavior or genetics. 307

cuboidal epithelium Type of epithelial tissue with cube-shaped cells. 588

cutaneous receptor Sensory receptors of the dermis that are activated by touch, pain, pressure, and temperature. 730

cuticle Waxy layer covering the epidermis of plants that protects the plant against water loss and disease-causing organisms. 422

cyanobacterium (pl., cyanobacteria) Photosynthetic bacterium that contains chlorophyll and releases oxygen; formerly called a blue-green alga. 66, 377

cyanogenic glycoside Plant compound that contains sugar; produces cyanide. 497

cycad Type of gymnosperm with palmate leaves and massive cones; cycads are most often found in the tropics and subtropics. 432

cyclic adenosine monophosphate (cAMP) ATP-related compound that acts as the second messenger in peptide hormone transduction; it initiates activity of the metabolic machinery. 755

cyclin Protein that cycles in quantity as the cell cycle progresses; combines with and activates the kinases that function to promote the events of the cycle. 155

cyst In protists and invertebrates, resting structure that contains reproductive bodies or embryos. 384, 533

cystic fibrosis (CF) Genetic disease caused by a defect in the *CFTR* gene, which is responsible for the formation of a transmembrane chloride ion transporter; causes the mucus of the body to be viscous. 675

cytochrome Any of several iron-containing protein molecules that are members of the electron transport chain in photosynthesis and cellular respiration. 144

cytokine Type of protein secreted by a T lymphocyte that attacks viruses, virally infected cells, and cancer cells. 633

cytokinesis Division of the cytoplasm following mitosis or meiosis. 155

cytokinin Plant hormone that promotes cell division; often works in combination with auxin during organ development in plant embryos. 487

cytoplasm Region of a cell between the nucleus, or the nucleoid region of a bacterium, and the plasma membrane; contains the organelles of the cell. 65

cytoplasmic segregation Process that parcels out the maternal determinants, which play a role in development, during mitosis. 800

cytosine (C) One of four nitrogen-containing bases in the nucleotides composing the structure of DNA and RNA; pairs with guanine. 217

cytoskeleton Internal framework of the cell, consisting of microtubules, actin filaments, and intermediate filaments. 68

cytotoxic T cell T lymphocyte that attacks and kills antigen-bearing cells. 635

D

data (sing., datum) Facts or information collected through observation and/or experimentation. 12

day-neutral plant Plant whose flowering is not dependent on day length—e.g., tomato and cucumber. 496

deamination Removal of an amino group (—NH$_2$) from an amino acid or other organic compound. 147

deciduous Land plant that sheds its leaves annually. 445

decomposer Organism, usually a bacterium or fungus, that breaks down organic matter into inorganic nutrients that can be recycled in the environment. 871

deductive reasoning The use of general principles to predict specific outcomes. Often uses "if . . . then" statements. 11

dehydration reaction Chemical reaction in which a water molecule is released during the formation of a covalent bond. 40

delayed allergic response Allergic response initiated at the site of the allergen by sensitized T cells, involving macrophages and regulated by cytokines. 642

deletion Change in chromosome structure in which the end of a chromosome breaks off or two simultaneous breaks lead to the loss of an internal segment; often causes abnormalities—e.g., cri du chat syndrome. 187

demographic transition Due to industrialization, a decline in the birthrate following a reduction in the death rate so that the population growth rate is lowered. 851

demography Properties of the rate of growth and the age structure of populations. 840

denatured Loss of a protein's or enzyme's normal shape so that it no longer functions; usually caused by a less than optimal pH and temperature. 53, 111

dendrite Part of a neuron that sends signals toward the cell body. 697

dendritic cell Antigen-presenting cell of the epidermis and mucous membranes. 632

denitrification Conversion of nitrate or nitrite to nitrogen gas by bacteria in soil. 878

dense fibrous connective tissue Type of connective tissue containing many collagen fibers packed together; found in tendons and ligaments, for example. 590

density-dependent factor Biotic factor, such as disease or competition, that affects population size in a direct relationship to the population's density. 847

density-independent factor Abiotic factor, such as fire or flood, that affects population size independent of the population's density. 847

deoxyribose Pentose sugar found in DNA. 42

derived trait Structural, physiological, or behavioral trait that is present in a specific lineage and is not present in the common ancestor for several related lineages. 354

dermis In mammals, thick layer of the skin underlying the epidermis. 598

desert Ecological biome characterized by a limited amount of rainfall; deserts have hot days and cool nights. 896

desmosome(s) Intercellular junctions that connect cytoskeletons of adjacent cells. 99

detritivore Any organism that obtains most of its nutrients from the detritus in an ecosystem. 870

deuterostome Group of coelomate animals in which the second embryonic opening is associated with the mouth; the first embryonic opening, the blastopore, is associated with the anus. 523

development Process of regulated growth and differentiation of cells and tissues. 508, 797

diabetic retinopathy Complication of diabetes that causes the capillaries in the retina to become damaged, potentially causing blindness. 724

diagnostic traits Characteristics that distinguish species from one another. 307

diaphragm In mammals, dome-shaped muscularized sheet separating the thoracic cavity from the abdominal cavity; contraceptive device that prevents sperm from reaching the egg. 669

diapsid Characteristic of a vertebrate skull in which there are two openings in the skull behind the eye socket (orbit). 558

diarrhea Excessively frequent and watery bowel movements. 653

diastole Relaxation period of a heart chamber during the cardiac cycle. 612

diatom Golden-brown alga with a cell wall in two parts, or valves; significant component of phytoplankton. 391

differentiation Specialization in the structure or function of a cell typically caused by the activation of specific genes. 508

diffusion Movement of molecules or ions from a region of higher to lower concentration; it requires no energy and tends to lead to an equal distribution (equilibrium). 91

dihybrid cross Cross between parents that differ in two traits. 198

dikaryotic Having two haploid nuclei that stem from different parent cells; during sexual reproduction, sac and club fungi have dikaryotic cells. 406

dinoflagellate Photosynthetic unicellular protist with two flagella, one whiplash and the other located within a groove between protective cellulose plates; significant part of phytoplankton. 392

dinosaur General term used to describe the large reptiles that existed prior to the start of the Cretaceous period. 559

dioecious Having unisexual flowers or cones, with the male flowers or cones confined to certain land plants and the female flowers or cones of the same species confined to other different plants. 432

diploid (2n) Cell condition in which two of each type of chromosome are present. 157, 172

diplomonad Protist that has modified mitochondria, two equal-sized nuclei, and multiple flagella. 395

directional selection Outcome of natural selection in which an extreme phenotype is favored, usually in a changing environment. 296

disaccharide Sugar that contains two monosaccharide units; e.g., maltose. 42

disruptive selection Outcome of natural selection in which the two extreme phenotypes are favored over the average phenotype, leading to more than one distinct form. 297

distal convoluted tubule Final portion of a nephron that joins with a collecting duct; associated with tubular secretion. 686

diverge Process by which a new evolutionary path begins; on a phylogenetic tree this is indicated by branching lines. 354

DNA (deoxyribonucleic acid) Nucleic acid polymer produced from covalent bonding of nucleotide monomers that contain the sugar deoxyribose; the genetic material of nearly all organisms. 54

DNA ligase Enzyme that links DNA fragments; used during production of recombinant DNA to join foreign DNA to vector DNA. 255

DNA microarray Glass or plastic slide containing thousands of single-stranded DNA fragments arranged in an array (grid); used to detect and measure gene expression; also called gene chips. 263

DNA polymerase During replication, an enzyme that joins the nucleotides complementary to a DNA template. 220

DNA repair enzyme One of several enzymes that restore the original base sequence in an altered DNA strand. 248

DNA replication Synthesis of a new DNA double helix prior to mitosis and meiosis in eukaryotic cells and during prokaryotic fission in prokaryotic cells. 220

domain Largest of the categories, or taxa, used by taxonomists to group species; the three domains are Archaea, Bacteria, and Eukarya. 6, 349

domain Archaea One of the three domains of life; contains prokaryotic cells that often live in extreme habitats and have unique genetic, biochemical, and physiological characteristics; its members are sometimes referred to as *archaea*. 7, 351

domain Bacteria One of the three domains of life; contains prokaryotic cells that differ from archaea because they have their own unique genetic, biochemical, and physiological characteristics. 7, 351

domain Eukarya One of the three domains of life, consisting of organisms with eukaryotic cells; includes protists, fungi, plants, and animals. 7, 351

dominance hierarchy Organization of animals in a group that determines the order in which the animals have access to resources. 298

dominant allele Allele that exerts its phenotypic effect in the heterozygote; it masks the expression of the recessive allele. 196

dormancy In plants, a cessation of growth under conditions that seem appropriate for growth. 487

dorsal root ganglion Mass of sensory neuron cell bodies located in the dorsal root of a spinal nerve. 707

double fertilization In flowering plants, one sperm nucleus unites with the egg nucleus, and a second sperm nucleus unites with the polar nuclei of an embryo sac. 437, 505

double helix Double spiral; describes the three-dimensional shape of DNA. 218

doubling time Number of years it takes for a population to double in size. 851

dryopithecine Tree dwelling primate existing 12–9 MYA; ancestral to apes. 574

duodenum First part of the small intestine, where chyme enters from the stomach. 652

duplication Change in chromosome structure in which a particular segment is present more than once in the same chromosome. 187

E

ecdysozoan Protostome characterized by periodic molting of their exoskeleton. Includes the roundworms and arthropods. 538

echinoderm Invertebrates such as sea stars, sea urchins, and sand dollars; characterized by radial symmetry and a water vascular system. 544

ecological niche Role an organism plays in its community, including its habitat and its interactions with other organisms. 859

ecological pyramid Visual depiction of the biomass, number of organisms, or energy content of various trophic levels in a food web—from the producer to the final consumer populations. 873

ecological release In an ecosystem, the freedom of a species to expand its use of available resources due to an elimination of competition. 316

ecological succession The gradual replacement of communities in an area following a disturbance (secondary succession) or the creation of new soil (primary succession). 868

ecology Study of the interactions of organisms with other organisms and with the physical and chemical environment. 839

ecosystem Biological community together with the associated abiotic environment; characterized by a flow of energy and a cycling of inorganic nutrients. 9, 839, 870

ecosystem diversity Variety of species in a particular locale, dependent on the species interactions. 909

ectoderm Outermost primary tissue layer of an animal embryo; gives rise to the nervous system and the outer layer of the integument. 798

ectotherm Organism having a body temperature that varies according to the environmental temperature. 553

edge effect Phenomenon in which the edges around a landscape patch provide a slightly different habitat than the favorable habitat in the interior of the patch. 919

effector Muscle or gland that receives signals from motor fibers and thereby allows an organism to respond to environmental stimuli. 708

egg Also called an ovum; haploid cell that is usually fertilized by a sperm to form a diploid zygote. 779

elastic cartilage Type of cartilage composed of elastic fibers, allowing greater flexibility. 591

elastic fiber Yellow fiber in the matrix of connective tissue, providing flexibility. 589

electrocardiogram (ECG) Recording of the electrical activity associated with the heartbeat. 613

electron Negative subatomic particle, moving about in an energy level around the nucleus of an atom. 22

electronegativity The ability of an atom to attract electrons toward itself in a chemical bond. 27

electron shell The average location, or energy level, of an electron in an atom. Often drawn as concentric circles around the nucleus. 23

electron transport chain (ETC) Process in a cell that involves the passage of electrons along a series of membrane-bound electron carrier molecules from a higher to lower energy level; the energy released is used for the synthesis of ATP. 137, 144

element Substances that cannot be broken down into substances with different properties; composed of only one type atom. 22

El Niño–Southern Oscillation Warming of water in the Eastern Pacific equatorial region such that the Humboldt Current is displaced, with possible negative results such as reduction in marine life. 904

elongation Middle stage of translation in which additional amino acids specified by the mRNA are added to the growing polypeptide. 231

embryo Stage of a multicellular organism that develops from a zygote before it becomes free-living; in seed plants, the embryo is part of the seed. 797

embryo sac Female gametophyte (megagametophyte) of flowering plants. 504

embryonic development In humans, the first two months of development following fertilization during which the major organs are formed. 805

embryonic disk During human development, flattened area during gastrulation from which the embryo arises. 808

embryophyta Bryophytes and vascular plants; both of which produce embryos. 422

emergent property A function or trait that appears as biological complexity increases. 2

emerging viruses Newly identified viruses that are becoming more prominent usually because they cause serious disease. 369

emphysema Disorder of the respiratory system, specifically the lungs, that is characterized by damage to the alveoli, thus reducing the ability to exchange gases with the external environment. 675

endangered species A species that is in peril of immediate extinction throughout all or most of its range (e.g., California condor, snow leopard). 908

endergonic reaction Chemical reaction that requires an input of energy; opposite of exergonic reaction. 107

endocrine gland Ductless organ that secretes hormone(s) into the bloodstream. 589, 753

endocrine system Organ system involved in the coordination of body activities; uses hormones as chemical signals secreted into the bloodstream. 753

endocytosis Process by which substances are moved into the cell from the environment; includes phagocytosis, pinocytosis, and receptor-mediated endocytosis. 97

endoderm Innermost primary tissue layer of an animal embryo that gives rise to the linings of the digestive tract and associated structures. 798

endodermis Internal plant root tissue forming a boundary between the cortex and the vascular cylinder. 451

endomembrane system Cellular system that consists of the nuclear envelope, endoplasmic reticulum, Golgi apparatus, and vesicles. 72

endometrium Mucous membrane lining the interior surface of the uterus. 779

endoplasmic reticulum (ER) System of membranous saccules and channels in the cytoplasm, often with attached ribosomes. 72

endoskeleton Protective internal skeleton, as in vertebrates. 737

endosperm In flowering plants, nutritive storage tissue that is derived from the union of a sperm nucleus and polar nuclei in the embryo sac. 505

endospore Spore formed within a cell; certain bacteria form endospores. 375

endosymbiosis See endosymbiotic theory.

endosymbiotic theory Explanation of the evolution of eukaryotic organelles by phagocytosis of prokaryotes. 67, 335

endotherm Organism in which maintenance of a constant body temperature is independent of the environmental temperature. 563

energy Capacity to do work and bring about change; occurs in a variety of forms. 4, 105

energy of activation Energy that must be added in order for molecules to react with one another. 109

enhancer DNA sequence that acts as a regulatory element to increase the level of transcription when regulatory proteins, such as transcription activators, bind to it. 244

entropy Measure of disorder or randomness in a system. 106

enzymatic protein Protein that catalyzes a specific reaction; may be found in the plasma membrane or the cytoplasm of the cell. 89

enzyme Organic catalyst, usually a protein, that speeds a reaction in cells due to its particular shape. 41, 109

enzyme inhibition Means by which cells regulate enzyme activity; may be competitive or noncompetitive inhibition. 112

eosinophil White blood cell containing cytoplasmic granules that stain with acidic dye. 632

epidermal tissue Exterior tissue, usually one cell thick, of leaves, young stems, roots, and other parts of plants. 447

epidermis In mammals, the outer, protective layer of the skin; in plants, tissue that covers roots, leaves, and stems of nonwoody organisms. 447, 451, 598

epididymis Location of sperm maturation in an adult human; located in the testis. 775

epigenetic inheritance An inheritance pattern in which a nuclear gene has been modified but the changed expression of the gene is not permanent over many generations; the transmission of genetic information by means that are not based on the coding sequences of a gene. 241

epiglottis Structure that covers the glottis, the air-tract opening, during the process of swallowing. 668

epinephrine Hormone secreted by the adrenal medulla in times of stress; adrenaline. 762

epiphyte Plant that takes its nourishment from the air because its placement in other plants gives it an aerial position. 428, 892

epithelial tissue Tissue that lines hollow organs and covers surfaces. 588

erythropoietin (EPO) Hormone produced by the kidneys that speeds red blood cell formation. 688

esophagus Muscular tube for moving swallowed food from the pharynx to the stomach. 651

essential nutrient In plants, substance required for normal growth, development, or reproduction; in humans, a nutrient that cannot be produced in sufficient quantities by the body and must be obtained from the diet. 465

estrogen Female sex hormone that helps maintain sexual organs and secondary sex characteristics. 779

estuary Portion of the ocean located where a river enters and fresh water mixes with salt water. 900

ethylene Plant hormone that causes ripening of fruit and is also involved in abscission. 488

etiolate Changes to a plant, such as yellowing of leaves and increase in shoot length, that occur when it is grown in darkness. 495

euchromatin Chromatin with a lower level of compaction and therefore accessible for transcription. 234

eudicot (Eudicotyledone) Flowering plant group; members have two embryonic leaves (cotyledons), net-veined leaves, vascular bundles in a ring, flower parts in fours or fives and their multiples, and other characteristics. 433, 446

euglenid Flagellated and flexible freshwater unicellular protist that usually contains chloroplasts and has a semirigid cell wall. 395

eukaryotic cell (eukaryote) Type of cell that has a membrane-bounded nucleus and membranous organelles; found in organisms within the domain Eukarya. 7, 65

euploidy Condition in which a cell contains the correct number, and combinations, of chromosomes. 183

eutrophication Enrichment of water by inorganic nutrients used by phytoplankton. Often, overenrichment caused by human activities leads to excessive bacterial growth and oxygen depletion. 877, 898

evergreen Land plant that sheds leaves over a long period, so some leaves are always present. 445

evolution Genetic change in a species over time resulting in the development of genetic and phenotypic differences that are the basis of natural selection; descent of organisms from a common ancestor. 5, 272

evolutionary species concept Every species has its own evolutionary history, which is partly documented in the fossil record. 307

excavate Supergroup of eukaryotes that includes euglenids, kinetoplastids, parabasalids, and diplomonads. 395

excretion Elimination of metabolic wastes by an organism at exchange boundaries such as the plasma membrane of unicellular organisms and excretory tubules of multicellular animals. 681

exergonic reaction Chemical reaction that releases energy; opposite of endergonic reaction. 107

exocrine gland Gland that secretes its product to an epithelial surface directly or through ducts. 589, 753

exocytosis Process in which an intracellular vesicle fuses with the plasma membrane so that the vesicle's contents are released outside the cell. 95

exon Segment of mRNA containing the protein-coding portion of a gene that remains within the mRNA after splicing has occurred. 226

exophthalmos Condition associated with an enlargement of the thyroid gland accompanied by an abnormal protrusion of the eyes. 761

exoskeleton Protective external skeleton, as in arthropods. 736

exotic species Nonnative species that migrate or are introduced by humans into a new ecosystem; also called alien species. 913

experiment A test of a hypothesis that examines the influence of a single variable. Often involves both control and test groups. 11

experimental design Methodology by which an experiment will seek to support the hypothesis. 11

experimental variable Factor of the experiment being tested. 12

expiration Act of expelling air from the lungs; exhalation. 656

exponential growth Growth, particularly of a population, in which the increase occurs in the same manner as compound interest. 844

extant Species, or other levels of taxa, that are still living. 272

external respiration Exchange of oxygen and carbon dioxide between alveoli of the lungs and blood. 664

extinct; extinction Total disappearance of a species or higher group. 10, 338

extracellular matrix (ECM) Nonliving substance secreted by some animal cells; is composed of protein and polysaccharides. 87

extraembryonic membrane Membrane that is not a part of the embryo but is necessary to the continued existence and health of the embryo. 774, 805

ex vivo gene therapy Gene therapy in which cells are removed from an organism, and DNA is injected to correct a genetic defect; the cells are returned to the organism to treat a disease or disorder. 256

F

facilitated transport Passive transfer of a substance into or out of a cell along a concentration gradient by a process that requires a protein carrier. 94

facultative anaerobe Prokaryote that is able to grow in either the presence or the absence of gaseous oxygen. 374

FAD Flavin adenine dinucleotide; a coenzyme of oxidation-reduction that becomes $FADH_2$ as oxidation of substrates occurs in the mitochondria during cellular respiration, FAD then delivers electrons to the electron transport chain. 136

fall overturn Mixing process that occurs in fall in stratified lakes, whereby oxygen-rich top waters mix with nutrient-rich bottom waters. 898

family One of the categories, or taxa, used by taxonomists to group species; the taxon located above the genus level. 6, 349

family pedigree Chart of genetic relationship of family individuals across generations. 204

farsighted (hyperopic) Condition in which an individual cannot focus on objects closely; caused by the focusing of the image behind the retina of the eye. 724

fat Organic molecule that contains glycerol and three fatty acids; energy storage molecule. 46

fatty acid Molecule that contains a hydrocarbon chain and ends with an acid group. 46

fermentation Anaerobic breakdown of glucose that results in a gain of two ATP and end products such as alcohol and lactate; occurs in the cytoplasm of cells. 137, 140

fern Member of a group of land plants that have large fronds; in the sexual life cycle, the independent gametophyte produces flagellated sperm, and the vascular sporophyte produces windblown spores. 428

fertilization Fusion of sperm and egg nuclei, producing a zygote that develops into a new individual. 175, 796

fibroblast Cell found in loose connective tissue that synthesizes collagen and elastic fibers in the matrix. 590

fibrocartilage Cartilage with a matrix of strong collagenous fibers. 591

fibrous protein A protein that has only a secondary structure; generally insoluble; includes collagens, elastins, and keratins. 51

fibrous root system In most monocots, a mass of similarly sized roots that cling to the soil. 452

filament End-to-end chains of cells that form as cell division occurs in only one plane; in plants, the elongated stalk of a stamen. 388

fimbria (pl., fimbriae) Small, bristlelike fiber on the surface of a bacterial cell, which attaches bacteria to a surface; also fingerlike extension from the oviduct near the ovary. 66, 372

fin In fishes and other aquatic animals, membranous, winglike, or paddlelike process used to propel, balance, or guide the body. 553

first messenger Chemical signal such as a peptide hormone that binds to a plasma membrane receptor protein and alters the metabolism of a cell because a second messenger is activated. 755

fishes Aquatic, gill-breathing vertebrate that usually has fins and skin covered with scales; fishes were among the earliest vertebrates that evolved. 553

fitness Ability of an organism to reproduce and pass its genes to the next fertile generation; measured against the ability of other organisms to reproduce in the same environment. 297

five-kingdom system System of classification that contains the kingdoms Monera, Protista, Plantae, Animalia, and Fungi. 351

fixed action pattern (FAP) Innate behavior pattern that is stereotyped, spontaneous, independent of immediate control, genetically encoded, and independent of individual learning. 822

flagellum (pl., flagella) Long, slender extension used for locomotion by some bacteria, protozoans, and sperm. 66, 80, 372

flagship species Species that evoke a strong emotional response in humans; charismatic, cute, regal (e.g., lions, tigers, dolphins, pandas). 918

flatworm Invertebrates such as planarians and tapeworms with extremely thin bodies; a three-branched gastrovascular cavity and a ladder type nervous system. 531

flower Reproductive organ of a flowering plant, consisting of several kinds of modified leaves arranged in concentric rings and attached to a modified stem called the receptacle. 434, 502

fluid-mosaic model Model for the plasma membrane based on the changing location and pattern of protein molecules in a fluid phospholipid bilayer. 87

follicle Structure in the ovary of animals that contains an oocyte; site of oocyte production. 780

follicular phase First half of the ovarian cycle, during which the follicle matures and much estrogen (and some progesterone) is produced. 780

food chain The order in which one population feeds on another in an ecosystem, thereby showing the flow of energy from a detritivore (detrital food chain) or a producer (grazing food chain) to the final consumer. 873

food web In ecosystems, a complex pattern of interlocking and crisscrossing food chains. 873

foraminiferan A protist bearing a calcium carbonate test with many openings through which pseudopods extend. 398

formed elements Portion of the blood that consists of erythrocytes, leukocytes, and platelets (thrombocytes). 618

formula A group of symbols and numbers used to express the composition of a compound. 26

fossil Any past evidence of an organism that has been preserved in the Earth's crust. 280, 333

founder effect Cause of genetic drift due to colonization by a limited number of individuals who, by chance, have different genotype and allele frequencies than the parent population. 295

fovea centralis Region of the retina consisting of densely packed cones; responsible for the greatest visual acuity. 721

frameshift mutation Insertion or deletion of at least one base so that the reading frame of the corresponding mRNA changes. 248

free energy Energy in a system that is capable of performing work. 107

frond Leaf of a fern, palm, or cycad. 428

fruit Flowering plant structure consisting of one or more ripened ovaries that usually contain seeds. 437, 511

fruiting body Spore-producing and spore-disseminating structure found in sac and club fungi. 409

functional genomics Study of gene function at the genome level. It involves the study of many genes simultaneously and the use of DNA microarrays. 263

functional group Specific cluster of atoms attached to the carbon skeleton of organic molecules that enters into reactions and behaves in a predictable way. 39

fungus (pl., fungi) Eukaryotic saprotrophic decomposer; the body is made up of filaments called hyphae that form a mass called a mycelium. 8, 404

G

gallbladder Organ attached to the liver that serves to store and concentrate bile. 656

gametangia Cell or multicellular structure in which gametes are formed. 408

gamete Haploid sex cell; e.g., egg or sperm. 172, 785

gametogenesis Development of the male and female sex gametes. 180

gametophyte Haploid generation of the alternation of generations life cycle of a plant; produces gametes that unite to form a diploid zygote. 180, 422, 504

ganglion (pl., ganglia) Collection or bundle of neuron cell bodies usually outside the central nervous system. 531, 694

gap junction Junction between cells formed by the joining of two adjacent plasma membranes; it lends strength and allows ions, sugars, and small molecules to pass between cells. 100

gastropod Mollusc with a broad, flat foot for crawling (e.g., snails and slugs). 535

gastrovascular cavity Blind digestive cavity in animals that have a sac body plan. 528

gastrula Stage of animal development during which the germ layers form, at least in part, by invagination. 798

gastrulation Formation of a gastrula from a blastula; characterized by an invagination to form cell layers of a caplike structure. 798

gel electrophoresis Process that separates molecules, such as proteins and DNA, based on their size and electrical charge by passing them through a matrix. 257

gene Unit of heredity existing as alleles on the chromosomes; in diploid organisms, typically two alleles are inherited—one from each parent. 5, 70

gene cloning DNA cloning to produce many identical copies of the same gene. 255

gene flow Sharing of genes between two populations through interbreeding. 294

gene mutation Altered gene whose sequence of bases differs from the original sequence. 247

gene pharming Production of pharmaceuticals using transgenic organisms, usually agricultural animals. 258

gene pool Total of the alleles of all the individuals in a population. 290

gene therapy Correction of a detrimental mutation by the insertion of DNA sequences into the genome of a cell. 255, 260

genetically modified organisms (GMO) Organism whose genetic material has been altered or enhanced using DNA technology. 258

genetic code Universal code that has existed for eons and allows for conversion of DNA and RNA's chemical code to a sequence of amino acids in a protein. Each codon consists of three bases that stand for one of the 20 amino acids found in proteins or directs the termination of translation. 224

genetic diversity Variety among members of a population. 908

genetic drift Mechanism of evolution due to random changes in the allelic frequencies of a population; more likely to occur in small populations or when only a few individuals of a large population reproduce. 294

genetic profile An individual's genome, including any possible mutations. 263

genetic recombination Process in which chromosomes are broken and rejoined to form novel combinations; in this way offspring receive alleles in combinations different from their parents. 175

genomics Area of study that examines the genome of a species or group of species. 261

genotype Genes of an organism for a particular trait or traits; often designated by letters—for example, *BB* or *Aa*. 196

genus One of the categories, or taxa, used by taxonomists to group species; contains those species that are most closely related through evolution. 6, 349

geologic timescale History of the Earth based on the fossil record and divided into eras, periods, and epochs. 334

germ cell During zygote development, cells that are set aside from the somatic cells and that will eventually undergo meiosis to produce gametes. 773

germinate Beginning of growth of a seed, spore, or zygote, especially after a period of dormancy. 512

germ layer Primary tissue layer of a vertebrate embryo—namely, ectoderm, mesoderm, or endoderm. 522, 798

gibberellin Plant hormone promoting increased stem growth; also involved in flowering and seed germination. 486

gills Respiratory organ in most aquatic animals; in fish, an outward extension of the pharynx. 550, 664

ginkgo Member of phylum Ginkgophyte; maidenhair tree. 432

girdling Removing a strip of bark from around a tree. 478

gland Epithelial cell or group of epithelial cells that are specialized to secrete a substance. 589

glaucoma Condition in which the fluid in the eye (aqueous humor) accumulates, increasing pressure in the eye and damaging nerve fibers. 724

global warming Predicted increase in the Earth's temperature due to human activities that promote the greenhouse effect. 877

globular protein Most of the proteins in the body; soluble in water or salt solution; includes albumins, globulins, histones. 52

glomerular capsule Cuplike structure that is the initial portion of a nephron. 685

glomerular filtration Movement of small molecules from the glomerulus into the glomerular capsule due to the action of blood pressure. 686

glomerulus Capillary network within the glomerular capsule of a nephron. 686

glottis Opening for airflow in the larynx. 668

glucagon Hormone produced by the pancreas that stimulates the liver to break down glycogen, thus raising blood glucose levels. 765

glucocorticoid Type of hormone secreted by the adrenal cortex that influences carbohydrate, fat, and protein metabolism; *See also* cortisol. 763

glucose Six-carbon monosaccharide; used as an energy source during cellular respiration and as a monomer of the structural polysaccharides. 42

glycerol Three-carbon carbohydrate with three hydroxyl groups attached; a component of fats and oils. 46

glycocalyx Gel-like coating outside the cell wall of a bacterium. If compact, it is called a capsule; if diffuse, it is called a slime layer. 65

glycogen Storage polysaccharide found in animals; composed of glucose molecules joined in a linear fashion but having numerous branches. 43

glycolipid Lipid in plasma membranes that contains an attached carbohydrate chain; assembled in the Golgi apparatus. 88

glycolysis Anaerobic breakdown of glucose that results in a gain of two ATP and the production of pyruvate; occurs in the cytoplasm of cells. 137, 138

glycoprotein Protein in plasma membranes that has an attached carbohydrate chain; assembled in the Golgi apparatus. 88

gnathostome Term used to describe any vertebrates with jaws. 552

gnetophyte Member of one of the four phyla of gymnosperms; Gnetophyta has only three living genera, which differ greatly from one another—e.g., *Welwitschia* and *Ephedra*. 432

golden brown algae Unicellular organism that contains pigments, including chlorophyll *a* and *c* and carotenoids, that produce its color. 391

Golgi apparatus Organelle consisting of sacs and vesicles that processes, packages, and distributes molecules about or from the cell. 72

gonad Organ that produces gametes; the ovary produces eggs, and the testis produces sperm. 773

gonadotropic hormone Substance secreted by the anterior pituitary that regulates the activity of the ovaries and testes; principally, follicle-stimulating hormone (FSH) and luteinizing hormone (LH). 757

granular leukocyte Form of white blood cell that contains spherical vesicles (granules) in its cytoplasm. 620

granum (pl., grana) Stack of chlorophyll-containing thylakoids in a chloroplast. 76, 121

grassland Biome characterized by rainfall greater than 25 cm/yr, grazing animals, and warm summers; includes the prairie of the U.S. midwest and the African savanna. 894

gravitational equilibrium Maintenance of balance when the head and body are motionless. 728

gravitropism Growth response of roots and stems of plants to the Earth's gravity; roots demonstrate positive gravitropism, and stems demonstrate negative gravitropism. 491

gray matter Nonmyelinated axons and cell bodies in the central nervous system. 702

green algae Members of a diverse group of photosynthetic protists; contain chlorophylls *a* and *b* and have other biochemical characteristics like those of plants. 387

greenhouse effect Reradiation of solar heat toward the Earth, caused by an atmosphere that allows the Sun's rays to pass through but traps the heat in the same manner as the glass of a greenhouse. 877

greenhouse gases Gases in the atmosphere such as carbon dioxide, methane, water vapor, ozone, and nitrous oxide that are involved in the greenhouse effect. 876

ground tissue Tissue that constitutes most of the body of a plant; consists of parenchyma, collenchyma, and sclerenchyma cells that function in storage, basic metabolism, and support. 447

growth factor A hormone or chemical, secreted by one cell, that may stimulate or inhibit growth of another cell or cells. 155

growth hormone (GH) Substance secreted by the anterior pituitary; controls size of an individual by promoting cell division, protein synthesis, and bone growth. 758

guanine (G) One of four nitrogen-containing bases in nucleotides composing the structure of DNA and RNA; pairs with cytosine. 217

guard cell One of two cells that surround a leaf stoma; changes in the turgor pressure of these cells cause the stoma to open or close. 477

guttation Liberation of water droplets from the edges and tips of leaves. 475

gymnosperm Type of woody seed plant in which the seeds are not enclosed by fruit and are usually borne in cones, such as those of the conifers. 430

H

habitat Place where an organism lives and is able to survive and reproduce. 839, 859

hair follicle Tubelike depression in the skin in which a hair develops. 599

halophile Type of archaean that lives in extremely salty habitats. 379

haploid (n) Cell condition in which only one of each type of chromosome is present. 157, 172

Hardy-Weinberg principle (equilibrium) Mathematical law stating that the gene frequencies in a population remain stable if evolution does not occur due to nonrandom mating, selection, migration, and genetic drift. 290,291

heart Muscular organ whose contraction causes blood to circulate in the body of an animal. 610

heart attack Damage to the myocardium due to blocked circulation in the coronary arteries; myocardial infarction. 616

heat Type of kinetic energy associated with the random motion of molecules. 105

helper T cell Secretes lymphokines, which stimulate all kinds of immune cells. 635

heme Iron-containing group found in hemoglobin. 672

hemizygous Possessing only one allele for a gene in a diploid organism; males are hemizygous for genes on the X chromosome. 209

hemocoel Residual coelom found in arthropods, which is filled with hemolymph. 534

hemoglobin (Hb) Iron-containing respiratory pigment occurring in vertebrate red blood cells and in the blood plasma of some invertebrates. 49, 618, 672

hemolymph Circulatory fluid that is a mixture of blood and interstitial fluid; seen in animals that have an open circulatory system, such as molluscs and arthropods. 606

hepatitis Inflammation of the liver. Viral hepatitis occurs in several forms. 654

herbaceous stem Nonwoody stem; herbaceous plants tend to die back to ground level at the end of the growing season. 454

herbivore Primary consumer in a grazing food chain; a plant eater. 870

hermaphroditic Type of animal that has both male and female sex organs. 531, 773

heterochromatin Highly compacted chromatin that is not accessible for transcription. 234

heterosporous Seed plant that produces two types of spores—microspores and megaspores. A plant that produces only one type of spore is *homosporous*. 426

heterotrophy Organism that cannot synthesize needed organic compounds from inorganic substances and therefore must take in organic food. 120, 870

heterozygote advantage Situation in which individuals heterozygous for a trait have a selective advantage over those who are homozygous dominant or recessive; an example is sickle-cell disease. 302

heterozygous Possessing unlike alleles for a particular trait. 196

hexose Any monosaccharide that contains six carbons; examples are glucose and galactose. 42

hippocampus Region of the central nervous system associated with learning and memory; part of the limbic system. 705

histamine Substance, produced by basophils in blood and mast cells in connective tissue, that causes capillaries to dilate. 631

histone A group of proteins involved in forming the nucleosome structure of eukaryote chromatin. 157, 233

homeobox 180-nucleotide sequence located in all homeotic genes. 804

homeodomain Conserved DNA-binding region of transcription factors encoded by the homeobox of homeotic genes. 804

homeostasis Maintenance of normal internal conditions in a cell or an organism by means of self-regulating mechanisms. 4, 599

homeothermic Ability to regulate the internal body temperature around an optimum value. 599

homeotic genes Genes that control the overall body plan by controlling the fate of groups of cells during development. 803

hominid Taxon that includes humans, chimpanzees, gorillas, and orangutans. 572

hominin Taxon that includes humans and species very closely related to humans plus chimpanzees. 572

hominine Taxon that includes humans, chimpanzees, and gorillas. 572

hominoid Taxon that includes the hominids plus the gibbons. 572

homologous chromosome Member of a pair of chromosomes that are alike and come together in synapsis during prophase of the first meiotic division; a *homologue*. 172

homologous gene Gene that codes for the same protein, even if the base sequence may be different. 265

homologous, homologous structure A structure that is similar in different types of organisms because these organisms descended from a common ancestor. 283, 317, 357

homologue Member of a homologous pair of chromosomes. 172

homology Similarity of parts or organs of different organisms caused by evolutionary derivation from a corresponding part or organ in a remote ancestor, and usually having a similar embryonic origin. 357

homosporous A plant that produces only one type of asexual spore. 426

homozygous Possessing two identical alleles for a particular trait. 196

hormone Chemical messenger produced in one part of the body that controls the activity of other parts. 484, 753

hornwort A bryophyte of the phylum Anthocerophyta with a thin gametophyte and tiny sporophyte that resembles a broom handle. 424

horsetail A seedless vascular plant having only one genus (*Equisetum*) in existence today;

characterized by rhizomes, scalelike leaves, strobili, and tough, rigid stems. 427

host Organism that provides nourishment and/or shelter for a parasite. 865

human chorionic gonadotropin (HCG) Gonadotropic hormone produced by the chorion that functions to maintain the uterine lining. 782, 806

Human Genome Project (HGP) Initiative to determine the complete sequence of the human genome and to analyze this information. 261

human immunodeficiency virus (HIV) Virus responsible for the disease AIDS. 637, 787

humus Decomposing organic matter in the soil. 467

hunter-gatherer A hominin that hunted animals and gathered plants for food. 580

hyaline cartilage Cartilage whose cells lie in lacunae separated by a white translucent matrix containing very fine collagen fibers. 591

hybridization Interbreeding between two different species; typically prevented by prezygotic isolation mechanisms. 310

hydrogen bond Weak bond that arises between a slightly positive hydrogen atom of one molecule and a slightly negative atom of another molecule or between parts of the same molecule. 28

hydrogen ion (H⁺) Hydrogen atom that has lost its electron and therefore bears a positive charge. 32

hydrolysis reaction Splitting of a chemical bond by the addition of water, with the H⁺ going to one molecule and the OH⁻ going to the other. 41

hydrophilic Type of molecule, often polar, that interacts with water by dissolving in water and/or by forming hydrogen bonds with water molecules. 40

hydrophobic Type of molecule, that is typically nonpolar, and therefore does not interact easily with water. 40

hydroponics Technique for growing plants by suspending them with their roots in a nutrient solution. 465

hydrostatic skeleton Fluid-filled body compartment that provides support for muscle contraction resulting in movement; seen in cnidarians, flatworms, roundworms, and segmented worms. 528, 736

hydrothermal vent Hot springs in the seafloor along ocean ridges where heated sea water and sulfate react to produce hydrogen sulfide; here, chemosynthetic bacteria support a community of varied organisms. 902

hydroxide ion (OH⁻) One of two ions that results when a water molecule dissociates; it has gained an electron and therefore bears a negative charge. 32

hypersensitive response (HR) Plants respond to pathogens by selectively killing plant cells to block the spread of the pathogen. 498

hyperthyroidism Caused by the over-secretion of hormones from the thyroid gland; symptoms include hyperactivity, nervousness, and insomnia. 761

hypertonic solution Higher solute concentration (less water) than the cytoplasm of a cell; causes cell to lose water by osmosis. 94

hypha Filament of the vegetative body of a fungus. 405

hypertension Form of cardiovascular disease characterized by blood pressure over 140/95 (over 45 years of age) or 130/90 (under 45 years of age). 616

hypothalamic-inhibiting hormone One of many hormones produced by the hypothalamus that inhibits the secretion of an anterior pituitary hormone. 757

hypothalamic-releasing hormone One of many hormones produced by the hypothalamus that stimulates the secretion of an anterior pituitary hormone. 757

hypothalamus In vertebrates, part of the brain that helps regulate the internal environment of the body—for example, heart rate, body temperature, and water balance. 704, 757

hypothesis Supposition established by reasoning after consideration of available evidence; it can be tested by obtaining more data, often by experimentation. 11

hypothyroidism Caused by the under-secretion of hormones from the thyroid gland; symptoms include weight gain, lethargic behavior, and depression. 761

hypotonic solution Solution that contains a lower solute (more water) concentration than the cytoplasm of a cell; causes cell to gain water by osmosis. 93

I

immediate allergic response Allergic response that occurs within seconds of contact with an allergen; caused by the attachment of the allergen to IgE antibodies. 641

immune system System associated with protection against pathogens, toxins, and some cancerous cells; in humans this is an organ system. 627

immunity Ability of the body to protect itself from foreign substances and cells, including disease-causing agents. 630

immunization Strategy for achieving immunity to the effects of specific disease-causing agents. 639

immunoglobulin (Ig) Globular plasma protein that functions as an antibody. 634

implantation In placental mammals, the embedding of an embryo at the blastocyst stage into the endometrium of the uterus. 806

imprinting Learning to make a particular response to only one type of animal or object. 823

inbreeding Mating between closely related individuals; influences the genotype ratios of the gene pool. 295

inclusive fitness Fitness that results from personal reproduction and from helping nondescendant relatives reproduce. 834

incomplete digestive tract Digestive tract that has a single opening, usually called a mouth. 647

incomplete dominance Inheritance pattern in which an offspring has an intermediate phenotype, as when a red-flowered plant and a white-flowered plant produce pink-flowered offspring. 205

incomplete penetrance Dominant alleles that are either not always, or partially expressed. 206

independent assortment Alleles of unlinked genes segregate independently of each other during meiosis so that the gametes could contain all possible combinations of alleles. 175

index fossil Deposits found in certain layers of strata; similar fossils can be found in the same strata around the world. 334

induced fit model Change in the shape of an enzyme's active site that enhances the fit between the active site and its substrate(s). 110

induced mutation Mutation that is caused by an outside influence, such as organic chemicals or ionizing radiation. 247

inducer Molecule that brings about activity of an operon by joining with a repressor and preventing it from binding to the operator. 240

induction Ability of a chemical or a tissue to influence the development of another tissue. 800

inductive reasoning Using specific observations and the process of logic and reasoning to arrive at general scientific principles. 11

infertility Inability to have as many children as desired. 784

inflammatory response Tissue response to injury that is characterized by redness, swelling, pain, and heat. 631

ingroup In a cladistic study of evolutionary relationships among organisms, the group that is being analyzed. 356

inheritance of acquired characteristics Lamarckian belief that characteristics acquired during the lifetime of an organism can be passed on to offspring. 273

initiation First stage of translation in which the translational machinery binds an mRNA and assembles. 230

innate immunity An immune response that does not require a previous exposure to the pathogen. 627

inner ear Portion of the ear consisting of a vestibule, semicircular canals, and the cochlea where equilibrium is maintained and sound is transmitted. 727

inorganic chemistry Branch of science that studies the chemical reactions and properties of all of the elements, except hydrogen and carbon. 38

insect Type of arthropod. The head has antennae, compound eyes, and simple eyes; the thorax has three pairs of legs and often wings; and the abdomen has internal organs. 541

insight learning Ability to apply prior learning to a new situation without trial-and-error activity. 825

inspiration Act of taking air into the lungs; inhalation. 669

insulin Hormone released by the pancreas that serves to lower blood glucose levels by stimulating the uptake of glucose by cells, especially muscle and liver cells. 765

integration Summing up of excitatory and inhibitory signals by a neuron or by some part of the brain. 701

interferon Antiviral agent produced by an infected cell that blocks the infection of another cell. 633

intergenic sequence Region of DNA that lies between genes on a chromosome. 262

interkinesis Period of time between meiosis I and meiosis II during which no DNA replication takes place. 176

intermediate filament Ropelike assemblies of fibrous polypeptides in the cytoskeleton that provide support and strength to cells; so called because they are intermediate in size between actin filaments and microtubules. 78

internal respiration Exchange of oxygen and carbon dioxide between blood and tissue fluid. 664

interneuron Neuron located within the central nervous system that conveys messages between parts of the central nervous system. 698

internode In vascular plants, the region of a stem between two successive nodes. 445

interphase Stages of the cell cycle (G_1, S, G_2) during which growth and DNA synthesis occur when the nucleus is not actively dividing. 154

interspersed repeat Repeated DNA sequence that is spread across several regions of a chromosome or across multiple chromosomes. 262

intertidal zone Region along a coastline where the tide recedes and returns. 900

intron Intervening sequence found between exons in mRNA; removed by RNA processing before translation. 226

inversion Change in chromosome structure in which a segment of a chromosome is turned around 180°; this reversed sequence of genes can lead to altered gene activity and abnormalities. 187

invertebrate Animal without a vertebral column or backbone. 521

ion Charged particle that carries a negative or positive charge. 26

ionic bond Chemical bond in which ions are attracted to one another by opposite charges. 26

iris Muscular ring that surrounds the pupil and regulates the passage of light through this opening. 721

iron-sulfur world Hypothesis that ocean thermal vents provided all of the materials needed for abiotic synthesis of the first molecules. 329

island biogeography model Proposes that the biodiversity on an island is dependent on its distance from the mainland, with islands located a greater distance having a lower level of diversity. 859

isomer Molecules with the same molecular formula but a different structure, and therefore a different shape. 40

isotonic solution Solution that is equal in solute concentration to that of the cytoplasm of a cell; causes cell to neither lose nor gain water by osmosis. 93

isotope Atoms of the same element having the same atomic number but a different mass number due to a variation in the number of neutrons. 24

iteroparity Repeated production of offspring at intervals throughout the life cycle of an organism. 843

J

jaundice Yellowish tint to the skin caused by an abnormal amount of bilirubin (bile pigment) in the blood, indicating liver malfunction. 654

jawless fishes Type of fishes that lack jaws (agnathan); includes hagfishes and lampreys. 553

joint Articulation between two bones of a skeleton. 742

junction protein(s) Proteins in the cell membrane that assist in cell-to-cell communication. 89

K

karyotype Chromosomes arranged by pairs according to their size, shape, and general appearance in mitotic metaphase. 184

keystone species Species whose activities significantly affect community structure. 918

kidneys Paired organs of the vertebrate urinary system that regulate the chemical composition of the blood and produce a waste product called urine. 684

kinetic energy Energy associated with motion. 105

kinetochore An assembly of proteins that attaches to the centromere of a chromosome during mitosis. 157

kinetoplastid Unicellular, flagellate protist characterized by the presence in their single mitochondrion of a kinetoplast (a structure containing a large mass of DNA). 396

kingdom One of the categories, or taxa, used by taxonomists to group species; the taxon above phylum. 6, 349

kin selection Indirect selection; adaptation to the environment due to the reproductive success of an individual's relatives. 834

***K*-selection** Favorable life-history strategy under stable environmental conditions characterized by the production of a few offspring with much attention given to offspring survival. 849

L

lactation Secretion of milk by mammary glands, usually for the nourishment of an infant. 782

lacteal Lymphatic vessel in an intestinal villus; aids in the absorption of fats. 653

lacuna Small pit or hollow cavity, as in bone or cartilage, where a cell or cells are located. 590

lake Body of fresh water, often classified by nutrient status, such as oligotrophic (nutrient-poor) or eutrophic (nutrient-rich). 898

landscape A number of interacting ecosystems. 909

landscape diversity Variety of habitat elements within an ecosystem (e.g., plains, mountains, and rivers). 909

large intestine In vertebrates, portion of the digestive tract that follows the small intestine; in humans, consists of the cecum, colon, rectum, and anal canal. 653

larynx Cartilaginous organ located between the pharynx and the trachea; in humans, contains the vocal cords; sometimes called the voice box. 668

last universal common ancestor (LUCA) The first living organism on the planet, from which all life evolved. 328

lateral line Canal system containing sensory receptors that allow fishes and amphibians to detect water currents and pressure waves from nearby objects. 726

law Universal principle that describes the basic functions of the natural world. 12

law of independent assortment Mendelian principle that explains how combinations of traits appear in gametes; *see also* independent assortment. 198

law of segregation Mendelian principle that explains how, in a diploid organism, alleles separate during the formation of the gametes. 196

laws of thermodynamics Two laws explaining energy and its relationships and exchanges. The first, also called the "law of conservation," says that energy cannot be created or destroyed but can only be changed from one form to another; the second says that energy cannot be changed from one form to another without a loss of usable energy. 105

leaf Lateral appendage of a stem, highly variable in structure, often containing cells that carry out photosynthesis. 445

leaf vein Vascular tissue within a leaf. 450

learning Relatively permanent change in an animal's behavior that results from practice and experience. 823

lens Transparent, disclike structure found in the vertebrate eye behind the iris; brings objects into focus on the retina. 721

leptin Hormone produced by adipose tissue that acts on the hypothalamus to signal satiety (fullness). 768

less-developed country (LDC) Country that is becoming industrialized; typically, population growth is expanding rapidly, and the majority of people live in poverty. 851

leucoplasts Plastid, generally colorless, that synthesizes and stores starch and oils. 77

leukocyte Another name for a white blood cell (WBC). 620

lichen Symbiotic relationship between certain fungi and either cyanobacteria or algae, in which the fungi possibly provide inorganic food or water and the algae or cyanobacteria provide organic food. 377, 414

life cycle Recurring pattern of genetically programmed events by which individuals grow, develop, maintain themselves, and reproduce. 180

ligament Tough cord or band of dense fibrous tissue that binds bone to bone at a joint. 590, 742

light reaction Portion of photosynthesis that captures solar energy and takes place in thylakoid membranes of chloroplasts; it produces ATP and NADPH. 122

lignin Chemical that hardens the cell walls of land plants. 426, 449

limbic system In humans, functional association of various brain centers, including the amygdala and hippocampus; governs learning and memory and various emotions such as pleasure, fear, and happiness. 705

limiting factor Resource or environmental condition that restricts the abundance and distribution of an organism. 840

lineage Line of descent represented by a branch in a phylogenetic tree. 354

lipase Fat-digesting enzyme secreted by the pancreas. 657

lipid Class of organic compounds that tends to be soluble in nonpolar solvents; includes fats and oils. 45

liposome Droplet of phospholipid molecules formed in a liquid environment. 331

liver Large, dark red internal organ that produces urea and bile, detoxifies the blood, stores glycogen, and produces the plasma proteins, among other functions. 654

liverwort Bryophyte with a lobed or leafy gametophyte and a sporophyte composed of a stalk and capsule. 424

lobe-finned fishes Type of fishes with limblike fins; also called the sarcopterygians. 555

locus Physical location of a trait (or gene) on a chromosome. 196

logistic growth Population increase that results in an S-shaped curve; growth is slow at first, steepens, and then levels off due to environmental resistance. 844

long-day plant Plant that flowers when day length is longer than a critical length; e.g., wheat, barley, clover, and spinach. 496

loop of the nephron Portion of a nephron between the proximal and distal convoluted tubules; functions in water reabsorption. 686

loose fibrous connective tissue Tissue composed mainly of fibroblasts widely separated by a matrix containing collagen and elastic fibers. 590

lophophoran A general term to describe several groups of lophotrochozoans that have a feeding structure called a lophophore. 530

lophotrochozoa Main group of protostomes; widely diverse. Includes the flatworms, rotifers, annelids, and molluscs. 530

lumen Cavity inside any tubular structure, such as the lumen of the digestive tract. 651

lung cancer Uncontrolled cell growth that affects any component of the respiratory system. 675

lungs Internal respiratory organ containing moist surfaces for gas exchange. 665

lungfishes Type of lobe-finned fishes that utilize lungs in addition to gills for gas exchange. 555

luteal phase Second half of the ovarian cycle, during which the corpus luteum develops and much progesterone (and some estrogen) is produced. 781

lycophyte Club mosses, among the first vascular plants to evolve and to have leaves. The leaves of the lycophytes are microphylls. 426

lymph Fluid, derived from tissue fluid, that is carried in lymphatic vessels. 592, 622, 629

lymphatic capillary Smallest vessels of the lymphatic system; closed-ended; responsible for the uptake of fluids from the surrounding tissues. 629

lymphatic system Organ system consisting of lymphatic vessels and lymphatic organs; transports lymph and lipids, and aids the immune system. 628

lymphatic vessel Vessel that carries lymph. 628

lymph node Mass of lymphatic tissue located along the course of a lymphatic vessel. 629

lymphocyte Specialized white blood cell that functions in specific defense; occurs in two forms—T lymphocytes and B lymphocytes. 620

lymphoid (lymphatic) organ Organ other than a lymphatic vessel that is part of the lymphatic system; the lymphatic organs are the lymph nodes, tonsils, spleen, thymus gland, and bone marrow. 629

lysogenic cell Cell that contains a prophage (virus incorporated into DNA), which is replicated when the cell divides. 368

lysogenic cycle Bacteriophage life cycle in which the virus incorporates its DNA into that of a bacterium; occurs preliminary to the lytic cycle. 368

lysosome Membrane-bounded vesicle that contains hydrolytic enzymes for digesting macromolecules and bacteria; used to recycle worn-out cellular organelles. 73

lytic cycle Bacteriophage life cycle in which the virus takes over the operation of the bacterium immediately upon entering it and subsequently destroys the bacterium. 368

M

macroevolution Large-scale evolutionary change, such as the formation of new species. 307

macronutrient Essential element needed in large amounts for plant growth, such as nitrogen, calcium, or sulfur. 465

macrophage In vertebrates, large phagocytic cell derived from a monocyte that ingests microbes and debris. 632

macular degeneration Condition in which the capillaries supplying the retina of the eye become damaged resulting in reduced vision and blindness. 724

malignant The power to threaten life; cancerous. 163

maltase Enzyme produced in small intestine that breaks down maltose to two glucose molecules. 656

mammal Endothermic vertebrate characterized especially by the presence of hair and mammary glands. 564

mantle In molluscs, an extension of the body wall that covers the visceral mass and may secrete a shell. 534

marsh Soft wetland, that is treeless. 897

marsupial Member of a group of mammals bearing immature young nursed in a marsupium, or pouch—for example, kangaroo and opossum. 564

mass extinction Episode of large-scale extinction in which large numbers of species disappear in a few million years or less. 338

mass number Mass of an atom equal to the number of protons plus the number of neutrons within the nucleus. 23

mast cell Connective tissue cell that releases histamine in allergic reactions. 631

maternal determinant One of many substances present in the egg that influences the course of development. 800

matrix Unstructured semifluid substance that fills the space between cells in connective tissues or inside organelles. 77

matter Anything that takes up space and has mass. 22

mechanical energy A type of kinetic energy associated with the position, or motion (such as walking or running) of an object. 105

mechanoreceptor Sensory receptor that responds to mechanical stimuli, such as pressure, sound waves, or gravity. 717

medulla oblongata In vertebrates, part of the brain stem that is continuous with the spinal cord; controls heartbeat, blood pressure, breathing, and other vital functions. 705

medusa Among cnidarians, bell-shaped body form that is directed downward and contains much mesoglea. 528

megafauna Large animals, such as humans, bears, and deer. Often defined as those over 100 pounds (44 kg) in adult size. 341

megaphyll Large leaf with several to many veins. 427

megaspore One of the two types of spores produced by seed plants; develops into a female gametophyte (embryo sac). 426, 502

meiosis Type of nuclear division that reduces the chromosome number from 2n to n; daughter cells receive the haploid number of chromosomes in varied combinations; also called reduction division. 172

melanocyte Specialized cell in the epidermis that produces melanin, the pigment responsible for skin color. 598

melanocyte-stimulating hormone (MSH) Substance that causes melanocytes to secrete melanin in most vertebrates. 757

melatonin Hormone, secreted by the pineal gland, that is involved in biorhythms. 768

membrane-first hypothesis Proposes that the plasma membrane was the first component of the early cells to evolve. 331

memory Capacity of the brain to store and retrieve information about past sensations and perceptions; essential to learning. 707

memory B cell Forms during a primary immune response but enters a resting phase until a secondary immune response occurs. 634

memory T cell T cell that differentiated during an initial infection and responds rapidly during subsequent exposure to the same antigen. 636

meninges Protective membranous coverings around the central nervous system. 702

meningitis A condition that refers to inflammation of the brain or spinal cord meninges (membranes). 702

menopause Termination of the ovarian and uterine cycles in older women. 782, 813

menstruation Periodic shedding of tissue and blood from the inner lining of the uterus in primates. 781

meristem Undifferentiated embryonic tissue in the active growth regions of plants. 447

mesoderm Middle primary tissue layer of an animal embryo that gives rise to muscle, several internal organs, and connective tissue layers. 798

mesoglea Transparent jellylike substance located between the endoderm and ectoderm of some sponges and cnidarians. 528

mesophyll Inner, thickest layer of a leaf consisting of palisade and spongy mesophyll; the site of most of photosynthesis. 459

mesosome In a bacterium, plasma membrane that folds into the cytoplasm and increases surface area. 65

messenger RNA (mRNA) Type of RNA formed from a DNA template and bearing coded information for the amino acid sequence of a polypeptide. 223

metabolic pathway Series of linked reactions, beginning with a particular reactant and terminating with an end product. 109

metabolic pool Metabolites that are the products of and/or the substrates for key reactions in cells, allowing one type of molecule to be changed into another type, such as carbohydrates converted to fats. 147

metabolism The sum of the chemical reactions that occur in a cell. 4, 107

metamorphosis Change in shape and form that some animals, such as insects, undergo during development. 521, 540

metaphase Third phase of mitosis; chromosomes are aligned at the metaphase plate. 159

metaphase plate A disk formed during metaphase in which all of a cell's chromosomes lie in a single plane at right angles to the spindle fibers. 159

metapopulation Population subdivided into several small and isolated populations due to habitat fragmentation. 919

metastasis Spread of cancer from the place of origin throughout the body; caused by the ability of cancer cells to migrate and invade tissues. 164

methanogen Type of archaean that lives in oxygen-free habitats, such as swamps, and releases methane gas. 378

MHC (major histocompatibility complex) protein Protein marker that is a part of cell-surface markers anchored in the plasma membrane, which the immune system uses to identify "self." 635

micelle Single layer of fatty acids (or phospholipids) that orientate themselves in an aqueous environment. 331

microevolution Change in gene frequencies between populations of a species over time. 290

microglia Supportive cells of the nervous system that help remove bacteria and debris, thus supporting the activity of the neurons. 697

micronutrient Essential element needed in small amounts for plant growth, such as boron, copper, and zinc. 465

microphyll Small leaf with one vein. 426

microRNA (miRNA) Short sequences of RNA, usually less than 22 nucleotides, that are involved in posttranscriptional regulation of gene expression. These molecules either inhibit, or reduce, the expression of specific genes. 246

microsphere Formed from proteinoids exposed to water; has properties similar to those of today's cells. 330

microspore One of the two types of spores produced by seed plants; develops into a male gametophyte (pollen grain). 426, 502

microtubule Small, cylindrical organelle composed of tubulin protein around an empty central core; present in the cytoplasm, centrioles, cilia, and flagella. 78

midbrain In mammals, the part of the brain located below the thalamus and above the pons. 704

middle ear Portion of the ear consisting of the tympanic membrane, the oval and round windows, and the ossicles, where sound is amplified. 727

migration Regular back-and-forth movement of animals between two geographic areas at particular times of the year. 825

millipede More-or-less cylindrical arthropod characterized by having two pairs of short legs on most of its body segments; may have 13 to almost 200 pairs of legs. 541

mimicry Superficial resemblance of two or more species; a survival mechanism that avoids predation by appearing to be noxious. 864

mineral Naturally occurring inorganic substance containing two or more elements; certain minerals are needed in the diet. 465

mineralocorticoid Hormones secreted by the adrenal cortex that regulate salt and water balance, leading to increases in blood volume and blood pressure. 763

mitochondria (sing., mitochondrion) Membrane-bounded organelle in which ATP molecules are produced during the process of cellular respiration. 76, 142

mitosis The stage of cellular reproduction in which nuclear division occurs; process in which a parent nucleus produces two daughter nuclei, each having the same number and kinds of chromosomes as the parent nucleus. 155

mitotic spindle A complex of microtubules and associated proteins that assist in separating the chromatids during cell division. 155

mixotrophic Organism that can use autotrophic and heterotrophic means of gaining nutrients. 384

model Simulation of a process that aids conceptual understanding until the process can be studied firsthand; a hypothesis that describes how a particular process could possibly be carried out. 12

mold Various fungi whose body consists of a mass of hyphae (filaments) that grow on and receive nourishment from organic matter such as human food and clothing. 410

mole The molecular weight of a molecule expressed in grams; contains 6.023×10^{23} molecules. 107

molecular clock Idea that the rate at which mutational changes accumulate in certain genes is constant over time and is not involved in adaptation to the environment. 359

molecule Union of two or more atoms of the same element; also, the smallest part of a compound that retains the properties of the compound. 26

mollusc Invertebrates including squids, clams, snails, and chitons; characterized by a visceral mass, a mantle, and a foot. 534

monoclonal antibody One of many antibodies produced by a clone of hybridoma cells that all bind to the same antigen. 635

monocot (Monocotyledone) Flowering plant group; members have one embryonic leaf (cotyledon), parallel-veined leaves, scattered vascular bundles, flower parts in threes or multiples of three, and other characteristics. 433, 446

monocyte Type of agranular leukocyte that functions as a phagocyte, particularly after it becomes a macrophage, which is also an antigen-presenting cell. 620

monoecious Having male flowers or cones and female flowers or cones on a single plant. 430

monogamous Breeding pair of organisms that reproduce only with each other through their lifetime. 831

monohybrid cross Cross between parents that differ in only one trait. 195

monomer Small molecule that is a subunit of a polymer—e.g., glucose is a monomer of starch. 40

monophyletic Group of species including the most recent common ancestor and all its descendants. 308, 384

monosaccharide Simple sugar; a carbohydrate that cannot be broken down by hydrolysis—e.g., glucose; also, any monomer of the polysaccharides. 41

monosomy Chromosome condition in which a diploid cell has one less chromosome than normal; designated as 2n-1. 183

monotreme Egg-laying mammal—e.g., duckbill platypus or spiny anteater. 564

monsoon Climate in India and southern Asia caused by wet ocean winds that blow onshore for almost half the year. 885

montane coniferous forest Coniferous forest of a mountain. 887

more-developed country (MDC) Country that is industrialized; typically, population growth is low, and the people enjoy a good standard of living overall. 851

morphogenesis Emergence of shape in tissues, organs, or entire embryo during development. 800

morphological species concept Definition of a species that defines species by specific diagnostic traits. 307

morphology Physical characteristics that contribute to the appearance of an organism. 307

morula Spherical mass of cells resulting from cleavage during animal development prior to the blastula stage. 797

mosaic evolution Concept that human characteristics did not evolve at the same rate; for example, some body parts are more humanlike than others in early hominins. 577

moss Bryophyte that is typically found in moist habitats. 424

motor molecule Protein that moves along either actin filaments or microtubules and translocates organelles. 78

motor (efferent) neuron Nerve cell that conducts nerve impulses away from the central nervous system and innervates effectors (muscle and glands). 697

mouth In humans, organ of the digestive tract where food is chewed and mixed with saliva. 650

mRNA transcript mRNA molecule formed during transcription that has a sequence of bases complementary to a gene. 226

mucosa Epithelial membrane containing cells that secrete mucus; found in the inner cell layers of the digestive (first layer) and respiratory tracts. 651

multicellular Organism composed of many cells; usually has organized tissues, organs, and organ systems. 2

multiple alleles Inheritance pattern in which there are more than two alleles for a particular trait; each individual has only two of all possible alleles. 205

multiple sclerosis Disease of the central nervous system characterized by the breakdown of myelin in the neurons; considered to be an autoimmune disease. 705

muscularis Smooth muscle layer found in the digestive tract. 651

muscular (contractile) tissue Type of animal tissue composed of fibers that shorten and lengthen to produce movements. 592

mutagen Chemical or physical agent that increases the chance of mutation. 248

mutualism Symbiotic relationship in which both species benefit in terms of growth and reproduction. 375, 866

mycelium Tangled mass of hyphal filaments composing the vegetative body of a fungus. 405

mycorrhizae (sing., mycorrhiza) Mutualistic relationship between fungal hyphae and roots of vascular plants. 414, 472

myelin sheath White, fatty material—derived from the membrane of neurolemmocytes—that forms a covering for nerve fibers. 697

myofibril Specific muscle cell organelle containing a linear arrangement of sarcomeres, which shorten to produce muscle contraction. 745

myosin Muscle protein making up the thick filaments in a sarcomere; it pulls actin to shorten the sarcomere, yielding muscle contraction. 745

N

NAD⁺ (nicotinamide adenine dinucleotide) Coenzyme in oxidation-reduction reactions that accepts electrons and hydrogen ions to become NADH + H⁺ as oxidation of substrates occurs. During cellular respiration, NADH carries electrons to the electron transport chain in mitochondria. 114, 136

NADP⁺ (nicotinamide adenine dinucleotide phosphate) Coenzyme in oxidation-reduction reactions that accepts electrons and hydrogen ions to become NADPH + H⁺. During photosynthesis, NADPH participates in the reduction of carbon dioxide to a carbohydrate. 113

nail Flattened epithelial tissue from the stratum lucidum of the skin; located on the tips of fingers and toes. 598

natural killer (NK) cell Lymphocyte that causes an infected or cancerous cell to burst. 632

natural group In systematics, groups of organisms that possess a shared evolutionary history. 348

natural selection Mechanism of evolutionary change caused by environmental selection of organisms most fit to reproduce; results in adaptation to the environment. 8, 277

navigate To steer or manage a course by adjusting one's bearings and following the result of the adjustment. 825

Neandertal Hominin with a sturdy build that lived during the last Ice Age in Europe and the Middle East; hunted large game and left evidence of being culturally advanced. 580

nearsighted (myopic) Condition in which an individual cannot focus on objects at a distance; caused by the focusing of the image in front of the retina of the eye. 724

negative feedback Mechanism of homeostatic response by which the output of a system suppresses or inhibits activity of the system. 600, 757

nematocyst In cnidarians, a capsule that contains a threadlike fiber, the release of which aids in the capture of prey. 528

nephridium (pl., nephridia) Segmentally arranged, paired excretory tubules of many invertebrates, as in the earthworm. 536

nephron Microscopic kidney unit that regulates blood composition by glomerular filtration, tubular reabsorption, and tubular secretion. 685

nerve Bundle of long axons outside the central nervous system. 593, 707

nerve fiber Axon; conducts nerve impulses away from the cell. Axons are classified as either myelinated or unmyelinated based on the presence or absence of a myelin sheath. 697

nerve net Diffuse, noncentralized arrangement of nerve cells in cnidarians. 529, 694

nervous tissue Tissue that contains nerve cells (neurons), which conduct impulses, and neuroglia, which support, protect, and provide nutrients to neurons. 593

neural plate Region of the dorsal surface of the chordate embryo that marks the future location of the neural tube. 799

neural tube Tube formed by closure of the neural groove during development. In vertebrates, the neural tube develops into the spinal cord and brain. 799

neurodegenerative disease Disease, usually caused by a prion, virus, or bacterium, that damages or impairs the function of nervous tissue. 371

neuroglia Nonconducting nerve cells that are intimately associated with neurons and function in a supportive capacity. 593, 697

neuromuscular junction Region where an axon bulb approaches a muscle fiber; contains a presynaptic membrane, a synaptic cleft, and a postsynaptic membrane. 747

neuron Nerve cell that characteristically has three parts: dendrites, cell body, and an axon. 593, 697

neurotransmitter Chemical stored at the ends of axons that is responsible for transmission across a synapse. 700

neutron Neutral subatomic particle, located in the nucleus and assigned one atomic mass unit. 22

neutrophil Granular leukocyte that is the most abundant of the white blood cells; first to respond to infection. 631

nitrification Process by which nitrogen in ammonia and organic compounds is oxidized to nitrites and nitrates by soil bacteria. 878

N₂ (nitrogen) fixation Process whereby free atmospheric nitrogen is converted into compounds, such as ammonium and nitrates, usually by bacteria. 878

node In plants, the place where one or more leaves attach to a stem. 421, 445

nodes of Ranvier Gaps in the myelin sheath around a nerve fiber. 697

noncompetitive inhibition Form of enzyme inhibition where the inhibitor binds to an enzyme at a location other than the active site; while at this site, the enzyme shape changes, the inhibitor is unable to bind to its substrate, and no product forms. 112

noncyclic pathway Light-dependent photosynthetic pathway that is used to generate ATP and NADPH; because the pathway is noncyclic, the electrons must be replaced by the splitting of water (photolysis). 124

nondisjunction Failure of the homologous chromosomes or sister chromatids to separate during either mitosis or meiosis; produces cells with abnormal chromosome numbers. 183

nomenclature In systematics, the process of assigning names to taxonomic groups; usually determined by governing organizations. 350

nonpolar covalent bond Bond in which the sharing of electrons between atoms is fairly equal. 27

nonrandom mating Mating among individuals on the basis of their phenotypic similarities or differences, rather than mating on a random basis. 295

nonseptate Lacking cell walls; some fungal species have hyphae that are nonseptate. 405

nonvascular plants Bryophytes, such as mosses and liverworts, that have no vascular tissue and either occur in moist locations or have special adaptations for living in dry locations. 423

norepinephrine Neurotransmitter of the postganglionic fibers in the sympathetic division of the autonomic system; also, a hormone produced by the adrenal medulla. 762

notochord Cartilage-like supportive dorsal rod in all chordates at some time in their life cycle; replaced by vertebrae in vertebrates. 550, 799

nuclear envelope Double membrane that surrounds the nucleus in eukaryotic cells and is connected to the endoplasmic reticulum; has pores that allow substances to pass between the nucleus and the cytoplasm. 70

nucleariid Protist that may be related to fungi although nucleariids lack the same type of cell wall and have threadlike pseudopods. 398

nuclear envelope Name for the phospholipid double membrane that separates the contents of the nucleus from the cytoplasm. 71

nuclear pore Opening in the nuclear envelope that permits the passage of proteins into the nucleus and ribosomal subunits out of the nucleus. 71

nucleic acid Polymer of nucleotides; both DNA and RNA are nucleic acids. 54

nucleoid Region of prokaryotic cells where DNA is located; it is not bounded by a nuclear envelope. 65, 166, 372

nucleolus Dark-staining, spherical body in the nucleus that produces ribosomal subunits. 70

nucleoplasm Semifluid medium of the nucleus containing chromatin. 70

nucleosome In the nucleus of a eukaryotic cell, a unit composed of DNA wound around a core of eight histone proteins, giving the appearance of a string of beads. 233

nucleotide Monomer of DNA and RNA consisting of a 5-carbon sugar bonded to a nitrogenous base and a phosphate group. 54

nucleus Membrane-bounded organelle within a eukaryotic cell that contains chromosomes and controls the structure and function of the cell. 64

O

obligate anaerobe Prokaryote unable to grow in the presence of free oxygen. 374

observation Initial step in the scientific method that often involves the recording of data from an experiment or natural event. 11

octet rule The observation that an atom is most stable when its outer shell is complete and contains eight electrons; an exception is hydrogen which requires only two electrons in its outer shell to have a completed shell. 25

oil Triglyceride, usually of plant origin, that is composed of glycerol and three fatty acids and is liquid in consistency due to many unsaturated bonds in the hydrocarbon chains of the fatty acids. 46

oil gland Gland of the skin, associated with a hair follicle, that secretes sebum; sebaceous gland. 599

olfactory cell Modified neuron that is a sensory receptor for the sense of smell. 719

oligodendrocyte Type of glial cell that forms myelin sheaths around neurons in the CNS. 697

omnivore Organism in a food chain that feeds on both plants and animals. 870

oncogene Cancer-causing gene formed by a mutation in a proto-oncogene; code for proteins that stimulate the cell cycle and inhibit apoptosis. 164

oocyte Immature egg that is undergoing meiosis; upon completion of meiosis, the oocyte becomes an egg. 779

oogenesis Production of eggs in females by the process of meiosis and maturation. 180

open circulatory system Arrangement of internal transport in which blood bathes the

organs directly, and there is no distinction between blood and interstitial fluid. 606

operant conditioning Learning that results from rewarding or reinforcing a particular behavior. 824

operator In an operon, the sequence of DNA that serves as a binding site for a repressor, and thereby regulates the expression of structural genes. 239

operon Group of structural and regulating genes that function as a single unit. 238

opisthokont Supergroup of eukaryotes that choanoflagellates animals, nucleariids and fungi. 397

opposable thumb Fingers arranged in such a way that the thumb can touch the ventral surface of the fingertips of all four fingers. 571

optimal foraging model Analysis of behavior as a compromise of feeding costs versus feeding benefits. 831

order One of the categories, or taxa, used by taxonomists to group species; the taxon located above the family level. 6, 349

organ Combination of two or more different tissues performing a common function. 444, 596

organelle Small, membranous structures in the cytoplasm having a specific structure and function. 67

organic chemistry Branch of science that deals with organic molecules including those that are unique to living things. 38

organic molecule Molecule that always contains carbon and hydrogen, and often contains oxygen as well; organic molecules are associated with living things. 38

organ of Corti Structure in the vertebrate inner ear that contains auditory receptors (also called spiral organ). 727

organ system Group of related organs working together; examples are the digestive and endocrine systems. 596

orientation In birds, the ability to know present location by tracking stimuli in the environment. 825

osmoregulation Regulation of the salt water balance to maintain a normal balance within internal fluids. 681

osmosis Diffusion of water through a selectively permeable membrane. 92

osmotic pressure Measure of the tendency of water to move across a selectively permeable membrane; visible as an increase in liquid on the side of the membrane with higher solute concentration. 93

ossicle One of the small bones of the vertebrate middle ear—malleus, incus, and stapes. 727

osteoblast Bone-forming cell. 738

osteoclast Cell that is responsible for bone resorption. 738

osteocyte Mature bone cell located within the lacunae of bone. 738

osteoporosis Condition characterized by a loss of bone density; associated with levels of sex hormones and diet. 738

ostracoderm Earliest vertebrate fossils of the Cambrian and Devonian periods; these fishes were small, jawless, and finless. 553

otolith Calcium carbonate granule associated with sensory receptors for detecting movement of the head; in vertebrates, located in the utricle and saccule. 728

outer ear Portion of the ear consisting of the pinna and the auditory canal. 726

outgroup In a cladistic study of evolutionary relationships among organisms, a group that has a known relationship to, but is not a member of, the taxa being analyzed. 356

ovarian cycle Monthly changes occurring in the ovary that determine the level of sex hormones in the blood. 780

ovary In flowering plants, the enlarged, ovule-bearing portion of the carpel that develops into a fruit; female gonad in animals that produces an egg and female sex hormones. 434, 503, 773, 779

overexploitation When the number of individuals taken from a wild population is so great that the population becomes severely reduced in numbers. 916

oviparous Type of reproduction in which development occurs in an egg, laid by mother, in reptiles. 774

ovoviviparous Animals that produce eggs that develop internally and hatch at around the same time as they are released to the environment; mostly aquatic. 774

ovulation Bursting of a follicle when a secondary oocyte is released from the ovary; if fertilization occurs, the secondary oocyte becomes an egg. 780

ovule In seed plants, a structure that contains the female gametophyte and has the potential to develop into a seed. 430

oxidation Loss of one or more electrons from an atom or molecule; in biological systems, generally the loss of hydrogen atoms. 113

oxidation-reduction reaction A paired set of chemical reactions in which one molecule gives up electrons (oxidized) while another molecule accepts electrons (reduced); commonly called a redox reaction. 113

oxygen debt Amount of oxygen required to oxidize lactic acid produced anaerobically during strenuous muscle activity. 747

oxyhemoglobin Compound formed when oxygen combines with hemoglobin. 672

oxytocin Hormone released by the posterior pituitary that causes contraction of the uterus and milk letdown. 757

ozone shield Accumulation of O_3, formed from oxygen in the upper atmosphere; a filtering layer that protects the Earth from ultraviolet radiation. 335

P

p53 The protein produced from a tumor suppressor gene that (1) attempts to repair DNA damage or (2) stops the cell cycle, or (3) initiates apoptosis. 155

pacemaker Cells of the sinoatrial node of the heart; electrical device designed to mimic the normal electrical patterns of the heart. 613

paleontology Study of fossils that results in knowledge about the history of life. 272, 333

palisade mesophyll Layer of tissue in a plant leaf containing elongated cells with many chloroplasts. 459

pancreas Internal organ that produces digestive enzymes and the hormones insulin and glucagon. 654, 765

pancreatic amylase Enzyme that digests starch to maltose. 656

pancreatic islet Masses of cells that constitute the endocrine portion of the pancreas. 765

panoramic vision Vision characterized by having a wide field of vision; found in animals with eyes to the side. 721

parabasalid Unicellular protist that lacks mitochondria; possesses flagella in clusters near the anterior of the cell. 395

parasite Species that is dependent on a host species for survival, usually to the detriment of the host species. 865

parasitism Symbiotic relationship in which one species (the *parasite*) benefits in terms of growth and reproduction to the detriment of the other species (the *host*). 375, 865

parasympathetic division Division of the autonomic system that is active under normal conditions; uses acetylcholine as a neurotransmitter. 713

parathyroid gland Gland embedded in the posterior surface of the thyroid gland; it produces parathyroid hormone. 761

parathyroid hormone (PTH) Hormone secreted by the four parathyroid glands that increases the blood calcium level and decreases the phosphate level. 761

parenchyma Plant tissue composed of the least-specialized of all plant cells; found in all organs of a plant. 448

Parkinson disease (PD) Progressive deterioration of the central nervous system due to a deficiency in the neurotransmitter dopamine. 704

parsimony In systematics, the simplest solution in the analysis of evolutionary relationships. 356

partial pressure Pressure exerted by each gas in a mixture of gases. 671

parthenogenesis Development of an egg cell into a whole organism without fertilization. 773

passive immunity Protection against infection acquired by transfer of antibodies to a susceptible individual. 639

pathogen Disease-causing agent such as viruses, parasitic bacteria, fungi, and animals. 375, 592

pattern formation Positioning of cells during development that determines the final shape of an organism. 800

pectoral girdle Portion of the vertebrate skeleton that provides support and attachment for the upper (fore) limbs; consists of the scapula and clavicle on each side of the body. 742

peduncle Flower stalk; expands into the receptacle. 434

pelagic zone Open portion of the ocean. 901

pelvic girdle Portion of the vertebrate skeleton to which the lower (hind) limbs are attached; consists of the coxal bones. 742

penis Male copulatory organ; in humans, the male organ of sexual intercourse. 777

pentose Five-carbon monosaccharide. Examples are deoxyribose found in DNA and ribose found in RNA. 42

pepsin Enzyme secreted by gastric glands that digests proteins to peptides. 656

peptidase Intestinal enzyme that breaks down short chains of amino acids to individual amino acids that are absorbed across the intestinal wall. 657

peptide Two or more amino acids joined together by covalent bonding. 50

peptide bond Type of covalent bond that joins two amino acids. 50

peptide hormone Type of hormone that is a protein, a peptide, or derived from an amino acid. 755

peptidoglycan Polysaccharide that contains short chains of amino acids; found in bacterial cell walls. 44, 374

perennial Flowering plant that lives more than one growing season because the underground parts regrow each season. 444

pericycle Layer of cells surrounding the vascular tissue of roots; produces branch roots. 451

periderm Protective tissue that replaces epidermis; includes cork, cork cambium. 447

peripheral nervous system (PNS) Nerves and ganglia that lie outside the central nervous system. 696

peristalsis Wavelike contractions that propel substances along a tubular structure such as the esophagus. 651

permafrost Permanently frozen ground, usually occurring in the tundra, a biome of Arctic regions. 888

peroxisome Enzyme-filled vesicle in which fatty acids and amino acids are metabolized to hydrogen peroxide that is broken down to harmless products. 75

petal A flower part that occurs just inside the sepals; often conspicuously colored to attract pollinators. 434, 503

petiole The part of a plant leaf that connects the blade to the stem. 445

phagocytosis Process by which cells engulf large substances, forming an intracellular vacuole. 97, 396

pharyngitis Inflammation of the pharynx; often caused by viruses or bacteria. 673

pharynx In vertebrates, common passageway for both food intake and air movement; located between the mouth and the esophagus. 668

phenomenon Observable natural event or fact. 11

phenotype Visible expression of a genotype—e.g., brown eyes or attached earlobes. 197

pheromone Chemical messenger that works at a distance and alters the behavior of another member of the same species. 755, 826

phloem Vascular tissue that conducts organic solutes in plants; contains sieve-tube members and companion cells. 426, 449

phoronids Lophophorans of the phylum Phoronida that are characterized by a long protective tube formed of chitin. 530

phospholipid Molecule that forms the bilayer of the cell's membranes; has a polar, hydrophilic head bonded to two nonpolar, hydrophobic tails. 46

photoautotroph Organism able to synthesize organic molecules by using carbon dioxide as the carbon source and sunlight as the energy source. 375

photoperiod (photoperiodism) Relative lengths of daylight and darkness that affect the physiology and behavior of an organism. 494, 495

photoreceptor Sensory receptor that responds to light stimuli. 717, 720

photorespiration Series of reactions that occurs in plants when carbon dioxide levels are depleted but oxygen continues to accumulate, and the enzyme RuBP carboxylase fixes oxygen instead of carbon dioxide. 130

photosynthesis Process, usually occurring within chloroplasts, that uses solar energy to reduce carbon dioxide to carbohydrate. 4, 120

photosystem Photosynthetic unit where solar energy is absorbed and high-energy electrons are generated; contains a pigment complex and an electron acceptor; occurs as PS (photosystem) I and PS II. 124

phototropism Growth response of plant stems to light; stems demonstrate positive phototropism. 491

pH scale Measurement scale for hydrogen ion concentration. Based on the formula $-\log[H+]$. 32

phylogenetic species concept Definition of a species that is determined by analysis of a phylogenetic tree to determine a common ancestor. 308

phylogenetic tree Diagram that indicates common ancestors and lines of descent among a group of organisms. 348

phylogeny Evolutionary history of a group of organisms. 348

phylum One of the categories, or taxa, used by taxonomists to group species; the taxon located above the class level. 6, 349

phytochrome Photoreversible plant pigment that is involved in photoperiodism and other responses of plants, such as etiolation. 495

phytoplankton Part of plankton containing organisms that photosynthesize, releasing oxygen to the atmosphere and serving as food producers in aquatic ecosystems. 899

phytoremediation The use of plants to restore a natural area to its original condition. 470

pineal gland Gland—either at the skin surface (fish, amphibians) or in the third ventricle of the brain (mammals)—that produces melatonin. 704, 768

pinocytosis Process by which vesicle formation brings macromolecules into the cell. 98

pioneer species Early colonizer of barren or disturbed habitats that usually has rapid growth and a high dispersal rate. 868

pit Any depression or opening; usually in reference to the small openings in the cell walls of xylem cells that function in providing a continuum between adjacent xylem cells. 449

pith Parenchyma tissue in the center of some stems and roots. 452

pituitary gland Small gland that lies just inferior to the hypothalamus; consists of the anterior and posterior pituitary, both of which produce hormones. 757

placenta Organ formed during the development of placental mammals from the chorion and the uterine wall; allows the embryo, and then the fetus, to acquire nutrients and rid itself of wastes; produces hormones that regulate pregnancy. 564, 774, 809

placental mammal Also called the eutherians; species that rely on internal development whereby the fetus exchanges nutrients and wastes with its mother via a placenta. 564

placoderm First jawed vertebrates; heavily armored fishes of the Devonian period. 553

plankton Freshwater and marine organisms that are suspended on or near the surface of the water; includes phytoplankton and zooplankton. 384

plants Multicellular, photosynthetic, eukaryotes that increasingly became adapted to live on land. 8, 420

plasma In vertebrates, the liquid portion of blood; contains nutrients, wastes, salts, and proteins. 618

plasma cell Mature B cell that mass-produces antibodies. 634

plasma membrane Membrane surrounding the cytoplasm that consists of a phospholipid bilayer with embedded proteins; functions to regulate the entrance and exit of molecules from cell. 65

plasmid Extrachromosomal ring of accessory DNA in the cytoplasm of prokaryotes. 65, 255, 372

plasmodesmata In plants, cytoplasmic connections in the cell wall that connect two adjacent cells. 100

plasmodial slime mold Free-living mass of cytoplasm that moves by pseudopods on a forest floor or in a field, feeding on decaying plant material by phagocytosis; reproduces by spore formation. 396

plasmolysis Contraction of the cell contents due to the loss of water. 94

plastid Organelles of plants and algae that are bounded by a double membrane and contain internal membranes and/or vesicles (i.e., chloroplasts, chromoplasts, leucoplasts). 77

plate tectonics Concept that the Earth's crust is divided into a number of fairly rigid plates whose movements account for continental drift. 343

platelet Component of blood that is necessary to blood clotting. 592, 620

pleiotropy Inheritance pattern in which one gene affects many phenotypic characteristics of the individual. 206

pneumonia Condition of the respiratory system characterized by the filling of the bronchi and alveoli with fluid; caused by a viral, fungal, or bacterial pathogen. 674

poikilothermic Body temperature that varies depending on environmental conditions; informally termed "cold-blooded." 599

point mutation Change of only one base in the sequence of bases in a gene. 248

polar body Nonfunctional product of oogenesis produced by the unequal division of cytoplasm in females during meiosis; in humans three of the four cells produced by meiosis are polar bodies. 181

polar covalent bond Bond in which the sharing of electrons between atoms is unequal. 27

pollen grain In seed plants, structure that is derived from a microspore and develops into a male gametophyte. 430, 504

pollen tube In seed plants, a tube that forms when a pollen grain lands on the stigma and germinates. The tube grows, passing between the cells of the stigma and the style to reach the egg inside an ovule, where fertilization occurs. 430

pollination In gymnosperms, the transfer of pollen from pollen cone to seed cone; in angiosperms, the transfer of pollen from anther to stigma. 430, 505

pollution Any environmental change that adversely affects the lives and health of living things. 915

polyandrous Practice of female animals having several male mates; found in the New World monkeys where the males help in rearing the offspring. 831

polygamous Practice of males having several female mates. 831

polygenic inheritance Pattern of inheritance in which a trait is controlled by several allelic pairs. 207

polygenic trait Traits that are under the control of multiple genes as opposed to monogenic (single-gene) traits. 207, 296

polymer Macromolecule consisting of covalently bonded monomers; for example, a polypeptide is a polymer of monomers called amino acids. 40

polymerase chain reaction (PCR) Technique that uses the enzyme DNA polymerase to produce millions of copies of a particular piece of DNA. 256

polyp Among cnidarians, body form that is directed upward and contains much mesoglea; in anatomy: small, abnormal growth that arises from the epithelial lining. 528, 654

polypeptide Polymer of many amino acids linked by peptide bonds. 50

polyploidy Having a chromosome number that is a multiple greater than twice that of the monoploid number. 315

polyribosome String of ribosomes simultaneously translating regions of the same mRNA strand during protein synthesis. 71, 230

polysaccharide Polymer made from carbohydrate monomers; the polysaccharides starch and glycogen are polymers of glucose monomers. 41

pons Portion of the brain stem above the medulla oblongata and below the midbrain; assists the medulla oblongata in regulating the breathing rate. 704

population Group of organisms of the same species occupying a certain area and sharing a common gene pool. 9, 290, 839

population density The number of individuals per unit area or volume living in a particular habitat. 840

population distribution The pattern of dispersal of individuals living within a certain area. 840

population genetics The study of gene frequencies and their changes within a population. 290

portal system Pathway of blood flow that begins and ends in capillaries, such as the portal system located between the small intestine and liver. 614

positive feedback Mechanism of homeostatic response in which the output of the system intensifies and increases the activity of the system. 601, 757

posterior pituitary Portion of the pituitary gland that stores and secretes oxytocin and antidiuretic hormone produced by the hypothalamus. 757

posttranscriptional control Gene expression following transcription that regulates the way mRNA transcripts are processed. 244

posttranslational control Alternation of gene expression by changing a protein's activity after it is translated. 247

postzygotic isolating mechanism Anatomical or physiological difference between two species that prevents successful reproduction after mating has taken place. 312

potential energy Stored energy in a potentially usable form, as a result of location or spatial arrangement. 105

predation Interaction in which one organism (the predator) uses another (the prey) as a food source. 860

predator Organism that practices predation. 860

prediction Step of the scientific process that follows the formulation of a hypothesis and assists in creating the experimental design. 11

preparatory (prep) reaction Reaction that oxidizes pyruvate with the release of carbon dioxide; results in acetyl CoA and connects glycolysis to the citric acid cycle. 137, 142

pressure-flow model Explanation for phloem transport; osmotic pressure following active transport of sugar into phloem produces a flow of sap from a source to a sink. 478

prey Organism that provides nourishment for a predator. 860

prezygotic isolating mechanism Anatomical, physiological, or behavioral difference between two species that prevents the possibility of mating. 310

primary lymphoid organ Location in the lymphatic system where lymphocytes develop and mature; e.g., bone marrow and thymus. 629

primary root Original root that grows straight down and remains the dominant root of the plant; contrasts with fibrous root system. 452

primate Member of the order Primates; includes prosimians, monkeys, apes, and hominins, all of whom have adaptations for living in trees. 571

primordial soup hypothesis Another name for the hypothesis of abiogenesis proposed by Alexander Oparin and J.B.S. Haldane that the first cells evolved in the oceans from present in the early atmosphere. 329

principle Theory that is generally accepted by an overwhelming number of scientists; also called a law. 12

prion Infectious particle consisting of protein only and no nucleic acid. 53, 369

producer Photosynthetic organism at the start of a grazing food chain that makes its own food—e.g., green plants on land and algae in water. 870

product Substance that forms as a result of a reaction. 107

progesterone Female sex hormone that helps maintain sexual organs and secondary sex characteristics. 779

proglottid Segment of a tapeworm that contains both male and female sex organs and becomes a bag of eggs. 533

prokaryote Organism that lacks the membrane-bounded nucleus and the membranous organelles typical of eukaryotes. 7, 371

prokaryotic cell Cells that generally lack a membrane-bounded nucleus and organelles; the cell type within the domains Bacteria and Archaea. 65

prolactin (PRL) Hormone secreted by the anterior pituitary that stimulates the production of milk from the mammary glands. 757

prometaphase Second phase of mitosis; chromosomes are condensed but not fully aligned at the metaphase plate. 159

promoter In an operon, a sequence of DNA where RNA polymerase binds prior to transcription. 238

prophase First phase of mitosis; characterized by the condensation of the chromatin; chromosomes are visible, but scattered in the nucleus. 158

proprioceptor Class of mechanoreceptors responsible for maintaining the body's equilibrium and posture; involved in reflex actions. 730

prosimian Group of primates that includes lemurs and tarsiers, and may resemble the first primates to have evolved. 575

prostaglandin Hormone that has various and powerful local effects. 768

prostate gland Gland in male humans that secretes an alkaline, cloudy, fluid that increases the motility of sperm. 777

proteasome Cellular structure containing proteases that is involved in the destruction of tagged proteins; used by cells for posttranslational control of gene expression. 247

protease Enzyme that breaks the peptide bonds between amino acids in proteins, polypeptides, and peptides. 247

protein Polymer of amino acids; often consisting of one or more polypeptides and having a complex three-dimensional shape. 49

protein-first hypothesis In chemical evolution, the proposal that protein originated before other macromolecules and made possible the formation of protocells. 330

proteinoid Abiotically polymerized amino acids that, when exposed to water, become microspheres having cellular characteristics. 330

proteome Sum of the expressed proteins in a cell. 265

proteomics Study of the complete collection of proteins that a cell or organism expresses. 232, 265

protists The group of eukaryotic organisms that are not a plant, fungus, or animal. Protists are generally a microscopic complex single cell; they evolved before other types of eukaryotes in the history of Earth. 7, 384

protobiont (protocell) In biological evolution, a possible cell forerunner that became a cell once it acquired genes. 330

proton Positive subatomic particle located in the nucleus and assigned one atomic mass unit. 22

proto-oncogene Gene that promotes the cell cycle and prevents apoptosis; may become an oncogene through mutation. 164

protostome Group of coelomate animals in which the first embryonic opening (the blastopore) is associated with the mouth. 523

protozoan Heterotrophic, unicellular protist that moves by flagella, cilia, or pseudopodia. 374

proximal convoluted tubule Portion of a nephron following the glomerular capsule where tubular reabsorption of filtrate occurs. 686

pseudocoelom Body cavity lying between the digestive tract and body wall that is incompletely lined by mesoderm. 538

pseudopod Cytoplasmic extension of amoeboid protists; used for locomotion and engulfing food. 78, 396

pteridophyte Ferns and their allies (horsetail and whisk ferns). 427

pulmonary circuit Circulatory pathway between the lungs and the heart. 609

pulmonary fibrosis Respiratory condition characterized by the buildup of connective tissue in the lungs; typically caused by inhalation of coal dust, silica, or asbestos. 675

pulmonary tuberculosis Respiratory infection caused by the bacterium *Mycobacterium tuberculosis*. 674

pulse Vibration felt in arterial walls due to expansion of the aorta following ventricle contraction. 613

Punnett square Visual representation developed by Reginald Punnett that is used to calculate the expected results of simple genetic crosses. 200

pupil Opening in the center of the iris of the vertebrate eye. 721

R

radial symmetry Body plan in which similar parts are arranged around a central axis, like spokes of a wheel. 522

radiolarian Protist that has a glassy silicon test, usually with a radial arrangement of spines; pseudopods are external to the test. 398

rain shadow Leeward side (side sheltered from the wind) of a mountainous barrier, which receives much less precipitation than the windward side. 885

rate of natural increase (r) Growth rate dependent on the number of individuals that are born each year and the number of individuals that die each year. 841

ray-finned bony fishes Group of bony fishes with fins supported by parallel bony rays connected by webs of thin tissue. 554

RB The protein of a tumor suppressor gene; interprets growth signals and nutrient availability before allowing the cell cycle to proceed. 155

reactant Substance that participates in a reaction. 107

receptacle Area where a flower attaches to a floral stalk. 434, 484

receptor Type of membrane protein that binds to specific molecules in the environment, providing a mechanism for the cell to sense and adjust to its surroundings. 474

receptor-mediated endocytosis Selective uptake of molecules into a cell by vacuole formation after they bind to specific receptor proteins in the plasma membrane. 98

receptor protein Proteins located in the plasma membrane or within the cell; bind to a substance that alters some metabolic aspect of the cell. 89

recessive allele Allele that exerts its phenotypic effect only in the homozygote; its expression is masked by a dominant allele. 196

reciprocal altruism The trading of helpful or cooperative acts, such as helping at the nest, by individuals—the animal that was helped will repay the debt at some later time. 834

recombinant DNA (rDNA) DNA that contains genes from more than one source. 255

red algae Marine photosynthetic protists with a notable abundance of phycobilin pigments; includes coralline algae of coral reefs. 389

red blood cell Erythrocyte; contains hemoglobin and carries oxygen from the lungs or gills to the tissues in vertebrates. 591, 618

red bone marrow Vascularized, modified connective tissue that is sometimes found in the cavities of spongy bone; site of blood cell formation. 740

radiocarbon dating Process of radiometric dating the measures the decay of ^{14}C to ^{14}N. 334

red tide A population bloom of dinoflagellates that causes coastal waters to turn red. Releases a toxin that can lead to paralytic shellfish poisoning. 392

redox reaction A paired set of chemical reactions in which one molecule gives up electrons (oxidized) while another molecule accepts electrons (reduced); also called an oxidation-reduction reaction. 113

reduction Gain of electrons by an atom or molecule with a concurrent storage of energy; in biological systems, the electrons are accompanied by hydrogen ions. 113

reflex action Automatic, involuntary response of an organism to a stimulus. 702

refractory period Time following an action potential when a neuron is unable to conduct another nerve impulse. 700

regulator gene Gene that controls the expression of another gene or genes; in an operon, regulator genes code for repressor proteins. 238

reinforcement Connection between natural selection and reproductive isolation that occurs when two closely related species come back into contact after a period of isolation. 314

relative dating (of fossils) Determining the age of fossils by noting their sequential relationships in strata; *absolute dating* relies on radioactive dating techniques to assign an actual date. 334

renin Enzyme released by the kidneys that leads to the secretion of aldosterone and a rise in blood pressure. 690, 763

repetitive DNA element Sequence of DNA on a chromosome that is repeated several times. 262

replacement reproduction Population in which each person is replaced by only one child. 852

replication fork In eukaryotic DNA replication, the location where the two parental DNA strands separate. 222

repressor In an operon, protein molecule that binds to an operator, preventing transcription of structural genes. 238

reproduce To produce a new individual of the same kind. 5

reproductive cloning Used to create an organism that is genetically identical to the original individual. 161

reproductive isolation Model by which new species arise when gene flow is disrupted between two populations, genetic changes accumulate, and the populations are subsequently unable to mate and produce viable offspring. 310

reproductively isolated Descriptive term that indicates that a population is incapable of interbreeding with another population. 294

reptile Terrestrial vertebrate with internal fertilization, scaly skin, and an egg with a leathery shell; includes snakes, lizards, turtles, crocodiles, and birds. 558

resource Abiotic and biotic components of an environment that support or are needed by living organisms. 840

resource partitioning Mechanism that increases the number of niches by apportioning the supply of a resource such as food or living space between species. 860

respiration Sequence of events that results in gas exchange between the cells of the body and the environment. 674

respiratory center Group of nerve cells in the medulla oblongata that send out nerve impulses on a rhythmic basis, resulting in involuntary inspiration on an ongoing basis. 670

responding variable Result or change that occurs when an experimental variable is utilized in an experiment. 14

resting potential Membrane potential of an inactive neuron. 698

restriction enzyme Bacterial enzyme that stops viral reproduction by cleaving viral DNA; used to cut DNA at specific points during production of recombinant DNA. 255

reticular activating system (RAS) Area of the brain that contains the reticular formation; acts as a relay for information to and from the peripheral nervous system and higher processing centers of the brain. 705

reticular fiber Very thin collagen fiber in the matrix of connective tissue, highly branched and forming delicate supporting networks. 589

retina Innermost layer of the vertebrate eyeball containing the photoreceptors—rod cells and cone cells. 721

retinal detachment Condition characterized by the separation of the retina from the choroid layer of the eye. 724

retrovirus RNA virus containing the enzyme reverse transcriptase that carries out RNA/DNA transcription. 368

reverse transcriptase Viral enzyme found in retroviruses that is capable of converting their RNA genome into a DNA copy. 368

rhizarian Supergroup of eukaryotes that includes foraminiferans and radiolarians. 398

rhizoid Rootlike hair that anchors a plant and absorbs minerals and water from the soil. 424

rhizome Rootlike underground stem. 426, 458, 513

rhodopsin Light-absorbing molecule in rod cells and cone cells that contains a pigment and the protein opsin. 722

ribose Pentose sugar found in RNA. 42

ribosomal RNA (rRNA) Structural form of RNA found in the ribosomes. 223

ribosome Site of protein synthesis in a cell; composed of proteins and ribosomal RNA (rRNA). 66, 71

ribozyme RNA molecule that functions as an enzyme that can catalyze chemical reactions. 109, 227

RNA (ribonucleic acid) Nucleic acid produced from covalent bonding of nucleotide monomers that contain the sugar ribose; occurs in many forms, including: messenger RNA, ribosomal RNA, and transfer RNA. 54

RNA-first hypothesis In chemical evolution, the proposal that RNA originated before other macromolecules and allowed the formation of the first cell(s). 330

RNA interference Cellular process that utilizes miRNA and siRNA molecules to reduce, or inhibit, the expression of specific genes. 246

RNA polymerase During transcription, an enzyme that creates an mRNA transcript by joining nucleotides complementary to a DNA template. 225

rod cell Photoreceptor in vertebrate eyes that responds to dim light. 721

root apical meristem Meristem tissue located under the root cap; the site of the majority of cell division in the root. 450

root cap Protective cover of the root tip, whose cells are constantly replaced as they are ground off when the root pushes through rough soil particles. 450

root hair Extension of a root epidermal cell that collectively increases the surface area for the absorption of water and minerals. 447

root nodule Structure on plant root that contains nitrogen-fixing bacteria. 452, 471

root pressure Osmotic pressure caused by active movement of mineral into root cells; serves to elevate water in xylem for a short distance. 475

root system Includes the main root and any and all of its lateral (side) branches. 444

rotational equilibrium Maintenance of balance when the head and body are suddenly moved or rotated. 728

rotifer Microscopic invertebrates characterized by ciliated corona that when beating looks like a rotating wheel. 533

rough ER (endoplasmic reticulum) Membranous system of tubules, vesicles, and sacs in cells; has attached ribosomes. 72

roundworm Invertebrates with nonsegmented cylindrical body covered by a cuticle that molts; some forms are free-living in water and soil, and many are parasitic. 538

r-**selection** Favorable life history strategy under certain environmental conditions; characterized by a high reproductive rate with little or no attention given to offspring survival. 849

RuBP carboxylase An enzyme that starts the Calvin cycle reactions by catalyzing attachment of the carbon atom from CO_2 to RuBP. 128

rumen The first chamber in the digestive tract of a ruminant mammal; microorganisms here break down complex carbohydrates. 649

S

saccule Saclike cavity in the vestibule of the vertebrate inner ear; contains sensory receptors for gravitational equilibrium. 728

sac fungi (Ascomycota) Fungi that produce spores in fingerlike sacs called asci within a fuiting body; includes morels, truffles, yeasts, and molds. 408

salivary amylase In humans, enzyme in saliva that digests starch to maltose. 651, 656

salivary gland In humans, gland associated with the mouth that secretes saliva. 651

salt Solid substances formed by ionic bonds that usually dissociate into individual ions in water. 27

saltatory conduction Movement of nerve impulses from one node to another along a myelinated axon. 699

saprotroph Organism that secretes digestive enzymes and absorbs the resulting nutrients back across the plasma membrane. 375, 404

sarcolemma Plasma membrane of a muscle fiber; also forms the tubules of the T system involved in muscular contraction. 744

sarcomere One of many units, arranged linearly within a myofibril, whose contraction produces muscle contraction. 745

sarcoplasmic reticulum Smooth endoplasmic reticulum of skeletal muscle cells; surrounds the myofibrils and stores calcium ions. 744

saturated fatty acid Fatty acid molecule that lacks double bonds between the carbons of its hydrocarbon chain. The chain bears the maximum number of hydrogens possible. 46

savanna Terrestrial biome that is a grassland in Africa, characterized by few trees and a severe dry season. 895

Schwann cell Cell that surrounds a fiber of a peripheral nerve and forms the myelin sheath. 697

scientific method Process by which scientists formulate a hypothesis, gather data by observation and experimentation, and come to a conclusion. 11

scientific theory Concept, or a collection of concepts, widely supported by a broad range of observations, experiments, and data. 12

sclera White, fibrous, outer layer of the eyeball. 721

sclerenchyma Plant tissue composed of cells with heavily lignified cell walls; functions in support. 449

scolex Tapeworm head region; contains hooks and suckers for attachment to host. 533

sea star An echinoderm with noticeable 5-pointed radial symmetry; found along rocky coasts where they feed on bivalves. 544

secondary lymphoid organ Location in the lymphatic system where lymphocytes are activated by antigens; example is the lymph nodes. 629

secondary metabolite Molecule not directly involved in growth, development, or reproduction of an organism; in plants, these molecules, which include nicotine, caffeine, tannins, and menthols, can discourage herbivores. 496

secondary oocyte In oogenesis, the functional product of meiosis I; becomes the egg. 181

second messenger Chemical signal such as cyclic AMP that causes the cell to respond to the first messenger—a hormone bound to plasma membrane receptor protein. 755

secretion Release of a substance by exocytosis from a cell. 73

sediment Particulate material (sand, clay, etc.) that is carried by streams and rivers and deposited in areas of slow water movement. 333

sedimentation Process by which particulate material accumulates and forms a stratum. 333

seed Mature ovule that contains an embryo, with stored food enclosed in a protective coat. 430, 502

seed plant Vascular plants that disperse seeds; the gymnosperms and angiosperms. 430

seedless vascular plant Collective name for club mosses (lycophytes) and ferns (pteridophytes) Characterized by windblown spores. 426

segmentation Repetition of body units as seen in the earthworm. 536

selectively permeable Property of the plasma membrane that allows some substances to pass, but prohibits the movement of others. 89

semelparity Condition of having a single reproductive effort in a lifetime. 843

semen (seminal fluid) Thick, whitish fluid consisting of sperm and secretions from several glands of the male reproductive tract. 777

semicircular canal One of three half-circle-shaped canals of the vertebrate inner ear; contains sensory receptors for rotational equilibrium. 727

semiconservative replication Process of DNA replication that results in two double helix

molecules, each having one parental and one new strand. 220

semilunar valve Valve resembling a half moon located between the ventricles and their attached vessels. 611

seminal vesicles Glands in male humans that secrete a viscous fluid that provides nutrition to the sperm cells. 777

seminiferous tubule Long, coiled structure contained within chambers of the testis where sperm are produced. 776

senescence Sum of the processes involving aging, decline, and eventual death of a plant or plant part. 487

sensory (afferent) neuron Nerve cell that transmits nerve impulses to the central nervous system after a sensory receptor has been stimulated. 697

sensory receptor Structure that receives either external or internal environmental stimuli and is a part of a sensory neuron or transmits signals to a sensory neuron. 717

sensory transduction Process by which a sensory receptor converts an input to a nerve impulse. 717

sepal Outermost, leaflike covering of the flower; usually green in color. 434, 503

septate Having cell walls; some fungal species have hyphae that are septate. 405

septum Partition or wall that divides two areas; the septum in the heart separates the right half from the left half. 405, 610

serosa Outer embryonic membrane of birds and reptiles; chorion. 651

seta (pl., setae) A needlelike, chitinous bristle in annelids, arthropods, and others. 536

severe combined immunodeficiency (SCID) Immune disease characterized by the impairment, or lack of, T or B cells in the body. 641

sexual dimorphism Species that have distinct differences between the sexes resulting in male and female forms. 297

sexual reproduction Reproduction involving meiosis, gamete formation, and fertilization; produces offspring with chromosomes inherited from each parent with a unique combination of genes. 172

sexual selection Changes in males and females, often due to male competition and female selectivity, leading to increased fitness. 297, 831

shoot apical meristem Group of actively dividing embryonic cells at the tips of plant shoots. 453

shoot system Aboveground portion of a plant consisting of the stem, leaves, and flowers. 444

short-day plant Plant that flowers when day length is shorter than a critical length—e.g., cocklebur, poinsettia, and chrysanthemum. 495

short tandem repeat (STR) profiling Procedure of analyzing DNA in which PCR and gel electrophoresis are used to create a banding pattern; these are usually unique for each individual; process used in DNA barcoding. 257

shrubland Arid terrestrial biome characterized by shrubs and tending to occur along coasts that have dry summers and receive most of their rainfall in the winter. 894

sieve-tube member Member that joins with others in the phloem tissue of plants as a means of transport for nutrient sap. 450

sign stimulus The environmental trigger that causes a fixed action pattern or unchanging behavioral response. 822

signal Molecule that stimulates or inhibits an event in the cell. 155

signal peptide Sequence of amino acids that binds with a signal recognition particle (SRP), causing a ribosome to bind to the endoplasmic reticulum (ER). 71

signal transduction Process that occurs within a cell when a molecular signal (protein, hormone, etc.) initiates a response within the interior of the cell. 484

sink In the pressure-flow model of phloem transport, the location (roots) from which sugar is constantly being removed. Sugar will flow to the roots from the source. 479

sink population Population that is found in an unfavorable area where at best the birthrate equals the death rate; sink populations receive new members from source populations. 919

sister chromatid One of two genetically identical chromosomal units that are the result of DNA replication and are attached to each other at the centromere. 154

skeletal muscle Striated, voluntary muscle tissue that comprises skeletal muscles; also called striated muscle. 592

skin Outer covering of the body; can be called the integumentary system because it contains organs such as sense organs. 597

sliding filament model An explanation for muscle contraction based on the movement of actin filaments in relation to myosin filaments. 745

small interfering RNAs (siRNA) Short sequences of RNA, typically less than 25 nucleotides, that are involved in posttranscriptional control of gene expression through a process called RNA interference. 246

small intestine In vertebrates, the portion of the digestive tract that precedes the large intestine; in humans, consists of the duodenum, jejunum, and ileum. 652

smooth (visceral) muscle Nonstriated, involuntary muscles found in the walls of internal organs. 592

smooth ER (endoplasmic reticulum) Membranous system of tubules, vesicles, and sacs in eukaryotic cells; site of lipid synthesis; lacks attached ribosomes. 72

society Group in which members of species are organized in a cooperative manner, extending beyond sexual and parental behavior. 826

sodium-potassium pump Carrier protein in the plasma membrane that moves sodium ions out of and potassium ions into cells; important in the function of nerve and muscle cells in animals. 95

soil Accumulation of inorganic rock material and organic matter that is capable of supporting the growth of vegetation. 467

soil erosion Movement of topsoil to a new location due to the action of wind or running water. 468

soil horizon Major layer of soil visible in vertical profile; for example, topsoil is the A horizon. 468

soil profile Vertical section of soil from the ground surface to the unaltered rock below. 468

solute Substance that is dissolved in a solvent, forming a solution. 30, 91

solution Fluid (the solvent) that contains a dissolved solid (the solute). 30, 91

solvent Liquid portion of a solution that serves to dissolve a solute. 91

somatic cell Body cell; excludes cells that undergo meiosis and become sperm or eggs. 155

somatic system Portion of the peripheral nervous system containing motor neurons that control skeletal muscles. 708

source In the pressure-flow model of phloem transport, the location (leaves) of sugar production. Sugar will flow from the leaves to the sink. 479

source population Population that can provide members to other populations of the species because it lives in a favorable area, and the birthrate is most likely higher than the death rate. 919

speciation Origin of new species due to the evolutionary process of descent with modification. 307, 313

species Group of similarly constructed organisms capable of interbreeding and producing fertile offspring; organisms that share a common gene pool; the taxon at the lowest level of classification. 6, 349

species concepts Models that describe the process by which new species arise. 307

species diversity Variety of species that make up a community. 858

species richness Number of species in a community. 858

specific epithet In the binomial system of taxonomy, the second part of an organism's name; it may be descriptive. 349

sperm Male gamete having a haploid number of chromosomes and the ability to fertilize an egg, the female gamete. 777

spermatogenesis Production of sperm in males by the process of meiosis and maturation. 180

spicule Skeletal structure of sponges composed of calcium carbonate or silicate. 527

spinal cord In vertebrates, the nerve cord that is continuous with the base of the brain and housed within the vertebral column. 702

spinal nerve Nerve that arises from the spinal cord. 707

spirillum (pl., spirilla) Long, rod-shaped bacterium that is twisted into a rigid spiral; if the spiral is flexible rather than rigid, it is called a spirochete. 65

spirochete Long, rod-shaped bacterium that is twisted into a flexible spiral; if the spiral is rigid rather than flexible, it is called a spirillum. 65

spleen Large, glandular organ located in the upper left region of the abdomen; stores and filters blood. 629

sponge Invertebrates that are pore-bearing filter feeders whose inner body wall is lined by collar cells that resemble a unicellular choanoflagellate. 527

spongy bone Type of bone that has an irregular, meshlike arrangement of thin plates of bone. 591, 740

spongy mesophyll Layer of tissue in a plant leaf containing loosely packed cells, increasing the amount of surface area for gas exchange. 459

spontaneous mutation Mutation that arises as a result of anomalies in normal biological processes, such as mistakes made during DNA replication. 247

sporangium (pl., sporangia) Structure that produces spores. 422

spore Asexual reproductive or resting cell capable of developing into a new organism

without fusion with another cell, in contrast to a gamete. 176, 406, 422

sporophyll Modified leaf that bears a sporangium or many sporangia. 426

sporophyte Diploid generation of the alternation-of-generations life cycle of a plant; produces haploid spores that develop into the haploid generation. 180, 422

spring overturn Mixing process that occurs in spring in stratified lakes whereby oxygen-rich top waters mix with nutrient-rich bottom waters. 899

squamous epithelium Type of epithelial tissue that contains flat cells. 588

stabilizing selection Outcome of natural selection in which extreme phenotypes are eliminated and the average phenotype is conserved. 296

standard deviation A statistical analysis of data from an observation or experiment; measures how much the data varies. 12

stamen In flowering plants, the portion of the flower that consists of a filament and an anther containing pollen sacs where pollen is produced. 434, 503

starch Storage polysaccharide found in plants that is composed of glucose molecules joined in a linear fashion with few side chains. 43

statolith Sensors found in root cap cells that cause a plant to demonstrate gravitropism. 491

stem Usually the upright, vertical portion of a plant that transports substances to and from the leaves. 445

stereoscopic vision Vision characterized by depth perception and three-dimensionality. 572, 721

steroid Type of lipid molecule having a complex of four carbon rings—e.g., cholesterol, estrogen, progesterone, and testosterone. 46

steroid hormone Type of hormone that has the same complex of four carbon rings, but each one has different side chains. 755

stigma In flowering plants, portion of the carpel where pollen grains adhere and germinate before fertilization can occur. 434, 503

stolon Stem that grows horizontally along the ground and may give rise to new plants where it contacts the soil—e.g., the runners of a strawberry plant. 456, 513

stomach In vertebrates, muscular sac that mixes food with gastric juices to form chyme, which enters the small intestine. 652

stomata (sing., stoma) Small openings between two guard cells on the underside of leaf epidermis through which gases pass. 121, 422, 447

stramenopile Group of protists that includes water molds, diatoms, and golden brown algae and is characterized by a "hairy" flagellum. 390

strata (sing. stratum) Ancient layer of sedimentary rock; results from slow deposition of silt, volcanic ash, and other materials. 273, 333

striated Having bands; in cardiac and skeletal muscle, alternating light and dark bands produced by the distribution of contractile proteins. 592

strobilus In club mosses, terminal clusters of leaves that bear sporangia. 426

stroke Condition resulting when an arteriole in the brain bursts or becomes blocked by an embolism; cerebrovascular accident. 616, 704

stroma Region within a chloroplast that surrounds the grana; contains enzymes involved in the synthesis of carbohydrates during the Calvin cycle of photosynthesis. 76, 121

stromatolite Domed structure found in shallow seas consisting of cyanobacteria bound to calcium carbonate. 335

structural gene Gene that codes for the amino acid sequence of a peptide or protein. 238

structural genomics Study of the sequence of DNA bases and the amount of genes in organisms. 261

style Elongated, central portion of the carpel between the ovary and stigma. 434, 503

submucosa Tissue layer just under the epithelial lining of the lumen of the digestive tract (second layer). 651

substrate Reactant in an enzyme-controlled reaction. 109

substrate-level ATP synthesis Process in which ATP is formed by transferring a phosphate from a metabolic substrate to ADP. 138

supergroup Systematic term that refers to the major groups of eukaryotes. 384

surface-area-to-volume ratio Ratio of a cell's outside area to its internal volume; the relationship limits the maximum size of a cell. 62

surface tension Force that holds moist membranes together due to the attraction of water molecules through hydrogen bonds. 31

survivorship Probability of newborn individuals of a cohort surviving to particular ages. 841

suture Line of union between two nonarticulating bones, as in the skull. 740

swamp Wet, spongy land that is saturated and sometimes partially or intermittently covered with water. 897

sweat gland Skin gland that secretes a fluid substance for evaporative cooling; sudoriferous gland. 599

swim bladder In fishes, a gas-filled sac whose pressure can be altered to change buoyancy. 554

symbiosis Relationship that occurs when two different species live together in a unique way; it may be beneficial, neutral, or detrimental to one or both species. 865

symbiotic relationship *See also* symbiosis. 375

symmetry Pattern of similarity in an object. 522

sympathetic division Division of the autonomic system that is active when an organism is under stress; uses norepinephrine as a neurotransmitter. 709

sympatric speciation Origin of new species in populations that overlap geographically. 314

synapse Junction between neurons consisting of the presynaptic (axon) membrane, the synaptic cleft, and the postsynaptic (usually dendrite) membrane. 700

synapsid Characteristic of a vertebrate skull in which there is a single opening in the skull behind the eye socket (orbit). 558

synapsis Pairing of homologous chromosomes during meiosis I. 173

synaptic cleft Small gap between presynaptic and postsynaptic cells of a synapse. 700

synaptonemal complex Protein structure that forms between the homologous chromosomes of prophase I of meiosis; promotes the process of crossing-over. 173

synovial joint Freely moving joint in which two bones are separated by a cavity. 742

systematics Study of the diversity of life for the purpose of understanding the evolutionary relationships between species. 6, 348

systemic circuit Circulatory pathway of blood flow between the tissues and the heart. 609

systemin In plants, an 18-amino-acid peptide that is produced by damaged or injured leaves that leads to the wound response. 497

systole Contraction period of the heart during the cardiac cycle. 612

T

tactile communication Communication through touch; for example, when a chick pecks its mother for food, chimpanzees groom each other, and honeybees "dance." 829

taiga Terrestrial biome that is a coniferous forest extending in a broad belt across northern Eurasia and North America. 890

tandem repeat Repetitive DNA sequence in which the repeats occur one after another in the same region of a chromosome. 262

taproot Main axis of a root that penetrates deeply and is used by certain plants (such as carrots) for food storage. 452

taste bud Structure in the vertebrate mouth containing sensory receptors for taste; in humans, most taste buds are on the tongue. 718

taxon (pl., taxa) Group of organisms that fills a particular classification category. 348

taxonomist Scientist that investigates the identification and naming of new organisms. 307, 348

taxonomy Branch of biology concerned with identifying, describing, and naming organisms. 6, 348

T cell Lymphocyte that matures in the thymus and exists in four varieties, one of which kills antigen-bearing cells outright. 629

T-cell receptor (TCR) Molecule on the surface of a T cell that can bind to a specific antigen fragment in combination with an MHC molecule. 635

telomere Tip of the end of a chromosome that shortens with each cell division and may thereby regulate the number of times a cell can divide. 164, 221

telophase Final phase of mitosis; daughter cells are located at each pole. 160

temperate deciduous forest Forest found south of the taiga; characterized by deciduous trees such as oak, beech, and maple, moderate climate, relatively high rainfall, stratified plant growth, and plentiful ground life. 891

temperate grassland Grasslands characterized by very cold winters and dry, hot summers; examples are the North American prairies and Russian steppes. 894

temperate rain forest Coniferous forest—e.g., that running along the west coast of Canada and the United States—characterized by plentiful rainfall and rich soil. 890

template Parental strand of DNA that serves as a guide for the complementary daughter strand produced during DNA replication. 220

tendon Strap of fibrous connective tissue that connects skeletal muscle to bone. 590, 744

terminal bud Bud that develops at the apex of a shoot. 453

termination End of translation that occurs when a ribosome reaches a stop codon on the mRNA that it is translating, causing release of the completed protein. 231

territoriality Marking and/or defending a particular area against invasion by another species member; area often used for the purpose of feeding, mating, and caring for young. 299, 830

territory Area occupied and defended exclusively by an animal or group of animals. 299, 830

test Loose-fitting shell of a foraminiferan or a radiolarian; made of calcium carbonate or silicon, respectively. 398

testcross Cross between an individual with a dominant phenotype and an individual with a recessive phenotype to determine whether the dominant individual is homozygous or heterozygous. 200

testes Male gonads that produce sperm and the male sex hormones. 773, 775

testosterone Male sex hormone that helps maintain sexual organs and secondary sex characteristics. 778

tetany Severe spasm caused by involuntary contraction of the skeletal muscles due to a calcium imbalance. 744

tetrapod Four-footed vertebrate; includes amphibians, reptiles, birds, and mammals. 552

thalamus In vertebrates, the portion of the diencephalon that passes on selected sensory information to the cerebrum. 704

therapeutic cloning Used to create mature cells of various cell types. Facilitates study of specialization of cells and provide cells and tissue to treat human illnesses. 161

thermoacidophile Type of archaean that lives in hot, acidic, aquatic habitats, such as hot springs or near hydrothermal vents. 379

thermoreceptor Sensory receptor that detects heat. 717

thigmotropism In plants, unequal growth due to contact with solid objects, as the coiling of tendrils around a pole. 491

threatened species Species that is likely to become an endangered species in the foreseeable future (e.g., bald eagle, gray wolf, Louisiana black bear). 908

thylakoid Flattened sac within a granum of a chloroplast; membrane contains chlorophyll; location where the light reactions of photosynthesis occur. 66, 76, 121

thymine (T) One of four nitrogen-containing bases in nucleotides composing the structure of DNA; pairs with adenine. 217

thymus gland Lymphoid organ involved in the development and functioning of the immune system; T lymphocytes mature in the thymus gland. 629,768

thyroid gland Large gland in the neck that produces several important hormones, including thyroxine, triiodothyronine, and calcitonin. 762

thyroid-stimulating hormone (TSH) Substance produced by the anterior pituitary that causes the thyroid to secrete thyroxine and triiodothyronine. 760

thyroxine (T_4) Hormone secreted from the thyroid gland that promotes growth and development; in general, it increases the metabolic rate in cells. 760

tight junction Junction between cells when adjacent plasma membrane proteins join to form an impermeable barrier. 99

tissue Group of similar cells combined to perform a common function. 588

tissue culture Process of growing tissue artificially, usually in a liquid medium in laboratory glassware. 513

tissue fluid Fluid that surrounds the body's cells; consists of dissolved substances that leave the blood capillaries by filtration and diffusion. 591, 621

tone Continuous, partial contraction of muscle. 744

tonicity The solute concentration (osmolarity) of a solution compared to that of a cell. If the solution is isotonic to the cell, there is no net movement of water; if the solution is hypotonic, the cell gains water; and if the solution is hypertonic, the cell loses water. 93

topography Surface features of the Earth. 885

totipotent Cell that has the full genetic potential of the organism, including the potential to develop into a complete organism. 514, 800

toxin Poisonous substance produced by living cells or organisms. Toxins are often proteins that are capable of causing disease on contact with or absorption by body tissues. 376

trachea (pl., tracheae) In insects, air tubes located between the spiracles and the tracheoles. In tetrapod vertebrates, air tube (windpipe) that runs between the larynx and the bronchi. 542, 664, 668

tracheid In vascular plants, type of cell in xylem that has tapered ends and pits through which water and minerals flow. 449

tract Bundle of myelinated axons in the central nervous system. 703

trait A characteristic of an organism; may be based on the physiology, morphology, or the genetics of the organism. 348

transcription First stage of gene expression; process whereby a DNA strand serves as a template for the formation of mRNA. 223

transcription activator Protein that participates in the initiation of transcription by binding to the enhancer regulatory regions. 244

transcriptional control Control of gene expression by the use of transcription factors, and other proteins, that regulate either the initiation of transcription or the rate at which it occurs. 244

transcription factor In eukaryotes, protein required for the initiation of transcription by RNA polymerase. 244

transduction Exchange of DNA between bacteria by means of a bacteriophage. 373

transduction pathway Series of proteins or enzymes that change a signal to one understood by the cell. 484

trans-fats Unsaturated fatty acid chains in which the configuration of the carbon-carbon double bonds is such that the hydrogen atoms are across from each other, as opposed to being on the same side (cis). 658

transfer rate Amount of a substance that moves from one component of the environment to another within a specified period of time. 875

transfer RNA (tRNA) Type of RNA that transfers a particular amino acid to a ribosome during protein synthesis; at one end, it binds to the amino acid, and at the other end it has an anticodon that binds to an mRNA codon. 223

transformation Taking up of extraneous genetic material from the environment by bacteria. 373

transgenic organism An organism whose genome has been altered by the insertion of genes from another species. 255

transitional fossil Fossil that bears a resemblance to two groups that in the present day are classified separately. 280

translation During gene expression, the process whereby ribosomes use the sequence of codons in mRNA to produce a polypeptide with a particular sequence of amino acids. 224

translational control Gene expression regulated by influencing the interaction of the mRNA transcripts with the ribosome. 246

translocation Movement of a chromosomal segment from one chromosome to another nonhomologous chromosome, leading to abnormalities—e.g., Down syndrome. 187, 231

transpiration Plant's loss of water to the atmosphere, mainly through evaporation at leaf stomata. 476

transposon DNA sequence capable of randomly moving from one site to another in the genome. 262

trichocyst Found in ciliates; contains long, barbed threads useful for defense and capturing prey. 393

trichomes In plants, specialized outgrowths of the epidermis (e.g., root hairs). 447

triglyceride Neutral fat composed of glycerol and three fatty acids; typically involved in energy storage. 46

triplet code During gene expression, each sequence of three nucleotide bases stands for a particular amino acid. 224

trisomy Chromosome condition in which a diploid cell has one more chromosome than normal; designated as 2n + 1. 183

trochozoan Type of protostome that produces a trochophore larva; also has two bands of cilia around its middle. 530

trophic level Feeding level of one or more populations in a food web. 873

trophoblast Outer membrane surrounding the embryo in mammals; when thickened by a layer of mesoderm, it becomes the chorion, an extraembryonic membrane. 806

tropical rain forest Biome near the equator in South America, Africa, and the Indo-Malay regions; characterized by warm weather, plentiful rainfall, a diversity of species, and mainly tree-living animal life. 892

tropism In plants, a growth response toward or away from a directional stimulus. 490

true coelom Body cavity completely lined with mesoderm; found in certain protostomes and all deuterostomes. 523

trypsin Protein-digesting enzyme secreted by the pancreas. 656

tubular reabsorption Movement of primarily nutrient molecules and water from the contents of the nephron into blood at the proximal convoluted tubule. 686

tubular secretion Movement of certain molecules from blood into the distal convoluted tubule of a nephron so that they are added to urine. 686

tumor Cells derived from a single mutated cell that has repeatedly undergone cell division; benign tumors remain at the site of origin, while malignant tumors metastasize. 164

tumor suppressor gene Gene that codes for a protein that ordinarily suppresses the cell cycle; inactivity due to a mutation can lead to a tumor. 164

turgor movement In plant cells, pressure of the cell contents against the cell wall when the central vacuole is full. 493

turgor pressure Pressure of the cell contents against the cell wall; in plant cells, determined by the water content of the vacuole; provides internal support. 93

tympanic membrane Membranous region that receives air vibrations in an auditory organ; in humans, the eardrum. 727

U

umbilical cord Cord connecting the fetus to the placenta through which blood vessels pass. 809

unicellular An organism comprised of a single cell, as in the bacteria. 2

uniformitarianism Belief, supported by James Hutton, that geological forces act at a continuous, uniform rate. 274

unsaturated fatty acid Fatty acid molecule that contains double bonds between some carbons of its hydrocarbon chain; thus contains fewer hydrogens than a saturated hydrocarbon chain. 46

upwelling Upward movement of deep, nutrient-rich water along coasts; it replaces surface waters that move away from shore when the direction of prevailing wind shifts. 904

uracil (U) Pyrimidine base that occurs in RNA, replacing thymine. 223

urea Main nitrogenous waste of terrestrial amphibians and most mammals. 681

ureter Tubular structure conducting urine from the kidney to the urinary bladder. 684

urethra Tubular structure that receives urine from the bladder and carries it to the outside of the body. 684

uric acid Main nitrogenous waste of insects, reptiles, and birds. 681

urinary bladder Organ where urine is stored. 684

urine Liquid waste product made by the nephrons of the vertebrate kidney through the processes of glomerular filtration, tubular reabsorption, and tubular secretion. 682

urochordate Group of aquatic invertebrate chordates that consists of the tunicates (sea squirts). 551

uterine cycle Cycle that runs concurrently with the ovarian cycle; it prepares the uterus to receive a developing zygote. 781

uterus In mammals, expanded portion of the female reproductive tract through which eggs pass to the environment or in which an embryo develops and is nourished before birth. 779

utricle Cavity in the vestibule of the vertebrate inner ear; contains sensory receptors for gravitational equilibrium. 728

V

vacuole Membrane-bounded sac, larger than a vesicle; usually functions in storage and can contain a variety of substances. In plants, the central vacuole fills much of the interior of the cell. 75

vagina Component of the female reproductive system that serves as the birth canal; receives the penis during copulation. 779

valence shell The outer electron shell of an atom. Contains the valence electrons, which determine the chemical reactivity of the atom. 25

vas deferens Also called the ductus deferens; storage location for mature sperm before they pass into the ejaculatory duct. 775

vascular bundle In plants, primary phloem and primary xylem enclosed by a bundle sheath. 450

vascular cambium In plants, lateral meristem that produces secondary phloem and secondary xylem. 454

vascular cylinder In eudicots, the tissues in the middle of a root, consisting of the pericycle and vascular tissues. 450

vascular plant Plant that has xylem and phloem. 426

vascular tissue Transport tissue in plants, consisting of xylem and phloem. 423, 447

vector In genetic engineering, a means to transfer foreign genetic material into a cell—e.g., a plasmid. 66, 255, 394

vein Blood vessel that arises from venules and transports blood toward the heart. 608

vena cava Large systemic vein that returns blood to the right atrium of the heart in tetrapods; either the superior or inferior vena cava. 614

ventilation Process of moving air into and out of the lungs; breathing. 664

ventricle Cavity in an organ, such as a lower chamber of the heart or the ventricles of the brain. 611, 702

venule Vessel that takes blood from capillaries to a vein. 608

vertebral column Portion of the vertebrate endoskeleton that houses the spinal cord; consists of many vertebrae separated by intervertebral disks. 741

vertebrate Chordate in which the notochord is replaced by a vertebral column. 520

vertigo Equilibrium disorder that is often associated with problems in the receptors of the semicircular canals in the ear. 730

vesicle Small, membrane-bounded sac that stores substances within a cell. 67, 331

vessel element Cell that joins with others to form a major conducting tube found in xylem. 449

vestibule Space or cavity at the entrance to a canal, such as the cavity that lies between the semicircular canals and the cochlea. 727

vestigial structure Remnant of a structure that was functional in some ancestor but is no longer functional in the organism in question. 272

villus Small, fingerlike projection of the inner small intestinal wall. 652

viroid Infectious strand of RNA devoid of a capsid and much smaller than a virus. 369

virus Noncellular parasitic agent consisting of an outer capsid and an inner core of nucleic acid. 364

visual accommodation Ability of the eye to focus at different distances by changing the curvature of the lens. 722

visual communication Form of communication between animals using their bodies, includes various forms of display. 828

vitamin Organic nutrient that is required in small amounts for metabolic functions. Vitamins are often part of coenzymes. 112

viviparous Animal that gives birth after partial development of offspring within mother. 774

vocal cord In humans, fold of tissue within the larynx; creates vocal sounds when it vibrates. 668

W

water column In plants, water molecules joined together in xylem from the leaves to the roots. 476

water (hydrologic) cycle Interdependent and continuous circulation of water from the ocean, to the atmosphere, to the land, and back to the ocean. 875

water mold Filamentous organisms having cell walls of cellulose; typically decomposers of dead freshwater organisms, but some are parasites of aquatic or terrestrial organisms. 391

water potential Potential energy of water; a measure of the capability to release or take up water relative to another substance. 473

water vascular system Series of canals that takes water to the tube feet of an echinoderm, allowing them to expand. 544

wax Sticky, solid, water-repellent lipid consisting of many long-chain fatty acids usually linked to long-chain alcohols. 48

wetland Area that is covered by water at some point in the year. (*See also* bog, marsh, or swamp.) 897

whisk fern Common name for seedless vascular plant that consists only of stems and has no leaves, or roots. 428

white blood cell Leukocyte, of which there are several types, each having a specific function in protecting the body from invasion by foreign substances and organisms. 620

white matter Myelinated axons in the central nervous system. 703

whorl Cluster of branches, or other plant structures, that occurs in a circular pattern. 421

wobble hypothesis Ability of the tRNAs to recognize more than one codon; the codons differ in their third nucleotide. 229

wood Secondary xylem that builds up year after year in woody plants and becomes the annual rings. 456

X

X-linked Allele that is located on an X chromosome; not all X-linked genes code for sexual characteristics. 208

xylem Vascular tissue that transports water and mineral solutes upward through the plant body; it contains vessel elements and tracheids. 426, 449

Y

yeast Unicellular fungus that has a single nucleus and reproduces asexually by budding or fission, or sexually through spore formation. 408

yolk Dense nutrient material in the egg of a bird or reptile. 774, 797

yolk sac One of the extraembryonic membranes that, in shelled vertebrates, contains yolk for the nourishment of the embryo, and in placental mammals is the first site for blood cell formation. 805

Z

zero population growth No growth in population size. 852

zooflagellate Nonphotosynthetic protist that moves by flagella; typically zooflagellates enter into symbiotic relationships, and some are parasitic. 395

zooplankton Part of plankton containing protozoans and other types of microscopic animals. 899

zoospore Spore that is motile by means of one or more flagella. 407

zygospore Thick-walled resting cell formed during sexual reproduction of zygospore fungi. 408

zygospore fungi (Zygomycota) Fungi such as black bread mold that reproduces by forming windblown spores in sporangia; sexual reproduction involves a thick-walled zygospore. 407

zygote Diploid cell formed by the union of two gametes; the product of fertilization. 172, 310

Credits

PHOTOGRAPHS

DESIGN ELEMENTS Part opener(wolf): Corbis RF/Alamy; p. xv (crab): © Brand X Pictures/PunchStock RF; p. xvii (butterfly): © Wally Eberhart/Getty Images; p. xix (bird): © Brand X Pictures/PunchStock RF.

CHAPTER 1 Opener: © Bob Gibbons/Alamy; 1.1(bacteria): © Dr. Dennis Kunkel/Phototake; 1.1(paramecium): © M. Abbey/Visuals Unlimited; 1.1(morel): © Corbis RF; 1.1(sunflower): © Photodisc Green/Getty RF; 1.1(octopus): © Erica S. Leeds; 1.3a: © Purestock/SuperStock RF; 1.3b: © Photodisc Blue/Getty RF; 1.3c: © Charles Bush Photography; 1.3d: © Michael Abby/Visuals Unlimited; 1.3e: © Pat Pendarvis; 1.3f: Courtesy National Park Service; 1.4: © Photodisc/Getty RF; 1.6: © Ralph Robinson/Visuals Unlimited; 1.7: © A.B. Dowsett/SPL/Photo Researchers, Inc.; 1.8(protist): © Michael Abby/Visuals Unlimited; 1.8(plant): © Pat Pendarvis; 1.8(fungi): © Tinke Hamming/Ingram Publishing RF; 1.8(animal): © Corbis RF; 1.11a: © Getty RF; 1.11b: © Dr. Phillip Dustan; 1.12: Courtesy Leica Microsystems Inc.; 1A(left): © Corbis RF; 1A(right): © JupiterImages/BananaStock/Alamy RF; 1.13: © Dr. Jeremy Burgess/Photo Researchers, Inc.; 1.14(all): Courtesy Jim Bidlack; 1.15: © Erica S. Leeds.

CHAPTER 2 Opener: © Peter Weimann/Animals Animals; 2.1: © Gunter Ziesler/Peter Arnold/Photolibrary; 2.4a: © Biomed Commun./Custom Medical Stock Photo; 2.4b(left): © Mazzlota et al./Photo Researchers, Inc; 2.4b(right): Courtesy National Institutes of Health; 2.5a(both): © Tony Freeman/PhotoEdit; 2.5b: © Geoff Tompkinson/SPL/Photo Researchers, Inc.; 2.7b(crystals): © Evelyn Jo Johnson; 2.7b(shaker): © Evelyn Jo Johnson; 2.10b: © Grant Taylor/Stone/Getty Images.

CHAPTER 3 Opener: Courtesy Sylvia S. Mader; 3.1b: © Getty RF; 3.1c: © Corbis RF; 3.1d: © Brand X Pictures/PunchStock RF; 3.1e: © Ingram Publishing/Alamy RF; 3.1f: © JewelryStock/Alamy RF; 3.1g: © Ingram Publishing/SuperStock RF; 3.3: © The McGraw Hill Companies, Inc. John Thoeming, photographer; 3.5a: © Brand X Pictures/PunchStock RF; 3.5b: © Ingram Publishing/Alamy RF; 3.5c: © H. Pol/CNRI/SPL/Photo Researchers, Inc.; 3.6a: © Steve Bloom/Taxi/Getty; 3.8a: © Jeremy Burgess/SPL/Photo Researchers, Inc.; 3.8b: © Don W. Fawcett/Photo Researchers, Inc.; 3.9: © Science VU; 3.12: © Ernest A. Janes/Bruce Coleman/Photoshot; 3.13a: © Das Fotoarchiv/Peter Arnold/Photolibrary; 3.13b: © Martha Cooper/Peter Arnold/Photolibrary; p. 49(woman running): © Duomo/Corbis; 3.17a: © Gregory Pace/Corbis; 3.17b: © Ronald Siemoneit/Corbis Sygma; 3.17c: © Kjell Sandved/Visuals Unlimited; 3.20a: © Photodisk Red/Getty Images; 3.21c: © Jennifer Loomis/Animals Animals.

CHAPTER 4 Opener: © Oliver Meckes/Nicole Ottawa/Photo Researchers, Inc.; 4.1a: © Geoff Bryant/Photo Researchers, Inc.; 4.1b: Courtesy Ray F. Evert/University of Wisconsin Madison; 4.1c: © Barbara J. Miller/Biological Photo Service; 4.1d: Courtesy O. Sabatakou and E. Xylouri-Frangiadaki; 4Aa: © Robert Brons/Biological Photo Service; 4Ab: © M. Schliwa/Visuals Unlimited; 4Ac: © Kessel/Shih/Peter Arnold/Photolibrary; 4B(bright-field): © Ed Reschke; 4B(bright-field stained): © Biophoto Associates/Photo Researchers, Inc.; 4B(differential, phase contrast, dark-field): © David M. Phillips/Visuals Unlimited; 4.4: © Howard Sochurek/The Medical File/Peter Arnold/Photolibrary; 4.6: © Dr. Dennis Kunkel/Visuals Unlimited; 4.7: © Newcomb/Wergin/Biological Photo Service; 4.8(top right): Courtesy E.G. Pollock; 4.8(bottom left): Courtesy Ron Milligan/Scripps Research Institute; 4.10: © R. Bolender & D. Fawcett/Visuals Unlimited; 4.11: Courtesy Charles Flickinger, from Journal of Cell Biology 49:221-226, 1971, Fig. 1 page 224; 4.12a: Courtesy Daniel S. Friend; 4.12b: Courtesy Robert D. Terry/University of California, San Diego School of Medicine; 4.14: © S.E. Frederick & E.H. Newcomb/Biolog-

ical Photo Service; 4.15: © Newcomb/Wergin/Biological Photo Service; 4.16a: Courtesy Herbert W. Israel, Cornell University; 4.17a: Courtesy Dr. Keith Porter; 4.18a(actin): © M. Schliwa/Visuals Unlimited; 4.18a(Chara): © The McGraw-Hill Companies, Inc. Dennis Strete and Darrell Vodopich, photographers; 4.18b(intermediate): © K.G. Murti/Visuals Unlimited; 4.18b(peacock): © Vol. 86/Corbis RF; 4.18c(microtubules): © K.G. Murti/Visuals Unlimited; 4.18c(chameleon): © Photodisc/Vol. 6/Getty RF; 4.19(one centrosome): Courtesy Kent McDonald, University of Colorado Boulder; 4.19(two centrosomes): This image originally appeared in: McGill M, Highfield DP, Monahan TM, Brinkley BR. Effects of nucleic acid specific dyes on centrioles of mammalian cells. J Ultrastruct Res. 1976; 57:43-53. The image was created by Manley McGill in 1976 while he was a member of the William Brinkley laboratory at The University of Texas Medical Branch, Galveston. It has been made available by William Brinkley (Baylor University) for the memory of his former student.; 4.20(sperm): © David M. Phillips/Photo Researchers, Inc.; 4.20(flagellum, basal body): © William L. Dentler/Biological Photo Service.

CHAPTER 5 Opener: © Professors P. Motta & T. Naguro/Science Photo Library/Photo Researchers, Inc.; 5Aa(human egg): © Anatomical Travelogue/Photo Researchers, Inc.; 5Aa(embryo): © Neil Harding/Stone/Getty Images; 5Aa(baby): © Photodisc Collection/Getty RF; 5.12a: © Eric Grave/Phototake; 5.12b: © Don W. Fawcett/Photo Researchers, Inc.; 5.12c(both): Courtesy Mark Bretscher; 5.14a: From Douglas E. Kelly, J. Cell Biol. 28 (1966): 51. Reproduced by copyright permission of The Rockefeller University Press; 5.14b: © David M. Phillips/Visuals Unlimited; 5.14c: Courtesy Camillo Peracchia, M.D.; 5.15: © E.H. Newcomb/Biological Photo Service.

CHAPTER 6 Opener: © PhotoAlto/PunchStock RF; 6.8b: © James Watt/Visuals Unlimited; 6.8c: © Creatas/PunchStock/PunchStock RF; 6.9: © G.K. & Vikki Hart/The Image Bank/Getty Images; 6A: © AP Photo/Chikumo Chiaki.

CHAPTER 7 Opener: © David Muench/Corbis; 7.1(moss): © Steven P. Lynch; 7.1(trees): © Digital Vision/PunchStock RF; 7.1(Euglena): © T.E. Adams/Visuals Unlimited; 7.1(sunflower): © Corbis RF; 7.1(kelp): © Chuck Davis/Stone/Getty Images; 7.1(cyanobacteria): © Sherman Thomas/Visuals Unlimited; 7.1(diatoms): © Ed Reschke/Peter Arnold/Photolibrary; 7.2: © Dr. George Chapman/Visuals Unlimited; 7.3: © Nigel Cattlin/Photo Researchers, Inc.; 7.4: Courtesy Lawrence Berkeley National Lab/Roy Kaltschmidt, LBL Photographer; 7.10: © Herman Eisenbeiss/Photo Researchers, Inc.; 7.11a: © Brand X Pictures/PunchStock RF; 7.11b: Courtesy USDA/Doug Wilson, photographer; 7.12: © PhotoLink/Getty RF.

CHAPTER 8 Opener: © Daniel H. Bailey/Age RF; 8.1: © E. & P. Bauer/zefa/Corbis; 8A(bread, wine, cheese): © The McGraw Hill Companies, Inc./John Thoeming, photographer; 8A(yogurt; pickles, vinegar): © The McGraw Hill Companies, Inc. Bruce M. Johnson, photographer; 8.6: © Dr. Donald Fawcett and Dr. Porter/Visuals Unlimited; 8.10: © C Squared Studios/Getty RF.

CHAPTER 9 Opener: © SPL/Photo Researchers, Inc.; 9.2: Courtesy Douglas R. Green/LaJolla Institute for Allergy and Immunology; 9.3a: © Andrew Syred/Photo Researchers, Inc.; 9.4 Animal cell(early prophase, prophase, metaphase, anaphase, telophase): © Ed Reschke; 9.4 Animal cell(prometaphase): © Michael Abbey/Photo Researchers, Inc.; 9.4 Plant cell(early prophase, prometaphase): © Ed Reschke; 9.4 Plant cell(prophase, metaphase, anaphase): © R. Calentine/Visuals Unlimited; 9.4 Plant cell(telophase): © Jack M. Bostrack/Visuals Unlimited; 9.5(top): © Thomas Deerinck/Visuals Unlimited; 9.5(bottom): © Steve Gschmeissner/SPL/Getty RF; 9.6: © B.A. Palevitz and E.H. Newcomb/BPS/Tom Stack & Associates; 9.8d: © Biophoto Associates/Photo Researchers, Inc.; 9.9: © Dennis Kunkel Microscopy, Inc./Visuals Unlimited.

CHAPTER 10 Opener: © Barcroft Media; 10.1a: © L. Willatt/Photo Researchers, Inc.; 10.3a: Courtesy Dr. D. Von Wettstein; 10.5(all): © Ed Reschke; 10.10a: © Jose Carrilo/PhotoEdit; 10.10b: © CNRI/SPL/Photo Researchers, Inc.; 10.11a(girl): Courtesy UNC Medical Illustration and Photography; 10.11b(boy): Courtesy Stefan D. Schwarz, http://klinefeltersyndrome.org; 10.11a, b(karyotypes): © CNRI/SPL/Photo Researchers, Inc.; 10.13b: Courtesy The Williams Syndrome Association; 10.14b: © Fred, Herbert, and Hendrik van Dijk. "Images of Memorable Cases, 50 Years at the Bedside: Case 133."

CHAPTER 11 Opener: © Felicia Martinez Photography/Photedit; 11.1: © Ned M. Seidler/National Geographic Image Collection; 11.12: Courtesy Division of Medical Toxicology, University of Virginia; 11.13: © Pat Pendarvis; 11.15(man): © AP/Wide World Photos; 11.15(tissue): © Ed Reschke; p. 206(sickle cell): © Phototake, Inc./Alamy; 11.19(left, right): Courtesy Dr. Rabi Tawil, Director, Neuromuscular Pathology Laboratory, University of Rochester Medical Center; 11.19(center): Courtesy Muscular Dystrophy Association.

CHAPTER 12 Openers (hibiscus): © Rosemary Calvert/PunchStock RF; (leopard): © James Martin/Stone/Getty Images; (crab): © Tui DeRoy/Bruce Coleman/Photoshot; (protozoa): © A.M. Siegelman/Visuals Unlimited; 12.2(bacteria): © Martin Shields/Photo Researchers, Inc.; 12.2(jellyfish): © David Wrobel/Visuals Unlimited; 12.2(pigs): Courtesy Shinn-Chih Wu, Ph.D., Dept. of Animal Science and Technology, National Taiwan University, Taipei, Taiwan; 12.2(mouse): © Dr. Charles Mazel/Visuals Unlimited/Corbis; 12.2(plant): Courtesy of The Samuel Roberts Noble Foundation, Ardmore, Oklahoma; 12.4a: © Photo Researchers, Inc.; 12.4c: © Science Source/Photo Researchers, Inc.; 12.5a(DNA model): © Photodisk Red/Getty RF; 12.5d: © A. Barrington Brown/Photo Researchers; 12.12a: © Oscar L. Miller/Photo Researchers, Inc.; 12.15d: Courtesy Alexander Rich; 12.20a: © Ada L. Olins and Donald E. Olins/Biological Photo Service; 12.20b: Courtesy Dr. Jerome Rattner, Cell Biology and Anatomy, University of Calgary; 12.20c: Courtesy of Ulrich Laemmli and J.R. Paulson, Dept. of Molecular Biology, University of Geneva, Switzerland; 12.20d, e: © Peter Engelhardt/Department of Pathology and Virology, Haartman Institute/Centre of Excellence in Computational Complex Systems Research, Biomedical Engineering and Computational Science, Faculty of Information and Natural Sciences, Helsinki University of Technology, Helsinki, Finland.

CHAPTER 13 Opener: © Francois Paquet-Durand/Photo Researchers, Inc.; 13.5a: © Dennis Kunkel Microscopy, Inc./Visuals Unlimited; 13.6: © Chanan Photo 2004; 13A: © Fitzpatrick Color Atlas & Synopsis of Clinical Dermatology 5/e. The McGraw-Hill Companies, Inc.; 13.12b, c: © Stan Flegler/Visuals Unlimited.

CHAPTER 14 Opener: © Zacharias Pereira da Mata/Shutterstock; 14.5: © U.S. Coast Guard photo by Petty Officer 3rd Class Patrick Kelley; 14A: © Deco/Alamy RF; 14.1(human): © Image Source/JupiterImages RF; 14.1(plant): Courtesy USDA/Peggy Greb, photographer; 14.1(mouse): © Redmond Durrell/Alamy RF; 14.1(fruit fly): © Herman Eisenbeiss/Photo Researchers, Inc.; 14.1(roundworm): © CMSP/J.L. Carson/Getty Images; 14.1(yeast): © G. Wanner/Getty Images; 14.10: Courtesy of Dr. Michael Thompson.

CHAPTER 15 Opener: © Tyler Keillor, University of Chicago; 15.1a(Darwin): © Carolina Biological/Visuals Unlimited; 15.1c(rhea): © Wolfgang Kaehler/Corbis; 15.1d(desert): © Luiz C. Marigo/Peter Arnold/Photolibrary; 15.1e(strata): © Gary J. James/Biological Photo Service; 15.1f(waterfall): © Charles Benes/Index Stock Imagery/Photolibrary; 15.1g(iguana): © Galen Rowell/Corbis; 15.4(Lepus): © WILDLIFE/Peter Arnold/Photolibrary; 15.4(Dolichotis): © Juan & Carmecita Munoz/Photo Researchers, Inc.; 15.5a: © Kevin Schafer/Corbis; 15.5b: © Michael Dick/Animals Animals; 15.6: © Lisette

CHAPTER 38 Openers (snake): © Nutscode/T Service/Photo Researchers, Inc.; (mouse): © Ted Kinsman/ Photo Researchers, Inc.; p. 717(snake): © Corbis RF; 38.1b, c(all): © Omikron/SPL/Photo Researchers, Inc.; 38.3: © Farley Bridges; 38.4(both): © Heather Angel/Natural Visions; 38.7: © Lennart Nilsson/Scanpix; 38.8b: © Biophoto Associates/Photo Researchers, Inc.; 38.12: © P. Motta/SPL/Photo Researchers, Inc.; 38.15a, b: Courtesy Dr. Yeohash Raphael, the University of Michigan, Ann Arbor.

CHAPTER 39 Opener: © 2008 The Associated Press. All rights reserved; 39.2: © Michael Fogden/OSF/Animals Animals; 39.3: © E. R. Degginger/Photo Researchers, Inc.; 39.4(hyaline cartilage): © Ed Reschke; 39.4(osteocyte): © Biophoto Associates/Photo Researchers, Inc.; 39.4(compact bone): © Ed Reschke; 39.13(gymnast): © Corbis RF; 39.13(myofibril): © Biology Media/Photo Researchers, Inc.; 39A: © Thinkstock/Getty RF; 39.14a(right): © Victor B. Eichler, Ph.D.

CHAPTER 40 Openers (caterpillar): © Doug Wechsler/Animals Animals; (moth): © Richard Kolar/ Animals Animals; 40.7a: © AP/Wide World Photos; 40.7b: © General Photographic Agency/Getty Images; 40.8(all): These photos were published in Clinical Pathological Conference, "Acromegaly, Diabetes, Hypermetabolism, Proteinura and Heart Failure", American Journal of Medicine, 20 (1956) 133. Reprinted with permission from Excerpta Medica Inc.; 40.9a: © Bruce Coleman, Inc./Alamy; 40.9b: © Medical-on-Line/Alamy; 40.9c: © Dr. P. Marazzi/ Photo Researchers, Inc.; 40.13a: © Custom Medical Stock Photo; 40.13b: © NMSB/Custom Medical Stock Photo; 40.14(both): Courtesy Shannon Halverson; 40.15(pancreatic islet, top and bottom): © Peter Arnold, Inc./Alamy; 40A: © Bettmann/Corbis; 40.17: © The McGraw-Hill Companies, Inc. Evelyn Jo Johnson, photographer.

CHAPTER 41 Opener: © Rudy Kuiter/OSF/Animals Animals; 41.1: © Dr. Dennis Kunkel/Visuals Unlimited; 41.2: © Herbert Kehrer/zefa/Corbis; 41.3: © Anthony Mercieca/ Photo Researchers, Inc.; 41.5b: © Ed Reschke; 41.6b: © Anatomical Travelogue/Photo Researchers, Inc.; 41.90: © Ed Reschke/Peter Arnold/Photolibrary; 41.13a: © Saturn Stills/ Photo Researchers, Inc.; 41.13b: © Getty RF; 41.13c: © The McGraw-Hill Companies, Inc. Lars A. Niki, photographer; 41.14: © CC Studio/SPL/Photo Researchers; 41Aa,b(both): © Brand X/SuperStock RF; 41.15: © Scott Camazine/Photo Researchers; 41.16: © CDC/Peter Arnold/Photolibrary; 41.17: © G. W. Willis/Visuals Unlimited; 41B(left): © Vol 161/ Corbis RF; 41B(right): © Vol 178/Corbis RF; 41C: © Hartmut Schwarzbach/Peter Arnold/Photolibrary; 41.18: © Biomedical Imaging Unit, Southampton General Hospital, Photo Researchers, Inc.; 41.19: © CNR/SPL/Photo Researchers, Inc.; 41.20: © Melba Photo Agency/Alamy RF.

CHAPTER 42 Openers (2 days): Courtesy of the film, Building Babies, © ICAM/Mona Lisa/Phototake; (3 weeks): © Lennart Nilsson/Scanpix; (6 weeks): © Claude Edelmann/ Photo Researchers, Inc.; (5 months): © Derek Bromhall/ OSF/Animals Animals; p. 797(chick): © Photodisc/Getty RF; 42.1: © David M. Phillips/Visuals Unlimited; 42.2a: © William Jorgensen/Visuals Unlimited; 42.4b: Courtesy Kathryn Tosney; 42.9a–c: Courtesy Steve Paddock, Howard Hughes Medical Research Institute; 42.10a: Courtesy E.B. Lewis; 42.12(fertilization): © Don W. Fawcett/Photo Researchers, Inc.; 42.12(2-cell): © Rawlins-CMSP/Getty Images; 42.12(morula): © RBM Online/epa/Corbis; 42.12(implantation): © Bettmann/Corbis; 42.14a: © Lennart Nilsson/Scanpix; 42.17: © Georges Gobet/AFP/Getty Images.

CHAPTER 43 Opener: © Jennifer Jarvis/Visuals Unlimited; 43.1a: © Joe McDonald; 43.1b: Courtesy Refuge for Saving the Wildlife, Inc.; 43.2(coastal): © John Sullivan/Monica Rua/Ribbitt Photography; 43.2(inland): © R. Andrew Odum/Peter Arnold/Photolibrary; 43.3b, c(both): Reprinted from Cell, Vol 86, Issue 2, J.R. Brown, Hong Ye, R. T. Bronson, P. Dikkes, M. E. Greenberg, A Defect in Nurturing in Mice Lacking the Immediate Early Gene fosB, fig. 2, p. 297-309, figure 2, 26 July 1996, with permission from Elsevier. Link from website to the journal: http://sciencedirect.com/science/journal/00928674; p. 824(imprinting): © Nina Leen/Time Life Pictures/Getty Images; p. 826(raven): Courtesy Dr. Bernd Heinrich; 43.7: © Gregory G. Dimijian/Photo Researchers, Inc.; 43A(left): © Creatas/ PunchStock RF; 43A(right): © Tom McHugh/Photo Researchers, Inc.; 43.8a(main): © Arco Images/GmbH/ Alamy; 43.8a(inset): © Fritz Polking/Visuals Unlimited; 43.9: © Raul Arboleda/AFP/Getty Images; 43.10(firefly): © Phil Degginger/Alamy; 43.10(trees): © PhotoLink RF; 43.11a: © OSF/Animals Animals; 43.12: © Nicole Duplaix/ Peter Arnold/Photolibrary; 43.14: © Alamy RF; 43B: © T & P Gardner/Bruce Coleman/Photoshot; 43.15: © D. Robert & Lorri Franz/Corbis; 43.16: © Mark Moffett/Minden Pictures; 43.17: © J & B Photo/Animals Animals.

CHAPTER 44 Opener: © AP Photo/Illinois River Biological Station via the Detroit Free Press, Nerissa Michaels; 44.1: © David Hall/Photo Researchers; 44.2a: © The McGraw Hill Companies, Inc. Evelyn Jo Johnson, photographer; 44.3(mice): © E. R. Degginger/Photo Researchers, Inc.; 44.3(rhinos): © Corbis RF; 44.4b: © Holt Studios/Photo Researchers, Inc.; 44.4c: © Bruce M. Johnson; 44.4d: © Digital Vison/Getty RF; 44.6a: © Breck P. Kent/Animals Animals; 44.6b: © Doug Sokell/Visuals Unlimited; 44.9: © Paul Janosi/Valan Photos; 44.13a: © Michael Fogden/Animals Animals; 44.13b: © Michael Fogden/Animals Animals; 44.13c: © Tom McHugh/Photo Researchers, Inc.; 44.14(dandelions): © Ted Levin/Animals Animals; 44.14(bears): © Michio Hoshino/Minden Pictures; 44Aa: © Tony Campbell/ Shutterstock Images; 44Ab: © Altrendo Images/Getty Images; 44.15b(top): © The McGraw-Hill Companies, Inc. Jill Braaten, photographer; 44.15b(bottom): © Robert Harding/Robert Harding World Imagery/Corbis; 44.16c: © Still Pictures/Peter Arnold/Photolibrary.

CHAPTER 45 Opener: © NPS photo by Barry O'Neill; 45.1a(forest): © Charlie Ott/Photo Researchers, Inc.; 45.1a(squirrel): © Stephen Dalton/Photo Researchers, Inc.; 45.1a(wolf): © Renee Lynn/Photo Researchers, Inc.; 45.1b(rain forest): © Michael Graybill and Jan Hodder/ Biological Photo Service; 45.1b(kinkajou): © Alan & Sandy Carey/Photo Researchers, Inc.; 45.1b(sloth): © Studio Carlo Dani/Animals Animals; 45Ab: © Minden Pictures/ Superstock; 45Ac: © Corbis RF; 45.6a: © Creatas/PunchStock RF; 45.7a: © Gustav Verderber/Visuals Unlimited; 45.7b: © Zig Leszczynski/Animals Animals; 45.7c: © A. Cosmos Blank/Photo Researchers, Inc.; 45.8a, b, d: © Edward S. Ross; 45.8c: © James H. Robinson/Photo Researchers, Inc.; 45.9: Courtesy the University of Tennessee Parasitology Laboratory; 45.10: © Dave B. Fleetham/ Visuals Unlimited; 45.11: © C. C. Lockwood/Animals Animals; 45.12: © Bill Wood/Bruce Coleman/Photoshot; 45Ba: Courtesy Dr. Ian Wyllie; 45.13a(all): © Breck P. Kent/Animals Animals; 45.15a(left): © Ed Reschke/Peter Arnold/Photolibrary; 45.15a(right): © Herman Eisenbeiss/Photo Researchers, Inc.; 45.15b(left): © Corbis RF; 45.15b(right): © Gerald C. Kelley/Photo Researchers, Inc.; 45.15c(left): © Bill Beatty/Visuals Unlimited; 45.15c(right): © Joe McDonald/Visuals Unlimited; 45.15d(left): © SciMAT/Photo Researchers, Inc.; 45.15d(right): © Michael Beug; 45.17: © George D. Lepp/Photo Researchers, Inc.; 45Cb: © Jan Martin Will/Shutterstock Images.

CHAPTER 46 Opener: © Andrew Michael/Photolibrary; 46.6a: © John Shaw/Tom Stack & Assoc.; 46.6b: © Danita Delimont/Getty Images; 46Aa: © Porterfield/Chickering/Photo Researchers, Inc.; 46Ab: © Michio Hoshino/ Minden Pictures; 46.7a: © Creatas/Jupiterimages RF; 46.7b: © Bill Silliker, Jr./Animals Animals; 46.8a: © E. R. Degginger/Animals Animals; 46.8b(chipmunk): © Carmela Lesczynski/Animals Animals; 46.8b(bobcat): © Tom McHugh/Photo Researchers, Inc.; 46.8b(marigolds): © Virginia Neefus/Animals Animals; 46.10(frog): © National Geographic/Getty Images; 46.13(zebras and others): © Darla G. Cox; 46.13(cheetah): © Digital Vision/Getty RF; 46.10(katydid): © M. Fogden/OSF/Animals Animals; 46.10(jaguar, macaw, monkey): © Professor David F. Cox, Lincoln Land Community College; 46.10(butterfly, lizard): © Kjell Sandved/Butterfly Alphabet; 46.11a(chaparall): © Walt Anderson/Visuals Unlimited; 46.11b(roadrunner): © John Cancalosi/Peter Arnold/Photolibrary; 46.12(bison): © Eastcott Momatiuk/Getty RF; 46.12(prairie): © Jim Steinberg/Photo Researchers, Inc.; 46.14a: © Jonathan Blair/Corbis; 46.14b(rat): © Bob Calhoun/Bruce Coleman/ Photoshot; 46.14b(roadrunner): © Jack Wilburn/Animals Animals; 46.14b(kit fox): © Jeri Gleiter/Peter Arnold/Photolibrary; 46.15(mayfly): Courtesy Michael Windelspecht; 46.15(trout): © William H. Mullins/Photo Researchers, Inc.; 46.15(carp): © Robert Maier/Animals Animals; 46.15(crab): © Gerlach Nature Photography/Animals Animals; 46.16a: © Roger Evans/Photo Researchers, Inc.; 46.16b: © The McGraw-Hill Companies, Inc. Pat Watson, photographer; 46.18(pike): © Robert Maier/Animals Animals; 46.18(water strider): © G.I. Bernard/Animals Animals; 46.19a: © John Eastcott/Yva Momatiuk/Animals Animals; 46.19b: © Theo Allofs/Visuals Unlimited; 46.19c: © Steve Bower/Shutterstock Images; 46B(left): © Brian J. Skerry/Getty Images; 46B(right): © Blend Images LLC RF.

CHAPTER 47 Opener: © Allen Blake Sheldon/Animals Animals; 47.3(periwinkle): © Steven P. Lynch; 47.3(armadillo): © Photodisc/Getty RF; 47.3(boat): © Getty RF; 47.3(bat): © Merlin D. Tuttle/Bat Conservation International; 47.3(ladybug): © Masterfile RF; 47.3(rubber harvest): © Bryn Campbell/Stone/Getty Images; 47.4: © Blend Images/Getty RF; 47.6b: © IT Stock/PunchStock RF; 47.6c (top): Courtesy Woods Hole Research Center; 47.6c(center): Courtesy R.O. Bierregaard; 47.6c(bottom): Courtesy Thomas Stone, Woods Hole Research Center; 47.7a: © Chuck Pratt/Bruce Coleman/ Photoshot; 47.7b: © Chris Johns/National Geographic Image Collection; 47.8b: Courtesy Walter C. Jaap/Florida Fish & Wildlife Conservation Commission; 47A, B: Courtesy Peter Paul van Dijk; 47.9a: © Shane Moore/Animals Animals; 47.9b: © Peter Auster/University of Connecticut; 47.10a: © Gerard Lacz/Peter Arnold/Photolibrary; 47.10b(forest): © Art Wolfe/Artwolfe.com; 47.10b(owl): © Pat & Tom Leeson/ Photo Researchers, Inc.; 47.11b: © Jeff Foott Productions; 47D(otter): © Getty RF; 47D(birds): Courtesy Deborah M. Berman; 47D(paddlefish): © Russ Kinne/Animals Animals; 47D(vial with sample, researcher in water: © Marilyn Kok, University of Illinois Springfield.

LINE ART & TEXT

CHAPTER 2 Biological Systems: The Interconnectedness of Water, Plants, and People: Textual data from G. Tyler Miller, Living in the Environment, 1983, Wadsworth Publishing Company, Belmont, CA; and Lester R. Brown, State of the World, 1992, W.W. Norton & Company, Inc. New York, NY, p. 29. Graphical data from EPA, 2010.

CHAPTER 17 17.12: Reprinted by permission from Macmillan Publishers Ltd on behalf of Cancer Research UK: RC Albertson, TD Kocher (2006) Genetic and developmental basis of cichlid trophic diversity. Heredity vol. 97 (3) pp. 211-221.

CHAPTER 25 Nature of Science: Plants Can Clean Up Toxic Messess: From Joe Bower, *National Wildlife Magazine*, June/July 2000, Vol. 38, No. 4, Reprinted with permission of the author.

CHAPTER 43 43.2: Data from S.J. Arnold, "The Microevolution of Feeding Behavior" in *Foraging Behavior: Ecology, Ethological, and Psychology Approaches,* edited A. Kamil and T. Sargent, 1980, Garland Publishing Company, New York, NY. Evolution: Sexual Selection in Male Bowerbirds: Courtesy of Gail Patricelli, University of Maryland.

CHAPTER 44 44.4b: Data from W.K. Purves, et al., Life: *The Science of Biology,* 4/e, Sinaeur & Associates. 44.4d: Data from A.K. Hegazy, 1990, "Population Ecology & Implications for Conservation of Cleome Droserifolia: A Threatened Xerophyte, "*Journal of Arid Environments,* 19:269-82. 44.8: From Raymond Pearl, *The Biology of Population Growth.* Copyright 1925 The McGraw-Hill Companies. All Rights Reserved. 44.9: Data from Charles J. Krebs, *Ecology,* 3/e, 1984, Harper & Row; after Scheffer, 1951. 44.16: United Nations Population Division, 2002.

CHAPTER 45 45.6: Data from D.A. MacLulich, *Fluctuations in the Numbers of the Varying Hare (Lepus americanus)*, University of Toronto Press, Toronto, 1937, reprinted 1974.

CHAPTER 46 Nature of Science: Wildlife Conservation and DNA: From T. Friend, "DNA Fingerprinting: Power Tool," *National Wildlife Magazine*, October/ November 1995, Vol. 33, No. 6. Reprinted with permissioin of the author.

CHAPTER 47 47.2: Redrawn from "Shrimp Stocking, Salmon Collapse, and Eagle Displacement" by C.N. Spencer, B.R. McClelland and J.A. Stanford, Bioscience, 41(1):14-21. Copyright © 1991 American Institute of Biological Sciences. 47.8a: Data from David M. Gates, *Climate Change and Its Biological Consequences,* 1993, Sinauer & Associates, Inc., Sunderland, MA.

Index

Note: This text contains a substantial number of supporting graphics. To ease index use, regular page numbers indicate *both* discussion and figure(s) on the same page; page numbers with *f* indicate additional figures on a different page; *t* indicates a table

A

A. *See* Adenine (A)
A band, 745
A horizon soil, 468
ABA (abscisic acid), 477, 488, 489*t*
Abalone, 534
ABC Islands, brown bears of, 889
Abdomen
    crustacean, 541*f*
    insect, 542, 543*f*
Abdominal cavity organs, 596
Abiogenesis, 329
Abiotic agents, of natural selection, 8
Abiotic synthesis, 328*f*, 329
ABO blood types, 205, 619
Abortion, 783–84
Aboveground vertical stems, 457*f*, 458
Abscisic acid (ABA), 477, 488, 489*t*
Abscission, 488, 489*f*, 489*t*
Absolute dating of fossils, 334
Absorption spectrum, 124
Abstinence, 783
Abulocetus natans, 281
Abyssal plain, 902
Acacia tree, 498, 866, 895
Accessory organs, digestive tract, 654–55
Accessory pigments, red algae, 389
Accessory structures, skin, 598–99
Accidental transport of alien species, 914
Accommodation, visual, 722
Acetaminophen, 246
Acetic acid bacteria, 141
Acetobacter aceti, 141
Acetyl CoA, 142–43, 147
Acetylcholine (ACh), 700, 708, 713, 747
Acetylcholinesterase (AChE), 701, 748
Acetylene, 39
ACh (acetylcholine), 700, 708, 713, 747
AChE (acetylcholinesterase), 701, 748
Achromatopsia (color blindness), 209, 300–301, 723
Acid deposition, 878, 915
Acid rain, 33, 467, 878
Acid reflux, 651
Acid-base balance, 690
Acidosis, 33
Acids, chemistry of, 32–34
Acinonyx jubatus, 895*f*
Acne, 599
Acquired characteristics, inheritance of, 273–74
Acquired immunity. *See* Adaptive immunity
Acquired immunodeficiency syndrome. *See* AIDS
Acrocephalus, 867*f*
Acromegaly, 759
Acrosome, 775*f*, 776*f*, 777, 796–97
Acrylamide, 248
ACTH (adrenocorticotropic hormone), 754*f*, 757, 758*f*, 764, 810
Actin, 49, 108, 592, 745
Actin filaments
    cell structure and function, 68*f*–69*f*
    cytoskeleton, 78, 79*f*
    in extracellular matrix, 99*f*
    in muscle contraction, 745–46, 747*t*, 748–49
Actinopterygii, 554, 555*f*
Action potentials, 698*f*, 699–700, 744
Active immunity, 637–39
Active site, of enzyme, 110
Active transport, 95–98
    carrier proteins and energy needed, 95, 96*f*

defined, 89*t*, 95
endocytosis, 97–98
exocytosis, 95, 96*f*
phagocytosis, 97–98
plasma membrane, 91
Acute bronchitis, 673, 674*f*
Acute phase proteins, 631
AD (Alzheimer disease), 53, 184, 701, 707
Adam's apple, 668, 778
Adaptation
    behavioral, 830–34
    defined, 5, 278
    to diet, digestive tracts, 648–49
    evolutionary theory, 278
    Lamarck and, 273
    photosynthesis, 131
    plants, 420
Adaptive immunity, 633–39
    active vs. passive, 637–39
    antibody-mediated, 633–35
    cell-mediated, 633, 635–37
    defined, 630
    evolution of, 627–28
    generally, 633–34
Adaptive radiation mode of speciation, 316–17
Addiction to drugs, 710–11
Addison disease, 764
Adductor longus muscle, 744*f*
Adenine (A)
    complimentary base pairing, 55
    DNA structure, 55*f*, 217–20, 223, 223*t*
    molecular structure of, 217*f*
    mRNA codons, 224*f*
    as nucleotide, 54*f*
    RNA structure, 220, 223*t*
Adenosine, 55, 755. *See also* Ribose
Adenosine deaminase deficiency, gene therapy for, 260
Adenosine diphosphate (ADP), 56, 107, 748
Adenosine triphosphate. *See* ATP
Adenovirus, 364*f*
ADH (antidiuretic hormone), 688, 754*f*, 757, 758*f*
Adhesion, 30, 476
Adhesion junctions, 98, 99*f*, 588, 593
Adiantum pedatum, 428*f*
Adipocytes, 590
Adipose tissue
    aging process, 811
    blood glucose regulation, 765
    functions of, 590
    leptin produced by, 768
    skin, 597*f*
    subcutaneous layer, 598
ADP (adenosine diphosphate), 56, 107, 748
Adrenal cortex, 754*f*, 756, 758*f*, 762–64
Adrenal glands, 754*f*, 758*f*, 762–64
Adrenal medulla, 754*f*, 755, 762–63
Adrenaline, 613, 754*f*, 755–56, 762
Adrenocorticotropic hormone (ACTH), 754*f*, 757, 758*f*, 764, 810
Adrenoleukodystrophy (ALD), 75, 210
Adult stem cells. *See* Stem cells
Adventitious roots, 452, 457*f*
Aerial roots, 461*f*
Aerial stems, 428
Aerobic bacteria, 335
Aerobic reactions, in cellular respiration, 137
Aerotropism, 491
AF-2 (food additive), 248
Afferent (sensory) neurons, 697–98, 717
Aflatoxin, 248
Africa
    convergent evolution Rift Valley, 317
    evolutionary hypotheses, 579
    savanna, 895
African sleeping sickness, 396, 398–99

Afrocoricida, 566*t*
Agapornis, 820
Agar, 389
Agaricus bisporus, 406*t*
Age distribution, 842–43, 852–53
Age of Amphibians, 339
Age of Cycads, 340
Age of Fishes, 339
Age spots, 811
Agent Orange, 485
Agglutination, 619
Aggregate fruits, 510*f*, 511, 511*t*
Aging process, 811–14
    cardiovascular systems, 811–12
    digestive systems, 812
    endocrine systems, 813
    excretory systems, 812
    hypotheses about, 813–14
    immune systems, 812
    integumentary systems, 811
    musculoskeletal systems, 737, 738, 812–13
    nervous systems, 812
    organ systems, 811–13
    reproductive systems, 813
    respiratory systems, 812
    sensory systems, 722, 724, 728, 812
Agnathans (jawless fish), 339, 526*t*, 552, 553
Agranular leukocytes, 620
Agriculture
    alien species introduced by, 914
    biodiversity, 910–11
    evolution of, 581
    livestock as tapeworm host, 533
    pollution from, 14–15, 877, 878, 915
    swidden (slash-and-burn), 899
AID (artificial insemination), 785
AIDS (acquired immunodeficiency syndrome). *See also* HIV
    electron micrograph of, 289*f*
    epidemiology of, 787
    microevolution, 289
    opportunistic infections in, 626, 638, 674, 787
    retrovirus, 368
    stages and symptoms of, 787
    treatment of, 788
    virus, 364
Air
    movements of, 884–85, 902
    nutritional function of soil, 467
    pollution of, 33, 414, 878
    temperature of, 902
Air bladder, brown algae, 390*f*
Air sacs, 667, 670*f*
Alagille syndrome, 188
Albumin, 618
Alces americanus, 890*f*
Alcohol
    abuse, 710
    ADH inhibition by, 757
    alcoholic fermentation, 140
    cardiovascular disease, 617
    cirrhosis of liver, 656
    in functional groups, 39*t*
    during pregnancy, 807
ALD (adrenoleukodystrophy), 75, 210
Aldosterone, 690, 763
Alexandrium catanella, 392–93
Alexis (son of Tzar Nicolas I of Russia), 210
Algae
    blue-green. *See* Cyanobacteria
    brown, 385*t*, 386*f*, 390, 900
    classification of, 354
    colonial, 2, 385*t*, 388
    in coral reefs, 901
    as decomposers in soil, 468
    filamentous, 388
    in food chain, 383
    golden brown, 385*t*, 386*f*, 390, 391
    green. *See* Green algae
    multicellular, 388, 389, 390
    photosynthesis by, 76

protist diversity, 385*t*
    red, 385*t*, 386*f*, 389
    size variety of, 384
    sperm production in, 422
Algae blooms, 392–93
Algin, 390
Ali, Muhammad, 693
Alien species, 838, 861, 909, 913–15
Alkaloids, 497
Alkalosis, 33
Allantois, 805, 808*f*, 809
Alleles. *See also* Gene(s)
    defined, 172, 196
    dominant, 290
    frequencies, and microevolution, 290–95, 292*t*
    in homologous chromosomes, 196*f*
    multiple, 205
Allelic traits, multiple, 205
Allen's rule, 583
Allergy(ies), 641–42, 642*t*
Alligators, 558*f*, 561–62, 918
Allopatric speciation mode of speciation, 313–14
Alloploidy, 315–16
Allosteric site, 112
Alouatta pigra, 893*f*
α-helix, 51, 52*f*
Alpine tundra, 887, 888
ALS (amyotrophic lateral sclerosis), 703
Alternate leaves, 460
Alternation of generations, 176, 388, 407, 422
Alternative splicing, 237
    mRNA, 227, 244–46
Altitude, and biomes, 887
Altman, Sidney, 330
Altruism, 833–34
Alu sequence, 262
Aluminum, in plants, 465
Alvarez, Luis, 343
Alvarez, Walter, 343
Alveolates, 385*t*, 392–94
Alveoli (sing., alveolus)
    mammary, 782, 783*f*
    pulmonary, 668
Alyces obstetricans, 848*f*
Alzheimer disease (AD), 53, 184, 701, 707
AM fungi, 404*f*, 406*t*, 408
Amanita phalloides, 411
Amborella trichopoda, 433
Ambulocetus natans, 281*f*, 310*f*
Ambystoma tigrinum, 557*f*
Amebic dysentery, 396
Amebocytes, 622
Amenorrhea, 782
American Cancer Society, 675
American Dietetic Association, 658, 659
American goldfinch, 774*f*
American Heart Association, 659
Ames test, 248
Amine group, polarity of, 27
Amino acids. *See also* Adenine (A); Cytosine (C); Guanine (G); Thymine (T); *specific amino acid*
    cellular respiration, 147
    codons for, 224
    essential, 658–59
    with ionized R groups, 51*f*
    metabolism of, 681
    protein digestion, 657
    as protein monomer, 50
    sequencing, and phylogeny, 358
    synthesis of, 129
    and tRNA, 223, 228–29
Amino functional groups, 39*t*
Aminoacyl-tRNA synthetases, 229
Ammonia
    as atmospheric gas, 329
    as buffer, 690
    in nitrogenous waste, 681
    photosynthesis vs. cellular respiration, 148
    water as solvent for, 30

I

M